HUMAN ANATOMY

AUTHOR TEAM

Frederic H. Martini received his Ph.D. in comparative and functional anatomy from Cornell University. Dr. Martini's publications include numerous journal articles, technical reports, and magazine articles. He is the author of *Fundamentals of Anatomy and Physiology* (Fifth Edition © 2001) and coauthor of *Essentials of Anatomy and Physiology* (Second Edition © 2000), *Structure and Function of the Human Body* (First Edition © 1999), and the *Human Body in Health and Disease* (First Edition © 2000) with Edwin Bartholomew, and *Foundations of Anatomy and Physiology* (First Edition © 1998) with George Karleskint. Dr. Martini is a member of the American Association of Anatomists, the American Physiological Society, the Human Anatomy and Physiology Society, the National Association of Biology Teachers, the Society for Integrative and Comparative Biology, the Society for College Science Teachers, the Western Society of Naturalists, the American Association for the Advancement of Science, and the National Association of Underwater Instructors.

Dr. Martini has been involved in teaching undergraduate courses in anatomy and physiology (comparative and/or human) since 1970. During the 1980s he spent his winters teaching courses, including human anatomy and physiology, at Maui Community College and his summers teaching an upper-level field course in vertebrate biology and evolution at the Shoals Marine Laboratory (SML) for Cornell University. Dr. Martini now retains affiliations with the University of Hawaii and SML.

Robert B. Tallitsch received his Ph.D. in physiology with an anatomy minor from the University of Wisconsin—Madison at the rip old age of 24. Since then Bob has been on the biology faculty at Augustana College in Rock Island, Illinois. His teaching responsibilities include Human Anatomy, Neuroanatomy, Histology, Kinesiology and Cadaver Dissection. Bob is also a member of the Asian Studies faculty at Augustana, and teaches a course in Traditional Chinese Medicine in this program. Bob's previous research grants, journal articles and abstracts center around the effect of hypoxia on contractile mechanics, sodium and calcium transport in papillary muscles isolated from genetically hypertensive rats. His publications for Prentice Hall include *Histology: A Diagnostic Atlas* with Ron Guastaferri (Art Coordinator and Illustrator, First Edition © 2004). Bob is a member of the American Physiological Society, AsiaNetwork and the Human Anatomy and Physiology Society. In addition to his teaching responsibilities at Augustana Bob has served as a visiting faculty member in the Department of Cardiology at Johns Hopkins College of Medicine, the Foreign Languages Faculty at the Beijing University of Chinese Medicine and Pharmacology, the Foreign Languages Faculty at Central China Normal University (Wuhan, PRC), and a visiting scientist in the National Institutes of Health Gerontology Research Center at the National Institutes of Aging.

Michael J. Timmons received his degrees from Loyola University, Chicago. For more than three decades he has had a strong commitment to teaching human anatomy and physiology to nursing and preprofessional students at Moraine Valley Community College. Early in his teaching career, Professor Timmons became very interested in publishing in those fields. He is a coauthor of Human Anatomy Laboratory Guide and Dissection Manual, the lab manual that accompanies this text, and of three anatomy and physiology laboratory manuals. He is also the editor of the Prentice Hall Laser Disk with Bar Code Manual; author of Atlas, the laboratory learning system; and content editor for the Instructor's Presentation Manager 4.0 CD-ROM, which are supplements to this text. His special areas of interest are biomedical photography, crafting illustration programs, and developing instructional technology learning systems. He has authored a series of titles on the dissection of the cat, including *A Photographic Atlas of Cat Anatomy, Cat Anatomy Slides: A Visual Guide to Dissection*, and numerous study guides.

Professor Timmons is an active member of scientific and professional associations. He is also a national and regional presenter at the League for Innovation Conferences on Information Technology for Colleges and Universities and at Human Anatomy and Physiology Society meetings.

Dr. Kathleen Welch, clinical consultant for all Martini projects, received her M.D. from the University of Seattle and did her residency at the University of North Carolina in Chapel Hill. For two years, she served as Director of Maternal and Child Health at the LBJ Tropical Medical Center in American Samoa. She then joined the Department of Family Practice at the Kaiser Permanente Clinic in Lahaina, Hawaii. She has been in private practice since 1987. Dr. Welch is a Fellow of the American Academy of Family Practice. She is a member of the Hawaii Medical Association and the Human Anatomy and Physiology Society. Drs. Martini and Welch were married in 1979; they have one child, "P. K.," born in January 1995.

ILLUSTRATION TEAM

Dr. William C. Ober (art coordinator and illustrator) received his undergraduate degree from Washington and Lee University and his M.D. from the University of Virginia in Charlottesville. While in medical school, he also studied in the Department of Art as Applied to Medicine at Johns Hopkins University. After graduation Dr. Ober completed a residency in family practice and is currently on the faculty of the University of Virginia, where he teaches in the Department of Sports Medicine. He is also part of the Core Faculty at the Shoals Marine Laboratory, where he teaches biological illustration in the summer program. Dr. Ober now devotes his full attention to medical and scientific illustration.

Claire W. Garrison, R.N. (illustrator), practiced pediatric and obstetric nursing for nearly 20 years before turning to medical illustration as a full-time career. Following a five-year apprenticeship, she has worked as Dr. Ober's associate since 1986. Ms. Garrison is also a Core Faculty member of the Shoals Marine Laboratory.

Texts illustrated by Dr. Ober and Ms. Garrison have received national recognition and awards from the Association of Medical Illustrators (Award of Excellence), American Institute of Graphics Arts (Certificate of Excellence), Chicago Book Clinic (Award for Art and Design), Printing Industries of America (Award of Excellence), and Bookbuilders West. They are also recipients of the Art Directors Award.

Ralph T. Hutchings is a biomedical photographer who was associated with the Anatomy Department of the Royal College of Surgeons for 20 years. An engineer by training, Mr. Hutchings has focused for years on photographing the structure of the human body. The result has been a series of color atlases, including the *Color Atlas of Human Anatomy*, *The Color Atlas of Surface Anatomy*, and *The Human Skeleton* (all published by Mosby-Yearbook Publishing, St. Louis, Missouri, USA). Mr. Hutchings makes his home in North London, where he tries to balance the demands of his photographic assignments with his hobbies of early motor cars and airplanes.

Human Anatomy

Fourth Edition

Frederic H. Martini, Ph.D.
University of Hawaii

Michael J. Timmons, M.S.
Moraine Valley Community College

Robert B. Tallitsch
Augustana College

WITH

William C. Ober, M.D.
Art Coordinator and Illustrator

Claire W. Garrison, R.N.
Illustrator

Kathleen Welch, M.D.
Clinical Consultant

Ralph T. Hutchings
Biomedical Photographer

Prentice Hall

Pearson Education
Upper Saddle River, New Jersey 07458

Library of Congress Cataloging-in-Publication Data
Martini, Frederic.
Human anatomy / Frederic H. Martini, Michael J. Timmons, Robert B. Tallitsch, with William C. Ober art coordinator and illustrator ; Claire W. Garrison, illustrator ; Kathleen Welch, clinical consultant ; Ralph T. Hutchings, biomedical photographer.— 4th ed.
p. cm.
Includes index.
ISBN 0-13-061569-2
1. Human anatomy. 2. Human anatomy—Atlases. I. Timmons, Michael J. II. Tallitsch, Robert B. III. Title.
QM23.2 .M356 2003
611—dc21
2002005105

Editor in Chief, Life and Geosciences: Sheri L. Snavely
Senior Acquisition's Editor: Halee Dinsey
Production Editor: Shari Toron
Manager of Page Formatting: Jim Sullivan
Formatters: Karen Noferi, Joanne Del Ben
Assistant Vice President of Production and Manufacturing: David W. Riccardi
Executive Managing Editor: Kathleen Schiaparelli
Assistant Managing Editor: Beth Sweeten
Editor in Chief, Development: Ray Mullaney
Development Editor: Deena Cloud
Director of Creative Services: Paul Belfanti
Director of Design: Carole Anson
Art Director: Jonathan Boylan
Interior and Cover Designer: Ann DeMarinis
Cover Illustration: Vincent Perez
Executive Marketing Manager for Life Sciences: Jennifer Welchans
Senior Marketing Manager: Martha McDonald
Editorial Assistant: Susan Zeigler
Manufacturing Manager: Trudy Pisciotti
Assistant Manufacturing Manager: Michael Bell
Managing Editor, Audio Visual Assets and Production: Patricia Burns
Audio Visual Editor: Adam Velthaus
Photo Editor: Carolyn Gauntt
Director, Image Resource Center: Melinda Reo
Manager, Rights and Permissions: Zina Arabia
Interior Image Specialist: Beth Boyd-Brenzel
Cover Image Specialist: Karen Sanatar
Image Permission Coordinator: Carolyn Gauntt

Prentice Hall

Published by Pearson Education, Inc.
Upper Saddle River, New Jersey 07458

Printed in the United States of America
10 9 8 7 6 5 4 3 2 1

ISBN 0-13-061569-2

Pearson Education LTD., *London*
Pearson Education Australia Pty Limited, *Sydney*
Pearson Education Singapore, Pte. Ltd
Pearson Education North Asia Ltd, *Hong Kong*
Pearson Education Canada, Ltd., *Toronto*
Pearson Education de Mexico, *S.A. de C.V.*
Pearson Education—Japan, *Tokyo*
Pearson Education Malaysia, Pte. Ltd

To Kitty and P.K.;

Judy, Molly, Kelly, Patrick, and Katie;

and

Mary, Molly, and Steven:

We couldn't have done this without you.

Thank you for your continued encouragement and devotion.

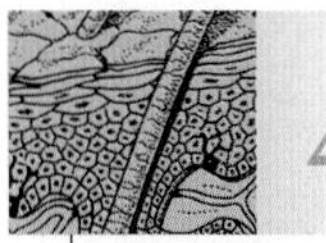

4

THE INTEGUMENTARY SYSTEM 87

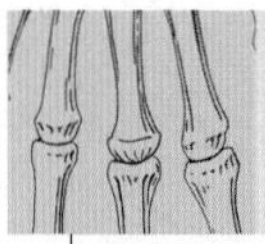

5

THE SKELETAL SYSTEM
Osseous Tissue and Skeletal Structure 112

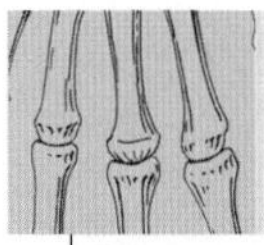

6

THE SKELETAL SYSTEM
Axial Division 133

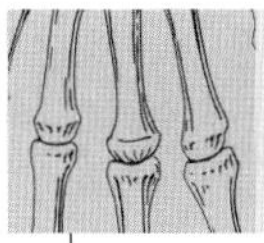

7

THE SKELETAL SYSTEM

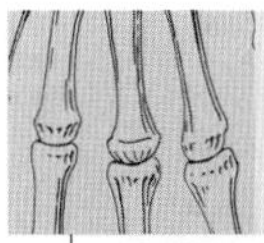

8

THE SKELETAL SYSTEM

9

THE MUSCULAR SYSTEM

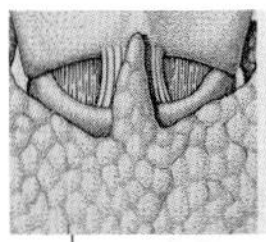

19

THE ENDOCRINE SYSTEM 506

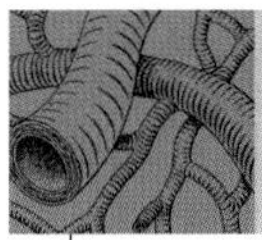

20

THE CARDIOVASCULAR SYSTEM Blood 532

PREFACE

Welcome to the new millenium! And, we might add, welcome to a new version of our Human Anatomy text. The Fourth Edition of Human Anatomy continues a tradition of excellence and innovation that began with the first edition. The book has long been known for its distinctive size and format, its rich pedagogical framework, the unsurpassed visual program, and its success at helping students understand and visualize anatomical structures.

The keys to the success of this text in classrooms around the world have always been our concern for accuracy and an approach that stresses the visual presentation of information. Anatomy is a visual science; it is much easier to visualize structures and grasp functional concepts through pictures and words. The visuals don't just provide occasional support for the narrative–they are part and parcel of the learning system that distinguishes this text from others. Over the years this text has received the Text and Academic Authors Association Award for Textbook Excellence and the Association of Medical Illustrators Award for Illustration.

What's New about the Fourth Edition

With this new edition of Human Anatomy, we welcome a new co-author, Dr. Robert Tallitsch of Augustana College. Dr. Tallitsch brings over years teaching experience and a specialty in. We believe Human Anatomy has been enhanced by his contribution and expertise.

With Bob's assistance, we've made a number of general organizational changes and improvements to the text and art programs. We have been helped in this process by the feedback from dozens of reviewers, as well as from adopting instructors, students, professionals in the field, and our own experiences in the classroom. As befits the overall theme of the text, we began the revision with a reassessment of the illustration program. This ultimately involved reorganizing, revising, adding, and replacing figures to improve and refine the visual presentation, cohesively and logically. This was simplifed greatly by the fact that we work with just a single pair of renowned medical illustrators, Bill Ober, M.D., and Claire Garrison, R.N., rather than with multiple artists and studios. Bill and Claire, with photos provided by the internationally respected biomedical photographer Ralph T. Hutchings, have played instrumental roles in the development of this text in each edition, the Fourth Edition being no exception.

Once the preliminary evaluation of the art program was completed, we began revising the narrative. Special attention was devoted to topics that students traditionally find to be the most difficult.

It is indeed a small world, and this book is used, in English or translated versions, almost everywhere. This has led us to confront one of the problems facing the anatomical world at large–the issue of standardizing anatomical terminology. In this edition, we have taken a large step toward incorporating the terminology endorsed by the International Federation on Anatomical Associations, as published in the 1998 *Terminologia Anatomica*. The terms used are internationally recognized, and the participating organizations, including the American Association of Anatomists, support the use of these terms in preference to the older, often idiosyncratic terms in common use. We have used our discretion as to whether the T.A. terms should be listed as primary or secondary, based on a survey of the current technical and medical literature. But we have made an effort to incorporate these terms wherever possible and practical.

■ A Emphasis on Digital Media

Human anatomy may evolve only slowly, but the technological options for teaching human anatomy change very quickly. The last three years have seen a revolution in the digital resources available to students and faculty. Computers are a part of daily life now, and digital media have become more important to teaching and learning as CD/DVD speeds and internet bandwidth have increased. In the field of human anatomy, the major changes have involved the increased ability to access sectional data and to create realistic three-dimensional computer models of anatomical structures. These digital graphics can now be brought into the classroom and used to enhance traditional lectures in ways that were either logistically or financially impossible just a few years ago.

In this revision we have begun the process of integrating these powerful media components into the text. One prominent example has been the creation of a series of 2-page media-based exercises called *Connections*. These exercises incorporate images derived from the Visible Human project of the National Library of Medicine, 3-D anatomical simulations from the student Interactive CD that is packaged with this text, and resources available at the text's Companion Website and at other websites around the world.

Connections evolved from the concerted efforts of anatomists, medical illustration, and software specialists. They are designed to help students visualize relationships in three-dimensions, and to reinforce key concepts more in an interactive way. In addition, the Human Anatomy Companion Web Site, offered via the World Wide Web, gives instructors the ability to create a custom syllabus, while it provides students with skill practice, lesson reinforcement, and topic exploration 24 hours a day.

Virtually every chapter in the text has been revised to a significant extent, and these revisions often involve changes to both the text and the artwork. In fact, only a small sampling of the changes made to the text and art will be cited here, and those in summary fashion only.

■ Changes and Additions to the Narrative

The introduction of alternative terminology has been discussed above and involved minor text and art changes too numerous to mention. The introduction of the Connections sections has also been described separately and will not be repeated here. In general terms, important topics have been updated and expanded as needed and those topics that students traditionally find most difficult have been augmented with greater explanation, description, and supporting illustrations. Additional subheadings have been added to aid students with their organization of anatomical hierarchy. Specifics include:

- Chapter 1 (Introduction): Material on body cavities recast to make it easier to follow the relationships among the ventral body cavities; added an introduction to the Visible Human project.
- Chapter 4 (Integumentary System): Added discussion of cutaneous anthrax; expanded treatment of injury and repair.
- Chapters 5-8 (Skeletal System): Reorganized material on endochondral ossification; reorganized descriptions of the temporal bones, the ossa coxae, the hip and knee joints; added material on aging effects on the skeletal system.
- Chapters 9-11 (Muscular System): Revised discussion of thin filament structure; reorganized discussions of contraction mechanism; improved presentations of muscles of the face, neck, thigh, leg, and

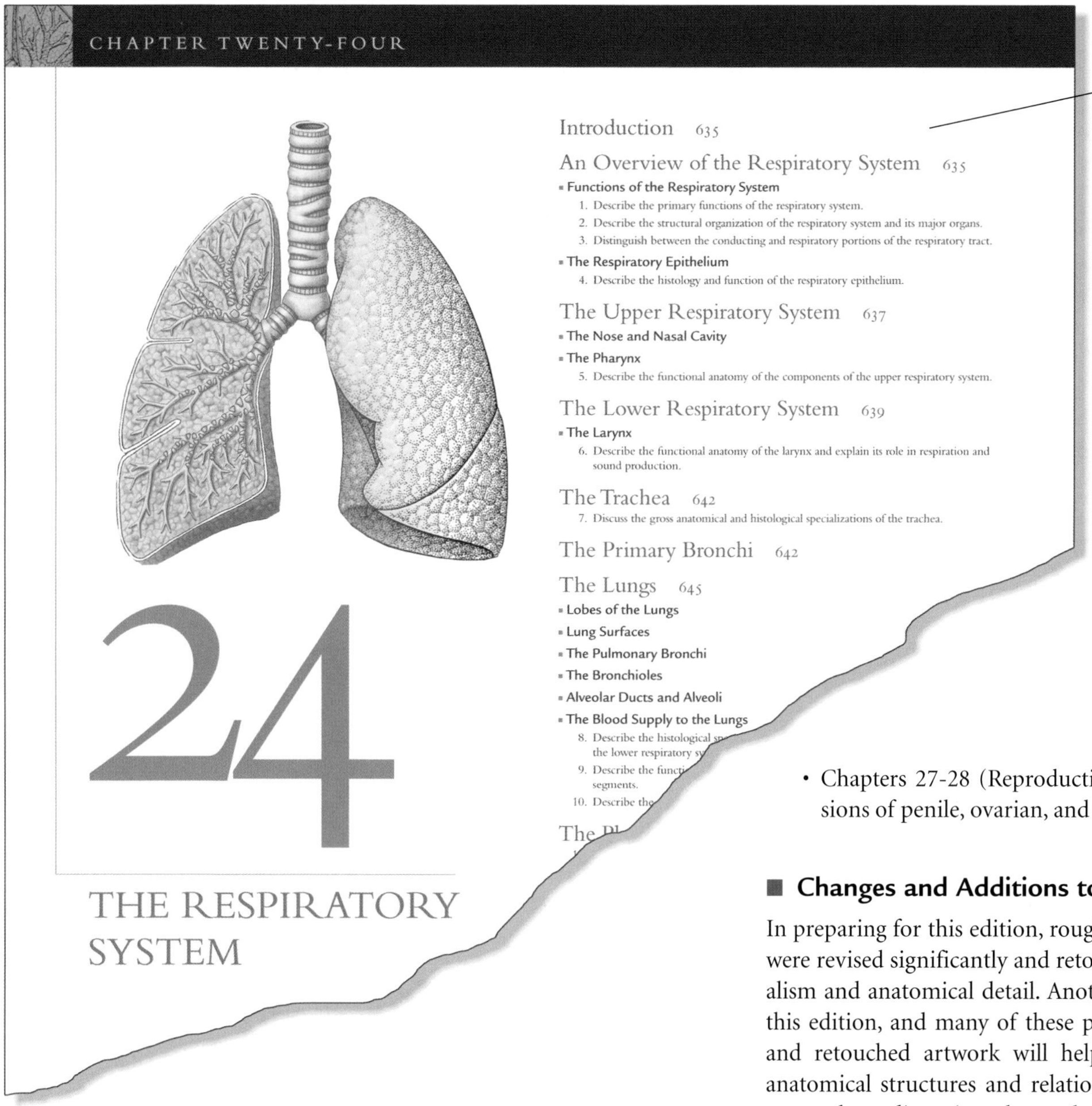
CHAPTER TWENTY-FOUR

Introduction 635

An Overview of the Respiratory System 635
- Functions of the Respiratory System
  1. Describe the primary functions of the respiratory system.
  2. Describe the structural organization of the respiratory system and its major organs.
  3. Distinguish between the conducting and respiratory portions of the respiratory tract.
- The Respiratory Epithelium
  4. Describe the histology and function of the respiratory epithelium.

The Upper Respiratory System 637
- The Nose and Nasal Cavity
- The Pharynx
  5. Describe the functional anatomy of the components of the upper respiratory system.

The Lower Respiratory System 639
- The Larynx
  6. Describe the functional anatomy of the larynx and explain its role in respiration and sound production.

The Trachea 642
  7. Discuss the gross anatomical and histological specializations of the trachea.

The Primary Bronchi 642

The Lungs 645
- Lobes of the Lungs
- Lung Surfaces
- The Pulmonary Bronchi
- The Bronchioles
- Alveolar Ducts and Alveoli
- The Blood Supply to the Lungs
  8. Describe the histological ... the lower respiratory ...
  9. Describe the ... segments.
  10. Describe the ...

24

THE RESPIRATORY SYSTEM

Chapter Outline and Objectives
Before you begin using a chapter, it is important to know where you are going and what is expected of you. Therefore, every chapter opens with an outline of that chapter's contents. Notice that each section of the outline corresponds to a heading in the text. Furthermore, the learning objectives are integrated into this outline so that you can preview the content and objectives for that chapter quickly. The addition of page references makes the feature useful as a guide for review.

foot; revised Tables throughout to provide additional information on origins, actions, and innervations..

- Chapters 13-18 (nervous system): Improved coverage of nerve plexuses and peripheral nerve distribution; added coverage on the blood supply to the brain; revised coverage of the basal nuclei; added/modified coverage of the corticospinal, medial, and lateral pathways; many small enhancements in cranial nerve coverage; revised tables throughout; revised treatment of ear anatomy.
- Chapter 19 (Endocrine System): Revised coverage of pituitary gland, pancreas.
- Chapters 20-22 (Cardiovascular System): Revised and enhanced coverage of coronary artery distribution, cardiac cycle, vascular walls, the branches of the abdominal aorta, arteries of the pelvis and lower limb and veins of the brain and abdominopelvic region.
- Chapter 24 (Respiratory System): revised discussions of lung anatomy
- Chapter 25 (Digestive System): Revised treatment of dental anatomy; added discussion of vascular supply to the large intestine.
- Chapter 26 (Urinary System): Reorganized material on nephron structure and circulation; revised material on segments of the nephron.
- Chapters 27-28 (Reproduction and Development): Revised discussions of penile, ovarian, and uterine anatomy.

■ Changes and Additions to the Illustrations

In preparing for this edition, roughly 60% of the illustrations in the book were revised significantly and retouched for greater three-dimensional realism and anatomical detail. Another 20% of the illustrations are new to this edition, and many of these pieces are full color paintings. This new and retouched artwork will help increase students' understanding of anatomical structures and relationships. You will find new artwork and new cadaver dissection photos throughout the text, and many new orientation icons have been created to help students orient on the photographs. In addition, we have created a Companion Atlas that contains radiological scans and anatomical photographs that supplement those in the textbook.

■ Enhanced Pedagogy

The teaching and learning framework of Human Anatomy has been designed to help students organize, interpret, and apply anatomical information. Features that characterized previous editions have been retained, including the end of section concept checks, figure-locator dots, icons identifying information in the Clinical Appendix (Appendix I) and the three-level end of chapter review system. Specific new pedagogy for the Fourth Edition includes:

- *Connections* sections follow chapters dealing with major topics and systems. These interactive, media-based activities give students the chance to explore relevant anatomical or clinical material on CD or the Internet. The new pedagogical features–new and old–can be surveyed easily by consulting the Student Walkthrough.
- A new icon appears in the *Connections* section of the text whenever there are associated animations or exercises on the Interactive CD that comes with the text.

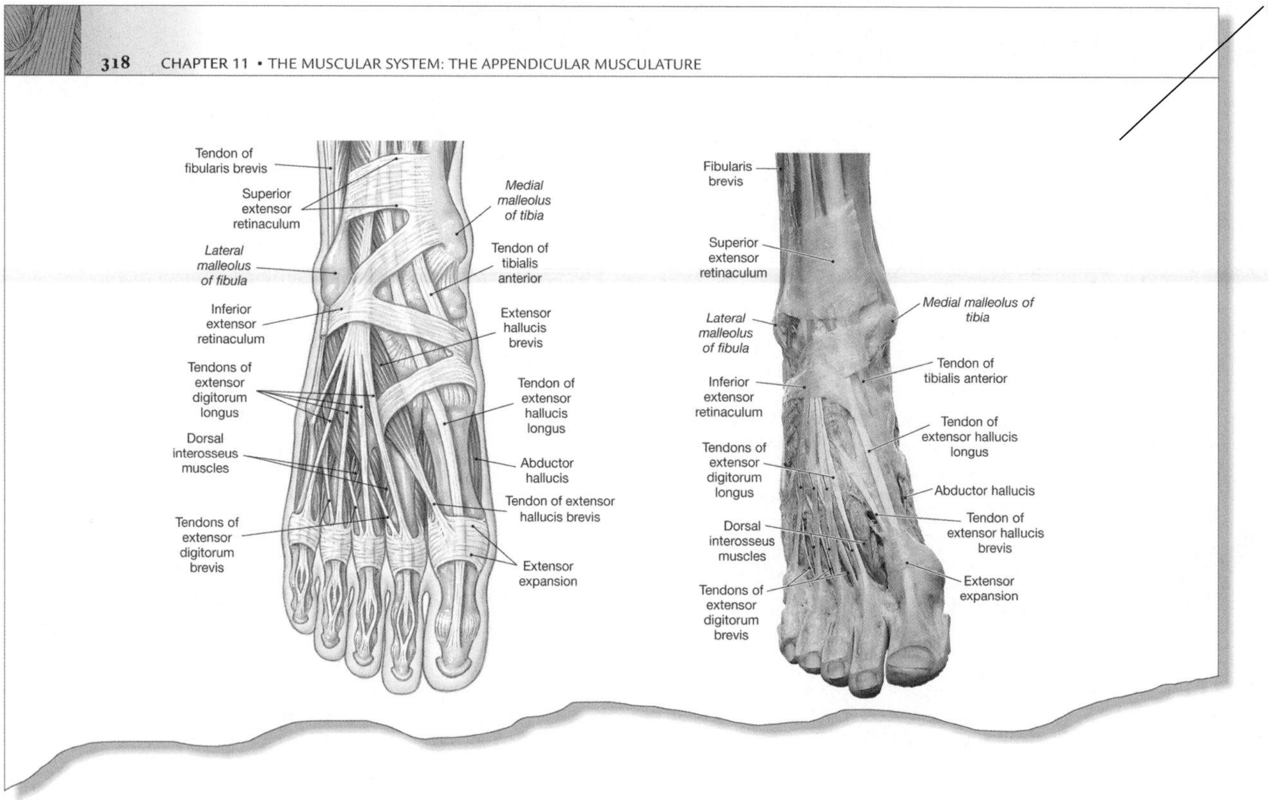

Illustrations Combined with Cadaver Dissection Photos An effectively rendered piece of art can communicate a lot, but we have tried to take your view of anatomy a step further by providing high quality cadaver photos, allowing you to compare a medical illustrator's interpretation with a photo of the actual structure. In addition to cadaver photos, some illustrations also include X-rays or MRI scans. Labels in plain type identify structures that are detailed in the accompanying text. Italicized labels identify anatomical landmarks that are not part of the system under discussion.

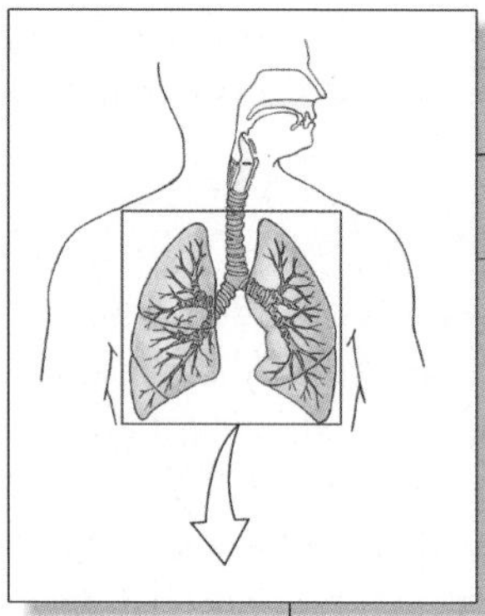

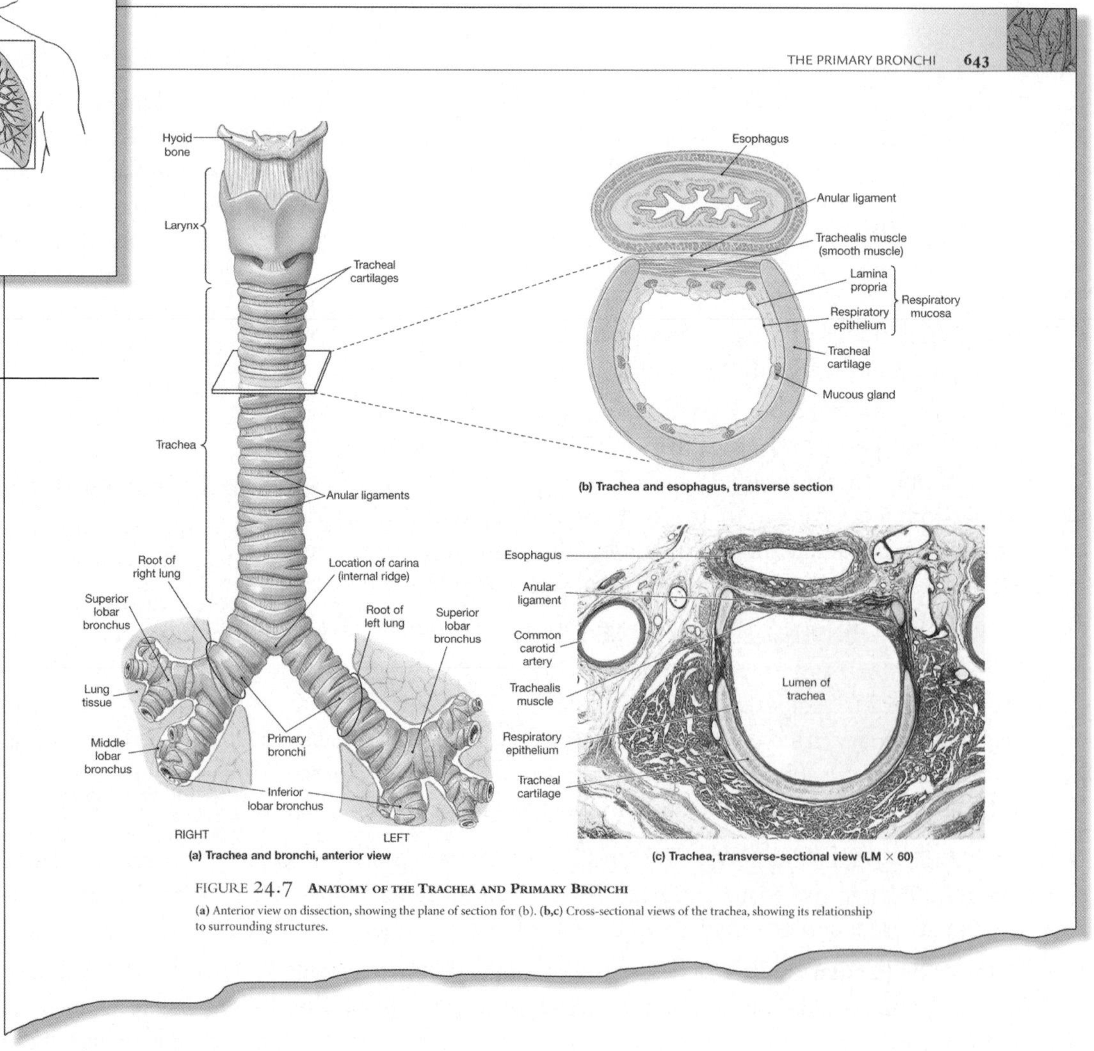

Macro-to-Micro Illustrations One of the challenges of learning anatomy is learning to "see"anatomical structures. Throughout this book you will find illustrations that provide an orientation icon indicating where in the human body a particular organ or structure is located. The orientation icon is then followed by (1) a large, clear painting of that structure, (2) a sectional view, and (3) a photomicrograph. In this fashion, the illustration provides you with a "macro-to-micro" view that should aid greatly in your understanding of the structure.

Major Text Headings Include Associated Figure Numbers
Because the text should dovetail with the illustrations, we have placed all associated figure references under the corresponding text heading. This is an organizational aid that allows you to group key illustrations with the relevant section of narrative. We think this feature will be especially important as you both preview and review the material in a chapter.

Pronunciation Guides
Key terms encountered for the first time include (parenthetically) pronunciation guides to help you with the correct pronunciation of the terms. The accented syllable is shown in all capitals. For vowels, an overbar indicates a long-vowel sound. When appropiate, the foreign word roots and combining forms are indicated. This should help you build your anatomical vocabulary and enable you to decipher new terms without assistance.

182 CHAPTER 7 • THE SKELETAL SYSTEM: APPENDICULAR DIVISION

muscles or openings for nerves and blood vessels that supply the bones or other organs of the body.

There are direct anatomical connections between the skeletal and muscular systems. As noted in Chapter 5, the connective tissue of the deep fascia that surrounds a skeletal muscle is continuous with that of its tendon, which continues into the periosteum and becomes part of the bone matrix at its attachment site. *p. 117* Muscles and bones are also physiologically linked, because muscle contractions can occur only when the extracellular concentration of calcium remains within relatively narrow limits. The skeleton contains most of the body's calcium, and these reserves are vital to calcium homeostasis.

The Pectoral Girdle And Upper Limb

[FIGURE 7.2]

Each arm articulates with the trunk at the **pectoral girdle**. The pectoral girdle consists of the S-shaped *clavicle (collarbone)* and a broad, flat *scapula (shoulder blade),* as seen in Figure 7.2●. The clavicle articulates with the manubrium of the sternum, and this is the *only* direct connection between the pectoral girdle and the axial skeleton. Skeletal muscles support and position the scapula, which has no direct bony or ligamentous connections to the thoracic cage. Each upper limb consists of the *brachium* (arm), the *antebrachium* (forearm), the wrist, and the hand. The skeleton of the upper limb consists of the *humerus* of the arm, the *ulna* and *radius* of the forearm, the *carpal bones* of the wrists, and the *metacarpal bones* and *phalanges* of the hand.

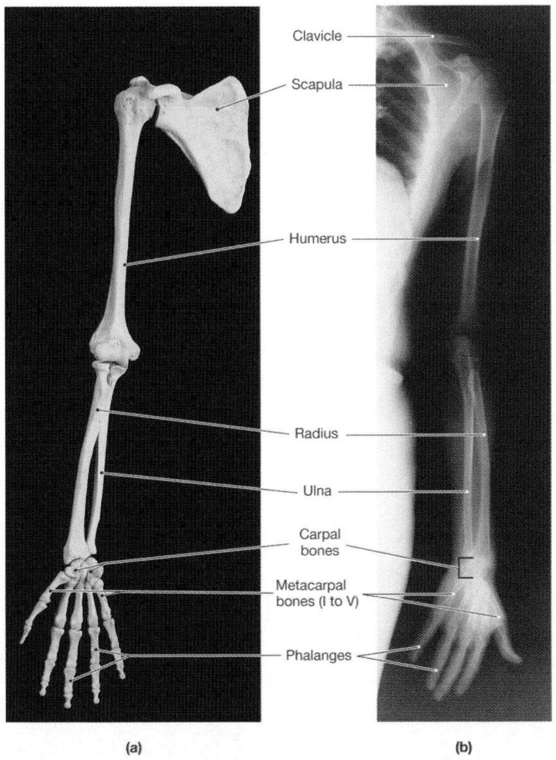

FIGURE 7.2 **The Pectoral Girdle and Upper Limb**
Each upper limb articulates with the axial skeleton at the trunk through the pectoral girdle. **(a)** Right upper limb, anterior view. **(b)** X-ray of right pectoral girdle and upper limb, posterior view.

The Pectoral Girdle

Movements of the clavicle and scapula position the shoulder joint and provide a base for arm movement. Once the shoulder joint is in position, muscles that originate on the pectoral girdle help to move the upper limb. The surfaces of the scapula and clavicle are therefore extremely important as sites for muscle attachment. Where major muscles attach, they leave their marks, creating bony ridges and flanges. Other bone markings, such as grooves or foramina, indicate the position of nerves or blood vessels that control the muscles and nourish the muscles and bones.

THE CLAVICLE [FIGURES 7.3/7.4]

The **clavicle** (KLAV-i-kl) (Figure 7.3●) connects the pectoral girdle and the axial skeleton. Each clavicle originates at the craniolateral border of the manubrium of the sternum, lateral to the jugular notch (see Figures 6.27a, p. 173 and Figure 7.4●). From the roughly pyramidal **sternal end**, the clavicle curves in an S-shape laterally and dorsally until it articulates with the acromion of the scapula. The **acromial end** is broader and flatter than the sternal end.

The smooth superior surface of the clavicle lies just deep to the skin; the rough inferior surface of the acromial end is marked by prominent lines and tubercles that indicate the attachment sites for muscles and ligaments. The **conoid tubercle** is on the inferior surface at the acromial end, and the **costal tuberosity** is at the sternal end. These are attachment sites for ligaments of the shoulder.

You can explore the interaction between scapulae and clavicles. With your fingers in the *jugular notch*, locate the clavicle to either side. *p. 175* When you move your shoulders you can feel the clavicles change their positions. Because the clavicles are so close to the skin, you can trace one laterally until it articulates with the scapula. Shoulder movements are limited by the position of the clavicle at the *sternoclavicular joint*, as shown in Figure 7.4●. The structure of this joint will be described in Chapter 8. Fractures of the medial portion of the clavicle are common because a fall on the palm of the hand of an outstretched arm produces compressive forces that are conducted to the clavicle and its articulation with the manubrium. Fortunately, these fractures usually heal rapidly without a cast.

THE SCAPULA [FIGURES 7.4/7.5]

The **body** of the **scapula** (SCAP-ū-lah) forms a broad triangle with many surface markings reflecting the attachment of muscles, tendons, and ligaments (Figure 7.5a,d●). The three sides of the scapular triangle are the **superior border**; the **medial**, or *vertebral*, **border**; and the **lateral**, or *axillary*, **border** (*axilla*, armpit). Muscles that position the scapula attach along these edges. The corners of the scapular triangle are called the **superior angle**, the **inferior angle**, and the **lateral angle**. The lateral angle, or *head* of the scapula, forms a broad process that supports the cup-shaped **glenoid cavity**, or *glenoid fossa*. At the glenoid cavity, the scapula articulates with the proximal end of the *humerus*, the bone of the arm. This articulation is the **glenohumeral joint**, or **shoulder joint**. The lateral angle is separated from the body of the scapula by the rounded **neck**. The relatively smooth, concave **subscapular fossa** forms most of the anterior surface of the scapula.

Figure Reference Locators
All textbooks have figure references that connect the running narrative with important illustrations. This text goes a step further by including visible but unobtrusive red dots next to every figure reference. These figure reference locators are designed to "mark your spot" in the narrative, facilitating an easier, more seamless return to the narrative.

Concept Links
Thoroughly understanding human anatomy requires a combination of rote memorization, three-dimensional visualizaton, and the integration of structural and functional concepts. Concept links provide a quick visual indicator and a page reference to signal you that it may be helpful to review a related concept from an earlier chapter.

Tables

Tables are used throughout the text to summarize concepts or organize important information in a format suitable for review.

THE CRANIAL NERVES **419**

TABLE 15.12 THE CRANIAL NERVES

Cranial Nerve (#)	Sensory Ganglion	Branch	Primary Function	Foramen	Innervation
Olfactory (I)			Special sensory	Cribriform plate	Olfactory epithelium
Optic (II)			Special sensory	Optic canal	Retina of eye
Oculomotor (III)			Motor	Superior orbital fissure	Inferior, medial, superior rectus, inferior oblique, and levator palpebrae muscles; intrinsic muscles of eye
Trochlear (IV)			Motor	Superior orbital fissure	Superior oblique muscle
Trigeminal (V)	Semilunar		Mixed		Areas associated with the jaws
		Ophthalmic	Sensory	Superior orbital fissure	Orbital structures, nasal cavity, skin of forehead, upper eyelid, eyebrows, nose (part)
		Maxillary	Sensory	Foramen rotundum	Lower eyelid; upper lip, gums, and teeth; cheek, nose (part), palate and pharynx (part)
		Mandibular	Mixed	Foramen ovale	*Sensory* to lower gums, teeth, lips; palate (part) and tongue (part); *motor* to muscles of mastication
Abducens (VI)			Motor	Superior orbital fissure	Lateral rectus muscle
Facial (VII)	Geniculate		Mixed	Internal acoustic meatus to facial canal; exits at stylomastoid foramen	*Sensory* to taste receptors on anterior 2/3 of tongue; *motor* to muscles of facial expression, lacrimal gland, submandibular salivary gland, sublingual salivary glands
Vestibulocochlear (Acoustic) (VIII)		Cochlear	Special sensory	Internal acoustic meatus	Cochlea (receptors for hearing)
		Vestibular	Special sensory	As above	Vestibule (receptors for motion and balance)
Glossopharyngeal (IX)	Superior (jugular) and inferior (petrosal)		Mixed	Jugular foramen	*Sensory* from posterior 1/3 of tongue; pharynx and palate (part); carotid body (monitors blood pressure, pH, and levels of respiratory gases); *motor* to pharyngeal muscles, parotid salivary gland
Vagus (X)	Superior (jugular) and inferior (nodose)		Mixed	Jugular foramen	*Sensory* from pharynx; auricle and external acoustic meatus; diaphragm; visceral organs in thoracic and abdominopelvic cavities; *motor* to palatal and pharyngeal muscles, and visceral organs in thoracic and abdominopelvic cavities
Accessory (XI)		Internal branch	Motor	Jugular foramen	Skeletal muscles of palate, pharynx and larynx (with branches of the vagus nerve)
		External branch	Motor	Jugular foramen	Sternocleidomastoid and trapezius muscles
Hypoglossal (XII)			Motor	Hypoglossal canal	Tongue musculature

Concept Check

This learning aid is designed to stop you for a moment and assess your understanding of the basic concepts just addressed in the previous few pages. Because concepts build on one another, you need to master important material sequentially. Although these concept checks can easily be skipped, we encourage you to pause and assess your understanding.

Answers to Concept Check Questions are found at the end of every chapter.

✓ CONCEPT CHECK

- Trace the path of a drop of blood from the renal artery to a glomerulus and back to a renal vein.
- Trace the path taken by filtrate in traveling from a glomerulus to a minor calyx.
- Explain why filtration alone is not sufficient for urine production.
- What is the function of the loop of Henle?

✓ ANSWERS TO CONCEPT CHECK QUESTIONS

p. 276 **1.** The muscles of facial expression originate on the surface of the skull. **2.** The muscles of mastication move the mandible at the temporomandibular joint during chewing. **3.** The contraction of the extra-ocular muscles causes the eye to look up, look down, rotate laterally, rotate medially, roll and look up to the side, or roll and look down to the side. **4.** The pharyngeal muscles are important in the initiation of swallowing.
p. 283 **1.** Damage to the external intercostal muscles would interfere with the process of breathing. **2.** A blow to the rectus abdominis muscle would cause the muscle to contract forcefully, resulting in flexion of the torso. In other words, you would "double up." **3.** The muscles of the pelvic floor have the following functions: (1) support the organs of the pelvic cavity; (2) flex the joints of the sacrum and coccyx; and (3) control the movement of materials through the urethra and anus. **4.** The diaphragm is a major muscle of respiration.

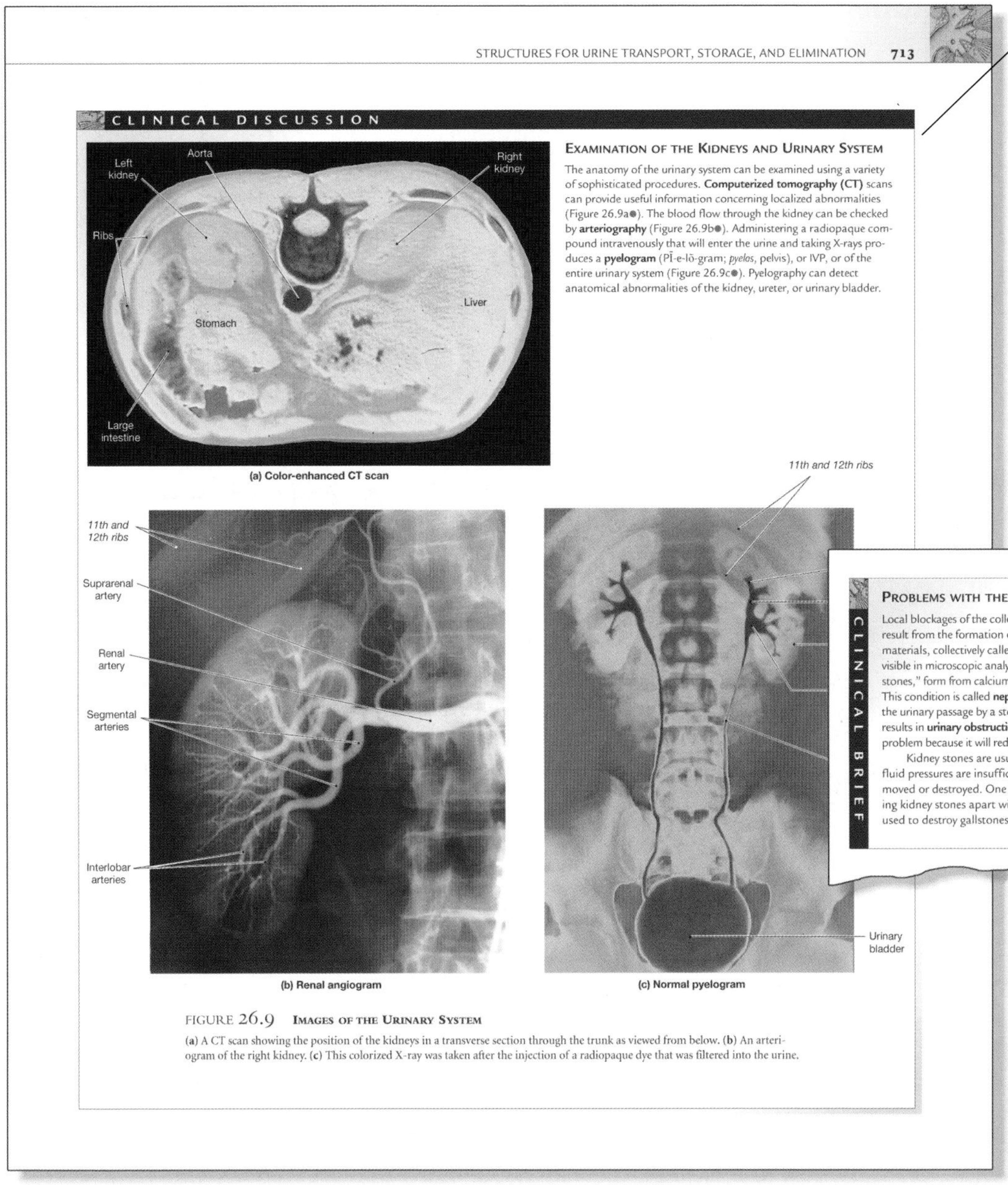
STRUCTURES FOR URINE TRANSPORT, STORAGE, AND ELIMINATION 713

CLINICAL DISCUSSION

EXAMINATION OF THE KIDNEYS AND URINARY SYSTEM

The anatomy of the urinary system can be examined using a variety of sophisticated procedures. **Computerized tomography (CT)** scans can provide useful information concerning localized abnormalities (Figure 26.9a●). The blood flow through the kidney can be checked by **arteriography** (Figure 26.9b●). Administering a radiopaque compound intravenously that will enter the urine and taking X-rays produces a **pyelogram** (PĪ-e-lō-gram; *pyelos*, pelvis), or IVP, or of the entire urinary system (Figure 26.9c●). Pyelography can detect anatomical abnormalities of the kidney, ureter, or urinary bladder.

FIGURE 26.9 IMAGES OF THE URINARY SYSTEM

(a) A CT scan showing the position of the kidneys in a transverse section through the trunk as viewed from below. (b) An arteriogram of the right kidney. (c) This colorized X-ray was taken after the injection of a radiopaque dye that was filtered into the urine.

Clinical Discussions

Large, illustrated clinical discussions are framed as a unit to distinguish them from the running narrative of the main text.

Clinical Briefs

Clinical Briefs are brief overviews designed to enhance your understanding of normal anatomy by focusing on disease, dysfunction, or injury. The topics are relevant and interesting, especially to those preparing for careers in the allied-health sciences.

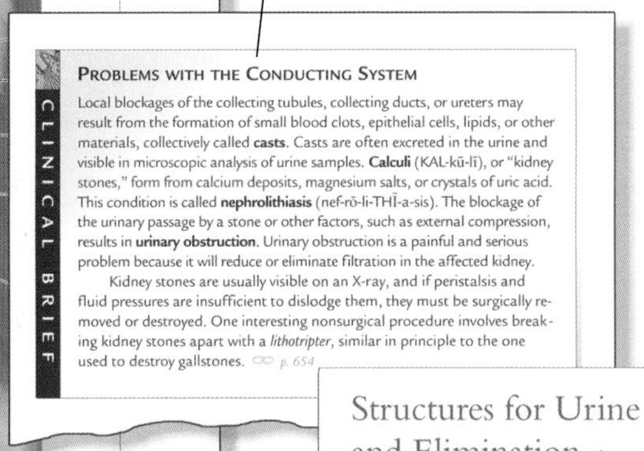
CLINICAL BRIEF

PROBLEMS WITH THE CONDUCTING SYSTEM

Local blockages of the collecting tubules, collecting ducts, or ureters may result from the formation of small blood clots, epithelial cells, lipids, or other materials, collectively called **casts**. Casts are often excreted in the urine and visible in microscopic analysis of urine samples. **Calculi** (KAL-kū-lī), or "kidney stones," form from calcium deposits, magnesium salts, or crystals of uric acid. This condition is called **nephrolithiasis** (nef-rō-li-THĪ-a-sis). The blockage of the urinary passage by a stone or other factors, such as external compression, results in **urinary obstruction**. Urinary obstruction is a painful and serious problem because it will reduce or eliminate filtration in the affected kidney.

Kidney stones are usually visible on an X-ray, and if peristalsis and fluid pressures are insufficient to dislodge them, they must be surgically removed or destroyed. One interesting nonsurgical procedure involves breaking kidney stones apart with a *lithotripter*, similar in principle to the one used to destroy gallstones. p. 654

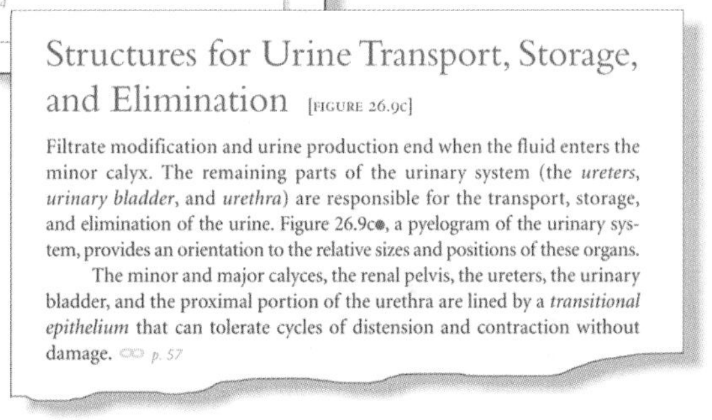
Structures for Urine Transport, Storage, and Elimination [FIGURE 26.9C]

Filtrate modification and urine production end when the fluid enters the minor calyx. The remaining parts of the urinary system (the *ureters*, *urinary bladder*, and *urethra*) are responsible for the transport, storage, and elimination of the urine. Figure 26.9c●, a pyelogram of the urinary system, provides an orientation to the relative sizes and positions of these organs.

The minor and major calyces, the renal pelvis, the ureters, the urinary bladder, and the proximal portion of the urethra are lined by a *transitional epithelium* that can tolerate cycles of distension and contraction without damage. p. 57

Related Clinical Terms

The end of each chapter contains a list of related clinical terms and their definitions. These terms appeared in the Clinical Brief or Clinical Discussions in this chapter, or they are found in the related section of the Clinical Issues appendix (indicated by the caduceus icon). Page references help you find the terms in context for further review or reference.

Clinical Appendix Reference

The Clinical Issues Appendix contains a substantial amount of clinical material, far more than you will find in other undergraduate Anatomy texts. This material will be especially interesting and useful for students planning careers in the medical or allied-health sciences. Within the running text and here in the end-of-chapter material, we have used the caduceus icon along with the title of the clinical topic to signal that the appendix contains additional relevant clinical information.

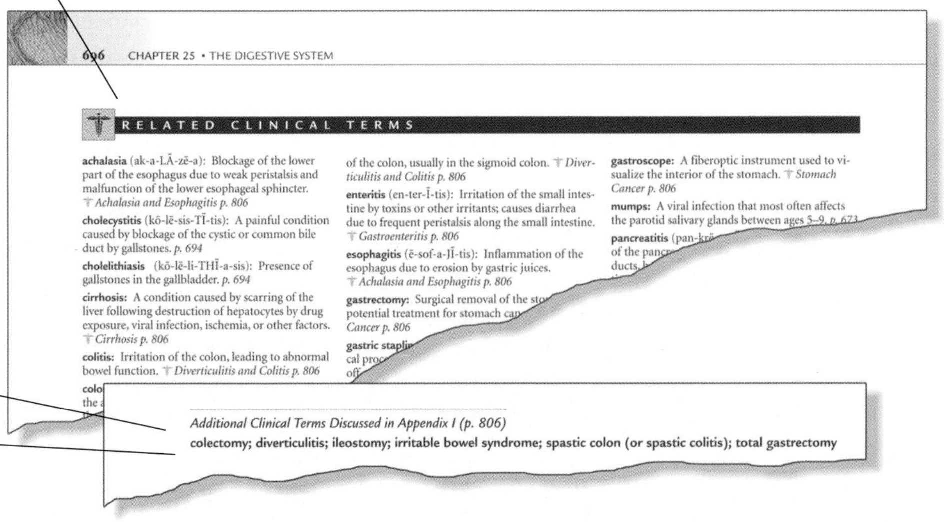
696 CHAPTER 25 • THE DIGESTIVE SYSTEM

RELATED CLINICAL TERMS

achalasia (ak-a-LĀ-zē-a): Blockage of the lower part of the esophagus due to weak peristalsis and malfunction of the lower esophageal sphincter. *Achalasia and Esophagitis p. 806*

cholecystitis (kō-lē-sis-TĪ-tis): A painful condition caused by blockage of the cystic or common bile duct by gallstones. *p. 694*

cholelithiasis (kō-lē-li-THĪ-a-sis): Presence of gallstones in the gallbladder. *p. 694*

cirrhosis: A condition caused by scarring of the liver following destruction of hepatocytes by drug exposure, viral infection, ischemia, or other factors. *Cirrhosis p. 806*

colitis: Irritation of the colon, leading to abnormal bowel function. *Diverticulitis and Colitis p. 806*

...of the colon, usually in the sigmoid colon. *Diverticulitis and Colitis p. 806*

enteritis (en-ter-Ī-tis): Irritation of the small intestine by toxins or other irritants; causes diarrhea due to frequent peristalsis along the small intestine. *Gastroenteritis p. 806*

esophagitis (ē-sof-a-JĪ-tis): Inflammation of the esophagus due to erosion by gastric juices. *Achalasia and Esophagitis p. 806*

gastrectomy: Surgical removal of the sto... potential treatment for stomach ca... *Cancer p. 806*

gastroscope: A fiberoptic instrument used to visualize the interior of the stomach. *Stomach Cancer p. 806*

mumps: A viral infection that most often affects the parotid salivary glands between ages 5–9. *p. 673*

Additional Clinical Terms Discussed in Appendix I (p. 806)

colectomy; diverticulitis; ileostomy; irritable bowel syndrome; spastic colon (or spastic colitis); total gastrectomy

Additional Clinical Terms Discussed in the Appendix

This listing gives you an overview of additional clinical terms that are discussed and defined in the Clinical Issues appendix.

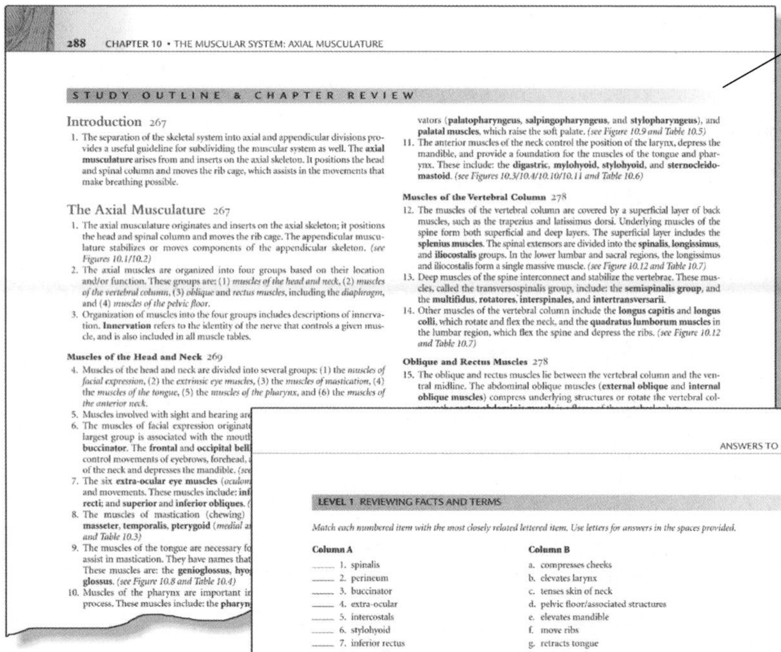

Study Outline and Chapter Review

The chapter summary and review section of each chapter begins with a detailed study outline. This outline includes page and figure references to help you quickly find the material you want to review in more depth.

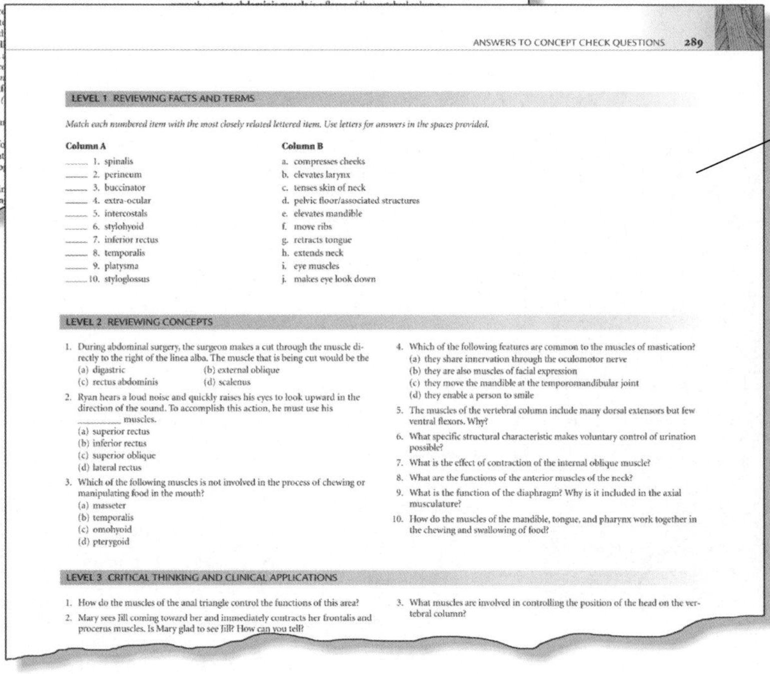

End-of-Chapter Three-Level Review System

Each chapter ends with a three-level questioning system. **Level One, Reviewing Facts and Terms**, tests your memory of specific details in the chapter. **Level Two, Reviewing Concepts**, requires slightly more advanced thinking skills since you are asked to synthesize or apply concepts. **Level Three, Critical Thinking and Clinical Applications**, asks you to address real-world or clinical scenarios. The three-level review system has been designed to help you grow intellectually from a basic level in which you master terms and concepts to more advanced levels in which critical-thinking skills are promoted. Students who desire a greater number and variety of questions should consider visiting our Web Site or obtaining a copy of the optional Study Guide.

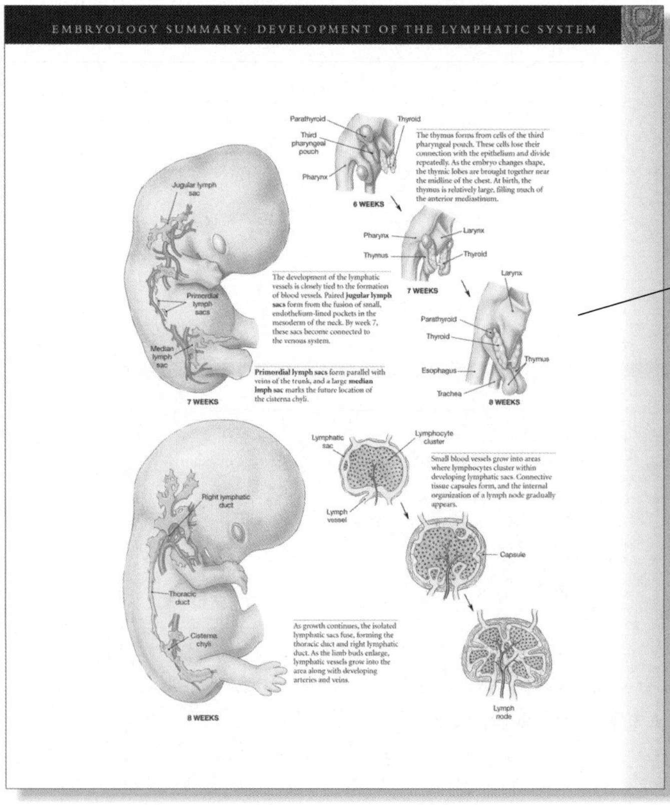

Embryology Summaries

Located throughout the text, these summaries highlight the developmental stages of significant organs, structures, and systems. They are boxed and are designed to be relevant but isolated from the running text. Your instructor may cover this material as part of each chapter or wait until Development is addressed in Chapter 28. In either case, these represent an important resource that can help you understand normal anatomy and explain the origins of both normal variation and congenital abnormalities.

New to this edition are the **Connection Spreads** that appear after every system. These Connections are meant to help students not only assimilate the material dealing with individual systems but also integrate information across system boundaries. Each includes two challenges, two critical thinking exercises, and a thought-provoking topic for discussion or further research.

CONNECTIONS

REVIEW IT

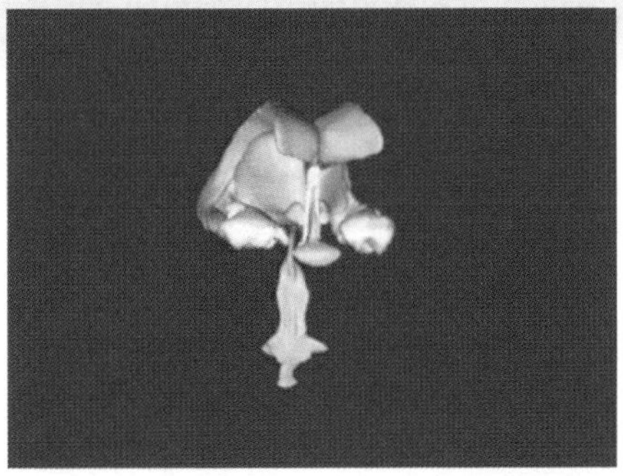

Challenge 1

Learning the anatomy of the brain is a challenging task. There are many layers to this organ and structures are found embedded within others. The Interactive CD has an animation that takes you through the portions of the brain in successive layers. To view this animation, go to Animations/Nervous system/Brain and brainstem. The image to the left has been taken from that animation. Determine first where in the brain these structures are found. What figure in the text is comparable to this image? Identify the structures that are visible here, keeping in mind that this image was re-constructed from transverse sections of an actual brain. Structures are not idealized in this image, but have the shapes and proportions characteristic of one particular individual. Some structures seen on the comparable text figure may not be visible here, or may appear differently.

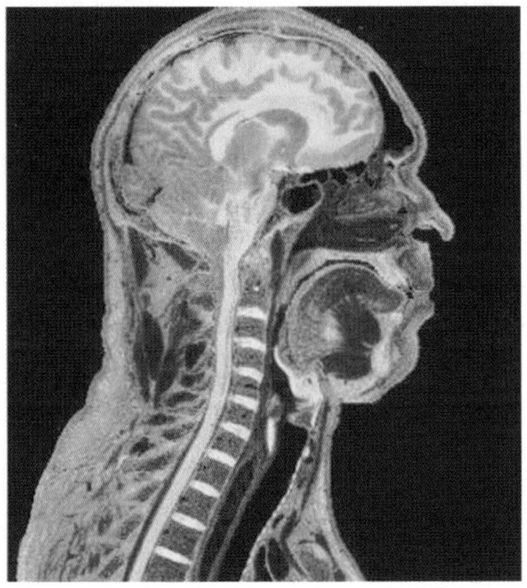

Challenge 2

Although it is emphasized in the text and portrayed properly in the figures, students are often surprised by the size of the human spinal cord when they first encounter one during dissection or surgical procedure. The sagittal image to the left is a reconstruction of the cord using the data set of transverse images from the Visible Human project. Note the relative size of the spinal cord. Find and label the meninges, the spinal nerves, the lumbar vertebrae, and discs. Do you notice anything else of interest in this image? What happens to the spinal cord as you approach the pelvis?

Images provided by the

504

The **Challenges** direct the student to compare and contrast different views of the body, identifying structures both on the page and on the accompanying Interactive CD. Designed to improve visualization ability, this activity helps students translate two-dimensional images into three-dimensional anatomical relationships. Images are taken from the Interactive CD and from an image bank created from the Visible Human data set. The student is asked to locate specific structures and compare these images to figures in the text. Such exercises provide valuable practice in mentally reconstructing the physical reality of the human body.

APPLY IT

Below are two separate exercises that provide more information on a topic presented in the preceding chapters. Each one is designed to take approximately 10 minutes, and will help you gain a better appreciation for the material presented in Chapters 13–18.

Critical Linking 1

There are many ways to view neuroanatomy, including sophisticated neuro-imaging techniques such as CTs and MRIs. These radiological procedures give very different pictures of the brain than any you have seen thus far. Go to the Companion Website and click on the key word "brain anatomy" to move to a site prepared by Dr. Keith Johnson of the Harvard Medical School. Here you will find a series of images of the brain, each with an explanation of the structures visible. To familiarize yourself with what you are viewing, first read the short introduction to neuro-imaging. Then look at the Top 100 Brain Structures. Can you identify all these? Compare the images you see there with those in your text. Which gives you a clearer understanding of the structures of the brain? Why?

Critical Linking 2

The peripheral nervous system is comprised of many nerves of clinical importance that anatomists should be able to identify and trace. For practice identifying these nerves, visit the Companion Website and choose the key word "nerves." This will take you to the Lumen "learn'em" Website prepared by Loyola University Medical School (Chicago). There you will find two sites of interest; "cutaneous innervation" and "nerves." Choose either of these to continue your studies. "Nerves" will lead you through practice identification of the cervical, brachial, lumbar, and sacral plexuses as well as the cranial nerves. "Cutaneous innervation" will provide practice in naming the sensory nerves.

FURTHER STUDIES

A study of the nervous system usually leads to interesting questions that lie just outside the realm of neuroanatomy. Many times these investigations assume a thorough understanding of the anatomy of the nervous system prior to fully understanding the issues. One such question that is currently in
news is in the area of neural development. A recently published text "*From Neurons to Neighbo*
of Early Childhood Development" is a synopsis of 2 1/2 years of research
strong scientific emphasis. This work was conduct
including neuroscience, psycholo
first scrutinized by
rigo

The **Critical Linking** feature of the Connections introduces students to authentic problem-solving situations, a technique known to increase both critical-thinking skills and academic motivation. A short case related to the preceding system (or integrated systems) is presented, followed by a ten to fifteen minute web-based activity. The activity involves both the material in the text and additional information provided by a specified Internet site (accessible through the Media Lab portion of the Companion Website). Completion of these activities requires students to integrate text-derived and web-derived knowledge.

The final portion of each Connection section presents an area for **Further Studies**, loosely related to the material presented in the preceding chapters. Included are such intriguing topics as the human body in art and poetry, the history of anatomy, science and ethics, stress and our lives, famine, and gender equity in societies. A short discussion introduces each subject, while web addresses are provided to encourage students in further independent investigation.

A Comprehensive Supplements Package

Like the textbook itself, each component of the ancillary package for Human Anatomy has been carefully designed to be part of an integrated teaching and learning component, designed to meet the needs of the instructor and student.

■ For Instructors

- ***Human Anatomy, Fourth Edition* Acetates** Contains over 450 full-color illustrations and photos—160 of which are new to the Fourth Edition—boxed and organized chapter-by-chapter to the text. 2D and 3D images include anatomical structures, cadaver dissections, specimen photos, MRI and CT scans, histology images, embryology summaries, and more.
- **Instructor's Presentation CD-ROM for *Human Anatomy, Fourth Edition*** Contains all content contained on the student CD-ROM PLUS an electronic editor to customize lecture and laboratory presentations using artwork—including text illustrations, dissection and specimen photos, MRI and CT scans, X-rays, and histology slides—from Human Anatomy, Fourth Edition. New features include electronic quizzes and PowerPoint presentations and fully interactive 3D animations and audio tutorials for use in class and lab.
- **NEW! Primal Pictures™ 7 CD-ROM Set** This award winning *7 CD-ROM* set from *Primal Pictures™* is available to instructors free upon adoption of Human Anatomy fourth edition. This Anatomical and Clinical Pathology software features 3-D computer graphic models of human anatomy derived from scan data. The model can rotated and layers stripped way. Movies illustrate muscle actions and slides highlight key pathology of anatomical structures. Anthroscopic materials provide fantastic views and stunning biomechanical animations demonstrate and describe the limits of motion and stresses. Radiology sections are also included. All materials can be imported into PowerPoint presentations.
- **Instructor's Guide to *Human Anatomy, Fourth Edition*** This instructor's guide, By Linda Banta and Kathleen Flickinger, provides the pedagogical tools and insights gained from the authors' 30-plus years of anatomy instruction. Each chapter contains ready-made Lecture Outlines and numerous teaching Strategies to aid, plan, and enhance instruction. Also provides Chapter Previews with emphasis on key concepts, listings of Chapter Goals and Chapter Objectives, and a General Remarks section with lesson specific insights.
- **Computerized Test Bank and Custom Test Manager** A fully customizable test back with thousands of questions—from multiple choice, labeling, true/false, and more—organized chapter by chapter. New labeling exercises using art figure from the text broaden your assessment capabilities.
- **Instructor's Companion Website for *Human Anatomy, Fourth Edition*** Contains all the resources of the Students' Companion Website PLUS an Instructor's Resource Center with links to related websites, student labeling exercises, and test item file. Website also contains a Syllabus Manager—an online syllabus creation and management tool that provides instructors with an easy, step-by-step editing function to help you adapt Human Anatomy, Fourth Edition and its comprehensive supplements package to your existing syllabus. New to this edition's website are embryology and histology atlases as well as a section highlighting important historical events.
- **Anatomy and Physiology Laser Disk** An encyclopedic visual resource to enrich your lecture and lab presentations. Contains animations, videos dissections, and still images.
- **Instructor's Guide and Resource Manual to the *Human Anatomy Laboratory Manual and Dissection Guide, Fourth Edition*** *By Michael Timmons and Ralph Hutchings* This companion pedagogical aide to Timmons and Hutchings' laboratory manual and dissection guide provides dozens of helpful teaching strategies, outlines, test questions, and lab resources.

■ For Students

- **Human Anatomy Atlas** This new 80 page full four colour photographic atlas can be shrink-wrapped with the fourth edition at no additional charge. It includes: images produced by radiological procedures, such as MRI and CT scans; histology images; skeletal system and cadaver dissection images, including some that appear in the text (at reduced size) and others that provide views not provided in the text.
- **Student CD-ROM for *Human Anatomy, Fourth Edition*** An Interactive CD is bound into every new copy of the fourth edition! This powerful multimedia tool contains visual aids to help students understand and appreciate anatomy. It includes enhanced rotatable, layered 3-D visualizations with "hot-spot" roll-over labeling to help your students learn the names of anatomical structures. Animations, tutorials, an audio glossary, and links to relevant tutorial and clinical websites round out this product. Seven 3-D "fly-throughs" guide your students through an endoscopic view of specific anatomical structures such as the stomach, heart etc.
- **Students' Companion Website for *Human Anatomy, Fourth Edition* www.prenhall.com/martini/ha4.** This comprehensive website, designed specifically for *Human Anatomy, Fourth Edition*, enhances the your students' learning experience. Site content correlates to every chapter in the book, providing students with a variety of self-testing questions, including multiple choice questions, labeling exercises, matching problems, and more. "Hot links" to related websites expand the realm of information available to your students. Contains many new labeling and matching exercises, new embryology, teratology, and histology sections, new images from the Visible Human Project with corresponding exercises in the *Connections* sections of the textbook, and much more.
- **A.D.A.M. *Interactive Anatomy Dissection Manual* by Martha DePecol Sanner and Harry Greer** This manual guides students through the A.D.A.M. Software's *Interactive Anatomy*. This software program has thousands of images (illustrations, cadaver images, X-ray photographs), advanced search features, and pinned images. Students can literally dissect the human body on the computer. This manual guides the student through this dissection. As students "cut" through the layers, this manual directs them to appropriate structures.
- **Correlation Guide for *A.D.A.M. Interactive Dissection Manual with Human Anatomy, Fourth Edition*.** This online supplement is the perfect complement to Timmons and Hutchings' *Human Anatomy Laboratory Manual, Fourth Edition*. It provided helpful

cross-references of images between *A.D.A.M. Interactive Anatomy* and *Human Anatomy, Fourth Edition*.

- ***Human Anatomy Laboratory Manual and Dissection Guide, Fourth Edition*** *By Michael Timmons and Ralph Hutchings* Designed specifically as the lab component of Human Anatomy, Fourth Edition, this manual is the ideal learning companion for majors and non-majors alike. It combines the proven features of a traditional anatomy dissection manual with the over-sized visual orientation of an anatomy atlas. All anatomical structures are grouped by system with precise labeling and are organized to correspond chapter-by-chapter with Human Anatomy, Fourth Edition. Its unparalleled art program includes photos and illustrations with instructional captions for myriad specimens—including prosected cadavers, isolated organs, histology slides, laboratory models, and more. Includes 175 labeled color and black-and-white photographs of cadaver and organ specimens; 60 labeled color photographs of anatomical models used in teaching laboratories; over 150 illustrations; and over 550 end of chapter assessment questions. An online correlation guide is available for this supplement on the *Human Anatomy, Fourth Edition* website.
- **Prentice Hall's Anatomy and Physiology Video Tutor** This highly regarded 75-minute video focuses on the concepts that instructors and students consistently identify as the most challenging. Processes are demonstrated through high-quality animations and video. On-camera narration and the accompanying frame-referenced study booklet allow for repeated concept review.

Acknowledgments

We would like to acknowledge the many users, reviewers, survey respondents, and focus group members whose advice, comments, and collective wisdom helped shape this text into its final form. Their passion for the subject, their concern for accuracy and method of presentation, and their experience with students of widely varying abilities and backgrounds has made the review process much more fruitful.

Several faculty members took the time to send us suggestions for improvements to this edition without being asked and without expecting compensation for their work. We would like to express our gratitude to these individuals here:

Dr. Chris Nicolay, Dept of Biology, University of North Carolina at Asheville

Dr. Abelhamid Zir, Faculty of Health Professions, Alquds University, Israel

Dr. David Seibel, Dept of Biology, Johnson County Community College. Dr. Ray Ochs, Department of Chemistry St. John's University

We would also like to thank Kevin Petti, Department of Biology, San Diego Miramar College and Mark Terrell, Indiana University—Purdue University/Indianapolis, for their outstanding technical reviews. Thanks to the focus group of student reviewers at Brigham Young University under the direction of Dr. Bob Seegmiller for their collective comments.

Their keen eyes and attention to detail helped maintain the highest standards of accuracy and currency in this new edition.

Dr. Steve Senger, University of Wisconsin/La Crosse, and Dr. Lewis Sadler and the talented team at Visible Productions, Ft. Collins, CO, worked with us on both the Interactive CD and on the images that appear in the Connections sections. We greatly appreciate their contributions. We would also like to thank Kelly Johnson, Department of Molecular Biosciences, University of Kansas, for his technical reviewing of pages, and for his indomitable spirit and tremendous efforts on a myriad of key projects, including the website and student and instructor CD-ROM that support this title. Many thanks also to John F. Neas (University of Kansas) and Brian Wisenden (Minnesota State University Moorhead) for their work and contributions to the *Human Anatomy* 4th edition website. Last but by no means least, we are grateful to Kate Flickinger for her invaluable work on the media plan for this text, as well as her creation of the Connections sections that are a major enhancement to this edition.

The creative talents brought to this project by our artist team, William Ober, M.D., and Claire Garrison, RN, are inspiring and very much appreciated. Bill and Claire worked intimately and tirelessly with us, imparting a unity of vision to the book as a whole while making it both clear and beautiful. Their superb art program is greatly enhanced by the incomparable bone and cadaver photographs of Ralph T. Hutchings, formerly of The College of Surgeons of England, and co-author of the best selling McMinns *Color Atlas of Human Anatomy*. In addition, Dr. Pietro Motta, Professor of Anatomy, University of Roma, La Sapienza, provided several superb SEM images for use in the text.

We are deeply indebted to the Prentice Hall production staff, whose efforts were so vital to the creation of this edition. Kudos go to Karen Noferi and Jim Sullivan for the indispensable roles they played in paging the text. A special note of thanks and gratitude to our production editor, Shari Toron for her skillful management of the project through the entire production process. Art director Jon Boylan and Carole Anson, Director of Design, oversaw the development of the beautiful internal text design by Anne Demaris, as well as an outstanding cover design. We must also express our appreciation to Crissy Statuto, Project Manager, for her work on the Companion Atlas (new to this edition), and to Susan Zeigler, Editorial Assistant, who greatly supported our efforts in countless ways.

We are grateful to Paul Corey, President of the Engineering, Science, and Mathematics Division of Prentice Hall, and Sheri Snavely, Vice President, Editor-in-Chief of Life and Geosciences for their continued enthusiastic support of this project, as well as of my other books at Prentice Hall. We would also like to acknowledge the contributions of Marty McDonald, Marketing Manager, and Jennifer Welchans, Executive Marketing Manager, who keep their fingers on the pulse of the market and help us meet the needs of our users. Above all, a heartfelt thanks to our editor, Halee Dinsey, for her patience in nurturing this project, her tireless efforts to coordinate the various components of the package, and her work on the media supplements and Interactive CD-ROM.

Finally, we would like to thank our families for their love and support during the revision process. We could not have accomplished this without the help of our wives—Kitty, Judy, and Mary—and the patience of our children—P. K., Molly, Kelly, Patrick, Katie, Ryan, Molly, and Steven.

No three people could expect to produce a flawless textbook of this scope and complexity. Any errors or oversights are strictly our own rather than those of the reviewers, artists, or editors. In an effort to improve future editions, we ask that readers with pertinent information, suggestions, or comments concerning the organization or content of this textbook send their remarks to us directly, by email, or care of Halee Dinsey, Senior Editor, Applied Biology, Prentice Hall, One Lake Street, Upper Saddle River, NJ,07458. Any and all comments and suggestions will be deeply appreciated and carefully considered in the preparation of the next edition.

Frederic H. Martini, Haiku, HI (martini@maui.net)
Michael J. Timmons, Orland Park, IL
Robert B. Tallitsch, Rock Island, IL

FIRST EDITION REVIEWERS

Cynthia Battie, University of Missouri-Kansas City
Annalisa Berta, San Diego State University
Paul Biersuck, Nassau Community College
Leann Blem, Virginia Commonwealth University
Debbie Borosh, Mount San Antonio Community College
Neil Cumberlidge, Northern Michigan University
Mary Jo Fourier, Johnson County Community College
Glenn Gorelick, Citrus College
Edward Lutsch, Northeastern Illinois University
Bryan Miller, Eastern Illinois University
Sherwin Mizell, Indiana University
Virginia Naples, Northern Illinois University
Gail Platt, Boston University
Ron Plakke, University of Northern Colorado
Carl Sievert, University of Wisconsin

First Edition Focus Group Participants
Mary Jo Fourier, Johnson County Community College
Edward Lutsch, Northeastern Illinois University
Bryan Miller, Eastern Illinois University
Virginia Naples, Northern Illinois University
Carl Sievert, University of Wisconsin
Reuben Barrett, Chicago State University

Technical Reviewers
Norman Lieska, University of Illinois School of Medicine
Lissa Little, Northwestern Medical School
Diane Merlos, Grossmont College
Larry Olver, Northwestern Medical School
Randy Perkins, Northwestern Medical School

SECOND EDITION REVIEWERS

Joan Aloi, Saddleback College
Cynthia Battie, University of Missouri, Kansas City
Annatisa Berla, San Diego University
Paul Biersuck, Nassau Community College
Leann Blem, Virginia Commonwealth University
Debbie Borosh, Mount San Antonio Community College
Neil Cumberlidge, Northern Michigan University
Mary Jo Fourier, Johnson County Community College
Anthony J. Gaudin, Ph.D., California State University-Northridge
Glenn Gornick, Citrus College
Glenn E. Kietzmann, Wayne State College
Tom Linder, University of Washington
Edward Lutsch, Northeastern Illinois University
Bryan Miller, Eastern Illinois University
Sherwin Mizell, Indiana University
Virginia Naples, Northern Illinois University
Mary Oreluk, R.N., Little Company of Mary Hospital
Gail Platt, Boston University
Annie Peterman, University of Washington
Ron Plakke, University of Northern Colorado
Carl Sievert, University of Wisconsin
James Smith, California State University, Fullerton
Thomas M. Vollberg, Sr., Creighton University

Technical Reviewers
Norman Lieska, University of Illinois School of Medicine
Lissa Little, Northwestern Medical School
Diane Merlos, Grossmont College
Larry Olver, Northwestern Medical School
Randy Perkins, Northwestern Medical School
Mike McKinley, Glendale Community College

THIRD EDITION REVIEWERS

Natalie Connors, Saint Louis University School of Medicine
Alan Dietsche, University of Rochester
Alexa Doing, Awakening Spirit Massage School
Wesley Hanson, John Brown University
Susan Kordas, Moraine Valley Community College
Shawn Miller, Utah College of Massage Therapy
Mary Oreluk, Little Company of Mary Hospital
Robert Seegmiller, Brigham Young University
Walter Sulskis, Moraine Valley Community College
Judith Tamburlin, SUNY Buffalo
Annette Tomaska, University of Illinois-Chicogo
Thomas Vollberg, Creighton University School of Medicine
Mark Zelman, William Rainey Harper College

Student Focus Group
Thanks to the focus group of student reviewers at Brigham Young University under the direction of Dr. Bob Seegmiller for their collective comments.

FOURTH EDITION REVIEWERS

Brian Wisenden, Minnesota State University at Moorhead
Wesley Hanson, Southern Nazarene University
Sidney L. Palmer, Ricks College
Lucy G. Andrews, University of Alabama at Birmingham
Robert Seegmiller, Brigham Young University
James D. Fawcett, University of Nebraska at Omaha
Judi Lindsley Nath, Lourdes College
Les MacKenzie, Queens's University
Ada Kelly Houston, AMEDDC'S
Judith Tamburlin, SUNY Buffalo
Emily Gay Williamson, Mississippi State University
Charles A. Ferguson, University of Colorado at Denver
Steven K. Itaya, University of South Alabama
Cheryl A. D. Wilga, University of Rhode Island
Rebecca M. Peterson, Pensylvannia State University
Patrick B. Fulks, Bakersfield College
Tony Yates, Seminole State College
Kent R. Thomas, Wichita State University
Tamara A. Stein, University of Michigan Medical School

Technical Reviewers for Human Anatomy Fourth Edition
Kelly Johnson, University of Kansas
Kevin Petti, San Diego Miramar
Mark Terrell, Indiana University, Purdue University Indianapolis

MEDIA ADVISORY BOARD FOR PRENTICE HALL

Cecil Hampton, Jefferson College
Joyce Ono, California State University
Betsy Ott, Tyler Junior College
Ed Pivorum, Clemson University
Anil Rao, Metropolitan State College
Chris Riegle, Irvine Valley College
Sharon Simpson, Broward College
Sandy Stewart, Vincennes University
Bob Tallitsch, Augustana College
Mary Pat Wenderoth, University of Washington
Lew Sadler, Visible Productions

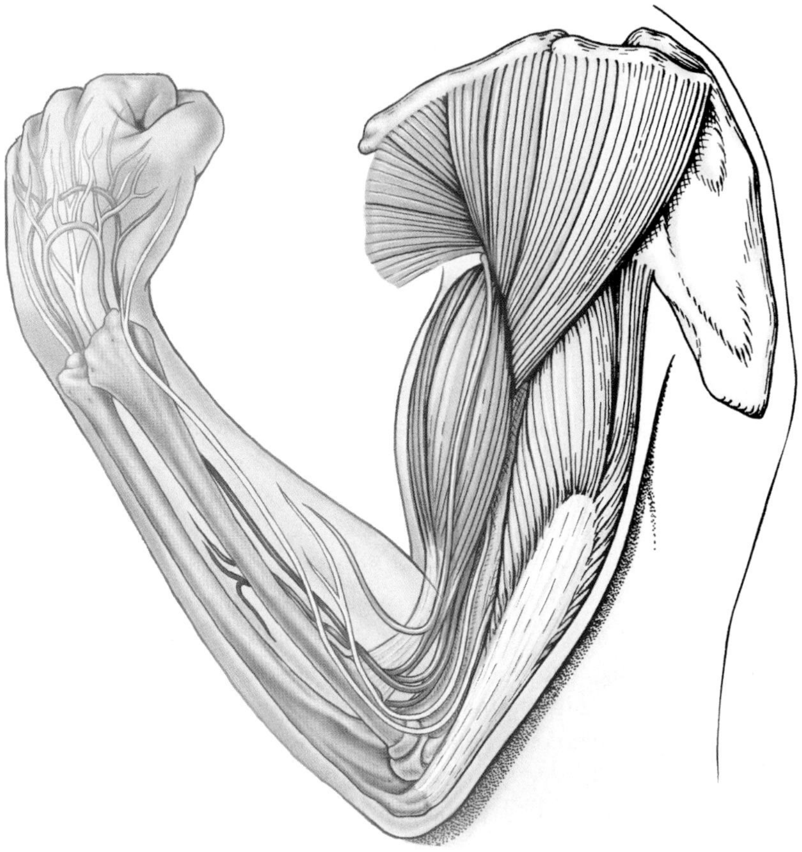

1

AN INTRODUCTION TO ANATOMY

We are all anatomists in our daily lives, if not in the classroom. For example, we rely on our memories of specific anatomical features to identify our friends and family, and we watch for subtle changes in body movement or position that give clues to what others are thinking or feeling. To be precise, anatomy is the study of external and internal structures and the physical relationships among body parts. But in practical terms, anatomy is the careful observation of the human body. Anatomical information provides clues about probable functions; physiology is the study of function, and physiological mechanisms can be explained only in terms of the underlying anatomy. *All specific functions are performed by specific structures.* For instance, filtering, warming, and humidifying inspired air are functions of the nasal cavity. The shapes of the bones projecting into the nasal cavity cause turbulence in the inhaled air, making it swirl against the moist lining. This contact warms and humidifies the air, and any suspended particles stick to the moist surfaces. In this way, the air is conditioned and filtered before it reaches the lungs.

The link between structure and function is always present, but not always understood. For example, the superficial anatomy of the heart was clearly described in the fifteenth century, but almost 200 years passed before the pumping action of the heart was demonstrated. On the other hand, many important cell functions were recognized decades before the electron microscope revealed the anatomical basis for those functions.

This text will discuss the anatomical structures and functions that make human life possible. The goals are to prepare you for more advanced courses in anatomy, physiology, and related subjects and to help you make informed decisions about your personal health.

Microscopic Anatomy [FIGURE 1.1]

Microscopic anatomy considers structures that cannot be seen without magnification. The boundaries of microscopic anatomy, or *fine anatomy,* are established by the limits of the equipment used (Figure 1.1•). A simple hand lens shows details that barely escape the naked eye, while an electron microscope demonstrates structural details that are at least one million times smaller. As we proceed through the text, we will be considering details at all levels, from macroscopic to microscopic. (Readers unfamiliar with the terms used to describe measurements and weights over this size range should consult the reference tables in Appendix II.)

Microscopic anatomy can be subdivided into specialties that consider features within a characteristic range of sizes. **Cytology** (sī-TOL-o-jē) analyzes the internal structure of **cells**, the smallest units of life. Living cells are composed of complex chemicals in various combinations, and our lives depend on the chemical processes occurring in the trillions of cells that form our body.

Histology (his-TOL-o-jē) takes a broader perspective and examines **tissues**, groups of specialized cells and cell products that work together to perform specific functions. The cells in the human body can be assigned to four major tissue types, and these tissues are the focus of Chapter 3.

Tissues in combination form **organs** such as the heart, kidney, liver, or brain. Organs are anatomical units that have multiple functions. Many organs are examined easily without a microscope, and at the organ level we cross the boundary from microscopic anatomy into gross anatomy.

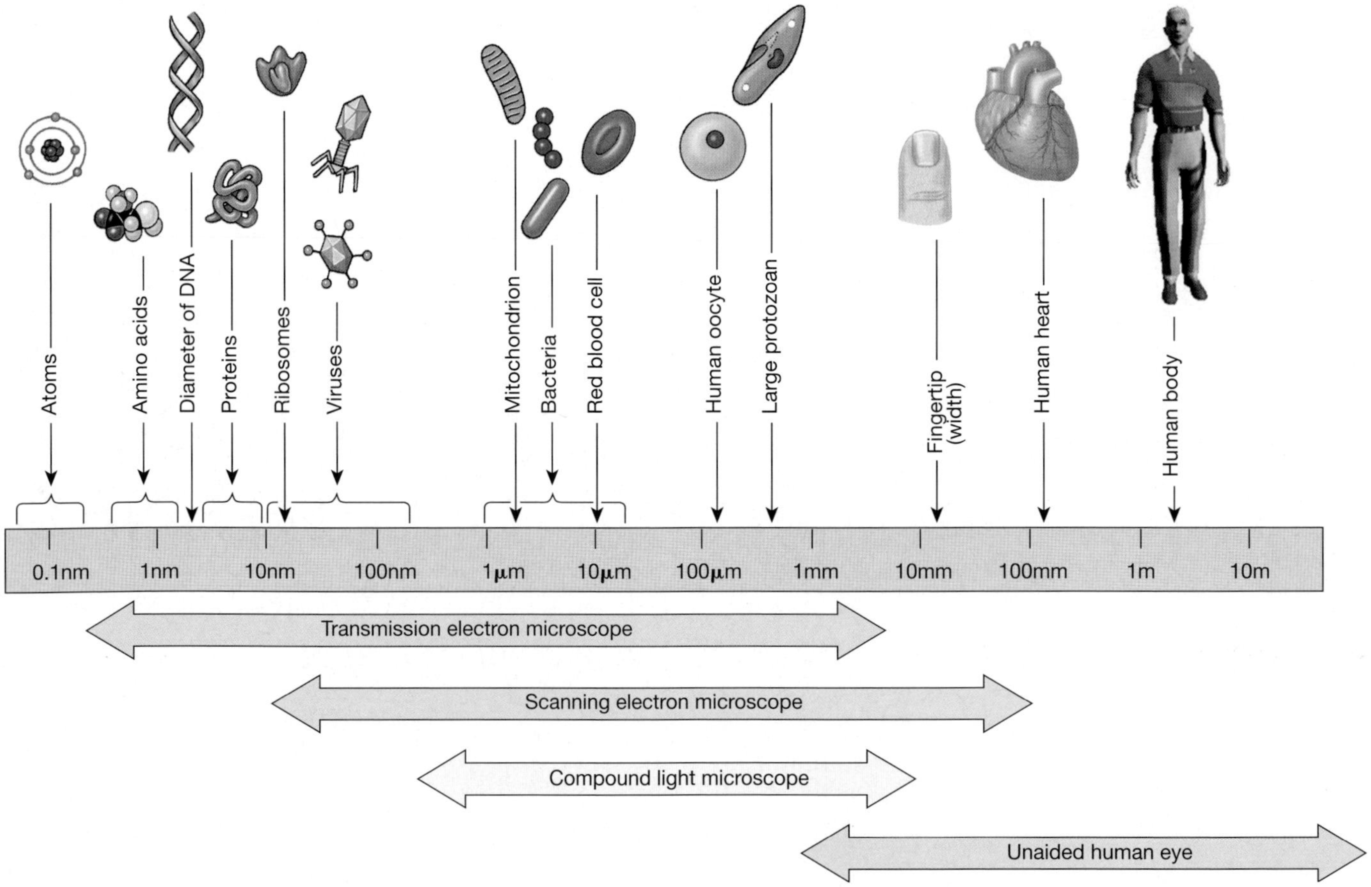

FIGURE 1.1 **The Study of Anatomy at Different Scales**
The amount of detail recognized depends on the method of study and the degree of magnification.

Gross Anatomy

Gross anatomy, or **macroscopic anatomy**, considers relatively large structures and features visible to the unaided eye. There are many ways to approach gross anatomy:

- *Surface anatomy* refers to the study of general form, or *morphology*, and superficial anatomical markings.
- *Regional anatomy* considers all of the superficial and internal features in a specific area of the body, such as the head, neck, or trunk. Advanced courses in anatomy often stress a regional approach because it emphasizes the spatial relationships among structures already familiar to the students.
- *Systemic anatomy* considers the structure of major *organ systems*, such as the skeletal or muscular system. Organ systems are groups of organs that function together to produce coordinated effects. For example, the heart, blood, and blood vessels form the *cardiovascular system*, which distributes oxygen and nutrients throughout the body. There are 11 organ systems in the human body, and they will be introduced later in the chapter. Introductory texts in anatomy, including this one, use a systemic approach because it provides a framework for organizing information about important structural and functional patterns.

Other Perspectives on Anatomy [FIGURE 1.2]

There are other anatomical specialties that will be encountered in this text.

- *Developmental anatomy* examines the changes in form that occur during the period between conception and physical maturity. Because it considers anatomical structures over such a broad range of sizes (from a single cell to an adult human), developmental anatomy involves the study of both microscopic and gross anatomy. Developmental anatomy is important in medicine because many structural abnormalities can result from errors that occur during development. The most extensive structural changes occur during the first two months of development. **Embryology** (em-brē-OL-o-jē) is the study of these early developmental processes.
- *Comparative anatomy* considers the anatomical organization of different types of animals. Observed similarities may reflect evolutionary relationships. Humans, lizards, and sharks are all called *vertebrates* because they share a combination of anatomical features that is not found in any other group of animals. All vertebrates have a spinal column composed of individual elements called *vertebrae.* Comparative anatomy uses techniques of gross and microscopic anatomy. Information on developmental anatomy is also very important, because related animals typically go through very similar developmental stages (Figure 1.2●).

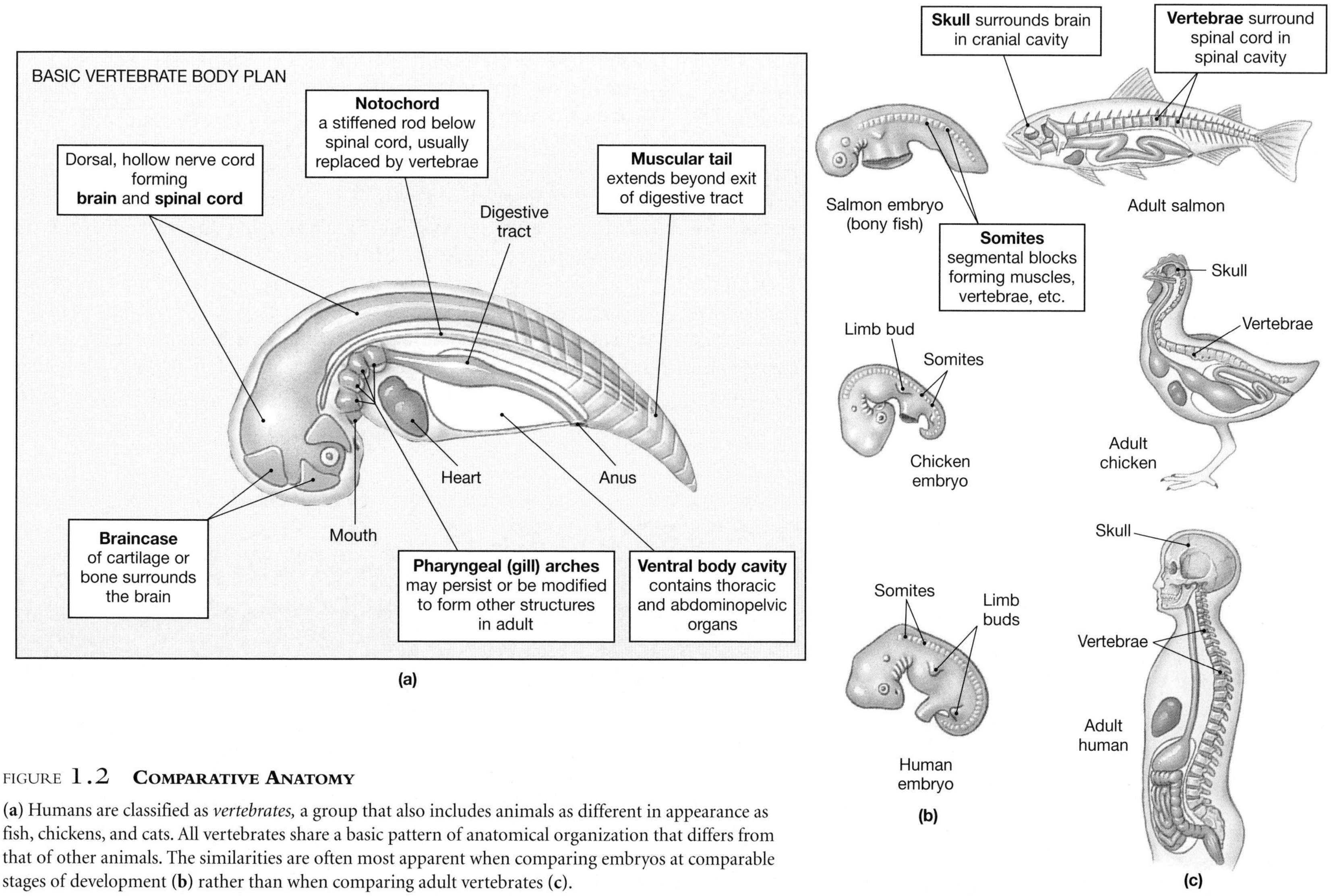

FIGURE 1.2 **Comparative Anatomy**

(**a**) Humans are classified as *vertebrates,* a group that also includes animals as different in appearance as fish, chickens, and cats. All vertebrates share a basic pattern of anatomical organization that differs from that of other animals. The similarities are often most apparent when comparing embryos at comparable stages of development (**b**) rather than when comparing adult vertebrates (**c**).

Several other gross anatomical specialties are important in medical diagnosis.

- *Medical anatomy* focuses on anatomical features that may undergo recognizable pathological changes during illness.
- *Radiographic anatomy* involves the study of anatomical structures as they are visualized by X-rays, ultrasound scans, or other specialized procedures performed on an intact body.
- *Surgical anatomy* studies anatomical landmarks important for surgical procedures.

✓ CONCEPT CHECK

- A histologist investigates structures at what level of organization?
- Which level(s) of organization would a gross anatomist investigate?
- How does the study of regional anatomy differ from the study of systemic anatomy?

Levels of Organization [FIGURES 1.3/1.4]

Our study of the human body will begin with an overview of cellular anatomy and then proceed to the gross and microscopic anatomy of each organ system. When considering events from the microscopic to macroscopic scales, we are examining several interdependent *levels of organization.*

We begin at the *chemical* or *molecular level of organization*. The human body consists of over a dozen different elements, but four of them (hydrogen, oxygen, carbon, and nitrogen) account for more than 99% of the total number of atoms (Figure 1.3a●). At the chemical level, atoms interact to form compounds with distinctive properties. The major classes of compounds in the human body are indicated in (Figure 1.3b●).

Figure 1.4● presents an example of the relationships between the chemical level and higher levels of organization. *Cells* are the smallest living units in the body. The *cellular level of organization* includes *cells*, the smallest living units in the body. Cells contain internal structures called *organelles*. Cells and their organelles are made up of complex chemicals. Cell structure and the function of the major organelles will be presented in Chapter 2. In Figure 1.4●, chemical interactions produce complex proteins within a *muscle cell* in the heart. Muscle cells are unusual because they can contract powerfully, shortening along their longitudinal axis.

Heart muscle cells are connected to form a distinctive *muscle tissue*, an example of the *tissue level of organization*. Layers of muscle tissue form the bulk of the wall of the heart, a hollow, three-dimensional organ. We are now at the *organ level* of organization.

Normal functioning of the heart depends on interrelated events at the chemical, cellular, tissue, and organ levels of organization. Coordinated contractions in the adjacent muscle cells of cardiac muscle tissue produce a heartbeat. When that beat occurs, the internal anatomy of the organ enables it to function as a pump. Each time it contracts, the heart pushes blood into the *circulatory system*, a network of blood vessels. Together the heart, blood, and circulatory system form an *organ system*, the *cardiovascular system (CVS)*.

Each level of organization is totally dependent on the others. For example, damage at the cellular, tissue, or organ level may affect the entire system. Thus, a chemical change in heart muscle cells may cause abnormal contractions or even stop the heartbeat. Physical damage to the muscle tissue, as in a chest wound, can make the heart ineffective even when most of the heart muscle cells are intact and uninjured. An inherited abnormality in heart structure can make it an ineffective pump, although the muscle cells and muscle tissue are perfectly normal.

Finally, it should be noted that something that affects the system will ultimately affect all of its components. For example, the heart may not be able to pump blood effectively after a massive blood loss. If the heart cannot pump and blood cannot flow, oxygen and nutrients cannot be distributed. In a very short time, the tissue begins to break down as heart muscle cells die from oxygen and nutrient starvation.

Of course the changes that occur when the heart is not pumping effectively will not be restricted to the cardiovascular system; all of the cells, tissues, and organs in the body will be damaged. This observation brings us to another higher level of organization, that of the *organism*; in this case a human being. This level reflects the interactions among organ systems. All are vital; every system must be working properly and in harmony with every other system, or survival will be impossible. When those systems are functioning normally, the characteristics of the internal environment will be relatively stable at all levels. This vital state of affairs is called **homeostasis** (hō-mē-ō-STĀ-sis; *homeo*, unchanging + *stasis*, standing).

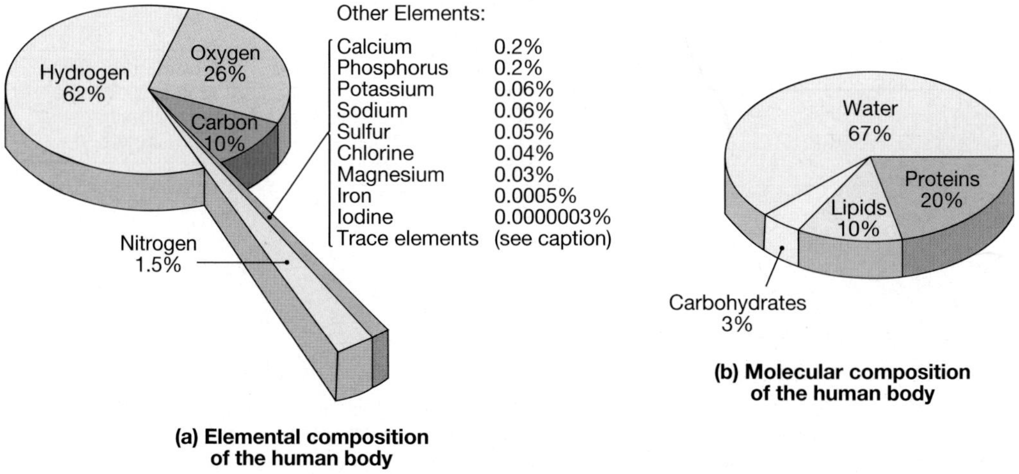

FIGURE 1.3 **COMPOSITION OF THE BODY AT THE CHEMICAL LEVEL OF ORGANIZATION**

The percent composition of elements and major molecules. (**a**) Elemental composition of the body. Trace elements include silicon, fluorine, copper, manganese, zinc, selenium, cobalt, molybdenum, cadmium, chromium, tin, aluminum, and boron. (**b**) Molecular composition of the body.

FIGURE 1.4 **Levels of Organization**

Interacting atoms form molecules that combine in the protein fibers of heart muscle cells. These cells interlock, forming heart muscle tissue that constitutes most of the walls of a three-dimensional organ, the heart. The heart is one component of the cardiovascular system, which also includes the blood and blood vessels. All of the organ systems must work together for the person to remain alive and healthy.

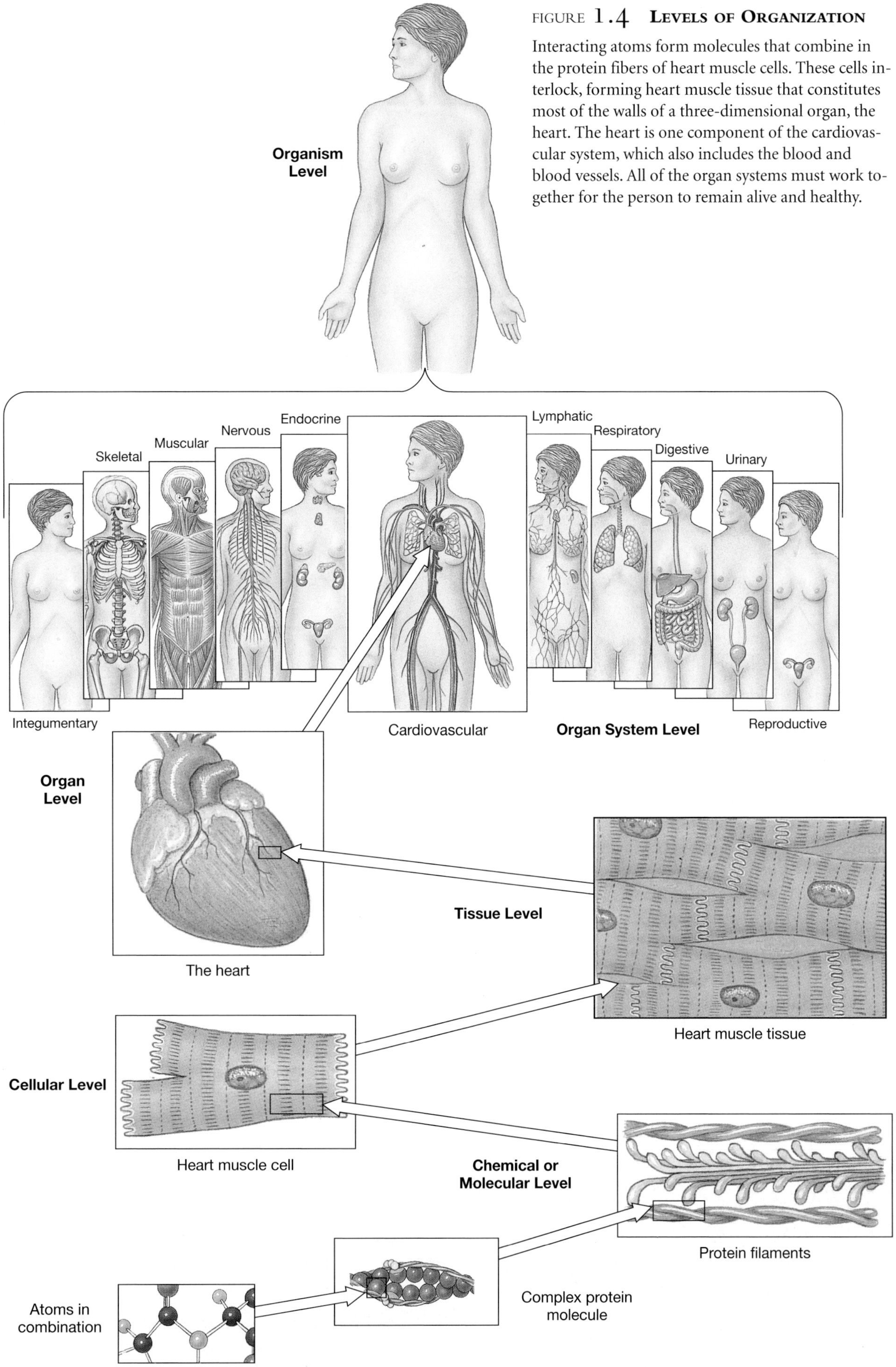

An Introduction To Organ Systems [FIGURES 1.5/1.6]

All living organisms share vital properties and processes:

- *Responsiveness*: Organisms respond to changes in their immediate environment; this property is also called *irritability.* You move your hand away from a hot stove; your dog barks at approaching strangers; fish are scared by loud noises; and amoebas glide toward potential prey. Organisms also make longer lasting changes as they adjust to their environments. For example, as winter approaches an animal may grow a heavier coat, or migrate to a warmer climate. The capacity to make such adjustments is termed *adaptability.*
- *Growth and Differentiation*: Over a lifetime, organisms grow larger, increasing in size through an increase in the size or number of their cells. In multicellular organisms, the individual cells become specialized to perform particular functions. This specialization is called **differentiation**. Growth and differentiation often produce changes in form and function. For example, the anatomical proportions and physiological capabilities of an adult human are quite different from those of an infant.
- *Reproduction*: Organisms reproduce, creating subsequent generations of their own kind, whether unicellular or multicellular.
- *Movement*: Organisms are capable of producing movement, which may be internal (transporting food, blood, or other materials inside the body) or external (moving through the environment).
- *Metabolism and Excretion*: Organisms rely on complex chemical reactions to provide the energy for responsiveness, growth, reproduction, and movement. They must also synthesize complex chemicals, such as proteins. The term **metabolism** refers to all the chemical operations under way in the body: *catabolism* is the breakdown of complex molecules into simple ones, and *anabolism* is the synthesis of complex molecules from simple ones. Normal metabolic operations require the **absorption** of materials from the environment. To generate energy efficiently, most cells require various nutrients, as well as oxygen, an atmospheric gas. The term **respiration** refers to the absorption, transport, and use of oxygen by cells. Metabolic operations often generate unneeded or potentially harmful waste products that must be removed through the process of **excretion**.

For very small organisms, absorption, respiration, and excretion involve the movement of materials across exposed surfaces. But creatures larger than a few millimeters seldom absorb nutrients directly from their environment. For example, human beings cannot absorb steaks, apples, or ice cream directly—they must first alter the foods' chemical structure. That processing, called **digestion**, occurs in specialized areas where complex foods are broken down into simpler components that can be absorbed easily. Respiration and excretion are also more complicated for large organisms, and we have specialized organs responsible for gas exchange (the lungs) and waste excretion (the kidneys). Finally, because absorption, respiration, and excretion are performed in different portions of the body, there must be an internal transportation system, or **cardiovascular system**.

Figure 1.5● provides an overview of the 11 organ systems in the human body. Figure 1.6● introduces the major organs in each system.

Organ System		*Major Functions*
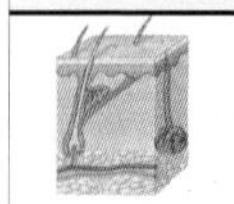	**Integumentary system**	Protection from environmental hazards; temperature control
	**Skeletal system**	Support, protection of soft tissues; mineral storage; blood formation
	**Muscular system**	Locomotion, support, heat production
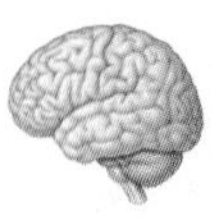	**Nervous system**	Directing immediate responses to stimuli, usually by coordinating the activities of other organ systems
	**Endocrine system**	Directing long-term changes in the activities of other organ systems
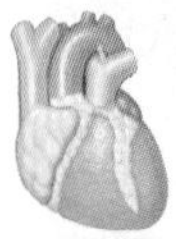	**Cardiovascular system**	Internal transport of cells and dissolved materials, including nutrients, wastes, and gases
	**Lymphatic system**	Defense against infection and disease
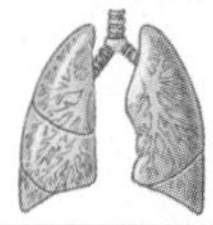	**Respiratory system**	Delivery of air to sites where gas exchange can occur between the air and circulating blood
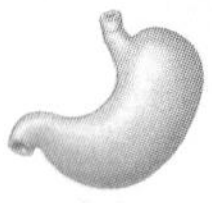	**Digestive system**	Processing of food and absorption of organic nutrients, minerals, vitamins, and water
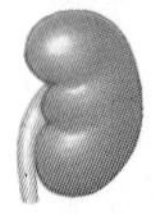	**Urinary system**	Elimination of excess water, salts, and waste products; control of pH
	**Reproductive system**	Production of sex cells and hormones

FIGURE 1.5 **AN INTRODUCTION TO ORGAN SYSTEMS**
An overview of the 11 organ systems and their major functions.

✓ **CONCEPT CHECK**

- What system includes the following structures: sweat glands, nails, and hair follicles?
- What system has structures with the following functions: production of hormones and ova, site of embryonic development?
- What is differentiation?

FIGURE 1.6 **THE ORGAN SYSTEMS OF THE BODY**

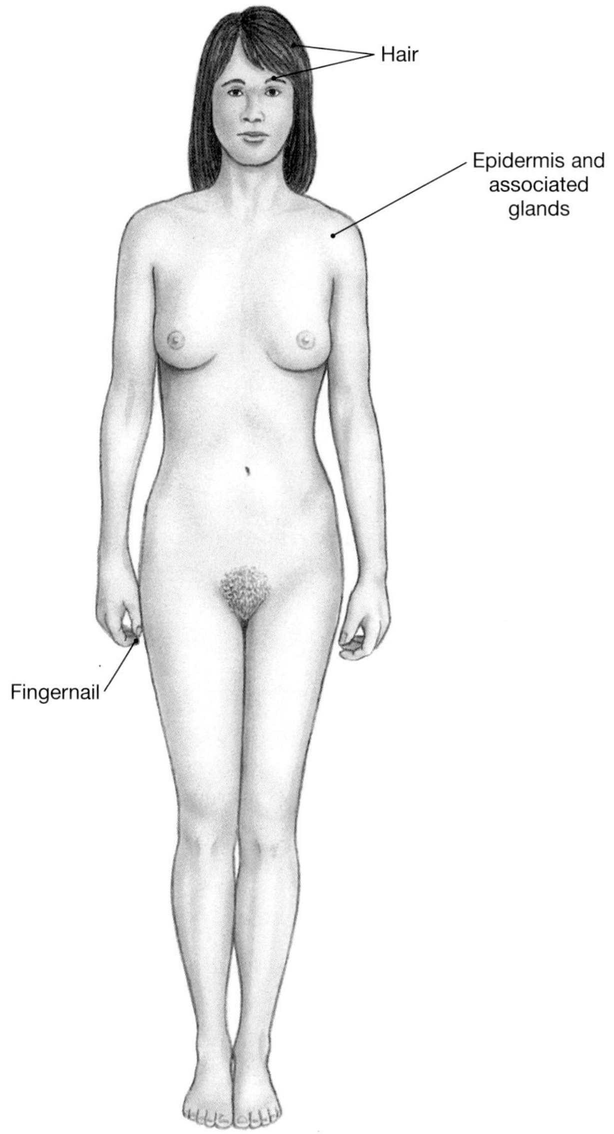

(a) THE INTEGUMENTARY SYSTEM

Protects against environmental hazards: helps control body temperature

Organ/Component	Primary Functions
CUTANEOUS MEMBRANE	
Epidermis	Covers surface; protects deeper tissues
Dermis	Nourishes epidermis; provides strength; contains glands
HAIR FOLLICLES	Produce hair; innervation provides sensation
Hairs	Provide some protection for head
Sebaceous glands	Secrete lipid coating that lubricates hair shaft and epidermis
SWEAT GLANDS	Produce perspiration for evaporative cooling
NAILS	Protect and stiffen distal tips of digits
SENSORY RECEPTORS	Provide sensations of touch, pressure, temperature, pain
SUBCUTANEOUS LAYER	Stores lipids; attaches skin to deeper structures

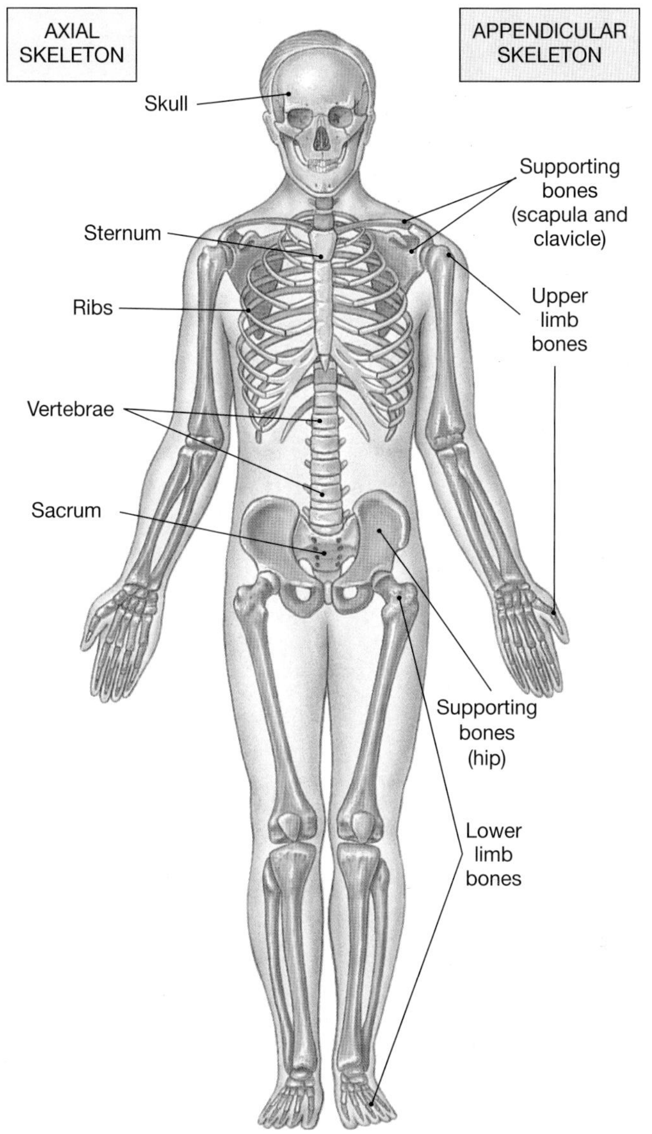

(b) THE SKELETAL SYSTEM

Provides support; protects tissues; stores minerals; forms blood

Organ/Component	Primary Functions
BONES, CARTILAGES, AND JOINTS	Support, protect soft tissues; bones store minerals
Axial skeleton (skull, vertebrae, ribs, sternum, sacrum, cartilages, and ligaments)	Protects brain, spinal cord, sense organs, and soft tissues of thoracic cavity; supports the body weight over the lower limbs
Appendicular skeleton (limbs and supporting bones and ligaments)	Provides internal support and positioning of the limbs; supports and moves axial skeleton
BONE MARROW	Acts as primary site of blood cell production (red blood cells, white blood cells)

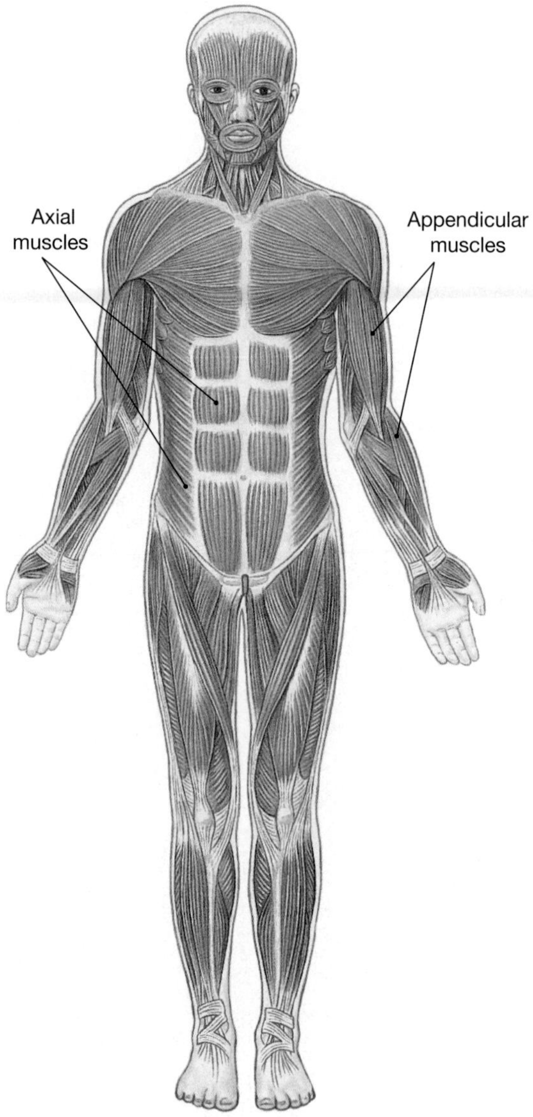

(c) THE MUSCULAR SYSTEM

Allows for locomotion; provides support; produces heat

Organ/Component	Primary Functions
SKELETAL MUSCLES (700)	Provide skeletal movement; control entrances to digestive and respiratory tracts and exits of digestive, urinary, and reproductive tracts; produce heat; support skeletal protect soft tissues
Axial muscles	Support and position axial skeleton
Appendicular muscles	Support, move, and brace limbs
TENDONS, APONEUROSES	Harness forces of contraction to perform specific tasks

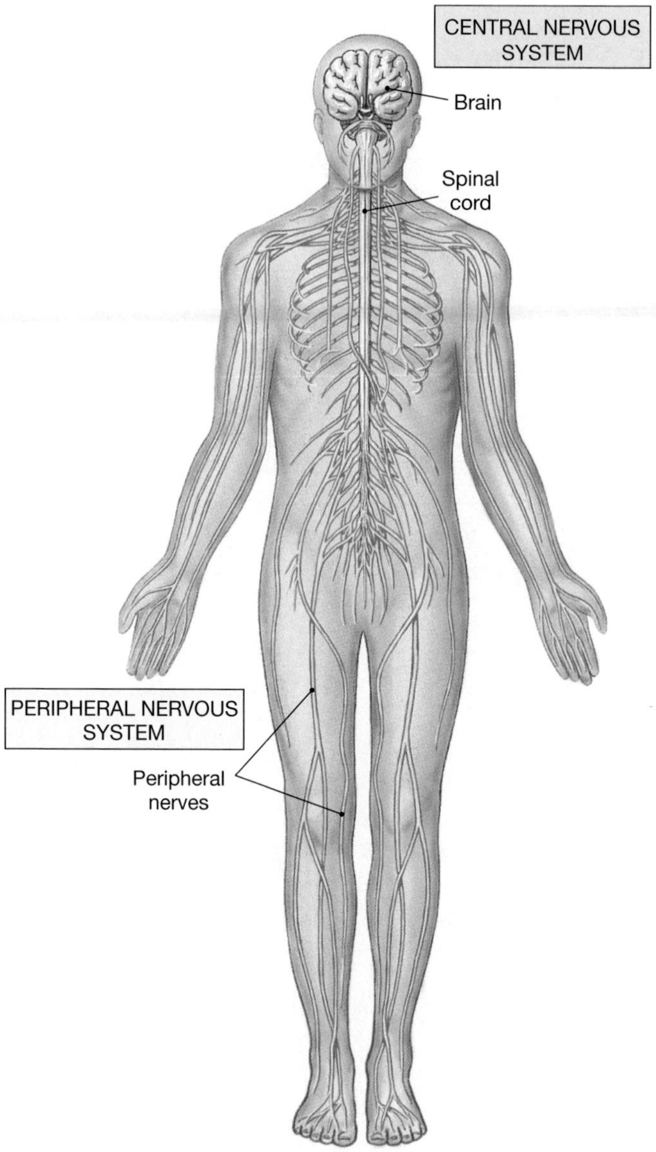

(d) THE NERVOUS SYSTEM

Directs immediate responses to stimuli, usually by coordinating the activities of other organ systems

Organ/Component	Primary Functions
CENTRAL NERVOUS SYSTEM (CNS)	Acts as control center for nervous system: processes information; provides short-term control over activities of other systems
Brain	Performs complex integrative functions; controls both voluntary and autonomic activities
Spinal cord	Relays information to and from brain; performs less-complex integrative functions; directs many simple involuntary activities
PERIPHERAL NERVOUS SYSTEM (PNS)	Links CNS with other systems and with sense organs

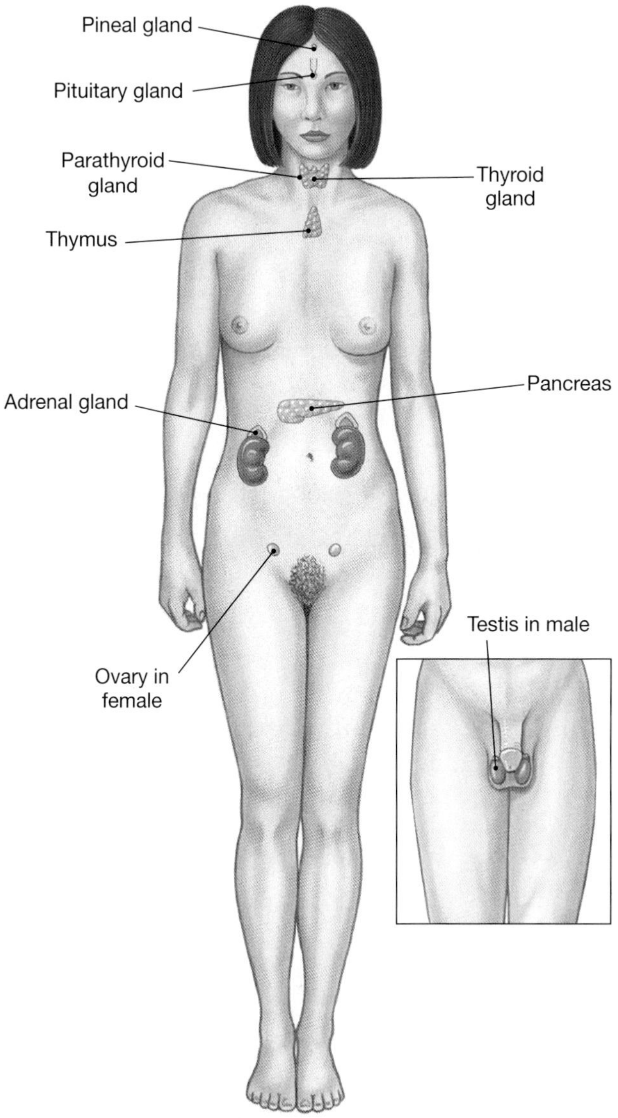

(e) THE ENDOCRINE SYSTEM

Directs long-term changes in activities of other organ systems

Organ/Component	Primary Functions
PINEAL GLAND	May control timing of reproduction and set day-night rhythms
PITUITARY GLAND	Controls other endocrine glands; regulates growth and fluid balance
THYROID GLAND	Controls tissue metabolic rate; regulates calcium levels
PARATHYROID GLANDS	Regulate calcium levels (with thyroid)
THYMUS	Controls maturation of lymphocytes
ADRENAL GLANDS	Adjust water balance, tissue metabolism, cardiovascular and respiratory activity
KIDNEYS	Control red blood cell production and elevate blood pressure
PANCREAS	Regulates blood glucose levels
GONADS	
Testes	Support male sexual characteristics and reproductive functions *(see Figure 1.6k)*
Ovaries	Support female sexual characteristics and reproductive functions *(see Figure 1.6l)*

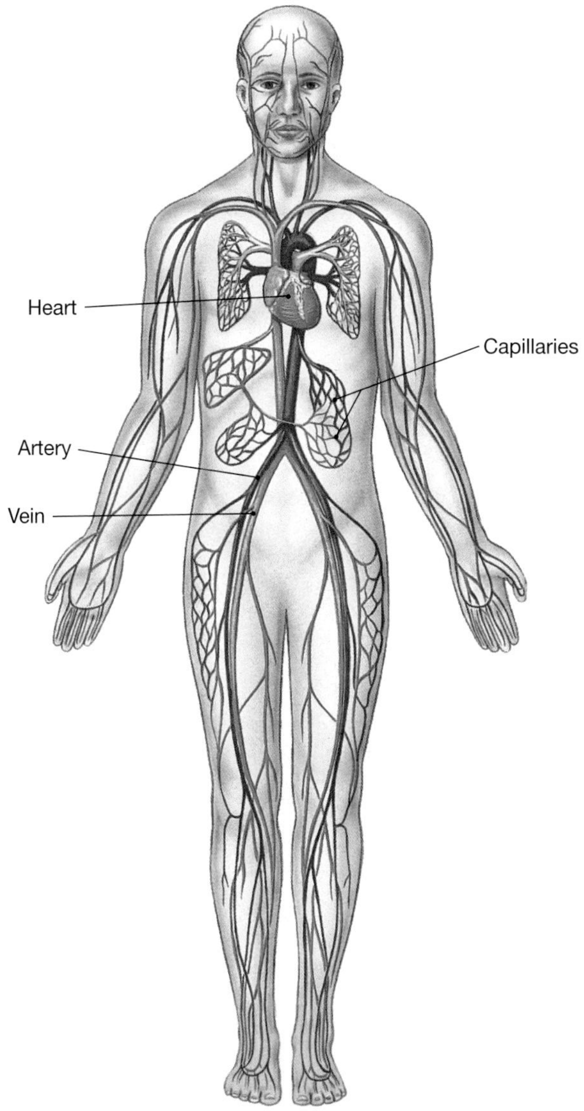

(f) THE CARDIOVASCULAR SYSTEM

Transports cells and dissolved materials, including nutrients, wastes, and gases

Organ/Component	Primary Functions
HEART	Propels blood; maintains blood pressure
BLOOD VESSELS	Distribute blood around the body
Arteries	Carry blood from heart to capillaries
Capillaries	Permit diffusion between blood and interstitial fluids
Veins	Return blood from capillaries to the heart
BLOOD	Transports oxygen, carbon dioxide, and blood cells; delivers nutrients and hormones; removes waste products; assists in temperature regulation and defense against disease

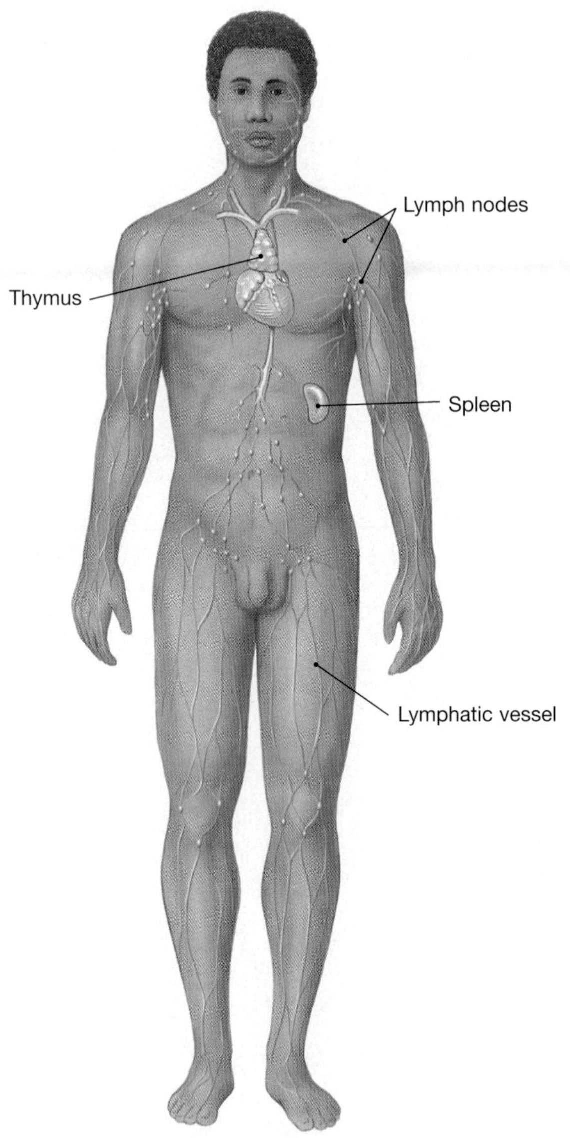

(g) The Lymphatic System

Defends against infection and disease; returns tissue fluid to the bloodstream

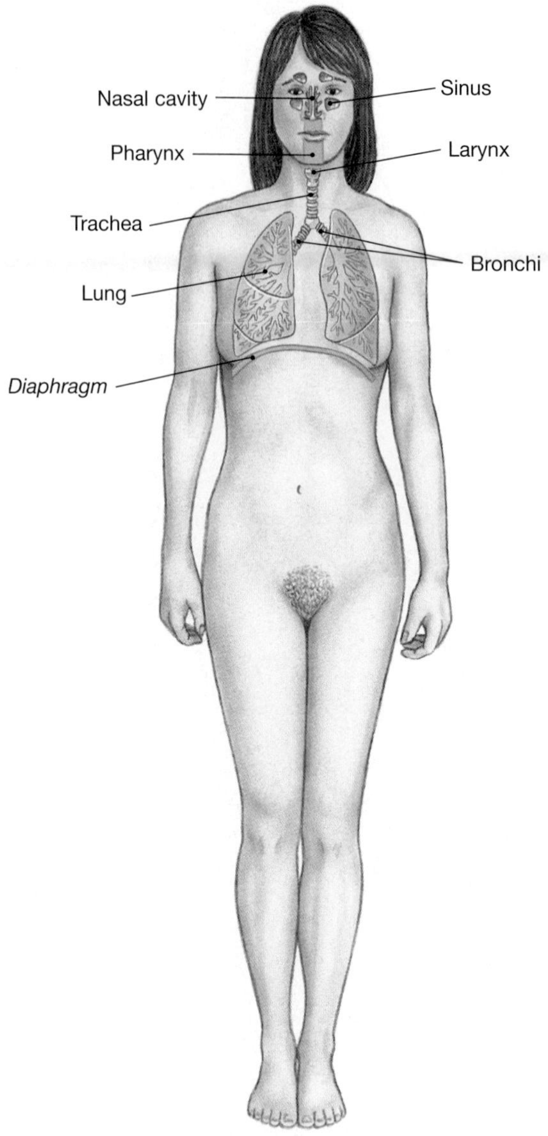

(h) The Respiratory System

Delivers air to sites where gas exchange can occur between the air and circulating blood

Organ/Component	Primary Functions
LYMPHATIC VESSELS	Carry lymph (water and proteins) and lymphocytes from peripheral tissues to veins of the cardiovascular system
LYMPH NODES	Monitor the composition of lymph; engulf pathogens; stimulates immune response
SPLEEN	Monitors circulating blood; engulfs pathogens; stimulates immune response
THYMUS	Controls development and maintenance of one class of lymphocytes (T cells)

Organ/Component	Primary Functions
NASAL CAVITIES, PARANASAL SINUSES	Filter, warm, humidify air; detect smells
PHARYNX	Conducts air to larynx; a chamber shared with the digestive tract *(see Figure 1.6i)*
LARYNX	Protects opening to trachea and contains vocal cords
TRACHEA	Filters air, traps particles in mucus; cartilages keep airway open
BRONCHI	(Same functions as trachea) through volume changes
LUNGS	Responsible for air movement during movements of ribs and diaphragm; include airways and alveoli
Alveoli	Act as sites of gas exchange between air and blood

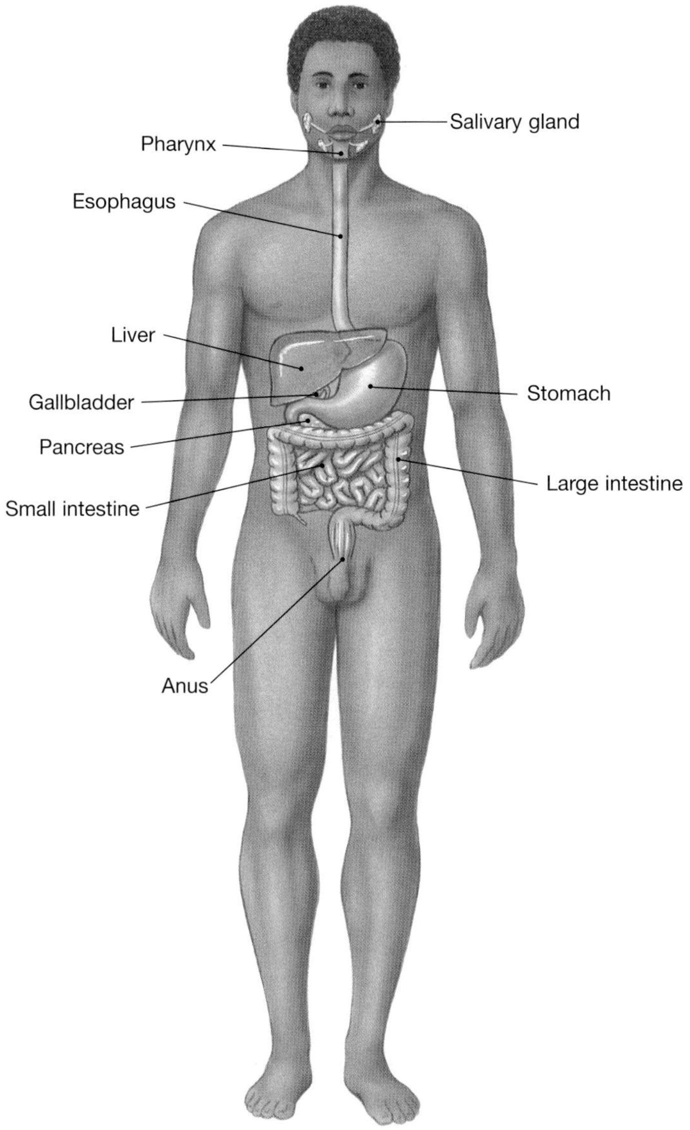

(i) THE DIGESTIVE SYSTEM
Processes food and absorbs nutrients

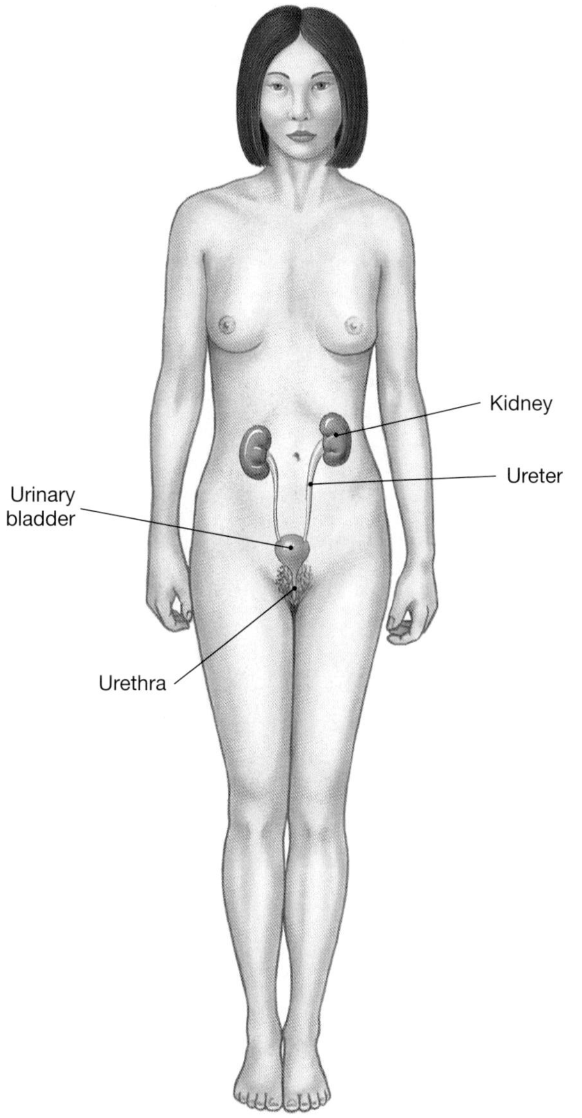

(j) THE URINARY SYSTEM
Eliminates excess water, salts, and waste products

Organ/Component	Primary Functions
SALIVARY GLANDS	Provide buffers and lubrication; produce enzymes that begin digestion
PHARYNX	Conducts solid food and liquids to esophagus; chamber shared with respiratory tract *(see Figure 1.6h)*
ESOPHAGUS	Delivers food to stomach
STOMACH	Secretes acids and enzymes
SMALL INTESTINE	Secretes digestive enzymes, buffers, and hormones; absorbs nutrients
LIVER	Secretes bile; regulates nutrient composition of blood
GALLBLADDER	Stores bile for release into small intestine
PANCREAS	Secretes digestive enzymes and buffers; contains endocrine cells (*see Figure 1.6e*)
LARGE INTESTINE	Removes water from fecal material; stores wastes

Organ/Component	Primary Functions
KIDNEYS	Form and concentrate urine; regulate blood pH and ion concentrations; perform endocrine functions (*see Figure 1.6e*)
URETERS	Conduct urine from kidneys to urinary bladder
URINARY BLADDER	Stores urine for eventual elimination
URETHRA	Conducts urine to exterior

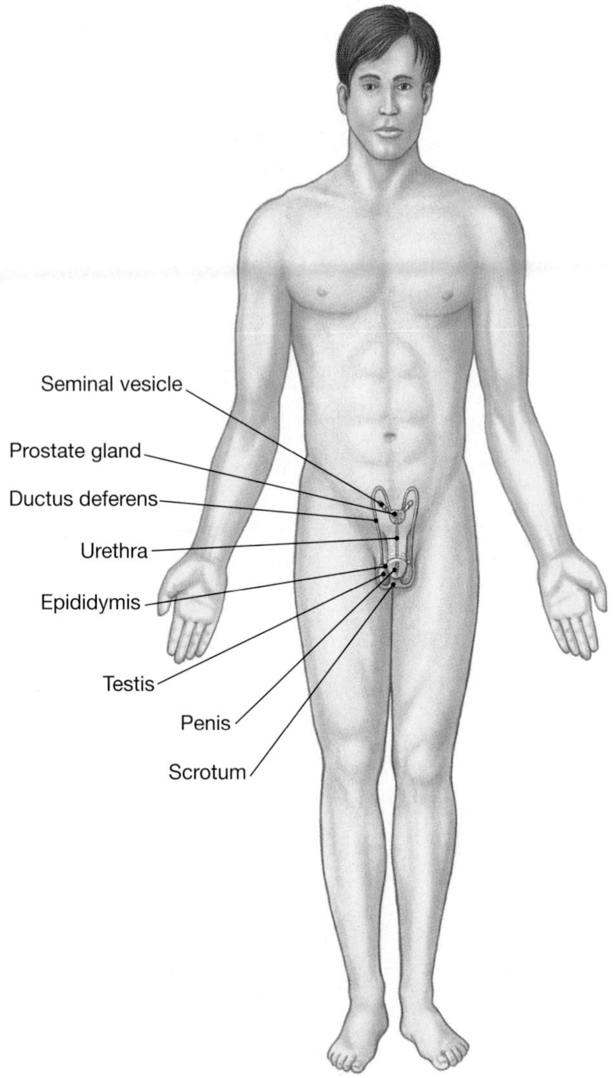

(k) The Male Reproductive System
Produces sex cells and hormones

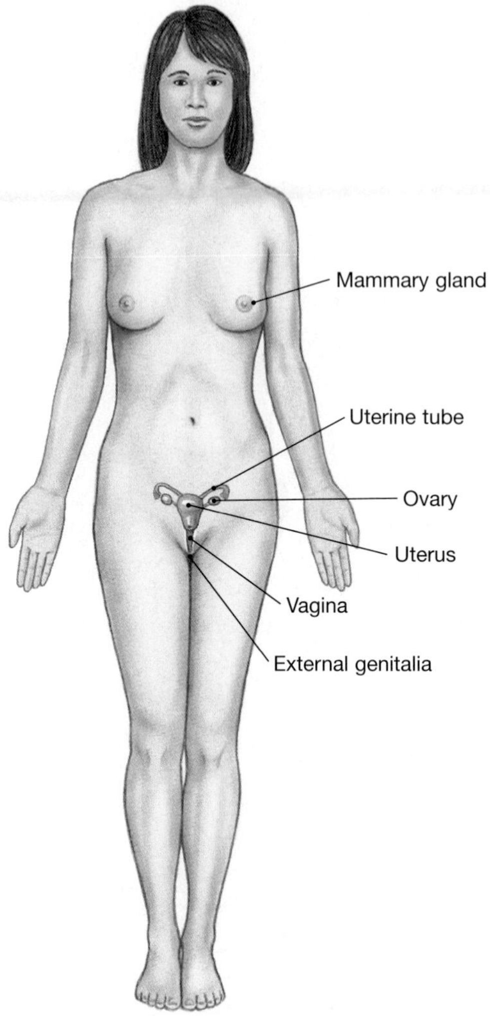

(l) The Female Reproductive System
Produces sex cells and hormones

Organ/Component	Primary Functions
TESTES	Produce sperm and hormones (*see Figure 1.6e*)
ACCESSORY ORGANS	
Epididymis	Acts as site of sperm maturation
Ductus deferens (sperm duct)	Conducts sperm between epididymis and prostate gland
Seminal vesicles	Secrete fluid that make up much of the volume of semen
Prostate gland	Secretes fluid and enzymes
Urethra	Conducts semen to exterior
EXTERNAL GENTALIA	
Penis	Contains erectile tissue; deposits sperm in vagina of female; produces pleasurable sensations during sexual activities
Scrotum	Surrounds the tests and controls their temperature

Organ/Component	Primary Functions
OVARIES	Produce oocytes and hormones (*see Figure 1.6e*)
UTERINE TUBES	Deliver oocyte or embryo to uterus; normal site of fertilization
UTERUS	Site of embryonic development and exchange between material and embryonic bloodstreams
VAGINA	Site of sperm deposition; acts as birth canal at delivery; provides passageway for fluids during menstruation
EXTERNAL GENITALIA	
Clitoris	Contains erectile tissue; produces pleasurable sensations during sexual activities
Labia	Contain glands that lubricate entrance to vagina
MAMMARY GLANDS	Produce milk that nourishes newborn infant

The Language of Anatomy [FIGURE 1.7]

If you discovered a new continent, how would you begin collecting information so that you could report your findings? You would have to construct a detailed map of the territory. The completed map would contain (1) prominent landmarks, such as mountains, valleys, or volcanoes; (2) the distance between them; and (3) the direction you traveled to get from one place to another. The distances might be recorded in miles, and the directions recorded as compass bearings (north, south, northeast, southwest, and so on). With such a map, anyone could go directly to a specific location on that continent.

Early anatomists faced similar communication problems. Stating that a bump is "on the back" does not give very precise information about its location. So anatomists created maps of the human body. The landmarks are prominent anatomical structures, and distances are measured in centimeters or inches. In effect, anatomy uses a special language that must be learned at the start. It will take some time and effort but is absolutely essential if you want to avoid a situation like that shown in Figure 1.7●.

New anatomical terms continue to appear as technology advances, but many of the older words and phrases remain in use. As a result, the vocabulary of this science represents a form of historical record. Latin and Greek words and phrases form the basis for an impressive number of anatomical terms. For example, many of the Latin names assigned to specific structures 2000 years ago are still in use today.

A familiarity with Latin roots and patterns makes anatomical terms more understandable, and the notes included on word derivation are intended to assist you in that regard. In English, when you want to indicate more than one of something, you usually add an *s* to the name—girl/girls or doll/dolls. Latin words change their endings. Those ending in *-us* convert to *-i*, and other conversions involve changing from *-um* to *-a*, and from *-a* to *-ae*. Additional information on foreign word roots, prefixes, suffixes, and combining forms can be found in Appendix IV on p. 823.

Latin and Greek terms are not the only foreign terms imported into the anatomical vocabulary over the centuries. Many anatomical structures and clinical conditions were initially named after either the discoverer or, in the case of diseases, the most famous victim. The major problem with this practice is that it is difficult for someone to remember a connection between the structure or disorder and the name. Over the last 100 years most of these commemorative names, or *eponyms*, have been replaced by more precise terms. For those interested in historical details, the section titled "Eponyms in Common Use" at the beginning of the Glossary provides information about the commemorative names in occasional use today.

Superficial Anatomy

A familiarity with major anatomical landmarks and directional references will make subsequent chapters more understandable, since none of the organ systems except the integument can be seen from the body surface. You must create your own mental maps and extract information from the anatomical illustrations that accompany this discussion.

ANATOMICAL LANDMARKS [FIGURE 1.8]

Important anatomical landmarks are presented in Figure 1.8●. You should become familiar with the adjectival form as well as the anatomical term. Understanding the terms and their origins will help you to remember the location of a particular structure, as well as its name. For example, the term **brachium** refers to the arm, and later chapters discuss the *brachialis muscle* and branches of the *brachial artery.*

Standard anatomical illustrations show the human form in the **anatomical position**. In the anatomical position, the person stands with the legs together and the feet flat on the floor. The hands are at the sides, and the palms face forward. The individual shown in Figure 1.8● is in the anatomical position as seen from the front (Figure 1.8a●) and back (Figure 1.8b●). Unless otherwise noted, all of the descriptions given in this text refer to the body in the anatomical position. A person lying down in the anatomical position is said to be **supine** (soo-PĪN) when lying face up and **prone** when lying face down.

Drawing by Ed Fisher; ©1990 The New Yorker Magazine, Inc.

FIGURE 1.7 **The Importance of Precise Vocabulary**

Would you want to be this patient?

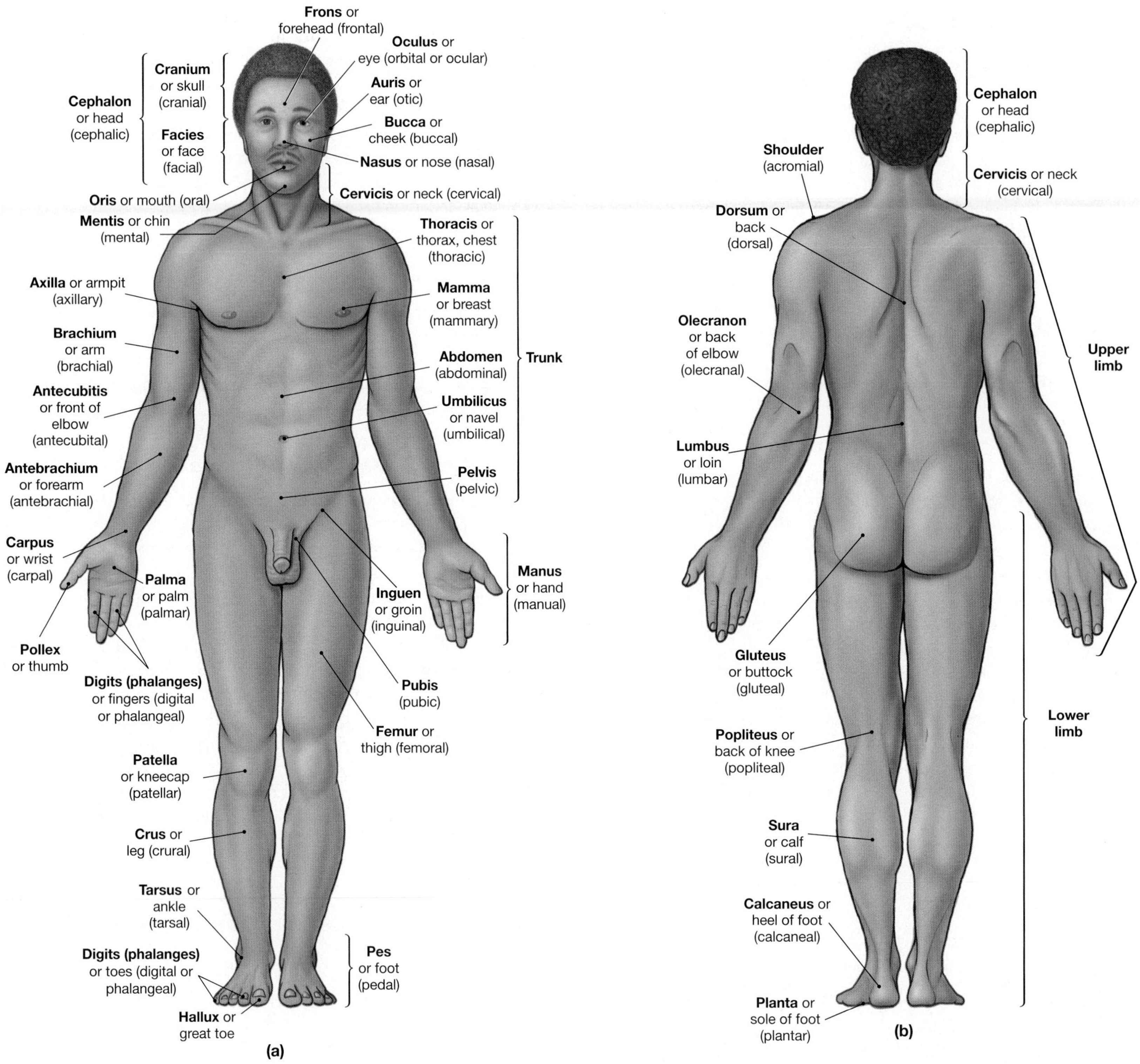

FIGURE 1.8 **ANATOMICAL LANDMARKS**

The anatomical terms are shown in boldface type, the common names are in plain type, and the anatomical adjectives are in parentheses. (**a**) Anterior view in the anatomical position. (**b**) Posterior view in the anatomical position.

ANATOMICAL REGIONS [FIGURES 1.8/1.9 AND TABLE 1.1]

Major regions of the body are indicated in Table 1.1. These and additional regions and anatomical landmarks are noted in Figure 1.8●. Anatomists and clinicians often use specialized regional terms to indicate a specific area of the abdominal or pelvic regions. There are two different methods in use. Clinicians refer to the **abdominopelvic quadrants**. The abdominopelvic surface is divided into four segments using a pair of imaginary lines (one horizontal and one vertical) that intersect at the *umbilicus* (navel). This simple method, shown in Figure 1.9a●, provides useful references for the description of aches, pains, and injuries. The location can assist the doctor in deciding the possible cause; for example, tenderness in the right lower quadrant (RLQ) is a symptom of appendicitis, whereas tenderness in the right upper quadrant (RUQ) may indicate gallbladder or liver problems. *Anatomy and Observation p. 784*

Anatomists tend to use more precise regional distinctions to describe the location and orientation of internal organs. They recognize nine **ab-**

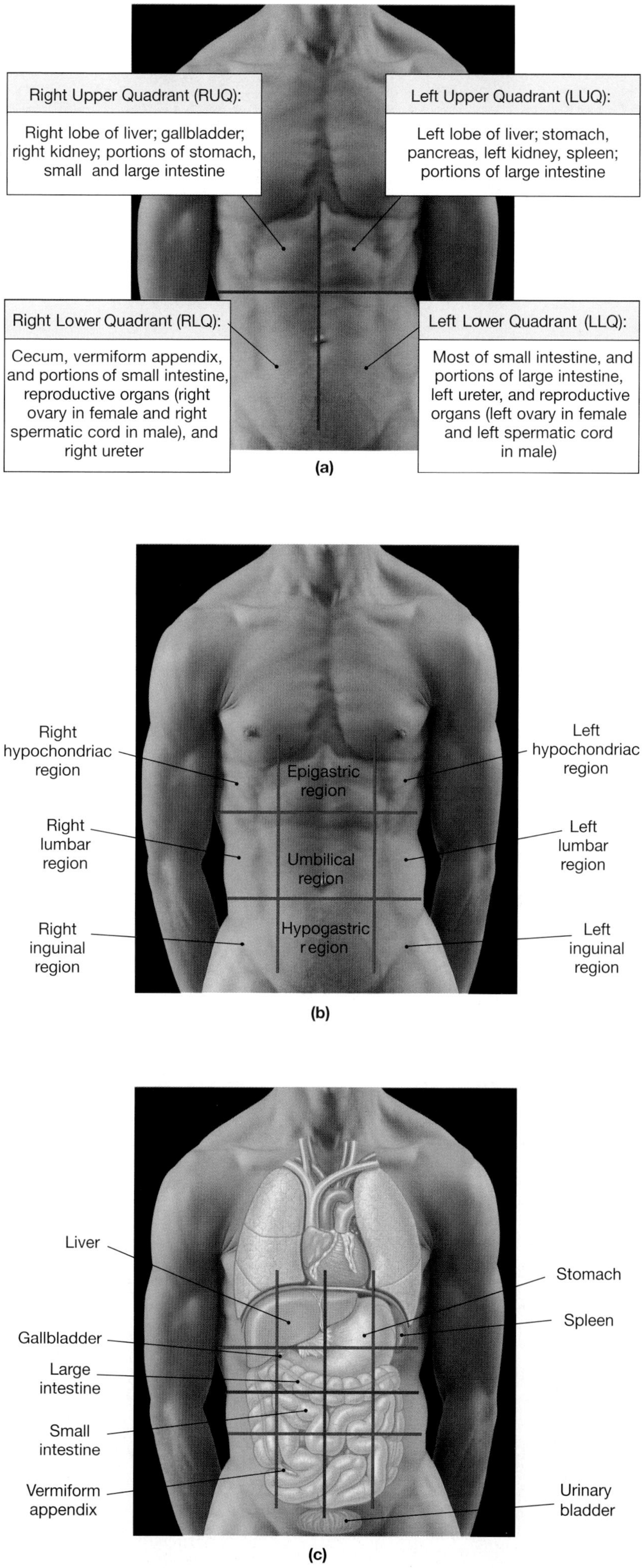

TABLE 1.1 **REGIONS OF THE HUMAN BODY***

Anatomical Name	Anatomical Region	Area Indicated
Cephalon	Cephalic	Area of head
Cervicis	Cervical	Area of neck
Thoracis	Thoracic	The chest
Brachium	Brachial	The segment of the upper limb closest to the trunk; the arm
Antebrachium	Antebrachial	The forearm
Carpus	Carpal	The wrist
Manus	Manual	The hand
Abdomen	Abdominal	The abdomen
Pelvis	Pelvic	The pelvis (in general)
Pubis	Pubic	The anterior pelvis
Inguen	Inguinal	The groin (crease between thigh and trunk)
Lumbus	Lumbar	The lower back
Gluteus	Gluteal	The buttock
Femur	Femoral	The thigh
Patella	Patellar	The kneecap
Crus	Crural	The leg, from knee to ankle
Sura	Sural	The calf
Tarsus	Tarsal	The ankle
Pes	Pedal	The foot
Planta	Sole	Plantar region of foot

*See Figures 1.8 and 1.9.

dominopelvic regions, shown in Figure 1.9b●. Figure 1.9c● shows the relationship between quadrants, regions, and internal organs.

ANATOMICAL DIRECTIONS [FIGURE 1.10 AND TABLE 1.2]

Figure 1.10● and Table 1.2 show the principal directional terms and examples of their use. There are many different terms, and some can be used interchangeably. For example, *anterior* refers to the front of the body, when viewed in the anatomical position; in humans, this term is equivalent to *ventral,* which actually refers to the belly side. Although your instructor may have additional terminology, the terms that appear frequently in later chapters have been emphasized in Table 1.2. When following anatomical descriptions, you will find it useful to remember that the terms *left* and *right* always refer to the left and right sides of the subject, not the observer. You should also note that although some reference terms are equivalent—*posterior* and *dorsal*, or *anterior* and *ventral*—anatomical descriptions use them in opposing pairs. For example, a discussion will give directions with reference either to posterior versus anterior, or dorsal versus ventral. Finally, you should be aware that some of the reference terms listed in Table 1.2 are not useful or have different meanings in veterinary anatomy.

FIGURE 1.9 **ABDOMINOPELVIC QUADRANTS AND REGIONS**

The abdominopelvic surface is separated into sections to identify anatomical landmarks more clearly and to define the location of contained organs more precisely. (**a**) Abdominopelvic quadrants divide the area into four sections. These terms, or their abbreviations, are most often used in clinical discussions. (**b**) More precise anatomical descriptions are provided by reference to the appropriate abdominopelvic region. (**c**) Quadrants or regions are useful because there is a known relationship between superficial anatomical landmarks and underlying organs.

FIGURE 1.10 **DIRECTIONAL REFERENCES**

Important directional references used in this text are indicated by arrows; definitions and descriptions are included in Table 1.2. (**a**) Lateral view. (**b**) Anterior view.

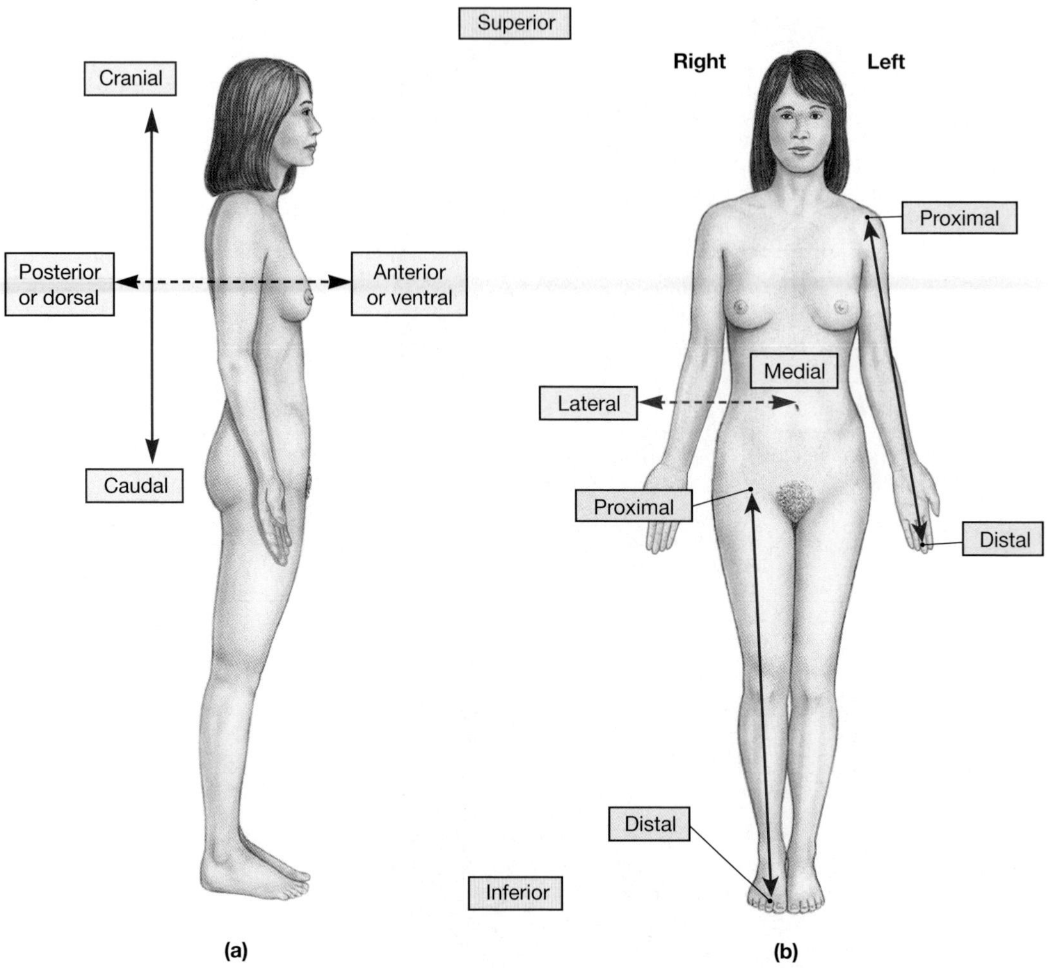

Sectional Anatomy

A presentation in sectional view is sometimes the only way to illustrate the relationships between the parts of a three-dimensional object. An understanding of sectional views has become increasingly important since the development of electronic imaging techniques that enable us to see inside the living body without resorting to surgery.

PLANES AND SECTIONS [FIGURES 1.11/1.12 AND TABLE 1.3]

Any slice through a three-dimensional object can be described with reference to three **sectional planes**, indicated in Table 1.3 and Figure 1.11•. The **transverse plane** lies at right angles to the long axis of the body, dividing it into **superior** and **inferior** sections. A division along this plane is called a **transverse section**, or *cross section*. The **frontal plane**, or **coronal plane**,

TABLE 1.2 REGIONAL AND DIRECTIONAL TERMS (SEE FIGURE 1.10)

Term	Region or Reference	Example
Anterior	The front; before	The navel is on the *anterior* surface of the trunk.
Ventral	The belly side (equivalent to anterior when referring to human body)	In humans, the navel is on the ventral surface.
Posterior	The back; behind	The shoulder blade is located *posterior* to the rib cage.
Dorsal	The back (equivalent to posterior when referring to human body)	The *dorsal* body cavity encloses the brain and spinal cord.
Cranial	Toward the head	The *cranial*, or *cephalic*, border of the pelvis is *superior* to the thigh.
Cephalic	Same as cranial	
Superior	Above; at a higher level (in human body, toward the head)	
Caudal	Toward the tail (coccyx in humans)	The hips are *caudal* to the waist.
Inferior	Below; at a lower level; toward the feet	The knees are *inferior* to the hips.
Medial	Toward the midline (the longitudinal axis of the body)	The *medial* surfaces of the thighs may be in contact.
Lateral	Away from the midline (the longitudinal axis of the body)	The femur articulates with the *lateral* surface of the pelvis.
Proximal	Toward an attached base	The thigh is *proximal* to the foot.
Distal	Away from an attached base	The fingers are *distal* to the wrist.
Superficial	At, near, or relatively close to the body surface	The skin is *superficial* to underlying structures.
Deep	Toward the interior of the body; farther from the surface	The bone of the thigh is *deep* to the surrounding skeletal muscles.

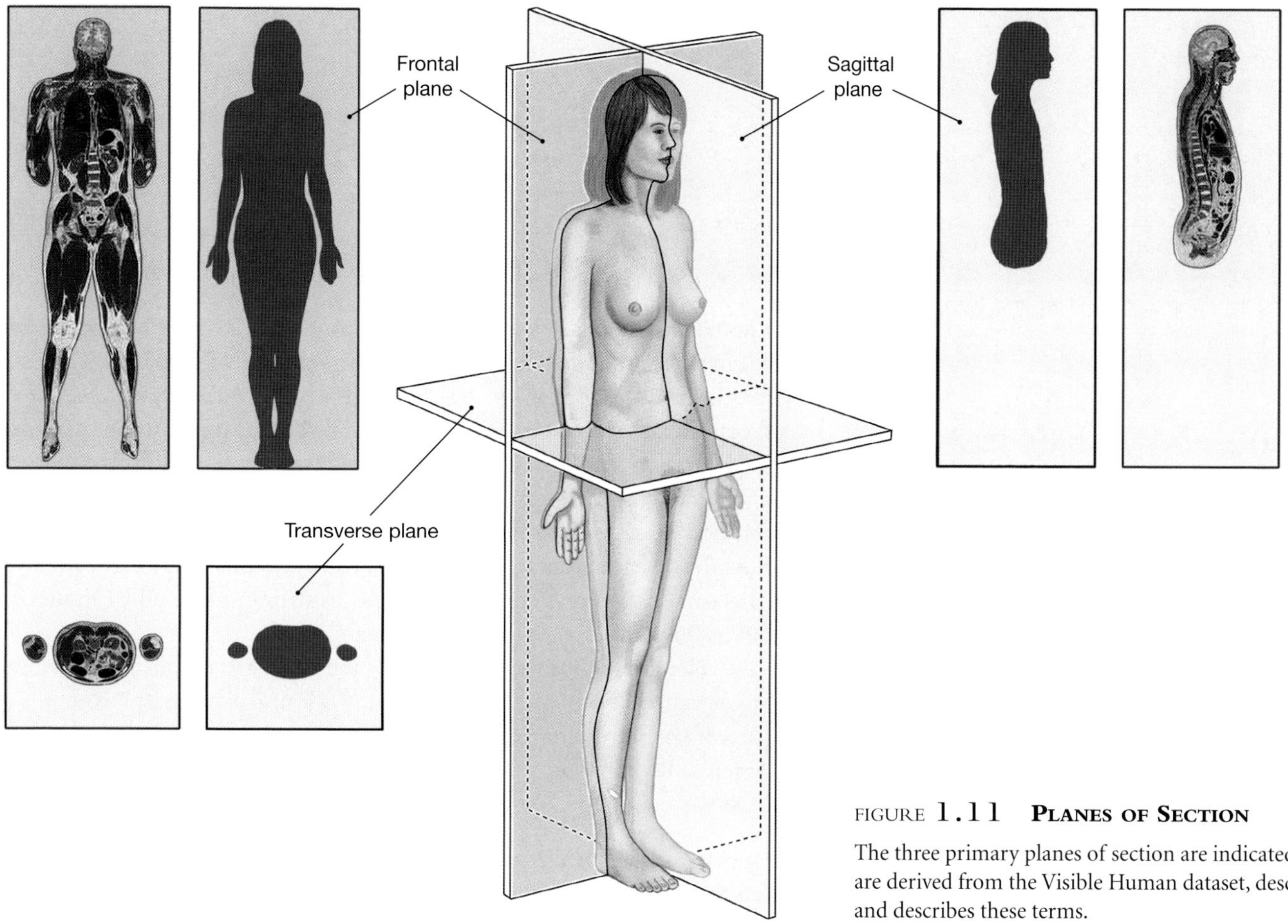

FIGURE 1.11 **PLANES OF SECTION**

The three primary planes of section are indicated here. The photographic images are derived from the Visible Human dataset, described on p. 20. Table 1.3 defines and describes these terms.

and the **sagittal plane** parallel the longitudinal axis of the body. The frontal plane extends from side to side, dividing the body into **anterior** and **posterior** sections. The sagittal plane extends from anterior to posterior, dividing the body into *left* and *right* sections. A section that passes along the midline and divides the body into left and right halves is a **midsagittal section**, or a **median sagittal section**; a section parallel to the midsagittal line is a **parasagittal section**.

Sometimes it is helpful to compare the information provided by sections made along different planes. Each sectional plane provides a different perspective on the structure of the body; when combined with observations on the external anatomy, they create a reasonably complete picture (see the Clinical Discussion on p. 21). You could develop a more accurate and complete picture by choosing one sectional plane and making a series of sections at small intervals. This process, called **serial reconstruction**, permits the analysis of relatively complex structures. Figure 1.12● shows the serial reconstruction of a simple bent tube, a piece of elbow macaroni. The procedure could be used to visualize the path of a small blood vessel or to follow a loop of the intestine. Serial reconstruction is an important method for studying histological structure and for analyzing the images produced by sophisticated clinical procedures (see the Clinical Discussion on p. 23).

TABLE 1.3 **TERMS THAT INDICATE PLANES OF SECTION (SEE FIGURE 1.10)**

Orientation of Plane	Adjective	Directional Term	Description
Perpendicular to long axis	Transverse or horizontal or cross-sectional	Transversely or horizontally	A *transverse*, or *horizontal*, *section* separates superior and inferior portions of the body; sections typically pass through head and trunk regions.
Parallel to long axis	Sagittal	Sagittally	A *sagittal section* separates right and left portions. You examine a sagittal section, but you section sagittally.
	Midsagittal		In a *midsagittal* section the plane passes through the midline, dividing the body in half and separating right and left sides.
	Parasagittal		A *parasagittal section* misses the midline, separating right and left portions of unequal size.
	Frontal or coronal	Frontally or coronally	A *frontal*, or *coronal*, *section* separates anterior and posterior portions of the body; coronal usually refers to sections passing through the skull.

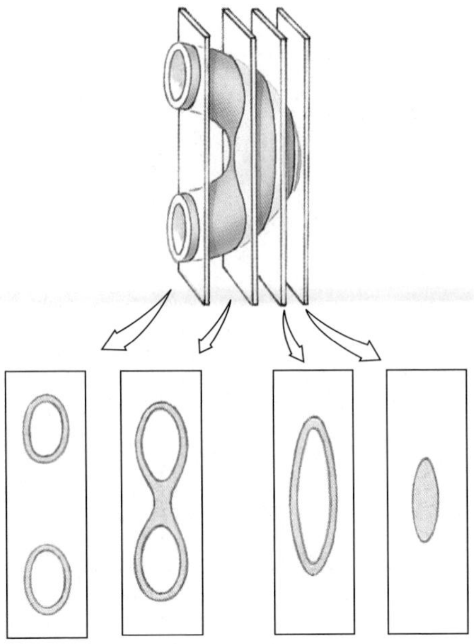

FIGURE 1.12 **SECTIONAL PLANES AND VISUALIZATION**

Here we are serially sectioning a bent tube, like a piece of elbow macaroni. Notice how the sectional views change as one approaches the curve; the effects of sectioning must be kept in mind when looking at slides under the microscope. They also affect the appearance of internal organs when seen in a sectional view, through a CT or MRI scan (see p. 22). For example, although it is a simple tube, the small intestine can look like a pair of tubes, a dumbbell, an oval, or a solid, depending on where the section was taken.

✓ CONCEPT CHECK

- What type of section would separate the two eyes?
- You fall and break your antebrachium. What part of the body is affected?
- What is the anatomical name for each of the following areas: groin, buttock, thigh?

BODY CAVITIES [FIGURE 1.13/1.14]

Viewed in sections, the human body is not a solid object, and many vital organs are suspended in internal chambers called **body cavities**. These cavities have two essential functions: (1) they protect delicate organs, such as the brain and spinal cord, from accidental shocks, and cushion them from the thumps and bumps that occur during walking, jumping, and running; and (2) they permit significant changes in the size and shape of body (visceral) organs. For example, because they are situated within body cavities, the lungs, heart, stomach, intestines, urinary bladder, and many other organs can expand and contract without distorting surrounding tissues and disrupting the activities of nearby organs.

The **dorsal body cavity** contains the brain and spinal cord; the much larger **ventral body cavity**, or **coelom** (SĒ-lōm; *koila*, cavity), contains organs of the respiratory, cardiovascular, digestive, urinary, and reproductive systems. Relationships between the dorsal and ventral body cavities and their various subdivisions are diagrammed in Figure 1.13●.

DORSAL BODY CAVITY [FIGURE 1.14a,c] The dorsal body cavity (Figure1.14a●) is a fluid-filled space whose limits are established by the *cranium*, the bones

FIGURE 1.13 **BODY CAVITIES**

Relationships, contents, and some selected functions of body cavities.

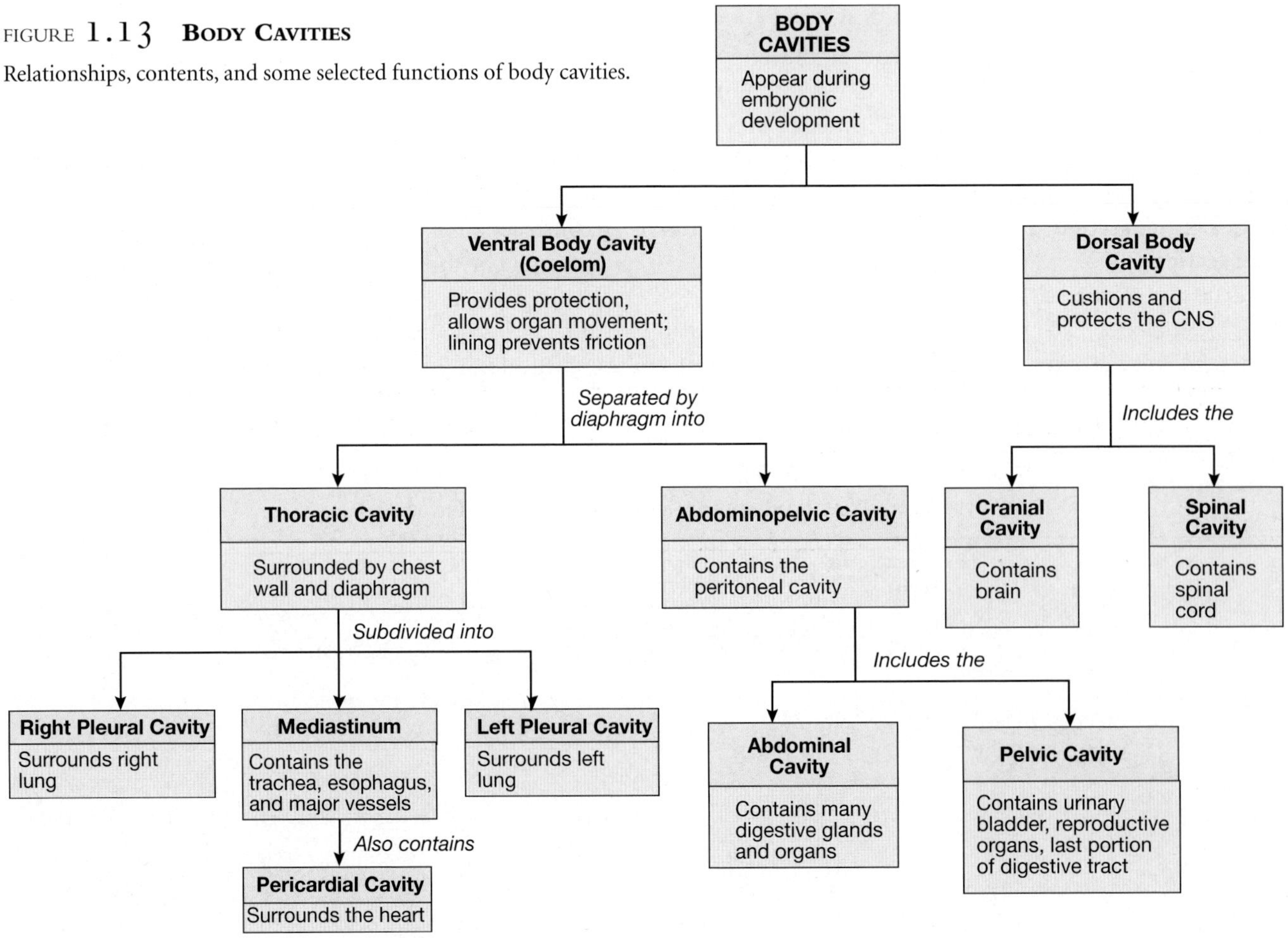

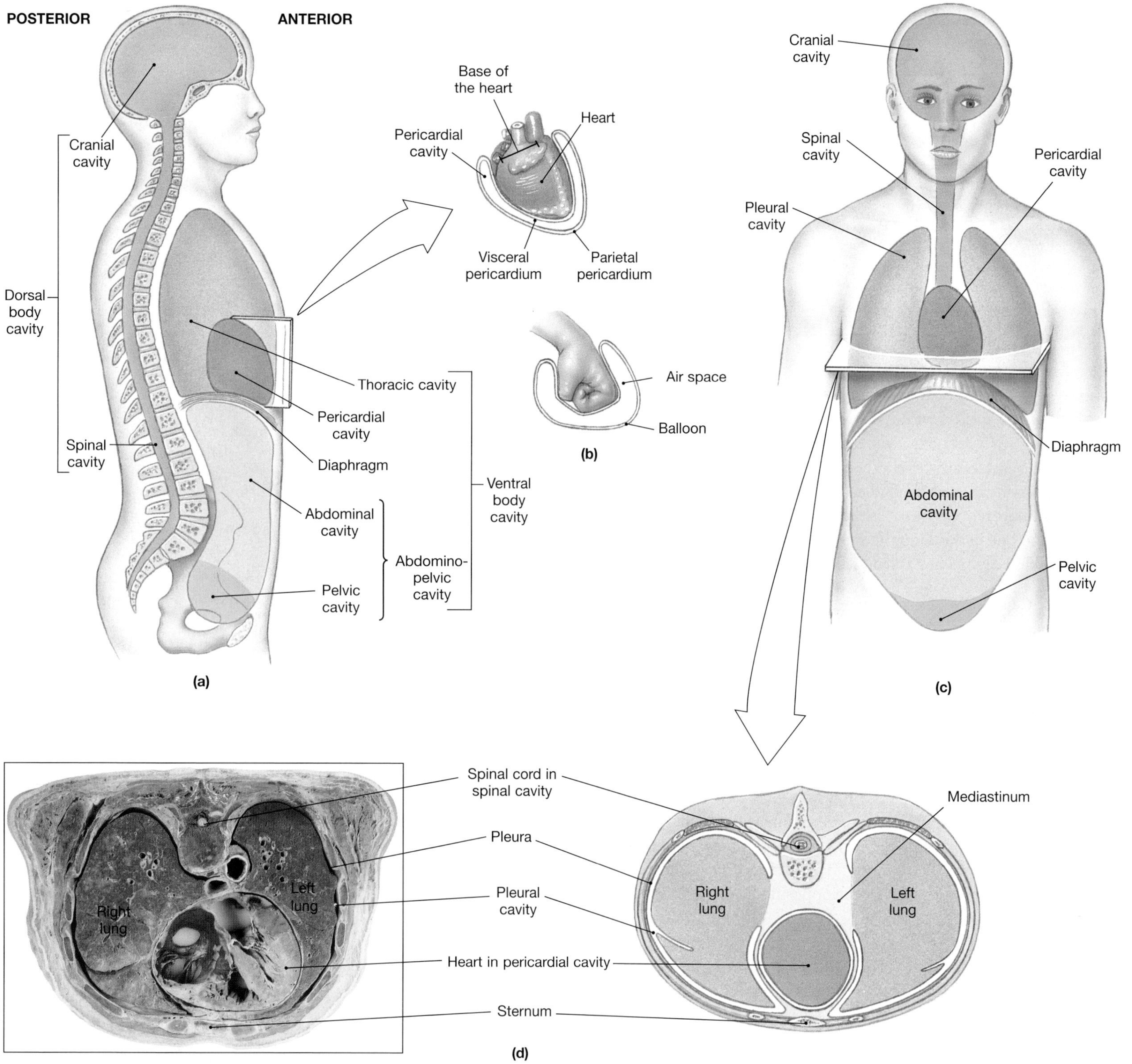

FIGURE 1.14 **Body Cavities**

Many vital organs are suspended in either the dorsal or the ventral body cavity. (**a**) Lateral view of dorsal and ventral body cavities. The dorsal body cavity is bounded by the bones of the skull and vertebral column. The muscular diaphragm divides the ventral body cavity into a superior thoracic (chest) cavity and an inferior abdominopelvic cavity. (**b**) The heart projects into the pericardial cavity like a fist pushed into a balloon. (**c**) Dorsal and ventral body cavities. (**d**) Sectional view of the thoracic cavity.

of the skull that surround the brain, and the *spinal canal*, enclosed by the spinal vertebrae. The dorsal body cavity is subdivided into the **cranial cavity**, which encloses the brain, and the **spinal (vertebral) cavity**, which surrounds the spinal cord.

Ventral Body Cavity [FIGURE 1.14a] As development proceeds, internal organs grow and change their relative positions. These changes lead to the subdivision of the ventral body cavity. The **diaphragm** (DĪ-a-fram), a flat muscular sheet, separates the ventral body cavity into a superior *thoracic*

CLINICAL DISCUSSION

THE VISIBLE HUMAN

The goal of the Visible Human project, funded by the U.S. National Library of Medicine, has been to create an accurate computerized human body that can be studied and manipulated in ways that would be impossible using a real body. The dataset in its current form consists of the digital images of cross-sections painstakingly prepared (by Dr. Victor Spitzer and colleagues at the University of Colorado Health Sciences Center) at 1 mm intervals for the visible male and 0.33 mm for the visible female. Even the relatively "low resolution" datasets are enormous—the male sections total 14 GB and the female sections total 40 GB. These images can be viewed on the web at **http://www.nlm.nih.gov**. These data have subsequently been used to generate a variety of enhanced images, such as those appearing in Figure 1.11●, and for interactive educational projects, such as the Digital Cadaver™. You will find images derived from the Visible Human project scattered throughout this text, and they form the basis for the anatomical animations found on the CD-ROM that accompanies this text.

cavity, enclosed by the chest wall, and an inferior *abdominopelvic cavity*, enclosed by the abdominal wall and pelvis.

Many of the organs within these cavities change size and shape as they perform their functions. For example, the stomach swells at each meal, and the heart contracts and expands with each beat. These organs project into moist internal chambers that permit expansion and limited movement, but prevent friction. There are three such chambers in the thoracic cavity and one in the abdominopelvic cavity. The internal organs that project into these cavities are called **viscera** (VIS-e-ra).

The Thoracic Cavity The **thoracic cavity** contains the lungs and heart, associated organs of the respiratory, cardiovascular, and lymphatic systems, the inferior portions of the esophagus, and the thymus. The boundaries of the thoracic cavity are established by the muscles and bones of the chest wall and the diaphragm, a muscular sheet that separates the thoracic cavity from the abdominopelvic cavity (see Figure 1.14a,c●). The thoracic cavity is subdivided into the left and right *pleural cavities* separated by the **mediastinum** (mē-dē-as-TĪ-num or mē-dē-AS-ti-num) (Figure 1.14a, c,d●).

Each **pleural cavity** contains a lung. The cavity is lined by a shiny, slippery *serous membrane*, which reduces friction as the lung expands and recoils during respiration. The serous membrane lining a pleural cavity is called a *pleura* (PLOO-rah). The *visceral pleura* covers the outer surfaces of a lung, and the *parietal pleura* covers the opposing mediastinal surface and the inner body wall.

The mediastinum consists of a mass of connective tissue that surrounds, stabilizes, and supports the esophagus, trachea, thymus, and the major blood vessels that originate or end at the heart. It also contains the **pericardial cavity**, a small chamber that surrounds the heart (Figure 1.14d●). The relationship between the heart and the pericardial cavity resembles that of a fist pushing into a balloon (Figure 1.14b●). The wrist corresponds to the base (attached portion) of the heart, and the balloon corresponds to the serous membrane that lines the pericardial cavity. The serous membrane covering the heart is called the **pericardium** (*peri-*, around + *kardia*, heart). The layer covering the heart is the *visceral pericardium*, and the opposing surface is the *parietal pericardium*. During each beat, the heart changes in size and shape. The pericardial cavity permits these changes, and the slippery pericardial lining prevents friction between the heart and adjacent structures in the mediastinum.

The Abdominopelvic Cavity The **abdominopelvic cavity** can be divided into a superior *abdominal cavity* and an inferior *pelvic cavity* (Figures 1.13 and 1.14a,c●). The abdominopelvic cavity contains the **peritoneal** (per-i-tō-NĒ-al) **cavity**, an internal chamber lined by a serous membrane known as the **peritoneum** (per-i-tō-NĒ-um). The *parietal peritoneum* lines the body wall. A narrow, fluid-filled space separates the parietal peritoneum from the *visceral peritoneum* that covers the enclosed organs. Organs such as the stomach, small intestine, and portions of the large intestine are suspended within the peritoneal cavity by double sheets of peritoneum, called **mesenteries** (MES-en-ter-ēs). Mesenteries provide support and stability while permitting limited movement.

- The **abdominal cavity** extends from the inferior surface of the diaphragm to an imaginary plane extending from the inferior surface of the lowest spinal vertebra to the anterior and superior margins of the pelvic girdle. The abdominal cavity contains the liver, stomach, spleen, kidneys, pancreas, small intestine, and most of the large intestine. (The positions of many of these organs can be seen in Figure 1.9c●, p. 15). These organs project partially or completely into the peritoneal cavity, much as the heart or lungs project into the pericardial or pleural cavities.
- The portion of the ventral body cavity inferior to the abdominal cavity is the **pelvic cavity**. The pelvic cavity, enclosed by the bones of the pelvis, contains the last segments of the large intestine, the urinary bladder, and various reproductive organs. For example, the pelvic cavity of a female contains the ovaries, uterine tubes, and uterus; in a male, it contains the prostate gland and seminal vesicles. The inferior portion of the peritoneal cavity extends into the pelvic cavity. The superior portion of the urinary bladder in both sexes, and the uterine tubes, the ovaries, and the superior portion of the uterus in females are covered by peritoneum.

✓ CONCEPT CHECK

- What is the general function of the mesenteries?
- If a surgeon makes an incision just inferior to the diaphragm, which body cavity will be opened?
- Use a directional term to describe the following:
  a. The toes are ____________ to the tarsus.
  b. The hips are ____________ to the head.

This chapter provided an overview of the locations and functions of the major components of each organ system, and it introduced the anatomical vocabulary needed to follow more detailed anatomical descriptions in later chapters. Modern methods of visualizing anatomical structures in living individuals are summarized in the Clinical Discussion and Figure 1.17●. Many of the figures in later chapters contain sectional images, and the Companion Atlas provides additional reference plates produced by the procedures outlined in that section. However, a true understanding of anatomy involves integrating the information provided by sectional images, interpretive artwork based on sections and dissections, and direct observation. This text will give you the basic information and show you interpretive illustrations, sectional views, and "real life" photos. But it will be up to you to integrate these views and develop your ability to observe and visualize anatomical structures. As you proceed, don't forget that every structure you encounter has a specific function. The goal of anatomy isn't simply to identify and catalog structural details, but to understand how those structures interact to perform the many and varied functions of the human body.

CLINICAL DISCUSSION

SECTIONAL ANATOMY AND CLINICAL TECHNOLOGY

Radiological procedures include various noninvasive techniques that use radioisotopes, radiation, and magnetic fields to produce images of internal structures. Physicians who specialize in the performance and analysis of these diagnostic images are called **radiologists**. Radiological procedures can provide detailed information about internal systems and structures. Figures1.15● through 1.17● compare the views provided by several different techniques; other examples will be found in later chapters and in the Companion Atlas. Most of the procedures produce black-and-white images on film sheets. Colors can be added by computer to illustrate subtle variations in contrast and shading. Note that when anatomical diagrams or scans present cross-sectional views, the sections are presented as though the observer were standing at the feet and looking toward the head of the subject.

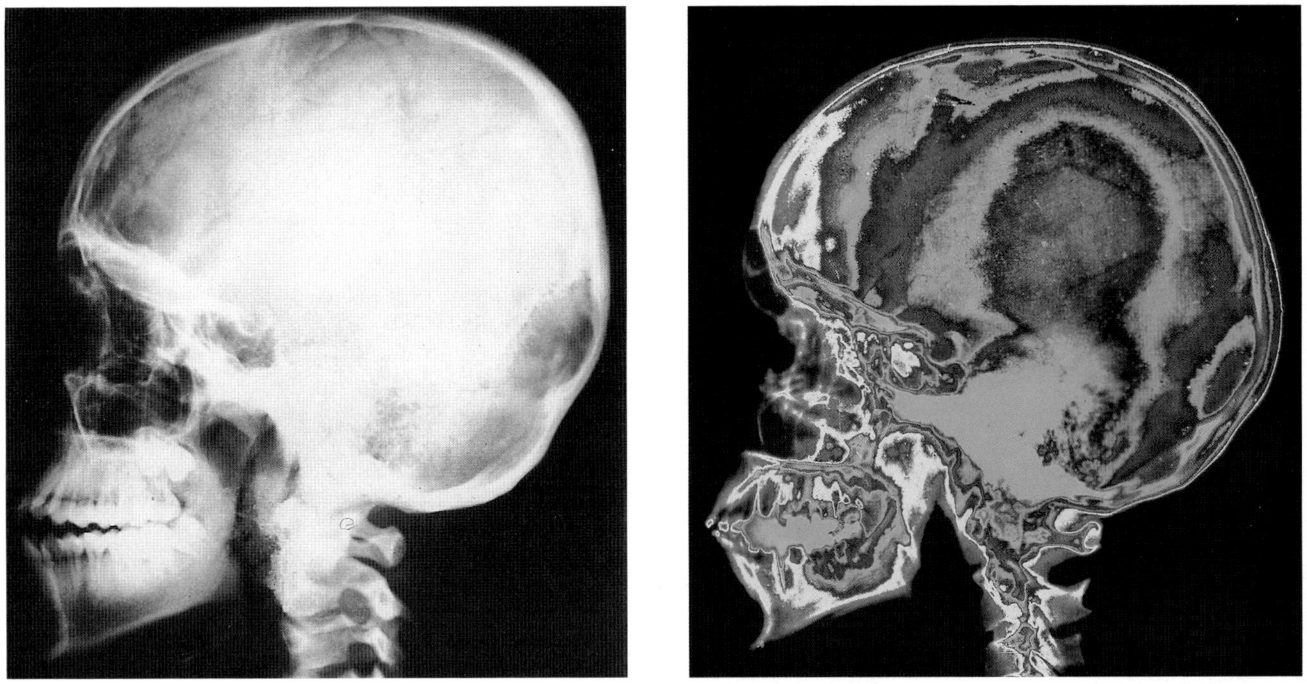

(a)

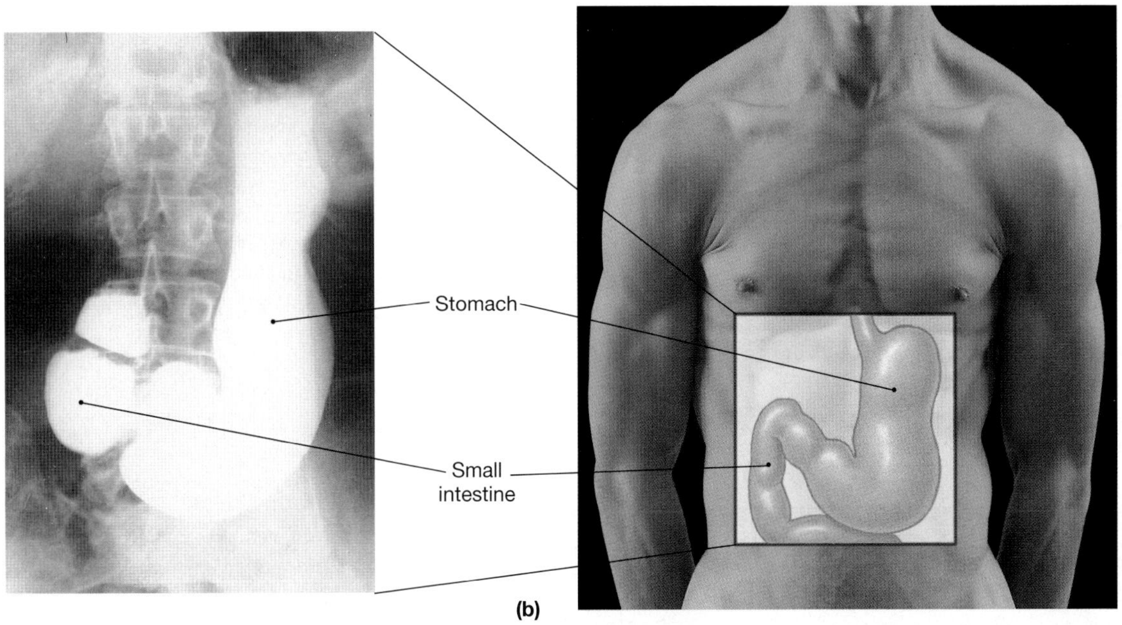

(b)

FIGURE 1.15 **X-RAYS**

Body images created by radiological imaging techniques. **(a)** An X-ray of the skull and color-enhanced X-ray of the skull, taken from the left side. X-rays are a form of high-energy radiation that can penetrate living tissues. In the most familiar procedure, a beam of X-rays travels through the body and strikes a photographic plate. All of the projected X-rays do not arrive at the film; some are absorbed or deflected as they pass through the body. The resistance to X-ray penetration is called **radiodensity**. In the human body, the order of increasing radiodensity is as follows: air, fat, liver, blood, muscle, bone. The result is an image with radiodense tissues, such as bone, appearing white, while less dense tissues are seen in shades of gray to black. (The image to the right has been scanned and digitally color enhanced.) A typical x-ray is a two-dimensional image of a three-dimensional object; it is usually difficult to decide whether a particular feature is on the left side (toward the viewer) or on the right side (away from the viewer). **(b)** A **barium-contrast X-ray** of the upper digestive tract. Barium is very radiodense, and the contours of the gastric and intestinal lining can be seen outlined against the white of the barium solution.

CLINICAL DISCUSSION (CONTINUED)

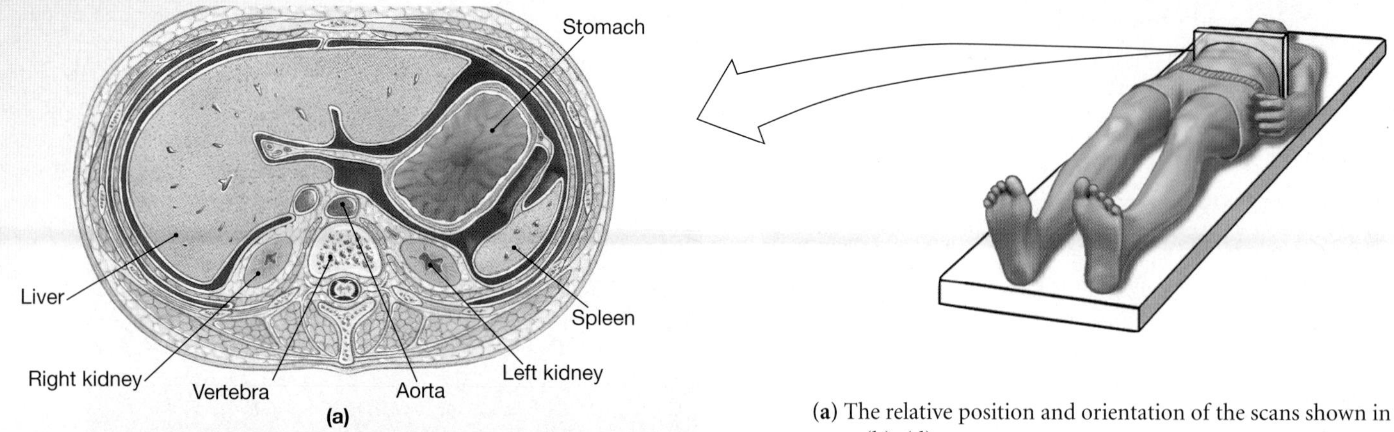

(a)

(a) The relative position and orientation of the scans shown in parts (b)–(d).

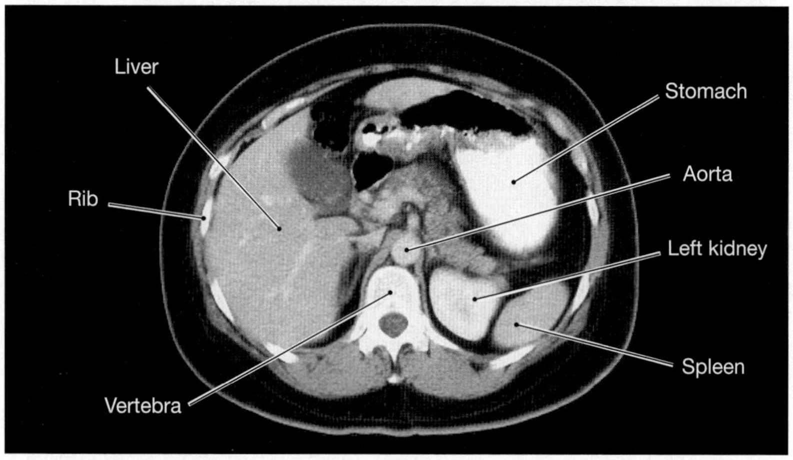

(b)

(b) A CT scan of the abdomen. **CT** (**C**omputerized **T**omography), formerly called **CAT** (Computerized Axial Tomography), uses computers to reconstruct sectional views. A single X-ray source rotates around the body, and the X-ray beam strikes a sensor monitored by the computer. The source completes one revolution around the body every few seconds; it then moves a short distance and repeats the process. By comparing the information obtained at each point in the rotation, the computer reconstructs the three-dimensional structure of the body. The result is usually displayed as a sectional view in black and white, but it can be colorized. CT scans show three-dimensional relationships and soft tissue structure more clearly than standard X-rays.

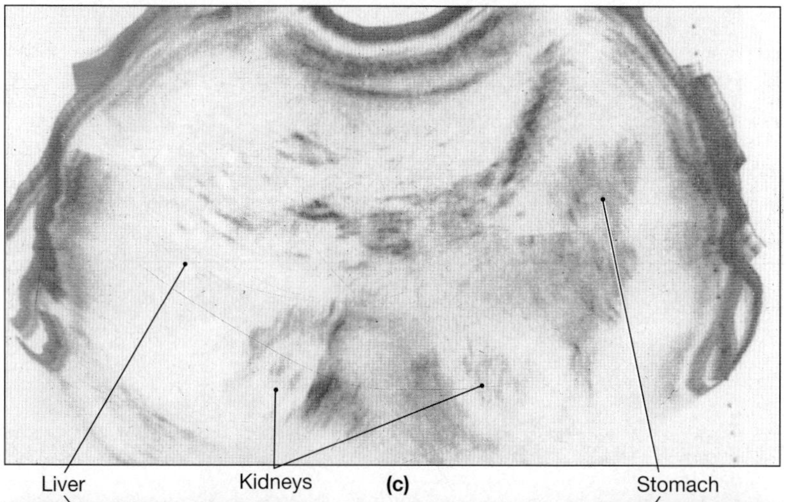

(c)

(c) An ultrasound scan of the abdomen. In **ultrasound** procedures, a small transmitter contacting the skin broadcasts a brief, narrow burst of high-frequency sound and then picks up the echoes. The sound waves are reflected by internal structures, and a picture, or **echogram**, can be assembled from the pattern of echoes. These images lack the clarity of other procedures, but no adverse effects have been attributed to the sound waves, and fetal development can be monitored without a significant risk of birth defects. Special methods of transmission and processing permit structural analysis of the beating heart, without the complications that can accompany dye injections.

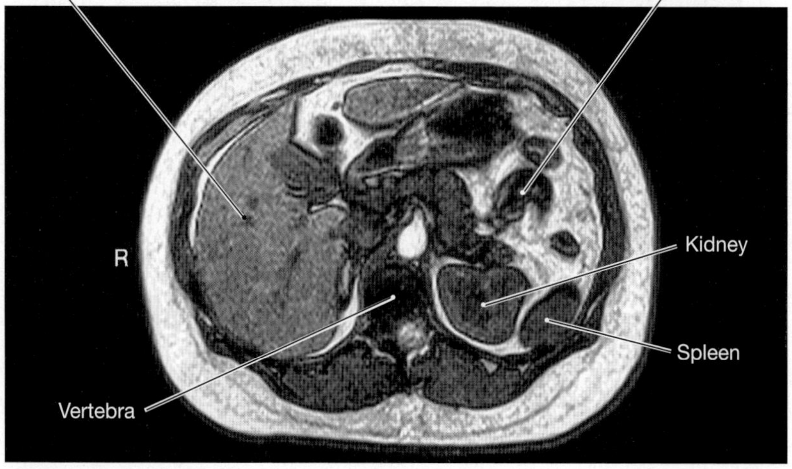

(d)

(d) An **MRI** (**M**agnetic **R**esonance **I**maging) scan of the same region. MRI surrounds part or all of the body with a magnetic field about 3000 times as strong as that of the earth. This field affects protons within atomic nuclei throughout the body, which line up along the magnetic lines of force like compass needles in the earth's magnetic field. When struck by a radio wave of the proper frequency, a proton will absorb energy. When the pulse ends, that energy is released, and the energy source of the radiation is detected by the MRI computers. Each element differs in terms of the radio frequency required to affect its protons. Note the differences in detail between this image, the CT scan, and the X-rays in Figure 1.15●.

FIGURE 1.16 **SCANNING TECHNIQUES**

CLINICAL DISCUSSION (CONTINUED)

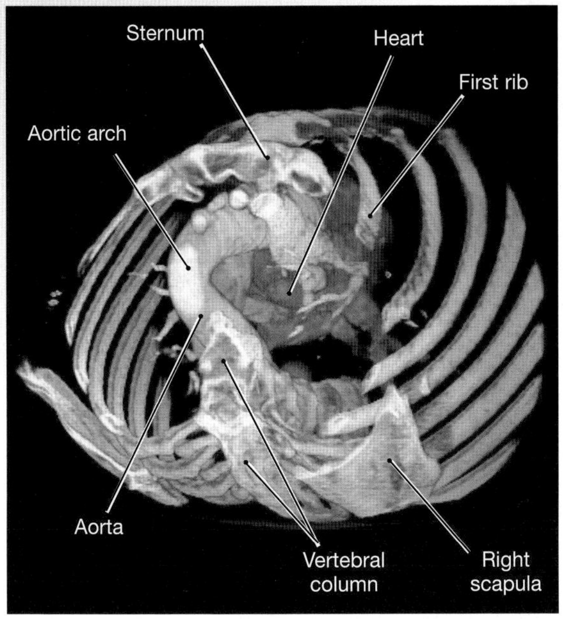

(a)

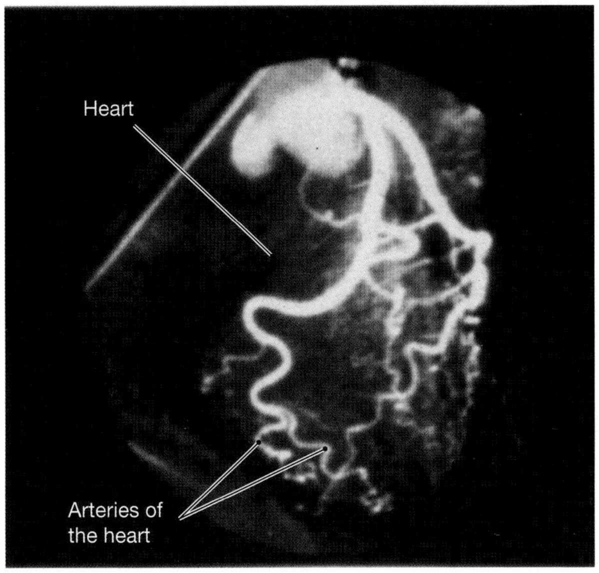

(b)

FIGURE 1.17 **SPECIAL SCANNING METHODS**

(a) A **spiral-CT** scan of the chest. Such an image is created by special processing of CT data. It permits rapid three-dimensional visualization of internal organs. Spiral-CT scans are becoming increasingly important in clinical settings. (b) **D**igital **S**ubtraction **A**ngiography (**DSA**) is used to monitor blood flow through specific organs, such as the brain, heart, lungs, or kidneys. X-rays are taken before and after radiopaque dye is administered, and a computer "subtracts" details common to both images. The result is a high-contrast image showing the distribution of the dye.

RELATED CLINICAL TERMS

abdominopelvic quadrant: One of four divisions of the anterior abdominal surface. *p. 15*

abdominopelvic region: One of nine divisions of the abdominal surface. *p. 15*

CT, CAT (computerized [axial] tomography): An imaging technique that reconstructs the three-dimensional structure of the body. *p. 22*

MRI (magnetic resonance imaging): An imaging technique that employs a magnetic field and radio waves to portray subtle structural differences. *p. 22*

radiologist: A physician who specializes in performing and analyzing diagnostic imaging procedures. *p. 21*

ultrasound: An imaging technique that uses brief bursts of high-frequency sound reflected by internal structures. *p. 22*

X-rays: High-energy radiation that can penetrate living tissues. *p. 21*

Additional Clinical Terms Discussed in Appendix I (p. 784)
auscultation; inspection; palpation; percussion

STUDY OUTLINE & CHAPTER REVIEW

Introduction 2

1. **Anatomy** is the study of both internal and external structures, and the physical relationships between body parts. Specific anatomical structures perform specific functions.

Microscopic Anatomy 2

1. The boundaries of **microscopic anatomy** are established by the equipment used. **Cytology** is the study of the internal structure of individual cells, the smallest units of life. **Histology** examines **tissues**, groups of cells that work together to perform specific functions. Specific arrangements of tissues form **organs**, anatomical units with multiple functions. A group of organs that function together forms an **organ system**. *(see Figure 1.1)*

Gross Anatomy 3

1. **Gross (macroscopic) anatomy** considers features relatively visible without a microscope. It includes **surface anatomy** (general form and superficial markings); **regional anatomy** (superficial and internal features in a specific area of the body); and **systemic anatomy** (structure of major organ systems).

Other Perspectives on Anatomy 3

1. **Developmental anatomy** examines the changes in form that occur continuously between conception and physical maturity. **Embryology** studies the processes that occur during the first 2 months of development.
2. **Comparative anatomy** considers both the similarities and relationships in anatomical organization of different animals. *(see Figure 1.2)*

3. Anatomical specialties important to clinical practice include **medical anatomy** (anatomical features that undergo characteristic changes during illness), **radiographic anatomy** (anatomical structures that are visualized by specialized procedures performed on an intact body), and **surgical anatomy** (landmarks important for surgical procedures). *(see Figures 1.15 to 1.17)*

Levels of Organization 4

1. Anatomical structures are arranged in a series of interacting levels of organization ranging from the chemical/molecular level, through cell/tissue levels, to the organ/system/organism level. *(see Figures 1.3/1.4)*

An Introduction to Organ Systems 6

1. All living organisms are recognized by a set of vital properties and processes: They **respond** to changes in their environment; they show *adaptability* to their environment; they **grow**, **differentiate**, and **reproduce** to create future generations; they are capable of producing **movement**; and they absorb materials from the environment, and use them in **metabolism**. Organisms absorb and consume oxygen during **respiration**, and discharge waste products during **excretion**. **Digestion** breaks down complex foods for use by the body. The **cardiovascular system** forms an internal transportation system between areas of the body. *(see Figures 1.5/1.6)*
2. The 11 organ systems of the human body perform these vital functions to maintain **homeostasis**. *(see Figure 1.4)*

The Language of Anatomy 13

1. Anatomy utilizes a special language that includes many terms and phrases derived from foreign languages, especially Latin and Greek. *(see Figures 1.7 to 1.14)*

Superficial Anatomy 13

2. Standard anatomical illustrations show the body in the **anatomical position**. Here, a person stands with the legs together and the feet flat on the floor. The hands are at the sides, and the palms face forward. *(see Figures 1.8/1.10)*
3. A person lying down in the anatomical position may be **supine** (face up) or **prone** (face down).
4. Specific terms identify specific anatomical regions. For example, *cephalic* (area of head), *cervical* (area of neck), and *thoracic* (area of chest). Other terms, including *abdominal, pelvic, lumbar, gluteal, pubic, brachial, antebrachial, manual, femoral, patellar, crural, sural, and pedal*, are applied to specific regions of the body. *(see Figure 1.8 and Table 1.1)*
5. **Abdominopelvic quadrants** and **abdominopelvic regions** represent two different approaches to describing locations in the abdominal and pubic areas of the body. *(see Figure 1.9)*
6. There are specific directional terms used to indicate relative location on the body. For example, *anterior* (front, before), *posterior* (back, behind), and *dorsal* (back). Other directional terms encountered throughout the text include: *ventral, superior, inferior, medial, lateral, cranial, cephalic, caudal, proximal, and distal. (see Figure 1.10 and Table 1.2)*

Sectional Anatomy 16

7. There are three **sectional planes**: **frontal plane** or **coronal plane** (anterior vs. posterior), **sagittal plane** (right vs. left sides), and **transverse plane** (superior vs. inferior). These sectional planes and related reference terms describe relationships between the parts of the three-dimensional human body. *(see Figure 1.11)*
8. **Serial reconstruction** is an important technique for studying histological structure and analyzing images produced by radiological procedures. *(see Figure 1.12)*
9. **Body cavities** protect delicate organs and permit changes in the size and shape of visceral organs. The **dorsal body cavity** includes the **cranial cavity** (enclosing the brain) and **spinal cavity** (surrounding the spinal cord). The **ventral body cavity**, or **coelom**, surrounds organs of the respiratory, cardiovascular, digestive, urinary, and reproductive systems. *(see Figures 1.13/1.14)*
10. The **diaphragm** divides the ventral body cavity into the superior **thoracic** and inferior **abdominopelvic cavities**. *(see Figures 1.13/1.14)*
11. The **abdominal cavity** extends from the inferior surface of the diaphragm to an imaginary line drawn from the inferior surface of the most inferior spinal vertebra to the anterior and superior margin of the pelvic girdle. The portion of the ventral body cavity inferior to this imaginary line is the **pelvic cavity**. *(see Figures 1.13/1.14)*
12. The ventral body cavity contains narrow, fluid-filled spaces lined by a *serous membrane*. The thoracic cavity contains two **pleural cavities** (each surrounding a lung) separated by the **mediastinum**. *(see Figures 1.13/1.14)*
13. The mediastinum contains the thymus, trachea, esophagus, blood vessels, and the **pericardial cavity**, which surrounds the heart. The membrane lining the pleural cavities is called the **pleura**; the membrane lining the pericardial cavity is called the pericardium. *(see Figures 1.13/1.14)*
14. The **abdominopelvic cavity** contains the **peritoneal cavity**, which is lined by the **peritoneum**. Many digestive organs are supported and stabilized by **mesenteries**.
15. Important **radiological procedures**, which can provide detailed information about internal systems, include **X-rays**, **CT scans**, **MRI**, and **ultrasound**. Physicians who perform and analyze these procedures are called **radiologists**. *(see Figures 1.15 to 1.17)*

LEVEL 1 REVIEWING FACTS AND TERMS

Match each numbered item with the most closely related lettered item. Use letters for answers in the spaces provided.

Column A	Column B
_____ 1. supine	a. study of tissues
_____ 2. cytology	b. face down
_____ 3. homeostasis	c. thoracic and abdominopelvic
_____ 4. lumbar	d. all chemical activity in body
_____ 5. prone	e. study of cells
_____ 6. metabolism	f. face up
_____ 7. ventral body cavity	g. constant internal environment
_____ 8. histology	h. lower back

9. A plane that passes perpendicular to the long axis of the body, dividing the body into a superior and inferior sections, is
   (a) sagittal (b) coronal
   (c) transverse (d) frontal
10. The study of the changes in form that occur during the period between conception and physical maturity is
    (a) systemic anatomy (b) comparative anatomy
    (c) developmental anatomy (d) microscopic anatomy
11. The process by which an organism increases the size and/or number of cells is
    (a) reproduction (b) metabolism
    (c) adaptation (d) growth
12. The mediastinum is the part of the thoracic cavity that lies between the
    (a) heart and pericardium
    (b) two pleural cavities
    (c) lungs and heart
    (d) pleural cavities and diaphragm

13. Making a sagittal section results in the separation of
(a) anterior and posterior portions of the body
(b) superior and inferior portions of the body
(c) dorsal and ventral portions of the body
(d) right and left portions of the body

14. Support, protection of soft tissue, mineral storage, and blood cell formation are functions of the
(a) skeletal system (b) lymphatic system
(c) integumentary system (d) endocrine system

15. Which of the following regions corresponds to the arm?
(a) cervical (b) brachial
(c) femoral (d) pedal

LEVEL 2 REVIEWING CONCEPTS

1. From the following selections, identify the directional terms equivalent to *ventral, posterior, superior*, and *inferior* in the correct sequence.
(a) anterior, dorsal, cephalic, caudal
(b) dorsal, anterior, caudal, cephalic
(c) caudal, cephalic, anterior, posterior
(d) cephalic, caudal, posterior, anterior

2. What properties and processes are associated with all living things?

3. Describe the position of the body in the anatomical position.

4. The body system that performs crisis management by directing rapid, short-term, and very specific responses is the
(a) lymphatic system (b) nervous system
(c) cardiovascular system (d) endocrine system

5. Which sectional plane could divide the body so that the face remains intact?
(a) sagittal section (b) coronal section
(c) midsagittal section (d) none of the above

6. What is the role of serous membranes in the body?

LEVEL 3 CRITICAL THINKING AND CLINICAL APPLICATIONS

1. How might events that follow a chemical imbalance in a heart muscle cell support the view that all levels of organization within an organism are interdependent?

2. A child born with a severe cleft palate may require surgery to repair the nasal cavity and reconstruct the roof of the mouth. What body systems are affected by the cleft palate? Also, studies of other mammals that develop cleft palates have helped us to understand the origins and treatment of such problems. What anatomical specialties are involved in identifying and correcting a cleft palate?

✓ ANSWERS TO CONCEPT CHECK QUESTIONS

p. 4 **1.** A histologist investigates the structure and properties of tissue. **2.** A gross anatomist investigates organ systems and their relationships to the body as a whole. **3.** Regional anatomy considers all of the superficial and internal features in a specific area of the body, such as the head, neck, or trunk. Systemic anatomy considers the structure of major organ systems, such as the skeletal or muscular system.
p. 6 **1.** Integumentary system. **2.** Reproductive system of the female. **3.** The gradual appearance of characteristic cellular specializations during development, as the result of gene activation or repression. **4.** Individual cells become specialized to perform specific functions.
p. 18 **1.** The two eyes would be separated by a midsagittal section. **2.** The fall would affect your forearm. **3.** Groin = inguen, buttock = gluteus, and thigh = femur.
p. 20 **1.** Mesenteries are double sheets of serous membranes in the peritoneal cavity that provide support and stability for organs [stomach, small intestine, and parts of the large intestine] while permitting limited movement. **2.** The body cavity inferior to the diaphragm is the abdominopelvic cavity. **3.** (a) distal, (b) inferior

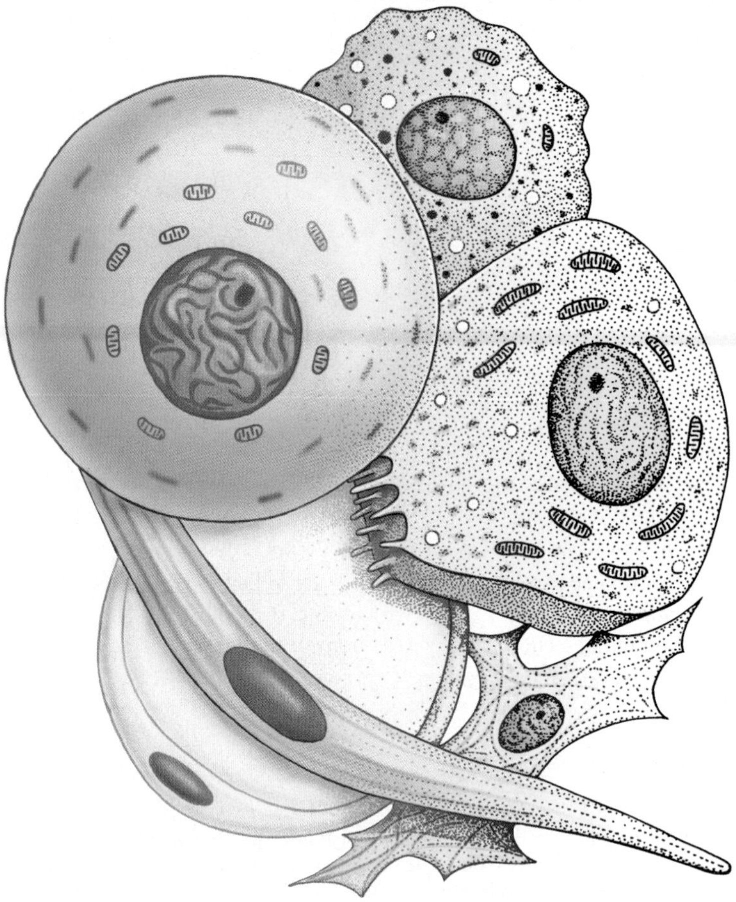

2

THE CELL

If you walk through a building supply store, you see many individual items —bricks, floor tiles, wall paneling, and a large assortment of lumber. Each item by itself is unremarkable and of very limited use. Yet, if you have all of them in sufficient quantity, you can build a functional unit, in this case a house. The human body is also composed of a multitude of individual components called *cells*. Much as individual bricks and lumber collectively form a wall of a house, individual cells work together to form *tissues*, such as the muscular wall of the heart.

Cells were first described by the English scientist Robert Hooke, around 1665. Hooke used an early light microscope to examine dried cork. He observed thousands of tiny empty chambers, which he named *cells*. Later, other scientists observed cells in living plants and realized that these spaces were filled with a gelatinous material. Research over the next 175 years led to the *cell theory*, the concept that cells are the fundamental units of all living things. Since the 1830s, when it was first proposed, the cell theory has been expanded to incorporate several basic concepts relevant to our discussion of the human body. These basic concepts are:

1. Cells are the structural "building blocks" of all plants and animals.
2. Cells are produced by the division of preexisting cells.
3. Cells are the smallest structural units that perform all vital functions.

The human body contains trillions of cells. All our activities, from running to thinking, result from the combined and coordinated responses of millions or even billions of cells. Yet each individual cell remains unaware of its role in the "big picture"—it is simply responding to changes in its local environment. Because cells form all of the structures in the body, and perform all vital functions, our exploration of the human body must begin with basic cell biology.

Two types of cells are contained in the body: sex cells and somatic cells. **Sex cells** (*germ cells* or *reproductive cells*) are either the sperm of males or the oocytes of females. **Somatic cells** (*soma*, body) include all the other cells in the body. In this chapter we will discuss somatic cells and in the reproductive system (Chapter 27) we will discuss sex cells.

The Study of Cells [FIGURE 2.1/2.2]

Cytology is the study of the structure and function of cells. Over the past 40 years we have learned a lot about cellular physiology and the mechanisms of homeostatic control. The two most common methods used to study cell and tissue structure are *light microscopy* and *electron microscopy*.

Before the 1950s most information about cells was provided by light microscopy. A photograph taken through a light microscope is called a light micrograph (LM) (see Figure 2.1a●). Light microscopy can magnify cellular structures about 1000 times and show details as fine as 0.25 μm. (The symbol μm stands for micrometer; 1 μm = 0.001 mm, or 0.0004 in.) With a light microscope one can identify cell types, such as cells lining the intestinal tract (Figure 2.1a●), and see large intracellular structures. Cells have a variety of sizes and shapes, as indicated in Figure 2.2●. The relative proportions of the cells in Figure 2.2● are correct, but all have been magnified roughly 500 times.

Individual cells are relatively transparent and difficult to distinguish from their neighbors. They become easier to see if they are treated with dyes that stain specific intracellular structures. Although special staining techniques can show the general distribution of protein, lipid, carbohydrate, or nucleic acids in the cell, many fine details of intracellular structure remained a mystery until investigators began using electron microscopy. This technique uses a focused beam of electrons, rather than a beam of light, to examine cell structure. In **transmission electron microscopy**, electrons penetrate an ultrathin section of tissue to strike a photographic plate. The result is a transmission electron micrograph (TEM). Transmission electron microscopy shows the fine structure of cell membranes and details of intracellular structures (Figure 2.1b●). In **scanning electron microscopy**, electrons bouncing off exposed surfaces create a scanning electron micrograph (SEM). Although scanning microscopy provides less magnification than transmission electron microscopy, it provides a three-dimensional perspective on cell structure (Figure 2.1c●).

Many other methods can be used to examine cell and tissue structure, and examples will be found in the pages that follow and throughout the book. This chapter describes the structure of a typical cell, some of the ways in which cells interact with their environment, and how cells reproduce.

Cellular Anatomy [FIGURES 2.3/2.4 AND TABLE 2.1]

The "typical" cell is like the "average" person. Any description can be thought of only in general terms because enormous individual variations occur. Our typical model cell will share features with most cells of the body without being identical to any. Figure 2.3● shows such a cell, and Table 2.1 summarizes the major structures and functions of its parts.

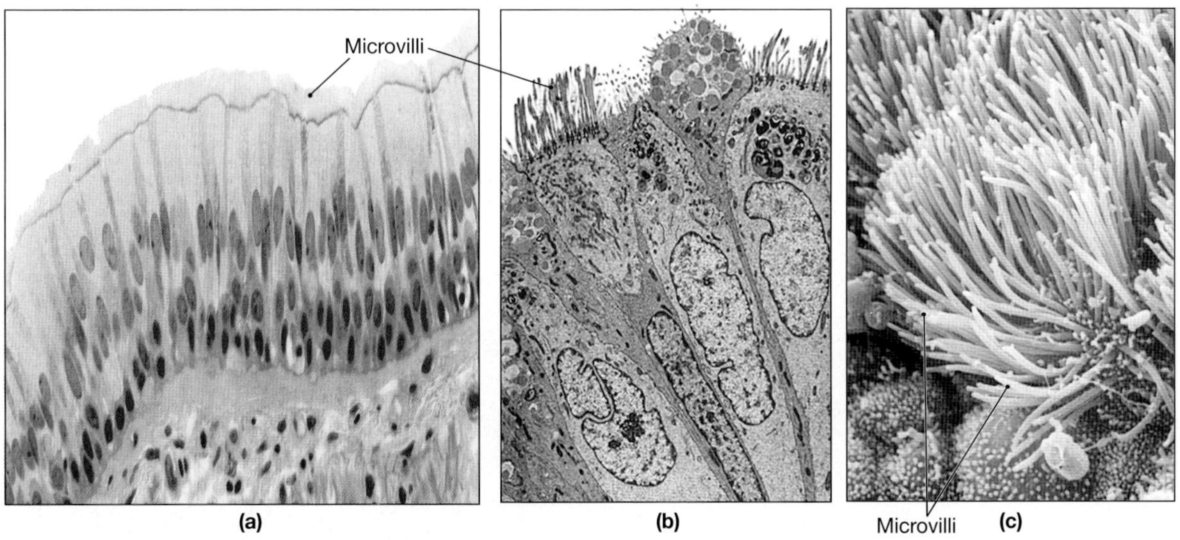

FIGURE 2.1 **DIFFERENT TECHNIQUES, DIFFERENT PERSPECTIVES**

Cells lining the intestine as seen in **(a)** light microscopy, **(b)** transmission electron microscopy, and **(c)** scanning electron microscopy. In (a), note that the fingerlike microvilli appear as a fuzzy border; only the nucleus and cell membrane are easily observed. Compare the appearance of the microvilli in (b) and (c) to (a).

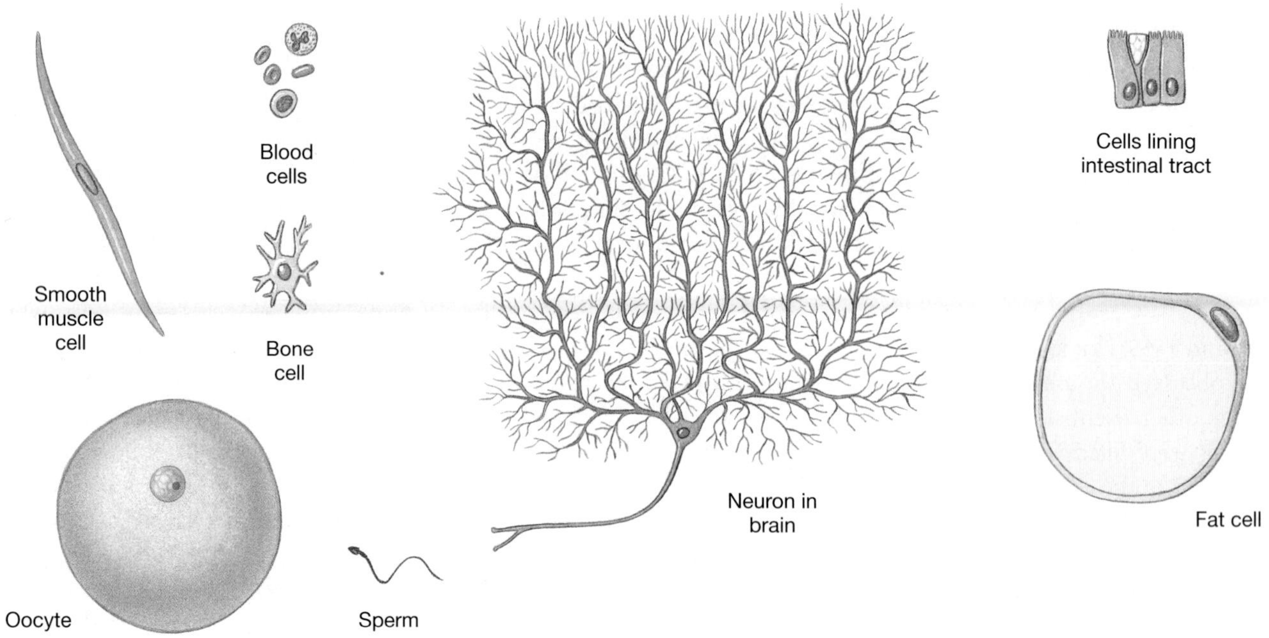

FIGURE 2.2 **The Diversity of Cells in the Body**

The cells of the body have many different shapes and a variety of special functions. These examples give an indication of the range of forms and sizes; all of the cells are shown with the dimensions they would have if magnified approximately 500 times.

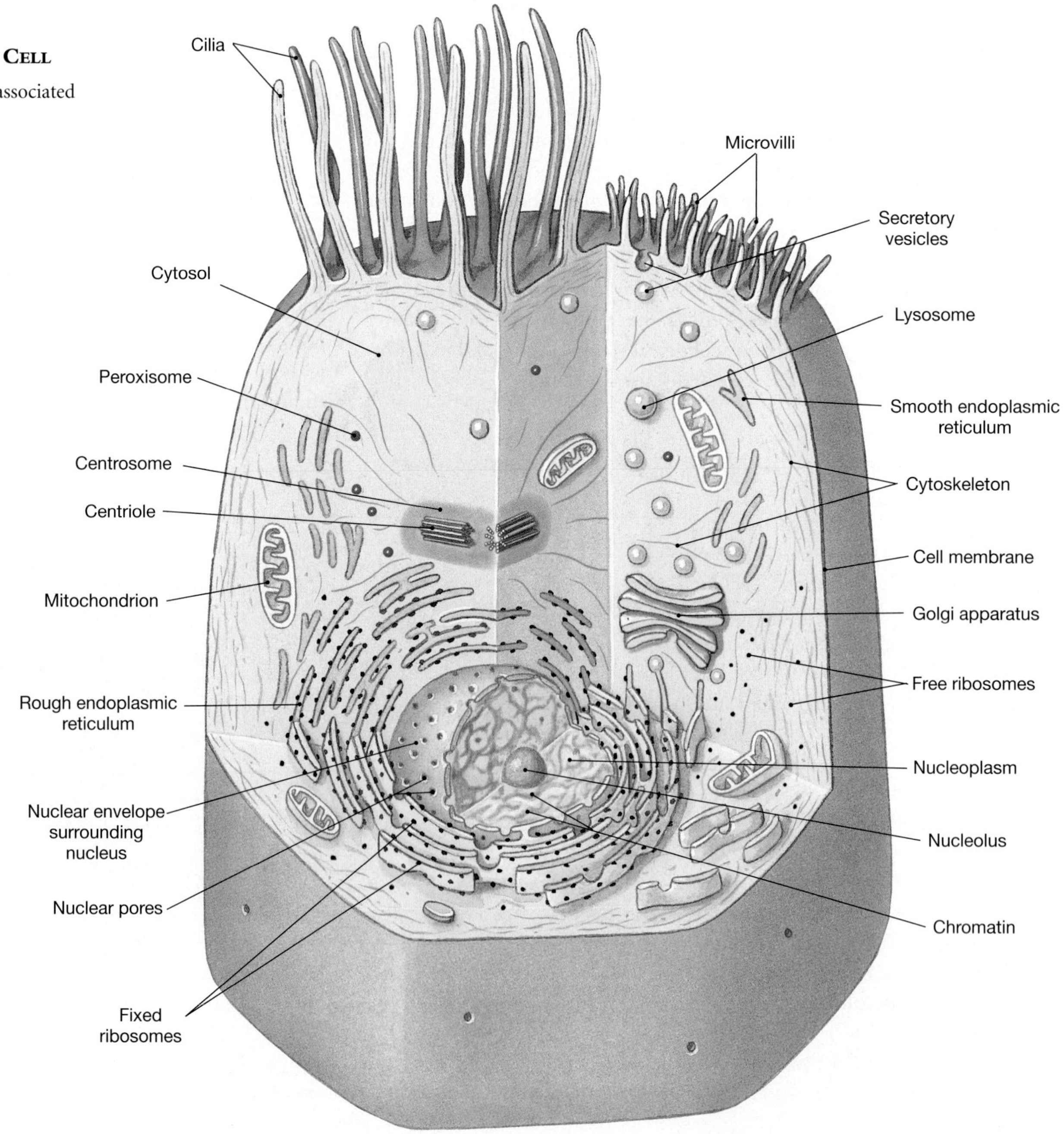

FIGURE 2.3 **Anatomy of a Typical Cell**

See Table 2.1 for a summary of the functions associated with the various cell structures.

TABLE 2.1

ANATOMY OF A REPRESENTATIVE CELL

Appearance	Structure	Composition	Function(s)
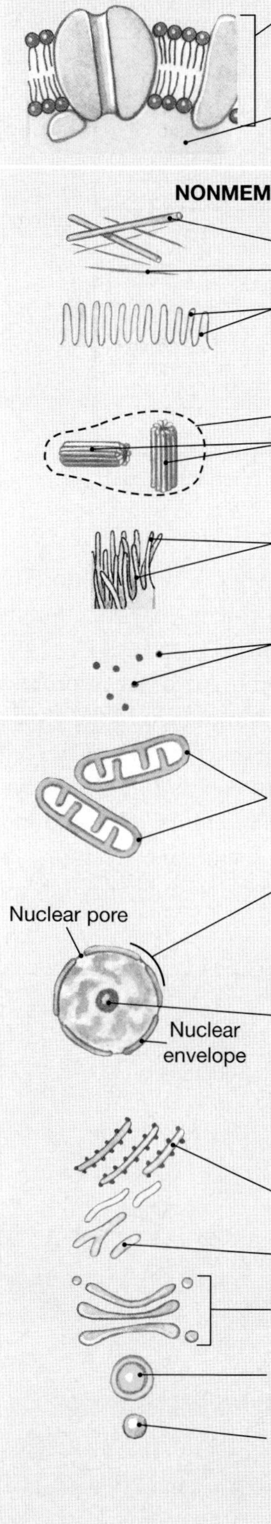	**CELL MEMBRANE**	Lipid bilayer, containing phospholipids, steroids, proteins, and carbohydrates	Isolation; protection; sensitivity; support; control of entrance/exit of materials
	CYTOSOL	Fluid component of cytoplasm; may contain inclusions of insoluble materials	Distributes materials by diffusion; stores glygogen pigments and other materials
	NONMEMBRANOUS ORGANELLES		
	Cytoskeleton Microtubule Microfilament	Proteins organized in fine filaments or slender tubes	Strength and support; movement of cellular structures and materials
	Microvilli	Membrane extensions containing microfilaments	Increase surface area to facilitate absorption of extracellular materials
	Centrosome Centrioles	Cytoplasm containing two centrioles, at right angles; each centriole is composed of 9 microtubule triplets	Essential for movement of chromosomes during cell division; organization of microtubules in cytoskeleton
	Cilia	Membrane extensions containing microtubule doublets in a 9 + 2 array	Movement of materials over cell surface
	Ribosomes	RNA + proteins; fixed ribosomes bound to rough endoplasmic reticulum, free ribosomes scattered in cytoplasm	Protein synthesis
	MEMBRANOUS ORGANELLES		
	Mitochondria	Double membrane, with inner membrane folds (cristae) enclosing metabolic enzymes	Produce 95% of the ATP required by the cell
	Nucleus	Nucleoplasm containing nucleotides, enzymes, nucleoproteins, and chromatin; surrounded by double membrane (nuclear envelope) containing nuclear pores	Control of metabolism; storage and processing of genetic information; control of protein synthesis
	Nucleolus	Dense region in nucleoplasm containing DNA and RNA	Site of rRNA synthesis and assembly of ribosomal subunits
	Endoplasmic reticulum (ER)	Network of membranous channels extending throughout the cytoplasm	Synthesis of secretory products; intracellular storage and transport
	Rough ER	Has ribosomes bound to membranes	Modification and packaging of newly synthesized proteins
	Smooth ER	Lacks attached ribosomes	Lipid and carbohydrate synthesis; calcium ion storage
	Golgi apparatus	Stacks of flattened membranes (cisternae) containing chambers	Storage, alteration, and packaging of secretory products and lysosomal enzymes
	Lysosome	Vesicles containing digestive enzymes	Intracellular removal of damaged organelles or of pathogens
	Peroxisome	Vesicles containing degradative enzymes	Catabolism of fats and other organic compounds; neutralization of toxic compounds generated in the process

Figure 2.4● previews the organization of this chapter. Our representative cells float in a watery medium known as the **extracellular fluid.** A *cell membrane* separates the cell contents, or *cytoplasm*, from the extracellular fluid. The cytoplasm can be further subdivided into a fluid, the *cytosol*, and intracellular structures collectively known as *organelles* (or-gan-ELS, "little organs").

The Cell Membrane [FIGURE 2.5]

The **cell membrane**, also called the **plasma membrane** or *plasmalemma*, forms the outer boundary of the cell. It is extremely thin and delicate, ranging from 6 to 10 nm (1 nm = 0.1μm) in thickness. Nevertheless, it has a complex structure composed of phospholipids, proteins, glycolipids, and cholesterol. The structure of the cell membrane is shown in Figure 2.5●.

The cell membrane is called a **phospholipid bilayer** because its phospholipids form two distinct layers. In each layer the phospholipid molecules lie so that the heads are at the surface and the tails are on the inside. Dissolved ions and water-soluble compounds cannot cross the lipid portion of a cell membrane because the lipid tails will not associate with water molecules. This feature makes the membrane very effective in isolating the

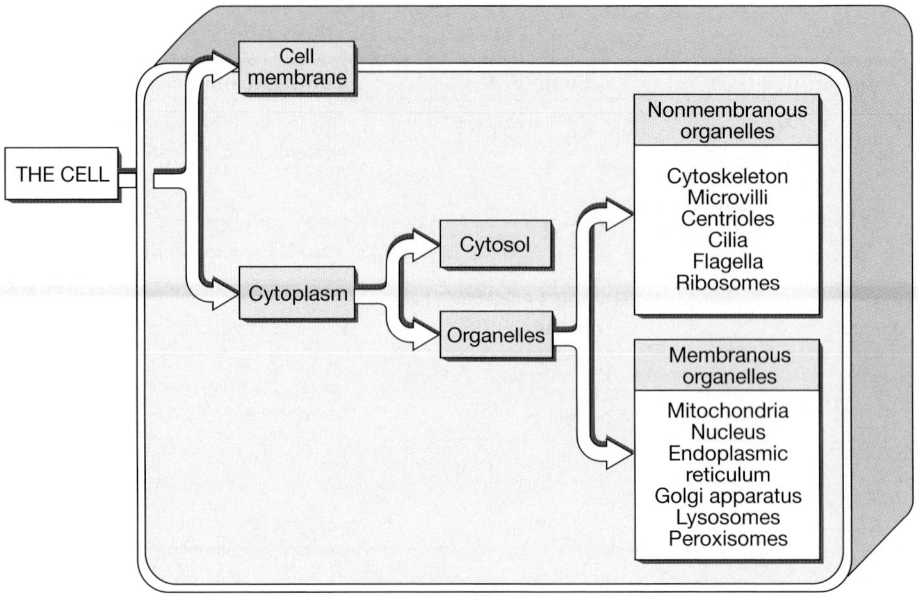

FIGURE 2.4 **A Flow Chart for the Study of Cell Structure**

Cytoplasm is subdivided into cytosol and organelles. Organelles are subdivided into nonmembranous organelles and membranous organelles.

cytoplasm from the surrounding fluid environment. Such isolation is important because the composition of the cytoplasm is very different from that of the extracellular fluid, and that difference must be maintained.

Peripheral proteins are attached to either the inner or the outer membrane surface. **Integral proteins** are embedded in the membrane. Most integral proteins span the entire width of the membrane one or more times, and are therefore called *transmembrane proteins*. Some of the integral proteins form **channels** that let water molecules, ions, and small water-soluble compounds into or out of the cell. Most of the communication between the interior and exterior of the cell occurs through these channels. Some of the channels are called **gated** because they can open or close to regulate the passage of materials. Other integral proteins may function as catalysts or receptor sites or in cell–cell recognition.

The inner and outer surfaces of the cell membrane differ in protein and lipid composition. The carbohydrate (*glyco-*) component of the glycolipids and glycoproteins that extend away from the outer surface of the cell membrane form a viscous, superficial coating known as the **glycocalyx** (*calyx*, cup). Some of these molecules function as receptors: When bound to a specific molecule in the extracellular fluid, a membrane receptor can trigger a change in cellular activity. For example, cytoplasmic enzymes on

FIGURE 2.5 **The Cell Membrane**

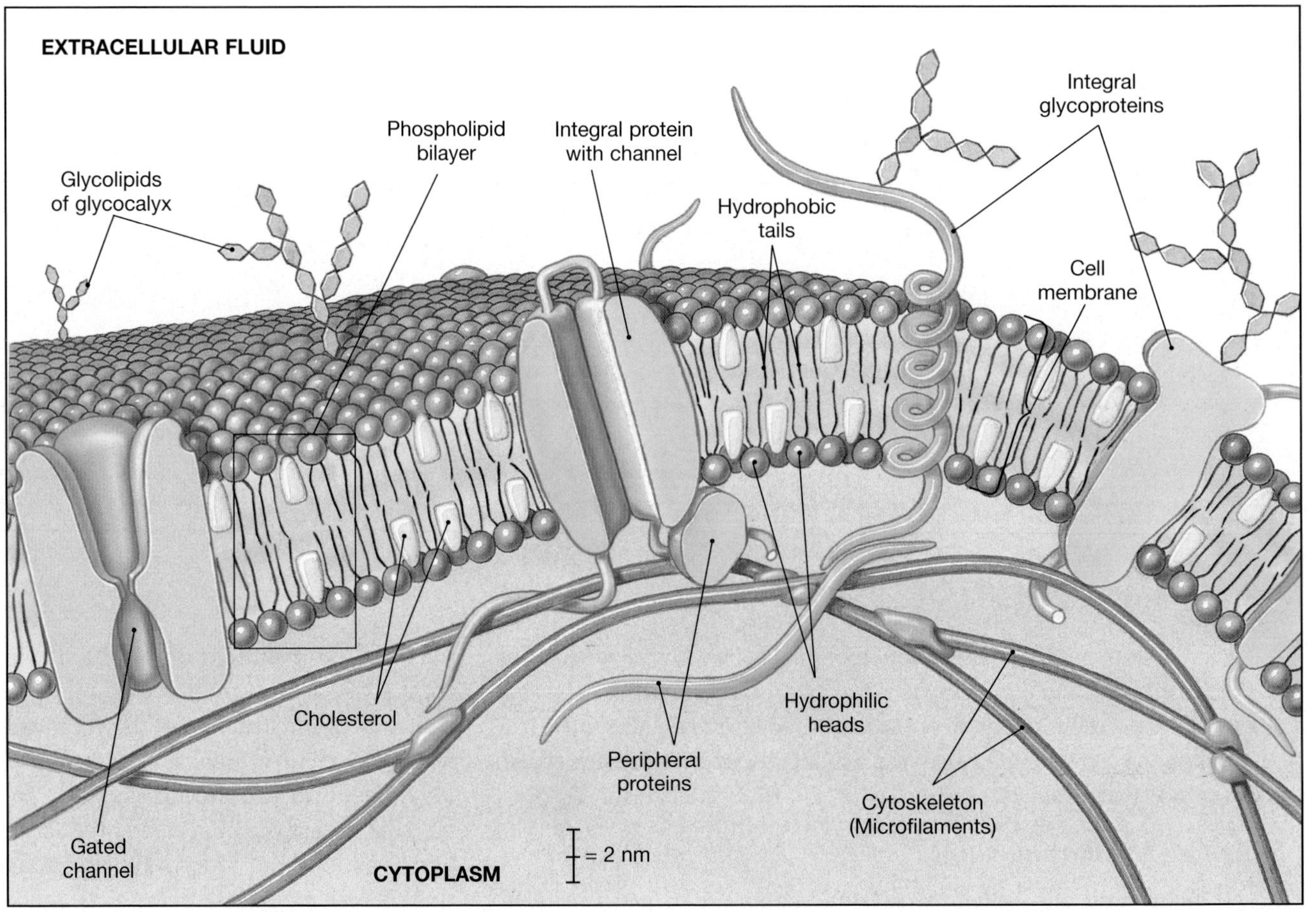

the inner surface of the cell membrane may be bound to integral proteins, and the activities of these enzymes may be affected by events on the membrane surface.

The general functions of the cell membrane include:

1. *Physical isolation*: The lipid bilayer of the cell membrane forms a physical barrier that separates the inside of the cell from the surrounding extracellular fluid.
2. *Regulation of exchange with the environment*: The cell membrane controls the entry of ions and nutrients, the elimination of wastes, and the release of secretory products.
3. *Sensitivity*: The cell membrane is the first part of the cell affected by changes in the extracellular fluid. It also contains a variety of receptors that allow the cell to recognize and respond to specific molecules in its environment, and to communicate with other cells. Any alteration in the cell membrane may affect all cellular activities.
4. *Structural support*: Specialized connections between cell membranes or between membranes and extracellular materials give tissues a stable structure.

Membrane structure is fluid. Cholesterol helps to stabilize the membrane structure and maintain its fluidity. Integral proteins can move within the membrane like ice cubes drifting in a bowl of punch. In addition, the composition of the cell membrane can change over time, through the removal and replacement of membrane components.

MEMBRANE PERMEABILITY: PASSIVE PROCESSES

The **permeability** of a membrane is a property that determines its effectiveness as a barrier. The greater the permeability, the easier it is for substances to cross the membrane. If nothing can cross a membrane, it is described as **impermeable**. If any substance can cross without difficulty, the membrane is **freely permeable**. Cell membranes fall somewhere in between and are said to be **selectively permeable**. A selectively permeable membrane permits the free passage of some materials and restricts the passage of others. The distinction may be on the basis of size, electrical charge, molecular shape, solubility, or some combination of factors.

The permeability of a cell membrane varies depending on the organization and characteristics of membrane lipids and proteins. Passage across the membrane may be active or passive. Active processes, discussed later in this chapter, require that the cell draw on an energy source, usually *adenosine triphosphate*, or ATP. Passive processes move ions or molecules across the cell membrane without any energy expenditure by the cell. Passive processes include *diffusion, osmosis, filtration*, and *facilitated diffusion*.

DIFFUSION [FIGURE 2.6] Ions and molecules in solution are in constant motion, bouncing off one another and colliding with water molecules. The result of the continual collisions and rebounds that occur is the process called diffusion. **Diffusion** can be defined as the net movement of material from an area where its concentration is relatively high to an area where its concentration is relatively low. The difference between the high and low concentrations represents a **concentration gradient**, and diffusion continues until that gradient has been eliminated. Because diffusion occurs from a region of higher concentration to one of relatively lower concentration, it is often described as proceeding "down a concentration gradient." When a concentration gradient has been eliminated, an equilibrium exists. Although molecular motion continues, there is no longer a net movement in any particular direction.

Diffusion is important in body fluids because it tends to eliminate local concentration gradients. For example, an active cell generates carbon dioxide and absorbs oxygen. As a result, the extracellular fluid around the cell develops a relatively high concentration of CO_2 and a relatively low concentration of O_2. Diffusion then distributes the carbon dioxide through the tissue and into the bloodstream. At the same time, oxygen diffuses out of the blood and into the tissue.

In the extracellular fluids of the body, water and dissolved solutes (substances dissolved in water) diffuse freely. A cell membrane, however, acts as a barrier that selectively restricts diffusion. Some substances can pass through easily, whereas others cannot penetrate the membrane at all. There are only two ways for an ion or molecule to cross a cell membrane: diffuse through one of the membrane channels or diffuse across the lipid portion of the membrane. The size of the ion or molecule and any electrical charge it might carry determine its ability to pass through membrane channels. To cross the lipid portion of the membrane, the molecule must be lipid-soluble. These mechanisms are summarized in Figure 2.6●.

OSMOSIS Cell membranes are very permeable to water molecules. The diffusion of water across a membrane from a region of high water concentration to a region of low water concentration is so important that it is given a special name, **osmosis** (oz-MŌ-sis; *osmos*, thrust). For convenience, we will always use the term *osmosis* when considering water movement and restrict use of the term *diffusion* to the movement of solutes.

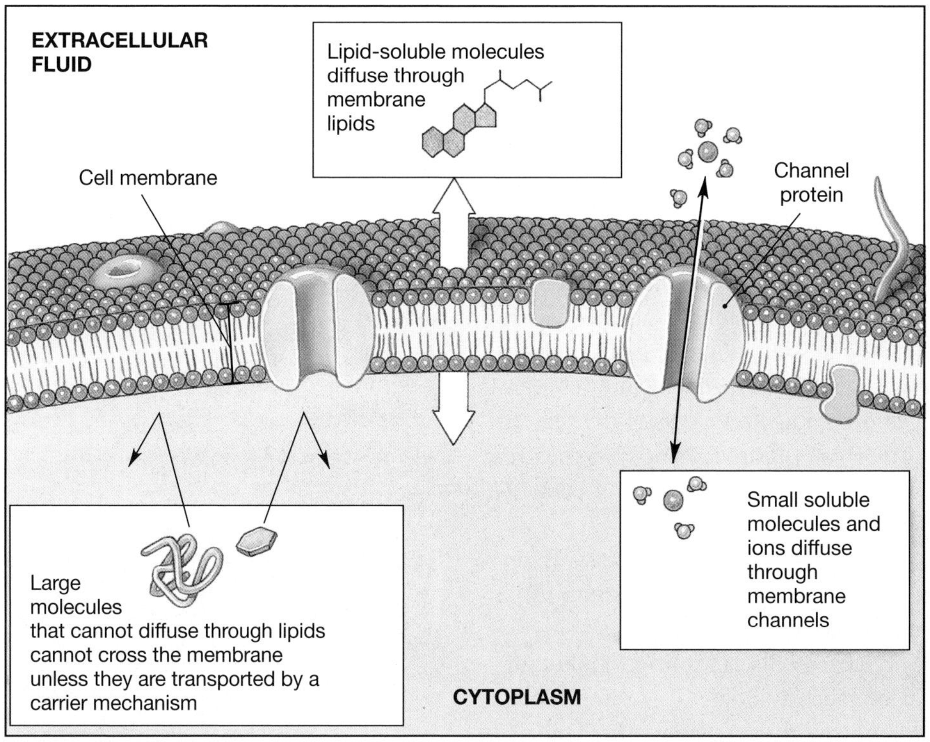
EXTRACELLULAR FLUID
Lipid-soluble molecules diffuse through membrane lipids
Cell membrane
Channel protein
Small soluble molecules and ions diffuse through membrane channels
Large molecules that cannot diffuse through lipids cannot cross the membrane unless they are transported by a carrier mechanism
CYTOPLASM

FIGURE 2.6 **DIFFUSION ACROSS CELL MEMBRANES**

Small ions and water-soluble molecules diffuse through membrane channels. Lipid-soluble molecules can cross the membrane by diffusing through the phospholipid bilayer. Large molecules that are not lipid-soluble cannot diffuse through the membrane at all.

Whenever an osmotic gradient exists, water molecules will diffuse rapidly across the cell membrane until the osmotic gradient is eliminated. Water molecules crossing membranes move in groups held together by hydrogen bonds. So, while solute molecules usually diffuse through membrane channels one at a time, water molecules move together in large clusters. This process is called *bulk flow*.

Filtration In **filtration**, hydrostatic pressure forces water across a membrane, and solute molecules are selected on the basis of size. If the membrane pores are large enough, molecules of solute will be carried along with the water. We can see filtration in action in a coffee machine. Gravity forces hot water through the filter, and the water carries with it a variety of dissolved compounds. The large coffee grounds never reach the pot because they cannot fit through the fine pores in the filter. In the body, the heart pushes blood through the circulatory system and generates *hydrostatic pressure*. Filtration occurs across the walls of small blood vessels, pushing water and dissolved nutrients into the tissues of the body.

Facilitated Diffusion Many essential nutrients, such as glucose and amino acids, are insoluble in lipids and too large to fit through membrane channels. These compounds can be passively transported across the membrane by special **carrier proteins** in a process called **facilitated diffusion**. The molecule to be transported first binds to a **receptor site** on an integral membrane protein. It is then moved to the inside of the cell membrane and released into the cytoplasm. No ATP is expended in facilitated diffusion or simple diffusion; in each case, molecules move from an area of higher concentration to one of lower concentration.

MEMBRANE PERMEABILITY: ACTIVE PROCESSES

All **active membrane processes** require energy. By spending energy, usually in the form of ATP, the cell can transport substances *against their concentration gradients*. We will consider two active processes: *active transport* and *endocytosis*.

Active Transport In **active transport** the high-energy bond in ATP provides the energy needed to move ions or molecules across the membrane. The process is complex, and specific enzymes must be present in addition to carrier proteins. Although it has an energy cost, active transport offers one great advantage: It is not dependent on a concentration gradient. As a result the cell can import or export specific materials *regardless of their intracellular or extracellular concentrations.*

All living cells show active transport of sodium (Na^+), potassium (K^+), calcium (Ca^{2+}), and magnesium (Mg^{2+}). Specialized cells can transport additional ions such as iodide (I^-) or iron (Fe^{2+}). Many of these carrier mechanisms, known as **ion pumps**, move a specific cation or anion in one direction, either into or out of the cell. If one ion moves in one direction while another moves in the opposite direction, the carrier is called an **exchange pump**. The energy demands of these pumps are impressive; a resting cell may use up to 40% of the ATP it produces to power its exchange pumps.

Endocytosis **Endocytosis** (EN-dō-sī-TŌ-sis) is the packaging of extracellular materials into a vesicle at the cell surface for importation into the cell. This process, which involves relatively large volumes of extracellular material, is sometimes called *bulk transport*. There are three major types of endocytosis: *pinocytosis, phagocytosis,* and *receptor-mediated endocytosis.* All three require energy in the form of ATP and so are classified as active processes. The mechanism is presumed to be the same in all three cases, but the mechanism itself remains unknown.

All forms of endocytosis produce small, membrane-bound compartments called *vesicles*. Once a vesicle has formed through endocytosis, its contents will enter the cytosol only if they can pass through the vesicle wall. This passage may involve active transport, simple or facilitated diffusion, or the destruction of the vesicle membrane.

Pinocytosis [FIGURE 2.7a] **Pinocytosis** (PIN-ō-sī-TŌ-sis), or "cell drinking," is the formation of vesicles filled with extracellular fluid. In this process, a deep groove or pocket forms in the cell membrane and then pinches off (Figure 2.7a●). Nutrients, such as lipids, sugars, or amino acids, then enter the cytoplasm by diffusion or active transport from the enclosed fluid. The membrane of the pinocytotic vesicle then returns to the cell surface.

Virtually all cells perform pinocytosis in this manner. In a few specialized cells, the vesicles form on one side of the cell and travel through the cytoplasm to the opposite side. There they fuse with the cell membrane

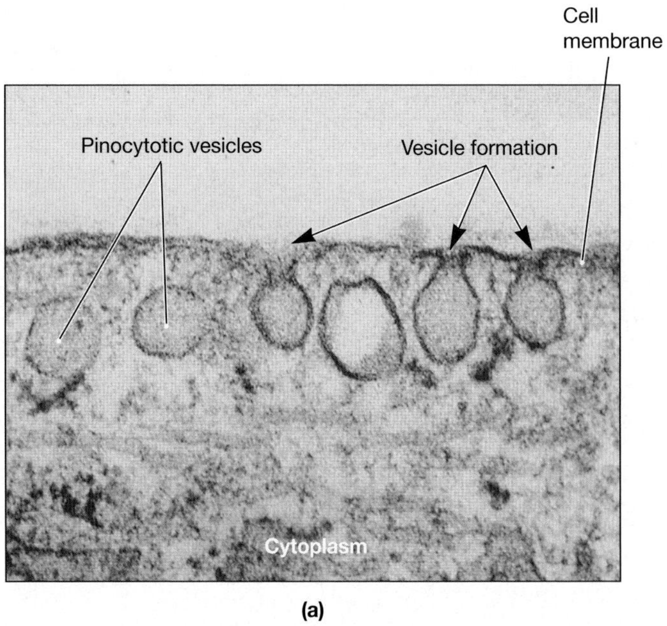

(a)

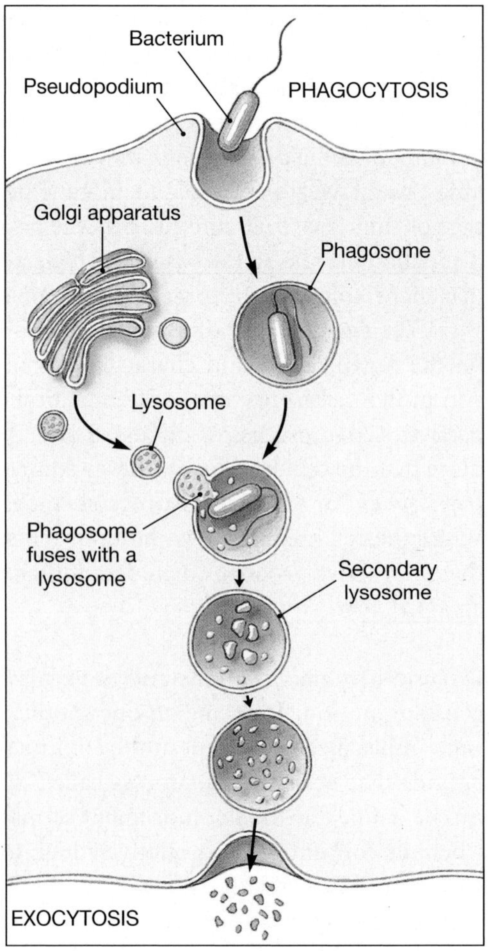

(b)

FIGURE 2.7 **Pinocytosis and Phagocytosis**

(a) An electron micrograph showing pinocytosis. (b) Material brought into the cell through phagocytosis is enclosed in a vesicle and subsequently exposed to lysosomal enzymes. After absorption of nutrients from the vesicle, the residue is discharged through exocytosis.

and discharge their contents through the process of *exocytosis*, described further on p. 40. This method of bulk transport is found in cells lining capillaries, the most delicate blood vessels. These cells use pinocytosis to transfer fluid and solutes from the bloodstream to the surrounding tissues.

Phagocytosis [FIGURE 2.7b] **Phagocytosis** (FA-gō-sī-TŌ-sis), or "cell eating," produces vesicles containing solid objects that may be as large as the cell itself. This process is shown in Figure 2.7b●. Cytoplasmic extensions called **pseudopodia** (soo-dō-PŌ-dē-a; *pseudo-*, false + *podon*, foot) surround the object, and their membranes fuse to form a vesicle. The vesicle may then fuse with a lysosome, whereupon its contents are digested by lysosomal enzymes.

Most cells display pinocytosis, but phagocytosis, especially the entrapment of living or dead cells, is performed only by specialized cells of the immune system. The phagocytic activity of these cells will be considered in chapters dealing with blood cells (Chapter 20) and the lymphatic system (Chapter 23).

Receptor-Mediated Endocytosis [FIGURE 2.8] **Receptor-mediated endocytosis** resembles pinocytosis, but it is far more selective (Figure 2.8●). Pinocytosis produces vesicles filled with extracellular fluid; receptor-mediated endocytosis produces vesicles that contain a specific target molecule in high concentrations. The target substances, called *ligands*, are bound to receptors on the membrane surface. Many important substances, including cholesterol and iron ions (Fe^{2+}), are distributed through the body attached to special transport proteins. The proteins are too large to pass through membrane pores, but they can enter the cell through receptor-mediated endocytosis. The vesicle eventually returns to the cell surface and fuses with the cell membrane. As the vesicle fuses with the cell membrane, the vesicle contents are released into the extracellular fluid. This release is another example of the process of exocytosis. A summary and comparison of the mechanisms involved in movement across cell membranes is presented in Table 2.2.

✓ **CONCEPT CHECK**

- What term is used to describe the permeability of cell membranes?
- Describe the processes of osmosis and diffusion. How do they differ?
- What are the three major types of endocytosis? How do they differ?

The Cytoplasm

Cytoplasm is a general term for all of the material inside the cell. Cytoplasm contains many more proteins than the extracellular fluid; proteins account for 15–30% of the weight of the cell. The cytoplasm includes two major subdivisions:

1. *Cytosol*, or intracellular fluid. The cytosol contains dissolved nutrients, ions, soluble and insoluble proteins, and waste products. The cell membrane separates the cytosol from the surrounding extracellular fluid.
2. *Organelles* (or-gan-ELS) are intracellular structures that perform specific functions.

THE CYTOSOL

Cytosol is significantly different from extracellular fluid. Three important differences are:

1. The cytosol contains a high concentration of potassium ions, whereas extracellular fluid contains a high concentration of sodium ions.

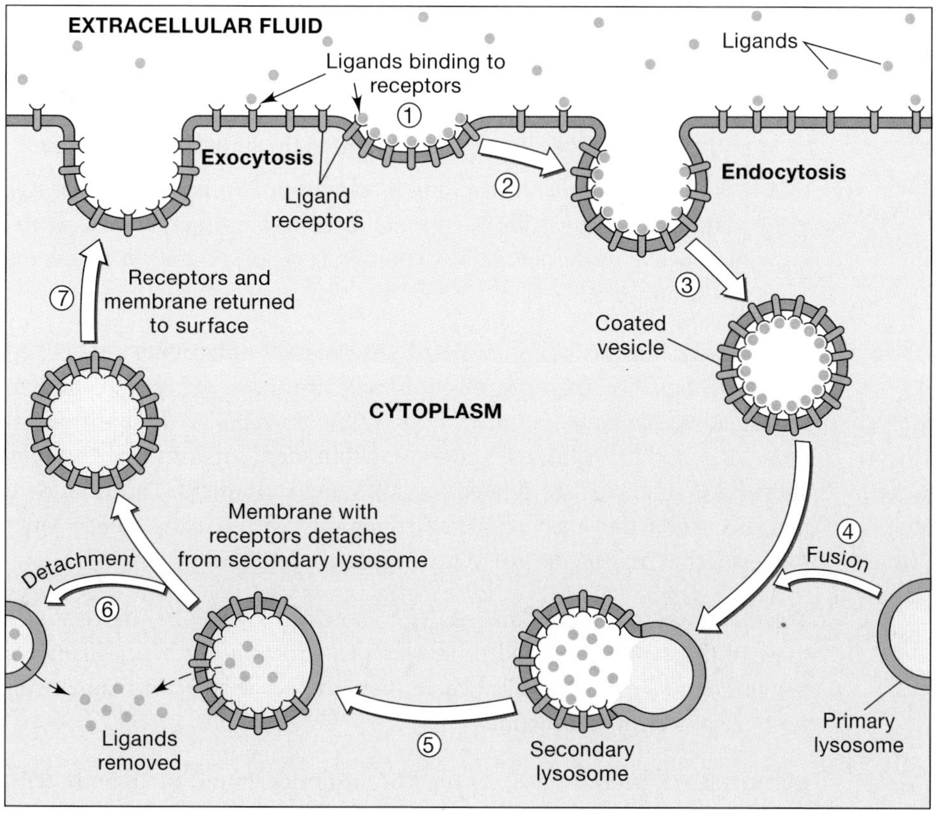

FIGURE 2.8 **Receptor-Mediated Endocytosis**

In this process, ① specific target molecules called ligands bind to receptors, generally gycoproteins, in the membrane surface. ② Membrane areas coated with ligands pinch off to form ③ vesicles that ④ fuse with primary lysosomes. ⑤ The ligands are freed from the receptors and, if necessary, broken down by enzymes before diffusing or being transported into the surrounding cytoplasm. ⑥ The membrane containing the receptor molecules detaches from the membrane of the lysosome and ⑦ returns to the cell surface to bind additional ligands.

TABLE 2.2

SUMMARY OF MECHANISMS INVOLVED IN MOVEMENT ACROSS CELL MEMBRANES

Mechanism	Process	Factors Affecting Rate	Substances Involved
PASSIVE			
Diffusion	Molecular movement of solutes; direction determined by relative concentrations	Size of gradient, molecular size, charge, lipid protein solubility, temperature	Small inorganic ions, lipid-soluble materials (all cells)
Osmosis	Movement of water (solvent) molecules toward high solute concentrations; requires membrane	Concentration gradient; opposing pressure	Water only (all cells)
Filtration	Movement of water; usually with solute, by hydrostatic pressure; requires membrane filter	Amount of pressure; size of pores in membrane filter	Water and small ions (blood vessels)
Facilitated diffusion	Carrier molecules transport materials down a concentration gradient; requires membrane	As above, plus availability of carrier protein	Glucose and amino acids (all cells)
ACTIVE			
Active transport	Carrier molecules work despite opposing concentration gradients	Availability of carrier, substrate, and ATP	Na^+, K^+, Ca^{2+}, Mg^{2+} (all cells); probably other solutes in special cases
Endocytosis	Formation of membranous vesicles containing fluid or solid material at the cell membrane	Stimulus and mechanism not understood; requires ATP	Fluids, nutrients (all cells); debris, pathogens (special cells)
Exocytosis	Fusion of vesicles containing fluids and/or solids with the cell membrane	Stimulus and mechanism incompletely understood; requires ATP and calcium ions	Fluid and wastes (all cells)

The numbers of positive and negative ions are not in balance across the membrane; the outside has a net excess of positive charges, and the inside a net excess of negative charges. The separation of unlike charges creates a *transmembrane potential*, like a miniature battery. The significance of the transmembrane potential will become clear in Chapter 13.

2. The cytosol contains a relatively high concentration of dissolved and suspended proteins. Many of these proteins are enzymes that regulate metabolic operations, while others are associated with the various organelles. These proteins give the cytosol a consistency that varies between that of thin maple syrup and almost-set gelatin.
3. The cytosol contains relatively small quantities of carbohydrates and large reserves of amino acids and lipids. The carbohydrates are broken down to provide energy, and the amino acids are used to manufacture proteins. The lipids stored in the cell are used primarily as an energy source when carbohydrates are unavailable.

The cytosol of cells contains masses of insoluble materials known as **inclusions**, or *inclusion bodies*. Among the most common inclusions are stored nutrients: for example, glycogen granules in liver or skeletal muscle cells, and lipid droplets in fat cells.

ORGANELLES [FIGURE 2.3]

Organelles are found in all body cells (Figure 2.3•, p. 28) although the types and numbers of organelles differ among the various cell types. Each organelle performs specific functions that are essential to normal cell structure, maintenance, and/or metabolism. Cellular organelles can be divided into two broad categories (Table 2.1, p. 29): (1) **nonmembranous organelles**, which are always in contact with the cytosol; and (2) **membranous organelles** surrounded by membranes that isolate their contents from the cytosol, just as the cell membrane isolates the cytosol from the extracellular fluid.

Nonmembranous Organelles

Nonmembranous organelles include the *cytoskeleton, microvilli, centrioles, cilia, flagella*, and *ribosomes*.

THE CYTOSKELETON [FIGURE 2.9]

The **cytoskeleton** is an internal protein framework that gives the cytoplasm strength and flexibility. It has four major components: *microfilaments, intermediate filaments, thick filaments*, and *microtubules*. None of these structures can be seen with the light microscope.

MICROFILAMENTS [FIGURE 2.9] **Microfilaments** are slender strands, composed primarily of the protein **actin**. In most cells microfilaments are scattered throughout the cytosol and form a dense network under the cell membrane. Figure 2.9a,b• shows the superficial layers of microfilaments in an intestinal cell.

Microfilaments have two major functions:

1. Microfilaments anchor the cytoskeleton to integral proteins of the cell membrane. This function stabilizes the position of the membrane proteins, provides additional mechanical strength to the cell, and firmly attaches the cell membrane to the underlying cytoplasm.
2. Actin microfilaments can interact with microfilaments or larger structures composed of the protein **myosin**. This interaction can produce active movement of a portion of a cell, or a change in the shape of the entire cell.

INTERMEDIATE FILAMENTS [FIGURE 2.9a] **Intermediate filaments** are defined chiefly by their size; their composition varies from one cell type to another. Intermediate filaments (1) provide strength, (2) stabilize the positions of organelles, and (3) transport materials within the cytoplasm. For example, specialized intermediate filaments, called **neurofilaments**, are found in neurons, where they provide structural support within *axons*, long cellular processes that may be up to a meter in length.

THICK FILAMENTS **Thick filaments**, not shown in Figure 2.9, are relatively massive filaments composed of myosin protein subunits. Thick filaments are abundant in muscle cells, where they interact with actin filaments to produce powerful contractions.

MICROTUBULES [FIGURES 2.9a,c/2.10] **Microtubules**, found in all body cells, are hollow tubes built from the globular protein **tubulin**. Figure 2.9a,c•

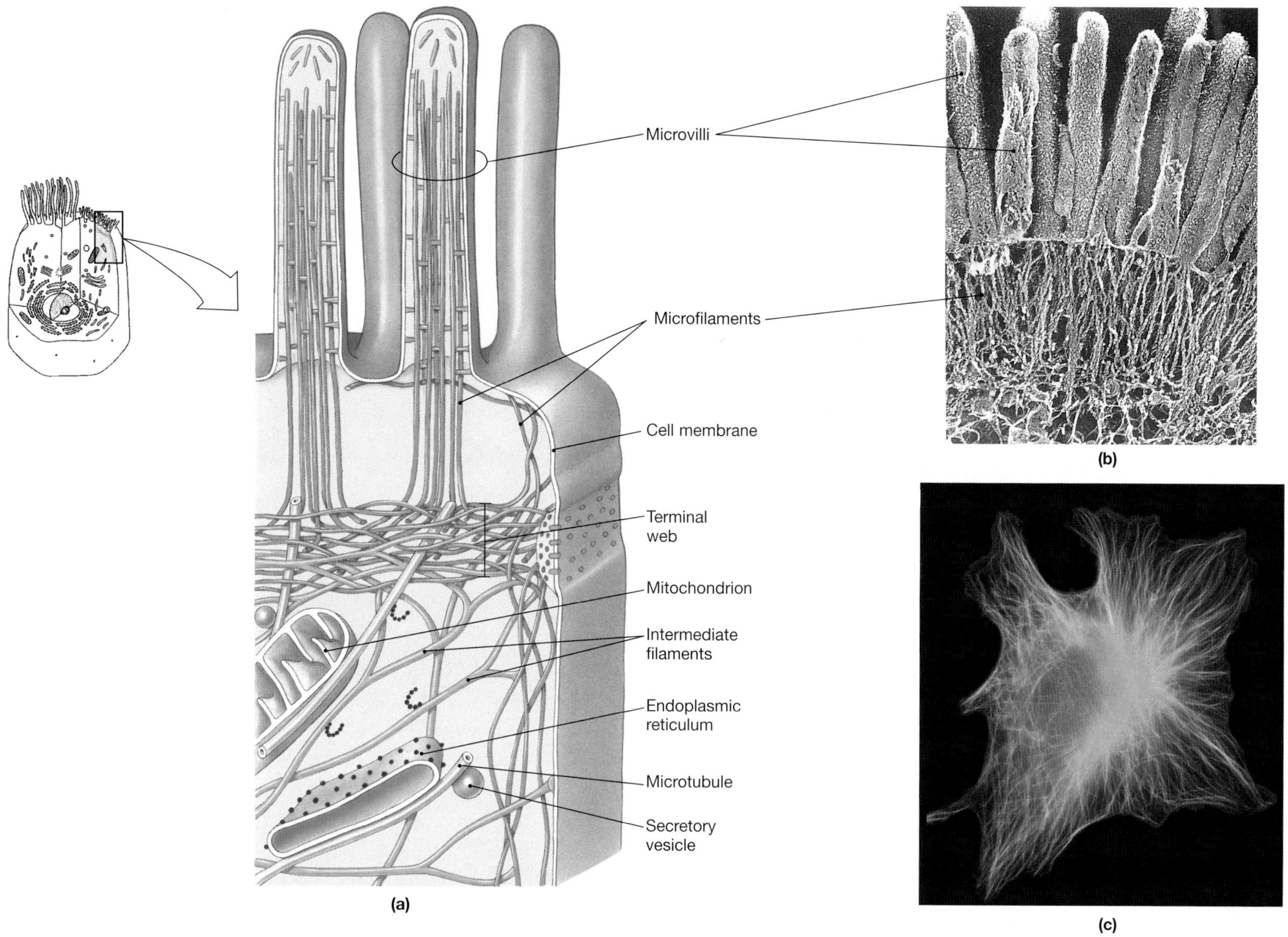

FIGURE 2.9 **THE CYTOSKELETON**

(**a**) The cytoskeleton provides strength and structural support for the cell and its organelles. Interactions between cytoskeletal elements are also important in moving organelles and in changing the shape of the cell. (**b**) An SEM image of the microfilaments and microvilli of an intestinal cell. (**c**) Microtubules in a living cell, as seen after special fluorescent labeling. (LM × 3200)

and Figure 2.10● show microtubules in the cytoplasm of representative cells. A microtubule forms through the aggregation of tubulin molecules; it persists for a time and then disassembles into individual tubulin molecules once again. The microtubular array is centered near the nucleus of the cell, in a region known as the *centrosome*. Microtubules radiate outward from the centrosome into the periphery of the cell.

Microtubules have a variety of functions:

1. Microtubules form the primary components of the cytoskeleton, giving the cell strength and rigidity and anchoring the positions of major organelles.
2. The assembly and/or disassembly of microtubules provides a mechanism for changing the shape of the cell, perhaps assisting in cell movement.
3. Microtubules can attach to organelles and other intracellular materials and move them around within the cell.
4. During cell division, microtubules form the *spindle apparatus* that distributes the duplicated chromosomes to opposite ends of the dividing cell. This process will be considered in more detail in a later section.
5. Microtubules form structural components of organelles such as *centrioles, cilia*, and *flagella*. Although these organelles are associated with the cell membrane, they are considered among the nonmembranous organelles because they do not have their own enclosing membrane.

The cytoskeleton as a whole incorporates microfilaments, intermediate filaments, and microtubules into a network that extends throughout the cytoplasm. The organizational details are as yet poorly understood, because the network is extremely delicate and difficult to study in an intact state.

MICROVILLI [FIGURE 2.9a,b]

Microvilli are small, finger-shaped projections of the cell membrane. They are found in cells that are actively engaged in absorbing materials from the

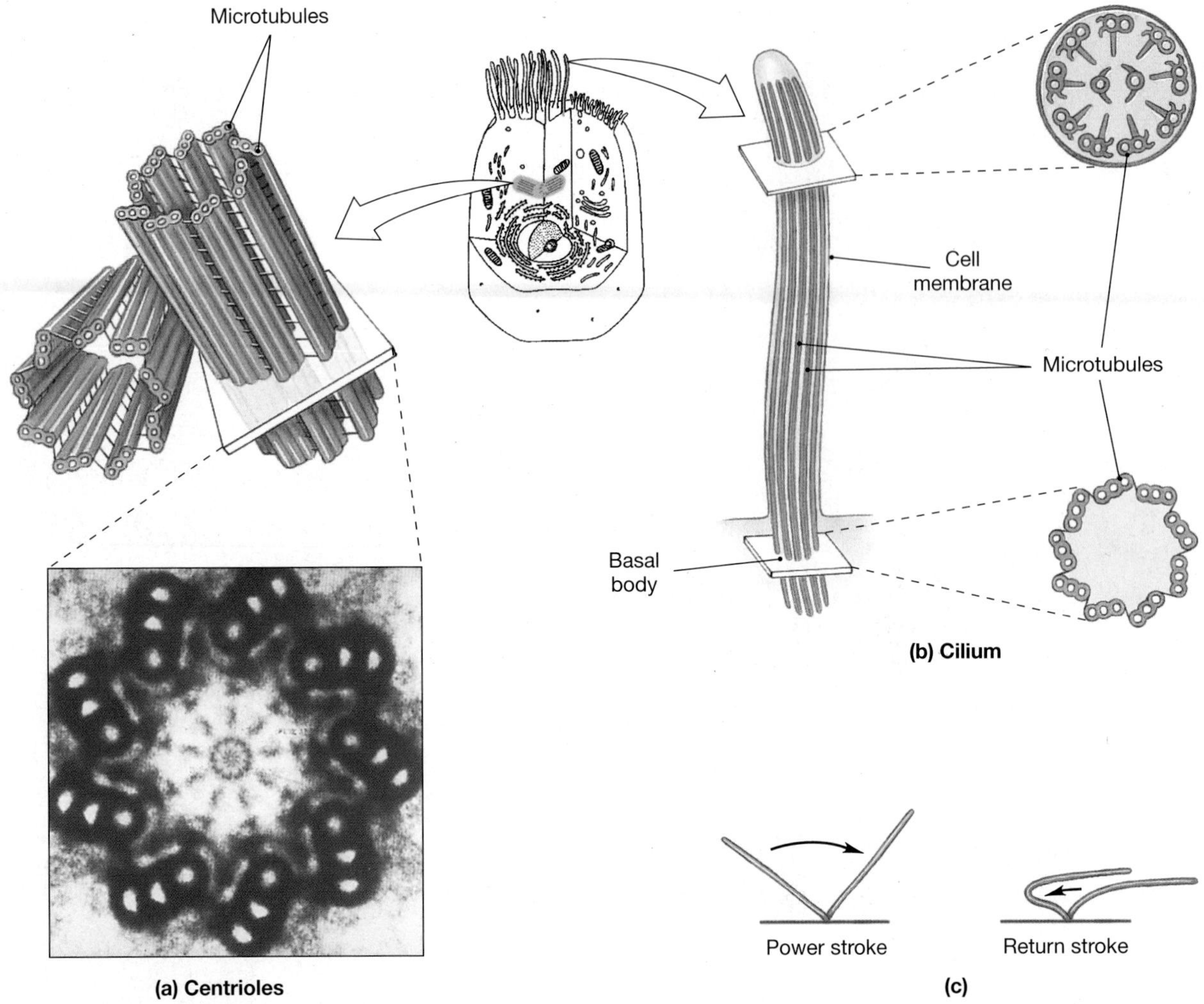

FIGURE 2.10 **Centrioles and Cilia**

(**a**) A centriole consists of nine microtubule triplets (9 + 0 array). The centrosome contains a pair of centrioles oriented at right angles to one another. (**b**) A cilium contains nine pairs of microtubules surrounding a central pair (9 + 2 array). (**c**) A single cilium swings forward and then returns to its original position. During the power stroke, the cilium is relatively stiff, but during the return stroke, it bends and moves parallel to the cell surface.

extracellular fluid, such as the cells of the small intestine and kidneys (Figure 2.9a,b●). Microvilli are important because they increase the surface area exposed to the extracellular environment for increased absorption. A network of microfilaments stiffens each microvillus and anchors it to the *terminal web*, a dense supporting network within the underlying cytoskeleton. Interactions between these microfilaments and the cytoskeleton can produce a waving or bending action. Their movements help to circulate fluid around the microvilli, bringing dissolved nutrients into contact with receptors on the membrane surface.

CENTRIOLES, CILIA, AND FLAGELLA [FIGURE 2.10]

The cytoskeleton contains numerous microtubules that function individually. Groups of microtubules form *centrioles, cilia*, and *flagella*. These structures are summarized in Table 2.3.

Centrioles [Figure 2.10a] A **centriole** is a cylindrical structure composed of short microtubules (Figure 2.10a●). There are nine groups of microtubules and each group is a triplet of microtubules. Because there are no central microtubules, this is called a *9 + 0 array*. This identification reflects the number of peripheral groups of microtubules oriented in a ring, with the number of microtubules at the center of the ring. However, some preparations show an axial structure that runs parallel to the long axis of the centriole, with radial spokes extending outward toward the microtubule groups. The function of this complex is not known. Cells capable of cell division contain a pair of centrioles arranged at right angles to each other. Centrioles direct the movement of chromosomes during cell division (discussed later in this chapter). Cells that do not divide, such as mature red blood cells and skeletal muscle cells, lack centrioles. The **centrosome** is the region of cytoplasm surrounding

TABLE 2.3 **A Comparison of Centrioles, Cilia, and Flagella**

Structure	Microtubule Organization	Location	Function
Centriole	Nine groups of microtubule triplets form a short cylinder	In centrosome near nucleus	Organizes microtubules in the spindle to move chromosomes during cell division
Cilium	Nine groups of long microtubule doublets form a cylinder around a central pair	At cell surface	Propels fluids or solids across cell surface
Flagellum	Same as cilium	At cell surface	Propels sperm cells through fluid

this pair of centrioles. It directs the organization of the microtubules of the cytoskeleton.

CILIA [FIGURE 2.10b,c] **Cilia** (singular *cilium*) contain nine groups of microtubule doublets surrounding a central pair (Figure 2.10b●). This is known as a *9 + 2 array*. Cilia are anchored to a compact **basal body** situated just beneath the cell surface. The structure of the basal body resembles that of a centriole. The exposed portion of the cilium is completely covered by the cell membrane. Cilia "beat" rhythmically, as depicted in Figure 2.10c●, and their combined efforts move fluids or secretions across the cell surface. Cilia lining the respiratory tract beat in a synchronized manner to move sticky mucus and trapped dust particles toward the throat and away from delicate respiratory surfaces. If the cilia are damaged or immobilized by heavy smoking or some metabolic problem, the cleansing action is lost, and the irritants will no longer be removed. As a result, chronic respiratory infections develop.

FLAGELLA **Flagella** (fla-JEL-ah; singular *flagellum*, "whip") resemble cilia but are much longer. A flagellum moves a cell through the surrounding fluid, rather than moving the fluid past a stationary cell. The sperm cell is the only human cell that has a flagellum, and it is used to move the cell along the female reproductive tract. If sperm flagella are paralyzed or otherwise abnormal, the individual will be sterile because immobile sperm cannot reach and fertilize an oocyte (egg).

RIBOSOMES [FIGURE 2.11]

Ribosomes are small, dense structures that cannot be seen with the light microscope. In an electron micrograph, ribosomes are dense granules roughly 25 nm in diameter (Figure 2.11a●). They are found in all cells, but their number varies depending on the type of cell and its activities. Each ribosome consists of roughly 60% RNA and 40% protein. At least 80 ribosomal proteins have been identified. These organelles are intracellular factories that manufacture proteins, using information provided by the DNA of the nucleus. A ribosome consists of two subunits that interlock as protein synthesis begins. When protein synthesis is complete, the subunits separate.

There are two major types of ribosomes: free ribosomes and fixed ribosomes (Figure 2.11a●). **Free ribosomes** are scattered throughout the cytoplasm; the proteins they manufacture enter the cytosol. **Fixed ribosomes** are attached to the *endoplasmic reticulum*, a membranous organelle. Proteins manufactured by fixed ribosomes enter the *lumen*, or internal cavity, of the endoplasmic reticulum, where they are modified and packaged for export. These processes are detailed later in this chapter.

✓ **CONCEPT CHECK**

- Cells lining the small intestine have numerous fingerlike projections on their free surfaces. What are these structures, and what is their function?
- How would the absence of a flagellum affect a sperm cell?
- Identify the two major subdivisions of the cytoplasm and the function of each.

Membranous Organelles

Each membranous organelle is completely surrounded by a phospholipid bilayer membrane similar in structure to the cell membrane. The membrane isolates the contents of a membranous organelle from the surrounding cytosol. This isolation allows the organelle to manufacture or store secretions, enzymes, or toxins that could adversely affect the cytoplasm in general. Table 2.1 on p. 29 includes six types of membranous organelles: *mitochondria, the nucleus, the endoplasmic reticulum, the Golgi apparatus, lysosomes*, and *peroxisomes*.

MITOCHONDRIA [FIGURE 2.12]

Mitochondria (mī-tō-KON-drē-ah; singular *mitochondrion*; *mitos*, thread + *chondrion*, small granules) are organelles that have an unusual double membrane (Figure 2.12●). An outer membrane surrounds the entire organelle, and a second, inner membrane contains numerous folds, called **cristae**. Cristae increase the surface area exposed to the fluid contents,

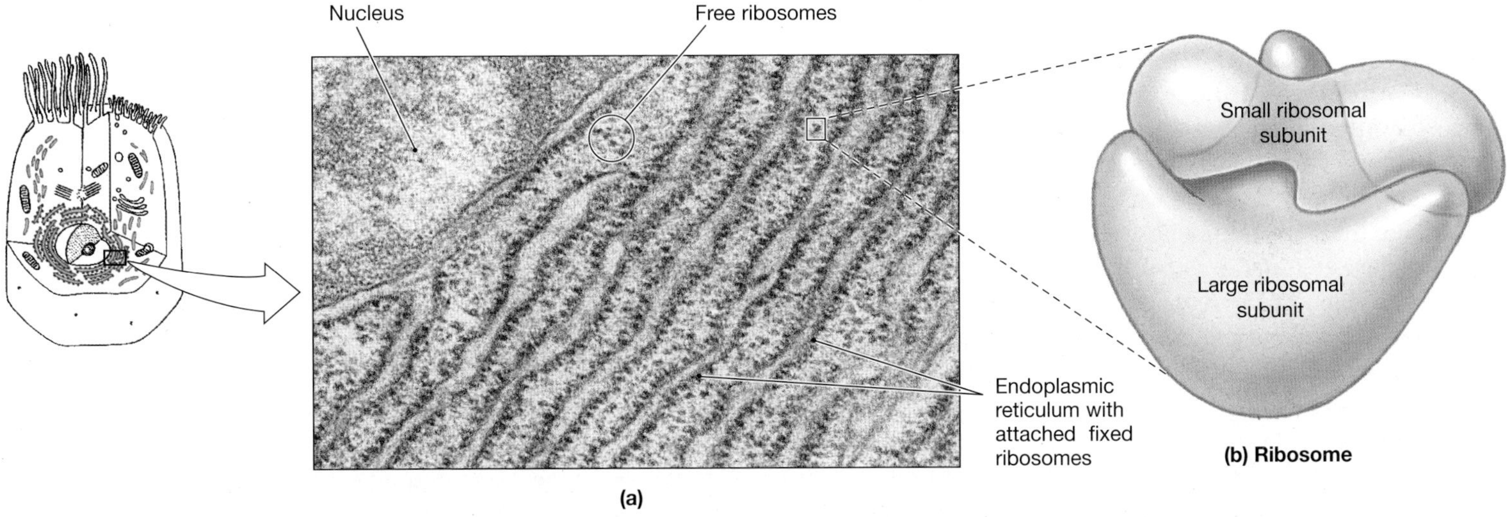

FIGURE 2.11 **RIBOSOMES**

These small, dense structures are involved in protein synthesis. (**a**) Both free and fixed ribosomes can be seen in the cytoplasm of this cell. (TEM × 73,600). (**b**) An individual ribosome, consisting of small and large subunits.

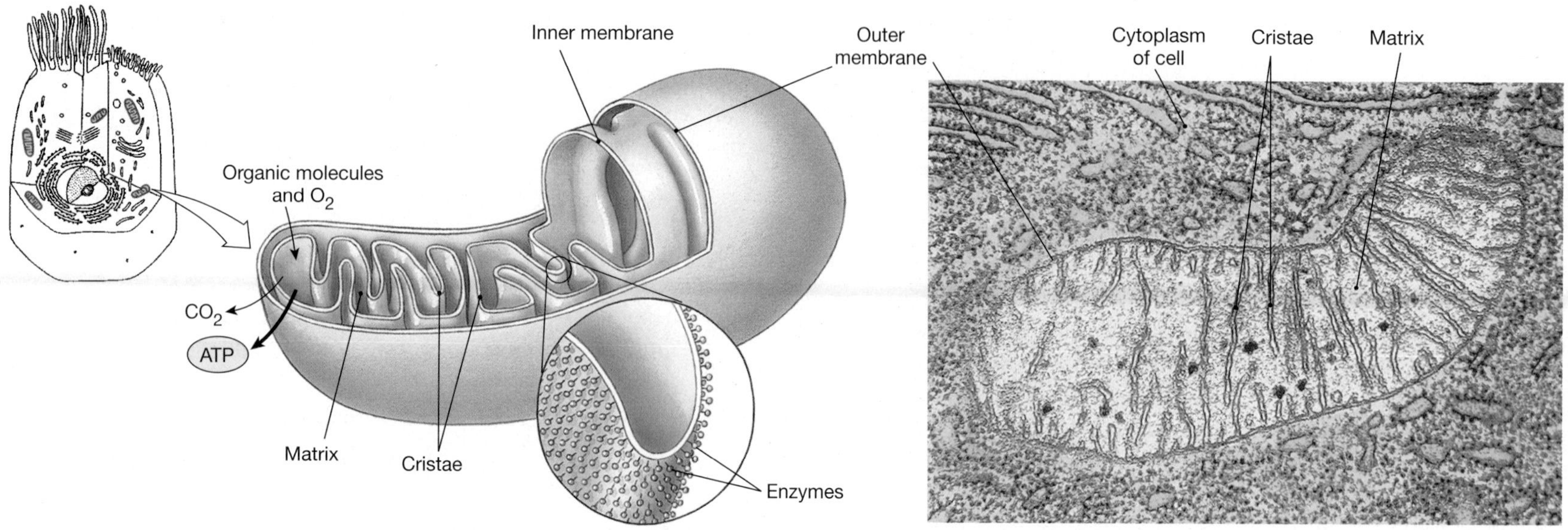

FIGURE 2.12 **MITOCHONDRIA**

The three-dimensional organization of a mitochondrion, and a color-enhanced TEM showing a typical mitochondrion in section. (TEM × 61,776)

or **matrix**, of the mitochondrion. The matrix contains metabolic enzymes that perform the reactions that provide energy for cellular functions.

Enzymes attached to the cristae produce most of the ATP generated by mitochondria. Mitochondrial activity produces about 95% of the energy needed to keep a cell alive. Mitochondria produce ATP through the breakdown of organic molecules in a series of reactions that also consume oxygen (O_2) and generate carbon dioxide (CO_2).

Mitochondria have various shapes, from long and slender to short and fat. Mitochondria control their own maintenance, growth, and reproduction. The number of mitochondria in a particular cell varies depending on the cell's energy demands. Red blood cells lack mitochondria—they obtain energy in other ways—but liver and skeletal muscle cells typically contain as many as 300 mitochondria. Muscle cells have high rates of energy consumption, and over time the mitochondria respond to the increased energy demands by reproducing. The increased numbers of mitochondria can provide energy faster and in greater amounts, improving muscular performance.

THE NUCLEUS [FIGURES 2.13/2.14]

The **nucleus** is the control center for cellular operations. A single nucleus stores all the information needed to control the synthesis of the approximately 100,000 different proteins in the human body. The nucleus determines the structural and functional characteristics of the cell by controlling what proteins are synthesized, and in what amounts. Most cells contain a single nucleus, but there are exceptions. For example, skeletal muscle cells are called multinucleate (*multi-*, many) because they have many nuclei, whereas mature red blood cells are called anucleate (*a-*, without) because they lack nuclei. A cell without a nucleus could be compared to a car without a driver. However, a car can sit idle for years, but a cell without a nucleus will survive only 3–4 months.

Figure 2.13● details the structure of a typical nucleus. A **nuclear envelope** surrounds the nucleus and separates it from the cytosol. The nuclear envelope is a double membrane enclosing a narrow **perinuclear space** (*peri-*, around). At several locations, the nuclear envelope is connected to the rough endoplasmic reticulum, as shown in Figure 2.3●, p. 28.

The nucleus directs processes that take place in the cytosol and must in turn receive information about conditions and activities in the cytosol. Chemical communication between the nucleus and cytosol occurs through **nuclear pores**, a complex of proteins that regulates movement of macromolecules into and out of the nucleus. These pores, which account for about 10% of the surface of the nucleus, permit the movement of water, ions, and small molecules but regulate the passage of large proteins, RNA, and DNA.

The term **nucleoplasm** refers to the fluid contents of the nucleus. The nucleoplasm contains ions, enzymes, RNA and DNA nucleotides, proteins, small amounts of RNA, and DNA. The DNA strands form complex structures known as *chromosomes* (*chroma*, color). The nucleoplasm also contains a network of fine filaments, the **nuclear matrix**, that provides structural support and may be involved in the regulation of genetic activity. Each **chromosome** contains DNA strands bound to special proteins called **histones.** The nucleus of each of your cells contains 23 pairs of chromosomes; one member of each pair was derived from your mother and one from your father. The structure of a typical chromosome is diagrammed in Figure 2.14●.

At intervals the DNA strands wind around the histones, forming a complex known as a **nucleosome**. The entire chain of nucleosomes may coil around other histones. The degree of coiling determines whether the chromosome is long and thin or short and fat. Chromosomes in a dividing cell are very tightly coiled, and so can be seen clearly as separate structures in light or electron micrographs. In cells that are not dividing, the chromosomes are loosely coiled, forming a tangle of fine filaments known as **chromatin** (KRŌ-ma-tin). Each chromosome may have some coiled regions, and only the coiled areas stain clearly. As a result, the nucleus has a clumped, grainy appearance.

The chromosomes also have direct control over the synthesis of RNA. Most nuclei contain one to four dark-staining areas called **nucleoli** (noo-KLĒ-ō-lī; singular *nucleolus*). Nucleoli are nuclear organelles that synthesize the components of ribosomes. A nucleolus contains histones and enzymes as well as RNA, and it forms around a chromosomal region containing the genetic instructions for producing ribosomal proteins and RNA. Nucleoli are most prominent in cells that manufacture large amounts of proteins, such as liver cells and muscle cells, because these cells need large numbers of ribosomes.

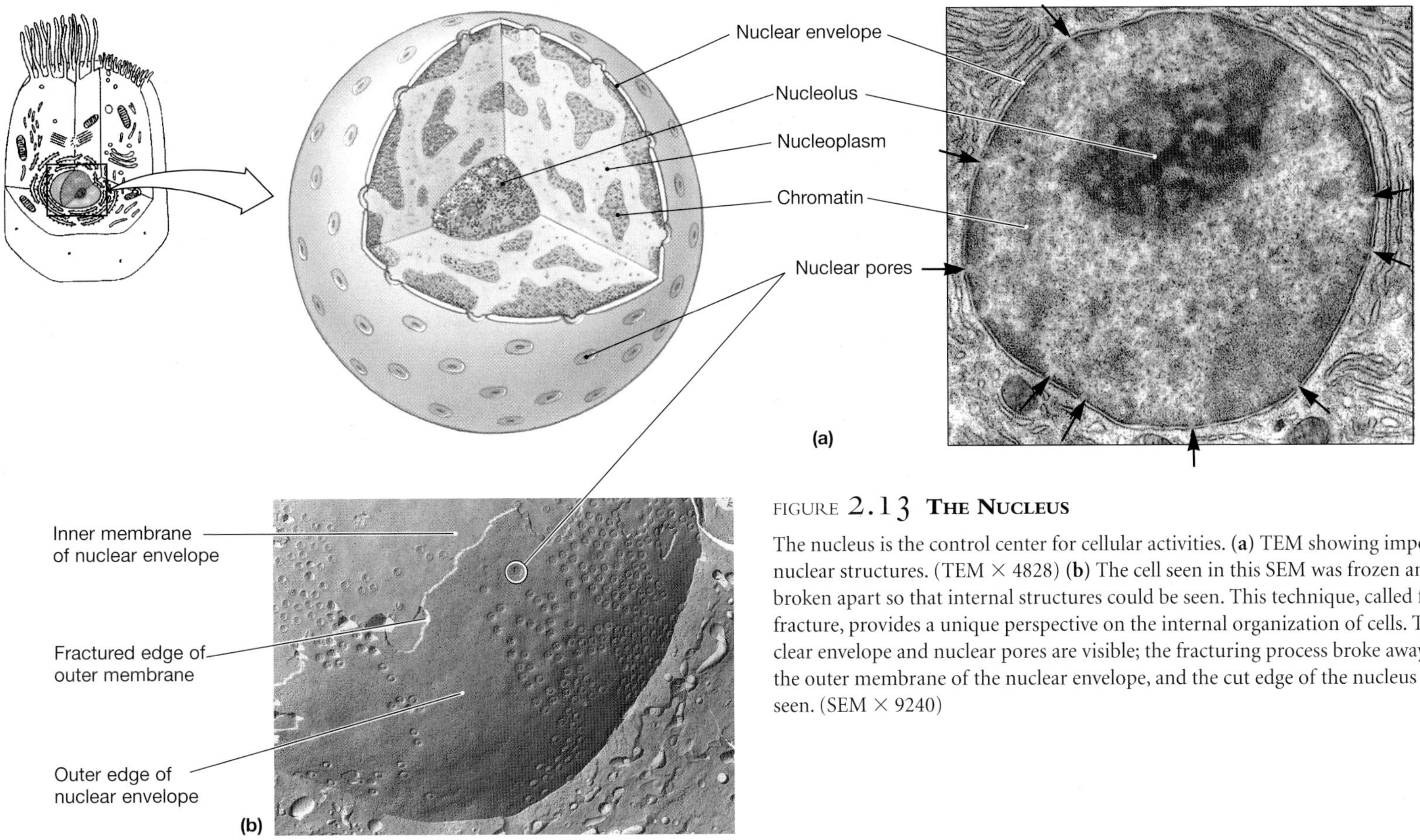

FIGURE 2.13 **THE NUCLEUS**

The nucleus is the control center for cellular activities. (**a**) TEM showing important nuclear structures. (TEM × 4828) (**b**) The cell seen in this SEM was frozen and then broken apart so that internal structures could be seen. This technique, called freeze-fracture, provides a unique perspective on the internal organization of cells. The nuclear envelope and nuclear pores are visible; the fracturing process broke away part of the outer membrane of the nuclear envelope, and the cut edge of the nucleus can be seen. (SEM × 9240)

THE ENDOPLASMIC RETICULUM [FIGURE 2.15]

The **endoplasmic reticulum** (en-dō-PLAZ-mik re-TIK-ū-lum), or **ER**, is a network of intracellular membranes that forms hollow tubes, flattened sheets, and round chambers (Figure 2.15●). The chambers are called **cisternae** (sis-TUR-nē; singular *cisterna*, a reservoir for water).

The ER has three major functions:

1. *Synthesis*: The membrane of the endoplasmic reticulum contains enzymes that manufacture carbohydrates and lipids; areas with fixed ribosomes synthesize proteins. These manufactured products are stored in the cisternae of the ER.
2. *Storage*: The ER can hold synthesized molecules or substances absorbed from the cytosol without affecting other cellular operations.
3. *Transport*: Substances can travel from place to place within the cell inside the endoplasmic reticulum.

The ER thus functions as a combination workshop, storage area, and shipping depot. It is where many newly synthesized proteins undergo chemical modification and where they are packaged for export to their next destination, the *Golgi apparatus*. There are two distinct types of endoplasmic reticulum, **rough endoplasmic reticulum (RER)** and **smooth endoplasmic reticulum (SER)**.

The outer surface of the rough endoplasmic reticulum contains fixed ribosomes. Ribosomes synthesize proteins using instructions provided by a strand of RNA. As the polypeptide chains grow, they enter the cisternae of the endoplasmic reticulum, where they may be further modified. Most of the proteins and glycoproteins produced by the RER are packaged into

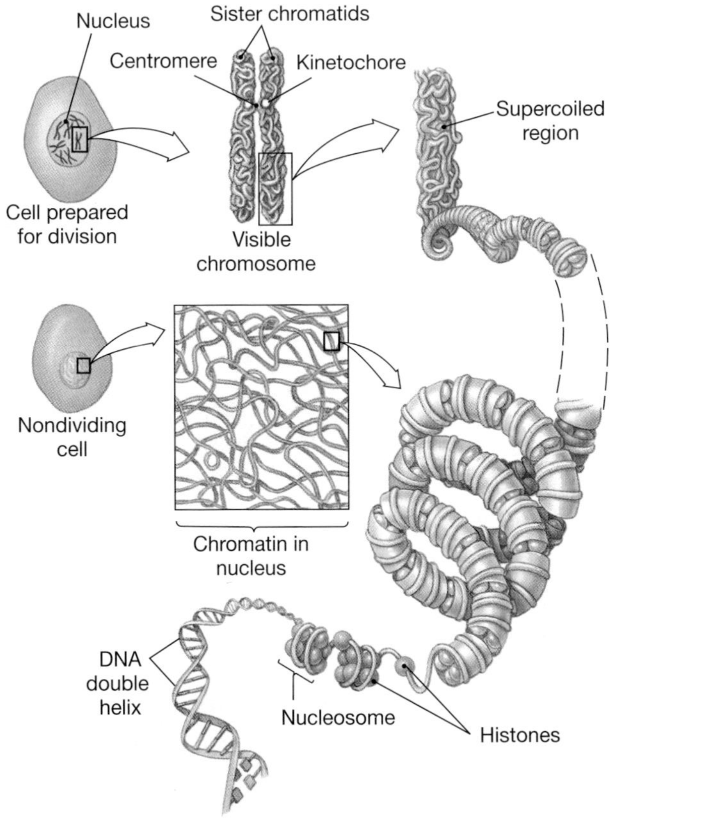

FIGURE 2.14 **CHROMOSOME STRUCTURE**

DNA strands are coiled around histones to form nucleosomes. Nucleosomes form coils that may be very tight or rather loose. In cells that are not dividing, the DNA is loosely coiled, forming a tangled network known as chromatin. When the coiling becomes tighter, as it does in preparation for cell division, the DNA becomes visible as distinct structures called chromosomes.

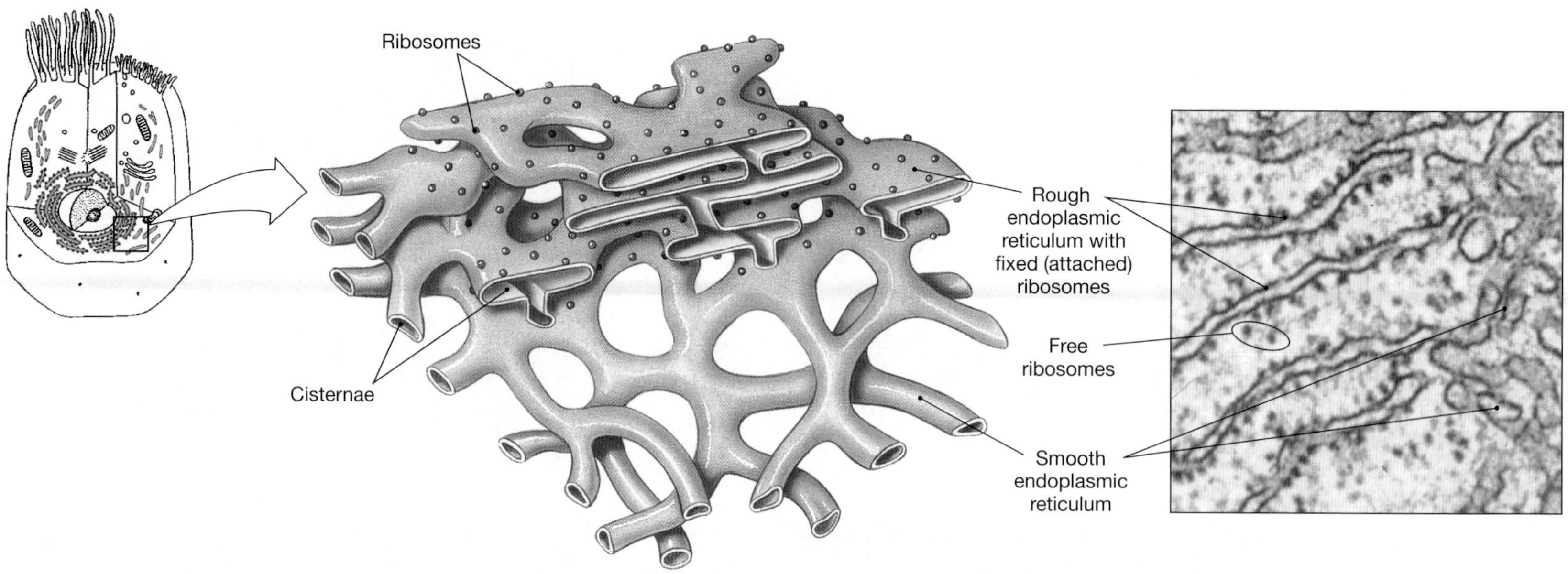

FIGURE 2.15 **THE ENDOPLASMIC RETICULUM**

This organelle is a network of intracellular membranes. Here, a diagrammatic sketch shows the three-dimensional relationships between the rough and smooth endoplasmic reticulum.

small membrane sacs that pinch off the edges or surfaces of the ER. These **transport vesicles** deliver the proteins to the Golgi apparatus.

No ribosomes are associated with smooth endoplasmic reticulum. The SER has a variety of functions that center around the synthesis of lipids and carbohydrates, the storage of calcium ions, and the removal and inactivation of toxins.

The amount of endoplasmic reticulum and the proportion of RER to SER vary depending on the type of cell and its ongoing activities. For example, pancreatic cells that manufacture digestive enzymes contain an extensive RER, and the SER is relatively small. The situation is reversed in cells that synthesize steroid hormones in reproductive organs.

THE GOLGI APPARATUS [FIGURE 2.16]

The **Golgi** (GŌL-jē) **apparatus**, or *Golgi complex*, consists of flattened membrane discs called **cisternae**. A typical Golgi apparatus, shown in Figure 2.16●, consists of five to six cisternae. Cells that are actively secreting have larger and more numerous cisternae than resting cells. The most actively secreting cells contain several sets of cisternae, each resembling a stack of dinner plates. Most often these stacks lie near the nucleus of the cell.

The major functions of the Golgi apparatus are:

1. Synthesis and packaging of secretions, such as mucins or enzymes.
2. Packaging of special enzymes for use in the cytosol.
3. Renewal or modification of the cell membrane.

The Golgi cisternae communicate with the ER and with the cell surface. This communication involves the formation, movement, and fusion of vesicles.

VESICLE TRANSPORT, TRANSFER, AND SECRETION [FIGURE 2.17] The role played by the Golgi apparatus in packaging secretions is illustrated in Figure 2.17a●. Protein and glycoprotein synthesis occurs in the RER, and transport vesicles (packages) then move these products to the Golgi apparatus. The vesicles usually arrive at a convex cisternae known as the *forming face* (or *cis face*). The transport vesicles then fuse with the Golgi membrane, emptying their contents into the cisternae, where enzymes modify the arriving proteins and glycoproteins.

Material moves between cisternae by means of small **transfer vesicles**. Ultimately the product arrives at the *maturing face* (or *trans face*). At the maturing face, vesicles form that carry materials away from the Golgi. Vesicles containing secretions that will be discharged from the cell are called **secretory vesicles**. Secretion occurs as the membrane of a secretory vesicle fuses with the cell membrane. This discharge process is called **exocytosis** (eks-ō-sī-TŌ-sis) (Figure 2.17b●).

MEMBRANE TURNOVER Because the Golgi apparatus continually adds new membrane to the cell surface, it has the ability to change the properties of the cell membrane over time. Such changes can profoundly alter the sensitivity and functions of the cell. In an actively secreting cell, the Golgi membranes may undergo a complete turnover every 40 minutes. The membrane lost from the Golgi is added to the cell surface, and that addition is balanced by the formation of vesicles at the membrane surface. As a result, an area equal to the entire membrane surface may be replaced each hour.

LYSOSOMES [FIGURE 2.18]

Many of the vesicles produced at the Golgi apparatus never leave the cytoplasm. The most important of these are lysosomes. **Lysosomes** (LĪ-sō-sōms; *lyso-*, dissolution + *soma*, body) are vesicles filled with digestive enzymes. Refer to Figure 2.18● as we describe the types of lysosomes and lysosomal functions. *Primary lysosomes* contain inactive enzymes. Activation occurs when the lysosome fuses with the membranes of damaged organelles, such as mitochondria or fragments of the endoplasmic reticulum. This fusion creates a *secondary lysosome*, which contains active enzymes. These enzymes then break down the lysosomal contents. Nutrients reenter the cytosol, and the remaining waste material is eliminated by exocytosis.

Lysosomes also function in the defense against disease. By the process of endocytosis, cells may remove bacteria, as well as fluids and organic debris, from their surroundings and isolate them within vesicles.

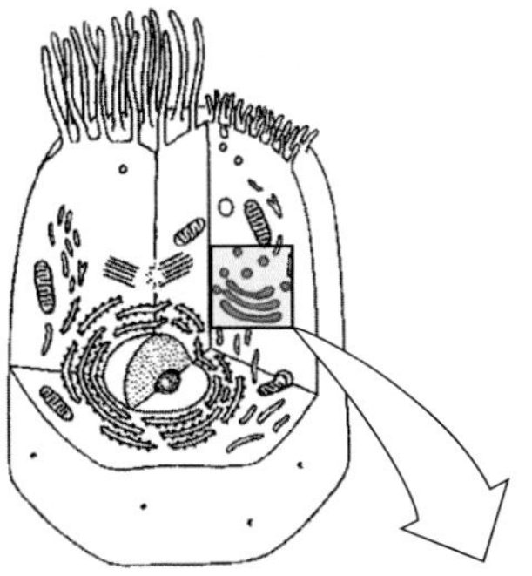

FIGURE 2.16 **The Golgi Apparatus**

(**a**) A sectional view of the Golgi apparatus of an active secretory cell. (TEM × 83,520) (**b**) A three-dimensional view of the Golgi apparatus with a cut edge corresponding to part (a).

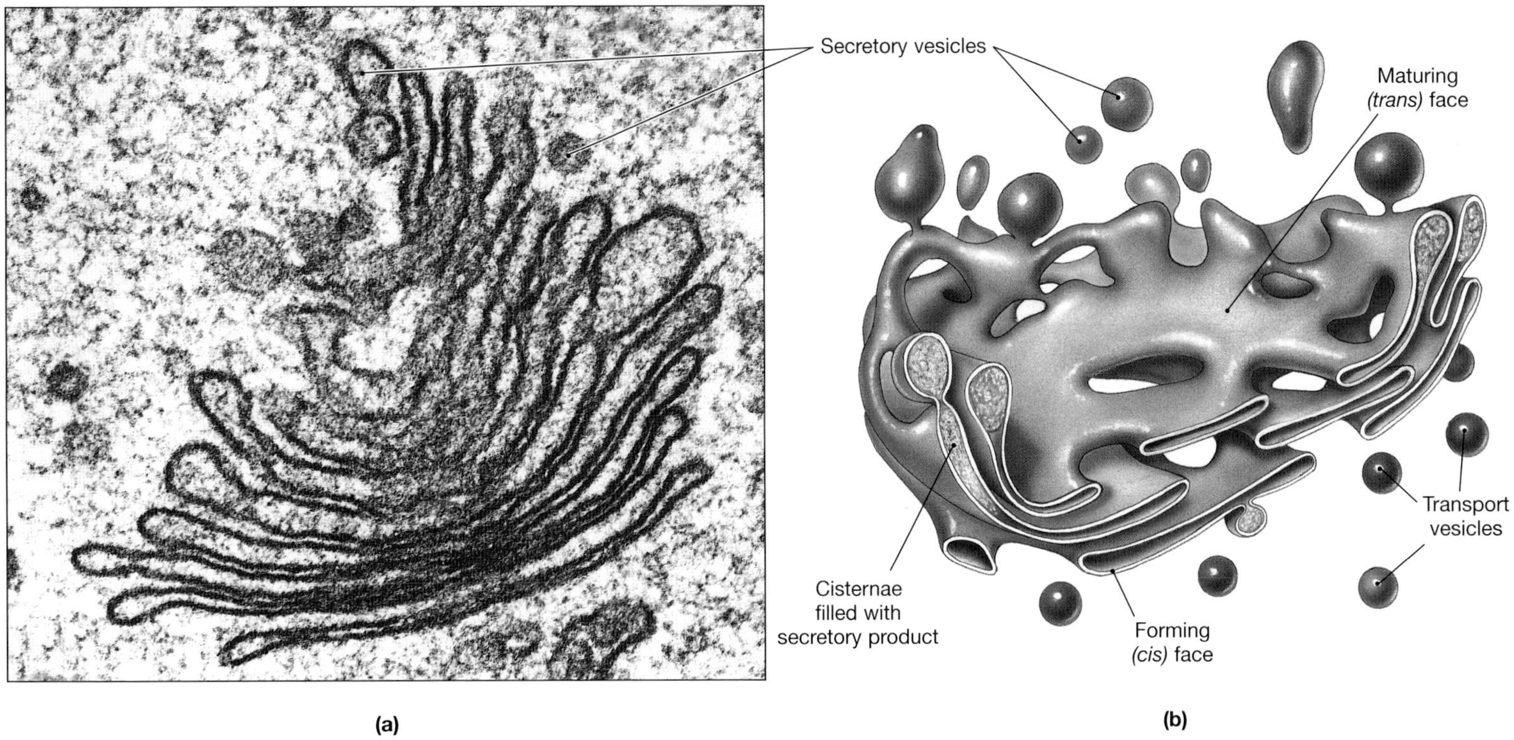

FIGURE 2.17 **The Function of the Golgi Apparatus**

(**a**) This diagram shows the functional link between the ER and the Golgi apparatus. Golgi structure has been simplified to clarify the relationships between the membranes. Transport vesicles carry the secretory product from the endoplasmic reticulum to the Golgi apparatus, and transfer vesicles move membrane and materials between the Golgi cisternae. At the maturing face, three functional categories of vesicles develop. Secretory vesicles carry the secretion from the Golgi to the cell surface, where exocytosis releases the contents into the extracellular fluid. Other vesicles add surface area and integral proteins to the cell membrane. Lysosomes, which remain in the cytoplasm, are vesicles filled with enzymes. (**b**) Exocytosis at the surface of a cell.

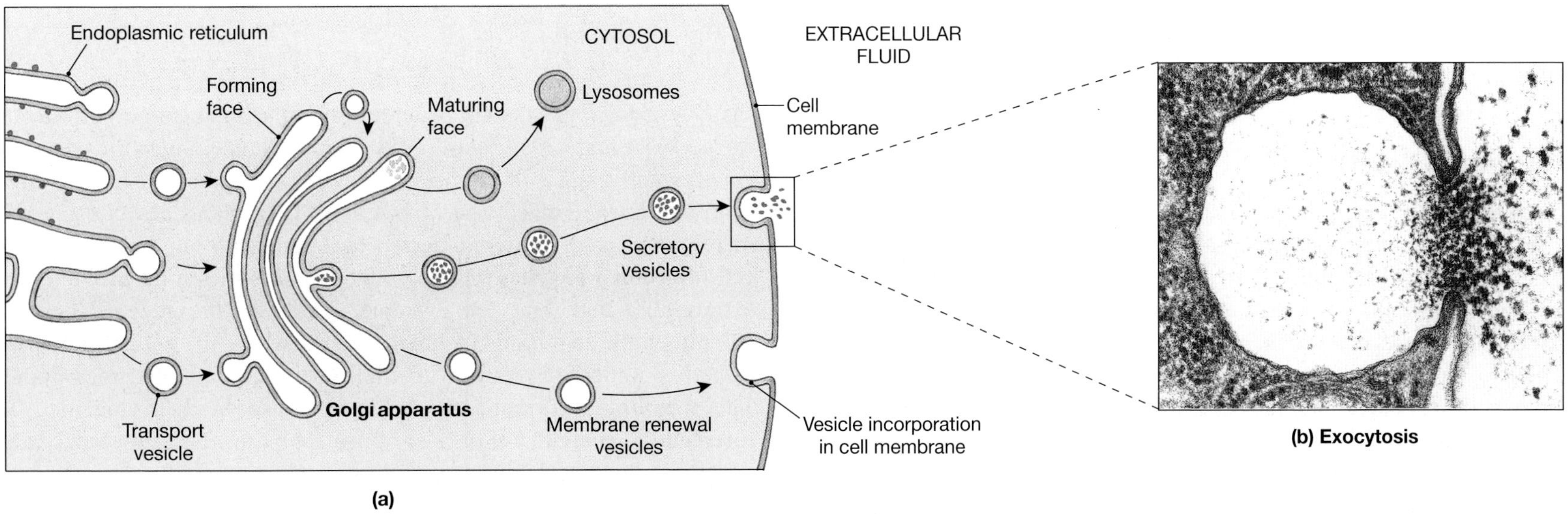

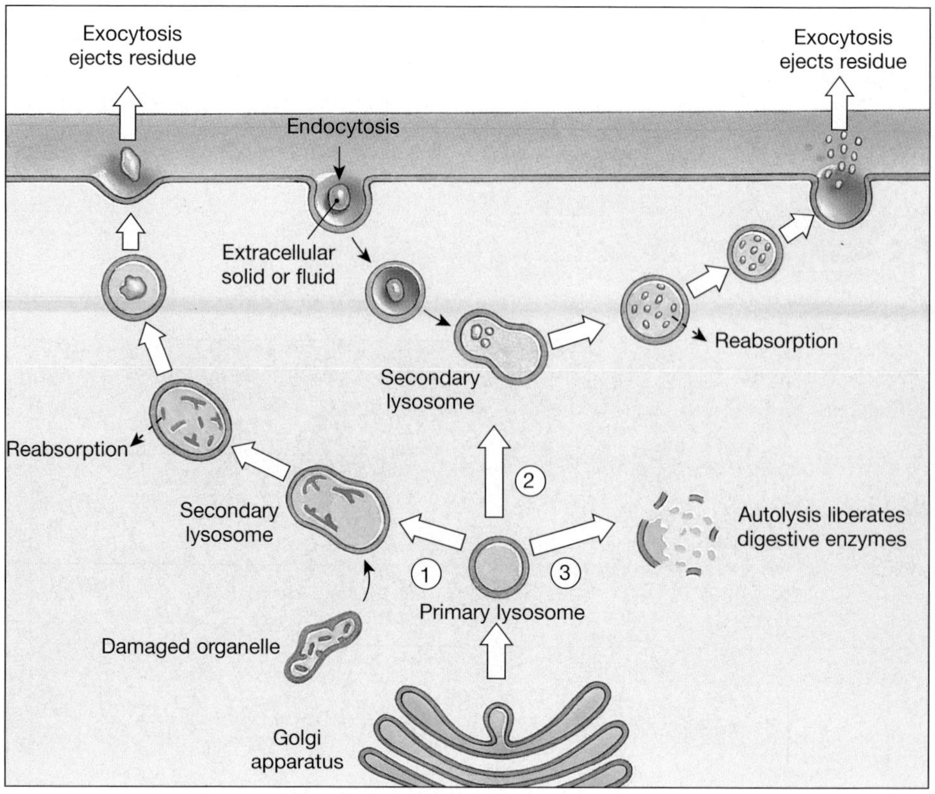

FIGURE 2.18 **LYSOSOMAL FUNCTIONS**

Primary lysosomes, formed at the Golgi apparatus, contain inactive enzymes. Activation may occur under three basic conditions: (**1**) when the primary lysosome fuses with the membrane of another organelle, such as a mitochondrion; (**2**) when the primary lysosome fuses with an endocytotic vesicle containing fluid or solid materials from outside the cell; or (**3**) in autolysis, when the lysosomal membrane breaks down following death or injury to the cell.

Lysosomes may fuse with vesicles created in this way, and the digestive enzymes within the secondary lysosome then break down the contents and release usable substances such as sugars or amino acids. In this way the cell not only protects itself against pathogenic organisms but obtains valuable nutrients.

Lysosomes also perform essential cleanup and recycling functions inside the cell. For example, when muscle cells are inactive, lysosomes gradually break down their contractile proteins; if the cells become active once again, this destruction ceases. This regulatory mechanism fails in a damaged or dead cell. Lysosomes then disintegrate, releasing active enzymes into the cytosol. These enzymes rapidly destroy the proteins and organelles of the cell, a process called **autolysis** (aw-TOL-i-sis; *auto-*, self). Because the breakdown of lysosomal membranes can destroy a cell, lysosomes have been called cellular "suicide packets." We do not know how to control lysosomal activities, or why the enclosed enzymes do not digest the lysosomal membranes unless the cell is damaged. Problems with lysosomal enzyme production cause more than 30 serious diseases affecting children. In these conditions, called *lysosomal storage diseases*, the lack of a specific lysosomal enzyme results in the buildup of waste products and debris normally removed and recycled by lysosomes. Affected individuals may die when vital cells, such as those of the heart, can no longer continue to function.

PEROXISOMES

Peroxisomes are smaller than lysosomes and carry a different group of enzymes. Peroxisomes are thought to originate at the RER, whereas lysosomes are produced at the Golgi apparatus. Peroxisomes absorb and break down fatty acids and other organic compounds. Enzymatic activity within a peroxisome may form toxins such as hydrogen peroxide as a byproduct; other enzymes then convert the hydrogen peroxide to water. Peroxisomes are most abundant in liver cells, which remove and neutralize toxins absorbed in the digestive tract.

Membrane Flow

With the exception of mitochondria, all the membranous organelles in the cell are either interconnected or in communication through the movement of vesicles. The RER and SER are continuous and connected to the nuclear envelope. Transport vesicles connect the ER with the Golgi apparatus, and secretory vesicles link the Golgi apparatus with the cell membrane. Finally, vesicles forming at the exposed surface of the cell remove and recycle segments of the cell membrane. This continual movement and exchange is called **membrane flow**.

Membrane flow is another example of the dynamic nature of cells. It provides a mechanism for cells to change the characteristics of their cell membranes—lipids, receptors, channels, anchors, and enzymes—as they grow, mature, or respond to a specific environmental stimulus.

✓ CONCEPT CHECK

- Microscopic examination of a cell reveals that it contains many mitochondria. What does this observation imply about the cell's energy requirements?
- Cells in the ovaries and testes contain large amounts of smooth endoplasmic reticulum (SER). Why?
- What occurs if lysosomes disintegrate in a damaged cell?

Intercellular Attachment [FIGURE 2.19]

Many cells form permanent or temporary attachments to other cells or extracellular materials (Figure 2.19●). Intercellular connections may involve extensive areas of opposing cell membranes, or they may be concentrated at specialized attachment sites. Large areas of opposing cell membranes may be interconnected by transmembrane proteins called **cell adhesion molecules (CAMs)**, which bind to each other and to other extracellular materials. For example, CAMs on the attached base of an epithelium help bind the basal surface (where the epithelium is attached to underlying tissues) to the underlying basement membrane. The membranes of adjacent cells may also be held together by **intercellular cement**, a thin layer of proteoglycans. These proteoglycans contain polysaccharide derivates known as *glycosaminoglycans*, most notably **hyaluronan** *(hyaluronic acid)*.

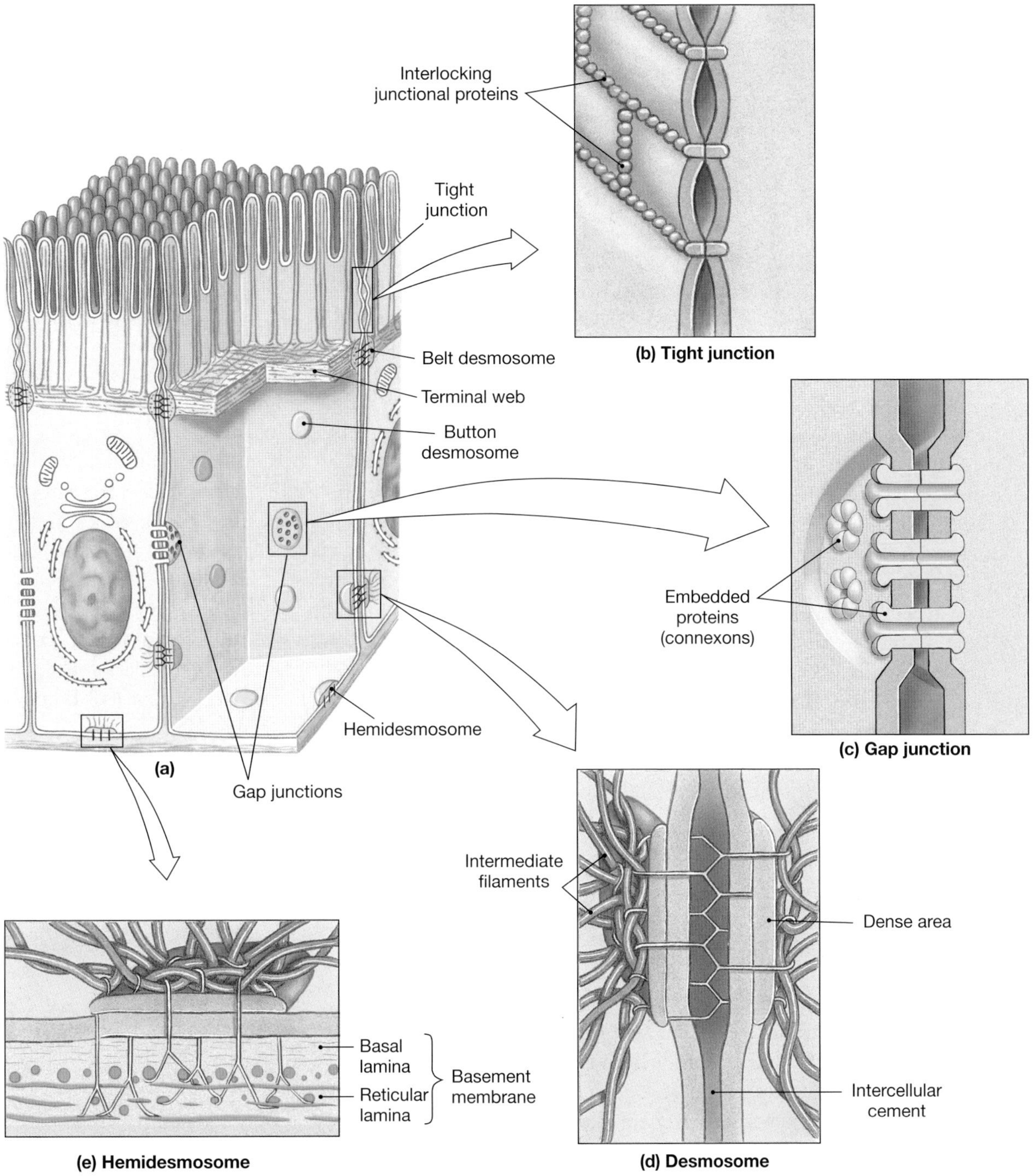

FIGURE 2.19 **CELL ATTACHMENTS**

Cells may attach to other cells or extracellular materials. (**a**) Diagrammatic view of an epithelial cell, showing the major types of intercellular connections. (**b**) A tight junction is formed by interlocking membrane proteins that lock together the two cell membranes. Bands of tight junctions encircle the apical portion of cuboidal and columnar epithelial cells, preventing the diffusion of fluids and solutes between the cells. (**c**) At a gap junction, binding of membrane proteins creates a cytoplasmic connection between two cells. (**d**) A desmosome has a more organized network of intermediate filaments. Desmosomes attach one cell to another. A continuous belt of desmosomes lies deep to the tight junctions. This belt is tied to the microfilaments of the terminal web. (**e**) Hemidesmosomes attach a cell to extracellular structures, such as the protein fibers in the basement membrane.

There are three major types of **cell junctions**: (1) *tight junctions*, (2) *gap junctions*, and (3) *desmosomes*.

- At a **tight junction**, the lipid portions of the two cell membranes are tightly bound together by interlocking membrane proteins (Figure 2.19b●). Because the membranes are tightly bound together, tight junctions block the passage of water and solutes between the cells. For example, tight junctions are found near the apical surfaces (the exposed surfaces) of cells that line the digestive tract, thereby keeping enzymes, acids, and wastes from damaging delicate underlying tissues.
- At a **gap junction**, two cells are held together by membrane proteins called *connexons* (Figure 2.19c●). Because these are channel proteins,

the result is a narrow passageway that lets small molecules and ions pass from cell to cell. Gap junctions are common among epithelial cells, where they help coordinate functions such as the beating of cilia. Gap junctions are also abundant in cardiac muscle and smooth muscle tissue, where they are essential to the coordination of muscle cell contractions.

- At a **desmosome** (DEZ-mō-sōm; *desmos*, ligament + *soma*, body), there are CAMs and proteoglycans linking the opposing cell membranes. Each cell contains a layered protein complex known as a *dense area* on the inside of the cell membrane. The cytoskeleton components of the cell are bound to this dense area (Figure 2.19d●). A *belt desmosome* forms a band that encircles the cell. This band is attached to the microfilaments of the *terminal web* (p. 35). *Button desmosomes* are small discs connected to bands of intermediate fibers (Figure 2.19a●). The intermediate fibers function like cross-braces to stabilize the shape of the cell. Desmosomes are very strong, and they can resist stretching and twisting. These connections are most abundant between cells in the superficial layers of the skin, where desmosomes create links so strong that dead skin cells are shed in thick sheets rather than individually. This attachment helps stabilize the position of the epithelial cell and anchors the cell to the underlying tissues. Desmosomes are also common in cardiac muscle tissue, interconnecting the muscle cells. *Hemidesmosomes* resemble half of a button desmosome. Rather than attaching one cell to another, a hemidesmosome attaches a cell to extracellular filaments and fibers in the basement membrane (Figure 2.19e●). This attachment helps stabilize the position of the epithelial cell and anchors the cell to adjacent tissues.

The Cell Life Cycle [FIGURE 2.20]

Between fertilization and physical maturity a human being increases in complexity from a single cell to roughly 75 trillion cells. This amazing increase in number occurs through a form of cellular reproduction called **cell division**. The division of a single cell produces a pair of *daughter cells*, each half the size of the original. Thus, two new cells have replaced the original one.

Even when development has been completed, cell division continues to be essential to survival. Although cells are highly adaptable, they can be damaged by physical wear and tear, toxic chemicals, temperature changes, or other environmental hazards. Cells are also subject to aging. The life span of a cell varies from hours to decades, depending on the type of cell and the environmental stresses involved. A typical cell does not live nearly as long as a typical person, so over time cell populations must be maintained by cell division.

The most important step in cell division is the accurate duplication of the cell's genetic material, a process called *DNA replication*, and the distribution of one copy of the genetic information to each of the two new daughter cells. The distribution process is called **mitosis** (mī-TŌ-sis). Mitosis occurs during the division of somatic (*soma*, body) cells. Somatic cells include all of the cells in the body other than the reproductive cells, which give rise to sperm or oocytes. Sperm and oocytes are called *gametes*; they are specialized cells containing half the number of chromosomes present in somatic cells. Production of gametes involves a distinct process, *meiosis* (mī-Ō-sis), which will be described in Chapter 28. An overview of the life cycle of a typical somatic cell is presented in Figure 2.20●.

Interphase [FIGURES 2.20/2.21/2.22a]

Most cells spend only a small part of their time actively engaged in cell division. Somatic cells spend the majority of their functional lives in *interphase*. During **interphase** the cell is performing all of its normal functions and, if necessary, preparing for division. In a cell preparing for division, interphase can be divided into the G_1, *S*, and G_2 phases (Figure 2.20●). An interphase cell in the **G_0 phase** is not preparing for mitosis, but is performing all other normal cell functions. Some mature cells, such as skeletal muscle cells and most neurons, remain in G_0 indefinitely and may never undergo mitosis. In contrast, *stem cells*, which divide repeatedly with very brief interphase periods, never enter G_0.

In the **G_1 phase** the cell manufactures enough mitochondria, centrioles, cytoskeletal elements, endoplasmic reticulum, ribosomes, Golgi membranes, and cytosol to make two functional cells. In cells dividing at top speed, G_1 may last as little as 8–12 hours. Such cells pour all of their energy into mitosis, and all other activities cease. If G_1 lasts for days, weeks, or months, preparation for mitosis occurs as the cells perform their normal functions. When G_1 preparations have been completed, the cell enters the **S phase**. Over the next 6–8 hours, the cell replicates its

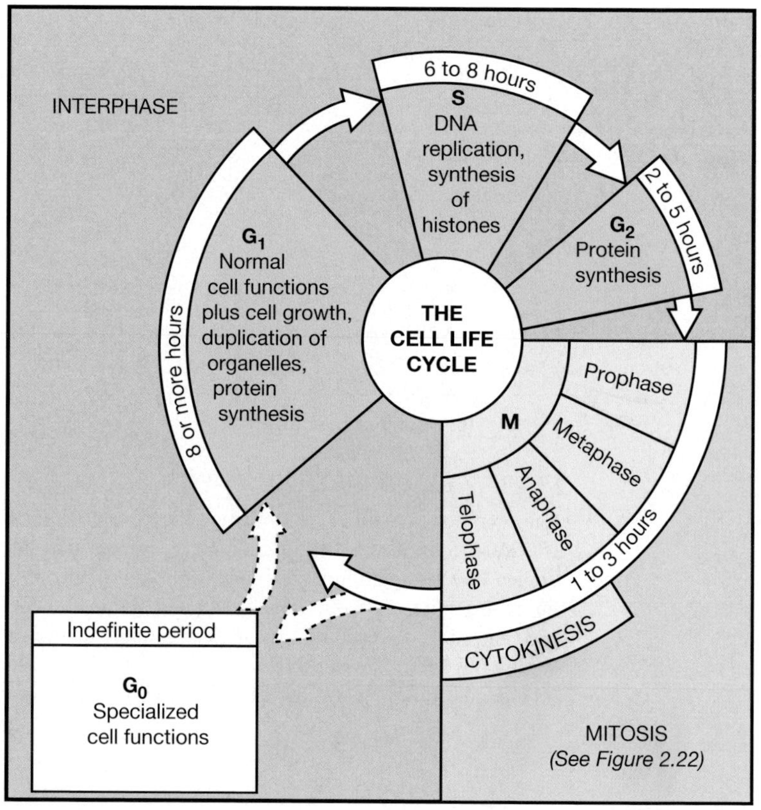

FIGURE 2.20 THE CELL LIFE CYCLE

The cell cycle is divided into interphase, comprising G_1, S, and G_2 stages, and the G_M phase, which includes mitosis and cytokinesis. The result is the production of two identical daughter cells.

CLINICAL BRIEF

Cell Division and Cancer

In normal tissue the rate of cell division balances cell loss or destruction. When that balance breaks down, the tissue begins to enlarge. A **tumor**, or **neoplasm**, is a mass or swelling produced by abnormal cell growth and division. In a **benign tumor** the cells remain within a connective tissue capsule. Such a tumor seldom threatens an individual's life. Surgery can usually remove the tumor if its size or position disturbs adjacent tissue function.

Cells in a **malignant tumor** are no longer responding to normal control mechanisms. These cells divide rapidly, spreading into the surrounding tissues, and they may also spread to other tissues and organs. This spread is called **metastasis** (me-TAS-ta-sis). Metastasis is dangerous and difficult to control. Once in a new location, the metastatic cells produce secondary tumors.

The term **cancer** refers to an illness characterized by malignant cells. Cancer cells gradually lose their resemblance to normal cells. They change size and shape, often becoming unusually large or abnormally small. Organ function begins to deteriorate as the number of cancer cells increases. The cancer cells may not perform their original functions at all, or they may perform normal functions in an unusual way. They also compete for space and nutrients with normal cells. They do not use energy very efficiently, and they grow and multiply at the expense of normal tissues. This activity accounts for the starved appearance of many patients in the late stages of cancer. *Causes of Cancer p. 784*

chromosomes, a process that involves the synthesis of both DNA and the associated histones.

Throughout the life of a cell, the DNA strands in the nucleus remain intact. DNA synthesis, or **DNA replication**, occurs in cells preparing to undergo mitosis or meiosis. The goal of replication is to copy the genetic information in the nucleus so that one set of chromosomes can be given to each of the two cells produced. Several different enzymes are needed for the process.

DNA REPLICATION

Each DNA molecule consists of a pair of nucleotide strands held together by hydrogen bonds between complementary nitrogen bases. Figure 2.21● diagrams the process of DNA replication. It starts when the weak bonds between the nitrogenous bases are disrupted, and the strands unwind. As they do so, molecules of the enzyme **DNA polymerase** bind to the exposed nitrogenous bases. This enzyme promotes bonding between the nitrogenous bases of the DNA strand and complementary DNA nucleotides suspended in the nucleoplasm.

Many molecules of DNA polymerase are working simultaneously along different portions of each DNA strand. This process produces short complementary nucleotide chains that are then linked together by enzymes called **ligases** (LĪ-gās-ez; *liga*, to tie). The final result is a pair of identical DNA molecules.

Once DNA replication has been completed, there is a brief (2–5 hours) **G_2 phase** devoted to last-minute protein synthesis. The cell then enters the **G_M phase**, and mitosis begins (Figures 2.20 and 2.22●).

Mitosis [FIGURE 2.22]

Mitosis consists of four stages, but the transitions from stage-to-stage are seamless. The stages are detailed in Figure 2.22●.

STEP 1 ***Prophase.*** (PRŌ-fāz; *pro*, before; Figure 2.22b,c●) Prophase begins when the chromosomes coil so tightly that they become visible as individual structures. As a result of DNA replication during the S phase, there are two copies of each chromosome, called **chromatids** (KRŌ-ma-tids), connected at a single point, the **centromere** (SEN-trō-mēr). The centrioles replicated in the G_1 phase; the two pairs of centrioles move apart during prophase. **Spindle fibers** extend between the centriole pairs; smaller microtubules called *astral rays* radiate into the surrounding cytoplasm. Prophase

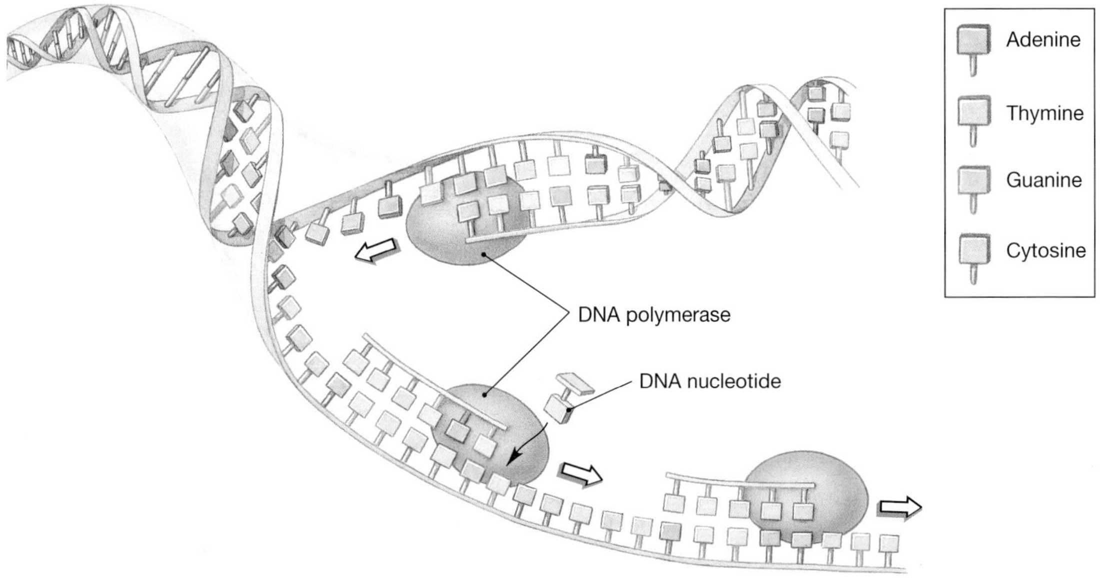

FIGURE 2.21 **DNA Replication**

In DNA replication the original paired strands unwind, and DNA polymerase begins attaching complementary DNA nucleotides along each strand. This process produces two identical copies of the original DNA molecule.

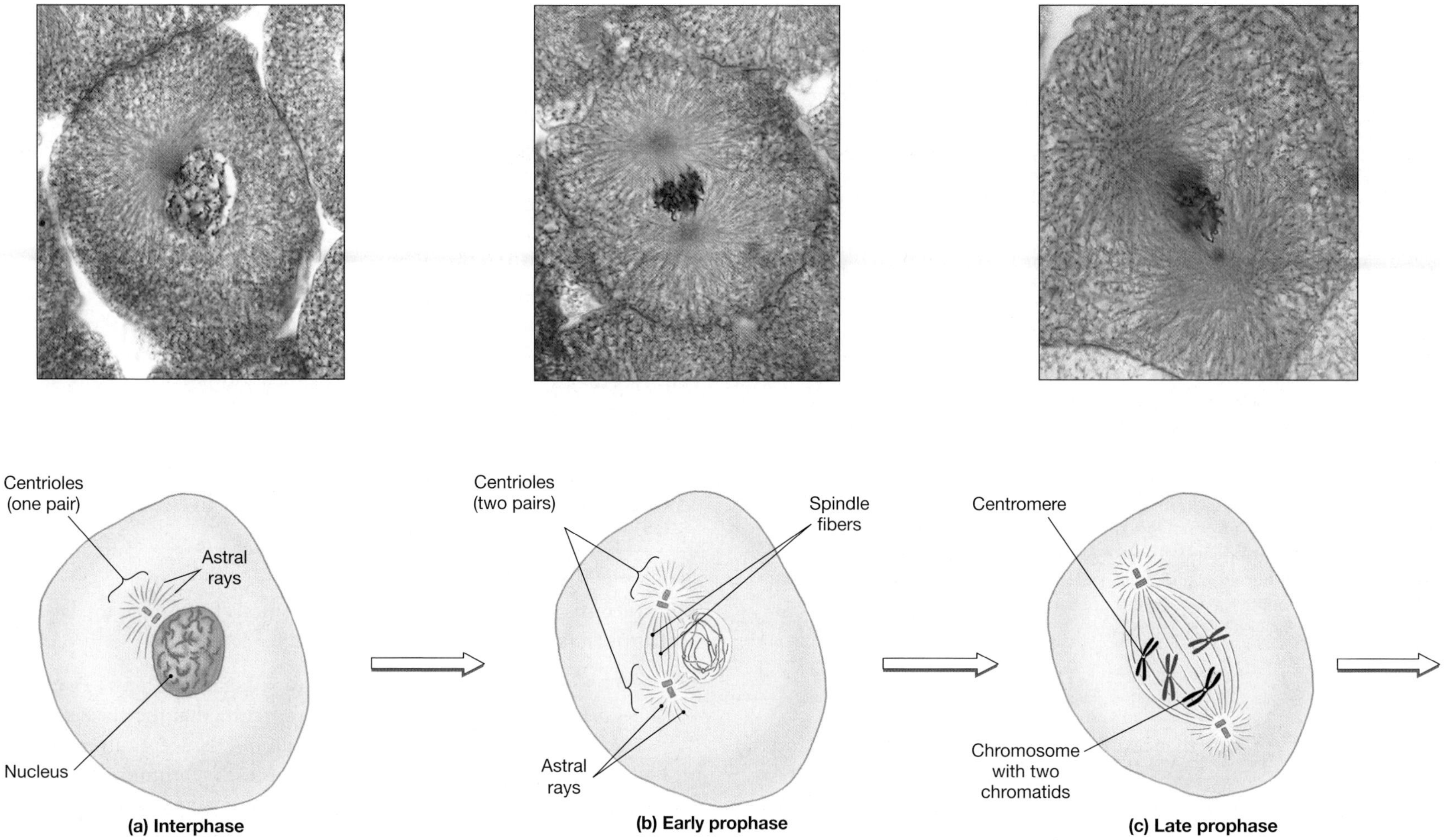

FIGURE 2.22 **Interphase and Mitosis**

The appearance of a cell in interphase and at the various stages of mitosis. (**a**) Interphase. (**b**) Early prophase. (**c**) Late prophase. (**d**) Metaphase. (**e**) Anaphase. (**f**) Telophase. (**g**) Daughter cells produced by cytokinesis. (LMs × 775)

ends with the disappearance of the nuclear envelope. The spindle fibers now form among the chromosomes and the kinetochore of each chromatid becomes attached to a spindle fiber called a *chromosomal microtubule.*

STEP 2 ***Metaphase.*** (MET-a-fāz; *meta*, after; Figure 2.22d●) Spindle fibers now pass among the chromosomes and the kinetochore of each chromatid becomes attached to a spindle fiber called a *chromosomal microtubule.* The chromosomes composed of chromatid pairs now move to a narrow central zone called the **metaphase plate**. A microtubule of the spindle apparatus attaches to each centromere.

STEP 3 ***Anaphase.*** (AN-uh-fāz; *ana*, back; Figure 2.22e●) As if responding to a single command, the chromatid pairs separate, and the **daughter chromosomes** move toward opposite ends of the cell. Anaphase ends as the daughter chromosomes arrive near the centrioles at opposite ends of the dividing cell.

STEP 4 ***Telophase.*** (TEL-ō-fāz; *telo*, end; Figure 2.22f●) This stage is in many ways the reverse of prophase, for in it the cell prepares to return to the interphase state. The nuclear membranes form and the nuclei enlarge as the chromosomes gradually uncoil. Once the chromosomes disappear, nucleoli reappear and the nuclei resemble those of interphase cells.

Telophase marks the end of mitosis proper, but the daughter cells have yet to complete their physical separation. This separation process, called **cytokinesis** (sī-tō-ki-NĒ-sis; *cyto-*, cell + *kinesis*, motion), usually begins in late anaphase. As the daughter chromosomes near the ends of the spindle apparatus, the cytoplasm constricts along the plane of the metaphase plate. This process continues through telophase, and the completion of cytokinesis (Figure 2.22g●) marks the end of cell division and the beginning of the next interphase period.

The frequency of cell division can be estimated by the number of cells in mitosis at any given time. As a result, the term **mitotic rate** is often used when discussing rates of cell division. In general, the longer the life expectancy of a cell type, the slower the mitotic rate. Relatively long-lived cells, such as muscle cells and neurons, either never divide or do so only under special circumstances. Other cells, like those lining the digestive tract survive only for days or even hours because they are constantly subjected to attack by chemicals, pathogens, and abrasion. Special cells called **stem cells** maintain these cell populations through repeated cycles of cell division.

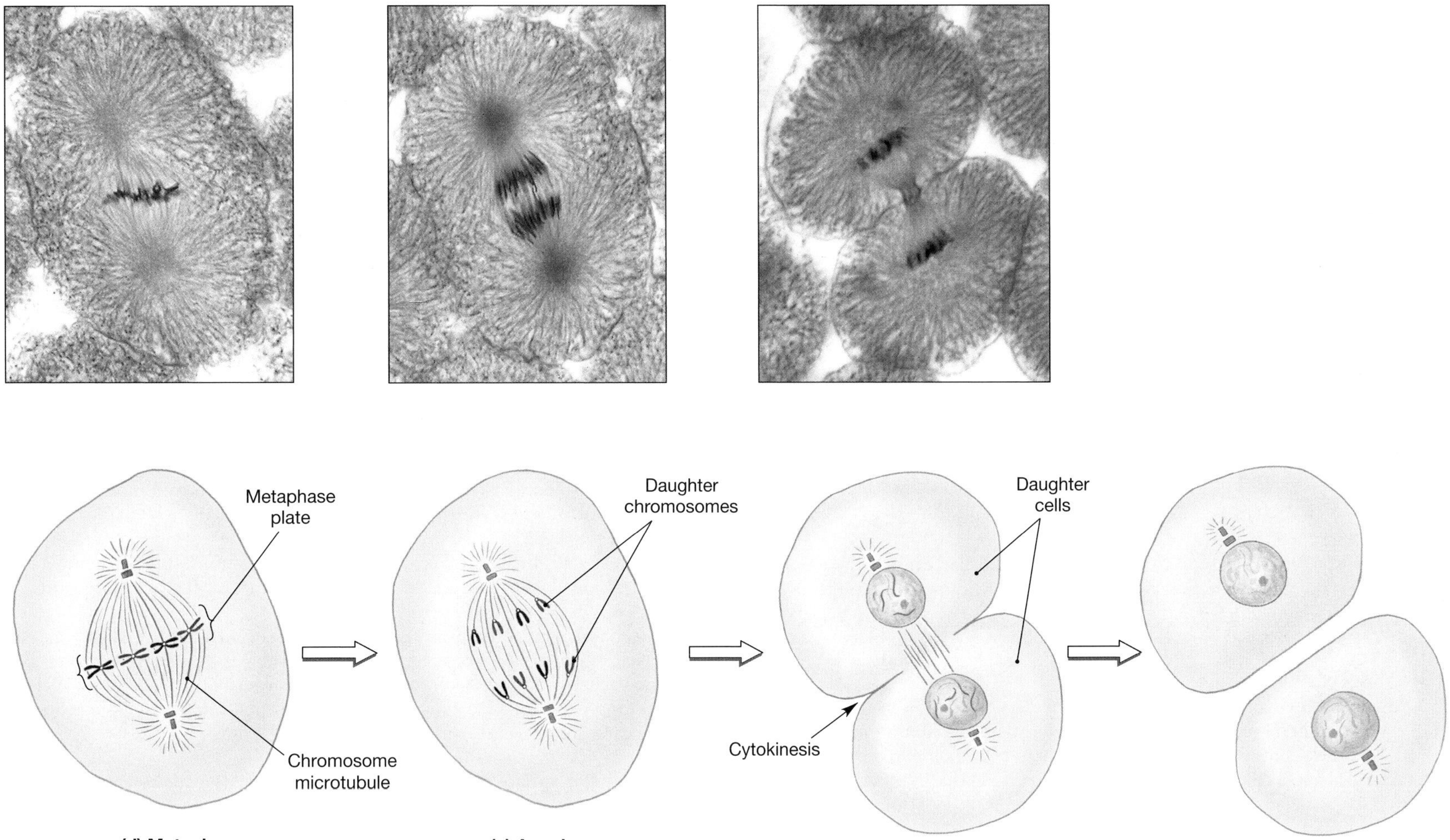

(d) Metaphase

(e) Anaphase

(f) Telophase

(g) Daughter cells separate

✓ CONCEPT CHECK

- What is cell division?
- Prior to cell division, mitosis must occur. What is mitosis?
- List in the order of appearance the stages of interphase and mitosis and the events that occur in each.

RELATED CLINICAL TERMS

benign tumor: A mass or swelling in which the cells remain within a connective tissue capsule; rarely life-threatening. *p. 45*

cancer: An illness characterized by malignant cells. *p. 45*

carcinogen (kar-SIN-ō-jen): An environmental factor that stimulates the conversion of a normal cell to a cancer cell. *Causes of Cancer p. 784*

hereditary predisposition: An individual born with genes that increase the likelihood of a specific disease. *Causes of Cancer p. 784*

malignant tumor: A mass or swelling in which the cells no longer respond to normal control mechanisms, but divide rapidly. *p. 45*

metastasis (me-TAS-ta-sis): The spread of malignant cells into surrounding and distant tissues and organs. *p. 45*

mutagen (MŪ-ta-jen): A factor that can damage DNA strands and sometimes cause chromosomal breakage, stimulating the development of cancer cells. *Causes of Cancer p. 784*

oncogene (ON-kō-jēn): A cancer-causing gene created by a somatic mutation in a normal gene (proto-oncogene) involved with growth, differentiation, or cell division. *Causes of Cancer p. 784*

tumor (neoplasm): A mass or swelling produced by abnormal cell growth and division. *p. 45*

tumor-suppressing genes (TSG) or **anti-oncogenes:** These genes suppress mitosis and growth in normal cells. *Causes of Cancer p. 784*

STUDY OUTLINE & CHAPTER REVIEW

Introduction 27

1. All living things are composed of **cells**, and contemporary *cell theory* incorporates several basic concepts: (1) cells are the building blocks of all plants and animals; (2) cells are produced by the division of preexisting cells; (3) cells are the smallest units that perform all vital functions.
2. The body contains two types of cells: **somatic cells** (body cells) and sex cells (**reproductive cells**)

The Study Of Cells 27

1. *Cytology* is the study of both the internal structure and function of individual cells.
2. *Light microscopy* uses light to permit magnification and viewing of cellular structures up to 1000 times their natural size. *(see Figure 2.1)*
3. *Electron microscopy* uses a focused beam of electrons to permit the examination of cell ultrastructure up to 1000 times smaller than what is possible by light microscopy. *(see Figure 2.1b,c)*

Cellular Anatomy 27

1. A cell is surrounded by a thin layer of **extracellular fluid**. The cell's outer boundary is the **cell membrane**, or **plasma membrane**. It is a **phospholipid bilayer** containing proteins and cholesterol. *Table 2.1* summarizes the anatomy of a typical cell. *(see Figures 2.3/2.4)*

The Cell Membrane 29

2. **Integral proteins** are embedded in the phospholipid bilayer of the membrane, while **peripheral proteins** are attached to the membrane but can separate from it. **Channels** allow water and ions to move across the membrane; some channels are called **gated channels** because they can open or close. *(see Figures 2.5/2.6)*
3. Cell membranes are **selectively permeable**; that is, they permit the free passage of some materials.
4. **Diffusion** is the net movement of material from an area where its concentration is high to an area where its concentration is lower. Diffusion occurs until the **concentration gradient** is eliminated. *(see Figure 2.6 and Table 2.2)*
5. Diffusion of water across a membrane in response to differences in water concentration is called **osmosis**. *(see Table 2.2)*
6. *Hydrostatic pressure* forces water across a membrane by **filtration**; if membrane pores are large enough, molecules of solute will be carried along. During *bulk flow* water molecules stream across a membrane in large groups. *(see Table 2.2)*
7. **Facilitated diffusion** is a passive transport process that requires the presence of **carrier proteins**. *(see Table 2.2)*
8. All **active membrane processes** require energy in the form of *adenosine triphosphate*, or ATP. Two important active processes are active transport and endocytosis. *(see Table 2.2)*
9. **Active transport** mechanisms consume ATP and are independent of concentration gradients. Some **ion pumps** are **exchange pumps**. *(see Table 2.2)*
10. **Endocytosis** is movement into a cell and is an active process that occurs in one of three forms: **pinocytosis** (cell drinking), **phagocytosis** (cell eating), or **receptor-mediated endocytosis** (selective movement). A summary of mechanisms involved in movement of substances across cell membranes is presented in *Table 2.2. (see Figures 2.7/2.8)*

The Cytoplasm 33

11. The **cytoplasm** contains an intracellular fluid, termed **cytosol**, which surrounds structures that perform specific functions, called **organelles**. *(see Figure 2.3 and Table 2.1)*

Nonmembranous Organelles 34

12. **Nonmembranous organelles** are not enclosed by membranes, and are always in contact with the cytosol. These include: cytoskeleton, microvilli, centrioles, cilia, flagella, and ribosomes. *(see Figure 2.9 to 2.11 and Table 2.1)*
13. The **cytoskeleton** is an internal protein network that gives the cytoplasm strength and flexibility. It has four components: **microfilaments, intermediate filaments, thick filaments**, and **microtubules**. *(see Figure 2.9 and Table 2.1)*
14. **Microvilli** are small, finger-like projections of the cell membrane that increase the surface area exposed to the extracellular environment. *(see Figure 2.9 and Table 2.1)*
15. **Centrioles** are small, microtubule-containing cylinders that direct the movement of chromosomes during cell division. *(see Figure 2.10 and Table 2.1)*
16. **Cilia** are anchored by a **basal body**, are microtubules-containing, hair-like projections from the cell surface that beat rhythmically to move fluids or secretions across the cell surface. *(see Figure 2.10 and Table 2.1)*
17. A whip-like **flagellum** moves a cell through surrounding fluid, rather than moving the fluid past a stationary cell. *Table 2.3* presents a comparison of centrioles, cilia, and flagella.
18. **Ribosomes** are intracellular factories consisting of small and large subunits; together they manufacture proteins. Two types of ribosomes, **free** (within the cytosal) and **fixed** (bound to the endoplasmic *reticulum*), are found in cells. *(see Figure 2.11 and Table 2.1)*

Membranous Organelles 37

19. **Membranous organelles** are surrounded by lipid membranes that isolate them from the cytosol. They include: mitochondria, nucleus, endoplasmic reticulum (rough and smooth), Golgi apparatus, lysosomes, and peroxisomes.
20. **Mitochondria** are responsible for producing 95% of the ATP within a typical cell. *(see Figure 2.12 and Table 2.1)*
21. The **nucleus** is the control center for cellular operations. It is surrounded by a **nuclear envelope**, through which it communicates with the cytosol through **nuclear pores**. *(see Figure 2.13 and Table 2.1)*
22. The **endoplasmic reticulum (ER)** is a network of intracellular membranes involved in synthesis, storage, and transport. The ER forms hollow tubes, flattened sheets, and rounded chambers termed **cisternae.** There are two types of ER: rough and smooth. **Rough endoplasmic reticulum (RER)** has attached ribosomes; **smooth endoplasmic reticulum (SER)** does not. *(see Figure 2.15 and Table 2.1)*
23. The **Golgi apparatus** packages materials for **lysosomes**, **peroxisomes**, **secretory vesicles**, and membrane segments that are incorporated into the cell membrane. Secretory products are discharged from the cell through the process of **exocytosis.** *(see Figures 2.16/2.17 and Table 2.1)*
24. **Lysosomes** are vesicles filled with digestive enzymes. The process of endocytosis is important in ridding the cell of bacteria and debris. The endocytic vesicle fuses with a lysosome, resulting in the digestion of its contents. *(see Figure 2.18 and Table 2.1)*
25. **Peroxisomes** carry enzymes used to break down organic molecules and neutralize toxins.
26. There is a continuous movement of membrane among the nuclear envelope, Golgi apparatus, endoplasmic reticulum, vesicles, and the cell membrane. This is called **membrane flow.**

Intercellular Attachment 42

1. Cells attach to other cells or to extracellular protein fibers by four different types of cell junctions: gap junctions, tight junctions, intermediate junctions, and desmosomes.
2. Cells in some areas of the body are linked by combinations of cell junctions. *(see Figure 2.19)*
3. At a **tight junction**, bands of interlocking membrane proteins bind together the adjoining cell membranes; these are the strongest intercellular connections. *(see Figure 2.19b)*
4. In a **gap junction**, two cells are held together by interlocked membrane proteins. These are channel proteins, which form a narrow passageway. *(see Figure 2.19c)*
5. A **desmosome** has a very thin proteoglycan layer between the cell membranes, reinforced by a network of microfilaments. *(see Figure 2.19d)*
6. A *hemidesmosome* attaches a cell to extracellular filaments or fibers. *(see Figure 2.19e)*

The Cell Life Cycle 44

1. **Cell division** is the reproduction of cells. In a dividing cell, an interphase or growth period alternates with a nuclear division phase, termed **mitosis.** *(see Figure 2.20)*

Interphase 44

2. Most somatic cells spend most of their time in **interphase**, a time of growth. *(see Figure 2.20)*

Mitosis 45

3. **Mitosis** refers to the nuclear division of somatic cells. Reproductive cells produce *gametes* (sperm or oocytes) through the process of *meiosis. (see Figures 2.20/2.22)*
4. Mitosis proceeds in four distinct, contiguous stages: **prophase**, **metaphase**, **anaphase**, and **telophase**. *(see Figure 2.22)*
5. In general, the longer the life expectancy of a cell type, the slower the **mitotic rate**. **Stem cells** undergo frequent mitosis to replace other, more specialized cells.

LEVEL 1 REVIEWING FACTS AND TERMS

Match each numbered item with the most closely related lettered item. Use letters for answers in the spaces provided.

Column A	**Column B**
1. ribosomes	a. DNA replication
2. lysosomes	b. flattened membrane discs, packaging
3. integral proteins	c. adjacent cell membranes bound by bands of interlocking proteins
4. Golgi apparatus	d. packaging of materials for import into cell
5. endocytosis	e. RNA and protein; protein synthesis
6. cytoskeleton	f. control center; stores genetic information
7. tight junction	g. cell vesicles with digestive enzymes
8. nucleus	h. embedded in the cell membrane
9. S phase	i. internal protein framework in cytoplasm

10. All of the following membrane transport mechanisms are passive processes except
    (a) facilitated diffusion (b) vesicular transport
    (c) filtration (d) diffusion
11. Which of the following is not a function of intermediate filaments?
    (a) part of spindle apparatus
    (b) provide strength to the cell
    (c) stabilize position of organelles
    (d) transport materials within cytoplasm
12. The interphase of a cell's life cycle is divided into the following phases
    (a) prophase, metaphase, anaphase, and telophase
    (b) G_0, G_1, S, and G_2
    (c) mitosis and cytokinesis
    (d) replication, rest, division
13. Which organelle is prevalent in cells involved in many phagocytic events?
    (a) free ribosomes (b) lysosomes
    (c) peroxisomes (d) microtubules
14. The principle cations in body fluids are
    (a) calcium and magnesium (b) chloride and bicarbonate
    (c) sodium and potassium (d) sodium and chloride
15. Membrane flow provides a mechanism for
    (a) continual change in the characteristics of membranes
    (b) increase in the size of the cell
    (c) response of the cell to a specific environmental stimulus
    (d) all of the above
16. ___ ion concentrations are high in the extracellular fluids, and ___ ion concentrations are high in the cytoplasm.
    (a) Calcium; magnesium (b) Chloride; sodium
    (c) Potassium; sodium (d) Sodium; potassium
17. Some integral membrane proteins form gated channels that open or close to
    (a) regulate the passage of materials into or out of the cell
    (b) permit water movement into or out of the cell
    (c) transport large proteins into the cell
    (d) communicate with neighboring cells
18. The three major functions of the endoplasmic reticulum are
    (a) hydrolysis, diffusion, osmosis
    (b) detoxification, packaging, modification
    (c) synthesis, storage, transport
    (d) pinocytosis, phagocytosis, storage
19. The viscous, superficial coating on the outer surface of the cell membrane is called the
    (a) glycocalyx (b) pseudopodia
    (c) inclusions (d) tubulin
20. Facilitated diffusion differs from ordinary diffusion in that facilitated diffusion
    (a) expends no ATP
    (b) moves molecules from an area of higher to lower concentration
    (c) requires a carrier protein for transport
    (d) moves molecules from an area of lower to higher concentration

LEVEL 2 REVIEWING CONCEPTS

1. If an animal cell lacked centrioles, it would not be able to
   (a) move (b) replicate DNA
   (c) divide (d) synthesize proteins
2. List the three basic concepts that make up the modern-day cell theory.
3. By what four passive processes do substances get into and out of cells?
4. List three important differences between cytosol and extracellular fluid.
5. What three major factors determine whether a substance can diffuse across a cell membrane?
6. What are organelles? Identify the two broad categories into which organelles may be divided, and describe the main difference between these groups.
7. What is cytokinesis? What is its role in the cell cycle?
8. List the stages of mitosis in order, and briefly describe the events that occur in each.
9. What are four general functions of the cell membrane?
10. Discuss the two major functions of microfilaments.

LEVEL 3 CRITICAL THINKING AND CLINICAL APPLICATIONS

1. Why does the skin of your hands get swollen and wrinkled if you soak them in fresh water for a long time?
2. Solutions A and B are separated by a selectively permeable barrier. Over time, the level of fluid on side A increases. Which solution initially had the higher concentration of solute?
3. What is the benefit of having some organelles enclosed by a membrane similar to a cellular membrane?
4. Experimental evidence demonstrates that the transport of a certain molecule exhibits the following characteristics: (1) the molecule moves against its concentration gradient, and (2) cellular energy is required for transport to occur. What type of transport process is at work?

✓ ANSWERS TO CONCEPT CHECK QUESTIONS

p. 33 **1.** Cell membranes are selectively permeable. **2.** Diffusion is a general term that refers to the passive movement of materials from regions of high concentration to regions of low concentration. Osmosis is the diffusion of water across a membrane; water moves into the solution containing the higher solute concentration (the lower water concentration). **3.** There are three types of endocytosis: pinocytosis, phagocytosis, and receptor-mediated endocytosis. Pinocytosis is the formation of vesicles filled with extracellular fluid. Phagocytosis produces vesicles containing solid objects. Receptor-mediated endocytosis resembles pinocytosis but is very selective: vesicles contain a specific target molecule in high concentrations bound to receptors on the membrane surface.
p. 37 **1.** The structures are called microvilli. Their function is to increase surface area for absorption. **2.** The lack of a flagellum would render the sperm cell immobile. **3.** The two divisions of the cytoplasm are: the cytosol, which is the intracellular fluid containing dissolved nutrients, ions, proteins and waste products; and organelles, which are structures that perform specific functions within the cell.
p. 42 **1.** Mitochondria produce ATP. The number of mitochondria in a particular cell varies depending on the cell's energy demand. If a cell contains many mitochondria, the energy demands for the cell are very high. **2.** Cells in the ovaries and testes contain large amounts of smooth endoplasmic reticulum (SER). Here, the SER functions to synthesize steroid hormones. **3.** If lysosomes disintegrate in a damaged cell, they release active enzymes into the cytosol. These enzymes rapidly destroy the proteins and organelles of the cell, a process called autolysis.
p. 47 **1.** Cell division is a form of cellular reproduction that results in an increase in cell number. **2.** Mitosis is a process that occurs during the division of somatic cells. It is the accurate duplication of the cell's genetic material and the distribution of one copy to each of the two new daughter cells. **3.** Interphase can be divided into the G_1, S, and G_2 phases. In the G_1 phase, the cell manufactures organelles and cytosol to make two functional cells. During S phase, the cell replicates its chromosomes. The G_2 phase is a time for last-minute protein synthesis before mitosis. An interphase cell in the G_0 phase is not preparing for mitosis, but is performing all other normal cell functions. Mitosis consists of four stages: prophase, metaphase, anaphase, and telophase. In prophase, the spindle fibers form and the nuclear membrane disappears; in metaphase, the chromatids line up along the metaphase plate; in anaphase, the chromatids separate and move to opposite poles of the spindle apparatus; and in telophase, the nuclear membrane reappears and chromosomes uncoil as the daughter cells separate through cytokinessis.

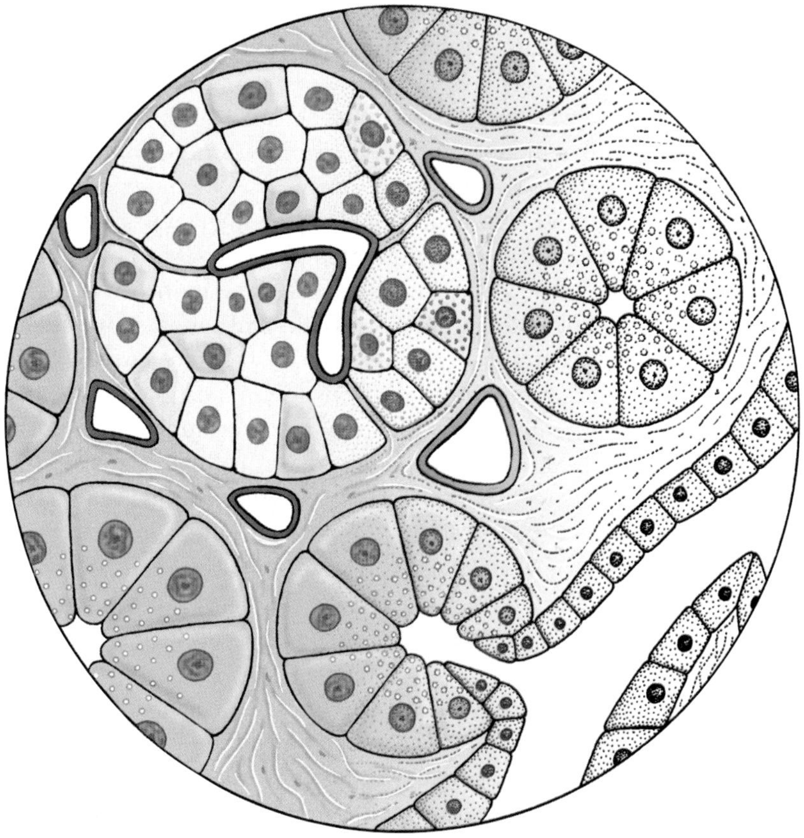

3

THE TISSUE LEVEL OF ORGANIZATION

A big corporation is a lot like a living organism, although it depends on its employees, rather than cells, to ensure its survival. It may take thousands of employees to keep the corporation going, and their duties vary—no one employee can do everything. So corporations usually have divisions with broad functions like marketing, production, maintenance, and so forth. The functions performed by the body are much more diverse than those of corporations, and no single cell contains the metabolic machinery and organelles needed to perform all of those functions. Instead, through the process of differentiation, each cell develops a characteristic set of structural features and a limited number of functions. These structures and functions can be quite distinct from those of nearby cells. Nevertheless, cells in a given location all work together. A detailed examination of the body reveals a number of patterns at the cellular level. Although the body contains trillions of cells, there are only about 200 types of cells. These cell types combine to form **tissues**, collections of specialized cells and cell products that perform a relatively limited number of functions. *Histology* is the study of tissues. There are four **primary tissue types**: *epithelial tissue, connective tissue, muscle tissue*, and *neural tissue*. The basic functions of these tissue types are introduced in Figure 3.1●.

This chapter will discuss the characteristics of each major tissue type, focusing on the relationship between cellular organization and tissue function. As noted in Chapter 2, histology is the study of groups of specialized cells and cell products that work together to perform specific functions.

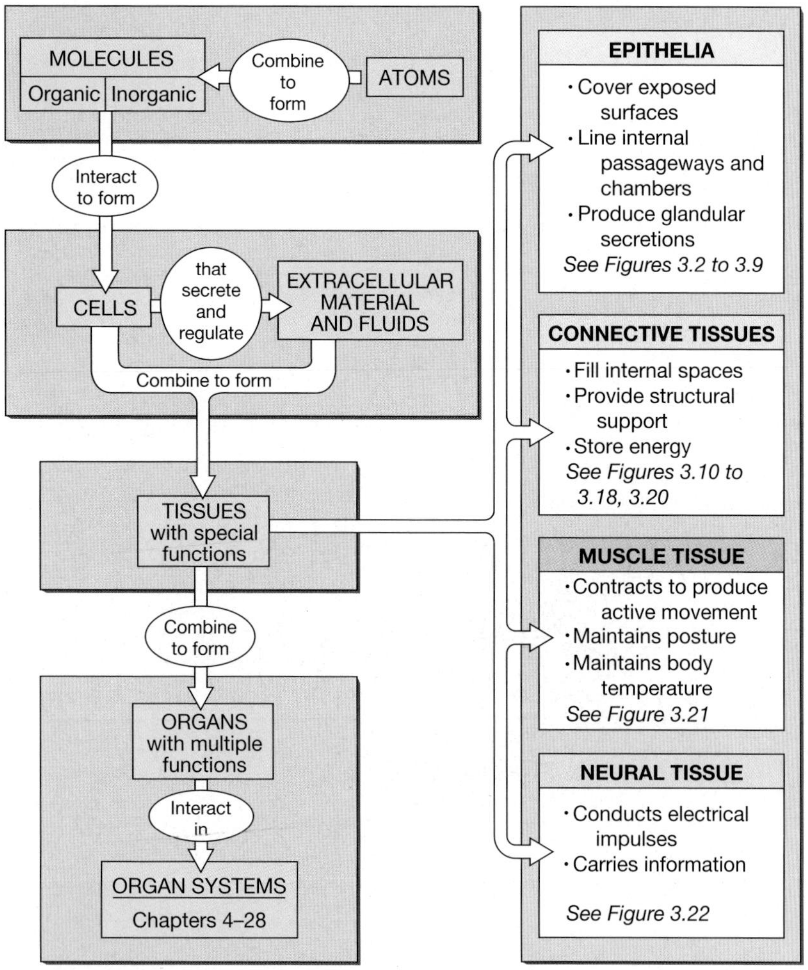

FIGURE 3.1 **AN ORIENTATION TO THE TISSUES OF THE BODY**

An overview of the levels of organization in the body, and an introduction to some of the functions of the four tissue types.

This chapter introduces the basic histological concepts needed to understand the patterns of tissue interaction in the organs and systems considered in later chapters.

Epithelial Tissue

Epithelial tissue includes epithelia and glands; glands are secretory structures derived from epithelia. An **epithelium** (epi-THĒ-lē-um; plural, *epithelia*) is a layer of cells that covers an exposed surface or lines an internal cavity or passageway. Each epithelium is a barrier with specific properties. Epithelia cover every exposed body surface. The surface of the skin is a good example, but epithelia also line the digestive, respiratory, reproductive, and urinary tracts—passageways that communicate with the outside world. Epithelia also line internal cavities and passageways, such as the chest cavity, fluid-filled chambers in the brain, eye, inner ear, and the inner surfaces of blood vessels and the heart.

Important characteristics of epithelia include:

1. *Cellularity*: Epithelia are composed almost entirely of cells bound closely together by cell junctions. This is very different from the situation in other tissues, where the individual cells are often widely separated by extracellular materials.
2. *Polarity*: An epithelium always has an exposed surface that faces the exterior of the body or some internal space. It also has an attached *base*, where the epithelium is attached to adjacent tissues. The surfaces differ in membrane structure and function. Whether the epithelium contains a single layer of cells or multiple layers, the organelles and other cytoplasmic components are not evenly distributed between the exposed and attached surfaces. **Polarity** is the term for this uneven distribution.
3. *Attachment*: The basal surface of a typical epithelium is bound to a thin **basement membrane**. The basement membrane is a complex structure produced by the basal surface of the epithelium and the underlying connective tissue.
4. *Avascularity*: Epithelia do not contain blood vessels. Because of this **avascular** (ā-VAS-kū-lar; *a*-, without + *vas*, vessel) condition, epithelial cells must obtain nutrients by diffusion or absorption across the apical or basal surfaces.
5. *Regeneration*: Epithelial cells damaged or lost at the surface are continually being replaced through the divisions of stem cells within the epithelium.

Functions of Epithelial Tissue

Epithelia perform several essential functions:

1. *Provide physical protection*: Epithelia protect exposed and internal surfaces from abrasion, dehydration, and destruction by chemical or biological agents.
2. *Control permeability*: Any substance that enters or leaves the body has to cross an epithelium. Some epithelia are relatively impermeable, whereas others are permeable to compounds as large as proteins. Many epithelia contain the molecular "machinery" needed for selective absorption or secretion. The epithelial barrier can be regulated and modified in response to various stimuli. For example, hormones can affect the transport of ions and nutrients through

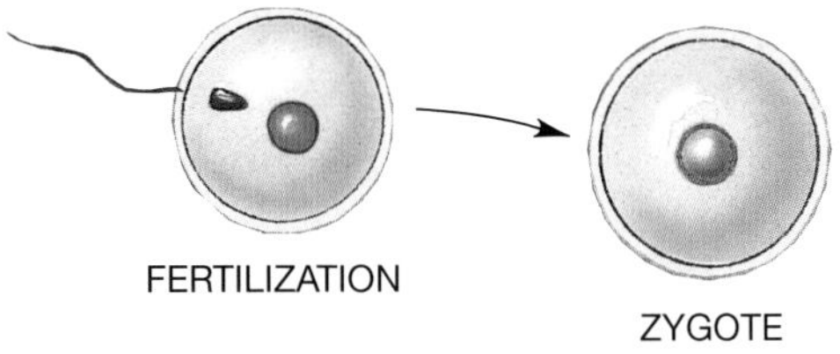

Fertilization produces a single cell, or **zygote** (ZĪ-gōt), that contains the normal number of chromosomes (46).

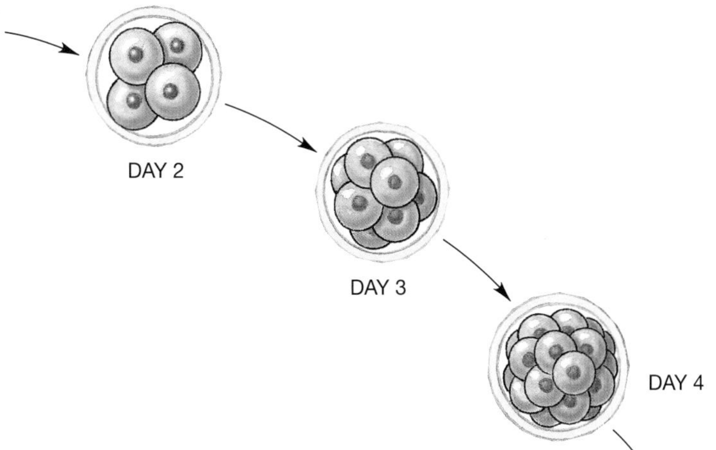

During **cleavage**, cell divisions produce a hollow ball of cells called a **blastocyst**. This process takes about a week to complete.

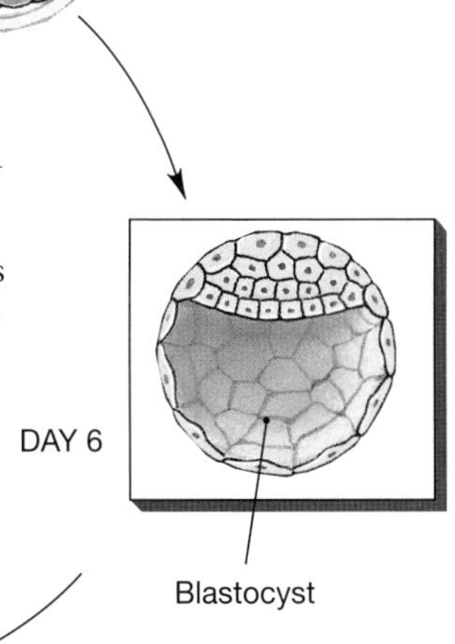

In section, the blastocyst contains two groups of cells with very different fates. The outer layer, or **trophoblast** (TRŌ-fō-blast; *trophos*, food + *blast*, precursor), will form the placenta, which nourishes the developing embryo. The **inner cell mass** will form the actual embryo.

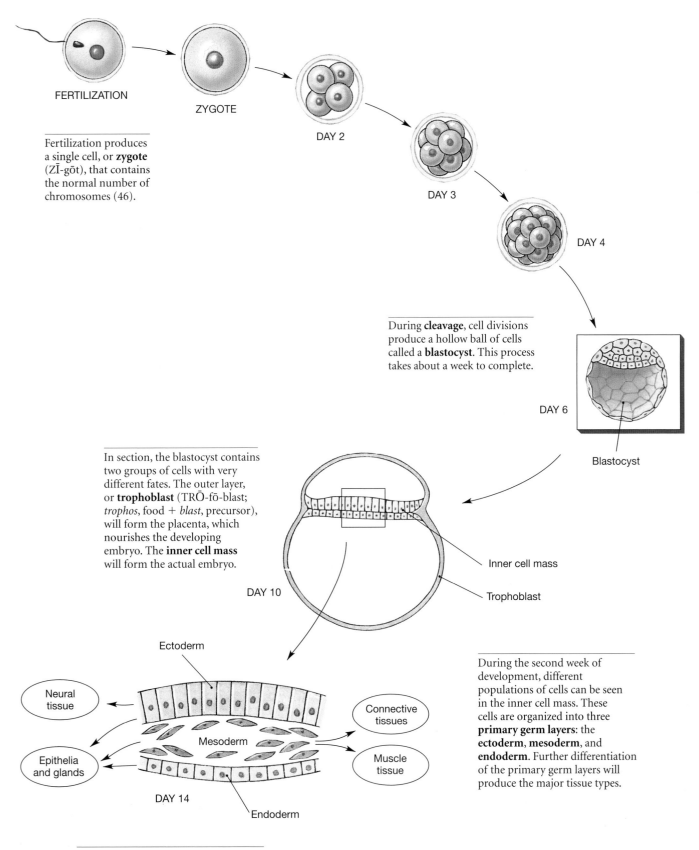

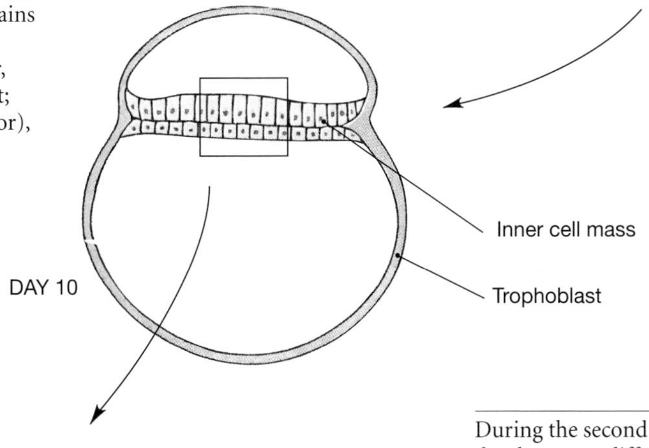

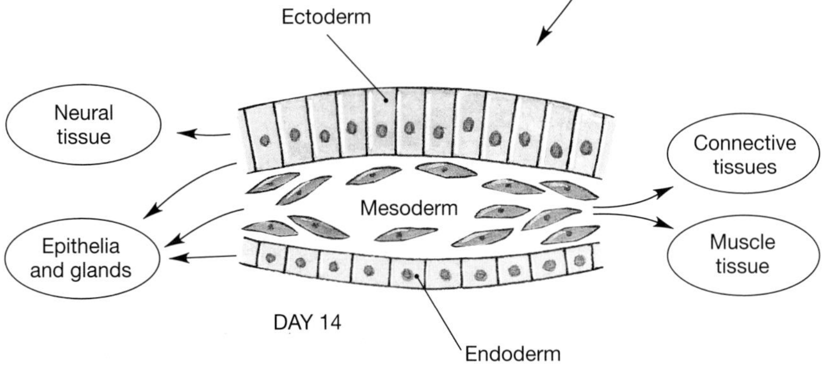

During the second week of development, different populations of cells can be seen in the inner cell mass. These cells are organized into three **primary germ layers**: the **ectoderm**, **mesoderm**, and **endoderm**. Further differentiation of the primary germ layers will produce the major tissue types.

All three germ layers participate in the formation of functional organs and organ systems. Their interactions will be detailed in later Embryology Summaries dealing with specific systems.

epithelial cells. Even physical stress can alter the structure and properties of epithelia—think of the calluses that form on your hands when you do rough work for a period of time.

3. *Provide sensations*: Most epithelia are extensively innervated by sensory nerves. Specialized epithelial cells can detect changes in the environment and convey information about such changes to the nervous system. For example, touch receptors in the deepest epithelial layers of the skin respond to pressure by stimulating adjacent sensory nerves. A **neuroepithelium** is a specialized sensory epithelium. Neuroepithelia are found in special sense organs that provide the sensations of smell, taste, sight, equilibrium, and hearing.
4. *Produce specialized secretions*: Epithelial cells that produce secretions are called **gland cells**. Individual gland cells are often scattered among other cell types in an epithelium. In a **glandular epithelium**, most or all of the epithelial cells produce secretions.

Specializations of Epithelial Cells [FIGURE 3.2]

Epithelial cells have several specializations that distinguish them from other body cells. Many epithelial cells are specialized for (1) the production of secretions, (2) the movement of fluids over the epithelial surface, or (3) the movement of fluids through the epithelium itself. These specialized epithelial cells usually show a definite polarity along the axis that extends from the *apical surface*, where the cell is exposed to an internal or external environment, to the *basolateral surface*, where the epithelium contacts the basement membrane and neighboring epithelial cells. This polarity means that (1) the intracellular organelles are unevenly distributed, and (2) the apical and basolateral cell membranes differ in terms of their associated proteins and functions. The actual arrangement of organelles varies depending on the functions of the individual cells.

Most epithelial cells have microvilli on their exposed apical surfaces; there may be just a few, or they may carpet the entire surface. Microvilli are especially abundant on epithelial surfaces where absorption and secretion take place, such as along portions of the digestive and urinary tracts. p. 35 The epithelial cells in these locations are transport specialists, and a cell with microvilli has at least 20 times the surface area of a cell without them. Increased surface area provides the cell with a much greater ability to absorb or secrete across the membrane. Microvilli are shown in Figure 3.2●. **Stereocilia** are very long microvilli (up to 250 μm) that are incapable of movement. Stereocilia are found along portions of the male reproductive tract and on receptor cells of the inner ear.

Figure 3.2b● shows the apical surface of a **ciliated epithelium**. A typical ciliated cell contains about 250 cilia that beat in a coordinated fashion. Substances are moved over the epithelial surface by the synchronized beating of cilia, like a continuously moving escalator. For example, the ciliated epithelium that lines the respiratory tract moves mucus from the lungs toward the throat. The mucus traps particles and pathogens and carries them away from more delicate surfaces deeper in the lungs.

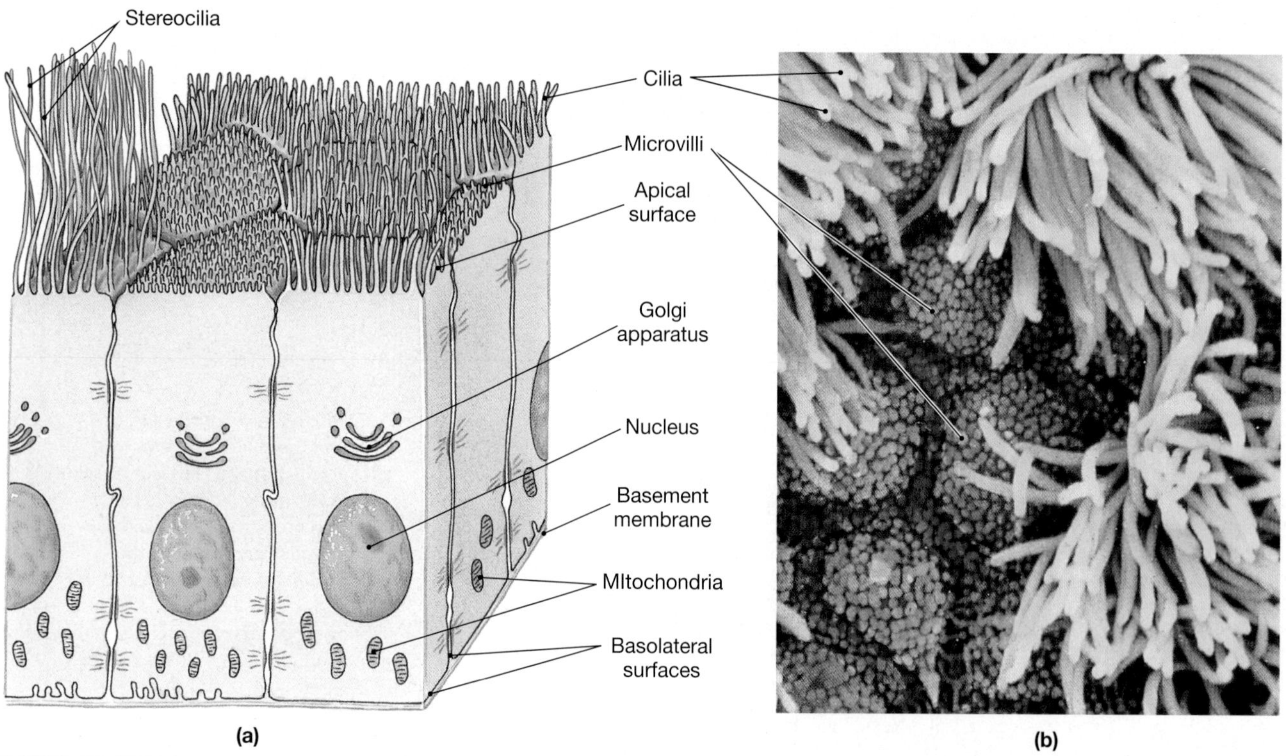

FIGURE 3.2 **POLARITY OF EPITHELIAL CELLS**

(**a**) Many epithelial cells differ in internal organization along an axis between the apical surface and the basement membrane. The apical surface frequently bears microvilli; less often, it may have cilia or (very rarely) stereocilia. A single cell typically has only one type of process; stereocilia and microvilli are shown together to highlight their relative proportions. (*All three would not normally be on the same group of cells but are depicted here for purposes of illustration.*) Tight junctions prevent movement of pathogens or diffusion of dissolved materials between the cells. Folds of membrane near the base of the cell increase the surface area exposed to the basement membrane. Mitochondria are typically concentrated at the basolateral region probably to provide energy for the cell's transport activities. (**b**) An SEM showing the surface of the epithelium that lines most of the respiratory tract. The small, bristly areas are microvilli found on the exposed surfaces of mucus-producing cells that are scattered among the ciliated epithelial cells. (SEM × 15,846) *Prof. P. Motta, Dept. of Anatomy, University "La Sapienza," Rome/Science Photo Library/ Photo Researchers, Inc.*

Maintaining the Integrity of the Epithelium

Three factors are involved in maintaining the physical integrity of an epithelium: (1) intercellular connections, (2) attachment to the basement membrane, and (3) epithelial maintenance and repair.

INTERCELLULAR CONNECTIONS [FIGURE 3.3]

Cells in epithelia are usually bound together by a variety of cell junctions, as detailed in Figure 2.19, p. 43•. There is often an extensive infolding of opposing cell membranes that both interlocks the cells and increases the surface area of the cell junctions. Note the degree of interlocking between cell membranes in Figure 3.3a,c•. The extensive connections between cells hold them together and may deny access to chemicals or pathogens that may contact their free surfaces. The combination of cell junctions, CAMs, intercellular cement, and physical interlocking gives the epithelium great strength and stability (Figure 3.3b•).

ATTACHMENT TO THE BASEMENT MEMBRANE [FIGURE 3.3b]

Epithelial cells not only hold onto one another, they also remain firmly connected to the rest of the body. The base of a typical epithelium is attached to a special two-part basement membrane (Figure 3.3b•). The layer closest to the epithelium, called the **basal lamina** (LA-mi-na; *lamina*, thin layer), contains glycoproteins and a network of microfilaments. The basal lamina, secreted by the adjacent layer of epithelial cells, provides a barrier that restricts the movement of proteins and other large molecules from the underlying connective tissue into the epithelium. The deeper portion of the basement membrane, the **reticular lamina**, contains bundles of coarse protein fibers produced by adjacent connective tissue cells. The reticular lamina gives the basement membrane its strength. Attachments between the protein fibers of the basal lamina and those of the reticular lamina hold the two layers together. Some specialized epithelia lack the reticular lamina, but all have a basal lamina.

EPITHELIAL MAINTENANCE AND RENEWAL

An epithelium must continually repair and renew itself. The rate of cell division varies depending upon the rate of loss of epithelial cells at the surface. Epithelial cells lead hard lives, for they may be exposed to disruptive enzymes, toxic chemicals, pathogenic bacteria, or mechanical abrasion. Under severe conditions, such as those encountered inside the small intestine, an epithelial cell may survive for just a day or two before it is destroyed. The only way the epithelium can maintain its integrity over time is through the continual division of stem cells. These stem cells, also known as **germinative cells**, are usually found close to the basement membrane.

✓ CONCEPT CHECK

- Identify the four primary tissue types.
- List four characteristics of epithelia.
- What are two specializations of epithelial cells?

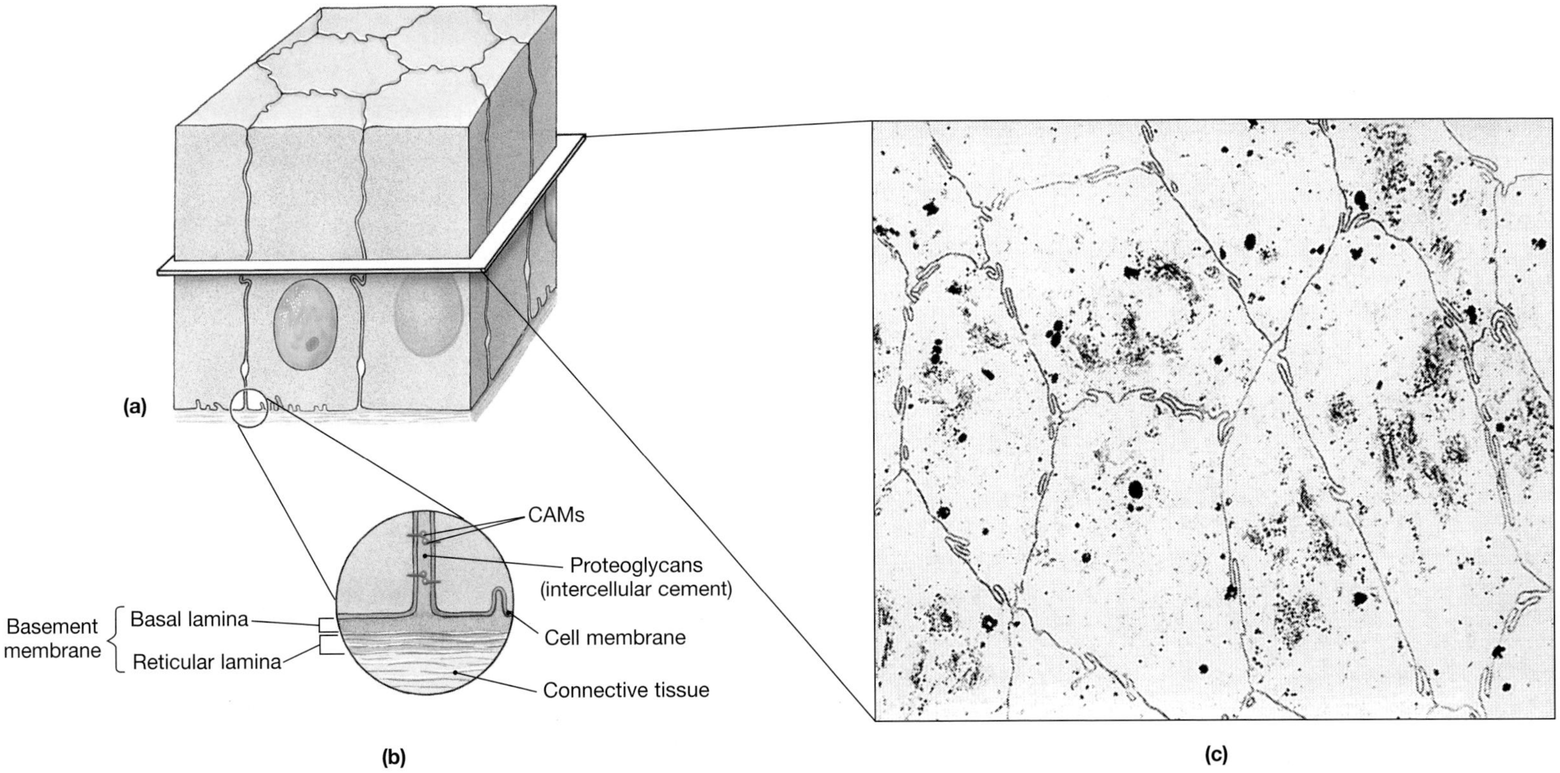

FIGURE 3.3 **EPITHELIA AND BASEMENT MEMBRANES**

The integrity of the epithelium depends on connections between adjacent epithelial cells and their attachment to the underlying basement membrane. **(a)** Epithelial cells are usually packed together and interconnected by intercellular attachments (*see Figure 2.19*). **(b)** At their basal surfaces, epithelia are attached to a basement membrane that forms the boundary between the epithelial cells and the underlying connective tissue. **(c)** Adjacent epithelial cell membranes are often interlocked. The TEM, magnified 2600 times, shows the degree of interlocking between columnar epithelial cells.

A Classification of Epithelia

Epithelia are classified according to the number of cell layers and the shape of the cells at the exposed surface. The classification scheme recognizes two types of layering—*simple* and *stratified*—and three cell shapes—*squamous, cuboidal,* and *columnar.*

If there is only a single layer of cells covering the basement membrane, the epithelium is a **simple epithelium**. Simple epithelia are relatively thin, and because all the cells have the same polarity, the nuclei form a row at roughly the same distance from the basement membrane. Because they are so thin, simple epithelia are also relatively fragile. A single layer of cells cannot provide much mechanical protection, and simple epithelia are found only in protected areas inside the body. They line internal compartments and passageways, including the ventral body cavities, the chambers of the heart, and all blood vessels.

Simple epithelia are also characteristic of regions where secretion, absorption, or filtration occurs, such as the lining of the intestines and the gas-exchange surfaces of the lungs. In these places the thin single layer of simple epithelia is an advantage, for it lessens the distance involved and thus the time required for materials to pass through or across the epithelial barrier.

A **stratified epithelium** has several layers of cells above the basement membrane. Stratified epithelia are usually found in areas subject to mechanical or chemical stresses, such as the surface of the skin and the lining of the mouth. The multiple layers of cells in a stratified epithelium make it thicker and sturdier than a simple epithelium. Regardless of whether an epithelium is simple or stratified, the epithelium must regenerate, replacing its cells over time. The germinative cells are always at or near the basement membrane. This means that in a simple epithelium, the germinative cells form part of the exposed epithelial surface, whereas in a stratified epithelium, the germinative cells are covered by more superficial cells.

Combining the two basic epithelial layouts (simple and stratified) and the three possible cell shapes (squamous, cuboidal, and columnar) enables one to describe almost every epithelium in the body.

SQUAMOUS EPITHELIA [FIGURE 3.4]

In a **squamous epithelium** (SKWĀ-mus; *squama,* plate or scale) the cells are thin, flat, and somewhat irregular in shape—like puzzle pieces (Figure 3.4a●). In a sectional view the nucleus occupies the thickest portion of each cell, and has a flattened shape similar to that of the cell as a whole; from the surface, the cells look like fried eggs laid side by side. A **simple squamous epithelium** is the most delicate type of epithelium in the body. This type of epithelium is found in protected regions where absorption

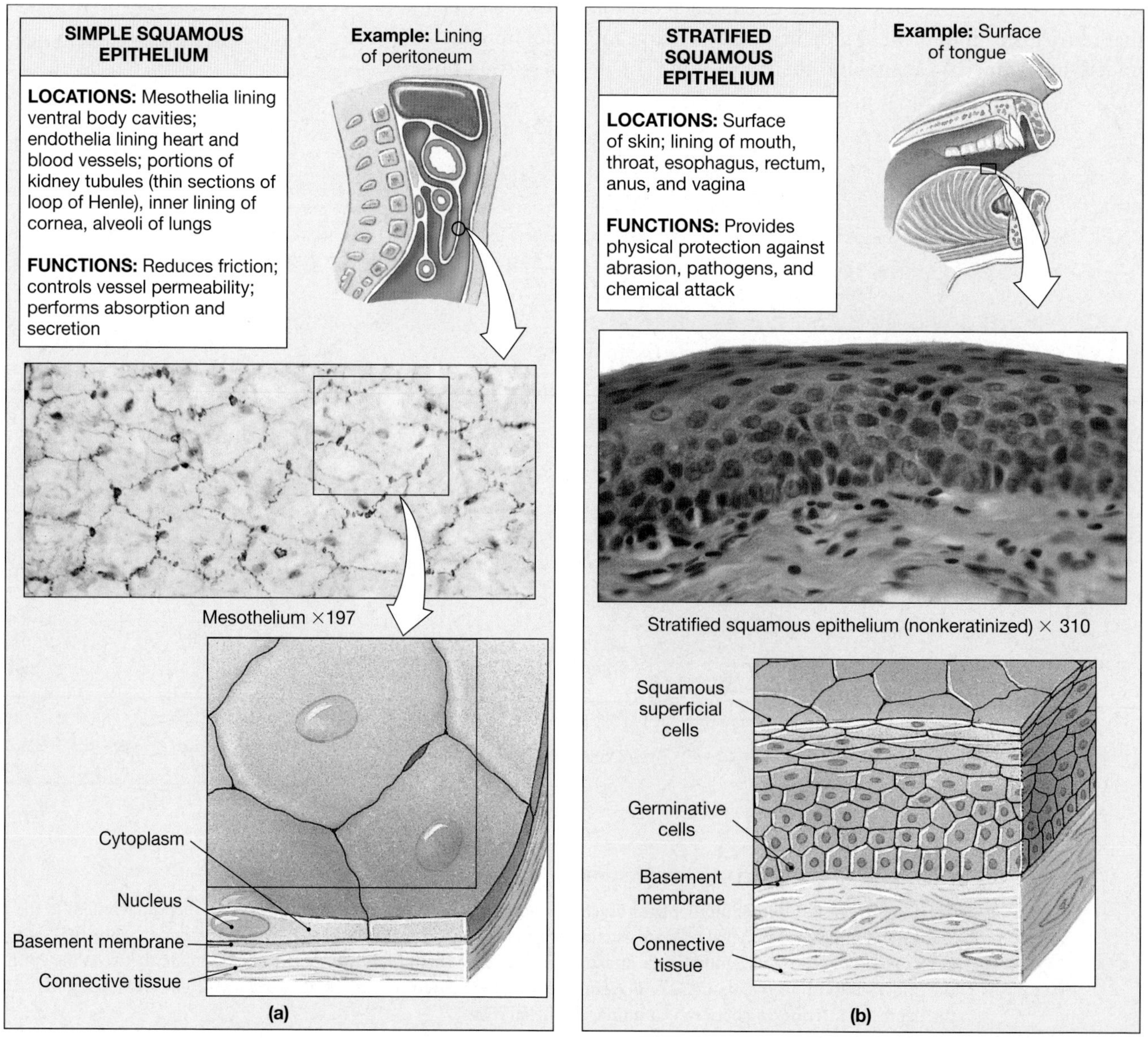
SIMPLE SQUAMOUS EPITHELIUM
LOCATIONS: Mesothelia lining ventral body cavities; endothelia lining heart and blood vessels; portions of kidney tubules (thin sections of loop of Henle), inner lining of cornea, alveoli of lungs
FUNCTIONS: Reduces friction; controls vessel permeability; performs absorption and secretion
Example: Lining of peritoneum
Mesothelium ×197
Cytoplasm
Nucleus
Basement membrane
Connective tissue
(a)
STRATIFIED SQUAMOUS EPITHELIUM
LOCATIONS: Surface of skin; lining of mouth, throat, esophagus, rectum, anus, and vagina
FUNCTIONS: Provides physical protection against abrasion, pathogens, and chemical attack
Example: Surface of tongue
Stratified squamous epithelium (nonkeratinized) × 310
Squamous superficial cells
Germinative cells
Basement membrane
Connective tissue
(b)

FIGURE 3.4 **SQUAMOUS EPITHELIA**

(a) Simple Squamous Epithelium. A superficial view of the simple squamous epithelium (*mesothelium*) that lines the peritoneal cavity. The three-dimensional drawing shows the epithelium in superficial and sectional view. **(b) Stratified Squamous Epithelium.** Sectional views of the stratified squamous epithelium that covers the tongue.

takes place or where a slick, slippery surface reduces friction. Examples include the respiratory exchange surfaces (*alveoli*) of the lungs, the serous membranes lining the ventral body cavities, and the inner surfaces of the circulatory system.

Special names have been given to simple squamous epithelia that line chambers and passageways that do not communicate with the outside world. The simple squamous epithelium that lines the ventral body cavities is known as a **mesothelium** (mez-ō-THĒ-lē-um; *mesos*, middle). The pleura, peritoneum, and pericardium each contain a superficial layer of mesothelium. The simple squamous epithelium lining the heart and all blood vessels is called an **endothelium** (en-dō-THĒ-lē-um).

A **stratified squamous epithelium** (Figure 3.4b●) is usually found where mechanical stresses are severe. Note how the cells form a series of layers, like a stack of plywood sheets. The surface of the skin and the lining of the mouth, throat, esophagus, rectum, vagina, and anus are areas where this epithelial type provides protection from physical and chemical attack. On exposed body surfaces, where mechanical stress and dehydration are potential problems, the apical layers of epithelial cells are packed with filaments of the protein *keratin*. As a result, the superficial layers are both tough and water-resistant, and the epithelium is said to be **keratinized**. A **nonkeratinized** stratified squamous epithelium provides resistance to abrasion, but will dry out and deteriorate unless kept moist. Nonkeratinized stratified squamous epithelia are found in the oral cavity, pharynx, esophagus, rectum, anus, and vagina.

CUBOIDAL EPITHELIA [FIGURE 3.5]

The cells of a **cuboidal epithelium** resemble little hexagonal boxes; they appear square in typical sectional views. Each nucleus is near the center of the cell, with the distance between adjacent nuclei roughly equal to the height of the epithelium. **Simple cuboidal epithelia** provide limited protection and occur in regions where secretion or absorption takes place. Such an epithelium lines portions of the kidney tubules, as seen in Figure 3.5a●. In the pancreas and salivary glands, simple cuboidal epithelia secrete enzymes and buffers and line the ducts that discharge those secretions. The thyroid gland contains chambers called *thyroid follicles* that are lined by a cuboidal secretory epithelium. Thyroid hormones, especially *thyroxine*, accumulate within the follicles before they are released into the bloodstream.

Stratified cuboidal epithelia are relatively rare; they are often found along the ducts of sweat glands (Figure 3.5b●) and in the larger ducts of some other exocrine glands, such as the mammary glands. A **transitional epithelium**, shown in Figure 3.5c,d●, lines the renal pelvis, the ureters, and the urinary bladder. This epithelium permits considerable stretching, and significant changes in volume occur at these locations. In an empty bladder (Figure 3.5c●), the epithelium seems to have many layers, and the outermost cells are typically plump cuboidal cells. The layered appearance results from overcrowding; the actual structure of the epithelium can be seen in the full bladder, when the pressure of the urine has stretched the lining (Figure 3.5d●).

COLUMNAR EPITHELIA [FIGURE 3.6]

Columnar epithelial cells, like cuboidal epithelial cells, are also hexagonal in cross section, but in contrast to cuboidal cells their height is much greater than their width. The nuclei are crowded into a narrow band close to the basement membrane, and the height of the epithelium is several times the distance between two nuclei (Figure 3.6a●). A **simple columnar epithelium** provides some protection and may also be encountered in areas where absorption or secretion occurs. This type of epithelium lines the stomach, intestinal tract, uterine tubes, and many excretory ducts.

Portions of the respiratory tract contain a specialized columnar epithelium, called a *respiratory epithelium*, that includes a mixture of cell types. Because their nuclei are situated at varying distances from the surface, the epithelium appears to be layered or stratified. But it is not truly stratified, because all of the cells contact the basement membrane. Since it looks stratified but isn't, it is known as a **pseudostratified columnar epithelium**. The exposed epithelial cells typically possess cilia, so this is often called a **pseudostratified ciliated columnar epithelium** (Figure 3.6b●). This type of epithelium lines most of the nasal cavity, the trachea (windpipe), bronchi, and also portions of the male reproductive tract.

Stratified columnar epithelia are relatively rare, providing protection along portions of the pharynx, urethra, and anus, as well as along a few large excretory ducts. The epithelium may have two layers (Figure 3.6c●) or multiple layers; when multiple layers exist, only the superficial cells have the classic columnar shape.

Glandular Epithelia

Many epithelia contain gland cells that produce secretions. *Exocrine glands* discharge their secretions onto an epithelial surface; *endocrine glands* are ductless glands that release their secretions into the surrounding extracellular fluid. Exocrine glands are classified by the type of secretions released, the structure of the gland, and the mode of secretion. Exocrine glands, which may be either unicellular or multicellular, secrete mucins, enzymes, water, and waste products. These secretions are released at the apical surfaces of the individual gland cells.

TYPES OF SECRETION

Exocrine (*exo-*, outside) secretions are discharged onto the surface of the skin or onto an epithelial surface lining one of the internal passageways that communicates with the exterior. There are many kinds of exocrine secretions, performing a variety of functions. Enzymes entering the digestive tract, perspiration on the skin, and the milk produced by mammary glands are examples of exocrine secretions. Exocrine cells often form pockets that are connected to the epithelial surface by tubes, called **ducts**.

Exocrine glands may be categorized according to the nature of the secretion produced:

- *serous glands* secrete a watery solution that usually contains enzymes, such as the salivary amylase in saliva;
- *mucous glands* secrete glycoproteins called **mucins** (MŪ-sins) that absorb water to form a slippery *mucus*, such as the mucus in saliva; and
- *mixed exocrine glands* contain more than one type of gland cell and may produce two different exocrine secretions, one serous and the other mucous. The submandibular gland, one of the salivary glands, is an example of a mixed exocrine gland.

Endocrine (*endo-*, inside) secretions are released by exocytosis from the gland cells into the fluid surrounding the cell. These secretions, called **hormones**, diffuse into the blood for distribution to other regions of the body, where they regulate or coordinate the activities of various tissues, organs, and organ systems. Endocrine cells may be part of an epithelial surface, such as the lining of the digestive tract, or they may be separate, as in

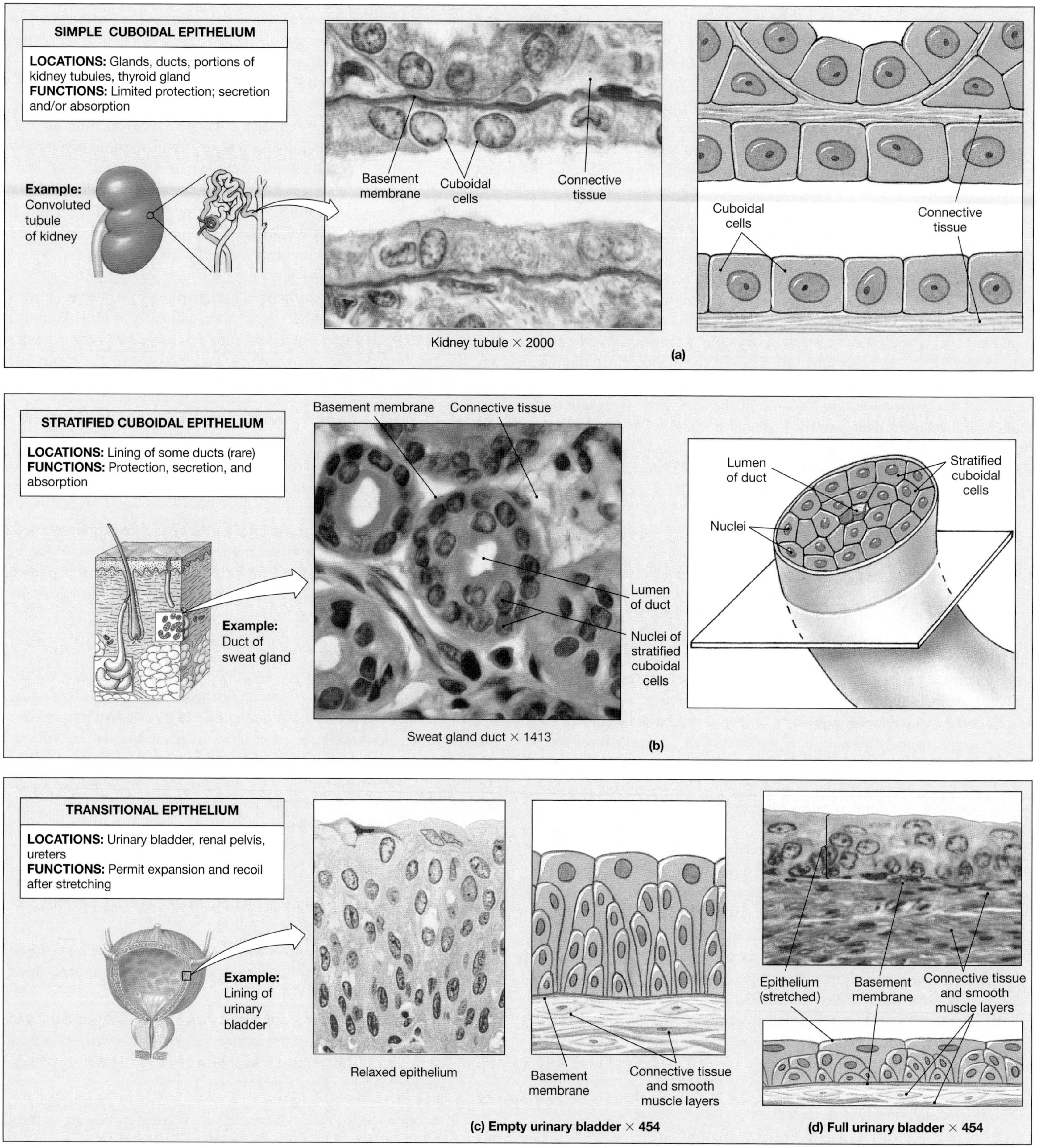

FIGURE 3.5 **CUBOIDAL AND TRANSITIONAL EPITHELIA**

(a) Simple Cuboidal Epithelium. A section through the simple cuboidal epithelium lining a kidney tubule. The diagrammatic view emphasizes structural details that permit the classification of an epithelium as cuboidal. **(b) Stratified Cuboidal Epithelium.** A sectional view of the stratified cuboidal epithelium lining a sweat gland duct in the skin. **(c) Transitional Epithelium, relaxed.** The lining of the empty urinary bladder, showing a transitional epithelium in the relaxed state. **(d) Transitional Epithelium, stretched.** The lining of the full urinary bladder, showing the effects of stretching on the arrangement of cells in the epithelium.

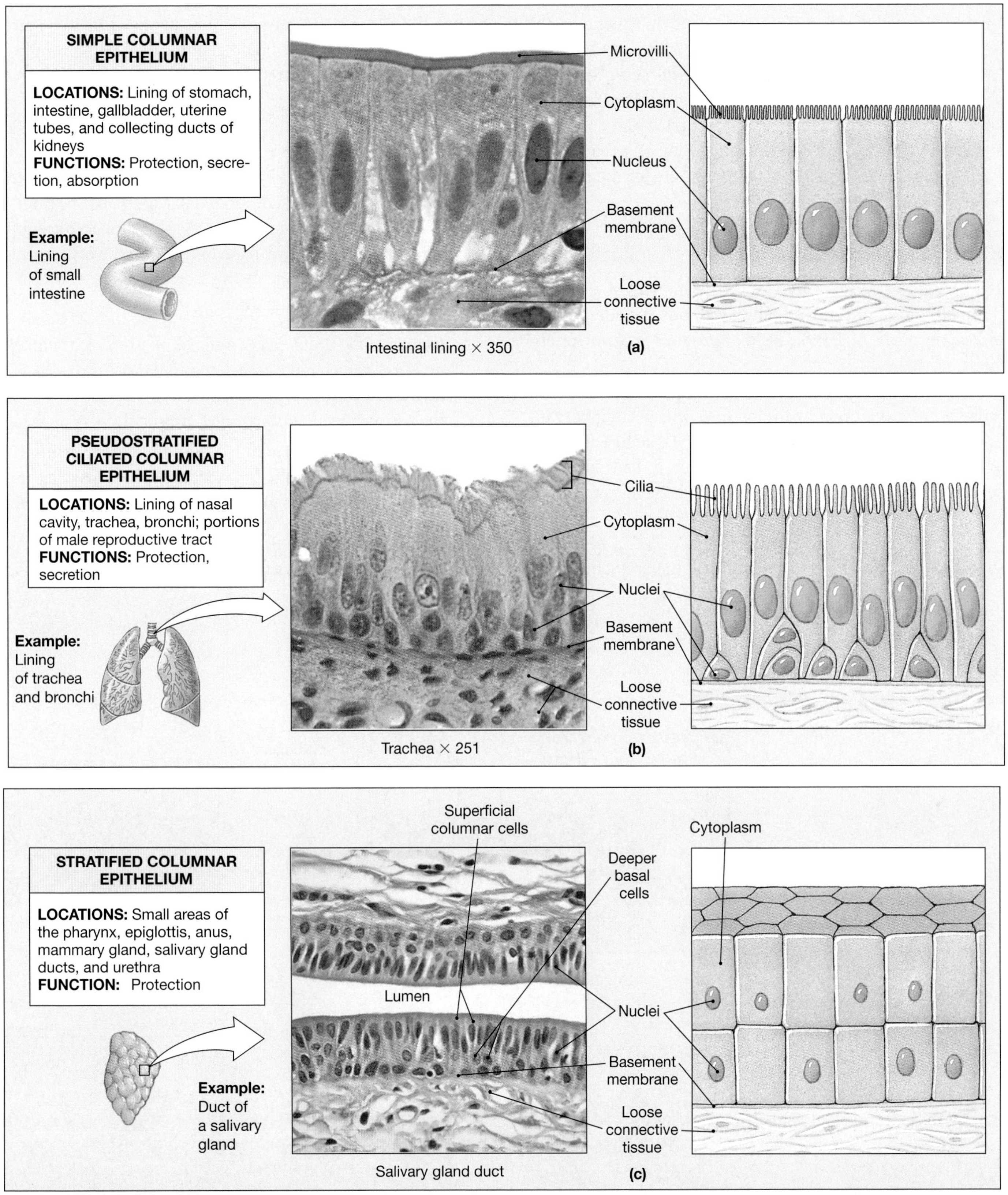

FIGURE 3.6 **COLUMNAR EPITHELIA**

(a) Simple Columnar Epithelium. A light micrograph showing the characteristics of simple columnar epithelium. In the diagrammatic sketch, note the relationships between the height and width of each cell; the relative size, shape, and location of nuclei; and the distance between adjacent nuclei. Contrast these observations with the corresponding characteristics of simple cuboidal epithelia. **(b) Pseudostratified Ciliated Columnar Epithelium**. The pseudostratified, ciliated, columnar epithelium of the respiratory tract. Note the uneven layering of the nuclei. **(c) Stratified Columnar Epithelium**. A stratified columnar epithelium is sometimes found along large ducts, such as this salivary gland duct. Note the overall height of the epithelium and the location and orientation of the nuclei.

the pancreas, thyroid gland, thymus, and pituitary gland. In either case, the gland cells release their hormones directly into body fluids, and there are no endocrine ducts; endocrine glands are often called *ductless glands*. Endocrine cells, tissues, organs, and hormones are considered further in Chapter 19. A few complex glands produce both exocrine and endocrine secretions. For example, the pancreas contains endocrine cells that secrete hormones, as well as exocrine cells and ducts responsible for the production of digestive enzymes and buffers.

GLAND STRUCTURE [FIGURES 3.7/3.8]

In epithelia that contain scattered gland cells, the individual secretory cells are called **unicellular glands**. **Multicellular glands** include glandular epithelia and aggregations of gland cells that produce exocrine or endocrine secretions.

Goblet cells are the only unicellular exocrine glands in the body. Goblet cells secrete *mucins* that upon hydration form *mucus*, a slippery lubricant. Goblet cells are scattered among other epithelial cells. For example, the pseudostratified ciliated columnar epithelium that lines the trachea and the columnar epithelium of the small and large intestines contain an abundance of goblet cells.

The simplest **multicellular exocrine gland** is called a **secretory sheet**. In a secretory sheet, glandular cells dominate the epithelium and release their secretions into an inner compartment (Figure 3.7a●). The mucus-secreting cells that line the stomach are an example of a secretory sheet. Their continual secretion protects the stomach from the acids and enzymes it contains.

Most other multicellular glands are found in pockets set back from the epithelial surface. Figure 3.7b● shows one example, a salivary gland that produces mucus and digestive enzymes. These multicellular exocrine glands have two epithelial components: a glandular portion that produces the secretion and a duct that carries the secretion to the epithelial surface.

Two characteristics are used to describe the organization of a multicellular gland: (1) the shape of the secretory portion of the gland and (2) the branching pattern of the duct.

1. Glands made up of cells arranged in a tube are **tubular**; those made up of cells in a blind pocket are **alveolar** (al-VĒ-ō-lar; *alveolus*, sac), or **acinar** (A-si-nar; *acinus*, chamber). Glands that have a combination of the two arrangements are called **tubuloalveolar** or *tubuloacinar*.
2. A duct is referred to as **simple** if it does not branch and **compound** if it branches repeatedly. Each glandular area may have its own duct; in the case of branched glands, several glands share a common duct.

Figure 3.8● diagrams this method of classification based on gland structure. Specific examples of each gland type will be discussed in later chapters.

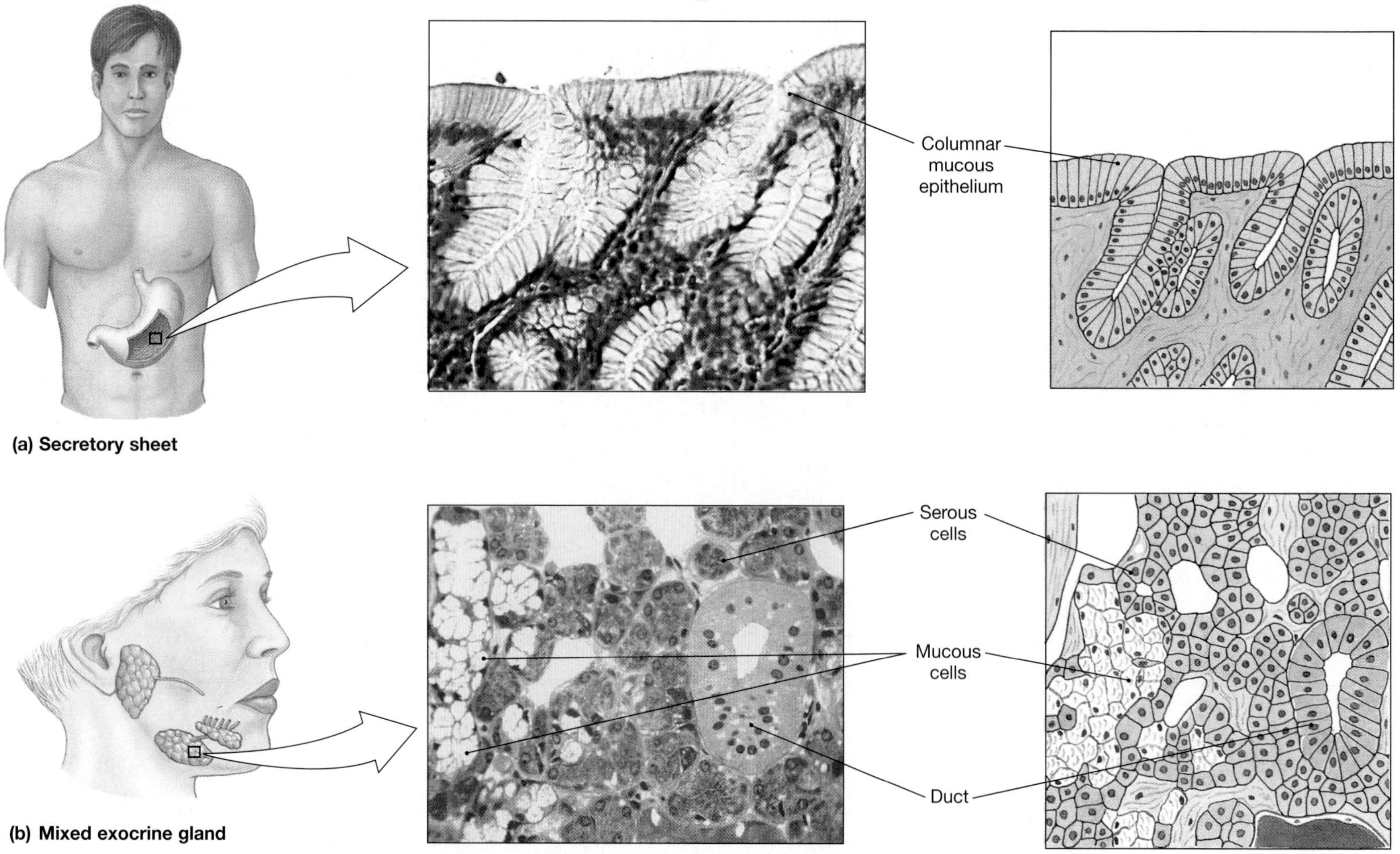

FIGURE 3.7 **MUCOUS AND MIXED GLANDULAR EPITHELIA**

(**a**) The interior of the stomach is lined by a secretory sheet whose secretions protect the walls from acids and enzymes. (The acids and enzymes are produced by glands that discharge their secretions onto the mucous epithelial surface.) (**b**) The submandibular salivary gland is a mixed gland containing cells that produce both serous and mucous secretions. The mucous cells contain large vesicles containing mucins, and they look pale and foamy. The serous cells secrete enzymes, and the proteins stain darkly. (LM × 252)

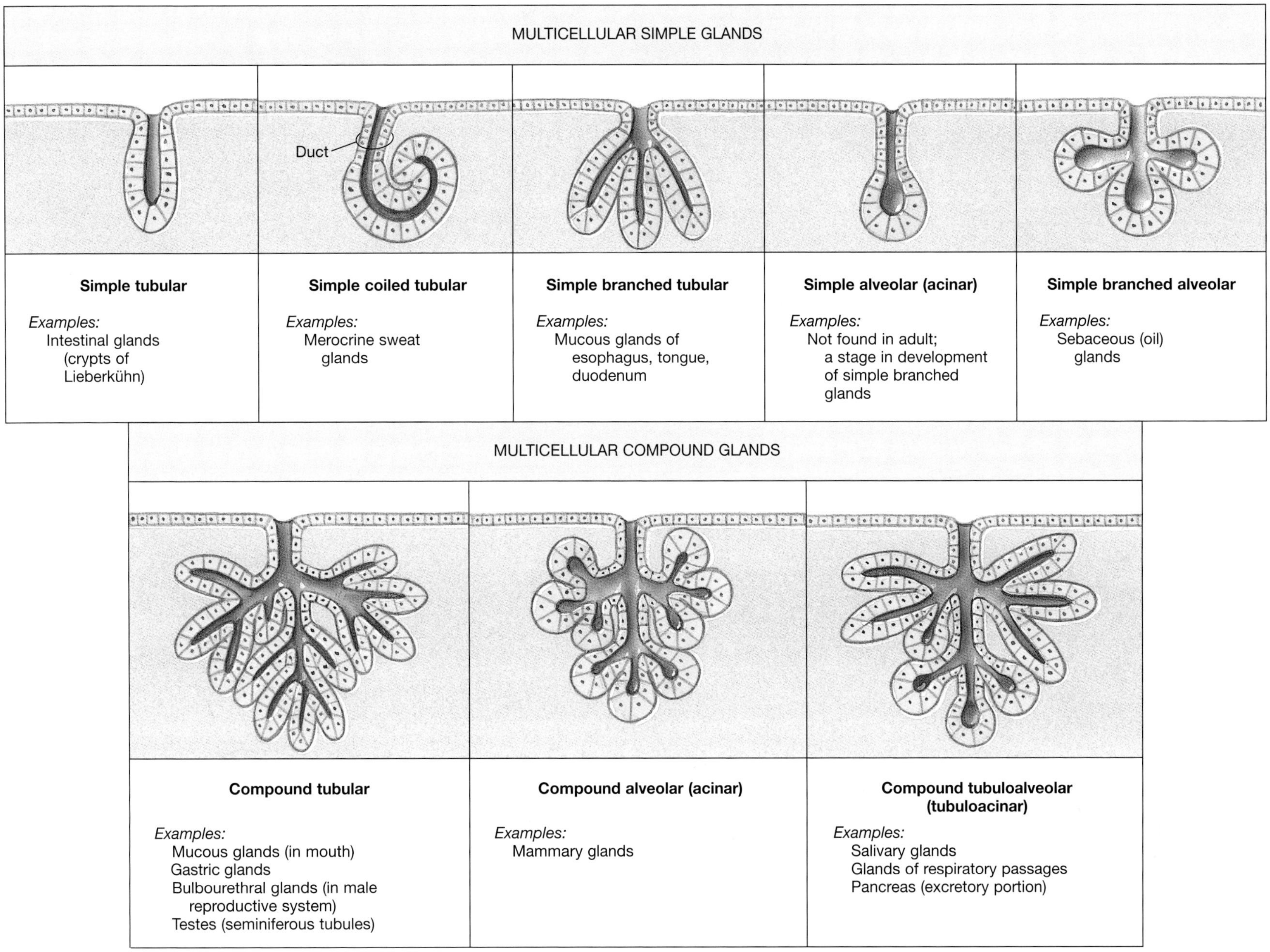

MULTICELLULAR SIMPLE GLANDS
Duct
Simple tubular
Examples:
Intestinal glands (crypts of Lieberkühn)
Simple coiled tubular
Examples:
Merocrine sweat glands
Simple branched tubular
Examples:
Mucous glands of esophagus, tongue, duodenum
Simple alveolar (acinar)
Examples:
Not found in adult; a stage in development of simple branched glands
Simple branched alveolar
Examples:
Sebaceous (oil) glands
MULTICELLULAR COMPOUND GLANDS
Compound tubular
Examples:
Mucous glands (in mouth)
Gastric glands
Bulbourethral glands (in male reproductive system)
Testes (seminiferous tubules)
Compound alveolar (acinar)
Examples:
Mammary glands
Compound tubuloalveolar (tubuloacinar)
Examples:
Salivary glands
Glands of respiratory passages
Pancreas (excretory portion)

FIGURE 3.8 **A Structural Classification of Simple and Compound Exocrine Glands**

MODES OF SECRETION [FIGURE 3.9]

A glandular epithelial cell may use one of three methods to release its secretions: *merocrine secretion*, *apocrine secretion*, or *holocrine secretion*. In **merocrine secretion** (MER-ō-krin; *meros*, part + *krinein*, to separate), the secretory product is released through exocytosis (Figure 3.9a●). This is the most common mode of secretion. For example, goblet cells release **mucus** through merocrine secretion. **Apocrine secretion** (AP-ō-krin; *apo-*, off) involves the loss of cytoplasm as well as the secretory product (Figure 3.9b●). The apical portion of the cytoplasm becomes packed with secretory vesicles before it is shed. Milk production by the lactiferous glands in the breasts involves a combination of merocrine and apocrine secretion.

Merocrine and apocrine secretions leave the nucleus and Golgi apparatus of the cell intact, so it can perform repairs and continue secreting. **Holocrine secretion** (HOL-ō-krin; *holos*, entire) destroys the gland cell. During holocrine secretion, the entire cell becomes packed with secretory products and then bursts apart (Figure 3.9c●). The secretion is released and the cell dies. Further secretion depends on gland cells being replaced by the division of stem cells. Sebaceous glands, associated with hair follicles, produce a waxy hair coating by means of holocrine secretion.

✓ CONCEPT CHECK

- You look at a tissue under a microscope and see a simple squamous epithelium. Can it be a sample of the skin surface?
- Why is epithelium regeneration a necessity in a gland that releases its product by holocrine secretion?
- The secretory cells of mammary glands release their products by apocrine secretion. What occurs in this mode of secretion?
- What functions are associated with a simple columnar epithelium?

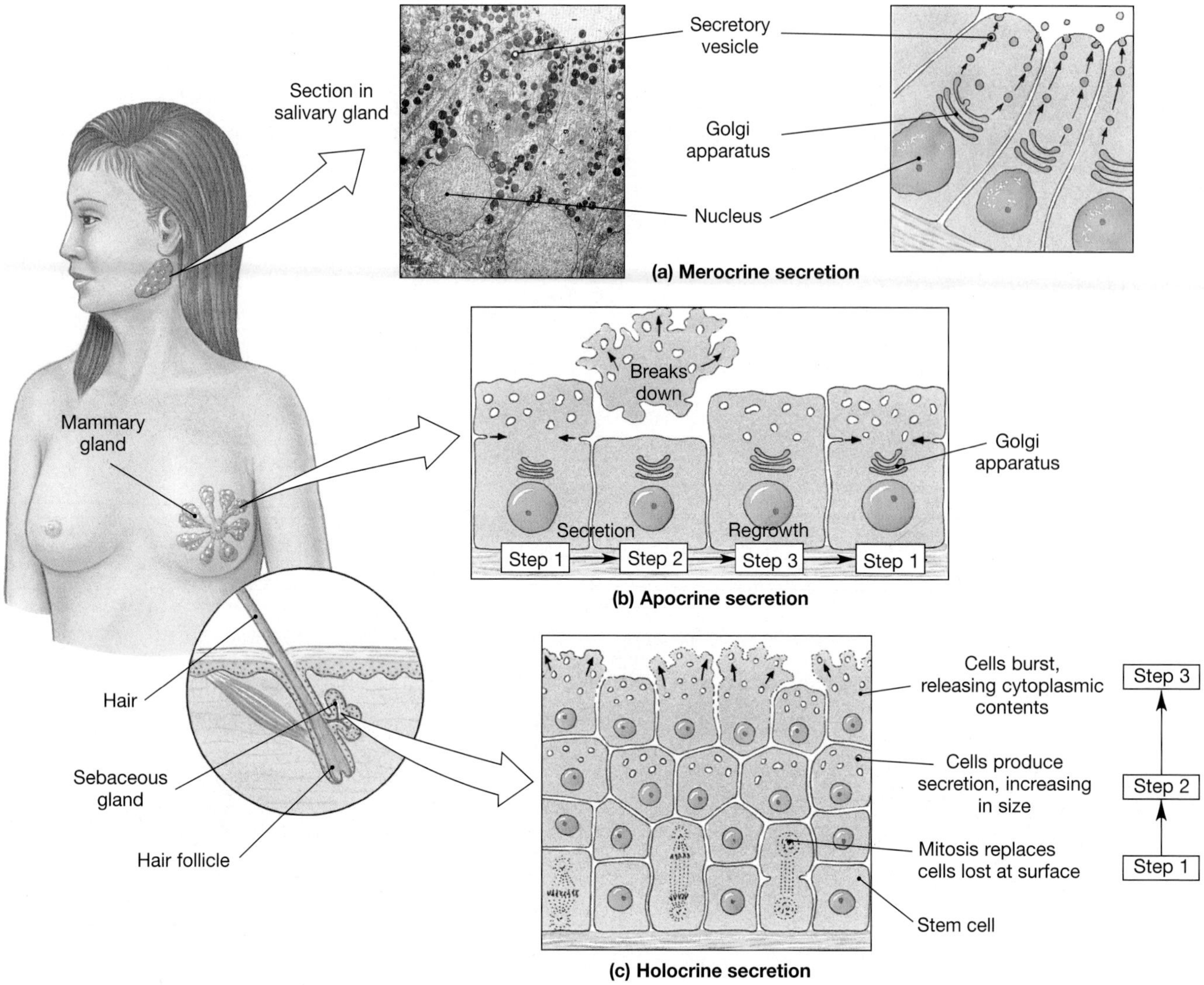

FIGURE 3.9 **Mechanisms of Glandular Secretion**

Diagrammatic representation of the mechanisms of exocrine gland secretion. (**a**) In **merocrine secretion**, secretory vesicles are discharged at the surface of the gland cell through exocytosis. (**b**) **Apocrine secretion** involves the loss of cytoplasm. Inclusions, secretory vesicles, and other cytoplasmic components are shed at the apical surface of the cell. The gland cell then undergoes a period of growth and repair before releasing additional secretions. (**c**) **Holocrine secretion** occurs as superficial gland cells break apart. Continued secretion involves the replacement of these cells through the mitotic division of underlying stem cells.

Connective Tissues

Connective tissues are found throughout the body but are never exposed to the environment outside the body. Connective tissues include bone, fat, and blood, tissues that are quite different in appearance and function. Nevertheless, all connective tissues have three basic components: (1) specialized cells, (2) extracellular protein fibers, and (3) a fluid known as the **ground substance**. The extracellular fibers and ground substance constitute the **matrix** that surrounds the cells. Although epithelial tissue consists almost entirely of cells, connective tissue consists mostly of extracellular matrix.

Connective tissues perform a variety of functions that involve far more than just connecting body parts together. Those functions include:

1. establishing a structural framework for the body;
2. transporting fluids and dissolved materials from one region of the body to another;
3. providing protection for delicate organs;
4. supporting, surrounding, and interconnecting other tissue types;
5. storing energy reserves, especially in the form of lipids; and
6. defending the body from invasion by microorganisms.

Although most connective tissues have multiple functions, no single connective tissue performs all of these functions.

All epithelia begin as simple epithelia that may later become stratified.

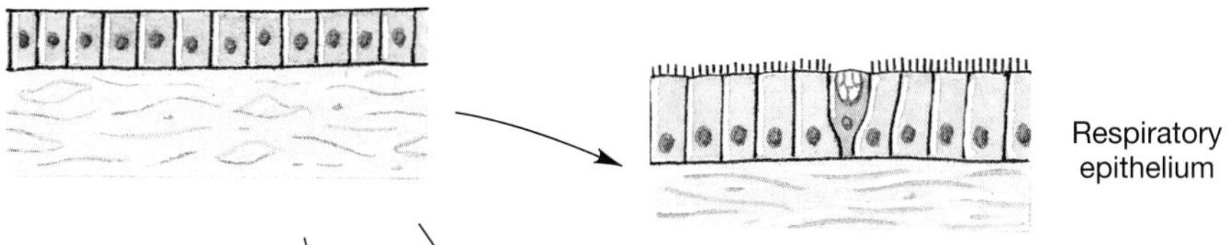

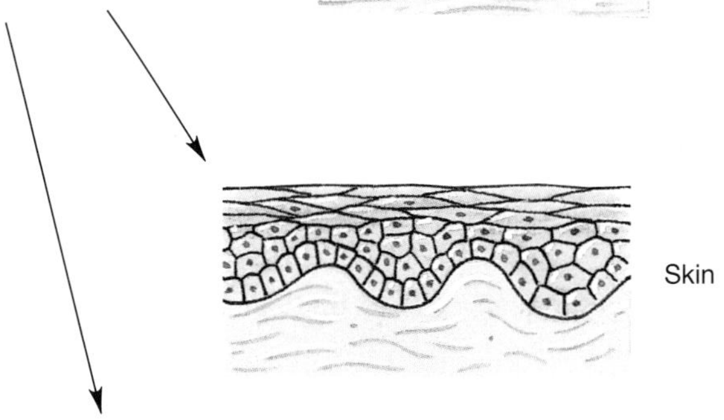

These cells differentiate into functional epithelial cells and gland cells that may have endocrine or exocrine functions.

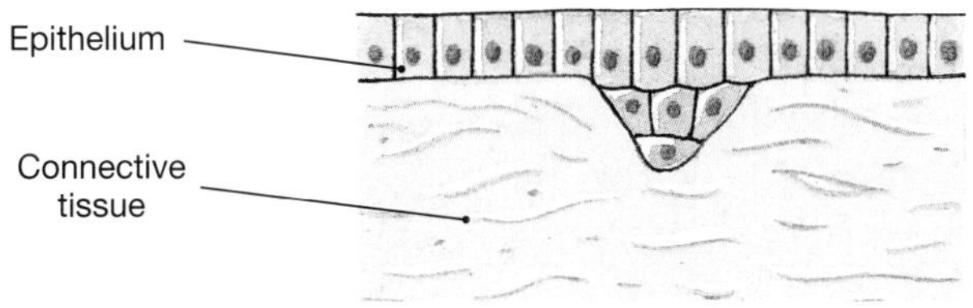

Complex glands begin to form as epithelial cells grow into the underlying connective tissue.

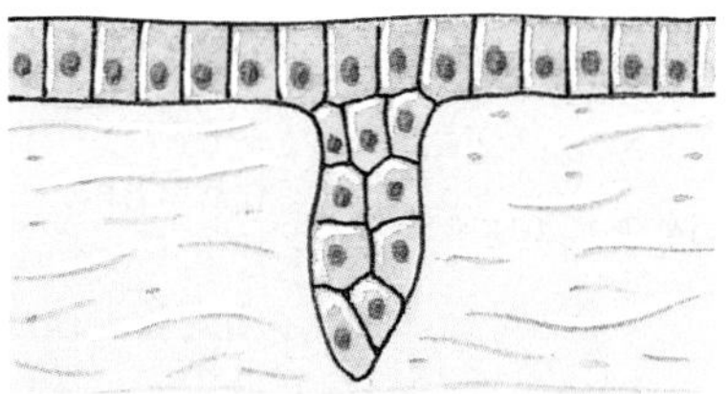

In the formation of an exocrine gland, the cells connecting the secretory cells to the surface form the duct that carries the secretions of the gland cells to the epithelial surface.

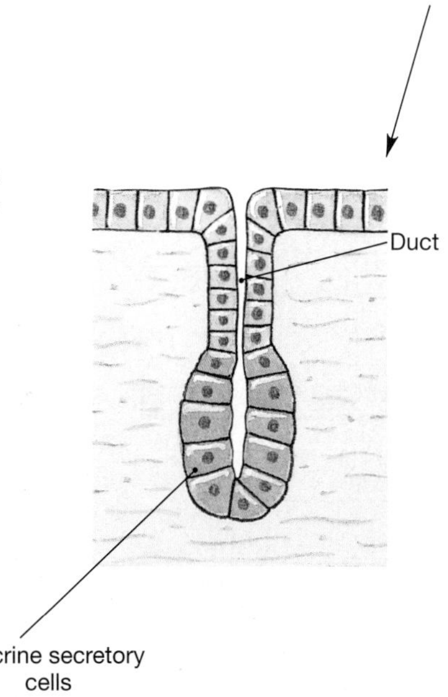

In the formation of an endocrine gland, the connecting cells disappear, and the gland cells secrete into blood vessels or into the surrounding tissue fluids.

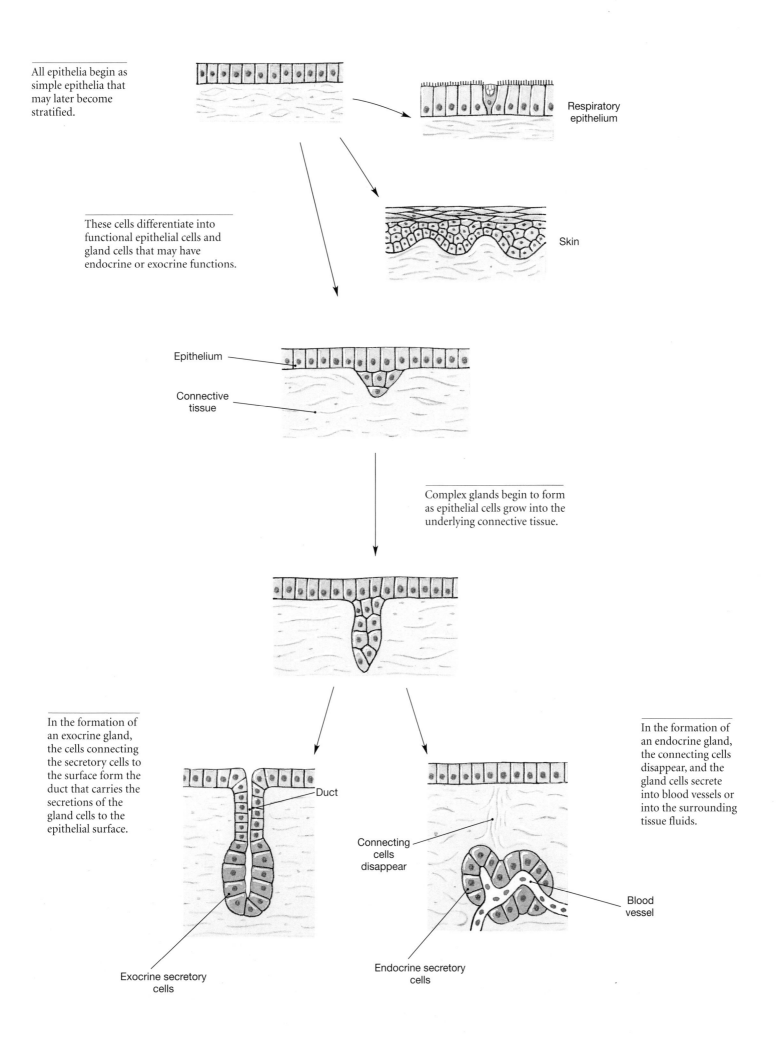

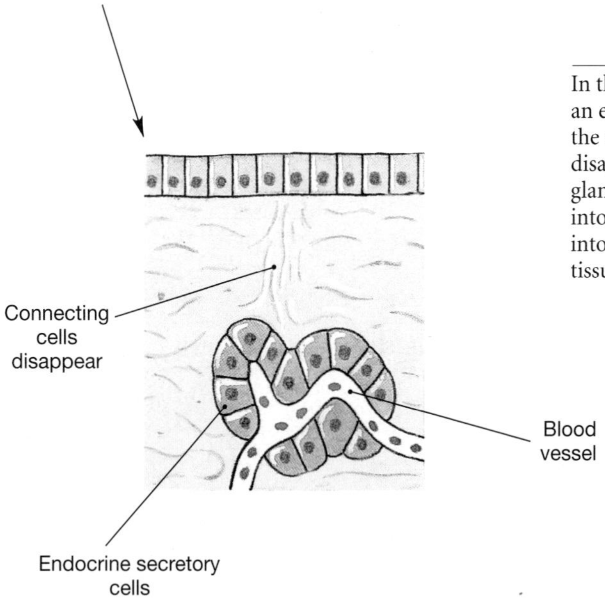

Classification of Connective Tissues [FIGURE 3.10]

Connective tissue can be classified into three categories: (1) *connective tissue proper*, (2) *fluid connective tissues*, and (3) *supporting connective tissues* (Figure 3.10●).

1. *Connective tissue proper* refers to connective tissues with many types of cells and extracellular fibers in a syrupy ground substance. These connective tissues may differ in terms of the number of cell types they contain and the relative properties and proportions of fibers and ground substance. *Adipose* (fat) *tissue* and *tendons* differ greatly, but both are examples of connective tissue proper.
2. *Fluid connective tissues* have a distinctive population of cells suspended in a watery matrix that contains dissolved proteins. There are two types of fluid connective tissues, *blood* and *lymph*.
3. *Supporting connective tissues* have a less diverse cell population than connective tissue proper and a matrix that contains closely packed fibers. There are two types of supporting connective tissues, *cartilage* and *bone*. The matrix of cartilage is a gel whose characteristics vary depending on the predominant fiber type. The matrix of bone is said to be **calcified** because it contains mineral deposits, primarily calcium salts. These minerals give the bone strength and rigidity.

Connective Tissue Proper [FIGURE 3.11 AND TABLE 3.1]

Connective tissue proper contains extracellular fibers, a viscous (syrupy) ground substance, and two classes of cells. **Fixed cells** are stationary and are involved primarily with local maintenance, repair, and energy storage. **Wandering cells** are concerned primarily with the defense and repair of damaged tissues. The number of cells at any given moment varies depending on local conditions. Refer to Figure 3.11● and Table 3.1 as we describe the cells and fibers of connective tissue proper.

CELLS OF CONNECTIVE TISSUE PROPER

FIXED CELLS Fixed cells include *fibroblasts, fixed macrophages, adipocytes, mesenchymal cells*, and, in a few locations, *melanocytes*.

- *Fibroblasts* (FĪ-brō-blasts) are the most abundant fixed cells in connective tissue proper and are the only cells always present. These slender or *stellate* (star-shaped) cells are responsible for the production and maintenance of all connective tissue fibers. Each fibroblast manufactures and secretes protein subunits that interact to form large extracellular fibers. In addition, fibroblasts secrete *hyaluronan*, which gives the ground substance its viscous consistency.
- *Fixed macrophages* (MAK-rō-fā-jez; *phagein*, to eat) are large, amoeboid cells that are scattered among the fibers. These cells engulf damaged cells or pathogens that enter the tissue. Although they are not abundant, they play an important role in mobilizing the body's defenses. When stimulated, they release chemicals that activate the immune system and attract large numbers of wandering cells involved in the body's defense mechanisms.
- *Adipocytes* (AD-i-pō-sīts) are also known as fat cells, or *adipose cells*. A typical adipocyte contains a single, enormous lipid droplet. The nucleus and other organelles are squeezed to one side, so that in section the cell resembles a class ring. The number of fat cells varies from one type of connective tissue to another, from one region of the body to another, and from individual to individual.
- *Mesenchymal* (MES-en-kī-mul) *cells* are stem cells that are present in many connective tissues. These cells respond to local injury or infection by dividing to produce daughter cells that differentiate into fibroblasts, macrophages, or other connective tissue cells.
- *Melanocytes* (MEL-an-ō-sīts or me-LAN-ō-sīts) synthesize and store a brown pigment, **melanin** (MEL-a-nin), that gives the tissue a dark color. Melanocytes are common in the epithelium of the skin, where they play a major role in determining skin color. They are also found in the underlying connective tissue (the *dermis*), although their distribution varies widely due to regional, individual, and racial factors. Melanocytes are also abundant in connective tissues of the eyes.

WANDERING CELLS Wandering cells include *free macrophages, mast cells, lymphocytes, plasma cells*, and *microphages*.

- *Free macrophages* are relatively large phagocytic cells that wander rapidly through the connective tissues of the body. When circulating within the blood, these cells are called *monocytes*. In effect, the few fixed macrophages in a tissue provide a "front-line" defense that is reinforced by the arrival of free macrophages and other specialized cells.
- *Mast cells* are small, mobile connective tissue cells often found near blood vessels. The cytoplasm of a mast cell is filled with secretory granules of **histamine** (HIS-ta-mēn) and **heparin** (HEP-a-rin).

FIGURE 3.10 **A CLASSIFICATION OF CONNECTIVE TISSUES**

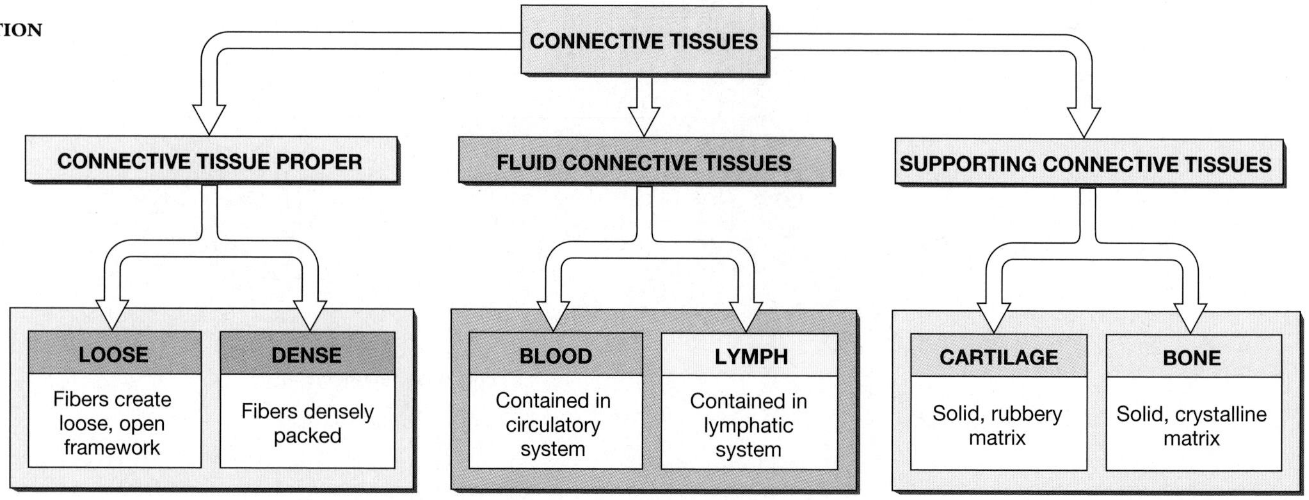

These chemicals, which are released after injury or infection, stimulate local inflammation.

- *Lymphocytes* (LIM-fō-sīts), like free macrophages, migrate throughout the body. Their numbers increase markedly wherever tissue damage occurs, and some may then develop into **plasma cells**. Plasma cells are responsible for the production of *antibodies*, proteins involved in defending the body against disease.
- *Microphages* are phagocytic blood cells that are smaller than monocytes. These cells, called *neutrophils* and *eosinophils*, migrate through connective tissues in small numbers. When an infection or injury occurs, chemicals released by macrophages and mast cells attract microphages in large numbers.

TABLE 3.1 A COMPARISION OF SOME FUNCTIONS OF FIXED CELLS AND WANDERING CELLS

Cell Types	Functions
FIXED CELLS	
Fibroblasts	Produce connective tissue fibers and maintain matrix
Fixed macrophages	Phagocytize pathogens and damaged cells
Adipocytes	Store lipid reserves
Mesenchymal cells	Connective tissue stem cells that can differentiate into other cell types
Melanocytes	Melanin synthesis
WANDERING CELLS	
Free macrophages	Mobile/traveling phagocytic cells (derived from monocytes of the blood)
Mast cells	Stimulate local inflammation
Lymphocytes	Participate in immune response
Microphages	Small, phagocytic cells (neutrophils, eosinophils) that mobilize during infection or tissue injury

CONNECTIVE TISSUE FIBERS [FIGURES 3.11/3.13/3.14]

Three types of fibers are found in connective tissue: *collagen, reticular*, and *elastic fibers*. Fibroblasts form all three types of fibers through the secretion of protein subunits that combine or aggregate within the matrix.

1. *Collagen fibers* are long, straight, and unbranched (Figure 3.11●). These are the most common, and the strongest, fibers in connective tissue proper. Each collagen fiber consists of three fibrous protein subunits wound together like the strands of a rope, and like a rope, a collagen fiber is flexible. Yet it is very strong when pulled from either end. This kind of applied force is called *tension*, and the ability to resist tension is called *tensile strength*. **Tendons** (Figure 3.14a●, p. 69) consist almost entirely of collagen fibers; they connect skeletal muscles to bones. Typical **ligaments** (LIG-a-ments) resemble tendons, but they connect one bone to another. The parallel alignment of collagen fibers in tendons and ligaments allows them to withstand tremendous forces; uncontrolled muscle contractions or skeletal movements are more likely to break a bone than to snap a tendon or ligament.
2. *Reticular fibers* (*reticulum*, network) contain the same protein subunits as collagen fibers, but the subunits interact in a different way. Reticular fibers are thinner than collagen fibers, and they form a

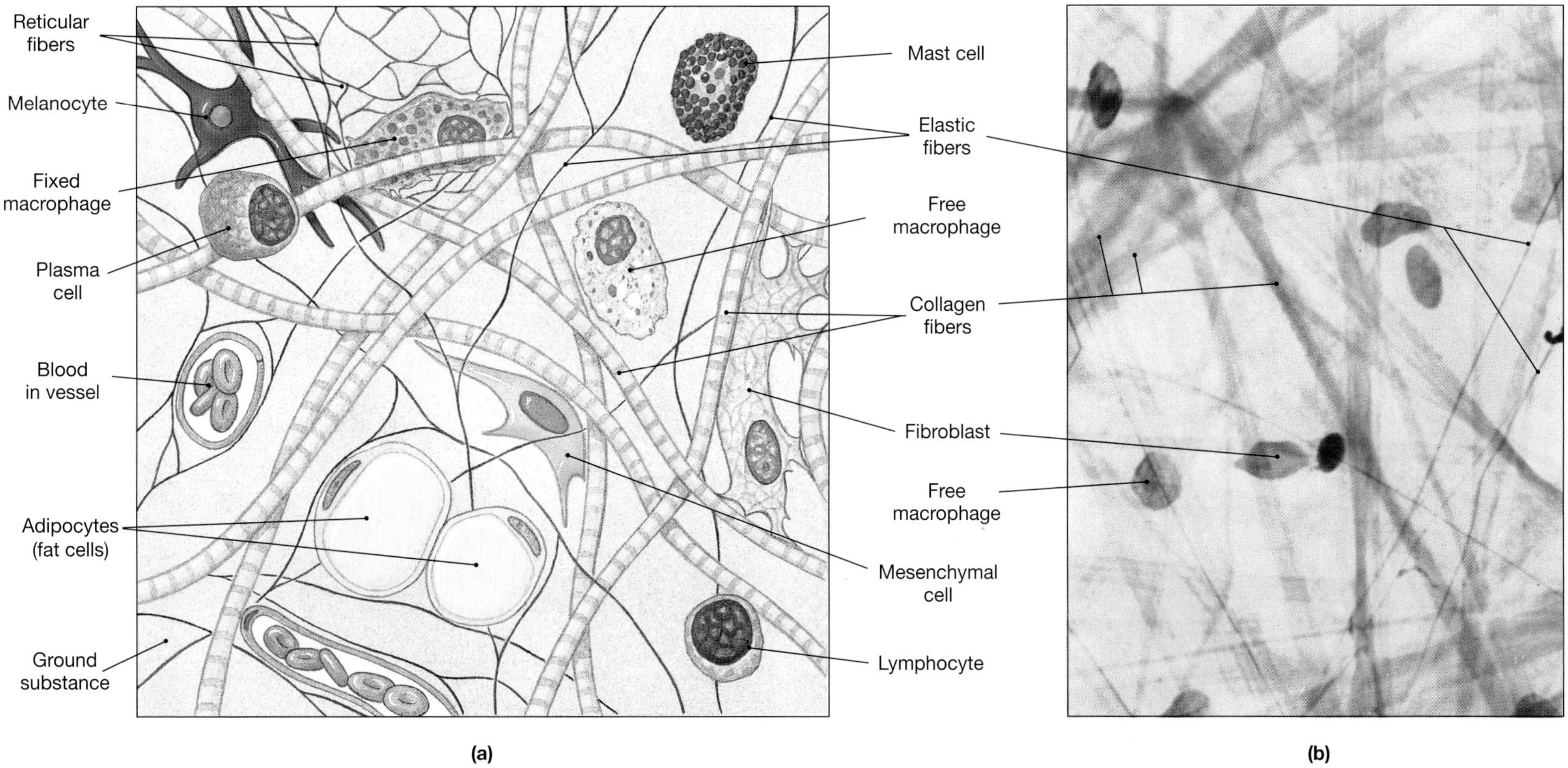

FIGURE 3.11 **THE CELLS AND FIBERS OF CONNECTIVE TISSUE PROPER**

(**a**) Diagrammatic view of the cells and fibers in areolar tissue, the most common type of connective tissue proper. (**b**) A light micrograph showing the areolar tissue that supports the mesothelium lining the peritoneum. (LM × 502) branching, interwoven framework that is tough but flexible. These fibers are especially abundant in organs such as the spleen and liver, where they create a complex three-dimensional network, or *stroma*, that supports the *parenchyma* (pa-RENG-ki-ma), or distinctive functional cells, of these organs (Figures 3.11a and 3.13c●). Because they form a network, rather than sharing a common alignment, reticular fibers can resist forces applied from many different directions. They are thus able to stabilize the relative positions of the organ's cells, blood vessels, and nerves despite changing positions and the pull of gravity.

3. *Elastic fibers* contain the protein *elastin*. Elastic fibers are branching and wavy, and after stretching up to 150% of their resting length, they recoil to their original dimensions. **Elastic ligaments** are dominated by elastic fibers. They are relatively rare, but have important functions, such as interconnecting the vertebrae (Figure 3.14b●, p. 69).

GROUND SUBSTANCE [FIGURE 3.11a]

The cellular and fibrous components of connective tissues are surrounded by a solution known as the ground substance (Figure 3.11a●). Ground substance in normal connective tissue proper is clear, colorless, and similar in consistency to maple syrup. In addition to hyaluronan, the ground substance contains a mixture of various proteoglycans and glycoproteins that interact to determine its consistency.

Connective tissue proper can be divided into *loose connective tissues* and *dense connective tissues* based on the relative proportions of cells, fibers, and ground substance.

EMBRYONIC TISSUES [FIGURE 3.12]

Mesenchyme is the first connective tissue to appear in the developing embryo. Mesenchyme contains star-shaped cells that are separated by a matrix that contains very fine protein filaments. This connective tissue (Figure 3.12a●) gives rise to all other connective tissues, including fluid connective tissues, cartilage, and bone. **Mucous connective tissue**, or *Wharton's jelly,* (Figure 3.12b●) is a loose connective tissue found in many regions of the embryo, including the umbilical cord.

Neither of these embryonic connective tissues is found in the adult. However, many adult connective tissues contain scattered mesenchymal (stem) cells that assist in repairs after the connective tissue has been injured or damaged.

LOOSE CONNECTIVE TISSUES

Loose connective tissues are the "packing material" of the body. These tissues fill spaces between organs, provide cushioning, and support epithelia. Loose connective tissues also surround and support blood vessels and nerves, store lipids, and provide a route for the diffusion of materials. There are three types of loose connective tissues: *areolar tissue, adipose tissue*, and *reticular tissue.*

AREOLAR TISSUE [FIGURE 3.13a] **Areolar tissue** (*areola*, a little space), is the least specialized connective tissue in the adult body. This tissue, shown in Figure 3.13a●, contains all of the cells and fibers found in any connective tissue proper. Areolar tissue has an open framework, and ground substance accounts for most of its volume. This viscous fluid cushions shocks, and because the fibers are loosely organized, areolar tissue can be distorted

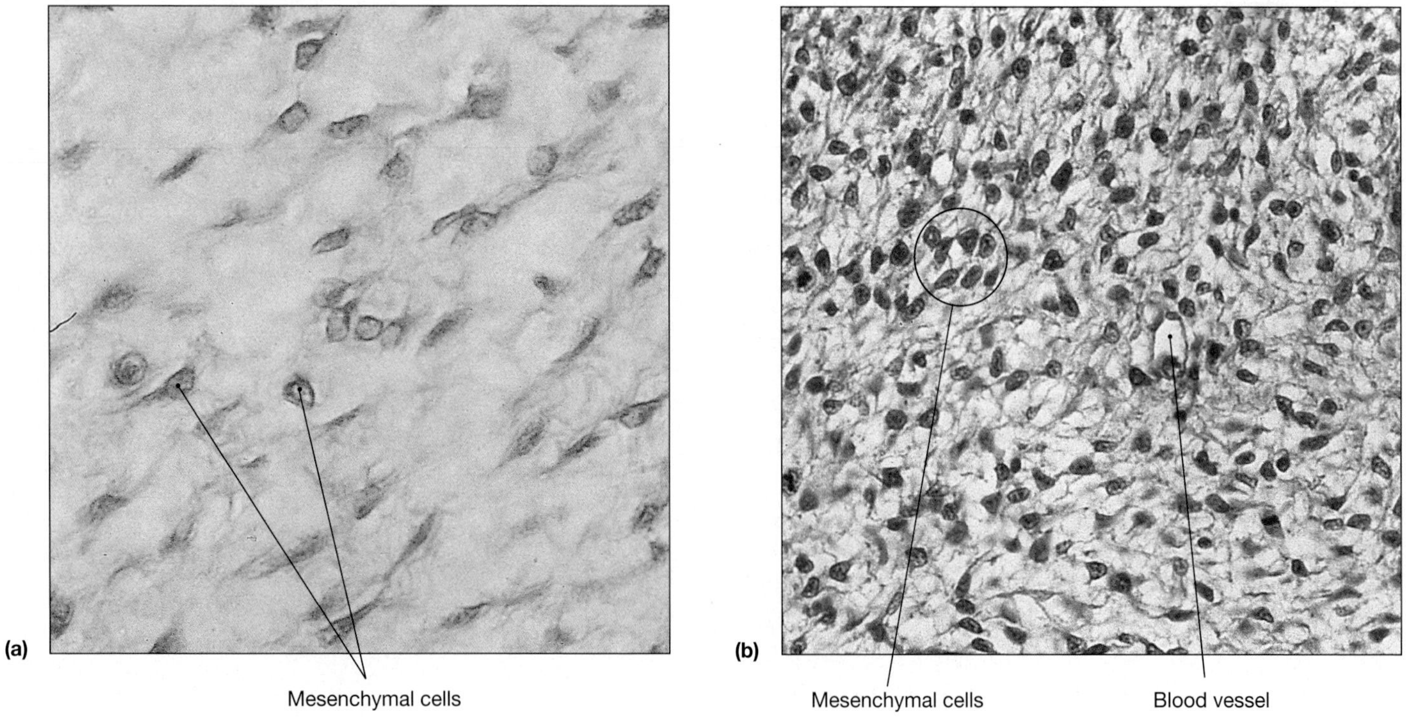

FIGURE 3.12 **EMBRYONIC CONNECTIVE TISSUES**

These connective tissue types give rise to all other connective tissue types. **(a) Mesenchyme**. This is the first connective tissue to appear in the embryo. (LM × 1036) **(b) Mucous Connective Tissue** (*Wharton's jelly*). This sample was taken from the umbilical cord of a fetus. (LM × 650)

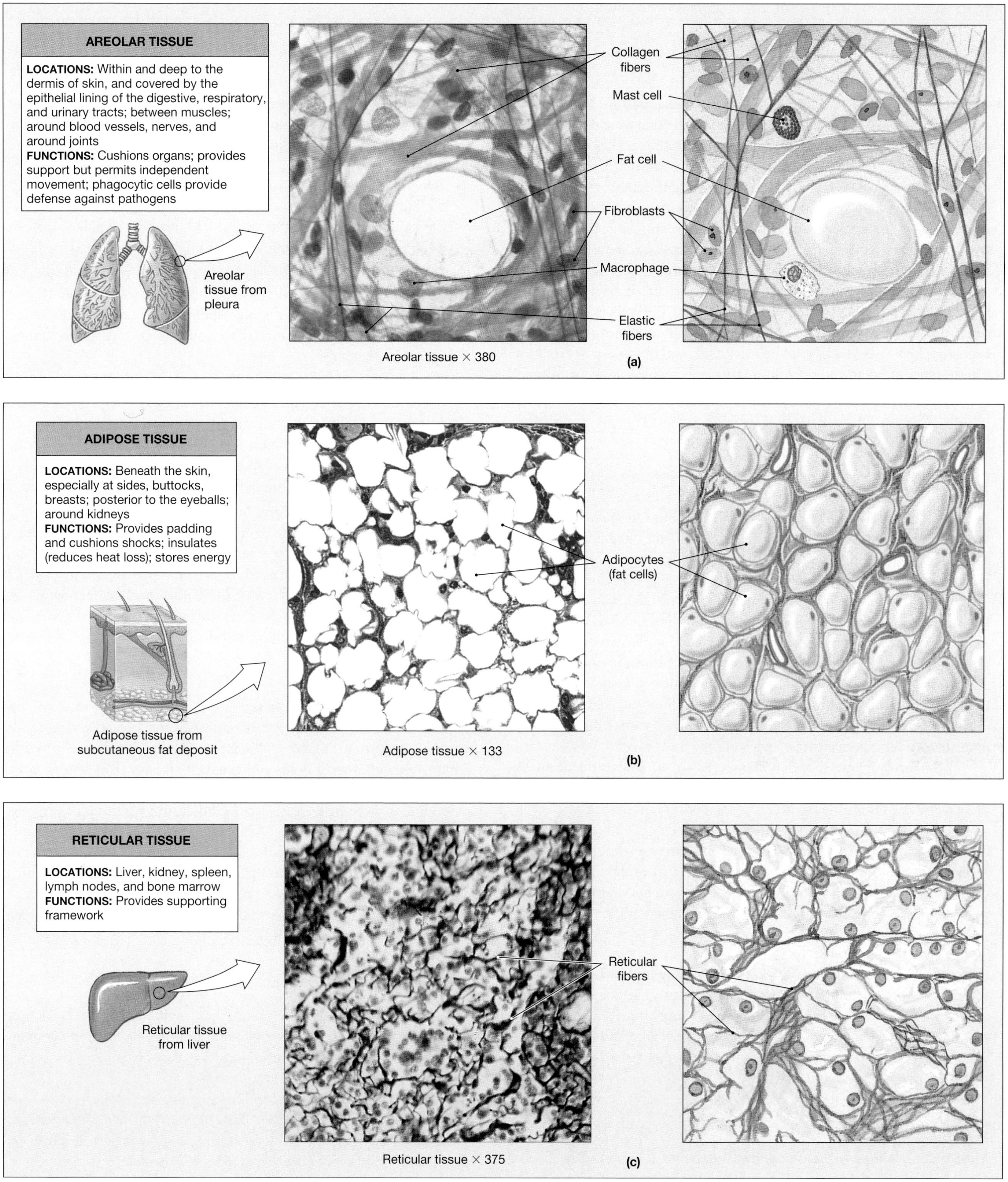

AREOLAR TISSUE
LOCATIONS: Within and deep to the dermis of skin, and covered by the epithelial lining of the digestive, respiratory, and urinary tracts; between muscles; around blood vessels, nerves, and around joints
FUNCTIONS: Cushions organs; provides support but permits independent movement; phagocytic cells provide defense against pathogens
Areolar tissue from pleura
Collagen fibers
Mast cell
Fat cell
Fibroblasts
Macrophage
Elastic fibers
Areolar tissue × 380
(a)
ADIPOSE TISSUE
LOCATIONS: Beneath the skin, especially at sides, buttocks, breasts; posterior to the eyeballs; around kidneys
FUNCTIONS: Provides padding and cushions shocks; insulates (reduces heat loss); stores energy
Adipose tissue from subcutaneous fat deposit
Adipocytes (fat cells)
Adipose tissue × 133
(b)
RETICULAR TISSUE
LOCATIONS: Liver, kidney, spleen, lymph nodes, and bone marrow
FUNCTIONS: Provides supporting framework
Reticular tissue from liver
Reticular fibers
Reticular tissue × 375
(c)

FIGURE 3.13 **Loose Connective Tissues**

This is the "packing material" of the body, filling spaces between other structures. (**a**) **Areolar Tissue.** Note the open framework; all the cells of connective tissue proper are found in aerolar tissue. (**b**) **Adipose Tissue.** Adipose tissue is a loose connective tissue dominated by adipocytes. In standard histological views, the cells look empty because their lipid inclusions dissolve during slide preparation. (**c**) **Reticular Tissue.** Reticular tissue consists of an open framework of reticular fibers. These fibers are usually very difficult to see because of the large numbers of cells organized around them.

without damage. The presence of elastic fibers makes it fairly resilient, so this tissue returns to its original shape after external pressure is relieved.

Areolar tissue forms a layer that separates the skin from deeper structures. In addition to providing padding, the elastic properties of this layer allow a considerable amount of independent movement. Thus, pinching the skin of the arm does not affect the underlying muscle. Conversely, contractions of the underlying muscles do not pull against the skin—as the muscle bulges, the areolar tissue stretches. Because this tissue has an extensive circulatory supply, drugs injected into the areolar tissue layer under the skin are quickly absorbed into the bloodstream.

In addition to delivering oxygen and nutrients and removing carbon dioxide and waste products, the capillaries (the smallest blood vessels) in areolar tissue carry wandering cells to and from the tissue. Epithelia usually cover a layer of areolar tissue, and fibroblasts are responsible for maintaining the reticular lamina of the basement membrane. The epithelial cells rely on diffusion across the basement membrane, and the capillaries in the underlying connective tissue provide the necessary oxygen and nutrients.

ADIPOSE TISSUE [FIGURE 3.13b] The distinction between areolar tissue and fat, or **adipose tissue**, is somewhat arbitrary. Adipocytes account for most of the volume of adipose tissue (Figure 3.13b●) but for only a fraction of the volume of areolar tissue. Adipose tissue provides padding, cushions shocks, acts as an insulator to slow heat loss through the skin, and serves as packing or filler around structures. Adipose tissue is common under the skin of the groin, sides, buttocks, and breasts. It fills the bony sockets behind the eyes, surrounds the kidneys, and dominates extensive areas of loose connective tissue in the pericardial and abdominal cavities. Most of the adipose tissue in the body is called **white fat**, because it has a pale, yellow-white color. In infants and young children, adipose tissue in the upper body is highly vascularized, and the individual adipocytes contain numerous mitochondria, which gives the tissue a deep, rich color from which the name **brown fat** is derived. † *Infants and Brown Fat p. 784*

Adipocytes are metabolically active cells—their lipids are continually being broken down and replaced. Although adipocytes are incapable of dividing, an excess of nutrients can cause the division of mesenchymal cells, which then differentiate into additional adipocytes. As a result, areas of aerolar tissue can become adipose tissue in someone who is overeating. When nutrients are scarce, as during a weight-loss program, adipocytes deflate like collapsing balloons. Because the cells are not destroyed, merely reduced in size, the lost weight can easily be regained in the same areas of the body. † *Liposuction p. 785*

RETICULAR TISSUE [FIGURE 3.13c] **Reticular tissue** consists of reticular fibers, macrophages, and fibroblasts (Figure 3.13c●). The fibers of reticular tissue form the stroma of the liver, the spleen, lymph nodes, and bone marrow. The fixed macrophages and fibroblasts of reticular tissue are seldom visible because they are vastly outnumbered by the parenchymal cells of these organs.

DENSE CONNECTIVE TISSUES

Most of the volume of **dense connective tissues** is occupied by fibers. Dense connective tissues are often called **collagenous** (ko-LA-jin-us) **tissues** because collagen fibers are the dominant fiber type. Two types of dense connective tissue are found in the body: (1) *dense regular connective tissue* and (2) *dense irregular connective tissue.*

DENSE REGULAR CONNECTIVE TISSUE [FIGURES 3.5c,d/3.14a,b] In dense regular connective tissue the collagen fibers are packed tightly and aligned parallel to applied forces. Four major examples of this tissue type are *tendons, aponeuroses, elastic tissue,* and *ligaments.*

1. *Tendons* (Figure 3.14a●) are cords of dense regular connective tissue that attach skeletal muscles to bones. The collagen fibers run along the longitudinal axis of the tendon and transfer the pull of the contracting muscle to the bone. Large numbers of fibroblasts are found between the collagen fibers.

2. *Aponeuroses* (ap-ō-noo-RŌ-sēz) are collagenous sheets or ribbons that resemble flat, broad tendons. Aponeuroses may cover the surface of a muscle and assist in attaching superficial muscles to another muscle or structure.

3. *Elastic tissue* contains large numbers of elastic fibers. Because elastic fibers outnumber collagen fibers, the tissue has a springy, resilient nature. This ability to stretch and rebound allows it to tolerate cycles of expansion and contraction. Elastic tissue often underlies transitional epithelia (Figure 3.5c,d●,p. 58); it is also found in the walls of blood vessels and surrounding the respiratory passageways.

4. *Ligaments* resemble tendons, but they usually connect one bone to another. Ligaments often contain significant numbers of elastic fibers as well as collagen fibers, and they can tolerate a modest amount of stretching. An even higher proportion of elastic fibers is found in **elastic ligaments**, which resemble tough rubber bands. Although uncommon elsewhere, elastic ligaments along the vertebral column are very important in stabilizing the positions of the vertebrae (Figure 3.14b●).

DENSE IRREGULAR CONNECTIVE TISSUE [FIGURE 3.14c] The fibers in **dense irregular connective tissue** form an interwoven meshwork and do not show any consistent pattern (Figure 3.14c●). This tissue provides strength and support to areas subjected to stresses from many directions. A layer of dense irregular connective tissue, the *dermis*, gives skin its strength; a piece of cured leather (the dermis of animal skin) provides an excellent illustration of the interwoven nature of this tissue. Except at joints, dense irregular connective tissue forms a sheath around cartilage (the perichondrium) and bone (the periosteum). Dense irregular connective tissue also forms the thick fibrous **capsule** that surrounds internal organs, such as the liver, kidneys, and spleen, and encloses the cavities of joints.

■ Fluid Connective Tissues

Blood and *lymph* are connective tissues that contain distinctive collections of cells in a fluid matrix. The watery matrix of blood and lymph contains cells and many types of suspended proteins that do not form insoluble fibers under normal conditions.

Blood contains blood cells and fragments of cells collectively known as *formed elements* (Figure 3.15●). Three types of formed elements exist: (1) red blood cells, (2) white blood cells, and (3) platelets. A single cell type, the **red blood cell**, or **erythrocyte** (e-RITH-rō-sīt; *erythros*, red), accounts for almost half the volume of blood. Red blood cells are responsible for the transport of oxygen and, to a lesser degree, of carbon dioxide in the blood. The watery matrix of blood, called **plasma**, also contains small numbers of **white blood cells**, or **leukocytes** (LOO-kō-sīts; *leukos*, white).

DENSE REGULAR CONNECTIVE TISSUE

LOCATIONS: Between skeletal muscles and skeleton (tendons and aponeuroses); between bones (ligaments); covering skeletal muscles; deep fasciae
FUNCTIONS: Provides firm attachment; conducts pull of muscles; reduces friction between muscles; stabilizes relative positions of bones

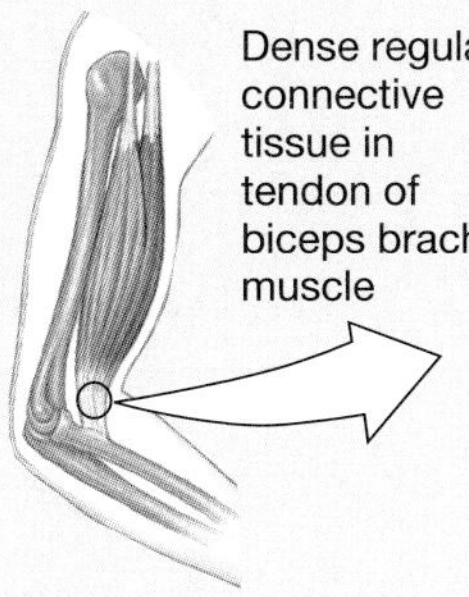

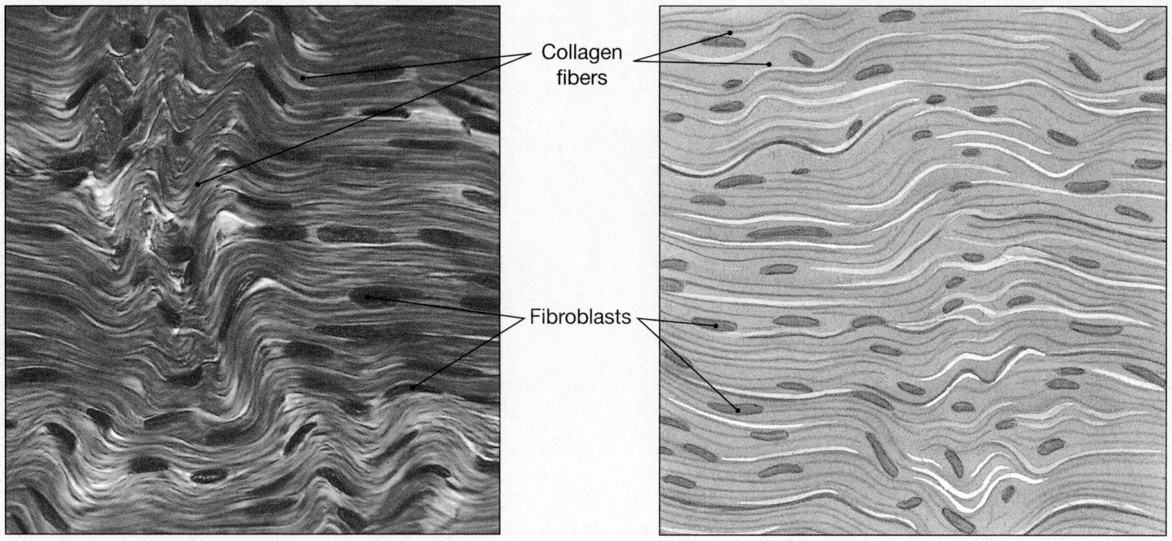

Tendon × 364

(a)

ELASTIC TISSUE

LOCATIONS: Between vertebrae of the spinal column (ligamentum flavum and ligamentum nuchae); ligaments supporting penis; underlying transitional epithelia; in walls of blood vessels
FUNCTIONS: Stabilize positions of vertebrae and penis; cushions shocks; allows expansion and recoil of organs

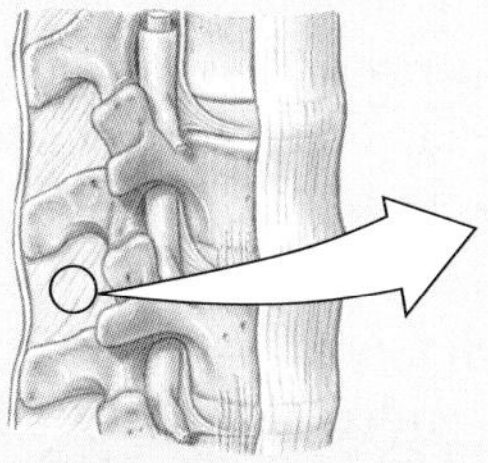

Elastic tissue in intervertebral ligaments

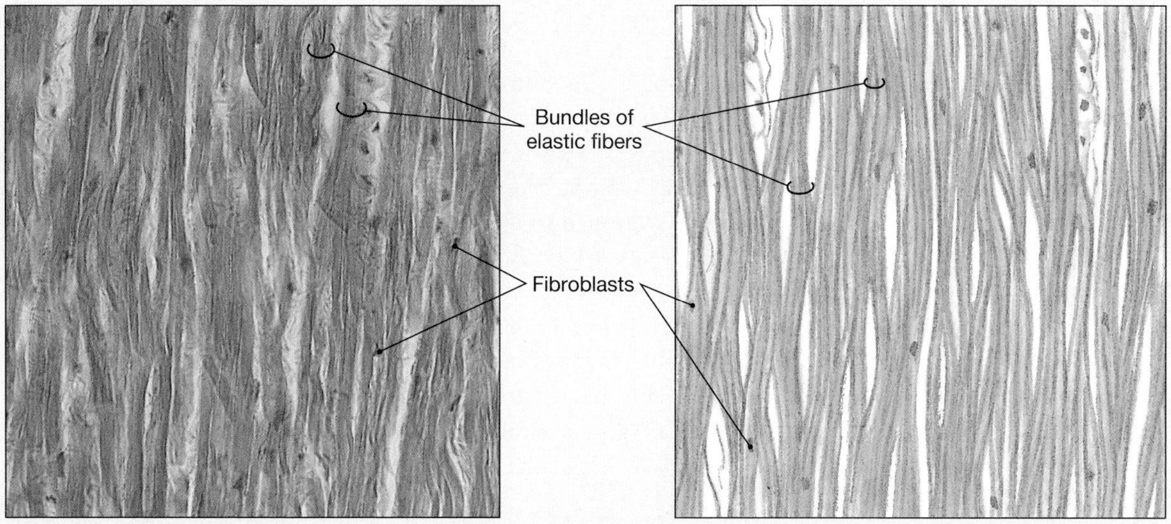

Elastic ligament × 887

(b)

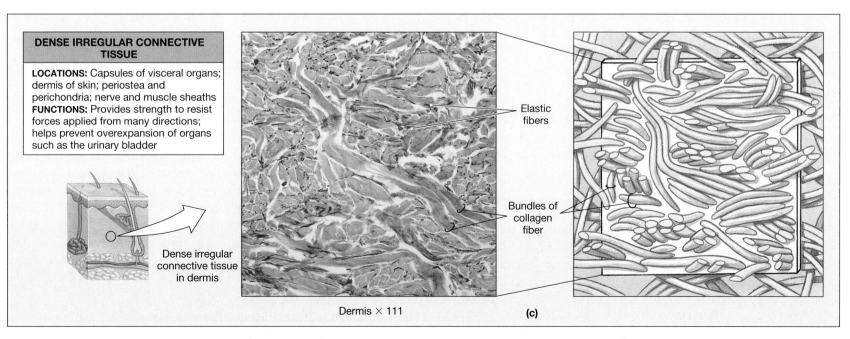

DENSE IRREGULAR CONNECTIVE TISSUE

LOCATIONS: Capsules of visceral organs; dermis of skin; periostea and perichondria; nerve and muscle sheaths
FUNCTIONS: Provides strength to resist forces applied from many directions; helps prevent overexpansion of organs such as the urinary bladder

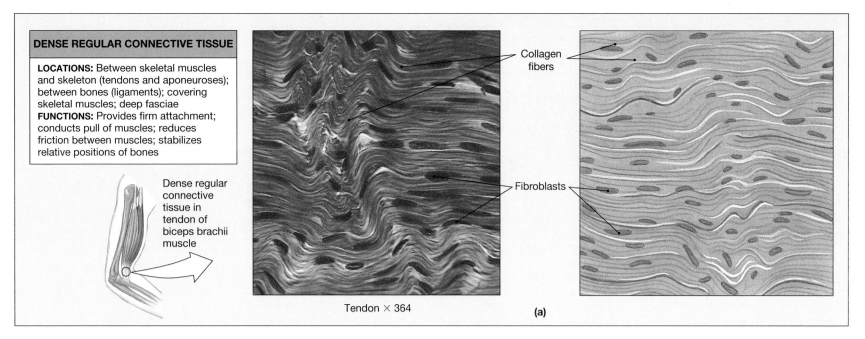

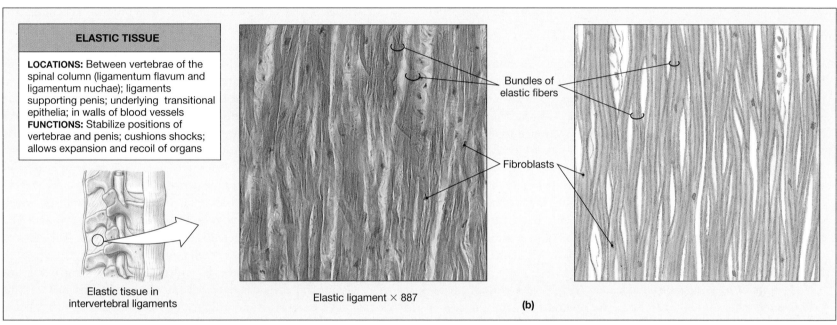

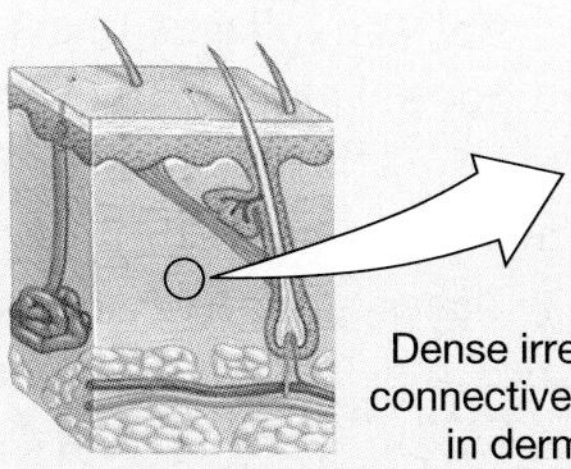

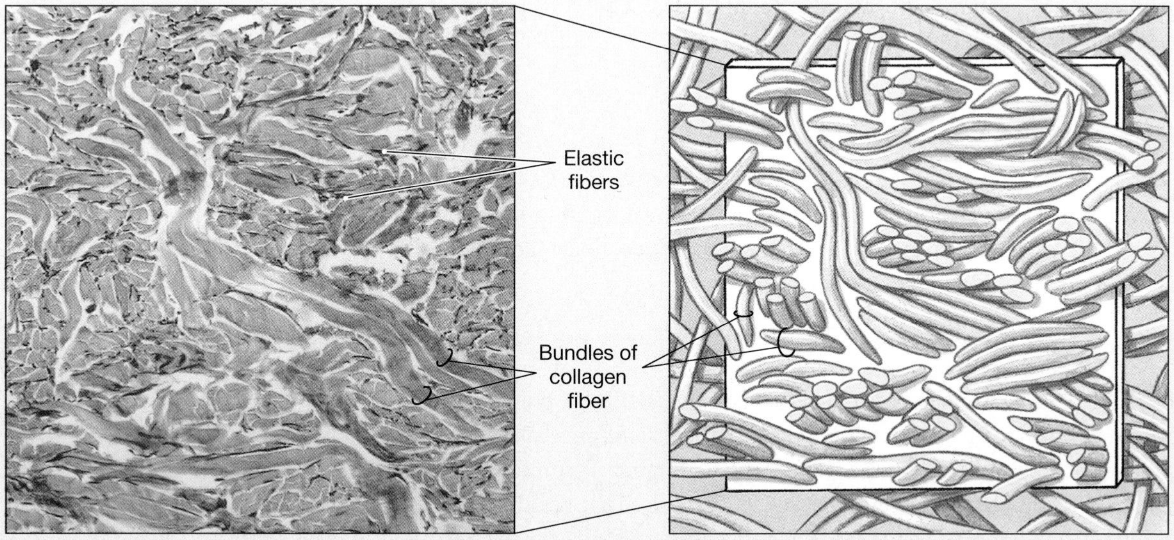

Dermis × 111

(c)

FIGURE 3.14 **DENSE CONNECTIVE TISSUES**

(a) Dense Regular Connective Tissue: Tendon. The dense regular connective tissue in a tendon consists of densely packed, parallel bundles of collagen fibers. The fibroblast nuclei can be seen flattened between the bundles. Most ligaments resemble tendons in their histological organization. **(b) Dense Regular Connective Tissue: Elastic Ligament**. Elastic ligaments extend between the vertebrae of the spinal column. The bundles of elastic fibers are fatter than the collagen fiber bundles of a tendon or typical ligament. **(c) Dense Irregular Connective Tissue**. The deep portion of the dermis of the skin consists of a thick layer of interwoven collagen fibers oriented in various directions.

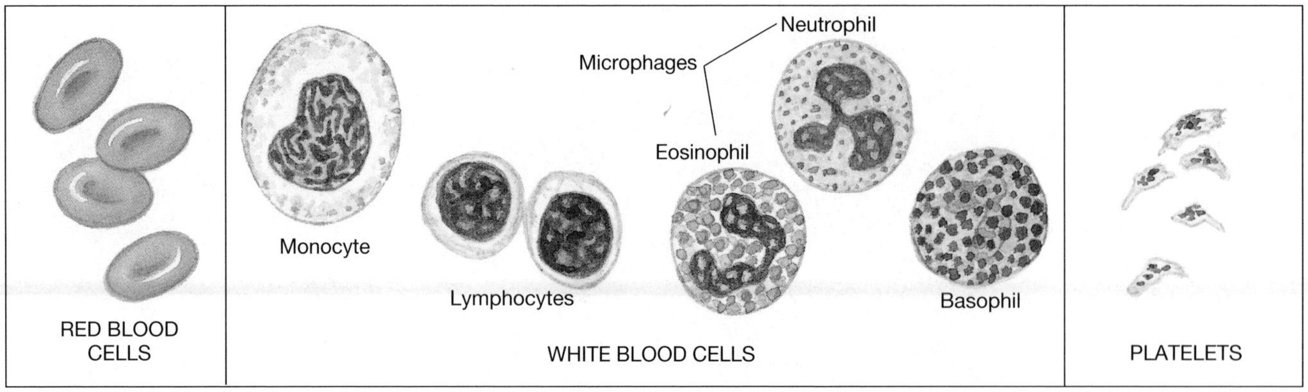

FIGURE 3.15 **FORMED ELEMENTS OF THE BLOOD**

White blood cells include the phagocytic microphages (*neutrophils* and *eosinophils*), *basophils*, *lymphocytes*, and *monocytes*. The white blood cells are important components of the immune system, which protects the body from infection and disease. Tiny membrane-enclosed packets of cytoplasm called **platelets** contain enzymes and special proteins. Platelets function in the clotting response that seals breaks in the vessel wall.

Extracellular fluid includes three major subdivisions: *plasma*, *interstitial fluid*, and *lymph*. *Plasma* is normally confined to the vessels of the circulatory system, and contractions of the heart keep it in motion. **Arteries** are vessels that carry blood away from the heart toward fine, thin-walled vessels called **capillaries.** **Veins** are vessels that drain the capillaries and return the blood to the heart, completing the circuit of blood. In tissues, filtration moves water and small solutes out of the capillaries and into the interstitial fluid, which bathes the body's cells. The major difference between the plasma and interstitial fluid is that plasma contains large number of suspended proteins.

Lymph forms as interstitial fluid enters **lymphatic vessels**, small passageways that return it to the cardiovascular system. Along the way, cells of the immune system monitor the composition of the lymph and respond to signs of injury or infection. The number of cells in lymph may vary, but ordinarily 99 percent of them are lymphocytes. The rest are primarily macrophages or microphages.

Supporting Connective Tissues

Cartilage and bone are called **supporting connective tissues** because they provide a strong framework that supports the rest of the body. In these connective tissues, the matrix contains numerous fibers and, in some cases, deposits of insoluble calcium salts.

CARTILAGE [FIGURE 3.16]

The matrix of **cartilage** is a firm gel that contains complex polysaccharides called **chondroitin sulfates** (kon-DROY-tin; *chondros*, cartilage). The chondroitin sulfates form complexes with proteins, forming proteoglycans. Cartilage cells, or **chondrocytes** (KON-drō-sīts), are the only cells found within the cartilage matrix (Figure 3.16•). These cells live in small chambers known as **lacunae** (la-KOO-nē; *lacus*, pool). The physical properties of cartilage depend on the nature of the matrix. Collagen fibers provide tensile strength, and the combined characteristics of the extracellular fibers and the ground substance give it flexibility and resilience.

Cartilage is avascular because chondrocytes produce a chemical that discourages the formation of blood vessels. All nutrient and waste-product exchange must occur by diffusion through the matrix. A cartilage is usually set apart from surrounding tissues by a fibrous **perichondrium** (per-ē-KON-drē-um; *peri*, around; Figure 3.16a•). The perichondrium contains two distinct layers: an outer, *fibrous layer* of dense irregular connective tissue and an inner, *cellular layer*. The fibrous layer provides mechanical support and protection, and attaches the cartilage to other structures. The cellular layer is important to the growth and maintenance of the cartilage.

Cartilages grow by two mechanisms (Figure 3.16b,c•). In **appositional growth**, cells of the inner layer of the perichondrium undergo repeated cycles of division. The innermost cells differentiate into chondroblasts, which begin producing cartilage matrix. After they are completely surrounded by matrix, the chondroblasts differentiate into chondrocytes. This growth mechanism gradually increases the dimensions of the cartilage by adding to its surface. Additionally, chondrocytes within the cartilage matrix can undergo division, and their daughter cells produce additional matrix. This cycle enlarges the cartilage from within, much like the inflation of a balloon; the process is called **interstitial growth**. Neither appositional nor interstitial growth occurs in adult cartilages, and most cartilages cannot repair themselves after a severe injury.

TYPES OF CARTILAGE [FIGURE 3.17] There are three major types of cartilage: (1) *hyaline cartilage*, (2) *elastic cartilage*, and (3) *fibrocartilage*.

1. **Hyaline cartilage** (HĪ-a-lin; *hyalos*, glass) is the most common type of cartilage. The matrix of hyaline cartilage contains closely packed collagen fibers. Although it is tough but somewhat flexible, this is the weakest type of cartilage. Because the collagen fibers of the matrix do not stain well, they are not always apparent in light microscopy (Figure 3.17a•). Examples of this type of cartilage in the adult body include: (1) the connections between the ribs and the sternum, (2) the supporting cartilages along the conducting passageways of the respiratory tract, and (3) the *articular cartilages* covering opposing bone surfaces within synovial joints, such as the elbow or knee.
2. **Elastic cartilage** contains numerous elastic fibers that make it extremely resilient and flexible. Among other structures elastic cartilage forms the external flap (*auricle* or *pinna*) of the external ear (Figure 3.17b•), the epiglottis, the airway to the middle ear (*auditory tube*),

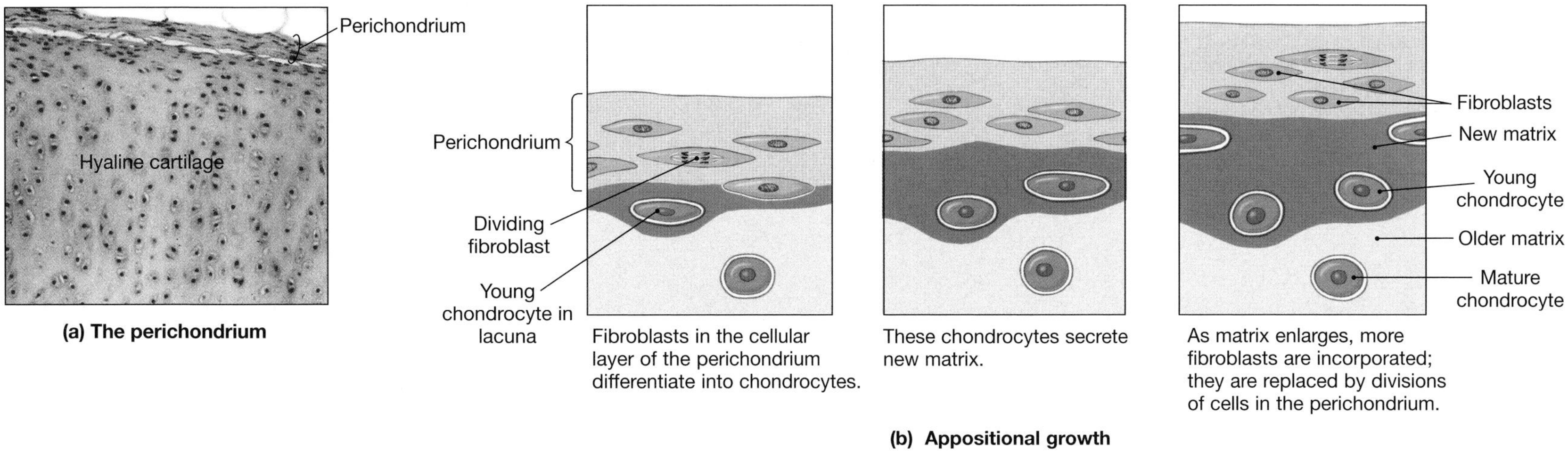

Perichondrium
Hyaline cartilage
(a) The perichondrium
Perichondrium
Dividing fibroblast
Young chondrocyte in lacuna
Fibroblasts in the cellular layer of the perichondrium differentiate into chondrocytes.
These chondrocytes secrete new matrix.
Fibroblasts
New matrix
Young chondrocyte
Older matrix
Mature chondrocyte
As matrix enlarges, more fibroblasts are incorporated; they are replaced by divisions of cells in the perichondrium.
(b) Appositional growth

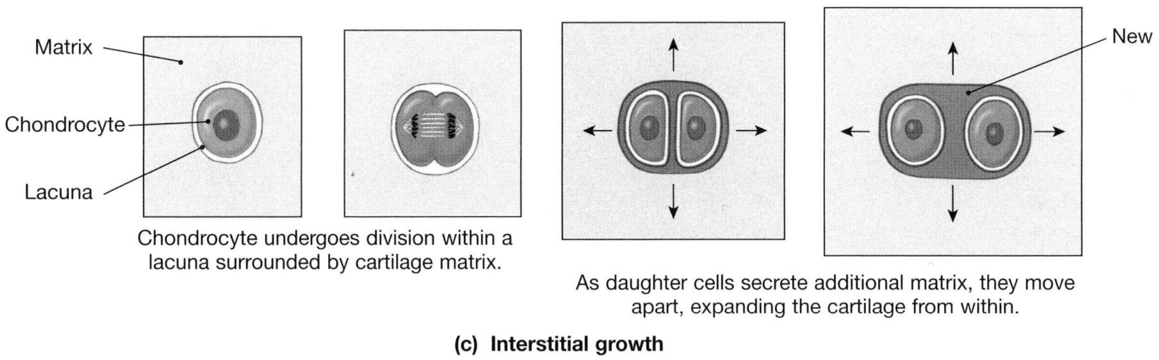

Matrix
Chondrocyte
Lacuna
Chondrocyte undergoes division within a lacuna surrounded by cartilage matrix.
New
As daughter cells secrete additional matrix, they move apart, expanding the cartilage from within.
(c) Interstitial growth

FIGURE 3.16 **THE FORMATION AND GROWTH OF CARTILAGE**

(**a**) This light microscope shows the organization of a small piece of hyaline cartilage and the surrounding perichondrium. (**b**) **Appositional Growth.** The cartilage grows at its external surface through the differentiation of fibroblasts into chondrocytes within the cellular layer of the perichondrium. (**c**) **Interstitial Growth.** The cartilage expands from within as chondrocytes in the matrix divide, grow, and produce new matrix.

and small (*cuneiform*) cartilages of the larynx. Although the cartilages at tip of the nose are very flexible, there is disagreement as to whether or not they should be classified as "true" elastic cartilages because their elastic fibers are not as abundant as they are at the auricle or epiglottis.

3. **Fibrocartilage** has little ground substance, and the matrix is dominated by collagen fibers (Figure 3.17c●). The collagen fibers are densely interwoven, making this tissue extremely durable and tough. Fibrocartilaginous pads lie between the spinal vertebrae, between the pubic bones of the pelvis, and around or within a few joints and tendons. In these positions they resist compression, absorb shocks, and prevent damaging bone-to-bone contact. Cartilages heal slowly and poorly, and damaged fibrocartilage in joints can interfere with the normal movements. *Cartilages and Knee Injuries p. 785*

BONE [FIGURE 3.18 AND TABLE 3.2]

Because the detailed histology of **bone**, or *osseous tissue* (OS-ē-us; *os*, bone), will be considered in Chapter 5, this discussion will focus on significant differences between cartilage and bone. Table 3.2 summarizes the similarities and differences between cartilage and bone. Roughly one-third of the matrix of bone consists of collagen fibers. The balance is a mixture of calcium salts, primarily calcium phosphate with lesser amounts of calcium carbonate. This combination gives bone truly remarkable properties. By themselves, calcium salts are strong but rather brittle. Collagen fibers are weaker, but relatively flexible. In bone, the minerals are organized around the collagen fibers. The result is a strong, somewhat flexible combination that is very resistant to shattering. In its overall properties, bone can compete with the best steel-reinforced concrete.

CLINICAL BRIEF

SHARKS AND THE FIGHT AGAINST CANCER ("JAWS" TO THE RESCUE)

One reason cartilages are avascular is that chondrocytes secrete a compound that inhibits blood vessel formation. This chemical has been named **antiangiogenesis factor** (*anti-*, against + *angeion*, vessel + *gennan*, to produce). One reason cancers can enlarge so rapidly is that blood vessels grow into areas where cells are crowded and active, improving nutrient and oxygen delivery. This growth could theoretically be prevented by antiangiogenesis factor, but the quantities produced in normal human cartilage are extremely small.

Sharks are highly successful marine predators, but their skeletons are cartilaginous, rather than bony. Researchers have long known that cancers are extremely rare in sharks and their cartilaginous relatives. The discovery of antiangiogenesis factor provided an explanation. For almost a decade, sharks were collected to obtain the antiangiogenesis factor from their cartilages. Because they contain so much cartilage, substantial quantities could be extracted from a single animal. Fortunately for shark populations, the gene for antiangiogenesis factor has now been identified, and large quantities can now be produced using recombinant DNA techniques.

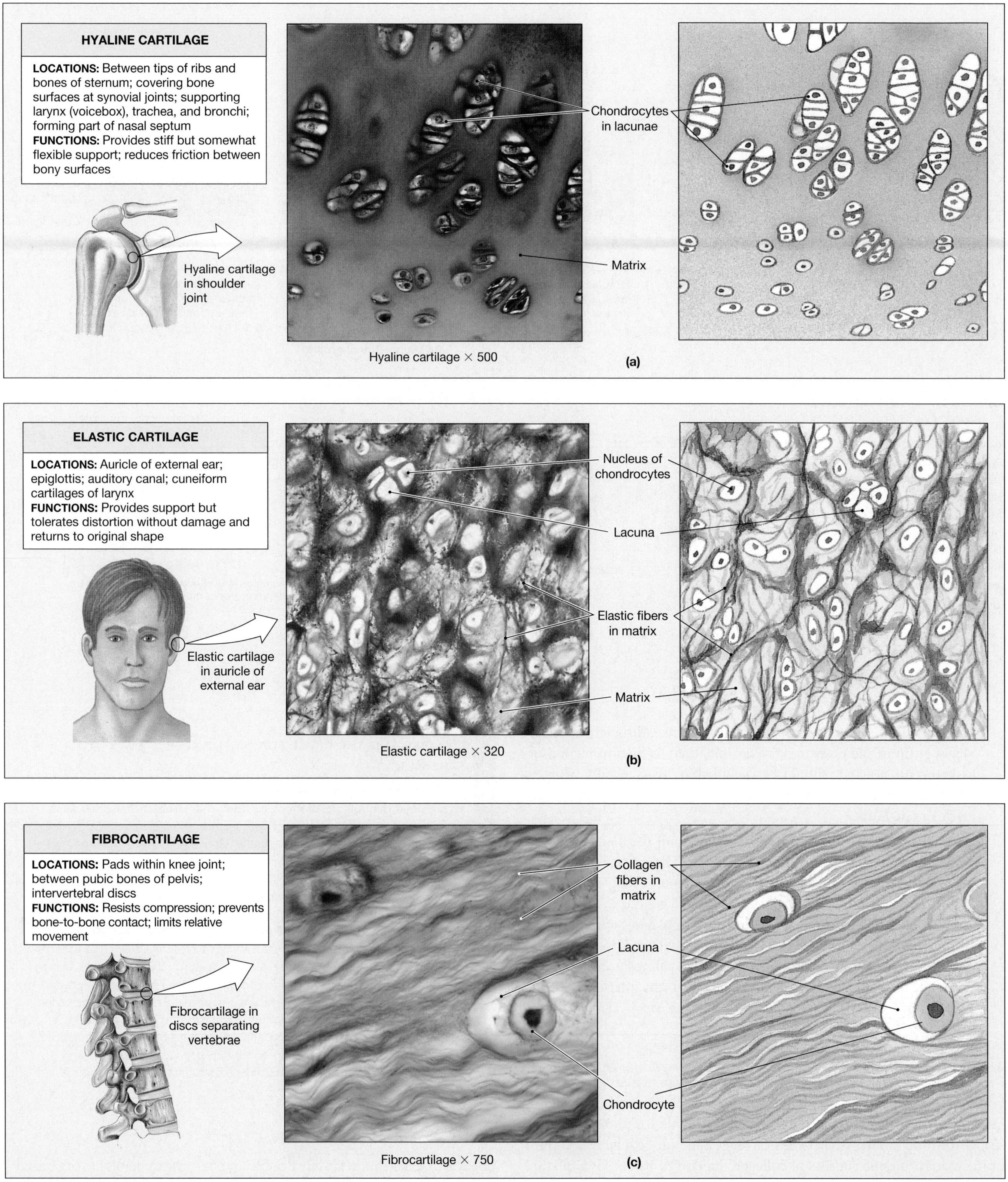
HYALINE CARTILAGE
LOCATIONS: Between tips of ribs and bones of sternum; covering bone surfaces at synovial joints; supporting larynx (voicebox), trachea, and bronchi; forming part of nasal septum
FUNCTIONS: Provides stiff but somewhat flexible support; reduces friction between bony surfaces
Hyaline cartilage in shoulder joint
Chondrocytes in lacunae
Matrix
Hyaline cartilage × 500
(a)
ELASTIC CARTILAGE
LOCATIONS: Auricle of external ear; epiglottis; auditory canal; cuneiform cartilages of larynx
FUNCTIONS: Provides support but tolerates distortion without damage and returns to original shape
Elastic cartilage in auricle of external ear
Nucleus of chondrocytes
Lacuna
Elastic fibers in matrix
Matrix
Elastic cartilage × 320
(b)
FIBROCARTILAGE
LOCATIONS: Pads within knee joint; between pubic bones of pelvis; intervertebral discs
FUNCTIONS: Resists compression; prevents bone-to-bone contact; limits relative movement
Fibrocartilage in discs separating vertebrae
Collagen fibers in matrix
Lacuna
Chondrocyte
Fibrocartilage × 750
(c)

FIGURE 3.17 **Types of Cartilage**

Cartilage is a supporting connective tissue with a firm, gelatinous matrix. (a) **Hyaline Cartilage.** Note the translucent matrix and the absence of prominent fibers. (b) **Elastic Cartilage.** The closely packed elastic fibers are visible between the chondrocytes. (c) **Fibrocartilage.** The collagen fibers are extremely dense, and the chondrocytes are relatively far apart.

TABLE 3.2 A COMPARISON OF CARTILAGE AND BONE

Characteristic	Cartilage	Bone
Structural Features		
Cells	Chondrocytes in lacunae	Osteocytes in lacunae
Matrix	Chondroitin sulfates with proteins, forming hydrated proteoglycans	Insoluble crystals of calcium phosphate and calcium carbonate
Fibers	Collagen, elastic, reticular fibers (proportions vary)	Collagen fibers predominate
Vascularity	None	Extensive
Covering	Perichondrium, two layers	Periosteum, two layers
Strength	Limited: bends easily but hard to break	Strong: resists distortion until breaking point is reached
Growth	Interstitial and appositional	Appositional only
Repair capabilities	Limited ability	Extensive ability
Oxygen demands	Low	High
Nutrient delivery	By diffusion through matrix	By diffusion through cytoplasm and fluid in canaliculi

The general organization of osseous tissue can be seen in Figure 3.18●. **Lacunae** within the matrix contain bone cells, or **osteocytes** (OS-tē-ō-sīts). The lacunae are often organized around blood vessels that branch through the bony matrix. Although diffusion cannot occur through the calcium salts, osteocytes communicate with blood vessels and with one another through slender cytoplasmic extensions. These extensions run through long, slender passages in the matrix. These passageways, called **canaliculi** (kan-a-LIK-ū-lē; "little canals"), form a branching network for the exchange of materials between the blood vessels and the osteocytes. There are two types of bone: *compact bone*, which contains blood vessels trapped within the matrix, and *spongy bone*, which does not.

Almost all bone surfaces are sheathed by a **periosteum** (per-ē-OS-tē-um) composed of a fibrous outer layer and a cellular inner layer. The periosteum is incomplete only at joints, where bones articulate. The periosteum assists in the attachment of a bone to surrounding tissues and to associated tendons and ligaments. The cellular layer functions in bone growth and participates in repairs after an injury. Unlike cartilage, bone undergoes extensive remodeling on a regular basis, and complete repairs can be made even after severe damage has occurred. Bones also respond to the stresses placed upon them, growing thicker and stronger with exercise, and thin and brittle with inactivity (Table 3.2).

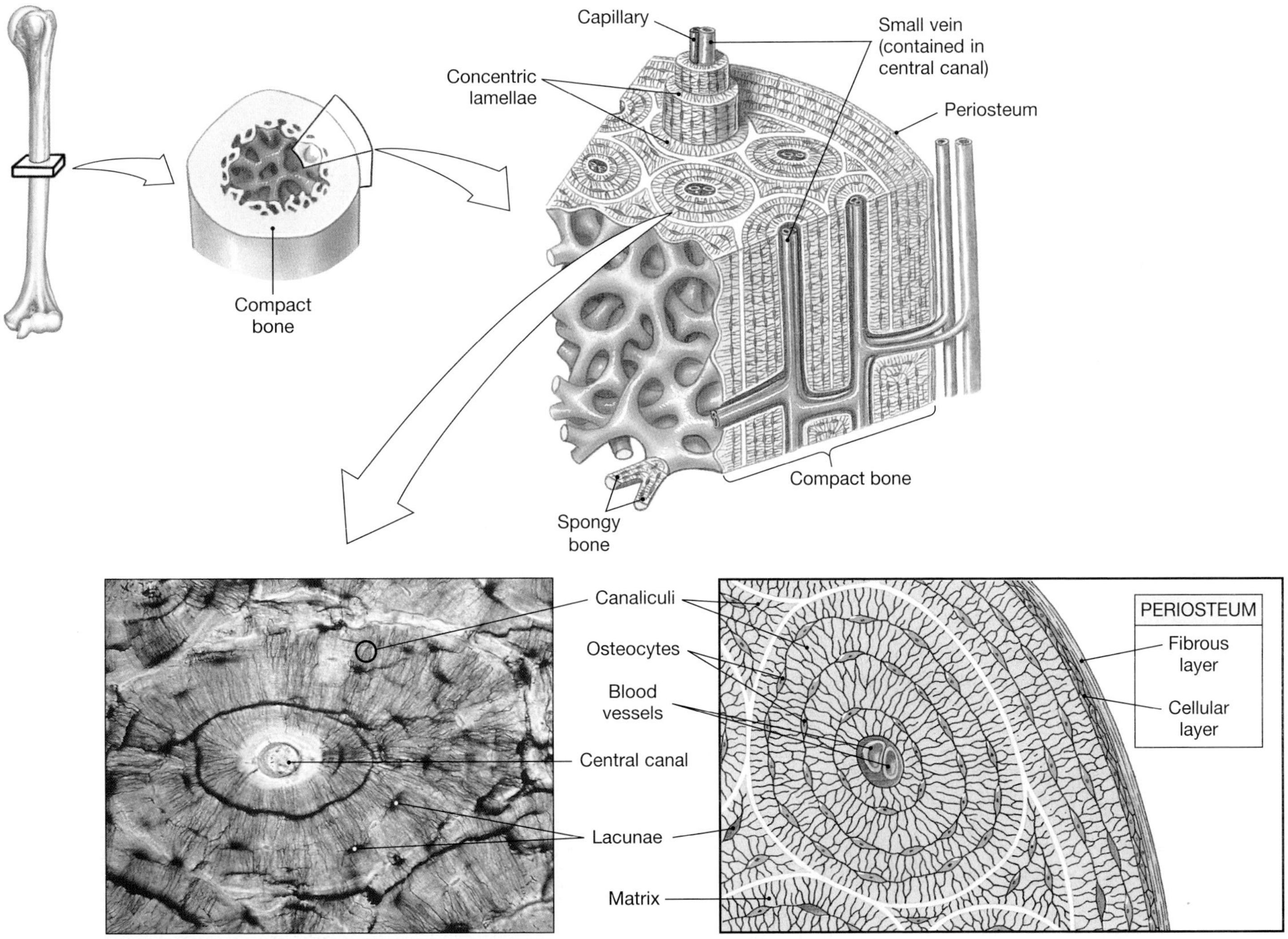

FIGURE 3.18 **BONE**

Bone is a supporting connective tissue with a hardened matrix. The osteocytes in compact bone are usually organized in groups around a central space that contains blood vessels. For the photomicrograph, a sample of bone was ground thin enough to become transparent. Bone dust produced during the grinding filled the lacunae and the central canal, making them appear dark.

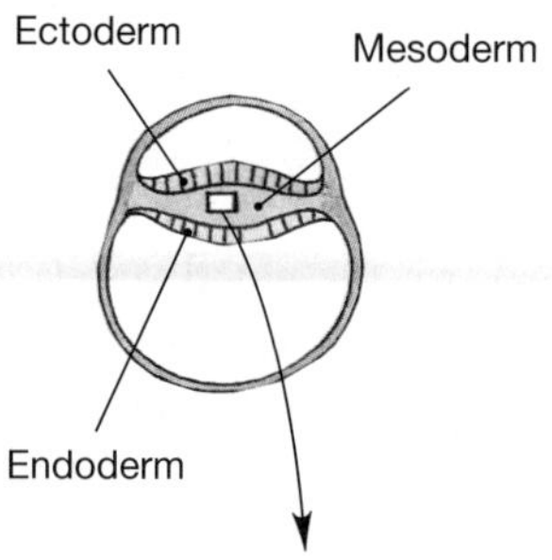

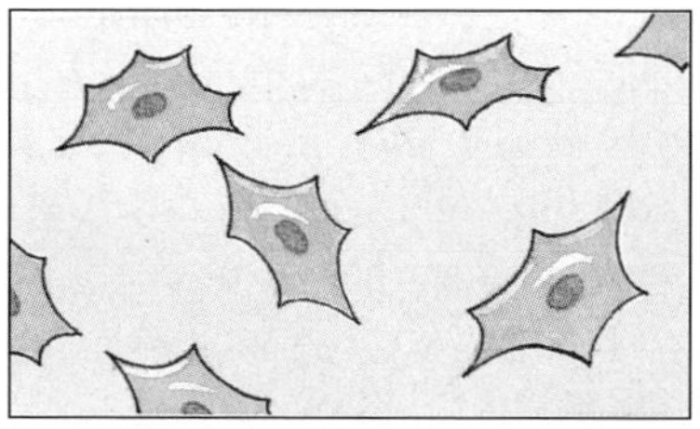

Mesenchyme is the first connective tissue to appear in the developing embryo. Mesenchyme contains star-shaped cells that are separated by a ground substance that contains fine protein filaments. Mesenchyme gives rise to all other forms of connective tissue, and scattered mesenchymal cells in adult connective tissues participate in their repair after injury.

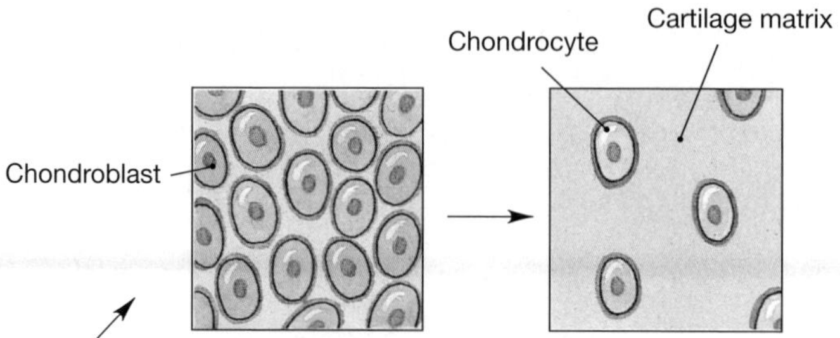

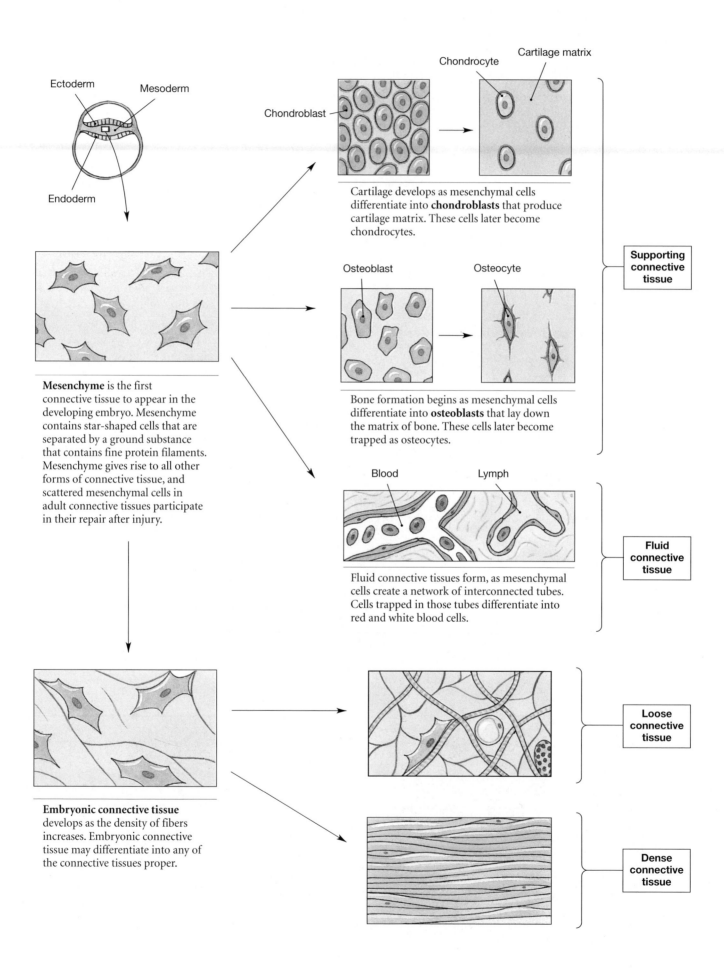

Cartilage develops as mesenchymal cells differentiate into **chondroblasts** that produce cartilage matrix. These cells later become chondrocytes.

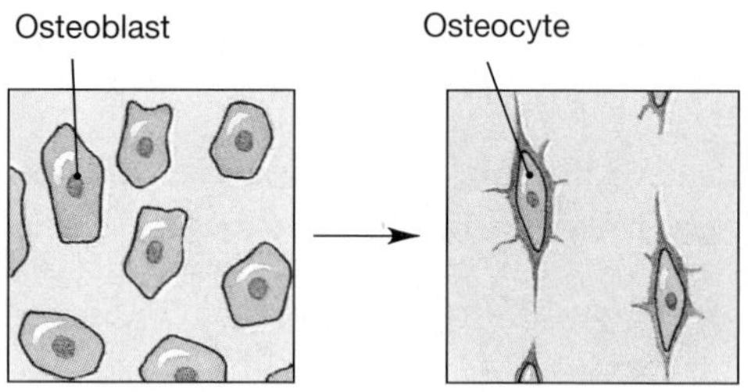

Bone formation begins as mesenchymal cells differentiate into **osteoblasts** that lay down the matrix of bone. These cells later become trapped as osteocytes.

Supporting connective tissue

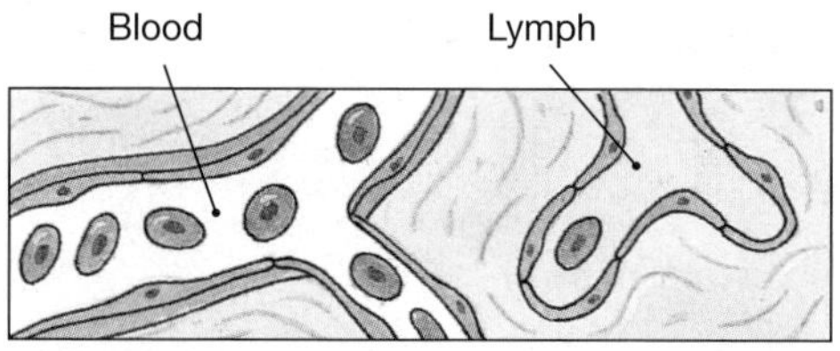

Fluid connective tissues form, as mesenchymal cells create a network of interconnected tubes. Cells trapped in those tubes differentiate into red and white blood cells.

Fluid connective tissue

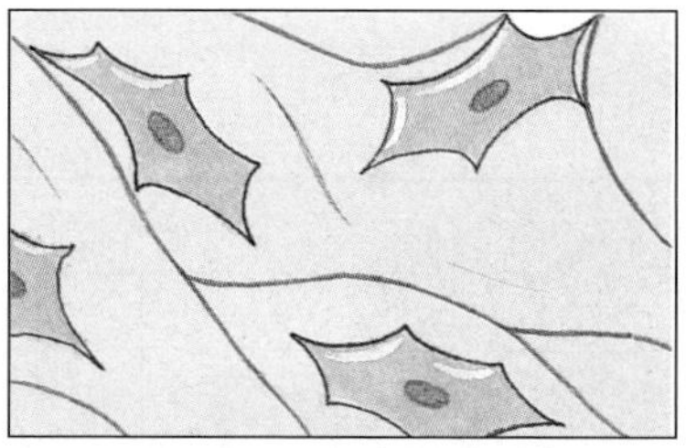

Embryonic connective tissue develops as the density of fibers increases. Embryonic connective tissue may differentiate into any of the connective tissues proper.

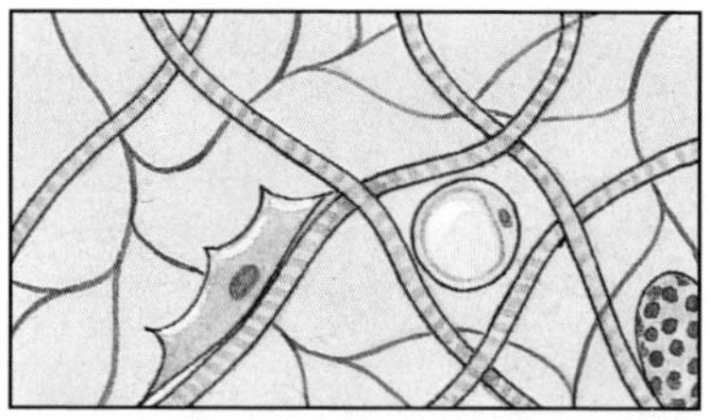

Loose connective tissue

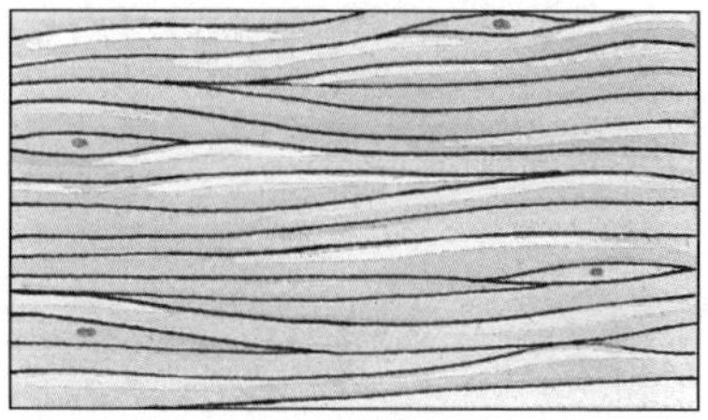

Dense connective tissue

✓ CONCEPT CHECK

- Identify the three basic components of all connective tissues.
- What is a major difference between connective tissue proper and supporting connective tissue?
- What are the two general classes of cells in connective tissue proper? What cells are found in each class?
- Lack of vitamin C in the diet interferes with the ability of fibroblasts to produce collagen. What effect might this limited ability to produce collagen have on connective tissue?

Membranes

Epithelia and connective tissues combine to form **membranes**. Each membrane consists of an epithelial sheet and an underlying connective tissue layer. Membranes cover and protect other structures and tissues in the body. There are four types of membranes: (1) *mucous membranes*, (2) *serous membranes*, (3) *the cutaneous membrane (skin)*, and (4) *synovial membranes*.

Mucous Membranes [FIGURE 3.19a]

Mucous membranes line passageways that communicate with the exterior, including the digestive, respiratory, reproductive, and urinary tracts (Figure 3.19a●). Mucous membranes, or mucosae (mū-KŌ-sē; singular, *mucosa*), form a barrier that resists the entry of pathogens. The epithelial surfaces are kept moist at all times; they may be lubricated by mucus or other glandular secretions or by exposure to fluids such as urine or semen. The areolar tissue component of a mucous membrane is called the **lamina propria** (PRŌ-prē-a). The lamina propria forms a bridge that connects the epithelium to underlying structures. It also provides support for the blood vessels and nerves that supply the epithelium. We will consider the organization of specific mucous membranes in greater detail in later chapters.

Many mucous membranes are lined by simple epithelia that perform absorptive or secretory functions. One example is the simple columnar epithelium of the digestive tract. However, other types of epithelia may be involved. For example, the mucous membrane of the mouth contains a stratified squamous epithelium, and the mucous membrane along most of the urinary tract has a transitional epithelium.

Serous Membranes [FIGURE 3.19b]

Serous membranes line the subdivisions of the ventral body cavity. There are three serous membranes, each consisting of a mesothelium (p. 57) supported by

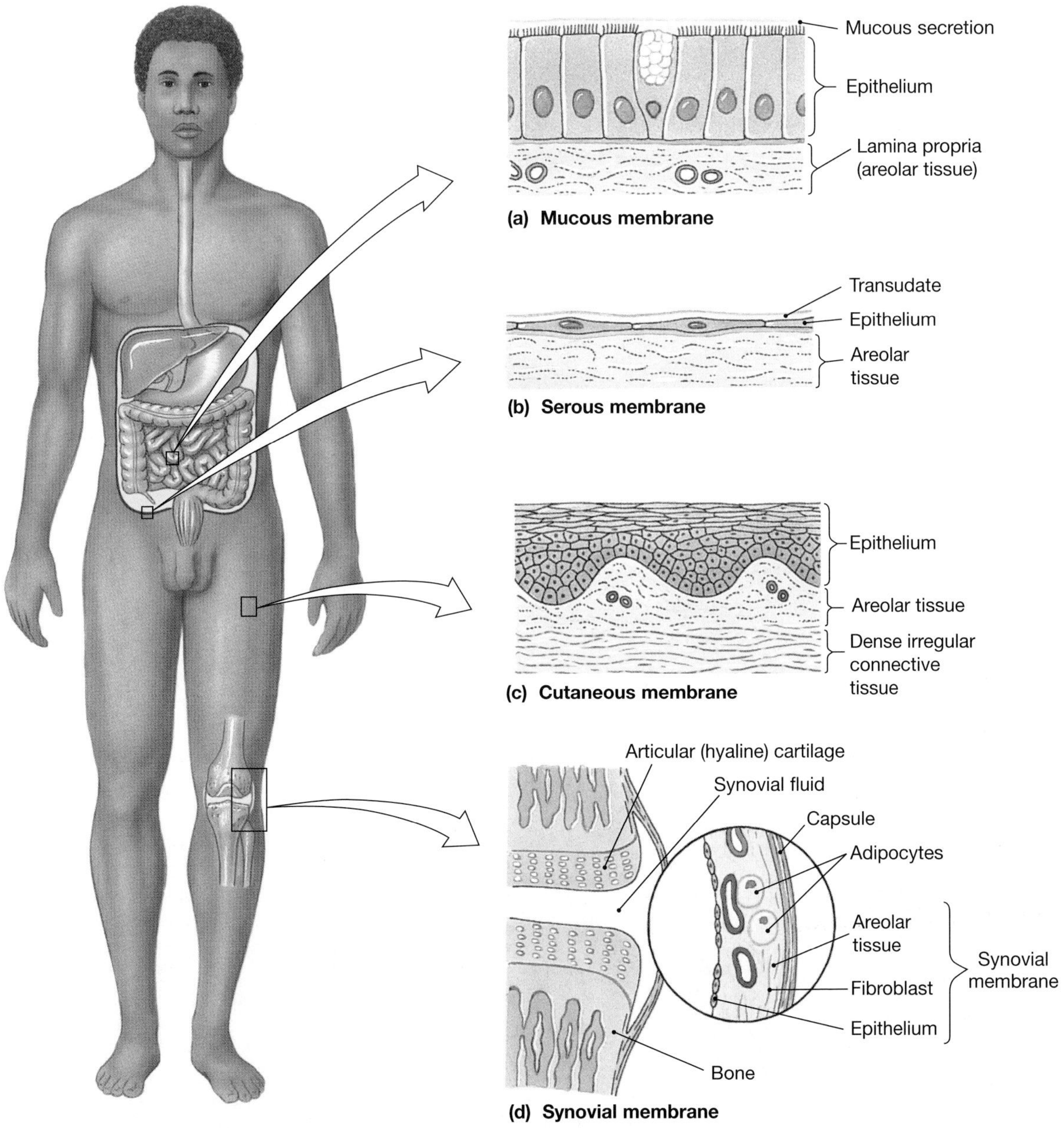

FIGURE 3.19 **MEMBRANES**

Membranes are composed of epithelia and connective tissues, which act to cover and protect other tissues and structures. **(a) Mucous membranes** are coated with the secretions of mucous glands. Mucous membranes line most of the digestive and respiratory tracts and portions of the urinary and reproductive tracts. **(b) Serous membranes** line the ventral body cavities (the peritoneal, pleural, and pericardial cavities). **(c)** The **cutaneous membrane**, the skin, covers the outer surface of the body. **(d) Synovial membranes** line joint cavities and produce the fluid within the joint.

areolar tissue (Figure 3.19b●). These membranes were introduced in Chapter 1: (1) The *pleura* lines the pleural cavities and covers the lungs; (2) the *peritoneum* lines the peritoneal cavity and covers the surfaces of the enclosed organs; and (3) the *pericardium* lines the pericardial cavity and covers the heart. *p. 76* Serous membranes are very thin, and they are firmly attached to the body wall and to the organs they cover. When you are looking at an organ, such as the heart or stomach, you are really seeing the tissues of the organ through a transparent serous membrane.

The parietal and visceral portions of a serous membrane are in close contact at all times. Minimizing friction between these opposing surfaces is the primary function of serous membranes. Because the mesothelia are very thin, serous membranes are relatively permeable, and tissue fluids diffuse onto the exposed surface, keeping it moist and slippery.

The fluid formed on the surfaces of a serous membrane is called a **transudate** (TRANS-ū-dāt; *trans-*, across). Specific transudates are called *pleural fluid, peritoneal fluid*, or *pericardial fluid*, depending on their source. In normal healthy individuals, the total volume of transudate at any given time is extremely small, just enough to prevent friction between the walls of the cavities and the surfaces of internal organs. But after an injury or in certain disease states, the volume of transudate may increase dramatically, complicating existing medical problems or producing new ones. *Problems with Serous Membranes p. 785*

The Cutaneous Membrane [FIGURE 3.19c]

The **cutaneous membrane**, or the skin, covers the surface of the body. It consists of a keratinized stratified squamous epithelium and an underlying layer of aerolar tissue reinforced by a layer of dense connective tissue (Figure 3.19c●). In contrast to serous or mucous membranes, the cutaneous membrane is thick, relatively waterproof, and usually dry. (The skin is discussed in detail in Chapter 4.)

Synovial Membranes [FIGURE 3.19d]

A **synovial membrane** (sin-Ō-vē-al) consists of extensive areas of areolar tissue bounded by an incomplete superficial layer of squamous or cuboidal cells (Figure 3.19d●). Bones contact one another at joints, or *articulations.* Joints that permit significant movement are surrounded by a fibrous capsule and contain a joint cavity lined by a synovial membrane. Although usually called an epithelium, it develops within connective tissue and differs from other epithelia in three respects: (1) there is no basal lamina or reticular lamina, (2) the cellular layer is incomplete, with gaps between adjacent cells, and (3) the "epithelial cells" are derived from macrophages and fibroblasts of the adjacent connective tissue. Some of the lining cells are phagocytic and others are secretory. The phagocytic cells remove cell debris or pathogens that could disrupt joint function. The secretory cells regulate the composition of the **synovial fluid** within the joint cavity. The synovial fluid lubricates the cartilages in the joint, distributes oxygen and nutrients, and cushions shocks at the joint.

The Connective Tissue Framework of the Body [FIGURE 3.20]

Connective tissues create the internal framework of the body. Layers of connective tissue connect the organs within the dorsal and ventral body cavities with the rest of the body. These layers (1) provide strength and stability, (2) maintain the relative positions of internal organs, and (3) provide a route for the distribution of blood vessels, lymphatics, and nerves. The connective tissue layers and wrappings can be divided into three major components: the superficial fascia, the deep fascia, and the subserous fascia. The functional anatomy of these layers is illustrated in Figure 3.20●.

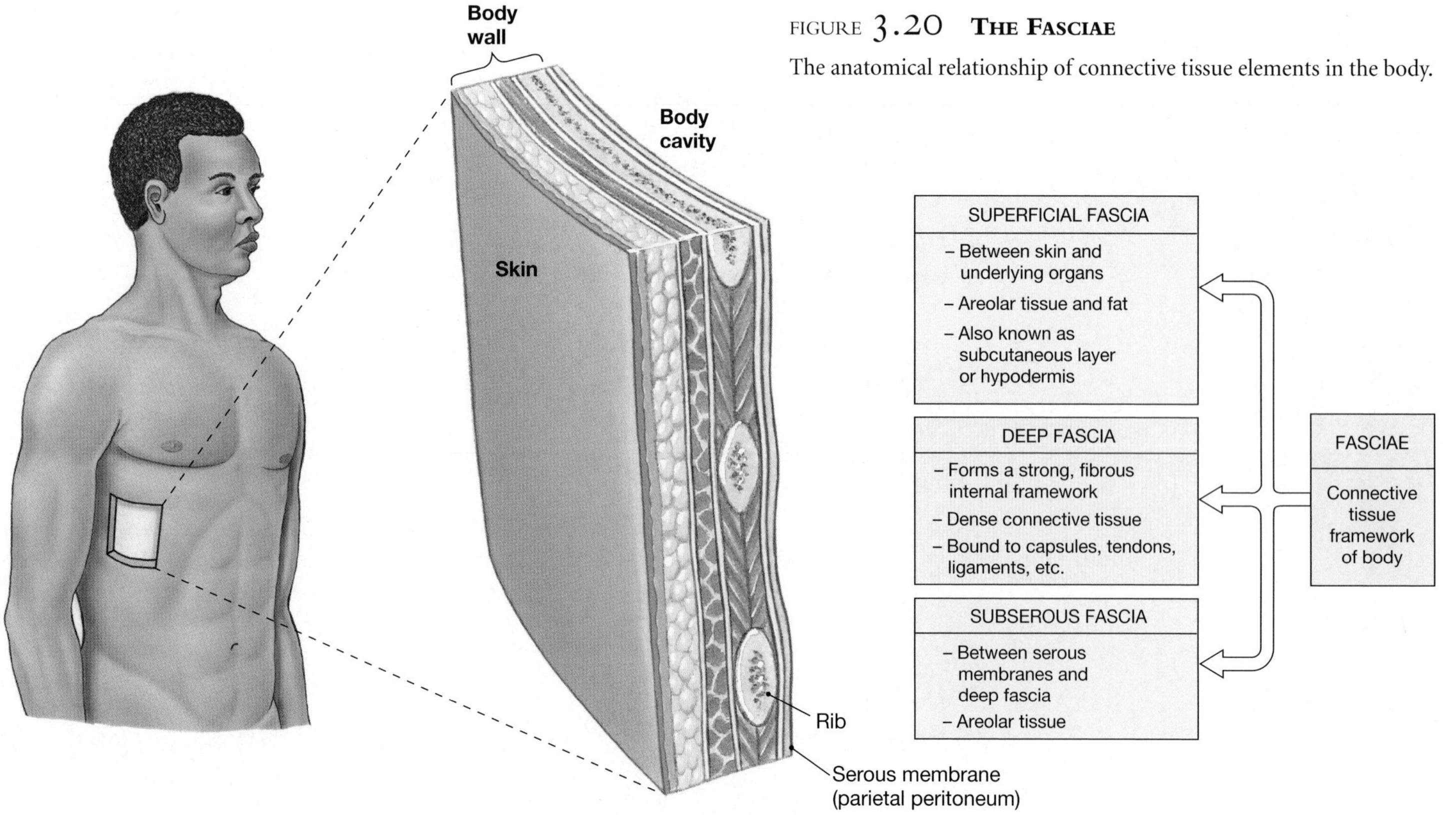

FIGURE 3.20 **The Fasciae**
The anatomical relationship of connective tissue elements in the body.

- The **superficial fascia**, or **subcutaneous layer** (*sub*, below + cutis, skin) is also termed the *hypodermis* (*hypo*, below + *derma*, skin). This layer of loose connective tissue separates the skin from underlying tissues and organs. It provides insulation and padding and lets the skin or underlying structures move independently.
- The **deep fascia** consists of dense connective tissue. The fiber organization resembles that of plywood: All the fibers in an individual layer run in the same direction, but the orientation of the fibers changes from one layer to another. This variation helps the tissue resist forces applied from many different directions. The tough *capsules* that surround most organs, including the kidneys and the organs in the thoracic and peritoneal cavities, are bound to the deep fascia. The perichondrium around cartilages, the periosteum around bones, and the connective tissue sheaths of muscle are also connected to the deep fascia. The dense connective tissue components are interwoven; for example, the deep fascia around a muscle blends into the tendon, whose fibers intermingle with those of the periosteum. This arrangement creates a strong, fibrous network for the body and ties structural elements together.
- The **subserous fascia** is a layer of loose connective tissue that lies between the deep fascia and the serous membranes that line body cavities. Because this layer separates the serous membranes from the deep fascia, movements of muscles or muscular organs do not severely distort the delicate lining.

✓ CONCEPT CHECK

- What type of membrane lines passageways of the respiratory or digestive system? Why is this type of membrane suited to these areas?
- Provide another name for the superficial fascia. What does it do?
- You are asked to locate the pericardium. What type of membrane is this, and where would you find it?
- What are the functions of the cutaneous membrane?

Muscle Tissue [FIGURE 3.21]

Muscle tissue is specialized for contraction (Figure 3.21●). Muscle cells possess organelles and properties distinct from those of other cells. They are capable of powerful contractions that shorten the cell along its longitudinal axis. Because they are different from "typical" cells, the term **sarcoplasm** is used to refer to the cytoplasm of a muscle cell, and **sarcolemma** is used to refer to the cell membrane.

Three types of muscle tissue are found in the body: (1) *skeletal*, (2) *cardiac*, and (3) *smooth*. The contraction mechanism is similar in all three, but they differ in their internal organization. We will describe each muscle type in greater detail in later chapters (skeletal muscle in Chapter 9, cardiac muscle in Chapter 21, and smooth muscle in Chapter 25). This discussion will focus on general characteristics rather than specific details.

Skeletal Muscle Tissue [FIGURE 3.21a]

Skeletal muscle tissue contains very large muscle cells. Because individual skeletal muscle cells are relatively long and slender, they are usually called **muscle fibers**. Skeletal muscle fibers are very unusual because they may be a foot (0.3 meter) or more in length, and each cell is **multinucleated**, containing hundreds of nuclei. Nuclei lie just under the surface of the **sarcolemma** (Figure 3.21a●). Skeletal muscle fibers are incapable of dividing, but new muscle fibers can be produced through the division of **satellite cells**, mesenchymal cells that persist in adult skeletal muscle tissue. As a result, skeletal muscle tissue can at least partially repair itself after an injury.

Skeletal muscle fibers contain *actin* and *myosin* filaments arranged in parallel within organized functional groups. As a result, skeletal muscle fibers appear to have a banded, or *striated*, appearance (Figure 3.21a●). Normally, skeletal muscle fibers will not contract unless stimulated by nerves, and the nervous system provides voluntary control over their activities. Thus, skeletal muscle is called **striated voluntary muscle**.

Skeletal muscle tissue is bound together by areolar connective tissue. The collagen and elastic fibers surrounding each cell and group of cells blend into those of a tendon or aponeurosis that conducts the force of contraction, usually to a bone of the skeleton. When the muscle tissue contracts, it pulls on the bone, and the bone moves.

Cardiac Muscle Tissue [FIGURE 3.21b]

Cardiac muscle tissue is found only in the heart. A typical cardiac muscle cell, or **cardiocyte**, is smaller than a skeletal muscle fiber, and it usually has one centrally placed nucleus. The prominent striations, seen in Figure 3.21b●, resemble those of skeletal muscle. Cardiac muscle cells form extensive connections with one another; these connections occur at specialized regions known as **intercalated discs**. As a result, cardiac muscle tissue consists of a branching network of interconnected muscle cells. The interconnections help channel the forces of contraction, and gap junctions at the intercalated discs help coordinate the activities of individual cardiac muscle cells. Like skeletal muscle fibers, cardiac muscle cells are incapable of dividing, and because this tissue lacks satellite cells, cardiac muscle tissue damaged by injury or disease cannot regenerate.

Cardiac muscle cells do not rely on nerve activity to start a contraction. Instead, specialized cardiac muscle cells called **pacemaker cells** establish a regular rate of contraction. Although the nervous system can alter the rate of pacemaker activity, it does not provide voluntary control over individual cardiac muscle cells. Therefore, cardiac muscle is called **striated involuntary muscle**.

Smooth Muscle Tissue [FIGURE 3.21c]

Smooth muscle tissue can be found in the walls of blood vessels; around hollow organs such as the urinary bladder; and in layers around the respiratory, circulatory, digestive, and reproductive tracts. A smooth muscle cell is a small cell with tapering ends, containing a single oval nucleus (Figure 3.21c●). Smooth muscle cells can divide, and smooth muscle tissue can regenerate after an injury. The actin and myosin filaments in smooth muscle cells are organized differently from those of skeletal and cardiac muscle, and as a result there are no striations; it is the only *nonstriated* muscle tissue. Smooth muscle cells usually contract on their own, through the action of *pacesetter cells*. Although smooth muscle contractions may be triggered by neural activity, the nervous system does not usually provide voluntary control over those contractions. Consequently, smooth muscle is called **nonstriated involuntary muscle**.

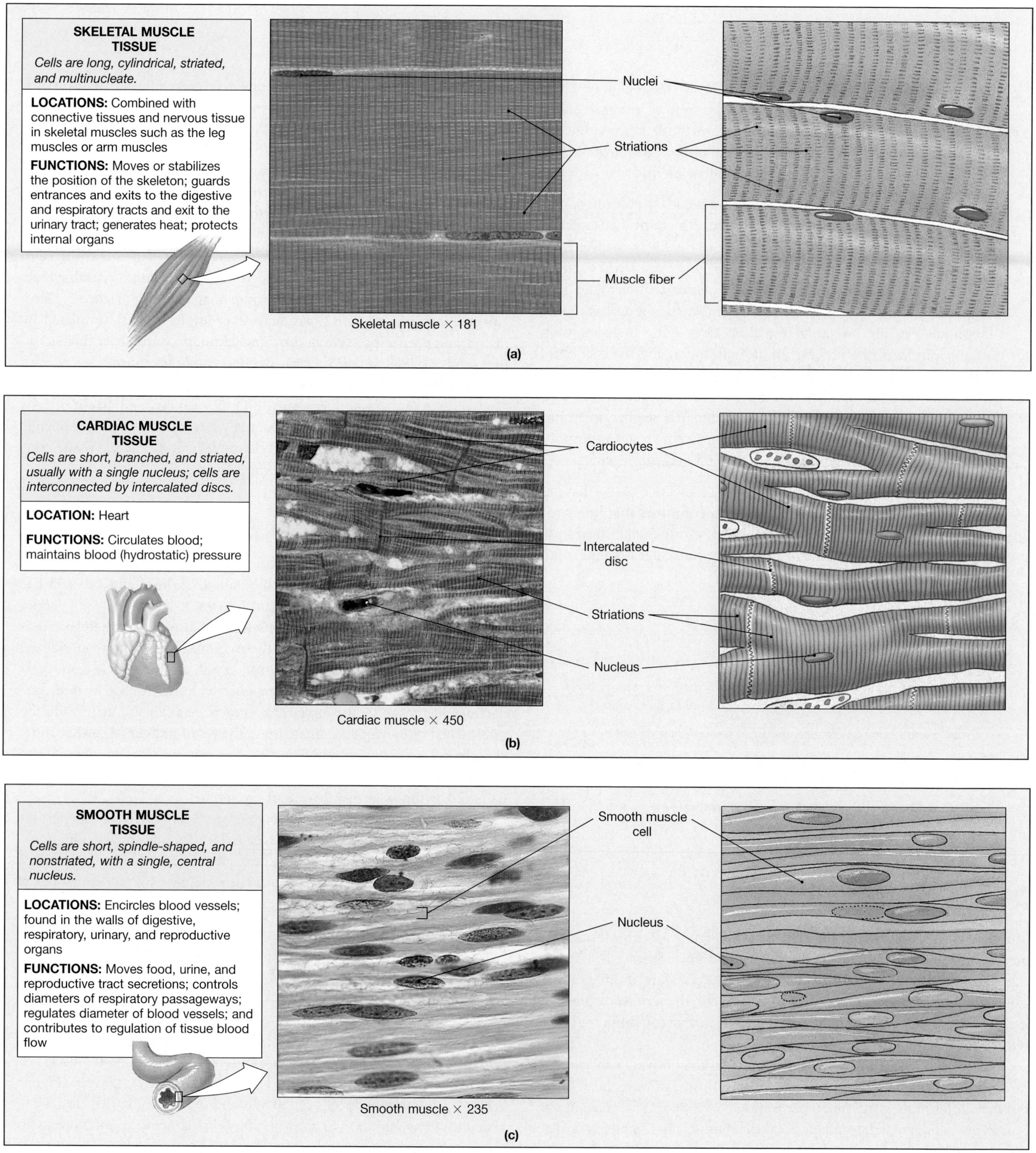

SKELETAL MUSCLE TISSUE
Cells are long, cylindrical, striated, and multinucleate.
LOCATIONS: Combined with connective tissues and nervous tissue in skeletal muscles such as the leg muscles or arm muscles
FUNCTIONS: Moves or stabilizes the position of the skeleton; guards entrances and exits to the digestive and respiratory tracts and exit to the urinary tract; generates heat; protects internal organs
Nuclei
Striations
Muscle fiber
Skeletal muscle × 181
(a)
CARDIAC MUSCLE TISSUE
Cells are short, branched, and striated, usually with a single nucleus; cells are interconnected by intercalated discs.
LOCATION: Heart
FUNCTIONS: Circulates blood; maintains blood (hydrostatic) pressure
Cardiocytes
Intercalated disc
Striations
Nucleus
Cardiac muscle × 450
(b)
SMOOTH MUSCLE TISSUE
Cells are short, spindle-shaped, and nonstriated, with a single, central nucleus.
LOCATIONS: Encircles blood vessels; found in the walls of digestive, respiratory, urinary, and reproductive organs
FUNCTIONS: Moves food, urine, and reproductive tract secretions; controls diameters of respiratory passageways; regulates diameter of blood vessels; and contributes to regulation of tissue blood flow
Smooth muscle cell
Nucleus
Smooth muscle × 235
(c)

FIGURE 3.21 **Muscle Tissue**

(a) Skeletal Muscle Fibers. Note the large fiber size, prominent banding pattern, multiple nuclei, and unbranched arrangement. **(b) Cardiac Muscle Cells**. Cardiac muscle cells differ from skeletal muscle fibers in three major ways: size (cardiac muscle cells are smaller), organization (cardiac muscle cells branch), and number of nuclei (a typical cardiac muscle cell has one centrally placed nucleus). Both contain actin and myosin filaments in an organized array that produces the striations seen in both types of muscle cell. **(c) Smooth Muscle Cells**. Smooth muscle cells are small and spindle-shaped, with a central nucleus. They do not branch, and there are no striations.

Neural Tissue [FIGURE 3.22]

Neural tissue, also known as **nervous tissue** or *nerve tissue*, is specialized for the conduction of electrical impulses from one region of the body to another. Most of the neural tissue in the body (roughly 96%) is concentrated in the brain and spinal cord, the control centers for the nervous system. Neural tissue contains two basic types of cells: **neurons** (NOO-rons; *neuro*, nerve), or *nerve cells*, and several different kinds of supporting cells, collectively called **neuroglia** (noo-ROG-lē-a; *glia*, glue). Neurons transmit electrical impulses along their cell membranes. All of the functions of the nervous system involve changes in the pattern and frequency of the impulses carried by individual neurons. Neuroglia have varied functions, such as providing a supporting framework for neural tissue, regulating the composition of the interstitial fluid, and providing nutrients to neurons.

Neurons are the longest cells in the body, many reaching a meter in length. Most neurons are incapable of dividing under normal circumstances, and they have a very limited ability to repair themselves after injury. A typical neuron has a **cell body**, or *soma*, that contains a large prominent nucleus (Figure 3.22●). Typically, the cell body is attached to several branching processes, called **dendrites** (DEN-drīts; *dendron*, tree), and a single **axon**. Dendrites receive incoming messages; axons conduct outgoing messages. It is the length of the axon that can make a neuron so long; because axons are very slender, they are also called **nerve fibers**. In Chapter 13 we will discuss the properties of neural tissue and provide additional histological and cytological details.

Tissues, Nutrition, and Aging

Tissues change with age. In general, repair and maintenance activities grow less efficient, and a combination of hormonal changes and alterations in lifestyle affect the structure and chemical composition of many tissues. Epithelia get thinner, and connective tissues more fragile. Individuals bruise easily and bones become brittle; joint pains and broken bones are common complaints. Because cardiac muscle cells and neurons cannot be replaced, over time, cumulative losses from relatively minor damage can contribute to major health problems such as cardiovascular disease or deterioration in mental function.

In future chapters we will consider the effects of aging on specific organs and systems. Some of these changes are genetically programmed. For example, the chondrocytes of older individuals produce a slightly different form of proteoglycan than those of younger people. The difference probably accounts for the observed changes in the thickness and resilience of cartilage. In other cases the tissue degeneration may be temporarily slowed or even reversed. The age-related reduction in bone strength in women, a condition called **osteoporosis**, is often caused by a combination of inactivity, low dietary calcium levels, and a reduction in circulating estrogens (female sex hormones). A program of exercise, calcium supplements, and hormonal replacement therapies can usually maintain normal bone structure for many years.

In this chapter we have introduced the four basic types of tissue found in the human body. In combination these tissues form all of the organs and systems that will be discussed in subsequent chapters. *Tissue Structure and Disease p. 785*

✓ CONCEPT CHECK

- What type of muscle tissue has small, tapering cells with single nuclei and no obvious striations?
- Why is skeletal muscle also called striated voluntary muscle?
- Which tissue is specialized for the conduction of electrical impulses from one body region to another?

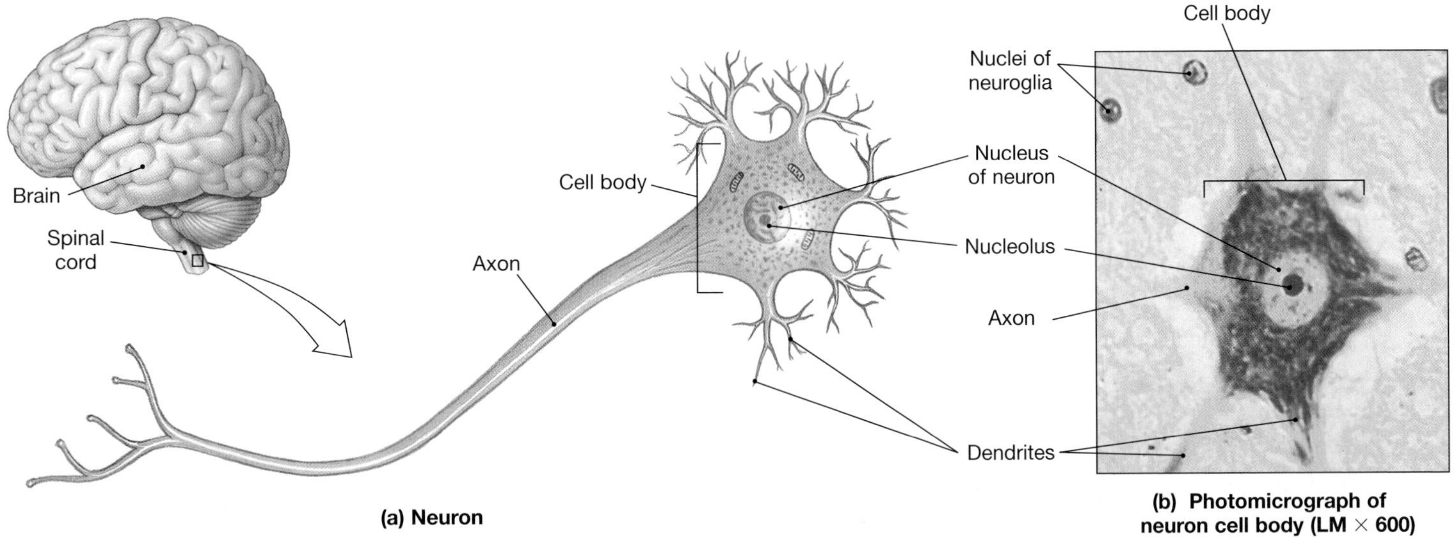

FIGURE 3.22 **NEURAL TISSUE**

Diagrammatic (**a**) and histological (**b**) views of a representative neuron. Neurons are specialized for conduction of electrical impulses over relatively long distances within the body.

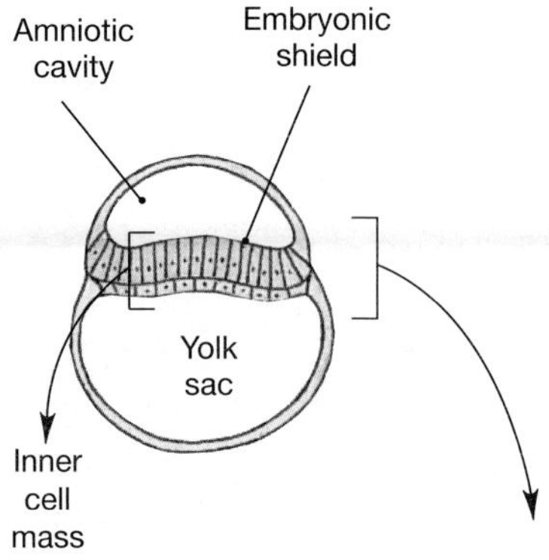

Many different organ systems show similar patterns of organization. For example, the digestive, respiratory, urinary, and reproductive systems each include passageways lined by epithelia and surrounded by layers of smooth muscle. These patterns are the result of developmental processes under way in the first 2 months of embryonic life.

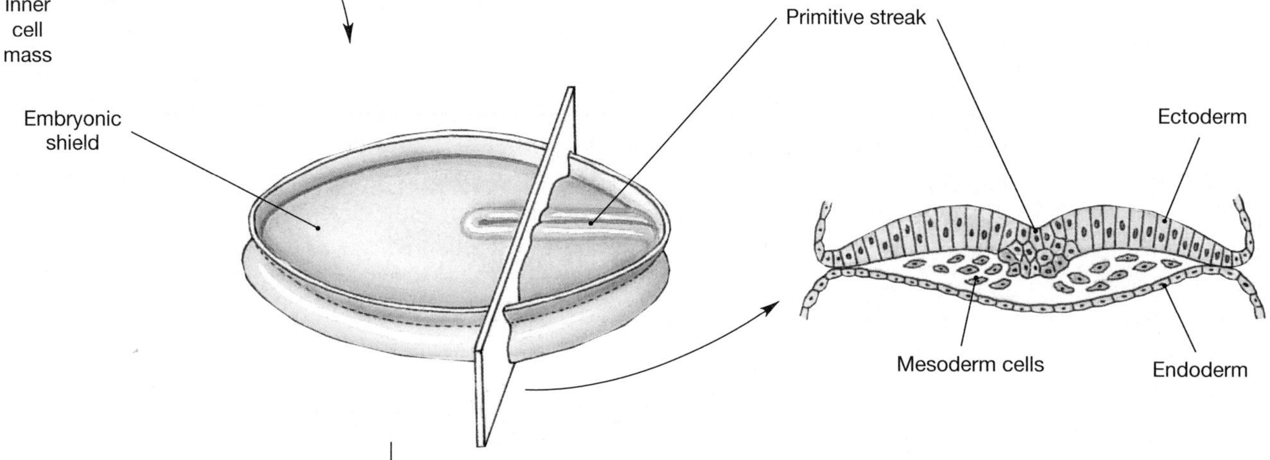

DAY 14

After roughly 2 weeks of development, the inner cell mass is only a millimeter in length. The region of embryonic development is called the **embryonic shield**. It contains a pair of epithelial layers: an upper ectoderm and an underlying endoderm. At a region called the **primitive streak**, superficial cells migrate between the two, adding to an intermediate layer of mesoderm.

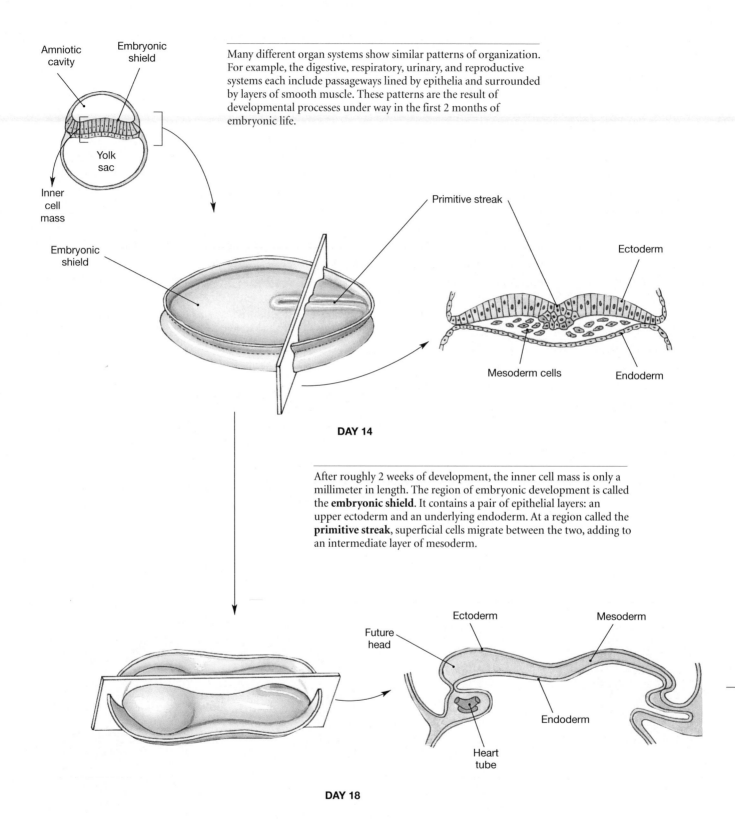

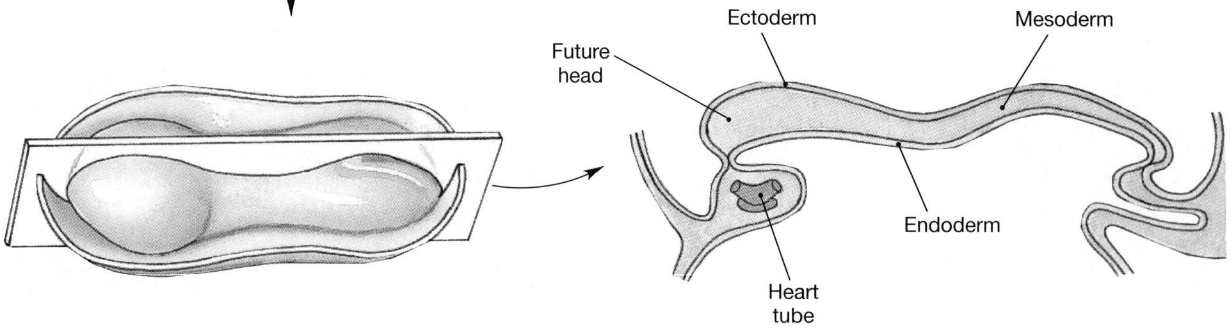

DAY 18

By day 18, the embryo has begun to lift off the surface of the embryonic shield. The heart and many blood vessels have already formed, well ahead of the other organ systems. Unless otherwise noted, discussions of organ system development in later chapters will begin at this stage.

Derivatives of Primary Germ Layers	
Ectoderm Forms:	Epidermis and epidermal derivatives of the integumentary system, including hair follicles, nails, and glands communicating with the skin surface (sweat, milk, and sebum) Lining of the mouth, salivary glands, nasal passageways, and anus Nervous system, including brain and spinal cord Portions of endocrine system (pituitary gland and parts of adrenal glands) Portions of skull, pharyngeal arches, and teeth
Mesoderm Forms:	Lining of the body cavities (pleural, pericardial, peritoneal) Muscular, skeletal, cardiovascular, and lymphatic systems Kidneys and part of the urinary tract Gonads and most of the reproductive tracts Connective tissues supporting all organ systems Portions of endocrine system (parts of adrenal glands and endocrine tissues of the reproductive tracts)
Endoderm Forms:	Most of the digestive system: epithelium (except mouth and anus), exocrine glands (except salivary glands), the liver and pancreas Most of the respiratory system: epithelium (except nasal passageways) and mucous glands Portions of urinary and reproductive systems (ducts and the stem cells that produce gametes) Portions of endocrine system (thymus, thyroid gland, parathyroid glands, and pancreas)

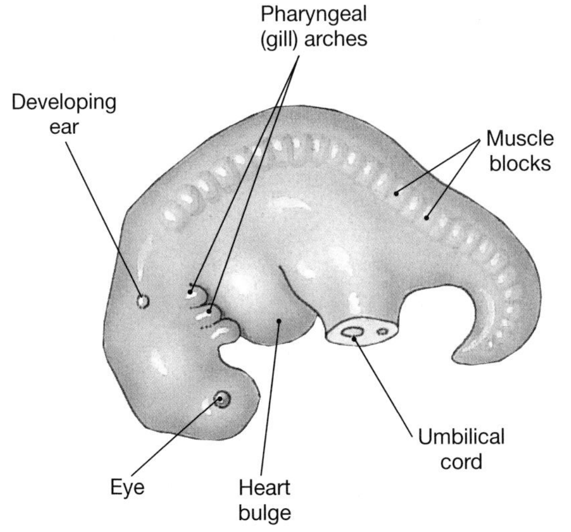

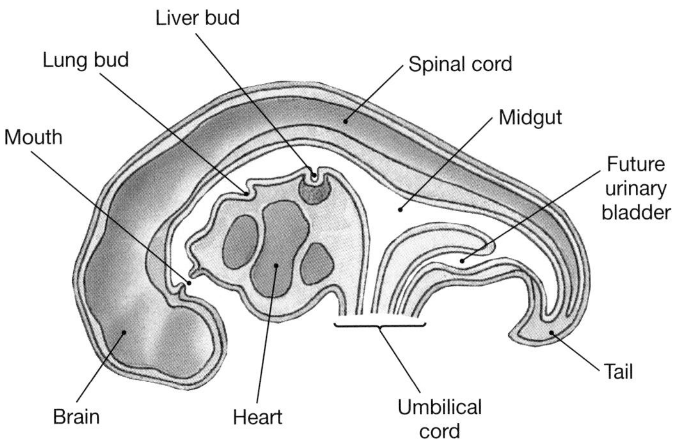

DAY 28

After 1 month, you can find the beginnings of all major organ systems. The role of each of the primary germ layers in the formation of organs is summarized in the accompanying table; details are given in later Embryology Summaries.

CLINICAL DISCUSSION

Tumor Formation and Growth

Physicians who specialize in the identification and treatment of cancers are called **oncologists** (on-KOL-ō-jists; *onkos*, mass). Pathologists and oncologists classify cancers according to their cellular appearance and their sites of origin. Over a hundred kinds have been described, but broad categories are used to indicate the usual location of the primary tumor. Table 3.3 summarizes information concerning benign and malignant tumors (cancers) associated with the tissues discussed in this chapter.

Cancer develops in a series of steps, diagrammed in Figure 3.23●. Initially the cancer cells are restricted to a single location, called the **primary tumor** or **primary neoplasm**. All of the cells in the tumor are usually the daughter cells of a single malignant cell. At first the growth of the primary tumor simply distorts the tissue, and the basic tissue organization remains intact. Metastasis begins as tumor cells "break out" of the primary tumor and invade the surrounding tissue. When this invasion is followed by penetration of nearby blood vessels, the cancer cells begin circulating throughout the body.

Responding to cues that are as yet unknown, these cells later escape from the circulatory system and establish **secondary tumors** at other sites. These tumors are extremely active metabolically, and their presence stimulates the growth of blood vessels into the area. The increased circulatory supply provides additional nutrients and further accelerates tumor growth and metastasis. Death may occur because vital organs have been compressed, because nonfunctional cancer cells have killed or replaced the normal cells in vital organs, or because the voracious cancer cells have starved normal tissues of essential nutrients. *Cancer Treatment and Statistics p. 785*

TABLE 3.3 **Benign and Malignant Tumors in the Major Tissue Types**

Tissue	Description
EPITHELIA	
Carcinomas	Any cancer of epithelial origin
Adenocarcinomas	Cancers of glandular epithelia
Angiosarcomas	Cancers of endothelial cells
Mesotheliomas	Cancers of mesothelial cells
CONNECTIVE TISSUES	
Fibromas	Benign tumors of fibroblast origin
Lipomas	Benign tumors of adipose tissue
Liposarcomas	Cancers of adipose tissue
Leukemias, lymphomas	Cancers of blood-forming tissues
Chondromas	Benign tumors in cartilage
Chondrosarcomas	Cancers of cartilage
Osteomas	Benign tumors in bone
Osteosarcomas	Cancers of bone
MUSCLE TISSUES	
Myomas	Benign muscle tumors
Myosarcomas	Cancers of skeletal muscle tissue
Cardiac sarcomas	Cancers of cardiac muscle tissue
Leiomyomas	Benign tumors of smooth muscle tissue
Leiomyosarcomas	Cancers of smooth muscle tissue
NEURAL TISSUES	
Gliomas, neuromas	Cancers of neuroglial origin

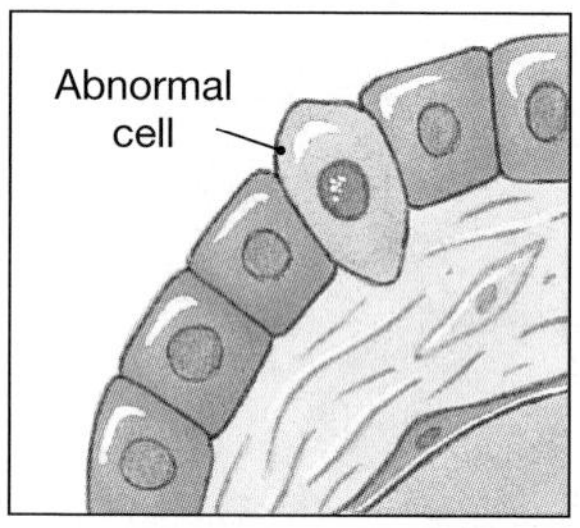

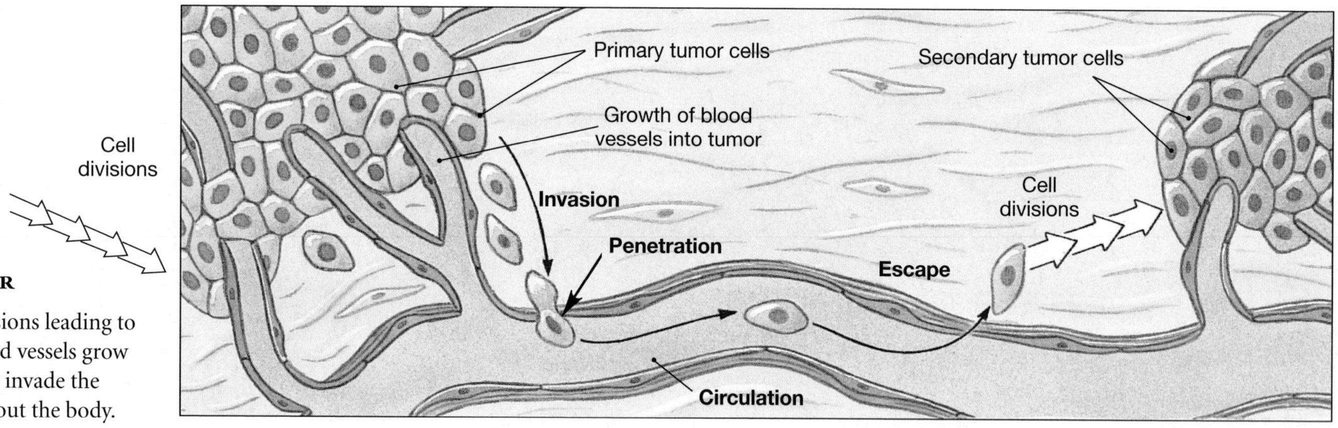

FIGURE 3.23 **The Development of Cancer**

Diagram of abnormal cell divisions leading to the formation of a tumor. Blood vessels grow into the tumor and tumor cells invade the blood vessels to travel throughout the body.

RELATED CLINICAL TERMS

adhesions: Restrictive fibrous connections that can result from surgery, infection, or other injuries to serous membranes. *Problems with Serous Membranes p. 785*

anaplasia (a-na-PLĀ-zē-a): An irreversible change in the size and shape of tissue cells. *Tissue Structure and Disease p. 785*

ascites (a-SĪ-tēz): An accumulation of peritoneal fluid that creates a characteristic abdominal swelling. *Problems with Serous Membranes p. 785*

chemotherapy: Administering drugs that either kill cancerous tissues or prevent mitotic divisions. *Cancer Treatment and Statistics p. 785*

dysplasia (dis-PLĀ-zē-a): A change in the normal shape, size, and organization of tissue cells. *Tissue Structure and Disease p. 785*

effusion: The accumulation of fluid in body cavities. *Problems with Serous Membranes p. 785*

immunotherapy: Administering drugs that help the immune system recognize and attack cancer cells. *Cancer Treatment and Statistics p. 785*

liposuction: A surgical procedure to remove unwanted adipose tissue by sucking it through a tube. *Liposuction p. 785*

metaplasia (me-ta-PLĀ-zē-a): A structural change that alters the character of a tissue. *Tissue Structure and Disease p. 785*

oncologists (on-KOL-ō-jists): Physicians who specialize in identifying and treating cancers. *p. 82*

pathologists (pa-THOL-ō-jists): Physicians who specialize in the diagnosis of diseases, primarily from an examination of body fluids, tissue samples, and other anatomical clues. *Tissue Structure and Disease p. 785*

pericarditis: An inflammation of the pericardium. *Problems with Serous Membranes p. 785*

peritonitis: An inflammation of the peritoneum. *Problems with Serous Membranes p. 785*

pleuritis (pleurisy): An inflammation of the lining of the pleural cavities. *Problems with Serous Membranes p. 785*

primary tumor (primary neoplasm): The site at which a cancer initially develops. *p. 82*

remission: A stage in which a tumor stops growing or grows smaller; the goal of cancer treatment. *Cancer Treatment and Statistics p. 785*

secondary tumor: A colony of cancerous cells formed by metastasis, the spread of cells from a primary tumor. *p. 82*

Additional Clinical Terms Discussed in Appendix 1 (pp. 784–786)

brown fat; cancer incidence; cancer survival rates; liposuction; pericardial effusion; pleural effusion; pleural rub; remission

STUDY OUTLINE & CHAPTER REVIEW

Introduction 52

1. **Tissues** are collections of specialized cells and cell products that are organized to perform a relatively limited number of functions. There are four **primary tissue types**: *epithelia, connective tissue, muscle tissue*, and *neural tissue*. *(see Figure 3.1)*
2. **Histology** is the study of tissues.

Epithelial Tissue 52

1. **Epithelial tissues** include *epithelia*, which cover surfaces, and *glands*, which are secretory structures derived from epithelia. An **epithelium** is an **avascular** layer of cells that forms a surface, lining, or covering. Epithelia consist mainly of tightly bound cells, rather than extracellular materials. (*see Figures 3.2* to *3.9*)
2. Epithelia cells are being replaced continually through stem cell activity.

Functions of Epithelial Tissue 52

3. Epithelia provide physical protection, control permeability, provide sensation, and produce specialized secretions. **Gland cells** are epithelial cells (or derived from them) that produce secretions.

Specializations of Epithelial Cells 54

4. Epithelial cells are specialized to both maintain the physical integrity of the epithelium and perform secretory or transport functions.
5. Epithelia may show **polarity** from the *basal* to *apical* surface; cells connect neighbor cells on their *lateral surfaces*; some epithelial cells have microvilli on their apical surfaces. There are often structural and functional differences between the apical surface and the basolateral surfaces of individual epithelial cells. *(see Figure 3.2)*
6. The coordinated beating of the cilia on a **ciliated epithelium** moves materials across the epithelial surface. *(see Figure 3.2)*

Maintaining the Integrity of the Epithelium 55

7. The inner surface of each epithelium interfaces with a two-part **basement membrane** consisting of a **basal lamina**, produced by the epithelium, and a **reticular lamina**, produced by the connective tissue on which the epithelium rests. In areas exposed to extreme chemical or mechanical stresses, divisions by **germinative cells** replace the short-lived epithelial cells. *(see Figure 3.3a)*

Classification of Epithelia 56

8. Epithelia are classified both on the basis of the number of cell layers in the epithelium and the shape of the exposed cells at the surface of the epithelium. *(see Figures 3.4 to 3.6)*
9. A **simple epithelium** has a single layer of cells covering the basement membrane. A **stratified epithelium** has several layers. In a **squamous epithelium** the surface cells are thin and flat; in a **cuboidal epithelium** the cells resemble short hexagonal boxes; in a **columnar epithelium** the cells are also hexagonal, but they are relatively tall and slender. A **transitional epithelium** is characterized by a mixture of what appears to be both cuboidal and squamous cells arranged to permit stretching. **Pseudostratified columnar epithelium** contains columnar cells, some of which possess cilia and goblet (secreting) cells that appear stratified, but are not. *(see Figures 3.4 to 3.6)*

Glandular Epithelia 57

10. Glands may be classified by the type of secretion produced, the structure of the gland or by their mode of secretion. *(see Figures 3.7 to 3.9)*
11. **Exocrine** secretions are discharged through **ducts** onto the skin or an epithelial surface that communicates with the exterior; **endocrine** secretions, known as **hormones**, are released by gland cells into the interstitial fluid surrounding the cell.
12. Exocrine glands may be classified as **serous** (producing a watery solution usually containing enzymes), **mucous** (producing a viscous, sticky mucus), or **mixed** (producing both types of secretions).
13. In epithelia that contain scattered gland cells, the individual secretory cells are called **unicellular glands**. **Multicellular glands** are glandular epithelia or aggregations of gland cells that produce exocrine or endocrine secretions. *(see Figures 3.7/3.8)*
14. A glandular, epithelial cell may release its secretions through merocrine, apocrine, or holocrine mechanism. *(see Figure 3.9)*
15. In **merocrine secretion**, the most common method of secretion, the product is released by exocytosis. **Apocrine secretion** involves the loss of both secretory product and some cytoplasm. Unlike the other two methods, **holocrine secretion** destroys the cell, which had become packed with secretory product before bursting. *(see Figure 3.9)*

Connective Tissues 62

1. All connective tissues have three components: specialized cells, extracellular **protein fibers**, and **ground substance**. The combination of protein fibers and ground substance forms the **matrix** of the tissue.
2. Whereas epithelia consist almost entirely of cells, the extracellular matrix accounts for most of the volume of a connective tissue.
3. **Connective tissue** is an internal tissue with many important functions. These include: establishing a structural framework; transporting fluids and dissolved materials; protecting delicate organs; supporting, surrounding, and interconnecting tissues; storing energy reserves; and defending the body from microorganisms.

Classification of Connective Tissues 64

4. **Connective tissue proper** refers to all connective tissue that contain varied cell populations and fiber types suspended in a viscous ground substance. *(see Figure 3.10)*

5. **Fluid connective tissues** have a distinctive population of cells suspended in a watery ground substance containing dissolved proteins. *Blood* and *lymph* are examples of fluid connective tissues. *(see Figure 3.10)*
6. **Supporting connective tissues** have a less diverse cell population than connective tissue proper. Additionally, they have a dense matrix that contains closely packed fibers. The two types of supporting connective tissues are *cartilage* and *bone. (see Figure 3.10)*

Connective Tissue Proper 64

7. Connective tissue proper is composed of extracellular fibers, a viscous ground substance, and two categories of cells: **fixed cells** and **wandering cells.** *(see Figure 3.11 and Table 3.1)*
8. There are three types of fibers in connective tissue: **collagen fibers, reticular fibers,** and **elastic fibers.** *(see Figures 3.11/3.13/3.14)*
9. All connective tissues are derived from embryonic **mesenchyme.** *(see Figure 3.12)*
10. Connective tissue proper includes **loose** and **dense connective tissues.** There are three types of loose connective tissues: **areolar tissue, adipose tissue,** and **reticular tissue.** Most of the volume of loose connective tissue is ground substance, a viscous fluid that cushions shocks. Most of the volume in dense connective tissue consists of extracellular protein fibers. There are two types of dense connective tissue: **dense regular connective tissue,** in which fibers are parallel and aligned along lines of stress, and **dense irregular connective tissue,** in which fibers form an interwoven meshwork. *(see Figures 3.13/3.14)*

Fluid Connective Tissues 68

11. **Blood** and **lymph** are examples of fluid connective tissues, each with a distinctive collection of cells in a watery matrix *(see Figure 3.15).* Both blood and lymph contain cells and many different types of dissolved proteins that do not form insoluble fibers under normal conditions.
12. Extracellular fluid includes the **plasma** of blood, the **interstitial fluid** within other connective tissues and other tissue types, and lymph. Lymph, which is confined to vessels of the lymphatic system.

Supporting Connective Tissues 70

13. Cartilage and bone are called **supporting connective tissues** because they support the rest of the body. A tough **perichondrium** covers the surface of most cartilages. *(see Figures 3.16 to 3.18)*
14. The matrix of **cartilage** is a firm gel that contains **chondroitin sulfates.** It is produced by immature cells called **chondroblasts,** and maintained by mature cells called **chondrocytes.** A fibrous covering called the **perichondrium** separates cartilage from surrounding tissues. Cartilage grows by two different mechanisms, **appositional growth** (growth at the surface) and **interstitial growth** (growth from within). *(see Figure 3.16)*
15. There are three types of cartilage: **hyaline cartilage, elastic cartilage,** and **fibrocartilage.** *(see Figure 3.17 and Table 3.2)*
16. **Bone (osseous tissue)** has a matrix consisting of collagen fibers and calcium salts, giving it unique properties. *(see Figure 3.18)*
17. **Osteocytes** in **lacunae** depend on diffusion through intercellular connections or **canaliculi** for nutrient intake. *(see Figure 3.18 and Table 3.2)*
18. All bone surfaces except those inside joint cavities are covered by a **periosteum** that has fibrous and cellular layers. The periosteum assists in attaching the bone to surrounding tissues, tendons, and ligaments, and it participates in the repair of bone after an injury.

Membranes 75

1. Membranes form a barrier or interface. Epithelia and connective tissues combine to form membranes that cover and protect other structures and tissues. There are four types of membranes: *mucous, serous, cutaneous,* and *synovial. (see Figure 3.19)*

Mucous Membranes 75

2. **Mucous membranes** line passageways that communicate with the exterior, such as the digestive and respiratory tracts. These surfaces are usually moistened by mucous secretions. They contain areolar tissue called the **lamina propria.** *(see Figure 3.19a)*

Serous Membranes 75

3. **Serous membranes** line internal cavities and are delicate, moist, and very permeable. Examples include peritoneal, pericardial, and pleural membranes. Each serous membrane forms a fluid call a **transudate.** *(see Figure 3.19b)*

The Cutaneous Membrane 76

4. The **cutaneous membrane,** or **skin** covers the body surface. Unlike other membranes, it is relatively thick, waterproof, and usually dry. *(see Figure 3.19c)*

Synovial Membranes 76

5. The **synovial membrane,** located within the cavity of synovial joints, produces **synovial fluid** that fills joint cavities. Synovial fluid helps lubricate the joint and promotes smooth movement in joints such as the knee. *(see Figure 3.19d)*

The Connective Tissue Framework of the Body 76

1. All organ systems are interconnected by a network of connective tissue proper that includes the **superficial fascia** (the **subcutaneous layer** or **hypodermis,** separating the skin from underlying tissues and organs), the **deep fascia** (dense connective tissue), and the **subserous fascia** (the layer between the deep fascia and the serous membranes that line body cavities). *(see Figure 3.20)*

Muscle Tissue 77

1. **Muscle tissue** consists primarily of cells that are specialized for contraction. There are three different types of muscle tissue: *skeletal muscle, cardiac muscle,* and *smooth muscle. (see Figure 3.21)*

Skeletal Muscle Tissue 77

2. **Skeletal muscle tissue** contains very large cylindrical **muscle fibers** interconnected by collagen and elastic fibers. Skeletal muscle fibers have striations due to the organization of their contractile proteins. Because we can control the contraction of skeletal muscle fibers through the nervous system, skeletal muscle is classified as **striated voluntary muscle.** New muscle fibers are produced by the division of **satellite cells.** *(see Figure 3.21a)*

Cardiac Muscle Tissue 77

3. **Cardiac muscle tissue** is found only in the heart. It is composed of unicellular, branched short cells called **cardiocytes.** The nervous system does not provide voluntary control over cardiac muscle cells. Thus, cardiac muscle is classified as **striated involuntary muscle.** *(see Figure 3.21b)*

Smooth Muscle Tissue 77

4. **Smooth muscle tissue** is composed of short, tapered, single nuclei–containing cells. It is found in the walls of blood vessels, around hollow organs, and in layers around various tracts. It is classified as **nonstriated involuntary muscle.** Smooth muscle cells can divide and therefore regenerate after injury. *(see Figure 3.21c)*

Neural Tissue 79

1. **Neural tissue** or **nervous tissue** (*nerve tissue*) is specialized to conduct electrical impulses from one area of the body to another.
2. Neural tissue consists of two cell types, neurons and neuroglia. **Neurons** transmit information as electrical impulses. There are different kinds of **neuroglia,** and among their other functions these cells provide a supporting framework for neural tissue and play a role in providing nutrients to neurons. *(see Figure 3.22)*
3. Neurons have a **cell body,** or **soma,** that contains a large prominent nucleus. Various branching processes termed **dendrites** and a single **axon** or **nerve fiber** extend from the cell body. Dendrites receive incoming messages; axons conduct messages toward other cells. *(see Figure 3.22)*

Tissues, Nutrition, and Aging 79

1. Tissues change with age. Repair and maintenance grow less efficient, and the structure and chemical composition of many tissues are altered.

LEVEL 1 REVIEWING FACTS AND TERMS

Match each numbered item with the most closely related lettered item. Use letters for answers in the spaces provided.

Column A	Column B
_____ 1. skeletal muscle	a. all epithelia
_____ 2. mast cell	b. single cell layer
_____ 3. avascular	c. urinary bladder
_____ 4. transitional	d. cell destroyed
_____ 5. goblet cell	e. connective tissue component
_____ 6. collagen	f. unicellular, exocrine gland
_____ 7. cartilage	g. tendon
_____ 8. simple epithelium	h. wandering cell
_____ 9. ground substance	i. lacunae
_____10. holocrine secretion	j. striated

11. Tissue that is specialized for the conduction of electrical impulses is called
    (a) muscle tissue (b) neural tissue
    (c) areolar tissue (d) osseous tissue
12. Which of the following refers to the dense connective tissue that forms the capsules that surround many organs?
    (a) superficial fascia (b) hypodermis
    (c) deep fascia (d) subserous fascia
13. The reduction of friction between the parietal and visceral surfaces of an internal cavity is the function of
    (a) cutaneous membranes (b) mucous membranes
    (c) serous membranes (d) synovial membranes
14. Which of the following is not a correct statement about simple epithelia?
    (a) afford little mechanical protection
    (b) characteristic of regions where secretion or absorption occurs
    (c) line internal compartments and passageways
    (d) cover surfaces subjected to mechanical or chemical stress
15. Functions of connective tissue include each of the following, except
    (a) establishing a structural framework for the body
    (b) transporting fluids and dissolved materials
    (c) storing energy reserves
    (d) provides sensations
16. Satellite cells would be found in association with
    (a) skeletal muscle (b) smooth muscle
    (c) cardiac muscle (d) neural tissue
17. Tissue changes with age include
    (a) decreased ability to repair
    (b) less efficient tissue maintenance
    (c) thinner epithelia
    (d) all of the above
18. The most common type of cartilage is
    (a) ligamentous cartilage
    (b) hyaline cartilage
    (c) elastic cartilage
    (d) fibrocartilage
19. An epithelium is connected to underlying connective tissue by
    (a) a basement membrane
    (b) canaliculi
    (c) stereocilia
    (d) proteoglycans
20. The basic shapes of epithelial cells include all but which of the following?
    (a) stratified (b) squamous
    (c) cuboidal (d) columnar

LEVEL 2 REVIEWING CONCEPTS

How does the role of a tissue in the body differ from that of a single cell?

1. A layer of glycoproteins and a network of fine protein filaments that perform limited functions, together act as a barrier that restricts the movement of proteins and other large molecules from the connective tissue to epithelium. This describes the structure and function of
   (a) interfacial canals (b) the reticular lamina
   (c) the basal lamina (d) areolar tissue
2. Connective tissue cells that respond to injury or infection by dividing to produce daughter cells that differentiate into other cell types are
   (a) mast cells (b) fibroblasts
   (c) plasma cells (d) mesenchymal cells
3. What are the main structural differences between bone and cartilage?
4. What is the relationship between the thickness of the epithelium and its function at a specific location in the body?
5. What is the difference between exocrine and endocrine secretions?
6. What is the significance of the cilia on the respiratory epithelium?
7. In what regions of the body would you expect to find dense, irregular connective tissue, and why would this tissue be located in these regions?
8. A significant structural feature in the digestive system is the presence of tight junctions near the exposed surfaces of cells lining the digestive tract. Why are these junctions so important?
9. What are germinative cells, and what is their function?

LEVEL 3 CRITICAL THINKING AND CLINICAL APPLICATIONS

1. Analysis of a glandular secretion indicates that it contains some DNA, RNA, and membrane components such as phospholipids. What kind of secretion is this and why?
2. During a laboratory examination, a student examines a tissue section that is composed of many parallel, densely packed protein fibers. There are no nuclei or striations, and there is no evidence of other cellular structures. The student identifies the tissue as skeletal muscle. Why is the student's choice wrong, and what tissue is being observed?
3. What type of epithelium would you expect to find lining the alveoli (air sacs) in the lungs?
4. Develop a two-step scheme that can be used to identify the three different types of muscle tissue.

✓ ANSWERS TO CONCEPT CHECK QUESTIONS

p. 55 **1.** There are four primary tissue types: epithelial tissue, connective tissue, muscle tissue, and neural tissue. **2.** Characteristics of epithelia include: cellularity, polarity, attachment, avascularity, and regeneration. **3.** Specializations of epithelial cells include: production of secretions, movement of fluids over the surface, and movement of fluids through the epithelium itself.
p. 61 **1.** No. A simple squamous epithelium does not provide enough protection against infection, abrasion, and dehydration, and is not found in the skin surface. **2.** In holocrine secretion, the entire gland cell is destroyed during the secretory process. Further secretion by the gland requires regeneration of cells to replace those lost during secretion. **3.** During apocrine secretion, the secretory product and the apical portion of the cell cytoplasm are shed. **4.** A simple columnar epithelium provides protection and may be involved in both absorption and secretion.
p. 75 **1.** The three basic components of connective tissue include: specialized cells, extracellular protein fibers, and a fluid called ground substance. **2.** Connective tissue proper refers to connective tissues with many types of cells and extracellular fibers in a syrupy ground substance. Supporting connective tissue (cartilage and bone) have a less diverse cell population and a matrix that contains closely packed fibers. The matrix is either gel-like (cartilage) or calcified (bone). **3.** There are two types in connective tissue proper. Fixed cells include: fibroblasts, fixed macrophages, adipocytes, mesenchymal cells, and sometimes melanocytes. Wandering cells include: free macrophages, mast cells, lymphocytes, plasma cells, and microphages. **4.** Collagen fibers add strength to connective tissue. We would expect vitamin C deficiency to result in the production of connective tissue that is weaker and more prone to damage.
p. 77 **1.** Mucous membranes line passageways that communicate with the exterior. They form a barrier that resists the entry of pathogens; they must remain moist at all times. **2.** Another name for superficial fascia is subcutaneous layer or hypodermis. It separates the skin from underlying tissues and organs. It provides insulation and padding and lets the skin or underlying structures move independently. **3.** The pericardium is a serous membrane that lines the pericardial cavity and covers the heart. **4.** The cutaneous membrane of the skin covers the surface of the body. It consists of a stratified squamous epithelium and an underlying layer of areolar tissue reinforced by a layer of dense connective tissue.
p. 79 **1.** Since both cardiac and skeletal muscle cells are striated (banded), this must be smooth muscle. **2.** Skeletal muscle cells have a banded or striated appearance because of the organization of actin and myosin filaments within the cells. **3.** Neural tissue.

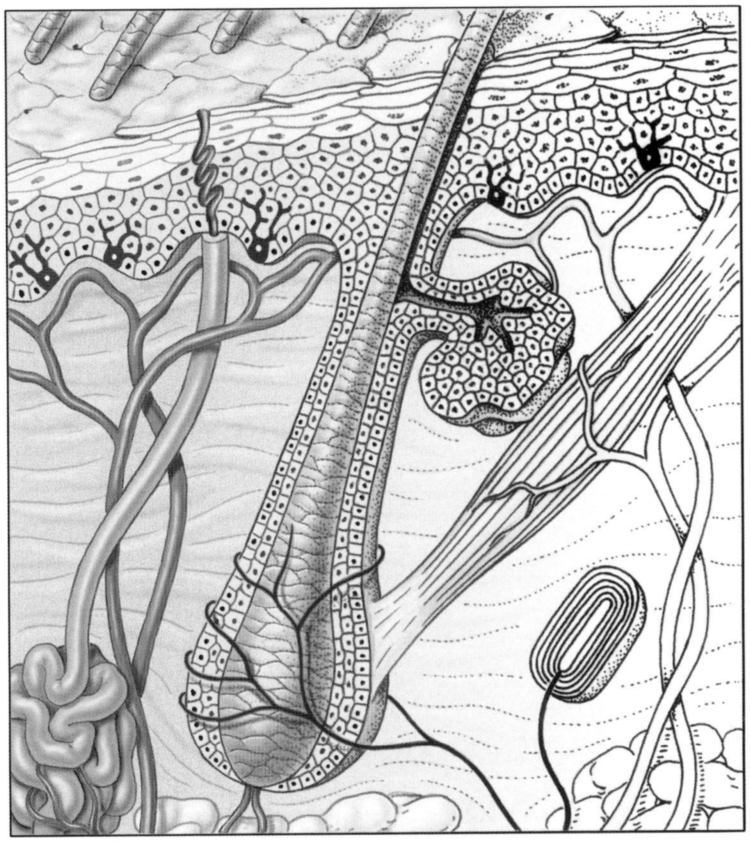

4

THE INTEGUMENTARY SYSTEM

The **integumentary system**, or *integument*, is probably the most closely watched yet underappreciated organ system. It is the only system we see every day, almost in its entirety. Because others see this system as well, we devote a lot of time to improving the appearance of the integument and associated structures. Washing the face, brushing and trimming hair, taking showers, and applying makeup are activities that modify the appearance or properties of the integument. Most people use the general appearance of the skin to estimate the overall health and age of a new acquaintance—healthy skin has a smooth sheen, and young skin has few wrinkles. The skin also gives clues to your emotional state, as when you blush with embarrassment or flush with rage.

When something goes wrong with the skin, the effects are immediately apparent. Even a relatively minor condition or blemish will be noticed at once, whereas more serious problems in other systems are often ignored. (That's probably why TV advertising devotes so much time to the control of minor acne, a temporary skin condition that is publicly displayed, rather than to the control of blood pressure, a potentially fatal cardiovascular problem that is easier to ignore.) The skin also mirrors the general health of other systems, and clinicians can use the appearance of the skin to detect signs of underlying disease. For example, the skin color changes from the presence of liver disease.

The skin has more than a cosmetic role, however. It protects you from the surrounding environment; its receptors tell you a lot about the outside world; and it helps to regulate your body temperature. You will encounter several more important functions as we examine the functional anatomy of the integumentary system in this chapter.

Integumentary Structure and Function

[FIGURE 4.1]

The integument covers the entire body surface, including the anterior surfaces of the eyes and the tympanic membranes (eardrums) at the ends of the external auditory canals. At the nostrils, lips, anus, urethral opening, and vaginal opening the integument turns inward, meeting the mucous membranes lining the respiratory, digestive, urinary, and reproductive tracts, respectively. At these sites the transition is seamless, and the epithelial defenses remain intact and functional.

All four tissue types are found within the integument. An epithelium covers its surface, and underlying connective tissues provide strength and resiliency. Blood vessels within the connective tissue nourish the epidermal cells. Smooth muscle tissue within the integument controls the diameters of the blood vessels and adjusts the positions of the hairs that project above the body surface. Neural tissue controls these smooth muscles and monitors sensory receptors providing sensations of touch, pressure, temperature, and pain.

The integument has numerous functions, including: physical protection, regulation of body temperature, excretion (secretion), nutrition (synthesis), sensation, and immune defense. Figure 4.1● shows the functional

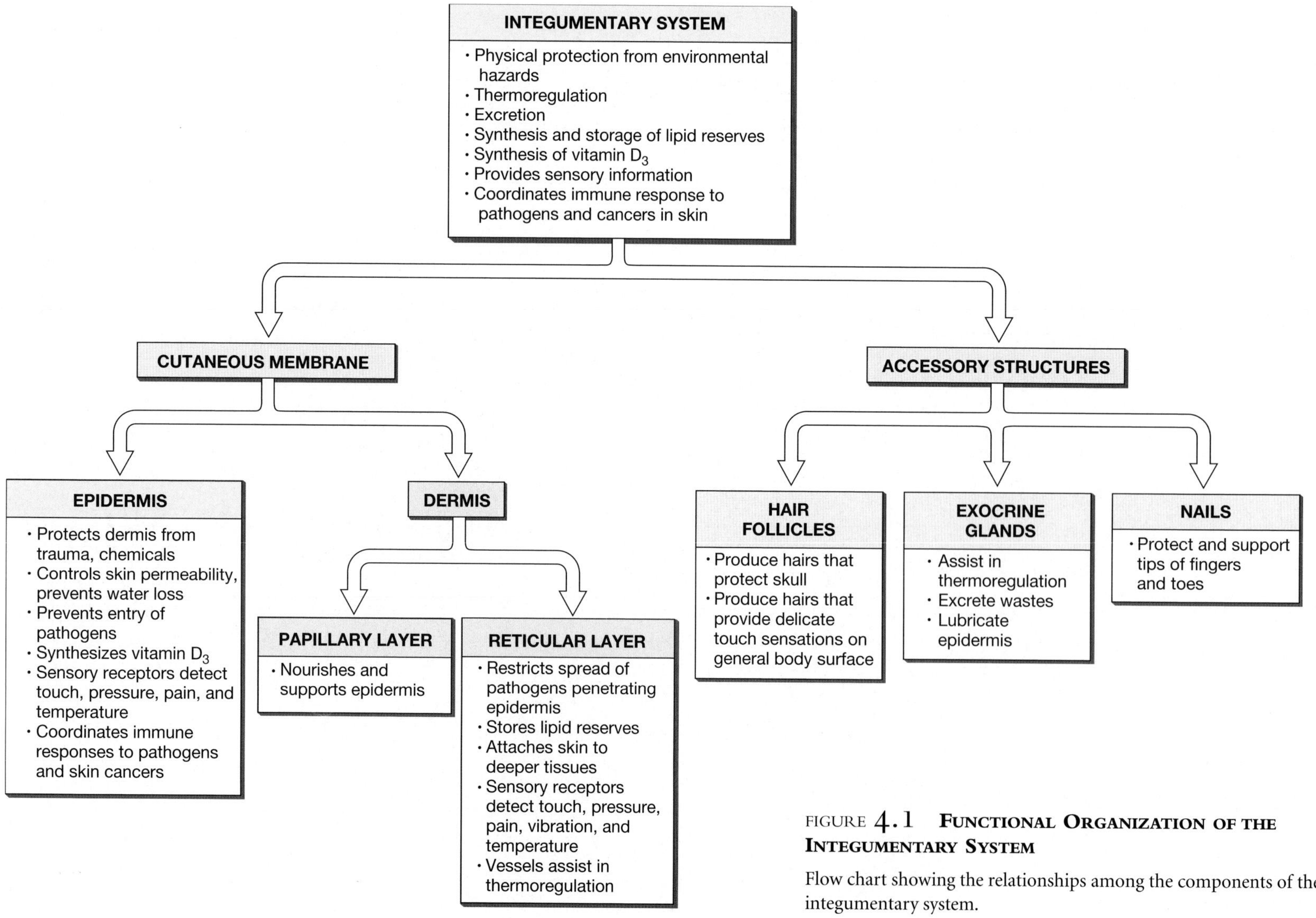

FIGURE 4.1 **Functional Organization of the Integumentary System**

Flow chart showing the relationships among the components of the integumentary system.

organization of the integumentary system. The integument has two major components, the *cutaneous membrane* and the *accessory structures.*

1. The **cutaneous membrane**, or skin, has two components, the superficial epithelium, termed the **epidermis** (*epi-*, above + *derma*, skin) and the underlying connective tissues of the **dermis**. Deep to the dermis, the loose connective tissue of the subcutaneous layer, also known as the superficial fascia, or *hypodermis*, separates the integument from the deep fascia around other organs, such as muscles or bones. p. 77 Although it is not usually considered to be a part of the integument, we will consider the subcutaneous layer in this chapter because of its extensive interconnections with the dermis.
2. The **accessory structures** include hair, nails, and a variety of multicellular exocrine glands. These structures are located in the dermis and protrude through the epidermis to the surface.

The Epidermis [FIGURE 4.2]

The **epidermis** consists of a stratified squamous epithelium, as seen in Figure 4.2•. There are four cell types in the epidermis: *keratinocytes, melanocytes, Merkel cells,* and *Langerhans cells.* The most abundant epithelial cells, the **keratinocytes** (ke-RAT-in-ō-sīts), form several different layers. The precise boundaries between them are often difficult to see in a light micrograph. In *thick skin,* found on the palms of the hands and soles of the feet, five layers can be distinguished. Only four layers can be distinguished in the *thin skin* that covers the rest of the body. Melanocytes are pigment-producing cells in the epidermis. Merkel cells have a role in detecting sensation, whereas Langerhans cells are phagocytic cells. All of these cell types are scattered among keratinocytes.

Layers of the Epidermis [FIGURE 4.3 AND TABLE 4.1]

Refer to Figure 4.3 and Table 4.1• as we describe the layers in a section of thick skin. Beginning at the basement membrane and traveling toward the outer epithelial surface, we find the *stratum germinativum,* the *stratum spinosum,* the *stratum granulosum,* the *stratum lucidum,* and the *stratum corneum.*

STRATUM GERMINATIVUM

The innermost epidermal layer is the **stratum germinativum** (STRĀ-tum jer-mi-na-TĒ-vum), or *stratum basale.* This layer is firmly attached to the basal lamina of the basement membrane that separates the epidermis from the loose connective tissue of the adjacent dermis. Large stem cells, or *basal cells,* dominate the stratum germinativum. The divisions of stem cells replace the more superficial keratinocytes that are lost or shed at the epithelial surface. The brown tones of the skin result from the synthetic activities

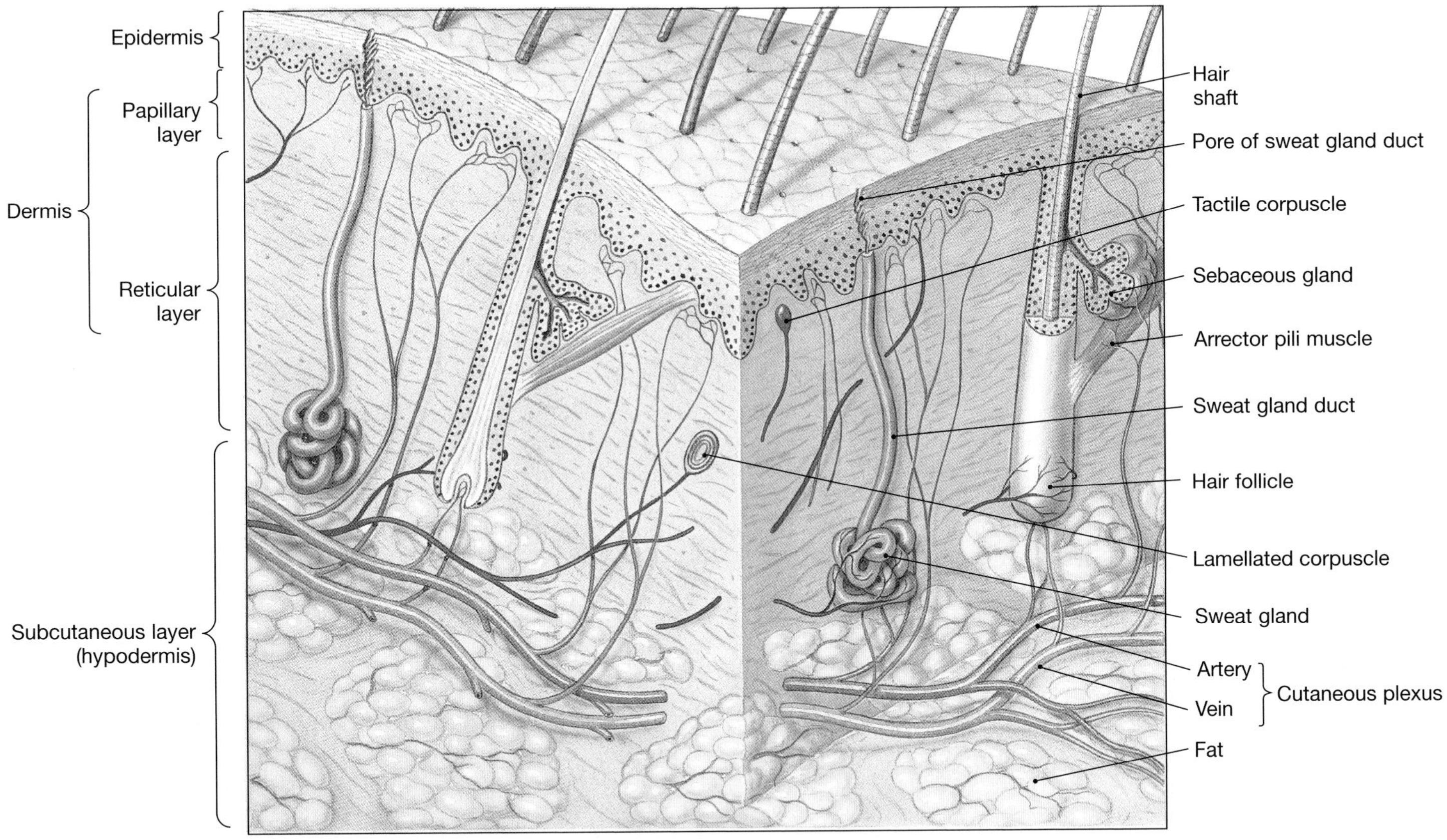

FIGURE 4.2 **COMPONENTS OF THE INTEGUMENTARY SYSTEM**

Relationships among the major components of the integumentary system (with the exception of nails, shown in Figure 4.15). The epidermis is a keratinized stratified squamous epithelium that overlies the dermis, a connective tissue region containing glands, hair follicles, and sensory receptors. Underlying the dermis is the subcutaneous layer, which contains fat and blood vessels supplying the dermis.

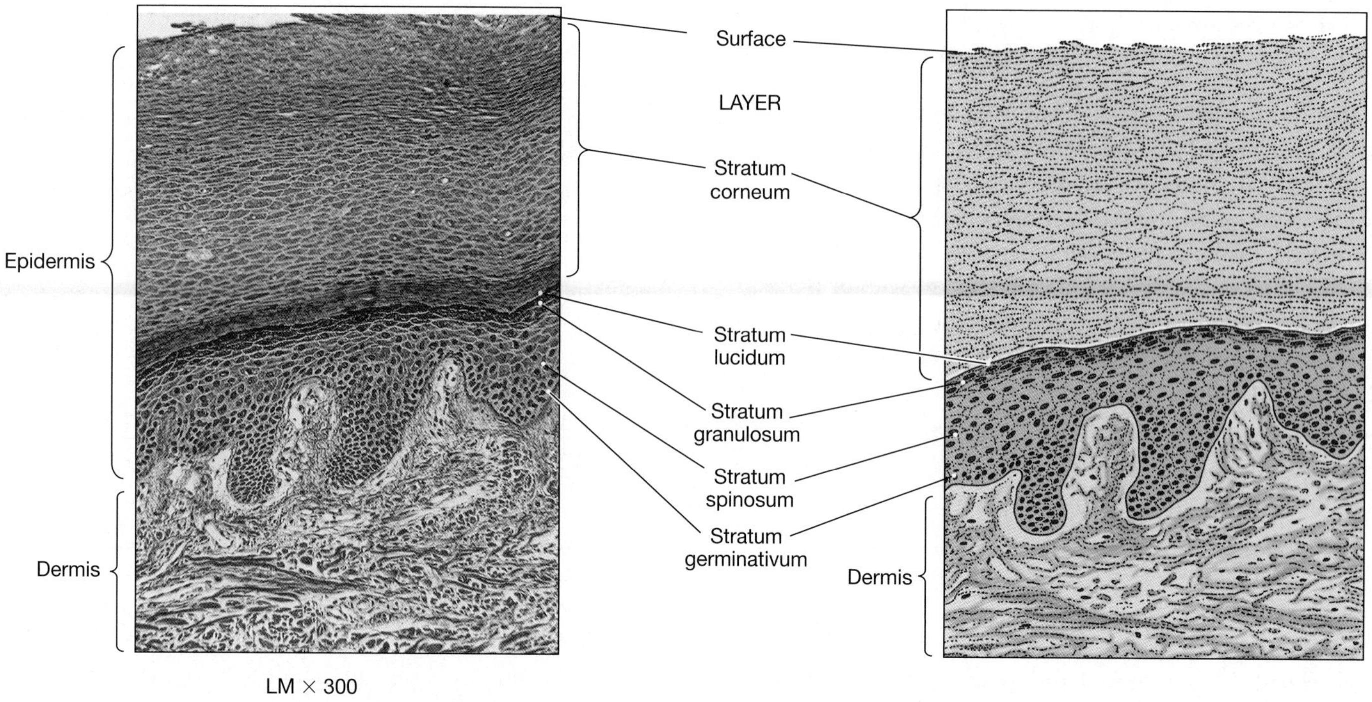

FIGURE 4.3 **The Structure of the Epidermis**
A light micrograph and corresponding diagrammatic sketch through a portion of the epidermis, showing the major stratified layers of epidermal cells.

of *melanocytes*, pigment cells introduced in Chapter 3. p. 64 Melanocytes are scattered among the basal cells of the stratum germinativum. They have numerous cytoplasmic processes that inject *melanin*, a black, yellow-brown, or brown pigment, into the keratinocytes in this layer and in more superficial layers. The ratio of melanocytes to germinative cells ranges between 1:4 and 1:20, depending on the region examined. They are most abundant in the cheeks and forehead, in the nipples, and in the genital region. Individual and racial differences in skin color result from different levels of melanocyte activity, not different numbers of melanocytes. Even *albino* individuals have normal numbers of melanocytes. (Albinism is an inherited condition in which melanocytes are incapable of producing melanin; it affects approximately 1 person in 10,000.)

Skin surfaces that lack hair contain specialized epithelial cells known as **Merkel cells**. These cells are found among the deepest cells of the stratum germinativum. They are sensitive to touch, and when compressed, Merkel cells release chemicals that stimulate sensory nerve endings, providing information about objects touching the skin. (There are many other kinds of touch receptors, but they are located in the dermis and will be introduced in later sections. All of the integumentary receptors are described in Chapter 18.)

TABLE 4.1 **Epidermal Layers**

Layer	Characteristics
Stratum Germinativum	Innermost, basal layer Attached to basement membrane Contains epidermal stem cells, melanocytes, and Merkel cells
Stratum Spinosum	Keratinocytes are bound together by desmosomes attached to tonofibrils of the cytoskeleton Some keratinocytes divide in this layer Langerhans cells and melanocytes are often present
Stratum Granulosum	Keratinocytes produce keratohyalin and keratin Keratin fibers develop as cells become thinner and flatter Gradually the cell membranes thicken, the organelles disintegrate, and the cells die
Stratum Lucidum	Appears as a "glassy" layer in thick skin only
Stratum Corneum	Multiple layers of flattened, dead, interlocking keratinocytes Typically relatively dry Water-resistant, but not waterproof Permits a slow water loss by insensible perspiration

STRATUM SPINOSUM

Each time a stem cell divides, one of the daughter cells is pushed superficial to the stratum germinativum into the next layer, the **stratum spinosum** ("spiny layer"), where it begins to differentiate into a keratinocyte. The stratum spinosum is several cells thick. Each keratinocyte in the stratum spinosum contains bundles of protein filaments that extend from one side of the cell to the other. These bundles, called *tonofibrils*, begin and end at desmosomes that connect the keratinocyte to its neighbors. The tonofibrils thus act as cross-braces, strengthening and supporting the cell junctions. All of the keratinocytes in the stratum spinosum are tied together by this network of interlocked desmosomes and tonofibrils. Standard histological procedures, used to prepare tissue for microscopic examination, shrink the cytoplasm but leave the tonofibrils and desmosomes intact. This makes the cells look like miniature pincushions, and that's why early histologists used the term "spiny layer" in their descriptions.

Some of the cells entering this layer from the stratum germinativum continue to divide, further increasing the thickness of the epithelium. Melanocytes are common in this layer, as are **Langerhans cells**, although the latter cells cannot be distinguished in standard histological preparations. Langerhans cells, which account for 3–8% of the cells in the epidermis, are most common in the superficial portion of the stratum spinosum. These cells play an important role in initiating an immune response against (1) pathogens that have penetrated the superficial layers of the epidermis and (2) epidermal cancer cells.

STRATUM GRANULOSUM

The layer of cells superficial to the stratum spinosum is the **stratum granulosum** ("granular layer"). The stratum granulosum consists of keratinocytes displaced from the stratum spinosum. By the time cells reach this layer, they have begun to manufacture large quantities of the proteins **keratohyalin** (ker-a-tō-HĪ-a-lin) and **keratin** (KER-a-tin; *keros*, horn). Keratohyalin accumulates in electron-dense granules called *keratohyalin granules*. These granules form an intracellular matrix that surrounds the keratin filaments. As large keratin filaments are developing within them, the keratinocytes gradually become thinner and flatter. The cell membranes thicken and become less permeable. The nuclei and other organelles then disintegrate, the cells die, and their subsequent dehydration creates a tightly interlocked layer of keratin fibers surrounded by keratohyalin and sandwiched between phospholipid membranes.

In humans, keratin forms the basic structural component of hair and nails. It's a very versatile material, however, and in other vertebrates it forms the claws of dogs and cats, the horns of cattle and rhinos, the feathers of birds, the scales of snakes, the baleen of whales, and a variety of other interesting epidermal structures.

STRATUM LUCIDUM

In the thick skin of the palms and soles, a glassy **stratum lucidum** ("clear layer") covers the stratum granulosum. The cells in this layer are flattened, densely packed, and filled with keratin, but they do not stain well in standard histological preparations.

STRATUM CORNEUM

The **stratum corneum** (KOR-nē-um; *cornu*, horn) is found at the surface of both thick and thin skin. It consists of 15–30 layers of flattened, dead, and interlocking cells. Because the interconnections established in the stratum spinosum remain intact, the cells are usually shed in large groups or sheets, rather than individually.

An epithelium containing large amounts of keratin is said to be **keratinized** (KER-a-tin-īz'd), or *cornified* (KOR-ni-fīd; *cornu*, horn + *facere*, to make). Normally the stratum corneum is relatively dry, which makes the surface unsuitable for the growth of many microorganisms. Maintenance of this barrier involves coating the surface with the secretions of integumentary glands (sebaceous and sweat glands, discussed in a later section). The process of **keratinization** (*cornification*) occurs everywhere on exposed skin surfaces except over the anterior surfaces of the eyes.

Although the stratum corneum is water-resistant, it is not waterproof, and water from the interstitial fluids slowly penetrates the surface, to be evaporated into the surrounding air. This process, called insensible perspiration, accounts for a loss of roughly 500 ml (about 1 pt) of water per day. † *Psoriasis and Xerosis p. 786*

It takes 15–30 days for a cell to move from the stratum germinativum to the stratum corneum. The dead cells usually remain in the exposed stratum corneum layer for an additional 2 weeks before they are shed or washed away. Thus the deeper portions of the epithelium—and all underlying tissues—are always protected by a barrier composed of dead, durable, and expendable cells.

✓ CONCEPT CHECK

- Excessive shedding of cells from the outer layer of skin in the scalp causes dandruff. What is the name of this layer of skin?
- As you pick up a piece of lumber, a splinter pierces the palm of your hand and lodges in the third layer of the epidermis. Identify this layer.
- What are the two major subdivisions of the integumentary system, and what are the components of each subdivision?
- What is keratinization? What are the stages of this process?

CLINICAL BRIEF

TRANSDERMAL DRUG ADMINISTRATION

Drugs in oils or other lipid-soluble carriers can penetrate the epidermis. The movement is slow, particularly through the layers of cell membranes in the stratum corneum, but once a drug reaches the underlying tissues, it will be absorbed into the circulation. A useful technique involves placing a sticky patch containing a drug over an area of thin skin. To overcome the relatively slow rate of diffusion, the patch must contain an extremely high concentration of the drug. This procedure, called *transdermal drug administration*, has the advantage that a single patch may work for several days, making daily pills unnecessary. *Scopolamine*, a drug that affects the nervous system, is administered transdermally to control the nausea associated with motion sickness. Transdermal *nitroglycerin* can be used to improve blood flow within heart muscle and prevent a heart attack. Transdermal *estrogen* can be used to reduce bone loss in postmenopausal women. Transdermal nicotine can be used to control the craving for tobacco and make it easier to quit smoking.

Thick and Thin Skin [FIGURE 4.4]

In descriptions of the skin, the terms *thick* and *thin* refer to the relative thickness of the epidermis, not to the integument as a whole. Most of the body is covered by **thin skin**. In a sample of thin skin, only four layers are present because the stratum lucidum is typically absent. Here the epidermis is a mere 0.08 mm thick, and the stratum corneum is only a few cell layers deep (Figure 4.4a,b●). **Thick skin** on the palms of the hands may be covered by 30 or more layers of keratinized cells. As a result, the epidermis in these locations exhibit all five layers and may be as much as six times thicker than the epidermis covering the general body surface (Figure 4.4c●).

Epidermal Ridges [FIGURES 4.4/4.5]

The stratum germinativum layer of the epidermis forms **epidermal ridges** that extend into the dermis, increasing the area of contact between the two regions. Projections from the dermis toward the epidermis, called **dermal papillae** (singular *papilla*, "nipple-shaped mound") extend between adjacent ridges, as indicated in Figure 4.4a,c●.

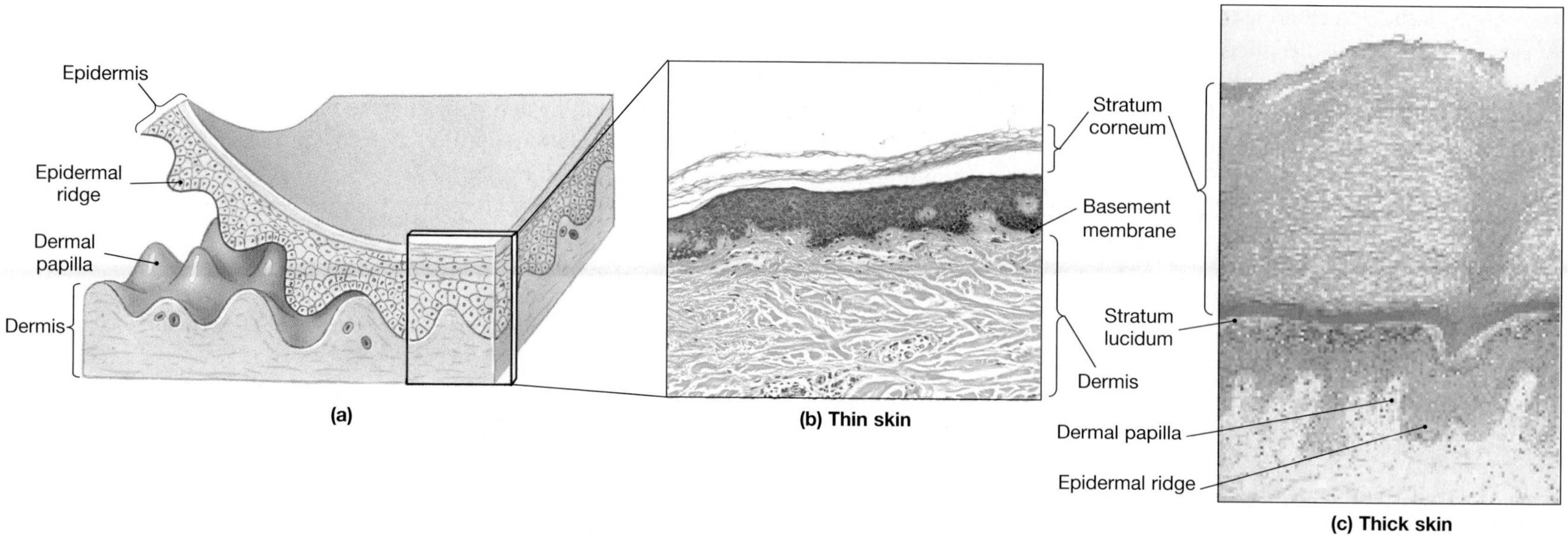

FIGURE 4.4 **THIN AND THICK SKIN**

The epidermis is a stratified squamous epithelium, which varies in thickness. **(a)** The basic organization of the epidermis. The thickness of the epidermis, especially the thickness of the stratum corneum, changes radically depending on the location sampled. **(b)** Thin skin covers most of the exposed body surface (LM × 154). (During sectioning the stratum corneum has pulled away from the rest of the epidermis.) **(c)** Thick skin covers the surfaces of the palms and soles (LM × 154).

The contours of the skin surface follow the ridge patterns, which vary from small conical pegs (in thin skin) to the complex whorls seen on the thick skin of the palms and soles. Ridges on the palms and soles increase the surface area of the skin and increase friction, ensuring a secure grip. Ridge shapes are genetically determined: Those of each person are unique and do not change in the course of a lifetime. Fingerprint-ridge patterns on the tips of the fingers (Figure 4.5●) can therefore be used to identify individuals, and they have been so used in criminal investigation for over a century.

SKIN COLOR [FIGURE 4.6]

The color of the epidermis is due to a combination of (1) the dermal blood supply and (2) variable quantities of two pigments, *carotene* and *melanin.* Blood contains red blood cells that carry the protein hemoglobin. When bound to oxygen, hemoglobin has a bright red color, giving blood vessels in the dermis a reddish tint that is seen most easily in lightly pigmented individuals. When those vessels are dilated, as during inflammation, the red tones become much more pronounced.

DERMAL BLOOD SUPPLY When the circulatory supply is temporarily reduced, the skin becomes relatively pale; a frightened Caucasian may "turn white" because of a sudden drop in blood supply to the skin. During a sustained reduction in circulatory supply, the blood in the superficial vessels loses oxygen and the hemoglobin changes color to a much darker red tone. Seen from the surface, the skin takes on a bluish coloration called **cyanosis** (sī-a-NŌ-sis; *kyanos,* blue). In individuals of any skin color, cyanosis is most apparent in areas of thin skin, such as the lips or beneath the nails. It can be a response to extreme cold or a result of circulatory or respiratory disorders, such as heart failure or severe asthma.

EPIDERMAL PIGMENT CONTENT **Carotene** (KAR-ō-tēn) is an orange-yellow pigment that is found in various orange-colored vegetables, such as carrots, corn, and squashes. It can be converted to vitamin A, which is required for epithelial maintenance and the synthesis of visual pigments by the photoreceptors of the eye. Carotene normally accumulates inside keratinocytes, and it becomes especially evident in the dehydrated cells of the stratum corneum and in the subcutaneous fat. **Melanin** (MEL-a-nin) is produced and stored in melanocytes (Figure 4.6●). The black, yellow-brown, or brown melanin forms in intracellular vesicles called *melanosomes.* These vesicles, which are transferred in-

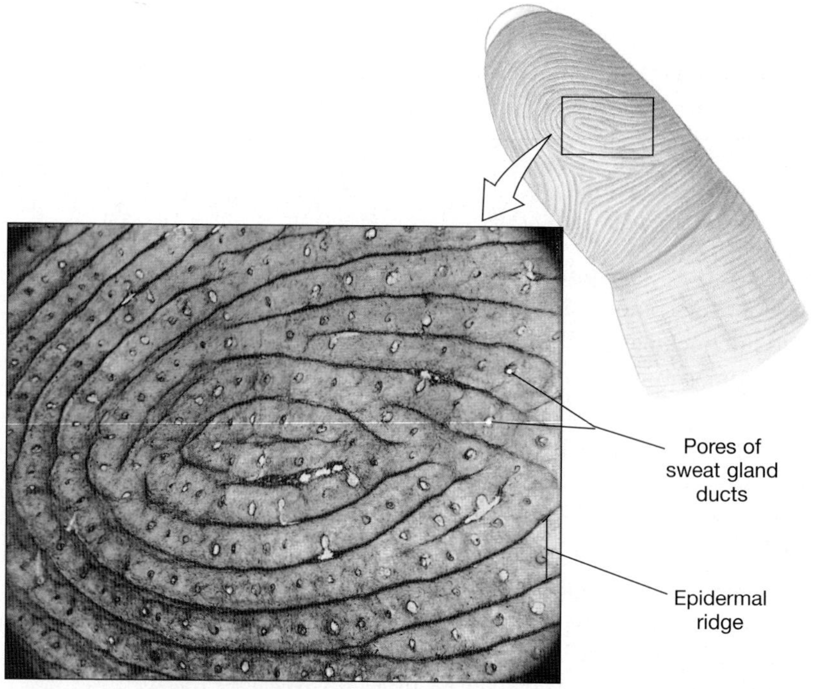

FIGURE 4.5 **THE EPIDERMAL RIDGES OF THICK SKIN**

Fingerprints reveal the pattern of epidermal ridges in thick skin. This scanning electron micrograph shows the ridges on a fingertip. The pits are the pores of sweat gland ducts. (SEM × 25) [Reproduced from R. G. Kessel and R. H. Kardon, *Tissues and Organs: A Text-Atlas of Scanning Electron Microscopy,* W. H. Freeman & Co., 1979.]

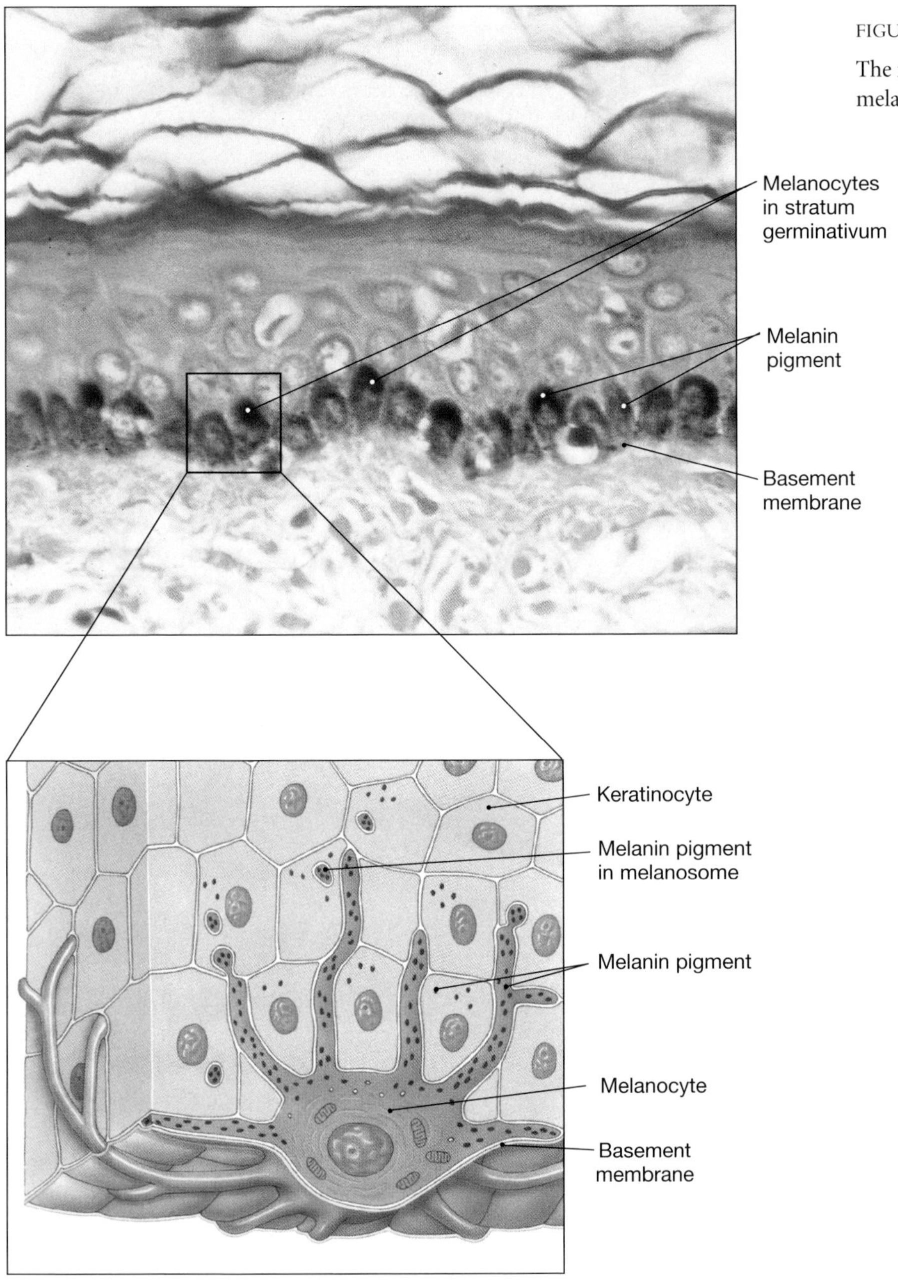

FIGURE 4.6 **MELANOCYTES**

The micrograph and accompanying drawing indicate the location and orientation of melanocytes in the stratum germinativum of a black person.

tact to keratinocytes, color the keratinocytes temporarily, until the melanosomes are destroyed by lysosomes. The cells in more superficial layers gradually lighten in color as the number of intact melanosomes decreases. In Caucasians, melanosome transfer occurs in the stratum germinativum and stratum spinosum, and the cells of more superficial layers lose their pigmentation. In blacks, the melanosomes are larger and the transfer may occur in the stratum granulosum as well; the pigmentation is thus darker and more persistent. Melanin pigments help prevent skin damage by absorbing **ultraviolet [UV] radiation** in sunlight. A little ultraviolet is necessary because the skin requires it to convert a cholesterol-related steroid precursor into a member of the family of hormones collectively known as vitamin D.[1] Vitamin D is required for normal calcium and phosphorus absorption by the small intestine; inadequate supplies of vitamin D lead to impaired bone maintenance and growth. However, too much UV radiation may damage chromosomes and cause widespread tissue damage similar to that caused by mild to moderate burns. Melanin in the epidermis as a whole protects the underlying dermis. Within each keratinocyte, the melanosomes are most abundant around the cell's nucleus. This increases the likelihood that the UV radiation will be absorbed before it can damage the nuclear DNA.

Melanocytes respond to UV exposure by increasing their rates of melanin synthesis and transfer. Tanning then occurs, but the response is not quick enough to prevent a sunburn on the first day at the beach; it takes about 10 days. Anyone can get a sunburn, but dark-skinned individuals have greater initial protection against the effects of UV radiation. Repeated UV exposure sufficient to stimulate tanning can result in long-term skin damage in the dermis and epidermis. In the dermis, damage to fibroblasts causes abnormal connective tissue structure and premature wrinkling. In the epidermis, skin cancers can develop from chromosomal damage in germinative cells or melanocytes. *Skin Cancers p. 786*

[1]Specifically, vitamin D_3 or *cholecalciferol*, which undergoes further modification in the liver and kidneys before circulating as the active hormone *calcitriol*.

✓ **CONCEPT CHECK**

- Describe the primary difference between thick and thin skin.
- Some criminals sand the tips of their fingers so as not to leave recognizable fingerprints. Would this practice permanently remove fingerprints? Why or why not?
- Identify the sources of color of the epidermis.
- Describe the relationship between epidermal ridges and dermal papillae.

The Dermis [FIGURE 4.2]

The dermis lies deep to the epidermis (Figure 4.2● p. 89). It has two major components, a superficial *papillary layer* and a deeper **reticular layer**.

Dermal Organization [FIGURES 4.4/4.7]

The superficial **papillary layer** consists of loose connective tissue (Figure 4.7a●). This region contains the capillaries supplying the epidermis and the axons of sensory neurons that monitor receptors in the papillary layer and the epidermis. The papillary layer derives its name from the dermal papillae that project between the epidermal ridges (Figure 4.4●).

The deeper **reticular layer** consists of fibers in an interwoven meshwork of dense irregular connective tissue that surrounds blood vessels, hair follicles, nerves, sweat glands, and sebaceous glands (Figure 4.7b●). The name of the layer derives from the interwoven arrangement of collagen fiber bundles in this region (*reticulum*, a little net). Some of the collagen fibers in the reticular

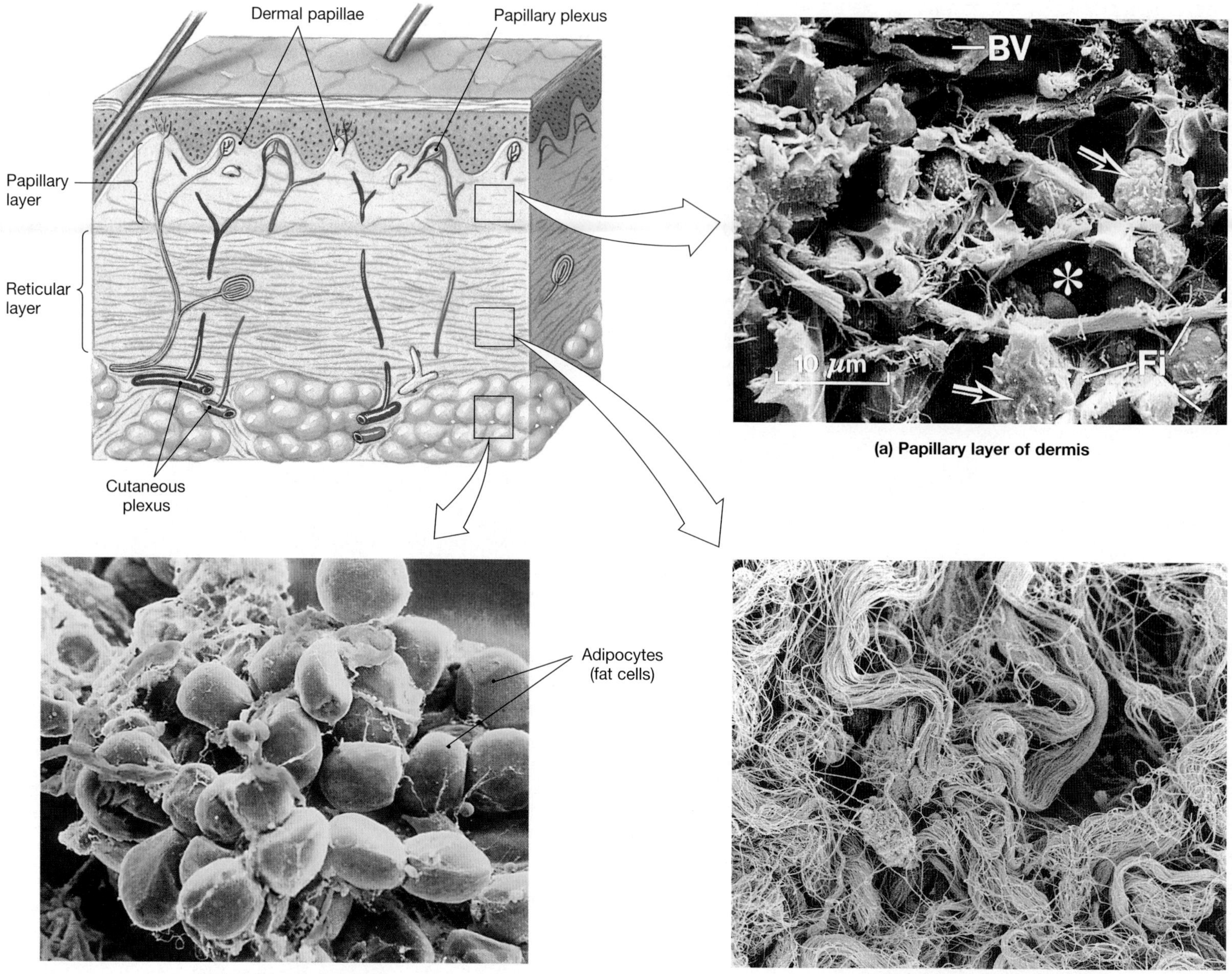

(a) Papillary layer of dermis

(c) Subcutaneous layer

(b) Reticular layer of dermis

FIGURE 4.7 **The Structure of the Dermis and the Subcutaneous Layer**

The dermis is a connective tissue layer deep to the epidermis; the subcutaneous layer (hypodermis) is a connective tissue layer deep to the dermis. **(a)** The papillary layer of the dermis consists of loose connective tissue that contains numerous blood vessels (BV), fibers (Fi), and macrophages (arrows). Open spaces, such as the one marked by an asterisk, would be filled with fluid ground substance. (SEM × 649) **(b)** The reticular layer of the dermis contains dense, irregular connective tissue. (SEM × 1340) **(c)** The subcutaneous layer contains large numbers of adipocytes in a framework of loose connective tissue fibers. (SEM × 268) [(a) Reproduced from R. G. Kessel and R. H. Kardon, *Tissues and Organs: A Text-Atlas of Scanning Electron Microscopy,* W. H. Freeman & Co., 1979. (c) Prof. P. Motta, Dept. of Anatomy, University "La Sapienza," Rome/Science Photo Library/Photo Researchers, Inc.]

layer extend into the papillary layer, tying the two layers together. The boundary line between these layers is therefore indistinct. Collagen fibers of the reticular layer also extend into the deeper subcutaneous layer (Figure 4.7c●). *Dermatitis p. 787*

WRINKLES, STRETCH MARKS, AND LINES OF CLEAVAGE [FIGURE 4.8]

The interwoven collagen fibers of the reticular layer provide considerable tensile strength, and the extensive array of elastic fibers enables the dermis to stretch and recoil repeatedly during normal movements. Age, hormones, and the destructive effects of ultraviolet radiation reduce the thickness and flexibility of the dermis, producing wrinkles and sagging skin. The extensive distortion of the dermis that occurs over the abdomen during pregnancy or after a substantial weight gain often exceeds the elastic capabilities of the skin. Elastic and collagen fibers then break, and although the skin stretches, it does not recoil to its original size after delivery or a rigorous diet. The skin then wrinkles and creases, creating a network of **stretch marks**.

Tretinoin (*Retin-A*) is a derivative of vitamin A that can be applied to the skin as a cream or gel. This drug was originally developed to treat acne, but it also increases blood flow to the dermis and stimulates dermal repairs. As a result, the rate of wrinkle formation decreases, and existing wrinkles become smaller. The degree of improvement varies from individual to individual.

At any one location, the majority of the collagen and elastic fibers are arranged in parallel bundles. The orientation of these bundles depends on the stress placed on the skin during normal movement; the bundles are aligned to resist the applied forces. The resulting pattern of fiber bundles establishes the **lines of cleavage** of the skin. Lines of cleavage, shown in Figure 4.8●, are clinically significant because a cut parallel to a cleavage line will usually remain closed, whereas a cut at right angles to a cleavage line will be pulled open as cut elastic fibers recoil. Surgeons choose their incision patterns accordingly, since an incision parallel to the cleavage lines will heal fastest and with minimal scarring.

Other Dermal Components [FIGURES 4.2/4.9]

In addition to extracellular protein fibers, the dermis contains all of the cells of connective tissue proper. p. 64 Accessory organs of epidermal origin, such as hair follicles and sweat glands, extend into the dermis (Figure 4.9●). In addition, the reticular and papillary layers of the dermis contain networks of blood vessels, lymph vessels, and nerve fibers (Figure 4.2●, p. 89).

The Blood Supply to the Skin [FIGURES 4.2/4.7]

Arteries supplying the skin form a network in the subcutaneous layer along the border with the reticular layer. This network is called the **cutaneous plexus** (Figure 4.2●, p. 89). Branches of these arteries supply the adipose tissues of the subcutaneous layer as well as the tissues of the skin. As small arteries travel toward the epidermis, branches supply the hair follicles, sweat glands, and other structures in the dermis. Upon reaching the papillary layer, these small arteries form another branching network, the **papillary plexus**, which provides arterial blood to capillary loops that follow the contours of the epidermal–dermal boundary (Figure 4.7a●). These capillaries empty into a network of delicate veins (*venules*) that merge to form larger veins that descend through the dermis to the subcutaneous layer.

There are two reasons why the circulation to the skin must be tightly regulated. First, it plays a key role in *thermoregulation*, the control of body temperature. When body temperature increases, increased circulation to the skin permits the loss of excess heat, whereas when body temperature decreases, reduced circulation to the skin promotes retention of body heat. Second, because the total blood volume is relatively constant, increased blood flow to the skin means a decreased blood flow to some other organ(s). The nervous, cardiovascular, and endocrine systems interact to regulate blood flow to the skin while maintaining adequate blood flow to other organs and systems. *Tumors in the Dermis p. 787*

THE NERVE SUPPLY TO THE SKIN

Nerve fibers in the skin control blood flow, adjust gland secretion rates, and monitor sensory receptors in the dermis and the deeper layers of the epidermis. We have already noted the presence of Merkel cells in the deeper layers of the epidermis. These cells are monitored by sensory terminals known as *Merkel's discs*. The epidermis also contains the dendrites of sensory nerves that probably respond to pain and temperature. The dermis contains similar receptors as well as other, more specialized receptors. Examples discussed in Chapter 18 include receptors sensitive to light touch (*tactile corpuscles*, located in dermal papillae and the *root hair plexus* surrounding each hair follicle), stretch (*Ruffini corpuscles*, in the reticular layer), and deep pressure and vibration (*lamellated corpuscles*, in the reticular layer).

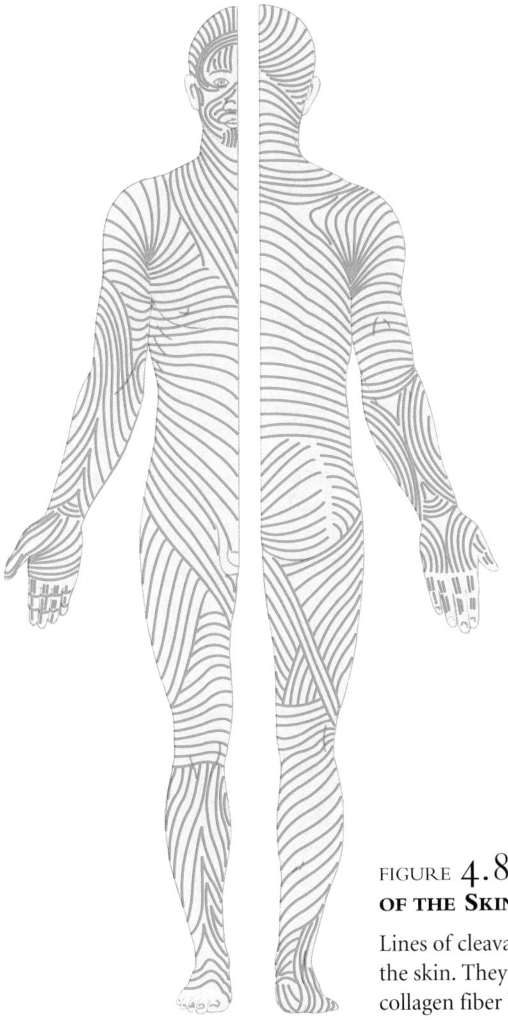

FIGURE 4.8 **LINES OF CLEAVAGE OF THE SKIN**

Lines of cleavage follow lines of tension in the skin. They reflect the orientation of collagen fiber bundles in the dermis.

The Subcutaneous Layer [FIGURES 4.2/4.7c]

The connective tissue fibers of the reticular layer are extensively interwoven with those of the **subcutaneous layer**, also referred to as *hypodermis*, or the *superficial fascia* p. 77, and the boundary between the two layers is usually indistinct (Figure 4.2●, p. 89). Although the subcutaneous layer is sometimes not considered to be a part of the integument, it is important in stabilizing the position of the skin in relation to underlying tissues, such as skeletal muscles or other organs, while still permitting independent movement.

The subcutaneous layer consists of loose connective tissue with abundant fat cells (Figure 4.7c●). Infants and small children usually have extensive "baby fat," p. 68 which helps reduce heat loss. Subcutaneous fat also serves as a substantial energy reserve and a shock absorber for the rough-and-tumble activities of our early years.

As we grow, the distribution of subcutaneous fat changes. Men accumulate subcutaneous fat at the neck, upper arms, along the lower back, and over the buttocks. In women the breasts, buttocks, hips, and thighs are the primary sites of subcutaneous fat storage. In adults of either sex, the subcutaneous layer of the backs of the hands and the upper surfaces of the feet contain few fat cells, whereas distressing amounts of adipose tissue can accumulate in the abdominal region, producing a prominent "pot belly."

The subcutaneous layer is quite elastic. Only the superficial region of the subcutaneous layer contains large arteries and veins; the remaining areas contain a limited number of capillaries and no vital organs. This last

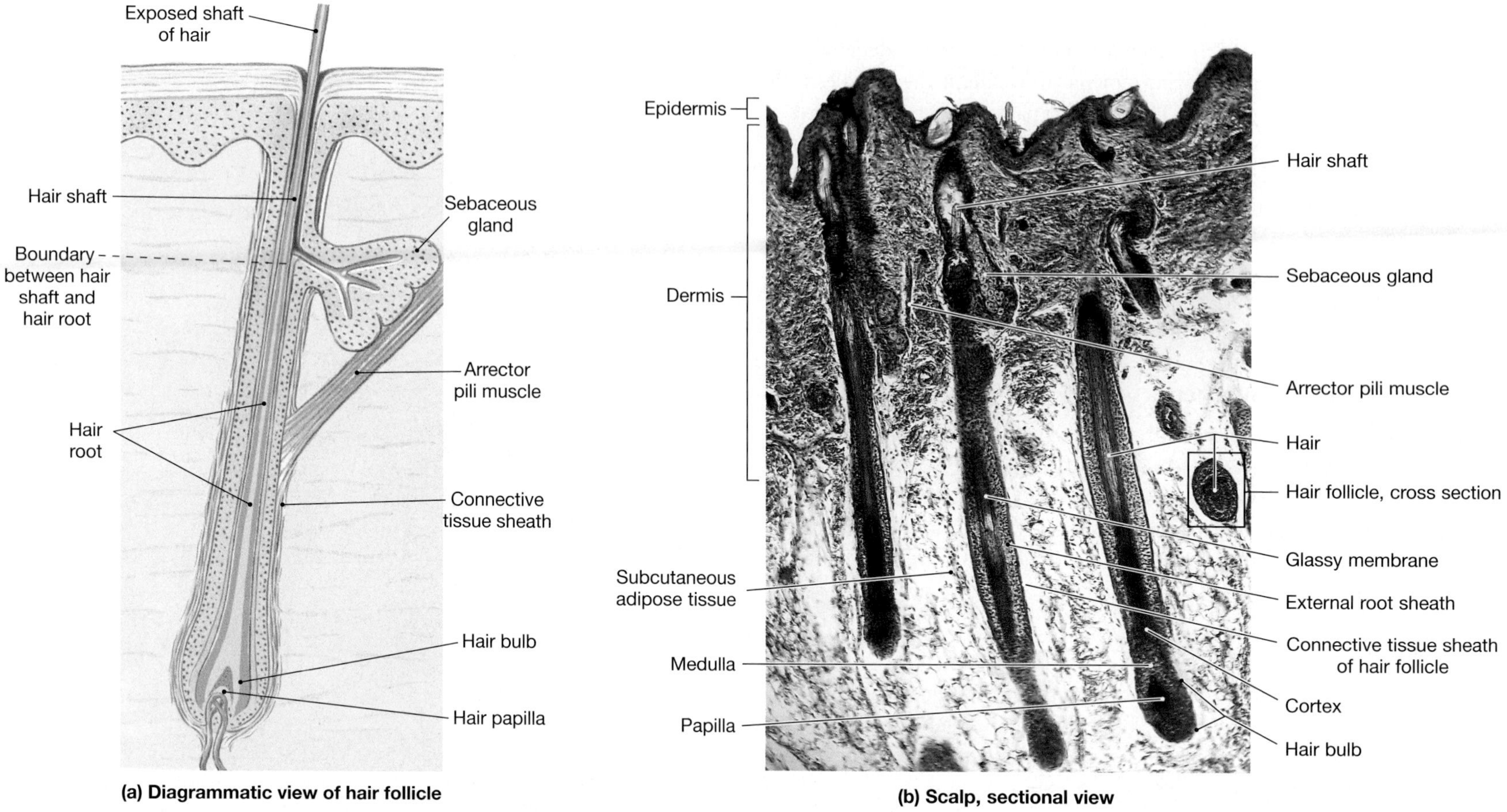

FIGURE 4.9 **Accessory Structures of the Skin**

(a) A diagrammatic view of a single hair follicle. **(b)** A light micrograph showing the sectional appearance of the skin of the scalp. Note the abundance of hair follicles and the way they extend into the dermis. (LM × 66)

characteristic makes *subcutaneous injection* a useful method for administering drugs. The familiar term *hypodermic needle* refers to the region targeted for injection.

Accessory Structures [FIGURE 4.2]

The accessory structures of the integument include *hair follicles*, *sebaceous glands*, *sweat glands*, and *nails* (Figure 4.2●, p. 89). During embryological development, these structures form through invagination of the epidermis (see the Embryology Summary on pp. 102–103).

Hair Follicles and Hair

Hairs project beyond the surface of the skin almost everywhere except over the sides and soles of the feet, the palms of the hands, the sides of the fingers and toes, the lips, and portions of the external genitalia.[2] There are about 5 million hairs on the human body, and 98% of them are on the general body surface, not the head. Hairs are nonliving structures that are formed in organs called **hair follicles**.

HAIR PRODUCTION [FIGURES 4.9/4.10]

Hair follicles extend deep into the dermis, often projecting into the underlying subcutaneous layer. The epithelium at the follicle base surrounds a small **hair papilla**, a peg of connective tissue containing capillaries and nerves. The **hair bulb** consists of epithelial cells that surround the papilla.

Hair production involves a specialization of the keratinization process. The **hair matrix** is the epithelial layer involved in hair production. When the superficial basal cells divide, they produce daughter cells that are pushed toward the surface as part of the developing hair. Most hairs have an inner *medulla* and an outer *cortex*. The medulla contains relatively soft and flexible **soft keratin**. Matrix cells closer to the edge of the developing hair form the relatively hard **cortex** (Figures 4.9b and 4.10●). The cortex contains **hard keratin** that gives hair its stiffness. A single layer of dead, keratinized cells at the outer surface of the hair overlap and form the **cuticle** that coats the hair.

The hair **root** extends from the hair bulb to the point where the internal organization of the hair is complete. The hair root attaches the hair to the hair follicle. The **shaft**, the part we see on the surface, extends from this point, usually halfway to the skin surface, to the exposed tip of the hair. The size, shape, and color of the hair shaft are highly variable.

FOLLICLE STRUCTURE [FIGURE 4.10]

The cells of the follicle walls are organized into concentric layers (Figure 4.10a●). Beginning at the hair cuticle, these layers include:

- *The internal root sheath*: This layer surrounds the hair root and the deeper portion of the shaft. It is produced by the cells at the periphery

[2]The glans penis and prepuce of the male; the clitoris, labia minora, and inner surfaces of the labia majora in the female.

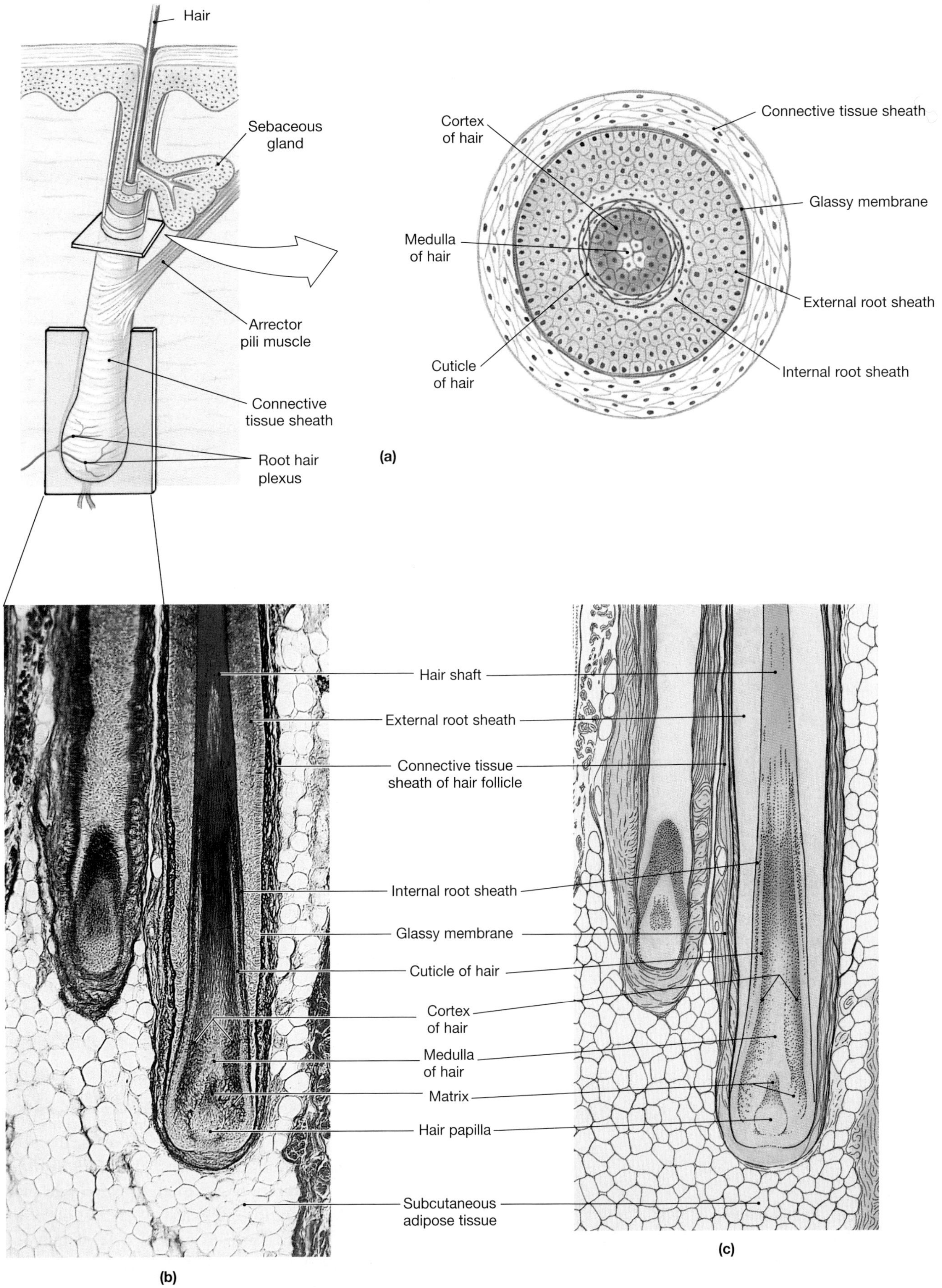

FIGURE 4.10 **HAIR FOLLICLES**

Hairs originate in hair follicles, which are complex organs. (**a**) A longitudinal section and a cross section through a hair follicle. Histological (**b**) and diagrammatic (**c**) sections along the longitudinal axis of a hair follicle. (LM × 60) of the hair matrix. Because the cells of the internal root sheath disintegrate relatively quickly, this layer does not extend the entire length of the follicle.

- *The external root sheath*: This layer extends from the skin surface to the hair matrix. Over most of that distance it has all of the cell layers found in the superficial epidermis. However, where the external root sheath joins the hair matrix, all of the cells resemble those of the stratum germinativum.
- *The glassy membrane*: This is a thickened basement membrane, wrapped in a dense connective tissue sheath.

FUNCTIONS OF HAIR [FIGURE 4.10a]

The 5 million hairs on the human body have important functions. The roughly 100,000 hairs on the head protect the scalp from ultraviolet light, cushion a blow to the head, and provide insulation for the skull. The hairs guarding the entrances to the nostrils and external auditory canals help prevent the entry of foreign particles and insects, and eyelashes perform a similar function for the surface of the eye. A **root hair plexus** of sensory nerves surrounds the base of each hair follicle (Figure 4.10a●). As a result, the movement of the shaft of even a single hair can be felt at a conscious level. This sensitivity provides an early-warning system that may help to prevent injury. For example, you may be able to swat a mosquito before it reaches the skin surface.

A ribbon of smooth muscle, called the **arrector pili** (a-REK-tor, PĪ-lī; plural, *arrectores pilorum*) extends from the papillary dermis to the connective tissue sheath surrounding the hair follicle (Figures 4.9a,b and 4.10a●). When stimulated, the arrector pili pulls on the follicle and elevates the hair. Contraction may be caused by emotional states, such as fear or rage, or as a response to cold, producing characteristic "goose bumps." In a furry mammal, this action increases the thickness of the insulating coat, rather like putting on an extra sweater. Although we do not receive any comparable insulating benefits, the reflex persists.

TYPES OF HAIRS

Hairs first appear after roughly 3 months of embryonic development. These hairs, collectively known as *lanugo* (la-NOO-gō), are extremely fine and unpigmented. Most lanugo hairs are shed before birth. They are replaced by one of the three types of hairs in the adult. The three major types of hairs in the integument of an adult are *vellus hairs, intermediate hairs*, and *terminal hairs*.

- *Vellus hairs* are the fine "peach fuzz" hairs found over much of the body surface.
- *Intermediate hairs* are those hairs that change in their distribution, such as the hairs of the upper and lower limbs.
- *Terminal hairs* are heavy, more deeply pigmented, and sometimes curly. The hairs on your head, including your eyebrows and eyelashes, are examples of terminal hairs.

The description of hair structure earlier in the chapter was based on an examination of terminal hairs. Vellus and intermediate hairs are similar, though neither has a distinct medulla. Hair follicles may alter the structure of the hairs they produce in response to circulating hormones. Thus a follicle that produces a vellus hair today may produce an intermediate hair tomorrow; this accounts for many of the changes in hair distribution that begin at puberty.

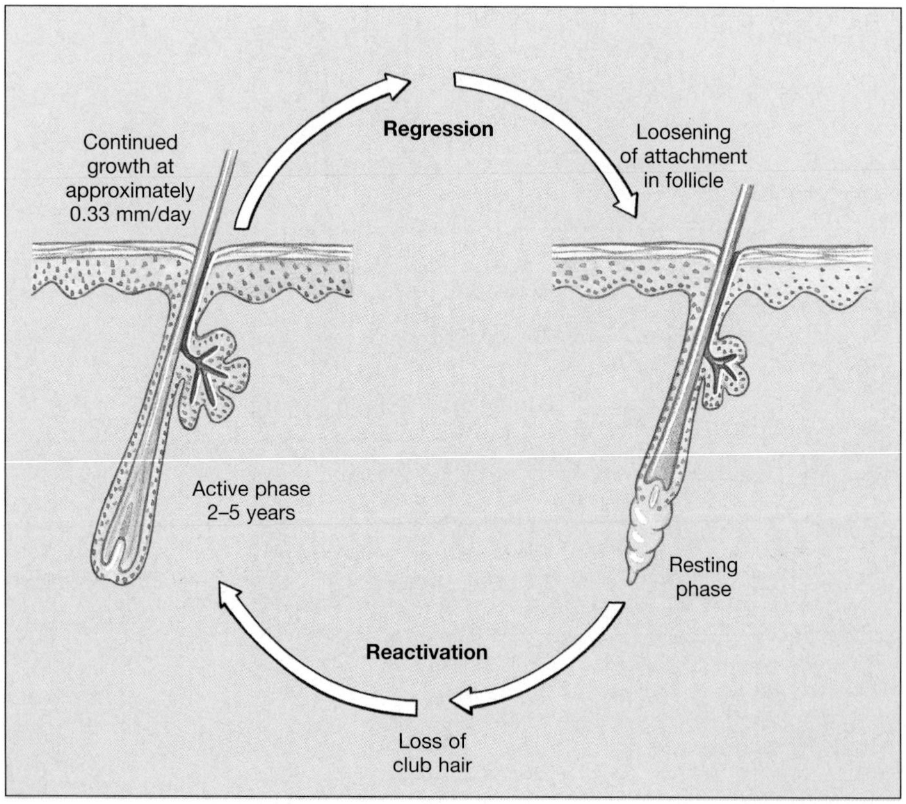

FIGURE 4.11 **THE HAIR GROWTH CYCLE**

Each hair follicle goes through growth cycles involving active and resting stages.

HAIR COLOR

Variations in hair color reflect differences in hair structure and variations in the pigment produced by melanocytes at the papilla. These characteristics are genetically determined, but the condition of your hair may be influenced by hormonal or environmental factors. Whether your hair is black or brown depends on the density of melanin in the cortex. Red hair results from the presence of a biochemically distinct form of melanin. As pigment production decreases with age, the hair color lightens toward gray. White hair results from the combination of a lack of pigment and the presence of air bubbles within the medulla of the hair shaft. Because the hair itself is dead and inert, changes in coloration are gradual; your hair can't "turn white overnight," as some horror stories suggest.

GROWTH AND REPLACEMENT OF HAIR [FIGURE 4.11]

A hair in the scalp grows for 2–5 years, at a rate of around 0.33 mm/day (about 1/64th inch). Variations in the hair growth rate and in the duration of the **hair growth cycle**, illustrated in Figure 4.11●, account for individual differences in uncut hair length.

While hair growth is under way, the root of the hair is firmly attached to the matrix of the follicle. At the end of the growth cycle, the follicle becomes inactive, and the hair is now termed a **club hair** (Figure 4.11●). The follicle gets smaller, and over time the connections between the hair matrix and the root of the club hair break down. When another growth cycle begins, the follicle produces a new hair, and the old club hair gets pushed toward the surface.

In healthy adults, about 50 hairs are lost each day, but several factors may affect this rate. Sustained losses of over 100 hairs per day usually indicate that something is wrong. Temporary increases in hair loss can result from drugs, dietary factors, radiation, high fever, stress, and hormonal factors related to pregnancy. Collecting hair samples can be helpful in diagnosing several disorders. For example, hairs of individuals with lead poisoning or other heavy-metal poisoning contain high quantities of those metal ions. In males, changes in the level of the sex hormones circulating in the blood can affect the scalp, causing a shift from terminal hair to vellus hair production. This alteration is called **male pattern baldness.** *Baldness and Hirsutism p. 787*

✓ CONCEPT CHECK

- What happens to the dermis when it is excessively stretched, as in pregnancy or obesity?
- What condition is produced by the contraction of the arrector pili?
- Describe the major features of a hair.

Glands in the Skin [FIGURE 4.12]

The skin contains two types of exocrine glands: *sebaceous glands* and *sweat (sudoriferous) glands*. Sebaceous glands produce an oily lipid that coats hair shafts and the epidermis. Sweat glands produce a watery solution and perform other special functions. Figure 4.12● summarizes the functional classification of the exocrine glands of the skin.

SEBACEOUS GLANDS [FIGURE 4.13]

Sebaceous (se-BĀ-shus) **glands**, or *oil glands*, discharge a waxy, oily secretion into hair follicles (Figure 4.13●). The gland cells manufacture large quantities of lipids as they mature, and the lipid product is released through holocrine secretion. p. 61 The ducts are short, and several sebaceous glands may open into a single follicle. Depending on whether the glands share a common duct, they may be classified as *simple alveolar glands* (each gland has its own duct) or *simple branched alveolar glands* (several glands empty into a single duct). p. 60

The lipids released by sebaceous gland cells enter the open passageway, or *lumen*, of the gland. Contraction of the arrector pili muscle that elevates the hair squeezes the sebaceous gland, forcing the waxy secretions into the follicle and onto the surface of the skin. This secretion, called

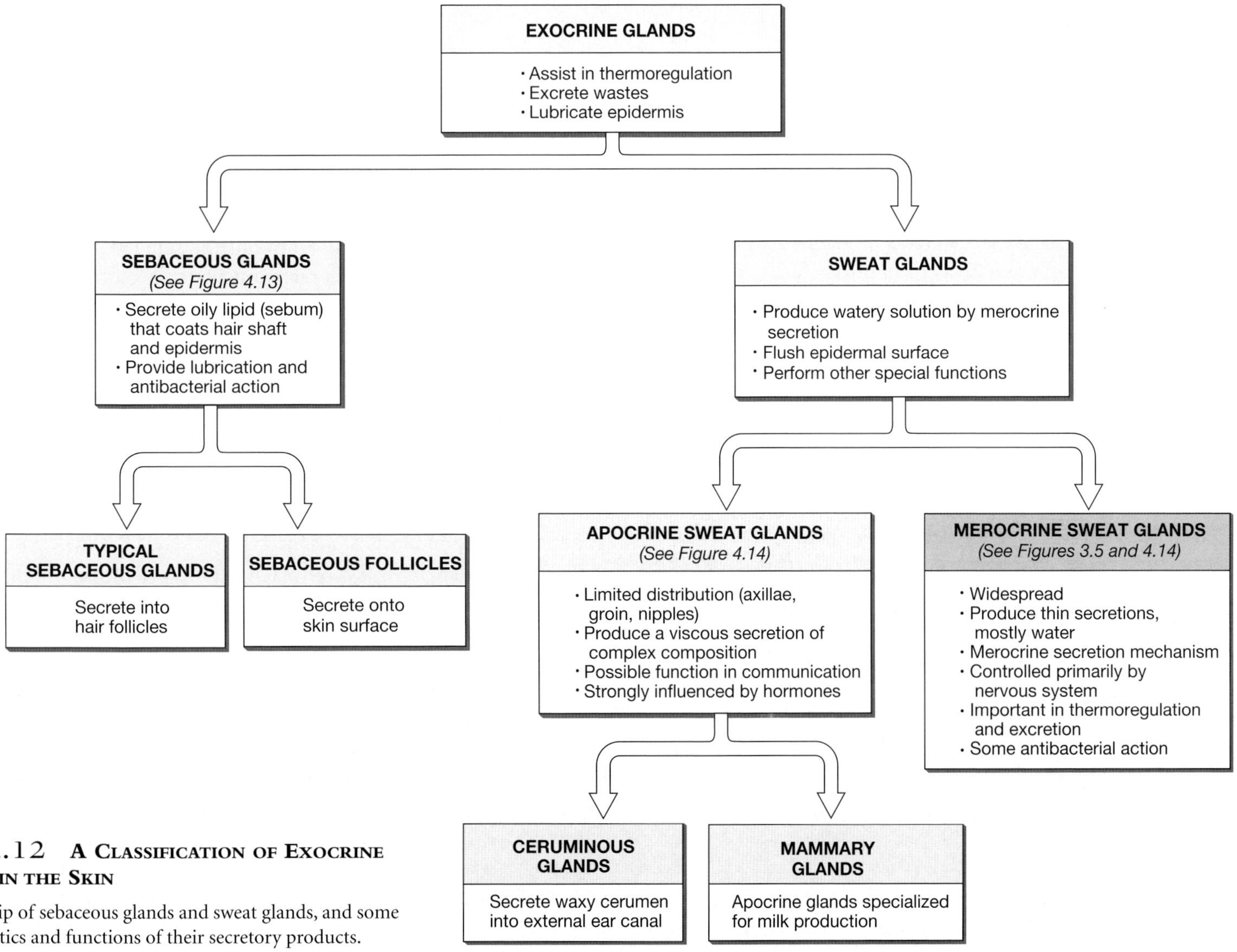

FIGURE 4.12 **A CLASSIFICATION OF EXOCRINE GLANDS IN THE SKIN**

Relationship of sebaceous glands and sweat glands, and some characteristics and functions of their secretory products.

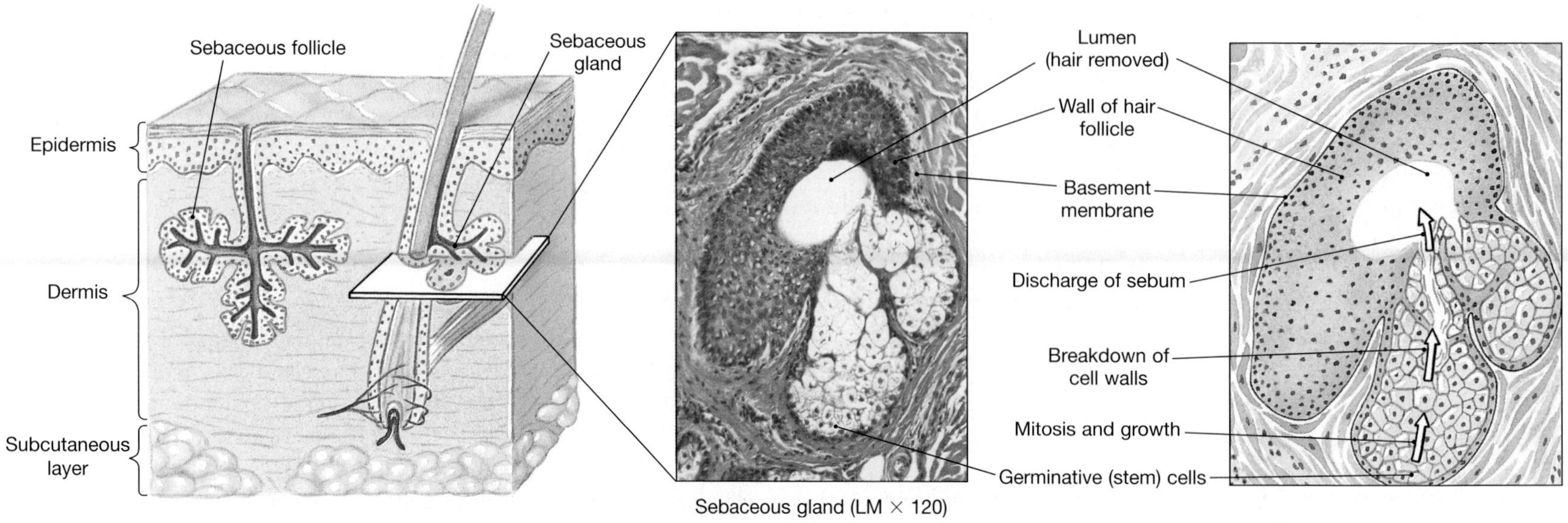

FIGURE 4.13 **SEBACEOUS GLANDS AND FOLLICLES**
The structure of sebaceous glands and sebaceous follicles in the skin.

sebum (SĒ-bum), provides lubrication and inhibits the growth of bacteria. Keratin is a tough protein, but dead, keratinized cells become dry and brittle once exposed to the environment. Sebum lubricates and protects the keratin of the hair shaft and conditions the surrounding skin. Shampooing removes the natural oily coating, and excessive washing can make hairs stiff and brittle.

Sebaceous follicles are large sebaceous glands that communicate directly with the epidermis. These follicles, which never produce hairs, are found on the integument covering the face, back, chest, nipples, and male sex organs. Although sebum has bactericidal (bacteria-killing) properties, under some conditions bacteria can invade sebaceous glands or follicles. The presence of bacteria in glands or follicles can produce a local inflammation known as **folliculitis** (fo-lik-ū-LĪ-tis). If the duct of the gland becomes blocked, a distinctive abscess called a **furuncle** (FUR-ung-kl), or "boil," develops. The usual treatment for a furuncle is to cut it open, or "lance" it, so that normal drainage and healing can occur.

CLINICAL BRIEF

ACNE AND SEBORRHEIC DERMATITIS

Sebaceous glands and sebaceous follicles are very sensitive to changes in the concentrations of sex hormones, and their secretory activities accelerate at puberty. For this reason an individual with large sebaceous glands may be especially prone to develop acne during adolescence. In **acne**, sebaceous ducts become blocked and secretions accumulate, causing inflammation and providing a fertile environment for bacterial infection. *Acne p. 787*

Seborrheic dermatitis is an inflammation around abnormally active sebaceous glands. The affected area becomes red, and there is usually some epidermal scaling. Sebaceous glands of the scalp are most often involved. In infants, mild cases are called "cradle cap." Adults know this condition as "dandruff." Anxiety, stress, and food allergies can increase the severity of the inflammation, as can a concurrent fungal infection.

SWEAT GLANDS [FIGURES 4.12/4.14]

The skin contains two different groups of sweat glands: *apocrine sweat glands* and *merocrine sweat glands* (Figures 4.12 and 4.14●). Both gland types contain **myoepithelial cells** (*myo-*, muscle), specialized epithelial cells located between the gland cells and the underlying basement membrane. Myoepithelial cell contractions squeeze the gland and discharge the accumulated secretions. The secretory activities of the gland cells and the contractions of myoepithelial cells are controlled by both the autonomic nervous system and by circulating hormones.

APOCRINE SWEAT GLANDS [FIGURES 4.9a/4.14a] **Apocrine sweat glands** are found in the axillae (armpits), around the nipples (*areolae*), and in the groin, where they release their secretion into hair follicles (Figures 4.9a and 4.14a●). The name *apocrine* was originally chosen because it was thought that the gland cells used an apocrine method of secretion. p. 61 Although we now know that their secretory products are produced through merocrine secretion, the name has not changed. These are coiled tubular glands that produce a viscous, cloudy, and potentially odorous secretion. Apocrine sweat glands begin secreting at puberty; the sweat produced may be acted upon by bacteria causing a noticeable odor. Apocrine gland secretions may also contain *pheromones*, chemicals that communicate information to other individuals at a subconscious level. The apocrine secretions of mature women have been shown to alter the menstrual timing of other women. The significance of these pheromones, and the role of apocrine secretions in males, remain unknown.

MEROCRINE SWEAT GLANDS [FIGURES 4.9a/4.14b] **Merocrine** (MER-ō-krin) **sweat glands**, also known as *eccrine sweat glands*, are far more numerous and widely distributed than apocrine sweat glands (Figures 4.9a and 4.14b●). The adult integument contains around 3 million merocrine glands. They are smaller than apocrine sweat glands, and they do not extend as far into the dermis. Palms and soles have the highest numbers; estimates are that the palm of the hand has about 500 per cm^2 (3000 glands per square inch). Merocrine sweat glands are coiled

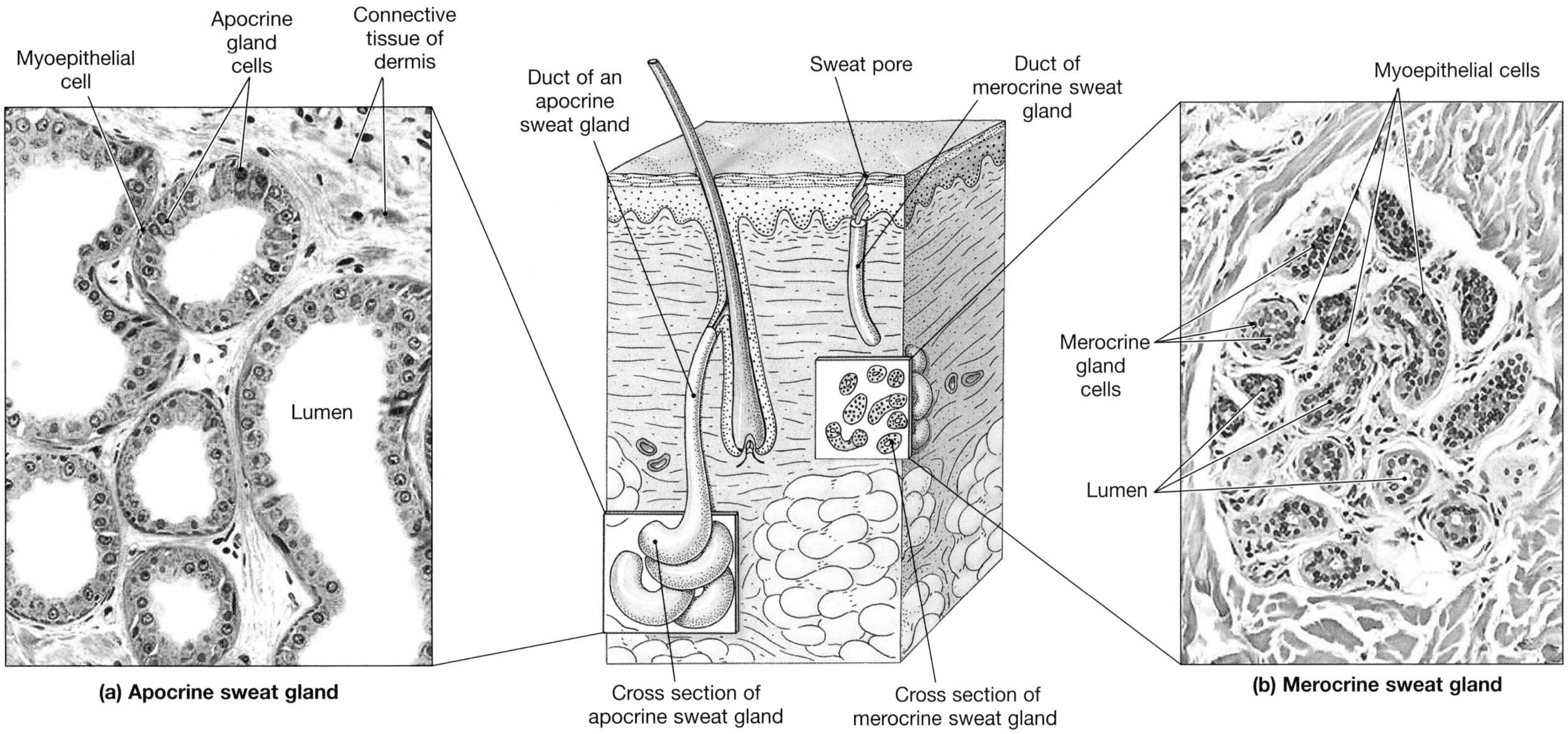

FIGURE 4.14 **SWEAT GLANDS**

(**a**) Apocrine sweat glands are found in the axillae (armpits), groin, and nipples. They produce a thick, potentially odorous fluid. (LM × 459) (**b**) Merocrine sweat glands produce a watery fluid commonly called sensible perspiration or sweat. (LM × 243)

tubular glands that discharge their secretions directly onto the surface of the skin.

The clear secretion produced by merocrine glands is termed **sweat**, or **sensible perspiration**. Sweat is mostly water (99%), but it does contain some electrolytes (chiefly sodium chloride), metabolites, and waste products. It is the presence of the sodium chloride that gives sweat a salty taste. The functions of merocrine sweat gland activity include:

- *Thermoregulation.* Sweat cools the surface of the skin and reduces body temperature. This cooling is the primary function of sensible perspiration, and the degree of secretory activity is regulated by neural and hormonal mechanisms. When all of the merocrine sweat glands are working at maximum, the rate of perspiration may exceed a gallon per hour, and dangerous fluid and electrolyte losses can occur. For this reason athletes in endurance sports must pause frequently to drink fluids.
- *Excretion.* Merocrine sweat gland secretion can also provide a significant excretory route for water and electrolytes, as well as for a number of prescription and nonprescription drugs.
- *Protection.* Merocrine sweat gland secretion provides protection from environmental hazards by diluting harmful chemicals and discouraging the growth of microorganisms.

CONTROL OF GLANDULAR SECRETIONS

Sebaceous glands and apocrine sweat glands can be turned on or off by the autonomic nervous system, but no regional control is possible—this means that when one sebaceous gland is activated, so are all the other sebaceous glands in the body. Merocrine sweat glands are much more precisely controlled, and the amount of secretion and the area of the body involved can be varied independently. For example, when you are nervously awaiting an anatomy exam, your palms may begin to sweat.

OTHER INTEGUMENTARY GLANDS

Sebaceous glands and merocrine sweat glands are found over most of the body surface. Apocrine sweat glands are found in relatively restricted areas. The skin also contains a variety of specialized glands that are restricted to specific locations. Many will be encountered in later chapters; two important examples will be noted here.

1. The **mammary glands** of the breasts are anatomically related to apocrine sweat glands. A complex interaction between sexual and pituitary hormones controls their development and secretion. Mammary gland structure and function will be discussed in Chapter 27.
2. **Ceruminous** (se-ROO-mi-nus) **glands** are modified sweat glands located in the external auditory canal. They differ from merocrine sweat glands in that they have a larger lumen and their gland cells contain pigment granules and lipid droplets not found in other sweat glands. Their secretions combine with those of nearby sebaceous glands, forming a mixture called **cerumen**, or simply "ear wax." Ear wax, together with tiny hairs along the ear canal, probably helps trap foreign particles or small insects and keeps them from reaching the eardrum.

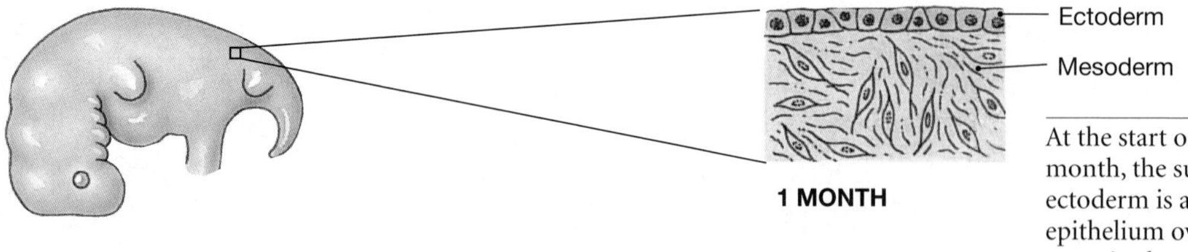

1 MONTH

At the start of the second month, the superficial ectoderm is a simple epithelium overlying loosely organized mesenchyme.

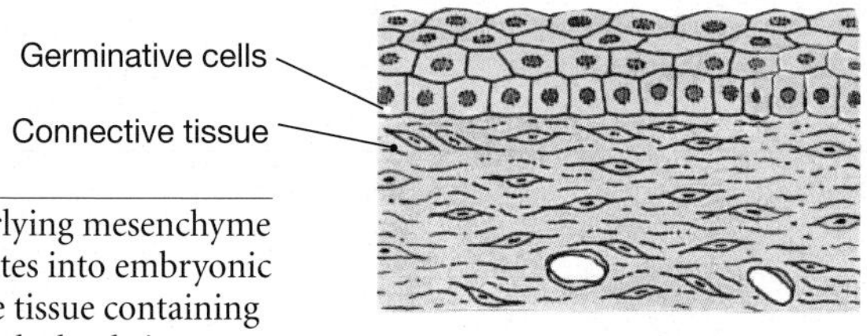

3 MONTHS

Over the following weeks, the epithelium becomes stratified through repeated divisions of the basal or *germinative* cells.

The underlying mesenchyme differentiates into embryonic connective tissue containing blood vessels that bring nutrients to the region.

SKIN

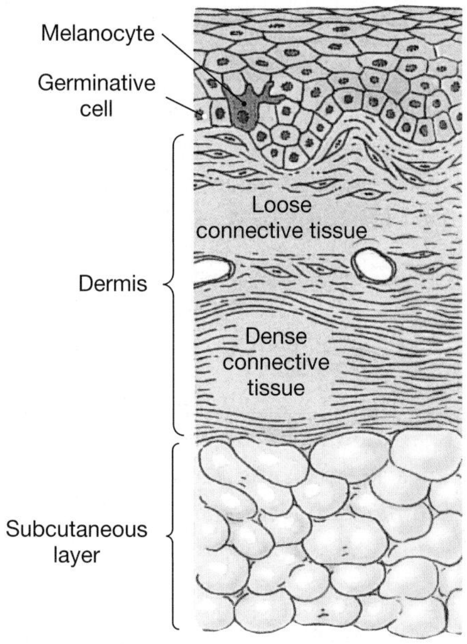

4 MONTHS

As basal cell divisions continue, the epithelial layer thickens and the basement membrane is thrown into irregular folds. Pigment cells called *melanocytes* migrate into the area and squeeze between the germinative cells. The epithelium now resembles the *epidermis* of the adult.

The embryonic connective tissue differentiates into the *dermis*. Fibroblasts and other connective tissue cells form from mesenchymal cells or migrate into the area. The density of fibers increases. Loose connective tissue extends into the ridges, but a deeper, less vascular region is dominated by a dense, irregular collagen fiber network. Below the dermis the embryonic connective tissue develops into the *subcutaneous layer*, a layer of loose connective tissue.

NAILS

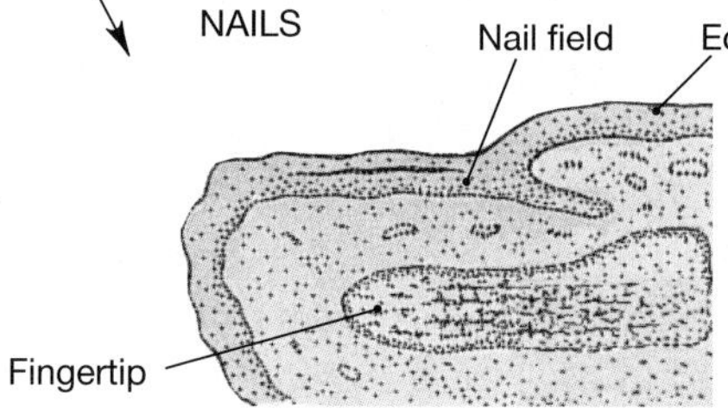

4 MONTHS

Nails begin as thickenings of the epidermis near the tips of the fingers and toes. These thickenings settle into the dermis, and the borderline with the general epidermis becomes distinct. Initially, nail production involves all of the germinative cells of the *nail field*.

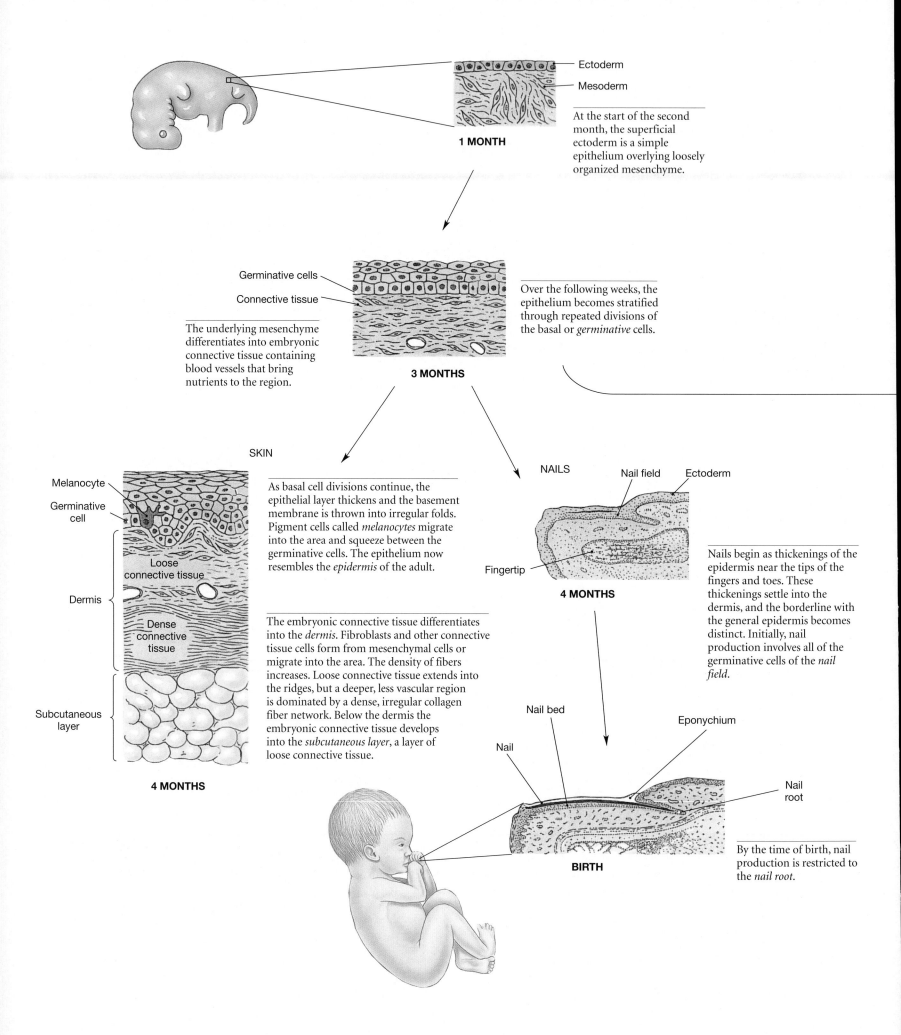

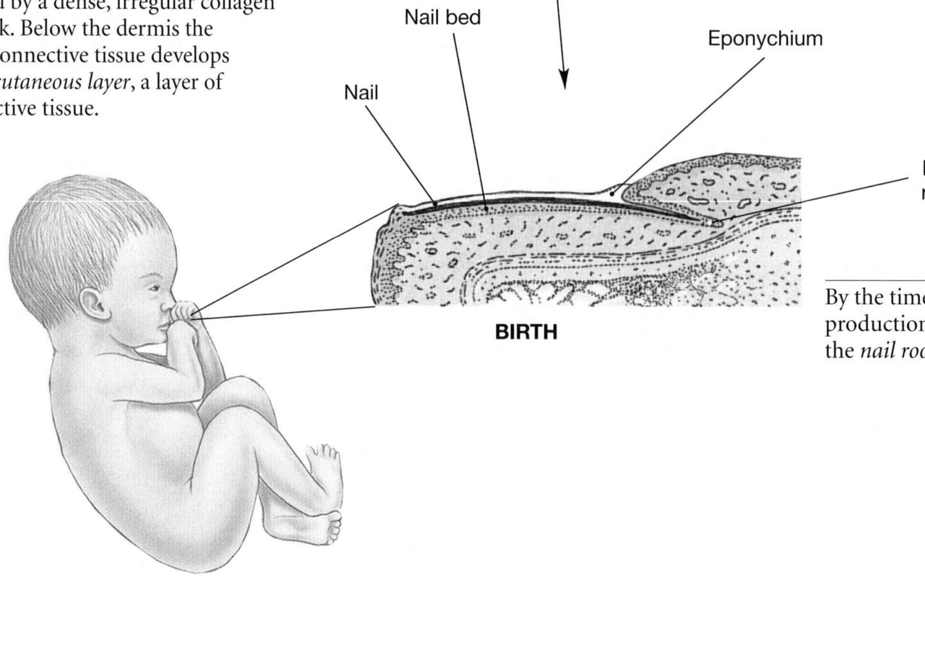

BIRTH

By the time of birth, nail production is restricted to the *nail root*.

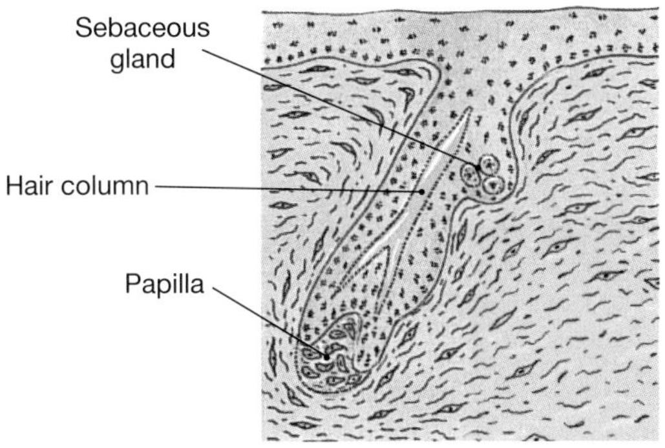

5 MONTHS

A hair follicle develops as a deep column surrounds a *papilla*, a small mass of connective tissue. Hair growth will occur in the epithelium covering the papilla. An outgrowth from the epithelial column forms a *sebaceous gland*.

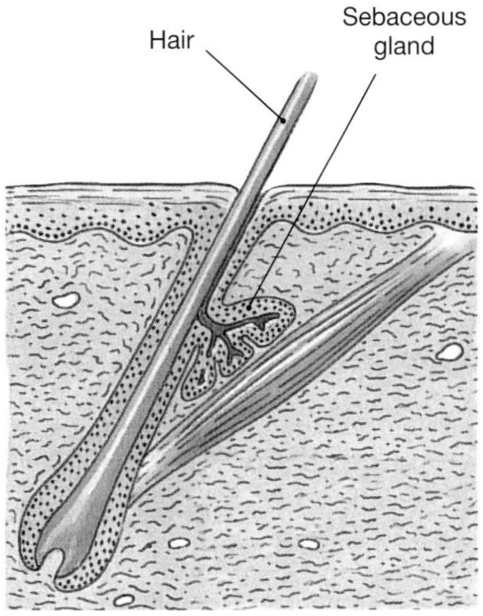

BIRTH

At birth a hair projects from the follicle, and the secretions of the sebaceous gland lubricate the hair shaft.

HAIR FOLLICLES AND EXOCRINE GLANDS

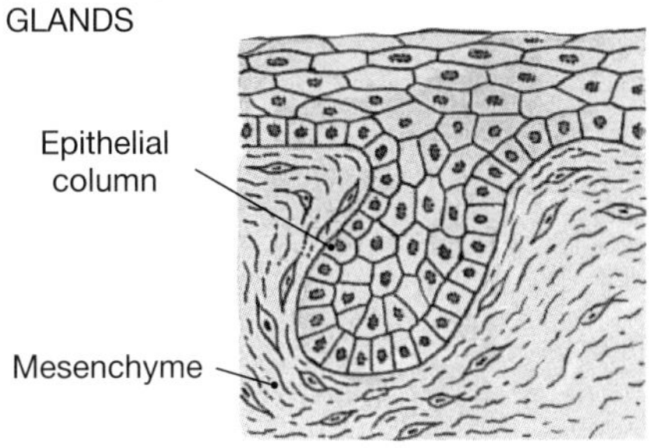

4 MONTHS

During the third and fourth months, small areas of epidermis undergo extensive divisions and form cords of cells that grow into the dermis. These are **epithelial columns**. Mesenchymal cells surround the columns as they extend deeper and deeper into the dermis. Hair follicles, sebaceous glands, and sweat glands develop from these columns.

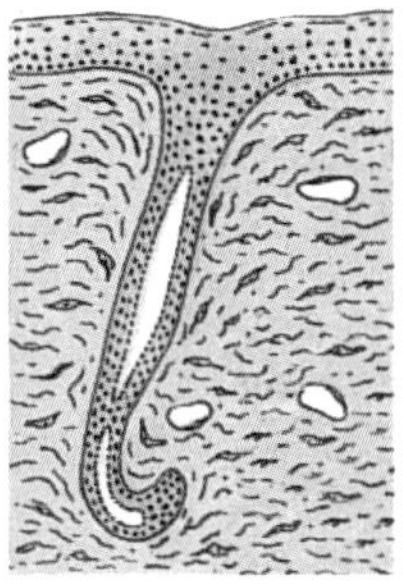

5 MONTHS

A sweat gland develops as an epithelial column elongates, coils, and becomes hollow.

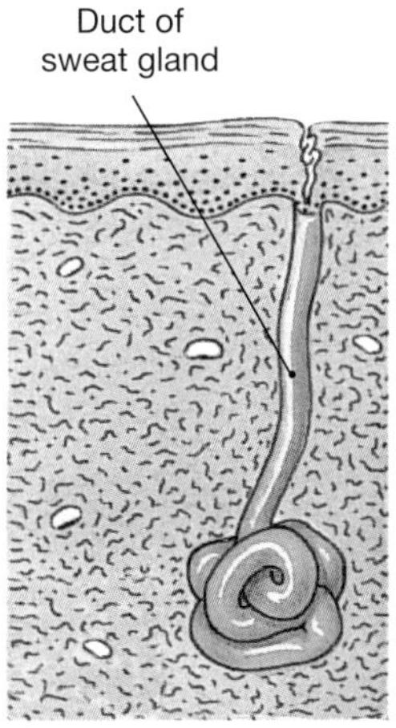

BIRTH

At birth, sweat gland ducts carry the secretions of the gland cells to the skin surface.

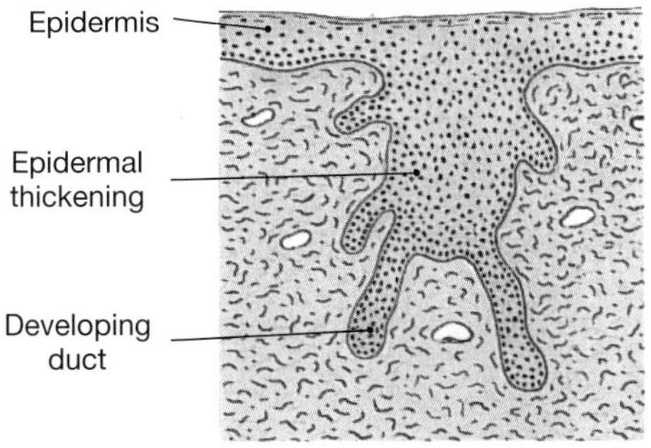

5 MONTHS

Mammary glands develop in a comparable fashion, but the epidermal thickenings are much broader and extensive branching occurs.

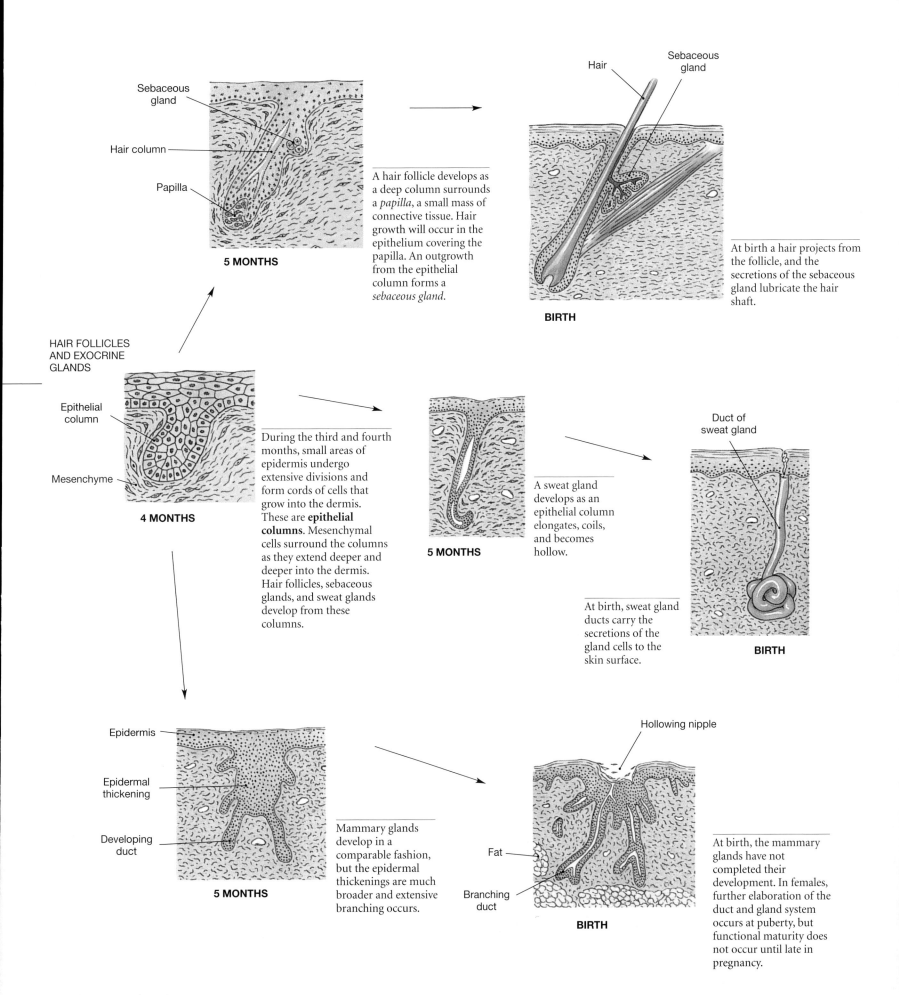

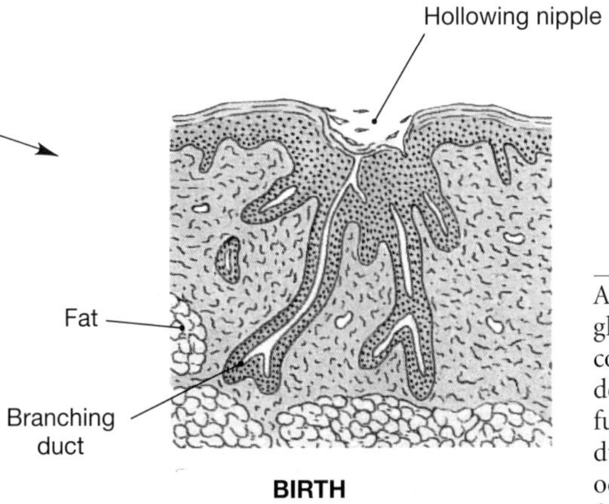

BIRTH

At birth, the mammary glands have not completed their development. In females, further elaboration of the duct and gland system occurs at puberty, but functional maturity does not occur until late in pregnancy.

✓ CONCEPT CHECK

- Compare the secretion of apocrine and merocrine sweat glands. Which produces secretions targeted by the deodorant industry?
- What is sensible perspiration?
- How does the control of merocrine gland secretion differ from the control of sebaceous and apocrine gland secretion?

Nails [FIGURE 4.15]

Nails form on the dorsal surfaces of the tips of the fingers and toes. The nails protect the exposed tips of the fingers and toes, and help limit distortion when the digits are subjected to mechanical stress—for example, in grasping objects or running. The structure of a nail can be seen in Figure 4.15●. The **nail body** covers the **nail bed**, but nail production occurs at the **nail root**, an epithelial fold not visible from the surface. The deepest portion of the nail root lies very close to the periosteum of the bone of the fingertip.

The nail body is recessed beneath the level of the surrounding epithelium, and it is bounded by **nail grooves** and **nail folds.** A portion of the stratum corneum of the nail fold extends over the exposed nail nearest the root, forming the **eponychium**, or *cuticle* (ep-ō-NIK-ē-um; *epi-*, over + *onyx*, nail). Underlying blood vessels give the nail its characteristic pink color, but near the root these vessels may be obscured, leaving a pale crescent known as the **lunula** (LOO-nū-la; *luna*, moon). The free edge of the nail body extends over a thickened stratum corneum, the **hyponychium** (hī-pō-NIK-ē-um).

Changes in the shape, structure, or appearance of the nails are clinically significant. A change may indicate the existence of a disease process affecting metabolism throughout the body. For example, the nails may turn yellow in patients who have chronic respiratory disorders, thyroid gland disorders, or AIDS. They may become pitted and distorted in psoriasis and concave in some blood disorders.

Local Control of Integumentary Function

The integumentary system displays a significant degree of functional independence. It responds directly and automatically to local influences without the involvement of the nervous or endocrine systems. For example, when the skin is subjected to mechanical stresses, stem cells in the stratum germinativum divide more rapidly, and the depth of the epithelium increases. That is why *calluses* form on your palms when you perform manual labor. A more dramatic display of local regulation can be seen after an injury to the skin. The skin can regenerate effectively even after considerable damage has occurred, because stem cells persist in both the epithelial and connective tissue components. Germinative cell divisions replace epidermal cells, and mesenchymal cell divisions replace lost fibroblasts and other dermal cells. This process can be slow, and when large surface areas are involved, problems of infection and fluid loss complicate the situation. † *Inflammation of the Skin, Complications of Inflammation, Burns and Grafts, Scar Tissue Formation*, and *Synthetic Skin pp. 787–789.*

After severe damage, the repair process does not return the integument to its original condition. The injury site contains an abnormal density of collagen fibers and relatively few blood vessels. Damaged hair follicles, sebaceous or sweat glands, muscle cells, and nerves are seldom repaired, and they

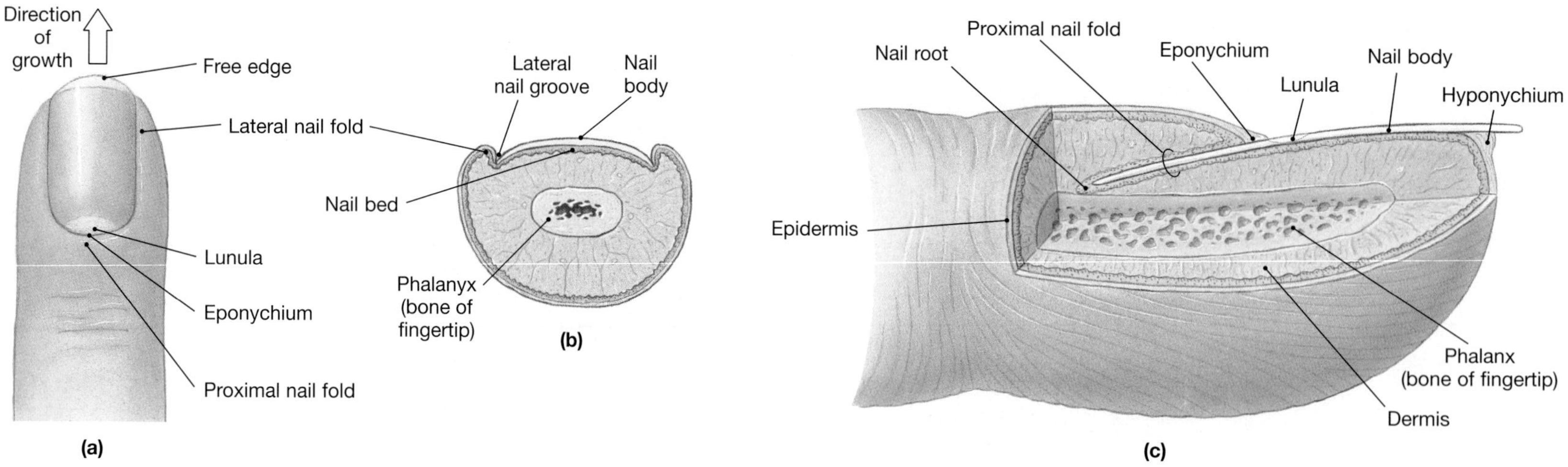

FIGURE 4.15 **STRUCTURE OF A NAIL**

These drawings illustrate the prominent features of a typical fingernail as viewed (**a**) from the surface, (**b**) in cross-section, and (**c**) in longitudinal section.

CLINICAL BRIEF

ANTHRAX

Anthrax is a bacteria that usually infects and kills hoofed animals such as goats, sheep, cows, pigs, and horses. It is spread by infected animals. In soil, these bacteria form spores; each spore consists of the cell nucleus surrounded by a bit of cytoplasm and a relatively thick, protective cell wall. Spores have no metabolic activity, and they are highly resistant to heat, dehydration, chemicals, and radiation. Spores can survive for years in soil or on animal hides or wool, but they are killed by burning, boiling for 10 minutes, autoclaving (a form of medical sterilization), treatment with potassium permanganate or hydrogen peroxide, or exposure to dilute formaldehyde.

Although it usually affects domesticated animals, on rare occasions the spores can infect humans. A skin infection called **cutaneous anthrax** is the most common form of human anthrax infection. If viable spores enter the body through a break in the skin they are engulfed by macrophages. The spores aren't damaged, and they then activate (germinate) to become bacteria that multiply and secrete toxins. This kills the macrophage and the release of toxins produces a localized firm edema. At the site of exposure, a pustule develops over 1-7 days with progressive redness and swelling. A fluid-filled vesicle containing bacteria may form, followed by a dry black scab, or *eschar* (ES-kar) develops. It is this black crust that gives this disease its name: in Greek, *anthrax* means charcoal. The eschar is not painful, but swollen and painful lymph nodes may occur as the bacteria spread. This transition from localized to systemic infection occurs in 10-20% of untreated cases, but these will probably result in fatalities.

Inhalation anthrax is much less common and much more dangerous. When inhaled and engulfed by macrophages, the spores germinate within the mediastinal lymph nodes. After a period of 1-3 days, vague, flu-like symptoms develop. Rapid spread of bacteria through the bloodstream can lead to death within 24 hours.

Both forms of anthrax infections can be treated with common antibiotics such as penicillin, tetracycline, or Cipro™. However, the window of time for successful treatment of inhalation anthrax is very short, due to its rapid spread into the circulatory system. Vaccination of agricultural workers, veterinarians, and soldiers at risk of exposure provides incomplete protection. However, until recently this was not a concern for the U.S. or Europe. Vaccination of farm animals, which began in the 1950s, essentially eliminated anthrax from these areas. This situation changed suddenly and dramatically after September 11, 2001 with the distribution of anthrax through the mail system.

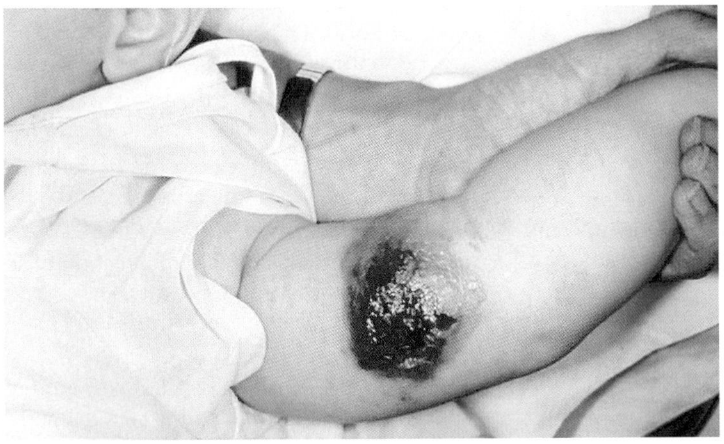

too are replaced by fibrous tissue. The formation of this rather inflexible, fibrous, noncellular **scar tissue** is a practical limit to the healing process.

Skin repairs proceed most rapidly in young, healthy individuals. For example, it takes 3–4 weeks to complete the repairs to a blister site in a young adult. The same repairs at age 65–75 take 6–8 weeks. However, this is just one example of the changes that occur in the integumentary system as a result of the aging process.

Aging and the Integumentary System

[FIGURE 4.16]

Aging affects all of the components of the integumentary system. These changes are summarized in Figure 4.16●.

1. *The epidermis thins* as germinative cell activity declines, making older people more prone to injury and skin infections.
2. *The number of Langerhans cells decreases* to around 50% of levels seen at maturity (approximately age 21). This decrease may reduce the sensitivity of the immune system and further encourage skin damage and infection.
3. *Vitamin D production declines* by around 75%. The result can be muscle weakness and a reduction in bone strength.
4. *Melanocyte activity declines*, and in Caucasians the skin becomes very pale. With less melanin in the skin, older persons are more sensitive to sun exposure and more likely to experience sunburn.
5. *Glandular activity declines.* The skin becomes dry and often scaly because sebum production is reduced; merocrine sweat glands are also less active. With impaired perspiration, older people cannot lose heat as fast as younger people can. Thus, the elderly are at greater risk of overheating in warm environments.
6. *The blood supply to the dermis is reduced* at the same time that sweat glands become less active. This combination makes the elderly less able to lose body heat, and overexertion or overexposure to warm temperatures (such as a sauna or hot-tub) can cause dangerously high body temperatures.
7. *Hair follicles stop functioning* or produce thinner, finer hairs. With decreased melanocyte activity, these hairs are gray or white.
8. *The dermis becomes thinner*, and the elastic fiber network decreases in size. The integument therefore becomes weaker and less resilient; sagging and wrinkling occur. These effects are most pronounced in areas exposed to the sun.
9. *Secondary sexual characteristics in hair and body fat distribution begin to fade* as the result of changes in levels of sex hormones. In consequence, people age 90–100 of both sexes and all races look very much alike.
10. *Skin repairs proceed relatively slowly*, and recurring infections may result.

FIGURE 4.16 **The Skin during the Aging Process**

Characteristic changes in the skin during aging; some causes and some effects.

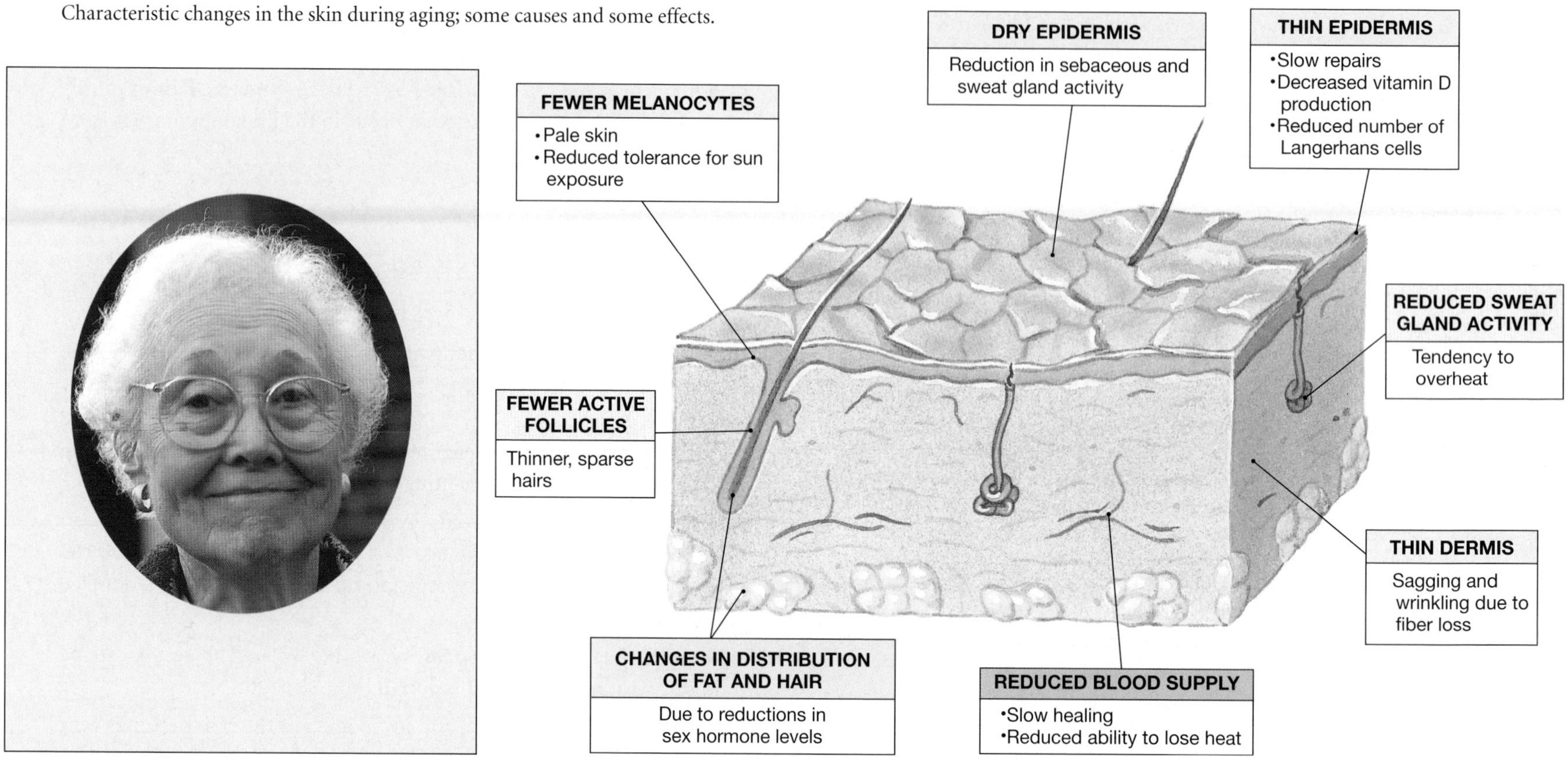

RELATED CLINICAL TERMS

acne: A sebaceous gland inflammation caused by an accumulation of secretions. *Acne p. 787*

basal cell carcinoma: A malignant cancer that originates in the stratum germinativum. This is the most common skin cancer, and roughly two-thirds of these cancers appear in areas subjected to chronic UV exposure. Metastasis rarely occurs. *Skin Cancers p. 786*

capillary hemangioma: A birthmark caused by a tumor in the capillaries of the papillary layer of the dermis. It usually enlarges after birth, but subsequently fades and disappears. *Tumors in the Dermis p. 787*

cavernous hemangiomas: ("port-wine stains") A birthmark caused by a tumor affecting larger vessels in the dermis. Such birthmarks usually last a lifetime. *Tumors in the Dermis p. 787*

contact dermatitis: A dermatitis usually caused by strong chemical irritants. It produces an itchy rash that may spread to other areas if scratching distributes the chemical agent; poison ivy is an example. *Dermatitis p. 787*

contraction: A pulling together of the edges of a wound during the healing process. *Inflammation of the Skin p. 787*

decubitis ulcers ("bedsores"): Ulcers that occur in areas subjected to restricted circulation, especially common in bedridden persons. *Complications of Inflammation p. 788*

dermatitis: An inflammation of the skin that involves primarily the papillary region of the dermis. *Dermatitis p. 787*

diaper rash: A localized dermatitis caused by a combination of moisture, irritating chemicals from fecal or urinary wastes, and flourishing microorganisms. *Dermatitis p. 787*

eczema (EK-se-ma): A dermatitis that can be triggered by temperature changes, fungus, chemical irritants, greases, detergents, or stress, and that can be related to hereditary or environmental factors. *Dermatitis p. 787*

erysipelas (er-i-SIP-e-las): A widespread inflammation of the dermis caused by bacterial infection. *Complications of Inflammation p. 788*

granulation tissue: A combination of fibrin, fibroblasts, and capillaries that forms during tissue repair following inflammation. *Inflammation of the Skin p. 787*

hypodermic needle: A needle used to administer drugs via subcutaneous injection. *p. 96*

keloid (KĒ-loyd): A thickened area of scar tissue covered by a shiny, smooth epidermal surface. Keloids most often develop on the upper back, shoulders, anterior chest, and earlobes in black or dark-skinned individuals. *Scar Tissue Formation p. 789*

malignant melanoma (mel-a-NŌ-ma): A skin cancer originating in malignant melanocytes. A potentially fatal metastasis often occurs. *Skin Cancers p. 786*

psoriasis (sō-RĪ-a-sis): A painless condition characterized by rapid stem cell divisions in the stratum germinativum of the scalp, elbows, palms, soles, groin, and nails. Affected areas appear dry and scaly. *Psoriasis and Xerosis p. 786*

scab: A fibrin clot that forms at the surface of a wound to the skin. *Inflammation of the Skin p. 788*

seborrheic dermatitis: An inflammation around abnormally active sebaceous glands. *p. 102*

sepsis: A dangerous, widespread bacterial infection. Sepsis is the leading cause of death in burn patients. *Burns and Grafts p. 788*

skin graft: Transplantation of a section of skin (partial thickness or full thickness) to cover an extensive injury site, such as a third-degree burn. *Burns and Grafts p. 788*

squamous cell carcinoma: A less common form of skin cancer almost totally restricted to areas of sun-exposed skin. Metastasis seldom occurs except in advanced tumors. *Skin Cancers p. 786*

urticaria (ur-ti-KAR-ē-a): or hives An extensive dermatitis resulting from an allergic reaction to food, drugs, an insect bite, infection, stress, or other stimulus. *Dermatitis p. 787*

xerosis (zē-RŌ-sis): "Dry skin," a common complaint of older persons and almost anyone living in an arid climate. *Psoriasis and Xerosis p. 786*

Additional Clinical Terms Discussed in Appendix I (pp. 767–789)

abscess; alopecia areata; athlete's foot; cellulitis; cold sores; comedos; first-degree burn; hirsutism; hyperkeratosis; inflammation; necrosis; pus; regeneration; second-degree burn; third-degree burn; ulcer; warts

STUDY OUTLINE & CHAPTER REVIEW

Integumentary Structure and Function 88

1. The **integumentary system**, or **integument**, consists of the **cutaneous membrane** or *skin*, which includes the superficial **epidermis** and deeper **dermis**, and the **accessory structures**, including: **hair follicles**, **exocrine glands**, and **nails**. The subcutaneous layer is deep to the cutaneous membrane. *(see Figures 4.1/4.2)*
2. The cutaneous membrane covers the outside of the body and consists of a stratified squamous epithelium, the epidermis, and an underlying layer of connective tissue, the dermis. *(see Figure 4.2)*

The Epidermis 89

1. There are four cell types in the epidermis: **keratinocytes**, the most abundant epithelial cells; **melanocytes**, pigment-producing cells; **Merkel cells**, involved in detecting sensation; and **Langerhans cells**, which are phagocytic cells of the immune system. Melanocytes, Merkel cells, and Langerhans cells are scattered among the keratinocytes.
2. The epidermis is a *stratified squamous epithelium*. There are five layers of keratinocytes in the epidermis in **thick skin** and four layers in **thin skin**. *(see Figure 4.4)*

Layers of the Epidermis 89

3. Division of basal cells in the **stratum germinativum** produces new keratinocytes, which replace more superficial cells. *(see Figures 4.2 to 4.6)*
4. As new committed epidermal cells differentiate, they pass through the **stratum spinosum**, the **stratum granulosum**, the **stratum lucidum** (of thick skin), and the **stratum corneum**. The keratinocytes move toward the surface, and through the process of keratinization the cells accumulate large amounts of **keratin**. Ultimately the cells are shed or lost at the epidermal surface. *(see Figure 4.3)*

Thick and Thin Skin 91

5. **Thin skin** covers most of the body; **thick skin** covers only the heavily abraded surfaces, such as the palms of the hands and the soles of the feet. *(see Figure 4.4)*
6. **Epidermal ridges**, such as those on the palms and soles, improve our gripping ability and increase the skin's sensitivity. Their pattern is determined genetically. The ridges interlock with **dermal papillae** of the underlying dermis. *(see Figures 4.4/4.5)*
7. The color of the epidermis depends on a combination of two factors: blood supply to the skin and composition and concentration of two pigments, **carotene** and **melanin**. Melanin helps to protect the skin from the damaging effects of excessive **ultraviolet radiation**. *(see Figure 4.6)*

The Dermis 93

Dermal Organization 93

1. Two layers comprise the dermis: the superficial **papillary layer** and the deeper **reticular layer**. *(see Figures 4.2/4.4/4.7 to 4.9)*
2. The papillary layer derives its name from its association with the dermal papillae. It contains blood vessels, lymphatics, and sensory nerves. This layer supports and nourishes the overlying epidermis. *(see Figures 4.4/4.7)*
3. The reticular layer consists of a meshwork of collagen and elastic fibers oriented in all directions to resist tension in the skin. *(see Figure 4.8)*

Other Dermal Components 95

4. An extensive blood supply to the skin includes the cutaneous and papillary arterial plexuses. The papillary layer contains abundant capillaries. *(see Figure 4.2)*
5. Sensory nerves innervate the skin. They monitor touch, temperature, pain, pressure, and vibration. *(see Figure 4.2)*

The Subcutaneous Layer 95

1. The **subcutaneous layer** is also referred to as the **hypodermis** or **superficial fascia**. Although not part of the integument, it stabilizes the skin's position against underlying organs and tissues yet permits limited independent movement. *(see Figures 4.2/4.7)*

Accessory Structures 96

Hair Follicles and Hair 96

1. Hairs originate in complex organs called **hair follicles**, which extend into the dermis. Each hair has a **bulb**, **root**, and **shaft**. Hair production involves a special keratinization of the epithelial cells of the **hair matrix**. At the center of the matrix, the cells form a soft core, or **medulla**; cells at the edge of the hair form a hard **cortex**. The **cuticle** is a hard layer of dead, keratinized cells that coats the hair. *(see Figures 4.2/4.9 to 4.11)*
2. The lumen of the follicle is lined by an **internal root sheath** produced by the hair matrix. An **external root sheath** surrounds the internal root sheath, between the skin surface and hair matrix. The **glassy membrane** is the thickened basement membrane external to the external root sheath; it is wrapped by a dense connective tissue layer. *(see Figure 4.10)*
3. A **root hair plexus** of sensory nerves surrounds the base of each hair follicle and detects the movement of the shaft. Contraction of the **arrector pili muscles** elevates the hairs by pulling on the follicles. *(see Figures 4.9/4.10a)*
4. **Vellus hairs** ("peach fuzz"), heavy **terminal hairs**, and **intermediate hairs** make up the hair population on our bodies. *(see Figure 4.11)*
5. Hairs grow and are shed according to the **hair growth cycle**. A single hair grows for 2–5 years and is subsequently shed. *(see Figure 4.11)*

Glands in the Skin 99

6. **Sebaceous oil glands** discharge a waxy, oily **sebum** into hair follicles. **Sebaceous follicles** are large sebaceous glands that produce no hair; they communicate directly with the epidermis. *(see Figure 4.13)*
7. **Apocrine sweat glands** produce an odorous secretion; the more numerous **merocrine sweat glands**, or *eccrine sweat glands*, produce a thin, watery secretion known as **sensible perspiration**, or **sweat**. *(see Figures 4.12/4.14)*
8. The **mammary glands** of the breast resemble larger and more complex apocrine sweat glands. Active mammary glands secrete milk. **Ceruminous glands** in the ear canal are modified sweat glands, which produce a waxy **cerumen**.

Nails 104

9. The **nails** protect the exposed tips of the fingers and toes, and help limit their distortion when they are subjected to mechanical stress.
10. The **nail body** covers the **nail bed**, with nail production occurring at the **nail root**. The **cuticle**, or **eponychium**, is formed by a fold of the stratum corneum, the **nail fold**, extending from the **nail root** to the exposed nail. *(see Figure 4.15)*

Local Control of Integumentary Function 104

1. The skin can regenerate effectively even after considerable damage, such as severe cuts or moderate burns. *(see Table A.3 in the Clinical Issues Appendix)*
2. Severe damage to the dermis and accessory glands cannot be completely repaired, and fibrous **scar tissue** remains at the injury site.

Aging and the Integumentary System 105

1. Aging affects all layers and accessory structures of the integumentary system. *(see Figure 4.16)*

LEVEL 1 REVIEWING FACTS AND TERMS

Match each numbered item with the most closely related lettered item. Use letters for answers in the spaces provided.

Column A	Column B
____ 1. hypodermis	a. fibrous, noncellular
____ 2. dermis	b. holocrine; oily secretion
____ 3. stem cell	c. pigment cells
____ 4. keratinized/cornified	d. stratum germinativum
____ 5. melanocytes	e. superficial fascia
____ 6. epidermis	f. papillary layer
____ 7. sebaceous gland	g. stratum corneum
____ 8. sweat gland	h. stratified squamous epithelium
____ 9. scar tissue	i. merocrine; clear secretion

10. The most abundant cells in the epidermis are
    (a) adipocytes (b) keratinocytes
    (c) Merkel cells (d) melanocytes
11. The effects of aging on the skin include
    (a) a decline in the activity of sebaceous glands
    (b) increased production of vitamin D
    (c) thickening of the epidermis
    (d) an increased blood supply to the dermis
12. Nail production occurs at the nail
    (a) body (b) bed
    (c) root (d) hyponychium
13. Sensible perspiration is produced by
    (a) ceruminous glands (b) apocrine sweat glands
    (c) merocrine sweat glands (d) sebaceous glands
14. The layer of the skin that contains both interwoven bundles of collagen fibers, and the protein elastin, and which is responsible for the strength of the skin, is the
    (a) papillary layer (b) reticular layer
    (c) epidermal layer (d) hypodermal layer
15. The layer of the epidermis that contains cells undergoing division is the
    (a) stratum corneum (b) stratum germinativum
    (c) stratum granulosum (d) stratum lucidum
16. Water loss due to penetration of interstitial fluid through the surface of the skin is termed
    (a) sensible perspiration (b) insensible perspiration
    (c) latent perspiration (d) active perspiration
17. Sweat glands that communicate with hair follicles in the axillae (armpits) and produce an odorous secretion are
    (a) apocrine glands (b) merocrine glands
    (c) sebaceous glands (d) both a and b
18. Each of the following is a function of the integumentary system except one. Identify the exception.
    (a) protection of underlying tissue
    (b) excretion
    (c) synthesis of vitamin C
    (d) thermoregulation
19. Carotene is
    (a) an orange-yellow pigment that accumulates inside epidermal cells
    (b) another name for melanin
    (c) deposited in stratum granulosum cells to protect the epidermis
    (d) a pigment that gives the characteristic color to hemoglobin
20. The cells in a hair follicle that are responsible for forming hair are
    (a) papillary cells (b) matrix cells
    (c) cortex cells (d) medullary cells

LEVEL 2 REVIEWING CONCEPTS

1. Epidermal ridges
   (a) are at the surface of the epidermis only
   (b) cause ridge patterns on the surface of the skin
   (c) produce patterns that are determined by the environment
   (d) interconnect with desmosomes of the stratum spinosum
2. What is the importance of the secretion of sebum?
3. Compare and contrast sensible and insensible perspiration.
4. Stretch marks may result from pregnancy. What makes stretch marks occur?
5. How does the protein keratin affect the appearance and function of the integument?
6. What characteristic(s) make(s) the subcutaneous layer a region frequently targeted for hypodermic injection?
7. Why do washing the skin and applying deodorant reduce the odor of apocrine sweat glands?
8. What is happening to an individual who is cyanotic, and what body structures would show this condition most easily?
9. Scar tissue is the result of
   (a) increased numbers of epidermal layers in the area of injury
   (b) a thickened stratum germinativum in the area of injury
   (c) increased numbers of fibroblasts and mast cells in the injured area
   (d) an abnormally large number of collagen fibers and relatively few blood vessels at the repair site
10. Skin can regenerate effectively even after considerable damage has occurred because
    (a) the epidermis of the skin has a rich supply of small blood vessels
    (b) fibroblasts in the dermis give rise to new epidermal germinal cells
    (c) contraction in the injured area brings cells of adjacent strata together
    (d) stem cells persist in both the epithelial and connective tissue components of the skin even after injury

LEVEL 3 CRITICAL THINKING AND CLINICAL APPLICATIONS

1. In a condition called sunstroke, the victim appears flushed, the skin is warm and dry, and the body temperature rises dramatically. Explain these observations based on what you know concerning the role of the skin in thermoregulation.
2. Exposure to optimum amounts of sunlight is necessary for proper bone maintenance and growth in children. How does sunlight promote bone maintenance and growth?
3. Many medications can be administered transdermally by applying patches that contain the medication to the surface of the skin. These patches can be attached anywhere on the skin except the palms of the hands and the soles of the feet. Why?

✓ ANSWERS TO CONCEPT CHECK QUESTIONS

p. 91 **1.** Cells are shed constantly from the outer layers of the stratum corneum. **2.** The splinter is lodged in the stratum granulosum. **3.** The two major subdivisions of the integumentary system are the cutaneous membrane and the accessory structures. The cutaneous membrane has two parts: the superficial epidermis and the deeper dermis. The accessory structures include hair follicles, exocrine glands, and nails. **4.** Keratinization is the production of keratin by epidermal cells. It occurs in the stratum granulosum of the epidermis. Keratin fibers develop within cells of the stratum granulosum. As keratin fibers are produced, these cells become thinner and flatter, and their cell membranes become thicker and less permeable. As these cells die, they form the densely packed layers of the stratum lucidum and stratum corneum.

p. 93 **1.** These terms refer to the relative thickness of the epidermis, not to the integument as a whole. Thick skin occurs on the palms of the hands (epidermal thickness may be as much as 0.50 mm thick) whereas thin skin covers most of the body (epidermal thickness averages 0.08 mm). **2.** Sanding the tips of one's fingers will not permanently remove fingertips. Since the ridges of the fingertips are formed in layers of the skin that are constantly regenerated, they will eventually reappear. The actual pattern of the ridges is determined by arrangement of tissue in the dermis, which is not affected by the sanding. **3.** The color of the epidermis is due to a combination of the dermal blood supply and variable quantities of two pigments, carotene and melanin. **4.** Epidermal ridges are formed by the deeper layers of the epidermis that extend into the dermis, increasing the contact area between the two regions. Dermal papillae are projections of the dermis that extend between adjacent ridges.

p. 99 **1.** When the dermis is stretched excessively, the elastic fibers are overstretched and are not able to recoil. The skin then forms folds or wrinkles, called stretch marks, in the affected areas. **2.** Contraction of the arrector pili pulls the hair follicle erect, depressing the area at the base of the hair and making the surrounding skin appear higher. The combined activity of the arrectores pilorum produces "goose bumps" or "goose pimples." **3.** Each hair has a medulla, produced by the central portion of the hair matrix, surrounded by a cortex and covered by a cuticle. The shaft of the hair begins where its internal organization is complete (roughly halfway toward the surface).

p. 104 **1.** Apocrine sweat glands produce a viscous, cloudy, and potentially odorous secretion. The secretion contains several kinds of organic compounds. Some of these have an odor, and others produce an odor when metabolized by skin bacteria. Deodorants are used to mask the resulting odor. Merocrine sweat glands produce a watery secretion that chiefly contains sodium chloride, metabolites, and waste products. Merocrine secretions are watery, and are generally known as sweat. **2.** Sensible perspiration is the sweat produced by merocrine sweat glands. **3.** Sebaceous and apocrine sweat glands can be collectively turned on or off by the autonomic nervous system, but no local or regional control is possible. The ANS controls the amount of merocrine sweat gland secretion and the region of the body involved.

REVIEW IT

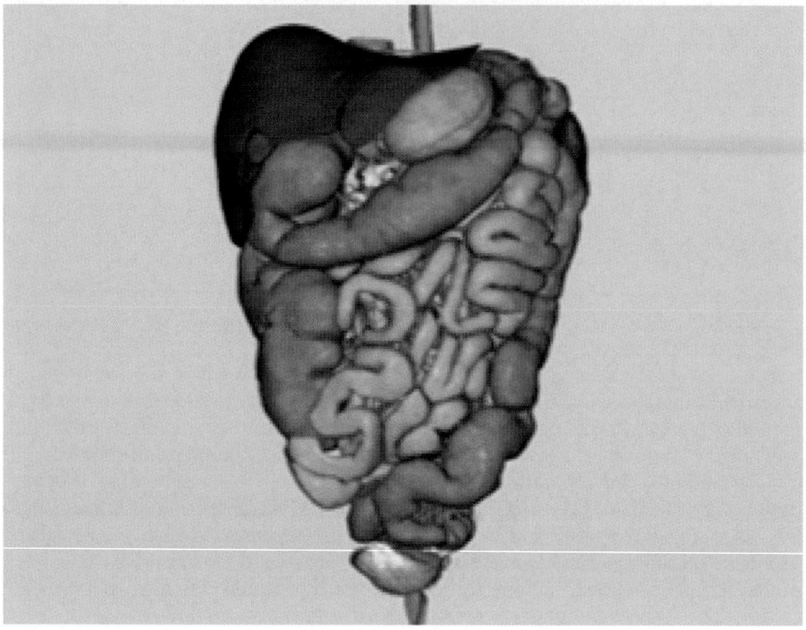

Challenge 1

It is usually challenging to compare three dimensional images with standard illustrations. Keep in mind that illustrations are often idealized views of the body. Practice your visualization skills by working with both the images in the chapter and those depicted on the enclosed Interactive CD.

The diagram to the left is taken from the animations on the CD (Regional view/Trunk). Using this diagram and your text, identify first what area of the body we are looking at using medical terms. Next correctly identify the organs indicated. Finally, compare this view with that seen in Figure 1.9●, p. 15. Indicate the quadrant where you would find each organ identified above.

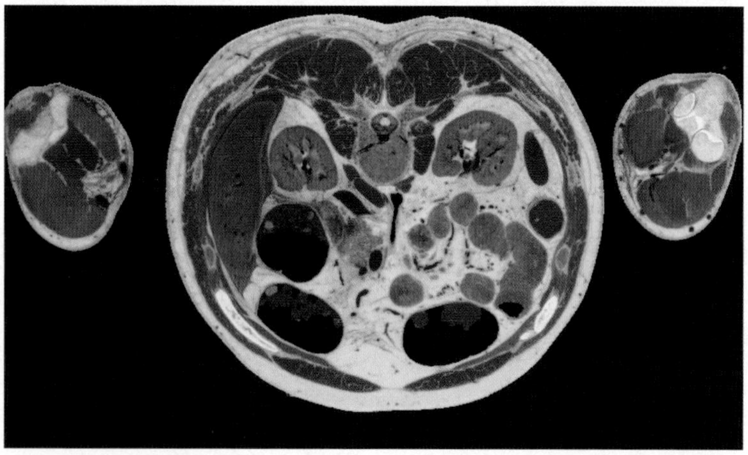

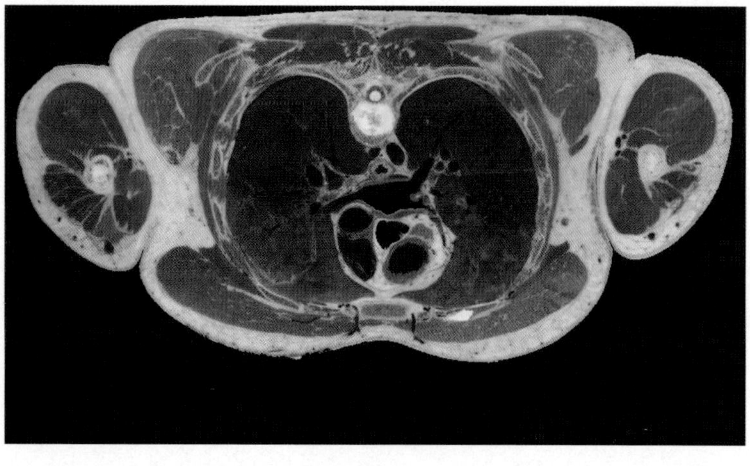

Challenge 2

What examples of each of the four tissue types do you see in section A? Copy the chart below onto a separate sheet of paper and complete it, making it longer as necessary. Compare this section with the one below it. What new tissues do you see? Add these to your chart.

Epithelial Tissue

Label number ______ – organ name: ______________________

Function of tissue: ______________________

Specific tissue type: ______________________

Connective Tissue

Label number ______ – organ name: ______________________

Function of tissue: ______________________

Specific tissue type: ______________________

Nervous Tissue

Label number ______ – organ name: ______________________

Function of tissue: ______________________

Specific tissue type: ______________________

Images provided by the Digital Cadaver™ project, courtesy of Visible Productions, Inc.

APPLY IT

Below are two separate exercises that provide more information on a topic presented in the preceding chapters. Each one is designed to take approximately 10 minutes, and will help you gain a better appreciation for the material presented in Chapters 1–4.

Critical Linking 1

Marfan syndrome is an inherited genetic disorder that affects connective tissue. The abnormal gene, which is located on chromosome 15, alters the structure of *fibrillin*, a glycoprotein important for strength, elasticity and cohesion within connective tissues. Fibrillin is composed of elastic fibers; in Marfan syndrome things that should be distensible and elastic become brittle. The older name for this disease is *arachnodactyly*, or spider fingers, because one of the outward signs of this disorder is extremely long fingers and toes. Individuals are usually very tall and slender, which can be unusual or inconvenient but not dangerous. However, Marfan syndrome can be fatal due to its effects on internal connective tissues. Deaths usually result from complications involving the cardiovascular system. Reflect on what you know of elastic connective tissues and their potential role in the cardiovascular system. Come up with a hypothesis as to how abnormal fibrillin could lead to fatal cardiovascular problems. Write your hypothesis on a sheet of paper or type it into your computer. Your thoughts should include the usual functions for connective tissue as well as your prediction of why fibrillin would be vital to this normal functioning.

After predicting why Marfan syndrome could lead to death, jump to the Companion Website critical linking section. Click on the keyword "Marfan" to read information on the disease from the National Marfan Foundation. Refine your predictions and prepare a short paragraph explaining the pathology of Marfan syndrome.

Critical Linking 2

Electron microscopy reveals yet another level of anatomical studies. Many times the images obtained using scanning electron microscopy have served to enhance the microanatomy already known.

Visit the Companion Website and move to the critical linking section for chapter 4. Click on the keyword "*Electron Microscopy*". This will take you to an electronic textbook of Dermatology, Anatomy of the Skin, by Rhett J. Drugge, MD, Internet Dermatology Society, Stamford, Connecticut. Move to the photomicrographs and compare what you see there to the images of skin in your text. What do these electron microscope images show you that differs from what you have seen in the text? Do the images on the website depict structures as you imagined them to appear, or do the structures have surprising surfaces etc? Come up with three new observations based on these images and present them in outline form to share with your classmates. Include these new observations in your notes to help in your studies of this material.

FURTHER STUDIES

Anatomy has been recognized as a science for ages. The history of anatomy is an interesting study of its own. Scholarly writings in anatomy can be found as far back as Mesopotamia. There are 660 medical tablets from the library of Asshurbanipal, the last great king of Assyria, indicating the extent of medical knowledge at the time. In studying these tablets, it is interesting to note that Mesopotamian practitioners blamed disease not only on divine spirits but also on the malfunctioning of the organ itself. This indicates a relatively advanced understanding of functional anatomy. Surgeries requiring a knowledge of anatomy were performed at this early date, as indicated by tablets that describe the draining of the pleural cavity, the use of a knife to scrape the skull of a patient, and an entire tablet dedicated to the dressing and care of post-operative surgical wounds. More information on this segment of history can be obtained by visiting *http://ww.indiana.edu/~ancmed/meso.HTM*. There are other good sites for information on the history of anatomy. For an overview of this topic, visit *http://www.english.upenn.edu/~jlynch/Frank/Contexts/anatomy.html*.

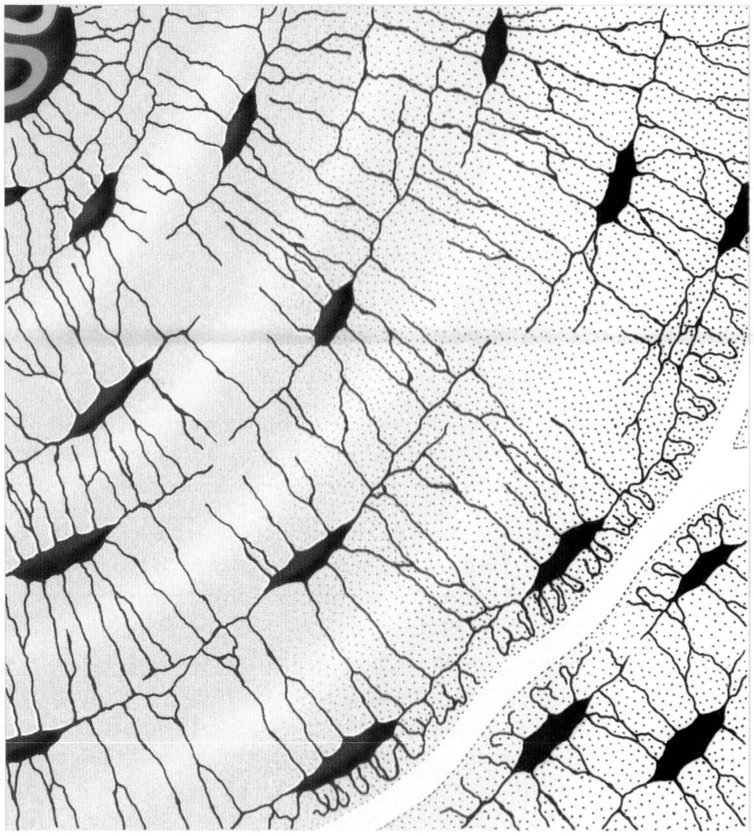

5

THE SKELETAL SYSTEM

Osseous Tissue and Skeletal Structure

The skeletal system includes the varied bones of the skeleton and the cartilages, ligaments, and other connective tissues that stabilize or interconnect them. Bones are more than racks that muscles hang from; they support our weight and work together with muscles to produce controlled, precise movements. Without a framework of bones to hold onto, contracting muscles would just get shorter and fatter. Our muscles must pull against the skeleton to make us sit, stand, walk, or run. The skeleton has many other vital functions; some may be unfamiliar to you, so we will begin this chapter by summarizing the major functions of the skeletal system.

1. *Support*: The skeletal system provides structural support for the entire body. Individual bones or groups of bones provide a framework for the attachment of soft tissues and organs.
2. *Storage of Minerals and Lipids*: The calcium salts of bone represent a valuable mineral reserve that maintains normal concentrations of calcium and phosphate ions in body fluids. Calcium is the most abundant mineral in the human body. A typical human body contains 1–2 kg (2.2–4.4 lb) of calcium, with more than 98% of it deposited in the bones of the skeleton. Additionally, energy reserves may be stored within bones in the form of lipids in areas of *yellow marrow*.
3. *Blood Cell Production*: Red blood cells, white blood cells, and platelets are produced in the *red marrow,* which fills the internal cavities of many bones. The role of the bone marrow in blood cell formation will be described in later chapters that discuss the cardiovascular and lymphatic systems (Chapters 20 and 23).
4. *Protection*: Delicate tissues and organs are often surrounded by skeletal elements. The ribs protect the heart and lungs, the skull encloses the brain, the vertebrae shield the spinal cord, and the pelvis cradles delicate digestive and reproductive organs.
5. *Leverage*: Many bones of the skeleton function as levers. They can change the magnitude and direction of the forces generated by skeletal muscles. The movements produced range from the delicate motion of a fingertip to powerful changes in the position of the entire body.

This chapter describes the structure, development, and growth of bone. The two chapters that follow organize bones into two divisions, the *axial skeleton* (consisting of the bones of the skull, vertebral column, sternum and ribs) and the *appendicular skeleton* (consisting of the bones of the limbs and the associated bones that connect the limbs to the trunk at the shoulders and pelvis). The final chapter in this group examines articulations or joints, structures where the bones meet and may move with respect to each other.

The bones of the skeleton are actually complex, dynamic organs that contain osseous tissue, other connective tissues, smooth muscle tissue, and neural tissue. We will now consider the internal organization of a typical bone.

Structure of Bone

Bone, or **osseous tissue**, is one of the supporting connective tissues. (You should review the sections on dense connective tissues, cartilage, and bone at this time.) p. 68–75 Like other connective tissues, osseous tissue contains specialized cells and an extracellular matrix consisting of protein fibers and a ground substance. The matrix of bone tissue is solid and sturdy due to the deposition of calcium salts around the protein fibers.

Osseous tissue is usually separated from surrounding tissues by a fibrous *periosteum*. When osseous tissue surrounds another tissue, the inner bony surfaces are lined by a cellular *endosteum*.

The Histological Organization

The basic organization of bone tissue was introduced in Chapter 3. pp. 71–73 We will now take a closer look at the organization of the matrix and cells of bone.

THE MATRIX OF BONE

Calcium phosphate, $Ca_3(PO_4)_2$, accounts for almost two-thirds of the weight of bone. The calcium phosphate interacts with calcium hydroxide $[Ca(OH)_2]$ to form crystals of **hydroxyapatite**, (hī-DROK-sē-ap-a-tīt) $Ca_{10}(PO_4)_6(OH)_2$. As they form, these crystals also incorporate other calcium salts, such as calcium carbonate, and ions such as sodium, magnesium, and fluoride. These inorganic components provide compressional strength to bone. Roughly one-third of the weight of bone is from collagen fibers, which contribute tensile strength to bone. Osteocytes and other cell types account for only 2% of the mass of a typical bone.

Calcium phosphate crystals are very strong, but relatively inflexible. They can withstand compression, but the crystals are likely to shatter when exposed to bending, twisting, or sudden impacts. Collagen fibers are tough and flexible. They can easily tolerate stretching, twisting, and bending, but when compressed, they simply bend out of the way. In bone, the collagen fibers provide an organic framework for the formation of mineral crystals. The hydroxyapatite crystals form small plates that lie alongside the collagen fibers. The result is a protein–crystal combination with properties intermediate between those of collagen and those of pure mineral crystals.

THE CELLS OF BONE [FIGURE 5.1]

Osseous tissue contains a distinctive population of cells, including *osteocytes, osteoblasts, osteoclasts*, and *osteoprogenitor cells* (Figure 5.1a●).

OSTEOCYTES **Osteocytes** (*osteon*, bone) are mature bone cells. They maintain and monitor the protein and mineral content of the surrounding matrix. As you will see in a later section, the minerals in the matrix are continually being recycled. Each osteocyte directs both the release of calcium from bone to blood and the deposition of calcium salts in the surrounding matrix. Osteocytes occupy small chambers, called **lacunae**, that are sandwiched between layers of calcified matrix. These matrix layers are known as **lamellae** (la-MEL-lē; singular: *lamella*, a thin plate) (Figure 5.1b–d●). Channels, called **canaliculi** (kan-a-LIK-ū-lī; "little canals"), radiate through the matrix from lacuna to lacuna and toward free surfaces and adjacent blood vessels. The canaliculi, which contain cytoplasmic processes and ground substance, interconnect the lacunae and provide a route for the diffusion of nutrients and waste products, either through the ground substance or from cell to cell across gap junctions.

OSTEOBLASTS **Osteoblasts** (OS-tē-ō-blasts; *blast*, precursor) are cuboidal cells that are found on the inner or outer surfaces of a bone. These cells secrete the organic components of the bone matrix. This material, called **osteoid** (OS-tē-oyd), later becomes mineralized through an unknown mechanism. Osteoblasts are responsible for the production of new bone, a process called **osteogenesis** (os-tē-ō-JEN-e-sis; *gennan*, to produce). If an osteoblast becomes surrounded by matrix, it differentiates into an osteocyte.

OSTEOPROGENITOR CELLS Bone tissue also contains small numbers of mesenchymal cells called **osteoprogenitor cells** (os-tē-ō-prō-JEN-i-tor; *progenitor*, ancestor). Osteoprogenitor cells can divide to produce daughter cells that differentiate into osteoblasts. The ability to produce additional

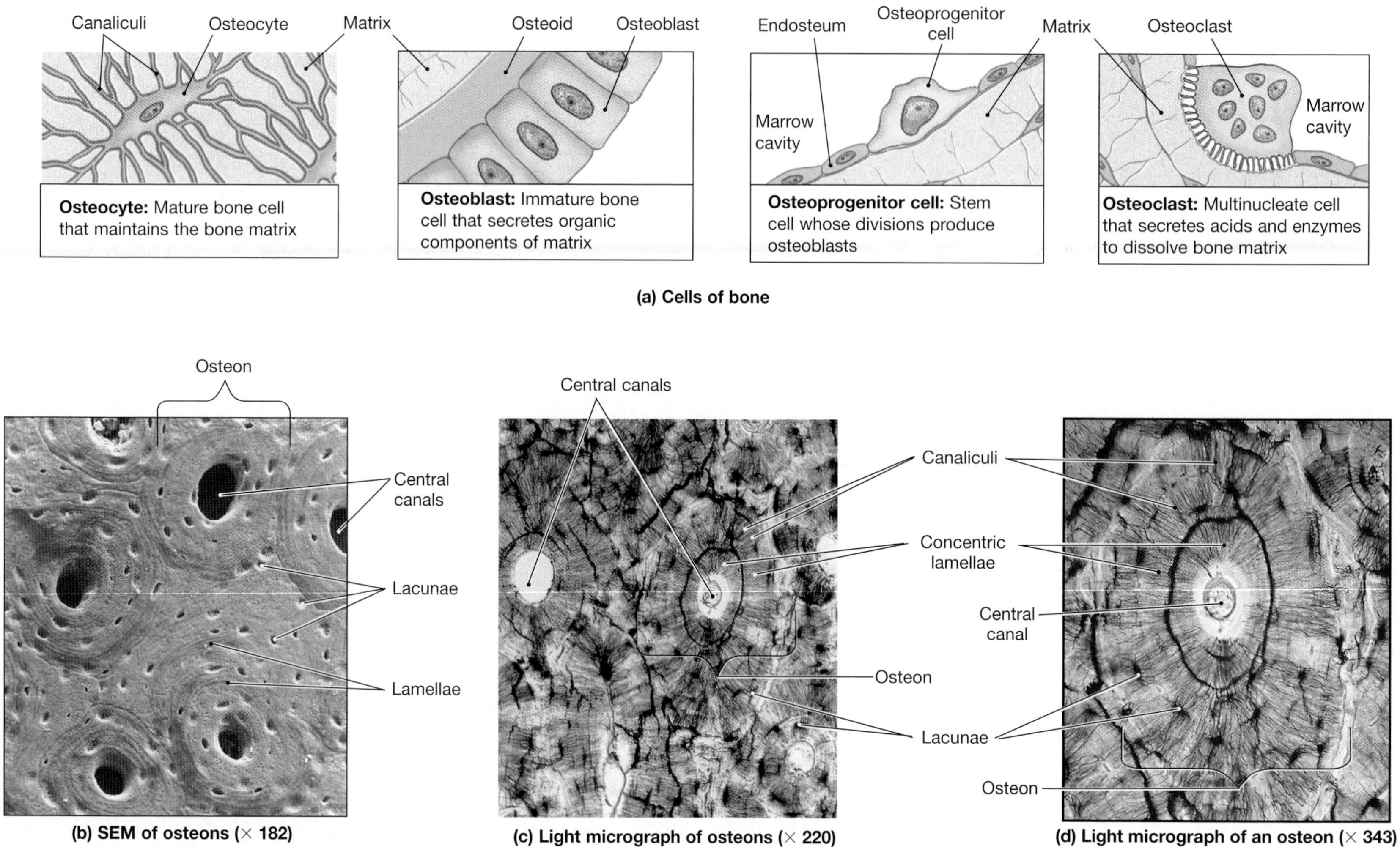

FIGURE 5.1 **STRUCTURE OF A TYPICAL BONE**

Osseous tissue contains specialized cells and a dense extracellular matrix containing calcium salts. **(a)** The cells of bone. **(b)** A scanning electron micrograph of several osteons in compact bone. (SEM × 182; reproduced from R. G. Kessel and R. H. Kardon, *Tissues and Organs: A Text-Atlas of Scanning Electron Microscopy*, W. H. Freeman & Co., 1979) **(c)** A thin section through compact bone; in this procedure the intact matrix and central canals appear white, and the lacunae and canaliculi are shown in blacks. **(d)** A single osteon at higher magnification.

osteoblasts becomes extremely important after a bone is cracked or broken. We will consider the repair process further in a later section.

OSTEOCLASTS **Osteoclasts** (OS-tē-ō-klasts; *clast*, to break) are giant cells with 50 or more nuclei. They secrete acids through the exocytosis of lysosomes. The acids dissolve the bony matrix and release the stored calcium and phosphate. This erosion process, called **osteolysis** (os-tē-OL-i-sis), increases the calcium and phosphate concentrations in body fluids. Osteoclasts are always removing matrix and releasing minerals, and osteoblasts are always producing matrix that quickly binds minerals. The balance between the activities of osteoblasts and osteoclasts is very important; when osteoclasts remove calcium salts faster than osteoblasts deposit them, bones become weaker. When osteoblast activity predominates, bones become stronger and more massive.

Compact and Spongy Bone [FIGURE 5.2]

There are two types of osseous tissue: *compact bone*, or dense bone; and *spongy bone*, or *cancellous* (KAN-se-lus) *bone*. **Compact bone** is relatively dense and solid, whereas **spongy bone** forms an open network of struts and plates. Both compact and spongy bone are present in typical bones of the skeleton, such as the *humerus*, the proximal bone of the upper limb, and the *femur*, the proximal bone of the lower limb. Compact bone tissue forms the walls, and an internal layer of spongy bone surrounds the **marrow cavity** (Figure 5.2a●). The marrow cavity contains **bone marrow**, a loose connective tissue that may be dominated by adipocytes (**yellow marrow**) or by a mixture of mature and immature red and white blood cells, and the stem cells that produce them (**red marrow**).

STRUCTURAL DIFFERENCES BETWEEN COMPACT AND SPONGY BONE

The matrix composition in compact bone is the same as that of spongy bone, but they differ in the three-dimensional arrangement of osteocytes, canaliculi, and lamellae.

COMPACT BONE [FIGURES 5.1b–d/5.2] The basic functional unit of mature compact bone is the cylindrical **osteon** (OS-tē-on), or *Haversian system* (Figure 5.1b–d●). Within an osteon the osteocytes are arranged in concentric layers around a **central canal**, or *Haversian canal*, which contains the blood vessels that supply the osteon. Central canals usually run parallel to the surface of the bone (Figure 5.2a●). Other passageways, known as **perforating canals**, or the *canals of Volkmann*, extend roughly perpendicular to the surface (Figure 5.2b●). Blood vessels in the perforating canals deliver blood to osteons deeper in the bone and service the interior marrow cavity. The lamellae of each osteon are cylindrical and aligned parallel to the long axis of the bone. These are known as *concentric lamellae.* Collectively, concentric lamellae form a series of

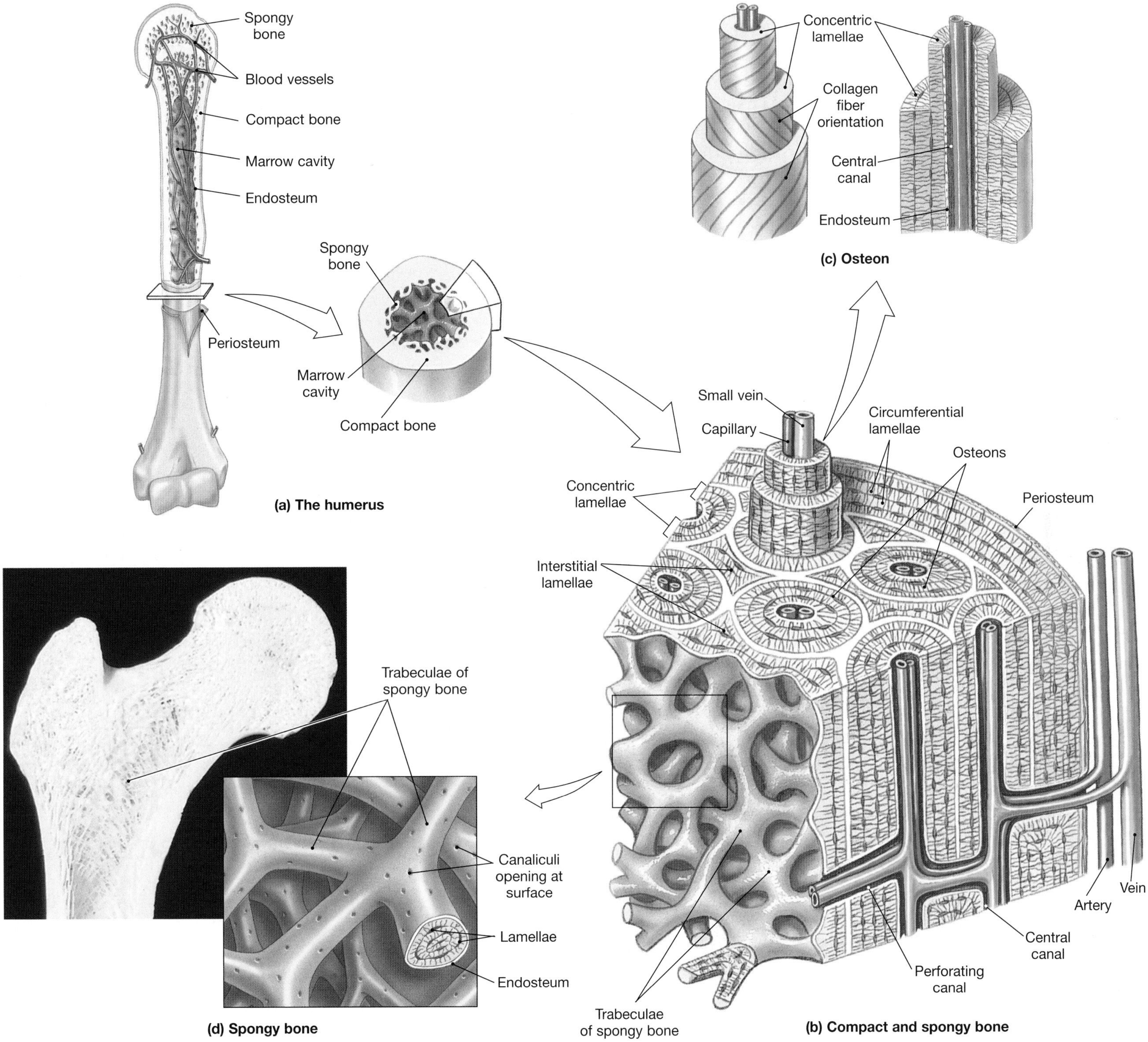

FIGURE 5.2 **THE INTERNAL ORGANIZATION IN REPRESENTATIVE BONES**

The structural relationship of compact and spongy bone in representative bones. (**a**) Gross anatomy of the humerus. (**b**) Diagrammatic view of the histological organization of compact and spongy bone. (**c**) The organization of collagen fibers within concentric lamellae. (**d**) Location and structure of spongy bone. The photo shows a sectional view of the head of the femur.

concentric rings, resembling a "bull's-eye" target, around the central canal (Figure 5.2b,c●). The collagen fibers spiral along the length of each lamella, and variations between the direction of spiraling in adjacent lamellae strengthen the osteon as a whole. Canaliculi interconnect the lacunae of an osteon and form a branching network that reaches the central canal. *Interstitial lamellae* fill in the spaces between the osteons in compact bone. Depending on their location, these lamellae either may have been produced during the growth of the bone, or they may represent remnants of osteons whose matrix components have been recycled by osteoclasts. A third type of lamellae, the *circumferential lamellae*, occur at the external and internal surfaces of the bone. In a limb bone such as the humerus or femur, the circumferential lamellae form the outer and inner surfaces of the shaft (Figure 5.2b●).

SPONGY BONE [FIGURE 5.2d] The primary differences between spongy bone and compact bone are: (1) in spongy bone, parallel lamellae form struts or thin, branching plates called **trabeculae** (tra-BEK-ū-lē; "a little beam");

and (2) there are no osteons in spongy bone, and nutrients reach the osteocytes by diffusion along canaliculi that open onto the surfaces of the trabeculae. In terms of the associated cells and the structure and composition of the lamellae, spongy bone is no different from compact bone. Spongy bone forms an open framework (Figure 5.2d●), and as a result it is much lighter than compact bone. However, the branching trabeculae give spongy bone considerable strength despite its relatively light weight. Thus, the presence of spongy bone reduces the weight of the skeleton and makes it easier for muscles to move the bones. Spongy bone is thus found wherever bones are not stressed heavily or where stresses arrive from many directions.

FUNCTIONAL DIFFERENCES BETWEEN COMPACT AND SPONGY BONE [FIGURE 5.3]

A layer of compact bone covers bone surfaces; the thickness of this layer varies from region to region and from one bone to another. This superficial layer of compact bone is in turn covered by the *periosteum*, a connective tissue wrapping that is connected to the deep fascia. The periosteum is complete everywhere except within a joint, where the edges or ends of two bones contact one another. In some joints, the two bones are interconnected by collagen fibers or a block of cartilage. In more mobile fluid-filled (*synovial*) joints, hyaline *articular cartilages* cover the opposing bony surfaces.

Compact bone is thickest where stresses arrive from a limited range of directions. Figure 5.3a● shows the general anatomy of the *femur*, the proximal bone of the lower limb. The compact bone of the **cortex** surrounds the marrow cavity, also known as the *medullary cavity* (*medulla*, innermost part). The bone has two ends, or **epiphyses** (e-PIF-i-sēs; singular, *epiphysis*; *epi*, above + *physis*, growth), separated by a tubular **diaphysis** (dī-AF-i-sis; "a growing between"), or **shaft**. The diaphysis is connected to the epiphysis at a narrow zone known as the **metaphysis** (me-TAF-i-sis). Figure 5.3● shows the organization of compact and spongy bone within the femur. The shaft of compact bone normally conducts applied stresses from one epiphysis to another. For example, when you are standing, the shaft of

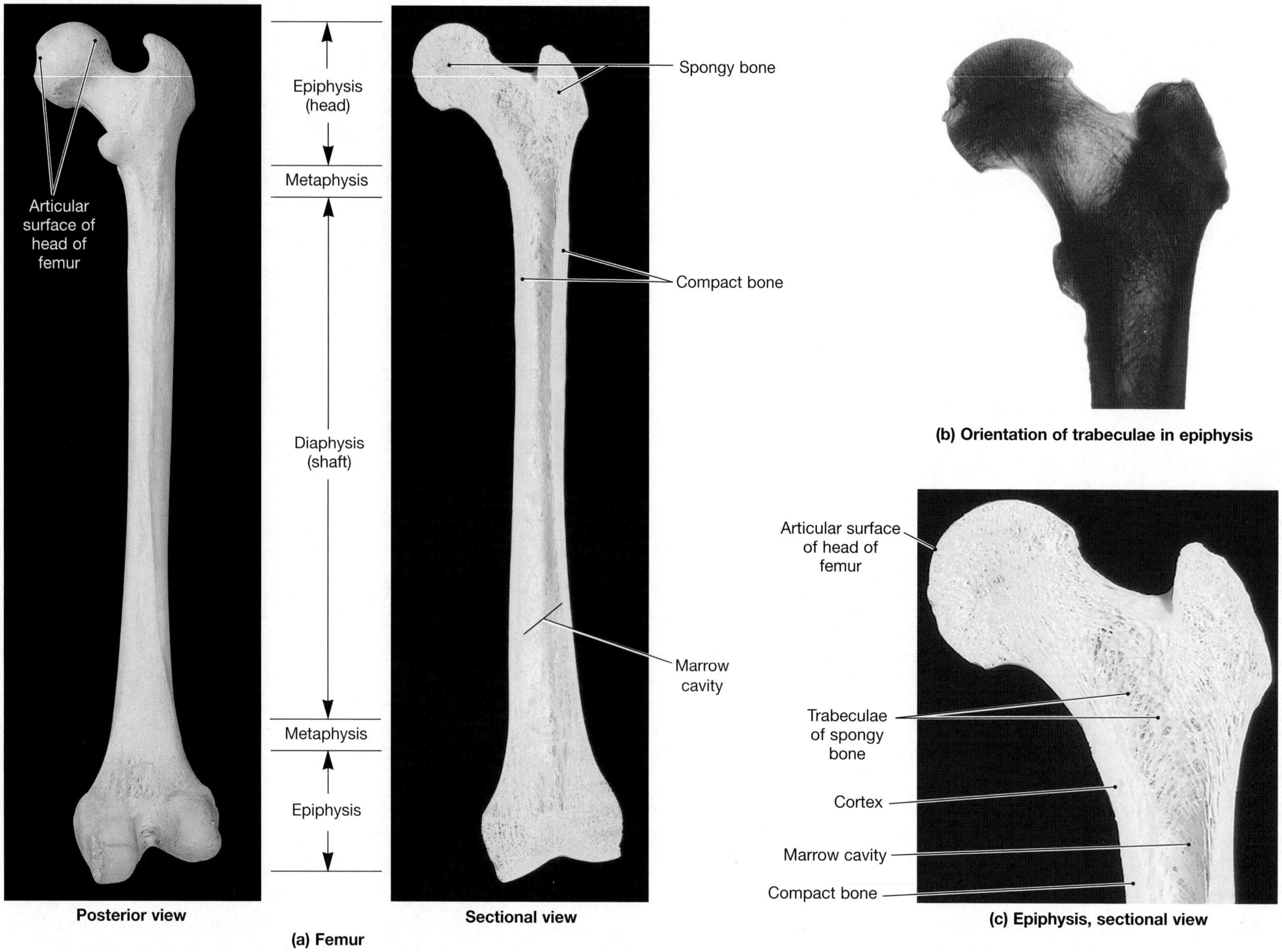

FIGURE 5.3 **ANATOMY OF A REPRESENTATIVE BONE**

(a) The femur, or thigh bone, in superficial and sectional views. The femur has a diaphysis (shaft) with walls of compact bone and epiphyses (heads) filled with spongy bone. A metaphysis separates the diaphysis and epiphysis at each end of the shaft. The body weight is transferred to the femur at the hip joint. Because the hip joint is off-center relative to the axis of the shaft, the body weight is distributed along the bone so that the medial portion of the shaft is compressed and the lateral portion is stretched. **(b)** An intact femur chemically cleared to show the orientation of the trabeculae in the epiphysis. **(c)** A photograph showing the epiphysis after sectioning.

the femur conducts your body weight from your hip to your knee. The osteons within the shaft are parallel to its long axis, and as a result the femur is very strong when stressed along that axis. You might envision a single osteon as a drinking straw with very thick walls. When you try to push the ends of a straw together it seems quite strong. However, if you hold the ends and push the side of the straw, it will break easily. Similarly, a long bone does not bend when forces are applied to either end, but an impact to the side of the shaft can easily cause a break, or *fracture.*

Spongy bone is not as massive as compact bone, but it is much more capable of resisting stresses applied from many different directions. The epiphyses of the femur are filled with spongy bone, and the trabecular alignment of the proximal epiphysis is shown in Figure 5.3b,c●. The trabeculae are oriented along the stress lines, but with extensive cross-bracing. At the proximal epiphysis, the trabeculae transfer forces from the hip across the metaphysis to the femoral shaft; at the distal epiphysis, the trabeculae direct the forces across the knee joint to the leg. In addition to reducing weight and handling stress from many directions, the open trabecular framework provides support and protection for the cells of the bone marrow. Yellow marrow, often found in the marrow cavity of the shaft, is an important energy reserve. Extensive areas of red marrow, such as that found in the spongy bone of the femoral epiphyses, are important sites of blood cell formation.

The Periosteum and Endosteum [FIGURE 5.4]

The outer surface of a bone is covered by a **periosteum** that consists of a fibrous outer layer and a cellular inner layer (Figure 5.4a●). The periosteum (1) isolates and protects the bone from surrounding tissues, (2) provides a route and a place of attachment for circulatory and nervous supply,

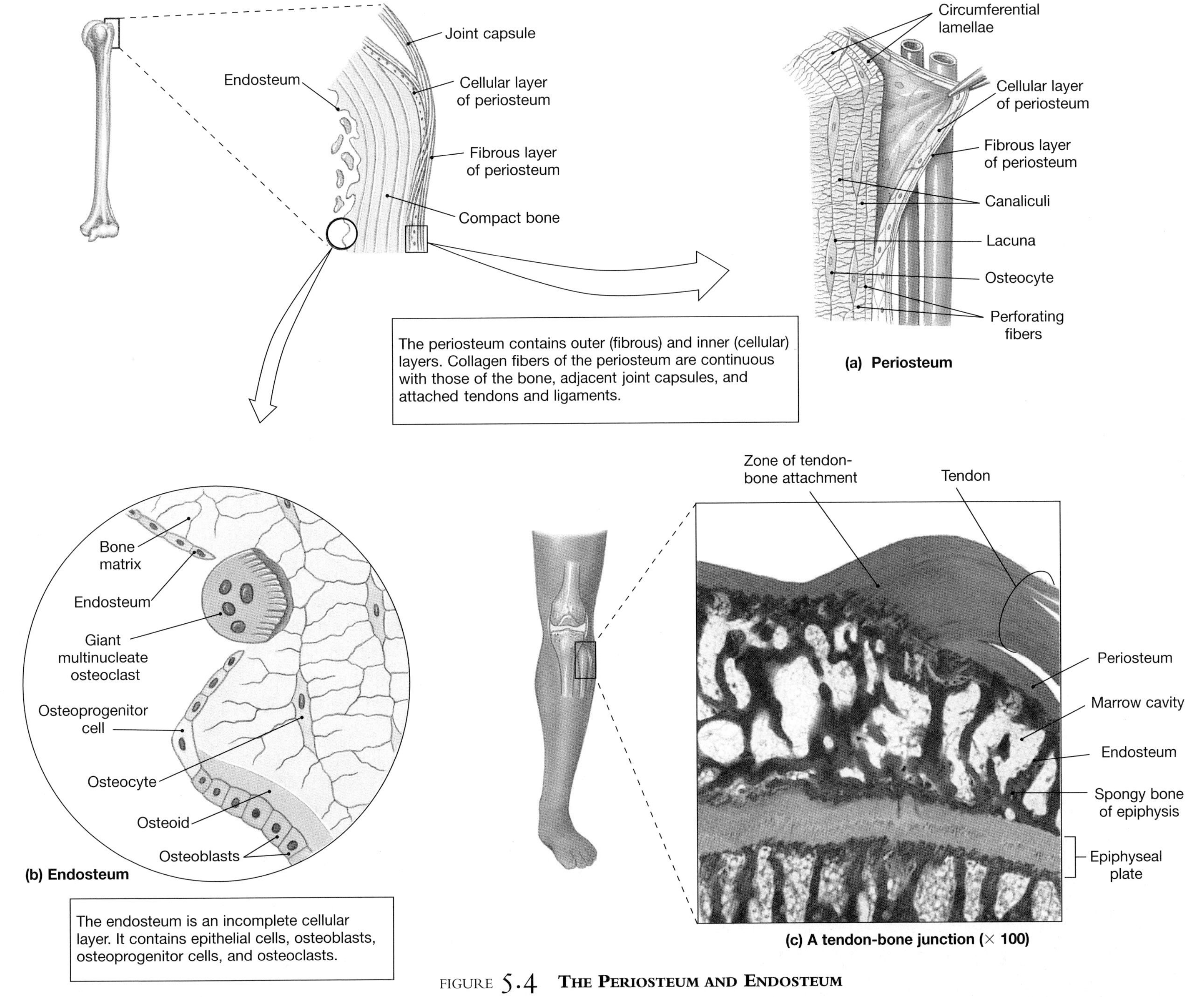

FIGURE 5.4 **THE PERIOSTEUM AND ENDOSTEUM**

Diagrammatic representation of periosteum and endosteum locations and their association with other bone structures; histology section shows both periosteum and endosteum. **(a)** The periosteum. **(b)** The endosteum. **(c)** A tendon–bone junction. (LM × 100)

(3) actively participates in bone growth and repair, and (4) attaches the bone to the connective tissue network of the deep fascia.

Near joints, the periosteum becomes continuous with the connective tissue network that surrounds and helps stabilize the joint. At a fluid-filled (*synovial*) joint, the periosteum is continuous with the *joint capsule* that encloses the joint complex. The fibers of the periosteum are also interwoven with those of the tendons attached to the bone (Figure 5.4c●). As the bone grows, these tendon fibers are cemented into the superficial lamellae by osteoblasts from the cellular layer of the periosteum. The collagen fibers incorporated into bone tissue from tendons and from the superficial periosteum are called *perforating fibers*, or *Sharpey's fibers* (Figure 5.4a●). The cementing process makes the tendon fibers part of the general structure of the bone, providing a much stronger bond than would otherwise be possible. An extremely powerful pull on a tendon or ligament will usually break the bone rather than snap the collagen fibers at the bone surface.

Inside the bone, a cellular **endosteum** lines the marrow cavity (Figure 5.4b●). This layer covers the trabeculae of spongy bone and lines the inner surfaces of the central canals. The endosteum is active during the growth of bone and whenever repair or remodeling is under way. The endosteum is not a complete epithelial layer, and the bone matrix is occasionally exposed. At these exposed sites, osteoclasts and osteoblasts have access to the mineralized surfaces.

CLINICAL BRIEF

RICKETS

The hip joint consists of the head of the femur and a corresponding socket on the lateral surface of the hip bone. The femoral head projects medially, and body weight compresses the medial side of the diaphysis. Because the force is applied off-center, the bone has a tendency to bend into a lateral bow. The other side of the shaft, which resists this bending, is placed under a stretching load, or *tension.* Bending does occur in disorders that reduce the amount of calcium salts in the skeleton. **Rickets** usually develops in children as a result of a vitamin D deficiency; this vitamin is essential to normal calcium absorption and deposition in the skeleton. In rickets, the bones are poorly mineralized, and they become very flexible. Affected individuals develop a bowlegged appearance as the thigh and leg bones bend under the weight of the body.

✓ CONCEPT CHECK

- How would the strength of a bone be affected if the ratio of collagen to calcium salts (hydroxyapatite) increased?
- A sample of bone shows concentric lamellae surrounding a central canal. Is the sample from the cortex or the medullary cavity of a long bone?
- If the activity of osteoclasts exceeds the activity of osteoblasts in a bone, how is the mass of the bone affected?
- If the osteoprogenitor cells in bone tissue were selectively destroyed by a poison, what future, normal process may be impeded?

Bone Development and Growth

The growth of the skeleton determines the size and proportions of our body. The bony skeleton begins to form about 6 weeks after fertilization, when the embryo is approximately 12 mm (1/2 in.) long. (Before this time all of the skeletal elements are cartilaginous.) During subsequent development, the bones undergo a tremendous increase in size. Bone growth continues through adolescence, and portions of the skeleton usually do not stop growing until age 25. The entire process is carefully regulated, and a breakdown in regulation will ultimately affect all of the body systems. In this section we will consider the physical process of *osteogenesis* (bone formation) and bone growth. The next section will examine the maintenance and replacement of mineral reserves in the adult skeleton.

During embryonic development, either mesenchyme or cartilage is replaced by bone. This process of replacing other tissues with bone is called **ossification**. The process of **calcification** refers to the deposition of calcium salts within a tissue. Any tissue can be calcified, but only ossification results in the formation of bone. There are two major forms of ossification. In *intramembranous ossification*, bone develops from mesenchyme or fibrous connective tissue. In *endochondral ossification*, bone replaces an existing cartilage model.

Intramembranous Ossification [FIGURES 5.5/5.6]

Intramembranous (in-tra-MEM-bra-nus) **ossification**, also called *dermal ossification*, begins when mesenchymal cells differentiate into osteoblasts within embryonic or fibrous connective tissue. This type of ossification normally occurs in the deeper layers of the dermis, and the bones that result are often called **dermal bones**, or *membrane bones.* Examples of dermal bones include the roofing bones of the skull, the *mandible* (lower jaw), and the *clavicle* (collar bone). *Sesamoid bones* form within tendons; the *patella* (kneecap) is an example of a sesamoid bone. Membrane bone may also develop in other connective tissues subjected to chronic mechanical stresses. For example, cowboys in the nineteenth century sometimes developed small bony plates in the dermis on the insides of their thighs, from friction and impact against their saddles. In some disorders affecting calcium ion metabolism or excretion, intramembranous bone formation occurs in many areas of the dermis and deep fascia. Bones in abnormal locations are called *heterotopic bones* (*heteros*, different + *topos*, place).

The steps in the process of intramembranous ossification are illustrated in Figure 5.5● and may be summarized as:

STEP 1. Mesenchymal cells differentiate into osteoblasts. The osteoblasts then cluster together and start to secrete the organic components of the matrix. The resulting mixture of collagen fibers and osteoid then becomes mineralized through the crystallization of calcium salts. The location in a bone where ossification begins is called an **ossification center**. As ossification proceeds, it traps some osteoblasts inside bony pockets; these cells differentiate into osteocytes.

STEP 2. The developing bone grows outward from the ossification center in small struts called **spicules**. Although osteoblasts are still being trapped in the expanding bone, mesenchymal cell divisions continue to produce additional osteoblasts. Bone growth is an active process, and osteoblasts require oxygen and a reliable supply of nutrients. As blood vessels branch within the region and grow between the spicules, the rate of bone growth accelerates.

STEP 3. Over time, the bone assumes the structure of spongy bone. However, subsequent remodeling around the trapped blood vessels can produce compact bone.

Figure 5.6a● shows skull bones forming through intramembranous ossification in the head of a 10-week fetus.

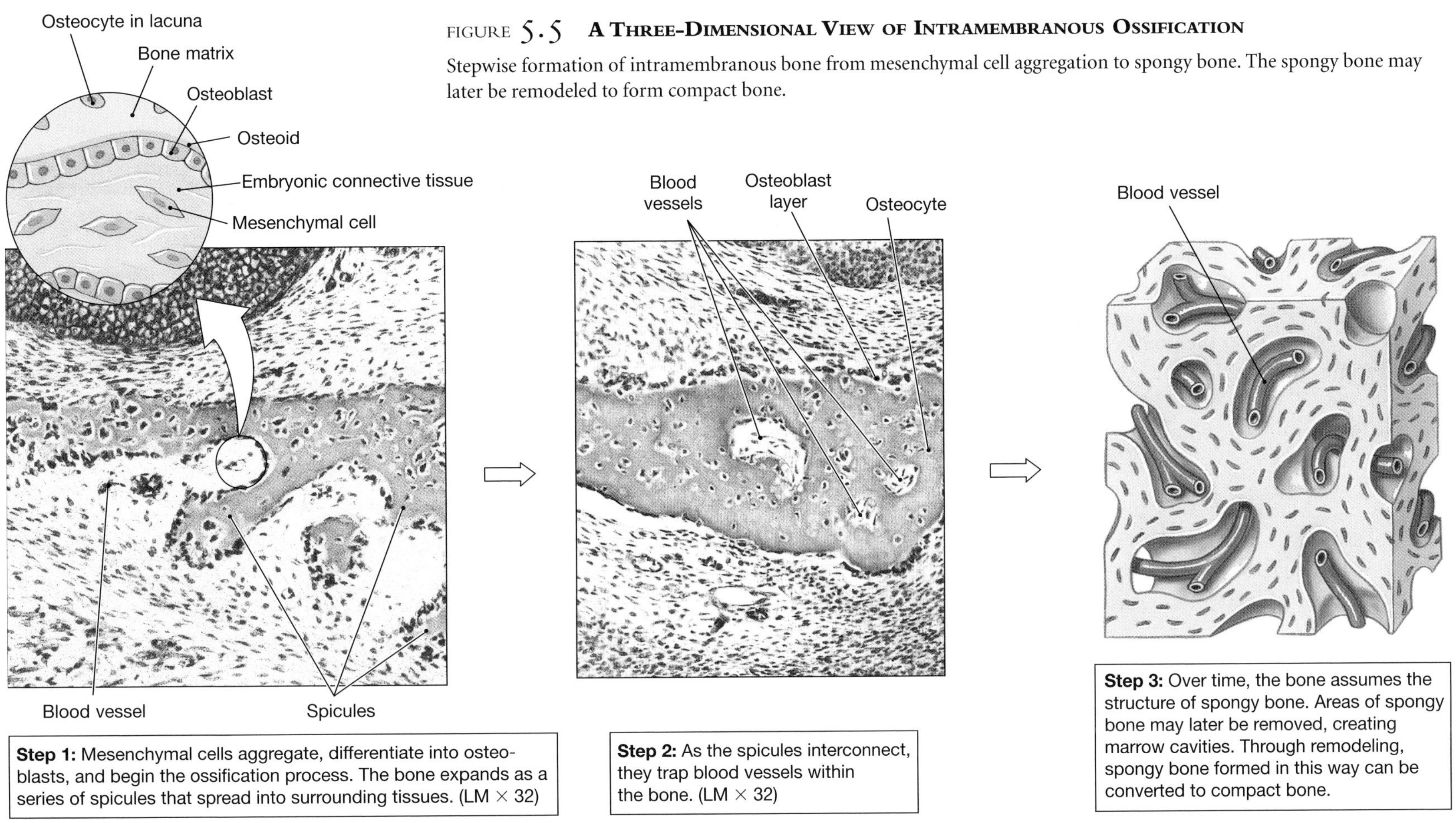

Osteocyte in lacuna
Bone matrix
Osteoblast
Osteoid
Embryonic connective tissue
Mesenchymal cell
Blood vessel
Spicules
Step 1: Mesenchymal cells aggregate, differentiate into osteoblasts, and begin the ossification process. The bone expands as a series of spicules that spread into surrounding tissues. (LM × 32)
Blood vessels
Osteoblast layer
Osteocyte
Step 2: As the spicules interconnect, they trap blood vessels within the bone. (LM × 32)
Blood vessel
Step 3: Over time, the bone assumes the structure of spongy bone. Areas of spongy bone may later be removed, creating marrow cavities. Through remodeling, spongy bone formed in this way can be converted to compact bone.

FIGURE 5.5 **A Three-Dimensional View of Intramembranous Ossification**

Stepwise formation of intramembranous bone from mesenchymal cell aggregation to spongy bone. The spongy bone may later be remodeled to form compact bone.

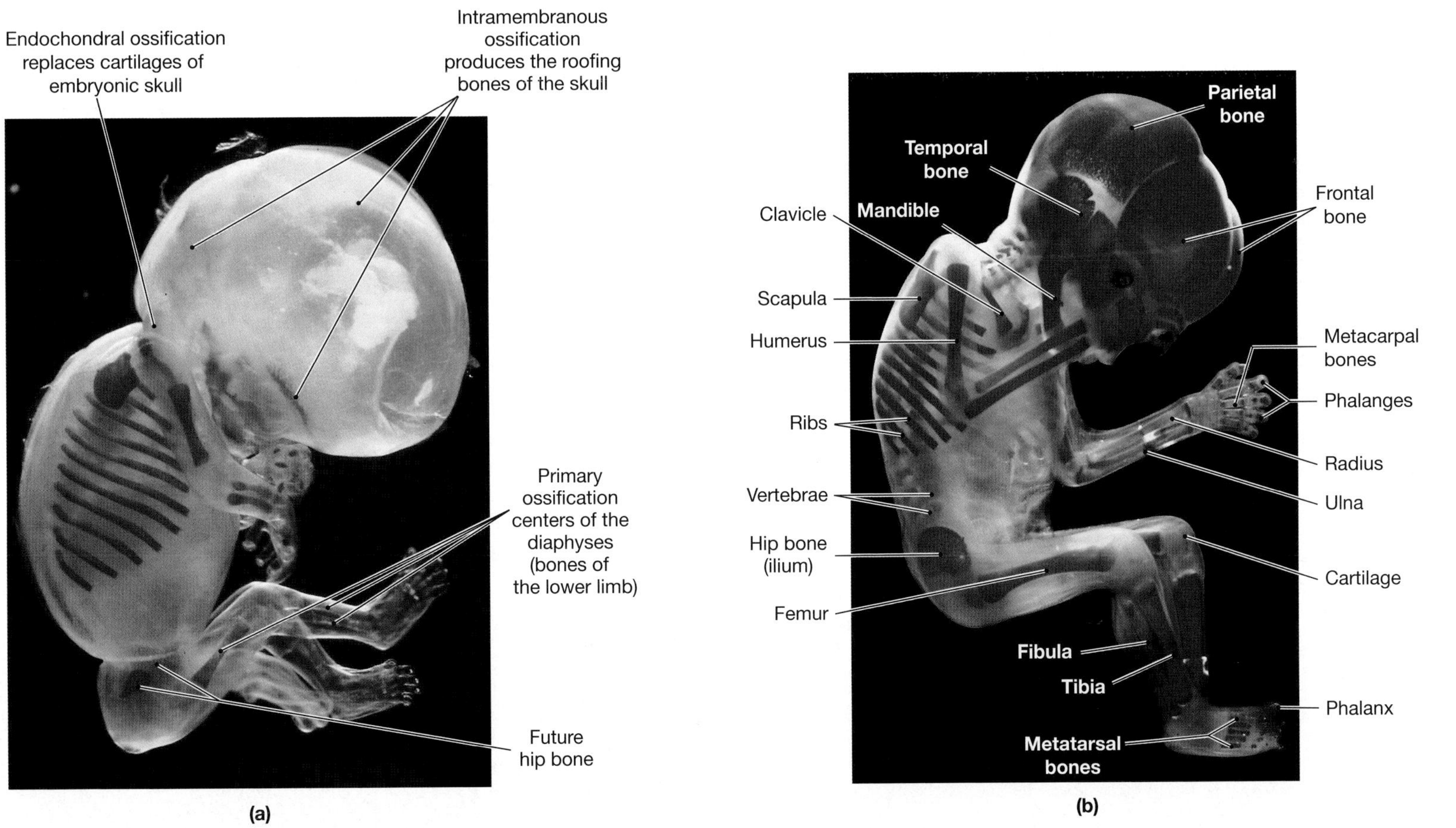

Endochondral ossification replaces cartilages of embryonic skull
Intramembranous ossification produces the roofing bones of the skull
Primary ossification centers of the diaphyses (bones of the lower limb)
Future hip bone
(a)
Parietal bone
Temporal bone
Frontal bone
Clavicle
Mandible
Scapula
Humerus
Metacarpal bones
Phalanges
Ribs
Radius
Ulna
Vertebrae
Hip bone (ilium)
Cartilage
Femur
Fibula
Tibia
Phalanx
Metatarsal bones
(b)

FIGURE 5.6 **Fetal Intramembranous and Endochondral Ossification**

These 10- and 16-week human fetuses have been specially stained (with alizarin red) and cleared to show developing skeletal elements. (**a**) At 10 weeks the fetal skull clearly shows both membrane and cartilaginous bone, but the boundaries that indicate the limits of future skull bones have yet to be established. (**b**) At 16 weeks the fetal skull shows the irregular margins of the future skull bones. Most elements of the appendicular skeleton form through endochondral ossification. Note the appearance of the wrist and ankle bones at 16 weeks versus at 10 weeks.

Endochondral Ossification [FIGURES 5.6/5.7]

Endochondral ossification (en-dō-KON-dral; *endo*, inside + *chondros*, cartilage) begins with the formation of a hyaline cartilage model. Limb bone development is a good example of this process. By the time an embryo is 6 weeks old, the proximal bone of the limb, either the *humerus* (upper limb) or *femur* (lower limb), has formed, but it is composed entirely of cartilage. This model continues to grow by expansion of the cartilage matrix (*interstitial growth*) and the production of more cartilage at the outer surface (*appositional growth*). These growth mechanisms were introduced in Chapter 3. ∞ pp. 70–71 Figure 5.6b● shows the extent of endochondral ossification occurring in the limb bones of a 16-week fetus. Steps in the growth and ossification of one of the limb bones are diagrammed in Figure 5.7a●.

STEP 1. As the cartilage enlarges, chondrocytes near the center of the shaft increase greatly in size, and the surrounding matrix begins to calcify. Deprived of nutrients, these chondrocytes die and disintegrate.

STEP 2. Cells of the perichondrium surrounding this region of the cartilage differentiate into osteoblasts. The perichondrium has now been converted into a periosteum, and the inner **osteogenic layer** (os-tē-ō-JEN-ik) soon produces a *bone collar*, a thin layer of bone around the shaft of the cartilage.

STEP 3. While these changes are under way, the blood supply to the periosteum increases, and capillaries and osteoblasts migrate into the heart of the cartilage, invading the spaces left by the disintegrating chondrocytes. The calcified cartilaginous matrix then breaks down, and osteoblasts replace it with spongy bone. Bone development proceeds from this **primary ossification center** in the shaft, toward both ends of the cartilaginous model.

STEP 4. While the diameter is small, the entire diaphysis is filled with spongy bone, but as it enlarges, osteoclasts erode the central portion and create a marrow cavity. Further growth involves two distinct processes: an increase in *length* and an enlargement in *diameter*.

INCREASING THE LENGTH OF A DEVELOPING BONE [FIGURES 5.7/5.8]

During the initial stages of osteogenesis, osteoblasts move away from the primary ossification center toward the epiphyses. But they do not manage to complete the ossification of the model immediately, because the cartilages of the epiphyses continue to grow. The region where the cartilage is being replaced by bone lies at the metaphysis, the junction between the diaphysis (shaft) and epiphyses (heads) of the bone. On the shaft side of the metaphysis, osteoblasts are continually invading the cartilage and replacing it with bone. But on the epiphyseal side, new cartilage is being produced at

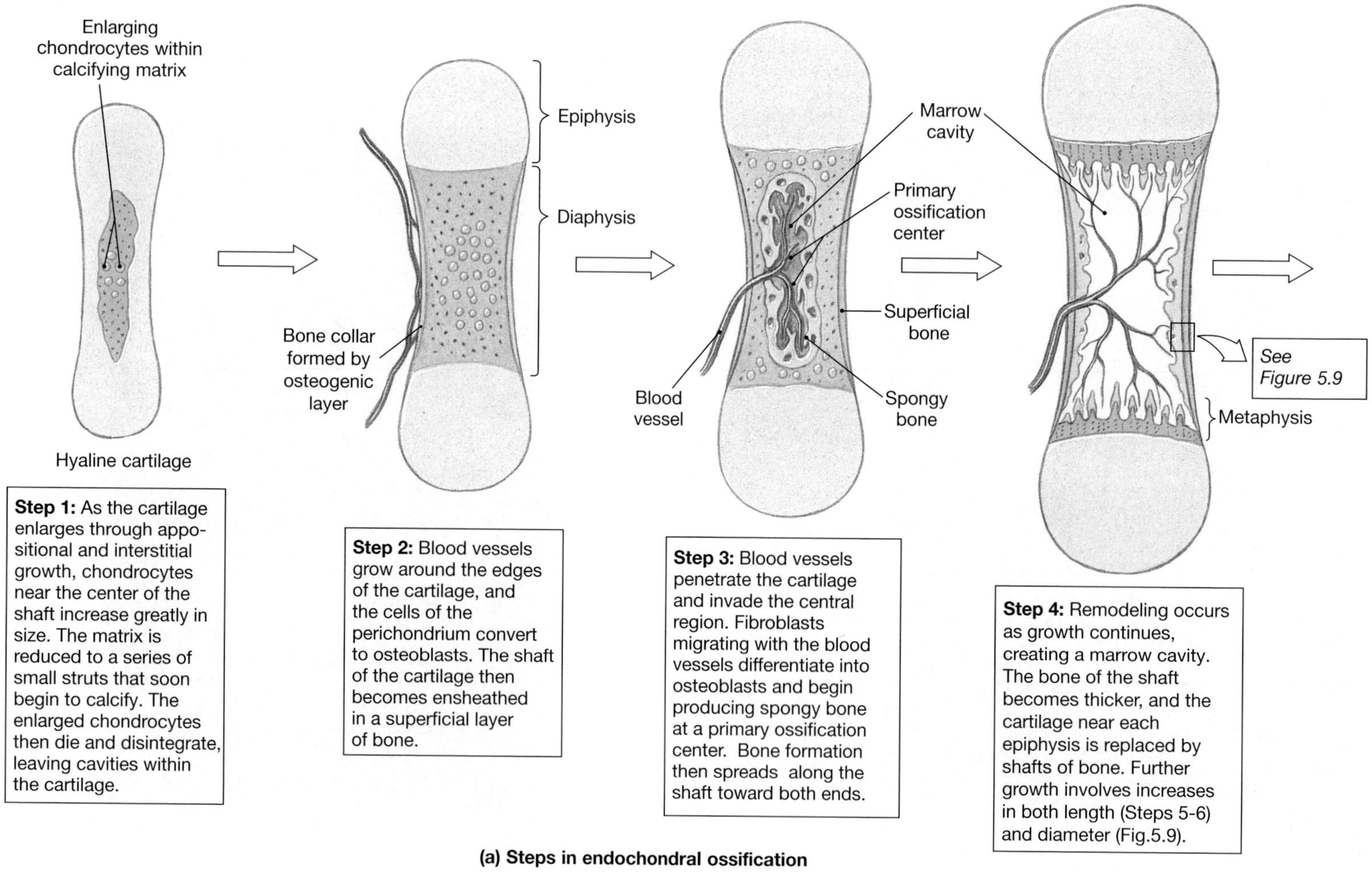

Enlarging chondrocytes within calcifying matrix
Hyaline cartilage
Step 1: As the cartilage enlarges through appositional and interstitial growth, chondrocytes near the center of the shaft increase greatly in size. The matrix is reduced to a series of small struts that soon begin to calcify. The enlarged chondrocytes then die and disintegrate, leaving cavities within the cartilage.
Epiphysis
Diaphysis
Bone collar formed by osteogenic layer
Step 2: Blood vessels grow around the edges of the cartilage, and the cells of the perichondrium convert to osteoblasts. The shaft of the cartilage then becomes ensheathed in a superficial layer of bone.
Marrow cavity
Primary ossification center
Superficial bone
Spongy bone
Blood vessel
Step 3: Blood vessels penetrate the cartilage and invade the central region. Fibroblasts migrating with the blood vessels differentiate into osteoblasts and begin producing spongy bone at a primary ossification center. Bone formation then spreads along the shaft toward both ends.
See Figure 5.9
Metaphysis
Step 4: Remodeling occurs as growth continues, creating a marrow cavity. The bone of the shaft becomes thicker, and the cartilage near each epiphysis is replaced by shafts of bone. Further growth involves increases in both length (Steps 5-6) and diameter (Fig.5.9).
(a) Steps in endochondral ossification

FIGURE 5.7 **ENDOCHONDRAL OSSIFICATION**

(**a**) Steps in the formation of a long bone from a hyaline cartilage model. (**b**) Light micrograph showing the zones of cartilage and the advancing osteoblasts at an epiphyseal cartilage.

the same rate. The situation is like a pair of joggers, one in front of the other. As long as they are running at the same speed, they can run for miles without colliding. In this case, the osteoblasts and the epiphysis are both "running away" from the primary ossification center. As a result, the osteoblasts never catch up with the epiphysis, although the skeletal element continues to grow longer and longer.

STEP 5. The next major change occurs when the centers of the epiphyses begin to calcify. Capillaries and osteoblasts then migrate into these areas, creating **secondary ossification centers**. The time of appearance of secondary ossification centers varies from one bone to another and from individual to individual. Secondary ossification centers may occur at birth in both ends of the humerus (arm), femur (thigh), and tibia (leg), but the ends of some other bones remain cartilaginous through childhood.

STEP 6. The epiphysis eventually become filled with spongy bone. A thin cap of the original cartilage model remains exposed to the joint cavity as the **articular cartilage.** This cartilage prevents damaging bone-to-bone contact within the joint. At the metaphysis, a relatively narrow cartilaginous region called the **epiphyseal cartilage,** or *epiphyseal plate,* now separates the epiphysis from the diaphysis. Figure 5.7b● shows the interface between the degenerating cartilage and the advancing osteoblasts. As long as the rate of cartilage growth keeps pace with the rate of osteoblast invasion, the shaft grows longer but the epiphyseal cartilage survives.

Within the epiphyseal cartilage, the chondrocytes are organized into zones (Figure 5.7b●). Chondrocytes at the epiphyseal side of the cartilage continue to divide and enlarge, while cartilage at the diaphyseal side of the cartilage is gradually replaced by bone. Overall, the thickness of the epiphyseal cartilage does not change. The continual expansion of the epiphyseal cartilage forces the epiphysis farther from the shaft. As the daughter cells mature, they become enlarged, and the surrounding matrix becomes calcified. On the shaft side of the epiphyseal cartilage, osteoblasts and capillaries continue to invade these lacunae and replace the dead cartilage with living bone organized in osteons.

Figure 5.8a● (p. 122) shows an X-ray of epiphyseal cartilages in the hand of a young child. At maturity, the rate of epiphyseal cartilage production slows, and the rate of osteoblast activity accelerates. As a result, the epiphyseal cartilage gets narrower and narrower, until it ultimately disappears. This event is called *epiphyseal closure.* The former location of the epiphyseal cartilage can often be detected in X-rays as a distinct *epiphyseal line* that remains after epiphyseal growth has ended (Figure 5.8b●). ☩ *Inherited Abnormalities in Skeletal Development p. 789; Hyperostosis p. 789*

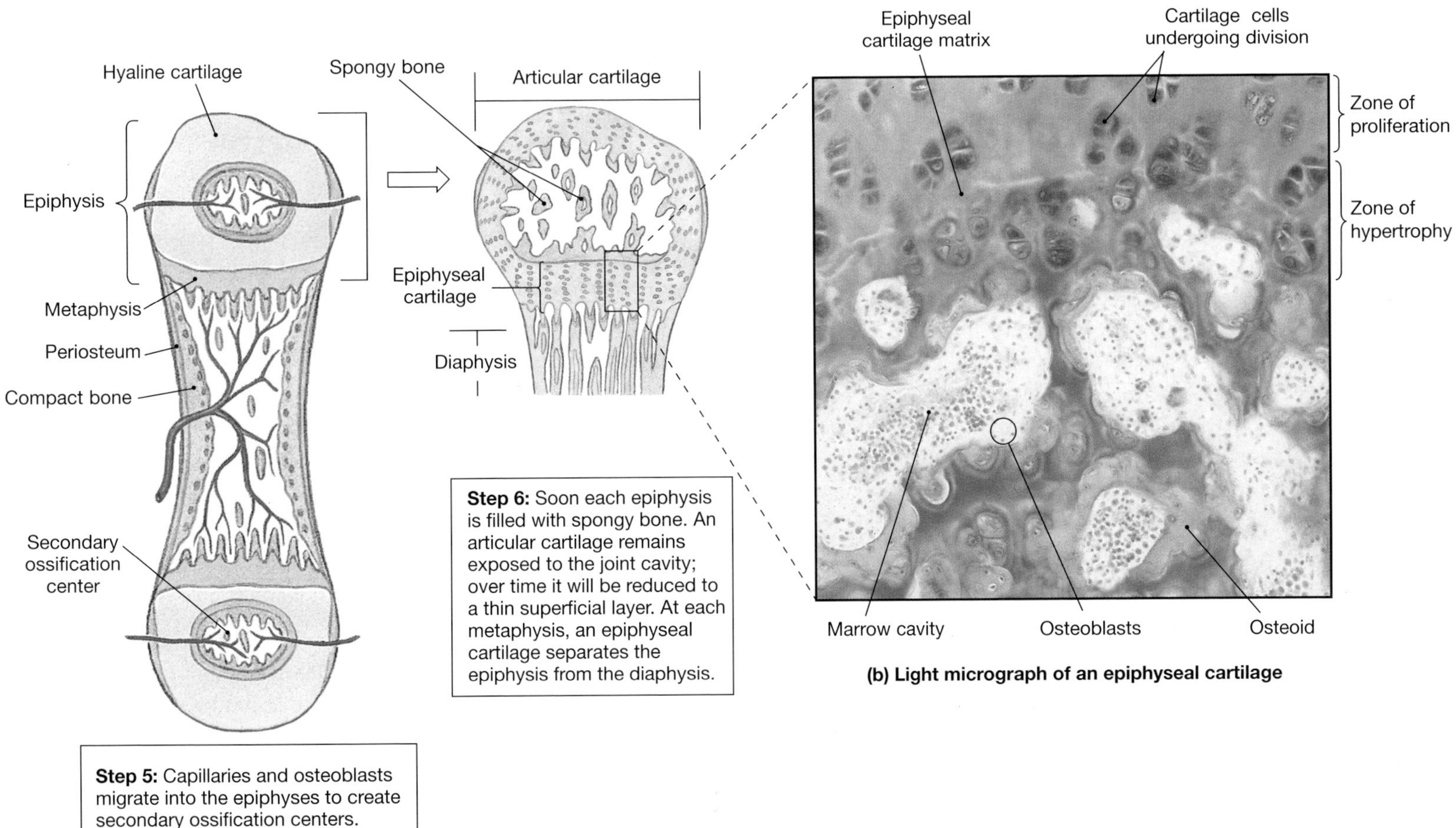
Hyaline cartilage
Spongy bone
Articular cartilage
Epiphysis
Metaphysis
Periosteum
Compact bone
Secondary ossification center
Epiphyseal cartilage
Diaphysis
Step 6: Soon each epiphysis is filled with spongy bone. An articular cartilage remains exposed to the joint cavity; over time it will be reduced to a thin superficial layer. At each metaphysis, an epiphyseal cartilage separates the epiphysis from the diaphysis.
Step 5: Capillaries and osteoblasts migrate into the epiphyses to create secondary ossification centers.
Epiphyseal cartilage matrix
Cartilage cells undergoing division
Zone of proliferation
Zone of hypertrophy
Marrow cavity
Osteoblasts
Osteoid

(b) Light micrograph of an epiphyseal cartilage

FIGURE 5.7 *(continued)*

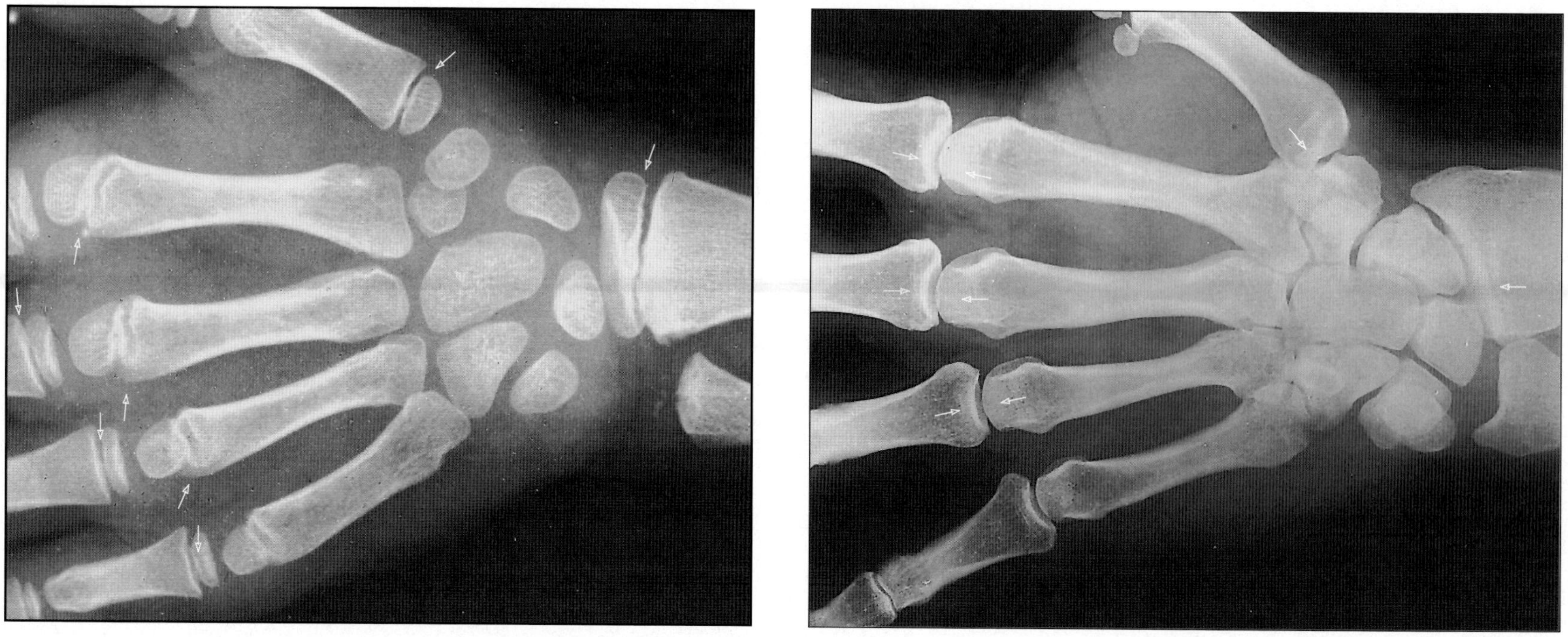

(a) (b)

FIGURE 5.8 **Epiphyseal Cartilages and Lines**

The epiphyseal cartilage is the location of long bone growth in length prior to maturity; the epiphyseal line marks the former location of the epiphyseal cartilage after growth has ended. **(a)** X-ray of the hand of a young child. The arrows indicate the locations of the epiphyseal cartilages. **(b)** X-ray of the hand of an adult. The arrows indicate the locations of epiphyseal lines.

INCREASING THE DIAMETER OF A DEVELOPING BONE [FIGURE 5.9]

The diameter of a bone enlarges through appositional growth at the outer surface. The mechanism of appositional growth is detailed in Figure 5.9a●. Cells of the inner layer of the periosteum differentiate into osteoblasts and add bone matrix to the surface. Eventually they become surrounded by matrix and differentiate into osteocytes. Over much of the surface, the bone is deposited in a series of layers that form circumferential lamellae. Over time, the deeper lamellae are recycled and replaced with the osteons typical of compact bone. However, blood vessels and collagen fibers of the periosteum can become enclosed within the matrix. As indicated in Figure 5.9a●, osteons then form around the smaller vessels. While bone is being added to the outer surface, osteoclasts are removing bone matrix at the inner surface. As a result, the marrow cavity gradually enlarges as the bone increases in diameter (Figure 5.9b●).

Formation of the Blood and Lymphatic Supply

[FIGURES 5.2b/5.9/5.10]

Osseous tissue is very vascular, and the bones of the skeleton have an extensive blood supply. In a typical bone such as the humerus, three major sets of blood vessels develop (Figure 5.10●).

1. *The nutrient artery and vein*: These vessels form as blood vessels invade the cartilage model at the start of endochondral ossification. There is usually only one **nutrient artery** and one **nutrient vein** entering the diaphysis through a nutrient foramen, although a few bones, including the femur, have two or more. These vessels penetrate the shaft to reach the marrow cavity. As they penetrate the shaft, branches of these vessels extend along the central canal to supply the osteons of the compact bone. (Figure 5.2b●, p. 115).
2. *Metaphyseal vessels*: These vessels supply blood to the inner (diaphyseal) surface of each epiphyseal cartilage, where bone is replacing cartilage.
3. *Periosteal vessels*: Blood vessels from the periosteum are incorporated into the developing bone surface as described and illustrated in Figure 5.9●. These vessels provide blood to the superficial osteons of the shaft. During endochondral bone formation, branches of periosteal vessels enter the epiphyses, providing blood to the secondary ossification centers. The periosteum also contains an extensive network of lymphatic vessels, and many of these have branches that enter the bone and reach individual osteons through numerous perforating canals.

Following the closure of the epiphyses, all three sets of blood vessels become extensively interconnected, as indicated in Figure 5.10●.

Bone Innervation

Bones are innervated by sensory nerves, and injuries to the skeleton can be very painful. Sensory nerve endings branch throughout the periosteum, and sensory nerves penetrate the cortex with the nutrient artery to innervate the endosteum, marrow cavity, and epiphyses.

Factors Regulating Bone Growth

Normal bone growth depends on a combination of nutritional and hormonal factors:

- Normal bone growth cannot occur without a constant dietary source of calcium and phosphate salts, as well as other ions such as magnesium, citrate, carbonate, and sodium.
- *Vitamins A* and *C* are essential for normal bone growth and remodeling. These vitamins must be obtained from the diet.

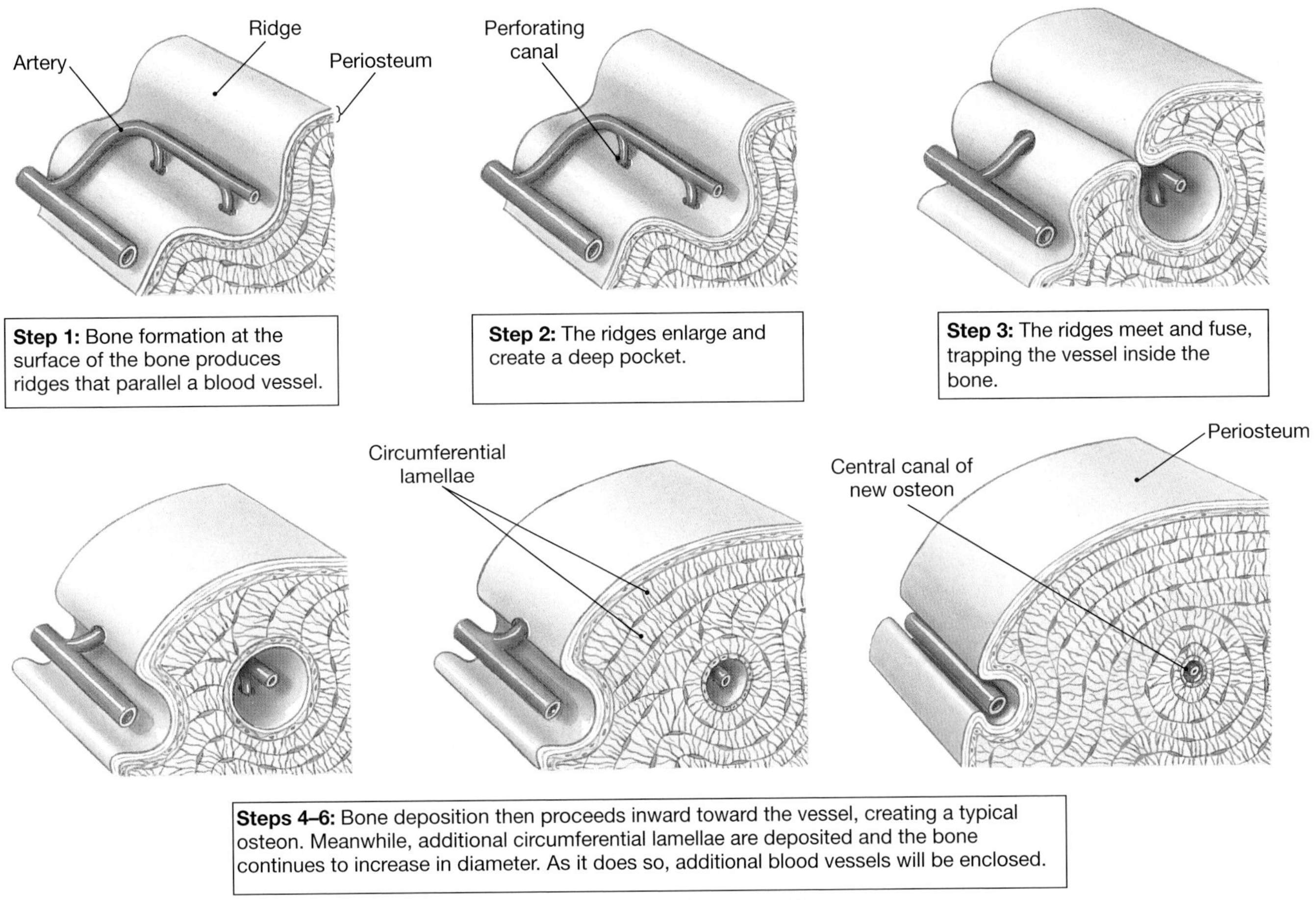

Artery
Ridge
Periosteum
Perforating canal
Step 1: Bone formation at the surface of the bone produces ridges that parallel a blood vessel.
Step 2: The ridges enlarge and create a deep pocket.
Step 3: The ridges meet and fuse, trapping the vessel inside the bone.
Circumferential lamellae
Central canal of new osteon
Periosteum
Steps 4–6: Bone deposition then proceeds inward toward the vessel, creating a typical osteon. Meanwhile, additional circumferential lamellae are deposited and the bone continues to increase in diameter. As it does so, additional blood vessels will be enclosed.

(a) Steps in appositional bone growth

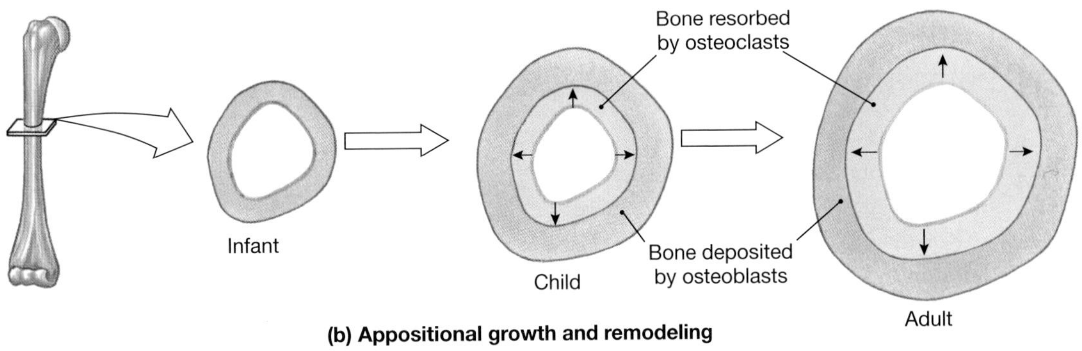

(b) Appositional growth and remodeling

FIGURE 5.9 **APPOSITIONAL BONE GROWTH**

(**a**) Three-dimensional diagrams illustrating the mechanism responsible for increasing the diameter of a growing bone. (**b**) A bone grows in diameter as new bone is added to the outer surface. At the same time, osteoclasts resorb bone on the inside, enlarging the marrow cavity.

- The group of related steroids collectively known as **vitamin D** plays an important role in normal calcium metabolism by stimulating the absorption and transport of calcium and phosphate ions into the blood. The active form of vitamin D, *calcitriol*, is synthesized in the kidneys; this process ultimately depends on the availability of a related steroid, *cholecalciferol*, that may be absorbed from the diet or synthesized in the skin in the presence of UV radiation. p. 93 *Osteomalacia p. 789*

Hormones regulate the pattern of growth by changing the rates of osteoblast and osteoclast activity:

- The parathyroid glands release **parathyroid hormone**, which stimulates osteoclast and osteoblast activity, increases the rate of calcium absorption along the small intestine, and reduces the rate of calcium loss in the urine. The action of parathyroid hormone on the intestine requires the presence of *calcitriol*, a hormone produced at the kidneys. p. 93
- The thyroid glands of children and pregnant women secrete the hormone **calcitonin** (kal-si-TŌ-nin), which inhibits osteoclasts and increases the rate of calcium loss in the urine. Calcitonin is of uncertain significance in the healthy nonpregnant adult.
- *Growth hormone*, produced by the pituitary gland, and **thyroxine**, from the thyroid gland, stimulate bone growth. In proper balance, these hormones maintain normal activity at the epiphyseal cartilages until roughly the time of puberty.

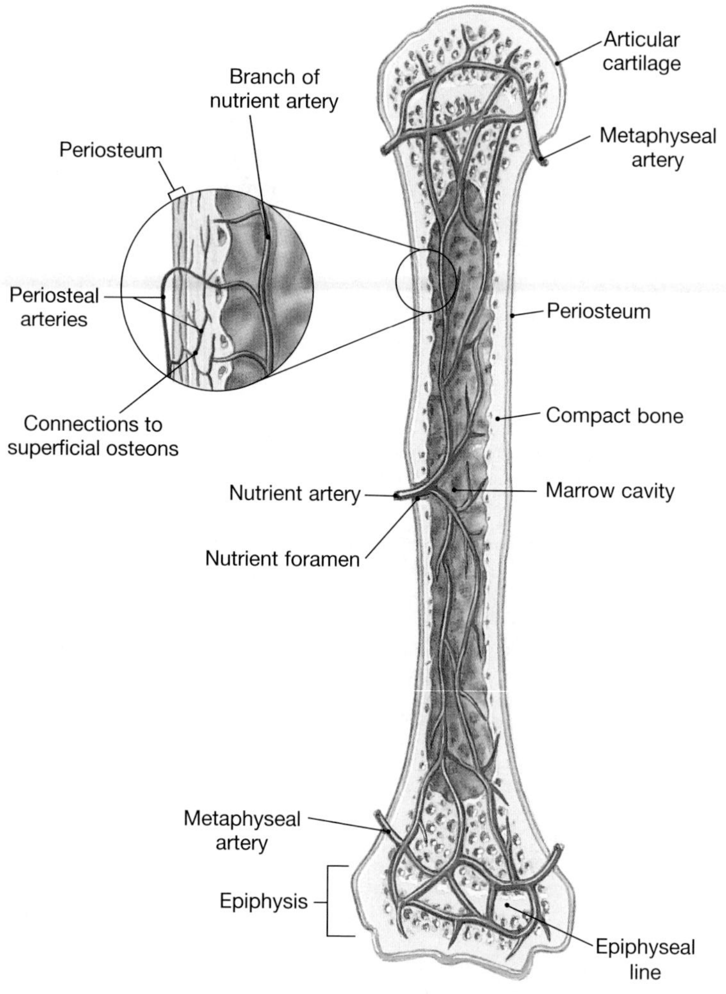

FIGURE 5.10 **CIRCULATORY SUPPLY TO A MATURE BONE**
Arrangement and association of blood vessels supplying the humerus.

- At puberty, bone growth accelerates dramatically. The **sex hormones** (*estrogen* and *testosterone*) stimulate osteoblasts to produce bone faster than the rate of epiphyseal cartilage expansion. Over time, the epiphyseal cartilages narrow and eventually ossify, or "close."

There are differences from bone to bone and individual to individual as to the timing of epiphyseal cartilage closure. The toes may complete their ossification by age 11, whereas portions of the pelvis or the wrist may continue to enlarge until age 25. Differences in the male and female sex hormones account for the variation between the sexes and for related variations in body size and proportions.

✓ CONCEPT CHECK

- How can X-rays of the femur be used to determine whether a person has reached full height?
- Briefly describe the major steps in the process of intramembranous ossification.
- Describe how bones increase in diameter.
- What is the epiphyseal cartilage? Where is it located? Why is it significant?

Bone Maintenance, Repair, and Remodeling

Bone growth occurs when osteoblasts are creating more bone matrix than osteoclasts are removing. Bone remodeling and repair may involve a change in the shape or internal architecture of a bone or a change in the total amount of minerals deposited in the skeleton. In the adult, osteocytes are continually removing and replacing the surrounding calcium salts. But osteoblasts and osteoclasts also remain active throughout life, not just during the growth years. In young adults osteoblast activity and osteoclast activity are in balance, and the rate of bone formation is equal to the rate of bone reabsorption. As one osteon forms through the activity of osteoblasts, another is destroyed by osteoclasts. The rate of mineral turnover is quite high; each year almost one-fifth of the adult skeleton is demolished and then rebuilt or replaced. Every part of every bone may not be affected, as there are regional and even local differences in the rate of turnover. For example, the spongy bone in the head of the femur may be replaced two or three times each year, whereas the compact bone along the shaft remains largely untouched. This high turnover rate continues into old age, but in older individuals osteoblast activity decreases faster than osteoclast activity. As a result bone resorption exceeds bone deposition, and the skeleton gradually gets weaker and weaker.

Changes in Bone Shape

The turnover and recycling of minerals give each bone the ability to adapt to new stresses. Osteoblast sensitivity to electrical events has been theorized as the mechanism that controls the internal organization and structure of bone. Whenever a bone is stressed, the mineral crystals generate minute electrical fields. Osteoblasts are apparently attracted to these electrical fields, and once in the area they begin to produce bone. (Electrical fields may also be used to stimulate the repair of severe fractures.)

Because bones are adaptable, their shapes and surface features reflect the forces applied to them. For example, bumps and ridges on the surface of a bone mark the sites where tendons attach to the bone. If muscles become more powerful, the corresponding bumps and ridges enlarge to withstand the increased forces. Heavily stressed bones become thicker and stronger, whereas bones not subjected to ordinary stresses will become thin and brittle. Regular exercise is therefore important as a stimulus that maintains normal bone structure, especially in growing children, postmenopausal women, and elderly men.

Degenerative changes in the skeleton occur after relatively brief periods of inactivity. For example, using a crutch while wearing a cast takes weight off the injured limb. After a few weeks, the unstressed bones will lose up to about a third of their mass. However, the bones rebuild just as quickly when normal loading resumes.

Injury and Repair [FIGURE 5.11]

Despite its mineral strength, bone may crack or even break if subjected to extreme loads, sudden impacts, or stresses from unusual directions. The damage produced constitutes a **fracture**. Healing of a fracture usually occurs even after severe damage, provided the blood supply and the cellular components of the endosteum and periosteum survive. Steps in the repair of a fracture are illustrated in Figure 5.11●. The final repair will be slightly

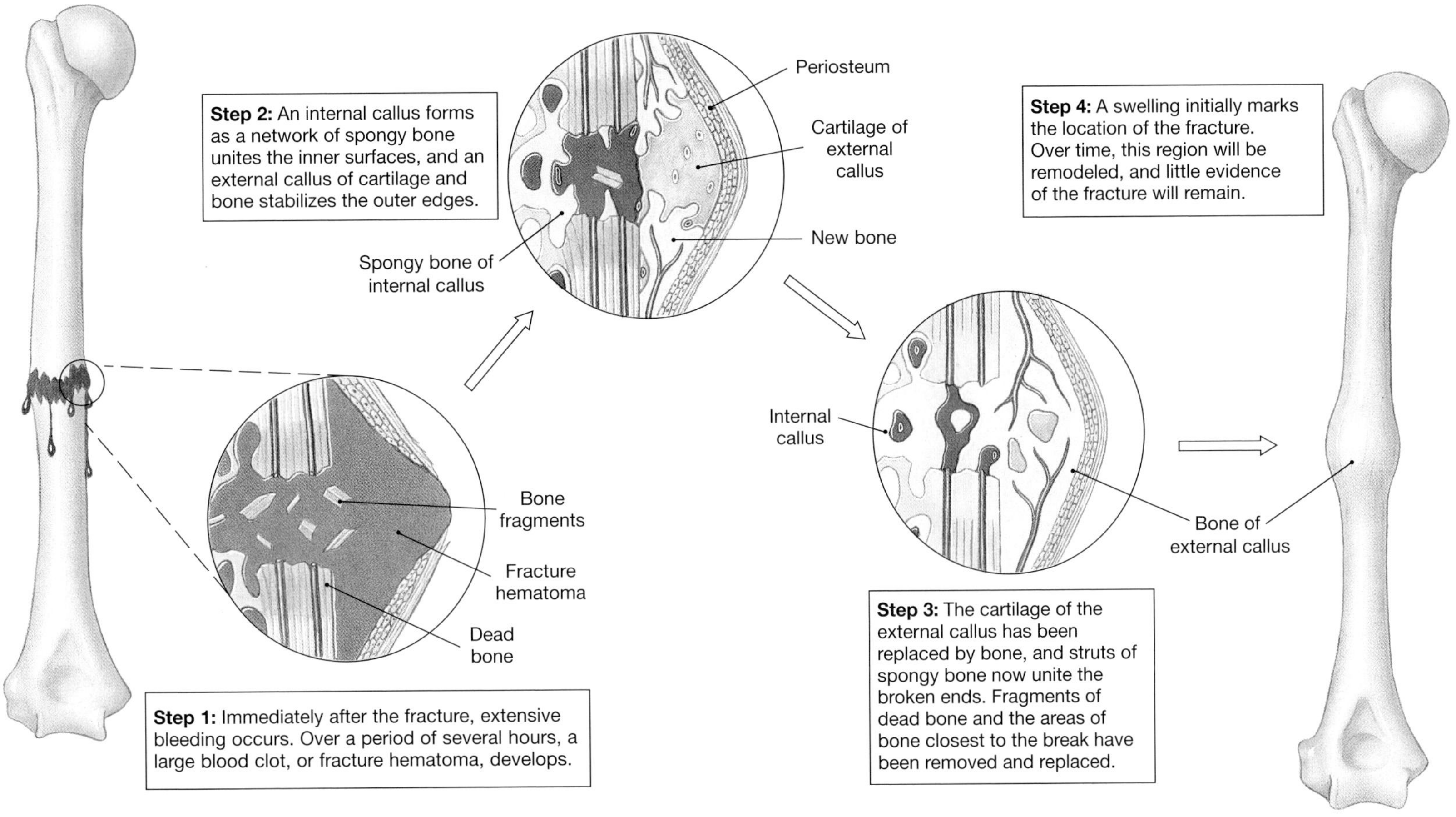

Step 2: An internal callus forms as a network of spongy bone unites the inner surfaces, and an external callus of cartilage and bone stabilizes the outer edges.
Periosteum
Cartilage of external callus
New bone
Spongy bone of internal callus
Step 4: A swelling initially marks the location of the fracture. Over time, this region will be remodeled, and little evidence of the fracture will remain.
Internal callus
Bone fragments
Fracture hematoma
Dead bone
Bone of external callus
Step 3: The cartilage of the external callus has been replaced by bone, and struts of spongy bone now unite the broken ends. Fragments of dead bone and the areas of bone closest to the break have been removed and replaced.
Step 1: Immediately after the fracture, extensive bleeding occurs. Over a period of several hours, a large blood clot, or fracture hematoma, develops.

FIGURE 5.11 **FRACTURE REPAIR**
Steps involved in the repair of a fracture.

thicker and probably stronger than the original bone; under comparable stresses, a second fracture will usually occur at a different site.

Aging and the Skeletal System [FIGURE 5.12]

The bones of the skeleton become thinner and relatively weaker as a normal part of the aging process. Inadequate ossification is called **osteopenia** (os-tē-ō-PĒ-nē-a; *penia*, lacking), and everyone becomes slightly osteopenic as they age. This reduction in bone mass occurs between the ages of 30 and 40. Over this period, osteoblast activity begins to decline while osteoclast activity continues at previous levels. Once the reduction begins, women lose roughly 8% of their skeletal mass every decade; the skeletons of men deteriorate at the slower rate of about 3% per decade. All parts of the skeleton are not equally affected. Epiphyses, vertebrae, and the jaws lose more than their fair share, resulting in fragile limbs, a reduction in height, and the loss of teeth. A significant percentage of older women and a smaller proportion of older men suffer from **osteoporosis** (os-tē-ō-po-RŌ-sis; *porosus*, porous). This condition is characterized by a reduction in bone mass sufficient to compromise normal function (Figure 5.12b●). Osteoporosis is discussed further in the Clinical Issues appendix. *Osteoporosis and Other Skeletal Abnormalities Associated with Aging p. 789*

Remodeling of Bone

Although bone is hard and dense, it is able to change its shape in response to environmental conditions. Bone remodeling involves the simultaneous process of adding new bone and removing previously formed bone. For example, bone remodeling occurs following the realignment of teeth by an orthodontist. As the teeth are moved the shape of the tooth socket changes by the resorption of old bone and the deposition of new bone according to the tooth's new position. In addition, increased muscular development (as in weight training) will involve the remodeling of bones to meet the new stress imposed at the site of muscular and tendon attachment.

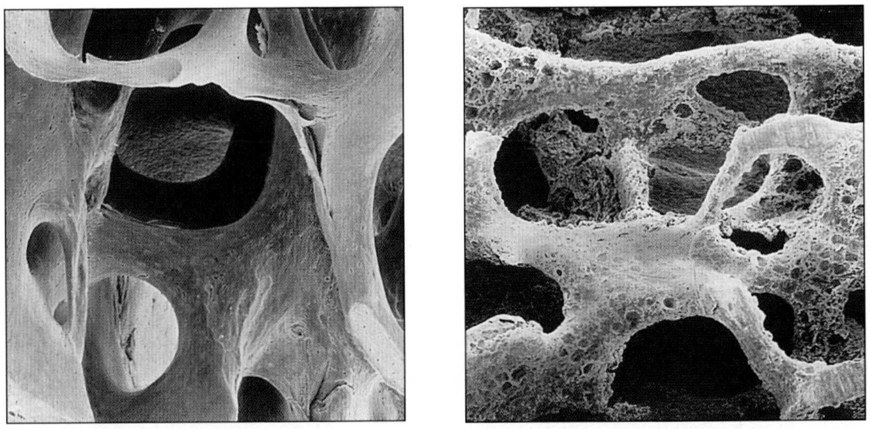
(a) Normal spongy bone **(b) Spongy bone in osteoporosis**

FIGURE 5.12 **THE EFFECTS OF OSTEOPOROSIS**
(a) Normal spongy bone from the epiphysis of a young adult. (SEM × 25)
(b) Spongy bone from a person with osteoporosis (SEM × 21)

CLINICAL DISCUSSION

A Classification of Fractures

Fractures are classified according to their external appearance, the site of the fracture, and the nature of the crack or break in the bone. Important fracture types are indicated below, with representative X-rays. Many fractures fall into more than one category. For example, a Colles' fracture is a transverse fracture, but depending on the injury, it may also be a comminuted fracture that can be either open or closed. **Closed**, or *simple*, fractures are completely internal; they do not involve a break in the skin. **Open**, or *compound*, fractures project through the skin; they are more dangerous because of the possibility of infection or uncontrolled bleeding.

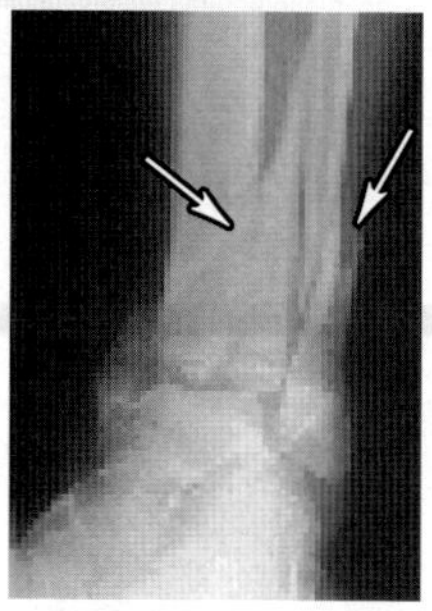

A **Pott's fracture** occurs at the ankle and affects both bones of the leg.

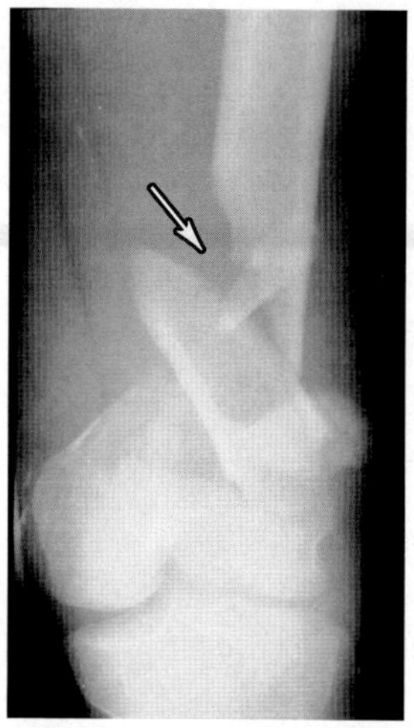

Comminuted fractures, such as this fracture of the femur, shatter the affected area into a multitude of bony fragments.

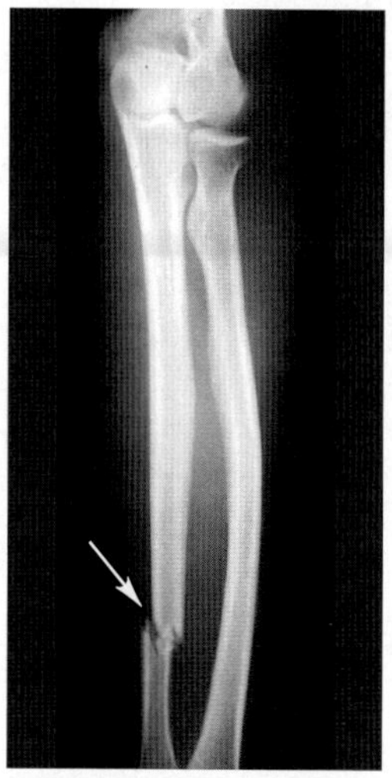

Transverse fractures, such as this fracture of the radius, break a shaft bone across its long axis.

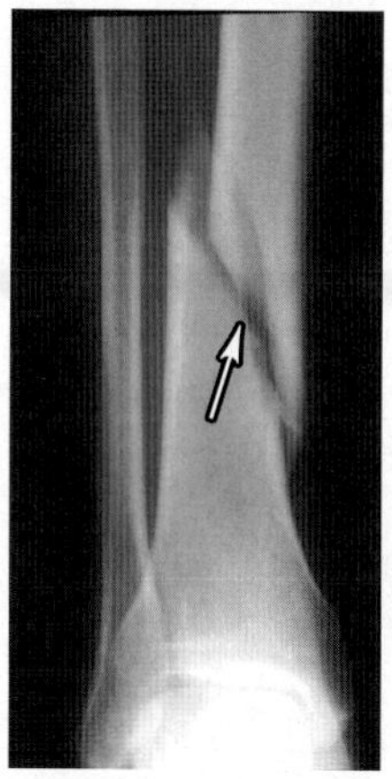

Spiral fractures, such as this fracture of the tibia, are produced by twisting stresses that spread along the length of the bone.

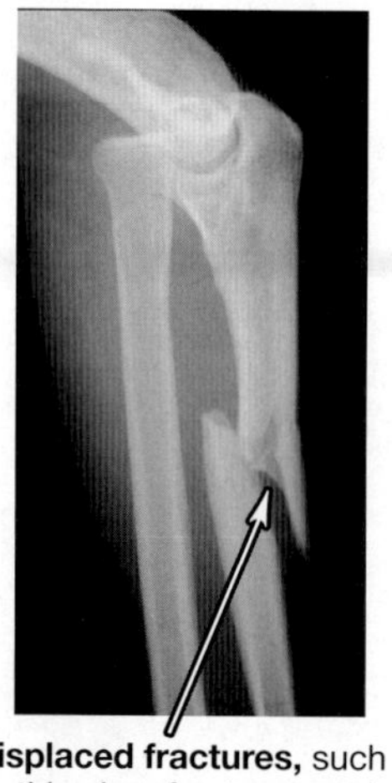

Displaced fractures, such as this ulnar fracture, produce new and abnormal bone arrangements; **nondisplaced fractures** retain the normal alignment of the bones or fragments.

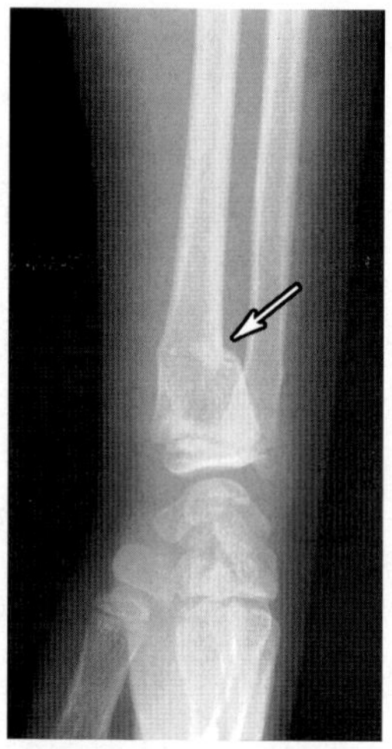

A **Colles' fracture**, a break in the distal portion of the radius, is typically the result of reaching out to cushion a fall.

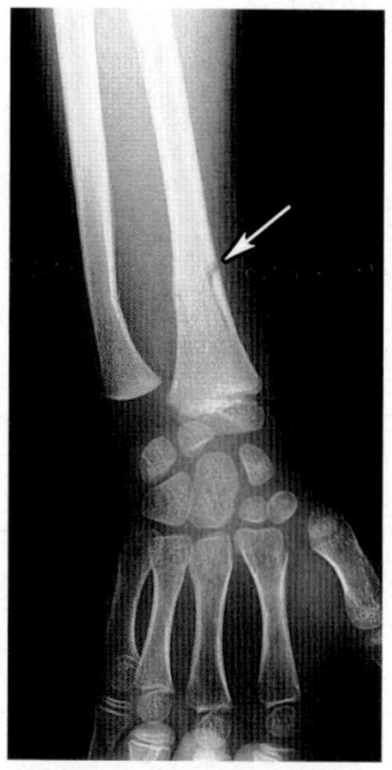

In a **greenstick fracture**, such as this fracture of the radius, only one side of the shaft is broken, and the other is bent. This type generally occurs in children, whose long bones have yet to ossify fully.

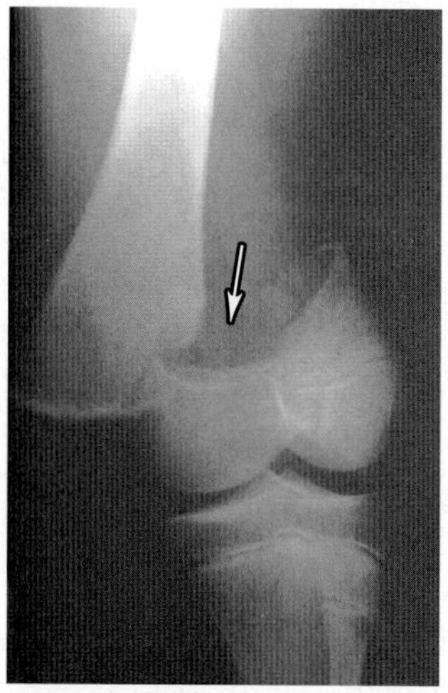

Epiphyseal fractures, such as this fracture of the femur, tend to occur where the bone matrix is undergoing calcification and chondrocytes are dying. A clean transverse fracture along this line generally heals well. Unless carefully treated, fractures between the epiphysis and the epiphyseal cartilage can permanently stop growth at this site.

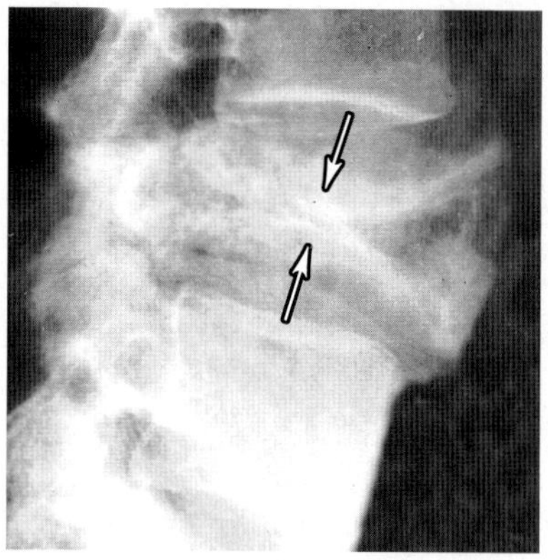

Compression fractures occur in vertebrae subjected to extreme stresses, as when you land on your seat in a fall. They are more common when bones are weakened by osteoporosis.

✓ CONCEPT CHECK

- Would you expect to see any difference in the bones of an athlete before and after extensive training to increase muscle mass? Why or why not?
- Which vitamins and hormones regulate bone growth?
- What major difference might we expect to find when comparing bone growth in a 15-year-old and that of a 30-year-old?

Anatomy of Skeletal Elements

The human skeleton contains 206 major bones. We can divide these bones into six broad categories according to their individual shapes.

Classification of Bones [FIGURES 5.3/5.13]

Refer to Figure 5.13● as we describe the anatomical classification of bones.

1. *Long bones* are relatively long and slender (Figure 5.13a●). They have a diaphysis, two metaphyses, two epiphyses, and a marrow cav-

ity, as detailed in Figure 5.3●, p. 116. Long bones are found in the upper and lower limbs. Examples include the *humerus*, *radius*, *ulna*, *femur*, *tibia*, and *fibula*.

2. *Short bones* are box-like in appearance (Figure 5.13b●). Their external surfaces are covered by compact bone, but the interior contains spongy bone. Examples of short bones include the *carpal bones* (wrists) and *tarsal bones* (ankles).
3. *Flat bones* have thin, roughly parallel surfaces of compact bone. In structure a flat bone resembles a spongy bone sandwich; such bones are strong but relatively light. Flat bones form the roof of the skull (Figure 5.13c●), the sternum, the ribs, and the scapulae. They provide protection for underlying soft tissues and offer an extensive surface area for the attachment of skeletal muscles. Special terms are used when describing the flat bones of the skull, such as the parietal bones. Their relatively thick layers of compact bone are called the internal and external tables, and the layer of spongy bone between the tables is called the diploë (DIP-lō-ē).
4. *Irregular bones* have complex shapes with short, flat, notched, or ridged surfaces (Figure 5.13d●). Their internal structure is equally varied. The vertebrae that form the spinal column and several bones in the skull are examples of irregular bones.
5. *Sesamoid bones* are usually small, round, and flat (Figure 5.13e●). They develop inside tendons and are most often encountered near joints at the knee, the hands, and the feet. Few individuals have sesamoid bones at every possible location, but everyone has sesamoid *patellae* (pa-TEL-ē), or kneecaps.
6. *Sutural (Wormian) bones* are small, flat, oddly shaped bones found between the flat bones of the skull in the suture line (Figure 5.13f●). There are individual variations in the number, shape, and position of the sutural bones. Their borders are like puzzle pieces and may range in size from a grain of sand to the size of a quarter.

Bone Markings (Surface Features) [FIGURE 5.14 AND TABLE 5.1]

Each bone in the body has a distinctive shape and characteristic external and internal features. Elevations or projections form where tendons and ligaments attach and where adjacent bones articulate. Depressions, grooves, and tunnels in bone indicate sites where blood vessels and nerves lie alongside or penetrate the bone. Detailed examination of these **bone markings**, or *surface features*, can yield an abundance of anatomical information. For example, anthropologists, criminologists, and pathologists can often determine the size, weight, sex, and general appearance of an individual on the basis of incomplete skeletal remains. (This topic will be discussed further in Chapter 6.) Bone marking terminology is presented in Table 5.1 and illustrated in Figure 5.14●.

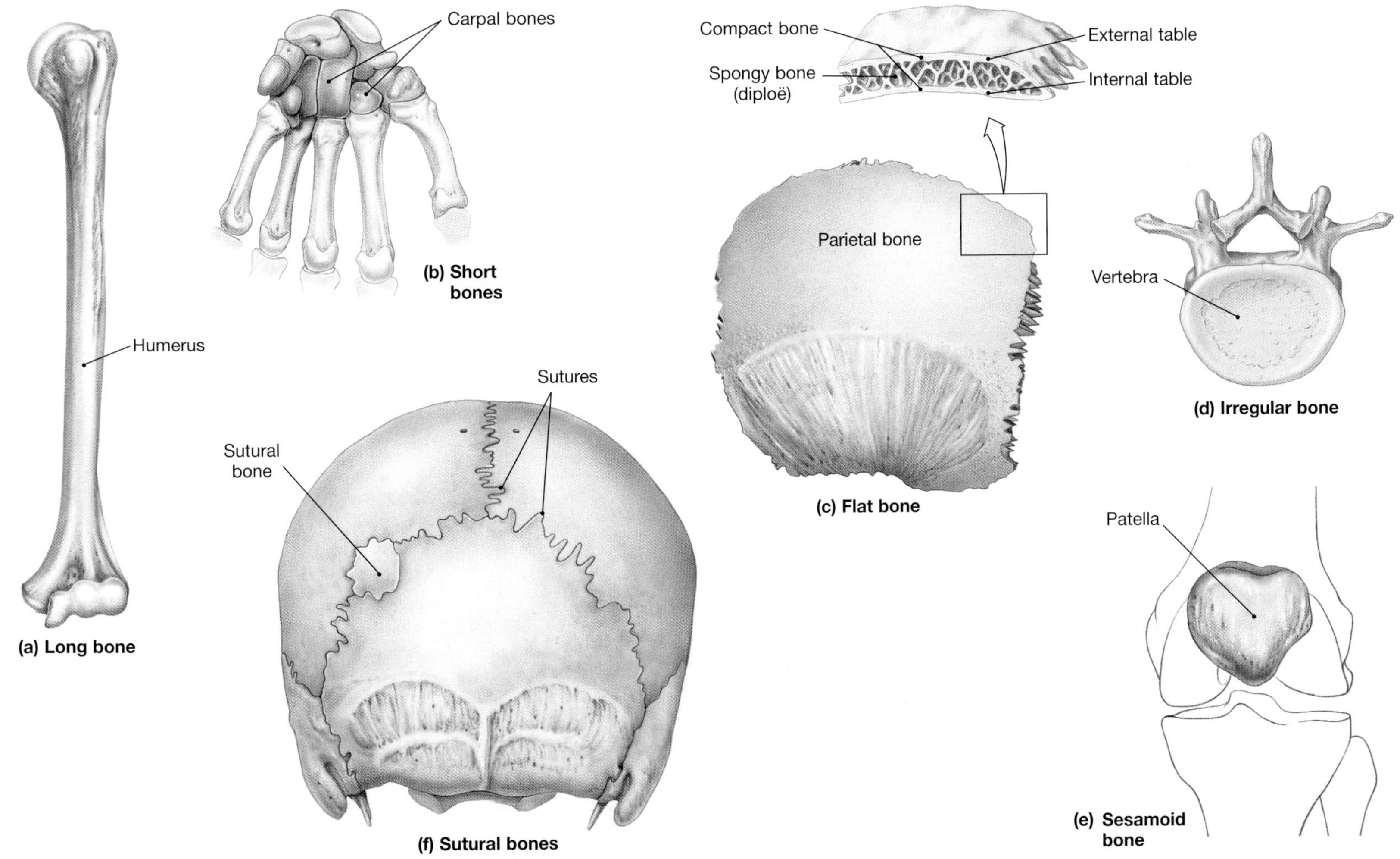

FIGURE 5.13 **SHAPES OF BONES**

Classification of bones depends on shape comparison. (**a**) Long bone. (**b**) Short bones. (**c**) Flat bone, in surface and sectional views. (**d**) Irregular bone. (**e**) Sesamoid bone. (**f**) Sutural bone.

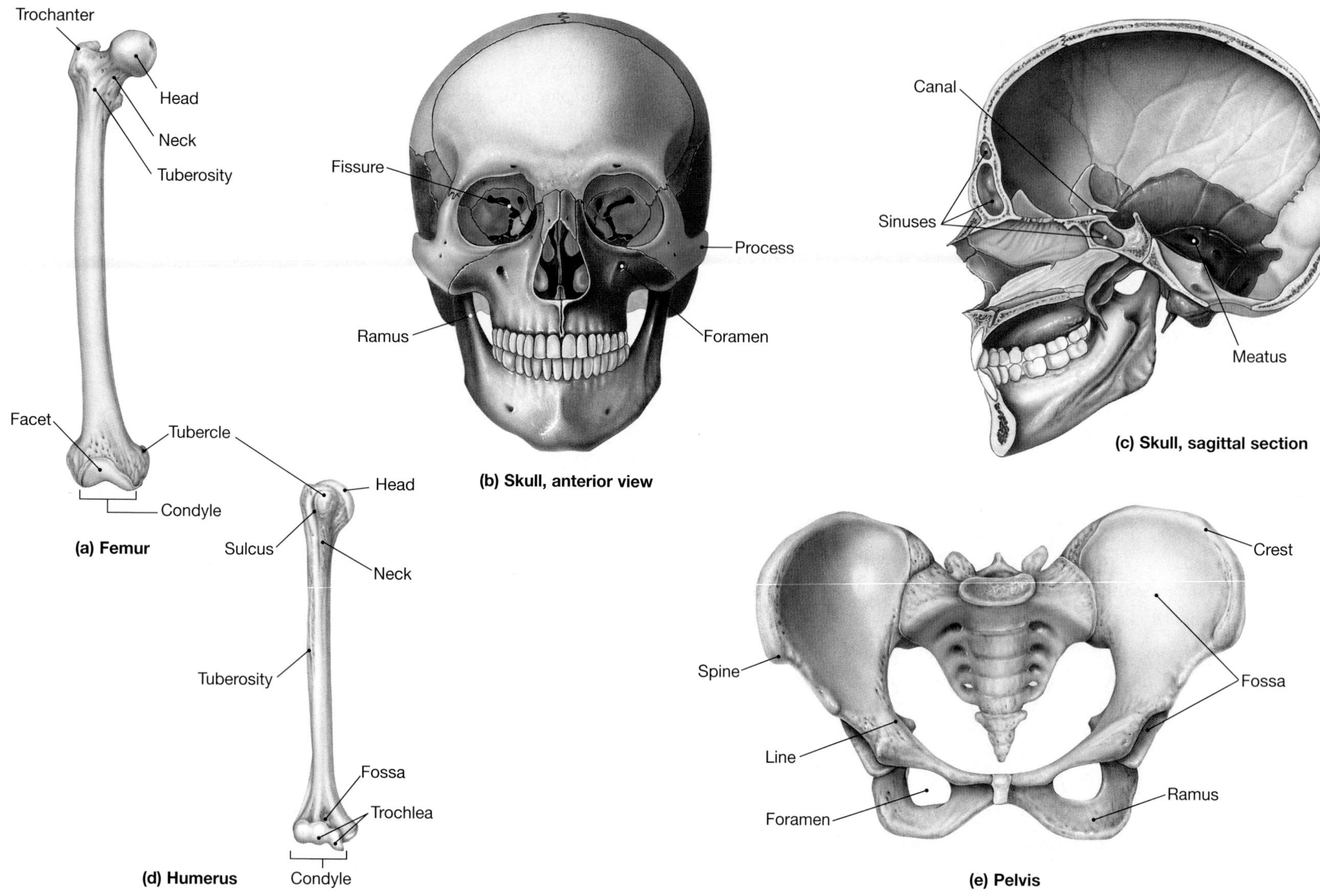

FIGURE 5.14 **EXAMPLES OF BONE MARKINGS (SURFACE FEATURES)**

Bone markings provide distinct and characteristic landmarks for orientation and identification of bones and associated structures.

TABLE 5.1

COMMON BONE MARKING TERMINOLOGY

General Description	Anatomical Term	Example (See Figure 5.14)	Definition
Elevations and projections (general)	Process	(b)	Any projection or bump
	Ramus	(b, e)	An extension of a bone making an angle to the rest of the structure
Processes formed where tendons or ligaments attach	Trochanter	(a)	A large, rough projection
	Tuberosity	(a, d)	A relatively smaller, rough projection
	Tubercle	(a, d)	A small, rounded projection
	Crest	(e)	A prominent ridge
	Line	(e)	A low ridge
	Spine	(e)	A pointed process
Processes formed for articulation with adjacent bones	Head	(a, d)	The expanded articular end of an epiphysis, separated from the shaft by a narrower neck
	Neck	(a, d)	A narrower connection between the epiphysis and diaphysis
	Condyle	(a, d)	A smooth, rounded articular process
	Trochlea	(d)	A smooth, grooved articular process shaped like a pulley
	Facet	(a)	A small, flat articular surface
Depressions	Fossa	(d, e)	A shallow depression
	Sulcus	(d)	A narrow groove
Openings	Foramen	(b, e)	A rounded passageway for blood vessels and/or nerves
	Fissure	(b)	An elongated cleft
	Meatus or canal	(c)	A large-diameter passageway through the substance of a bone
	Sinus or antrum	(c)	A chamber within a bone, normally filled with air

Our discussion will focus on prominent features that are useful in identifying a bone. These markings are also useful because they provide fixed landmarks that can help in determining the position of the soft tissue components of other systems. Specific anatomical terms are used to describe the various elevations and depressions.

✓ CONCEPT CHECK

- Why is a working knowledge of bone markings important in a clinical setting?
- What is the primary difference between sesamoid and irregular bones?
- Where would you look for sutural bones in a skeleton?

Integration with other Systems

Although bones may seem inert, you should now realize that they are quite dynamic structures. The entire skeletal system is intimately associated with other systems. Bones are attached to the muscular system, extensively connected to the cardiovascular and lymphatic systems, and largely under the physiological control of the endocrine system. Also, the digestive and excretory systems play important roles in providing the calcium and phosphate minerals needed for bone growth. In return, the skeleton represents a reserve of calcium, phosphate, and other minerals that can compensate for changes in the dietary supply of these ions.

RELATED CLINICAL TERMS

achondroplasia (a-kon-drō-PLĀ-sē-a): A condition resulting from abnormal epiphyseal plate activity; the epiphyseal plates grow unusually slowly, and the individual develops short, stocky limbs. The trunk is normal in size, and sexual and mental development remain unaffected. *Inherited Abnormalities in Skeletal Development p. 789*

acromegaly: A condition caused by excessive secretion of growth hormone after puberty and closure of epiphyseal cartilage. Skeletal abnormalities develop, affecting the cartilages and various small bones. *Hyperostosis p. 789*

external callus: A toughened layer of connective tissue that encircles and stabilizes a bone at a fracture site. *p. 125*

fracture: A crack or break in a bone. *p. 124*

fracture hematoma: A large blood clot that closes off the injured vessels and leaves a fibrous meshwork in the damaged area. *p. 125*

gigantism: A condition resulting from an overproduction of growth hormone before puberty. *p. 123*

hyperostosis (hī-per-os-TŌ-sis): The excessive formation of bone tissue. *Hyperostosis p. 789*

internal callus: A bridgework of trabecular bone that unites the broken ends of a bone on the marrow side of the fracture. *p. 125*

Marfan's syndrome: An inherited condition linked to defective production of a connective tissue glycoprotein. Extreme height and long, slender limbs are the most obvious physical indications of this disorder. *Inherited Abnormalities in Skeletal Development p. 789*

osteoclast-activating factor: A compound released by cancers of the bone marrow, breast, or other tissues. It produces a severe osteoporosis. *Osteoporosis and Other Skeletal Abnormalities Associated with Aging p. 789*

osteogenesis imperfecta (im-per-FEK-ta): An inherited condition affecting the organization of collagen fibers. Osteoblast function is impaired, growth is abnormal, and the bones are very fragile, leading to progressive skeletal deformation and repeated fractures. *Inherited Abnormalities in Skeletal Development p. 789*

osteomalacia (os-tē-ō-ma-LĀ-shē-ah): A softening of bone due to a decrease in the mineral content. *Osteomalacia p. 789*

osteomyelitis (os-tē-ō-mī-e-LĪ-tis): A painful infection in a bone, usually caused by bacteria. *Osteoporosis and Other Skeletal Abnormalities Associated with Aging p. 789*

osteopenia (os-tē-ō-PĒ-nē-a): Inadequate ossification, leading to thinner, weaker bones. *p. 125*

osteopetrosis (os-tē-ō-pe-TRŌ-sis): A condition caused by a decrease in osteoclast activity, causing increased bone mass and various skeletal deformities. *Hyperostosis p. 789*

osteoporosis (os-tē-ō-po-RŌ-sis): A reduction in bone mass to a degree that compromises normal function. *Osteoporosis and Other Skeletal Abnormalities Associated with Aging p. 789*

Paget's disease (osteitis deformans) (os-tē-Ī-tis de-FOR-mans): A condition characterized by gradual deformation of the skeleton. *Osteoporosis and Other Skeletal Abnormalities Associated with Aging p. 789*

pituitary growth failure: A type of dwarfism caused by inadequate growth hormone production. *p. 123*

rickets: A disorder that reduces the amount of calcium salts in the skeleton; often characterized by a "bowlegged" appearance. *p. 118*

STUDY OUTLINE & CHAPTER REVIEW

Introduction 113

1. The skeletal system includes the bones of the skeleton, the cartilages, ligaments, and other connective tissues that stabilize or interconnect bones. Its functions include structural support, storage of minerals and lipids, blood cell production, protection of delicate tissues and organs, and leverage.

Structure of Bone 113

1. **Osseous (bone) tissue** is a supporting connective tissue with specialized cells and a solid, extracellular **matrix** of protein fibers and a ground substance.

The Histological Organization 113

2. Bone matrix consists largely of crystals of **hydroxyapatite,** accounting for almost two-thirds of the weight of bone. The remaining third is dominated by collagen fibers and small amounts of other calcium salts; bone cells and other cell types contribute only about 2% to the volume of bone tissue.
3. **Osteocytes** are mature bone cells that are completely surrounded by hard bone matrix. Osteocytes reside in spaces, termed **lacunae.** Osteocytes in lacunae are interconnected by small, hollow channels called **canaliculi. Lamellae** are layers of calcified matrix. *(see Figure 5.1)*

4. **Osteoblasts** are immature, bone-forming cells. By the process of **osteogenesis**, osteoblasts synthesize **osteoid**, the matrix of bone prior to its calcification. *(see Figure 5.1)*
5. **Osteoprogenitor cells** are mesenchymal cells that play a role in the repair of bone fractures. *(see Figure 5.1)*
6. **Osteoclasts** are large, multinucleated cells that help to dissolve the bony matrix through the process of **osteolysis**. They are important in the regulation of calcium and phosphate concentrations in body fluids. *(see Figure 5.1)*

Compact and Spongy Bone 114

7. There are two types of bone: **compact**, or *dense*, bone, and **spongy**, or *cancellous*, bone. The matrix composition in compact bone is the same as that of spongy bone, but they differ in the three-dimensional arrangement of osteocytes, canaliculi, and lamellae. *(see Figures 5.1/5.2)*
8. The basic functional unit of compact bone is the **osteon**, or *Haversian system*. Osteocytes in an osteon are arranged in concentric layers around a **central canal**. *(see Figures 5.1b–d, 5.2)*
9. Spongy bone contains struts or plates called **trabeculae**, often in an open network. *(see Figure 5.2)*
10. Compact bone covers bone surfaces. It is thickest where stresses come from a limited range of directions. Spongy bone is located internally in bones. It is found where stresses are few or come from many different directions. *(see Figure 5.3)*

The Periosteum and Endosteum 117

11. A bone is covered externally by a two-layered **periosteum** (outer fibrous, inner cellular) and lined internally by a cellular **endosteum**. *(see Figure 5.4)*

Bone Development and Growth 118

1. **Ossification** is the process of replacing other tissue by bone; **calcification** is the process of deposition of calcium salts within a tissue.

Intramembranous Ossification 118

2. **Intramembranous ossification**, also called **dermal ossification**, begins when osteoblasts differentiate within a mesenchymal or fibrous connective tissue. This process can ultimately produce spongy or compact bone. Such ossification begins at an **ossification center**. *(see Figures 5.5, 5.6)*

Endochondral Ossification 120

3. **Endochondral ossification** begins with the formation of a cartilaginous model. This hyaline cartilage model is gradually replaced by osseous tissue. *(see Figures 5.6/5.7)*
4. The length of a developing bone increases at the **epiphyseal cartilage**, which separates the epiphysis from the diaphysis. Here, new cartilage is added at the epiphyseal side, while osseous tissue replaces older cartilage at the diaphyseal side. The time of closure of the epiphyseal cartilage differs among bones and among individuals. *(see Figure 5.8)*
5. The diameter of a bone enlarges through appositional growth at the outer surface. *(see Figure 5.9)*

Formation of the Blood and Lymphatic Supply 122

6. A typical bone formed through endochondral ossification has three major sets of vessels: the *nutrient vessels*, *metaphyseal vessels*, and *periosteal vessels*. Lymphatic vessels are distributed in the periosteum and enter the osteons through the nutrient and perforating canals. *(see Figures 5.7/5.10)*

Bone Innervation 122

7. Sensory nerve endings branch throughout the periosteum, and sensory nerves penetrate the cortex with the nutrient artery to innervate the endosteum, marrow cavity, and epiphyses.

Factors Regulating Bone Growth 122

8. Normal osteogenesis requires a continual and reliable source of minerals, vitamins, and hormones.
9. **Parathyroid hormone**, secreted by the parathyroid glands, stimulates osteoclast and osteoblast activity. In contrast, **calcitonin**, secreted by the thyroid gland, inhibits osteoclast activity and increases calcium loss in the urine. These hormones control the rate of mineral deposition in the skeleton and regulate the calcium ion concentrations in body fluids.
10. **Growth hormone**, **thyroxine**, and **sex hormones** stimulate bone growth by changing osteoblast activity.
11. There are differences between individual bones and between individuals with respect to the timing of epiphyseal cartilage closure.

Bone Maintenance, Repair, and Remodeling 124

1. The turnover rate for bone is quite high. Each year almost one-fifth of the adult skeleton is broken down and then rebuilt or replaced.

Changes in Bone Shape 124

2. Mineral turnover and recycling allows bone to adapt to new stresses.
3. Calcium is the most common mineral in the human body, with more than 98% of it located in the skeleton.

Injury and Repair 124

4. A **fracture** is a crack or break in a bone. Healing of a fracture can usually occur if portions of the blood supply, endosteum, and periosteum remain intact. *(see Figure 5.11)* For a classification of fracture types, see the Clinical Discussion on p. 126.

Aging and the Skeletal System 125

5. The bones of the skeleton become thinner and relatively weaker as a normal part of the aging process. **Osteopenia** usually develops to some degree, but in some cases this process progresses to **osteoporosis** and the bones become dangerously weak and brittle. *(see Figure 5.12)*

Remodeling of Bone 125

6. Bone remodeling involves the simultaneous process of adding new bone and removing previously formed bone.

Anatomy of Skeletal Elements 126

Classification of Bones 126

1. Categories of bones are based on anatomical classification and include: *long bones*, *short bones*, *flat bones*, *irregular bones*, *sesamoid bones*, and *sutural bones* (*Wormian bones*). *(see Figure 5.13)*

Bone Markings (Surface Features) 127

2. **Bone markings** (or *surface features*) can be used to identify specific elevations, depressions, and openings of bones. *(see Figure 5.14)* Common bone marking terminology is presented in *Table 5.1*.

Integration with other Systems 129

1. The skeletal system is anatomically and physiologically linked to other body systems and represents a reservoir for calcium, phosphate, and other minerals.

LEVEL 1 REVIEWING FACTS AND TERMS

1. Mature bone cells are termed
   (a) osteocytes (b) osteoblasts
   (c) osteoclasts (d) osteons
2. Spongy bone is formed of
   (a) osteons (b) struts and plates
   (c) concentric lamellae (d) spicules only
3. The narrow passageways that contain cytoplasmic extensions of osteocytes are termed
   (a) lamellae (b) lacunae
   (c) canaliculi (d) marrow cavities
4. Endochondral ossification begins with the formation of
   (a) a fibrous connective tissue model
   (b) a hyaline cartilage model
   (c) a membrane model
   (d) a calcified model
5. When sexual hormone production increases, bone production
   (a) slows down (b) accelerates rapidly
   (c) increases slowly (d) is not affected
6. The presence of an epiphyseal line indicates
   (a) epiphyseal growth has ended
   (b) epiphyseal growth is just beginning
   (c) growth in bone diameter is just beginning
   (d) the bone is fractured at that location
7. The inadequate ossification which occurs with aging is called
   (a) osteopenia (b) osteomyelitis
   (c) osteitis (d) osteoporosis
8. Decreased levels of calcium ions in the blood stimulate the secretion of the hormone
   (a) calcitonin (b) thyroid hormone
   (c) parathyroid hormone (d) growth hormone
9. The sternum is an example of a(n)
   (a) flat bone (b) long bone
   (c) irregular bone (d) sesamoid bone
10. A small, rough projection of a bone is termed a
    (a) ramus (b) tuberosity
    (c) trochanter (d) tubercle

LEVEL 2 REVIEWING CONCEPTS

1. How would decreasing the proportion of organic molecules to inorganic components in the bony matrix affect the physical characteristics of bone?
   (a) the bone would be less flexible
   (b) the bones would be stronger
   (c) the bones would be more brittle
   (d) the bones would be more flexible
2. Premature closure of the epiphyseal cartilages could be caused by
   (a) elevated levels of sex hormones
   (b) high levels of vitamin D
   (c) too little parathyroid hormone
   (d) an excess of growth hormone
3. What factors determine the type of ossification that occurs in a specific bone?
4. What kind of bone would most likely be found in a tissue sample taken from the area of a long bone immediately internal to the periosteum?
5. What are the advantages of spongy bone over compact bone in an area such as the expanded ends of long bones?
6. Describe the function of the periosteum.
7. Why is a healed area of bone less likely to fracture in the same place again from similar stresses?
8. Of what significance is the presence of epiphyseal cartilage in a long bone?
9. What properties are used to distinguish a sesamoid bone from a sutural bone?
10. Contrast the processes of ossification and calcification.

LEVEL 3 CRITICAL THINKING AND CLINICAL APPLICATIONS

1. A small child falls off a bicycle and breaks an arm. The bone is set correctly and heals well. After the cast is removed, there remains an enlarged bony bump at the region of the fracture. After several months this enlargement disappears, and the arm is essentially normal in appearance. What happened during this healing process?
2. John is 14 and lives in an urban apartment. He spends most of his time watching TV and eating "junk" food. One afternoon, during recess, he falls on the playground and breaks his leg. Although he appears to be healthy, his leg takes longer to heal than expected. What might be the cause of the longer healing time?
3. An otherwise healthy but inactive 78-year-old woman falls when she gets up suddenly. What is the most likely diagnosis as to why the fracture occurred? Activity of what type(s) of bone cells is implicated in this result? How might these conditions be improved?

✓ ANSWERS TO CONCEPT CHECK QUESTIONS

p. 118 **1.** If the ratio of collagen to hydroxyapatite in a bone increased, the bone would be more flexible and less strong. **2.** Concentric layers of bone around a central canal are indicative of an osteon or Haversian system. Osteons are found in compact bone. Since the ends (epiphyses) of long bones are primarily cancellous (spongy) bone, this sample most likely came from the shaft (diaphysis) of a long bone. **3.** Since osteoclasts function in breaking down or demineralizing bone, the bone would have less mineral content and as a result would be weaker. **4.** Fracture repair would be impeded.

p. 124 **1.** Long bones of the body, like the femur, have an epiphyseal cartilage separating the epiphysis from the diaphysis, as long as the bone is still growing in length. An X-ray would indicate whether or not the epiphyseal cartilage was still present. If it was, then growth was still occurring, and if not, the bone had reached its adult length. **2.** (1) In a fibrous connective tissue, osteoblasts secrete matrix components in an ossification center. (2) Growth occurs outward from the ossification center in small struts called spicules. (3) Over time, the bone assumes the shape of spongy bone; subsequent remodeling can produce compact bone. **3.** The diameter of a bone enlarges through appositional growth at the outer surface. In this process, periosteal cells differentiate into osteoblasts and contribute to the growth of the bone matrix. **4.** The epiphyseal cartilage is a relatively narrow band of cartilage separating the epiphysis from the diaphysis. It is located at the metaphysis. The continued growth of chondrocytes at the epiphyseal side and their subsequent replacement by bone at the diaphyseal side allow for an increase in the length of a developing bone.

p. 126 **1.** Bones increase in thickness in response to physical stress. One common type of stress that is applied to a bone is that produced by muscles. We would expect the bones of an athlete to be thicker after the addition of the extra muscle mass because of the greater stress that the muscle would apply to the bone. **2.** Vitamin D plays an important role in normal calcium metabolism by stimulating the absorption and transport of calcium and phosphate ions into the blood. Calcitonin inhibits osteoclasts and increases the rate of calcium loss in the urine. Parathyroid hormone stimulates osteoclasts and osteoblasts, increases calcium absorption at the intestines (this action requires another hormone, calcitriol), and decreases the rate of urinary calcium loss. Vitamins A and C are also important, as are growth hormone and sex hormones. **3.** In a 15-year-old, the rate of bone formation would be expected to exceed the rate of bone reabsorption, whereas in the 30-year-old, these rates would be expected to be approximately the same.

p. 129 **1.** Bone markings are often palpable at the surface; they provide reference points for orientation on associated soft tissues. For pathologists, bone markings can be used to estimate the size, weight, sex, and general appearance of an individual on the basis of incomplete remains. **2.** A sesamoid bone is usually round, small, and flat. Irregular bones may have complex shapes with short, flat, notched, or ridged surfaces. **3.** Sutural bones, also called *Wormian bones*, are small, flat, oddly shaped bones found between the flat bones of the skull in the suture line.

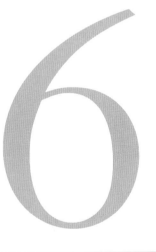

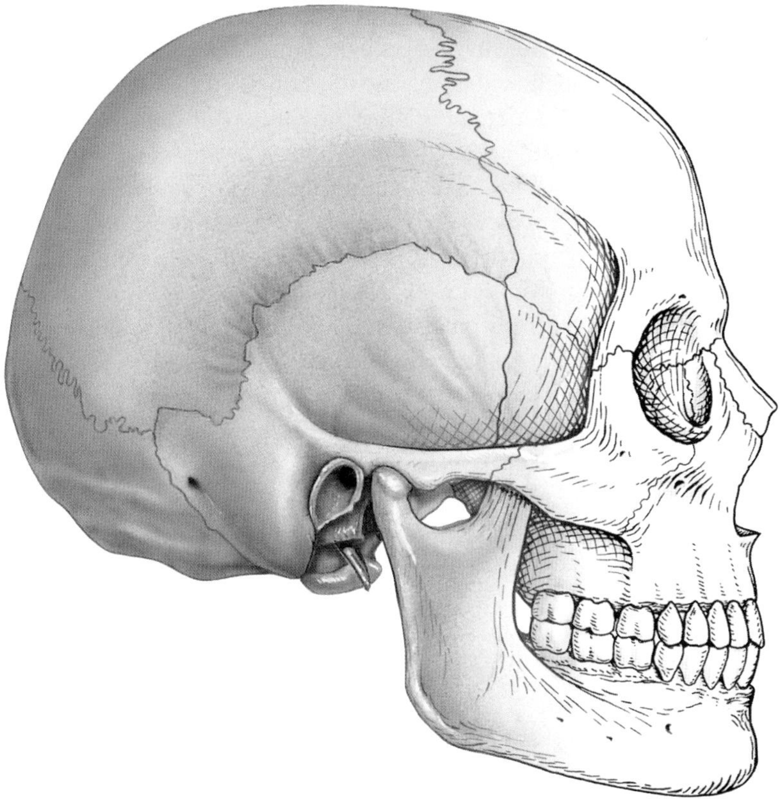

6

THE SKELETAL SYSTEM
Axial Division

The basic features of the human skeleton have been shaped by evolution, but because no two people have exactly the same combination of age, diet, activity pattern, and hormone levels, the bones of each individual are unique. As discussed in Chapter 5, bones are continually remodeled and reshaped, and your skeleton changes throughout your lifetime. Examples include the proportional changes at puberty and the gradual osteoporosis of aging. This chapter provides other examples of the dynamic nature of the human skeleton, such as the changes in the shape of the vertebral column during the transition from crawling to walking.

The skeletal system is divided into *axial* and *appendicular divisions*; the axial division components are shown in yellow and blue in Figure 6.1●. The skeletal system includes 206 separate bones and a number of associated cartilages. The **axial skeleton** consists of the bones of the skull, thorax, and vertebral column. These elements form the longitudinal axis of the

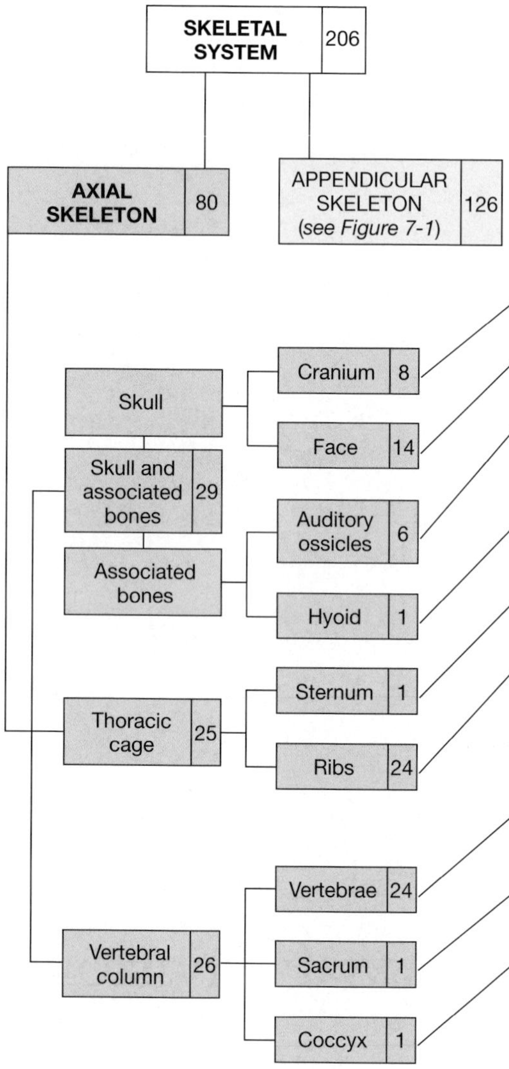

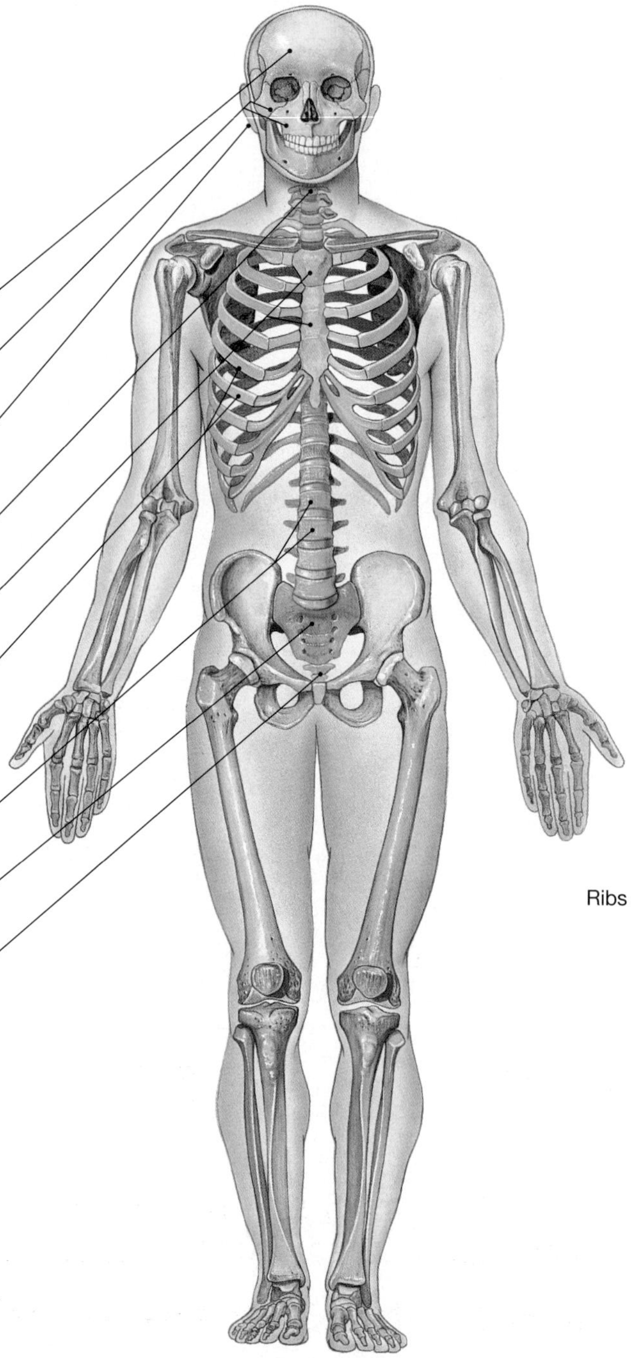

(a) Skeletal system, axial components highlighted

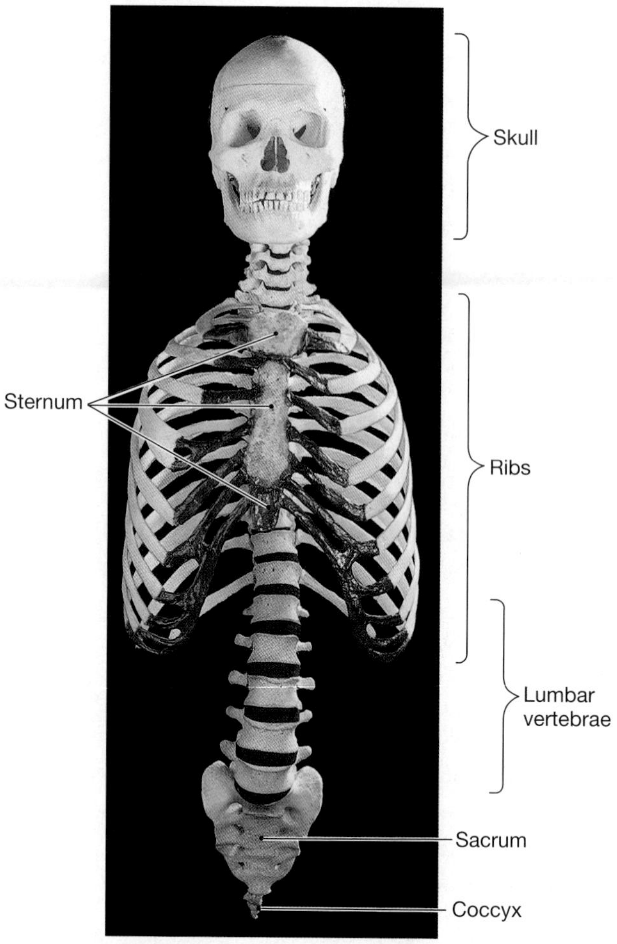

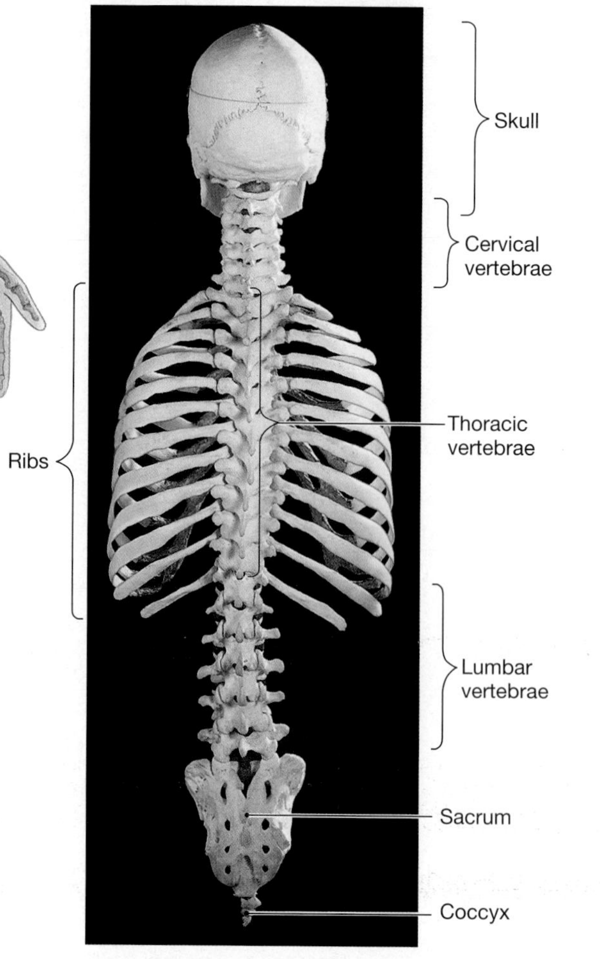

(b) Axial skeleton

FIGURE 6.1 **THE AXIAL SKELETON**

(a) Anterior view of the skeleton, highlighting components of the axial skeleton; the flow chart indicates relationships among the axial components. **(b)** Anterior (above) and posterior (below) views of the bones of the axial skeleton.

body. There are 80 bones in the axial skeleton, roughly 40% of the bones in the human body. The axial components include:

- the **skull** (22 bones),
- bones associated with the skull (6 auditory ossicles and 1 hyoid bone),
- the **vertebral column** (24 vertebrae, 1 sacrum, and 1 coccyx), and
- the **thoracic cage** (24 ribs and 1 sternum).

The axial skeleton functions as a framework that supports and protects organs in the dorsal and ventral body cavities. It houses special sense organs for taste, smell, hearing, balance, and sight. Additionally, it provides an extensive surface area for the attachment of muscles that (1) adjust the positions of the head, neck, and trunk, (2) perform respiratory movements, and (3) stabilize or position structures of the appendicular skeleton. The joints of the axial skeleton permit limited movement, but they are very strong and often heavily reinforced with ligaments. Finally, some parts of the axial skeleton, including portions of the vertebrae, sternum, ribs, and many long bones, contain red marrow for blood cell formation. p. 114

This chapter describes the structural anatomy of the axial skeleton, and we will begin with the skull. Before proceeding, you may find it helpful to review the directional references included in Tables 1.1 and 1.2 and the terms introduced in Table 5.1. p. 128 The remaining 126 bones of the human skeleton constitute the **appendicular skeleton**. This division includes the bones of the limbs and the **pectoral** and **pelvic girdles** that attach the limbs to the trunk. The appendicular skeleton will be examined in Chapter 7.

The Skull and Associated Bones

[FIGURES 6.2 TO 6.5/6.7a]

The bones of the skull protect the brain and guard the entrances to the digestive and respiratory systems. The skull contains 22 bones: 8 form the **cranium**, or "*braincase*," and 14 are associated with the face (Figures 6.2 to 6.5●).

The cranium consists of the *occipital*, *parietal*, *frontal*, *temporal*, *sphenoid*, and *ethmoid* bones. The superficial facial bones, the *maxillae*, *palatine*, *nasal*, *zygomatic*, *lacrimal*, *vomer*, and *mandible* (Figure 6.2●), provide areas for the attachment of muscles that control facial expressions and assist in the manipulation of food.

The cranial bones enclose the **cranial cavity**, a fluid-filled chamber that cushions and supports the brain. Blood vessels, nerves, and membranes that stabilize the position of the brain are attached to the inner surface of the cranium. Its outer surface provides an extensive area for the attachment of muscles that move the eyes, jaws, and head. A specialized joint between the occipital bone and the first spinal vertebra stabilizes the positions of the cranium and vertebral column while permitting a considerable range of head movements.

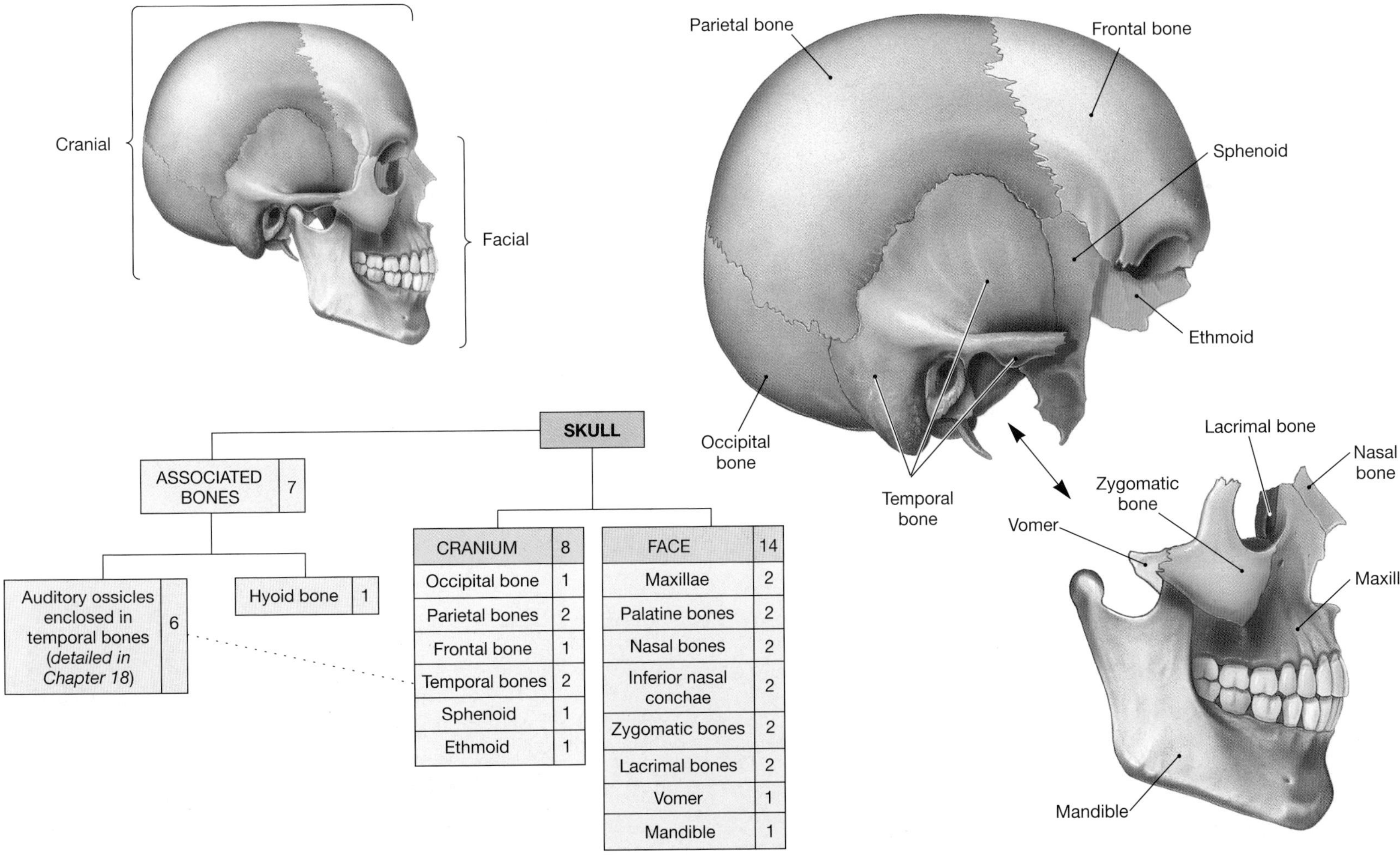

FIGURE 6.2 **CRANIAL AND FACIAL SUBDIVISIONS OF THE SKULL**

The skull can be divided into the cranial and the facial divisions. The palatine bones and the inferior nasal conchae of the facial division are not visible from this perspective. The seven associated bones are not shown.

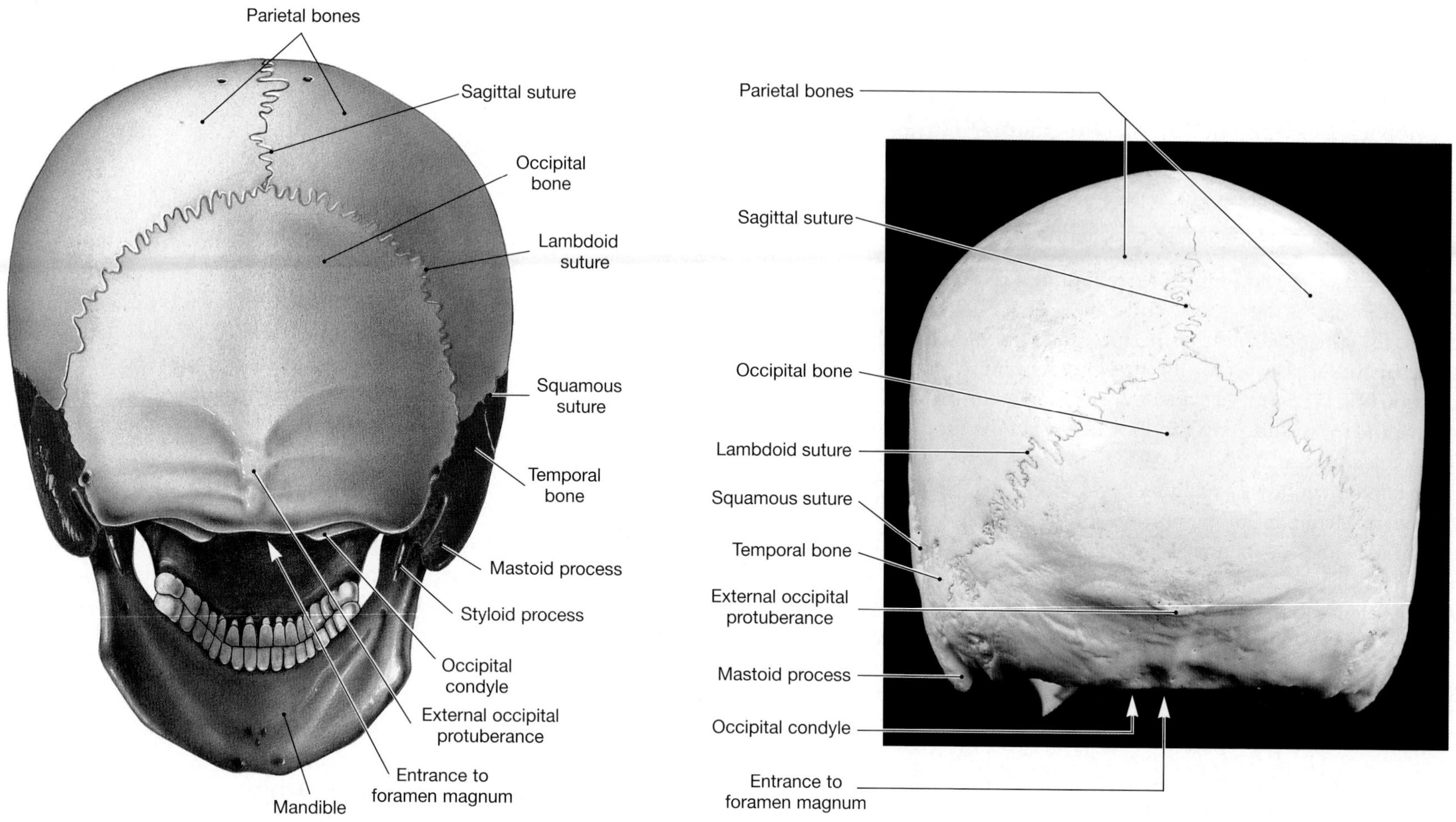

(a) Posterior view

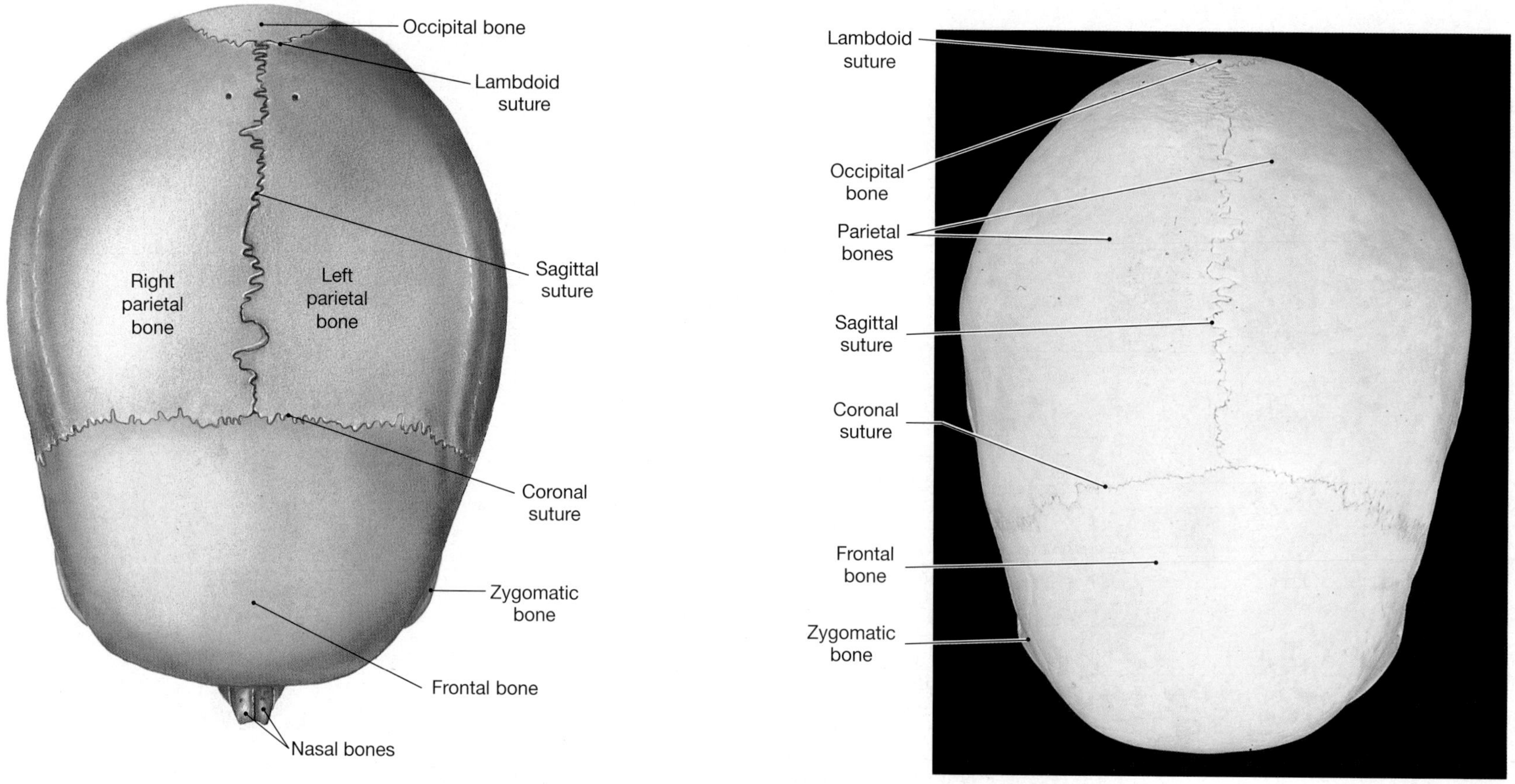

(b) Superior view

FIGURE 6.3 **The Adult Skull**

The bones in an adult skull are shown in (**a**) posterior view, (**b**) superior view, (**c**) lateral view, (**d**) anterior view, and (**e**) inferior view.

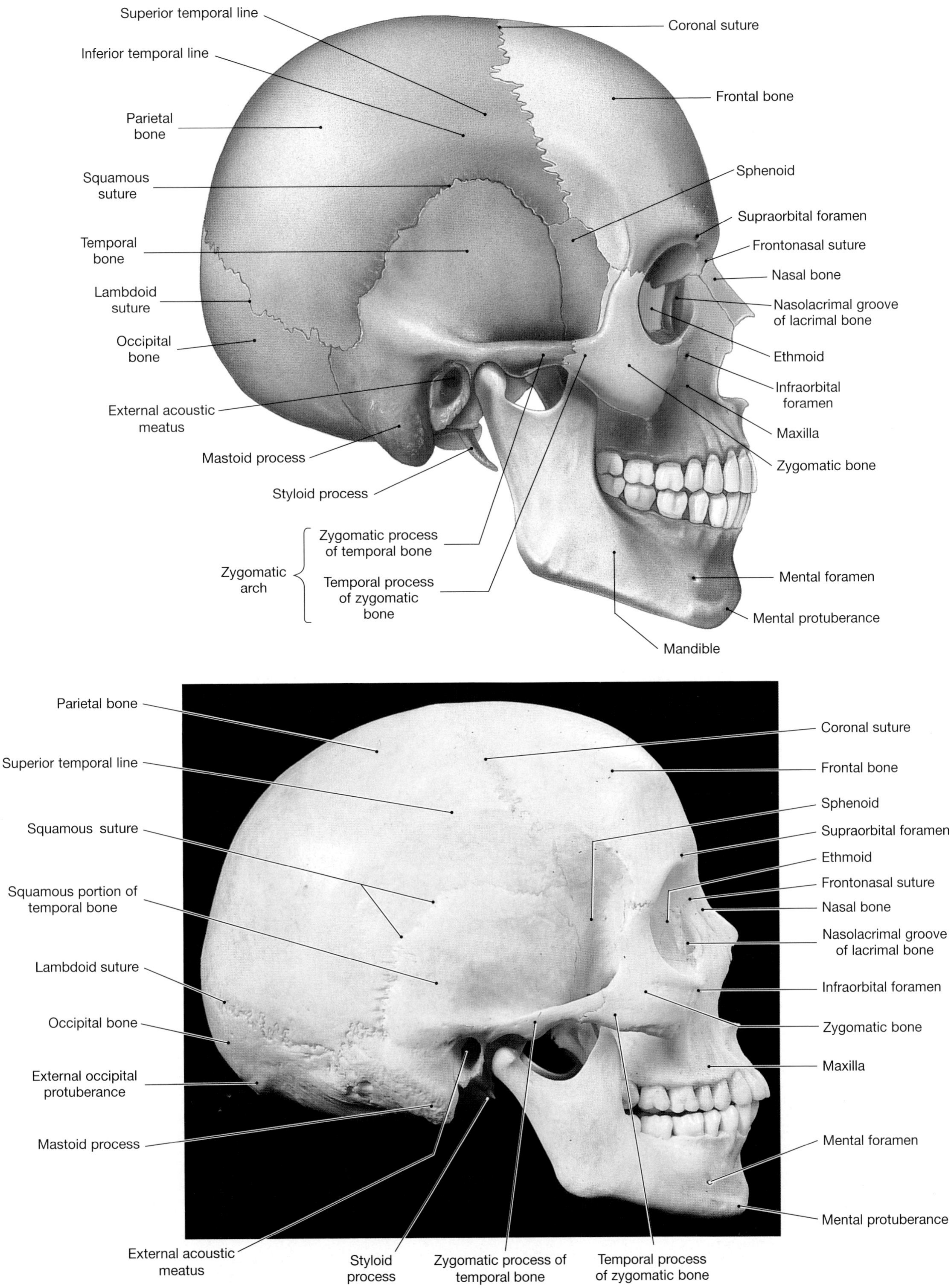

FIGURE 6.3 *(continued)*

(c) Lateral view

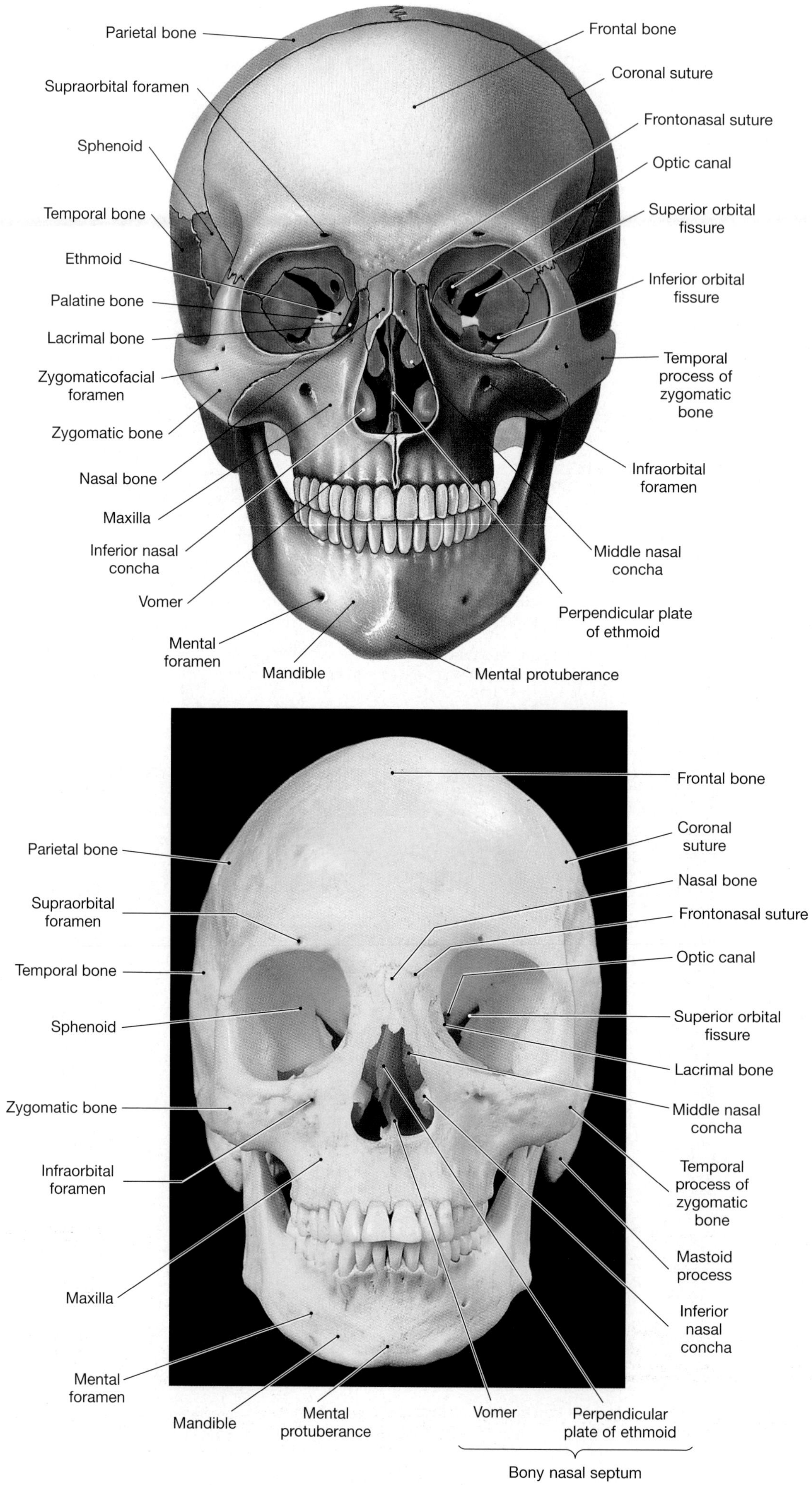

FIGURE 6.3 *(continued)*

(d) Anterior view

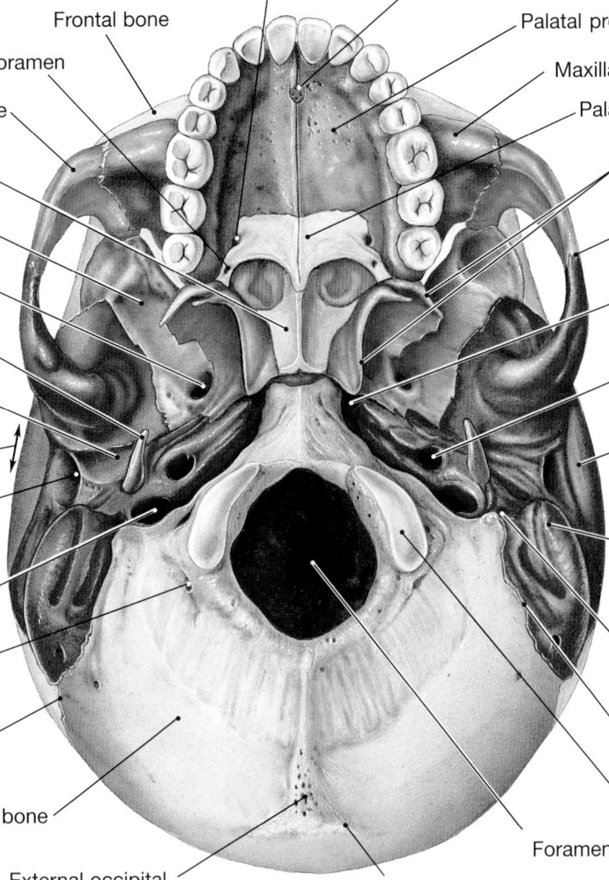

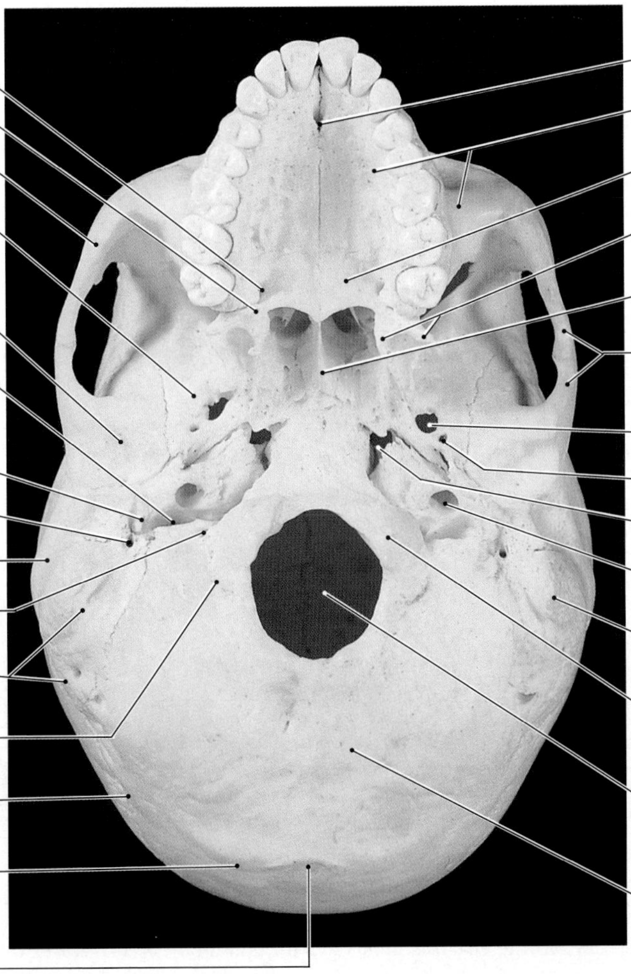

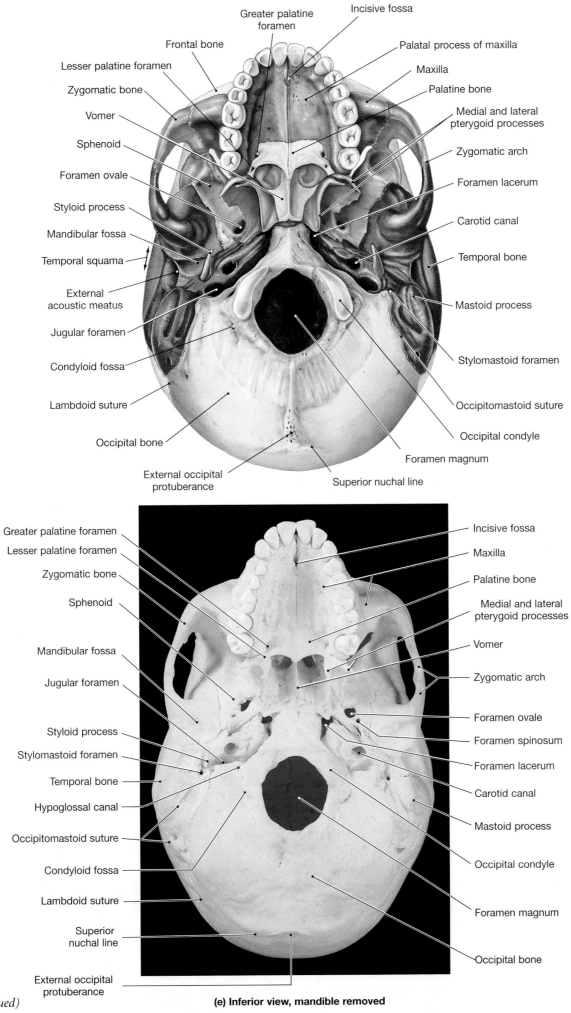

FIGURE 6.3 *(continued)* **(e) Inferior view, mandible removed**

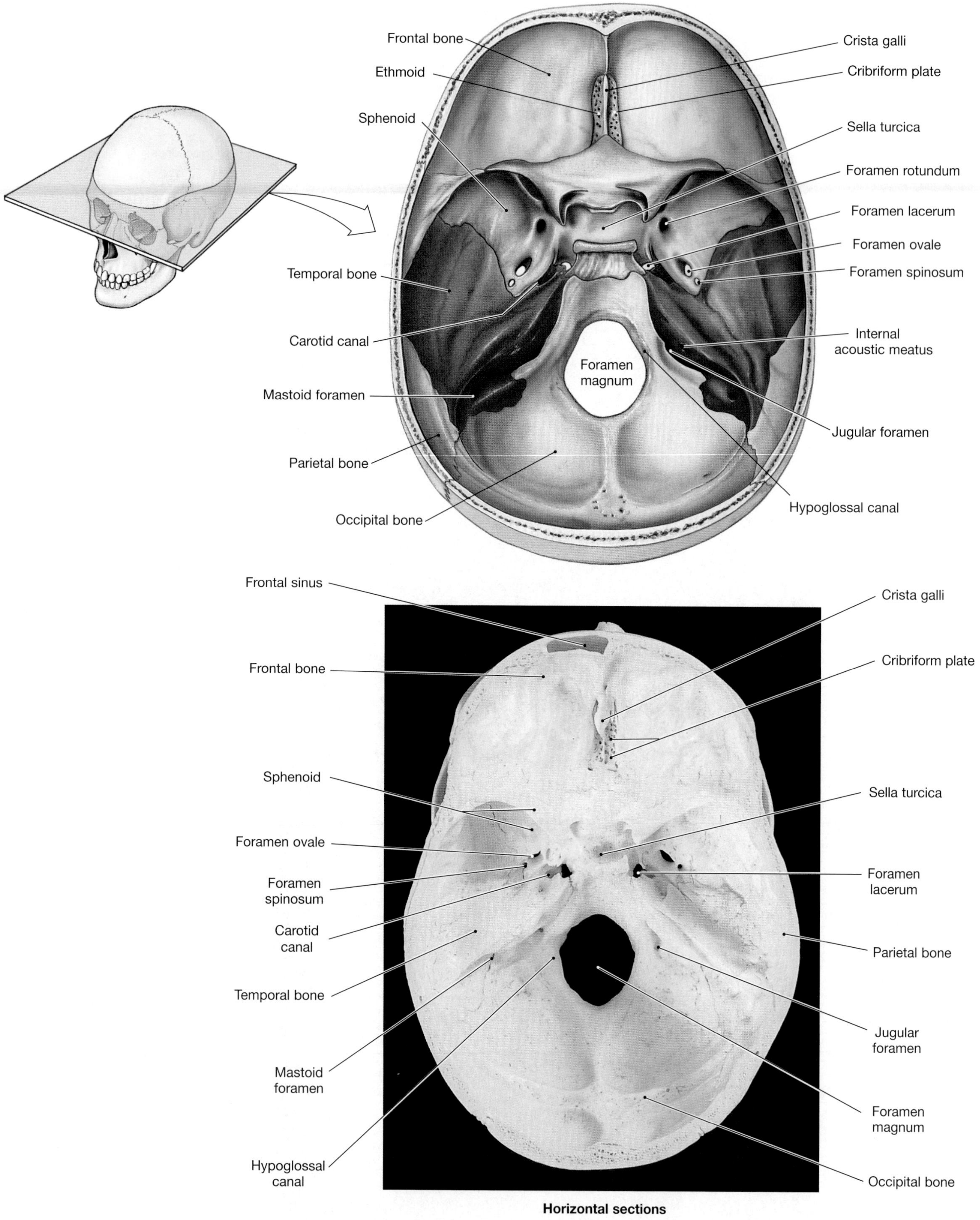

FIGURE 6.4 **SECTIONAL ANATOMY OF THE SKULL, PART I**

Horizontal section: A superior view showing major landmarks in the floor of the cranial cavity. *Compare with Figure 6.3e and MRI Scans 1a–c in the Companion Atlas.*

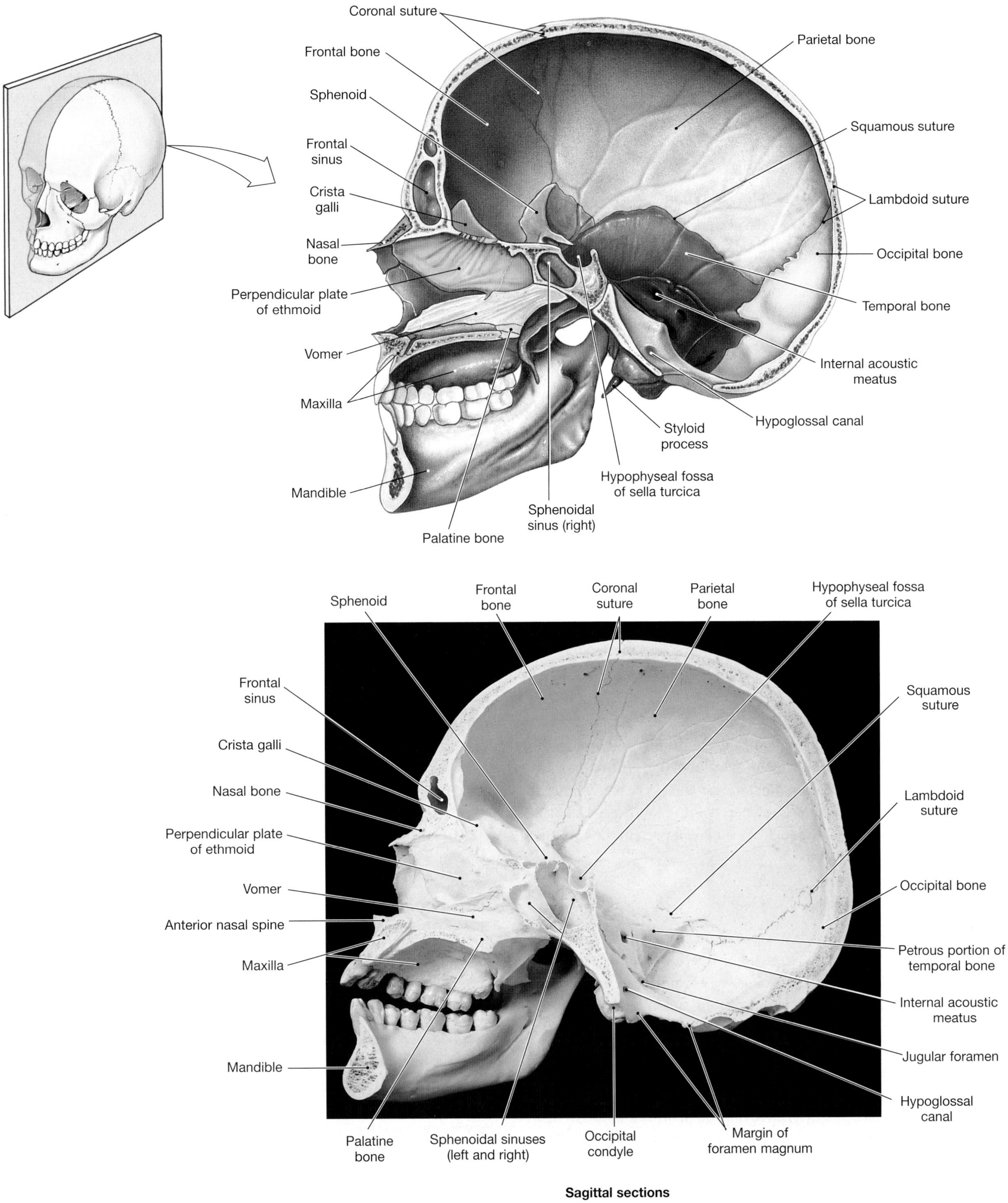

FIGURE 6.5 **SECTIONAL ANATOMY OF THE SKULL, PART II**

Sagittal section: A medial view of the right half of the skull. Because the bony nasal septum is intact, the right nasal cavity cannot be seen. *See MRI Scans 1d–e in the Companion Atlas.*

If the cranium is the house where the brain resides, the *facial complex* is the front porch. **Facial bones** protect and support the entrances to the digestive and respiratory tracts.

The boundaries between skull bones are immovable joints called **sutures**. At a suture, the bones are joined firmly together with dense fibrous connective tissue. Each of the sutures of the skull has a name, but you need to know only five major sutures at this time: the *lambdoid, coronal, sagittal, squamous*, and *frontonasal* sutures.

- *Lambdoid* (lam-DOYD) *suture*. The **lambdoid suture** arches across the posterior surface of the skull (Figure 6.3a●), separating the *occipital bone* from the *parietal bones*. One or more **sutural bones** (*Wormian bones*) may be found along this suture; they range from a bone the size of a grain of sand to one as large as a quarter. p. 127
- *Sagittal suture*. The **sagittal suture** begins at the superior midline of the lambdoid suture and extends anteriorly between the parietal bones to the coronal suture (Figure 6.3b●).
- *Coronal suture*. Anteriorly, the sagittal suture ends when it intersects the coronal suture. The **coronal suture** crosses the superior surface of the skull, separating the anterior *frontal bone* from the more posterior parietal bones (Figure 6.3b●). The occipital, parietal, and frontal bones form the **calvaria** (kal-VAR-ē-a), also called the *cranial vault* or "skullcap."
- *Squamous suture*. A **squamous suture** on each side of the skull marks the boundary between the *temporal bone* and the parietal bone of that side. The squamous sutures can be seen in Figure 6.3a●, where they intersect the lambdoid suture. The path of the squamous suture on the right side of the skull can be seen in Figure 6.3c●
- *Frontonasal suture*. The **frontonasal suture** is the boundary between the superior aspects of the two nasal bones and the frontal bone (Figure 6.3c,d●).

Bones of the Cranium [FIGURES 6.3 TO 6.5]

We will now consider each of the bones of the cranium. As we proceed, use the figures provided to develop a three-dimensional perspective on the individual bones. Ridges and foramina that are detailed here mark either the attachment of muscles or the passage of nerves and blood vessels that will be studied in later chapters. Figures 6.3, 6.4, and 6.5● present the adult skull in superficial and sectional views.

OCCIPITAL BONE [FIGURES 6.3a–c,e/6.6a,b]

The **occipital bone** contributes to the posterior, lateral, and inferior surfaces of the cranium (Figure 6.3a,b,c,e●). The inferior surface of the occipital bone contains a large circular opening, the **foramen magnum** (Figure 6.3e●), which connects the cranial cavity with the spinal cavity enclosed by the vertebral column. At the adjacent **occipital condyles**, the skull articulates with the first cervical vertebra. The posterior, external surface of the occipital bone (Figure 6.6a●) bears a number of prominent ridges. The **occipital crest** extends posteriorly from the foramen magnum, ending in a small midline bump called the **external occipital protuberance**. Two horizontal ridges intersect the crest, the **inferior** and **superior nuchal** (NOO-kal) **lines**. These lines mark the attachment of muscles and ligaments that stabilize the articulation at the occipital condyles and balance the weight of the head over the vertebrae of the neck. The occipital bone forms part of the wall of the large **jugular foramen** (Figure 6.3e●). The *internal jugular vein* passes through this foramen to drain venous blood from the brain. The **hypoglossal canals** begin at the lateral base of each occipital condyle, just superior to the condyles (Figure 6.6a●). The *hypoglossal nerves*, cranial nerves that control the tongue muscles, pass through these canals.

Inside the skull, the hypoglossal canals begin on the inner surface of the occipital bone near the foramen magnum (Figure 6.6b●). Note the concave internal surface of the occipital bone, which closely follows the contours of the brain. The grooves follow the path of major vessels, and the ridges mark the attachment site of membranes (the *meninges*) that stabilize the position of the brain.

PARIETAL BONES [FIGURES 6.3b,c/6.5/6.6c]

The paired **parietal** (pa-RĪ-e-tal) **bones** contribute to the superior and lateral surfaces of the cranium and form the major part of the calvaria (Figure 6.3b,c●). The external surface of each parietal bone (Figure 6.6c●) bears a pair of low ridges, the **superior** and **inferior temporal lines**. These lines mark the attachment of the *temporalis muscle*, a large muscle that closes the mouth. The smooth parietal surface superior to these lines is called the **parietal eminence**. The internal surfaces of the parietal bones retain the impressions of cranial veins and arteries that branch inside the cranium (Figure 6.5●).

FRONTAL BONE [FIGURES 6.3b,c,d/6.5/6.7]

The **frontal bone** forms the forehead and roof of the orbits (Figure 6.3b,c,d●). During development, the bones of the cranium form through the fusion of separate centers of ossification, and at birth the fusions have not been completed. At this time there are two frontal bones that articulate along the **metopic suture**. Although the suture usually disappears by age 8 with the fusion of the bones, the frontal bone of an adult often retains traces of the suture line.

The metopic suture, or what remains of it, runs down the center of the **frontal part** of the frontal bone (Figure 6.7a●). The convex anterior surface of the frontal part is called the *frontal squama*, or forehead. The lateral surfaces contain the anterior continuations of the superior temporal lines. The frontal part of the frontal bone ends at the **supraorbital margins** that mark the superior limits of the orbits, the bony recesses that support and protect the eyeballs. Above the supraorbital margins are thickened ridges, the **superciliary arches**, which support the eyebrows. The center of each margin is perforated by a single **supraorbital foramen** or **notch**.

The **orbital part** of the frontal bone forms the roughly horizontal roof of each orbit. The inferior surface of the orbital part is relatively smooth, but it contains small openings for blood vessels and nerves heading to or from structures in the orbit. This is often called the *orbital surface* of the frontal bone. The shallow **lacrimal fossa** marks the location of the *lacrimal* (tear) *gland* that lubricates the surface of the eye (Figure 6.7b●).

The internal surface of the frontal bone roughly conforms to the shape of the anterior portion of the brain (Figure 6.7c●). The inner surface of the frontal part bears a prominent **frontal crest** that marks the attachment of membranes that, among their other functions, prevent contact between the delicate brain tissues and the bone of the cranium.

The **frontal sinuses** (Figure 6.5a and 6.7b●) are variable in size and in time of appearance. They usually develop after age 6, but some people never develop them at all. The frontal sinuses and other sinuses will be described in a later section.

TEMPORAL BONES [FIGURES 6.3c,d,e/6.8]

The paired **temporal bones** contribute to the lateral and inferior walls of the cranium; contribute to the *zygomatic arches* of the cheek; form the only articulations with the mandible; and protect the sense organs of the inner

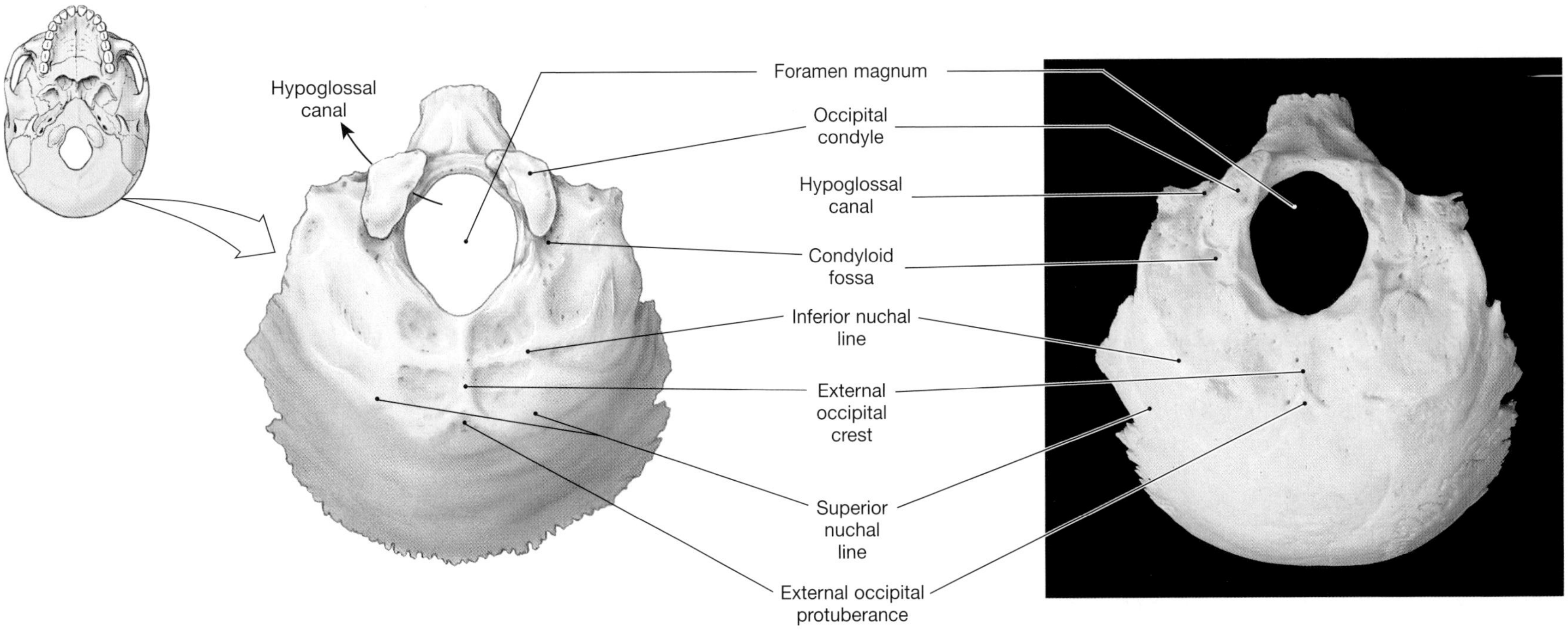

(a) Occipital bone, inferior (external) view

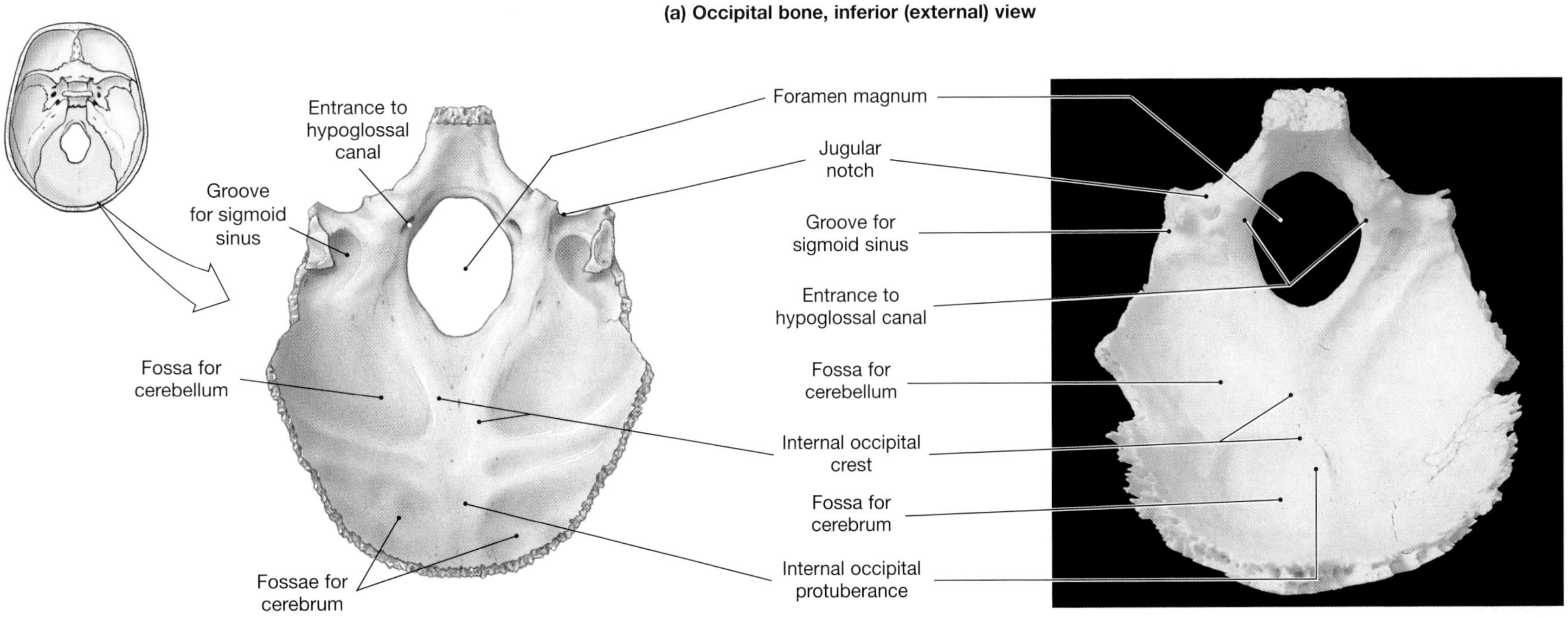

(b) Occipital bone, superior (internal) view

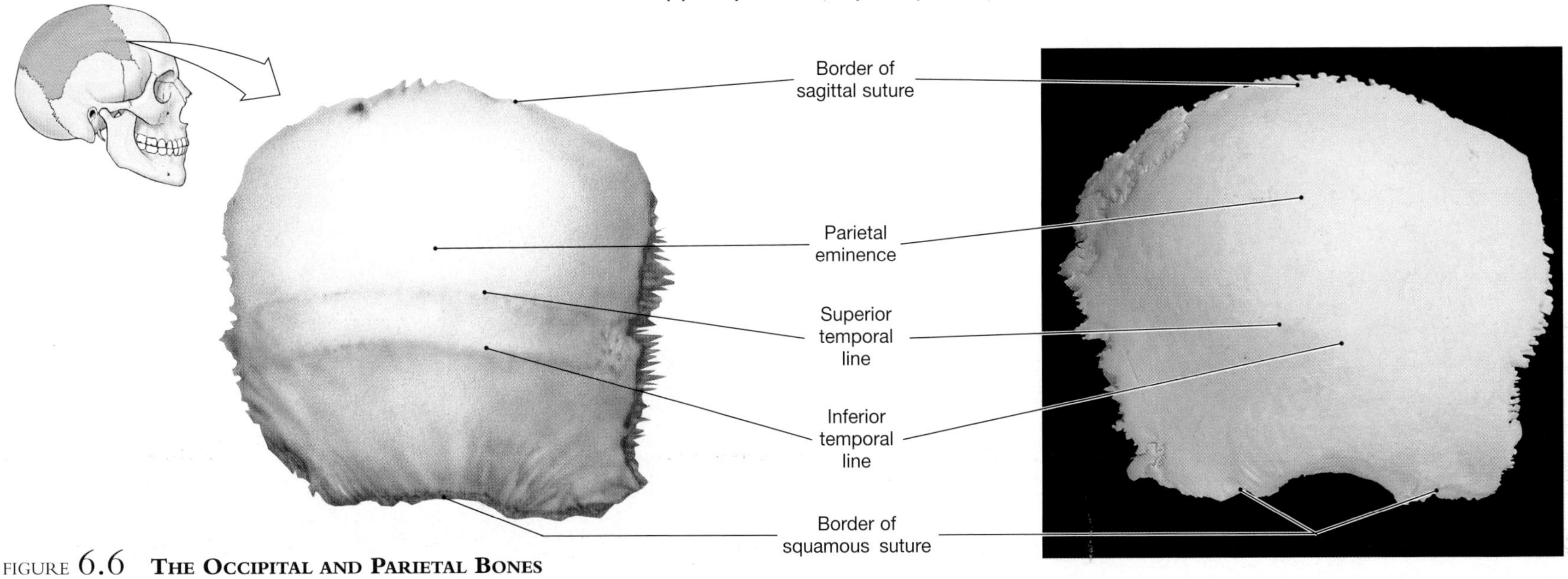

(c) Parietal bone, external surface

FIGURE 6.6 **THE OCCIPITAL AND PARIETAL BONES**

The occipital bone is shown in (**a**) inferior (external) view and (**b**) superior (internal) view. (**c**) A lateral view of the right parietal bone; for a medial view, see Figure 6.5.

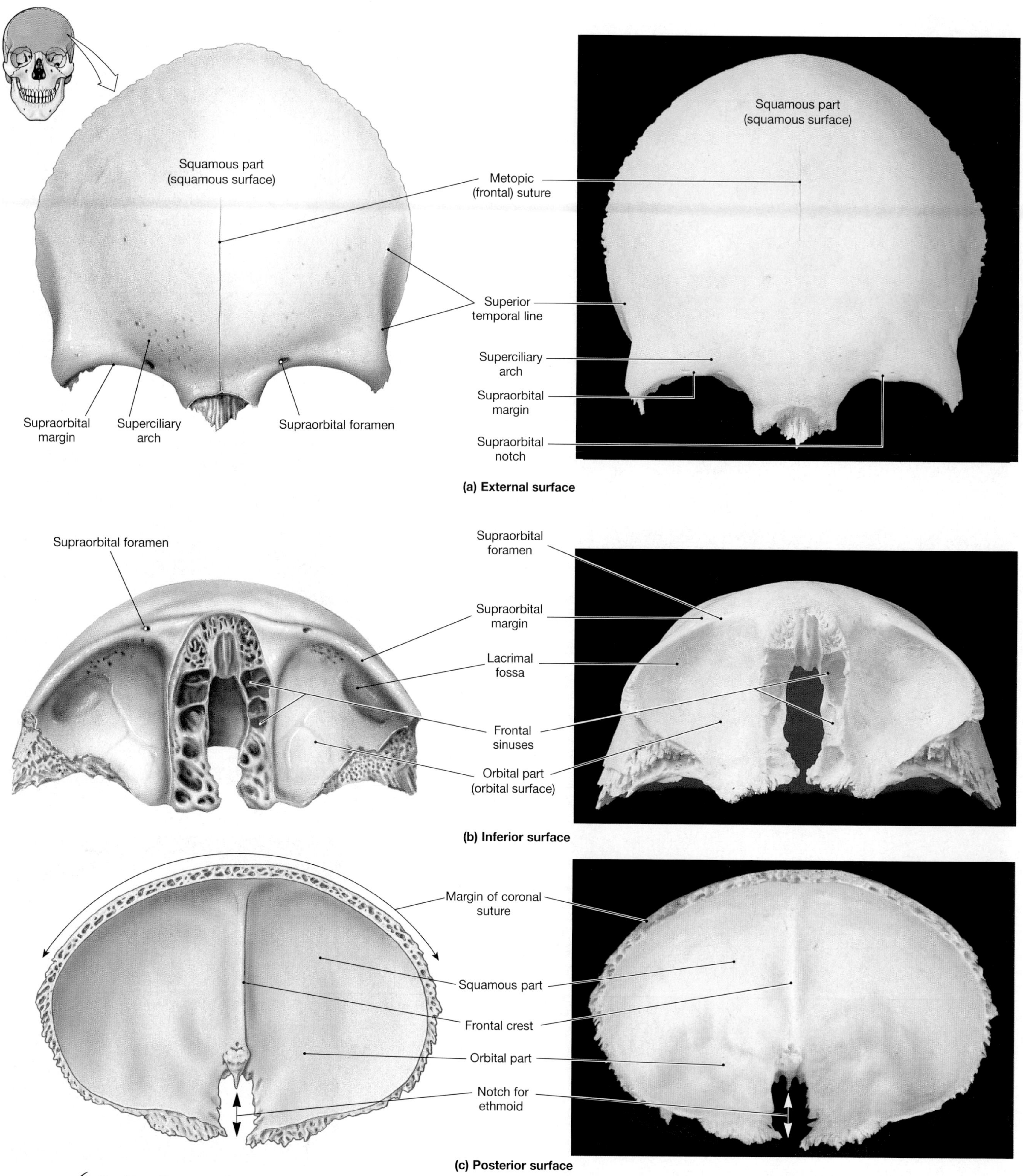

FIGURE 6.7 **The Frontal Bone**

The frontal bone is shown in (**a**) anterior view, (**b**) inferior view, and (**c**) posterior (internal) view.

ear. In addition, the convex surfaces inferior to each parietal bone form an extensive area for the attachment of muscles that close the jaws and move the head (Figure 6.3c●, p. 137). The temporal bones articulate with the zygomatic, parietal, and occipital bones and with the sphenoid and mandible. Each temporal bone has squamous, tympanic, and petrous parts.

The **squamous part** of the temporal bone is the lateral surface bordering the squamous suture (Figure 6.8a,d●). The convex external surface of the squamous part is the *squama;* the concave internal surface, whose curvature parallels the surface of the brain, is the *cerebral surface.* The inferior margin of the squamous part is formed by the prominent **zygomatic process**. The zygomatic process curves laterally and anteriorly to meet the **temporal process** of the *zygomatic bone.* Together these processes form the **zygomatic arch**, or cheekbone. Inferior to the base of the zygomatic process, the temporal bone articulates with the mandible. A depression called the **mandibular fossa** and an elevated **articular tubercle** mark this site (Figure 6.8a,c●).

Immediately posterior and lateral to the mandibular fossa is the **tympanic part** of the temporal bone (Figure 6.8b●). This region surrounds the entrance to the **external acoustic meatus**, or *external auditory canal.* In life, this passageway ends at the delicate **tympanic membrane**, or eardrum, but this membrane disintegrates during the preparation of a dried skull.

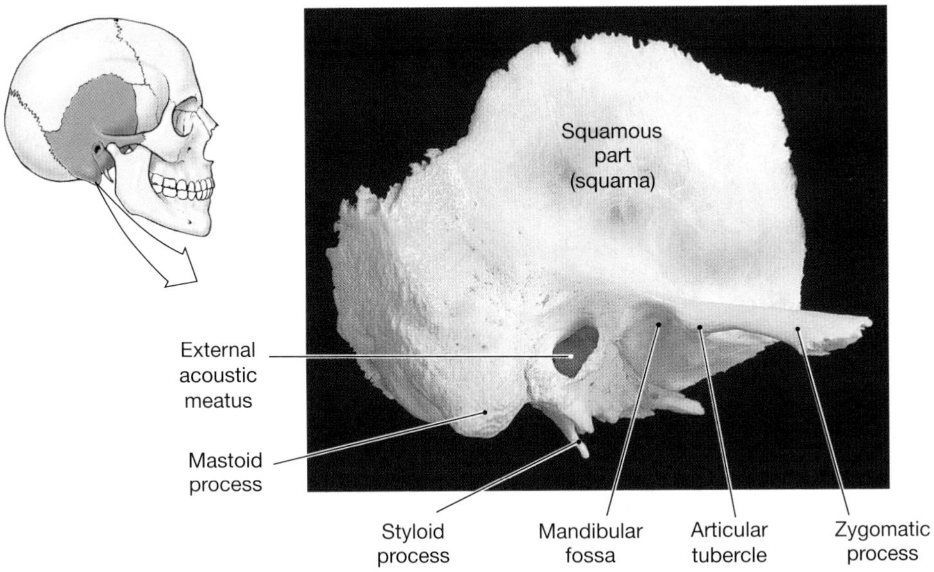

(a) Right temporal bone, lateral view

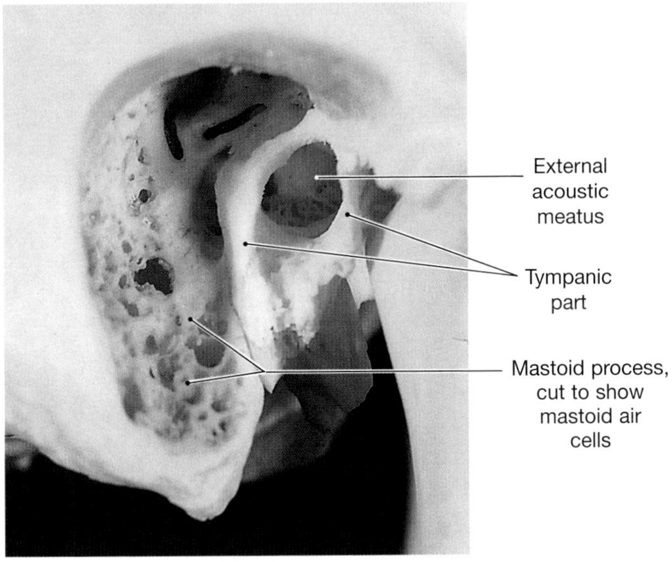

(b) The mastoid air cells

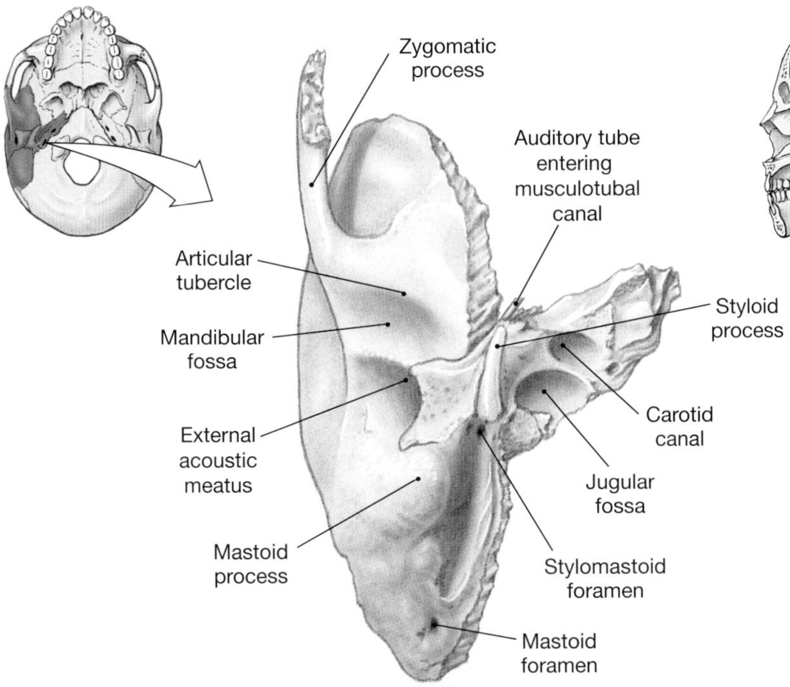

(c) Right temporal bone, inferior view

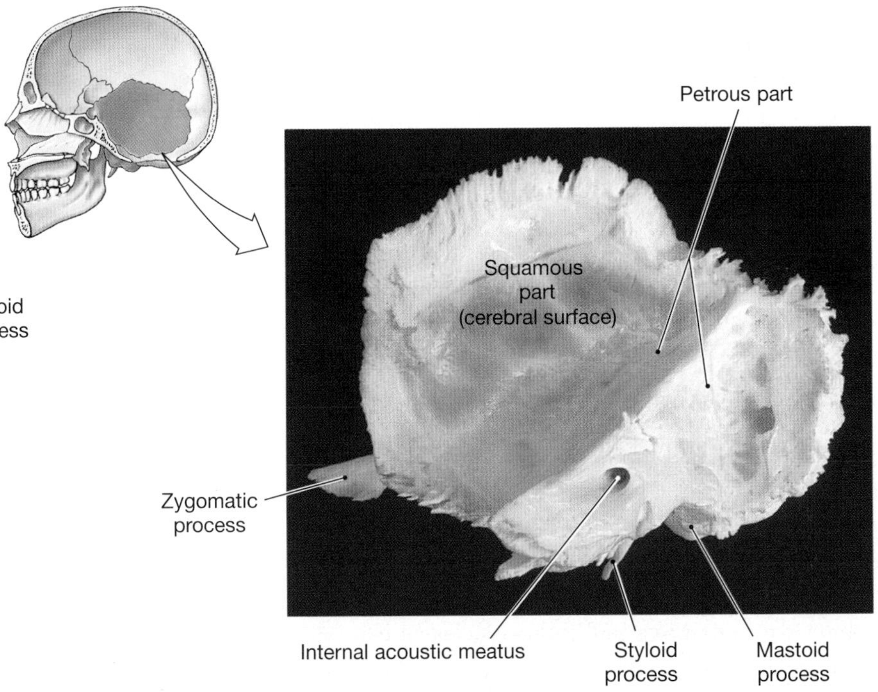

(d) Right temporal bone, medial view

FIGURE 6.8 **THE TEMPORAL BONE**

Major anatomical landmarks are shown on a right temporal bone: (**a**) lateral view, (**b**) a cutaway view of the mastoid air cells, (**c**) an inferior view, (**d**) a medial view.

The most massive portion of the temporal bone is the **petrous part** (*petrous,* stone). The petrous part of the temporal bone surrounds and protects sense organs providing our senses of hearing and balance. On the lateral surface, the bulge just posterior and inferior to the external acoustic meatus is the **mastoid process** (Figure 6.8a–c●). This process provides an attachment site for muscles that rotate or extend the head. Numerous interconnected mastoid sinuses, termed *mastoid air cells,* are contained within the mastoid process (Figure 6.8b●). Infections in the respiratory tract may spread to these air cells, and such an infection is called *mastoiditis.*

Several other landmarks on the petrous part of the temporal bone can be seen on its inferior surface (Figure 6.8c●). Near the base of the mastoid process, the **mastoid foramen** penetrates the temporal bone. Blood vessels travel through this passageway to reach the membranes surrounding the brain. Ligaments that support the hyoid bone attach to the sharp **styloid process** (STĪ-loyd; *stylos,* pillar), as do some of the tongue muscles. The **stylomastoid foramen** lies posterior to the base of the styloid process. The *facial nerve* passes through this foramen to control the facial muscles. Medially, the **jugular foramen** is bounded by the temporal and occipital bones (Figure 6.3e●, p. 139). Anterior and slightly medial to the jugular foramen is the entrance to the **carotid canal**. The *internal carotid artery,* a major artery that supplies blood to the brain, penetrates the skull through this passageway. Anterior and medial to the carotid canal, a jagged slit, the **foramen lacerum** (LA-se-rum; *lacerare,* to tear), extends between the occipital and temporal bones. In life, this space contains hyaline cartilage and small arteries supplying the inner surface of the cranium.

Lateral and anterior to the carotid foramen, the temporal bone articulates with the sphenoid. A small canal begins at that articulation and ends inside the mass of the temporal bone (Figure 6.8c●). This is the *musculotubal canal,* which surrounds the **auditory tube**, an air-filled passageway. The auditory tube, also known as the *Eustachian* (ū-STĀ-kē-an) *tube,* or *pharyngotympanic tube,* begins at the pharynx. It ends at the **tympanic cavity**, a chamber inside the temporal bone. The tympanic cavity, or *middle ear,* contains the **auditory ossicles**, or ear bones. These tiny bones transfer sound vibrations from the eardrum toward the receptor complex in the inner ear, which provides the sense of hearing.

The petrous part dominates the medial surface of the temporal bone (Figure 6.8d●). The

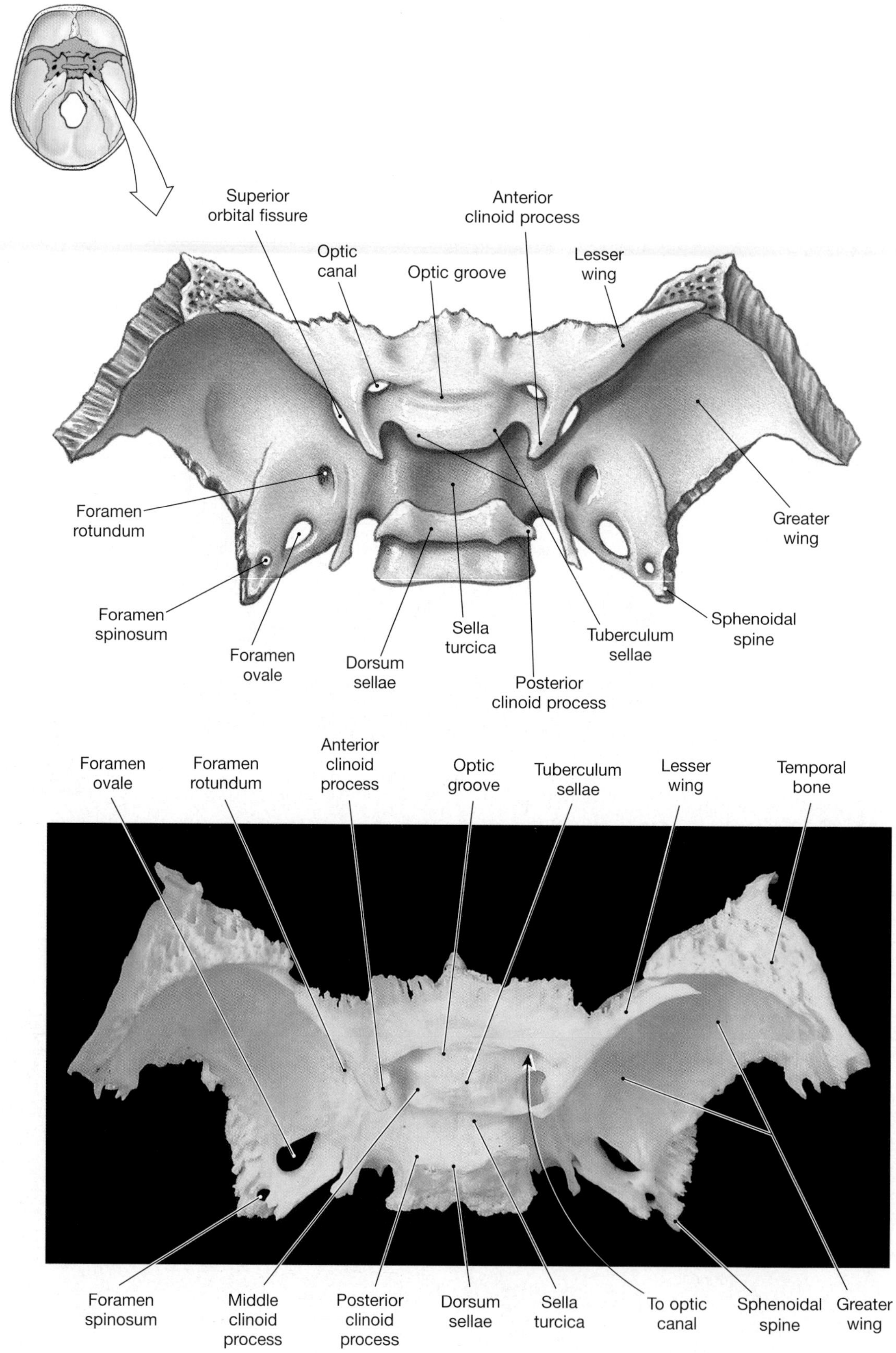

(a) Superior surface

FIGURE 6.9 **The Sphenoid**

Views of the sphenoid showing major anatomical landmarks on the (**a**) superior surface, and (**b**) anterior surface.

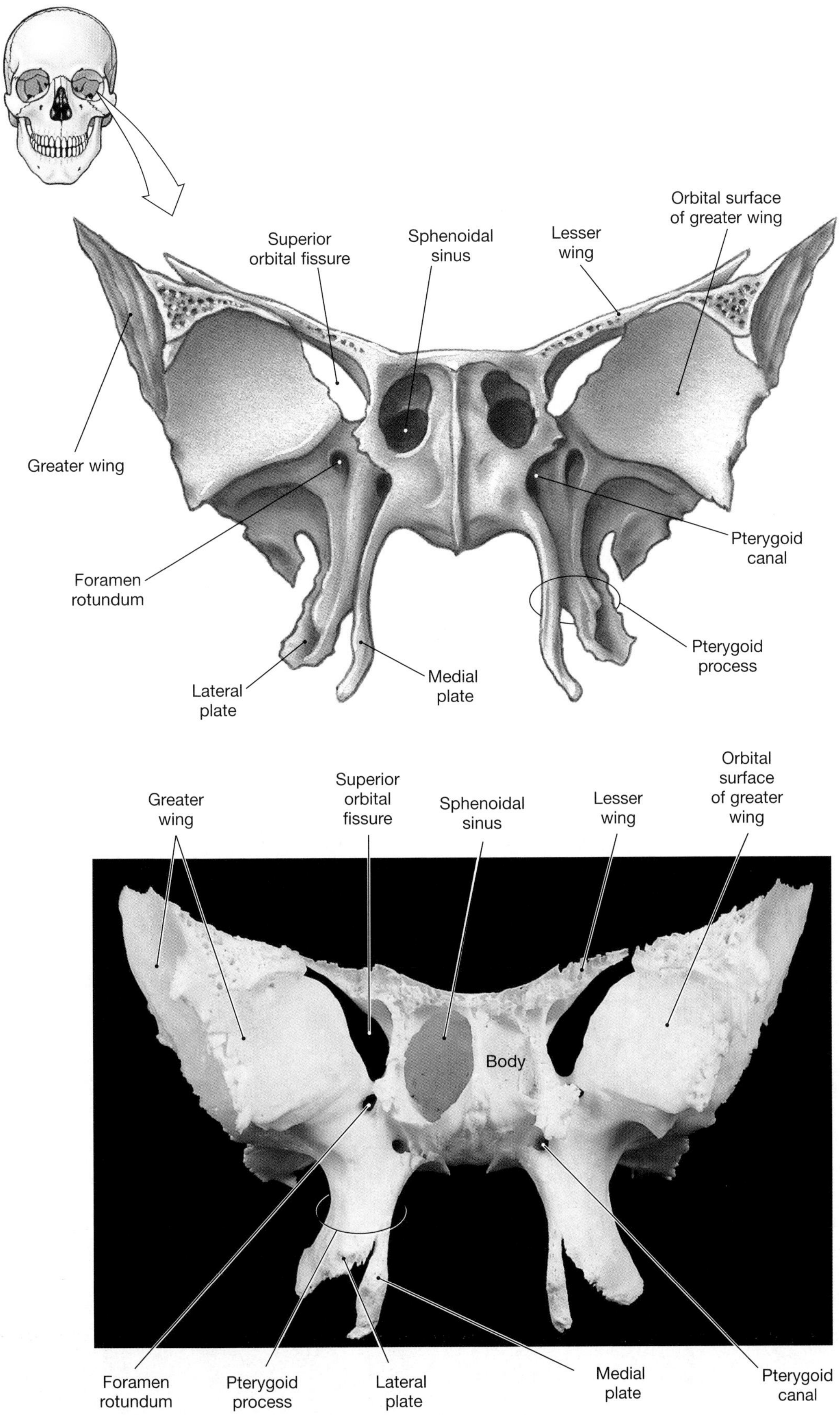

(b) Anterior surface

internal acoustic meatus carries blood vessels and nerves to the inner ear and the facial nerve to the stylomastoid foramen. The entire medial surface of the temporal bone is marked by grooves that mark the location of blood vessels passing along the inner surface of the cranium. The sharp ridge on the inner surface of the petrous part marks the attachment of a membrane that helps stabilize the position of the brain.

SPHENOID [FIGURES 6.3c,d,e/6.4/6.9]

The **sphenoid**, or *sphenoidal bone*, articulates with every other cranial bone and extends from one side to the other across the floor of the cranium. Although it is relatively large, much of the sphenoid is hidden by more superficial bones. It acts as a bridge uniting the cranial and facial bones; it articulates with the frontal, occipital, parietal, ethmoid, and temporal bones of the cranium, and the palatine bones, zygomatic bones, maxillae, and vomer of the facial complex (Figure 6.3c,d,e●, pp. 137–139). The sphenoid also acts as a brace, strengthening the sides of the skull. The **body** forms the central portion of the bone.

The general shape of the sphenoid has been compared to a giant bat, with its wings extended. The wings can be seen most clearly on the superior surface (Figures 6.4 and 6.9a●). A prominent central depression between the wings cradles the pituitary gland below the brain. This recess is called the **hypophyseal** (hī-pō-FIZ-ē-al) **fossa**, and the bony enclosure is called the **sella turcica** (TUR-si-ka) because it supposedly resembles a "Turkish saddle." If the rider faced forward, he could grasp the **anterior clinoid processes** on either side. The anterior clinoid processes are posterior projections of the **lesser wings** of the sphenoid. The **tuberculum sellae** forms the anterior border of the sella turcica; the **dorsum sellae** forms the posterior border. A **posterior clinoid** (KLĪ-noyd) **process** extends laterally on either side of the dorsum sellae.

The transverse groove that crosses to the front of the saddle, above the level of the seat, is the **optic groove**. At either end of this groove is an **optic canal**. The *optic nerves* that carry visual information from the eyes to the brain travel through these canals. On either side of the sella turcica, the **superior orbital fissure**, the **foramen rotundum**, the **foramen ovale** (ō-VAH-lā), and the **foramen spinosum** penetrate the **greater wings** of the sphenoid. These passages carry blood vessels and cranial nerves to structures of the orbit, face, and jaws. Posterior and lateral to these foramina the greater wings end at a sharp **sphenoidal spine**. The superior orbital fissures and the left and right foramen rotundum can also be seen in an anterior view (Figure 6.9b●).

The **pterygoid processes** (TER-i-goyd; *pterygion*, wing) of the sphenoid are vertical projections that begin at the boundary between the greater and lesser wings. Each process forms a pair of *plates* that are important sites for the attachment of muscles that move the lower jaw and soft palate. At the base of each pterygoid process, the **pterygoid canal** provides a route for a small nerve and an artery that supply the soft palate and adjacent structures.

ETHMOID [FIGURES 6.3d/6.4/6.5/6.10]

The **ethmoid**, or *ethmoidal bone*, is an irregularly shaped bone that forms part of the orbital wall (Figure 6.3d●) the anteromedial floor of the cranium (Figure 6.4●), the roof of the nasal cavity, and part of the nasal septum (Figure 6.5●). The ethmoid has three parts: the *cribriform plate*, the paired *lateral masses*, and the *perpendicular plate* (Figure 6.10●).

The superior surface of the ethmoid (Figure 6.10a●) contains the **cribriform plate**, an area perforated by the *cribriform foramina*. These openings allow passage of the branches of the *olfactory nerves*, which provide the sense of smell. A prominent ridge, the **crista galli** (*crista*, crest + *gallus*, chicken; "cock's comb") separates the right and left sides of the cribriform plate. The *falx cerebri*, a membrane that stabilizes the position of the brain, attaches to this bony ridge.

The **lateral masses**, dominated by the **superior nasal conchae** (KON-kē; singular *concha*, "a snail shell") and the **middle nasal conchae**, are best viewed from the anterior and posterior surfaces of the ethmoid (Figure 6.10b,c●). The lateral masses contain the **ethmoidal labyrinth**, an interconnected network of air cells. The labyrinth opens into the nasal cavity on each side. Mucous secretions from these air cells flush the surfaces of the nasal cavities.

The nasal conchae are thin scrolls of bone that project into the nasal cavity on either side of the perpendicular plate. The projecting conchae break up the airflow, creating swirls and eddies. This mechanism slows air movement, but provides additional time for warming, humidification, and dust removal before the air reaches more delicate portions of the respiratory tract.

The **perpendicular plate** forms part of the *nasal septum*, a partition that also includes the vomer and a piece of hyaline cartilage. Olfactory receptors are located in the epithelium covering the inferior surfaces of the cribriform plate, the medial surfaces of the superior nasal conchae, and the superior portion of the perpendicular plate.

THE CRANIAL FOSSAE [FIGURE 6.11]

The contours of the cranium closely follow the shape of the brain. Proceeding from anterior to posterior, the floor of the cranium is not horizontal; it descends in two steps (Figure 6.11a●). Viewed from the superior surface

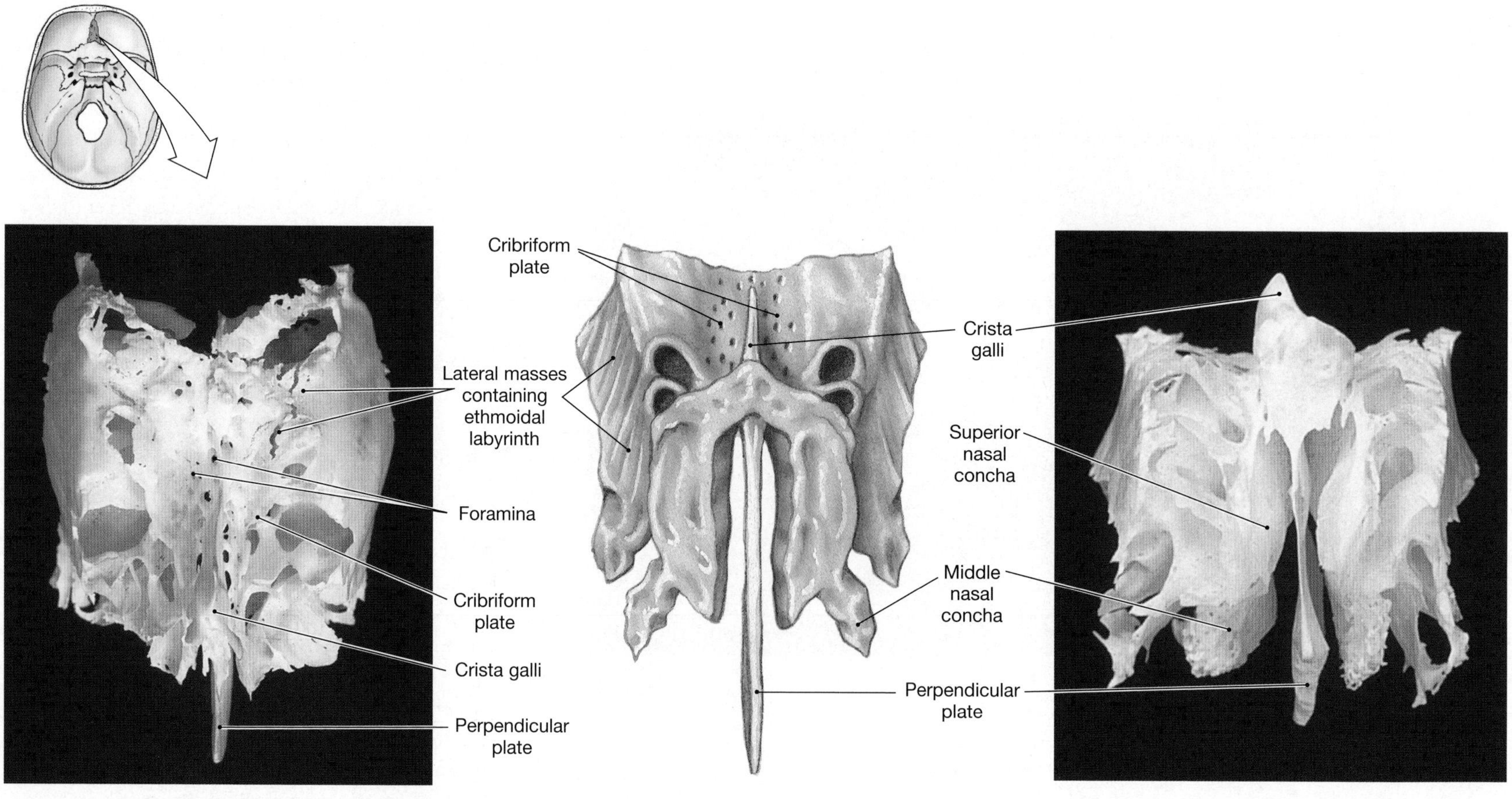

(a) Superior view **(b) Anterior view** **(c) Posterior view**

FIGURE 6.10 **The Ethmoid**

Views of the ethmoid showing major anatomical landmarks on the **(a)** superior surface, **(b)** anterior surface, and **(c)** posterior surface.

FIGURE 6.11 **The Cranial Fossae**

Cranial fossae are curved depressions in the floor of the cranium. (**a**) A sagittal section through the skull, showing the relative positions of the cranial fossae. (**b**) Horizontal sections, superior view. The superior portion of the brain has been removed, but portions of the brain stem and associated nerves and blood vessels remain. *See MRI Scans 1a–e in the Companion Atlas.*

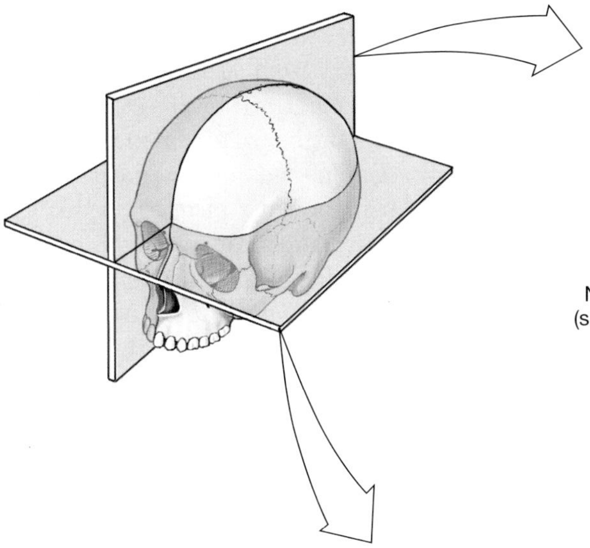

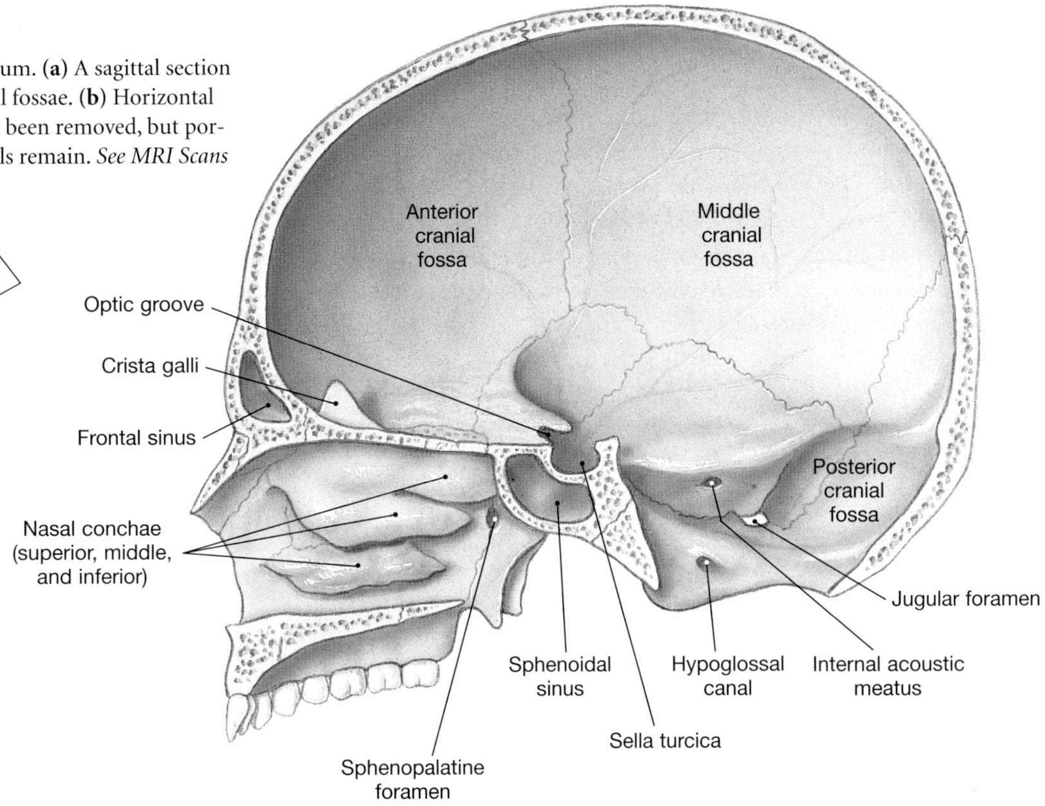

(a) Skull, sagittal section

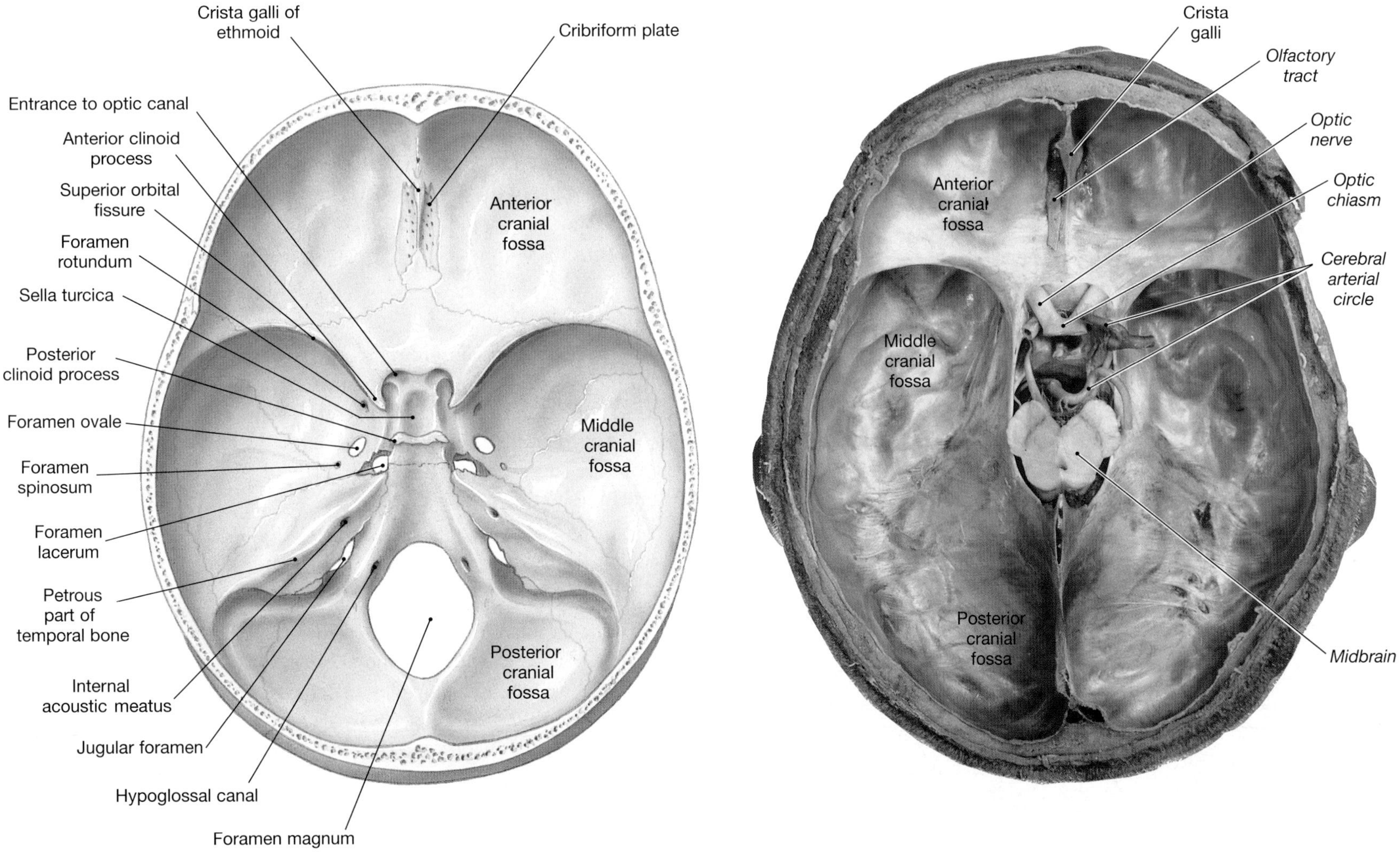

(b) Horizontal sections, superior view

(Figure 6.11b●), the cranial floor at each level forms a curving depression known as a **cranial fossa.** The **anterior cranial fossa** is formed by the frontal bone, the ethmoid, and the lesser wings of the sphenoid. The anterior cranial fossa cradles the frontal lobes of the cerebral hemispheres. The **middle cranial fossa** extends from the "step" at the lesser wings to the petrous portion of the temporal bone. The sphenoid, temporal, and parietal bones form this fossa, which cradles the temporal lobes of the cerebral hemispheres, the *diencephalon*, and the anterior portion of the brain stem (*mesencephalon*). The more inferior **posterior cranial fossa** extends from the petrous parts of the temporal bones to the posterior surface of the skull. The posterior fossa is formed primarily by the occipital bone, with contributions from the temporal and parietal bones. The posterior cranial fossa supports the occipital lobes of the cerebral hemispheres, the cerebellum, and the posterior brain stem (*pons* and *medulla oblongata*).

✓ CONCEPT CHECK

- The internal jugular veins are important blood vessels of the head. Through what opening do these blood vessels pass?
- What bone contains the depression called the sella turcica? What is located in the depression?
- Which of the five senses would be affected if the cribriform plate of the ethmoid failed to form?
- Identify the bones of the cranium.

Bones of the Face

The facial bones are the paired *maxillae, palatine bones, nasal bones, inferior nasal conchae, zygomatic bones*, and *lacrimal bones*, and the single *vomer* and *mandible*.

THE MAXILLAE [FIGURES 6.3d/6.5/6.12a,b,c]

The left and right **maxillae** (singular *maxilla*), or *maxillary bones,* are the largest facial bones, and together they form the upper jaw. The maxillae articulate with all other facial bones except the mandible (Figure 6.3d●, p. 138). The **orbital rim** (Figure 6.12a●) provides protection for the eye and other structures in the orbit. The **frontal process** of each maxilla articulates with the frontal bone of the cranium and with a nasal bone. The oral margins of the maxillae form the **alveolar processes** that contain the upper teeth. An elongated **inferior orbital fissure** within each orbit lies between the maxillae and the sphenoid (Figure 6.3d●). The **infraorbital foramen** that penetrates the orbital rim marks the path of a major sensory nerve from the face. In the orbit, it runs along the *infraorbital groove* (shown in Figure 6.15●) before passing through the inferior orbital fissure and the foramen rotundum to reach the brain stem.

The large **maxillary sinuses** are evident in medial view and in horizontal section (Figure 6.12b,c●). These are the largest sinuses in the skull; they lighten the portion of the maxillae superior to the teeth and produce mucous secretions that flush the inferior surfaces of the nasal cavities. The sectional view also shows the extent of the **palatine processes** that form most of the bony roof, or **hard palate**, of the mouth. The **incisive fossa** on the inferior midline of the palatal process marks the openings of the *incisive canals* (Figure 6.5●, p. 141), which contain small arteries and nerves.

THE PALATINE BONES [FIGURES 6.3e/6.12c/6.13/6.15]

The **palatine bones** are small, L-shaped bones (Figure 6.13●). The *horizontal plates* articulate with the maxillae to form the posterior portions of the hard palate (Figure 6.12c●). On its inferior surface, a *greater palatine foramen* lies between the palatine bone and the maxilla on each side (Figure 6.3e●, p. 139).One or more *lesser palatine foramina* are usually present as well. The *nasal crest,* a ridge that forms where the left and right palatine bones interconnect, marks the articulation with the vomer. The vertical portion of the "L" is formed by the *perpendicular plate* of the palatine bone. This portion of the palatine bone articulates with the maxillae, sphenoid, and ethmoid, and with the inferior nasal concha. The medial surface of the perpendicular plate has two ridges: (1) the *conchal crest*, marking the articulation with the inferior nasal concha, and (2) the *ethmoidal crest*, marking the articulation with the middle nasal concha of the ethmoid. The *orbital process*, based on the perpendicular plate, forms a small portion of the posterior floor of the orbit (Figure 6.15●, p. 153).

THE NASAL BONES [FIGURES 6.3c,d/6.15]

The paired **nasal bones** articulate with the frontal bone at the midline of the face at the *frontonasal suture* (Figure 6.3c,d●, p. 137-138). The nasal bones extend to the superior border of the **external nares** (NA-rēz), or nasal openings. Cartilage attached to the nasal bones forms the flexible portion of the nose. The lateral edge of each nasal bone articulates with the frontal process of a maxillae (Figures 6.3c and 6.15●).

THE INFERIOR NASAL CONCHAE [FIGURES 6.3d/6.16]

The **inferior nasal conchae** are paired scroll-like bones that resemble the superior and middle conchae of the ethmoid. One inferior concha is located on each side of the nasal septum, attached to the lateral wall of the nasal cavity (Figures 6.3d and 6.16●). They perform the same functions as the conchae of the ethmoid.

THE ZYGOMATIC BONES [FIGURES 6.3c,d/6.15]

As noted above, the temporal process of the **zygomatic bone** articulates with the zygomatic process of the temporal bone to form the zygomatic arch (Figure 6.3c,d●). A **zygomaticofacial foramen** on the anterior surface of each zygomatic bone carries a sensory nerve innervating the cheek. The zygomatic bone also forms the lateral rim of the orbit (Figure 6.15●) and contributes to the inferior orbital wall.

THE LACRIMAL BONES [FIGURES 6.3c,d/6.15]

The paired **lacrimal bones** (*lacrima*, tear) are the smallest bones in the skull. The lacrimal bone is situated in the medial portion of each orbit, where it articulates with the frontal bone, maxilla, and ethmoid (Figures 6.3c,d and 6.15●). A shallow depression, the **lacrimal groove**, or *lacrimal sulcus*, leads to a narrow passageway, the **nasolacrimal canal**, formed by the lacrimal bone and the maxilla. This canal encloses the tear duct as it passes toward the nasal cavity.

THE VOMER [FIGURES 6.3d,e/6.5]

The **vomer** forms the inferior portion of the nasal septum (Figure 6.5●, p. 141). It is based on the floor of the nasal cavity and articulates with

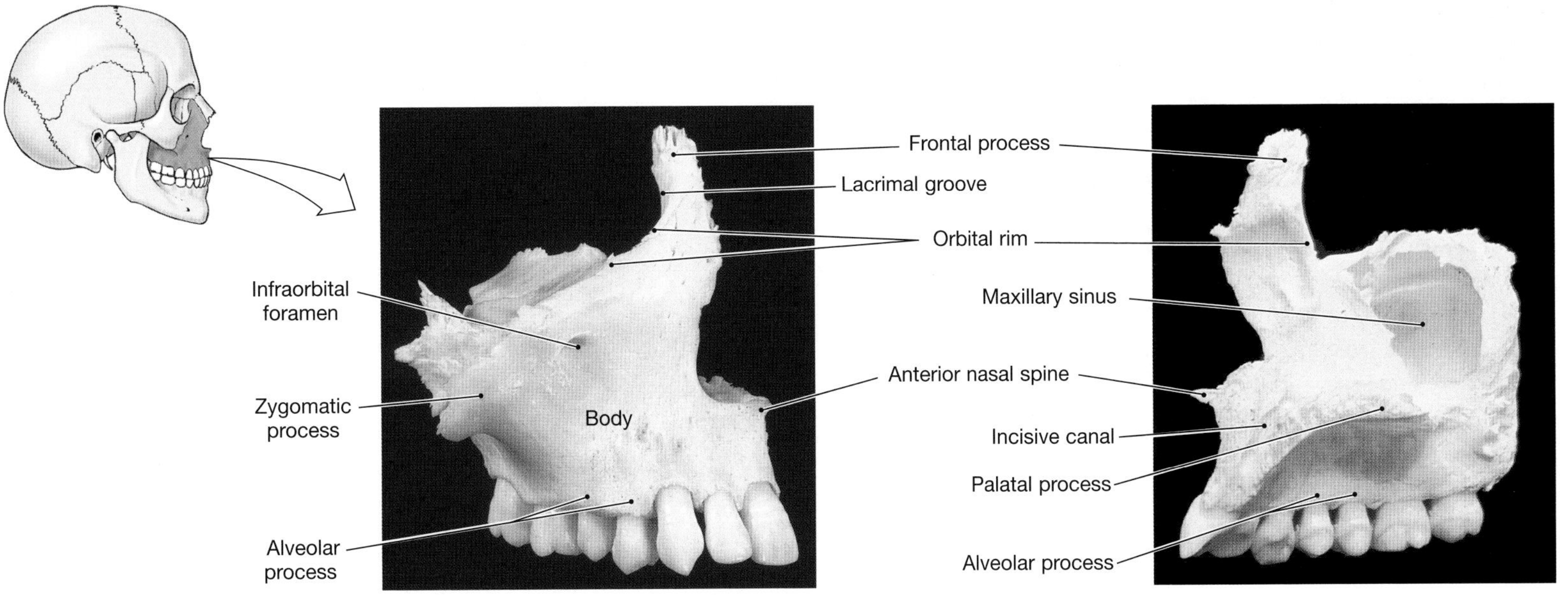

(a) Right maxilla, lateral view

(b) Medial view

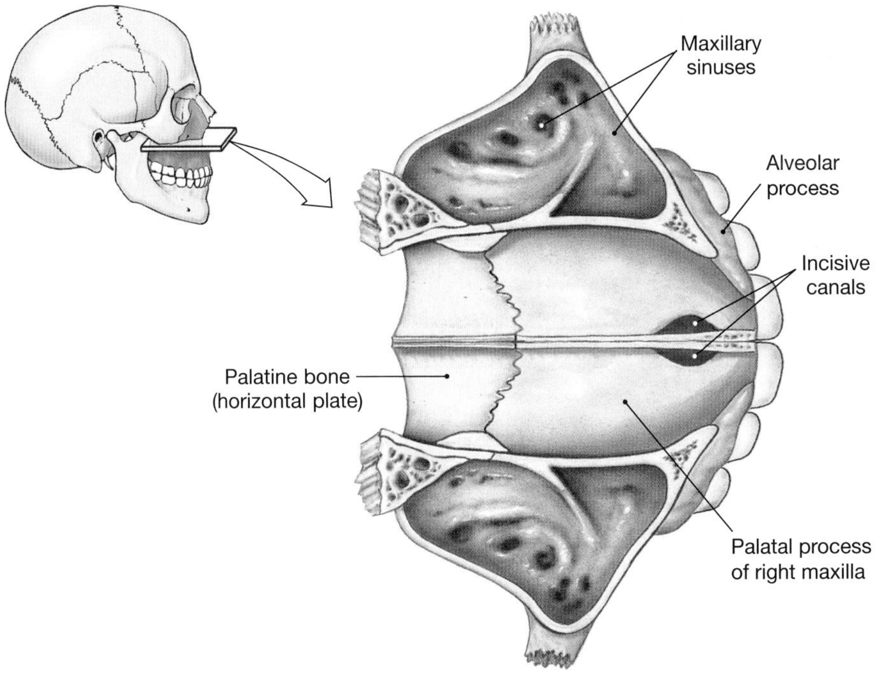

(c) Maxillae and palatine bones, sectional view

FIGURE 6.12 **THE MAXILLAE**

Views of the right maxilla showing major anatomical landmarks on (**a**) the anterior and lateral surfaces, and (**b**) the medial surface. (**c**) Superior view of a horizontal section through both maxillae, showing the orientation of the maxillary sinuses and the structure of the hard palate.

FIGURE 6.13 **THE PALATINE BONES**

Views of the palatine bones, showing major anatomical landmarks on (**a**) the anterior surfaces of the palatine bones, (**b**) the medial surface of the right palatine bone, and (**c**) the lateral surface of the right palatine bone.

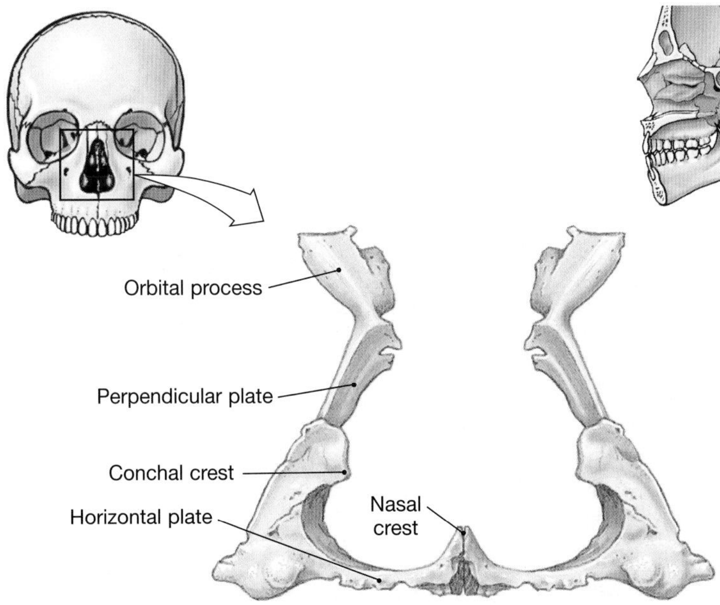

(a) Palatine bones, anterior view

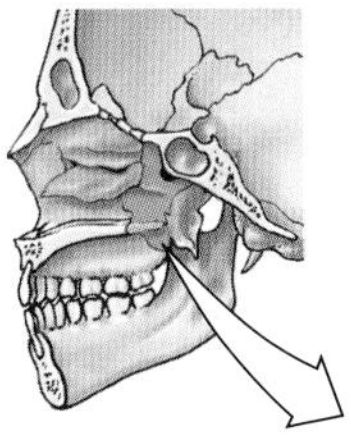

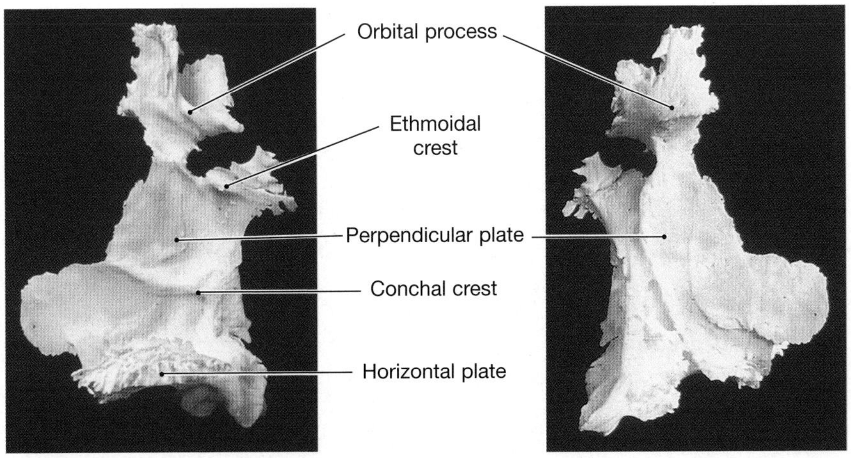

(b) Palatine bone, medial view

(c) Palatine bone, lateral view

both the maxillae and palatine bones along the midline. The vertical portion of the vomer is thin. Its curving superior surface artic-ulates with the sphenoid and the perpendicular plate of the ethmoid, forming a bony **nasal septum** (*septum*, wall) that separates the right and left nasal cavities (Figure 6.3d,e●, pp. 138–139). Anteriorly, the vomer supports a cartilaginous extension of the nasal septum that continues into the fleshy portion of the nose and separates the external nares.

THE MANDIBLE [FIGURES 6.3c,d/6.14]

The **mandible** forms the entire lower jaw (Figures 6.3c,d and 6.14●). This bone can be subdivided into the horizontal **body** and the ascending **rami** (singular, *ramus*, "branch"). The teeth are supported by the mandibular body. Each ramus meets the body at the mandibular **angle**. The **condylar processes** extend to the smooth articular surface of the *head* of the mandible. The head articulates with the mandibular fossae of the temporal bone at the *temporomandibular joint (TMJ)*. This joint is quite mobile, as evidenced by the jaw movements during chewing or talking. The disadvantage of such mobility is that the jaw can easily be dislocated by forceful movement of the mandible anteriorly or laterally.

At the **coronoid** (kor-Ō-noyd) **processes**, the *temporalis muscle* inserts onto the mandible. This is one of the most forceful muscles involved in closing the mouth. Anteriorly, the **mental foramina** (*mentalis*, chin) penetrate the body on each side of the chin. Nerves pass through these foramina, carrying sensory information from the lips and chin back to the brain. The **mandibular notch** is the depression that lies between the condylar and coronoid processes.

The **alveolar part** of the mandible is a thickened area that contains the alveoli and the roots of the teeth (Figure 6.14b●). A **mylohyoid line** lies on the medial aspect of each ramus. It marks the insertion of the *mylohyoid muscle* that supports the floor of the mouth and tongue. The *submandibular salivary gland* nestles in the **submandibular fossa**, a depression inferior to the mylohyoid line. Near the posterior, superior end of the mylohyoid line, a prominent **mandibular foramen** leads into the **mandibular canal**. This is a passageway for blood vessels and nerves that service the lower teeth. The nerve that uses this passage carries sensory information from the teeth and gums; dentists typically anesthetize this nerve before working on the lower teeth.

The Orbital and Nasal Complexes

Several of the facial bones articulate with cranial bones to form the *orbital complex* surrounding each eye and the *nasal complex* that surrounds the nasal cavities.

THE ORBITAL COMPLEX [FIGURE 6.15]

The **orbits** are the bony recesses that enclose and protect the eyes. In addition to the eye, each orbit also contains a lacrimal gland, adipose tissue, muscles that move the eye, blood vessels, and nerves. Seven bones fit together to create the **orbital complex** that forms each orbit (Figure 6.15●). The frontal bone

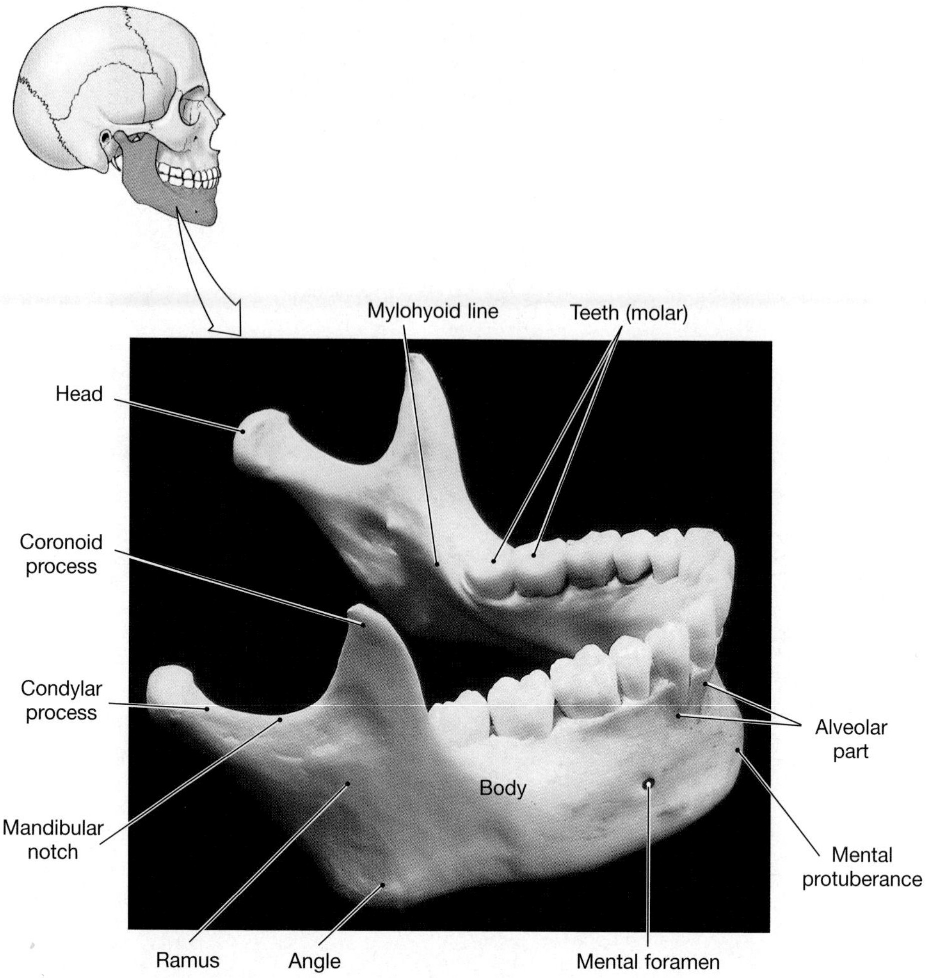

(a) Lateral view

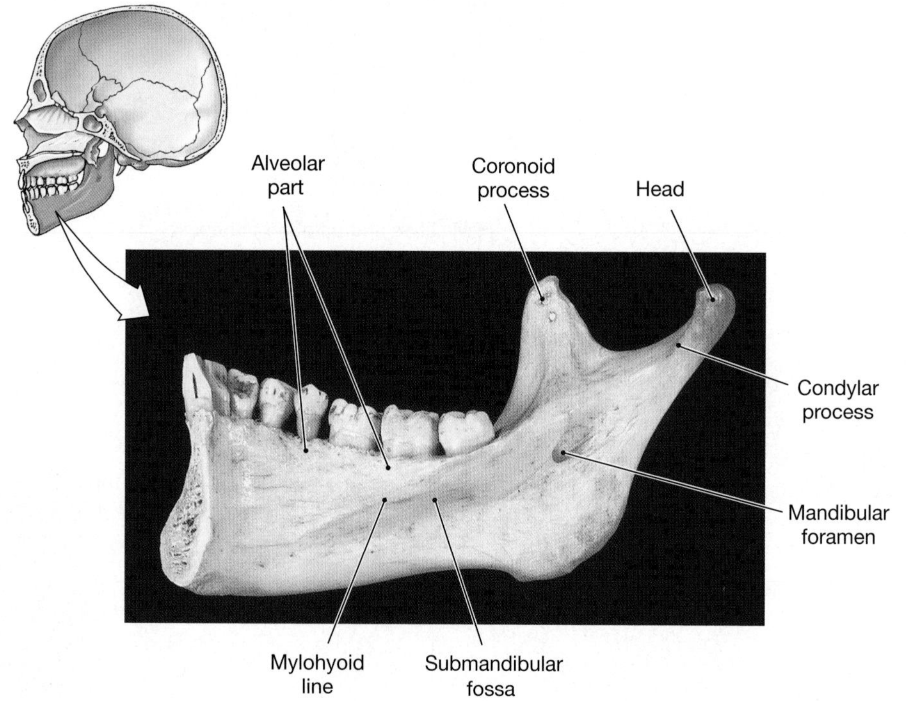

(b) Medial view

FIGURE 6.14 **THE MANDIBLE**

Views of the mandible showing major anatomical landmarks on (**a**) the superior and lateral surfaces and (**b**) the medial surface of the right half of the mandible.

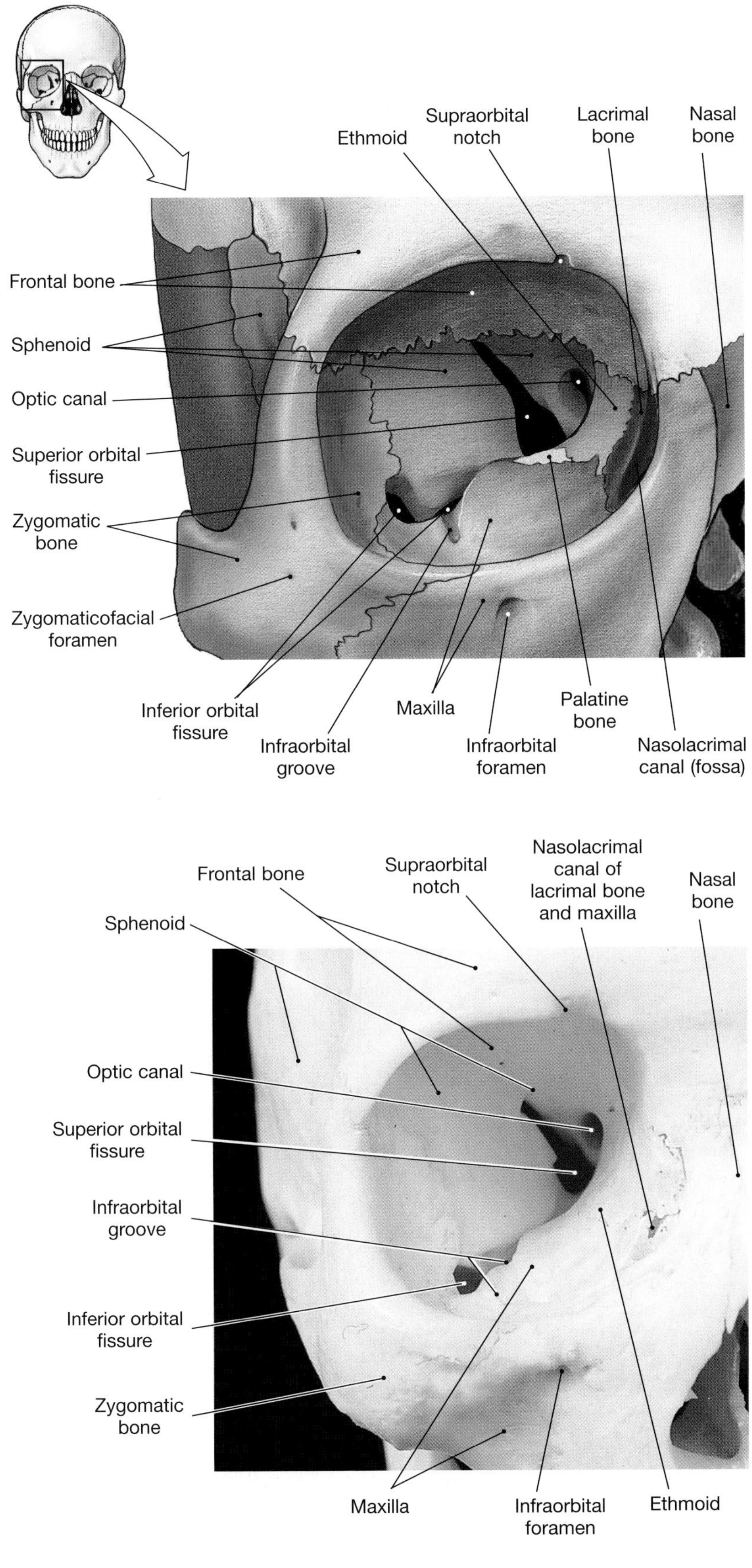

FIGURE 6.15 **The Orbital Complex**

The structure of the orbital complex on the right side. Seven bones form the bony orbit that encloses and protects the right eye.

forms the roof, and the maxilla forms most of the orbital floor. Proceeding from medial to lateral, the orbital rim and the first portion of the wall are contributed by the maxilla, the lacrimal bone, and the lateral mass of the ethmoid, which articulates with the sphenoid and a small process of the palatine bone. The sphenoid forms most of the posterior orbital wall. Several prominent foramina and fissures penetrate the sphenoid or lie between the sphenoid and maxilla. Laterally, the sphenoid and maxilla articulate with the zygomatic bone, which forms the lateral wall and rim of the orbit.

THE NASAL COMPLEX [FIGURES 6.5/6.16]

The **nasal complex** (Figure 6.16●) includes the bones and cartilage that enclose the nasal cavities and the *paranasal sinuses*, airspaces connected to the nasal cavities. The frontal bone, sphenoid, and ethmoid form the superior wall of the nasal cavities. The perpendicular plate of the ethmoid and the vomer form the bony portion of the nasal septum (see Figures 6.5●, p. 141, and 6.16a●). The lateral walls are primarily formed by the maxillae, the lacrimal bones, the ethmoid, and the inferior nasal conchae (Figure 6.16b–d●). The bridge of the nose is supported by the maxillae and nasal bones. The soft tissues of the nose enclose anterior extensions of the nasal cavities. These are supported by cartilaginous extensions of the bridge of the nose and the nasal septum.

Paranasal Sinuses [FIGURE 6.16] The frontal bone, sphenoid, ethmoid, and maxilla contain the **paranasal sinuses**, air-filled chambers that act as extensions of and open into the nasal cavities. Figure 6.16● shows the location of the **frontal** and **sphenoidal sinuses**. The **ethmoidal** and **maxillary sinuses** are shown in Figure 6.16c,d●. These sinuses lighten skull bones, produce mucus, and resonate during sound production. The mucous secretions are released into the nasal cavities, and the ciliated epithelium passes the mucus back toward the throat, where it is eventually swallowed. Incoming air is humidified and warmed as it flows across this carpet of mucus. Foreign particulate matter, such as dust or microorganisms, becomes

CLINICAL BRIEF

Sinus Problems and Septal Defects

The mucous membrane of the paranasal sinuses responds to environmental stress by accelerating the production of mucus. The mucus flushes irritants off the walls of the nasal cavities. A variety of stimuli produce this result, including sudden changes in temperature or humidity, irritating vapors, and bacterial or viral infections. The flushing action often succeeds in removing a mild irritant. But a viral or bacterial infection produces an inflammation of the mucous membrane of the nasal cavity. As swelling occurs, the communicating passageways narrow. Mucus drainage slows, congestion increases, and the victim experiences headaches and a feeling of pressure within the facial bones. This condition of sinus inflammation and congestion is called *sinusitis*. The maxillary sinuses are often involved, because gravity does little to assist mucus drainage from these sinuses.

Temporary sinus problems may accompany allergies or exposure of the mucous epithelium to chemical irritants or invading microorganisms. Chronic sinusitis may occur as the result of a *deviated* (nasal) *septum*. In this condition the nasal septum has a bend in it, most often at the junction between the bony and cartilaginous regions. Septal deviation often blocks drainage of one or more sinuses, producing chronic cycles of infection and inflammation. A deviated septum can result from developmental abnormalities or injuries to the nose, and the condition can usually be corrected or improved by surgery.

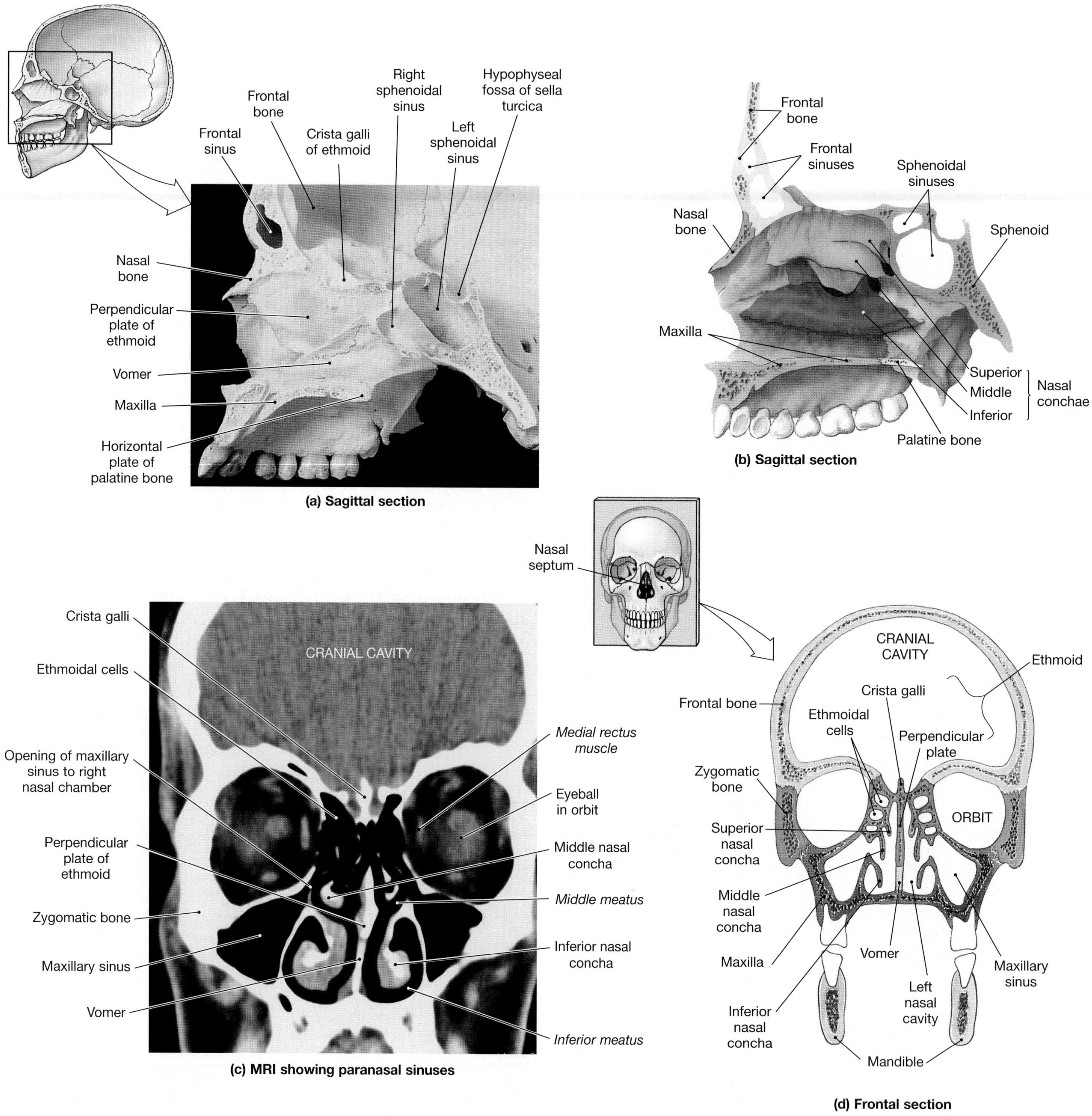

FIGURE 6.16 **THE NASAL COMPLEX**

Sections through the skull showing relationships among the bones of the nasal complex. **(a)** Sagittal section with the nasal septum in place. **(b)** Diagrammatic sagittal section with the nasal septum removed to show major features of the wall of the right nasal cavity. **(c)** An MRI scan showing the location of several paranasal sinuses. This individual has an abnormal (deviated) nasal septum that interferes with normal airflow. (d) A diagrammatic frontal section showing the positions of the paranasal sinuses. *See MRI Scan 2a in the Companion Atlas.*

trapped in this sticky mucus and then swallowed. This mechanism helps protect the delicate exchange surfaces of the fragile lung tissue portions of the respiratory tract.

The Hyoid Bone

The hyoid bone is associated with the skull, but does not articulate with any of the cranial or facial bones.

THE HYOID BONE [FIGURE 6.17]

The **hyoid bone** lies inferior to the skull, suspended by the **stylohyoid ligaments**, but not in direct contact with any other bone of the skeleton (Figure 6.17●). The body of the hyoid serves as a base for several muscles concerned with movements of the tongue and larynx. Because muscles and ligaments form the only connections between the hyoid and other skeletal structures, the entire complex is quite mobile. The larger processes on the hyoid are the **greater horns**, which help support the larynx and serve as the base for muscles that move the tongue. The **lesser horns** are connected to the stylohyoid ligaments, and from these ligaments the hyoid and larynx hang beneath the skull like a swing from the limb of a tree.

Many superficial bumps and ridges in the axial skeleton are associated with the skeletal muscles described in Chapter 10; learning the names now will help you organize the material in that chapter. Tables 6.1 and 6.2 summarize information concerning the foramina and fissures introduced thus far. Use Table 6.1 (p. 158) as a reference for foramina and fissures of the skull and Table 6.2 (p. 159) as a reference for surface features and foramina of the skull. These references will be especially important in later chapters dealing with the nervous and cardiovascular systems.

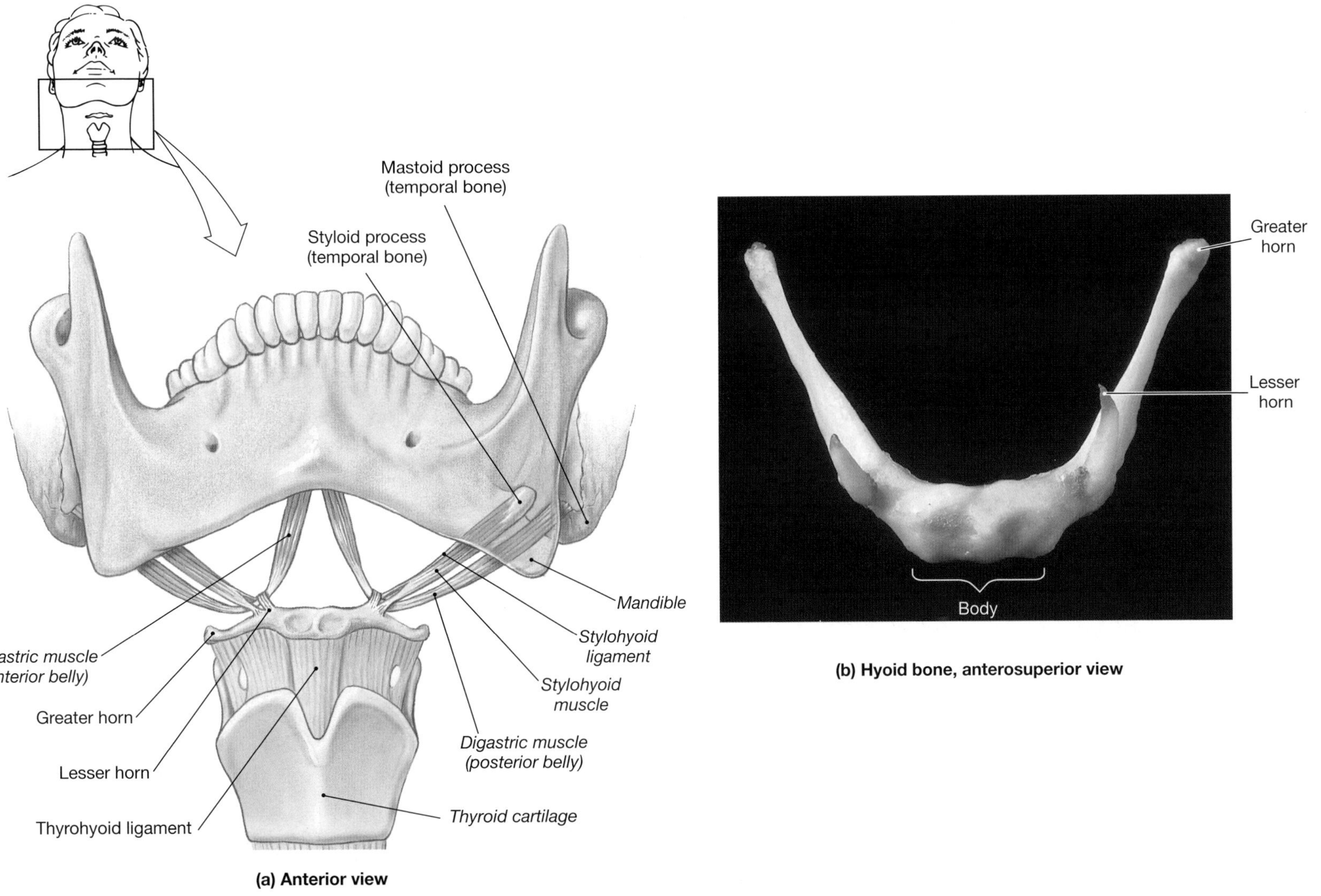

FIGURE 6.17 **THE HYOID BONE**

Anterior views showing (**a**) the relationship of the hyoid bone to the skull, the larynx, and selected skeletal muscles; (**b**) the isolated hyoid bone.

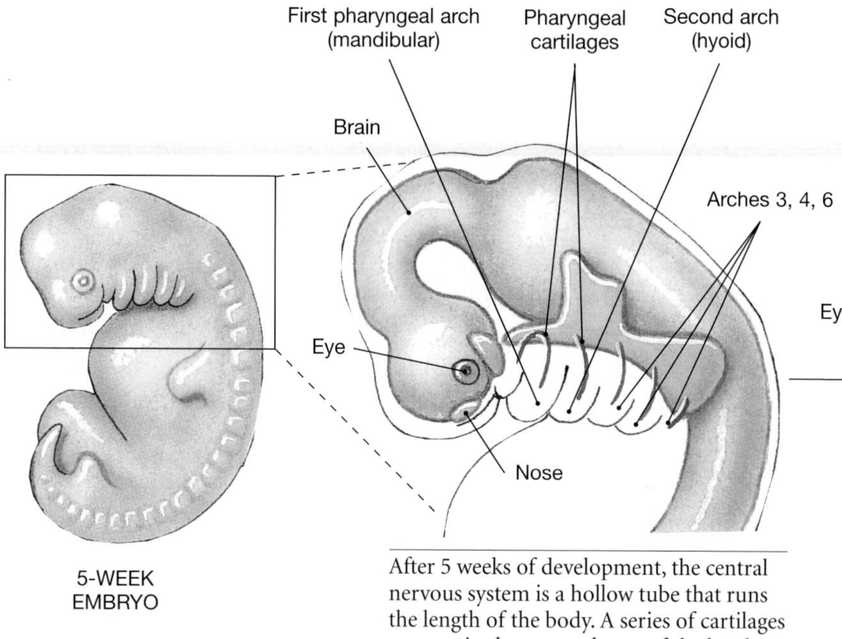

After 5 weeks of development, the central nervous system is a hollow tube that runs the length of the body. A series of cartilages appears in the mesenchyme of the head beneath and alongside of the expanding brain and around the developing nose, eyes, and ears. These cartilages are shown in light blue. Five additional pairs of cartilages develop in the walls of the pharynx. These cartilages, shown in dark blue, are located within the **pharyngeal**, or **branchial, arches**. (*Branchial* refers to gills—in fish the caudal arches develop into skeletal supports for the gills.) The first arch, or **mandibular arch**, is the largest.

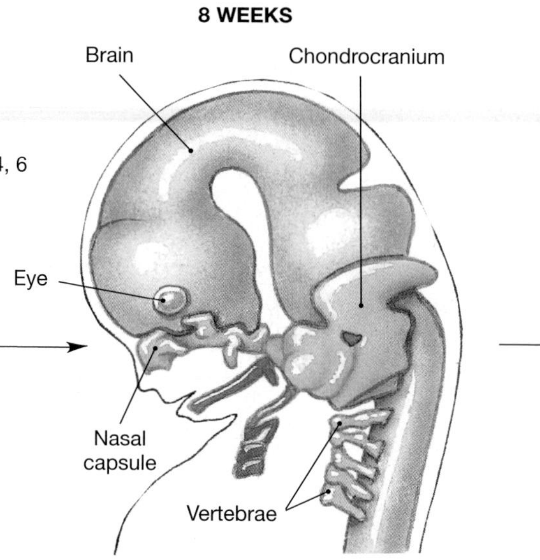

The cartilages associated with the brain enlarge and fuse, forming a cartilaginous **chondrocranium** (kon-drō-KRĀ-nē-yum; *chondros*, cartilage + *cranium*, skull) that cradles the brain and sense organs. At 8 weeks its walls and floor are incomplete, and there is no roof.

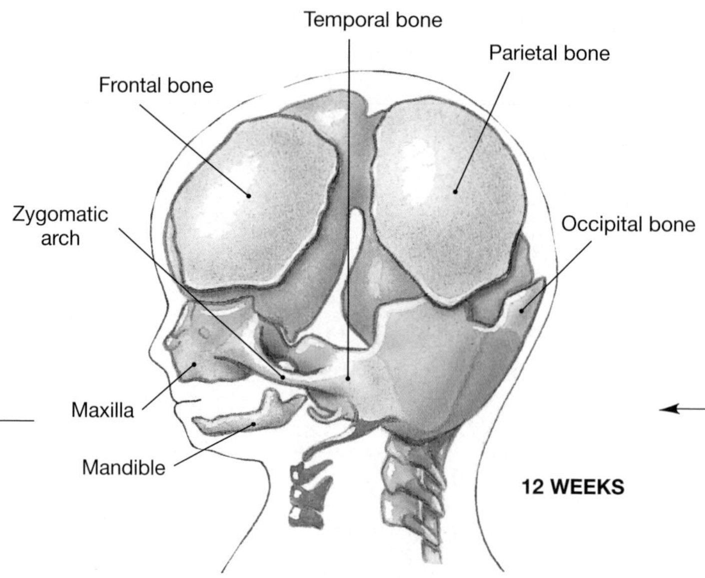

After 12 weeks ossification is well under way in the cranium and face. Compare with Figure 5.6, p. 119.

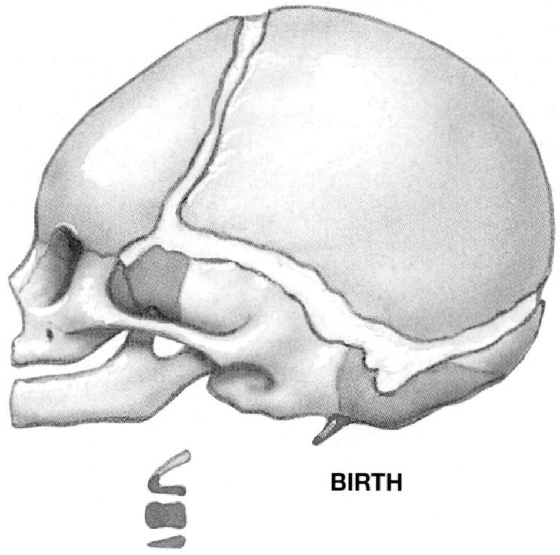

The skull at birth; compare with the situation at 12 weeks. Extensive fusions have occurred, but the cranial roof remains incomplete. (For further details, see Figure 6.18, p. 162.)

During the ninth week, numerous centers of endochondral ossification appear within the chondrocranium. These centers are shown in pink. Gradually, the frontal and parietal bones of the cranial roof appear as intramembranous ossification begins in the overlying dermis. As these centers (beige) enlarge and expand, extensive fusions occur.

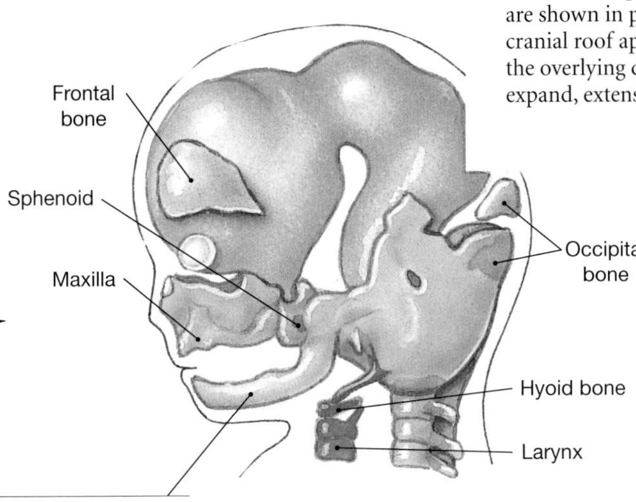

The mandible forms as dermal bone develops around the inferior portion of the mandibular arch.

The dorsal portion of the mandibular arch fuses with the chondrocranium. The fused cartilages do not ossify; instead, osteoblasts begin sheathing them in dermal bone. On each side this sheath fuses with a bone developing at the entrance to the nasal cavity, producing the two maxillae. Ossification centers in the roof of the mouth spread to form the palatine processes and later fuse with the maxillae.

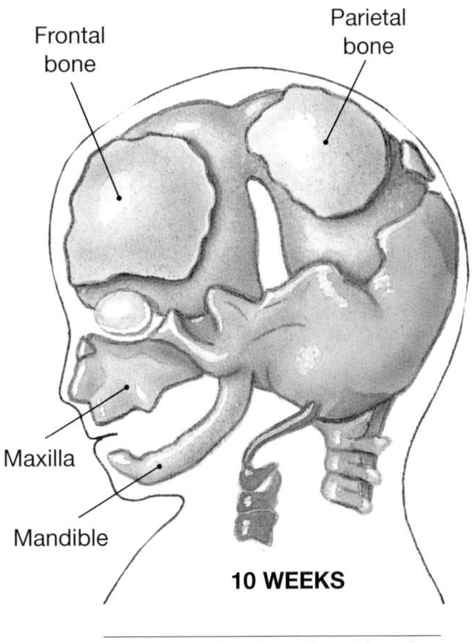

The second arch, or **hyoid arch**, forms near the temporal bones. Fusion of the superior tips of the hyoid with the temporals forms the styloid processes. The ventral portion of the hyoid arch ossifies as the hyoid bone. The third arch fuses with the hyoid, and the fourth and sixth arches form laryngeal cartilages.

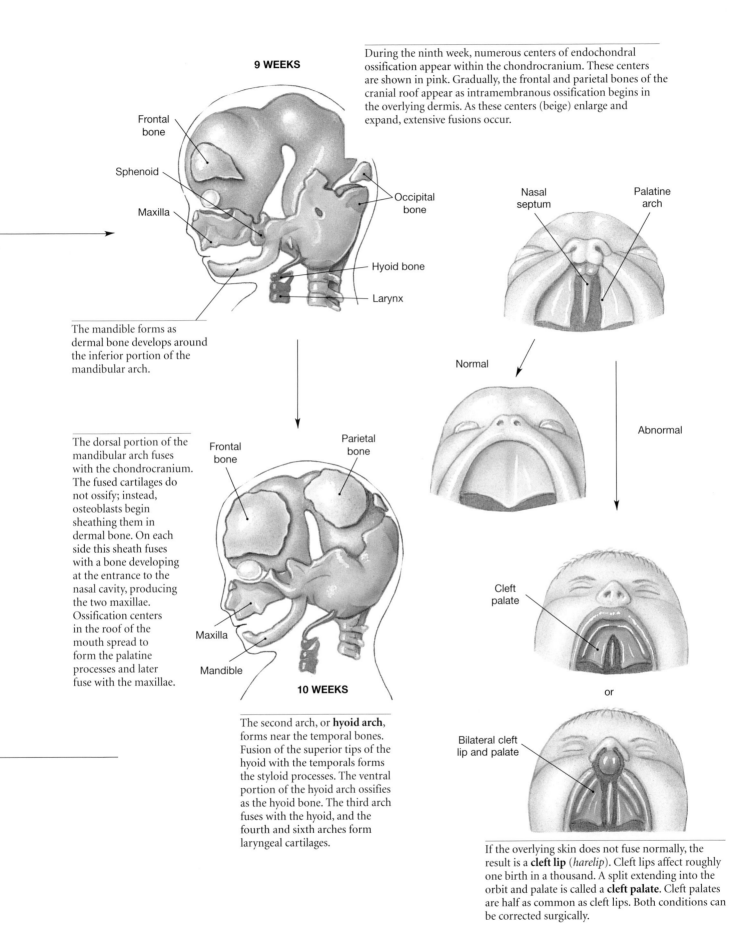

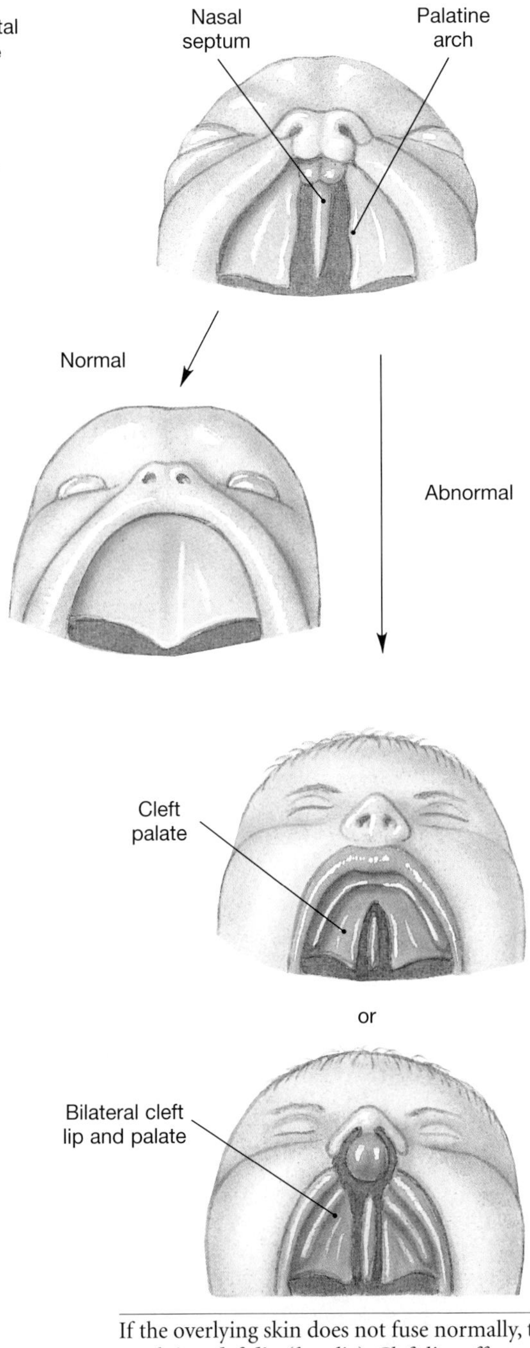

If the overlying skin does not fuse normally, the result is a **cleft lip** (*harelip*). Cleft lips affect roughly one birth in a thousand. A split extending into the orbit and palate is called a **cleft palate.** Cleft palates are half as common as cleft lips. Both conditions can be corrected surgically.

TABLE 6.1

A Key to the Foramina and Fissures of the Skull

		Major Structures Using Passageway	
Bone	**Foramen/Fissure**	**Neural Tissue**	**Vessels and Other Structures**
OCCIPITAL BONE	Foramen magnum	Medulla oblongata (last portion of brain) and accessory nerve (XI) controlling several muscles of the back, pharynx, and larynx	Vertebral arteries to brain and supporting membranes around CNS
	Hypoglossal canal	Hypoglossal nerve (XII) provides motor control to muscles of the tongue	
With temporal bone	Jugular foramen	Glossopharyngeal nerve (IX), vagus nerve (X), accessory nerve (XI). Nerve IX provides taste sensation; X is important for visceral functions; XI innervates important muscles of the back and neck	Internal jugular vein; important vein returning blood from brain to heart
FRONTAL BONE	Supraorbital foramen (or notch)	Supraorbital nerve, sensory branch of the ophthalmic nerve, innervating the eyebrow, eyelid, and frontal sinus	Supraorbital artery delivers blood to same region
TEMPORAL BONE	Mastoid foramen		Vessels to membranes around CNS
	Stylomastoid foramen	Facial nerve (VII) provides motor control of facial muscles	
	Carotid canal		Internal carotid artery; major arterial supply to the brain
	External acoustic meatus		Air conducts sound to eardrum
	Internal acoustic meatus	Vestibulocochlear nerve (VIII) from sense organs for hearing and balance. Facial nerve (VII) enters here, exits at stylomastoid foramen	Internal acoustic artery to inner ear
SPHENOID BONE	Optic canal	Optic nerve (II) brings information from the eye to the brain	Ophthalmic artery brings blood into orbit
	Superior orbital fissure	Oculomotor nerve (III), trochlear nerve (IV), ophthalmic branch of trigeminal nerve (V), abducens nerve (VI). Ophthalmic nerve provides sensory information about eye and orbit; other nerves control muscles that move the eye	Ophthalmic vein returns blood from orbit
	Foramen rotundum	Maxillary branch of trigeminal nerve (V) provides sensation from the face	
	Foramen ovale	Mandibular branch of trigeminal nerve (V) controls the muscles that move the lower jaw and provides sensory information from that area	
With temporal and occipital bones	Foramen spinosum		Vessels to membranes around CNS
	Foramen lacerum		Internal carotid artery leaves carotid canal, enters cranium through foramen lacerum
With maxillae	Inferior orbital fissure	Maxillary branch of trigeminal nerve (V). See *Foramen rotundum*	
ETHMOID	Cribriform foramina	Olfactory nerve (I) provides sense of smell	
MAXILLA	Infraorbital foramen	Infraorbital nerve, maxillary branch of trigeminal nerve (V) from the inferior orbital fissure to face	Infraorbital artery with the same distribution
	Incisive canals	Nasopalatine nerve	Small arteries to the palatal surface
ZYGOMATIC BONE	Zygomaticofacial foramen	Zygomaticofacial nerve, sensory branch of mandibular nerve to cheek	
LACRIMAL BONE	Lacrimal sulcus, nasolacrimal canal (with maxilla)		Tear duct drains into the nasal cavity
MANDIBLE	Mental foramen	Mental nerve, sensory nerve branch of the mandibular nerve, provides sensation for the chin and lips.	Mental vessels to chin and lips
	Mandibular foramen	Inferior alveolar nerve, sensory branch of the mandibular nerve, provides sensation for the gums, teeth	Inferior alveolar vessels supply same region

TABLE 6.2

Surface Features of the Skull

Region	Bone	Articulates with:	Surface Features			
			Structures	**Functions**	**Foramina**	**Functions**
CRANIUM (8)						
	Occipital bone (1) *(Figure 6.6)*	Parietal bone, temporal bone, sphenoid	**External:**			
			Occipital condyles	Articulate with first cervical vertebra	Jugular foramen (with temporal)	Carries blood from smaller veins in the cranial cavity
			Occipital crest, external occipital protuberance, and inferior and superior nuchal lines	Attachment of muscles and ligaments that move the head and stabilize the atlanto-occipital joint	Hypoglossal canal	Passageway for hypoglossal nerve that controls tongue muscles
			Internal:			
			Internal occipital crest	Attachment of membranes that stabilize position of the brain		
	Parietal bones (2) *(Figure 6.6)*	Occipital, frontal, temporal bones, sphenoid	**External:**			
			Superior and inferior temporal lines	Attachment of major jaw-closing muscle		
			Parietal eminence	Attachment of scalp to skull		
	Frontal bone (1) *(Figure 6.7)*	Parietal, nasal, zygomatic bones sphenoid, ethmoid, maxillae	Metopic suture	Marks fusion of frontal bones in development	Supraorbital foramina	Passageways for sensory branch of ophthalmic nerve and supraorbital artery to the eyebrow and eyelid
			Frontal squama	Attachment of muscles of scalp		
			Supraorbital margin	Protects eye		
			Lacrimal fossae	Recesses enclose lacrimal glands		
			Frontal sinuses	Lighten bone and produce mucous secretions		
			Frontal crest	Attachment of stabilizing membranes (meninges) within the cranium		
	Temporal bones (2) *(Figure 6.8)*	Occipital, parietal, frontal, zygomatic bones, sphenoid and mandible; encloses auditory ossicles and suspends hyoid bone by stylohyoid ligaments	**External:**		**External:**	
			Squamous part:		Carotid canal	Entryway for carotid artery bringing blood to the brain
			Squama	Attachment of jaw muscles		
			Mandibular fossa and articular tubercle	Form articulation with mandible	Stylomastoid foramen	Exit for nerve that controls facial muscles
			Zygomatic process	Articulates with zygomatic bone	Jugular foramen (with occipital bone)	Carries blood from smaller veins in the cranial cavity
			Petrous part:		External acoustic meatus	Entrance and passage to tympanum
			Mastoid process	Attachment of muscles that extend or rotate head		
			Styloid process	Attachment of stylohyoid ligament and muscles attached to hyoid bone	Mastoid foramen	Passage for blood vessels to membranes of brain
			Internal:		**External:**	
			Mastoid air cells	Lighten mastoid process	Foramen lacerum between temporal and occipital bones	Cartilage and small arteries to the inner surface of the cranium
					Internal:	
			Petrous part	Protects middle and inner ear	Auditory tube	Connects airspace of middle ear with pharynx
					Internal acoustic meatus	Passage for blood vessels and nerves to the inner ear and stylomastoid foramen

TABLE 6.2

SURFACE FEATURES OF THE SKULL *(continued)*

Region	Bone	Articulates with:	Surface Features			
			Structures	**Functions**	**Foramina**	**Functions**
	Sphenoid (1) *(Figure 6.9)*	Occipital, frontal, temporal, zygomatic, palatine bones, maxillae, ethmoid and vomer	**Internal:** Sella turcica	Protects pituitary gland	Optic foramen	Passage of optic nerve
			Anterior and posterior clinoid processes, optic groove	Protect pituitary gland and optic nerve	Superior orbital fissure	Entrance for nerves that control eye movements
			External: Pterygoid processes and spines	Attachment of jaw muscles	Foramen rotundum	Passage for sensory nerves from face
					Foramen ovale	Passage for nerves that control jaw movement
					Foramen spinosum	Passage of vessels to membranes around brain
	Ethmoid (1) *(Figure 6.10)*	Frontal, nasal, palatine, lacrimal bones, sphenoid, maxillae and vomer	Crista galli	Attachment of membranes that stabilize position of brain	Cribriform foramina	Passage of olfactory nerves
			Ethmoidal labyrinth	Lighten bone and site of mucus production		
			Superior and middle conchae	Create turbulent airflow		
			Perpendicular plate	Separates nasal cavities (with vomer and nasal cartilage)		
FACE (14)						
	Maxillae (2) *(Figure 6.12)*	Frontal, zygomatic, palatine, lacrimal bones, sphenoid, ethmoid and inferior nasal concha	Orbital margin	Protects eye	Inferior orbital fissure	Exit for nerves entering skull at foramen rotundum
			Palatal process	Forms most of the bony palate	Infraorbital foramen	Passage of sensory nerves from face
			Maxillary sinus	Lightens bone, secretes mucus		
			Alveolar process	Surrounds articulations with teeth		
	Palatine bones (2) *(Figure 6.13)*	Sphenoid, maxillae, and vomer		Contribute to bony palate and orbit		
	Nasal bones (2) *(Figures 6.3c,d, 6.15)*	Frontal bone, ethmoid, maxillae		Support bridge of nose		
	Vomer (1) *(Figures 6.3d,e, 6.5, 6.16)*	Ethmoid, maxillae, palatine bones		Forms inferior and posterior part of nasal septum		
	Inferior nasal conchae (2) *(Figures 6.3d, 6.16)*	Maxillae and palatine bones		Create turbulent airflow		
	Zygomatic bones (2) *(Figures 6.3c,d, 6.15)*	Frontal and temporal bones, sphenoid, maxillae	Temporal process	With zygomatic process of temporal, completes zygomatic arch for attachment of jaw muscles		
	Lacrimal bones (2) *(Figures 6.3c,d, 6.15)*				Lacrimal foramen	Drains tears from orbit to nasal cavity

TABLE 6.2

SURFACE FEATURES OF THE SKULL *(continued)*

Region	Bone	Articulates with:	Surface Features: Structures	Functions	Foramina	Functions
	Mandible (1) *(Figure 6.14)*	Temporal bones	Condylar process	Articulates with temporal bone	Mental foramen	Passage for sensory nerve from chin and lips
			Coronoid process	Attachment of temporalis muscle from parietal surface	Mandibular foramen	Passage of sensory nerve from teeth and gums
			Alveolar process	Protects articulations with teeth		
			Mylohoid line	Attachment of muscle supporting floor of mouth		
			Submandibular fossa	Protects submandibular salivary gland		
ASSOCIATED BONES (7)	**Hyoid bone (1)** *(Figure 6.17)*	Suspended by ligaments from styloid process of temporal bone; connected by ligaments to larynx	Greater horns	Attachment of tongue muscles and ligaments to larynx		
			Lesser horns	Attachment of stylohyoid ligaments		
	Auditory ossicles (6)	3 are enclosed by the petrous part of each temporal bone		Conduct sound vibrations from tympanic membrane to fluid-filled chambers of inner ear		

The Skulls of Infants, Children, and Adults [FIGURE 6.18]

Many different centers of ossification are involved in the formation of the skull, but as development proceeds, fusion of the centers produces a smaller number of composite bones. For example, the sphenoid begins as 14 separate ossification centers. At birth, fusion has not been completed, and there are two frontal bones, four occipital bones, and a number of sphenoid and temporal elements.

The skull organizes around the developing brain, and as the time of birth approaches, the brain enlarges rapidly. Although the bones of the skull are also growing, they fail to keep pace, and at birth the cranial bones are connected by areas of fibrous connective tissue. These connections are quite flexible, and the skull can be distorted without damage. Such distortion normally occurs during delivery and eases the passage of the infant along the birth canal. The largest fibrous regions between the cranial bones are known as **fontanels** (fon-tah-NELS; sometimes spelled *fontanelles*) (Figure 6.18●):

- The *anterior fontanel* is the largest fontanel. It lies at the intersection of the frontal, sagittal, and coronal sutures.
- The *occipital fontanel* is at the junction between the lambdoid and sagittal sutures.
- The *sphenoidal fontanels* are at the junctions between the squamous sutures and the coronal suture.
- The *mastoid fontanels* are at the junctions between the squamous sutures and the lambdoid suture.

The skulls of infants and adults differ in terms of the shape and structure of cranial elements, and this difference accounts for variations in proportions as well as in size. The most significant growth in the skull occurs before age 5; at that time the brain stops growing and the cranial sutures develop. As a result, when compared with the skull as a whole, the cranium of a young child is relatively larger than that of an adult.

CLINICAL BRIEF

PROBLEMS WITH GROWTH OF THE SKULL

The growth of the cranium is usually coordinated with the expansion of the brain. Unusual distortions of the skull result from the premature closure of one or more fontanels, a condition called **craniostenosis** (krā-nē-ō-sten-Ō-sis; *stenosis*, narrowing). As the brain continues to enlarge, the rest of the skull accommodates it. A long and narrow head will be produced by early closure of the sagittal suture, whereas a very broad skull results if the coronal suture forms prematurely. Closure of all of the cranial sutures restricts the development of the brain, and surgery must be performed to prevent brain damage. However, if the brain enlargement stops because of genetic or developmental abnormalities, skull growth ceases as well. This condition, which results in a very undersized head, is called **microcephaly** (mī-krō-SEF-a-lē).

✓ CONCEPT CHECK

- What are the names and functions of the facial bones?
- Identify the functions of the paranasal sinuses.
- What bones form the orbital complex?
- What are fontanels? Where are they found?

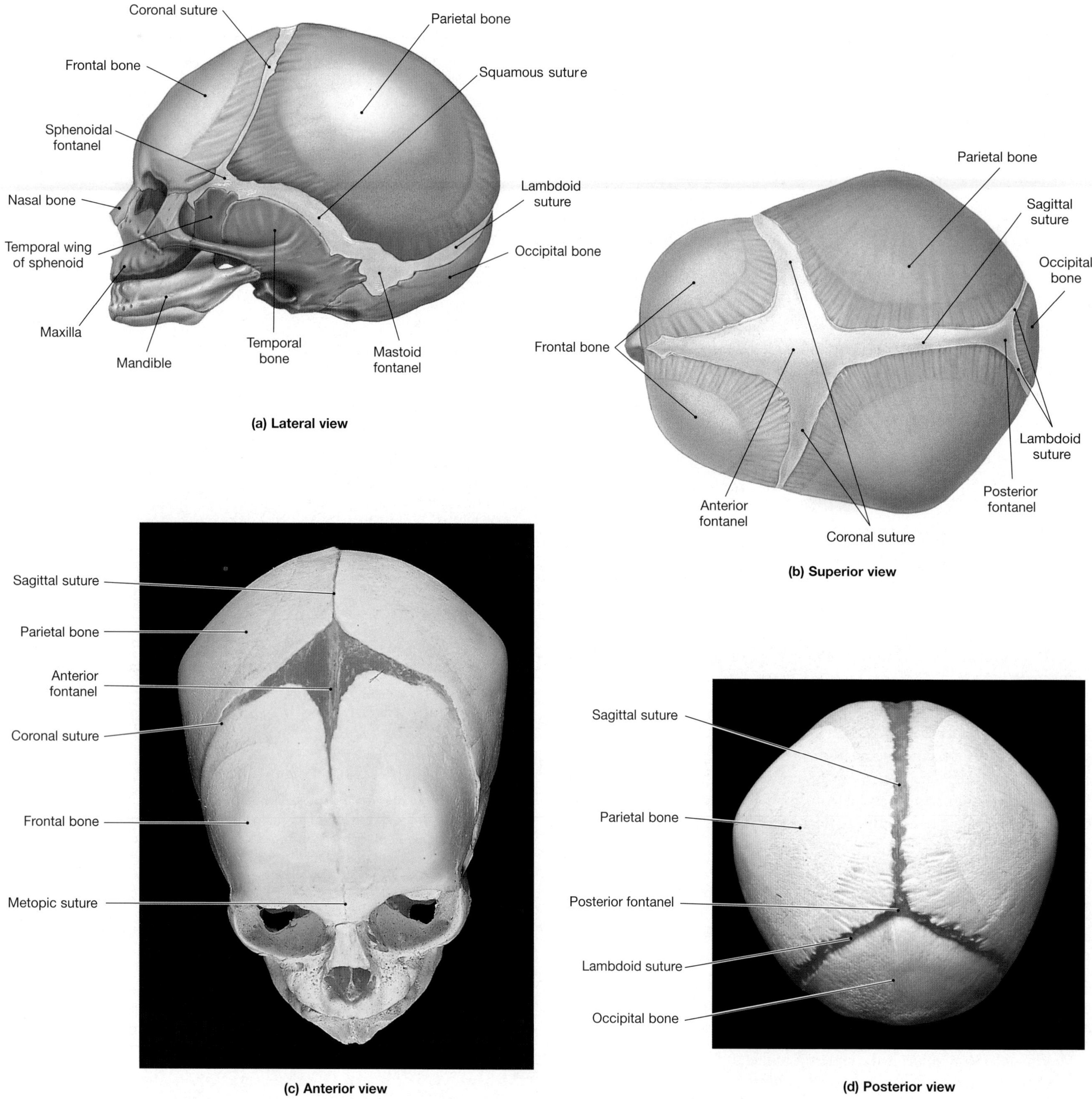

FIGURE 6.18 **The Skull of an Infant**

The flat bones in the infant skull are separated by fontanels, which allow for cranial expansion and the distortion of the skull during birth. By about age 4 these areas will disappear, and skull growth will be completed. (**a**) Lateral view. (**b**) Superior view. (**c**) Anterior view. (**d**) Posterior view.

The Vertebral Column [FIGURE 6.19]

The rest of the axial skeleton is subdivided into the vertebral column and the thoracic cage. The adult **vertebral column** consists of 26 bones, including the vertebrae (24), the sacrum, and the coccyx. The vertebrae provide a column of support, bearing the weight of the head, neck, and trunk, and ultimately transferring that weight to the appendicular skeleton of the lower limbs. They also protect the spinal cord, provide a passageway for spinal nerves that begin or end at the spinal cord, and help maintain an upright body position, as in sitting or standing.

The vertebral column is divided into regions. Beginning at the skull, the regions are *cervical*, *thoracic*, *lumbar*, *sacral*, and *coccygeal* (Figure 6.19●). Seven *cervical vertebrae* constitute the neck and extend inferiorly to the trunk. The first cervical vertebra forms a pair of joints, or articulations, with the occipital condyles of the skull. The seventh cervical vertebra articulates with the first thoracic vertebra. Twelve *thoracic vertebrae* form the midback

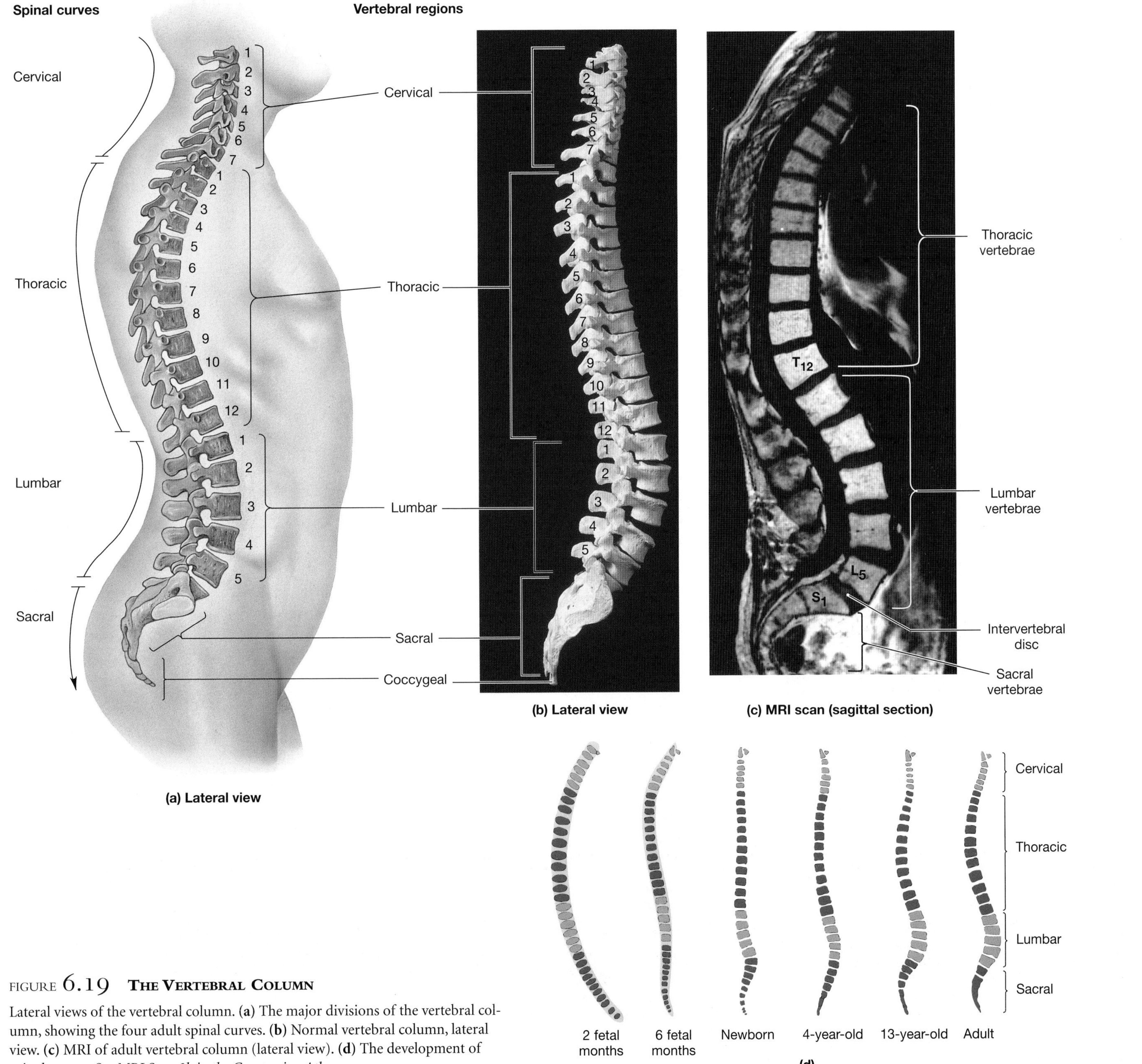

FIGURE 6.19 **THE VERTEBRAL COLUMN**

Lateral views of the vertebral column. (**a**) The major divisions of the vertebral column, showing the four adult spinal curves. (**b**) Normal vertebral column, lateral view. (**c**) MRI of adult vertebral column (lateral view). (**d**) The development of spinal curves. *See MRI Scan 3b in the Companion Atlas.*

regions, and each articulates with one or more pairs of ribs. The twelfth thoracic vertebra articulates with the first lumbar vertebra. Five *lumbar vertebrae* form the lower back; the fifth articulates with the sacrum, which in turn articulates with the coccyx. The cervical, thoracic, and lumbar regions consist of individual vertebrae. During development, the *sacrum* originates as a group of five vertebrae, and the *coccyx* (KOK-siks), or "tailbone," begins as three to five very small vertebrae. The vertebrae of the sacrum usually complete their fusion by age 25. The distal coccygeal vertebrae do not complete their ossification before puberty, and thereafter fusion occurs at a variable pace. The total length of the vertebral column of an adult averages 71 cm (28 inches).

Spinal Curvature [FIGURE 6.19]

The vertebral column does not form a straight and rigid structure. A side view of the adult vertebral column shows four **spinal curves** (Figure 6.19a–c●): (1) **cervical curve**, (2) **thoracic curve**, (3) **lumbar curve**, and (4) **sacral curve**.

The sequence of appearance of the spinal curves from fetus, to newborn, to child, and to adult is illustrated in Figure 6.19d●. The thoracic and sacral curves are called **primary curves** because they appear late in fetal development. These are also called **accommodation curves** because they accommodate the thoracic and abdominopelvic viscera. The vertebral column in the newborn is C-shaped in contrast to the reversed S-shape of the adult, because only the primary curves are present. The lumbar and cervical curves, known as **secondary curves**, do not appear until several months after birth. These are also called **compensation curves** because they help shift the trunk weight over the legs. They become accentuated as the toddler learns to walk and run. All four curves are fully developed by the time a child is 10 years old.

When standing, the weight of the body must be transmitted through the vertebral column to the hips and ultimately to the lower limbs. Yet most of the body weight lies in front of the vertebral column. The various curves bring that weight in line with the body axis. Consider what people do automatically when they stand holding a heavy object. To avoid toppling forward, they exaggerate the lumbar curve, bringing the weight closer to the body axis. This posture can lead to discomfort at the base of the spinal column. Similarly, women in the last 3 months of pregnancy often develop chronic back pain from the changes in lumbar curvature that adjust for the increasing weight of the fetus. No doubt you have seen pictures of African or South American people carrying heavy objects balanced on their heads. Such a practice increases the load on the vertebral column, but because the weight is aligned with the axis of the spine, the spinal curves are not affected and strain is minimized.

CLINICAL BRIEF

Kyphosis, Lordosis, Scoliosis

There are several abnormal distortions of the normal spinal curvature (Figure 6.20●). In **kyphosis** (kī-FŌ-sis), the normal thoracic curve becomes exaggerated posteriorly, producing a "roundback" appearance. This can be caused by (1) osteoporosis or a compression fracture affecting the anterior portions of vertebral bodies, (2) chronic contractions in muscles that insert on the vertebrae, or (3) abnormal vertebral growth. In **lordosis** (lor-DŌ-sis), or "swayback," the abdomen and buttocks protrude because of an anterior exaggeration of the lumbar curve.

Scoliosis (skō-lē-Ō-sis) involves an abnormal lateral curvature. This lateral deviation may occur in one or more of the movable vertebrae. Scoliosis is the most common distortion of the vertebral column. It may result from developmental problems, such as incomplete vertebral formation, or from muscular paralysis affecting one side of the back. In four out of five cases, it is impossible to determine the structural or functional cause of the abnormal spinal curvature. Scoliosis usually appears in girls during adolescence, when periods of growth are most rapid. Treatment consists of a combination of exercises, braces, and sometimes surgical modifications of the affected vertebrae. Early detection greatly improves the chances for successful treatment.

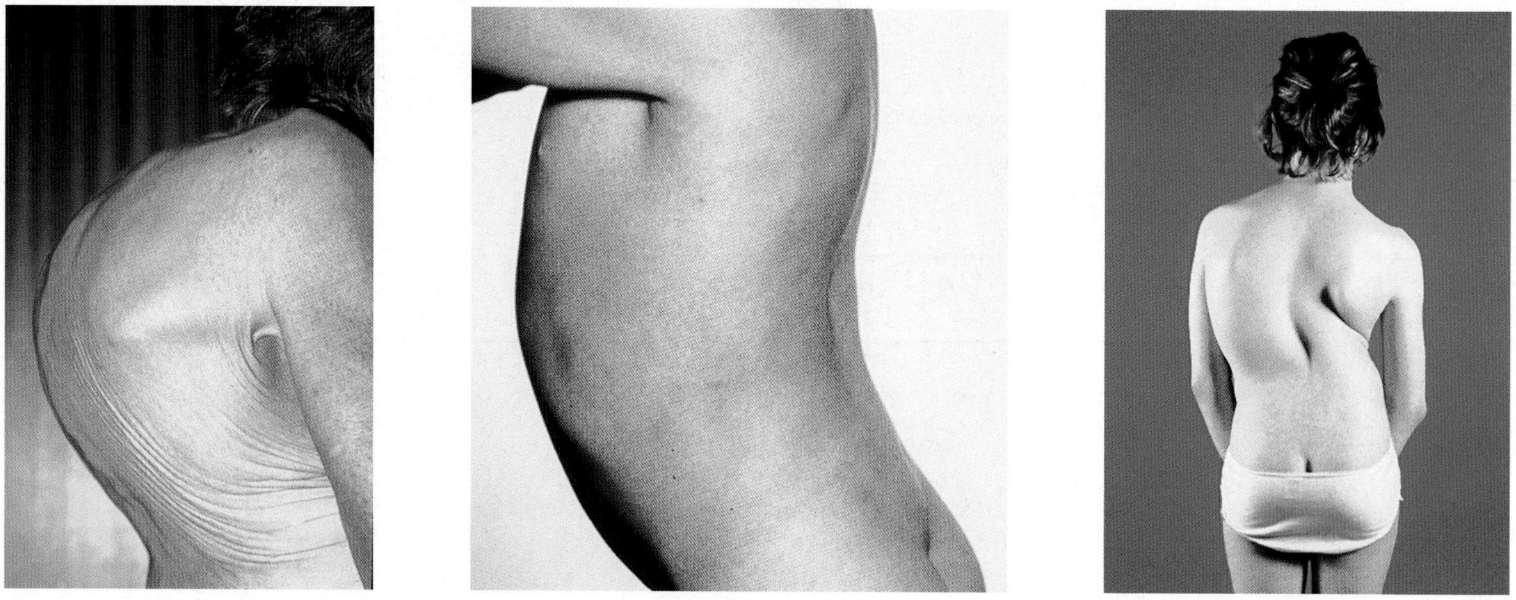

(a) Kyphosis **(b) Lordosis** **(c) Scoliosis**

FIGURE 6.20 **SPINAL DEFORMITIES**

(**a**) Kyphosis. (**b**) Lordosis. (**c**) Scoliosis

Vertebral Anatomy [FIGURE 6.21]

Generally, vertebrae have a common structural plan (Figure 6.21●). Anteriorly, each vertebra has a relatively thick, spherical to oval *body*, from which a *vertebral arch* extends posteriorly. Various processes either for muscle attachment or for rib articulation extend from the vertebral arch. Paired *articular processes* on both the superior and inferior surfaces project from the vertebral arch. These points represent the articulation between adjacent vertebrae (Figure 6.21●).

THE VERTEBRAL BODY [FIGURE 6.21]

The **vertebral body**, or *centrum* (plural, *centra*), is the part of a vertebra that transfers weight along the axis of the vertebral column (Figure 6.21●). Each vertebra articulates with neighboring vertebrae; the bodies are interconnected by ligaments and separated by pads of fibrocartilage, the **intervertebral discs**.

THE VERTEBRAL ARCH [FIGURE 6.21]

The **vertebral arch** (Figure 6.21●), also called the **neural arch**, forms the lateral and posterior margins of the **vertebral foramen** that in life surrounds a portion of the spinal cord. The vertebral arch has a floor (the posterior surface of the body), walls (the *pedicles*), and a roof (the *laminae*) (Lam-i-nē; singular *lamina*, "a thin plate"). The **pedicles** (ped-i-sels) arise along the posterolateral (posterior and lateral) margins of the body. The **laminae** on either side extend dorsomedially (dorsally and medially) to complete the roof. From the fusion of the laminae, a **spinous process**, also known as a *spinal process*, projects dorsally and posteriorly from the midline. These processes can be seen and felt through the skin of the back. **Transverse processes** project laterally or dorsolaterally on both sides from the point where the laminae join the pedicles. These processes are sites of muscle attachment, and they may also articulate with the ribs.

THE ARTICULAR PROCESSES [FIGURE 6.21]

The **articular processes** also arise at the junction between the pedicles and laminae. There is a superior and inferior articular process on each side of the vertebra. The **superior articular processes** project cranially; the **inferior articular processes** project caudally (Figure 6.21●).

VERTEBRAL ARTICULATION [FIGURE 6.21]

The inferior articular processes of one vertebra articulate with the superior articular processes of the more caudal vertebra. Each articular process has a polished surface called an **articular facet**. The superior processes have articular facets on their dorsal surfaces, whereas the inferior processes articulate along their ventral surfaces.

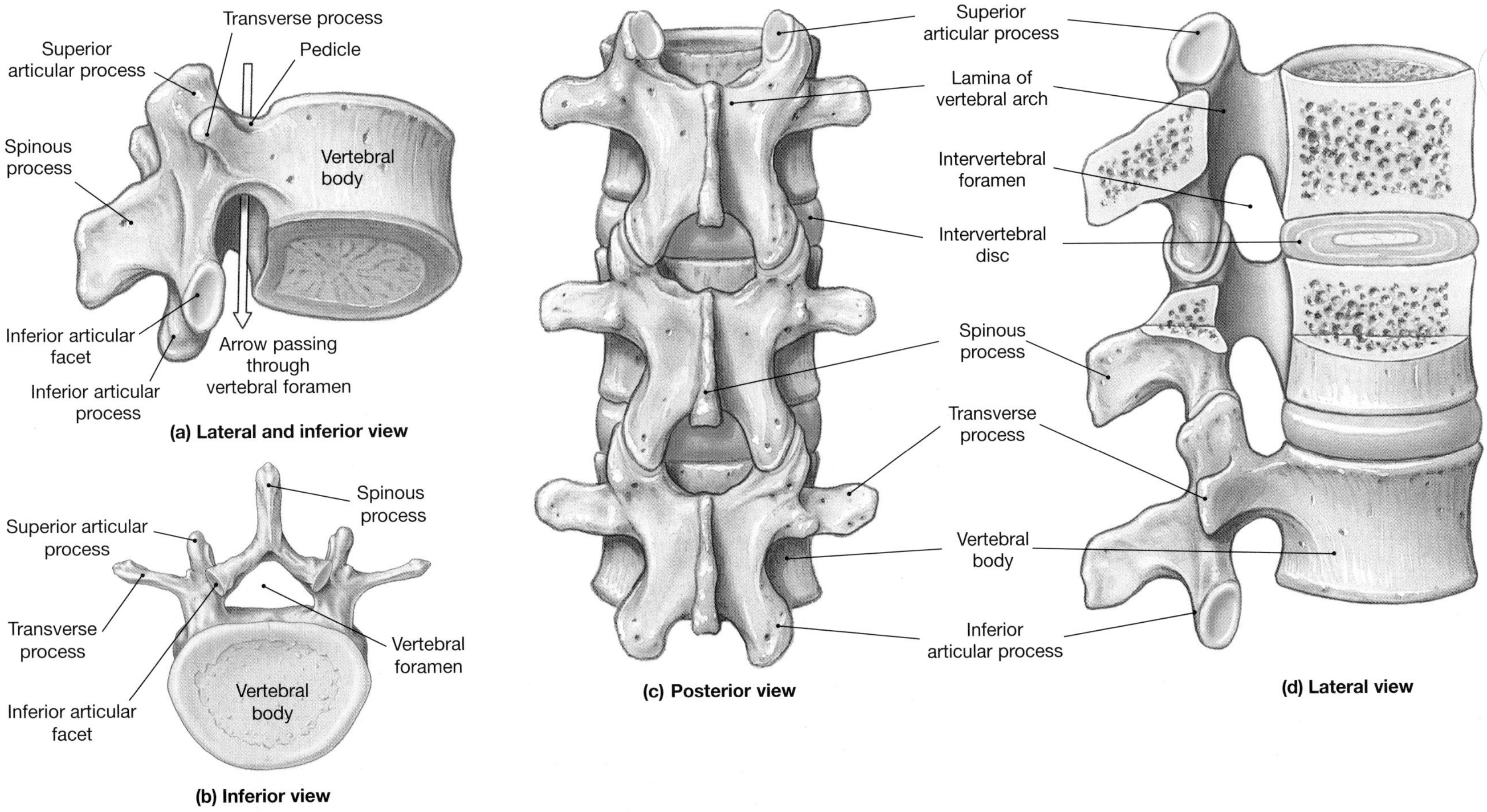

FIGURE 6.21 **VERTEBRAL ANATOMY**

The anatomy of a typical vertebra and the arrangement of articulations between vertebrae.**(a)** A lateral and slightly inferior view of a vertebra. **(b)** An inferior view of a vertebra. **(c)** A posterior view of three articulated vertebrae. **(d)** A lateral and sectional view of three articulated vertebrae.

The vertebral arches of the vertebral column together form the **vertebral canal**, a space that encloses the spinal cord. However, the spinal cord is not completely encased in bone. The vertebral bodies are separated by the intervertebral discs, and there are gaps between the pedicles of successive vertebrae. These **intervertebral foramina** (Figure 6.21●) permit the passage of nerves running to or from the enclosed spinal cord.

Vertebral Regions [FIGURE 6.19a AND TABLE 6.3]

In references to the vertebrae, a capital letter indicates the vertebral region, and a subscript number indicates the vertebra in question, starting with the cervical vertebra closest to the skull. For example, C_3 refers to the third cervical vertebra, with C_1 in contact with the skull; L_4 is the fourth lumbar vertebra, with L_1 in contact with the last thoracic vertebra (Figure 6.19a●). This shorthand will be used throughout the text.

Although each vertebra bears characteristic markings and articulations, focus on the general characteristics of each region and how the regional variations determine the vertebral group's basic function. Table 6.3 compares typical vertebrae from each region of the vertebral column.

CERVICAL VERTEBRAE [FIGURE 6.22 AND TABLE 6.3]

The seven **cervical vertebrae** are the smallest of the vertebrae (Figure 6.22●). They extend from the occipital bone of the skull to the thorax. Notice that the body of a cervical vertebra is relatively small as compared with the size of the triangular vertebral foramen. At this level the spinal cord still contains most of the nerves that connect the brain to the rest of the body. As you continue along the vertebral canal, the diameter of the spinal cord decreases, and so does the diameter of the vertebral arch. On the other hand, cervical vertebrae support only the weight of the head, so the vertebral bodies can be relatively small and light. As you continue caudally along the vertebral column, the loading increases and the vertebral bodies gradually enlarge.

In a typical cervical vertebra (C_1-C_7), the superior surface of the body is concave from side to side, and it slopes, with the anterior edge inferior to the posterior edge. The spinous process is relatively stumpy, usually shorter than the diameter of the vertebral foramen. The tip of each process other than C_7 bears a prominent notch. A notched spinous process is described as *bifid* (BĪ-fid; *bifidus*, cut into two parts). Laterally, the transverse processes are fused to the **costal processes** that originate near the ventrolateral portion of the body. *Costal* refers to rib, and these processes represent the fused remnants of cervical ribs. The costal and transverse processes encircle prominent, round, **transverse foramina**. These passageways protect the *vertebral arteries* and *vertebral veins*, important blood vessels supplying the brain.

This description would be adequate to identify all but the first two cervical vertebrae. When cervical vertebrae C_3–C_7 articulate, their interlocking vertebral bodies permit a relatively greater degree of flexibility than do those of other regions. The first two cervical vertebrae are unique and the seventh is modified. Table 6.3 summarizes the features of cervical vertebrae.

THE ATLAS (C_1) [FIGURE 6.23a,b] The **atlas** (C_1) holds up the head, articulating with the occipital condyles of the skull at the superior articular facet of the superior articular process (Figure 6.23a, b●). It is named after Atlas, a figure in Greek mythology who held up the world. The articulation between the occipital condyles and the atlas is a joint that permits nodding (as when indicating "yes") but prevents twisting. The atlas can be distinguished from the other vertebrae by the following features: (1) the lack of a body, (2) the possession of semicircular **anterior** and **posterior vertebral arches**, each containing **anterior** and **posterior tubercles**, and (3) the presence of oval **superior articular facets** and round **inferior articular facets**.

The atlas articulates with the second cervical vertebra, the *axis*. This articulation permits rotation (as when shaking the head to indicate "no").

THE AXIS (C_2) [FIGURE 6.23c–f] During development, the body of the atlas fuses to the body of the second cervical vertebra, called the **axis** (C_2) (Figure 6.23c,d●). This fusion creates the prominent **dens** (*denz*, tooth), or *odontoid process* (ō-DON-toyd; *odontos*, tooth) of the axis. Thus, there is no intervertebral disc between the atlas and the axis. A *transverse ligament* binds the dens to the inner surface of the atlas, forming a pivot for rotation of the atlas and skull relative to the rest of the vertebral column. This permits the turning of the head from side to side (as when indicating "no"; Figure 6.23e,f●). Important muscles controlling the position of the head and neck attach to the especially robust spinous process of the axis.

In a child the fusion between the dens and axis is incomplete, and impacts or even severe shaking can cause dislocation of the dens and severe damage to the spinal cord. In the adult, a blow to the base of the skull can be equally dangerous because a dislocation of the axis–atlas joint can force the dens into the base of the brain, with fatal results.

TABLE 6.3 REGIONAL DIFFERENCES IN VERTEBRAL STRUCTURE AND FUNCTION

Type (Number)	Location	Vertebral Body	Vertebral Foramen	Spinous Process	Transverse Process	Functions
Cervical vertebrae (7) *(see Figure 6.22)*	Neck	Small; oval; curved faces	Large	Long; split; tip points inferiorly	Has transverse foramen	Support skull, stabilize relative positions of brain and spinal cord, allow controlled head movement
Thoracic vertebrae (12) *(see Figure 6.24)*	Chest	Medium; heart-shaped; flat faces; facets for rib articulations	Smaller	Long; slender; not split; tip points inferiorly	All but two (T_{11}, T_{12}) have facets for rib articulations	Support weight of head, neck, upper limbs, organs of thoracic cavity; articulate with ribs to allow changes in volume of thoracic cage
Lumbar vertebrae (5) *(see Figure 6.25)*	Lower back	Massive; oval; flat faces	Smallest	Blunt; broad tip points posteriorly	Short; no articular facets or transverse foramen	Support weight of head, neck, upper limbs, organs of thoracic and abdominal cavities

VERTEBRA PROMINENS (C_7) [FIGURES 6.22a, 6.24a] The transition from one vertebral region to another is not abrupt, and the last vertebra of one region usually resembles the first vertebra of the next. The **vertebra prominens** (C_7) has a long, slender spinous process that ends in a broad tubercle that can be felt beneath the skin at the base of the neck. This vertebra, shown in Figures 6.22a● and 6.24a●, is the interface between the cervical curve, which arches anteriorly, and the thoracic curve, which arches posteriorly. The transverse processes are large, providing additional surface area for muscle attachment, and the transverse foramina are either reduced or absent. A large elastic ligament, the **ligamentum nuchae** (li-ga-MEN-tum NOO-kā; *nucha*, nape) begins at the vertebra prominens and extends cranially to an insertion along the external occipital crest. Along the way, it attaches to the spinous processes of the other cervical vertebrae. When the head is upright, this ligament acts like the string on a bow, maintaining the cervical curvature without muscular effort. If the neck has been bent forward, the elasticity in this ligament helps return the head to an upright position.

The head is relatively massive, and it sits atop the cervical vertebrae like a soup bowl on the tip of a finger. With this arrangement, small muscles can produce significant effects by tipping the balance one way or another. But if the body suddenly changes position, as in a fall or during rapid acceleration (a jet taking off) or deceleration (a car crash), the balancing muscles are not strong enough to stabilize the head. A dangerous partial or complete dislocation of the cervical vertebrae can result, with injury to muscles and ligaments and potential injury to the spinal cord. The term **whiplash** is used to describe such an injury, because the movement of the head resembles the cracking of a whip.

THORACIC VERTEBRAE [FIGURE 6.24 AND TABLE 6.3]

There are 12 **thoracic vertebrae**. A typical thoracic vertebra (Figure 6.24●) has a distinctive heart-shaped body that is more massive than that of a cervical vertebra. The round vertebral foramen is relatively smaller, and the

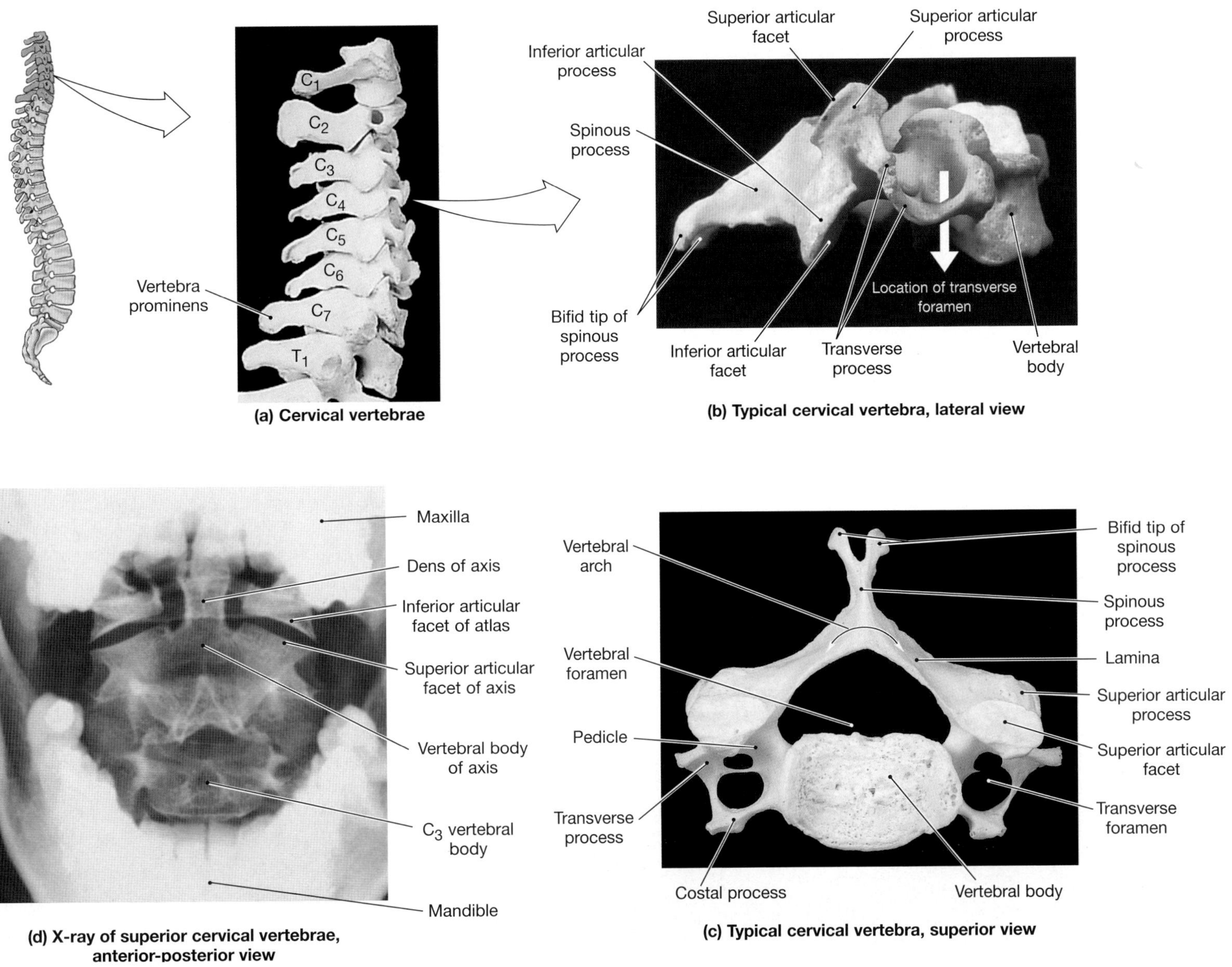

FIGURE 6.22 **CERVICAL VERTEBRAE**

These are the smallest and most superior vertebrae. (**a**) Lateral view of the cervical vertebrae. (**b**) Lateral view of a typical (C_3–C_6) cervical vertebra. (**c**) Superior view of the same vertebra. Note the characteristic features listed in Table 6.3. (**d**) An X-ray of the superior cervical vertebrae, anterior-posterior view. The mouth is open and the lower teeth are visible. *See MRI Scan 3a in the Companion Atlas.*

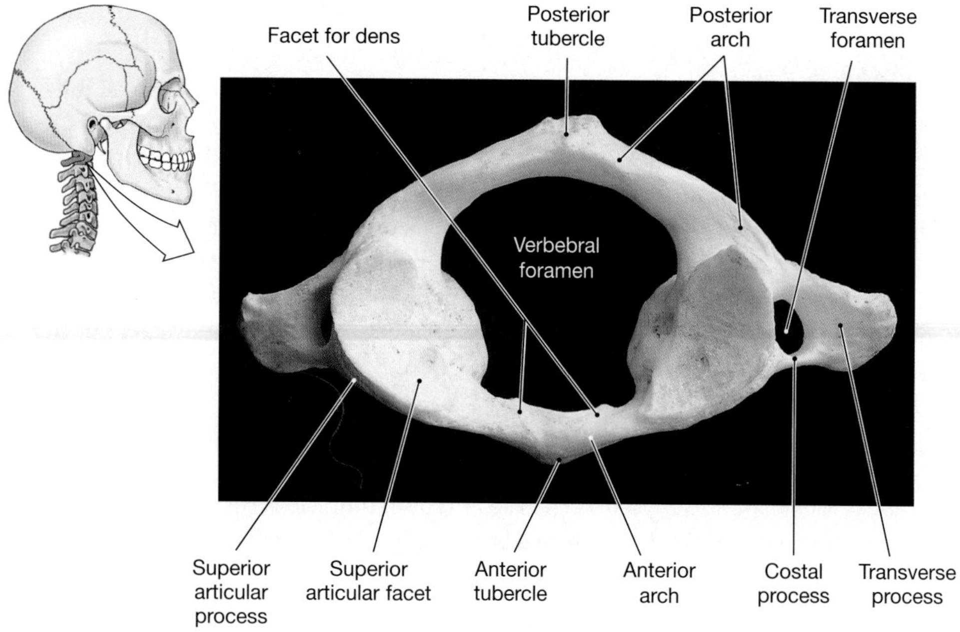

(a) Atlas, superior view

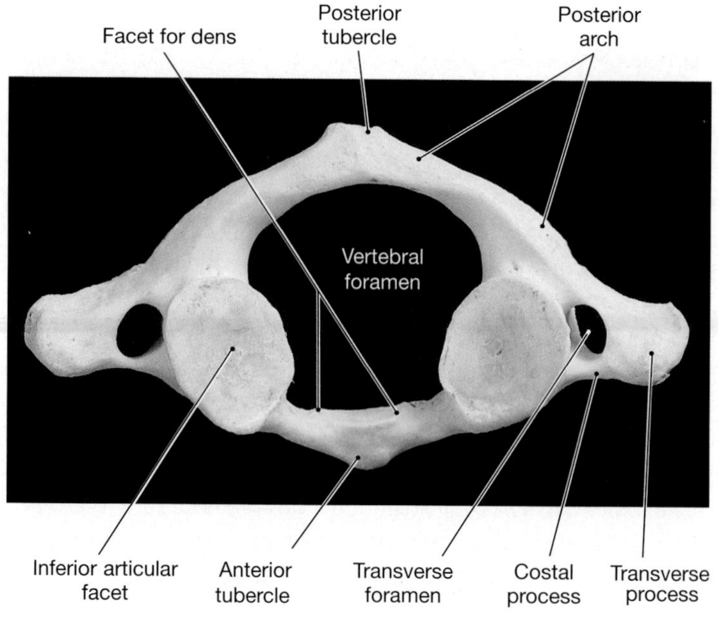

(b) Atlas, inferior view

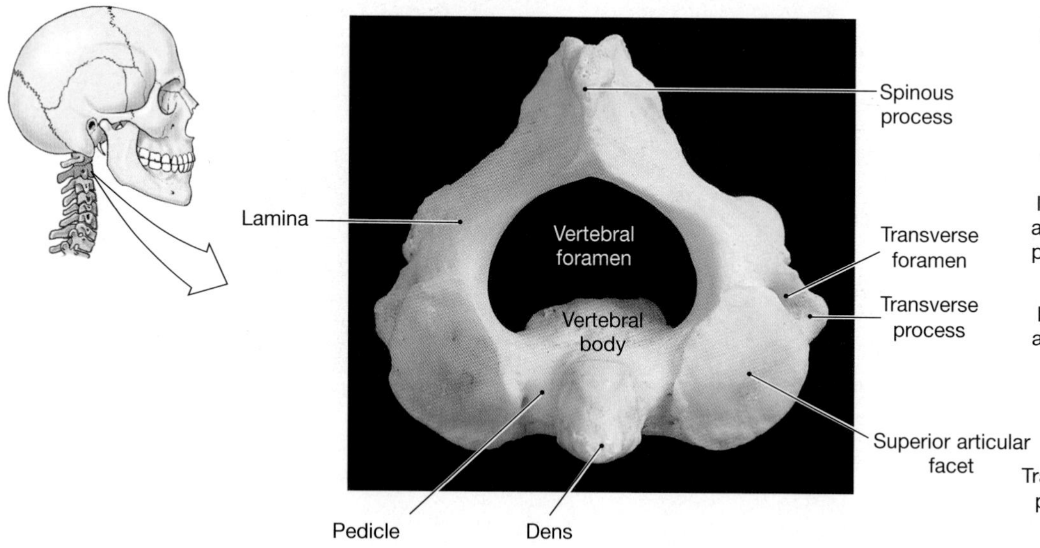

(c) Axis, superior view

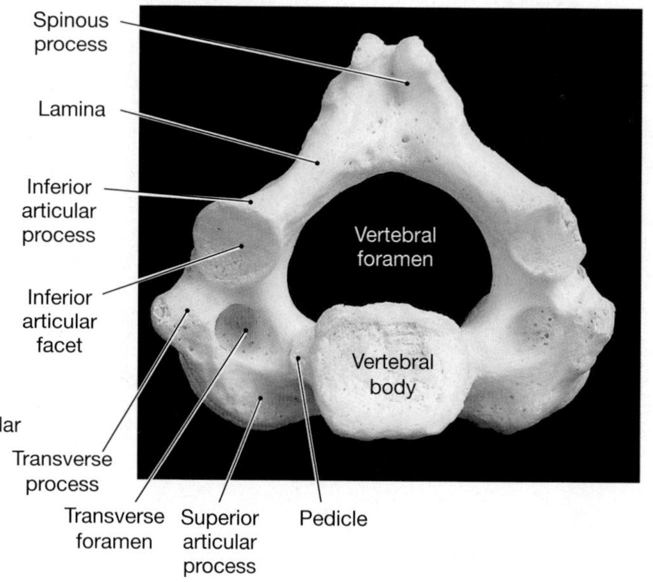

(d) Axis, inferior view

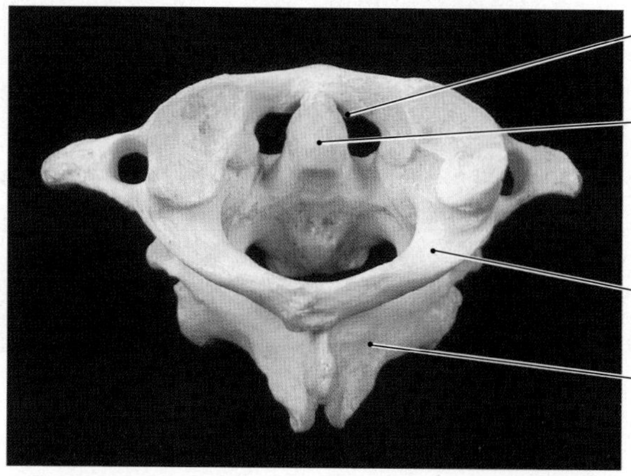

(e) Articulated atlas and axis, superior and posterior view

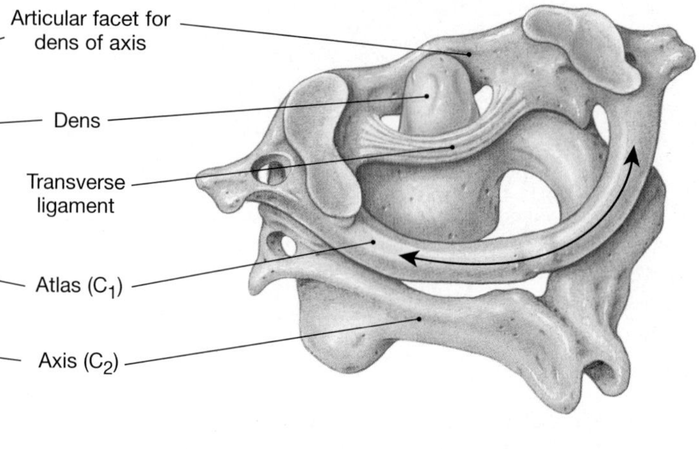

(f) The articulated atlas and axis; note the location and orientation of the transverse ligament.

FIGURE 6.23 **ATLAS AND AXIS**

Unique anatomical characteristics of vertebrae C_1 (atlas) and C_2 (axis). The atlas is seen in **(a)** superior view and **(b)** inferior view. The axis is seen in **(c)** superior view and **(d)** inferior view. **(e)** The articulated atlas and axis, in superior and posterior view. **(f)** The articulated atlas (C_1) and axis (C_2), showing the transverse ligament that holds the dens of the axis in position at the articular facet of the atlas.

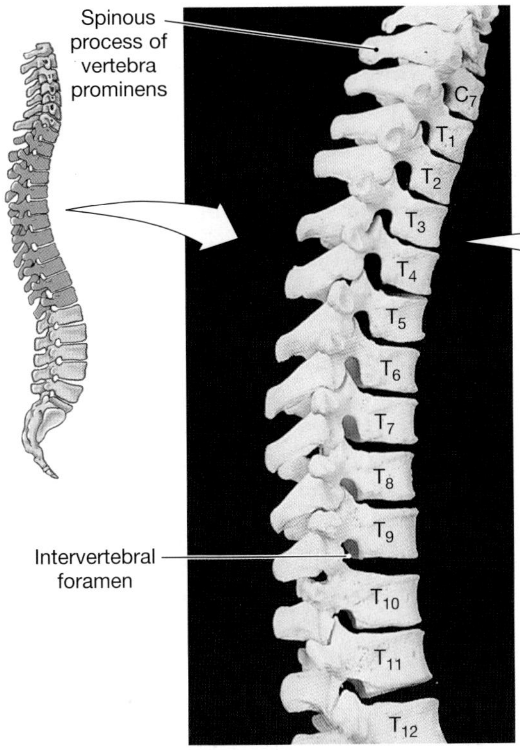

(a) Thoracic vertebrae, lateral view

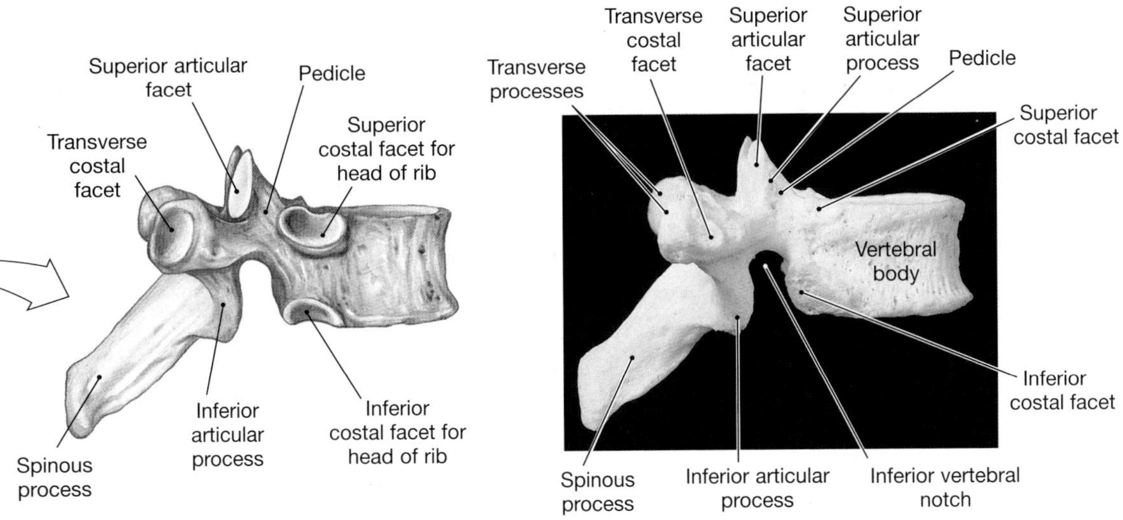

(b) Thoracic vertebra, lateral view

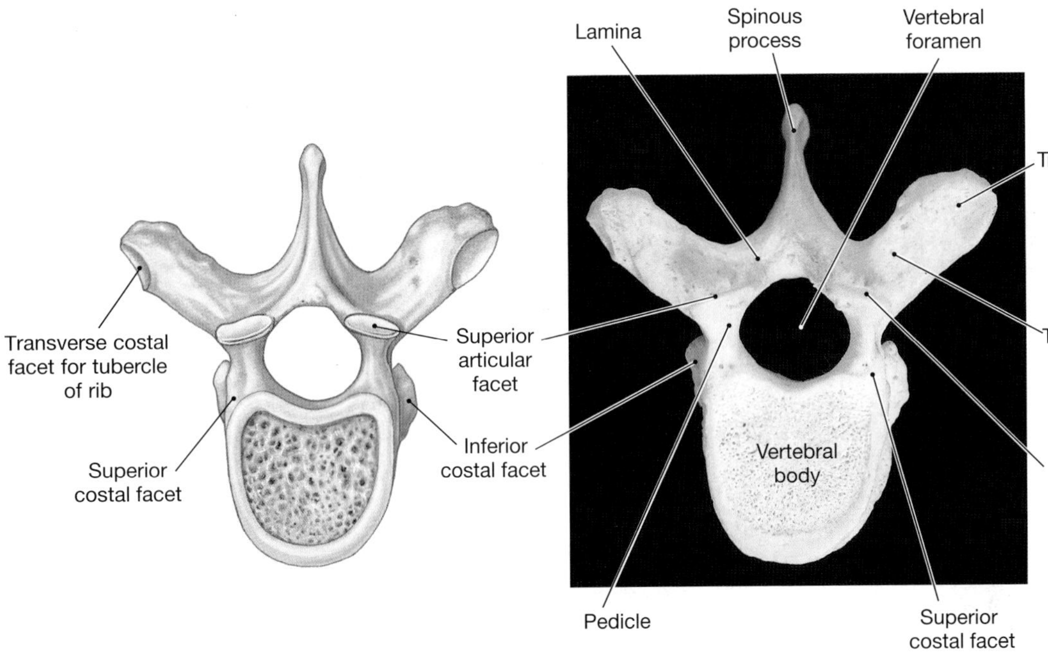

(c) Thoracic vertebra, superior view

FIGURE 6.24 **THORACIC VERTEBRAE**

The body of each thoracic vertebra articulates with ribs. Note the characteristic features listed in Table 6.3. **(a)** Lateral view of the thoracic region of the vertebral column. The vertebra prominens (C_7) resembles T_1, but it lacks facets for rib articulation. Vertebra T_{12} resembles the first lumbar vertebra (L_1), but it has a facet for rib articulation. A representative thoracic vertebra is shown in **(b)** lateral view, **(c)** superior view, and **(d)** posterior view. *See MRI Scan 3b in the Companion Atlas.*

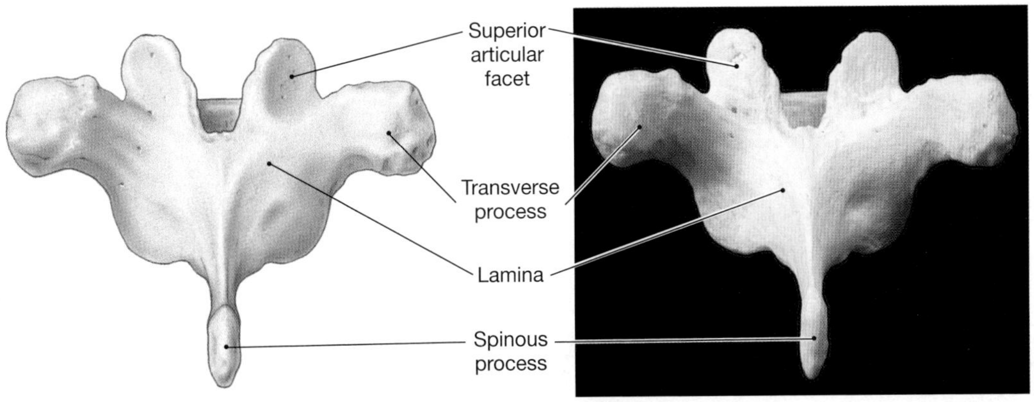

(d) Thoracic vertebra, posterior view

long, slender spinous process projects posterocaudally. The spinous processes of T_{10}, T_{11}, and T_{12} increasingly resemble those of the lumbar series, as the transition between the thoracic and lumbar curvatures approaches. Because of the weight carried by the lower thoracic and lumbar vertebrae, it is difficult to stabilize the transition between the thoracic and lumbar curves. As a result, compression fractures or compression-dislocation fractures after a hard fall most often involve the last thoracic and first two lumbar vertebrae.

Each thoracic vertebra articulates with ribs along the dorsolateral surfaces of its body. The location and structure of the articulations vary somewhat from vertebra to vertebra (Figure 6.24b,c●). Thoracic vertebrae T_1 to T_8 have **superior** and **inferior costal facets**, as they articulate with two pairs of ribs. Vertebrae T_9 to T_{12} have only a single costal facet on either side.

The transverse processes of vertebrae T_1 to T_{10} are relatively thick, and their anterolateral surfaces contain **transverse costal facets** for articulation with the tubercles of ribs. Thus, ribs 1 through 10 contact their vertebrae at two points, at a costal facet and at a transverse costal facet. This dual articulation with the ribs limits the mobility of the thoracic vertebrae. Table 6.3, p. 166, summarizes the features of the thoracic vertebrae.

LUMBAR VERTEBRAE [FIGURE 6.25 AND TABLE 6.3]

The **lumbar vertebrae** are the largest of the vertebrae. The body of a typical lumbar vertebra (Figure 6.25●) is thicker than that of a thoracic vertebra, and the superior and inferior surfaces are oval rather than heart-shaped. There are no articular facets on either the body or the transverse processes, and the vertebral foramen is triangular. The transverse processes are slender and project dorsolaterally, and the stumpy spinous processes project dorsally.

The lumbar vertebrae bear the most weight. Thus a compression injury to the vertebrae or intervertebral discs most often occurs in this region. The most common injury is a tear or rupture in the connective tissues of the intervertebral disc; this condition is known as a herniated

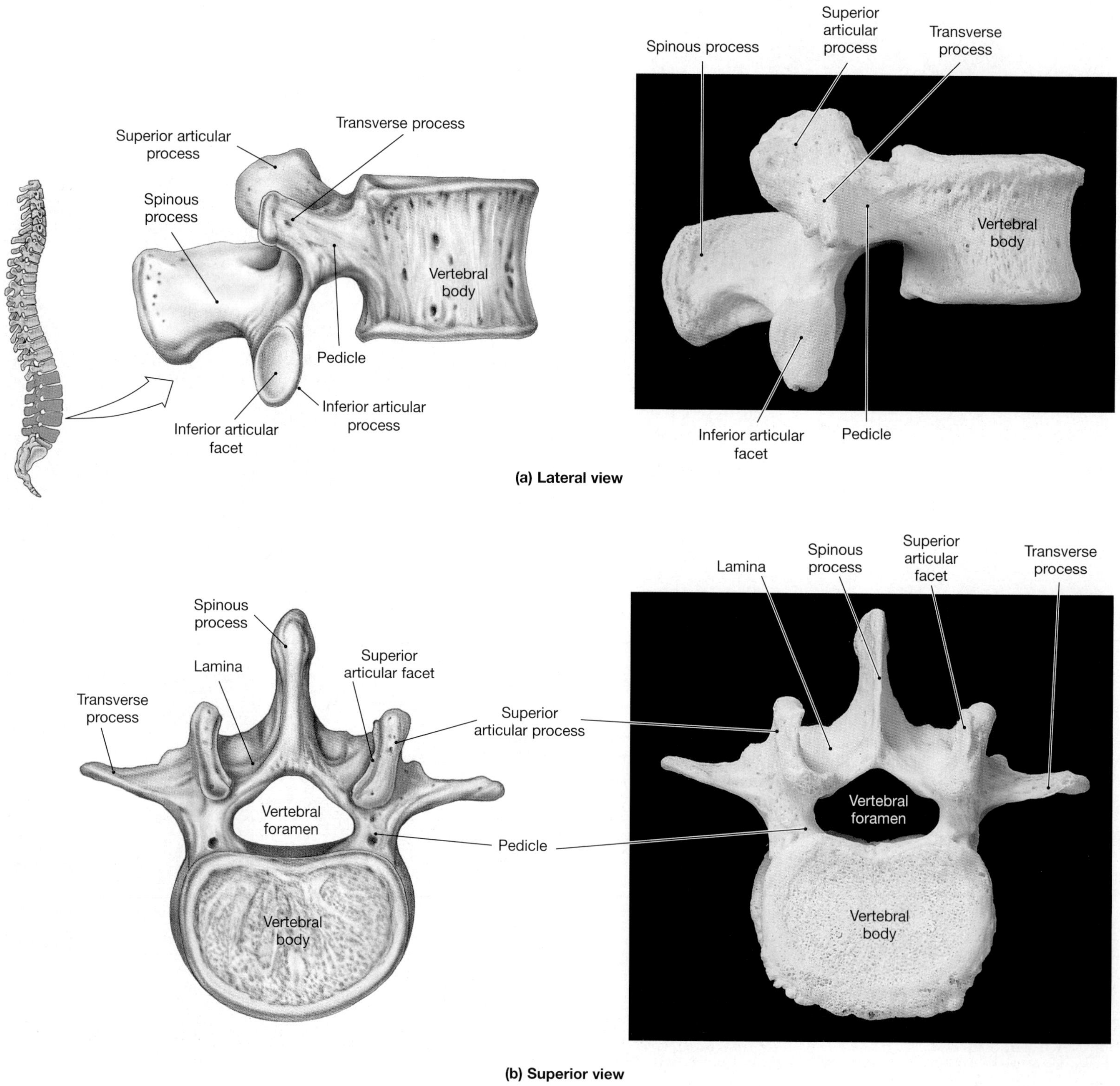

FIGURE 6.25 **LUMBAR VERTEBRAE**

These are the largest vertebrae and they bear the most weight. A representative lumbar vertebra is shown in (**a**) lateral views and (**b**) superior views. *See MRI Scans 3b, and 10a in the Companion Atlas.*

disc. The massive spinous processes of the lumbar vertebrae provide surface area for the attachment of lower back muscles that reinforce or adjust the lumbar curvature. Table 6.3, p. 166, summarizes the characteristics of lumbar vertebrae.

THE SACRUM [FIGURE 6.26]

The **sacrum** (Figure 6.26●) consists of the fused components of five sacral vertebrae. These vertebrae begin fusing shortly after puberty and are usually completely fused between ages 25 and 30. Once this fusion is complete, prominent *transverse lines* mark the former boundaries of individual vertebrae. This composite structure protects reproductive, digestive, and excretory organs and, via paired articulations, attaches the axial skeleton to the pelvic girdle of the appendicular skeleton. The broad surface area of the sacrum provides an extensive area for the attachment of muscles, especially those responsible for movement of the thigh.

The sacrum is curved, with a convex dorsal surface (Figure 6.26a●). The narrow, caudal portion is the sacral **apex**, whereas the broad superior surface forms the **base**. The **sacral promontory**, a prominent bulge at the anterior tip of the base, is an important landmark in females during pelvic examinations and during labor and delivery. The **superior articular processes** form synovial articulations with the last lumbar vertebra. The **sacral canal** begins between those processes and extends the length of the sacrum. Nerves and membranes that line the vertebral canal in the spinal cord continue into the sacral canal.

The spinous processes of the five fused sacral vertebrae form a series of elevations along the **median sacral crest**. The laminae of the fifth sacral vertebra fail to contact one another at the midline, and they form the **sacral cornua**. These ridges establish the margins of the **sacral hiatus** (hī-Ā-tus), the end of the sacral canal. In life, this opening is covered by connective tissues. On either side of the median sacral crest are the **sacral foramina**. The intervertebral foramina, now enclosed by the fused sacral bones, open into these passageways. A broad sacral *wing*, or **ala**, extends laterally from each **lateral sacral crest**. The median and lateral sacral crests provide surface area for the attachment of muscles of the lower back and hip.

Viewed laterally (Figure 6.26b●), the *sacral curve* is more apparent. The degree of curvature is greater in males than in females (see Table 7-1, p. 209). Laterally, the **auricular surface** of the sacrum articulates with the pelvic girdle at the **sacroiliac joint**. Dorsal to the auricular surface is a roughened area, the **sacral tuberosity**, which marks the attachment of a ligament that stabilizes this articulation. The anterior surface, or *pelvic surface,* of the sacrum is concave (Figure 6.26c●). At the apex, a flattened area marks the site of articulation with the *coccyx*. The wedge-like shape of the mature sacrum provides a strong foundation for transferring the weight of the body from the axial skeleton to the pelvic girdle.

THE COCCYX [FIGURE 6.26]

The small **coccyx** consists of three to five (most often four) coccygeal vertebrae that have usually begun fusing by age 26 (Figure 6.26●). The coccyx provides an attachment site for a number of ligaments and for a muscle that constricts the anal opening. The first two coccygeal vertebrae have transverse processes and unfused vertebral arches. The prominent laminae of the first coccygeal vertebra are known as the **coccygeal cornua**; they curve to meet the cornua of the sacrum. The coccygeal vertebrae do not complete their fusion until late in adulthood. In males, the adult coccyx points anteriorly, whereas in females, it points inferiorly. In very elderly people, the coccyx may fuse with the sacrum.

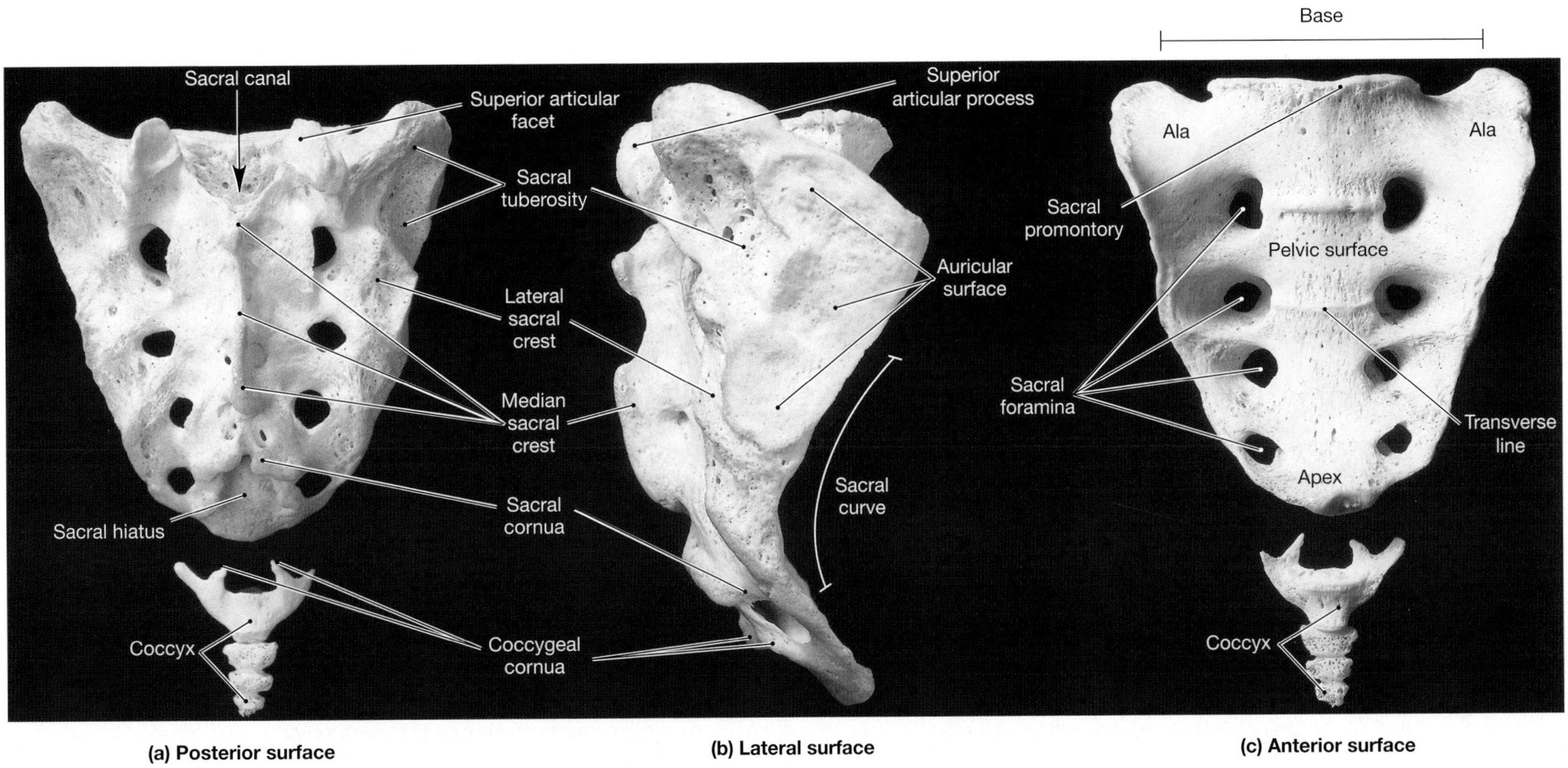

FIGURE 6.26 **THE SACRUM AND COCCYX**

Fused vertebrae form the adult sacrum and coccyx. These bones are shown in (**a**) posterior view, (**b**) lateral view from the right side, and (**c**) anterior view.

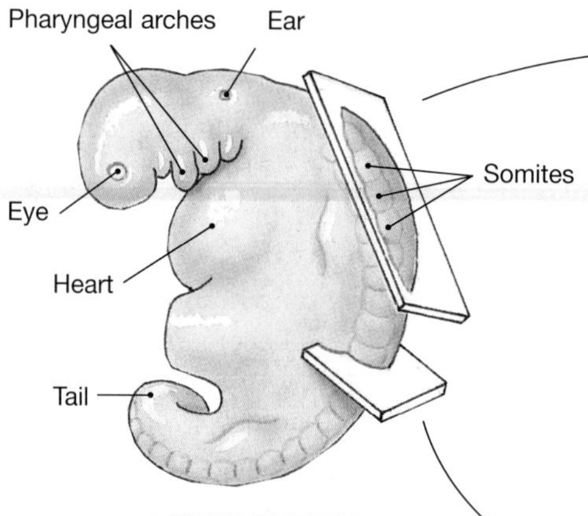

4-WEEK EMBRYO

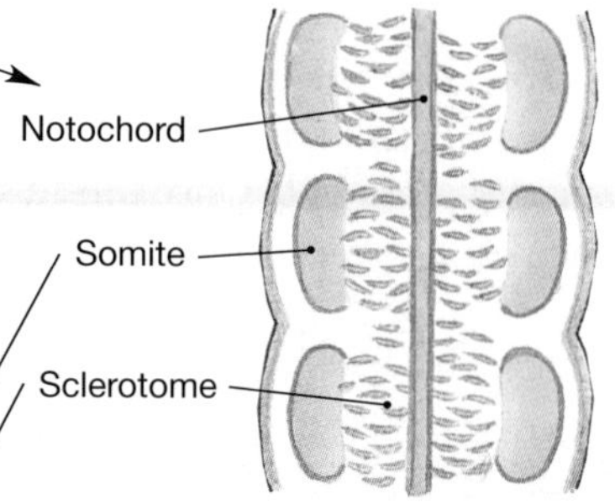

Cells of the sclerotomal segments migrate away from the somites and cluster around the notochord.

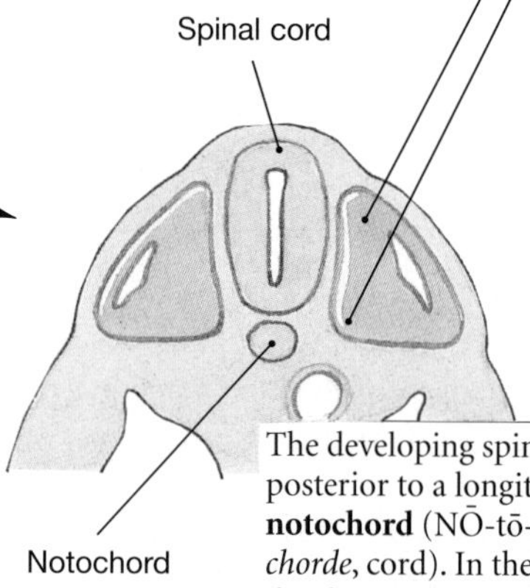

The developing spinal cord lies posterior to a longitudinal rod, the **notochord** (NŌ-tō-kōrd; *noton*, back + *chorde*, cord). In the fourth week of development, mesoderm on either side of the spinal cord and notochord forms a series of mesenchymal blocks called **somites** (SŌ-mīts). Mesenchyme in the medial portions of each somite, a region known as the **sclerotome** (SKLE-ro-tōme; *skleros*, hard), will produce the vertebral column and contribute to the floor of the cranium.

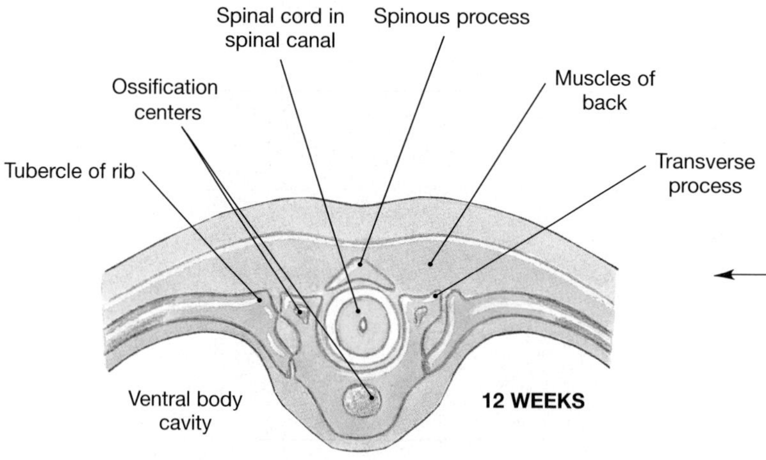

12 WEEKS

About the time the ribs separate from the vertebrae, ossification begins. Only the shortest ribs undergo complete ossification. In the rest, the distal portions remain cartilaginous, forming the costal cartilages. Several ossification centers appear in the sternum, but fusion gradually reduces the number.

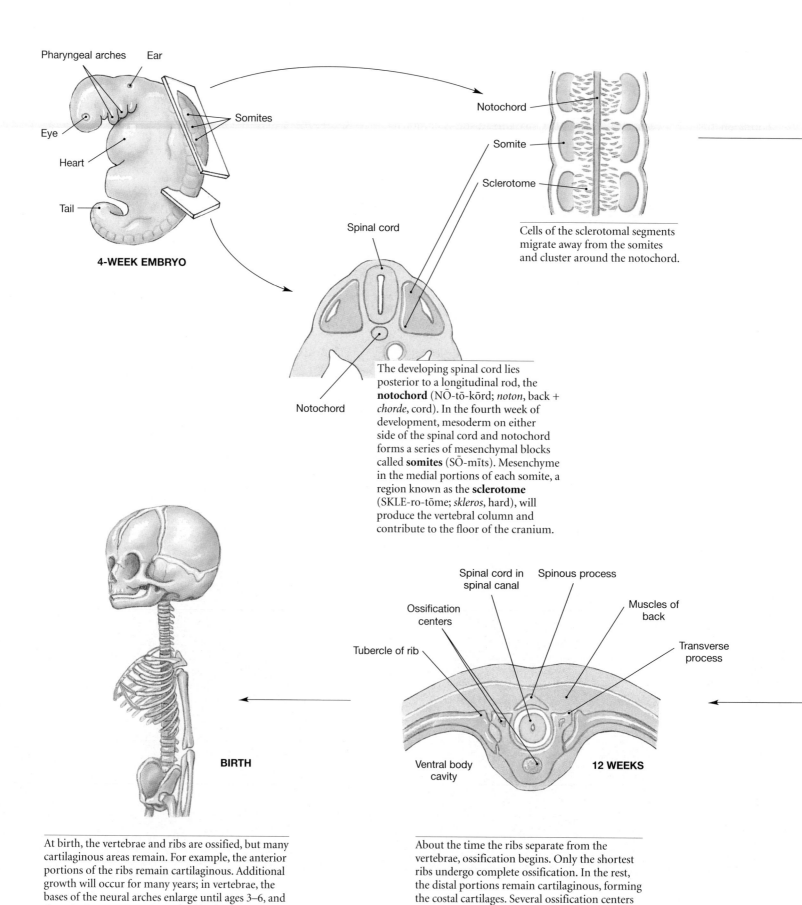

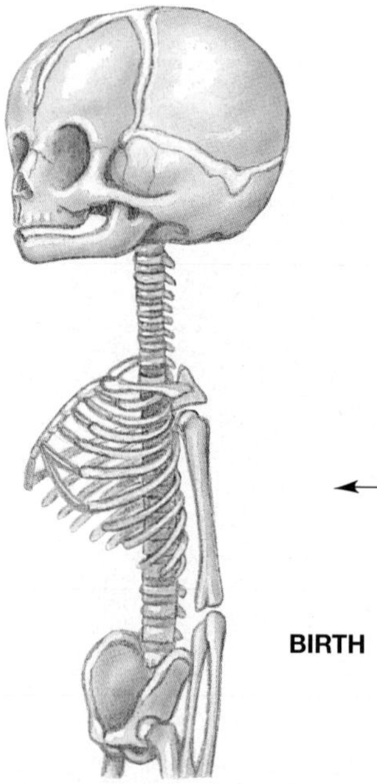

BIRTH

At birth, the vertebrae and ribs are ossified, but many cartilaginous areas remain. For example, the anterior portions of the ribs remain cartilaginous. Additional growth will occur for many years; in vertebrae, the bases of the neural arches enlarge until ages 3–6, and the spinal processes and vertebral bodies grow until ages 18–25.

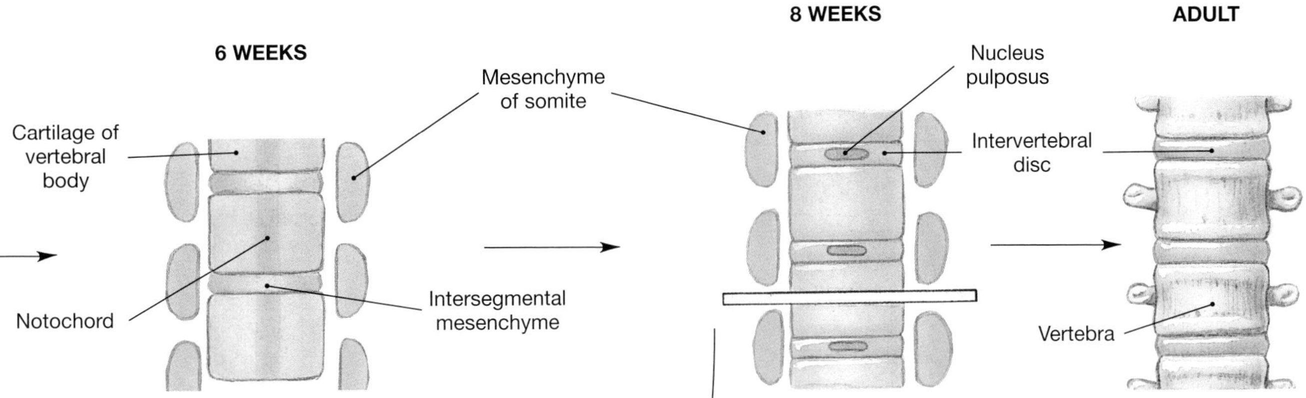

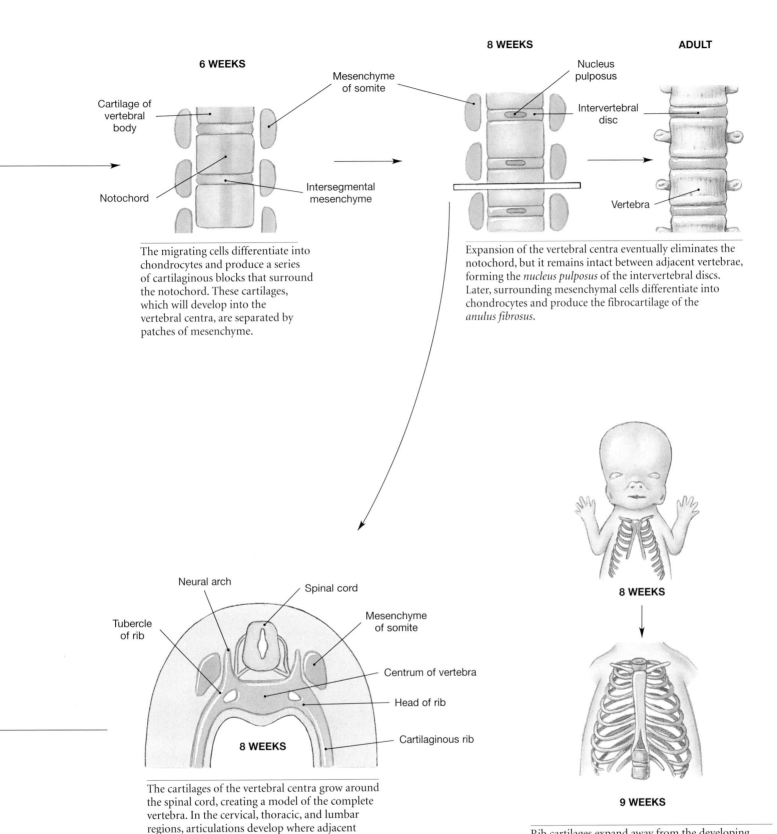

The migrating cells differentiate into chondrocytes and produce a series of cartilaginous blocks that surround the notochord. These cartilages, which will develop into the vertebral centra, are separated by patches of mesenchyme.

Expansion of the vertebral centra eventually eliminates the notochord, but it remains intact between adjacent vertebrae, forming the *nucleus pulposus* of the intervertebral discs. Later, surrounding mesenchymal cells differentiate into chondrocytes and produce the fibrocartilage of the *anulus fibrosus*.

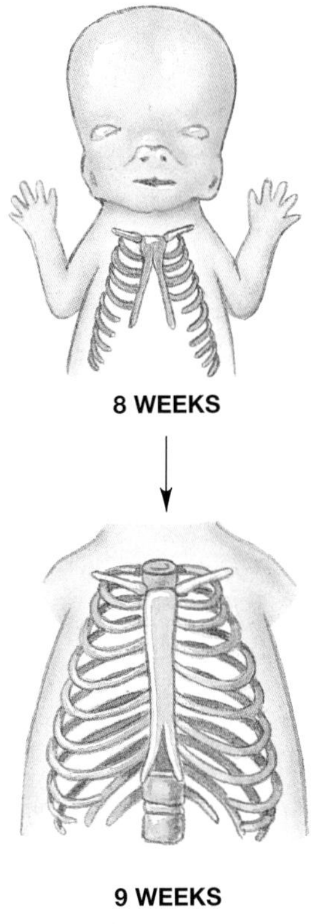

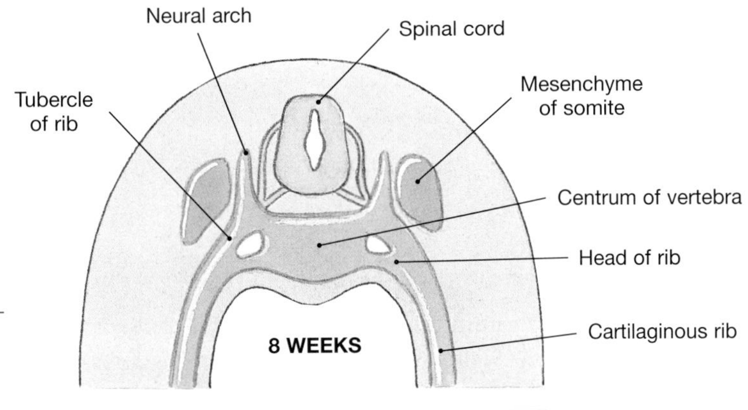

The cartilages of the vertebral centra grow around the spinal cord, creating a model of the complete vertebra. In the cervical, thoracic, and lumbar regions, articulations develop where adjacent cartilaginous blocks come into contact. In the sacrum and coccyx, the cartilages fuse together.

Rib cartilages expand away from the developing transverse processes of the vertebrae. At first they are continuous, but by week 8 the ribs have separated from the vertebrae. Ribs form at every vertebra, but in the cervical, lumbar, sacral, and coccygeal regions, they remain small and later fuse with the growing vertebrae. The ribs of the thoracic vertebrae continue to enlarge, following the curvature of the body wall. When they reach the ventral midline, they fuse with the cartilages of the sternum.

CLINICAL BRIEF

Spina Bifida

Spina bifida (SPĪ-na BI-fi-da) results when the vertebral laminae fail to unite during development. The vertebral arch is incomplete, and the membranes that line the dorsal body cavity bulge outward. In mild cases, most often involving the sacral and lumbar regions, the condition may pass unnoticed. In severe cases, the entire vertebral column and skull are affected. This condition is often associated with developmental abnormalities of the brain and spinal cord. (See the Embryology Summary: Development of Spinal Cord and Nerves, in Chapter 14.)

The Thoracic Cage [FIGURE 6.27]

The skeleton of the chest, or **thoracic cage**, consists of the thoracic vertebrae, the *ribs*, and the *sternum* (Figure 6.27a,c●). The *ribs*, or **costae**, and the sternum form the **rib cage** and support the walls of the thoracic cavity. This cavity is narrow superiorly, broad inferiorly, and somewhat flattened in an anterior-posterior direction. The thoracic cage serves two functions:

- It protects the heart, lungs, thymus, and other structures in the thoracic cavity; and
- It serves as an attachment point for muscles involved with (1) respiration, (2) the position of the vertebral column, and (3) movements of the pectoral girdle and upper limbs.

The Ribs [FIGURES 6.24/6.27]

Ribs, or *costae*, are elongated, curved, flattened bones that (1) originate on or between thoracic vertebrae and (2) end in the wall of the thoracic cavity. There are 12 pairs of ribs (Figure 6.27●). The first seven pairs are called **true ribs**, or *vertebrosternal ribs.* At the anterior body wall the true ribs are connected to the sternum by separate cartilages, the **costal cartilages**. Beginning with the first rib, the vertebrosternal ribs gradually increase in length and in the radius of curvature.

Ribs 8–12 are called **false ribs** or *vertebrochondral ribs*, because they do not attach directly to the sternum. The costal cartilages of ribs 8–10 fuse together before reaching the sternum (Figure 6.27a●). The last two pairs of ribs are called **floating ribs** because they have no connection with the sternum.

Figure 6.27b● shows the superior surface of the *vertebral end* of a representative rib. The **head**, or *capitulum* (ka-PIT-ū-lum) of each rib articulates with the body of a thoracic vertebra or between adjacent vertrebral bodies. After a short **neck**, the **tubercle**, or *tuberculum* (too-BER-kū-lum), projects dorsally. The inferior portion of the tubercle contains an articular facet that contacts the transverse process of the thoracic vertebra. When the rib articulates between adjacent vertebrae, the articular surface is divided into **superior** and **inferior articular facets** by the **interarticular crest** (Figure 6.27c,d●). Ribs 1 through 10 originate at costal facets on the bodies of vertebrae T_1 to T_{10}, and their tubercular facets articulate with the transverse costal facets of their respective vertebrae. Ribs 11 and 12 originate at costal facets on T_{11} and T_{12}. These ribs do not have tubercular facets and they do not articulate with transverse processes. The difference in rib orientation can be seen by comparing Figures 6.24●, p. 169, and 6.27c,d●.

The bend, or **angle**, of the rib indicates the site where the tubular **body**, or *shaft*, begins curving toward the sternum. The internal rib surface is concave, and a prominent **costal groove** along its inferior border marks the path of nerves and blood vessels. The superficial surface is convex and provides an attachment site for muscles of the pectoral girdle and trunk. The *intercostal muscles* that move the ribs are attached to the superior and inferior surfaces.

With their complex musculature, dual articulations at the vertebrae, and flexible connection to the sternum, the ribs are quite mobile. Note how the ribs curve away from the vertebral column to angle downward. Functionally, a typical rib acts as if it were the handle on a bucket, lying just below the horizontal plane. Pushing it down forces it inward; pulling it up swings it outward. In addition, because of the curvature of the ribs, the same movements change the position of the sternum. Depressing the ribs moves the sternum posteriorly (inward), whereas elevation moves it anteriorly (outward). As a result, movements of the ribs affect both the width and the depth of the thoracic cage, increasing or decreasing its volume accordingly. *The Thoracic Cage and Surgical Procedures. p. 789*

CLINICAL BRIEF

Cracked Ribs

A hockey player is checked into the boards; a basketball player flies out of bounds after a loose ball, slamming into the first row of seats; a wide receiver is hit hard after catching a pass over the middle. Sudden impacts in the chest such as these are relatively common, and the ribs usually take the full force of the contact.

The ribs are composed of spongy bone with a thin outer covering of compact bone. They are firmly bound in connective tissues and are interconnected by layers of muscle. As a result, displaced fractures are uncommon, and rib injuries usually heal swiftly and effectively. In extreme injuries, a broken rib can be forced into the thoracic cavity and can damage internal organs. The entry of air into one of the pleural cavities, a condition known as a **pneumothorax** (noo-mō-THOR-aks), may lead to a collapsed lung. Damage to a blood vessel or even the heart can cause bleeding into the thoracic cavity, a condition called a hemothorax. A **hemothorax** can also impair lung function because fluid may accumulate and compress a lung.

The Sternum [FIGURE 6.27a]

The adult **sternum** is a flat bone that forms in the anterior midline of the thoracic wall (Figure 6.27a●). The sternum has three components:

- The broad, triangular **manubrium** (ma-NOO-brē-um) articulates with the *clavicles* (or collarbones) of the appendicular skeleton and the costal cartilages of the first pair of ribs. The manubrium is the widest and most superior portion of the sternum. The **jugular notch** is the shallow indentation on the superior surface of the manubrium. It is located between the clavicular articulations.
- The tongue-shaped **body** attaches to the inferior surface of the manubrium and extends caudally along the midline. Individual costal cartilages from rib pairs 2–7 are attached to this portion of the sternum. The rib pairs 8–10 are also attached to the body, but by a single pair of cartilages shared with rib pair 7.

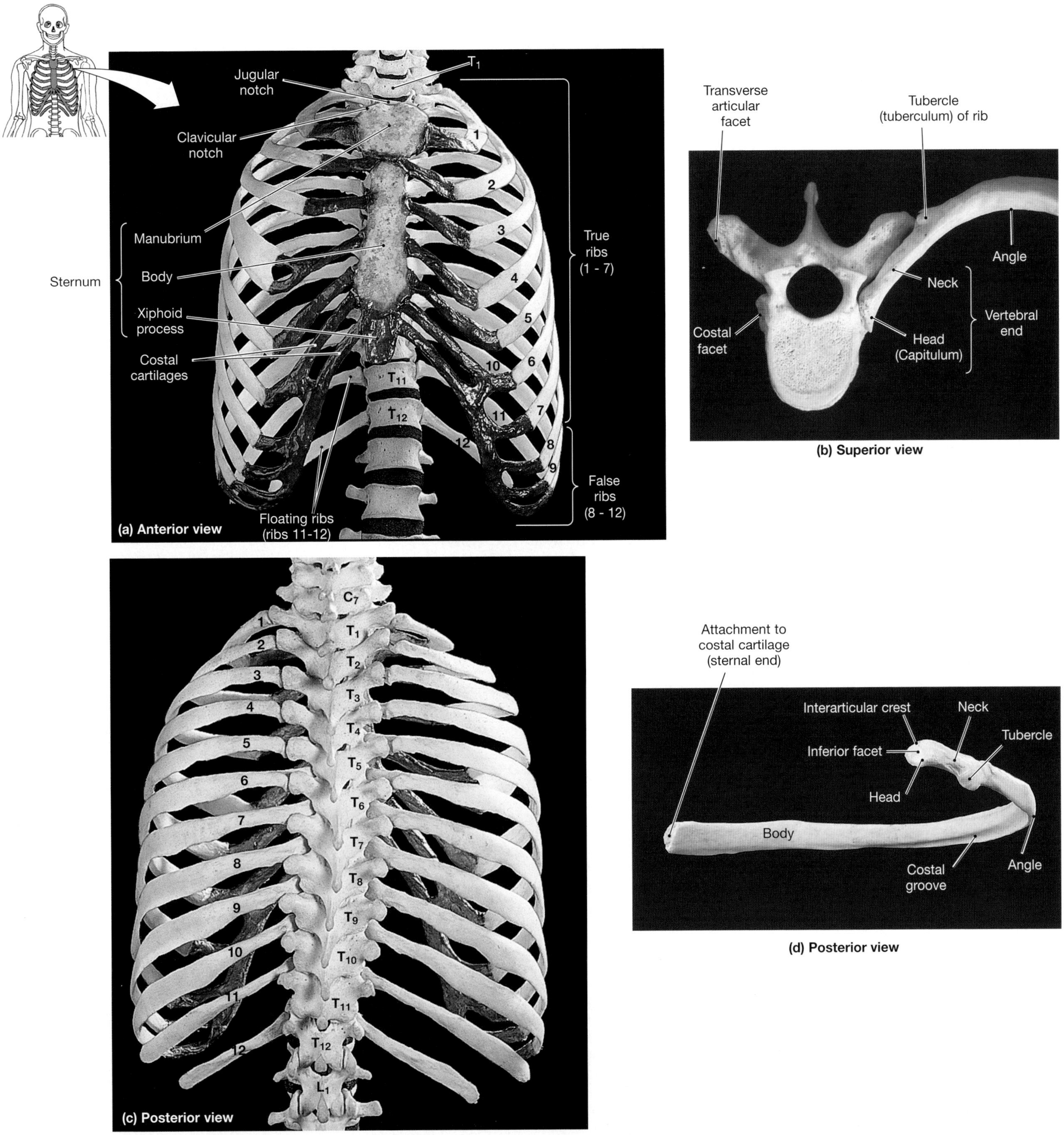

FIGURE 6.27 **The Thoracic Cage**

(**a**) Anterior view of the rib cage and sternum. (**b**) A superior view of the articulation between a thoracic vertebra and the vertebral end of a right rib. (**c**) Posterior view of the rib cage. (**d**) A posterior and medial view showing major anatomical landmarks on an islolated right rib (rib 10). *See MRI Scan 10c in the Companion Atlas.*

- The **xiphoid** (ZĪ-foyd) **process**, the smallest part of the sternum, is attached to the inferior surface of the body. The muscular *diaphragm* and the *rectus abdominis muscle* attach to the xiphoid process.

Ossification of the sternum begins at six to ten different centers, and fusion is not completed until at least age 25. Before age 25, the sternal body consists of four separate bones. Their boundaries can be detected as a series of transverse lines crossing the adult sternum. The xiphoid process is usually the last of the sternal components to undergo ossification and fusion. Its connection to the body of the sternum can be broken by an impact or strong pressure, creating a spear of bone that can severely damage the liver. To reduce the chances of that happening, strong emphasis is placed on the proper positioning of the hand during cardiopulmonary resuscitation (CPR) training.

✓ CONCEPT CHECK

- Joe suffered a hairline fracture at the base of the odontoid process. What bone is fractured, and where would you find it?
- Improper administration of CPR (cardiopulmonary resuscitation) could result in a fracture of what bone?
- What are the five vertebral regions? What are the identifying features of each region?
- List the spinal curves in order from superior to inferior.

RELATED CLINICAL TERMS

chest tube: A drain installed after thoracic surgery to permit removal of blood and pleural fluid. *The Thoracic Cage and Surgical Procedures p. 789*

craniostenosis (krā-nē-ō-sten-Ō-sis): Premature closure of one or more fontanels, which can lead to unusual distortions of the skull. *p. 161*

deviated nasal septum: A bent nasal septum that may slow or prevent sinus drainage. *p. 153*

hemothorax: Bleeding into the thoracic cavity. *p. 174*

kyphosis (kī-FŌ-sis): Abnormal exaggeration of the thoracic curvature that produces a "roundback" appearance. *p. 164*

lordosis (lōr-DŌ-sis): Abnormal lumbar curvature giving a "swayback" appearance. *p. 164*

microcephaly (mī-krō-SEF-a-lē): An undersized head resulting from genetic or developmental abnormalities. *p. 161*

pneumothorax (noo-mō-THŌR-aks): The entry of air into a pleural cavity. *p. 174*

scoliosis (skō-lē-Ō-sis): Abnormal lateral curvature of the spine. *p. 164*

sinusitis: Inflammation and congestion of the paranasal sinuses. *p. 153*

spina bifida (SPĪ-na BI-fi-da): A condition resulting from failure of the vertebral laminae to unite during development; it is often associated with developmental abnormalities of the brain and spinal cord. *p. 174*

thoracentesis (thō-ra-sen-TĒ-sis) or **thoracocentesis:** The penetration of the thoracic wall along the superior border of one of the ribs. *The Thoracic Cage and Surgical Procedures p. 789*

whiplash: An injury resulting from a sudden change in body position that can injure the cervical vertebrae. *p. 167*

STUDY OUTLINE & CHAPTER REVIEW

Introduction 134

1. The skeletal system consists of the axial skeleton and the appendicular skeleton. The **axial skeleton** can be subdivided into the **skull** and associated bones (the **auditory ossicles** and **hyoid bone**), the **vertebral column**, and the **thoracic cage** composed of the **ribs** and **sternum**. *(see Figure 6.1)*
2. The **appendicular skeleton** includes the **pectoral** and **pelvic girdles** that support and attach the upper and lower limbs to the trunk. *(see Figure 6.1)*

The Skull and Associated Bones 135

1. The **skull** consists of the *cranium* and the *bones of the face*. Skull bones protect the brain and guard entrances to the digestive and respiratory systems. Eight skull bones form the **cranium**, which encloses the **cranial cavity**, a division of the dorsal body cavity. The **facial bones** protect and support the entrances to the respiratory and digestive systems. *(see Figures 6.2 to 6.15 and Tables 6.1/6.2)*
2. Prominent superficial landmarks on the skull include the **lambdoid**, **sagittal**, **coronal**, **squamous**, and **frontonasal** sutures. **Sutures** are immovable joints that form boundaries between skull bones. *(see Figures 6.3a,b,c,d and Tables 6.1/6.2)*

Bones of the Cranium 142

3. For articulations of cranial bones with other cranial bones and/or facial bones, *see Table 6.2.*
4. The **occipital bone** forms part of the base of the skull. It surrounds the **foramen magnum** and forms part of the wall of the **jugular foramen**. *(see Figures 6.3a,c,e/6.6a,b)*
5. The **parietal bones** form part of the superior and lateral surfaces of the cranium. *(see Figures 6.3b,c/6.5/6.6c)*
6. The **frontal bone** forms the forehead and roof of the orbit. *(see Figures 6.3b,c,d/6.5a/6.7)*
7. The **temporal bone** forms part of the wall of the jugular foramen and houses the **carotid canal**. The thick **petrous part** of the temporal bone houses the tympanic cavity containing the **auditory ossicles**. The auditory ossicles transfer sound vibrations from the tympanic membrane to a fluid-filled chamber in the inner ear. *(see Figures 6.3c,e/6.8/18.10)*
8. The **sphenoid** contributes to the floor of the cranium. It is a bridge between the cranial and facial bones. **Optic nerves** pass through the **optic canal** in the sphenoid to reach the brain. **Pterygoid processes** form **plates** that serve as sites for attachment of muscles that move the mandible and soft palate. *(see Figures 6.3c,d,e/6.4/6.9)*
9. The **ethmoid** is an irregularly shaped bone that forms part of the orbital wall and the roof of the nasal cavity. The **cribriform plate** of the ethmoid contains perforations for olfactory nerves. The **perpendicular plate** forms part of the *nasal septum*. *(see Figures 6.3d/6.4/6.5/6.10)*
10. **Cranial fossae** are curving depressions in the cranial floor that closely follow the shape of the brain. The **anterior cranial fossa** is formed by the frontal bone, the ethmoid, and the **lesser wings** of the sphenoid. The **middle cranial fossa** is created by the sphenoid, temporal, and parietal bones. The **posterior cranial fossa** is primarily formed by the occipital bone, with contributions from the temporal and parietal bones. *(see Figure 6.11)*

Bones of the Face 150

11. For articulations of facial bones with other facial bones and/or cranial bones, *see Table 6.2.*
12. The left and right **maxillae**, or *maxillary bones,* are the largest facial bones and form the upper jaw. *(see Figures 6.3d/6.12)*
13. The **palatine bones** are small, L-shaped bones that form the posterior portions of the hard palate, and contribute to the floor of the orbit. *(see Figures 6.3e/6.13)*
14. The paired **nasal bones** articulate with the frontal bone at the midline of the face and extend to the superior border of the **external nares.** *(see Figures 6.3c,d/6.15/6.16)*
15. The **vomer** forms the inferior portion of the nasal septum. It is based on the floor of the nasal cavity and articulates with both the maxillae and the palatines along the midline. *(see Figures 6.3c,d/6.5/6.16)*
16. One **inferior nasal concha** is located on each side of the **nasal septum**, attached to the lateral wall of the nasal cavity. They increase the epithelial surface area and create turbulence in the inspired air. The superior and middle conchae of the ethmoid perform the same functions. *(see Figures 6.3d/6.16)*
17. The **temporal process** of the **zygomatic bone** articulates with **zygomatic process** of the temporal bone to form the **zygomatic arch** (*cheekbone*). *(see Figures 6.3c,d/6.16)*
18. The paired **lacrimal bones** are the smallest bones in the skull. They are situated in the medial portion of each orbit. Each lacrimal bone forms a **nasolacrimal groove** with the adjacent maxilla, and this groove leads to a **nasolacrimal canal** that delivers tears to the nasal cavity. *(see Figures 6.3c,d/6.16)*
19. The **mandible** is the entire lower jaw. It articulates with the temporal bone at the *temporomandibular joint* (TMJ). *(see Figures 6.3c,d/6.14)*

The Orbital and Nasal Complexes 152

20. Seven bones form the **orbital complex**, a bony recess that contains an eye: frontal, lacrimal, palatine and zygomatic bones and the ethmoid, sphenoid, and maxillae. *(see Figure 6.16)*
21. The **nasal complex** includes the bones and cartilage that enclose the nasal cavities and the **paranasal sinuses.** Paranasal sinuses are hollow airways that interconnect with the nasal passages. Large paranasal sinuses are present in the frontal bone and the sphenoid, ethmoid, and maxillae. *(see Figures 6. 5/6.15/6.16)*

The Hyoid Bone 155

22. The **hyoid bone,** suspended by **stylohyoid ligaments**, consists of a **body**, the **greater horns** and the **lesser horns**. The hyoid bone serves as a base for several muscles concerned with movements of the tongue and larynx. *(see Figure 6.17)*

The Skulls of Infants, Children, and Adults 161

1. Fibrous connections at **fontanels** permit the skulls of infants and children to continue growing. *(see Figure 6.18)*

The Vertebral Column 163

1. The adult **vertebral column** consists of a total of 26 vertebrae (24 individual *vertebrae,* the *sacrum,* and the *coccyx.*) There are *7 cervical vertebrae* (the first articulates with the occipital bone), *12 thoracic vertebrae* (which articulate with the ribs), and *5 lumbar vertebrae* (the fifth articulates with the *sacrum*). The sacrum and coccyx consist of fused vertebrae. *(see Figures 6.19 to 6.25)*

Spinal Curvature 164

2. The spinal column has four **spinal curves**: the **thoracic** and **sacral curvatures** are called **primary**, or **accommodation, curves**; the **lumbar** and **cervical curvatures** are known as **secondary**, or **compensation, curves.** *(see Figure 6.19)*

Vertebral Anatomy 165

3. A typical vertebra has a thick, supporting **body**, or *centrum*; a **vertebral arch** (**neural arch**) formed by walls (**pedicles**) and a roof (**lamina**) that provide a space for the spinal cord; and it articulates with other vertebrae at the **superior** and **inferior articular processes.** *(see Figure 6.21)*
4. Adjacent vertebrae are separated by **intervertebral discs**. Spaces between successive pedicles form the **intervertebral foramina** through which nerves pass to and from the spinal cord. *(see Figure 6.21)*

Vertebral Regions 166

5. **Cervical vertebrae** are distinguished by the shape of the vertebral body, the relative size of the vertebral foramen, the presence of **costal processes** with **transverse foramina**, and bifid **spinous processes.** *(see Figures 6.19/6.22/6.23 and Table 6.3)*
6. **Thoracic vertebrae** have distinctive heart-shaped bodies, long, slender spinous processes, and articulations for the ribs. *(see Figures 6.24/6.19)*
7. The **lumbar vertebrae** are the most massive and least mobile; they are subjected to the greatest strains. *(see Figures 6.19/6.25)*
8. The **sacrum** protects reproductive, digestive, and excretory organs. It has an **auricular surface** for articulation with the pelvic girdle. The sacrum articulates with the fused elements of the **coccyx.** *(see Figure 6.26)*

The Thoracic Cage 174

1. The skeleton of the **thoracic cage** consists of: the *thoracic vertebrae*, the *ribs*, and the *sternum*. The ribs and sternum form the **rib cage.** *(see Figure 6.27a,c)*

The Ribs 174

2. **Ribs** 1–7 are **true**, or *vertebrosternal,* **ribs**. Ribs 8–12 are called **false**, or *vertebrochondral*, **ribs**. The last two pairs of ribs are **floating ribs.** The *vertebral end* of a typical rib articulates with the vertebral column at the **head**, or **capitulum**. After a short **neck**, the **tubercle**, or *tuberculum*, projects dorsally. A bend, or **angle**, of the rib indicates the site where the tubular **body**, or **shaft**, begins curving toward the sternum. A prominent, inferior **costal groove** marks the path of nerves and blood vessels. *(see Figures 6.24/6.27)*

The Sternum 174

3. The **sternum** consists of a **manubrium**, a **body**, and a **xiphoid process.** *(see Figure 6.27a)*

LEVEL 1 REVIEWING FACTS AND TERMS

Match each numbered item with the most closely related lettered item. Use letters for answers in the spaces provided.

Column A	Column B
____ 1. suture	a. mandible
____ 2. foramen magnum	b. boundary between skull bones
____ 3. mastoid process	c. maxillae
____ 4. optic canal	d. cervical vertebrae
____ 5. crista galli	e. occipital bone
____ 6. condylar process	f. sternum
____ 7. transverse foramen	g. thoracic vertebrae
____ 8. costal facets	h. temporal bone
____ 9. manubrium	i. ethmoid
____10. upper jaw	j. sphenoid

11. Which of the following is not a function of the axial skeleton?
    (a) supports and protects organs in the dorsal and ventral body cavities
    (b) provides attachments for muscles that move head, neck, and trunk
    (c) provides an attachment for muscles involved in respiration
    (d) provides an attachment for muscle involved in limb movement
12. The bony portion of the nasal septum is formed by the
    (a) nasal bones
    (b) perpendicular plate of the ethmoid
    (c) perpendicular plate of the ethmoid and vomer
    (d) vomer and sphenoid
13. The lower jaw articulates with the temporal bone at the
    (a) mandibular fossa
    (b) mastoid process
    (c) superior clinoid process
    (d) cribriform plate
14. The hyoid bone
    (a) serves as a base of attachment for muscles that move the tongue
    (b) is part of the mandible
    (c) is located inferior to the larynx
    (d) articulates with the maxillae
15. An exaggerated lateral curvature of the vertebral column is termed
    (a) kyphosis
    (b) lordosis
    (c) scoliosis
    (d) gomphosis
16. The role of fontanels is to
    (a) allow for compression of the skull during childbirth
    (b) serve as ossification centers for the facial bones
    (c) serve as the final bony plates of the skull
    (d) lighten the weight of the skull bones
17. The sacrum
    (a) provides protection for reproductive, digestive, and excretory organs
    (b) bears the most weight in the vertebral column
    (c) articulates with the pectoral girdle
    (d) is composed of vertebrae that are completely fused by puberty
18. The side walls of the vertebral foramen are formed by the
    (a) body of the vertebra
    (b) spinous process
    (c) pedicles
    (d) laminae
19. The portion of the sternum that articulates with the clavicles is the
    (a) manubrium
    (b) body
    (c) xiphoid process
    (d) angle
20. A point of attachment for muscles that rotate or extend the head is the
    (a) styloid process
    (b) mastoid process
    (c) posterior clinoid process
    (d) articular tubercle

LEVEL 2 REVIEWING CONCEPTS

1. The primary spinal curves
   (a) are also called compensation curves
   (b) include the lumbar curvature
   (c) develop several months after birth
   (d) accommodate the thoracic and abdominopelvic viscera
2. As you proceed from the head inferiorly along the vertebral column
   (a) the vertebrae become smaller
   (b) the spinous process becomes larger
   (c) the bodies of the vertebrae become lighter
   (d) the size of the neural arch increases
3. What is the relationship between the pituitary gland and the sphenoid bone?
4. What properties of sutures make them unique to the skull?
5. Describe the relationship between the ligamentum nuchae and the axial skeleton with respect to holding the head in the upright position.
6. Discuss factors that can cause increased mucus production by the mucous membranes of the paranasal sinuses.
7. In addition to the functions served by all vertebrae, the first two cervical vertebrae perform some special functions. What unique roles do these vertebrae perform in spinal mobility?
8. What is the relationship between the temporal bone and the ear?
9. What is the purpose of the many small openings in the cribriform plate of the ethmoid bone?
10. Why are the largest vertebral bodies found in the lumbar region?

LEVEL 3 CRITICAL THINKING AND CLINICAL APPLICATIONS

1. Elise is in her last month of pregnancy and is suffering from lower back pains. Since she is carrying her excess weight in front of her, she wonders why her back hurts. What would you tell her?
2. Jeff gets into a brawl at a sports event and receives a broken nose. After the nose heals, he starts to have sinus headaches and discomfort in the area of his maxillae. What is the probable cause of Jeff's discomfort?
3. Some of the symptoms of the common cold or flu include an ache in all of the teeth in the maxillae, even though there is nothing wrong with them, as well as a heavy feeling in the front of the head. What anatomical response to the infection causes these unpleasant sensations?
4. The skull of a newborn baby often appears to be deformed, but after a short period of time, the skull assumes a normal shape. What happened during the birth process, and why did this change occur?

✓ ANSWERS TO CONCEPT CHECK QUESTIONS

p. 150 **1.** Each internal jugular vein passes through the jugular foramen, an opening between the occipital bone and the temporal bone. **2.** The sella turcica is located in the sphenoid, and it contains the pituitary gland. **3.** Nerve fibers to the olfactory bulb, which is involved with the sense of smell, pass through the cribriform plate from the nasal cavity. If the cribriform plate failed to form, these sensory nerves could not reach the olfactory bulbs and the sense of smell (olfaction) would be lost. **4.** Eight bones of the skull form the cranium, or "braincase": the frontal bone, parietal bones (2), occipital bone, temporal bones (2), sphenoid, and ethmoid.
p. 161 **1.** The 14 facial bones are: the maxillae (2), zygomatic bones (2), nasal bones (2), lacrimal bones (2), interior nasal conchae (2), palatine bones (2), vomer, and mandible. These bones protect and support the entrances to the digestive and respiratory tracts, and provide extensive areas for skeletal muscle attachment. **2.** The paranasal sinuses function to make some of the skull bones lighter, to produce mucus, and to resonate during sound production. **3.** The orbital complex consists of portions of 7 bones: palatine, zygomatic, frontal, and lacrimal bones and the maxillae, sphenoid, and ethmoid. **4.** Fontanels are the relatively soft, flexible, fibrous region between flat bones in the developing skull.
p. 176 **1.** The odontoid process, or dens, is found on the second cervical vertebra, or axis, which is located in the neck. **2.** Improper compression of the chest during CPR could result in a fracture of the sternum, especially at the xiphoid process, or the ribs. **3.** The vertebral column is divided into cervical, thoracic, lumbar, sacral, and coccygeal regions. Distinguishing features are as follows: cervical—triangular foramen, bifid spinous process, transverse foramina; thoracic—round foramen, heart-shaped body, transverse facets, costal facets; lumbar—triangular foramen, oval-shaped, large robust body; sacral—5 fused vertebrae; coccygeal: 3–5 small fused vertebrae. **4.** The spinal curves from superior to inferior are: (1) cervical curve, (2) thoracic curve, (3) lumbar curve, and (4) sacral curve.

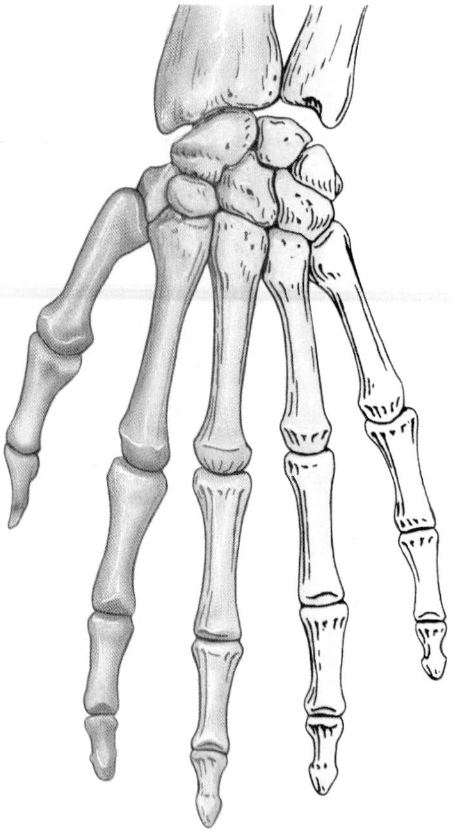

7

THE SKELETAL SYSTEM
Appendicular Division

If you make a list of the things you've done today, you will see that your appendicular skeleton plays a major role in your life. Standing, walking, writing, eating, dressing, shaking hands, and turning the pages of a book—the list goes on and on. Your axial skeleton protects and supports internal organs, and it participates in vital functions, such as respiration. But it is your appendicular skeleton that gives you control over your environment, changes your position in space, and provides mobility.

The **appendicular skeleton** includes the bones of the upper and lower limbs and the supporting elements, called *girdles*, that connect them to the trunk (Figure 7.1●). This chapter describes the bones of the appendicular skeleton. As in Chapter 6, the descriptions emphasize surface features that have functional importance and highlight the interactions among the skeletal system and other systems. For example, many of the anatomical features noted in this chapter are attachment sites for skeletal

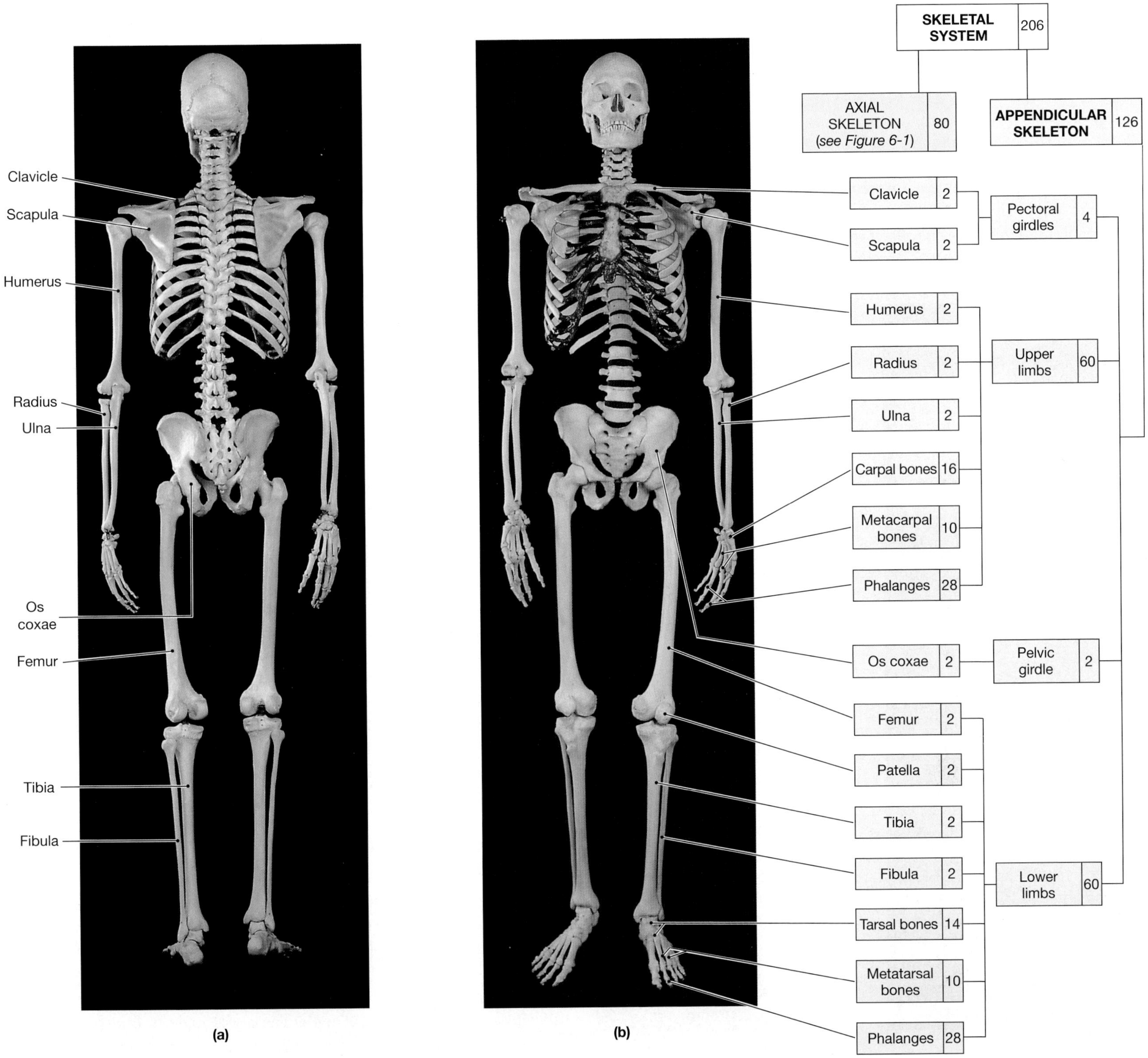

FIGURE 7.1 **THE APPENDICULAR SKELETON**

A flowchart showing the relationship of the components of the appendicular skeleton: pectoral and pelvic girdles, and upper and lower limbs. (**a**) Posterior view of the skeleton. (**b**) Anterior view of the skeleton, highlighting the appendicular components. The numbers in the boxes indicate the total number of bones of that type or category in the adult skeleton.

muscles or openings for nerves and blood vessels that supply the bones or other organs of the body.

There are direct anatomical connections between the skeletal and muscular systems. As noted in Chapter 5, the connective tissue of the deep fascia that surrounds a skeletal muscle is continuous with that of its tendon, which continues into the periosteum and becomes part of the bone matrix at its attachment site. p. 117 Muscles and bones are also physiologically linked, because muscle contractions can occur only when the extracellular concentration of calcium remains within relatively narrow limits. The skeleton contains most of the body's calcium, and these reserves are vital to calcium homeostasis.

The Pectoral Girdle And Upper Limb

[FIGURE 7.2]

Each arm articulates with the trunk at the **pectoral girdle**. The pectoral girdle consists of the S-shaped *clavicle (collarbone)* and a broad, flat *scapula (shoulder blade),* as seen in Figure 7.2●. The clavicle articulates with the manubrium of the sternum, and this is the *only* direct connection between the pectoral girdle and the axial skeleton. Skeletal muscles support and position the scapula, which has no direct bony or ligamentous connections to the thoracic cage. Each upper limb consists of the *brachium* (arm), the *antebrachium* (forearm), the wrist, and the hand. The skeleton of the upper limb consists of the *humerus* of the arm, the *ulna* and *radius* of the forearm, the *carpal bones* of the wrists, and the *metacarpal bones* and *phalanges* of the hand.

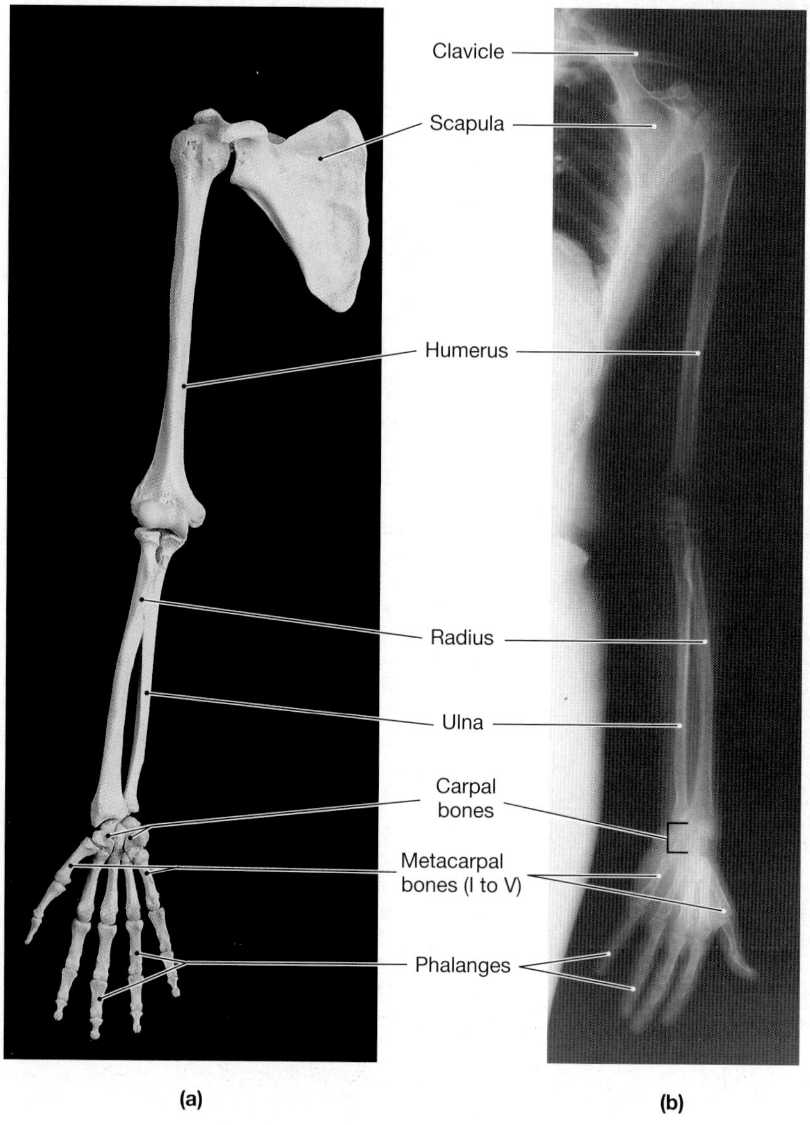

FIGURE 7.2 **The Pectoral Girdle and Upper Limb**

Each upper limb articulates with the axial skeleton at the trunk through the pectoral girdle. **(a)** Right upper limb, anterior view. **(b)** X-ray of right pectoral girdle and upper limb, posterior view.

The Pectoral Girdle

Movements of the clavicle and scapula position the shoulder joint and provide a base for arm movement. Once the shoulder joint is in position, muscles that originate on the pectoral girdle help to move the upper limb. The surfaces of the scapula and clavicle are therefore extremely important as sites for muscle attachment. Where major muscles attach, they leave their marks, creating bony ridges and flanges. Other bone markings, such as grooves or foramina, indicate the position of nerves or blood vessels that control the muscles and nourish the muscles and bones.

THE CLAVICLE [FIGURES 7.3/7.4]

The **clavicle** (KLAV-i-kl) (Figure 7.3●) connects the pectoral girdle and the axial skeleton. Each clavicle originates at the craniolateral border of the manubrium of the sternum, lateral to the jugular notch (see Figures 6.27a, p. 173 and Figure 7.4●). From the roughly pyramidal **sternal end**, the clavicle curves in an S-shape laterally and dorsally until it articulates with the acromion of the scapula. The **acromial end** is broader and flatter than the sternal end.

The smooth superior surface of the clavicle lies just deep to the skin; the rough inferior surface of the acromial end is marked by prominent lines and tubercles that indicate the attachment sites for muscles and ligaments. The **conoid tubercle** is on the inferior surface at the acromial end, and the **costal tuberosity** is at the sternal end. These are attachment sites for ligaments of the shoulder.

You can explore the interaction between scapulae and clavicles. With your fingers in the *jugular notch*, locate the clavicle to either side. p. 175 When you move your shoulders you can feel the clavicles change their positions. Because the clavicles are so close to the skin, you can trace one laterally until it articulates with the scapula. Shoulder movements are limited by the position of the clavicle at the *sternoclavicular joint*, as shown in Figure 7.4●. The structure of this joint will be described in Chapter 8. Fractures of the medial portion of the clavicle are common because a fall on the palm of the hand of an outstretched arm produces compressive forces that are conducted to the clavicle and its articulation with the manubrium. Fortunately, these fractures usually heal rapidly without a cast.

THE SCAPULA [FIGURES 7.4/7.5]

The **body** of the **scapula** (SCAP-ū-lah) forms a broad triangle with many surface markings reflecting the attachment of muscles, tendons, and ligaments (Figure 7.5a,d●). The three sides of the scapular triangle are the **superior border**; the **medial**, or *vertebral*, **border**; and the **lateral**, or *axillary*, **border** (*axilla*, armpit). Muscles that position the scapula attach along these edges. The corners of the scapular triangle are called the **superior angle**, the **inferior angle**, and the **lateral angle**. The lateral angle, or *head* of the scapula, forms a broad process that supports the cup-shaped **glenoid cavity**, or *glenoid fossa*. At the glenoid cavity, the scapula articulates with the proximal end of the *humerus*, the bone of the arm. This articulation is the **glenohumeral joint**, or **shoulder joint**. The lateral angle is separated from the body of the scapula by the rounded **neck**. The relatively smooth, concave **subscapular fossa** forms most of the anterior surface of the scapula.

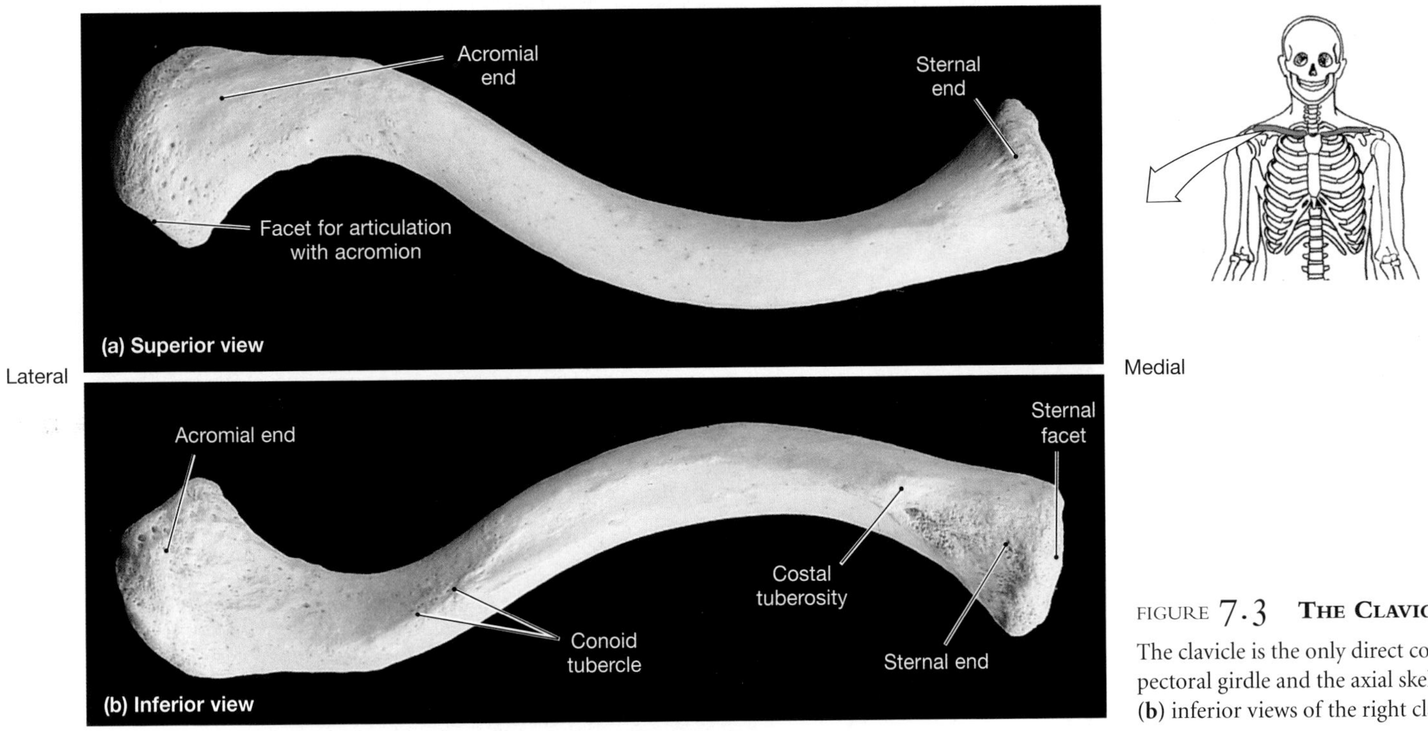

FIGURE 7.3 **THE CLAVICLE**

The clavicle is the only direct connection between the pectoral girdle and the axial skeleton. (**a**) Superior and (**b**) inferior views of the right clavicle.

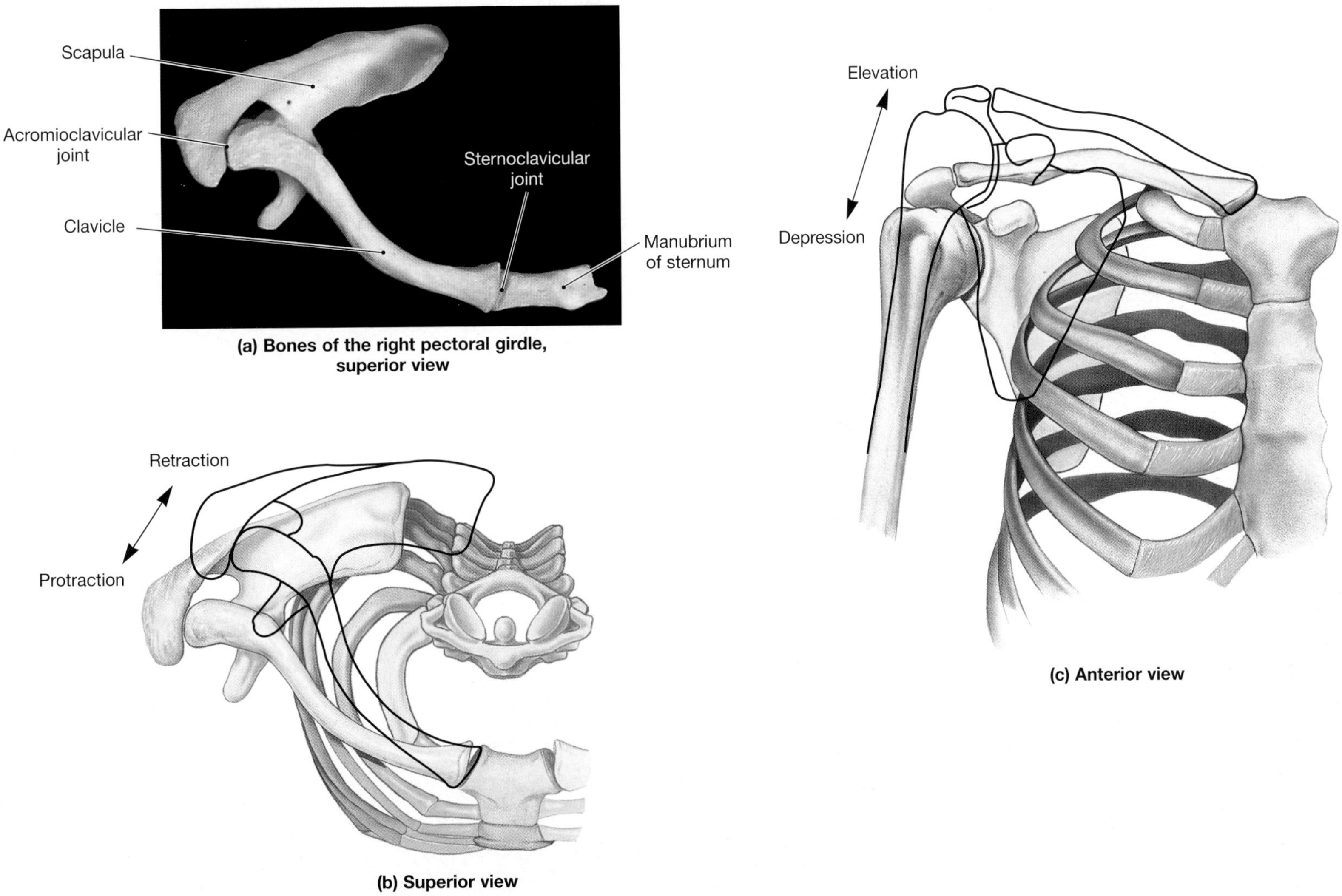

FIGURE 7.4 **MOBILITY OF THE PECTORAL GIRDLE**

Diagrammatic representation of normal movements of the pectoral girdle. (**a**) Bones of the right pectoral girdle. (**b**) Alterations in the position of the right shoulder during *protraction* (movement anteriorly) and *retraction* (movement posteriorly). (**c**) Alterations in the position of the right shoulder that occur during *elevation* (superior movement) and *depression* (inferior movement). In each instance, note that the clavicle is responsible for limiting the range of motion. (*see Figure 8.5d,f*)

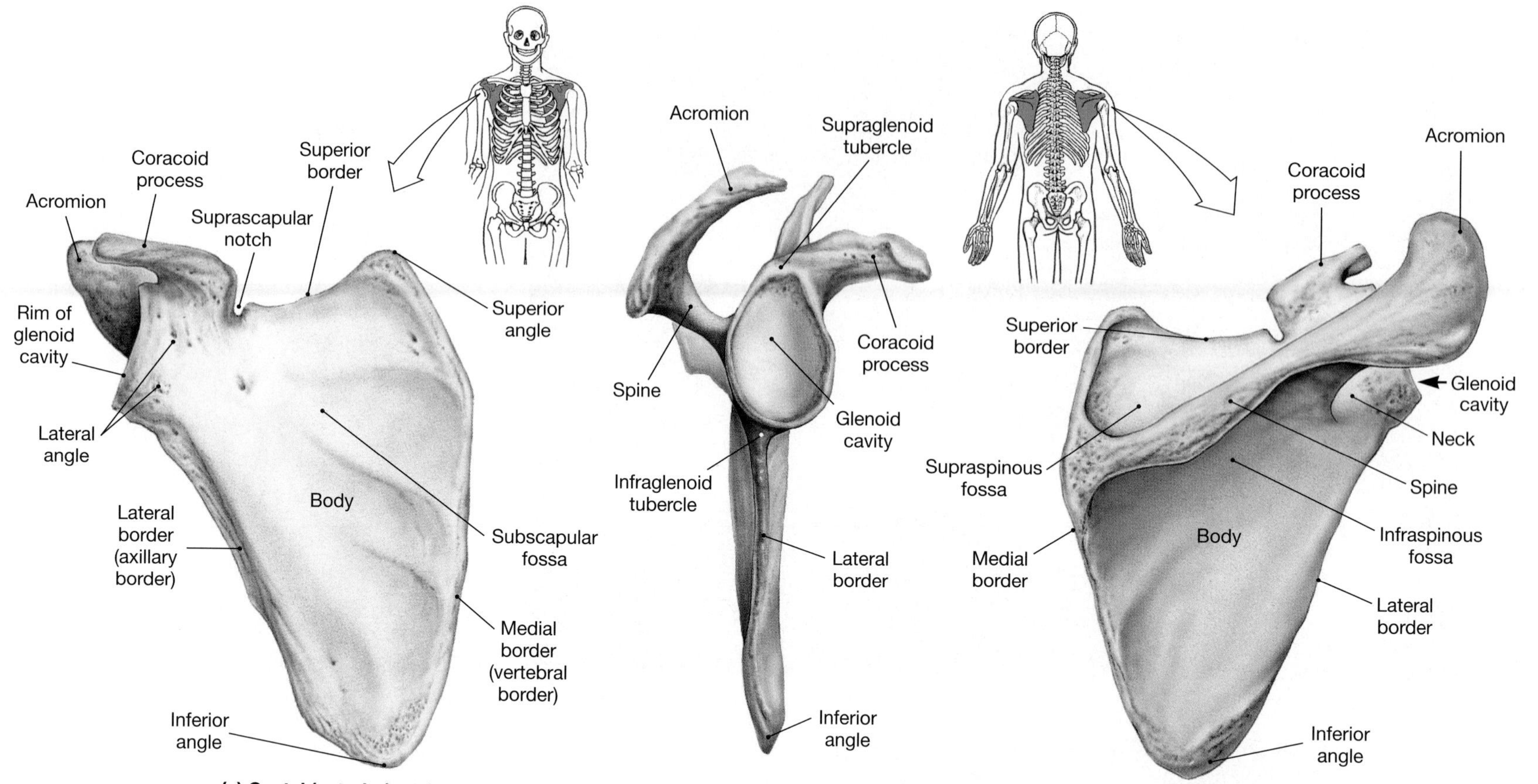

(a) Costal (anterior) surface **(b) Lateral view** **(c) Posterior surface**

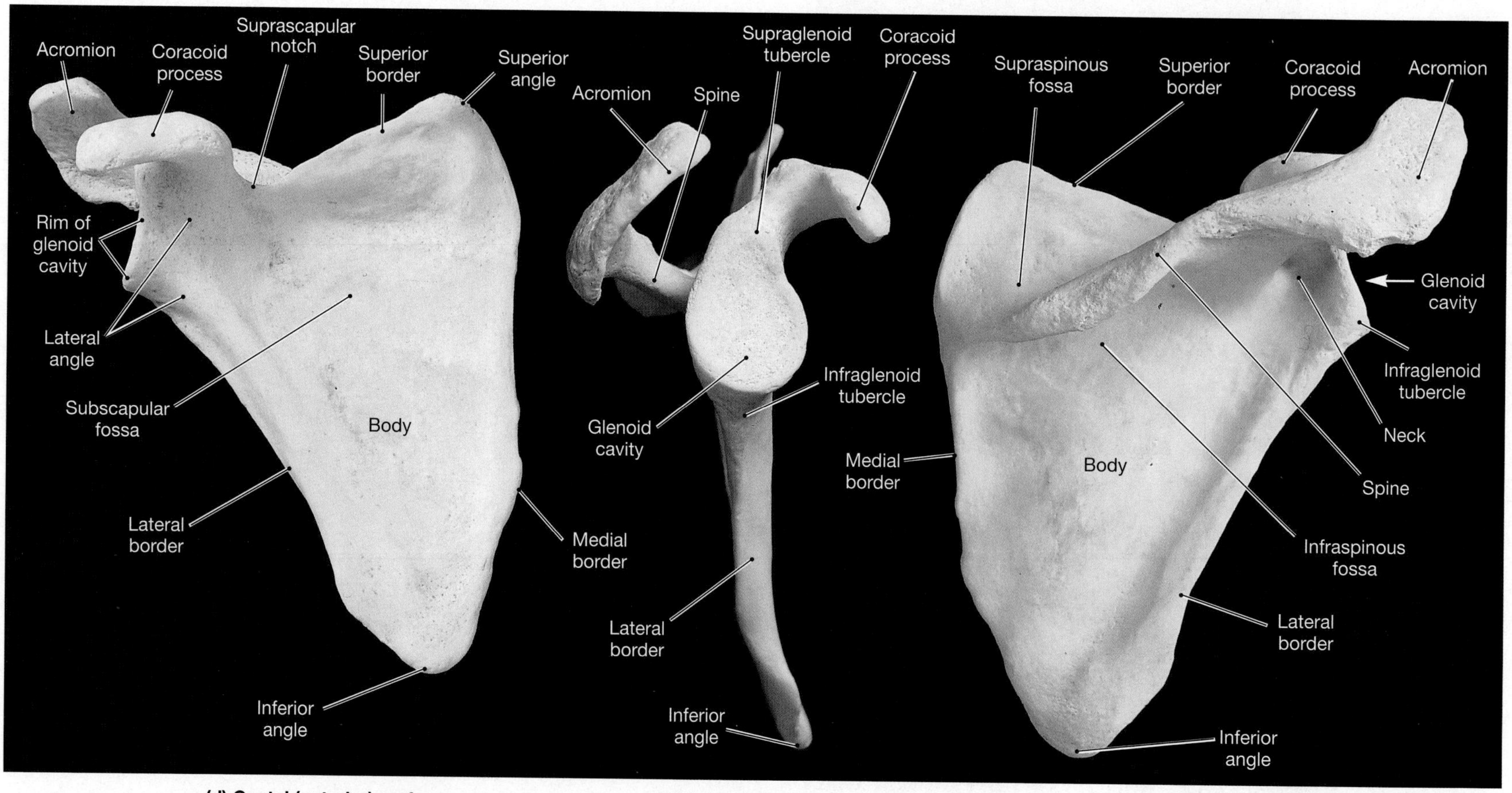

(d) Costal (anterior) surface **(e) Lateral view** **(f) Posterior surface**

FIGURE 7.5 **The Scapula**

This part of the pectoral girdle articulates with the upper limb. (**a,d**) Anterior, (**b,e**) lateral, and (**c,f**) posterior views of the right scapula.

Two large scapular processes extend over the superior margin of the glenoid cavity, superior to the head of the humerus. The smaller, anterior projection is the **coracoid** (KOR-a-koyd; *korakodes*, like a crow's beak) **process**. This process serves as an attachment site for the *biceps brachii muscle*, a prominent muscle on the anterior surface of the arm. The **suprascapular notch** is an indentation lateral to the base of the coracoid process. The **acromion** (a-KRŌ-mē-on; *akron*, tip + *omos*, shoulder), the larger, posterior process, serves as an attachment point for part of the *trapezius muscle* of the back. If you run your fingers along the superior surface of the shoulder joint, you will feel this process. The acromion articulates with the clavicle at the **acromioclavicular joint** (Figure 7.4a●). Both the acromion and the coracoid process are attached to ligaments and tendons associated with the shoulder joint, which will be described further in Chapter 8.

Most of the surface markings of the scapula represent the attachment sites for muscles that position the shoulder and arm. For example, the **supraglenoid tubercle** marks the origin of a second portion of the *biceps brachii muscle*. The **infraglenoid tubercle** marks the origin of one portion of the *triceps brachii muscle*, an equally prominent muscle on the posterior surface of the arm. The acromion is continuous with the **scapular spine**. This ridge crosses the scapular body before ending at the medial border. The scapular spine divides the convex dorsal surface of the body into two regions. The area superior to the spine constitutes the **supraspinous fossa** (*supra*, above); the region inferior to the spine is the **infraspinous fossa** (*infra*, beneath). The spine is an attachment site for two large muscles (the *supraspinatus muscle* and the *infraspinatus muscle*), and the entire posterior surface is marked by small ridges and lines where smaller muscles attach to the scapula.

The Upper Limb [FIGURE 7.2]

The bones of each upper limb consist of a humerus, an ulna and radius, the carpal bones of the wrist, and the metacarpal bones and phalanges of the hand (Figure 7.2●, p. 182).

THE HUMERUS [FIGURE 7.6]

The **humerus** is the proximal bone of the upper limb. The superior, medial portion of the proximal epiphysis is smooth and round. This is the **head** of the humerus, which articulates with the scapula at the glenoid cavity. The lateral edge of the epiphysis bears a large projection, the **greater tubercle** of the humerus (Figure 7.6a,b●). The greater tubercle forms the lateral margin of the shoulder; you can find it by feeling for a bump situated a few centimeters anterior and inferior to the tip of the acromion. Three muscles that originate on the scapula (the *supraspinatus*, *infraspinatus*, and *teres minor muscles*) are attached to the humerus at the greater tubercle. The **lesser tubercle** lies on the anterior and medial surface of the epiphysis. The lesser tubercle marks the insertion point of another scapular muscle, the *subscapularis*. The lesser tubercle and greater tubercle are separated by the **intertubercular groove**, or *intertubercular sulcus*. A tendon of the *biceps brachii muscle* runs along this groove from its origin at the supraglenoid tubercle of the scapula. The **anatomical neck**, a constriction inferior to the head of the humerus, marks the distal limit of the articular capsule of the shoulder joint. It lies between the tubercles and the smooth articular surface of the head. Distal to the tubercles, the narrow **surgical neck** corresponds to the metaphysis of the growing bone. This name reflects the fact that fractures often occur at this site.

The proximal **shaft**, or *body*, of the humerus is round in section. The elevated **deltoid tuberosity** runs along the lateral border of the shaft, extending more than halfway down its length. The deltoid tuberosity is named after the *deltoid muscle* that attaches to it. On the anterior surface of the shaft, the intertubercular groove continues alongside the deltoid tuberosity.

The articular **condyle** dominates the distal, inferior surface of the humerus (Figure 7.6a,c●). A low ridge crosses the condyle, dividing it into two distinct articular regions. The **trochlea** (*trochlea*, pulley) is the spool-shaped medial portion that articulates with the *ulna*, the medial bone of the forearm. The trochlea extends from the base of the **coronoid fossa** (KOR-o-noyd; *corona*, crown) on the anterior surface to the **olecranon fossa** on the posterior surface (Figure 7.6a,d●). These depressions accept projections from the surface of the ulna as the elbow approaches full *flexion* (elbow bent) or full *extension* (elbow straight). The rounded **capitulum** forms the lateral surface of the condyle. The capitulum articulates with the head of the *radius*, the smaller bone of the forearm. A shallow **radial fossa** superior to the capitulum accommodates a small part of the radial head as the forearm approaches the humerus.

On the posterior surface (Figure 7.6d●), the **radial groove** runs along the posterior margin of the deltoid tuberosity. This depression marks the path of the *radial nerve*, a large nerve that provides sensory information from the back of the hand and motor control over large muscles that straighten the elbow. The radial groove ends at the inferior margin of the deltoid tuberosity, where the nerve turns toward the anterior surface of the arm. Near the distal end of the humerus, the shaft expands to either side, forming a broad triangle. *Epicondyles* are processes that develop proximal to an articulation and provide additional surface area for muscle attachment. The **medial** and **lateral epicondyles** project to either side of the distal humerus at the elbow joint (Figure 7.6c,d●). The *ulnar nerve* crosses the posterior surface of the medial epicondyle. Bumping the humeral side of the elbow joint can strike this nerve and produce a temporary numbness and paralysis of muscles on the anterior surface of the forearm. It causes an odd sensation, so this area is sometimes called the *funny bone*.

THE ULNA [FIGURES 7.2/7.7]

The **ulna** and **radius** are parallel bones that support the forearm (Figure 7.2●). In the anatomical position, the ulna lies medial to the radius (Figure 7.7a●).The **olecranon** (ō-LEK-ra-non), or *olecranon process*, of the ulna forms the point of the elbow (Figure 7.7b●). This process is the superior and posterior portion of the proximal epiphysis. On its anterior surface, the **trochlear notch** (or *semilunar notch*) interlocks with the trochlea of the humerus (Figure 7.7c–e●). The olecranon forms the superior lip of the trochlear notch, and the **coronoid process** forms its inferior lip. When the elbow is straightened (*extension*), the olecranon projects into the olecranon fossa on the posterior surface of the humerus. When the elbow is bent (*flexion*), the coronoid process projects into the coronoid fossa on the anterior humeral surface. Lateral to the coronoid process, a smooth **radial notch** (Figure 7.7d,e●) accommodates the head of the radius at the *proximal radioulnar joint*.

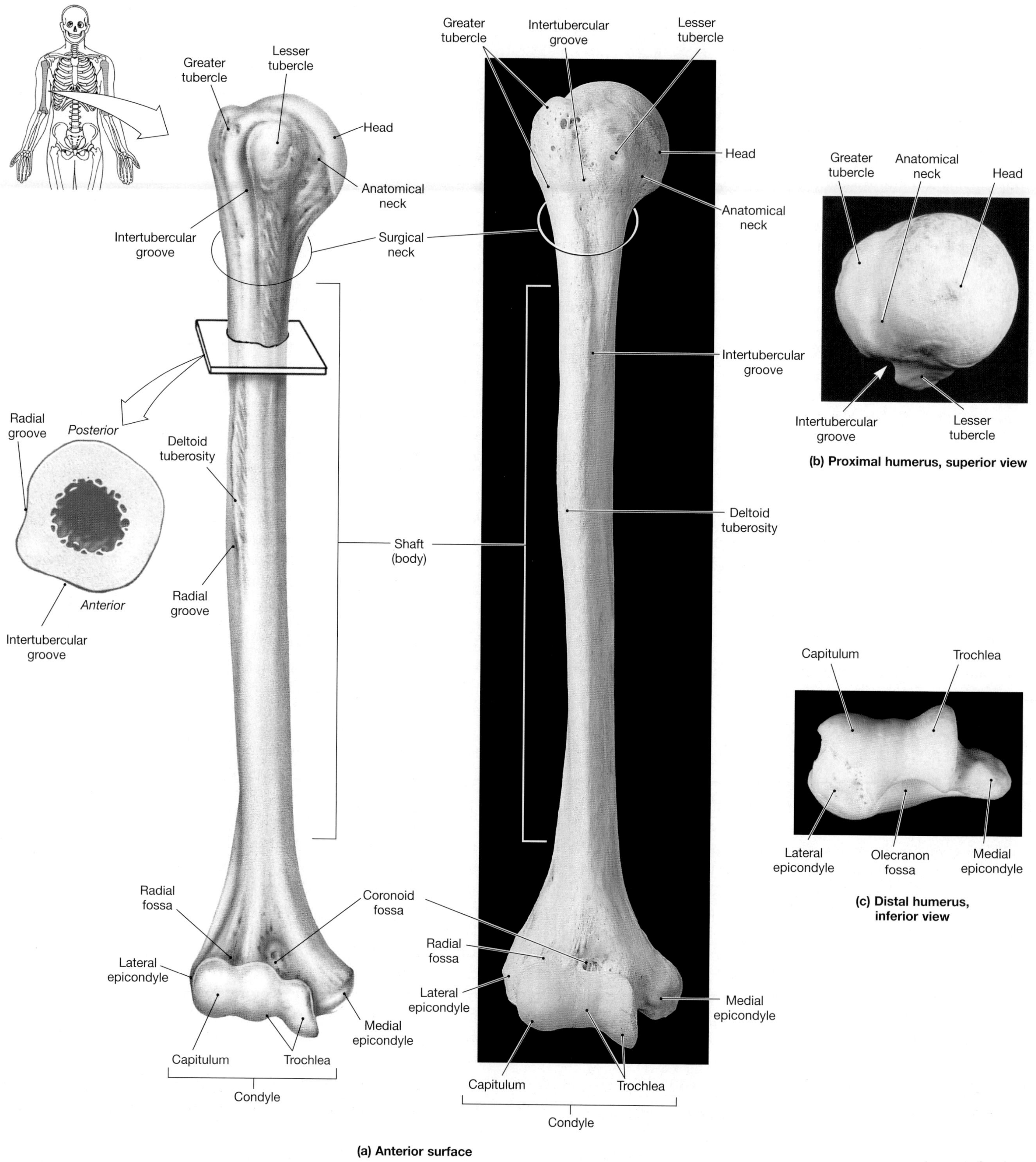

FIGURE 7.6 **The Humerus**

(**a**) Anterior views. (**b**) Superior view of the head of the humerus. (**c**) Inferior view of the distal humerus. (**d**) Posterior views.

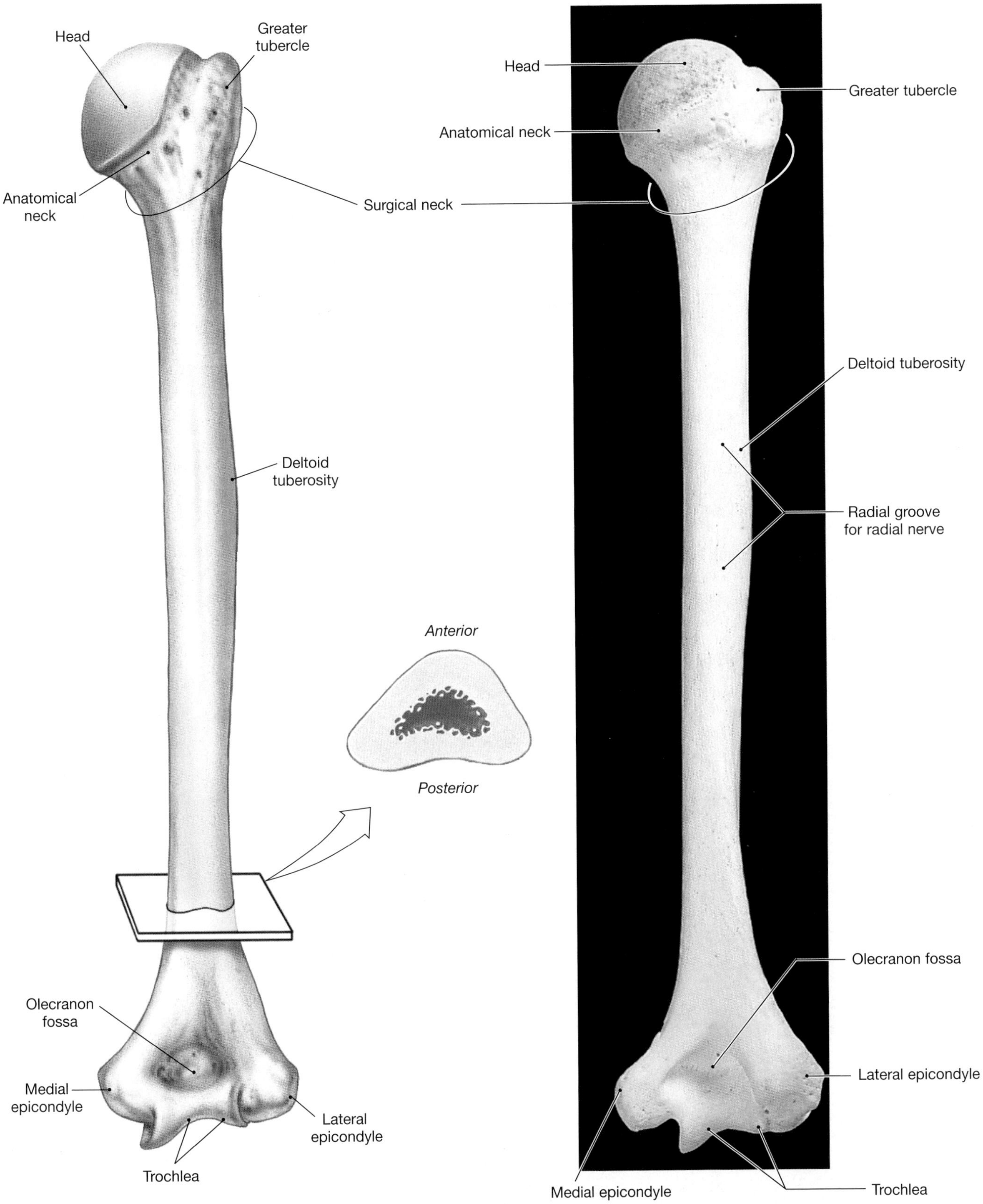

(d) Posterior surface

FIGURE 7.6 *(continued)*

The shaft of the ulna is roughly triangular in cross section, with the smooth medial surface at the base of the triangle and the lateral margin at the apex. A fibrous sheet, the **antebrachial interosseous membrane**, connects the lateral margin of the ulna to the medial margin of the radius and provides additional surface area for muscle attachment (Figure 7.7a,d●). Distally, the ulnar shaft narrows before ending at a disc-shaped **ulnar head** whose posterior margin supports a short **styloid process** (styloid, long and pointed). A triangular articular cartilage attaches to the styloid process, isolating the ulnar head from the bones of the wrist. The *distal radioulnar joint* lies near the lateral border of the ulnar head.

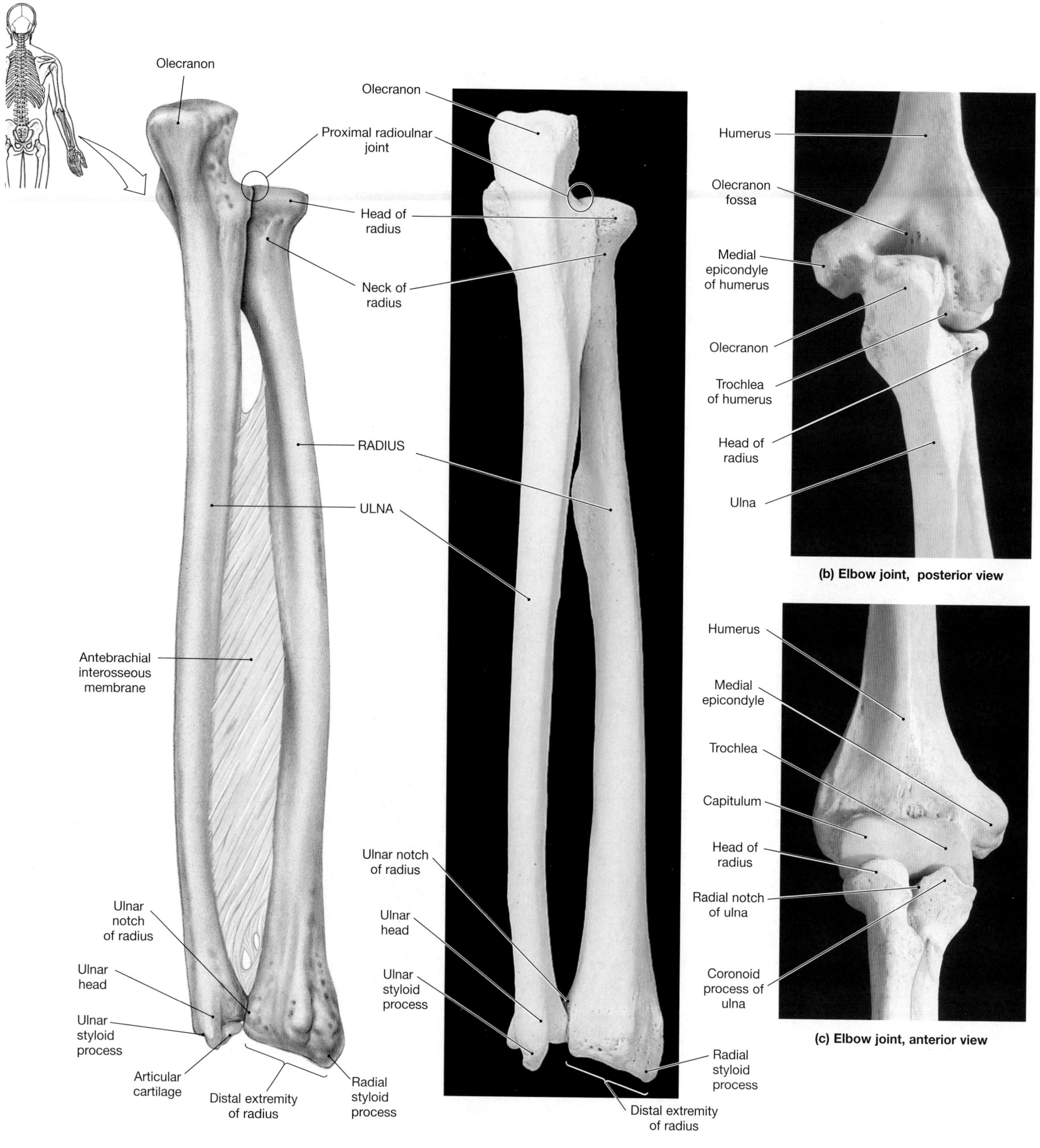

(a) Posterior view

(b) Elbow joint, posterior view

(c) Elbow joint, anterior view

FIGURE 7.7 **THE RADIUS AND ULNA**

The radius and ulna are the bones of the forearm. (**a**) Posterior view of the right radius and ulna. (**b**) A posterior view of elbow joint, showing the interlocking of the participating bones. (**c**) An anterior view of the elbow joint. (**d**) Anterior view of the radius and ulna. (**e**) A lateral view of the proximal end of the ulna.

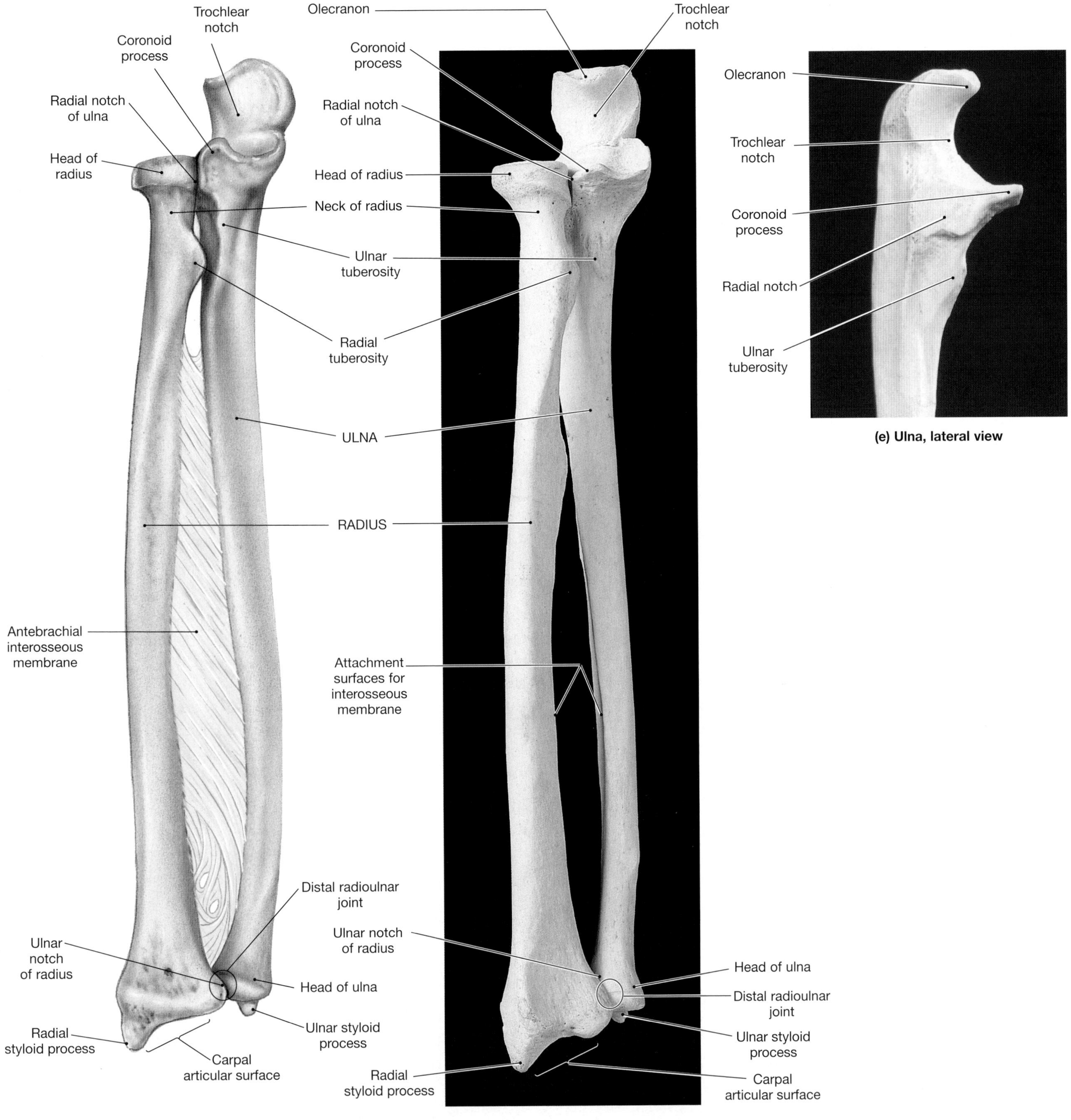

(e) Ulna, lateral view

(d) Anterior view

FIGURE 7.7 *(continued)*

The *elbow joint* is a stable, two-part joint that functions like a hinge (Figure 7.7b,c●). Much of the stability comes from the interlocking of the trochlea of the humerus with the trochlear notch of the ulna; this is the *humeroulnar joint*. The other portion of the elbow joint consists of the *humeroradial joint* formed by the capitulum of the humerus and the flat superior surface of the head of the radius. We will examine the structure of the elbow joint in Chapter 8.

THE RADIUS [FIGURE 7.7]

The radius is the lateral bone of the forearm (Figure 7.7●). The disc-shaped **head** of the radius, or *radial head*, articulates with the capitulum of the humerus. A narrow *neck* extends from the radial head to a prominent **radial tuberosity** that marks the attachment site of the *biceps brachii muscle*. This muscle bends (flexes) the elbow, swinging the forearm toward the arm. The shaft of the radius curves along its length, and the **distal extremity** is considerably larger than the distal portion of the ulna. Because the articular cartilage and an articulating disc separate the ulna from the wrist, only the distal extremity of the radius participates in the wrist joint. The **styloid process** on the lateral surface of the distal extremity helps stabilize the wrist.

The medial surface of the distal extremity articulates with the ulnar head at the **ulnar notch**, forming the distal radioulnar joint. The proximal radioulnar joint permits rotation of the radial head; when this movement occurs, the ulnar notch rolls across the rounded surface of the ulnar head. This movement is called **pronation**; the reverse movement, which returns the radius to the anatomical position, is called **supination**.

THE CARPAL BONES [FIGURE 7.8a]

The wrist, or *carpus*, is formed by the eight **carpal bones**. The bones form two rows, with four **proximal carpal bones** and four **distal carpal bones**. The proximal carpal bones are the *scaphoid bone*, the *lunate bone*, the *triquetrum*, and the *pisiform* (PĪS-i-form) *bone*. The distal carpal bones are the *trapezium*, the *trapezoid bone*, the *capitate bone*, and the *hamate bone* (Figure 7.8●). The carpal bones are linked with one another at joints that permit limited sliding and twisting movements. Ligaments interconnect the carpal bones and help stabilize the wrist.

The Proximal Carpal Bones

- The **scaphoid bone** is the proximal carpal bone located on the lateral border of the wrist adjacent to the styloid process of the radius.
- The comma-shaped **lunate** (*luna*, moon) **bone** lies medial to the scaphoid bone. Like the scaphoid bone, the lunate bone articulates with the radius.
- The **triquetrum** (*triangular bone*) is medial to the lunate bone. It has the shape of a small pyramid. The triquetrum articulates with the cartilage that separates the ulnar head from the wrist.
- The small, pea-shaped **pisiform bone** lies anterior to the triquetrum and extends farther medially than any other carpal bone in the proximal or distal rows.

The Distal Carpal Bones

- The **trapezium** is the lateral bone of the distal row. It forms a proximal articulation with the scaphoid bone.
- The wedge-shaped **trapezoid bone** lies medial to the trapezium; it is the smallest distal carpal bone. Like the trapezium, it has a proximal articulation with the scaphoid bone.
- The **capitate bone** is the largest carpal bone. It sits between the trapezoid and the hamate bone.
- The **hamate** (*hamatum*, hooked) **bone** is a hook-shaped bone that is the medial distal carpal bone.

A phrase to help you remember the names of the carpal bones in the order given is: "**S**am **l**ikes **t**o **p**ush **t**he **t**oy **c**ar **h**ard." The first letter of each word is the first letter of the bone; the first four are proximal, the last four distal.

THE HAND [FIGURE 7.8b,c]

Five **metacarpal** (met-a-KAR-pal) **bones** articulate with the distal carpal bones and support the palm of the hand (Figure 7.8b,c●). Roman numerals I–V are used to identify the metacarpal bones, beginning with the lateral metacarpal bone that articulates with the trapezium. Each metacarpal bone has a wide, concave, proximal *base*, a small *body*, and a distal *head*. Distally, the metacarpal bones articulate with the finger bones, or **phalanges** (fa-LAN-jēz; singular, *phalanx*). There are 14 phalangeal bones in each hand. The thumb, or **pollex** (POL-eks), has two phalanges (proximal and distal), and each of the other fingers has three phalanges (proximal, middle, and distal).

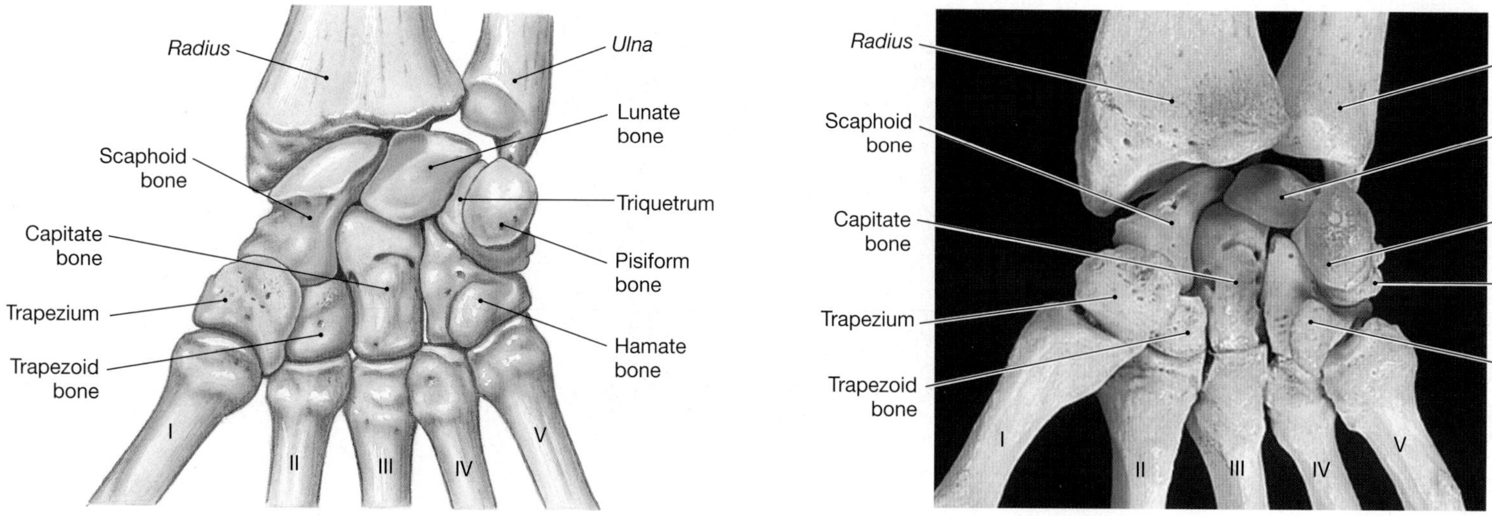

(a) Right wrist, anterior (palmar) view

FIGURE 7.8 **The Bones of the Wrist and Hand**

Carpal bones form the wrist; metacarpal bones and phalanges form the hand. (**a**) Anterior (palmar) view of the bones of the right wrist. (**b**) Anterior (palmar) view of the bones of the right wrist and hand. (**c**) Posterior (dorsal) view of the bones of the right wrist and hand.

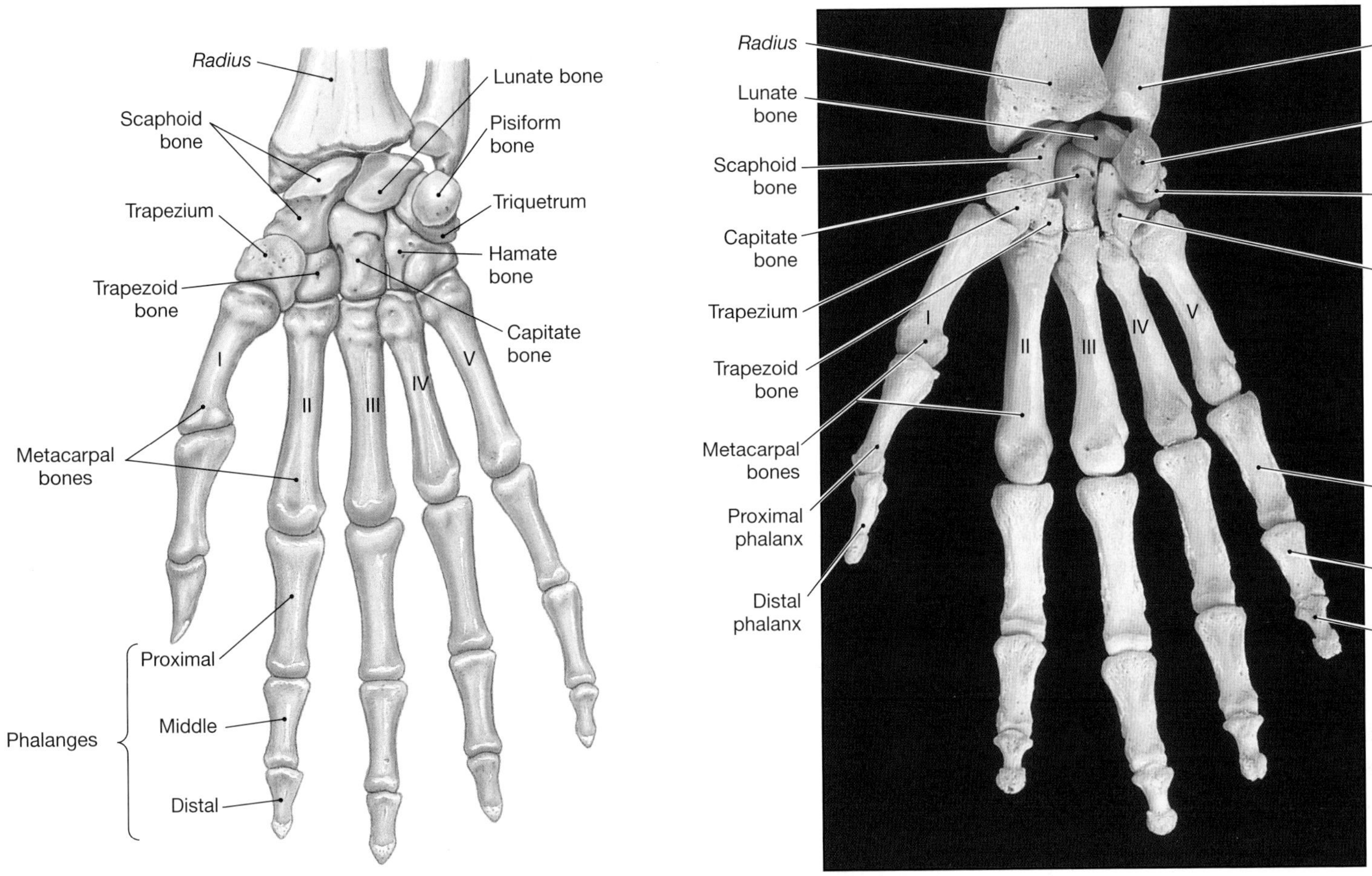

(b) Right hand, anterior (palmar) view

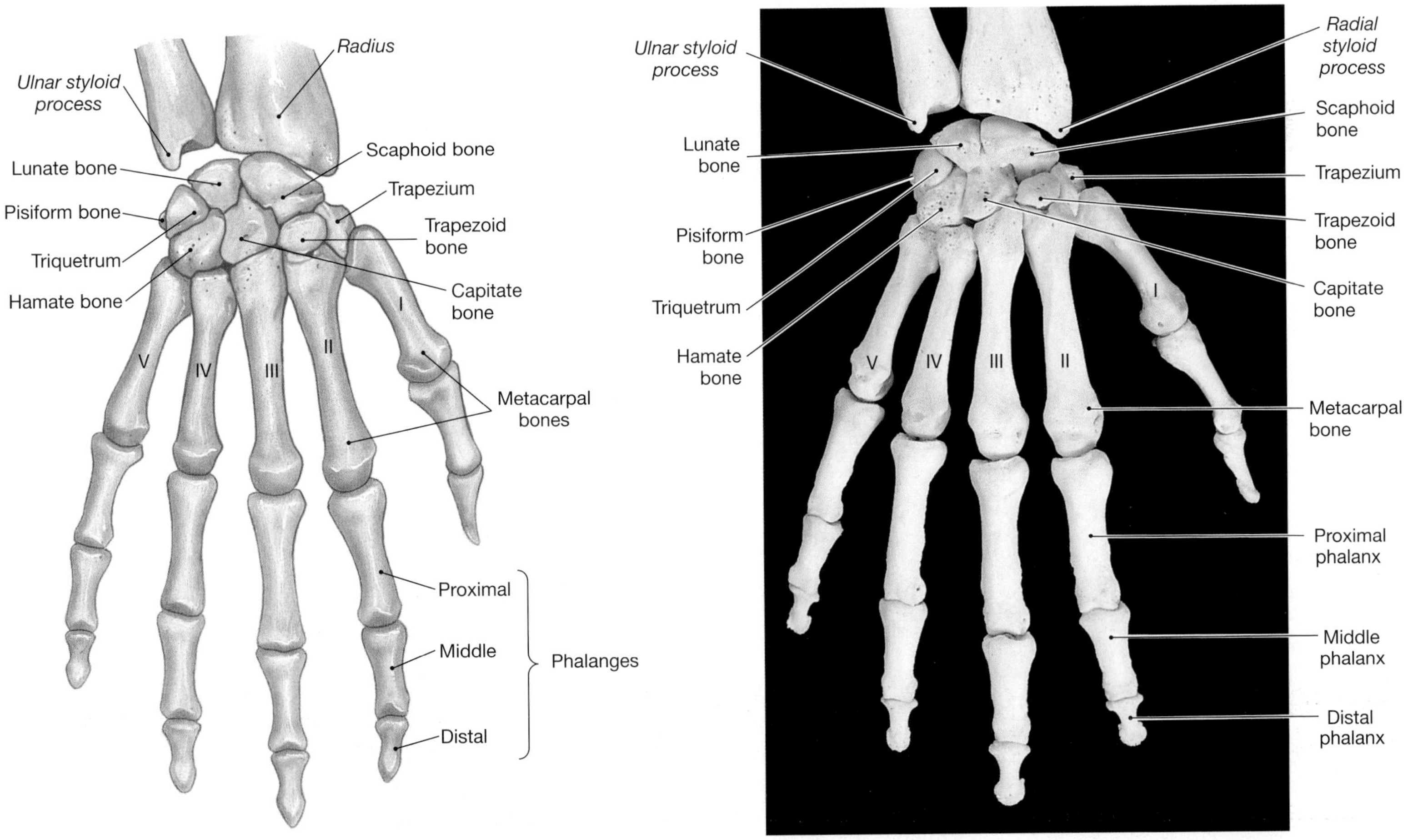

(c) Right hand, posterior (dorsal) view

FIGURE 7.8 *(continued)*

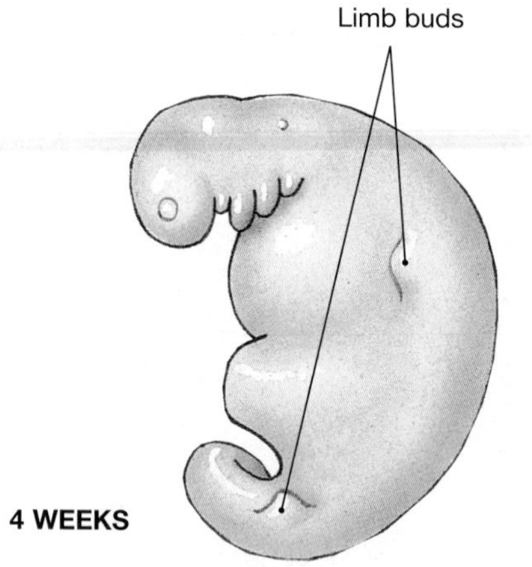

In the fourth week of development, ridges appear along the flanks of the embryo, extending from just behind the throat to just before the anus. These ridges form as mesodermal cells congregate beneath the ectoderm of the flank. Mesoderm gradually accumulates at the end of each ridge, forming two pairs of limb buds.

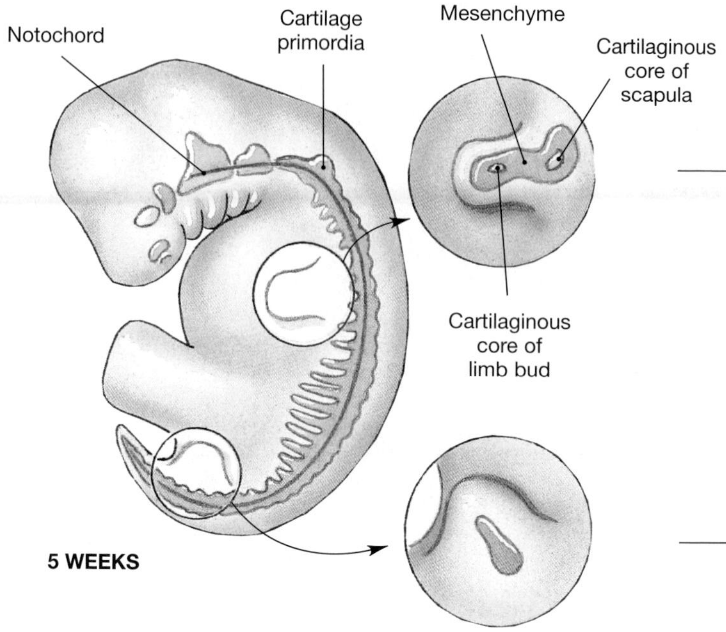

After 5 weeks of development, the pectoral limb buds have a cartilaginous core and scapular cartilages are developing in the mesenchyme of the trunk.

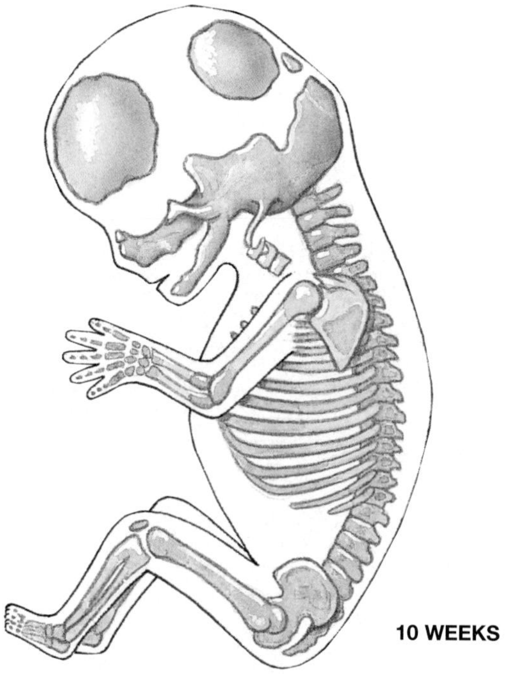

Ossification in the embryonic skeleton after approximately 10 weeks of development. The shafts of the limb bones are undergoing rapid ossification, but the distal bones of the carpus and tarsus remain cartilaginous.

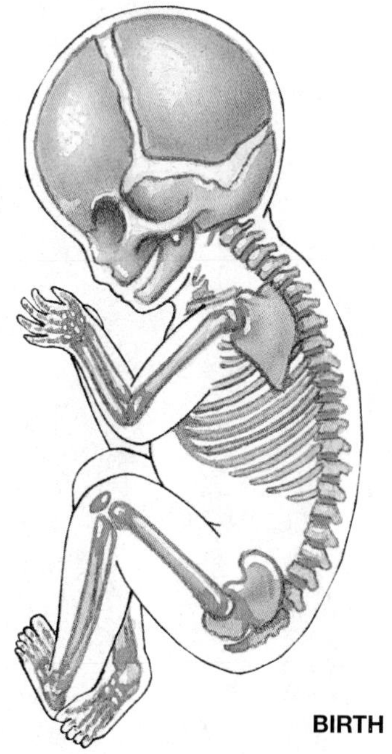

The skeleton of a newborn infant. Note the extensive areas of cartilage (blue) in the humeral head, in the wrist, between the bones of the palm and fingers, and in the coxae. Notice the appearance of the axial skeleton, with reference to the Embryology Summaries in Chapter 6.

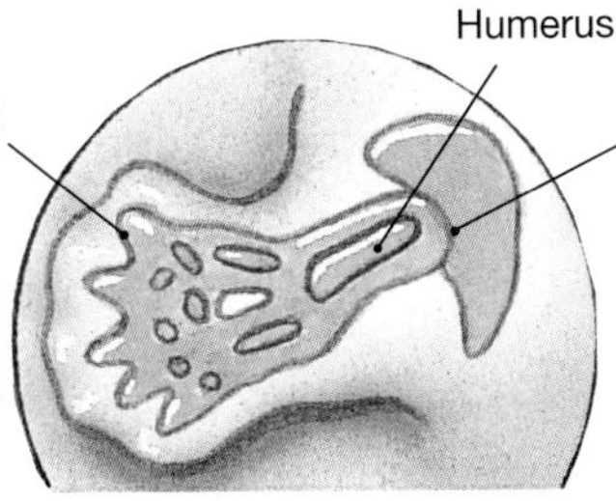

As the limb bud enlarges, bends develop at the future locations of the shoulder and elbow joints. Two cartilages form in the forearm, and a lateral rotation of the **apical ridge** places the elbow in its proper orientation.

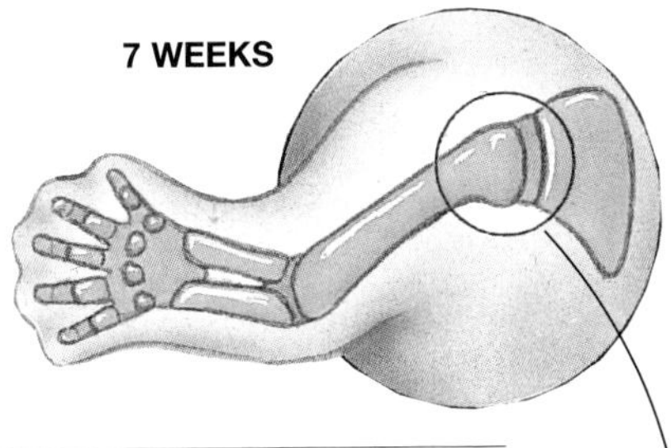

The hands originate as paddles, but the death of cells between the phalangeal cartilages produces individual fingers.

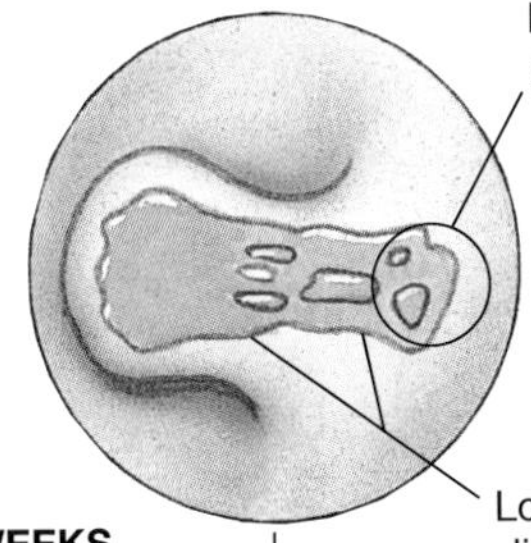

The formation of the pelvic girdle and legs closely parallels that of the pectoral complex. But as the pelvic limb bud enlarges, the apical ridge rotates medially rather than laterally. As a result, the knee joint faces posteriorly, while the elbow faces anteriorly.

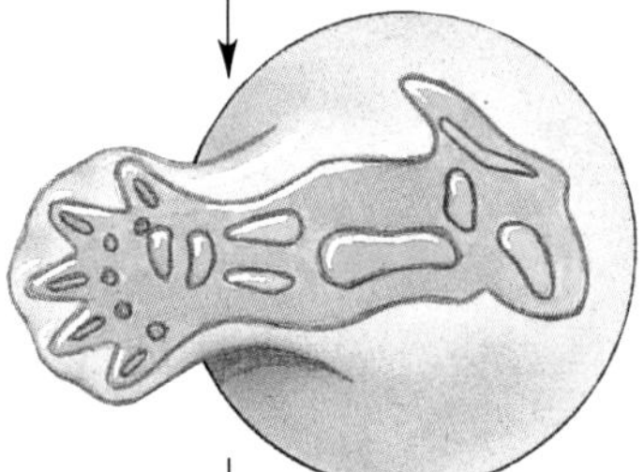

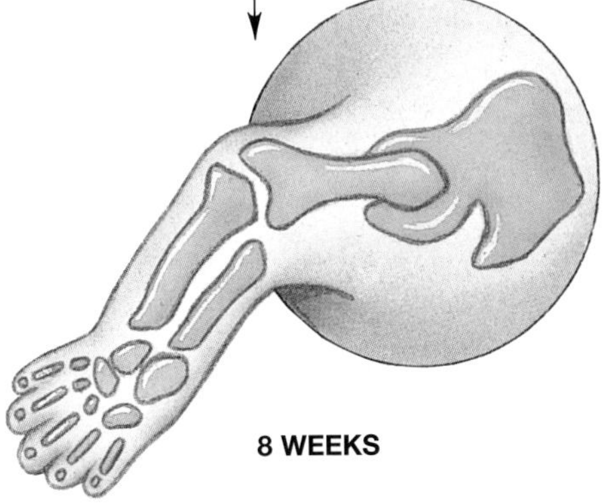

By week 8, cartilaginous models of all of the major skeletal components are well formed, and endochondral ossification begins in the future limb bones. Ossification of the coxal bones begins at three separate centers that gradually enlarge.

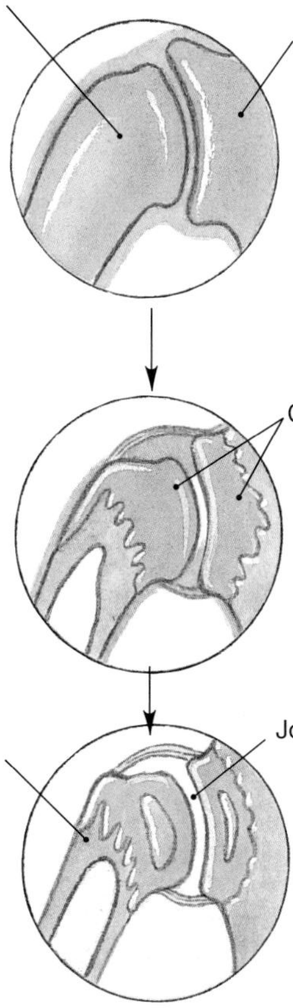

Joints form where two cartilages are in contact. The surfaces within the joint cavity remain cartilaginous, while the rest of the bones undergo ossification.

✓ CONCEPT CHECK

- Why would a broken clavicle affect the mobility of the scapula?
- Which antebrachial bone is lateral in the anatomical position?
- What is the function of the olecranon?
- Which bone is the only direct connection between the pectoral girdle and the axial skeleton?

The Pelvic Girdle And Lower Limb

[FIGURE 7.9]

The bones of the **pelvic girdle** support and protect the lower viscera, including the reproductive organs and developing fetus in females. The pelvic bones are more massive than those of the pectoral girdle because of the stresses involved in weight bearing and locomotion. The bones of the lower limbs are more massive than those of the upper limbs, for similar reasons. The pelvic girdle consists of two **ossa coxae** (ossa, bone + *coxa*, hip; singular, *os coxae*), also called *hip bones*, or *innominate bones*. The *pelvis* is a com-

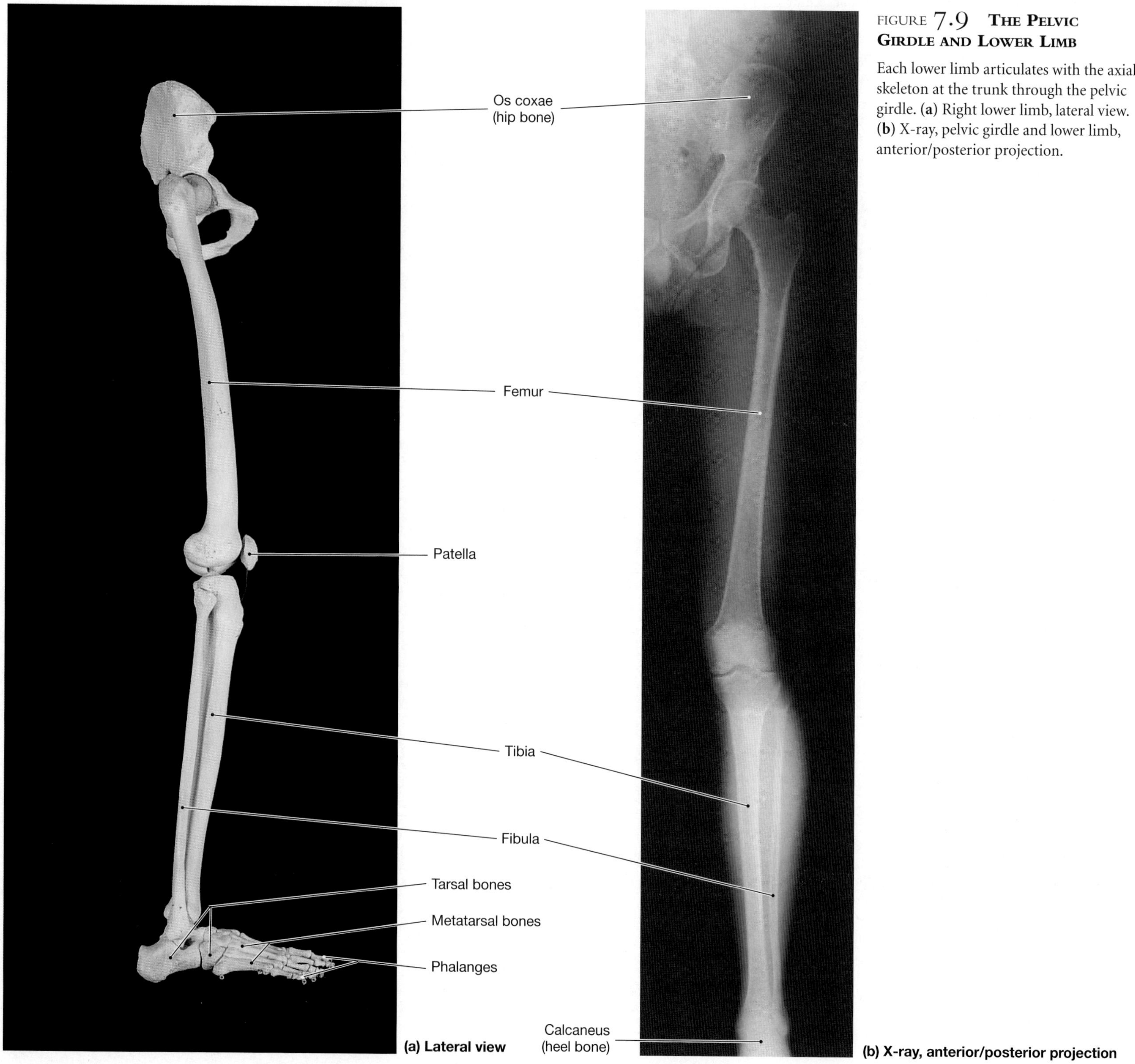

FIGURE 7.9 **THE PELVIC GIRDLE AND LOWER LIMB**

Each lower limb articulates with the axial skeleton at the trunk through the pelvic girdle. (**a**) Right lower limb, lateral view. (**b**) X-ray, pelvic girdle and lower limb, anterior/posterior projection.

posite structure that includes the ossa coxae of the appendicular skeleton and the sacrum and coccyx of the axial skeleton. The skeleton of each lower limb includes the *femur* (thigh), *patella* (kneecap), the *tibia* and *fibula* (leg), and the bones of the ankle (*tarsal bones*) and foot *(metatarsal bones* and *phalanges*) (Figure 7.9●). In anatomical terms, *leg* refers only to the distal portion of the limb, not to the entire lower limb. We use *thigh* for *upper leg* and *leg* for *lower leg*.

The Pelvic Girdle [FIGURE 7.10]

Each os coxae of the adult pelvic girdle forms through the fusion of three bones: an *ilium* (IL-ē-um), an *ischium* (IS-kē-um), and a *pubis* (PŪ-bis) (Figure 7.10●). At birth the three bones are separated by hyaline cartilage. Growth and fusion of the three bones into a single os coxae is usually completed by age 25. The articulation between an os coxae and the auricular surfaces of the sacrum occurs at the posterior and medial aspect of the ilium, forming the *sacroiliac joint*. The anterior and medial portions of the ossa coxae are connected by a pad of fibrocartilage at the *pubic symphysis*. On the lateral surface of each os coxae, the head of the femur articulates with the curved surface of the **acetabulum** (as-e-TAB-ū-lum; *acetabulum*, a vinegar cup).

The acetabulum lies inferior and anterior to the center of the os coxae (Figure 7.10a●). The space enclosed by the walls of the acetabulum is the **acetabular fossa**, which has a diameter of approximately 5 cm (2 inches). The acetabulum contains a smooth curved surface that forms the shape of the letter C. This is the **lunate surface**, which articulates with the head of the femur. A ridge of bone forms the lateral and superior margins of the acetabulum. There is no ridge marking the anterior and inferior margins. This gap is called the *acetabular notch*.

THE OS COXAE [FIGURES 7.10/7.11a]

The ilium, ischium, and pubis meet inside the acetabular fossa, as if it were a pie sliced into three pieces. The **ilium** (plural, *ilia*), the largest of the bones, provides the superior slice that includes around two-fifths of the acetabular surface. Superior to the acetabulum, the broad, curved, lateral surface of the ilium provides an extensive area for the attachment of muscles, tendons, and ligaments (Figure 7.10a●). The **anterior**, **posterior**, and **inferior gluteal lines** mark the attachment sites for the *gluteal muscles* that move the femur. The iliac expansion begins superior to the **arcuate** (AR-kū-āt) **line**. The anterior border includes the **anterior inferior iliac spine**, superior to the **inferior iliac notch**, and continues anteriorly to the **anterior superior iliac spine**. Curving posteriorly, the superior border supports the **iliac crest**, a ridge marking the attachments of both ligaments and muscles. The iliac crest ends at the **posterior superior iliac spine**. Inferior to the spine, the ilial margin continues inferiorly to the rounded **posterior inferior iliac spine** that is superior to the **greater sciatic** (sī-AT-ik) **notch**, through which the sciatic nerve passes into the lower limb.

Near the superior and posterior margin of the acetabulum, the ilium fuses with the **ischium**, which accounts for the posterior two-fifths of the acetabular surface. The ischium is the strongest of the coxal bones. Posterior to the acetabulum, the prominent **ischial spine** projects superior to the **lesser sciatic notch**. The rest of the ischium forms a sturdy process that turns medially and inferiorly. A roughened projection, the **ischial tuberosity**, forms its posterolateral border. When seated, the body weight is borne by the ischial tuberosities. The narrow **ischial ramus** (branch) of the ischium continues toward its anterior fusion with the **pubis** (PŪ-bis).

At the point of fusion, the ramus of the ischium meets the **inferior ramus** of the pubis. Anteriorly, the inferior ramus ends at the **pubic tubercle**, where it meets the **superior ramus** of the pubis. The anterior, superior surface of the superior ramus bears a roughened ridge, the **pubic crest**, which extends medially from the pubic tubercle. The pubic and ischial rami encircle the **obturator** (OB-too-rā-tor) **foramen**. In life, this space is closed by a sheet of collagen fibers whose inner and outer surfaces provide a firm base for the attachment of muscles, blood vessels, and nerves. The superior ramus originates at the anterior margin of the acetabulum. Inside the acetabulum, the pubis contacts the ilium and ischium.

Figures 7.10b and 7.11a● show additional features visible on the medial and anterior surfaces of the right os coxae:

- The concave medial surface of the **iliac fossa** helps support the abdominal organs and provides additional surface area for muscle attachment. The arcuate line marks the inferior border of the iliac fossa.
- The anterior and medial surface of the pubis contains a roughened area that marks the site of articulation with the pubis of the opposite side. At this articulation, the **pubic symphysis**, the two pubic bones are attached to a median fibrocartilage pad.
- The **pectineal** (pek-TIN-ē-al) **line** begins near the symphysis and extends diagonally across the pubis to merge with the arcuate line, which continues toward the **auricular surface** of the ilium. Auricular surfaces of the ilium and sacrum unite to form the *sacroiliac joint*. Ligaments arising at the **iliac tuberosity** stabilize this joint.
- On the medial surface of the superior ramus of the pubis lies the **obturator groove**, for the obturator blood vessels and nerves.

THE PELVIS [FIGURES 7.11 TO 7.13]

Figure 7.11● shows anterior and posterior views of the **pelvis**, which consists of four bones: the two ossa coxae, the sacrum, and the coccyx. The pelvis is a ring of bone, with the hip bones forming the anterior and lateral parts, the sacrum and coccyx the posterior part. An extensive network of ligaments connects the lateral borders of the sacrum with the iliac crest, the ischial tuberosity, the ischial spine, and the iliopectineal line. Other ligaments bind the ilia to the posterior lumbar vertebrae. These interconnections increase the stability of the pelvis.

The pelvis may be subdivided into the **false** *(greater)* **pelvis** and the **true** *(lesser)* **pelvis**. The boundaries of each are indicated in Figure 7.12●. The false pelvis consists of the expanded, bladelike portions of each ilium superior to the iliopectineal line. The false pelvis encloses organs within the inferior portion of the abdominal cavity. Structures inferior to the iliopectineal line form the true pelvis, which forms the boundaries of the pelvic cavity. ∞ p. 19 These pelvic structures include the inferior portions of each ilium, both pubic bones, the ischia, the sacrum, and the coccyx. In lateral view (Figure 7.12b●), the superior limit of the true pelvis is a line that extends from either side of the base of the sacrum, along the iliopectineal lines to the superior margin of the pubic symphysis. The bony edge of the true pelvis is called the **pelvic brim**. The space enclosed by the pelvic brim is the **pelvic inlet**.

The **pelvic outlet** is the opening bounded by the inferior margins of the pelvis (Figure 7.12b,c●), specifically the coccyx, the ischial tuberosities, and the inferior border of the pubic symphysis. In life, the region of the pelvic outlet is called the *perineum* (per-i-NĒ-um). Perineal muscles form the floor of the pelvic cavity and support the enclosed organs. These muscles are described in Chapter 10.

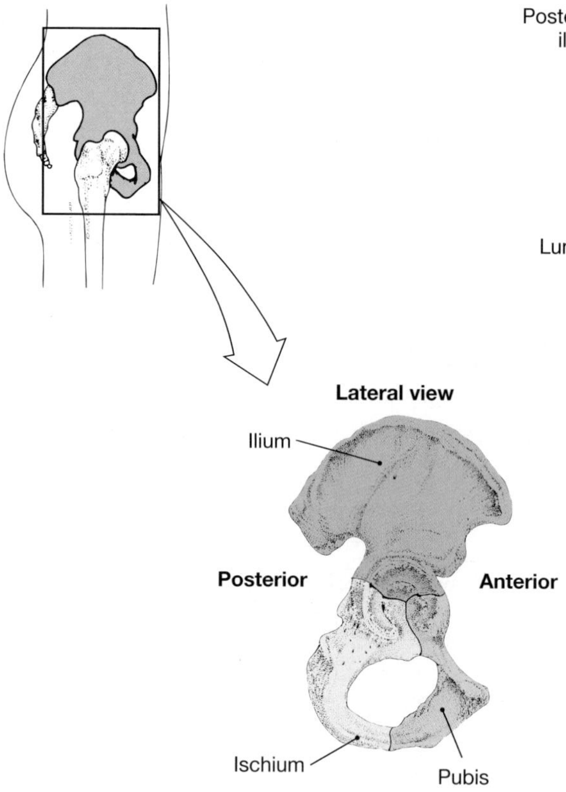

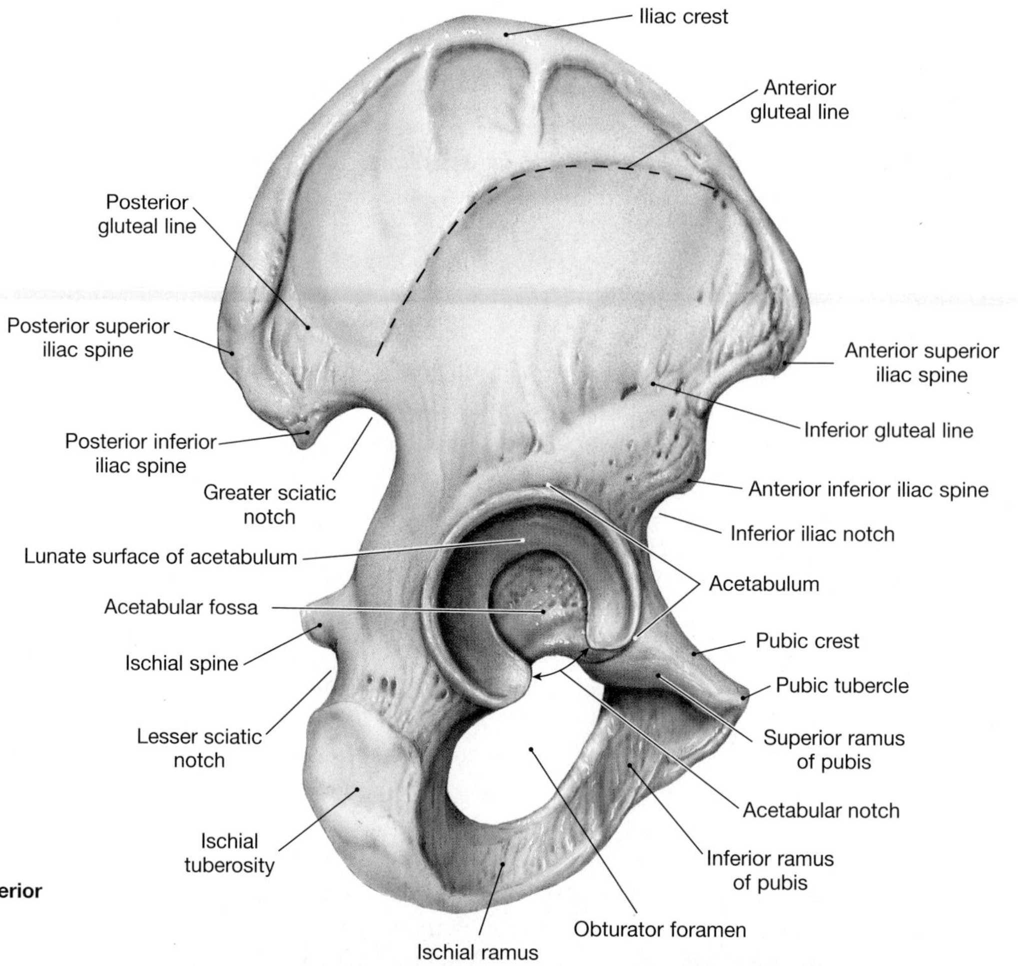

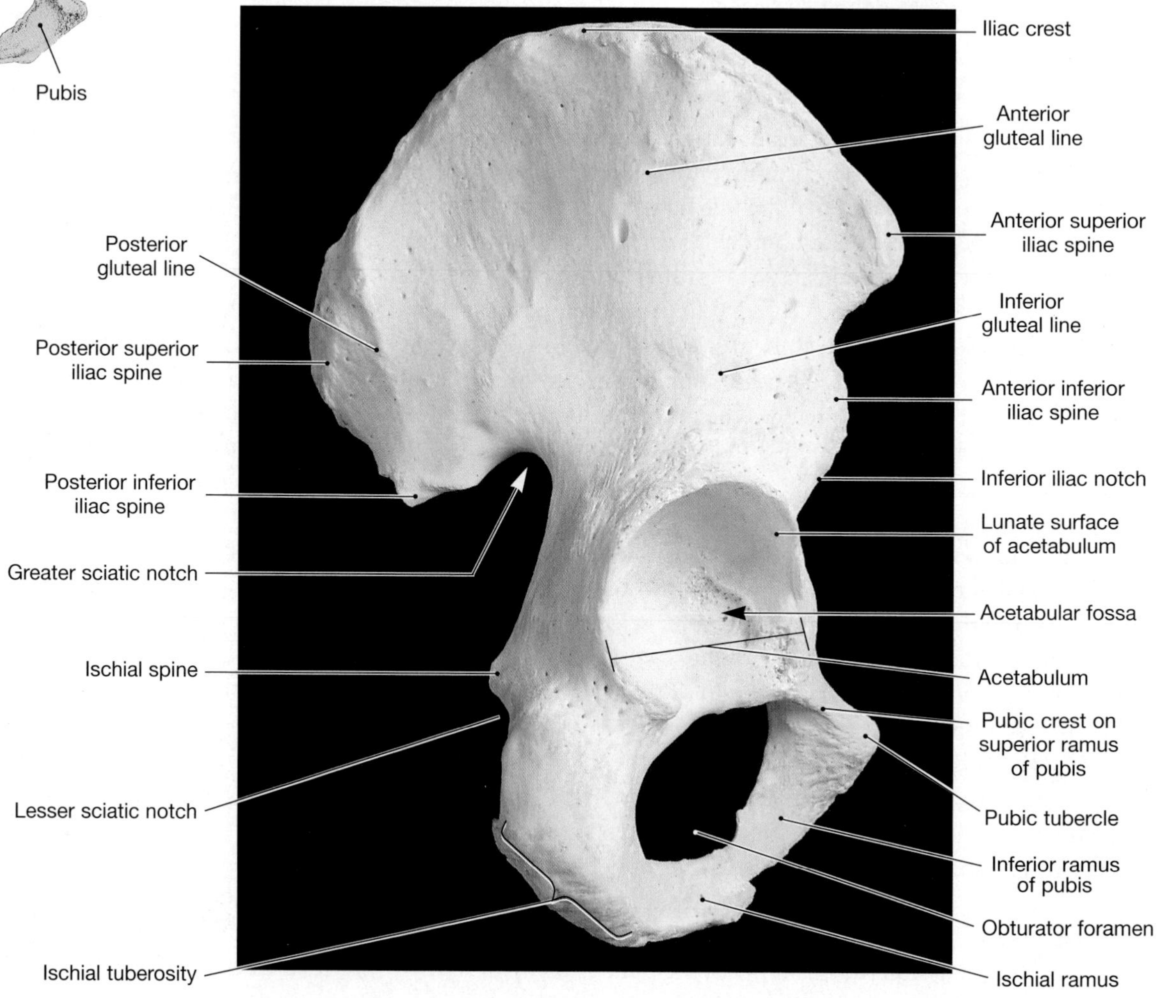

(a) Lateral view

FIGURE 7.10 **The Pelvic Girdle**

The pelvic girdle consists of the two ossa coxae. Each os coxae forms as a result of the fusion of an ilium, an ischium, and a pubis. (**a**) Lateral view. (**b**) Medial view.

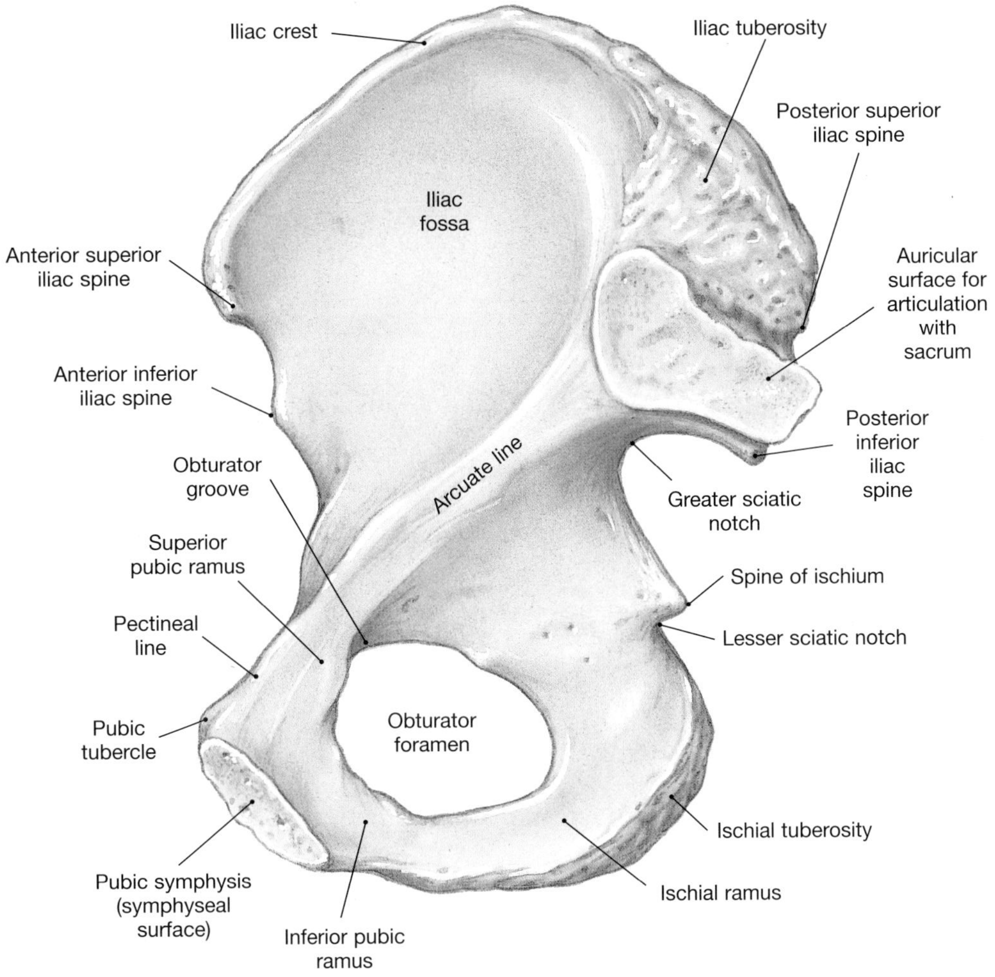

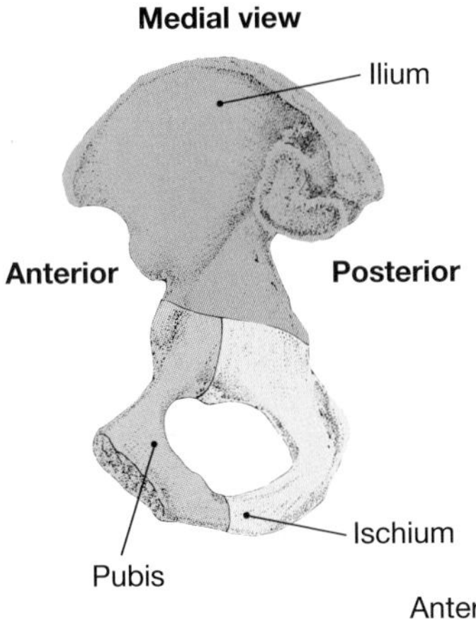

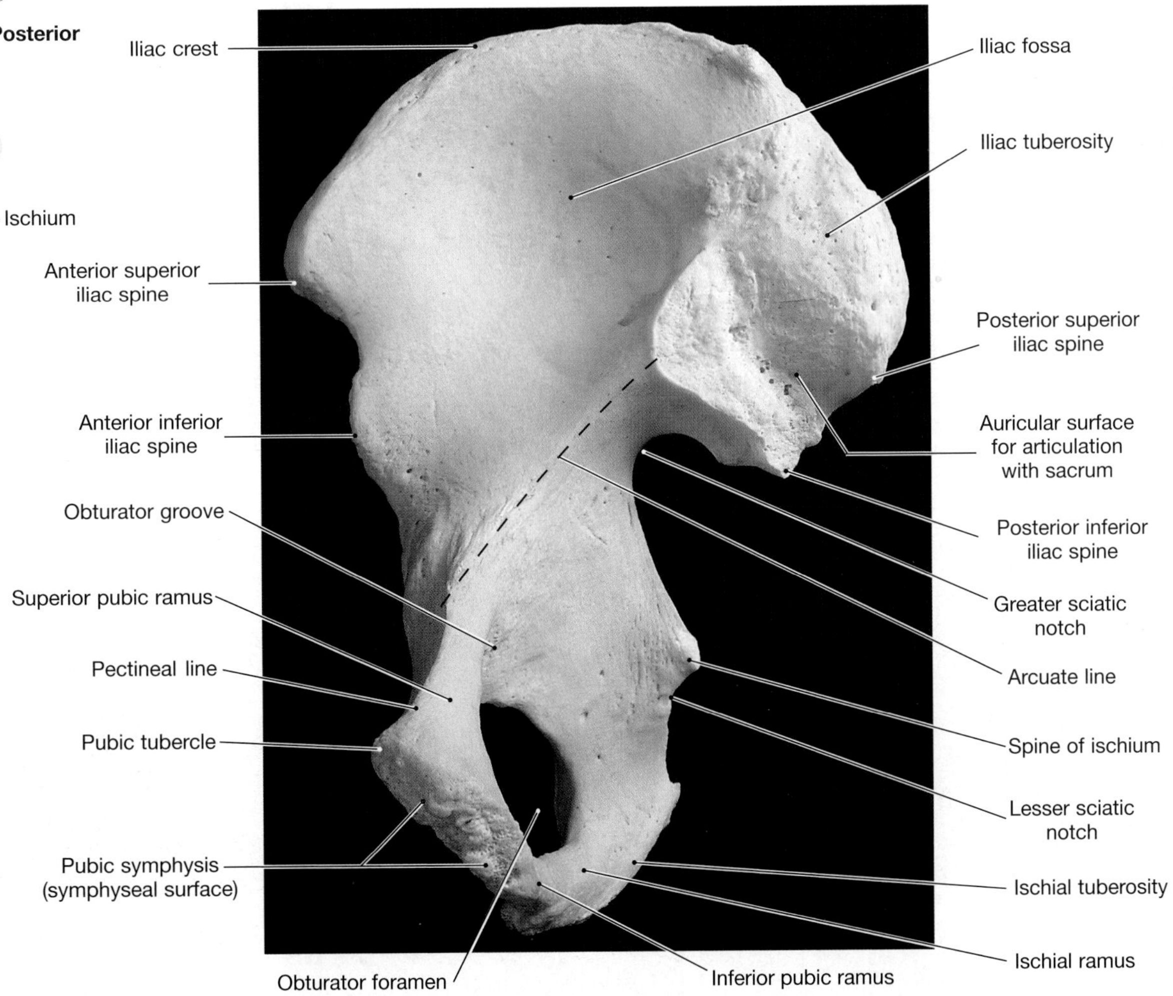

(b) Medial view

FIGURE 7.10 *(continued)*

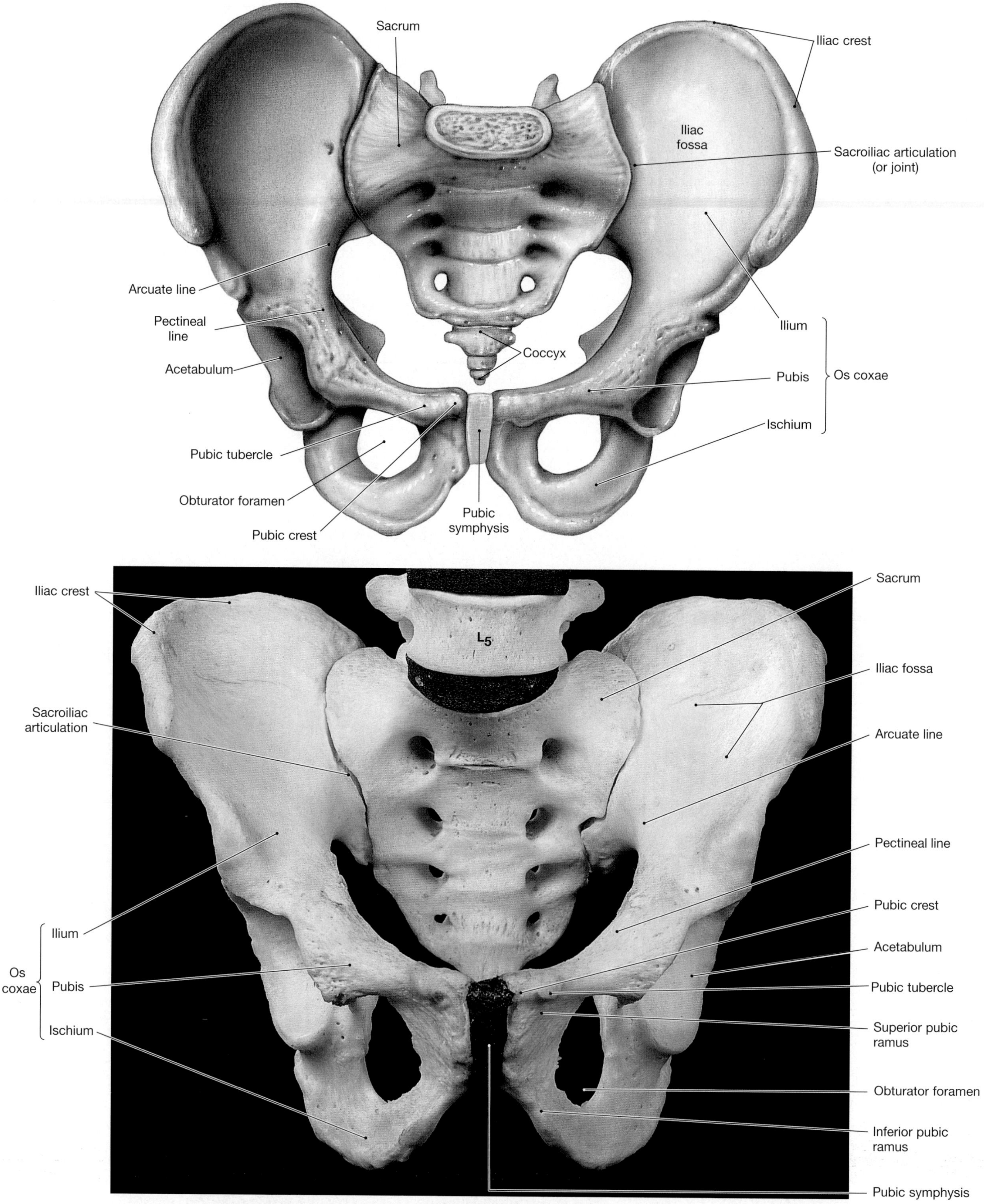

(a) Anterior view

FIGURE 7.11 **THE PELVIS**

A pelvis consists of two ossa coxae, the sacrum, and the coccyx. (**a**) Anterior view of the pelvis of adult male. (**b**) Posterior view.

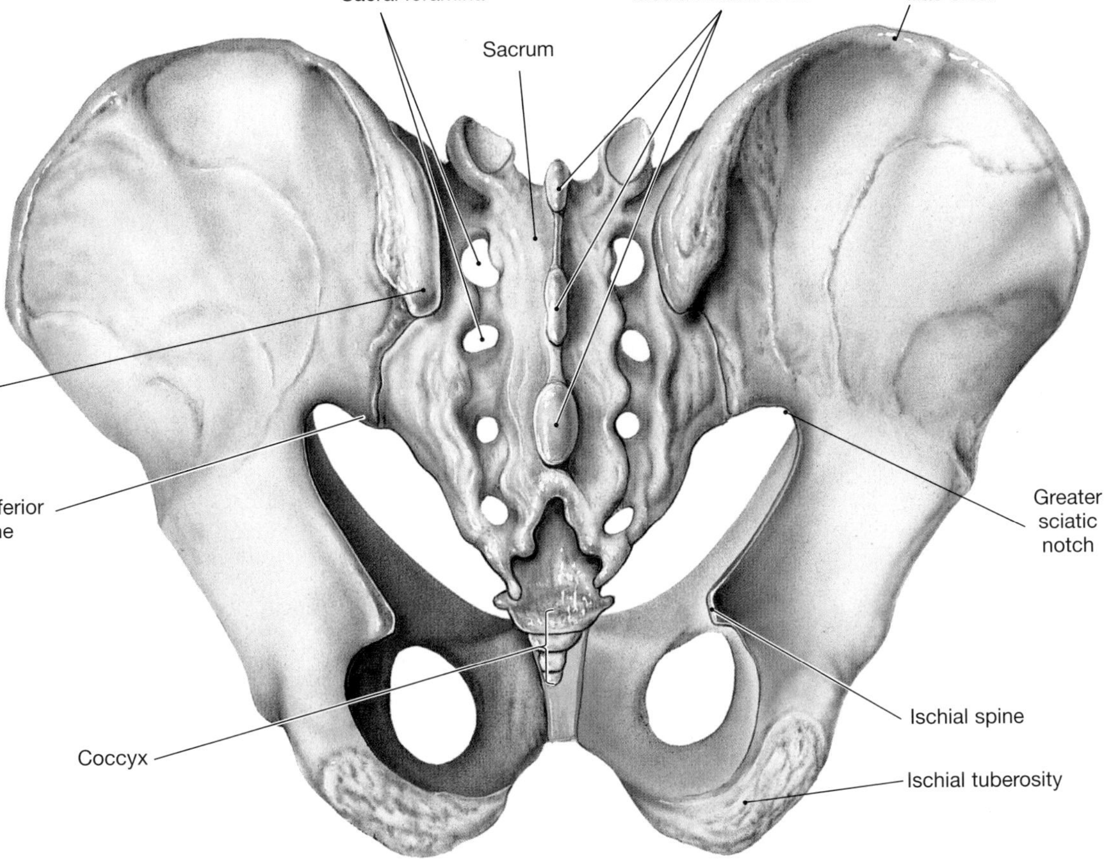
Sacral foramina
Median sacral crest
Iliac crest
Sacrum
Posterior superior iliac spine
Posterior inferior iliac spine
Greater sciatic notch
Ischial spine
Coccyx
Ischial tuberosity

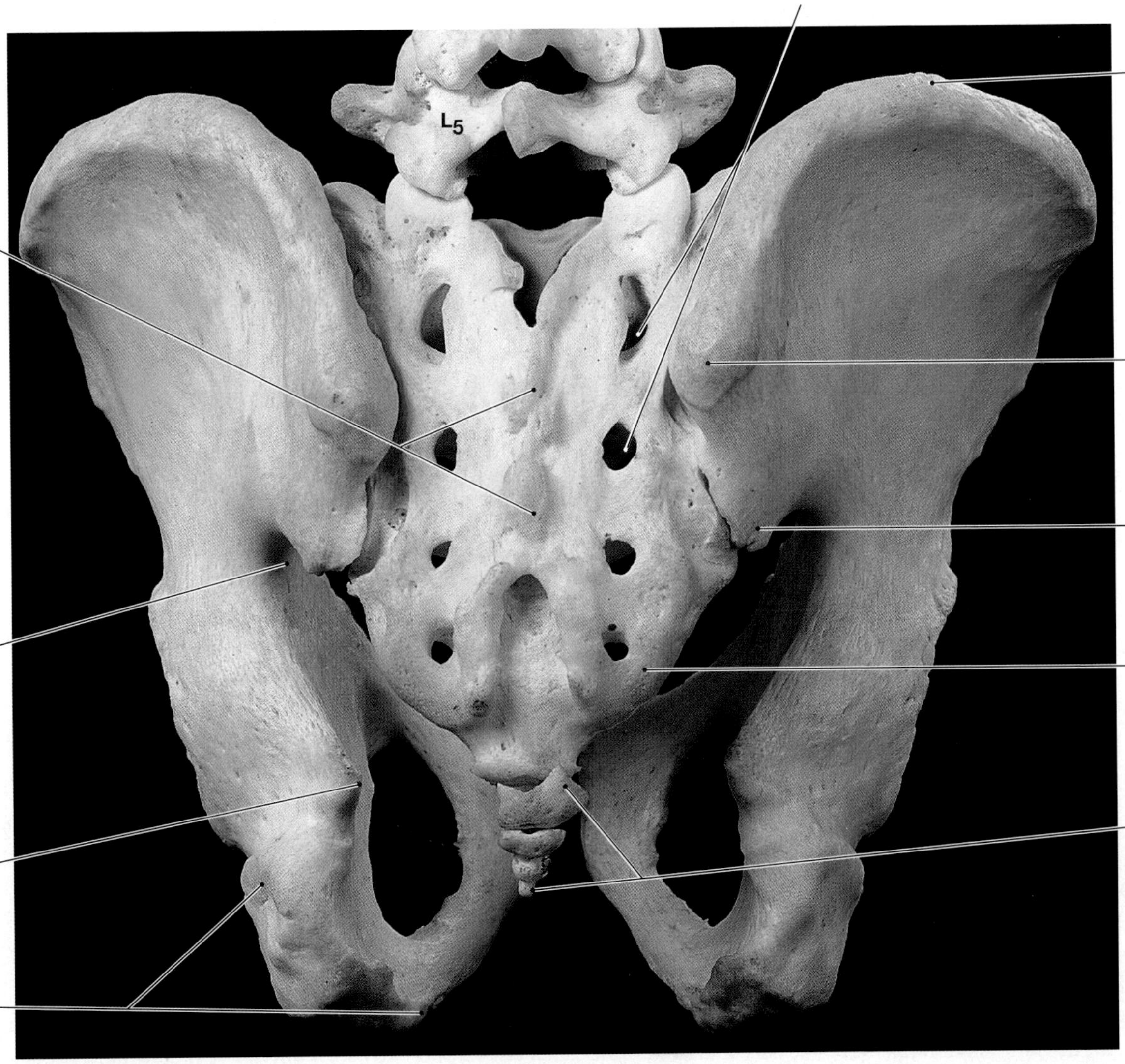
Sacral foramina
Iliac crest
L5
Median sacral crest
Posterior superior iliac spine
Posterior inferior iliac spine
Greater sciatic notch
Sacrum
Coccyx
Ischial spine
Ischial tuberosity

FIGURE 7.11 *(continued)*

(b) Posterior view

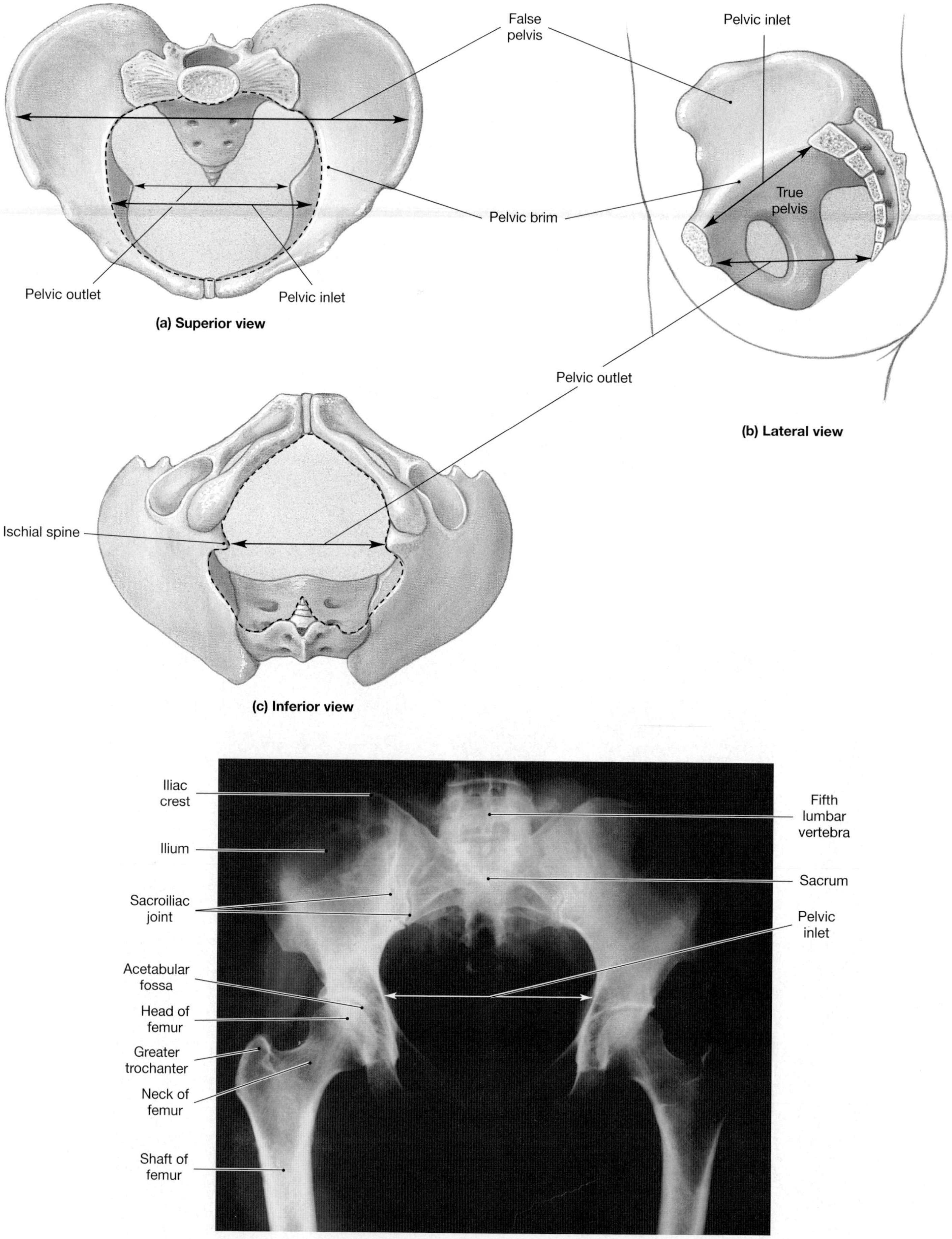

(d) X-ray of pelvis and proximal femora, anterior/posterior projection

FIGURE 7.12 **Divisions of the Pelvis**

A pelvis is subdivided into the true (lesser) and false (greater) pelvis. (**a**) Superior view, showing the pelvic brim and pelvic inlet of a male. (**b**) Lateral view, showing the boundaries of the true (lesser) and false (greater) pelvis. (**c**) Inferior view, showing the limits of the pelvic outlet. (**d**) X-ray (anterior/posterior projection) of the pelvis and femora.

Figure 7.12d● shows the appearance of the pelvis in anterior view. The shape of the female pelvis is somewhat different from that of the male pelvis (Figure 7.13●). Some of these differences are the result of variations in body size and muscle mass. For example, the female pelvis is usually smoother, lighter, and has less prominent markings. Other differences are adaptations for childbearing, including:

- an enlarged pelvic outlet, due in part to greater separation of the ischial spines,
- less curvature on the sacrum and coccyx, which in the male arc into the pelvic outlet,
- a wider, more circular pelvic inlet,
- a relatively broad, low pelvis,
- ilia that project farther laterally, but do not extend as far superior to the sacrum, and
- a broader *pubic angle*, with the inferior angle between the pubic bones greater than 100°.

These adaptations are related to (1) the support of the weight of the developing fetus and uterus and (2) easing the passage of the newborn through the pelvic outlet at the time of delivery. In addition, a hormone produced during pregnancy loosens the pubic symphysis, allowing relative movement between the coxae that can further increase the size of the pelvic inlet and outlet and thus facilitate delivery.

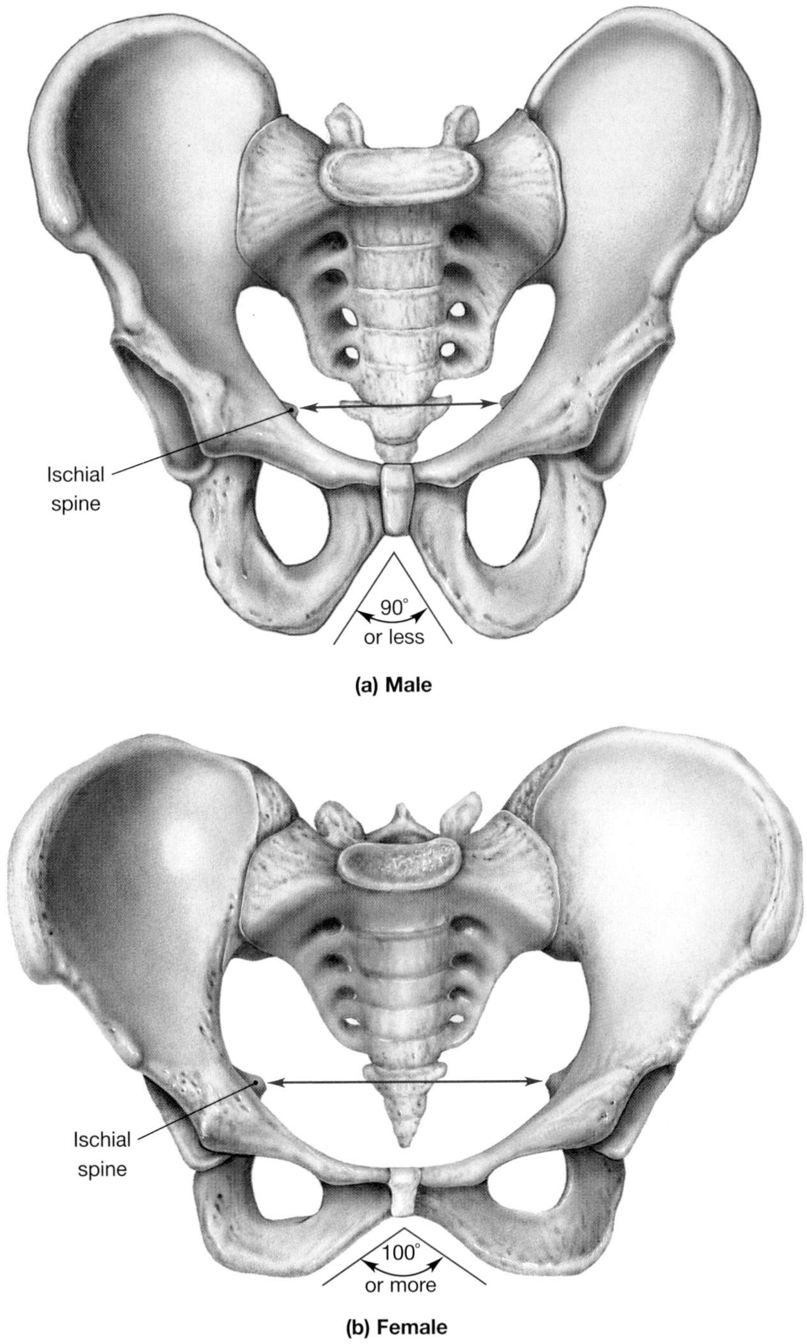

FIGURE 7.13 **ANATOMICAL DIFFERENCES IN THE MALE AND FEMALE PELVIS**

The black arrows indicate the pubic angle. Note the much sharper pubic angle in the pelvis of a **(a)** male compared to a **(b)** female. The red arrows indicate the width of the pelvic outlet *(see Figure 7.12)*. The female pelvis has a much wider pelvic outlet.

The Lower Limb [FIGURE 7.9]

The skeleton of the lower limb consists of the femur, patella (kneecap), tibia and fibula, and tarsal bones of the ankle and metatarsal bones and phalanges of the foot (Figure 7.9●, p. 194). The functional anatomy of the lower limb is very different from that of the upper limb, primarily because the lower limb must transfer the body weight to the ground.

THE FEMUR [FIGURES 7.9/7.12a/7.14]

The **femur** (Figure 7.14●) is the longest and heaviest bone in the body. Distally, the femur articulates with the tibia of the leg at the knee joint. Proximally, the rounded **head** of the femur articulates with the pelvis at the acetabulum (Figure 7.9, p. 194, and 7.12a●). A stabilizing ligament (the *ligament of the head*) attaches to the femoral head at a depression, the **fovea** (Figure 7.14b●). Distal to the head, the **neck** joins the shaft at an angle of about 125°. The shaft is solid and massive, but curves along its length (Figure 7.14a,d,e●). This lateral bow becomes greatly exaggerated if the skeleton weakens; a bowlegged stance is characteristic of rickets, a metabolic disorder discussed in Chapter 5. p. 118

The **greater trochanter** (trō-KAN-ter) projects laterally from the junction of the neck and shaft. The **lesser trochanter** originates on the posteromedial surface of the femur. Both trochanters develop where large tendons attach to the femur. On the anterior surface of the femur, the raised **intertrochanteric** (in-ter-trō-kan-TER-ik) **line** marks the distal edge of the articular capsule (Figure 7.14a,c●). This line continues around to the posterior surface, passing inferior to the trochanters as the **intertrochanteric crest** (Figure 7.14b,d●). Inferior to the intertrochanteric crest, the **pectineal line** (medial) and the **gluteal tuberosity** (lateral) mark the attachment of the *pectineus muscle* and the *gluteus maximus muscle*, respectively. A prominent elevation, the **linea aspera** (*aspera*, rough), runs along the center of the posterior surface of the femoral shaft. This ridge marks the attachment site of other powerful hip muscles (the *adductor muscles*). Distally, the linea aspera divides into a **medial** and **lateral supracondylar ridge** to form a flattened triangular area, the **popliteal surface**. The medial supracondylar ridge terminates in a raised, rough projection, the **adductor tubercle**, on the **medial epicondyle**. The lateral supracondylar ridge ends at the **lateral epicondyle**. The smoothly rounded **medial** and **lateral condyles** are primarily distal to the epicondyles. The condyles continue across the inferior surface of the femur to the anterior surface, but the intercondylar fossa does not. As a result,

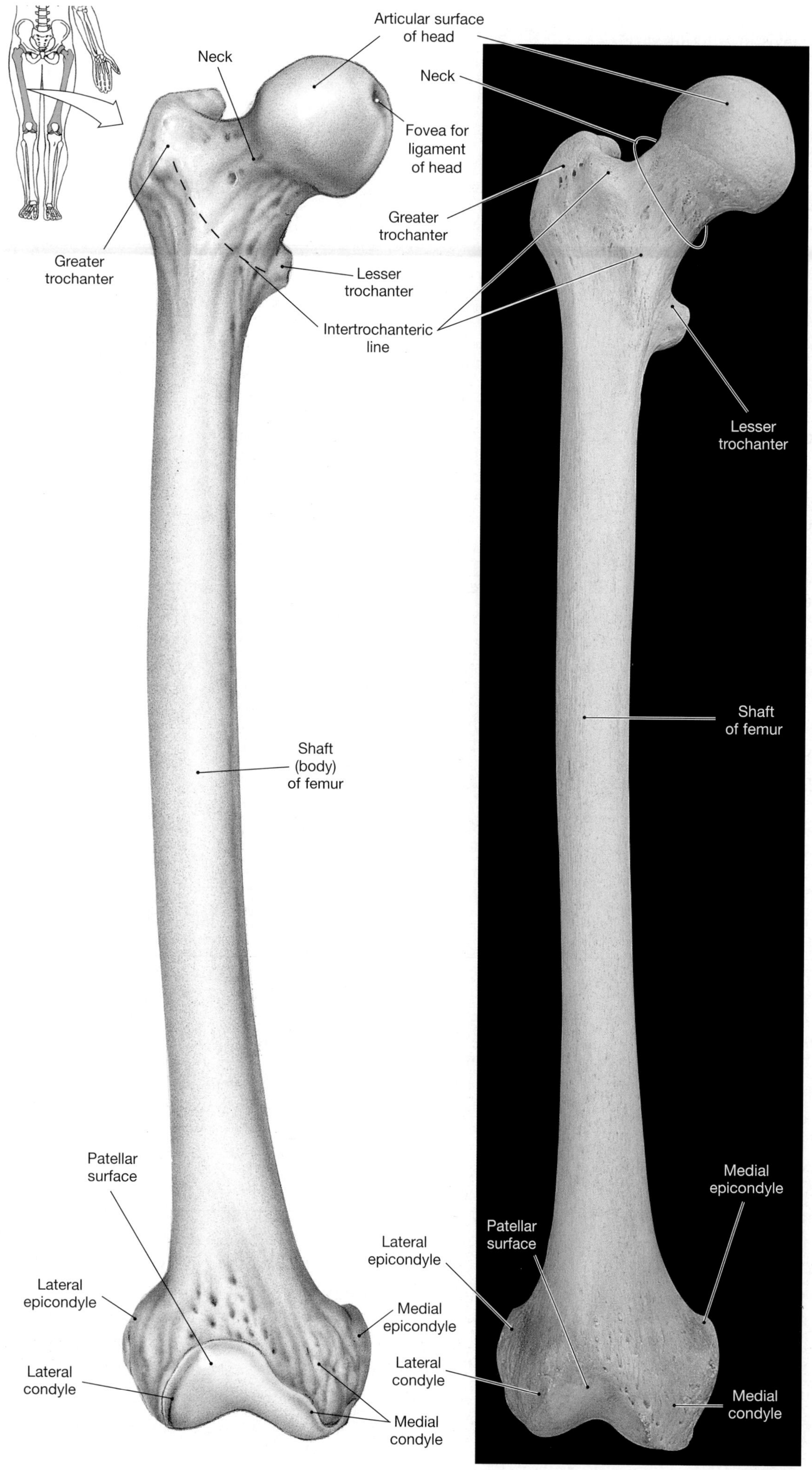

(a) Anterior surface

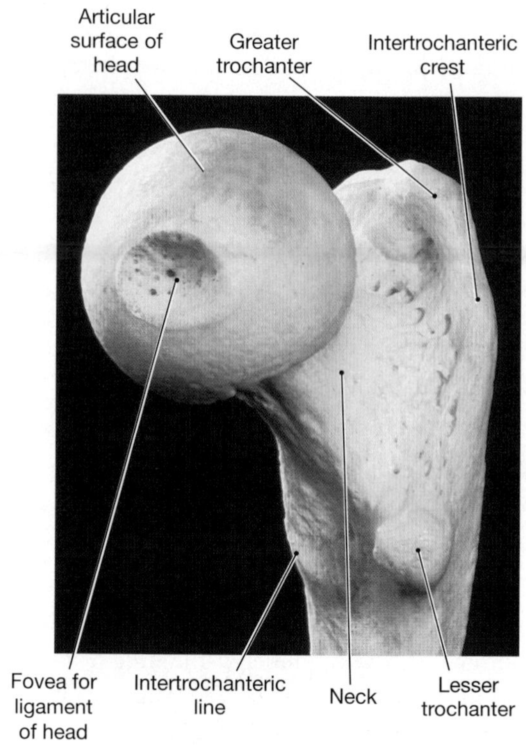

(b) Femoral head, medial view

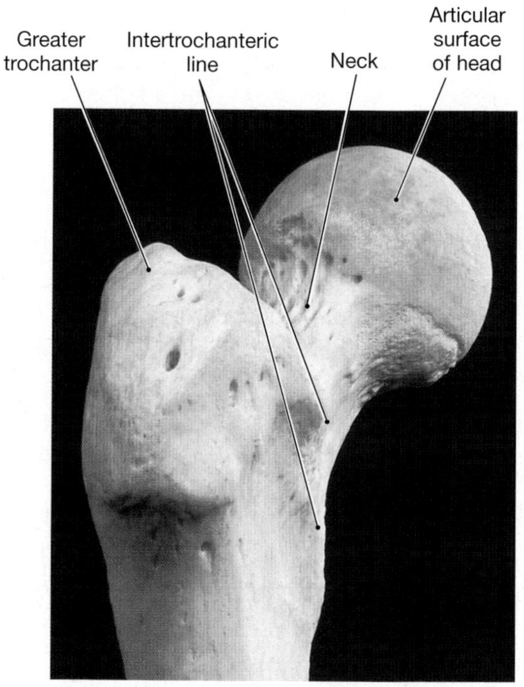

(c) Femoral head, lateral view

FIGURE 7.14 **THE FEMUR**

(a) Landmarks on the anterior surface of the right femur. **(b)** A medial view of the femoral head. **(c)** A lateral view of the femoral head. **(d)** Landmarks on the posterior surface of the right femur. **(e)** A superior view of the femur. **(f)** An inferior view of the right femur, showing the articular surfaces that participate in the knee joint.

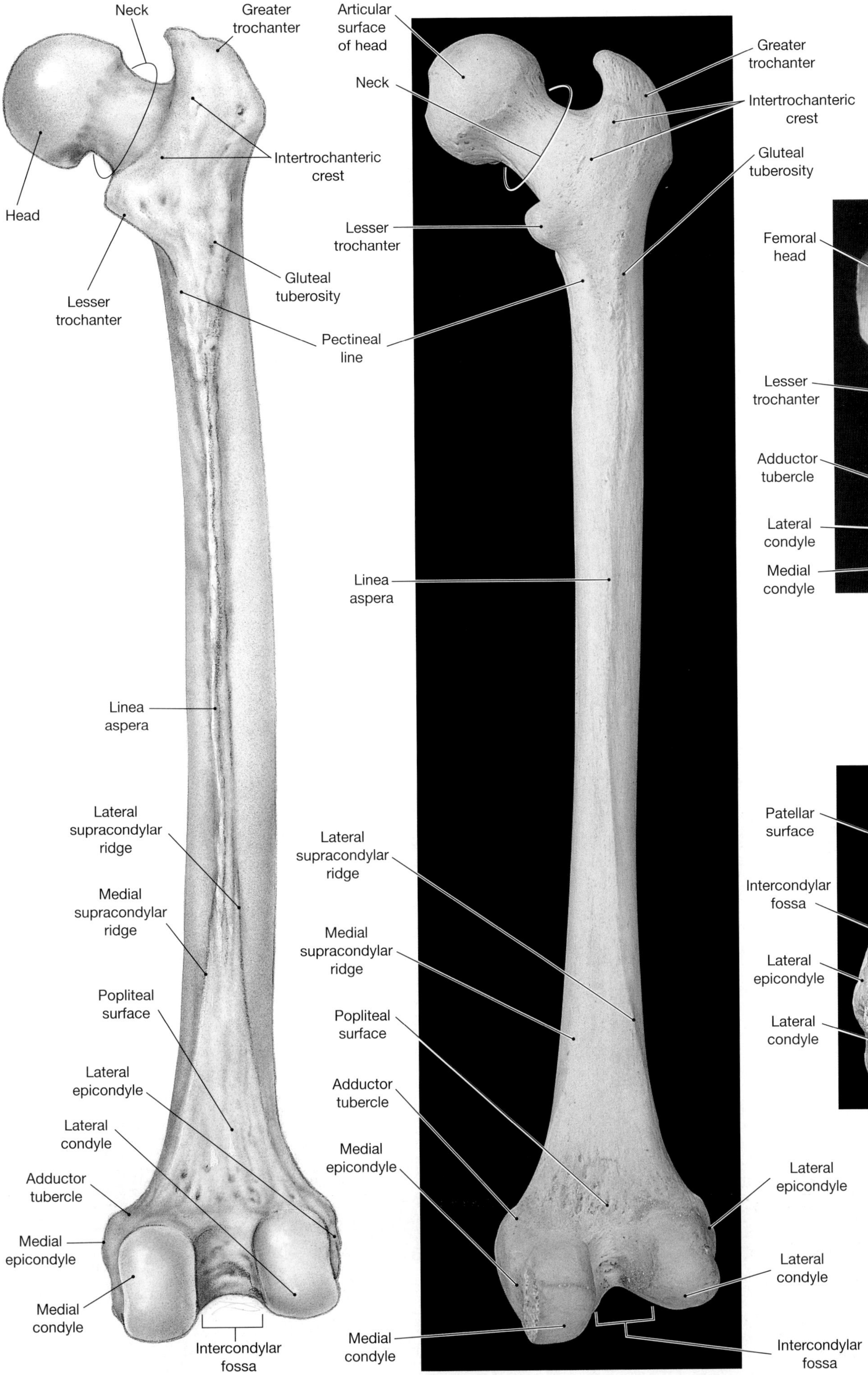

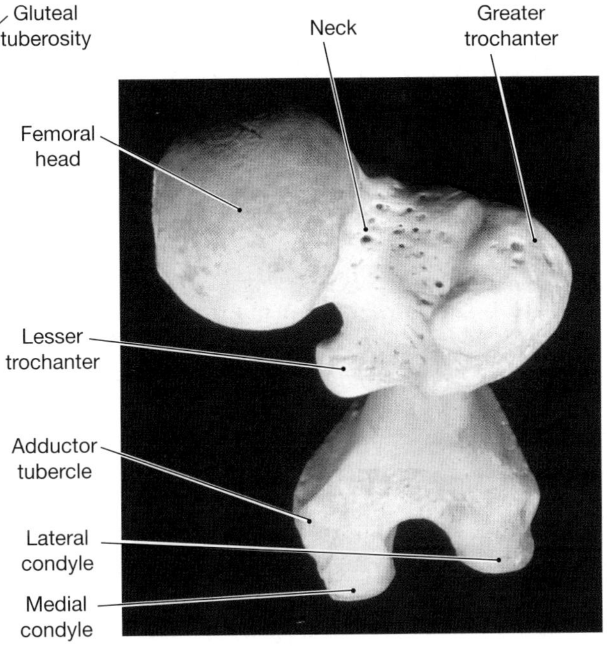

(e) Femur, superior view

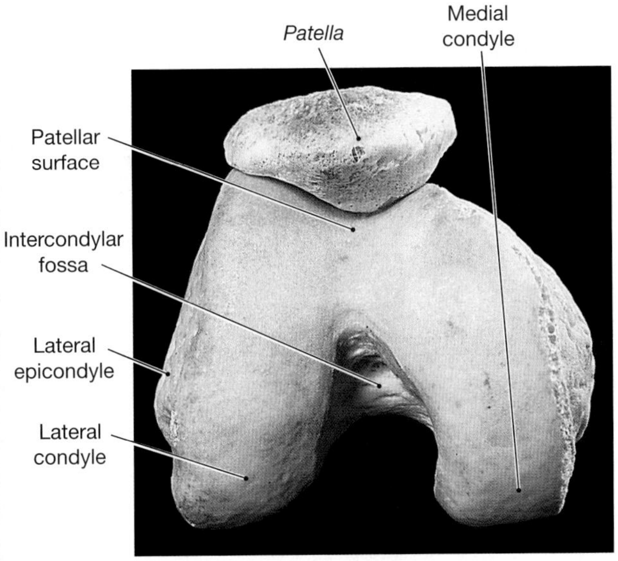

(f) Femur, inferior view

FIGURE 7.14 *(continued)*

(d) Posterior surface

the smooth articular faces merge, producing an articular surface with elevated lateral borders. This is the **patellar surface** over which the patella glides (Figure 7.14d,f●). On the posterior surface, the two condyles are separated by a deep **intercondylar fossa**.

THE PATELLA [FIGURES 7.14f/7.15]

The **patella** (pa-TEL-a) is a large sesamoid bone that forms within the tendon of the *quadriceps femoris*, a group of muscles that straightens (extends) the knee. This bone both strengthens the *quadriceps tendon* and protects the anterior surface of the knee joint. The triangular patella has a rough, convex anterior surface (Figure 7.15a●). It has a broad, superior **base** and a roughly pointed inferior **apex**. The roughened surface and broad **base** reflect the attachment of the quadriceps tendon (along the anterior and superior surfaces) and the *patellar ligament* (along the anterior and inferior surfaces). The patellar ligament extends from the apex of the patella to the tibia. The posterior patellar surface (Figure 7.15b●) presents two concave **facets** (*medial* and *lateral*) for articulation with the medial and lateral condyles of the femur (Figure 7.14f●).

THE TIBIA [FIGURE 7.16]

The **tibia** (TIB-ē-a) is the large medial bone of the leg (Figure 7.16●). The medial and lateral condyles of the femur articulate with the **medial** and **lateral condyles** of the proximal end of the tibia. A ridge, the **intercondylar eminence**, separates the medial and lateral condyles of the tibia (Figure 7.16b,d●). There are two **tubercles** (**medial** and **lateral**) on the intercondylar eminence. The anterior surface of the tibia near the condyles bears a prominent, rough **tibial tuberosity** that can easily be felt beneath the skin of the leg. This tuberosity marks the attachment of the stout patellar ligament.

The **anterior margin**, or *border*, is a ridge that begins at the distal end of the tibial tuberosity and extends distally along the anterior tibial surface. The anterior margin of the tibia can be felt through the skin. The lateral margin of the shaft is the **interosseous border**; from here, a collagenous sheet extends to the medial margin of the fibula. Distally, the tibia narrows, and the medial border ends in a large process, the **medial malleolus** (ma-LĒ-ō-lus; *malleolus*, hammer). The inferior surface of the tibia (Figure 7.16c●) forms a hinge joint with the *talus*, the proximal bone of the ankle. Here the tibia passes the weight of the body, received from the femur at the knee, to the foot across the ankle joint, or *talocrural joint*. The medial malleolus provides medial support for this joint, preventing lateral sliding of the tibia across the talus. The posterior surface of the tibia bears a prominent **popliteal line**, or *soleal line* (Figure 7.16d●). This marks the attachment of several leg muscles, including the *popliteus* and the *soleus*.

THE FIBULA [FIGURE 7.16]

The slender **fibula** (FIB-ū-la) parallels the lateral border of the tibia (Figure 7.16●). The **head** of the fibula, or *fibular head*, articulates along the lateral margin of the tibia, inferior and slightly posterior to the lateral tibial condyle. The medial border of the thin shaft is bound to the tibia by the *crural interosseous membrane*, which extends from the **interosseous border** of the fibula to that of the tibia. A sectional view through the shafts of the tibia and fibula (Figure 7.16e●) shows the locations of the tibial and fibular interosseous borders and the fibrous *crural interosseous membrane* that extends between them. This membrane helps stabilize the positions of the two bones and provides additional surface area for muscle attachment.

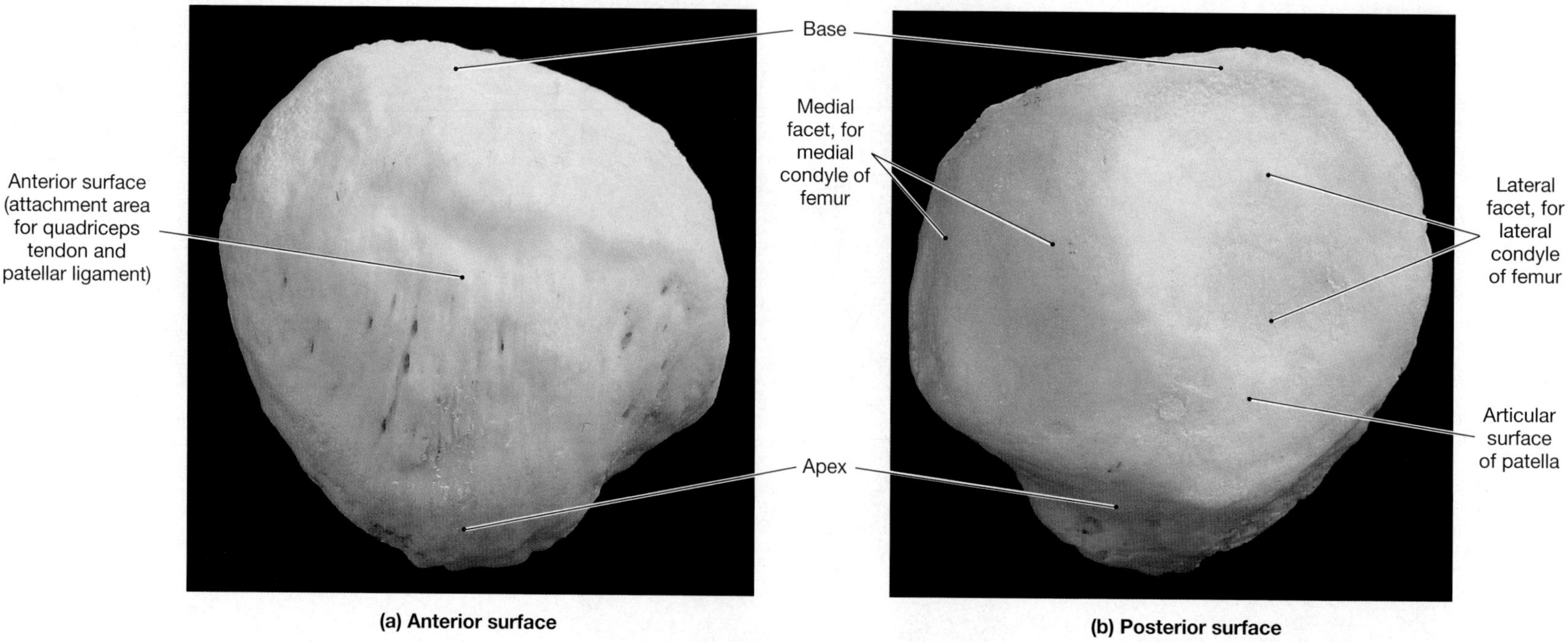

FIGURE 7.15 **THE PATELLA**
This sesamoid bone forms within the tendon of the *quadriceps femoris*. (**a**) Anterior surface of the right patella. (**b**) Posterior surface.

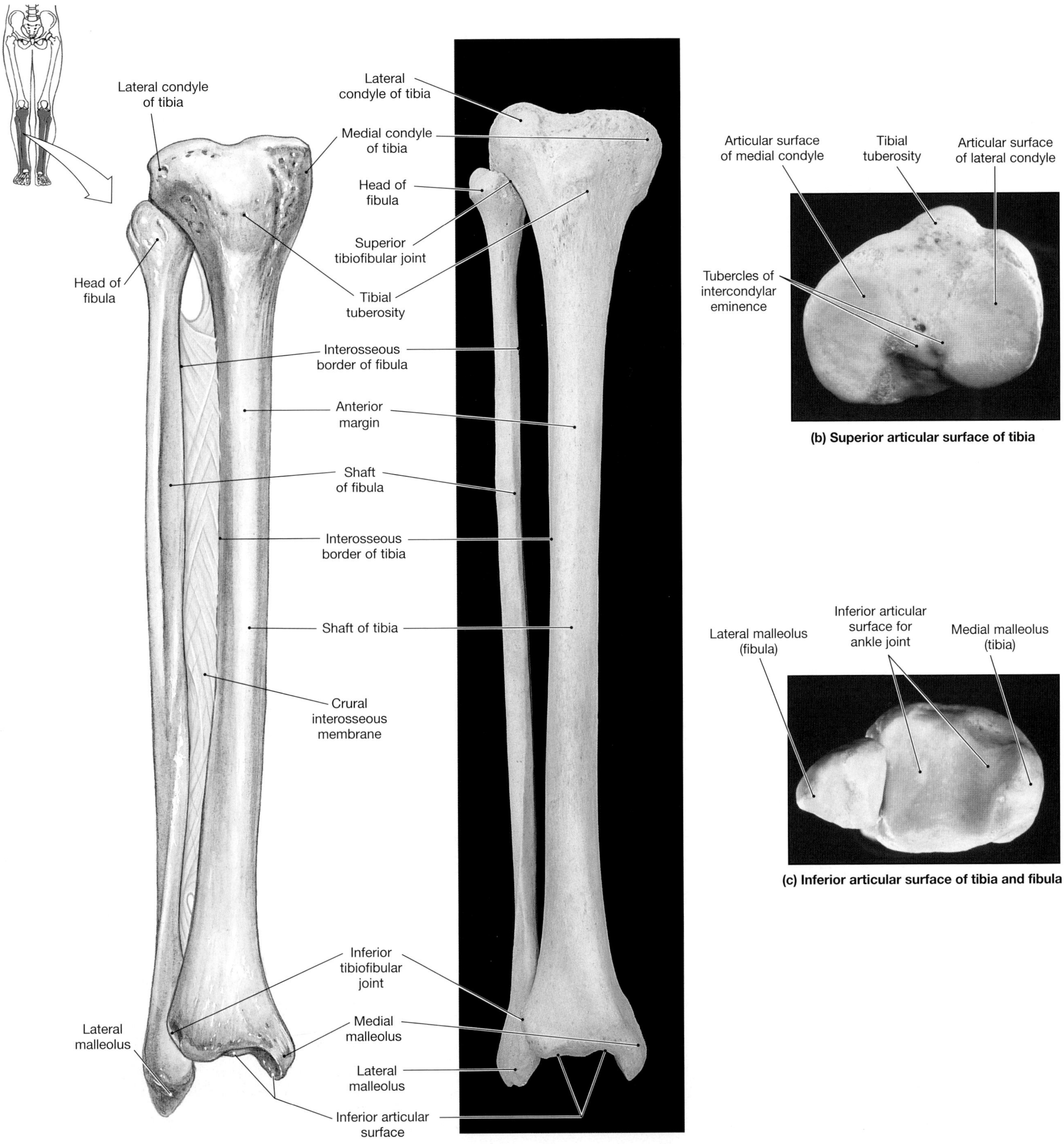

(a) Anterior views

(b) Superior articular surface of tibia

(c) Inferior articular surface of tibia and fibula

FIGURE 7.16 **THE TIBIA AND FIBULA**

(**a**) Anterior views of the right tibia and fibula. (**b**) Superior view of the proximal end of the tibia. (**c**) Inferior view of the distal surfaces of the tibia and fibula, showing the surfaces that participate in the ankle joint. (**d**) Posterior views of the right tibia and fibula. (**e**) A cross-sectional view at the plane indicated in (d).

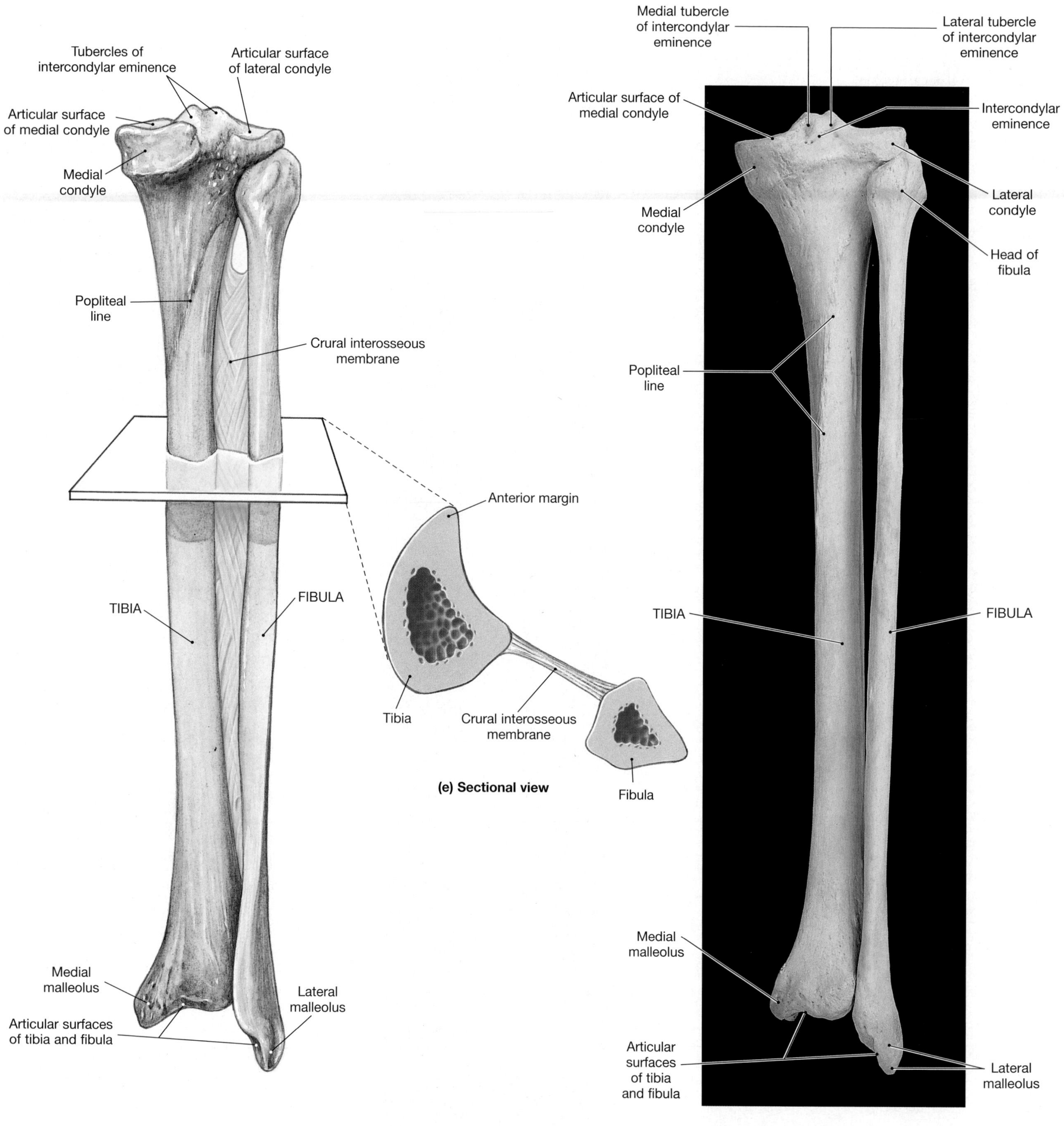
Tubercles of intercondylar eminence
Articular surface of lateral condyle
Articular surface of medial condyle
Medial condyle
Popliteal line
Crural interosseous membrane
TIBIA
FIBULA
Medial malleolus
Lateral malleolus
Articular surfaces of tibia and fibula
Anterior margin
Tibia
Crural interosseous membrane
Fibula
Medial tubercle of intercondylar eminence
Lateral tubercle of intercondylar eminence
Articular surface of medial condyle
Intercondylar eminence
Medial condyle
Lateral condyle
Head of fibula
Popliteal line
TIBIA
FIBULA
Medial malleolus
Articular surfaces of tibia and fibula
Lateral malleolus

(e) Sectional view

(d) Posterior views

FIGURE 7.16 *(continued)*

The fibula is excluded from the knee joint and does not transfer weight to the ankle and foot. However, it is an important site for muscle attachment. In addition, the distal tip of the fibula provides lateral support to the ankle joint. This fibular process, the **lateral malleolus**, provides stability to the ankle joint by preventing medial sliding of the tibia across the surface of the talus.

THE TARSUS [FIGURE 7.17]

The ankle, or **tarsus**, contains seven **tarsal bones**: the *talus*, the *calcaneus*, the *cuboid*, the *navicular*, and three *cuneiform bones* (Figures 7.17 and 7.18●).

- The **talus** is the second largest foot bone. It transmits the weight of the body from the tibia anteriorly, toward the toes. The primary tibial articulation is between the talus and the tibia; this involves the smooth superior surface of the **trochlea** of the talus. The trochlea has lateral and medial extensions that articulate with the lateral malleolus (fibula) and medial malleolus (tibia). The lateral surfaces of the talus are roughened where ligaments connect it to the tibia and fibula, further stabilizing the ankle joint.
- The **calcaneus** (kal-KĀ-nē-us), or heel bone, is the largest of the tarsal bones and may be easily palpated. When standing normally, most of your weight is transmitted from the tibia to the talus to the calcaneus, and then to the ground. The posterior surface of the calcaneus is a rough, knob-shaped projection. This is the attachment site for the *calcaneal tendon* (*calcanean tendon* or *Achilles tendon*) that arises at strong calf muscles. These muscles raise the heel and depress the sole of the foot, as when standing on tiptoe. The superior and anterior surfaces of the calcaneus bear smooth facets for articulation with other tarsal bones.
- The **cuboid bone** articulates with the anterior, lateral surface of the calcaneus.
- The **navicular bone**, located on the medial side of the ankle, articulates with the anterior surface of the talus. The distal surface of the navicular bone articulates with the three cuneiform bones.
- The three **cuneiform bones** are wedge-shaped bones arranged in a row, with articulations between them, located anterior to the navicular bone. They are named according to their position: **medial cuneiform**, **intermediate cuneiform**, and **lateral cuneiform bones**. Proximally, the cuneiform bones articulate with the anterior surface of the navicular bone. The lateral cuneiform bone also articulates with the medial surface of the cuboid bone. The distal surfaces of the cuboid bone and the cuneiform bones articulate with the metatarsal bones of the foot.

THE FOOT [FIGURES 7.17/7.18]

The **metatarsal bones** are five long bones that form the *metatarsus* (or distal portion) of the foot (Figures 7.17 and 7.18●). The metatarsal bones are identified with Roman numerals I–V, proceeding from medial

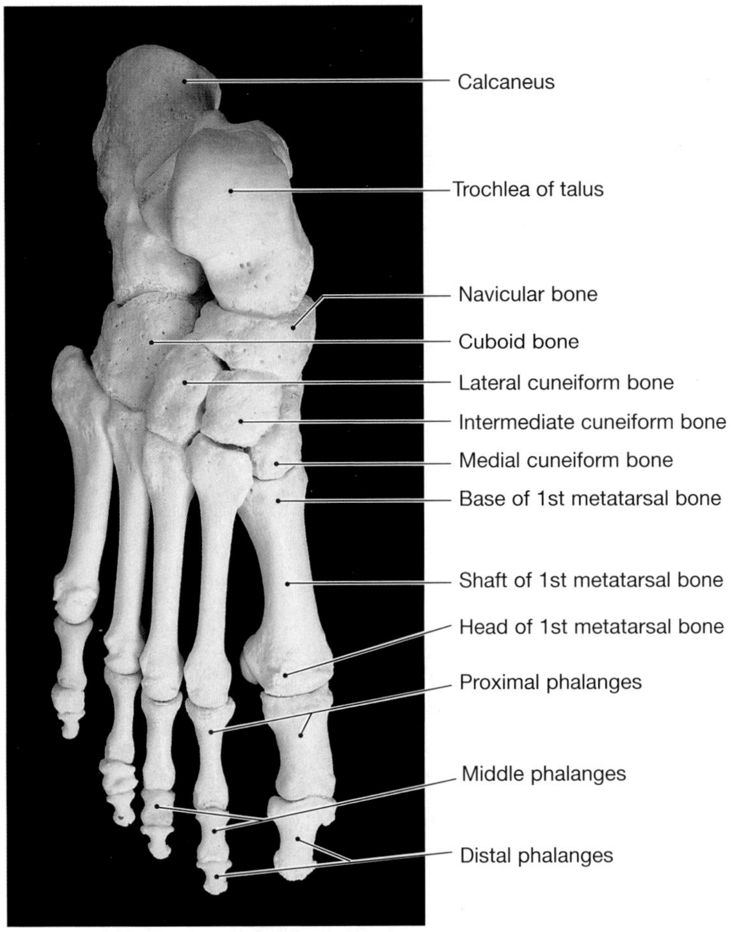

(a) Superior (dorsal) view

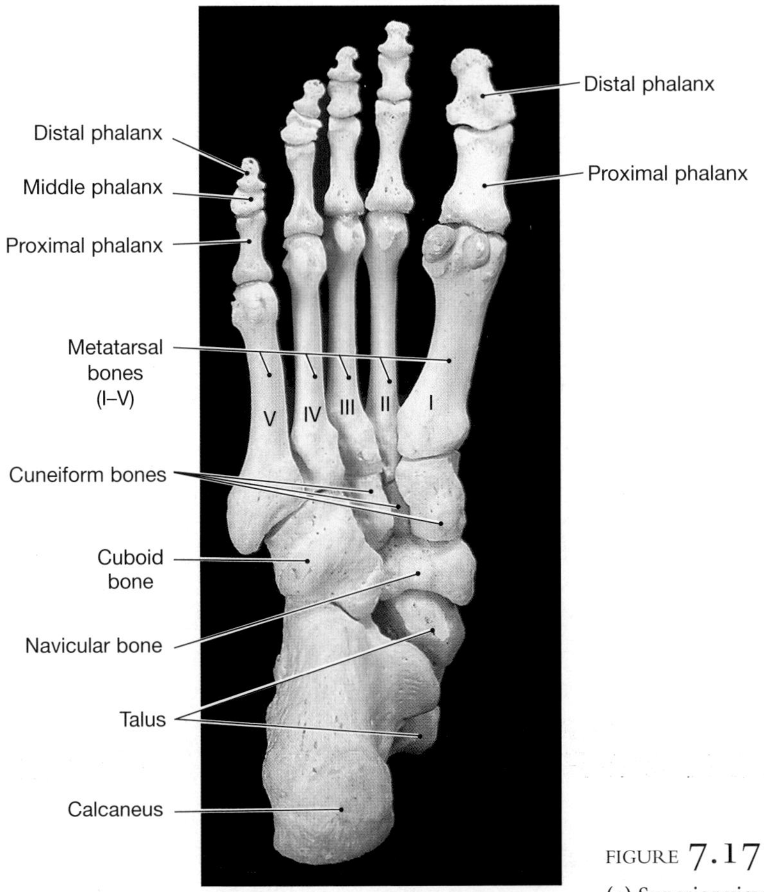

(b) Inferior (plantar) view

FIGURE 7.17 **BONES OF THE ANKLE AND FOOT, PART I**

(a) Superior view of the bones of the right foot. Note the orientation of the tarsal bones that convey the weight of the body to both the heel and the plantar surfaces of the foot. **(b)** Inferior (plantar) view.

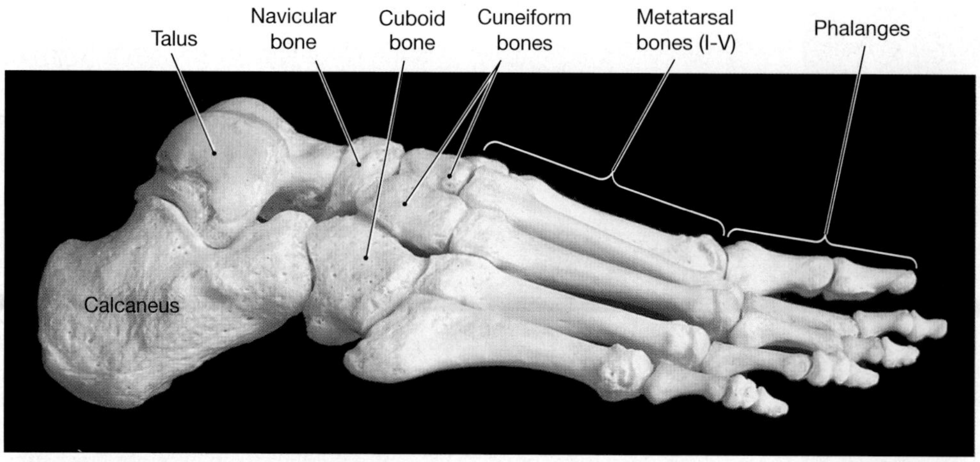

(a) Lateral view

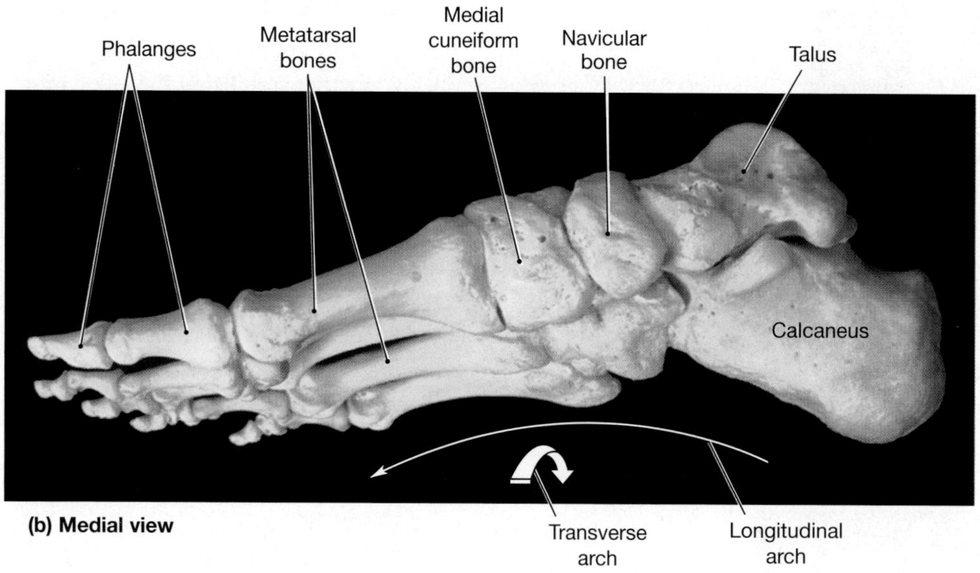

(b) Medial view

FIGURE 7.18 **BONES OF THE ANKLE AND FOOT, PART II**

(**a**) Lateral view. (**b**) Medial view, showing the relative positions of the tarsal bones and the orientation of the transverse and longitudinal arches. *See MRI Scan 8a in the Companion Atlas.*

to lateral across the sole. Proximally, the first three metatarsal bones articulate with the three cuneiform bones, and the last two articulate with the cuboid bone. Distally, each metatarsal bone articulates with a different proximal phalanx. The first metatarsal helps support the weight of the body.

The 14 **phalanges**, or toe bones have the same anatomical organization as the phalanges of the fingers. The great toe, or **hallux**, has two phalanges (proximal and distal), and the other four toes have three phalanges apiece (proximal, middle, and distal).

ARCHES OF THE FOOT [FIGURE 7.18b] Weight transfer occurs along the **longitudinal arch** of the foot (Figure 7.18b●). Ligaments and tendons maintain this arch by tying the calcaneus to the distal portions of the metatarsal bones. The lateral side of the foot carries most of the weight of the body while standing normally. This calcaneal portion of the arch has much less curvature than the medial, talar portion. The talar portion also has considerably more elasticity than the calcaneal portion of the longitudinal arch. As a result, the medial, plantar (sole) surface remains elevated, and the muscles, nerves, and blood vessels that supply the inferior surface of the foot are not squeezed between the metatarsal bones and the ground. This elasticity also helps absorb the shocks that accompany sudden changes in weight loading. For example, the stresses involved with running or ballet dancing on the toes are cushioned by the elasticity of this portion of the longitudinal arch. Because the degree of curvature changes from the medial to the lateral borders of the foot, a **transverse arch** also exists. In the condition known as *flat feet*, normal aches are lost ("fall") or never form. Individuals with this condition cannot walk long distances.

When you stand normally, your body weight is distributed evenly between the calcaneus and the distal ends of the metatarsal bones. The amount of weight transferred forward depends on the position of the foot and the placement of body weight. During dorsiflexion of the foot, as when "digging in the heels," all of the body weight rests on the calcaneus. During plantar flexion and "standing on tiptoe," the talus and calcaneus transfer the weight to the metatarsal bones and phalanges through more anterior tarsal bones.

✓ CONCEPT CHECK

- What three bones make up the os coxae?
- The fibula does not participate in the knee joint nor does it bend; but when it is fractured, it is difficult to walk. Why?
- While jumping off the back steps of his house, 10-year-old Mark lands on his right heel and breaks his foot. What foot bone is most likely broken?
- Describe at least three differences between the female and male pelvis.
- Where does the weight of the body rest during dorsiflexion? During plantar flexion?

Individual Variation In The Skeletal System [TABLES 7.1/7.2]

A comprehensive study of a human skeleton can reveal important information about the individual. For example, there are characteristic racial differences in portions of the skeleton, especially the skull and pelvis, and the development of various ridges and general bone mass can permit an estimation of muscular development and body weight. Details such as the condition of the teeth or the presence of healed fractures can provide information about the individual's medical history. Two important details, gender and age, can be determined or closely estimated on the basis of measurements indicated in Tables 7.1 and 7.2 on p. 210. Table 7.1 identifies characteristic differences between the skeletons of males and females, but not every skeleton shows every feature in classic detail. Many differences, including markings on the skull, cranial capacity, and general skeletal features, reflect differences in average body size, muscle mass, and muscular strength. The general changes in the skeletal system that take place with age are summarized in Table 7.2. Note how these changes begin at age 3 and continue throughout life. For example, fusion of the epiphyseal plates begins at about age 3, while degenerative changes in the normal skeletal system, such as a reduction in mineral content in the bony matrix, do not begin until age 45.

TABLE 7.1 GENDER DIFFERENCES IN THE HUMAN SKELETON

Region/Feature	Male	Female
SKULL		
General appearance	Heavier; rougher surface	Lighter; smoother surface
Forehead	More sloping	More vertical
Sinuses	Larger	Smaller
Cranium	About 10% larger (average)	About 10% smaller
Mandible	Larger, more robust	Lighter, smaller
Teeth	Larger	Smaller
PELVIS		
General appearance	Narrow; robust; heavier; rougher surface	Broad; light; smoother surface
Pelvic inlet	Heart-shaped	Oval to round
Iliac fossa	Deeper	Shallower
Ilium	Extends farther superior to sacroiliac joint	More vertical; less extension superior to the sacroiliac joint
Angle inferior to pubic symphysis	Less than 90°	100° or more
Acetabulum	Directed laterally	Faces slightly anteriorly as well as laterally
Obturator foramen	Oval	Triangular
Ischial spine	Points medially	Points posteriorly
Sacrum	Long, narrow triangle with pronounced sacral curvature	Broad, short triangle with less curvature
Coccyx	Points anteriorly	Points inferiorly
OTHER SKELETAL ELEMENTS		
Bone weight	Heavier	Lighter
Bone markings	More prominent	Less prominent

TABLE 7.2 AGE-RELATED CHANGES IN THE SKELETON

Region/Structure	Event(s)	Age (Years)
GENERAL SKELETON		
Bony matrix	Reduction in mineral content	Begins at ages 30–45; values differ for males versus females between ages 45 and 65; similar reductions occur in both sexes after age 65
Markings	Reduction in size, roughness	Gradual reduction with increasing age and decreasing muscular strength and mass
SKULL		
Fontanels	Closure	Completed by age 2
Metopic suture	Fusion	2–8
Occipital bone	Fusion of ossification centers	1–4
Styloid process	Fusion with temporal bone	12–16
Hyoid bone	Complete ossification and fusion	25–30
Teeth	Loss of "baby teeth"; appearance of secondary dentition; eruption of posterior molars	Detailed in Chapter 25 (Digestive System)
Mandible	Loss of teeth; reduction in bone mass; change in angle at mandibular notch	Accelerates in later years (age 60±)
VERTEBRAE		
Curvature	Appearance of major curves	3 months – 10 years (see Figure 6.19, p. 163)
Intervertebral discs	Reduction in size, percentage contribution to height	Accelerates in later years (age 60±)
LONG BONES		
Epiphyseal cartilages	Fusion	Ranges vary according to specific bone under discussion, but general analysis permits determination of approximate age (3–7, 15–22, etc.)
PECTORAL AND PELVIC GIRDLES		
Epiphyseal cartilages	Fusion	Overlapping ranges are somewhat narrower than the above, including 14–16, 16–18, 22–25 years

STUDY OUTLINE & CHAPTER REVIEW

Introduction 181

1. The **appendicular skeleton** includes the bones of the upper and lower limbs and the pectoral and pelvic girdles that support the limbs and connect them to the trunk. *(see Figure 7.1)*

The Pectoral Girdle and Upper Limb 182

1. Each upper limb articulates with the trunk through the **pectoral girdle**, or *shoulder girdle*, which consists of the **clavicle** (collarbone) and the **scapula** (shoulder blade). *(see Figures 7.2 to 7.5)*

The Pectoral Girdle 182

2. The clavicle and scapula position the shoulder joint, help move the upper limb, and provide a base for muscle attachment. *(see Figures 7.3/7.4)*
3. The **clavicle** is an S-shaped bone that extends between the manubrium of the sternum and the **acromion process** of the scapula. This bone provides the only direct connection between the pectoral girdle and the axial skeleton.
4. The **scapula** articulates with the round head of the humerus at the **glenoid cavity** of the scapula, the **glenohumeral joint** *(shoulder joint)*. Two scapular processes, the **coracoid process** and the **acromion process**, are attached to ligaments and tendons associated with the shoulder joint. The acromion articulates with the clavicle at the **acromioclavicular joint**. The acromion is continuous with the **scapular spine**, which crosses the posterior surface of the scapular body. *(see Figure 7.5)*

The Upper Limb 185

5. The **humerus** articulates with the glenoid cavity of the scapula. The articular capsule of the shoulder attaches distally to the humerus at its **anatomical neck**. Two prominent projections, the **greater tubercle** and **lesser tubercle**, are important sites for muscle attachment. Other prominent surface features include: the **deltoid tuberosity**, site of *deltoid muscle* attachment; the **radial groove**, marking the path of the *radial nerve*; the **medial** and **lateral epicondyles** for other muscle attachment; and the articular **condyle**, divided into two articular regions, the **trochlea** (medial) and **capitulum** (lateral). *(see Figures 7.2/7.6 to 7.8)*
6. Distally the humerus articulates with the ulna (at the trochlea) and the radius (at the capitulum). The trochlea extends from the **coronoid fossa** to the **olecranon fossa**. *(see Figure 7.6)*
7. The **ulna** and **radius** are the parallel bones of the forearm. The olecranon fossa of the humerus accommodates the **olecranon** of the ulna during straightening (extension) of the elbow joint (**olecranal joint**). The coronoid fossa accommodates the **coronoid process** of the ulna during bending (flexion) of the elbow joint. *(see Figures 7.2/7.7)*
8. The **carpal bones** of the wrist form two rows, **proximal** and **distal**. From lateral to medial, the proximal row consists of the **scaphoid bone, lunate bone, triquetrum** and **pisiform bone**. From lateral to medial, the distal row consists of the **trapezium, trapezoid bone, capitate bone,** and **hamate bone**. *(see Figure 7.8a)*
9. Five **metacarpal bones** articulate with the distal carpal bones. Distally, the metacarpal bones articulate with the phalanges. Four of the fingers contain three **phalanges**; the **pollex** (thumb) has only two. *(see Figure 7.8)*

The Pelvic Girdle and Lower Limb 194

The Pelvic Girdle 195

1. The pelvic girdle consists of two **ossa coxae** *(hip bones* or *innominate bones)*; each os coxae forms through the fusion of three bones – an ilium, an ischium, and a pubis. *(see Figures 7.9/7.10)*
2. The **ilium** is the largest of the hip bones. Inside the **acetabulum**, the fossa on the lateral surface of the os coxae that accommodates the **head** of the femur, the ilium is fused to the **ischium** (posteriorly) and the **pubis** (anteriorly). The **pubic symphysis** limits movement between the pubic bones of the left and right hip bones. *(see Figures 7.11/7.13)*
3. The **pelvis** consists of the two hip bones, the sacrum, and coccyx. It may be subdivided into the **false** *(greater)* **pelvis** and the **true** *(lesser)* **pelvis**. The true pelvis encloses the **pelvic cavity**. *(see Figures 7.11 to 7.13)*

The Lower Limb 201

4. The **femur** is the longest bone in the body. At its rounded **head**, it articulates with the pelvis at the acetabulum, and at its distal end its **medial** and **lateral condyles** articulate with the tibia at the knee joint. The **greater** and **lesser trochanters** are projections near the head where large tendons attach to the femur. *(see Figures 7.9/7.12d/7.14)*
5. The **patella** is a large sesamoid bone that forms within the tendon of the *quadriceps femoris* muscle group. The patellar ligament extends from the patella to the **tibial tuberosity.** *(see Figures 7.14f/7.15)*
6. The **tibia** is the largest medial bone of the leg. The prominent rough surface markings of the tibia include the **tibial tuberosity**, the **anterior margin**, the **interosseous border**, and the **medial malleolus**. The medial malleolus is a large process that provides medial support for the **talocrural joint** (ankle). *(see Figure 7.16)*
7. The **fibula** is the slender leg bone lateral to the tibia. The **head** articulates with the tibia inferior to the knee, inferior and slightly posterior to the lateral tibial condyle. A fibular process, the **lateral malleolus**, stabilizes the ankle joint by preventing medial movement of the tibia across the talus. *(see Figure 7.16/7.17)*
8. The **tarsus**, or ankle, includes seven **tarsal bones**; only the smooth superior surface of the trochlea of the talus articulates with the tibia and fibula. It has lateral and medial extensions that articulate with the lateral and medial malleoli of the fibula and tibia, respectively. When standing normally, most of the body weight is transferred to the calcaneus, and the rest is passed on to the **metatarsal bones**. Weight transfer occurs along the **longitudinal arch** and **transverse arches** of the foot. *(see Figures 7.17/7.18)*
9. The basic organizational pattern of the **metatarsal bones** and **phalanges** of the foot is the same as that of palm and phalanges of the hand. *(see Figures 7.17/7.18)*

Individual Variation in the Skeletal System 209

1. Studying a human skeleton can reveal important information such as race, medical history, weight, sex, body size, muscle mass, and age. *(see Tables 7.1/7.2)*
2. A number of age-related changes and events take place in the skeletal system. These changes begin at about age 3 and continue throughout life. *(see Tables 7.1/7.2)*

LEVEL 1 REVIEWING FACTS AND TERMS

Match each numbered item with the most closely related lettered item. Use letters for answers in the spaces provided.

Column A	Column B
______ 1. shoulder	a. tibia
______ 2. hip	b. pectoral girdle
______ 3. scapula	c. radius
______ 4. trochlea	d. phalanges
______ 5. ulnar notch	e. pelvic girdle
______ 6. one coxal bone	f. femur
______ 7. greater trochanter	g. infraspinous fossa
______ 8. medial malleolus	h. calcaneus
______ 9. heel bone	i. ilium
______10. toes	j. humerus

11. Structural characteristics of the pectoral girdle that adapt it to a wide range of movement include
    (a) heavy bones
    (b) relatively weak joints
    (c) limited range of motion at the shoulder joint
    (d) joints stabilized by ligaments and tendons to the thoracic cage
12. The depression on the anterior surface at the distal end of the humerus is the
    (a) olecranon fossa (b) coronoid fossa
    (c) intercondylar fossa (d) intertubercular groove
13. What bone articulates with the os coxae at the acetabulum?
    (a) sacrum (b) humerus
    (c) femur (d) tibia
14. The fibula
    (a) forms an important part of the knee joint
    (b) articulates with the femur
    (c) helps to bear the weight of the body
    (d) provides lateral stability to the ankle
15. Structural characteristics of the pelvic girdle that adapt it to the role of bearing weight of the body include
    (a) heavy bones (b) stable joints
    (c) limited range of movement at some joints (d) all of the above
16. Which of the following is a characteristic of the male pelvis?
    (a) triangular obturator foramen
    (b) coccyx points into the pelvic outlet
    (c) sacrum broad and short
    (d) ischial spine points posteriorly
17. The sole of the foot is partially supported by the
    (a) metacarpal bones (b) metatarsal bones
    (c) carpal bones (d) all of the above
18. The ___ of the radius assists in the stabilization of the wrist joint.
    (a) olecranon (b) coronoid process
    (c) styloid process (d) radial tuberosity
19. The olecranon is found on the
    (a) humerus (b) radius
    (c) ulna (d) femur
20. The small, anterior projection of the scapula that extends over the superior margin of the glenoid cavity is the
    (a) scapular spine (b) acromion
    (c) coracoid process (d) supraspinous process

LEVEL 2 REVIEWING CONCEPTS

1. The only fixed support for the pectoral girdle is the
   (a) scapula (b) clavicle
   (c) humerus (d) sternum
2. The pelvis
   (a) is a composite structure
   (b) contains bones from both the axial and appendicular skeleton
   (c) protects abdominal organs
   (d) all of the above
3. Lee fractured her pisiform bone in an accident. What part of her body was injured?
   (a) leg (b) ankle
   (c) wrist (d) shoulder
4. In determining the age of a skeleton, what pieces of information would be helpful?
5. What is the importance of maintaining the correct amount of curvature of the longitudinal arch of the foot?
6. Why are fractures of the clavicle so common?
7. Why is the tibia, but not the fibula, involved in the transfer of weight to the ankle and foot?
8. What is the function of the olecranon of the ulna?
9. How is body weight passed to the metatarsal bones?

LEVEL 3 CRITICAL THINKING AND CLINICAL APPLICATIONS

1. Why would a person suffering from osteoporosis be more likely to suffer a broken hip than a broken shoulder?
2. Archaeologists find the pelvis of a primitive human and are able to tell the sex, relative age, and some physical characteristics of the individual. How is this possible from the pelvis only?
3. Due to a development defect, Joe is born without clavicles. What effect might this have on his normal activities?
4. The condition of lower than normal longitudinal arches is known as "flat feet." What structural problem causes flat feet?

✓ ANSWERS TO CONCEPT CHECK QUESTIONS

p. 194 **1.** The clavicle attaches the scapula to the sternum and thus restricts the scapula's range of movement. If the clavicle is broken, the scapula will have a greater range of movement and will be less stable. **2.** The radius is in a lateral position when the forearm is in the anatomical position. **3.** The olecranon is the point of the elbow. During extension of the elbow, the olecranon swings into the olecranon fossa on the posterior surface of the humerus to prevent overextension. **4.** The clavicle articulates with the manubrium of the sternum, and this provides the only direct connection between the pectoral girdle and the axial skeleton.
p. 209 **1.** The three bones that make up the os coxae are the ilium, ischium, and pubic bones. **2.** Although the fibula is not part of the knee joint and does not bear weight, it is an important point of attachment for many leg muscles. When the fibula is fractured, these muscles cannot function properly to move the leg and walking is difficult and painful. The fibula also helps stabilize the ankle joint. **3.** Mark has most likely fractured his calcaneus (heel bone). **4.** There are six differences that are adaptations for childbearing, including: an enlarged pelvic outlet; less curvature on the sacrum and coccyx, which in the male arc anteriorly into the pelvic outlet; a wider, more circular pelvic inlet; a relatively broad, low pelvis; ilia that project farther laterally, but do not extend as far superior to the sacrum; and a broader pubic arch, with the inferior angle between the pubic bones greater than 100°. **5.** During dorsiflexion of the foot, as when "digging in the heels," all of the body weight rests on the calcaneus. During plantar flexion and "standing on tiptoe," the talus and calcaneus transfer the weight to the metatarsal bones and phalanges through more anterior tarsal bones.

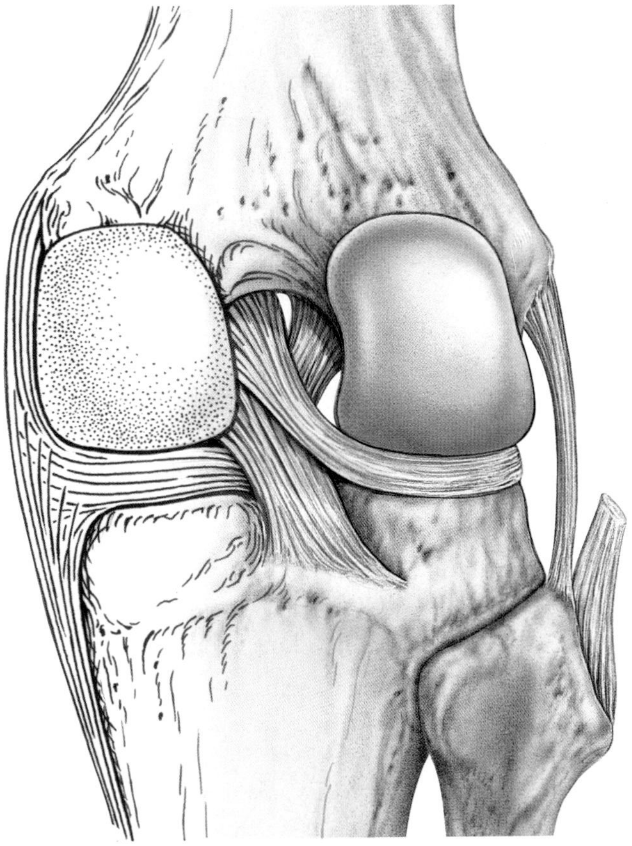

8

THE SKELETAL SYSTEM

Articulations

We depend upon our bones for support, but support without mobility would leave us little better than statues. Body movements must conform to the limits of the skeleton. For example, you cannot bend the shaft of the humerus or femur; movements are restricted to joints. Joints, or **articulations** (ar-tik-ū-LĀ-shuns), exist wherever two bones meet; they may be in direct contact or separated by fibrous tissue, cartilage, or fluid. Each joint tolerates a specific range of motion, and a variety of bony surfaces, cartilages, ligaments, tendons, and muscles work together to keep movement within the normal range. In this chapter we will focus on how bones are linked together to give us freedom of movement. The function and mobility of each joint depends on its anatomical design. Some joints are interlocking and completely prohibit movement, whereas other joints permit either slight movement or extensive movement. Immovable and slightly movable joints are more common in the axial skeleton, whereas the freely movable joints are more common in the appendicular skeleton.

Classification of Joints [TABLES 8.1/8.2]

Three functional categories of joints are based on the range of motion permitted (Table 8.1). An immovable joint is a **synarthrosis** (sin-ar-THRŌ-sis; *syn*, together + *arthros*, joint); a slightly movable joint is an **amphiarthrosis** (am-fē-ar-THRŌ-sis; *amphi*, on both sides); and a freely movable joint is a **diarthrosis** (dī-ar-THRŌ-sis; *dia*, through). Subdivisions within each functional category indicate significant structural differences. Synarthrotic or amphiarthrotic joints are classified as fibrous or cartilaginous, and diarthrotic joints are subdivided according to the degree of movement permitted. An alternative classification scheme is based on joint structure only (bony fusion, fibrous, cartilaginous, or synovial). This classification scheme is presented in Table 8.2. We will use the functional classification here, as our focus will be on the degree of motion permitted, rather than the histological structure of the articulation.

Synarthroses (Immovable Joints)

At a synarthrosis the bony edges are quite close together and may even interlock. A **suture** (*sutura*, a sewing together) is a synarthrotic joint found only between the bones of the skull. The edges of the bones are interlocked and bound together at the suture by dense connective tissue. A different type of fibrous synarthrosis binds each tooth to the surrounding bony socket. This fibrous connection is the **periodontal ligament** (per-ē-ō-DON-tal; *peri*, around + *odontos*, tooth), and the articulation is a **gomphosis** (gom-FŌ-sis; *gomphosis*, a bolting together).

In a growing bone, the diaphysis and each epiphysis are bound together by an epiphyseal plate, an example of a cartilaginous synarthrosis. This rigid connection is called a **synchondrosis** (sin-kon-DRŌ-sis; *syn*, together + *chondros*, cartilage). Sometimes two separate bones actually fuse together, and the boundary between them disappears. This creates a **synostosis** (sin-os-TŌ-sis), a totally rigid, immovable joint.

Amphiarthroses (Slightly Movable Joints)

An amphiarthrosis permits very limited movement, and the bones are usually farther apart than they are at a synarthrosis. The bones may be connected by collagen fibers or cartilage. At a **syndesmosis** (sin-dez-MŌ-sis; *desmo*, band or ligament), the articulating bones are connected by a ligament. Examples include the distal articulation between the tibia and fibula. At a **symphysis** the bones are separated by a wedge or pad of fibrocartilage. The articulations between adjacent vertebral bodies (via the *intervertebral disc*) and the anterior connection between the two pubic bones (the *pubic symphysis*) are examples of this type of joint.

Diarthroses (Freely Movable Joints) [FIGURE 8.1]

Diarthroses, or **synovial** (si-NŌ-vē-al) **joints**, permit a wide range of motion. Synovial joints are typically found at the ends of long bones, such as those of the upper and lower limbs. Under normal conditions, the bony surfaces cannot contact one another, because the articulating surfaces are covered by special **articular cartilages**. These cartilages act as shock absorbers and also help reduce friction. Although articular cartilages otherwise resemble hyaline cartilages, they have no perichondrium, and the matrix contains much more fluid. Figure 8.1● introduces the structure of synovial joints.

TABLE 8.1

A Functional Classification of Articulations

Functional Category	Structural Category	Description	Example
Synarthrosis (no movement)	**Fibrous**		
	Suture	Fibrous connections plus extensive interlocking	Between the bones of the skull
	Gomphosis	Fibrous connections plus insertion in alveolar process	Periodontal ligaments between the teeth and jaws
	Cartilaginous		
	Synchondrosis	Interposition of cartilage plate	Epiphyseal cartilages
	Bony fusion		
	Synostosis	Conversion of other articular form to solid mass of bone	Portions of the skull, such as along the metopic suture of the frontal bone
AMPHIARTHROSIS (little movement)	**Fibrous**		
	Syndesmosis	Ligamentous connection	Between the tibia and fibula
	Cartilaginous Symphysis	Connection by a fibrocartilage pad	Between right and left ossa coxae of pelvis; between adjacent vertebral bodies
DIARTHROSIS (free movement)	**Synovial**	Complex joint bounded by joint capsule and containing synovial fluid	Numerous; subdivided by range of movement *(see Figures 8.3 to 8.6)*
	Monaxial	Permits movement in one plane	Elbow, ankle
	Biaxial	Permits movement in two planes	Ribs, wrist
	Triaxial	Permits movement in all three planes	Shoulder, hip

TABLE 8.2 **A STRUCTURAL CLASSIFICATION OF ARTICULATIONS**

Structure	Type	Functional Category	Example*
BONY FUSION	Synostosis	Synarthrosis	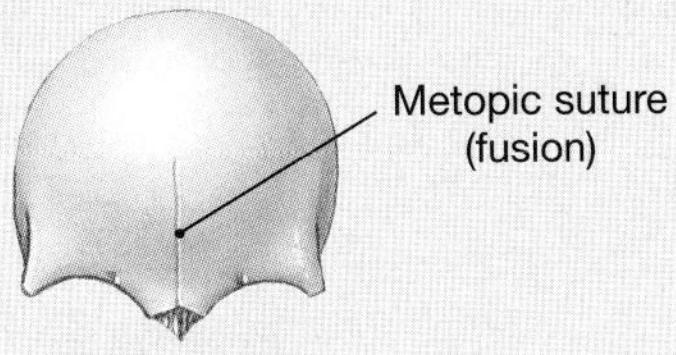
FIBROUS JOINT	Suture Gomphosis Syndesmosis	Synarthrosis Synarthrosis Amphiarthrosis	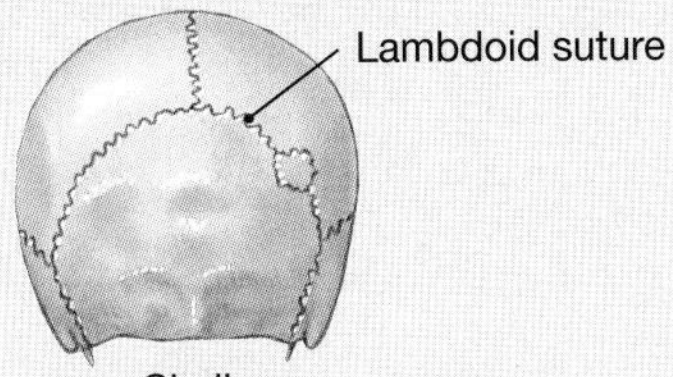
CARTILAGINOUS JOINT	Synchondrosis Symphysis	Synarthrosis Amphiarthrosis	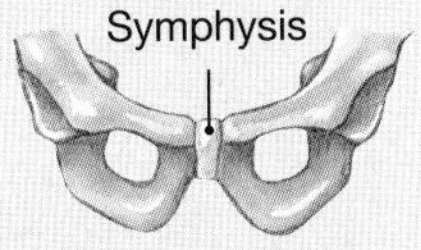
SYNOVIAL JOINT	Monaxial Biaxial Triaxial	All diarthroses	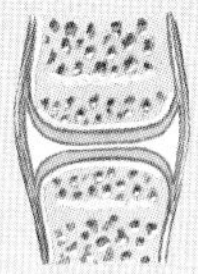

*For other examples, see Table 8.1

SYNOVIAL FLUID

A synovial joint is surrounded by a **joint capsule**, or **articular capsule**, composed of a thick layer of dense connective tissue. A *synovial membrane* lines the joint cavity but stops at the edges of the articular cartilages. p. 76 Synovial membranes produce the **synovial fluid** that fills the joint cavity. Synovial fluid serves three functions:

1. *Provides lubrication*: When a portion of the cartilage is compressed, some of the fluid is squeezed out of the cartilage matrix and into the space between the opposing surfaces. This thin layer of fluid reduces friction between moving surfaces in a joint to around one-fifth of that between two pieces of ice. The articular cartilages act like sponges, for when the compression stops, fluid is sucked back into them.
2. *Nourishes the chondrocytes*: The total quantity of synovial fluid in a joint is normally less than 3 ml, even in a large joint such as the knee. This relatively small volume of fluid must be circulated continually to provide nutrients and a route for waste disposal for the chondrocytes of the articular cartilages. The synovial fluid circulates whenever the joint moves, and the compression and reexpansion of the articular cartilages pump synovial fluid into and out of the cartilage matrix.
3. *Acts as a shock absorber*: Synovial fluid cushions shocks in joints that are subjected to compression. For example, the hip, knee, and ankle joints are compressed during walking, and they are severely compressed during jogging or running. When the pressure suddenly increases, the synovial fluid absorbs the shock and distributes it evenly across the articular surfaces. *Rheumatism, Arthritis, and Synovial Function p. 790*

ACCESSORY STRUCTURES [FIGURE 8.1]

Synovial joints may have a variety of accessory structures, including pads of cartilage or fat, ligaments, tendons, and bursae (Figure 8.1●).

CARTILAGES AND FAT PADS [FIGURE 8.1] In complex joints such as the knee (Figure 8.1●), accessory structures may lie between the opposing articular surfaces and modify the shapes of the joint surfaces. These include:

- *Menisci* (men-IS-kē; *meniscus*, crescent), or **articular discs**, are fibrocartilage pads that may subdivide a synovial cavity, channel the flow of synovial fluid, allow for variations in the shapes of the articular surfaces, or restrict movements at the joint.
- *Fat pads* are often found around the periphery of the joint, lightly covered by a layer of synovial membrane. Fat pads provide protection for the articular cartilages and serve as packing material for the joint as a whole. Fat pads fill spaces created when bones move and the joint cavity changes shape.

LIGAMENTS [FIGURE 8.1b] The joint capsule that surrounds the entire joint is continuous with the periostea of the articulating bones. **Accessory ligaments** are localized thickening of the capsule; these ligaments reinforce and strengthen the capsule. They include **extracapsular ligaments** on the outer surface of the capsule and **intracapsular ligaments** on the inner surface of the capsule (Figure 8.1b●).

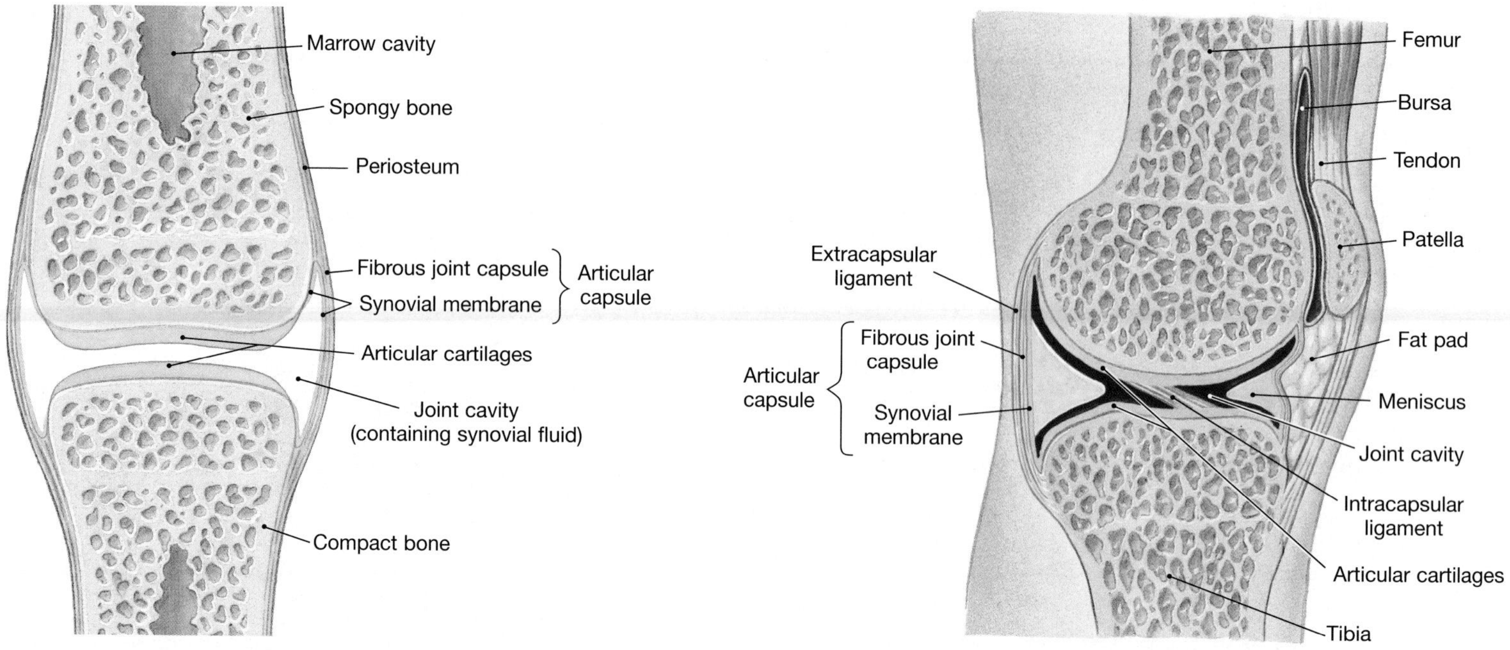

(a) Synovial joint, sagittal section **(b) Knee joint, sagittal section**

FIGURE 8.1 **STRUCTURE OF A SYNOVIAL JOINT**

Synovial joints are diarthrotic joints that permit a wide range of motion. **(a)** Diagrammatic view of a simple articulation. **(b)** A simplified sectional view of the knee joint.

TENDONS [FIGURE 8.1b] While not part of the articulation itself, tendons (Figure 8.1b●) usually pass across or around a joint. Normal muscle tone keeps them taut, and their presence may limit the range of motion and provide additional mechanical support.

BURSAE [FIGURE 8.1b] **Bursae** (Figure 8.1b●) are small, fluid-filled pockets in connective tissue. They are filled with synovial fluid and lined by a synovial membrane. Bursae may be connected to the joint cavity, or they may be completely separate from it. Bursae form where a tendon or ligament rubs against other tissues. Their function is to reduce friction and act as a shock absorber. Bursae are found around most synovial joints, such as the shoulder joint. **Synovial tendon sheaths** are tubular bursae that surround tendons where they pass across bony surfaces. Bursae may also appear beneath the skin covering a bone or within other connective tissues exposed to friction or pressure. Bursae that develop in abnormal locations, or due to abnormal stresses, are called *adventitious bursae.*

STRENGTH VERSUS MOBILITY

A joint cannot be both highly mobile and very strong. The greater the range of motion at a joint, the weaker it becomes. A synarthrosis, the strongest type of joint, does not permit any movement, whereas any mobile diarthrosis may be damaged by movement beyond its normal range of motion. Several factors combine to limit mobility and reduce the chance of injury:

- the presence of accessory ligaments and the collagen fibers of the joint capsule;
- the shapes of the articulating surfaces that prevent movement in specific directions;
- the presence of other bones, skeletal muscles, or fat pads around the joint;
- tension in tendons attached to the articulating bones. When a skeletal muscle contracts and pulls on a tendon, it may either encourage or oppose movement in a specific direction.

✓ CONCEPT CHECK

- Distinguish between a synarthrosis and an amphiarthrosis.
- What is the main advantage of a synovial joint?
- Identify two functions of synovial fluid.
- What are bursae? What is their function?

CLINICAL BRIEF

DISLOCATION OF A SYNOVIAL JOINT

When a **dislocation**, or **luxation** (luks-Ā-shun), occurs, the articulating surfaces are forced out of position. This displacement can damage the articular cartilages, tear ligaments, or distort the joint capsule. Although the *inside* of a joint has no pain receptors, nerves that monitor the capsule, ligaments, and tendons are quite sensitive, and dislocations are very painful. The damage accompanying a partial dislocation, or **subluxation** (sub-luks-Ā-shun), is less severe. People who are "double-jointed" have joints that are weakly stabilized. Although their joints permit a greater range of motion than those of other individuals, they are more likely to suffer partial or complete dislocations.

Articular Form and Function

To *understand* human movement you must become aware of the relationship between structure and function at each articulation. To *describe* human movement you need a frame of reference that permits accurate and precise communication. The synovial joints can be classified according to their anatomical and functional properties. To demonstrate the basis for that classification, we will describe the movements that can occur at a typical synovial joint, using a simplified model.

Describing Dynamic Motion [FIGURE 8.2]

Take a pencil (or pen) as your model, and stand it upright on the surface of a desk or table, as shown in Figure 8.2a●. The pencil represents a bone, and the desk is an articular surface. A little imagination and a lot of twisting, pushing, and pulling will demonstrate that there are only three ways to move the model. Considering them one at a time will provide a frame of reference for analyzing any complex movement.

Possible Movement 1: Moving the point.

If you hold the pencil upright but do not secure the point, you can push the pencil across the surface. This kind of motion is called *gliding* (Figure 8.2b●), and it is an example of **linear motion**. You could slide the point forward or backward, from one side to the other, or diagonally. However you choose to move the pencil, the motion can be described using two lines of reference. One line represents forward/backward motion, and the other left/right movement. For example, a simple movement along one axis could be described as "forward 1 cm" or "left 2 cm." A diagonal movement could be described using both axes, as in "backward 1 cm and to the right 2.5 cm."

Possible Movement 2: Changing the angle of the shaft.

While holding the tip in position, you can still move the free (eraser) end forward and backward or from side to side. These movements, which change the angle between the shaft and the articular surface, are examples of **angular motion** (Figure 8.2c●).

Any angular movement can be described with reference to the same two axes (forward/backward, left/right) and the angular change (in degrees). However, in one instance a special term is used to describe a complex angular movement. Grasp the free end of the pencil, and move it until the shaft is no longer vertical. Now with the point held firmly in place, move the free end through a complete circle (Figure 8.2d●). This movement is very difficult to describe. Anatomists avoid the problem entirely by using a special term, **circumduction** (ser-kum-DUK-shun; *circum*, around), for this type of angular motion.

Possible Movement 3: Rotating the shaft.

If you prevent movement of the base and keep the shaft vertical, you can still spin the shaft around its longitudinal axis. This movement is called **rotation** (Figure 8.2e●). Several articulations will permit partial rotation, but none can rotate freely; such a movement would hopelessly tangle the blood vessels, nerves, and muscles that cross the joint.

An articulation that permits movement along only one axis is called **monaxial** (mon-AKS-ē-al), or *uniaxial* (unē-AKS-ē-al). In the above model, if an articulation permits angular movement only in the forward/backward plane, or prevents any movement other than rotation around its longitudinal axis, it is monaxial. If movement can occur along two axes, the articulation is **biaxial** (bī-AKS-ē-al). If the pencil could undergo angular motion in the forward/backward or left/right plane, but not in some combination of the two, it would be biaxial. **Triaxial** (trī-AKS-ē-al) joints permit a combination of rotational and angular motion.

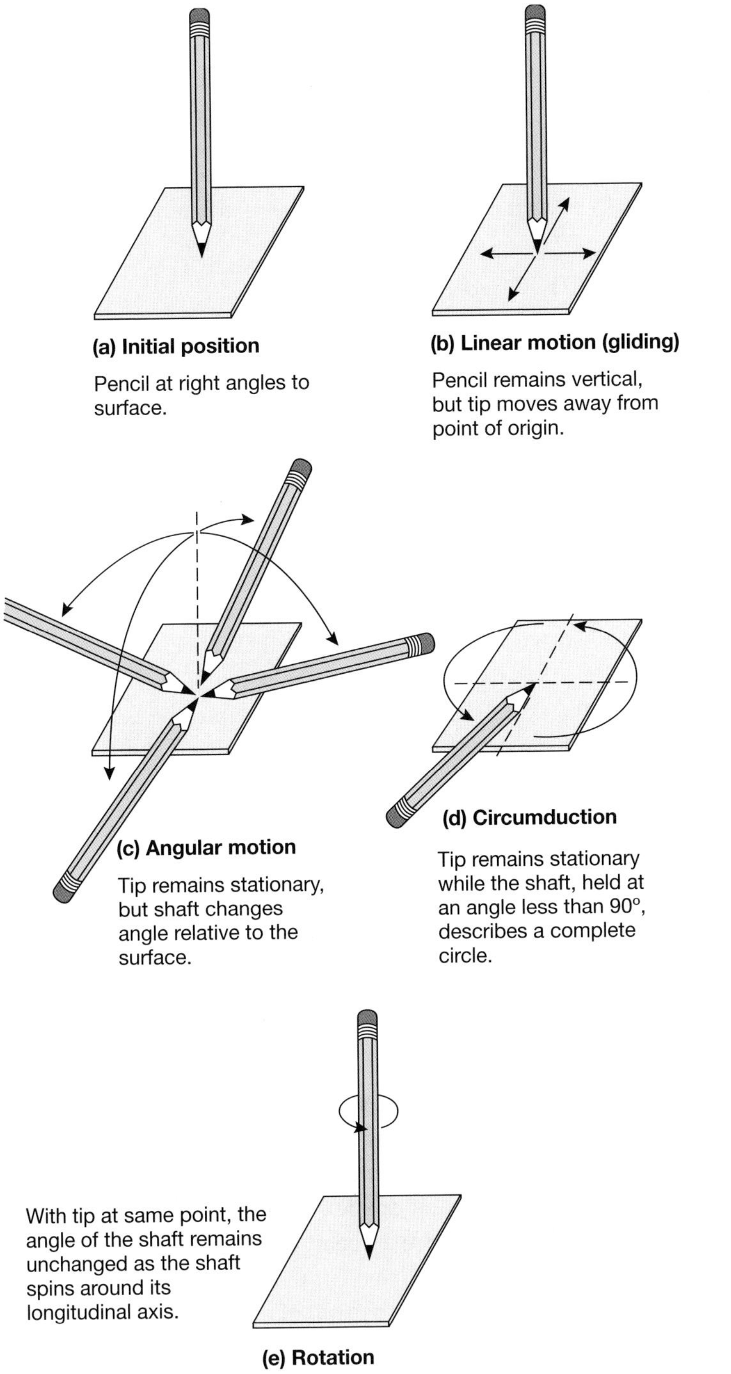

(a) Initial position
Pencil at right angles to surface.
(b) Linear motion (gliding)
Pencil remains vertical, but tip moves away from point of origin.
(c) Angular motion
Tip remains stationary, but shaft changes angle relative to the surface.
(d) Circumduction
Tip remains stationary while the shaft, held at an angle less than 90°, describes a complete circle.
With tip at same point, the angle of the shaft remains unchanged as the shaft spins around its longitudinal axis.
(e) Rotation

FIGURE 8.2 **A Simple Model of Articular Motion**

Three types of dynamic motion are described: **(a)** initial position of the model, **(b)** possible movement 1, showing gliding, an example of linear motion, **(c)** possible movement 2, showing angular motion, **(d)** possible movement 2, showing a special type of angular motion called circumduction, and **(e)** possible movement 3, showing rotation.

Types of Movements

All movements, unless otherwise indicated, are described with reference to a figure in the anatomical position. In descriptions of motion at synovial joints, anatomists use descriptive terms that have specific meanings. We will consider these movements with regard to the basic categories of movement considered in the previous section.

LINEAR MOTION (GLIDING) [FIGURE 8.2b]

In **gliding**, two opposing surfaces slide past one another (Figure 8.2b●). Gliding occurs between the surfaces of articulating carpal bones and tarsal bones and between the clavicles and the sternum. The movement can occur in almost any direction, but the amount of movement is slight, and rotation is usually prevented by the capsule and associated ligaments.

ANGULAR MOTION [FIGURE 8.3]

Examples of angular motion include *abduction, adduction, flexion*, and *extension*. The description of each movement is based on reference to an individual in the anatomical position (Figure 8.3●).

- **Abduction** (*ab*, from) is movement *away from the longitudinal axis of the body* in the frontal plane. For example, swinging the upper limb to the side is abduction of the limb; moving it back constitutes **adduction** (*ad*, to). Abduction of the wrist moves the heel of the hand away from the body, whereas adduction moves it toward the body. Spreading the fingers or toes apart abducts them, because they move *away* from a central digit (finger or toe). Bringing them together constitutes adduction. Abduction and adduction always refer to movements of the appendicular skeleton (Figure 8.3a,c●).
- **Flexion** (FLEK-shun) can be defined as movement in the anterior-posterior plane that *reduces the angle between the articulating elements*. **Extension** occurs in the same plane, but it *increases the angle between articulating elements* (Figure 8.3b●). When you bring your head toward your chest, you flex the intervertebral articulations of the neck. When you bend down to touch your toes, you flex the intervertebral articulations of the entire vertebral column. Extension reverses these movements, returning the limb or body axis to the anatomical position.

  Flexion at the shoulder or hip swings the limbs anteriorly, whereas extension moves them posteriorly. Flexion at the wrist moves the palm forward, and extension moves it back. In each of these examples, extension can be continued past the anatomical position, in which case **hyperextension** occurs. You can also hyperextend the joints of the neck, a movement that allows you to gaze at the ceiling (Figure 8.3b●). Hyperextension of other joints is usually prevented by ligaments, bony processes, or soft tissues.
- A special type of angular motion, **circumduction** (Figure 8.3d●), was also introduced in our model. A familiar example of circumduction is moving your arm in a loop, as when drawing a large circle on a chalkboard.

ROTATION [FIGURE 8.4]

Rotation of the head may involve **left rotation** or **right rotation**, as in shaking the head "no." In analysis of movements of the limbs, if the anterior aspect of the limb rotates *inward*, toward the ventral surface of the body, you have **internal rotation**, or **medial rotation**. If it turns outward, you have **external rotation**, or **lateral rotation**. These rotational movements are illustrated in Figure 8.4●.

The articulations between the radius and ulna permit the rotation of the distal end of the radius from the anatomical position across the anterior surface of the ulna. This moves the wrist and hand from palm-facing-front to palm-facing-back. This motion is called **pronation** (prō-NĀ-shun); the opposing movement, which turns the palm forward, is **supination** (soo-pi-NĀ-shun).

SPECIAL MOVEMENTS [FIGURE 8.5]

There are a number of special terms that apply to specific articulations or unusual types of movement (Figure 8.5●).

- **Eversion** (ē-VER-shun; *e*, out + *vertere*, to turn) is a twisting motion of the foot that turns the sole outward (Figure 8.5a●). The opposite movement, turning the sole inward, is called **inversion** (*in*, into).
- **Dorsiflexion** and **plantar flexion** (*planta*, sole) also refer to movements of the foot (Figure 8.5b●). Dorsiflexion is flexion of the ankle and elevation of the sole, as when "digging in the heels." Plantar flexion, the opposite movement, extends the ankle and elevates the heel, as when standing on tiptoes.
- **Lateral flexion** occurs when the vertebral column bends to the side. This movement is most pronounced in the cervical and thoracic regions (Figure 8.5c●).
- **Protraction** entails moving a part of the body anteriorly in the horizontal plane. **Retraction** is the reverse movement (Figure 8.5d●). You protract your jaw when you grasp your upper lip with your lower teeth, and you protract your clavicles when you cross your arms.
- **Opposition** is the special movement of the thumb that produces pad-to-pad contact of the thumb with any other finger. (Figure 8.5e●).
- **Elevation** and **depression** occur when a structure moves in a superior or inferior direction. You depress your mandible when you open your mouth and elevate it as you close it (Figure 8.5f●). Another familiar elevation occurs when you shrug your shoulders.

A Structural Classification of Synovial Joints [FIGURE 8.6]

Synovial joints are freely movable diarthrotic joints. Since they permit a wide range of motion, they are classified according to the type and range of movement permitted. The structure of the joint defines its movement.

- *Gliding joints*: **Gliding joints**, also called *planar joints*, have flattened or slightly curved faces (Figure 8.6a●). The relatively flat articular surfaces slide across one another, but the amount of movement is very slight. Ligaments usually prevent or restrict rotation. Gliding joints are found at the ends of the clavicles, between the carpal bones, between the tarsal bones, and between the articular facets of adjacent vertebrae. Gliding joints may be *nonaxial*, which means that they permit only small sliding movements, or *multiaxial*, which means that they permit sliding in any direction.
- *Hinge joints*: **Hinge joints** permit angular movement in a single plane, like the opening and closing of a door (Figure 8.6b●). A hinge joint is an example of a monaxial joint. Examples of hinged joints would be the elbow and knee.
- *Pivot joints*: **Pivot joints** are also monaxial, but they permit only rotation (Figure 8.6c●). A pivot joint between the atlas and axis allows you to rotate your head to either side.
- *Ellipsoidal joints*: In an **ellipsoidal joint**, or *condyloid joint*, an oval articular face nestles within a depression on the opposing surface (Figure 8.6d●). With such an arrangement, angular motion occurs in two planes, along or across the length of the oval. It is thus an example of a biaxial joint. Ellipsoidal joints connect the fingers and toes with the metacarpal bones and metatarsal bones, respectively.
- *Saddle joints*: **Saddle joints** (Figure 8.6e●) have complex articular faces. Each one resembles a saddle because it is concave on one axis

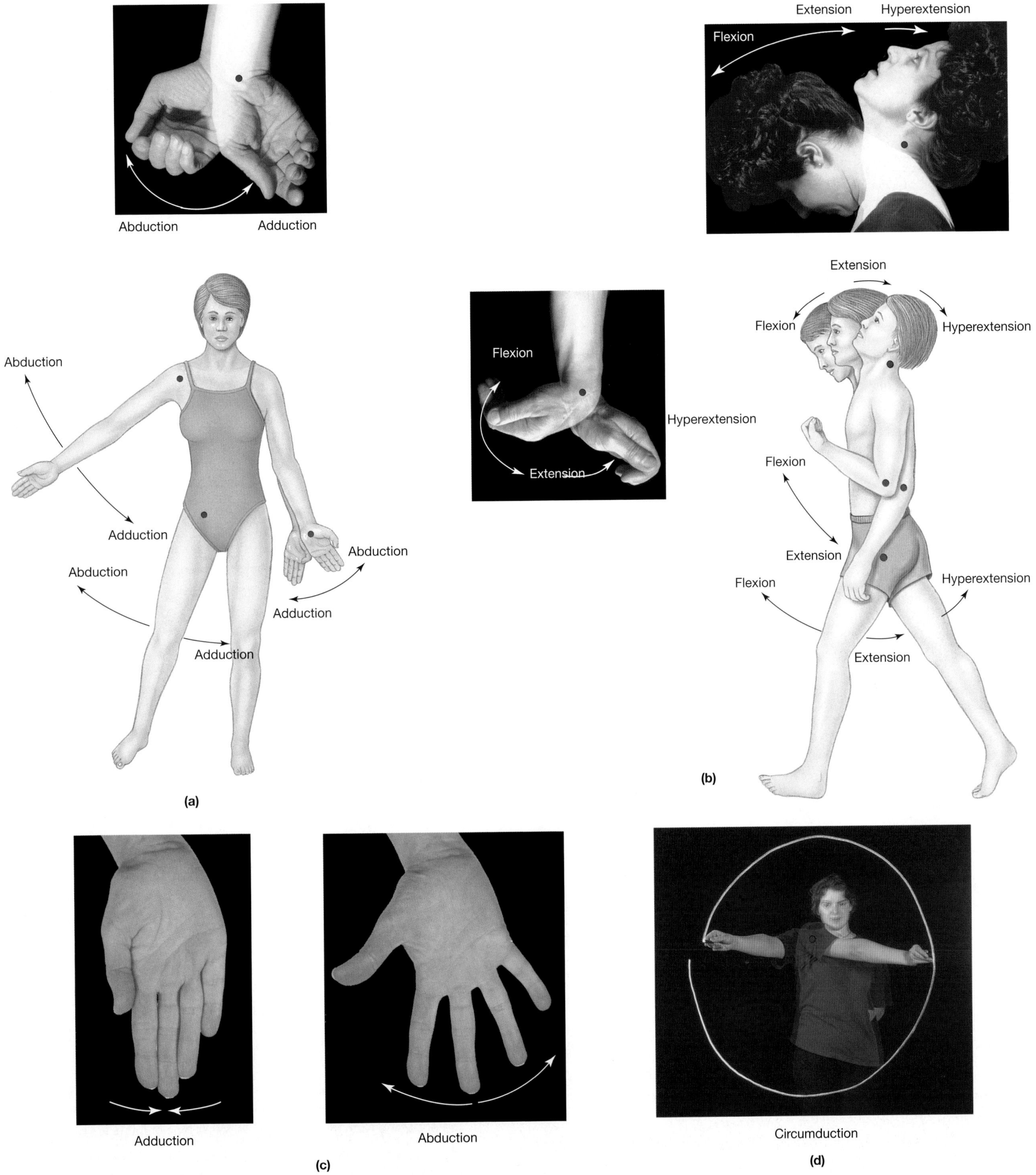

FIGURE 8.3 **Angular Movements**

Examples of movements that change the angle between the shaft and the articular surface. The red dots indicate the locations of the joints involved in the illustrated movement. (**a**) abduction/adduction, (**b**) flexion/extension, (**c**) abduction/adduction, and (**d**) circumduction.

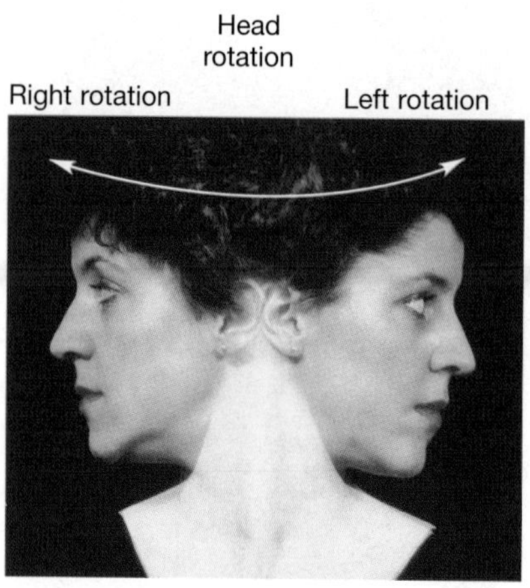

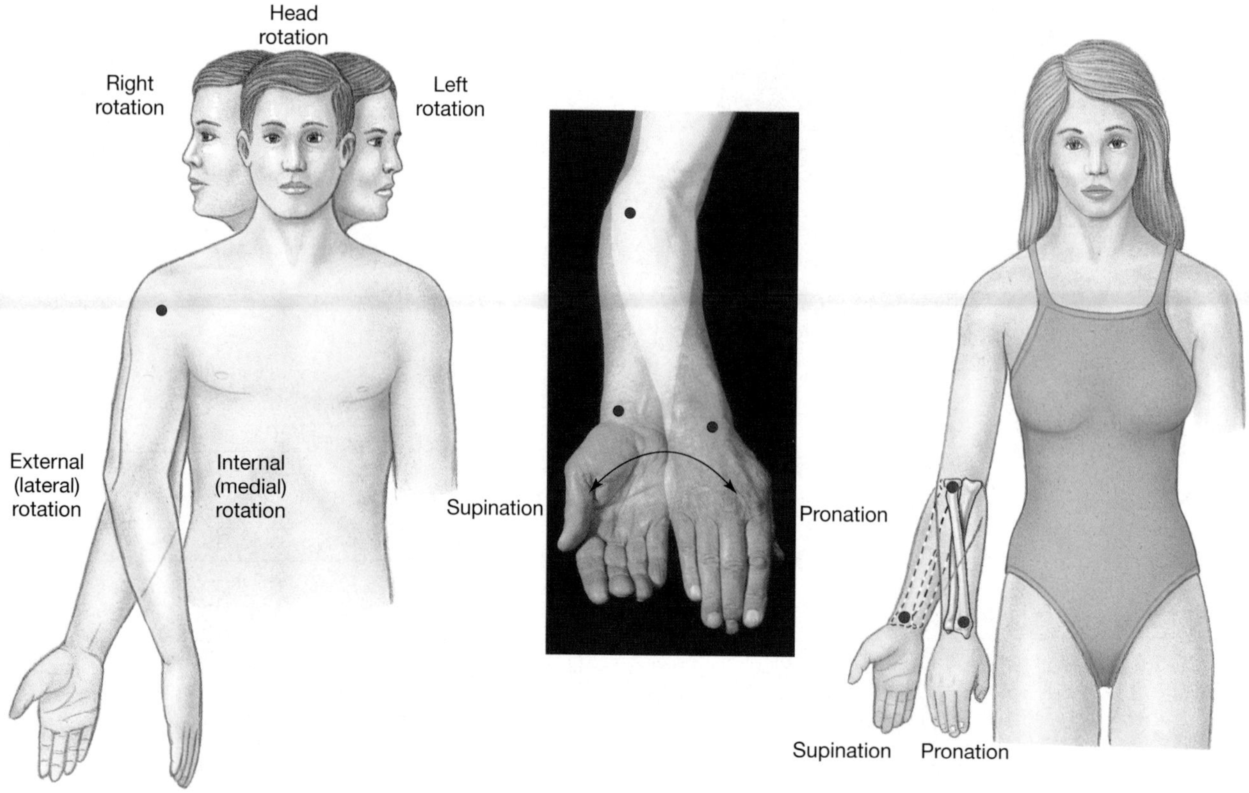

FIGURE 8.4 **Rotational Movements**

Examples of motion in which the shaft of the bone rotates.

FIGURE 8.5 **Special Movements**

Examples of special terms used to describe movement at specific joints or unique directions of movement. (**a**) eversion/inversion, (**b**) dorsiflexion/plantar flexion, (**c**) lateral flexion, (**d**) retraction/protraction, (**e**) opposition, and (**f**) elevation/depression.

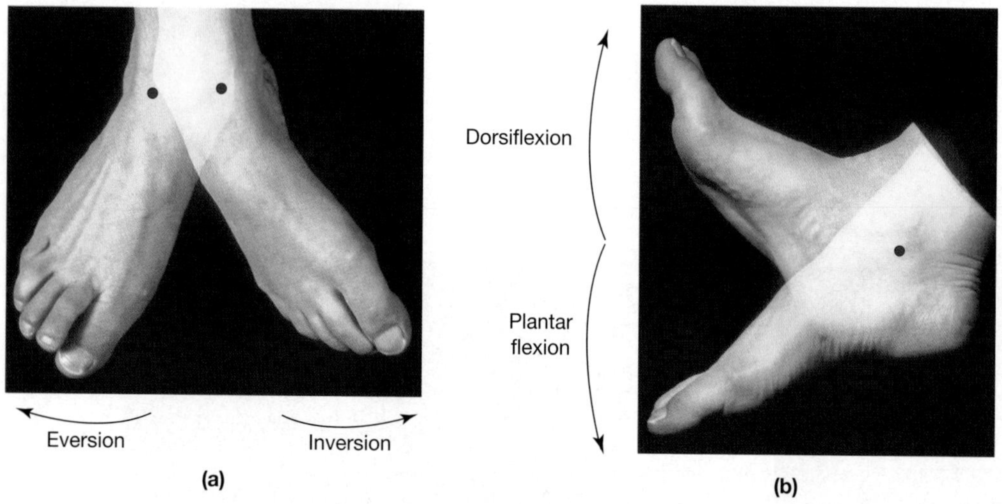

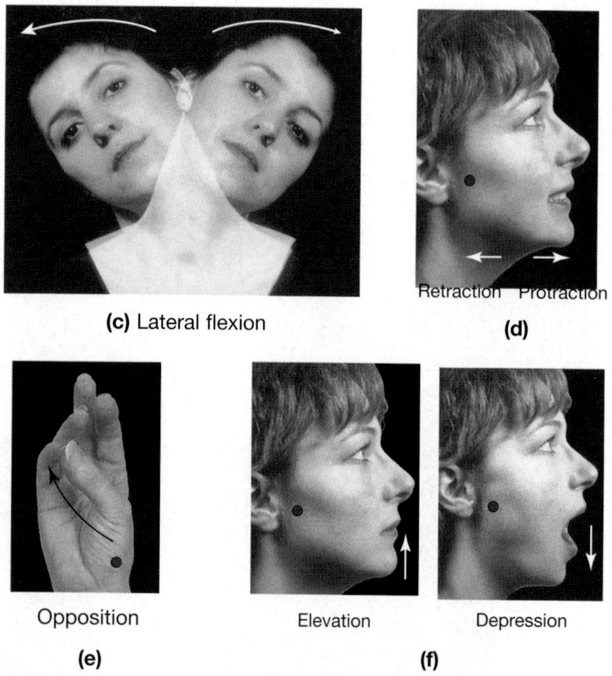

and convex on the other. Saddle joints are extremely mobile, allowing extensive angular motion without rotation. They are usually classified as biaxial joints. Moving the saddle joint at the base of your thumb is an excellent demonstration that also provides an excuse for twiddling your thumbs during a lecture.

- *Ball-and-socket joints*: In a **ball-and-socket joint** (Figure 8.6f●), the round head of one bone rests within a cup-shaped depression in another. All combinations of movements, including rotation, can be performed at ball-and-socket joints. These are triaxial joints, and examples include the shoulder and hip joints.

✓ CONCEPT CHECK

- In a newborn infant, the large bones of the skull are joined by fibrous connective tissue. What type of joint is this? These bones later grow, interlock, and form immovable joints. What type of joints are these?
- Give the proper term for each of the following types of motion: (a) moving the humerus away from the midline of the body; (b) turning the palms so that they face forward; (c) bending the elbow.

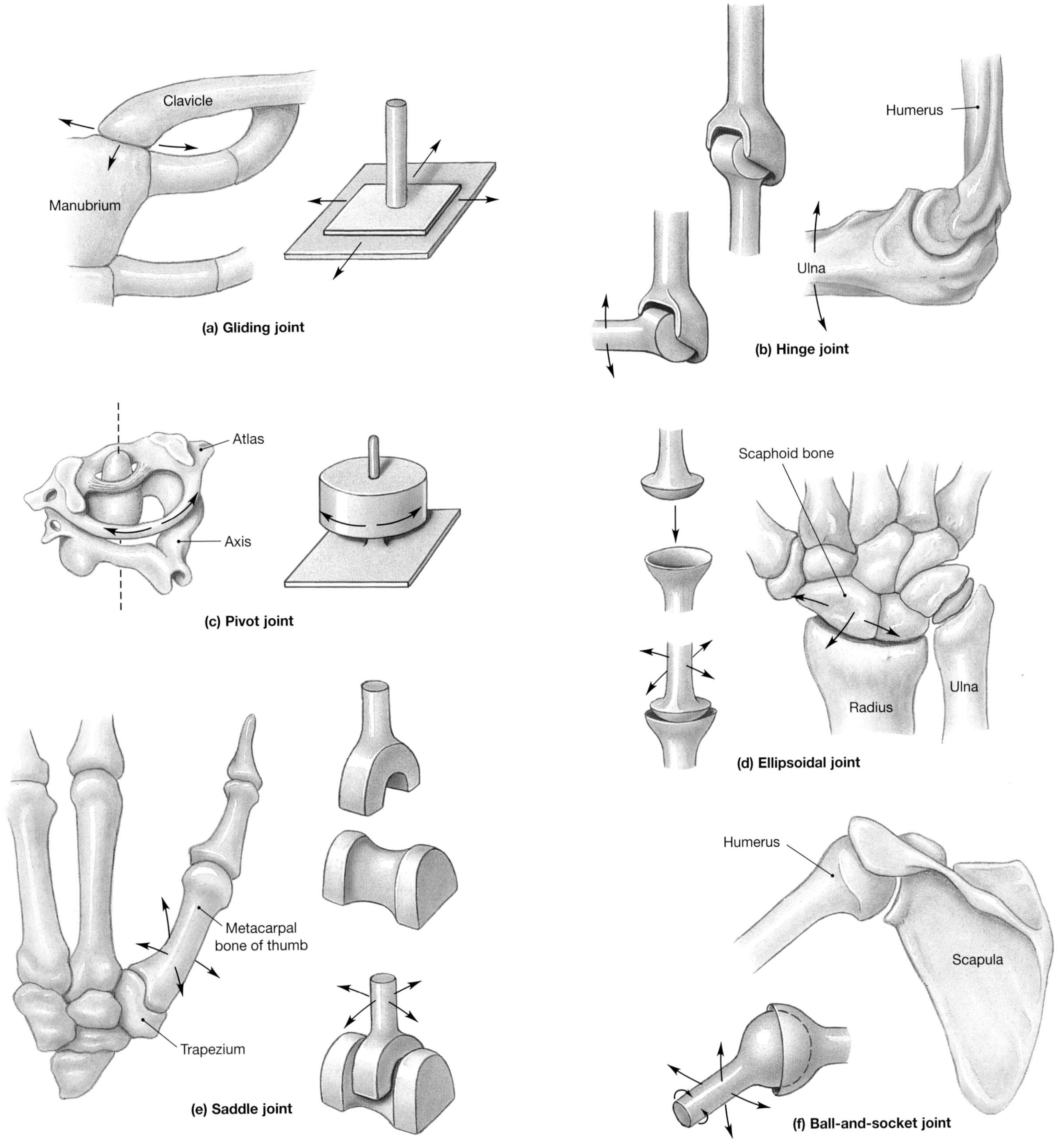

FIGURE 8.6 **A Structural Classification of Synovial Joints**
This classification scheme is based on the amount of movement permitted.

Representative Articulations

This section considers examples of articulations that demonstrate important functional principles. We will first consider several articulations of the axial skeleton: (1) the *temporomandibular joint* (TMJ), between the mandible and the temporal bone, (2) the *intervertebral articulations* between adjacent vertebrae, and (3) the *sternoclavicular joint* between the clavicle and the sternum. Next, we will examine synovial joints of the appendicular skeleton. The shoulder has great mobility, the elbow has great strength, and the wrist makes fine adjustments in the orientation of the palm and fingers. The functional requirements of the joints in the lower limb are very different from those of the upper limb. Articulations at the hip, knee, and ankle must transfer the body weight to the ground, and during movements such as running, jumping, or twisting, the applied forces are considerably greater than the weight of the body. Although this section considers representative articulations, Tables 8.3, 8.4, and 8.5 summarize information concerning the majority of articulations in the body.

The Temporomandibular Joint [FIGURE 8.7]

The **temporomandibular joint** (Figure 8.7●) is a small but complex articulation between the mandibular fossa of the temporal bone and the condylar process of the mandible. p.152 The articulating bones are separated by a thick fibrocartilage pad. This cartilage, which extends horizontally, divides the joint cavity into two separate chambers. As a result, the temporomandibular joint is really two synovial joints, one between the temporal bone and the articular disc, and the second between the articular disc and the mandible.

The articular capsule surrounding this joint complex is relatively loose, and this permits an extensive range of motion. However, because the joint is poorly stabilized, a forceful lateral or anterior movement of the mandible can result in a partial or complete dislocation.

The lateral portion of the articular capsule, which is relatively thick, is called the **lateral** (*temporomandibular*) **ligament**. There are also two extracapsular ligaments:

- the **stylomandibular ligament**, which extends from the styloid process to the posterior margin of the angle of the mandibular ramus;
- the **sphenomandibular ligament**, which extends from the sphenoidal spine to the medial surface of the mandibular ramus. Its insertion covers the posterior portion of the mylohyoid line.

The temporomandibular joint is primarily a hinge joint, but the loose capsule and relatively flat articular surfaces also permit small gliding and rotational movements. These secondary movements are important when positioning food on the occlusal surfaces of the teeth.

Intervertebral Articulations [FIGURE 8.8]

The articulations between the superior and inferior articular processes of adjacent vertebrae are gliding joints that permit small movements associated with flexion and rotation of the vertebral column. Little gliding occurs between adjacent vertebral bodies. Figure 8.8● illustrates the structure of the intervertebral joints. From axis to sacrum, the vertebrae are separated and cushioned by pads of fibrocartilage called **intervertebral discs**. Intervertebral discs are not found in the sacrum and coccyx, where vertebrae have fused, nor are they found between the first and second cervical vertebrae. The articulation between C_1 and C_2 was described in Chapter 6. p.166

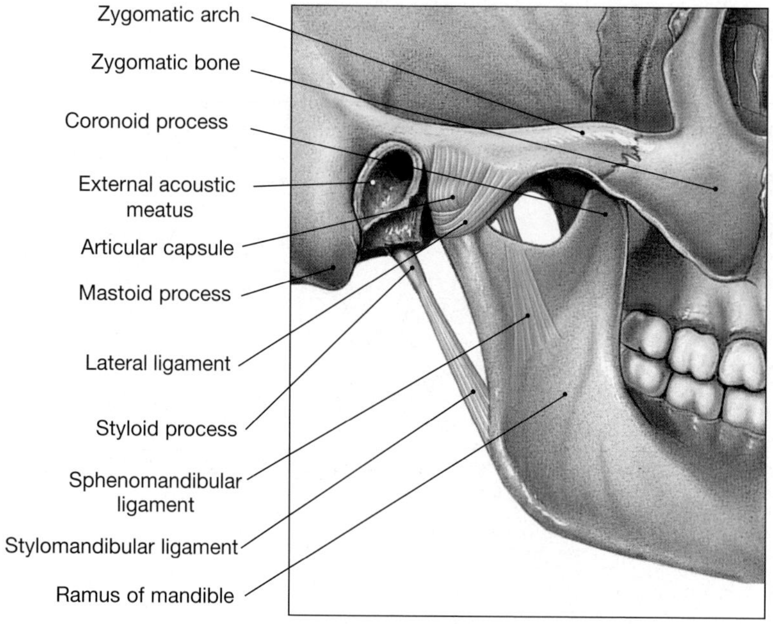

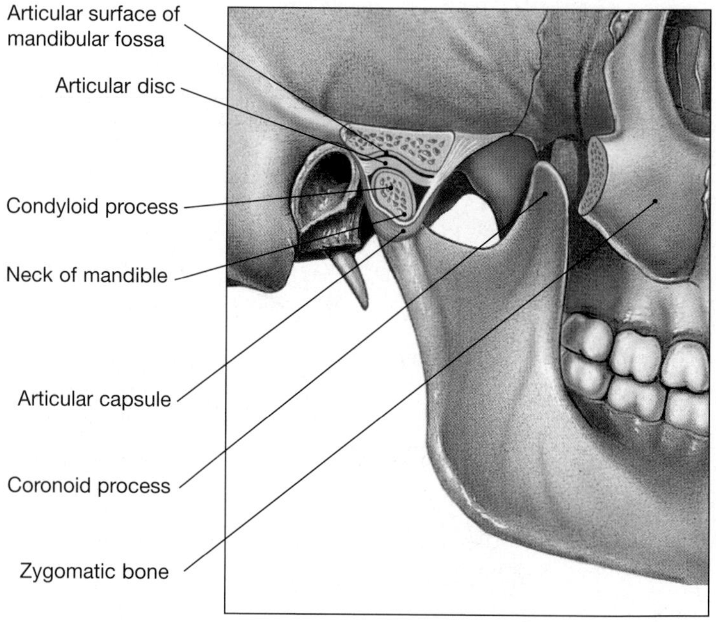

FIGURE 8.7 **THE TEMPOROMANDIBULAR JOINT**

This hinge joint forms between the condylar process of the mandible and the mandibular fossa of the temporal bone. (**a**) Lateral view of the right temporomandibular joint. (**b**) Sectional view of the same joint.

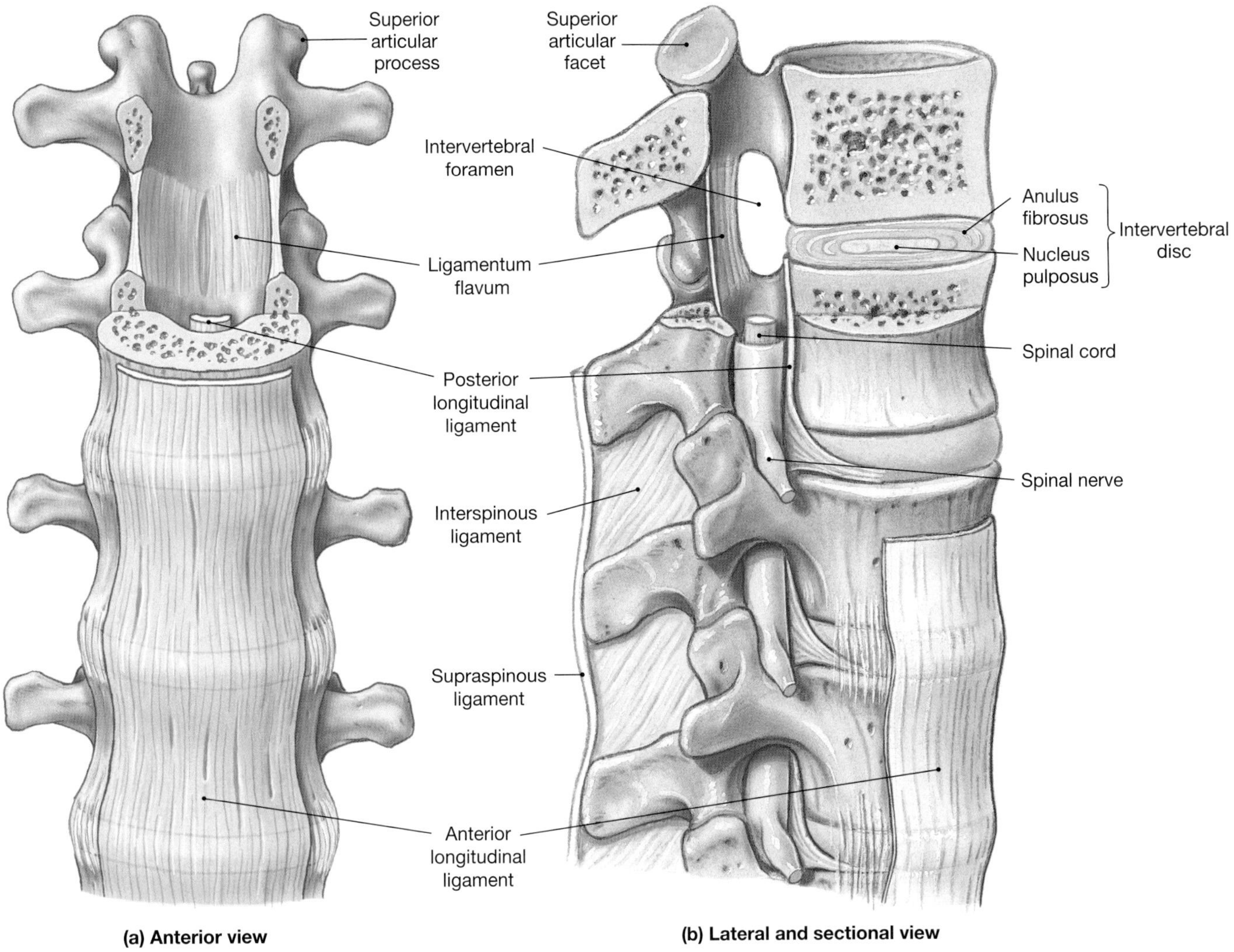

FIGURE 8.8 **Intervertebral Articulations**

Adjacent vertebrae articulate at their superior and inferior articular processes; their bodies are separated by intervertebral discs. (**a**) Anterior view. (**b**) Lateral and sectional views.

THE INTERVERTEBRAL DISCS [FIGURES 8.8/8.9a]

Each intervertebral disc (Figures 8.8 and 8.9a●) has a tough outer layer of fibrocartilage, the **anulus fibrosus** (AN-ū-lus fī-BRŌ-sus). The collagen fibers of the anulus fibrosus attach the disc to the bodies of adjacent vertebrae. The anulus also surrounds a soft, elastic, and gelatinous core, the **nucleus pulposus** (pul-PŌ-sus). The nucleus pulposus is composed primarily of water (about 75%) with scattered reticular and elastic fibers. The nucleus pulposus gives the disc resiliency and enables it to act as a shock absorber.

Movements of the vertebral column compress the nucleus pulposus and displace it in the opposite direction. This displacement permits smooth gliding movements by each vertebra while still maintaining the alignment of all the vertebrae. The discs make a significant contribution to an individual's height; they account for roughly one-quarter of the length of the vertebral column above the sacrum. As we grow older, the water content of the nucleus pulposus within each disc decreases. The discs gradually become less effective as a cushion, and the chances for vertebral injury increase. Loss of water by the discs also causes shortening of the vertebral column; this shortening accounts for the characteristic decrease in height with advanced age.

INTERVERTEBRAL LIGAMENTS [FIGURE 8.8]

Numerous ligaments are attached to the bodies and processes of all vertebrae to bind them together and stabilize the vertebral column (Figure 8.8●). Ligaments interconnecting adjacent vertebrae include: the *anterior longitudinal ligament*, the *posterior longitudinal ligament*, the *ligamentum flavum*, the *interspinous ligament*, and the *supraspinous ligament*.

- The *anterior longitudinal ligament* connects the anterior surfaces of each vertebral body.
- The *posterior longitudinal ligament* parallels the anterior longitudinal ligament but passes across the posterior surfaces of each body.
- The *ligamentum flavum* (plural, *ligamenta flava*) connects the laminae of adjacent vertebrae.
- The *interspinous ligament* connects the spinous processes of adjacent vertebrae.
- The *supraspinous ligament* interconnects the tips of the spinous processes from C_7 to the sacrum. The *ligamentum nuchae*, discussed in Chapter 6, is a supraspinous ligament that extends from C_7 to the base of the skull. p. 168

CLINICAL DISCUSSION

Problems with the Intervertebral Discs

An intervertebral disc compressed beyond its normal limits may become temporarily or permanently damaged. If the posterior longitudinal ligaments are weakened, as often occurs with advancing age, the compressed nucleus pulposus may distort the anulus fibrosus, partially forcing it into the vertebral canal. This condition is often called a **slipped disc** (Figure 8.9a●), although disc slippage does not actually occur. The most common sites for disc problems are at C_5–C_6, L_4–L_5, and at L_5–S_1.

Under severe compression the nucleus pulposus may break through the anulus fibrosus and enter the vertebral canal. This condition is called a **herniated disc** (Figure 8.9b●). When a disc herniates, sensory nerves are distorted, producing pain; the protruding mass can also compress the nerves passing through the intervertebral foramen. **Sciatica** (sī-AT-i-ka) is the painful result of compression of the roots of the sciatic nerve. The acute initial pain in the lower back is sometimes called **lumbago** (lum-BĀ-gō).

Most lumbar disc problems can be treated successfully with some combination of rest, back braces, analgesic (pain-killing) drugs, and physical therapy. Surgery to relieve the symptoms is required in only about 10% of cases involving lumbar disc herniation. In this procedure, the disc is removed, and the vertebral bodies fused together to prevent movement. To access the offending disc, the surgeon may remove the nearest vertebral arch by shaving away the laminae. For this reason, the procedure is known as a **laminectomy** (la-mi-NEK-tō-mē).

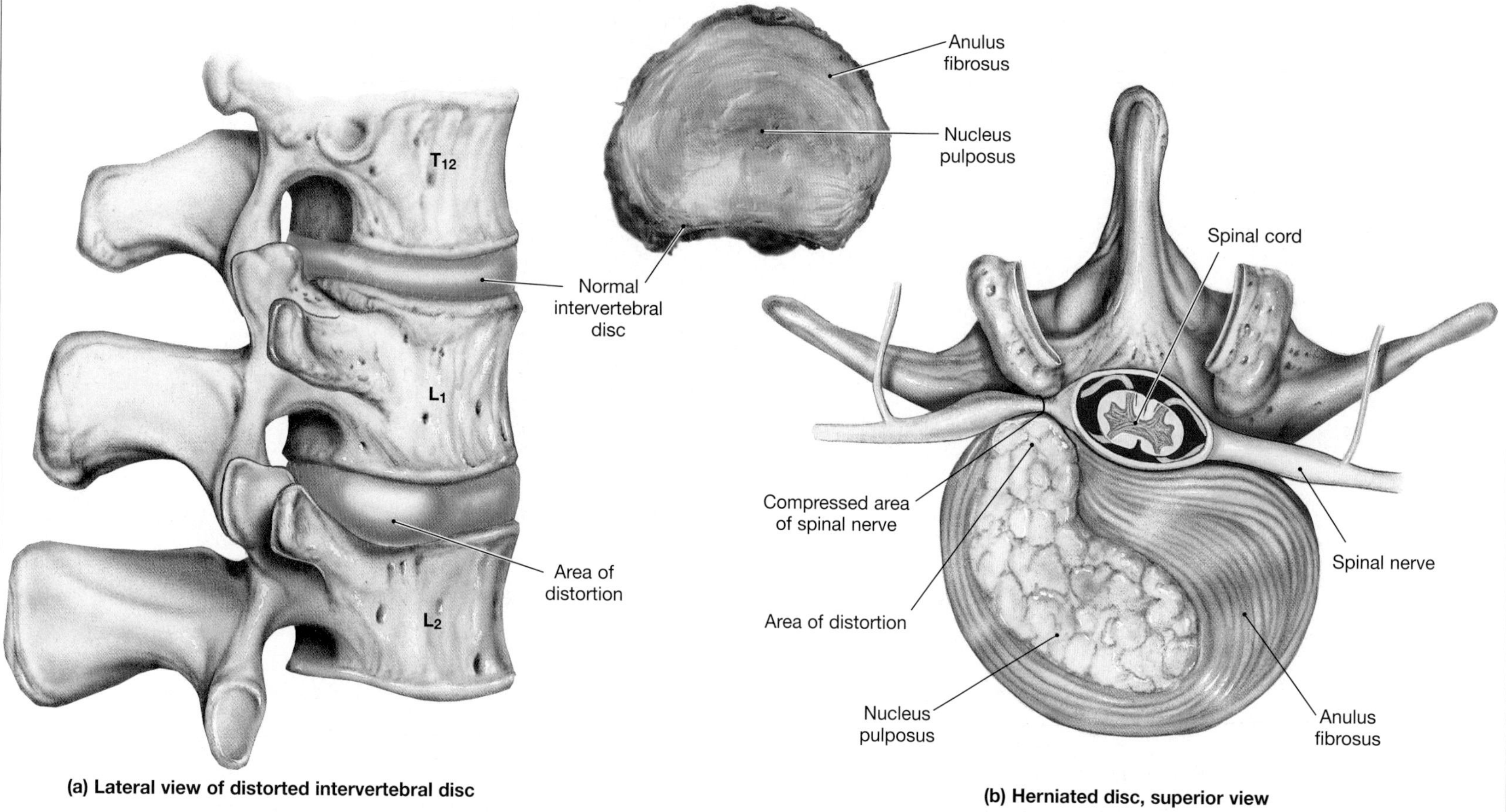

FIGURE 8.9 **Damage to the Intervertebral Discs**

(**a**) Lateral view of the lumbar region of the spinal column, showing normal and distorted ("slipped") intervertebral discs. The superior surface of an isolated normal intervertebral disc is shown for comparison with the sectional view in (**b**). Sectional view through a herniated disc, showing displacement of the nucleus pulposus and its effect on the spinal cord and adjacent nerves.

VERTEBRAL MOVEMENTS [TABLE 8.3]

The following movements of the vertebral column are possible: (1) **anterior flexion**, bending forward; (2) **extension**, bending backward; (3) **lateral flexion**, bending to the side; and (4) **rotation**, or twisting.

Table 8.3 summarizes information concerning the articulations and movements of the axial skeleton.

The Sternoclavicular Joint [FIGURE 8.10]

The **sternoclavicular joint** is a synovial joint between the medial end of the clavicle and the manubrium of the sternum. As at the temporomandibular joint (p. 222), an articular disk divides the sternoclavicular joint and separates two synovial cavities (Figure 8.10●). The articular capsule is both tense and dense, providing stability but limiting movement. The capsule is reinforced by two accessory ligaments, the **anterior sternoclavicular ligament** and the **posterior sternoclavicular ligament**. There are also two extracapsular ligaments:

- The **interclavicular ligament** interconnects the clavicles and reinforces the superior portions of the adjacent articular capsules. This ligament, which is also firmly attached to the superior border of the manubrium, prevents dislocation when the shoulder is depressed.

TABLE 8.3 ARTICULATIONS OF THE AXIAL SKELETON

Element	Joint	Type of Articulation	Movements
SKULL			
Cranial and facial bones of skull	Various	Synarthroses (suture or synostosis)	None
Maxillae/teeth	Alveolar	Synarthrosis (gomphosis)	None
Mandible/teeth	Alveolar	As above	None
Temporal bone/mandible	Temporomandibular	Combined gliding joint and hinge diarthrosis	Elevation/depression, lateral gliding, limited protraction/retraction
VERTEBRAL COLUMN			
Occipital bone/atlas	Atlanto-occipital	Ellipsoidal diarthrosis	Flexion/extension
Atlas/axis	Atlanto-axial	Pivot diarthrosis	Rotation
Other vertebral elements	Intervertebral (between vertebral bodies)	Amphiarthrosis (symphysis)	Slight movement
	Intervertebral (between articular processes)	Gliding diarthrosis	Slight rotation and flexion/extension
Thoracic vertebrae/ribs	Vertebrocostal	Gliding diarthrosis	Elevation/depression
Ribs/sternum	Sternocostal	Synarthrosis (synchondrosis)	None
L_5 /sacrum	Between body of L_5 and sacral body	Amphiarthrosis (symphysis)	Slight movement
	Between inferior articular processes of L_5 and articular processes of sacrum	Gliding diarthrosis	Slight flexion/extension
Sacrum/os coxae	Sacroiliac	Gliding diarthrosis	Slight gliding movement
Sacrum/coccyx	Sacrococcygeal	Gliding diarthrosis (may become fused)	Slight movement
Coccygeal bones		Synarthrosis (synostosis)	None

- The broad **costoclavicular ligament** extends from the costal tuberosity of the clavicle, near the inferior margin of the articular capsule, to the superior and medial borders of the first rib and the first costal cartilage. This ligament prevents dislocation when the shoulder is elevated.

The sternoclavicular joint is primarily a gliding joint, but the capsular fibers permit a slight rotation and circumduction of the clavicle.

The Shoulder Joint [FIGURE 8.11]

The **shoulder joint**, or **glenohumeral joint**, is a loose and shallow joint that permits the greatest range of motion of any joint in the body. Because it is also the most frequently dislocated joint, it provides an excellent demonstration of the principle that strength and stability must be sacrificed to obtain mobility.

This joint is a ball-and-socket type, formed by the articulation of the head of the humerus with the glenoid cavity of the scapula (Figure 8.11●).

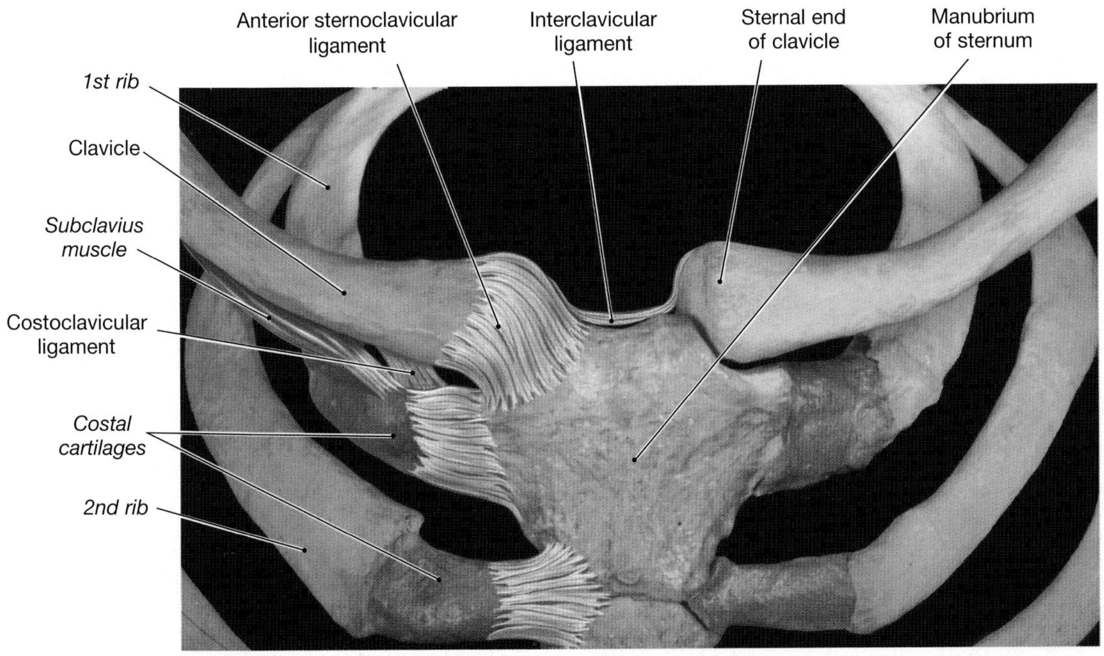

FIGURE 8.10 **THE STERNOCLAVICULAR JOINT**

A superior view of the thorax showing the bones and ligaments of the sternoclavicular joint. This joint is classified as a stable, heavily reinforced gliding diarthrosis.

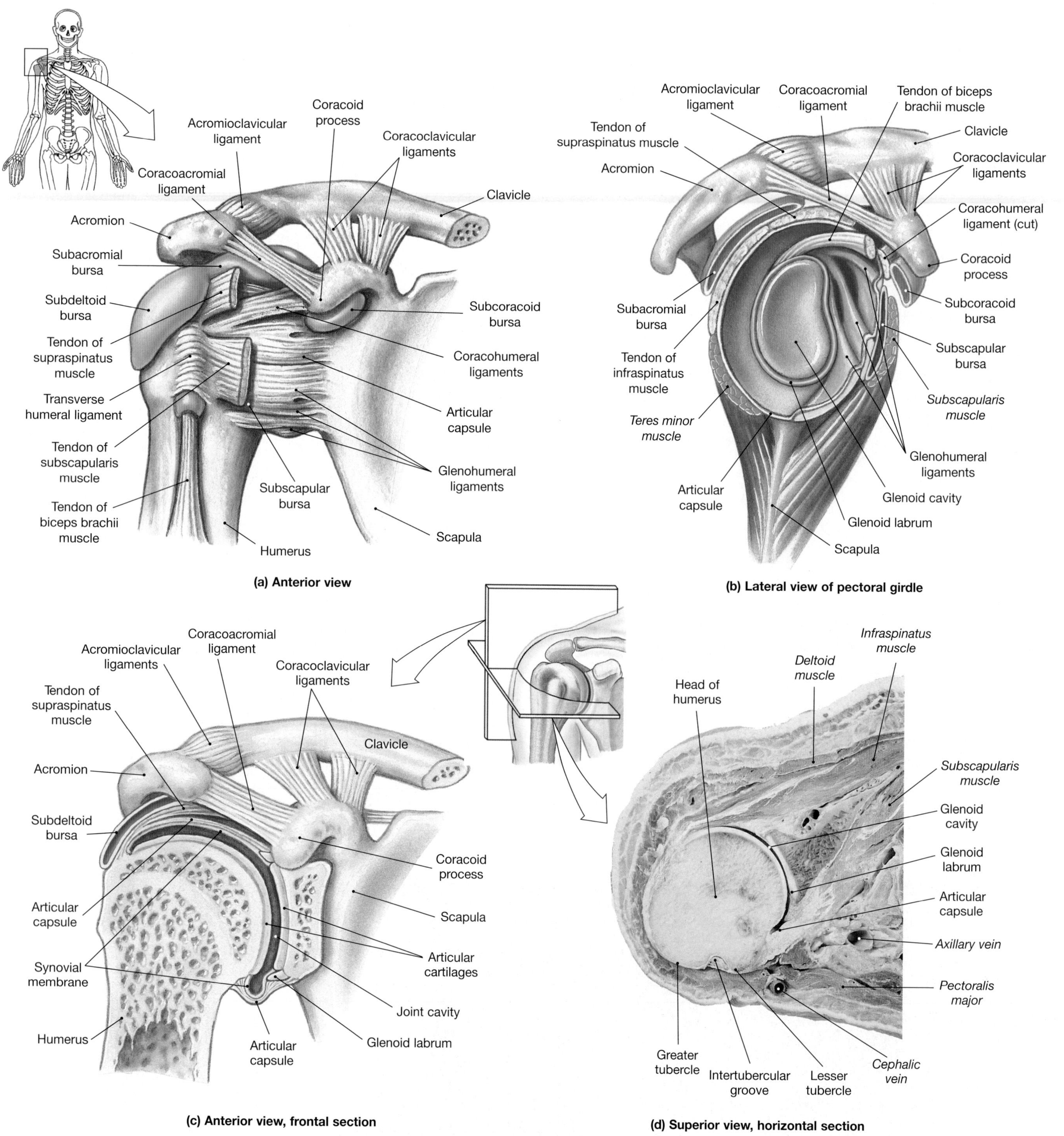

FIGURE 8.11 **THE (SHOULDER) GLENOHUMERAL JOINT**

A ball-and-socket joint formed between the humerus and the scapula. (**a**) Anterior view of the right shoulder joint. (**b**) Lateral view, right shoulder joint (humerus removed). (**c**) A frontal section through the right shoulder joint. (**d**) Horizontal section of the right shoulder joint, superior view.

In life, the surface of the glenoid cavity is covered by a fibrocartilaginous **glenoid labrum** (*labrum*, lip or edge) (Figure 8.11c,d•), which deepens the joint slightly. The relatively loose articular capsule extends from the scapular neck to the humerus. It is a relatively oversized capsule that permits an extensive range of motion. The capsule is weakest at its inferior surface. The bones of the pectoral girdle provide some stability to the superior surface, because the acromion and coracoid processes project laterally superior to the humeral head. However, most of the stability at this joint is provided by (1) ligaments and (2) surrounding skeletal muscles and their associated tendons.

LIGAMENTS [FIGURE 8.11]

Major ligaments involved with stabilizing the glenohumeral joint are shown in Figure 8.11a,b,c• and described below.

- The capsule surrounding the shoulder joint is relatively thin, but it thickens anteriorly in regions known as the **glenohumeral ligaments**. Because the capsular fibers are usually loose, these ligaments participate in joint stabilization only as the humerus approaches, or exceeds, the limits of normal motion.
- The large **coracohumeral ligament** originates at the base of the coracoid process and inserts on the head of the humerus. This ligament strengthens the superior part of the articular capsule and helps support the weight of the upper limb.
- The **coracoacromial ligament** spans the gap between the coracoid process and the acromion, just superior to the capsule. This ligament provides additional support to the superior surface of the capsule.
- The strong **acromioclavicular ligament** binds the acromion to the clavicle, thereby restricting clavicular movement at the acromion end. A *shoulder separation* is a relatively common injury involving partial or complete dislocation of the acromioclavicular joint. This injury can result from a blow to the superior surface of the shoulder. The acromion is forcibly depressed, but the clavicle is held back by strong muscles.
- The **coracoclavicular ligaments** tie the clavicle to the coracoid process and help to limit the relative motion between the clavicle and scapula.
- The **transverse humeral ligament** extends between the greater and lesser tubercles, and holds down the tendon from the long head of the biceps brachii muscle in the intertubercular groove of the humerus.

SKELETAL MUSCLES AND TENDONS

Muscles that move the humerus do more to stabilize the glenohumeral joint than all the ligaments and capsular fibers combined. Muscles originating on the trunk, pectoral girdle, and humerus cover the anterior, superior, and posterior surfaces of the capsule. Tendons passing across the joint reinforce the anterior and superior portions of the capsule. The tendons of specific appendicular muscles support the shoulder and limit its movement range. These muscles, collectively called the *rotator cuff*, are a frequent site of sports injury.

BURSAE [FIGURE 8.11a,b,c]

As at other joints, *bursae* at the shoulder reduce friction where large muscles and tendons pass across the joint capsule. p. 216 The shoulder has a relatively large number of important bursae. The **subacromial bursa** and the **subcoracoid bursa** (Figure 8.11a,b•) prevent contact between the acromial and coracoid processes and the capsule. The **subdeltoid bursa** and the **subscapular bursa** (Figure 8.11a,b,c•) lie between large muscles and the capsular wall. Inflammation of one or more of these bursae can restrict motion and produce the painful symptoms of **bursitis.** *Bursitis p. 790*

The Elbow Joint [FIGURE 8.12]

The **elbow joint** is a complex hinge joint involving the humerus, radius, and ulna. The largest and strongest articulation at the elbow is the *humeroulnar joint*, where the trochlea of the humerus projects into the trochlear notch of the ulna. p. 189 At the smaller *humeroradial joint*, which lies lateral to the humeroulnar joint, the capitulum of the humerus articulates with the head of the radius (Figure 8.12•).

Muscles that extend the elbow attach to the rough surface of the olecranon. These muscles are primarily under the control of the radial nerve, which passes along the *radial groove* of the humerus. p. 185 The large *biceps brachii muscle* covers the anterior surface of the arm. Its tendon is attached to the radius at the *radial tuberosity*. p. 190 Contraction of this muscle produces flexion of the elbow and supination of the forearm.

The elbow joint is extremely stable because: (1) the bony surfaces of the humerus and ulna interlock to prevent lateral movement and rotation; (2) the articular capsule is very thick; and (3) the capsule is reinforced by strong ligaments. The medial surface of the joint is stabilized by the **ulnar collateral ligament**. This ligament extends from the medial epicondyle of the humerus anteriorly to the coronoid processes of the ulna, and posteriorly to the olecranon (Figure 8.12a,b•). The **radial collateral ligament** stabilizes the lateral surface of the joint. It extends between the lateral epicondyle of the humerus and the **annular ligament** that binds the proximal radial head to the ulna (Figure 8.12e•)

Despite the strength of the capsule and ligaments, the elbow joint can be damaged by severe impacts or unusual stresses. For example, when you fall on a hand with a partially flexed elbow, contractions of muscles that extend the elbow may break the ulna at the center of the trochlear notch. Less violent stresses can produce dislocations or other injuries to the elbow, especially if epiphyseal growth has not been completed. For example, parents in a hurry may drag a toddler along behind them, exerting an upward, twisting pull on the elbow joint that can result in a partial dislocation known as a "nursemaid's elbow."

CLINICAL BRIEF

SHOULDER INJURIES

When a head-on charge leads to a collision, such as a block (in football) or check (in hockey), the shoulder usually lies in the impact zone. The clavicle provides the only fixed support for the pectoral girdle, and it cannot resist large forces. Because the inferior surface of the shoulder capsule is poorly reinforced, a dislocation caused by an impact or violent muscle contraction most often occurs at this site. Such a dislocation can tear the inferior capsular wall and the glenoid labrum. The healing process often leaves a weakness and inherent instability of the joint that increases the chances for future dislocations.

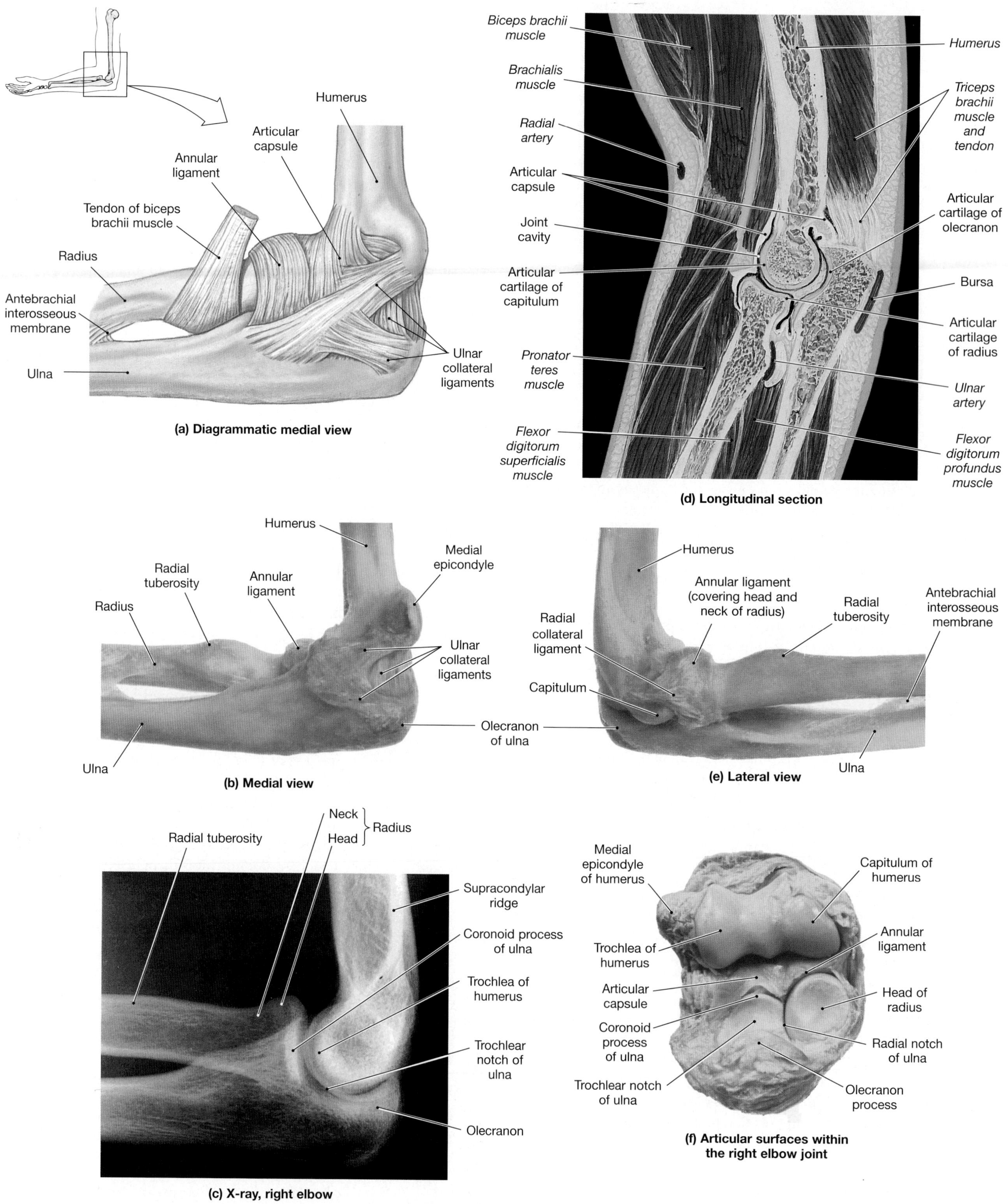

FIGURE 8.12 **THE ELBOW JOINT**

This is a complex hinge joint formed between the humerus and the ulna and radius. All views are of the right elbow joint. **(a)** Diagrammatic medial view. The radius is shown pronated; note the position of the biceps brachii tendon, which inserts on the radial tuberosity. **(b)** Medial view. **(c)** X-ray. **(d)** A plastic model based on an oblique section through the elbow region. **(e)** Lateral view. **(f)** A posterior view; the posterior portion of the capsule has been cut, and the joint cavity opened to show the opposing surfaces. *See also Figure 7.7 and Scan 10b in the Companion Atlas.*

The Joints of the Wrist [FIGURE 8.13]

The carpus, or wrist, contains the **wrist joint** (Figure 8.13●). The wrist joint consists of the **distal radioulnar joint**, the **radiocarpal joint**, and the **intercarpal joints**. The distal radioulnar joint permits pronation and supination. The radiocarpal joint involves the distal articular surface of the radius and three proximal carpal bones, the scaphoid bone, lunate bone, and triquetrum. The radiocarpal joint is an ellipsoidal articulation that permits flexion/extension, adduction/abduction, and circumduction.

The intercarpal joints are gliding joints that permit sliding and slight twisting movements.

STABILITY OF THE WRIST [FIGURE 8.13b,c]

Carpal surfaces that do not participate in articulations are roughened by the attachment of ligaments and for the passage of tendons. A tough connective tissue capsule, reinforced by broad ligaments, surrounds the wrist

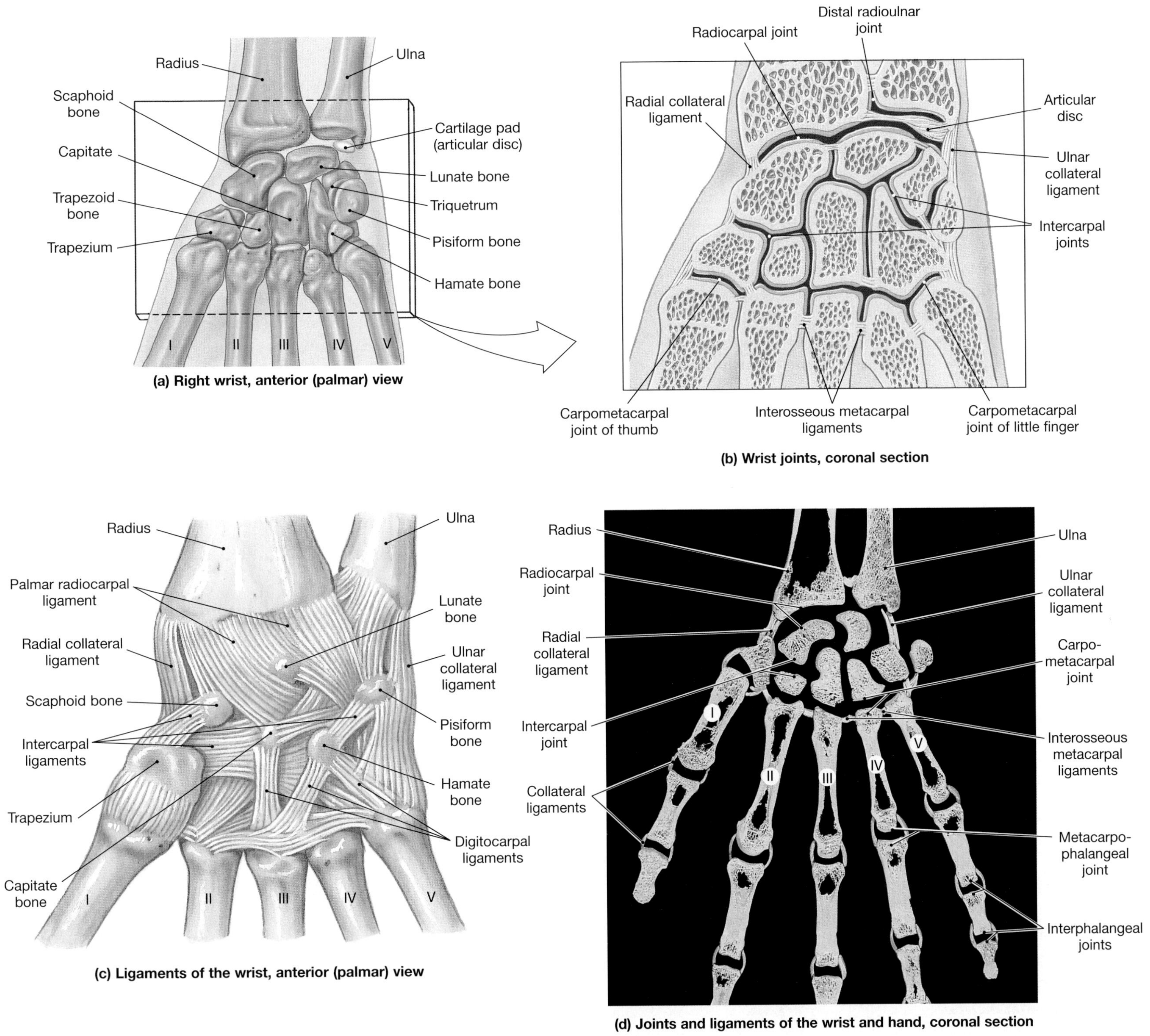

FIGURE 8.13 **JOINTS OF THE WRIST AND HAND**

(**a**) Anterior view of the right wrist, identifying the components of the wrist joint. (**b**) Sectional view through the wrist, showing the radiocarpal, intercarpal, and carpometacarpal joints. (**c**) Stabilizing ligaments on the anterior (palmar) surface of the wrist. (**d**) Sectional view of the bones that form the wrist and hand.

complex and stabilizes the positions of the individual carpal bones (Figure 8.13b,c●). The major ligaments include:

- the *palmar radiocarpal ligament*, which connects the distal radius to the anterior surfaces of the scaphoid bone, lunate bone, and triquetrum;
- the *dorsal radiocarpal ligament*, which connects the distal radius to the posterior surfaces of the same carpal bones (not seen from palmar surface);
- the *ulnar collateral ligament*, which extends from the styloid process of the ulna to the medial surface of the triquetrum; and
- the *radial collateral ligament*, which extends from the styloid process of the radius to the lateral surface of the scaphoid bone.

In addition to these prominent ligaments, a variety of *intercarpal ligaments* interconnect the carpal bones, and *digitocarpal ligaments* bind the distal carpal bones to the metacarpal bones (Figure 8.13c●). Tendons that pass across the wrist joint provide additional reinforcement. (There are many tendons involved, and they are not illustrated in the accompanying figure; we will examine these tendons and associated muscles in Chapter 11.) Tendons of muscles producing flexion of the wrist and finger joints pass over the anterior surface of the wrist joint superficial to the ligaments of the wrist joint. Tendons of muscles producing extension and hyperextension pass across the posterior surface in a similar fashion. A pair of broad transverse ligaments arch across the anterior and posterior surfaces of the wrist superficial to these tendons, holding the tendons in position.

The Joints of the Hand [FIGURE 8.13 AND TABLE 8.4]

The carpal bones articulate with the metacarpal bones of the palm (Figure 8.13a●). The first metacarpal bone has a saddle-type articulation at the wrist, the **carpometacarpal joint** of the thumb (Figure 8.13b,d●). All other carpal/metacarpal articulations are gliding joints. An **intercarpal joint** is formed by carpal/carpal articulation. The articulations between the metacarpal bones and the proximal phalanges (**metacarpophalangeal joints**) are ellipsoidal, permitting flexion/extension and adduction/abduction. The **interphalangeal joints** are hinge joints that allow flexion and extension (Figure 8.13d●).

Table 8.4 summarizes the characteristics of the articulations of the upper limb.

✓ CONCEPT CHECK

- Would a tennis player or a jogger be more likely to develop inflammation of the subscapular bursa? Why?
- Mary falls on the palm of her hands with her elbows slightly flexed. After the fall, she can't move her left arm at the elbow. If a fracture exists, what bone is most likely broken?

The Hip Joint [FIGURE 8.14]

Figure 8.14● introduces the structure of the **hip joint**. In this ball-and-socket joint, a fibrocartilage pad covers the articular surface of the acetabulum and extends like a horseshoe along the sides of the **acetabular notch** (Figure 8.14a●). A fat pad covered by a synovial membrane covers the central portion of the acetabulum. This pad acts as a shock absorber, and the adipose tissue stretches and distorts without damage.

THE ARTICULAR CAPSULE [FIGURE 8.14a–c]

The articular capsule of the hip joint is extremely dense, strong, and deep (Figure 8.14b,c●). It extends from the lateral and inferior surfaces of the pelvic girdle to the intertrochanteric line and trochanteric crest of the

TABLE 8.4 **ARTICULATIONS OF THE PECTORAL GIRDLE AND UPPER LIMB**

Element	Joint	Type of Articulation	Movements
Sternum/clavicle	Sternoclavicular	Gliding diarthrosis (a double "gliding joint," with two joint cavities separated by an articular cartilage)	Protraction/retraction, depression/elevation, slight rotation
Scapula/clavicle	Acromioclavicular	Gliding diarthrosis	Slight gliding movement
Scapula/humerus	Glenohumeral (shoulder)	Ball-and-socket diarthrosis	Flexion/extension, adduction/abduction, circumduction, rotation
Humerus/ulna and humerus/radius	Elbow (humeroulnar and humeroradial)	Hinge diarthrosis	Flexion/extension
Radius/ulna	Proximal radioulnar Distal radioulnar	Pivot diarthrosis Pivot diarthrosis	Rotation Pronation/supination
Radius/carpal bones	Radiocarpal	Ellipsoidal diarthrosis	Flexion/extension, adduction/abduction, circumduction
Carpal bone/carpal bone	Intercarpal	Gliding diarthrosis	Slight gliding movement
Carpal bone/first metacarpal bone	Carpometacarpal of thumb	Saddle diarthrosis	Flexion/extension, adduction/abduction, circumduction, opposition
Carpal bones/metacarpal bones II–V	Carpometacarpal	Gliding diarthrosis	Slight flexion/extension, adduction/abduction
Metacarpal bones/phalanges	Metacarpophalangeal	Ellipsoidal diarthrosis	Flexion/extension, adduction/abduction, circumduction
Phalanx/phalanx	Interphalangeal	Hinge diarthrosis	Flexion/extension

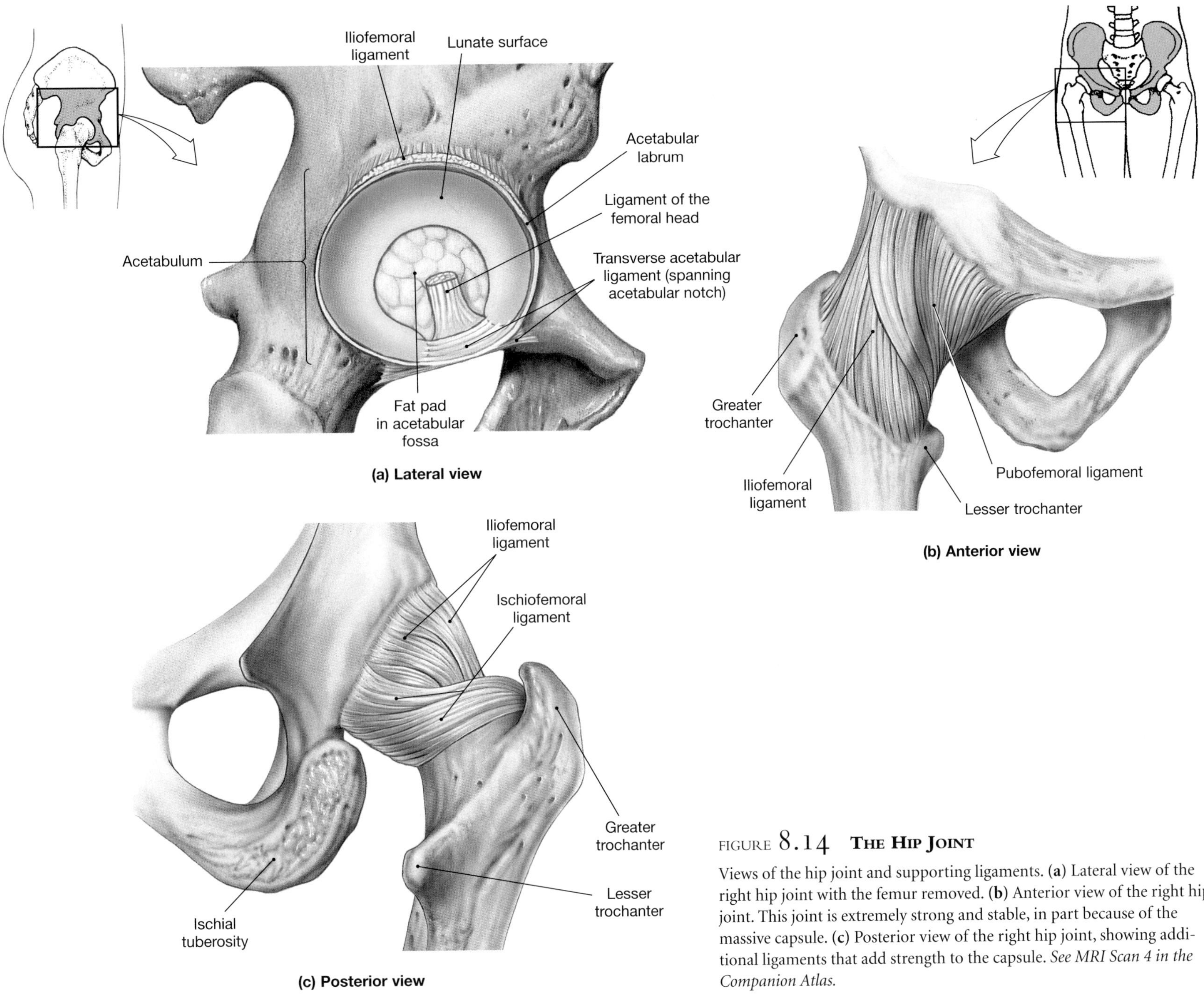

FIGURE 8.14 **THE HIP JOINT**

Views of the hip joint and supporting ligaments. (**a**) Lateral view of the right hip joint with the femur removed. (**b**) Anterior view of the right hip joint. This joint is extremely strong and stable, in part because of the massive capsule. (**c**) Posterior view of the right hip joint, showing additional ligaments that add strength to the capsule. *See MRI Scan 4 in the Companion Atlas.*

femur, enclosing both the femoral head and neck. This arrangement helps keep the head from moving away from the acetabulum. Additionally, a circular rim of fibrocartilage, called the **acetabular labrum** (Figure 8.14a●), increases the depth of the acetabulum.

STABILIZATION OF THE HIP [FIGURES 8.14b,c/8.15]

Four broad ligaments reinforce the articular capsule (Figure 8.14b,c●). Three of them are regional thickening of the capsule: the **iliofemoral**, **pubofemoral**, and **ischiofemoral ligaments**. The **transverse acetabular ligament** crosses the acetabular notch and completes the inferior border of the acetabular fossa. A fifth ligament, the **ligament of the femoral head**, or *ligamentum capitis femoris*, originates along the transverse acetabular ligament and attaches to the center of the femoral head (Figures 8.14a and 8.15●). This ligament tenses only when the thigh is flexed and undergoing external rotation. Much more important stabilization is provided by the bulk of the surrounding muscles. Although flexion, extension, adduction, abduction, and rotation are permitted, hip flexion is the most important normal movement. Movements are restricted by the combination of ligaments, capsular fibers, the depth of the bony socket, and the bulk of the surrounding muscles.

The almost complete bony socket enclosing the head of the femur, the strong articular capsule, the stout supporting ligaments, and the dense muscular padding make this an extremely stable joint. Fractures of the femoral neck or between the trochanters are actually more common than hip dislocations. *Hip Fractures and the Aging Process p.790*, and *A Case Study: Avascular Necrosis p. 790*

The Knee Joint

Although the knee functions as a hinge joint, the articulation is far more complex than that of the elbow or even the ankle. The rounded femoral condyles roll across the superior surface of the tibia, so the points of contact are constantly changing. The knee is much less stable than other hinge

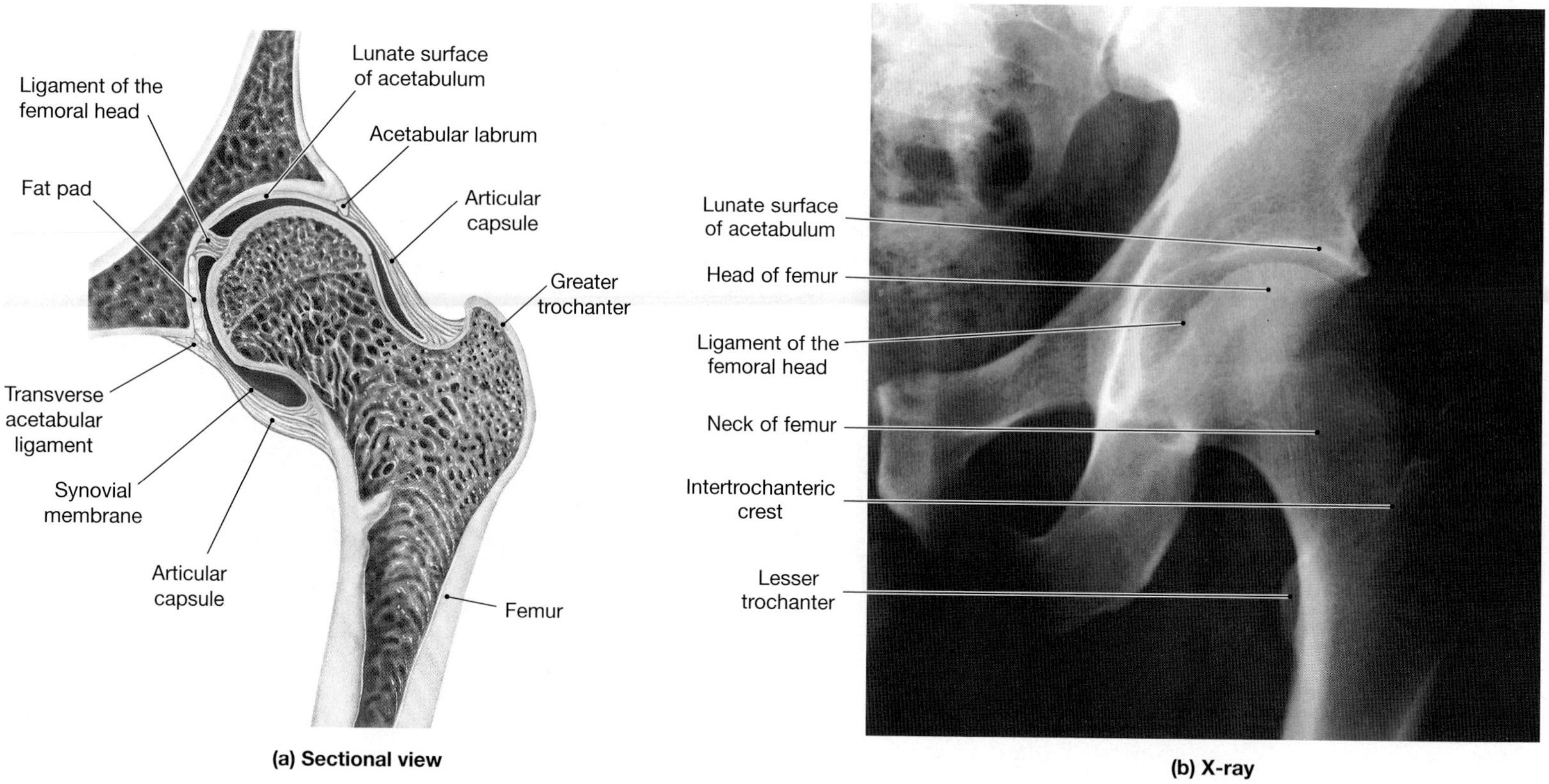

FIGURE 8.15 **ARTICULAR STRUCTURE OF THE HIP JOINT**

Coronal sectional views of the hip joint. **(a)** View showing the position and orientation of the ligament of the femoral head. **(b)** X-ray of right hip joint, anterior/posterior view. *See MRI Scan 4 in the Companion Atlas.*

joints, and some degree of rotation is permitted in addition to flexion and extension.

Structurally the knee resembles three separate joints, two between the femur and tibia (medial condyle to medial condyle and lateral condyle to lateral condyle) and one between the patella and the patellar surface of the femur. pp. 201–204

THE ARTICULAR CAPSULE [FIGURES 8.16/8.17b,c,d]

There is no single unified capsule in the knee, nor is there a common synovial cavity (Figure 8.16●). A pair of fibrocartilage pads, the **medial** and **lateral menisci**, lie between the femoral and tibial surfaces (Figure 8.17b,c,d●). The menisci (1) act as cushions, (2) conform to the shape of the articulating surfaces as the femur changes position, and (3) provide some lateral stability to the joint. Prominent **fat pads** provide padding around the margins of the joint and assist the bursae in reducing friction between the patella and other tissues (Figure 8.16a,b●).

SUPPORTING LIGAMENTS [FIGURES 8.16/8.17]

Seven major ligaments stabilize the knee joint, and a complete dislocation of the knee is an extremely rare event.

- The tendon from the muscles responsible for extending the knee passes over the anterior surface of the joint (Figure 8.16a●). The patella is embedded within this tendon, and the **patellar ligament** continues to its attachment on the anterior surface of the tibia. The patellar ligament provides support to the anterior surface of the knee joint (Figure 8.16b●), where there is no continuous capsule.

The remaining supporting ligaments are grouped either as extracapsular ligaments or intracapsular ligaments, depending on the location of the ligament with respect to the articular capsule. The extracapsular ligaments include:

- The **tibial collateral ligament** reinforcing the medial surface of the knee joint, and the **fibular collateral ligament** reinforcing the lateral surface (Figures 8.16a and 8.17●). These ligaments tighten only at full extension, and in this position they act to stabilize the joint.
- Two superficial **popliteal ligaments** extending between the femur and the heads of the tibia and fibula (Figure 8.17a●). These ligaments reinforce the back of the knee joint.
- The intracapsular ligaments include: The **anterior cruciate** and **posterior cruciate ligaments** attaching the intercondylar area of the tibia to the condyles of the femur. *Anterior* and *posterior* refer to their sites of origin on the tibia, and they cross one another as they proceed to their destinations on the femur (Figure 8.17b–d●). (The term *cruciate* is derived from the Latin word *crucialis*, meaning "a cross.") These ligaments limit the anterior and posterior movement of the femur and maintain the alignment of the femoral and tibial condyles.

LOCKING OF THE KNEE [FIGURE 8.17/TABLE 8.5]

The knee joint can "lock" in the extended position. At full extension a slight external rotation of the tibia tightens the anterior cruciate ligament and jams the meniscus between the tibia and femur. This mechanism allows you to stand for prolonged periods without using (and tiring) the extensor muscles. Unlocking the joint requires muscular contractions that produce internal rotation of the tibia or external rotation of the femur. *Knee Injuries p. 790*

Table 8.5 summarizes information about the articulations of the lower limb.

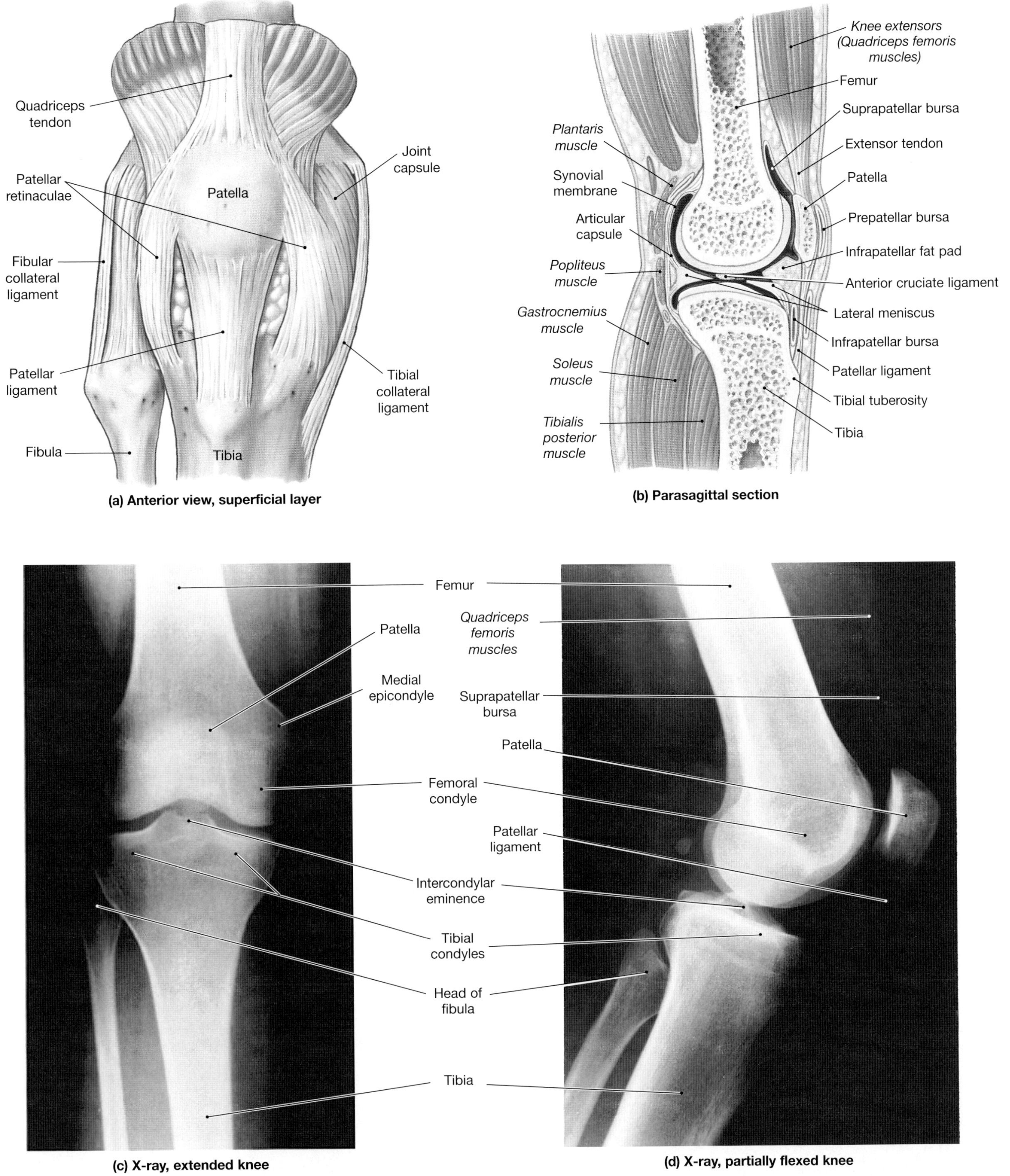

FIGURE 8.16 **The Knee Joint, Part I**

(a) Anterior view of a superficial dissection of the extended right knee. **(b)** A diagrammatic parasagittal section through the extended right knee. **(c)** X-ray of the right knee in extension, anteroposterior projection. **(d)** X-ray of the partly flexed right knee joint, lateral projection. *See MRI Scans 5a,b and 6a,b in the Companion Atlas.*

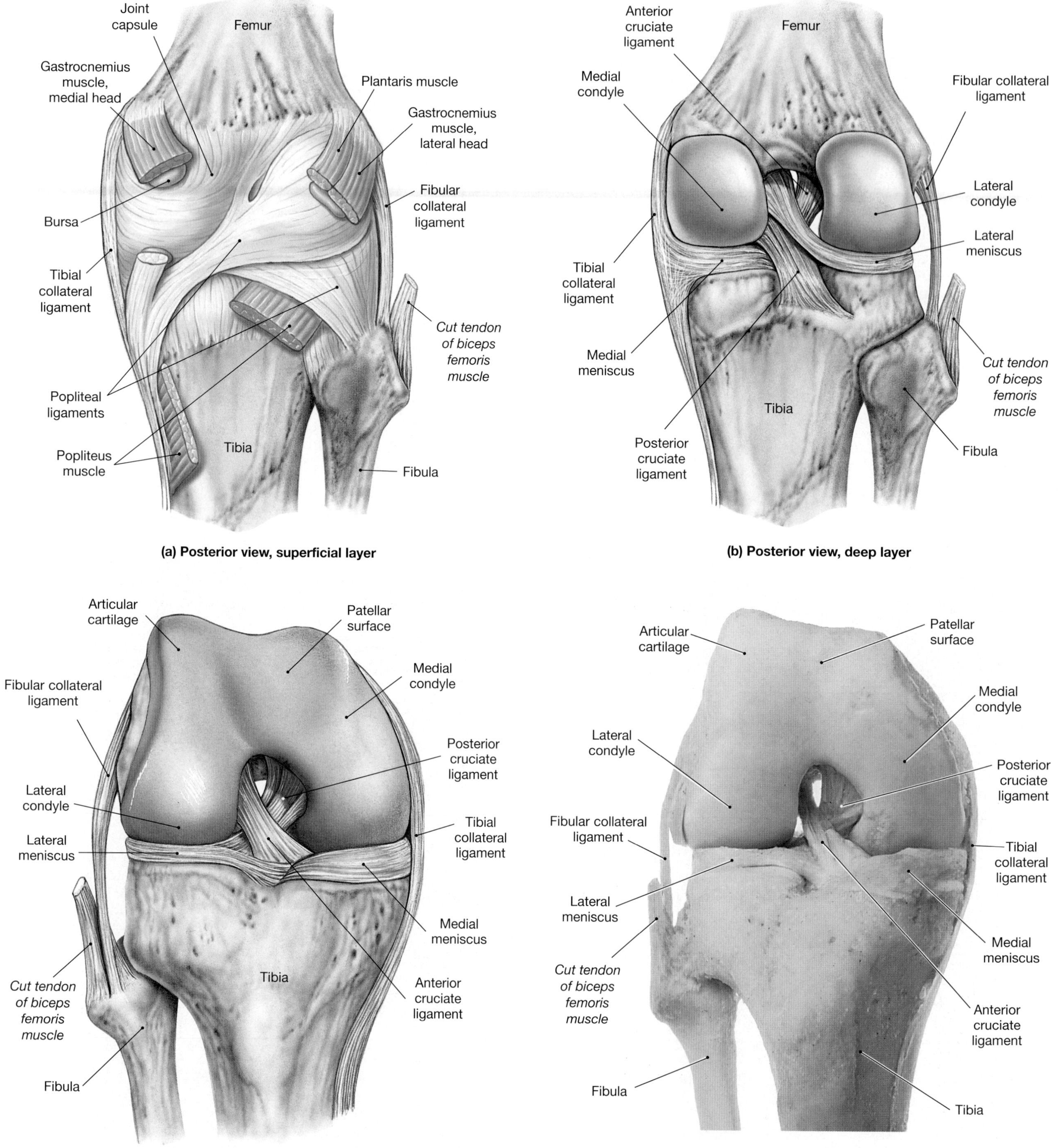

(d) Anterior view, flexed knee

FIGURE 8.17 **The Knee Joint, Part II**

(**a**) Posterior view of a dissection of the extended right knee, showing the ligaments supporting the capsule. (**b**) and (**c**) Posterior views of the right knee at full extension after removal of the joint capsule. (**d**) Anterior views of the right knee at full flexion after removal of the joint capsule, patella, and associated ligaments. *See MRI Scans 5a,b, and 7a,b in the Companion Atlas.*

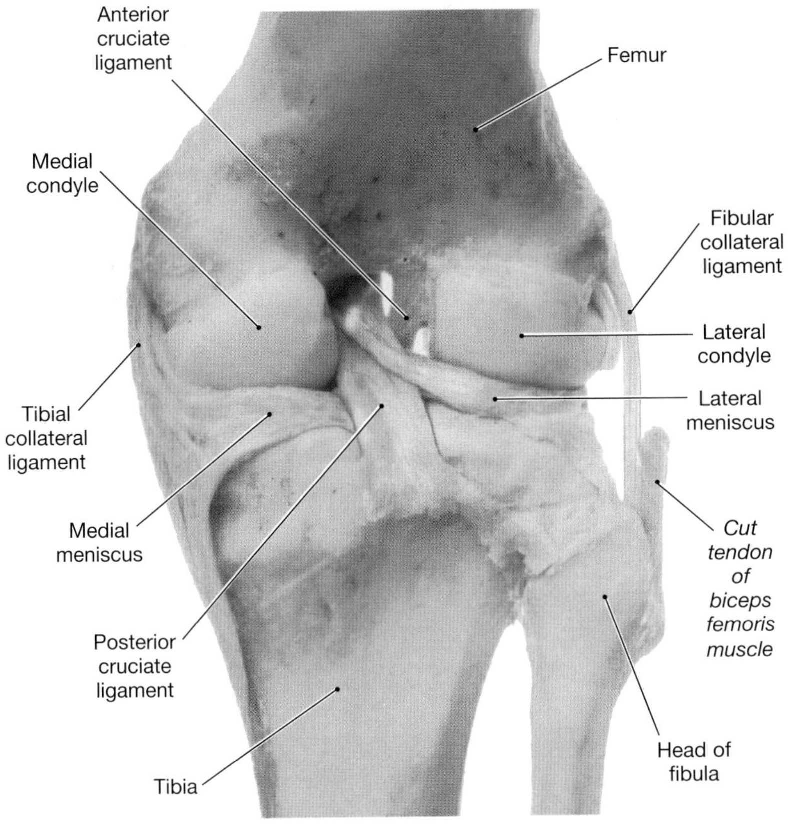

(c) Posterior view, deep layer

FIGURE 8.17 *(continued)*

✓ CONCEPT CHECK

- Where would you find the following ligaments: iliofemoral ligament, pubofemoral ligament, and ischiofemoral ligament?
- What symptoms would you expect to see in an individual who has damaged the menisci of the knee joint?
- How is the knee joint affected by damage to the patellar ligament?
- How do both the tibial and fibular collateral ligaments function to stabilize the knee joint?

The Joints of the Ankle and Foot

THE ANKLE JOINT [FIGURES 8.18/8.19]

The ankle joint, or **talocrural joint**, is a hinge joint formed by articulations among the tibia, the fibula, and the talus (Figures 8.18 and 8.19●). The primary weight-bearing articulation is the *tibiotalar joint*, between the distal articular surface of the tibia and the trochlea of the talus. The *distal tibiofibular joint* is a fibrous syndesmosis that binds the two bones together. The lateral malleolus of the fibula and the lateral articular surface of the talus form the *fibulotalar joint*. The ankle joint permits limited dorsiflexion (sole elevated) and plantar flexion (sole depressed). p. 220

The articular capsule of the ankle joint extends between the distal surfaces of the tibia and medial malleolus, the lateral malleolus, and the talus. The anterior and posterior portions of the articular capsule are thin, but the lateral and medial surfaces are strong and reinforced by stout ligaments shown in Figure 8.19b,c,d●. The major ligaments are the medial **deltoid ligament** and the three **lateral ligaments**. The malleoli, supported by these ligaments and bound together by the *tibiofibular ligaments*, prevent the ankle bones from sliding from side to side.

TABLE 8.5 ARTICULATIONS OF THE PELVIC GIRDLE AND LOWER LIMB

Element	Joint	Type of Articulation	Movements
Sacrum/os coxae	Sacroiliac	Gliding diarthrosis	Gliding movements
Os coxae/os coxae	Pubic symphysis	Amphiarthrosis	None[1]
Os coxae/femur	Hip	Ball-and-socket diarthrosis	Flexion/extension, adduction/abduction, circumduction, rotation
Femur/tibia	Knee	Complex, functions as hinge	Flexion/extension, limited rotation
Tibia/fibula	Tibiofibular (proximal) Tibiofibular (distal)	Gliding diarthrosis Gliding diarthrosis and amphiarthrotic syndesmosis	Slight gliding movements Slight gliding movements
Tibia and fibula with talus	Ankle, or talocrural	Hinge diarthrosis	Dorsiflexion/plantar flexion
Tarsal bone to tarsal bone	Intertarsal	Gliding diarthrosis	Slight gliding movements
Tarsal bones to metatarsal bones	Tarsometatarsal	Gliding diarthrosis	Slight gliding movements
Metatarsal bones to phalanges	Metatarsophalangeal	Ellipsoidal diarthrosis	Flexion/extension, adduction/abduction
Phalanx/phalanx	Interphalangeal	Hinge diarthrosis	Flexion/extension

[1]During pregnancy hormones weaken the symphysis and permit movement important to childbirth (see Chapter 28)

FIGURE 8.18 **JOINTS OF THE ANKLE AND FOOT, PART I**

(**a**) Longitudinal section of the left foot identifying major joints and associated structures. (**b**) A corresponding MRI scan of the left ankle and proximal portion of the foot. *See MRI Scan 8a in the Companion Atlas.*

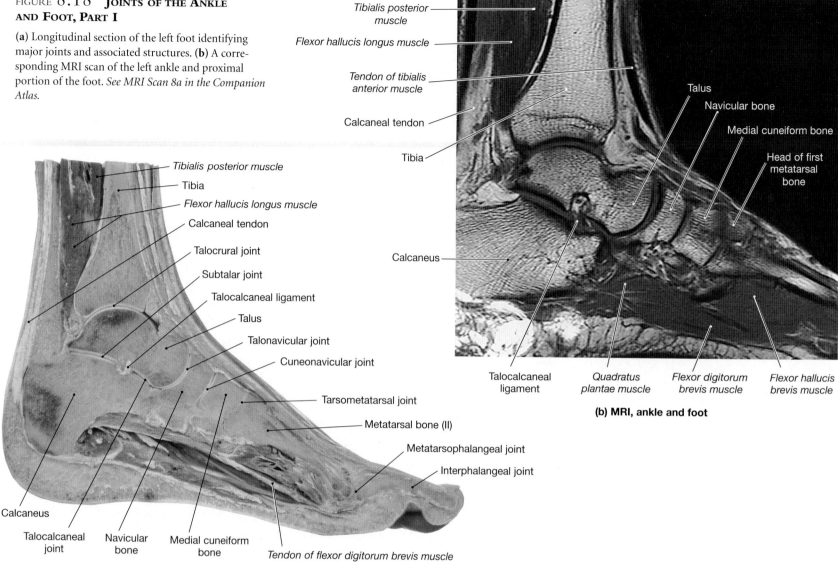

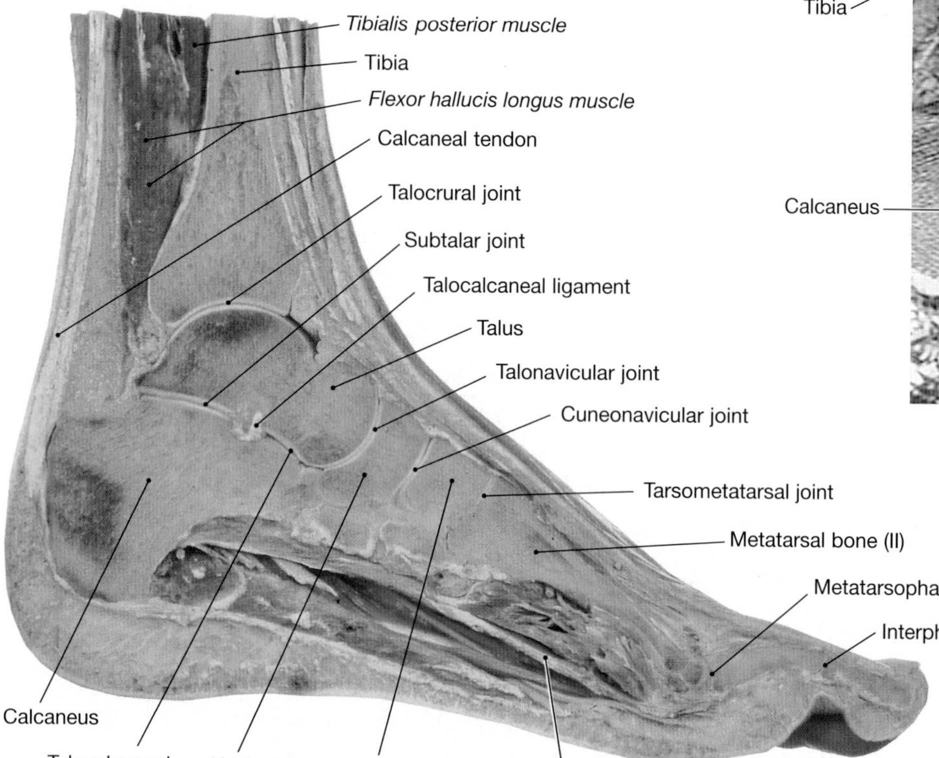

(a) Ankle and foot, longitudinal section

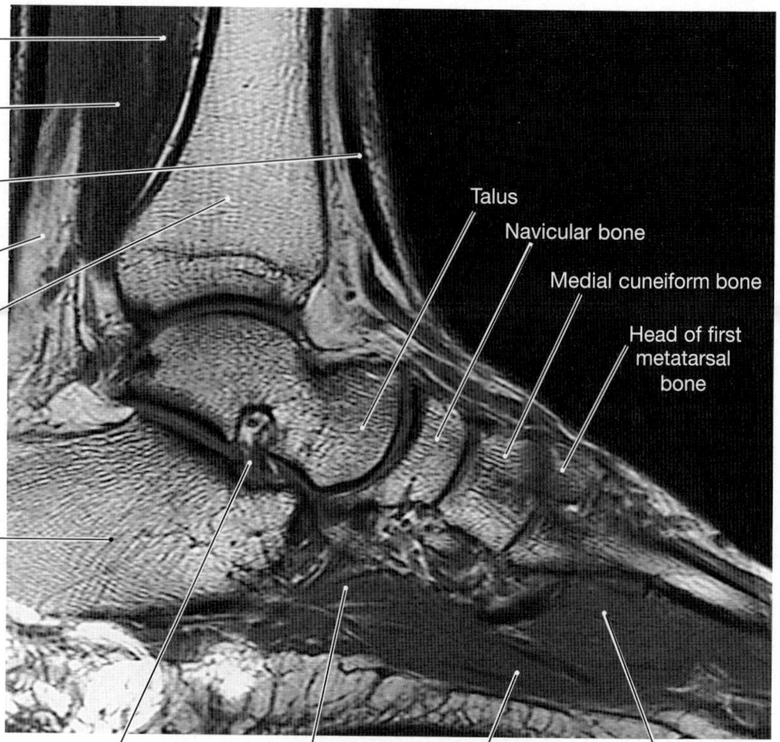

(b) MRI, ankle and foot

JOINTS OF THE FOOT [FIGURES 8.18/8.19]

Four groups of synovial joints are found in the foot (Figures 8.18 and 8.19●):

1. Tarsal bone to tarsal bone (**intertarsal joints**). These are gliding joints that permit limited sliding and twisting movements. The articulations between the tarsal bones are comparable to those between the carpal bones of the wrist.
2. Tarsal bone to metatarsal bone (**tarsometatarsal joints**). These are gliding articulations that allow limited sliding and twisting movements. The first three metatarsal bones articulate with the medial, intermediate, and lateral cuneiform bones. The fourth and fifth metatarsal bones articulate with the cuboid bone.
3. Metatarsal bone to phalanx (**metatarsophalangeal joints**). These are ellipsoidal joints that permit flexion/extension and adduction/abduction. Joints between the metatarsal bones and phalanges resemble those between the metacarpal bones and phalanges of the hand. Because the first metatarsophalangeal joint is ellipsoidal, rather than saddle-shaped like the first metatarsophalangeal joint of the hand, the great toe lacks the mobility of the thumb. A pair of sesamoid bones often forms in the tendons that cross the inferior surface of this joint, and their presence further restricts movement.
4. Phalanx to phalanx (**interphalangeal joints**). These are hinge joints that permit flexion and extension. *Problems with the Ankle and Foot p. 791*

Aging and Articulations

Joints are subjected to heavy wear and tear throughout our lifetime, and problems with joint function are relatively common, especially in older individuals. **Rheumatism** (ROO-muh-tizm) is a general term that indicates pain and stiffness affecting the skeletal system, the muscular system, or both. Several major forms of rheumatism exists. **Arthritis** (ar-THRĪ-tis) encompasses all the rheumatic diseases that affect synovial joints. Arthritis always involves damage to the articular cartilages, but the specific cause can vary. For example, arthritis can result from bacterial or viral infection, injury to the joint, metabolic problems, or severe physical stresses.

Osteoarthritis (os-tē-ō-ar-THRĪ-tis), also known as *degenerative arthritis* or *degenerative joint disease (DJD),* generally affects individuals age 60 or older. Osteoarthritis can result from cumulative wear and tear at the joint surfaces or from factors affecting collagen formation. In the U.S. population, 25 percent of women and 15 percent of men over age 60 show signs of this disease.

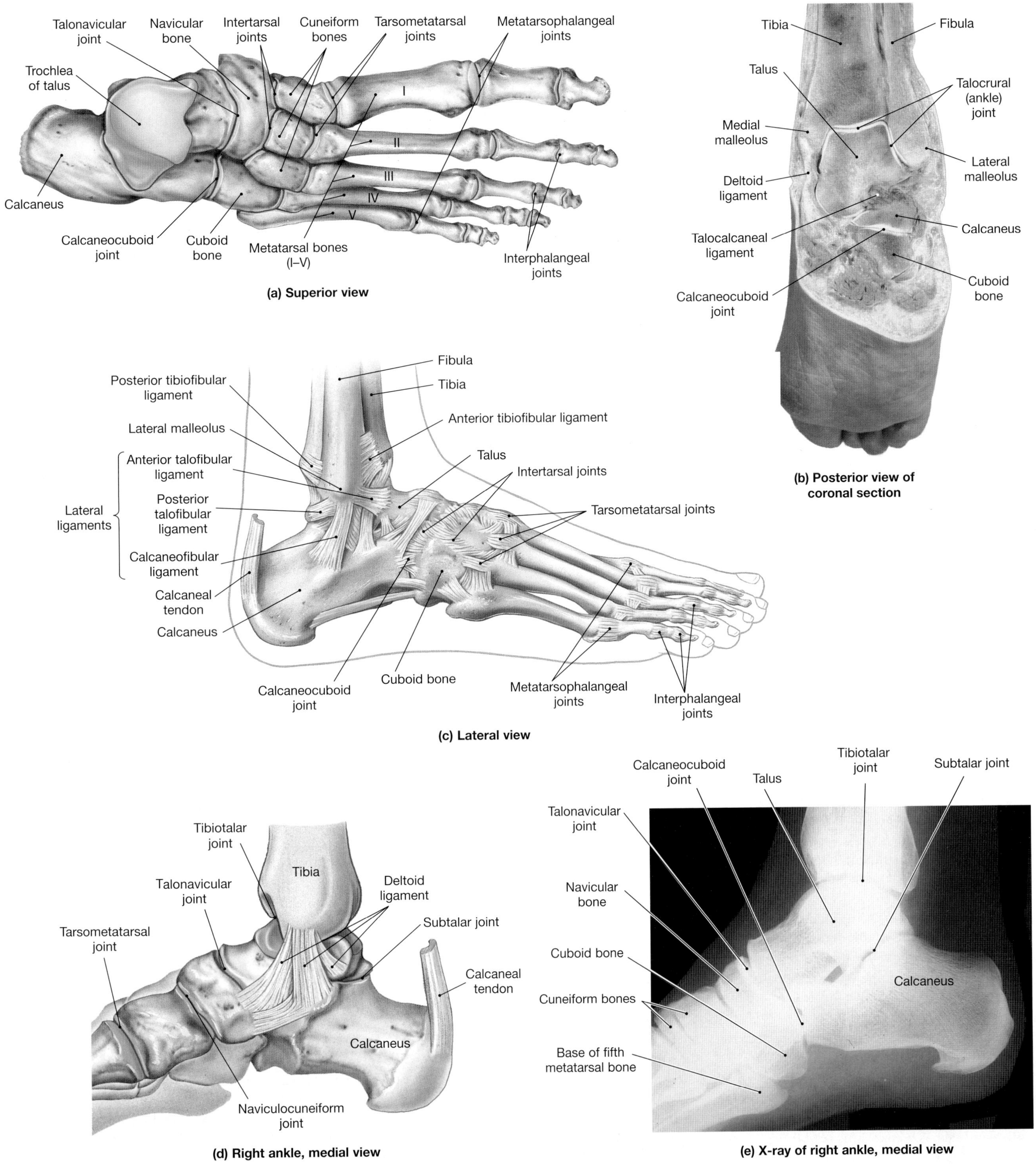

FIGURE 8.19 **JOINTS OF THE ANKLE AND FOOT, PART II**

(**a**) Superior view of bones and joints of the right foot. (**b**) Posterior view of a coronal section through the right ankle after plantar flexion.Note the placement of the medial and lateral malleoli. (**c**) Lateral view of the right foot, showing ligaments that stabilize the ankle joint. (**d**) Medial view of the right ankle, showing the lateral ligaments. (**e**) X-ray of right ankle, medial/lateral projection. *See MRI Scan 8b in the Companion Atlas.*

Rheumatoid arthritis is an inflammatory condition that affects roughly 2.5 percent of the adult population. At least some cases occur when the immune response mistakenly attacks the joint tissues. Such a condition, in which the body attacks its own tissues, is called an *autoimmune disease.* Allergies, bacteria, viruses, and genetic factors have all been proposed as contributing to or triggering the destructive inflammation.

Regular exercise, physical therapy, and drugs that reduce inflammation, such as aspirin, provide symptomatic relief. Anti-rheumatic medications that slow the progression of the disease show promise. Most seem to work by blocking stages in the abnormal inflammatory or immune process responsible for rheumatoid arthritis. Surgical procedures can realign or redesign the affected joint. In extreme cases involving the hip, knee, elbow, or shoulder, the defective joint can be replaced by an artificial one.

Degenerative changes comparable to those seen in arthritis may result from joint immobilization. When motion ceases, so does the circulation of synovial fluid, and the cartilages begin to suffer. **Continuous passive motion (CPM)** of any injured joint appears to encourage the repair process by improving the circulation of synovial fluid. The movement is often performed by a physical therapist or a machine during the recovery process.

With age, bone mass decreases and bones become weaker, so the risk of fractures increases. If osteoporosis develops, the bones may weaken to the point at which fractures occur in response to stresses that could easily be tolerated by normal bones. Hip fractures are among the most dangerous fractures seen in elderly persons. These fractures, most often involving individuals over age 60, may be accompanied by hip dislocation or by pelvic fractures.

Healing proceeds very slowly, and the powerful muscles that surround the hip joint often prevent proper alignment of the bone fragments. Fractures at the greater or lesser trochanter generally heal well if the joint can be stabilized; steel frames, pins, screws, or some combination of these devices may be needed to preserve alignment and to permit healing to proceed normally.

Although hip fractures are most common among those over age 60, in recent years the frequency of hip fractures has increased dramatically among young, healthy professional athletes.

Bones and Muscles

The skeletal and muscular systems are structurally and functionally interdependent; their interactions are so extensive that they are often considered to be parts of a single *musculoskeletal system.*

There are direct physical connections, for the connective tissues that surround the individual muscle fibers are continuous with those that establish the tissue framework of an attached bone. Muscles and bones are also physiologically linked, because muscle contractions can occur only when the extracellular concentration of calcium remains within relatively narrow limits, and most of the body's calcium reserves are held within the skeleton. The next three chapters will examine the structure and function of the muscular system and discuss how muscular contractions perform specific movements.

RELATED CLINICAL TERMS

ankylosis (an-ki-LŌ-sis): An abnormal fusion between articulating bones in response to trauma and friction within a joint. *Rheumatism, Arthritis, and Synovial Function p. 790*

arthritis (ar-THRĪ-tis): Rheumatic diseases that affect synovial joints. Arthritis always involves damage to the articular cartilages, but the specific cause may vary. The diseases of arthritis are usually classified as either **degenerative** or **inflammatory** in nature. *Rheumatism, Arthritis, and Synovial Function p. 790*

arthroscope: An instrument that uses fiber optics to explore a joint without major surgery. *Knee Injuries p. 790*

arthroscopic surgery: The surgical modification of a joint using an arthroscope. *Knee Injuries p. 790*

bunion: The most common pressure-related bursitis, involving a tender nodule formed around bursae over the base of the great toe. *Bursitis p. 790*

bursectomy: The surgical removal of inflamed bursae. *Bursitis p. 790*

bursitis: Inflammation of a bursa that causes pain whenever the associated tendon or ligament moves. *Bursitis p. 790*

congenital talipes equinovarus (clubfoot): A congenital deformity affecting one or both feet. It develops secondary to abnormalities in neuromuscular development. *Problems with the Ankle and Foot p. 791*

continuous passive motion (CPM): A therapeutic procedure involving passive movement of an injured joint to stimulate circulation of synovial fluid. The goal is to prevent degeneration of the articular cartilages. *Rheumatism, Arthritis, and Synovial Function p. 790*

dancer's fracture: A fracture of the fifth metatarsal, usually near its proximal articulation. *Problems with the Ankle and Foot p. 791*

flatfeet: The loss or absence of a longitudinal arch. *Problems with the Ankle and Foot p. 791*

herniated disc: A common name for a condition caused by distortion of an intervertebral disc. The distortion applies pressure to spinal nerves, causing pain and limiting range of motion. *p. 224*

laminectomy (la-mi-NEK-tō-mē): Removal of vertebral laminae; may be performed to access the vertebral canal and relieve symptoms of a herniated disc. *p. 224*

luxation (luks-Ā-shun): A dislocation; a condition in which the articulating surfaces are forced out of position. *p. 216*

meniscectomy: The surgical removal of an injured meniscus. *Knee Injuries p. 790*

osteoarthritis (os-tē-ō-ar-THRĪ-tis) (*degenerative arthritis,* or *degenerative joint disease (DJD)*): An arthritic condition resulting from (1) cumulative wear and tear on joint surfaces or (2) genetic predisposition. In the U.S. population, 25% of women and 15% of men over 60 years of age show signs of this disease. *Rheumatism, Arthritis, and Synovial Function p. 790*

rheumatism (ROO-ma-tizm): A general term that indicates pain and stiffness affecting the skeletal system, the muscular system, or both. *Rheumatism, Arthritis, and Synovial Function p. 790*

rheumatoid arthritis: An inflammatory arthritis that affects roughly 2.5% of the adult population. The cause is uncertain, although allergies, bacteria, viruses, and genetic factors have all been proposed. The primary symptom is **synovitis** (sin-ō-VĪ-tis), swelling and inflammation of the synovial membrane. *Rheumatism, Arthritis, and Synovial Function p. 790*

sciatica (sī-AT-i-ka): The painful result of compression of the roots of the sciatic nerve. The acute initial pain in the lower back is sometimes called **lumbago** (lum-BĀ-gō). *p. 224*

shoulder separation: The partial or complete dislocation of the acromioclavicular joint. *p. 227*

sprain: Condition caused when a ligament is stretched to the point where some of the collagen fibers are torn. Unless torn completely, the ligament remains functional, and the structure of the joint is not affected. *Problems with the Ankle and Foot p. 791*

subluxation (sub-luks-Ā-shun): A partial dislocation; displacement of articulating surfaces sufficient to cause discomfort, but resulting in less physical damage to the joint than during a complete dislocation. *p. 216*

Additional Clinical Terms Discussed in Appendix I (pp. 790–791)
avascular necrosis; synovitis; torn ligaments

STUDY OUTLINE & CHAPTER REVIEW

Introduction 214

1. **Articulations** (joints) exist wherever two bones interact. The function of a joint is dependent on its anatomical design. Joints may permit: (1) no movement, (2) slight movement, or (3) extensive movement.

Classification of Joints 214

1. Three categories of joints are based on range of movement. *Immovable joints* are **synarthroses,** *slightly movable joints* are **amphiarthroses,** and those that are *freely movable joints* are called **diarthroses.** Joints may be classified by function *(see Table 8.1)* or by structure. *(see Table 8.2)*

Synarthroses (Immovable Joints) 214

2. In a **synarthrosis,** bony edges are close together and may interlock. Examples of synarthroses include a **suture** between skull bones, a **gomphosis** between teeth and jaws, a **synchondrosis** between bone and cartilage in an epiphyseal plate, and a **synostosis** where two bones fuse and the boundary between them disappears.

Amphiarthroses (Slightly Movable Joints) 214

3. Very limited movements are permitted in an **amphiarthrosis.** Examples of amphiarthroses are a **syndesmosis,** where collagen fibers connect bones of the leg, and a **symphysis,** where bones are separated by a pad of cartilage.

Diarthroses (Freely Movable Joints) 214

4. A wide range of movement is permitted at a **diarthrosis,** or **synovial joint.** The bony surfaces at diarthroses are covered by **articular cartilages,** lubricated by **synovial fluid,** and enclosed within a **joint,** or **articular, capsule.** Other synovial and accessory structures can include: **menisci** or **articular discs; fat pads; ligaments; tendons; bursae;** and **tendon sheaths.** *(see Figure 8.1)*
5. A joint cannot have both great strength and great mobility at the same time. The greater the strength of a joint, the lesser its mobility, and vice versa.

Articular Form and Function 217

Describing Dynamic Motion 217

1. Possible movements of a bone at an articulation can be classified as **linear motion** (back and forth motion), **angular motion** (movement where the angle between the shaft and the articular surface changes), and **rotation** (spinning of the shaft on its longitudinal axis). *(see Figure 8.2)*
2. Joints are described as **monaxial, biaxial,** or **triaxial** depending on the number of axes along which they permit movement. *(see Figure 8.6)*

Types of Movements 217

3. In **gliding,** the opposing surfaces at an articulation slide past each other. *(see Figure 8.2b)*
4. Several important terms describe angular motion, including: **abduction** (movement away from the longitudinal axis of the body), **adduction** (movement toward the longitudinal axis of the body), **flexion** (reduction in angle between articulating elements), **extension** (increase in angle between articulating elements), **hyperextension** (extension beyond anatomical position), and **circumduction** (a special type of angular motion that includes flexion, abduction, extension, and adduction). *(see Figure 8.3)*
5. Description of rotational movements requires reference to a figure in the anatomical position. **Rotation** of the head to the left or right is observed with a "shaking the head no" movement. An **internal** or **external rotation** is observed in limb movements if the anterior aspect of the limb either turns toward or away from the ventral surface of the body. The bones in the forearm permit **pronation** (motion to bring palm facing back) and **supination** (motion to bring palm facing front). *(see Figure 8.4)*
6. Several special terms apply to specific articulations or unusual movement types. Movements of the foot include **eversion** (bringing the sole of the foot outward), and **inversion** (bringing the sole of foot inward). The ankle undergoes **dorsiflexion** ("digging in the heels") and **plantar flexion** ("standing on tiptoes"). **Lateral flexion** occurs when the vertebral column bends to the side. **Opposition** is the thumb movement that enables us to grasp objects. **Protraction** involves moving a body part anteriorly ("jutting out the lower jaw"), **retraction** involves moving it posteriorly ("pulling jaw back"). **Elevation** and **depression** occur when we move a structure inferiorly or superiorly (occurs with opening and closing of the mouth). *(see Figure 8.5)*

A Structural Classification of Synovial Joints 218

7. **Gliding joints** permit limited movement, usually in a single plane. *(see Figure 8.6 and Table 8.2)*
8. **Hinge joints** and **pivot joints** are monaxial joints that permit angular movement in a single plane. *(see Figure 8.6 and Table 8.2)*
9. Biaxial joints include **ellipsoidal joints** and **saddle joints.** They allow angular movement in two planes. *(see Figure 8.6 and Table 8.2)*
10. Triaxial, or **ball-and-socket joints,** permit all combinations of movement, including rotation. *(see Figure 8.6 and Table 8.2)*

Representative Articulations 222

The Temporomandibular Joint 222

1. The **temporomandibular joint (TMJ)** involves the mandibular fossa of the temporal bone and the condylar process of the mandible. This joint has a thick pad of fibrocartilage, the articular disc. Supporting structures include the dense capsule, the **temporomandibular ligament,** the **stylomandibular ligament,** and the **sphenomandibular ligament.** This is a relatively loose hinge joint that permits small amounts of gliding and rotation. *(see Figure 8.7)*

Intervertebral Articulations 222

2. The superior and inferior articular processes of vertebrae form gliding joints with those of adjacent vertebrae. The bodies form symphyseal joints. They are separated by **intervertebral discs** containing an inner soft, elastic gelatinous core, the **nucleus pulposus,** and an outer layer of fibrocartilage, the **anulus fibrosus.** (see Figures 8.8/8.9)
3. Numerous ligaments bind together the bodies and processes of all vertebrae. *(see Figure 8.8)*
4. The articulations of the vertebral column permit **flexion** and **extension** (anterior-posterior), **lateral flexion,** and **rotation.**
5. Articulations of the axial skeleton are summarized in *Table 8.3.*

The Sternoclavicular Joint 224

6. The **sternoclavicular joint** is a gliding joint that lies between the sternal end of each clavicle and the manubrium of the sternum. An articular disk separates the opposing surfaces. The capsule is reinforced by the **anterior** and **posterior sternoclavicular ligaments,** plus the **interclavicular** and **costoclavicular ligaments.** *(see Figure 8.10)*

The Shoulder Joint 225

7. The shoulder, or **glenohumeral joint,** formed by the glenoid fossa and the head of the humerus, is a loose, shallow joint that permits the greatest range of motion of any joint in the body. It is a ball-and-socket diarthrosis. Strength and stability are sacrificed to obtain mobility. The ligaments and surrounding muscles and tendons provide strength and stability. The shoulder has a large number of **bursae** that reduce friction as large muscles and tendons pass across the joint capsule. *(see Figure 8.11)*

The Elbow Joint 227

8. The **elbow** is a hinge joint that permits flexion and extension. It is really two joints, one between the humerus and ulna (*humeroulnar joint*) and one between the humerus and the radius (*humeroradial joint*). **Radial** and **ulnar collateral ligaments** and **annular ligaments** aid in stabilizing this joint. *(see Figure 8.12)*

The Joints of the Wrist 229

9. The **wrist joint** is formed by the **distal radioulnar joint**, the **radiocarpal joint**, and the **intercarpal joints.** The radioulnar articulation is a pivot diarthrosis that permits pronation and supination. The radiocarpal articulation is an ellipsoidal articulation that involves the distal articular surface of the radius and three proximal carpal bones (scaphoid bone, lunate bone, and triquetrum). The radiocarpal joint permits: flexion/extension, adduction/abduction, and circumduction. A connective tissue capsule and broad ligaments stabilize the positions of the individual carpal bones. The intercarpal joints are gliding joints. *(see Figure 8.13)*

The Joints of the Hand 230

10. Five types of diarthrotic joints are found in the hand: (1) carpal bone/carpal bone (**intercarpal joints**); gliding diarthrosis; (2) carpal bone/first metacarpal bone (**carpometacarpal joint of the thumb**); saddle diarthrosis, permitting flexion/extension, adduction/abduction, circumduction, opposition; (3) carpal bones/metacarpal bones II–V (**carpometacarpal joints**); gliding diarthrosis, permitting slight flexion/extension and adduction/abduction; (4) metacarpal bone/phalanx (**metacarpophalangeal joints**); ellipsoidal diarthrosis, permitting flexion/extension, adduction/abduction, and circumduction; and (5) phalanx/phalanx (**interphalangeal joints**); hinge diarthrosis, permitting flexion/extension. *(see Figure 8.13)*

The Hip Joint 230

11. The **hip joint** is a ball-and-socket diarthrosis that is formed by the union of the acetabulum of the os coxae with the head of the femur. The joint permits flexion/extension, adduction/abduction, circumduction, and rotation. *(see Figures 8.14/8.15)*
12. The articular capsule of the hip joint is reinforced and stabilized by four broad ligaments: **iliofemoral**, **pubofemoral**, **ischiofemoral**, and **transverse acetabular ligaments**. Another ligament, the **ligament of the femoral head** *(ligamentum capitis femoris)*, also helps to stabilize the hip joint. *(see Figures 8.14/8.15)*

The Knee Joint 231

13. The knee joint functions as a hinge joint, but is more complex than standard hinge joints such as the elbow. Structurally, the knee resembles three separate joints: (1) medial condyles of femur and tibia, (2) lateral condyles of femur and tibia, and (3) the patella and patellar surface of femur. The joint permits flexion/extension and limited rotation. *(see Figures 8.16/8.17)*
14. The articular capsule of the knee is not a single unified capsule with a common synovial cavity. It contains: (1) **fibrocartilage pads**, the **medial** and **lateral menisci**, and (2) **fat pads.** *(see Figures 8.16/8.17)*
15. Seven major ligaments bind and stabilize the knee joint: **patellar**, **popliteal** (two), **anterior** and **posterior cruciate**, **tibial collateral**, and **fibular collateral ligaments**. *(see Figures 8.16/8.17)*

The Joints of the Ankle and Foot 235

16. The ankle joint, or **talocrural joint**, is a hinge joint formed by the inferior surface of the tibia, the lateral malleolus of the fibula, and the trochlea of the talus. The primary joint is the tibiotalar joint. The tibia and fibula are bound together by **anterior** and **posterior tibiofibular ligaments.** With these stabilizing ligaments holding the bones together, the medial and lateral malleoli can prevent lateral or medial sliding of the tibia across the trochlear surface. The ankle joint permits dorsiflexion/plantar flexion. The medial **deltoid ligament** and three **lateral ligaments** further stabilize the ankle joint. *(see Figures 8.18/8.19)*
17. Four types of diarthrotic joints are found in the foot: (1) tarsal bone/tarsal bone (**intertarsal joints**, named after the participating bone); gliding diarthrosis; (2) tarsal bone/metatarsal bone (**tarsometatarsal joints**); gliding diarthrosis; (3) metatarsal bone/phalanx (**metatarsophalangeal joints**); ellipsoidal diarthrosis, permitting flexion/extension, adduction/abduction; and (4) phalanx/phalanx (**interphalangeal joints**); hinge diarthrosis, permitting flexion/extension. *(see Figures 8.18/8.19* and *Table 8.5)*

Aging and Articulations 236

1. Problems with joint function are relatively common, especially in older individuals. **Rheumatism** is a general term for pain and stiffness affecting the skeletal system, the muscular system, or both; several major forms exist. **Arthritis** encompasses all the rhematic diseases that affect synovial joints. Both conditions become increasingly common with age.

Bones and Muscles 238

1. The skeletal and muscular systems are structurally and functionally interdependent and comprise the *musculoskeletal system.*

LEVEL 1 REVIEWING FACTS AND TERMS

Match each numbered item with the most closely related lettered item. Use letters for answers in the spaces provided.

Column A	Column B
____ 1. no movement	a. wrist joint
____ 2. synovial	b. dorsiflexion
____ 3. increased angle	c. fluid-filled pockets
____ 4. bursae	d. diarthrosis
____ 5. palm facing front	e. knee
____ 6. digging in heels	f. intervertebral discs
____ 7. fibrocartilage	g. supination
____ 8. carpus	h. extension
____ 9. menisci	i. synarthrosis

10. The type of joint formed by the fusion of two bones is a
    (a) gomphosis (b) synostosis
    (c) synchondrosis (d) symphysis
11. In monaxial articulation
    (a) movement can occur in only one plane
    (b) movement can occur in two planes
    (c) movement can occur in all three planes
    (d) only circumduction is possible
12. Which of the following is not a function of synovial fluid?
    (a) shock absorption (b) increase osmotic pressure within joint
    (c) lubrication (d) provide nutrients
13. The joint that provides the greatest range of mobility of any joint in the body is the
    (a) knee joint (b) hip joint
    (c) elbow joint (d) shoulder joint
14. Which of the following ligaments is not associated with the hip joint?
    (a) iliofemoral ligament (b) pubofemoral ligament
    (c) ligament of the femoral head (d) ligamentum flavum
15. The back of the knee joint is reinforced by
    (a) tibial collateral ligaments (b) popliteal ligaments
    (c) posterior cruciate ligament (d) patellar ligaments

16. The shoulder joint is primarily stabilized by
    (a) ligaments and muscles that move the humerus
    (b) the scapula
    (c) glenohumeral ligaments only
    (d) the clavicle
17. A twisting motion of the foot that turns the sole inward is
    (a) dorsiflexion (b) eversion
    (c) inversion (d) protraction
18. Which of the following is not a function of the intervertebral discs?
    (a) act as shock absorbers
    (b) prevent bone–bone contact
    (c) contribute to the height of the individual
    (d) lubricate the joint
19. Luxations are painful due to stimulation of pain receptors in all locations except the following
    (a) inside the joint cavity
    (b) in the capsule
    (c) in the ligaments around the joint
    (d) in the tendons around the joint
20. The ligaments that limit the anterior and posterior movement of the femur and maintain the alignment of the femoral and tibial condyles are the
    (a) cruciate ligaments
    (b) fibular collateral ligaments
    (c) patellar ligaments
    (d) tibial collateral ligaments

LEVEL 2 REVIEWING CONCEPTS

1. Muscles that extend the elbow attach to the
   (a) coronoid process (b) radial tuberosity
   (c) olecranon (d) lateral epicondyle
2. Compare and contrast the strength and stability of a joint with respect to the amount of mobility in the joint.
3. What would be the expected result of the contraction of the muscle that is attached to the radial tuberosity?
4. How do the malleoli of the tibia and fibula function to retain the correct positioning of the tibiotalar joint?
5. How do articular cartilages differ from other cartilages in the body?
6. When you stand for a long period of time, why should you "lock" your knee in extended position? How does the knee lock?
7. What role is played by capsular ligaments in a complex synovial joint? Use the humeroulnar joint to illustrate your answer.
8. What is the common mechanism that holds together immovable joints such as skull sutures and the gomphoses, holding teeth in their alveoli?
9. How can pronation be distinguished from circumduction of a skeletal element?
10. What would you tell your grandfather about his decrease in height as he grows older?

LEVEL 3 CRITICAL THINKING AND CLINICAL APPLICATIONS

1. When a person involved in an automobile accident suffers from "whiplash," what structures have been affected and what movements could be responsible for this injury?
2. A marathon runner steps on an exposed tree root, causing a twisted ankle. After being examined, she was told the ankle was severely sprained, not broken. The ankle will probably take longer to heal than a broken bone would. Which structures were damaged, and why would they take so long to heal?
3. Lee injures his knee during a football practice such that the synovial fluid in the knee joint no longer circulates. The physician who examines him tells him that they have to reestablish circulation of the synovial fluid before the articular cartilages become damaged. Why?

✓ ANSWERS TO CONCEPT CHECK QUESTIONS

p. 216 1. No relative movement is permitted at a synarthrosis, whereas an amphiarthrosis permits slight movement. **2.** The main advantage is that it permits a broad range of motion without significant friction. **3.** Friction reduction and the distribution of dissolved gases, nutrients, and waste products. **4.** Bursae are pockets lined by synovial membrane and filled with synovial fluid. They reduce friction between adjacent structures, such as tendons and bones or muscles.

p. 220 1. Originally, the joint is a type of syndesmosis. When the bones fuse, the bones along the suture represent a synostosis. **2.** a. abduction, b. supination, c. flexion.

p. 230 1. Since the subscapular bursa is located in the shoulder joint, an inflammation of this structure (bursitis) would be found in the tennis player. The condition is associated with repetitive motion that occurs at the shoulder, such as swinging a tennis racket. The jogger would be more at risk for injuries to the knee joint. **2.** Mary has most likely fractured her ulna.

p. 235 1. The iliofemoral, pubofemoral, and ischiofemoral ligaments would all be found in the hip joint. **2.** Damage to the menisci in the knee joint would result in a decrease in the joints stability. The individual would have a harder time locking the knee in place while standing and would have to use muscle contractions to stabilize the joint. When standing for long periods, the muscles would fatigue and the knee would "give out." We would also expect the individual to experience pain. **3.** The patellar ligament provides support to the anterior surface of the knee joint. Damage to the patellar ligament would affect this support. **4.** The tibial collateral ligament reinforces the medial surface of the knee joint, and the fibular collateral ligament reinforces the lateral surface. These ligaments tighten only at full extension, and in this position they act to stabilize the joint.

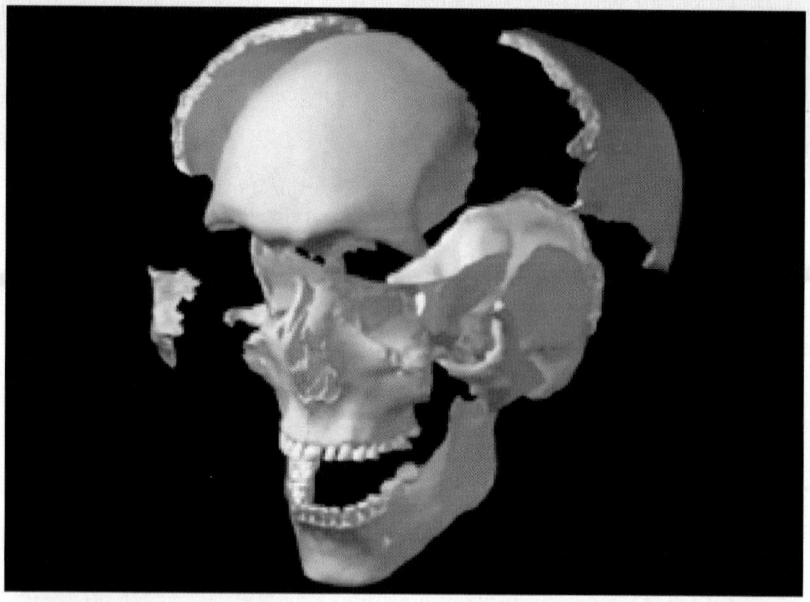

REVIEW IT

Challenge 1

In Chapter 6 we looked at the various bones of the skull using photographs and images of the skull as well as photos of the isolated bones.

On the Interactive CD go to Animations/ Skeletal System/ Skull/ Exploded Skull. This animation allows you to see the individual bones of the skull in relation to one another. The image here is taken from that animation. Identify the bones in this image using the photographs and illustrations from Chapter 6 as a guide. You can also go directly to the CD and manipulate this animation to gain further insights into how these bones articulate.

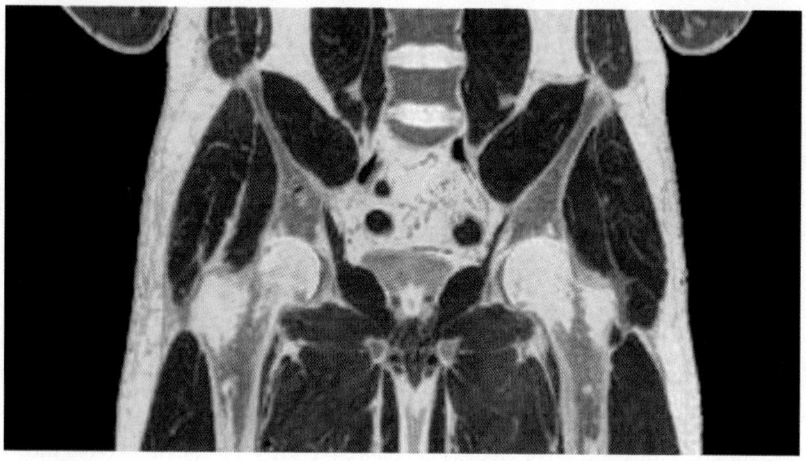

Challenge 2

This image comes from a digital reconstruction based on the Visible Human data set. You are looking at the pelvic girdle. What bones are visible in this section? Find the figure in your text (Chapter 8) which most closely corresponds to this image. Which areas contain spongy bone? Which areas contain compact bone? What specific bony landmarks are identifiable on this section? If you were to look at these structures on a skeleton, what clues could you use to determine the gender of the individual?

Images provided by the Digital Cadaver™ project, courtesy of Visible Productions, Inc.

APPLY IT

Below are two separate exercises that provide more information on a topic presented in the preceding chapters. Each one is designed to take approximately 10 minutes, and will help you gain a better appreciation for the material presented in Chapters 5–8.

Critical Linking 1

The best way to learn the bones and their processes is to practice, practice, practice! After studying the text and interacting with the animations on the CD, make sure you really know the structures by visiting the Bone Box.

Get to this site through the Critical Linking portion of the Companion Website for this text. Once there, click on the key word "Bone Box" to reach the Lumen Bone Box homepage. The site includes photographs of all 206 bones in the body. Each photograph includes numbered lead lines pointing to prominent features of the bone. Begin with the vertebrae, clicking on the atlas. On a separate sheet of paper, write the answers to each of the numbered structures. Check yourself by clicking on the number. The answers are provided in the upper left portion of the site. Continue to review the entire skeleton in this fashion.

Critical Linking 2

More than 120,000 artificial hips are being implanted each year in the United States. Based on your knowledge of the hip, what anatomical structures do you expect to be involved in this procedure? What structures will the prostheses replace? How are these prostheses created? Before reading the information provided on the Website indicated, try to answer these questions.

Write an outline of your thoughts then move to the companion website, Critical Linking section and click on the key word "Hip replacement". There you will find a comprehensive article prepared for the National Institutes of Health designed to answer questions such as this. How closely did your answers match current practices?

FURTHER STUDIES

Bone is one of the few structures in the human body that will remain intact long after death. Due to the inclusion of naturally occurring radioactive minerals as well as other trace elements trapped in the matrix, bone is used forensically to provide clues about the lifestyles and age of unearthed specimens. This opens up an entire field of study referred to as *paleoradiology*, combining the studies of anatomy, medicine, and history. Although this field can sometimes be a bit gruesome, the information obtained from bone located during archeological digs can be fascinating. One such discovery is the observation that ancient Egyptians suffered from osteoporosis and osteopenia just as we do today.

A complete description of this study can be found in the article prepared by Carol Haigh, Biomedical and Forensic Egyptologist at *http://www.shef.ac.uk/~assem/5/haigh.html* There is another, more recent example of the importance of this field of study. Controversy has erupted in the scientific community over the discovery that previously unearthed bones found on Arrlington ridge in the Channel Islands were misidentified. Apparently these bones were labeled as belonging to a male. Recent scientific studies have confirmed they actually belonged to a female, causing a re-evaluation of the cultural history of the region. Articles on this can be obtained by visiting *http://europe.cnn.com/NATURE/9906/08/ancient.woman/* or *http://www.sbnature.org/htmls/islandbones.htm*.

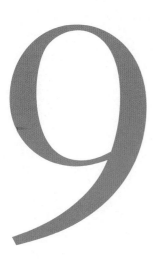

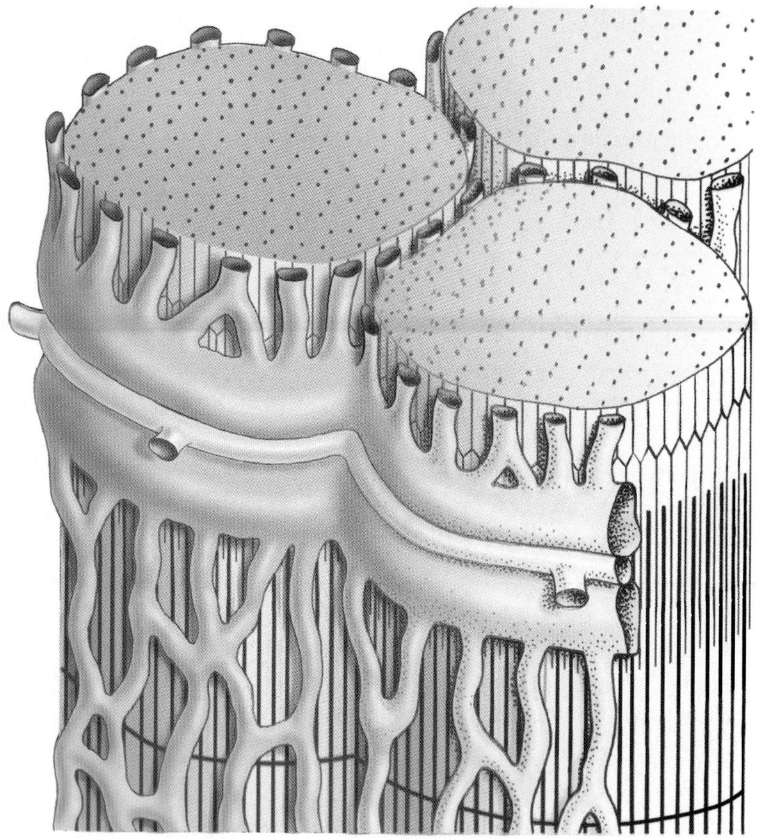

9

THE MUSCULAR SYSTEM

Skeletal Muscle Tissue and Muscle Organization

It is hard to imagine what life would be like without muscle tissue. We would be unable to sit, stand, walk, speak, or grasp objects. Blood would not circulate, because there would be no heartbeat to propel it through the vessels. The lungs could not rhythmically empty and fill nor could food move along the digestive tract. In fact, there would be practically no movement through any of your internal passageways.

This is not to say that all life depends on muscle tissue. There are large organisms that get by very nicely without it—we call them plants. But life as *we* live it would be impossible, for many of our physiological processes, and virtually all our dynamic interactions with the environment, involve muscle tissue. Muscle tissue, one of the four primary tissue types, consists chiefly of *muscle fibers*—elongate cells, each capable of contracting along its longitudinal axis. Muscle tissue also includes the connective tissue fibers that harness those contractions to perform useful work. There are three types of muscle tissue: *skeletal muscle*, *cardiac muscle*, and *smooth muscle*. p. 77 **Skeletal muscle tissue** moves the body by pulling on bones of the skeleton, making it possible for us to walk, dance, or play a musical instrument. **Cardiac muscle tissue** pushes blood through the arteries and veins of the circulatory system; **smooth muscle tissues** push fluids and solids along the digestive tract and perform varied functions in other systems. These muscle tissues share four basic properties:

1. *Excitability*: the ability to respond to stimulation. For example, skeletal muscles normally respond to stimulation by the nervous system, and some smooth muscles respond to circulating hormones.
2. *Contractility*: the ability to shorten actively and exert a pull or tension that can be harnessed by connective tissues.
3. *Extensibility*: the ability to continue to contract over a range of resting lengths. For example, a smooth muscle cell can be stretched to several times its original length and still contract when stimulated.
4. *Elasticity*: the ability of a muscle to rebound toward its original length after a contraction.

This chapter focuses attention on skeletal muscle tissue; cardiac muscle tissue will be considered in Chapter 21, which deals with the anatomy of the heart, and smooth muscle tissue will be considered in Chapter 25, in our discussion of the digestive system.

Skeletal muscles are organs that include all four basic tissue types but that consist primarily of skeletal muscle tissue. The **muscular system** of the human body consists of over 700 skeletal muscles and includes all of the skeletal muscles that can be controlled voluntarily. This system will be the focus of the next three chapters. This chapter considers the function, gross anatomy, microanatomy, and organization of skeletal muscles, as well as muscle terminology. Chapter 10 discusses the gross anatomy of the axial musculature, skeletal muscles associated with the axial skeleton; Chapter 11 discusses the gross anatomy of the appendicular musculature, skeletal muscles associated with the appendicular skeleton.

Functions of Skeletal Muscle

Skeletal muscles are contractile organs directly or indirectly attached to bones of the skeleton. Skeletal muscles perform the following functions:

1. *Produce skeletal movement.* Muscle contractions pull on tendons and move the bones of the skeleton. The effects range from simple motions such as extending the arm to the highly coordinated movements of swimming, skiing, or typing.
2. *Maintain posture and body position.* Contraction of specific muscles also maintains body posture—for example, holding the head in position when reading a book or balancing the weight of the body above the feet when walking involves the contraction of muscles that stabilize joints. Without constant muscular contraction, we could not sit upright without collapsing or stand without toppling over.
3. *Support soft tissues.* The abdominal wall and the floor of the pelvic cavity consist of layers of skeletal muscle. These muscles support the weight of visceral organs and protect internal tissues from injury.
4. *Regulate entering and exiting of material.* Openings, or **orifices**, of the digestive and urinary tracts are encircled by skeletal muscles. These muscles provide voluntary control over swallowing, defecation, and urination.
5. *Maintain body temperature.* Muscle contractions require energy, and whenever energy is used in the body, some of it is converted to heat. The heat lost by contracting muscles keeps our body temperature in the range required for normal functioning.

Anatomy of Skeletal Muscles

When naming structural features of muscles and their components, anatomists often used the Greek words *sarkos* (flesh) and *mys* (muscle). These root words should be kept in mind as our discussion proceeds. We will first discuss the gross anatomy of skeletal muscle and then describe the microstructure that makes contraction possible.

Gross Anatomy [FIGURE 9.1]

Figure 9.1● illustrates the appearance and organization of a typical skeletal muscle. We begin our study of the gross anatomy of muscle with a description of the connective tissues that bind and attach skeletal muscles to other structures.

CONNECTIVE TISSUE OF MUSCLE [FIGURE 9.1]

Each skeletal muscle has three concentric layers, or wrappings, of connective tissue: an outer *epimysium*, a central *perimysium*, and an inner *endomysium* (Figure 9.1●).

- The **epimysium** (ep-i-MIS-ē-um; *epi*, on + *mys*, muscle) is a dense irregular connective tissue layer that surrounds the entire skeletal muscle. The epimysium, which separates the muscle from surrounding tissues and organs, is connected to the *deep fascia*. p. 77
- The connective tissue fibers of the **perimysium** (per-i-MIS-ē-um; *peri-*, around) divide the muscle into a series of internal compartments, each containing a bundle of muscle fibers called a **fascicle** (FAS-i-kul; *fasciculus*, bundle). In addition to collagen and elastic fibers, the perimysium contains numerous blood vessels and nerves that branch to supply each individual fascicle.
- The **endomysium** (en-dō-MIS-ē-um; *endo*, inside + *mys*, muscle) surrounds each skeletal muscle fiber, binds each muscle fiber to its neighbor, and supports capillaries that supply individual fibers. The endomysium consists of a delicate network of reticular fibers. Scattered **satellite cells** lie between the endomysium and the muscle fibers. These cells function in the repair of damaged muscle tissue.

TENDONS AND APONEUROSES The connective tissue fibers of the endomysium and perimysium are interwoven, and those of the perimysium blend into the

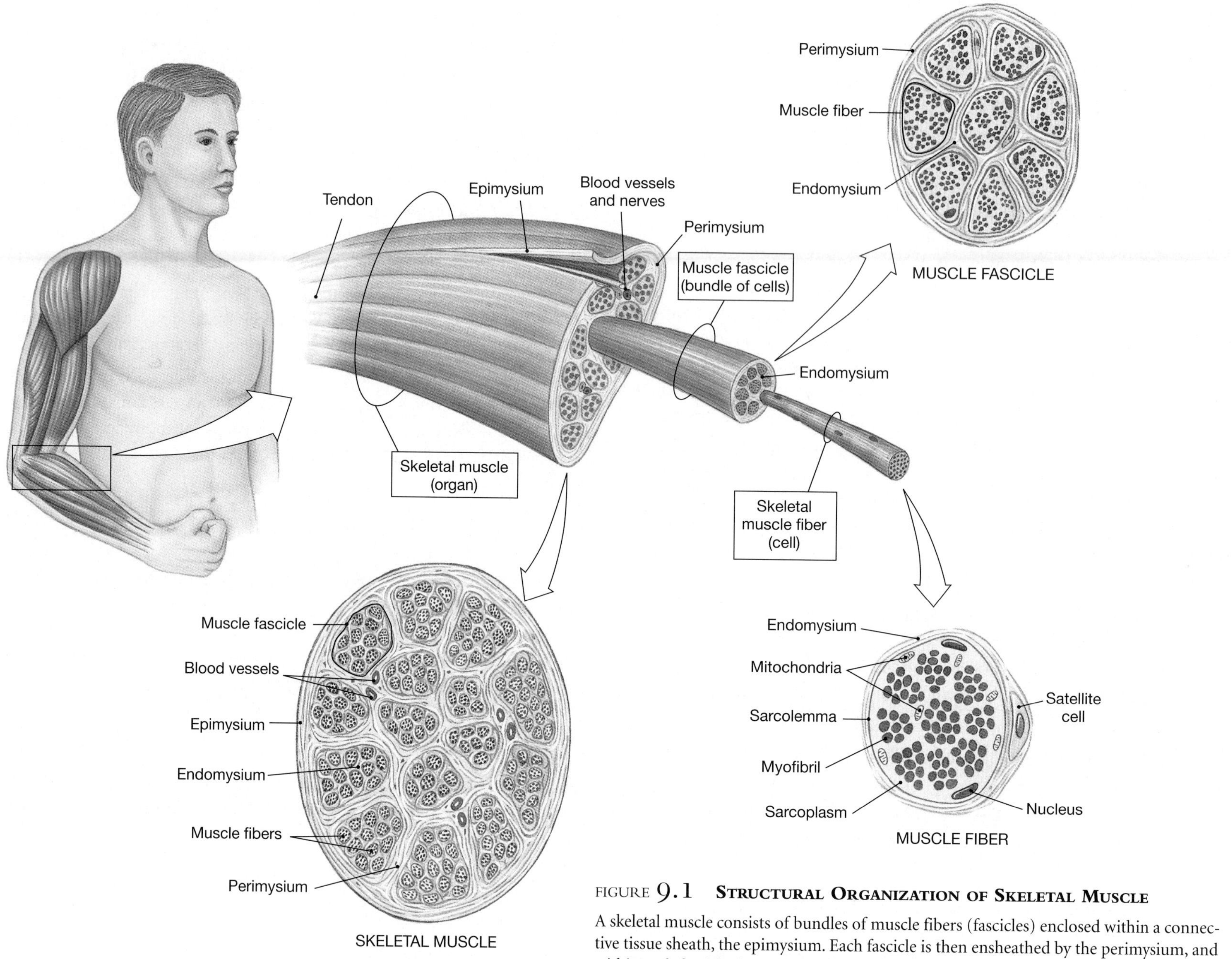

FIGURE 9.1 **Structural Organization of Skeletal Muscle**

A skeletal muscle consists of bundles of muscle fibers (fascicles) enclosed within a connective tissue sheath, the epimysium. Each fascicle is then ensheathed by the perimysium, and within each fascicle the individual muscle fibers are surrounded by the endomysium. Each muscle fiber has many nuclei as well as mitochondria and other organelles seen here and in Figure 9.3.

epimysium. At each end of the muscle, the collagen fibers of the epimysium, perimysium, and endomysium often converge to form a fibrous **tendon** that attaches the muscle to bone, skin, or another muscle. Tendons often resemble thick cords or cables. Tendons that form thick, flattened sheets are called **aponeuroses**. The structural characteristics of tendons and aponeuroses were considered in Chapter 3. p. 68

The tendon fibers are interwoven into the periosteum and matrix of the associated bone. This meshwork provides an extremely strong bond, and any contraction of the muscle exerts a pull on the attached bone.

NERVES AND BLOOD VESSELS [FIGURES 9.2]

The connective tissues of the epimysium and perimysium contain the nerves and blood vessels that supply the muscle fibers. Skeletal muscles are often called *voluntary muscles* because their contractions can be consciously controlled. This control is provided by the nervous system. Nerves, which are bundles of axons, penetrate the epimysium, branch through the perimysium, and enter the endomysium to innervate individual muscle fibers. Chemical communication between a synaptic terminal of the neuron and a skeletal muscle fiber occurs at a site called the **neuromuscular junction**, or *myoneural junction*. Several neuromuscular junctions are shown in Figure 9.2•. Each muscle fiber has one neuromuscular junction, usually located midway along its length. At a neuromuscular junction the synaptic terminal of the neuron is bound to the *motor end plate* of the skeletal muscle fiber. The motor end plate is a specialized area of the muscle cell membrane within a neuromuscular junction. (A later section will consider the structure of the motor end plate and its role in nerve–muscle communication.)

Muscle contraction requires tremendous quantities of energy, and an extensive vascular supply delivers the oxygen and nutrients needed for the production of ATP in active skeletal muscles. These blood vessels often enter the muscle alongside the associated nerves, and the vessels and nerves follow the same branching pattern through the perimysium. Once within the endomysium, the arteries supply an extensive capillary network around each muscle fiber. Because these capillaries are coiled, rather than straight, they are able to tolerate changes in the length of the muscle fiber.

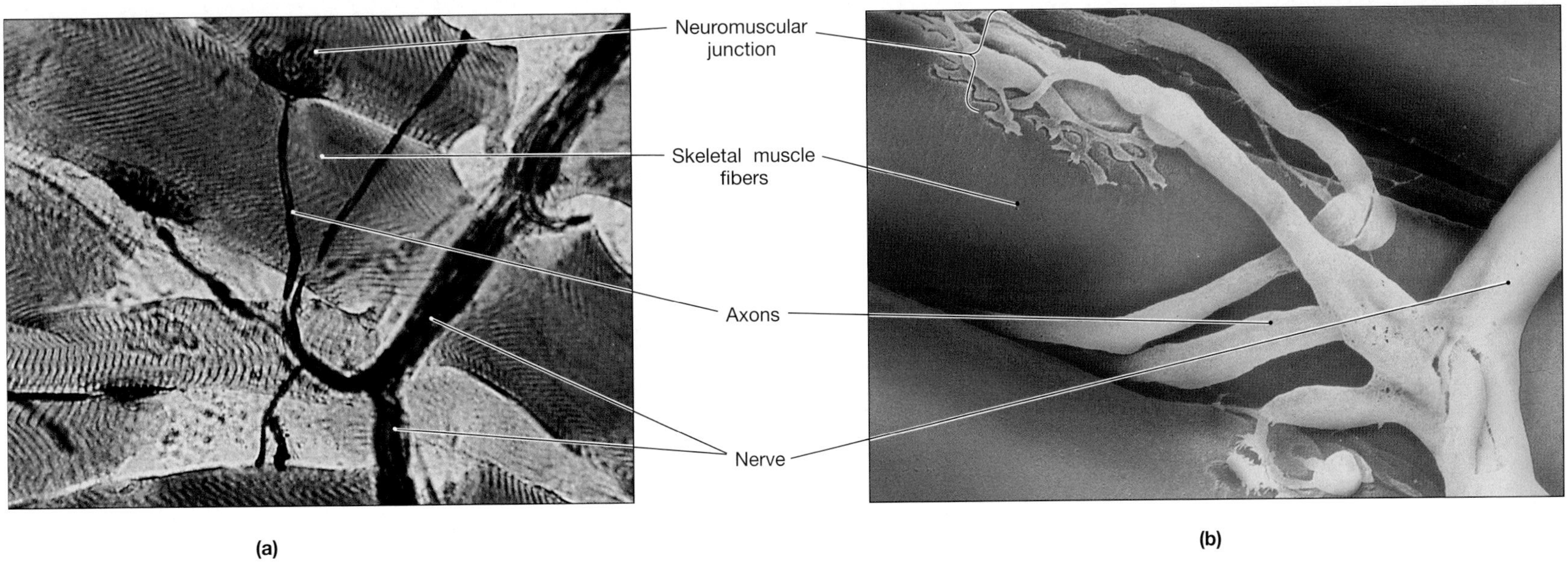

FIGURE 9.2 **Skeletal Muscle Innervation**

Each skeletal muscle fiber is stimulated by a nerve fiber at a neuromuscular junction. **(a)** Several neuromuscular junctions are seen on the muscle fibers of this fascicle. (LM × 230) **(b)** Colorized SEM of neuromuscular junction.

✓ **CONCEPT CHECK**

- What are the three types of muscle tissue, and what is the function of each?
- What is the perimysium? What structures would be located here?
- Describe the difference between a tendon and an aponeurosis.
- What is the difference between a myoneural junction and a motor end plate?

Microanatomy of Skeletal Muscle Fibers [FIGURES 9.1/9.3]

The cell membrane, or **sarcolemma** (sar-kō-LEM-a; *sarkos*, flesh + *lemma*, husk) of a skeletal muscle fiber surrounds the cytoplasm, or **sarcoplasm** (SAR-kō-plazm). Skeletal muscle fibers differ in several other respects from the "typical" cell described in Chapter 2.

- Skeletal muscle fibers are very large. A fiber from a leg muscle could have a diameter of 100 μm and a length equal to that of the entire muscle (30–40 cm, or 10–16 in.).
- Skeletal muscle fibers are *multinucleated*. During development, groups of embryonic cells called **myoblasts** fuse together to create individual skeletal muscle fibers (Figure 9.3a●). Each nucleus in a skeletal muscle fiber reflects the contribution of a single myoblast. Each skeletal muscle fiber contains hundreds of nuclei just inside the sarcolemma (Figure 9.3b,c●). This characteristic distinguishes skeletal muscle fibers from cardiac and smooth muscle fibers. Some myoblasts do not fuse with developing muscle fibers, but remain in adult skeletal muscle tissue as *satellite cells* (Figures 9.1 and 9.3a●). When a skeletal muscle is injured, satellite cells may differentiate and assist in the repair and regeneration of the muscle.
- Deep indentations in the sarcolemmal surface form a network of narrow tubules that extend into the sarcoplasm. Electrical impulses conducted by the sarcolemma and these **transverse tubules**, or **T-tubules**, help stimulate and coordinate muscle contractions.

MYOFIBRILS AND MYOFILAMENTS [FIGURE 9.3c,d]

The sarcoplasm of a skeletal muscle fiber contains hundreds to thousands of **myofibrils**. Each myofibril is a cylindrical structure 1–2 μm in diameter and as long as the entire cell (Figure 9.3c,d●). Myofibrils can shorten, and these are the structures responsible for skeletal muscle fiber contraction. Because the myofibrils are attached to the sarcolemma at each end of the cell, their contraction shortens the entire cell.

Surrounding each myofibril is a sleeve made up of membranes of the **sarcoplasmic reticulum (SR)**, a membrane complex similar to the smooth endoplasmic reticulum of other cells (Figure 9.3d●). This membrane network, which is closely associated with the transverse tubules, plays an essential role in controlling the contraction of individual myofibrils. On either side of a transverse tubule, the tubules of the SR enlarge, fuse, and form expanded chambers called **terminal cisternae**. The combination of a pair of terminal cisternae plus a transverse tubule is known as a **triad**. Although the membranes of the triad are in close contact and tightly bound together, there is no direct connection between them.

Mitochondria and glycogen granules are scattered among the myofibrils. The breakdown of glycogen and the activity of mitochondria provide the ATP needed to power muscular contractions. A typical skeletal muscle fiber has hundreds of mitochondria, more than most other cells in the body.

Myofibrils consist of bundles of **myofilaments**, protein filaments consisting primarily of the proteins *actin* and *myosin*. The actin filaments are found in *thin filaments,* and the myosin filaments are found in *thick filaments.* p. 34 The actin and myosin filaments are organized in repeating units called **sarcomeres** (SAR-kō-mērz; *sarkos*, flesh + *meros*, part).

SARCOMERE ORGANIZATION [FIGURES 9.2/9.3/9.4/9.5]

Thick and thin filaments within a myofibril are organized in the sarcomeres. The arrangement of thick and thin filaments within the sarcomere gives it a banded appearance. All of the myofibrils are arranged parallel to the long axis of the cell, with their sarcomeres lying side by side. As a result, the entire muscle fiber has a banded appearance corresponding to the bands of the individual sarcomeres (see Figures 9.2 and 9.3●).

Each myofibril consists of a linear series of approximately 10,000 sarcomeres. Sarcomeres are the smallest functional units of the muscle fiber—interactions between the thick and thin filaments of sarcomeres are responsible for skeletal muscle fiber contractions. Figure 9.4● diagrams the structure of an individual sarcomere. The thick filaments lie in the center of the sarcomere, linked by proteins of the **M-line**. Thin filaments at either end of the sarcomere, attached to interconnecting proteins that make up the **Z-lines**, extend toward the M-line. The Z-lines delineate the ends of the

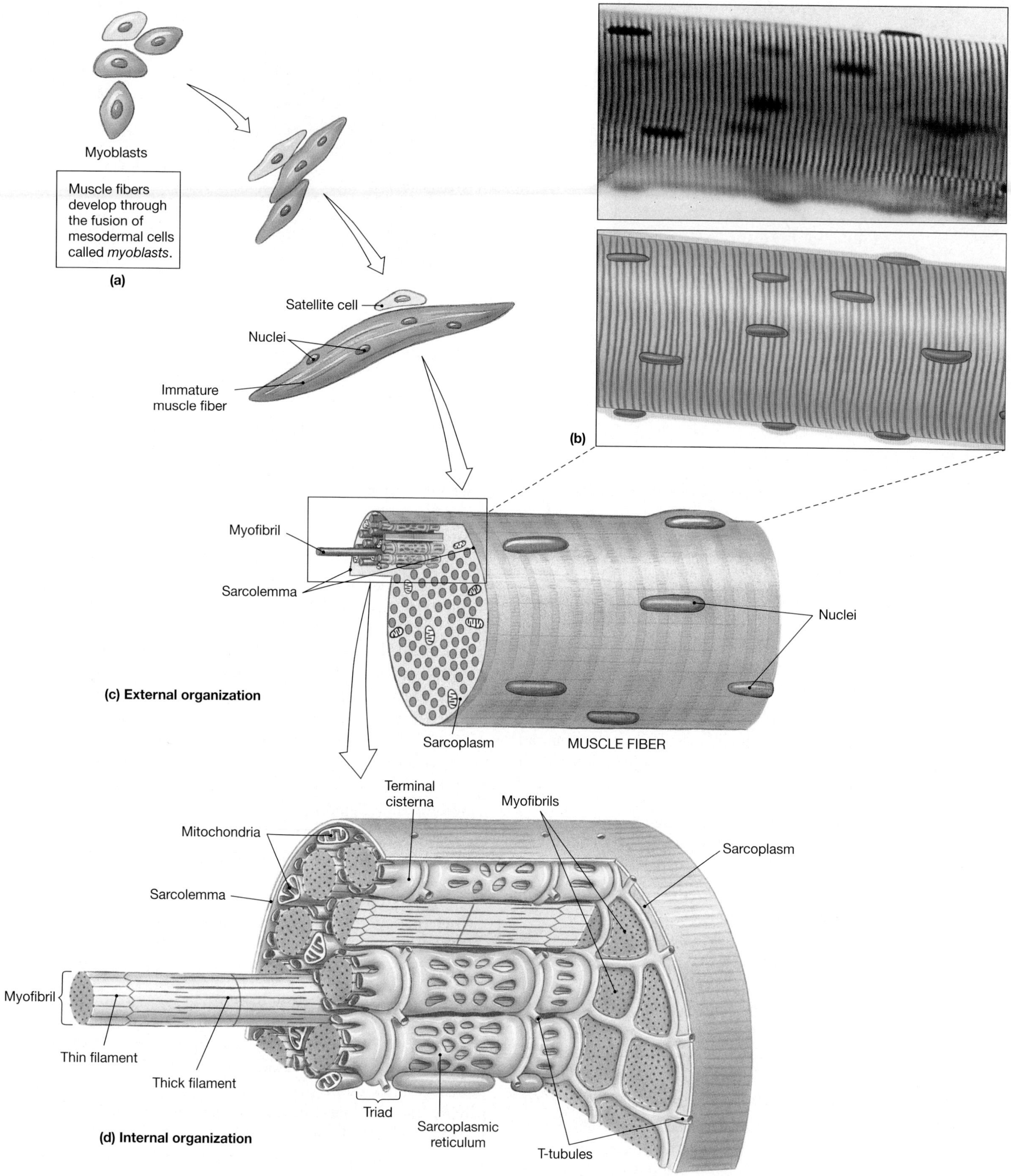

FIGURE 9.3 **THE FORMATION AND STRUCTURE OF A SKELETAL MUSCLE FIBER**

(**a**) Development of a skeletal muscle fiber. (**b**) External appearance and histological view (LM × 612). (**c**) The external organization of a muscle fiber. (**d**) The internal organization of a muscle fiber. Note the relationships among myofibrils, sarcoplasmic reticulum, mitochondria, triads, and thick and thin filaments.

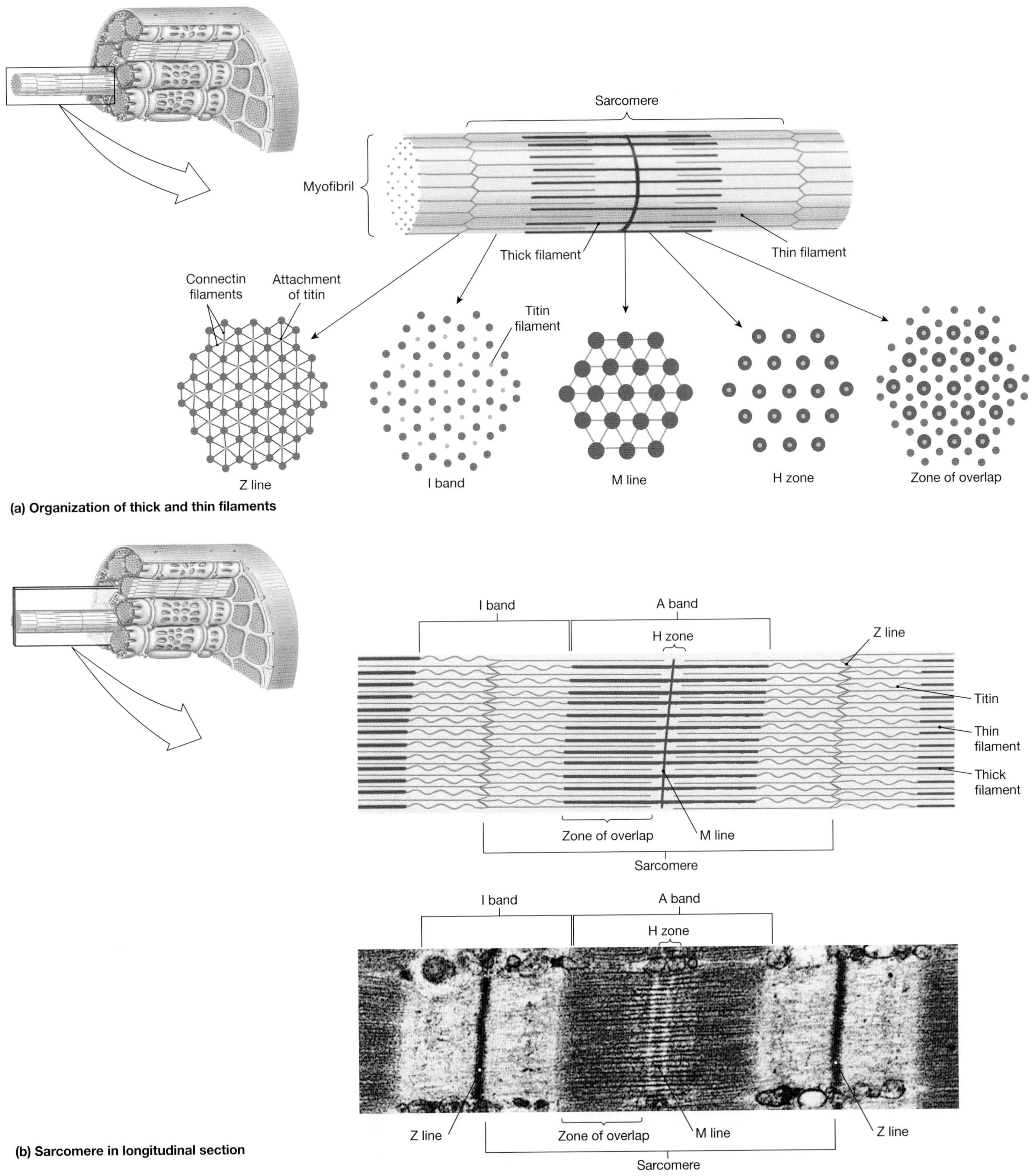

FIGURE 9.4 **SARCOMERE STRUCTURE**

(**a**) The basic arrangement of thick and thin filaments within a sarcomere and cross-sectional views of each region of the sarcomere. (**b**) A corresponding view of a sarcomere in a myofibril in the gastrocnemius muscle of the calf (TEM × 64,000) and a diagram showing the various components of this sarcomere.

sarcomere. In the **zone of overlap**, the thin filaments pass between the thick filaments. Figure 9.4a● shows cross sections through different portions of the sarcomere. Note the relative sizes and arrangement of thick and thin filaments at the zone of overlap. Each thin filament sits in a triangle formed by three thick filaments, and each thick filament is surrounded by six thin filaments.

The differences in the size and density of thick filaments and thin filaments account for the banded appearance of the sarcomere. The **A band** is the area containing thick filaments (Figure 9.4b●). The A band includes the M-line, the **H band** (thick filaments only), and the zone of overlap (thick and thin filaments). Between the A band and the Z-line is the **I band**, which contains only thin filaments. From the Z-lines at either end of the sarcomere, thin filaments extend into the zone of overlap toward the M-line. The terms *A band* and *I band* are derived from *anisotropic* and *isotropic*, which refer to the appearance of these bands when viewed under polarized light. You may find it helpful to remember that A bands are d*a*rk and I bands are l*i*ght. Figure 9.5● reviews the levels of organization we have considered thus far.

THIN FILAMENTS [FIGURE 9.6a,b] Each **thin filament** consists of a twisted strand 5–6 nm in diameter and 1 μm in length (Figure 9.6a,b●). This strand, called **F actin**, is composed of 300–400 globular *G actin* molecules. A slender strand of the protein *nebulin* holds the F actin strand together. Each molecule of G actin contains an **active site** that can bind to a thick filament in much the same way that a substrate molecule binds to the active site of an enzyme. A thin filament also contains the associated proteins **tropomyosin** (trō-pō-MĪ-ō-sin) and **troponin** (TRŌ-pō-nin; *trope*, turning). Tropomyosin molecules form a long chain that covers the active sites, preventing actin–myosin interaction. Troponin holds the tropomyosin strand in place. Before a contraction can begin, the troponin molecules must change position, moving the tropomyosin molecules and exposing the active sites; the mechanism will be detailed in a later section.

At either end of the sarcomere, the thin filaments are attached to the Z-line. Although called a *line* because it looks like a dark line on the surface of the myofibril, in sectional view the Z-line is more like an open meshwork created by proteins called *actinins*. For this reason, the Z-line is often called the *Z disc*.

THICK FILAMENTS [FIGURE 9.4a/9.6c,d] **Thick filaments** are 10–12 nm in diameter and 1.6 μm long (Figure 9.6c●). They are composed of a bundle of myosin molecules. Each of the roughly 500 myosin molecules within a thick filament consists of a double myosin strand with an attached, elongate *tail* and a free globular *head* (Figure 9.6d●). Adjacent thick filaments are interconnected midway along their length by proteins of the M-line. The myosin molecules are oriented away from the M-line, with the heads projecting outward toward the surrounding thin filaments. Myosin heads are also known as **cross-bridges**, because they connect thick filaments and thin filaments during a contraction.

Each thick filament has a core of *titin* (Figures 9.4a and 9.6c●). On either side of the M-line, a strand of titin extends the length of the filament and continues past the myosin portion of the thick filament to an attachment at the Z-line. The portion of the titin strand exposed within the I band is highly elastic and will recoil after stretching. In the normal resting sarcomere, the titin strands are completely relaxed; they become tense only when some external force stretches the sarcomere. When this occurs, the titin strands help maintain the normal alignment of the thick and thin filaments, and when the tension is removed, the recoil of the titin fibers helps return the sarcomere to its normal resting length.

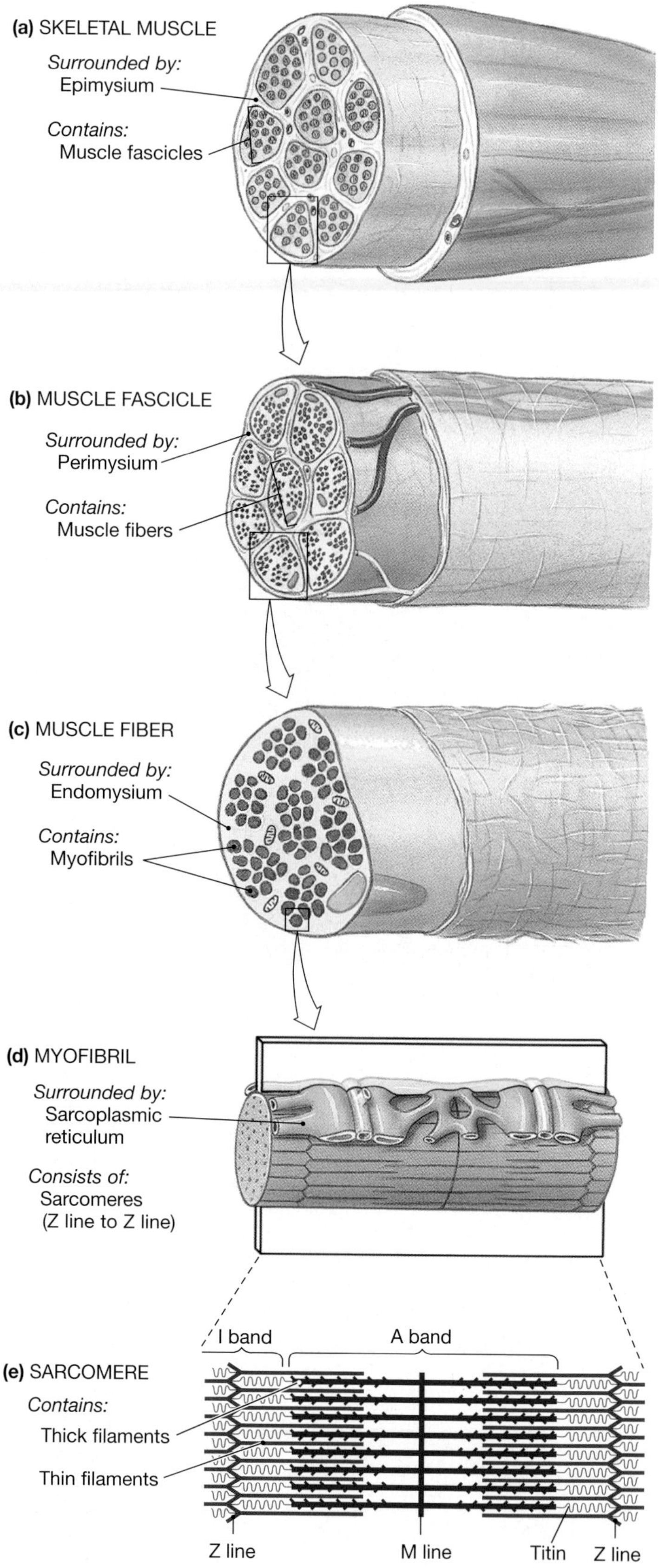

FIGURE 9.5 **LEVELS OF FUNCTIONAL ORGANIZATION IN A SKELETAL MUSCLE FIBER**

✓ CONCEPT CHECK

- Why does skeletal muscle appear striated when viewed with a microscope?
- What are myofibrils? Where are they found?
- Myofilaments consist primarily of what proteins?
- What is the functional unit of skeletal muscle?
- What two proteins help regulate actin and myosin interaction?

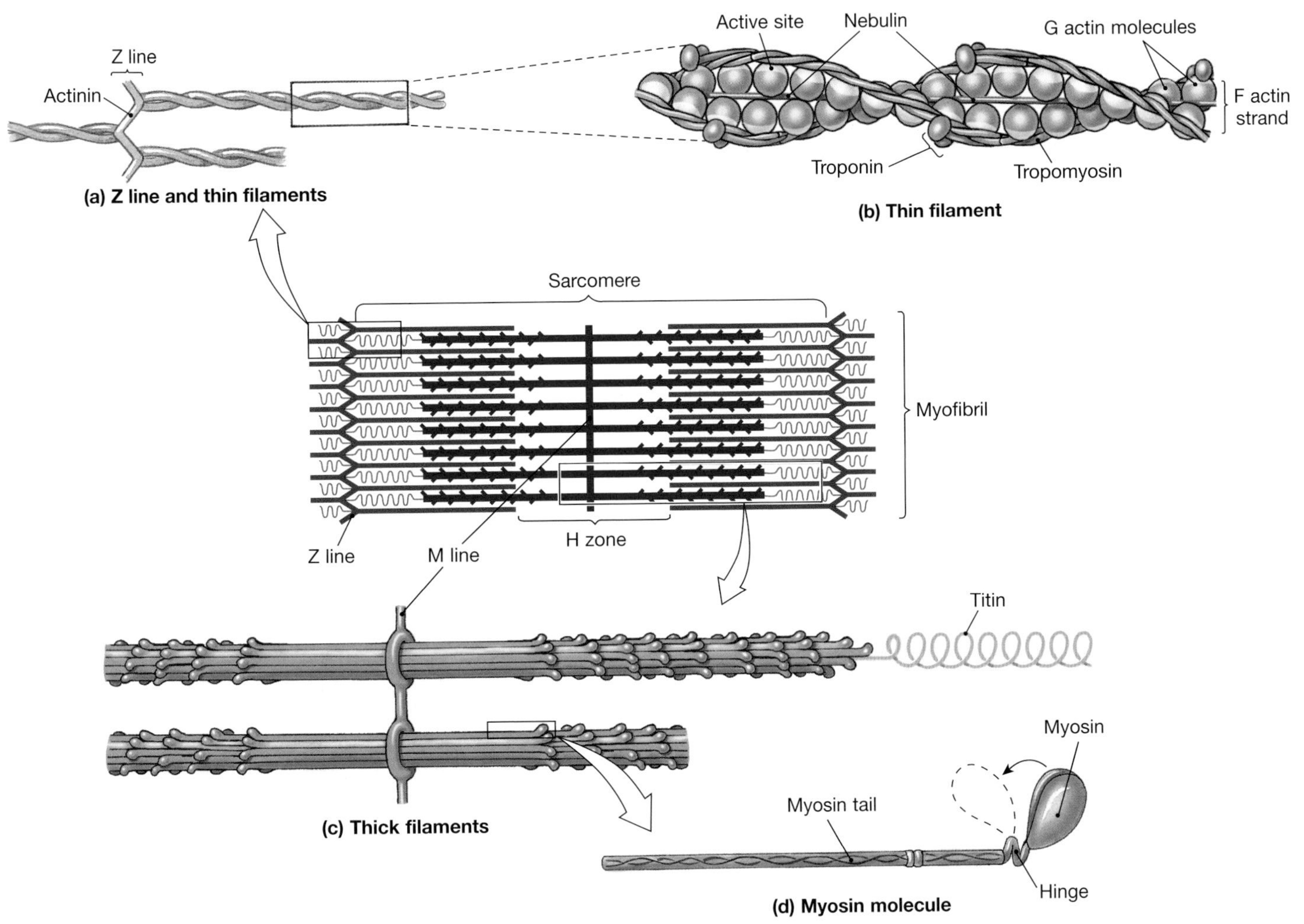

FIGURE 9.6 THIN AND THICK FILAMENTS

Myofilaments are bundles of thin and thick filament proteins. (**a**) The attachment of thin filaments to the Z-line. (**b**) The detailed structure of a thin filament, showing the organization of G actin, troponin, and tropomyosin. (**c**) The structure of thick filaments. (**d**) A single myosin molecule, detailing the structure and movement of the myosin head after cross-bridge binding occurs.

Muscle Contraction

A contracting muscle fiber exerts a pull, or **tension**, and shortens in length. Muscle fiber contraction results from interactions between the thick and thin filaments in each sarcomere. The mechanism for muscle contraction is explained by the *sliding filament theory*. The trigger for a contraction is the presence of calcium ions, and the contraction itself requires the presence of ATP.

The Sliding Filament Theory [FIGURES 9.7/9.8]

Direct observation of contracting muscle fibers indicates that in a contraction, (1) the H band and I band get smaller, (2) the zone of overlap gets larger, and (3) the Z-lines move closer together, but (4) the width of the A band remains constant throughout the contraction (Figure 9.7●). The explanation for these observations is known as the **sliding filament theory**. The sliding filament theory explains the physical changes that occur between thick and thin filaments during contraction.

Sliding occurs when the myosin heads of thick filaments bind to active sites on thin filaments. When cross-bridge binding occurs, the myosin head pivots toward the M-line, pulling the thin filament toward the center of the sarcomere. The cross-bridge then detaches and returns to its original position, ready to repeat the cycle of "attach, pivot, detach, and return." When the thick filaments pull on the thin filaments, the Z-lines move toward the M-line, and the sarcomere shortens.

When many people are pulling on a rope, the amount of tension produced is proportional to the number of people involved. In a muscle fiber, the amount of tension generated during a contraction depends on the number of cross-bridge interactions that occur in the sarcomeres of the myofibrils. The number of cross-bridges is in turn determined by the degree of overlap between thick and thin filaments. Only myosin heads within the zone of overlap can bind to active sites and produce tension. The tension produced by the muscle fiber can therefore be related directly to the structure of an individual sarcomere (Figure 9.8●). At optimal lengths the muscle fiber develops maximum tension (Figure 9.8c●). The normal range of sarcomere lengths is from 75 to 130% of this optimal length. During normal movements, our muscle fibers perform over a broad range of intermediate lengths, and the tension produced therefore varies from moment to moment. During an activity such as walking, in which muscles contract and relax in a cyclical fashion, muscle fibers are stretched to a length very close to optimal before they are stimulated to contract.

THE START OF A CONTRACTION [FIGURES 9.3d/9.9]

The immediate trigger for contraction is the appearance of free calcium ions in the sarcoplasm. The intracellular calcium ion concentration is usually very low. In most cells, this is because any calcium ions entering the cytoplasm are immediately pumped across their cell membranes and into the extracellular fluid. Although skeletal muscle fibers do pump Ca^{2+} out of

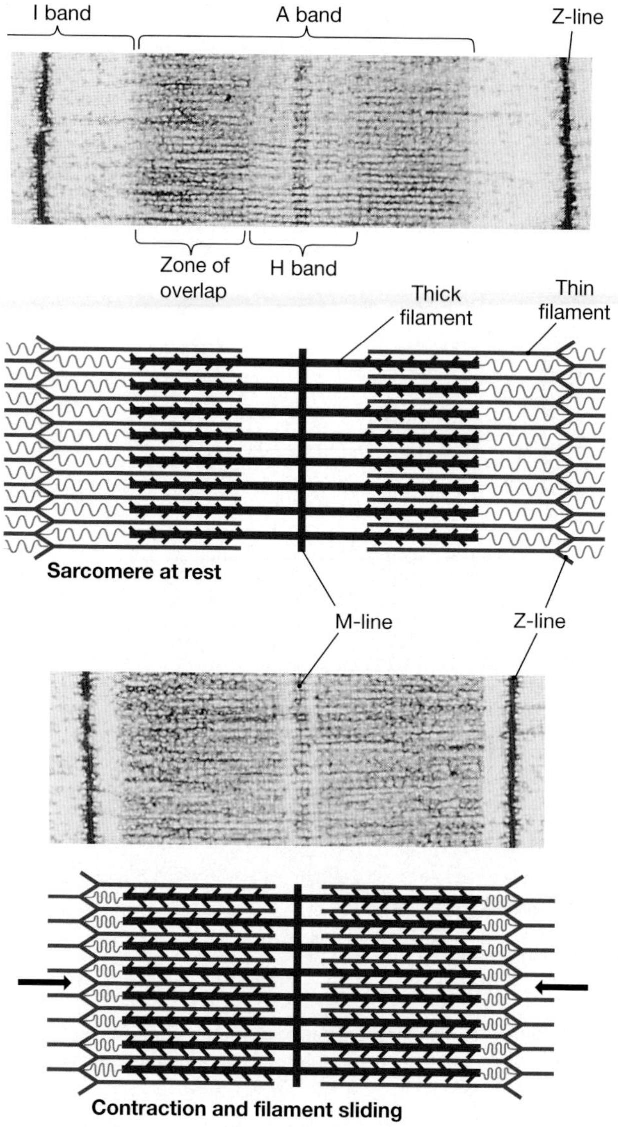

FIGURE 9.7 **CHANGES IN THE APPEARANCE OF A SARCOMERE DURING CONTRACTION OF A SKELETAL MUSCLE FIBER**

During a contraction, the A band stays the same width, but the Z-lines move closer together and the I band is reduced in width.

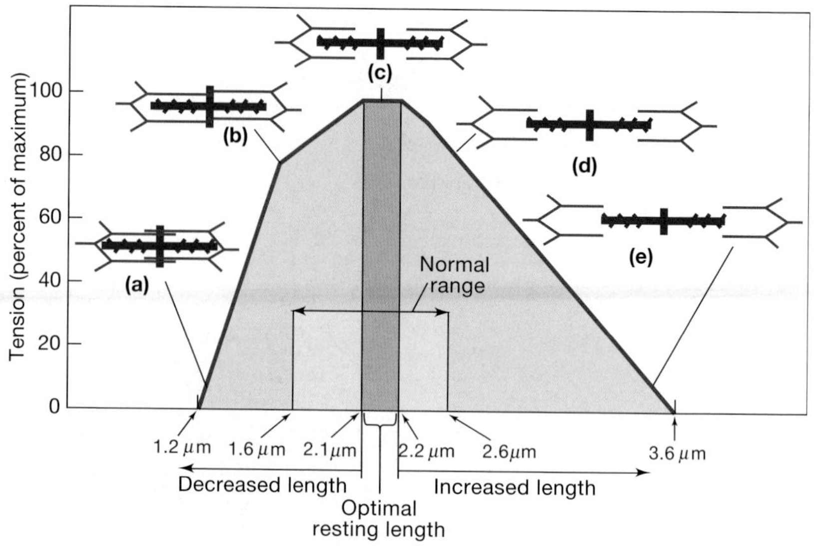

FIGURE 9.8 **THE EFFECT OF SARCOMERE LENGTH ON TENSION**

If sarcomeres are too short or too long the efficiency of contraction is affected. When the sarcomeres are too short, contraction cannot occur, either because **(a)** the thick filaments come in contact with the Z-lines or **(b)** the thin filaments overlap across the center of the sarcomere. **(c)** The tension produced by a contracting skeletal muscle fiber reaches a maximum when the zone of overlap is large, but the thin filaments do not extend across the center of the sarcomere. If the sarcomeres are stretched too far, the zone of overlap is reduced **(d)** or disappears **(e)**, and the cross-bridge interactions are reduced or cannot occur.

the cell in this way, they also transport them into the terminal cisternae of the sarcoplasmic reticulum (Figure 9.9●). The sarcoplasm of a resting skeletal muscle fiber contains very low concentrations of calcium ions, but the Ca^{2+} concentration inside the terminal cisternae may be as much as 40,000 times higher.

Electrical events at the sarcolemmal surface cause a contraction by triggering the release of calcium ions from the terminal cisternae. The electrical "message" or impulse, is distributed by the transverse tubules that extend deep into the sarcoplasm of the muscle fiber. A transverse tubule begins at the sarcolemma and travels inward at right angles to the membrane surface (Figure 9.3d●, p. 248). Along the way, branches from the transverse tubule encircle each of the individual sarcomeres at the boundary between the A band and the I band.

When an electrical impulse travels along a nearby T-tubule, the terminal cisternae become freely permeable to calcium ions. These calcium ions diffuse into the zone of overlap, where they bind to troponin. This results in a change in the shape of the troponin molecule, and this alters the position of the tropomyosin strand and exposes the active sites on the actin molecules. Cross-bridge binding then occurs, and the contraction begins.

THE END OF A CONTRACTION

The duration of the contraction usually depends on the duration of the electrical stimulation. The change in calcium permeability at the terminal cisternae is only temporary, so if the contraction is to continue, additional electrical impulses must be conducted along the T-tubules. If the electrical stimulation ceases, the sarcoplasmic reticulum will recapture the calcium

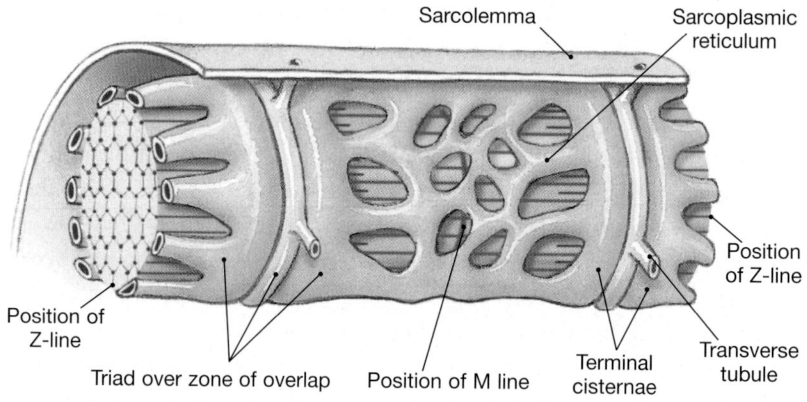

FIGURE 9.9 **THE ORIENTATION OF THE SARCOPLASMIC RETICULUM, T-TUBULES, AND INDIVIDUAL SARCOMERES**

A triad occurs where a T-tubule encircles a sarcomere between two terminal cisternae. Compare with *Figure 9.3d*; note that triads occur at the zones of overlap.

ions, the troponin-tropomyosin complex will cover the active sites, and the contraction will end.

Each time a cross-bridge detaches, an ATP molecule is broken down. Thus, even with continued stimulation, muscle fibers will eventually stop contracting as they run out of ATP. Each myosin head may cycle five times each second, and there are hundreds of myosin heads on each thick filament, hundreds of thick filaments in a sarcomere, thousands of sarcomeres in a myofibril, and hundreds to thousands of myofibrils in each muscle fiber. In other words, a muscle fiber contraction consumes enormous amounts of ATP!

Although muscle contraction is an active process, the return to resting length is entirely passive. Muscles cannot push; they can only pull. Factors that help return a shortened muscle to its normal resting length include elastic forces (such as the recoil of elastic fibers in the epimysium, perimysium, and endomysium), the pull of other muscles, or gravity.

The Neural Control of Muscle Fiber Contraction

[FIGURES 9.2/9.10]

The basic sequence of events in the process can be summarized as follows:

1. Chemicals released by the motor neuron at the neuromuscular junction alter the transmembrane potential of the sarcolemma. This change sweeps across the surface of the sarcolemma and into the transverse tubules.
2. The change in the transmembrane potential of the T-tubules triggers the release of calcium ions by the sarcoplasmic reticulum. This release initiates the contraction, as detailed above.

Each skeletal muscle fiber is controlled by a *motor neuron* whose cell body is located inside the central nervous system. The axon of this motor neuron extends into the periphery to reach the neuromuscular junction of that muscle fiber. The general appearance of a neuromuscular junction was shown in Figure 9.2●, p. 247. Figure 9.10● provides additional details. The expanded tip of the axon at the neuromuscular junction is called the **synaptic terminal**. The cytoplasm of the synaptic terminal contains numerous mitochondria and small secretory vesicles, called **synaptic vesicles**, filled with molecules of **acetylcholine (ACh)** (as-e-til-KŌ-lēn).

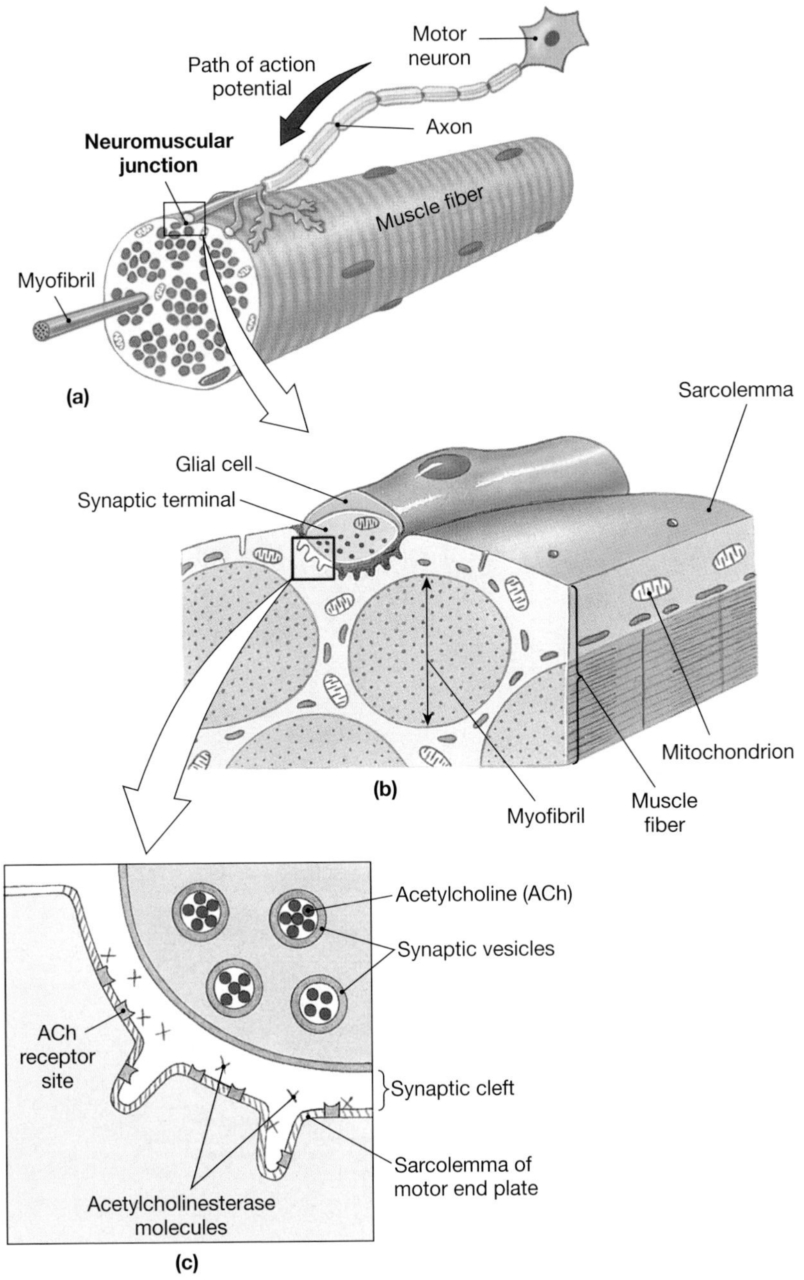

FIGURE 9.10 **THE NEUROMUSCULAR JUNCTION**

(a) A diagramatic view of a neuromuscular junction. **(b)** One portion of a neuromuscular junction. **(c)** Detailed view of a terminal, synaptic cleft, and motor end plate. *See also Figure 9.2.*

CLINICAL BRIEF

RIGOR MORTIS

When death occurs, circulation ceases and the skeletal muscles are deprived of nutrients and oxygen. Within a few hours, the skeletal muscle fibers have run out of ATP, and the sarcoplasmic reticulum becomes unable to remove calcium ions from the sarcoplasm. Calcium ions diffusing into the sarcoplasm from the extracellular fluid or leaking out of the sarcoplasmic reticulum then trigger a sustained contraction. Without ATP, the cross-bridges cannot detach from the active sites, and the muscle locks in the contracted position. All of the body's skeletal muscles are involved, and the individual becomes "stiff as a board." This physical state, called **rigor mortis**, lasts until the lysosomal enzymes released by autolysis break down the myofilaments 15–25 hours later.

Acetylcholine is an example of a *neurotransmitter*, a chemical released by a neuron to communicate with another cell. That communication takes the form of a change in the transmembrane potential of that cell. A narrow space, the **synaptic cleft**, separates the synaptic terminal from the motor end plate of the skeletal muscle fiber. The synaptic cleft contains the enzyme **acetylcholinesterase (AChE)**, or *cholinesterase*, which breaks down molecules of ACh.

When an electrical impulse arrives at the synaptic terminal, ACh is released into the synaptic cleft. The ACh released then binds to receptor sites on the motor end plate, initiating a change in the local transmembrane potential. This change results in the generation of an electrical impulse, or **action potential**, that sweeps over the surface of the sarcolemma and into each T-tubule. Action potentials will continue to be generated, one after another, until acetylcholinesterase removes the bound ACh. *Duchenne's Muscular Dystrophy* and *Myasthenia Gravis p. 791*

■ Muscle Contraction: A Summary [FIGURE 9.11]

The entire sequence of events from neural activation to the completion of a contraction is visually summarized in Figure 9.11●.

✓ CONCEPT CHECK

- What happens to the A bands and I bands of a myofibril during a contraction?
- List the sequence of activities during a contraction.
- How do terminal cisternae and transverse tubules interact to cause a skeletal muscle contraction?
- What is a neurotransmitter? What does it do?

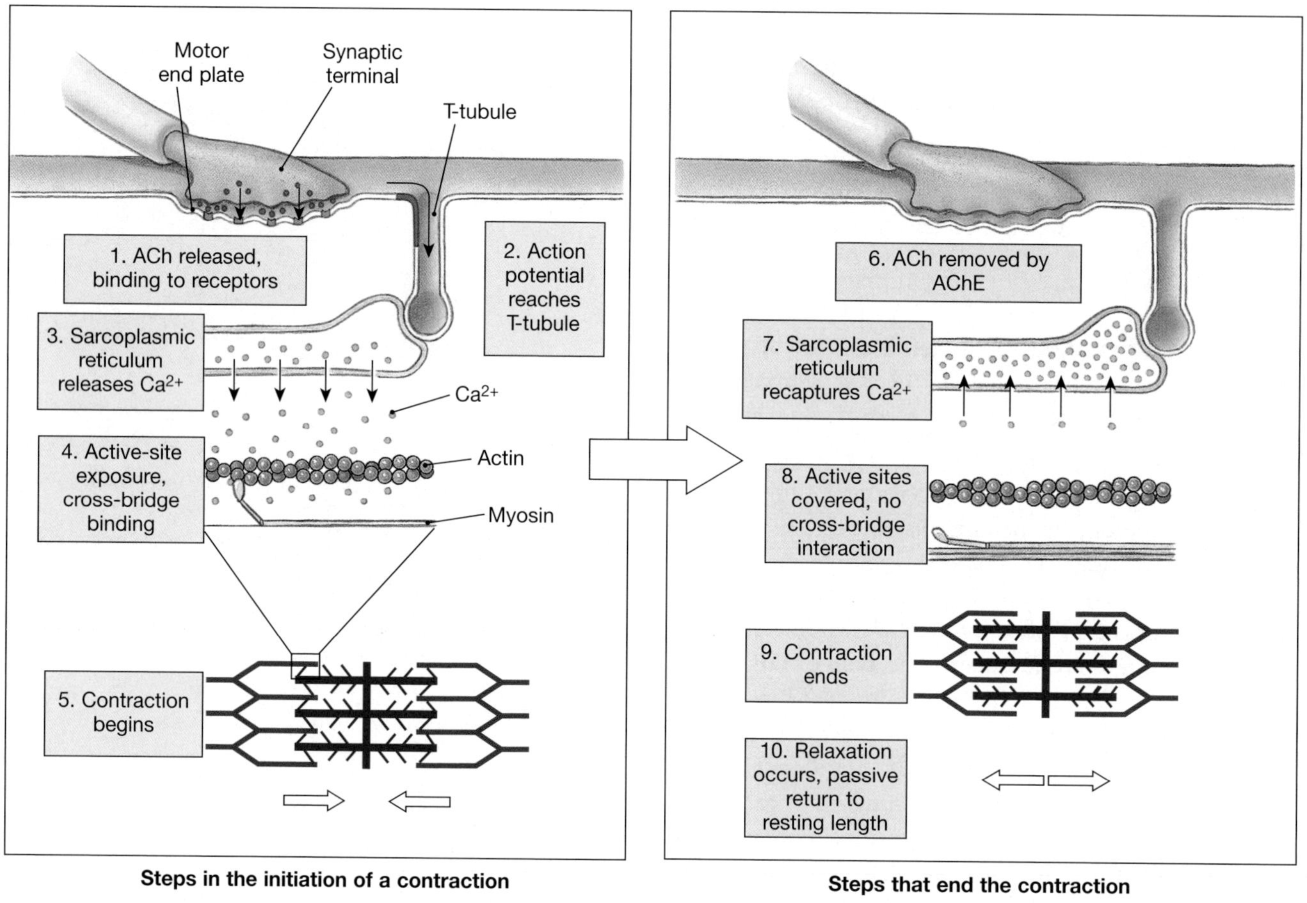

FIGURE 9.11 **THE EVENTS IN MUSCLE CONTRACTION**
A summary of the sequence of events in a muscle contraction.

Key steps in the initiation of a contraction include the following:

1. At the neuromuscular junction (NMJ), ACh released by the synaptic terminal binds to receptors on the sarcolemma.
2. The resulting change in the transmembrane potential of the muscle fiber leads to the production of an action potential that spreads across its entire surface and along the T-tubules.
3. The sarcoplasmic reticulum (SR) releases stored calcium ions, increasing the calcium concentration of the sarcoplasm in and around the sarcomeres.
4. Calcium ions bind to troponin, producing a change in the orientation of the troponin–tropomyosin complex that exposes active sites on the thin (actin) filaments. Myosin cross-bridges form when myosin heads bind to active sites.
5. Repeated cycles of cross-bridge binding, pivoting, and detachment occur, powered by the hydrolysis of ATP. These events produce filament sliding, and the muscle fiber shortens.

This process continues for a brief period, until:

6. Action potential generation ceases as ACh is broken down by acetylcholinesterase (AChE).
7. The SR reabsorbs calcium ions, and the concentration of calcium ions in the sarcoplasm declines.
8. When calcium ion concentrations approach normal resting levels, the troponin–tropomyosin complex returns to its normal position. This change covers the active sites and prevents further cross-bridge interaction.
9. Without cross-bridge interactions, further sliding will not take place, and the contraction will end.
10. Muscle relaxation occurs, and the muscle fiber returns passively to resting length.

Motor Units and Muscle Control

[FIGURE 9.12]

All of the muscle fibers controlled by a single motor neuron constitute a **motor unit**. A typical skeletal muscle contains thousands of muscle fibers. Although some motor neurons control a single muscle fiber, most control hundreds. The size of a motor unit is an indication of how fine the control of movement can be. In the muscles of the eye, where precise control is extremely important, a motor neuron may control two or three muscle fibers. We have much less precise control over power-generating muscles, such as our leg muscles, where up to 2000 muscle fibers may be controlled by a single motor neuron.

A skeletal muscle contracts when its motor units are stimulated. The amount of tension produced depends on two factors: (1) the frequency of stimulation and (2) the number of motor units involved. A single, momentary contraction is called a **muscle twitch**. A twitch is the response to a single stimulus. As the rate of stimulation increases, tension production will rise to a peak and plateau at maximal levels. Most muscle contractions involve this type of stimulation.

Each muscle fiber either contracts completely or does not contract at all. This characteristic is called the *all or none principle*. All of the fibers in a motor unit contract at the same time, and the amount of force exerted by the muscle as a whole thus depends on how many motor units are activated. By varying the number of motor units activated at any one time, the nervous system provides precise control over the pull exerted by a muscle.

When a decision is made to perform a movement, specific groups of motor neurons are stimulated. The stimulated neurons do not respond simultaneously, and over time, the number of activated motor units gradually increases. Figure 9.12● shows how the muscle fibers of each motor unit are intermingled with those of other units. Because of this intermingling, the direction of pull exerted on the tendon does not change as more motor units are activated, but the total amount of force steadily increases. The smooth but steady increase in muscular tension produced by increasing the number of active motor units is called **recruitment**, or **multiple motor unit summation**.

Peak tension occurs when all of the motor units in the muscle are contracting at the maximal rate of stimulation. However, such powerful contractions cannot last long, because the individual muscle fibers soon use up their available energy reserves. To lessen the onset of fatigue during periods of sustained contraction, motor units are activated on a rotating basis, so that some of them are resting and recovering while others are actively contracting.

■ Muscle Tone

Even when a muscle is at rest, some motor units are always active. Their contractions do not produce enough tension to cause movement, but they do tense the muscle. This resting tension in a skeletal muscle is called **muscle tone**. Motor units are randomly stimulated, so that there is a constant tension in the attached tendon but individual muscle fibers can have some time to relax. Resting muscle tone stabilizes the position of bones and joints. For example, in muscles involved with balance and posture, enough motor units are stimulated to produce the tension needed to maintain body position. Specialized muscle cells called **muscle spindles** are monitored by sensory nerves that control the muscle tone in the surrounding muscle tissue. Reflexes triggered by activity in these sensory nerves play an important role in the reflex control of position and posture, a topic that we will discuss in Chapter 14.

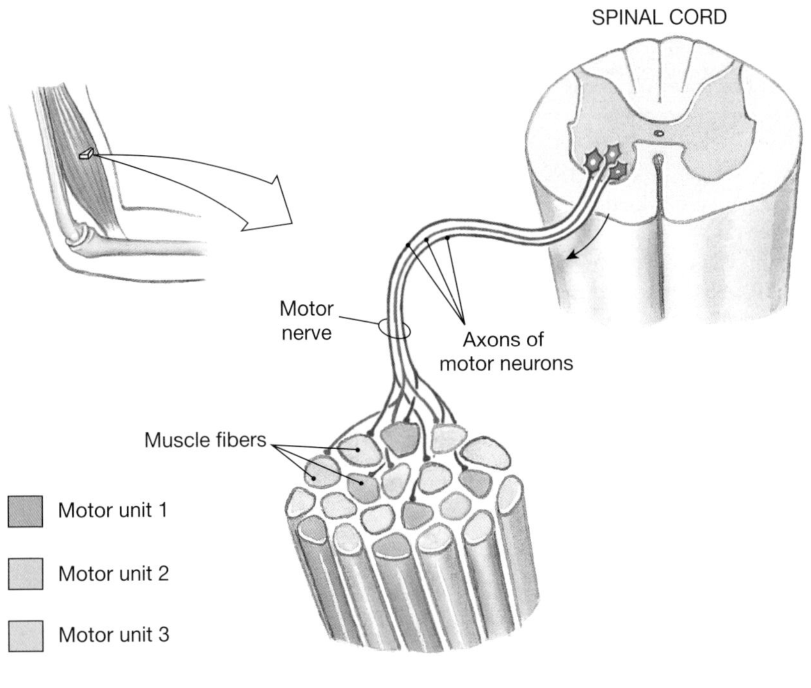

FIGURE 9.12 **THE ARRANGEMENT OF MOTOR UNITS IN A SKELETAL MUSCLE**

Muscle fibers of different motor units are intermingled, so that the net distribution of force applied to the tendon remains constant even when individual muscle groups cycle between contraction and relaxation. The number of muscle fibers in a motor unit ranges from as few as one to more than 2000.

■ Muscle Hypertrophy

Exercise increases the activity of muscle spindles and may enhance muscle tone. As a result of repeated, exhaustive stimulation, muscle fibers develop a larger number of mitochondria, a higher concentration of glycolytic enzymes, and larger glycogen reserves. These muscle fibers have more myofibrils, and each myofibril contains a larger number of thick and thin filaments. The net effect is an enlargement, or **hypertrophy** (hī-PER-trō-fē), of the stimulated muscle. Hypertrophy occurs in muscles that have been repeatedly stimulated to produce near-maximal tension; the intracellular changes that occur increase the amount of tension produced when these muscles contract. A champion weight lifter or bodybuilder is an excellent example of hypertrophied muscular development.

■ Muscle Atrophy

When a skeletal muscle is not stimulated by a motor neuron on a regular basis, it loses muscle tone and mass. The muscle becomes flaccid, and the muscle fibers become smaller and weaker. This reduction in muscle size, tone, and power is called **atrophy**. Individuals paralyzed by spinal injuries or other damage to the nervous system will gradually lose muscle tone and size in the areas affected. Even a temporary reduction in muscle use can lead to muscular atrophy; this loss of tone and size may easily be seen by comparing limb muscles before and after a cast has been worn. Muscle atrophy is initially reversible, but dying muscle fibers are not replaced, and in extreme atrophy the functional losses are permanent. That is why physical therapy is crucial in cases where people are temporarily unable to move normally. *Polio p. 791*

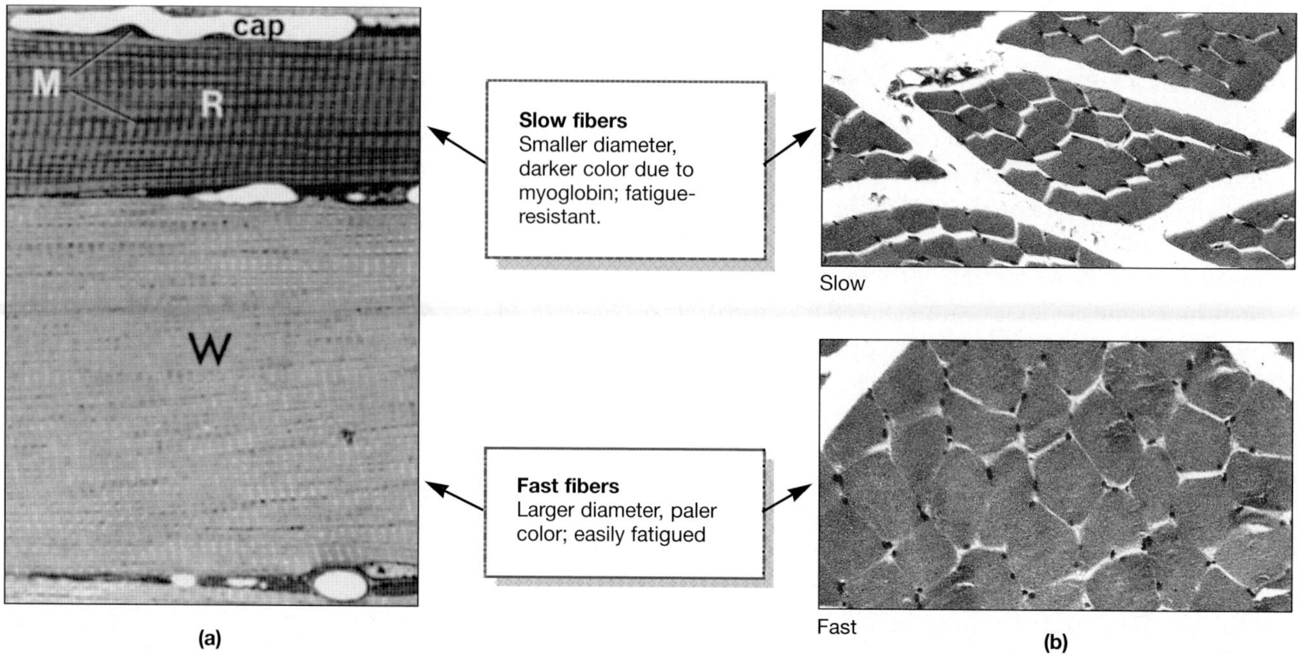

FIGURE 9.13 **TYPES OF SKELETAL MUSCLE FIBERS**

Fast fibers are for rapid contractions and slow fibers are for slower, but extended contractions. **(a)** The relatively slender slow muscle fiber (R) has more mitochondria (M) and a more extensive capillary supply (cap) than the fast muscle fiber (W). (LM × 1044) **(b)** Note the difference in the size of slow muscle fibers (above) and fast muscle fibers (below). (LMs × 229)

Slow fibers
Smaller diameter, darker color due to myoglobin; fatigue-resistant.

Slow

Fast fibers
Larger diameter, paler color; easily fatigued

Fast

(a) (b)

Types of Skeletal Muscle Fibers [FIGURE 9.13]

Skeletal muscles are designed for various actions. The types of fibers that comprise a muscle will, in part, determine its action. There are three major types of skeletal muscle fibers in the body: *fast*, *slow*, and *intermediate*. Fast and slow muscle fibers are shown in Figure 9.13●. The differences among these groups reflect differences in the way they obtain the ATP to support their contractions.

Fast fibers, or *white fibers*, are large in diameter; they contain densely packed myofibrils, large glycogen reserves, and relatively few mitochondria. Most of the skeletal muscle fibers in the body are called fast fibers because they can contract in 0.01 seconds or less following stimulation. The tension produced by a muscle fiber is directly proportional to the number of sarcomeres, so fast-fiber muscles produce powerful contractions. However, these contractions use enormous amounts of ATP, and their mitochondria are unable to meet the demand. As a result, their contractions are supported primarily by *anaerobic* (*an*, without + *aer*, air + *bios*, life) *glycolysis*. This reaction pathway, which does not require oxygen, converts stored glycogen to lactic acid. Fast fibers fatigue rapidly, both because their glycogen reserves are limited and because lactic acid builds up, and the acid pH interferes with the contraction mechanism.

Slow fibers, or *red fibers*, are only about half the diameter of fast fibers, and they take three times as long to contract after stimulation. Slow fibers are specialized to continue contracting for extended periods, long after a fast muscle would have become fatigued. They can do so because their mitochondria are able to continue producing ATP throughout the contraction. As you will recall from Chapter 2, mitochondria absorb oxygen and generate ATP. The reaction pathway involved is called *aerobic metabolism*. The oxygen required comes from two sources:

1. Skeletal muscles containing slow muscle fibers have a more extensive network of capillaries than do muscles dominated by fast muscle fibers. This means that there is greater blood flow, and the red blood cells can deliver more oxygen to the active muscle fibers.

2. Slow fibers are red because they contain the red pigment **myoglobin** (MĪ-ō-glō-bin). This globular protein is structurally related to hemoglobin, the oxygen-binding pigment found in red blood cells. Myoglobin binds oxygen molecules as well. Thus, resting slow muscle fibers contain substantial oxygen reserves that can be mobilized during a contraction.

Slow muscles also contain a relatively larger number of mitochondria than do fast muscle fibers. Whereas fast muscle fibers must rely on their glycogen reserves during peak levels of activity, the mitochondria in slow muscles can break down carbohydrates, lipids, or even proteins. They can therefore continue to contract for extended periods; for example, the leg muscles of marathon runners are dominated by slow muscle fibers.

Intermediate fibers have properties intermediate between those of fast fibers and slow fibers. For example, intermediate fibers contract faster than slow fibers but slower than fast fibers. Histologically, intermediate fibers are very similar to fast fibers, although they have more mitochondria, a slightly increased capillary supply, and a greater resistance to fatigue.

The properties of the various types of skeletal muscles are detailed in Table 9.1.

✓ CONCEPT CHECK

- Why does a sprinter experience muscle fatigue after a few minutes, while a marathon runner can run for hours?

- What type of muscle fibers would you expect to predominate in the large leg muscles of someone who excels at endurance activities such as cycling or long-distance running?

- Why do some motor units control only a few muscle fibers, whereas others control many fibers?

- What is recruitment?

■ Distribution of Fast, Slow, and Intermediate Fibers

The percentage of fast, slow, and intermediate muscle fibers varies from one skeletal muscle to another. Most muscles contain a mixture of fiber types, although all of the fibers within one motor unit are of the same type.

TABLE 9.1 PROPERTIES OF SKELETAL MUSCLE FIBER TYPES

Property	Slow	Intermediate	Fast
Cross-sectional diameter	Small	Intermediate	Large
Tension	Low	Intermediate	High
Contraction speed	Slow	Fast	Fast
Fatigue resistance	High	Intermediate	Low
Color	Red	White	White
Myoglobin content	High	Low	Low
Capillary supply	Dense	Intermediate	Scarce
Mitochondria	Many	Intermediate	Few
Glycolytic enzyme concentration in sarcoplasm	Low	High	High
Substrates used for ATP generation during contraction	Lipids, carbohydrates, amino acids (aerobic)	Primarily carbohydrates (anaerobic)	Carbohydrates (anaerobic)
Alternative names	Type I, S (slow), red, SO (slow oxidizing), slow-twitch oxidative	Type II-B, FR (fast resistant), fast-twitch oxidative	Type II-A, FF (fast fatigue), white, fast-twitch glycolytic

However, there are no slow fibers in muscles of the eye and hand, where swift but brief contractions are required. Many back and calf muscles are dominated by slow fibers; these muscles contract almost continually to maintain an upright posture.

The percentage of fast versus slow fibers in each muscle is genetically determined, and there are significant individual differences. These variations have an effect on endurance. A person with more slow muscle fibers in a particular muscle will be better able to perform repeated contractions under aerobic conditions. For example, marathon runners with high proportions of slow muscle fibers in their leg muscles outperform those with more fast muscle fibers. For brief periods of intense activity, such as a sprint or a weight-lifting event, the individual with a higher percentage of fast muscle fibers will have the advantage.

The characteristics of the muscle fibers changes with physical conditioning. Repeated, intense workouts promote the enlargement of fast muscle fibers and muscular hypertrophy. Training for endurance events, such as cross-country or marathon running, increases the proportion of intermediate fibers in the active muscles. This occurs through the gradual conversion of fast fibers to intermediate fibers.

Endurance training does not promote hypertrophy, and many athletes train using a combination of aerobic activity, such as swimming, with anaerobic activities such as weight lifting or sprinting. This combination, known as *interval training*, enlarges muscles and improves strength and endurance.

The Organization of Skeletal Muscle Fibers [FIGURES 9.1/9.14]

Although most skeletal muscle fibers contract at comparable rates and shorten to the same degree, variations in microscopic and macroscopic organization can dramatically affect the power, range, and speed of movement produced when a muscle contracts.

Muscle fibers within a skeletal muscle form bundles called *fascicles* (Figure 9.1●, p. 246). The muscle fibers of each fascicle lie parallel to one another, but the organization of the fascicles in the skeletal muscle can vary, as can the relationship between the fascicles and the associated tendon. Four different patterns of fascicle arrangement or organization produce *parallel muscles*, *convergent muscles*, *pennate muscles*, and *circular muscles*. Figure 9.14● illustrates the fascicle organization of skeletal muscle fibers.

Parallel Muscles [FIGURE 9.14a]

In a **parallel muscle**, the fascicles are parallel to the long axis of the muscle. The functional characteristics of a parallel muscle resemble those of an individual muscle fiber. Consider the skeletal muscle shown in Figure 9.14a●. It has a firm attachment by a tendon that extends from the free tip to a movable bone of the skeleton. Most of the skeletal muscles in the body are parallel muscles. Some form flat bands, with broad aponeuroses at each end; others are spindle-shaped, with cordlike tendons at one or both ends. Such a muscle has a central **body**, also known as the *belly*, or *gaster* (GAS-ter; *gaster*, stomach). When this muscle contracts, it gets shorter and the body increases in diameter. The *biceps brachii muscle* of the arm is an example of a parallel muscle with a central body. The bulge of the contracting biceps can be seen on the anterior surface of the arm when the elbow is flexed.

A skeletal muscle cell can contract effectively until it has been shortened by roughly 30%. Because the muscle fibers are parallel to the long axis of the muscle, when they contract together, the entire muscle shortens by the same amount. For example, if the skeletal muscle is 10 cm long, the end of the tendon will move 3 cm when the muscle contracts. The tension developed by the muscle during this contraction depends on the total number of myofibrils it contains. Because the myofibrils are distributed evenly through the sarcoplasm of each cell, the tension can be estimated on the basis of the cross-sectional area of the resting muscle. A parallel skeletal muscle 6.45 cm^2 (1 $in.^2$) in cross-sectional area can develop approximately 23 kg (50 lb) of tension.

Convergent Muscles [FIGURE 9.14b]

In a **convergent muscle**, the muscle fibers are based over a broad area, but all the fibers come together at a common attachment site. They may pull on

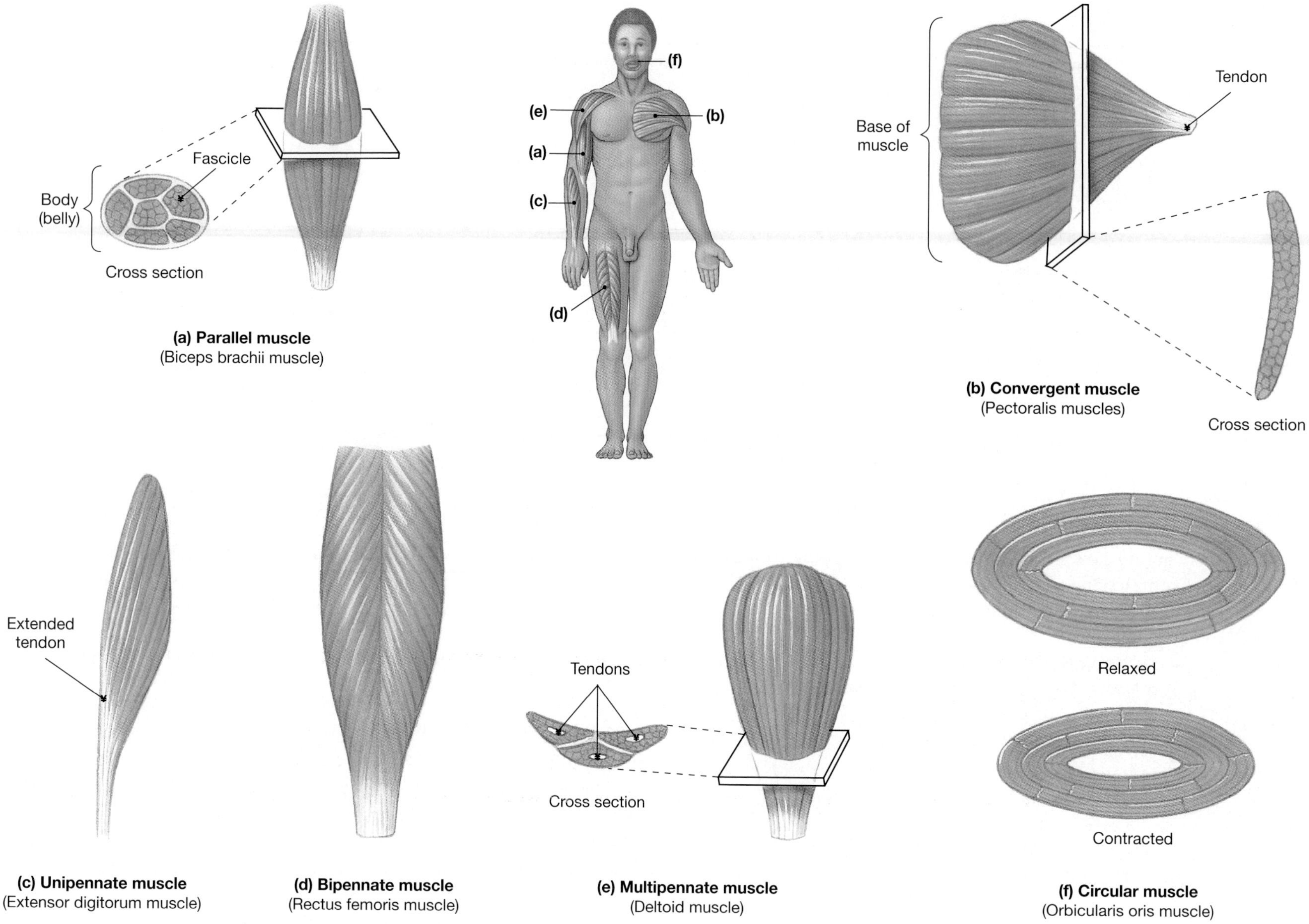

FIGURE 9.14 **SKELETAL MUSCLE FIBER ORGANIZATION**

Four different arrangements of muscle fiber patterns are observed: (**a**) parallel, (**b**) convergent, (**c,d,e**) pennate, and (**f**) circular.

a tendon, a tendinous sheet, or a slender band of collagen fibers known as a **raphe** (RĀ-fē; seam). The muscle fibers often spread out, like a fan or a broad triangle, with a tendon at the tip as shown in Figure 9.14b●. The prominent *pectoralis muscles* of the chest have this shape. This type of muscle has versatility; the direction of pull can be changed by stimulating only one group of muscle cells at any one time. However, when they all contract at once, they do not pull as hard on the tendon as a parallel muscle of the same size because the muscle fibers on opposite sides of the tendon pull in different directions rather than all working together.

Pennate Muscles [FIGURE 9.14c,d,e]

In a **pennate muscle** (*penna*, feather), one or more tendons run through the body of the muscle, and the fascicles form an oblique angle to the tendon. Because they pull at an angle, contracting pennate muscles do not move their tendons as far as parallel muscles do. However, a pennate muscle will contain more muscle fibers than a parallel muscle of the same size, and as a result, the contraction of the pennate muscle generates more tension than that of a parallel muscle of the same size.

If all of the muscle cells are found on the same side of the tendon, the muscle is **unipennate** (Figure 9.14c●). A long muscle that extends the fingers, the *extensor digitorum muscle* is an example of a unipennate muscle. More commonly, there are muscle fibers on both sides of the tendon. A prominent muscle of the thigh, the *rectus femoris muscle*, is a **bipennate muscle** (Figure 9.14d●) that helps to extend the knee. If the tendon branches within the muscle, the muscle is multipennate (Figure 9.14e●). The triangular *deltoid muscle* that covers the superior surface of the shoulder joint is an example of a **multipennate muscle.**

Circular Muscles [FIGURE 9.14f]

In a **circular muscle**, or **sphincter** (SFINK-ter), the fibers are concentrically arranged around an opening or recess (Figure 9.14f●). When the muscle contracts, the diameter of the opening decreases. Circular muscles guard entrances and exits of internal passageways such as the digestive and urinary tracts. An example is the *orbicularis oris muscle* of the mouth.

Levers: A Systems Design for Movement [FIGURE 9.15]

Skeletal muscles do not work in isolation. When a muscle is attached to the skeleton, the nature and site of the connection will determine the force, speed, and range of the movement produced. These characteristics are interdependent, and the relationships can explain a great deal about the general organization of the muscular and skeletal systems.

The force, speed, or direction of movement produced by contraction of a muscle can be modified by attaching the muscle to a lever. The applied force is the effort produced by the muscle contraction. This effort is opposed by a resistance, which is a load or weight. A **lever** is a rigid structure—such as a board, a crowbar, or a bone—that moves on a fixed point called the **fulcrum**. In the body, each bone is a lever and each joint a fulcrum. The teeter-totter, or seesaw, at the park provides a more familiar example of lever action. Levers can change (1) the direction of an applied force, (2) the distance and speed of movement produced by a force, and (3) the strength of a force.

Classes of Levers [FIGURE 9.15]

Three classes of levers are found in the human body:

1. *First-Class Levers*: The seesaw is an example of a **first-class lever**—one in which the fulcrum lies between the applied force and the resistance as seen in Figure 9.15a●. There are not many examples of first-class levers in the body. One, involving the muscles that extend the neck, is shown in this figure.
2. *Second-Class Levers*: In a **second-class lever**, the resistance is located between the applied force and the fulcrum. A familiar example of such a lever is a loaded wheelbarrow. The weight of the load is the resistance, and the upward lift on the handle is the applied force. Because in this arrangement the force is always farther from the fulcrum than the resistance, a small force can balance a larger weight. In other words, the force is magnified. Notice, however, that when a force moves the handle, the resistance moves more slowly and covers a shorter distance. There are few examples of second-class levers in the body. In performing plantar flexion, the calf muscles act across a second-class lever (Figure 9.15b●).
3. *Third-Class Levers*: In a **third-class lever** system, a force is applied between the resistance and the fulcrum (Figure 9.15c●). Third-class levers are the most common levers in the body. The effect of this arrangement is the reverse of that produced by a second-class lever: Speed and distance traveled are increased at the expense of force. In the example illustrated (the biceps brachii muscle, which flexes the elbow), the resistance is six times farther away from the fulcrum than the applied force. The biceps brachii muscle can develop an effective force of 180 kg, which now will be reduced from 180 kg to 30 kg. However, the distance traveled and the speed of movement are *increased* by the same ratio (6 to 1): the resistance travels 45 cm while the insertion point moves only 7.5 cm.

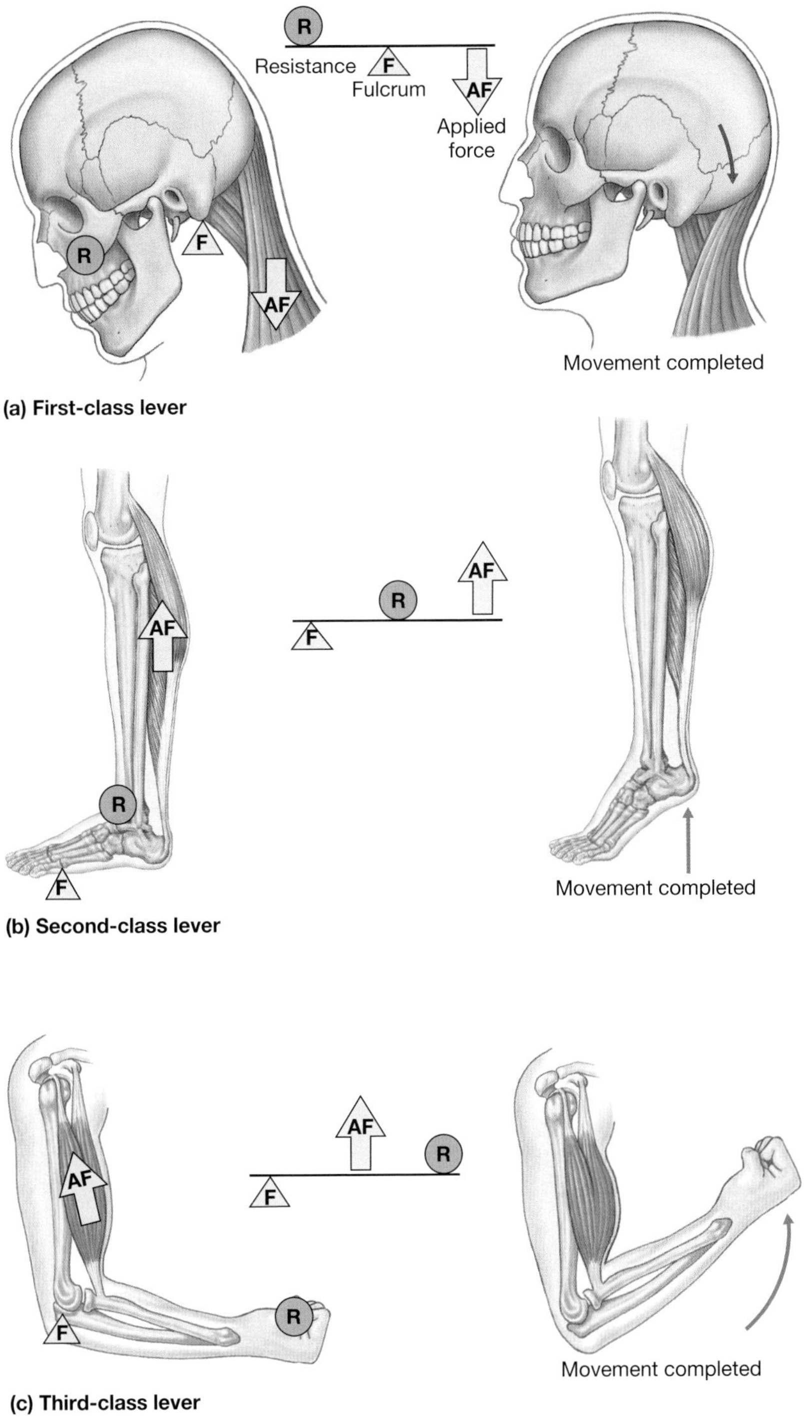

FIGURE 9.15 **The Three Classes of Levers**

Levers are rigid structures that move on a fixed point called a fulcrum. **(a)** In a **first-class lever**, the applied force and the resistance are on opposite sides of the fulcrum. This lever can change the amount of force transmitted to the resistance and alter the direction and speed of movement. **(b)** In a **second-class lever**, the resistance lies between the applied force and the fulcrum. This arrangement magnifies force at the expense of distance and speed; the direction of movement remains unchanged. **(c)** In a **third-class lever**, the force is applied between the resistance and the fulcrum. This arrangement increases speed and distance moved but requires a larger applied force.

Although every muscle does not operate as part of a lever system, the presence of levers provides speed and versatility far in excess of what we would predict on the basis of muscle physiology alone. Skeletal muscle cells resemble one another closely, and their abilities to contract and generate tension are quite similar. Consider a skeletal muscle that can contract in 500 ms and shorten 1 cm while exerting a 10-kg pull. Without using a lever, this muscle would be performing efficiently only when moving a 10-kg weight a distance of 1 cm. But by using a lever, the same muscle operating at the same efficiency could move 20 kg a distance of 0.5 cm, 5 kg a distance of 2 cm, or 1 kg a distance of 10 cm. Thus, the lever system design produces the maximum movements with the greatest efficiency.

Muscle Terminology [TABLE 9.2]

Each muscle begins at an **origin**, ends at an **insertion**, and contracts to produce a specific **action**. Terms indicating the actions of muscles, specific regions of the body, and structural characteristics of muscle are presented in Table 9.2.

Origins and Insertions

Typically, the origin remains stationary and the insertion moves, or the origin is proximal to the insertion. For example, the triceps inserts on the olecranon process and originates closer to the shoulder. Such determinations are made during normal movement with the individual in the anatomical position. Part of the fun of studying the muscular system is that you can actually do the movements and think about the muscles involved. (Laboratory discussions of the muscular system often resemble a poorly organized aerobics class.)

When the origins and insertions cannot be determined easily on the basis of movement or position, other criteria are used. If a muscle extends between a broad aponeurosis and a narrow tendon, the aponeurosis is considered to be the origin, and the tendon is attached to the insertion. If there are several tendons at one end and just one at the other, there are multiple origins and a single insertion. These simple rules cannot cover every situation, and knowing which end is the origin and which is the insertion is ultimately less important than knowing where the two ends attach and what the muscle does when it contracts.

Actions

Almost all skeletal muscles either originate or insert upon the skeleton. When a muscle moves a portion of the skeleton, that movement may in-

TABLE 9.2 MUSCLE TERMINOLOGY

Terms Indicating Direction Relative to Axes of the Body	Terms Indicating Specific Regions of the Body*	Terms Indicating Structural Characteristics of the Muscle	Terms Indicating Actions
Anterior (front)	Abdominis (abdomen)	**ORIGIN**	**GENERAL**
Externus (superficial)	Anconeus (elbow)	Biceps (two heads)	Abductor
Extrinsic (outside)	Auricularis (auricle of ear)	Triceps (three heads)	Adductor
Inferioris (inferior)	Brachialis (brachium)	Quadriceps (four heads)	Depressor
Internus (deep, internal)	Capitis (head)		Extensor
Intrinsic (inside)	Carpi (wrist)	**SHAPE**	Flexor
Lateralis (lateral)	Cervicis (neck)	Deltoid (triangle)	Levator
Medialis/medius (medial, middle)	Cleido/clavius (clavicle)	Orbicularis (circle)	Pronator
Oblique	Coccygeus (coccyx)	Pectinate (comb-like)	Rotator
Posterior (back)	Costalis (ribs)	Piriformis (pear-shaped)	Supinator
Profundus (deep)	Cutaneous (skin)	Platys- (flat)	Tensor
Rectus (straight, parallel)	Femoris (femur)	Pyramidal (pyramid)	
Superficialis (superficial)	Genio- (chin)	Rhomboideus (rhomboid)	**SPECIFIC**
Superioris (superior)	Glosso/glossal (tongue)	Serratus (serrated)	Buccinator (trumpeter)
Transversus (transverse)	Hallucis (great toe)	Splenius (bandage)	Risorius (laugher)
	Ilio- (ilium)	Teres (long and round)	Sartorius (like a tailor)
	Inguinal (groin)	Trapezius (trapezoid)	
	Lumborum (lumbar region)		
	Nasalis (nose)	**OTHER STRIKING FEATURES**	
	Nuchal (back of neck)	Alba (white)	
	Oculo- (eye)	Brevis (short)	
	Oris (mouth)	Gracilis (slender)	
	Palpebrae (eyelid)	Lata (wide)	
	Pollicis (thumb)	Latissimus (widest)	
	Popliteus (behind knee)	Longissimus (longest)	
	Psoas (loin)	Longus (long)	
	Radialis (radius)	Magnus (large)	
	Scapularis (scapula)	Major (larger)	
	Temporalis (temples)	Maximus (largest)	
	Thoracis (thoracic region)	Minimus (smallest)	
	Tibialis (tibia)	Minor (smaller)	
	Ulnaris (ulna)	-tendinosus (tendinous)	
	Uro- (urinary)	Vastus (great)	

*For other regional terms, refer to Figure 1.8, p. 14, which identifies anatomical landmarks.

volve *abduction, adduction, flexion, extension, hyperextension, circumduction, rotation, pronation, supination, eversion, inversion, dorsiflexion, plantar flexions, lateral flexion, opposition, protraction, retraction, elevation*, and *depression*. Before proceeding, consider reviewing the discussion of planes of motion and Figures 8.3 to 8.5●. pp. 219–220

There are two methods of describing actions. The first references the bone region affected. Thus the biceps brachii muscle is said to perform "flexion of the forearm." The second method specifies the joint involved. Thus, the action of the biceps brachii muscle is described as "flexion of (or at) the elbow." Both methods are valid, and each has its advantages, but we will primarily use the latter method when describing muscle actions in later chapters.

Muscles can be grouped according to their **primary actions** into three types:

1. *Prime Movers (Agonists)*: A **prime mover**, or **agonist**, is a muscle whose contraction is chiefly responsible for producing a particular movement, such as flexion at the elbow. The biceps brachii muscle is an example of a prime mover or agonist producing flexion at the elbow.
2. *Synergists*: When a **synergist** (*syn*-, together + *ergon*, work) contracts, it assists the prime mover in performing that action. Synergists may provide additional pull near the insertion or stabilize the point of origin. Their importance in assisting a particular movement may change as the movement progresses; in many cases they are most useful at the start, when the prime mover is stretched and its power is relatively low. For example, the *latissimus dorsi muscle* and the *teres major muscle* pull the arm inferiorly. With the arm pointed at the ceiling, the muscle fibers of the massive latissimus dorsi muscle are at maximum stretch, and they are aligned parallel to the humerus. The latissimus dorsi muscle cannot develop much tension in this position. However, the orientation of the teres major muscle, which originates on the scapula, can contract more efficiently, and it assists the latissimus dorsi muscle in starting an inferior movement. The importance of this smaller "assistant" decreases as the inferior movement proceeds. In this example, the latissimus dorsi muscle is the agonist and the teres major muscle the synergist. Synergists may also assist an agonist by preventing movement at a joint and thereby stabilizing the origin of the agonist. These muscles are called *fixators*.
3. *Antagonists*: **Antagonists** are muscles whose actions oppose that of the agonist; if the agonist produces flexion, the antagonist will produce extension. When an agonist contracts to produce a particular movement, the corresponding antagonist will be stretched, but it will usually not relax completely. Instead, its tension will be adjusted to control the speed of the movement and ensure its smoothness. For example, the biceps brachii muscle acts as an agonist when it contracts, thereby producing flexion of the elbow. The triceps brachii muscle, located on the opposite side of the humerus, acts as an antagonist to stabilize the flexion movement and to produce the opposing action, extension of the elbow.

Names of Skeletal Muscles [TABLE 9.2]

You will not need to learn every one of the nearly 700 muscles in the human body, but you will have to become familiar with the most important ones. Fortunately, the names of most skeletal muscles provide clues to their identification (Table 9.2). Skeletal muscles are named according to several criteria, including specific body regions, orientation of muscle fibers, specific or unusual features, identification of origin and insertion, and primary functions. The name may indicate a specific region such as the *brachialis muscle* of the arm, the shape of the muscle (*trapezius* or *piriformis muscles*), or some combination of the two (*biceps femoris muscle*).

Some names include reference to the orientation of the muscle fibers within a particular skeletal muscle. For example, **rectus** means "straight," and rectus muscles are parallel muscles whose fibers generally run along the long axis of the body. Because there are several rectus muscles, the name usually includes a second term that refers to a precise region of the body. The *rectus abdominis muscle* is found on the abdomen, and the *rectus femoris muscle* on the thigh. Other directional indicators include **transversus** and **oblique** for muscles whose fibers run across or at an oblique angle to the longitudinal axis of the body.

Other muscles were named after specific and unusual structural features. A *biceps* muscle has two tendons of origin (*bi*-, two + *caput*, head), the names *triceps* has three, and the *quadriceps* four. Shape is sometimes an important clue to the name of a muscle. For example, the names *trapezius* (tra-PĒ-zē-us), *deltoid, rhomboideus* (rom-BOY-dē-us), and *orbicularis* (or-bik-ū-LAR-is) refer to prominent muscles that look like a trapezoid, a triangle, a rhomboid, and a circle, respectively. Long muscles are called **longus** (long) or **longissimus** (longest), and **teres** muscles are both long and round. Short muscles are called **brevis**; large ones are called **magnus** (big), **major** (bigger), or **maximus** (biggest); and small ones are called **minor** (smaller) or **minimus** (smallest).

Muscles visible at the body surface are external and often called **externus** or **superficialis** (superficial), whereas those lying beneath are internal, termed **internus** or **profundus**. Superficial muscles that position or stabilize an organ are called **extrinsic** muscles; those that operate within the organ are called **intrinsic** muscles.

The names of many muscles identify their origins and insertions. In such cases, the first part of the name indicates the origin and the second part the insertion. For example, the *genioglossus muscle* originates at the chin (*geneion*) and inserts in the tongue (*glossus*).

Names that include flexor, extensor, retractor, and so on indicate the primary function of the muscle. These are such common actions that the names almost always include other clues concerning the appearance or location of the muscle. For example, the *extensor carpi radialis longus* is a long muscle found along the radial (lateral) border of the forearm. When it contracts, its primary function is extension at the wrist.

A few muscles are named after the specific movements associated with special occupations or habits. For example, the *sartorius* (sar-TŌR-ē-us) *muscle* is active when crossing the legs. Before sewing machines were invented, a tailor would sit on the floor cross-legged, and the name of the muscle was derived from *sartor*, the Latin word for "tailor." On the face, the *buccinator* (BUK-si-nā-tor) *muscle* compresses the cheeks, as when pursing the lips and blowing forcefully. *Buccinator* translates as "trumpet player." Finally, another facial muscle, the *risorius* (ri-SOR-ē-us) *muscle*, was supposedly named after the mood expressed. However, the Latin term *risor* means "laughter," while a more appropriate description for the effect would be "grimace."

Except for the *platysma* and the *diaphragm*, the complete names of all skeletal muscles include the term "muscle." Although we will generally use the full name of the muscle in the text, to save space and reduce clutter we will use only the descriptive portion of the name in the accompanying figures (triceps brachii, instead of triceps brachii muscle).

Aging and the Muscular System

As the body ages, there is a general reduction in the size and power of all muscle tissues. The effects of aging on the muscular system can be summarized as follows:

1. *Skeletal muscle fibers become smaller in diameter.* This reduction in size reflects primarily a decrease in the number of myofibrils. In addition, the muscle fibers contain less ATP, glycogen reserves, and myoglobin. The overall effect is a reduction in muscle strength and endurance and a tendency to fatigue rapidly. Because cardiovascular performance also decreases with age, blood flow to active muscles does not increase with exercise as rapidly as it does in younger people.
2. *Skeletal muscles become smaller in diameter and less elastic.* Aging skeletal muscles develop increasing amounts of fibrous connective tissue, a process called **fibrosis.** Fibrosis makes the muscle less flexible, and the collagen fibers can restrict movement and circulation.
3. *Tolerance for exercise decreases.* A lower tolerance for exercise results in part from the tendency for rapid fatigue and in part from the reduction in the ability to eliminate the heat generated during muscular contraction.
4. *Ability to recover from muscular injuries decreases.* The number of satellite cells steadily decreases with age, and the amount of fibrous tissue increases. As a result, when an injury occurs, repair capabilities are limited, and scar tissue formation is the usual result.

The rate of decline in muscular performance is the same in all individuals, regardless of their exercise patterns or lifestyle. Therefore, to be in good shape late in life, one must be in very good shape early in life. Regular exercise helps control body weight, strengthens bones, and generally improves the quality of life at all ages. Extremely demanding exercise is not as important as regular exercise. In fact, extreme exercise in the elderly may lead to problems with tendons, bones, and joints. Although it has obvious effects on the quality of life, there is no clear evidence that exercise prolongs life expectancy.

✓ CONCEPT CHECK

- What does the name *flexor digitorum longus* tell you about this muscle?
- Describe the difference between the origin and insertion of a muscle.
- What type of a muscle is a synergist?
- What is the difference between *major* and *minor* designations for a muscle?

RELATED CLINICAL TERMS

fibrosis: A process in which increasing amounts of fibrous connective tissue develop, making muscles less flexible. *p. 262*

muscular dystrophy: Several congenital conditions characterized by a generalized muscular weakness most evident in the upper limbs, head, and chest; caused by a reduction in the number of ACh receptors on motor end plates. *Duchenne's Muscular Dystrophy p. 791*

myasthenia gravis: A disease that produces progressive muscular weakness due to the loss of ACh receptors at the motor end plate. *Myasthenia Gravis p. 791*

polio: Progressive paralysis due to destruction of CNS motor neurons by the polio virus. *Polio p. 791*

rigor mortis: A state following death during which muscles are locked in the contracted position, making the body extremely stiff. *p. 253*

Additional Clinical Terms Discussed in Appendix I (p. 791)
Duchenne's muscular dystrophy (DMD)

STUDY OUTLINE & CHAPTER REVIEW

Introduction 245

1. There are three types of muscle tissue: **skeletal muscle**, **cardiac muscle**, and **smooth muscle**. The muscular system includes all the skeletal muscle tissue that can be controlled voluntarily.

Functions of Skeletal Muscle 245

1. **Skeletal muscles** attach to bones directly or indirectly and perform these functions: (1) produce skeletal movement, (2) maintain posture and body position, (3) support soft tissues, (4) regulate the entering and exiting of materials, (5) maintain body temperature.

Anatomy of Skeletal Muscles 245

Gross Anatomy 245

1. Each muscle fiber is wrapped by three concentric layers of connective tissue: an **epimysium**, a **perimysium**, and an **endomysium**. At the ends of the muscle are **tendons** or **aponeuroses** that attach the muscle to other structures. *(see Figure 9.1)*
2. Communication between a neuron and a muscle fiber occurs across the **neuromuscular** (*myoneural*) **junction**. *(see Figure 9.2)*

Microanatomy of Skeletal Muscle Fibers 247

3. A skeletal muscle cell has a cell membrane, or **sarcolemma**; cytoplasm, or **sarcoplasm**; and an internal membrane system, or **sarcoplasmic reticulum (SR)**, similar to the endoplasmic reticulum of other cells. *(see Figure 9.3)*
4. A skeletal muscle cell is large and multinucleated. Invaginations of the sarcolemma into the sarcoplasm of the skeletal muscle cell are called **transverse (T) tubules**. The transverse tubules carry the electrical impulse that stimulates contraction into the sarcoplasm, which contains numerous **myofibrils.** Protein filaments inside a myofibril are organized into repeating functional units called **sarcomeres**.
5. **Myofilaments** form myofibrils, which consist of **thin filaments** (actin and accessory proteins) and **thick filaments** (myosin). *(see Figures 9.3 to 9.5)*

✓ ANSWERS TO CONCEPT CHECK QUESTIONS

p. 247 **1.** Skeletal muscle tissue moves the body by pulling on bones of the skeleton. Cardiac muscle tissue pushes blood through the arteries and veins of the circulatory system. Smooth muscle tissues push fluids and solids along the digestive tract and perform varied functions in other systems. **2.** The perimysium is the connective tissue partition that separates adjacent fasciculi in a skeletal muscle. The perimysium contains blood vessels and nerves that supply each individual fascicle. **3.** Tendons are collagenous bands that connect skeletal muscle to the skeleton. Aponeuroses are thick, flattened tendinous sheets. **4.** A myoneural junction (also called a neuromuscular junction) is the site where the axon meets the muscle cell membrane. A motor end plate is the region of the sarcolemma at the neuromuscular junction.
p. 250 **1.** Skeletal muscle appears striated when viewed under the microscope because it is composed of the myofilaments actin and myosin, which are arranged in such a way as to produce a banded appearance in the muscle. **2.** A myofibril is a cylindrical collection of myofilaments within a cardiac or skeletal muscle cell. **3.** Myofilaments consist primarily of actin and myosin, along with accessory proteins of the thin filament (tropomyosin and troponin). **4.** The functional unit of skeletal muscle is the sarcomere. **5.** The proteins tropomyosin and troponin help regulate the actin and myosin interactions.
p. 254 **1.** During contraction, the width of the A band remains the same and the I band gets smaller. **2.** Stimulation of a motor neuron triggers the release of chemicals at the neuromuscular junction, which alter the transmembrane potential at the sarcolemma. This change sweeps across the surface of the sarcolemma and into the T-tubules. The change in the transmembrane potential of the T-tubules triggers the release of calcium ions by the sarcoplasmic reticulum. This release initiates the contraction, which proceeds as myosin heads go through repeated cycles of attach-pivot-detach-return. **3.** Terminal cisternae are expanded chambers of the sarcoplasmic reticulum—they store calcium ions, which are required to initiate contractile activities within skeletal muscle cells. Transverse tubules are sandwiched between terminal cisternae. A transverse tubule is an invagination of the sarcolemma, which conducts the stimulation into the interior of the cell. On arrival, it causes the release of calcium ions from the terminal cisternae, and this results in a muscle contraction. **4.** A neurotransmitter is a chemical compound released by one neuron to affect the transmembrane potential of another.
p. 256 **1.** The sprinter requires large amounts of energy for a relatively short burst of activity. To supply this demand for energy, the muscles switch to anaerobic metabolism. Anaerobic metabolism is not as efficient in producing energy as is aerobic metabolism, and the process also produces acidic waste products. The combination of less energy and the waste products contributes to fatigue. Marathon runners, on the other hand, derive most of their energy from aerobic metabolism, which is more efficient and does not produce the level of waste products that anaerobic metabolism does. **2.** Individuals who are naturally better at endurance types of activities such as cycling or marathon running have a higher percentage of slow-twitch muscle fibers, which are physiologically better adapted to this type of activity than the fast-twitch fibers, which are less vascular and fatigue faster. **3.** It depends on the number of branches in the motor neurons. A lesser number of branches (collaterals) results in the activation in only a few motor neurons. **4.** Recruitment is the smooth but steady increase in muscular tension produced by increasing the number of active motor units.
p. 262 **1.** This is a long muscle that flexes the joints of the finger. **2.** Each muscle begins at an origin, which typically remains stationary, and ends at an insertion, which is the part of the muscle that moves during a contraction. **3.** A synergist contracts to assist the prime mover in performing a specific action. **4.** Major is used to describe muscles that are bigger, whereas minor is used to describe muscles that are smaller.

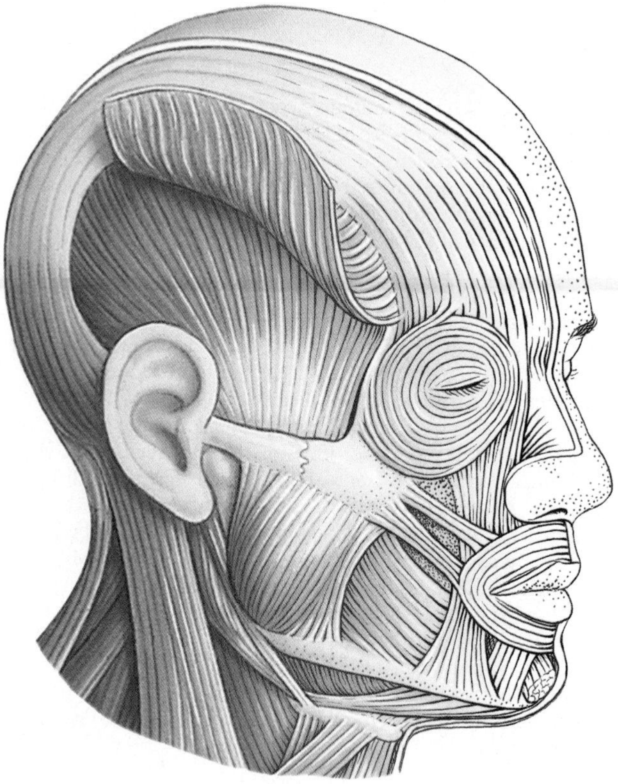

1. Identify and locate the principal axial muscles of the body, together with their origins and insertions, and describe their innervations and actions.

10

THE MUSCULAR SYSTEM

Axial Musculature

The separation of the skeletal system into axial and appendicular divisions provides a useful guideline for subdividing the muscular system as well. The **axial musculature** arises on the axial skeleton. It positions the head and vertebral column and moves the rib cage, which assists in breathing. Axial muscles do not play a role in the movement or stabilization of the pectoral or pelvic girdles or the limbs. Roughly 60% of the skeletal muscles in the body are axial muscles. The **appendicular musculature** stabilizes or moves components of the appendicular skeleton. The major axial and appendicular muscles are illustrated in Figures 10.1 and 10.2●. Although in almost all cases the word "muscle" is officially a part of each name, it has not been included in the figure labels.

The Axial Musculature

[FIGURES 10.1/10.2]

The axial musculature is involved in movements of the head and spinal column. Because our discussion of axial musculature relies heavily on an understanding of skeletal anatomy, you may find it helpful to review appropriate figures in Chapters 6 and 7 as we proceed. The relevant figures in those chapters are noted in the figure captions throughout this chapter.

The axial muscles fall into four logical groups based on location and/or function. The groups do not always have distinct anatomical boundaries. For example, a function such as the extension of the vertebral column involves muscles along its entire length.

1. The first group includes the *muscles of the head and neck* that are not associated with the vertebral column. These muscles include those that move the face, tongue, and larynx. They are responsible for verbal and nonverbal communication—laughing, talking, frowning, smiling, and whistling are examples. This group of muscles also performs movements associated with feeding, such as sucking, chewing, or swallowing, as well as contractions of the eye muscles that help us look around for something else to eat.
2. The second group, the *muscles of the vertebral column*, includes numerous flexors and extensors of the axial skeleton.

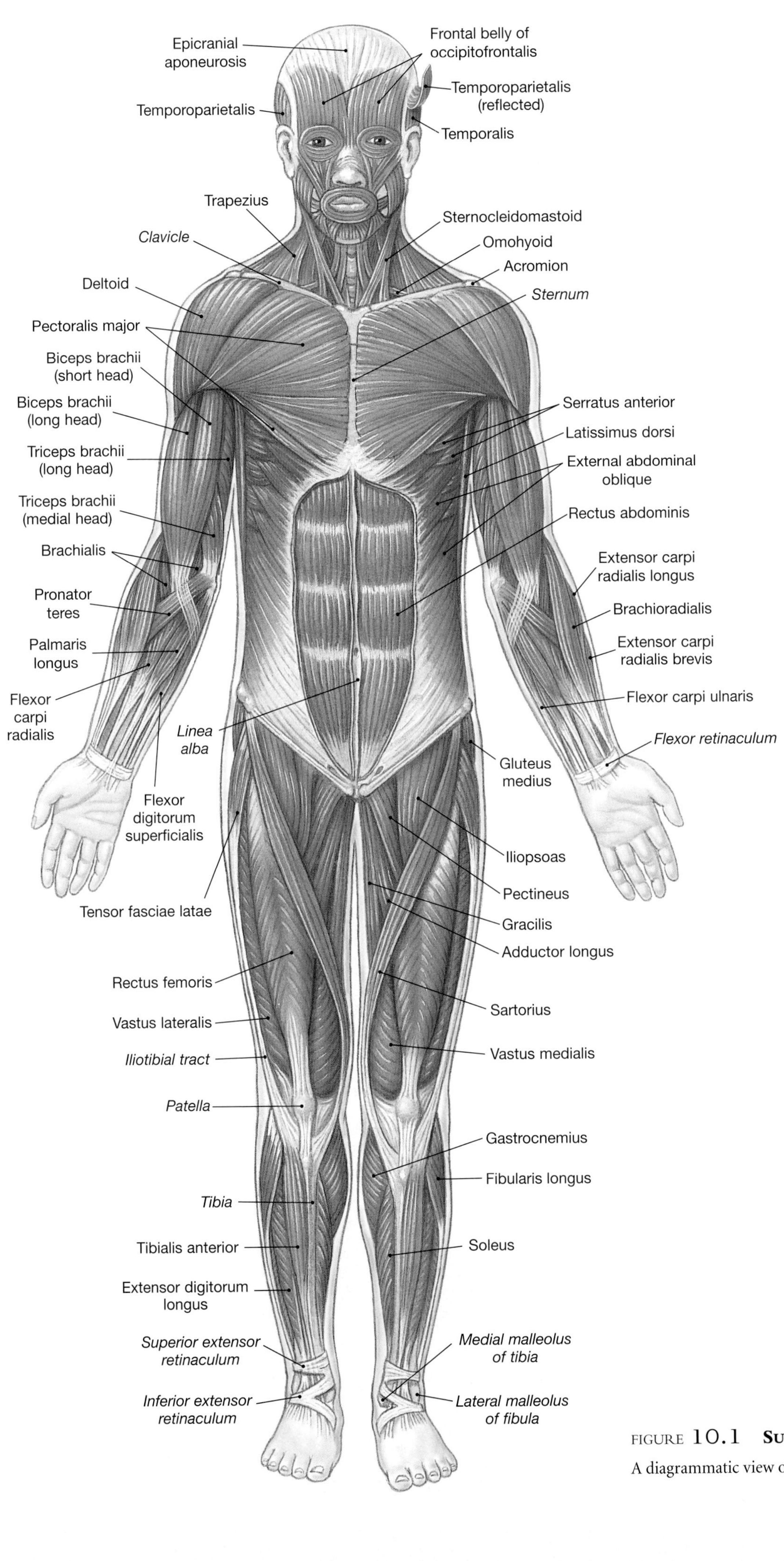

FIGURE 10.1 **Superficial Skeletal Muscles, Anterior View**

A diagrammatic view of the major axial and appendicular muscles.

3. The third group, the *oblique* and *rectus muscles*, form the muscular walls of the thoracic and abdominopelvic cavities between the first thoracic vertebra and the pelvis. In the thoracic area, these muscles are partitioned by the ribs, but over the abdominal surface, they form broad muscular sheets. There are also oblique and rectus muscles in the neck. Although they do not form a complete muscular wall, they are included in this group because they share a common developmental origin. The diaphragm is placed within this group because it is developmentally linked to other muscles of the chest wall.
4. The fourth group, the *muscles of the pelvic floor*, extend between the sacrum and pelvic girdle, forming the muscular *perineum*, which closes the pelvic outlet. ∞ *p. 200*

Figures 10.1 and 10.2● provide an overview of the major axial and appendicular muscles of the human body. These are the superficial muscles, which tend to be relatively large. The superficial muscles cover deeper, smaller muscles that cannot be seen unless the overlying muscles are either removed or *reflected*—that is, cut and pulled out of the way. Later figures that show deep muscles in specific regions will indicate whether superficial muscles have been removed or reflected for the sake of clarity.

To facilitate the review process, information concerning the origin, insertion, and action of each muscle has been summarized in tables. These tables also contain information about the *innervation* of individual muscles. The term **innervation** refers to the nerve supply to a particular structure or organ, and each skeletal muscle is controlled by one or more motor nerves. The names of the nerves provide clues to the distribution of the nerve or the site at which the nerve leaves the cranial or spinal cavities. For example, the *facial nerve* innervates the facial musculature, and the various *spinal nerves* leave the vertebral canal by way of the intervertebral foramina. ∞ *pp. 142, 166* To help you understand the relationships between the skeletal muscles and the bones of the skeleton, we have included skeletal icons showing the origins and insertions of representative muscles in each group. On each icon, the areas where muscles originate are shown in red, and the areas where muscles insert are shown in blue.

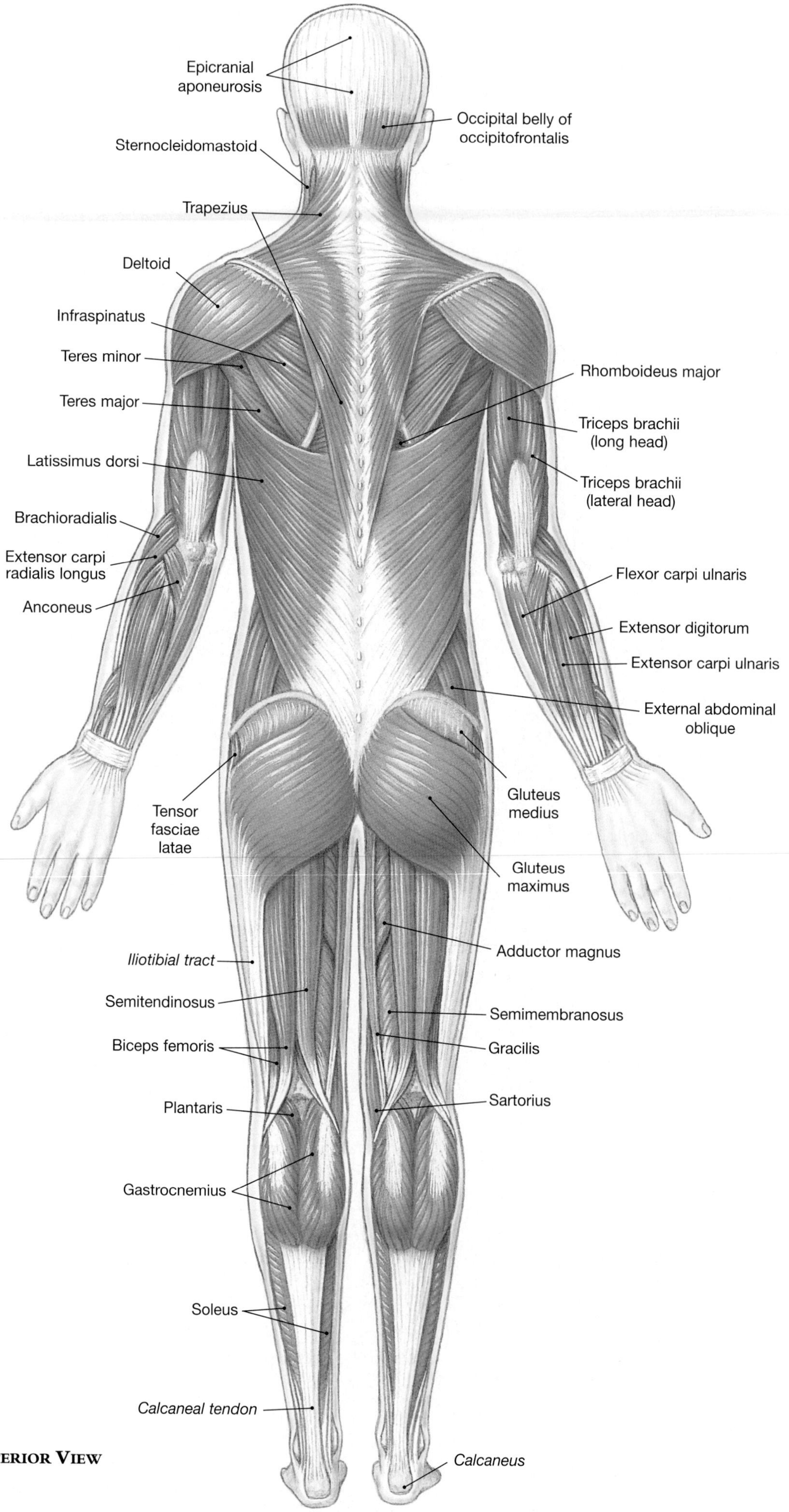

FIGURE 10.2 **Superficial Skeletal Muscles, Posterior View**
A diagrammatic view of the major axial and appendicular muscles.

Muscles of the Head and Neck

The muscles of the head and neck can be subdivided into several groups. The *muscles of facial expression*, the *extra-ocular muscles*, the *muscles of mastication*, the *muscles of the tongue*, and the *muscles of the pharynx* originate on the skull or hyoid bone. Other muscles involved with sight and hearing originate on the skull. These muscles are discussed in Chapter 18 (general and special senses), along with muscles associated with the ear and hearing. The *anterior muscles of the neck* are concerned primarily with altering the position of the larynx, hyoid bone, and floor of the mouth.

MUSCLES OF FACIAL EXPRESSION [FIGURES 10.2/10.3/10.4 AND TABLE 10.1]

The muscles of facial expression originate on the surface of the skull. View Figures 10.3 and 10.4● as we describe their structure. Table 10.1 provides a detailed summary of their characteristics. At their insertions the epimysial fibers are woven into those of the superficial fascia and the dermis of the skin; when they contract, the skin moves. These muscles are innervated by the seventh cranial nerve, the *facial nerve*.

The largest group of facial muscles is associated with the mouth (Figure 10.3●). The **orbicularis oris** (OR-is) **muscle** constricts the opening, while other muscles move the lips or the corners of the mouth. The **buccinator muscle** has two functions related to feeding (in addition to its importance to musicians). During chewing it cooperates with the masticatory muscles by moving food back across the teeth from the space inside the cheeks. In infants, the buccinator is responsible for producing the suction required for suckling at the breast.

Smaller groups of muscles control movements of the eyebrows and eyelids, the scalp, the nose, and the external ear. The *epicranius* (ep-i-KRĀ-nē-us; *epi-*, on + *kranion*, skull), or scalp, consists of the **occipitofrontalis**

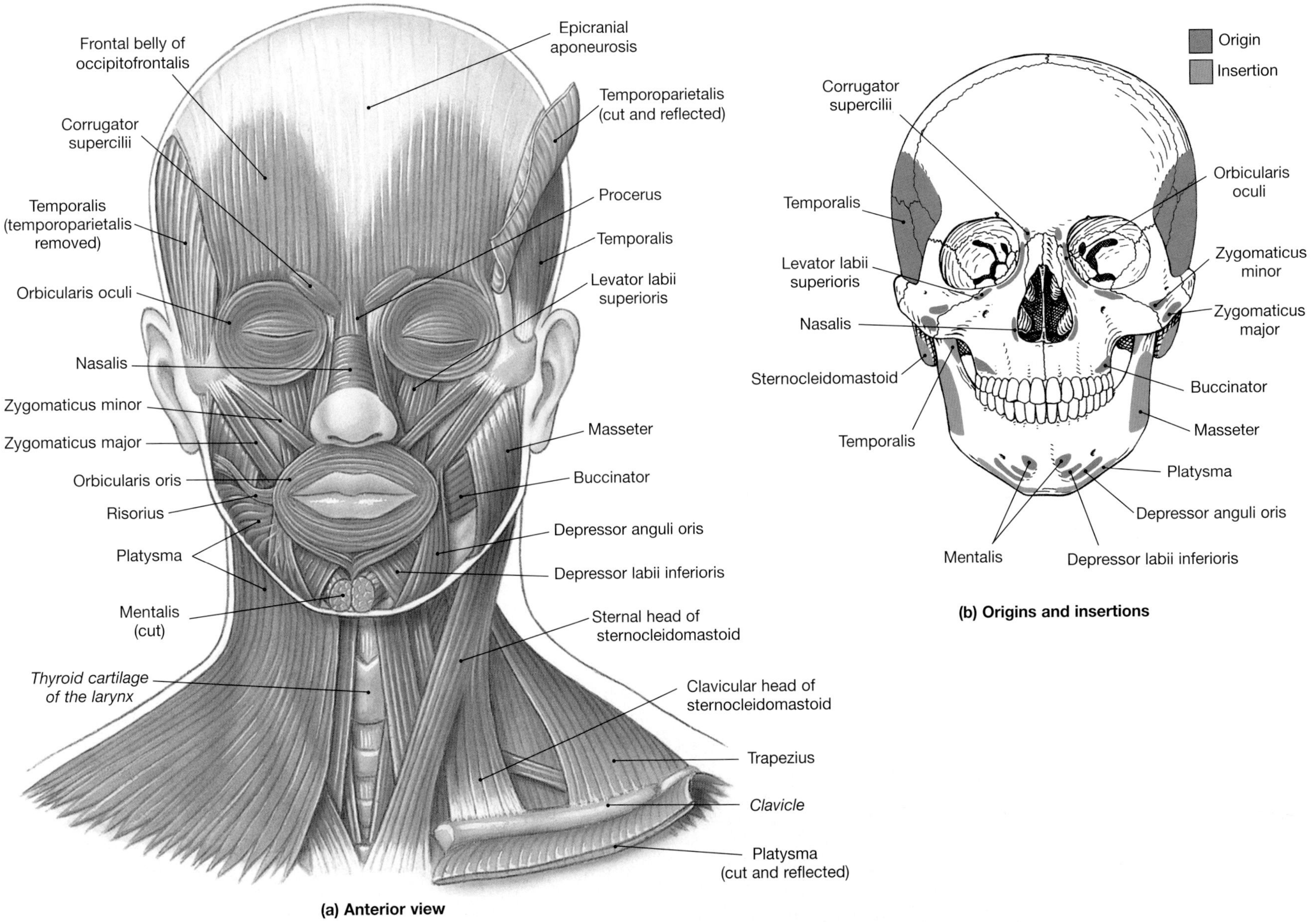

FIGURE 10.3 **MUSCLES OF THE HEAD AND NECK, PART I**

(**a**) Anterior view. (**b**) Origins and insertions of selected muscles.

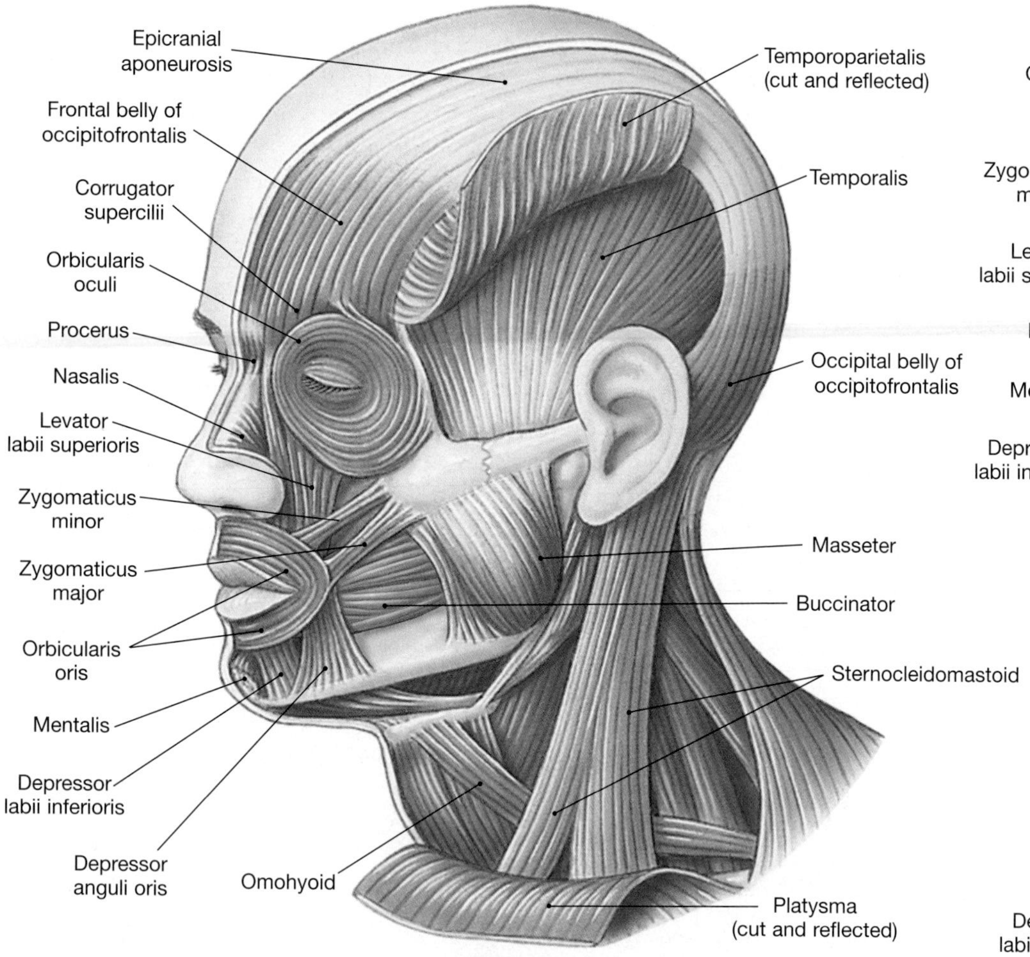

(a) Lateral view

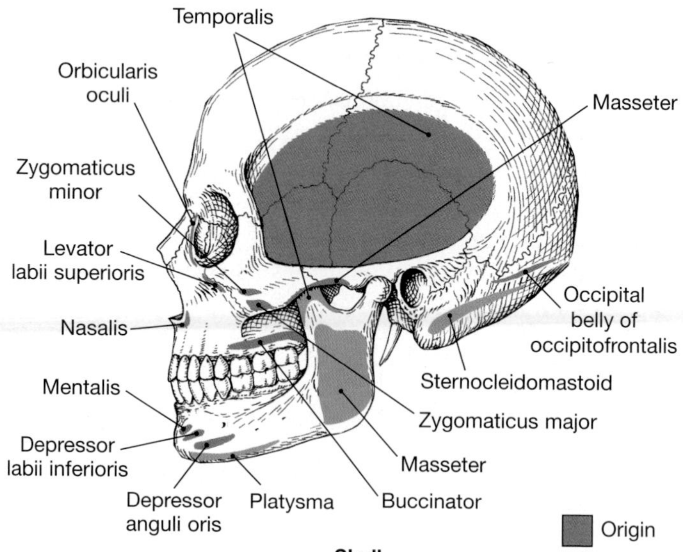

Skull

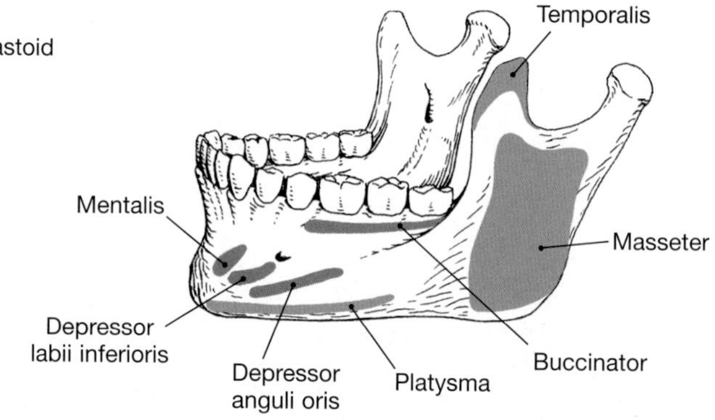

Mandible

(c) Origins and insertions

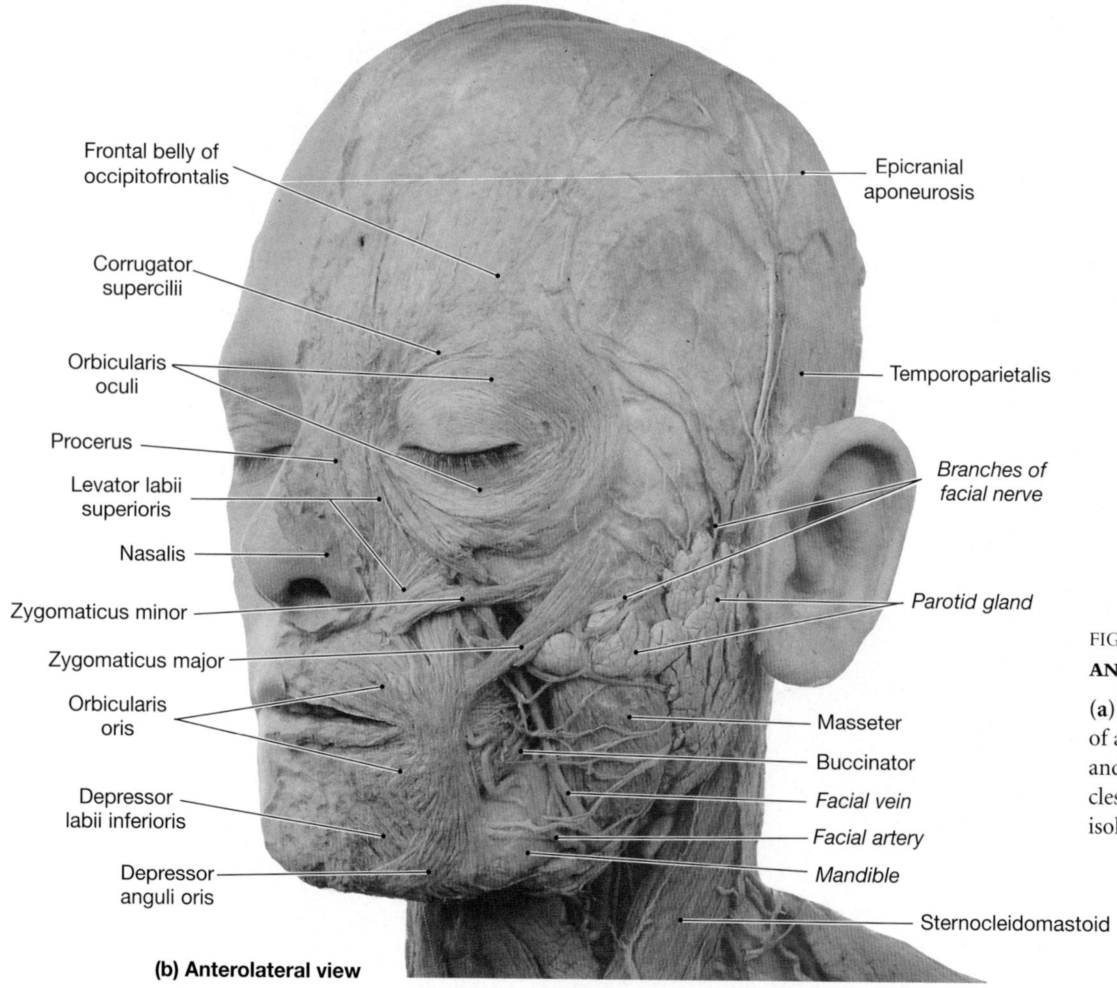

(b) Anterolateral view

FIGURE 10.4 **Muscles of the Head and Neck, Part II**

(**a**) A diagrammatic lateral view. (**b**) A corresponding view of a dissection showing many of the muscles of the head and neck. (**c**) Origins and insertions of representative muscles on the lateral surface of the entire skull (above) and the isolated mandible (below). *See also Figure 6.3.*

TABLE 10.1 MUSCLES OF FACIAL EXPRESSION

Region/Muscle	Origin	Insertion	Action	Innervation
MOUTH				
Buccinator	Alveolar processes of maxilla and mandible	Blends into fibers of orbicularis oris	Compresses cheeks	Facial nerve (N VII)
Depressor labii inferioris	Mandible between the anterior midline and the mental foramen	Skin of lower lip	Depresses lower lip	As above
Levator labii superioris	Inferior margin of orbit, superior to the infraorbital foramen	Orbicularis oris	Elevates upper lip	As above
Mentalis	Incisive fossa of mandible	Skin of chin	Elevates and protrudes lower lip	As above
Orbicularis oris	Maxilla and mandible	Lips	Compresses, purses lips	As above
Risorius	Fascia surrounding parotid salivary gland	Angle of mouth	Draws corner of mouth to the side	As above
Depressor anguli oris	Anterolateral surface of mandibular body	Skin at angle of mouth	Depresses corner of mouth	As above
Zygomaticus major	Zygomatic bone near the zygomaticomaxillary suture	Angle of mouth	Retracts and elevates corner of mouth	As above
Zygomaticus minor	Zygomatic bone posterior to zygomaticotemporal suture	Upper lip	Retracts and elevates upper lip	As above
EYE				
Corrugator supercilii	Orbital rim of frontonasal bone near nasal suture	Eyebrow	Pulls skin inferiorly and anteriorly; wrinkles brow	As above
Levator palpebrae superioris	Tendinous band around optic foramen	Upper eyelid	Elevates upper eyelid	Oculomotor nerve (N III)[a]
Orbicularis oculi	Medial margin of orbit	Skin around eyelids	Closes eye	Facial nerve (N VII)
NOSE				
Procerus	Nasal bones and lateral nasal cartilages	Aponeurosis at bridge of nose and skin of forehead	Moves nose, changes position, shape of nostrils	As above
Nasalis	Maxilla and alar cartilage of nose	Bridge of nose	Compresses bridge, depresses tip of nose; elevates corners of nostrils	As above
EAR (EXTRINSIC)				
Temporoparietalis	Fascia around external ear	Epicranial aponeurosis	Tenses scalp, moves auricle of ear	As above
SCALP (EPICRANIUS)[b]				
Occipitofrontalis				
Frontal belly	Epicranial aponeurosis	Skin of eyebrow and bridge of nose	Raises eyebrows, wrinkles forehead	As above
Occipital belly	Superior nuchal line	Epicranial aponeurosis	Tenses and retracts scalp	As above
NECK				
Platysma	Superior thorax between cartilage of second rib and acromion of scapula	Mandible and skin of cheek	Tenses skin of neck, depresses mandible	As above

[a] This muscle originates in association with the extra-ocular muscles, so its innervation is unusual. (*See Figure 15.24.*)
[b] Epicranial aponeurosis, frontal belly of occipitofrontalis, and occipital belly of occipitofrontalis muscle collectively comprise the epicranius muscle.

muscle, which has two bellies, the **frontal belly** and the **occipital belly**, separated by a collagenous sheet, *epicranial aponeurosis*, or *galea aponeurotica* (GĀ-lē-a ap-Ō-nū-RŌ-ti-ka), (Figures 10.2 and 10.3●). The superficial **platysma** (pla-TIZ-ma; *platys*, flat) covers the ventral surface of the neck, extending from the base of the neck to the periosteum of the mandible and the fascia at the corners of the mouth (Figures 10.3 and 10.4●).

EXTRA-OCULAR MUSCLES [FIGURE 10.5 AND TABLE 10.2]

Six **extra-ocular muscles**, sometimes called the *oculomotor* (ok-ū-lō-MŌ-ter), or *extrinsic eye muscles*, originate on the surface of the orbit and control the position of each eye. These muscles are the **inferior rectus, medial rectus**, **superior rectus**, **lateral rectus**, **inferior oblique**, and **superior**

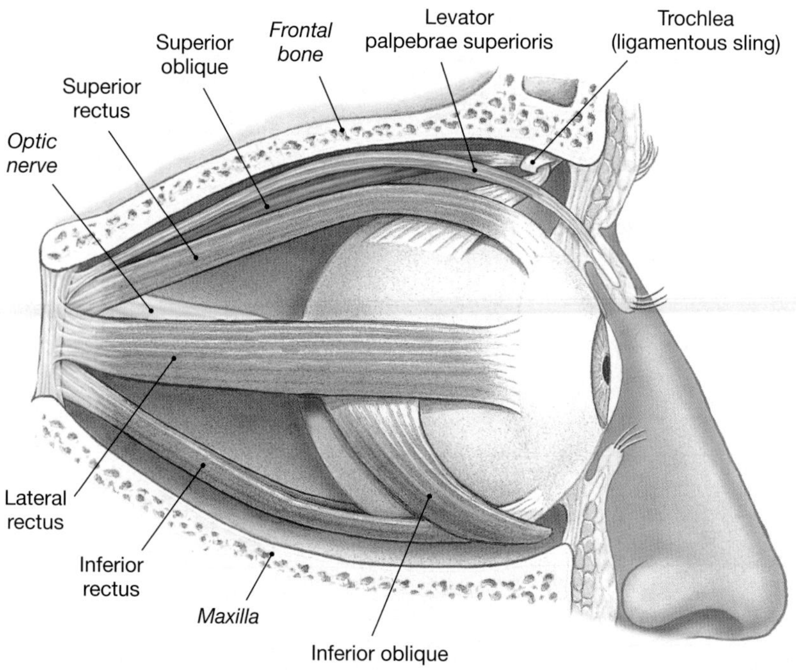

(a) Lateral surface, right eye

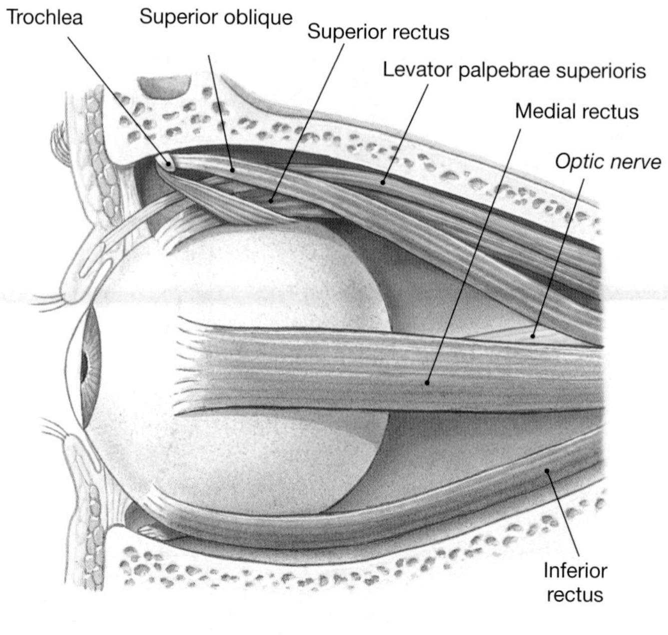

(b) Medial surface, right eye

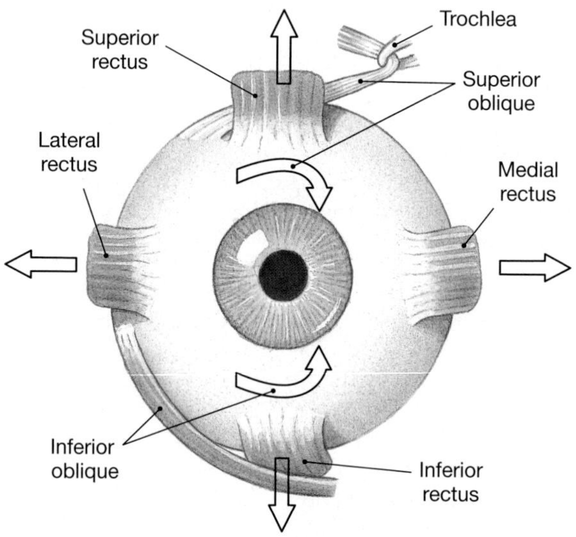

(c) Anterior view, right eye

FIGURE 10.5 **EXTRA-OCULAR MUSCLES**

(a) Muscles on the lateral surface of the right eye. **(b)** Muscles on the medial surface of the right eye. **(c)** Anterior view of the right eye, showing the orientation of the extra-ocular muscles and the directions of eye movement produced by contractions of the individual muscles. *See also Figure 6.3.*

oblique muscles (Figure 10.5• and Table 10.2). The rectus muscles move the eyes in the direction indicated by their names. Additionally, the superior and inferior rectus muscles also cause a slight movement of the eye medially, whereas superior and inferior oblique muscles cause a slight lateral movement. Thus, to roll the eye straight up, one contracts the superior rectus and the inferior oblique muscles; to roll the eye straight down requires the inferior rectus and the inferior oblique muscles. The extra-oular muscles are innervated by the third (*oculomotor*), fourth (*trochlear*), and sixth (*abducens*) cranial nerves. The *intrinsic eye muscles*, which are smooth muscles inside the eyeball, control pupil diameter and lens shape. These muscles are discussed in Chapter 18.

TABLE 10.2 EXTRAO-CULAR MUSCLES

Muscle	Origin	Insertion	Action	Innervation
Inferior rectus	Sphenoid around optic canal	Inferior, medial surface of eyeball	Eye looks down	Oculomotor nerve (N III)
Medial rectus	As above	Medial surface of eyeball	Eye looks medially	As above
Superior rectus	As above	Superior surface of eyeball	Eye looks up	As above
Lateral rectus	As above	Lateral surface of eyeball	Eye looks laterally	Abducens nerve (N VI)
Inferior oblique	Maxilla at anterior portion of orbit	Inferior, lateral surface of eyeball	Eye rolls, looks up and medially	Oculomotor nerve (N III)
Superior oblique	Sphenoid around optic canal	Superior, lateral surface of eyeball	Eye rolls, looks down and medially	Trochlear nerve (N IV)

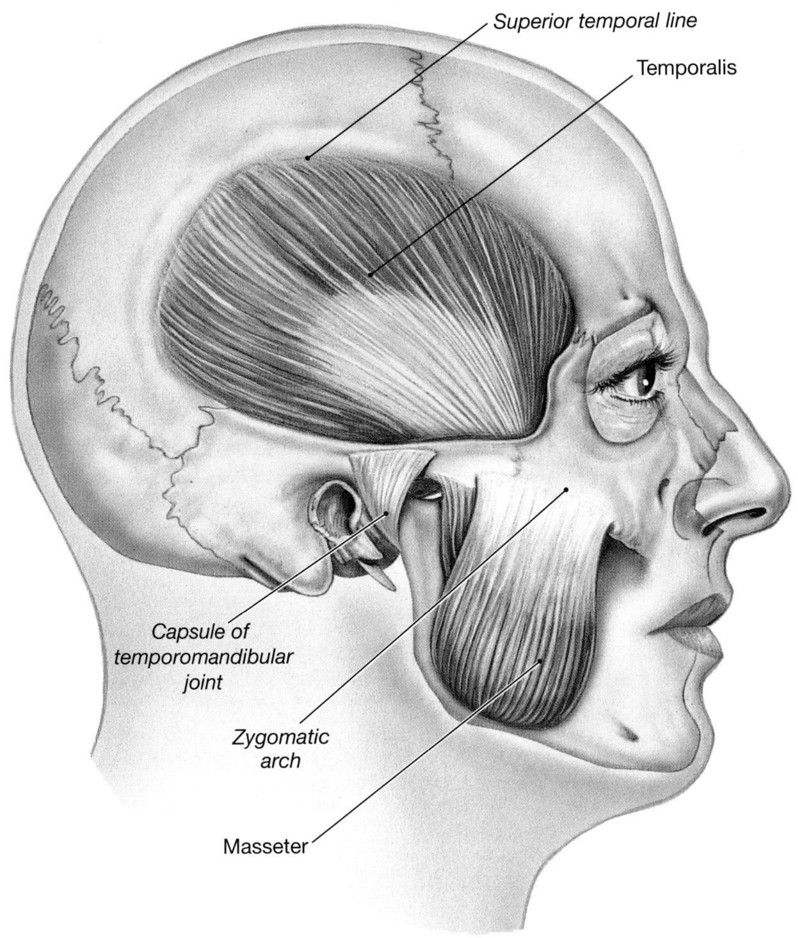

(a) Lateral view

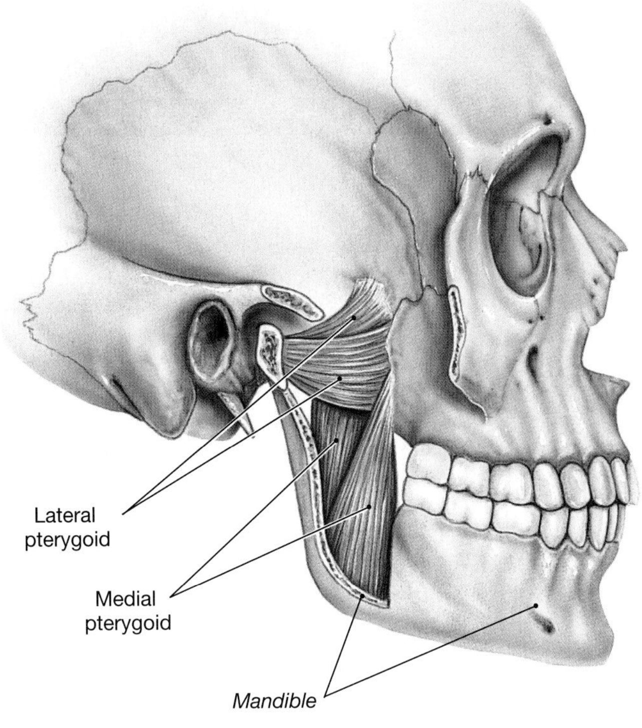

(b) Lateral view, pterygoid muscles exposed

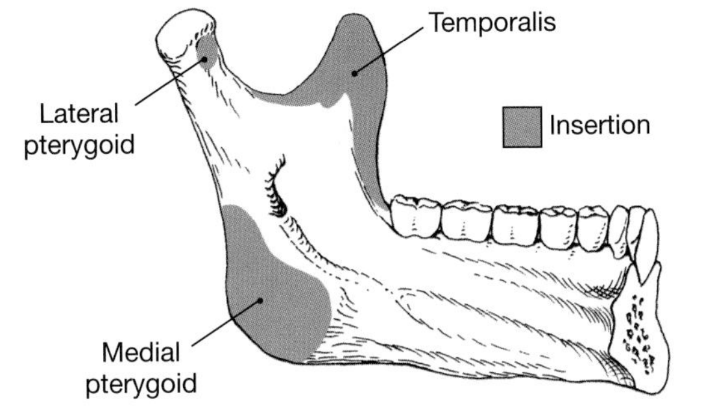

(c) Insertions, medial view of left mandibular ramus

FIGURE 10.6 **MUSCLES OF MASTICATION**

These muscles move the mandible during chewing. **(a)** The temporalis and masseter are prominent muscles on the lateral surface of the skull. The temporalis passes medial to the zygomatic arch to insert on the coronoid process of the mandible. The masseter inserts on the angle and lateral surface of the mandible. **(b)** The location and orientation of the pterygoid muscles can be seen after removing the overlying muscles, along with a portion of the mandible. **(c)** Selected insertions on the medial surface of the mandible. *See also Figures 6.3 and 6.14.*

MUSCLES OF MASTICATION (FIGURE 10.6 AND TABLE 10.3)

The muscles of mastication (Figure 10.6● and Table 10.3) move the mandible at the temporomandibular joint. p. 222 The large **masseter** (ma-SĒ-ter) **muscle** elevates the mandible and is the most powerful and important of the masticatory muscles. The **temporalis** (tem-po-RĀ-lis) **muscle** assists in elevation of the mandible, whereas the medial and lateral **pterygoid** (TER-i-goyd) **muscles**, when used in various combinations can elevate, protract, or slide the mandible from side to side; such movement is called *lateral excursion*. These movements are important in maximizing the efficient use of the teeth while chewing or grinding foods of various consistencies. The muscles of mastication are innervated by the fifth cranial nerve, the *trigeminal nerve*.

TABLE 10.3 **MUSCLES OF MASTICATION**

Muscle	Origin	Insertion	Action	Innervation
Masseter	Zygomatic arch	Lateral surface and angle of mandibular ramus	Elevates mandible and closes jaws	Trigeminal nerve (N V), mandibular branch
Temporalis	Along temporal lines of skull	Coronoid process of mandible	As above	As above
Pterygoids	Lateral pterygoid plate	Medial surface of mandibular ramus		
Medial pterygoid	Lateral pterygoid plate and adjacent portions of palatine bone and maxilla	Medial surface of mandibular ramus	Elevates the mandible and closes the jaws, or moves mandible side to side	As above
Lateral pterygoid	Lateral pterygoid plate and greater wing of sphenoid	Anterior part of the neck of the mandibular condyle	Opens jaws, protrudes mandible, or moves mandible side to side	As above

CLINICAL DISCUSSION

What's New?

You may have heard someone comment that "There's nothing new in anatomy." That statement reflects a popular opinion that every anatomical structure in the human body was described centuries ago. Many people were therefore surprised when, in 1996, anatomical researchers at the University of Maryland documented the existence of a "new" skeletal muscle. This muscle, the *sphenomandibularis muscle* (Figure 10.7●), assists the muscles of mastication; it extends from the lateral surface of the sphenoid to the mandible. The work was begun through computer analysis of the Visible Human database, a digitized photographic atlas of cross-sectional anatomy. The initial work was then supported by careful cadaver dissections. Although there remains some controversy about this muscle (for example, it may have been described previously as a portion of the temporalis muscle), this is a good example of how modern technologies are providing new perspectives on the human body.

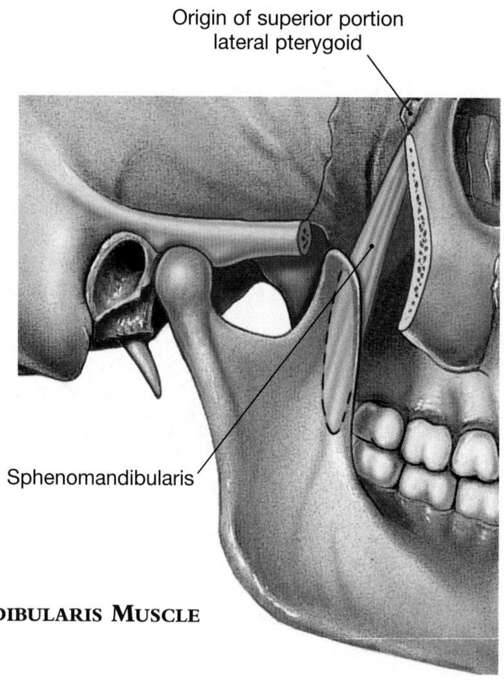

FIGURE 10.7
The Sphenomandibularis Muscle

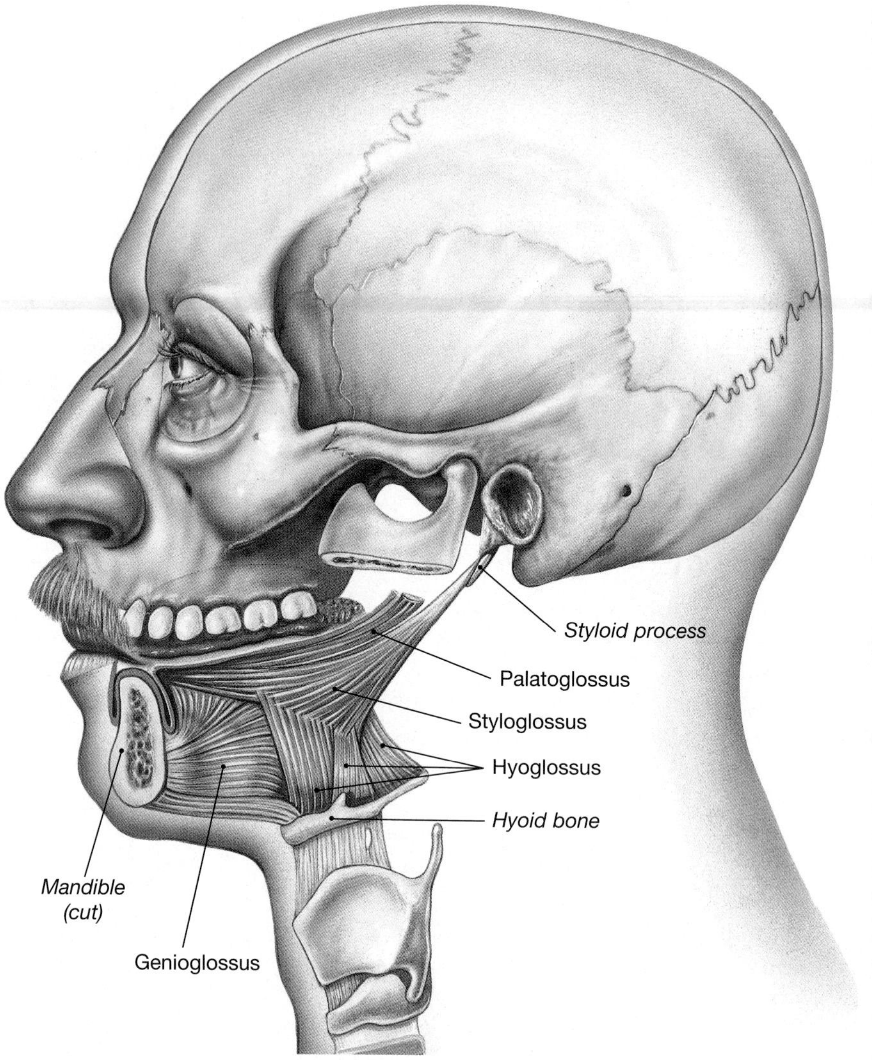

FIGURE 10.8 **Muscles of the Tongue**
The left mandibular ramus has been removed to show the muscles on the left side of the tongue.

MUSCLES OF THE TONGUE (FIGURE 10.8 AND TABLE 10.4)

These muscles have names ending in *-glossus*, meaning "tongue." Once you can recall the structures referred to by *genio-*, *hyo-*, *palato-*, and *stylo*, you shouldn't have much trouble with this group. The **genioglossus muscle** originates at the chin, the **hyoglossus muscle** at the hyoid bone, the **palatoglossus muscle** at the palate, and the **styloglossus muscle** at the styloid process (Figure 10.8●). These muscles, the extrinsic tongue muscles, are used in various combinations to move the tongue in the delicate and complex patterns necessary for speech. They also manipulate food within the mouth in preparation for swallowing. The intrinsic tongue muscles, located entirely within the tongue, assist in these activities. Most of these muscles are innervated by the twelfth cranial nerve, the *hypoglossal nerve*; its name indicates its function as well as its location (Table 10.4).

TABLE 10.4 **Muscles of the Tongue**

Muscle	Origin	Insertion	Action	Innervation
Genioglossus	Medial surface of mandible around chin	Body of tongue, hyoid bone	Depresses and protracts tongue	Hypoglossal nerve (N XII)
Hyoglossus	Body and greater horn of hyoid bone	Side of tongue	Depresses and retracts tongue	As above
Palatoglossus	Anterior surface of soft palate	As above	Elevates tongue, depresses soft palate	Cranial root of accessory nerve (N XI)
Styloglossus	Styloid process of temporal bone	Along the side to tip and base of tongue	Retracts tongue, elevates sides	Hypoglossal nerve (N XII)

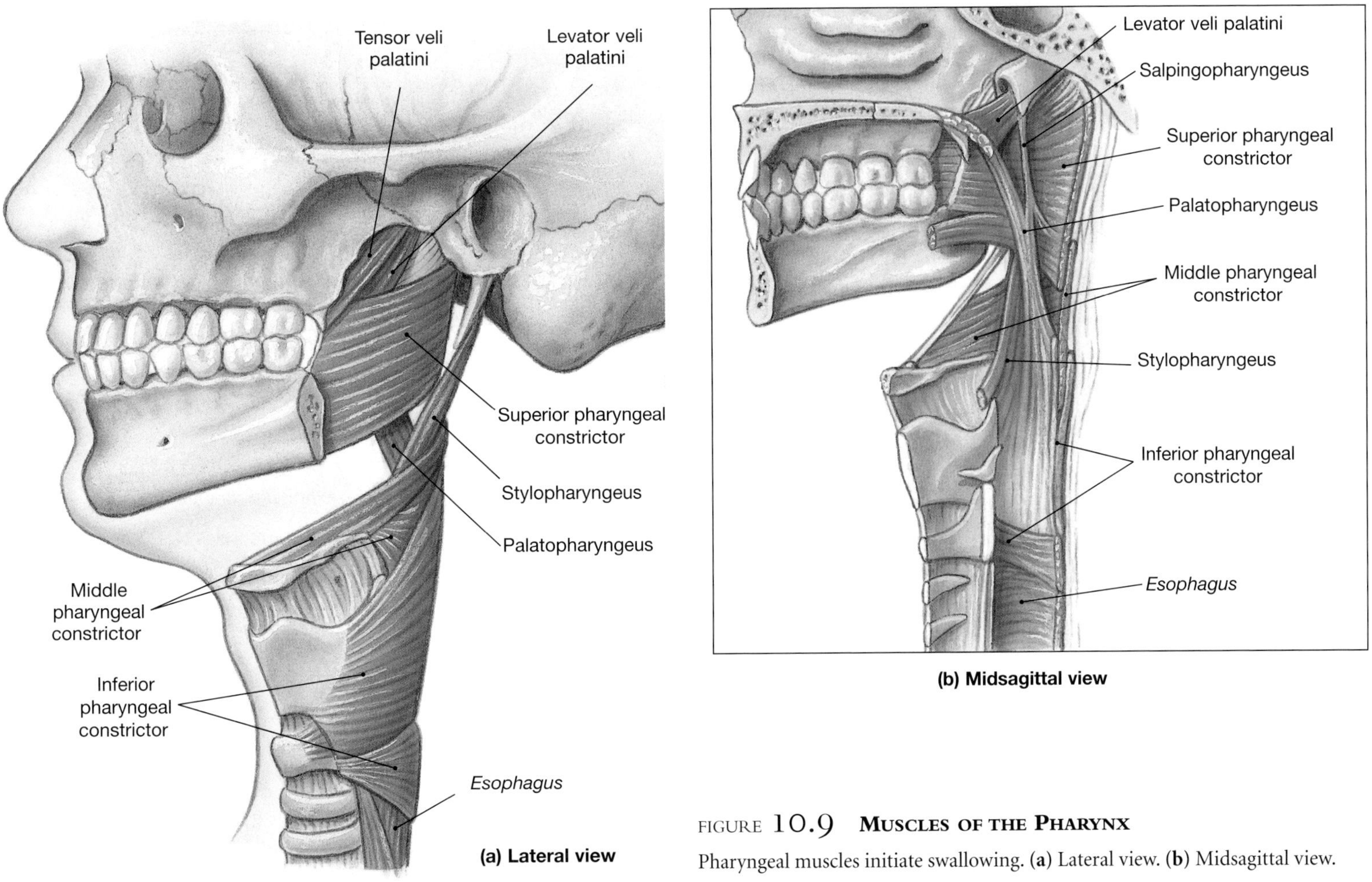

FIGURE 10.9 **MUSCLES OF THE PHARYNX**

Pharyngeal muscles initiate swallowing. (**a**) Lateral view. (**b**) Midsagittal view.

MUSCLES OF THE PHARYNX (FIGURE 10.9 AND TABLE 10.5)

The paired pharyngeal muscles are important in the initiation of swallowing. The **pharyngeal constrictors** begin the process of moving a *bolus*, or chewed mass of food, into the esophagus. The **palatopharyngeus** (pal-āt-ō-far-IN-jē-us), **salpingopharyngeus** (sal-pin-gō-far-IN-jē-us), and **stylopharyngeus** (stī-lō-far-IN-jē-us) **muscles** elevate the larynx and are grouped together as *laryngeal elevators.* The **palatal muscles,** the **tensor veli palatini** and **levator veli palatini,** raise the soft palate and adjacent portions of the pharyngeal wall. The latter muscles also pull open the entrance to the auditory tube. As a result, swallowing repeatedly can help one adjust to pressure changes when flying or diving. Pharyngeal muscles are innervated by the ninth (*glossopharyngeal*) and tenth (*vagus*) cranial nerves. These muscles are illustrated in Figure 10.9●, and additional information can be found in Table 10.5.

TABLE 10.5 **MUSCLES OF THE PHARYNX**

Muscle	Origin	Insertion	Action	Innervation
PHARYNGEAL CONSTRICTORS			Constrict pharynx to propel bolus into esophagus	Branches of pharyngeal plexus (N X)
Superior constrictor	Pterygoid process of sphenoid, medial surfaces of mandible	Median raphe attached to occipital bone		N X
Middle constrictor	Horns of hyoid bone	Median raphe		N X
Inferior constrictor	Cricoid and thyroid cartilages of larynx	Median raphe		N X
LARYNGEAL ELEVATORS*			Elevate larynx	Branches of pharyngeal plexus (N IX & X)
Palatopharyngeus	Soft palate	Thyroid cartilage		N X
Salpingopharyngeus	Cartilage around the inferior portion of the auditory tube	Thyroid cartilage		N X
Stylopharyngeus	Styloid process of temporal bone	Thyroid cartilage		N IX
PALATAL MUSCLES				
Levator veli palatini	Petrous part of temporal bone, tissues around the auditory tube	Soft palate	Elevates soft palate	N XI
Tensor veli palatini	Sphenoidal spine and tissues around the auditory tube	Soft palate	As above	N V

* Assisted by the thyrohyoid, geniohyoid, stylohyoid, and hyoglossus muscles, discussed in Tables 10.4 and 10.6.

ANTERIOR MUSCLES OF THE NECK (FIGURES 10.10/10.11 AND TABLE 10.6)

The anterior muscles of the neck control the position of the larynx, depress the mandible, tense the floor of the mouth, and provide a stable foundation for muscles of the tongue and pharynx (Figures 10.10 and 10.11•, and Table 10.6). The anterior neck muscles that position the larynx are called *extrinsic* muscles, while those that affect the vocal cords are termed *intrinsic*. (The vocal cords will be discussed in Chapter 24.) Additionally, the muscles of the neck are either *suprahyoid* or *infrahyoid* based on their location relative to the hyoid bone. The **digastric** (dī-GAS-trik) **muscle** has two bellies, as the name implies (*di-*, two + *gaster*, stomach). One belly extends from the chin to the hyoid bone, and the other continues from the hyoid bone to the mastoid portion of the temporal bone. This muscle opens the mouth by depressing the mandible. The anterior belly overlies the broad, flat **mylohyoid** (mī-lō-HĪ-oyd) **muscle**, which provides muscular support to the floor of the mouth. The **geniohyoid muscles**, which lie superior to the mylohyoid muscle, provide additional support. The **stylohyoid** (stī-lō-HĪ-oyd) **muscle** forms a muscular connection between the hyoid bone and the styloid process of the skull. The **sternocleidomastoid** (ster-nō-klī-dō-MAS-toid) **muscle** extends from the clavicle and the sternum to the mastoid region of the skull. It originates at two heads, a *sternal head* and a *clavicular head* (Table 10.6). These extensive muscles are innervated by more than one nerve, and specific regions can be made to contract independently. As a result, their actions are quite varied. The other members of this group are straplike muscles that run between the sternum and the larynx (*sternothyroid*), or hyoid bone (*sternohyoid*) and between the larynx and hyoid bone (*thyrohyoid*).

✓ CONCEPT CHECK

- Where do muscles of facial expression originate?
- What is the general function of the muscles of mastication?
- Describe the general function(s) of the extra-ocular muscles.
- What is the importance of the pharyngeal muscles?

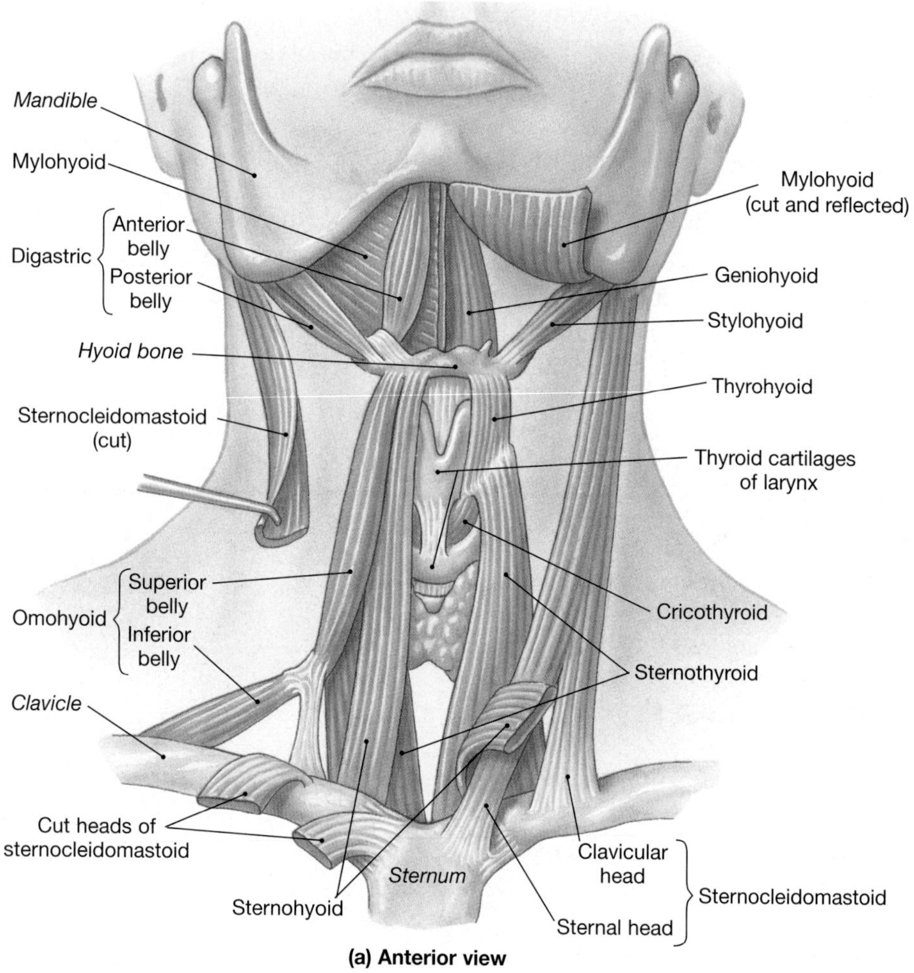

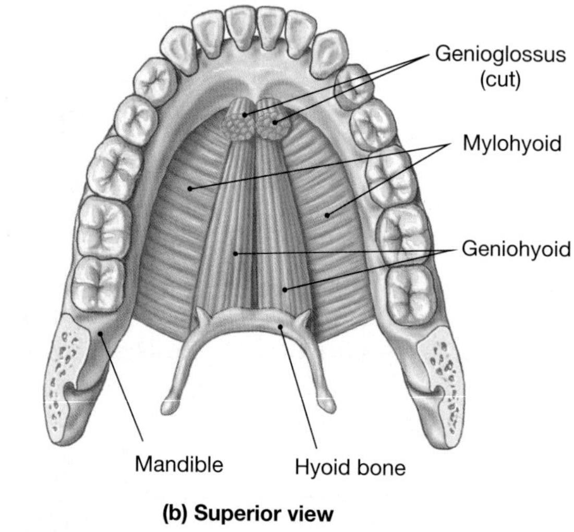

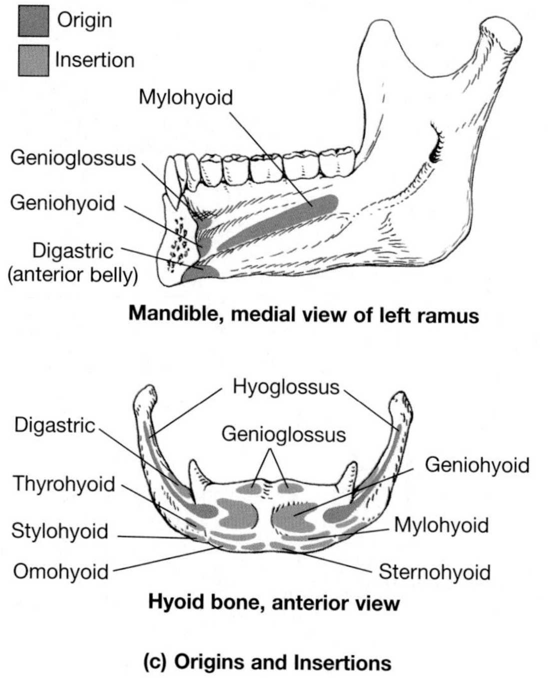

FIGURE 10.10 **ANTERIOR MUSCLES OF THE NECK, PART I**

These muscles adjust the position of the larynx, mandible, and floor of the mouth, and establish a foundation for attachment of both tongue and pharyngeal muscles. (**a**) Anterior view of neck muscles. (**b**) Muscles that form the floor of the oral cavity, superior view. (**c**) Origins and insertions on the mandible and hyoid. *See also Figures 6.3, 6.4, and 6.17.*

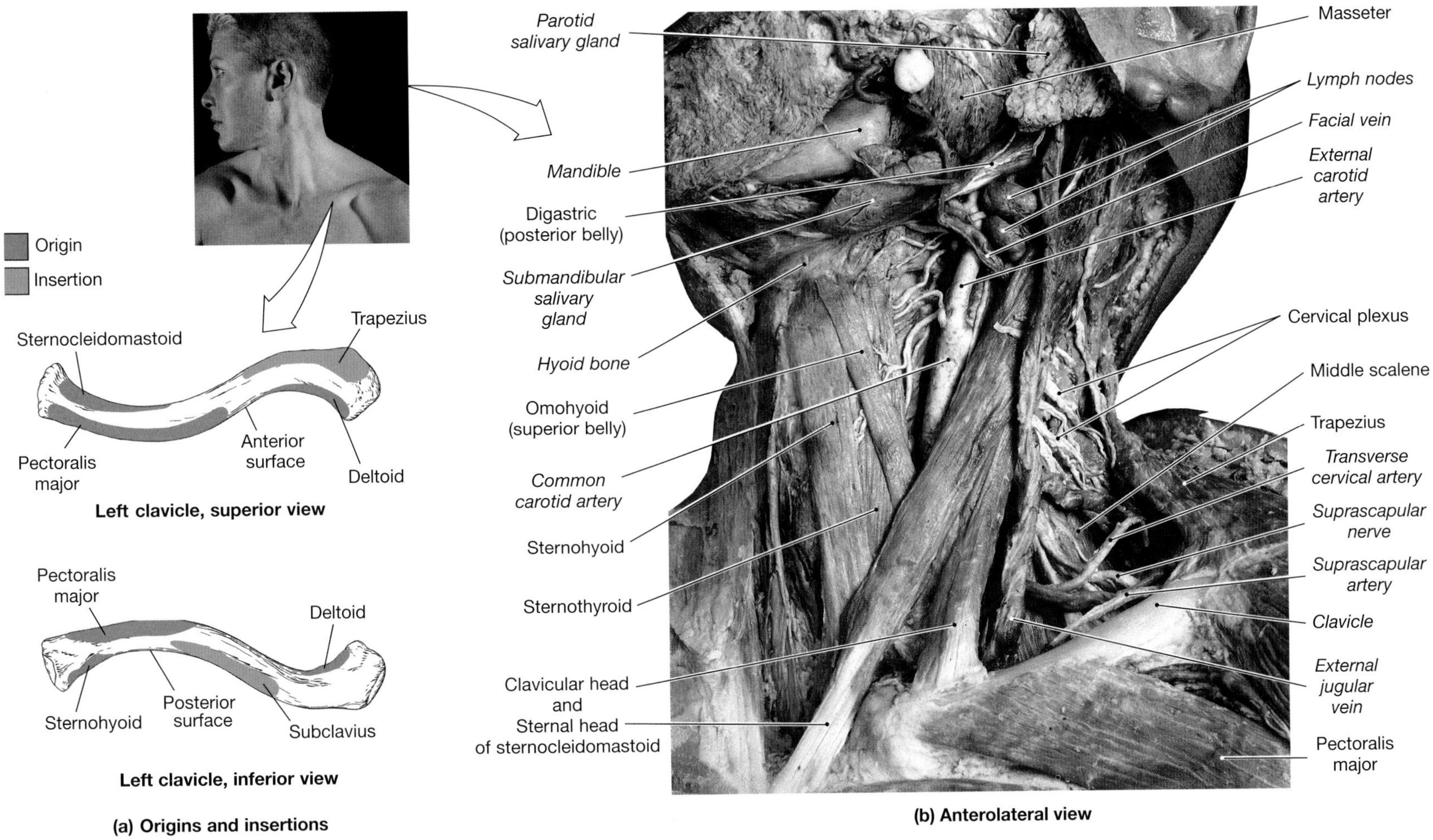

FIGURE 10.11 **Anterior Muscles of the Neck, Part II**

(**a**) Origins and insertions on the clavicle. (**b**) An anterolateral view of a dissection of the neck, showing neck muscles and adjacent structures. *See also Figures 6.17 and 7.3.*

TABLE 10.6 **Anterior Muscles of the Neck**

Muscle	Origin	Insertion	Action	Innervation
Digastric		Hyoid bone	Depresses mandible, opening mouth, and/or elevates larynx	
Anterior belly	From inferior surface of mandible at chin			Facial nerve (N VII)
Posterior belly	From mastoid region of temporal bone			Trigeminal nerve (N V), mandibular branch
Geniohyoid	Medial surface of mandible at chin	Hyoid bone	As above and retracts hyoid bone	Cervical nerve C_1 via hypoglossal nerve (N XIII)
Mylohyoid	Mylohyoid line of mandible	Median connective tissue band (raphe) that runs to hyoid bone	Elevates floor of mouth, elevates hyoid bone, and/or depresses mandible	Trigeminal nerve (N V), mandibular branch
Omohyoid	Central tendon attaches to clavicle and first rib	Two bellies: superior attaches to hyoid bone; inferior to superior margin of scapula	Depresses hyoid bone and larynx	Cervical spinal nerves C_2–C_3
Sternohyoid	Clavicle and manubrium	Hyoid bone	As above	Cervical spinal nerves C_1–C_3
Sternothyroid	Dorsal surface of manubrium and first costal cartilage	Thyroid cartilage of larynx	As above	As above
Stylohyoid	Styloid process of temporal bone	Hyoid bone	Elevates larynx	Facial nerve (N VII)
Thyrohyoid	Thyroid cartilage of larynx	Hyoid bone	Elevates larynx, depresses hyoid bone	Cervical spinal nerves C_1–C_2 via hypoglossal nerve (N XII)
Sternocleidomastoid		Mastoid region of skull and lateral portion of superior nuchal line	Together, they flex the neck; alone, one side bends neck toward shoulder and turns face to opposite side	Accessory nerve (N XI) and cervical spinal nerves (C_2–C_3) of cervical plexus
Clavicular head	Attaches to sternal end of clavicle			
Sternal head	Attaches to manubrium			

Muscles of the Vertebral Column [FIGURES 10.2/10.12 AND TABLE 10.7]

The muscles of the vertebral column (Figure 10.12● and Table 10.7) are covered by the more superficial back muscles, such as the trapezius and latissimus dorsi (see Figure 10.2●, p. 268). The spinal extensors, or **erector spinae**, extend the vertebral column. The names of the individual muscles within these groups provide useful information about their insertions. For example, a muscle with the name *capitis* inserts on the skull, whereas *cervicis* indicates an insertion on the upper cervical vertebrae and *thoracis* an insertion on the lower cervical and upper thoracic vertebrae. The erector spinae are subdivided into **spinalis**, **longissimus**, and **iliocostalis** muscle groups (Figure 10.12a●). These divisions are based on proximity to the vertebral column, with the spinalis group being the closest and the iliocostalis the farthest away. In the inferior lumbar and sacral regions, the distinction between the longissimus and iliocostalis muscles becomes indistinct. When contracting together, the erector spinae extend the vertebral column. When the muscles on only one side contract, there is lateral flexion of the vertebral column.

Deep to the spinalis muscles, deep muscles of the spine interconnect and stabilize the vertebrae. These muscles, sometimes called the *transversospinalis muscles*, include the **semispinalis group**, and the **multifidus**, **rotatores**, **interspinales**, and **intertransversarii muscles** (Figure 10.12a,b●). These are all relatively short muscles that work in various combinations to produce slight extension or rotation of the vertebral column. They are also important in making delicate adjustments in the positions of individual vertebrae and stabilizing adjacent vertebrae. If injured, these muscles can start a cycle of pain → muscle stimulation → contraction → pain. This can lead to pressure on adjacent spinal nerves, leading to sensory losses as well as limiting mobility. Many of the warmup and stretching exercises recommended before athletic events are intended to prepare these small but very important muscles for their supporting roles.

The muscles of the vertebral column include many dorsal extensors but few ventral flexors. The vertebral column does not need a massive series of flexor muscles because (1) many of the large trunk muscles flex the vertebral column when they contract, and (2) most of the body weight lies anterior to the vertebral column, and gravity tends to flex the spine. However, there are a few spinal flexors associated with the anterior surface of the vertebral column. In the neck (Figure 10.12c●) the **longus capitis** and the **longus colli** rotate or flex the neck, depending on whether the muscles of one or both sides are contracting. In the lumbar region, the large **quadratus lumborum** muscles flex the vertebral column and depress the ribs.

Oblique and Rectus Muscles (FIGURES 10.12 TO 10.14 AND TABLE 10.8)

The muscles of the oblique and rectus groups (Figures 10.12 to 10.14● and Table 10.8) lie between the vertebral column and the ventral midline. The oblique muscles can compress underlying structures or rotate the spinal column, depending on whether one or both sides are contracting. The rectus muscles are important flexors of the spinal column, acting in opposition to the erector spinae. The oblique and rectus muscles of the trunk and the diaphragm that subdivides the ventral body cavity are united by their common embryological origins. The oblique and rectus muscles can be divided into cervical, thoracic, and abdominal groups.

The oblique group includes the **scalene** (SKĀ-lēn) **muscles** of the cervical region and the **intercostal** (in-ter-KOS-tul) and **transversus muscles** of the thoracic region. In the neck, the *anterior, middle*, and *posterior scalene muscles* elevate the first two ribs and assist in flexion of the neck (Figures 10.12a,c●). In the thorax, the oblique muscles lie between the ribs, and the **external intercostal muscles** cover the **internal intercostal muscles** (Figure 10.13a●). Both intercostal muscles are important in respiratory movements of the ribs. A small **transversus thoracis muscle** crosses the inner surface of the rib cage and is covered by the serous membrane (*pleura*) that lines the pleural cavities.

In the abdomen, the same basic pattern of musculature extends unbroken across the abdominopelvic surface. The cross-directional arrangement of muscle fibers in these muscles strengthens the wall of the abdomen. These muscles are the **external** and **internal oblique muscles** (also called the *abdominal obliques*), the **transversus abdominis muscles** (ab-DOM-i-nus), and the **rectus abdominis muscle** (Figure 10.13a,d●). An excellent way to observe the relationship of these muscles is to view them in horizontal section (Figure 10.13b●). The rectus abdominis muscle begins at the xiphoid process and ends near the pubic symphysis. This muscle is divided longitudinally by a median collagenous partition, the **linea alba** (white line). The transverse **tendinous inscriptions** are bands of fibrous tissue that divide this muscle into four repeated segments (Figure 10.13a,d●). The surface anatomy of the oblique and rectus muscles of the thorax and abdomen is shown in Figure 10.13c●.

THE DIAPHRAGM (FIGURE 10.14)

The term *diaphragm* refers to any muscular sheet that forms a wall. When used without a modifier, however, the **diaphragm**, or *diaphragmatic muscle*, specifies the muscular partition that separates the abdominopelvic and thoracic cavities (Figure 10.14●). The diaphragm is a major muscle of respiration: Its contraction increases the volume of the thoracic cavity to promote inspiration; its relaxation decreases the volume to facilitate expiration (the muscles of respiration will be examined in Chapter 24). † *Hernias p. 791*

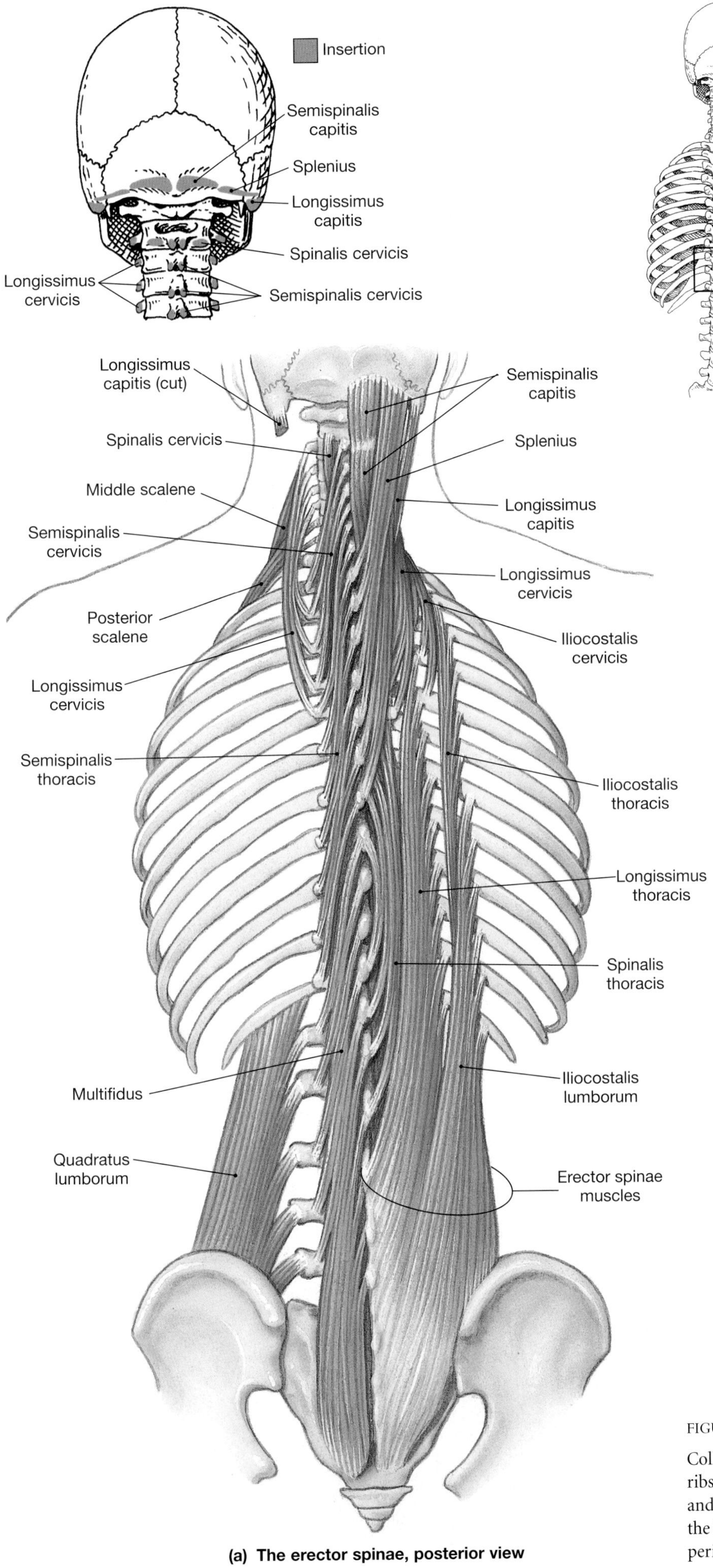

(a) The erector spinae, posterior view

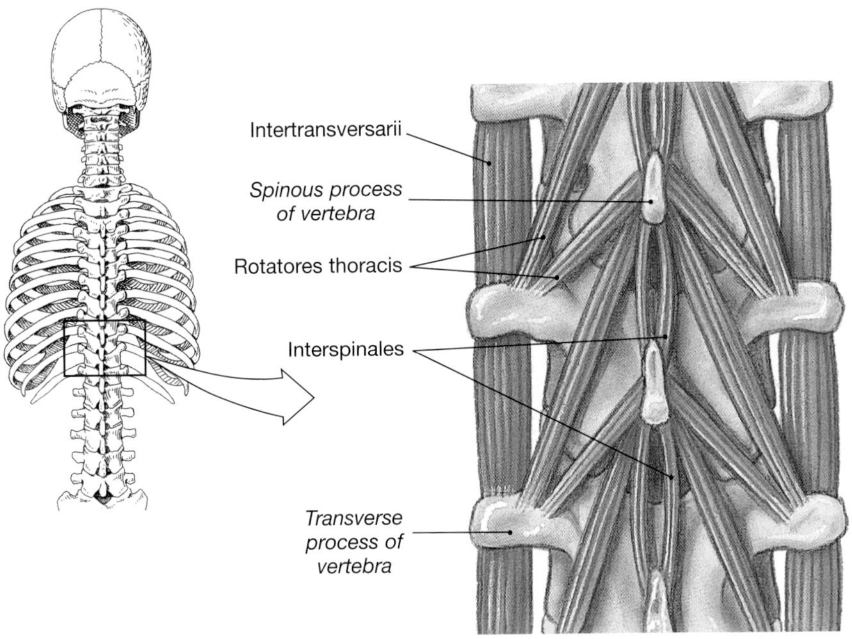

(b) Intervertebral muscles, posterior view

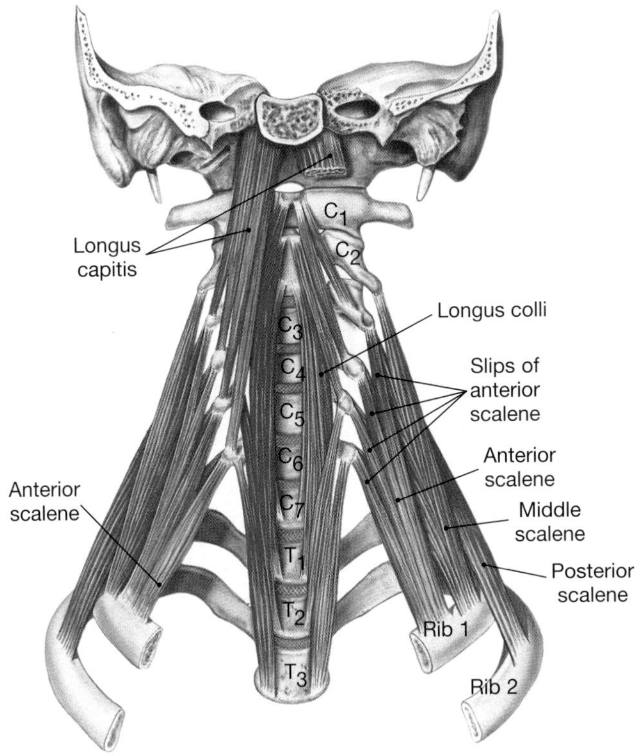

(c) Muscles on the anterior surfaces of the cervical and upper thoracic vertebrae

FIGURE 10.12 **MUSCLES OF THE VERTEBRAL COLUMN**

Collectively, these muscles adjust the position of the vertebral column, head, neck, and ribs. Selected origins and insertions are shown. **(a)** Posterior view of superficial (right) and deep (left) muscles collectively known as the erector spinae. **(b)** Posterior view of the intervertebral muscles. **(c)** Muscles on the anterior surfaces of the cervical and superior thoracic vertebrae. *See also Figures 6.19, 6.20, and 6.26c.*

TABLE 10.7

MUSCLES OF THE VERTEBRAL COLUMN

Group/Muscle	Origin	Insertion	Action	Innervation
SUPERFICIAL LAYER				
Splenius (Splenius capitis, splenius cervicis)	Spinous processes and ligaments connecting inferior cervical and superior thoracic vertebrae	Mastoid process, occipital bone of skull, superior cervical vertebrae	The two sides act together to extend neck; either alone rotates and laterally flexes neck to that side	Cervical spinal nerves
Erector Spinae				
SPINALIS GROUP				
Spinalis cervicis	Inferior portion of ligamentum nuchae and spinous process of C_7	Spinous process of axis	Extends neck	As above
Spinalis thoracis	Spinous processes of inferior thoracic and superior lumbar vertebrae	Spinous processes of superior thoracic vertebrae	Extends vertebral column	Thoracic and lumbar spinal nerves
LONGISSIMUS GROUP				
Longissimus capitis	Transverse processes of inferior cervical and superior thoracic vertebrae	Mastoid process of temporal bone	The two sides act together to extend neck; either alone rotates and laterally flexes neck to that side	Cervical and thoracic spinal nerves
Longissimus cervicis	Transverse processes of superior thoracic vertebrae	Transverse processes of middle and superior cervical vertebrae	As above	As above
Longissimus thoracis	Broad aponeurosis and at transverse processes of inferior thoracic and superior lumbar vertebrae; joins iliocostalis	Transverse processes of superior thoracic and lumbar vertebrae and inferior surfaces of lower 10 ribs	Extension of vertebral column; alone, each produce lateral flexion to that side	Thoracic and lumbar spinal nerves
ILIOCOSTALIS GROUP				
Iliocostalis cervicis	Superior borders of vertebrosternal ribs near the angles	Transverse processes of middle and inferior cervical vertebrae	Extends or laterally flexes neck, elevates ribs	Cervical and superior thoracic spinal nerves
Iliocostalis thoracis	Superior borders of ribs 6-12 medial to the angles	Superior ribs and transverse processes of last cervical vertebra	Stabilizes thoracic vertebrae in extension	Thoracic spinal nerves
Iliocostalis lumborum	Iliac crest, sacral crests, and lumbar spinous processes	Inferior surfaces of ribs 6-12 near their angles	Extends vertebral column, depresses ribs	Inferior thoracic nerves and lumbar spinal nerves
DEEP LAYER				
Transversospinalis				
SEMISPINALIS GROUP				
Semispinalis capitis	Processes of inferior cervical and superior thoracic vertebrae	Occipital bone, between nuchal lines	Together the two sides extend neck; alone each extends and laterally flexes neck	Cervical spinal nerves
Semispinalis cervicis	Transverse processes of T_1–T_5 or T_6	Spinous processes of C_2–C_5	Extends vertebral column and rotates toward opposite side	As above
Semispinalis thoracis	Transverse processes of T_6–T_{10}	Spinous processes of C_5–T_4	As above	Thoracic spinal nerves
Multifidus	Sacrum and transverse process of each vertebra	Spinous processes of the third or fourth more-superior vertebra	As above	Cervical, thoracic, and lumbar spinal nerves
Rotatores (cervicis, thoracis, and lumborum)	From the articular processes of cervical, transverse processes of thoracic, and mamillary processes of lumbar vertebrae	Spinous process of adjacent, more superior vertebra	As above	As above
Interspinales	Spinous process of each vertebra	Spinous processes of more superior vertebra	Extends vertebral column	As above
Intertransversarii	Transverse processes of each vertebra	Transverse process of more superior vertebra	Lateral flexion of vertebral column	As above
SPINAL FLEXORS				
Longus capitis	Transverse processes of cervical vertebrae	Base of the occipital bone	Together the two sides flex the neck; alone each rotates head to that side	Cervical spinal nerves
Longus colli	Anterior surfaces of cervical and superior thoracic vertebrae	Transverse processes of superior cervical vertebrae	Flexes and/or rotates neck; limits hyperextension	As above
Quadratus lumborum	Iliac crest and iliolumbar ligament	Last rib and transverse processes of lumbar vertebrae	Together they depress ribs; alone each side produces lateral flexion of vertebral column	Thoracic and lumbar spinal nerves

FIGURE 10.13 **The Oblique and Rectus Muscles**

Oblique muscles compress underlying structures between the vertebral column and the ventral midline; rectus muscles are flexors of the vertebral column. **(a)** Anterior view of the trunk, showing superficial and deep members of the oblique and rectus groups, and the sectional plane shown in part (b). **(b)** Diagrammatic horizontal section through the abdominal region. **(c)** Surface anatomy of the abdominal wall, anterior view. The *serratus anterior muscle*, seen in parts (a) and (c), is an appendicular muscle detailed in Chapter 11. **(d)** Cadaver, anterior superficial view of the abdominal wall. *See also Figures 6.19, 6.26, and 7.11.*

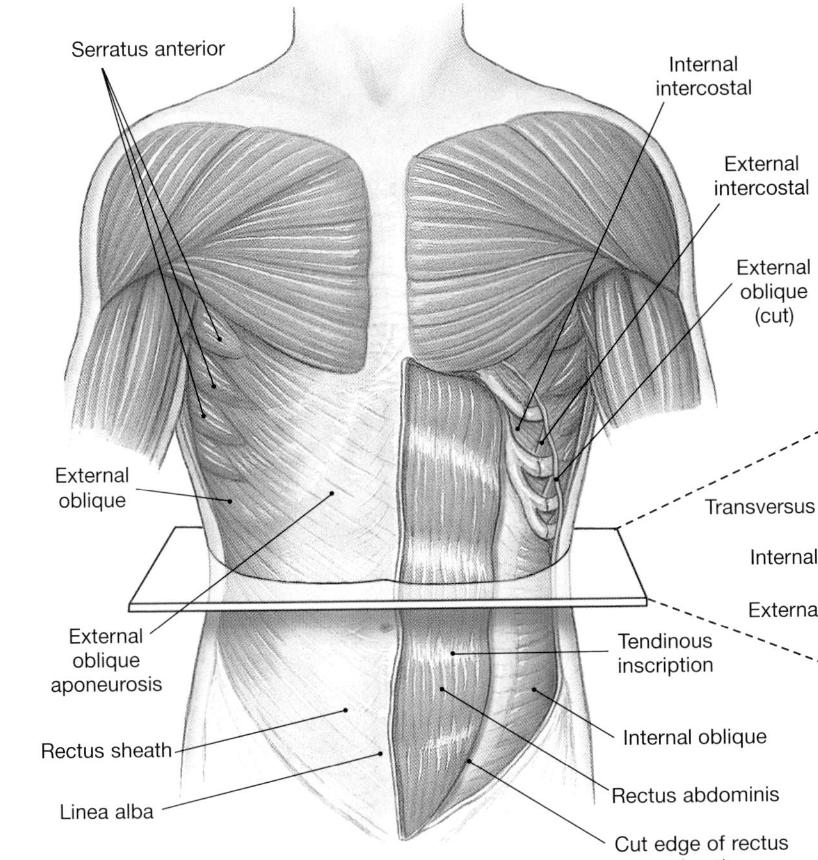

(a) Anterior view

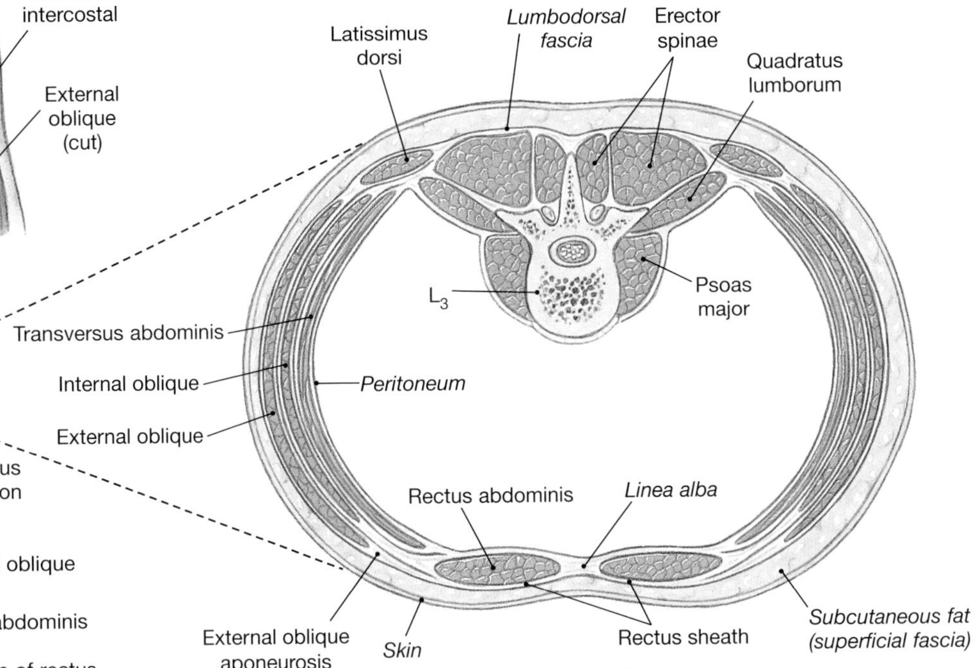

(b) Horizontal section view

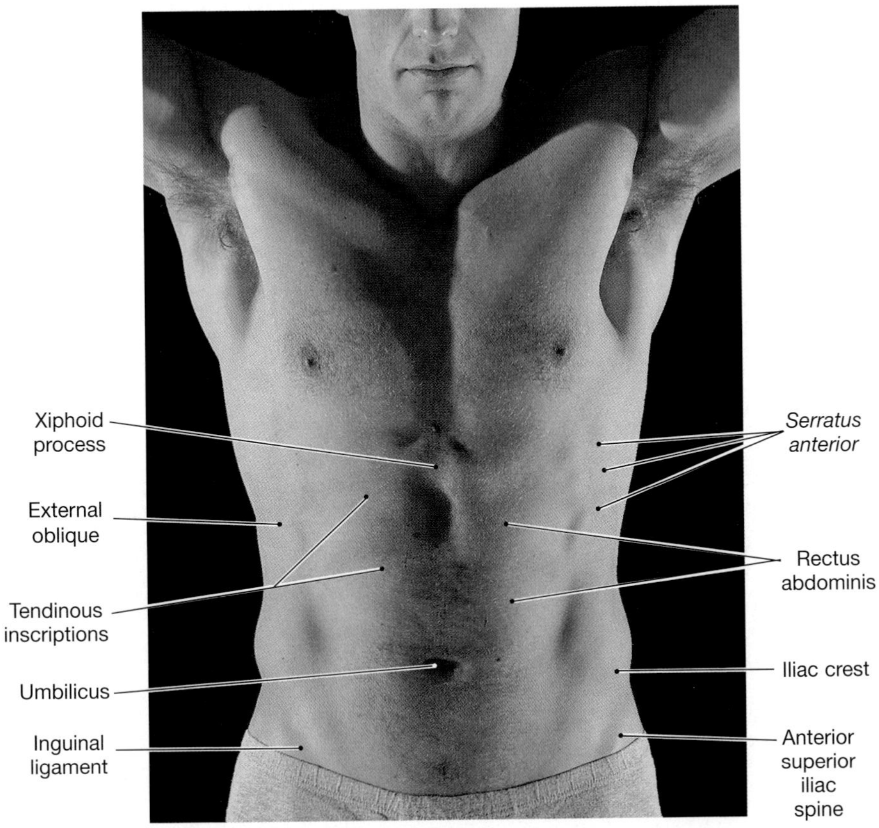

(c) Anterior view, surface anatomy

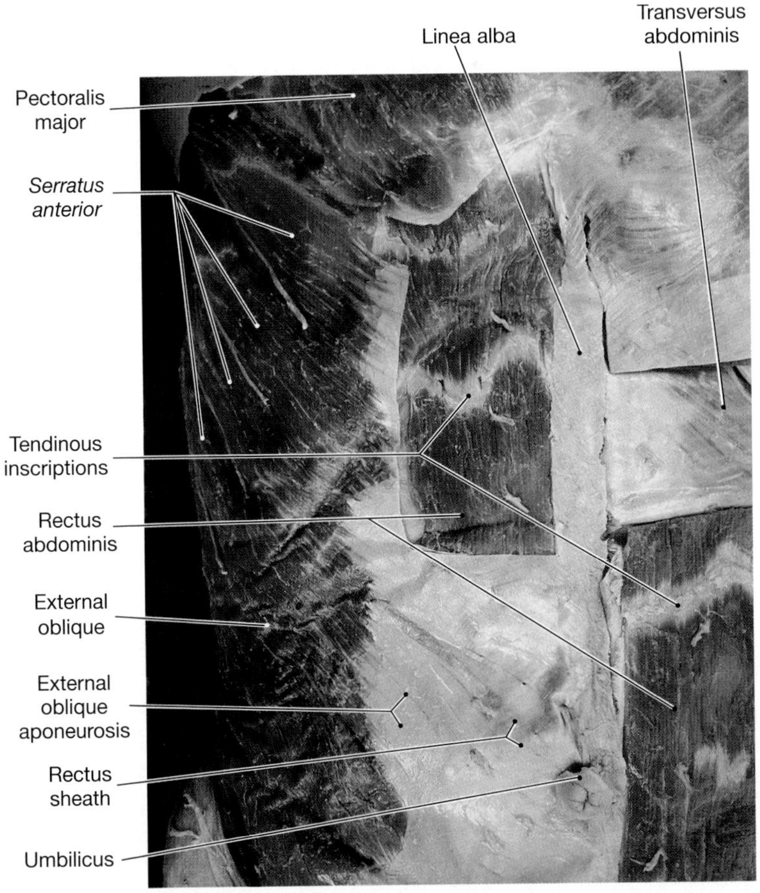

(d) Anterior view

FIGURE 10.14 **The Diaphragm**

This muscular sheet separates the thoracic cavity from the abdominopelvic cavity. (**a**) Inferior view. (**b**) Diagrammatic superior view. (**c**) Superior view of a transverse section through the thorax, with other organs removed to show the location and orientation of the diaphragm.

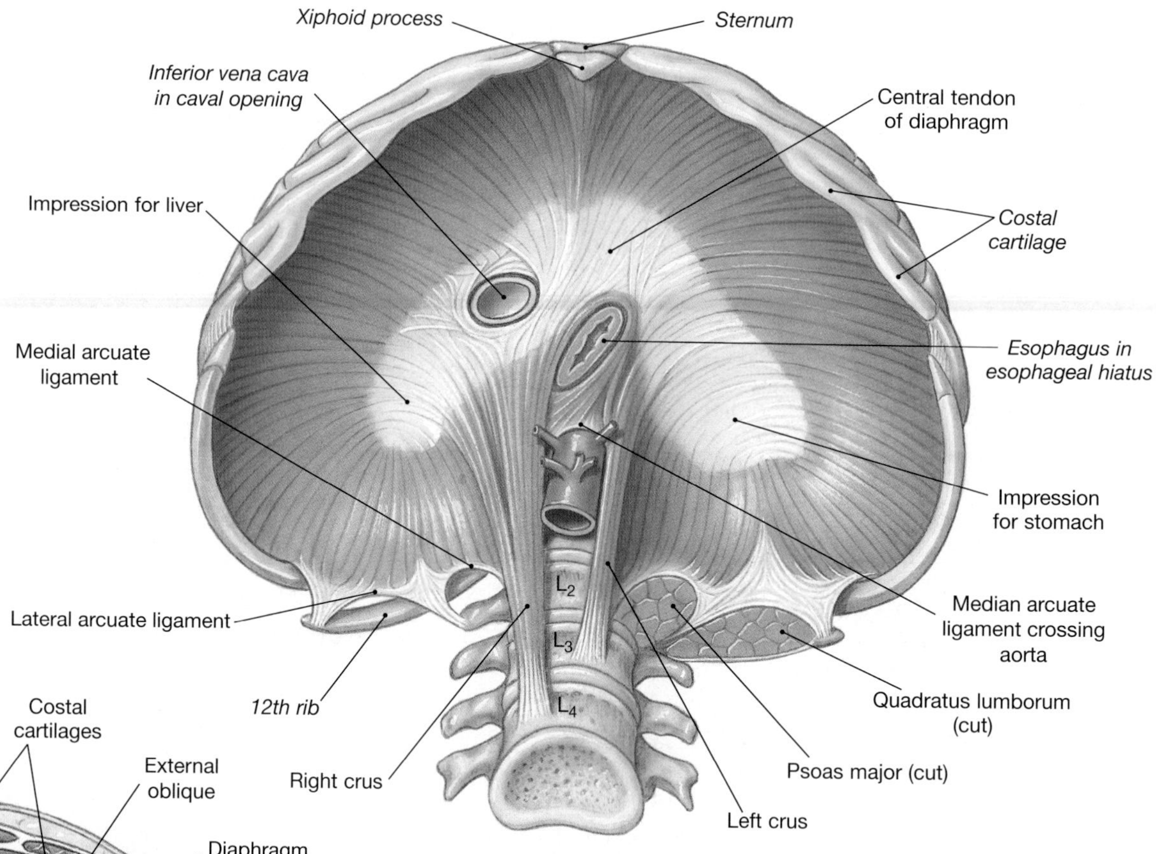

(a) Inferior view

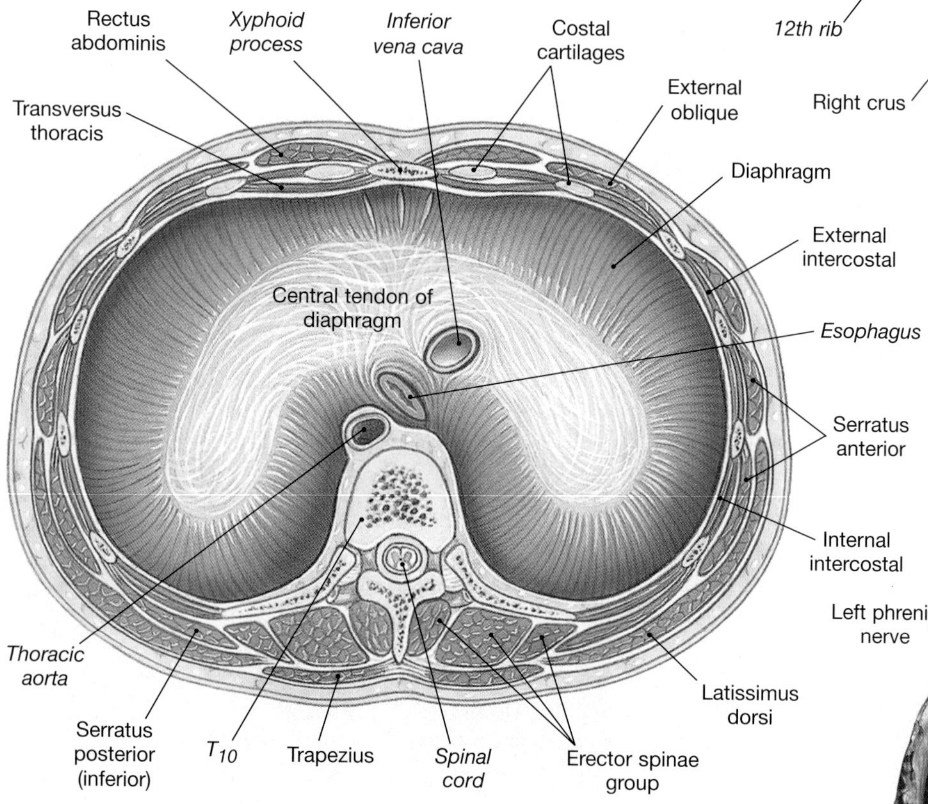

(b) Superior view

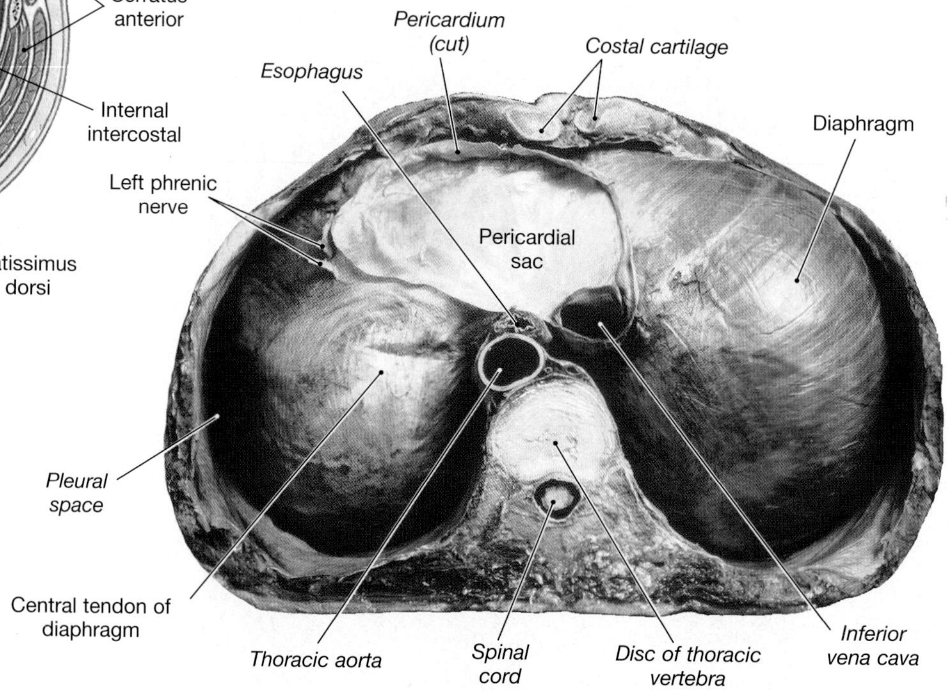

(c) Diaphragm, superior view

TABLE 10.8 OBLIQUE AND RECTUS MUSCLES

Group/Muscle	Origin	Insertion	Action	Innervation
OBLIQUE GROUP				
Cervical region				
Scalenes (anterior, middle, and posterior)	Transverse and costal processes of cervical vertebrae C_2–C_7	Superior surface of first two ribs	Elevate ribs, and/or flex neck; one side bends neck and rotates head and neck to opposite side	Cervical spinal nerves
Thoracic region				
External intercostals	Inferior border of each rib	Superior border of more-inferior rib	Elevate ribs	Intercostal nerves (branches of thoracic spinal nerves)
Internal intercostals	Superior border of each rib	Inferior border of the more-superior rib	Depress ribs	As above
Transverse thoracis	Posterior surface of sternum	Cartilages of ribs	As above	As above
Abdominal region				
External oblique	External and inferior borders of ribs 5–12	External oblique aponeuroses extending to linea alba and iliac crest	Compresses abdomen; depresses ribs; flexes, laterally flexes, or rotates vertebral column	Intercostal, iliohypogastric, iliohypogastric, and ilioinguinal nerves
Internal oblique	Lumbodorsal fascia and iliac crest	Inferior surfaces of ribs 9–12, costal cartilages 8–10, linea alba and pubis	As above	As above
Transverse abdominis	Cartilages of ribs 6–12, iliac crest, and lumbodorsal fascia	Linea alba and pubis	Compresses abdomen	As above
Serratus posterior				
superior	Spinous processes of C_7–T_3 and ligmentum nuchae	Superior borders of ribs 2–5 near angles	Elevates ribs, enlarges thoracic cavity	Thoracic nerves (T_1–T_4)
inferior	Aponeurosis from spinous processes of T_{10}–L_3	Inferior borders of ribs 8–12	Pulls ribs inferiorly; also pulls outward, opposing diaphragm	Thoracic nerves (T_9–T_{12})
RECTUS GROUP				
Cervical region	*See muscles in Table 10.6 (except sternocleidomastoid)*			
Thoracic region				
Diaphragm	Xiphoid process, ribs 7–12 and associated costal cartilages, and anterior surfaces of lumbar vertebrae	Central tendinous sheet	Contraction expands thoracic cavity, compresses abdominopelvic cavity	Phrenic nerves (C_3–C_5)
Abdominal region				
Rectus abdominis	Superior surface of pubis around symphysis	Inferior surfaces of cartilages (ribs 5–7) and xiphoid process of sternum	Depresses ribs, flexes vertebral column and compresses abdomen	Intercostal nerves (T_7–T_{12})

Muscles of the Pelvic Floor (FIGURE 10.15 AND TABLE 10.9)

The muscles of the pelvic floor extend from the sacrum and coccyx to the ischium and pubis. These muscles (1) support the organs of the pelvic cavity, (2) flex the joints of the sacrum and coccyx, and (3) control the movement of materials through the urethra and anus (Figure 10.15● and Table 10.9).

The boundaries of the **perineum** (the pelvic floor and associated structures) are established by the inferior margins of the pelvis. If you draw a line between the ischial tuberosities, you will divide the perineum into two triangles: an anterior or **urogenital triangle**, and a posterior or **anal triangle** (Figure 10.15b●). The superficial muscles of the anterior triangle are the muscles of the external genitalia. They overlie deeper muscles that strengthen the pelvic floor and encircle the urethra. These deep muscles constitute the **urogenital diaphragm**, a muscular layer that extends between the pubic bones.

An even more extensive muscular sheet, the **pelvic diaphragm**, forms the muscular foundation of the anal triangle. This layer extends anteriorly superior to the urogenital diaphragm as far as the pubic symphysis.

The urogenital and pelvic diaphragms do not completely close the pelvic outlet, because the urethra, vagina, and anus pass through them to open on the external surface. Muscular sphincters surround their openings and permit voluntary control of urination and defecation. Muscles, nerves, and blood vessels also pass through the pelvic outlet as they travel to or from the lower limbs.

✓ CONCEPT CHECK

- Damage to the external intercostal muscles would interfere with what important process?
- If someone hit you in your rectus abdominis muscle, how would your body position change?
- What is the function of the muscles of the pelvic floor?
- What is the function of the diaphragm?

FIGURE 10.15 **Muscles of the Pelvic Floor**

These muscles form the urogenital triangle and anal triangle to support organs of the pelvic cavity, flex the sacrum and coccyx, and control material movement through the urethra and anus. **(a)** Inferior view, female. **(b)** Inferior view, male. **(c)** Selected origins and insertions. *See also Figures 7.10 to 7.12.*

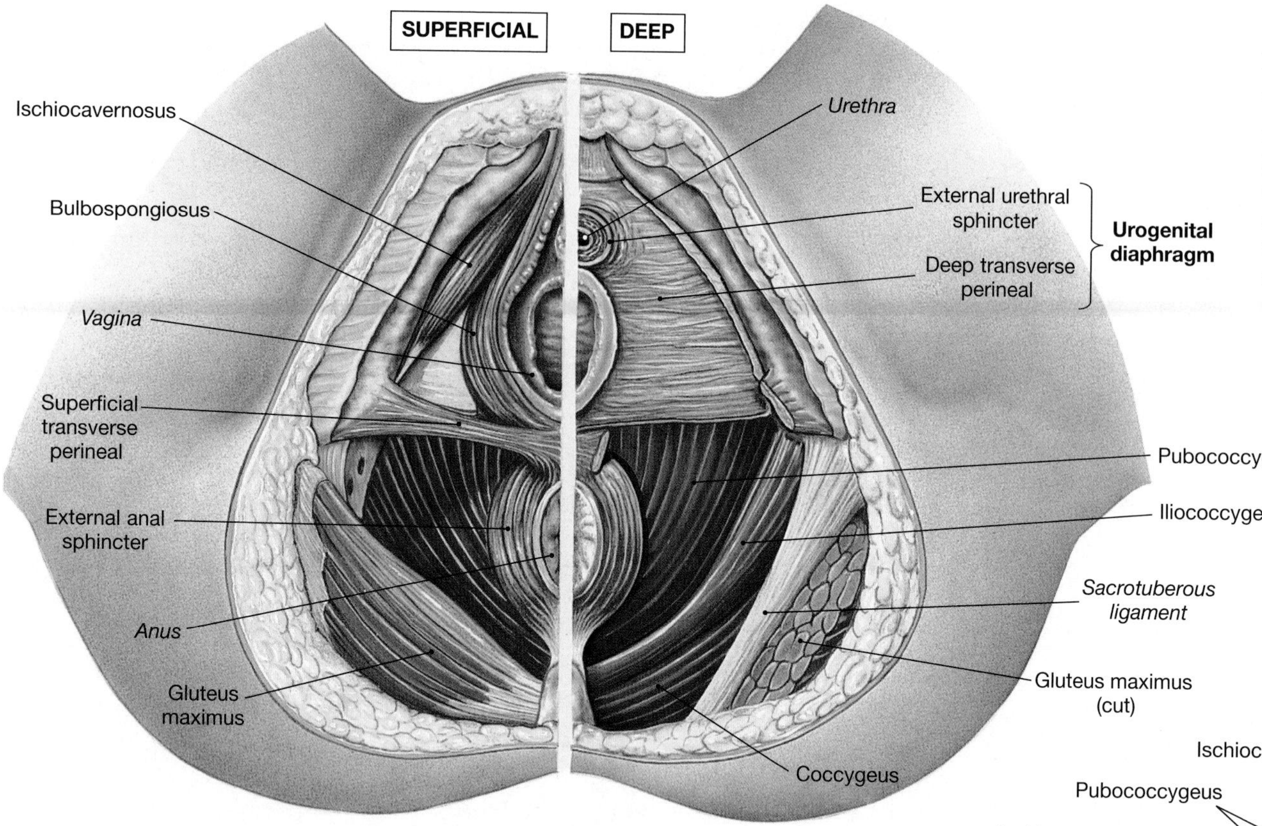

(a) Female

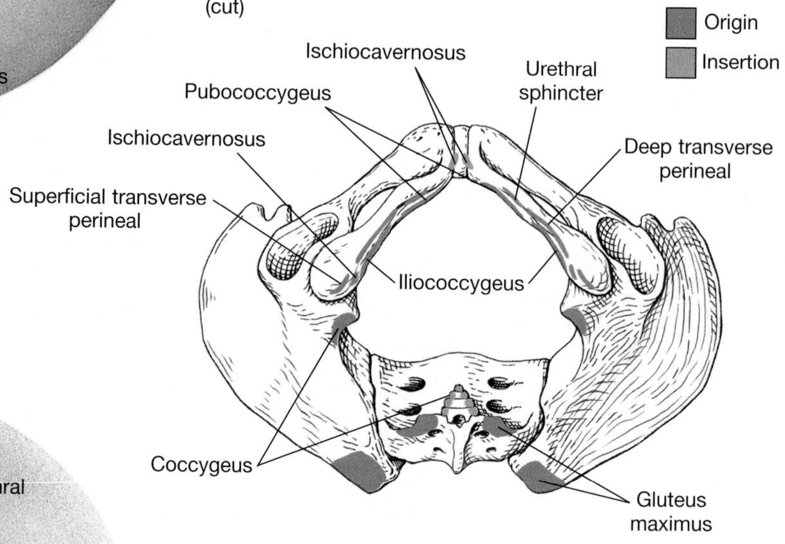

(c) Origins and insertions, inferior view of pelvis

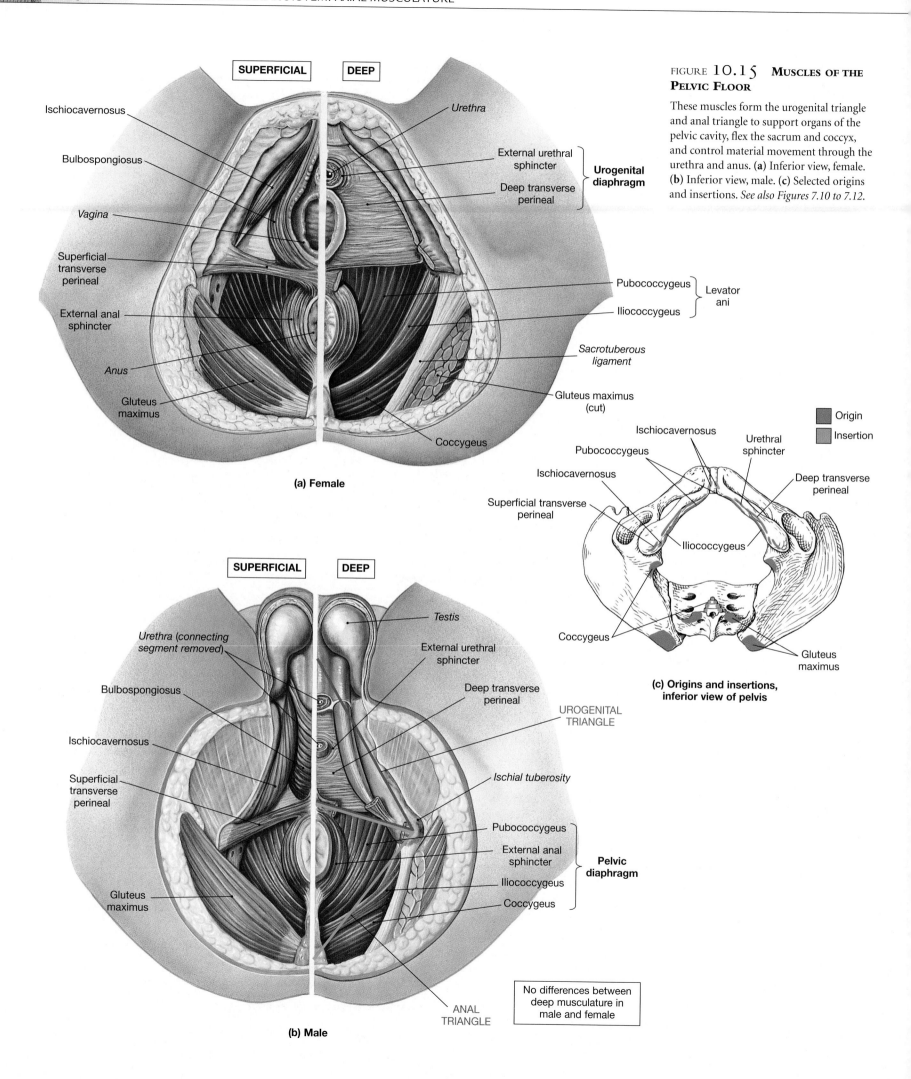

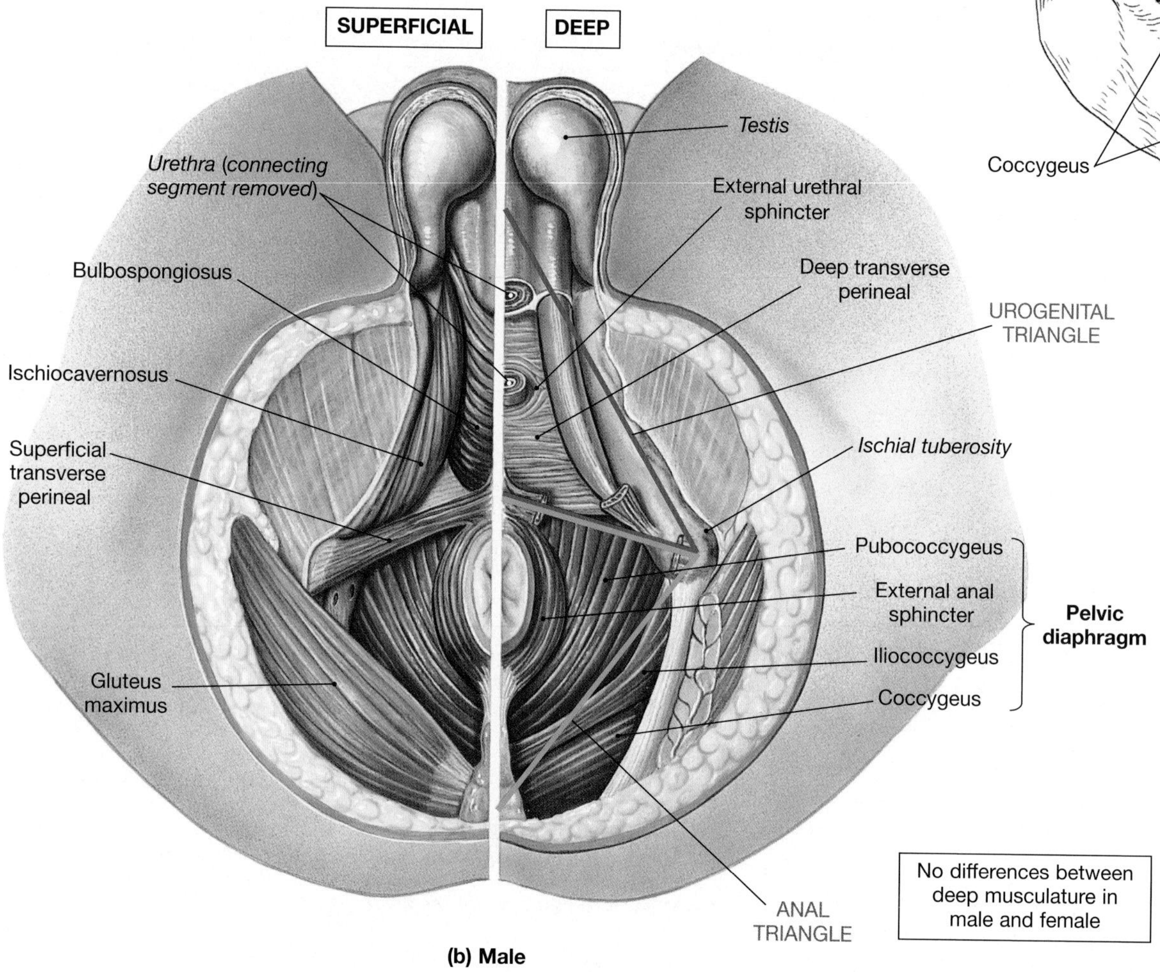

(b) Male

TABLE 10.9

MUSCLES OF THE PELVIC FLOOR

Group/Muscle	Origin	Insertion	Action	Innervation
UROGENITAL TRIANGLE				
Superficial muscles				
Bulbospongiosus				
Male	Collagen sheath at base of penis; fibers cross over urethra	Median raphe and central tendon of perineum	Compresses base, stiffens penis, ejects urine or semen	Pudendal nerve, perineal branch (S_2–S_4)
Female	Collagen sheath at base of clitoris; fibers run on either side of urethral and vaginal openings	Central tendon of perineum	Compresses and stiffens clitoris, narrows vaginal opening	As above
Ischiocavernosus	Ramus and tuberosity of ischium	Symphysis pubis anterior to base of penis or clitoris	Compresses and stiffens penis or clitoris, helping to maintain erection	As above
Superficial transverse perineal	Ischial ramus	Central tendon of perineum	Stabilizes central tendon of perineum	As above
Deep muscles: urogenital diaphragm				
Deep transverse perineal	Ischial ramus	Median raphe of urogenital diaphragm	As above	As above
Urethral sphincter:				
Male	Ischial and pubic rami	To median raphe at base of penis; inner fibers encircle urethra	Closes urethra; compresses prostate and bulbourethral glands	As above
Female	Ischial and pubic rami	To median raphe; inner fibers encircle urethra	Closes urethra; compresses vagina and greater vestibular glands	As above
ANAL TRIANGLE				
Pelvic diaphragm				
Coccygeus	Ischial spine	Lateral, inferior borders of the sacrum and coccyx	Flexes coccygeal joints	Inferior sacral nerves (S_4–S_5)
Levator ani:				
Iliococcygeus	Ischial spine, pubis	Coccyx and median raphe	Tenses floor of pelvis, supports pelvic organs, flexes coccygeal joints, elevates and retracts anus	Pudendal nerve (S_2–S_4)
Pubococcygeus	Inner margins of pubis	As above	As above	As above
External anal sphincter	Via tendon from coccyx	Encircles anal opening	Closes anal opening	Pudendal nerve; hemorrhoidal branch(S_2–S_4)

RELATED CLINICAL TERMS

diaphragmatic hernia (hiatal hernia): A hernia that occurs when abdominal organs slide into the thoracic cavity through an opening in the diaphragm. *Hernias p. 792*

hernia: A condition involving an organ or body part that protrudes through an abnormal opening. *Hernias p. 791*

inguinal hernia: A condition in which the inguinal canal enlarges and abdominal contents are forced into the inguinal canal. *Hernias p. 791*

Additional Clinical Term Discussed in Appendix I (p. 792)
esophageal hiatus

Near the head, mesoderm forms skeletal muscle associated with the *pharyngeal arches.*

Mesoderm from the parietal portion of the lateral plate and the adjacent myotome forms the *limb buds.*

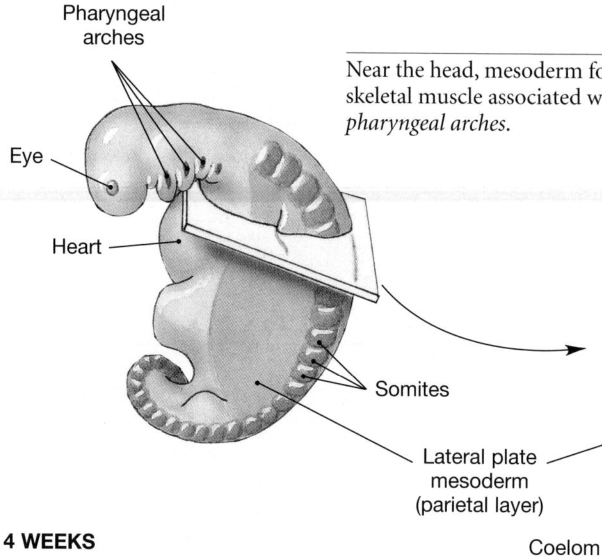

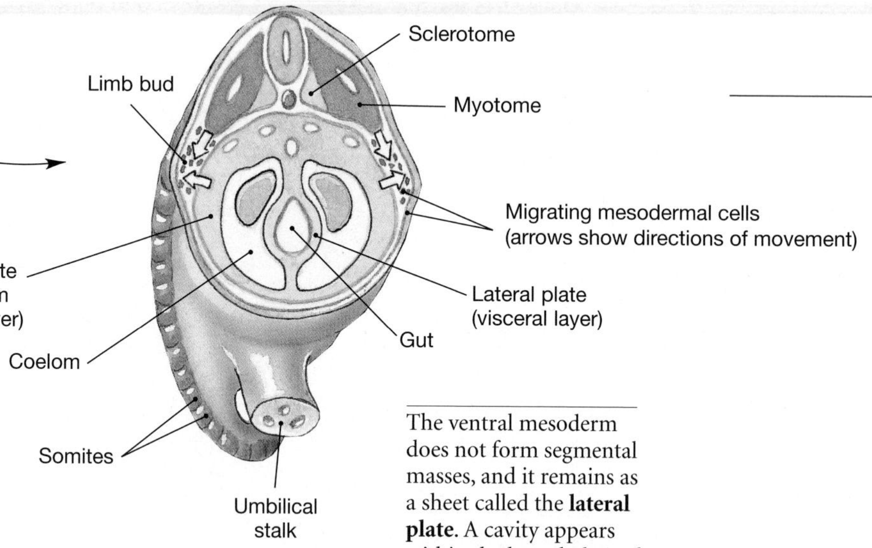

After 4 weeks of development, mesoderm on either side of the notochord has formed *somites*. The medial portion of each somite will form skeletal muscles; this region is called the **myotome**.

The ventral mesoderm does not form segmental masses, and it remains as a sheet called the **lateral plate**. A cavity appears within the lateral plate of the chest and abdomen; this cavity is the *coelom*. Formation of the coelom divides the lateral plate into an inner **visceral layer** and an outer **parietal layer**.

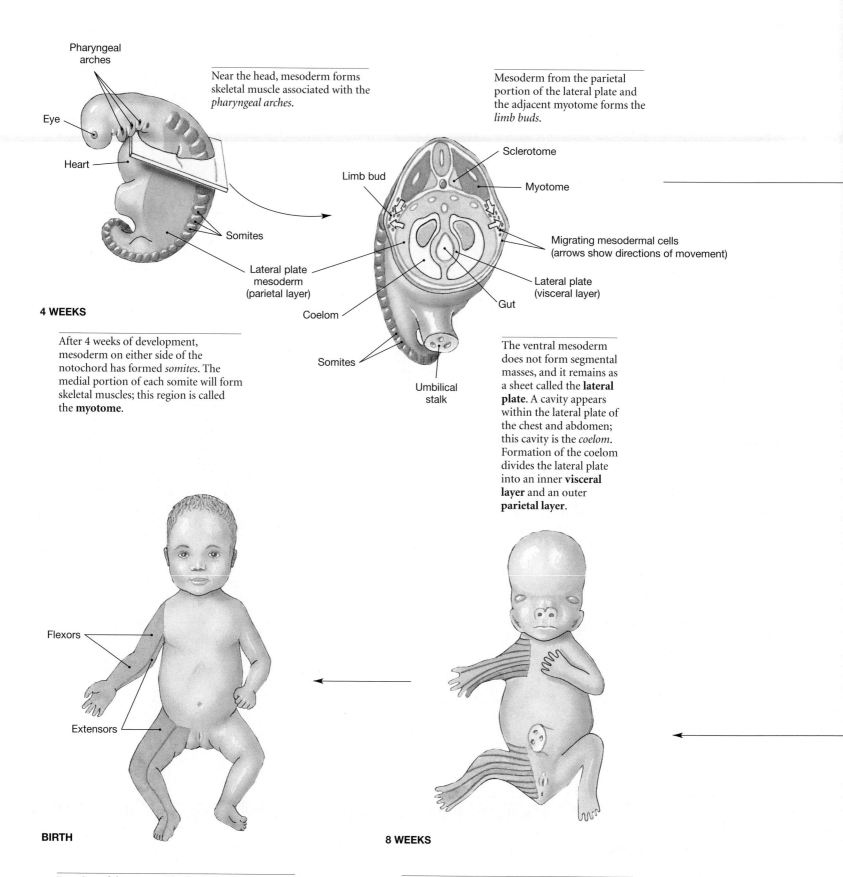

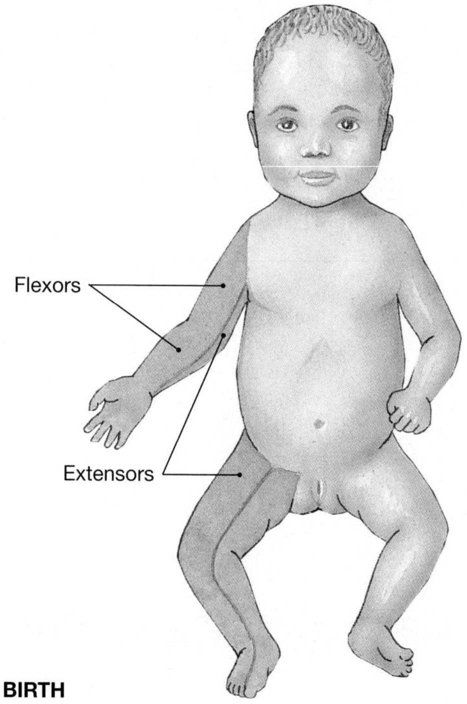

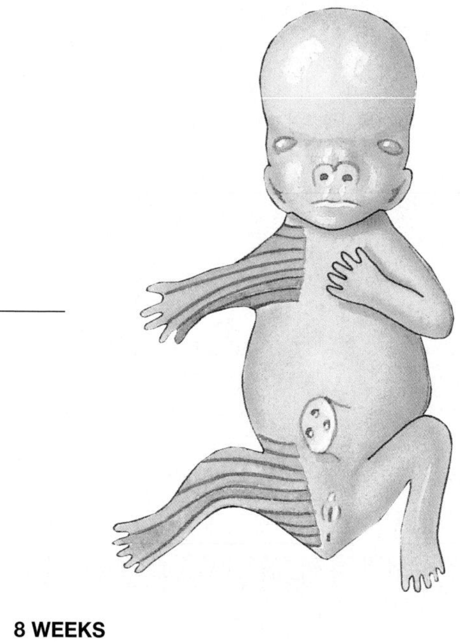

Rotation of the arm and leg buds produces a change in the position of these masses relative to the body axis.

While the limb buds enlarge, additional myoblasts invade the limb from myotomal segments nearby. Lines indicate the boundaries between myotomes providing myoblasts to the limb.

Hypaxial mesoderm in the trunk grows around the body wall toward the sternum in company with the ribs. This creates a mesodermal layer that extends from the chin to the pelvic girdle.

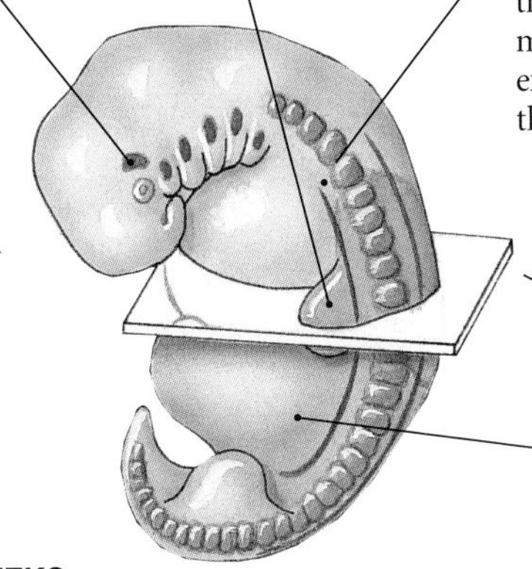

6 WEEKS

The hypaxial mesoderm near the sacrum migrates caudally to produce the **muscles of the pelvic floor**.

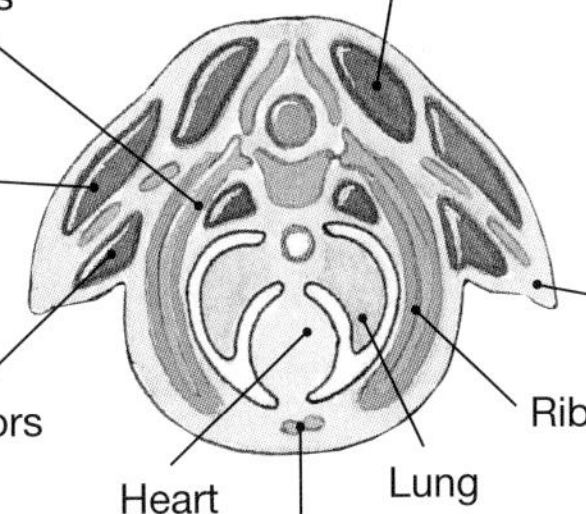

Each limb bud has a flattened distal tip, with a thickened **apical ridge**. As cartilages appear in the limb buds, surrounding mesodermal cells from the lateral plate and myotomes differentiate into *myoblasts*.

Myotomal muscles organize around the developing vertebral column in two groups, one dorsal (**epaxial muscles**) and the other ventral (**hypaxial muscles**).

Epaxial muscles remain arranged in segments. These deep muscles include the **intervertebral muscles**. Superficial epaxial muscles form the major muscles of the **erector spinae group**.

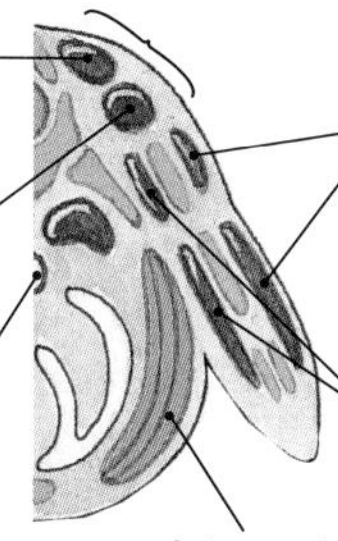

Muscles forming at the pharyngeal arches are associated with the head and neck. The **muscles of mastication** develop from the mesoderm surrounding the *mandibular arch*.

Mesoderm of the *hyoid* (second) *arch* migrates over the lateral and ventral surfaces of the neck and the surfaces of the skull to form the **muscles of facial expression.**

Mesoderm of the third, fourth and sixth pharyngeal arches form the pharyngeal and intrinsic laryngeal muscles.

Pharyngeal myoblasts form a superficial layer that later subdivides to create the *trapezius* and *sternocleidomastoid muscles*.

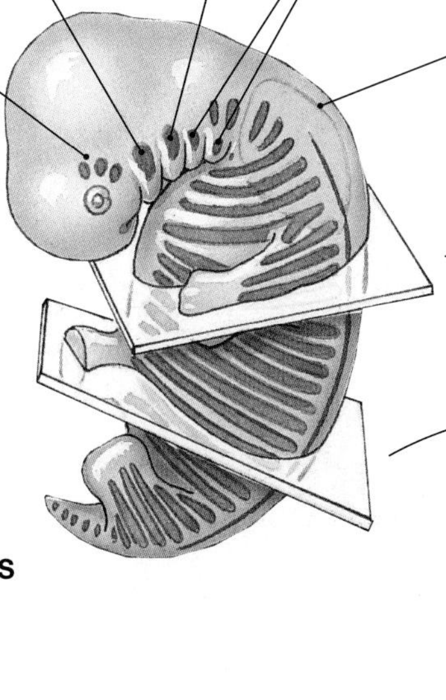

7 WEEKS

Migration of myoblasts over the dorsal surface of the trunk creates limb extensors; migration of ventral myoblasts produces the flexors.

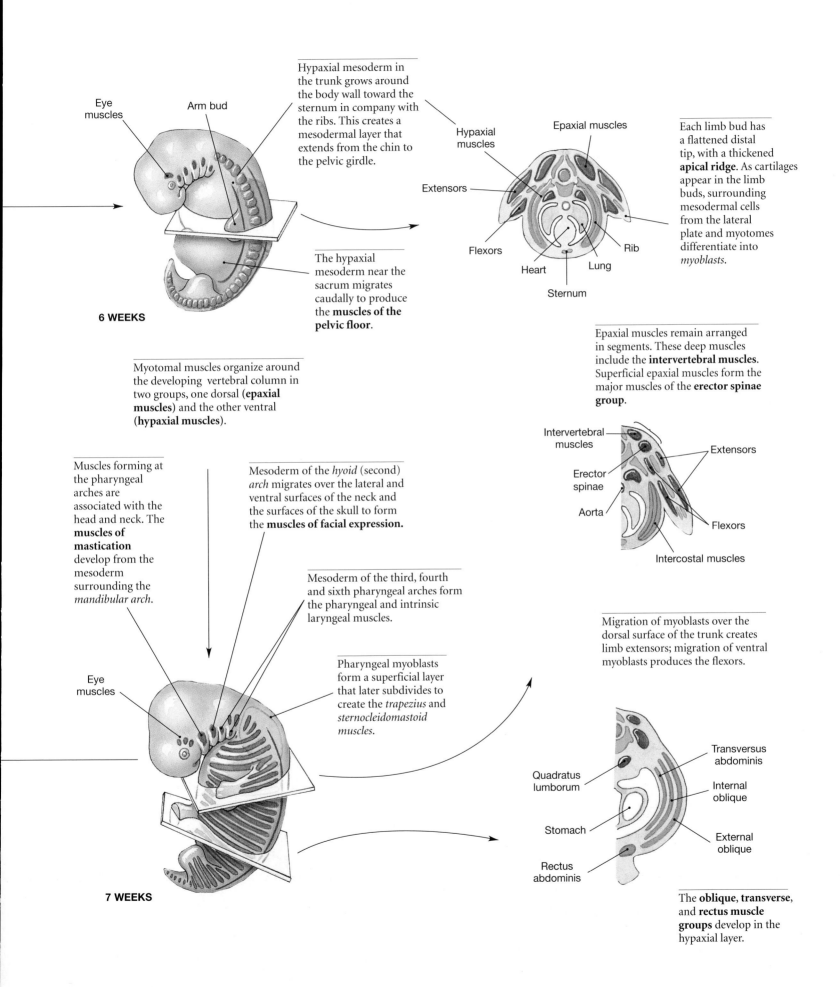

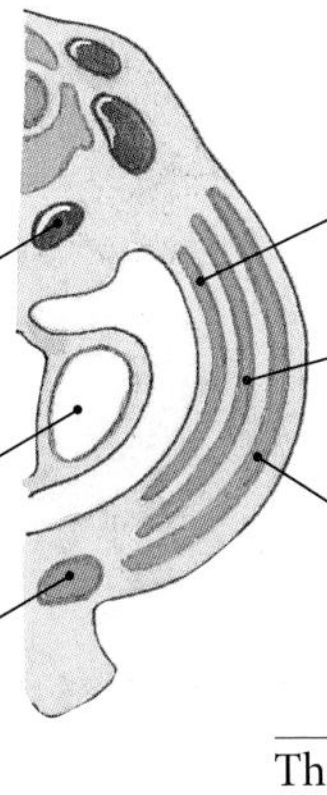

The **oblique**, **transverse**, and **rectus muscle groups** develop in the hypaxial layer.

STUDY OUTLINE & CHAPTER REVIEW

Introduction 267

1. The separation of the skeletal system into axial and appendicular divisions provides a useful guideline for subdividing the muscular system as well. The **axial musculature** arises from and inserts on the axial skeleton. It positions the head and spinal column and moves the rib cage, which assists in the movements that make breathing possible.

The Axial Musculature 267

1. The axial musculature originates and inserts on the axial skeleton; it positions the head and spinal column and moves the rib cage. The appendicular musculature stabilizes or moves components of the appendicular skeleton. *(see Figures 10.1/10.2)*
2. The axial muscles are organized into four groups based on their location and/or function. These groups are: (1) *muscles of the head and neck*, (2) *muscles of the vertebral column*, (3) *oblique* and *rectus muscles*, including the *diaphragm*, and (4) *muscles of the pelvic floor*.
3. Organization of muscles into the four groups includes descriptions of innervation. **Innervation** refers to the identity of the nerve that controls a given muscle, and is also included in all muscle tables.

Muscles of the Head and Neck 269

4. Muscles of the head and neck are divided into several groups: (1) the *muscles of facial expression*, (2) the *extrinsic eye muscles*, (3) the *muscles of mastication*, (4) the *muscles of the tongue*, (5) the *muscles of the pharynx*, and (6) the *muscles of the anterior neck*.
5. Muscles involved with sight and hearing are based on the skull.
6. The muscles of facial expression originate on the surface of the skull. The largest group is associated with the mouth. It includes: **orbicularis oris** and **buccinator**. The **frontal** and **occipital bellies** of the **occipitofrontalis** muscle control movements of eyebrows, forehead, and scalp. The **platysma** tenses skin of the neck and depresses the mandible. *(see Figures 10.3 to 10.6 and Table 10.1)*
7. The six **extra-ocular eye muscles** (*oculomotor muscles*) control eye position and movements. These muscles include: **inferior**, **lateral**, **medial**, and **superior recti**; and **superior** and **inferior obliques**. *(see Figure 10.5 and Table 10.2)*
8. The muscles of mastication (chewing) act on the mandible. They are: **masseter**, **temporalis**, **pterygoid** (*medial* and *lateral*) muscles. *(see Figure 10.6 and Table 10.3)*
9. The muscles of the tongue are necessary for speech and swallowing, and they assist in mastication. They have names that end in *-glossus*, meaning "tongue." These muscles are: the **genioglossus**, **hyoglossus**, **palatoglossus**, and **styloglossus**. *(see Figure 10.8 and Table 10.4)*
10. Muscles of the pharynx are important in the initiation of the swallowing process. These muscles include: the **pharyngeal constrictors**, the laryngeal elevators (**palatopharyngeus**, **salpingopharyngeus**, and **stylopharyngeus**), and **palatal muscles**, which raise the soft palate. *(see Figure 10.9 and Table 10.5)*
11. The anterior muscles of the neck control the position of the larynx, depress the mandible, and provide a foundation for the muscles of the tongue and pharynx. These include: the **digastric**, **mylohyoid**, **stylohyoid**, and **sternocleidomastoid**. *(see Figures 10.3/10.4/10.10/10.11 and Table 10.6)*

Muscles of the Vertebral Column 278

12. The muscles of the vertebral column are covered by a superficial layer of back muscles, such as the trapezius and latissimus dorsi. Underlying muscles of the spine form both superficial and deep layers. The superficial layer includes the **splenius muscles**. The spinal extensors are divided into the **spinalis**, **longissimus**, and **iliocostalis** groups. In the lower lumbar and sacral regions, the longissimus and iliocostalis form a single massive muscle. *(see Figure 10.12 and Table 10.7)*
13. Deep muscles of the spine interconnect and stabilize the vertebrae. These muscles, called the transversospinalis group, include: the **semispinalis group**, and the **multifidus**, **rotatores**, **interspinales**, and **intertransversarii**.
14. Other muscles of the vertebral column include the **longus capitis** and **longus colli**, which rotate and flex the neck, and the **quadratus lumborum muscles** in the lumbar region, which flex the spine and depress the ribs. *(see Figure 10.12 and Table 10.7)*

Oblique and Rectus Muscles 278

15. The oblique and rectus muscles lie between the vertebral column and the ventral midline. The abdominal oblique muscles (**external oblique** and **internal oblique muscles**) compress underlying structures or rotate the vertebral column; the **rectus abdominis muscle** is a flexor of the vertebral column.
16. The oblique muscles of the neck and thorax include the **scalenes**, the **intercostals**, and the **transversus** muscles. The **external intercostals** and **internal intercostals** are important in respiratory movements of the ribs. *(see Figure 10.13 and Table 10.8)*
17. The **diaphragm** (*diaphragmatic muscle*) is also important in respiration. It separates the abdominopelvic and thoracic cavities. *(see Figure 10.14)*

Muscles of the Pelvic Floor 283

18. Pelvic floor muscles extend from the sacrum and coccyx to the ischium and pubis. These muscles (1) support the organs of the pelvic cavity, (2) flex the joints of the sacrum and coccyx, and (3) control the movement of materials through the urethra and anus.
19. The **perineum** (the pelvic floor and associated structures) can be divided into an anterior, or **urogenital triangle** and a posterior, or **anal triangle**. The pelvic floor consists of the **urogenital diaphragm** and the **pelvic diaphragm**. *(see Figure 10.15 and Table 10.9)*

LEVEL 1 REVIEWING FACTS AND TERMS

Match each numbered item with the most closely related lettered item. Use letters for answers in the spaces provided.

Column A	Column B
____ 1. spinalis	a. compresses cheeks
____ 2. perineum	b. elevates larynx
____ 3. buccinator	c. tenses skin of neck
____ 4. extra-ocular	d. pelvic floor/associated structures
____ 5. intercostals	e. elevates mandible
____ 6. stylohyoid	f. move ribs
____ 7. inferior rectus	g. retracts tongue
____ 8. temporalis	h. extends neck
____ 9. platysma	i. eye muscles
____10. styloglossus	j. makes eye look down

11. Which of the following muscles compresses the abdomen?
(a) diaphragm (b) internal intercostal
(c) external oblique (d) rectus abdominis

12. The muscle that inserts on the pubis is the
(a) internal oblique (b) rectus abdominis
(c) transversus abdominis (d) scalene

13. The iliac crest is the origin of the
(a) quadratus lumborum (b) iliocostalis cervicis
(c) longissimus cervicis (d) splenius

14. Which of the following describes the action of the digastric muscle?
(a) elevates the larynx
(b) elevates the larynx and depresses the mandible
(c) depresses the larynx
(d) elevates the mandible

15. Which of the following muscles has its insertion on the cartilages of the ribs?
(a) diaphragm
(b) external intercostal
(c) transversus thoracis
(d) scalene

16. Some of the muscles of the tongue are innervated by the
(a) hypoglossal nerve (N XII) (b) trochlear nerve (N IV)
(c) abducens nerve (N VII) (d) b and c

17. Which of the following is not a spinal flexor?
(a) iliocostalis lumborum (b) longus capitis
(c) longus cervicis (d) quadratus lumborum

18. The muscular partition that separates the abdominopelvic and thoracic cavities is the
(a) masseter (b) perineum
(c) diaphragm (d) transversus abdominis

19. The scalenes have their origin on the
(a) transverse and costal processes of cervical vertebrae
(b) inferior border of the previous rib
(c) cartilages of the ribs
(d) lumbodorsal fascia and iliac crest

20. Which of the following is not a muscle in the urogenital triangle?
(a) ischiocavernosus (b) perineus group
(c) bulbospongiosus (d) coccygeus

LEVEL 2 REVIEWING CONCEPTS

1. During abdominal surgery, the surgeon makes a cut through the muscle directly to the right of the linea alba. The muscle that is being cut would be the
(a) digastric (b) external oblique
(c) rectus abdominis (d) scalenus

2. Ryan hears a loud noise and quickly raises his eyes to look upward in the direction of the sound. To accomplish this action, he must use his __________ muscles.
(a) superior rectus
(b) inferior rectus
(c) superior oblique
(d) lateral rectus

3. Which of the following muscles is not involved in the process of chewing or manipulating food in the mouth?
(a) masseter
(b) temporalis
(c) omohyoid
(d) pterygoid

4. Which of the following features are common to the muscles of mastication?
(a) they share innervation through the oculomotor nerve
(b) they are also muscles of facial expression
(c) they move the mandible at the temporomandibular joint
(d) they enable a person to smile

5. The muscles of the vertebral column include many dorsal extensors but few ventral flexors. Why?

6. What specific structural characteristic makes voluntary control of urination possible?

7. What is the effect of contraction of the internal oblique muscle?

8. What are the functions of the anterior muscles of the neck?

9. What is the function of the diaphragm? Why is it included in the axial musculature?

10. How do the muscles of the mandible, tongue, and pharynx work together in the chewing and swallowing of food?

LEVEL 3 CRITICAL THINKING AND CLINICAL APPLICATIONS

1. How do the muscles of the anal triangle control the functions of this area?

2. Mary sees Jill coming toward her and immediately contracts her frontalis and procerus muscles. Is Mary glad to see Jill? How can you tell?

3. What muscles are involved in controlling the position of the head on the vertebral column?

✓ ANSWERS TO CONCEPT CHECK QUESTIONS

p. 276 **1.** The muscles of facial expression originate on the surface of the skull. **2.** The muscles of mastication move the mandible at the temporomandibular joint during chewing. **3.** The contraction of the extra-ocular muscles causes the eye to look up, look down, rotate laterally, rotate medially, roll and look up to the side, or roll and look down to the side. **4.** The pharyngeal muscles are important in the initiation of swallowing.
p. 283 **1.** Damage to the external intercostal muscles would interfere with the process of breathing. **2.** A blow to the rectus abdominis muscle would cause the muscle to contract forcefully, resulting in flexion of the torso. In other words, you would "double up." **3.** The muscles of the pelvic floor have the following functions: (1) support the organs of the pelvic cavity; (2) flex the joints of the sacrum and coccyx; and (3) control the movement of materials through the urethra and anus. **4.** The diaphragm is a major muscle of respiration.

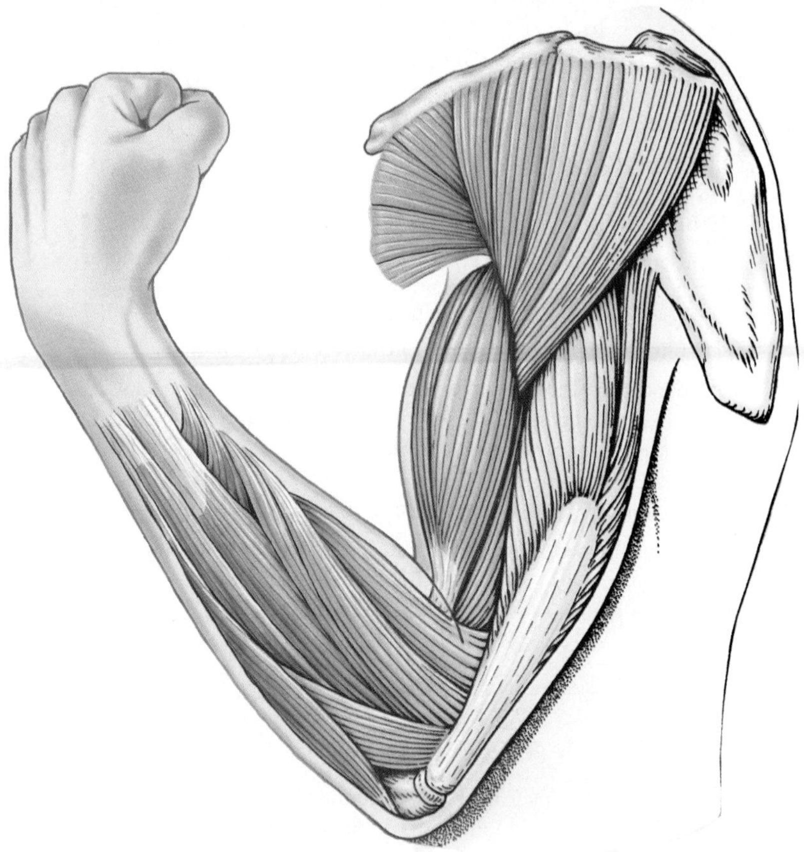

11

THE MUSCULAR SYSTEM

The Appendicular Musculature

In this chapter we will describe the **appendicular musculature**. These muscles are responsible for stabilizing the pectoral and pelvic girdles and for moving the upper and lower limbs. Appendicular muscles account for roughly 40% of the skeletal muscles in the body.

This discussion assumes an understanding of skeletal anatomy, and you may find it helpful to review figures in Chapters 6 and 7 as you proceed. The appropriate figures are referenced in the figure captions throughout this chapter as are related images appearing in the Companion Atlas.

There are two major groups of appendicular muscles: (1) the muscles of the pectoral girdle and upper limbs, and (2) the muscles of the pelvic girdle and lower limbs. The functions and required ranges of motion differ greatly between these groups. The muscular connections between the pectoral girdle and the axial skeleton increase upper limb mobility because the skeletal elements are not locked in position relative to the axial skeleton. The muscular connections also act as shock absorbers. For example, people can jog and still perform delicate hand movements because the appendicular muscles absorb the shocks and jolts, smoothing the bounces in their stride. In contrast, the pelvic girdle has evolved a strong skeletal connection to transfer weight from the axial to the appendicular skeleton. The emphasis is on strength rather than versatility, and the very features that strengthen the joints limit the range of movement.

Muscles of the Pectoral Girdle and Upper Limbs [FIGURES 11.1/11.4]

Muscles associated with the pectoral girdle and upper limbs can be divided into four groups: (1) *muscles that position the pectoral girdle*, (2) *muscles that move the arm*, (3) *muscles that move the forearm and hand*, and (4) *muscles that move the hand and fingers*. As we describe the various muscles of the pectoral girdle and upper limbs, refer first to Figure 11.1●, then to Figure 11.4● for the general location of the muscle under study.

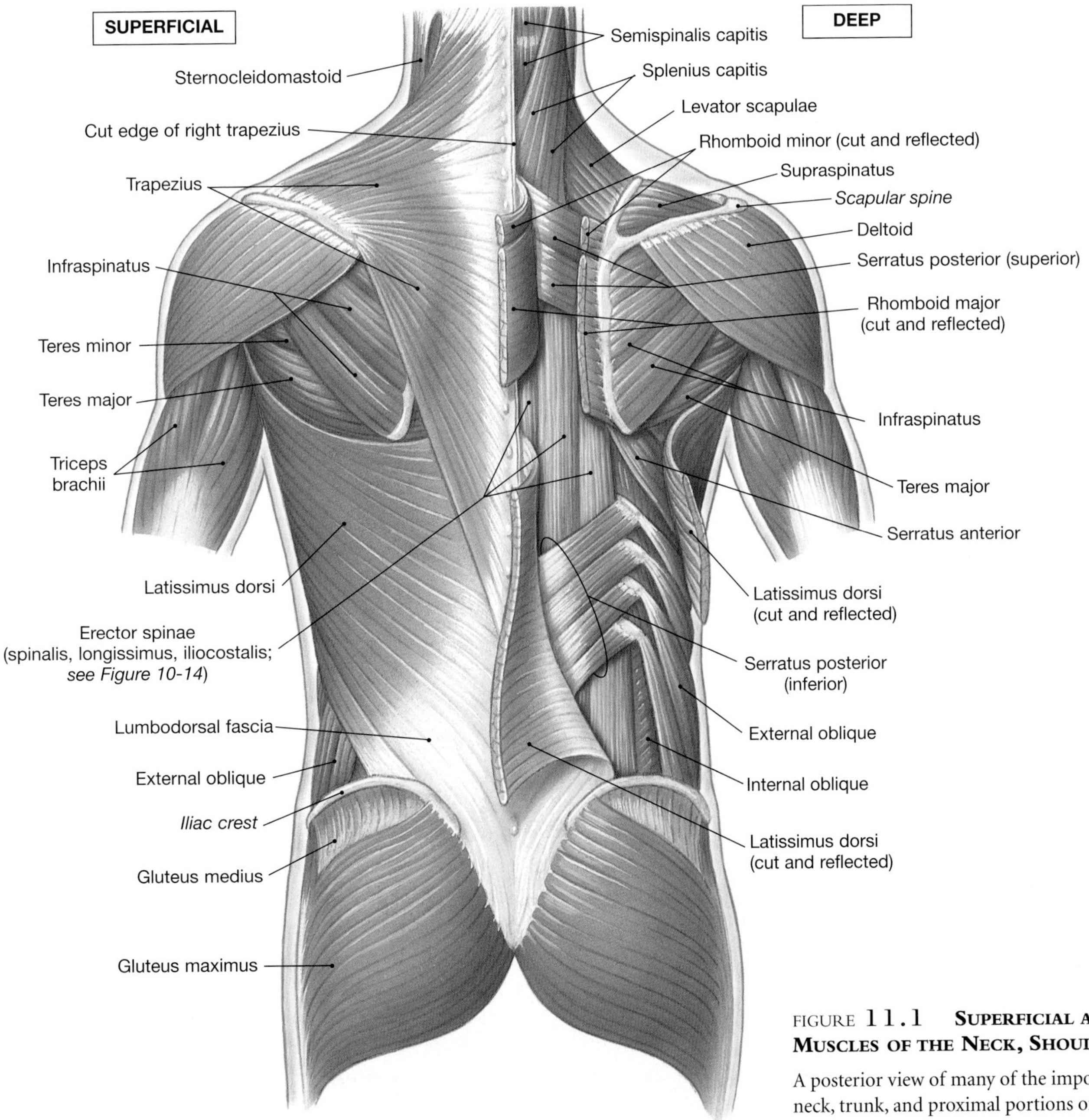

FIGURE 11.1 **SUPERFICIAL AND DEEP MUSCLES OF THE NECK, SHOULDER, AND BACK**

A posterior view of many of the important muscles of the neck, trunk, and proximal portions of the upper limbs.

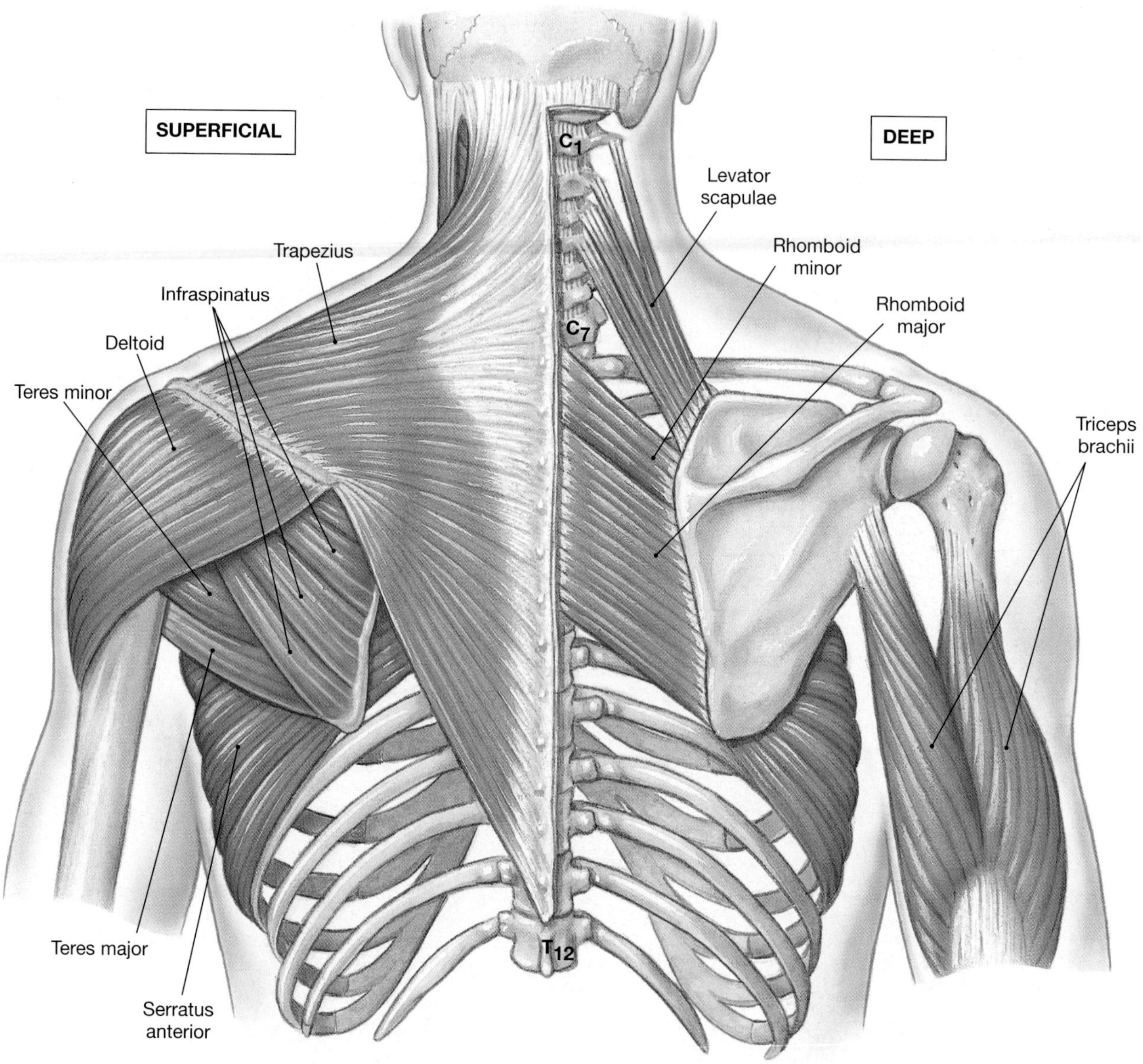

FIGURE 11.2 **MUSCLES THAT POSITION THE PECTORAL GIRDLE, PART I**

Posterior view, showing superficial muscles and deep muscles of the pectoral girdle. *See also Figures 6.26, 7.4, and 8.11. See Figure 11.5c for insertions of some of the muscles shown in this figure.*

Muscles That Position the Pectoral Girdle [FIGURES 11.1 TO 11.4 AND TABLE 11.1]

The large **trapezius** (tra-PĒ-zē-us) **muscles** cover the back and portions of the neck, extending to the base of the skull. These muscles originate along the middle of the neck and back and insert upon the clavicles and the scapular spines. Together, these triangular muscles form a broad diamond (Figures 11.1 and 11.2●). The trapezius muscles are innervated by more than one nerve (Table 11.1). Because specific regions of the trapezius can be made to contract independently, its actions are quite varied.

Removing the trapezius reveals the **rhomboid** (rom-BOYD) and the **levator scapulae** (SKAP-ū-lē) **muscles** (Figures 11.1 and 11.2●). These muscles are attached to the dorsal surfaces of the cervical and thoracic vertebrae. They insert along the vertebral border of each scapula, between the superior and inferior angles. Contraction of the rhomboid muscles adducts (retracts) the scapula, pulling it toward the center of the back. The levator scapulae elevates the scapula, as in shrugging the shoulders.

On the chest, the **serratus** (se-RĀ-tus) **anterior muscle** originates along the anterior and superior surfaces of several ribs (Figures 11.3 and 11.4●). This fan-shaped muscle inserts along the anterior margin of the vertebral border of the scapula. When the serratus anterior contracts, it abducts (protracts) the scapula and swings the shoulder anteriorly.

Two deep chest muscles arise along the ventral surfaces of the ribs. The **subclavius** (sub-KLĀ-vē-us; *sub*, below + *clavius*, clavicle) **muscle** inserts upon the inferior border of the clavicle (Figures 11.3 and 11.4●). When it contracts, it depresses and protracts the scapular end of the clavicle. Because ligaments connect this end to the shoulder joint and scapula, those structures move as well. The **pectoralis minor** (pek-tō-RĀ-lis) **muscle** attaches to the coracoid process of the scapula (Figures 11.3 and 11.4●). Its contraction usually complements that of the subclavius. Table 11.1 identifies the muscles that move the pectoral girdle and the nerves that innervate those muscles.

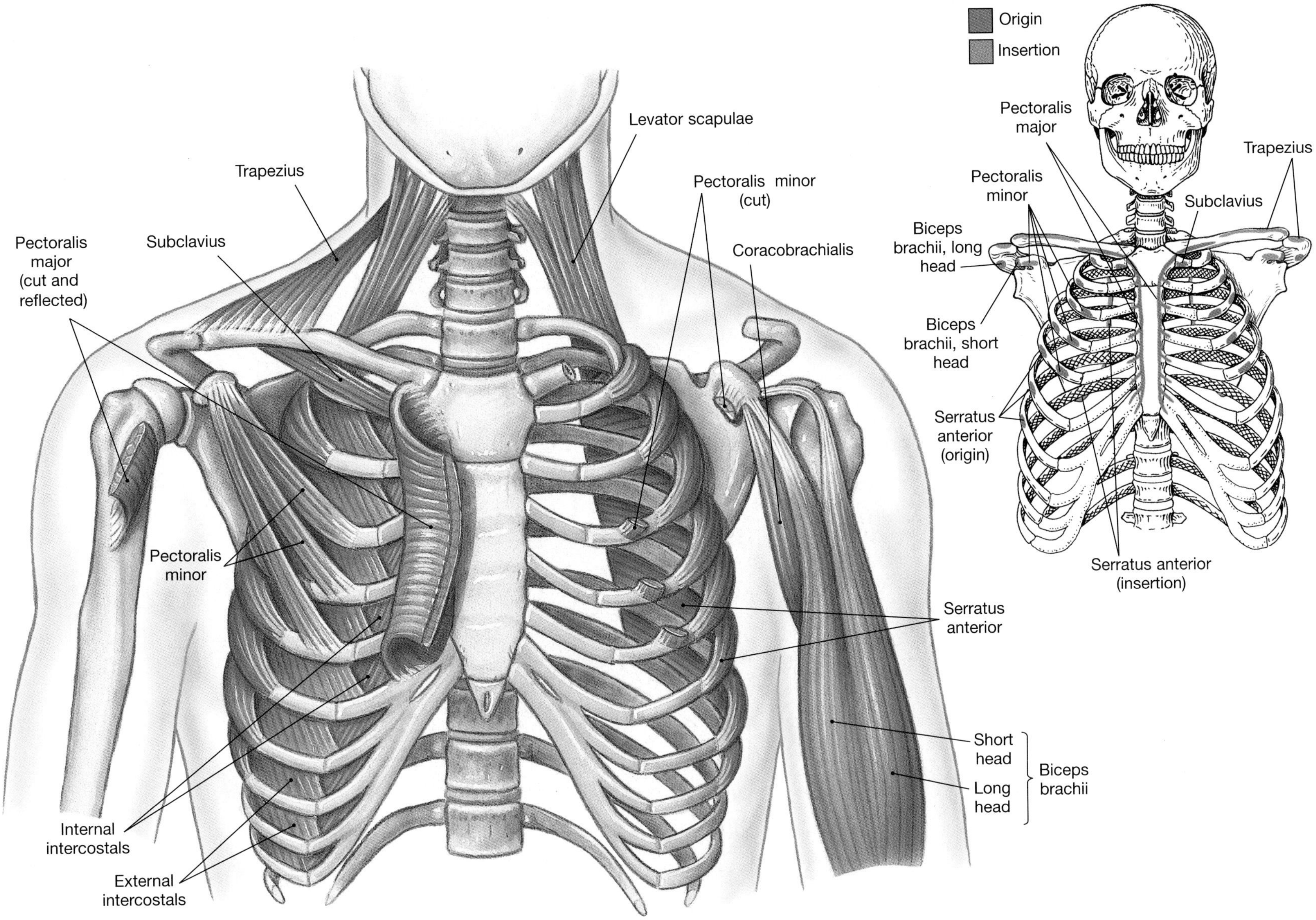

FIGURE 11.3 **Muscles That Position the Pectoral Girdle, Part II**

Anterior view, showing superficial muscles and deep muscles of the pectoral girdle. Selected origins and insertions are detailed.

TABLE 11.1 **Muscles That Position the Pectoral Girdle**

Muscle	Origin	Insertion	Action	Innervation
Levator scapulae	Transverse processes of first 4 cervical vertebrae	Vertebral border of scapula near superior angle	Elevates scapula	Cervical nerves C_3–C_4 and dorsal scapular nerve (C_5)
Pectoralis minor	Anterior surfaces and superior margins of ribs 3–5	Coracoid process of scapula	Depresses and protracts shoulder; rotates scapula so glenoid cavity moves inferiorly (downward rotation); elevates ribs if scapula is stationary	Medial pectoral nerve (C_8, T_1)
Rhomboid major	Spinous processes of superior thoracic vertebrae	Vertebral border of scapula from spine to inferior angle	Adducts and performs downward rotation of the scapula	Dorsal scapular nerve (C_5)
Rhomboid minor	Spinous process of vertebrae C_7–T_1	Vertebral border of scapula	As above	As above
Serratus anterior	Anterior and superior margins of ribs 1-8 or 1–9	Anterior surface of vertebral border of scapula	Protracts shoulder; rotates scapula so glenoid cavity moves superiorly (upward rotation)	Long thoracic nerve (C_5–C_7)
Subclavius	First rib	Clavicle (inferior border)	Depresses and protracts shoulder	Nerve to subclavius (C_5–C_6)
Trapezius	Occipital bone, ligamentum nuchae, and spinous processes of thoracic vertebrae	Clavicle and scapula (acromion and scapular spine)	Depends on active region and state of other muscles; may elevate, retract, depress, or rotate scapula upward and/or elevate clavicle; can also extend neck	Accessory nerve (N XI) and cervical spinal nerves (C_3–C_4)

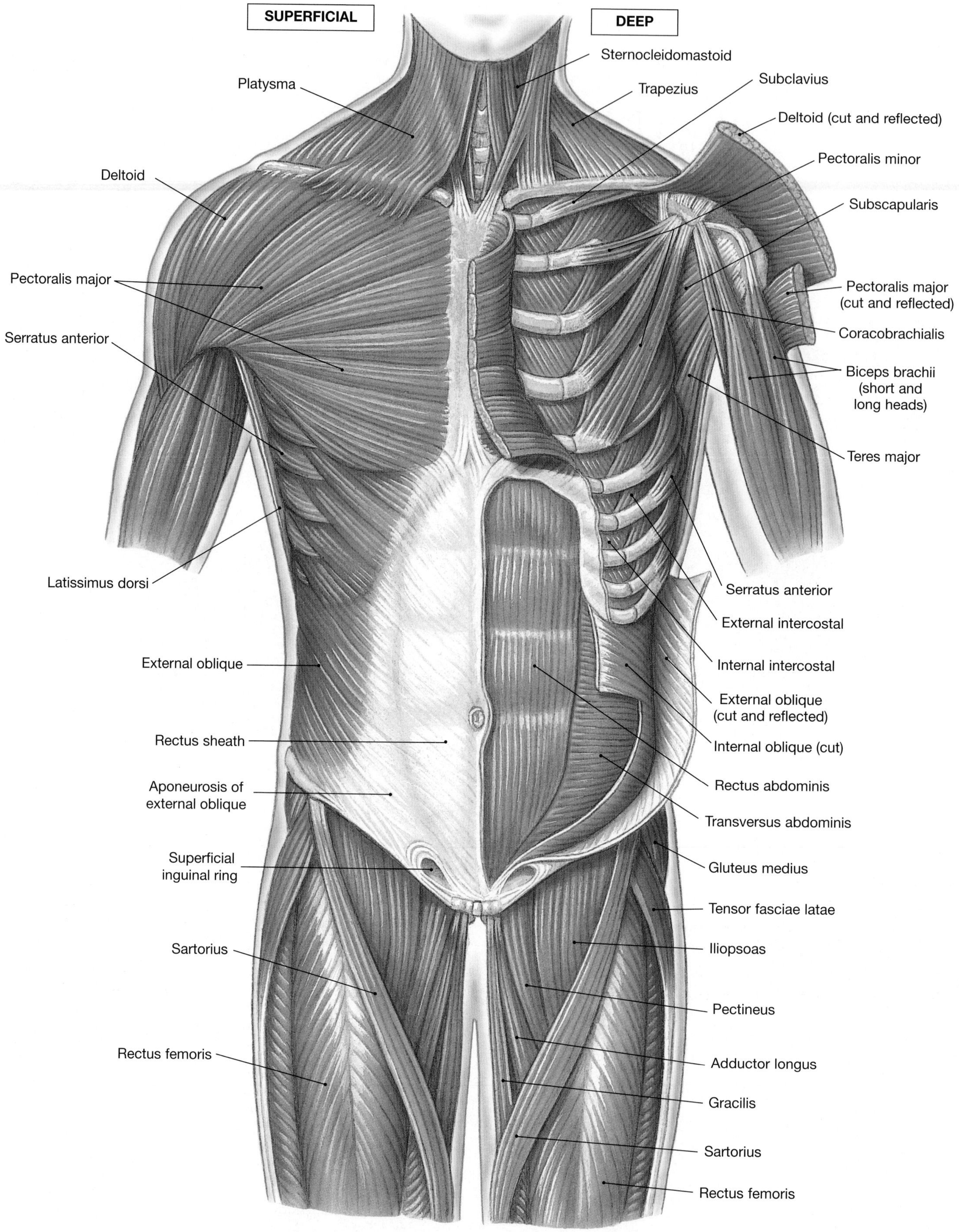

FIGURE 11.4 **Superficial and Deep Muscles of the Trunk and Proximal Limbs**

Anterior view of the axial muscles of the trunk and the appendicular musculature associated with the pectoral and pelvic girdles and the proximal portions of the limbs.

TABLE 11.2 MUSCLES THAT MOVE THE ARM

Muscle	Origin	Insertion	Action	Innervation
Coracobrachialis	Coracoid process	Medial margin of shaft of humerus	Adduction and flexion at shoulder	Musculocutaneous nerve (C_5–C_7)
Deltoid	Clavicle and scapula (acromion and adjacent scapular spine)	Deltoid tuberosity of humerus	*Whole muscle*: abduction of shoulder; *anterior part*: flexion and medical rotation of humerus; *posterior part*: extension and lateral rotation of humerus	Axillary nerve (C_5–C_6)
Supraspinatus	Supraspinous fossa of scapula	Greater tubercle of humerus	Abduction at the shoulder	Suprascapular nerve (C_5)
Infraspinatus	Infraspinous fossa of scapula	Greater tubercle of humerus	Lateral rotation at shoulder	Suprascapular nerve (C_5–C_6)
Subscapularis	Subscapular fossa of scapula	Lesser tubercle of humerus	Medial rotation at shoulder	Subscapular nerve (C_5–C_6)
Teres major	Inferior angle of scapula	Medial lip of intertubercular groove of humerus	Extension, adduction, and medial rotation at shoulder	Lower subscapular nerve (C_5–C_6)
Teres minor	Lateral border of scapula	Greater tubercle of humerus	Lateral rotation at shoulder	Axillary nerve (C_5)
Triceps brachii (long head)	*See Table 11.3*			
Biceps brachii	*See Table 11.3*			
Latissimus dorsi	Spinous process of inferior thoracic and all lumbar vertebrae, ribs 8–12, and lumbodorsal fascia	Floor of intertubercular groove of the humerus	Extension, adduction, and medial rotation at shoulder	Thoracodorsal nerve (C_6–C_8)
Pectoralis major	Cartilages of ribs 2–6, body of sternum, and inferior, medial portion of clavicle	Crest of greater turbercle and lateral lip of intertubercular groove of humerus	Flexion, adduction, and medial rotation at shoulder	Pectoral nerves (C_5–T_1)

■ Muscles That Move the Arm [FIGURES 11.1/11.4 TO 11.6 AND TABLE 11.2]

The muscles that move the arm are easiest to remember when grouped by primary actions. Some of these muscles are best seen in posterior view (Figure 11.1●) and others in anterior view (Figure 11.4●). Information on the muscles that move the arm is summarized in Table 11.2. The **deltoid muscle** is the major abductor of the arm, but the **supraspinatus** (soo-pra-spī-NĀ-tus) **muscle** assists at the start of this movement. The **subscapularis** and **teres** (TER-ēz) **major muscles** rotate the arm medially, whereas the **infraspinatus** (in-fra-spī-NĀ-tus) and **teres minor muscles** perform lateral rotation. All of these muscles originate on the scapula. The small **coracobrachialis** (kor-a-kō-brā-kē-AL-is) **muscle** (Figure 11.5a●) is the only muscle attached to the scapula that produces flexion and adduction at the shoulder joint.

The **pectoralis major muscle** extends between the anterior portion of the chest and the crest of the greater tubercle of the humerus. The **latissimus dorsi** (la-TIS-i-mus DOR-sē) **muscle** extends between the thoracic vertebrae at the posterior midline and the floor of the intertubercular groove of the humerus (Figures 11.1, 11.4, and 11.5a,b●). The pectoralis major muscle flexes the shoulder joint, and the latissimus dorsi muscle extends it. These two muscles can also work together to produce adduction and medial rotation of the humerus at the shoulder joint.

The surrounding muscles provide substantial support for the highly mobile but relatively weak shoulder joint. p. 225 The tendons of the supraspinatus, infraspinatus, subscapularis, and teres minor muscles support the joint capsule and limit the range of movement. These muscles form the *rotator cuff*, a frequent site of sports injuries. Powerful, repetitive arm movements (such as pitching a fastball at 96 mph for many innings) can place intolerable strains on the muscles of the rotator cuff, leading to muscle strains, bursitis, and other painful injuries. *Sports Injuries p. 792*

✓ CONCEPT CHECK

- Sometimes baseball pitchers will suffer from rotator cuff injuries. What muscles are involved in this type of injury?
- Identify the fan-shaped muscle that inserts along the anterior margin of the scapula at its vertebral border and acts to protract the scapula.
- What is the primary muscle producing abduction at the shoulder joint?
- What muscle produces extension, adduction, and medial rotation at the shoulder joint?

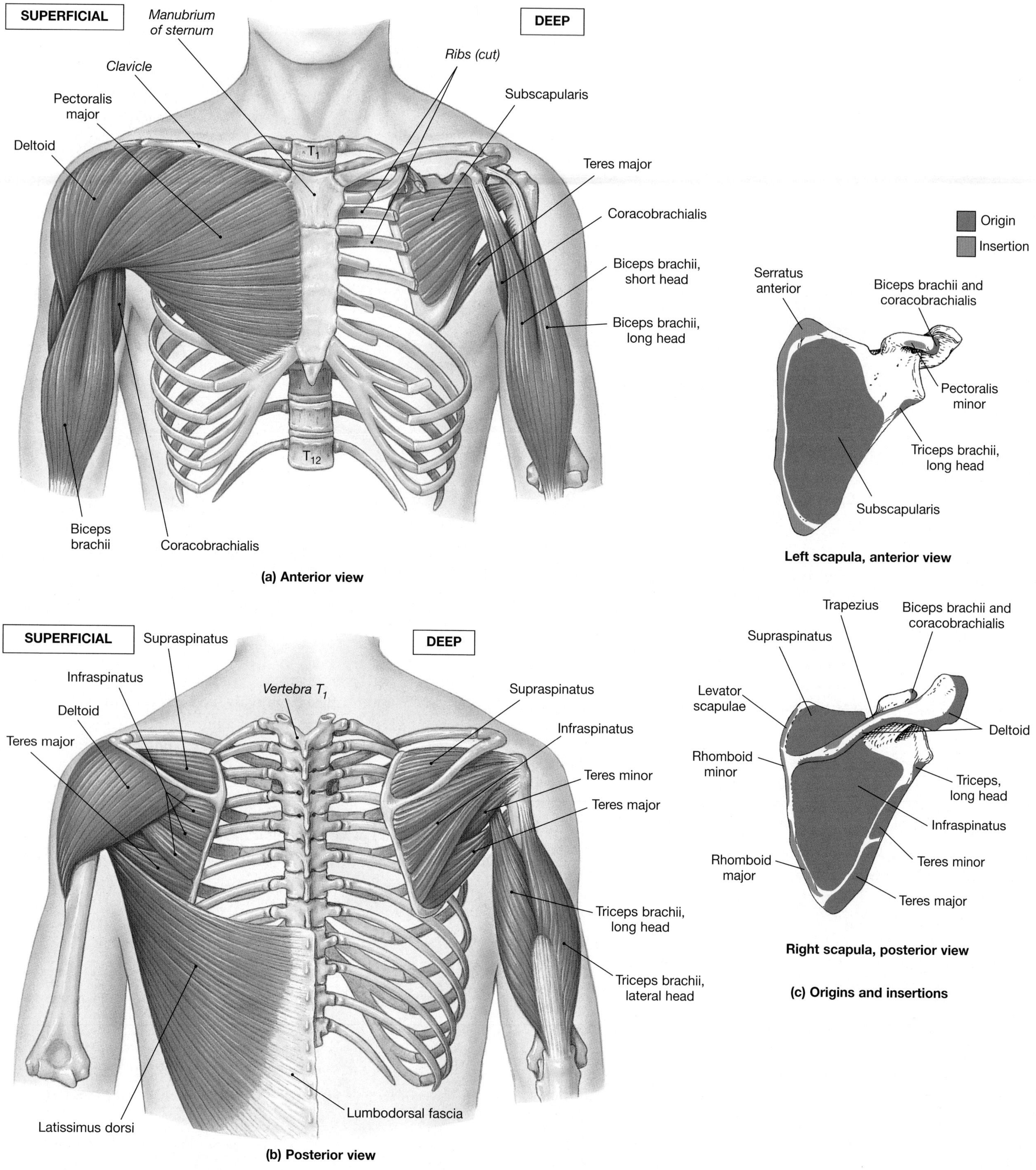

FIGURE 11.5 **Muscles That Move the Arm**

(**a**) Anterior view. (**b**) Posterior view. (**c**) Anterior and posterior views of the scapula, showing selected origins and insertions. *See also Figures 7.4 to 7.6, and 8.11.*

Muscles That Move the Forearm and Hand

[FIGURES 11.4 TO 11.7 AND TABLE 11.3]

Most of the muscles that move the forearm and hand originate on the humerus and insert upon the forearm and wrist. There are two noteworthy exceptions: The *long head* of the **triceps brachii** (TRĪ-seps BRĀ-kē-ī) **muscle** originates on the scapula and inserts on the olecranon; the *long head* of the **biceps brachii muscle** originates on the scapula and inserts on the radial tuberosity of the radius (Figures 11.4 to 11.7●). Although their contractions can have a secondary effect on the shoulder, their primary actions are at the elbow joint. The triceps brachii muscle extends the elbow when, for example, we do push-ups. The biceps brachii muscle both flexes the elbow and supinates the forearm. With the forearm pronated the biceps brachii muscle cannot function effectively due to the position of its muscular insertion. As a result, we are strongest when flexing the elbow with the forearm supinated; the biceps brachii muscle then makes a prominent bulge. Muscles that move the forearm and hand along with their nerve innervations are detailed in Table 11.3.

TABLE 11.3 MUSCLES THAT MOVE THE FOREARM AND HAND

Muscle	Origin	Insertion	Action	Innervation
ACTION AT THE ELBOW				
FLEXORS				
Biceps brahii	*Short head* from the coracoid process; *long head* from the supraglenoid tubercle (both on the scapula)	Radial tuberosity	Flexion at elbow and shoulder; supination	Musculocutaneous nerve (C_5–C_6)
Brachialis	Anterior, distal surface of humerus	Ulnar tuberosity	Flexion at elbow	As above and radial nerve (C_7–C_8)
Brachioradialis	Ridge superior to the lateral epicondyle of humerus	Lateral aspect of styloid process of radius	As above	Radial nerve (C_6–C_8)
EXTENSORS				
Anconeus	Posterior surface of lateral humerus	Lateral margin of olecranon and ulnar shaft	Extension at elbow	Radial nerve (C_6–C_8)
Triceps brachii				
lateral head	Superior, lateral margin of humerus	Olecranon of ulna	As above	Radial nerve (C_6–C_8)
long head	Infraglenoid tubercle of scapula	As above	As above	As above
medial head	Posterior surface of humerus, inferior to radial groove	As above	As above	As above
PRONATORS/ SUPINATORS				
Pronator quadratus	Medial surface of distal portion of ulna	Anterolateral surface of distal portion of radius	Pronates forearm and hand by medial rotation of radius at radioulnar joints	Median nerve (C_8–T_1)
Pronator teres	Medial epicondyle of humerus and coronoid process of ulna	Distal lateral surface of radius	As above	Median nerve (C_6–C_7)
Supinator	Lateral epicondyle of humerus and ridge near radial notch of ulna	Anterolateral surface of radius distal to the radial tuberosity	Supinates forearm and hand by lateral rotation of radius at radioulnar joints	Deep radial nerve (C_6–C_8)
ACTION AT THE WRIST				
FLEXORS				
Flexor carpi radialis	Medial epicondyle of humerus	Bases of 2nd and 3rd metacarpal bones	Flexion and abduction at wrist	Median nerve (C_6–C_7)
Flexor carpi ulnaris	Medial epicondyle of humerus; adjacent medial surface of olecranon and anteromedial portion of ulna	Pisiform bone, hamate bone, and base of 5th metacarpal bone	Flexion and adduction at wrist	Ulnar nerve (C_8–T_1)
Palmaris longus	Medial epicondyle of humerus	Palmar aponeurosis and flexor retinaculum	Flexion at wrist	Median nerve (C_6–C_7)
EXTENSORS				
Extensors carpi radialis longus	Lateral supracondylar ridge of humerus	Base of 2nd metacarpal bone	Extension and abduction at wrist	Radial nerve (C_6–C_7)
Extensor carpi radalis brevis	Lateral epicondyle of humers	Base of 3rd metacarpal bone	As above	As above
Extensor carpi ulnaris	Lateral epicondyle of humerus; adjacent dorsal surface of ulna	Base of 5th metacarpal bone	Extension and adduction at wrist	Deep radial nerve (C_6–C_8)

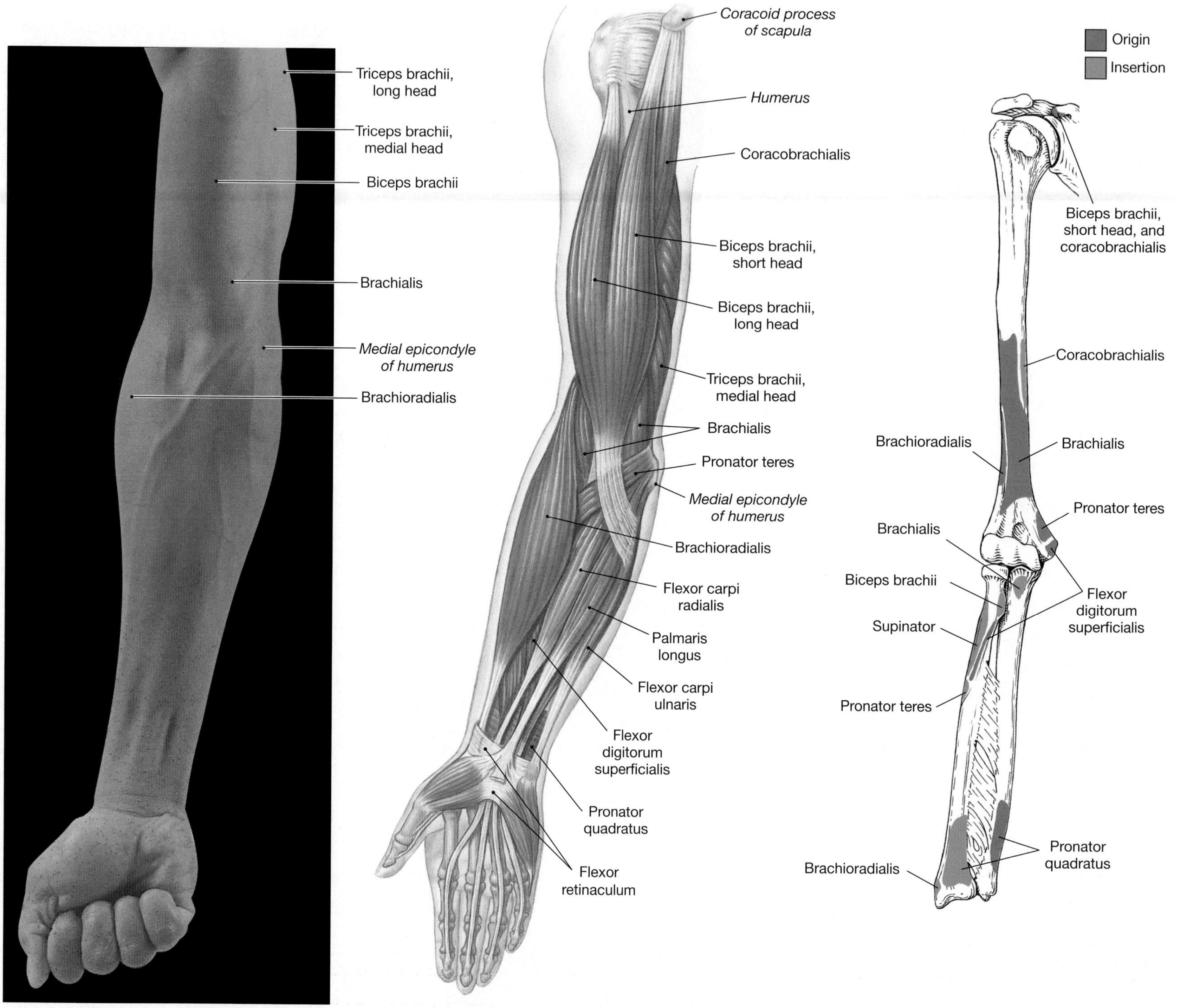

(a) Surface anatomy, anterior view **(b) Superficial muscles, anterior view** **(c) Origins and insertions, anterior view**

FIGURE 11.6 **Muscles That Move the Forearm and Hand, Part I**

Relationships among the muscles of the right upper limb are shown. **(a)** Surface anatomy of the right upper limb, anterior view. **(b)** Superficial muscles of the right upper limb, anterior view. **(c)** Anterior view of bones of the right upper limb, showing selected muscle origins and insertions. **(d)** Anterior view of a dissection of the muscles of the right upper limb. The palmaris longus and flexor carpi muscles (radialis and ulnaris) have been partly removed, and the flexor retinaculum has been cut. **(e)** The relationships among the deeper muscles of the arm are best seen in the sectional view. **(f)** Anterior view of the deep muscles of the supinated forearm. *See also Figures 7.6, 7.7, and 7.8.*

The **brachialis** (brā-kē-ā-lis) and **brachioradialis** (brā-kē-ō-rā-dē-ā-lis) **muscles** also flex the elbow; they are opposed by the **anconeus** (an-KŌ-nē-us) and the triceps brachii muscles. The **flexor carpi ulnaris**, the **flexor carpi radialis**, and the **palmaris longus** are superficial muscles that work together to produce flexion of the wrist (Figures 11.6b–e and 11.7b–e●). Because of differences in their sites of origin, the flexor carpi radialis muscle flexes and abducts while the flexor carpi ulnaris muscle flexes and adducts. The **extensor carpi radialis muscle** and the **extensor carpi ulnaris muscle** have a similar relationship; the former produces extension and abduction, the latter extension and adduction.

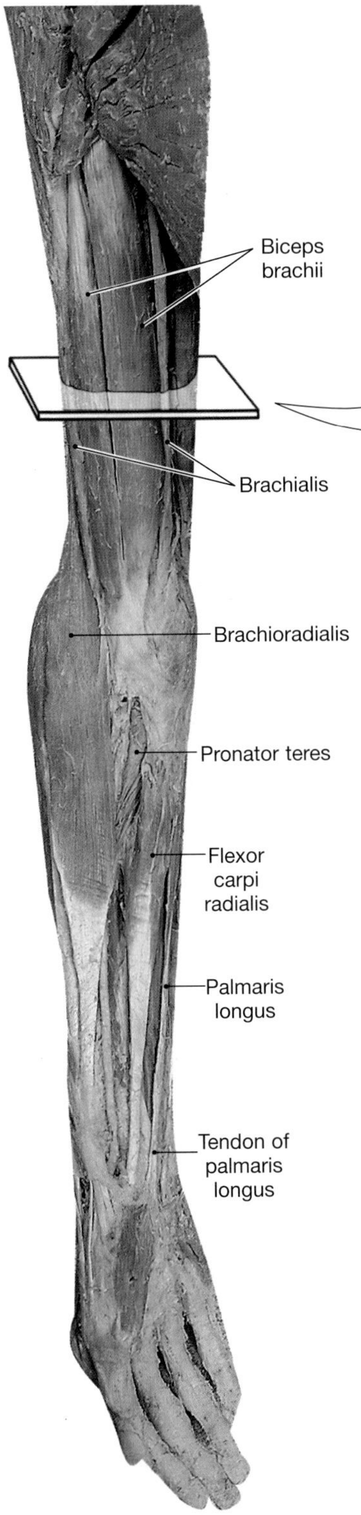

(d) Anterior view

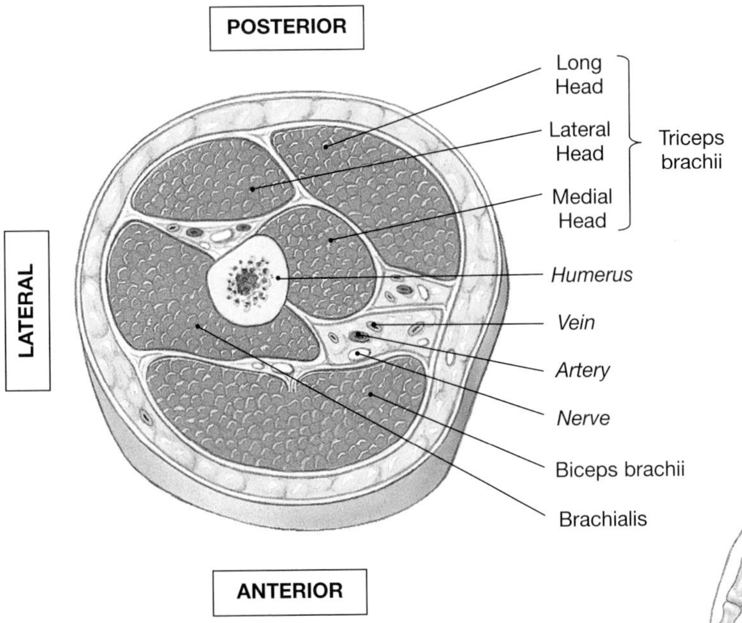

(e) Arm, cross-sectional view

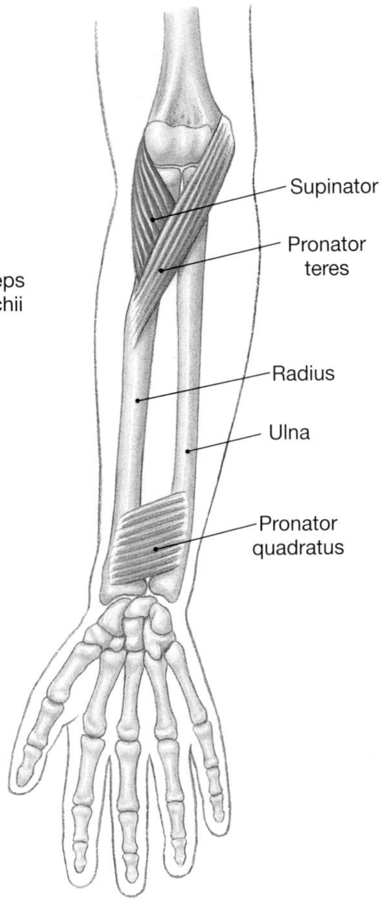

(f) Anterior view of forearm, deep muscles

FIGURE 11.6 *(continued)*

The **pronator teres muscle** and the **supinator muscle** are antagonistic muscles that originate on both the humerus and the ulna. They insert on the radius and cause rotation without flexing or extending the elbow. The **pronator quadratus muscle** originates on the ulna and assists the pronator teres muscle in opposing the actions of the supinator muscle or biceps brachii muscle. The muscles involved in pronation and supination can be seen in Figures 11.6f and 11.7f•. Note the changes in orientation that occur as the pronator teres and pronator quadratus muscles contract. During pronation the tendon of the biceps brachii muscle rolls under the radius, and a bursa prevents abrasion against the tendon. p. 227

As you study the muscles in Table 11.3, note that in general, extensor muscles lie along the posterior and lateral surfaces of the forearm, and flexors are found on the anterior and medial surfaces. Many of the muscles that move the forearm and hand can be seen from the body surface (Figures 11.6a and 11.7a•).

Muscles That Move the Hand and Fingers [FIGURES 11.6 TO 11.9 AND TABLES 11.4/11.5]

Several superficial and deep muscles of the forearm (Table 11.4) perform flexion and extension at the joints of the fingers. Only the tendons of these muscles extend across the wrist joint. These are relatively large muscles (Figures 11.6, 11.7, and 11.8•), and keeping them clear of the joints ensures maximum mobility at both the wrist and hand. The tendons that cross the dorsal and ventral surfaces of the wrist pass through **synovial tendon sheaths**, elongated bursae that reduce friction. These muscles and their

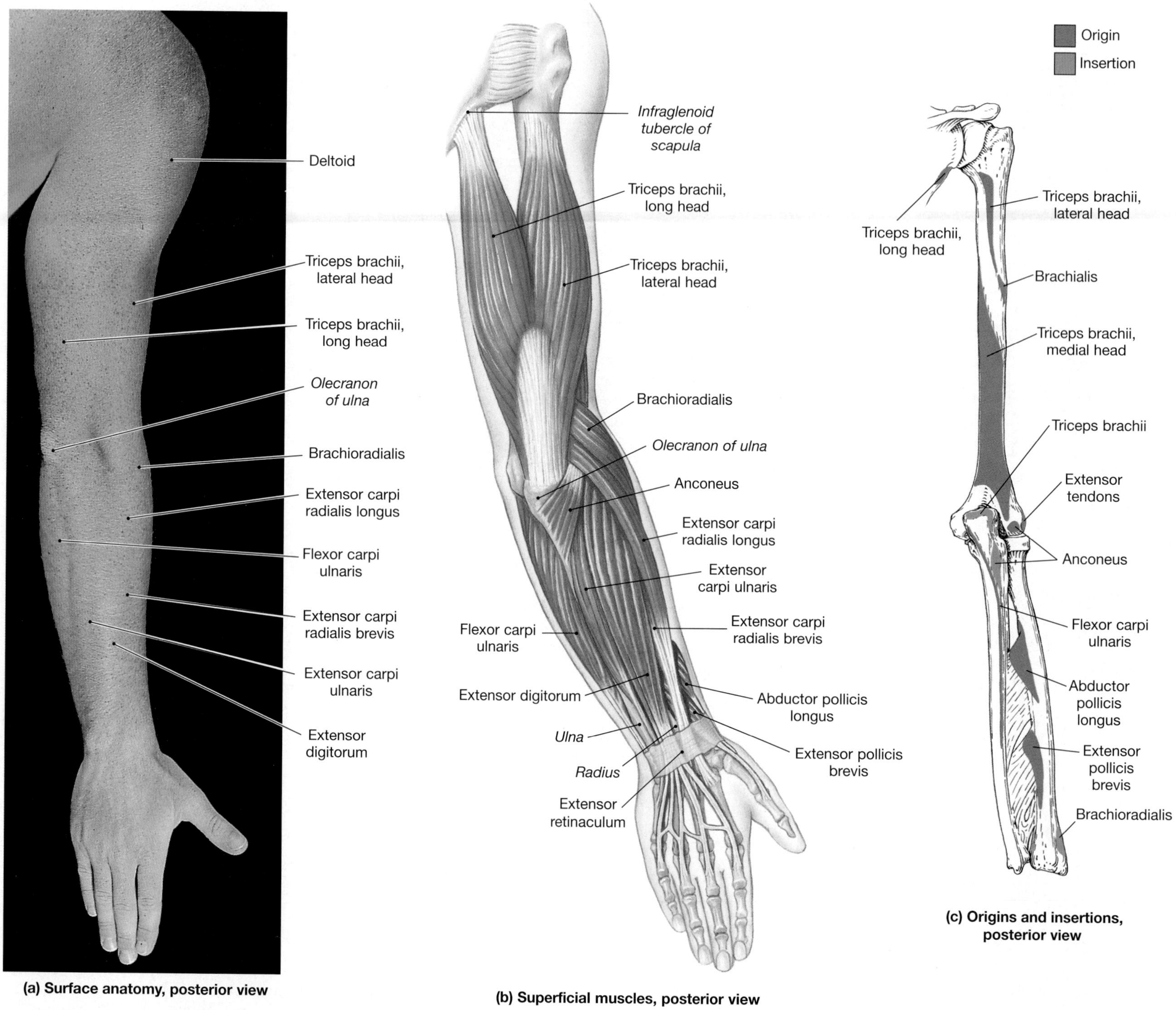

FIGURE 11.7 **MUSCLES THAT MOVE THE FOREARM AND HAND, PART II**

Relationships among the muscles of the right upper limb are shown. **(a)** Surface anatomy of the right upper limb, posterior view. **(b)** A diagrammatic view of a dissection of the superficial muscles. **(c)** Posterior view of the bones of the upper limb, showing the origins and insertions of selected muscles. **(d)** A posterior view of superficial dissection of the forearm. **(e)** Relationships among deeper muscles are best seen in this sectional view. The deep digital extensors and flexors are shown in *Figure 11.8*. **(f)** Deep muscles involved with pronation and supination. *See also Figure 7.7.*

tendons are shown in anterior view in Figures 11.6b,d, and 11.9d,f●, and in posterior view in Figures 11.7b,d, 11.8d–f, and 11.9a,e●. The fascia of the forearm thickens on the posterior surface of the wrist to form a wide band of connective tissue, the **extensor retinaculum** (ret-i-NAK-ū-lum) (Figure 11.9a,e●). The extensor retinaculum holds the tendons of the extensor muscles in place. On the anterior surface, the fascia also thickens to form another wide band of connective tissue, the **flexor retinaculum**, which retains the tendons of the flexor muscles (Figure 11.9d,f●). Inflammation of the retinacula and tendon sheaths can restrict movement and irritate the *median nerve*, a sensory and motor nerve that innervates the hand. This condition, known as *carpal tunnel syndrome*, causes chronic pain.

The muscles of the forearm that provide strength and crude control of the hand and fingers are called the *extrinsic muscles of the hand*. Fine control of the hand involves small *intrinsic muscles* that originate on the carpal and metacarpal bones (Figure 11.9a,d,f●). No muscles originate on the phalanges, and only tendons extend across the distal joints of the fingers. The intrinsic muscles of the hand are detailed in Table 11.5.

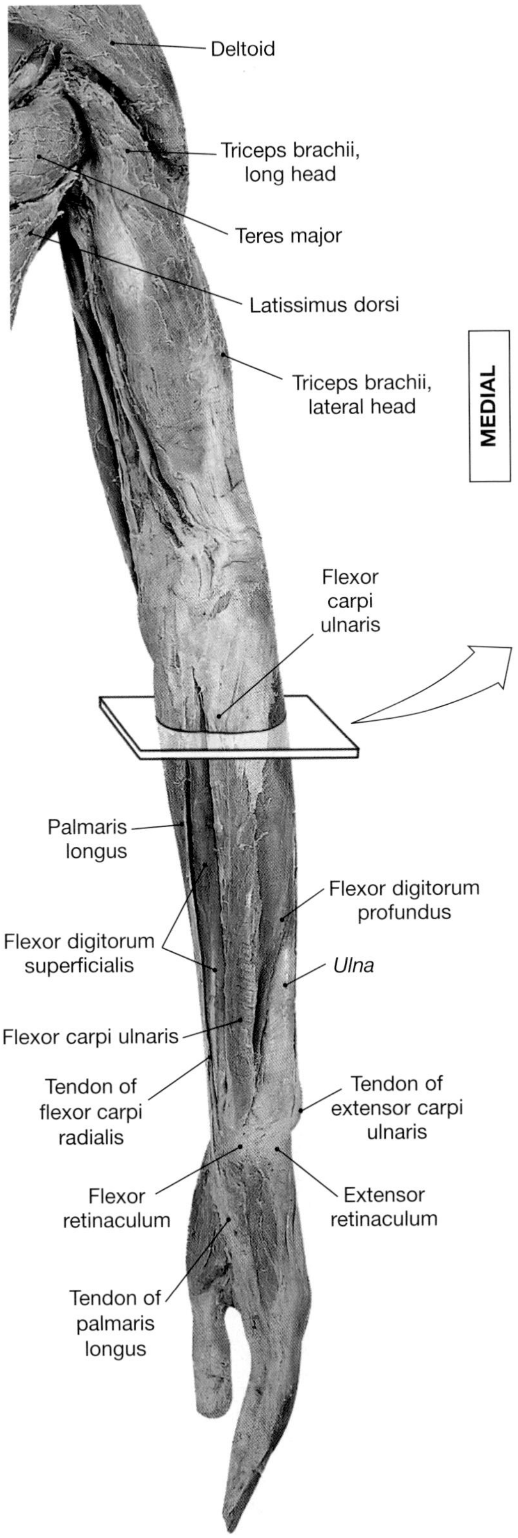

(d) Superficial muscles, posterior view

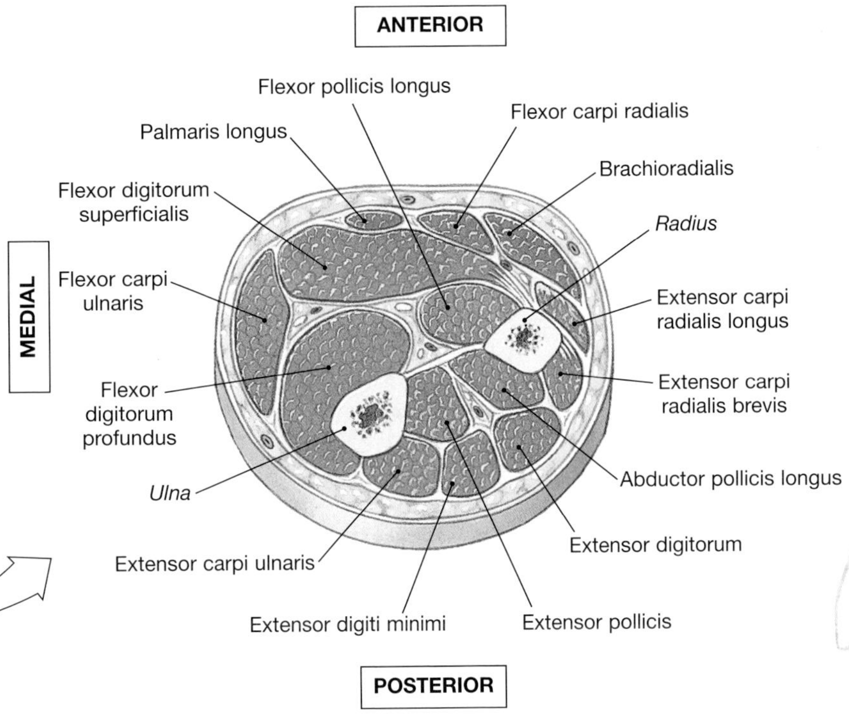

(e) Forearm, cross-sectional view

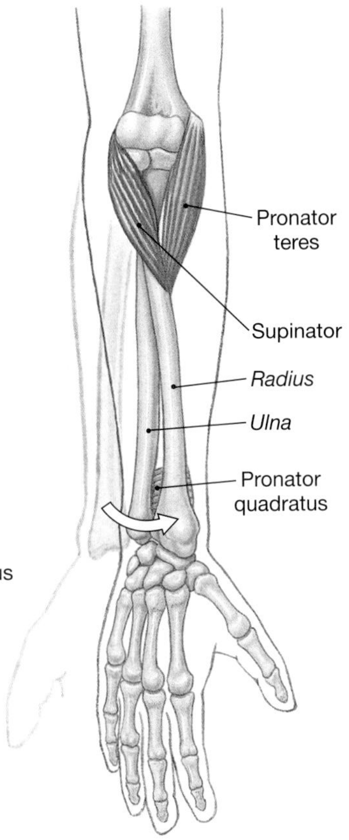

(f) Pronation

FIGURE 11.7 *(continued)*

CLINICAL BRIEF

Carpal Tunnel Syndrome

In **carpal tunnel syndrome**, inflammation within the sheath surrounding the flexor tendons of the palm leads to compression of the *median nerve*, a mixed (sensory and motor) nerve that innervates the palm and palmar surfaces of the thumb, index, and middle fingers. Symptoms include pain, especially on wrist flexion, a tingling sensation or numbness on the palm, and weakness in the abductor pollicis brevis. This condition is fairly common and often affects those involved in hand movements that place repetitive stress on the tendons crossing the wrist. Activities commonly associated with this syndrome include typing at a computer keyboard, playing the piano, or as in the case of carpenters, repeated use of a hammer. Treatment involves administration of antiinflammatory drugs, such as aspirin, and use of a splint to prevent wrist movement and stabilize the region. A number of specially designed computer keyboards are available to reduce the stresses associated with typing.

✓ CONCEPT CHECK

- Injury to the flexor carpi ulnaris muscle would impair what two movements?
- Identify the muscles that rotate the radius without flexing or extending the elbow.
- What structure do the tendons that cross the dorsal and ventral surfaces of the wrist pass through before reaching the point of insertion?
- Identify the thickened fascia on the posterior surface of the wrist that forms a wide band of connective tissue.

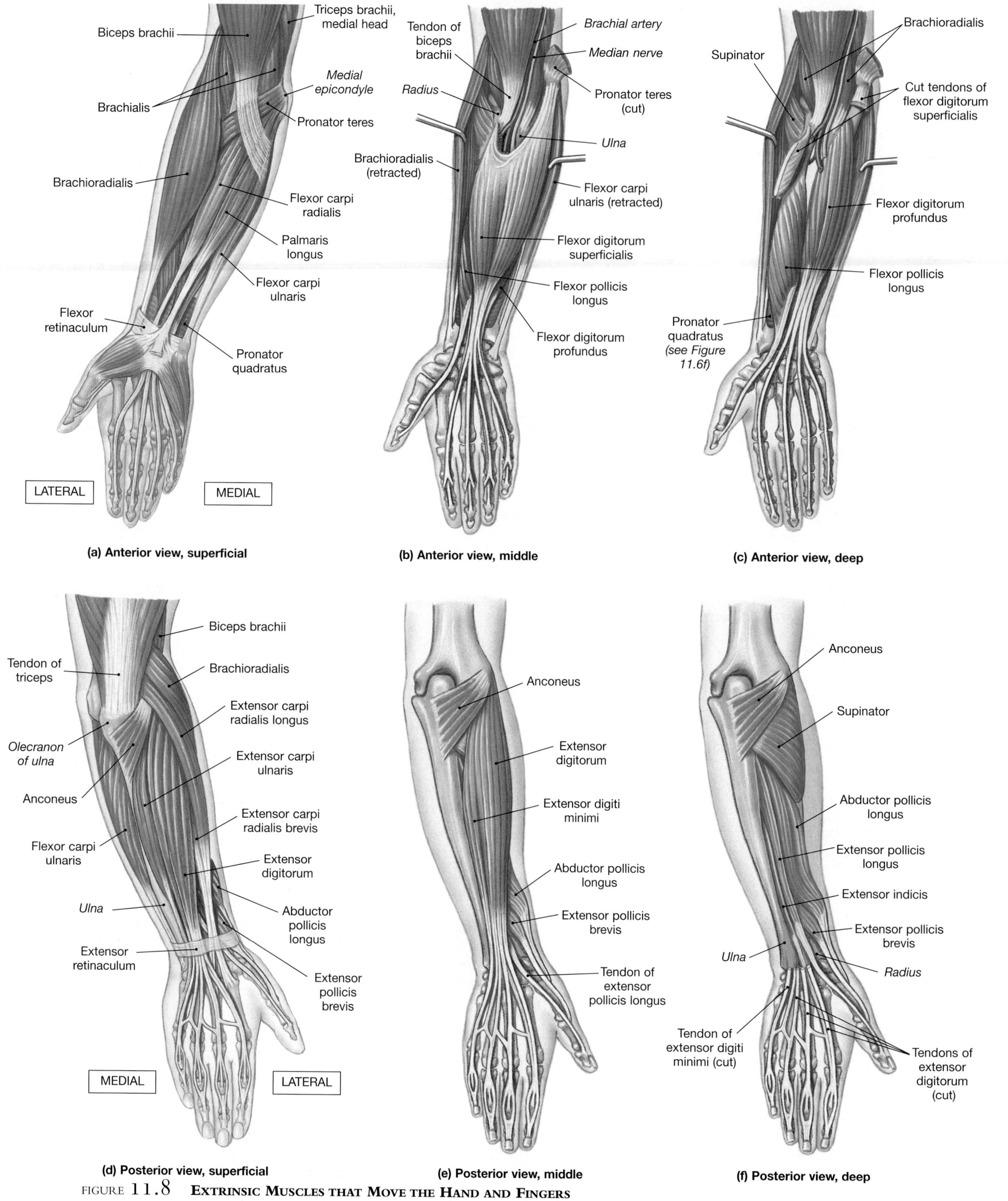

FIGURE 11.8 **EXTRINSIC MUSCLES THAT MOVE THE HAND AND FINGERS**

(**a**) Anterior view, showing superficial muscles of the right forearm. (**b**) Anterior view of middle layer of muscles. The flexor carpi radialis muscle and palmaris longus muscle have been removed. (**c**) Anterior view of the deep layer of muscles. (**d**) Posterior view, showing superficial muscles of the right forearm. (**e**) Posterior view of middle layer of muscles. (**f**) Posterior view of the deep layer of muscles. *See also Figure 7.7, 7.8, and 11.7.*

TABLE 11.4

Muscles That Move the Fingers and Hand

Muscle	Origin	Insertion	Action	Innervation
Abductor pollicis longus	Proximal dorsal surfaces of ulna and radius	Lateral margin of 1st metacarpal bone	Abduction at joints of of thumb and wrist	Deep radial nerve (C_6–C_7)
Extensor digitorum	Lateral epicondyle of humerus	Posterior surfaces of the phalanges, digits 2-5	Extension at finger joints and wrist	Deep radial nerve (C_6–C_8)
Extensor pollicis brevis	Shaft of radius distal to origin of adductor pollicis longus	Base of proximal phalanx of thumb	Extension at joints of thumb; abduction at wrist	Deep radial nerve (C_6–C_7)
Extensor pollicis longus	Posterior and lateral surfaces of ulna and interosseous membrane	Base of distal phalanx of thumb	As above	Deep radial nerve (C_6–C_8)
Extensor indicis	Posterior surface of ulna and interosseous membrane	Posterior surface of proximal phalanx of little finger (2), with tendon of extensor digitorum	Extension and adduction at joints of index finger	As above
Extensor digiti minimi	Via extensor tendon to lateral epicondyle of humerus and from intermuscular septa	Posterior surface of proximal phalanx of little finger	Extension at joints of little finger	As above
Flexor digitorum superficialis	Medial epicondyle of humerus; adjacent anterior surfaces of ulna and radius	Midlateral surfaces of middle phalanges of digits 2–5	Flexion at proximal interphalangeal, metacarpophalangeal, and wrist joints	Median nerve (C_7–T_1)
Flexor digitorum profundus	Medial and posterior surfaces of ulna, medial surfaces of coronoid process, and interosseus membrane	Bases of distal phalanges of digits 2–5	Flexion at distal interphalangeal joints, and, to lesser degree, proximal interphalangeal joints and wrist	Palmar interosseous nerve, from median nerve and ulnar nerve (C_8–T_1)
Flexor pollicis longus	Anterior shaft of radius, interosseous membrane	Base of distal phalanx of thumb	Flexion at joints of thumb	Median nerve (C_8–T_1)

TABLE 11.5

Intrinsic Muscles of the Hand

Muscle	Origin	Insertion	Action	Innervation
Adductor pollicis	Metacarpal and carpal bones	Proximal phalanx of thumb	Adduction of thumb	Ulnar nerve, deep branch (C_8–T_1)
Opponens pollicis	Trapezium	First metacarpal bone	Opposition of thumb	Median nerve (C_6–C_7)
Palmaris brevis	Palmar aponeurosis	Skin of medial border of hand	Moves skin on medial border toward midline of palm	Ulnar nerve, superficial branch (C_8)
Abductor digiti minimi	Pisiform bone	Proximal phalanx of little finger	Abduction of little finger and flexion at its metacarpophalangeal joint	Ulnar nerve, deep branch (C_6–T_1)
Abductor pollicis brevis	Transverse carpal ligament, scaphoid and trapezium bones	Radial side of base of proximal phalanx of thumb	Abduction of thumb	Median nerve (C_6–C_7)
Flexor pollicis brevis*	Flexor retinaculum, trapezium, capitate bone and ulnar side of 1st metacarpal bone	Ulnar side of proximal phalanx of thumb	Flexion and adduction of thumb	Branches of median and ulnar nerves
Flexor digiti minimi brevis	Hamate bone	Proximal phalanx of little finger	Flexion at joints of little finger	Ulnar nerve, deep branch (C_8–T_1)
Opponens digiti minimi	As above	5th metacarpal bone	Opposition of 5th metacarpal bone	As above
Lumbricals (4)	The four tendons of flexor digitorum profundus	Tendons of extensor digitorum to digits 2-5	Flexion at metacarpophalangeal joints; extension at proximal and distal interphalangeal joints	No. 1 and no. 2 by median nerve; no. 3 and no. 4 by ulnar nerve, deep branch
Dorsal interossei (4)	Each originates from opposing faces of two metacarpal bones (I and II, II and III, III and IV, IV and V)	Bases of proximal phalanes of digits 2–4	Abduction at metacarpophalangeal joints of digits 2–4, flexion at metacarpophalangeal joints; extension at interphalangeal joints	Ulnar nerve, deep branch (C_8–T_1)
Palmar interossei	Sides of metacarpal bones II, IV and V	Bases of proximal phalanges of digits 2,4, and 5	Adduction at metacarpophalangeal joints of digits 2, 4, and 5; flexion at metacarpophalangeal joints; extension at interphalangeal joints	As above

*The portion of the flexor pollicis brevis originating on the first metacarpal bone is sometimes called the *first palmar interosseus muscle*

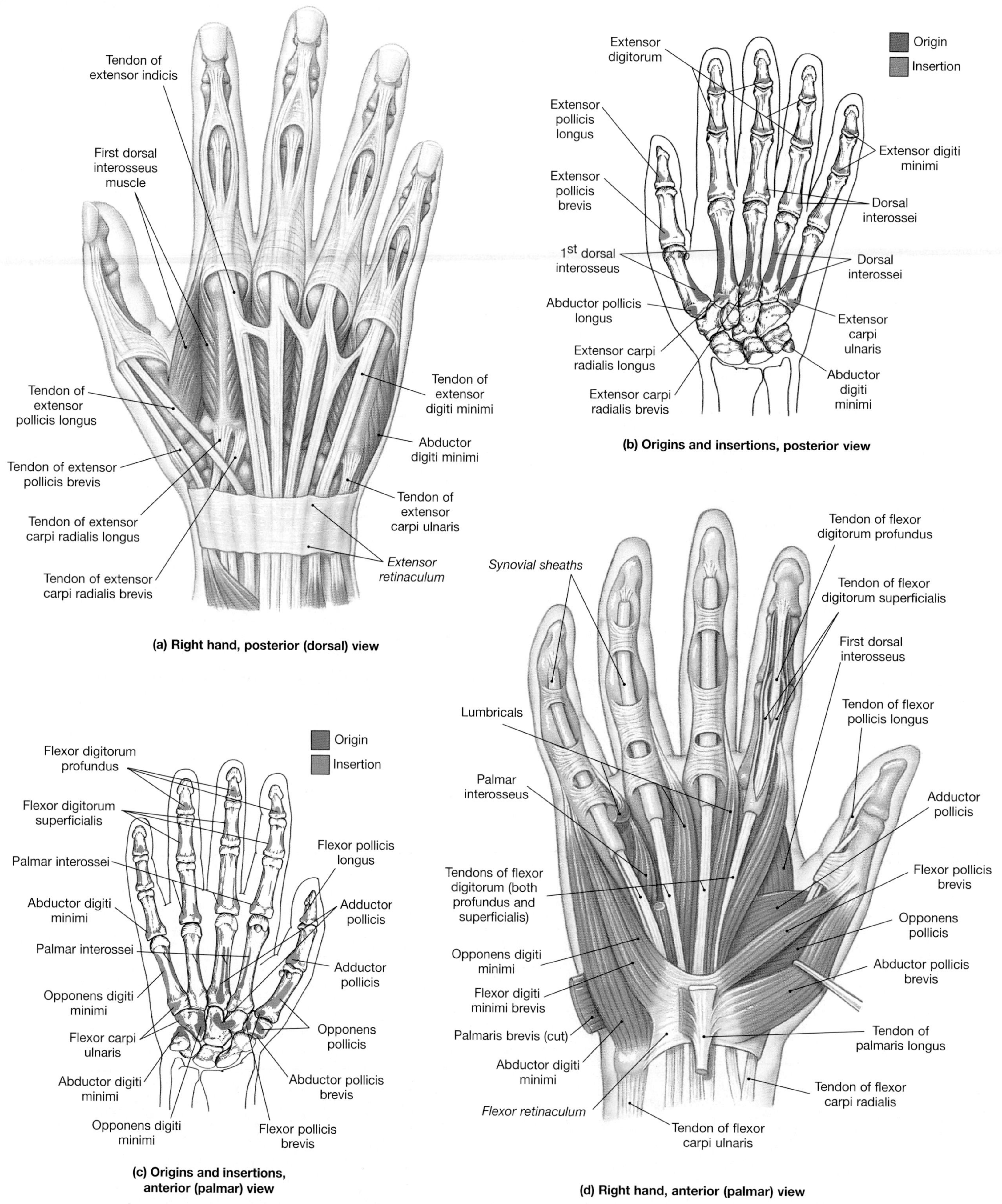

FIGURE 11.9 **INTRINSIC MUSCLES, TENDONS, AND LIGAMENTS OF THE HAND**

Anatomy of the right wrist and hand. **(a)** Posterior (dorsal) view. **(b)** Posterior view of the bones of the right hand, showing the origins and insertions of selected muscles. **(c)** Anterior view of the bones of the right hand, showing the origins and insertions of selected muscles. **(d)** Anterior (palmar) view. **(e)** Right hand, transverse sectional view through metacarpal bones. **(f)** Anterior view of a deep palmar dissection of the right hand.

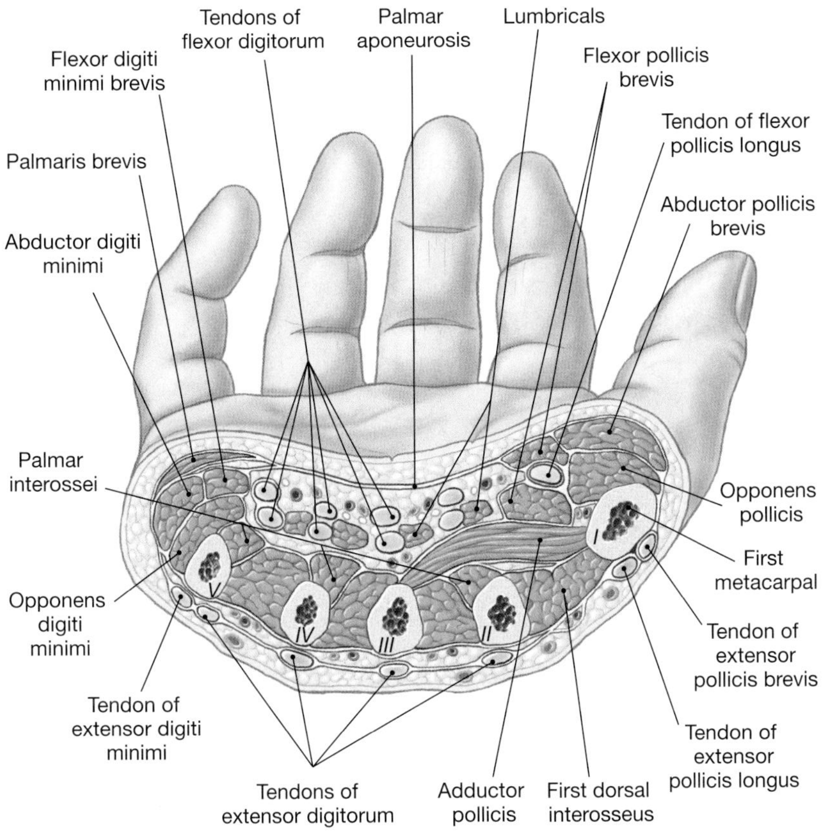

(e) Right hand, sectional view through metacarpal bones

FIGURE 11.9 *(continued)*

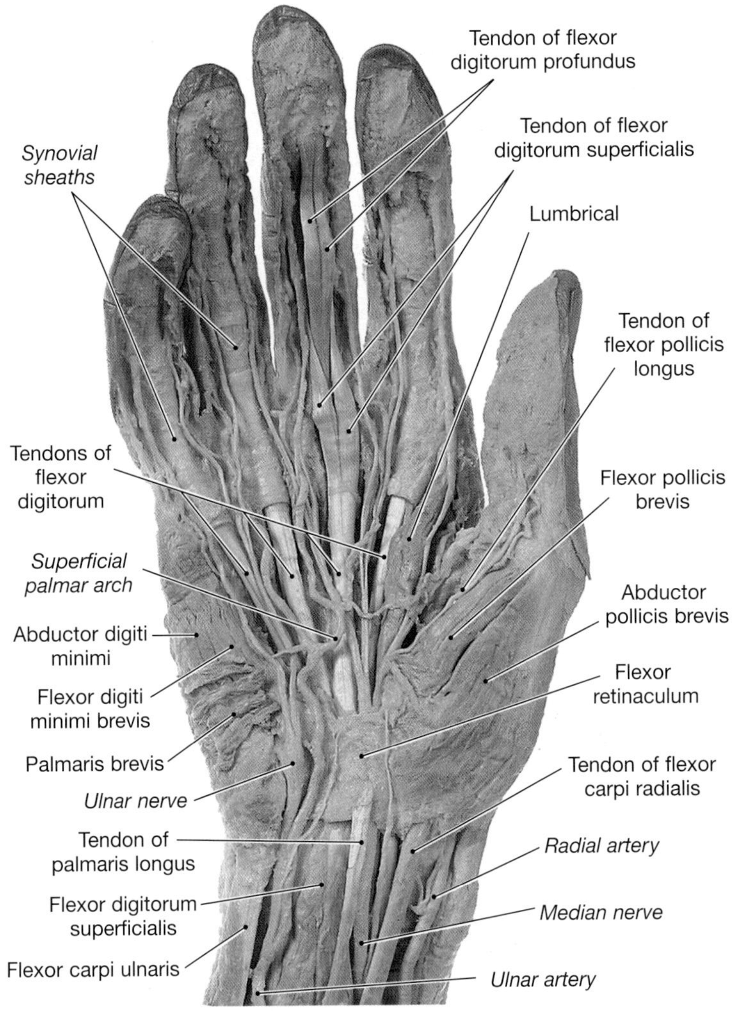

(f) Right hand, anterior (palmar) view

Muscles of the Pelvic Girdle and Lower Limbs

The pelvic girdle is tightly bound to the axial skeleton, and relatively little movement is permitted. The few muscles that can influence the position of the pelvis were considered in Chapter 10, during the discussion of the axial musculature. p. 283 The muscles of the lower limbs are larger and more powerful than those of the upper limb. These muscles can be divided into three groups: (1) muscles that move the thigh, (2) muscles that move the leg, and (3) muscles that move the foot and toes.

Muscles That Move the Thigh [FIGURES 11.1/11.4/11.10 TO 11.11 AND TABLE 11.6]

One method for organizing the diverse muscles that originate on the pelvis considers the orientation of each muscle around the hip joint. Muscles originating on the surface of the pelvis and inserting on the femur will produce characteristic movements determined by their position relative to the acetabulum (Table 11.6).

Gluteal muscles cover the lateral surface of the ilium (Figures 11.1, 11.4, and 11.10•). The **gluteus maximus muscle** is the largest and most superficial of the gluteal muscles. It originates along the posterior gluteal line and adjacent portions of the iliac crest; the sacrum, coccyx, and associated ligaments; and the lumbodorsal fascia. Acting alone, this massive muscle produces extension and lateral rotation at the hip. The gluteus maximus muscle shares an insertion with the **tensor fasciae latae** (TEN-sor FASH-ē-ē LĀ-tē) **muscle**, which originates on the iliac crest and lateral surface of the anterior superior iliac spine. Together these muscles pull on the **iliotibial** (il-ē-ō-TIB-ē-al) **tract**, a band of collagen fibers that extends along the lateral surface of the thigh and inserts upon the tibia. This tract provides a lateral brace for the knee that becomes particularly important when a person balances on one foot.

The **gluteus medius** and **gluteus minimus** (Figure 11.10•) **muscles** originate anterior to the gluteus maximus and insert upon the greater trochanter of the femur. Both produce abduction and medial rotation at the hip joint. The anterior gluteal line on the lateral surface of the ilium marks the boundary between these muscles. p. 195

The six **lateral rotators** (Figures 11.10a,c and 11.11•) originate at or are inferior to the horizontal axis of the acetabulum and insert on the femur. All cause lateral rotation of the thigh; additionally, the **piriformis** (pir-i-

TABLE 11.6

Muscles That Move the Thigh

Muscle	Origin	Insertion	Action	Innervation
GLUTEAL GROUP				
Gluteus maximus	Iliac crest, posterior gluteal line, and lateral surface of ilium; sacrum, coccyx, and lumbodorsal facia	Iliotibial tract and gluteal tuberosity of femur	Extension and lateral rotation at hip; helps stabilize the extended knee	Inferior gluteal nerve (L_5–S_2)
Gluteus medius	Anterior iliac crest, lateral surface of ilium between posterior and anterior gluteal lines	Greater trochanter of femur	Abduction and medial rotation at hip	Superior gluteal nerve (L_4–S_1)
Gluteus minimus	Lateral surface of ilium between inferior and anterior gluteal lines	As above	As above	As above
Tensor fasciae latae	Iliac crest and lateral surface of anterior superior iliac spine	Iliotibial tract	Flexion and medial rotation at hip; tenses fasciae latae, which laterally supports the knee	As above
LATERAL ROTATOR GROUP				
Obturators (externus and internus)	Lateral and medial margins of obturator foramen	Trochanteric fossa of femur (externus); medial surface of greater trochanter (internus)	Lateral rotation of hip	Obturator nerve (externus: L_3–L_4) and special nerve from sacral plexus (internus: L_5–S_2)
Piriformis	Anterolateral surface of sacrum	Greater trochanter of femur	Lateral rotation and abduction at hip	Branches of sacral nerves (S_1–S_2)
Gemelli (superior and inferior)	Ischial spine and ischial tuberosity	Medial surface of greater trochanter	Lateral rotation at hip	Nerves to obturator internus and quadratus femoris
Quadratus femoris	Lateral border of ischial tuberosity	Intertrochanteric crest of femur	As above	Special nerves from sacral plexus (L_4–S_1)
ADDUCTOR GROUP				
Adductor brevis	Inferior ramus of pubis	Linea aspera of femur	Adduction, flexion, and medial rotation at hip	Obturator nerve (L_3–L_4)
Adductor longus	Inferior ramus of pubis, anterior to adductor brevis	As above	As above	As above
Adductor magnus	Inferior ramus of pubis posterior to adductor brevis and ischial tuberosity	Linea aspera and adductor tubercle of femur	Whole muscle produces adduction at the hip; anterior part produces flexion and medial rotation; posterior part produces extension and lateral rotation	Obturator and sciatic nerves
Pectineus	Superior ramus of pubis	Pectineal line inferior to lesser trochanter of femur	Flexion, medial rotation, and adduction at hip	Femoral nerve (L_2–L_4)
Gracilis	Inferior ramus of pubis	Medial surface of tibia inferior to medial condyle	Flexion at knee; adduction and medial rotation at hip	Obturator nerve (L_3–L_4)
ILIOPSOAS GROUP				
Iliacus	Iliac fossa	Femur distal to lesser trochanter; tendon fused with that of psoas major	Flexion at hip	Femoral nerve (L_2–L_3)
Psoas major	Anterior surfaces and transverse processes of vertebrae T_{12}–L_5	Lesser trochanter in company with iliacus	Flexion at hip and/or lumbar intervertebral joints	Branches of the lumbar plexus (L_2–L_3)

FOR-mis) **muscle** produces abduction at the hip. The piriformis and the **obturator muscles** (*externus* and *internus*) are the dominant lateral rotators.

The **adductors** are located inferior to the acetabular surface. The adductors include the **adductor magnus**, **adductor brevis**, **adductor longus**, **pectineus** (pek-TIN-ē-us), and **gracilis** (GRAS-i-lis) **muscles** (Figure 11.11●). All originate on the pubis; all of the adductors except the gracilis muscle insert on the linea aspera, a ridge along the posterior surface of the femur. (The gracilis inserts on the tibia.) Their actions are varied. All of the adductors except the adductor magnus muscle originate both anterior and inferior to the hip joint, so they produce hip flexion as well as adduction. They also produce medial rotation at the hip. The adductor magnus muscle can produce either adduction and flexion or adduction and extension, depending on the region stimulated. It may also produce either medial or lateral rotation. When an athlete suffers a *pulled groin,* the problem is a *strain*—a muscle tear or break—in one of these adductor muscles.

The medial surface of the pelvis is dominated by a single pair of muscles. The **psoas** (SŌ-us) **major muscle** originates alongside the inferior

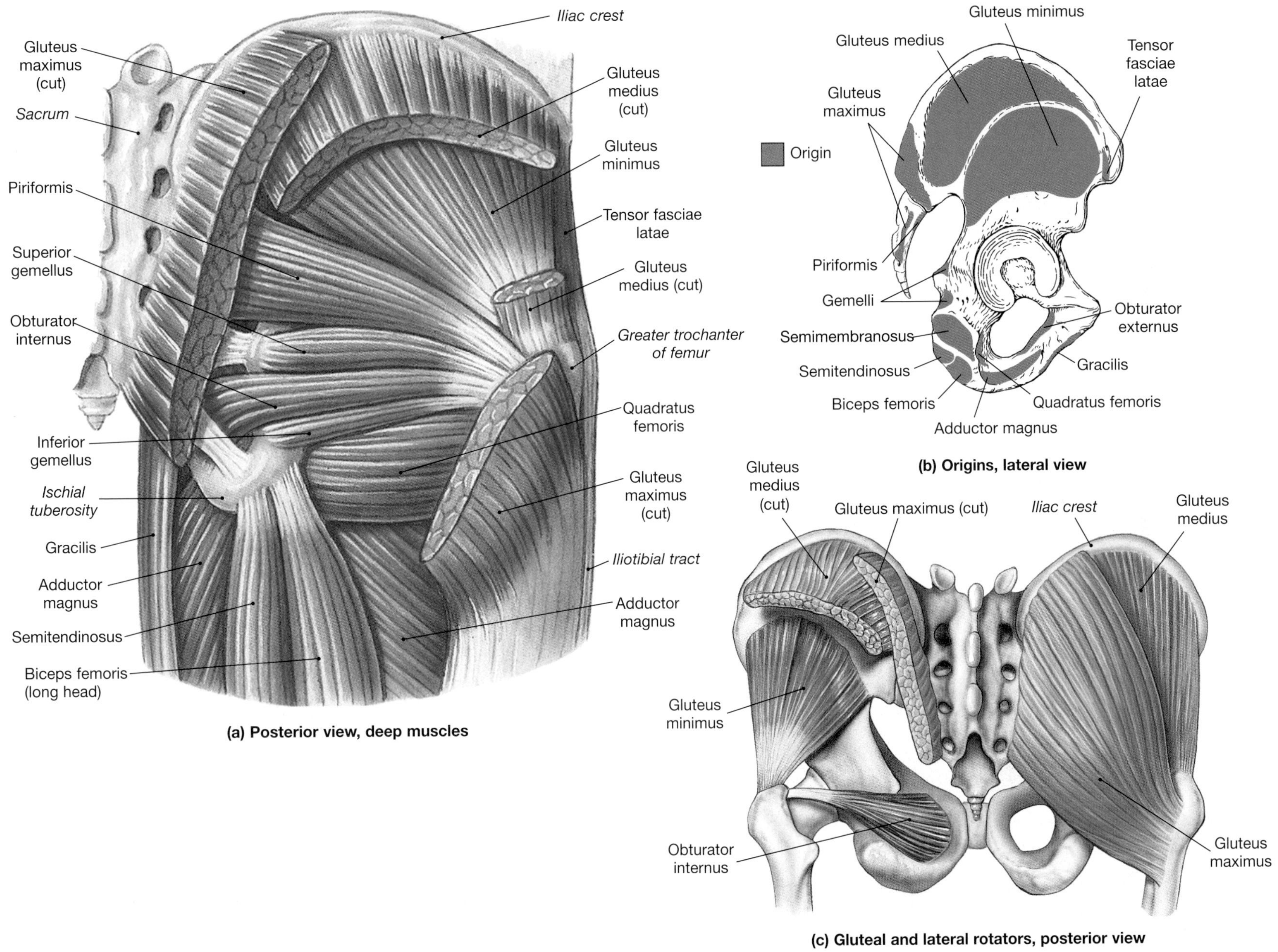

FIGURE 11.10 **Muscles That Move the Thigh, Part I**

The gluteal and lateral rotator muscles of the right hip. (**a**) Posterior view of pelvis, showing deep dissections of the gluteal muscles and lateral rotators. For a superficial view of the gluteal muscles, see Figures 11.1 and 11.13. (**b**) Lateral view of the right pelvis, showing the origins of selected muscles. (**c**) Posterior view of the gluteal and lateral rotator muscles; the gluteus maximus muscle has been removed to show the deeper muscles. *See also Figures 7.10, 7.11, and 7.14.*

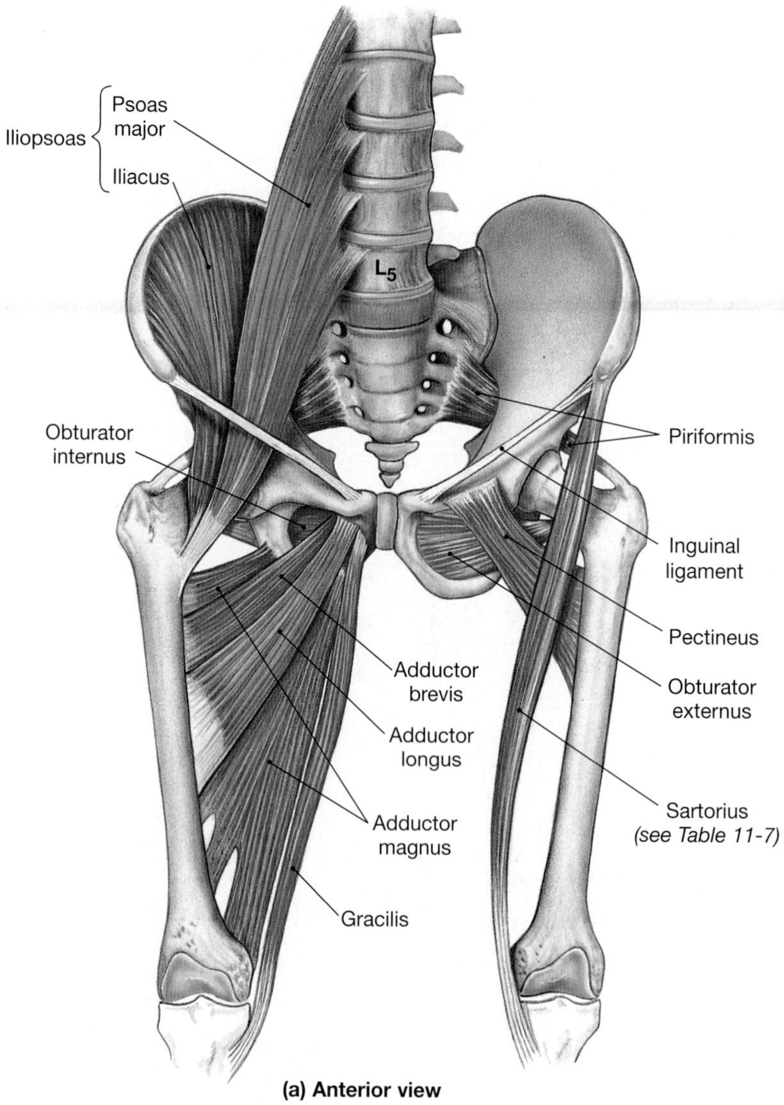

(a) Anterior view

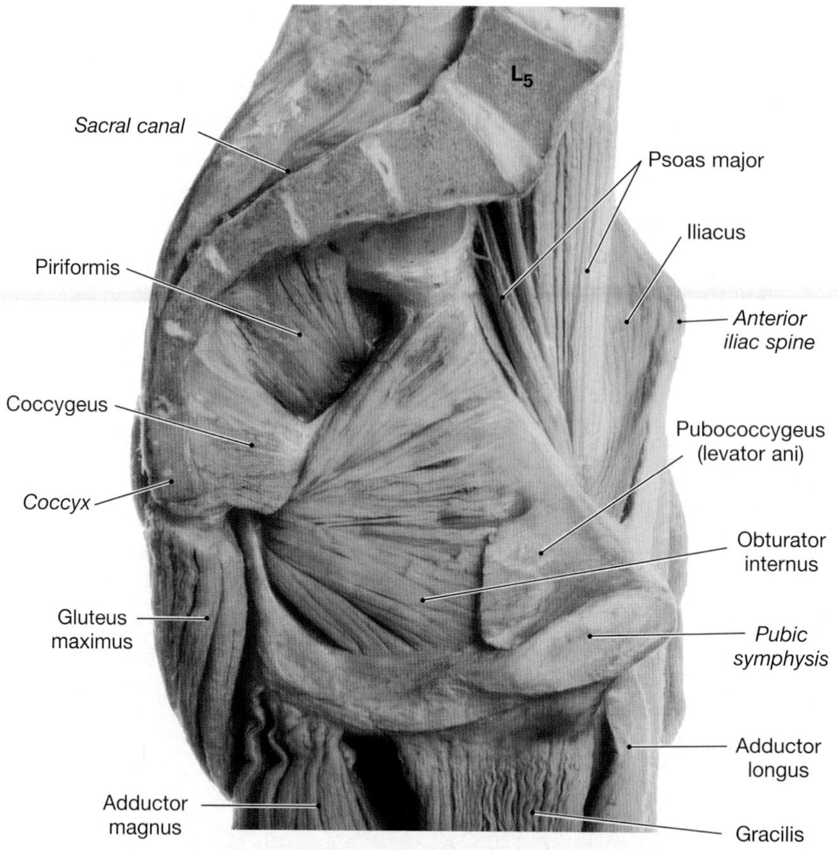

(b) Sagittal section through the pelvis

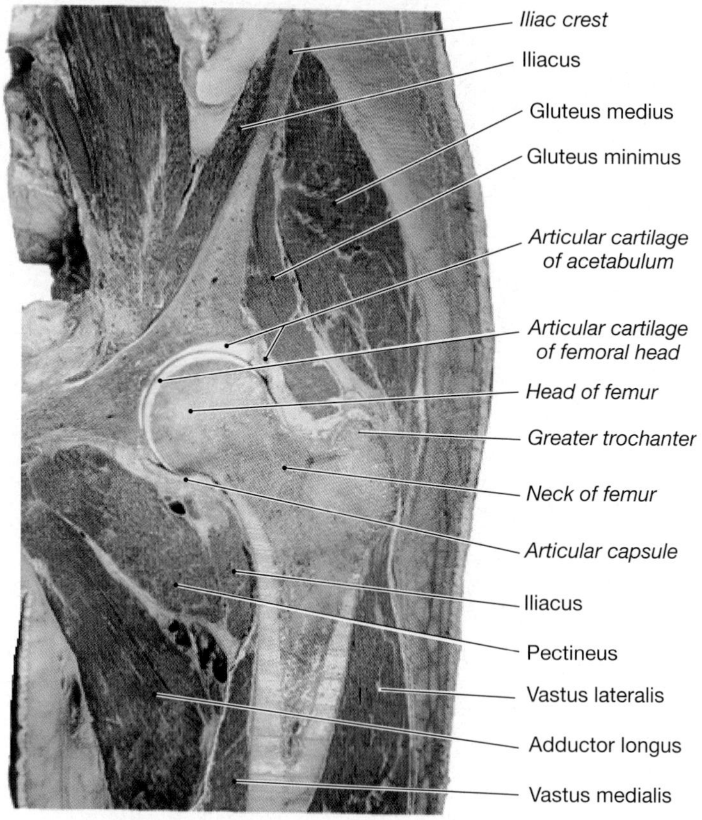

(c) Coronal section through the pelvis

FIGURE 11.11 **Muscles That Move the Thigh, Part II**

The iliopsoas muscle and adductors of the right hip. (**a**) Anterior view of the iliopsoas muscle and the adductor group. (**b**) Muscles and associated structures seen in a sagittal section through the pelvis. *See also Figures 7.10, 7.11, and 7.14.* (**c**) Coronal section through the hip, showing the hip joint in relation to surrounding muscles. *See MRI Scan 4 in the Companion Atlas.*

thoracic and lumbar vertebrae, and its insertion lies on the lesser trochanter of the femur. Before reaching this insertion, its tendon merges with that of the **iliacus** (i-LĒ-ak-us) **muscle**, which lies nestled within the iliac fossa. These two muscles, which are powerful flexors of the hip, pass deep to the *inguinal ligament,* and are often referred to as the **iliopsoas** (i-lē-ō-SŌ-us) **muscle** (Figure 11.11●).

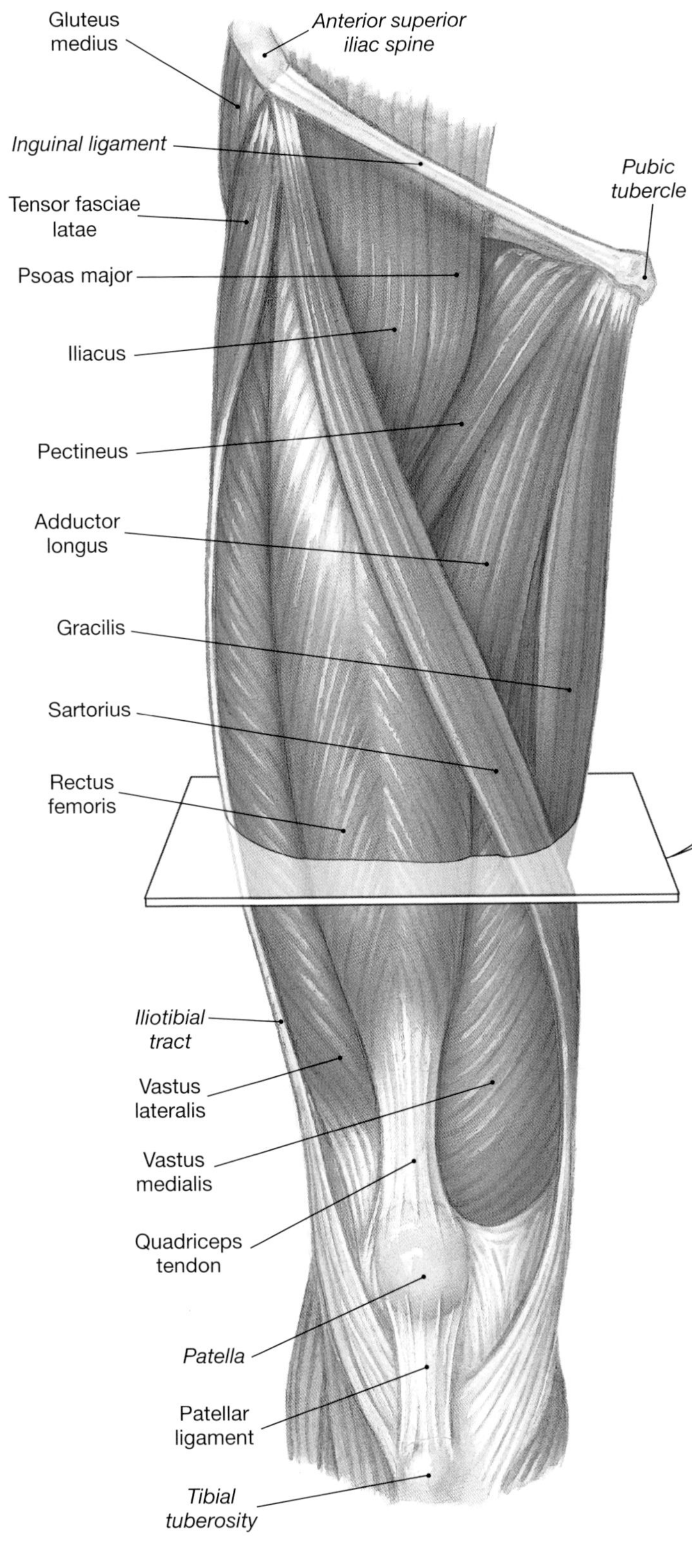

(a) Quadriceps and thigh muscles, anterior view

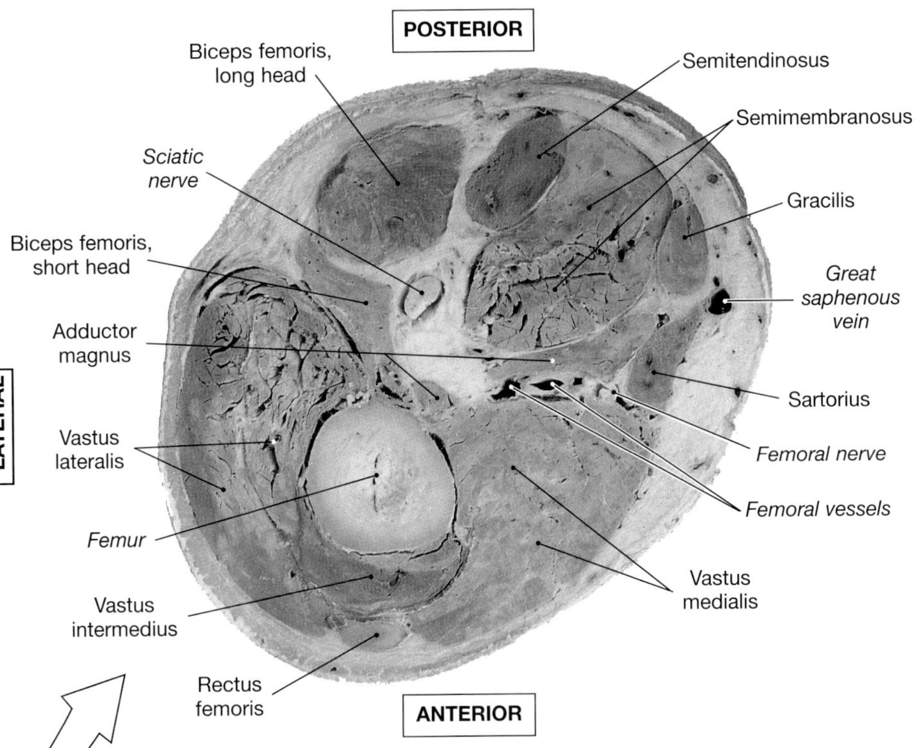

(b) Thigh, transverse section

FIGURE 11.12 **MUSCLES THAT MOVE THE LEG, PART I**

(**a**) Superficial anterior view of the muscles of the right thigh. (**b**) Transverse section of the right thigh. *See Figures 7.10, 7.11, 7.14, and 8.12; also Scans 5–7 in the Companion Atlas.*

Muscles That Move the Leg [FIGURES 11.12 TO 11.15 AND TABLE 11.7]

Muscles that move the leg are detailed in Figures 11.12 to 11.15● and Table 11.7. Most of the extensors originate on the femoral surface and extend along the anterior and lateral surfaces of the thigh (Figures 11.12 and 11.13●). Flexors originate on the pelvic girdle and extend along the posterior and medial surfaces of the thigh (Figure 11.14●). Collectively the knee extensors (Figures 11.12 and 11.13●) are called the *quadriceps muscles* or the *quadriceps femoris.* The three **vastus muscles** (**vastus lateralis**, **vastus medialis**, and **vastus intermedius**) originate along the body of the femur, and they cradle the **rectus femoris muscle** the way a bun surrounds a hot dog. All four muscles insert upon the patella and reach the tibial tuberosity via the patellar ligament. The rectus femoris muscle originates on the anterior inferior iliac spine, so in addition to producing extension of the knee, it can assist in flexion of the hip.

The flexors of the knee include: the **biceps femoris**, **semimembranosus** (sem-ē-mem-bra-NŌ-sus), **semitendinosus** (sem-ē-ten-di-NŌ-sus), and **sartorius** (sar-TOR-ē-us) **muscles** (Figures 11.12a, 11.13a,b

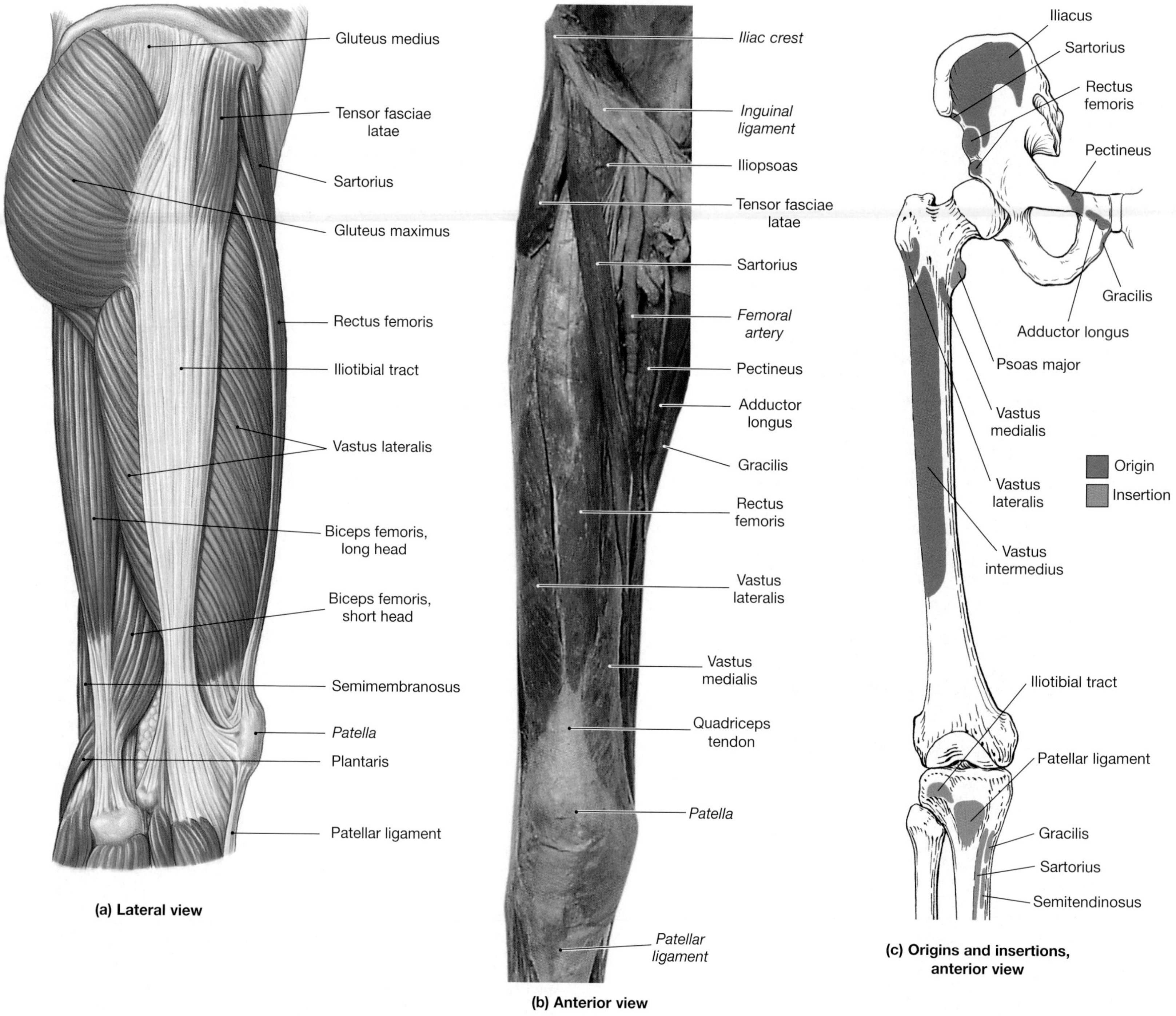

FIGURE 11.13 **MUSCLES THAT MOVE THE LEG, PART II**

(**a**) Lateral view of the muscles of the right thigh. (**b**) Anterior view of a dissection of the muscles of the right thigh. (**c**) Anterior view of the bones of the right lower limb showing the origins and insertions of selected muscles.

and 11.14●). These muscles originate along the edges of the pelvis and insert upon the tibia and fibula. Their contractions produce flexion at the knee. Because the biceps femoris, semimembranosus, and semitendinosus muscles originate on the pelvis inferior and posterior to the acetabulum, their contractions also produce extension at the hip. These muscles are often called the *hamstrings*. The sartorius muscle is the only knee flexor that originates superior to the acetabulum, and its insertion lies along the medial aspect of the tibia. When it contracts, it produces flexion and lateral rotation at the hip, as when crossing the legs. In Chapter 8 we noted that the knee joint can be locked at full extension by a slight lateral rotation of the tibia. ∞ p. 232 The small **popliteus** (pop-LI-tē-us) **muscle** originates on the femur near the lateral condyle and inserts on the posterior tibial shaft (Figure 11.15●). When knee flexion is initiated, this muscle contracts to produce a slight medial rotation of the tibia that unlocks the joint. Figure 11.14d● shows the surface anatomy of the posterior surface of the thigh and the landmarks associated with some of the knee flexors.

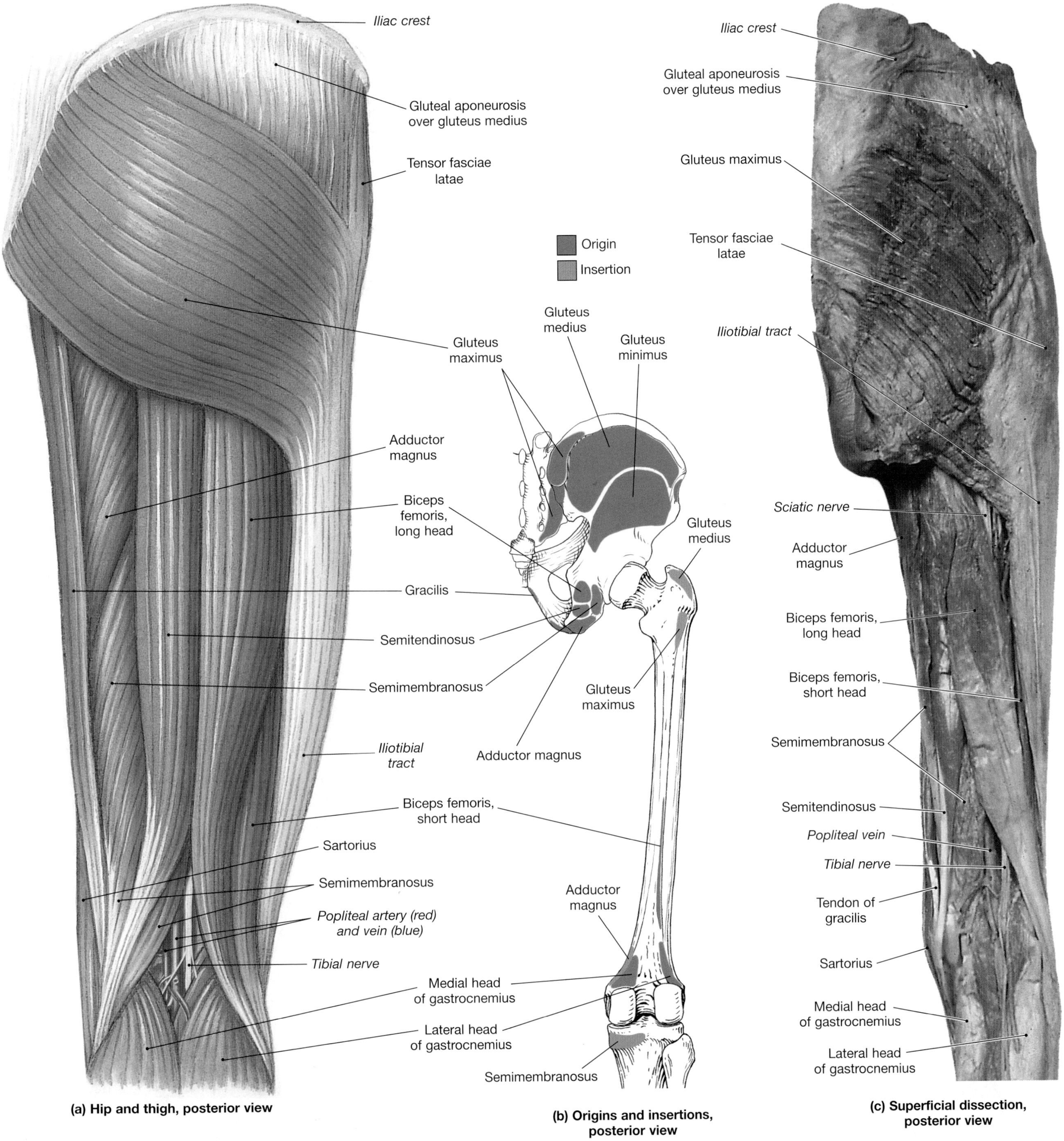

FIGURE 11.14 **MUSCLES THAT MOVE THE LEG, PART III**

(**a**) Posterior view of superficial muscles of the right thigh. (**b**) Posterior view of the bones of the right hip, thigh, and proximal leg, showing the origins and insertions of selected muscles (part 1). (**c**) Posterior view of a dissection of the muscles of the thigh and proximal leg. (**d**) Surface anatomy of the right thigh, posterior view. (**e**) Deep muscles of the posterior thigh. (**f**) Posterior view of the bones of the right hip, thigh, and proximal leg, showing the origins and insertions of selected muscles (part 2). *See also Scans 4–7 in the Companion Atlas.*

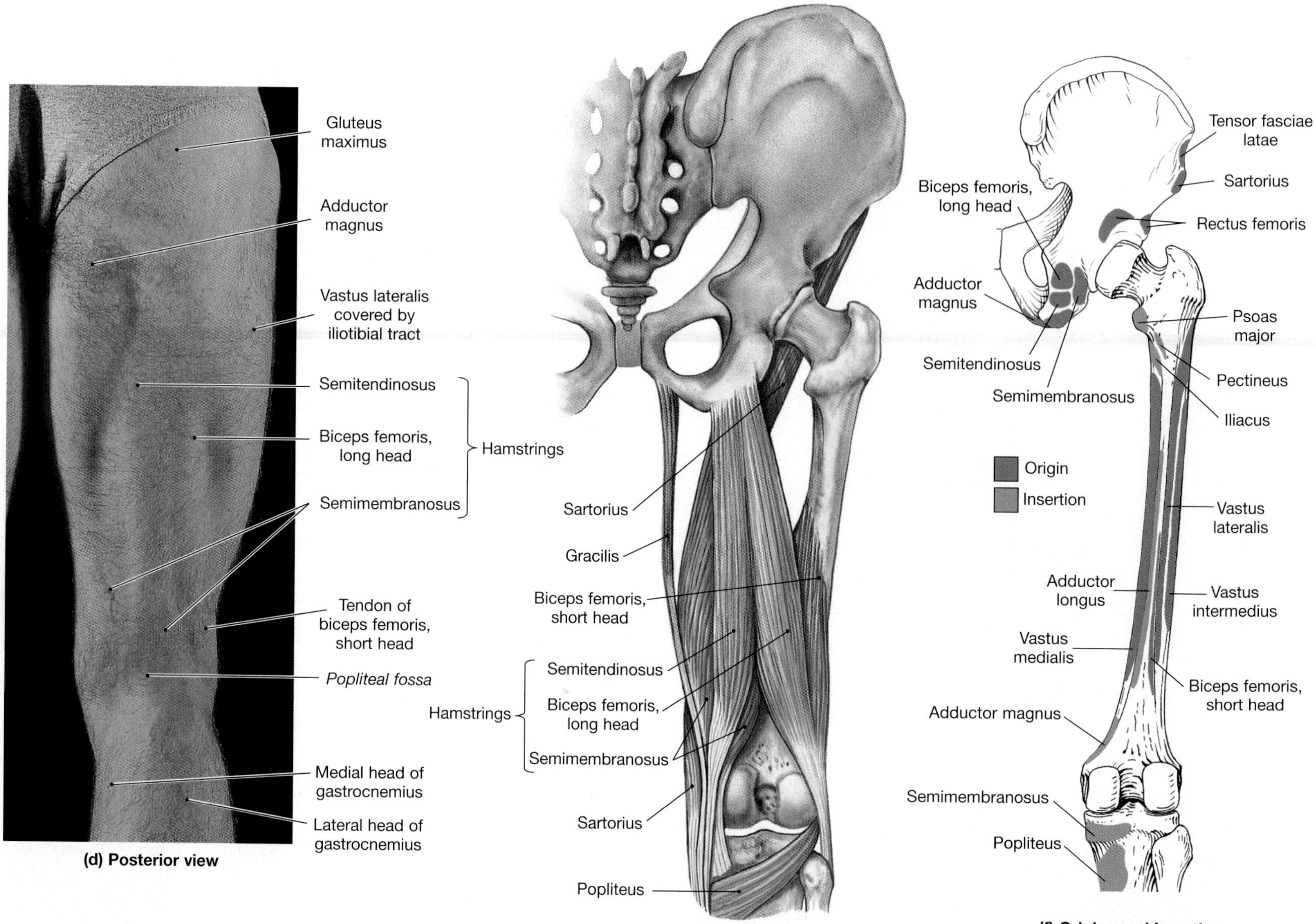

(d) Posterior view

(e) Posterior view

(f) Origins and Insertions, posterior view

FIGURE 11.14 *(continued)*

TABLE 11.7

Muscles That Move the Leg

Muscle	Origin	Insertion	Action	Innervation
FLEXORS OF THE KNEE				
Biceps femoris	Ischial tuberosity and linea aspera of femur	Head of fibula, lateral condyle of tibia	Flexion at knee; extension and lateral rotation at hip	Sciatic nerve; tibial portion (S_1–S_3 to long head) and common fibular branch (L_5–S_2 to short head)
Semimembranosus	Ischial tuberosity	Posterior surface of medial condyle of tibia	Flexion at knee; extension and medial rotation at hip	Sciatic nerve (tibial portion L_5–S_2)
Semitendinosus	As above	Proximal, medial surface of tibia near insertion of gracilis	As above	As above
Sartorius	Anterior superior iliac spine	Medial surface of tibia near tibial tuberosity	Flexion at knee; flexion and lateral rotation at hip	Femoral nerve (L_2–L_3)
Popliteus	Lateral condyle of femur	Posterior surface of proximal tibial shaft	Medial rotation of tibia (or lateral rotation of femur) at knee	Tibial nerve (L_4–S_1)
EXTENSORS OF THE KNEE				
Rectus femoris	Anterior inferior iliac spine and superior acetabular rim of ilium	Tibial tuberosity via patellar tendon	Extension at knee; flexion at hip	Femoral nerve (L_2–L_4)
Vastus intermedius	Anterolateral surface of femur and linea aspera (distal half)	As above	Extension at knee	As above
Vastus lateralis	Anterior and inferior to greater trochanter of femur and along linea aspera (proximal half)	As above	As above	As above
Vastus medialis	Entire length of linea aspera of femur	As above	As above	As above

Muscles That Move the Foot and Toes [FIGURES 11.15 TO 11.18 AND TABLES 11.8/11.9]

Extrinsic and intrinsic muscles that move the foot and toes (Figures 11.15 to 11.18●) are detailed in Tables 11.8 and 11.9. Most of the muscles that move the ankle produce the plantar flexion involved with walking and running movements.

The large **gastrocnemius** (gas-trok-NĒ-mē-us; *gaster*, stomach + *kneme*, knee) **muscle** of the calf is an important plantar flexor, but the slow muscle fibers of the underlying **soleus** (SŌ-lē-us) **muscle** make this the more powerful muscle. These muscles are best seen from the posterior and lateral views (Figures 11.15 and 11.16b,c●). The gastrocnemius muscle arises from two heads located on the medial and lateral epicondyles of the femur just proximal to the knee. A sesamoid bone, the *fabella*, is usually found within the gastrocnemius muscle. The gastrocnemius and soleus muscles share a common tendon, the **calcaneal tendon**. This tendon may also be called the *calcanean tendon* or the *Achilles tendon*.

Deep to the gastrocnemius and soleus muscles lie the two **fibularis muscles** (Figure 11.15b,c,d●). These muscles, also known as the *peroneus muscles*, produce eversion of the foot as well as plantar flexion of the ankle. Inversion of the foot is caused by contraction of the **tibialis** (tib-ē-Ā-lis) **muscles**; the large **tibialis anterior muscle** opposes the gastrocnemius muscle and dorsiflexes the ankle (Figures 11.16 and 11.17●).

Important muscles that move the toes originate on the surface of the tibia, the fibula, or both (Figures 11.15, 11.16, and 11.17●). Large tendon sheaths surround the tendons of the tibialis anterior, extensor digitorum longus, and extensor hallucis longus muscles where they cross the ankle joint. The positions of these sheaths are stabilized by the **superior** and **inferior extensor retinacula** (Figures 11.16, 11.17a, and 11.18a●).

The small intrinsic muscles that move the toes originate on the bones of the tarsus and foot (Figure 11.18● and Table 11.9). Some of the flexor muscles originate at the anterior border of the calcaneus; their muscle tone contributes to maintenance of the longitudinal arch of the foot. As in the hand, small **interosseus muscles** (plural, *interossei*) originate on the lateral and medial surfaces of the metatarsal bones.

✓ CONCEPT CHECK

- What leg movement would be impaired by injury to the obturator muscle?
- Often one hears of athletes suffering a "pulled hamstring." Describe the muscles in such an injury.
- To which group of muscles do the pectineus and gracilis belong?
- What is the collective name for the knee extensors?

TABLE 11.8 EXTRINSIC MUSCLES THAT MOVE THE FOOT AND TOES

Muscle	Origin	Insertion	Action	Innervation
ACTION AT THE ANKLE				
DORSIFLEXORS				
Tibialis anterior	Lateral condyle and proximal shaft of tibia	Base of 1st metatarsal bone and medial cuneiform bone	Dorsiflexion at ankle; inversion of foot	Deep fibular nerve (L_4–S_1)
PLANTAR FLEXORS				
Gastrocnemius	Femoral condyles	Calcaneus via calcaneal tendon	Plantar flexion at ankle; inversion and adduction of foot; flexion at knee	Tibial nerve (S_1–S_2)
Fibularis brevis	Midlateral margin of fibula	Base of 5th metatarsal bone	Eversion of foot and plantar flexion at ankle	Superficial fibular nerve (L_4–S_1)
Fibularis longus	Head and proximal shaft of fibula	Base of 1st metatarsal bone and medial cuneiform bone	Eversion of foot and plantar flexion at ankle; supports ankle; supports longitudinal arch	As above
Plantaris	Lateral supracondylar ridge	Posterior portion of calcaneus	Plantar flexion at ankle; flexon at knee	Tibial nerve (L_4–S_1)
Soleus	Head and proximal shaft of fibula, and adjacent posteromedial shaft of tibia	Calcaneus via calcaneal tendon (with gastrocnemius)	Plantar flexion at ankle; postural muscle when standing	Sciatic nerve, tibial branch (S_1–S_2)
Tibialis posterior	Interosseous membrane and adjacent shafts of tibia and fibula	Navicular, all 3 cuneiforms, cuboid, 2nd, 3rd, and 4th metatarsal bones	Adduction and inversion of foot; plantar flexion at ankle	As above
ACTION AT THE TOES				
DIGITAL FLEXORS				
Flexor digitorum longus	Posteromedial surface of tibia	Inferior surface of distal phalanges, toes 2–5	Flexion of joints of toes 2–5; plantar flexes ankle; inverts foot	Sciatic nerve, tibial branch (L_5–S_1)
Flexor hallucis longus	Posterior surface of fibula	Inferior surface, distal phalanx of great toe	Flexion at joints of great toe; plantar flexes ankle; inverts foot	As above
DIGITAL EXTENSORS				
Extensor digitorum longus	Lateral condyle of tibia, anterior surface of fibula	Superior surfaces of phalanges, toes 2–5	Extension of toes 2–5; dorsiflexes ankle and everts foot	Deep fibular nerve (L_4–S_1)
Extensor hallucis longus	Anterior surface of fibula	Superior surface, distal phalanx of great toe	Extension at joints of great toe; dorsiflexes ankle and everts foot	As above

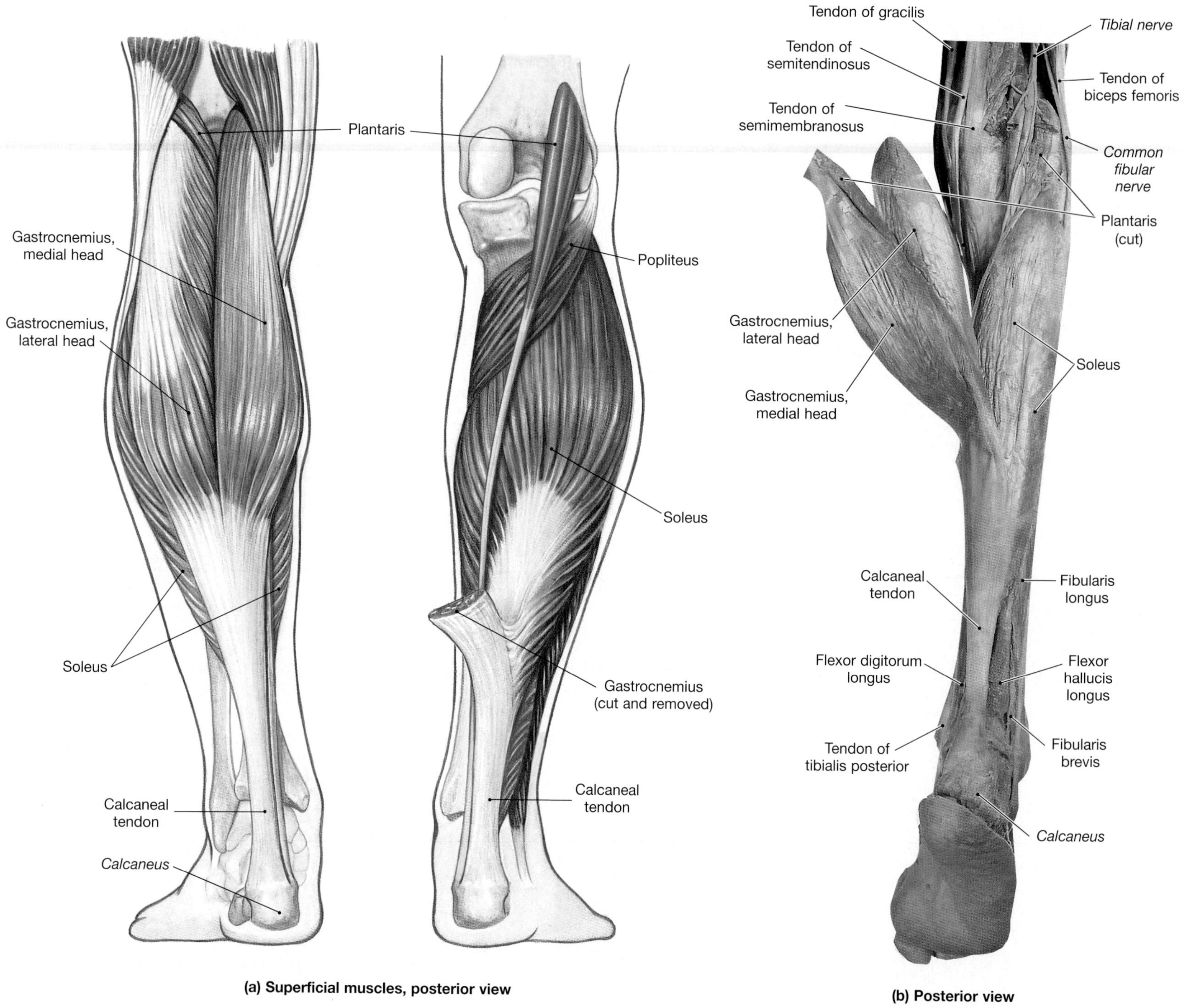

FIGURE 11.15 **Extrinsic Muscles That Move the Foot and Toes, Part I**

(**a**) Superficial muscles of the posterior surface of the legs; these large muscles are primarily responsible for plantar flexion. (**b**) Posterior view of a dissection of the superficial muscles of the right leg. (**c**) Posterior view of deeper muscles of the leg. (**d**) A posterior view of the bones of the right leg and foot, showing the origins and insertions of selected muscles. *See also Figures 7.16 and 7.17 and MRI Scans 5–7 in the Companion Atlas.*

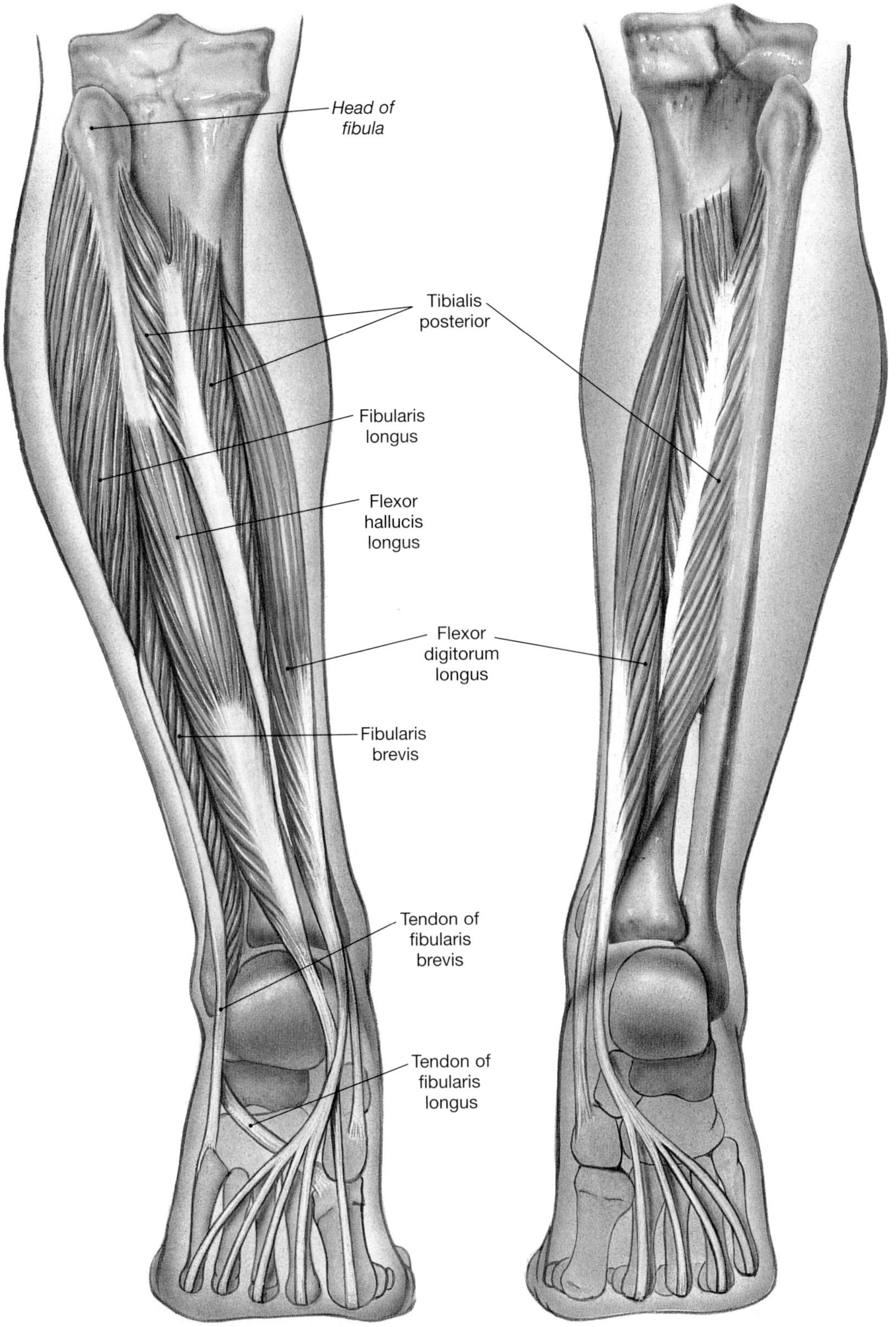

(c) Deep muscles, posterior view

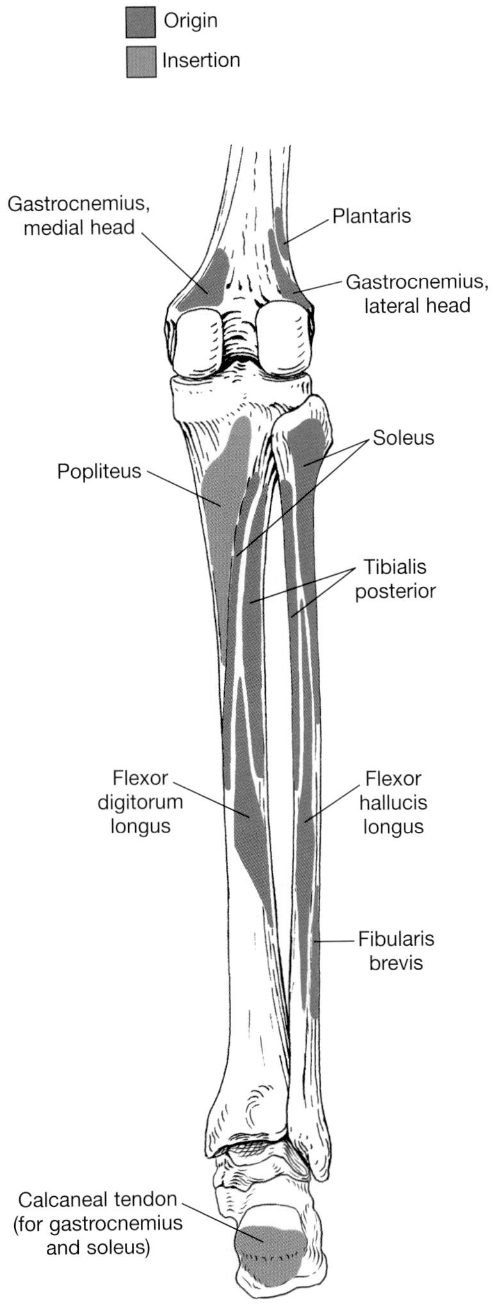

(d) Origins and insertions, posterior view

FIGURE 11.15 *(continued)*

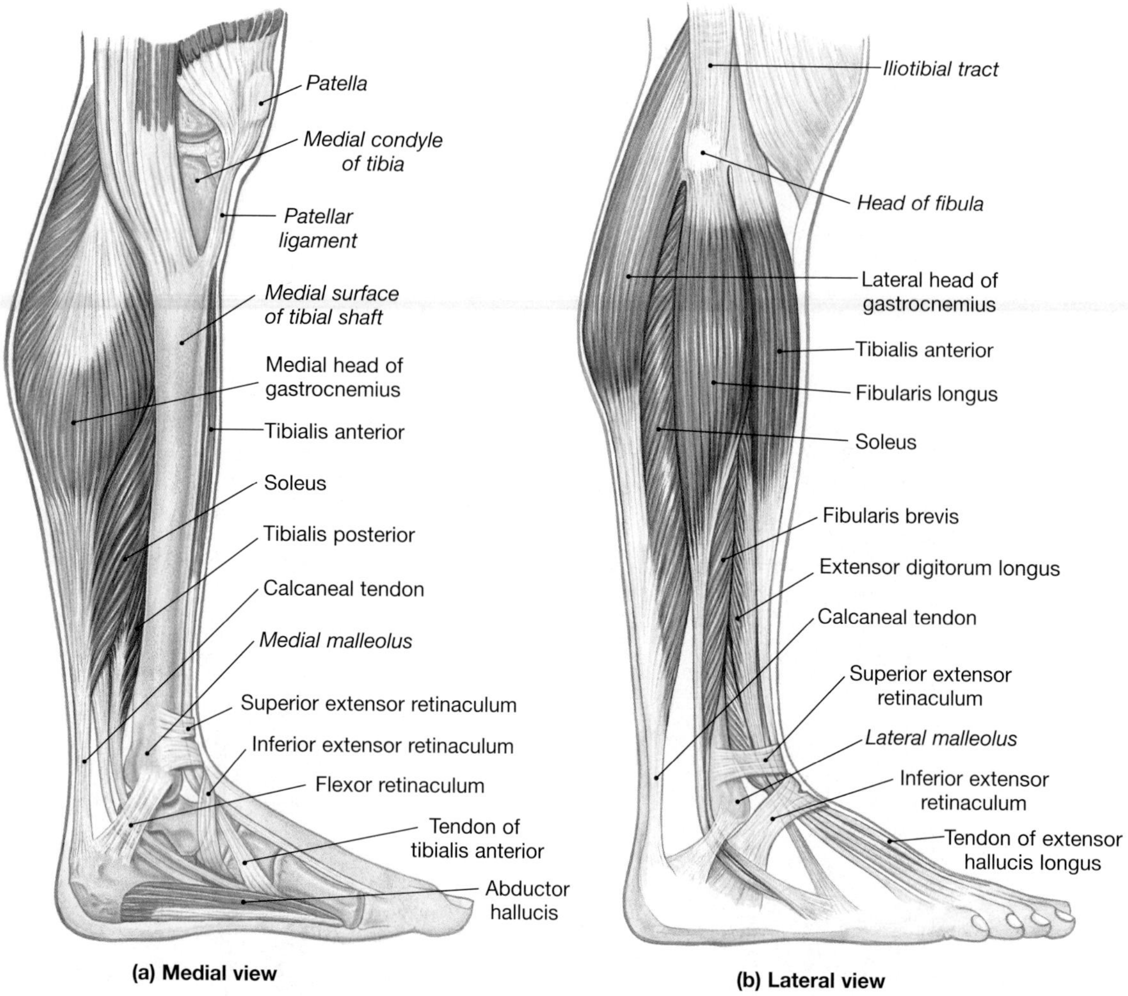

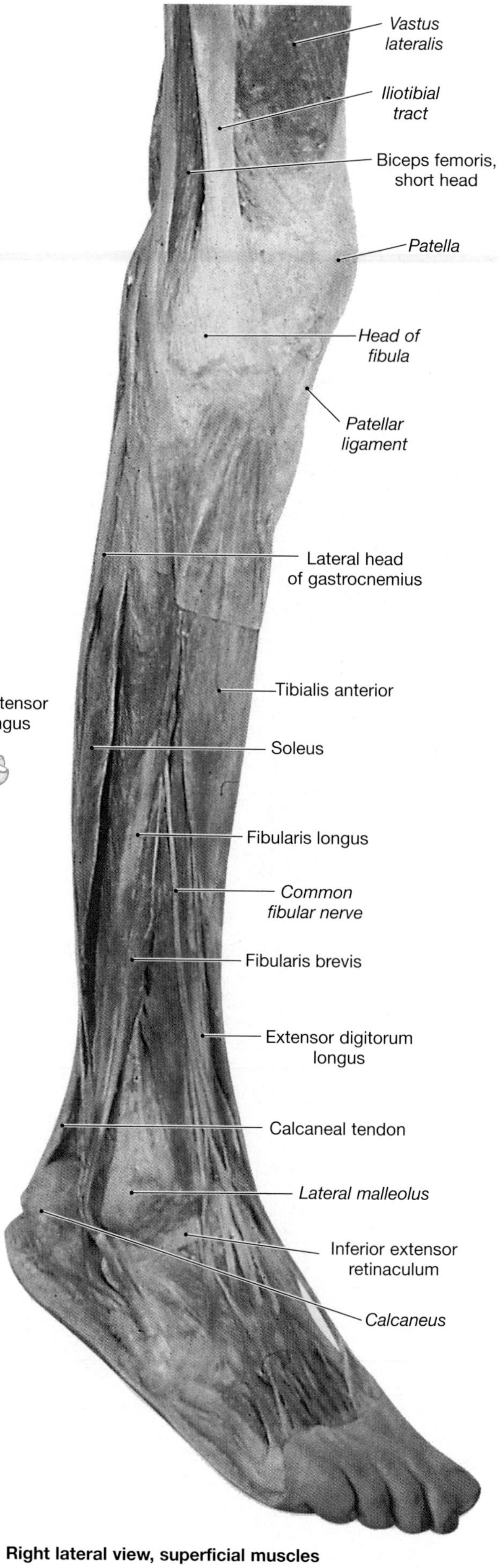

FIGURE 11.16 **Extrinsic Muscles That Move the Foot and Toes, Part II**

(**a**) Medial view of the superficial muscles of the left leg. (**b**) Lateral view of the superficial muscles of the right leg. (**c**) Lateral view of a dissection of the superficial muscles of the right leg. *See also Scans 5–7 in the Companion Atlas.*

CLINICAL BRIEF

Intramuscular Injections

Drugs are frequently injected into tissues rather than directly into the circulation. This method enables the physician to introduce a large amount of a drug at one time, yet have it enter the circulation gradually. In an **intramuscular (IM) injection**, the drug is introduced into the mass of a large skeletal muscle. Uptake is usually faster and accompanied by less tissue irritation than when drugs are administered *intradermally* or *subcutaneously* (injected into the dermis or subcutaneous layers). p. 95. Up to 5 ml of fluid may be injected at one time, and multiple injections are possible.

The most common complications involve accidental injection into a blood vessel or piercing of a nerve. The sudden entry of massive quantities of a drug into the bloodstream can have unpleasant or even fatal consequences, while damage to a nerve can cause pain, motor paralysis, or sensory loss. As a result, the site of injection must be selected with care. Bulky muscles that contain few large vessels or nerves make ideal targets; thus, the gluteus medius or the posterior, lateral, superior portion of the gluteus maximus is often selected. The deltoid muscle of the arm, about 2.5 cm (1 in.) distal to the acromion, is another popular site. Probably the most satisfactory from an anatomical point of view is the vastus lateralis of the thigh, for an injection into this thick muscle will not encounter large vessels or nerves. However, because these musles are used in walking, there can be lingering pain after an injection. This is the preferred site in infants whose gluteal and deltoid muscles are relatively small and who have yet to learn to walk.

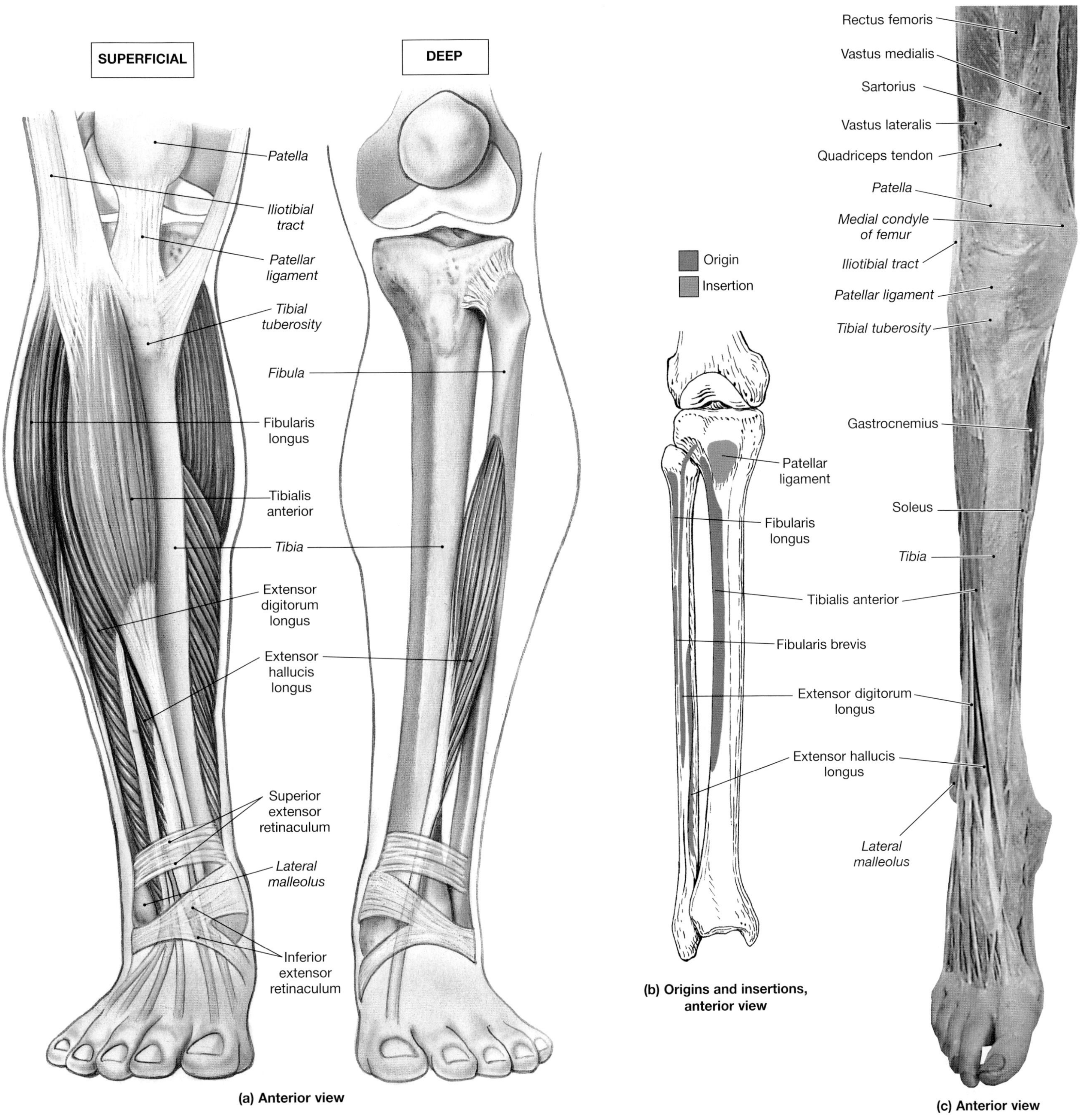

FIGURE 11.17 **EXTRINSIC MUSCLES THAT MOVE THE FOOT AND TOES, PART III**

(**a**) Anterior views showing superficial and deep muscles of the right leg and left leg, respectively. (**b**) Anterior view of the bones of the right leg, showing the origins and insertions of selected muscles. (**c**) Anterior view of a dissection of the superficial muscles of the right leg.

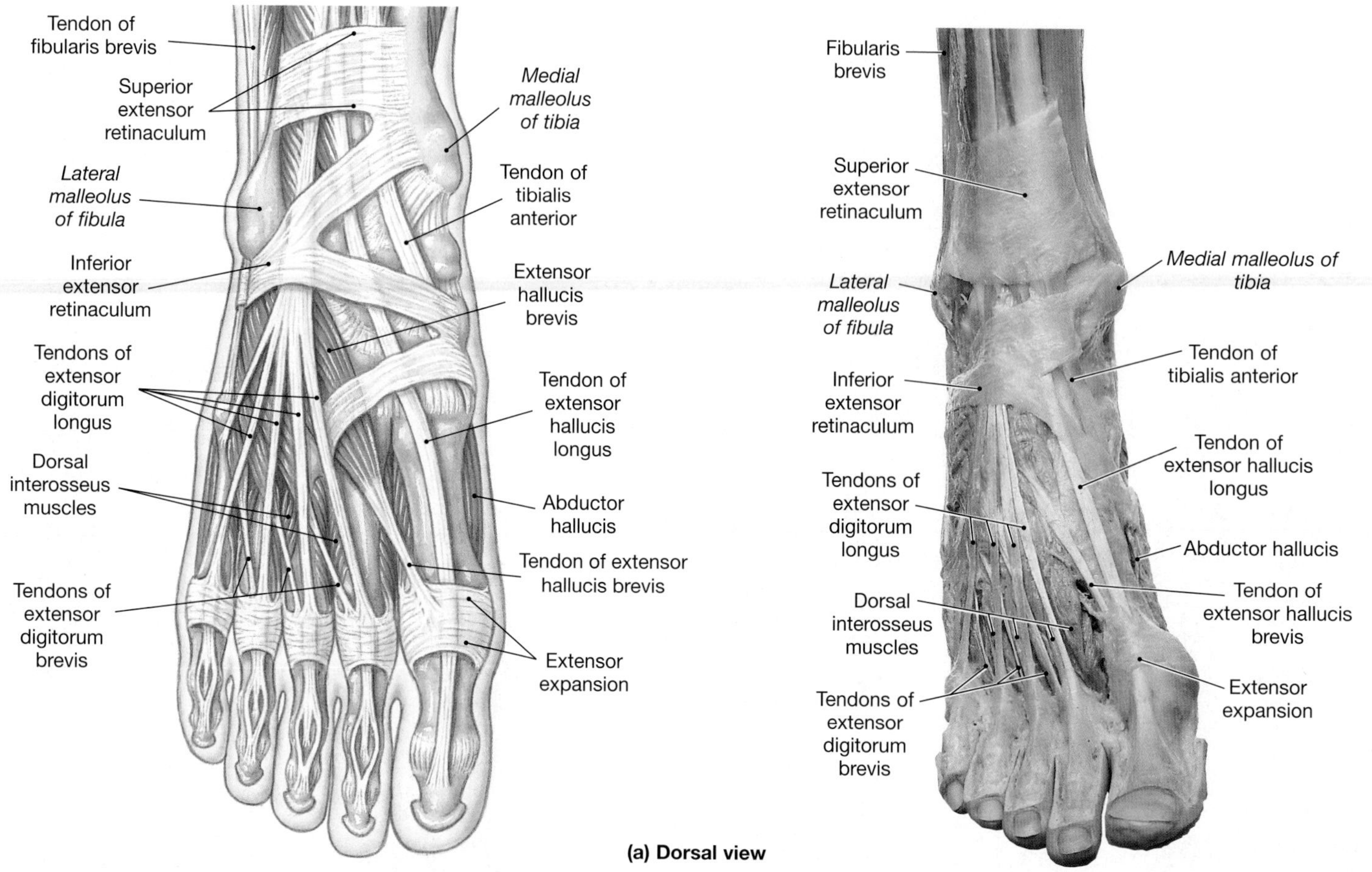

(a) Dorsal view

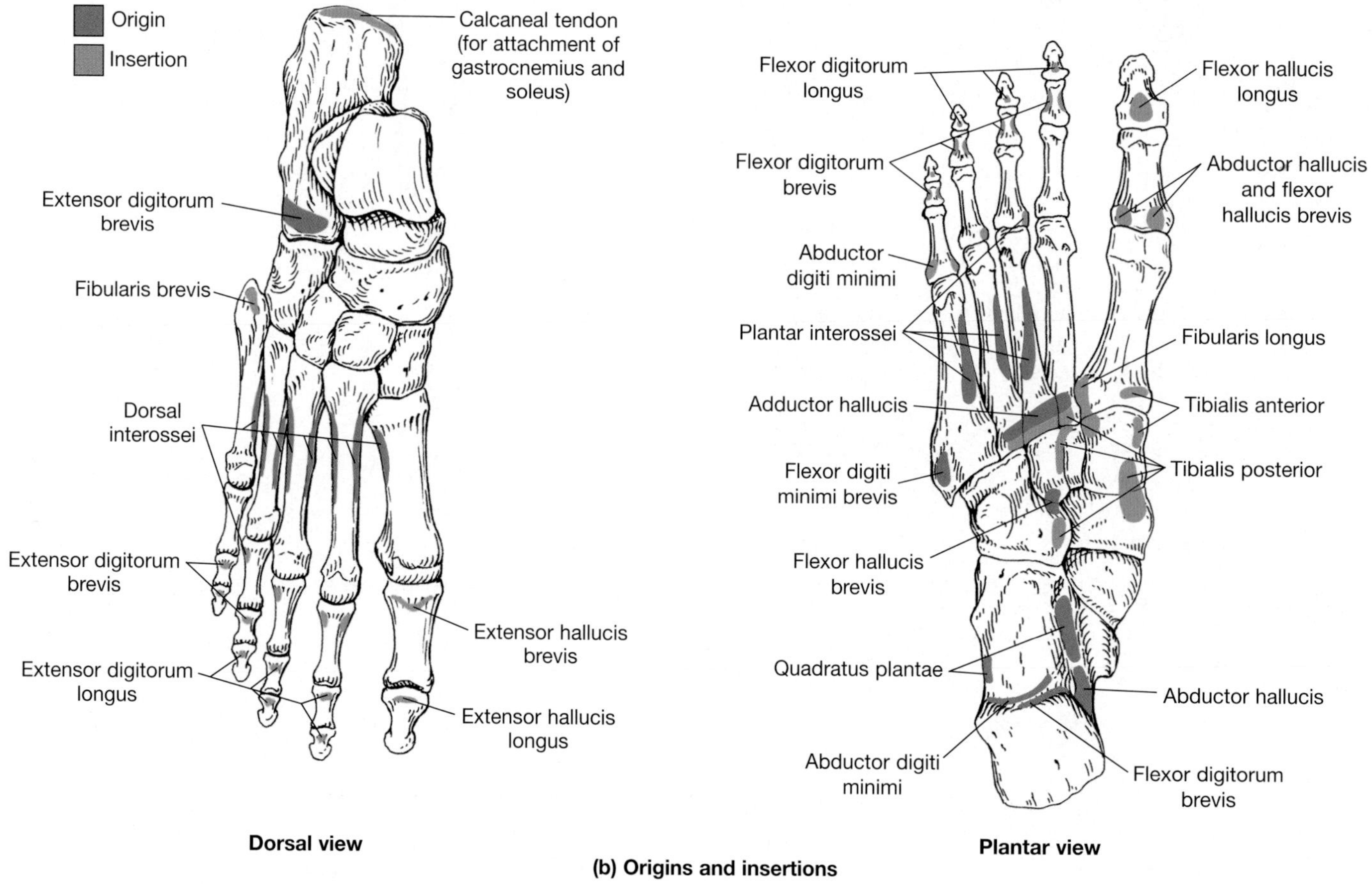

(b) Origins and insertions

FIGURE 11.18 **Intrinsic Muscles That Move the Foot and Toes**

(a) Dorsal views of right foot. **(b)** Dorsal (superior) and plantar (inferior) views of the bones of the right foot, showing the origins and insertions of selected muscles. **(c)** Right foot, sectional view through the metatarsal bones. **(d)** Plantar (inferior) view, superficial layer of the right foot. **(e)** Plantar (inferior) view, deep layer of the right foot **(f)** Plantar (inferior) view, deepest layer of the right foot.

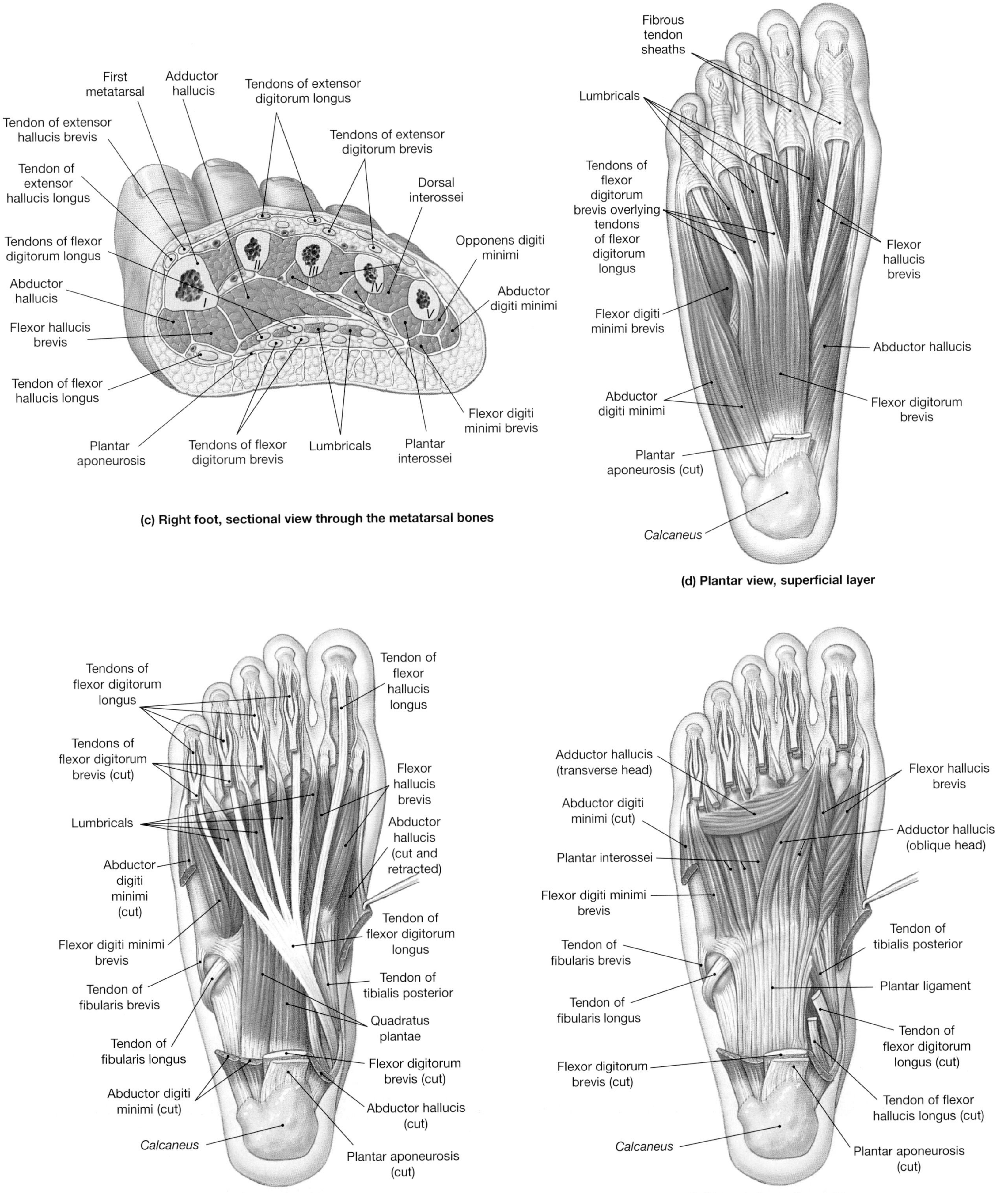

(c) Right foot, sectional view through the metatarsal bones

(d) Plantar view, superficial layer

(e) Plantar view, deep layer

(f) Plantar view, deepest layer

FIGURE 11.18 *(continued)*

TABLE 11.9

INTRINSIC MUSCLES OF THE FOOT

Muscle	Origin	Insertion	Action	Innervation
Extensor digitorum brevis	Calcaneus (superior and lateral surfaces)	Dorsal surface of toes 1–4	Extension at metatarsophalangeal joints of toes 1–4	Deep fibular nerve (L_5–S_1)
Abductor hallucis	Calcaneus (tuberosity on inferior surface)	Medial side of proximal phalanx of great toe	Abduction at metatarsophalangeal jointof great toe	Medial plantar nerve (L_4–L_5)
Flexor digitorum brevis	As above	Sides of middle phalanges, toes 2–5	Flexion of proximal interphalangeal joints of toes 2–5	As above
Abductor digiti minimi	As above	Lateral side of proximal phalanx, toe 5	Abduction at metatarsophalangeal joint of toe 5	Lateral plantar nerve (L_4–L_5)
Quadratus plantae	Calcaneus (medial, inferior surfaces)	Tendon of flexor digitorum longus	Flexion at joints of toes 2–5	As above
Lumbricals (4)	Tendons of flexor digitorum longus	Insertions of extensor digitorum longus	Flexion at metatarsophalaneal joints; extension at proximal interphalangeal joints of toes 2–5	Medial plantar nerve (1), lateral plantar nerve (2–4)
Flexor hallucis brevis	Cuboid and lateral cuneiform bones	Proximal phalanx of great toe	Flexion at metatarsophalangeal joint of great toe	Medial plantar nerve (L_4–S_5)
Adductor hallucis	Bases of metatarsal bones II–IV and plantar ligaments	As above	Adduction at metatarsophalangeal joint of great toe	Lateral plantar nerve (S_1–S_2)
Flexor digiti minimi brevis	Base of metatarsal bone V	Lateral side of proximal phalanx of toe 5	Flexion at metatarsophalangeal joint of toe 5	As above
Dorsal interossei (4)	Sides of metatarsal bones	Medial and lateral sides of toe 2; lateral sides of toes 3 and 4	Abduction at metatarsophalangeal joints of toes 3 and 4	As above
Plantar interossei (3)	Bases and medial sides of metatarsal bones	Medial sides of toes 3–5	Adduction of metatarsophalangeal joints of toes 3–5	As above

CLINICAL DISCUSSION

COMPARTMENT SYNDROME

In the limbs, the interconnections between the superficial fascia, the deep fascia of the muscles, and the periostea of the appendicular skeleton are quite substantial. The muscles within a limb are effectively isolated in **compartments** formed by dense collagenous sheets, as shown in Figure 11.19●. Blood vessels and nerves traveling to specific muscles within the limb enter and branch within the appropriate compartments. When a crushing injury, severe contusion, or strain occurs, the blood vessels within one or more compartments may be damaged. When damaged, these compartments become swollen with tissue, fluid, and blood that has leaked from damaged blood vessels. Because the connective tissue partitions are very strong, the accumulated fluid cannot escape, and pressures rise within the affected compartments. Eventually compartment pressures may become so high that they compress the regional blood vessels and eliminate the circulatory supply to the muscles and nerves of the compartment. This compression produces a condition of **ischemia** (is-KĒ-mē-a), or "blood starvation," known as **compartment syndrome**. Slicing into the compartment along its longitudinal axis or implanting a drain are emergency measures used to relieve the pressure. If such steps are not taken, the contents of the compartment will suffer severe damage. Nerves in the affected compartment will be destroyed after 2–4 hours of ischemia, although they can regenerate to some degree if the circulation is restored. After six hours or more, the muscle tissue will also be destroyed, and no regeneration will occur. The muscles will be replaced by scar tissue, and shortening of the connective tissue fibers may result in *contracture*, a permanent reduction in muscle length.

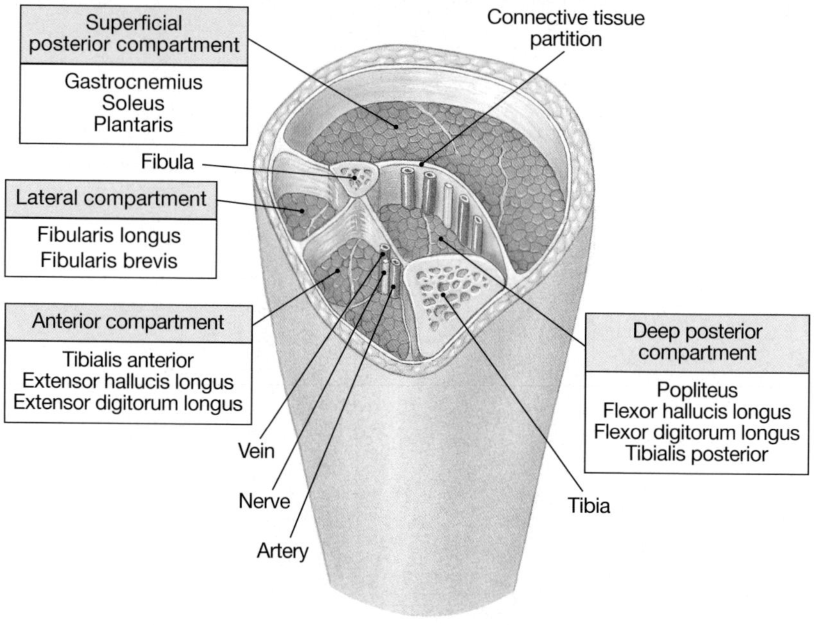

FIGURE 11.19 **MUSCULOSKELETAL COMPARTMENTS**
A diagrammatic section through the right leg, with the muscles removed to show the arrangement of the compartments.

RELATED CLINICAL TERMS

bone bruise: Bleeding within the periosteum of a bone. *Sports Injuries p. 792*

bursitis: Inflammation of the bursae around one or more joints. *Sports Injuries p. 792*

carpal tunnel syndrome: An inflammation within the sheath surrounding the flexor tendons of the palm. *p. 301*

compartment syndrome: Ischemia resulting from accumulated blood and fluid trapped within a musculoskeletal compartment. *p. 320*

intramuscular (IM) injection: Injection into the mass of a large skeletal muscle. *p. 316*

ischemia (is-KĒ-mē-a): A condition of "blood starvation" resulting from compression of regional blood vessels. *p. 320*

muscle cramps: Prolonged, involuntary, painful muscular contractions. *Sports Injuries p. 792*

rotator cuff: The muscles that surround the shoulder joint; a frequent site of sports injuries. *p. 295*

sprains: Tears or breaks in ligaments or tendons. *Sports Injuries p. 792*

strains: Tears or breaks in muscles. *Sports Injuries p. 792*

stress fractures: Cracks or breaks in bones subjected to repeated stress or trauma. *Sports Injuries p. 792*

tendinitis: Inflammation of the connective tissue surrounding a tendon. *Sports Injuries p. 792*

STUDY OUTLINE & CHAPTER REVIEW

Introduction 291

1. The **appendicular musculature** is responsible for stabilizing the pectoral and pelvic girdles and for moving the upper and lower limbs.

Muscles of the Pectoral Girdle and Upper Limbs 291

1. There are four groups of muscles associated with the pectoral girdle and upper limbs: (1) *muscles that position the pectoral girdle*, (2) *muscles that move the arm*, (3) *muscles that move the forearm and hand*, and (4) *muscles that move the hand and fingers*.

Muscles That Position the Pectoral Girdle 292

2. The **trapezius muscles** cover the back and parts of the neck, to the base of the skull. The trapezius muscle affects the position of the pectoral (shoulder) girdle, head, and neck. *(see Figures 11.1 to 11.4, and Table 11.1)*
3. Deep to the trapezius, the **rhomboid muscles** adduct the scapula, and the **levator scapulae muscle** elevates the scapula. Both insert on the scapula. *(see Figures 11.1 to 11.4, and Table 11.1)*
4. The **serratus anterior muscle**, which abducts the scapula and swings the shoulder anteriorly, originates along the ventrosuperior surfaces of several ribs. *(see Figures 11.1 to 11.4, and Table 11.1)*
5. Two deep chest muscles arise along the ventral surfaces of the ribs. Both the **subclavius** and the **pectoralis minor muscles** depress and protract the shoulder. *(see Figures 11.1 to 11.4, and Table 11.1)*

Muscles That Move the Arm 295

6. Muscles that move the arm are best remembered when they are grouped by primary actions.
7. The **deltoid** and the **supraspinatus muscles** produce abduction at the shoulder. The **subscapularis** and the **teres major** rotate the arm medially, whereas the **infraspinatus** and **teres minor muscles** rotate the arm laterally. The supraspinatus, infraspinatus, subscapularis, and teres minor are known as the muscles of the **rotator cuff**. The **coracobrachialis muscle** produces flexion and adduction at the shoulder. *(see Figures 11.1/11.4/11.5, and Table 11.2)*
8. The **pectoralis major muscle** flexes the shoulder, while the **latissimus dorsi** muscle extends it. Additionally, both adduct and medially rotate the arm. *(see Figures 11.1/11.4/11.5, and Table 11.2)*

Muscles That Move the Forearm and Hand 297

9. The primary actions of the **biceps brachii muscle** and the **triceps brachii muscle** (*long head*) affect the elbow joint. The biceps brachii flexes the elbow and supinates the forearm, while the triceps brachii extends the elbow. Additionally, both have a secondary effect on the pectoral girdle.
10. The **brachialis** and **brachioradialis muscles** flex the elbow. This action is opposed by the **anconeus muscle** and the **triceps brachii muscle**. The **flexor carpi ulnaris**, the **flexor carpi radialis**, and the **palmaris longus muscles** are superficial muscles of the forearm that cooperate to flex the wrist. Additionally, the flexor carpi ulnaris muscle adducts the wrist, while the flexor carpi radialis muscle abducts it. Extension of the wrist is provided by the **extensor carpi radialis muscle**, which also abducts the wrist, and the **extensor carpi ulnaris muscle**, which also adducts the wrist. The **pronator teres** and **pronator quadratus muscles** pronate the forearm without flexion or extension at the elbow; their action is opposed by the **supinator muscle**. *(see Figures 11.4/11.5/11.6/11.7, and Table 11.3)*

Muscles That Move the Hand and Fingers 299

11. Muscles that perform flexion and extension of the finger joints are illustrated in *Figures 11.7 to 11.9* and detailed in *Tables 11.4/11.5*.

Muscles of the Pelvic Girdle and Lower Limbs 305

1. There are three groups of muscles associated with the pelvis and lower limbs: (1) *muscles that move the thigh*, (2) *muscles that move the leg*, and (3) *muscles that move the foot and toes*.

Muscles That Move the Thigh 305

2. Muscles originating on the surface of the pelvis and inserting on the femur produce characteristic movements determined by their position relative to the acetabulum. *(see Figure 11.10 and Table 11.6)*
3. **Gluteal muscles** cover the lateral surface of the ilium. The largest is the **gluteus maximus muscle**, which produces extension and lateral rotation at the hip. It shares an insertion with the **tensor fasciae latae muscle**, which produces flexion, abduction, and medial rotation at the hip. Together these muscles pull on the **iliotibial tract** to provide a lateral brace for the knee. *(see Figures 11.1/11.4/11.10 to 11.14, and Table 11.6)*
4. The **piriformis** and the **obturator muscles** are the most important lateral rotators.
5. The **adductor group** (**adductor magnus**, **adductor brevis**, **adductor longus**, **pectineus**, and **gracilis muscles**) produce adduction at the hip. Individually, they can produce various other movements, such as medial or lateral rotation and flexion or extension at the hip. *(see Figures 11.10 to 11.14, and Table 11.6)*
6. The **psoas major** and the **iliacus** merge to form the **iliopsoas muscle**, a powerful flexor of the hip. *(see Figures 11.11/11.12 and Table 11.6)*

Muscles That Move the Leg 309

7. *Extensor muscles of the knee* are found along the anterior and lateral surfaces of the thigh; flexor muscles lie along the posterior and medial surfaces of the thigh. Flexors and adductors originate on the pelvic girdle, whereas most extensors originate at the femoral surface.
8. The *flexors of the knee* include: the **biceps femoris**, **semimembranosus**, and **semitendinosus muscles** (together they also produce extension of the hip and are termed "hamstrings"), and the **sartorius muscle**. The **popliteus muscle** medially rotates the tibia (or laterally rotates the femur) to unlock the knee joint. *(see Figures 11.13/11.14, and Table 11.7)*
9. Collectively, the *knee extensors* are known as the **quadriceps femoris**. This group includes the **vastus intermedius**, **vastus lateralis** and **vastus medialis muscles** and the **rectus femoris muscle**. *(see Figures 11.12/11.13, and Table 11.7)*

Muscles That Move the Foot and Toes 313

10. Extrinsic and intrinsic muscles that move the foot and toes are illustrated in *Figures 11.15 to 11.18* and listed in *Tables 11.8 and 11.9*.
11. The **gastrocnemius** and **soleus muscles** produce plantar flexion. The large **tibialis anterior muscle** opposes the gastrocnemius and dorsiflexes the ankle. A pair of **fibularis muscles** produce eversion as well as plantar flexion. *(see Figures 11.16/11.17, and Table 11.8)*
12. Smaller muscles of the calf and shin position the foot and move the toes. Precise control of the phalanges is provided by muscles originating on the tarsal and metatarsal bones. *(see Figure 11.18, and Table 11.9)*

LEVEL 1 REVIEWING FACTS AND TERMS

Match each numbered item with the most closely related lettered item. Use letters for answers in the spaces provided.

Column A	Column B
____ 1. rhomboid muscles	a. abducts the toes
____ 2. latissimus dorsi	b. flexes hip and/or lumbar spine
____ 3. infraspinatus	c. adduct (retract) scapula
____ 4. brachialis	d. connective tissue bands
____ 5. supinator	e. plantar flexes, inverts, abducts foot
____ 6. flexor retinaculum	f. origin—surface of ilium
____ 7. gluteal muscles	g. flexes elbow
____ 8. iliacus	h. dorsiflexes ankle and inverts foot
____ 9. gastrocnemius	i. lateral rotation of humerus at shoulder
____ 10. tibialis anterior	j. supinates forearm
____ 11. interossei	k. extends, adducts, medially rotates humerus at shoulder

12. The powerful flexors of the thigh are the
    (a) piriformis (b) obturators
    (c) pectineus (d) iliopsoas
13. Which of the following is not a muscle of the rotator cuff?
    (a) supraspinatus (b) subclavius
    (c) subscapularis (d) teres minor
14. Which of the following does not originate on the humerus?
    (a) anconeus (b) biceps brachii
    (c) brachialis (d) triceps brachii, lateral head
15. Knee extensors, known as the quadriceps femoris, include
    (a) three vastus muscles and the rectus femoris
    (b) biceps femoris, gracilis, sartorius
    (c) popliteus, iliopsoas, gracilis
    (d) gastrocnemius, tibialis, fibularis
16. The muscle that causes opposition of the thumb is the
    (a) adductor pollicis (b) extensor digitorum
    (c) abductor pollicis (d) opponens pollicis

LEVEL 2 REVIEWING CONCEPTS

1. Damage to the pectoralis major muscle would interfere with the ability to
   (a) extend the elbow (b) abduct the humerus
   (c) adduct the humerus (d) elevate the scapula
2. Which of the following muscles does not produce adduction at the hip?
   (a) pectineus (b) psoas
   (c) obturator internus (d) piriformis
3. Which of the following muscles belongs to the group known as "hamstrings"?
   (a) biceps femoris (b) semitendinosus
   (c) semimembranosus (d) all of the above
4. If you bruised your gluteus maximus muscle, you would expect to experience discomfort when performing
   (a) flexion at the knee (b) extension at the knee
   (c) abduction at the hip (d) all of the above
5. What is the anatomical name of an appendicular muscle that makes it possible to do a push-up?
6. What muscle supports the knee laterally and becomes greatly enlarged in ballet dancers because of the need for flexion and abduction at the hip?
7. What is the most important muscle involved in sitting cross-legged?
8. What is the function of the intrinsic muscles of the hand?
9. How does the tensor fasciae latae muscle act synergistically with the gluteus maximus muscle?
10. What are the main functions of the flexor and extensor retinacula of the wrist and ankle?

LEVEL 3 CRITICAL THINKING AND CLINICAL APPLICATIONS

1. What muscles are the most important in typing?
2. While playing soccer, Jerry pulls his hamstring muscle. As a result of the injury, he has difficulty flexing and medially rotating his thigh. Which muscle(s) of the hamstring group did he probably injure?
3. While unloading her car trunk, Linda pulls a muscle and, as a result, has difficulty moving her arm. The doctor in the emergency room tells her that she pulled her pectoralis major muscle. Linda tells you that she thought the pectoralis major was a chest muscle and doesn't understand what that has to do with her arm. What would you tell her?

✓ ANSWERS TO CONCEPT CHECK QUESTIONS

p. 295 **1.** The rotator cuff muscles include the supraspinatus, infraspinatus, subscapularis, and teres minor muscles. The tendons of these muscles help to enclose and stabilize the shoulder joint. **2.** The fan-shaped muscle that inserts along the anterior margin of the vertebral border of the scapula is the serratus anterior muscle. **3.** The deltoid muscle is the major abductor working at the shoulder. **4.** The teres major muscle produces extension, adduction, and medial rotation at the shoulder.

p. 301 **1.** Injury to the flexor carpi ulnaris muscle would prevent flexion and adduction at the wrist. **2.** The pronator teres and the supinator muscles arise on both the humerus and forearm. They rotate the radius without flexing or extending the elbow. **3.** The tendons that cross the dorsal and ventral surfaces of the wrist pass through tendon sheaths, elongated bursae that reduce friction. **4.** The thickened fascia of the forearm on the posterior surface of the wrist is the extensor retinaculum.

p. 313 **1.** Injury to the obturator muscles would interfere with your ability to perform lateral rotation at the hip. **2.** The hamstring refers to a group of three muscles that collectively function in flexing the knee. These muscles are the biceps femoris, semimembranosus, and semitendinosus, muscles. **3.** The pectineus and the gracilis muscles belong to the adductor group of muscles that move the thigh. **4.** Collectively, knee extensors are known as the quadriceps femoris.

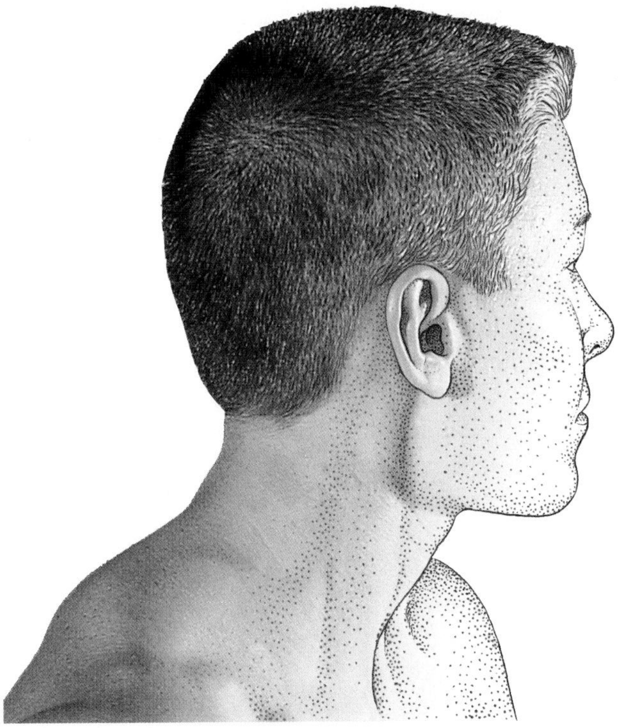

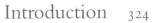

12

SURFACE ANATOMY

This chapter focuses attention on anatomical structures that can be identified from the body surface. **Surface anatomy** is the study of anatomical landmarks on the exterior of the human body. The photographs in this chapter survey the entire body, providing a visual tour that highlights skeletal landmarks and muscle contours. Chapter 1 provided an overview of surface anatomy. pp. 14–15 Now that you are familiar with the basic anatomy of the skeletal and muscular systems, a detailed examination of surface anatomy will help demonstrate the structural and functional relationships between those systems. Many of the figures in earlier chapters included views of surface anatomy; those figures will be referenced throughout this chapter.

Surface anatomy has many practical applications. For example, an understanding of surface anatomy is crucial to medical examination in a clinical setting. In the laboratory, a familiarity with surface anatomy is essential for both invasive and noninvasive laboratory procedures. *Anatomy and Observation p. 784*

A Regional Approach to Surface Anatomy

Surface anatomy is best studied using a regional approach. The regions are: *head and neck, thorax, abdomen, upper limb*, and *lower limb*. This chapter presents the information in pictorial fashion, using photographs of the living human body. These models have well-developed muscles and very little body fat. Because many anatomical landmarks can be hidden by a layer of subcutaneous fat, you may not find it as easy to locate these structures on your own body. In practice, anatomical observation often involves estimating the location and then palpating for specific structures. In the sections that follow, identify through visual observation and palpation the surface anatomy of the regions of the body, using the labeled photographs for reference.

The Head and Neck [FIGURE 12.1]

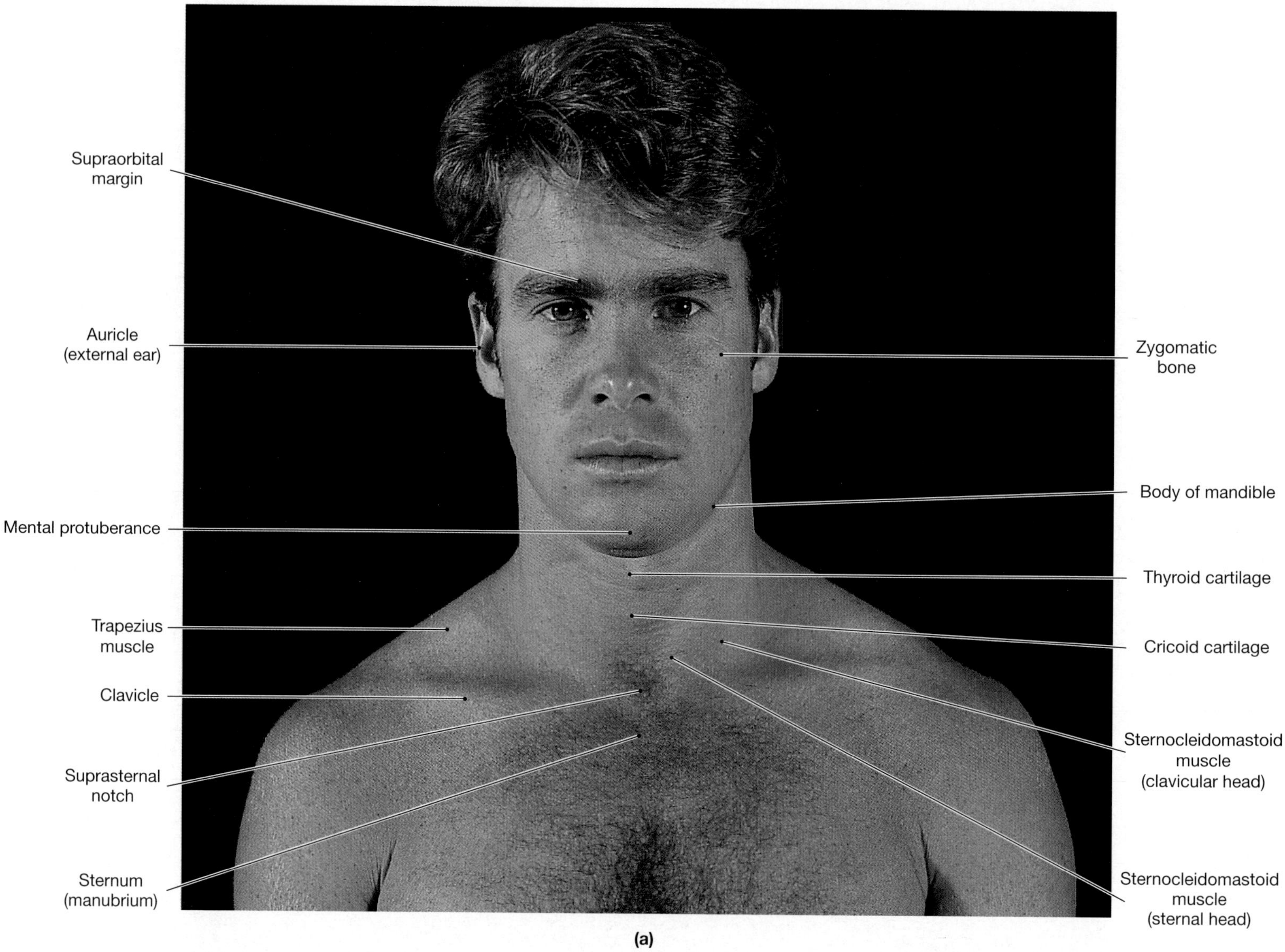

FIGURE 12.1 **Head and Neck**

(**a**) Anterior view. *For details concerning the musculature of this region, see Figures 10.3 and 10.4.* (**b**) Posterior cervical triangle. (**c**) Detail of anterior and posterior cervical triangles. The **anterior triangle** of the neck (b,c) extends from the anterior midline to the anterior border of the sternocleidomastoid muscle. The **posterior triangle** (b) extends between the posterior border of the sternocleidomastoid and the anterior border of the trapezius.

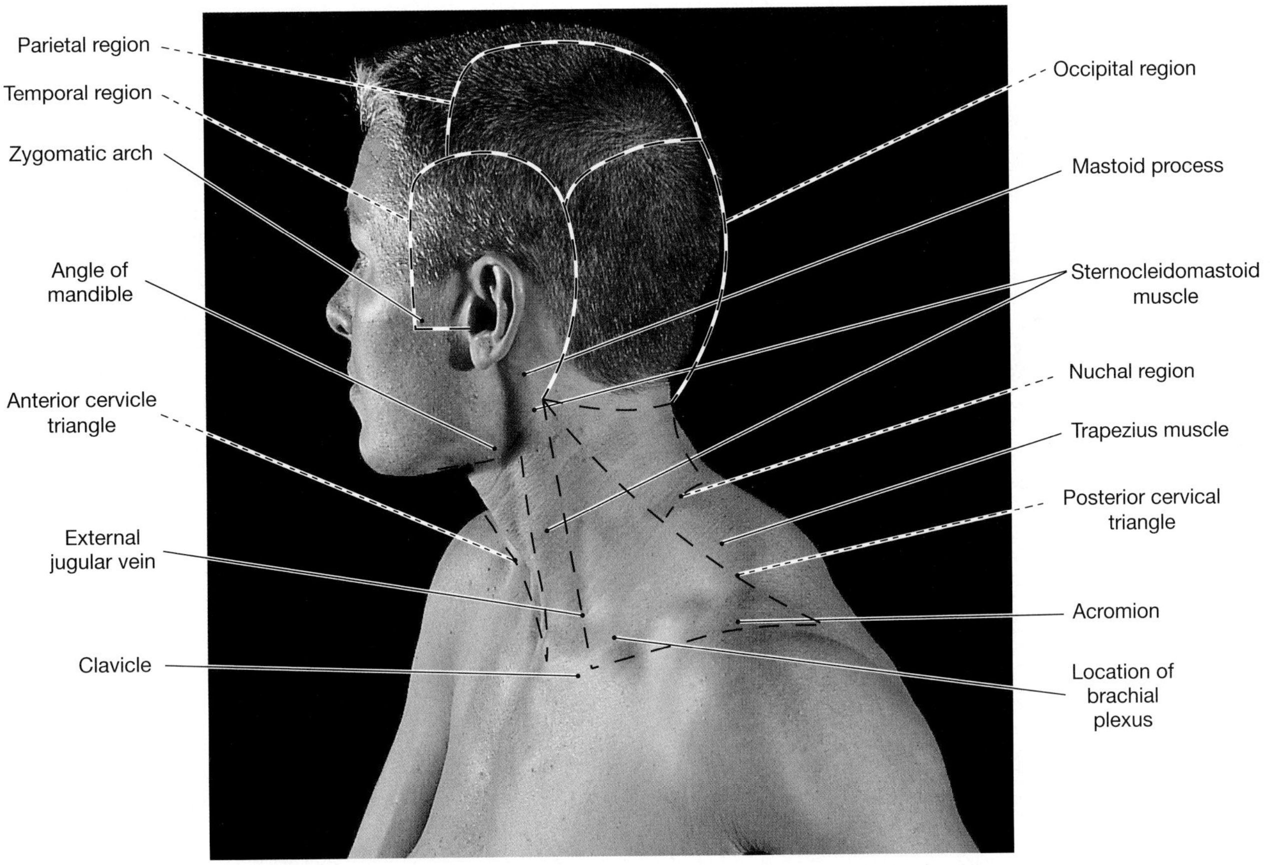

(b) The posterior cervical triangles and the larger regions of the head and neck

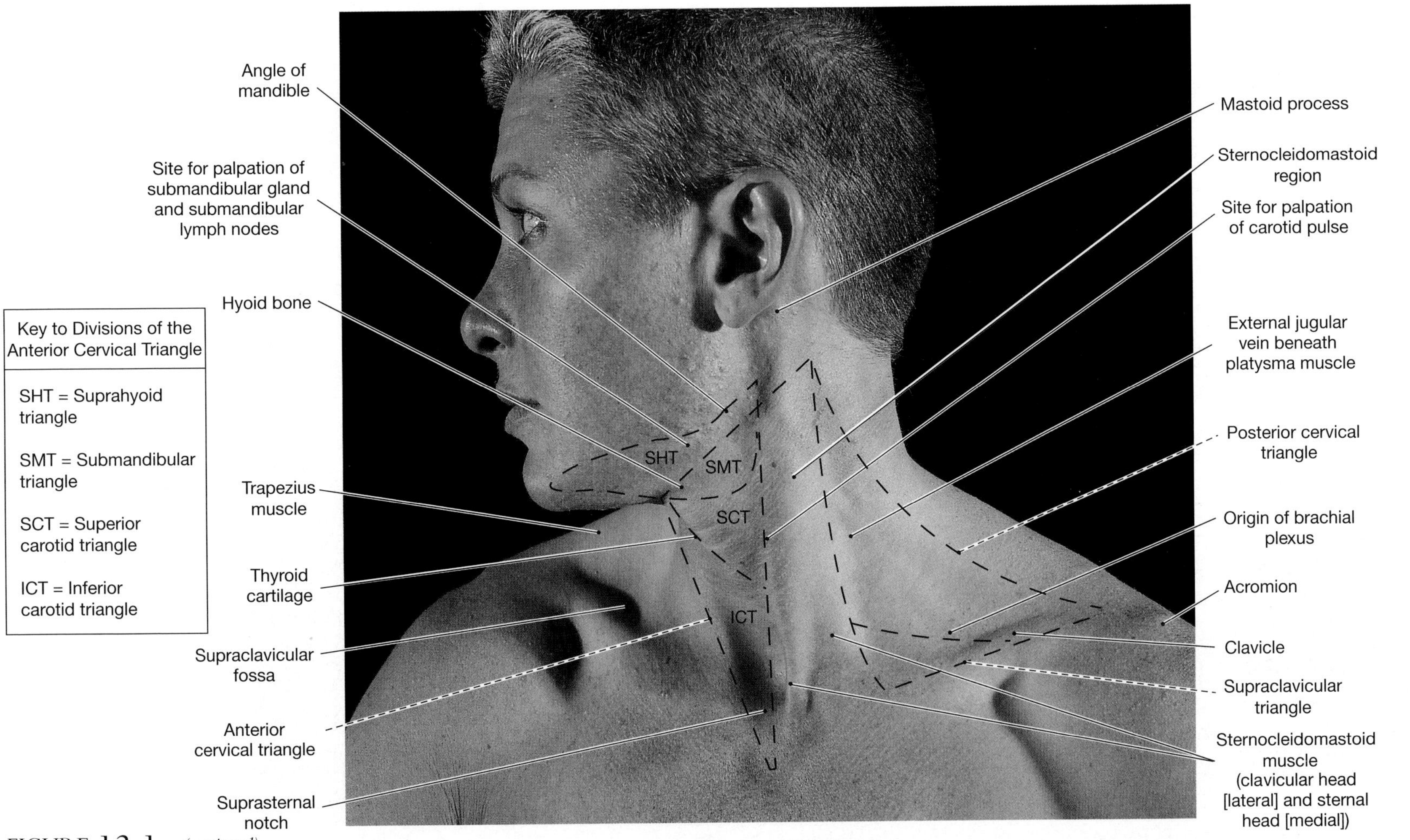

FIGURE 12.1 *(continued)*

(c) The cervical triangles

The Thorax

THE ANTERIOR THORAX [FIGURE 12.2a]

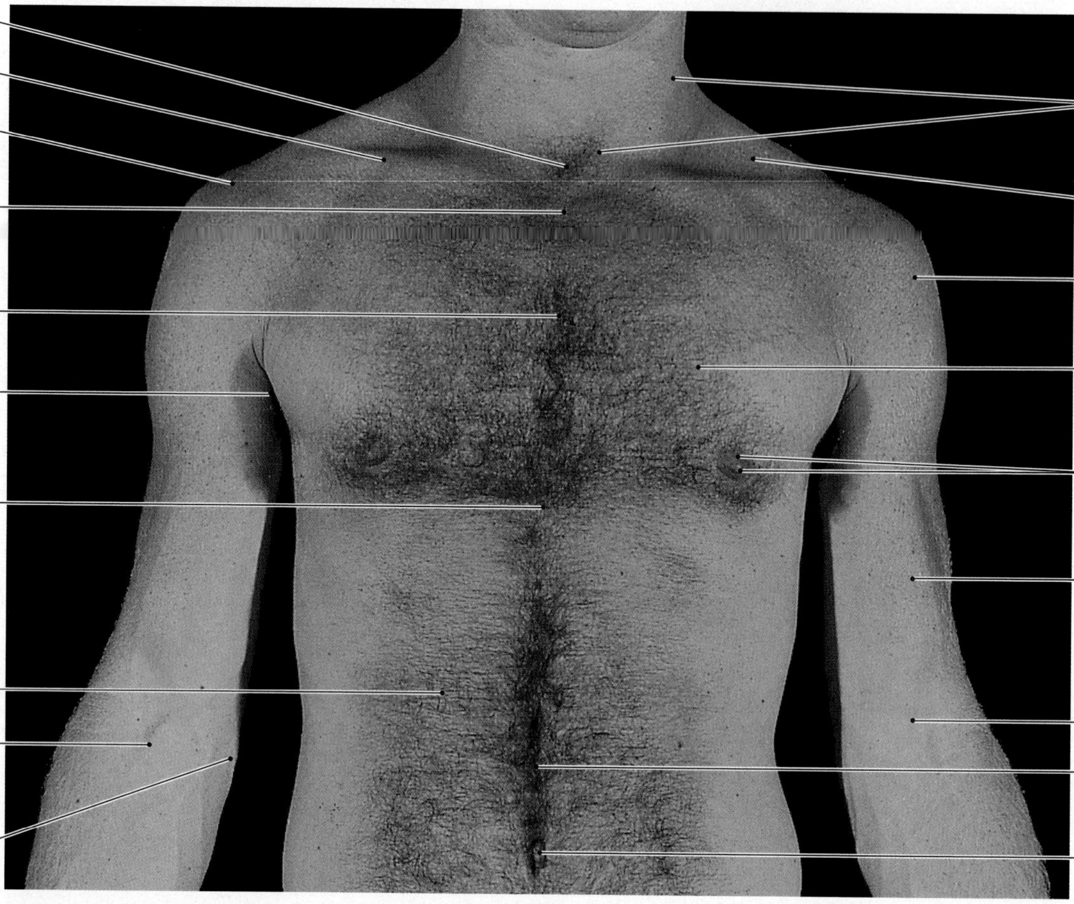

(a)

THE BACK AND SHOULDER REGIONS [FIGURE 12.2b]

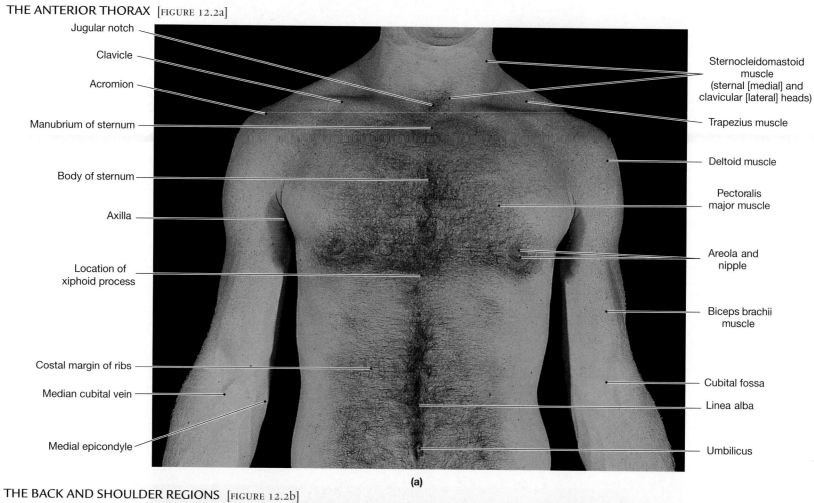

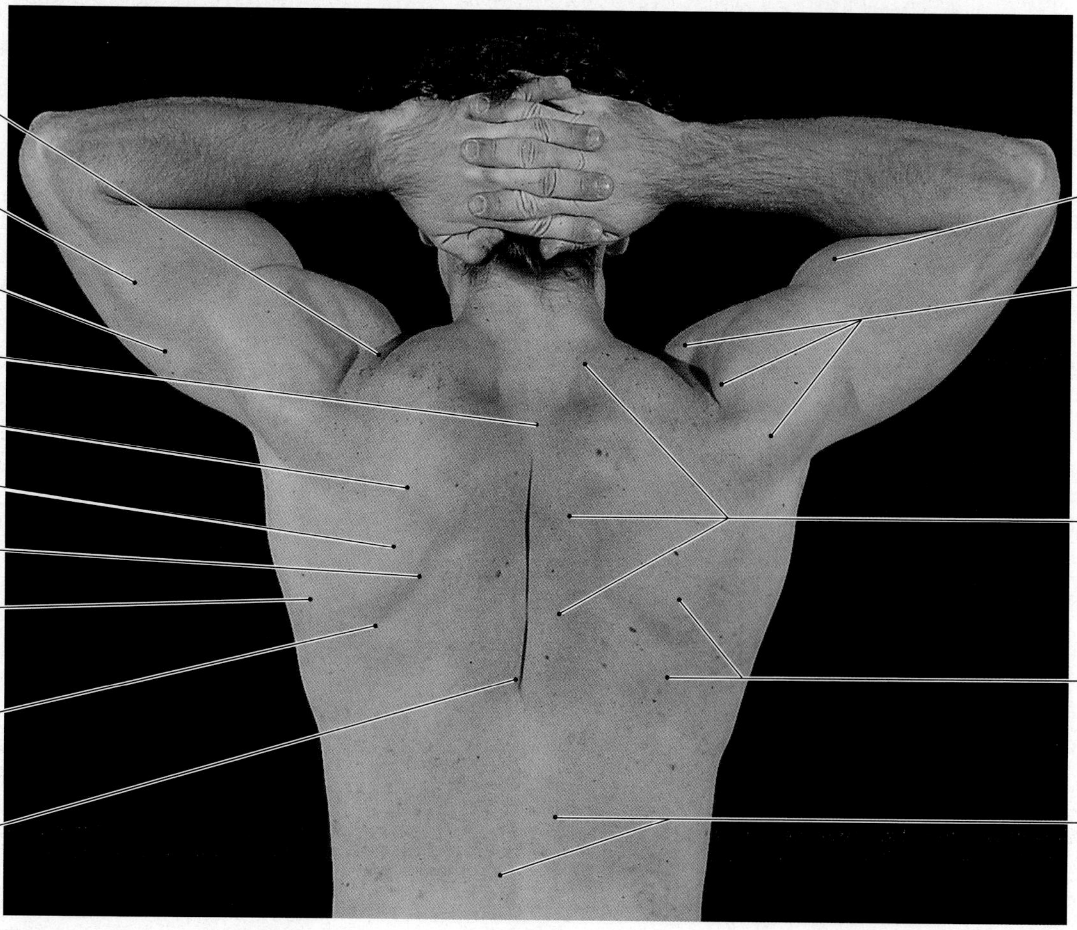

(b)

FIGURE 12.2 **The Thorax**
(**a**) Anterior view. (**b**) Posterior view.

The Abdomen [FIGURE 12.3]

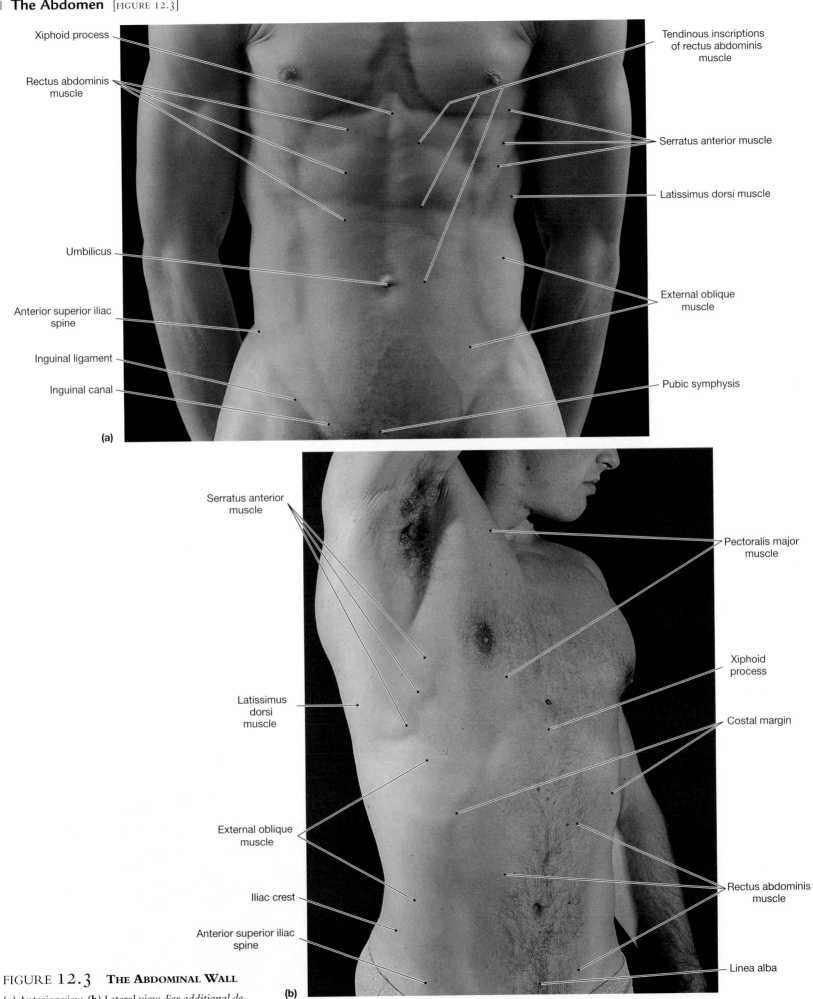

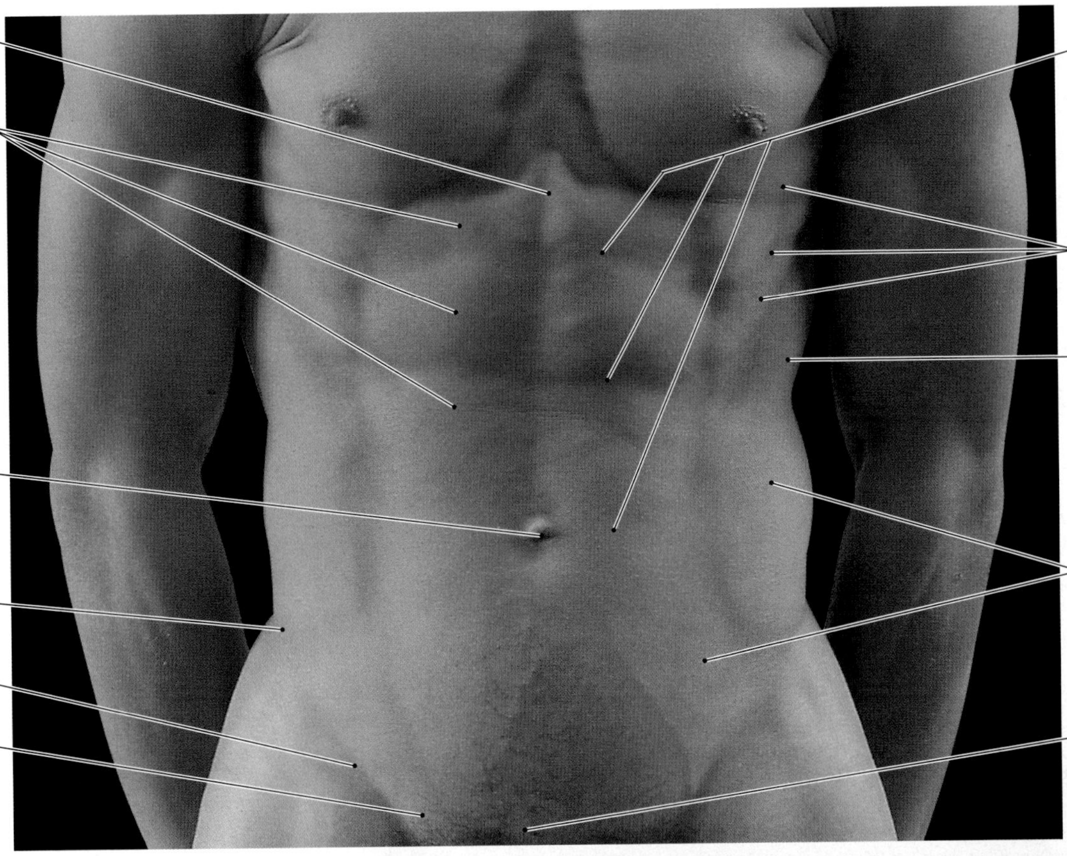

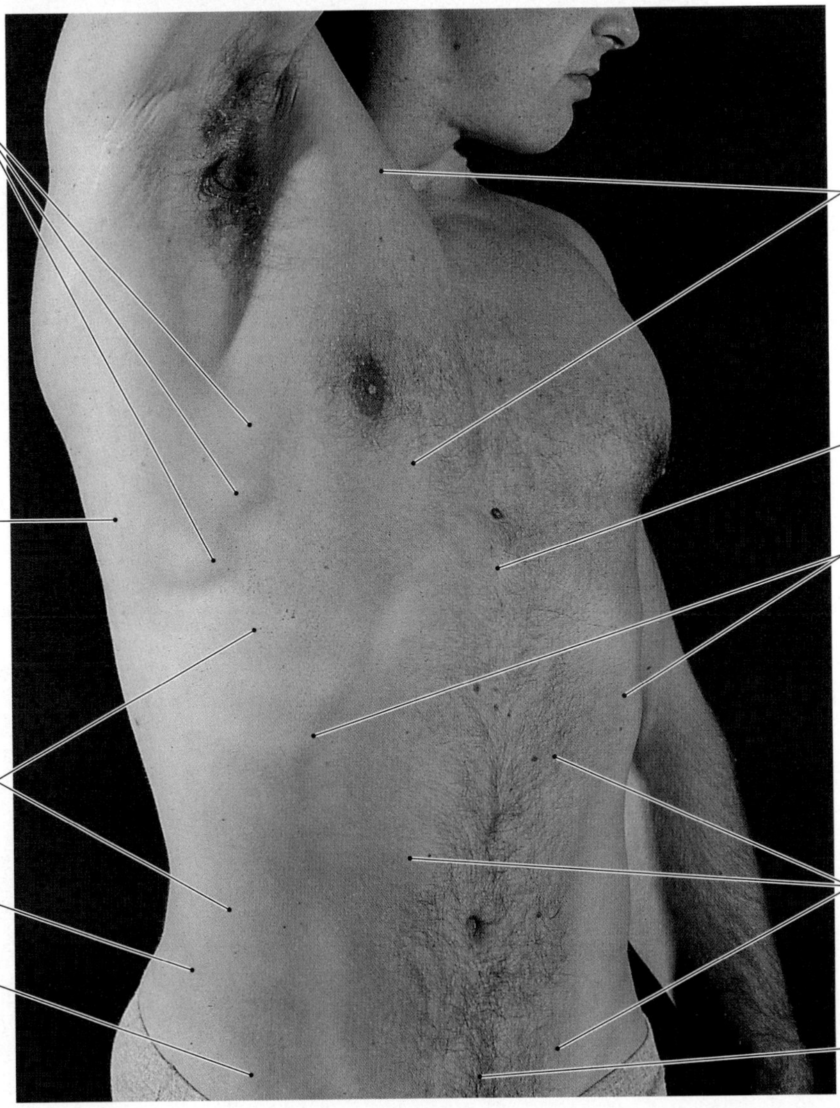

FIGURE 12.3 **The Abdominal Wall**

(**a**) Anterior view. (**b**) Lateral view. *For additional details of the abdominal wall, see Figure 10.13.*

The Upper Limb [FIGURES 12.4/12.5]

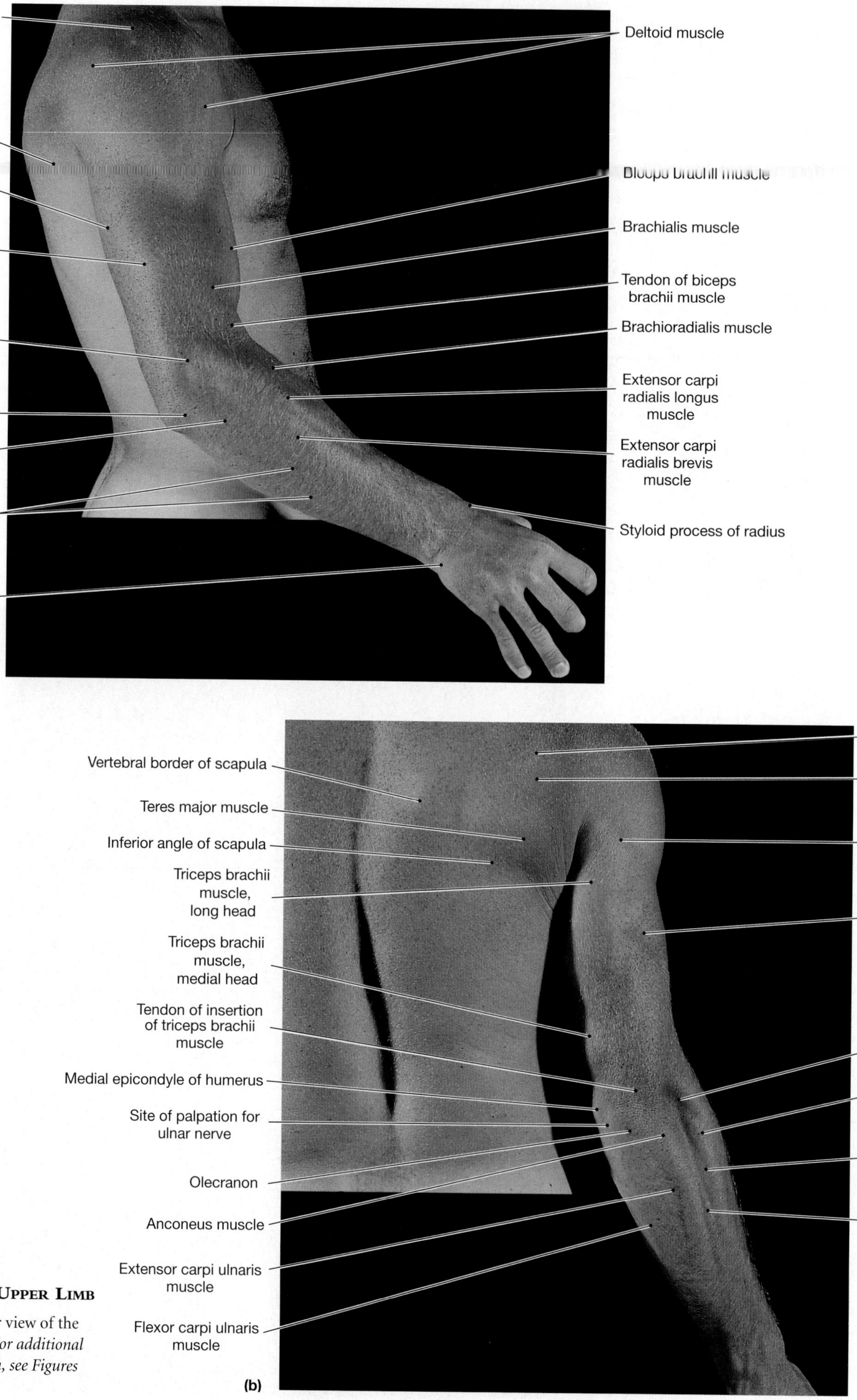

FIGURE 12.4 **The Upper Limb**

(**a**) Lateral view. (**b**) Posterior view of the chest and right upper limb. *For additional details of the arm and forearm, see Figures 11.7 and 11.8.*

THE FOREARM AND WRIST [FIGURE 12.5]

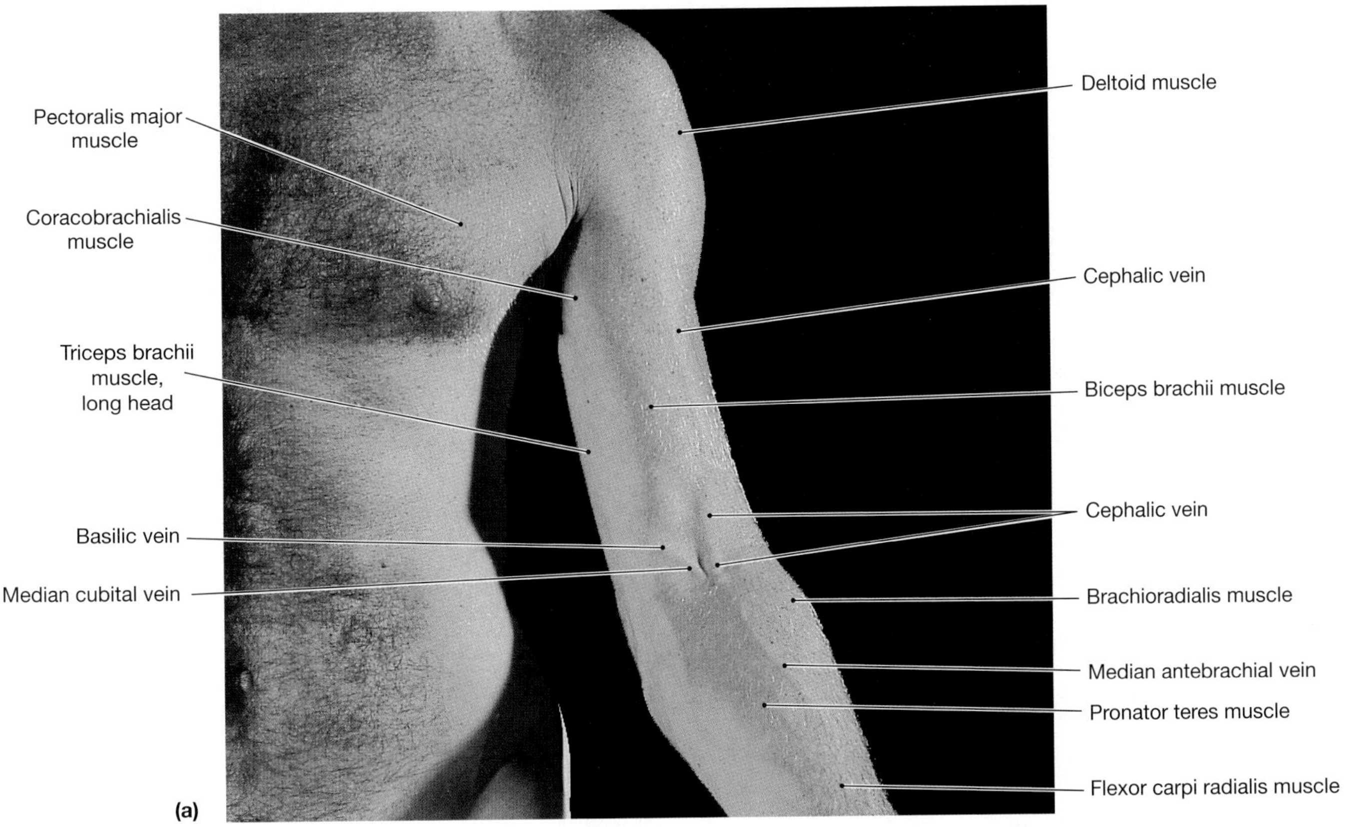

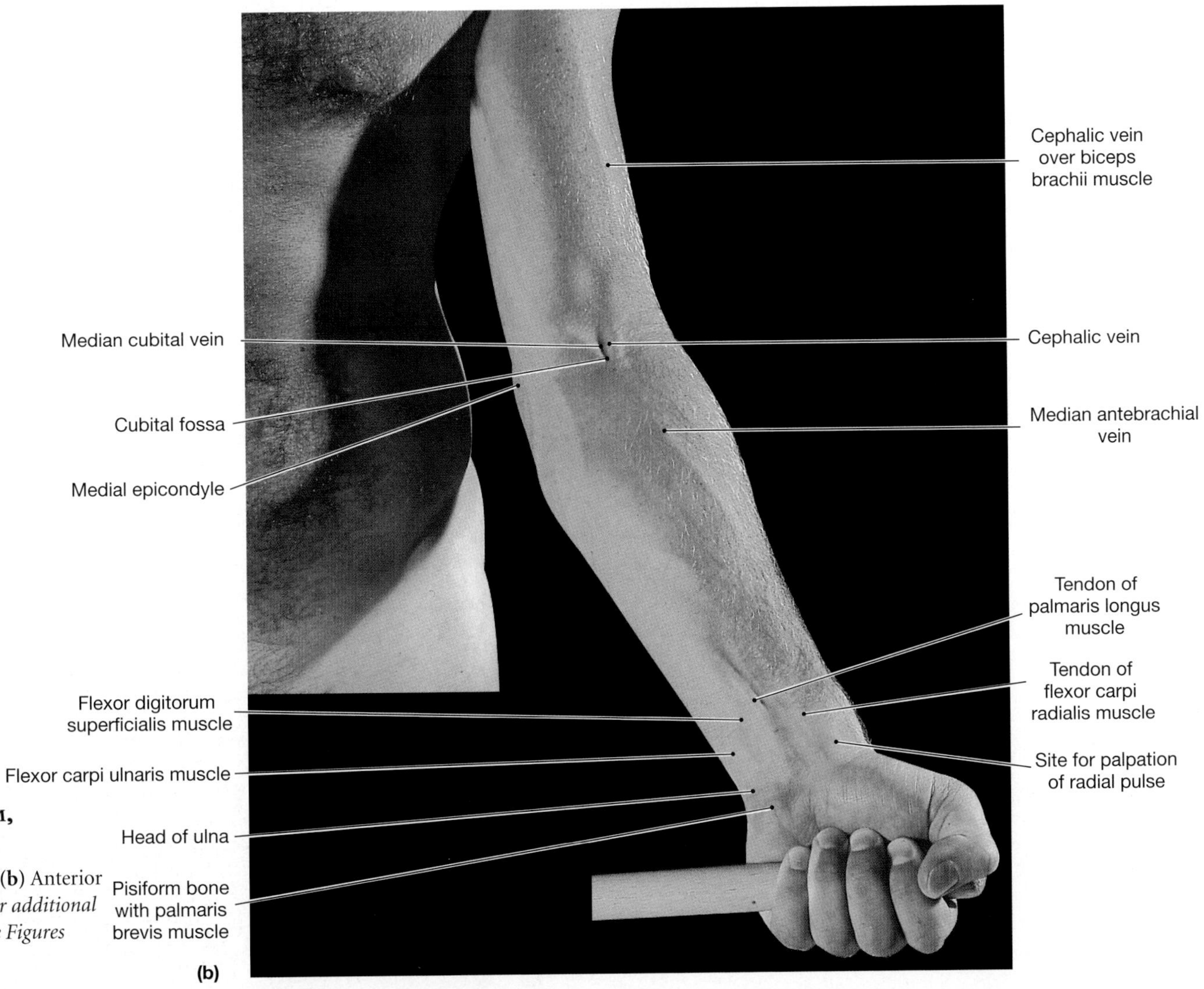

FIGURE 12.5 **THE ARM, FOREARM, AND WRIST**

(**a**) Anterior view of the left arm. (**b**) Anterior view of left forearm and wrist. *For additional details of the arm and forearm, see Figures 11.5, 11.6, and 11.8.*

The Pelvis and Lower Limb [FIGURES 12.6/12.7]

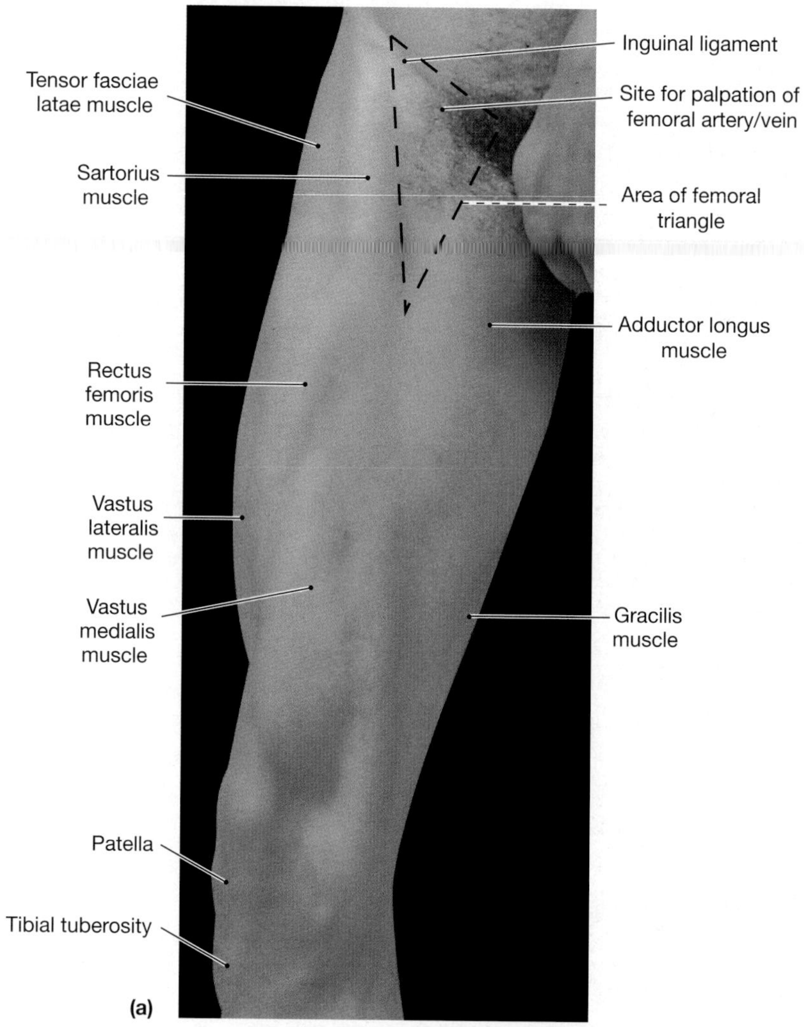

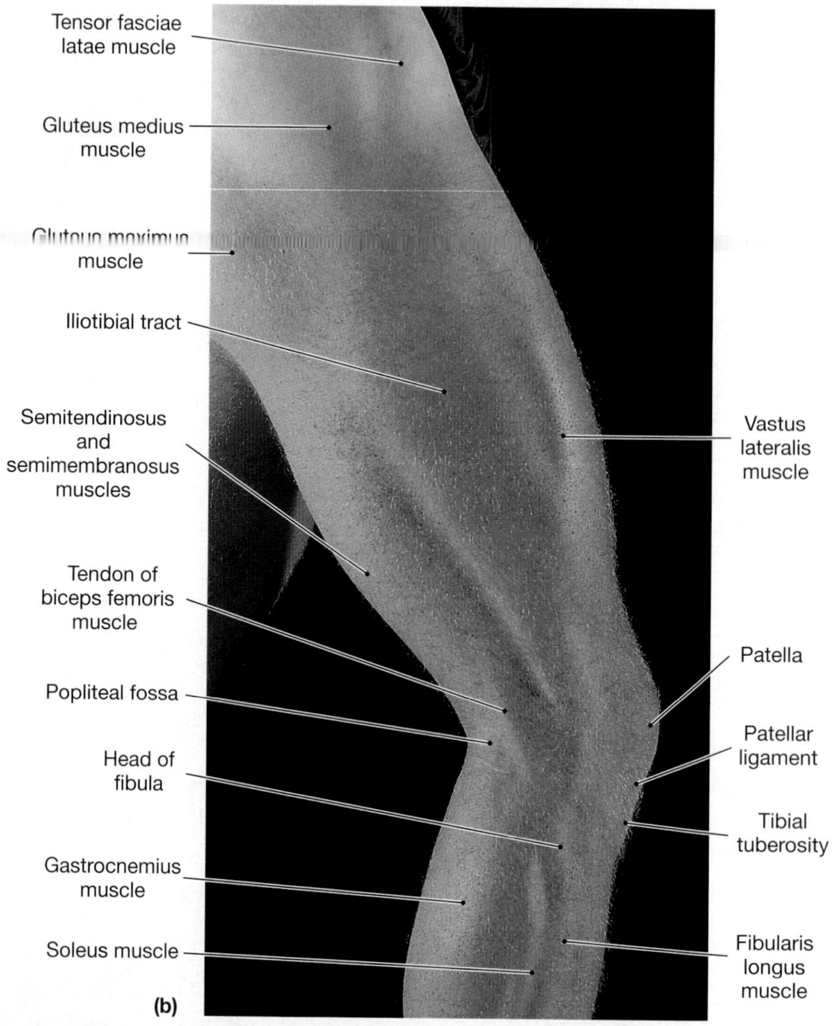

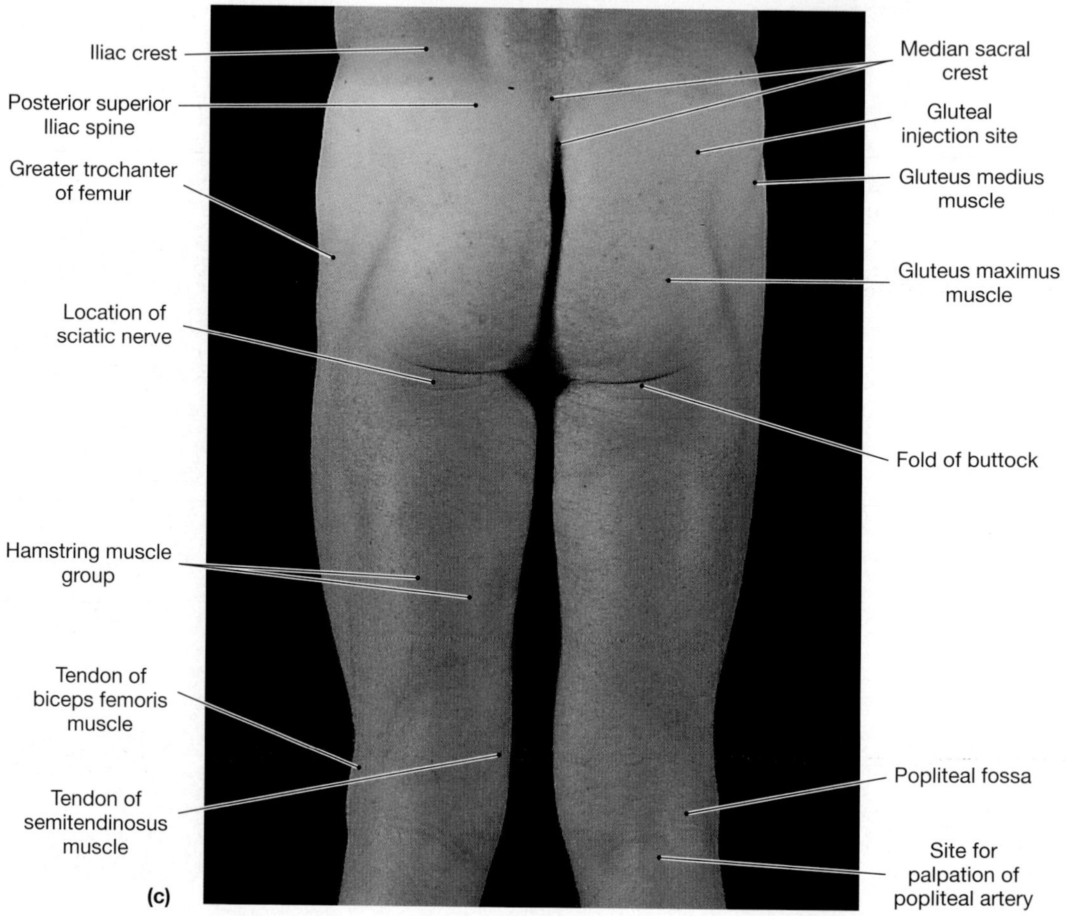

FIGURE 12.6 **The Pelvis and Lower Limb**

(**a**) Anterior surface of right thigh showing the femoral region and several adductor muscles. The boundaries of the *femoral triangle* are the inguinal ligament, medial border of the sartorius, and medial border of the adductor longus muscles. (**b**) Lateral view showing right gluteal region and thigh. (**c**) Posterior view of gluteal region and posterior surface of thigh. *For additional details of the thigh, see Figures 11.11 to 11.15.*

THE KNEE, LEG, ANKLE, AND FOOT [FIGURES 12.6/12.7]

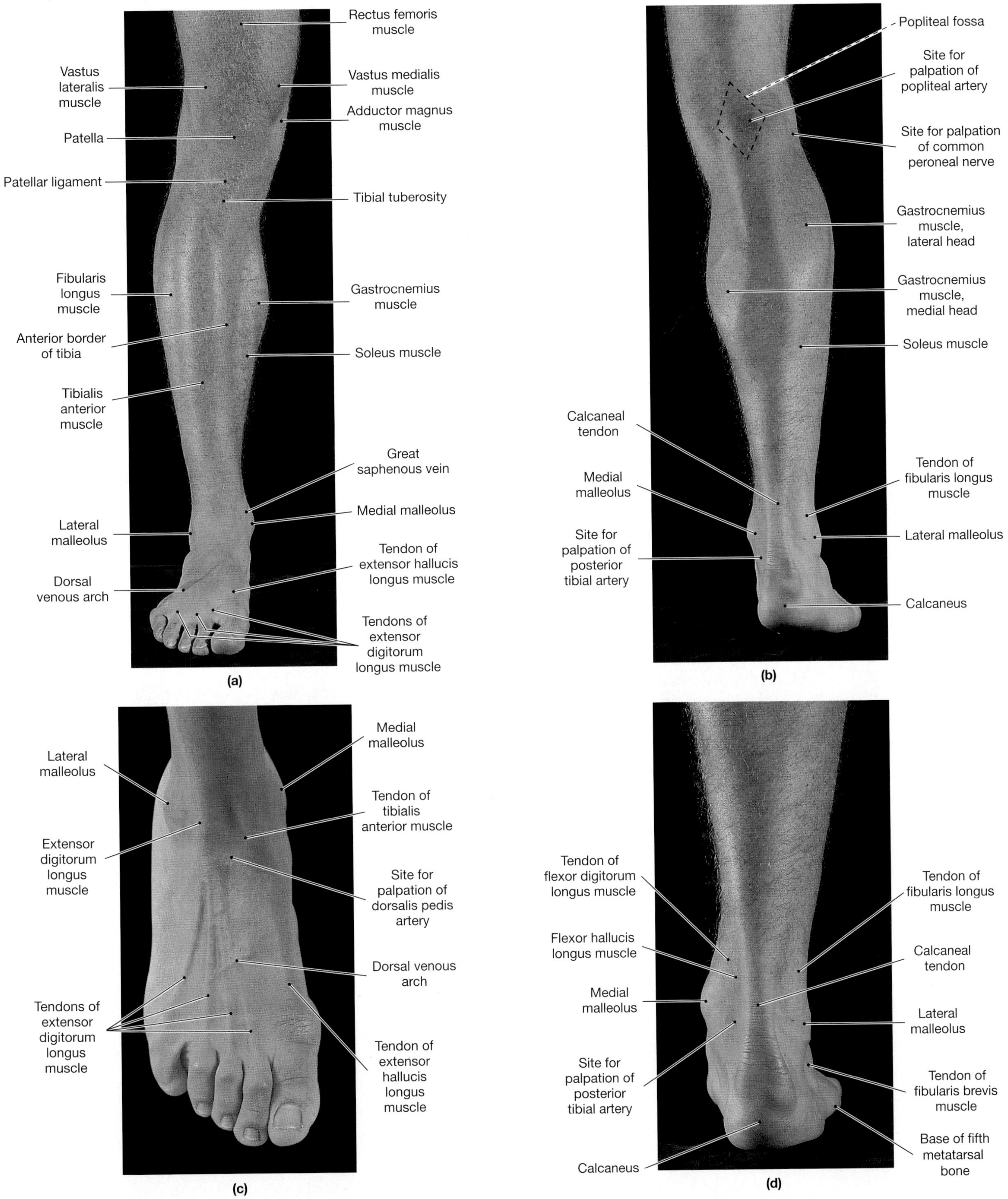

FIGURE 12.7 **THE LEG AND FOOT**

(**a**) Knee and leg, anterior view. (**b**) Knee and leg, posterior view. (**c**) Ankle and foot, anterior view. (**d**) Ankle and foot, posterior view. *For other views of the ankle and foot, see Figures 7.17 and 7.18 and 11.17 to 11.19.*

REVIEW IT

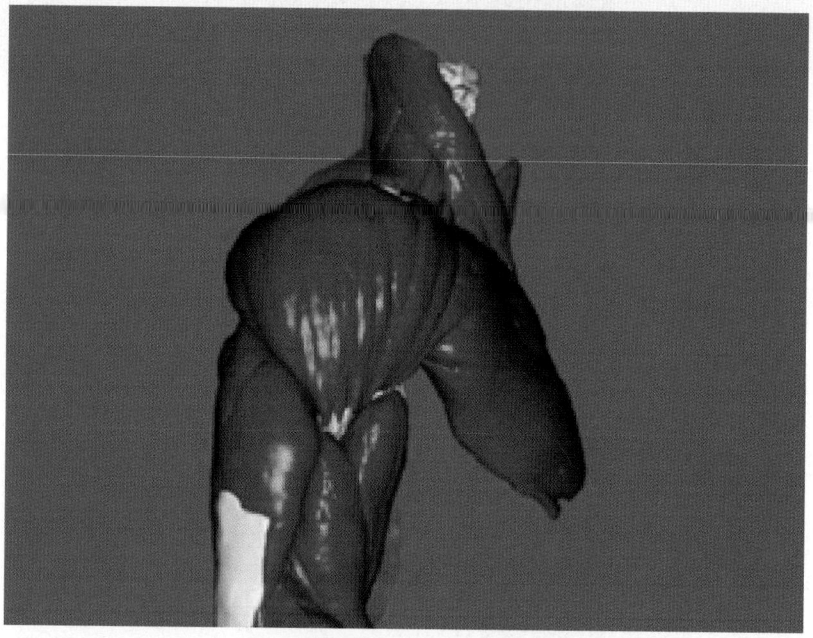

Challenge 1

The illustrations of the arm presented in the text are depicted with the subject in the anatomical position, as this is the best way to begin a study of the anatomy of the muscles. This is not always the view from which you will be working, however. The image seen here was taken from the Interactive CD/ Muscular system/Shoulder. The view is rotated to show the superficial muscles of the arm and the pectoral girdle. Identify these muscles by comparing this view with those presented in the text. Figures 11.5 and 11.6, pp. 296 and 298, will prove useful in this task.

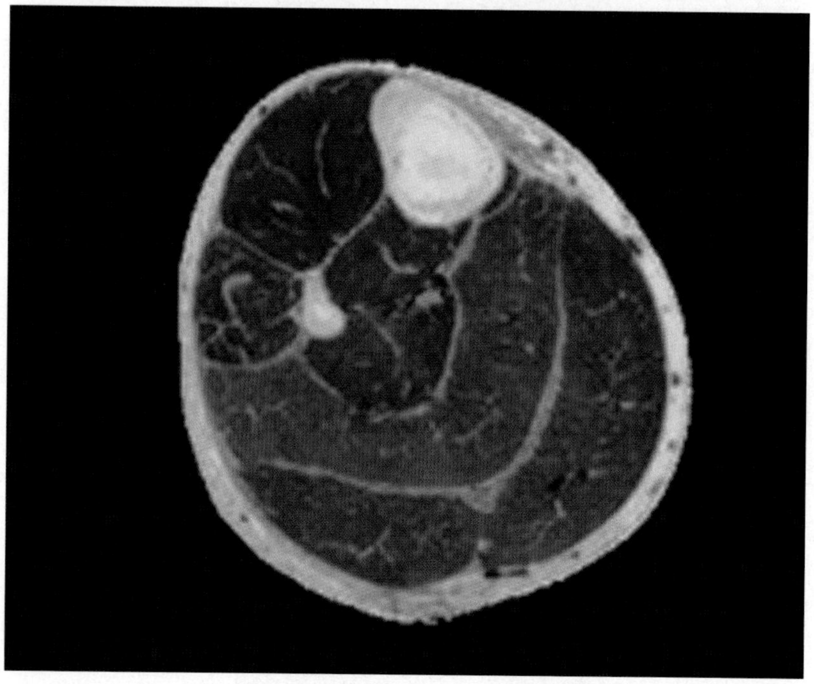

Challenge 2

One of the most challenging aspects of the study of anatomy is identifying structures as they appear in the body as opposed to identifying them in artists renderings or neatly prosected cadavers. You have been studying the musculature of the leg through animations on the Interactive CD, drawings in the anatomical position, and cadaver photographs. Presented here is a cross section of the right leg, immediately inferior to the knee. Using the text as a reference, identify the gastrocnemius, tibialis anterior, fibularis longus, tibialis posterior and soleus muscles. Can you identify any other structures?

Images provided by the Digital Cadaver™ project, courtesy of Visible Productions, Inc.

INTEGRATE IT

Below are two separate exercises that provide more information on the preceding topics. These exercises differ from the Challenge exercises in that these are designed to help you understand the interactions between systems, rather than hone your knowledge of one particular system. As before, each activity is designed to take approximately 10 minutes, and will help you gain a better appreciation for the integration of the material presented in the text.

Critical Linking 1

Muscles put stresses on the skeletal system, causing enlarged areas on the bones they contact. The skeletal system supports the muscles, providing the levers through which movement becomes functional and efficient. This interaction is taken for granted as we live out our lives here on Earth. What happens to these two systems and their interaction when we move the body to the environment of space? Without gravity, the support function of the skeletal system is less important. Levers are not required to function with as great an efficiency to propel the body. What do you predict will be the effects of space travel on these two systems and their interactions? Jot down your initial thoughts, then visit the critical linking portion of the Companion Website. Click on the key word "Space travel" to read the findings of Richard W. Lymn, Ph.D., Chief, Muscle Biology Program, National Institute of Arthritis and Musculoskeletal and Skin Diseases (NIAMS), and Stephen L. Gordon, Ph.D., Chief, Musculoskeletal Diseases Branch, NIAMS, in collaboration with Frank D. Sultzman, Ph.D., Chief, Life Support Branch, National Aeronautics and Space Administration (NASA). Did your thoughts match theirs?

Critical Linking 2

Pathologies that affect the joints or bones also affect the associated muscles. This close interaction of the two systems is easily seen in radiographs of affected bone and muscle. Go to the Companion Website Critical Linking area, and click on the key word "Musculoskeletal Disorders". You will be taken to a website provided by the University of Washington, in Seattle. From the list of disorders on the left, choose *Osteomyelitis of the arm*. Look at the radiographs presented (clicking on the small thumbnails on the left will provide an enlarged radiograph on the right) and determine how this infection of the bone is affecting the muscular system. What symptoms would you expect this patient to develop? How many of these symptoms are skeletal and how many are muscular? Follow the same line of questioning as you investigate fractures (the knee fractures are quite interesting) and rotator cuff injuries.

FURTHER STUDIES

"I think any artist who is serious about his art cannot ignore the human form, it is too essential to artistic expression to ignore," said Jim Fields, Artist and English professor at the University of Kentucky. "It forces us to take a look at ourselves and shows us who we are and I think it helps us understand that there is beauty in the body." (anatomy, *http://www.studentadvantage.lycos.com*) Surface anatomy has long been a subject for artistic expression. As early as 10,000–15,000 b.c. artists have depicted the human body in various forms. This fascination with the human form can be traced through history, with examples abounding in all cultures and all periods. During the 2nd International Symposium of the German Society of Aesthetic Plastic Surgeons held in 1998 in Kloster, Germany, G. W. Cichon-Hollander prepared a short slide show "The ideal beauty of the human body in art". This is an excellent overview of surface anatomy in art, and can be viewed at *http://www.cichon.de/ideal-beauty/*. For a more modern view of anatomical art, open a recent edition of Gray's Anatomy—there are some spectacular images in there. A less clinical, more artistic approach to anatomy can be found in this very text, on the chapter opening pages. These images are designed to depict the evolution of anatomical art from pen and ink line drawings to the sophisticated images generated via computers today. A short search of the Internet will reveal a wealth of information on the human form in art, and is well worth a few minutes time from your day!

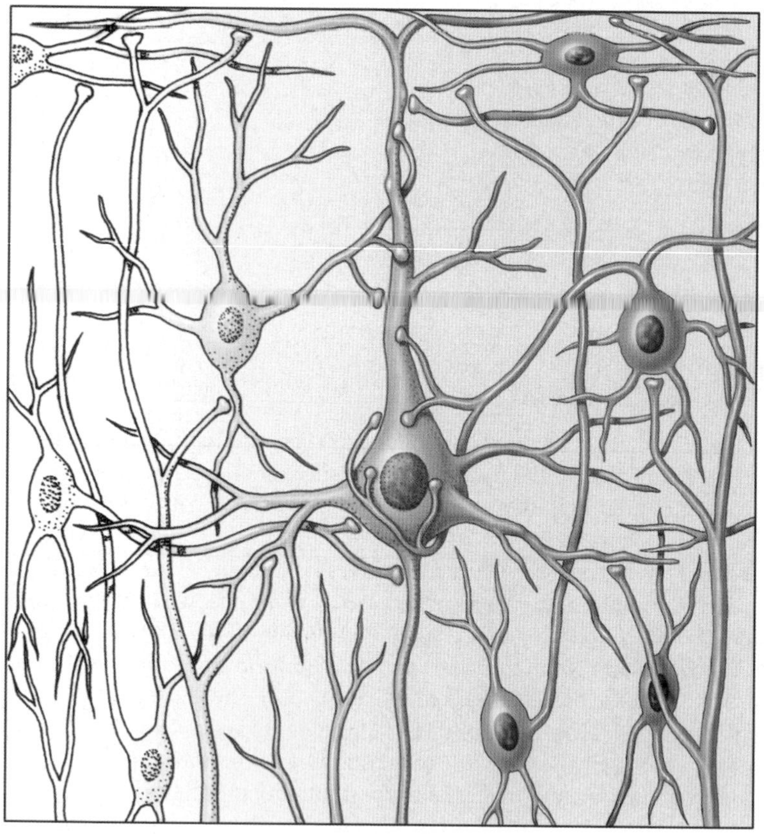

13

THE NERVOUS SYSTEM
Neural Tissue

The *nervous system* is among the smallest organ systems in terms of body weight, yet it is by far the most complex. Although it is often compared to a computer, the nervous system is much more complicated and versatile than any electronic device. Yet as in a computer, the rapid flow of information and high processing speed depend on electrical activity. Unlike a computer, however, portions of the brain can rework their electrical connections as new information arrives—that's part of the learning process.

Along with the *endocrine system*, discussed in Chapter 19, the nervous system controls and adjusts the activities of other systems. These two systems share important structural and functional characteristics. Both rely on some form of chemical communication with targeted tissues and organs, and they often act in a complementary fashion. The nervous system usually provides relatively swift but brief responses to stimuli by temporarily modifying the activities of other organ systems. The response may appear almost immediately—in a few milliseconds—but the effects disappear quickly after neural activity ceases. In contrast, endocrine responses are typically slower to develop than neural responses, but they often last much longer. The endocrine system adjusts the metabolic activity of other systems in response to changes in nutrient availability and energy demands. It also coordinates processes that continue for extended periods (months to years), such as growth and development. Chapters 13–18 detail the various components and functions of the nervous system. This chapter begins the series by considering the structure and function of neural tissue and basic principles of neural function. Subsequent chapters will build on this foundation as they explore the functional organization of the brain, spinal cord, higher-order functions, and sense organs.

An Overview of the Nervous System

[FIGURES 13.1/13.2 AND TABLE 13.1]

The **nervous system** includes all of the **neural tissue** in the body. p. 79 Table 13.1 provides an overview of the most important concepts and terms introduced in this chapter.

The nervous system has two anatomical subdivisions: the *central nervous system* and the *peripheral nervous system* (Figure 13.1●). The **central nervous system (CNS)** consists of the *brain* and *spinal cord*. The CNS is responsible for integrating, processing, and coordinating sensory data and motor commands. It is also the seat of higher functions, such as intelligence, memory, learning, and emotion. Early in development, the CNS begins as a mass of neural tissue organized into a hollow tube. As development continues, the central cavity decreases in relative size, but the thickness of the walls and the diameter of the enclosed space vary from one region to another. The narrow central cavity that persists within the spinal cord is called the *central canal*; the *ventricles* are expanded chambers, continuous with the central canal, found in specific regions of the brain. *Cerebrospinal fluid* (*CSF*) fills the central canal and ventricles and surrounds the CNS, which is suspended within the dorsal body cavity. p. 19

The **peripheral nervous system (PNS)** includes all of the neural tissue outside the CNS. The PNS provides sensory information to the CNS and carries motor commands from the CNS to peripheral tissues and systems. The PNS is subdivided into two divisions (Figure 13.2●). The **afferent division** of the PNS brings sensory information to the CNS, and the **efferent division** carries motor commands to muscles and glands. The afferent division begins at **receptors** that monitor specific characteristics of the environment. A receptor may be a *dendrite* (a sensory process of a neuron), a specialized cell or cluster of cells, or a complex sense organ (such as the eye). Whatever its structure, the stimulation of a receptor provides information that may be carried to the CNS. The efferent division begins inside the CNS and ends at an **effector**: a muscle cell, gland cell, or other cell specialized to perform specific functions. Both divisions have somatic and

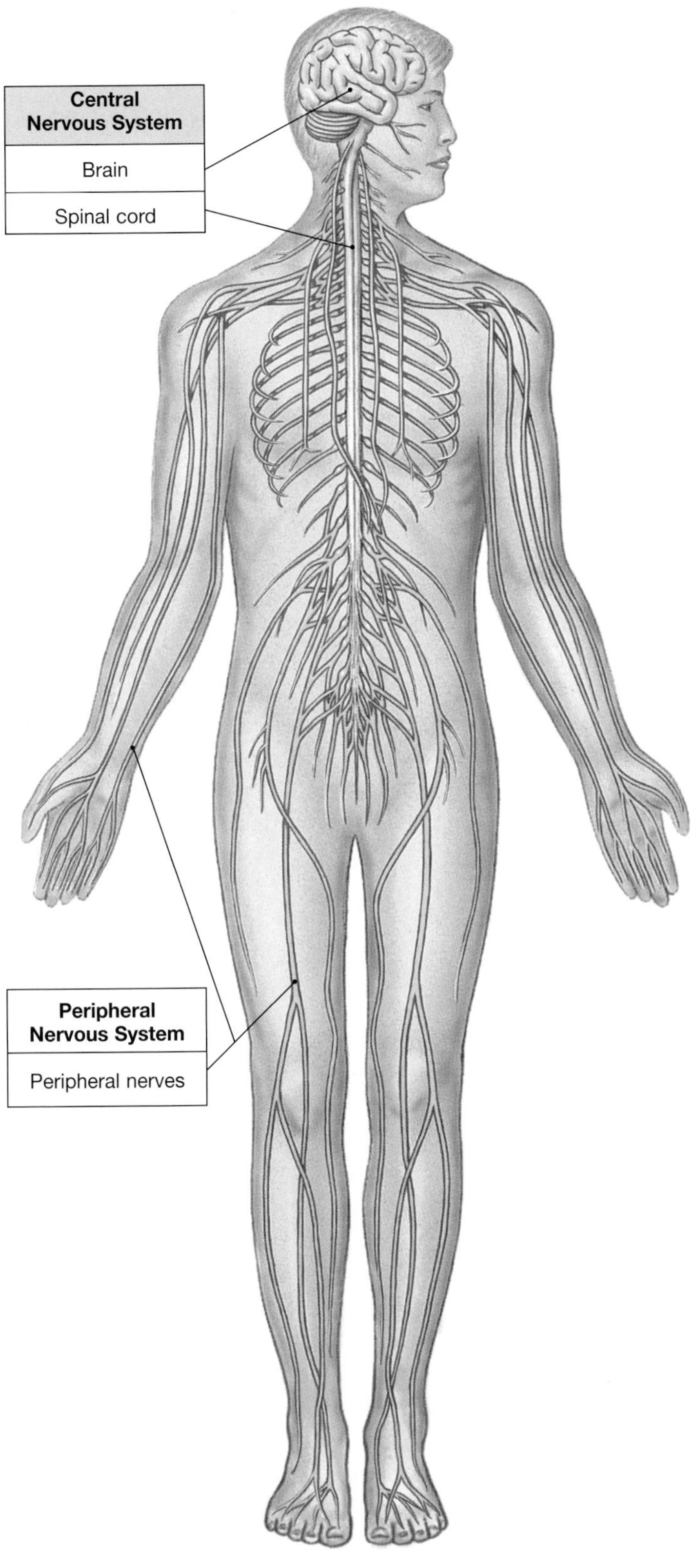

FIGURE 13.1 **THE NERVOUS SYSTEM**

The nervous system includes all of the neural tissue in the body. Its components include the brain, the spinal cord, sense organs such as the eye and ear, and the nerves that interconnect those organs and link the nervous system with other systems.

visceral components. The afferent division carries information from *somatic* sensory receptors, which monitor skeletal muscles and joints, and *visceral* sensory receptors that monitor other internal tissues, such as smooth muscle, cardiac muscle, and glands. The afferent division also delivers information provided by special sense organs, such as the eye and ear. The efferent division includes the **somatic nervous system (SNS)**, which controls skeletal muscle contractions; and the **autonomic nervous system (ANS)**, or *visceral motor system*, which regulates smooth muscle, cardiac muscle, and glandular activity.

The activities of the somatic nervous system may be *voluntary* or *involuntary*. Voluntary contractions of our skeletal muscles are under conscious control; you exert voluntary control over your arm muscles as you raise a full glass of water to your lips. Involuntary contractions are directed outside of your conscious awareness; if you accidentally place your hand on a hot stove, it will be withdrawn immediately, usually before you even notice the pain. The activities of the autonomic nervous system are usually outside of our conscious awareness or control.

The organs of the CNS and PNS are complex, with numerous blood vessels and layers of connective tissue that provide physical protection and mechanical support. Nevertheless, all of the varied and essential functions of the nervous system are performed by individual neurons that must be kept safe, secure, and fully functional. Our discussion of the nervous system will begin at the cellular level, with the histology of neural tissue.

Cellular Organization in Neural Tissue [FIGURE 13.3]

Neural tissue contains two distinct cell types: nerve cells, or *neurons*, and supporting cells, or *neuroglia*. **Neurons** (*neuro*, nerve) are responsible for the transfer and processing of information in the nervous system. Neuron structure was introduced in Chapter 3. p. 79 A representative neuron (Figure 13.3●) has a **cell body**, or *soma*. The region around the nucleus is called the **perikaryon** (per-i-KAR-ē-on; *karyon*). The cell body usually has several branching **dendrites.** Typical dendrites are highly branched, and each branch bears fine processes called *dendritic spines*. In the CNS, a neuron receives information from other neurons primarily at the **dendritic spines**, which represent 80–90% of the neuron's total surface area.

The cell body is attached to an elongated **axon** that ends at one or more **synaptic terminals.** At each synaptic terminal, the neuron communicates with another cell. The soma contains the organelles responsible for energy production and the biosynthesis of organic molecules, such as enzymes.

Supporting cells, or **neuroglia** (noo-RŌ-glē-a; *glia*, glue), isolate the neurons, provide a supporting framework for the neural tissue, and act as phagocytes. The neural tissue of the body contains approximately 100 billion neuroglia, or **glial cells**, which is roughly 5 times the number of neurons. Glial cells are smaller than neurons, and they retain the ability to divide—an ability lost by most neurons. Collectively, neuroglia account for roughly half of the volume of the nervous system. There are significant organizational differences between the neural tissue of the CNS and that of the PNS, primarily due to differences in the glial cell populations.

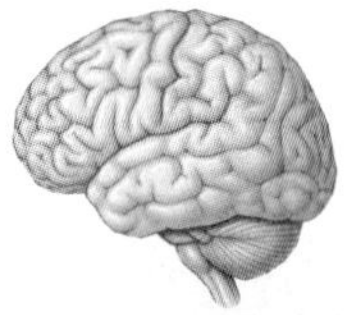

TABLE 13.1 AN INTRODUCTORY GLOSSARY FOR THE NERVOUS SYSTEM

MAJOR ANATOMICAL AND FUNCTIONAL DIVISIONS	
Central Nervous System (CNS)	The brain and spinal cord, which contain control centers responsible for processing and integrating sensory information, planning and coordinating responses to stimuli, and providing short-term control over the activities of other systems.
Peripheral Nervous System (PNS)	Neural tissue outside of the CNS, whose function is to link the CNS with sense organs and other systems.
Autonomic Nervous System (ANS)	Components of the CNS and PNS that are concerned with the control of visceral functions.
GROSS ANATOMY	
Nucleus:	A CNS center with discrete anatomical boundaries (p. 350).
Center:	A group of neuron cell bodies in the CNS that share a common function (p. 350).
Tract:	A bundle of axons within the CNS that share a common origin, destination, and function (p. 350).
Column:	A group of tracts found within a specific region of the spinal cord (p. 350).
Pathways:	Centers and tracts that connect the brain with other organs and systems in the body (p. 350).
Ganglia:	An anatomically distinct collection of sensory or motor neuron cell bodies within the PNS (p. 339).
Nerve:	A bundle of axons in the PNS (p. 340).
HISTOLOGY:	
Gray Matter:	Neural tissue dominated by neuron cell bodies (p. 338).
White Matter:	Neural tissue dominated by myelinated axons (p. 338).
Neural Cortex:	A layer of gray matter at the surface of the brain (p. 350).
Neuron:	The basic functional unit of the nervous system; a highly specialized cell; a nerve cell (pp. 334, 340).
Sensory Neuron:	A neuron whose axon carries sensory information from the PNS toward the CNS (p. 344).
Motor Neuron:	A neuron whose axon carries motor commands from the CNS toward effectors in the PNS (p. 344).
Soma:	The cell body of a neuron (p. 336).
Dendrites:	Neuronal processes that are specialized to respond to specific stimuli in the extracellular environment (p. 336).
Axon:	A long, slender cytoplasmic process of a neuron; axons are capable of conducting nerve impulses (action potentials) (p. 336).
Myelin:	A membranous wrapping, produced by glial cells, that coats axons and increases the speed of action potential propagation; axons coated with myelin are said to be *myelinated* (p. 338).
Neuroglia or Glial Cells:	Supporting cells that interact with neurons and regulate the extracellular environment, provide defense against pathogens, and perform repairs within neural tissue (p. 336).
FUNCTIONAL CATEGORIES:	
Receptor:	A specialized cell, dendrite, or organ that responds to specific stimuli in the extracellular environment and whose stimulation alters the level of activity in a sensory neuron (pp. 336, 344).
Effector:	A muscle, gland, or other specialized cell or organ that responds to neural stimulation by altering its activity and producing a specific effect (p. 336).
Reflex:	A rapid stereotyped response to a specific stimulus.
Somatic:	Pertaining to the control of skeletal muscle activity (*somatic motor*) or sensory information from skeletal muscles, tendons, and joints (*somatic sensory*) (pp. 343–344).
Visceral:	Pertaining to the control of functions, such as digestion, circulation, etc. (*visceral motor*) or sensory information from visceral organs (*visceral sensory*) (pp. 343–344).
Voluntary:	Under direct conscious control (p. 336).
Involuntary:	Not under direct conscious control (p. 336).
Subconscious:	Pertaining to centers in the brain that operate outside a person's conscious awareness (p. 336).
Action Potentials:	Sudden, transient changes in the membrane potential that are propagated along the surface of an axon or sarcolemma (p. 346).

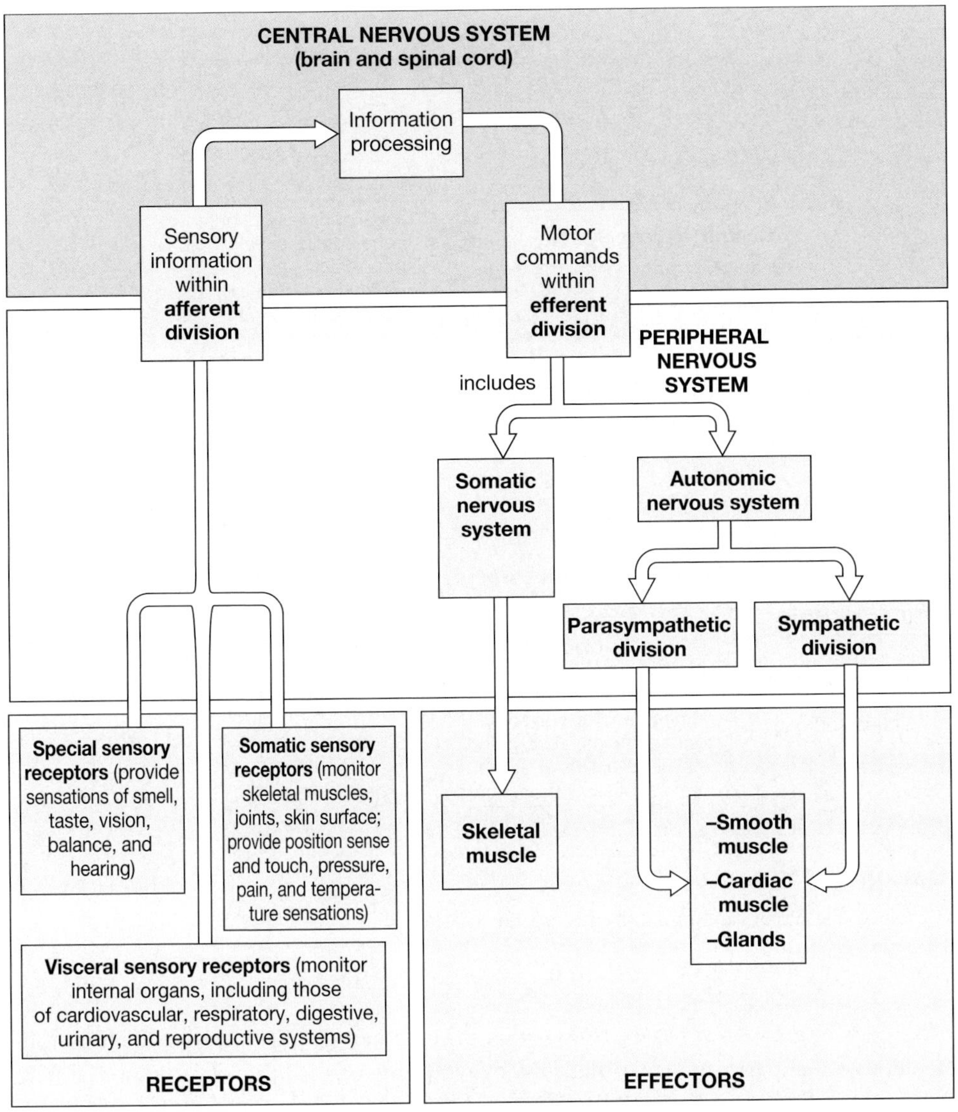

FIGURE 13.2 **A Functional Overview of the Nervous System**

This diagram shows the relationship between the CNS and PNS and the functions and components of the afferent and efferent divisions.

Neuroglia [FIGURE 13.4]

The greatest variety of glial cells is found within the central nervous system. Figure 13.4● compares the functions of the major glial cell populations in the CNS and PNS.

NEUROGLIA OF THE CNS [FIGURES 13.4/13.5]

Four types of glial cells are found in the central nervous system: *astrocytes*, *oligodendrocytes*, *microglia*, and *ependymal cells*. These cell types can be distinguished on the basis of size, intracellular organization, and the presence of specific cytoplasmic processes (Figures 13.4/13.5●).

Astrocytes (FIGURE 13.5) The largest and most numerous glial cells are the **astrocytes** (AS-trō-sīts; *astro-*, star + *cyte*, cell). Their processes contact the surfaces of neuron cell bodies and axons and the walls of capillaries. Astrocytes act both to shield neurons from direct contact with other neurons and limit their exposure to the surrounding interstitial fluid. Astrocytes (Figure 13.5●) have a variety of functions, but many are poorly understood. These functions can be summarized as:

- *Maintaining the blood-brain barrier*: Neural tissue must be physically and biochemically isolated from the general circulation because hormones or other chemicals normally present in the blood could have disruptive effects on neuron function. The endothelial cells lining CNS capillaries have very restricted permeability characteristics that control the chemical exchange between blood and interstitial fluid. They are responsible for the **blood-brain barrier (BBB)**, which isolates the CNS from the general circulation. Chemicals secreted by astrocytes are essential for the maintenance of the blood-brain barrier. The slender cytoplasmic extensions of astrocytes end in expanded "feet" that wrap around capillaries. Astrocyte feet form a com-

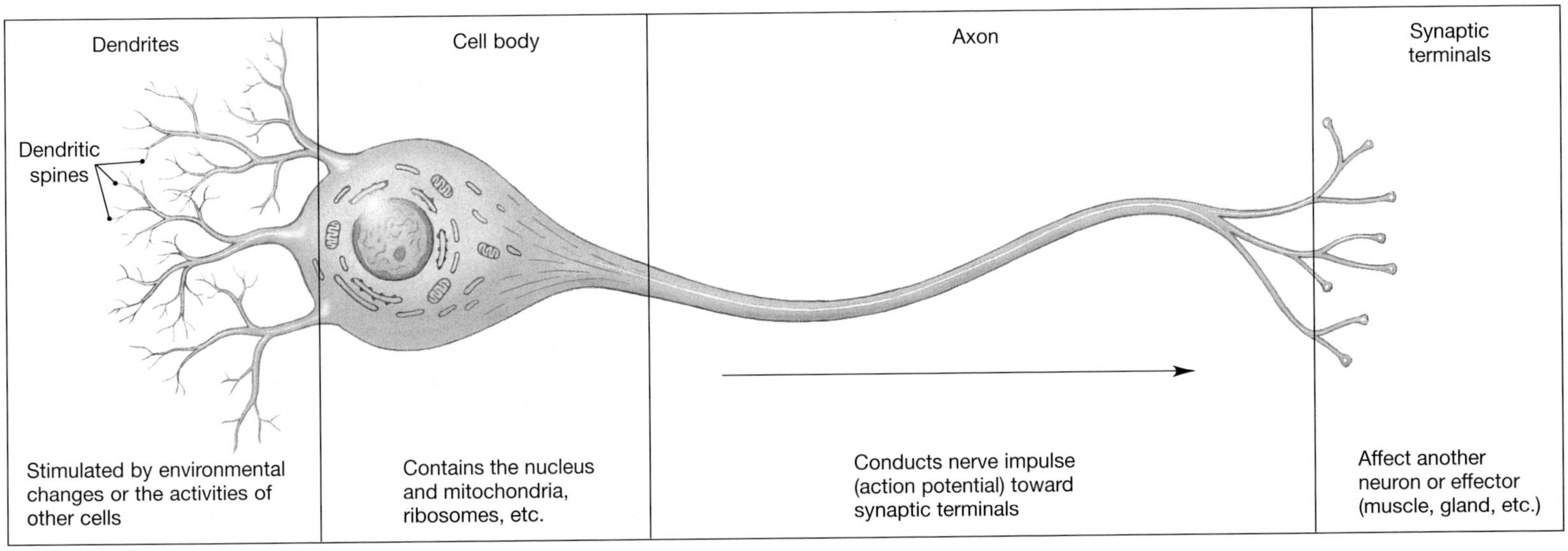

FIGURE 13.3 **A Review of Neuron Structure**

The relationship of the four parts of a neuron (dendrites, cell body, axon, and synaptic terminals): the functional activities of each part and the normal direction of action potential conduction are shown.

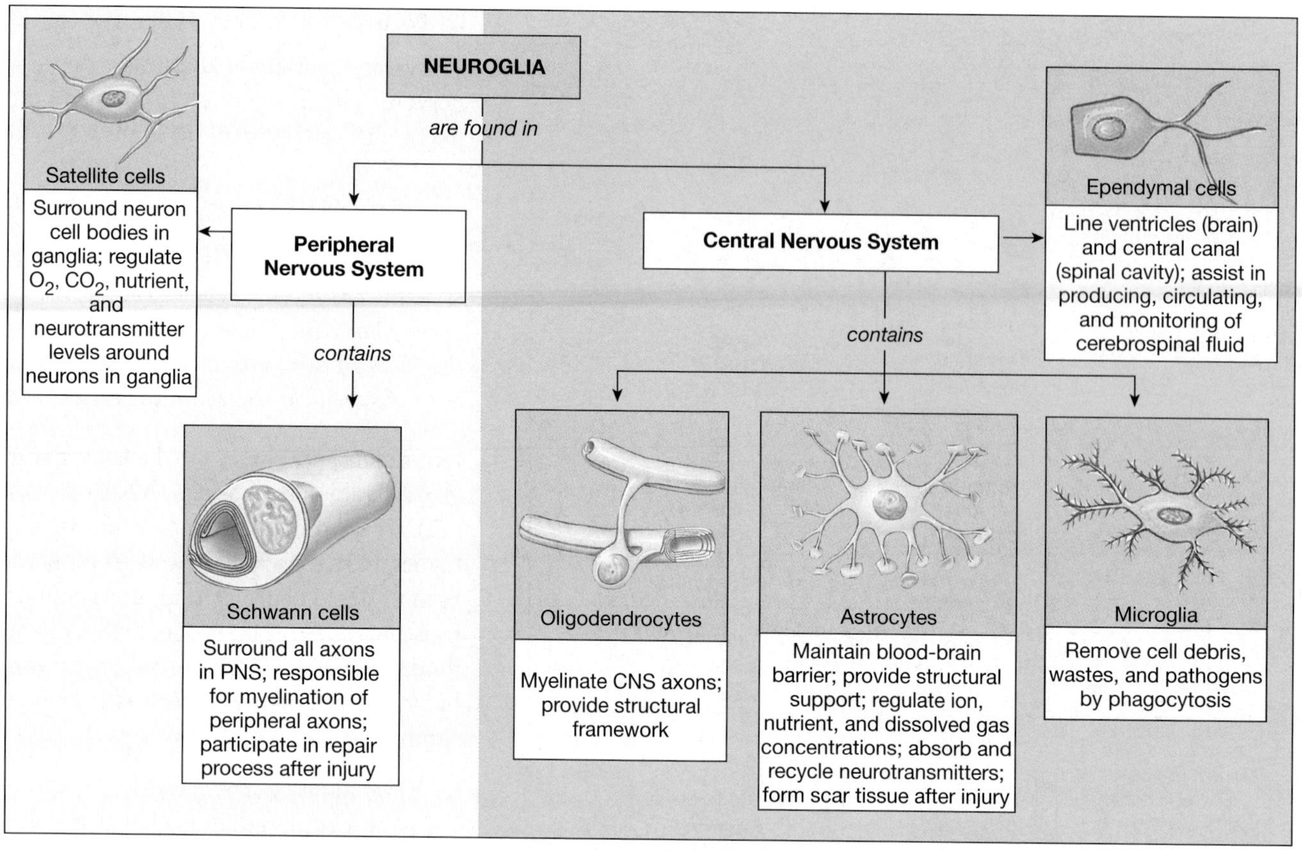

FIGURE 13.4 **THE CLASSIFICATION OF NEUROGLIA**
The categories and functions of the various glial cell types are summarized in this flowchart.

plete cytoplasmic blanket around the capillaries of the CNS, interrupted only where other glial cells contact the capillary walls. (The blood-brain barrier will be discussed further in Chapter 15.)

- *Creating a three-dimensional framework for the CNS*: Astrocytes are packed with microfilaments that extend across the breadth of the cell. This reinforcement provides mechanical strength, and astrocytes form a structural framework that supports the neurons of the brain and spinal cord.
- *Performing repairs in damaged neural tissue*: After damage to the CNS, astrocytes make structural repairs that stabilize the tissue and prevent further injury by producing scar tissue at the injury site.
- *Guiding neuron development*: In the embryonic brain, astrocytes appear to be involved in directing the growth and interconnection of developing neurons through the secretion of chemicals known as *neurotropic factors*.
- *Controlling the interstitial environment*: Astrocytes adjust the interstitial fluid composition by providing a rapid-transit system for the transport of nutrients, ions, and dissolved gases between capillaries and neurons. Additionally, they absorb and recycle neurotransmitters released at active synaptic terminals.

OLIGODENDROCYTES [FIGURE 13.5] **Oligodendrocytes** (ō-li-gō-DEN-drō-sīts; *oligo*, few) resemble astrocytes only in that they both possess slender cytoplasmic extensions. However, oligodendrocytes have smaller cell bodies and fewer processes (Figure 13.5●). Oligodendrocyte processes usually contact the axons or cell bodies of neurons. Oligodendrocyte processes (1) tie clusters of axons together, and (2) improve the functional performance of neurons by wrapping axons in *myelin*, a material with insulating properties. The functions of processes ending at the cell bodies have yet to be determined.

Many axons in the CNS are completely sheathed by the processes of oligodendrocytes. Near the tip of each process, the cell membrane expands to form a flattened pad that wraps around the axon (Figure 13.5●). This creates a multilayered membrane sheath composed primarily of phospholipids. This membranous coating is called **myelin** (MĪ-e-lin), and the axon is said to be **myelinated**. Myelin improves the speed at which an action potential, or *nerve impulse*, is conducted along an axon. Not all axons in the CNS are myelinated. **Unmyelinated** axons may be incompletely covered by glial cell processes.

Many oligodendrocytes cooperate in the formation of the myelin sheath along the entire length of a myelinated axon. The relatively large areas wrapped in myelin are called **internodes** (*inter*, between). Small gaps, called **nodes**, or the *nodes of Ranvier* (rahn-vē-Ā), exist between the myelin sheaths produced by adjacent oligodendrocytes. When dissected, myelinated axons appear a glossy white, primarily because of the lipids present. Regions dominated by myelinated axons constitute the **white matter** of the CNS. In contrast, regions dominated by neuron cell bodies, dendrites, and unmyelinated axons are called **gray matter** because of their dusky gray color.

MICROGLIA [FIGURE 13.5] Roughly 5% of the CNS glial cells are **microglia** (mī-KRŌ-glē-a). These cells are smaller than the other glial cells, and their slender cytoplasmic processes have many fine branches (Figure

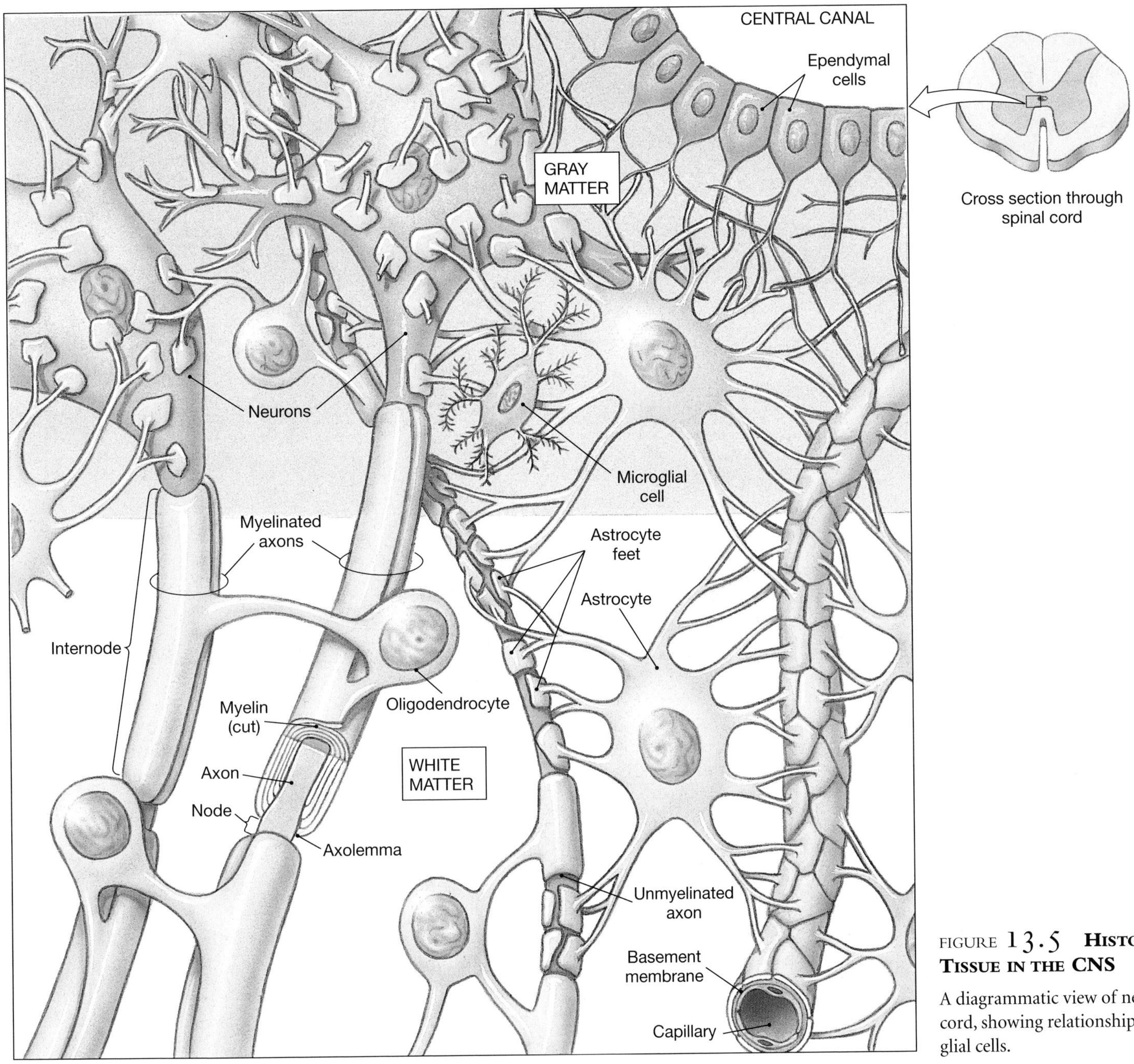

FIGURE 13.5 **HISTOLOGY OF NEURAL TISSUE IN THE CNS**

A diagrammatic view of neural tissue in the spinal cord, showing relationships between neurons and glial cells.

13.5●). Microglia appear early in embryonic development through the division of mesodermal stem cells. The stem cells that produce microglia are related to those that produce tissue macrophages and monocytes of the blood. The microglia migrate into the CNS as it forms, and thereafter they remain within the neural tissue, acting as a roving security force. Microglia engulf cellular debris, waste products, and pathogens. In times of infection or injury, the number of microglia increases dramatically.

EPENDYMAL CELLS [FIGURE 13.6] The ventricles of the brain and central canal of the spinal cord are lined by a cellular layer called the **ependyma** (ep-EN-di-mah). These chambers and passageways are filled with **cerebrospinal fluid (CSF)**. This fluid, which also surrounds the brain and spinal cord, provides a protective cushion and transports dissolved gases, nutrients, wastes, and other materials. The composition, formation, and circulation of CSF will be discussed in Chapter 15.

Ependymal cells are cuboidal to columnar in form. Unlike typical epithelial cells, ependymal cells have slender processes that branch extensively and make direct contact with glial cells in the surrounding neural tissue (Figure 13.6a●). Experimental evidence suggests that ependymal cells may act as receptors that monitor the composition of the CSF. The cilia may also assist in the circulation of CSF. During development and early childhood, the free surfaces of ependymal cells are covered with cilia. In the adult, cilia may persist on ependymal cells lining the ventricles of the brain (Figure 13.6b●), but the ependyma elsewhere usually has only scattered microvilli. Within the ventricles, specialized ependymal cells participate in the secretion of cerebrospinal fluid.

NEUROGLIA OF THE PNS

Neuron cell bodies in the PNS are usually clustered together in masses called **ganglia** (singular, *ganglion*). Axons are bundled together and

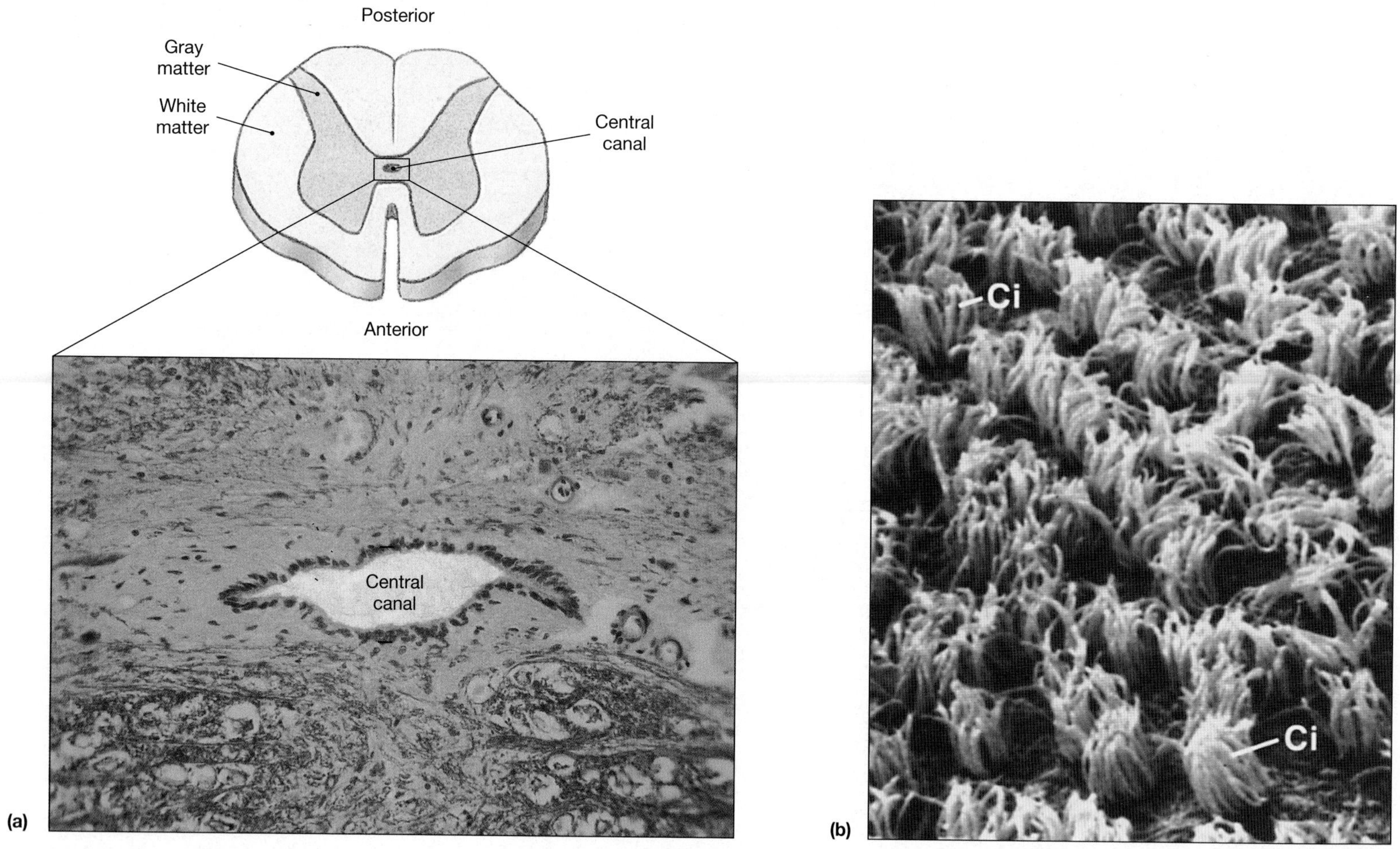

FIGURE 13.6 **The Ependyma**

The ependyma is a cellular layer that lines brain ventricles and the central canal of the spinal cord. (**a**) Light micrograph showing the ependymal lining of the central canal. (LM × 257) The diagrammatic view of ependymal organization is based on information provided by electron microscopy. (**b**) An SEM of the ciliated surface of the ependyma from one of the ventricles. (Ci = cilia; SEM × 1825) (Reproduced from R. G. Kessel and R. H. Kardon, *Tissues and Organs: A Text-Atlas of Scanning Electron Microscopy*, W. H. Freeman & Co., 1979.)

wrapped in connective tissue, forming **peripheral nerves**, or simply *nerves*. All neuron cell bodies and axons in the PNS are completely insulated from their surroundings by the processes of glial cells. The two glial cell types involved are called *satellite cells* and *Schwann cells*.

Satellite Cells (FIGURE 13.7) **Satellite cells**, or *amphicytes* (AM-fi-sīts), surround the neuron cell bodies in peripheral ganglia (Figure 13.7●). Satellite cells regulate the exchange of nutrients and waste products between the neuron cell body and the extracellular fluid. They also help isolate the neuron from stimuli other than those provided at synapses.

Schwann Cells [FIGURES 13.5/13.8] **Schwann cells**, or *neurolemmocytes*, produce a covering around every peripheral axon, whether it is unmyelinated or myelinated. The cell membrane of an axon is called the **axolemma** (*lemma*, husk); the superficial cytoplasmic covering provided by the Schwann cells is known as the **neurilemma** (noo-ri-LEM-ma). The

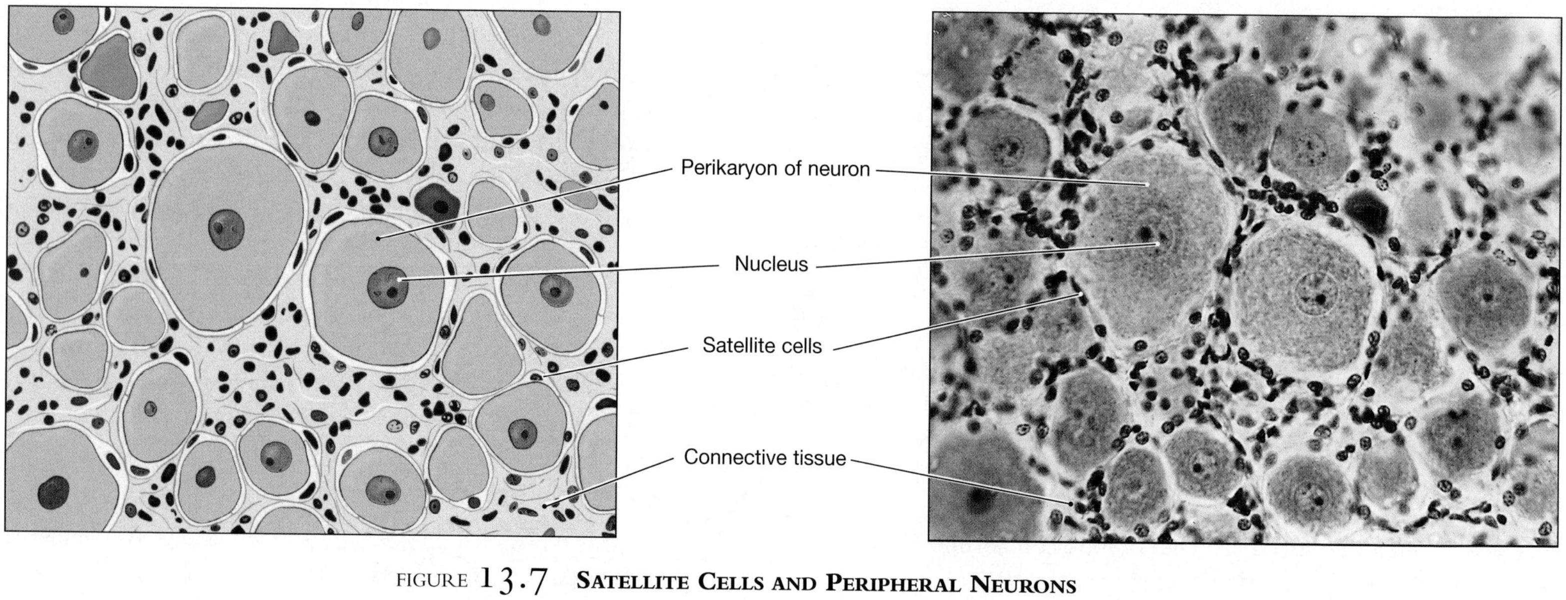

FIGURE 13.7 **Satellite Cells and Peripheral Neurons**

Satellite cells surround neuron cell bodies in peripheral ganglia. (LM × 20)

physical relationship between a Schwann cell and a myelinated peripheral axon differs from that of an oligodendrocyte and a myelinated axon in the CNS. A Schwann cell can myelinate only about 1 mm along the length of a single axon. In contrast, an oligodendrocyte can myelinate portions of several axons (compare Schwann cells, Figure 13.8a, with oligodendrocytes, Figure 13.5●, p. 339). Although the mechanism of myelination differs, myelinated axons in both the CNS and PNS have nodes and internodes, and the presence of myelin—however formed—increases the rate of nerve impulse conduction. Unmyelinated axons are enclosed by the processes of Schwann cells, but the relationship is simple and no myelin forms. A single Schwann cell may surround several different unmyelinated axons, as indicated in Figure 13.8b●.

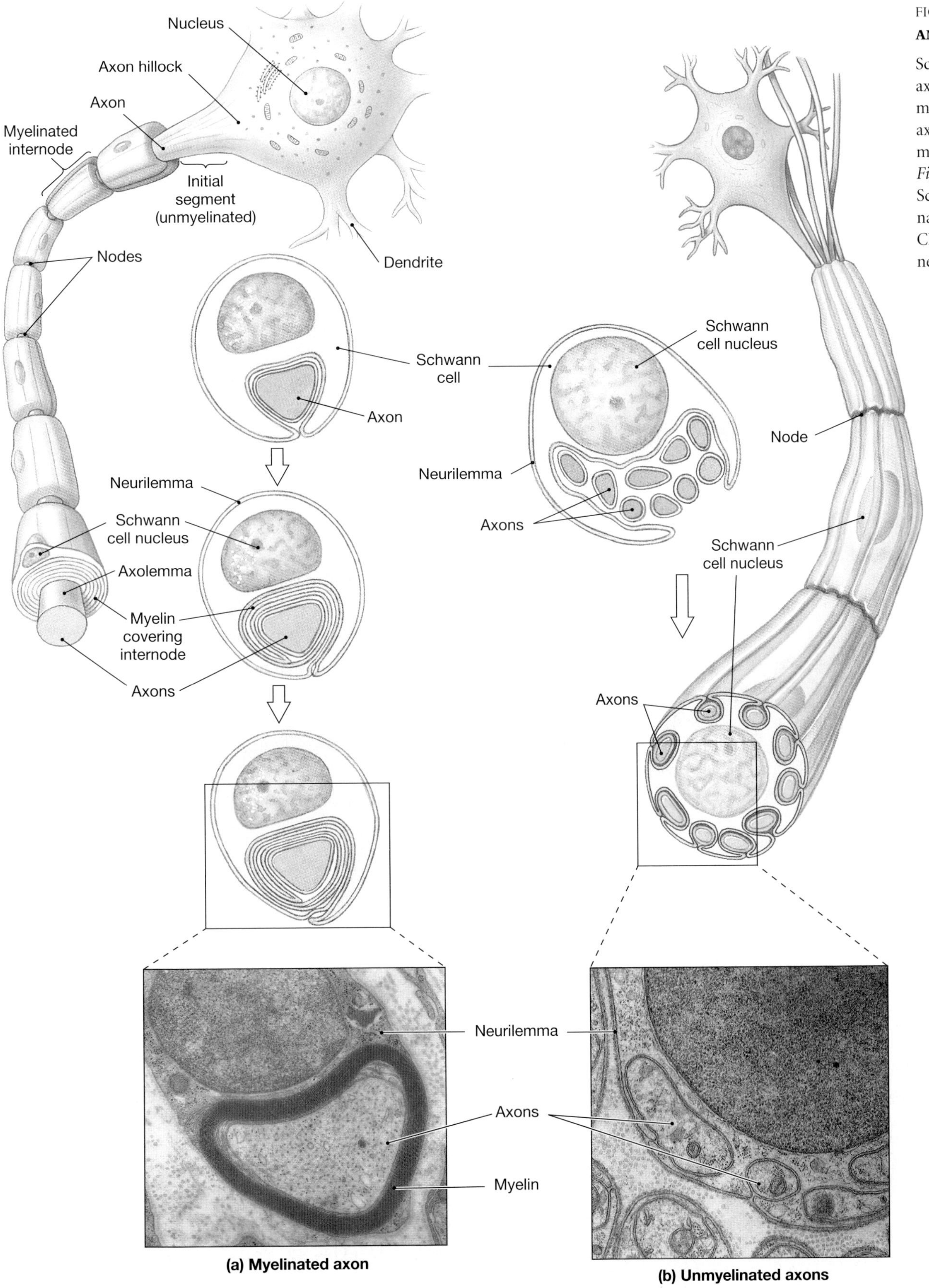

FIGURE 13.8 **Schwann Cells and Peripheral Axons**

Schwann cells ensheath every peripheral axon. **(a)** A single Schwann cell forms the myelin sheath around a portion of a single axon. This situation differs from the way myelin forms inside the CNS. *Compare with Figure 13.5.* (TEM × 20,603) **(b)** A single Schwann cell can encircle several unmyelinated axons. Unlike the situation inside the CNS, every axon in the PNS has a complete neurilemmal sheath. (LM × 27,627)

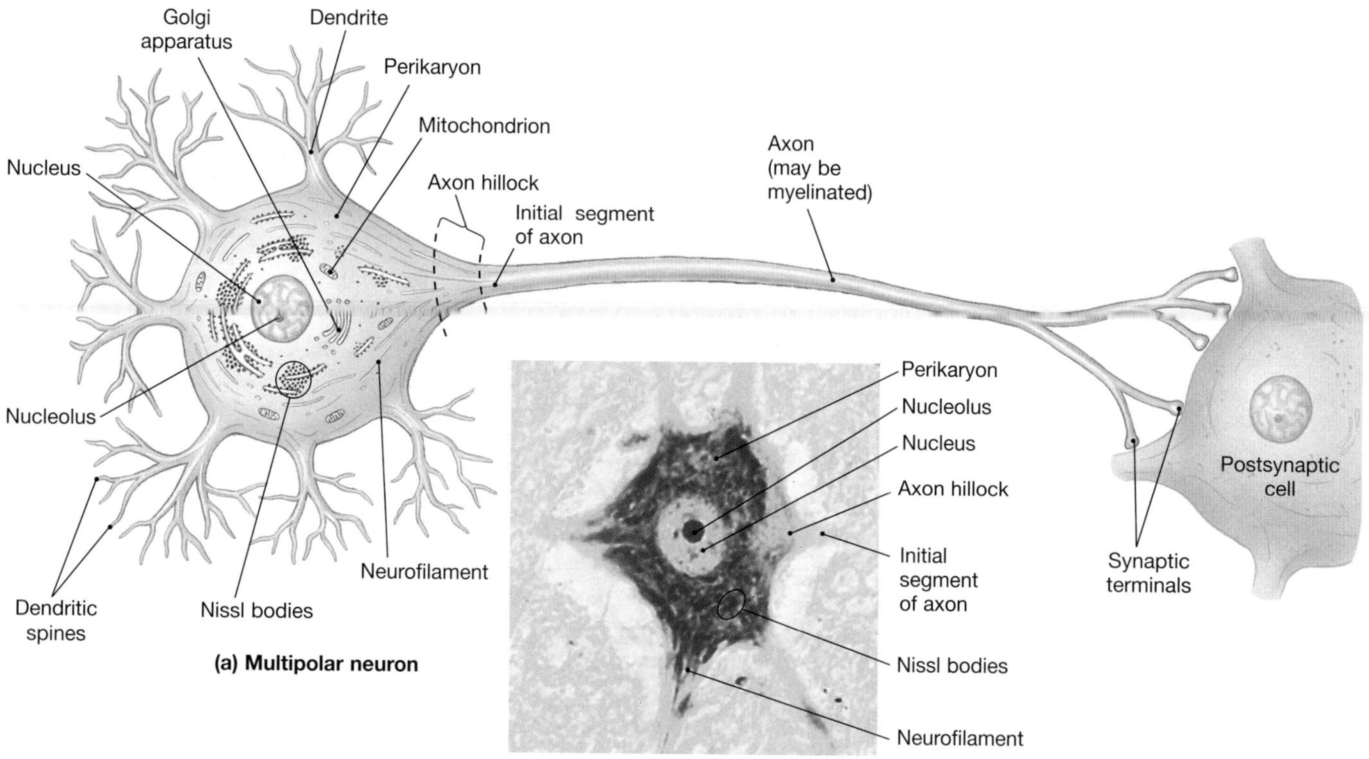

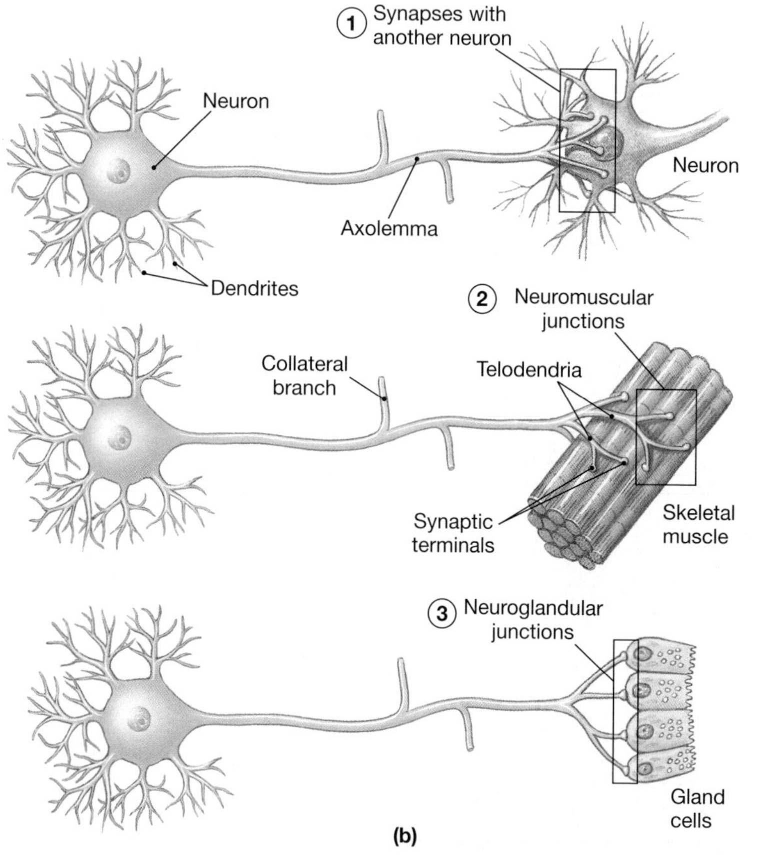

FIGURE 13.9 **ANATOMY OF A REPRESENTATIVE NEURON**

A neuron has a cell body (soma), some branching dendrites, and a single axon. (**a**) Detailed organization of the cell body. (LM × 1600) (**b**) A neuron may innervate (1) other neurons, (2) skeletal muscle fibers, or (3) gland cells. Synapses are shown in boxes for each example. A single neuron would not innervate all three.

✓ **CONCEPT CHECK**

- Identify the two anatomical subdivisions of the nervous system.
- What two terms are used to refer to the supporting cells in neural tissue?
- Specifically, what cells help maintain the blood-brain barrier?
- What is the name of the membranous coating formed around axons by oligodendrocytes?
- Identify the cells in the peripheral nervous system that form a covering around axons.

Neurons [FIGURE 13.9]

The cell body of a representative neuron contains a relatively large, round nucleus with a prominent nucleolus (Figure 13.9a●). The surrounding cytoplasm constitutes the **perikaryon** (per-i-KAR-ē-on; *karyon,* nucleus). The cytoskeleton of the perikaryon contains **neurofilaments** and **neurotubules.** Bundles of neurofilaments, called **neurofibrils,** are cytoskeletal elements that extend into the dendrites and axon.

The perikaryon contains organelles that provide energy and perform biosynthetic activities. The numerous mitochondria, free and fixed ribosomes, and membranes of the rough endoplasmic reticulum (RER) give the perikaryon a coarse, grainy appearance. Mitochondria generate ATP to meet the high energy demands of an active neuron. The ribosomes and RER synthesize peptides and proteins. Groups of fixed and free ribosomes are present in large numbers. These ribosomal clusters are called *Nissl bodies* because they were first described by the German microscopist Franz Nissl. Nissl bodies account for the gray color of areas that contain neuron cell bodies—the *gray matter* seen in gross dissection of the brain or spinal cord.

Most neurons lack the *centrosome* complex. p. 36 In other cells, the centrioles of the centrosome form the spindle fibers that move chromosomes during cell division. Neurons usually lose their centrioles during differentiation, and they become incapable of undergoing cell division. If these specialized neurons are later lost to injury or disease, they cannot be replaced.

The membrane permeability of the dendrites and cell body can be changed by exposure to chemical, mechanical, or electrical stimuli. One of the primary functions of glial cells is to limit the number or types of stimuli affecting individual neurons. Glial cell processes cover most of the surfaces of the cell body and dendrites, except where synaptic terminals exist or where dendrites function as sensory receptors, monitoring conditions in the extracellular environment. Exposure to appropriate stimuli can produce a localized change in the transmembrane potential and lead to the generation of an action potential at the axon. The transmembrane potential is a property resulting from the unequal distribution of ions across the cell membrane. We will examine transmembrane potentials and action potentials later in this chapter.

An axon, or *nerve fiber*, is a long cytoplasmic process capable of propagating an action potential. In a multipolar neuron, a specialized region, the **axon hillock**, connects the **initial segment** of the axon to the soma. The **axoplasm** (AK-sō-plazm), or cytoplasm of the axon, contains neurofibrils, neurotubules, numerous small vesicles, lysosomes, mitochondria, and various enzymes. An axon may branch along its length, producing side branches called **collaterals**. The main trunk and the collaterals end in a series of fine terminal extensions, called **telodendria** (tel-ō-DEN-drē-a; *telo-*, end + *dendron*, tree) (Figure 13.9b●). The telodendria end in a **synaptic terminal**, where the neuron contacts another neuron or effector. *Axoplasmic transport* is the movement of organelles, nutrients, synthesized molecules, and waste products between the cell body and the synaptic terminals. This is a complex process that consumes energy and relies on movement along the neurofibrils of the axon and its branches. *Axoplasmic Transport and Disease p. 792*

Each synaptic terminal is part of a **synapse**, a specialized site where the neuron communicates with another cell (Figure 13.9b●). The structure of the synaptic terminal varies with the type of postsynaptic cell. A relatively simple, round **synaptic knob**, or *terminal bouton*, is found where one neuron synapses on another. The synaptic terminal found at a *neuromuscular junction*, where a neuron contacts a skeletal muscle fiber, is much more complex. p. 253 Synaptic communication most often involves the release of specific chemicals called **neurotransmitters**. The release of these chemicals is triggered by the arrival of a nerve impulse; additional details are provided in a later section.

NEURON CLASSIFICATION

The billions of neurons in the nervous system are quite variable in form. Neurons may be classified based on: (1) structure or (2) function.

STRUCTURAL CLASSIFICATION OF NEURONS [FIGURE 13.10] The structural classification is based on the number of processes that project from the cell body (Figure 13.10●).

- **Anaxonic** (an-ak-SON-ik) **neurons** are small, and there are no anatomical clues to distinguish dendrites from axons (Figure 13.10a●). Anaxonic neurons are found only in the CNS and in special sense organs, and their functions are poorly understood.
- **Bipolar neurons** have a single dendrite and a single axon, with the cell body between them (Figure 13.10b●). Bipolar neurons are relatively rare but play an important role in relaying information concerning sight, smell, and hearing. Their axons are not myelinated.
- **Unipolar neurons**, or *pseudounipolar neurons*, have continuous dendritic and axonal processes, and the cell body lies off to one side. In these neurons, the initial segment lies at the base of the dendritic branches (Figure 13.10c●), and the rest of the process is considered

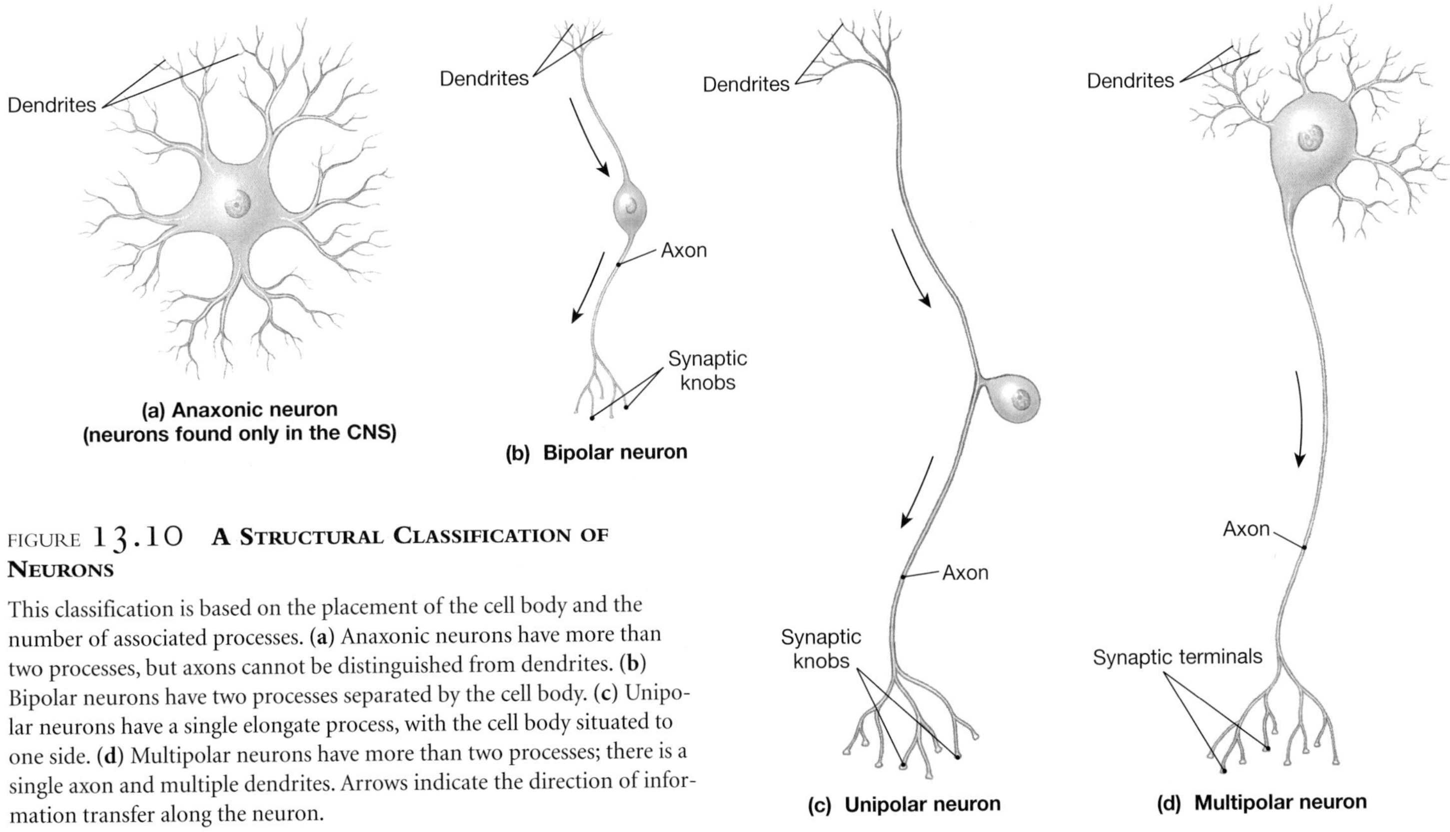

FIGURE 13.10 **A STRUCTURAL CLASSIFICATION OF NEURONS**

This classification is based on the placement of the cell body and the number of associated processes. **(a)** Anaxonic neurons have more than two processes, but axons cannot be distinguished from dendrites. **(b)** Bipolar neurons have two processes separated by the cell body. **(c)** Unipolar neurons have a single elongate process, with the cell body situated to one side. **(d)** Multipolar neurons have more than two processes; there is a single axon and multiple dendrites. Arrows indicate the direction of information transfer along the neuron.

an axon on both structural and functional grounds. Sensory neurons of the peripheral nervous system are usually unipolar, and their axons may be myelinated.

- **Multipolar neurons** have several dendrites and a single axon that may have one or more branches (Figure 13.10d●). Multipolar neurons are the most common type of neuron in the CNS. For example, all of the motor neurons that control skeletal muscles are multipolar neurons with myelinated axons.

Functional Classification of Neurons [Figure 13.11] Neurons can be categorized into three functional groups: (1) *sensory neurons,* (2) *motor neurons,* and (3) *interneurons,* or *association neurons.* Their relationships are diagrammed in Figure 13.11●.

Sensory Neurons Almost all sensory **neurons** are unipolar neurons with their cell bodies located outside the CNS in peripheral sensory ganglia. They form the *afferent division* of the PNS. Their function is to deliver information to the CNS. The axons of sensory neurons, called **afferent fibers**, extend between a sensory receptor and the spinal cord or brain. Sensory neurons collect information concerning the external or internal environment. There are about 10 million sensory neurons. **Somatic sensory neurons** transmit information about the outside world and our position within it. **Visceral sensory neurons** transmit information about internal conditions and the status of other organ systems.

Receptors may be either the processes of specialized sensory neurons or cells monitored by sensory neurons. Receptors are broadly categorized as:

- **Exteroceptors** (*extero-*, outside) provide information about the external environment in the form of touch, temperature, and pressure sensations and the more complex *special senses* of sight, smell, and hearing.
- **Proprioceptors** (prō-prē-ō-SEP-torz) monitor the position and movement of skeletal muscles and joints.
- **Interoceptors** (*intero-*, inside) monitor the digestive, respiratory, cardiovascular, urinary, and reproductive systems and provide sensations of deep pressure and pain as well as taste, another special sense.

Data from exteroceptors and proprioceptors are carried by somatic sensory neurons. Interoceptive information is carried by visceral sensory neurons.

Motor Neurons **Motor neurons** are multipolar neurons that form the efferent division of the nervous system. A motor neuron stimulates or modifies the activity of a peripheral tissue, organ, or organ system. About half a million motor neurons are found in the body. Axons traveling away from the CNS are called **efferent fibers.** The two efferent divisions of the PNS—the somatic nervous system (SNS) and the autonomic nervous system (ANS)—differ in the way they innervate peripheral effectors. The SNS includes all of the somatic motor neurons that innervate skeletal muscles. The cell bodies of these motor neurons lie inside the CNS, and their axons extend to the neuromuscular junctions that control skeletal muscles. Most of the activities of the SNS are consciously controlled.

The autonomic nervous system includes all the **visceral motor neurons** that innervate peripheral effectors other than skeletal muscles. There are two groups of visceral motor neurons—one group has cell bodies inside the CNS, and the other has cell bodies in peripheral ganglia. The neurons inside the CNS control the neurons in the peripheral ganglia, and these neurons

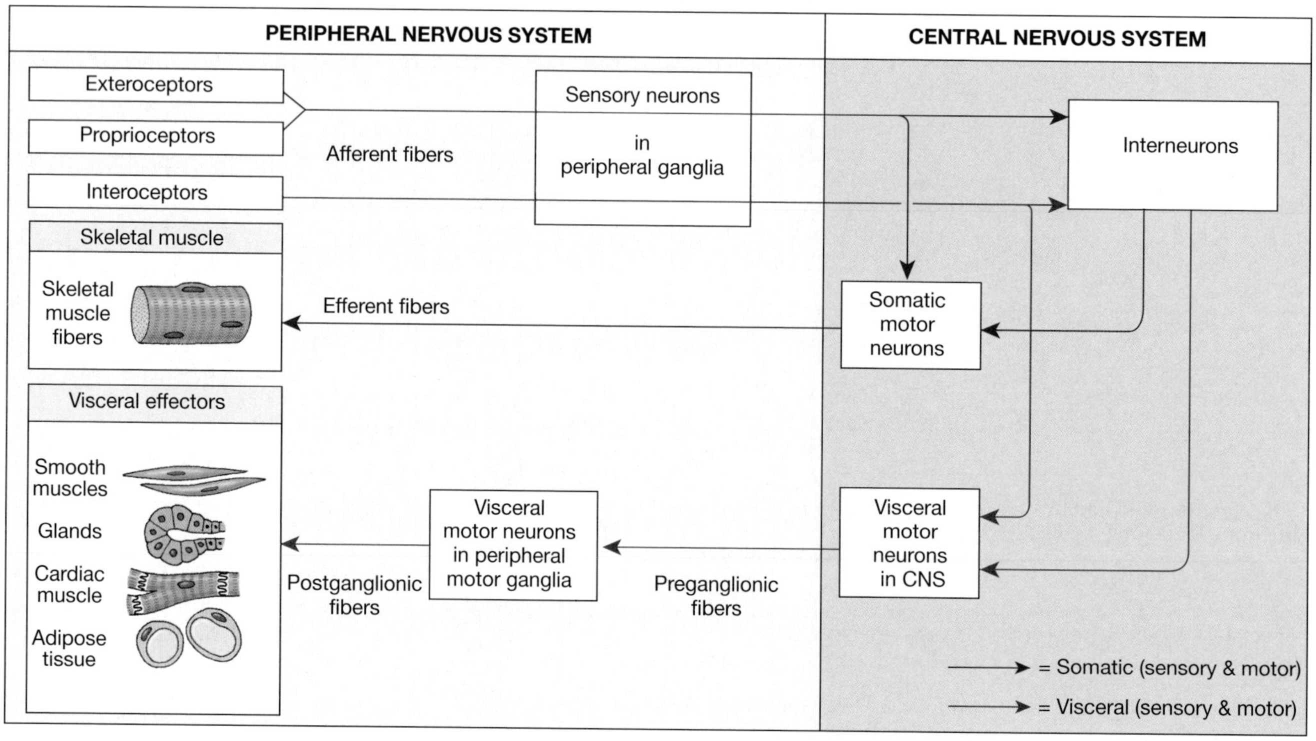

FIGURE 13.11 **A Functional Classification of Neurons**

Neurons are classified functionally into three categories: (1) sensory neurons that detect stimuli in the PNS and send information to the CNS, (2) motor neurons to carry instructions from the CNS to peripheral effectors, and (3) interneurons in the CNS that process sensory information and coordinate motor activity.

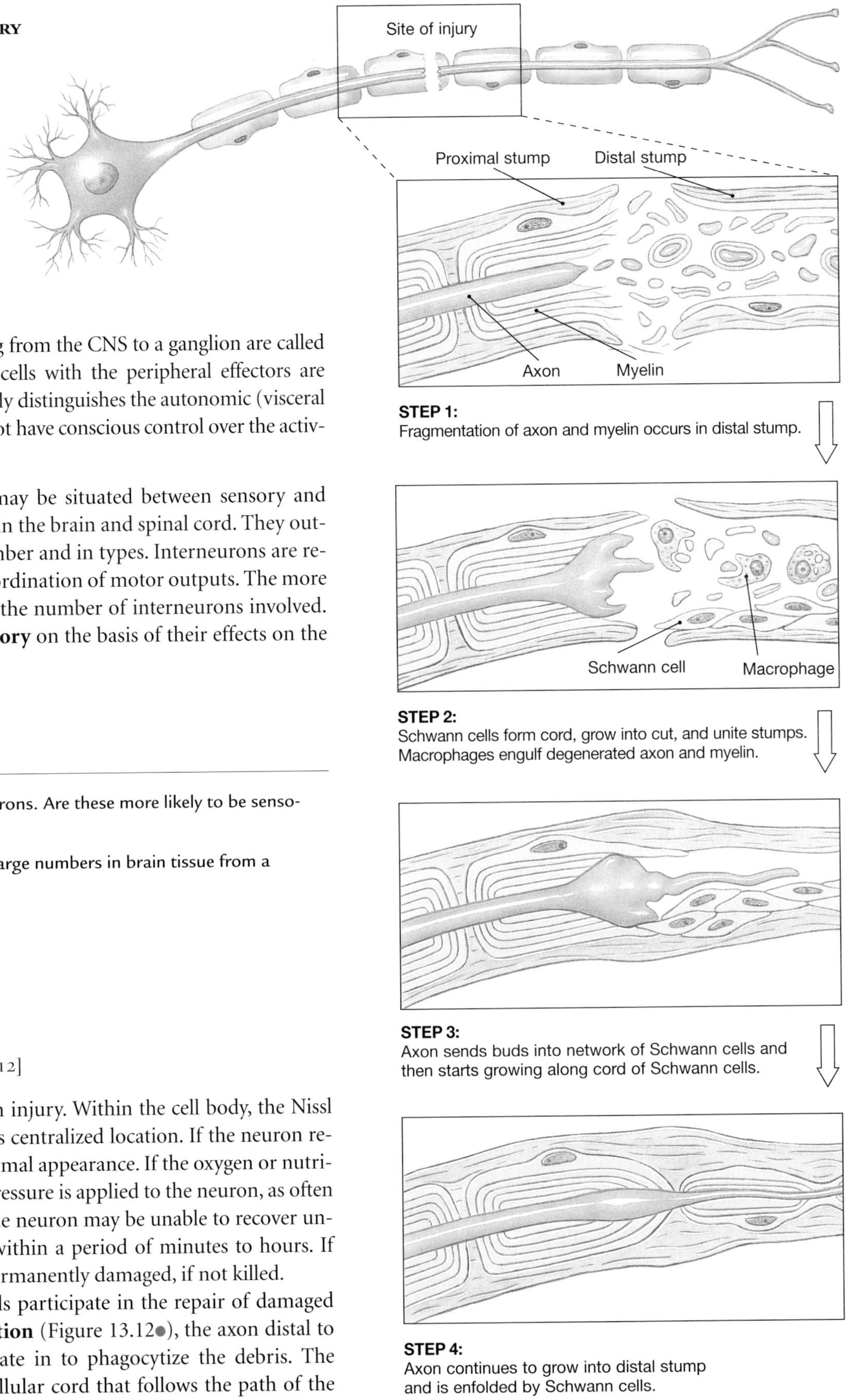
Site of injury
Proximal stump
Distal stump
Axon
Myelin
STEP 1:
Fragmentation of axon and myelin occurs in distal stump.
Schwann cell
Macrophage
STEP 2:
Schwann cells form cord, grow into cut, and unite stumps.
Macrophages engulf degenerated axon and myelin.
STEP 3:
Axon sends buds into network of Schwann cells and
then starts growing along cord of Schwann cells.
STEP 4:
Axon continues to grow into distal stump
and is enfolded by Schwann cells.

FIGURE 13.12 **NERVE REGENERATION AFTER INJURY**
Steps involved in the repair of a peripheral nerve by the process of Wallerian degeneration.

in turn control the peripheral effectors. Axons extending from the CNS to a ganglion are called **preganglionic fibers**. Axons connecting the ganglion cells with the peripheral effectors are known as **postganglionic fibers**. This arrangement clearly distinguishes the autonomic (visceral motor) system from the somatic motor system. We do not have conscious control over the activities of the ANS.

Interneurons Interneurons **(association neurons)**, may be situated between sensory and motor neurons. Interneurons are located entirely within the brain and spinal cord. They outnumber all other neurons combined both in total number and in types. Interneurons are responsible for the analysis of sensory inputs and the coordination of motor outputs. The more complex the response to a given stimulus, the greater the number of interneurons involved. Interneurons can be classified as **excitatory** or **inhibitory** on the basis of their effects on the postsynaptic membranes of other neurons.

✓ CONCEPT CHECK

- Examination of a tissue sample shows unipolar neurons. Are these more likely to be sensory neurons or motor neurons?
- What type of glial cell would you expect to find in large numbers in brain tissue from a person suffering from a CNS infection?

Neural Regeneration [FIGURE 13.12]

A neuron has a very limited ability to recover after an injury. Within the cell body, the Nissl bodies disappear and the nucleus moves away from its centralized location. If the neuron regains normal function, it will gradually return to a normal appearance. If the oxygen or nutrient supply is restricted, as in a stroke, or mechanical pressure is applied to the neuron, as often happens in spinal cord or peripheral nerve injuries, the neuron may be unable to recover unless circulation is restored or the pressure removed within a period of minutes to hours. If these stresses continue, the affected neurons will be permanently damaged, if not killed.

In the peripheral nervous system, Schwann cells participate in the repair of damaged nerves. In the process known as **Wallerian degeneration** (Figure 13.12●), the axon distal to the injury site deteriorates, and macrophages migrate in to phagocytize the debris. The Schwann cells in the area divide and form a solid cellular cord that follows the path of the original axon. Additionally, these Schwann cells release growth factors to promote axonal regrowth. As the neuron recovers, its axon grows into the injury site, and the Schwann cells wrap around it.

If the axon continues to grow into the periphery alongside the appropriate cord of Schwann cells, it may eventually reestablish its normal synaptic contacts. If it stops growing, or wanders off in some new direction, normal function will not return. The growing axon is most likely to arrive at its appropriate destination if the cut proximal and distal stumps remain in contact after the injury. When an entire peripheral nerve is severed, only a relatively small number of axons will successfully reestablish normal synaptic contacts. As a result, nerve function will be permanently impaired.

Limited regeneration can occur inside the central nervous system, but the situation is more complicated because (1) many more axons are likely to be involved, (2) astrocytes produce scar tissue that can prevent axon growth across the damaged area, and (3) astrocytes release chemicals that block the regrowth of axons. Additional information concerning neural regeneration and surgical repairs will be presented in the next chapter.

The Nerve Impulse

Excitability is the ability of a cell membrane to conduct electrical impulses. Membranes of skeletal muscle fibers, cardiac muscle cells, some gland cells, and the axolemma of most neurons (including all multipolar and unipolar neurons) are examples of excitable membranes. An electrical impulse, or **action potential**, develops after the membrane is stimulated to a level known as the *threshold*. After the threshold level has been reached, the membrane permeability to sodium and potassium ions changes. The ion movements that result produce a sudden change in the transmembrane potential, and this change constitutes an action potential. The permeability changes are temporary and initially confined to the point of stimulation. However, the change in ion distribution almost immediately triggers changes in the permeability of adjacent portions of the cell membrane. In this way, the action potential is conducted along the membrane surface. For example, in a skeletal muscle fiber, action potentials begin at the neuromuscular junction and sweep across the entire surface of the sarcolemma. p. 246 In the nervous system, action potentials traveling along axons are known as **nerve impulses**.

Before a nerve impulse can occur, a stimulus of sufficient strength must be applied to the membrane of the neuron. Once initiated, the rate of impulse conduction depends on the properties of the axon, specifically:

1. *The presence or absence of a myelin sheath*: A myelinated axon conducts impulses five to seven times faster than an unmyelinated axon.
2. *The diameter of the axon*: The larger the diameter, the more rapidly the impulse will be conducted.

The largest myelinated axons, with diameters ranging from 4 to 20 mm, conduct nerve impulses at speeds of up to 140 m/s (300 mph). In contrast, small unmyelinated fibers (less than 2 mm in diameter) conduct impulses at speeds below 1 m/s (2 mph). *Demyelination Disorders p. 792*

✓ CONCEPT CHECK

- What effect would cutting the axon have in transmitting the action potential?
- Two axons are tested for conduction velocities. One conducts action potentials at 50 m/s, the other at 1 m/s. Which axon is myelinated?
- Define excitability.
- What term is used to identify conducted changes in transmembrane potential?

Synaptic Communication [FIGURE 13.9b]

A synapse between neurons may involve a synaptic terminal and (1) a dendrite (*axodendritic*), (2) the cell body (*axosomatic*), or (3) an axon (*axoaxonic*). A synapse may also permit communication between a neuron and another cell type; such synapses are called **neuroeffector junctions**. The *neuromuscular junction* described in Chapter 9 was an example of a neuroeffector junction. p. 246 Neuroeffector junctions involving other cell types are shown in Figure 13.9b●, p. 342.

At a synaptic terminal, a nerve impulse triggers events at a synapse that transfers the information either to another neuron or to an effector cell. A synapse may be **chemical**, involving the passage of a neurotransmitter substance between cells, or **electrical**, with gap junctions permitting ion flow between the cells.

Chemical Synapses [FIGURE 13.13]

Chemical synapses are by far the most abundant; there are several different types. Most interactions between neurons and all communications between neurons and peripheral effectors involve chemical synapses. At a chemical synapse between neurons (Figure 13.13●), a neurotransmitter released at the *presynaptic membrane* of a synaptic knob binds to receptor proteins on the *postsynaptic membrane* and triggers a transient change in the transmembrane potential of the receptive cell. Only the presynaptic membrane releases a neurotransmitter. As a result, communication occurs in one direction only: from the presynaptic neuron to the postsynaptic neuron.

The neuromuscular junction described in Chapter 9 is a chemical synapse that releases the neurotransmitter *acetylcholine* (*ACh*). p. 253 Over 50 different neurotransmitters have been identified, but ACh is the best known. All neuromuscular junctions utilize ACh; it is also released at many chemical synapses in the CNS and PNS. The general sequence of events is similar, regardless of the location of the synapse or the nature of the neurotransmitter.

- Arrival of the action potential at the synaptic knob triggers release of neurotransmitter from secretory vesicles, through exocytosis at the presynaptic membrane.
- The neurotransmitter diffuses across the synaptic cleft and binds to receptors on the postsynaptic membrane.
- Receptor binding results in a change in the permeability of the postsynaptic cell membrane. Depending on the identity and abundance of the receptor proteins on the postsynaptic membrane, the result may be excitatory or inhibitory. In general, excitatory effects promote the generation of action potentials, whereas inhibitory effects reduce the ability to generate an action potential.
- If the degree of excitation is sufficient, receptor binding may lead to the generation of an action potential in the axon (if the postsynaptic cell is a neuron) or sarcolemma (if the postsynaptic cell is a skeletal muscle fiber).
- The effects of one action potential on the postsynaptic membrane are short-lived, because the neurotransmitter molecules are either enzymatically broken down or reabsorbed. To prolong or enhance the effects, additional action potentials must arrive at the synaptic terminal, and additional molecules of ACh must be released into the synaptic cleft.

Examples of neurotransmitters other than ACh will be presented in later chapters. There may be thousands of synapses on the cell body of a

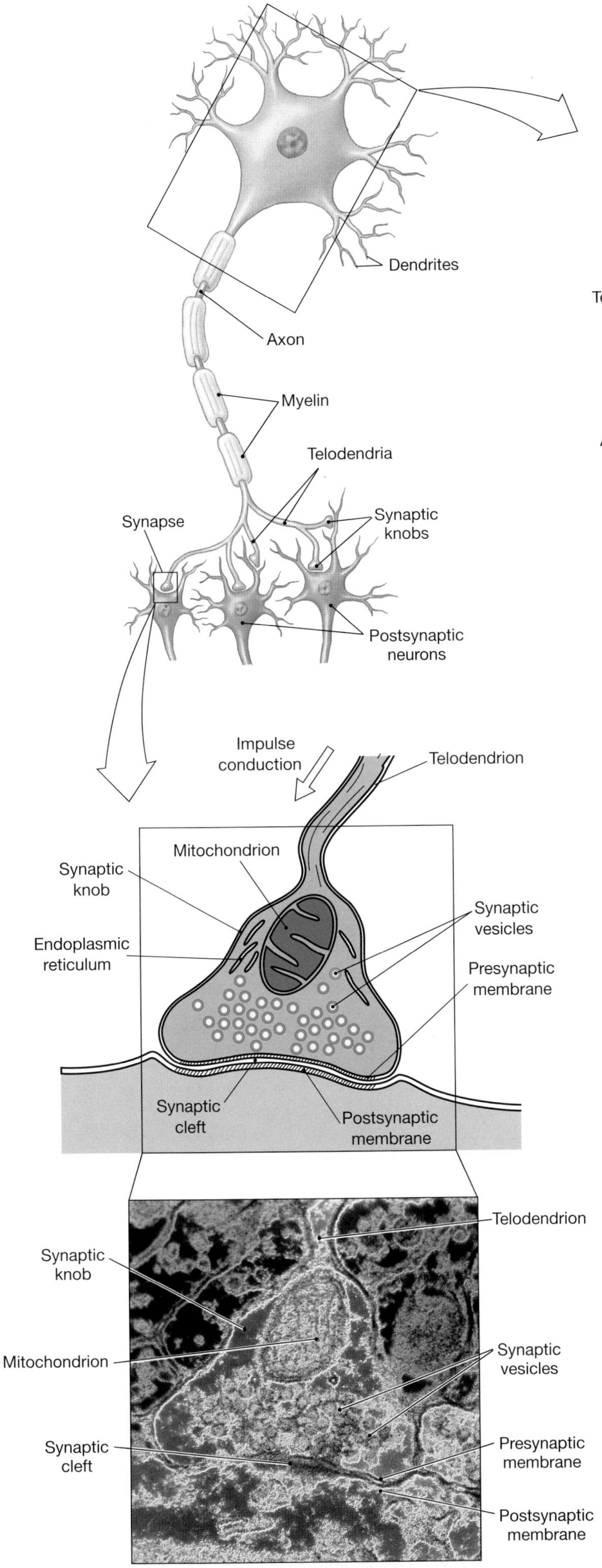

(a) A chemical synapse

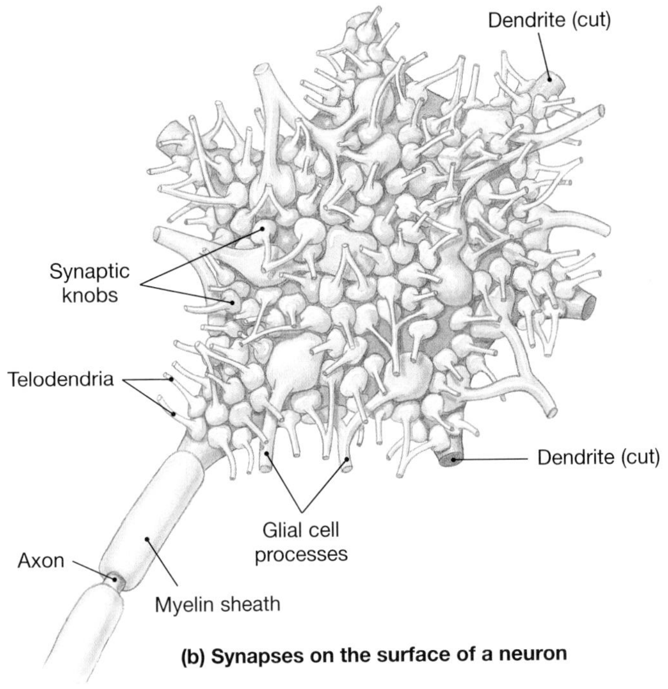

(b) Synapses on the surface of a neuron

FIGURE 13.13 **THE STRUCTURE OF A SYNAPSE**

A synapse is the site of communication between a neuron and another cell. **(a)** Diagrammatic view of a chemical synapse between two neurons, paired with a color-enhanced TEM of a chemical synapse. (TEM × 186,480) **(b)** There may be thousands of chemical synapses on the surface of a single neuron. Many of these synapses may be active at any one moment.

single neuron (Figure 13.13b●). Many of these will be active at any given moment, releasing a variety of different neurotransmitters. Some will have excitatory effects, others inhibitory effects. The activity of the receptive neuron depends on the sum of all of the excitatory and inhibitory stimuli influencing the axon hillock at any given moment.

Electrical Synapses

Chemical synapses dominate the nervous system. **Electrical synapses** are found between neurons in the CNS and PNS, but they are relatively rare. At an electrical synapse, the presynaptic and postsynaptic cell membranes are tightly bound together, and *gap junctions* permit the passage of ions between the cells. *p. 43* Because the two cells are linked in this way, they function as if they shared a common cell membrane, and the nerve impulse crosses from one cell membrane to the next without delay.

Neuron Organization and Processing

[FIGURE 13.14]

Neurons are the basic building blocks of the nervous system. The billions of interneurons within the CNS are organized into a much smaller number of **neuronal pools**. A neuronal pool is a group of interconnected neurons with specific functions. Neuronal pools are defined on the basis of function rather than on anatomical grounds. A pool may be diffuse, involving neurons in several different regions of the brain, or localized, with all of the neurons restricted to one specific location in the brain or spinal cord. Estimates concerning the actual number of neuronal pools range between a few hundred and a few thousand. Each has a limited number of input sources and output destinations, and the pool may contain both excitatory and inhibitory neurons.

The basic "wiring pattern" found in a neuronal pool is called a *neural circuit*. A neural circuit may have one of the following functions:

1. **Divergence** is the spread of information from one neuron to several neurons, as in Figure 13.14a●, or from one pool to multiple pools. Divergence permits the broad distribution of a specific input. Considerable divergence occurs when sensory neurons bring information into the CNS, for the information is distributed to neuronal pools throughout the spinal cord and brain.
2. In **convergence**, several neurons synapse on the same postsynaptic neuron (Figure 13.14b●). Several different patterns of activity in the presynaptic neurons can have the same effect on the postsynaptic neuron. Convergence permits the variable control of motor neurons by providing a mechanism for their voluntary and involuntary control. For example, the movements of your diaphragm and ribs are now being controlled by respiratory centers in the brain that operate outside of your conscious awareness. But the same motor neurons also can be controlled voluntarily, as when you take a deep breath and hold it. Two different neuronal pools are involved, both synapsing on the same motor neurons.
3. Information may be relayed in a stepwise sequence, from one neuron to another or from one neuronal pool to the next. This pattern, called **serial processing**, is shown in Figure 13.14c●. Serial processing occurs as sensory information is relayed from one processing center to another in the brain.
4. **Parallel processing** occurs when several neurons or neuronal pools are processing the same information at one time (Figure 13.14d●). Thanks to parallel processing, many different responses occur simultaneously. For example, stepping on a sharp object stimulates sensory neurons that distribute the information to a number of neuronal pools. As a result of parallel processing, you might withdraw your foot, shift your weight, move your arms, feel the pain, and shout "Ouch!" all at the same time.
5. Some neural circuits utilize positive feedback to produce **reverberation**. In this arrangement, collateral axons extend back toward the source of an impulse and further stimulate the presynaptic neurons. Once a reverberating circuit has been activated, it will continue to function until synaptic fatigue or inhibitory stimuli break the cycle. As with convergence or divergence, reverberation can occur within a single neuronal pool, or it may involve a series of interconnected pools. An example of reverberation is shown in Figure 13.14e●; much more complicated examples of reverberation between neuronal pools in the brain may be involved in the maintenance of consciousness, muscular coordination, and normal breathing patterns. We will discuss these and other "wiring patterns" as we consider the organization of the spinal cord and brain in subsequent chapters.

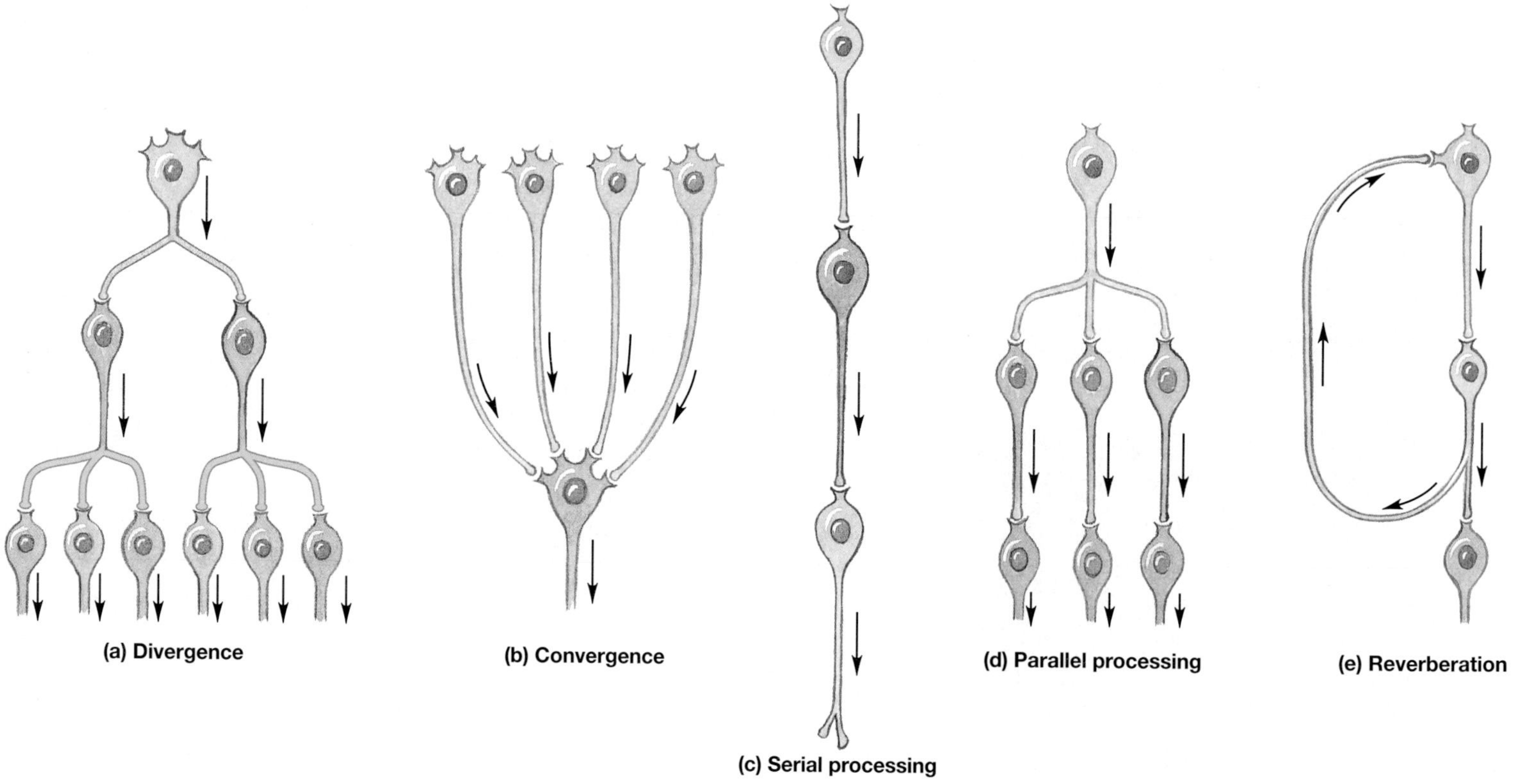

FIGURE 13.14 **ORGANIZATION OF NEURONAL CIRCUITS**

(**a**) Divergence, a mechanism for spreading stimulation to multiple neurons or neuronal pools in the CNS. (**b**) Convergence, a mechanism providing input to a single neuron from multiple sources. (**c**) Serial processing, in which neurons or pools work in a sequential manner. (**d**) Parallel processing, in which individual neurons or neuronal pools process information simultaneously. (e) Reverberation, a feedback mechanism that may be excitatory or inhibitory.

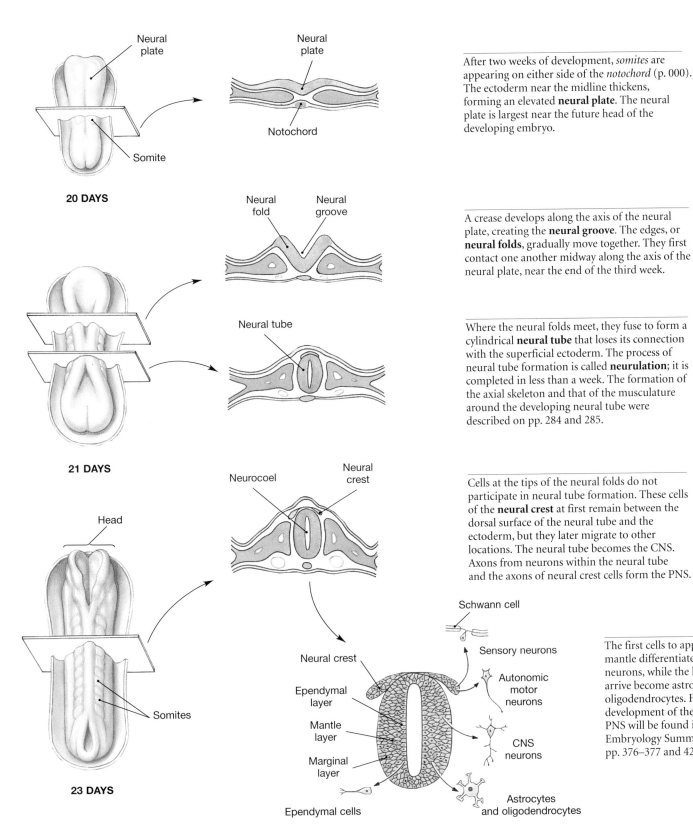

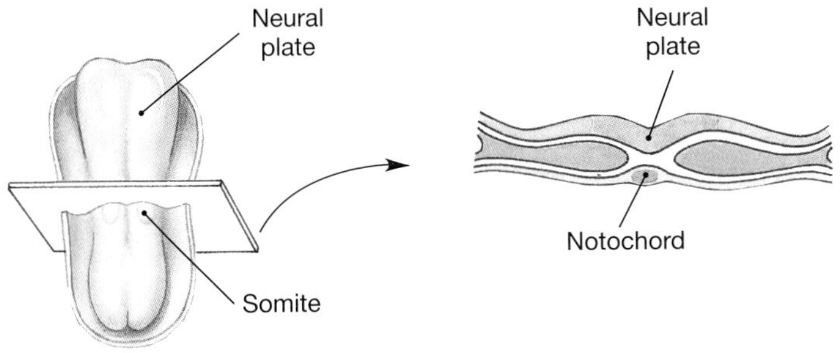

20 DAYS

After two weeks of development, *somites* are appearing on either side of the *notochord* (p. 000). The ectoderm near the midline thickens, forming an elevated **neural plate**. The neural plate is largest near the future head of the developing embryo.

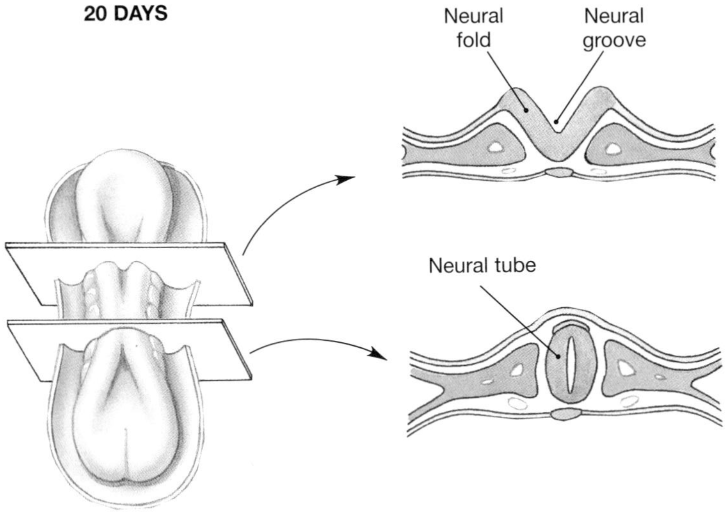

21 DAYS

A crease develops along the axis of the neural plate, creating the **neural groove**. The edges, or **neural folds**, gradually move together. They first contact one another midway along the axis of the neural plate, near the end of the third week.

Where the neural folds meet, they fuse to form a cylindrical **neural tube** that loses its connection with the superficial ectoderm. The process of neural tube formation is called **neurulation**; it is completed in less than a week. The formation of the axial skeleton and that of the musculature around the developing neural tube were described on pp. 284 and 285.

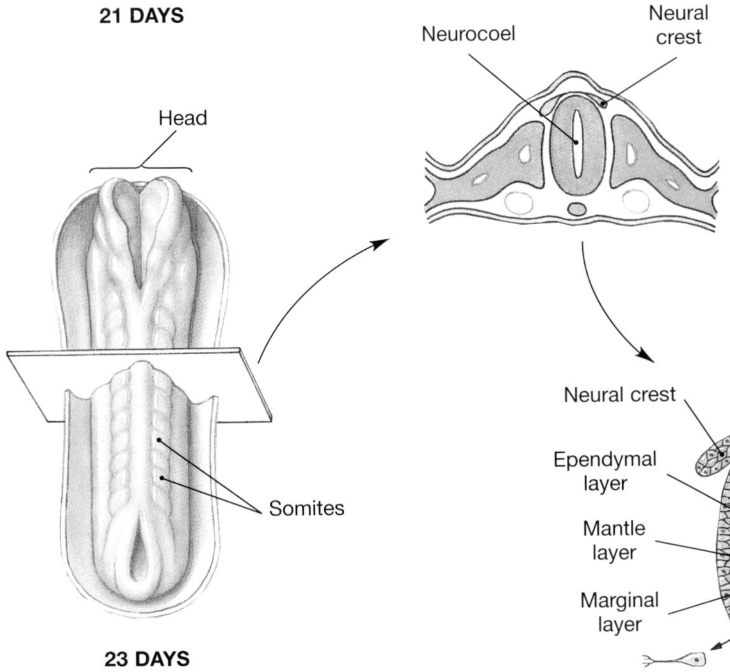

23 DAYS

Cells at the tips of the neural folds do not participate in neural tube formation. These cells of the **neural crest** at first remain between the dorsal surface of the neural tube and the ectoderm, but they later migrate to other locations. The neural tube becomes the CNS. Axons from neurons within the neural tube and the axons of neural crest cells form the PNS.

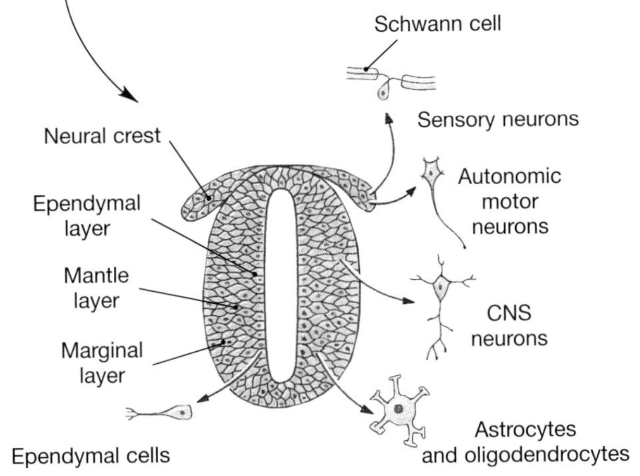

The first cells to appear in the mantle differentiate into neurons, while the last cells to arrive become astrocytes and oligodendrocytes. Further development of the CNS and PNS will be found in the Embryology Summaries on pp. 376–377 and 420–421.

The neural tube increases in thickness as its epithelial lining undergoes repeated mitoses. By the middle of the fifth developmental week, there are three distinct layers. The **ependymal layer** lines the enclosed cavity, or **neurocoel**. The ependymal cells continue their mitotic activities, and daughter cells create the surrounding **mantle layer**. Axons from developing neurons form a superficial **marginal layer**.

Anatomical Organization of the Nervous System [FIGURE 13.15]

The functions of the nervous system depend on the interactions between neurons in neuronal pools, with the most complex neural processing steps occurring in the spinal cord and brain (CNS). The arriving sensory information and the outgoing motor commands are carried by the peripheral nervous system (PNS). Axons and cell bodies in the CNS and PNS are not randomly scattered. Instead, they form masses or bundles with distinct anatomical boundaries. The anatomical organization of the nervous system is depicted in Figure 13.15●.

In the PNS:

- The cell bodies of sensory neurons and visceral motor neurons are found in *ganglia*.
- Axons are bundled together in nerves, with *spinal nerves* connected to the spinal cord, and *cranial nerves* connected to the brain.

In the CNS:

- A collection of neuron cell bodies with a common function is called a **center**. A center with a discrete anatomical boundary is called a **nucleus**. Portions of the brain surface are covered by a thick layer of gray matter, called the **neural cortex**. The term *higher centers* refers to the most complex integration centers, nuclei, and cortical areas of the brain.
- The white matter of the CNS contains bundles of axons that share common origins, destinations, and functions. These bundles are called **tracts**. Tracts in the spinal cord form larger groups, called **columns**.
- The centers and tracts that link the brain with the rest of the body are called **pathways**. For example, **sensory pathways**, or *ascending pathways*, distribute information from peripheral receptors to processing centers in the brain, and **motor pathways**, or *descending pathways*, begin at CNS centers concerned with motor control and end at the effectors they control. ✝ *Growth and Myelination of the Nervous System p. 793*

✓ CONCEPT CHECK

- Identify the two types of synapses.
- In general, how do excitatory and inhibitory synapses differ?
- Distinguish between a neuronal pool whose function is divergence and a neuronal pool whose function is convergence.
- Describe the following anatomical structures that occur within the central nervous system: center, tract, and pathway.

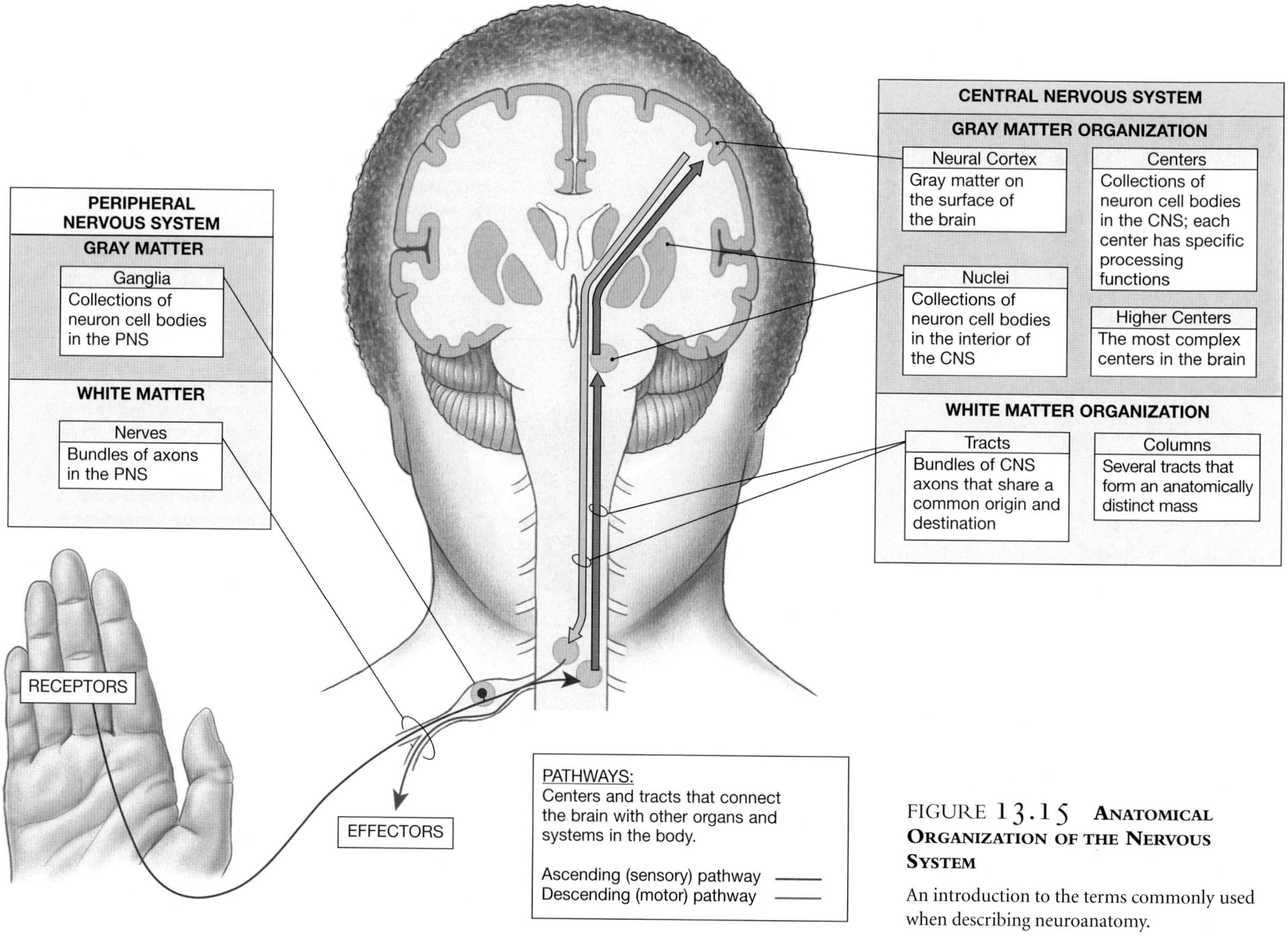

PERIPHERAL NERVOUS SYSTEM
GRAY MATTER
Ganglia
Collections of neuron cell bodies in the PNS
WHITE MATTER
Nerves
Bundles of axons in the PNS
RECEPTORS
EFFECTORS
CENTRAL NERVOUS SYSTEM
GRAY MATTER ORGANIZATION
Neural Cortex
Gray matter on the surface of the brain
Centers
Collections of neuron cell bodies in the CNS; each center has specific processing functions
Nuclei
Collections of neuron cell bodies in the interior of the CNS
Higher Centers
The most complex centers in the brain
WHITE MATTER ORGANIZATION
Tracts
Bundles of CNS axons that share a common origin and destination
Columns
Several tracts that form an anatomically distinct mass
PATHWAYS:
Centers and tracts that connect the brain with other organs and systems in the body.
Ascending (sensory) pathway
Descending (motor) pathway

FIGURE 13.15 **ANATOMICAL ORGANIZATION OF THE NERVOUS SYSTEM**

An introduction to the terms commonly used when describing neuroanatomy.

RELATED CLINICAL TERMS

demyelination: The progressive destruction of myelin sheaths in the CNS and PNS, leading to a loss of sensation and motor control. Demyelination is associated with *heavy metal poisoning, diphtheria, multiple sclerosis*, and *Guillain-Barré syndrome.* *Demyelination Disorders p. 793*

rabies: An acute viral disease of the central nervous system usually transmitted by the bite of an infected mammal. The virus reaches the CNS by transport along the axons of neurons innervating the region of the bite. *Axoplasmic Transport and Disease p. 792*

Additional Clinical Terms Discussed in Appendix I (pp. 792-793)
diphtheria; Guillain-Barré syndrome; heavy metal poisoning; plasmapheresis

STUDY OUTLINE & CHAPTER REVIEW

Introduction 335

1. Two organ systems—the nervous and endocrine systems—coordinate and direct the activities of other organ systems. The nervous system provides swift, brief responses to stimuli; the endocrine system adjusts metabolic operations and directs long-term changes.

An Overview of the Nervous System 335

1. The **nervous system** encompasses all the **neural tissue** in the body. Its anatomical subdivisions are the **central nervous system (CNS)** (the brain and spinal cord) and the **peripheral nervous system (PNS)** (all of the neural tissue outside the CNS).
2. Functionally, the nervous system is subdivided into an **afferent division**, which transmits sensory information from somatic and visceral receptors to the CNS, and an **efferent division**, which carries motor commands to muscles and glands.
3. The efferent division includes both the **somatic nervous system (SNS)** (voluntary control over skeletal muscle contractions) and the **autonomic nervous system (ANS)** (automatic, involuntary regulation of smooth muscle, cardiac muscle, and glandular activity). *(see Figures 13.1/13.2 and Table 13.1)*

Cellular Organization in Neural Tissue 336

1. There are two types of cells in neural tissue: **neurons**, which are responsible for information transfer and processing, and **neuroglia**, or **glial cells**, which are supporting cells in the nervous system. A typical neuron has a **cell body (soma)**, an **axon**, and several **dendrites**. *(see Figures 13.3/13.4)*

Neuroglia 337

2. There are four types of neuroglia in the CNS: (1) *astrocytes*; (2) *oligodendrocytes*; (3) *microglia*; and (4) *ependymal cells (see Figures 13.4 to 13.8).*
3. **Astrocytes** are the largest, most numerous glial cell. They maintain the blood-brain barrier to isolate the CNS from the general circulation, provide structural support for the CNS, regulate ion and nutrient concentrations, and perform repairs to stabilize the tissue and prevent further injury. *(see Figures 13.4/13.5)*
4. **Oligodendrocytes** wrap CNS axons in a membrane sheath termed **myelin.** Gaps between the myelin wrappings along an axon are called **nodes of Ranvier**, whereas the large areas wrapped in myelin are called **internodes.** Regions primarily containing myelinated axons appear glossy white and are termed **white matter.** *(see Figures 13.4/13.5)*
5. **Microglia** are small cells with many fine cytoplasmic processes. These are phagocytic cells that engulf cellular debris, waste products, and pathogens. They increase in number as a result of infection or injury. *(see Figures 13.4/13.5)*
6. **Ependymal cells** are atypical epithelial cells that line chambers and passageways filled with **cerebrospinal fluid (CSF)** in the brain and spinal cord. They assist in producing, circulating, and monitoring CSF. *(see Figures 13.4/13.6)*
7. Neuron cell bodies in the PNS are clustered into **ganglia**, and their axons form **peripheral nerves.** *(see Figures 13.7/13.8)*
8. The PNS glial cell types are *satellite cells* and *Schwann cells. (see Figures 13.7/13.8)*
9. **Satellite cells** enclose neuron cell bodies in ganglia. *(see Figure 13.7)*
10. **Schwann cells** *(neurolemmocytes)* cover all peripheral axons, whether myelinated or unmyelinated. *(see Figure 13.8)*

Neurons 342

11. The **perikaryon** of a neuron is the cytoplasm surrounding the nucleus. It contains organelles, including: **neurofilaments, neurotubules,** and bundles of neurofilaments, termed **neurofibrils,** which extend into the dendrites and axon. The **axon hillock** is a specialized region of an axon. It connects the **initial segment** of the axon to the cell body. The cytoplasm of the axon, the **axoplasm,** contains numerous organelles. *(see Figure 13.9)*
12. **Collaterals** are side branches from an axon. **Telodendria** are a series of fine, terminal extensions branching from the axon tip. *(see Figure 13.9)*
13. Telodendria end at *synaptic terminals.* A **synapse** is a site of intercellular communication between a neuron and another cell. A **synaptic knob** is located where one neuron synapses on another. Synaptic communication usually involves the release of specific chemicals, called neurotransmitters. *(see Figure 13.9)*
14. Structurally, neurons may be classified on the basis of the number of processes that project from the cell body. These include; (1) **anaxonic** (no distinguishable axon); (2) **bipolar** (one dendrite and one axon); (3) **unipolar** (dendrite and axon are continuous at one side of cell body); and (4) **multipolar** (several dendrites and one axon). *(see Figure 13.10)*
15. There are three functional categories of neurons: *sensory neurons, motor neurons,* and *interneurons* (association neurons). *(see Figure 13.11)*
16. **Sensory neurons** form the afferent division of the PNS and deliver information from sensory receptors to the CNS. Receptors are categorized as: **exteroceptors** (information from external environment), **proprioceptors** (position and movement of joints), and **interoceptors** (monitor digestive, respiratory, cardiovascular, urinary, and reproductive systems). *(see Figure 13.11)*
17. **Motor neurons** form the efferent pathways that stimulate or modify the activity of a peripheral tissue, organ, or organ system. **Somatic motor neurons** innervate skeletal muscle. **Visceral motor neurons** innervate all peripheral effectors other than skeletal muscles. Axons of visceral motor neurons from the CNS (**preganglionic fibers**) synapse on neurons in ganglia; these ganglion cells project axons (**postganglionic fibers**) to control the peripheral effectors. *(see Figure 13.11)*
18. **Interneurons (association neurons)** may be located between sensory and motor neurons; they analyze sensory inputs and coordinate motor outputs. Interneurons are classified as **excitatory** or **inhibitory** on the basis of their effects on postsynaptic neurons. *(see Figure 13.11)*

Neural Regeneration 345

1. Neurons have a very limited ability to regenerate after an injury. When an entire peripheral nerve is severed, only a relatively small number of axons within the nerve will successfully reestablish normal synaptic contacts. As a result, complete nerve function is impaired permanently. *(see Figure 13.12)*
2. Schwann cells participate in the repair of damaged peripheral nerves. This process is known as **Wallerian degeneration.** *(see Figure 13.12)*
3. Limited regeneration can occur inside the central nervous system, but the situation is more complicated because: (1) many more axons are likely to be involved, (2) astrocytes produce scar tissue that can prevent axon growth across the damaged area, and (3) astrocytes release chemicals that block the regrowth of axons. *(see Figure 13.12)*

The Nerve Impulse 346

1. **Excitability** is the ability of a cell membrane to conduct electrical impulses; the cell membranes of skeletal muscle fibers and most neurons are excitable.
2. The conducted changes in the transmembrane potential occur as a result of changes in the flow of sodium and potassium ions when the membrane *threshold* is reached and are called **action potentials**. An action potential traveling along an axon is called a **nerve impulse**.
3. The rate of impulse conduction depends on the properties of the axon, specifically the presence or absence of a myelin sheath (a myelinated axon conducts impulses five to seven times faster than an unmyelinated axon) and the diameter of the axon (the larger the diameter, the faster the rate of conduction).

Synaptic Communication 346

1. Synapses occur on dendrites, the soma, or along axons. Synapses permit communication between neurons and other cells at **neuroeffector junctions**. *(see Figure 13.9b)*
2. A synapse may be **chemical** (involving a neurotransmitter) or **electrical** (with direct physical contact between cells). **Chemical synapses** are more common. *(see Figure 13.13a)*

Chemical Synapses 346

3. At a **chemical synapse** between two neurons, a special relationship is established. Only the *presynaptic membrane* releases a neurotransmitter, which binds to receptor proteins on the *postsynaptic membrane*, causing a change in the transmembrane potential of the receptive cell. Thus, communication can occur in only one direction across a synapse: from the presynaptic neuron to the postsynaptic neuron. *(see Figure 13.13)*
4. More than 50 neurotransmitters have been identified. All neuromuscular junctions utilize ACh as a neurotransmitter; ACh is also released at many chemical synapses in both the CNS and PNS.
5. The general sequence of events at a chemical synapse is as follows: (1) neurotransmitter release is triggered by the arrival of an action potential at the synaptic knob of the presynaptic membrane, (2) the neurotransmitter binds to receptors on the postsynaptic membrane after it diffuses across the synaptic cleft, (3) binding of neurotransmitter causes a change in the permeability of the postsynaptic cell membrane, resulting in either excitatory or inhibitory effects, depending on the identity and abundance of receptor proteins, (4) the initiation of an action potential depends upon the degree of excitation, and (5) the effects on the postsynaptic membrane fade rapidly as the neurotransmitter molecules are degraded by enzymes.
6. A single neuron may have thousands of synapses on its cell body. The activity of the neuron depends upon the summation of all of the excitatory and inhibitory stimuli arriving at any given moment at the axon hillock.

Electrical Synapses 347

7. **Electrical synapses** are found between neurons in the CNS and PNS, although they are rare. At this synapse, the membranes of the presynaptic and postsynaptic cells are tightly bound together, and the cells function as if they shared a common cell membrane.

Neuron Organization And Processing 347

1. The roughly 20 billion interneurons can be classified into **neuronal pools**. The neural circuits of these neuronal pools may show: (1) *divergence*, (2) *convergence*, (3) *serial processing*, (4) *parallel processing*, and (5) *reverberation. (see Figure 13.14)*
2. **Divergence** is the spread of information from one neuron to several neurons or from one pool to several pools. This facilitates the widespread distribution of a specific input. *(see Figure 13.14a)*
3. **Convergence** is the presence of synapses from several neurons on one postsynaptic neuron. It permits the variable control of motor neurons. *(see Figure 13.14b)*
4. **Serial processing** is a pattern of stepwise information processing, from one neuron to another or from one neuronal pool to the next. This is the way sensory information is relayed between processing centers in the brain. *(see Figure 13.14c)*
5. **Parallel processing** is a pattern that processes information by several neurons or neuronal pools at one time. Many different responses occur at the same time. *(see Figure 13.14d)*
6. **Reverberation** occurs when neural circuits utilize positive feedback to continue the activity of the circuit. Collateral axons establish a circuit to continue to stimulate presynaptic neurons. *(see Figure 13.14e)*

Anatomical Organization of the Nervous System 350

1. Nervous system functions depend on interactions between neurons in neuronal pools. Almost all complex processing steps occur inside the brain and spinal cord. *(see Figure 13.15)*
2. Neuronal cell bodies and axons in both the PNS and CNS are organized into masses or bundles with distinct anatomical boundaries. *(see Figure 13.15)*
3. In the PNS, *ganglia* contain the cell bodies of sensory and visceral motor neurons. Axons in nerves occur within *spinal nerves* to the spinal cord, and *cranial nerves* to the brain. *(see Figure 13.11)*
4. In the CNS, cell bodies are organized into **centers**; a center with discrete boundaries is called a **nucleus**. The **neural cortex** is the gray matter that covers portions of the brain. It is termed *high center* to reflect its involvement in complex activities. White matter has bundles of axons called **tracts**. Tracts organize into larger units, called **columns**. The centers and tracts that link the brain and body are **pathways**. Sensory *(ascending)* pathways carry information from peripheral receptors to the brain; motor *(descending)* pathways extend from CNS centers concerned with motor control to the associated skeletal muscles. *(see Figure 13.15)*

LEVEL 1 REVIEWING FACTS AND TERMS

Match each numbered item with the most closely related lettered item. Use letters for answers in the spaces provided.

Column A	Column B
____ 1. afferent division	a. positive feedback
____ 2. effector	b. connects initial segment to soma
____ 3. astrocyte	c. sensory information
____ 4. oligodendrocyte	d. monitor position/movement of joints
____ 5. axon hillock	e. myelin
____ 6. collaterals	f. one dendrite
____ 7. bipolar neurons	g. neuron cell bodies in PNS
____ 8. proprioceptors	h. blood-brain barrier
____ 9. reverberation	i. side branches of axons
____ 10. ganglia	j. skeletal muscle cells

11. Which of the following is not a function of the neuroglia?
    (a) support
    (b) information processing
    (c) secretion of cerebrospinal fluid
    (d) phagocytosis
12. Glial cells found surrounding the cell bodies of peripheral neurons are
    (a) astrocytes
    (b) ependymal cells
    (c) microglia
    (d) satellite cells
13. Cells responsible for information processing and transfer are the
    (a) neuroglia
    (b) neurons
    (c) Schwann cells
    (d) astrocytes
14. Axons terminate in a series of fine extensions known as
    (a) telodendria
    (b) synapses
    (c) collaterals
    (d) hillocks
15. Which of the following activities or sensations are not monitored by interoceptors?
    (a) urinary activities
    (b) digestive system activities
    (c) visual activities
    (d) cardiovascular activities
16. Neurons in which dendritic and axonal processes are continuous and the cell body lies off to one side are called
    (a) anaxonic
    (b) unipolar
    (c) bipolar
    (d) multipolar
17. Which of the following does not influence the rate of transmission of a nerve impulse?
    (a) presence or absence of myelin
    (b) diameter of the axon
    (c) presence or absence of nodes
    (d) whether the axon is sensory or motor
18. At an electrical synapse, presynaptic and postsynaptic membranes are bound together by
    (a) neurotransmitters (b) gap junctions
    (c) telodendria (d) internodes
19. In neuron pools, parallel processing occurs when
    (a) several neurons synapse on the same postsynaptic neuron
    (b) information is relayed stepwise from one neuron to another
    (c) several neurons process the same information at the same time
    (d) neurons utilize positive feedback
20. A column is a
    (a) collection of neuron cell bodies
    (b) group of tracts in the spinal cord
    (c) bundle of white matter with a common origin and destination
    (d) none of the above

LEVEL 2 REVIEWING CONCEPTS

1. In the peripheral nervous system, Schwann cells participate in the repair of damaged nerves by
    (a) producing new axons
    (b) regenerating cell bodies for the neurons
    (c) forming a cellular tube that can direct growth of new axons
    (d) all of the above
2. Most neurons lack centrosomes. This observation explains
    (a) why neurons grow such long axons
    (b) why neurons cannot regenerate
    (c) the conducting ability of neurons
    (d) the ability of neurons to produce axoplasmic flow
3. Each of the following is an example of a neuroeffector junction except
    (a) skeletal muscle (b) endocrine gland
    (c) smooth muscle (d) another neuron
4. What purpose do collaterals serve in the nervous system?
5. How does exteroceptor activity differ from interoceptor activity?
6. What is the purpose of the blood-brain barrier?
7. Differentiate between CNS and PNS functions.
8. Distinguish between the somatic nervous system and the autonomic nervous system.
9. Why is an electrical synapse more efficient than a chemical synapse? Why is it less versatile?
10. Differentiate between serial and parallel processing.

LEVEL 3 CRITICAL THINKING AND CLINICAL APPLICATIONS

1. In multiple sclerosis, there is progressive and intermittent damage to the myelin sheath of peripheral nerves. This results in poor motor control of the affected area. Why does destruction of the myelin sheath affect motor control?
2. An 8-year-old girl was cut on the elbow when she fell into a window while skating. This injury caused only minor muscle damage but partially severed a nerve in her arm. What is likely to happen to the severed axons of this nerve, and will the little girl regain normal function of the nerve and the muscles it controls?
3. Eve's father suffers a stroke that leaves him partially paralyzed on his right side. What type of glial cell would you expect to find in increased numbers in the area of the brain that is affected by the stroke?

✓ ANSWERS TO CONCEPT CHECK QUESTIONS

p. 342 **1.** The two anatomical subdivisions of the nervous system are the central nervous system and the peripheral nervous system. **2.** The supporting cells in neural tissue are called either glial cells or neuroglia. **3.** Astrocytes help maintain the blood-brain barrier. **4.** Oligodendrocytes produce a membranous coating that wraps around axons and is called myelin. **5.** In the peripheral nervous system, Schwann cells form a myelin covering around axons.
p. 345 **1.** Sensory neurons of the peripheral nervous system are usually unipolar; thus, this tissue is most likely associated with a sensory organ. **2.** Microglial cells are small phagocytic cells that are found in increased number in damaged and diseased areas of the CNS.
p. 346 **1.** Cutting the axon of a neuron prevents the transmission of the nerve impulse along the length of the axon. **2.** Myelinated fibers conduct action potentials much faster than nonmyelinated fibers, so the axon conducting at 50 m/s is myelinated. **3.** Excitability is the ability of a cell membrane to conduct electrical impulses. **4.** The conducted changes in the transmembrane potential are called action potentials.
p. 350 **1.** A synapse may be either chemical, involving a neurotransmitter substance, or electrical, with gap junctions providing direct physical contact between the cells. **2.** Excitatory synapses promote the generation of nerve impulses in the postsynaptic cell, whereas inhibitory synapses oppose the generation of nerve impulses in the postsynaptic cell. **3.** Divergence is the spread of information from one neuron to several neurons, or from one pool to multiple pools. Convergence occurs when several neurons synapse on the same postsynaptic neuron, or several neuronal pools synapse on one neuronal pool. **4.** A center is a collection of neuron cell bodies with a common function. Bundles of axons in the CNS that share common origins, destinations, and functions are called tracts. The centers and tracts that link the brain with the rest of the body are called pathways.

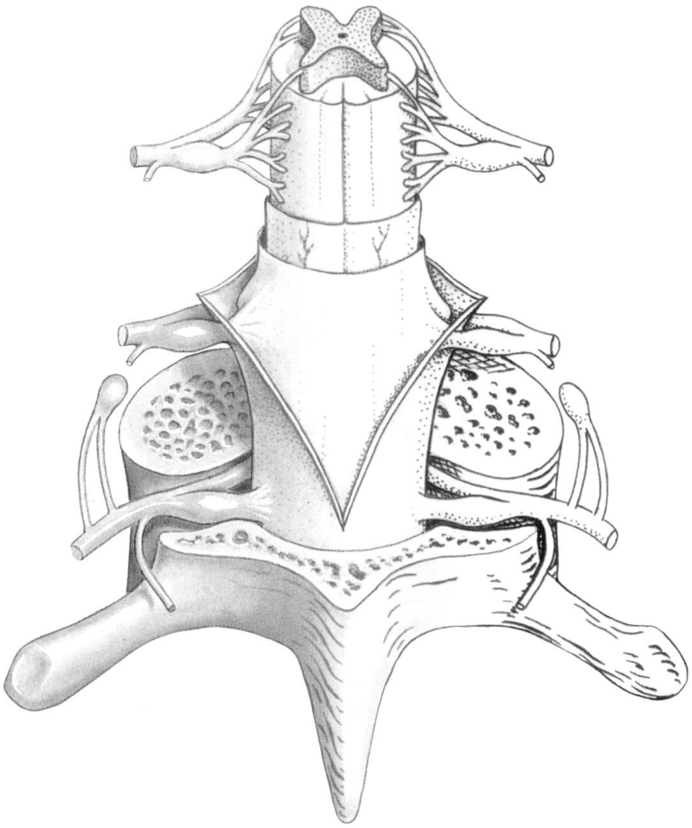

14

THE NERVOUS SYSTEM

The Spinal Cord and Spinal Nerves

The **central nervous system (CNS)** consists of the *spinal cord* and *brain.* Despite the fact that the two are anatomically connected, the spinal cord and brain show significant degrees of functional independence. The spinal cord is far more than just a highway for information traveling to or from the brain. Although most sensory data is relayed to the brain, the spinal cord also integrates and processes information on its own. This chapter describes the anatomy of the spinal cord and examines the integrative activities that occur in this portion of the CNS.

Gross Anatomy of the Spinal Cord

[FIGURES 14.1 TO 14.3]

The adult spinal cord (Figure 14.1a●) measures approximately 45 cm (18 in.) in length. The dorsal surface of the spinal cord bears a shallow longitudinal groove, the **posterior median sulcus**. The deep crease along the ventral surface is the **anterior median fissure**. Each region of the spinal cord (cervical, thoracic, lumbar, and sacral) contains tracts involved with that particular segment and those inferior to it. For example, the thoracic region of the spinal cord contains all the tracts involved with thoracic, lumbar, and sacral segments. In contrast, the sacral spinal cord, the narrowest of all, consists only of tracts that begin or end in that region. Figure 14.1d● provides a series of sectional views that demonstrate the variations in the relative mass of gray matter versus white matter along the length of the spinal cord.

The amount of gray matter is increased substantially in segments of the spinal cord concerned with the sensory and motor innervation of the limbs. These areas contain interneurons responsible for relaying arriving sensory information and coordinating the activities of the somatic motor neurons that control the complex muscles of the limbs. These areas of the spinal cord are expanded to form the **enlargements** of the spinal cord seen in Figure 14.1a●. The **cervical enlargement** supplies nerves to the pectoral girdle and upper limbs; the **lumbar enlargement** provides innervation to structures of the pelvis and lower limbs. Inferior to the lumbar enlargement, the spinal cord tapers to a conical tip, called the **conus medullaris**, at or inferior to the level of the first lumbar vertebra. A slender strand of fibrous tissue, the **filum terminale** ("terminal thread"), extends from the inferior tip of the conus medullaris along the length of the vertebral canal as far as the second or third sacral vertebra (Figure 14.1a,c●). There it provides longitudinal support to the spinal cord as a component of the *coccygeal ligament.*

The entire spinal cord can be divided into 31 segments. Each segment is identified by a letter and number designation. For example, C_3 is the third cervical segment (Figure 14.1a●).

Every spinal segment is associated with a pair of **dorsal root ganglia** that contain the cell bodies of sensory neurons. These sensory ganglia lie between the pedicles of adjacent vertebrae. p. 165 On either side of the spinal cord, a **dorsal root** contains the axons of the sensory neurons in the dorsal root ganglion (Figure 14.1b,c●). Anterior to the dorsal root, a **ventral root** leaves the spinal cord. The ventral root contains the axons of both somatic and visceral motor neurons that control peripheral effectors. The dorsal and ventral roots of each segment enter and leave the vertebral canal between adjacent vertebrae at the *intervertebral foramina.* p. 166 Distal to each dorsal root ganglion, the sensory and motor fibers form a single **spinal nerve** (Figures 14.1d, 14.2c, and 14.3●). Spinal nerves are classified as **mixed nerves**, because they contain both afferent (sensory) and efferent (motor) fibers. Figure 14.3● shows the spinal nerves as they emerge from intervertebral foramina.

The spinal cord continues to enlarge and elongate until an individual is approximately 4 years old. Up to that time, enlargement of the spinal cord keeps pace with the growth of the vertebral column. Throughout this period the ventral and dorsal roots are short, and they leave the vertebral canal through the adjacent intervertebral foramina. After age 4 the vertebral column continues to grow, but the spinal cord does not. This vertebral growth carries the dorsal root ganglia and spinal nerves farther and farther away from their original position relative to the spinal cord. As a result, the dorsal and ventral roots gradually elongate. The adult spinal cord extends only to the level of the first or second lumbar vertebra.

When seen in gross dissection, the filum terminale and the long ventral and dorsal roots that extend caudal to the conus medullaris reminded early anatomists of a horse's tail. With this in mind the complex was called the **cauda equina** (KAW-da ek-WĪ-na; *cauda*, tail + *equus*, horse) (Figure 14.1a,c●).

CLINICAL BRIEF

SHINGLES

In **shingles**, or *Herpes zoster*, the *Herpes varicella-zoster virus* attacks neurons within ganglia of the dorsal roots and cranial nerves. This disorder produces a painful, blistered rash whose distribution corresponds to that of the affected sensory nerves. Shingles usually develops in adults who were first exposed to the virus as children. The initial infection produces symptoms known as *chickenpox*. After this encounter the virus remains dormant within neurons of the spinal cord. It is not known what triggers reactivation of this pathogen. The rash heals over several weeks but pain in the affected nerve may persist for months.

Most people suffer only a single episode of shingles in their adult lives. However, the problem may recur in people with weakened immune systems, including those with AIDS or some forms of cancer. Treatment typically involves large doses of the antiviral drug acyclovir (*Zovirax*™).

Spinal Meninges [FIGURES 14.1b,c/14.2]

The vertebral column and its surrounding ligaments, tendons, and muscles isolate the spinal cord from the external environment. p. 166 The delicate neural tissues also must be protected against damaging contacts with the surrounding bony walls of the vertebral canal. Specialized membranes, collectively known as the **spinal meninges** (men-IN-jēz), provide protection, physical stability, and shock absorption (Figure 14.1b,c●). The spinal meninges cover the spinal cord and surround the spinal nerve roots (Figure 14.2●). Blood vessels branching within these layers also deliver oxygen and nutrients to the spinal cord. There are three meningeal layers: the *dura mater*, the *arachnoid*, and the *pia mater*. At the foramen magnum of the skull, the spinal meninges are continuous with the **cranial meninges** that surround the brain. (The cranial meninges, which have the same three layers, will be described in Chapter 15.)

The Dura Mater [FIGURES 14.1b,c/14.2]

The tough, fibrous **dura mater** (DOO-ra MĀ-ter; *dura*, hard + *mater*, mother) forms the outermost covering of the spinal cord and brain (Figure 14.1b,c●). The dura mater of the spinal cord consists of a layer of dense irregular connective tissue whose outer and inner surfaces are covered by a simple squamous epithelium. The outer epithelium is not bound

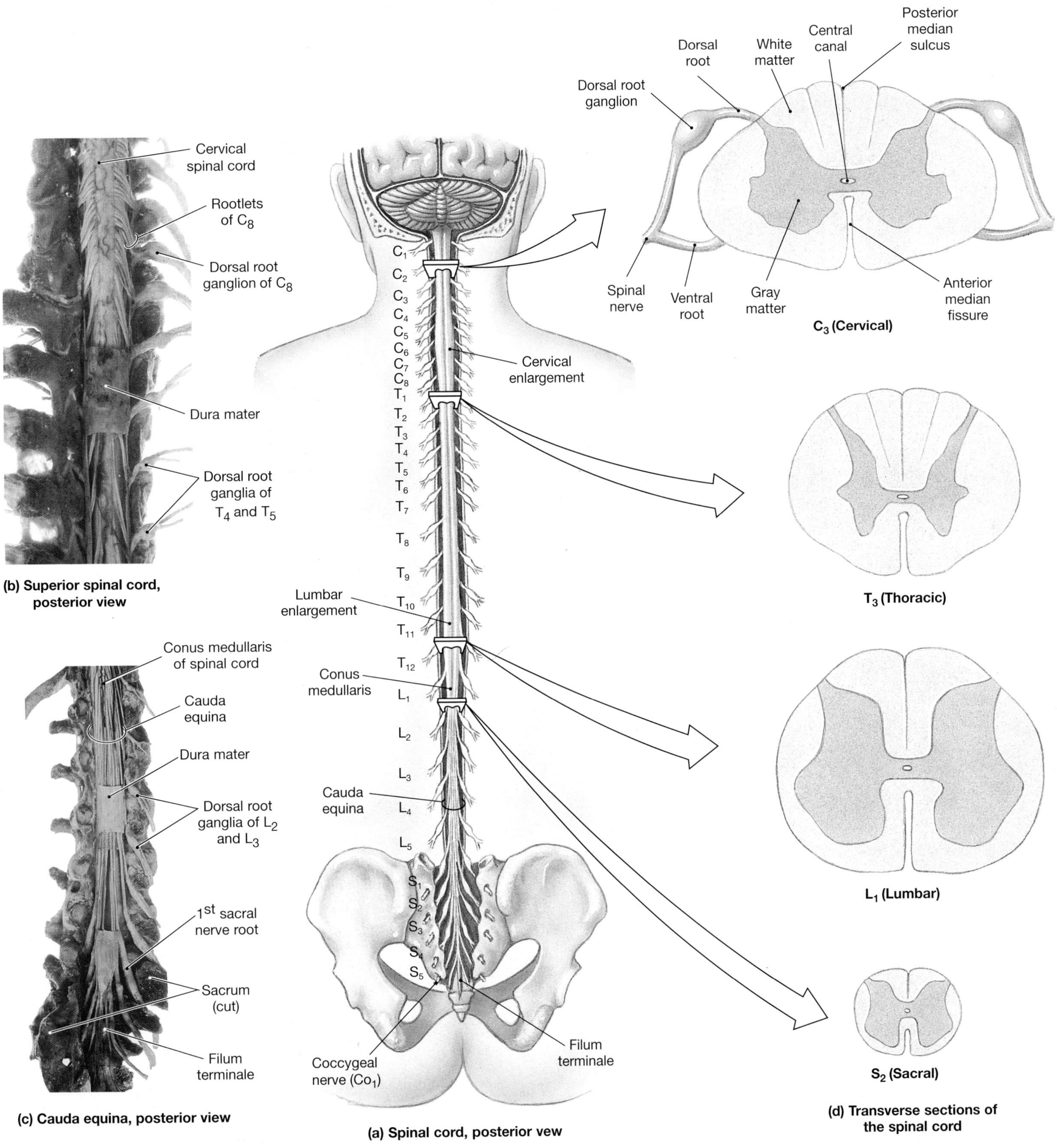

FIGURE 14.1 **Gross Anatomy of the Spinal Cord**

The spinal cord extends inferiorly from the base of the brain along the vertebral canal. (**a**) Superficial anatomy and orientation of the adult spinal cord. The numbers to the left identify the spinal nerves and indicate where the nerve roots leave the vertebral canal. The spinal cord, however, extends from the brain only to the level of vertebrae L_1–L_2 (**b**) Posterior view of a dissection of the cervical spinal cord. (**c**) Posterior view of a dissection of the conus medullaris, cauda equina, filum terminale, and associated spinal nerve roots. (**d**) Inferior views of cross sections through representative regions of the spinal cord, showing the arrangement of gray and white matter.

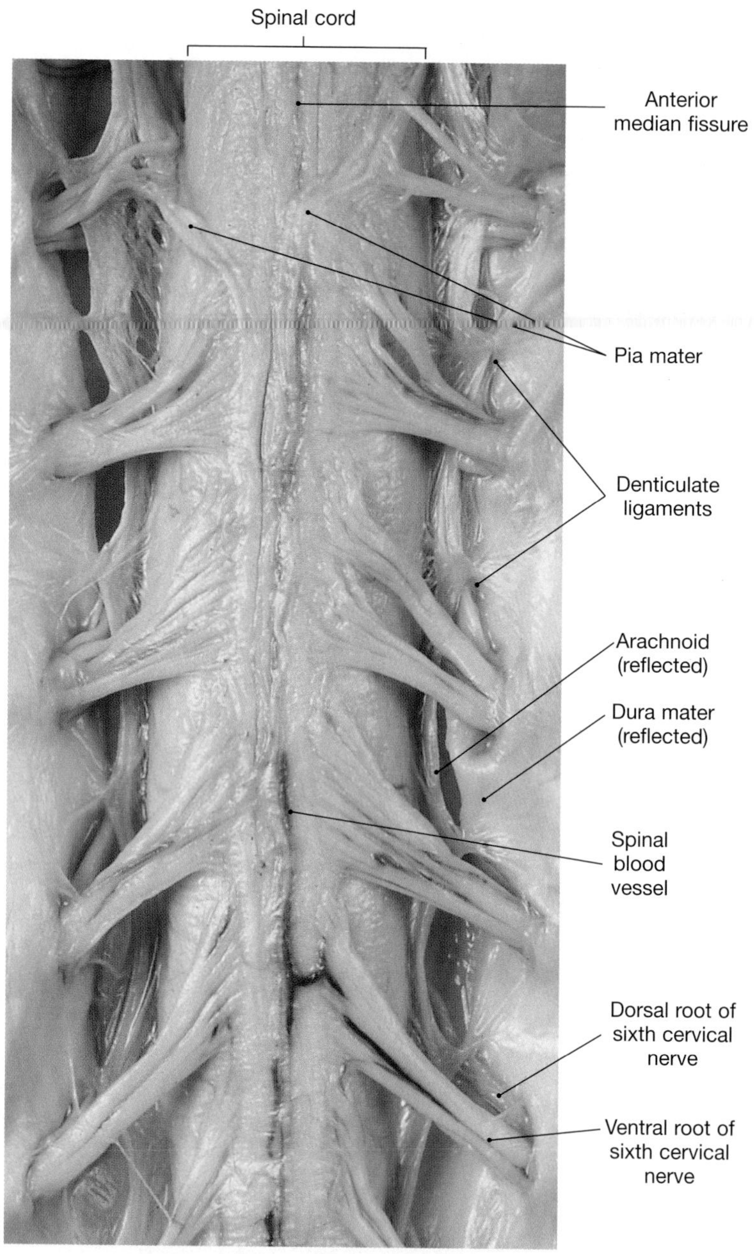

(a) Anterior view

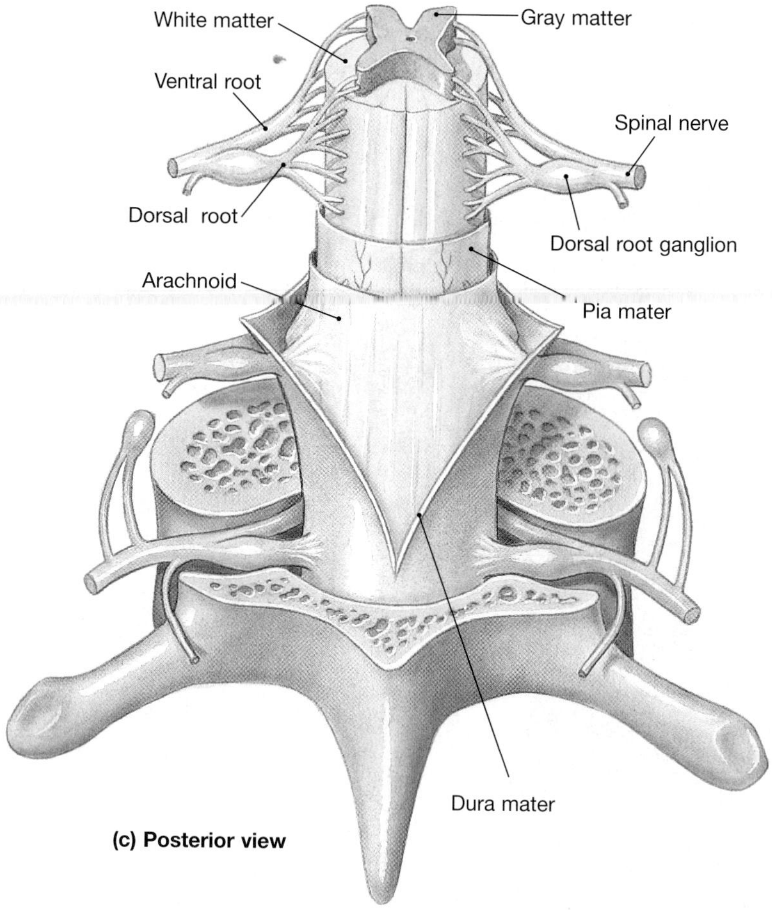

(c) Posterior view

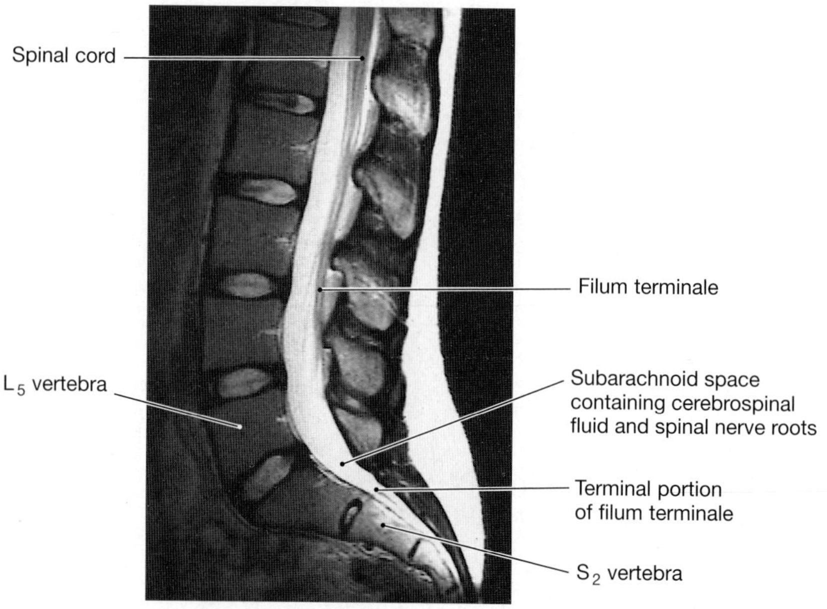

(b) MRI sectional view

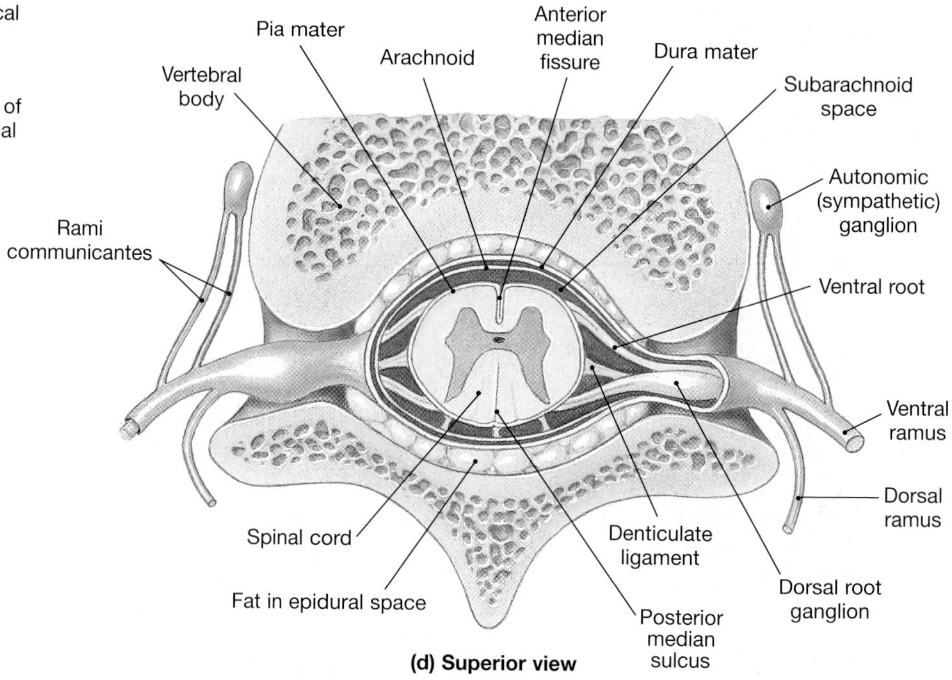

(d) Superior view

FIGURE 14.2 **THE SPINAL CORD AND SPINAL MENINGES**

(a) Anterior view of spinal cord showing meninges and spinal nerves. For this view, the dura and arachnoid membranes have been cut longitudinally and retracted (pulled aside); notice the blood vessels that run in the subarachnoid space, bound to the outer surface of the delicate pia meter. **(b)** An MRI scan of the inferior portion of the spinal cord, showing its relationship to the vertebral column. **(c)** Posterior view of the spinal cord, showing the meningeal layers, superficial landmarks, and distribution of gray and white matter. **(d)** Sectional view through the spinal cord and meninges, showing the peripheral distribution of the spinal nerves.

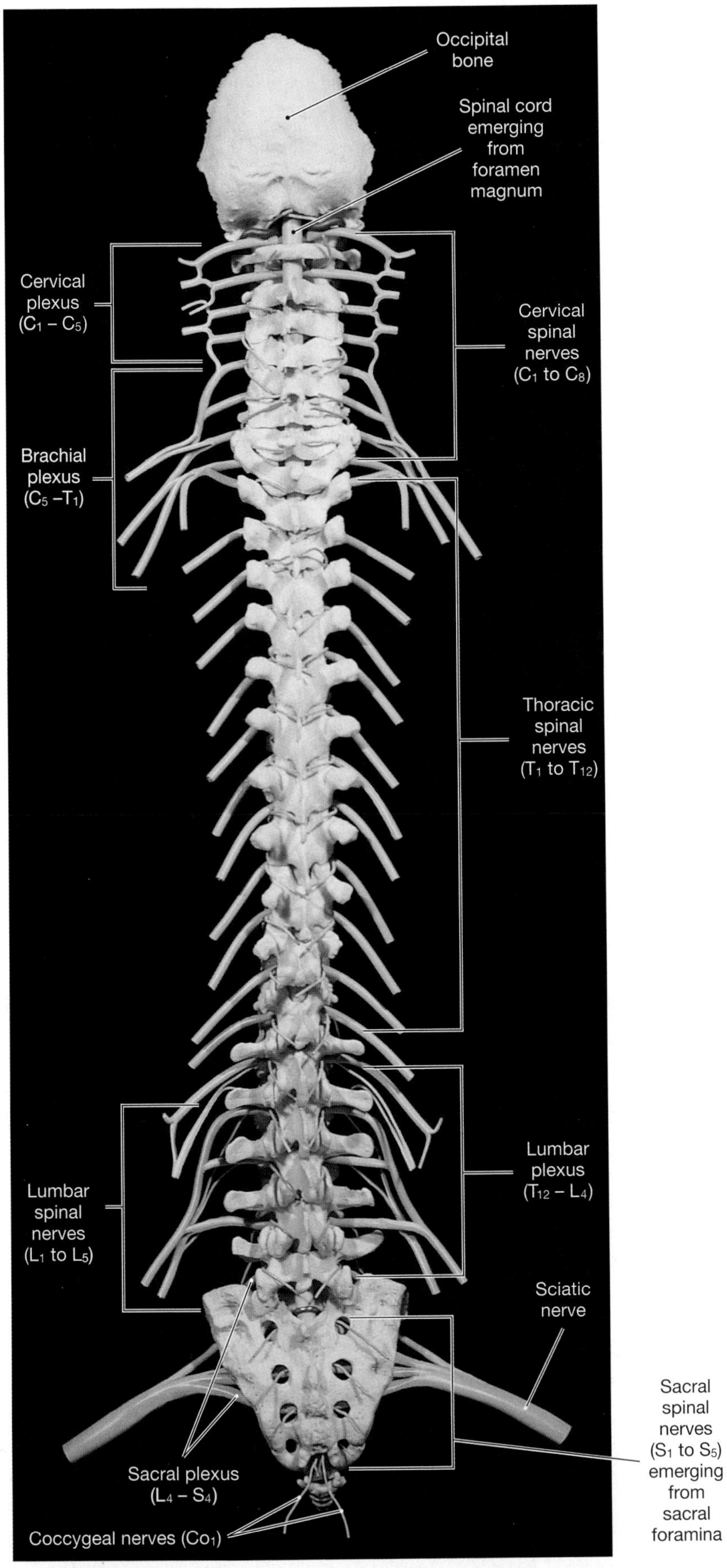

FIGURE 14.3 **POSTERIOR VIEW OF VERTEBRAL COLUMN AND SPINAL NERVES**

This model of the vertebral column and spinal nerves extends from the occipital bone to the coccyx. Note the spinal nerves as they emerge from the intervertebral foramina.

to the bony walls of the vertebral canal, and the intervening **epidural space** contains loose connective tissue, blood vessels, and adipose tissue (Figure 14.2d●).

Localized attachment of the dura mater to the skull, the sacrum, and each vertebra stabilize the position of the spinal cord within the vertebral canal. Cranially, the spinal dura mater fuses with the periosteum of the cranial cavity at the margins of the foramen magnum. Caudally, the spinal dura mater tapers from a sheath to a dense cord of collagen fibers that ultimately blend with components of the filum terminale to form the **coccygeal ligament**. The coccygeal ligament extends along the sacral canal and is interwoven into the periosteum of the sacrum and coccyx. The cranial and sacral attachments provide longitudinal stability. Lateral support is provided by the connective tissues within the epidural space and by the extensions of the dura mater that accompany the spinal nerve roots as they pass through the intervertebral foramina. Distally, the connective tissue of the spinal dura mater is continuous with the connective tissue sheath that surrounds each spinal nerve (Figure 14.2a,c,d●).

The Arachnoid [FIGURES 14.2a,c,d/14.4]

In most anatomical and histological preparations, a narrow **subdural space** separates the dura mater from deeper meningeal layers. It is likely, however, that in life no such space exists, and the inner surface of the dura is in contact with the outer surface of the **arachnoid** (a-RAK-noyd; *arachne*, spider) (Figure 14.2a,c●). The arachnoid, the middle meningeal layer, consists of a simple squamous epithelium. It is separated from the innermost layer, the *pia mater*, by the **subarachnoid space**. This space contains **cerebrospinal fluid (CSF)** that acts as a shock absorber as well as a diffusion medium for dissolved gases, nutrients, chemical messengers, and waste products. The cerebrospinal fluid flows through a meshwork of collagen and elastin fibers produced by modified fibroblasts. Bundles of fibers, known as *arachnoid trabeculae*, extend from the inner surface of the arachnoid to the outer surface of the pia mater. The subarachnoid space and the role of cerebrospinal fluid will be discussed in Chapter 15. The subarachnoid space of the spinal meninges can be accessed easily between L_3 and L_4 (Figure 14.4●) for the clinical examination of cerebrospinal fluid or for the administration of anesthetics. *Spinal Meningitis* and *Spinal Anesthesia p. 793*

The Pia Mater [FIGURE 14.2]

The subarachnoid space bridges the gap between the arachnoid epithelium and the innermost meningeal layer, the **pia mater** (*pia*, delicate + *mater*, mother) as seen in Figure 14.2a,c,d●. The elastic and collagen fibers of the pia mater are interwoven with those of the arachnoid trabeculae. The blood vessels supplying the spinal cord are found here. The pia mater is firmly bound to the underlying neural tissue, conforming to its bulges and fissures. The surface of the spinal cord consists of a thin layer of astrocytes, and cytoplasmic extensions of these glial cells lock the collagen fibers of the spinal pia mater in place.

Along the length of the spinal cord, paired **denticulate ligaments** are extensions of the spinal pia mater that connect the pia mater and spinal arachnoid to the dura mater (Figure 14.2a,d●). These ligaments originate along either side of the spinal cord, between the ventral and dorsal roots. They begin at the foramen magnum of the skull, and collectively they help prevent side-to-side movement and inferior movement of the spinal cord. The connective tissue fibers of the spinal pia mater continue from the inferior tip of the conus medullaris as the filum terminale. As noted above, the filum terminale blends into the coccygeal ligament; this arrangement prevents superior movement of the spinal cord.

The spinal meninges surround the dorsal and ventral roots within the intervertebral foramina. As seen in Figure 14.2c,d●, the meningeal membranes are continuous with the connective tissues surrounding the spinal nerves and their peripheral branches.

✓ CONCEPT CHECK

- Damage to which root of a spinal nerve would interfere with motor function?
- Identify the location of the cerebrospinal fluid that surrounds the spinal cord.
- What are the two spinal enlargements? Why are these regions of the cord increased in diameter?
- What is found within a dorsal root ganglion?

CLINICAL DISCUSSION

Spinal Taps and Myelography

Tissue samples, or *biopsies*, are taken from many organs to assist in diagnosis. For example, when a liver or skin disorder is suspected, small plugs of tissue or tissue fluid are removed and examined for signs of cell damage or are used to identify the microorganisms causing an infection. Unlike many other tissues, however, neural tissue consists largely of cells rather than extracellular fluids or fibers. Tissue samples are seldom removed for analysis, because any extracted or damaged neurons will not be replaced. Instead, small volumes of cerebrospinal fluid (CSF) are collected and analyzed. CSF is intimately associated with the neural tissue of the CNS, and pathogens, cell debris, or metabolic wastes in the CNS are detectable in the CSF.

The withdrawal of cerebrospinal fluid, known as a **spinal tap**, must be done with care to avoid injuring the spinal cord. The adult spinal cord extends only as far as vertebra L_1 or L_2. Between vertebra L_2 and the sacrum, the meningeal layers remain intact, but they enclose only the relatively sturdy components of the cauda equina and a significant quantity of CSF. With the vertebral column flexed, a needle can be inserted between the lower lumbar vertebrae and into the subarachnoid spaces with minimal risk to the cauda equina. In this procedure, known as a **lumbar puncture (LP)**, 3–9 ml of fluid are taken from the subarachnoid space between vertebrae L_3 and L_4 (Figure 14.4a●). Spinal taps are performed when CNS infection is suspected or when diagnosing severe headaches, disc problems, some types of strokes and other altered mental states.

Myelography involves the introduction of radiopaque dyes into the CSF of the subarachnoid space. Because the dyes are opaque to X-rays, the CSF appears white on an X-ray photograph (Figure 14.4b●). Any tumors, inflammation, or adhesions that distort or divert CSF circulation will be shown in silhouette. Pain medication and/or local anesthetics can be injected into the subarachnoid space. In the event of severe infection, inflammation, or leukemia (cancer of the white blood cells), antibiotics, steroids, or anticancer drugs can be injected as well.

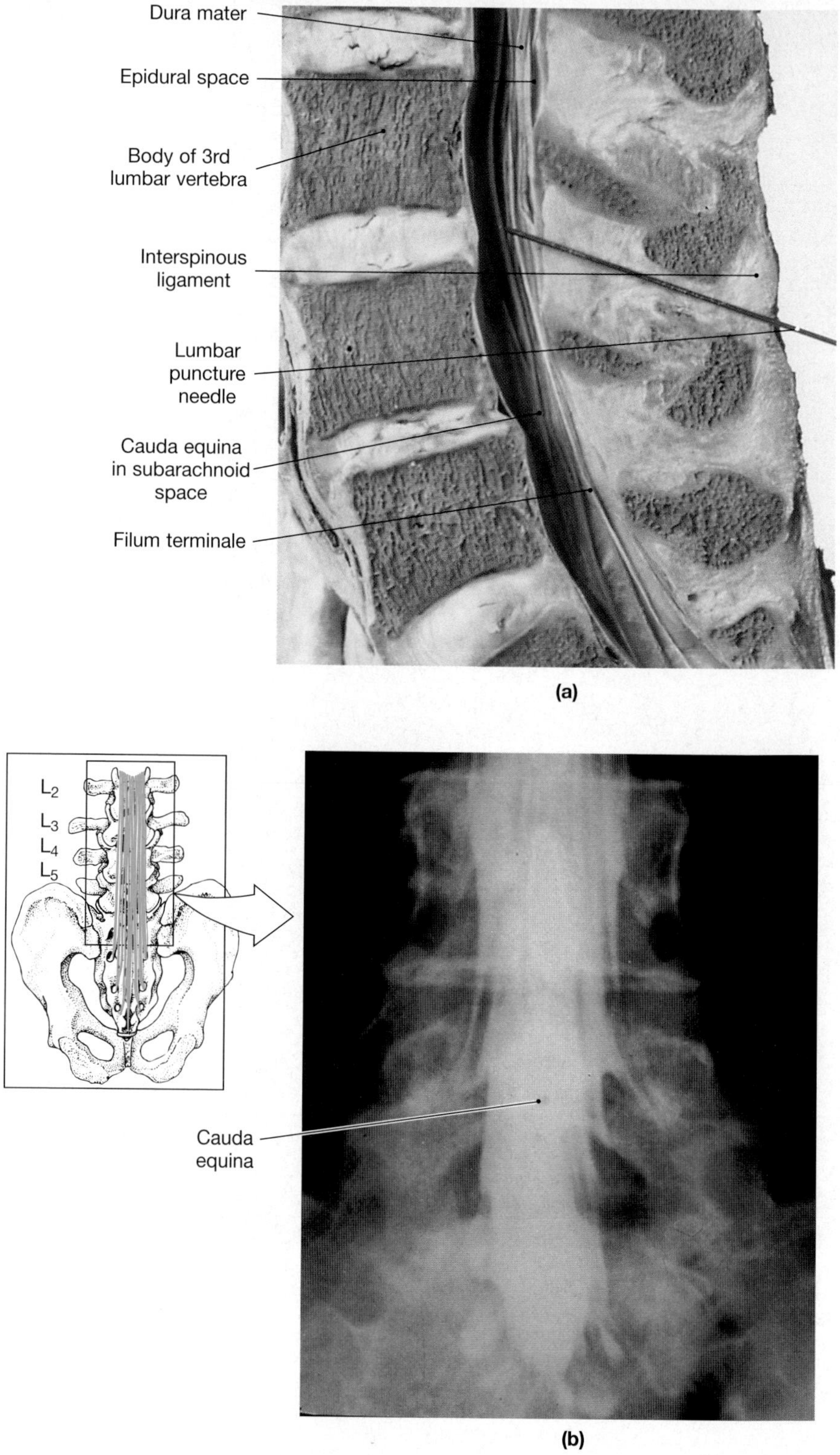

FIGURE 14.4 **Spinal Taps and Myelography**

(**a**) The position of the lumbar puncture needle is in the subarachnoid space, near the nerves of the cauda equina. The needle has been inserted in the midline between the third and fourth lumbar vertebral spines, pointing at a superior angle toward the umbilicus. Once the needle correctly punctures the dura and enters the subarachnoid space, a sample of CSF may be obtained. (**b**) A myelogram—an X-ray photograph of the spinal cord after introduction of a radiopaque dye into the CSF—showing the cauda equina in the lower lumbar region.

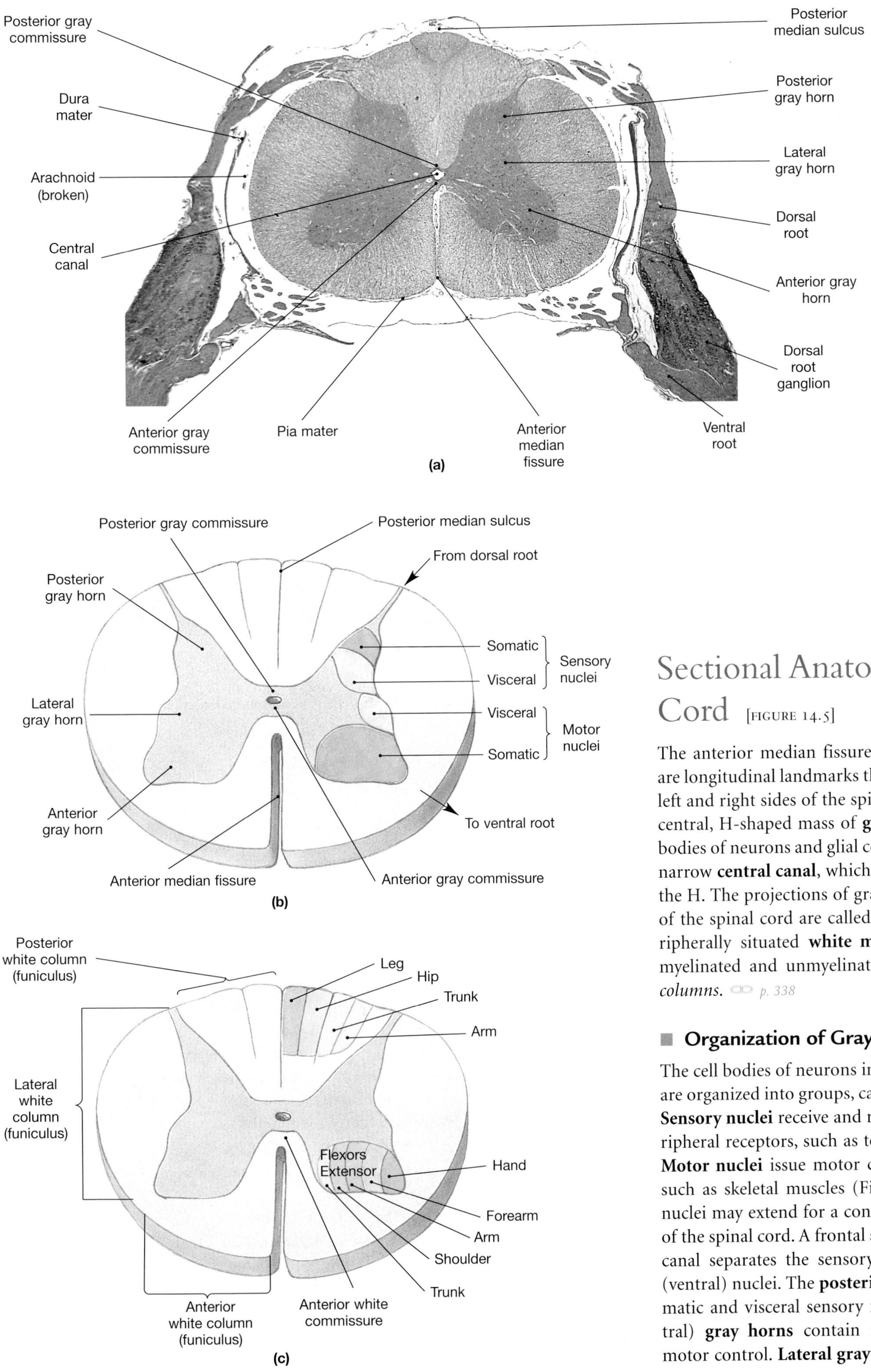

FIGURE 14.5 **SECTIONAL ORGANIZATION OF THE SPINAL CORD**

(**a**) Transverse section of the spinal cord. (**b**) The left half of this sectional view shows important anatomical landmarks; the right half indicates the functional organization of the gray matter in the anterior, lateral, and posterior gray horns. (**c**) The left half of this sectional view shows the major columns of white matter. The right half indicates the anatomical organization of sensory tracts in the posterior white column for comparison with the organization of motor nuclei in the anterior gray horn. Note that both sensory and motor components of the spinal cord have a definite regional organization.

Sectional Anatomy of the Spinal Cord [FIGURE 14.5]

The anterior median fissure and the posterior median sulcus are longitudinal landmarks that follow the division between the left and right sides of the spinal cord (Figure 14.5●). There is a central, H-shaped mass of **gray matter**, dominated by the cell bodies of neurons and glial cells. The gray matter surrounds the narrow **central canal**, which is located in the horizontal bar of the H. The projections of gray matter toward the outer surface of the spinal cord are called **horns** (Figure 14.5a,b●). The peripherally situated **white matter** contains large numbers of myelinated and unmyelinated axons organized in *tracts* and *columns.* p. 338

Organization of Gray Matter [FIGURE 14.5b,c]

The cell bodies of neurons in the gray matter of the spinal cord are organized into groups, called *nuclei*, with specific functions. **Sensory nuclei** receive and relay sensory information from peripheral receptors, such as touch receptors located in the skin. **Motor nuclei** issue motor commands to peripheral effectors, such as skeletal muscles (Figure 14.5b●). Sensory and motor nuclei may extend for a considerable distance along the length of the spinal cord. A frontal section along the axis of the central canal separates the sensory (dorsal) nuclei from the motor (ventral) nuclei. The **posterior** (dorsal) **gray horns** contain somatic and visceral sensory nuclei, whereas the **anterior** (ventral) **gray horns** contain neurons concerned with somatic motor control. **Lateral gray horns** (*intermediate horns*), found

CLINICAL BRIEF

SPINAL CORD INJURIES

Injuries affecting the spinal cord produce symptoms of sensory loss or motor paralysis that reflect the specific nuclei and tracts involved. At the outset, any severe injury to the spinal cord produces a period of sensory and motor paralysis termed **spinal shock**. The skeletal muscles become flaccid; neither somatic nor visceral reflexes function; and the brain no longer receives sensations of touch, pain, heat, or cold. The location and severity of the injury determine the extent and duration of these symptoms and how much recovery takes place.

Violent jolts, such as those associated with blows or gunshot wounds, may cause **spinal concussion** without visibly damaging the spinal cord. Spinal concussion produces a period of spinal shock, but the symptoms are only temporary and recovery may be complete in a matter of hours. More serious injuries, such as whiplash or falls, usually involve physical damage to the spinal cord. In a **spinal contusion**, hemorrhages occur in the meninges and within the spinal cord, pressure rises in the cerebrospinal fluid, and the white matter of the spinal cord may degenerate at the site of injury. Gradual recovery over a period of weeks may leave some functional losses. Recovery from a **spinal laceration** by vertebral fragments or other foreign bodies will usually be far slower and less complete. **Spinal compression** occurs when the spinal cord becomes physically squeezed or distorted within the vertebral canal. In a **spinal transection** the spinal cord is completely severed. Current surgical procedures cannot repair a severed spinal cord, but experimental techniques have restored partial function in laboratory rats. *Technology and Motor Paralysis p. 793*

Spinal injuries often involve some combination of compression, laceration, contusion, and partial transection. Relieving pressure and stabilizing the affected area through surgery may prevent further damage and allow the injured spinal cord to recover as much as possible.

Extensive damage at or above the fourth or fifth cervical vertebra will eliminate sensation and motor control of the upper and lower limbs. The extensive paralysis produced is called **quadriplegia**. If the damage extends from C_3 to C_5, the motor paralysis will include all of the major respiratory muscles, and the patient will usually need mechanical assistance in breathing. **Paraplegia**, the loss of motor control of the lower limbs, may follow damage to the thoracic vertebrae and spinal cord. Injuries to the inferior lumbar vertebrae may compress or distort the elements of the cauda equina, causing problems with peripheral nerve function.

in thoracic and superior lumbar segments, contain visceral motor neurons. The **gray commissures** (*commissura*, a joining together) contain axons crossing from one side of the cord to the other before reaching a destination within the gray matter (Figure 14.5b●). There are two gray commissures, one posterior to and one anterior to the central canal.

Figure 14.5b● shows the relationship between the function of a particular nucleus (sensory or motor) and its relative position within the gray matter of the spinal cord. The nuclei within each gray horn are also highly organized. Figure 14.5b,c● illustrates the distribution of somatic motor nuclei in the anterior gray horns of the cervical enlargement. The size of the anterior horns varies with the number of skeletal muscles innervated by that segment. Thus, the anterior horns are largest in cervical and lumbar regions, which control the muscles associated with the limbs.

Organization of White Matter [FIGURE 14.5]

The white matter can be divided into regions, or **columns** (*funiculi*, singular, *funiculus*) (Figure 14.5c●). The **posterior white columns** are sandwiched between the posterior gray horns and the posterior median sulcus. The **anterior white columns** lie between the anterior gray horns and the anterior median fissure; they are interconnected by the **anterior white commissure**. The white matter on either side between the anterior and posterior columns represents the **lateral white columns**.

Each column contains **tracts**, or *fasciculi*, whose axons share functional and structural characteristics (specific tracts are detailed in Chapter 16). A specific tract conveys either sensory data or motor commands, and the axons are relatively uniform with respect to diameter, myelination, and conduction speed. All of the axons within a tract relay information in the same direction. Small commissural tracts carry sensory or motor signals between segments of the spinal cord; other, larger tracts connect the spinal cord with the brain. **Ascending tracts** carry sensory information toward the brain, and **descending tracts** convey motor commands into the spinal cord. Within each column the tracts show a regional organization comparable to that found in the nuclei of the gray matter (Figure 14.5b,c●). The identities of the major CNS tracts will be discussed when we consider sensory and motor pathways in Chapter 16.

✓ **CONCEPT CHECK**

- A patient suffering from polio has lost the use of his leg muscles. In what area of the spinal cord would you expect to locate the virally infected motor neurons in this individual?
- How is white matter organized within the spinal cord?
- What is the term used to describe the projections of gray matter toward the outer surface of the spinal cord?
- What is the difference between ascending tracts and descending tracts in the white matter?

Spinal Nerves [FIGURES 14.1/14.6]

There are 31 pairs of spinal nerves, and each can be identified by its association with adjacent vertebrae. Every spinal nerve has a regional number, as indicated in Figure 14.1●, p. 357.

In the cervical region the first pair of spinal nerves, C_1, exits between the skull and the first cervical vertebra. For this reason, cervical nerves take their names from the vertebra immediately *following* them. In other words, cervical nerve C_2 *precedes* vertebra C_2, and the same system is used for the rest of the cervical series. The transition from this identification method occurs between the last cervical and first thoracic vertebrae. The spinal nerve lying between these two vertebrae has been designated C_8, and is shown in Figure 14.1b●. Thus, there are seven cervical vertebrae but *eight* cervical nerves. Spinal nerves caudal to the first thoracic vertebra take their names from the vertebra immediately preceding them. Thus, the spinal nerve T_1 emerges immediately caudal to vertebra T_1, spinal nerve T_2 follows vertebra T_2, and so forth.

Each peripheral nerve has three layers of connective tissue: an outer *epineurium*, a central *perineurium*, and an inner *endoneurium* (Figure 14.6●). These are comparable to the connective tissue layers associated with skeletal muscles. p. 357 The **epineurium** is a tough fibrous sheath that consists of a dense network of collagen fibers. At each intervertebral

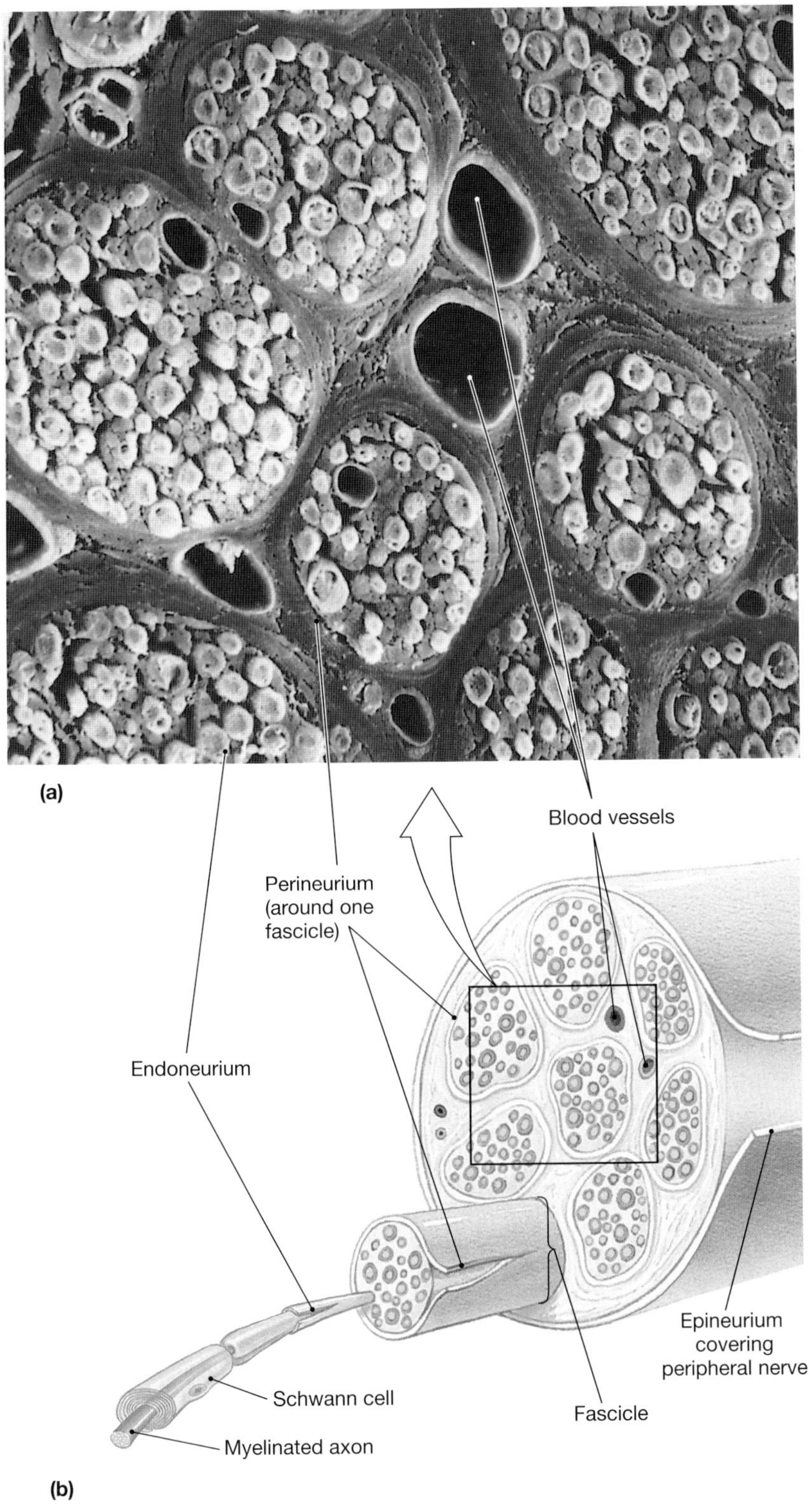

FIGURE 14.6 **ANATOMY OF A PERIPHERAL NERVE**

A peripheral nerve consists of an outer epineurium enclosing a variable number of fascicles (bundles of nerve fibers). The fascicles are wrapped by the perineurium, and within each fascicle the individual axons, which are ensheathed by Schwann cells, are surrounded by the endoneurium. **(a)** A scanning electron micrograph showing the various layers in great detail. (SEM × 425) (Reproduced from R. G. Kessel and R. H. Kardon, *Tissues and Organs: A Text-Atlas of Scanning Electron Microscopy*, W. H. Freeman & Co., 1979.) **(b)** A typical peripheral nerve and its connective tissue wrappings.

foramen, the epineurium of a spinal nerve becomes continuous with the dura mater of the spinal cord. The fibers of the **perineurium** divide the nerve into a series of compartments that contain bundles of axons. A single bundle of axons is known as a **fascicle**, or **fasciculus**. Arteries and veins penetrate the epineurium and branch within the perineurium. The **endoneurium** consists of delicate connective tissue fibers that surround individual axons. Capillaries leaving the perineurium branch in the endoneurium and provide oxygen and nutrients to the axons and Schwann cells of the nerve. † *Multiple Sclerosis p. 794*

Peripheral Distribution of Spinal Nerves

[FIGURES 14.2d/14.7/14.8]

Each spinal nerve forms through the fusion of dorsal and ventral nerve roots as those roots pass through an intervertebral foramen. Distally, the spinal nerve divides into several branches. In the thoracic and superior lumbar regions, the first branch of each spinal nerve carries visceral motor fibers to a nearby **autonomic ganglion** (Figure 14.7a●); this ganglion is associated with the *sympathetic division* of the ANS. (We will examine this division in Chapter 17.) Because preganglionic axons are myelinated, this branch has a light color and is known as the **white ramus** (*ramus*, branch). Two groups of unmyelinated postganglionic fibers leave the ganglion. Those innervating glands and smooth muscles in the body wall or limbs form the **gray ramus** that rejoins the spinal nerve. The gray ramus is typically proximal to the white ramus. The gray and white rami are collectively termed the **rami communicantes** (Figure 14.7a●), or "communicating branches." Preganglionic or postganglionic fibers that innervate internal organs do not rejoin the spinal nerves. Instead, they form a series of separate autonomic nerves, such as the *splanchnic nerves*, involved with regulating the activities of organs in the abdominopelvic cavity.

The **dorsal ramus** of each spinal nerve provides sensory innervation from, and motor innervation to, a specific segment of the skin and muscles of the back. The region innervated resembles a horizontal band that begins at the origin of the spinal nerve. The relatively large **ventral ramus** supplies the ventrolateral body surface, structures in the body wall, and the limbs.

The distribution of the sensory nerves within the dorsal rami illustrates the segmental division of labor along the length of the spinal cord (Figure 14.7b●). Each pair of spinal nerves monitors a specific region of the body surface, an area known as a **dermatome** (Figure 14.8●). Dermatomes are clinically important because damage to either a spinal nerve or dorsal root ganglion will produce a characteristic loss of sensation in specific areas of the skin.

Nerve Plexuses [FIGURES 14.3/14.7/14.9]

The distribution pattern illustrated in Figure 14.7● applies to spinal nerves T_2–T_{12}. But in segments controlling the skeletal musculature of the neck and the upper and lower limbs, the situation is more complicated. During development, small skeletal muscles fuse with their neighbors to form larger muscles with compound origins. Although the anatomical distinctions may disappear, ventral rami from the associated spinal segments continue to provide innervation and motor control. As they converge, the ventral

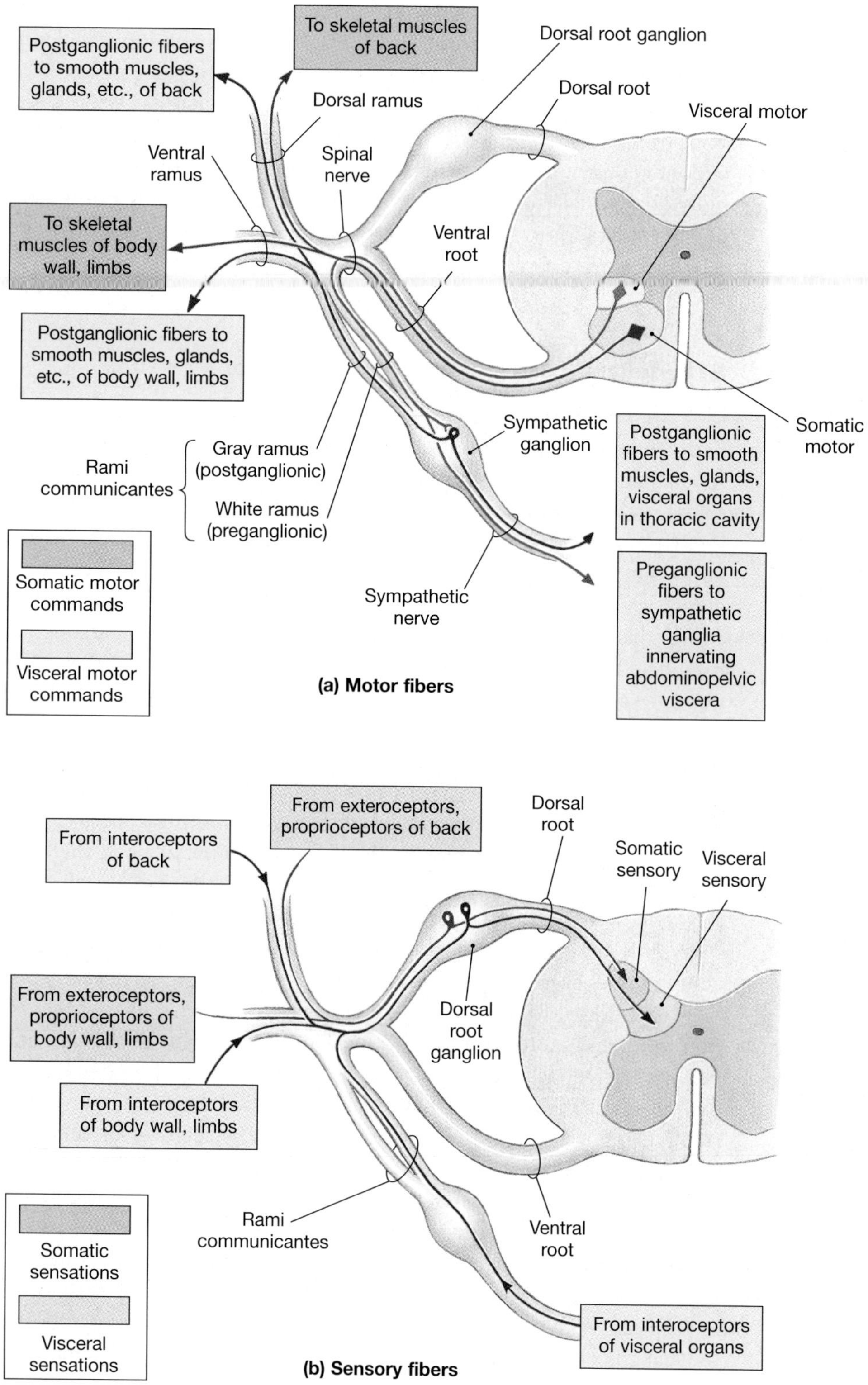

FIGURE 14.7 **PERIPHERAL DISTRIBUTION OF SPINAL NERVES**

Diagrammatic view illustrating the distribution of fibers in the major branches of a representative spinal nerve. **(a)** The distribution of motor neurons in the spinal cord and motor fibers within the spinal nerve and its branches. Although the gray ramus is typically proximal to the white ramus, this simplified diagrammatic view makes it easier to follow the relationships between preganglionic and postganglionic fibers. **(b)** A comparable view, detailing the distribution of sensory neurons and sensory fibers.

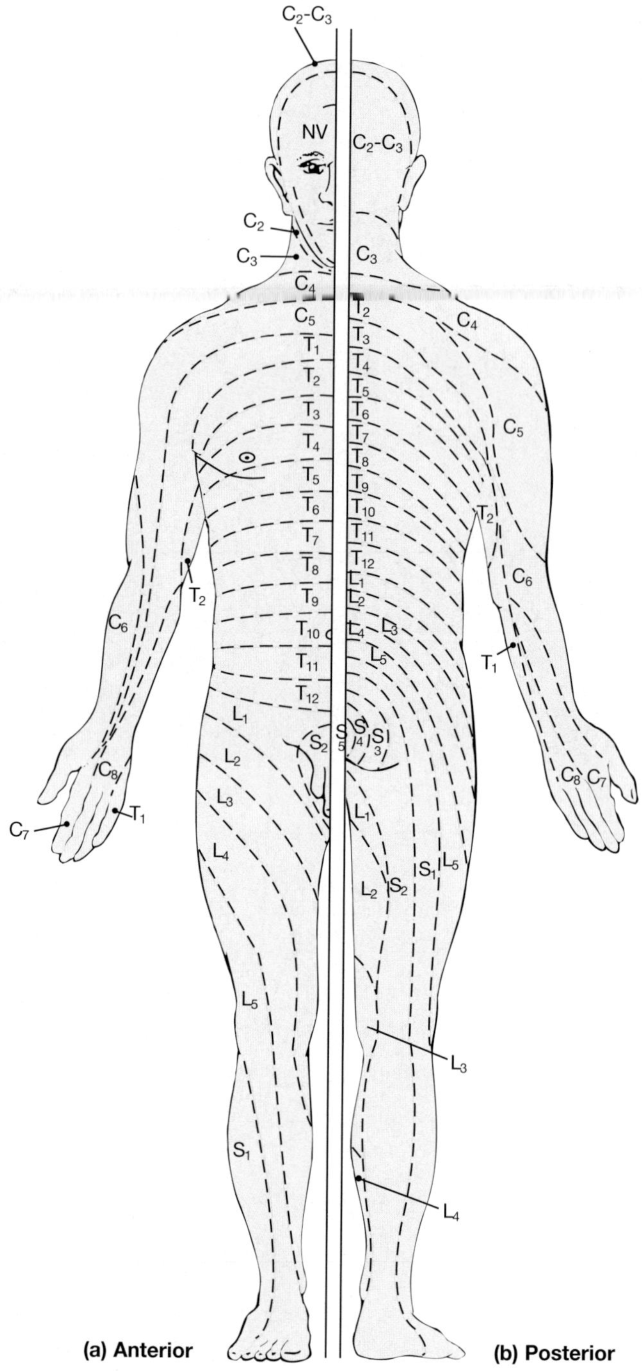

FIGURE 14.8 **DERMATOMES**

Anterior and posterior distribution of dermatomes; the related spinal nerves are indicated for each dermatome.

rami of adjacent spinal nerves blend their fibers to produce a series of compound nerve trunks. Such a complex interwoven network of nerves is called a **nerve plexus** (PLEK-sus, "braid"). The four major nerve plexuses are the *cervical plexus, brachial plexus, lumbar plexus, and sacral plexus* (Figures 14.3, p. 358 and 14.9●).

THE CERVICAL PLEXUS [FIGURES 14.9/14.10 AND TABLE 14.1]

Branches from the **cervical plexus** innervate the muscles of the neck and extend into the thoracic cavity to control the diaphragmatic muscles (Table 14.1). The cervical plexus (Figures 14.9 and 14.10●) consists of muscular and cutaneous branches in the ventral rami of spinal nerves C_1–C_4 and some nerve fibers from C_5. The **phrenic nerve**, the major nerve of this plexus, provides the entire nerve supply to the diaphragm. Other branches are distributed to the skin of the neck, shoulder, and superior portion of the chest. Figures 14.9 and 14.10● identify the nerves responsible for the control of axial and appendicular skeletal muscles considered in Chapters 10 and 11. p. 279

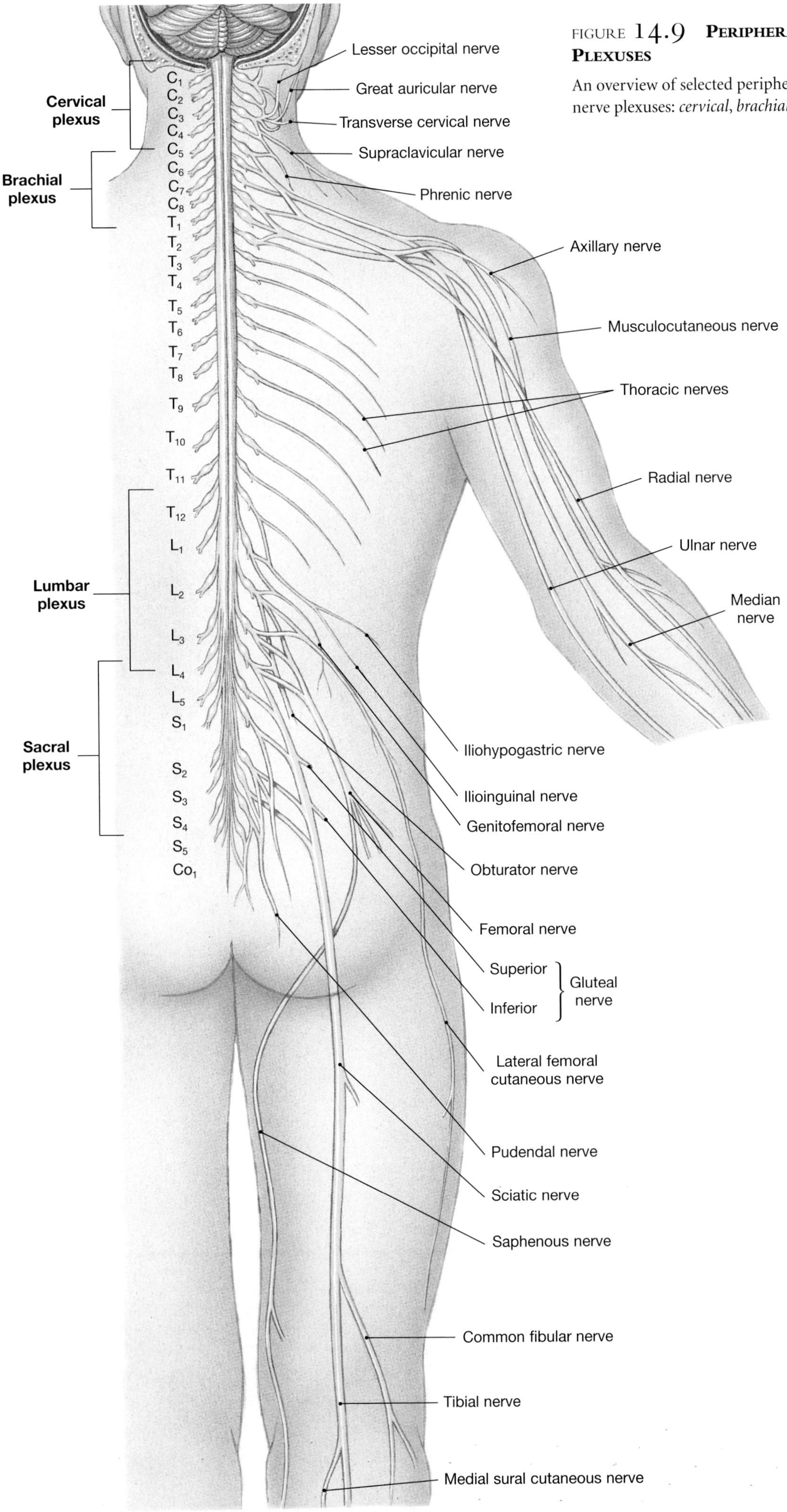

FIGURE 14.9 **PERIPHERAL NERVES AND NERVE PLEXUSES**

An overview of selected peripheral nerves and the four major nerve plexuses: *cervical*, *brachial*, *lumbar*, and *sacral*.

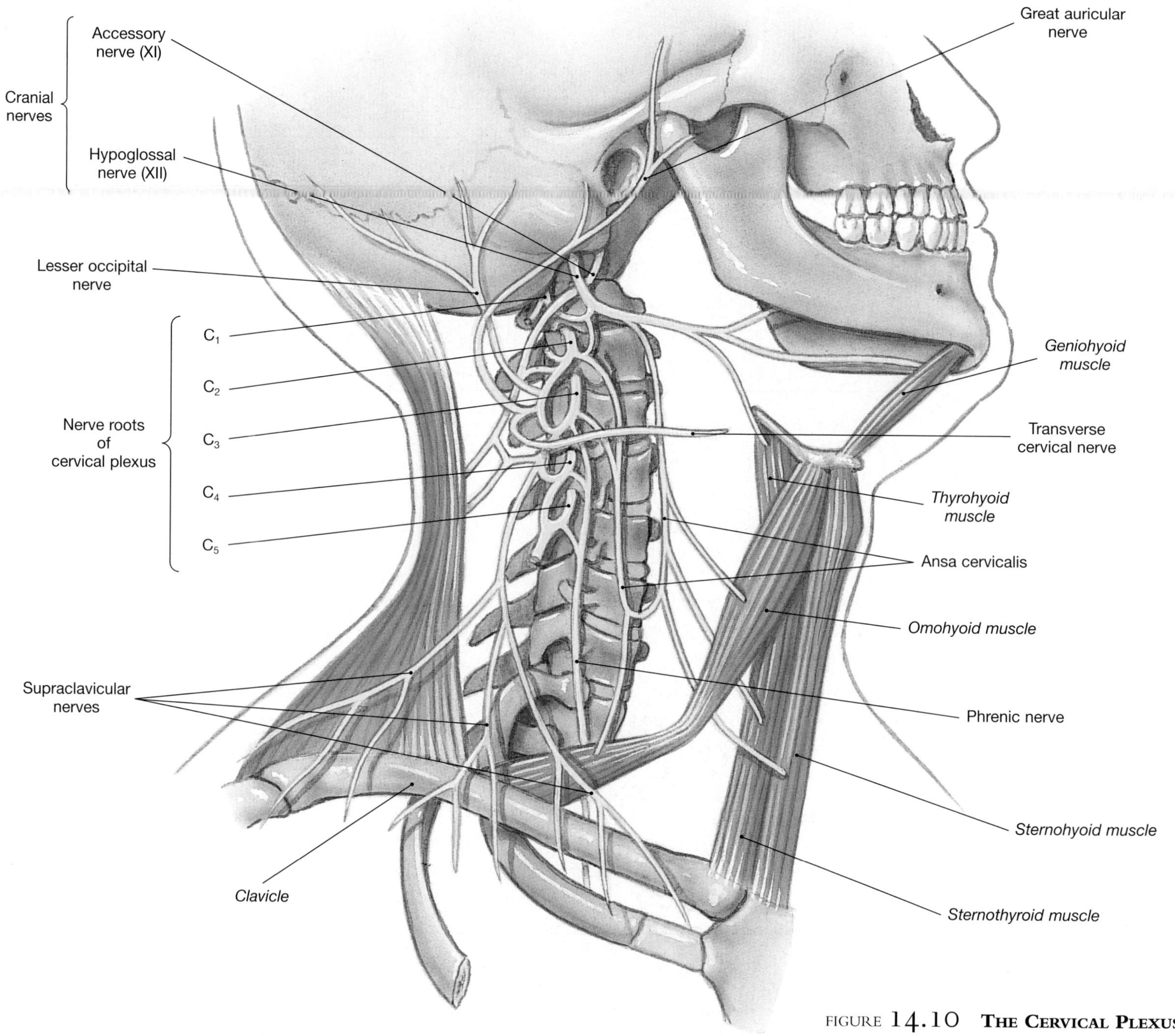

FIGURE 14.10 **THE CERVICAL PLEXUS**

TABLE 14.1

THE CERVICAL PLEXUS

Spinal Segments	Nerves	Distribution
C_1–C_4	**Ansa cervicalis (superior and inferior branches)**	Five of the extrinsic laryngeal muscles (sternothyroid, sternohyoid, omohyoid, geniohyoid and thyrohyoid) by way of N XII
C_2–C_3	**Lesser occipital, transverse cervical, supraclavicular, and great auricular nerves**	Skin of upper chest, shoulder, neck, and ear
C_3–C_5	**Phrenic nerve**	Diaphragm
C_1–C_5	**Cervical nerves**	Levator scapulae, scalenes, sternocleidomastoid, and trapezius muscles (with N XI)

THE BRACHIAL PLEXUS

[FIGURES 14.9/14.11/14.12 AND TABLE 14.2]

The **brachial plexus** (Table 14.2) is larger and more complex than the cervical plexus. It innervates the pectoral girdle and upper limb, with contributions from the ventral rami of spinal nerves C_5–T_1 (Figures 14.9, 14.11, and 14.12●). The spinal nerves converge to form the **superior, middle**, and **inferior trunks**. Branches from these large trunks interconnect to form the **lateral, medial**, and **posterior cords**. The nerves of the brachial plexus arise from one or more trunks or cords; their names indicate their positions relative to the axillary artery.

FIGURE 14.11 **THE BRACHIAL PLEXUS**

(**a**) The trunks and cords of the brachial plexus.

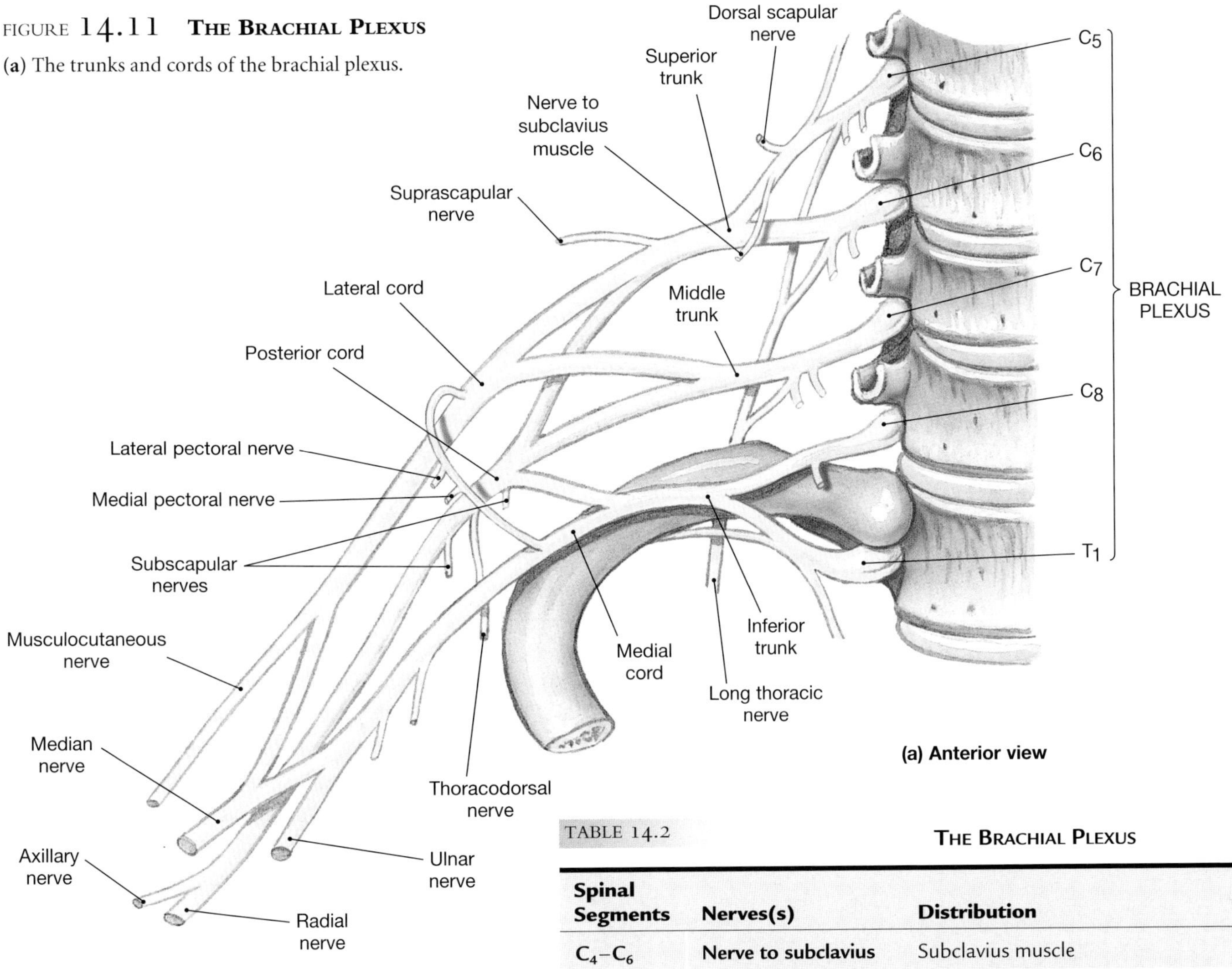

(a) Anterior view

TABLE 14.2 **THE BRACHIAL PLEXUS**

Spinal Segments	Nerves(s)	Distribution
C_4–C_6	**Nerve to subclavius**	Subclavius muscle
C_5	**Dorsal scapular nerve**	Rhomboid muscles and levator scapulae muscle
C_5–C_7	**Long thoracic nerve**	Serratus anterior muscle
C_5, C_6	**Suprascapular nerve**	Supraspinatus and infraspinatus muscles
C_5–T_1	**Pectoral nerves (medial and lateral)**	Pectoralis muscles
C_5, C_6	**Subscapular nerves**	Subscapularis and teres major muscles
C_6–C_8	**Thoracodorsal nerve**	Latissimus dorsi muscle
C_5, C_6	**Axillary nerve**	Deltoid and teres minor muscles; skin of shoulder
C_5–T_1	**Radial nerve**	Extensor muscles of the arm and forearm (triceps brachii, extensor carpi radialis, and extensor carpi ulnaris muscles) and brachioradialis muscle; digital extensors and abductor pollicis muscle; skin over the posterolateral surface of the arm
C_5–C_7	**Musculocutaneous nerve**	Flexor muscles on the arm (biceps brachii, brachialis, and coracobrachialis muscles); skin over lateral surface of forearm
C_6–T_1	**Median nerve**	Flexor muscles on the forearm (flexor carpi radialis and palmaris longus muscles); pronator quadratus and pronator teres muscles; digital flexors (through the *palmar interosseous nerve*); skin over anterolateral surface of hand
C_8, T_1	**Ulnar nerve**	Flexor carpi ulnaris muscle; adductor pollicis muscle and small digital muscles; skin over medial surface of the hand

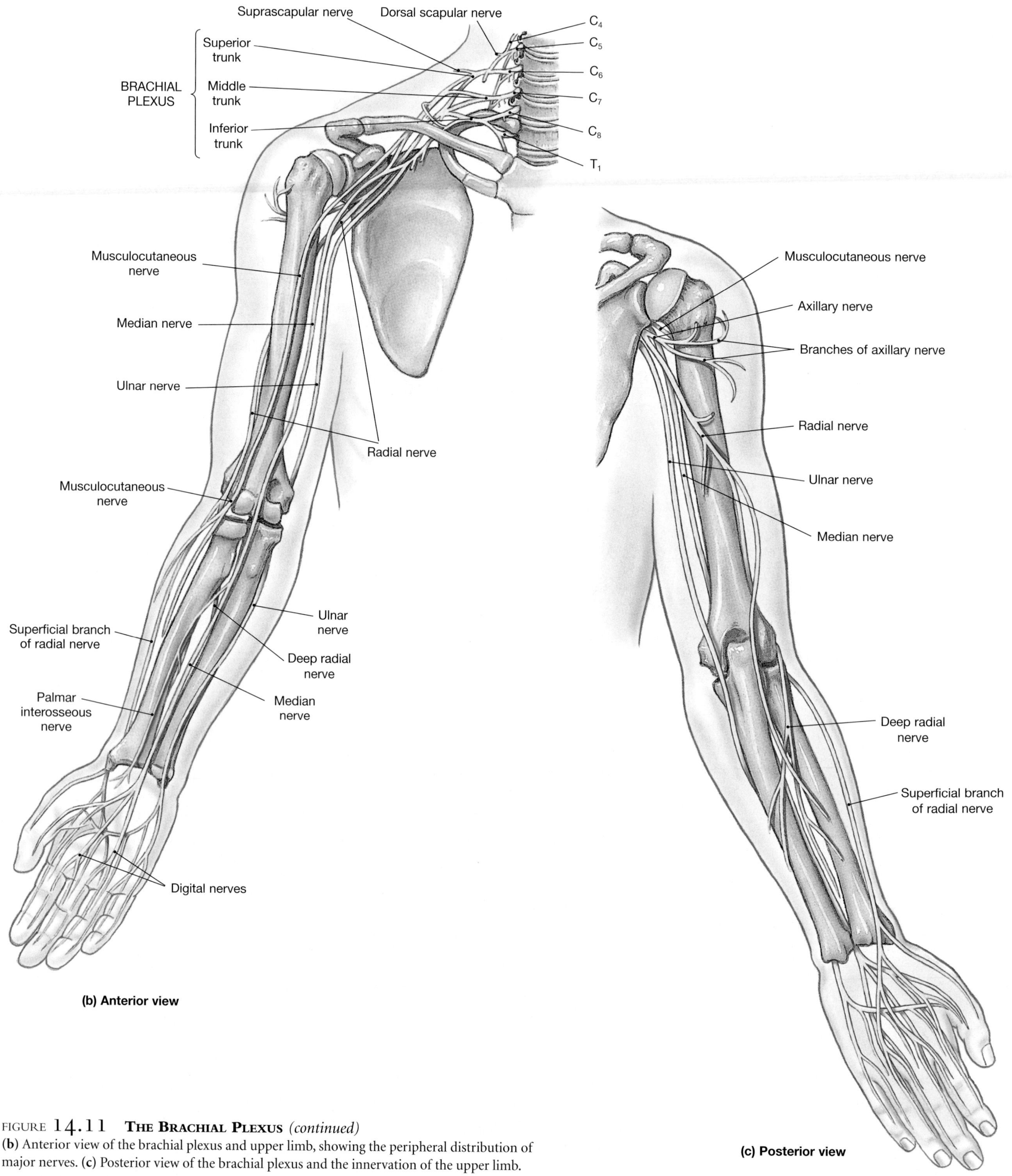

FIGURE 14.11 **The Brachial Plexus** *(continued)*
(**b**) Anterior view of the brachial plexus and upper limb, showing the peripheral distribution of major nerves. (**c**) Posterior view of the brachial plexus and the innervation of the upper limb.

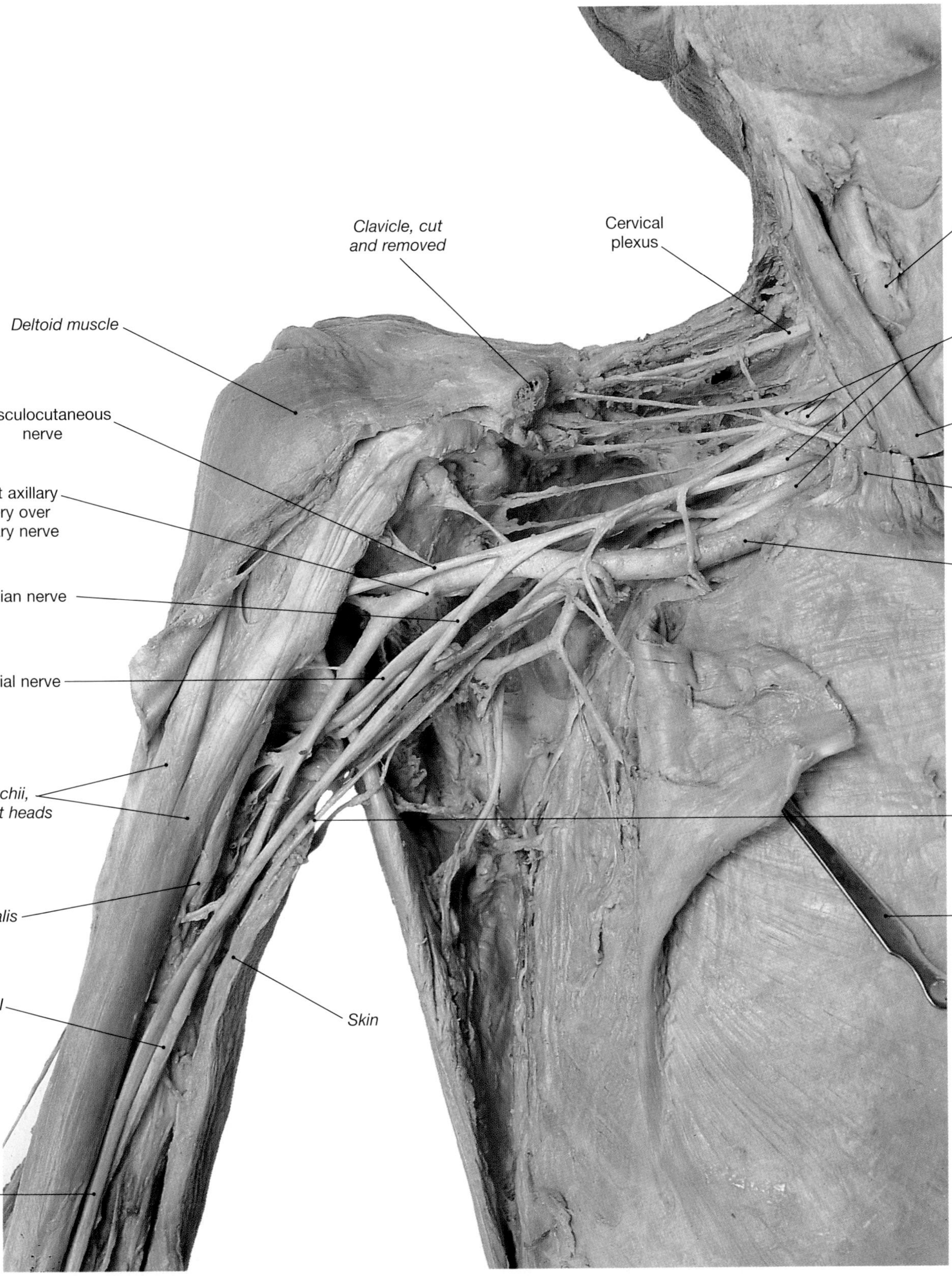

FIGURE 14.12 **THE CERVICAL AND BRACHIAL PLEXUSES**

This dissection shows the major nerves arising from the cervical and brachial plexuses.

The lateral cord forms the **musculocutaneous nerve** exclusively, and contributes to the **median nerve** together with the medial cord. The medial cord contributes to the median nerve and exclusively forms the **ulnar nerve**. The posterior cord forms both the **axillary nerve** and the **radial nerve**.

Figures 14.9 and 14.11● identify these nerves as well as the smaller nerves responsible for the control of axial and appendicular skeletal muscles considered in Chapters 10 and 11. p. 296

THE LUMBAR AND SACRAL PLEXUSES

[FIGURES 14.9/14.13/14.14 AND TABLE 14.3]

The **lumbar plexus** and the **sacral plexus** arise from the lumbar and sacral segments of the spinal cord. The ventral rami of these nerves supply the pelvic girdle and lower limb (Figures 14.9, p. 365, and 14.13●). Because the ventral rami of both plexuses are distributed to the lower limb, they are often collectively referred to as the *lumbosacral plexus*. The nerves that form the lumbar and sacral plexuses are detailed in Table 14.3.

The lumbar plexus is formed by the ventral rami of T_{12}–L_4. The major nerves of the lumbar plexus are the **genitofemoral nerve, lateral femoral cutaneous nerve,** and **femoral nerve**.

The sacral plexus contains the ventral rami from spinal nerves L_4–S_4. The ventral rami of L_4 and L_5 form the **lumbosacral trunk**, which contributes to the sacral plexus along with the ventral rami of S_1–S_4 (Figure 14.13a,b●). The major nerves of the sacral plexus are the **sciatic nerve** and the **pudendal nerve**. The sciatic nerve passes posterior to the femur and deep to the long head of the biceps femoris muscle. As it approaches the popliteal fossa, the sciatic nerve divides into two branches: the **common fibular** (*peroneal*) **nerve** and the **tibial nerve** (Figures 14.9, p. 365, and 14.14●).

Figures 14.9, 14.13, and 14.14● show these nerves as well as the smaller nerves responsible for controlling the axial and appendicular muscles detailed in Chapters 10 and 11. p. 283

Although dermatomes can provide clues to the location of injuries along the spinal cord, the loss of sensation at the skin does not provide precise information concerning the site of injury, because the boundaries of dermatomes are not precise, clearly defined lines. More exact conclusions can be drawn from the loss of motor control on the basis of the origin and distribution of the peripheral nerves originating at nerve plexuses. In the assessment of motor performance, a distinction is made between the conscious ability to control motor activities and the performance of automatic, involuntary motor responses. These latter, programmed motor patterns, called *reflexes*, will be described now.

✓ CONCEPT CHECK

- Injury to which of the nerve plexuses would interfere with the ability to breathe?
- Describe in order from outermost to innermost the three connective tissue layers surrounding each peripheral nerve.
- Distinguish between a white ramus and a gray ramus.
- Which nerve plexus may have been damaged if motor activity in the arm and forearm are affected by injury?

TABLE 14.3 The Lumbar and Sacral Plexuses

Spinal Segment(s)	Nerves(s)	Distribution
LUMBAR PLEXUS		
T_{12}–L_1	**Iliohypogastric nerve**	Abdominal muscles (external and internal oblique muscles, transverse abdominis muscles); skin over inferior abdomen and buttocks
L_1	**Ilioinguinal nerve**	Abdominal muscles (with *iliohypogastric nerve*); skin over superior, medial thigh and portions of external genitalia
L_1, L_2	**Genitofemoral nerve**	Skin over anteromedial surface of thigh and portions of external genitalia
L_2, L_3	**Lateral femoral cutaneous nerve**	Skin over anterior, lateral, and posterior surfaces of thigh
L_2–L_4	**Femoral nerve**	Anterior muscles of thigh (sartorius muscle and quadriceps group); adductors of thigh (pectineus and iliopsoas muscles); skin over anteromedial surface of thigh, medial surface of leg and foot
L_2–L_4	**Obturator nerve**	Adductors of thigh (adductors magnus, brevis, and longus); gracilis muscle; skin over medial surface of thigh
L_2–L_4	**Saphenous nerve**	Skin over medial surface of leg
SACRAL PLEXUS		
L_4–S_2	**Gluteal nerves:**	
	Superior	Abductors of thigh (gluteus minimus, gluteus medius, and tensor fasciae latae)
	Inferior	Extensor of thigh (gluteus maximus)
S_1–S_3	**Posterior femoral cutaneous nerve**	Skin of perineum and posterior surface of thigh and leg
L_4–S_3	**Sciatic nerve:**	Two of the hamstrings (semimembranosus and semitendinosus); adductor magnus (with *obturator nerve*)
	Tibial nerve	Flexors of knee and plantar flexors of ankle (popliteus, gastrocnemius, soleus, and tibialis posterior muscles and long head of biceps femoris muscle); flexors of toes; skin over posterior surface of leg, plantar surface of foot
	Fibular nerve	Biceps femoris muscle (short head); fibularis (brevis and longus) and tibialis anterior muscles; extensors of toes; skin over anterior surface of leg and dorsal surface of foot; skin over lateral portion of foot (through the *sural nerve*)
S_2–S_4	**Pudendal nerve**	Muscles of perineum, including urogenital diaphragm and external anal and urethral sphincter muscles; skin of external genitalia and related skeletal muscles (bulbospongiosus, ischiocavernosus muscles)

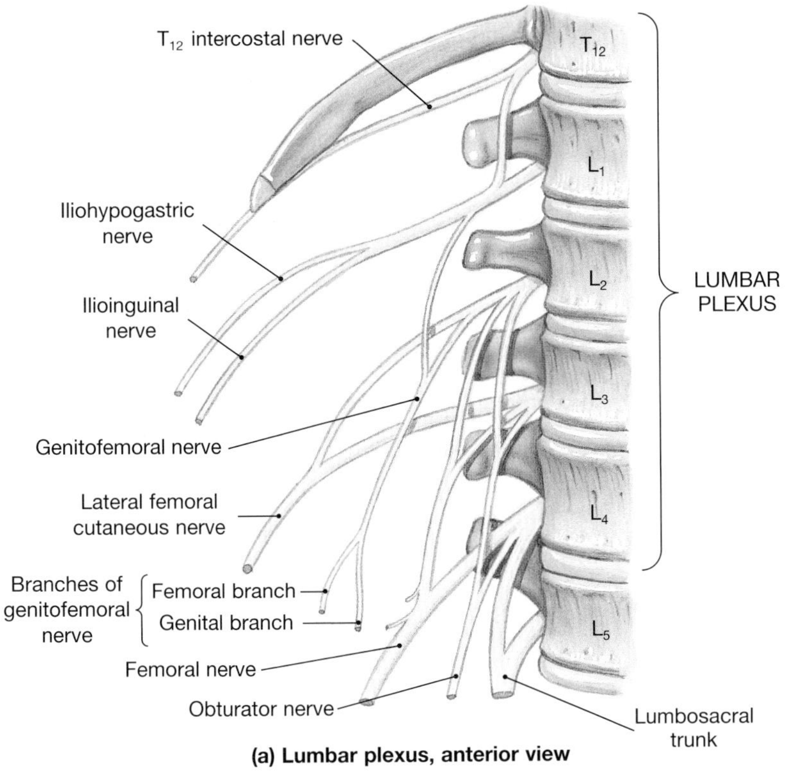

(a) Lumbar plexus, anterior view

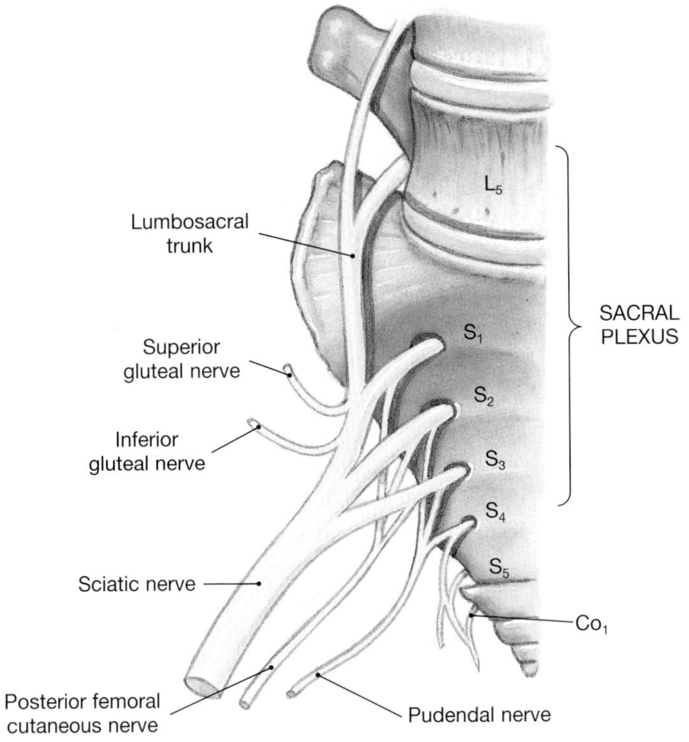

(b) Sacral plexus, anterior view

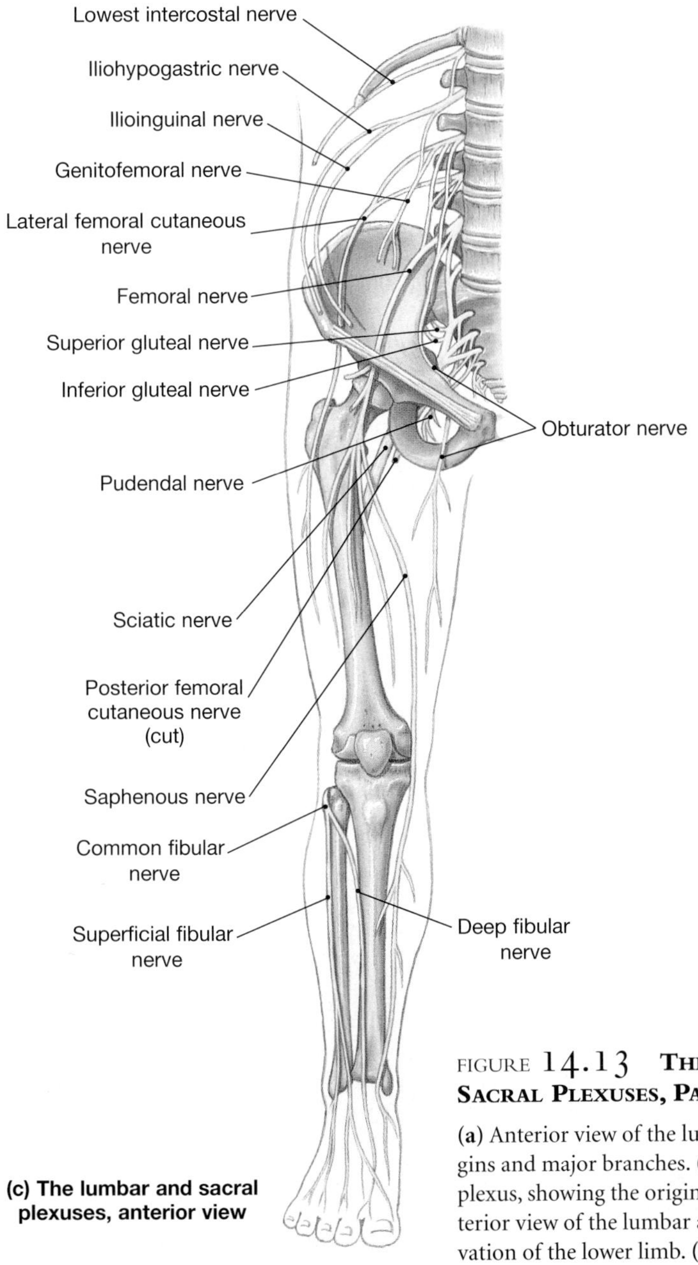

(c) The lumbar and sacral plexuses, anterior view

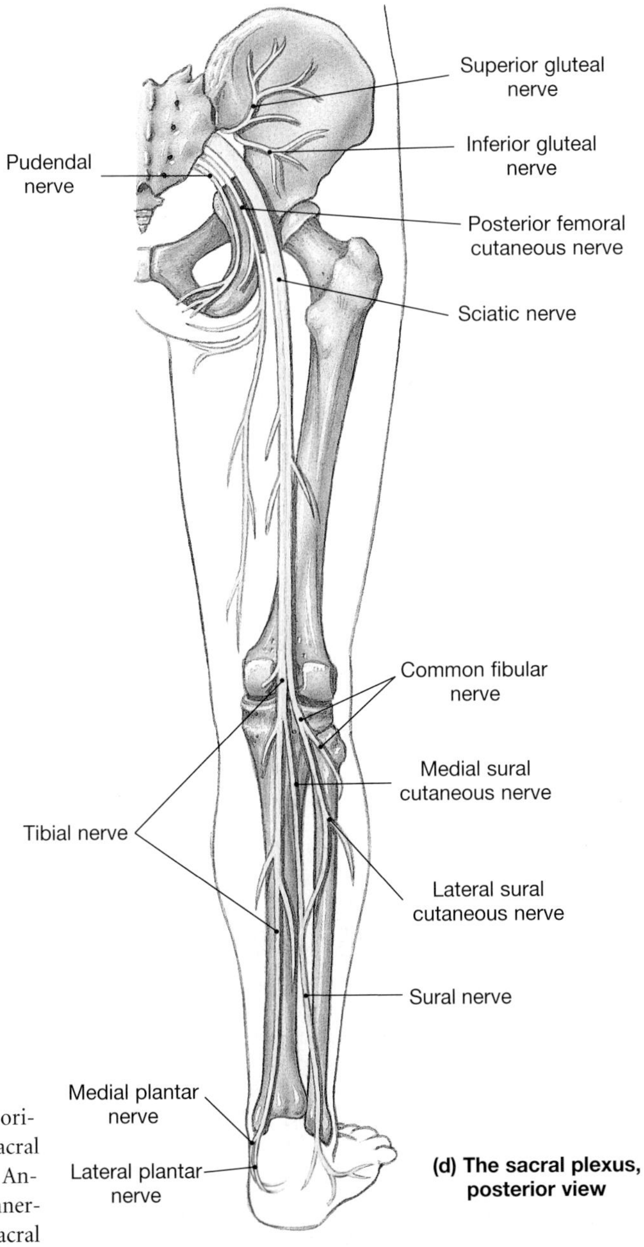

(d) The sacral plexus, posterior view

FIGURE 14.13 **The Lumbar and Sacral Plexuses, Part I**

(**a**) Anterior view of the lumbar plexus, showing its origins and major branches. (**b**) Anterior view of the sacral plexus, showing the origins of the major nerves. (**c**) Anterior view of the lumbar and sacral plexuses and innervation of the lower limb. (**d**) Posterior view of the sacral plexus and the associated innervation of the lower limb.

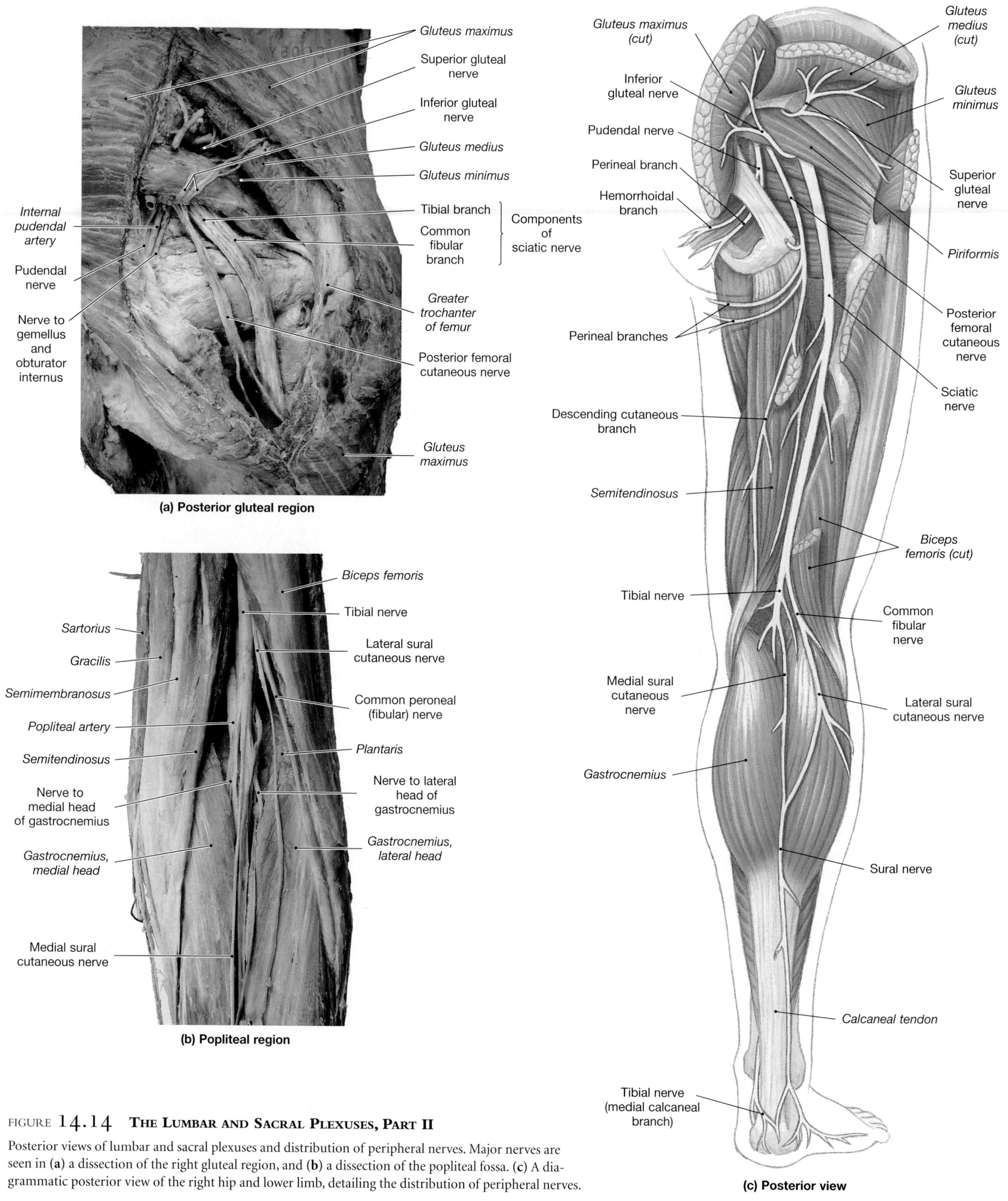

FIGURE 14.14 **THE LUMBAR AND SACRAL PLEXUSES, PART II**

Posterior views of lumbar and sacral plexuses and distribution of peripheral nerves. Major nerves are seen in (**a**) a dissection of the right gluteal region, and (**b**) a dissection of the popliteal fossa. (**c**) A diagrammatic posterior view of the right hip and lower limb, detailing the distribution of peripheral nerves.

Reflexes [FIGURE 14.15]

Conditions inside or outside the body can change rapidly and unexpectedly. A **reflex** is an immediate involuntary motor response to a specific stimulus. Reflexes help preserve homeostasis by making rapid adjustments in the function of organs or organ systems. The response shows little variability—activation of a particular reflex always produces the same motor response. The neural "wiring" of a single reflex is called a **reflex arc**. A reflex arc begins at a receptor and ends at a peripheral effector, such as a muscle or gland cell. Figure 14.15● illustrates the five steps involved in a neural reflex:

STEP 1 *Arrival of a Stimulus and Activation of a Receptor.* There are many types of sensory receptors, and general categories were introduced in Chapter 13. p. 344 Each receptor has a characteristic range of sensitivity; some receptors, such as pain receptors, respond to almost any stimulus. These receptors, the dendrites of sensory neurons, are stimulated by pressure, temperature extremes, physical damage, or exposure to abnormal chemicals. Other receptors, such as those providing visual, auditory, or taste sensations, are specialized cells that respond to only a limited range of stimuli.

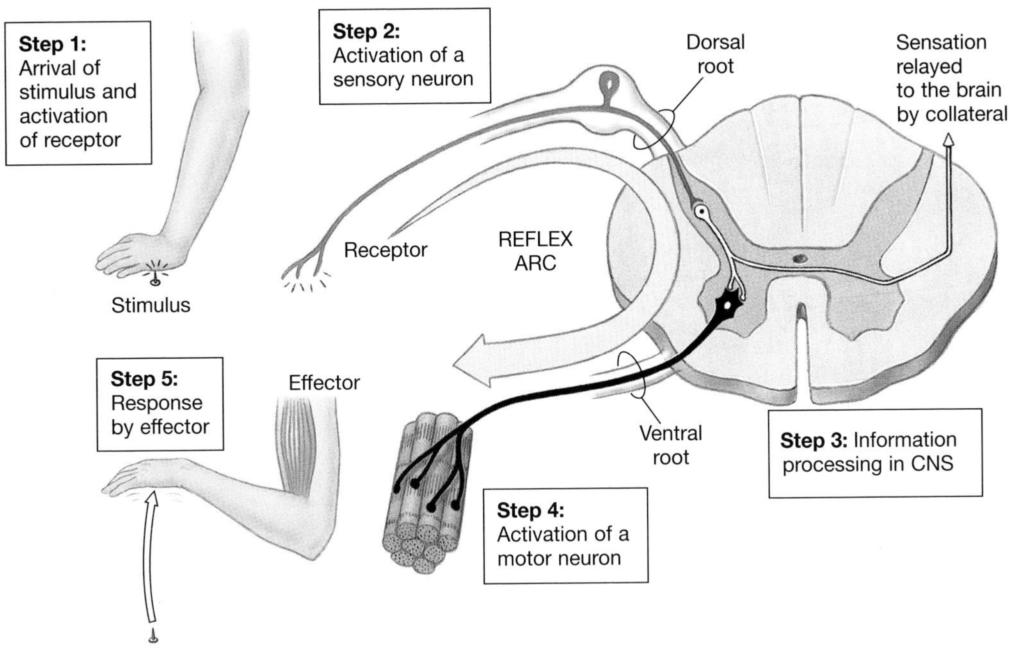

FIGURE 14.15 **A Reflex Arc**
This diagram illustrates the five steps involved in a neural reflex.

STEP 2 *Relay of Information to the CNS.* Information is carried in the form of action potentials along an afferent fiber. In this case, the axon conducts the action potentials into the spinal cord via one of the dorsal roots.

STEP 3 *Information Processing.* Information processing begins when a neurotransmitter released by synaptic terminals of the sensory neuron reaches the postsynaptic membrane of either a motor neuron or an interneuron. p. 346 In the simplest reflexes, such as the one diagrammed in Figure 14.15●, this processing is performed by the motor neuron that controls peripheral effectors. In more complex reflexes, several pools of interneurons are interposed between the sensory and motor neurons, and both serial and parallel processing occur. p. 348 The goal of this information processing is the selection of an appropriate motor response through the activation of specific motor neurons.

STEP 4 *Activation of a Motor Neuron.* A motor neuron stimulated to threshold conducts action potentials along its axon into the periphery; in this example, through the ventral root of a spinal nerve.

STEP 5 *Response of a Peripheral Effector.* Activation of the motor neuron causes a response by a peripheral effector, such as a skeletal muscle or gland. In general, this response is aimed at removing or counteracting the original stimulus. Reflexes play an important role in opposing potentially harmful changes in the internal or external environment.

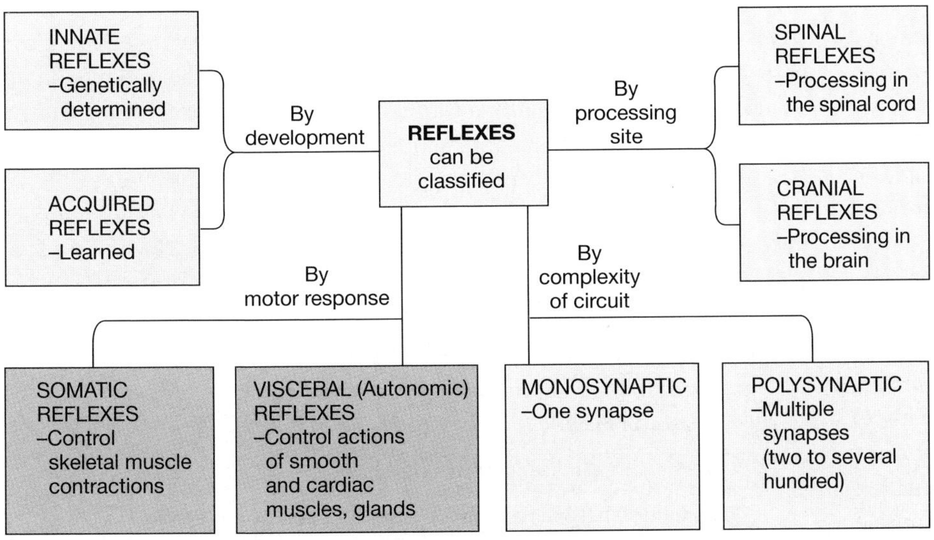

FIGURE 14.16 **The Classification of Reflexes**
Four different methods are used to classify reflexes.

Classification of Reflexes [FIGURES 14.16/14.17]

Reflexes can be classified according to (1) their development (**innate** and **acquired reflexes**), (2) the site where information processing occurs (**spinal** and **cranial reflexes**), (3) the nature of the resulting motor response (**somatic** and **visceral**, or **autonomic, reflexes**), or (4) the complexity of the neural circuit involved (*monosynaptic* and *polysynaptic reflexes*). These categories, presented in Figure 14.16●, are not mutually exclusive; they represent different ways of describing a single reflex.

In the simplest reflex arc, a sensory neuron synapses directly on a motor neuron. Such a reflex is termed a **monosynaptic reflex** (Figure 14.17a●). Transmission across a chemical synapse always involves a synaptic delay, but with only one synapse, the delay between stimulus and response is minimized.

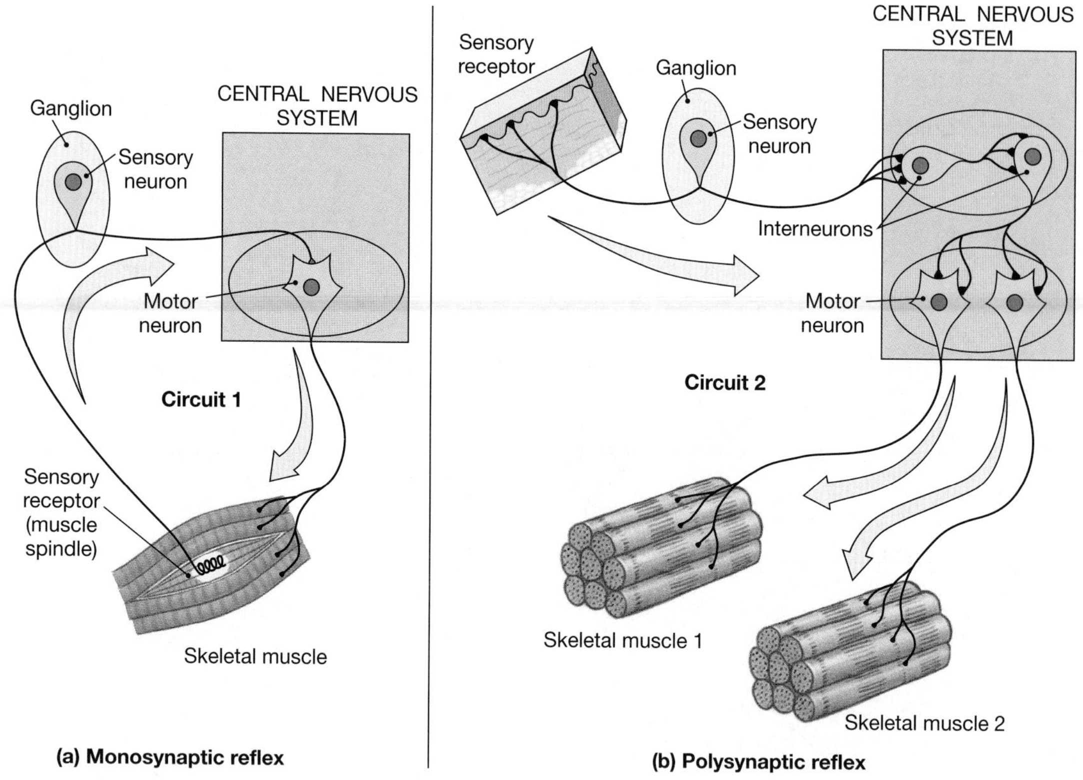

FIGURE 14.17 **Neural Organization and Simple Reflexes**

A comparison of monosynaptic and polysynaptic reflexes. **(a)** A monosynaptic reflex circuit involves a peripheral sensory neuron and a central motor neuron. In this example, stimulation of the receptor will lead to a reflexive contraction in a skeletal muscle. **(b)** A polysynaptic reflex circuit involves a sensory neuron, interneurons, and motor neurons. In this example, the stimulation of the receptor leads to the coordinated contractions of two different skeletal muscles.

Polysynaptic reflexes (Figure 14.17b●) have a longer delay between stimulus and response, the length of the delay being proportional to the number of synapses involved. Polysynaptic reflexes can produce far more complicated responses because the interneurons can control several different muscle groups. Many of the motor responses are extremely complicated; for example, stepping on a sharp object not only causes withdrawal of the foot, but triggers all of the muscular adjustments needed to prevent a fall. Such complicated responses result from the interactions between multiple interneuron pools.

Spinal Reflexes [FIGURES 14.17a/14.18]

The neurons in the gray matter of the spinal cord participate in a variety of reflex arcs. These *spinal reflexes* range in complexity from simple monosynaptic reflexes involving a single segment of the spinal cord to polysynaptic reflexes that integrate motor output from many different spinal cord segments to produce a coordinated motor response.

The best-known spinal reflex is the **stretch reflex**. It is a simple monosynaptic reflex that provides automatic regulation of skeletal muscle length (Figure 14.18a●). The stimulus stretches a relaxed muscle, thus activating a sensory neuron and triggering the contraction of that muscle. The stretch reflex also provides for the automatic adjustment of muscle tone, increasing or decreasing it in response to information provided by the stretch receptors of *muscle spindles* (Figure 14.17a●). Muscle spindles, which will be considered in Chapter 18, consist of specialized muscle fibers whose lengths are monitored by sensory neurons.

The most familiar stretch reflex is probably the *knee jerk*, or **patellar reflex**. In this reflex, a sharp rap on the patellar ligament stretches muscle spindles in the quadriceps muscles (Figure 14.18b●). With so brief a stimulus, the reflexive contraction occurs unopposed and produces a noticeable kick. Physicians often test this reflex to check the status of the lower segments of the spinal cord. A normal patellar reflex indicates that spinal nerves and spinal segments L_2–L_4 are undamaged.

The stretch reflex is an example of a **postural reflex**, a reflex that maintains normal upright posture. Postural muscles usually have a firm muscle tone and extremely sensitive stretch receptors. As a result, very fine adjustments are continually being made, and you are not aware of the cycles of contraction and relaxation that occur.

Higher Centers and Integration of Reflexes

Reflexive motor activities occur automatically, without instructions from higher centers in the brain. However, higher centers can have a profound effect on reflex performance. For example, processing centers in the brain can enhance or suppress spinal reflexes via descending tracts that synapse on interneurons and motor neurons throughout the spinal cord. Motor control therefore involves a series of interacting levels. At the lowest level are monosynaptic reflexes that are rapid but stereotyped and relatively inflexible. At the highest level are centers in the brain that can modulate or build upon reflexive motor patterns. *Reflexes and Diagnostic Testing p. 794*

✓ CONCEPT CHECK

- What is a reflex?
- In order, list the five steps in a reflex arc.
- Distinguish between a monosynaptic and polysynaptic reflex.
- What are the four methods of classifying reflexes?

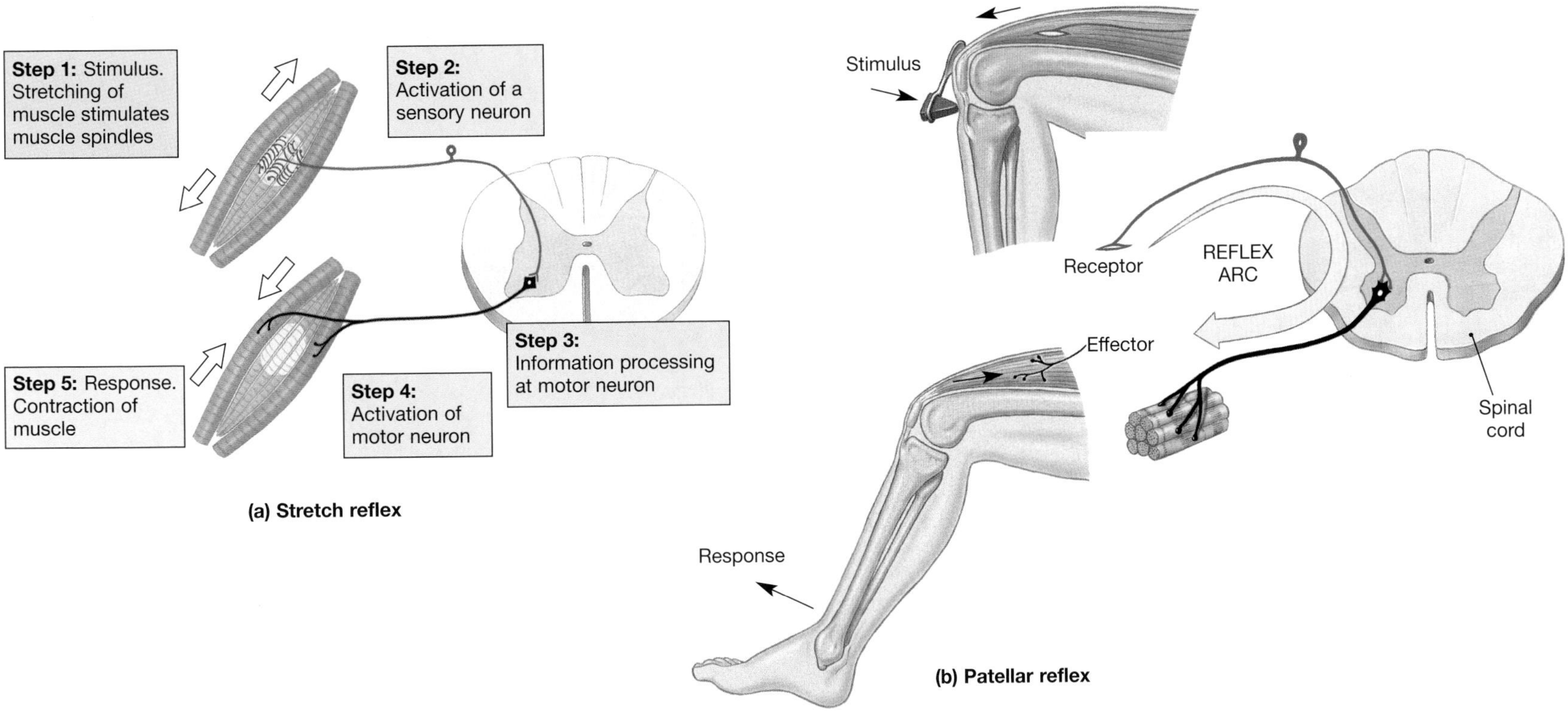

FIGURE 14.18 **STRETCH REFLEXES**

(**a**) Steps common to all stretch reflexes. (**b**) The patellar reflex is controlled by muscle spindles in the quadriceps group. The stimulus is a reflex hammer striking the muscle tendon, stretching the spindle fibers. This results in a sudden increase in the activity of the sensory neurons, which synapse on spinal motor neurons. The response occurs upon the activation of motor units in the quadriceps group, which produces an immediate increase in muscle tone and a reflexive kick.

RELATED CLINICAL TERMS

areflexia (ā-re-FLEK-sē-a): The lack of normal reflex responses to stimuli. *Reflexes and Diagnostic Testing p. 794*

Babinski sign (positive Babinski reflex): A spinal reflex in infants, consisting of a fanning of the toes, produced by stroking the foot on the side of the sole; in adults, a sign of CNS injury. *Reflexes and Diagnostic Testing p. 794*

caudal anesthesia: Injection of anesthetics into the epidural space of the sacrum to paralyze and reduce sensation of lower abdominal and perineal structures. *Spinal Anesthesia p. 793*

epidural block: Regional anesthesia produced by the injection of an anesthetic into the epidural space near targeted spinal nerve roots. *Spinal Anesthesia p. 793*

functional electrical stimulation (FES): A technique for stimulating specific muscles and muscle groups using electrodes controlled by a computer. *Technology and Motor Paralysis p. 793*

hyperreflexia: Exaggerated reflex responses that may develop in some pathological states or following stimulation of spinal and cranial nuclei by higher centers. *Reflexes and Diagnostic Testing p. 794*

hyporeflexia: A condition where normal spinal reflexes are present but weak. *Reflexes and Diagnostic Testing p. 794*

lumbar puncture: A spinal tap performed between adjacent lumbar vertebrae. *p. 360*

meningitis: An inflammation of the meningeal membranes. *Spinal Meningitis p. 793*

multiple sclerosis (skler-Ō-sis) **(MS):** A disease of the nervous system characterized by recurrent, often progressive incidents of demyelination affecting tracts in the brain and/or spinal cord. Common symptoms include partial loss of vision and problems with speech, balance, and general motor coordination. *Multiple Sclerosis p. 794*

myelography: A diagnostic procedure in which a radiopaque dye is introduced into the cerebrospinal fluid in order to obtain an X-ray of the spinal cord. *p. 360*

nerve graft: Insertion of an intact section from a different peripheral nerve to bridge the gap between the cut ends of a damaged nerve and to provide a route for axonal regeneration. *Technology and Motor Paralysis p. 793*

paraplegia: Paralysis involving loss of motor control of the lower limbs. *p. 362*

patellar reflex: The "knee jerk" reflex; often used to provide information about the related spinal segments. *p. 374*

quadriplegia: Paralysis involving loss of sensation and motor control of the upper and lower limbs. *p. 362*

spinal shock: A period of sensory and motor paralysis following any severe injury to the spinal cord. *p. 362*

spinal tap: A procedure in which fluid is extracted from the subarachnoid space through a needle inserted between the vertebrae. *p. 360*

Additional Clinical Terms Discussed in Appendix I (pp. 793–794)

abdominal reflex; ankle jerk reflex; biceps reflex; clonus; mass reflex; plantar reflex (negative Babinski reflex); triceps reflex

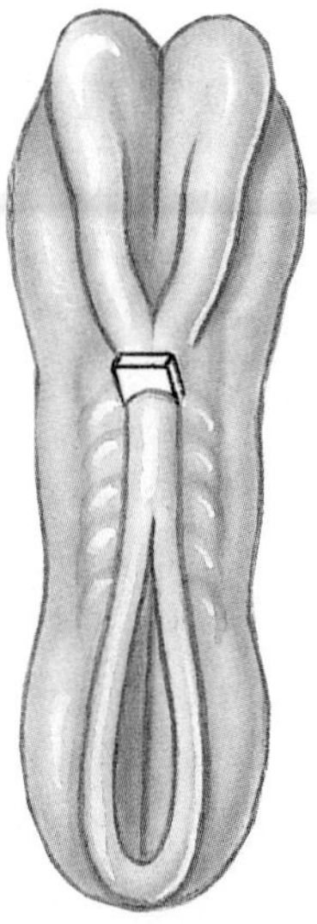

22 DAYS

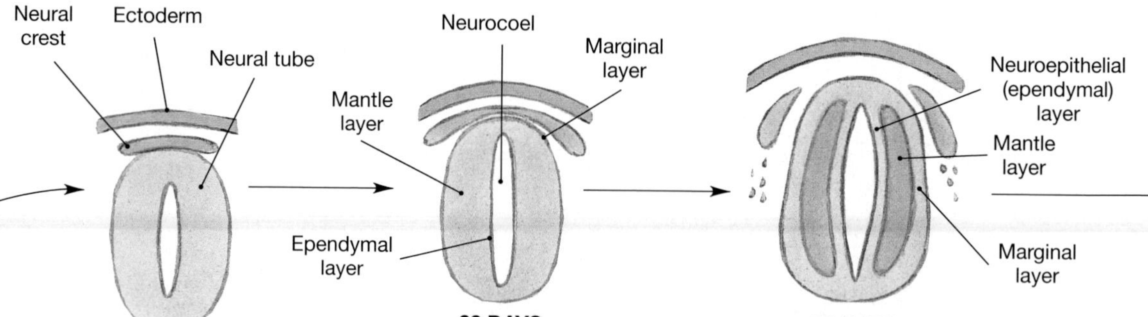

By the end of the fifth developmental week, the *neural tube* (p. 349) is almost completely closed. In the spinal cord the *mantle layer* that contains developing neurons and neuroglial cells will produce the gray matter that surrounds the *neurocoel*. As neurons develop in the mantle layer, their axons grow toward central or peripheral destinations. The axons leave the mantle layer and travel toward synaptic targets within a peripheral *marginal layer*.

Eventually, the growing axons will form bundles, or tracts, in the marginal layer, and these tracts will crowd together in the columns that form the white matter of the spinal cord.

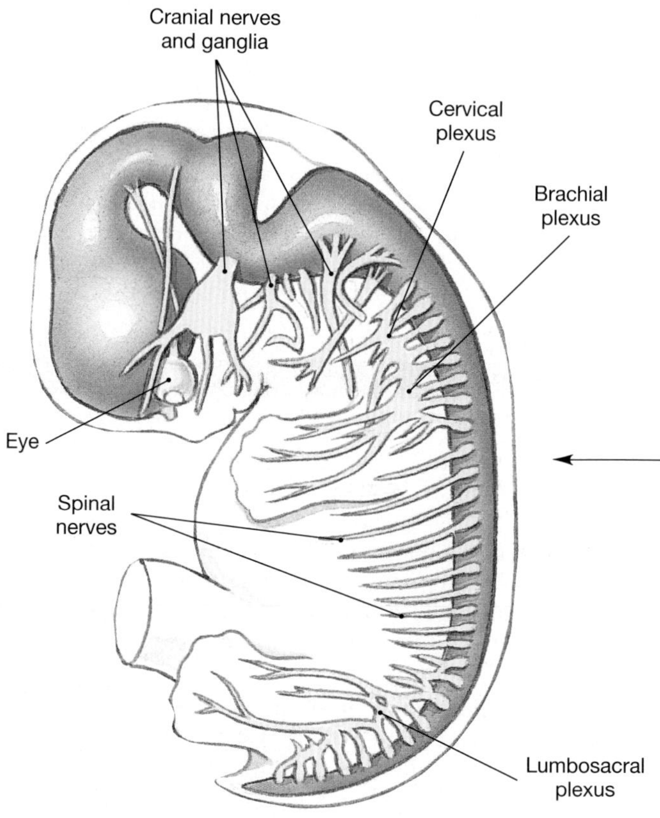

7 WEEKS
(Peripheral nerve distribution)

Several spinal nerves innervate each developing limb. When embryonic muscle cells migrate away from the myotome, the nerves grow right along with them. If a large muscle in the adult is derived from several myotomal blocks, connective tissue partitions will often mark the original boundaries, and the innervation will always involve more than one spinal nerve.

DEVELOPMENTAL ABNORMALITIES

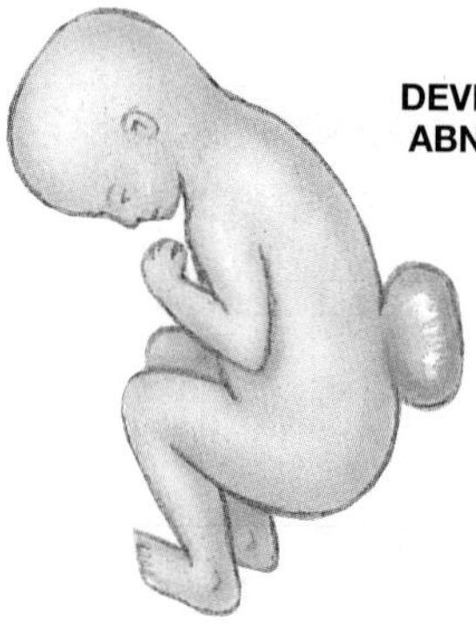

Spina bifida

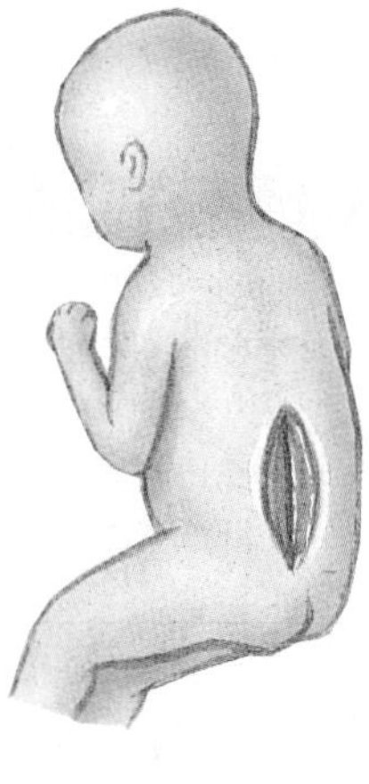

Neural tube defect

Spina bifida (BĪ-fi-da) results when the developing vertebral laminae fail to unite due to abnormal neural tube formation at that site. The neural arch is incomplete, and the meninges bulge outward beneath the skin of the back. The extent of the abnormality determines the severity of the defects. In mild cases, the condition may pass unnoticed; extreme cases involve much of the length of the vertebral column.

A **neural tube defect (NTD)** is a condition that is secondary to a developmental error in the formation of the spinal cord. Instead of forming a hollow tube, a portion of the spinal cord develops as a broad plate. This is often associated with spina bifida. Neural tube defects affect roughly one individual in 1000; prenatal testing can detect the existence of these defects with an 80–85% success rate.

As the mantle enlarges, the neurocoel becomes laterally compressed and relatively narrow. The relatively thin **roof plate** and **floor plate** will not thicken substantially, but the **dorsolateral** and **ventrolateral plates** enlarge rapidly. Neurons developing within the dorsolateral plate will receive and relay sensory information, while those in the ventrolateral region will develop into motor neurons.

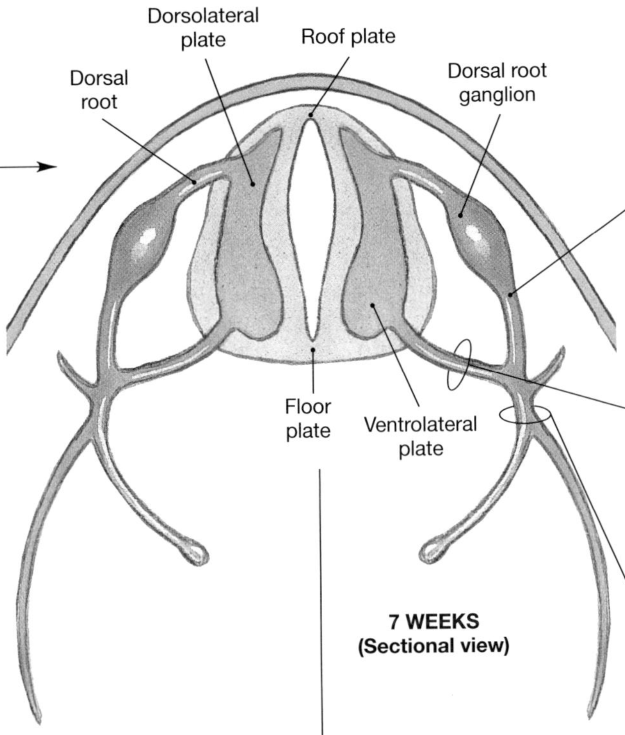

7 WEEKS
(Sectional view)

By this time, cells of the *neural crest* have migrated to either side of the spinal cord and have formed the dorsal root ganglia. The neural crest cells become sensory neurons and glial cells (Schwann cells and satellite cells). Processes from these sensory neurons grow both into the periphery, to contact receptors, and into the CNS via the dorsal roots.

In each segment the axons of developing motor neurons form a pair of ventral roots that grow away from the spinal cord.

Distal to each dorsal root ganglion, the motor efferents of the ventral root and the sensory afferents of the dorsal root are bound together into a single spinal nerve. Along much of the spinal cord, these nerves share a stereotyped pattern of peripheral branches, and the pattern accounts for the distribution of dermatomes.

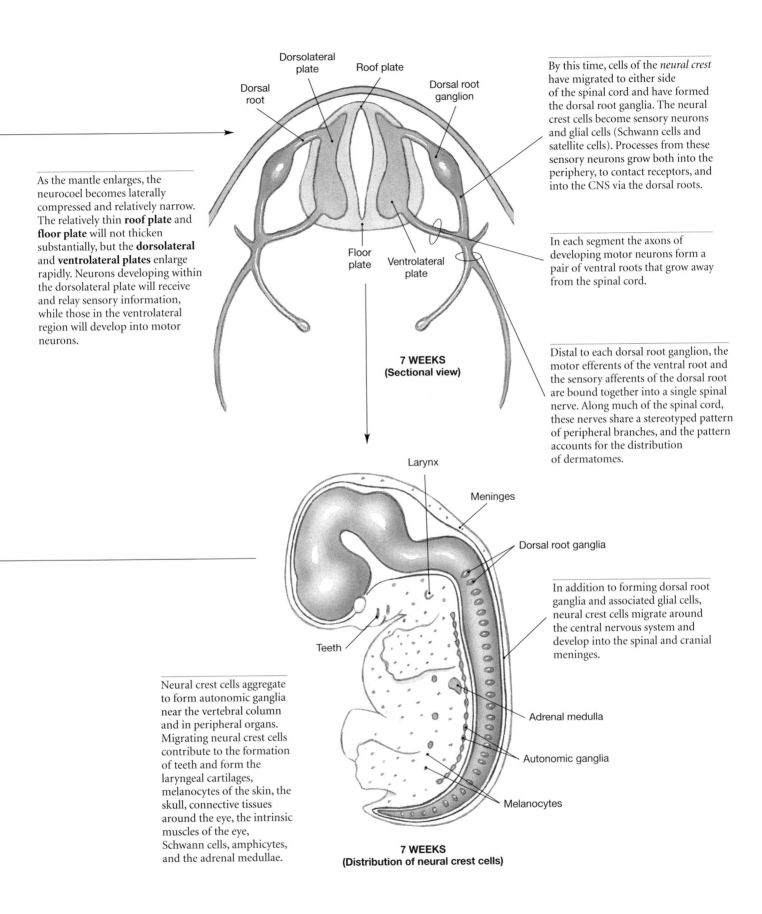

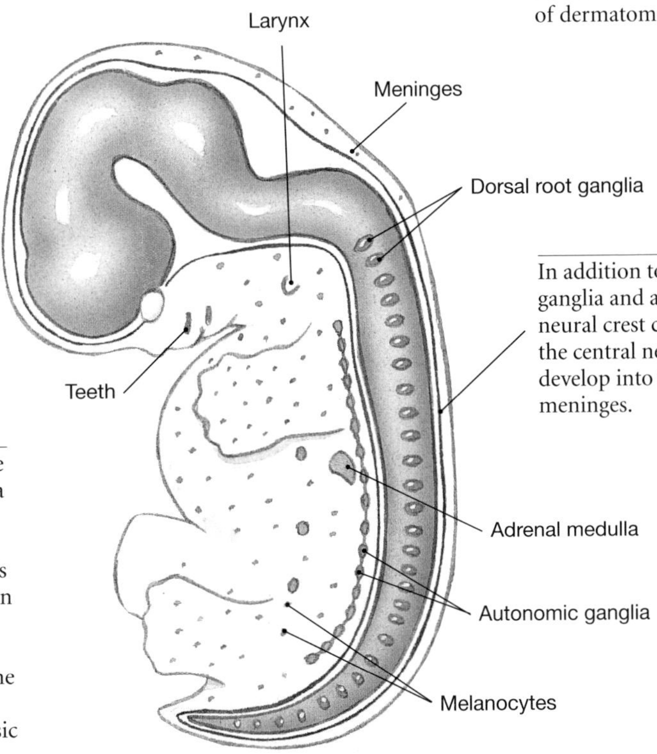

7 WEEKS
(Distribution of neural crest cells)

In addition to forming dorsal root ganglia and associated glial cells, neural crest cells migrate around the central nervous system and develop into the spinal and cranial meninges.

Neural crest cells aggregate to form autonomic ganglia near the vertebral column and in peripheral organs. Migrating neural crest cells contribute to the formation of teeth and form the laryngeal cartilages, melanocytes of the skin, the skull, connective tissues around the eye, the intrinsic muscles of the eye, Schwann cells, amphicytes, and the adrenal medullae.

STUDY OUTLINE & CHAPTER REVIEW

Introduction 356

1. The **central nervous system (CNS)** consists of the *spinal cord* and *brain.* Although they are connected, they have some functional independence. The spinal cord integrates and processes information on its own, in addition to relaying information to and from the brain.

Gross Anatomy of the Spinal Cord 356

1. The adult spinal cord has a **posterior median sulcus** (shallow) and an **anterior median fissure** (wide). It includes localized **enlargements** (*cervical* and *lumbar*), which are expanded regions where there is increased gray matter to provide innervation of the limbs. *(see Figures 14.1 to 14.3)*
2. The spinal cord tapers to a conical tip, the **conus medullaris.** The **filum terminale** (a strand of fibrous tissue) originates at this tip and extends through the vertebral canal to the second sacral vertebra, ultimately becoming part of the *coccygeal ligament. (see Figures 14.1 to 14.4)*
3. The spinal cord has 31 segments, each associated with a pair of **dorsal root ganglia** (containing sensory neuron cell bodies), and pairs of **dorsal roots** and **ventral roots.** *(see Figures 14.1 to 14.3)*
4. Sensory and motor fibers unite as a single **spinal nerve** distal to each dorsal root ganglion. Spinal nerves emerge from **intervertebral foramina** and are **mixed nerves,** since they contain both sensory and motor fibers. *(see Figures 14.1 to 14.3)*
5. The **cauda equina** is the inferior extension of the ventral and dorsal roots and the filum terminale in the vertebral canal. *(see Figures 14.1/14.4)*

Spinal Meninges 356

1. The **spinal meninges** are a series of specialized membranes that provide physical stability and shock absorption for neural tissues of the spinal cord; the **cranial meninges** are membranes that surround the brain (Chapter 15). There are three meningeal layers: the *dura mater*, the *arachnoid*, and the *pia mater. (see Figure 14.2)*

The Dura Mater 356

2. The spinal **dura mater** is the tough, fibrous outermost layer that covers the spinal cord; caudally it forms the **coccygeal ligament** with the filum terminale. The **epidural space** separates the dura mater from the inner walls of the vertebral canal. *(see Figures 14.1b,c/14.2/14.4a)*

The Arachnoid 359

3. Internal to the inner surface of the dura mater is the **subdural space.** It separates the dura mater from the middle meningeal layer, the **arachnoid.** Internal to the arachnoid is the **subarachnoid space,** which has a network of collagen and elastic fibers, the *arachnoid trabeculae.* This space also contains **cerebrospinal fluid,** which acts as a shock absorber and a diffusion medium for dissolved gases, nutrients, chemical messengers, and waste products. *(see Figures 14.2/14.4a)*

The Pia Mater 359

4. The **pia mater** is the innermost meningeal layer. It is bound firmly to the underlying neural tissue. Paired **denticulate ligaments** are supporting fibers extending laterally from the spinal cord surface, binding the spinal pia mater and spinal arachnoid to the dura mater to prevent either side-to-side or inferior movement of the spinal cord. *(see Figure 14.2)*

Sectional Anatomy of the Spinal Cord 361

1. The central **gray matter** surrounds the **central canal** and contains cell bodies of neurons and glial cells. The gray matter projections toward the outer surface of the spinal cord are called **horns.** The peripheral **white matter** contains myelinated and unmyelinated axons in *tracts* and *columns. (see Figure 14.5)*

Organization of Gray Matter 361

2. Neuron cell bodies in the spinal cord gray matter are organized into groups, termed *nuclei.* The **posterior gray horns** contain somatic and visceral sensory nuclei, while nuclei in the **anterior gray horns** are involved with somatic motor control. The **lateral gray horns** contain visceral motor neurons. The **gray commissures** are posterior and anterior to the central canal. They contain the axons of interneurons that cross from one side of the cord to the other. *(see Figure 14.5)*

Organization of White Matter 362

3. The white matter can be divided into six **columns** (*funiculi*), each of which contains **tracts** *(fasciculi).* **Ascending tracts** relay information from the spinal cord to the brain, and **descending tracts** carry information from the brain to the spinal cord. *(see Figure 14.5)*

Spinal Nerves 362

1. There are 31 pairs of spinal nerves; each is identified through its association with an adjacent vertebra (*cervical, thoracic, lumbar,* and *sacral*). *(see Figures 14.1/14.3)*
2. Each spinal nerve is ensheathed by a series of connective tissue layers. The outermost layer, **epineurium,** is a dense network of collagen fibers; the middle layer, **perineurium,** partitions the nerve into a series of bundles *(fascicles)* and conveys blood vessels into each individual fiber; and the inner layer, **endoneurium,** is composed of delicate connective tissue fibers that surround individual axons. *(see Figure 14.6)*

Peripheral Distribution of Spinal Nerves 363

3. The first branch of each spinal nerve in the thoracic and upper lumbar regions is the **white ramus,** which contains myelinated axons going to an **autonomic ganglion.** Two groups of unmyelinated fibers exit this ganglion: a **gray ramus,** carrying axons that innervate glands and smooth muscles in the body wall or limbs back to the spinal nerve, and an autonomic nerve carrying fibers to internal organs. Collectively, the white and gray rami are termed the **rami communicantes.** *(see Figures 14.2/14.7)*
4. Each spinal nerve has both a **dorsal ramus** (sensory/motor innervation to the skin and muscles of the back) and a **ventral ramus** (supplies ventrolateral body surface, body wall structures, and limbs). Each pair of spinal nerves monitors a region of the body surface, an area called a **dermatome.** *(see Figures 14.2/14.7/14.8)*

Nerve Plexuses 363

5. A complex, interwoven network of nerves is called a **nerve plexus.** The four major plexuses are: the *cervical plexus*, the *brachial plexus*, the *lumbar plexus*, and the *sacral plexus. (see Figures 14.3/14.9 to 14.14, and Tables 14.1 to 14.3)*
6. The **cervical plexus** consists of C_1-C_4 ventral rami and some fibers from C_5. Muscles of the neck are innervated; some branches extend into the thoracic cavity to the diaphragm. The **phrenic nerve** is the major nerve in this plexus. *(see Figures 14.3/14.9/14.10/14.12, and Table 14.1)*
7. The **brachial plexus** innervates the pectoral girdle and upper limbs by ventral rami of C_5-T_1. The nerves in this plexus originate from cords or trunks: **superior, middle,** and **inferior trunks** give rise to the **lateral cord, medial cord,** and **posterior cord.** *(see Figures 14.3/14.9/14.11/14.12, and Table 14.2)*
8. Collectively the **lumbar plexus** and **sacral plexus** originate from the posterior abdominal wall and ventral rami of nerves supplying the pelvic girdle and lower limb. The lumbar plexus contains fibers from spinal segments T_{12}–L_4 and the sacral plexus contains fibers from spinal segments L_4-S_4. *(see Figures 14.3/14.9/14.13/14.14, and Table 14.3)*

Reflexes 373

1. A neural **reflex** is a rapid, automatic, involuntary motor response to stimuli. Reflexes help preserve homeostasis by rapidly adjusting the functions of organs or organ systems. *(see Figure 14.15)*

2. A **reflex arc** is the neural "wiring" of a single reflex. *(see Figure 14.15)*
3. A *receptor* is a specialized cell that monitors conditions in the body or external environment. Each receptor has a characteristic range of sensitivity.
4. There are five steps involved in a neural reflex: (1) arrival of a stimulus/receptor activation; (2) relay of information to the CNS; (3) information processing; (4) activation of a motor neuron; (5) response by a peripheral effector. *(see Figure 14.15)*

Classification of Reflexes 373

5. Reflexes are classified by: (1) their development *(innate, acquired)*; (2) where information is processed *(spinal, cranial)*; (3) motor response *(somatic, visceral (autonomic))*; and (4) complexity of the neural circuit *(monosynaptic, polysynaptic)*. *(see Figure 14.16)*
6. **Innate reflexes** are genetically determined. **Acquired reflexes** are learned following repeated exposure to a stimulus. *(see Figure 14.16)*
7. Reflexes processed in the brain are **cranial reflexes**. In a **spinal reflex** the important interconnections and processing occur inside the spinal cord. *(see Figure 14.16)*
8. **Somatic reflexes** control skeletal muscle contractions, and **visceral** *(autonomic)* **reflexes** control the activities of smooth and cardiac muscles and glands. *(see Figure 14.16)*
9. A **monosynaptic reflex** is the simplest reflex arc. A sensory neuron synapses directly on a motor neuron that acts as the processing center. **Polysynaptic reflexes** have at least one interneuron placed between the sensory afferent and the motor efferent. Thus, they have a longer delay between stimulus and response. *(see Figures 14.16/14.17)*

Spinal Reflexes 374

10. **Spinal reflexes** range from simple monosynaptic reflexes (involves only one segment of the cord) to more complex polysynaptic reflexes (in which many segments of the cord interact to produce a coordinated motor response). *(see Figure 14.17)*
11. The **stretch reflex** is a monosynaptic reflex that automatically regulates skeletal muscle length and muscle tone. The sensory receptors involved are stretch receptors of *muscle spindles*. *(see Figure 14.18a)*
12. A **patellar reflex** is the familiar *knee jerk*, wherein a rap on the patellar ligament stretches the **muscle spindles** in the quadriceps muscles. *(see Figure 14.18b)*
13. A **postural reflex** is a stretch reflex that maintains normal upright posture.

Higher Centers and Integration of Reflexes 374

14. Higher centers in the brain can enhance or inhibit reflex motor patterns based in the spinal cord.

LEVEL 1 REVIEWING FACTS AND TERMS

Match each numbered item with the most closely related lettered item. Use letters for answers in the spaces provided.

Column A	Column B
_____ 1. ventral root	a. tracts and columns
_____ 2. epidural space	b. specific region of body surface
_____ 3. white matter	c. cervical plexus
_____ 4. fascicle	d. motor neuron axons
_____ 5. dermatome	e. sacral plexus
_____ 6. phrenic nerve	f. lumbar plexus
_____ 7. brachial plexus	g. single bundle of axons
_____ 8. obturator nerve	h. involuntary motor response
_____ 9. reflex	i. loose connective tissue, adipose tissue
_____ 10. pudendal nerve	j. pectoral girdle/upper extremity

11. The ___ is a strand of fibrous tissue that provides longitudinal support as a component of the coccygeal ligament.
    (a) conus medullaris (b) filum terminale
    (c) cauda equina (d) dorsal root
12. Axons crossing from one side of the spinal cord to the other within the gray matter are found in the
    (a) anterior gray horns (b) white commissures
    (c) gray commissures (d) lateral gray horns
13. The dorsal root of a spinal nerve contains
    (a) axons of sensory neurons (b) axons of motor neurons
    (c) cell bodies of sensory neurons (d) interneurons
14. The tough, fibrous outermost covering of the spinal cord is the
    (a) arachnoid (b) dura mater
    (c) pia mater (d) epidural block
15. Sensory and motor innervations of the skin of the lateral and ventral surfaces of the body are provided by the
    (a) white rami communicantes (b) gray rami communicantes
    (c) dorsal ramus (d) ventral ramus
16. The ulnar nerve is found in the ___ plexus.
    (a) cranial (b) brachial
    (c) sacral (d) cervical
17. The middle layer of connective tissue that surrounds each peripheral nerve is the
    (a) epineurium
    (b) perineurium
    (c) endoneurium
    (d) endomysium
18. The expanded area of the spinal cord that supplies nerves to the pectoral girdle and upper limbs is the
    (a) conus medullaris
    (b) filum terminale
    (c) lumbar enlargement
    (d) cervical enlargement
19. Spinal nerves are called mixed nerves because
    (a) they contain sensory and motor fibers
    (b) they exit at intervertebral foramina
    (c) they are associated with a pair of dorsal root ganglia
    (d) they are associated with dorsal and ventral roots
20. The gray matter of the spinal cord is dominated by
    (a) myelinated axons only
    (b) cell bodies of neurons and glial cells
    (c) unmyelinated axons only
    (d) Schwann cells and satellite cells

LEVEL 2 REVIEWING CONCEPTS

1. The epidural space contains
   (a) cerebrospinal fluid (b) connective tissue and blood vessels
   (c) lymph (d) denticulate ligaments
2. Polysynaptic reflexes can produce far more complicated responses than can monosynaptic reflexes because
   (a) the response time is quicker
   (b) the response is initiated by highly sensitive receptors
   (c) motor neurons carry impulses at a rate faster than sensory neurons
   (d) involved interneurons can control several different muscle groups
3. Proceeding inward from the outermost layer, number the list in the correct sequence:
   ___ walls of vertebral canal ___ subdural space
   ___ pia mater ___ subarachnoid space
   ___ dura mater ___ epidural space
   ___ arachnoid membrane ___ spinal cord
4. What is the role of the meninges in protecting the spinal cord?
5. How does a reflex differ from a voluntary muscle movement?
6. If the dorsal root of the spinal cord were damaged, what would be affected?
7. Why is response time in a monosynaptic reflex much faster than the response time in a polysynaptic reflex?
8. Why are there eight cervical spinal nerves but only seven cervical vertebrae?
9. What prevents side-to-side movements of the spinal cord?
10. Why is it important that a spinal tap be done between the third and fourth lumbar vertebrae?

LEVEL 3 CRITICAL THINKING AND CLINICAL APPLICATIONS

1. Bowel and bladder control involve spinal reflex arcs that are located in the sacral region of the spinal cord. In both instances two sphincter muscles, an inner sphincter of smooth muscle and an outer sphincter of skeletal muscle, control the passageway out of the body. How would a transection of the spinal cord at the L_1 level affect an individual's bowel and urinary bladder control?
2. Cindy is in an automobile accident and injures her spinal cord. She has lost feeling in her right hand, and her doctor tells her that it is the result of swelling compressing a portion of her spinal cord. Which part of her cord is likely to be compressed?
3. Karen falls down a flight of stairs and suffers spinal cord damage due to hyperextension of the cord during the fall. The injury results in edema of the central cord with resulting compression of the anterior horn cells of the lumbar region. What symptoms would you expect to observe as a result of this injury?

✓ ANSWERS TO CONCEPT CHECK QUESTIONS

p. 360 **1.** The ventral root of spinal nerves is composed of visceral and somatic motor fibers. Damage to this root would interfere with motor functions. **2.** The cerebrospinal fluid that surrounds the spinal cord is found in the subarachnoid space, which lies between the arachnoid membrane and the pia mater. **3.** The two spinal enlargements are the cervical enlargement and the lumbar enlargement. These regions are increased in size because of the increase in neuron cell bodies in the gray matter here. These segments of the spinal cord are concerned with innervation of the limbs. **4.** Each dorsal root ganglion contains cell bodies of sensory neurons.

p. 362 **1.** Since the poliovirus would be located in the somatic motor neurons, we would find it in the anterior gray horns of the spinal cord where the cell bodies of these neurons are located. **2.** White matter is organized in columns (anterior, lateral, and posterior) around the periphery of the spinal cord. **3.** Projections of gray matter toward the surface of the spinal cord are called horns. **4.** Ascending tracts carry sensory information toward the brain. Descending tracts carry motor commands into the spinal cord.

p. 370 **1.** The phrenic nerves that innervate the diaphragm originate in the cervical plexus. Damage to this plexus or, more specifically, to the phrenic nerves, would greatly interfere with the ability to breathe and possibly result in death by suffocation. **2.** The outermost layer is called the epineurium. It surrounds the entire nerve. The middle layer, or perineurium, divides the nerve into a series of compartments that contain bundles of axons. A single bundle is called a fascicle. The endoneurium is the innermost layer, and it surrounds individual axons. **3.** The white and gray ramus connect the spinal nerve to a nearby autonomic ganglion. The white ramus carries preganglionic axons that are myelinated from the nerve to the ganglion. The gray ramus carries postganglionic, unmyelinated axons from the ganglion back to the spinal nerve. **4.** The brachial plexus may have been damaged.

p. 374 **1.** A reflex is an immediate, involuntary motor response to a specific stimulus. **2.** (1) Arrival of a stimulus and activation of a receptor. (2) Relay of information to the CNS. (3) Information processing. (4) Activation of a motor neuron in the CNS. (5) Response of a peripheral effector. **3.** A monosynaptic reflex has a sensory neuron synapsing directly on a motor neuron. A polysynaptic reflex has more than one synapse between stimulus and response. **4.** Reflexes are classified according to (1) their development, (2) the site where information processing occurs, (3) the nature of the resulting motor response, or (4) the complexity of the neural circuit involved.

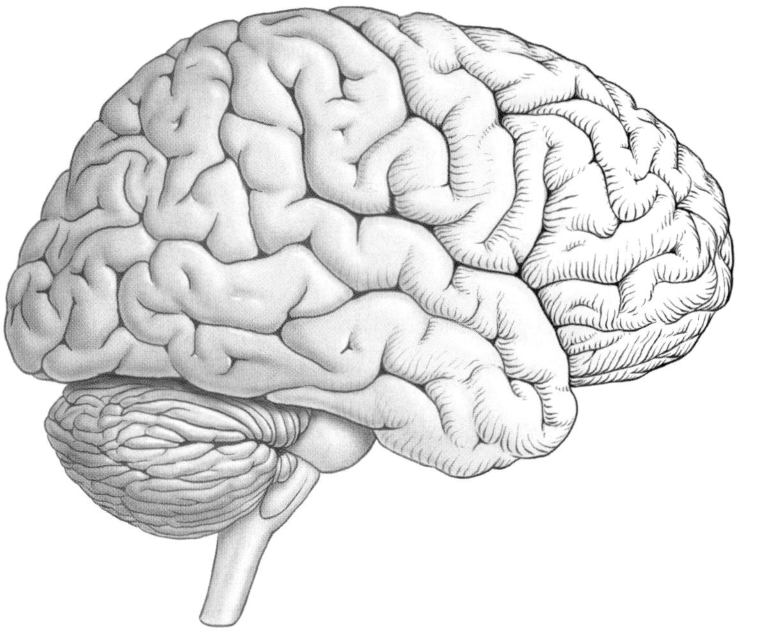

15

THE NERVOUS SYSTEM

The Brain and Cranial Nerves

The brain is probably the most fascinating organ in the body. It has a complex three-dimensional structure and performs a bewildering array of functions. Often the brain is likened to an organic computer, with its individual neurons compared to silicon "chips." Like the brain, a computer receives enormous amounts of incoming information, files and processes this information, and directs appropriate output responses. However, any direct comparison between your brain and a computer is misleading, because even the most sophisticated computer lacks the versatility and adaptability of a single neuron. One neuron may process information from 100,000 different sources at the same time and there are tens of billions of neurons in the nervous system. Rather than continuing to list the number of activities that can be performed by the brain, it is more appropriate to appreciate that this incredibly complex organ is the source of all of our dreams, passions, plans, memories, and behaviors. Everything we do and everything we are results from its activity.

The brain is far more complex than the spinal cord, and it can respond to stimuli with greater versatility. That versatility results from the tremendous number of neurons and neuronal pools in the brain and the complexity of their interconnections. The brain contains roughly 20 billion neurons, each of which may receive information across thousands of synapses at one time. Excitatory and inhibitory interactions among the extensively interconnected neuronal pools ensure that the response can vary to meet changing circumstances. But adaptability has a price: A response cannot be immediate, precise, and adaptable all at the same time. Adaptability requires multiple processing steps, and every synapse adds to the delay between stimulus and response. One of the major functions of spinal reflexes is to provide an *immediate* response that can be fine-tuned or elaborated on by more versatile but slower processing centers in the brain.

We now begin a detailed examination of the brain. This chapter focuses attention on the major structures of the brain and their relationships with the cranial nerves.

An Introduction to the Organization of the Brain [FIGURE 15.1]

The adult human brain (Figure 15.1●) contains almost 98% of the neural tissue in the body. An average adult brain weighs 1.4 kg (3 lb) and has a volume of 1200 cc (71 in.^{3}). There is considerable individual variation, and the brains of males are on average about 10% larger than those of females, owing to differences in average body size. Its relatively unimpressive external appearance gives few clues to its real complexity and importance. An adult brain can be held easily in both hands; no correlation exists between brain size and intelligence, and individuals with the smallest (750 cc) and largest (2100 cc) brains are functionally normal. A freshly removed brain is gray externally, and its internal tissues are tan to pink. Overall, the brain has the consistency of medium-firm tofu or chilled jello.

Early in development, the brain resembles the spinal cord—hollow, and with a narrow central passageway filled with cerebrospinal fluid. As development proceeds, this simple passageway becomes subdivided, and in several regions it expands to form enlarged chambers called *ventricles*. We will consider the anatomy of these ventricles in a later section.

Embryology of the Brain [TABLE 15.1]

The development of the brain is detailed in the Embryology Summary on pp. 420–421. However, a brief overview will help you understand adult brain structure and organization. The central nervous system begins as a hollow *neural tube*, with a fluid-filled internal cavity called the *neurocoel*. In the fourth week of development, three areas in the cephalic portion of the neural tube enlarge rapidly through expansion of the neurocoel. This enlargement creates three prominent **primary brain vesicles** named for their relative positions: the **prosencephalon** (prō-zen-SEF-a-lon; *proso*, forward + *enkephalos*, brain), or "forebrain"; the **mesencephalon** (mez-en-SEF-a-lon; *mesos*, middle), or midbrain; and the **rhombencephalon** (rom-ben-SEF-a-lon), or "hindbrain."

The fate of the three primary divisions of the brain is summarized in Table 15.1. The prosencephalon and rhombencephalon are subdivided further, forming **secondary brain vesicles** The prosencephalon forms the **telencephalon** (tel-en-SEF-a-lon; *telos*, end) and the *diencephalon*. The telencephalon forms the *cerebrum*, the paired cerebral hemispheres that dominate the superior and lateral surfaces of the adult brain. The hollow diencephalon has a roof (the *epithalamus*), walls (the left and right *thalamus*), and a floor (the *hypothalamus*). By the time the posterior end of the neural tube closes, secondary bulges, the *optic vesicles*, have extended laterally from the sides of the diencephalon. Additionally, the developing brain bends, forming creases that mark the boundaries between the ventricles. The mesencephalon does not subdivide, but its walls thicken and the neurocoel becomes a relatively narrow passageway with a diameter comparable to that of the central canal of the spinal cord. The portion of the rhombencephalon closest to the mesencephalon forms the **metencephalon** (met-en-SEF-a-lon; *meta*, after). The ventral portion of the metencephalon develops into the *pons*, and the dorsal portion becomes the *cerebellum*. The portion of the rhombencephalon closer to the spinal cord becomes the **myelencephalon** (mī-el-en-SEF-a-lon; *myelon*, spinal cord), which will form the *medulla oblongata*. We will now examine each of these structures in the adult brain.

Major Regions and Landmarks [FIGURE 15.1]

There are six major divisions in the adult brain: (1) the *cerebrum*, (2) the *diencephalon*, (3) the *mesencephalon*, (4) the *pons*, (5) the *cerebellum*, and (6) the *medulla oblongata*. Refer to Figure 15.1● as we provide an overview of each division.

THE CEREBRUM

The **cerebrum** (ser-Ē-brum or SER-e-brum) is divided into large, paired **cerebral hemispheres** separated by the **longitudinal fissure**. Conscious thought processes, intellectual functions, memory storage and retrieval, and complex motor patterns originate in the cerebrum. Motor activities are adjusted automatically by the cerebellum and other brain regions on the basis of sensory information and memories of learned patterns of movement.

DIENCEPHALON

The deep portion of the brain attached to the cerebrum is called the **diencephalon** (dī-en-SEF-a-lon; *dia*, through). The diencephalon can be divided into three regions:

- The **epithalamus**, or roof of the diencephalon, contains the hormone-secreting *pineal gland*.
- The right **thalamus** and left thalamus (THAL-a-mus; plural, *thalami*) form the walls of the diencephalon. Each thalamus is a sensory information relay and processing center.
- The floor of the diencephalon is the **hypothalamus** (*hypo-*, below), a visceral control center. A narrow stalk connects the hypothalamus to the **pituitary gland**, or *hypophysis* (*phyein*, to generate). The hypothalamus contains centers involved with emotions, autonomic function, and hormone production. It is the primary link between the nervous and endocrine systems.

The remaining regions of the brain are collectively known as the *brain stem*. The **brain stem** consists of the mesencephalon, pons, and medulla oblongata[1]. The brain stem contains important processing centers and also relays information to or from the cerebrum or cerebellum. Refer to Figure 15.1● as we describe the structure of the brain stem.

[1] Some sources consider the brain stem to include the diencephalon. We will use the more restrictive definition here.

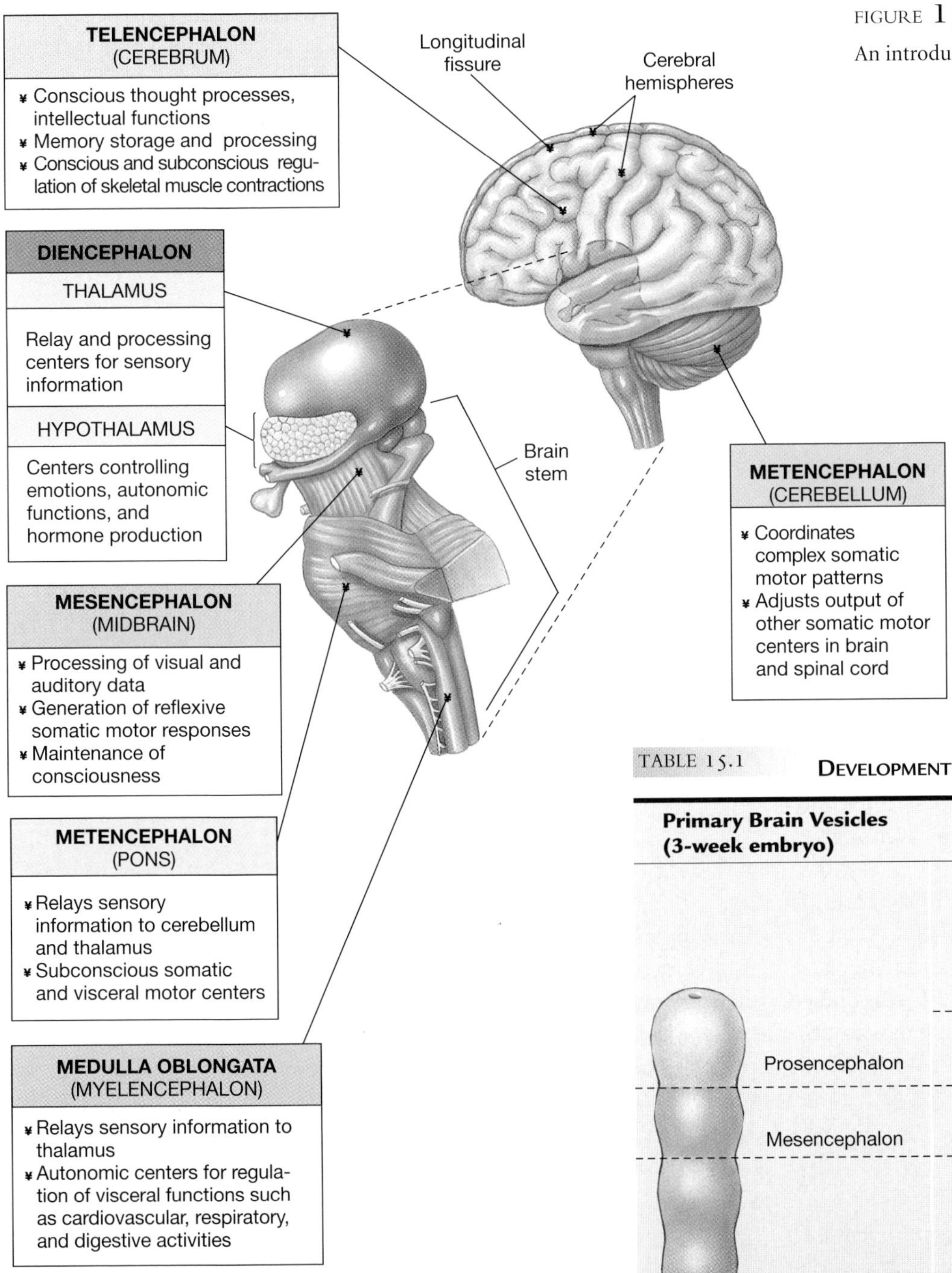

TELENCEPHALON
(CEREBRUM)
Conscious thought processes, intellectual functions
Memory storage and processing
Conscious and subconscious regulation of skeletal muscle contractions
Longitudinal fissure
Cerebral hemispheres
DIENCEPHALON
THALAMUS
Relay and processing centers for sensory information
HYPOTHALAMUS
Centers controlling emotions, autonomic functions, and hormone production
Brain stem
METENCEPHALON
(CEREBELLUM)
Coordinates complex somatic motor patterns
Adjusts output of other somatic motor centers in brain and spinal cord
MESENCEPHALON
(MIDBRAIN)
Processing of visual and auditory data
Generation of reflexive somatic motor responses
Maintenance of consciousness
METENCEPHALON
(PONS)
Relays sensory information to cerebellum and thalamus
Subconscious somatic and visceral motor centers
MEDULLA OBLONGATA
(MYELENCEPHALON)
Relays sensory information to thalamus
Autonomic centers for regulation of visceral functions such as cardiovascular, respiratory, and digestive activities

FIGURE 15.1 **Major Divisions of the Brain**
An introduction to brain regions and their major functions.

TABLE 15.1 **Development of the Human Brain** *(See also pp. 420–421 for embryological summary)*

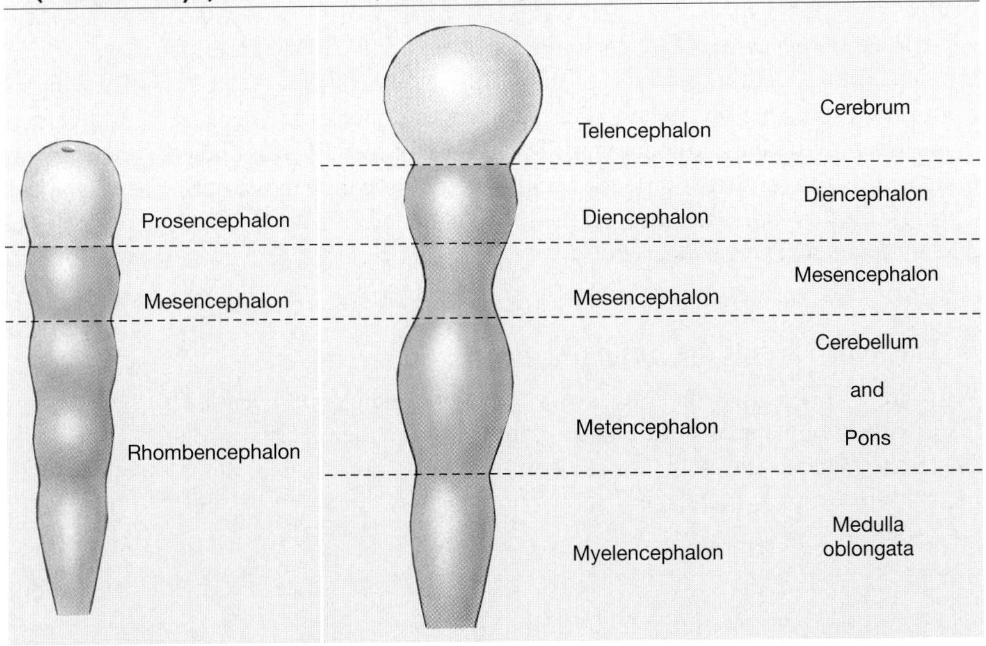

Primary Brain Vesicles (3-week embryo)	Secondary Brain Vesicles (6-week embryo)	Brain Regions at Birth
Prosencephalon	Telencephalon	Cerebrum
	Diencephalon	Diencephalon
Mesencephalon	Mesencephalon	Mesencephalon
Rhombencephalon	Metencephalon	Cerebellum and Pons
	Myelencephalon	Medulla oblongata

MESENCEPHALON

Nuclei in the **mesencephalon**, or midbrain, process visual and auditory information and generate involuntary somatic motor responses to these stimuli. This region also contains centers involved with the maintenance of consciousness.

THE PONS AND CEREBELLUM

The **pons** is immediately inferior to the mesencephalon. It contains nuclei involved with both somatic and visceral motor control. The term *pons* refers to a bridge, and the pons connects the cerebellum to the brain stem. The relatively small hemispheres of the **cerebellum** (ser-e-BEL-um) lie posterior to the pons and inferior to the cerebral hemispheres. The cerebellum automatically adjusts motor activities on the basis of sensory information and memories of learned patterns of movement.

MEDULLA OBLONGATA

The spinal cord connects to the brain at the **medulla oblongata**. The superior portion of the medulla oblongata has a thin, membranous roof, whereas the inferior portion resembles the spinal cord. The medulla oblongata relays sensory information to the thalamus and to other brain stem centers. In addition, it contains major centers concerned with the regulation of autonomic function, such as heart rate, blood pressure, and digestive activities.

■ Gray Matter and White Matter Organization

The general distribution of gray matter in the brain stem resembles that in the spinal cord; there is an inner region of gray matter surrounded by tracts of white matter. The gray matter surrounds the fluid-filled ventricles and passageways that correspond to the central canal of the spinal cord. The gray matter forms *nuclei*—spherical, oval, or irregularly shaped clusters of neuron cell bodies. Although tracts of white matter surround these nuclei, the arrangement is not as predictable as it is in the spinal cord. For example, the tracts may begin, end, merge, or branch as they pass around or through nuclei in their path. In the cerebrum and cerebellum the white matter is covered by **neural cortex** (*cortex*, rind), a superficial layer of gray matter.

The term *higher centers* refers to nuclei, centers, and cortical areas of the cerebrum, cerebellum, diencephalon, and mesencephalon. Output from these processing centers modifies the activities of nuclei and centers in the lower brain stem and spinal cord. The nuclei and cortical areas of the brain can receive sensory information and issue motor commands to peripheral effectors indirectly, through the spinal cord and spinal nerves, or directly through the cranial nerves.

■ The Ventricles of the Brain [FIGURE 15.2]

Ventricles (VEN-tri-kls) are fluid-filled cavities within the brain. They are filled with cerebrospinal fluid and lined by ependymal cells. p. 339 There are four ventricles in the adult brain: one within each cerebral hemisphere, a third within the diencephalon, and a fourth that lies between the pons and cerebellum and extends into the superior portion of the medulla oblongata. The fourth ventricle is continuous with the central canal of the spinal cord. Figure 15.2● shows the position and orientation of the ventricles.

The ventricles in the cerebral hemispheres have a complex shape. A thin medial partition, the **septum pellucidum**, separates this pair of **lateral ventricles**. There is no direct connection between the two lateral ventricles, but each communicates with the ventricle of the diencephalon through an **interventricular foramen** (*foramen of Monro*). Because there are two lateral ventricles (first and second), the cavity within the diencephalon is called the **third ventricle**.

The mesencephalon has a slender canal known as the **aqueduct of the midbrain** (*aqueduct of Sylvius* or *cerebral aqueduct*). This passageway connects the third ventricle with the **fourth ventricle**, which begins between the pons and cerebellum. In the inferior portion of the medulla oblongata, the fourth ventricle narrows and becomes continuous with the central canal of the spinal cord. There is a circulation of cerebrospinal fluid from the ventricles and central canal into the subarachnoid space through foramina in the roof of the fourth ventricle. However, before you can understand the origin and circulation of cerebrospinal fluid, you will need to know more about the organization of the cranial meninges and how they differ from the spinal meninges introduced in Chapter 14. p. 356

✓ CONCEPT CHECK

- List the six major divisions in the adult brain.
- What are the three major structures of the brain stem?
- What are the ventricles? What type of epithelial cell lines them?
- List the secondary brain vesicles and the brain regions associated with each at birth.

Protection and Support of the Brain

The human brain is an extremely delicate organ that must be protected from injury yet remain in touch with the rest of the body. It also has a high demand for nutrients and oxygen and thus an extensive blood supply, yet it must be isolated from compounds in the blood that could interfere with its complex operations. Protection, support, and nourishment of the brain involves (1) the bones of the skull, which were detailed in Chapter 6 (pp. 135–150), (2) the cranial meninges, (3) the cerebrospinal fluid, and (4) the blood-brain barrier.

■ The Cranial Meninges [FIGURE 15.3]

The brain lies cradled within the cranium of the skull, and there is an obvious correspondence between the shape of the brain and that of the cranial

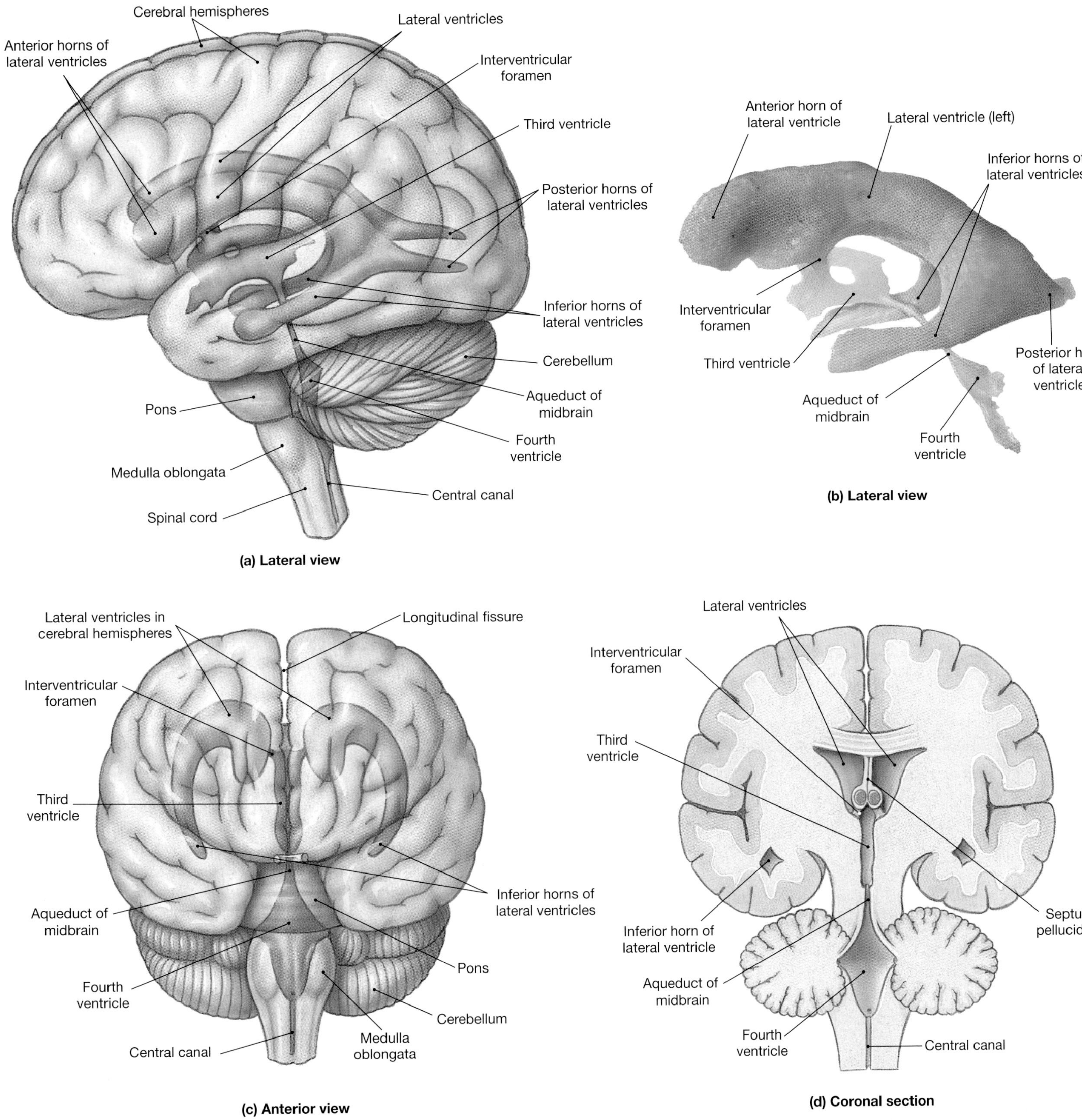

FIGURE 15.2 **Ventricles of the Brain**

These chambers contain cerebrospinal fluid, which transports nutrients, chemical messengers, and waste products. (**a**) Orientation and extent of the ventricles as seen in a lateral view of a transparent brain. (**b**) A lateral view of a plastic cast of the ventricles. (**c**) Anterior view of the ventricles as if seen through a transparent brain. (**d**) Diagrammatic coronal section, showing the interconnections between the ventricles. *See MRI Scans 1 and 2 in the Companion Atlas.*

cavity (Figure 15.3●). The massive cranial bones provide mechanical protection, but they also pose a threat. The brain is like a person driving a car. If the car hits a tree, the car protects the driver from contact with the tree, but serious injury will occur unless a seat belt or airbag protects the driver from contact with the car. *Cranial Trauma p. 794*

Within the cranial cavity, the **cranial meninges** that surround the brain provide this protection, acting as shock absorbers that prevent contact with surrounding bones (Figure 15.3a●). The cranial meninges are continuous with the spinal meninges, and they have the same three layers: *dura mater* (outermost), *arachnoid* (middle), and *pia mater* (innermost). p. 356. However, the cranial meninges have distinctive specializations and functions.

THE DURA MATER [FIGURES 15.3/15.4a,b]

The cranial **dura mater** consists of two fibrous layers. The outermost layer, or *endosteal layer*, is fused to the periosteum lining the cranial bones (Figure 15.3a●). The innermost layer is called the *meningeal layer;* in many areas the endosteal and meningeal layers are separated by a slender gap that contains interstitial fluid and blood vessels, including the large veins known as **dural sinuses**. The veins of the brain open into these sinuses, which in turn deliver that blood to the internal jugular vein of the neck.

At four locations the meningeal layer of the cranial dura mater extends deep into the cranial cavity, providing additional stabilization and support to the brain (Figures 15.3b,c and 15.4●):

- The **falx cerebri** (falks ser-Ē-brē; *falx*, curving, or sickle-shaped) is a fold of dura mater that projects between the cerebral hemispheres in the longitudinal fissure. Its inferior portions attach to the crista galli (anteriorly) and the internal occipital crest and *tentorium cerebelli* (posteriorly). Two large venous sinuses, the **superior sagittal sinus** and the **inferior sagittal sinus**, travel within this dural fold.
- The **tentorium cerebelli** (ten-TOR-ē-um ser-e-BEL-ē; *tentorium*, covering) separates and protects the cerebellar hemispheres from those of the cerebrum. It extends across the cranium at right angles to the falx cerebri. The **transverse sinus** lies within the tentorium cerebelli.
- The **falx cerebelli** extends in the midsagittal line inferior to the tentorium cerebelli, dividing the two cerebellar hemispheres.
- The **diaphragma sellae** is a continuation of the dural sheet that lines the *sella turcica* of the sphenoid (Figure 15.3b●). p. 146 The diaphragma sellae anchors the dura mater to the sphenoid and ensheathes the base of the pituitary gland.

THE ARACHNOID [FIGURE 15.4]

In most anatomical preparations a narrow **subdural space** separates the opposing epithelia of the dura mater and the cranial **arachnoid** (Figure 15.4b●). It is likely, however, that in life no such space exists. The cranial arachnoid covers the brain, providing a smooth surface that does not follow the underlying neural convolutions or *sulci*. Deep to the arachnoid is the **subarachnoid space**, which contains a delicate, weblike meshwork of collagen and elastic fibers that link the arachnoid to the underlying pia mater. Externally, along the axis of the superior sagittal sinus, finger-like extensions of the cranial arachnoid penetrate the dura mater. At these projections, called **arachnoid granulations**, cerebrospinal fluid flows past bundles of fibers (the *arachnoid trabeculae*), crosses the arachnoid membrane, and enters the venous circulation (Figure 15.4b,c●). The cranial arachnoid acts as a roof over the cranial blood vessels, and the underlying pia mater forms a floor. Cerebral arteries and veins are supported by the arachnoid trabeculae and surrounded by cerebrospinal fluid. Blood vessels, surrounded and suspended by arachnoid trabeculae, penetrate the substance of the brain within channels lined by pia mater.

THE PIA MATER [FIGURE 15.4]

The cranial **pia mater** is tightly attached to the surface contours of the brain, anchored by the processes of astrocytes. The cranial pia mater is a highly vascular membrane that acts as a floor to support the large cerebral blood vessels as they branch over the surface of the brain, invading the neural contours to supply superficial areas of neural cortex (Figure 15.4●). An extensive circulatory supply is vital, because the brain requires a constant supply of nutrients and oxygen.

CLINICAL BRIEF

EPIDURAL AND SUBDURAL HEMORRHAGES

A severe head injury may damage meningeal vessels and cause bleeding into the epidural or subdural spaces. The most common cases of epidural bleeding, or **epidural hemorrhage**, involve an arterial break. The arterial blood pressure rapidly forces considerable quantities of blood into the epidural space, distorting the underlying soft tissues of the brain. The individual loses consciousness from minutes to hours after the injury, and death follows in untreated cases.

An epidural hemorrhage involving a damaged vein does not produce massive symptoms immediately, and the individual may become unconscious from several hours to several days or even weeks after the original incident. Consequently, the problem may not be noticed until the nervous tissue has been severely damaged by distortion, compression, and secondary hemorrhaging. Epidural hemorrhages are rare, occurring in fewer than 1% of head injuries. This rarity is rather fortunate, for the mortality rate is 100% in untreated cases and over 50% even after removal of the blood pool and closure of the damaged vessels.

The term **subdural hemorrhage** is somewhat misleading, because blood actually enters the inner layer of the dura, flowing beneath the epithelium that contacts the arachnoid membrane. Subdural hemorrhages are roughly twice as common as epidural hemorrhages. The most common source of blood is a small vein or one of the dural sinuses. Because the blood pressure is somewhat lower than in a typical epidural hemorrhage the extent and effects of the condition may be quite variable. The hemorrhage produces a mass of clotted and partially clotted blood; this mass is called a *hematoma* (hē-ma-TŌ-ma). *Acute subdural hematomas* become symptomatic in minutes to hours after injury. *Chronic subdural hematomas* may produce symptoms weeks, months, or even years after a head injury.

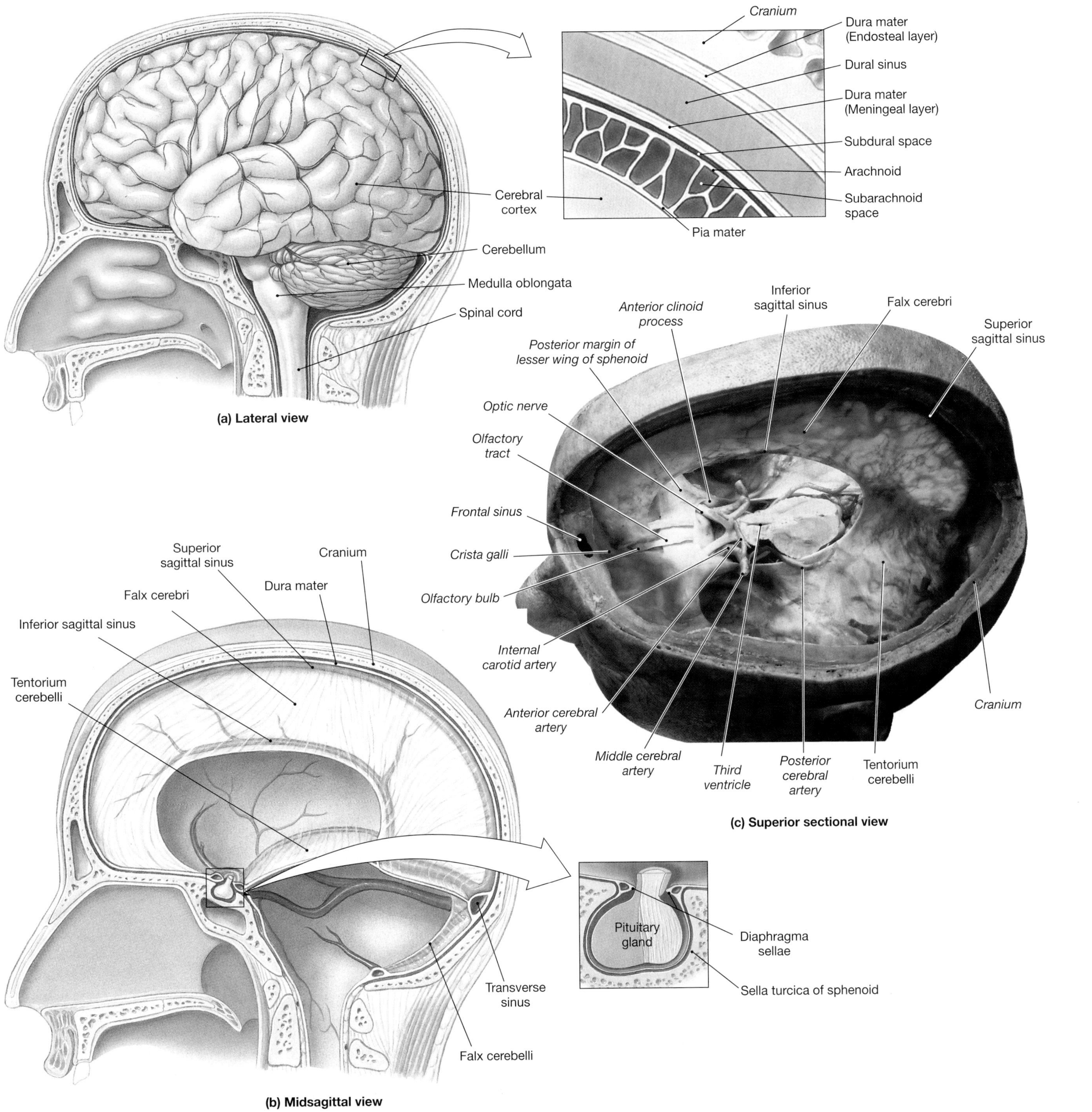

FIGURE 15.3 **Relationships among the Brain, Cranium, and Meninges**

(**a**) Lateral view of the brain, showing its position in the cranium and the organization of the meningeal coverings. (**b**) A corresponding view of the cranial cavity with the brain removed, showing the orientation and extent of the falx cerebri and tentorium cerebelli. (**c**) Superior view of the open cranial cavity with the cerebrum and diencephalon removed. *See MRI Scan 2 in the Companion Atlas.*

FIGURE 15.4 **The Cranial Meninges**

(**a**) A superior view of a dissection of the cranial meninges. (**b**) Organization and relationship of the cranial meninges to the brain. (**c**) A detailed view of the arachnoid membrane, the subarachnoid space, and the pia mater. Note the relationship between the cerebral vein and the subarachnoid space.

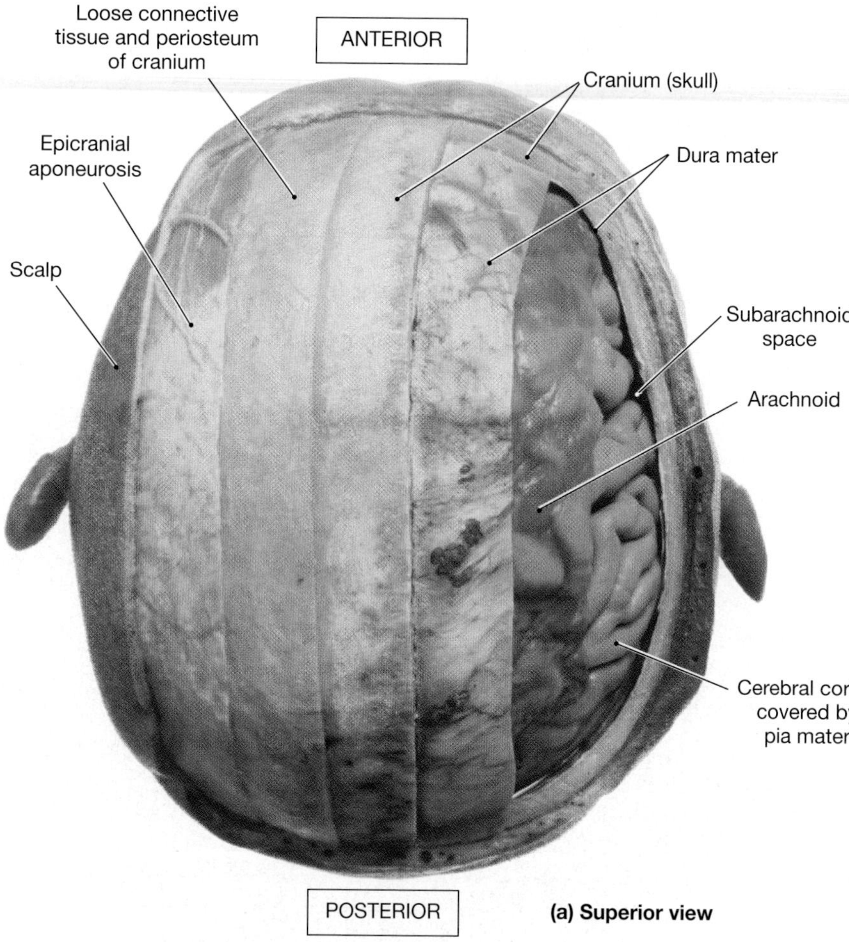

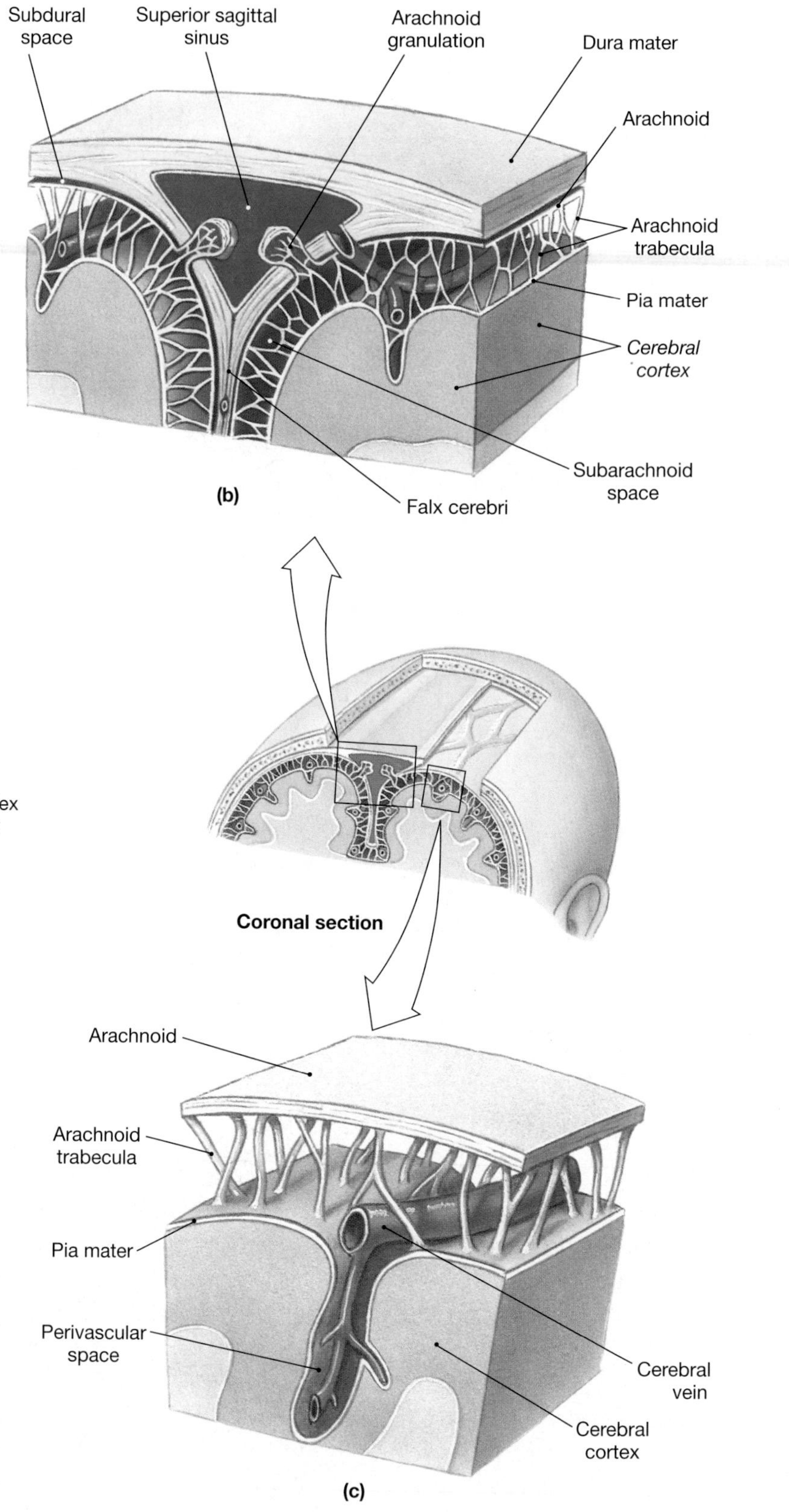

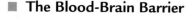

The Blood-Brain Barrier

Neural tissue in the CNS has an extensive blood supply, yet it is isolated from the general circulation by the **blood-brain barrier (BBB)**. This barrier provides a means to maintain a constant environment, which is necessary for both control and proper functioning of CNS neurons. The barrier exists because the endothelial cells lining the capillaries of the CNS are extensively interconnected by tight junctions that prevent the diffusion of materials between adjacent endothelial cells. *p. 43* In general, only lipid-soluble compounds can diffuse across the endothelial cell membranes and into the interstitial fluid of the brain and spinal cord. Water-soluble compounds can cross the capillary walls only through passive or active transport mechanisms. *pp. 31–32* Many different transport proteins are involved, and their activities are quite specific. For example, the transport system that handles glucose is different from those transporting large amino acids. The restricted permeability characteristics of the endothelial lining of brain capillaries is in some way dependent on chemicals secreted by astrocytes. These cells, which are in close contact with CNS capillaries, were described in Chapter 13. *p. 337*

Endothelial transport across the blood-brain barrier is selective and directional. Neurons have a constant need for glucose that must be met regardless of the relative concentrations in the blood and interstitial fluid. Even when circulating glucose levels are low, endothelial cells continue to transport glucose from the blood to the interstitial fluid of the brain. In contrast, the amino acid *glycine* is a neurotransmitter, and its concentration in neural tissue must be kept much lower than that in the circulating blood. Endothelial cells actively absorb this compound from the interstitial fluid of the brain and secrete it into the blood.

The blood-brain barrier remains intact throughout the CNS, with three noteworthy exceptions:

1. In portions of the hypothalamus, the capillary endothelium has an increased permeability, which both exposes hypothalamic nuclei in

the anterior and tuberal regions to circulating hormones, and permits the diffusion of hypothalamic hormones into the circulation.

2. Capillaries in the *pineal gland* are also very permeable. The pineal gland, an endocrine structure, is located in the roof of the diencephalon. The capillary permeability allows pineal secretions into the general circulation.
3. In the membranous roof of both the third and fourth ventricles, the pia mater supports extensive capillary networks that project into the ventricles of the brain. These capillaries are unusually permeable. Substances do not have free access to the CNS, however, because the capillaries are covered by modified ependymal cells that are interconnected by tight junctions. This complex, the *choroid plexus*, is the site of cerebrospinal fluid production.

Cerebrospinal Fluid

Cerebrospinal fluid completely surrounds and bathes the exposed surfaces of the central nervous system. It has several important functions, including:

1. *Cushioning* delicate neural structures.
2. *Supporting the brain*: In essence, the brain is suspended inside the cranium floating in the cerebrospinal fluid. A human brain weighs about 1400 g in air, but it is only a little denser than water; when supported by the cerebrospinal fluid, it weighs only about 50 g.
3. *Transporting nutrients, chemical messengers, and waste products*: Except at the choroid plexus, the ependymal lining is freely permeable, and the CSF is in constant chemical communication with the interstitial fluid of the CNS.

Because free exchange occurs between the interstitial fluid and CSF, changes in CNS function may produce changes in the composition of the CSF. As noted in Chapter 14, a spinal tap can provide useful clinical information concerning CNS injury, infection, or disease. p. 360

FORMATION OF CSF [FIGURE 15.5]

The **choroid plexus** (*choroid*, vascular coat + *plexus*, network) consists of a combination of specialized ependymal cells and permeable capillaries. Two extensive folds of the choroid plexus originate in the roof of the third ventricle and extend through the interventricular foramina into the lateral ventricles. These folds cover the floors of the lateral ventricles (Figure 15.5●). In the lower brain stem, a region of the choroid plexus in the roof of the fourth ventricle projects between the cerebellum and the pons.

The choroid plexus is responsible for the production of cerebrospinal fluid (CSF). The capillaries are highly permeable, but large, highly specialized ependymal cells cover the capillaries and prevent free exchange between those capillaries and the CSF of the ventricles. The ependymal cells use both active and passive transport mechanisms to secrete cerebrospinal fluid into the ventricles. The regulation of CSF composition involves transport in both directions, and the choroid plexus removes waste products from the CSF and fine-tunes its composition over time. There are many differences in composition between cerebrospinal fluid and blood plasma (blood with the cellular elements removed). For example, the blood contains high concentrations of suspended proteins, but the CSF does not. There are also differences in the concentrations of individual ions and in the levels of amino acids, lipids, and waste products. Thus, although CSF is derived from plasma, it is not merely a simple filtrate of blood.

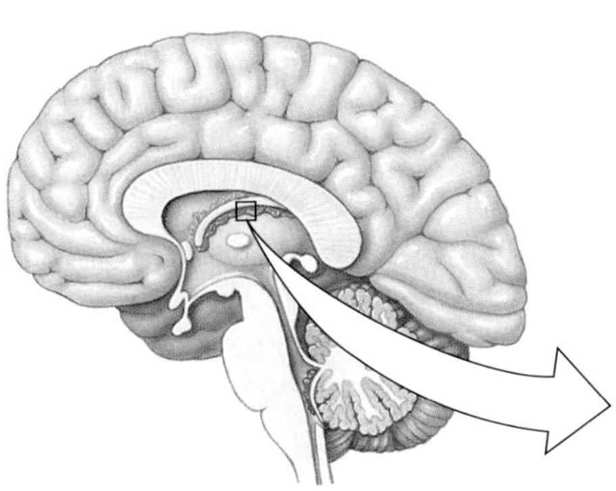

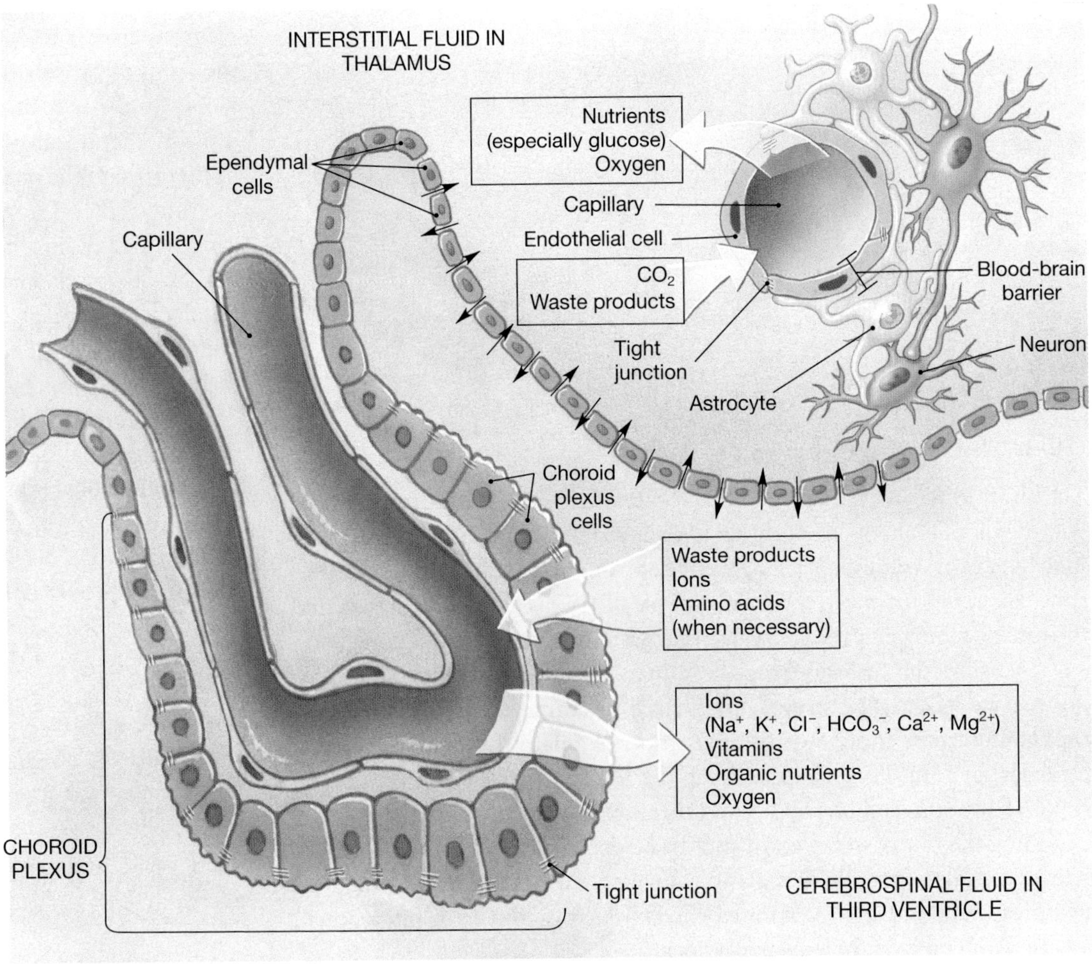

FIGURE 15.5 **The Choroid Plexus and Blood-Brain Barrier**

The choroid plexus consists of a combination of specialized ependymal cells and permeable capillaries in the pia mater. The ependymal cells act as the selective barrier, actively transporting nutrients, vitamins, and ions into the CSF. When necessary, these cells also actively remove ions or compounds from the CSF to stabilize its composition.

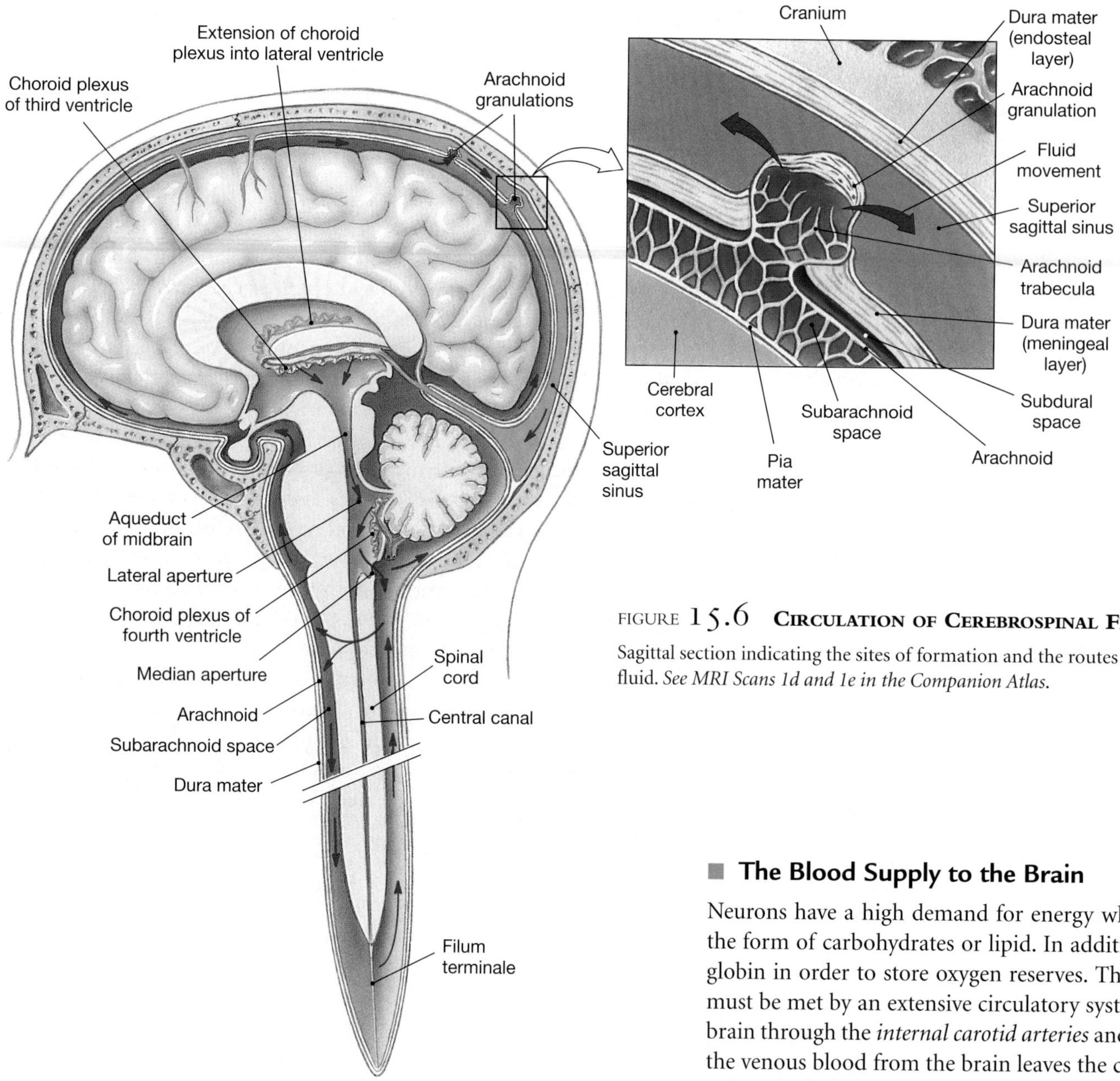

FIGURE 15.6 **CIRCULATION OF CEREBROSPINAL FLUID**

Sagittal section indicating the sites of formation and the routes of circulation of cerebrospinal fluid. *See MRI Scans 1d and 1e in the Companion Atlas.*

CIRCULATION OF CSF [FIGURES 15.4b/15.6]

The choroid plexus produces CSF at a rate of about 500 ml/day. The total volume of CSF at any given moment is approximately 150 ml; this means that the entire volume of CSF is replaced roughly every eight hours. Despite this rapid turnover, the composition of CSF is closely regulated, and the rate of removal normally keeps pace with the rate of production.

Most of the CSF reaching the fourth ventricle enters the subarachnoid space by passing through the paired **lateral apertures** and a single **median aperture** in its membranous roof. (A relatively small quantity of cerebrospinal fluid circulates between the fourth ventricle and the central canal of the spinal cord.) CSF flows through the subarachnoid space surrounding the brain, spinal cord, and cauda equina (Figure 14-4a●, p. 360). It eventually reenters the circulation through the arachnoid granulations (Figures 15.4b and 15.6●). If the normal circulation of CSF is interrupted, a variety of clinical problems may appear.

The Blood Supply to the Brain

Neurons have a high demand for energy while lacking energy reserves in the form of carbohydrates or lipid. In addition, neurons are lacking myoglobin in order to store oxygen reserves. Therefore these energy demands must be met by an extensive circulatory system. Arterial blood reaches the brain through the *internal carotid arteries* and the *vertebral arteries.* Most of the venous blood from the brain leaves the cranium in the *internal jugular veins,* which drain the dural sinuses. A head injury that damages the cerebral blood vessels may cause bleeding into the dura mater, either near the dural epithelium or between the outer layer of the dura mater and the bones of the skull. These are serious conditions, because the blood entering these spaces compresses and distorts the relatively soft tissues of the brain.

Cerebrovascular diseases are circulatory disorders that interfere with the normal blood supply to the brain. The particular distribution of the vessel involved determines the symptoms, and the degree of oxygen or nutrient starvation determines the severity. A **cerebrovascular accident (CVA)**, or *stroke,* occurs when the blood supply to a portion of the brain is shut off. Affected neurons begin to die in a matter of minutes.

✓ CONCEPT CHECK

- Identify the four extensions of the innermost layer of the dura mater into the cranial cavity that provide stabilization and support to the brain.
- Discuss the structure and function of the pia mater.
- What is the function of the blood-brain barrier?
- What is the function of the cerebrospinal fluid? Where is it formed?

CLINICAL DISCUSSION

HYDROCEPHALUS

The adult brain is surrounded by the inflexible bones of the cranium. The cranial cavity contains two fluids—blood and cerebrospinal fluid—and the relatively firm tissues of the brain. Because the total volume cannot change, when the volume of blood or CSF increases, the volume of the brain must decrease. In a subdural or epidural hemorrhage, the fluid volume increases as blood collects within the cranial cavity. p. 384 The rising intracranial pressure compresses the brain, leading to neural dysfunction that often ends in unconsciousness and death.

Any alteration in the rate of cerebrospinal fluid production is normally matched by an increase in the rate of removal at the arachnoid granulations. If this equilibrium is disturbed, clinical problems appear as the intracranial pressure changes. The volume of cerebrospinal fluid will increase if the rate of formation accelerates or the rate of removal decreases. In either event the increased fluid volume leads to compression and distortion of the brain. Increased rates of formation may accompany head injuries, but the most common problems arise from masses, such as tumors or abscesses, or from developmental abnormalities. These conditions have the same effect: They restrict the normal circulation and reabsorption of CSF. Because CSF production continues, the ventricles gradually expand, distorting the surrounding neural tissues and causing the deterioration of brain function.

Infants are especially sensitive to alterations in intracranial pressure, because the arachnoid granulations do not appear until roughly 3 years of age. (Over the interim, CSF is reabsorbed into small vessels within the subarachnoid space and underlying the ependyma.) As in an adult, if intracranial pressure becomes abnormally high, the ventricles will expand. But in an infant, the cranial sutures have yet to fuse, and the skull can enlarge to accommodate the extra fluid volume. This enlargement produces an enormously expanded skull, a condition called **hydrocephalus**, or "water on the brain." Infant hydrocephalus (Figure 15.7●) often results from some interference with normal CSF circulation, such as blockage of the aqueduct of the midbrain or constriction of the connection between the subarachnoid spaces of the cranial and spinal meninges. Untreated infants often suffer some degree of mental retardation. Successful treatment usually involves the installation of a **shunt**, a tube that either bypasses the blockage site or drains the excess cerebrospinal fluid. In either case, the goal is reduction of the intracranial pressure. The shunt may be removed if (1) further growth of the brain eliminates the blockage or (2) the intracranial pressure decreases following the development of the arachnoid granulations at 3 years of age.

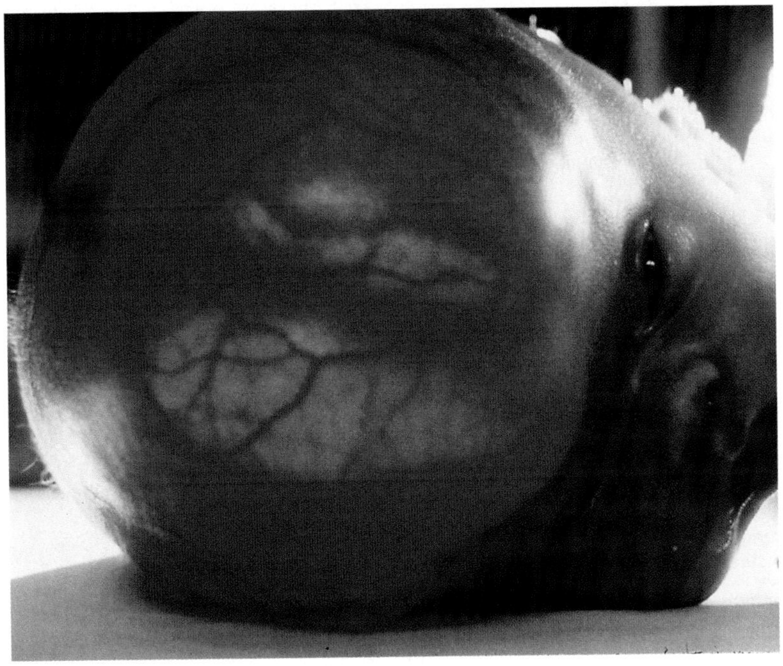

FIGURE 15.7 **HYDROCEPHALUS**

This infant has severe hydrocephalus, a condition usually caused by impaired circulation and removal of cerebrospinal fluid. CSF buildup leads to distortion of the brain and enlargement of the cranium.

The Cerebrum [FIGURES 15.1/15.8/15.9]

The cerebrum is the largest region of the brain. It consists of the paired *cerebral hemispheres*, which rest on the diencephalon and brain stem. Conscious thought processes and all intellectual functions originate in the *cerebral hemispheres*. Much of the cerebrum is involved in the processing of somatic sensory and motor information. Somatic sensory information relayed to the cerebrum reaches our conscious awareness, and cerebral neurons exert direct (voluntary) or indirect (involuntary) control over somatic motor neurons. Most visceral sensory processing and visceral motor (autonomic) control occur at centers elsewhere in the brain, usually outside our conscious awareness. Figures 15.1, p. 383, 15.8, and 15.9● provide additional perspective on the cerebrum and its relationships with other regions of the brain.

The Cerebral Hemispheres [FIGURES 15.8/15.9]

A thick blanket of neural cortex (superficial gray matter) covers the paired **cerebral hemispheres** that form the superior and lateral surfaces of the cerebrum (Figures 15.8 and 15.9●). The cortical surface forms a series of elevated ridges, or **gyri** (JĪ-rī), separated by shallow depressions, called **sulci** (SUL-sī), or deeper grooves, called **fissures.** These features increase the surface area of the cerebral hemispheres and provide space for additional cortical neurons. The cerebral cortex performs the most complicated neural functions, and analytical and integrative activities require large numbers of neurons. The brain and cranium have both enlarged in the course of human evolution, but the cerebral cortex has grown out of proportion to the rest of the brain. The total surface area of the cerebral hemispheres is roughly equivalent to 2200 cm^2 (2.5 ft^2) of flat surface, and that large an area can be packed into the skull only when folded, like a crumpled piece of paper.

THE CEREBRAL LOBES [FIGURES 15.8/15.9]

The two cerebral hemispheres are separated by a deep **longitudinal fissure** (Figure 15.8a,b●), and each hemisphere can be divided into **lobes** named after the overlying bones of the skull (Figure 15.9a●). There are individual differences in the appearance of the sulci and gyri of each brain, but the boundaries between lobes are reliable landmarks. A deep groove, the **central sulcus**, extends laterally from the longitudinal fissure. The area anterior to the central sulcus is the **frontal lobe**, and the **lateral sulcus** marks its inferior border. The region inferior to the lateral sulcus is the **temporal lobe.** Reflecting this lobe to the side (Figure 15.9b●) exposes the **insula** (IN-sū-la), an "island" of cortex that is otherwise hidden. The **parietal lobe** extends posteriorly from the central sulcus to the **parieto-occipital sulcus.** The region posterior to the parieto-occipital sulcus is the **occipital lobe.**

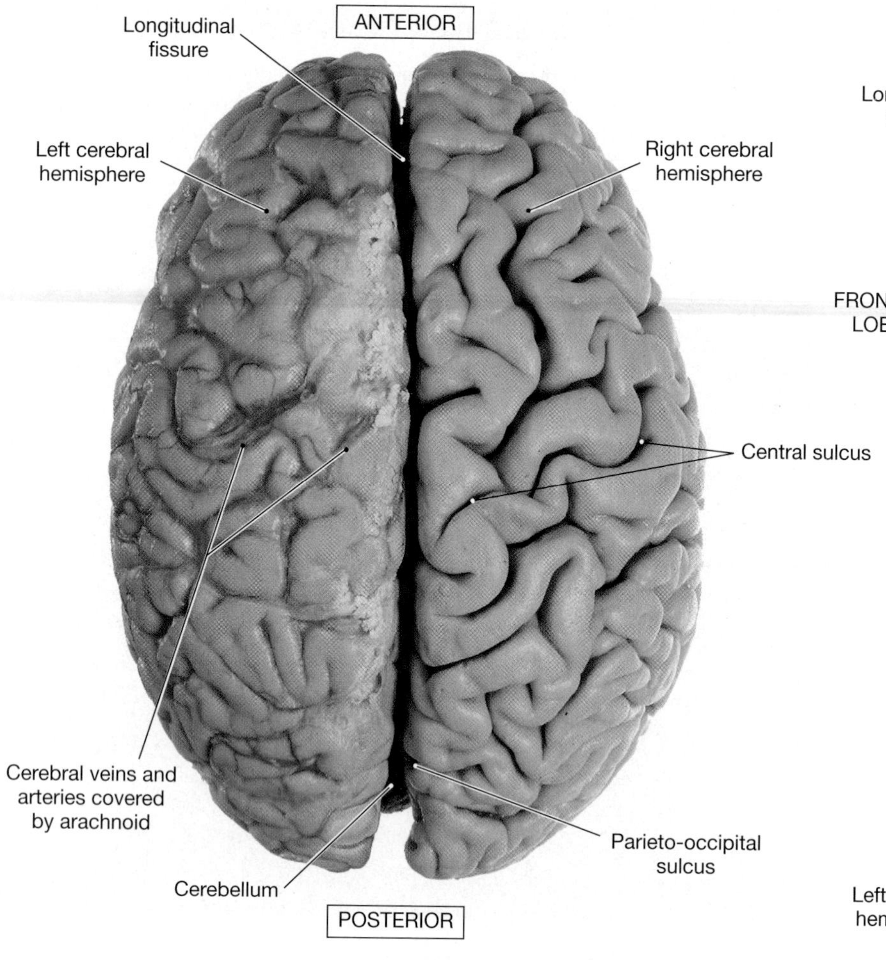

(a) Superior view

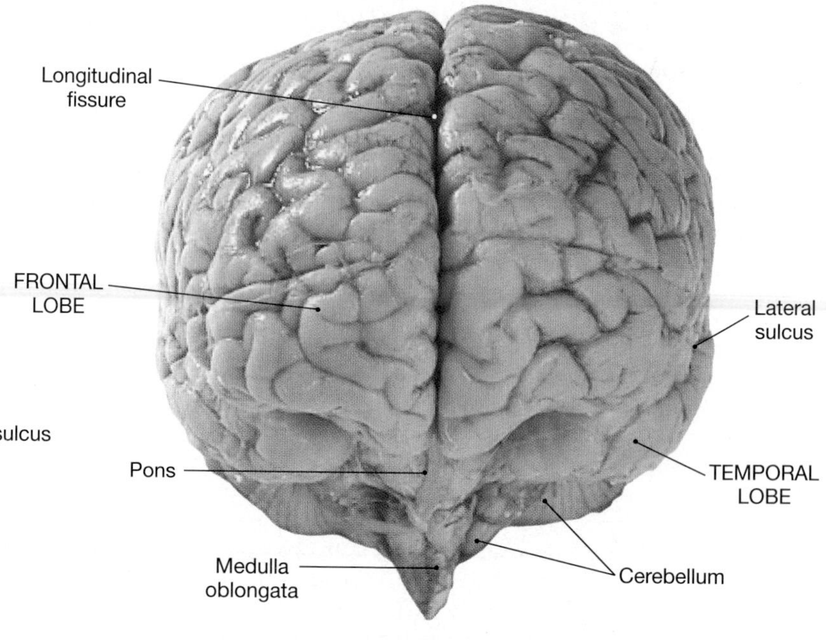

(b) Anterior view

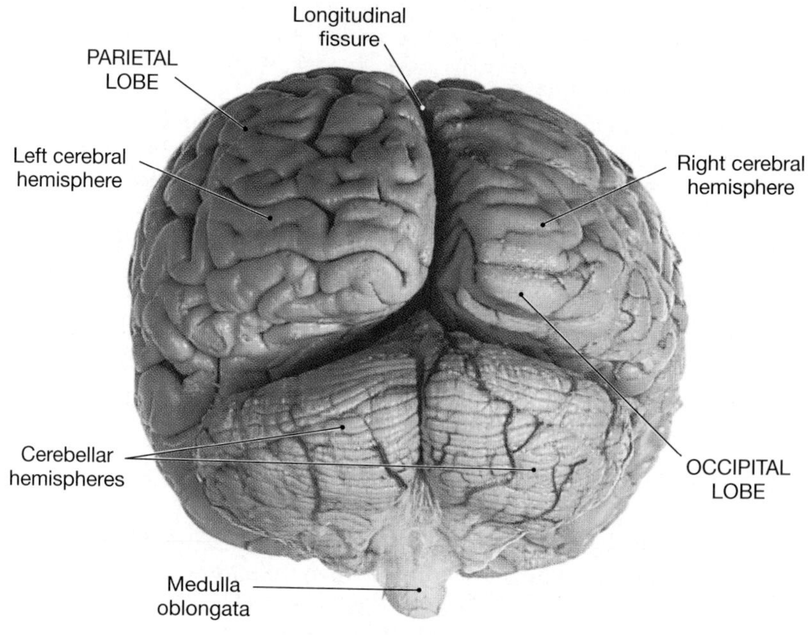

(c) Posterior view

FIGURE 15.8 **THE CEREBRAL HEMISPHERES, PART I**

The cerebral hemispheres are the largest part of the adult brain. (**a**) Superior view. (**b**) Anterior view. (**c**) Posterior view. Note the relatively small size of the cerebellar hemispheres.

Each lobe contains functional regions whose boundaries are less clearly defined. Some of these functional regions process sensory information, while others are responsible for motor commands. Three points about the cerebral lobes should be kept in mind:

1. *Each cerebral hemisphere receives sensory information from and generates motor commands to the opposite side of the body.* The left hemisphere controls the right side, and the right hemisphere controls the left side. This crossing over has no known functional significance.
2. *The two hemispheres have some functional differences, although anatomically they appear to be identical.* These differences affect primarily higher-order functions, a topic that will be discussed in Chapter 16.
3. *The assignment of a specific function to a specific region of the cerebral cortex is imprecise.* Because the boundaries are indistinct, with considerable overlap, any one region may have several different functions. Some aspects of cortical function, such as consciousness, cannot easily be assigned to any single region.

Our understanding of brain function is still incomplete, and not every anatomical feature has a known function. However, it is clear from studies on metabolic activity and blood flow that all portions of the brain are used in a normal individual.

MOTOR AND SENSORY AREAS OF THE CEREBRAL CORTEX

[FIGURE 15.9b AND TABLE 15.2]

Conscious thought processes and all intellectual functions originate in the cerebral hemispheres. However, much of the cerebrum is involved with the processing of somatic sensory and motor information. The major motor and sensory regions of the cerebral cortex are detailed in Figure 15.9b● and Table 15.2. The central sulcus separates the motor and sensory portions of the cortex. The **precentral gyrus** of the frontal lobe forms the anterior

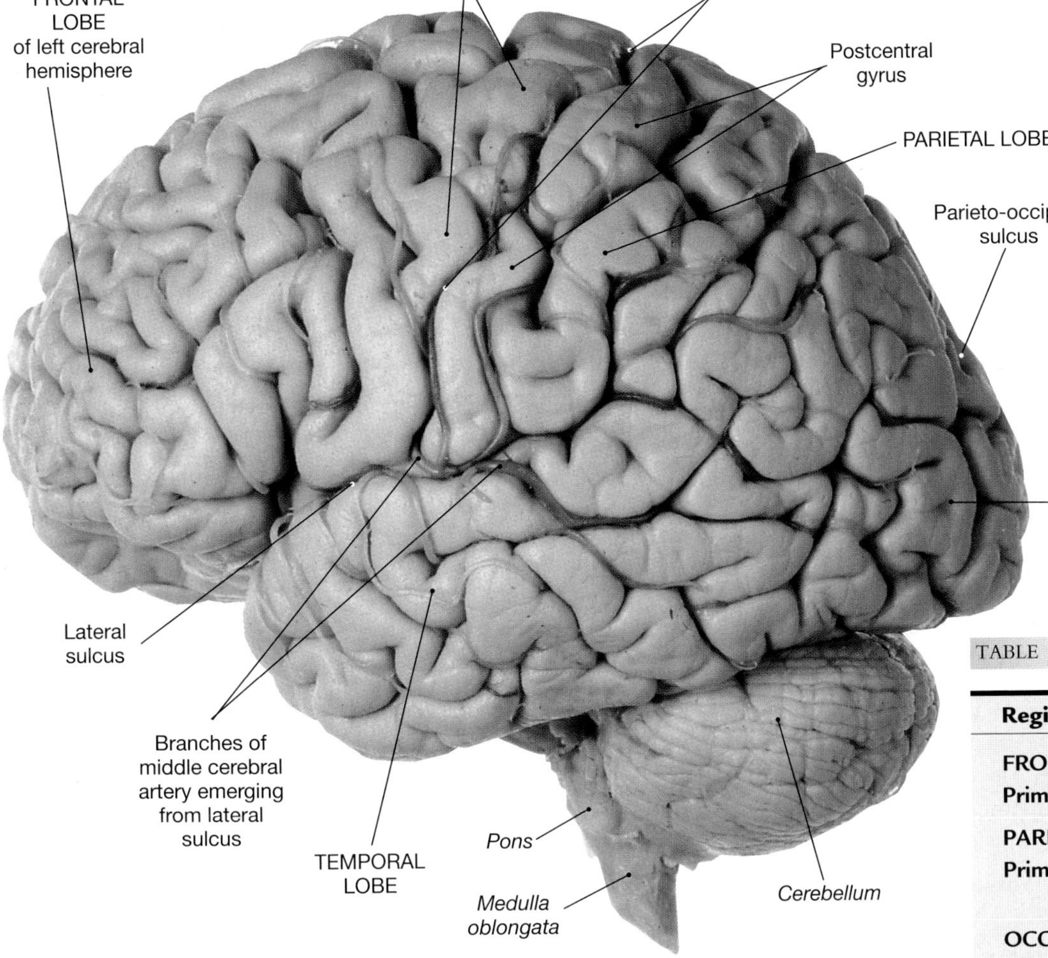

(a) Lateral view

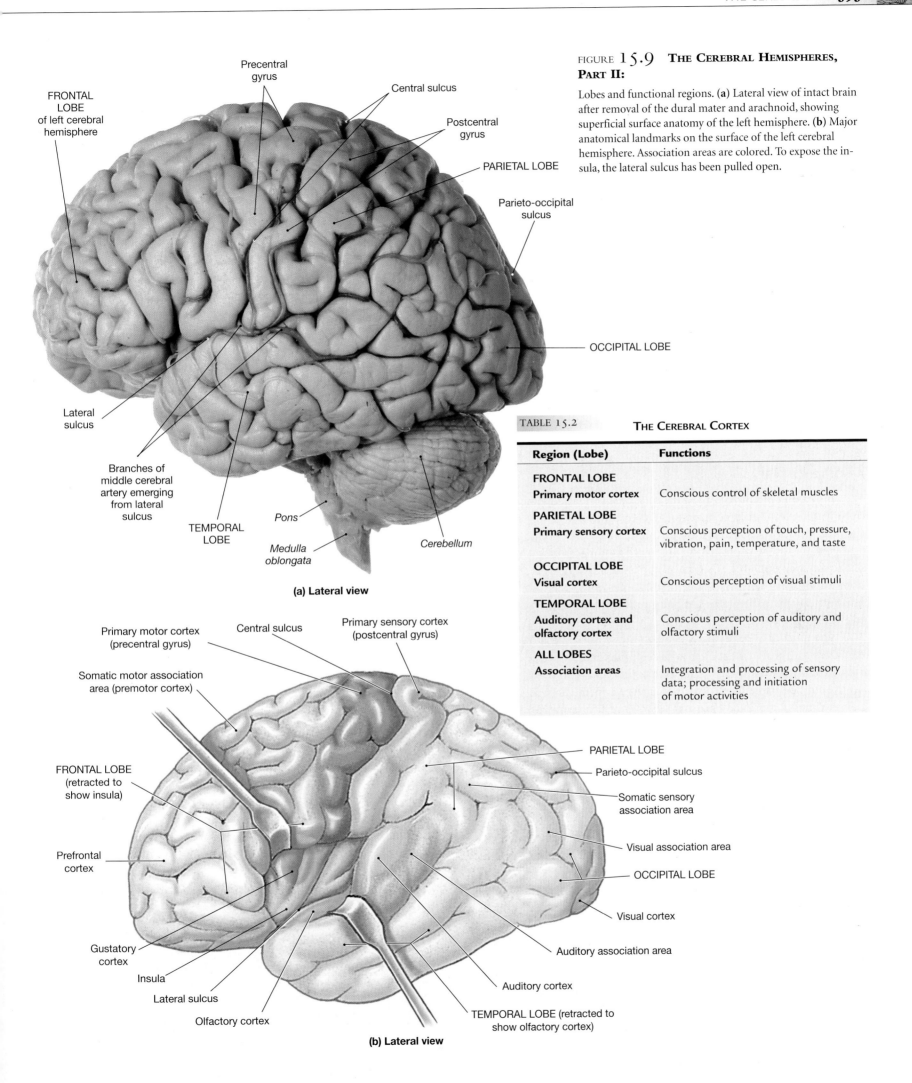

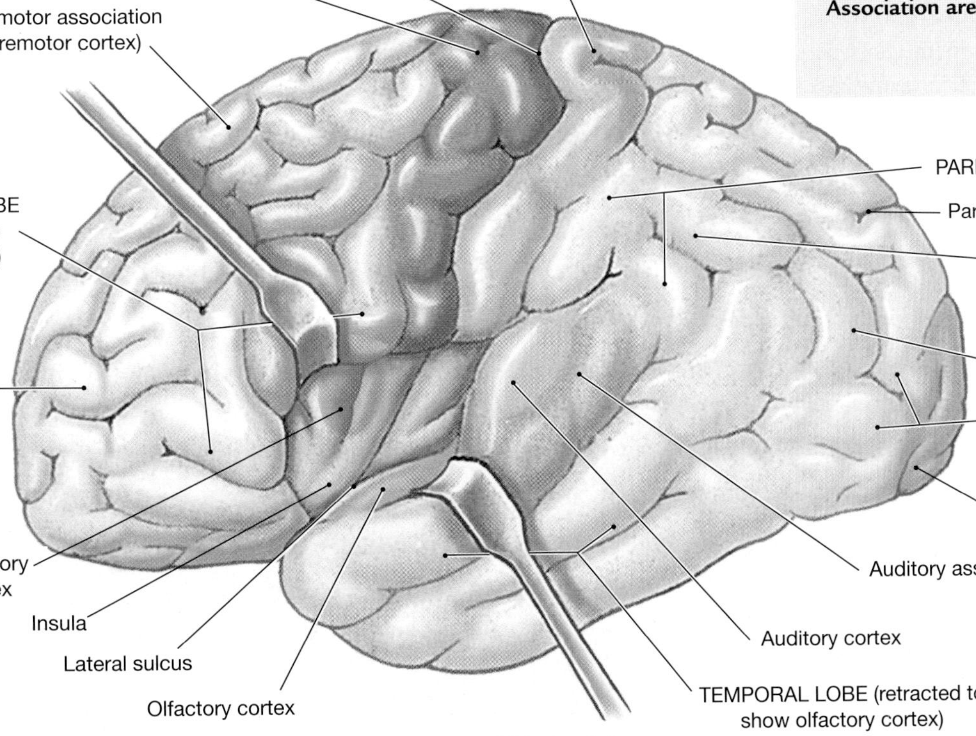

(b) Lateral view

FIGURE 15.9 **THE CEREBRAL HEMISPHERES, PART II:**

Lobes and functional regions. **(a)** Lateral view of intact brain after removal of the dural mater and arachnoid, showing superficial surface anatomy of the left hemisphere. **(b)** Major anatomical landmarks on the surface of the left cerebral hemisphere. Association areas are colored. To expose the insula, the lateral sulcus has been pulled open.

TABLE 15.2 THE CEREBRAL CORTEX

Region (Lobe)	Functions
FRONTAL LOBE **Primary motor cortex**	Conscious control of skeletal muscles
PARIETAL LOBE **Primary sensory cortex**	Conscious perception of touch, pressure, vibration, pain, temperature, and taste
OCCIPITAL LOBE **Visual cortex**	Conscious perception of visual stimuli
TEMPORAL LOBE **Auditory cortex and olfactory cortex**	Conscious perception of auditory and olfactory stimuli
ALL LOBES **Association areas**	Integration and processing of sensory data; processing and initiation of motor activities

margin of the central sulcus. The surface of this gyrus is the **primary motor cortex**. Neurons of the primary motor cortex direct voluntary movements by controlling somatic motor neurons in the brain stem and spinal cord. The neurons of the primary motor cortex are called **pyramidal cells**, and the pathway that provides voluntary motor control is known as the **corticospinal pathway** or *pyramidal system*.

The **postcentral gyrus** of the parietal lobe forms the posterior margin of the central sulcus, and its surface contains the **primary sensory cortex**. Neurons in this region receive somatic sensory information from touch, pressure, pain, taste, and temperature receptors. We are consciously aware of these sensations because the sensory information has been relayed to the primary sensory cortex. At the same time, collaterals deliver information to the basal nuclei and other centers. As a result, sensory information is monitored at both conscious and unconscious levels.

Sensory information concerning sensations of sight, sound, and smell arrive at other portions of the cerebral cortex. The **visual cortex** of the occipital lobe receives visual information, and the **auditory cortex** and **olfactory cortex** of the temporal lobe receive information concerned with hearing and smelling, respectively. The **gustatory cortex** lies in the anterior portion of the insula and adjacent portions of the frontal lobe. This region receives information from taste receptors of the tongue and pharynx. The regions of the cerebral cortex involved with special sensory information are shown in Figure 15.9b•.

ASSOCIATION AREAS [FIGURE 15.9b]

The sensory and motor regions of the cortex are connected to nearby **association areas** that interpret incoming data or coordinate a motor response (Figure 15.9b•). The **somatic motor association area**, or **premotor cortex**, is responsible for the coordination of learned motor activities. The **somatic sensory association area** posterior to the postcentral gyrus receives input from the primary sensory cortex, the thalamus, and other brain regions. This cortical region integrates and interprets sensations concerning the size and shape of objects.

The functional distinctions between the sensory and motor association areas are most evident after localized brain damage. For example, an individual with a damaged **visual association area** may see letters quite clearly, but be unable to recognize or interpret them. This person would scan the lines of a printed page and see rows of clear symbols that convey no meaning. Someone with damage to the area of the premotor cortex concerned with coordination of eye movements can understand written letters and words but cannot read, because his or her eyes cannot follow the lines on a printed page.

INTEGRATIVE CENTERS [FIGURE 15.9b]

Integrative centers receive and process information from many different association areas. These regions direct extremely complex motor activities and perform complicated analytical functions. For example, the **prefrontal cortex** of the frontal lobe (Figure 15.9b•) integrates information from sensory association areas and performs abstract intellectual functions, such as predicting the consequences of possible responses.

These lobes and cortical areas are found on both cerebral hemispheres. Higher-order integrative centers concerned with complex processes, such as speech, writing, mathematical computation, or understanding spatial relationships, are restricted to the left or right hemisphere. These centers and their functions are described in Chapter 16.

The Central White Matter [FIGURE 15.10 AND TABLE 15.3]

The **central white matter** is covered by the gray matter of the cerebral cortex (Figure 15.10•). It contains myelinated fibers that form bundles that extend from one cortical area to another or that connect areas of the cortex to other regions of the brain. These bundles include: (1) *association fibers*, tracts that interconnect areas of neural cortex within a single cerebral hemisphere; (2) *commissural fibers*, tracts that connect the two cerebral hemispheres; and (3) *projection fibers*, tracts that link the cerebrum with other regions of the brain and the spinal cord. The names and functions of these groups are summarized in Table 15.3.

Association fibers interconnect portions of the cerebral cortex within the same cerebral hemisphere. The shortest association fibers are called **arcuate** (AR-kū-āt) **fibers** because they curve in an arc to pass from one gyrus to another. The longer association fibers are organized into discrete bundles. The **longitudinal fasciculi** connect the frontal lobe to the other lobes of the same hemisphere.

A dense band of **commissural fibers** (kom-MIS-ū-ral; *commissura*, a crossing over) permit communication between the two hemispheres. Prominent commissural bundles linking the cerebral hemispheres include the **corpus callosum** and the **anterior commissure**.

Projection fibers link the cerebral cortex to the diencephalon, brain stem, cerebellum, and spinal cord. All ascending and descending axons must pass through the diencephalon on their way to or from sensory, motor, or association areas of the cerebral cortex. In gross dissection the afferent fibers and efferent fibers look alike, and the entire collection of fibers is known as the **internal capsule**.

The Basal Nuclei [FIGURE 15.11 AND TABLE 15.4]

The **basal nuclei** are paired masses of gray matter within the cerebral hemispheres[2]. These nuclei lie within each hemisphere inferior to the floor of the lateral ventricle (Figure 15.11•). They are embedded within the central white matter, and the radiating projection and commissural fibers travel around or between these nuclei.

The **caudate nucleus** has a massive head and a slender, curving tail that follows the curve of the lateral ventricle. At the tip of the tail there is a separate nucleus, the **amygdaloid body** (ah-MIG-da-loyd; *amygdale*, almond). Three masses of gray matter lie between the bulging surface of the insula and the lateral wall of the diencephalon. These are the **claustrum** (KLAWS-trum), the **putamen** (pū-TĀ-men), and the **globus pallidus** (GLŌ-bus PAL-i-dus; pale globe).

Several additional terms are used to designate specific anatomical or functional subdivisions of the cerebral nuclei. The putamen and globus pallidus are often considered as subdivisions of a larger **lentiform** (lens-shaped) **nucleus**, for when exposed on gross dissection, they form a rather compact, rounded mass (Figure 15.11b,d,e•). The term *corpus striatum* is sometimes used to refer to the caudate *and* lentiform nuclei, or to the caudate nucleus and the putamen. Table 15.4 summarizes these relationships and the functions of the basal nuclei.

FUNCTIONS OF THE BASAL NUCLEI

The basal nuclei are involved with the subconscious control of skeletal muscle tone and the coordination of learned movement patterns. Under

[2] These have also been called the cerebral nuclei or the basal ganglia.

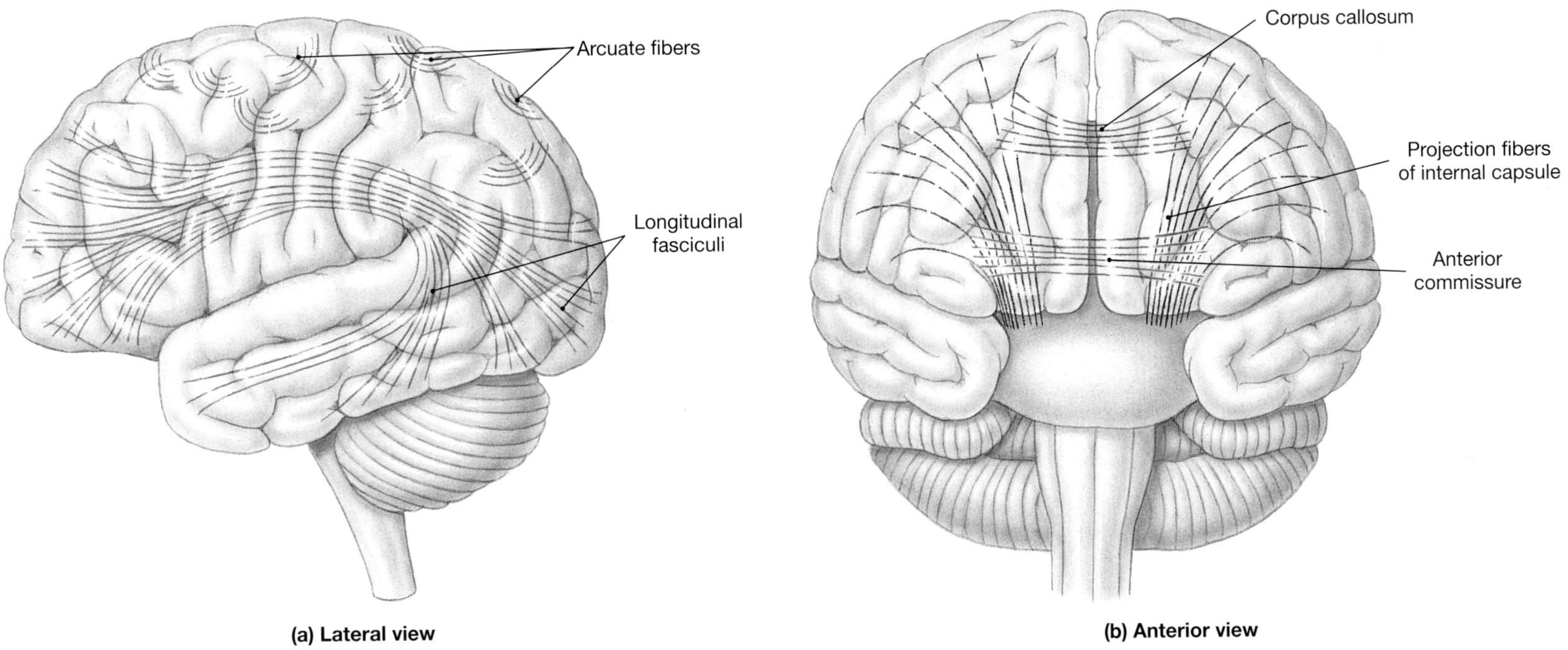

FIGURE 15.10 **THE CENTRAL WHITE MATTER**

Shown are the major groups of axon fibers and tracts of the central white matter. (**a**) Lateral aspect of the brain, showing arcuate fibers and longitudinal fasciculi. (**b**) Anterior view of the brain, showing orientation of the commissural and projection fibers.

TABLE 15.3 WHITE MATTER OF THE CEREBRUM

Fibers/Tracts	Functions
Association fibers	Interconnect cortical areas within the same hemisphere
Arcuate fibers	Interconnect gyri within a lobe
Longitudinal fasciculi	Interconnect the frontal lobe with other cerebral lobes
Commissural fibers (anterior commissure and corpus callosum)	Interconnect corresponding lobes of different hemispheres
Projection fibers	Connect cerebral cortex to diencephalon, brain stem, cerebellum, and spinal cord

normal conditions, these nuclei do not initiate particular movements. But once a movement is under way, the basal nuclei provide the general pattern and rhythm, especially for movements of the trunk and proximal limb muscles. (This system will be discussed further in Chapter 16.) Some functions assigned to specific basal nuclei are detailed below.

CAUDATE NUCLEUS When walking, the caudate nucleus and putamen control the cycles of arm and leg movements that occur between the time the decision is made to "start walking" and the time the "stop" order is given.

CLAUSTRUM AND AMYGDALOID BODY The claustrum appears to be involved in the processing of visual information at the subconscious level. Evidence suggests that it focuses attention on specific patterns or relevant features. The amygdaloid body is an important component of the *limbic system* and will be considered in the next section. The functions of other basal nuclei are poorly understood.

GLOBUS PALLIDUS The globus pallidus controls and adjusts muscle tone, particularly in the appendicular muscles, to set body position in preparation for a voluntary movement. For example, when you decide to pick up a pencil, the globus pallidus positions the shoulder and stabilizes the arm as you consciously reach and grasp with the forearm, wrist, and hand.

The Limbic System [FIGURES 15.11/15.12 AND TABLE 15.5]

The **limbic system** (LIM-bik; *limbus*, border) includes nuclei and tracts along the border between the cerebrum and diencephalon. The functions of the limbic system include: (1) establishment of emotional states and related behavioral drives, (2) linking the conscious, intellectual functions of the cerebral cortex with the unconscious and autonomic functions of other portions of the brain, and (3) facilitating memory storage and retrieval. This system is a functional grouping rather than an anatomical one, and the limbic system includes components of the cerebrum, diencephalon, and mesencephalon (Table 15.5).

The amygdaloid body (Figures 15.11a,b and 15.12●) appears to act as an integration center between the limbic system, the cerebrum, and various sensory systems. The **limbic lobe** of the cerebral hemisphere consists of the gyri and deeper structures that are adjacent to the diencephalon. The **cingulate gyrus** (SIN-gū-lāt; *cingulum*, girdle or belt) sits superior to the corpus callosum. The **dentate gyrus** and the adjacent **parahippocampal** (pa-ra-hip-ō-KAM-pal) **gyrus** conceal an underlying nucleus, the **hippocampus**, which lies beneath the floor of the lateral ventricle (see Figures 15.11e, 15.12b, and 15.13a●). Early anatomists thought this nucleus resembled a seahorse (*hippocampus*); it plays an essential role in learning and the storage of long-term memories.

FIGURE 15.11 **The Basal Nuclei**

(**a**) Lateral view showing the relative positions of the basal nuclei. Compare this three-dimensional representation with the frontal sections (**b, c**) and the horizontal section (**d**). The frontal section in (c) was taken slightly anterior to the plane of the diagrammatic section. As a result, the amygdaloid body cannot be seen. (**e**) Diagrammatic view of the brain dissected to show the orientation of cerebral and thalamic structures. *See MRI Scans 1a, 1b, and 2 in the Companion Atlas.*

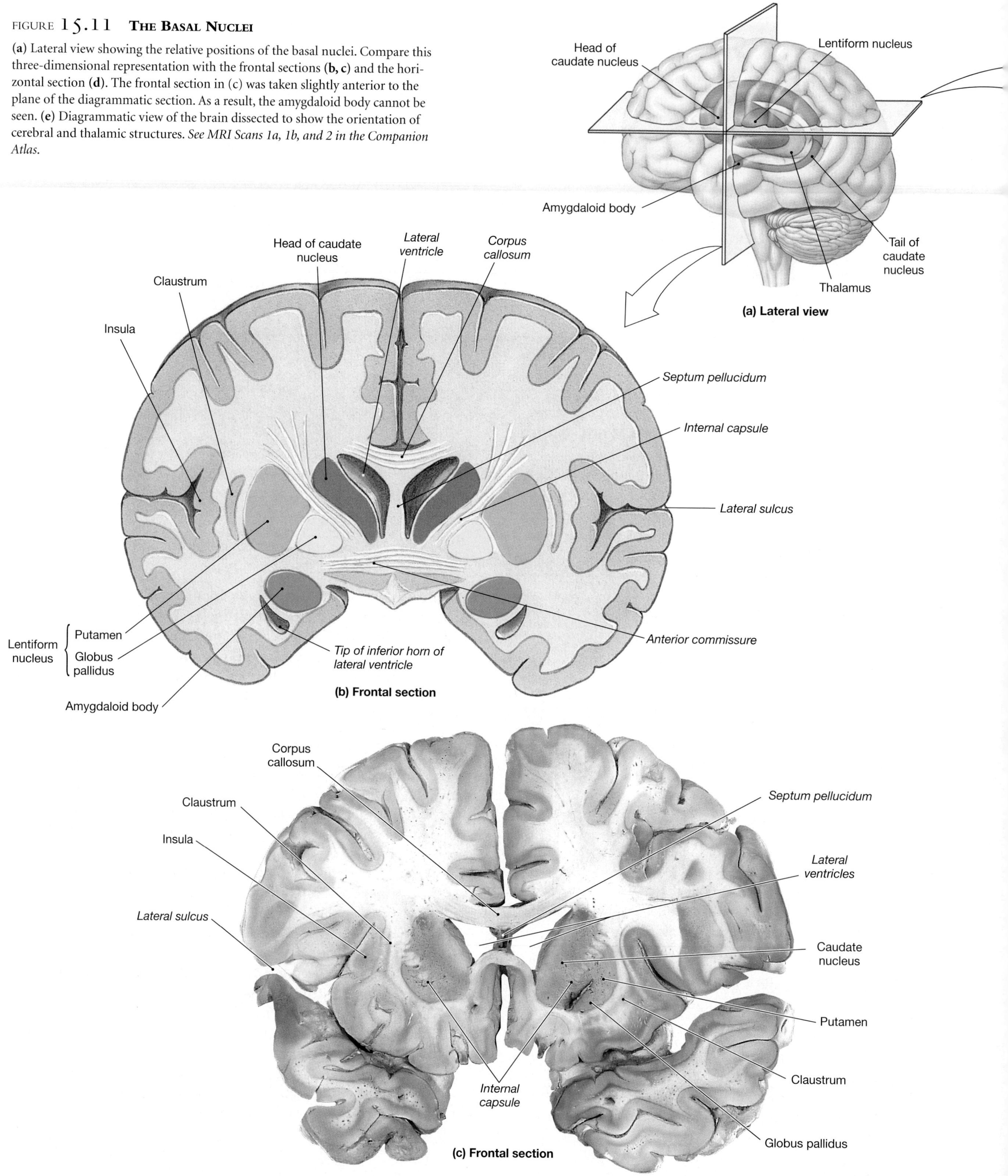

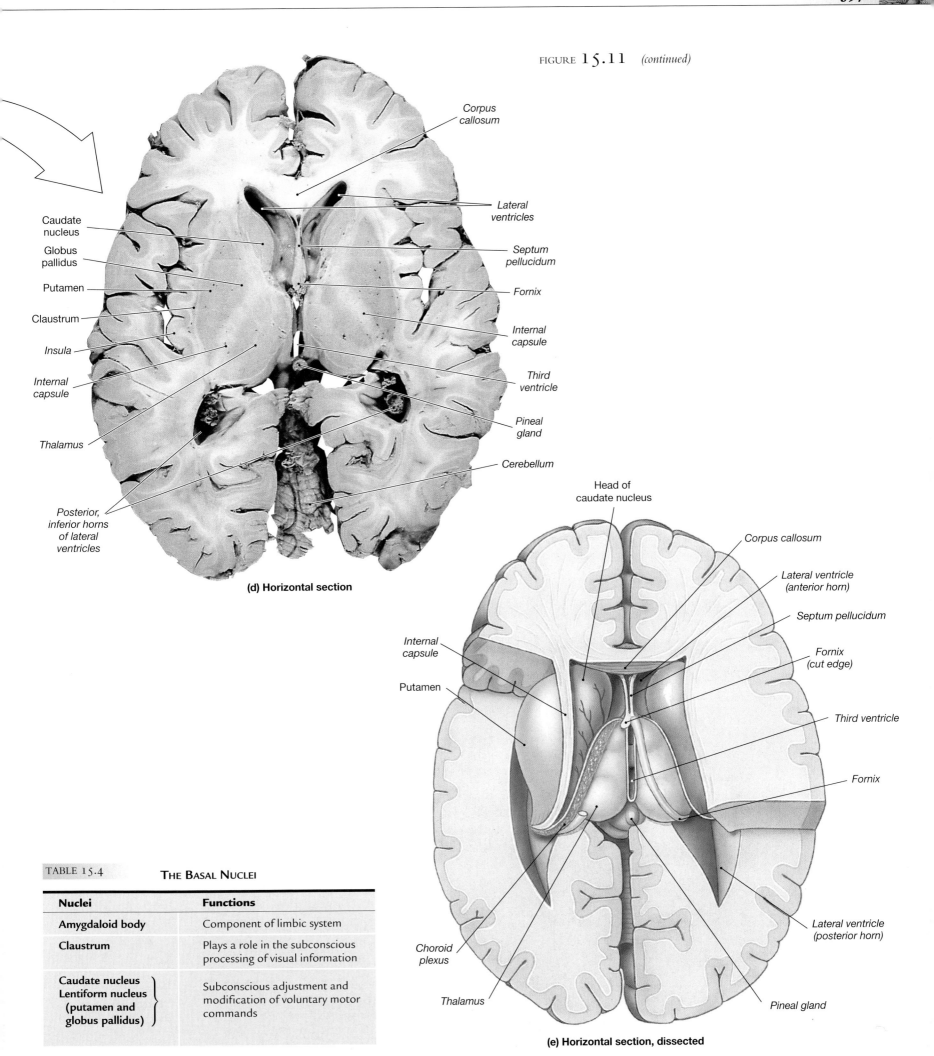

FIGURE 15.11 *(continued)*

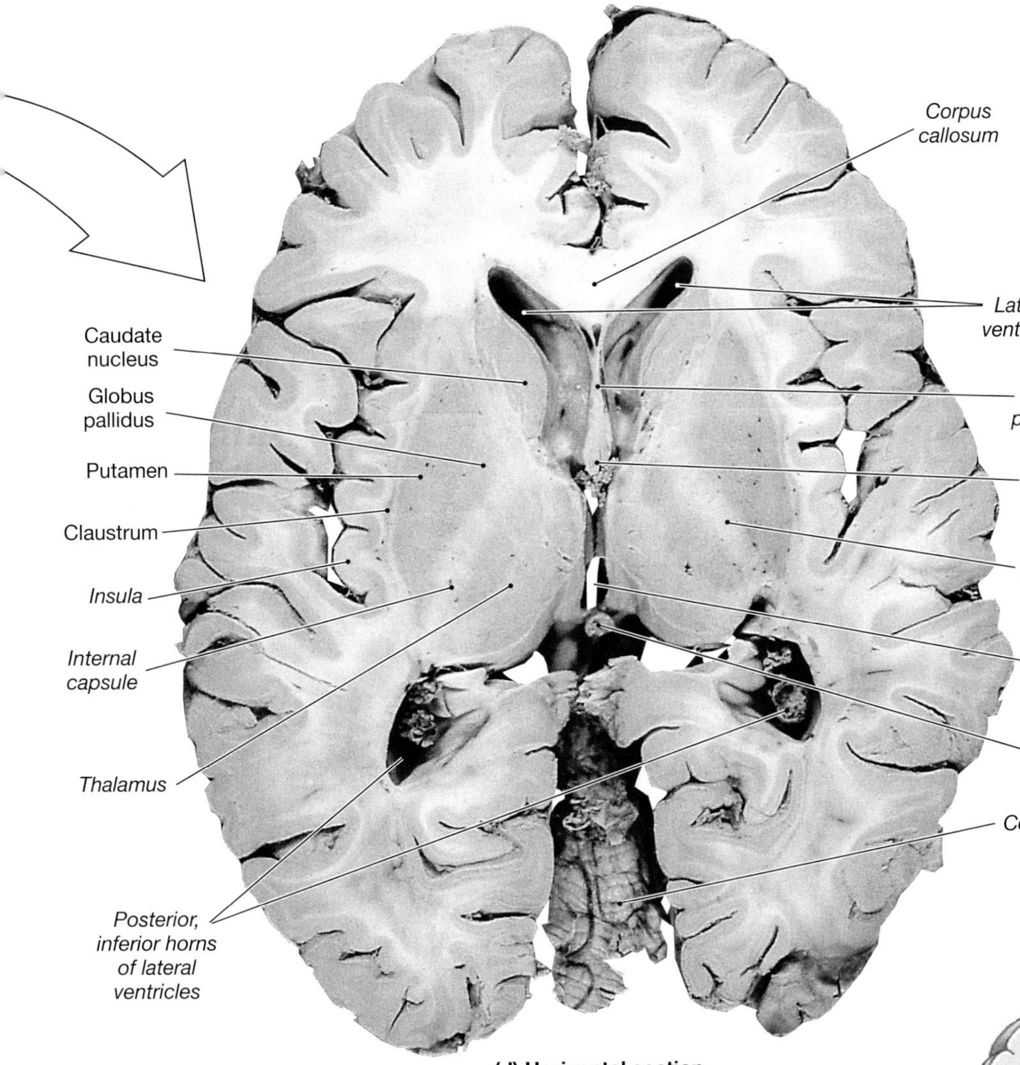

(d) Horizontal section

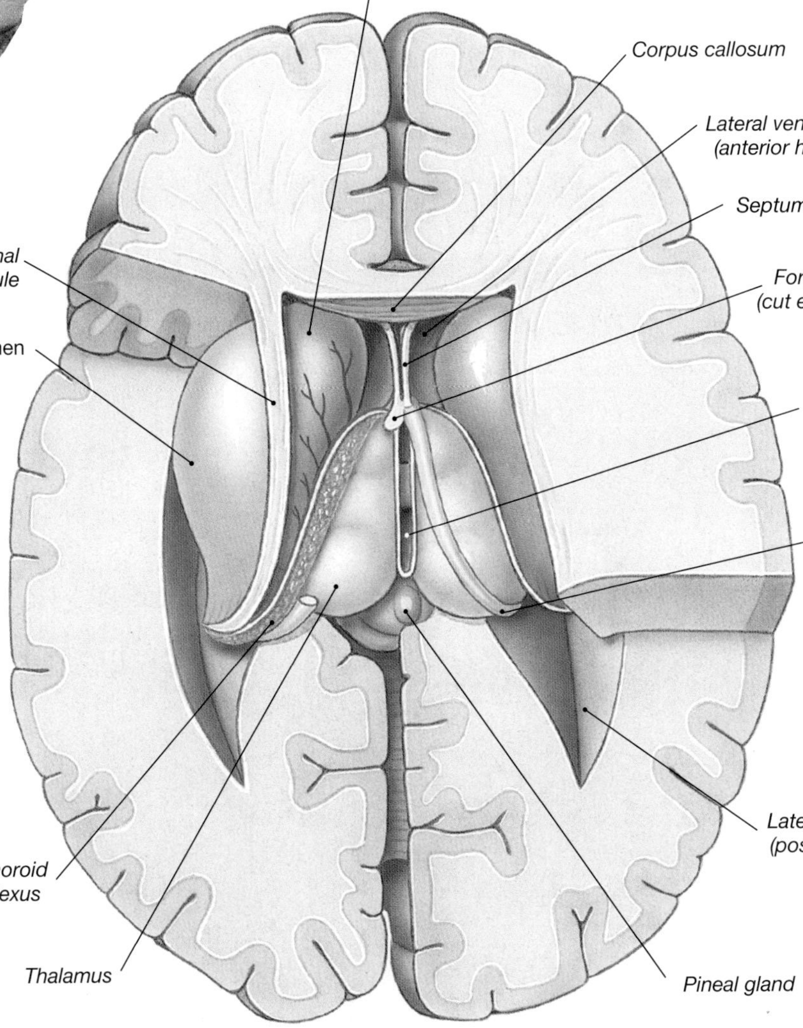

(e) Horizontal section, dissected

TABLE 15.4 **THE BASAL NUCLEI**

Nuclei	Functions
Amygdaloid body	Component of limbic system
Claustrum	Plays a role in the subconscious processing of visual information
Caudate nucleus **Lentiform nucleus (putamen and globus pallidus)**	Subconscious adjustment and modification of voluntary motor commands

The **fornix** (FŌR-niks) is a tract of white matter that connects the hippocampus with the hypothalamus. From the hippocampus the fornix curves medially and superiorly to the corpus callosum, and then forms an arch that curves anteriorly, ending in the hypothalamus. Many of the fibers end in the **mamillary bodies** (MAM-i-lar-ē; *mamilla*, or *mammilla*, breast), prominent nuclei in the floor of the hypothalamus. The mamillary bodies contain motor nuclei that control reflex movements associated with eating, such as chewing, licking, and swallowing.

Several other nuclei in the wall (thalamus) and floor (hypothalamus) of the diencephalon are components of the limbic system. Among its other functions, the **anterior nucleus** of the thalamus relays visceral sensations from the hypothalamus to the cingulate gyrus. Experimental stimulation of the hypothalamus has outlined a number of important centers responsible for the emotions of rage, fear, pain, sexual arousal, and pleasure.

Stimulation of the hypothalamus can also produce heightened alertness and a generalized excitement. This response is caused by widespread stimulation of the **reticular formation**, an interconnected network of brain stem nuclei whose dominant nuclei lie within in the mesencephalon. Stimulation of adjacent portions of the hypothalamus or thalamus will depress reticular activity, resulting in generalized lethargy or actual sleep.

TABLE 15.5 **The Limbic System**

FUNCTIONS	Processing of memories, creation of emotional states, drives, and associated behaviors
CEREBRAL COMPONENTS	
Cortical areas	Limbic lobe (cingulate gyrus, dentate gyrus, and parahippocampal gyrus)
Nuclei	Hippocampus, amygdaloid body
Tracts	Fornix
DIENCEPHALIC COMPONENTS	
Thalamus	Anterior nuclear group
Hypothalamus	Centers concerned with emotions, appetites (thirst, hunger), and related behaviors (Table 15.7)
OTHER COMPONENTS	
Reticular formation	Network of interconnected nuclei throughout brain stem

✓ **CONCEPT CHECK**

- Each cerebral hemisphere is subdivided into lobes. Identify the lobes and their general functions.
- What are gyri and sulci?
- List and describe the three major groups of axons in central white matter.

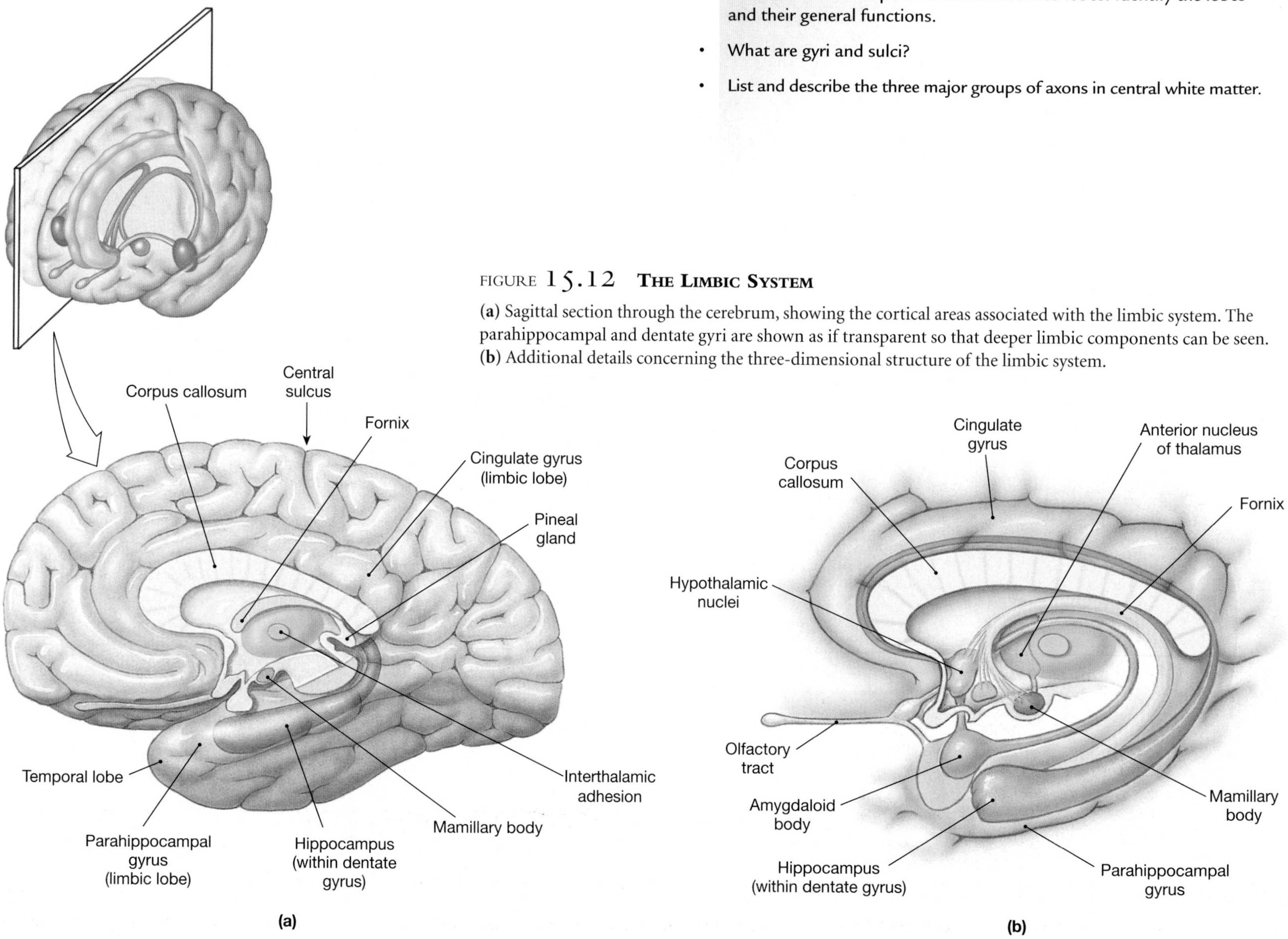

FIGURE 15.12 **The Limbic System**

(**a**) Sagittal section through the cerebrum, showing the cortical areas associated with the limbic system. The parahippocampal and dentate gyri are shown as if transparent so that deeper limbic components can be seen. (**b**) Additional details concerning the three-dimensional structure of the limbic system.

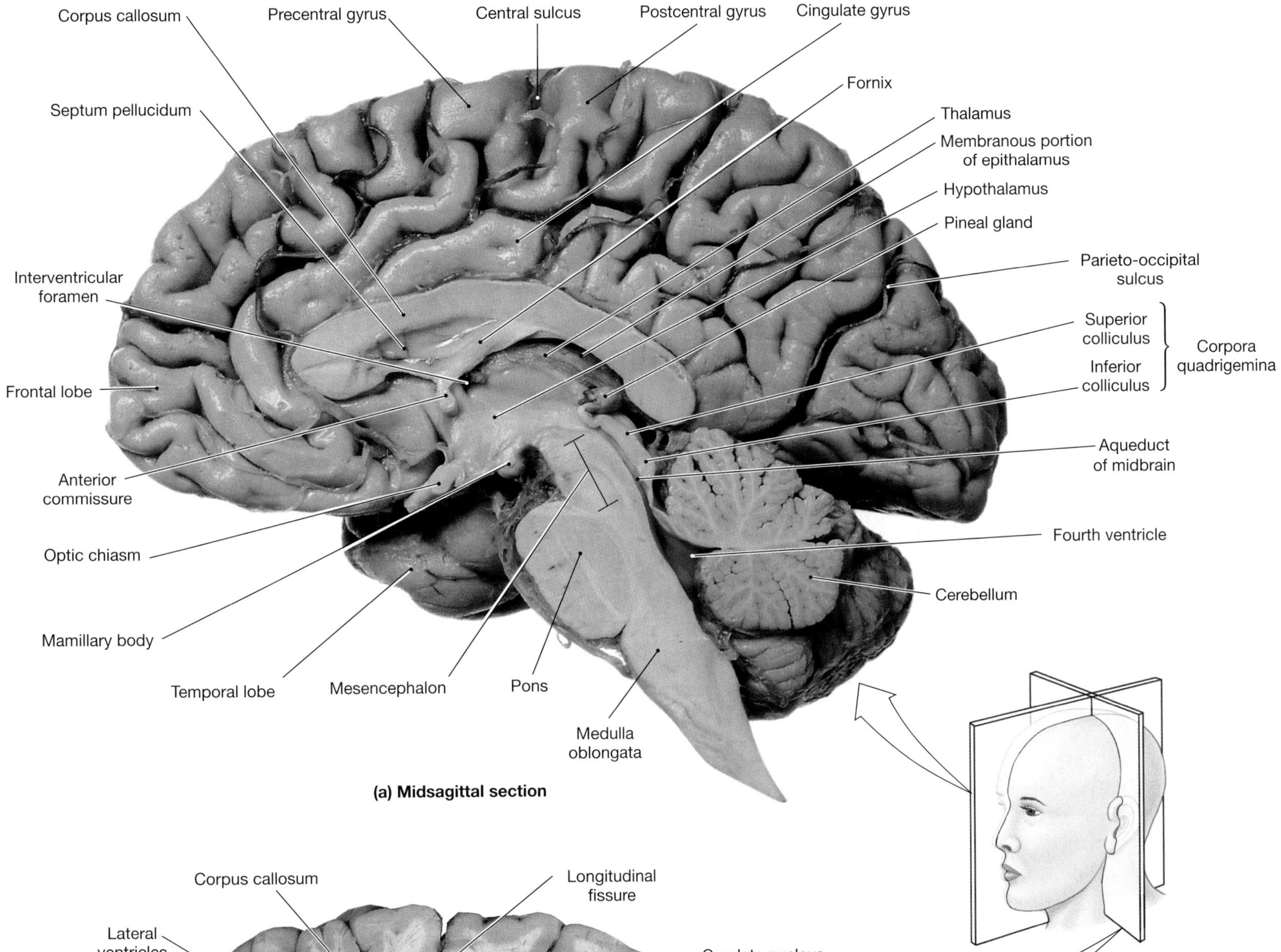

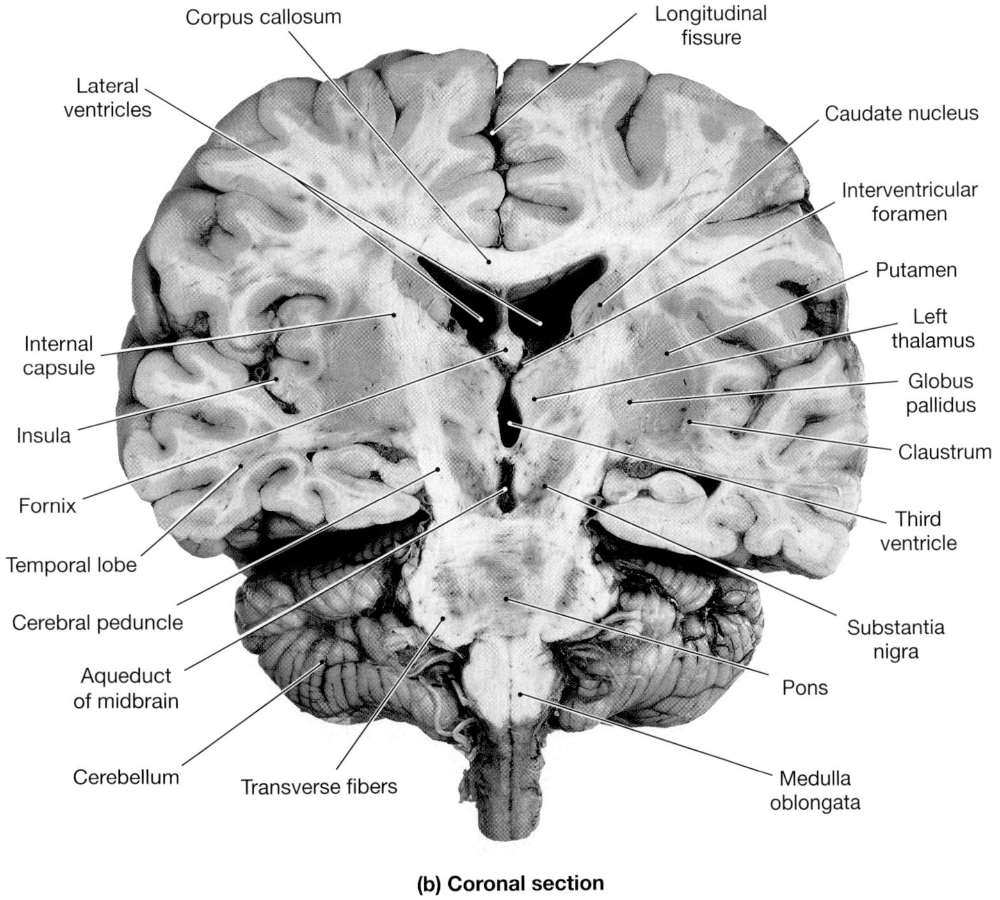

FIGURE 15.13 **SECTIONAL VIEWS OF THE BRAIN**

(**a**) A sagittal section through the brain. (**b**) A coronal section through the brain. *See MRI Scans 1 and 2 in the Companion Atlas.*

The Diencephalon [FIGURES 15.1/15.12/15.13]

The diencephalon connects the cerebral hemispheres to the brain stem. It consists of the *epithalamus*, the left and right *thalamus*, and the *hypothalamus*. Figures 15.1, p. 383, 15.12, and 15.13● show the position of the diencephalon and its relationship to other landmarks in the brain.

The Epithalamus [FIGURE 15.13a]

The **epithalamus** is the roof of the third ventricle (Figure 15.13a●). Its membranous anterior portion contains an extensive area of choroid plexus that extends through the interventricular foramina into the lateral ventricles. The posterior portion of the epithalamus contains the **pineal gland**, an endocrine structure that secretes the hormone **melatonin**. Melatonin is important in the regulation of day-night cycles, with secondary effects on reproductive function. (The role of melatonin will be described in Chapter 19.)

The Thalamus [FIGURES 15.11e/15.12/15.14]

Most of the neural tissue in the diencephalon is concentrated in the left thalamus and right thalamus. These two egg-shaped bodies form the walls of the diencephalon and surround the third ventricle (Figure 15.11e, p. 397 and 15.13b●). The thalamic nuclei provide the switching and relay centers for both sensory and motor pathways. Ascending sensory information from the spinal cord and cranial nerves (other than the olfactory nerve) is processed in the thalamic nuclei before the information is relayed to the cerebrum or brain stem. The thalamus is thus the final relay point for ascending sensory information that will be projected to the primary sensory cortex. It acts as an information filter, passing on only a small portion of the arriving sensory information. The thalamus also acts as a relay station that coordinates motor activities at the conscious and subconscious levels.

The two thalami are separated by the third ventricle. Viewed in midsagittal section, the thalamus extends from the anterior commissure to the inferior base of the pineal gland (Figure 15.13a●). A medial projection of gray matter, the **interthalamic adhesion**, or *massa intermedia*, extends into the ventricle from the thalamus on either side (Figure 15.12a●). In roughly 70% of the population, the two intermediate masses fuse in the midline, interconnecting the two thalami.

The thalamus on each side bulges laterally, away from the third ventricle, and anteriorly toward the cerebrum (Figures 15.11e, 15.13b, and 15.14●). The lateral border of each thalamus is established by the fibers of the internal capsule. p. 394 Embedded within each thalamus is a rounded mass composed of several interconnected *thalamic nuclei*.

FUNCTIONS OF THALAMIC NUCLEI [FIGURE 15.14 AND TABLE 15.6]

The thalamic nuclei are concerned primarily with the relay of sensory information to the basal nuclei and cerebral cortex. The five major groups of thalamic nuclei, detailed in Figure 15.14● and Table 15.6, are (1) the *anterior group*, (2) the *medial group*, (3) the *ventral group*, (4) the *lateral group*, and (5) the *posterior group*.

1. The **anterior nuclei** are part of the limbic system, and they play a role in emotions, memory, and learning. They relay information from the hypothalamus and hippocampus to the cingulate gyrus.
2. The **medial nuclei** provide a conscious awareness of emotional states by connecting the basal nuclei and emotional centers in the hypothalamus with the prefrontal cortex of the cerebrum. These nuclei also integrate sensory information arriving at other portions of the thalamus for relay to the frontal lobes.
3. The **ventral nuclei** relay information to and from the basal nuclei and cerebral cortex. Two of the nuclei (*ventral anterior* and *ventral lateral*) relay information concerning somatic motor commands from the basal nuclei and cerebellum to the primary motor cortex and premotor cortex. They are part of a feedback loop that helps plan a movement and then fine-tunes it. The *ventral posterior nuclei* relay sensory information concerning touch, pressure, pain, temperature, and proprioception from the spinal cord and brain stem to the primary sensory cortex of the parietal lobe.
4. The **posterior nuclei** include the *pulvinar* and the *geniculate nuclei*. The **pulvinar** integrates sensory information for projection to the association areas of the cerebral cortex. The **lateral geniculate** (je-NIK-ū-lāt; *genicula*, little knee) **nucleus** of each thalamus receives visual information from the eyes, brought by the **optic tract**. Efferent fibers project to the visual cortex and descend to the mesencephalon. The **medial geniculate nuclei** relay auditory information to the auditory cortex from the specialized receptors of the inner ear.
5. The **lateral nuclei** are relay stations in feedback loops that adjust activity in the cingulate gyrus and parietal lobe. They thus have an impact on emotional states and the integration of sensory information.

The Hypothalamus [FIGURES 15.13a/15.15]

The hypothalamus contains centers involved with emotions and visceral processes that affect the cerebrum as well as other components of the brain stem. It also controls a variety of autonomic functions and forms the link between the nervous and endocrine systems. The hypothalamus, which forms the floor of the third ventricle, extends from the area superior to the **optic chiasm**, where the *optic tracts* from the eyes arrive at the brain, to the posterior margins of the mamillary bodies (Figure 15.15●). (The mamillary bodies were introduced in the discussion of the limbic system on p. 395.) Posterior to the optic chiasm, the **infundibulum** (in-fun-DIB-ū-lum; *infundibulum*, funnel) extends inferiorly, connecting the hypothalamus to the pituitary gland. In life, the diaphragma sellae (p. 387) surrounds the infundibulum as it enters the hypophyseal fossa of the sphenoid.

Viewed in midsagittal section (Figures 15.13a, and 15.15●), the floor of the hypothalamus between the infundibulum and the mamillary bodies is the **tuberal area** (*tuber*, swelling). The tuberal area contains nuclei involved with the control of pituitary gland function.

FUNCTIONS OF THE HYPOTHALAMUS [FIGURE 15.15b, TABLE 15.7]

The hypothalamus contains a variety of important control and integrative centers, in addition to those associated with the limbic system. These centers and their functions are summarized in Figure 15.15b● and Table 15.7. Hypothalamic centers are continually receiving sensory information from the cerebrum, brain stem, and spinal cord. Hypothalamic neurons also detect and respond to changes in the CSF and interstitial fluid composition; they also respond to stimuli in the circulating blood because of the high permeability of capillaries in this region. Hypothalamic functions include:

1. *Subconscious control of skeletal muscle contractions:* By stimulation of appropriate centers in other portions of the brain, hypothalamic nuclei direct somatic motor patterns associated with the emotions of rage, pleasure, pain, and sexual arousal.
2. *Control of autonomic function*: Hypothalamic centers adjust and coordinate the activities of autonomic centers in other parts of the brain stem concerned with regulating heart rate, blood pressure, respiration, and digestive functions.

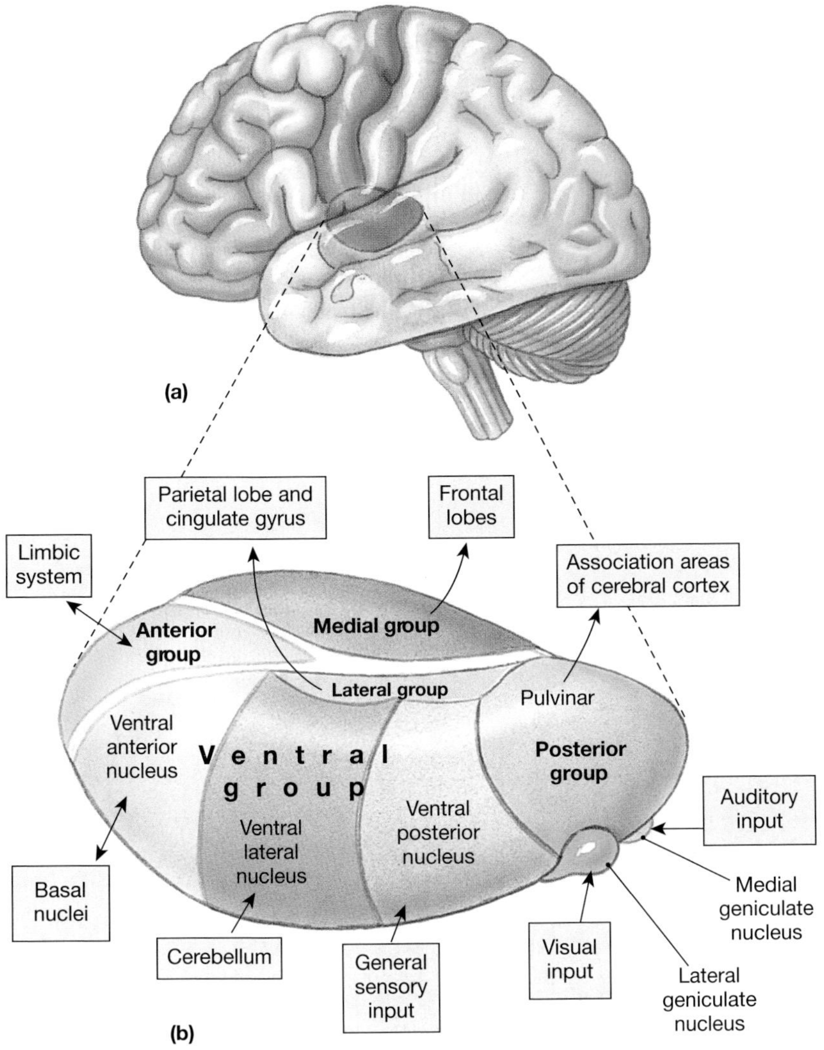

FIGURE 15.14 **THE THALAMUS**

(**a**) Lateral view of the brain, showing the positions of the major thalamic structures. Functional areas of cerebral cortex are also indicated, with colors corresponding to those of the associated thalamic nuclei. (**b**) Enlarged view of the thalamic nuclei of the left side. The color of each nucleus or group of nuclei matches the color of the associated cortical region. The boxes either provide examples of the types of sensory input relayed to the basal nuclei and cerebral cortex or indicate the existence of important feedback loops involved with emotional states, learning, and memory.

3. *Coordination of activities of the nervous and endocrine systems*: Much of the regulatory control is exerted through inhibition or stimulation of endocrine cells within the pituitary gland.
4. *Secretion of hormones*: The hypothalamus secretes two hormones: (1) *antidiuretic hormone*, produced by the **supraoptic nucleus**, restricts water loss at the kidneys, and (2) *oxytocin*, produced by the **paraventricular nucleus**, stimulates smooth muscle contractions in the uterus and prostate gland, and myoepithelial cell contractions in the mammary glands. Both hormones are transported along axons down the infundibulum for release into the circulation at the posterior portion of the pituitary gland.
5. *Production of emotions and behavioral drives*: Specific hypothalamic centers produce sensations that lead to changes in voluntary or involuntary behavior patterns. For example, stimulation of the **feeding center** produces the desire to eat, and stimulation of the **thirst center** produces the desire to drink.

TABLE 15.6 **THE THALAMUS**

Structure/Nuclei	Functions
ANTERIOR GROUP	Part of the limbic system
MEDIAL GROUP	Integrates sensory information and other data arriving at the thalamus and hypothalamus for projection to the frontal lobes of the cerebral hemispheres
VENTRAL GROUP	Projects sensory information to the primary sensory cortex of the parietal lobe; relays information from cerebellum and basal nuclei to motor areas of cerebral cortex
POSTERIOR GROUP	
Pulvinar	Integrates sensory information for projection to association areas of cerebral cortex
Lateral geniculate nuclei	Project visual information to the visual cortex of occipital lobe
Medial geniculate nuclei	Project auditory information to the auditory cortex of temporal lobe
LATERAL GROUP	Form feedback loops involving the cingulate gyrus (emotional states) and the parietal lobe (integration of sensory information)

6. *Coordination between voluntary and autonomic functions*: When facing a stressful situation, your heart rate and respiratory rate go up and your body prepares for an emergency. These autonomic adjustments are made because cerebral activities are monitored by the hypothalamus. Recall the autonomic nervous system (ANS) is a division of the peripheral nervous system introduced in Chapter 13. p. 335 The ANS consists of two divisions: (1) *sympathetic* and (2) *parasympathetic*. The sympathetic division stimulates tissue metabolism, increases alertness, and prepares the body to respond to emergencies; whereas, the parasympathetic division promotes sedentary activities and conserves body energy. These divisions and their relationships will be discussed in Chapter 17.
7. *Regulation of body temperature*: The **preoptic area** of the hypothalamus controls the physiological responses to changes in body temperature. In doing so, it coordinates the activities of other CNS centers and regulates other physiological systems.
8. *Control of circadian rhythms*: The **suprachiasmatic nucleus** coordinates daily cycles of activity that are linked to the day-night cycle. This nucleus receives direct input from the retina of the eye, and its output adjusts the activities of other hypothalamic nuclei, the pineal gland, and the reticular formation.

Table 15.7 lists the major nuclei and centers of the hypothalamus and provides greater detail regarding their known functions.

✓ CONCEPT CHECK

- What area of the diencephalon is stimulated by changes in body temperature?
- Which region of the diencephalon helps coordinate somatic motor activities?
- What endocrine structure in the diencephalon secretes melatonin?
- What hormones are produced by the hypothalamus and released at the pituitary gland?

FIGURE 15.15 **THE HYPOTHALAMUS**

(**a**) Midsagittal section through the brain, showing major features of the diencephalon and adjacent portions of the brain stem. (**b**) Enlarged view of the hypothalamus, showing the locations of major nuclei and centers. Functions for these centers are summarized in Table 15.7. *See MRI Scan 1e in the Companion Atlas.*

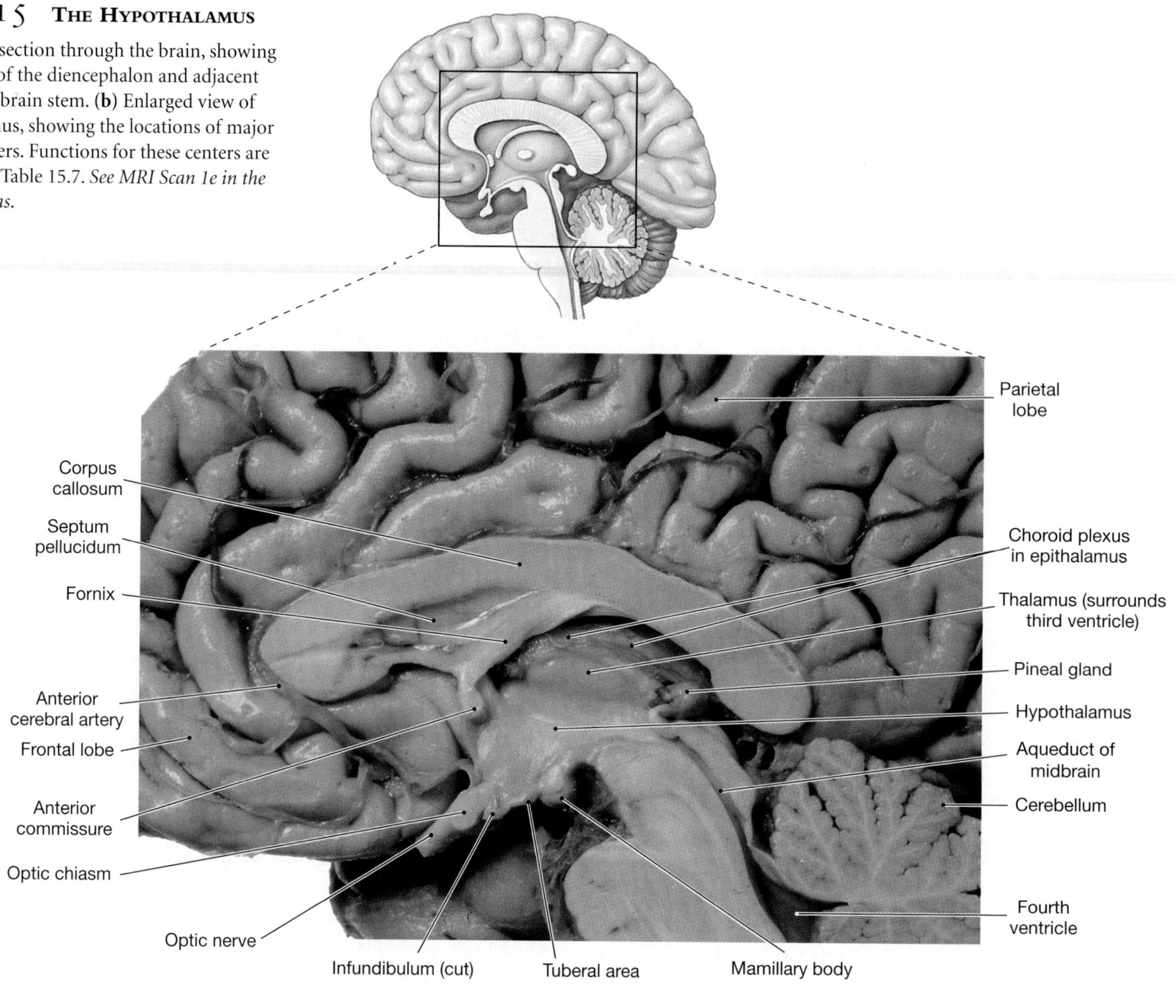

(a) Midsagittal section

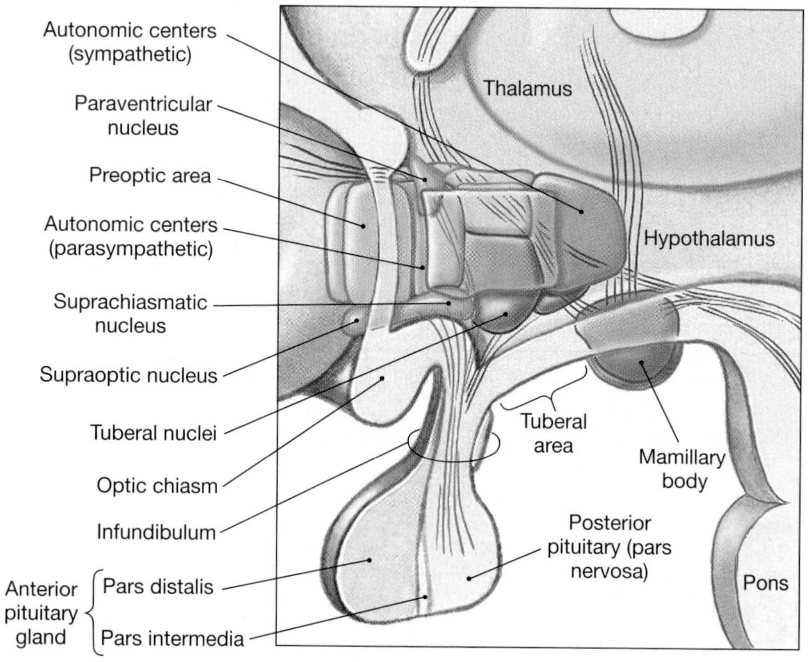

(b) Hypothalamus

TABLE 15.7 **THE HYPOTHALAMUS**

Region/Nucleus	Functions
Hypothalamus in general	Controls autonomic functions; sets appetitive drives (thirst, hunger, sexual desire) and behaviors; sets emotional states (with limbic system); integrates with endocrine system (see Chapter 19)
Supraoptic nucleus	Secretes antidiuretic hormone, restricting water loss at the kidneys
Suprachiasmatic nucleus	Regulates daily (circadian) rhythms
Paraventricular nucleus	Secretes oxytocin, stimulating smooth muscle contractions in uterus and mammary glands
Preoptic area	Regulates body temperature via control of autonomic centers in the medulla oblongata
Tuberal area	Produces inhibitory and releasing hormones that target endocrine cells of the anterior lobe of the pituitary gland
Autonomic centers	Control heart rate and blood pressure via regulation of autonomic centers in the medulla oblongata
Mamillary bodies	Control feeding reflexes (licking, swallowing, etc.)

The Mesencephalon [FIGURES 15.16/15.17 AND TABLE 15.8]

The mesencephalon, or midbrain, contains nuclei that process visual and auditory information and generate reflexive responses to these stimuli.

The external anatomy of the mesencephalon can be seen in Figure 15.16•, and the major nuclei are detailed in Figure 15.17• and Table 15.8. The surface of the midbrain posterior to the aqueduct of the midbrain is called the roof, or **tectum**, of the mesencephalon. This region contains two pairs of sensory nuclei known collectively as the **corpora quadrigemina** (KOR-pō-ra quad-ri-JEM-i-na). These nuclei are relay stations concerned with the processing of visual and auditory sensations. Each **superior colliculus** (kol-IK-ū-lus; *colliculus*, small hill) receives visual inputs from the lateral geniculate of the thalamus on that side. The **inferior colliculus** receives auditory data from nuclei in the medulla oblongata; some of this information may be forwarded to the medial geniculate on the same side.

The mesencephalon also contains the major nuclei of the reticular formation. Specific patterns of stimulation in this region can produce a variety of involuntary motor responses. Each side of the mesencephalon contains a pair of nuclei, the **red nucleus** and the *substantia nigra* (Figure 15.17•). The **red nucleus** is provided with numerous blood vessels, giving it a rich red coloration. This nucleus integrates information from the cerebrum and cerebellum and issues involuntary motor commands concerned with the maintenance of muscle tone and limb position. The **substantia nigra** (NĪ-grah, "black") lies lateral to the red nucleus. The gray matter in this region contains darkly pigmented cells, giving it a black color. The substantia nigra plays an important role in regulating the motor output of the basal nuclei.

The nerve fiber bundles on the ventrolateral surfaces of the mesencephalon (Figures 15.16 and 15.17b•) are the **cerebral peduncles** (*peduncles*, little feet). They contain (1) ascending fibers that synapse in the thalamic nuclei and (2) descending fibers of the corticospinal pathway that carry voluntary motor commands from the primary motor cortex of each cerebral hemisphere.

CLINICAL BRIEF

The Substantia Nigra and Parkinson's Disease

The basal nuclei contain two discrete populations of neurons. One group stimulates motor neurons by releasing acetylcholine (ACh), and the other inhibits motor neurons by the release of the neurotransmitter *gamma-aminobutyric acid*, or *GABA*. Under normal conditions, the excitatory neurons remain inactive, and the descending tracts are responsible primarily for inhibiting motor neuron activity. The excitatory neurons are quiet because they are continually exposed to the inhibitory effects of the neurotransmitter *dopamine*. This compound is manufactured by neurons in the substantia nigra and transported along axons to synapses in the basal nuclei. If the ascending tract or the dopamine-producing neurons are damaged, this inhibition is lost, and the excitatory neurons become increasingly active. This increased activity produces the motor symptoms of **Parkinson's disease**, or *paralysis agitans*.

Parkinson's disease is characterized by a pronounced increase in muscle tone. Voluntary movements become hesitant and jerky, for a movement cannot occur until one muscle group manages to overpower its antagonists. Individuals with Parkinson's disease show **spasticity** during voluntary movement and a continual **tremor** when at rest. A tremor represents a tug of war between antagonistic muscle groups that produces a background shaking of the limbs. Individuals with Parkinson's disease also have difficulty starting voluntary movements. Even changing one's facial expression requires intense concentration, and the individual acquires a blank, static expression. Finally, the positioning and preparatory adjustments normally performed automatically no longer occur. Every aspect of each movement must be voluntarily controlled, and the extra effort requires intense concentration that may prove tiring and extremely frustrating. In the late stages of this condition, other CNS effects, such as depression and hallucinations, often appear.

Providing the basal nuclei with dopamine can significantly reduce the symptoms for two-thirds of Parkinson's patients. Dopamine cannot cross the blood-brain barrier, and the most common treatment involves the oral administration of the drug L-DOPA (*levodopa*), a related compound that crosses the cerebral capillaries and is converted to dopamine. Surgery to control Parkinson's symptoms focuses on the destruction of large areas within the basal nuclei or thalamus to control the motor symptoms of tremor and rigidity. Transplantation of tissues that produce dopamine or related compounds directly into the basal nuclei is one method attempted as a cure. The transplantation of fetal brain cells into the basal nuclei of adult brains has slowed or even reversed the course of the disease in a significant number of patients, although problems with involuntary muscle contractions later developed in many cases.

TABLE 15.8 The Mesencephalon

Subdivision	Region/Nucleus	Functions
GRAY MATTER		
Tectum (roof)	Superior colliculi	Integrate visual information with other sensory inputs; initiate reflex responses to visual stimuli
	Inferior colliculi	Relay auditory information to medial geniculate nuclei; initiate reflex responses to auditory stimuli
Walls and floor	Red nuclei	Involuntary control of background muscle tone and limb position
	Substantia nigra	Regulates activity in the basal nuclei
	Reticular formation	Automatic processing of incoming sensations and outgoing motor commands; can initiate motor responses to stimuli; helps maintain consciousnes (RAS)
	Other nuclei/centers	Nuclei associated with two cranial nerves (N III, N IV)
WHITE MATTER	Cerebral peduncles	Connect primary motor cortex with motor neurons in brain and spinal cord; carry ascending sensory information to thalamus

FIGURE 15.16 **THE DIENCEPHALON AND BRAIN STEM**

(**a**) Diagrammatic view of the diencephalon and brain stem, seen from the left side. (**b**) Lateral view of the brain stem, with a portion of the cerebellum sectioned and removed. (**c**) Posterior diagrammatic view of the diencephalon and brain stem. (**d**) Brain stem, posterior view.

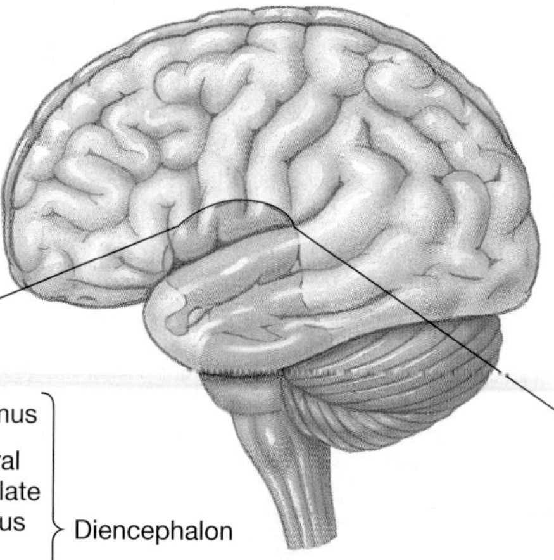

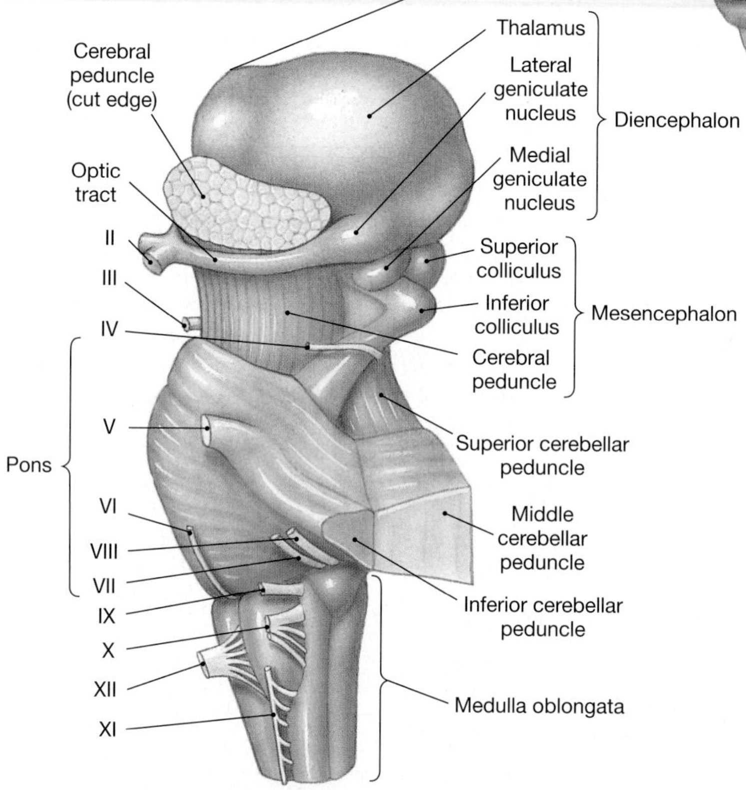

(a) Lateral view

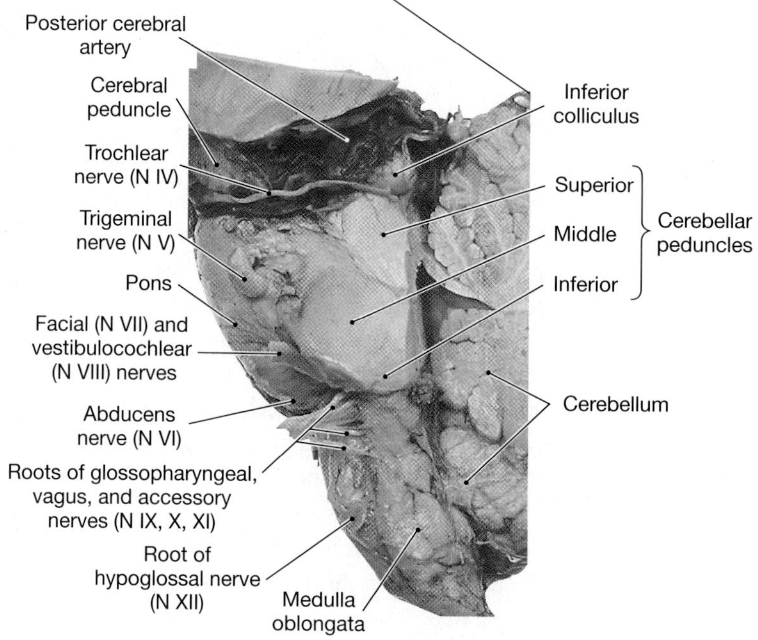

(b) Sagittal section

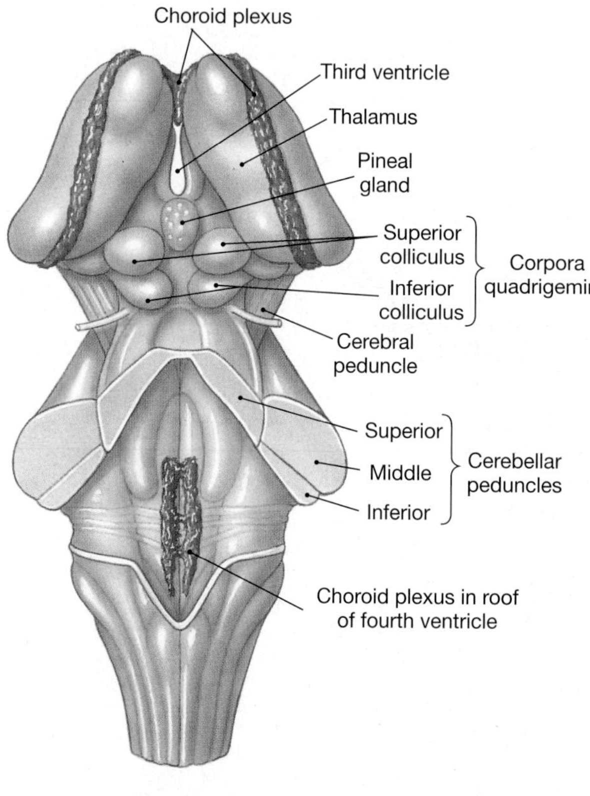

(c) Posterior view

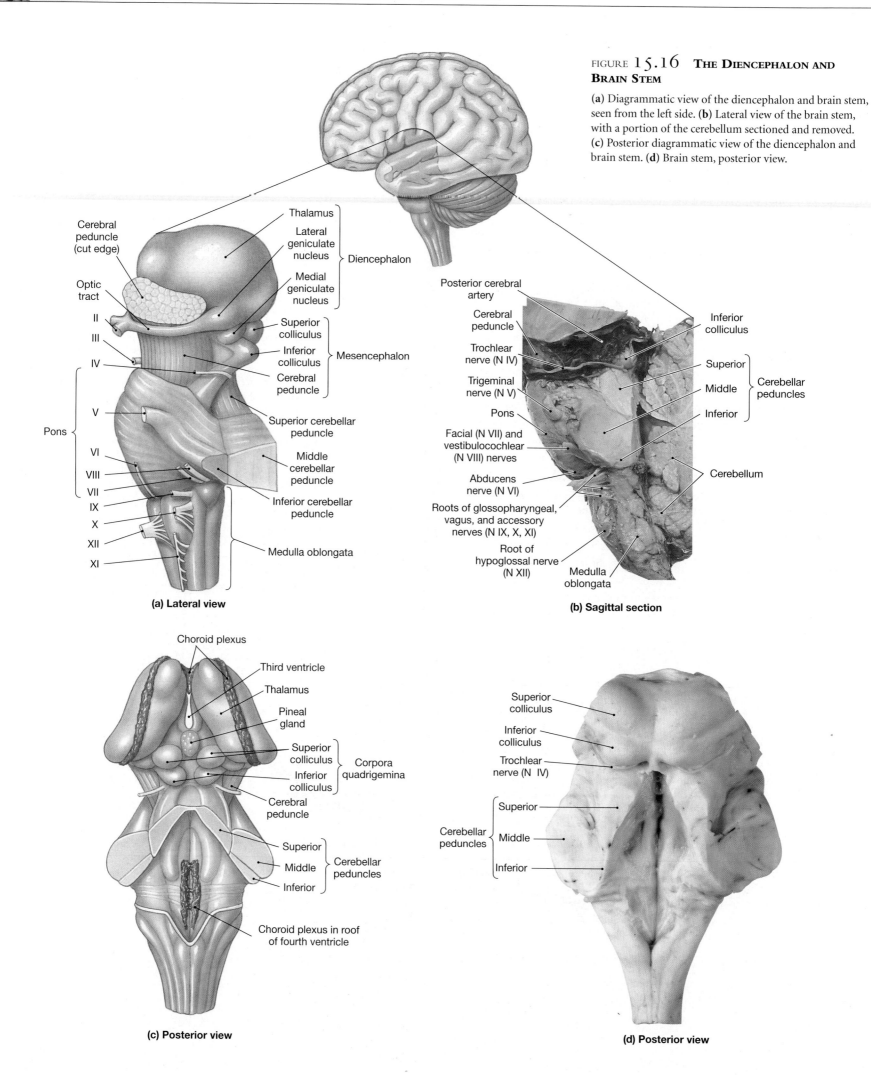

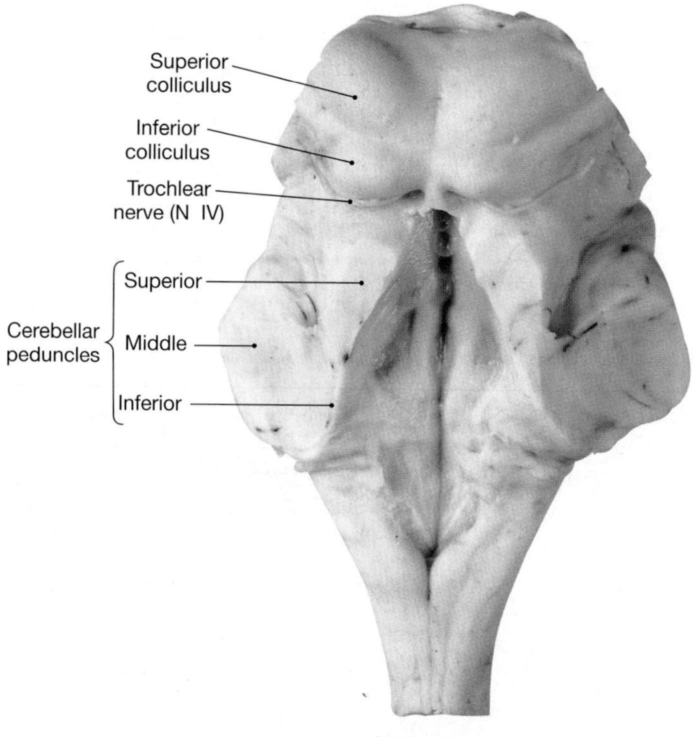

(d) Posterior view

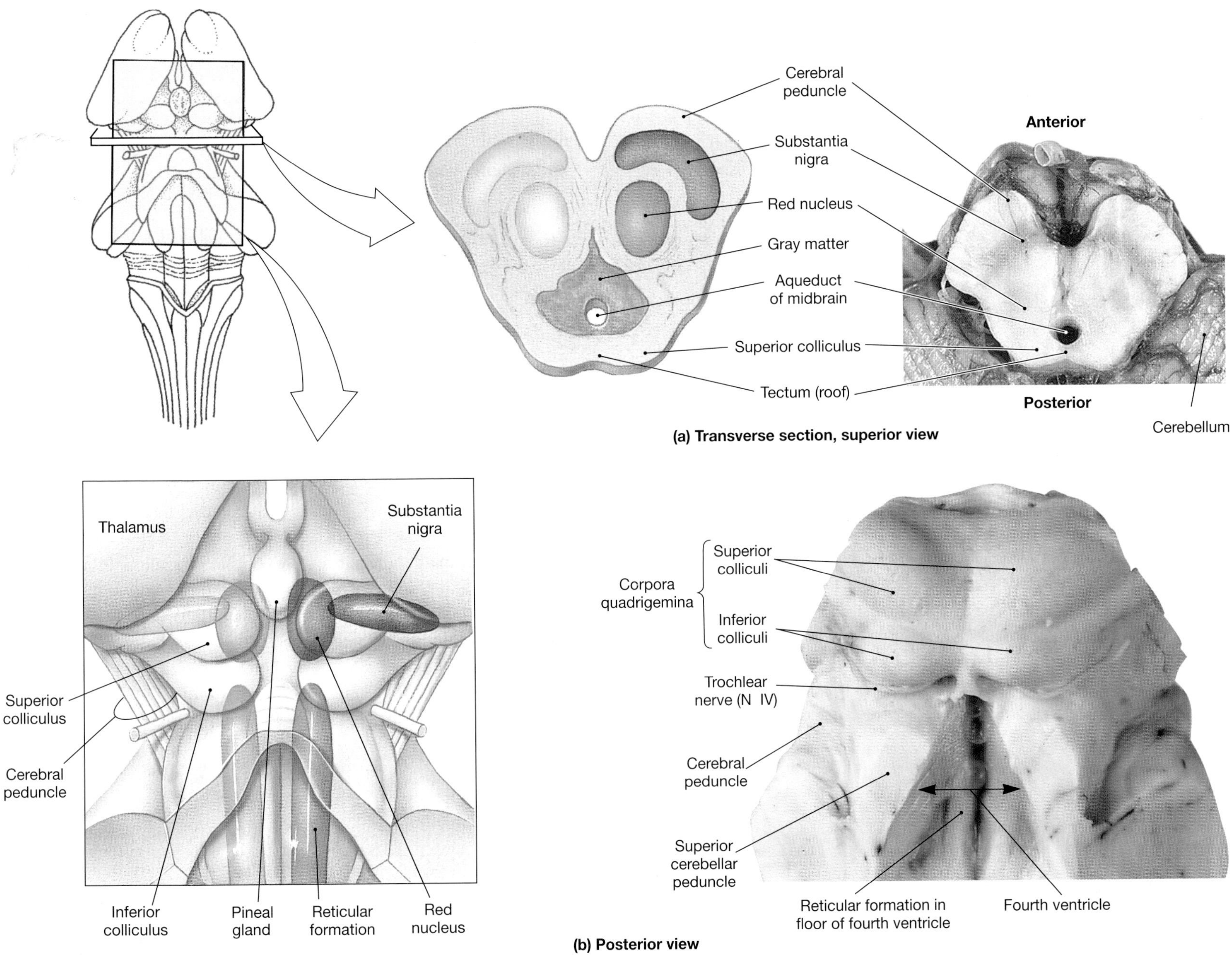

FIGURE 15.17 **THE MESENCEPHALON**

(**a**) Diagrammatic view and sectioned brain stem, with the sections taken at the level indicated in the icon. (**b**) Diagrammatic and posterior views of the diencephalon and brain stem. The diagrammatic view is drawn as if transparent, to show the positions of important nuclei.

The Pons [FIGURES 15.13/15.16/15.18 AND TABLE 15.9]

The pons extends inferiorly from the mesencephalon to the medulla oblongata. It forms a prominent bulge on the anterior surface of the brain stem. The cerebellar hemispheres lie posterior to the pons; the two are partially separated by the fourth ventricle. On either side, the pons is attached to the cerebellum by three *cerebellar peduncles.* Important features and regions are indicated in Figures 15.13, 15.16, and 15.18•; structures are detailed in Table 15.9. The pons contains:

- *Sensory and motor nuclei for four cranial nerves* (N V, N VI, N VII, and N VIII). These cranial nerves innervate the jaw muscles, the anterior surface of the face, one of the extra-ocular muscles (the lateral rectus), and organs of hearing and equilibrium in the inner ear.
- *Nuclei concerned with the involuntary control of respiration.* On each side of the brain, the reticular formation in this region contains two respiratory centers, the *apneustic center* and the *pneumotaxic center.* These centers modify the activity of the *respiratory rhythmicity center* in the medulla oblongata.
- *Nuclei that process and relay cerebellar commands arriving over the middle cerebellar peduncles.*
- *Ascending, descending, and transverse tracts.* The longitudinal tracts interconnect other portions of the CNS. The middle cerebellar peduncles are connected to the **transverse fibers** of the pons that cross its anterior surface. These fibers permit communication between the cerebellar hemispheres of opposite sides.

The Cerebellum [FIGURES 15.8/15.13/15.19, AND TABLE 15.10]

The cerebellum has two **cerebellar hemispheres**, each with a highly convoluted surface composed of neural cortex (Figures 15.8, p. 392, and 15.19●). These folds, or **folia** (FŌ-lē-a), of the surface are less prominent than the gyri of the cerebral hemispheres. Each hemisphere consists of two **lobes, anterior** and **posterior**, which are separated by the **primary fissure**. Along the midline a narrow band of cortex known as the **vermis** (VER-mis; "worm") separates the cerebellar hemispheres. Slender **flocculonodular** (flok-ū-lō-NOD-ū-lar) **lobes** lie anterior and inferior to the cerebellar hemisphere. The anterior and posterior lobes assist in the planning, execution, and coordination of limb and trunk movements. The flocculonodular lobe is important in the maintenance of balance and the control of eye movements. The structures of the cerebellum and their functions are summarized in Table 15.10.

The cerebellar cortex contains huge, highly branched **Purkinje** (pur-KIN-jē) **cells** (Figure 15.19b●). Purkinje cells have large pear-shaped somas, which have large, numerous dendrites fanning out into the gray matter (neural cortex) of the cerebellar cortex. Axons project from the basal portion of the cell into the white matter to reach the cerebellar nuclei. Internally, the white matter of the cerebellum forms a branching array that in sectional view resembles a tree. Anatomists call it the **arbor vitae**, or "tree of life." The cerebellum receives proprioceptive information, indicating body position, (position sense) from the spinal cord and monitors all proprioceptive, visual, tactile, balance, and auditory sensations received by the brain. Information concerning motor commands issued by the cerebral cortex reaches the cerebellum indirectly, relayed from nuclei in the pons. A relatively small portion of the afferent fibers synapse within **cerebellar nuclei** before projecting to the cerebellar cortex. Most axons carrying sensory information do not synapse in the cerebellar nuclei but pass through the deeper layers of the cerebellar cortex to end near the cortical surface. There they synapse with the dendritic processes of the Purkinje cells. Tracts containing the axons of Purkinje cells then relay motor commands to nuclei within the cerebrum and brain stem. *Cerebellar Dysfunction p. 794*

Tracts that link the cerebellum with the brain stem, cerebrum, and spinal cord leave the cerebellar hemispheres as the *superior*, *middle*, and *inferior cerebellar peduncles* (Figures 15.16 and 15.19b●). The **superior cerebellar peduncles** link the cerebellum with nuclei in the mesen-

TABLE 15.10 **THE CEREBELLUM**

Subdivision	Region/Nucleus	Functions
Gray matter	Cerebellar cortex	Subconscious coordination and control of ongoing movements of body parts
	Cerebellar nuclei	As above
White matter	Arbor vitae	Connects cerebellar cortex and nuclei with cerebellar peduncles
	Cerebellar peduncles	
	Superior	Link the cerebellum with mesencephalon, diencephalon, and cerebrum
	Middle	Contain transverse fibers and carry communications between the cerebellum and pons
	Inferior	Link the cerebellum with the medulla oblongata and spinal cord

FIGURE 15.18 **THE PONS**

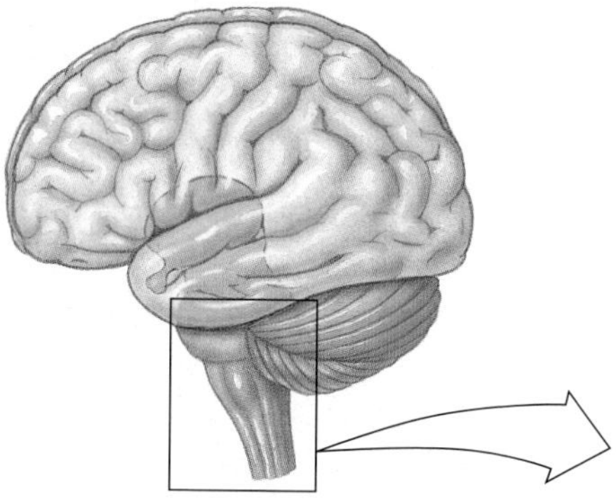

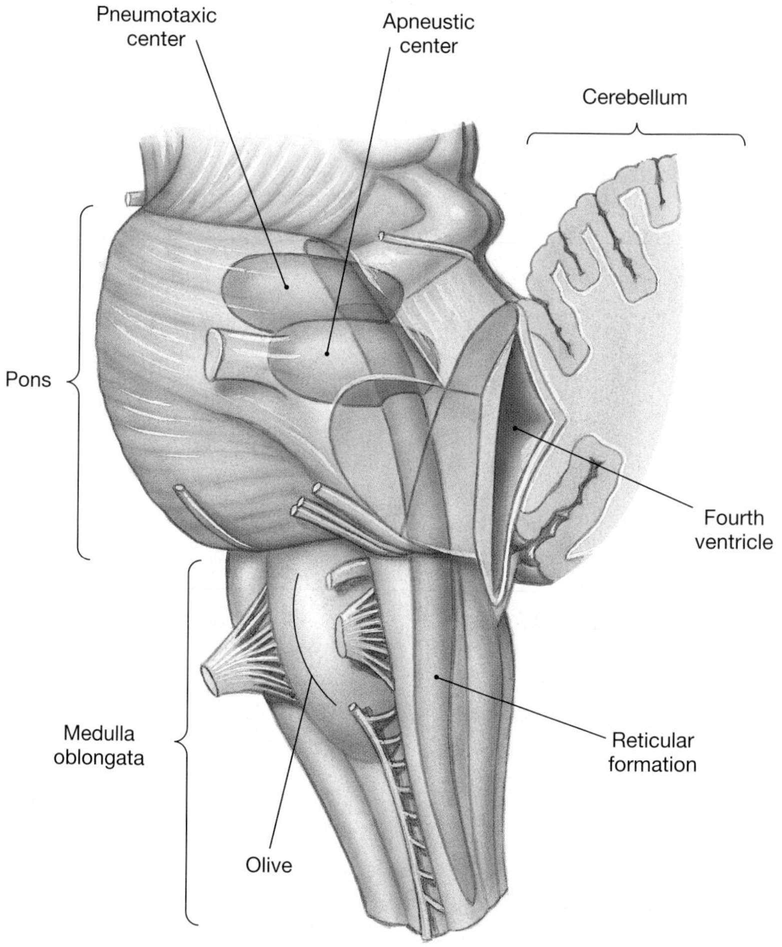

TABLE 15.9 **THE PONS**

Subdivision	Region/Nucleus	Functions
Gray matter	Respiratory centers	Modify output of respiratory centers in the medulla oblongata
	Other nuclei/centers	Nuclei associated with four cranial nerves and cerebellum
White matter	Ascending and descending tracts	Interconnect other portions of CNS
	Transverse fibers	Interconnect cerebellar hemispheres; interconnect pontine nuclei with the cerebellar hemispheres on the opposite side

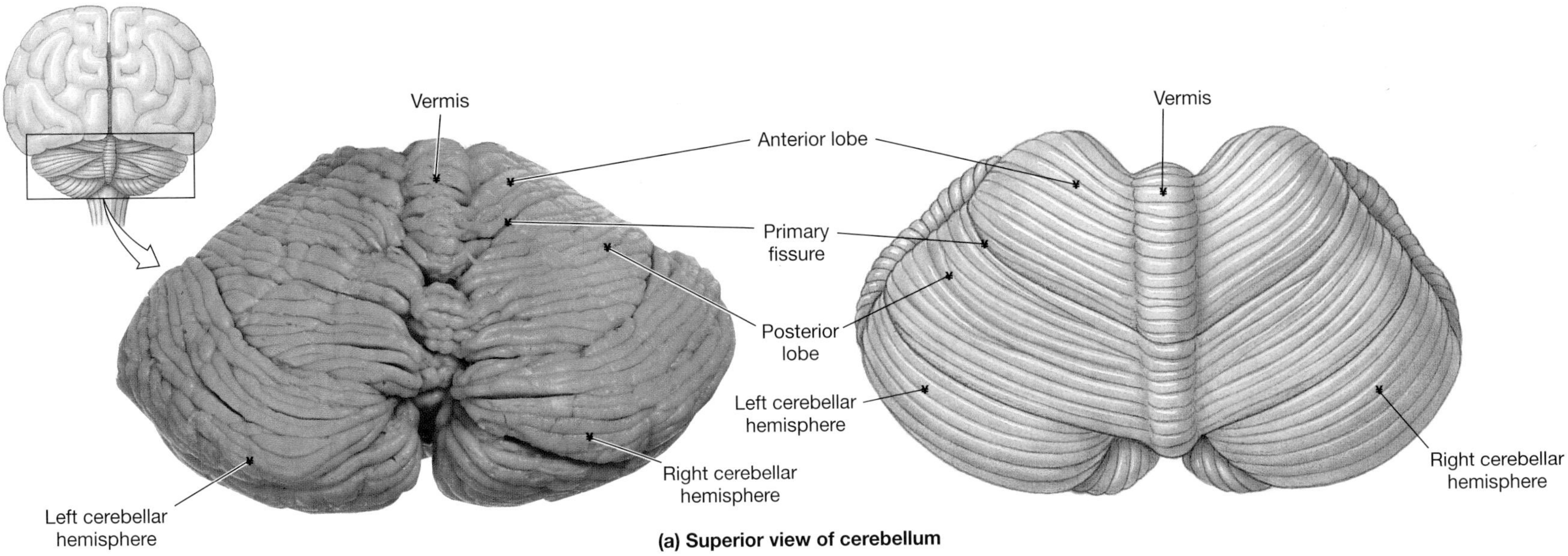

FIGURE 15.19 **The Cerebellum**

(**a**) Superior surface of the cerebellum, showing major anatomical landmarks and regions. (**b**) Sagittal view of the cerebellum, showing the arrangement of gray matter and white matter. Purkinje cells are seen in the photograph; these large neurons are found in the cerebellar cortex. (LM × 120)

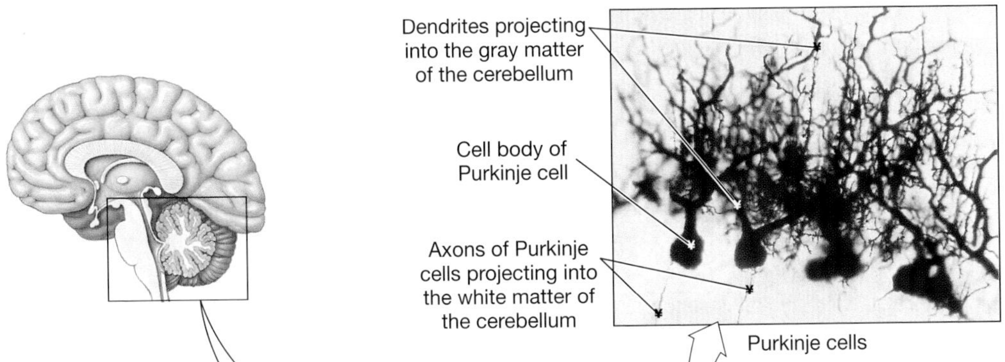

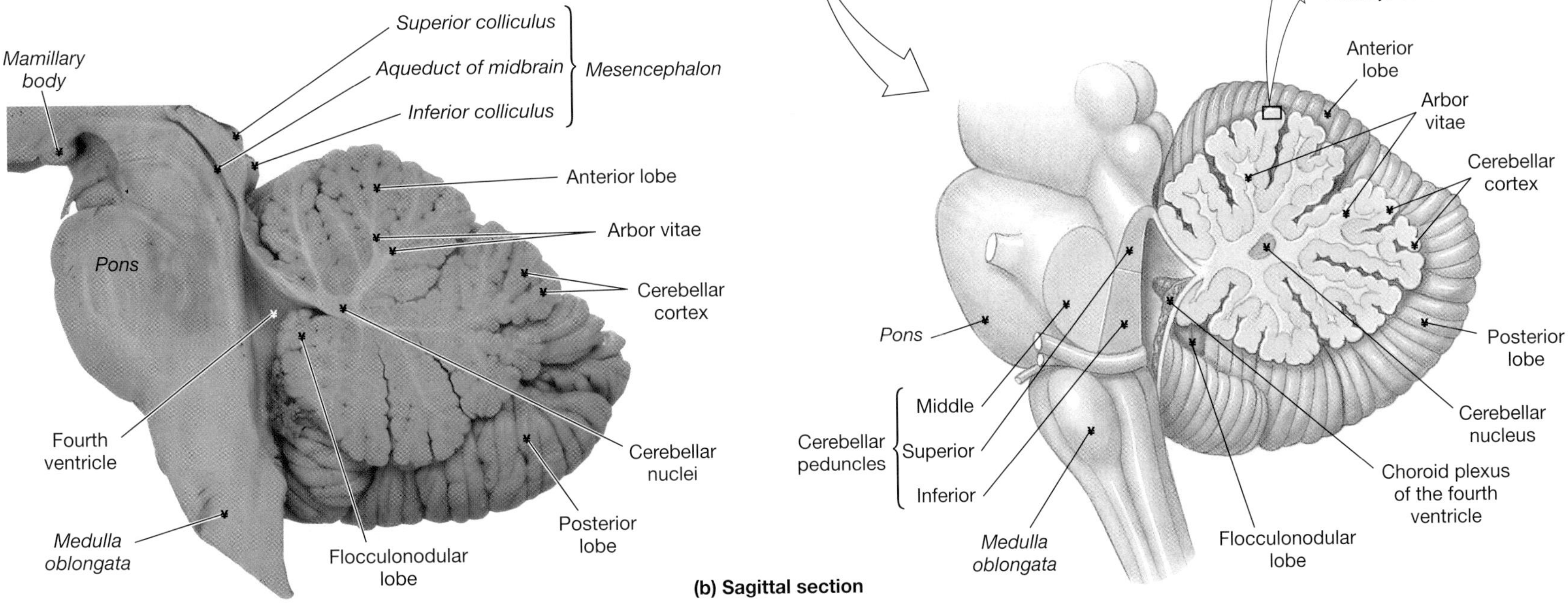

cephalon, diencephalon, and cerebrum. The **middle cerebellar peduncles** are connected to a broad band of fibers that cross the ventral surface of the pons at right angles to the axis of the brain stem. The middle cerebellar peduncles also connect the cerebellar hemispheres with sensory and motor nuclei in the pons. The **inferior cerebellar peduncles** permit communication between the cerebellum and nuclei in the medulla oblongata and carry ascending and descending cerebellar tracts from the spinal cord.

The cerebellum is an automatic processing center that has two primary functions:

- *Adjusting the postural muscles of the body*: The cerebellum coordinates rapid, automatic adjustments that maintain balance and equilibrium. These alterations in muscle tone and position are made by modifying the activity of the red nucleus.
- *Programming and fine-tuning voluntary and involuntary movements*: The cerebellum stores memories of learned movement patterns. These functions are performed indirectly, by regulating activity along motor pathways involving the cerebral cortex, basal nuclei, and motor centers in the brain stem.

The Medulla Oblongata [FIGURES 15.9/15.13/15.16/15.20 AND TABLE 15.11]

The spinal cord connects to the brain at the **medulla oblongata**, which corresponds to the embryonic myelencephalon. The medulla oblongata, or *medulla*, is continuous with the spinal cord. The external appearance of the medulla oblongata should be examined in Figures 15.9a, p. 393, and 15.16●. The important nuclei and centers are diagrammed in Figure 15.20● and detailed in Table 15.11.

Figure 15.13a●, p. 399, shows the medulla oblongata in midsagittal section. The caudal portion resembles the spinal cord in having a rounded shape and a narrow central canal. Closer to the pons, the central canal becomes enlarged and continuous with the fourth ventricle.

The medulla oblongata physically connects the brain with the spinal cord, and many of its functions are directly related to this connection. For example, all communication between the brain and spinal cord involves tracts that ascend or descend through the medulla oblongata.

Nuclei in the medulla oblongata may be (1) relay stations along sensory or motor pathways, (2) sensory or motor nuclei associated with cranial nerves connected to the medulla oblongata, or (3) nuclei associated with the autonomic control of visceral activities.

1. *Relay stations*: Ascending tracts may synapse in sensory or motor nuclei that act as relay stations and processing centers. For example, the **nucleus gracilis** and the **nucleus cuneatus** pass somatic sensory information to the thalamus, and the **olivary nuclei** relay information from the spinal cord, the cerebral cortex, diencephalon, and brain stem to the cerebellar cortex. The bulk of these nuclei create the **olives**, prominent bulges along the ventrolateral surface of the medulla oblongata (Figure 15.18●).
2. *Nuclei of cranial nerves*: The medulla oblongata contains sensory and motor nuclei associated with five of the cranial nerves (N VIII, N IX, N X, N XI, and N XII). These cranial nerves innervate muscles of the pharynx, neck, and back, as well as visceral organs of the thoracic and peritoneal cavities.
3. *Autonomic nuclei*: The reticular formation in the medulla oblongata contains nuclei and centers responsible for the regulation of vital autonomic functions. These **reflex centers** receive input from cranial nerves, the cerebral cortex, the diencephalon, and the brain stem, and their output controls or adjusts the activities of one or more peripheral systems. Major centers include:
   - The **cardiovascular centers**, which adjust heart rate, the strength of cardiac contractions, and the flow of blood through peripheral tissues. On functional grounds, the cardiovascular centers may be subdivided into **cardiac** (*kardia*, heart) and **vasomotor** (*vas*, canal) centers, but their anatomical boundaries are difficult to determine.
   - The **respiratory rhythmicity centers**, which set the basic pace for respiratory movements, and their activity is regulated by inputs from the apneustic and pneumotaxic centers of the pons.

✓ CONCEPT CHECK

- In what part of the brain would you find a worm (vermis) and a tree (arbor vitae)?
- The medulla oblongata is one of the smallest sections of the brain, yet damage there can cause death, whereas similar damage in the cerebrum might go unnoticed. Why?
- What are the functions of the nuclei in the tectum of the mesencephalon?
- When the substantia nigra loses its "dark" color, neurons here fail to make dopamine. What might be some clinical signs of this condition?

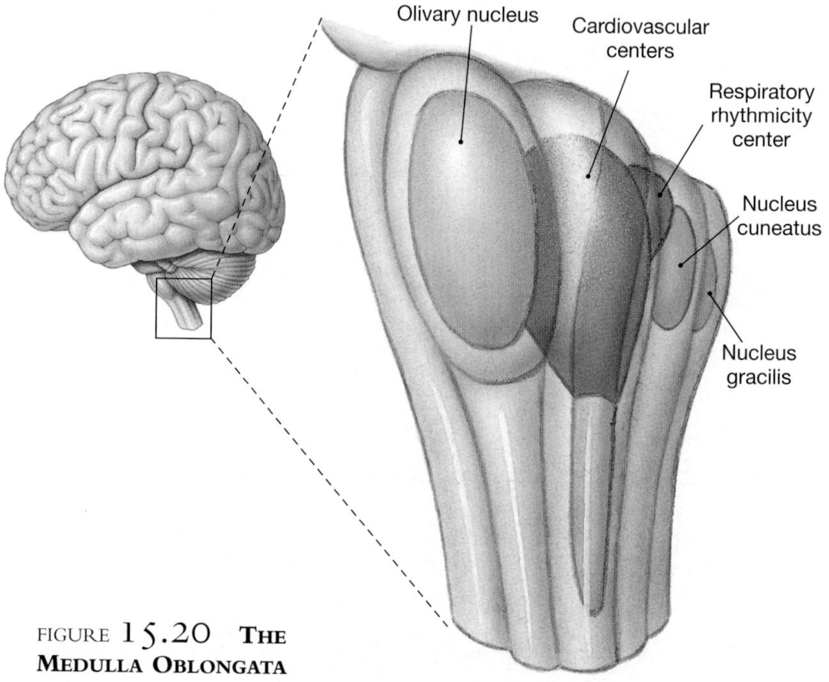

FIGURE 15.20 THE MEDULLA OBLONGATA

TABLE 15.11

THE MEDULLA OBLONGATA

Subdivision	Region/Nucleus	Functions
Gray matter	Nucleus gracilis Nucleus cuneatus	Relay somatic sensory information to the ventral posterior nuclei of the thalamus
	Olivary nuclei	Relay information from the red nucleus, other midbrain centers, and the cerebral cortex to the vermis of cerebellum
	Reflex centers	
	Cardiac centers	Regulate heart rate and force of contraction
	Vasomotor centers	Regulate distribution of blood flow
	Respiratory rhythmicity centers	Set the pace of respiratory movements
	Other nuclei/centers	Sensory and motor nuclei of five cranial nerves Nuclei relaying ascending sensory information from the spinal cord to higher centers
White matter	Ascending and descending tracts	Link the brain with the spinal cord

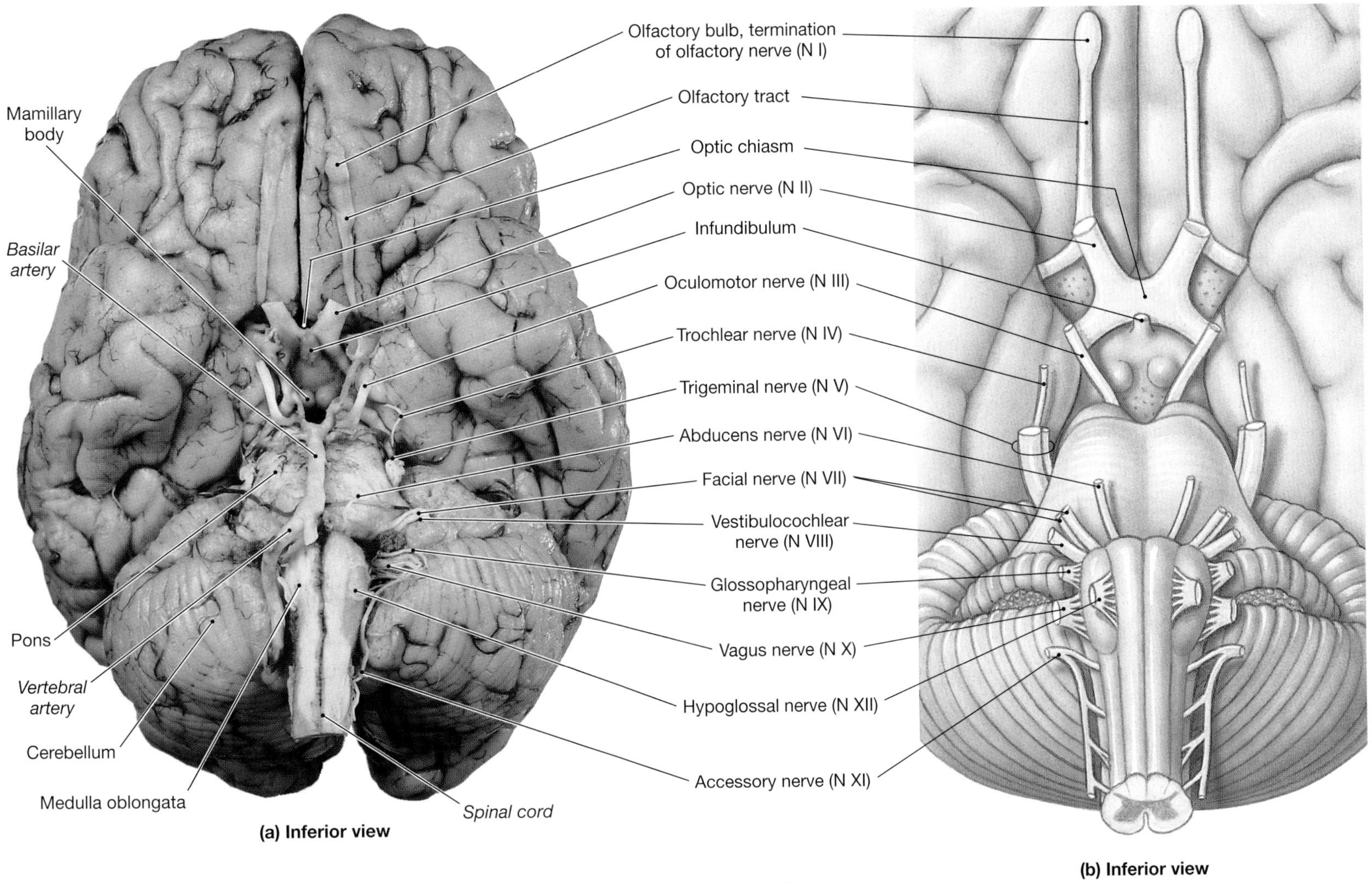

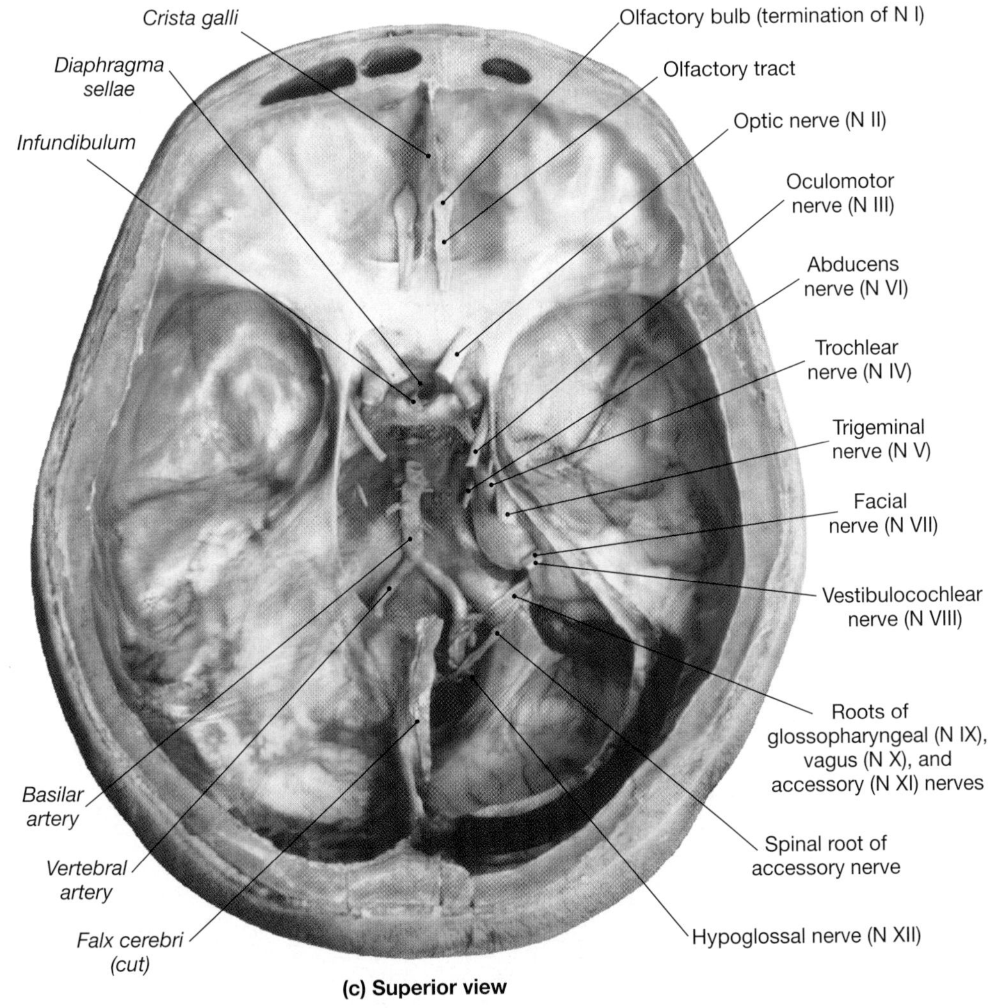

FIGURE 15.21 **ORIGINS OF THE CRANIAL NERVES**

(**a**) The inferior surface of the brain as it appears on gross dissection. The roots of the cranial nerves are clearly visible. (**b**) Diagrammatic inferior view of the human brain; compare with (a). (**c**) Superior view of the cranial fossae, with brain and right half of tentorium cerebelli removed. Portions of several cranial nerves are visible.

The Cranial Nerves [FIGURE 15.21]

Cranial nerves are components of the peripheral nervous system that connect to the brain rather than to the spinal cord. Twelve pairs of cranial nerves can be found on the ventrolateral surface of the brain (Figure 15.21●), each with a name related to its appearance or function.

Cranial nerves are numbered according to their position along the longitudinal axis of the brain, beginning at the cerebrum. Roman numerals are usually used, either alone or with the prefix N or CN. We will use the abbreviation N, which is generally preferred by neuroanatomists and clinical neurologists. Comparative anatomists prefer CN, an equally valid abbreviation.

Each cranial nerve attaches to the brain near the associated sensory or motor nuclei. The sensory nuclei act as switching centers, with the postsynaptic neurons relaying the information either to other nuclei or to processing centers within the cerebral or cerebellar cortex. Similarly, the motor nuclei receive convergent inputs from higher centers or from other nuclei along the brain stem.

The next section classifies cranial nerves as primarily sensory, special sensory, motor, or mixed (sensory and motor). This is a useful method of classification, but it is based on the primary function, and a cranial nerve can have important secondary functions. Two examples are worth noting:

1. As elsewhere in the PNS, a nerve containing tens of thousands of motor fibers to a skeletal muscle will also contain sensory fibers from proprioceptors in that muscle. These sensory fibers are assumed to be present but are ignored in the primary classification of the nerve.
2. Regardless of their other functions, several cranial nerves (N III, N VII, N IX, and N X) distribute autonomic fibers to peripheral ganglia, just as spinal nerves deliver them to ganglia along the spinal cord. The presence of small numbers of autonomic fibers will be noted (and discussed further in Chapter 17) but ignored in the classification of the nerve.

The Olfactory Nerve (N I) [FIGURES 15.21/15.22]

Primary function: Special sensory (smell)
Origin: Receptors of olfactory epithelium
Passes through: Cribriform plate of ethmoid p. 149
Destination: Olfactory bulbs

The first pair of cranial nerves (Figure 15.22•) carries special sensory information responsible for the sense of smell. The olfactory receptors are specialized neurons in the epithelium covering the roof of the nasal cavity, the superior nasal conchae of the ethmoid and the superior parts of the nasal septum. Axons from these sensory neurons collect to form 20 or more bundles that penetrate the cribriform plate of the ethmoid. These bundles are components of the **olfactory nerves (N I)**. Almost at once these bundles enter the **olfactory bulbs,** neural masses on either side of the crista galli. The olfactory afferents synapse within the olfactory bulbs. The axons of the postsynaptic neurons proceed to the cerebrum along the slender **olfactory tracts** (Figures 15.21 and 15.22•).

Because the olfactory tracts look like typical peripheral nerves, anatomists about one hundred years ago misidentified these tracts as the first cranial nerve. Later studies demonstrated that the olfactory tracts and bulbs are part of the cerebrum, but by then the numbering system was already firmly established. Anatomists were left with a forest of tiny olfactory nerve bundles lumped together as N I.

The olfactory nerves are the only cranial nerves attached directly to the cerebrum. The rest originate or terminate within nuclei of the diencephalon or brain stem, and the ascending sensory information synapses in the thalamus before reaching the cerebrum.

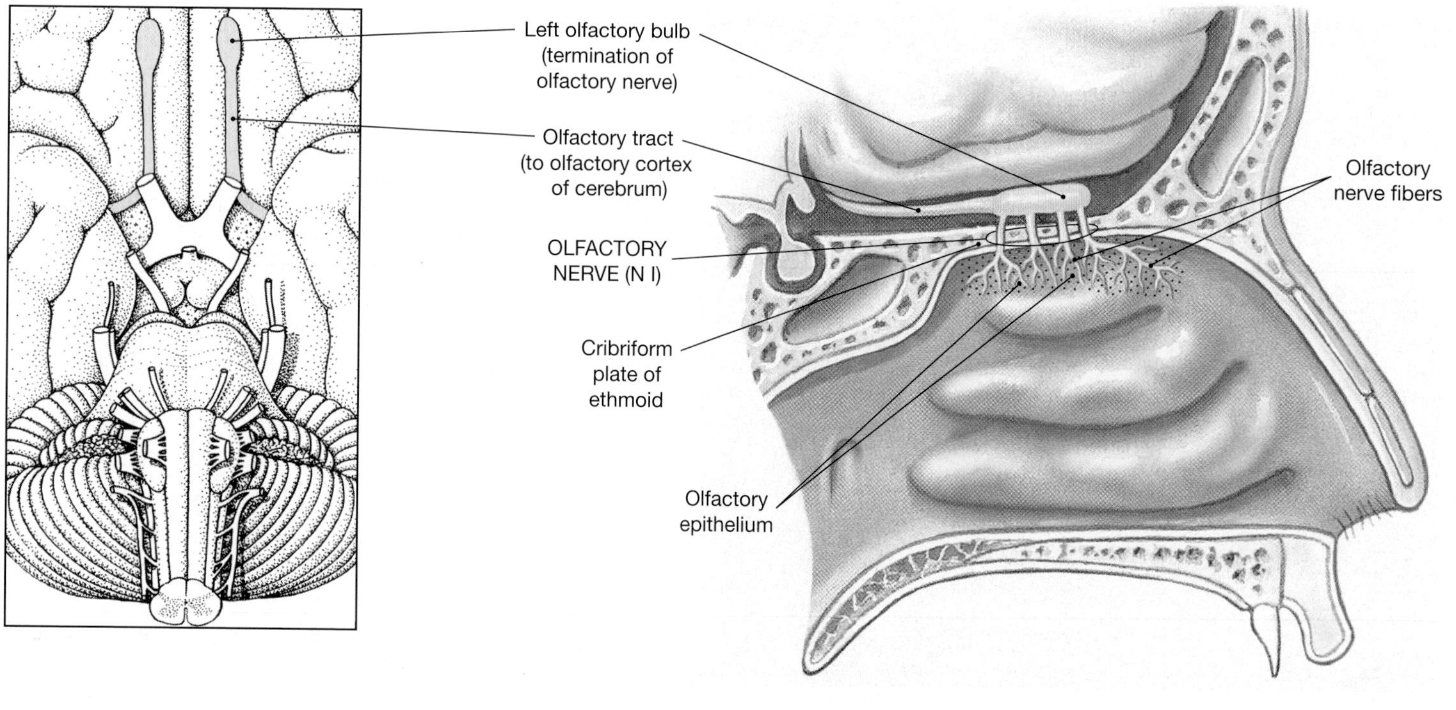

FIGURE 15.22 **THE OLFACTORY NERVE**

The Optic Nerve (N II) [FIGURES 15.21/15.23]

Primary function: Special sensory (vision)
Origin: Retina of eye
Passes through: Optic canal of sphenoid p. 149
Destination: Diencephalon by way of the optic chiasm

The **optic nerves** (**N II**) carry visual information from special sensory ganglia in the eyes. These nerves, diagrammed in Figure 15.23●, contain about 1 million sensory nerve fibers. These nerves pass through the optic foramina of the sphenoid before converging at the ventral and anterior margin of the diencephalon, at the **optic chiasm** (*chiasma*, a crossing). At the optic chiasm, some of the fibers from each optic nerve cross over to the opposite side of the brain. The reorganized axons continue toward the lateral geniculate nuclei of the thalamus as the **optic tracts** (Figures 15.21 and 15.23●). After synapsing in the lateral geniculate nuclei, projection fibers deliver the information to the occipital lobe of the brain. This arrangement results in each cerebral hemisphere receiving visual information from the lateral half of the retina of the eye on that side and from the medial half of the retina of the eye on the opposite side. A relatively small number of axons in the optic tracts bypass the lateral geniculate nuclei and synapse in the superior colliculi of the mesencephalon. This pathway will be considered in Chapter 18.

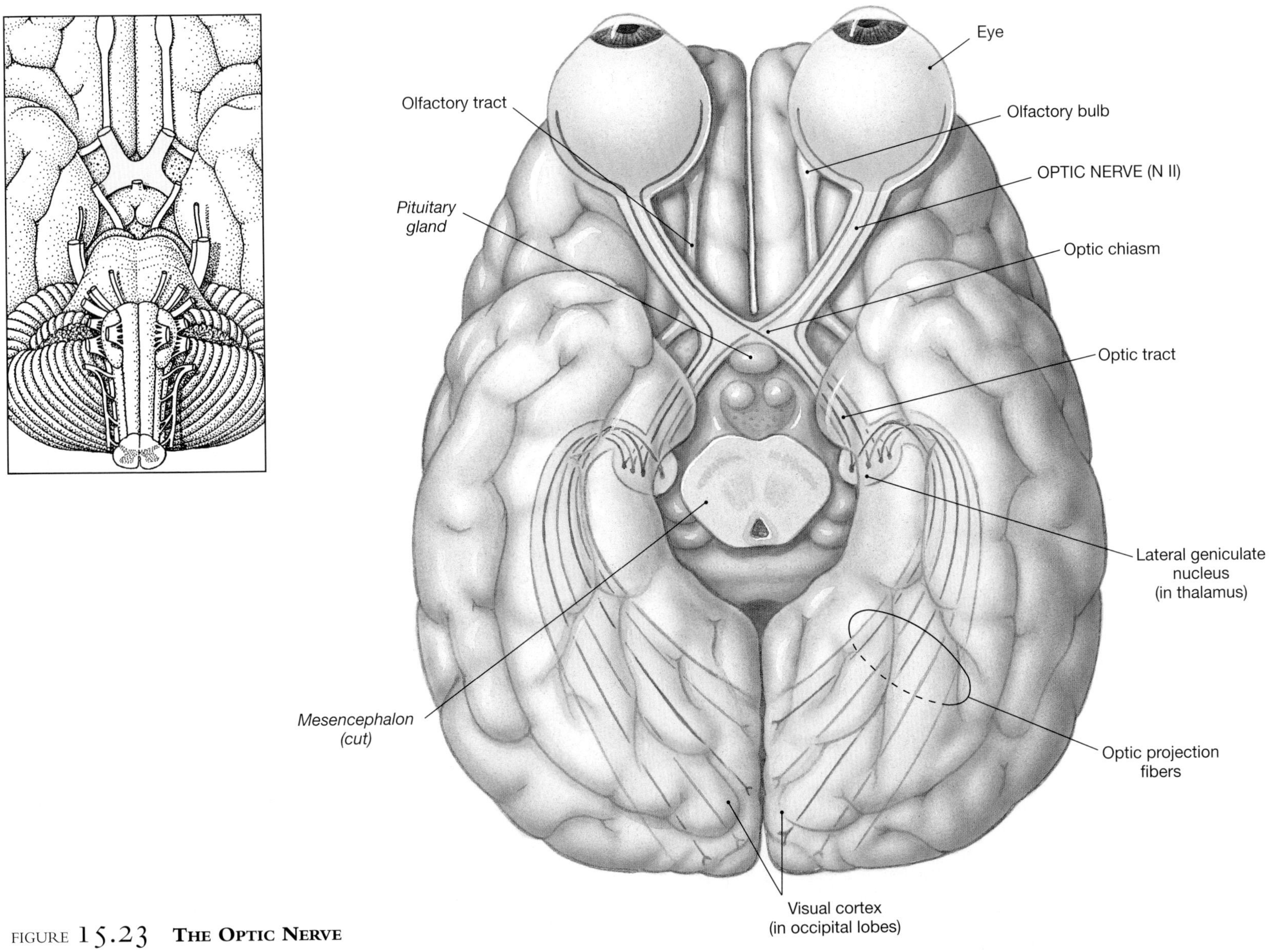

FIGURE 15.23 **The Optic Nerve**

The Oculomotor Nerve (N III) [FIGURES 15.21/15.24]

Primary function: Motor, eye movements
Origin: Mesencephalon
Passes through: Superior orbital fissure of sphenoid p. 149
Destination: *Somatic motor*: superior, inferior, and medial rectus muscles; the inferior oblique muscle; the levator palpebrae superioris muscle p. 272 *Visceral motor*: intrinsic eye muscles

The mesencephalon contains the motor nuclei controlling the third and fourth cranial nerves. The **oculomotor nerves (N III)** innervate all but two of the extra-ocular muscles. These cranial nerves emerge from the ventral surface of the mesencephalon (Figure 15.21●) and penetrate the posterior orbital wall at the superior orbital fissure. The oculomotor nerve (Figure 15.24●) controls four of the six extra-ocular muscles and the levator palpebrae superioris muscle, which raises the upper eyelid.

The oculomotor nerve also delivers preganglionic autonomic fibers to neurons of the **ciliary ganglion**. The ganglionic neurons control intrinsic eye muscles. These muscles change the diameter of the pupil, adjusting the amount of light entering the eye, and change the shape of the lens to focus images on the retina.

The Trochlear Nerve (N IV) [FIGURES 15.21/15.24]

Primary function: Motor, eye movements
Origin: Mesencephalon
Passes through: Superior orbital fissure of sphenoid p. 149
Destination: Superior oblique muscle p. 272

The **trochlear nerve** (TRŌK-lē-ar; *trochlea*, pulley), smallest of the cranial nerves, innervates the superior oblique muscle of the eye (Figure 15.24●). The motor nucleus lies in the ventrolateral portion of the mesencephalon, but the fibers emerge from the surface of the tectum to enter the orbit through the superior orbital fissure (Figure 15.21●). The name "trochlear nerve" should remind you that the innervated muscle passes through a ligamentous sling, or *trochlea*, on its way to its insertion on the superior surface of the eye.

The Trigeminal Nerve (N V) [FIGURES 15.21/15.25]

Primary function: Mixed (sensory and motor); ophthalmic and maxillary branches sensory, mandibular branch mixed
Origin: *Ophthalmic branch* (sensory): orbital structures, nasal cavity, skin of forehead, superior eyelid, eyebrow, and part of the nose
Maxillary branch (sensory): inferior eyelid, upper lip, gums, and teeth; cheek; nose, palate, and part of the pharynx
Mandibular branch (mixed): sensory from lower gums, teeth, and lips; palate and tongue (part); motor from motor nuclei of pons (Figure 15.21●)
Passes through: Ophthalmic branch through superior orbital fissure, maxillary branch through foramen rotundum, mandibular branch through foramen ovale p. 149
Destination: Ophthalmic and maxillary branches to sensory nuclei in pons; mandibular branch innervates muscles of mastication p. 273

FIGURE 15.24 **CRANIAL NERVES CONTROLLING THE EXTRA-OCULAR MUSCLES**

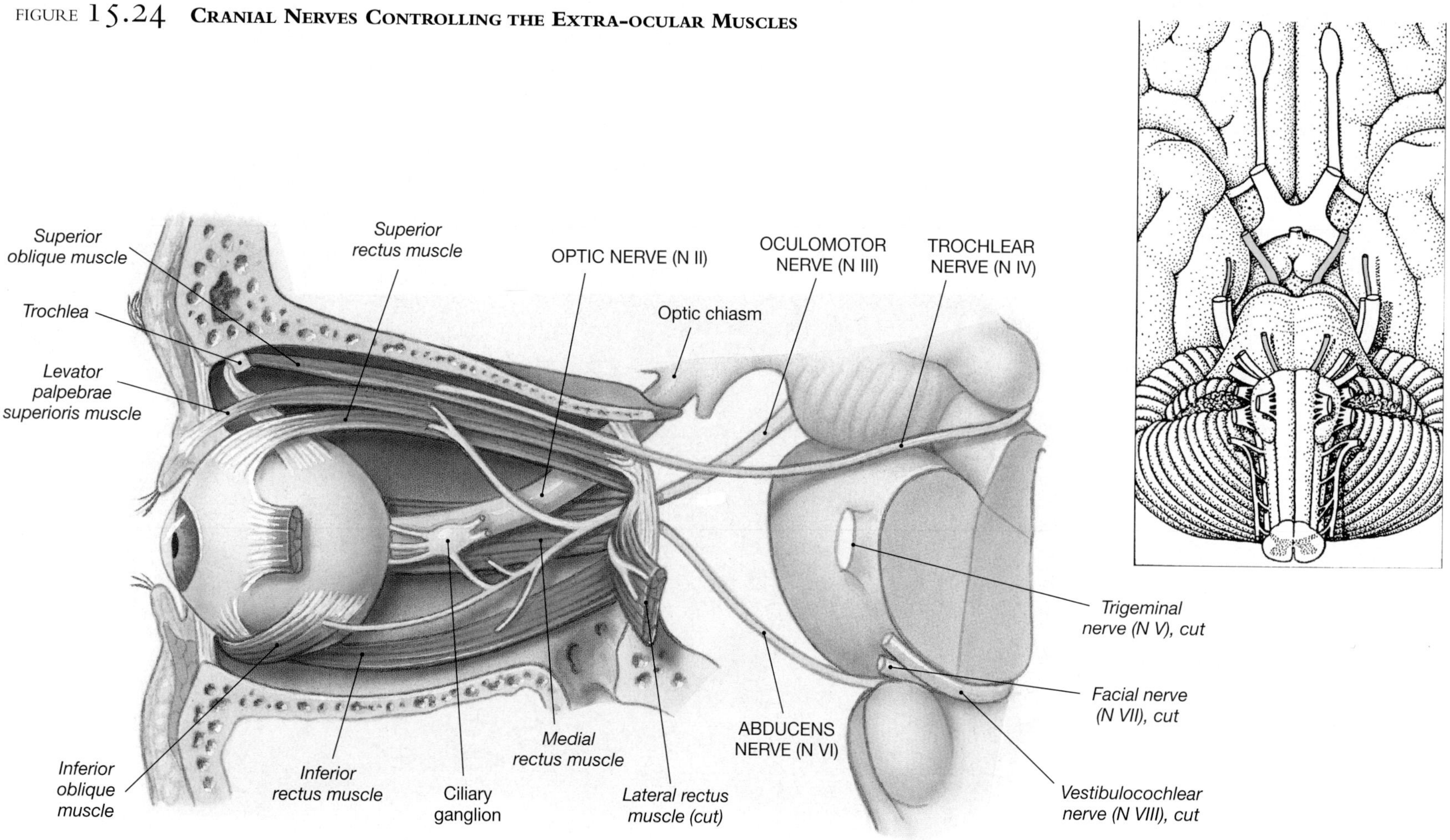

The pons contains the nuclei associated with three cranial nerves (N V, N VI, and N VII) and contributes to the control of a fourth (N VIII). The **trigeminal** (trī-JEM-i-nal) **nerve** (Figure 15.25●) is the largest cranial nerve. This mixed nerve provides sensory information from the head and face and motor control over the muscles of mastication. Sensory (dorsal) and motor (ventral) roots originate on the lateral surface of the pons. The sensory branch is larger, and the enormous **semilunar ganglion** (*trigeminal ganglion*) contains the cell bodies of the sensory neurons. As the name implies, the trigeminal has three major branches; the relatively small motor root contributes to only one of the three.

Branch 1. The **ophthalmic branch** of the trigeminal nerve is purely sensory. This nerve innervates orbital structures, the nasal cavity and sinuses, and the skin of the forehead, eyebrows, eyelids, and nose. It leaves the cranium through the superior orbital fissure, then branches within the orbit.

Branch 2. The **maxillary branch** of the trigeminal nerve is also purely sensory. It supplies the lower eyelid, upper lip, cheek, and nose. Deeper sensory structures of the upper gums and teeth, the palate, and portions of the pharynx are also innervated by the maxillary nerve branch. The maxillary branch leaves the cranium at the foramen rotundum, entering the floor of the orbit through the inferior orbital fissure. A major branch of the maxillary, the *infraorbital nerve*, passes through the infraorbital foramen to supply adjacent portions of the face.

Branch 3. The **mandibular branch** is the largest branch of the trigeminal nerve, and it carries all of the fibers of the motor root. This branch exits the cranium through the foramen ovale. The motor components of the mandibular nerve innervate the muscles of mastication. The sensory fibers carry proprioceptive information from those muscles and monitor (1) the skin of the temples; (2) the lateral surfaces, gums, and teeth of the mandible; (3) the salivary glands; and (4) the anterior portions of the tongue.

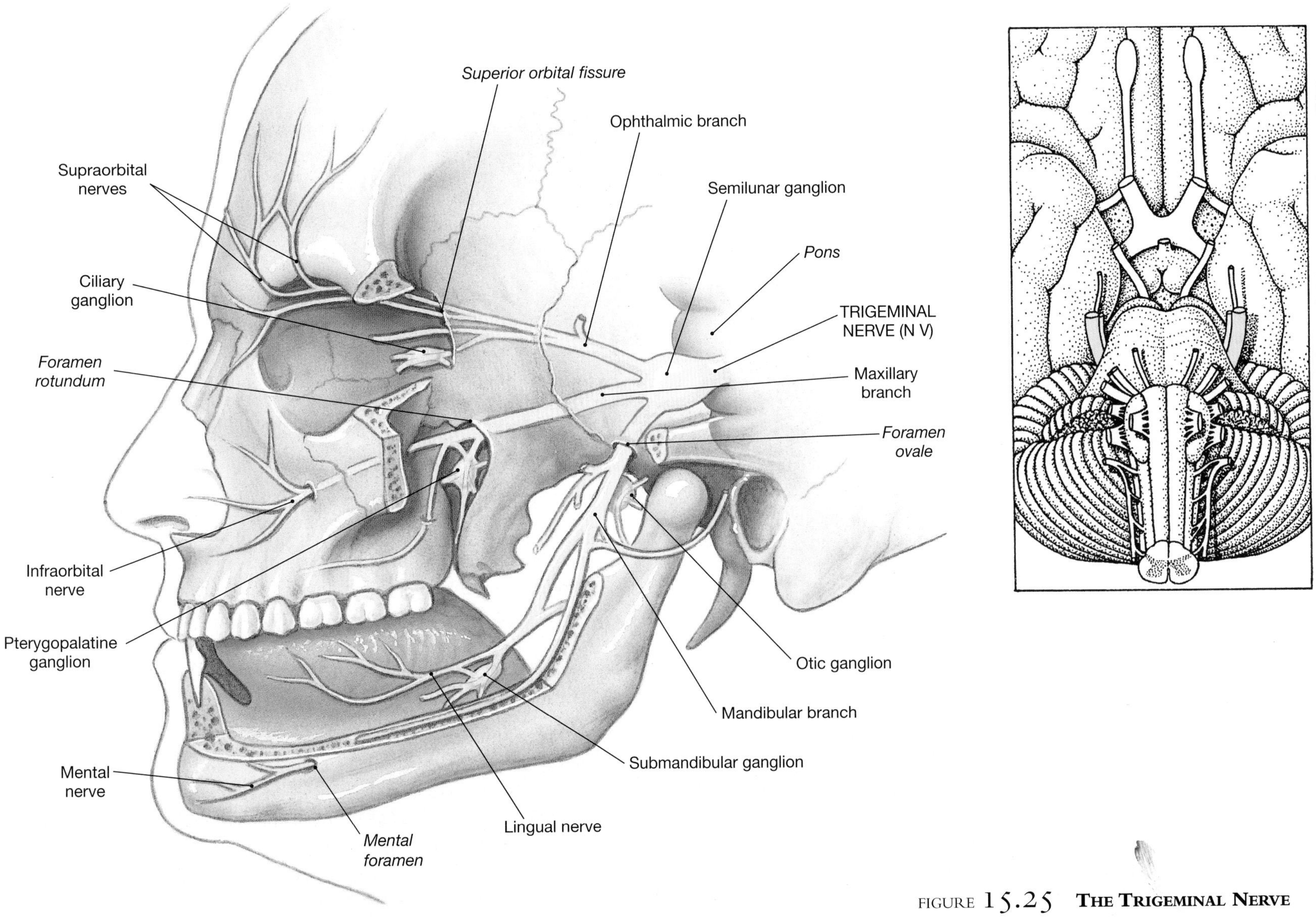

FIGURE 15.25 **THE TRIGEMINAL NERVE**

The trigeminal nerve branches are associated with the *ciliary*, *pterygopalatine*, *submandibular*, and *otic ganglia*. These are autonomic ganglia whose neurons innervate structures of the face. The trigeminal nerve does not contain visceral motor fibers, and all of its fibers pass through these ganglia without synapsing. However, branches of other cranial nerves, such as the *facial nerve*, can be bound to the trigeminal nerve; these branches may innervate the ganglion, and the postganglionic autonomic fibers may then travel with the trigeminal nerve to peripheral structures. The ciliary ganglion was discussed earlier (p. 412), and the other ganglia will be detailed below, with the branches of the *facial nerve* (N VII).

CLINICAL BRIEF

Tic Douloureux

Tic douloureux (doo-loo-ROO; *douloureux*, painful) affects one individual out of every 25,000. Sufferers complain of severe, almost totally debilitating pain triggered by contact with the lip, tongue, or gums. The pain arrives with a sudden, shocking intensity and then disappears. Usually only one side of the face is involved. Another name for this condition is **trigeminal neuralgia**, for it is the maxillary and mandibular branches of N V that innervate the sensitive areas. This condition usually affects adults over 40 years of age; the cause is unknown. The pain can often be temporarily controlled by drug therapy, but surgical procedures may eventually be required. The goal of the surgery is the destruction of the sensory nerves carrying the pain sensations. They can be destroyed by actually cutting the nerve, a procedure called a **rhizotomy** (*rhiza*, root), or by injecting chemicals such as alcohol or phenol into the nerve at the foramina ovale and rotundum. The sensory fibers may also be destroyed by inserting an electrode and cauterizing the sensory nerve trunks as they leave the semilunar ganglion.

The Abducens Nerve (N VI) [FIGURES 15.21/15.24]

Primary function: Motor, eye movements
Origin: Pons
Passes through: Superior orbital fissure of sphenoid p. 149
Destination: Lateral rectus muscle p. 272

The **abducens** (ab-DŪ-senz) **nerve** innervates the lateral rectus, the sixth of the extrinsic eye muscles. Innervation of this muscle makes lateral movements of the eyeball possible. The nerve emerges from the inferior surface of the brain at the border between the pons and the medulla oblongata (Figure 15.21●). It reaches the orbit through the superior orbital fissure in company with the oculomotor and trochlear nerves (Figure 15.24●).

The Facial Nerve (N VII) [FIGURES 15.21/15.26]

Primary function: Mixed (sensory and motor)
Origin: Sensory from taste receptors on anterior two-thirds of tongue; motor from motor nuclei of pons
Passes through: Internal acoustic meatus of temporal bone, along facial canal to reach stylomastoid foramen p. 149
Destination: Sensory to sensory nuclei of pons; *Somatic motor*: muscles of facial expression: p. 271 *Visceral motor*: lacrimal (tear) gland and nasal mucous glands via pterygopalatine ganglion; submandibular and sublingual salivary glands via submandibular ganglion

The **facial nerve** is a mixed nerve. The cell bodies of the sensory neurons are located in the **geniculate ganglion**, and the motor nuclei are in the pons (Figure 15.21●). The sensory and motor roots combine to form a large nerve that passes through the internal acoustic meatus of the temporal bone (Figure 15.26●). The nerve then passes through the facial canal to reach the face through the stylomastoid foramen. p. 139 The sensory neurons monitor proprioceptors in the facial muscles, provide deep pressure sensations over the face, and receive taste information from receptors along the anterior two-thirds of the tongue. Somatic motor fibers control the superficial muscles of the scalp and face and deep muscles near the ear.

The facial nerve carries preganglionic autonomic fibers to the sphenopalatine and submandibular ganglia.

- *Pterygopalatine ganglion*: The *greater petrosal nerve* innervates the pterygopalatine ganglion. Postganglionic fibers from this ganglion innervate the lacrimal gland and small glands of the nasal cavity and pharynx.
- *Submandibular ganglion*: To reach the submandibular ganglion, autonomic fibers leave the facial nerve and travel along the mandibular branch of the trigeminal nerve. Postganglionic fibers from this ganglion innervate the *submandibular* and *sublingual* (*sub*, under + *lingua*, tongue) *salivary glands*.

CLINICAL BRIEF

Bell's Palsy

Bell's palsy results from an inflammation of the facial nerve that is probably related to viral infection. Involvement of the facial nerve (N VII) can be deduced from symptoms of paralysis of facial muscles on the affected side and loss of taste sensations from the anterior two-thirds of the tongue. The individual does not show prominent sensory deficits, and the condition is usually painless. In most cases, Bell's palsy "cures itself" after a few weeks or months but this process can be accelerated by early treatment with corticosteroids and antiviral drugs.

The Vestibulocochlear Nerve (N VIII) [FIGURES 15.21/15.27]

Primary function: Special sensory: balance and equilibrium (vestibular branch) and hearing (cochlear branch)
Origin: Monitors receptors of the inner ear (vestibule and cochlea)
Passes through: Internal acoustic meatus of the temporal bone p. 149
Destination: Vestibular and cochlear nuclei of pons and medulla oblongata (Figure 15.21●)

The **vestibulocochlear nerve** is also known as the *acoustic nerve*, the *auditory nerve*, and the *statoacoustic nerve*. We will use the term vestibulocochlear because it indicates the names of its two major branches: the *vestibular branch* and the *cochlear branch*. The vestibulocochlear nerve lies posterior to the origin of the facial nerve, straddling the boundary between the pons and the medulla oblongata (Figure 15.27●). This nerve reaches the sensory receptors of the inner ear by entering the internal acoustic meatus in company with the facial nerve. There are two distinct bundles of

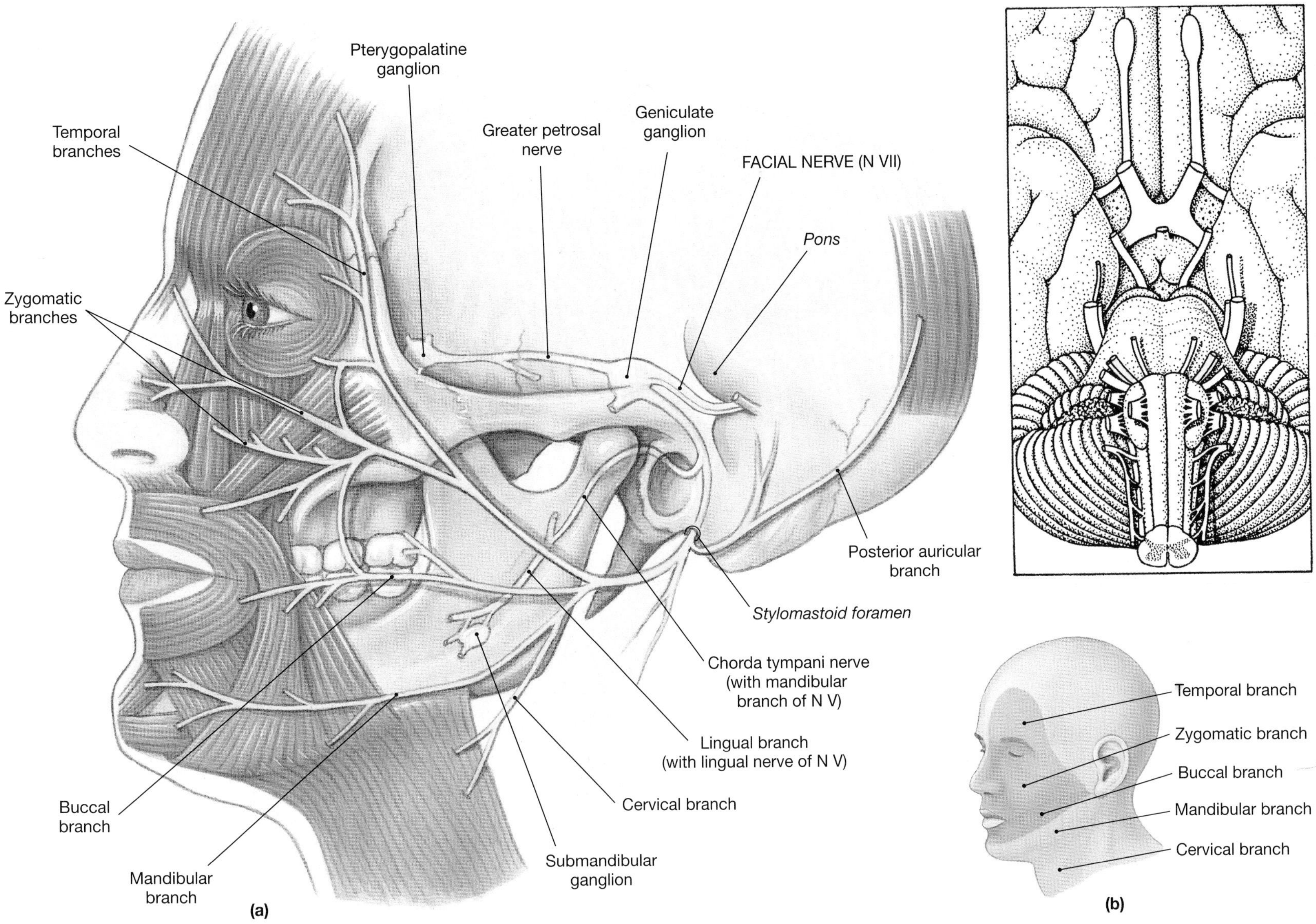

FIGURE 15.26 **The Facial Nerve**

(**a**) Origin and branches of the facial nerve. (**b**) The superficial distribution of the five major branches of the facial nerve.

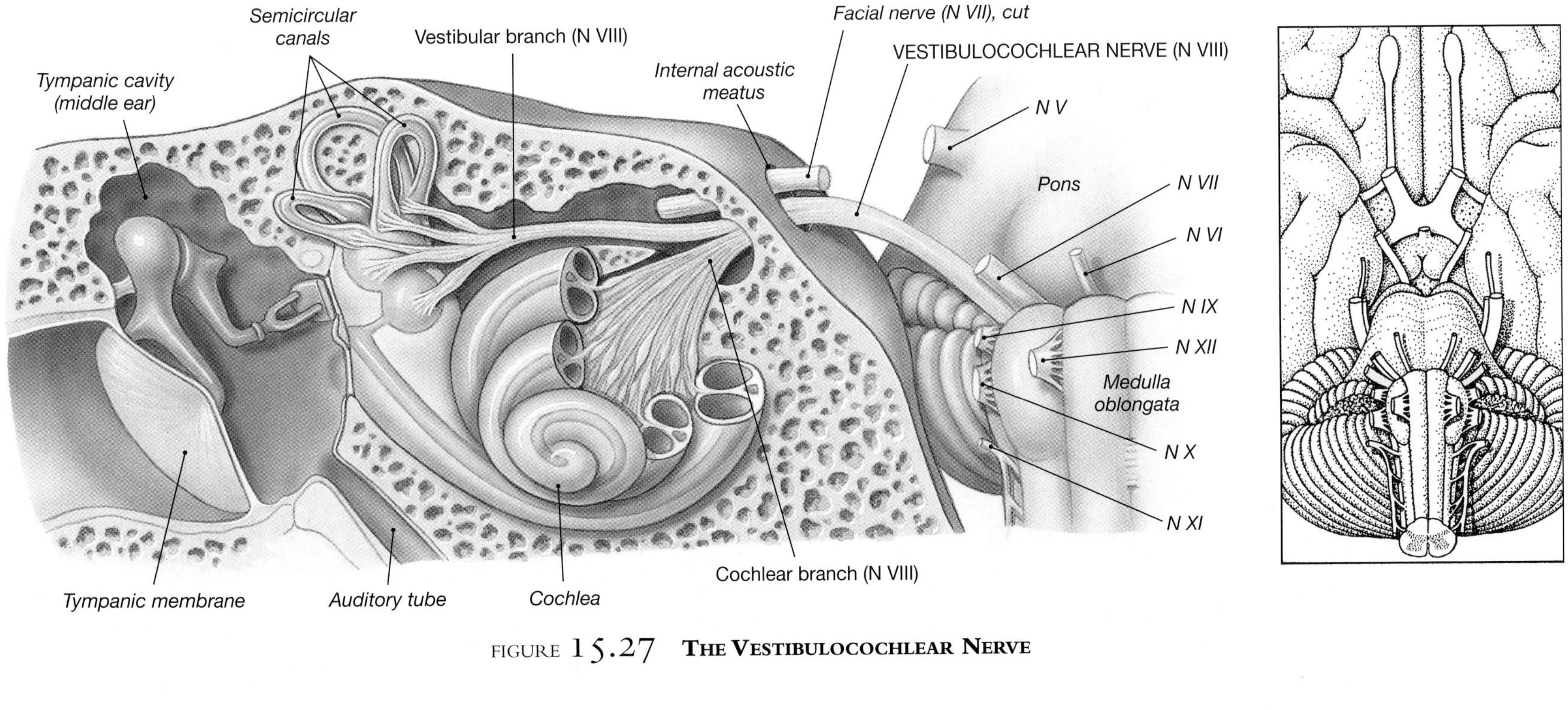

FIGURE 15.27 **The Vestibulocochlear Nerve**

sensory fibers within the vestibulocochlear nerve. The **vestibular nerve** (*vestibulum*, cavity) originates at the receptors of the *vestibule*, the portion of the inner ear concerned with balance sensations. The sensory neurons are located within an adjacent sensory ganglion, and their axons target the **vestibular nuclei** of the medulla oblongata. These afferents convey information concerning position, movement, and balance. The **cochlear nerve** (KOK-lē-ar; *cochlea*, snail shell) monitors the receptors in the cochlea that provide the sense of hearing. The nerve cells are located within a peripheral ganglion, and their axons synapse within the **cochlear nuclei** of the medulla oblongata. Axons leaving the vestibular and cochlear nuclei relay the sensory information to other centers or initiate reflexive motor responses. Balance and the sense of hearing will be discussed in Chapter 18.

The Glossopharyngeal Nerve (N IX) [FIGURES 15.21/15.28]

Primary function: Mixed (sensory and motor)
Origin: Sensory from posterior one-third of the tongue, part of the pharynx and palate, the carotid arteries of the neck; motor from motor nuclei of medulla oblongata
Passes through: Jugular foramen between occipital and temporal bones p. 149
Destination: Sensory fibers to sensory nuclei of medulla oblongata (Figure 15.21●); *Somatic motor*: pharyngeal muscles involved in swallowing; *Visceral motor*: parotid salivary gland, after synapsing in the otic ganglion

In addition to the vestibular nucleus of N VIII, the medulla oblongata contains the sensory and motor nuclei for the ninth, tenth, eleventh, and twelfth cranial nerves. The **glossopharyngeal nerve** (glos-ō-fah-RIN-jē-al; *glossum*, tongue) innervates the tongue and pharynx. The glossopharyngeal nerve passes through the cranium through the jugular foramen in company with N X and N XI (Figure 15.28●).

The glossopharyngeal is a mixed nerve, but sensory fibers are most abundant. The sensory neurons are in the **superior ganglion** (*jugular ganglion*) and the **inferior ganglion** (*petrosal ganglion*)[3]. The afferent fibers carry general sensory information from the lining of the pharynx and the soft palate to a nucleus in the medulla oblongata. The glossopharyngeal nerve also provides taste sensations from the posterior third of the tongue and has special receptors monitoring the blood pressure and dissolved gas concentrations within major blood vessels.

The somatic motor fibers control the pharyngeal muscles involved in swallowing. Visceral motor fibers synapse in the otic ganglion, and postganglionic fibers innervate the parotid salivary gland of the cheek.

[3] The names of the ganglia associated with N IX and N X vary from reference to reference. N IX has a *superior ganglion*, also called the *jugular ganglion*, and an *inferior ganglion*, also called the *petrosal* (or *petrous*) *ganglion*. N X also has two major ganglia, a *superior ganglion*, or *jugular ganglion*, and an *inferior ganglion*, or *nodose ganglion*. *Superior* and *inferior* are the names recommended by the Terminologia Anatomica.

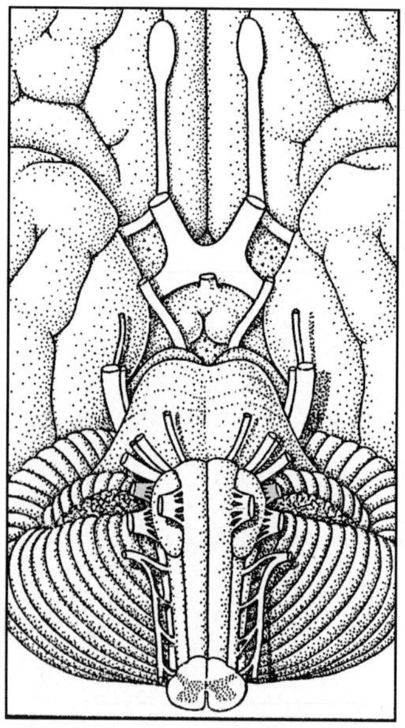

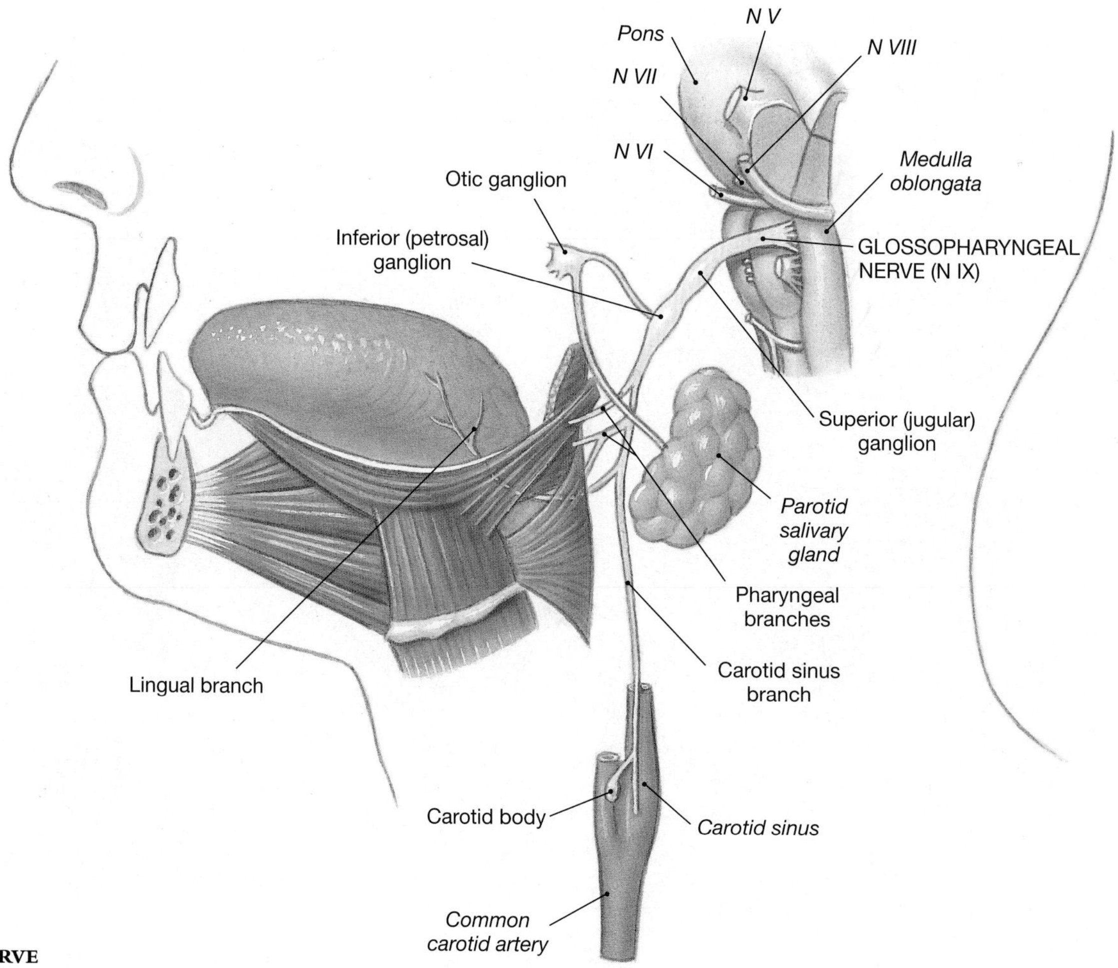

FIGURE 15.28 **The Glossopharyngeal Nerve**

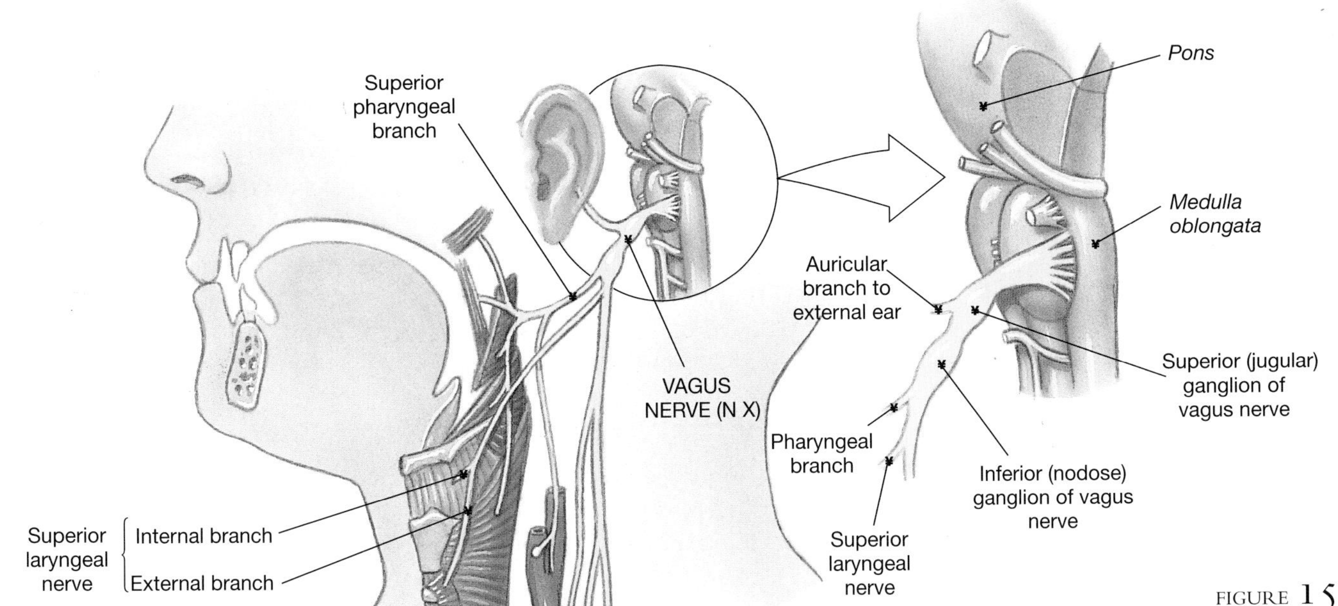

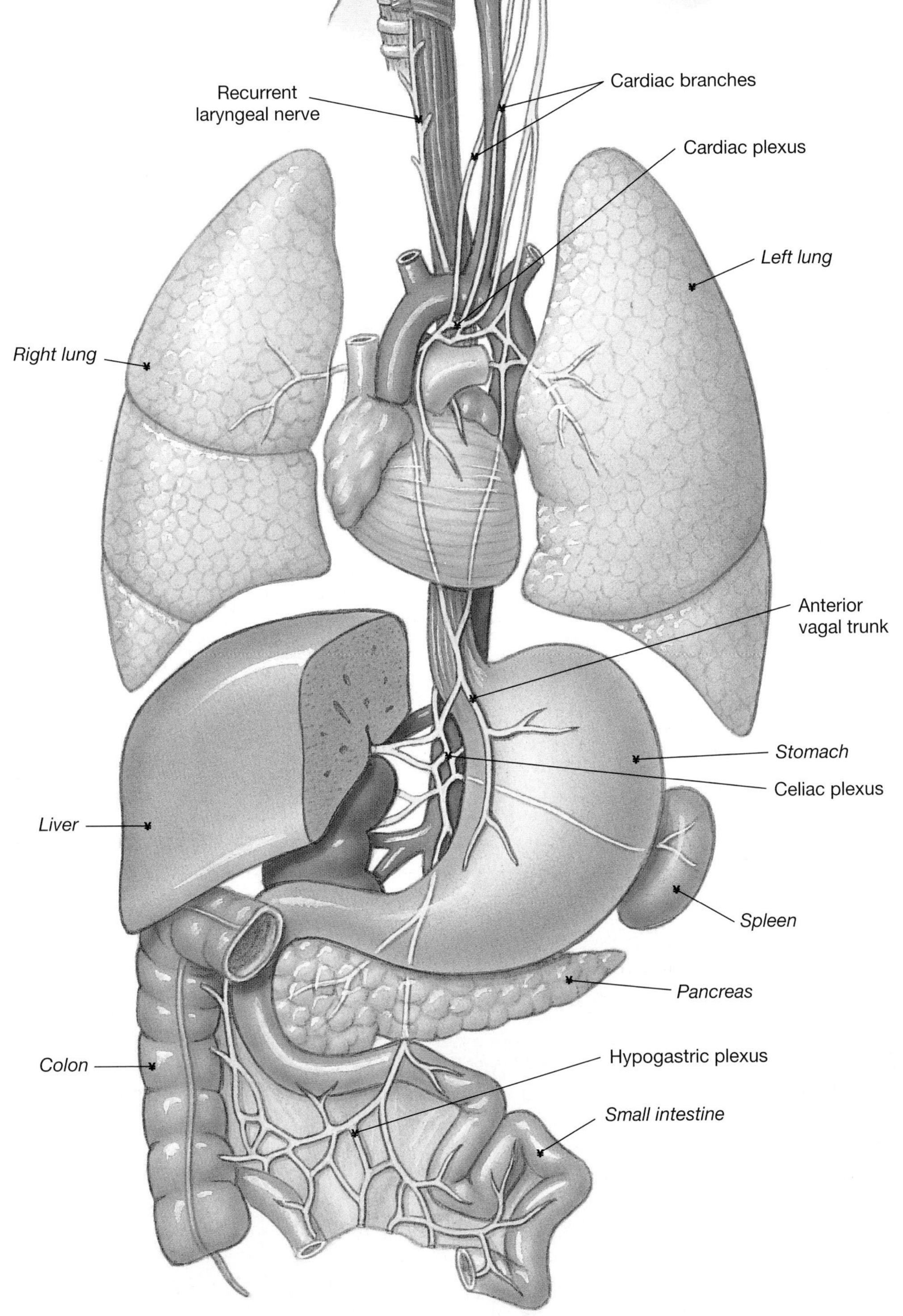

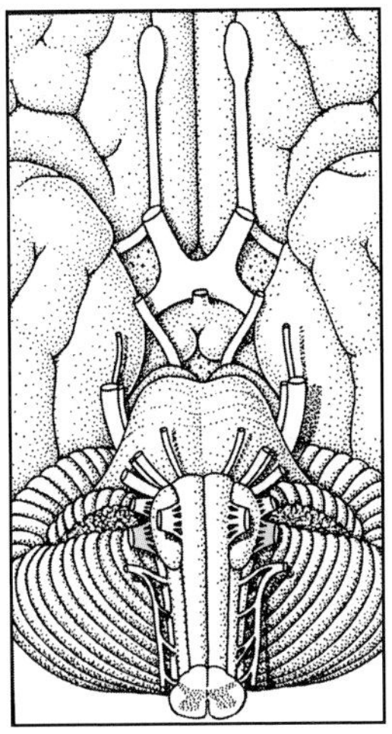

FIGURE 15.29 **THE VAGUS NERVE**

The Vagus Nerve (N X) [FIGURES 15.21/15.29]

Primary function: Mixed (sensory and motor)
Origin: *Visceral sensory* from pharynx (part), auricle, external acoustic meatus, diaphragm, and visceral organs in thoracic and abdominopelvic cavities. *Visceral motor* from motor nuclei in the medulla oblongata
Passes through: Jugular foramen between occipital and temporal bones p. 149
Destination: Sensory fibers to sensory nuclei and autonomic centers of medulla oblongata (Figure 15.21●); *Visceral motor* fibers to muscles of the palate, pharynx, digestive, respiratory, and cardiovascular systems in the thoracic and abdominal cavities

The **vagus** (VĀ-gus) **nerve** arises immediately posterior to the glossopharyngeal nerve. Many small rootlets contribute to its formation, and developmental studies indicate that this nerve probably represents the fusion of several smaller cranial nerves during our evolution. As its name suggests (*vagus*, wanderer), the vagus nerve branches and radiates extensively. Figure 15.29● shows only the general pattern of distribution.

Sensory neurons are located within the **superior ganglion**, or *jugular ganglion*, and the **inferior ganglion**, or *nodose ganglion*. The vagus nerve provides somatic sensory information concerning the external acoustic meatus, a portion of the ear, and the diaphragm, and special sensory information from pharyngeal taste receptors. But the majority of the vagal afferents provide visceral sensory information from receptors along the esophagus, respiratory tract, and abdominal viscera as distant as the terminal segments of the large intestine. Vagal afferents are vital to the autonomic control of visceral function, but because the information often fails to reach the cerebral cortex, we are seldom aware of the sensations they provide.

The motor components of the vagus nerve are equally diverse. The vagus nerve carries preganglionic autonomic fibers that affect the heart and control smooth muscles and glands within the areas monitored by its sensory fibers, including the stomach, intestines, and gallbladder. The vagus nerve also distributes somatic motor fibers to muscles of the palate and pharynx, but these are actually branches of the *accessory nerve*, described below.

The Accessory Nerve (N XI) [FIGURES 15.21/15.30]

Primary function: Motor
Origin: Motor nuclei of spinal cord and medulla oblongata
Passes through: Jugular foramen between occipital and temporal bones p. 149
Destination: Medullary branch innervates voluntary muscles of palate, pharynx, and larynx; spinal branch controls sternocleidomastoid and trapezius muscles

The **accessory nerve** differs from other cranial nerves in that some of its motor fibers originate in the lateral gray horns of the first five cervical segments of the spinal cord (Figures 15.21 and 15.30●). These fibers form the *spinal root*, which enters the cranium through the foramen magnum, uniting with the motor fibers of the *cranial root*, which originates at a nucleus in the medulla oblongata, and leave the cranium through the jugular foramen. The accessory nerve consists of two branches:

1. The **internal branch** joins the vagus nerve and innervates the voluntary swallowing muscles of the soft palate and pharynx and the intrinsic muscles that control the vocal cords.
2. The **external branch** controls the sternocleidomastoid and trapezius muscles of the neck and back. pp. 277, 293 The motor fibers of this branch originate in the lateral gray horns of C_1 to C_5.

The Hypoglossal Nerve (N XII) [FIGURES 15.21/15.30]

Primary function: Motor, tongue movements
Origin: Motor nuclei of the medulla oblongata (Figure 15.21●)
Passes through: Hypoglossal canal of occipital bone p. 149
Destination: Muscles of the tongue p. 274

The **hypoglossal** (hī-pō-GLOS-al) **nerve** leaves the cranium through the hypoglossal canal of the occipital bone. It then curves inferiorly, anteriorly, and then superiorly to reach the skeletal muscles of the tongue (Figure 15.30●). This nerve provides voluntary motor control over movements of the tongue.

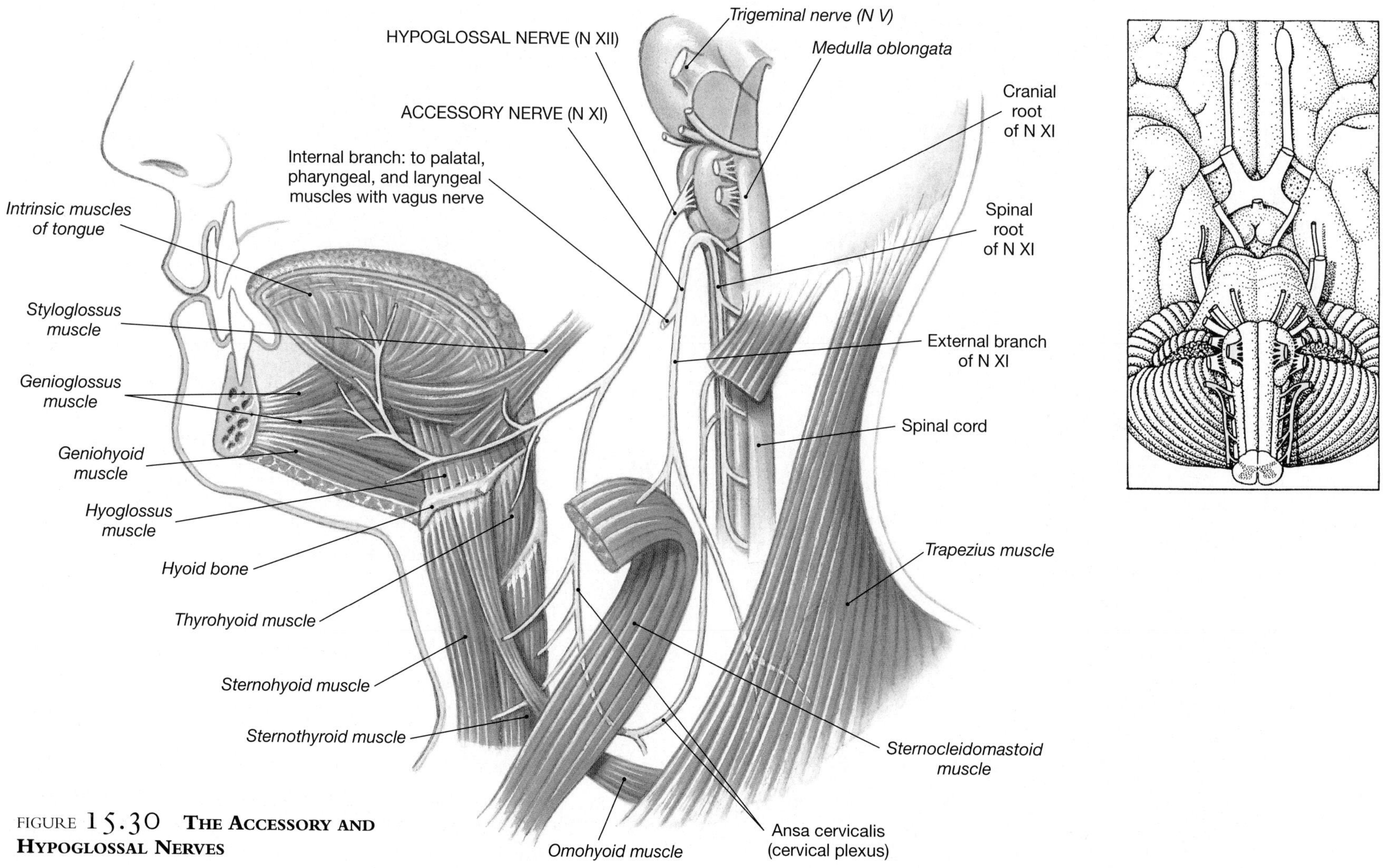

FIGURE 15.30 **THE ACCESSORY AND HYPOGLOSSAL NERVES**

TABLE 15.12 THE CRANIAL NERVES

Cranial Nerve (#)	Sensory Ganglion	Branch	Primary Function	Foramen	Innervation
Olfactory (I)			Special sensory	Cribriform plate	Olfactory epithelium
Optic (II)			Special sensory	Optic canal	Retina of eye
Oculomotor (III)			Motor	Superior orbital fissure	Inferior, medial, superior rectus, inferior oblique, and levator palpebrae muscles; intrinsic muscles of eye
Trochlear (IV)			Motor	Superior orbital fissure	Superior oblique muscle
Trigeminal (V)	Semilunar		Mixed		Areas associated with the jaws
		Ophthalmic	Sensory	Superior orbital fissure	Orbital structures, nasal cavity, skin of forehead, upper eyelid, eyebrows, nose (part)
		Maxillary	Sensory	Foramen rotundum	Lower eyelid; upper lip, gums, and teeth; cheek, nose (part), palate and pharynx (part)
		Mandibular	Mixed	Foramen ovale	*Sensory* to lower gums, teeth, lips; palate (part) and tongue (part); *motor* to muscles of mastication
Abducens (VI)			Motor	Superior orbital fissure	Lateral rectus muscle
Facial (VII)	Geniculate		Mixed	Internal acoustic meatus to facial canal; exits at stylomastoid foramen	*Sensory* to taste receptors on anterior 2/3 of tongue; *motor* to muscles of facial expression, lacrimal gland, submandibular salivary gland, sublingual salivary glands
Vestibulocochlear (Acoustic) (VIII)		Cochlear	Special sensory	Internal acoustic meatus	Cochlea (receptors for hearing)
		Vestibular	Special sensory	As above	Vestibule (receptors for motion and balance)
Glossopharyngeal (IX)	Superior (jugular) and inferior (petrosal)		Mixed	Jugular foramen	*Sensory* from posterior 1/3 of tongue; pharynx and palate (part); carotid body (monitors blood pressure, pH, and levels of respiratory gases); *motor* to pharyngeal muscles, parotid salivary gland
Vagus (X)	Superior (jugular) and inferior (nodose)		Mixed	Jugular foramen	*Sensory* from pharynx; auricle and external acoustic meatus; diaphragm; visceral organs in thoracic and abdominopelvic cavities; *motor* to palatal and pharyngeal muscles, and visceral organs in thoracic and abdominopelvic cavities
Accessory (XI)		Internal branch	Motor	Jugular foramen	Skeletal muscles of palate, pharynx and larynx (with branches of the vagus nerve)
		External branch	Motor	Jugular foramen	Sternocleidomastoid and trapezius muscles
Hypoglossal (XII)			Motor	Hypoglossal canal	Tongue musculature

A Summary of Cranial Nerve Branches and Functions

Few people are able to remember the names, numbers, and functions of the cranial nerves without a struggle. Mnemonic devices may prove useful. The most famous and oft-repeated is *On Old Olympus's Towering Top A Finn And German Viewed Some Hops.* (The *And* refers to the acoustic nerve, an alternative name for N VIII and the *Some* refers to the spinal accessory nerve, an alternative name for N XI.) A more modern one, *Oh, Once One Takes The Anatomy Final, Very Good Vacations Are Heavenly*, may be a bit easier to remember. A summary of the basic distribution and function of each cranial nerve is detailed in Table 15.12.

✓ CONCEPT CHECK

- John is experiencing problems in moving his tongue. His doctor tells him the problems are due to pressure on a cranial nerve. Which cranial nerve is involved?
- What symptoms would you associate with damage to the abducens nerve (N VI)?
- A blow to the head has caused Julie to lose her balance. Which cranial nerve and what branch of that nerve are probably involved?
- Bruce has lost the ability to detect tastes on the tip of his tongue. What cranial nerve is involved?

Before proceeding, briefly review the summaries of skull formation (pp. 156–157) and vertebral column development (pp. 172–173).

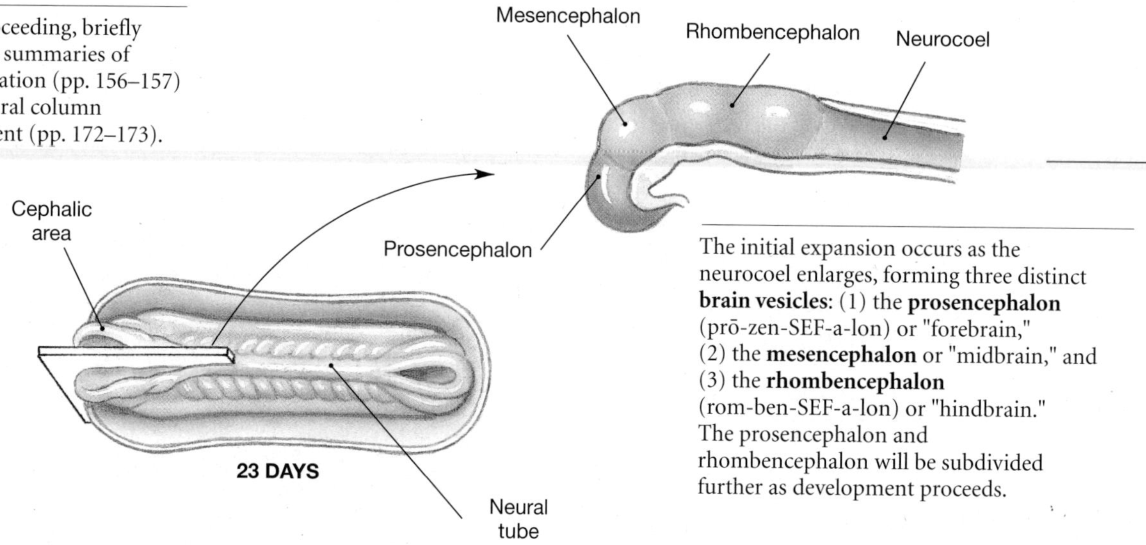

The initial expansion occurs as the neurocoel enlarges, forming three distinct **brain vesicles**: (1) the **prosencephalon** (prō-zen-SEF-a-lon) or "forebrain," (2) the **mesencephalon** or "midbrain," and (3) the **rhombencephalon** (rom-ben-SEF-a-lon) or "hindbrain." The prosencephalon and rhombencephalon will be subdivided further as development proceeds.

Even before **neural tube** formation (p. 349) has been completed, the cephalic portion begins to enlarge. Major differences in brain versus spinal cord development include: (1) early breakdown of mantle (gray matter) and marginal (white matter) organization; (2) appearance of areas of neural cortex; (3) differential growth between and within specific regions; (4) appearance of characteristic bends and folds; and (5) loss of obvious segmental organization.

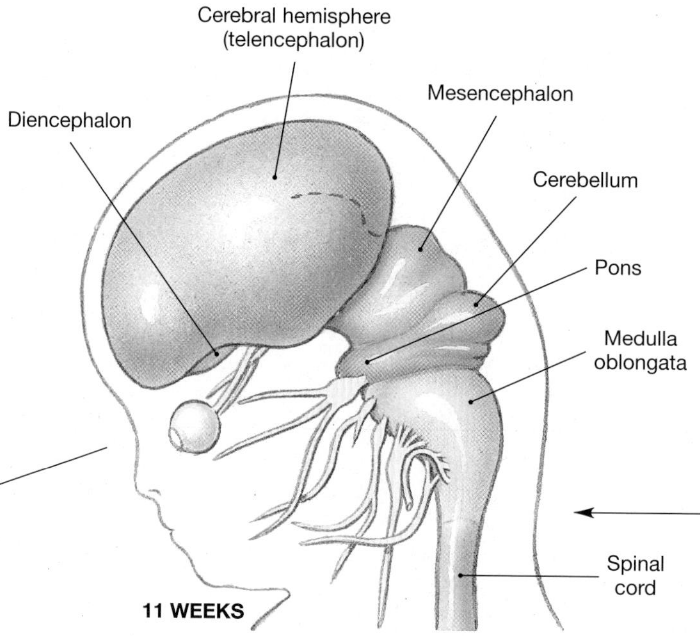

After 11 weeks, the expanding cerebral hemispheres have overgrown the diencephalon. At the metencephalon, cortical formation and expansion produce the cerebellum, which overlies the nuclei and tracts of the pons.

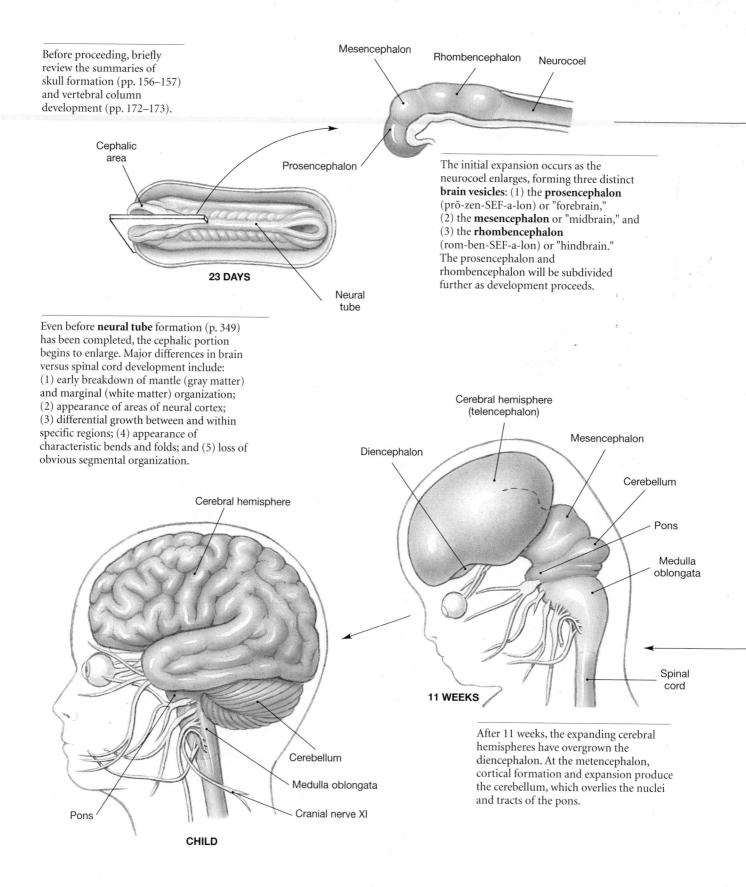

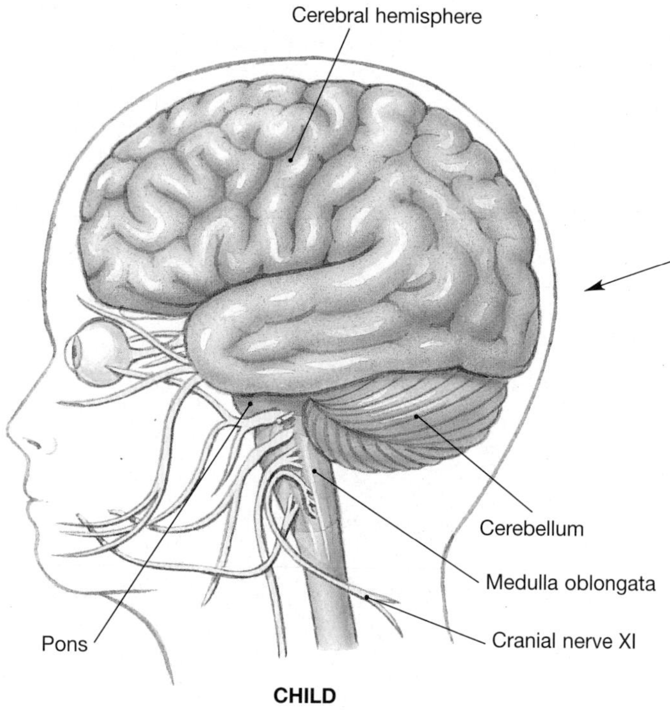

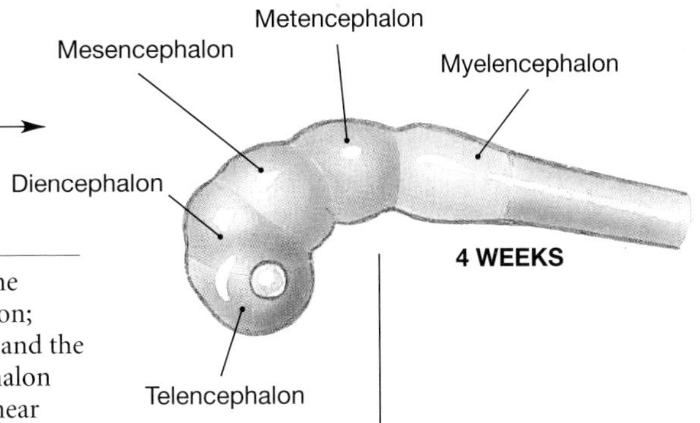

The rhombencephalon first subdivides into the **metencephalon** (met-en-SEF-a-lon; *meta* after) and the **myelencephalon** (mī-el-en-SEF-a-lon; *myelon* spinal cord).

The prosencephalon forms the **telecephalon** (tel-en-SEF-a-lon; *telos* end + *enkiphalos* brain) and the **diencephalon**. The telencephalon begins as a pair of swellings near the rostral, dorsolateral border of the prosencephalon.

Cranial nerves develop as sensory ganglia, link peripheral receptors with the brain, and motor fibers grow out of developing cranial nuclei. Special sensory neurons of cranial nerves I, II, and VIII develop in association with the developing receptors. The somatic motor nerves III, IV, and VI grow to the eye muscles; the mixed nerves (V, VII, IX, and X) innervate the **pharyngeal arches** (p. 156).

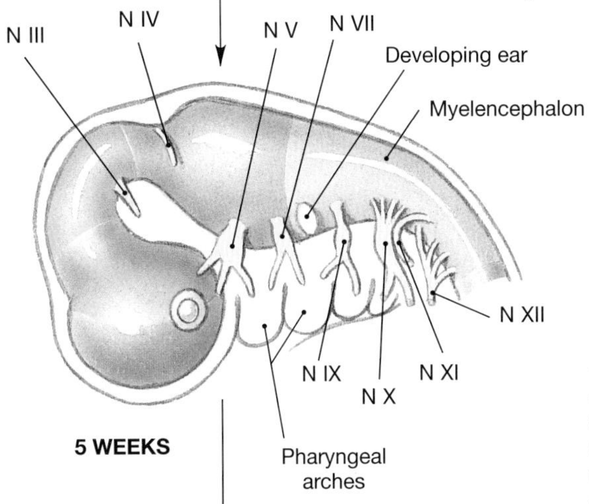

Development of the **mesencephalon** produces a small mass of neural tissue with a constricted neurocoel, the aqueduct of the midbrain.

As differential growth proceeds and the position and orientation of the embryo change, a series of bends, or **flexures** (FLEK-sherz), appears along the axis of the developing brain.

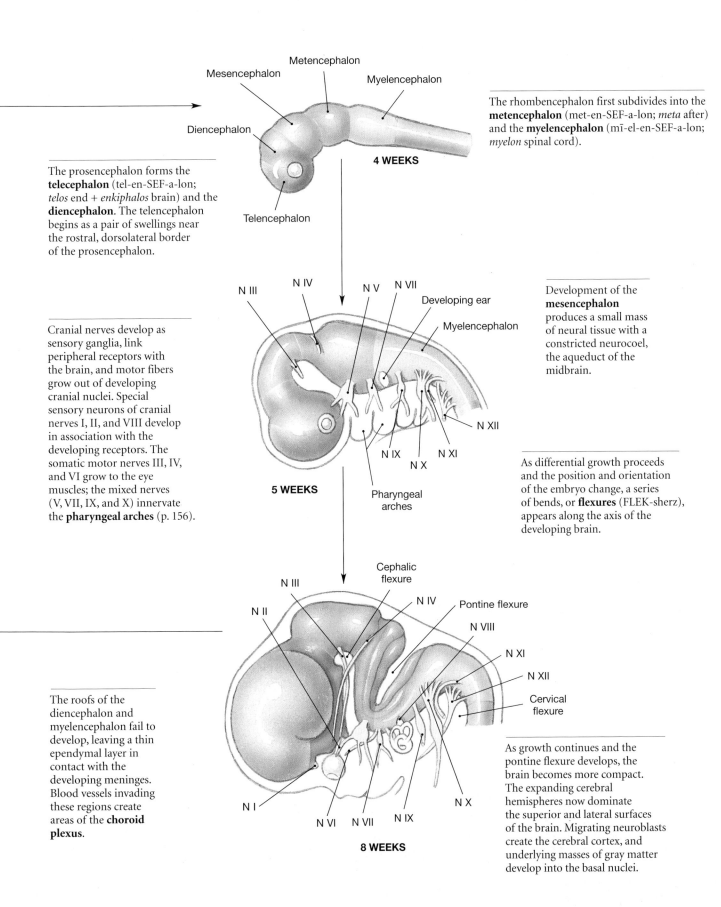

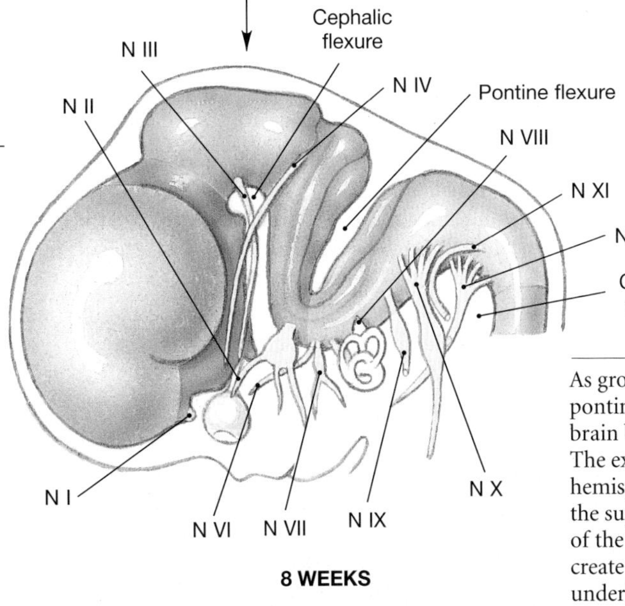

The roofs of the diencephalon and myelencephalon fail to develop, leaving a thin ependymal layer in contact with the developing meninges. Blood vessels invading these regions create areas of the **choroid plexus**.

As growth continues and the pontine flexure develops, the brain becomes more compact. The expanding cerebral hemispheres now dominate the superior and lateral surfaces of the brain. Migrating neuroblasts create the cerebral cortex, and underlying masses of gray matter develop into the basal nuclei.

CLINICAL DISCUSSION

Cranial Reflexes

Cranial reflexes are reflex arcs that involve the sensory and motor fibers of cranial nerves. Examples of cranial reflexes are discussed in later chapters, and this section will simply provide an overview and general introduction.

Table 15.13 lists representative examples of cranial reflexes and their functions. These reflexes are clinically important because they provide a quick and easy method for observing the condition of cranial nerves and specific nuclei and tracts in the brain.

Cranial somatic reflexes are seldom more complex than the somatic reflexes of the spinal cord. This table includes four somatic reflexes: the *corneal reflex*, the *tympanic reflex*, the *auditory reflex*, and the *vestibulo-ocular reflex*. These reflexes are often used to check for damage to the cranial nerves or processing centers involved. The brain stem contains many reflex centers that control visceral motor activity. Many of these reflex centers are in the medulla oblongata, and they can direct very complex visceral motor responses to stimuli. These visceral reflexes are essential to the control of respiratory, digestive, and cardiovascular functions.

TABLE 15.13 **Cranial Reflexes**

Reflex	Stimulus	Afferents	Central Synapse	Efferents	Response
SOMATIC REFLEXES					
Corneal reflex	Contact with corneal surface	N V (trigeminal)	Motor nucleus for N VII (facial nerve)	N VII	Blinking of eyelids
Tympanic reflex	Loud noise	N VIII (vestibulocochlear)	Inferior colliculus (midbrain)	N VII	Reduced movement of auditory ossicles
Auditory reflexes	Loud noise	N VIII	Motor nuclei of brain stem and spinal cord	N III, IV, VI, VII, X, cervical nerves	Eye and/or head movements triggered by sudden sounds
Vestibulo-ocular reflexes	Rotation of head	N VIII	Motor nuclei controlling extra-ocular muscles	N III, IV, VI	Opposite movement of eyes to stabilize field of vision
VISCERAL REFLEXES					
Direct light reflex	Light striking photoreceptors	N II (optic)	Superior colliculus (midbrain)	N III (oculomotor)	Constriction of ipsilateral pupil
Consensual light reflex	Light striking photoreceptors	N II	Superior colliculus	N III	Constriction of contralateral pupil

RELATED CLINICAL TERMS

ataxia: A disturbance of balance that in severe cases leaves the individual unable to stand without assistance; caused by problems affecting the cerebellum. *Cerebellar Dysfunction p. 794*

Bell's palsy: A condition resulting from an inflammation of the facial nerve; symptoms include paralysis of facial muscles on the affected side and loss of taste sensations from the anterior two-thirds of the tongue. *p. 414*

cranial trauma: A head injury resulting from violent contact with another object. Cranial trauma may cause a **concussion**, a condition characterized by a temporary loss of consciousness and a variable period of amnesia. *Cranial Trauma p. 794*

decerebrate rigidity: A generalized state of muscular contraction resulting from loss of CNS inhibitory control.

dysmetria: (dis-MET-rē-a) An inability to stop a movement at a precise, predetermined position; it often leads to an *intention tremor* in the affected individual; usually reflects cerebellar dysfunction. *Cerebellar Dysfunction p. 794*

epidural hemorrhage: A condition involving bleeding into the epidural spaces. *p. 386*

hydrocephalus: Also known as "water on the brain;" a condition in which the skull expands to accommodate extra fluid. *p. 391*

Parkinson's disease (paralysis agitans): A condition characterized by a pronounced increase in muscle tone, resulting from loss of inhibitory control over neurons in the basal nuclei. *p. 403*

spasticity: A condition characterized by hesitant, jerky, voluntary movements and increased muscle tone. *p. 403*

subdural hemorrhage: A condition in which blood accumulates between the dura and the arachnoid. *p. 386*

tic douloureux (doo-loo-ROO), or **trigeminal neuralgia:** A disorder of the maxillary and mandibular branches of N V characterized by severe, almost totally debilitating pain triggered by contact with the lip, tongue, or gums. *p. 414*

tremor: A background shaking of the limbs resulting from a "tug of war" between antagonistic muscle groups. *p. 403*

STUDY OUTLINE & CHAPTER REVIEW

Introduction 382

1. The brain is far more complex than the spinal cord; its complexity makes it adaptable but slower in response than spinal reflexes.

An Introduction to the Organization of the Brain 382

Embryology of the Brain 382

1. The brain forms from three swellings at the superior tip of the developing neural tube: the **prosencephalon**, **mesencephalon**, and **rhombencephalon**. *(see Table 15.1 and Embryology Summary, pp. 391, 420–421)*

Major Regions and Landmarks 382

2. There are six regions in the adult brain: cerebrum, diencephalon, mesencephalon, pons, cerebellum, and medulla oblongata. *(see Figure 15.1)*
3. Conscious thought, intellectual functions, memory, and complex involuntary motor patterns originate in the **cerebrum**. *(see Figure 15.1)*
4. The roof of the **diencephalon** is the **epithalamus**, the walls are the **thalami**, which contain relay and processing centers for sensory data. The floor is the **hypothalamus**, which contains centers involved with emotions, autonomic function, and hormone production. *(see Figure 15.1)*
5. The **mesencephalon** processes visual and auditory information and generates involuntary somatic motor responses. *(see Figure 15.1)*
6. The **pons** connects the cerebellum to the brain stem and is involved with somatic and visceral motor control. The **cerebellum** adjusts voluntary and involuntary motor activities on the basis of sensory data and stored memories. *(see Figure 15.1)*
7. The spinal cord connects to the brain at the **medulla oblongata**, which relays sensory information and regulates autonomic functions. *(see Figure 15.1)*

Gray Matter and White Matter Organization 384

8. The brain contains extensive areas of **neural cortex**, a layer of gray matter on the surfaces of the cerebrum and cerebellum that covers underlying white matter.

The Ventricles of the Brain 384

9. The central passageway of the brain expands to form chambers called **ventricles**. *Cerebrospinal fluid (CSF)* continually circulates from the ventricles and central canal of the spinal cord into the subarachnoid space of the meninges that surround the CNS. *(see Figures 15.2/15.13)*

Protection and Support of the Brain 384

The Cranial Meninges 384

1. The **cranial meninges** —the **dura mater**, **arachnoid**, and **pia mater** —are continuous with the same spinal meninges layers surrounding the spinal cord. However, they have anatomical and functional differences. *(see Figures 15.3/15.4)*
2. Folds of dura mater stabilize the position of the brain within the cranium and include: the **falx cerebri**, **tentorium cerebelli**, **falx cerebelli**, and **diaphragma sellae**. *(see Figures 15.3/15.4)*

The Blood-Brain Barrier 388

3. The **blood-brain barrier** isolates neural tissue from the general circulation.
4. The blood-brain barrier remains intact throughout the CNS except in portions of the hypothalamus, the pineal gland, and at the choroid plexus in the membranous roof of the diencephalon and medulla.

Cerebrospinal Fluid 389

5. The **choroid plexus** is the site of cerebrospinal fluid production. *(see Figure 15.5)*
6. Cerebrospinal fluid (CSF) (1) cushions delicate neural structures, (2) supports the brain, and (3) transports nutrients, chemical messengers, and waste products. Cerebrospinal fluid reaches the subarachnoid space via the **lateral apertures** and a **median aperture**. Diffusion across the **arachnoid granulations** into the superior sagittal sinus returns CSF to the venous circulation. *(see Figures 15.4/15.6)*
7. The blood-brain barrier isolates neural tissue from the general circulation.

The Blood Supply to the Brain 390

8. Arterial blood reaches the brain through the *internal carotid arteries* and the *vertebral arteries*. Venous blood leaves primarily in the *internal jugular veins*.

The Cerebrum 391

The Cerebral Hemispheres 391

1. The cortical surface contains **gyri** (elevated ridges) separated by **sulci** (shallow depressions) or deeper grooves (**fissures**). The **longitudinal fissure** separates the two **cerebral hemispheres**. The **central sulcus** marks the boundary between the **frontal lobe** and the **parietal lobe**. Other sulci form the boundaries of the **temporal lobe** and the **occipital lobe**. *(see Figures 15.1/15.8/15.9)*
2. Each cerebral hemisphere receives sensory information from and generates motor commands to the opposite side of the body. There are significant functional differences between the two; thus, the assignment of a specific function to a specific region of the cerebral cortex is imprecise.
3. The **primary motor cortex** of the **precentral gyrus** directs voluntary movements. The **primary sensory cortex** of the **postcentral gyrus** receives somatic sensory information from touch, pressure, pain, taste, and temperature receptors. *(see Figure 15.9 and Table 15.2)*
4. **Association areas**, such as the **visual association area** and **somatic motor association area** (**premotor cortex**), control our ability to understand sensory information and coordinate a motor response. "Higher-order" integrative centers receive information from many different association areas and direct complex motor activities and analytical functions. *(see Figure 15.9 and Table 15.2)*

The Central White Matter 394

5. The **central white matter** contains three major groups of axons: (1) **association fibers** (tracts that interconnect areas of neural cortex within a single cerebral hemisphere); (2) **commissural fibers** (tracts connecting the two cerebral hemispheres); and (3) **projection fibers** (tracts that link the cerebrum with other regions of the brain and spinal cord). *(see Figure 15.10 and Table 15.3)*

The Basal Nuclei 394

6. The **basal nuclei** within the central white matter include the **caudate nucleus**, **amygdaloid body**, **claustrum**, **globus pallidus**, and **putamen**. The basal nuclei control muscle tone and the coordination of learned movement patterns and other somatic motor activities. *(see Figure 15.11 and Table 15.4)*

The Limbic System 395

7. The **limbic system** includes the amygdaloid body, **cingulate gyrus**, **dentate gyrus**, **parahippocampal gyrus**, **hippocampus**, and **fornix**. The **mamillary bodies** control reflex movements associated with eating. The functions of the limbic system involve emotional states and related behavioral drives. *(see Figures 15.12/15.13 and Table 15.5)*
8. The **anterior nucleus** relays visceral sensations, and stimulating the **reticular formation** produces heightened awareness and a generalized excitement.

The Diencephalon 400

1. The diencephalon provides the switching and relay centers necessary to integrate the sensory and motor pathways. *(see Figures 15.12/15.13)*

The Epithalamus 400

2. The epithalamus forms the roof of the diencephalon. It contains the hormone-secreting *pineal gland*.

The Thalamus 400

3. The thalamus is the principal and final relay point for ascending sensory information and coordinates voluntary and involuntary somatic motor activities. *(see Figures 15.11d,e/15.12/15.13, and Table 15.6)*

The Hypothalamus 400

4. The hypothalamus contains important control and integrative centers. It can (1) control involuntary somatic motor activities; (2) control autonomic function; (3) coordinate activities of the nervous and endocrine systems; (4) secrete hormones; (5) produce emotions and behavioral drives; (6) coordinate voluntary and autonomic functions; (7) regulate body temperature; and (8) control of circadian cycles of activity. *(see Figures 15.12/15.14, and Table 15.7)*

The Mesencephalon 403

1. The **tectum** (roof) of the mesencephalon contains two pairs of nuclei, the **corpora quadrigemina**. On each side, the **superior colliculus** receives visual inputs from the thalamus, and the **inferior colliculus** receives auditory data from the medulla oblongata. The **red nucleus** integrates information from the cerebrum and issues involuntary motor commands related to muscle tone and limb position. The **substantia nigra** regulates the motor output of the basal nuclei. The **cerebral peduncles** contain ascending fibers headed for thalamic nuclei and descending fibers of the corticospinal pathway that carry voluntary motor commands from the primary motor cortex of each cerebral hemisphere. *(see Figures 15.13/15.16/15.17b, and Table 15.8)*

The Pons 405

1. The pons contains: (1) sensory and motor nuclei for four cranial nerves; (2) nuclei concerned with involuntary control of respiration; (3) nuclei that process and relay cerebellar commands arriving over the middle cerebellar peduncles; and (4) ascending and descending tracts. *(see Figures 15.1/15.13/15.16/15.18, and Table 15.9)*

The Cerebellum 406

1. The cerebellum oversees the body's postural muscles, and programs and tunes voluntary and involuntary movements. The **cerebellar hemispheres** consist of neural cortex formed into folds, or **folia**. The surface can be divided into the **anterior** and **posterior lobes**, the **vermis**, and the **flocculonodular lobes**. *(see Figures 15.1/15.8/15.9/15.13/15.19, and Table 15.10)*

The Medulla Oblongata 408

1. The medulla oblongata connects the brain to the spinal cord. It contains the **nucleus gracilis** and the **nucleus cuneatus**, which are processing centers, and the **olivary nuclei**, which relay information from the spinal cord, the cerebral cortex, and brain stem to the cerebellar cortex. Its **reflex centers**, including the **cardiovascular centers** and the **respiratory rhythmicity center**, control or adjust the activities of peripheral systems. *(see Figures 15.1/15.8/15.13/15.16/15.18, and Table 15.11)*

Cranial Nerves 409

1. There are 12 pairs of cranial nerves. Each nerve attaches to the brain near the associated sensory or motor nuclei on the ventrolateral surface of the brain. *(see Figure 15.21)*

The Olfactory Nerve (N I) 410

2. The **olfactory tract** (nerve) (N I) carries sensory information responsible for the sense of smell. The olfactory afferents synapse within the **olfactory bulbs**. *(see Figure 15.22)*

The Optic Nerve (N II) 411

3. The **optic nerve** (N II) carries visual information from special sensory receptors in the eyes. *(see Figure 15.23)*

The Oculomotor Nerve (N III) 412

4. The **oculomotor nerve** (N III) is the primary source of innervation for the extra-ocular muscles that move the eyeball. *(see Figure 15.24)*

The Trochlear Nerve (N IV) 412

5. The **trochlear nerve** (N IV), the smallest cranial nerve, innervates the superior oblique muscle of the eye. *(see Figure 15.24)*

The Trigeminal Nerve (N V) 412

6. The **trigeminal nerve** (N V), the largest cranial nerve, is a mixed nerve with **ophthalmic, maxillary**, and **mandibular branches.** *(see Figure 15.25)*

The Abducens Nerve (N VI) 414

7. The **abducens nerve** (N VI) innervates the sixth extrinsic oculomotor muscle, the lateral rectus. *(see Figure 15.24)*

The Facial Nerve (N VII) 414

8. The **facial nerve** (N VII) is a mixed nerve that controls muscles of the scalp and face. It provides pressure sensations over the face and receives taste information from the tongue. *(see Figure 15.26)*

The Vestibulocochlear Nerve (N VIII) 414

9. The **vestibulocochlear nerve** (N VIII) contains the **vestibular nerve**, which monitors sensations of balance, position, and movement, and the **cochlear nerve**, which monitors hearing receptors. *(see Figure 15.27)*

The Glossopharyngeal Nerve (N IX) 416

10. The **glossopharyngeal nerve** (N IX) is a mixed nerve that innervates the tongue and pharynx and controls the action of swallowing. *(see Figure 15.28)*

The Vagus Nerve (N X) 417

11. The **vagus nerve** (N X) is a mixed nerve that is vital to the autonomic control of visceral function and has a variety of motor components. *(see Figure 15.29)*

The Accessory Nerve (N XI) 418

12. The **accessory nerve** (N XI) has an **internal branch**, which innervates voluntary swallowing muscles of the soft palate and pharynx, and an **external branch**, which controls muscles associated with the pectoral girdle. *(see Figure 15.30)*

The Hypoglossal Nerve (N XII) 418

13. The **hypoglossal nerve** (N XII) provides voluntary motor control over tongue movements. *(see Figure 15.30)*

A Summary of Cranial Nerve Branches and Functions 419

14. The branches and functions of the cranial nerves are summarized in *Table 15.12.*

LEVEL 1 REVIEWING FACTS AND TERMS

Match each numbered item with the most closely related lettered item. Use letters for answers in the spaces provided.

Column A	Column B
____ 1. mesencephalon	a. visual cortex
____ 2. myelencephalon	b. learned motor patterns
____ 3. tentorium cerebelli	c. midbrain
____ 4. abducens nerve	d. motor, tongue movements
____ 5. diencephalon	e. third ventricle
____ 6. occipital lobe	f. motor, eye movements
____ 7. hypoglossal nerve	g. sensory information relay
____ 8. basal nuclei	h. medulla oblongata
____ 9. thalamus	i. Purkinje cells
____10. cerebellum	j. separate cerebrum/cerebellum

11. Conscious thought processes and all intellectual functions originate in the
(a) cerebellum (b) corpus callosum
(c) cerebral hemispheres (d) medulla oblongata

12. The primary link between the nervous and the endocrine systems is the
(a) hypothalamus (b) pons
(c) mesencephalon (d) medulla oblongata

13. The primary purpose of the blood-brain barrier is to
(a) provide the brain with oxygenated blood
(b) drain venous blood via the internal jugular vein
(c) isolate neural tissue in the CNS from the general circulation
(d) all of the above

14. The only cranial nerves that are attached to the cerebrum are the
(a) optic (b) oculomotor
(c) trochlear (d) olfactory

15. The anterior nuclei of the thalamus
(a) are part of the limbic system
(b) are connected to the pituitary gland
(c) produce the hormone melatonin
(d) receive impulses from the optic nerve

16. The cortex inferior to the lateral sulcus is the
(a) parietal lobe (b) temporal lobe
(c) frontal lobe (d) occipital lobe

17. Lying within each hemisphere inferior to the floor of the lateral ventricles are the
(a) anterior commissures (b) motor association areas
(c) auditory cortex (d) basal nuclei

18. Nerve fiber bundles on the ventrolateral surface of the mesencephalon are the
(a) tegmenta (b) corpora quadrigemina
(c) cerebral peduncles (d) superior colliculi

19. Efferent tracts from the hypothalamus
(a) control involuntary motor activities
(b) control autonomic function
(c) coordinate activities of the nervous and endocrine systems
(d) all of the above

20. The limbic system components include all except
(a) amygdaloid body (b) cingulate gyrus
(c) globus pallidus (d) hippocampus

LEVEL 2 REVIEWING CONCEPTS

1. Damage to the pyramidal cells of the cerebral cortex would directly affect
(a) perception of pain (b) voluntary motor activity
(c) the ability to see (d) the ability to hear

2. Damage to the superior colliculi would interfere with the ability to
(a) express rage (b) voluntarily move the arm
(c) react to the movement of a car with the eyes (d) maintain proper posture

3. Damage to the basal nuclei would lead to
(a) loss of sight (b) loss of hearing
(c) inability to sense pain (d) difficulty in maintaining balance

4. Which lobe and specific area of the brain would be affected if one could no longer cut designs from construction paper?

5. Impulses from proprioceptors must pass through specific nuclei before arriving at their destination in the brain. What are the nuclei, and what is the destination of this information?

6. Which nuclei are more likely involved in the coordinated movement of the head in the direction of a loud noise?

7. Which cranial nerves are responsible for all aspects of eye function?

8. If an individual has poor emotional control and difficulty in remembering past events, what area of the brain might be damaged or have a lesion?

9. Which region of the brain provides links between the cerebellar hemispheres and the mesencephalon, diencephalon, cerebrum, and spinal cord?

10. Why is the blood-brain barrier less intact in the hypothalamus?

LEVEL 3 CRITICAL THINKING AND CLINICAL APPLICATIONS

1. Shortly after birth, the head of an infant begins to enlarge rapidly. What is occurring, why, and is there a clinical explanation and solution to this problem?

2. Rose awakened one morning and discovered that her face was paralyzed on the left side and she had no sensation of taste from the anterior two-thirds of the tongue on the same side. What is the cause of these symptoms, and what can be done to help Rose with this situation?

3. After suffering a stroke, Alexandra finds that she cannot move her right arm. Where might one expect the stroke damage to have occurred?

✓ ANSWERS TO CONCEPT CHECK QUESTIONS

p. 384 **1.** The six major divisions in the adult brain are: the cerebrum; the diencephalon; the mesencephalon or midbrain; the pons and the cerebellum; and the medulla oblongata. **2.** The three major structures of the brain stem include: the mesencephalon, the pons, and the medulla oblongata. **3.** The ventricles are fluid-filled chambers inside the telencephalon, the diencephalon, the metencephalon, and the superior portion of the myelencephalon. The ventricles are lined by ependymal cells. **4.** The secondary brain vesicles and the brain regions associated with each at birth are: telencephalon (cerebrum), diencephalon (epithalamus, thalamus, and hypothalamus), mesencephalon (midbrain), metencephalon (cerebellum and pons), and myelencephalon (medulla oblongata).

p. 390 **1.** The four extensions of the dura mater include: falx cerebri, tentorium cerebelli, falx cerebelli, and diaphragma sellae. **2.** The pia mater is a highly vascular membrane composed primarily of areolar connective tissue. It acts as a floor to support the large cerebral blood vessels as they branch over the surface of the brain, invading the neural contours to supply superficial areas of neural cortex. **3.** The blood-brain barrier functions to isolate the CNS from the general circulation. **4.** Cerebrospinal fluid (CSF) has several functions: cushioning delicate neural structures; supporting the brain; and transporting nutrients, chemical messengers, and waste products. CSF is produced by the choroid plexus, a combination of specialized ependymal cells and permeable capillaries in the ventricle lining.

p. 398 **1.** The frontal lobe contains the primary motor cortex for voluntary muscular activities. Additionally, it contains the prefrontal cortex, which integrates information and performs abstract intellectual functions. The parietal lobe contains the primary sensory cortex, which receives somatic sensory information. The occipital lobe contains the visual cortex for conscious perception of visual stimuli. The temporal lobe contains the auditory cortex and olfactory cortex for the conscious perception of auditory and olfactory stimuli. **2.** Gyri are the elevated ridges on the surface of the cerebrum. Sulci are the shallow depressions that separate neighboring gyri. **3.** The three major groups of axons in the central white matter are: (1) association fibers, tracts that interconnect areas of neural cortex within a single cerebral hemisphere; (2) commissural fibers, tracts that connect the two cerebral hemispheres; and (3) projection fibers, tracts that link the cerebrum with other regions of the brain and spinal cord.
p. 401 **1.** Changes in body temperature would stimulate the preoptic area of the hypothalamus, a division of the diencephalon. **2.** The thalamus coordinates the activities at the conscious and subconscious levels. **3.** The posterior epithalamus contains the pineal gland, which secretes melatonin. **4.** The hypothalamus produces and secretes two hormones: antidiuretic hormone, produced by the supraoptic nucleus, and oxytocin, produced by the paraventricular nucleus.
p. 408 **1.** The vermis and arbor vitae are structures associated with the cerebellum. **2.** The medulla oblongata is small, but it contains many vital reflex centers, including those that control breathing and regulate the heart and blood pressure. Damage to the medulla oblongata can result in a cessation of breathing, or changes in heart rate and blood pressure that are incompatible with life. In contrast, damage to a small region of the cerebrum may affect a specific area/function. It should not be fatal. **3.** These nuclei, collectively called the corpora quadrigemina, are relay stations concerned with the processing of visual and auditory sensations. **4.** Voluntary movements become hesitant and jerky. A continual tremor may occur during rest.
p. 419 **1.** The cranial nerve responsible for tongue movements is the hypoglossal nerve. **2.** Since the abducens (N VI) controls lateral movements of the eyes, we would expect an individual with damage to this nerve to be unable to move their eyes laterally. **3.** The vestibulocochlear nerve (N VIII) is concerned with balance. The branch would be the vestibular branch. **4.** The facial nerve (N VII) is involved with the sensory detection of taste.

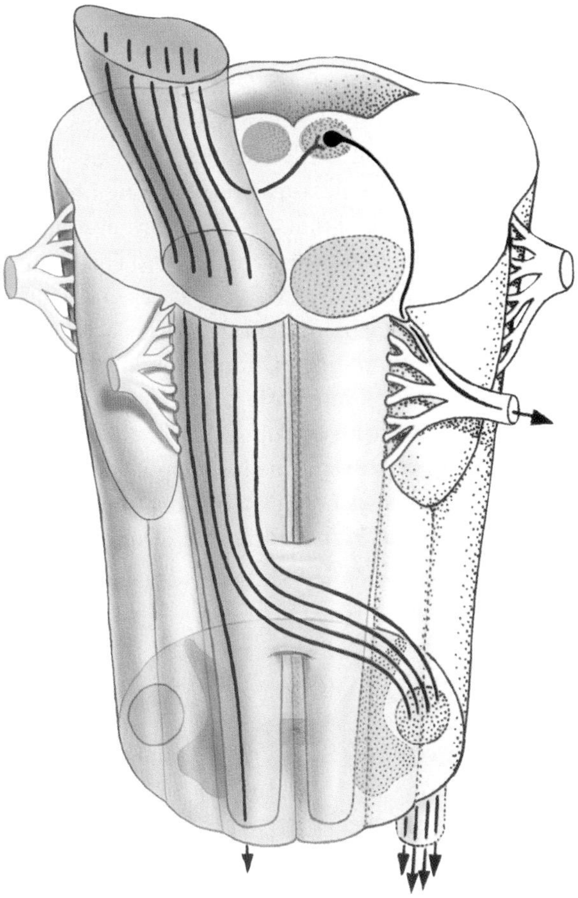

16

THE NERVOUS SYSTEM

Pathways and Higher-Order Functions

There is a saying that "big cities never sleep." In Chicago or Los Angeles at 3:00 A.M., shops are open, deliveries are made, people are on the streets, and traffic moves briskly. The central nervous system is much more complex than any city, and far busier. There is a continuous flow of information among the brain, spinal cord, and peripheral nerves. At any given moment, millions of sensory neurons are delivering information to processing centers in the CNS, and millions of motor neurons are controlling or adjusting the activities of peripheral effectors. This process continues around the clock, whether we are awake or asleep. Your conscious mind may sleep soundly, but many brain stem centers are active throughout our lives, performing vital autonomic functions at the subconscious level.

Many subtle forms of interaction, feedback, and regulation link higher centers with the various components of the brain stem. Only a few are understood in any detail. In this chapter, we will focus on the anatomical design that allows neural structures to perform sensory and motor operations as well as higher-order brain functions, such as learning and memory.

Sensory and Motor Pathways

Communication between the CNS, the PNS, and peripheral organs and systems involves pathways that relay sensory and motor information between the periphery and higher centers of the brain. Each ascending (sensory) or descending (motor) pathway consists of a chain of tracts and associated nuclei. Processing usually occurs at several points along a pathway, wherever synapses relay signals from one neuron to another. The number of synapses varies from one pathway to another. For example, a sensory pathway ending in the cerebral cortex involves three neurons whereas a sensory pathway ending in the cerebellum involves two neurons. Our attention will focus on pathways involving the major sensory and motor tracts of the spinal cord. In general, (1) these tracts are paired (bilaterally and symmetrically along the spinal cord), and (2) the axons within each tract are grouped according to the body region innervated. All tracts involve the brain and spinal cord, and a tract name often indicates its origin and destination. If the name begins with *spino-* the tract must *start* in the spinal cord and *end* in the brain; it must therefore carry sensory information. The last part of the name indicates a major nucleus or region of the brain near the end of the pathway. For example, the *spinocerebellar tract* begins in the spinal cord and ends in the cerebellum. If the name ends in *-spinal* the tract must *start* in the brain and *end* in the spinal cord; it bears motor commands. Once again, the start of the name usually indicates the origin of the tract. For example, the *vestibulospinal tract* starts in the vestibular nucleus and ends in the spinal cord.

Sensory Pathways [FIGURES 16.1/16.2 AND TABLE 16.1]

Sensory receptors monitor conditions both inside the body and in the external environment. When stimulated, a receptor passes information to the central nervous system. This information, called a **sensation**, arrives in the form of action potentials in an afferent (sensory) fiber. The complexity of the response to a particular stimulus depends in part on where processing occurs and where the motor response is initiated. For example, processing in the spinal cord can produce a very rapid, stereotyped motor response, such as a stretch reflex. p. 374 Processing within the brain stem may result in more complex motor activities, such as coordinated changes in the position of the eyes, head, neck, or trunk. Most of the processing of sensory information occurs in the spinal cord or brain stem; only about 1% of the information provided by afferent fibers reaches the cerebral cortex and our conscious awareness. However, the information arriving at the sensory cortex is organized so that we can determine the source and nature of the stimulus with great precision. Chapter 17 describes the distribution of visceral sensory information and considers reflexive responses to visceral sensations. Chapter 18 examines the origins of sensations and the pathways involved in relaying special sensory information, such as olfaction (smell) or vision, to conscious and subconscious processing centers in the brain.

We will describe three sensory pathways that deliver somatic sensory information to the sensory cortex of the cerebral or cerebellar hemispheres. These pathways involve a chain of neurons.

- A **first-order neuron** is the sensory neuron that delivers the sensations to the CNS.
- A **second-order neuron** is the interneuron upon which the axon of the first-order neuron synapses. The second-order neuron may be located either in the spinal cord or the brain stem.
- In pathways ending at the cerebral cortex, the second-order neuron synapses on a **third-order neuron** in the thalamus. The axon of the third-order neuron carries the sensory information from the thalamus to the appropriate sensory area of the cerebral cortex.

In most cases, the axon of either the first-order or second-order neuron crosses over to the opposite side of the spinal cord or brain stem as it ascends. As a result of this crossover, or *decussation*, sensory information from the left side of the body is delivered to the right side, and vice versa. The functional or evolutionary significance of this decussation is unknown. In two of the sensory pathways (the *posterior column pathway* and the *spinothalamic pathway*), the axons of the third-order neurons ascend within the internal capsule to synapse on neurons of the primary sensory cortex of the cerebral hemisphere. Because deccussation occurred at the level of the first-order or second-order neurons, the right cerebral hemisphere receives sensory information from the left side of the body, and vice versa.

Table 16.1 identifies and summarizes the three major somatic sensory pathways, also called *somatosensory pathways*: (1) the *posterior column pathway*, (2) the *spinothalamic pathway*, and (3) the *spinocerebellar pathway*. Figure 16.1● indicates their relative positions in the spinal cord. For clarity, the figure dealing with spinal pathways (Figure 16.2●) shows how sensations originating on one side of the body are relayed to the cerebral cortex. Keep in mind, however, that these pathways are present on *both* sides of the body.

THE POSTERIOR COLUMN PATHWAY [FIGURE 16.2a]

The **posterior column pathway** (Figure 16.2a●) carries highly localized proprioceptive (position), "fine" touch, pressure, and vibration sensations. The axons of the first-order neurons reach the CNS through the dorsal roots of spinal nerves and the sensory roots of cranial nerves. Axons from the dorsal roots of spinal nerves ascend within the **fasciculus gracilis** or the **fasciculus cuneatus**, synapsing at the nucleus gracilis or the nucleus cuneatus of the medulla oblongata. The second-order neurons then relay the information to the thalamus of the opposite side of the brain along a tract called the **medial lemniscus** (*lemniskos*, ribbon). The decussation occurs as the axons of the second-order neurons leave the nuclei to enter the medial lemniscus. As it travels toward the thalamus, the medial lemniscus incorporates the same classes of sensory information (fine touch, pressure, and vibration) collected by cranial nerves V, VII, IX, and X.

TABLE 16.1

PRINCIPAL ASCENDING (SENSORY) TRACTS IN THE SPINAL CORD AND THE SENSORY INFORMATION THEY PROVIDE

		Location of Neurons				
Pathway/Tract	**Sensations**	**First-Order**	**Second-Order**	**Third-Order**	**Final Destination**	**Site of Cross-over**
POSTERIOR COLUMN PATHWAY						
Fasciculus gracilis	Proprioception and fine touch, pressure, and vibration from the inferior half of the body	Dorsal root ganglia of lower body; axons enter CNS in dorsal roots and ascend within fasciculus gracilis	Nucleus gracilis of medulla oblongata; axons cross over before entering medial lemniscus	Ventral posterolateral nucleus of thalamus	Primary sensory cortex on side opposite stimulus	Axons of second-order neurons, before joining medial lemniscus
Fasciculus cuneatus	Proprioception and fine touch, pressure, and vibration from the superior half of the body	Dorsal root ganglia of upper body; axons enter CNS in dorsal roots and ascend within fasciculus cuneatus	Nucleus cuneatus of medulla oblongata; axons cross over before entering medial lemniscus	Ventral posterolateral and ventral posteromedial nuclei of thalamus	As above	As above
SPINOTHALAMIC PATHWAY						
Lateral spinothalamic tracts	Pain and temperature sensations	Dorsal root ganglia; axons enter CNS in dorsal roots on opposite side	Interneurons in posterior gray horn; axons enter lateral spinothalamic tract	Ventral posterolateral nucleus of thalamus	Primary sensory cortex on side opposite stimulus	Axons of second-order neurons, at level of entry
Anterior spinothalamic tracts	Crude touch and pressure sensations	As above	Interneurons in posterior gray horn; axons enter anterior spinothalamic tract on opposite side	As above	As above	As above
SPINOCEREBELLAR PATHWAY						
Posterior spinocerebellar tracts	Proprioception	Dorsal root ganglia; axons enter CNS in dorsal roots	Interneurons in posterior gray horn; axons enter posterior spinocerebellar tract on same side	Not present	Cerebellar cortex on side of stimulus	None
Anterior spinocerebellar tracts	Proprioception	As above	Interneurons in same spinal segment; axons enter anterior spinocerebellar tract on same or opposite side	Not present	Cerebellar cortex, primarily on side opposite (and side of) stimulus	Axons of most second-order neurons cross before entering tract

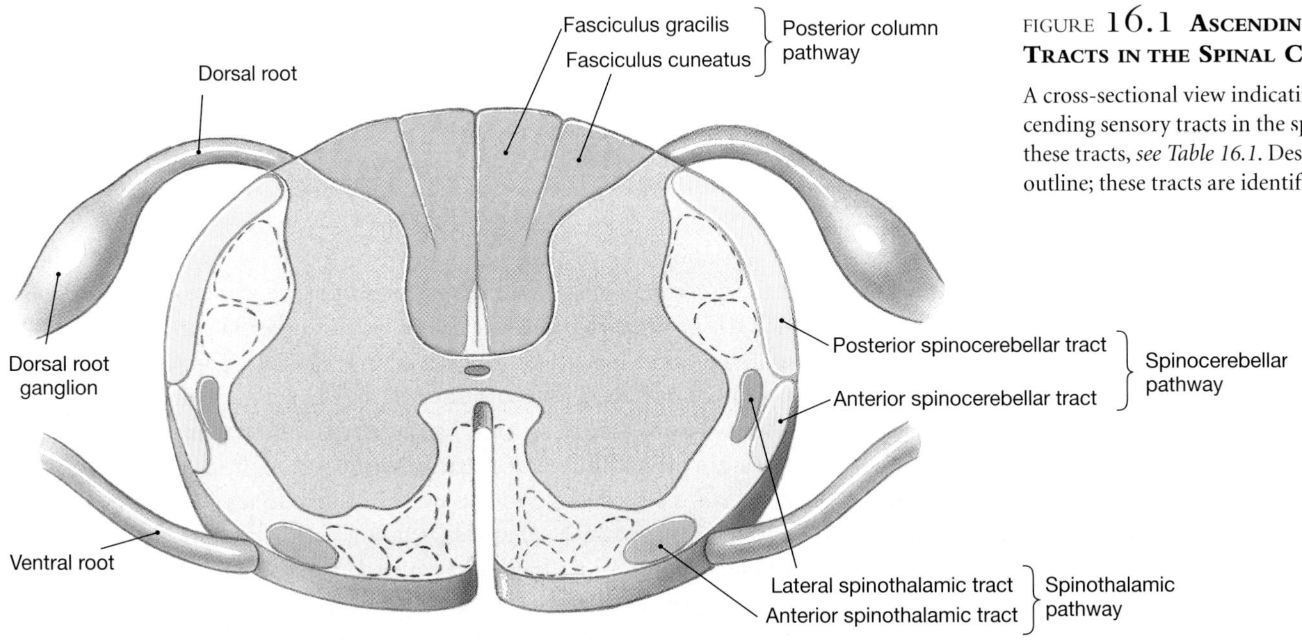

FIGURE 16.1 **ASCENDING SENSORY PATHWAYS AND TRACTS IN THE SPINAL CORD**

A cross-sectional view indicating the locations of the major ascending sensory tracts in the spinal cord. For information about these tracts, *see Table 16.1*. Descending motor tracts are shown in outline; these tracts are identified in *Figure 16.4*.

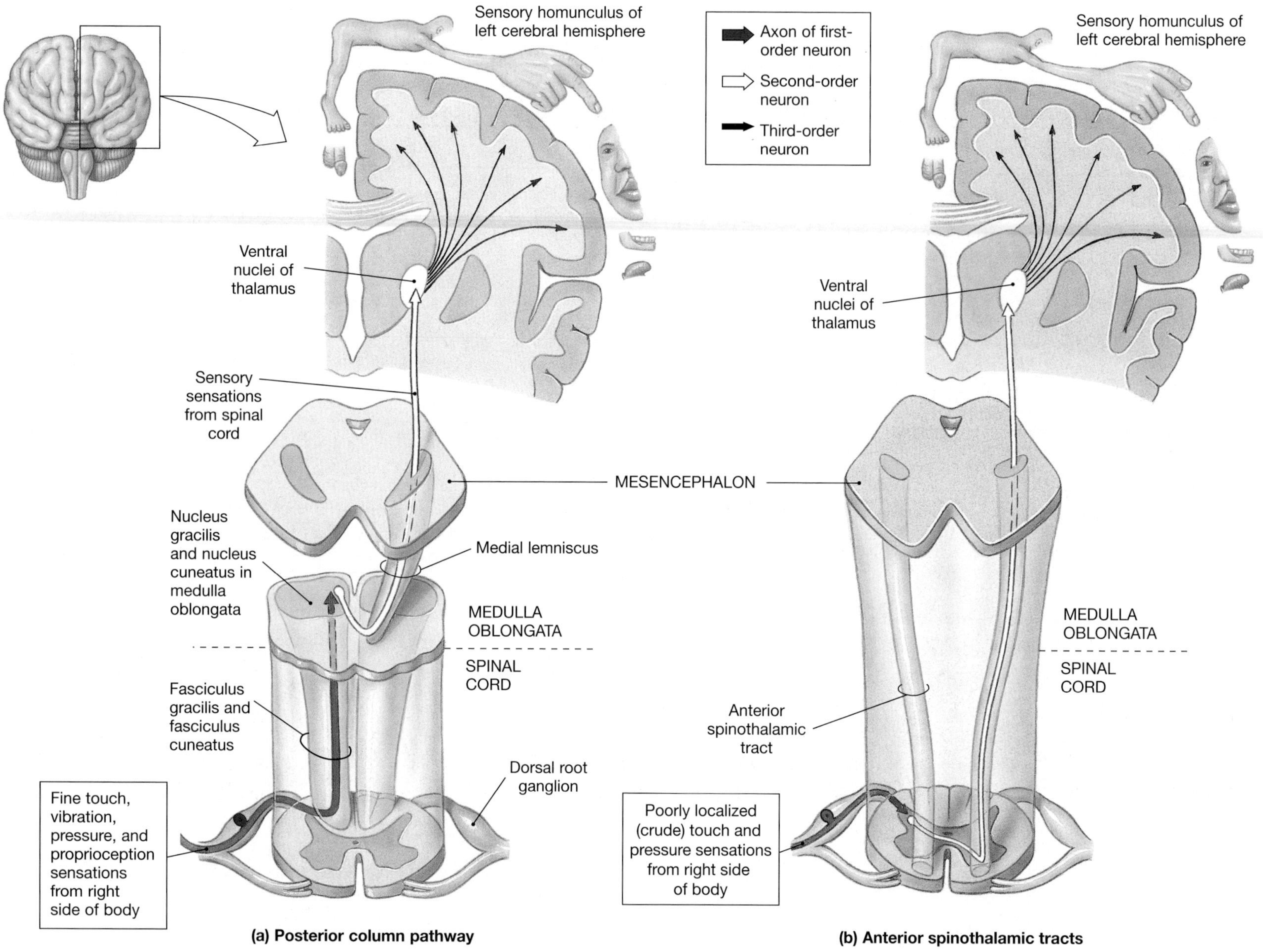

FIGURE 16.2 **The Posterior Column, Spinothalamic, and Spinocerebellar Sensory Pathways**

Diagrammatic comparison of first-, second-, and third-order neurons in ascending pathways. For clarity, this figure shows only the pathway for sensations originating on the right side of the body. **(a)** The posterior column pathway delivers fine-touch, vibration, and proprioception information to the primary sensory cortex of the cerebral hemisphere on the opposite side of the body. The crossover occurs in the medulla, after a synapse in the nucleus gracilis or nucleus cuneatus. **(b)** The anterior spinothalamic pathway carries crude touch and pressure sensations to the primary sensory cortex on the opposite side of the body. The crossover occurs in the spinal cord at the level of entry. **(c)** The lateral spinothalamic tract carries sensations of pain and temperature to the primary sensory cortex on the opposite side of the body. The crossover occurs in the spinal cord, at the level of entry. **(d)** The spinocerebellar pathway carries proporiceptive information to the cerebellum. (Only one tract is detailed on each side, although each side has both tracts.)

Ascending sensory information maintains a strict regional organization along the pathway from center to center. *p. 362* Data arriving over the posterior column pathway are integrated by the ventral posterolateral nucleus of the thalamus, which sorts data according to the region of the body involved and projects it to specific regions of the primary sensory cortex. The axons carrying the information from the thalamus to the sensory cortex are collectively known as *projection fibers*. The sensations arrive with sensory information from the toes at one end of the primary sensory cortex and information from the head at the other.

The individual "knows" the nature of the stimulus and its location because the information has been projected to a specific portion of the primary sensory cortex. If it is relayed to another part of the sensory cortex, the sensation will be perceived as having originated in a different part of the body. For example, the pain of a heart attack is often felt in the left arm; this is an example of referred pain, a topic addressed in Chapter 18. Our perception of a given sensation as touch, rather than as temperature or pain, depends on processing in the thalamus. If the cerebral cortex were damaged or the projection fibers cut, a person could still be aware of a light touch because the thalamic nuclei remain intact. The individual, however, would be unable to determine its source, because localization is provided by the primary sensory cortex.

If a site on the primary sensory cortex is electrically stimulated, the individual reports feeling sensations in a specific part of the body. By elec-

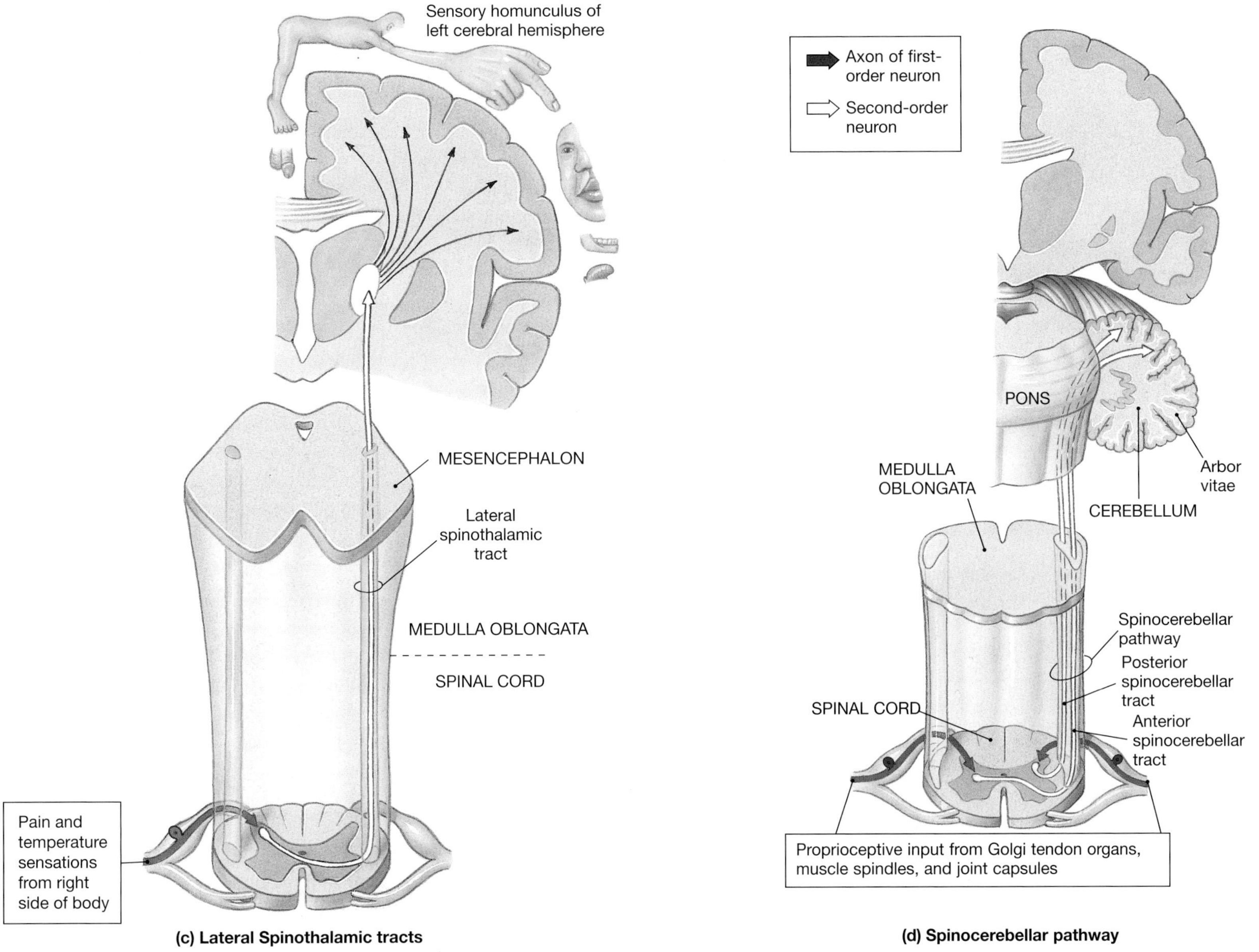

FIGURE 16.2 *(continued)*

trically stimulating the cortical surface, investigators have been able to create a functional map of the primary sensory cortex (Figure 16.2a●). This sensory map is called a **sensory homunculus** ("little man"). The proportions of the homunculus are obviously very different from those of the individual. For example, the face is huge and distorted, with enormous lips and tongue, whereas the back is relatively tiny. These distortions occur because the area of sensory cortex devoted to a particular region is proportional not to its absolute size but rather to *the number of sensory receptors* the region contains. In other words, it takes many more cortical neurons to process sensory information arriving from the tongue, which has tens of thousands of taste and touch receptors, than it does to analyze sensations originating on the back, where touch receptors are few and far between.

THE SPINOTHALAMIC PATHWAY [FIGURE 16.2b,c]

The **spinothalamic pathway** (Figure 16.2b,c●) begins as axons of first-order neurons carrying "crude" sensations of touch, pressure, pain, and temperature enter the spinal cord and synapse within the posterior gray horns. The second-order neurons are interneurons whose axons cross to the opposite side of the spinal cord before ascending within the **anterior** and **lateral spinothalamic tracts**. These tracts converge on the ventral posterolateral nuclei of the thalamus. Projection fibers of third-order neurons then carry the information to the primary sensory cortex. Table 16.1 summarizes the origin and destination of these tracts and the associated sensations. For clarity, Figure 16.2● shows the distribution route for crude touch and pressure sensations and pain and temperature sensations from the right side of the body. However, both sides of the spinal cord have anterior and lateral spinothalamic tracts.

THE SPINOCEREBELLAR PATHWAY [FIGURE 16.2d]

The **spinocerebellar pathway** carries proprioceptive information concerning the position of muscles, tendons, and joints to the cerebellum, which is responsible for fine coordination of body movements. The axons of first-order sensory neurons synapse on interneurons in the posterior gray horns of the spinal cord. The axons of these second-order neurons ascend in the **anterior** and **posterior spinocerebellar tracts** (Figure 16.2d●).

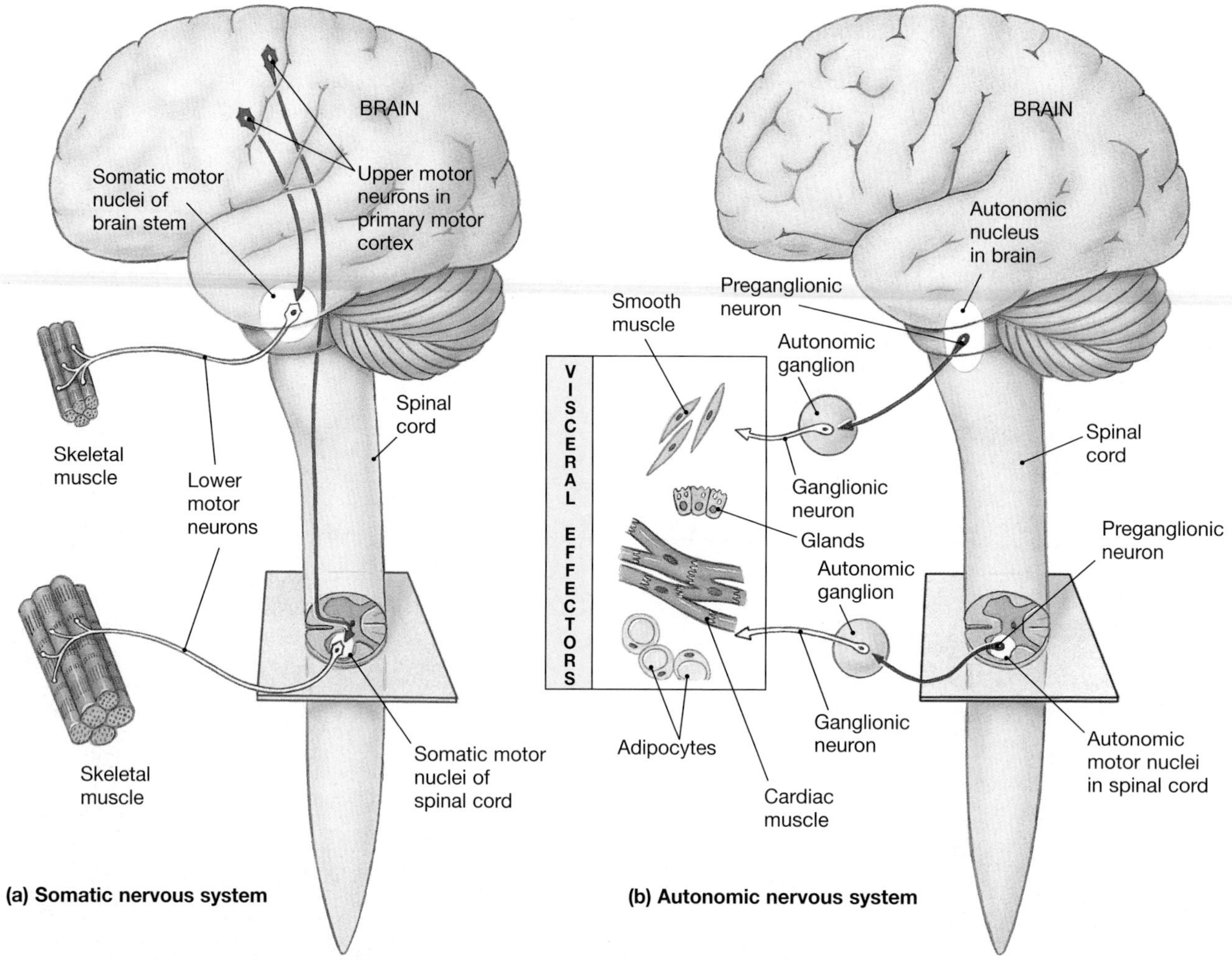

FIGURE 16.3 **MOTOR PATHWAYS IN THE CNS AND PNS**

Organization of the somatic and autonomic nervous systems. (**a**) In the SNS, an upper motor neuron in the CNS controls a lower motor neuron in the brain stem or spinal cord. The axon of the lower motor neuron has direct control over skeletal muscle fibers. Stimulation of the lower motor neuron always has an excitatory effect on the skeletal muscle fibers. (**b**) In the ANS, the axon of a preganglionic neuron in the CNS controls ganglionic neurons in the periphery. Stimulation of the ganglionic neurons may lead to excitation or inhibition of the visceral effector innervated.

- Axons that cross-over to the opposite side of the spinal cord enter the anterior spinocerebellar tract and ascend to the cerebellar cortex by way of the superior cerebellar peduncle.[1]
- The posterior spinocerebellar tract carries axons that do not cross over to the opposite side of the spinal cord. These axons ascend to the cerebellar cortex by way of the inferior cerebellar peduncle.

Table 16-1 summarizes the origin and destination of these tracts and the associated sensations.

Motor Pathways [FIGURE 16.3/16.4]

The central nervous system issues motor commands in response to information provided by sensory systems. These commands are distributed by the somatic nervous system and the autonomic nervous system. The *somatic nervous system (SNS)* issues somatic motor commands that direct the contractions of skeletal muscles. The *autonomic nervous system (ANS)*, or *visceral motor system*, innervates visceral effectors, such as smooth muscles, cardiac muscle, and glands.

[1] The anterior spinocerebeller tract also contains a relatively small number of uncrossed axons from interneurons on the same side as the stimulus.

The motor neurons of the SNS and ANS are organized in different ways. Somatic motor pathways (Figure 16.3a●) always involve at least two motor neurons: an **upper motor neuron**, whose cell body lies in a CNS processing center, and a **lower motor neuron** located in a motor nucleus of the brain stem or spinal cord. Activity in the upper motor neuron can excite or inhibit the lower motor neuron, but only the axon of the lower motor neuron extends to skeletal muscle fibers. Destruction of or damage to a lower motor neuron produces a flaccid paralysis of the innervated motor unit. Damage to an upper motor neuron may produce muscle rigidity, flaccidity, or uncoordinated contractions.

At least two neurons are involved in ANS pathways, and one of them is always located in the periphery (Figure 16.3b●). Autonomic motor control involves a **preganglionic neuron** whose cell body lies within the CNS, and a **ganglionic neuron** in a peripheral ganglion. Higher centers in the hypothalamus and elsewhere in the brain stem may stimulate or inhibit the preganglionic neuron. Motor pathways of the ANS will be described in Chapter 17.

Conscious and subconscious motor commands control skeletal muscles by traveling over three integrated motor pathways: the *corticospinal pathway*, the *medial pathway*, and the *lateral pathway*. Figure 16.4● indicates the positions of the associated motor tracts in the spinal cord. Activity within these motor pathways is monitored and adjusted by the basal

FIGURE 16.4 **The Corticospinal Pathway and Descending Motor Tracts in the Spinal Cord**

(**a**) The corticospinal pathway originates at the primary motor cortex. The corticobulbar tracts end at the motor nuclei of cranial nerves on the opposite side of the brain. Most fibers in this pathway cross-over in the medulla and enter the lateral corticospinal tracts; the rest descend in the anterior corticospinal tracts and cross-over after reaching target segments in the spinal cord. (**b**) A cross-sectional view indicating the locations of the major descending motor tracts that contain the axons of upper motor neurons. The origins and destinations of these tracts are listed in *Table 16.2*. Sensory tracts (shown in *Figure 16.1*) appear in dashed outline.

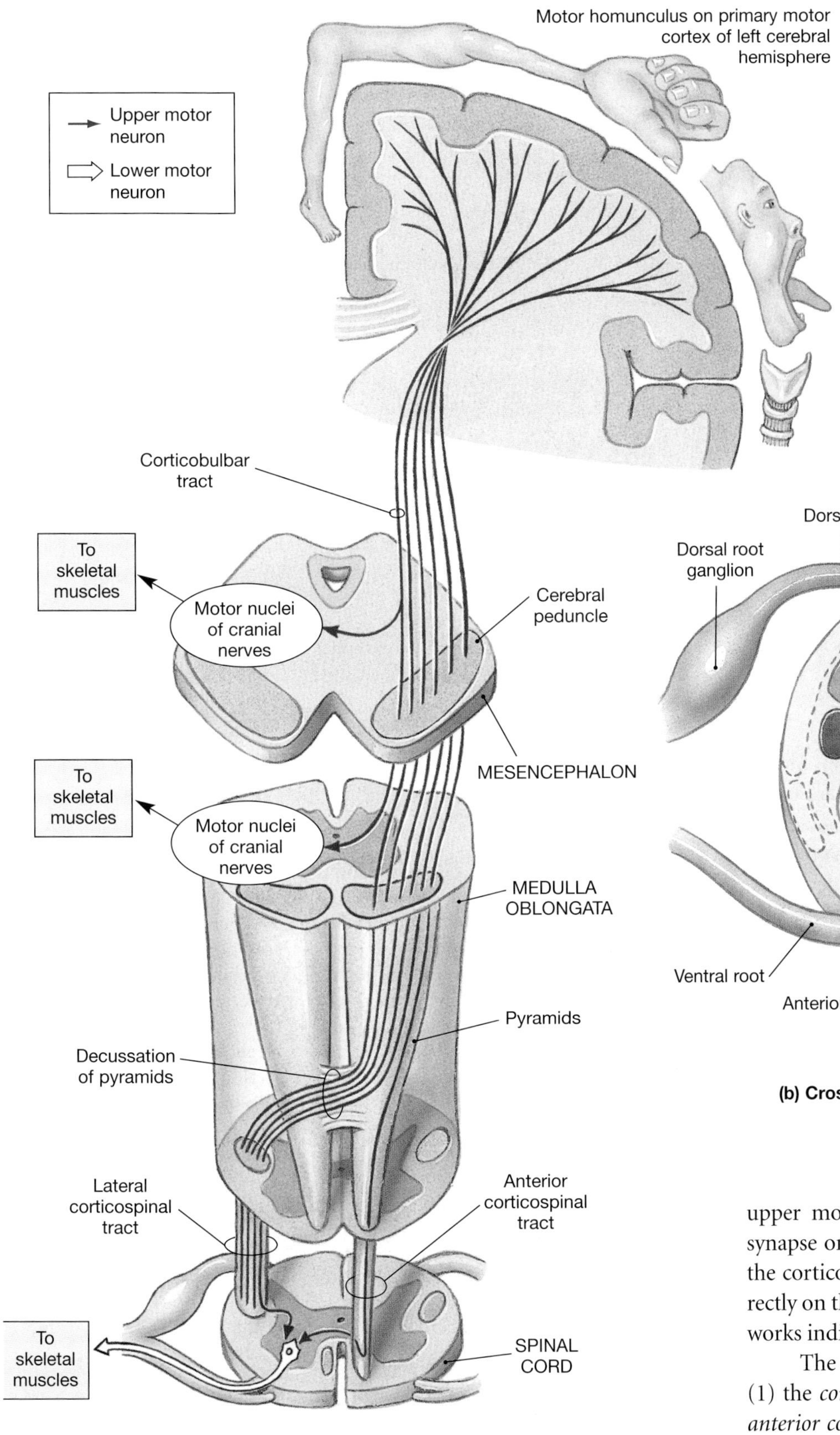

(a) Corticospinal pathway

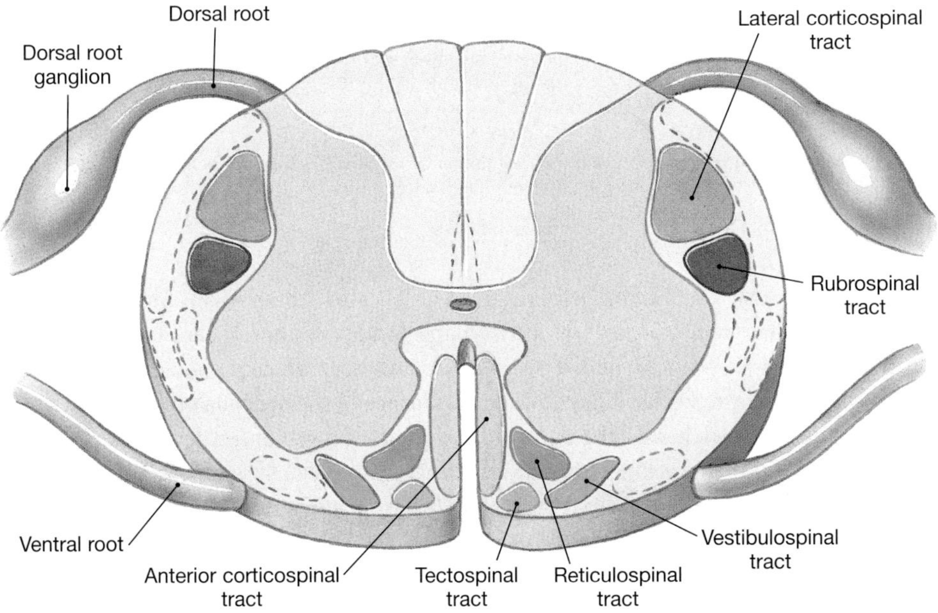

(b) Cross-sectional view of descending motor tracts in the spinal cord

nuclei and cerebellum. Their input stimulates or inhibits the activity of either (1) motor nuclei or (2) the primary motor cortex.

THE CORTICOSPINAL PATHWAY [FIGURE 16.4]

The **corticospinal pathway**, sometimes called the *pyramidal system* (Figure 16.4a●) provides voluntary control over skeletal muscles. This system begins at the *pyramidal cells* of the primary motor cortex. The axons of these upper motor neurons descend into the brain stem and spinal cord to synapse on lower motor neurons that control skeletal muscles. In general, the corticospinal pathway is direct: The upper motor neurons synapse directly on the lower motor neurons. However, the corticospinal pathway also works indirectly, as it innervates centers of the medial and lateral pathways.

The corticospinal pathway contains three pairs of descending tracts: (1) the *corticobulbar tracts*, (2) the *lateral corticospinal tracts*, and (3) the *anterior corticospinal tracts*. These tracts enter the white matter of the internal capsule, descend into the brain stem, and emerge on either side of the mesencephalon as the *cerebral peduncles*. p. 403

The Corticobulbar Tracts [FIGURE 16.4a AND TABLE 16.2] Axons in the **corticobulbar tracts** (kor-ti-kō-BUL-bar; *bulbar*, brain stem) (Figure 16.4a● and Table 16.2) synapse on lower motor neurons in the motor nuclei of cranial nerves III, IV, V, VI, VII, IX, XI, and XII. The corticobulbar tracts provide conscious control over skeletal muscles that move the eye, jaw, and face and some muscles of the neck and pharynx. The corticobulbar tracts also innervate the motor centers of the medial and lateral pathways. p. 419

TABLE 16.2

Principal Descending (Motor) Tracts in the Spinal Cord and the General Functions of the Associated Nuclei in the Brain

Tract	Location of Upper	Destination	Site of Cross-Over	Action
CORTICALSPINAL PATHWAY				
Corticobulbar tracts	Primary motor cortex (cerebral hemisphere)	Lower motor neurons of cranial nerve nuclei in brain stem	Brain stem	Conscious motor control of skeletal muscles
Lateral corticospinal tracts	As above	Lower motor neurons of anterior gray horns of spinal cord	Pyramids of medulla oblongata	As above
Anterior corticospinal tracts	As above	As above	Level of lower motor neuron	As above
MEDIAL PATHWAY				
Vestibulospinal tract	Vestibular nucleus (at border of pons and medulla oblongata)	As above	Uncrossed	Subconscious regulation of balance and muscle tone
Tectospinal tract	Tectum (mesencephalon: superior and inferior colliculi)	Lower motor neurons of anterior gray horns (cervical spinal cord only)	Brain stem (mesencephalon)	Subconscious regulation of eye, head, neck, and upper limb position in response to visual and auditory stimuli
Reticulospinal tracts	Reticular formation (network of nuclei in brain stem)	Lower motor neurons of anterior gray horns of spinal cord	Uncrossed	Subconscious regulation of reflex activity
LATERAL PATHWAY				
Rubrospinal tracts	Red nuclei of mesencephalon	As above	Brain stem (mesencephalon)	Subconscious regulation of upper limb muscle tone and movement

The Corticospinal Tracts [FIGURE 16.4 AND TABLE 16.2] Axons in the **corticospinal tracts** (Figure 16.4b●) synapse on lower motor neurons in the anterior gray horns of the spinal cord. As they descend, the corticospinal tracts are visible along the ventral surface of the medulla oblongata as a pair of thick bands, the **pyramids**. Along the length of the pyramids, roughly 85% of the axons cross the midline (decussate) to enter the descending **lateral corticospinal tracts** on the opposite side of the spinal cord. The other 15% continue uncrossed along the spinal cord as the **anterior corticospinal tracts**. At the spinal segment it targets, an axon in the anterior corticospinal tract crosses over to the opposite side of the spinal cord in the anterior white commissure before synapsing on lower motor neurons in the anterior gray horns. Information concerning these tracts and their associated functions is summarized in Table 16.2.

The Motor Homunculus The activity of pyramidal cells in a specific portion of the primary motor cortex will result in the contraction of specific peripheral muscles. The identities of the stimulated muscles depend on the region of motor cortex that is active. As in the primary sensory cortex, the primary motor cortex corresponds point by point with specific regions of the body. The cortical areas have been mapped out in diagrammatic form, creating a **motor homunculus**. Figure 16.4a● shows the motor homunculus of the left cerebral hemisphere and the corticospinal pathway controlling skeletal muscles on the right side of the body.

The proportions of the motor homunculus are quite different from those of the actual body (Figure 16.4a●), because the motor area devoted to a specific region of the cortex is proportional to the number of motor units involved in the region's control rather than its actual size. As a result, the homunculus provides an indication of the degree of fine motor control available. For example, the hands, face, and tongue, all of which are capable of varied and complex movements, appear very large, whereas the trunk is relatively small. These proportions are similar to those of the sensory homunculus (Figure 16.2a●, p. 430). The sensory and motor homunculi differ in other respects because some highly sensitive regions, such as the sole of the foot, contain few motor units, and some areas with an abundance of motor units, such as the eye muscles, are not particularly sensitive.

> **CLINICAL BRIEF**
>
> **Cerebral Palsy**
>
> The term **cerebral palsy** refers to a number of disorders affecting voluntary motor performance that appear during infancy or childhood and persist throughout the life of the affected individual. The cause may be CNS trauma associated with premature or unusually stressful birth, maternal exposure to drugs, including alcohol, or a genetic defect that causes the improper development of motor pathways. Problems with labor and delivery may cause compression or interruption of placental circulation or oxygen supplies. If the oxygen concentration in fetal blood declines significantly for as little as 5–10 minutes, CNS function can be permanently impaired. The cerebral cortex, cerebellum, basal nuclei, hippocampus, and thalamus are particularly sensitive, and abnormalities in motor skills, posture and balance, memory, speech, and learning abilities can result. *Tay-Sachs Disease p. 795*

THE MEDIAL AND LATERAL PATHWAYS [FIGURES 16.4b/16.5 AND TABLE 16.3]

Several centers in the cerebrum, diencephalon, and brain stem may issue somatic motor commands as a result of processing performed at a subconscious level. *pp. 394, 403, 408* These centers and their associated tracts were long known as the *extrapyramidal system (EPS)*, because it was thought that they operated independent of, and in parallel to, the *pyramidal system* (corticospinal pathway). This classification scheme is both inaccurate and misleading, because motor control is integrated at all levels through extensive feedback loops and interconnections. It is more appropriate to group these nuclei and tracts in terms of their primary functions: the components of the **medial pathway** help control gross movements of the trunk and proximal limb muscles, whereas those of the **lateral pathway** help control the distal limb muscles that perform more-precise movements.

The medial and lateral pathways can modify or direct skeletal muscle contractions by stimulating, facilitating, or inhibiting lower motor neurons. It is important to note that the axons of upper motor neurons in the medial and lateral pathways synapse on the same lower motor neurons innervated by the corticospinal pathway. This means that the various motor pathways interact not only within the brain, through interconnections between the primary motor cortex and motor centers in the brain stem, but also through excitatory or inhibitory interactions at the level of the lower motor neurons.

THE MEDIAL PATHWAY The medial pathway is primarily concerned with the control of muscle tone and gross movements of the neck, trunk, and proximal limb muscles. The upper motor neurons of the medial pathway are located in the *vestibular nuclei*, the *superior* and *inferior colliculi*, and the *reticular formation* (Figure 16.5●).

The vestibular nuclei receive information, over the vestibulocochlear nerve (N VIII), from receptors in the inner ear that monitor the position and movement of the head. These nuclei respond to changes in the orientation of the head, sending motor commands that alter the muscle tone, extension, and position of the neck, eyes, head, and limbs. The primary goal is to maintain posture and balance. The descending fibers in the spinal cord constitute the **vestibulospinal tracts** (Figure 16.4b●).

The superior and inferior colliculi are located in the *tectum*, or roof, of the mesencephalon. pp. 405 The colliculi receive visual (superior) and auditory (inferior) sensations. Axons of upper motor neurons in the colliculi descend in the **tectospinal tracts**. These axons cross to the opposite side immediately, before descending to synapse on lower motor neurons in the brain stem or spinal cord. Axons in the tectospinal tracts direct reflexive changes in the position of the head, neck, and upper limbs in response to bright lights, sudden movements, or loud noises.

The reticular formation is a loosely organized network of neurons that extends throughout the brain stem. The reticular formation receives input from almost every ascending and descending pathway. It also has extensive interconnections with the cerebrum, the cerebellum, and brain stem nuclei. Axons of upper motor neurons in the reticular formation descend in the **reticulospinal tracts** without crossing to the opposite side. The effects of reticular formation stimulation are determined by the region stimulated. For example, the stimulation of upper motor neurons in one portion of the reticular formation produces eye movements, whereas the stimulation of another portion activates respiratory muscles.

THE LATERAL PATHWAY The lateral pathway is primarily concerned with the control of muscle tone and the more-precise movements of the distal parts of the limbs. The upper motor neurons of the lateral pathway lie within the *red nuclei* of the mesencephalon. pp. 405 Axons of upper motor neurons in the red nuclei cross to the opposite side of the brain and descend into the spinal cord in the **rubrospinal tracts** (*ruber*, red). In humans, the rubrospinal tracts are small and extend only to the cervical spinal cord. There they provide motor control over distal muscles of the upper limbs; normally, their role is insignificant as compared with that of the lateral corticospinal tracts. However, the rubrospinal tracts can be important in maintaining motor control and muscle tone in the upper limbs if the lateral corticospinal tracts are damaged.

Table 16.2 reviews the major motor tracts we discussed in this section.

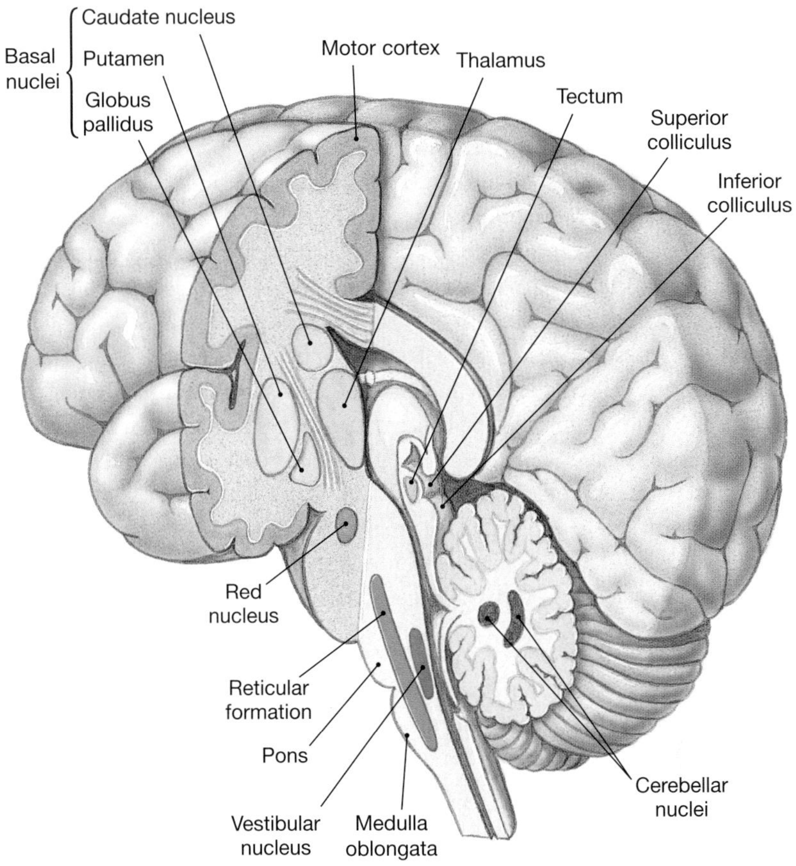

FIGURE 16.5 **NUCLEI OF THE MEDIAL AND LATERAL PATHWAYS**

Cutaway view showing the location of major nuclei whose motor output is carried by the medial and lateral pathways. *See also Figure 15.11.*

THE BASAL NUCLEI AND CEREBELLUM [FIGURE 16.5]

The basal nuclei are within the cerebrum lateral to the thalamus (Figure 16.5●). The basal nuclei and cerebellum are responsible for coordination and feedback control over muscle contractions, whether those contractions are consciously or subconsciously directed.

THE BASAL NUCLEI The basal nuclei provide the background patterns of movement involved in voluntary motor activities. For example, they may control muscles that determine the background position of the trunk and limbs, or they may direct rhythmic cycles of movement, as in walking or running. The basal nuclei do not exert direct control over lower motor neurons; instead they adjust the activities of upper motor neurons in the various motor pathways. The basal nuclei receive input from all portions of the cerebral cortex as well as from the substantia nigra.

The basal nuclei use two major pathways to adjust or establish patterns of movement:

1. One group of axons synapses on thalamic neurons, which then send their axons to the premotor cortex, the motor association area that directs activities of the primary motor cortex. This arrangement creates a feedback loop that changes the sensitivity of the pyramidal cells and alters the pattern of instructions carried by the corticospinal tracts.
2. A second group of axons synapses in the reticular formation, altering the excitatory or inhibitory output of the reticulospinal tracts.

Two distinct populations of neurons exist: one that stimulates neurons by releasing acetylcholine (ACh) and another that inhibits neurons by the release of gamma aminobutyric acid (GABA). Under normal conditions, the excitatory interneurons are kept inactive and the tracts leaving the basal nuclei have an inhibitory effect on upper motor neurons. In *Parkinson's disease*, the excitatory neurons become more active, leading to problems with the voluntary control of movement. p. 403

If the primary motor cortex is damaged, the individual loses the ability to exert fine control over skeletal muscles. However, some voluntary movements can still be controlled by the basal nuclei. In effect, the medial and lateral pathways function as they usually do, but the corticospinal pathway cannot fine-tune the movements. For example, after damage to the primary motor cortex, the basal nuclei can still receive information about planned movements from the prefrontal cortex and can perform preparatory movements of the trunk and limbs. But, because the corticospinal pathway is inoperative, precise movements of the forearms, wrists, and hands cannot occur. An individual in this condition can stand, maintain balance, and even walk, but all movements are hesitant, awkward, and poorly controlled.

The Cerebellum The cerebellum monitors proprioceptive (position) sensations, visual information from the eyes, and vestibular (balance) sensations from the inner ear as movements are under way. pp. 406–407 Axons relaying proprioceptive information reach the cerebellar cortex in the spinocerebellar tracts. Visual information is relayed from the superior colliculi, and balance information is relayed from the vestibular nuclei. The output of the cerebellum affects upper motor neuron activity in the corticospinal, medial, and lateral pathways.

All motor pathways send information to the cerebellum when motor commands are issued. As the movement proceeds, the cerebellum monitors proprioceptive and vestibular information and compares the arriving sensations with those experienced during previous movements. It then adjusts the activities of the upper motor neurons involved. In general, any voluntary movement begins with the activation of far more motor units than are required—or even desirable. The cerebellum provides the necessary inhibition, reducing the number of motor commands to an efficient minimum. As the movement proceeds, the pattern and degree of inhibition change, producing the desired result.

The patterns of cerebellar activity are learned by trial and error, over many repetitions. Many of the basic patterns are established early in life; examples include the fine balancing adjustments you make while standing and walking. The ability to fine-tune a complex pattern of movement improves with practice, until the movements become fluid and automatic. Consider the relaxed, smooth movements of acrobats, golfers, and sushi chefs. These people move without thinking about the details of their movements. This ability is important, because when you concentrate on voluntary control, the rhythm and pattern of the movement usually fall apart as your primary motor cortex starts overriding the commands of the basal nuclei and cerebellum.

Levels of Somatic Motor Control [FIGURE 16.6]

Ascending information is relayed from one nucleus or center to another in a series of steps. For example, somatic sensory information from the spinal cord goes from a nucleus in the medulla oblongata to a nucleus in the thalamus before it reaches the primary sensory cortex. Information processing occurs at each step along the way. As a result, conscious awareness of the stimulus may be blocked, reduced, or heightened.

These processing steps are important, but they take time. Every synapse means another delay, and between conduction time and synaptic delays it takes several milliseconds to relay information from a peripheral receptor to the primary sensory cortex. Additional time will pass before the primary motor cortex orders a voluntary motor response.

This delay is not dangerous, because interim motor commands are issued by relay stations in the spinal cord and brain stem. While the conscious mind is still processing the information, neural reflexes provide an immediate response that can later be "fine-tuned." For example, if you touch a hot stovetop, in the few milliseconds it takes for you to become consciously aware of the danger, you could be severely burned. But that doesn't happen, because your response (withdrawing your hand) occurs almost immediately, through a withdrawal reflex coordinated in the spinal cord. Voluntary motor responses, such as shaking the hand, stepping back, and crying out, occur somewhat later. In this case the initial reflexive response, directed by neurons in the spinal cord, was supplemented by a voluntary response controlled by the cerebral cortex. The spinal reflex provided a rapid, automatic, preprogrammed response that preserved homeostasis. The cortical response was more complex, but it required more time to prepare and execute.

Nuclei in the brain stem also are involved in a variety of complex reflexes. Some of these nuclei receive sensory information and generate appropriate motor responses. These motor responses may involve direct control over motor neurons or the regulation of reflex centers in other parts of the brain. Figure 16.6• illustrates the various levels of somatic motor control from simple spinal reflexes to complex patterns of movement.

All of the levels of somatic motor control affect the activity of lower motor neurons. Reflexes coordinated in the spinal cord and brain stem are the simplest mechanisms of motor control. Higher levels perform more elaborate processing; as one moves from the medulla oblongata to the cerebral cortex, the motor patterns become increasingly complex and variable. For example, the respiratory rhythmicity center of the medulla oblongata sets a basic breathing rate. Centers in the pons adjust that rate in response to commands received from the hypothalamus (subconscious) or cerebral cortex (conscious).

The basal nuclei, cerebellum, mesencephalon, and hypothalamus control the most complicated involuntary motor patterns. Examples include motor patterns associated with eating and reproduction (hypothalamus), walking and body positioning (basal nuclei), learned movement patterns (cerebellum), and movements in response to sudden visual or auditory stimuli (mesencephalon).

At the highest level are the complex, variable, and voluntary motor patterns dictated by the cerebral cortex. Motor commands may be given to specific motor neurons directly, or they may be given indirectly by altering the activity of a reflex control center. Figure 16.6b,c• provides a simple diagram of the steps involved in the planning and execution of a voluntary movement.

During development, the control levels appear in sequence, beginning with the spinal reflexes. More complex reflexes develop as the neurons grow and interconnect. The process proceeds relatively slowly, as billions of neurons establish trillions of synaptic connections. At birth neither the cerebral nor the cerebellar cortex is fully functional, and their abilities take years to mature. A number of anatomical factors, noted in earlier chapters, contribute to this maturation:

1. Cortical neurons continue to increase in number until at least age 1.
2. The brain grows in size and complexity until at least age 4.
3. Myelination of CNS axons continues at least until age 1–2; peripheral myelination may continue through puberty.

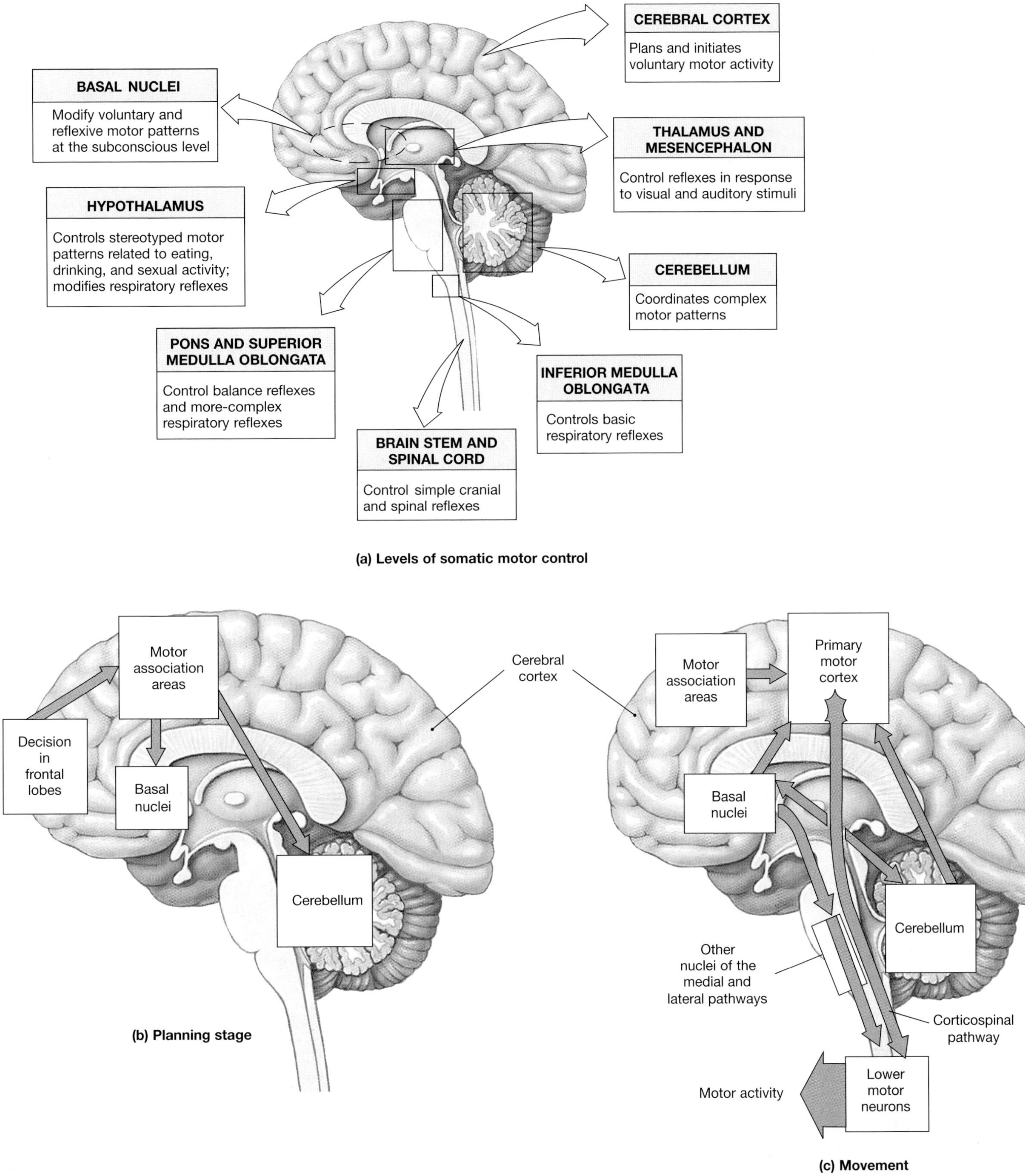
CEREBRAL CORTEX
Plans and initiates voluntary motor activity
BASAL NUCLEI
Modify voluntary and reflexive motor patterns at the subconscious level
THALAMUS AND MESENCEPHALON
Control reflexes in response to visual and auditory stimuli
HYPOTHALAMUS
Controls stereotyped motor patterns related to eating, drinking, and sexual activity; modifies respiratory reflexes
CEREBELLUM
Coordinates complex motor patterns
PONS AND SUPERIOR MEDULLA OBLONGATA
Control balance reflexes and more-complex respiratory reflexes
INFERIOR MEDULLA OBLONGATA
Controls basic respiratory reflexes
BRAIN STEM AND SPINAL CORD
Control simple cranial and spinal reflexes
(a) Levels of somatic motor control
Motor association areas
Decision in frontal lobes
Basal nuclei
Cerebellum
Cerebral cortex
(b) Planning stage
Motor association areas
Primary motor cortex
Basal nuclei
Cerebellum
Other nuclei of the medial and lateral pathways
Corticospinal pathway
Motor activity
Lower motor neurons
(c) Movement

FIGURE 16.6 **Somatic Motor Control**

(**a**) Somatic motor control involves a series of levels, with simple spinal and cranial reflexes at the bottom and complex voluntary motor patterns at the top. (**b**) The planning stage: When a conscious decision is made to perform a specific movement, information is relayed from the frontal lobes to motor association areas. These areas in turn relay the information to the cerebellum and basal nuclei. (**c**) Movement: As the movement begins, the motor association areas send instructions to the primary motor cortex. Feedback from the basal nuclei and cerebellum modifies those commands, and output along the medial and lateral pathways directs involuntary adjustments in position and muscle tone.

CLINICAL BRIEF

ANENCEPHALY

Although it may sound strange, physicians usually take a newborn infant into a dark room and shine a light against the skull. They are checking for **anencephaly** (an-en-SEF-a-lē), a rare condition in which the brain fails to develop at levels above the mesencephalon or lower diencephalon. The cranium typically does not develop at all, making diagnosis easy; if the cranial vault does develop it is empty and translucent enough to transmit light.

Unless the condition is discovered right away, the parents may take the infant home, totally unaware of the problem. All the normal behavior patterns expected of a newborn are present, including suckling, stretching, yawning, crying, kicking, sticking fingers in the mouth, and tracking movements with the eyes. However, death will occur naturally over a period of days to months.

This tragic condition provides a striking demonstration of the role of the brain stem in controlling complex involuntary motor patterns. During normal development these patterns become incorporated into variable and versatile behaviors as control and analytical centers appear in the cerebral cortex.

As these events occur, cortical neurons continue to establish new interconnections that will have a long-term effect on the functional capabilities of the individual.

✓ CONCEPT CHECK

- As a result of pressure on her spinal cord, Jill cannot feel touch or pressure on her legs. What spinal tract is being compressed?
- What is the anatomical reason for the left side of the brain controlling motor function on the right side of the body?
- An injury to the superior portion of the motor cortex would affect what part of the body?
- Through which of the motor tracts would the following commands travel: (a) reflexive change of head position due to bright lights, (b) automatic alterations in limb position to maintain balance?

Higher-Order Functions

Higher-order functions have the following characteristics:

1. They are performed by the cerebral cortex.
2. They involve complex interconnections and communication between areas within the cerebral cortex and between the cerebral cortex and other areas of the brain.
3. They involve both conscious and unconscious information processing.
4. They are not part of the programmed "wiring" of the brain; therefore, the functions are subject to modification and adjustment over time.

Our discussion of higher-order functions begins by identifying the cortical areas involved and considering functional differences between the left and right cerebral hemispheres. We will then briefly consider the mechanisms of memory, learning, and consciousness.

Integrative Regions of the Cerebral Cortex [FIGURE 16.7]

The sensory, motor, and association areas of the cerebral hemispheres were introduced in Chapter 15. p. 394 Figure 16.7a● reviews the major cortical regions of the left cerebral hemisphere. Scanning data, electrical monitoring, and clinical observation have shown that several cortical areas integrate complex sensory stimuli and motor responses. These centers (Figure 16.7b●) include the *general interpretive area*, the *speech center*, and the *prefrontal cortex*.

THE GENERAL INTERPRETIVE AREA

The **general interpretive area**, or *gnostic area*, receives information from all the sensory association areas. This analytical center is present in only one hemisphere, usually the left. Damage to the general interpretive area affects the ability to interpret what is read or heard, even though the words are understood as individual entities. For example, an individual might understand the meaning of the words "sit" and "here" but be totally bewildered by the request "sit here."

THE SPEECH CENTER

Efferents from the general interpretive area target the **speech center**, or *Broca's area*. This center lies along the edge of the premotor cortex in the same hemisphere as the general interpretive area. The speech center is a motor center that regulates the patterns of breathing and vocalization needed for normal speech. The corresponding regions on the opposite hemisphere are not "inactive," but their functions are less well defined. Damage to the speech center can manifest itself in various ways. Some individuals have difficulty speaking, although they know exactly what words to use; others talk constantly but use all the wrong words.

THE PREFRONTAL CORTEX

The **prefrontal cortex**, or *prefrontal association area*, of the frontal lobe is the most complex brain area. It has extensive connections with other cortical areas and with other portions of the brain, such as the limbic system. The prefrontal cortex performs complicated learning and reasoning functions. Through its interconnections with the limbic system, it also provides emotional context and motivation. The prefrontal cortex performs such abstract intellectual functions as predicting the future consequences of events or actions. Damage to the prefrontal cortex leads to difficulties in estimating the temporal relationships between events; questions such as "How long ago did this happen?" or "What happened first?" become difficult to answer. *Huntington's Disease p. 795*

Feelings of frustration, tension, and anxiety are generated at the prefrontal cortex as it interprets ongoing events and makes predictions about future situations or consequences. If the connections between the prefrontal cortex and other brain regions are severed, the tensions, frustrations, and anxieties are removed. Through the mid-twentieth century, this rather drastic procedure, called a **prefrontal lobotomy**, was used to "cure" a variety of mental illnesses, especially those associated with violent or antisocial behavior.

After a lobotomy, the patient would no longer be concerned about what had previously been a major problem, whether psychological (hallucinations) or physical (severe pain). However, the individual was often equally unconcerned about tact, decorum, and toilet training. Now that drugs have been developed to target specific pathways and regions of the CNS, lobotomies are no longer used to control behavior.

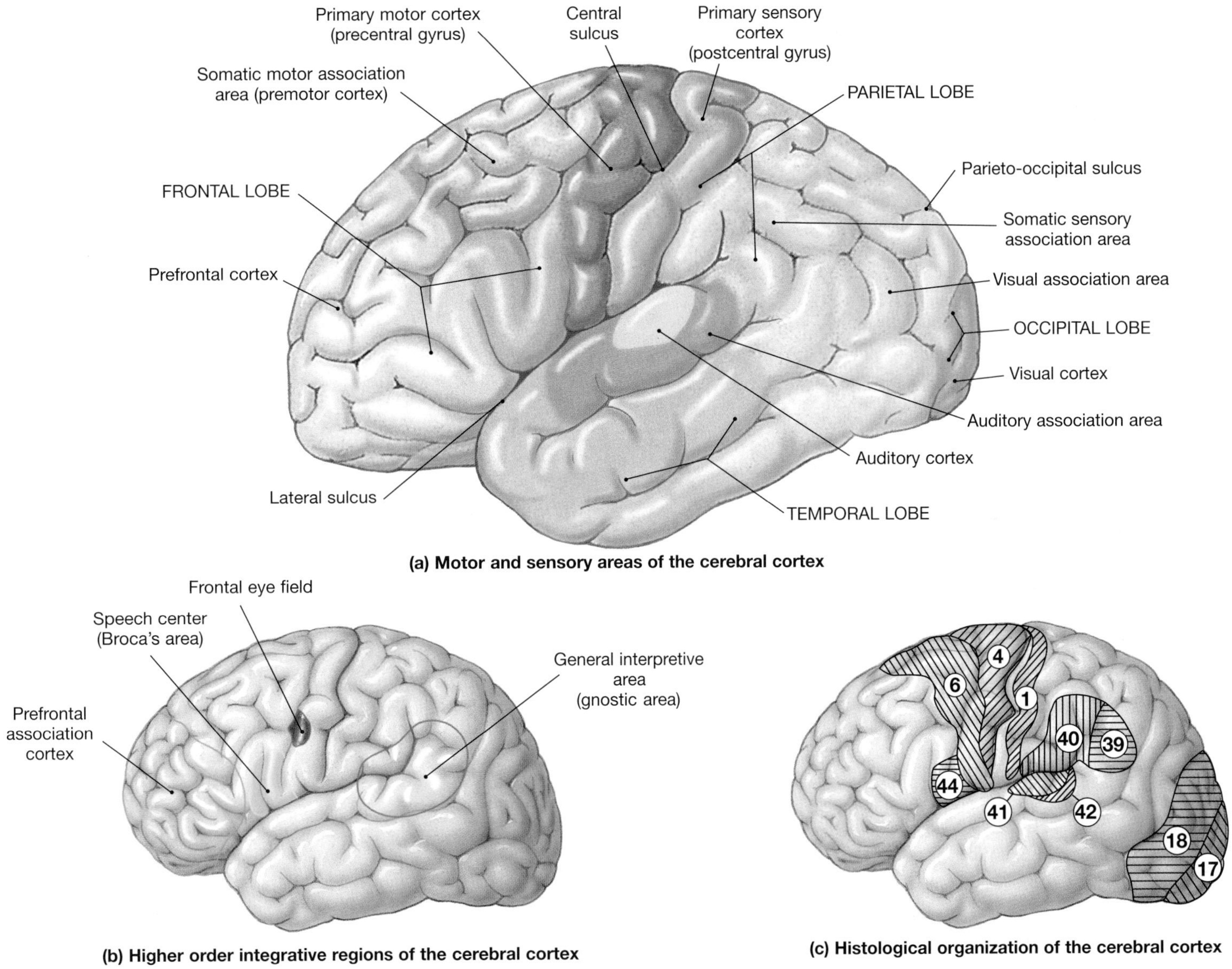

FIGURE 16.7 **FUNCTIONAL AREAS OF THE CEREBRAL CORTEX**

(**a**) Motor, sensory, and special sensory areas of the cerebral cortex that are found in both cerebral hemispheres. (**b**) The left hemisphere usually contains the general interpretive area and the speech center. Other specializations of the two cerebral hemispheres are shown in *Figure 16.8*. (**c**) These are some of the Brodman areas of the cerebral cortex, each characterized by a distinctive pattern of cellular organization. Compare these anatomical regions with the functional zones shown in (a) and (b).

BRODMAN AREAS AND CORTICAL FUNCTION [FIGURE 16.7a]

In the early 1900s, several attempts were made to describe and classify regional differences in the histological organization of the cerebral cortex. It was hoped that the patterns of cellular organization could be correlated to specific sensory, motor, and integrative functions. By 1919, at least 200 different patterns had been described, and most of the classification schemes have since been abandoned. However, the cortical map prepared by Brodman in 1909 has proved useful to neuroanatomists. Brodman described 47 different patterns of cellular organization in the cerebral cortex. Several of these *Brodman areas* are shown in Figure 16.7c●. Some correspond to known functional areas. For example, Brodman's area 44 corresponds to the motor speech center, and area 4 follows the contours of the primary motor cortex. In other cases, the correspondence is less precise. For example, Brodman's area 1 lies within the primary sensory cortex.

Hemispheric Specialization [FIGURES 16.7a/16.8]

The regions seen in Figure 16.7a● are present on both hemispheres, but the higher-order functions are not equally distributed. Figure 16.8● indicates the major functional differences between the hemispheres. Higher-order centers in the left and right hemispheres have different but complementary functions. In the majority of people in the United States, the left hemisphere contains the general interpretive and speech centers. This is the hemisphere responsible for language-based skills; reading, writing, and speaking are dependent on processing done in the left cerebral hemisphere. It is also important in performing analytical tasks, such as mathematical calculations and logical decision making. Although the left hemisphere was formerly called the *dominant hemisphere,* the term **categorical hemisphere** is more appropriate, because the right hemisphere has many important functions.

CLINICAL BRIEF

Damage to the Integrative Centers

Aphasia (*a-*, without + *phasia*, speech) is a disorder affecting the ability to speak or read. Extreme aphasia, or **global aphasia**, results from extensive damage to the general interpretive area or to the associated sensory tracts. Affected individuals are totally unable to speak, to read, or to understand or interpret the speech of others. Global aphasia often accompanies a severe stroke or tumor that affects a large area of cortex including the speech and language areas. Recovery is possible when the condition results from edema or hemorrhage, but the process often takes months or even years.

Dyslexia (*lexis,* diction) is a disorder affecting the comprehension and use of words. **Developmental dyslexia** affects children; there are estimates that up to 15% of children in the United States suffer from some degree of dyslexia. These children have difficulty reading and writing, although their other intellectual functions may be normal or above normal. Their writing looks uneven and disorganized with letters reversed or written in the wrong order more frequently than in the writing of unaffected children. Recent evidence suggests that at least some forms of dyslexia result from problems in processing and sorting visual information.

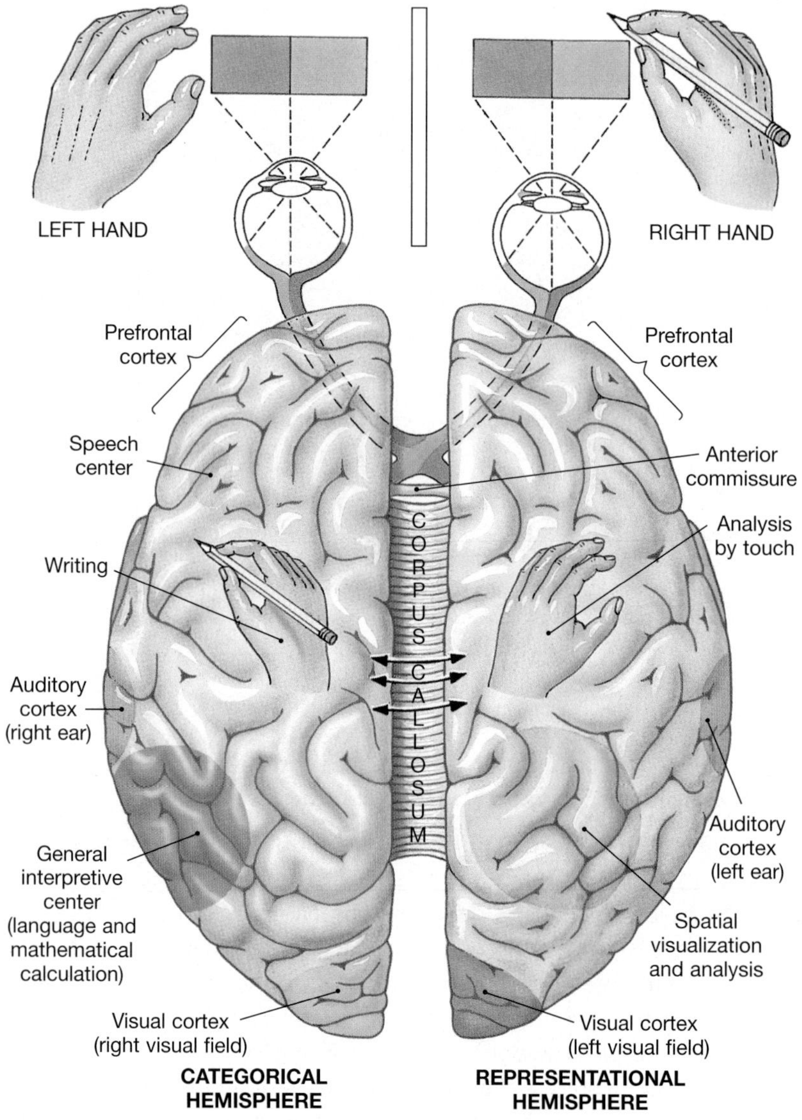

FIGURE 16.8 **Hemispheric Specialization**

Functional differences between the left and right cerebral hemispheres. Notice that special sensory information is relayed to the cerebral hemisphere on the opposite side of the body. Chapter 18 will provide additional details on these pathways.

The right cerebral hemisphere analyzes sensory information and relates the body to the sensory environment. Interpretive centers in this hemisphere permit the identification of familiar objects by touch, smell, taste, or feel. Because it is concerned with spatial relationships and analysis, the term **representational hemisphere** is used to refer to the right hemisphere.

The designation of a hemisphere as categorical rather than representational depends on the location of the major functional centers, especially the general interpretive and speech centers. Because the left hemisphere is categorical in the majority of both left-handed and right-handed individuals, there is probably a genetic basis for the distribution of functions. An estimated 90% of the population have an enlarged left hemisphere at birth. (The remaining 10% may have hemispheres of equal size or an enlarged right hemisphere.)

The functional significance (if any) of having the categorical hemisphere on the left side is unknown. Interestingly, there may be a link between being right- or left-handed and sensory and spatial abilities. An unusually high percentage of musicians and artists are left-handed; the complex motor activities performed by these individuals are directed by the primary motor cortex and association areas on the right (representational) hemisphere.

Hemispheric specialization does not mean that the two hemispheres are independent, merely that certain centers have evolved to process information gathered by the system as a whole. The intercommunication occurs over commissural fibers, especially those of the corpus callosum. p. 394 The corpus callosum alone contains over 200 million axons, carrying an estimated 4 billion impulses per second!

Memory

Memory is a higher-order function that involves considerable interaction between the cerebral cortex and other areas of the brain. Memory is the process of accessing stored bits of information gathered through experience; these bits of information are called *memories*, or *memory engrams*. Some memories can be retrieved voluntarily and expressed verbally, as when you recall and recite a phone number. Other memories are retrieved subconsciously; for example, when hungry, you may salivate at the smell of food. There are *short-term memories* that last seconds to hours, and *long-term memories* that can last for years. Each type of memory involves different anatomical structures. The conversion from a short-term to a long-term memory is called *memory consolidation.*

Two components of the limbic system, the amygdaloid body and the hippocampus, are essential to memory consolidation. p. 395 Damage to either of those areas will interfere with normal memory consolidation. Damage to the hippocampus leads to an immediate loss of short-term memory, although long-term memories remain intact and accessible. Tracts leading from the amygdaloid body to the hypothalamus may link memories to specific emotions. A cerebral nucleus near the diencephalon, the *nucleus basalis*, plays an uncertain role in memory storage and retrieval. Tracts connect this nucleus with the hippocampus, amygdaloid body, and all areas of the cerebral cortex. Damage to this nucleus is associated with changes in emotional states, memory, and intellectual function. (See the discussion of Alzheimer's disease later in this chapter.)

Long-term memories are stored in the cerebral cortex. Conscious motor and sensory memories are referred to the appropriate association areas. For example, visual memories are stored in the visual association area, and memories of voluntary motor activity in the premotor cortex. Special portions of the occipital and temporal lobes retain the memories of facial images, the sounds of voices, and the pronunciation of words.

CLINICAL BRIEF

Disconnection Syndrome

Communication across the corpus callosum permits the integration of sensory information and motor commands. Yet the two hemispheres are significantly different in terms of their ongoing processing activities. Otherwise untreatable seizures may sometimes be "cured" by severing the corpus callosum. This surgery produces symptoms of **disconnection syndrome**. In this condition the two hemispheres function independently, each remaining "unaware" of stimuli or motor commands involving their counterpart. The result is a number of rather interesting changes in the individual's abilities. For example, objects touched by the left hand can be recognized but not verbally identified, because the sensory information arrives at the right hemisphere and the speech center is on the left. The object can be verbally identified if felt with the right hand, but the person will not be able to say whether or not it is the same object previously touched with the left hand. This problem with cross-referencing sensory information applies to all incoming sensations.

Two years after a surgical sectioning of the corpus callosum, the most striking behavioral abnormalities have disappeared, and the individual may test normally. In addition, individuals born without a functional corpus callosum do not show obvious sensory or motor deficits. In some way, the CNS adapts to the situation, probably by increasing the amount of information transferred across the anterior commissure.

✓ Concept Check

- After suffering a head injury in an automobile accident, David has difficulty comprehending what he hears or reads. This problem might indicate damage to what portion of the brain?
- Kyle has difficulty speaking; he talks continuously, but uses incorrect words. What center or area of the brain is probably involved?
- What is memory consolidation?
- Gina cannot perform abstract intellectual functions such as predicting the future consequences of her actions. What part of her brain is affected?

Consciousness: The Reticular Activating System

[FIGURE 16.9 AND TABLE 16.3]

A conscious individual is alert and attentive; an unconscious individual is not. The difference is obvious, but there are many gradations of both the conscious and unconscious states (Table 16.3). The state of consciousness experienced by an individual is determined by complex interactions between the brain stem and the cerebral cortex. One of the most important brain stem components is the **reticular activating system (RAS)**, a poorly defined network in the reticular formation. The RAS extends from the mesencephalon to the medulla oblongata (Figure 16.9●). The output of the RAS projects throughout the cerebral cortex. When the RAS is inactive, so is the cerebral cortex; stimulation of the RAS produces a widespread activation of the cerebral cortex. The primary function of this activation is to keep the individual alert and conscious; if the level of cortical activity starts declining, the person gradually becomes increasingly lethargic and eventually becomes unconscious.

The mesencephalic portion of the RAS appears to be the control center of the system, and stimulation of this area produces the most pronounced and long-lasting effects on the cerebral cortex. Stimulating other portions of the RAS seems to have an effect only insofar as it changes the activity of the mesencephalic region. The greater the stimulation to the mesencephalic region of the RAS, the more alert and attentive the individual will be to incoming sensory information. Associated nuclei in the thalamus play a supporting role by focusing attention on specific mental processes.

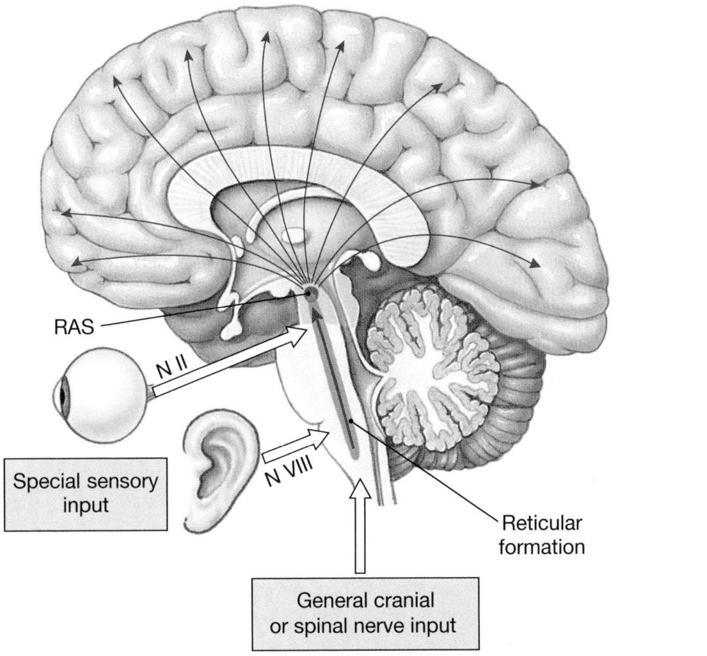

FIGURE 16.9 **The Reticular Activating System**

The mesencephalic control center of the reticular formation is the reticular activating system (RAS). It receives collateral inputs from a variety of sensory pathways. Stimulation of this region produces arousal and heightened states of attentiveness.

CLINICAL BRIEF

Amnesia

Amnesia refers to the loss of memory from disease or trauma. The type of memory loss depends upon the specific regions of the brain affected. Damage to sensory association areas produces memory loss of sensations arriving at the adjacent sensory cortex. Damage to thalamic and limbic structures, especially the hippocampus, will affect memory storage and consolidation. Amnesia may occur suddenly or progressively, and recovery may be complete, partial, or nonexistent, depending on the nature of the problem.

In **retrograde amnesia** (*retro-*, behind), the individual loses memories of past events. Some degree of retrograde amnesia often follows a head injury, and accident victims are frequently unable to remember the moments preceding a car wreck. In **anterograde amnesia** (*antero-*, ahead), an individual may be unable to store additional memories, but earlier memories are intact and accessible. The problem appears to involve an inability to generate long-term memories. At least two drugs—diazepam (*Valium*™) and *Halcion*™—have been known to cause brief periods of anterograde amnesia. A person with permanent anterograde amnesia lives in surroundings that are always new. Magazines can be read, chuckled over, and then reread a few minutes later with equal pleasure, as if they had never been seen before. Physicians and nurses must introduce themselves at every meeting, even if they have been treating the patient for years.

Post-traumatic amnesia (PTA) may develop after a head injury. The duration and extent of the amnesia varies depending on the severity of the injury. PTA combines the characteristics of retrograde and anterograde amnesia; the individual can neither remember parts of the past nor consolidate memories of the present. After recovery, a stable, permanent memory gap persists.

CLINICAL BRIEF

Cerebrovascular Diseases

Cerebrovascular diseases are circulatory disorders that interfere with the normal circulatory supply to the brain. The particular distribution of the vessel involved will determine the symptoms, and the degree of oxygen or nutrient starvation will determine their severity. A stroke, or **cerebrovascular accident (CVA)**, occurs when the blood supply to a portion of the brain is shut off by a vascular blockage or hemorrhage. Affected neurons begin to die in a matter of minutes.

The symptoms of a stroke provide an indication of the vessel and region of the brain involved. For example, the carotid artery enters the skull via the carotid canal. One major branch of the carotid, the *middle cerebral artery*, is the most common site of a stroke. Superficial branches deliver blood to the temporal lobe and to large portions of the frontal and parietal lobes; deep branches supply the basal nuclei and portions of the thalamus. If a stroke blocks the middle cerebral artery on the left side of the brain, aphasia and a sensory and motor paralysis of the right side result. In a stroke affecting the middle cerebral artery on the right side, the individual experiences a loss of sensation and motor control over the left side and has difficulty drawing or interpreting spatial relationships. Strokes affecting vessels supplying the brain stem also produce distinctive symptoms; those affecting the lower brain stem are often fatal. (Further information on the causes, diagnosis, and treatment of strokes will be found in Chapter 22.).

Aging and the Nervous System

The aging process affects all bodily systems, and the nervous system is no exception. Anatomical changes begin shortly after maturity (probably by age 30) and accumulate over time. Although an estimated 85% of the elderly (above age 65) lead relatively normal lives, there are noticeable changes in mental performance and CNS functioning.

Some common, age-related anatomical changes in the nervous system include:

1. *A reduction in brain size and weight.* This reduction results primarily from a decrease in the volume of the cerebral cortex. The brains of elderly individuals have narrower gyri and wider sulci than those of young persons, and the subarachnoid space and ventricles are enlarged.
2. *A reduction in the number of neurons.* Reduced brain size has been linked to a loss of cortical neurons. However, neuronal loss neither occurs to the same degree in all individuals nor in all brain stem nuclei.
3. *A decrease in blood flow to the brain.* With age, fatty deposits gradually accumulate in the walls of blood vessels and reduce the rate of arterial blood flow. (This process, called *atherosclerosis*, may affect arteries throughout the body; it is described further in Chapter 22.) Reduced blood flow does not cause a cerebral crisis, but it does increase the probability that the individual will suffer a stroke.
4. *Changes in synaptic organization of the brain.* The number of dendritic branches and interconnections appears to decrease. As synaptic connections are lost, the rate of neurotransmitter production declines.
5. *Intracellular and extracellular changes in CNS neurons.* Many neurons in the brain begin accumulating abnormal intracellular deposits. *Plaques* are extracellular accumulations of an unusual fibrillar protein, *amyloid*, surrounded by abnormal dendrites and axons. *Neurofibrillary tangles* are masses of neurofibrils that form dense mats inside the soma. The significance of these cellular and extracellular abnormalities remains to be determined. There is evidence that they appear in all aging brains (see the discussion of Alzheimer's disease on the next page), but when present in excess, they seem to be associated with clinical abnormalities.

CLINICAL DISCUSSION

Levels of Consciousness

Normal individuals cycle between the alert, conscious state and the asleep state each day. For reference purposes, this table indicates the entire range of conscious and unconscious states, ranging from *delirium* through *coma*. It is important to realize that these states are external indications of the level of ongoing CNS activity. When CNS function becomes abnormal, the state of consciousness can be affected. As a result, clinicians are quick to note any abnormalities in the state of consciousness in their patients.

TABLE 16.3 States of Awareness

Level or State	Description
CONSCIOUS STATES	
Delirium	Disorientation, restlessness, confusion, hallucinations, agitation, alternating with other conscious states; develops quickly
Dementia	Progressive decline in spatial orientation, memory, behavior, and language
Confusion	Reduced awareness, easily distracted, easily startled by sensory stimuli, alternates between drowsiness and excitability; resembles minor form of delirium state
Normal consciousness	Aware of self and external environment, well-oriented, responsive
Somnolence	Extreme drowsiness, but will respond normally to stimuli
Chronic vegetative state	Conscious but unresponsive, no evidence of cortical function
UNCONSCIOUS STATES	
Asleep	Can be aroused by normal stimuli (light touch, sound, etc.)
Stupor	Can be aroused by extreme and/or repeated stimuli
Coma	Cannot be aroused and does not respond to stimuli (coma states can be further subdivided according to the effect on reflex responses to stimuli)

These anatomical changes are linked to a series of functional alterations. In general, neural processing becomes less efficient. For example, memory consolidation often becomes more difficult, and the sensory systems of the elderly, notably hearing, balance, vision, smell, and taste, become less acute. Light must be brighter, sounds louder, and smells stronger before they are perceived. Reaction times are slowed, and reflexes—even some monosynaptic reflexes—become weaker or even disappear. There is a decrease in the precision of motor control, and it takes longer to perform a given motor pattern than it did 20 years earlier. For the majority of the elderly population, these changes do not interfere with their abilities to function in society. But for as yet unknown reasons, many elderly individuals become incapacitated by progressive CNS changes. By far the most common such incapacitating condition is *Alzheimer's disease*.

✓ CONCEPT CHECK

- What would happen to a sleeping individual if his or her reticular activating system (RAS) were suddenly stimulated?
- Which major anatomical structures in the CNS are affected by aging?

CLINICAL BRIEF

Alzheimer's Disease

Alzheimer's disease is a complex disease resulting from a combination of genetic mutations, susceptibility genes, and environmental factors. It is a progressive disorder characterized by the loss of higher cerebral functions and causes roughly 75 percent of cases of **senile dementia**, commonly termed "senility." The first symptoms usually appear at 50–60 years of age. An estimated 10 percent of people over age 65 in the United States suffer from some form of Alzheimer's disease; the incidence rises to almost 50% for those over age 85.

Alzheimer's disease produces a gradual deterioration of mental organization. The afflicted individual loses memories, verbal and reading skills, and emotional control. Initial symptoms are subtle—moodiness, loss of social skills, irritability, and depression. As the condition progresses, however, it becomes more difficult to ignore or accommodate. The victim has difficulty making decisions, even minor ones. Mistakes—sometimes dangerous ones—are made, through either bad judgment or simple forgetfulness. The memory losses continue, and the problems become more severe. The affected person may not recognize relatives or know how to use the telephone. The memory loss often starts with a loss of short term memory, and progresses to the loss of recently stored memories, and eventually the loss of basic long-term memories, such as the sound of the person's own name. The loss of memory affects both intellectual and motor abilities, and a person with severe Alzheimer's disease has difficulty in performing even the simplest motor tasks.

Individuals with Alzheimer's disease show a pronounced decrease in the number of cortical neurons, especially in the frontal and temporal lobes. There are also unusually large concentrations of plaques and neurofibrillary tangles in the basal nucleus, hippocampus, and parahippocampal gyrus. These plaques and tangles contain deposits of several peptides, primarily 2 forms of **amyloid β (Aβ)** protein, and appear in brain regions such as the hippocampus, specifically associated with memory processing.

In Down Syndrome and in some inherited cases of Alzheimer's disease, abnormal genes lead to excessive production and deposition of Aβ protein and increased incidence of the disease. There is no cure but a few medications and supplements slow disease progression in many patients. Diagnosis involves excluding metabolic conditions that can mimic dementia, a detailed history and physical, and evaluation of mental functioning.

RELATED CLINICAL TERMS

Alzheimer's disease: A progressive disorder marked by the loss of higher-order cerebral functions. *p. 443*

amnesia: The temporary or permanent loss of memory from disease or trauma. *p. 441*

anencephaly (an-en-SEF-a-lē): A rare condition in which the brain fails to develop at levels above the mesencephalon or lower diencephalon. *p. 438*

aphasia: A disorder affecting the ability to speak or read. *p. 440*

cerebral palsy: A number of disorders affecting voluntary motor performance that appear during infancy or childhood and persist throughout the life of the affected individual. *p. 434*

cerebrovascular accident (CVA): A condition in which the blood supply to a portion of the brain is blocked off. *p. 442*

cerebrovascular diseases: Vascular disorder that interferes with the arterial supply to the brain. *p. 442*

delirium: A conscious state involving confusion and wild oscillations in the level of consciousness. *p. 442*

dementia: A chronic state of consciousness characterized by deficits in memory, spatial orientation, language, or personality. *p. 442*

disconnection syndrome: Syndrome caused by severing the corpus callosum and separating the two cerebral hemispheres. Each hemisphere continues to function independently, and the right hand literally doesn't know what the left hand is doing. *p. 441*

dyslexia: A disorder affecting the comprehension and use of words. *p. 440*

Tay-Sachs disease: A disease resulting from a genetic abnormality involving the metabolism of *gangliosides*. Affected infants seem normal at birth, but the progress of symptoms typically includes muscular weakness, blindness, seizures, and death, usually before age 4. *Tay-Sachs Disease p. 795*

Additional Clinical Term Discussed in Appendix I (p. 795)

Huntington's disease

STUDY OUTLINE & CHAPTER REVIEW

Introduction 428

1. Information passes continually between the brain, spinal cord, and peripheral nerves. Sensory information is delivered to CNS processing centers, and motor neurons control and adjust peripheral effector activities.

Sensory and Motor Pathways 428

1. Pathways relay sensory and motor information between the CNS, the PNS, and peripheral organs and systems. Ascending (sensory) and descending (motor) pathways contain a chain of tracts and associated nuclei.

Sensory Pathways 428

2. Sensory receptors detect changes in the body or external environment and pass this information to the CNS. This information, called a **sensation**, arrives as action potentials in an afferent *(sensory)* fiber. The response to the stimulus depends on where the processing occurs.
3. Sensory neurons that deliver the sensations to the CNS are termed **first-order neurons. Second-order neurons** are the CNS interneurons on which the first-order neurons synapse. These neurons synapse on a **third-order neuron** in the thalamus. The axon of either the first-order or second-order neuron crosses to the opposite side of the CNS, in a process called **decussation.** Thus, the right cerebral hemisphere receives sensory information from the left side of the body, and vice versa. *(see Table 16.1)*
4. The **posterior column pathway** carries fine touch, pressure, and proprioceptive (position) sensations. The axons ascend within the **fasciculus gracilis** and **fasciculus cuneatus**, and synapse in the nucleus gracilis and nucleus cuneatus within the medulla oblongata. This information is then relayed to the thalamus via the **medial lemniscus.** Decussation occurs as the second-order neurons enter the medial lemniscus. *(see Figures 16.1/16.2 and Table 16.1)*
5. The nature of any stimulus and its location is known because the information projects to a specific portion of the primary sensory cortex. Perceptions of sensations such as touch depend on processing in the thalamus. The precise localization

is provided by the primary sensory cortex. A functional map of the primary sensory cortex is called the **sensory homunculus**. *(see Figure 16.2)*

6. The **spinothalamic pathway** carries poorly localized sensations of touch, pressure, pain, and temperature. The axons of the interneurons decussate in the spinal cord and ascend in the **anterior** and **lateral spinothalamic tracts** to the ventral posterolateral nuclei of the thalamus. *(see Figure 16.2 and Table 16.1)*
7. The **spinocerebellar pathway**, including the **posterior** and **anterior spinocerebellar tracts**, carries sensations to the cerebellum concerning the position of muscles, tendons, and joints. *(see Figures 16.1/16.2, and Table 16.1)*

Motor Pathways 432

8. Motor commands from the CNS are issued in response to sensory system information. These commands are distributed by either the *somatic nervous system (SNS)* for skeletal muscles or the *autonomic nervous system (ANS)* for visceral effectors. *(see Figure 16.3)*
9. Somatic motor pathways always involve an **upper motor neuron** (whose cell body lies in a CNS processing center) and a **lower motor neuron** (located in a motor nucleus of the brain stem or spinal cord). Autonomic motor control requires a **preganglionic neuron** (in the CNS) and a **ganglionic neuron** (in a peripheral ganglion). *(see Figures 16.3 to 16.5)*
10. The neurons of the primary motor cortex are **pyramidal cells**; the **corticospinal pathway** provides a rapid, direct mechanism for voluntary skeletal muscle control. The corticospinal pathway consist of three pairs of descending motor tracts: (1) the *corticobulbar tracts*, (2) the *lateral corticospinal tracts*, and (3) the *anterior corticospinal tracts*. A functional map of the primary motor cortex is called the **motor homunculus**. *(see Figure 16.4 and Table 16.2)*
11. The **corticobulbar tracts** end at the motor nuclei of cranial nerves controlling eye movements, facial muscles, tongue muscles, and neck and superficial back muscles. *(see Figure 16.4a)*
12. **Corticospinal tracts** synapse on motor neurons in the anterior gray horns of the spinal cord and control movement in the neck, trunk, and some coordinated movements in the axial skeleton. They are visible along the ventral side of the medulla oblongata as a pair of thick elevations, the **pyramids**, where most of the axons decussate to enter the descending **lateral corticospinal tracts**. The remaining axons are uncrossed here and enter the **anterior corticospinal tracts**. These fibers will cross inside the anterior gray commissure before they synapse on motor neurons in the anterior gray horns. *(see Figure 16.4 and Table 16.2)*
13. The **medial and lateral pathways** consist of several centers that may issue motor commands as a result of processing performed at an unconscious, involuntary level. These pathways can modify or direct somatic motor patterns. Their outputs may descend in (1) the *vestibulospinal*, (2) the *tectospinal*, (3) the *reticulospinal* or (4) the *rubrospinal, tracts. (see Figures 16.4/16.5 and Tables 16.2/16.3)*
14. The vestibular nuclei receive sensory information from inner ear receptors through N VIII. These nuclei issue motor commands to maintain posture and balance. The fibers descend through the **vestibulospinal tracts**. *(see Figure 16.5 and Table 16.2)*
15. Commands carried by the **tectospinal tracts** change the position of the eyes, head, neck, and arms in response to bright lights, sudden movements, or loud noises. *(see Figure 16.5 and Table 16.2)*
16. Motor commands carried by the **reticulospinal tracts** vary according to the region stimulated. The reticular formation receives inputs from almost all ascending and descending pathways and from numerous interconnections with the cerebrum, cerebellum, and brain stem nuclei. *(see Figure 16.5 and Table 16.2)*
17. Basal nuclei, the cerebellum, and the reticular formation give inputs to the red nucleus. The **rubrospinal tracts** carry motor commands to lower motor neurons that provide supplemental control over distal muscles of the upper limbs. In humans these tracts are small and of little significance under normal circumstances. *(see Figure 16.5 and Table 16.2)*
18. The **basal nuclei** adjust the motor commands issued in other processing centers. They do not initiate specific movements, but provide a background pattern once movement is initiated. *(see Figure 16.5 and Table 16.2)*
19. The **cerebellum** regulates the activity along both conscious (corticospinal) and subconscious (medial and lateral) pathways. The integrative activities performed by neurons in the cerebellar cortex and cerebellar nuclei are essential for precise control of voluntary and involuntary movements. After voluntary movement is initiated, the cerebellum triggers inhibition of unnecessary motor units to complete the pattern of activity. *(see Figure 16.6)*

Levels of Somatic Motor Control 436

20. Ascending sensory information is relayed from one nucleus or center to another in a series of steps. Information processing occurs at each step along the way. Processing steps are important but time-consuming. Nuclei in the spinal cord, brain stem, and the cerebrum work together in various complex reflexes. *(see Figure 16.6)*

Higher-Order Functions 438

1. Higher-order functions have certain characteristics: (1) performed by the cerebral cortex; (2) involve complex fiber connections between areas of the cerebral cortex and between the cortex and other areas of the brain; (3) involve conscious and unconscious information processing; and (4) subject to modification and adjustment over time. *(see Figures 16.7 to 16.9)*

Integrative Regions of the Cerebral Cortex 438

2. Several cortical areas act as higher-order integrative centers for complex sensory stimuli and motor responses. These centers include: (1) the *general interpretive area*; (2) the *speech center*; and (3) the *prefrontal cortex*. *(see Figure 16.7)*
3. The **general interpretive area** *(gnostic area)* receives information from all the sensory association areas. It is present in only one hemisphere, usually the left. *(see Figures 16.7b/16.8)*
4. The **speech center** *(Broca's area)* regulates the patterns of breathing and vocalization needed for normal speech. *(see Figure 16.7b)*
5. The **prefrontal cortex** coordinates information from the secondary and special association areas of the entire cortex and performs abstract intellectual functions. *(see Figure 16.7b)*
6. Many functional areas of the cerebral cortex show a characteristic pattern of cellular organization, as described by Brodman. *(see Figure 16.7c)*

Hemispheric Specialization 439

7. The left hemisphere is usually the **categorical hemisphere**; it contains the general interpretive and speech centers and is responsible for language-based skills. The right hemisphere, or **representational hemisphere**, is concerned with spatial relationships and analysis. *(see Figures 16.7/16.8)*

Memory 440

8. **Memory** is the process of accessing stored bits of information gathered through experience. It involves considerable interaction between the cerebral cortex and other areas of the brain. *Short-term memories* last for seconds to hours; *long-term memories* can last for years. Conversion of short-term memory to a long-term memory is called *memory consolidation*. The amygdaloid body and the hippocampus are essential to memory consolidation.

Consciousness: The Reticular Activating System 441

9. Consciousness is determined by interactions between the brain stem and cerebral cortex. One of the most important brain stem components is a network in the reticular formation called the **reticular activating system** (**RAS**). *(see Figure 16.9 and Table 16.3)*

Aging and the Nervous System 442

1. Age-related changes in the nervous system include: (1) *reduction in brain size and weight*, (2) *reduction in number of neurons*, (3) *decrease in blood flow to the brain*, (4) *changes in synaptic organization of the brain*, and (5) *intracellular and extracellular changes in CNS neurons*.

LEVEL 1 REVIEWING FACTS AND TERMS

Match each numbered item with the most closely related lettered item. Use letters for answers in the spaces provided.

Column A	Column B
_____ 1. decussation	a. second-order
_____ 2. sensory	b. pain, temperature, crude touch, pressure
_____ 3. interneuron	c. voluntary-control skeletal muscle
_____ 4. posterior column	d. general interpretive
_____ 5. spinothalamic	e. afferent
_____ 6. spinocerebellar	f. proprioceptive information
_____ 7. corticospinal system	g. speech
_____ 8. tectospinal tracts	h. cross-over
_____ 9. medial pathway	i. position change—noise related
_____ 10. gnostic area	j. subconscious motor command
_____ 11. Broca's area	k. fasciculi gracilis and cuneatus

12. The spinal tracts that regulate voluntary motor control of skeletal muscles on the opposite side of the body are the
    (a) corticospinal tracts (b) rubrospinal tracts
    (c) reticulospinal tracts (d) vestibulospinal tracts
13. The area of the motor cortex that is devoted to a particular region of the body is relative to
    (a) the size of the body area
    (b) the number of motor units in the area of the body
    (c) the number of sensory receptors in the area of the body
    (d) the size of the nerves that serve the area of the body
14. Which of the following is a spinal tract of the medial and lateral pathways?
    (a) vestibulospinal tracts (b) tectospinal tracts
    (c) reticulospinal tracts (d) all of the above
15. Axons of the corticospinal tract synapse at
    (a) motor nuclei of cranial nerves
    (b) motor neurons in the anterior horns of the spinal cord
    (c) motor neurons in the posterior horns of the spinal cord
    (d) motor neurons in ganglia near the spinal cord
16. We can distinguish between sensations that originate in different areas of the body because
    (a) sensory neurons carry only one type of information
    (b) sensory neurons from each body region synapse in specific brain regions
    (c) incoming sensory information is first assessed by the thalamus
    (d) different sensory-receptor types produce different action-potential types
17. The cerebellum adjusts voluntary and involuntary motor activity in response to each of the following except
    (a) proprioceptive data
    (b) visual information
    (c) information from the cerebral cortex
    (d) information from the "vital centers"
18. The conscious state depends upon the proper functioning of the
    (a) prefrontal lobes (b) general interpretive area
    (c) reticular activating system (d) limbic system
19. Which of the following is essential for memory consolidation?
    (a) Brodman's area 4 (b) basal nuclei
    (c) hippocampus (d) prefrontal lobe
20. Each of the following describes some aspect of Alzheimer's disease except
    (a) it is the most common cause of senile dementia
    (b) it occurs more frequently in patients with a CVA
    (c) it is associated with the formation of plaques and neurofibrillary tangles in regions of the brain that are involved with memory
    (d) it is characterized by a progressive loss of memory

LEVEL 2 REVIEWING CONCEPTS

1. The general interpretative area
   (a) is the speech center of the brain
   (b) is responsible for predicting future consequences
   (c) allows us to interpret what is read or heard
   (d) all of the above
2. Abstract intellectual functions are performed in the
   (a) prefrontal cortex
   (b) general interpretive area
   (c) premotor cortex
   (d) speech center
3. What symptoms would you associate with damage to the nucleus gracilis on the right side of the medulla oblongata?
   (a) inability to perceive fine touch from the left lower limb
   (b) inability to perceive fine touch from the right lower limb
   (c) inability to direct fine motor activities involving the left shoulder
   (d) inability to direct fine motor activities involving the right shoulder
4. A cerebrovascular accident occurs when
   (a) the reticular activating system fails to function
   (b) the prefrontal lobe is damaged
   (c) the blood supply to a portion of the brain is cut off
   (d) a descending tract in the spinal cord is severed
5. Describe the function of first-order neurons in the CNS.
6. Why do the proportions of the sensory homunculus differ from those of the body?
7. What is the primary role of the basal nuclei in the function of the medial and lateral pathways?
8. How does the cerebellum influence overall motor activities in the body?
9. Where is the speech center and what are its functions?
10. Compare the actions directed by motor commands in the vestibulospinal tracts with those in the tectospinal tracts.

LEVEL 3 CRITICAL THINKING AND CLINICAL APPLICATIONS

1. Cindy has a biking accident and injures her back. She is examined by a doctor who notices that Cindy cannot feel pain sensations (a pinprick) from her left hip and lower limb, but she has normal sensation elsewhere and has no problems with the motor control of her limbs. The physician tells Cindy that he thinks a portion of the spinal cord may be compressed and that this is responsible for her symptoms. Where might the problem be located?

2. Evan suffered a fractured skull when he fell while roller-blading. Diagnostic tests indicate severe damage to the motor cortex. His mother is anxious to know if he will ever be able to move or walk again. What would you tell her?

✓ ANSWERS TO CONCEPT CHECK QUESTIONS

p. 438 **1.** The fasciculus gracilis in the posterior column of the spinal cord is responsible for carrying information about touch and pressure from the lower part of the body to the brain. **2.** The anatomical basis for motor control occurring on the opposite side is that crossing over (decussation) occurs, and the corticospinal motor fibers innervate lower motor neurons on the opposite side of the body. **3.** The superior portion of the motor cortex exercises control over the upper limb and superior portion of the lower limb. An injury to this area would affect the ability to control the muscles in those regions of the body. **4.** (a) tectospinal tracts, (b) vestibulospinal tracts.

p. 441 **1.** An inability to comprehend the written or spoken word indicates a problem with the general interpretive area of the brain, which in most individuals is located in the left temporal lobe of the cerebrum. **2.** Difficulty speaking indicates that the speech center, also called Broca's area, is involved. **3.** Memory consolidation is the conversion from short-term memory to long-term memory. **4.** The prefrontal cortex makes predictions about future events.

p. 442 **1.** The reticular activating system (RAS) is responsible for rousing the cerebrum to a state of consciousness. If a sleeping individual's RAS were stimulated, she would certainly wake up. **2.** The brain is affected by age. The common age-related anatomical changes in the CNS are (1) reduction in brain size and weight, (2) the loss of cortical neurons, (3) decrease in blood flow to the brain due to the accumulation of fatty deposits, (4) decrease in the number and organization of dendritic branching and synaptic connections, and (5) abnormal accumulation of intercellular deposits, including lipofuscin, plaques, and neurofibrillary tangles.

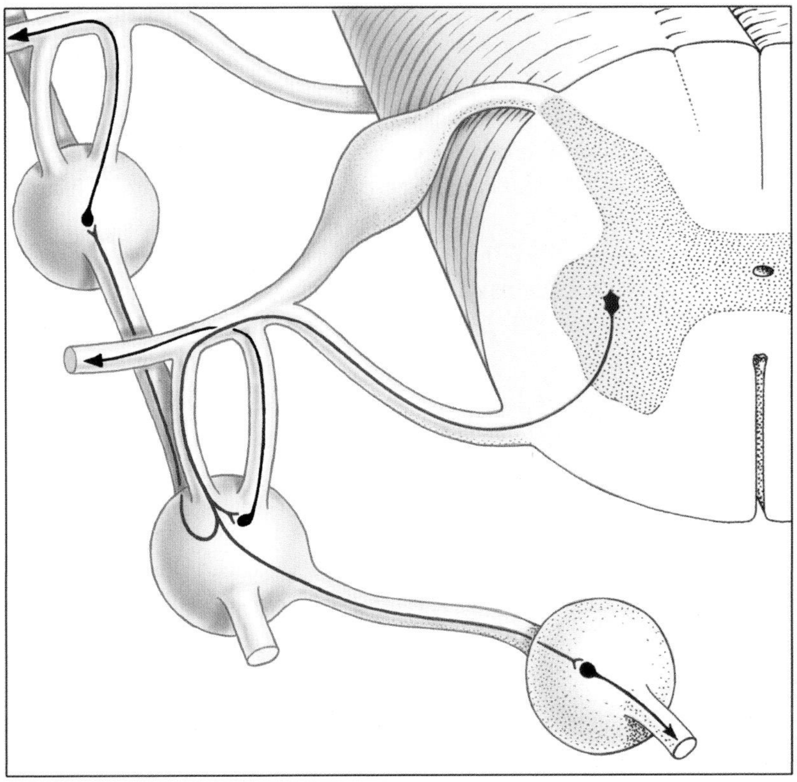

17

THE NERVOUS SYSTEM

Autonomic Division

Our conscious thoughts, plans, and actions represent only a tiny fraction of the activities of the nervous system. If all consciousness were eliminated, vital physiological processes would continue virtually unchanged—a night's sleep is not a life-threatening event. Longer, deeper states of unconsciousness are not necessarily more dangerous, as long as nourishment is provided. People who have suffered severe brain injuries have survived in a coma for decades. Survival under these conditions is possible because routine adjustments in physiological systems are made by the autonomic nervous system (ANS). It is the ANS that regulates body temperature and coordinates cardiovascular, respiratory, digestive, excretory, and reproductive functions. In doing so, it adjusts internal water, electrolyte, nutrient, and dissolved gas concentrations in body fluids outside of our conscious awareness.

This chapter examines the anatomical structure and subdivisions of the autonomic nervous system. Each subdivision has a characteristic anatomical and functional organization. Our examination of the ANS will begin with a description of the sympathetic and parasympathetic divisions. Then, we will briefly examine the way these divisions maintain and adjust various organ systems to meet the body's ever changing physiological needs.

An Overview of the Autonomic Nervous System (ANS)

It is useful to compare the organization of the ANS, which innervates visceral effectors, with the somatic nervous system (SNS), which was discussed in Chapter 16. We will focus on (1) the pathways involved in visceral motor output, and (2) the subdivisions of the ANS, based on structural and functional patterns of peripheral innervation.

Pathways for Visceral Motor Output

The axons of lower motor neurons of the SNS extend from the CNS to contact and exert direct control over skeletal muscles. p. 432 In the ANS, the axon of a visceral motor neuron within the CNS innervates a second neuron located in a peripheral ganglion. It is this second neuron that controls the peripheral effector. Visceral motor neurons in the CNS, known as **preganglionic neurons**, send their axons, called *preganglionic fibers*, to synapse on **ganglionic neurons**, whose cell bodies are located outside the CNS, in autonomic ganglia. Axons that leave the autonomic ganglia are relatively small and unmyelinated. These axons are called *postganglionic fibers* because they carry impulses away from the ganglion. Postganglionic fibers innervate peripheral organs, such as cardiac muscle, smooth muscle, glands, and adipose tissue.

Subdivisions of the ANS [FIGURE 17.1]

The ANS contains two subdivisions, the *sympathetic division* and the *parasympathetic division* (Figure 17.1●). Most often, the two divisions have opposing effects; if the sympathetic division causes excitation, the parasympathetic division causes inhibition. However, this is not always the case because (1) the two divisions may work independently, with some structures innervated by only one division, and (2) the two divisions may work together, each controlling one stage of a complex process. In general, the parasympahthetic division predominates under resting conditions, and the sympathetic division "kicks in" during times of exertion, stress, or emergency.

SYMPATHETIC (THORACOLUMBAR) DIVISION [FIGURE 17.1]

Preganglionic fibers from both the thoracic and upper lumbar spinal segments synapse in ganglia near the spinal cord. These axons and ganglia are part of the **sympathetic division**, or **thoracolumbar** (thor-a-kō-LUM-bar) **division** of the ANS (Figure 17.1●). This division is often called the "fight or flight" system because an increase in sympathetic activity generally stimulates tissue metabolism, increases alertness, and prepares the body to deal with emergencies.

PARASYMPATHETIC (CRANIOSACRAL) DIVISION [FIGURE 17.1]

Preganglionic fibers originating in either the brain stem or the sacral spinal cord are part of the **parasympathetic division**, or **craniosacral** (krā-nē-ō-SĀ-kral) **division** of the ANS (Figure 17.1●). The preganglionic fibers synapse on neurons of **terminal ganglia**, located close to the target organs, or **intramural ganglia** (*murus*, wall), within the tissues of the target organs. This division is often called the "rest and repose" system because it conserves energy and promotes sedentary activities, such as digestion.

INNERVATION PATTERNS

The sympathetic and parasympathetic divisions of the ANS affect their target organs through the controlled release of neurotransmitters by postganglionic fibers. Target organ activity may be either stimulated or inhibited, depending on the response of the membrane receptor to the presence of the neurotransmitter. Three general statements describe ANS neurotransmitters and their effects:

1. All preganglionic autonomic fibers release acetylcholine (ACh) at their synaptic terminals. The effects are always stimulatory.
2. Postganglionic parasympathetic fibers also release ACh, but the effects may be either stimulatory or inhibitory, depending on the nature of the receptor.
3. Most postganglionic sympathetic terminals release the neurotransmitter **norepinephrine (NE)**. The effects are usually stimulatory.

✓ CONCEPT CHECK

- Describe the difference(s) between preganglionic and ganglionic neurons.
- List the two subdivisions of the autonomic nervous system. What common name or term is applied to each?
- What neurotransmitter is released by most postganglionic sympathetic terminals?
- What organs are innervated by postganglionic fibers of the autonomic nervous system?

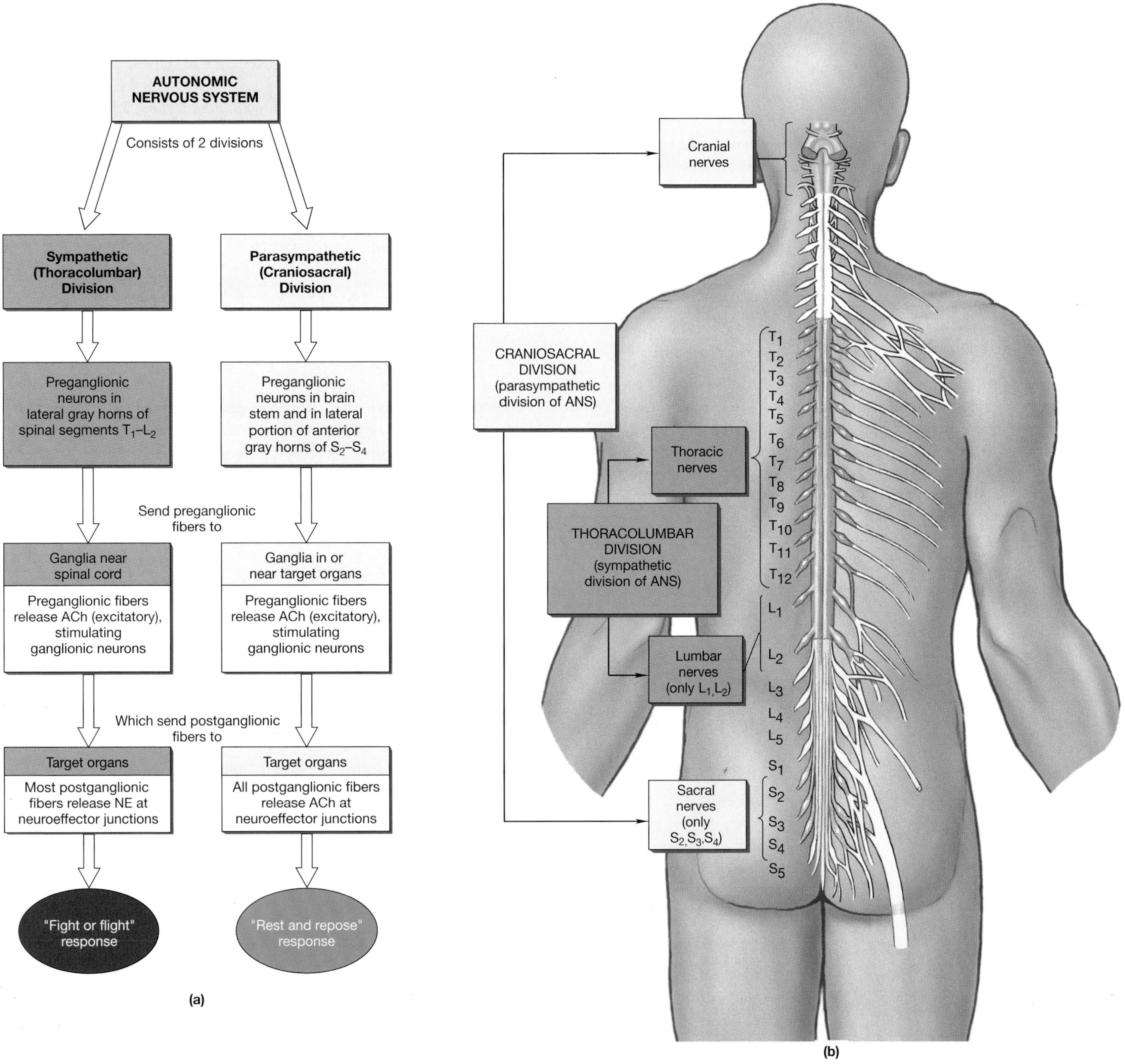

FIGURE 17.1 **COMPONENTS AND ANATOMIC SUBDIVISIONS OF THE ANS**

(**a**) Functional components. (**b**) Anatomical subdivisions. At the thoracic and lumbar levels, the visceral efferent fibers that emerge form the sympathetic division, *detailed in Figure 17.4*. At the cranial and sacral levels, the visceral efferent fibers from the CNS form the parasympathetic division, *detailed in Figure 17.8*.

The Sympathetic Division [FIGURE 17.2]

The sympathetic division (Figure 17.2●) consists of:

1. *Preganglionic neurons located between segments T_1 and L_2 of the spinal cord.* The cell bodies of these neurons occupy the lateral gray horns between T_1 and L_2, and their axons enter the ventral roots of those segments.
2. *Ganglionic neurons in ganglia near the vertebral column.* There are two types of ganglia in the sympathetic division:
   - *Sympathetic chain ganglia*, also called *paravertebral,* or *lateral, ganglia*, lie lateral to the vertebral column on each side. Neurons in these ganglia control effectors in the body wall, head and neck, limbs, and inside the thoracic cavity.
   - *Collateral ganglia*, also known as *prevertebral ganglia*, lie anterior to the vertebral column. Neurons in these ganglia innervate effectors in the abdominopelvic cavity.
3. *Specialized neurons in the interior of the adrenal gland.* The core of each adrenal gland, an area known as the *adrenal medulla*, is a modified sympathetic ganglion. The ganglionic neurons here have very short axons, and, when stimulated, they release neurotransmitters into the bloodstream for distribution throughout the body as hormones.

The Sympathetic Chain Ganglia [FIGURES 17.1a/17.2]

The ventral roots of spinal segments T_1 to L_2 contain sympathetic preganglionic fibers. The basic pattern of sympathetic innervation in these regions was described in Figure 17.1a●. Each ventral root joins the corresponding dorsal root, which carries afferent sensory fibers, to form a spinal nerve that passes through an intervertebral foramen. p. 356 As it clears the foramen, a *white ramus*, or *white ramus communicans*, branches from the spinal nerve (Figure 17.3a●). The white ramus carries myelinated preganglionic fibers into a nearby sympathetic chain ganglion. Fibers entering a sympathetic chain ganglion may have one of three destinations: (1) They may synapse within the sympathetic chain ganglion at the level of entry (Figure 17.2a●); (2) they may ascend or descend within the sympathetic chain and synapse with a ganglion at a different level; or (3) they may pass through the sympathetic chain without synapsing and proceed to one of the collateral ganglia (Figure 17.3b●) or the adrenal medullae (Figure 17.3c●). Extensive divergence occurs in the sympathetic division, with one preganglionic fiber synapsing on as many as 32 ganglionic neurons. Preganglionic fibers projecting between the sympathetic chain ganglia interconnect them, making the chain resemble a string of beads. Each ganglion in the sympathetic chain innervates a particular body segment or group of segments.

If a preganglionic fiber carries motor commands that target structures in the body wall or the thoracic cavity, it will synapse in one or more of the sympathetic chain ganglia. Unmyelinated postganglionic fibers then leave the sympathetic chain and proceed to their peripheral

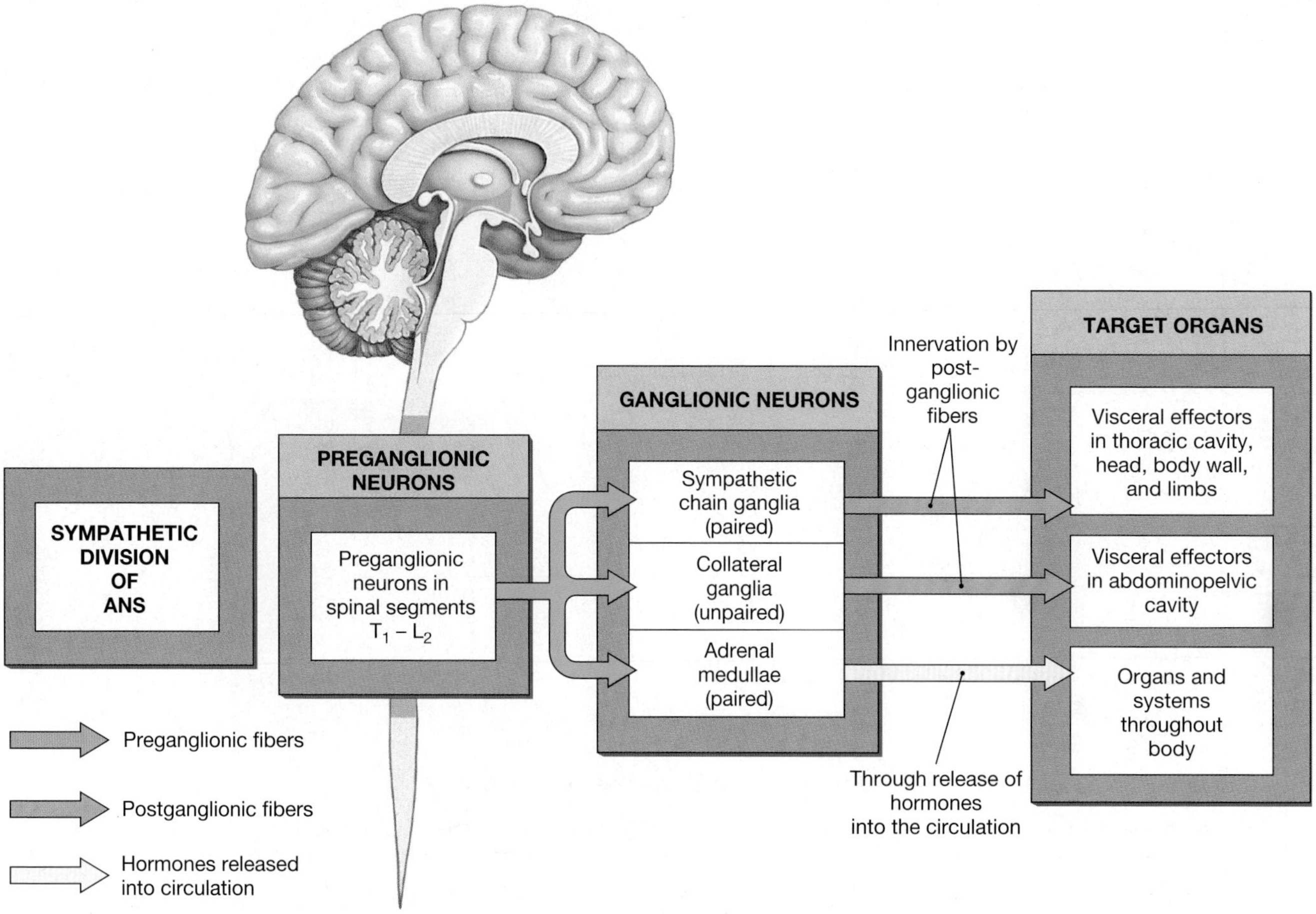

FIGURE 17.2 **ORGANIZATION OF THE SYMPATHETIC DIVISION OF THE ANS**

This diagram highlights the relationships between preganglionic and ganglionic neurons and between ganglionic neurons and target organs.

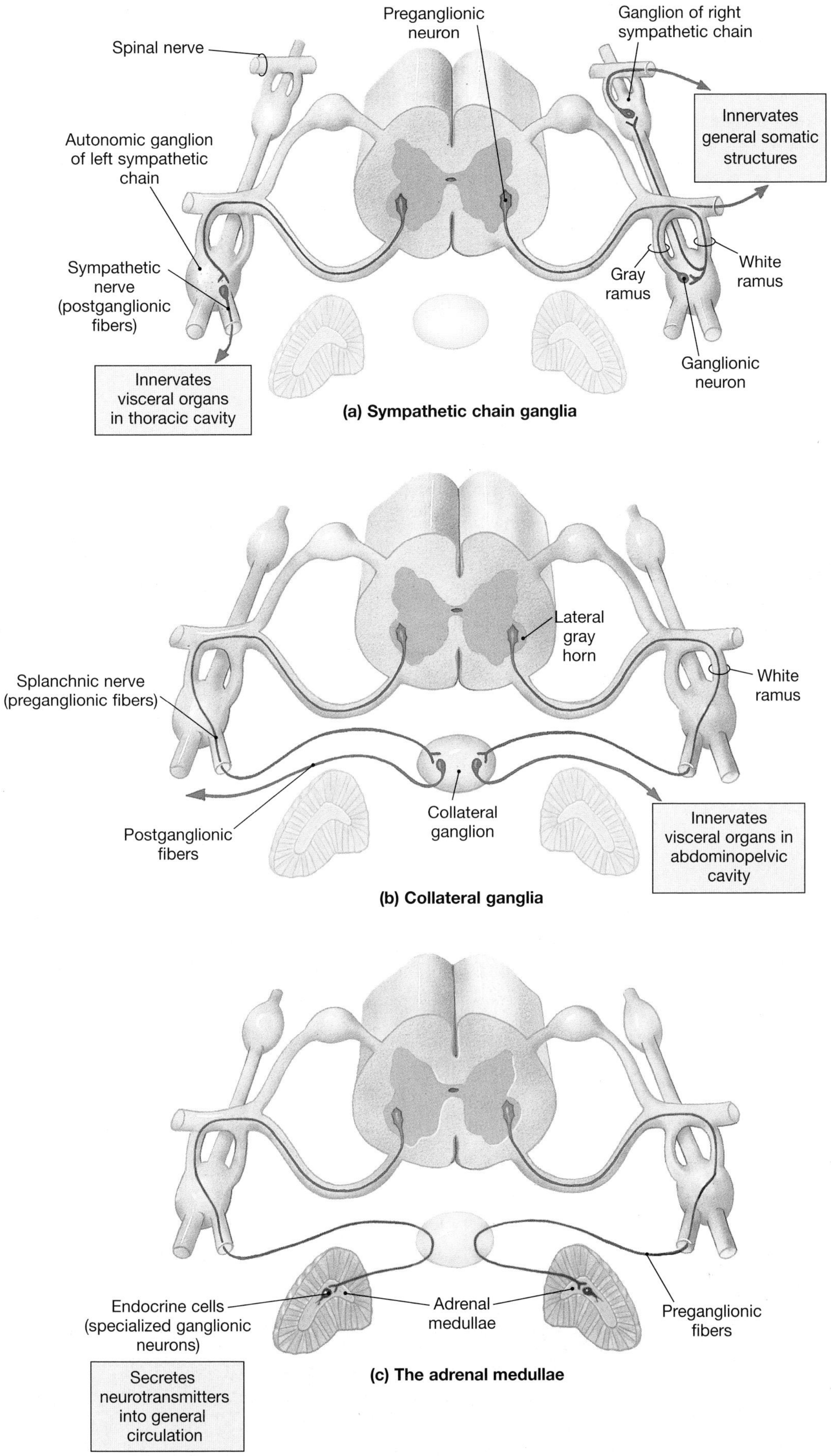

Major effects produced by sympathetic postganglionic fibers in spinal nerves:
- Constriction of cutaneous blood vessels, reduction in circulation to the skin and to most other organs in the body wall
- Acceleration of blood flow to skeletal muscles and brain
- Stimulation of energy production and use by skeletal muscle tissue
- Release of stored lipids from subcutaneous adipose tissue
- Stimulation of secretion by sweat glands
- Stimulation of arrector pili
- Dilation of the pupils and focusing for distant objects

Major effects produced by postganglionic fibers entering the thoracic cavity in sympathetic nerves:
- Acceleration of heart rate and increasing the strength of cardiac contractions
- Dilation of respiratory passageways

Major effects produced by preganglionic fibers innervating the collateral ganglia:
- Constriction of small arteries and reduction in the flow of blood to visceral organs
- Decrease in the activity of digestive glands and organs
- Stimulation of the release of glucose from glycogen reserves in the liver
- Stimulation of the release of lipids from adipose tissue
- Relaxation of the smooth muscle in the wall of the urinary bladder
- Reduction of the rate of urine formation at the kidneys
- Control of some aspects of sexual function, such as ejaculation in males

Major effect produced by preganglionic fibers innervating the adrenal medullae:
- Release of epinephrine and norepinephrine into the general circulation

FIGURE 17.3 **SYMPATHETIC PATHWAYS AND THEIR GENERAL FUNCTIONS**

Preganglionic fibers leave the spinal cord in the ventral roots of spinal nerves. They synapse on ganglionic neurons in (**a**) sympathetic chain ganglia, (**b**) in collateral ganglia, or (**c**) in the adrenal medullae. These sections are seen in inferior view, the standard format for radiographic scans and sectional views of the nervous system.

targets within spinal nerves and sympathetic nerves. Postganglionic fibers that innervate general somatic structures, such as the sweat glands of the skin or the smooth muscles in superficial blood vessels, enter the *gray ramus* (*gray ramus communicans*) and return to the spinal nerve for subsequent distribution. However, spinal nerves do not innervate structures in the ventral body cavities. Postganglionic fibers innervating visceral organs in the thoracic cavity, such as the heart and lungs, proceed directly to their peripheral targets as sympathetic nerves. These nerves are usually named after their primary targets, as in the case of the *cardiac nerves* and *esophageal nerves*.

FUNCTIONS OF THE SYMPATHETIC CHAIN [FIGURE 17.3a]

The primary results of increased activity along the postganglionic fibers leaving the sympathetic chain ganglia within spinal nerves and sympathetic nerves are summarized in Figure 17.3a●. In general, the target cell responses help prepare the individual for a crisis that will require sudden, intensive physical activity.

ANATOMY OF THE SYMPATHETIC CHAIN [FIGURE 17.4]

Each sympathetic chain has 3 cervical, 11–12 thoracic, 2–5 lumbar, 4–5 sacral, and 1 coccygeal sympathetic ganglia. Numbers of ganglia may vary because adjacent ganglia may fuse. For example, the coccygeal ganglia from both sides usually fuse to form a single median ganglion, the *ganglion impar*. Preganglionic sympathetic neurons are limited to segments T_1–L_2 of the spinal cord, and the spinal nerves of these segments have both white rami (preganglionic fibers) and gray rami (postganglionic fibers). The neurons in the cervical, inferior lumbar, and sacral sympathetic chain ganglia are innervated by preganglionic fibers extending along the axis of the chain. In turn, these chain ganglia provide postganglionic fibers, through gray rami, to the cervical, lumbar, and sacral spinal nerves. *Every spinal nerve has a gray ramus that carries sympathetic postganglionic fibers.* About 8% of the axons in each spinal nerve are sympathetic postganglionic fibers. The dorsal and ventral rami of the spinal nerves provide extensive sympathetic innervation to structures in the body wall and limbs. In the head, postganglionic fibers leaving the cervical chain ganglia supply the regions and structures innervated by cranial nerves N III, N VII, N IX, and N X (Figure 17.4●).

In summary: (1) Only the thoracic and superior lumbar ganglia receive preganglionic fibers from white rami; (2) The cervical, inferior lumbar, and sacral chain ganglia receive preganglionic innervation from the thoracic and superior lumbar segments through preganglionic fibers that ascend or descend along the sympathetic chain; and (3) Every spinal nerve receives a gray ramus from a ganglion of the sympathetic chain.

This anatomical arrangement has interesting functional consequences. If the ventral roots of thoracic spinal nerves are damaged, there will be no sympathetic motor function on the affected side of the head, neck, and trunk. Yet damage to the ventral roots of cervical spinal nerves will produce voluntary muscle paralysis on the affected side, *but leave sympathetic function intact* because the preganglionic fibers innervating the cervical ganglia originate in the white rami of thoracic segments, which are undamaged.

Collateral Ganglia [FIGURES 17.3b/17.4]

Preganglionic fibers that regulate the activities of the abdominopelvic viscera originate at preganglionic neurons in the inferior thoracic and superior lumbar segments. These fibers pass through the sympathetic chain without synapsing, and converge to form the **greater**, **lesser**, **lumbar,** and **sacral splanchnic** (SPLANK-nik) **nerves** in the dorsal wall of the abdominal cavity. Splanchnic nerves from both sides of the body converge on the collateral ganglia (Figures 17.3b and 17.4●). Collateral ganglia, which are variable in appearance, are located anterior and lateral to the descending aorta. These ganglia are most often single, rather than paired structures.

FUNCTIONS OF THE COLLATERAL GANGLIA [FIGURE 17.3b]

Postganglionic fibers that originate within the collateral ganglia extend throughout the abdominopelvic cavity, innervating visceral tissues and organs. A summary of the effects of increased sympathetic activity along these postganglionic fibers is included in Figure 17.3b●. The general pattern is (1) a reduction of blood flow, energy use, and activity by visceral organs that are not important to short-term survival (such as the digestive tract), and (2) the release of stored energy reserves.

ANATOMY OF THE COLLATERAL GANGLIA [FIGURES 17.4/17.9]

The splanchnic nerves (greater, lesser, lumbar and sacral) innervate three collateral ganglia. Preganglionic fibers from the seven inferior thoracic segments end at the **celiac** (SĒ-lē-ak) **ganglion** and the **superior mesenteric ganglion.** These ganglia are embedded in an extensive, web-like network of nerve fibers termed an *autonomic plexus*. Preganglionic fibers from the lumbar segments form splanchnic nerves that end at the **inferior mesenteric ganglion**. These ganglia are diagrammed in Figure 17.4● and detailed in Figure 17.9●. The sacral splanchnic nerves end in the *hypogastric plexus,* an autonomic network supplying pelvic organs and the external genitalia.

THE CELIAC GANGLION The celiac ganglion is variable in appearance. It most often consists of a pair of interconnected masses of gray matter situated at the base of the *celiac trunk*. The celiac ganglion may also form a single mass, or many small, interwoven masses. Postganglionic fibers from the celiac ganglion innervate the stomach, duodenum, liver, gallbladder, pancreas, and spleen.

THE SUPERIOR MESENTERIC GANGLION The superior mesenteric ganglion is located near the base of the *superior mesenteric artery*. Postganglionic fibers leaving the superior mesenteric ganglion innervate the small intestine and the initial segments of the large intestine.

THE INFERIOR MESENTERIC GANGLION The inferior mesenteric ganglion is located near the base of the *inferior mesenteric artery*. Postganglionic fibers from this ganglion provide sympathetic innervation to the terminal portions of the large intestine, the kidney and bladder, and the sex organs.

The Adrenal Medullae [FIGURES 17.3c/17.4/17.5]

Some preganglionic fibers originating between T_5 and T_8 pass through the sympathetic chain and the celiac ganglion without synapsing and proceed to the core of the **adrenal medulla** (Figures 17.3c, 17.4, and 17.5●). There, these preganglionic fibers synapse on modified neurons that perform an endocrine function. These neurons have very short axons. When stimulated, they release the neurotransmitters *epinephrine* (E) and *norepinephrine* (NE) into an extensive network of capillaries

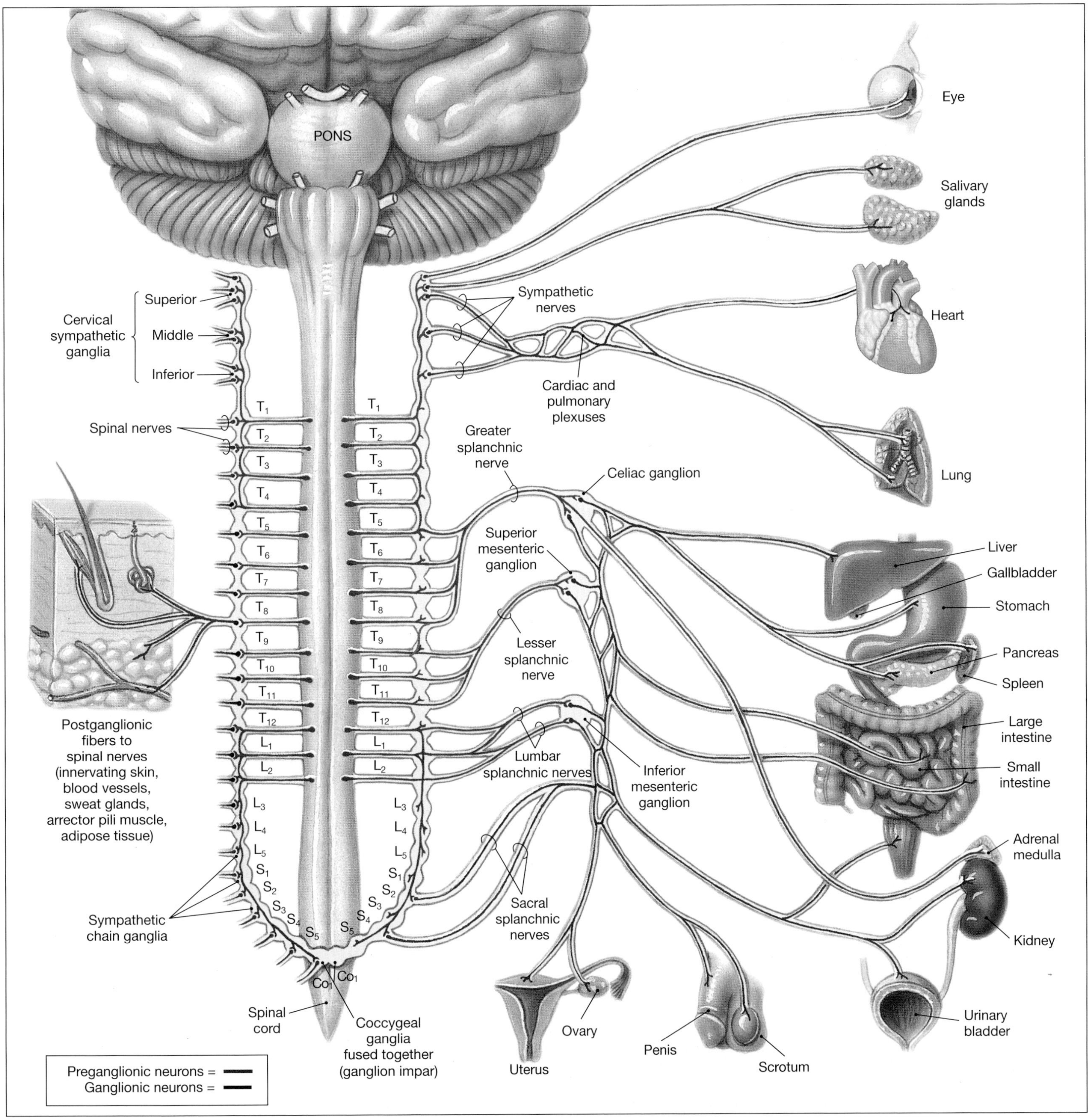

FIGURE 17.4 **Anatomical Distribution of Sympathetic Postganglionic Fibers**

The left side of this figure shows the distribution of sympathetic postganglionic fibers through the gray rami and spinal nerves. The right side shows the distribution of preganglionic and postganglionic fibers innervating visceral organs. However, *both* innervation patterns are found on *each* side of the body.

(Figure 17.5●). The neurotransmitters then function as hormones, exerting their effects in other regions of the body. Epinephrine, also called *adrenaline*, accounts for 75–80% of the secretory output; the rest is norepinephrine (*noradrenaline*).

The circulating blood then distributes these hormones throughout the body. This causes changes in the metabolic activities of many different cells. In general, the effects resemble those produced by the stimulation of sympathetic postganglionic fibers. But they differ in two respects: (1) Cells not innervated by sympathetic postganglionic fibers are affected by circulating levels of epinephrine and norepinephrine if they possess receptors for these molecules, and (2) The effects last much longer than those produced by direct sympathetic innervation because the released hormones continue to diffuse out of the circulating blood for an extended period.

Effects of Sympathetic Stimulation

The sympathetic division can change tissue and organ activities both by releasing norepinephrine at peripheral synapses and by distributing epinephrine and norepinephrine throughout the body in the bloodstream. The motor fibers that target specific effectors, such as smooth muscle fibers in blood vessels of the skin, can be activated in reflexes that do not involve other peripheral effectors. In a crisis, however, the entire division responds. This event, called **sympathetic activation**, affects peripheral tissues and alters CNS activity. Sympathetic activation is controlled by sympathetic centers in the hypothalamus.

When sympathetic activation occurs, an individual experiences:

1. Increased alertness, through stimulation of the reticular activating system, causing the individual to feel "on edge."

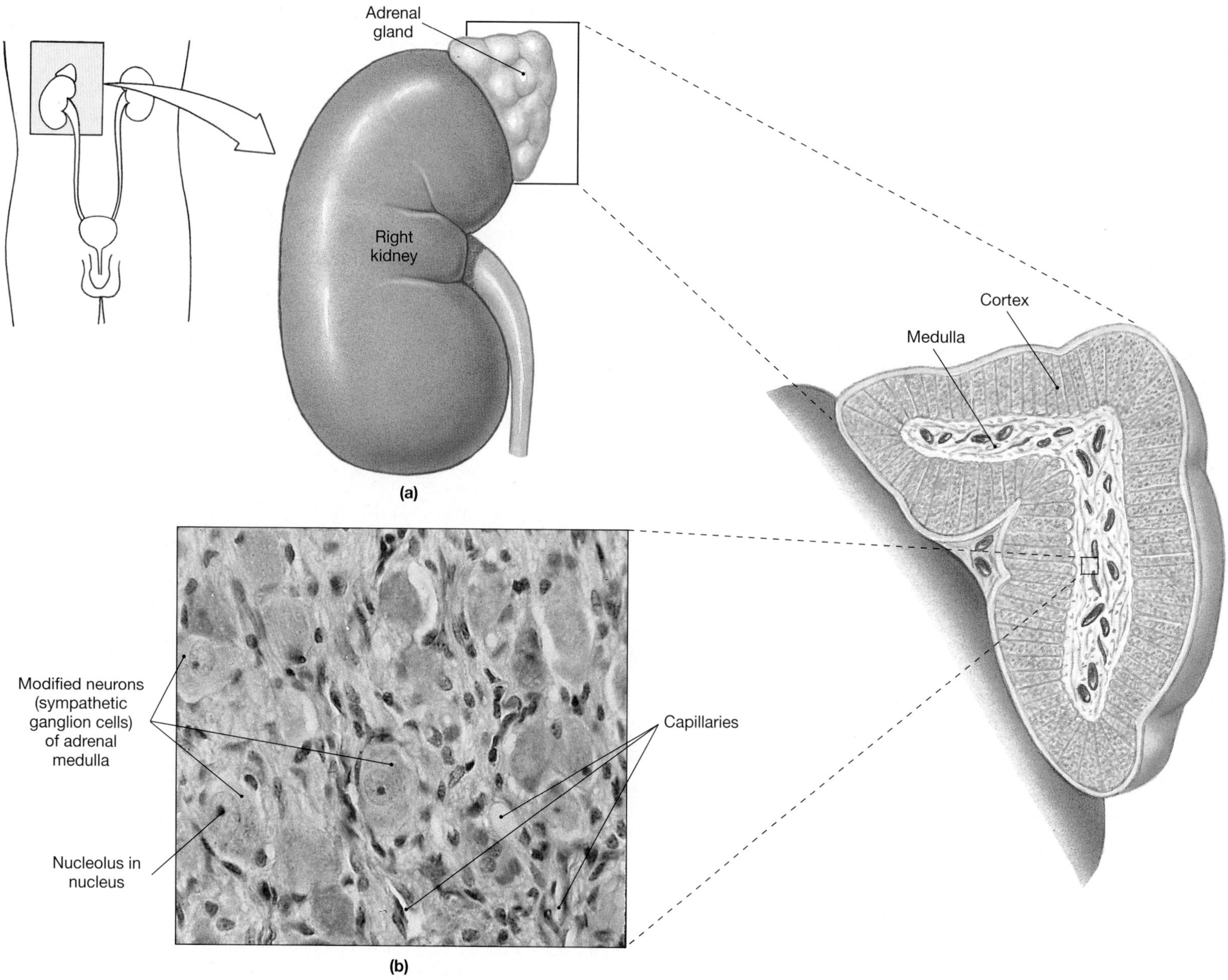

FIGURE 17.5 **ADRENAL MEDULLA**
(**a**) Relationship of an adrenal gland to a kidney. (**b**) The adrenal medulla, a modified sympathetic ganglion. (LM × 426)

2. A feeling of energy and euphoria, often associated with a disregard for danger and a temporary insensitivity to painful stimuli.
3. Increased activity in the cardiovascular and respiratory centers of the pons and medulla oblongata, leading to elevations in blood pressure, heart rate, breathing rate, and depth of respiration.
4. A general elevation in muscle tone through stimulation of the extrapyramidal system, so that the person *looks* tense and may even begin to shiver.
5. The mobilization of energy reserves, through the accelerated breakdown of glycogen in muscle and liver cells and the release of lipids by adipose tissues.

These changes, coupled with the peripheral changes already noted, complete the preparations necessary for the individual to cope with stressful and potentially dangerous situations. We will now consider the cellular basis for the general effects of sympathetic activation on peripheral organs.

CLINICAL BRIEF

Horner's Syndrome

In **Horner's syndrome**, the sympathetic postganglionic innervation to one side of the face becomes interrupted, possibly as the result of an injury, a tumor, or some progressive condition such as multiple sclerosis. p. 416 The affected side of the face becomes flushed because in the absence of sympathetic tone, the blood vessels dilate. Sweating stops in the affected region, and the pupil on that side becomes markedly constricted. Other symptoms include a drooping eyelid and an apparent retreat of the eye into the orbit.

The elimination of sympathetic innervation can have additional consequences that appear more gradually. Under normal conditions, sympathetic innervation provides the effectors with a background level of stimulation. After the disappearance of sympathetic stimulation, the effectors may become extremely sensitive to norepinephrine and epinephrine. This hypersensitivity can produce changes in facial blood flow and other functions when the adrenal medulla is stimulated.

Sympathetic Activation and Neurotransmitter Release

[FIGURE 17.6]

When they are active, sympathetic preganglionic fibers release ACh at their synapses with ganglionic neurons. These are *cholinergic* synapses. p. 346 The ACh released always stimulates the ganglionic neurons. This stimulation of ganglionic neurons usually leads to the release of norepinephrine at neuroeffector junctions. These synapses are called *adrenergic*. The sympathetic division also contains a small but significant number of ganglionic neurons that release ACh, rather than NE, at their neuroeffector junctions. For example, ACh is released at sympathetic neuroeffector junctions in the body wall, the skin, and in skeletal muscles.

Figure 17.6● details a representative sympathetic neuroeffector junction. Rather than ending at a single synaptic knob, the telodendria form an extensive branching network. Each branch resembles a string of pop-beads, and each pop-bead, or **varicosity**, is packed with mitochondria and neurotransmitter vesicles. These varicosities pass along or near the surfaces of many effector cells. A single axon may supply 20,000 varicosities that can affect dozens of surrounding cells. Receptor proteins are scattered across most cell membranes, and there are no specialized postsynaptic membranes.

The effects of neurotransmitter released by varicosities persist for at most a few seconds before the neurotransmitter is reabsorbed, broken down by enzymes, or removed by diffusion into the bloodstream. In contrast, the effects of the E and NE secreted by the adrenal medullae are considerably longer in duration because (1) the bloodstream does not contain the enzymes that break down epinephrine or norepinephrine, and (2) most tissues contain relatively low concentrations of these enzymes. As a result, the adrenal stimulation causes widespread effects that continue for a relatively long time. For example, tissue concentrations of epinephrine may remain elevated for as long as 30 seconds, and the effects may persist for several minutes.

Membrane Receptors and Sympathetic Function

The effects of sympathetic stimulation result primarily from interactions with membrane receptors sensitive to epinephrine and norepinephrine. (A few sympathetic neuroeffector junctions release ACh; these will be detailed below.) There are two classes of sympathetic receptors sensitive to E and NE: *alpha receptors* and *beta receptors*. Each of these

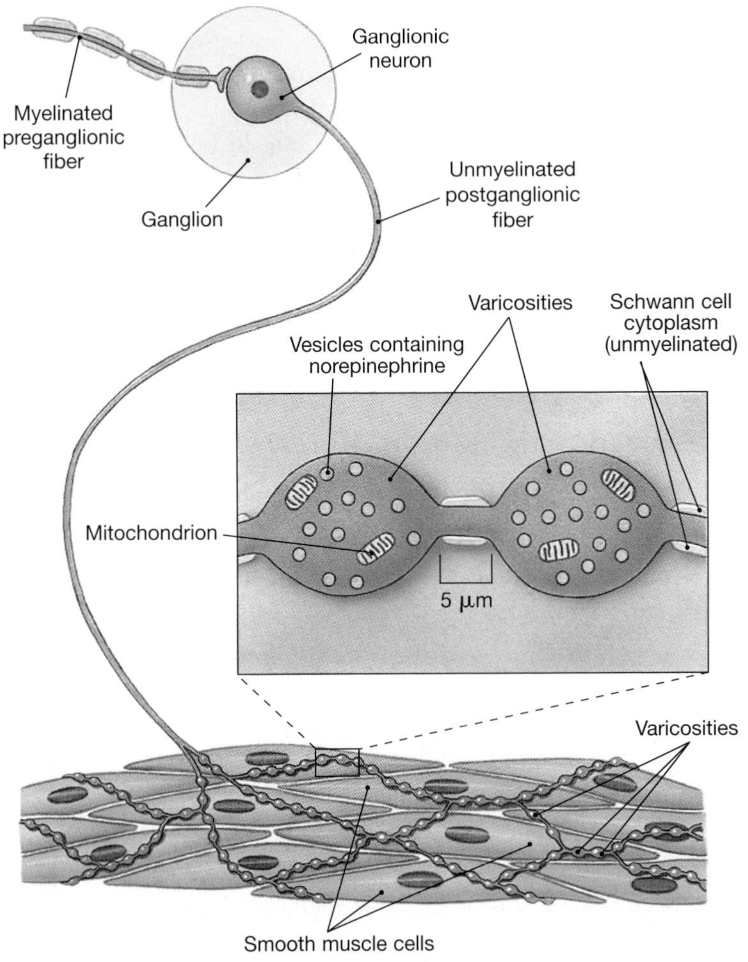

FIGURE 17.6 **Sympathetic Postganglionic Nerve Endings**
A diagrammatic view of sympathetic neuroeffector junctions.

classes of receptors has 2–3 subtypes. The diversity of receptors and their presence alone or in combination accounts for the variability of target organ responses to sympathetic stimulation. In general, epinephrine stimulates both classes of receptors, while norepinephrine primarily stimulates alpha receptors.

ALPHA AND BETA RECEPTORS

Stimulation of **alpha receptors** on the surfaces of smooth muscle cells causes constriction of peripheral blood vessels and the closure of sphincters along the digestive tract. **Beta receptors** are found in many organs, including skeletal muscles, the smooth muscle surrounding respiratory airways, the heart, and the liver. Stimulation of these beta receptors triggers changes in the metabolic activity of the target cells. The effects depend on which enzymes are involved. The most common response is an increase in metabolic activity. For example, skeletal muscles use energy at a faster rate, and the heart rate accelerates. However, in some target organs these neurotransmitters have an inhibitory effect. For example, the dilation of blood vessels supplying skeletal muscles and the enlargement of the respiratory passageways—two important responses to sympathetic activation—result from the relaxation of smooth muscle cells. This relaxation follows the stimulation of beta receptors on their cell membranes.

SYMPATHETIC STIMULATION AND ACh

Although the vast majority of sympathetic postganglionic fibers release NE and are adrenergic, a few postganglionic fibers are cholinergic. Activation of these sympathetic fibers stimulates sweat gland secretion and blood vessel dilation in skeletal muscle. Both the body wall and skeletal muscles are innervated by the sympathetic division of the ANS only. Thus, the distribution of sympathetic, postganglionic cholinergic fibers provides a method for regulating sweat gland secretion and selectively controlling blood flow to skeletal muscles, while norepinephrine released by adrenergic terminals reduces the blood flow to other tissues in the body wall.

■ A Summary of the Sympathetic Division [TABLE 17.1]

1. The sympathetic division of the ANS includes two sympathetic chains resembling a string of beads, one on each side of the vertebral column; three collateral ganglia anterior to the spinal column; and two adrenal medullae.
2. Preganglionic fibers are short because the ganglia are close to the spinal cord. The postganglionic fibers are relatively long and extend a considerable distance before reaching their target organs. (In the case of the adrenal medullae, very short axons from modified ganglionic neurons end at capillaries that carry their secretions to the bloodstream.)
3. The sympathetic division shows extensive divergence; a single preganglionic fiber may innervate as many as 32 ganglionic neurons in several different ganglia. As a result, a single sympathetic motor neuron inside the CNS can control a variety of peripheral effectors and produce a complex and coordinated response.
4. All preganglionic neurons release ACh at their synapses with ganglionic neurons. Most of the postganglionic fibers release norepinephrine, but a few release ACh.
5. The effector response depends on the function of the membrane receptor activated when epinephrine or norepinephrine binds to either alpha or beta receptors.

Table 17.1 (p. 461) summarizes the characteristics of the sympathetic division of the ANS.

✓ CONCEPT CHECK

- Where do the nerve fibers that synapse in the collateral ganglia originate?
- Individuals with high blood pressure may be given a medication that blocks beta receptors. How would this medication help their condition?
- What are the two types of sympathetic ganglia, and where are they located?

The Parasympathetic Division

[FIGURE 17.7]

The parasympathetic division of the ANS (Figure 17.7●) includes:

1. *Preganglionic neurons located in the brain stem and in sacral segments of the spinal cord.* In the brain, the mesencephalon, pons, and medulla oblongata contain autonomic nuclei associated with cranial nerves III, VII, IX, and X. In the sacral segments of the spinal cord, the autonomic nuclei lie in spinal segments S_2–S_4.
2. *Ganglionic neurons in peripheral ganglia located very close to—or even within—the target organs.* As noted earlier (p. 448), ganglionic neurons in the parasympathetic division are found in terminal ganglia (near the target organs) or intramural ganglia (within the tissues of the target organs). The preganglionic fibers of the parasympathetic division do not diverge as extensively as do those of the sympathetic division. A typical preganglionic fiber synapses on 6–8 ganglionic neurons. These neurons are all located in the same ganglion, and their postganglionic fibers influence the same target organ. As a result, *the effects of parasympathetic stimulation are more specific and localized than those of the sympathetic division.*

■ Organization and Anatomy of the Parasympathetic Division [FIGURE 17.8]

Parasympathetic preganglionic fibers leave the brain in cranial nerves III (oculomotor), VII (facial), IX (glossopharyngeal), and X (vagus) (Figure 17.8●). The fibers in N III, N VII, and N IX help control visceral structures in the head. These preganglionic fibers synapse in the **ciliary**, **pterygopalatine**, **submandibular**, and **otic ganglia.** pp. 410–416 Short postgan-

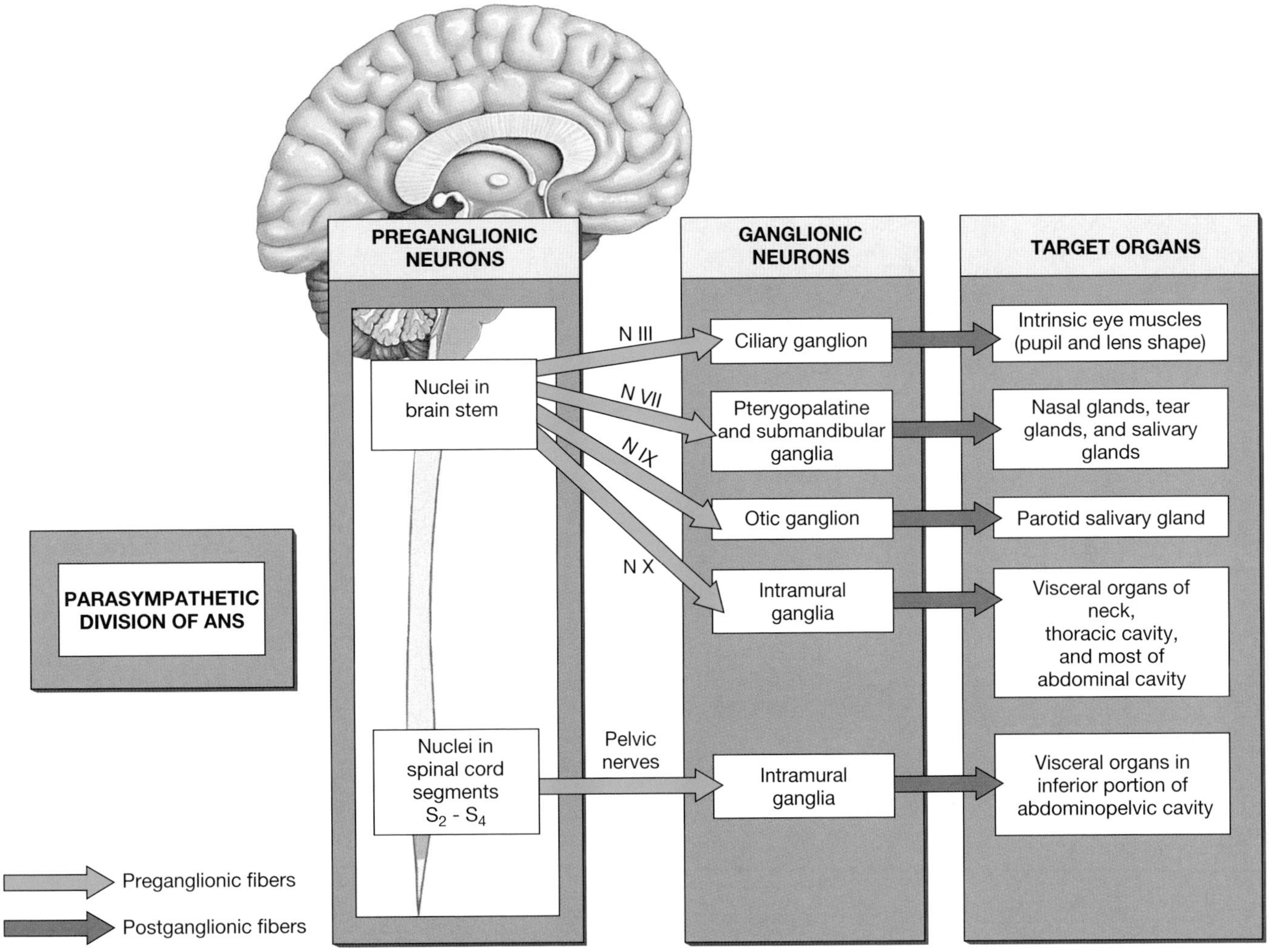

FIGURE 17.7 **ORGANIZATION OF THE PARASYMPATHETIC DIVISION OF THE ANS**
This diagram summarizes the relationships between preganglionic and ganglionic neurons and between ganglionic neurons and target organs.

glionic fibers then continue to their peripheral targets. The vagus nerve provides preganglionic parasympathetic innervation to intramural ganglia within structures in the thoracic cavity and in the abdominopelvic cavity as distant as the last segments of the large intestine. The vagus nerve alone provides roughly 75% of all parasympathetic outflow.

The sacral parasympathetic outflow does not join the ventral rami of the spinal nerves. pp. 364, 458 Instead, the preganglionic fibers form distinct **pelvic nerves** that innervate intramural ganglia in the kidney and urinary bladder, the terminal portions of the large intestine, and the sex organs.

General Functions of the Parasympathetic Division

A partial listing of the major effects produced by the parasympathetic division includes:

1. Constriction of the pupils to restrict the amount of light entering the eyes; assists in focusing on nearby objects.
2. Secretion by digestive glands, including salivary glands, gastric glands, duodenal and other intestinal glands, the pancreas, and the liver.
3. Secretion of hormones that promote nutrient absorption by peripheral cells.
4. Increased smooth muscle activity along the digestive tract.
5. Stimulation and coordination of defecation.
6. Contraction of the urinary bladder during urination.
7. Constriction of the respiratory passageways.
8. Reduction in heart rate and force of contraction.
9. Sexual arousal and stimulation of sexual glands in both sexes.

These functions center on relaxation, food processing, and energy absorption. The parasympathetic division has been called the *anabolic system* because stimulation leads to a general increase in the nutrient content of the blood. Cells throughout the body respond to this increase by absorbing nutrients and using them to support growth and other anabolic activities.

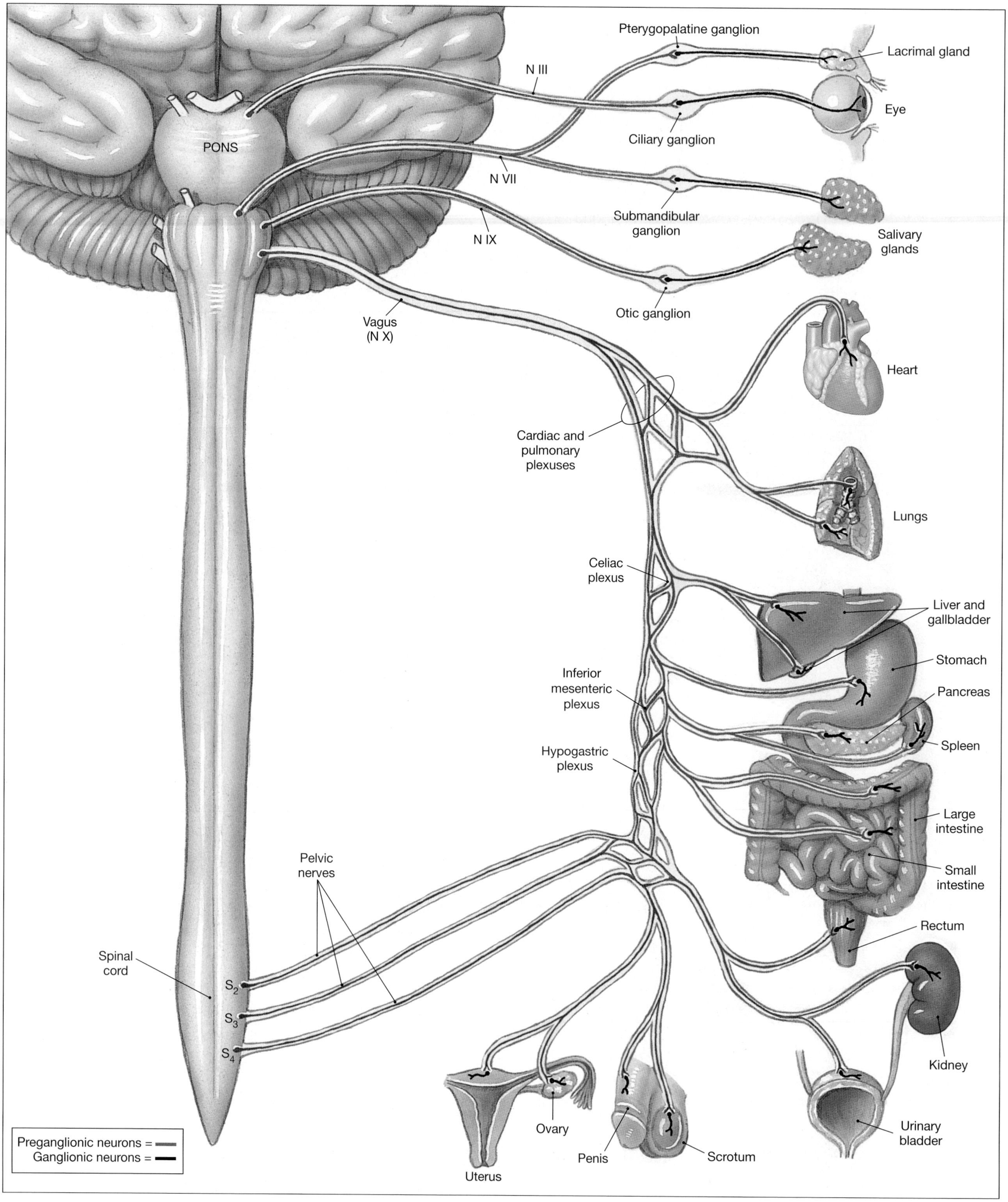

FIGURE 17.8 **Anatomical Distribution of the Parasympathetic Output**

Preganglionic fibers exit the CNS through either cranial nerves or pelvic nerves. The pattern of target-organ innervation is similar on each side of the body although only nerves on the left side are illustrated.

Parasympathetic Activation and Neurotransmitter Release

All of the preganglionic and postganglionic fibers in the parasympathetic division release ACh at their synapses and neuroeffector junctions. The neuroeffector junctions are small, with narrow synaptic clefts. The effects of stimulation are short-lived, because most of the ACh released is inactivated by acetylcholinesterase within the synapse. Any ACh diffusing into the surrounding tissues is deactivated by the enzyme *tissue cholinesterase.* As a result, the effects of parasympathetic stimulation are quite localized, and they last a few seconds at most.

MEMBRANE RECEPTORS AND RESPONSES

Although all the synapses (neuron-to-neuron) and neuroeffector junctions (neuron-to-effector) of the parasympathetic division use the same transmitter, acetylcholine, two different types of ACh receptors are found on the postsynaptic membranes:

1. **Nicotinic** (nik-ō-TIN-ik) **receptors** are found on the surfaces of ganglion cells of both the parasympathetic and sympathetic divisions, as well as at neuromuscular junctions of the SNS. Exposure to ACh always causes excitation of the ganglionic neuron or muscle fiber through the opening of membrane ion channels.
2. **Muscarinic** (mus-kar-IN-ik) **receptors** are found at cholinergic neuroeffector junctions in the parasympathetic division, as well as at the few cholinergic neuroeffector junctions in the sympathetic division. Stimulation of muscarinic receptors produces longer-lasting effects than does stimulation of nicotinic receptors. The response, which reflects the activation or inactivation of specific enzymes, may be either excitatory or inhibitory.

The names *nicotinic* and *muscarinic* indicate the chemical compounds that stimulate these receptor sites. Nicotinic receptors bind *nicotine*, a powerful component of tobacco smoke. Muscarinic receptors are stimulated by *muscarine*, a toxin produced by some poisonous mushrooms.

A Summary of the Parasympathetic Division [TABLE 17.1]

1. The parasympathetic division includes visceral motor nuclei in the brain stem associated with four cranial nerves (III, VII, IX, and X). In sacral segments S_2–S_4, autonomic nuclei lie in the lateral portions of the anterior gray horns.
2. The ganglionic neurons are situated in intramural ganglia or in ganglia closely associated with their target organs.
3. The parasympathetic division innervates structures in the head and organs in the thoracic and abdominopelvic cavities.
4. All parasympathetic neurons are cholinergic. Release of ACh by preganglionic neurons stimulates nicotinic receptors on ganglionic neurons, and the effect is always excitatory. The release of ACh at neuroeffector junctions stimulates muscarinic receptors, and the effects may be either excitatory or inhibitory, depending on the nature of the enzymes activated when ACh binds to the receptor.
5. The effects of parasympathetic stimulation are usually brief and restricted to specific organs and sites.

Table 17.1 (p. 461) summarizes the characteristics of the parasympathetic division of the ANS.

✓ CONCEPT CHECK

- Identify the neurotransmitter released by preganglionic fibers and by the postganglionic fibers in the parasympathetic division of the autonomic nervous system.
- What are the two different ACh receptors found on postsynaptic membranes in the parasympathetic division?
- What are intramural ganglia?
- Why does sympathetic stimulation have such widespread effects?

Relationships between the Sympathetic and Parasympathetic Divisions

The sympathetic division has widespread impact, reaching visceral organs as well as tissues throughout the body. The parasympathetic division modifies the activity of structures innervated by specific cranial nerves and pelvic nerves. This includes the visceral organs within the thoracic and abdominopelvic cavities. Although some of these organs are innervated by only one autonomic division, most vital organs receive **dual innervation**—that is, they receive instructions from both the sympathetic and parasympathetic divisions. Where dual innervation exists, the two divisions often have opposing or antagonistic effects. Dual innervation is most prominent in the digestive tract, the heart, and the lungs. For example, sympathetic stimulation decreases digestive tract motility, whereas parasympathetic stimulation increases its motility.

Anatomy of Dual Innervation [FIGURE 17.9]

In the head, parasympathetic postganglionic fibers from the ciliary, pterygopalatine, submandibular, and otic ganglia accompany the cranial nerves to their peripheral destinations. The sympathetic innervation reaches the same structures by traveling directly from the superior cervical ganglia of the sympathetic chain.

In the thoracic and abdominopelvic cavities, the sympathetic postganglionic fibers mingle with parasympathetic preganglionic fibers at a series of plexuses (Figure 17.9●). These are the *cardiac plexus*, the *pulmonary plexus*, the *esophageal plexus*, the *celiac plexus*, the *inferior mesenteric plexus*, and the *hypogastric plexus*. Nerves leaving these plexuses travel with the blood vessels and lymphatics that supply visceral organs.

Autonomic fibers entering the thoracic cavity intersect at the **cardiac plexus** and the **pulmonary plexus**. These plexuses contain both sympathetic and parasympathetic fibers bound for the heart and lungs, respectively, as well as the parasympathetic ganglia whose output affects those organs. The **esophageal plexus** contains descending branches of the vagus nerve and splanchnic nerves leaving the sympathetic chain on either side.

Parasympathetic preganglionic fibers of the vagus nerve enter the abdominopelvic cavity with the esophagus. There they join the network of the **celiac plexus**, also called the *solar plexus*. The celiac plexus and an associated smaller plexus, the **inferior mesenteric plexus**, innervate viscera down to the initial segments of the large intestine. The **hypogastric plexus** contains the parasympathetic outflow of the pelvic nerves, sympathetic postganglionic

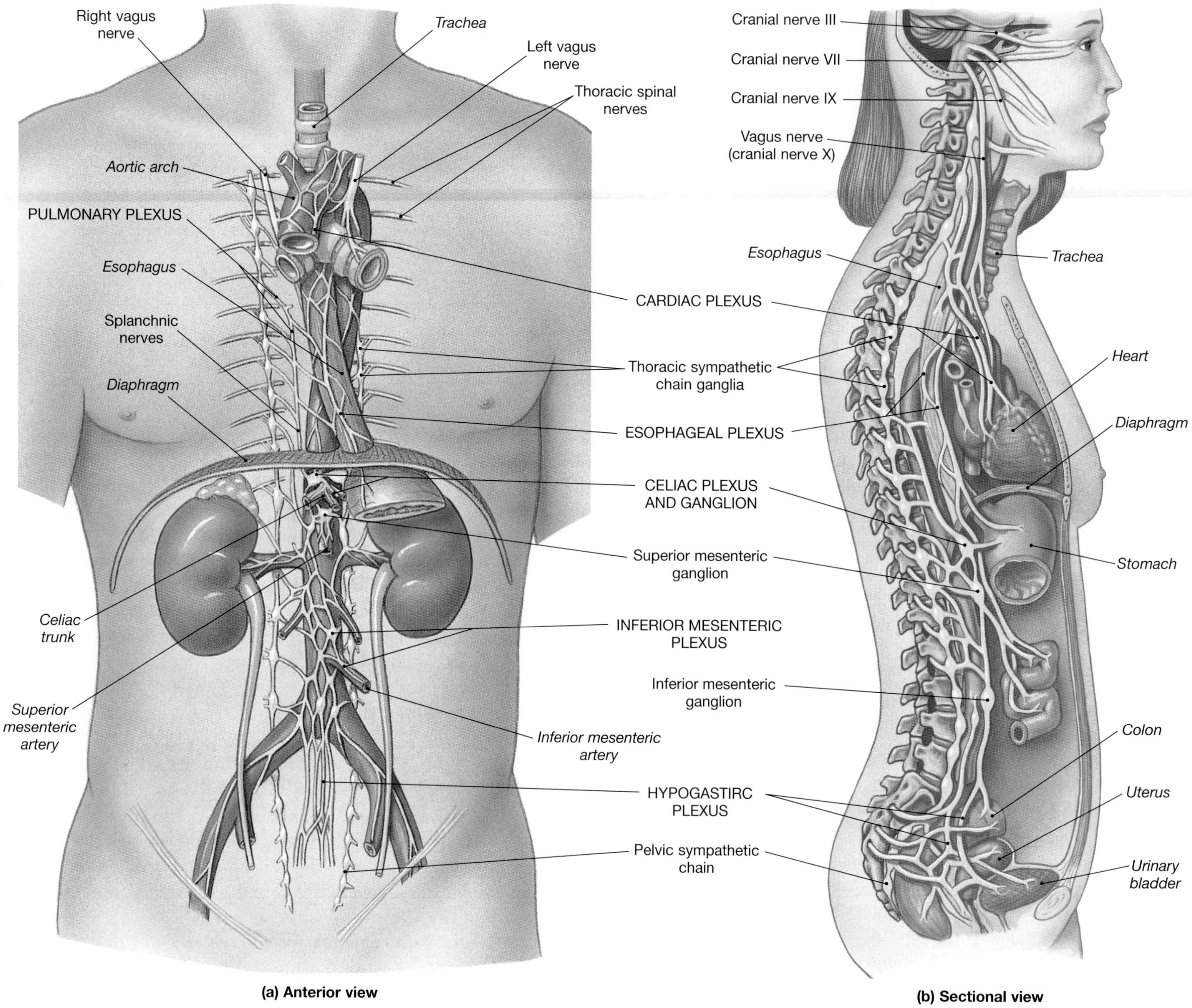

FIGURE 17.9 **THE PERIPHERAL AUTONOMIC PLEXUSES**

(a) This is a diagrammatic view of the distribution of ANS plexuses in the thoracic cavity (cardiac, esophageal, and pulmonary plexuses) and the abdominopelvic cavity (celiac, inferior mesenteric, and hypogastric plexuses). **(b)** A sectional view of the autonomic plexuses.

fibers from the inferior mesenteric ganglion, and sacral splanchnic nerves from the sympathetic chain. The hypogastric plexus innervates the digestive, urinary, and reproductive organs of the pelvic cavity.

A Comparison of the Sympathetic and Parasympathetic Divisions

Figure 17.10● and Table 17.1 compare key features of the sympathetic and parasympathetic divisions of the ANS.

Integration and Control of Autonomic Functions

The ANS, like the somatic nervous system (SNS), is organized into a series of interacting levels. At the first level you find the visceral motor neurons that participate in cranial and spinal *visceral reflexes.* The cell bodies of these motor neurons are located in the spinal cord and lower brain stem.

Visceral Reflexes [FIGURE 17.11 AND TABLE 17.2]

Visceral reflexes are the simplest functional units in the autonomic nervous system. They provide automatic motor responses that can be modified, facilitated, or inhibited by higher centers, especially those of the hypothalamus. All visceral reflexes are polysynaptic. p. 374 Each visceral reflex arc (Figure 17.11●) consists of a receptor, a sensory nerve, a processing center (interneuron or motor neuron), and two visceral motor neurons, preganglionic and ganglionic. Sensory nerves deliver information to the CNS along spinal nerves, cranial nerves, and the autonomic nerves that innervate peripheral effectors. For example, shining a light in the eye triggers a visceral reflex (the *consensual light reflex*) that constricts the pupils of both eyes. p. 373 In total darkness, the pupils dilate. The motor nuclei directing pupillary constriction or dilation are also controlled by hypothalamic centers concerned with emotional states. For example, when you are queasy or nauseated, your pupils constrict; when you are sexually aroused, your pupils dilate.

Visceral reflexs may be either long reflexes or short reflexes. **Long reflexes** are the autonomic equivalents of the polysynaptic reflexes introduced in Chapter 13. p. 374 Visceral sensory neurons deliver information to the CNS along the dorsal roots of spinal nerves, within the sensory branches of cranial nerves, and within the autonomic nerves that innervate visceral effectors. The processing steps involve interneurons within the CNS, and the motor neurons involved are located within the brain stem or spinal cord. The ANS carries the motor commands to the

FIGURE 17.10 **A Comparison of the Sympathetic and Parasympathetic Divisions**

This diagram compares fiber length (preganglionic and postganglionic), the general location of ganglia, and the primary neurotransmitter released by each division of the autonomic nervous system.

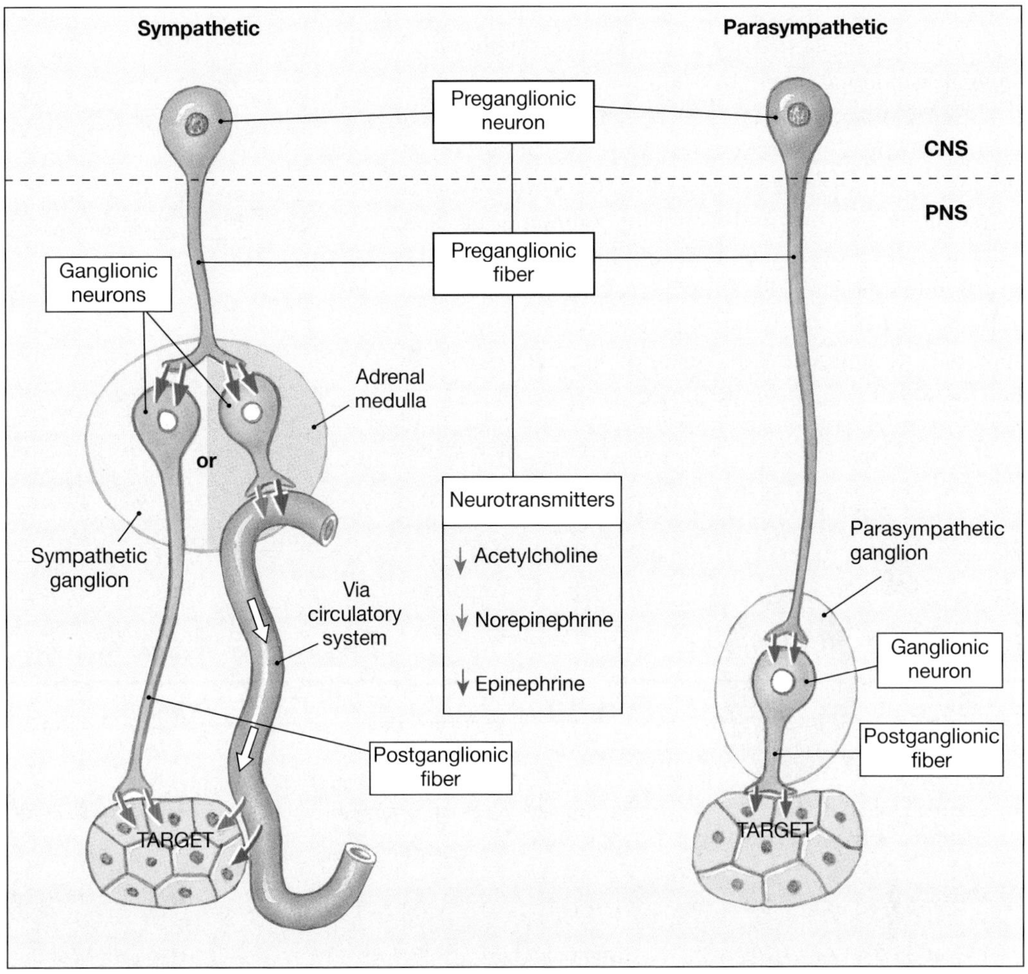

TABLE 17.1 A Comparison of the Sympathetic and Parasympathetic Divisions of the ANS

Characteristic	Sympathetic Division	Parasympathetic Division
Location of CNS Visceral Motor Neurons	Lateral gray horns of spinal segments T_1–L_2	Brain stem and spinal segments S_2–S_4
Location of PNS Ganglia	Paravertebral sympathetic chain; collateral ganglia (celiac, superior mesenteric, and inferior mesenteric) located anterior and lateral to the descending aorta	Intramural or terminal
Preganglionic Fibers:		
Length	Relatively short, myelinated	Relatively long, myelinated
Neurotransmitter released	Acetylcholine	Acetylcholine
Postganglionic Fibers:		
Length	Relatively long, unmyelinated	Relatively short, unmyelinated
Neurotransmitter released	Usually norepinephrine	Always acetylcholine
Neuroeffector Junction	Varicosities and enlarged terminal knobs that release transmitter near target cells	Neuroeffector junctions that release transmitter to special receptor surface
Degree of Divergence from CNS to Ganglion Cells	Approximately 1:32	Approximately 1:6
General Functions	Stimulate metabolism, increase alertness, prepare for emergency "fight or flight" response	Promote relaxation, nutrient uptake, energy storage ("rest and repose")

FIGURE 17.11 **VISCERAL REFLEXES**

Visceral reflexes have the same basic components as somatic reflexes, but all visceral reflexes are polysynaptic.

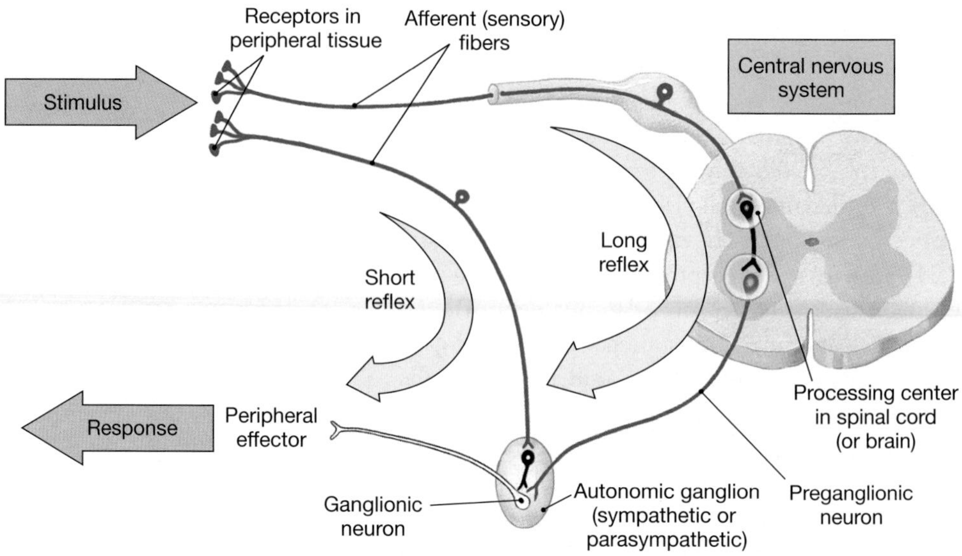

appropriate visceral effectors, after a synapse within a peripheral autonomic ganglion.

Short reflexes bypass the CNS entirely; they involve sensory neurons and interneurons whose cell bodies are located within autonomic ganglia. The interneurons synapse on ganglionic neurons, and the motor commands are then distributed by postganglionic fibers. Short reflexes control very simple motor responses with localized effects. In general, short reflexes may control patterns of activity in one small part of a target organ, whereas long reflexes coordinate the activities of the entire organ.

TABLE 17.2 **REPRESENTATIVE VISCERAL REFLEXES**

Reflex	Stimulus	Response	Comments
PARASYMPATHETIC REFLEXES			
Gastric and Intestinal reflexes *(see Chapter 25)*	Pressure and physical contact with food materials	Smooth muscle contractions that propel food materials and mix with secretions	Mediated by the vagus nerve (N X)
Defecation *(see Chapter 25)*	Distention of rectum	Relaxation of internal anal sphincter	Requires voluntary relaxation of external anal sphincter
Urination *(see Chapter 26)*	Distention of urinary bladder	Contraction of urinary bladder walls, relaxation of internal urethral sphincter	Requires voluntary relaxation of external urethral sphincter
Direct light and Consensual light reflexes *(see Chapter 18)*	Bright light shining in eye(s)	Constriction of pupils of both eyes	
Swallowing reflex *(see Chapter 25)*	Movement of food and drink into superior pharynx	Smooth muscle and skeletal muscle contractions	Coordinated by swallowing center in medulla oblongata
Vomiting reflex *(see Chapter 25)*	Irritation of digestive tract lining	Reversal of normal smooth muscle action to eject contents	Coordinated by vomiting center in medulla oblongata
Coughing reflex *(see Chapter 24)*	Irritation of respiratory tract lining	Sudden explosive ejection of air	Coordinated by coughing center in medulla oblongata
Barorecepter reflex *(see Chapter 21)*	Sudden rise in blood pressure in carotid artery	Reduction in heart rate and force of contraction	Coordinated in cardiac center in medulla oblongata
Sexual arousal *(see Chapter 27)*	Erotic stimuli (visual or tactile)	Increased glandular secretions, sensitivity	
SYMPATHETIC REFLEXES			
Cardioacceleratory reflex *(see Chapter 21)*	Sudden decline in blood pressure in carotid artery	Increase in heart rate and force of contraction	Coordinated in cardiac center in medulla oblongata
Vasomotor reflexes *(see Chapter 22)*	Changes in blood pressure in major arteries	Changes in diameter of peripheral blood vessels	Coordinated in vasomotor center in medulla oblongata
Pupillary reflex *(see Chapter 18)*	Low light level reaching visual receptors	Dilation of pupil	
Emission and ejaculation (in males) *(see Chapter 27)*	Erotic stimuli (tactile)	Contraction of seminal vesicles and prostate, and skeletal muscle contractions that eject semen	Ejaculation involves the contractions of the bulbospongiosus muscles

In most organs, long reflexes are most important in regulating visceral activities, but this is not the case with the digestive tract and its associated glands. The ganglia in the walls of the digestive tract contain the cell bodies of visceral sensory neurons, interneurons, and visceral motor neurons. Although parasympathetic innervation of the visceral motor neurons can stimulate and coordinate various digestive activities, the digestive system is quite capable of controlling its own functions independent of the central nervous system. For this reason it has been called the *enteric nervous system (ENS);* we shall consider its functions further in Chapter 25.

As we examine other body systems, many examples of autonomic reflexes involved in respiration, cardiovascular function, and other visceral activities will be examined. Table 17.2 summarizes information concerning important visceral reflexes. Note that the parasympathetic division participates in reflexes that affect individual organs and systems, reflecting the relatively specific and restricted pattern of innervation. In contrast, there are fewer sympathetic reflexes. This division is typically activated as a whole, in part because of the degree of divergence and in part because the release of hormones by the adrenal medullae also produces widespread peripheral effects.

CLINICAL BRIEF

Diabetic Neuropathy and the ANS

In the condition *diabetes mellitus*, blood glucose levels are elevated yet most cells are unable to absorb and use the glucose as an energy source. A variety of physiology problems result; these will be discussed further in Chapter 19. People with chronic, untreated or poorly managed cases of diabetes mellitus often develop peripheral nerve problems. The peripheral nerve dysfunction, a condition known a *diabetic neuropathy,* has widespread effects. We will defer considering specifics until Chapter 19 except to note that diabetic neuropathy has multiple effects on the ANS. Most notably, it interferes with normal visceral reflexes (Table 17.2). Symptoms often include delayed gastric emptying, reduced sympathetic control of the cardiovascular system, leading to a slow heart rate and low blood pressure when standing, difficulty with urination, and impotence.

Higher Levels of Autonomic Control [FIGURE 17.12]

The levels of activity in the sympathetic and parasympathetic divisions are controlled by centers in the brain stem concerned with specific visceral functions. Figure 17.12● diagrams the levels of autonomic control. As in the SNS, simple reflexes based in the spinal cord provide relatively rapid and automatic responses to stimuli. More complex sympathetic and parasympathetic reflexes are coordinated by processing centers in the medulla oblongata. In addition to the cardiovascular centers, the medulla oblongata contains centers and nuclei involved with respiration, digestive secretions, peristalsis, and urinary function. These medullary centers are in turn subject to regulation by the hypothalamus. In general, centers in the posterior and lateral hypothalamus are concerned with the coordination and regulation of sympathetic function, and portions of the anterior and medial hypothalamus control the parasympathetic division. p. 400

The term *autonomic* was originally applied to the visceral motor system because it was thought that the regulatory centers functioned without regard for other CNS activities. This view has been revised drastically in light of subsequent research. We now know that the hypothalamus interacts with all other portions of the brain, and that activity in the limbic system (memories, emotional states), thalamus (sensory information), or cerebral cortex (conscious thought processes) can have dramatic effects on autonomic function. Interconnections between the limbic system and the hypothalamus provide a direct link between emotional states and hypothalamic activity. For example, when you become angry, your heart rate accelerates and your blood pressure rises.

✓ CONCEPT CHECK

- What is meant by dual innervation?
- What are visceral reflexes?
- Name three plexuses in the abdominopelvic cavity.

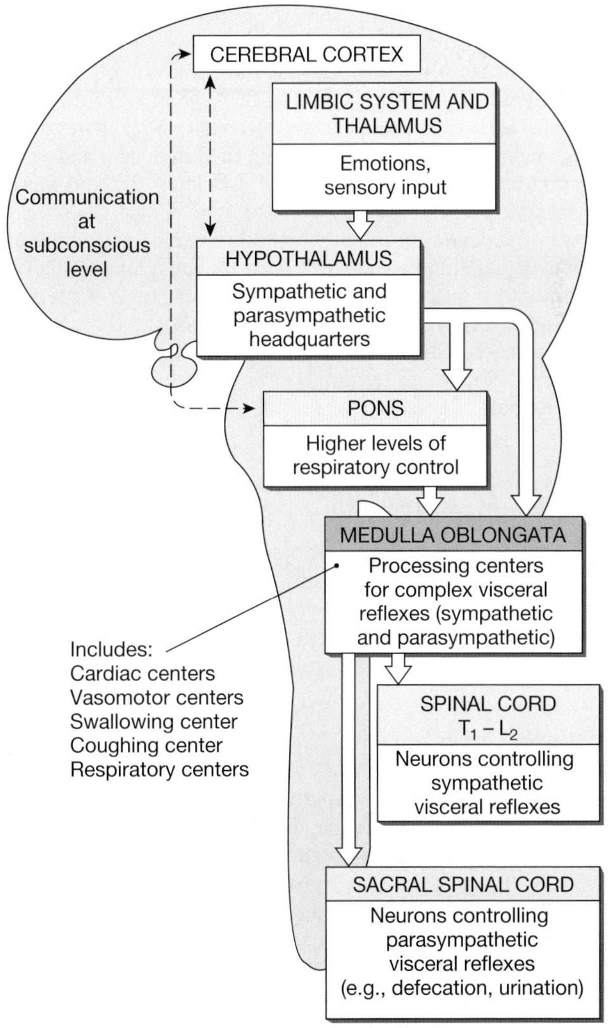

FIGURE 17.12 **Levels of Autonomic Control**

RELATED CLINICAL TERMS

diabetic neuropathy: A degenerative neurological disorder that may develop in persons with *diabetes mellitus*. *p. 463*

Horner's syndrome A condition characterized by unilateral loss of sympathetic innervation to the face. *p. 455*

STUDY OUTLINE & CHAPTER REVIEW

Introduction 448

1. The **autonomic nervous system (ANS)** regulates body temperature and coordinates cardiovascular, respiratory, digestive, excretory, and reproductive functions. Routine physiological adjustments to systems are made by the autonomic nervous system operating at the subconscious level.

An Overview of the Autonomic Nervous System (ANS) 448

Pathways for Visceral Motor Output 448

1. Visceral motor neurons in the CNS, termed **preganglionic neurons**, send axons (*preganglionic fibers*) to synapse on **ganglionic neurons**, whose cell bodies are located in autonomic ganglia outside the CNS. The axon of the ganglionic neuron is a **postganglionic fiber** that innervates peripheral organs.

Subdivisions of the ANS 448

2. There are two subdivisions in the ANS: the *sympathetic division* and the *parasympathetic division*. *(see Figure 17.1)*
3. Visceral efferents from the thoracic and lumbar segments form the **thoracolumbar (sympathetic) division** ("fight or flight" system) of the ANS. Generally, it stimulates tissue metabolism, increases alertness, and prepares the body to deal with emergencies. Visceral efferents leaving the brain stem and sacral segments form the **craniosacral (parasympathetic) division** ("rest and repose" system). Generally, it conserves energy and promotes sedentary activities. *(see Figure 17.1)*
4. Both divisions affect target organs via neurotransmitters. Membrane receptors determine whether the response will be stimulatory or inhibitory. Generally, neurotransmitter effects are as follows: (1) All preganglionic terminals release **acetylcholine (ACh)** and are excitatory; (2) All postganglionic parasympathetic terminals release ACh and effects may be excitatory or inhibitory; (3) Most postganglionic sympathetic terminals release **norepinephrine (NE)** and effects are usually excitatory.

The Sympathetic Division 450

1. The **sympathetic division** consists of preganglionic neurons between spinal cord segments T_1 and L_2; ganglionic neurons in ganglia near the vertebral column; and specialized neurons within the adrenal gland. *(see Figures 17.1a to 17.4/17.10)*
2. There are two types of sympathetic ganglia: **sympathetic chain ganglia** *(paravertebral ganglia* or *lateral ganglia)* and **collateral ganglia** *(prevertebral ganglia)*.

The Sympathetic Chain Ganglia 450

3. Between spinal segments T_1 and L_2 each ventral root gives off a *white ramus* with preganglionic fibers to a **sympathetic chain ganglion**. These preganglionic fibers tend to undergo extensive divergence before they synapse with the ganglionic neuron. The synapse occurs within the sympathetic chain ganglia, within one of the collateral ganglia, or within the adrenal medullae. Preganglionic fibers run between the sympathetic chain ganglia and interconnect them. Postganglionic fibers targeting visceral effectors in the body wall enter the *gray ramus* to return to the spinal nerve for distribution, whereas those that target thoracic cavity structures form autonomic nerves that go directly to their visceral destination. *(see Figures 17.1a,b/17.3)*
4. There are 3 cervical, 11–12 thoracic, 2–5 lumbar, 4–5 sacral, and 1 coccygeal sympathetic ganglia in each sympathetic chain. *Every spinal nerve has a gray ramus that carries sympathetic postganglionic fibers.* In summary: (1) Only thoracic and superior lumbar ganglia receive preganglionic fibers by way of white rami; (2) The cervical, inferior lumbar, and sacral chain ganglia receive preganglionic innervation from collateral fibers of sympathetic neurons; and (3) every spinal nerve receives a gray ramus from a ganglion of the sympathetic chain. *(see Figure 17.4)*

Collateral Ganglia 452

5. The abdominopelvic viscera receive sympathetic innervation via preganglionic fibers that pass through the sympathetic chain to synapse within collateral ganglia. The preganglionic fibers that innervate the collateral ganglia form the **splanchnic nerves** (greater, lesser, lumbar, and sacral). *(see Figures 17.3b/17.5/17.9)*
6. The splanchnic nerves innervate the hypogastric plexus and three collateral ganglia: (1) the *celiac ganglion*; (2) the *superior mesenteric ganglion*; and (3) the *inferior mesenteric ganglion*. *(see Figures 17.4/17.9)*
7. The **celiac ganglion** innervates the stomach, liver, pancreas, and spleen; the **superior mesenteric ganglion** innervates the small intestine and initial segments of the large intestine; and the **inferior mesenteric ganglion** innervates the kidney, bladder, sex organs, and terminal portions of the large intestine. *(see Figures 17.3b/17.4/17.9)*

The Adrenal Medullae 452

8. Some preganglionic fibers do not synapse as they pass through both the sympathetic chain and collateral ganglia. Instead, they enter one of the adrenal glands and synapse on modified neurons within the **adrenal medulla**. These cells release *norepinephrine* (NE) and *epinephrine* (E) into the circulation, causing a prolonged sympathetic innervation effect. *(see Figures 17.3c/17.4/17.5)*

Effects of Sympathetic Stimulation 454

9. In a crisis, the entire division responds, an event called **sympathetic activation**. Its effects include increased alertness, a feeling of energy and euphoria, increased cardiovascular and respiratory activity, general elevation in muscle tone, and mobilization of energy reserves.

Sympathetic Activation and Neurotransmitter Release 455

10. Stimulation of the sympathetic division has two distinctive results: the release of norepinephrine (or in some cases acetylcholine) at neuroeffector junctions, and the secretion of epinephrine and norepinephrine into the general circulation. *(see Figure 17.6)*

Membrane Receptors and Sympathetic Function 455

11. There are two classes of sympathetic receptors, which are stimulated by both norepinephrine and epinephrine: **alpha receptors** (which respond to stimulation by depolarizing the membrane) and **beta receptors** (which respond to stimulation by changing metabolic activity of the cells).
12. Most postganglionic fibers release norepinephrine (adrenergic), but a few release acetylcholine (cholinergic). Postganglionic fibers innervating sweat glands of the skin and blood vessels to skeletal muscles release ACh.

A Summary of the Sympathetic Division 456

13. The sympathetic division has the following characteristics: (1) Two segmentally arranged sympathetic chains lateral to the vertebral column, three collateral ganglia anterior to the vertebral column, and two adrenal medullae; (2) preganglionic fibers are relatively short, except for those of the adrenal medullae, while postganglionic fibers are quite long; (3) extensive divergence typically occurs, with a single preganglionic fiber synapsing with many ganglionic neurons in different ganglia; (4) all preganglionic fibers release ACh, while most postganglionic fibers release NE; and (5) effector response depends on the nature and activity of the receptor. *(see Table 17.1)*

The Parasympathetic Division 456

1. The parasympathetic division consists of preganglionic neurons in the brain stem and sacral segments of the spinal cord; and ganglionic neurons in peripheral ganglia located within or immediately next to target organs. *(see Figures 17.7/17.8 and Table 17.1)*

Organization and Anatomy of the Parasympathetic Division 456

2. Preganglionic fibers leave the brain in cranial nerves III (oculomotor), VII (facial), IX (glossopharyngeal), and X (vagus). *(see Figures 17.7/17.8)*
3. Parasympathetic fibers in the oculomotor, facial, and glossopharyngeal nerves help control visceral structures in the head, and they synapse in the **ciliary**, **pterygopalatine**, **submandibular**, and **otic ganglia**. Fibers in the vagus nerve supply preganglionic parasympathetic innervation to intramural ganglia within structures in the thoracic and abdominopelvic cavity. *(see Figures 17.7/17.8)*
4. Preganglionic fibers leaving the sacral segments form **pelvic nerves** that innervate intramural ganglia in the kidney, bladder, latter parts of the large intestine, and sex organs. *(see Figure 17.8)*

General Functions of the Parasympathetic Division 457

5. The effects produced by the parasympathetic division include: (1) pupil constriction, (2) digestive gland secretion, (3) hormone secretion for nutrient absorption, (4) increased digestive tract activity, (5) defecation activities, (6) urination activities, (7) reduced heart rate, (8) respiratory passageway constriction, and (9) sexual arousal. These general functions center on relaxation, food processing, and energy absorption.

Parasympathetic Activation and Neurotransmitter Release 459

6. All the parasympathetic preganglionic and postganglionic fibers release ACh at synapses and neuroeffector junctions. The effects are short-lived because of the actions of enzymes at the postsynaptic membrane and in the surrounding tissues.
7. Two different types of ACh receptors are found in postsynaptic membranes. **Nicotinic receptors** are located on ganglion cells of both divisions of the ANS and at neuromuscular junctions. Exposure to ACh causes excitation by opening membrane channels. **Muscarinic receptors** are located at neuroeffector junctions in the parasympathetic division and those cholinergic neuroeffector junctions in the sympathetic division. Stimulation of muscarinic receptors produces a longer-lasting effect than does stimulation of nicotinic receptors.

A Summary of the Parasympathetic Division 459

8. The parasympathetic division has the following characteristics: (1) Includes visceral motor nuclei associated with cranial nerves III, VII, IX, and X and sacral segments S_2–S_4; (2) ganglionic neurons are located in terminal or intramural ganglia near or within target organs, respectively; (3) innervates areas serviced by cranial nerves and organs in the thoracic and abdominopelvic cavities; (4) all parasympathetic neurons are cholinergic. The postganglionic neurons are also cholinergic, and are further subdivided as being either muscarinic or nicotinic receptors; and (5) effects are usually brief and restricted to specific sites. *(see Figure 17.10 and Table 17.1)*

Relationships Between The Sympathetic And Parasympathetic Divisions 459

1. The sympathetic division has widespread influence, reaching visceral and somatic structures throughout the body. *(see Figure 17.4 and Table 17.1)*
2. The parasympathetic division innervates only visceral structures serviced by cranial nerves or lying within the thoracic and abdominopelvic cavity. Organs with **dual innervation** receive instructions from both divisions. *(see Figure 17.10 and Table 17.1)*

Anatomy of Dual Innervation 459

3. In body cavities the parasympathetic and sympathetic nerves intermingle to form a series of characteristic nerve plexuses (nerve networks), which include the **cardiac**, **pulmonary**, **esophageal**, **celiac**, **inferior mesenteric**, and **hypogastric plexuses**. *(see Figure 17.9)*

A Comparison of the Sympathetic and Parasympathetic Divisions 460

4. Review *Figure 17.10* and *Table 17.1.*

Integration and Control of Autonomic Functions 460

Visceral Reflexes 461

1. **Visceral reflexes** are the simplest functions of the ANS. They provide automatic motor responses that can be modified, facilitated, or inhibited by higher centers, especially in the hypothalamus. *(see Figure 17.11 and Table 17.2)*

Higher Levels of Autonomic Control 463

2. In general, higher brain centers in the posterior and lateral hypothalamus are concerned with the coordination and regulation of sympathetic function, and portions of the anterior and medial hypothalamus control the parasympathetic division. *(see Figure 17.12)*

LEVEL 1 REVIEWING FACTS AND TERMS

Match each numbered item with the most closely related lettered item. Use letters for answers in the spaces provided.

Column A	**Column B**
_____ 1. preganglionic	a. all preganglionic fibers
_____ 2. thoracolumbar	b. preganglionic fibers to collateral ganglia
_____ 3. parasympathetic	c. first neuron
_____ 4. prevertebral	d. collateral ganglia
_____ 5. paravertebral	e. adrenal medulla
_____ 6. acetylcholine	f. sympathetic activation
_____ 7. norepinephrine	g. sympathetic division
_____ 8. sympathetic	h. terminal ganglia
_____ 9. splanchnic	i. sympathetic chain
_____10. crisis	j. long postganglionic fiber

11. In the autonomic nervous system,
    (a) the lower motor neurons directly innervate effector organs
    (b) there is always a synapse between the CNS and the effector organ
    (c) motor neurons do not synapse but are connected by gap junctions
    (d) some motor neurons release both NE and ACh

12. Splanchnic nerves
    (a) are formed by parasympathetic postganglionic fibers
    (b) include preganglionic fibers that go to collateral ganglia
    (c) control sympathetic function of structures in the head
    (d) connect one chain ganglion with another

13. Preganglionic fibers of the sympathetic nervous system that carry motor impulses to targets in the body wall or thoracic cavity synapse in a(n)
    (a) intramural ganglion (b) collateral ganglion
    (c) sympathetic chain ganglion (d) adrenal ganglion
14. Preganglionic fibers of the ANS sympathetic division originate in the
    (a) cerebral cortex of the brain (b) medulla oblongata
    (c) brain stem and sacral spinal cord (d) thoracolumbar spinal cord
15. The neurotransmitter at all synapses and neuroeffector junctions in the parasympathetic division of the ANS is
    (a) epinephrine (b) cyclic-AMP
    (c) norepinephrine (d) acetylcholine
16. The large cells in the adrenal medulla, which resemble neurons in sympathetic ganglia,
    (a) are located in the adrenal cortex
    (b) release acetylcholine into blood capillaries
    (c) release epinephrine and norepinephrine into blood capillaries
    (d) have no endocrine functions
17. Sympathetic preganglionic fibers are characterized as
    (a) being short in length and unmyelinated
    (b) being short in length and myelinated
    (c) being long in length and myelinated
    (d) being long in length and unmyelinated
18. All preganglionic autonomic fibers release _______ at their synaptic terminals, and the effects are always _______.
    (a) norepinephrine; inhibitory (b) norepinephrine; excitatory
    (c) acetylcholine; excitatory (d) acetylcholine; inhibitory
19. Postganglionic fibers of autonomic neurons are usually
    (a) myelinated (b) unmyelinated
    (c) larger than preganglionic fibers (d) located in the spinal cord
20. The autonomic division of the nervous system directs
    (a) voluntary motor activity
    (b) conscious control of skeletal muscles
    (c) processes that maintain homeostasis
    (d) all behavioral activities

LEVEL 2 REVIEWING CONCEPTS

1. Dual innervation refers to situations in which
    (a) vital organs receive instructions from both sympathetic and parasympathetic fibers
    (b) the atria and ventricles of the heart receive autonomic stimulation from the same nerves
    (c) sympathetic and parasympathetic fibers have similar effects
    (d) sympathetic fibers always stimulate organs and parasympathetic fibers always inhibit organs
2. Damage to the ventral roots of the first five thoracic spinal nerves on the right side of the body would interfere with
    (a) the ability to dilate the right pupil
    (b) the ability to dilate the left pupil
    (c) the ability to contract the right biceps brachii muscle
    (d) the ability to contract the left biceps brachii muscle
3. Under which set of circumstances would the diameter of peripheral blood vessels be the greatest?
    (a) increased sympathetic stimulation
    (b) decreased sympathetic stimulation
    (c) increased parasympathetic stimulation
    (d) decreased parasympathetic stimulation
4. Injury to the neurons of a collateral ganglion would affect the function of the
    (a) heart (b) pupils
    (c) sweat glands (d) digestive tract
5. The effects of epinephrine and norepinephrine released by the adrenal glands last longer than those of either chemical when released at neuroeffector junctions. Why?
6. Why are the effects of parasympathetic stimulation more specific and localized than those of the sympathetic division?
7. How is sweat gland function regulated by sympathetic activity?
8. How do sympathetic chain ganglia differ from both collateral ganglia and intramural ganglia?
9. Compare the general effects of the sympathetic and parasympathetic divisions of the ANS.
10. Describe the general organization of the pathway for visceral motor output.

LEVEL 3 CRITICAL THINKING AND CLINICAL APPLICATIONS

1. In some severe cases, a person suffering from stomach ulcers may need to have surgery to cut the branches of the vagus nerve that innervates the stomach. How would this help the problem?
2. A patient is about to undergo major surgery. Two hours before the surgery is to begin, the patient experiences "jitters," an elevated heart rate and blood pressure, increased rate of breathing, and cold sweats. What is happening?
3. Kassie is stung on the neck by a wasp. Because she is allergic to wasp venom, her throat begins to swell and her respiratory passages constrict. Would acetylcholine or epinephrine be more helpful in relieving her symptoms? Why?

✓ ANSWERS TO CONCEPT CHECK QUESTIONS

p. 448 **1.** Preganglionic neurons have their cell bodies in the CNS and their axons project to ganglia in the PNS; ganglionic neurons have their cell bodies in ganglia in the PNS and their axons innervate effector cells. **2.** Sympathetic division (thoracolumbar division) and parasympathetic division (craniosacral division). **3.** Norepinephrine is the neurotransmitter released by most postganglionic sympathetic terminals. **4.** Cardiac muscle in the heart and the smooth muscle, glands, and adipose tissues in all peripheral organs.
p. 456 **1.** The neurons that synapse in the collateral ganglia originate in the inferior thoracic and superior lumbar regions of the spinal cord and pass through the sympathetic chain ganglia without synapsing before reaching the collateral ganglia. **2.** Blocking the beta receptors on cells would decrease or prevent sympathetic stimulation of those tissues. This would result in decreased heart rate and force of contraction, and relaxation of the smooth muscle in the walls; the combination would lower blood pressure. **3.** Sympathetic chain ganglia (paravertebral ganglia) are lateral to each side of the vertebral column. Collateral ganglia (prevertebral ganglia) are anterior to the vertebral column.
p. 459 **1.** The neurotransmitter is acetylcholine. **2.** Nicotinic receptors and muscarinic receptors. **3.** These ganglia are located in the tissues of their target organs. **4.** (1) The extensive divergence of preganglionic fibers in the sympathetic division distributes sympathetic output to many different visceral organs and tissues simultaneously, and (2) the release of E and NE by the adrenal medullae affects tissues and organs throughout the body.
p. 463 **1.** It means that most vital organs receive instructions from both the sympathetic and parasympathetic divisions. **2.** Visceral reflexes are the simplest functional units in the autonomic nervous system. They provide automatic motor responses that can be modified, facilitated, or inhibited by higher centers, especially those of the hypothalamus. **3.** The celiac plexus, the inferior mesenteric plexus, and the hypogastric plexus.

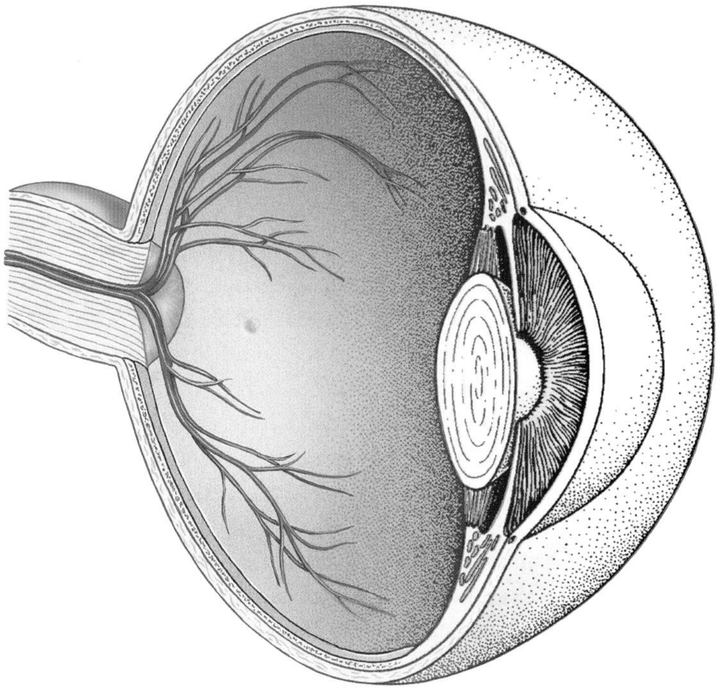

18

THE NERVOUS SYSTEM

General and Special Senses

Every cell membrane functions as a receptor for the cell, because it responds to changes in the extracellular environment. Cell membranes differ in their sensitivities to specific electrical, chemical, and mechanical stimuli. For example, a hormone that stimulates a neuron may have no effect on an osteocyte, because the cell membranes of neurons and osteocytes contain different receptor proteins. A **sensory receptor** is a specialized cell or cell process that monitors conditions in the body or the external environment. Stimulation of the receptor directly or indirectly alters the production of action potentials in a sensory neuron. p. 344 The sensory information arriving at the CNS is called a **sensation**; a **perception** is a conscious awareness of a sensation. The term **general senses** refers to sensations of temperature, pain, touch, pressure, vibration, and proprioception (body position). General sensory receptors are distributed throughout the body. These sensations arrive at the primary sensory cortex, or *somatosensory cortex*, via pathways previously described. p. 428

The **special senses** are smell (*olfaction*), taste (*gustation*), balance (*equilibrium*), hearing, and vision. The sensations are provided by specialized receptor cells that are structurally more complex than those of the general senses. These receptors are localized within complex **sense organs**, such as the eye or ear. The information provided by these receptors is distributed to specific areas of the cerebral cortex (such as the visual cortex) and to centers throughout the brain stem.

Sensory receptors represent the interface between the nervous system and the internal and external environments. The nervous system relies on accurate sensory data to control and coordinate relatively swift responses to specific stimuli. This chapter begins by summarizing receptor function and basic concepts in sensory processing. We will then apply this information to each of the general and special senses as we discuss their structure.

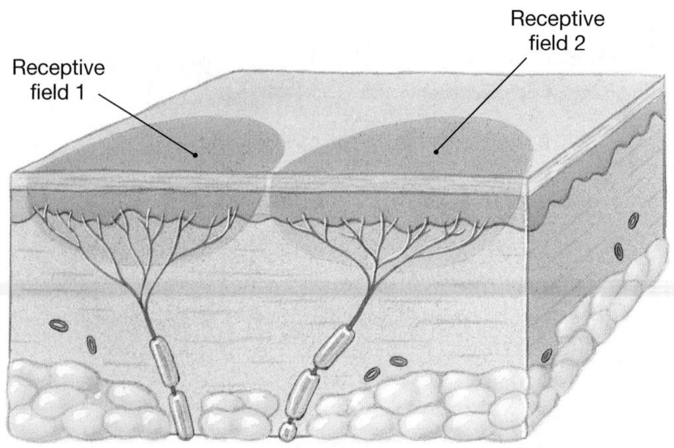

FIGURE 18.1 **RECEPTORS AND RECEPTIVE FIELDS**

Each receptor monitors a specific area known as the receptive field.

Receptors [FIGURE 18.1]

Each receptor has a characteristic sensitivity. For example, a touch receptor is very sensitive to pressure but relatively insensitive to chemical stimuli. This concept is called **receptor specificity**. Specificity results from the structure of the receptor cell itself or from the presence of accessory cells or structures that shield it from other stimuli. The simplest receptors are the dendrites of sensory neurons, called **free nerve endings**. They can be stimulated by many different stimuli. For example, free nerve endings that provide the sensation of pain may be responding to chemical stimulation, pressure, temperature changes, or physical damage. In contrast, the receptor cells of the eye are surrounded by accessory cells that normally prevent their stimulation by anything other than light. The area monitored by a single receptor cell is its **receptive field** (Figure 18.1●). Whenever a sufficiently strong stimulus arrives in the receptive field, the CNS receives the information. The larger the receptive field, the poorer our ability to localize a stimulus. For example, a touch receptor on the general body surface may have a receptive field 7 cm (2.5 in.) in diameter. As a result, a light touch can be described only as affecting a general area of this size. On the tongue, where the receptive fields are less than a millimeter in diameter, we can be very precise about the location of a stimulus.

An arriving stimulus can take many different forms; it may be a physical force, such as pressure, a dissolved chemical, a sound, or a beam of light. Regardless of the nature of the stimulus, however, sensory information must be sent to the CNS in the form of action potentials, which are electrical events. The arriving information is processed and interpreted by the CNS at the conscious and subconscious levels.

Interpretation of Sensory Information

When sensory information arrives at the CNS, it is routed according to the location and nature of the stimulus. Along sensory pathways, axons relay information from point A (the receptor) to point B (a neuron at a specific site in the cerebral cortex). The connection between receptor and cortical neuron is called a **labeled line**. Each labeled line carries information concerning a specific sensation (touch, pressure, vision, and so forth) from receptors in a specific part of the body. The identity of the active labeled line indicates the location and nature of the stimulus. *All other characteristics of the stimulus are conveyed by the pattern of action potentials in the afferent fibers*. This **sensory coding** provides information about the strength, duration, variation, and movement of the stimulus.

Some sensory neurons, called **tonic receptors**, are always active. The photoreceptors of the eye and various receptors that monitor body position are examples of tonic receptors. Other receptors are normally inactive, but become active for a short time whenever there is a change in the conditions they are monitoring. These are **phasic receptors** and provide information on the intensity and rate of change of a stimulus. Many touch and pressure receptors in the skin are examples of phasic receptors. Receptors that combine phasic and tonic coding convey extremely complicated sensory information; receptors that monitor the positions and movements of joints are in this category.

Central Processing and Adaptation

Adaptation is a reduction in sensitivity in the presence of a constant stimulus. **Peripheral (sensory) adaptation** occurs when the receptors or sensory neurons alter their levels of activity. The receptor responds strongly at first, but thereafter the activity along the afferent fiber gradually declines, in part because of synaptic fatigue. This response is characteristic of phasic receptors, which are also called **fast-adapting receptors**. Tonic receptors show little peripheral adaptation and so are called **slow-adapting receptors**.

Adaptation also occurs inside the CNS along the sensory pathways. For example, a few seconds after exposure to a new smell, conscious awareness of the stimulus virtually disappears, although the sensory neurons are still quite active. This process is known as **central adaptation**. Central adaptation usually involves the inhibition of nuclei along a sensory pathway. At the subconscious level, central adaptation further restricts the amount of detail arriving at the cerebral cortex. Most of the incoming sensory information is processed in centers along the spinal cord or brain stem, potentially triggering involuntary reflexes. Only about 1% of the information provided by afferent fibers reaches the cerebral cortex and our conscious awareness.

Sensory Limitations

Our sensory receptors provide a constant detailed picture of our bodies and our surroundings. This picture is, however, incomplete for several reasons:

1. Humans do not have receptors for every possible stimulus.
2. Our receptors have characteristic ranges of sensitivity.
3. A stimulus must be interpreted by the CNS. Our perception of a particular stimulus is an interpretation and not always a reality.

This discussion has introduced basic concepts of receptor function and sensory processing. We can now describe and discuss the receptors responsible for the general senses.

✓ CONCEPT CHECK

- What different types of stimuli may activate free nerve endings?
- Contrast tonic and phasic receptors.
- What is a sensation?
- Which sensations are grouped under the banner "general senses"?

The General Senses

Receptors for the general senses are scattered throughout the body and are relatively simple in structure. A simple classification scheme divides them into *exteroceptors* and *interoceptors*. **Exteroceptors** provide information about the external environment; **interoceptors** monitor conditions inside the body.

A more detailed classification system divides the general sensory receptors into four types according to the nature of the stimulus that excites them:

1. **Nociceptors** (nō-sē-SEP-torz; *noceo*, hurt) respond to a variety of stimuli usually associated with tissue damage. Receptor activation causes the sensation of pain.
2. **Thermoreceptors** respond to changes in temperature.
3. **Mechanoreceptors** are stimulated or inhibited by physical distortion, contact, or pressure on their cell membranes.
4. **Chemoreceptors** monitor the chemical composition of body fluids and respond to the presence of specific molecules.

Each class of receptors has distinct structural and functional characteristics. You will find that some tactile receptors and mechanoreceptors are identified by eponyms (commemorative names). Contemporary anatomists have proposed differing alternatives for these names, and, as yet, no standardization or consensus exists. More significantly, *none* of the alternative names has been widely accepted in the primary literature (professional, technical, or clinical journals or reports). To avoid later confusion, this chapter will use eponyms whenever there is no generally accepted or widely used alternative.

Nociceptors [FIGURES 18.2/18.3a]

Nociceptors, or pain receptors, are especially common in the superficial portions of the skin (Figure 18.3a●), in joint capsules, within the periostea of bones, and around the walls of blood vessels. There are few nociceptors in other deep tissues or in most visceral organs. Pain receptors are free nerve endings with large receptive fields. As a consequence, it is often difficult to determine the exact origin of a painful sensation.

There are three types of nociceptors: (1) receptors sensitive to extremes of temperature, (2) receptors sensitive to mechanical damage, and (3) receptors sensitive to dissolved chemicals, such as those released by injured cells. However, very strong temperature, pressure, or chemical stimuli will excite all three receptor types.

Sensations of **fast pain**, or *prickling pain*, are produced by deep cuts or similar injuries. These sensations reach the CNS very quickly, where they often trigger somatic reflexes. They are also relayed to the primary sensory cortex and so receive conscious attention. Painful sensations cease only after tissue damage has ended. However, central adaptation may reduce *perception* of the pain while the pain receptors are still stimulated. † *The Control of Pain p. 795*

Sensations of **slow pain**, or *burning and aching pain*, result from the same types of injuries as fast pain sensations. However, sensations of slow pain begin later and persist longer than sensations of fast pain. For example, a cut on the hand would produce an immediate awareness of fast pain, followed somewhat later by the ache of slow pain. Slow pain sensations cause a generalized activation of the reticular formation and thalamus. The individual is aware of the pain but has only a general idea of the area affected. A person experiencing slow pain sensations will often palpate the area in an attempt to locate the source of the pain.

Pain sensations from visceral organs are often perceived as originating in more superficial regions; generally, regions innervated by the same respective spinal nerves. The precise mechanism responsible for this **referred pain** remains to be determined, but several clinical examples are shown in Figure 18.2●. Cardiac pain, for example, is often perceived as originating in the upper chest and left arm.

Thermoreceptors

Temperature receptors are found in the dermis of the skin, in skeletal muscles, in the liver, and in the hypothalamus. Cold receptors are three or four times more numerous than warm receptors. The receptors are free nerve endings, and there are no known structural differences between cold and warm thermoreceptors.

Temperature sensations are conducted along the same pathways that carry pain sensations. They are sent to the reticular formation, the thalamus, and (to a lesser extent) the primary sensory cortex. Thermoreceptors

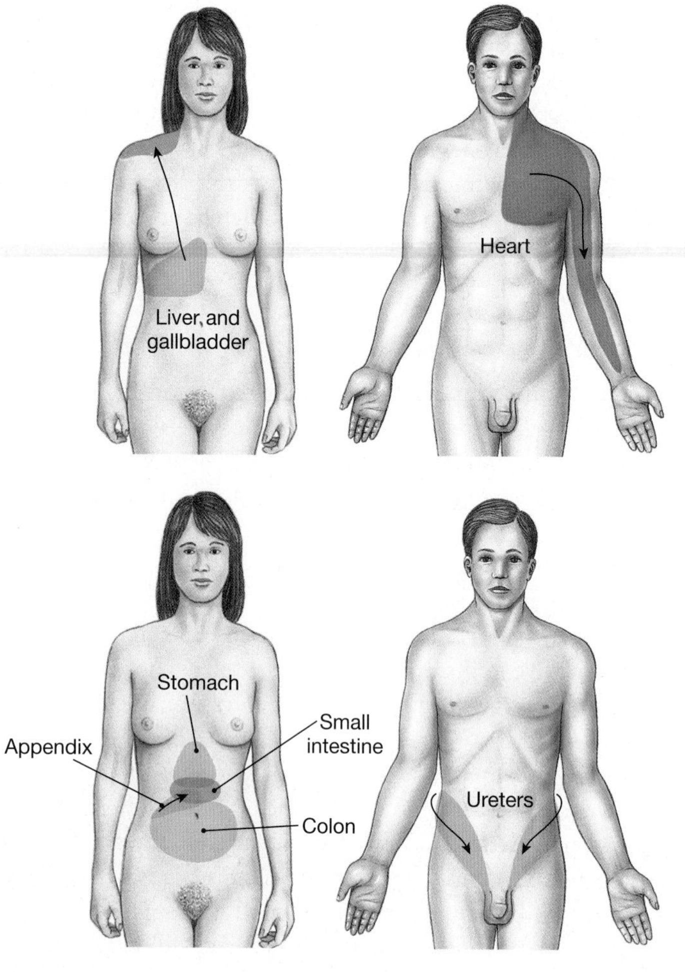

FIGURE 18.2 **Referred Pain**

Pain sensations originating in visceral organs are often perceived as involving specific regions of the body surface innervated by the same spinal nerves.

are phasic receptors. They are very active when the temperature is changing, but they quickly adapt to a stable temperature. When you enter an air-conditioned classroom on a hot summer day or a warm lecture hall on a brisk fall evening, the temperature seems unpleasant at first, but the discomfort fades as adaptation occurs.

Mechanoreceptors

Mechanoreceptors are sensitive to stimuli that stretch, compress, twist, or distort their cell membranes. There are three classes of mechanoreceptors: (1) *tactile receptors* provide sensations of touch, pressure, and vibration; (2) *baroreceptors* (bar-ō-rē-SEP-torz; *baro-*, pressure) detect pressure changes in the walls of blood vessels and in portions of the digestive, reproductive, and urinary tracts; and (3) *proprioceptors* monitor the positions of joints and muscles and are the most complex of the general sensory receptors.

TACTILE RECEPTORS [FIGURE 18.3/TABLE 18.1]

Tactile receptors range in structural complexity from the simple free nerve endings to specialized sensory complexes with accessory cells and supporting structures. **Fine touch** and **pressure receptors** provide detailed information about a source of stimulation, including its exact location, shape, size, texture, and movement. These receptors are extremely sensitive and have relatively narrow receptive fields. **Crude touch** and **pressure receptors** provide poor localization and little additional information about the stimulus.

Figure 18.3● shows six different types of tactile receptors in the skin. They can be subdivided into two groups: *unencapsulated receptors* (free nerve endings, tactile disc, and root hair plexus) and *encapsulated receptors* (tactile corpuscle, lamellated corpuscle, and Ruffini corpuscle).

Unencapsulated Receptors [FIGURE 18.3a–c] **Free nerve endings** are common in the papillary layer of the dermis (Figure 18.3a●). In sensitive areas, the dendritic branches penetrate the epidermis and contact *Merkel cells* in the stratum germinativum. p. 90 Each Merkel cell communicates with a sensory neuron across a chemical synapse that involves an expanded nerve terminal known as a **tactile disc** (also termed *Merkel's* (MER-kelz) *disc*) (Figure 18.3b●). Merkel cells are sensitive to fine touch and pressure. They are tonically active, extremely sensitive, and have narrow receptive fields. Free nerve endings are also associated with hair follicles.

The free nerve endings of the **root hair plexus** monitor distortions and movements across the body surface (Figure 18.3c●). When the hair is displaced, the movement of the follicle distorts the sensory dendrites and produces action potentials in the afferent fiber. These receptors adapt rapidly, so they are best at detecting initial contact and subsequent movements. For example, most people feel their clothing only when they move or when they consciously focus attention on tactile sensations from their skin.

Encapsulated Receptors [FIGURE 18.3d–f and TABLE 18.1] Large, oval **tactile corpuscles** (also called *Meissner's* (MĪS-nerz) *corpuscles*), are found where tactile sensitivities are extremely well developed (Figure 18.3d●). They are especially common at the eyelids, lips, fingertips, nipples, and external genitalia. The dendrites are highly coiled and interwoven, and they are surrounded by modified Schwann cells. A fibrous capsule surrounds the entire complex and anchors it within the dermis. Tactile corpuscles detect light touch, movement, and vibration; they adapt to stimulation within a second after contact.

Ruffini (ru-FĒ-nē) **corpuscles**, located in the dermis, are also sensitive to pressure and distortion of the skin, but they are tonically active and show little if any adaptation. The capsule surrounds a core of collagen fibers that are continuous with those of the surrounding dermis. Dendrites within the capsule are interwoven around the collagen fibers (Figure 18.3e●). Any tension or distortion of the dermis tugs or twists the fibers within the capsule, and this change stretches or compresses the attached dendrites and alters the activity in the myelinated afferent fiber.

Lamellated corpuscles (also called *pacinian* (pa-SIN-ē-an) *corpuscles*), are considerably larger encapsulated receptors (Figure 18.3f●). The dendritic process lies within a series of concentric cellular layers. These layers shield the dendrite from virtually every source of stimulation other than direct pressure. Lamellated corpuscles respond to heavy pressure but are most sensitive to pulsing or vibrating stimuli. Although both lamellated corpuscles and Ruffini corpuscles respond to pressure, the lamellated corpuscles adapt rapidly while the Ruffini corpuscles adapt quite slowly. These receptors are scattered throughout the dermis, notably in the fingers, breasts, and external genitalia. They are also encountered in the superficial and deep fasciae, in periostea and joint capsules, in mesenteries, in the pancreas, and in the walls of the urethra and urinary bladder.

Table 18.1 summarizes the functions and characteristics of the six tactile receptors discussed. The distribution of tactile sensations inside the CNS is via the posterior column and spinothalamic pathways. p. 428 Tactile sensitivities may be altered by peripheral infection, disease processes,

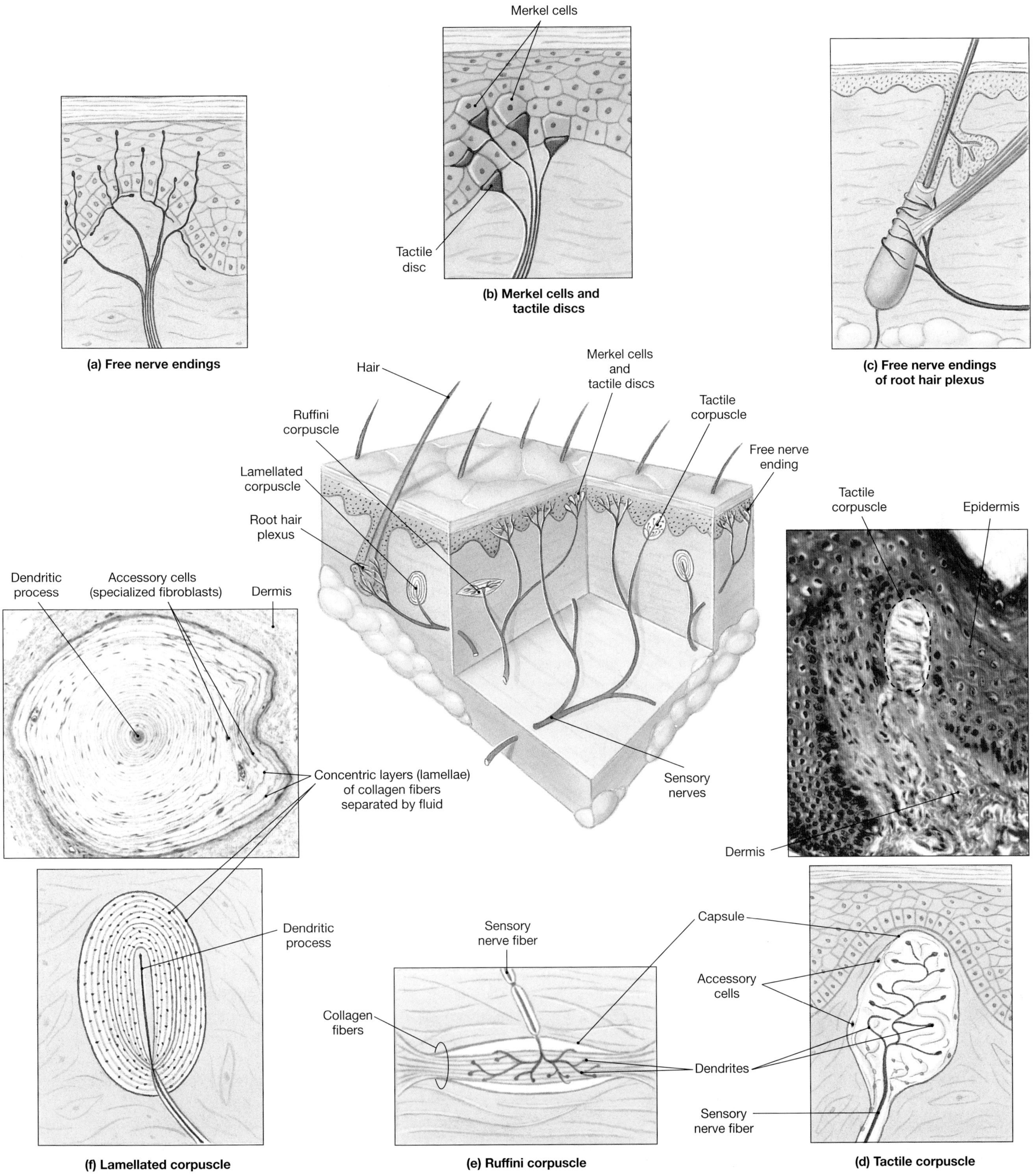

FIGURE 18.3 **TACTILE RECEPTORS IN THE SKIN**

The location and general appearance of six important tactile receptors. (**a**) Free nerve endings. (**b**) Merkel cells and tactile discs. (**c**) Free nerve endings of root hair plexus. (**d**) Tactile corpuscle; the capsule boundary in the micrograph is indicated by a dashed line. (LM × 550) (**e**) Ruffini corpuscle. (**f**) Lamellated corpuscle. (LM × 125) and damage to sensory afferents or central pathways, and there are important clinical tests that evaluate tactile sensitivity. *Assessment of Tactile Sensitivities p. 795*

✓ CONCEPT CHECK

- When the nociceptors in your hand are stimulated, what sensation do you perceive?
- What would happen to an individual if the information from proprioceptors in the lower limbs were blocked from reaching the CNS?
- What are the three classes of mechanoreceptors?
- What receptor type is being stimulated when carbon dioxide concentration is being monitored in the blood?

BARORECEPTORS [FIGURE 18.4]

Baroreceptors are stretch receptors that monitor changes in pressure. The receptor consists of free nerve endings that branch within the elastic tissues in the wall of a hollow organ, a blood vessel, or the respiratory, digestive, or urinary tracts. When the pressure changes, the elastic walls of these tubes or organs stretch or recoil. These changes in shape distort the dendritic branches and alters the rate of action-potential generation. Baroreceptors respond immediately to a change in pressure.

Baroreceptors monitor blood pressure in the walls of major vessels, including the carotid artery (at the *carotid sinus*) and the aorta (at the *aortic sinus*). The information provided by these baroreceptors plays a major role in regulating cardiac function and adjusting blood flow to vital tissues. Baroreceptors in the lungs monitor the degree of lung expansion. This information is relayed to the respiratory rhythmicity center, which sets the pace of respiration. Baroreceptors in the urinary and digestive tracts trigger a variety of visceral reflexes, such as urination. Figure 18.4● provides examples of baroreceptor locations and functions.

PROPRIOCEPTORS

Proprioceptors monitor the position of joints, the tension in tendons and ligaments, and the state of muscular contraction. Generally, proprioceptors do not adapt to constant stimulation. *Muscle spindles*, introduced in Chapter 9, are proprioceptors that monitor the length of skeletal muscles. p. 255 **Golgi tendon organs** monitor the tension in tendons during muscle contraction. Joint capsules are richly supplied with free nerve endings that detect tension, pressure, and movement at the joint. Your sense of body position results from the integration of information from these proprioceptors with information from the inner ear.

TABLE 18.1 TOUCH AND PRESSURE RECEPTORS

Sensation	Receptor	Responds to
Fine touch	Free nerve ending	Light contact with skin
	Tactile disc	As above
	Root hair plexus	Initial contact with hair shaft
Pressure and vibration	Tactile corpuscle	Initial contact and low-frequency vibrations
	Lamellated corpuscle	Initial contact (deep) and high-frequency vibrations
Deep pressure	Ruffini corpuscle	Stretching and distortion of the dermis

Chemoreceptors [FIGURE 18.5]

Chemoreceptors are specialized neurons that can detect small changes in the concentration of specific chemicals or compounds. In general, chemoreceptors respond only to water-soluble and lipid-soluble substances that are dissolved in the surrounding fluid.

The locations of important chemosensory receptors are indicated in Figure 18.5●. Neurons within the respiratory centers of the brain respond to the concentration of hydrogen ions (pH) and carbon dioxide (P_{CO_2}) in the cerebrospinal fluid. Chemoreceptive neurons are found within the **carotid bodies**, near the origin of the internal carotid arteries on each side of the neck, and in the **aortic bodies** between the major branches of the aortic arch. These receptors monitor the carbon dioxide (P_{CO_2}) and oxygen (P_{O_2}) concentration of arterial blood. The afferent fibers leaving the carotid and aortic bodies reach the respiratory centers by traveling along the ninth (glossopharyngeal) and tenth (vagus) cranial nerves. These chemoreceptors play an important role in the reflexive control of respiration and cardiovascular function.

Olfaction (Smell) [FIGURE 18.6]

The sense of smell, more precisely called **olfaction**, is provided by paired **olfactory organs**. These organs are located in the nasal cavity on either side of the nasal septum. The olfactory organs (Figure 18.6●) consist of:

- a specialized neuroepithelium, the **olfactory epithelium**, contains the **olfactory receptors**, **supporting cells**, and **basal cells** (*stem cells*);
- an underlying layer of loose connective tissue known as the *lamina propria*. This layer contains (1) **olfactory glands**, also called *Bowman's glands*, which produce a thick, pigmented mucus, (2) blood vessels, and (3) nerves.

The olfactory epithelium covers the inferior surface of the cribriform plate and the superior portions of the nasal septum and superior nasal conchae of the ethmoid. p. 148

When air is drawn in through the nose, the nasal conchae produce turbulent airflow that brings airborne compounds into contact with the olfactory organs. A normal, relaxed inhalation provides a small sample of the inhaled air (around 2%) to the olfactory organs. Sniffing repeatedly increases the flow of air across the olfactory epithelium, intensifying the stimulation of the olfactory receptors. Once compounds have reached the olfactory organs, water-soluble and lipid-soluble materials must diffuse into the mucus before they can stimulate the olfactory receptors.

Olfactory Receptors [FIGURE 18.6b]

The olfactory receptors are highly modified neurons. The apical, dendritic portion of each receptor forms a prominent knob that projects beyond the epithelial surface and into the nasal cavity (Figure 18.6b●). That projection provides a base for up to 20 cilia that extend into the surrounding mucus, exposing their considerable surface area to the dissolved chemical compounds. Somewhere between 10 and 20 million olfactory receptors are packed into an area of roughly 5 cm^2. Olfactory reception occurs on the surface of an olfactory cilium, through binding to specific membrane receptors. When the odorous substance binds to its

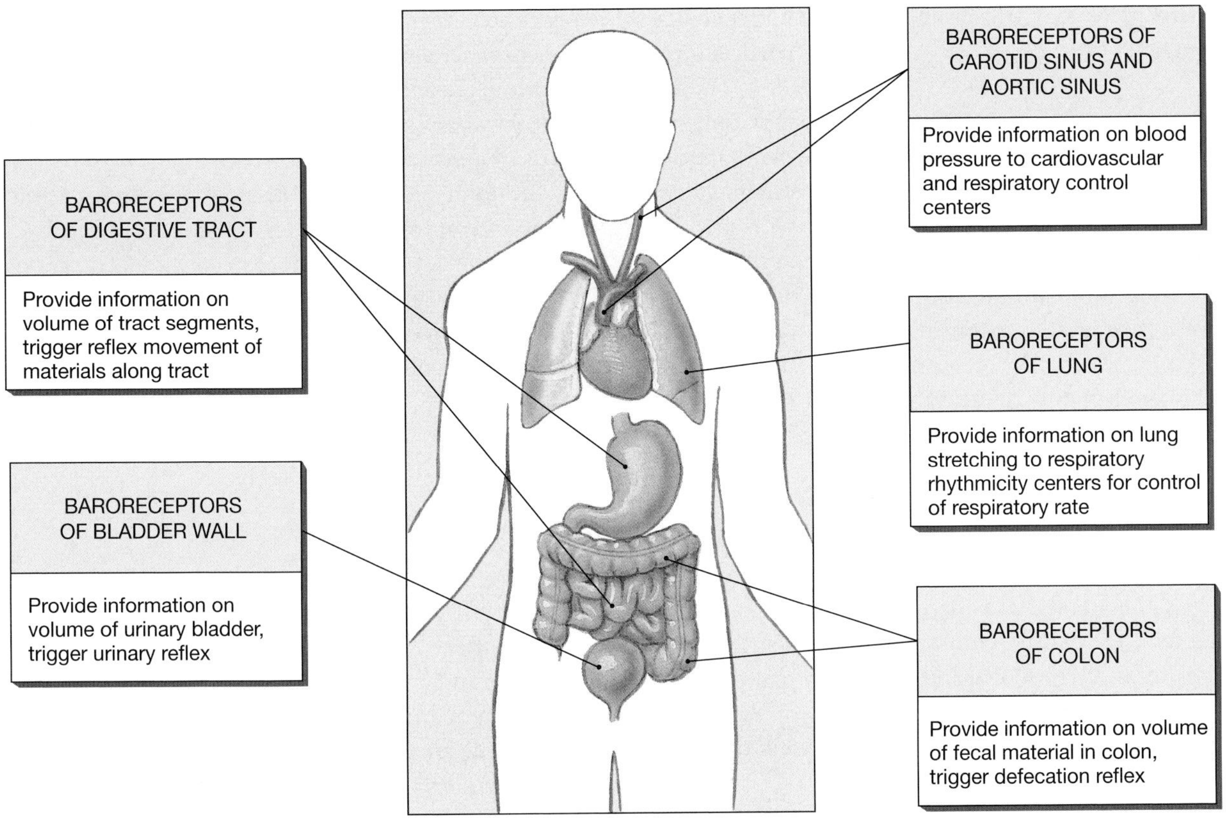
BARORECEPTORS OF CAROTID SINUS AND AORTIC SINUS
Provide information on blood pressure to cardiovascular and respiratory control centers
BARORECEPTORS OF DIGESTIVE TRACT
Provide information on volume of tract segments, trigger reflex movement of materials along tract
BARORECEPTORS OF LUNG
Provide information on lung stretching to respiratory rhythmicity centers for control of respiratory rate
BARORECEPTORS OF BLADDER WALL
Provide information on volume of urinary bladder, trigger urinary reflex
BARORECEPTORS OF COLON
Provide information on volume of fecal material in colon, trigger defecation reflex

FIGURE 18.4 **Baroreceptors and the Regulation of Autonomic Functions**

Baroreceptors provide information essential to the regulation of autonomic activities, including respiration, digestion, urination, and defecation.

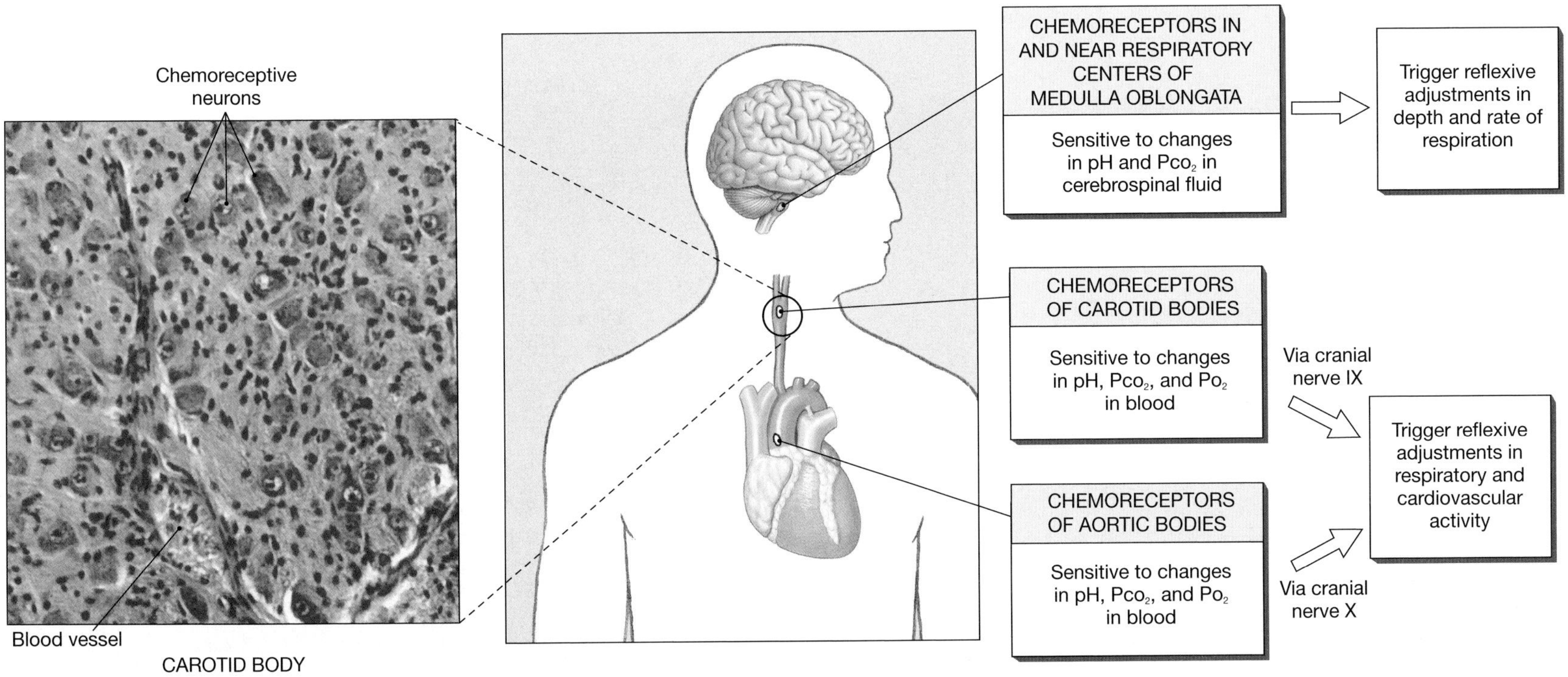
Chemoreceptive neurons
Blood vessel
CAROTID BODY
CHEMORECEPTORS IN AND NEAR RESPIRATORY CENTERS OF MEDULLA OBLONGATA
Sensitive to changes in pH and Pco₂ in cerebrospinal fluid
Trigger reflexive adjustments in depth and rate of respiration
CHEMORECEPTORS OF CAROTID BODIES
Sensitive to changes in pH, Pco₂, and Po₂ in blood
Via cranial nerve IX
Trigger reflexive adjustments in respiratory and cardiovascular activity
CHEMORECEPTORS OF AORTIC BODIES
Sensitive to changes in pH, Pco₂, and Po₂ in blood
Via cranial nerve X

FIGURE 18.5 **Chemoreceptors**

Chemoreceptors are found both inside the CNS, on the ventrolateral surfaces of the medulla oblongata, and in the aortic and carotid bodies. These receptors are involved in the autonomic regulation of respiratory and cardiovascular function. The micrograph shows the histological appearance of the chemoreceptive neurons in the carotid body. (LM × 1500) receptor, the receptor membrane depolarizes. This may trigger an action potential in the axon of the olfactory receptor.

Olfactory Pathways [FIGURE 18.6]

The olfactory system is very sensitive. As few as four molecules of an odorous substance can activate an olfactory receptor. However, the activation of an afferent fiber does not guarantee a conscious awareness of the stimulus. Considerable convergence occurs along the olfactory pathway, and inhibition at the intervening synapses can prevent the sensations from reaching the cerebral cortex.

Axons leaving the olfactory epithelium collect into 20 or more bundles that penetrate the cribriform plate of the ethmoid bone to synapse on neurons within the olfactory bulbs (Figure 18.6•). This collection of nerve bundles constitutes the first cranial nerve (N I). The axons of the second-order neurons in the olfactory bulb travel within the olfactory tract to reach the olfactory cortex, the hypothalamus, and portions of the limbic system.

Olfactory sensations are the only sensations that reach the cerebral cortex without first synapsing in the thalamus. The extensive limbic and hypothalamic connections help to explain the profound emotional and behavioral responses that can be produced by certain smells, such as perfumes.

Olfactory Discrimination

The olfactory system can make subtle distinctions between thousands of chemical stimuli. We know that there are at least 50 different "primary smells." No apparent structural differences exist among the olfactory cells, but the epithelium as a whole contains receptor populations with distinctly different sensitivities. The CNS interprets the smell on the basis of the particular pattern of receptor activity.

The olfactory receptor cells are the best known examples of neuronal replacement in the adult human. (Neuron replacement can also occur in the hippocampus, but the regulatory mechanisms are unknown.) Despite ongoing replacement, the total number of olfactory receptors declines with age, and the remaining receptors become less sensitive. As a result, elderly individuals have difficulty detecting odors in low concentrations. This decline in the number of receptors accounts for Grandmother's tendency to apply perfume in excessive quantities and explains why Grandfather's aftershave seems so overdone; they must apply more to be able to smell it themselves.

Gustation (Taste) [FIGURE 18.7a]

Gustation, or taste, provides information about the foods and liquids that we consume. The **gustatory** (GUS-ta-tō-rē) **receptors** (taste receptors) are distributed over the dorsal surface of the tongue (Figure 18.7a•) and adjacent portions of the pharynx and larynx. By adulthood the taste receptors on the epithelium of the pharynx and larynx have decreased in importance, and the *taste buds* of the tongue are the primary gustatory receptors.

Taste buds lie along the sides of epithelial projections called **papillae** (pa-PIL-lē; *papilla*, nipple-shaped mound). There are three types of papillae on the human tongue: **filiform** (*filum*, thread), **fungiform** (*fungus*, mushroom), and **circumvallate** (sir-kum-VAL-āt; *circum-*, around + *vallum*, wall). There are regional differences in the distribution of the papillae (Figure 18.7a•).

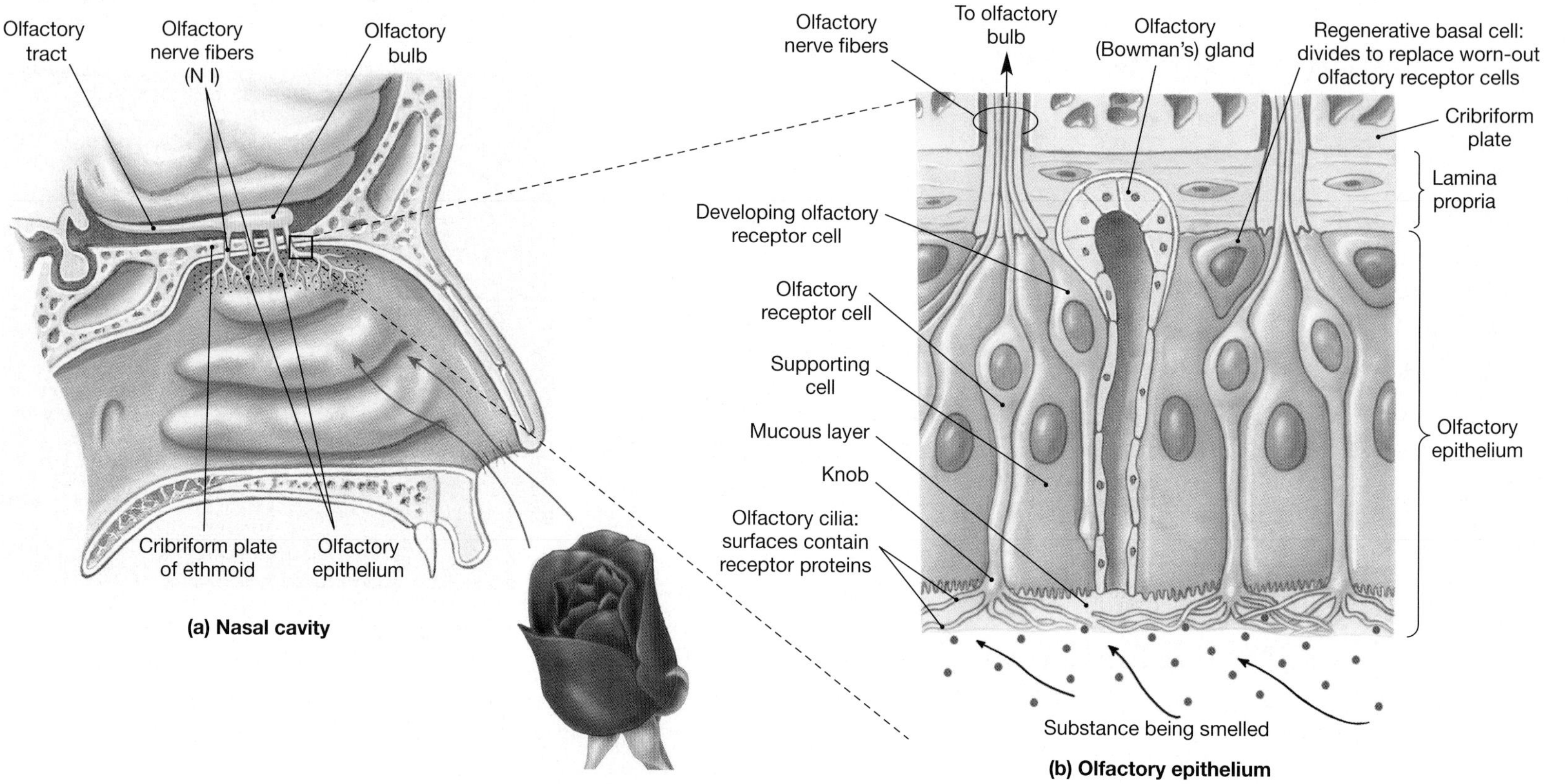

FIGURE 18.6 **THE OLFACTORY ORGANS**

(**a**) The distribution of the olfactory receptors on the left side of the nasal septum is shown by the shading. (**b**) A detailed view of the olfactory epithelium.

Gustatory Receptors [FIGURE 18.7b,c]

The taste receptors are clustered within individual **taste buds** (Figure 18.7b,c●). Each taste bud contains around 40 slender receptors, called **gustatory cells**. Each taste bud contains at least 3 different gustatory cell types, plus *basal cells* that are probably stem cells. A typical gustatory cell remains intact for only 10–12 days. Taste buds are recessed into the surrounding epithelium and isolated from the relatively unprocessed oral contents. Each gustatory cell extends slender microvilli, sometimes called *taste hairs*, into the surrounding fluids through a narrow opening, the **taste pore**.

The small fungiform papillae each contain about five taste buds, while the large circumvallate papillae, which form a V shape near the posterior margin of the tongue, contain as many as 100 taste buds per papilla. The typical adult has over 10,000 taste buds.

The mechanism of gustatory reception appears to parallel that of olfaction. Dissolved chemicals contacting the taste hairs provide the stimulus that produces a change in the transmembrane potential of the taste cell. Stimulation of the gustatory cell results in action potentials in the afferent fiber.

Gustatory Pathways [FIGURE 18.8]

Taste buds are monitored by cranial nerves VII (facial), IX (glossopharyngeal), and X (vagus) (Figure 18.8●). The sensory afferents synapse within the **nucleus solitarius** of the medulla oblongata; the axons of the postsynaptic neurons then enter the medial lemniscus. p. 428 After another synapse in the thalamus, the information is projected to the appropriate regions of the gustatory cortex.

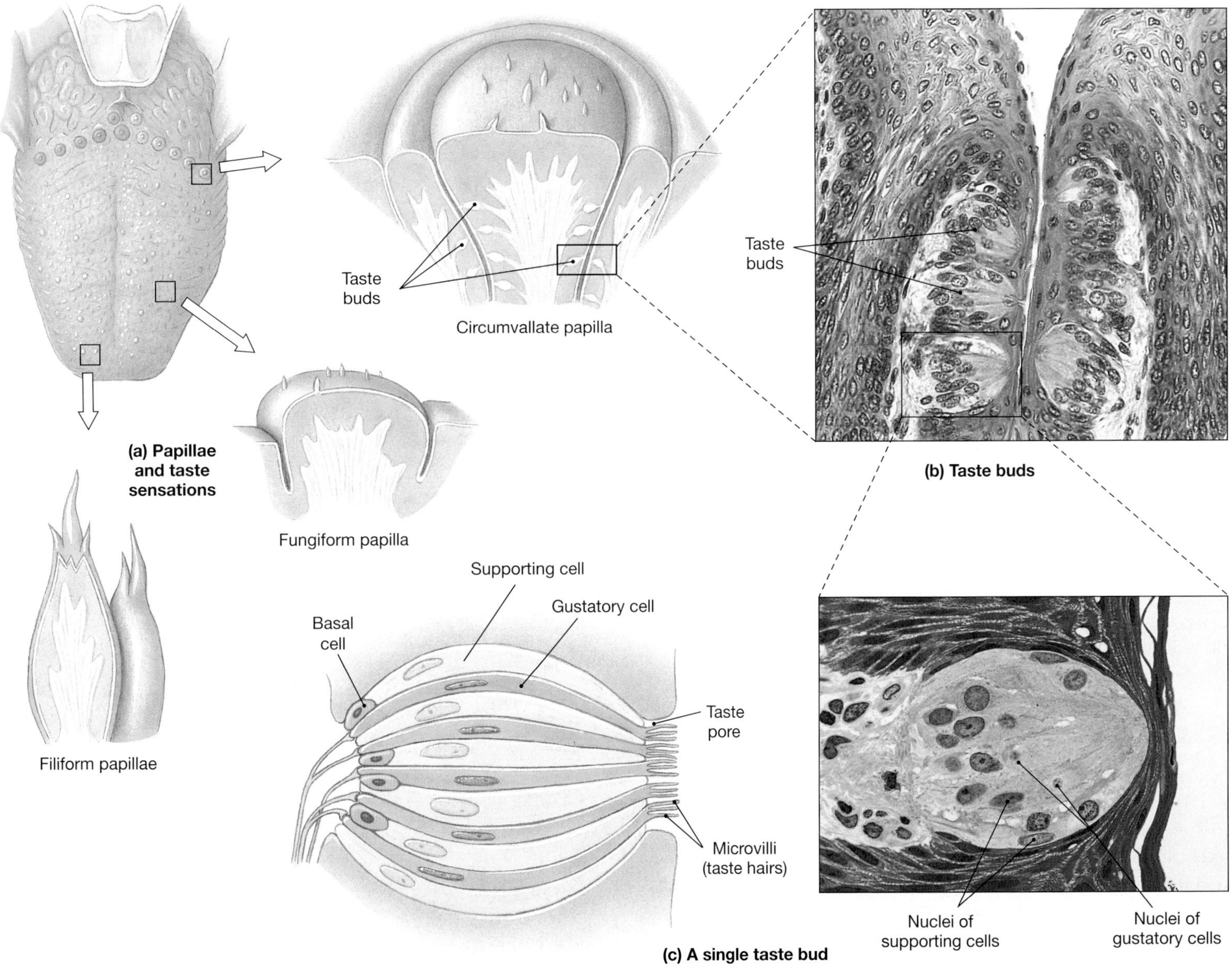

FIGURE 18.7 **GUSTATORY RECEPTION**

(**a**) Gustatory receptors are found in taste buds that form pockets in the epithelium of the fungiform and circumvallate papillae.
(**b**) Photomicrograph of taste buds in a circumvallate papilla. (LM × 280) (**c**) A taste bud, showing receptor cells and supporting cells. The diagrammatic view shows details of the taste pore not visible in the light micrograph. (LM × 650)

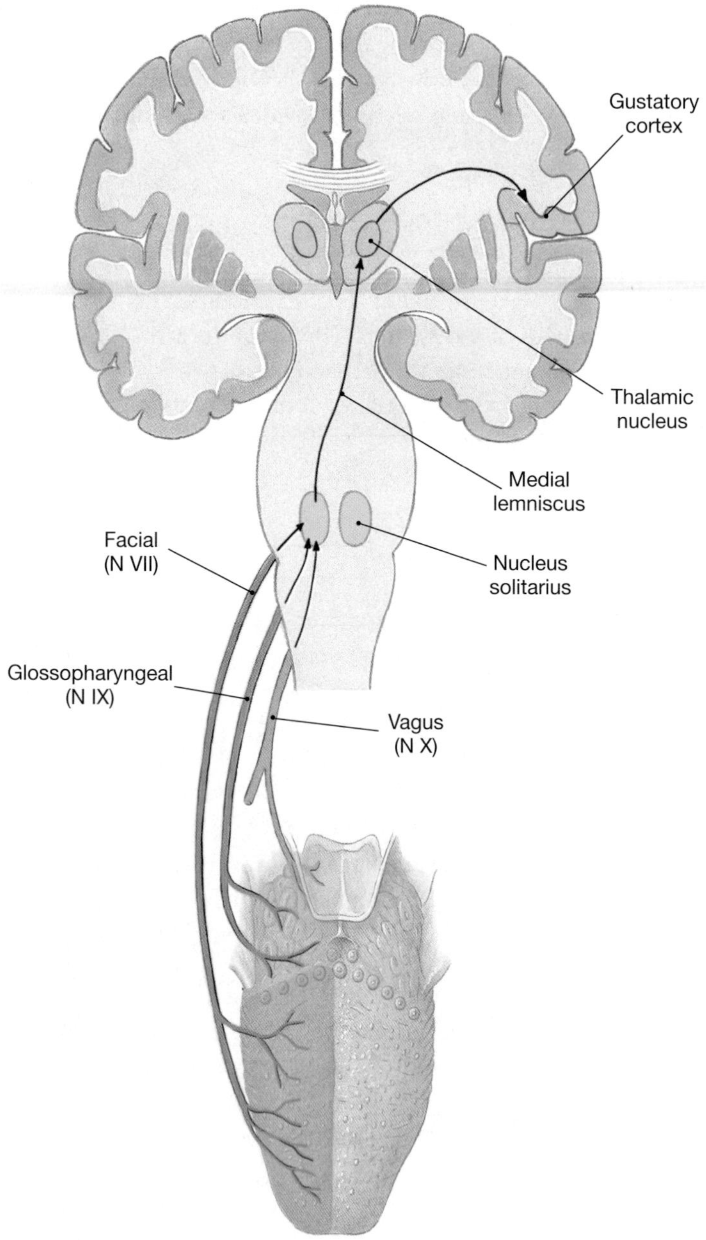

FIGURE 18.8 **Gustatory Pathways**

Three cranial nerves (VII, IX, and X) carry gustatory information to the gustatory cortex of the cerebrum.

A conscious perception of taste involves correlating the information received from the taste buds with other sensory data. Information concerning the general texture of the food, together with taste-related sensations of "peppery" or "burning hot," is provided by sensory afferents in the trigeminal nerve (N V). In addition, the level of stimulation from the olfactory receptors plays an overwhelming role in taste perception. We are several thousand times more sensitive to "tastes" when our olfactory organs are fully functional. When you have a cold, airborne molecules cannot reach the olfactory receptors, and meals taste dull and unappealing. This reduction in taste perception will occur even though the taste buds may be responding normally.

Gustatory Discrimination

You are probably already familiar with the four **primary taste sensations**: sweet, salt, sour, and bitter. At least two additional tastes have been discovered in humans:

- *Umami.* **Umami** (oo-MAH-mē) is a pleasant taste that is characteristic of beef broth and chicken broth. This taste is produced by receptors sensitive to the presence of amino acids, especially glutamate, small peptides, and nucleotides. The distribution of these receptors is not known in detail, but they are present in taste buds of the circumvallate papillae.
- *Water.* Most people say that water has no flavor. However, research on humans and other vertebrates has demonstrated the presence of **water receptors**, especially in the pharynx. Their sensory output is processed in the hypothalamus and affects several systems that deal with water balance and the regulation of blood pressure.

The threshold for receptor stimulation varies for each of the primary taste sensations, and the taste receptors respond most readily to unpleasant rather than to pleasant stimuli. For example, we are almost a thousand times more sensitive to acids, which give a sour taste, than to either sweet or salty chemicals, and we are a hundred times more sensitive to bitter compounds than to acids. This sensitivity has survival value, for acids can damage the mucous membranes of the mouth and pharynx, and many potent biological toxins produce an extremely bitter taste.

Our tasting abilities change with age. We begin life with over 10,000 taste buds, but the number begins declining dramatically by age 50. The sensory loss becomes especially significant because aging individuals also experience a decline in the population of olfactory receptors. As a result, many elderly people find that their food tastes bland and unappetizing, whereas children often find the same foods too spicy.

✓ CONCEPT CHECK

- What are the primary taste sensations?
- Why does food taste bland when you have a cold?
- Where are taste receptors located?
- List the three types of papillae on the tongue.

Equilibrium And Hearing [FIGURE 18.9]

The ear is divided into three anatomical regions; the *external ear*, the *middle ear*, and the *inner ear* (Figure 18.9●). The external ear is the visible portion of the ear, and it collects and directs sound waves to the *eardrum*. The *middle ear* is a chamber located within the petrous portion of the temporal bone. Structures within the middle ear amplify sound waves and transmit them to an appropriate portion of the inner ear. The *inner ear* contains the sensory organs for equilibrium and hearing.

The External Ear [FIGURES 18.9/18.10a,b]

The **external ear** includes the flexible **auricle**, or *pinna*, which is supported by elastic cartilage. The auricle of the ear surrounds the **external acoustic meatus**. The auricle protects the external acoustic meatus and provides directional sensitivity to the ear by blocking or facilitating the passage of sound into the external auditory canal. The external acoustic meatus is a passageway that ends at the eardrum, called the **tympanic membrane**, or **tympanum** (Figures 18.9 and 18.10a●). The tympanic membrane is a thin, semitransparent connective tissue sheet (Figure 18.10b●) that separates the external ear from the middle ear.

The tympanic membrane is very delicate. The auricle and the narrow external acoustic meatus provide some protection from accidental injury to the tympanic membrane. In addition, **ceruminous glands** distributed along the external auditory meatus secrete a waxy material, and many small, outwardly projecting hairs help deny access to foreign objects or insects. The waxy secretion of the ceruminous glands, called **cerumen**, also slows the growth of microorganisms in the external acoustic meatus and reduces the chances of infection.

The Middle Ear [FIGURES 18.9/18.10]

The **middle ear** consists of an air-filled space, the **tympanic cavity**, which contains the *auditory ossicles* (Figures 18.9 and 18.10●). The tympanic cavity is separated from the external acoustic meatus by the tympanic membrane, but it communicates with the nasopharynx through the auditory canal and with the mastoid sinuses through a number of small and variable connections. p. 145 The **auditory tube** is also called the *pharyngotympanic tube* or the *Eustachian tube.* This tube, about 4.0 cm in length, penetrates the petrous part of the temporal bone within the *musculotubal canal.* p. 145 The connection to the tympanic cavity is relatively narrow and supported by elastic cartilage. The opening into the nasopharynx is relatively broad and funnel-shaped. The auditory tube serves to equalize the pressure in the middle ear cavity with external, atmospheric pressure. Pressure must be equal on both sides of the tympanic membrane or there will be a painful distortion of the membrane. Unfortunately, the auditory tube can also allow microorganisms to travel from the nasopharynx into the tympanic cavity, resulting in an "ear infection." Such infections are especially common in children, because their auditory tubes are relatively short and broad, as compared to those of adults.

THE AUDITORY OSSICLES [FIGURES 18.9/18.10]

The tympanic cavity contains three tiny ear bones collectively called **auditory ossicles.** p. 146 These ear bones, the smallest bones in the body, connect the tympanic membrane with the receptor complex of the inner ear (Figures 18.9 and 18.10●). The three auditory ossicles are the *malleus*, the

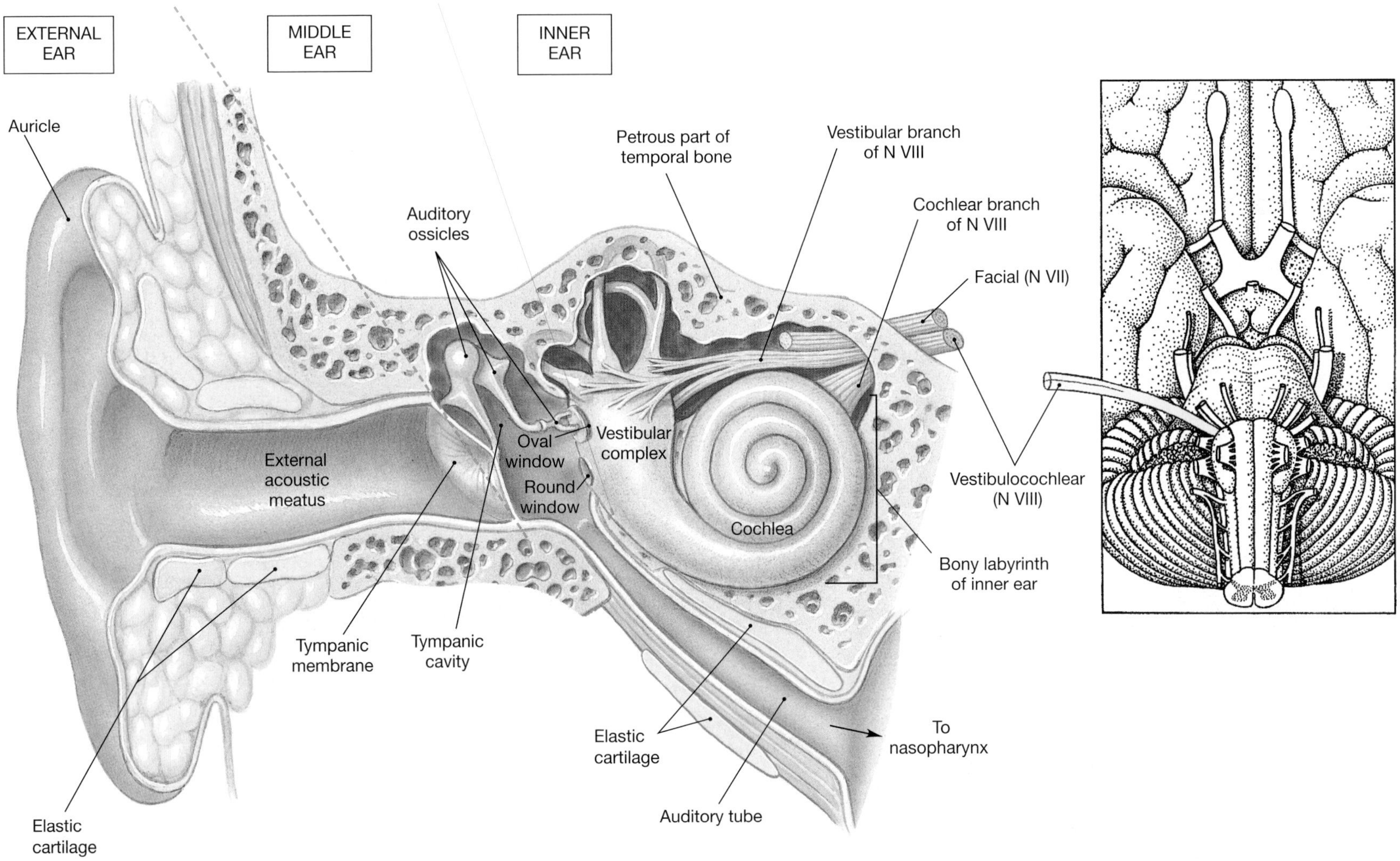

FIGURE 18.9 **ANATOMY OF THE EAR**

A general orientation to the external, middle, and inner ear.

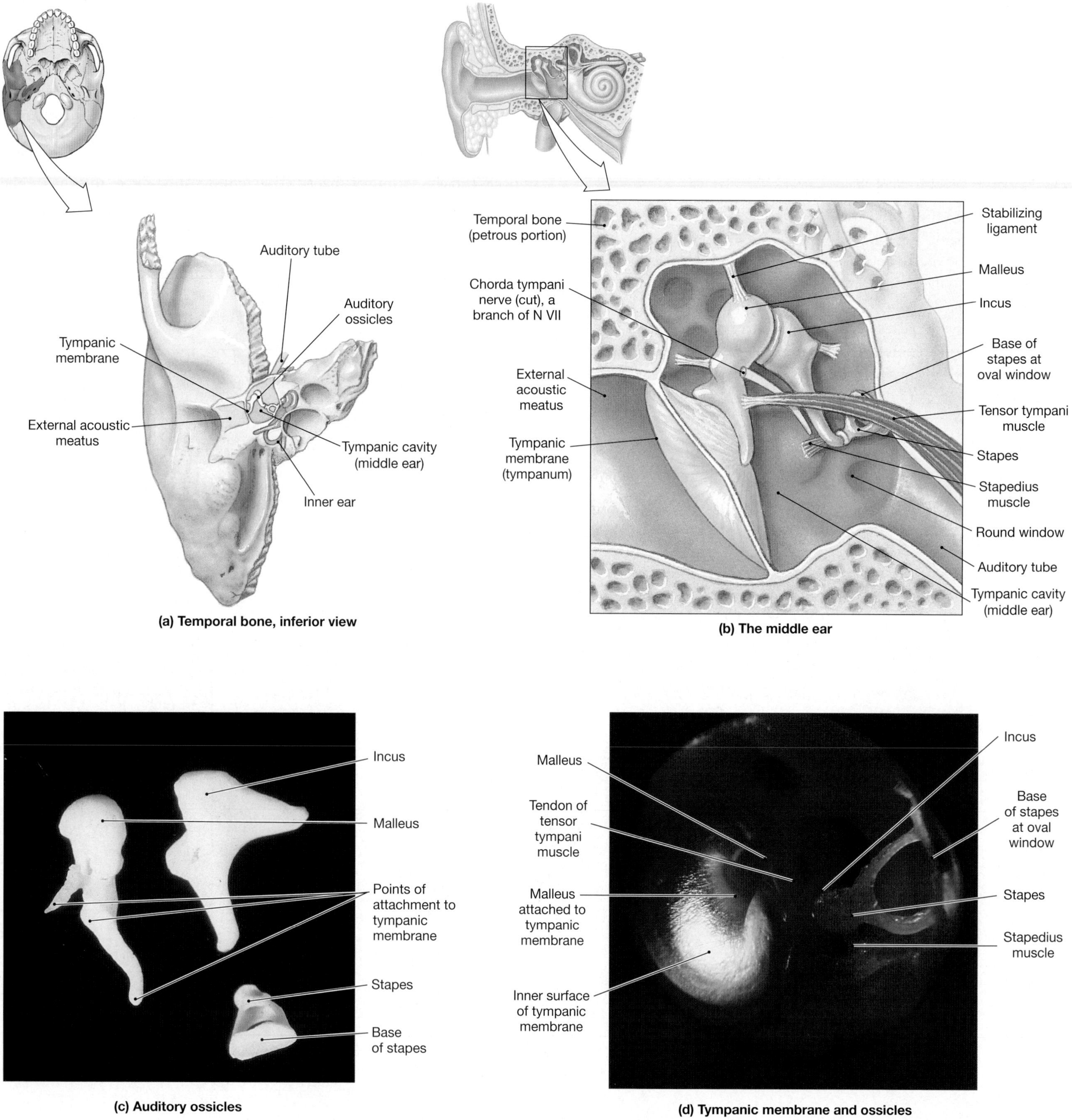

FIGURE 18.10 **THE MIDDLE EAR**

(**a**) Inferior view of the right temporal bone, drawn as if transparent, to show the location of the middle and inner ear. (**b**) Structures within the middle ear cavity. (**c**) The isolated auditory ossicles. (**d**) The tympanic membrane and auditory ossicles as seen through a fiber-optic tube inserted along the auditory canal and into the middle ear cavity.

CLINICAL BRIEF

Otitis Media and Mastoiditis

Otitis media is an infection of the middle ear, most often of bacterial origin. *Acute otitis media* typically affects infants and children and is less common in adults. The pathogens usually gain access via the auditory tube, usually during an upper respiratory infection. As the pathogen population rises in the tympanic cavity, white blood cells rush to the site, and the middle ear becomes filled with pus.

Untreated the tympanum may rupture, producing a characteristic oozing from the external acoustic meatus. Before this, the bacteria can usually be controlled by antibiotics, the pain reduced by analgesics, and the swelling reduced by decongestants. In the United States, it is rare for otitis media to progress to the stage at which tympanic rupture occurs.

Otitis media is extremely common in developing countries, where medical care and antibiotics are not readily available. Both children and adults in these countries often suffer from *chronic otitis media*, a condition characterized by recurring bouts of infection and chronic perforation, with draining ears. This condition produces scarring at the tympanic membrane, which often leads to some degree of hearing loss. Resulting damage to the inner ear or the auditory ossicles may further reduce auditory sensitivity.

If the pathogens leave the middle ear and invade the air cells within the mastoid process, **mastoiditis** develops. The connecting passageways are very narrow, and as the infection progresses, the subject experiences severe earaches, fever, and swelling behind the ear in addition to symptoms of otitis media. The same antibiotic treatment is used to deal with both conditions, the particular antibiotic selected depending on the identity of the bacterium involved. The major risk of mastoiditis is the spread of the infection to the brain via the connective tissue sheath of the facial nerve (N VII). Should mastoiditis occur, prompt antibiotic therapy is needed, and if the problem remains, the person may have to undergo **mastoidectomy** (opening and drainage of the mastoid sinuses) or **myringotomy** (drainage of the middle ear through a surgical opening in the tympanic membrane).

incus, and the *stapes*. These bones act as levers that transfer sound vibrations from the tympanum to a fluid-filled chamber within the inner ear.

The lateral surface of the **malleus** (*malleus*, a hammer) attaches to the interior surface of the tympanum at three points. The middle bone, the **incus** (*incus*, anvil), connects the medial surface of the malleus to the **stapes** (STĀ-pēz; *stapes*, stirrup). The *base*, or *footplate*, of the stapes almost completely fills the *oval window*, a hole in the bony wall of the middle ear cavity. An *anular ligament* extends between the base of the stapes and the bony margins of the oval window.

Vibration of the tympanum converts arriving sound waves into mechanical movements. The auditory ossicles then conduct those vibrations, and movement of the stapes sets up vibrations in the fluid contents of the inner ear. Because of the way these ossicles are connected, an in-out movement of the tympanic membrane produces a rocking motion at the stapes. The tympanic membrane is 22 times larger and heavier than the oval window, and the amount of force applied increases proportionally from the tympanic membrane to the oval window. This amplification process produces a relatively powerful deflection of the stapes at the oval window.

Because this amplification occurs, we can hear very faint sounds. But this degree of magnification can be a problem if we are exposed to very loud noises. Within the tympanic cavity, two small muscles serve to protect the eardrum and ossicles from violent movements under very noisy conditions.

- The **tensor tympani** (TEN-sor tim-PAN-ē) **muscle** is a short ribbon of muscle whose origin is the petrous part of the temporal bone, within the musculotubal canal and whose insertion is on the "handle" of the malleus (Figure 18.10b,d●). When the tensor tympani contracts, the malleus is pulled medially, stiffening the tympanum. This increased stiffness reduces the amount of possible movement. The tympani muscle is innervated by motor fibers of the mandibular branch of the trigeminal nerve (N V).
- The **stapedius** (sta-PĒ-dē-us) **muscle**, innervated by the facial nerve (N VII), originates from the posterior wall of the tympanic cavity and inserts on the stapes (Figure 18.10b,d●). Contraction of the stapedius pulls the stapes, reducing movement of the stapes at the oval window.

The Inner Ear [FIGURES 18.9 TO 18.12]

The senses of equilibrium and hearing are provided by the receptors of the **inner ear** (Figures 18.9 and 18.11●). The receptors are housed within a collection of fluid-filled tubes and chambers known as the **membranous labyrinth** (*labyrinthos*, network of canals). The membranous labyrinth contains a fluid called **endolymph** (EN-dō-limf). The receptor cells of the inner ear can function only when exposed to the unique ionic composition of the endolymph. Endolymph has a relatively high potassium ion concentration and a relatively low sodium ion concentration whereas typical extracellular fluids have high sodium and low potassium ion concentrations.

The **bony labyrinth** is a shell of dense bone that surrounds and protects the membranous labyrinth. Its inner contours closely follow the contours of the membranous labyrinth (Figure 18.12b●), while its outer walls are fused with the surrounding temporal bone. p. 146 Between the bony and membranous labyrinths flows the **perilymph** (PER-i-limf), a liquid whose properties closely resemble those of cerebrospinal fluid.

The bony labyrinth can be subdivided into the **vestibule** (VES-ti-būl), the **semicircular canals**, and the **cochlea** (KOK-lē-a; *cochlea*, snail shell) as seen in Figures 18.9 and 18.12b●. Receptors in the membranous labyrinth of the vestibule and semicircular canals provide the sense of equilibrium; those in the cochlea provide the sense of hearing. The structures and air spaces of the external ear and middle ear function in the capture and transmission of sound to the cochlea.

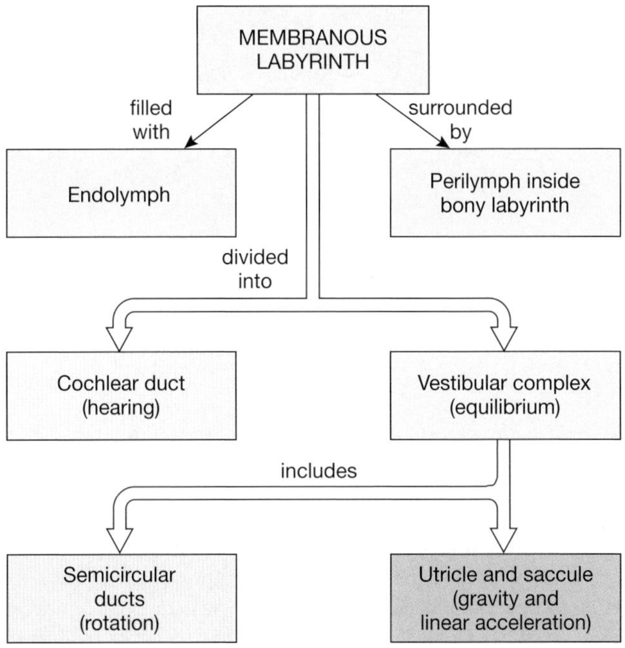

FIGURE 18.11 **STRUCTURAL RELATIONSHIPS OF THE INNER EAR**

Flow chart showing inner ear structures and spaces, their contained fluids, and what stimulates these receptors.

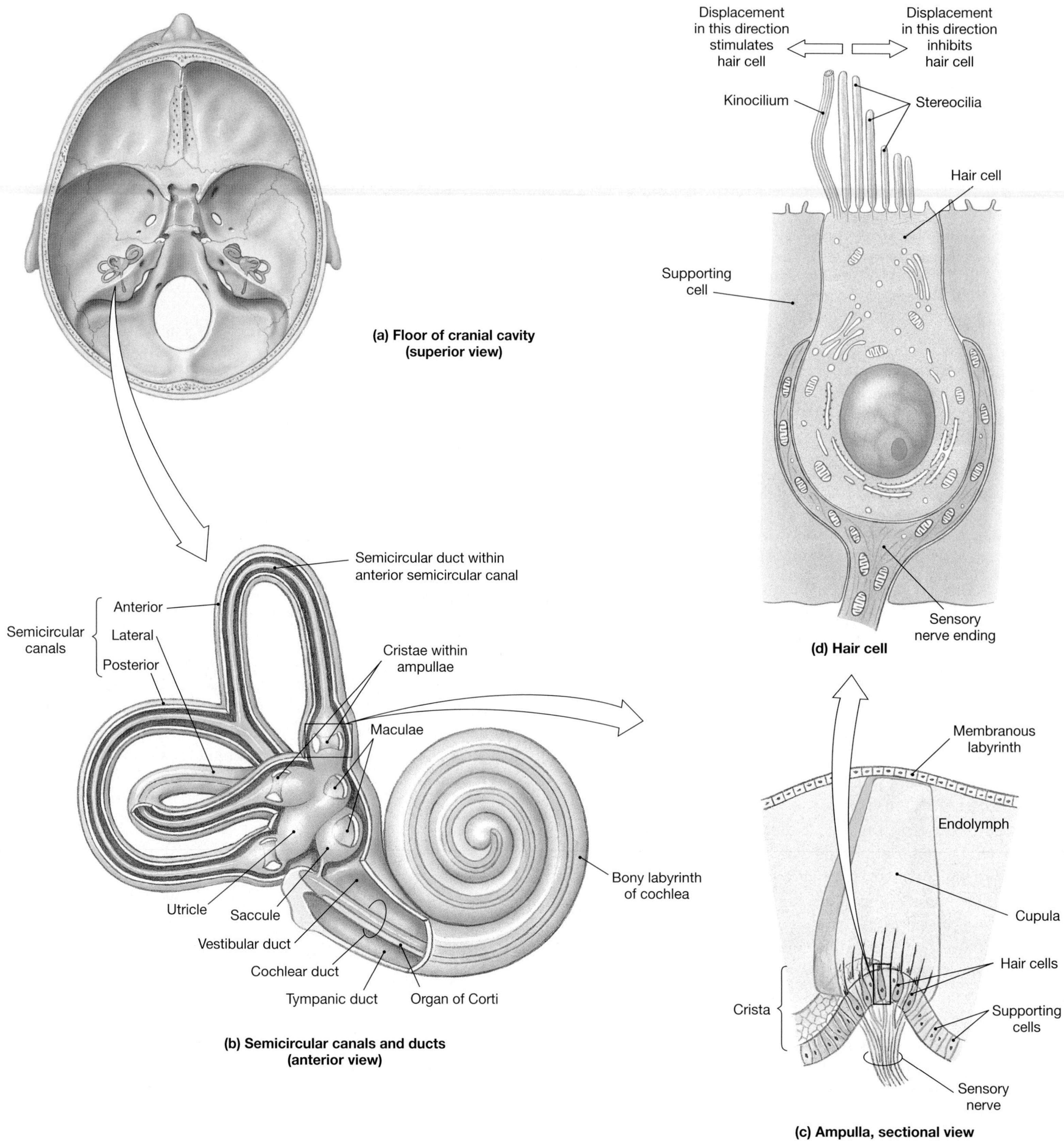

FIGURE 18.12 **Receptor Hair Cells of the Inner Ear**

(**a**) The orientation of the bony labyrinth within the petrous part of each temporal bone. (**b**) Anterior view of the bony labyrinth, cut away to show the enclosed membranous labyrinth. (**c**) A section through the ampulla of a semicircular canal. Displacement of the cupula stimulates the hair cells. (**d**) Structure of a typical hair cell, showing details revealed by electron microscopy. An individual hair cell has a kinocilium and numerous stereocilia. Each hair cell contacts a sensory neuron. Bending the stereocilia toward the kinocilium depolarizes the cell and stimulates the sensory neuron. Displacement in the opposite direction inhibits the sensory neuron.

The vestibule and semicircular canals together are called the *vestibular complex*, because the fluid-filled chambers of the vestibule are broadly continuous with those of the semicircular canals. The cavity within the vestibule contains a pair of membranous sacs, the **utricle** (Ū-tre-kl) and the **saccule** (SAK-ūl), or the *utriculus* and *sacculus*. Receptors in the utricle and saccule provide sensations of gravity and linear acceleration. Those in the semicircular canals are stimulated by rotation of the head.

The cochlea contains a slender, elongated portion of the membranous labyrinth known as the **cochlear duct** (Figure 18.12b●). The cochlear duct sits sandwiched between a pair of perilymph-filled chambers, and the entire complex makes turns around a central bony hub. In sectional view the spiral arrangement resembles that of a snail shell, *cochlea* in Latin.

The outer walls of the perilymphatic chambers consist of dense bone everywhere except at two small areas near the base of the cochlear spiral. The **round window**, or *cochlear window*, is the more inferior of the two openings. A thin, flexible membrane spans the opening and separates the perilymph in one of the cochlear chambers from the air in the middle ear (Figure 18.9●). The **oval window** is the more superior of the two openings in the cochlear wall (Figure 18.10b,c,d●). The base of the stapes almost completely fills the oval window. The anular ligament, which extends between the edges of the base and the margins of the oval window, completes the seal. When a sound vibrates the tympanum, the movements are conducted to the perilymph of the inner ear by the movements of the stapes. This process ultimately leads to the stimulation of receptors within the cochlear duct, and we "hear" the sound.

The sensory receptors of the inner ear are called **hair cells** (Figure 18.12d●). These receptor cells are surrounded by **supporting cells** and are monitored by sensory afferent fibers. The free surface of the hair cell supports 80–100 long **stereocilia**. Stereocilia are structurally similar to cilia, but they are several times longer and they do not beat.

Each hair cell in the vestibule also contains a **kinocilium**, a single large cilium. Hair cells do not actively move their kinocilia and stereocilia. However, when an external force pushes against these processes, the distortion of the cell membrane alters the rate of chemical transmitter released by the hair cell. Hair cells are thus highly specialized mechanoreceptors. Their abilities to provide equilibrium sensations in the vestibule and hearing in the cochlea depends on the presence of accessory structures that restrict the sources of stimulation. The importance of these accessory structures will become apparent as we consider the semicircular canals, utricle, and saccule.

THE VESTIBULAR COMPLEX AND EQUILIBRIUM

The **vestibular complex** is the part of the inner ear that provides equilibrium sensations by detecting rotation, gravity, and acceleration. It consists of the semicircular canals, the utricle, and the saccule.

The Semicircular Canals [FIGURES 18.12b,c/18.13] The **anterior, posterior**, and **lateral semicircular canals** are continuous with the vestibule (Figure 18.12b●). Each semicircular canal surrounds a **semicircular duct**. The duct contains a swollen region, the **ampulla**, which contains the sensory receptors. These receptors respond to rotational movements of the head. Hair cells attached to the wall of the ampulla form a raised structure known as a **crista** (Figure 18.12b,c●). The kinocilia and stereocilia of the hair cells are embedded in a gelatinous structure, the **cupula** (KŪ-pū-la). Because the cupula has a density very close to that of the surrounding endolymph, it essentially "floats" above the receptor surface, nearly filling the ampulla. When the head rotates in the plane of the duct, movement of the endolymph along the duct axis pushes the cupula and distorts the receptor processes. Fluid movement in one direction stimulates the hair cells, and movement in the opposite direction inhibits them. When the endolymph stops moving, the elastic nature of the cupula makes it "bounce back" to its normal position.

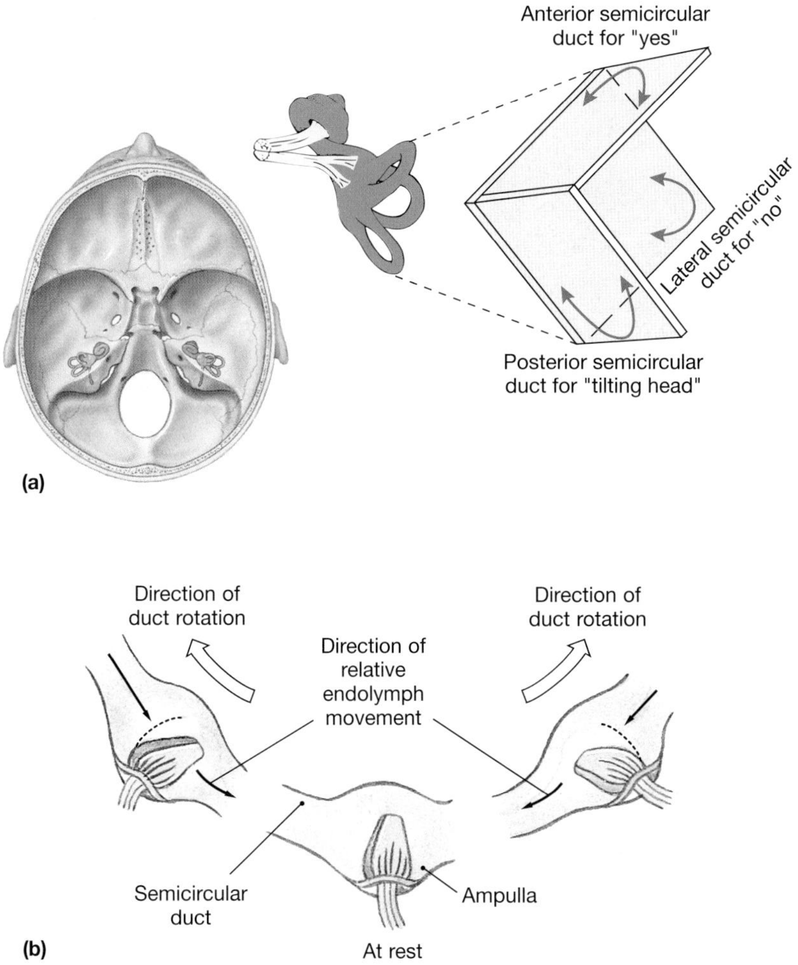

FIGURE 18.13 **Function of the Semicircular Ducts**

(**a**) A superior view showing the planes of sensitivity for the semicircular ducts. (**b**) Endolymph movement along the length of the duct moves the cupula and stimulates the hair cells.

Even the most complex movement can be analyzed in terms of motion in three rotational planes. The receptors within each semicircular duct responds to one of these rotational movements (Figure 18.13●). A horizontal rotation, as in shaking the head "no," stimulates the hair cells of the lateral semicircular duct. Nodding "yes" excites the anterior duct, while tilting the head from side to side activates the receptors in the posterior duct.

CLINICAL BRIEF

Nystagmus

Automatic eye movements occur in response to sensations of motion (whether real or illusory) under the direction of the *superior colliculi*. p. 403 These movements attempt to keep the gaze focused on a specific point in space. When you spin around, your eyes fix on one point for a moment, then jump ahead to another, in a series of short, rhythmic, jerky movements. These eye movements may appear in normal stationary individuals with extreme lateral gaze and after damage to or stimulation of the brain stem or inner ear. This condition is called **nystagmus**. Physicians often check for nystagmus by asking the subject to watch a small penlight as it is moved across the field of vision.

The Utricle and Saccule [FIGURES 18.12b/18.14] The utricle and saccule (Figures 18.12b and 18.14●) are connected by a slender passageway continuous with the narrow **endolymphatic duct**. The endolymphatic duct ends in a blind pouch, the **endolymphatic sac**, that projects through the dura mater lining the temporal bone and into the subdural space. Portions of the cochlear duct continually secrete endolymph, and at the endolymphatic sac, excess fluids return to the general circulation.

The hair cells of the utricle and saccule are clustered in the oval **maculae** (MAK-ū-lē; *macula*, spot) (Figures 18.12b and 18.14a●). As in the ampullae, the hair cell processes are embedded in a gelatinous mass. However, the surface of this gelatinous material contains densely packed calcium carbonate crystals known as **statoconia** (sta-tō-KŌ-nē-a; *conia*, dust). The complex as a whole (gelatinous matrix and statoconia), called an **otolith** (Ō-tō-lith; *oto-*, ear + *lithos*, stone) can be seen in Figure 18.14b●.

When the head is in the normal, upright position, the otoliths rest atop the maculae. Their weight presses down on the macular surfaces,

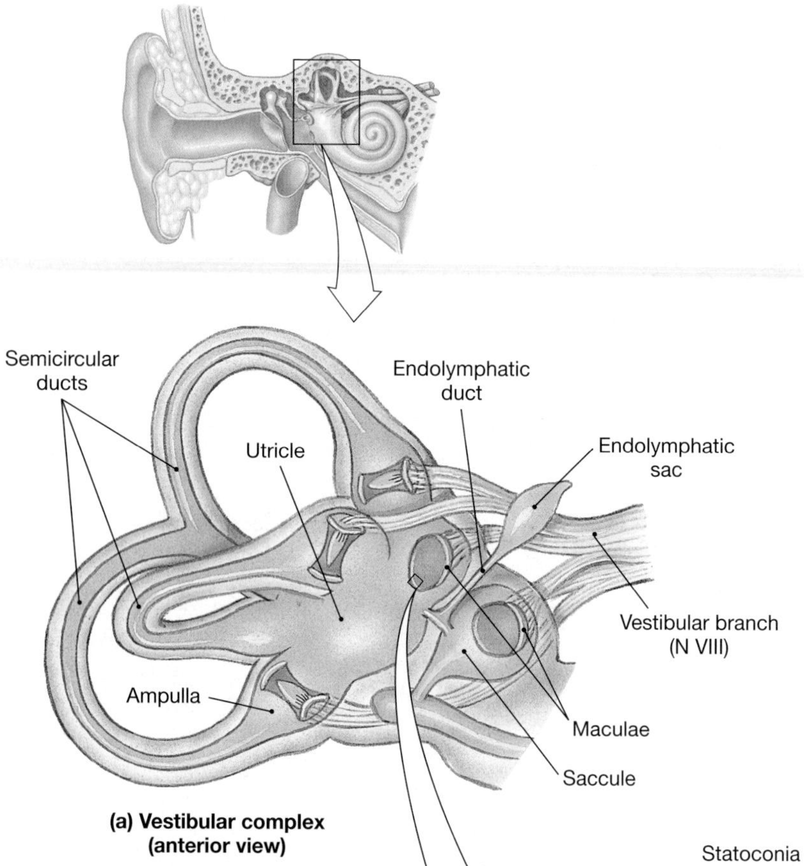

(a) Vestibular complex (anterior view)

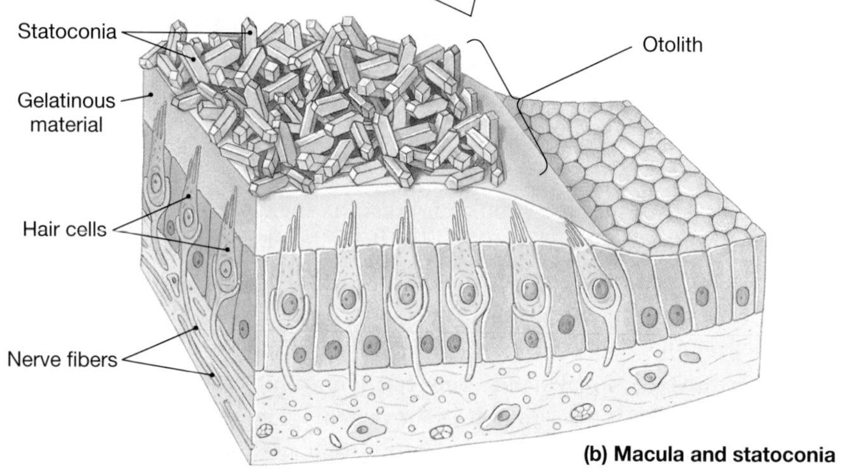

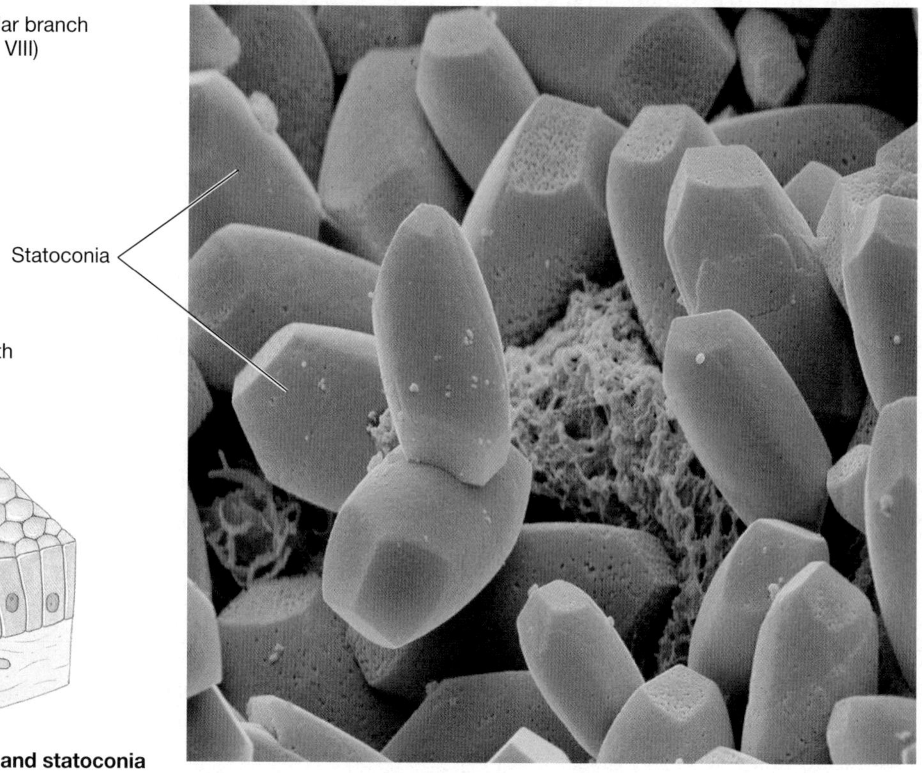

(b) Macula and statoconia

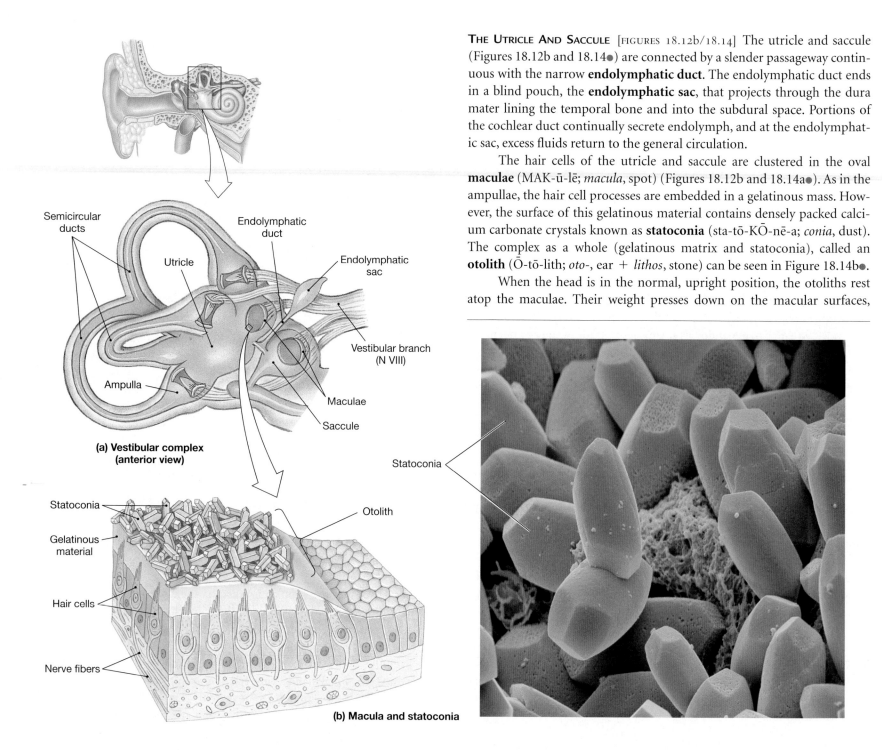

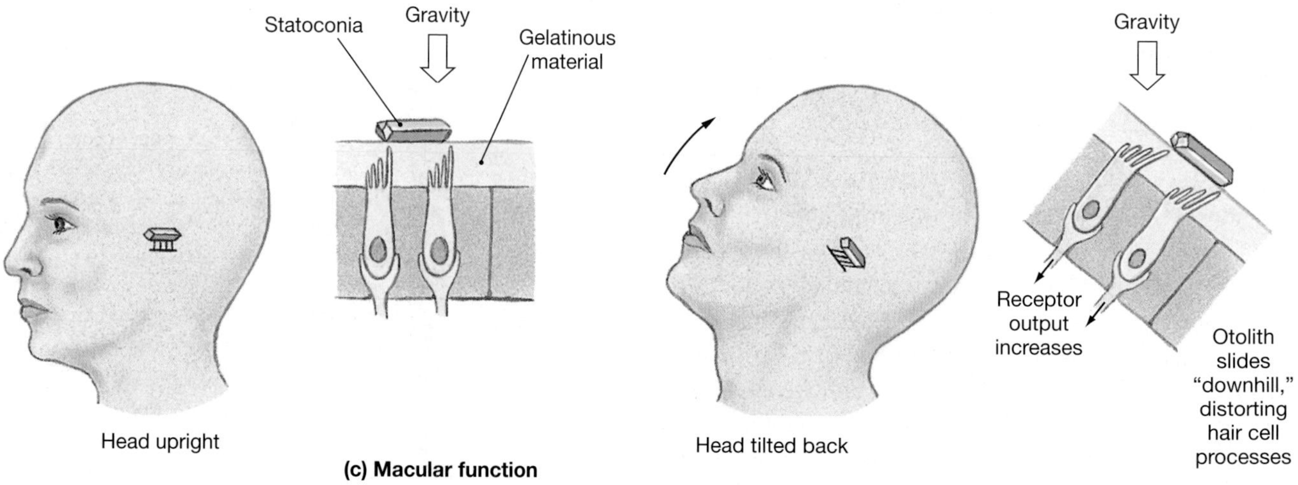

(c) Macular function

FIGURE 18.14 **The Maculae of the Vestibule**

(**a**) Location of the maculae within the utricle and saccule. (**b**) Detailed structure of a sensory macula and a scanning electron micrograph showing the crystalline structure of otoliths. (**c**) Diagrammatic view of changes in otolith position during tilting of the head.

pushing the sensory hairs down rather than to one side or another. When the head is tilted, the pull of gravity on the otoliths shifts them to the side. This shift distorts the sensory hairs, and the change in receptor activity tells the CNS that the head is no longer level (Figure 18.14c●).

For example, when an elevator starts its downward plunge, we are immediately aware of it because the otoliths no longer push so forcefully against the surface of the receptor cells. Once they catch up, we are no longer aware of any movement until the elevator brakes to a halt. As the body slows down, the otoliths press harder against the hair cells, and we "feel" the force of gravity increase. A similar mechanism accounts for our perception of linear acceleration in a car that speeds up suddenly. The otoliths lag behind, distorting the sensory hairs and changing the activity in the sensory neurons.

Pathways For Vestibular Sensations [FIGURE 18.15] Hair cells of the vestibule and semicircular ducts are monitored by sensory neurons located in adjacent **vestibular ganglia**. Sensory fibers from each ganglion form the **vestibular branch** of the vestibulocochlear nerve (N VIII). These fibers synapse on neurons within the vestibular nuclei at the boundary between the pons and medulla oblongata. The two vestibular nuclei:

1. integrate the sensory information concerning balance and equilibrium arriving from each side of the head;
2. relay information from the vestibular apparatus to the cerebellum;
3. relay information from the vestibular apparatus to the cerebral cortex, providing a conscious sense of position and movement; and
4. send commands to motor nuclei in the brain stem and spinal cord.

The reflexive motor commands issued by the vestibular nucleus are distributed to the motor nuclei for cranial nerves involved with eye, head, and neck movements (N III, N IV, N VI, and N XI). Descending instructions along the **vestibulospinal tracts** of the spinal cord adjust peripheral muscle tone to complement the reflexive movements of the head or neck. These pathways are illustrated in Figure 18.15●. ✝ *Vertigo, Dizziness, and Motion Sickness p. 795*

✓ CONCEPT CHECK

- You are exposed unexpectedly to very loud noises. What happens within the tympanic cavity to protect the tympanum from damage?
- Identify the auditory ossicles and describe their functions.
- What is perilymph? Where is it located?
- As you shake your head "no," you are aware of this head movement. How are these sensations detected?

Hearing

THE COCHLEA [FIGURE 18.16]

The bony cochlea (Figure 18.16●) coils around a central hub, or **modiolus** (mō-DĪ-ō-lus). There are usually 2.5 turns in the cochlear spiral. The modiolus encloses the **spiral ganglion,** which contains the cell bodies of the sensory neurons that monitor the receptors in the cochlear duct. In sectional

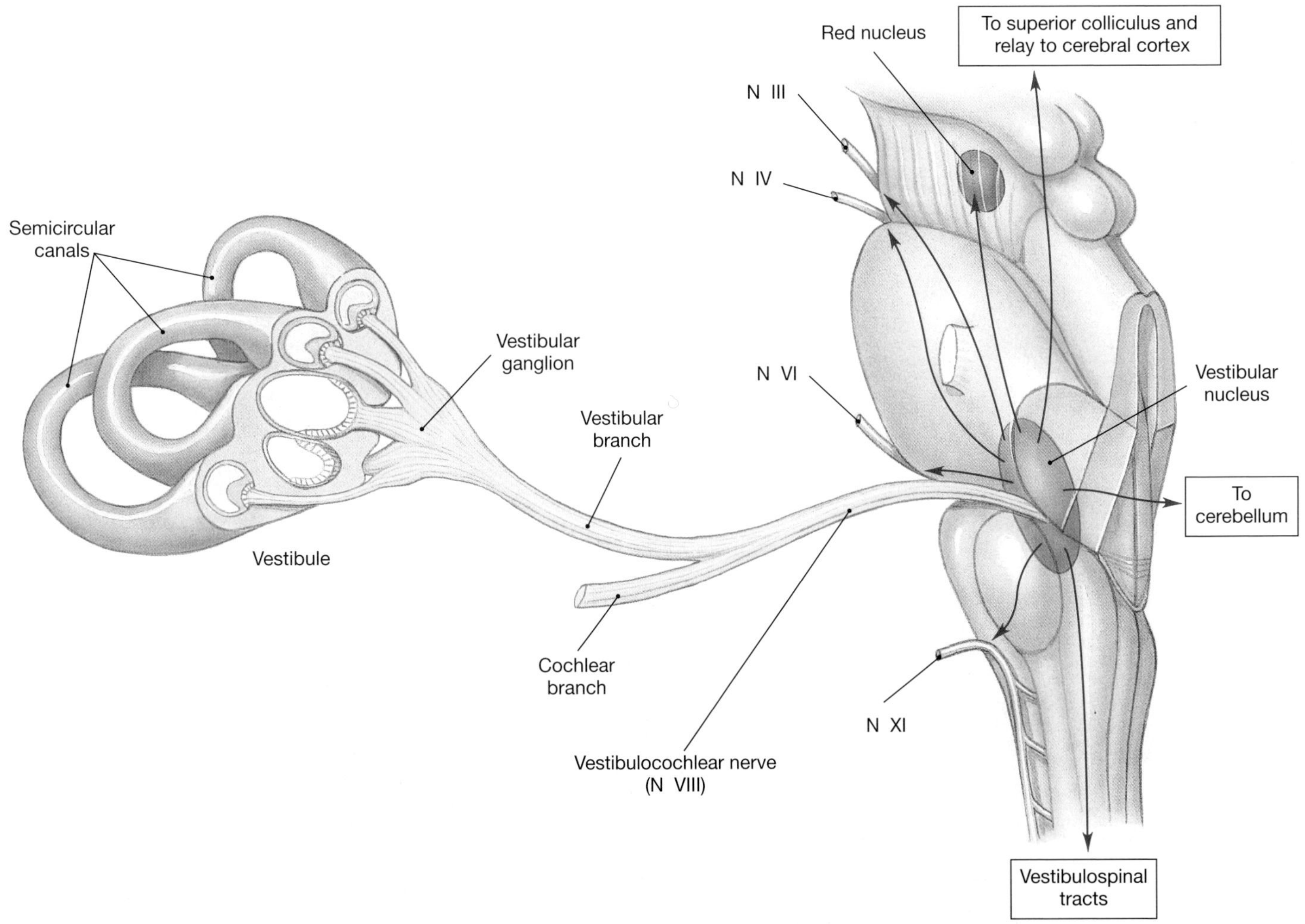

FIGURE 18.15 **Neural Pathways for Equilibrium Sensations**

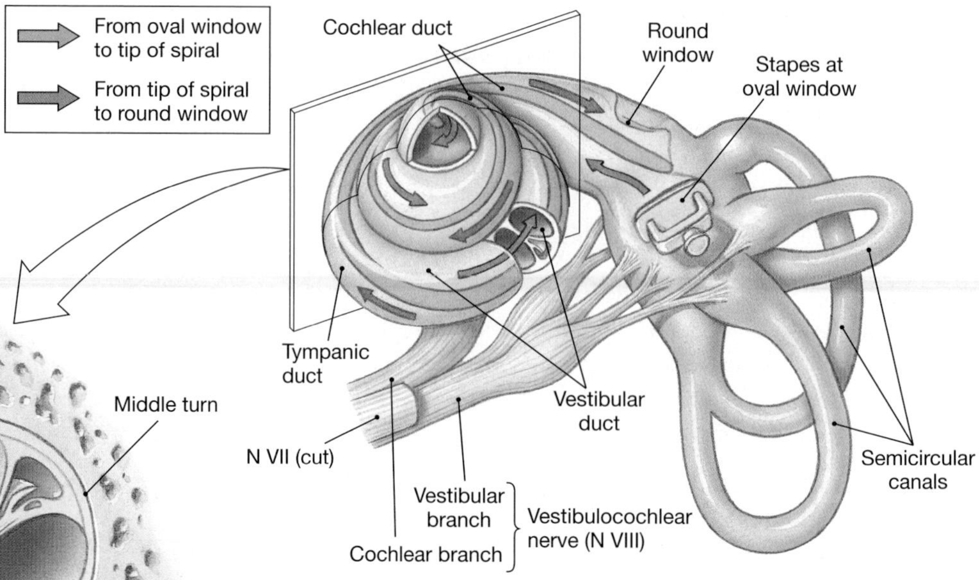

(a) Structure and orientation of the cochlea

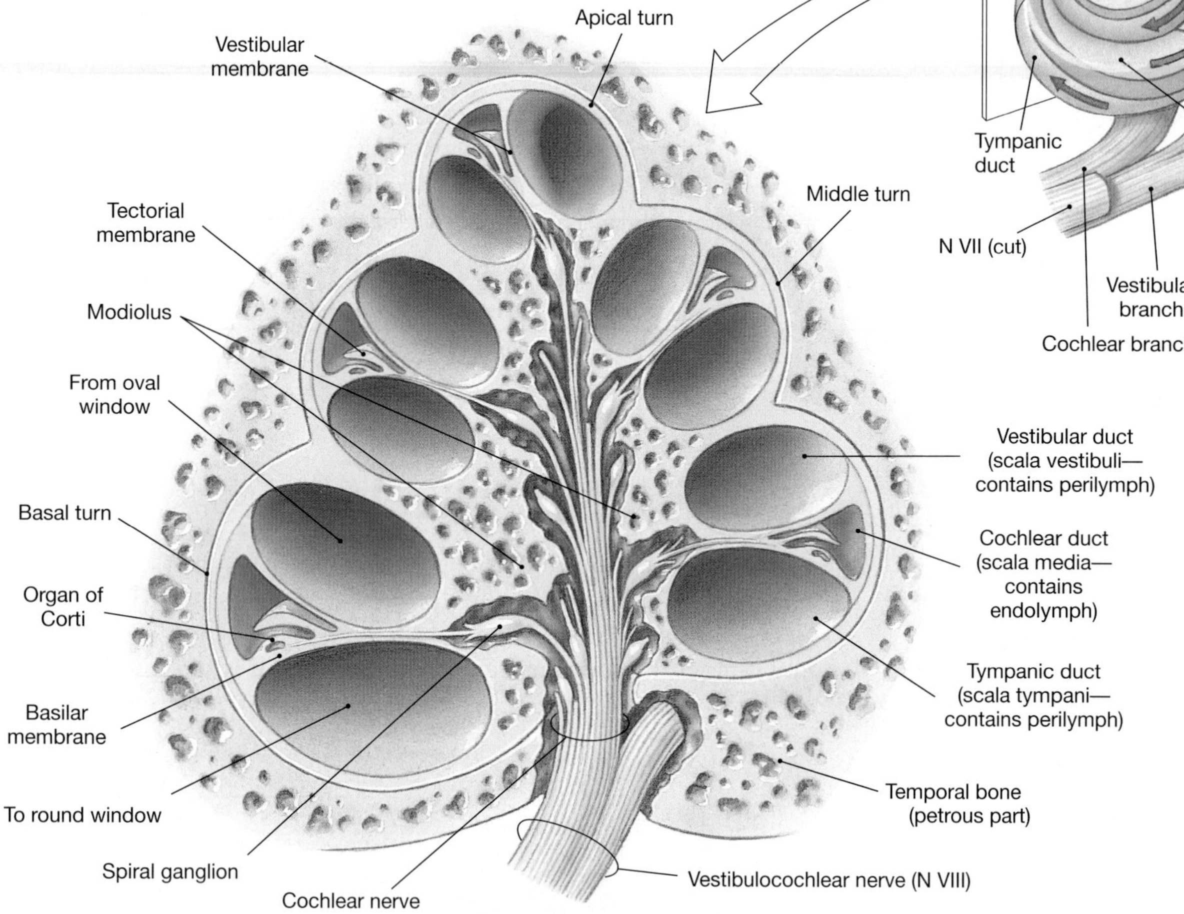

(b) Cochlear section, diagrammatic

FIGURE 18.16 **The Cochlea and Organ of Corti**

(**a**) Structure of the cochlea in partial section. (**b**) Structure of the cochlea within the temporal bone as seen in section, showing the turns of the vestibular duct, cochlear duct, and tympanic duct. (**c**) Photomicrograph of the cochlea showing many of the structures in (b).

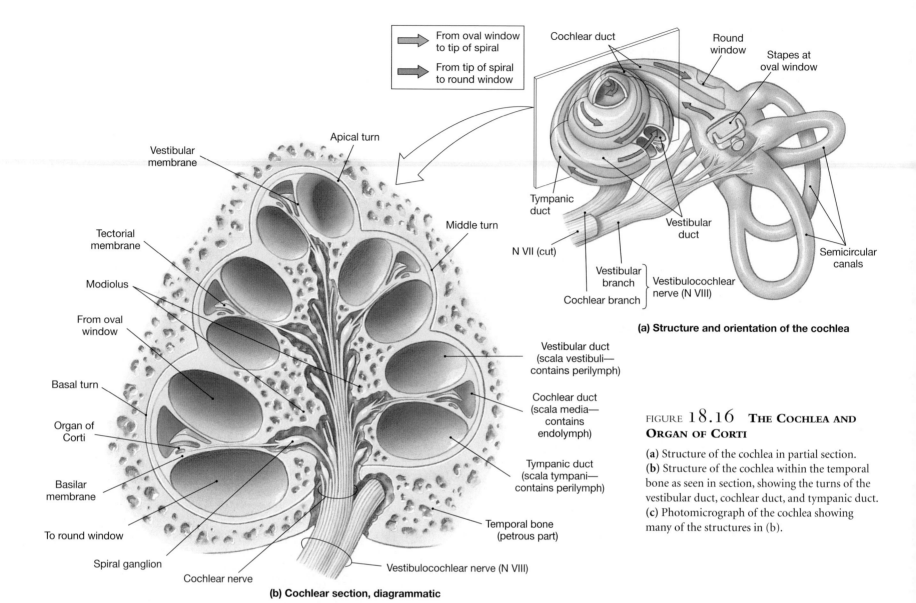

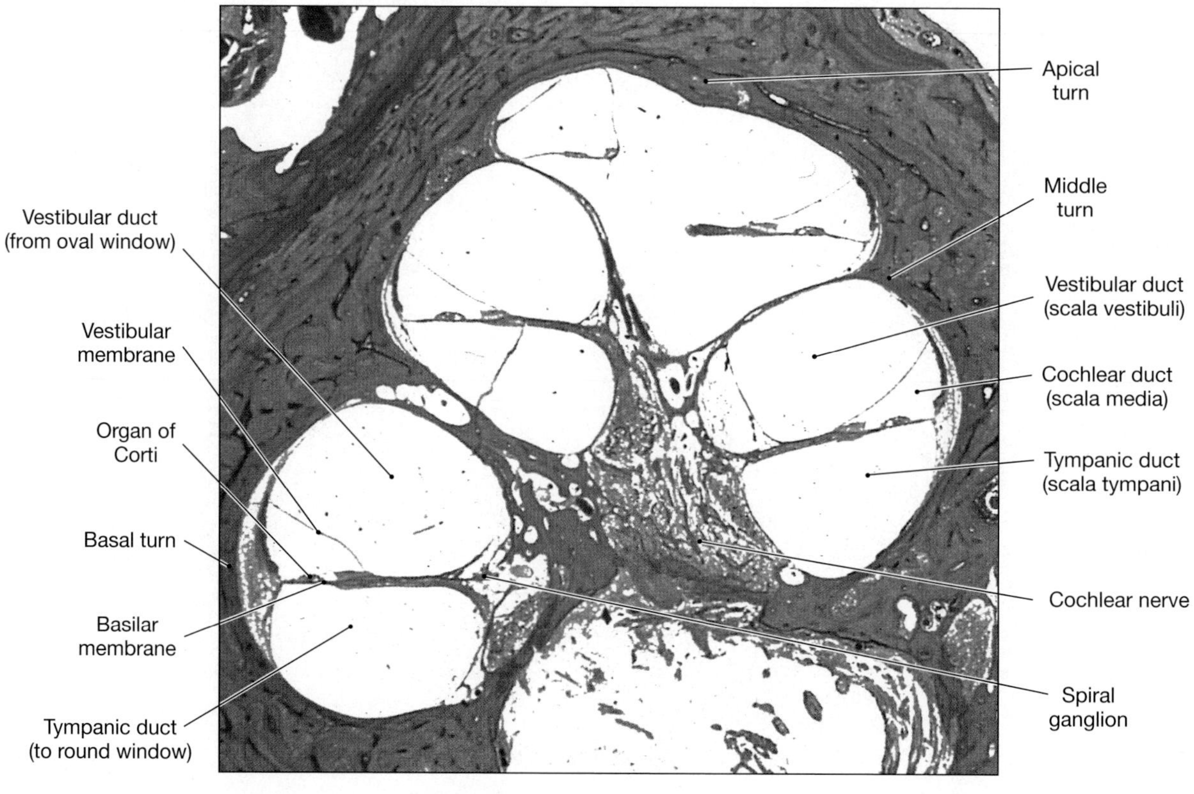

(c) Cochlear section, photomicrograph

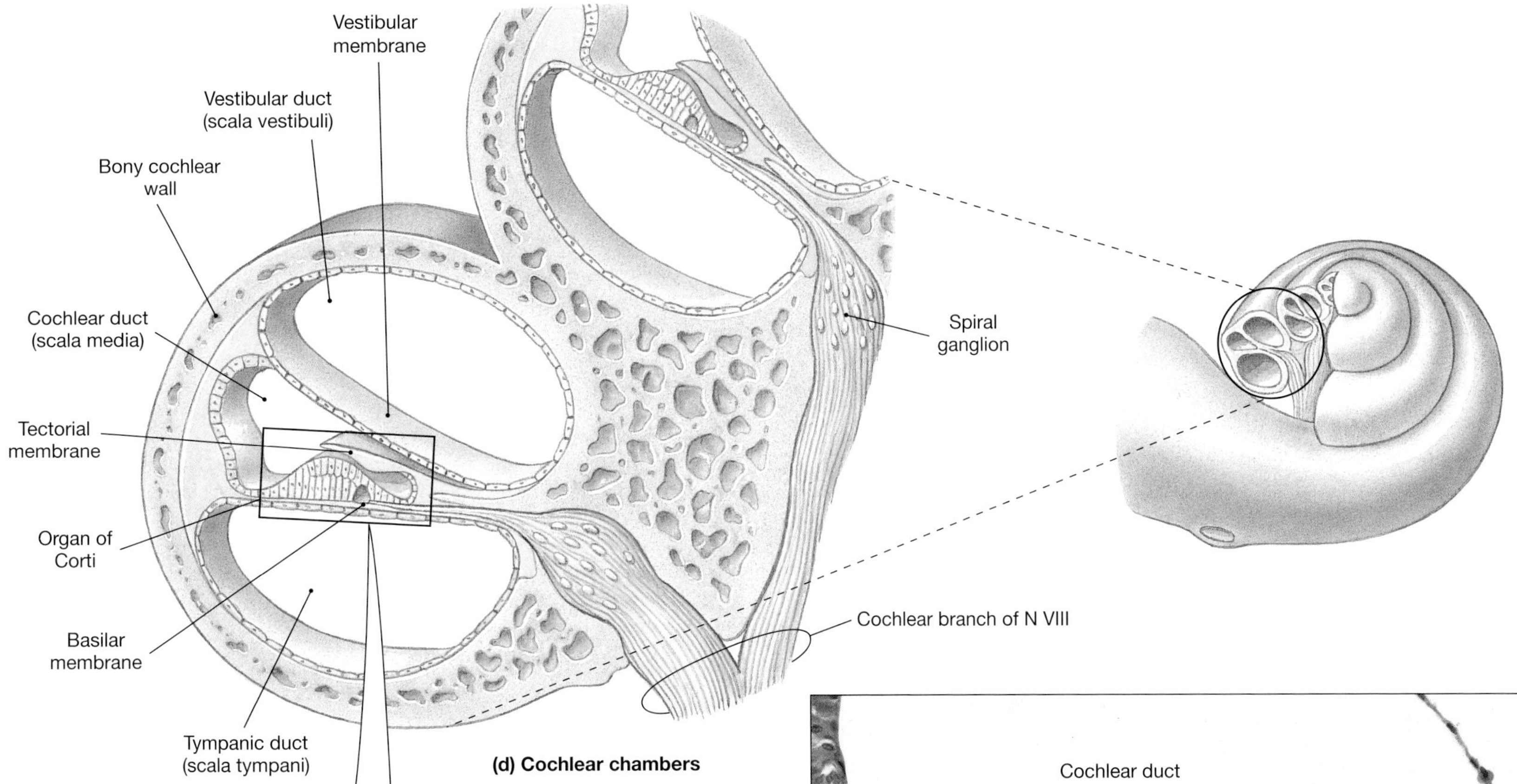

(d) Cochlear chambers

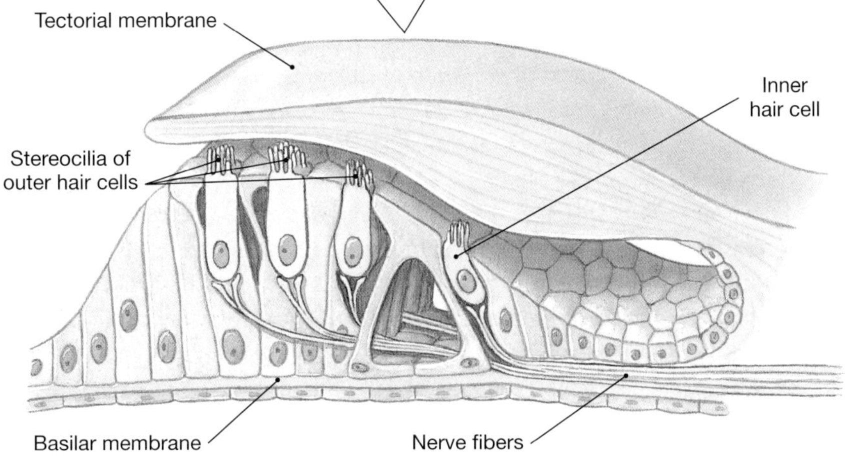

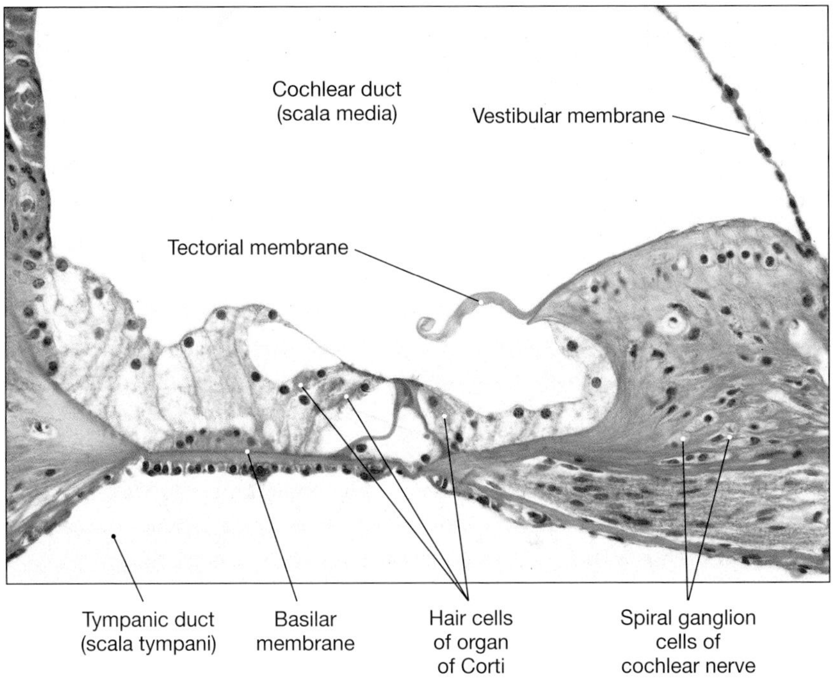

(e) Organ of Corti

(f) The receptor surface of the Organ of Corti (SEM × 1320)

FIGURE 18.16 **THE COCHLEA AND ORGAN OF CORTI** *(continued)*

(**d**) Three-dimensional section showing the detail of the cochlear chambers, tectorial membrane, and organ of Corti. (**e**) Diagrammatic and histological sections through the receptor hair cell complex of the organ of Corti. (LM × 2022) (**f**) A color-enhanced SEM showing a portion of the receptor surface of the organ of Corti. *Prof. P. Motta, Dept. of Anatomy, University "La Sapienza," Rome/Science Photo Library/Photo Researchers, Inc.*

view, the **cochlear duct**, or *scala media*, lies between a pair of perilymphatic chambers, the **vestibular duct** (*scala vestibuli*) and the **tympanic duct** (*scala tympani*). The two perilymphatic chambers are interconnected at the tip of the cochlear spiral. The oval window is at the base of the vestibular duct, and the round window is at the base of the tympanic duct.

The Organ Of Corti [FIGURE 18.16b–e] The hair cells of the cochlear duct are found in the **organ of Corti**, or **spiral organ** (Figure 18.16b–e●). This sensory structure rests upon the **basilar membrane** that separates the cochlear duct from the tympanic duct. The hair cells are arranged in inner and outer longitudinal rows. These hair cells lack kinocilia, and their stereocilia are in contact with the overlying **tectorial membrane** (tek-TOR-ē-al; *tectum*, roof). This membrane is firmly attached to the inner wall of the cochlear duct. When a portion of the basilar membrane bounces up and down, the stereocilia of the hair cells are distorted.

SOUND DETECTION [TABLE 18.2]

Hearing is the detection of sound, which consists of pressure waves conducted through air or water. Sound waves enter the external acoustic meatus and travel toward the tympanum. The tympanum provides the surface for sound collection, and it vibrates in response to sound waves with frequencies between approximately 20 and 20,000 Hz; this is the range in a young child, but with age the range decreases. As previously mentioned, the auditory ossicles transfer these vibrations in modified form to the oval window.

Movement of the stapes at the oval window applies pressure to the perilymph of the vestibular duct. A property of liquids is their inability to be compressed. For example, when you sit on a waterbed, you know that when you push down *here*, the waterbed bulges over *there*. Because the rest of the cochlea is sheathed in bone, pressure applied at the oval window can be relieved only at the round window. When the base of the stapes moves inward at the oval window, the membrane that spans the round window bulges outward.

Movement of the stapes sets up pressure waves in the perilymph. These waves distort the cochlear duct and the organ of Corti, stimulating the hair cells. The location of maximum stimulation varies depending on the frequency (pitch) of the sound. High-frequency sounds affect the basilar membrane near the oval window; the lower the frequency of the sound, the farther away from the oval window the distortion will be.

The actual amount of movement at a given location depends on the amount of force applied to the oval window. This relationship provides a mechanism for detecting the intensity (volume) of the sound. Very high-intensity sounds can produce hearing losses by breaking the stereocilia off the surfaces of the hair cells. The reflex contraction of the tensor tympani and stapedius in response to a dangerously loud noise occurs in less than 0.1 second, but this may not be fast enough to prevent damage and related hearing loss. Table 18.2 summarizes the steps involved in translating a sound wave into an auditory sensation.

TABLE 18.2 **Steps in the Production of an Auditory Sensation**

1. Sound waves arrive at the tympanic membrane.
2. Movement of the tympanic membrane causes displacement of the auditory ossicles.
3. Movement of the stapes at the oval window establishes pressure waves in the perilymph of the vestibular duct.
4. The pressure waves distort the basilar membrane on their way to the round window of the tympanic duct.
5. Vibration of the basilar membrane causes vibration of hair cells against the tectorial membrane resulting in hair cell stimulation and neurotransmitter release.
6. Information concerning the region and intensity of stimulation is relayed to the CNS over the cochlear branch of N VIII.

CLINICAL BRIEF

Hearing Loss

Conductive deafness results from conditions in the middle ear that block the normal transfer of vibration from the tympanic membrane to the oval window. A plugged external auditory canal from accumulated wax or trapped water may cause a temporary hearing loss. Scarring or perforation of the tympanum, fluid in the middle ear chamber, and immobilization of one or more of the auditory ossicles are more serious examples of conduction deafness.

In **nerve deafness** the problem lies within the cochlea or somewhere along the auditory pathway. The vibrations are reaching the oval window, but either the receptors cannot respond or their response cannot reach its central destinations. Also, certain drugs entering the endolymph may kill the receptors, and infections may damage the hair cells or affect the cochlear nerve. Hair cells can also be damaged by exposure to high doses of aminoglycoside antibiotics, such as neomycin or gentamicin; this potential side effect must be balanced against the severity of infection before these drugs are prescribed. *Testing and Treating Hearing Deficits p. 795*

Auditory Pathways [FIGURE 18.17]

Hair cell stimulation activates sensory neurons whose cell bodies are in the adjacent spiral ganglion. Their afferent fibers form the **cochlear branch** of the vestibulocochlear nerve (N VIII). These axons enter the medulla oblongata and synapse at the **cochlear nucleus** on the same side of the brain. The axons of the second-order neurons cross to the opposite side of the brain and ascend to the inferior colliculus of the midbrain (Figure 18.17●). This processing center coordinates a number of responses to acoustic stimuli, including auditory reflexes involving skeletal muscles of the head, face, and trunk. These reflexes automatically change the position of the head in response to a sudden loud noise.

Before reaching the cerebral cortex and our conscious awareness, ascending auditory sensations synapse in the thalamus. Projection fibers then deliver the information to the auditory cortex of the temporal lobe. In effect, the auditory cortex contains a map of the organ of Corti. High-frequency sounds activate one portion of the cortex, and low-frequency sounds affect another. If the auditory cortex is damaged, the individual will respond to sounds and have normal acoustic reflexes, but sound interpretation and pattern recognition will be difficult or impossible. Damage to the adjacent association area does not affect the ability to detect the tones and patterns, but produces an inability to comprehend their meaning.

✓ CONCEPT CHECK

- If the membrane spanning the round window were not able to bulge out with increased pressure in the perilymph, how would sound perception be affected?
- How would loss of stereocilia from the hair cells of the organ of Corti affect hearing?
- Distinguish between the cochlear and tympanic ducts.
- What structure stimulates hair cells in the organ of Corti?

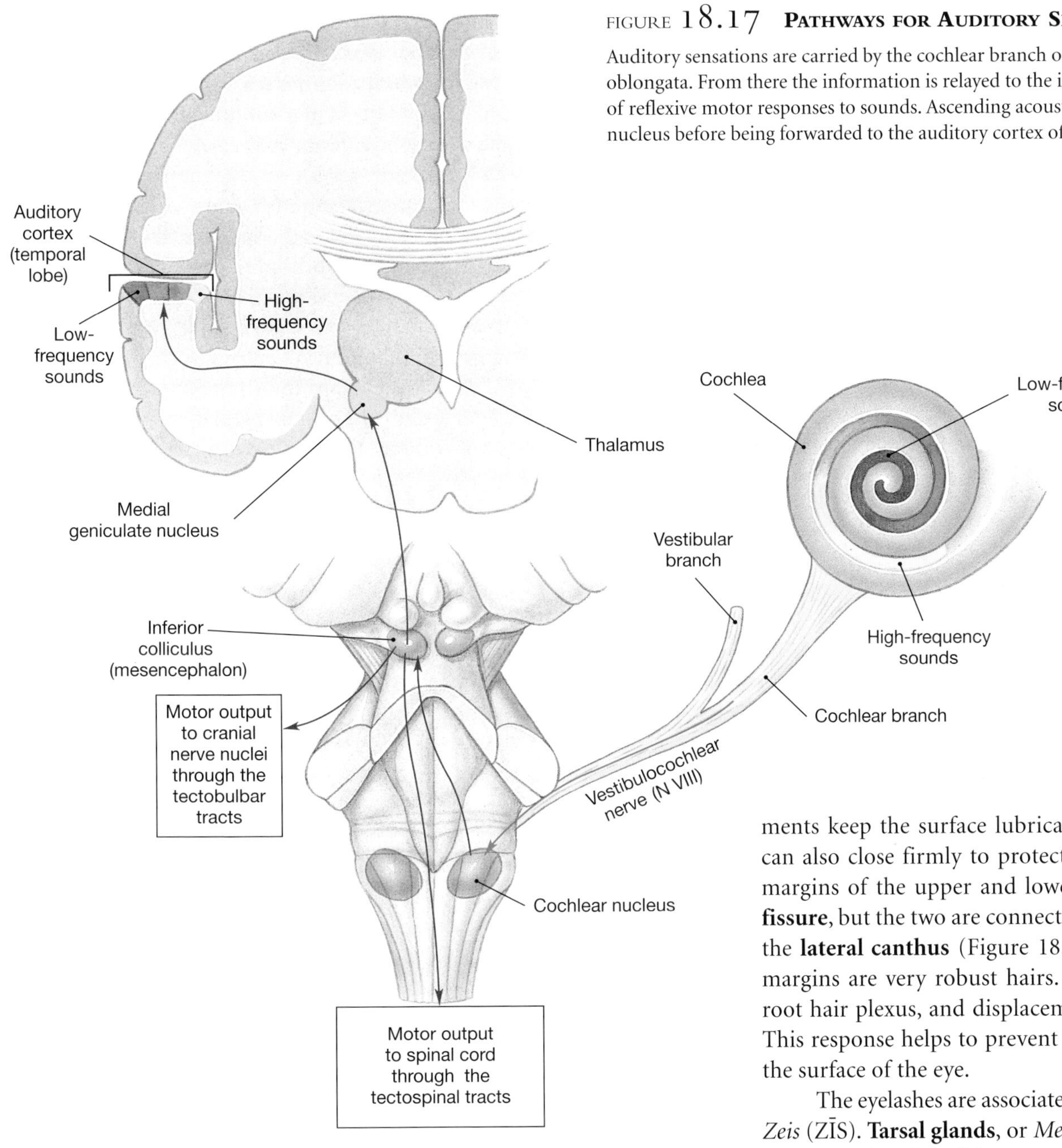

FIGURE 18.17 **Pathways for Auditory Sensations**

Auditory sensations are carried by the cochlear branch of N VIII to the cochlear nucleus of the medulla oblongata. From there the information is relayed to the inferior colliculus, a center that directs a variety of reflexive motor responses to sounds. Ascending acoustic information goes to the medial geniculate nucleus before being forwarded to the auditory cortex of the temporal lobe.

Vision [FIGURE 18.18]

Humans rely more on vision than on any other special sense, and the visual cortex is several times larger than the cortical areas devoted to other special senses. Our visual receptors are contained in elaborate structures, the eyes, which enable us not only to detect light but to create detailed visual images. We will begin our discussion with the *accessory structures* of the eye that provide protection, lubrication, and support. The superficial anatomy of the eye and the major accessory structures are illustrated in Figure 18.18•.

Accessory Structures of the Eye

The **accessory structures** of the eye include the eyelids, the superficial epithelium of the eye, and the structures associated with the production, secretion, and removal of tears.

EYELIDS [FIGURES 18.18/18.19/18.20b,e]

The eyelids, or **palpebrae** (pal-PĒ-brē) are a continuation of the skin. The eyelids act like windshield wipers; their continual blinking movements keep the surface lubricated and free from dust and debris. They can also close firmly to protect the delicate surface of the eye. The free margins of the upper and lower eyelids are separated by the **palpebral fissure**, but the two are connected at the **medial canthus** (KAN-thus) and the **lateral canthus** (Figure 18.18•). The **eyelashes** along the palpebral margins are very robust hairs. Each of the eyelashes is monitored by a root hair plexus, and displacement of the hair triggers a blinking reflex. This response helps to prevent foreign matter and insects from reaching the surface of the eye.

The eyelashes are associated with large sebaceous glands, the *glands of Zeis* (ZĪS). **Tarsal glands**, or *Meibomian* (mī-BŌ-mē-an) *glands*, along the inner margin of the lid secrete a lipid-rich product that helps to keep the eyelids from sticking together. At the medial canthus glands within the **lacrimal caruncle** (KAR-unk-ul) (Figure 18.18a•) produce the thick secretions that contribute to the gritty deposits occasionally found after a good night's sleep. These various glands are subject to occasional invasion and infection by bacteria. A cyst, or *chalazion* (kah-LĀ-zē-on; "small lump"), usually results from the infection of a tarsal gland. An infection in a sebaceous gland of an eyelash, a tarsal gland, or one of the many sweat glands that open to the surface between the follicles of the eyelashes, produces a painful localized swelling known as a *sty*.

The visible surface of the eyelid is covered by a thin layer of stratified squamous epithelium. Deep to the subcutaneous layer, the eyelids are supported and strengthened by broad sheets of connective tissue, collectively called the **tarsal plate** (Figure 18.18b•). The muscle fibers of the *orbicularis oculi muscle* and the *levator palpebrae superioris muscle* (Figures 18.18b and 18.19•) lie between the tarsal plate and the skin. These skeletal muscles are responsible for closing the eyelids (orbicularis oculi) and raising the upper eyelid (levator palpebrae superioris). pp. 271–272

The epithelium covering the inner surface of the eyelids and the outer surface of the eye is called the **conjunctiva** (kon-junk-TĪ-va; "uniting" or

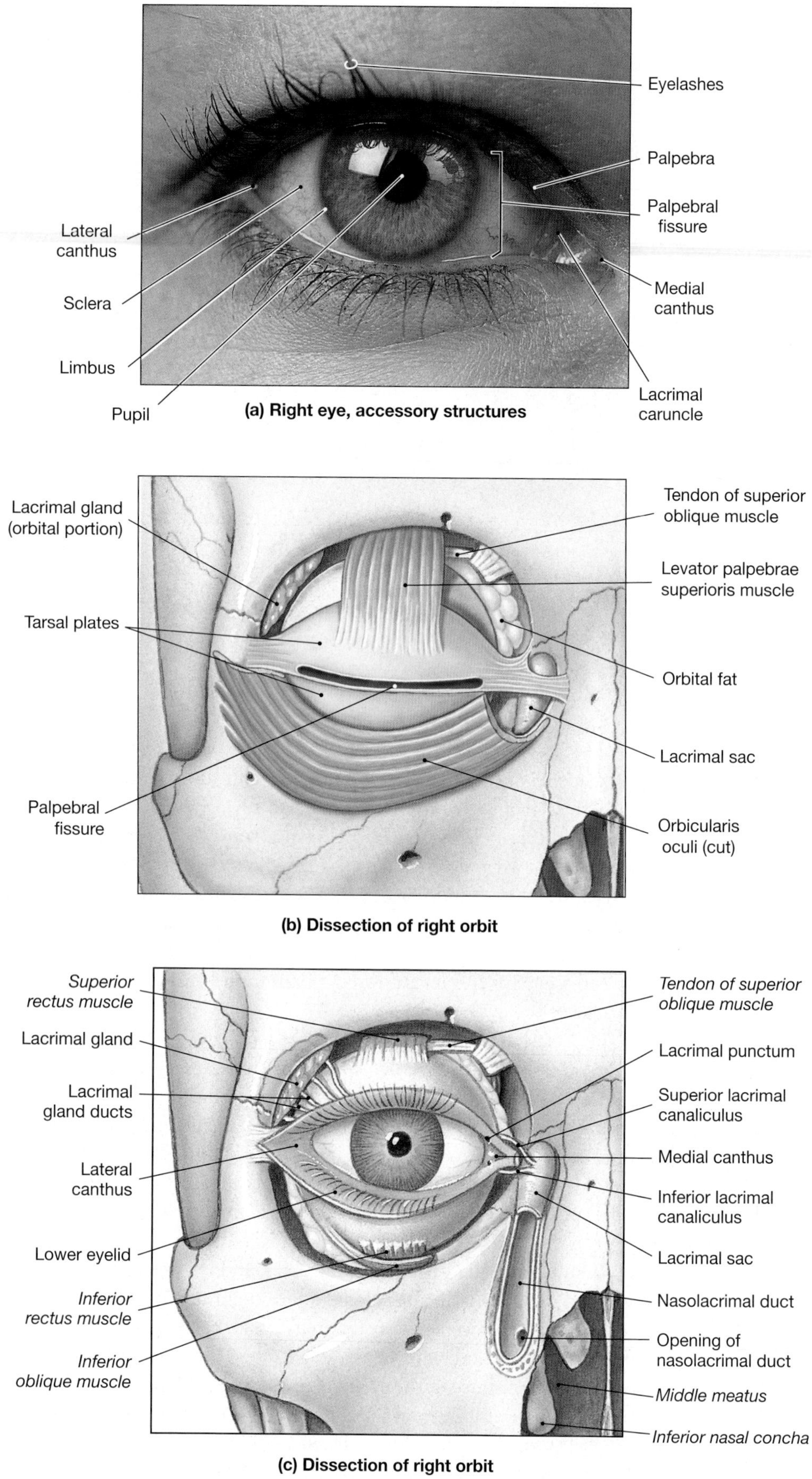

FIGURE 18.18 **ACCESSORY STRUCTURES OF THE EYE, PART I**

(a) Superficial anatomy of the right eye and its accessory structures. **(b)** Diagrammatic representation of a superficial dissection of the right orbit. **(c)** Diagrammatic representation of a deeper dissection of the right eye showing its position within the orbit and its relationship to accessory structures, especially the lacrimal apparatus.

"connecting") (Figure 18.20b,e●). It is a mucous membrane covered by a specialized stratified squamous epithelium. The **palpebral conjunctiva** covers the inner surface of the eyelids, and the **ocular conjunctiva**, or **bulbar conjunctiva**, covers the anterior surface of the eye. A continuous supply of fluid washes over the surface of the eyeball, keeping the conjunctiva moist and clean. Goblet cells within the epithelium assist the various accessory glands in providing a superficial lubricant that prevents friction and drying of the opposing conjunctival surfaces.

Over the transparent **cornea** (KOR-nē-a) of the eye, the relatively thick stratified epithelium changes to a very thin and delicate squamous epithelium 5–7 cells thick. Near the edges of the lids, the conjunctiva develops a more robust stratified squamous epithelium characteristic of exposed bodily surfaces. Although there are no specialized sensory receptors monitoring the surface of the eye, there are abundant free nerve endings with very broad sensitivities.

THE LACRIMAL APPARATUS [FIGURES 18.18b,c/18.19]

A constant flow of tears keeps conjunctival surfaces moist and clean. Tears reduce friction, remove debris, prevent bacterial infection, and provide nutrients and oxygen to portions of the conjunctival epithelium. The **lacrimal apparatus** produces, distributes, and removes tears. The lacrimal apparatus of each eye consists of: (1) a *lacrimal gland,* (2) *superior* and *inferior lacrimal canaliculi,* (3) a *lacrimal sac,* and (4) a *nasolacrimal duct* (Figures 18.18b,c and 18.19●).

The pocket created where the conjunctiva of the eyelid connects with that of the eye is known as the **fornix** (FOR-niks). The lateral portion of the superior fornix receives 10–12 ducts from the **lacrimal gland**, or tear gland. The lacrimal gland is about the size and shape of an almond, measuring roughly 12–20 mm (0.5–0.75 in.). It nestles within a depression in the frontal bone, within the orbit and superior and lateral to the eyeball (Figure 18.19●). ∞ p. 153 The lacrimal gland normally provides the key ingredients and most of the volume of the tears that bathe the conjunctival surfaces. Its secretions are watery, slightly alkaline, and contain the enzyme **lysozyme**, which attacks microorganisms.

The lacrimal gland produces tears at a rate of around 1 ml/day. Once the lacrimal secretions have reached the ocular surface, they mix with the products of accessory glands and the oily secretions of the tarsal glands and glands of Zeis. The latter contributions produce a superficial "oil slick" that assists in lubrication and slows evaporation.

The blinking of the eye sweeps the tears across the ocular surface, and they accumulate at the medial canthus in an area known as the *lacus lacrimalis,* or "lake of tears." Two small pores, the **superior** and **inferior lacrimal puncta** (singular, *punctum*), drain the lacrimal lake, emptying into the **lacrimal canaliculi** that run along grooves in the surface of the lacrimal bone. These passageways lead to the **lacrimal sac** which fills the lacrimal groove of the lacrimal bone. From there the **nasolacrimal duct** extends along the nasolacrimal canal formed by lacrimal bone and maxilla to deliver the tears to the inferior meatus on that side of the nasal cavity. p. 150

Blockage of the lacrimal puncta or oversecretion of the lacrimal glands can produce "watery eyes" that are constantly overflowing. Inadequate tear production, a more common condition, produces "dry eyes." Lubricating "artificial tears" in the form of eye drops are the usual answer, but more serious cases may be treated by surgically closing the lacrimal puncta.

The Eye [FIGURES 18.19/18.20a,e,f]

The eyes are slightly irregular spheroids with an average diameter of 24 mm (almost 1 in.), slightly smaller than a Ping-Pong ball. Each eye weighs around 8 g (0.28 oz). The eyeball shares space within the orbit with the extra-ocular muscles, the lacrimal gland, and the cranial nerves and blood vessels that supply the eye and adjacent portions of the orbit and face (Figures 18.19 and 18.20e,f●). p. 272 A mass of **orbital fat** provides padding and insulation.

The wall of the eye contains three distinct layers, or tunics (Figure 18.20a●): an outer *fibrous tunic*, an intermediate *vascular tunic*, and an inner *neural tunic*. The eyeball is hollow, and the interior is divided into two *cavities*. The large **posterior cavity** is also called the *vitreous chamber*, because it contains the gelatinous *vitreous body*. The smaller **anterior cavity** is subdivided into two chambers, *anterior* and *posterior*. The shape of the eye is stabilized in part by the vitreous body and the clear *aqueous humor* that fills the anterior cavity.

THE FIBROUS TUNIC [FIGURES 18.19/18.20a,b,c,e]

The **fibrous tunic**, the outermost layer of the eye, consists of the *sclera* and the *cornea* (Figure 18.20●). The fibrous tunic: (1) provides mechanical support and some degree of physical protection, (2) serves as an attachment site for the extra-ocular muscles, and (3) contains structures that assist in the focusing process.

Most of the ocular surface is covered by the **sclera** (SKLER-a). The sclera, or "white of the eye," consists of a dense fibrous connective tissue containing both collagen and elastic fibers. This layer is thickest at the posterior portion of the eye, near the exit of the optic nerve, and thinnest over the anterior surface. The six extra-ocular muscles insert upon the sclera, and the collagen fibers of their tendons are interwoven into the collagen fibers of the outer tunic (Figure 18.19●). p. 272

The anterior surface of the sclera contains small blood vessels and nerves that penetrate the sclera to reach internal structures. The network of small vessels that lie deep to the ocular conjunctiva usually does not carry enough blood to lend an obvious color to the sclera, but they are visible, on close inspection, as red lines against the white background of collagen fibers.

The transparent cornea of the eye is part of the fibrous tunic, and it is continuous with the sclera. The corneal surface is covered by a delicate stratified squamous epithelium continuous with the ocular conjunctiva. Deep to that epithelium, the cornea consists primarily of a dense matrix containing multiple layers of collagen fibers. The transparency of the cornea results from the precise alignment of the collagen fibers within these layers. A simple squamous epithelium separates the innermost layer of the cornea from the anterior chamber of the eye.

The cornea is structurally continuous with the sclera; the **limbus** is the border between the two. The cornea is avascular, and there are no blood vessels between the cornea and the overlying conjunctiva. As a result, the superficial epithelial cells must obtain oxygen and nutrients from the tears that flow across their free surfaces while the innermost epithelial layer receives its nutrients from the aqueous humor within the anterior chamber. There are numerous free nerve endings in the cornea, and this is the most sensitive portion of the eye. This sensitivity is important because corneal damage will cause blindness even though the rest of the eye—photoreceptors included—is perfectly normal.

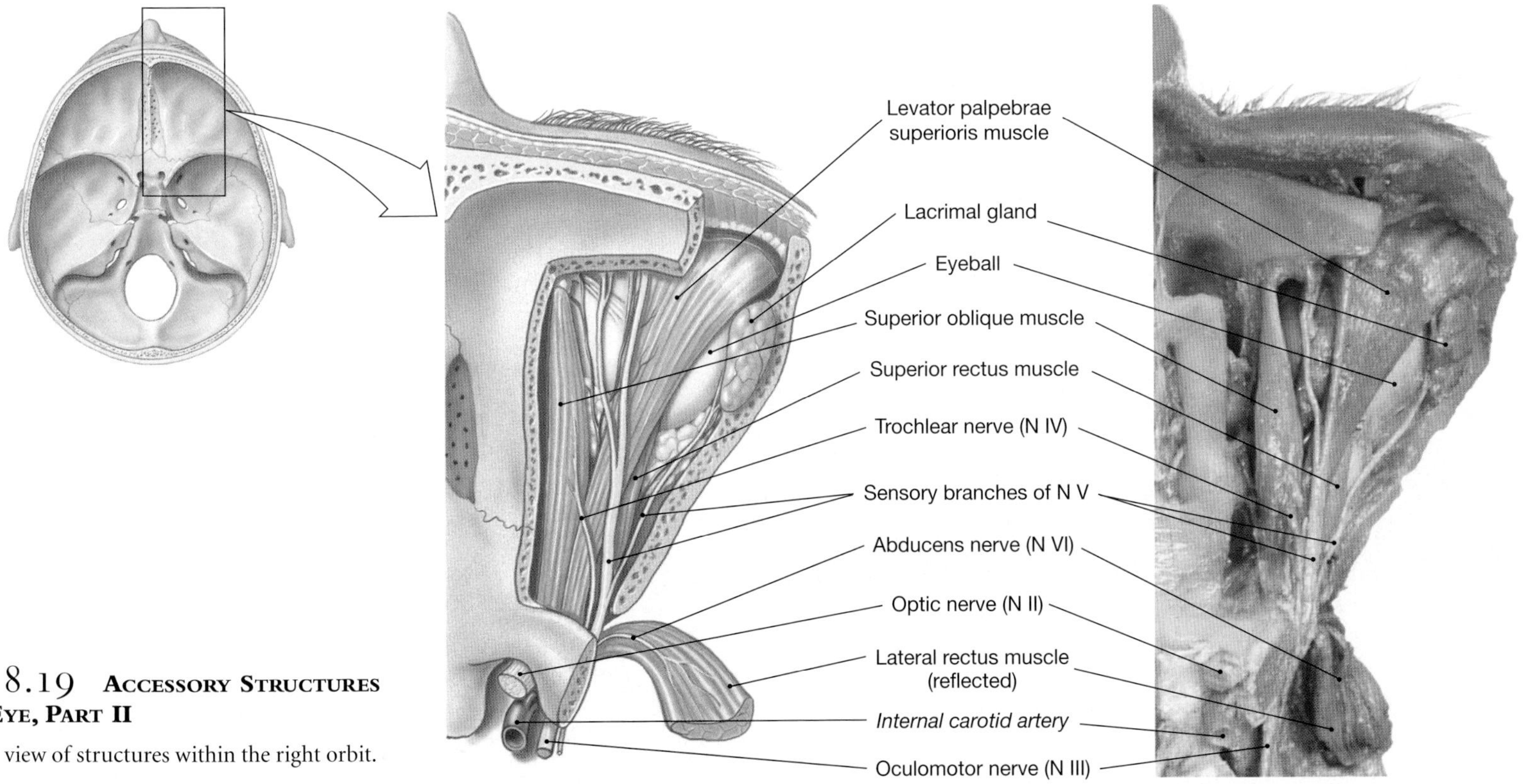

FIGURE 18.19 **ACCESSORY STRUCTURES OF THE EYE, PART II**

A superior view of structures within the right orbit.

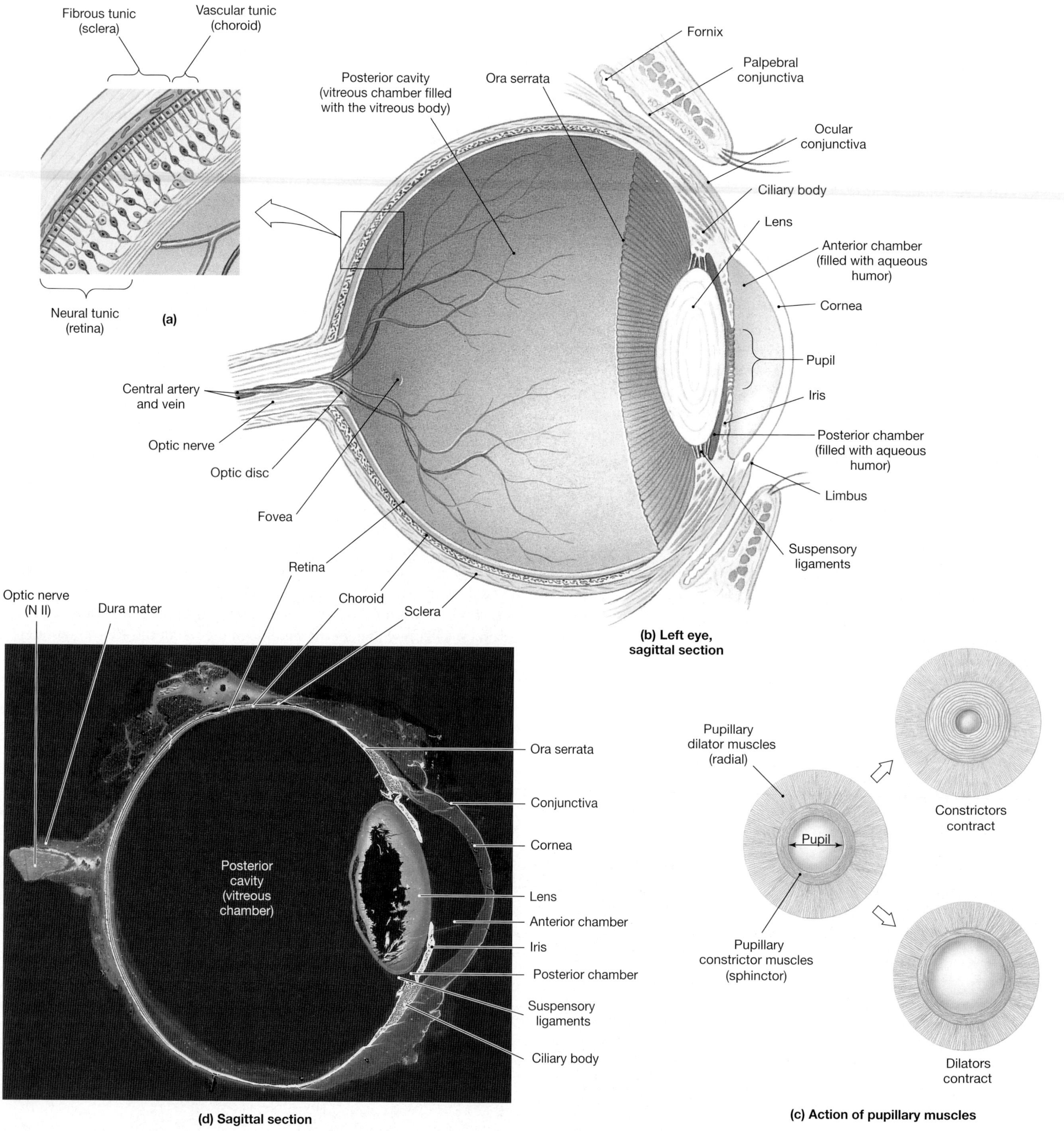

FIGURE 18.20 **Sectional Anatomy of the Eye**

(**a**) The three layers, or *tunics*, of the eye. (**b**) Major anatomical landmarks and features in a diagrammatic view of the left eye. (**c**) The action of pupillary muscles and changes in pupillary diameter. (**d**) Sagittal section through the eye. (**e**) Horizontal section through right eye. (**f**) Horizontal section through the top of the right eye and orbit.

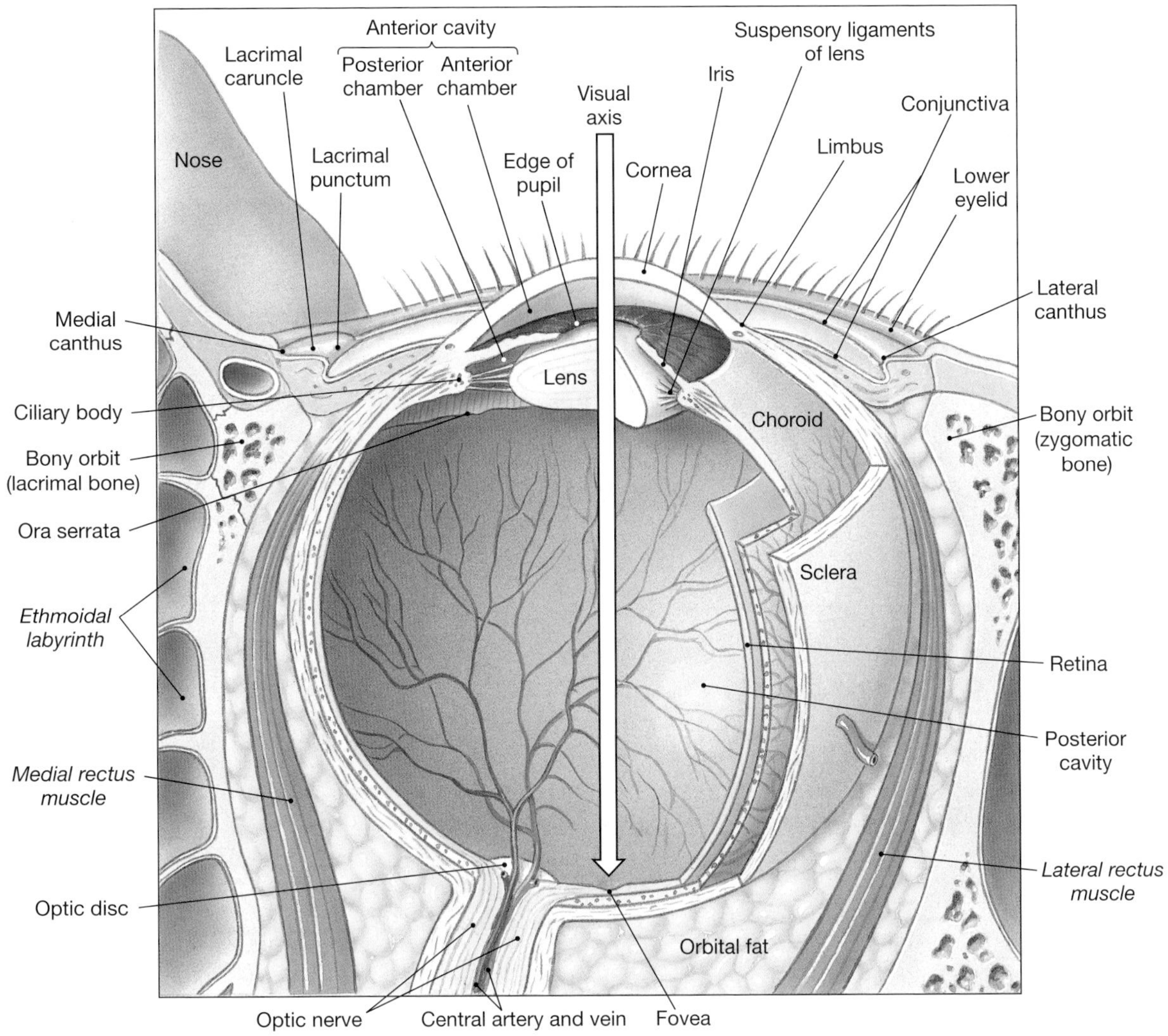

(e) Horizontal section, superior view

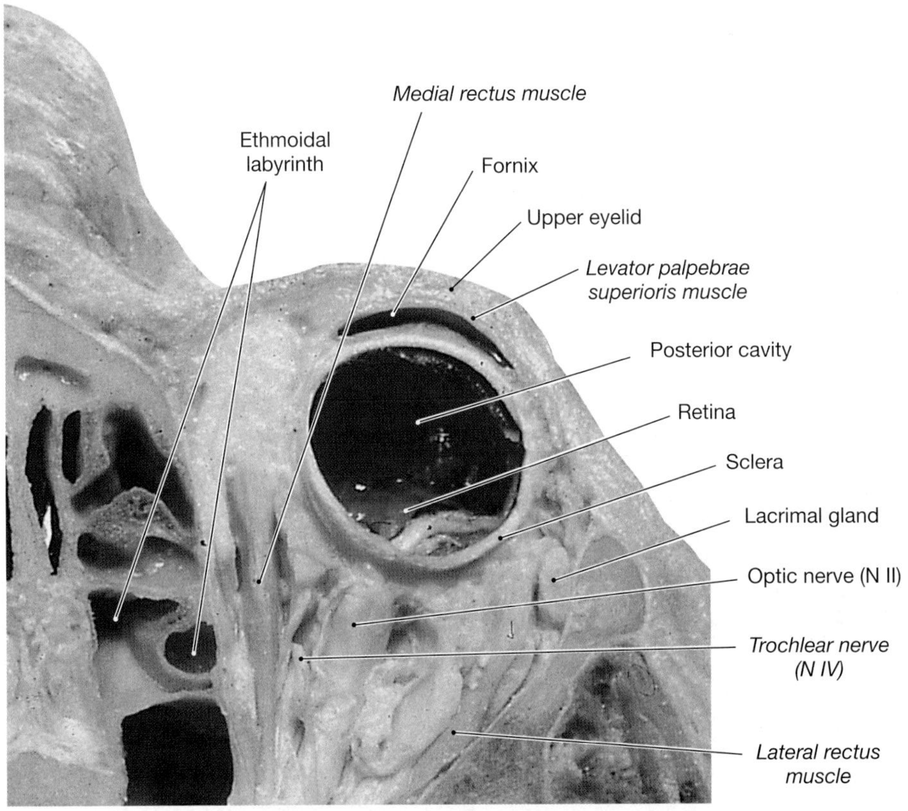

FIGURE 18.20 *(continued)*

(f) Horizontal section, superior view

CLINICAL BRIEF

Conjunctivitis

Conjunctivitis, or "pink-eye," results from damage to and irritation of the conjunctival surface. The most obvious symptom results from dilation of the blood vessels deep to the conjunctival epithelium. The term *conjunctivitis* is more useful as the description of a symptom than as a name for a specific disease. A great variety of pathogens, including bacteria, viruses, and fungi can cause conjunctivitis, and a temporary form of the condition may be produced by allergic, chemical, or physical irritation (including even such mundane experiences as prolonged crying or peeling an onion).

Chronic conjunctivitis, or **trachoma**, results from bacterial or viral invasion of the conjunctiva. Many of these infections are highly contagious, and severe cases may scar the corneal surface and affect vision. The bacteria most often involved is *Chlamydia trachomatis*. Trachoma is a relatively common problem in southwestern North America, North Africa, and the Middle East. The condition must be treated with topical and systemic antibiotics to prevent corneal damage and vision loss.

THE VASCULAR TUNIC (UVEA) [FIGURES 18.20d,e/18.21]

The **vascular tunic**, or **uvea**, contains numerous blood vessels, lymphatics, and the intrinsic eye muscles. The functions of this layer include: (1) providing a route for blood vessels and lymphatics that supply tissues of the eye, (2) regulating the amount of light entering the eye, (3) secreting and reabsorbing the *aqueous humor* that circulates within the eye, and (4) controlling the shape of the lens, an essential part of the focusing process. The vascular tunic includes the *iris*, the *ciliary body*, and the *choroid* (Figures 18.20b,d,e and 18.21●).

The Iris [FIGURES 18.20/18.21] The **iris** can be seen through the transparent corneal surface. The iris contains blood vessels, pigment cells, and the two layers of smooth muscle fibers that comprise the *intrinsic eye muscles*. Contraction of these muscles changes the diameter of the central opening of the iris, the **pupil**. One group of smooth muscle fibers forms a series of concentric circles around the pupil (Figure 18.20c●). The diameter of the pupil decreases when these **pupillary sphincter muscles** contract. A second group of smooth muscles extends radially from the edge of the pupil. Contraction of these **pupillary dilator muscles** enlarges the pupil. These antagonistic muscle groups are controlled by the autonomic nervous system; parasympathetic activation causes pupillary constriction, and sympathetic activation causes pupillary dilation. p. 462

The body of the iris consists of a connective tissue whose posterior surface is covered by an epithelium containing pigment cells. Pigment cells may also be present in the connective tissue of the iris and in the epithelium covering its anterior surface. Eye color is determined by the density and distribution of pigment cells. When there are no pigment cells in the body of the iris, light passes through it and bounces off the inner surface of the pigmented epithelium. The eye then appears blue. Individuals with gray, brown, and black eyes have more pigment cells, respectively, in the body and surface of the iris.

The Ciliary Body [FIGURES 18.20b,d,e/18.21] At its periphery the iris attaches to the anterior portion of the ciliary body. The **ciliary body** begins at the junction between the cornea and sclera and extends posteriorly to the **ora serrata** (Ō-ra ser-RĀ-ta; "serrated mouth") (Figures 18.20b,d,e and 18.21●). The bulk of the ciliary body consists of the **ciliary muscle**, a muscular ring that projects into the interior of the eye. The epithelium is thrown into numerous folds, called **ciliary processes**. The **suspensory ligaments**, or *zonular fibers*, of the lens attach to these processes. These connective tissue fibers hold the lens posterior to the iris and centered on the pupil. As a result, any light passing through the pupil and headed for the photoreceptors will pass through the lens.

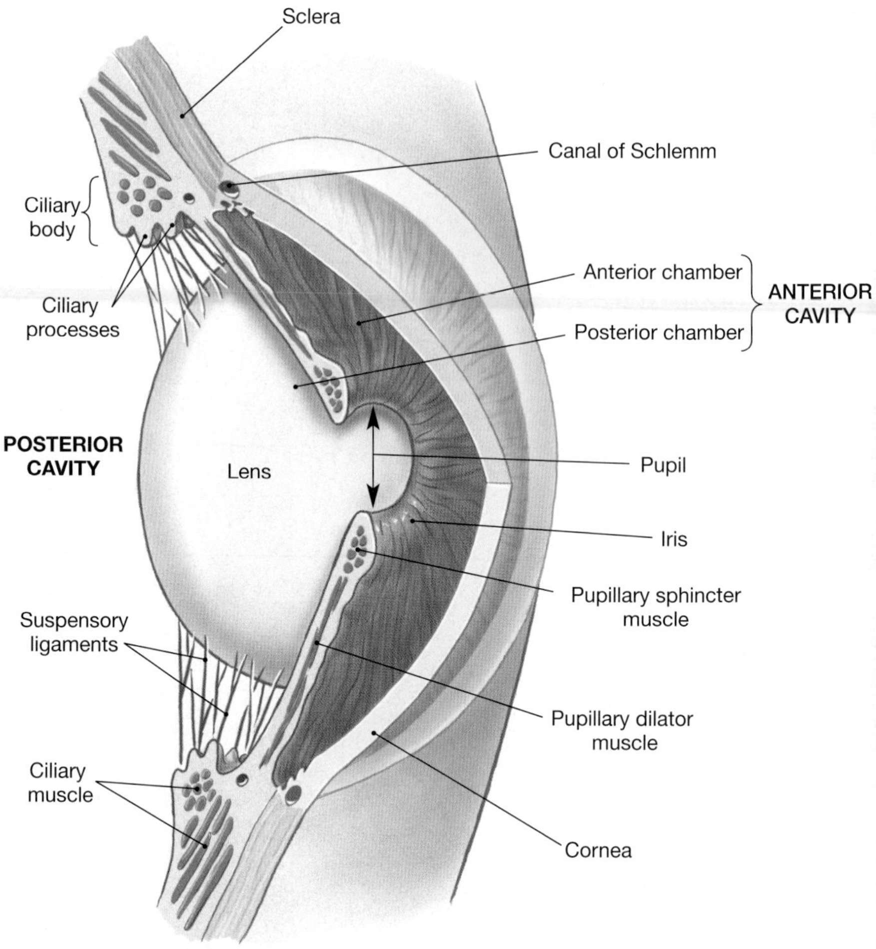

FIGURE 18.21 **The Lens and Chambers of the Eye**

The lens is suspended between the posterior cavity and the posterior chamber of the anterior cavity. Its position is maintained by the suspensory ligaments that attach the lens to the ciliary body.

The Choroid [FIGURE 18.20] The **choroid** contains an extensive capillary network that delivers oxygen and nutrients to the outer portion of the retina. It also contains scattered melanocytes, which are especially dense in the outermost portion of the choroid adjacent to the sclera (Figure 18.20a,b,d,e●). The innermost portion of the choroid attaches to the outer retinal layer.

CLINICAL BRIEF

Corneal Transplants

The cornea has a very restricted ability to repair itself, so corneal injuries must be treated immediately to prevent serious vision loss. To restore vision after corneal scarring, it is usually necessary to replace the cornea through a *corneal transplant*. Corneal replacements are probably the most common form of transplant surgery. They can be performed between unrelated individuals because there are no corneal blood vessels, and white blood cells that would otherwise reject the graft are unlikely to enter the area. Corneal tissues must be removed within five hours after the donor's death.

THE NEURAL TUNIC [FIGURES 18.20 TO 18.22]

The **neural tunic**, or **retina**, consists of two distinct layers, an outer **pigmented layer** and an inner **neural layer**, called the **neural retina**, which contains the visual receptors and associated neurons (Figures 18.20 to 18.22●). The pigment layer absorbs light after it passes through the retina and has important biochemical interactions with retinal photoreceptors. The neural retina contains (1) the photoreceptors that respond to light, (2) supporting cells and neurons that perform preliminary processing and integration of visual information, and (3) blood vessels supplying tissues lining the posterior cavity.

The neural retina and pigmented layers are normally very close together, but not tightly interconnected. The pigmented layer continues over the ciliary body and iris, although the neural retina extends anteriorly only as far as the ora serrata. The neural retina thus forms a cup that establishes the posterior and lateral boundaries of the posterior cavity (Figure 18.20b,d,e,f●).

RETINAL ORGANIZATION [FIGURES 18.20b,e/18.22] There are approximately 130 million photoreceptors in the retina, each monitoring a specific location on the retinal surface. A visual image results from the processing of information provided by the entire receptor population. In sectional view, the retina contains several layers of cells (Figure 18.22a,b●). The outermost layer, closest to the pigmented layer, contains the visual receptors. There are two types of **photoreceptors**: **rods** and **cones**. Rods do not discriminate between different colors of light. They are very light-sensitive and enable us to see in dimly lit rooms, at twilight, or in pale moonlight. Cones provide us with color vision. There are three types of cones, and their stimulation in various combinations provides the perception of different colors. Cones give us sharper, clearer images, but they require more intense light than rods. If you sit outside at sunset (or sunrise), you will probably be able to tell when your visual system shifts from cone-based vision (clear images in full color) to rod-based vision (relatively grainy images in black and white).

Rods and cones are not evenly distributed across the outer surface of the retina. Approximately 125 million rods form a broad band around the periphery of the retina. The posterior retinal surface is dominated by the presence of roughly 6 million cones. Most of these are concentrated in the area where a visual image arrives after passing through the cornea and lens. There are no rods in this region, which is known as the **macula lutea** (LOO-tē-a; "yellow spot"). The highest concentration of cones is found in the central portion of the macula lutea, at the **fovea** (FŌ-vē-a; "shallow depression"), or *fovea centralis*. The fovea is the site of sharpest vision; when you look directly at an object, its image falls upon this portion of the retina (Figures 18.20b,e and 18.22c●).

The rods and cones synapse with roughly 6 million **bipolar cells** (Figure 18.22a,b●). Stimulation of rods and cones alters their rates of neurotransmitter release, and this in turn alters the activity of the bipolar cells. **Horizontal cells** at this same level form a network that inhibits or facilitates communication between the visual receptors and bipolar cells. Bipolar cells in turn synapse within the layer of **ganglion cells** that faces the vitreous chamber. **Amacrine** (AM-a-krin) **cells** at this level modulate communication between bipolar and ganglion cells. The ganglion cells are the only cells in the retina that generate action potentials to the brain.

Axons from an estimated 1 million ganglion cells converge on the **optic disc**, penetrate the wall of the eye, and proceed toward the diencephalon as the optic nerve (N II) (Figure 18.20b,e●). The *central retinal artery* and *central retinal vein* that supply the retina pass through the center of the optic nerve and emerge on the surface of the optic disc (Figure 18.22c●). There are no photoreceptors or other retinal structures at the optic disc. Because light striking this area goes unnoticed, it is commonly called the **blind spot**. You do not "notice" a blank spot in the visual field, because involuntary eye movements keep the visual image moving and allow the brain to fill in the missing information. *Scotomas and Floaters p. 796*

THE CHAMBERS OF THE EYE

The chambers of the eye are the *anterior*, *posterior*, and *vitreous chambers*. The anterior and posterior chambers are filled with *aqueous humor*.

AQUEOUS HUMOR [FIGURE 18.23] **Aqueous humor** forms continuously as interstitial fluids pass between the epithelial cells of the ciliary processes and enter the posterior chamber (Figure 18.23●). The epithelial cells appear to regulate its composition, which resembles that of cerebrospinal fluid. The aqueous humor circulates so that in addition to forming a fluid cushion, it provides an important route for nutrient and waste transport.

Aqueous humor returns to the circulation in the anterior chamber near the edge of the iris. After diffusing through the local epithelium, it passes into the **canal of Schlemm**, or *scleral venous sinus*, which communicates with the veins of the eye.

The **lens** lies posterior to the cornea, held in place by the suspensory ligaments that originate on the ciliary body of the choroid (Figure 18.23●). The lens and its attached suspensory ligaments form the anterior boundary of the vitreous chamber. This chamber contains the **vitreous body,** a gelatinous mass sometimes called the *vitreous humor*. The vitreous body helps to maintain the shape of the eye, support the posterior surface of the lens, and give physical support to the retina by pressing the neural layer against the pigment layer. Aqueous humor produced in the posterior chamber freely diffuses through the vitreous body and across the retinal surface.

THE LENS [FIGURES 18.20/18.23]

The primary function of the lens is to focus the visual image on the retinal photoreceptors. It accomplishes this by changing its shape. The lens consists of concentric layers of cells that are precisely organized (Figures 18.20b,d,e and 18.23●). A dense, fibrous capsule covers the entire lens.

CLINICAL BRIEF

GLAUCOMA

Glaucoma affects roughly 2% of the population over 40. In this condition the aqueous humor no longer has free access to the canal of Schlemm. The primary factors responsible cannot be determined in 90% of all cases. Although drainage is impaired, production of aqueous humor continues unabated, and the intraocular pressure begins to rise. The fibrous scleral coat cannot expand significantly, so the increasing pressures begin to push against the surrounding intraocular soft tissues.

The optic nerve is not wrapped in connective tissue, for it penetrates all three tunics. When intraocular pressures have risen to roughly twice normal levels, distortion of the nerve fibers begins to affect visual perception. If this condition is not corrected, blindness eventually results.

Most eye exams include a glaucoma test. Intraocular pressure is tested by bouncing a tiny blast of air off the surface of the eye and measuring the deflection produced. Glaucoma may be treated by the application of drugs that constrict the pupil and tense the edge of the iris, making the surface more permeable to aqueous humor. Surgical correction involves perforating the wall of the anterior chamber to encourage drainage and is now performed by laser surgery on an outpatient basis.

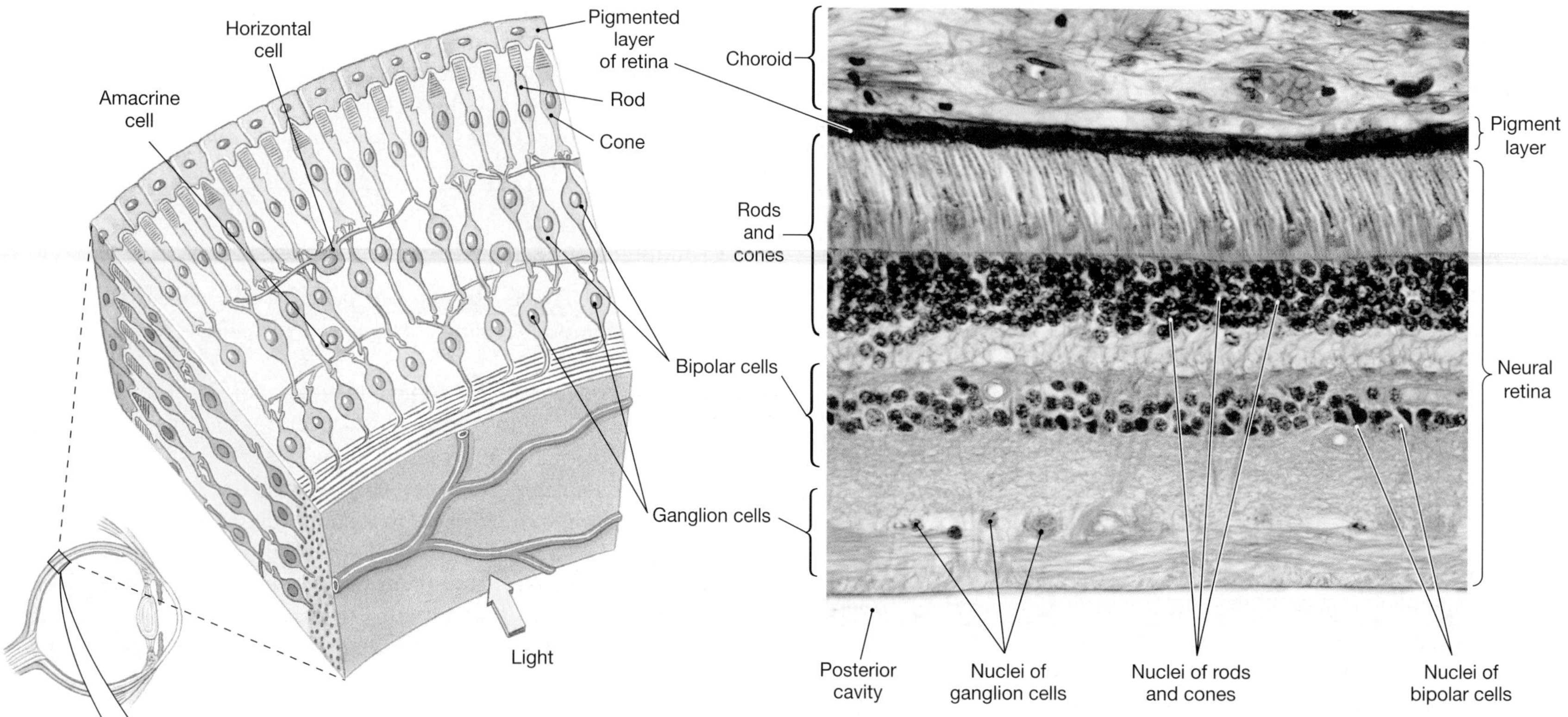

(a)

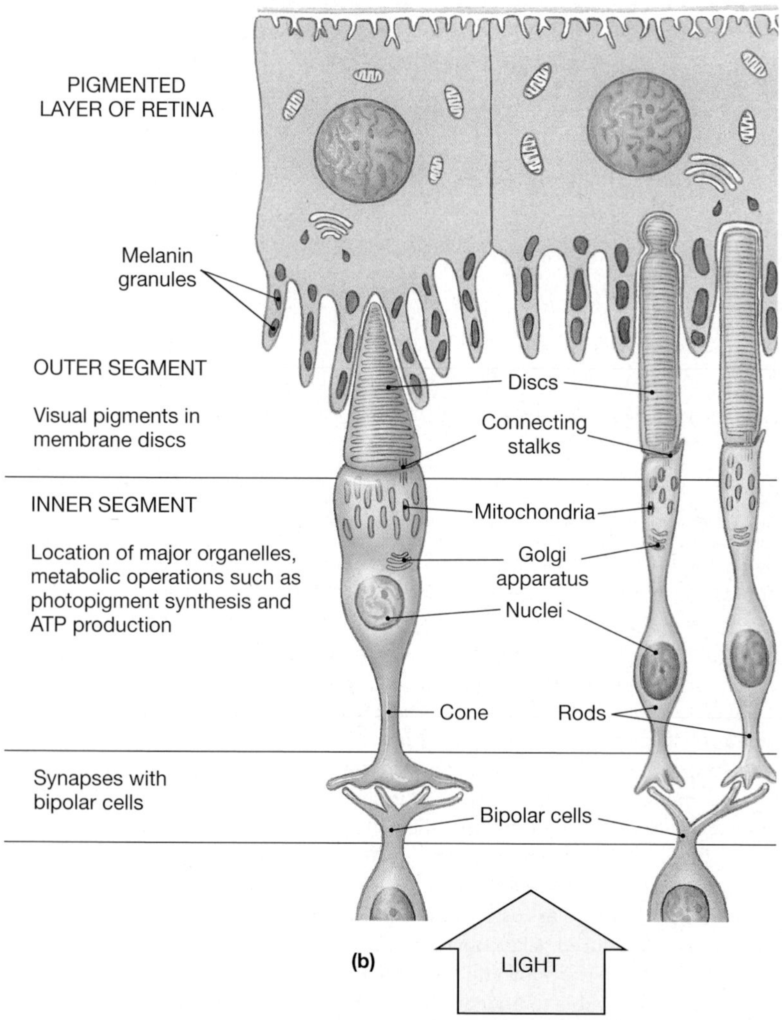

(b)

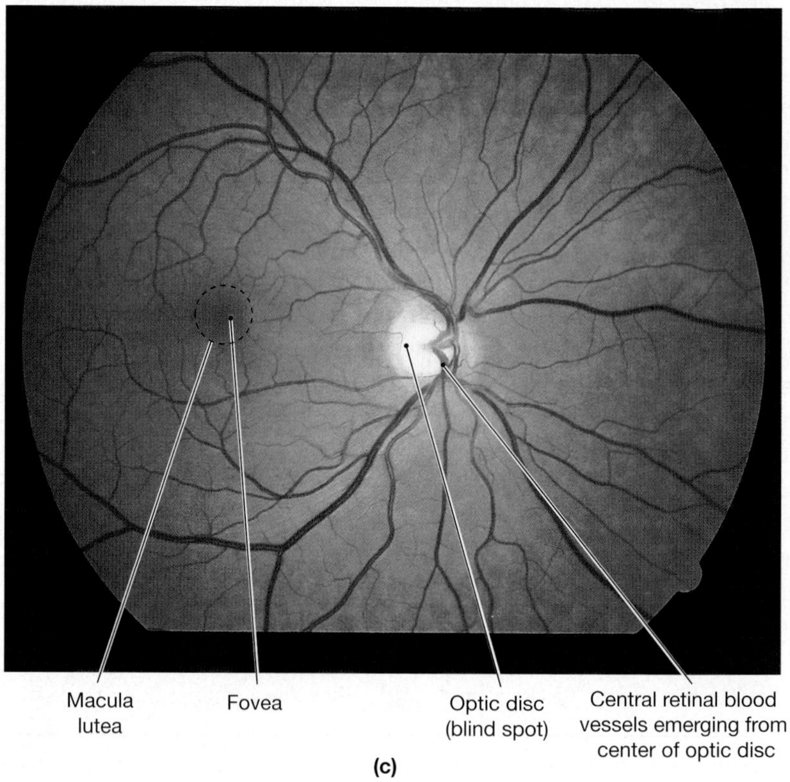

(c)

FIGURE 18.22 **Retinal Organization**

(a) Cellular organization of the retina. Note that the photoreceptors are located closest to the choroid rather than near the vitreous chamber. (LM × 73) **(b)** Diagrammatic view of the fine structure of rods and cones, based on data from electron microscopy. **(c)** A photograph taken through the pupil of the eye, showing the retinal blood vessels, the origin of the optic nerve, and the optic disc, or "blind spot."

Many of the capsular fibers are elastic, and unless an outside force is applied, they will contract and make the lens spherical. Around the edges of the lens, the capsular fibers intermingle with those of the suspensory ligaments.

At rest, tension in the suspensory ligaments overpowers the elastic capsule and flattens the lens. In this position the eye is focused for distant vision. When the ciliary muscles contract, the ciliary body moves toward the lens. This movement reduces the tension in the suspensory ligaments, and the elastic lens assumes a more spherical shape, which focuses the eye on nearby objects.

Visual Pathways [FIGURES 18.24/18.25]

Each rod and cone cell monitors a specific receptive field. A visual image results from the processing of information provided by the entire receptor population. A significant amount of processing occurs in the retina before the information is sent to the brain, because of interactions between the various cell types.

The two optic nerves, one from each eye, reach the diencephalon at the optic chiasm (Figure 18.24●). From this point, a partial decussation occurs: approximately one-half of the fibers proceed toward the lateral geniculate nucleus of the same side of the brain, while the other half cross over to reach the lateral geniculate

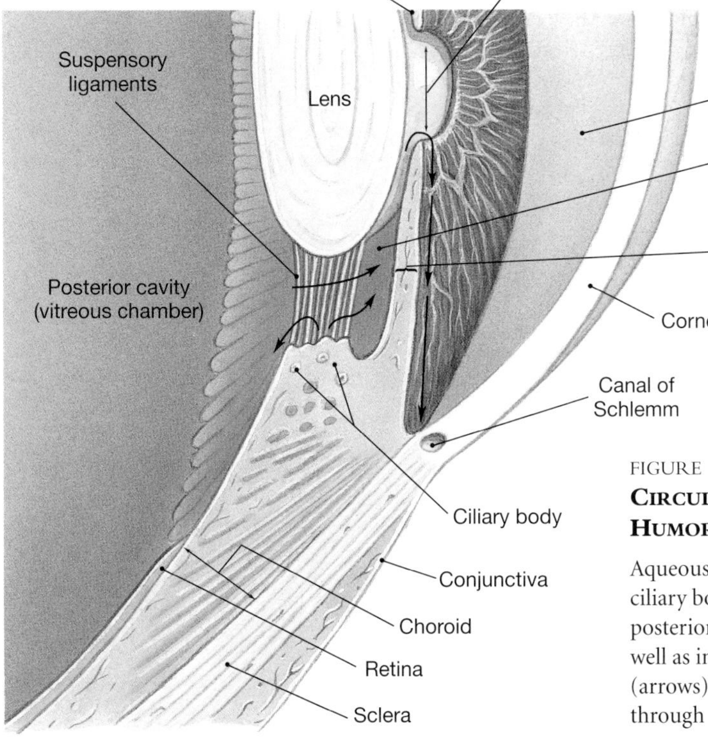

FIGURE 18.23 **THE CIRCULATION OF AQUEOUS HUMOR**

Aqueous humor secreted at the ciliary body circulates through the posterior and anterior chambers as well as into the posterior cavity (arrows) before it is reabsorbed through the canal of Schlemm.

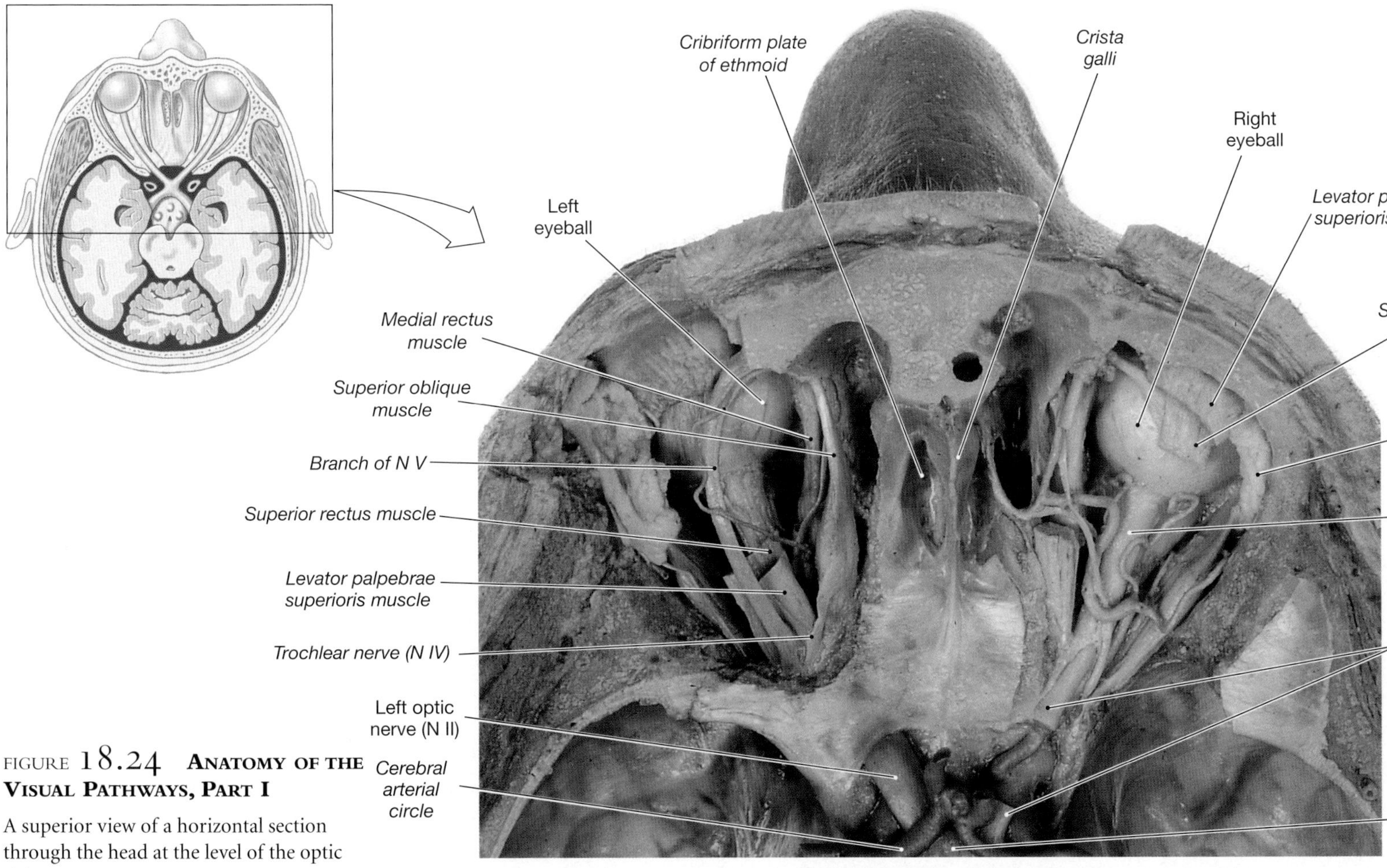

FIGURE 18.24 **ANATOMY OF THE VISUAL PATHWAYS, PART I**

A superior view of a horizontal section through the head at the level of the optic chiasm.

FIGURE 18.25 **Anatomy of the Visual Pathways, Part II**

(**a**) An inferior view of the brain. Much of the brain stem has been removed. Portions of the cerebral hemispheres have been dissected away to show the optic tract. (**b**) At the optic chiasm, a partial crossover of nerve fibers occurs. As a result, each hemisphere receives visual information from the lateral half of the retina of the eye on that side and from the medial half of the retina of the eye on the opposite side. Visual association areas integrate this information to develop a composite picture of the entire visual field.

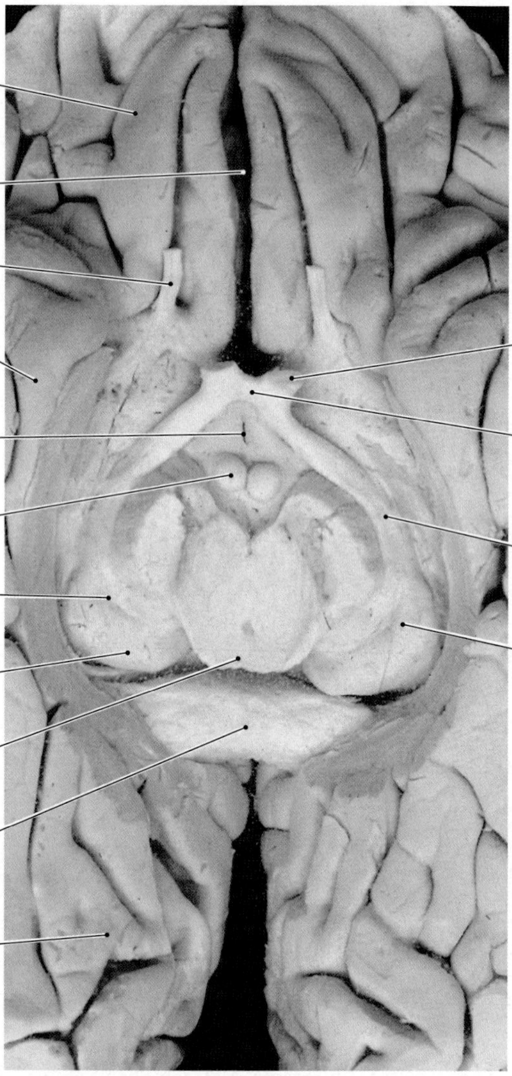

(a) Inferior view of the brain

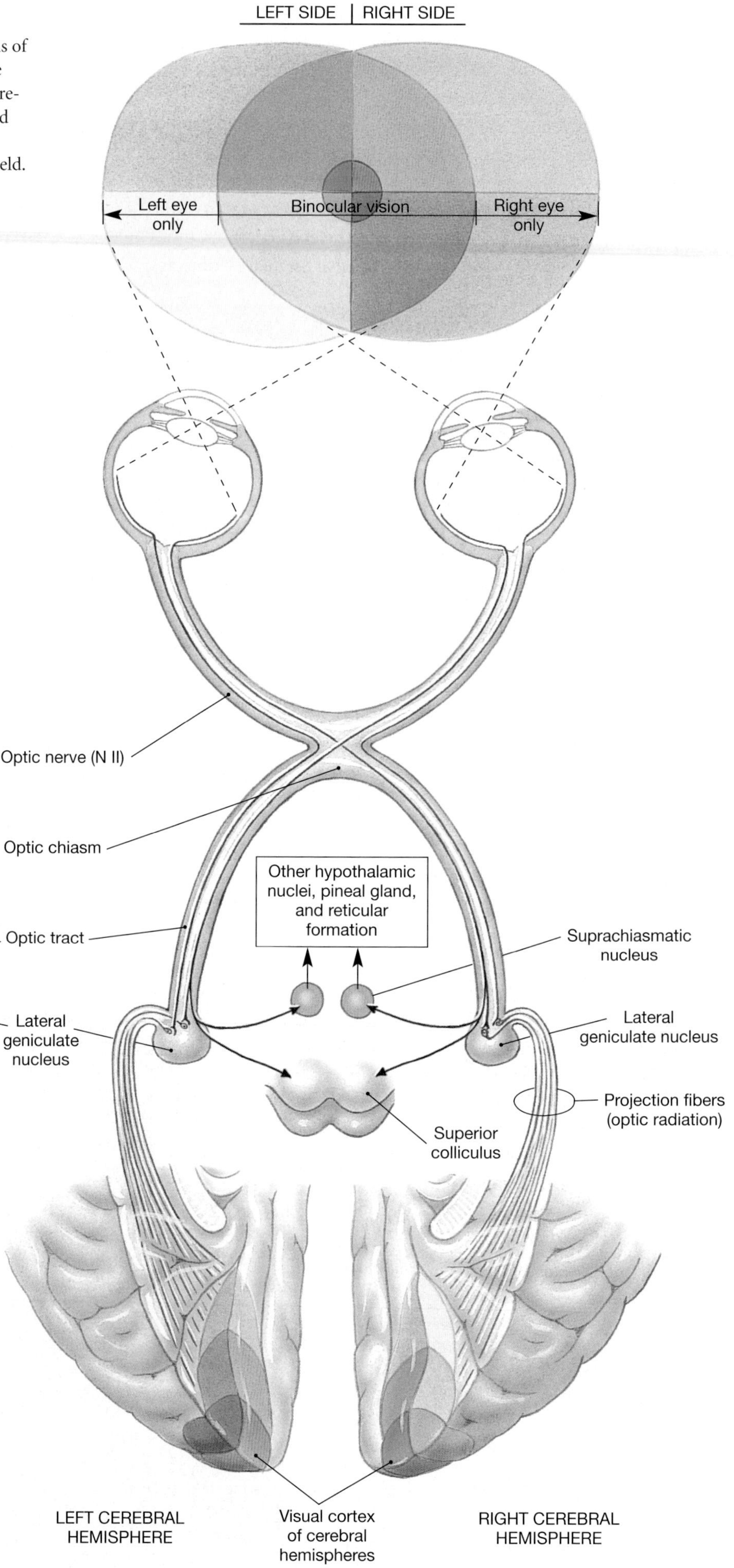

(b) Visual pathways

nucleus of the opposite side (Figure 18.25●). Visual information from the left half of each retina arrives at the lateral geniculate nucleus of the left side; information from the right half of each retina goes to the right side. The lateral geniculate nuclei act as a switching center that relays visual information to reflex centers in the brain stem as well as to the cerebral cortex. The reflexes that control eye movement are triggered by information that bypasses the lateral geniculate nuclei to synapse in the superior colliculi.

CORTICAL INTEGRATION [FIGURE 18.25]

The sensation of vision arises from the integration of information arriving at the visual cortex of the occipital lobes of the cerebral hemispheres. The visual cortex contains a sensory map of the entire field of vision. As in the case of the primary sensory cortex, the map does not faithfully duplicate the relative areas within the sensory field.

Each eye also receives a slightly different image, because (1) their foveae are 2–3 inches apart, and (2) the nose and eye socket block the view of the opposite side. The association and integrative areas of the cortex compare the two perspectives (Figure 18.25●) and use them to provide us with depth perception. The partial crossover that occurs at the optic chiasm ensures that the visual cortex receives a *composite* picture of the entire visual field.

THE BRAIN STEM AND VISUAL PROCESSING [FIGURE 18.25]

Many centers in the brain stem receive visual information, either from the lateral geniculate nuclei or via collaterals from the optic tracts. Collaterals that bypass the lateral geniculate nuclei synapse in the superior colliculus or hypothalamus (Figure 18.25b●). The superior colliculus of the midbrain issues motor commands controlling unconscious eye, head, or neck movements in response to visual stimuli. Visual inputs to the **suprachiasmatic** (soo-pra-kī-az-MA-tic) **nucleus** of the hypothalamus and the endocrine cells of the *pineal gland* affect the function of other brain stem nuclei. These nuclei establish a daily pattern of visceral activity that is tied to the day-night cycle. This **circadian rhythm** (*circa*, about + *dies*, day) affects metabolic rate, endocrine function, blood pressure, digestive activities, the awake-asleep cycle, and other physiological processes.

CLINICAL BRIEF

CATARACTS

The transparency of the lens depends on a precise combination of structural and biochemical characteristics. When that balance becomes disturbed, the lens loses its transparency, and the abnormal lens is known as a **cataract**. Cataracts may be congenital or result from drug reactions, injuries, or radiation, but **senile cataracts** are the most common form. As aging proceeds, the lens becomes less elastic, and the individual has difficulty focusing on nearby objects. (The person becomes "far-sighted.")

Over time, the lens takes on a yellowish hue, and eventually it begins to lose its transparency. As the lens becomes "cloudy," the individual needs brighter reading lights and larger type. Visual clarity begins to fade. If the lens becomes completely opaque, the person will be functionally blind, even though the retinal receptors are normal. Modern surgical procedures involve removing the lens, either intact or in pieces, after shattering it with high-frequency sound. The missing lens can be replaced by an artificial one placed behind the iris. Vision can then be fine-tuned with glasses or contact lenses.

✓ CONCEPT CHECK

- What layer of the eye would be the first to be affected by inadequate tear production?
- If the intraocular pressure becomes abnormally high, which structures of the eye are affected, and how are they affected?
- Would a person born without cones in her eyes be able to see? Explain.
- In anatomy laboratory, your partner asks, "What are ciliary processes, and what do they do?" How do you answer?

RELATED CLINICAL TERMS

anesthesia: A total loss of sensation. *Assessment of Tactile Sensitivities p. 795*

audiogram: A graphical record of a subject's performance during a hearing test. *Testing and Treating Hearing Deficits p. 795*

bone conduction test: A test for conductive deafness, usually involving placement of a vibrating tuning fork against the skull. *Testing and Treating Hearing Deficits p. 795*

cataract: An abnormal lens that has lost its transparency. *p. 497*

cochlear implant: Insertion of electrodes into the cochlear nerve to provide external stimulation that provides some sensitivity to sounds in the absence of a functional organ of Corti. *Testing and Treating Hearing Deficits p. 795*

conductive deafness: Deafness resulting from conditions in the middle ear that block the transfer of vibrations from the tympanic membrane to the oval window. *p. 486*

hypesthesia: A reduction in sensitivity. *Assessment of Tactile Sensitivities p. 795*

mastoidectomy: Surgical opening and draining of the mastoid sinuses. *p. 479*

Ménière's disease: Acute vertigo caused by the rupture of the wall of the membranous labyrinth. *Vertigo, Dizziness, and Motion Sickness p. 795*

myringotomy: Drainage of the middle ear through a surgical opening in the tympanum. *p. 479*

nerve deafness: Deafness resulting from problems within the cochlea or along the auditory pathway. *p. 486*

nystagmus: Short, jerky eye movements that sometimes appear after damage to the brain stem or inner ear. *p. 481*

paresthesia: Abnormal sensations. *Assessment of Tactile Sensitivities p. 795*

referred pain: Pain sensations from visceral organs, often perceived as originating in more superficial areas innervated by the same spinal nerves. *p. 469*

rhizotomy: Cutting the dorsal roots, which provide sensation from an area, to relieve pain. *The Control of Pain p. 795*

scotomas: Abnormal blind spots that are fixed in position. *Scotomas and Floaters p. 796*

tractotomy: Severing ascending spinal tracts to relieve pain. *The Control of Pain p. 795*

vertigo: An inappropriate sense of motion. *Vertigo, Dizziness, and Motion Sickness p. 795*

Additional Clinical Terms Discussed in Appendix I (pp. 795–796)

dizziness; motion sickness; two-point discrimination test

VISION

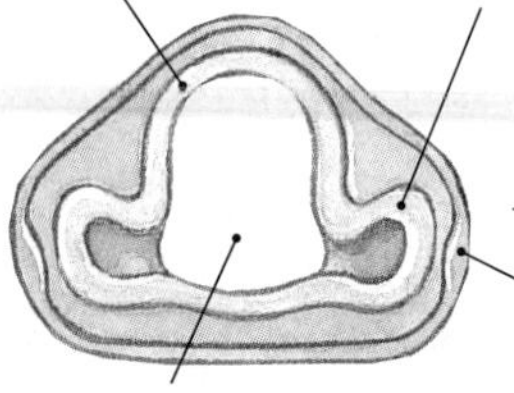

The first indication of optic development appears as a pair of bulges called **optic vesicles** in the lateral walls of the prosencephalon. These extend to either side like a pair of dumbbells, each containing a cavity continuous with the neurocoel.

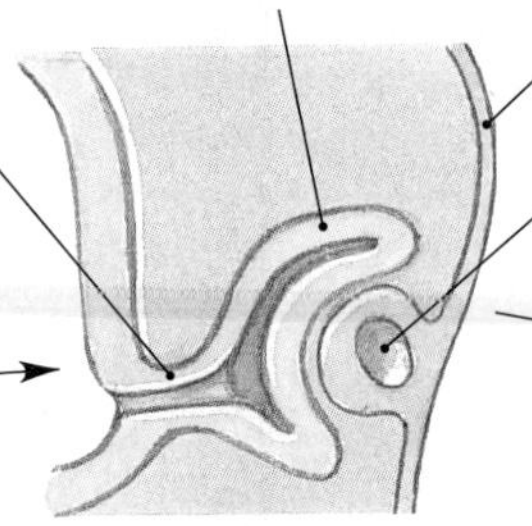

These bulges become indented, forming a pair of **optic cups**, which remain connected to the diencephalon by **optic stalks**. The epidermis overlying the optic cup responds by forming a **lens placode**, which thickens and creates another vesicle. This **lens vesicle** becomes the lens.

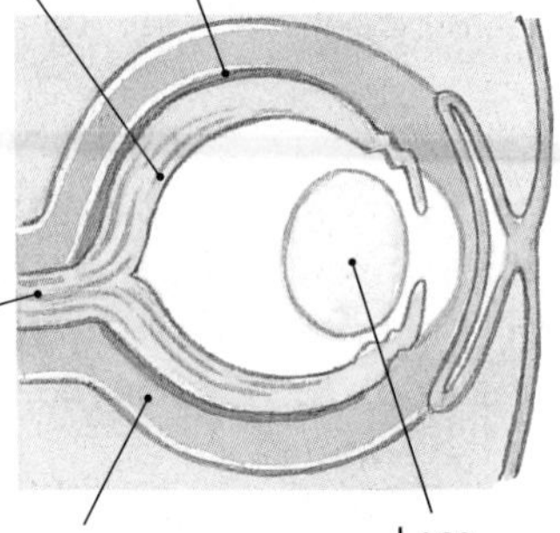

Mesoderm aggregating around this complex forms the choroid and scleral coats. The anterior and posterior chambers develop as cavities appear within the mesoderm.

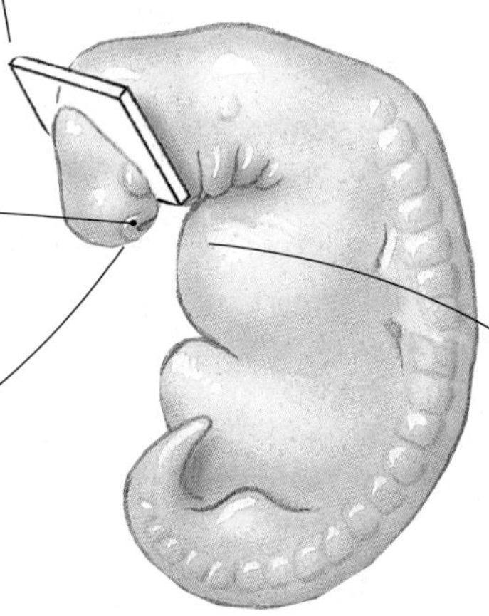

4½ WEEKS

All special sense organs develop from the epithelia of the embryo.

OLFACTION

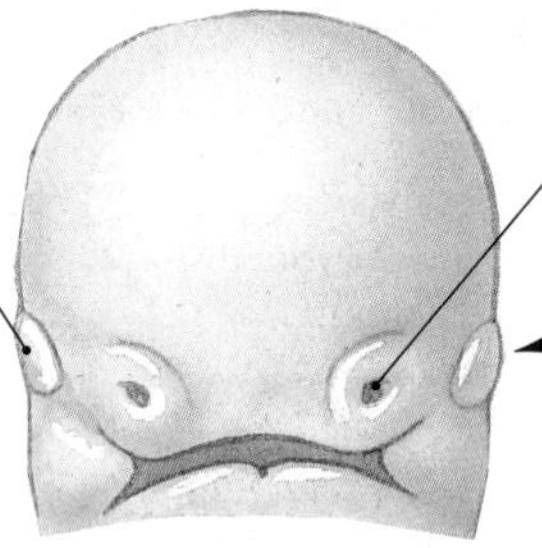

5 WEEKS

Olfactory receptors begin as a pair of thickened areas in front of the prosencephalon during the fifth developmental week. The thickenings are called **nasal placodes**.

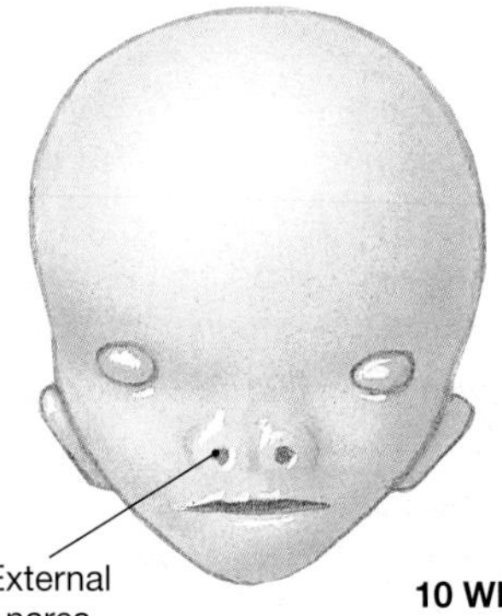

10 WEEKS

Over time, the nasal placodes are enfolded and protected by developing facial structures. (Development of the face was discussed in the Embryology Summary on pp. 156–157.)

GUSTATION

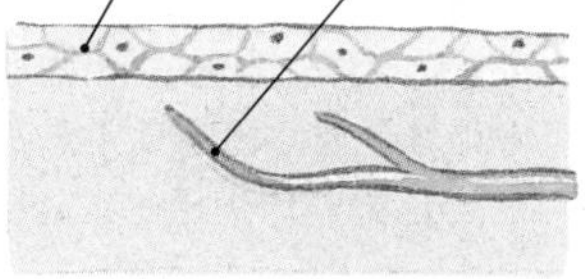

Gustatory receptors are the least specialized of any of the special sense organs. Taste buds develop as sensory fibers grow into the developing mouth and pharynx.

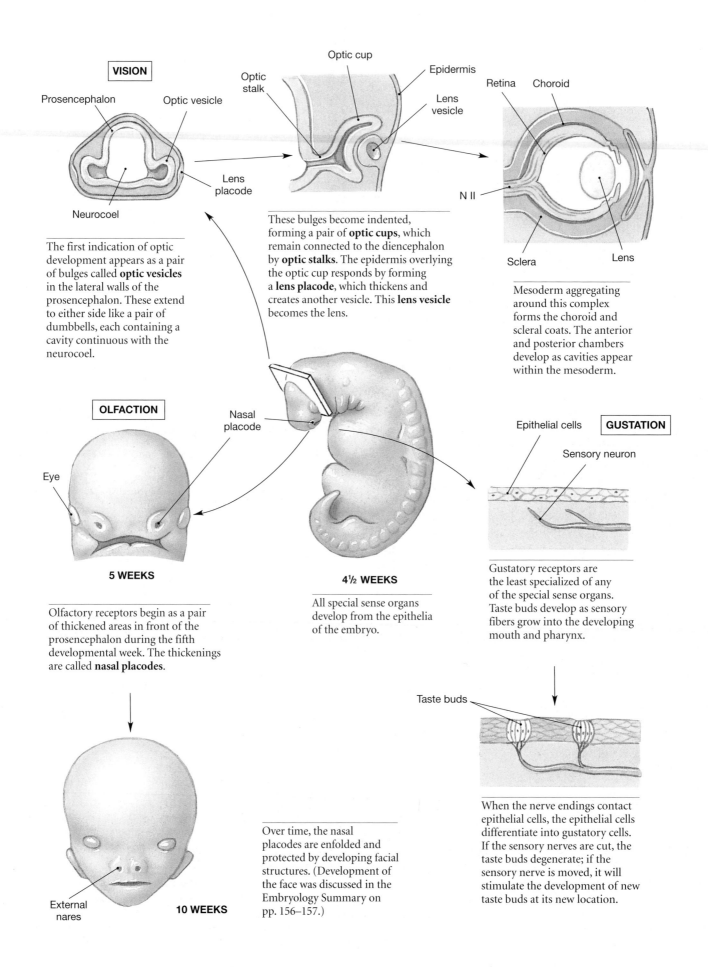

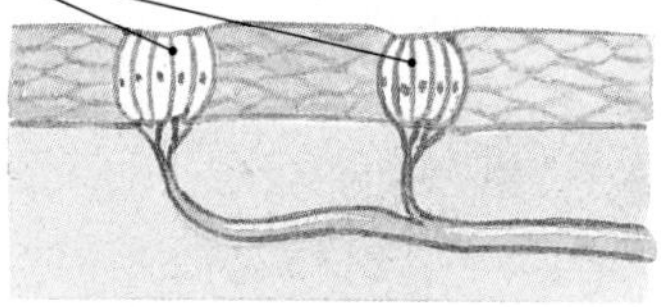

When the nerve endings contact epithelial cells, the epithelial cells differentiate into gustatory cells. If the sensory nerves are cut, the taste buds degenerate; if the sensory nerve is moved, it will stimulate the development of new taste buds at its new location.

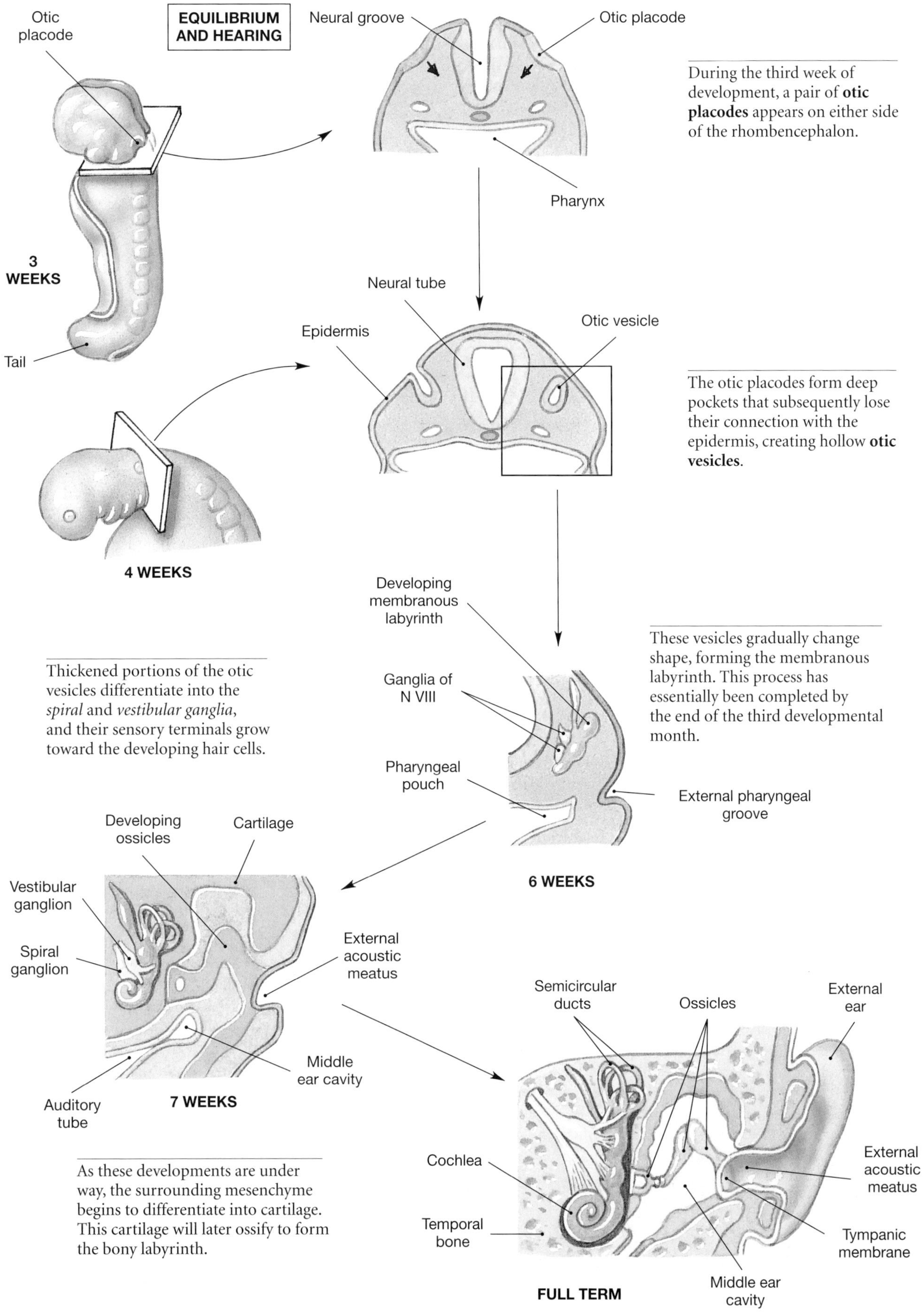

During the third week of development, a pair of **otic placodes** appears on either side of the rhombencephalon.

The otic placodes form deep pockets that subsequently lose their connection with the epidermis, creating hollow **otic vesicles**.

These vesicles gradually change shape, forming the membranous labyrinth. This process has essentially been completed by the end of the third developmental month.

Thickened portions of the otic vesicles differentiate into the *spiral* and *vestibular ganglia*, and their sensory terminals grow toward the developing hair cells.

As these developments are under way, the surrounding mesenchyme begins to differentiate into cartilage. This cartilage will later ossify to form the bony labyrinth.

STUDY OUTLINE & CHAPTER REVIEW

Introduction 468

1. The **general senses** are temperature, pain, touch, pressure, vibration, and proprioception; receptors for these sensations are distributed throughout the body. Receptors for the **special senses** (**olfaction**, **gustation**, **equilibrium**, **hearing**, and **vision**) are located in specialized areas, or **sense organs**. A **sensory receptor** is a specialized cell that when stimulated sends a **sensation** to the CNS.

Receptors 468

1. **Receptor specificity** allows each receptor to respond to particular stimuli. The simplest receptors are **free nerve endings**; the area monitored by a single receptor cell is the **receptive field**. *(see Figure 18.1)*

Interpretation of Sensory Information 468

2. **Tonic receptors** are always sending signals to the CNS; **phasic receptors** become active only when the conditions that they monitor change.

Central Processing and Adaptation 468

3. **Adaptation** (a reduction in sensitivity in the presence of a constant stimulus) may involve changes in receptor sensitivity (**peripheral**, or **sensory**, **adaptation**) or inhibition along the sensory pathways (**central adaptation**). **Fast-adapting receptors** are phasic; **slow-adapting receptors** are tonic.

Sensory Limitations 469

4. The information provided by our sensory receptors is incomplete because: (1) we do not have receptors for every stimulus; (2) our receptors have limited ranges of sensitivity; and (3) a stimulus produces a neural event that must be interpreted by the CNS.

The General Senses 469

1. Receptors are classified as **exteroceptors** if they provide information about the external environment and **interoceptors** if they monitor conditions inside the body.

Nociceptors 469

2. **Nociceptors** respond to a variety of stimuli usually associated with tissue damage. There are two types of these painful sensations: **fast** (*prickling*) **pain** and **slow** (*burning and aching*) **pain**. *(see Figures 18.2/18.3a)*

Thermoreceptors 469

3. **Thermoreceptors** respond to changes in temperature. They conduct sensations along the same pathways that carry pain sensations.

Mechanoreceptors 470

4. **Mechanoreceptors** respond to physical distortion, contact, or pressure on their cell membranes: **tactile receptors** to touch, pressure, and vibration; **baroreceptors** to pressure changes in the walls of blood vessels and the digestive, reproductive, and urinary tracts; and **proprioceptors** (*muscle spindles*) to positions of joints and muscles. *(see Figures 18.3/18.4)*
5. **Fine touch** and **pressure receptors** provide detailed information about a source of stimulation; **crude touch** and **pressure receptors** are poorly localized. Important tactile receptors include **free nerve endings**, the **root hair plexus**, **tactile discs** (*Merkel's discs*), **tactile corpuscles** (*Meissner's corpuscles*), **lamellated corpuscles** (*pacinian corpuscles*), and **Ruffini corpuscles**. *(see Figure 18.3)*
6. **Baroreceptors** (stretch receptors) monitor changes in pressure; they respond immediately but adapt rapidly. Baroreceptors in the walls of major arteries and veins respond to changes in blood pressure. Receptors along the digestive tract help coordinate reflex activities of digestion. *(see Figure 18.4)*
7. **Proprioceptors** monitor the position of joints, tension in tendons and ligaments, and the state of muscular contraction.

Chemoreceptors 472

8. In general, **chemoreceptors** respond to water-soluble and lipid-soluble substances that are dissolved in the surrounding fluid. They monitor the chemical composition of body fluids. *(see Figure 18.5)*

Olfaction (Smell) 472

1. The **olfactory organs** contain the **olfactory epithelium** with **olfactory receptors** (neurons sensitive to chemicals dissolved in the overlying mucus), **supporting cells**, and **basal** (*stem*) **cells**. Their surfaces are coated with the secretions of the **olfactory glands**. *(see Figure 18.6)*

Olfactory Receptors 472

2. The olfactory receptors are modified neurons. *(see Figure 18.6b)*

Olfactory Pathways 474

3. The olfactory system has extensive limbic and hypothalamic connections that help explain the emotional and behavioral responses that can be produced by certain smells. *(see Figure 18.6b)*

Olfactory Discrimination 474

4. The olfactory system can make subtle distinctions between thousands of chemical stimuli; the CNS interprets the smell.
5. The olfactory receptor population shows considerable turnover and is the only known example of neuronal replacement in the adult human. The total number of receptors declines with age.

Gustation (Taste) 474

1. **Gustation**, or taste, provides information about the food and liquids that we consume.

Gustatory Receptors 475

2. **Gustatory receptors** are clustered in **taste buds**, each of which contains **gustatory cells**, which extend *taste hairs* through a narrow **taste pore**. *(see Figure 18.7b,c)*
3. Taste buds are associated with epithelial projections (**papillae**). *(see Figure 18.7a)*

Gustatory Pathways 475

4. The taste buds are monitored by cranial nerves VII, IX, and X. The afferent fibers synapse within the **nucleus solitarius** before proceeding to the thalamus and cerebral cortex. *(see Figure 18.8)*

Gustatory Discrimination 476

5. The **taste sensations** are sweet, salty, sour, bitter, umami, and water.
6. There are individual differences in the sensitivity to specific tastes. The number of taste buds and their sensitivity decline with age. *(see Figure 18.8)*

Equilibrium And Hearing 476

The External Ear 477

1. The **external ear** includes the **auricle**, which surrounds the entrance to the **external acoustic meatus** that ends at the **tympanic membrane (tympanum)**, or eardrum. *(see Figures 18.9/18.10)*

The Middle Ear 477

2. In the **middle ear**, the **tympanic cavity** encloses and protects the **auditory ossicles**, which connect the tympanic membrane with the receptor complex of the inner ear. The tympanic cavity communicates with the nasopharynx via the **auditory tube**. *(see Figures 18.9/18.10)*
3. The **tensor tympani** and **stapedius muscles** contract to reduce the amount of motion of the tympanum when very loud sounds arrive. *(see Figures 18.9/18.10b,d)*

The Inner Ear 479

4. The senses of equilibrium and hearing are provided by the receptors of the **inner ear** (housed within fluid-filled tubes and chambers known as the **membranous labyrinth**). Its chambers and canals contain **endolymph**. The **bony labyrinth** surrounds and protects the membranous labyrinth. The bony labyrinth can be subdivided into the **vestibule** and **semicircular canals** (providing the sense of equilibrium) and the **cochlea** (providing the sense of hearing). *(see Figures 18.9/18.11 to 18.17)*
5. The vestibule includes a pair of membranous sacs, the **utricle** and **saccule**, whose receptors provide sensations of gravity and linear acceleration. The cochlea contains the **cochlear duct**, an elongated portion of the membranous labyrinth. *(see Figure 18.12)*
6. The basic receptors of the inner ear are **hair cells** whose surfaces support **stereocilia**. Hair cells provide information about the direction and strength of varied mechanical stimuli. *(see Figure 18.12d)*
7. The utricle and saccule are connected by a passageway continuous with the **endolymphatic duct**, which terminates in the **endolymphatic sac**. In the saccule and utricle, hair cells cluster within **maculae**, where their cilia contact **otoliths** consisting of densely packed mineral crystals (**statoconia**) in a gelatinous matrix. When the head tilts, the mass of each otolith shifts, and the resulting distortion in the sensory hairs signals the CNS. *(see Figures 18.12b/18.14)*
8. The **anterior**, **posterior**, and **lateral semicircular** ducts are continuous with the utricle. Each contains an **ampulla** with sensory receptors. Here the cilia contact a gelatinous **cupula**. *(see Figures 18.12/18.13)*
9. The vestibular receptors activate sensory neurons of the **vestibular ganglia**. The axons form the **vestibular branch** of the vestibulocochlear nerve (N VIII), synapsing within the **vestibular nuclei**. *(see Figure 18.15)*

Hearing 483

10. Sound waves travel toward the tympanum, which vibrates; the auditory ossicles conduct the vibrations to the base of the stapes at the oval window. Movement at the oval window applies pressure first to the perilymph of the **vestibular duct**. This pressure is passed on to the perilymph in the **tympanic duct**. *(see Figure 18.16)*
11. Pressure waves distort the **basilar membrane** and push the hair cells of the **organ of Corti (spiral organ)** against the **tectorial membrane**. *(see Figure 18.16 and Table 18.2)*

Auditory Pathways 486

12. The sensory neurons for hearing are located in the **spiral ganglion** of the cochlea. Their afferent fibers form the **cochlear branch** of the vestibulocochlear nerve (N VIII), synapsing at the **cochlear nucleus**. *(see Figure 18.17)*

Vision 487

Accessory Structures of the Eye 487

1. The **accessory structures** of the eye include the **palpebrae** (eyelids), which are separated by the **palpebral fissure**. The **eyelashes** line the palpebral margins. **Tarsal glands**, which secrete a lipid-rich product, line the inner margins of the eyelids. Glands at the **lacrimal caruncle** produce other secretions. *(see Figure 18.18)*
2. The secretions of the **lacrimal gland** bathe the conjunctiva; these secretions are slightly alkaline and contain **lysozymes** (enzymes that attack bacteria). Tears collect in the *lacus lacrimalis*. The tears reach the inferior meatus of the nose after passing through the **lacrimal puncta**, the **lacrimal canaliculi**, the **lacrimal sac**, and the **nasolacrimal duct**. Collectively, these structures constitute the **lacrimal apparatus**. *(see Figures 18.18 to 18.20)*

The Eye 489

3. The eye has three layers: an outer *fibrous tunic*, a *vascular tunic*, and an inner *neural tunic*.
4. The **fibrous tunic** includes most of the ocular surface, which is covered by the **sclera** (a dense, fibrous connective tissue of the fibrous tunic); the **limbus** is the border between the sclera and the cornea. *(see Figure 18.20)*
5. An epithelium called the **conjunctiva** covers most of the exposed surface of the eye; the **bulbar**, or **ocular**, **conjunctiva** covers the anterior surface of the eye, and the **palpebral conjunctiva** lines the inner surface of the eyelids. The **cornea** is transparent. *(see Figure 18.20)*
6. The **vascular tunic**, or **uvea**, includes the **iris**, the **ciliary body**, and the **choroid**. The iris forms the boundary between the anterior and posterior chambers. The ciliary body contains the **ciliary muscle** and the **ciliary processes**, which attach to the **suspensory ligaments** *(zonular fibers)* of the lens. *(see Figures 18.20/18.21)*
7. The **neural tunic (retina)** consists of an outer **pigmented layer** and an inner **neural retina**; the latter contains visual receptors and associated neurons. *(see Figures 18.20 to 18.22)*
8. There are two types of **photoreceptors** (visual receptors of the retina). **Rods** provide black and white vision in dim light; **cones** provide color vision in bright light. Cones are concentrated in the **macula lutea**; the **fovea** *(fovea centralis)* is the area of sharpest vision. *(see Figures 18.20/18.22)*
9. The direct line to the CNS proceeds from the photoreceptors to **bipolar cells**, then to **ganglion cells**, and to the brain via the optic nerve. **Horizontal cells** and **amacrine cells** modify the signals passed between other retinal components. *(see Figure 18.22a)*
10. The **aqueous humor** continuously circulates within the eye and reenters the circulation after diffusing through the walls of the anterior chamber and into the **canal of Schlemm** (*scleral venous sinus*). *(see Figure 18.23)*
11. The **lens**, held in place by the suspensory ligaments, lies posterior to the cornea and forms the anterior boundary of the vitreous chamber. This chamber contains the **vitreous body**, a gelatinous mass that helps to stabilize the shape of the eye and to support the retina. *(see Figures 18.20/18.23)*
12. The lens focuses a visual image on the retinal receptors.

Visual Pathways 495

13. Each photoreceptor monitors a specific receptive field. The axons of ganglion cells converge on the **optic disc** and proceed along the optic tract to the optic chiasm. *(see Figures 18.20b,e/18.22/18.24/18.25)*
14. From the optic chiasm, after a partial decussation, visual information is relayed to the lateral geniculate nuclei. From there the information is sent to the visual cortex of the occipital lobes. *(see Figure 18.25)*
15. Visual inputs to the **suprachiasmatic nucleus** and the pineal gland affect the function of other brain stem nuclei. These nuclei establish a visceral **circadian rhythm** that is tied to the day-night cycle and affects other metabolic processes. *(see Figure 18.25b)*

LEVEL 1 REVIEWING FACTS AND TERMS

Match each numbered item with the most closely related lettered item. Use letters for answers in the spaces provided.

Column A	Column B
____ 1. monitored area	a. lost transparency
____ 2. physical distortion	b. receptive field
____ 3. olfactory	c. similar to CSF
____ 4. gustatory	d. mechanoreceptors
____ 5. ceruminous glands	e. Bowman's glands
____ 6. perilymph	f. cochlear duct
____ 7. ampulla	g. anterior cavity
____ 8. organ of Corti	h. taste buds
____ 9. aqueous humor	i. semicircular duct
____10. cataract	j. external acoustic meatus

11. The anterior, transparent part of the fibrous tunic is known as the
    (a) cornea (b) sclera
    (c) iris (d) fovea
12. Nociceptors
    (a) are especially common in internal organs
    (b) usually have large receptive fields
    (c) are large, complex receptors
    (d) convey precise information about the site of stimulation
13. Fine touch and pressure receptors provide detailed information about
    (a) the source of the stimulus (b) the shape of the stimulus
    (c) the texture of the stimulus (d) all of the above
14. Receptors in the saccule and utricle provide sensations of
    (a) balance and equilibrium (b) hearing
    (c) vibration (d) gravity and linear acceleration
15. The lacrimal glands
    (a) are located in pockets in the lacrimal bones
    (b) produce less fluid than the ocular conjunctiva
    (c) produce watery, slightly alkaline secretions
    (d) function only during stress
16. The neural tunic
    (a) consists of three distinct layers (b) contains the photoreceptors
    (c) forms the iris (d) all of the above
17. The auditory ossicles connect the
    (a) tympanic membrane to the oval window
    (b) tympanic membrane to the round window
    (c) cochlea to the tympanic membrane
    (d) cochlea to the oval window
18. Mechanoreceptors that detect pressure changes in the walls of blood vessels as well as in portions of the digestive, reproductive, and urinary tracts are
    (a) tactile receptors (b) baroreceptors
    (c) proprioceptors (d) free nerve receptors
19. Pupillary muscle groups are controlled by the ANS. Parasympathetic activation causes pupillary ___, and sympathetic activation causes ___.
    (a) dilation, constriction (b) dilation, dilation
    (c) constriction, dilation (d) constriction, constriction
20. Auditory information about the region and intensity of stimulation is relayed to the CNS over the cochlear branch of cranial nerve
    (a) N IV (b) N VI
    (c) N VIII (d) N X

LEVEL 2 REVIEWING CONCEPTS

1. Damage to the fovea of the eye interferes with the ability to
   (a) focus an image (b) regulate the amount of light striking the retina
   (c) see color (d) bleach visual pigments
2. Receptors that combine phasic and tonic coding
   (a) do not send sensory information to the CNS
   (b) convey extremely complicated sensory information to the CNS
   (c) function in sensations such as crude touch
   (d) are most likely to be peripheral sense receptors
3. Damage to the cupula of the lateral semicircular duct would interfere with our perception of
   (a) the direction of gravitational pull
   (b) horizontal rotation of the head
   (c) vertical rotation of the head
   (d) linear acceleration
4. What is receptor specificity? What causes it?
5. What could stimulate the release of an increased quantity of neurotransmitter by a hair cell into the synapse with a sensory neuron?
6. What are the functions of hair cells in the inner ear?
7. What is the functional role of sensory adaptation?
8. What type of information about a stimulus does sensory coding provide?
9. What would be the consequence of damage to the lamellated corpuscles of the arm?
10. What is the structural relationship between the bony labyrinth and the membranous labyrinth?

LEVEL 3 CRITICAL THINKING AND CLINICAL APPLICATIONS

1. Beth has surgery to remove some polyps (growths) from her sinuses. After she heals from the surgery, she notices that her sense of smell is not as keen as it was before the surgery. Can you suggest a reason for this?
2. Jared is 10 months old, and his pediatrician diagnoses him with otitis media. What does the physician tell his mother?
3. Ruth suffers from conjunctivitis. What is wrong with her?

✓ ANSWERS TO CONCEPT CHECK QUESTIONS

p. 469 **1.** Free nerve endings may be stimulated by chemical stimulation, pressure, temperature changes, or physical damage. **2.** Tonic receptors are always active, whereas phasic receptors are normally inactive, but become active for a short time. **3.** A sensation is the sensory information arriving at the CNS. **4.** The general senses refer to sensations of temperature, pain, touch, pressure, vibration, and proprioception.
p. 472 **1.** Since nociceptors are pain receptors, if they are stimulated, you perceive a pain sensation in your affected hand. **2.** Proprioceptors relay information about limb position and movement to the central nervous system, especially the cerebellum. Lack of this information would result in uncoordinated movements, and the individual would probably be unable to walk. **3.** The three classes of mechanoreceptors are: tactile receptors, baroreceptors, and proprioceptors. **4.** Carbon dioxide concentration in the blood is being monitored by chemoreceptors in the carotid and aortic bodies.
p. 476 **1.** There are four primary taste sensations: sweet, salt, sour, and bitter. **2.** When you have a cold, airborne molecules cannot reach the olfactory receptors, and meals taste dull and unappealing. **3.** Taste receptors are clustered in individual taste buds. Papillae on the tongue contain taste buds. **4.** The tongue has three types of papillae: filiform, fungiform, and circumvallate.
p. 483 **1.** Two small muscles contract to protect the eardrum and ossicles from violent movements. These are the tensor tympani and stapedius muscles. **2.** The auditory ossicles are the three tiny ear bones that are located within the middle ear. They act as levers that transfer sound vibrations from the tympanum to a fluid-filled chamber within the inner ear. **3.** Perilymph is a liquid whose properties closely resemble those of cerebrospinal fluid. It fills the space between the bony and membranous labyrinths. **4.** Shaking the head "no" stimulates the hair cells of the lateral semicircular duct. This stimulation is interpreted by the brain as a movement of the head.
p. 486 **1.** Without the movement of the membrane spanning the round window, the perilymph would be moved by the vibration of the stapes at the oval window, and there would be little or no perception of sound. **2.** Loss of stereocilia (as a result of constant exposure to loud noises, for instance) would reduce hearing sensitivity and could eventually result in deafness. **3.** The cochlear duct (scala media) is sandwiched between the vestibular (scala vestibuli) and tympanic duct (scala tympani). The hair cells for hearing are located in the cochlear duct. **4.** The tectorial membrane overlies the hair cells in the organ of Corti. When the membrane containing these hair cells vibrates, the stereocilia of the hair cells are distorted and sound is detected.
p. 497 **1.** Inadequate tear production would affect the cornea first. Since the cornea is avascular, the cells of the cornea must obtain oxygen and nutrients from the tear fluid that passes over its surface. **2.** The two structures most affected by an abnormally high intraocular pressure are: (1) the canal of Schlemm (the aqueous humor no longer has free access to this structure), and (2) the optic nerve (the nerve fibers of this structure are distorted, which affects visual perception). **3.** An individual born without cones would be able to see only in black and white (monochromatic) and would have very poor visual acuity. **4.** Ciliary processes are folds in the epithelium of the ciliary body. The ciliary body consists of the ciliary muscle, which helps to control the shape of the lens for near and far vision.

REVIEW IT

Challenge 1

Learning the anatomy of the brain is a challenging task. There are many layers to this organ and structures are found embedded within others. The Interactive CD has an animation that takes you through the portions of the brain in successive layers. To view this animation, go to Animations/Nervous system/Brain and brainstem. The image to the left has been taken from that animation. Determine first where in the brain these structures are found. What figure in the text is comparable to this image? Identify the structures that are visible here, keeping in mind that this image was re-constructed from transverse sections of an actual brain. Structures are not idealized in this image, but have the shapes and proportions characteristic of one particular individual. Some structures seen on the comparable text figure may not be visible here, or may appear differently.

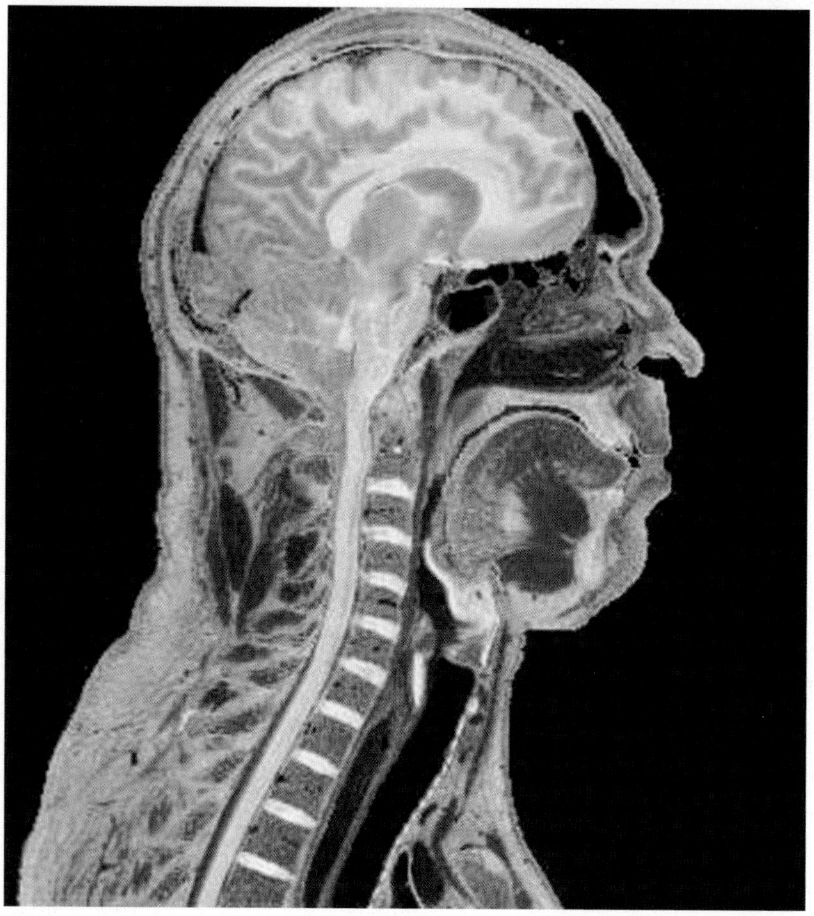

Challenge 2

Although it is emphasized in the text and portrayed properly in the figures, students are often surprised by the size of the human spinal cord when they first encounter one during dissection or surgical procedure. The sagittal image to the left is a reconstruction using the data set of transverse images from the Visible Human project. Note the relative sizes of the brain and spinal cord. Find and label the corpus callosum, thalamus, midbrain, cerebellum, pons, medulla oblongata, and cervical spinal cord. Identify the cervical vertebrae and intervertebral discs.

Images provided by the Digital Cadaver™ project, courtesy of Visible Productions, Inc.

APPLY IT

Below are two separate exercises that provide more information on a topic presented in the preceding chapters. Each one is designed to take approximately 10 minutes, and will help you gain a better appreciation for the material presented in Chapters 13–18.

Critical Linking 1

There are many ways to view neuroanatomy, including sophisticated neuro-imaging techniques such as CTs and MRIs. These radiological procedures give very different pictures of the brain than any you have seen thus far. Go to the Companion Website and click on the key word "brain anatomy" to move to a site prepared by Dr. Keith Johnson of the Harvard Medical School. Here you will find a series of images of the brain, each with an explanation of the structures visible. To familiarize yourself with what you are viewing, first read the short introduction to neuro-imaging. Then look at the Top 100 Brain Structures. Can you identify all these? Compare the images you see there with those in your text. Which gives you a clearer understanding of the structures of the brain? Why?

Critical Linking 2

The peripheral nervous system is comprised of many nerves of clinical importance that anatomists should be able to identify and trace. For practice identifying these nerves, visit the Companion Website and choose the key word "nerves." This will take you to the Lumen "learn'em" Website prepared by Loyola University Medical School (Chicago). There you will find two sites of interest; "cutaneous innervation" and "nerves." Choose either of these to continue your studies. "Nerves" will lead you through practice identification of the cervical, brachial, lumbar, and sacral plexuses as well as the cranial nerves. "Cutaneous innervation" will provide practice in naming the sensory nerves.

FURTHER STUDIES

A study of the nervous system usually leads to interesting questions that lie just outside the realm of neuroanatomy. Many times these investigations assume a thorough understanding of the anatomy of the nervous system prior to fully understanding the issues. One such question that is currently in the news is in the area of neural development. A recently published text "*From Neurons to Neighborhoods: The Science of Early Childhood Development*" is a synopsis of $2\frac{1}{2}$ years of research into early childhood development with a strong scientific emphasis. This work was conducted by a 17-member committee representing many disciplines including neuroscience, psychology, child development, pediatrics, and education. The resulting information was first scrutinized by both the National Research Council and the Institute of Medicine for scientific accuracy and rigorous application of scientific principles before being released. Consequently the findings of this text are being given serious consideration in many diverse disciplines. Some of the conclusions reached by this study include: (1) The "nature versus nurture" debate is too simplistic in its contentions. The factors that influence development of the mind are intertwined and connected in ways this theory ignores. (2) While it is agreed that the period from birth to age three is extremely important to healthy neural development, this period should be extended in both directions to include prenatal care as well as the years beyond age three. (3) The developing brain can be permanently harmed by many biological and chemical factors that interfere with the developing physiology. These factors include certain diseases, environmental toxins, poor nutrition, and chronic stress. (4) The notion that certain activities and stimulations given young children are better neural activators than others and lead to more advanced brain development in infants is not scientifically valid. You can read this entire text on-line by visiting *http://www.nap.edu/books/0309069882/html/*. Additionally, a good review of the material presented in the text can be found at *http://www4.nationalacademies.org/onpi/webextra.nsf/web/investing?OpenDocument*

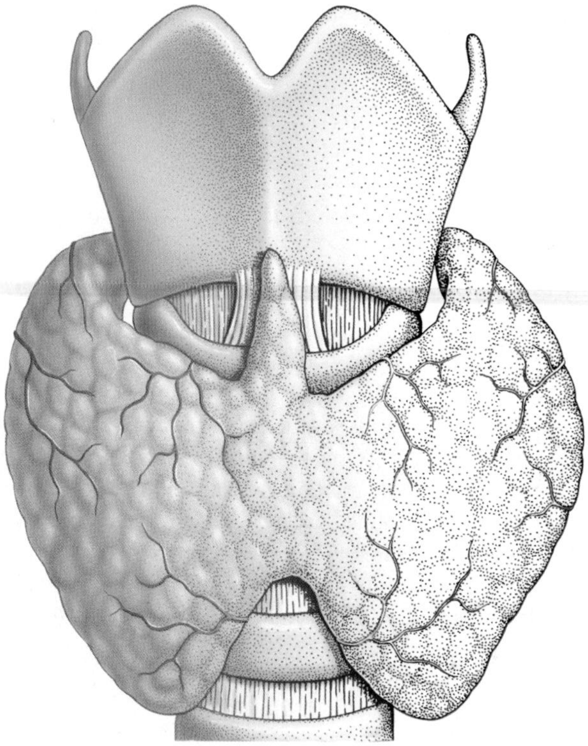

19

THE ENDOCRINE SYSTEM

Homeostatic regulation involves coordinating the activities of organs and systems throughout the body. At any given moment, the cells of both the nervous and endocrine systems are working together to monitor and adjust the physiological activities under way. The activities of these two systems are coordinated closely, and their effects are typically complementary. In general, the nervous system produces short-term (usually a few seconds) but very specific responses to environmental stimuli. In contrast, endocrine gland cells release chemicals into the bloodstream for distribution throughout the body. p. 57 These chemicals, called **hormones** (meaning "to excite"), alter the metabolic activities of many different tissues and organs simultaneously. The hormonal effects may not be apparent immediately, but when they appear, they often persist for days. This response pattern makes the endocrine system particularly effective in regulating ongoing processes such as growth and development.

The nervous and endocrine systems are compared in Table 19.1. From this general perspective, the two systems are distinguished easily. Yet when viewed in detail, there are instances in which the two systems are difficult to separate either anatomically or functionally. For example, the adrenal medulla is a modified sympathetic ganglion whose neurons secrete epinephrine and norepinephrine into the blood. p. 452 The adrenal medulla is thus an endocrine structure that is functionally and developmentally part of the nervous system.

Endocrinology is the study of ductless glands or tissues of the endocrine system and their hormonal products. This chapter describes the components and functions of the endocrine system and considers some important interactions between the nervous and endocrine systems.

An Overview of the Endocrine System

[FIGURES 19.1/19.2]

The endocrine system (Figure 19.1●) includes all of the endocrine cells and tissues of the body. Endocrine cells are glandular secretory cells that release hormones into the interstitial fluids. In contrast, secretions from exocrine glands are released onto an epithelial surface. p. 57

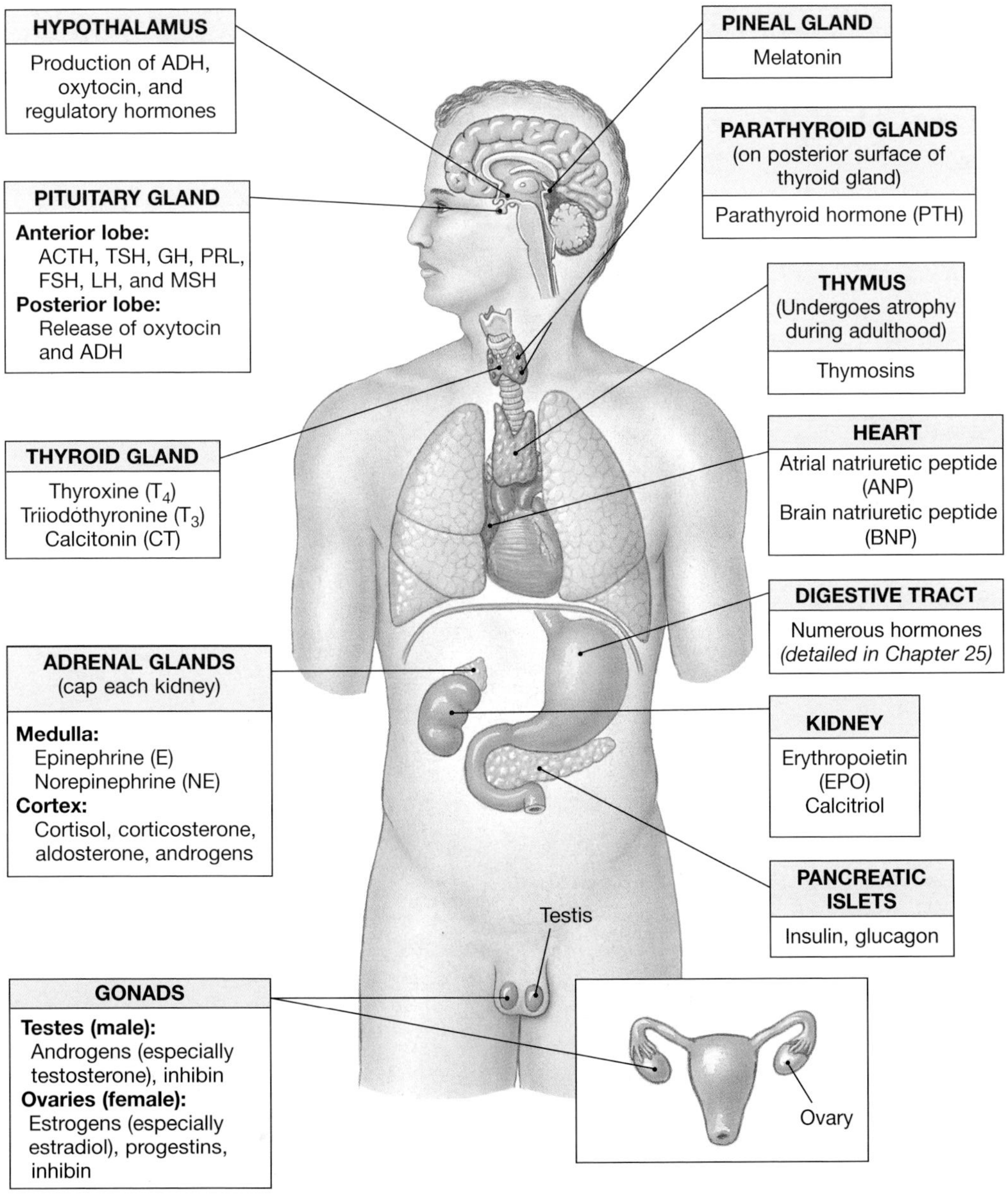

FIGURE 19.1 **The Endocrine System**

Location of endocrine glands and endocrine cells, and the major hormones produced by each gland.

KEY TO PITUITARY HORMONES:	
ACTH	Adrenocorticotropic hormone
TSH	Thyroid-stimulating hormone
GH	Growth hormone
PRL	Prolactin
FSH	Follicle-stimulating hormone
LH	Luteinizing hormone
MSH	Melanocyte-stimulating hormone
ADH	Antidiuretic hormone

TABLE 19.1 A COMPARISON OF THE NERVOUS AND ENDOCRINE SYSTEMS

	Nervous System	Endocrine System
METHOD OF INFORMATION TRANSFER	Release of neurotransmitters at specialized synapses at specific sites in the body	Release of hormones into general circulation for distribution throughout the body
EFFECTS		
SCOPE	Localized	Widespread/systemic
PRIMARY TARGETS	Neurons, gland cells, muscle cells, fat cells	All tissues
ONSET	Immediate (milliseconds)	Gradual (seconds to hours)
DURATION	Short-term (milliseconds to minutes)	Long-term (minutes to days)
RECOVERY	Immediately after stimulation ends	Slow, persisting after hormonal secretion stops

Hormones are organized into four groups based on chemical structure:

- *Amino acid derivatives*: The **amino acid derivatives** are relatively small molecules that are structurally similar to amino acids. Examples include (1) derivatives of tyrosine, such as the **thyroid hormones** released by the thyroid gland and the **catecholamines** (epinephrine, norepinephrine) released by the adrenal medullae, and (2) derivatives of tryptophan, such as **melatonin** synthesized by the pineal gland.
- *Peptide hormones*: **Peptide hormones** are chains of amino acids. This is the largest group of hormones; all pituitary gland hormones are peptide hormones.
- *Steroid hormones*: **Steroid hormones**, derived from cholesterol, are released by the reproductive organs and the adrenal glands.
- *Eicosanoids*: **Eicosanoids** are small molecules with a five-carbon ring at one end. These compounds coordinate cellular activities and affect enzymatic processes (such as blood clotting) that occur in extracellular fluids.

Enzymes control all cellular activities and metabolic reactions. Hormones influence cellular operations by changing the *types, activities, or quantities* of key cytoplasmic enzymes. In this way, a hormone can regulate the metabolic operations of its **target cells**—peripheral cells that respond to its presence. The first step in the process involves the release of the hormone by the endocrine cell. At the target cell, the hormone binds to a specific receptor protein in the cell membrane or cytoplasm. The presence or absence of the specific receptor determines each cell's sensitivities to a hormone. In turn, the binding of a hormone to its receptor starts a biochemical chain of events that changes the pattern of enzymatic activity within the cell. The different classes of hormones work by different mechanisms to achieve this result (Figure 19.2●).

The catecholamines, peptide hormones, and several eicosanoids must bind to receptors on the cell membrane because they cannot diffuse through a cell membrane. Because they cannot reach the cytoplasm, they cannot exert their effects directly. Instead, the hormone acts as a *first messenger* that causes the appearance of a *second messenger* in the cytoplasm of the target cell. The second messenger may function as an enzyme activator, inhibitor, or cofactor to change the direction and rate of the cell's metabolic reactions. Many hormones discussed later in this chapter, including calcitonin, parathyroid hormone, ADH, ACTH, FSH, LH, TSH, and glucagon, produce their effects in this way. The specific response of the target cell depends on which enzymes are affected.

The steroid hormones and thyroid hormones cross the cell membrane and bind to intracellular receptors. The lipid-based steroid hormones diffuse rapidly through the lipid portion of the cell membrane and bind to receptors in the cytoplasm or nucleus. The hormone-receptor complex then binds to specific DNA segments in the nucleus and triggers the activation or inactivation of specific genes. Thyroid hormones are passively transported across the cell membrane. As in the case of steroid hormones, they bind to specific DNA segments and trigger the activation of specific genes.

By activating or inactivating genes, these hormones alter the rate of mRNA transcription in the nucleus. This process gradually changes the

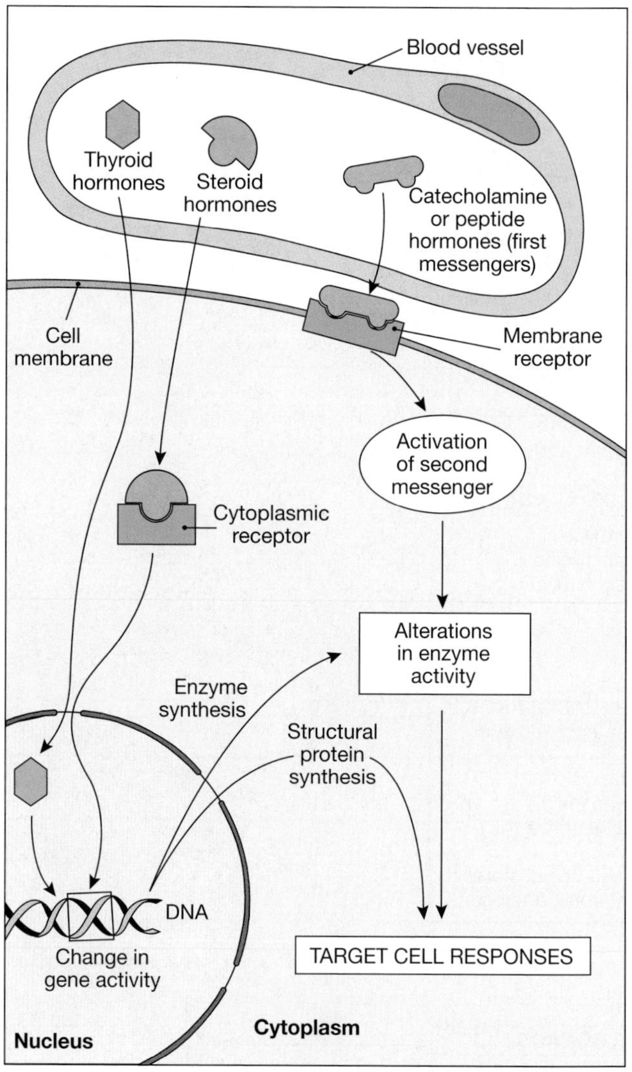

FIGURE 19.2 **A SUMMARY OF THE MECHANISMS OF HORMONE ACTION**

Catecholamines and peptide hormones act through second messengers produced when the hormones bind to receptors at the membrane surfaces of target cells. Steroid hormones bind to receptors in the cytoplasm or nucleus. Thyroid hormones proceed directly to the nucleus to reach hormonal receptors. The hormone-receptor complex activated by steroid or thyroid hormones then binds to DNA segments and alters gene activity.

pattern of protein synthesis in their target cells. As a result, cell structure may be altered or, if the affected proteins are enzymes, the metabolic activity of the target cell may change. For example, the hormone *testosterone* stimulates the production of enzymes and proteins in skeletal muscle fibers, causing an increase in muscle size and strength.

Endocrine activity is controlled by **endocrine reflexes** that are triggered by (1) *humoral stimuli* (changes in the composition of the extracellular fluid), (2) *hormonal stimuli* (arrival or removal of a specific hormone), or (3) *neural stimuli* (the arrival of neurotransmitters at neuroglandular junctions). In most cases, endocrine reflexes are regulated by some form of negative feedback. In direct negative-feedback control: (1) the endocrine cell responds to a disturbance in homeostasis (such as a change in the composition of the extracellular fluid) by releasing its hormone into the circulatory system, (2) the released hormone stimulates a target cell, and (3) the target cell response restores homeostasis and eliminates the source of stimulation of the endocrine cell. One example noted in Chapter 5 was the control of calcium levels by parathyroid hormone. p. 123 When circulating calcium levels decline, parathyroid hormone is released, and the responses of target cells (osteoclasts) elevate blood calcium levels. As calcium levels climb, the level of parathyroid stimulation declines and so does its rate of hormone secretion.

More complex endocrine reflexes involve one or more intermediary steps and often two or more hormones. These complicated chains of events may be controlled by complex negative-feedback loops or, rarely, through positive feedback. Complex negative-feedback loops are the most common regulatory mechanisms. In these cases the secretion of one hormone, such as the thyroid-stimulating hormone from the anterior lobe of the pituitary gland, triggers the secretion of a second hormone, such as the thyroid hormones produced by the thyroid gland. The second hormone may have multiple effects, one of which always includes the suppression of the release of the first hormone.

Hormone regulation through positive feedback is restricted to processes that must be rushed to completion. In these instances, the secretion of a hormone produces an effect that further stimulates hormone release. For example, the release of oxytocin during labor and delivery causes smooth muscle contractions in the uterus, and the uterine contractions further stimulate oxytocin release.

The Hypothalamus and Endocrine Regulation [FIGURE 19.3]

Coordinating centers in the hypothalamus regulate the activities of the nervous and endocrine systems by three different mechanisms (Figure 19.3●):

1. The hypothalamus contains autonomic centers that exert direct neural control over the endocrine cells of the adrenal medullae. p. 400 When the sympathetic division is activated, the adrenal medullae release hormones into the bloodstream.
2. The hypothalamus acts as an endocrine organ, releasing the hormones *ADH* and *oxytocin* into the circulation at the posterior lobe of the pituitary gland.
3. The hypothalamus secretes **regulatory hormones**, or *regulatory factors*, that control the activities of endocrine cells in the anterior lobe of the pituitary gland. There are two classes of regulatory hormones. (1) **Releasing hormones (RH)** stimulate production of one or more hormones at the anterior lobe of the pituitary gland, whereas (2) **inhibiting hormones (IH)** prevent the synthesis and secretion of specific pituitary hormones.

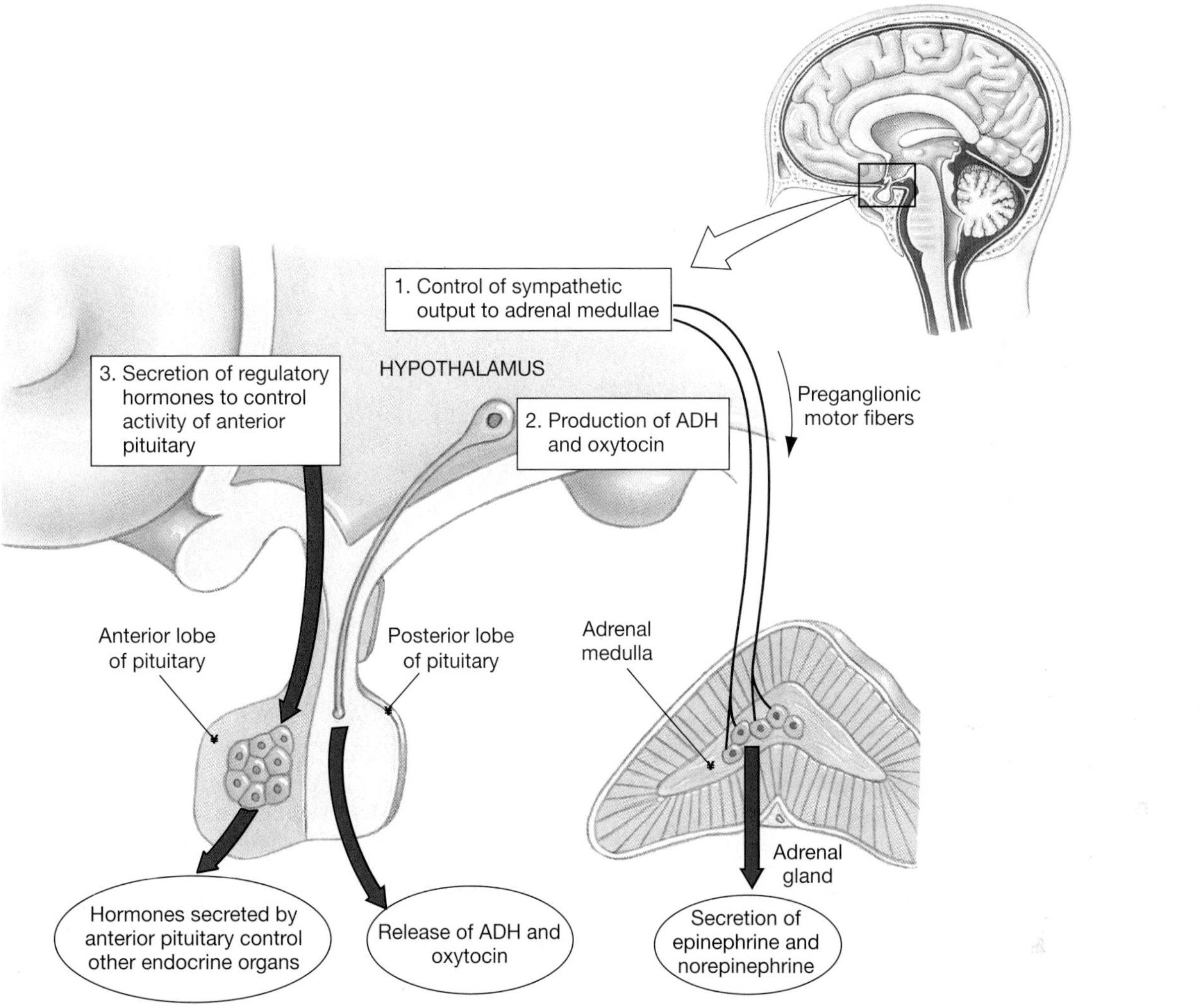

FIGURE 19.3 **HYPOTHALAMIC CONTROL OVER ENDOCRINE ORGANS**

A comparison of the three types of hypothalamic control. **(1)** The hypothalamus exerts direct neural control over the secretory activity of the adrenal medullae. **(2)** Hypothalamic neurons secrete ADH and oxytocin, hormones that produce specific responses in peripheral target organs. **(3)** Hypothalamic neurons release regulatory hormones that control the secretory activity of the anterior lobe of the pituitary gland.

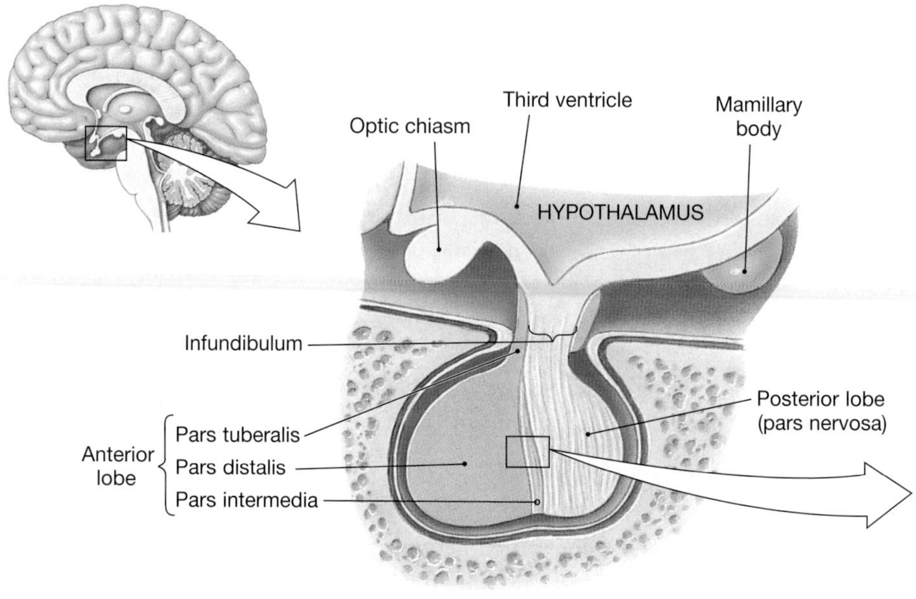

(a)

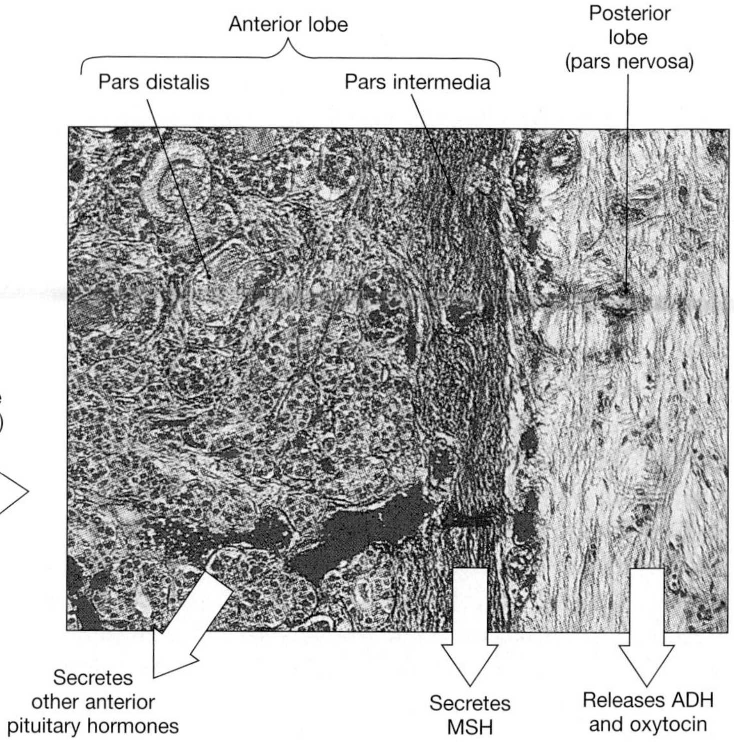

(b) Anterior and posterior pituitary tissues (LM × 77)

FIGURE 19.4 **Anatomy and Orientation of the Pituitary Gland and Its Subdivisions**

(**a**) Relationship of the pituitary gland to the hypothalamus. (**b**) Photomicrograph of pituitary gland showing anterior and posterior lobes.

The Pituitary Gland

[FIGURES 19.4 TO 19.6 AND TABLE 19.2]

The **pituitary gland**, or **hypophysis** (hī-POF-i-sis), weighs one-fifth of an ounce and is the most compact chemical factory in the body. This small, oval gland, about the size and weight of a small grape, lies inferior to the hypothalamus within the *sella turcica*, a depression in the sphenoid. p. 147 The **infundibulum** (in-fun-DIB-ū-lum) extends from the hypothalamus inferiorly to the posterior and superior surfaces of the pituitary gland (Figure 19.4a●). The *diaphragma sellae* encircles the stalk of the infundibulum and holds the pituitary gland in position within the sella turcica. p. 387

The pituitary gland has both *posterior* and *anterior lobes* based on anatomical and developmental grounds (Figure 19.4●). Nine important

TABLE 19.2 **The Pituitary Hormones**

Region/Area	Hormone	Targets	Hormonal Effects
ANTERIOR LOBE (Adenohypophysis)			
Pars distalis	Thyroid-stimulating hormone (TSH)	Thyroid gland	Secretion of thyroid hormones
	Adrenocorticotropic hormone (ACTH)	Adrenal cortex (zona fasciculata)	Glucocorticoid secretion
	Gonadotropins:		
	Follicle-stimulating hormone (FSH)	Follicle cells of ovaries in female	Estrogen secretion, follicle development
		Sustentacular cells of testes in male	Stimulation of sperm maturation
	Luteinizing hormone (LH)	Follicle cells of ovaries in female	Ovulation, formation of corpus luteum, progesterone secretion
		Interstitial cells of testes in male	Testosterone secretion
	Prolactin (PRL)	Mammary glands in female	Production of milk
	Growth hormone (GH)	All cells	Growth, protein synthesis, lipid mobilization and catabolism
Pars intermedia (not active in normal adults)	Melanocyte-stimulating hormone (MSH)	Melanocytes	Increased melanin synthesis in epidermis
POSTERIOR LOBE (Neurohypophysis, or Pars nervosa)	Antidiuretic hormone (ADH or vasopressin)	Kidneys	Reabsorption of water; elevation of blood volume and pressure
	Oxytocin (OT)	Uterus, mammary glands (females)	Labor contractions, milk ejection
		Ductus deferens and prostate gland (males)	Contractions of ductus deferens and prostate gland; ejection of secretions

peptide hormones are released by the pituitary gland, two by the posterior lobe and seven by the *pars distalis* and *pars intermedia* of the anterior lobe. Table 19.2 summarizes information about the pituitary gland hormones and their targets; representative target organs are diagrammed in Figure 19.5●.

The Posterior Lobe of the Pituitary Gland [FIGURES 19.4 TO 19.6]

The **posterior lobe** of the pituitary gland (Figure 19.4●) is also called the **neurohypophysis** (noo-rō-hī-POF-i-sis) or *pars nervosa* ("nervous part"). It contains the axons and axon terminals of roughly 50,000 hypothalamic neurons whose cell bodies are in either the **supraoptic** or **paraventricular nuclei** (Figure 19.6●). The axons extend from these nuclei through the infundibulum and end in synaptic terminals in the posterior lobe. The hypothalamic neurons manufacture ADH (supraoptic nuclei) and oxytocin (paraventricular nuclei). ADH and oxytocin are called *neurosecretions* because they are produced and released by neurons. Once released, these hormones enter local capillaries supplied by the **inferior hypophyseal artery** (Figure 19.6●). From there they will be transported into the general circulation.

Hormones released by the posterior lobe (Figure 19.5●) include the following:

1. **ADH: Antidiuretic hormone**, or *vasopressin*, is released in response to a variety of stimuli, most notably to a rise in the concentration of electrolytes in the blood or a fall in blood volume or pressure. The primary function of ADH is to decrease the amount of water lost at the kidneys. ADH also causes the constriction of peripheral blood vessels, which helps to elevate blood pressure. *Diabetes Insipidus p. 796*

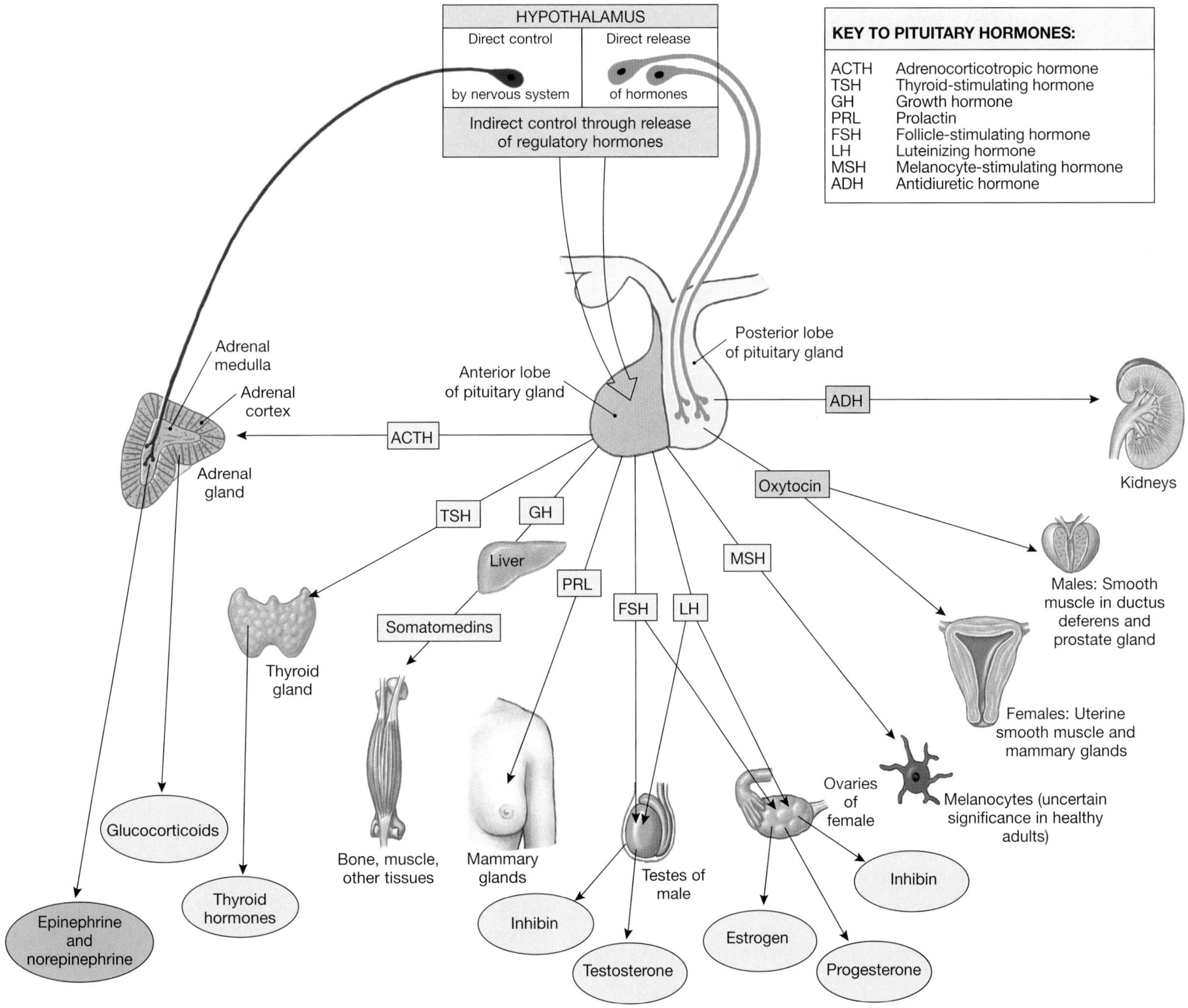

FIGURE 19.5 **Pituitary Hormones and Their Targets**

This schematic diagram shows the hypothalamic control of the pituitary gland, the pituitary hormones produced, and the responses of representative target tissues.

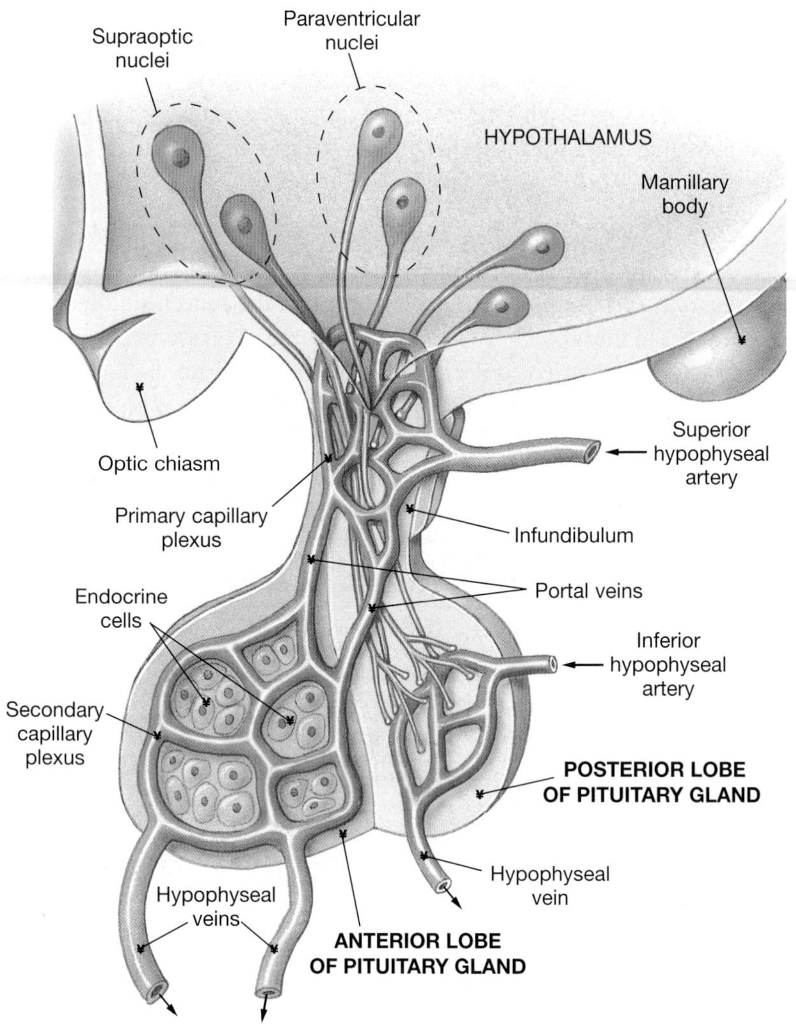

FIGURE 19.6 **THE PITUITARY GLAND AND THE HYPOPHYSEAL PORTAL SYSTEM**

This circulatory arrangement forms the hypophyseal portal system, which permits control of the anterior lobe by hypothalamic regulatory hormones.

2. **Oxytocin**: The functions of **oxytocin** (ok-sē-TŌ-sin; *oxy-*, quick + *tokos*, childbirth) are best known in women, where it stimulates the contractions of smooth muscle cells in the uterus and contractile (myoepithelial) cells surrounding the secretory cells of the mammary glands. The stimulation of uterine muscles by oxytocin in the last stage of pregnancy is required for normal labor and childbirth. After birth, the sucking of an infant at the breast stimulates the release of oxytocin into the blood. Oxytocin then stimulates contraction of the myoepithelial cells in the mammary glands, causing the discharge of milk from the nipple. In the human male, oxytocin causes smooth muscle contractions in the prostate gland. Animal studies suggest that its release may also trigger sexual behaviors by stimulating centers in the hypothalamus.

The Anterior Lobe of the Pituitary Gland [FIGURE 19.4]

The **anterior lobe** of the pituitary gland, or **adenohypophysis** (ad-e-nō-hī-POF-i-sis), contains a variety of different cell types that can be distinguished on the basis of size and staining properties. The adenohypophysis can be subdivided into three regions: (1) a large **pars distalis** (dis-TAL-is; "distal part"), which represents the major portion of the pituitary gland, (2) a slender **pars intermedia** (in-ter-MĒ-dē-a; "intermediate part"), which forms a narrow band adjacent to the neurohypophysis, and (3) an extension, called the **pars tuberalis**, which wraps around the adjacent portion of the infundibulum (Figure 19.4●). The entire adenohypophysis is richly vascularized with an extensive capillary network.

THE HYPOPHYSEAL PORTAL SYSTEM [FIGURE 19.6]

The production of hormones in the anterior lobe of the pituitary gland is controlled by the hypothalamus through the secretion of specific regulatory factors. Near the attachment of the infundibulum, hypothalamic neurons release regulatory factors into the surrounding interstitial fluids. The regulatory factors can easily enter the circulation in this region because the capillaries have a "Swiss cheese" appearance, with open spaces between adjacent endothelial cells. Such capillaries are called **fenestrated** (FEN-es-trāt-ed; *fenestra*, window), and they are found only where relatively large molecules enter or leave the circulatory system. This *primary capillary plexus* in the floor of the tuberal area receives blood from the **superior hypophyseal artery** (Figure 19.6●).

Before leaving the hypothalamus, the capillary network unites to form a series of larger vessels that spiral around the infundibulum to reach the anterior lobe of the pituitary gland. Once within the lobe, these vessels form a *secondary capillary plexus*, which branches among the endocrine cells (Figure 19.6●). This is an unusual vascular arrangement. Typically, an artery conducts blood from the heart to a capillary network, and a vein carries blood from a capillary network back to the heart. The vessels between the hypothalamus and the anterior lobe of the pituitary, however, carry blood from one capillary network to another. Blood vessels that link two capillary networks are called **portal vessels**, and the entire complex is termed a **portal system**. Portal systems provide an efficient means of chemical communication by ensuring that all of the blood entering the portal vessels will reach the intended target cells before returning to the general circulation. The communication is strictly one way, however, because any chemicals released by the cells "downstream" must do a complete tour of the cardiovascular system before reaching the capillaries at the start of the portal system. Portal vessels are named after their destinations, so this particular network of vessels is the **hypophyseal portal system**.

HORMONES OF THE ANTERIOR LOBE [FIGURE 19.5 AND TABLE 19.2]

We will restrict our discussion of anterior pituitary hormones to the seven whose functions and control mechanisms are reasonably well understood. All but one of these hormones are produced by the pars distalis, and five of them regulate the production of hormones by other endocrine glands. These hormones are termed *tropic* (*tropos*, turning). Their names indicate their activities; details are summarized in Table 19.2 and Figure 19.5●.

1. **Thyroid-stimulating hormone (TSH)** targets the thyroid gland and triggers the release of thyroid hormones. TSH is secreted by cells called *thyrotropes.*
2. **Adrenocorticotropic hormone (ACTH)** stimulates the release of steroid hormones by the adrenal gland. ACTH specifically targets cells producing hormones called **glucocorticoids (GC)** (gloo-kō-KOR-ti-koyds) that affect glucose metabolism. The cells that secrete ACTH are called *corticotropes.*

3. **Follicle-stimulating hormone (FSH)** promotes the development of oocytes (female gametes) within the ovaries of mature women. The process begins within structures called follicles, and FSH also stimulates the secretion of **estrogens** (ES-trō-jens) by follicle cells. Estrogens, which are steroids, are female sex hormones; *estradiol* is the most important estrogen. In men, FSH secretion supports sperm production in the testes.
4. **Luteinizing** (LOO-tē-in-ī-zing) **hormone (LH)** induces ovulation in women and promotes the ovarian secretion of **progestins** (prō-JES-tins), steroid hormones that prepare the body for possible pregnancy. *Progesterone* is the most important progestin. In men, LH, formerly called *interstitial cell-stimulating hormone (ICSH)*, stimulates the production of male sex hormones called **androgens** (AN-drō-jenz; *andros*, man) by the interstitial cells of the testes. *Testosterone* is the most important androgen. Because they regulate the activities of the male and female sex organs (gonads) FSH and LH are called **gonadotropins** (gō-nad-ō-TRŌ-pinz; *tropos*, turning). These hormones are produced by cells called *gonadotropes.*
5. **Prolactin** (prō-LAK-tin; *pro-*, before + *lac*, milk) **(PRL)** stimulates the development of the mammary glands and the production of milk. PRL exerts the dominant effect on the glandular cells, but the mammary glands are regulated by the interaction of a number of other hormones, including estrogen, progesterone, growth hormone, glucocorticoids, and hormones produced by the placenta. The functions of prolactin in males are poorly understood. PRL is secreted by cells called *lactotropes.*
6. **Growth hormone (GH)**, also called *human growth hormone (HGH)* or *somatotropin* (*soma*, body), stimulates cell growth and replication by accelerating the rate of protein synthesis. GH is secreted by cells called *somatotropes.* Although virtually every tissue responds to some degree, growth hormone has a particularly strong effect on skeletal and muscular development. Liver cells respond to GH by synthesizing and releasing somatomedins, which are peptide hormones. Somatomedins stimulate protein synthesis and cell growth in skeletal muscle fibers, cartilage cells, and other target cells. Children unable to produce adequate concentrations of growth hormone suffer from *pituitary growth failure*, sometimes called *pituitary dwarfism.* The steady growth and maturation that normally precede and accompany puberty do not occur in these individuals.
7. **Melanocyte-stimulating hormone (MSH)** is the only hormone released by the pars intermedia. As the name indicates, MSH stimulates the melanocytes of the skin, increasing their rates of melanin production and distribution. MSH is secreted only during fetal development, in young children, in pregnant women, and in some disease states.

✓ CONCEPT CHECK

- Which brain region controls production of hormones in the pituitary gland?
- What is a target cell? What is the relationship between a hormone and its target cell?
- Identify the two regions of the pituitary gland, and describe how hormone release is controlled for each.

The Thyroid Gland [FIGURE 19.7]

The **thyroid gland** curves across the anterior surface of the trachea (windpipe), just inferior to the **thyroid** ("shield-shaped") **cartilage** that dominates the anterior surface of the larynx (Figure 19.7a●). Because of its location, the thyroid gland easily can be felt with the fingers; when something goes wrong with the gland, it may even become prominent. The size of the thyroid gland is quite variable, depending on heredity, environment, and nutritional factors, but the average weight is about 34 g (1.2 oz). The gland has a deep red coloration because of the large number of blood vessels supplying the glandular cells. Blood supply to the gland is from two sources: (1) a *superior thyroid artery*, which is a branch from the external carotid artery, and (2) an *inferior thyroid artery*, a branch of the thyrocervical trunk. Venous drainage of the gland is through the *superior* and *middle thyroid veins*, which end in the internal jugular vein, and the *inferior thyroid veins*, which terminate at the brachiocephalic veins.

The thyroid gland has a "butterfly-like" appearance and consists of two main **lobes**. The superior portion of each lobe extends over the lateral surface of the trachea toward the inferior border of the thyroid cartilage. Inferiorly, the lobes of the thyroid gland extend to the level of the second or third ring of cartilage in the trachea. The two lobes are united by a slender connection, the **isthmus** (IS-mus). The thyroid gland is anchored to the tracheal rings by a thin capsule that is continuous with connective tissue partitions that segment the glandular tissue and surround the *thyroid follicles.*

Thyroid Follicles and Thyroid Hormones [FIGURES 19.7b,c/19.8 AND TABLE 19.3]

Thyroid follicles manufacture, store, and secrete thyroid hormones. Individual follicles are spherical, resembling miniature tennis balls, and they are lined by a simple cuboidal epithelium (Figure 19.7b,c●). These follicle cells surround a **follicle cavity**, which contains **colloid**, a viscous fluid containing large quantities of suspended proteins. A network of capillaries surrounds each follicle, delivering nutrients and regulatory hormones to the follicular cells and accepting their secretory products and metabolic wastes.

The follicle cells have abundant mitochondria and an extensive rough endoplasmic reticulum. As you would expect from that description, these cells are actively synthesizing proteins. Follicular cells synthesize a globular protein called **thyroglobulin** (thī-rō-GLOB-ū-lin) and secrete it into the colloid of the thyroid follicle. Thyroglobulin contains molecules of tyrosine, and some of these amino acids will be modified into thyroid hormones inside the follicle, through the attachment of iodine. The follicle cells actively transport iodide ions (I^-) into the cell from the interstitial fluid. The iodine is converted to a special ionized form (I^+) and attached to the tyrosine molecules of thyroglobulin by enzymes on the lumenal surfaces of the follicle cells. Two thyroid hormones are created in this way, and while in the colloid they remain part of the structure of the thyroglobulin. The hormone **thyroxine** (thī-ROKS-ēn), or **TX**, consists of two tyrosine molecules and four atoms of iodine; it is also known as tetraiodothyronine, or simply $\mathbf{T_4}$. **Triiodothyronine**, or $\mathbf{T_3}$, is a related molecule containing three iodine atoms. Eventually each molecule of thyroglobulin stored within a thyroid follicle will contain 4–8 molecules of T_3 and/or T_4 hormones. The thyroid is the only endocrine gland that stores its hormone product extracellularly.

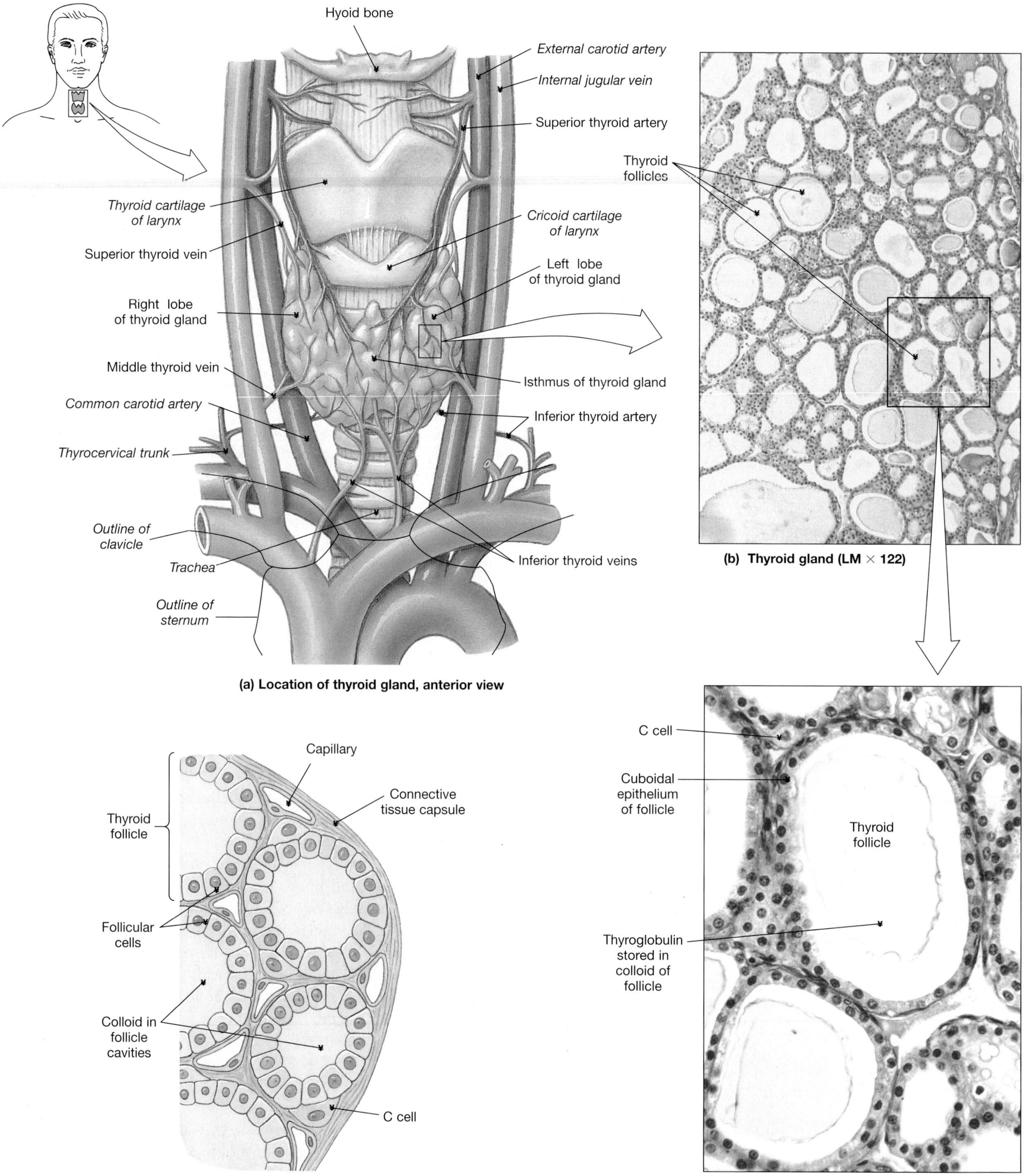

FIGURE 19.7 **The Thyroid Gland**

(**a**) Location and anatomy of the thyroid gland. (**b**) Histological organization of the thyroid. (**c**) Histological details of the thyroid gland showing thyroid follicles and both of the cell types in the follicular epithelium.

The major factor controlling the rate of thyroid hormone release is the concentration of TSH in the circulating blood (Figure 19.8●). Under the influence of TRH from the hypothalamus, the anterior pituitary releases TSH. Follicle cells respond by removing thyroglobulin from the lumen of the follicles by endocytosis. They next break down the protein through lysosomal activity, which releases molecules of both T_3 and T_4. These hormones then leave the cell, primarily by diffusion, and enter the circulation. Thyroxine (T_4) accounts for roughly 90% of all thyroid secretions. The two thyroid hormones, which have complementary effects, increase the rate of cellular metabolism and oxygen consumption in almost every cell in the body. These hormones are included in Table 19.3. ✝ *Thyroid Gland Disorders p. 796*

The C Cells of the Thyroid Gland [FIGURE 19.7c]

A second type of endocrine cell lies among the cuboidal follicle cells. Although in contact with the basement membrane, these cells do not reach the lumen. These are the **C cells**, or *parafollicular cells*. C cells occur singly or in small groups. They are larger than the cuboidal follicular cells and do not stain as clearly (Figure 19.7c●). C cells produce the hormone **calcitonin** (kal-si-TŌ-nin) **(CT)**. Calcitonin assists in the regulation of calcium ion concentrations in body fluids, especially under stresses such as starvation or pregnancy. Calcitonin lowers calcium ion concentrations by (1) inhibiting osteoclasts and (2) stimulating calcium ion excretion at the kidneys. The actions of calcitonin are opposed by those of *parathyroid hormone*, produced by the parathyroid glands.

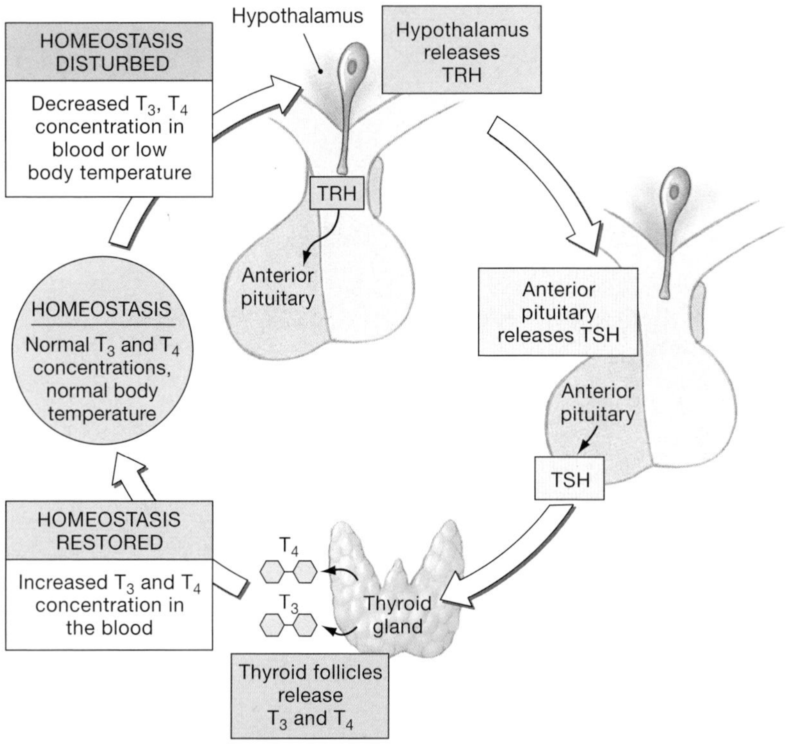

FIGURE 19.8 **THE REGULATION OF THYROID SECRETION**

This negative feedback loop is responsible for the homeostatic control of thyroid hormone release.

The Parathyroid Glands [FIGURES 19.7a/19.9]

There are typically four pea-sized, reddish brown **parathyroid glands** located on the posterior surfaces of the thyroid gland (Figure 19.9a●). The glands usually are attached at the surface of the thyroid gland by the thyroid capsule. Like the thyroid gland, the parathyroid glands are surrounded by a connective tissue capsule that invades the interior of the gland, forming separations and small irregular *lobules*. Blood is supplied to the superior pair by the *superior thyroid artery* and to the inferior pair by the *inferior thyroid artery* (Figure 19.7a●). The venous drainage is the same as that of the thyroid. All together the four parathyroid glands weigh a mere 1.6 g.

There are two types of cells in the parathyroid gland. The **principal cells**, or **chief cells**, (Figure 19.9b,c●) are glandular cells that produce the hormone **parathyroid hormone (PTH)**; the other major cell types (*oxyphil cells* and *transitional cells*) are probably immature or inactive principal cells. Like the C cells of the thyroid, the chief cells of the parathyroids monitor the circulating concentration of calcium ions. When the calcium concentration falls below normal, the chief cells secrete parathyroid hormone. PTH stimulates osteoclasts and osteoblasts (although osteoclast effects predominate), and reduces urinary excretion of calcium ions. It also stimulates the production of *calcitriol*, a kidney hormone that promotes intestinal absorption of calcium. PTH levels remain elevated until blood Ca^{2+} concentrations return to normal. PTH has been shown to be effective in reducing the progress of osteoporosis in the elderly. ✝ *Disorders of Parathyroid Function p. 796*

TABLE 19.3 **HORMONES OF THE THYROID GLAND, PARATHYROID GLANDS, AND THYMUS**

Gland/Cells	Hormones	Targets	Effects
THYROID			
Follicular epithelium	Thyroxine (T_4), triiodothyronine (T_3)	Most cells	Increase energy utilization, oxygen consumption, growth, and development
C cells	Calcitonin (CT)	Bone and kidneys	Decreases calcium ion concentrations in body fluids; uncertain significance in healthy nonpregnant adults
PARATHYROIDS			
Chief cells	Parathyroid hormone (PTH)	Bone and kidneys	Increases calcium ion concentrations in body fluids; increase bone mass
THYMUS	"Thymosins" (see Chapter 23)	Lymphocytes	Maturation and functional competence of immune system

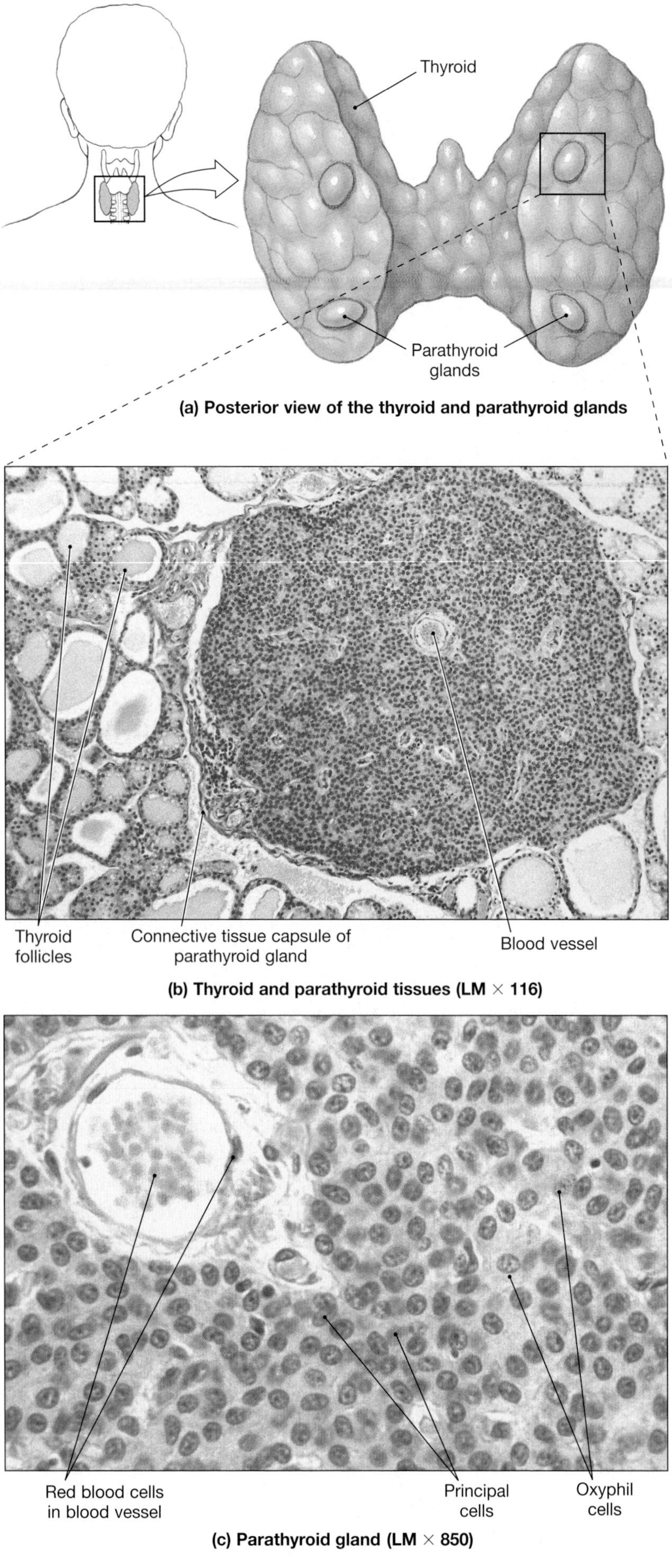

FIGURE 19.9 **The Parathyroid Glands**

There are usually four separate parathyroid glands bound to the posterior surface of the thyroid gland. **(a)** The location and size of the parathyroid glands on the posterior surface of the thyroid lobes. **(b)** This photomicrograph shows both parathyroid and thyroid tissues. **(c)** A photomicrograph showing principal and oxyphil cells of the parathyroid gland. (LM × 850)

The Thymus [FIGURE 19.1 AND TABLE 19.3]

The **thymus** is embedded in a mass of connective tissue inside the thoracic cavity, usually just posterior to the sternum (Figure 19.1●, p. 507). In newborn infants and young children, the thymus is relatively large, often extending from the base of the neck to the superior border of the heart. Although its relative size decreases as a child grows, the thymus continues to enlarge slowly, reaching its maximum size just before puberty, at a weight of around 40 g. After puberty it gradually diminishes in size; by age 50 the thymus may weigh less than 12 g.

The thymus produces several hormones important to the development and maintenance of normal immunological defenses (Table 19.3). **Thymosin** (thī-MŌ-sin) was the name originally given to a thymic extract that promoted the development and maturation of lymphocytes and thus increased the effectiveness of the immune system. It has since become apparent that "thymosin" is a blend of several different, complementary hormones (thymosin-1, thymopoietin, thymopentin, thymulin, thymic humoral factor, and IGF-1).

It has been suggested that the gradual decrease in the size and secretory abilities of the thymus may make the elderly more susceptible to disease. The histological organization of the thymus and the functions of the various "thymosins" will be discussed further in Chapter 23.

The Adrenal Gland [FIGURE 19.10]

A yellow, pyramid-shaped **adrenal gland**, or **suprarenal gland** (soo-pra-RĒ-nal; *supra-*, above + *renes*, kidneys), is firmly attached to the superior border of each kidney by a dense, fibrous **capsule** (Figure 19.10a●). The adrenal gland on each side nestles between the kidney, the diaphragm, and the major arteries and veins running along the dorsal wall of the abdominopelvic cavity. The adrenal glands project into the abdominopelvic cavity, but they remain separated from it by the peritoneal lining. Like the other endocrine glands, the adrenal glands are highly vascularized. Branches of the *renal artery*, the *inferior phrenic artery*, and a direct branch from the aorta (the *middle suprarenal artery*) supply blood to each adrenal gland. The *suprarenal veins* carry blood away from the adrenal glands.

A typical adrenal gland weighs about 7.5 g. It is generally heavier in men than in women, but adrenal size can vary greatly as secretory demands change. Structurally and functionally, the adrenal gland can be divided into two regions, each secreting different hormone types, but both aiding in managing stress: a superficial **adrenal cortex** and an inner **adrenal medulla** (Figure 19.10b,c●).

The Adrenal Cortex [FIGURE 19.10C AND TABLE 19.4]

The yellowish color of the adrenal cortex is due to the presence of stored lipids, especially cholesterol and various fatty acids. The adrenal cortex produces more than two dozen different steroid hormones, collectively called **adrenocortical steroids**, or simply **corticosteroids**. These hormones are vital; if the adrenal glands are destroyed or removed, corticosteroids must be administered or the person will not survive. The corticosteroids exert their effects on metabolic operations by determining which genes are transcribed in their target cells, and at what rates.

Deep to the adrenal capsule there are three distinct regions, or *zones*, in the adrenal cortex: (1) an outer, *zona glomerulosa*; (2) a middle, *zona fasciculata*; and (3) an inner, *zona reticularis* (Figure 19.10c●). Although each

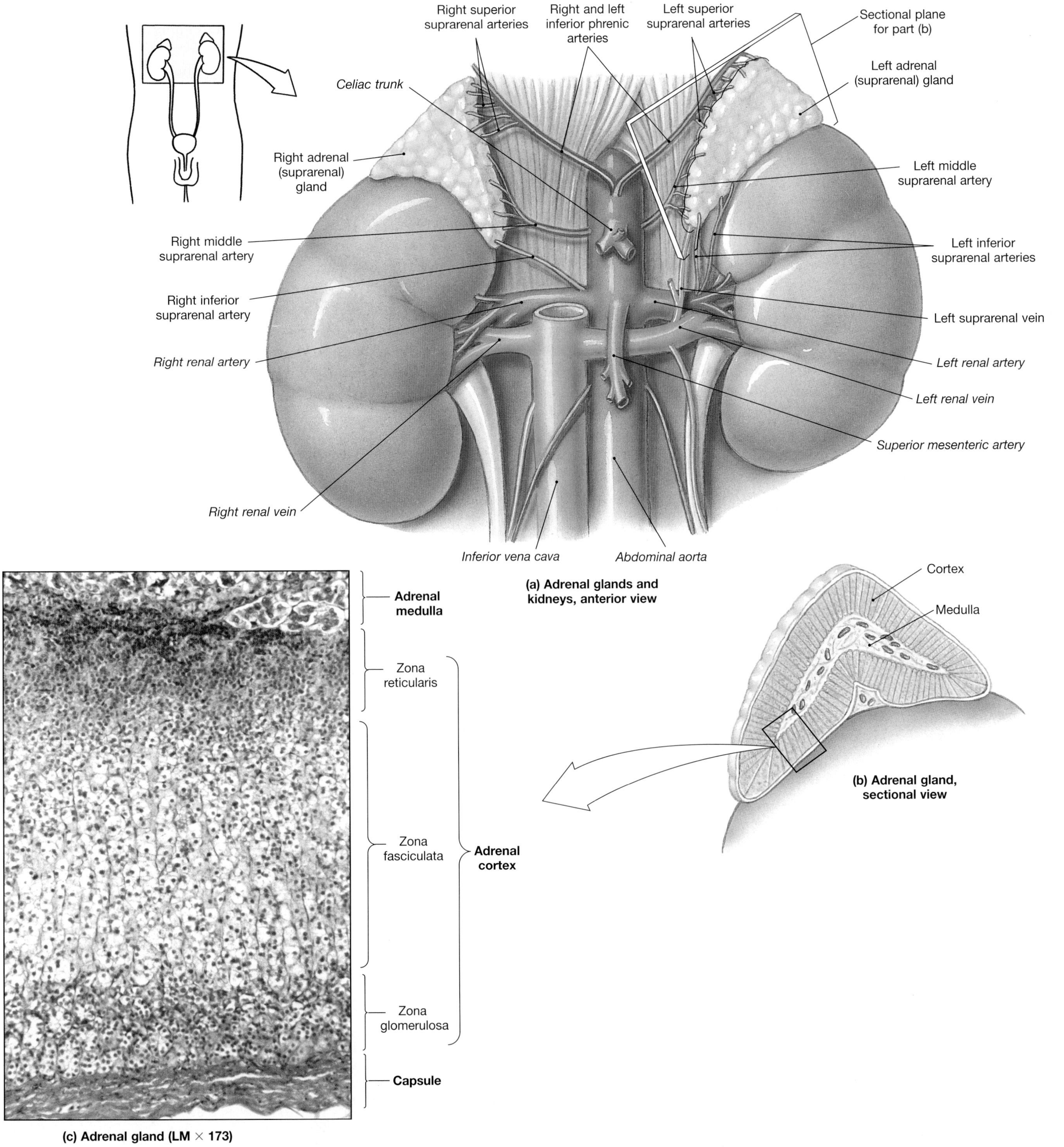

FIGURE 19.10 **THE ADRENAL GLAND.**

(**a**) Anterior view of the kidney and adrenal gland. Note the sectional plane for part (b). (**b**) An adrenal gland cut to show both the cortex and the medulla. Note the orientation of the section for part (c). (**c**) Photomicrograph, with the major regions identified.

zone synthesizes different steroid hormones (Table 19.4), all of the cortical cells have an extensive smooth ER for the manufacture of lipid-based steroids. This is in marked contrast to the abundant rough ER characteristic of protein-secreting gland cells, such as those of the anterior pituitary gland or thyroid gland.

THE ZONA GLOMERULOSA [FIGURE 19.10C]

The **zona glomerulosa** (glō-mer-ū-LŌ-sa), the outermost cortical region, accounts for about 15% of the cortical volume (Figure 19.10c●). This zone extends from the capsule to the radiating cords of the underlying zona fasciculata. A *glomerulus* is a little ball or knot, and here the endocrine cells form densely packed clusters.

The zona glomerulosa produces **mineralocorticoids (MC)**, steroid hormones that affect the electrolyte composition of body fluids. **Aldosterone** (al-DOS-ter-ōn) is the principal mineralocorticoid and targets kidney cells that regulate the ionic composition of the urine. It causes the retention of sodium ions (Na^+) and water, thereby reducing fluid losses in the urine. Aldosterone also reduces sodium and water losses at the sweat glands, salivary glands, and along the digestive tract. It also promotes the loss of potassium ions (K^+) in the urine and at other sites as well. Aldosterone secretion occurs when the zona glomerulosa is stimulated by a fall in blood Na^+ levels, a rise in blood K^+ levels, or arrival of the hormone *angiotensin II* (*angeion*, vessel + *teinein*, to stretch). *Disorders of the Adrenal Cortex p. 796*

THE ZONA FASCICULATA [FIGURE 19.10C]

The **zona fasciculata** (fa-sik-ū-LA-ta; *fasciculus*, little bundle) begins at the inner border of the zona glomerulosa and extends toward the medulla (Figure 19.10c●). It represents about 78% of the cortical volume. The cells are larger and contain more lipids than those of the zona glomerulosa, and the lipid droplets give the cytoplasm a pale, foamy appearance. The cells of the zona fasciculata form cords that radiate like a sunburst from the zona reticularis. Adjacent cords are separated by flattened vessels with fenestrated walls.

Steroid production in the zona fasciculata is stimulated by ACTH from the anterior pituitary. This zone produces steroid hormones collectively known as **glucocorticoids (GC)** because of their effects on glucose metabolism. **Cortisol** (KOR-ti-sol; also called *hydrocortisone*) and **corticosterone** (kor-ti-KOS-te-rōn) are the most important glucocorticoids secreted by the adrenal cortex; the liver converts some of the circulating cortisol to **cortisone**, another active glucocorticoid. These hormones speed up the rates of glucose synthesis and glycogen formation, especially within the liver.

THE ZONA RETICULARIS [FIGURE 19.10C]

The **zona reticularis** (re-tik-ū-LAR-is; *reticulum*, network) forms a narrow band between the zona fasciculata and the outer border of the adrenal medulla (Figure 19.10c●). The cells of the zona reticularis are much smaller than those of the medulla, and this makes the boundary relatively easy to distinguish. In total, the zona reticularis accounts for only around 7% of the total cell volume of the adrenal cortex. The endocrine cells of the zona reticularis form a folded, branching network with an extensive capillary supply. The zona reticularis normally secretes small amounts of sex hormones called *androgens*. Androgens are produced in large quantities by the testes, and the significance of the small adrenal androgen contribution—in both males *and* females—remains uncertain.

The Adrenal Medulla [FIGURE 19.10B,C AND TABLE 19.4]

The boundary between the adrenal cortex and medulla does not form a straight line (Figure 19.10b,c●), and the supporting connective tissues and blood vessels are extensively interconnected. The adrenal medulla has a reddish brown coloration due in part to the many blood vessels in this area. **Pheochromocytes**, or **chromaffin cells**, are large, rounded cells of the medulla that resemble the neurons in sympathetic ganglia. These cells are innervated by preganglionic sympathetic fibers; sympathetic activation, provided by the splanchnic nerves, triggers the secretory activity of these modified ganglionic neurons. p. 452

The adrenal medulla contains two populations of endocrine cells—one secreting epinephrine (adrenaline), and the other norepinephrine (noradrenaline). The adrenal medulla secretes roughly 3 times as much epinephrine as norepinephrine. p. 452 Their secretion triggers cellular energy utilization and the mobilization of energy reserves. This combination increases muscular strength and endurance (Table 19.4). The meta-

TABLE 19.4 **THE ADRENAL HORMONES**

Region/Zone	Hormones	Targets	Effects
CORTEX			
Zona glomerulosa	Mineralocorticoids (MC), primarily aldosterone	Kidneys	Increases renal reabsorption of sodium ions and water (especially in the presence of ADH) and accelerates urinary loss of potassium ions
Zona fasciculata	Glucocorticoids (GC): cortisol (hydrocortisone), corticosterone; cortisol converted to cortisone and released by the liver	Most cells	Releases amino acids from skeletal muscles, lipids from adipose tissues; promotes formation of liver glycogen and glucose; promotes peripheral utilization of lipids (glucose-sparing); anti-inflammatory effects
Zona reticularis	Androgens		Uncertain significance under normal conditions
MEDULLA	Epinephrine, norepinephrine	Most cells	Increased cardiac activity, blood pressure, glycogen breakdown, blood glucose; release of lipids by adipose tissue (see Chapter 17)

bolic changes that follow catecholamine release are at their peak 30 seconds after adrenal stimulation, and they linger for several minutes thereafter. As a result, the effects produced by stimulation of the adrenal medulla outlast the other signs of sympathetic activation. *Disorders of the Adrenal Medulla p. 796*

✓ CONCEPT CHECK

- When a person's thyroid gland is removed, signs of decreased thyroid hormone concentration do not appear until about one week later. Why?
- Removal of the parathyroid glands would result in a decrease in the blood of what important mineral?
- A disorder of the adrenal gland prevents Bill from retaining sodium ions in body fluids. Which region of the gland is affected, and what hormone is deficient?

Endocrine Functions of the Kidneys and Heart

The kidneys and heart produce several hormones, and most of them are involved with the regulation of blood pressure and blood volume. The kidneys produce *renin*, an enzyme (often called a hormone), and two hormones: *erythropoietin*, a peptide, and *calcitriol*, a steroid. Once in the circulation, **renin** converts circulating **angiotensinogen**, an inactive protein produced by the liver, to **angiotensin I**. In capillaries of the lungs, this compound is converted to **angiotensin II**, the hormone that stimulates the secretion of aldosterone by the adrenal cortex. **Erythropoietin** (e-rith-rō-POY-e-tin) **(EPO)** stimulates red blood cell production by the bone marrow. This hormone is released either when blood pressure or blood oxygen levels in the kidneys decline. EPO stimulates red blood cell production and maturation, thus increasing the blood volume and its oxygen-carrying capacity.

Calcitriol is a steroid hormone secreted by the kidney in response to the presence of parathyroid hormone (PTH). Calcitriol synthesis is dependent on the availability of a related steroid, *cholecalciferol* (vitamin D_3), which may be synthesized in the skin or absorbed from the diet. Cholecalciferol from either source is absorbed from the bloodstream by the liver and converted to an intermediary product that is released into the circulation and absorbed by the kidneys for conversion to calcitriol. The term *vitamin D* is used to indicate the entire group of related steroids, including calcitriol, cholecalciferol, and various intermediaries.

The best-known function of calcitriol is the stimulation of calcium and phosphate ion absorption along the digestive tract. PTH stimulates the release of calcitriol, and in this way, PTH has an indirect effect on intestinal calcium absorption. The effects of calcitriol on the skeletal system and kidney are not well understood.

Cardiac muscle cells in the heart produce **atrial natriuretic peptide (ANP)** and **brain natriuretic peptide (BNP)** in response to increased blood pressure or blood volume. ANP and BNP suppresses the release of ADH and aldosterone and stimulate water and sodium ion loss at the kidneys. These effects gradually reduce both blood pressure and blood volume.

The Pancreas and Other Endocrine Tissues of the Digestive System

The pancreas, the lining of the digestive tract, and the liver produce various exocrine secretions that are essential to the normal digestion of food. Although the pace of digestive activities can be affected by the autonomic nervous system, most digestive processes are controlled locally by the individual organs. The various digestive organs communicate with one another using hormones detailed in Chapter 25. This section focuses attention on one digestive organ, the pancreas, which produces hormones that affect metabolic operations throughout the body.

The Pancreas [FIGURE 19.11 AND TABLE 19.5]

The **pancreas** is a mixed gland with both exocrine and endocrine activities. It lies within the abdominopelvic cavity in the J-shaped loop between the stomach and small intestine (Figure 19.11a●). It is a slender, pink organ with a nodular or lumpy consistency. The adult pancreas ranges between 20 and 25 cm (8–10 in.) in length and weighs about 80 g (2.8 oz). Chapter 25 considers the detailed anatomy of the pancreas, because the **exocrine pancreas**, roughly 99% of the pancreatic volume, produces large quantities of a digestive enzyme-rich fluid that enters the digestive tract through a prominent secretory duct.

The **endocrine pancreas** consists of small groups of cells scattered throughout the gland, each group surrounded by exocrine cells. The groups, known as **pancreatic islets**, or the *islets of Langerhans* (LAN-ger-hanz), account for only about 1% of the pancreatic cell population (Figure 19.11b●). Nevertheless, there are roughly 2 million islets in the normal pancreas.

Like other endocrine tissue, the islets are surrounded by an extensive, fenestrated capillary network that carries its hormones into the circulation. Two major arteries supply blood to the pancreas, the *pancreaticoduodenal arteries* and *pancreatic arteries*. Venous blood returns to the *hepatic portal vein*. (Circulation to and from major organs will be considered in Chapter 22.) The islets are also innervated by the autonomic nervous system, through branches from the *celiac plexus*. *p. 459*

Each islet contains four major cell types.

1. **Alpha cells** produce the hormone **glucagon** (GLOO-ka-gon), which raises blood glucose levels by increasing the rates of glycogen breakdown and glucose release by the liver (Figure 19.11c●).
2. **Beta cells** produce the hormone **insulin** (IN-su-lin), which lowers blood glucose by increasing the rate of glucose uptake and utilization by most body cells (Figure 19.11c●).
3. **Delta cells** produce the hormone **somatostatin** (**growth-hormone-inhibiting hormone**), which inhibits the production and secretion of glucagon and insulin and slows the rates of food absorption and enzyme secretion along the digestive tract.

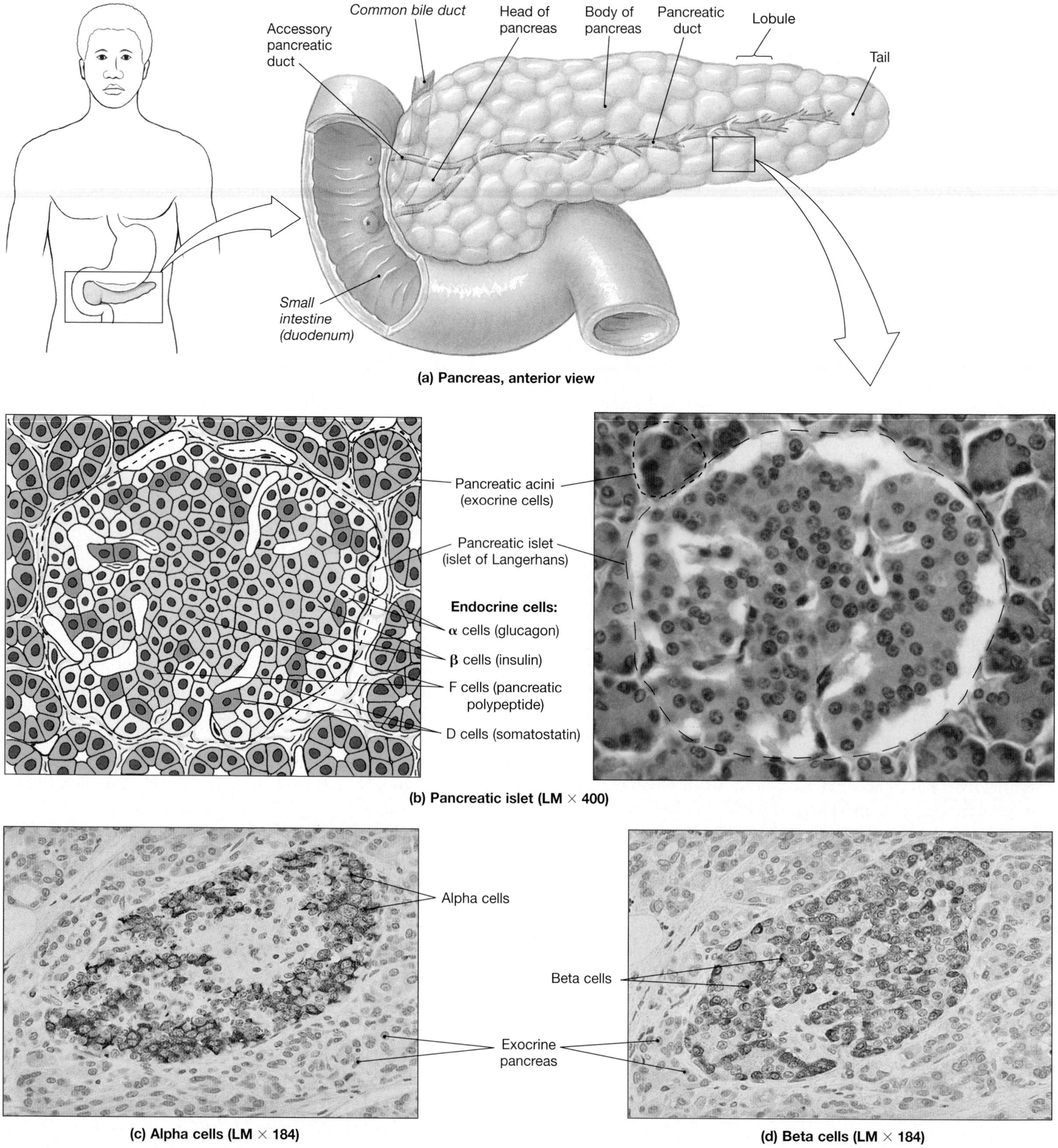

FIGURE 19.11 **THE PANCREAS**

This organ, which is dominated by exocrine cells, contains clusters of endocrine cells known as the pancreatic islets. **(a)** The gross anatomy of the pancreas. **(b)** The general histology of the pancreatic islets. **(c)** and **(d)** Special staining techniques can be used to differentiate between alpha cells (c) and beta cells (d) in pancreatic islets.

TABLE 19.5 HORMONES OF THE PANCREAS

Structure/Cells	Hormones	Primary Targets	Effects
PANCREATIC ISLETS			
Alpha cells	Glucagon	Liver, adipose tissues	Mobilization of lipid reserves; glucose synthesis and glycogen breakdown in liver; elevation of blood glucose concentrations
Beta cells	Insulin	All cells except those of brain, kidneys, digestive tract epithelium, and RBCs	Facilitation of uptake of glucose by cells; stimulation of lipid and glycogen formation and storage; decrease in blood glucose concentrations
Delta cells	Somatostatin	Alpha and beta cells, digestive epithelium	Inhibition of secretion of insulin and glucagon
F cells	Pancreatic polypeptide (PP)	Gallbladder and pancreas, possibly gastrointestinal tract	Inhibits gallbladder contractions; regulates production of some pancreatic enzymes; may control nutrient absorption

4. **F cells** produce the hormone **pancreatic polypeptide (PP)**. It inhibits gallbladder contractions and regulates the production of some pancreatic enzymes; it may help control the rate of nutrient absorption by the digestive tract.

Pancreatic alpha and beta cells are sensitive to blood glucose concentrations, and their regulatory activities are not under the direct control of other endocrine or nervous components. Yet because the islet cells are extremely sensitive to variations in blood glucose levels, any hormone that affects blood glucose concentrations will affect the production of insulin and glucagon indirectly. The major hormones of the pancreas are summarized in Table 19.5. *Diabetes Mellitus p. 797*

Endocrine Tissues of the Reproductive System

The endocrine tissues of the reproductive system are restricted primarily to the male and female gonads, the testes and ovaries respectively. The anatomy of the reproductive organs will be described in Chapter 27.

Testes [TABLE 19.6]

In the male, the **interstitial cells** of the testes produce androgens. The most important androgen is **testosterone** (tes-TOS-ter-ōn). This hormone promotes the production of functional sperm, maintains the secretory glands of the male reproductive tract, influences secondary sexual characteristics, and stimulates muscle growth (Table 19.6). During embryonic development, the production of testosterone affects the anatomical development of the hypothalamic nuclei of the CNS.

Sustentacular cells, which are directly associated with the formation of functional sperm, secrete an additional hormone, called **inhibin** (in-HIB-in). Inhibin production, which occurs under FSH stimulation, depresses the secretion of FSH by the anterior lobe of the pituitary gland. Throughout adult life, these two hormones interact to maintain sperm production at normal levels.

Ovaries [TABLE 19.6]

In the ovaries, oocytes begin their maturation into female gametes (sex cells) within specialized structures called **follicles**. The maturation process starts in response to stimulation by FSH. Follicle cells surrounding the oocytes produce estrogens, especially the hormone **estradiol**. These steroid hormones support the maturation of the oocytes and stimulate the growth of the uterine lining (Table 19.6). Under FSH stimulation, active follicles secrete inhibin, which suppresses FSH release through a feedback mechanism comparable to that described for males.

After ovulation has occurred, the remaining follicular cells reorganize into a **corpus luteum** (LOO-tē-um) that releases a mixture of estrogens and progestins, especially **progesterone** (prō-JES-ter-ōn). Progesterone accelerates the movement of the oocyte along the uterine tube and prepares the uterus for the arrival of the developing embryo. A summary of information concerning the reproductive hormones can be found in Table 19.6.

TABLE 19.6 HORMONES OF THE REPRODUCTIVE SYSTEM

Structure/Cells	Hormones	Primary Targets	Effects
TESTES			
Interstitial cells	Androgens	Most cells	Supports functional maturation of sperm; protein synthesis in skeletal muscles; male secondary sex characteristics and associated behaviors
Sustentacular cells	Inhibin	Anterior lobe of pituitary gland	Inhibits secretion of FSH
OVARIES			
Follicle cells	Estrogens (especially estradiol)	Most cells	Support follicle maturation; female secondary sex characteristics and associated behaviors
	Inhibin	Anterior lobe of pituitary gland	Inhibits secretion of FSH
Corpus luteum	Progestins (especially progesterone)	Uterus, mammary glands	Prepare uterus for implantation; prepare mammary glands for secretory functions
	Relaxin	Pubic symphysis, uterus, mammary glands	Loosens pubic symphysis; relaxes uterine (cervical) muscles; stimulates mammary gland development

As noted in Chapter 3, all secretory glands, whether exocrine or endocrine, are derived from epithelia. Endocrine organs develop from epithelia (1) covering the outside of the embryo, (2) lining the digestive tract, and (3) lining the coelomic cavity.

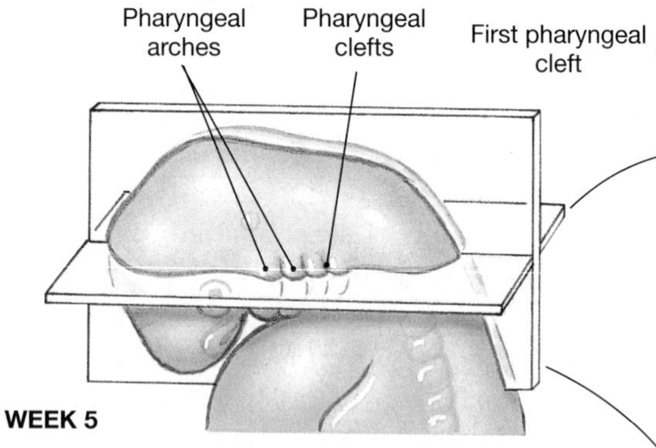

The pharyngeal region of the embryo plays a particularly important role in endocrine development. After 4–5 weeks of development, the *pharyngeal arches* are well formed. Human embryos develop five or six pharyngeal arches, not all visible from the exterior. (Arch 5 may not appear or may form and degenerate almost immediately.) The five major arches (I-IV, VI) are separated by *pharyngeal clefts*, deep ectodermal grooves.

PARATHYROID GLANDS AND THYMUS

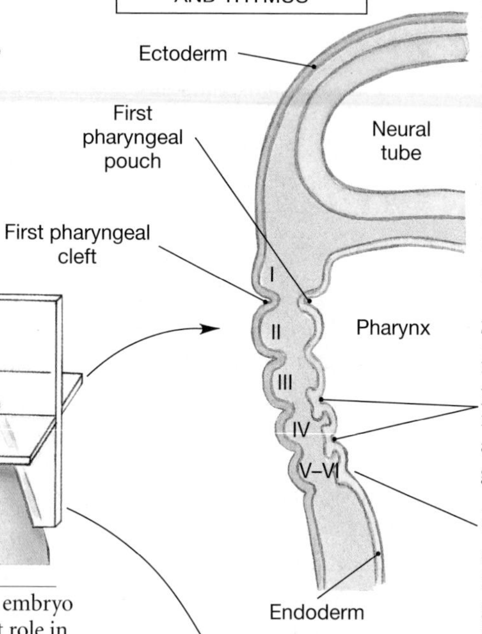

In sectional view, five **pharyngeal pouches** extend laterally toward the pharyngeal clefts. The first pouch lies caudal to the first (mandibular) arch. Pharyngeal pouches 5 and 6 are very small and are interconnected. Endoderm lining the third, fourth, and fifth pairs of pharyngeal pouches forms dorsal and ventral masses of cells that migrate beneath the endodermal epithelium.

The dorsal masses of the third and fourth pouches form the parathyroid glands. The ventral masses move toward the midline and fuse to create the thymus gland.

Cells originating in the walls of the small fifth pouch will be incorporated into the thyroid gland, where they will differentiate into C cells (see below).

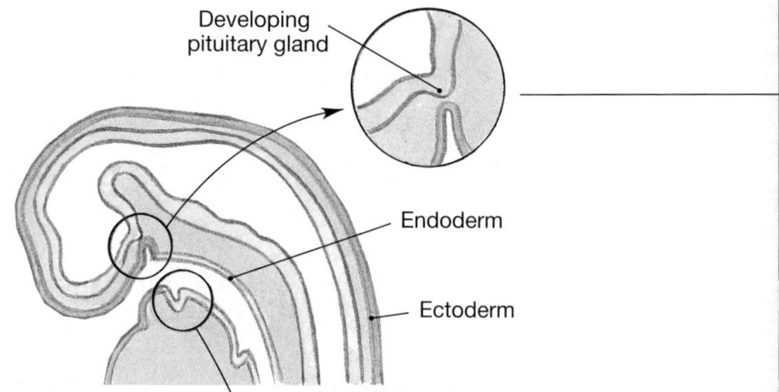

THYROID GLAND

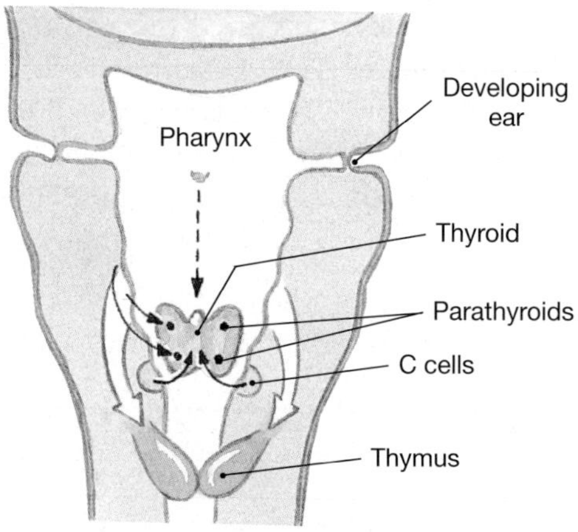

As the embryo enlarges and changes shape, the thyroid shifts caudally to a position near the thyroid cartilage of the larynx. On its way, the thyroid gland incorporates C cells from the walls of the fifth pouch.

The boundary between ectoderm and endoderm lies along the line formed by the circumvallate papillae of the tongue (*refer to Figure 18.7*). This line roughly corresponds to the middle of the mandibular (first) arch. The thyroid gland forms here in the ventral midline, and the pituitary gland forms in the dorsal midline.

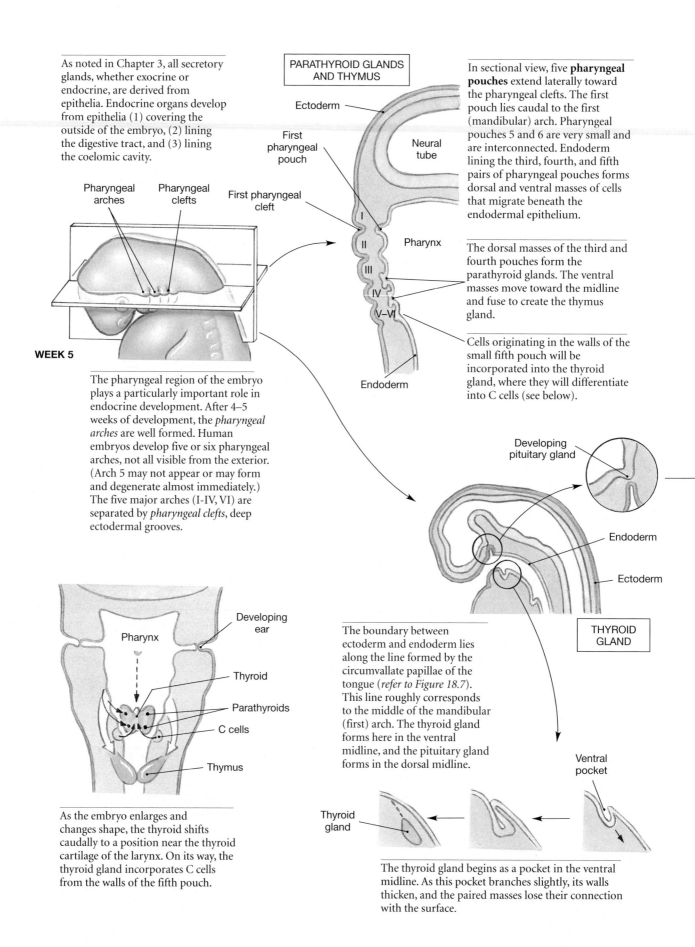

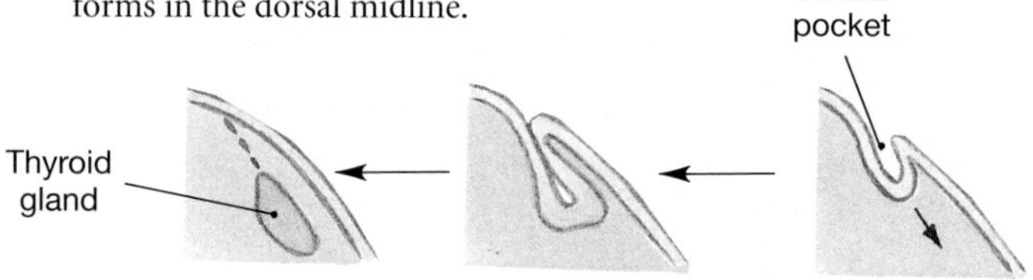

The thyroid gland begins as a pocket in the ventral midline. As this pocket branches slightly, its walls thicken, and the paired masses lose their connection with the surface.

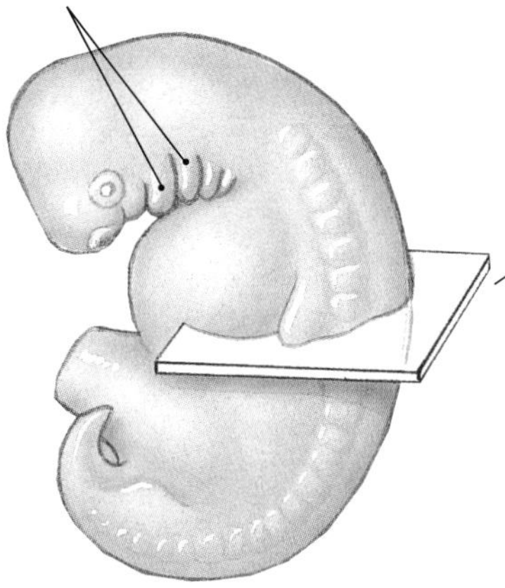

ADRENAL GLANDS

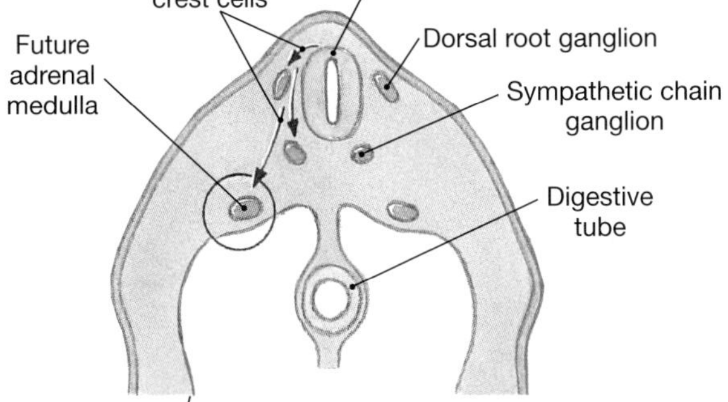

Each adrenal gland also has a compound origin. Shortly after the formation of the *neural tube*, neural crest cells migrate away from the CNS. This migration leads to the formation of the dorsal root ganglia and autonomic ganglia. On each side of the coelomic cavity, neural crest cells aggregate in a mass that will become an adrenal medulla.

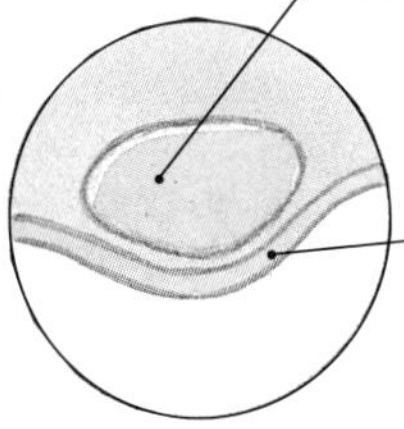

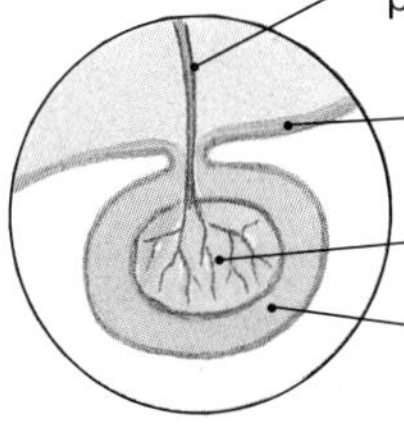

Overlying epithelial cells respond by undergoing division, and the daughter cells surround the neural crest cells to form a thick adrenal cortex.

PITUITARY GLAND

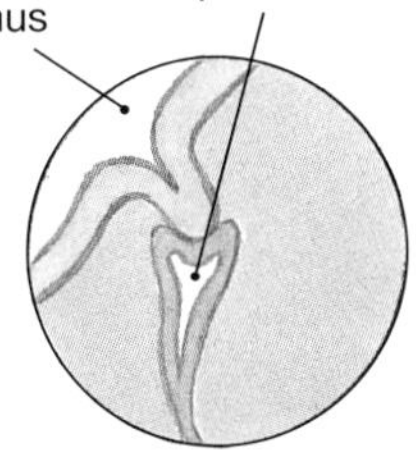

The pituitary gland has a compound origin. The first step is the formation of an ectodermal pocket in the dorsal midline of the pharynx. This pocket loses its connection to the pharynx, creating a hollow ball of cells that lies inferior to the floor of the diencephalon posterior to the optic chaism.

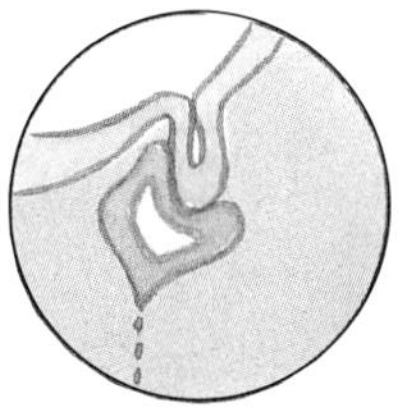

As these cells undergo division, the central chamber gradually disappears. This endocrine mass will become the anterior lobe of the pituitary gland. The posterior lobe of the pituitary gland begins as a depression in the hypothalamic floor and grows toward the developing anterior lobe.

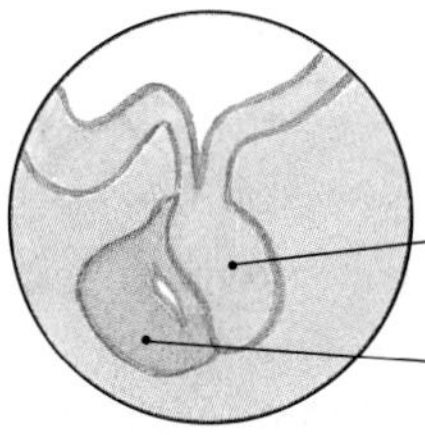

For additional details concerning the development of other endocrine organs refer to the Embryology Summaries in Chapters 23, 25, 26, and 27.

CLINICAL DISCUSSION

Endocrine Disorders

Endocrine disorders fall into two basic categories: They reflect either abnormal hormone production or abnormal cellular sensitivity. The symptoms are interesting because they highlight the significance of normally "silent" hormonal contributions. The characteristic features of many of these conditions are illustrated in Figure 19.12●.

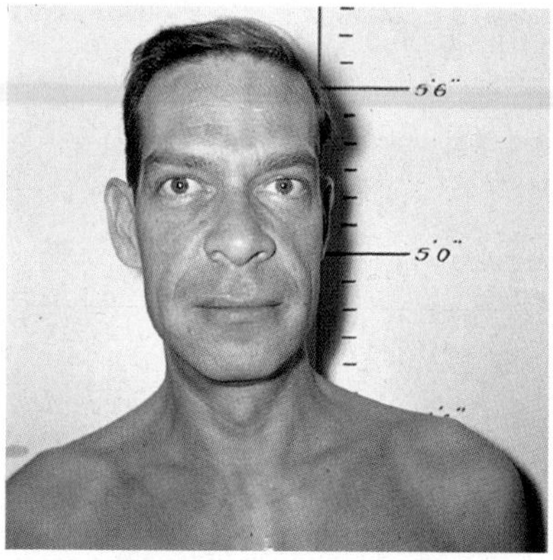

(a) *Acromegaly* results from the overproduction of growth hormone after the epiphyseal plates have fused. Bone shapes change and cartilaginous areas of the skeleton enlarge. Note the broad facial features and the enlarged lower jaw.

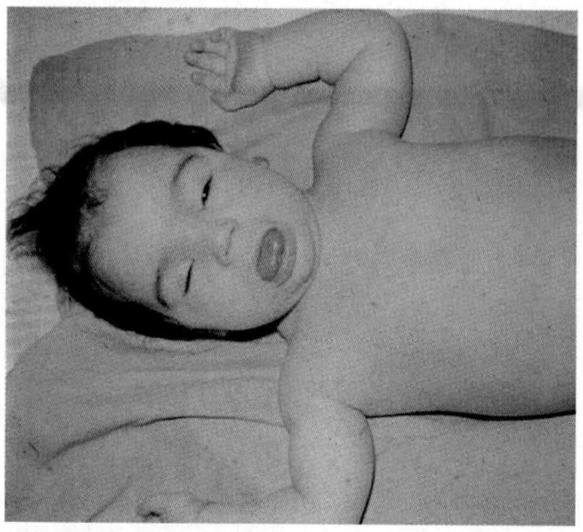

(b) *Cretinism* results from thyroid hormone insufficiency in infancy.

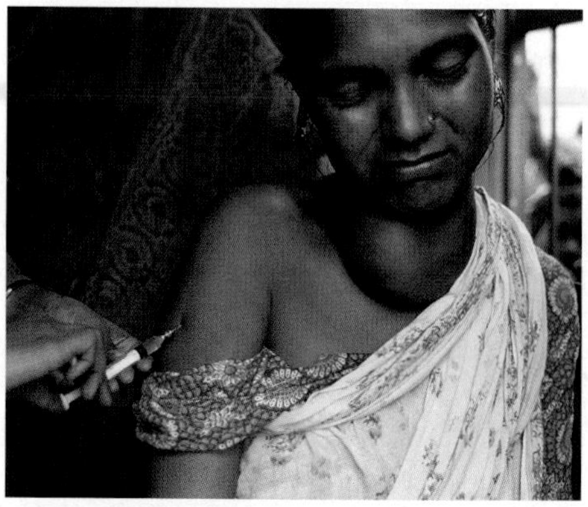

(c) An enlarged thyroid gland, or *goiter,* is usually associated with thyroid hyposecretion due to nutritional iodine insufficiency.

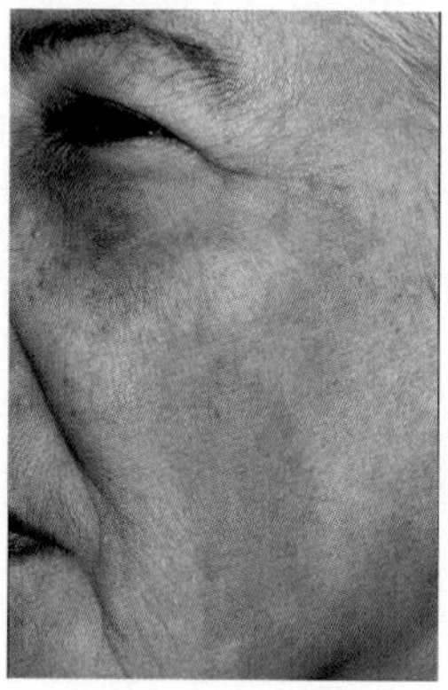

(d) *Addison's disease* is caused by hyposecretion of corticosteroids, especially glucocorticoids. Pigment changes result from stimulation of melanocytes by ACTH, which is structurally similar to MSH.

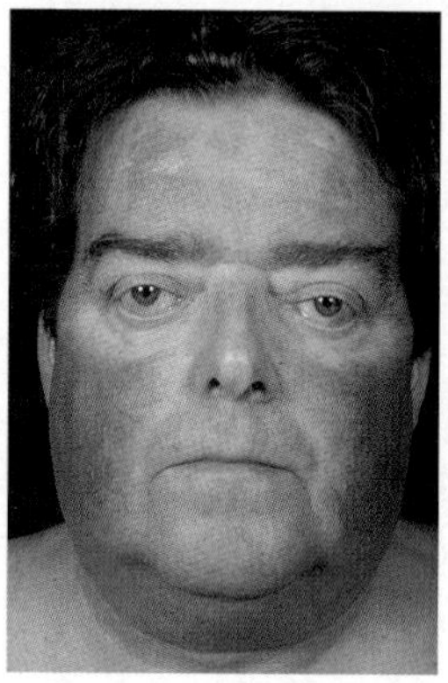

(e) *Cushing's disease* is caused by hypersecretion of glucocorticoids. Lipid reserves are mobilized, and adipose tissue accumulates in the cheeks and at the base of the neck.

FIGURE 19.12 **ENDOCRINE ABNORMALITIES**

Representative examples of conditions resulting from endocrine disorders.

The Pineal Gland [FIGURE 19.1]

The small, red, pinecone-shaped **pineal gland,** or *epiphysis* (e-PIF-e-sis) *cerebri,* is part of the epithalamus (Figure 19.1●, p. 507). p. 399 The pineal gland contains neurons, glial cells, and special secretory cells called **pinealocytes** (PĪN-ē-a-lō-sīts). Pinealocytes synthesize the hormone **melatonin** (mel-a-TŌ-nin), which is derived from molecules of the neurotransmitter *serotonin.* Melatonin slows the maturation of sperm, oocytes, and reproductive organs by inhibiting the production of a hypothalamic releasing factor that stimulates FSH and LH secretion. Collaterals from the visual pathways enter the pineal gland and affect the rate of melatonin production. Melatonin production rises at night and declines during the day. This cycle is apparently important in regulating *circadian rhythms,* our natural awake-asleep cycles. This hormone is also a powerful antioxidant that may help protect CNS tissues from toxins generated by active neurons and glial cells.

Hormones and Aging

The endocrine system shows relatively few functional changes with advancing age. The most dramatic exceptions are (1) the changes in reproduction hormone levels at puberty, and (2) the decline in the concentration of reproductive hormones at menopause in women. It is interesting to note that age-related changes in other tissues affect their abilities to respond to hormonal stimulation. As a result, most tissues may become less responsive to circulating hormones, even though hormone concentrations remain normal.

CLINICAL COMMENT Endocrine disorders may develop for a variety of reasons. For example, a hormone level may rise because its target organs are becoming less responsive, because a tumor has formed among the gland cells, or because something has interfered with the normal feedback control mechanism. When naming endocrine disorders, clinicians use the prefix *hyper-* when referring to excessive hormone production and *hypo-* when referring to inadequate hormone production. For a more complete overview of endocrine disorders, refer to the Clinical Appendix. *Endocrine Disorders p. 796*

✓ CONCEPT CHECK

- Where are the islets of Langerhans located? Name the hormones produced here and their functions.
- What is the function of inhibin, and where is it produced?
- Which hormone(s) of the endocrine system show the most dramatic decline in concentration as a result of aging?

RELATED CLINICAL TERMS

diabetes insipidus: A disorder that develops when the posterior pituitary no longer releases adequate amounts of ADH. *Diabetes Insipidus p. 796*

diabetes mellitus (mel-LĪ-tus): A disorder characterized by glucose concentrations high enough to overwhelm the kidneys' reabsorption capabilities. *Diabetes Mellitus p. 797*

exophthalmos (eks-ahf-THAL-mos): Protrusion of the eyes, a symptom of hyperthyroidism. *Thyroid Gland Disorders p. 796*

goiter: A diffuse enlargement of the thyroid gland. *Thyroid Gland Disorders p. 796*

insulin-dependent diabetes mellitus (IDDM) (also known as **Type 1 diabetes** or **juvenile-onset diabetes**): A type of diabetes mellitus; the primary cause is inadequate insulin production by the beta cells of the pancreatic islets. *Diabetes Mellitus p. 797*

ketoacidosis: A condition in which large numbers of ketone bodies in the blood lead to a dangerously low blood pH. *Diabetes Mellitus p. 797*

myxedema (miks-e-DĒ-ma): Symptoms of severe hypothyroidism, which include subcutaneous swelling, dry skin, hair loss, low body temperature, muscular weakness, and slowed reflexes. *Thyroid Gland Disorders p. 796*

non-insulin-dependent diabetes mellitus (NIDDM) (also known as **Type 2 diabetes** or **maturity-onset diabetes**): A type of diabetes mellitus in which insulin levels are normal or elevated but peripheral tissues no longer respond normally. *Diabetes Mellitus p. 797*

thyrotoxic crisis: A period when a subject with acute hyperthyroidism experiences an extremely high fever, rapid heart rate, and the malfunctioning of a variety of physiological systems. *Thyroid Gland Disorders p. 796*

Additional Clinical Terms Discussed in Appendix I (pp. 796–797)

Addison's disease; aldosteronism; androgenital syndrome; Cushing's disease; diabetic nephropathy; diabetic neuropathy; diabetic retinopathy; glycosuria; Graves' disease; gynecomastia; hyperthyroidism; hypoaldosteronism; hypoparathyroidism; hypothyroidism; ketone bodies; pheochromocytoma; polyuria; thyrotoxic crisis; thyrotoxicosis

STUDY OUTLINE & CHAPTER REVIEW

Introduction 507

1. The nervous and endocrine systems work together in a complementary way to monitor and adjust physiological activities for the regulation of homeostasis.
2. In general, the nervous system performs short-term "crisis management," while the endocrine system regulates longer-term, ongoing metabolic processes. *Endocrine cells* release chemicals called **hormones** that alter the metabolic activities of many different tissues and organs simultaneously. *(see Figure 19.1 and Table 19.1)*

An Overview of the Endocrine System 507

1. The endocrine system consists of all endocrine cells and tissues. They release their secretory products into interstitial fluids.
2. Hormones can be divided into three groups based on chemical structure: *amino acid derivatives*, *peptide hormones*, and *steroids*.
3. **Amino acid derivatives** are small molecules structurally similar to amino acids. They include: (1) derivatives of tyrosine—for example, **thyroid hormones** from the thyroid gland, and **catecholamines** (epinephrine and norepinephrine) from the adrenal medullae, and (2) derivatives of trytophan—for example, **melatonin**, from the pineal gland.
4. **Peptide hormones** are chains of amino acids. This is the largest group of hormones. All pituitary gland hormones belong to this group.
5. **Steroid hormones** are lipid derivatives, structurally similar to cholesterol. This type of hormone is produced by the adrenal cortex and reproductive organs.
6. Cellular activities and metabolic reactions are controlled by enzymes. Hormones exert their effects by modifying the activities of **target cells** (cells that are sensitive to that particular hormone). A hormone must interact with a specific receptor in the cell membrane or cytoplasm. The presence or absence of the receptor determines a cell's sensitivities to a hormone. Binding the hormone initiates biochemical events that change activity patterns within the cell. The different classes of hormones work by different mechanisms. *(see Figure 19.2)*
7. Receptors for catecholamines and peptide hormones are located on the cell membranes of target cells; in this case, the hormone acts as a *first messenger* that causes the appearance of a *second messenger* to appear in the cytoplasm. This second messenger may act to activate or inhibit enzymes, or as a cofactor to change the direction and rate of the cell's metabolic reactions.
8. Thyroid and steroid hormones cross the cell membrane and bind to intracellular receptors. This hormone-receptor complex binds to DNA segments and causes the activation or inactivation of specific genes. *(see Figure 19.2)*
9. Negative-feedback control usually regulates endocrine activity. In this control: (1) the endocrine cell releases its hormone in response to changes in the composition of the extracellular fluid; (2) target cell is stimulated; (3) target cell restores homeostasis. This response eliminates the source of stimulation of the endocrine cell.

10. Endocrine activity can be controlled by: (1) neural activity, (2) positive feedback (rare), or (3) complex negative feedback mechanisms.

The Hypothalamus and Endocrine Regulation 509

11. The hypothalamus regulates endocrine and neural activities. It (1) controls the output of the adrenal medulla, an endocrine component of the sympathetic division of the ANS, (2) produces two hormones of its own (ADH and oxytocin), which are released from the posterior pituitary, and (3) controls the activity of the anterior pituitary through the production of **regulatory hormones (releasing hormones**, or **RH**, and **inhibiting hormones**, or **IH**). *(see Figure 19.3)*

The Pituitary Gland 510

1. The **pituitary gland (hypophysis)** releases nine important peptide hormones. Two are synthesized in the hypothalamus and released at the posterior lobe of the pituitary and seven are synthesized in the anterior lobe of the pituitary. *(see Figures 19.4/19.5 and Table 19.2)*

The Posterior Lobe of the Pituitary Gland 511

2. The **posterior lobe of the pituitary gland (neurohypophysis**, or *pars nervosa*) contains the axons of some hypothalamic neurons. Neurons within the **supraoptic** and **paraventricular nuclei** manufacture **antidiuretic hormone (ADH)** and **oxytocin**, respectively. ADH decreases the amount of water lost at the kidneys. It is released in response to a rise in the concentration of electrolytes in the blood or a fall in blood volume. In women, oxytocin stimulates smooth muscle cells in the uterus and contractile cells in the mammary glands. It is released in response to stretched uterine muscles and/or suckling of an infant. In men, it stimulates prostatic smooth muscle contractions. *(see Figures 19.4 to 19.6 and Table 19.2)*

The Anterior Lobe of the Pituitary Gland 512

3. The **anterior lobe of the pituitary gland (adenohypophysis)** can be subdivided into the large **pars distalis** and the slender **pars intermedia**. The entire adenohypophysis is highly vascularized.
4. In the floor of the hypothalamus in the tuberal area, neurons release regulatory factors into the surrounding interstitial fluids. Endocrine cells in the anterior lobe of the pituitary gland are controlled by releasing factors, inhibiting factors (hormones), or some combination of the two. These secretions enter the circulation through **fenestrated** capillaries that contain open spaces between their epithelial cells. Blood vessels, called **portal vessels**, form an unusual vascular arrangement that connects the hypothalamus and anterior lobe of the pituitary gland. This complex is the **hypophyseal portal system.** It ensures that all of the blood entering the portal vessels will reach the intended target cells before returning to the general circulation. *(see Figures 19.4/19.6)*
5. Important hormones released by the pars distalis include: (1) **thyroid-stimulating hormone (TSH)** (triggers the release of thyroid hormones); (2) **adrenocorticotropic hormone (ACTH)** (stimulates the release of **glucocorticoids** by the adrenal gland); (3) **follicle-stimulating hormone (FSH)** (stimulates **estrogen** secretion (*estradiol*) and egg development in women and sperm production in men); (4) **luteinizing hormone (LH)**, or *interstitial cell-stimulating hormone (ICSH)* (causes ovulation and production of **progestins** (*progesterone*) in women and **androgens** (*testosterone*) in men). Together, FSH and LH are called **gonadotropins;** (5) **prolactin (PRL)** (stimulates the development of the mammary glands and the production of milk); and (6) **growth hormone (GH**, or **somatotropin)** (stimulates cells' growth and replication). *(see Figures 19.4/19.5)*
6. **Melanocyte-stimulating hormone (MSH)**, released by the pars intermedia, stimulates melanocytes to produce melanin. *(see Figure 19.5)*

The Thyroid Gland 513

1. The **thyroid gland** lies inferior to the **thyroid cartilage** of the larynx. It consists of two **lobes** connected by a narrow **isthmus.** *(see Figure 19.7)*

Thyroid Follicles and Thyroid Hormones 513

2. The thyroid gland contains numerous **thyroid follicles**. Cells of the follicles manufacture **thyroglobulin** and store it within the **colloid** (a viscous fluid containing suspended proteins) in the **follicle cavity**. The cells also transport iodine from the extracellular fluids into the cavity, where it complexes with tyrosine residues of the thyroglobulin molecules to form thyroid hormones. *(see Figure 19.7b,c and Table 19.3)*
3. When stimulated by TSH, the follicle cells reabsorb the thyroglobulin, break down the protein, and release the thyroid hormones, **thyroxine** (**TX** or T_4) and **triiodothyronine** (T_3), into the circulation. *(see Figure 19.8)*

The C Cells of the Thyroid Gland 515

4. The **C cells** of the thyroid follicles produce **calcitonin (CT)**, which helps to lower calcium ion concentrations in body fluids by inhibiting osteoclast activities and stimulating calcium ion excretion at the kidneys. *(see Figure 19.7c)*
5. Actions of calcitonin are opposed by those of the *parathyroid hormone* produced by the parathyroid glands. *(see Table 19.3)*

The Parathyroid Glands 515

1. Four **parathyroid glands** are embedded in the posterior surface of the thyroid gland. The **principal (chief) cells** of the parathyroid produce **parathyroid hormone (PTH)** in response to lower-than-normal concentrations of calcium ions. Oxyphil cells of the parathyroid have no known function. *(see Figure 19.9 and Table 19.3)*
2. PTH: (1) stimulates osteoclast activity, (2) stimulates osteoblast activity to a lesser degree, (3) reduces calcium loss in the urine, and (4) promotes calcium absorption in the intestine (by stimulating calcitriol production). *(see Table 19.3)*
3. The parathyroid glands and the C cells of the thyroid gland maintain calcium ion levels within relatively narrow limits. *(see Figure 19.9c and Table 19.3)*

The Thymus 516

1. The **thymus**, embedded in a connective tissue mass in the thoracic cavity, produces several hormones that stimulate the development and maintenance of normal immunological defenses. *(see Figure 19.1)*
2. **Thymosins** produced by the thymus promote the development and maturation of lymphocytes.

The Adrenal Gland 516

1. A single **adrenal** *(suprarenal)* **gland** rests on the superior border of each kidney. Each gland is surrounded by a fibrous **capsule** and is subdivided into a superficial **adrenal cortex** and an inner **adrenal medulla.** *(see Figure 19.10)*

The Adrenal Cortex 516

2. The adrenal cortex manufactures steroid hormones called **adrenocortical steroids (corticosteroids)**. The cortex can be subdivided into three separate areas: (1) The outer **zona glomerulosa** releases **mineralocorticoids (MC)**, principally **aldosterone**, which restrict sodium and water losses at the kidneys, sweat glands, digestive tract, and salivary glands. The zona glomerulosa responds to the presence of the hormone angiotensin II, which appears after the enzyme renin has been secreted by kidney cells exposed to a decline in blood volume and/or blood pressure; (2) The middle **zona fasciculata** produces **glucocorticoids (GC)**, notably **cortisol** and **corticosterone**. All of these hormones accelerate the rates of both glucose synthesis and glycogen formation, especially in liver cells; (3) The inner **zona reticularis** produces small amounts of sex hormones called androgens, whose significance is uncertain. *(see Figure 19.10c and Table 19.4)*

The Adrenal Medulla 518

3. Each adrenal medulla contains clusters of **chromaffin cells**, which resemble sympathetic ganglia neurons. They secrete either epinephrine (75–80%) or

norepinephrine (20–25%). These catecholamines trigger cellular energy utilization and the mobilization of energy reserves (see Chapter 17). *(see Figure 19.10b,c and Table 19.4)*

Endocrine Functions of the Kidneys and Heart 519

1. Endocrine cells in both the kidneys and heart produce hormones that are important for the regulation of blood pressure and blood volume, blood oxygen levels, and calcium and phosphate ion absorption.
2. The kidney produces the enzyme *renin* and the peptide hormone *erythropoietin* when blood pressure or blood oxygen levels in the kidneys decline, and it secretes the steroid hormone *calcitriol* when parathyroid hormone is present. **Renin** catalyzes the conversion of circulating **angiotensinogen** to **angiotensin I**. In lung capillaries, it is converted to **angiotensin II**, the hormone that stimulates the production of aldosterone in the adrenal cortex. **Erythropoietin (EPO)** stimulates red blood cell production by the bone marrow. **Calcitriol** stimulates the absorption of both calcium and phosphate in the digestive tract.
3. Specialized muscle cells of the heart produce **atrial natriuretic peptide (ANP)** and **brand natriuretic peptide (BNP)** when blood pressure or blood volume becomes excessive. These hormones stimulate water and sodium ion loss at the kidneys, eventually reducing blood volume.

The Pancreas and Other Endocrine Tissues of the Digestive System 519

1. The lining of the digestive tract, the liver, and the pancreas produce exocrine secretions that are essential to the normal breakdown and absorption of food.

The Pancreas 519

2. The **pancreas** is a nodular organ occupying a space between the stomach and small intestine. It contains both exocrine and endocrine cells. The **exocrine pancreas** secretes an enzyme-rich fluid into the lumen of the digestive tract. Cells of the **endocrine pancreas** form clusters called **pancreatic islets** (*islets of Langerhans*). Each islet contains four cell types: **alpha cells** produce **glucagon** to raise blood glucose levels; **beta cells** secrete **insulin** to lower blood glucose levels; **delta cells** secrete **somatostatin (growth hormone–inhibiting factor)** to inhibit the production and secretion of glucagon and insulin; and **F cells** secrete **pancreatic polypeptide (PP)** to inhibit gallbladder contractions and regulate the production of some pancreatic enzymes; it may also help to control the rate of nutrient absorption by the G.I. tract. *(see Figure 19.11 and Table 19.5)*
3. Insulin lowers blood glucose by increasing the rate of glucose uptake and utilization by most body cells; glucagon raises blood glucose levels by increasing the rates of glycogen breakdown and glucose synthesis in the liver. Somatostatin reduces the rates of hormone secretion by alpha and beta cells and slows food absorption and enzyme secretion in the digestive tract. *(see Table 19.5)*

Endocrine Tissues of the Reproductive System 521

Testes 521

1. The **interstitial cells** of the male testes produce androgens. **Testosterone** is the most important androgen. It promotes the production of functional sperm, maintains reproductive-tract secretory glands, influences secondary sexual characteristics, and stimulates muscle growth. *(see Table 19.6)*
2. The hormone **inhibin** produced by **sustentacular cells** in the testes, interacts with FSH from the anterior lobe of the pituitary gland to maintain sperm production at normal levels.

Ovaries 521

3. Oocytes develop in **follicles** in the female ovary; follicle cells surrounding the oocytes produce estrogens, especially **estradiol**. Estrogens support the maturation of the oocytes and stimulate the growth of the uterine lining. Active follicles secrete inhibin, which suppresses FSH release by negative feedback. *(see Table 19.6)*
4. After ovulation, the follicle cells remaining within the ovary reorganize into a **corpus luteum**, which produces a mixture of estrogens and progestins, especially **progesterone**. Progesterone facilitates the movement of a fertilized egg through the uterine tube to the uterus and stimulates the preparation of the uterus for implantation. *(see Table 19.6)*

The Pineal Gland 524

1. The **pineal gland** *(epiphysis cerebri)* contains secretory cells called **pinealocytes**, which synthesize **melatonin**. Melatonin slows the maturation of sperm, eggs, and reproductive organs by inhibiting the production of FSH- and LH-releasing factors from the hypothalamus. Additionally, melatonin may establish circadian rhythms. *(see Figure 19.1)*

Hormones and Aging 524

1. The endocrine system shows relatively few functional changes with advancing age. The most dramatic endocrine changes are the rise in reproductive hormone levels at puberty and the decline in reproductive hormone levels at menopause.

LEVEL 1 REVIEWING FACTS AND TERMS

Match each numbered item with the most closely related lettered item. Use letters for answers in the spaces provided.

Column A	Column B
____ 1. target cells	a. unknown function
____ 2. hypothalamus	b. stimulates milk production
____ 3. ADH	c. regulated by hormones
____ 4. prolactin	d. pineal gland
____ 5. FSH	e. norepinephrine release
____ 6. colloid	f. decreases water loss
____ 7. oxyphil	g. lymphocyte maturation
____ 8. thymosin	h. stimulates estrogen secretion
____ 9. chromaffin cells	i. produces releasing hormone
____10. melatonin	j. viscous fluid with stored hormones

11. Adrenocorticotropic hormone (ACTH) stimulates the release of
    (a) thyroid hormones by the hypothalamus
    (b) gonadotropins by the testes
    (c) somatotropins by the hypothalamus
    (d) steroid hormones by the adrenal glands
12. When a catecholamine or peptide hormone binds to receptors on the surface of a cell,
    (a) the hormone receptor complex moves into the cytoplasm
    (b) the cell membrane becomes depolarized
    (c) a second messenger appears in the cytoplasm
    (d) the hormone is transported to the nucleus to alter DNA activity
13. The two hormones released by the posterior lobe of the pituitary gland are
    (a) somatotropin and
    (b) estrogen and progesterone gonadotropin
    (c) ADH and oxytocin
    (d) GH and prolactin
14. Hormones can alter cellular operations by changing
    (a) the quantities of enzymes
    (b) the activities of enzymes in a cell
    (c) the types of enzymes in a cell
    (d) all of the above
15. Endocrine organs can be controlled by
    (a) hormones from other endocrine glands
    (b) direct neural stimulation
    (c) changes in the composition of extracellular fluid
    (d) all of the above
16. Reduced fluid losses in the urine due to retention of sodium ions and water is a result of the action of
    (a) antidiuretic hormone
    (b) calcitonin
    (c) aldosterone
    (d) cortisone
17. When blood glucose levels fall,
    (a) insulin is released
    (b) glucagon is released
    (c) peripheral cells quit taking up glucose
    (d) aldosterone is released to stimulate these cells
18. Hormones released by the kidneys include
    (a) calcitriol and erythropoietin
    (b) ADH and aldosterone
    (c) epinephrine and norepinephrine
    (d) cortisol and cortisone norepinephrine
19. The element required for normal thyroid function is
    (a) magnesium
    (b) potassium
    (c) iodine
    (d) calcium
20. A structure known as the corpus luteum secretes
    (a) testosterone
    (b) progesterone
    (c) aldosterone
    (d) cortisone

LEVEL 2 REVIEWING CONCEPTS

1. Diabetes insipidus is caused by
   (a) decreased levels of insulin
   (b) decreased numbers of insulin receptors
   (c) decreased levels of ADH
   (d) increased levels of aldosterone
2. Damage to the cells of the zona fasciculata of the adrenal cortex would result in
   (a) the disappearance of axillary and pubic hair
   (b) increased volume of urine formation
   (c) decreased ability to convert lipids to glucose
   (d) increased water retention
3. Decreased levels of parathyroid hormone could result in
   (a) spasms (b) profuse urination
   (c) depressed immune activity (d) increased sweating
4. Discuss the functional differences between the endocrine and the nervous system.
5. Hormones can be divided into three groups on the basis of chemical structure. What are these three groups?
6. Describe the primary target and effects of testosterone.
7. What effects do thyroid hormones have on body tissues?
8. Why is normal parathyroid function essential in maintaining normal calcium ion levels?
9. Describe the role of melatonin in regulating reproductive function.
10. What is the significance of the capillary network within the hypophysis?

LEVEL 3 CRITICAL THINKING AND CLINICAL APPLICATIONS

1. Steve is 50 years old and has a pituitary tumor that produces excess amounts of growth hormone. Describe the symptoms that you would expect to observe in this patient.
2. Ernesto has been suffering from extreme thirst; he drinks numerous glasses of water every day and urinates a great deal. Name two disorders that could produce these symptoms. What test could a clinician perform to determine which disorder is present?
3. Carl is suffering from hypothyroidism (his thyroid gland does not produce enough thyroid hormone). How could his physician determine if the problem is due to problems at the level of the hypothalamus and pituitary gland or at the level of the thyroid gland?
4. What are two benefits of having a portal system connect the median eminence of the hypothalamus with the anterior lobe of the pituitary gland?

✓ ANSWERS TO CONCEPT CHECK QUESTIONS

p. 513 **1.** The hypothalamus is the region of the brain responsible for regulating hormone secretion by the pituitary gland. **2.** It is a peripheral cell that responds to the presence of a specific hormone. Hormones change cellular metabolic activities. **3.** The posterior lobe of the pituitary contains axon terminals of neurons whose cell bodies are in the hypothalamus. When these neurons are stimulated, their axon terminals release neurosecretions (oxytocin or ADH). Most endocrine cells of the anterior lobe of the pituitary gland are controlled by the secretion of regulatory factors by the hypothalamus.
p. 519 **1.** Most of the thyroid hormone in the blood is bound to proteins called thyroid-binding globulins. This represents a large reservoir of thyroxine that guards against rapid fluctuations in the level of this important hormone. Because there is such a large amount stored in this way, it takes several days to deplete the supply of hormone, even after the thyroid gland has been removed. **2.** Removal of the parathyroid glands would result in a decrease in the blood levels of calcium ions. This can be counteracted by increasing the amount of vitamin D and calcium in the diet. **3.** The region of the gland affected is the zona glomerulosa. The deficient hormone is aldosterone.
p. 525 **1.** The islets of Langerhans are located within the pancreas. The primary hormones released there are glucagon, insulin, and somatostatin. **2.** It depresses the secretion of FSH by the anterior lobe of the pituitary as part of a negative-feedback control mechanism. Inhibin is produced by sustentacular cells of the testes and by follicular cells in the ovaries. **3.** The reproductive hormones.

REVIEW IT

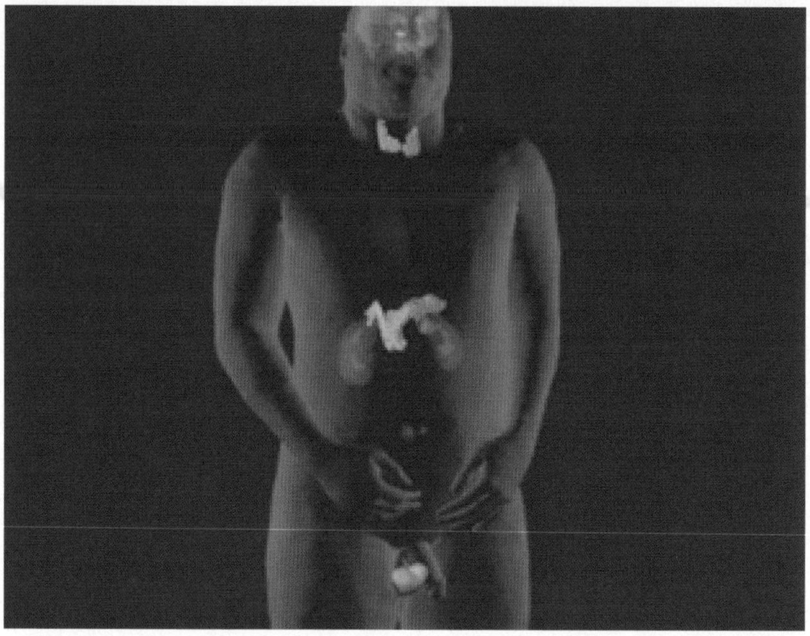

Challenge 1

As you have learned, the endocrine system is a series of glands and organs linked by their ability to secrete hormones. The components of this system communicate only via the cardiovascular system, making any depiction of the endocrine system appear more like a collection of individual organs than a unified system. Compare text Figure 19.1 to Figure 6.1 or 14.1● to fully appreciate the unique character of this system. Due to the inclusion in the endocrine system of many organs whose primary functions lie within other systems, overview illustrations of the endocrine system vary widely from one source to another.

The figure here is an anterior view of the endocrine system as depicted on the Interactive CD. It is based on the organs that can be digitally reconstructed from the Visible Human data set. List the organs presented, noting their relationship to one another. Compare this with Figure 19.1●. Which organs are not depicted on the CD? Why do you suppose they are not included in the endocrine image? Go to the CD and locate these additional hormone-producing organs.

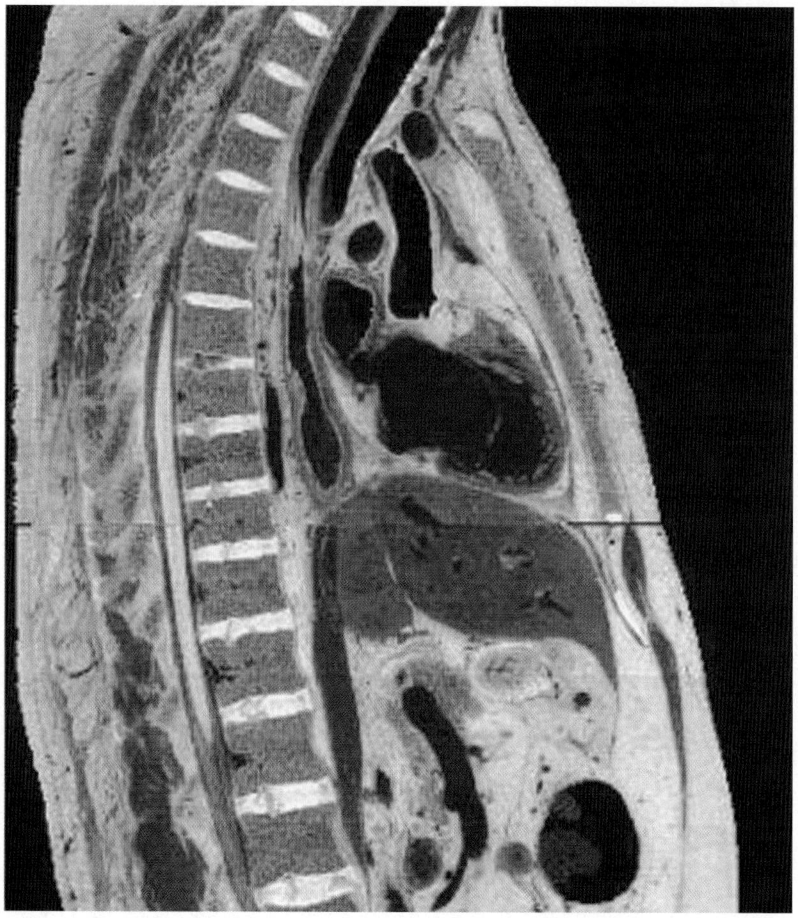

Challenge 2

Here we see a sagittal section of the chest and superior portion of the abdomen. Identify the heart, thymus, liver, pancreas, small intestine, adipose tissues, spinal cord, vertebrae, vertebral discs. Can you identify any other structures? Look again at the list of structures you have identified. How many of your structures are included as part of the endocrine system? How many of your listed structures are affected by the endocrine system? Prepare a table that provides the name of each identified structure, the hormones that affect this structure, what affect the hormone has on the structure, and where that hormone is produced.

Images provided by the Digital Cadaver™ project, courtesy of Visible Productions, Inc.

APPLY IT

Below are two separate exercises that provide more information on the preceding topics. These exercises differ from the usual critical linking exercises found in the media spreads in that these are designed to help you understand the interactions between systems, rather than hone your knowledge of one particular system. As before, each activity is designed to take approximately 10 minutes, and will help you gain a better appreciation for the integration of the material presented in the preceding chapters.

Critical Linking 1

The endocrine system and the nervous system function in concert to control both immediate and long-term processes in the body. In general the nervous system is responsible for immediate reactions to the environment, while the endocrine system works slowly, maintaining homeostasis during longer processes such as growth and sexual maturity. These two systems are intertwined to such a degree as to be dealt with in many medical institutions as one integrated system. The Harvard Medical School, Massachusetts General Hospital periodically publishes the Neuroendocrine bulletin, filled with clinical discussions of the interactions between these two systems. While it is generally accepted that these two control systems work in concert, are there interactions that extend beyond simple control?

Current research indicates there are links between these two systems beyond the obvious. Spend a few minutes listing those ways in which you believe the endocrine system could affect the nervous system. Once you have prepared this list, go to the Critical Linking portion of the Companion Website and click on the key word "Neurocognitive dysfunction." There you can read an article on how the pituitary gland is directly related to cognitive functioning, or thought processes. Read this article, then reflect on what else you might add to your list of neuroendocrine interactions. This site is filled with other articles on the topic, should you want to read further.

Critical Linking 2

There are numerous clinical indications of the interaction between the nervous system and the endocrine system. Many of the psychological troubles that were assigned to strictly nervous system malfunctions have recently been re-evaluated to include an endocrine involvement. One such illness is clinical depression. This disease affects millions of people each year, with severe complications for both the individual and society at large. Recent studies have uncovered a relationship between depression and the hypothalamus/pituitary interactions. Given what you know of the endocrine system and it's interaction with the nervous system, how might you expect these interactions to be altered in a clinically depressed patient?

Prepare a brief outline of the functions of the hypothalamus in regulating the pituitary gland. Visit the Companion Website, Critical Linking section and click on the key word "depression." Read the *Scientific American* article, paying special attention to the section titled "Hormonal Abnormalities." Are the functions you identified previously affected in depression? How does this information help you further your understanding of the interactions of the endocrine and nervous system in maintaining homeostasis?

FURTHER STUDIES

We constantly hear about this, read about it, and experience it ourselves on a daily basis. It seems stress is a normal part of daily existence in the twenty-first century. What is stress, from a biological perspective? Why does it have such a negative effect on our lives? Is there such a thing as "good stress"? This topic has interested everyone for years. Consequently there is a wealth of information, both reliable and questionable, to be found on the subject. The symptoms of a stressful experience are both immediate and long-term. The immediate effects include elevated heart rate, increased respirations, increased blood glucose levels, and a feeling of "butterflies in the stomach" or nervous energy. Long term effects of stress include changes in blood chemistry, depressed organ function, ulcers, and sleep disturbances. Which of these would you expect to fall under the control of the endocrine system versus the nervous system? Here again we see a coordinated effort between the two major control systems of the body. In this case however, they seem to be working in opposition to homeostatic maintenance of the body.

To gain a scientific perspective on stress, visit the following Internet address: *http://www.uams.edu/department_of_psychiatry/syllabus/STRESS/Stress97.htm*. Dr. Jeffrey Clothier has prepared an excellent introduction to the anatomy of stress. Further your investigations by visiting sites such as *http://www.virusmyth.com/aids/data/ahstress/htm*, or *http://www.esportmed.com/smu/content/viewsumm.cfm?sid=129*. At this site, you may have to locate the specific article by searching for "fitness and neuroendocrine immune function."

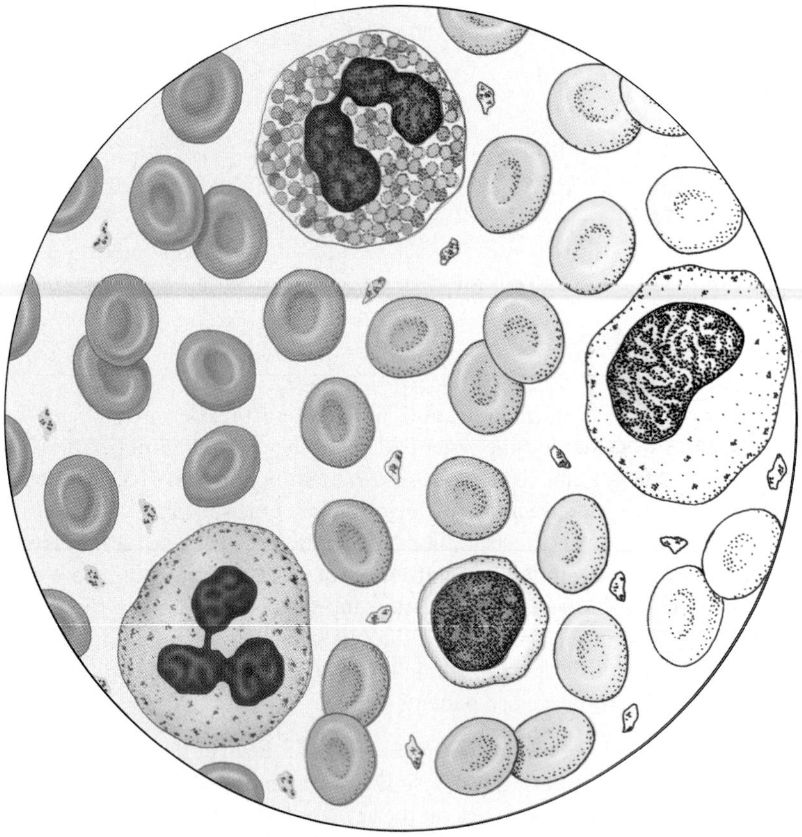

Introduction 533

Functions of the Blood 533

1. List and describe the functions of the blood.

Composition of the Blood 533

- Plasma
  2. Discuss the composition of blood and the physical characteristics of plasma.

Formed Elements 535

- Red Blood Cells (RBCs)
  3. List the structural characteristics and functions of red blood cells.
  4. Explain what determines a person's blood type and why blood types are important.
- White Blood Cells (WBCs)
  5. Categorize the various white blood cells on the basis of their structures and functions, and describe how white blood cells fight infection.
- Platelets
  6. Discuss the function of platelets.

Hemopoiesis 543

- Erythropoiesis
- Leukopoiesis
  7. Discuss the differentiation and life cycles of blood cells.
  8. Identify the locations where the components of blood are produced, and discuss the factors that regulate their production.

20

THE CARDIOVASCULAR SYSTEM

Blood

The living body is in constant chemical communication with its external environment. Nutrients are absorbed across the lining of the digestive tract, gases diffuse across the delicate epithelium of the lungs, and wastes are excreted in the feces and urine, as well as in saliva, bile, sweat, and other exocrine secretions. These chemical exchanges occur at specialized sites or organs because all parts of the body are linked by the *cardiovascular system* (*CVS*). The cardiovascular system can be compared to the cooling system of a car. The basic components include a circulating fluid (blood), a pump (the heart), and an assortment of conducting pipes (a network of blood vessels). The three chapters on the cardiovascular system examine those components individually: Chapter 20 discusses the nature of the circulating blood, Chapter 21 considers the structure and function of the heart, and Chapter 22 examines the network of blood vessels and the integrated functioning of the cardiovascular system. You will then be ready for Chapter 23, which considers the *lymphatic system*, whose vessels and organs are structurally and functionally linked to the CVS.

Functions of the Blood [TABLE 20.1]

Blood is a specialized fluid connective tissue p. 68 that (1) distributes nutrients, oxygen, and hormones to each of the roughly 75 trillion cells in the human body, (2) carries metabolic wastes to the kidneys for excretion, and (3) transports specialized cells that defend peripheral tissues from infection and disease. Table 20.1 contains a detailed listing of the functions of the blood. The services performed by the blood are absolutely essential; any body cells or region completely deprived of circulation may die in a matter of minutes.

Composition of the Blood [FIGURE 20.1]

Blood is a fluid connective tissue normally confined to the circulatory system. It has a characteristic and unique composition (Figure 20.1●). Blood consists of two components:

1. **Plasma** (PLAZ-mah), the liquid matrix of blood, has a density only slightly greater than that of water. It contains dissolved proteins, rather than the network of insoluble fibers found in loose connective tissues or cartilage, and numerous dissolved solutes.
2. **Formed elements** are blood cells and cell fragments that are suspended in the plasma. These elements are present in great abundance and are highly specialized. **Red blood cells (RBCs)** transport oxygen and carbon dioxide. The less numerous **white blood cells (WBCs)** are components of the immune system. **Platelets** (PLĀT-lets) are small, membrane-enclosed packets of cytoplasm that contain enzymes and other factors involved in *blood clotting*.

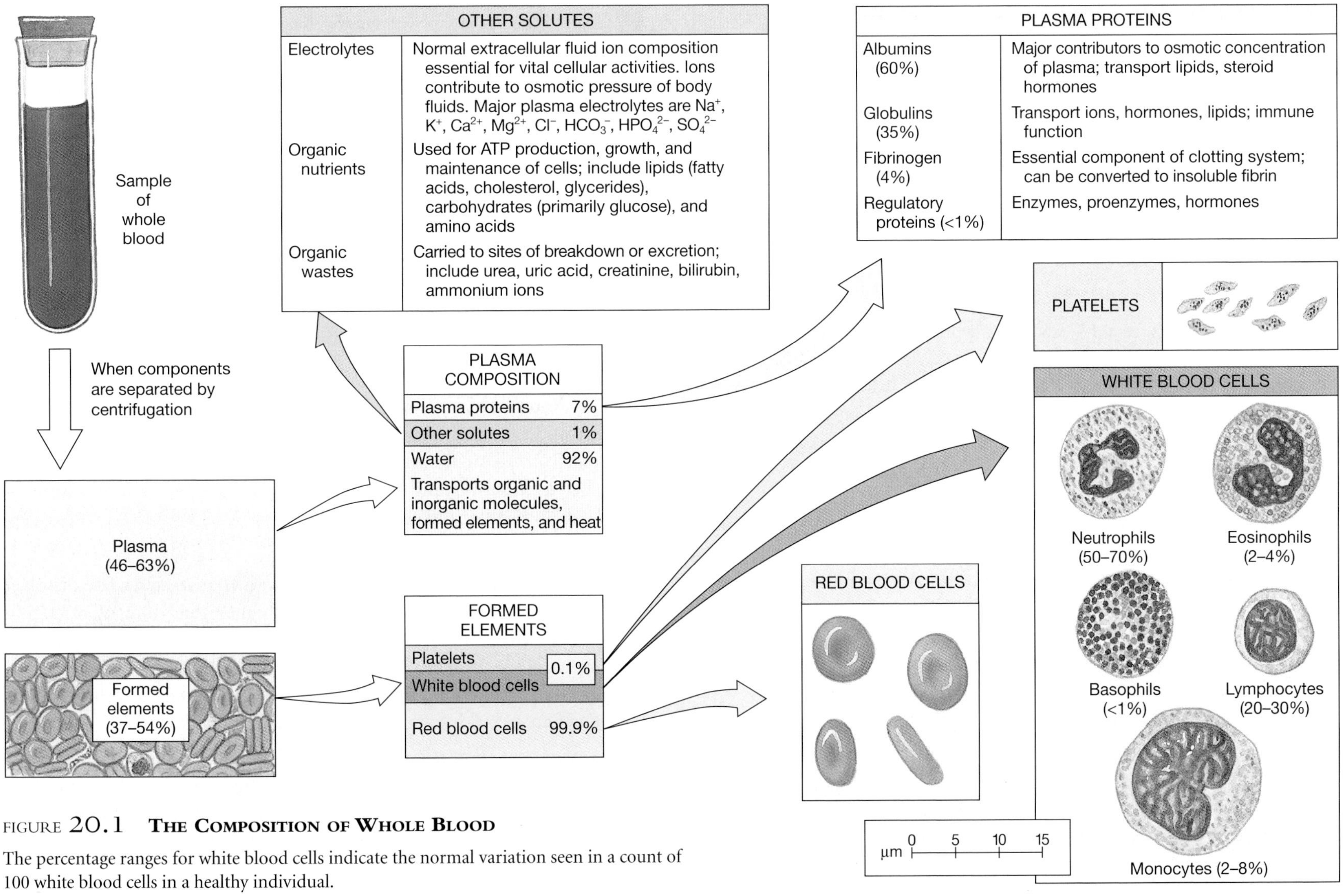

FIGURE 20.1 **THE COMPOSITION OF WHOLE BLOOD**

The percentage ranges for white blood cells indicate the normal variation seen in a count of 100 white blood cells in a healthy individual.

TABLE 20.1 **FUNCTIONS OF THE BLOOD**

1. *Transport of dissolved gases*, bringing oxygen from the lungs to the tissues and carrying carbon dioxide from the tissues to the lungs.
2. *Distribution of nutrients* absorbed at the digestive tract or released from storage in adipose tissue or the liver.
3. *Transport of metabolic wastes* from peripheral tissues to sites of excretion, especially the kidneys.
4. *Delivery of enzymes and hormones* to specific target tissues.
5. *Stabilization of the pH and electrolyte composition of interstitial fluids* throughout the body. By absorbing, transporting, and releasing ions as it circulates, blood helps prevent regional variations in the ion concentrations of body tissues. An extensive array of buffers enables the bloodstream to deal with the acids generated by tissues, such as the lactic acid produced by skeletal muscles.
6. *Prevention of fluid losses* through damaged vessels or at other injury sites. The **clotting reaction** seals the breaks in the vessel walls, preventing changes in blood volume that could seriously affect blood pressure and cardiovascular function.
7. *Defense against toxins and pathogens*. Blood transports *white blood cells*, specialized cells that migrate into peripheral tissues to fight infections or remove debris, and delivers *antibodies*, special proteins that attack invading organisms or foreign compounds. The blood also collects toxins, such as those produced by infection, and delivers them to the liver and kidneys, where they can be inactivated or excreted.
8. *Stabilization of body temperature* by absorbing and redistributing heat. Active skeletal muscles and other tissues generate heat, and the bloodstream carries it away. When body temperature is already high, that heat is lost across the surface of the skin. When body temperature is too low, warm blood is directed to the most temperature-sensitive organs. These changes in circulatory flow are controlled and coordinated by the *cardiovascular centers* in the medulla oblongata. p. 408

CLINICAL BRIEF

TRANSFUSIONS

In a transfusion, blood components are provided to an individual whose blood volume has been reduced or whose blood is deficient in some way. Transfusions of whole blood are most often used to restore blood volume after massive hemorrhaging has occurred. In an **exchange transfusion**, most of the blood volume of an individual is removed and simultaneously replaced with whole blood from another source. This may be necessary to treat acute drug poisoning or *hemolytic diseases of the newborn* (p. 546).

The blood is obtained under sterile conditions from carefully screened donors. It is tested for infectious bacteria and viruses and discarded if pathogens are detected. Whole blood is treated to prevent clotting and stabilize the red blood cells, and refrigerated. Chilled whole blood remains usable for around 3.5 weeks. For longer storage the blood must be fractionated. The red blood cells are separated from the plasma, and if necessary they may be frozen after treatment with a special antifreeze solution. The plasma can then be stored chilled, frozen, or freeze-dried. This procedure permits long-term storage of rare blood types that might not otherwise be available for emergency use.

Fractionated blood has many uses. **Packed red blood cells (PRBCs)**, with most of the plasma removed, are preferred for cases of anemia, in which the volume of blood may be close to normal but its oxygen-carrying capabilities have been reduced. Plasma may be administered when massive fluid losses are occurring, such as after severe burns. Plasma samples can be further fractionated to yield albumins, immunoglobulins, clotting factors, and plasma enzymes that can be administered separately. White blood cells and platelets can also be collected, sorted, and stored for subsequent transfusion.

Some 6 million units (3 million liters) of blood are used each year in the United States alone, and the demand for blood or blood components often exceeds the supply. Moreover, there has been increasing concern over the danger that transfusion recipients will become infected with hepatitis viruses or HIV (the virus that causes AIDS) from contaminated blood. The result has been a number of changes in transfusion practices over recent years. In general, fewer units of blood are now administered. There has also been an increase in **autologous transfusion**, in which blood is removed from a patient (or potential patient), stored, and later transfused back into the original donor when needed, such as after a surgical procedure. Moreover, new technology permits the reuse of blood "lost" during surgery. The blood is collected and filtered; the platelets are removed, and the remainder of the blood is reinfused into the patient.

Whole blood is a mixture of plasma and formed elements. Its components may be separated, or **fractionated**, for clinical purposes. Whole blood is sticky, cohesive, and resistant to flow, characteristics that determine the **viscosity** of a solution. Solutions are usually compared with pure water, which has a viscosity of 1.0. Plasma has a viscosity of 1.5, but the viscosity of whole blood is much greater (about 5.0) because of interactions between water molecules and the formed elements.

On average there are 5–6 liters of whole blood in the cardiovascular system of an adult man, and 4–5 liters in an adult woman. The blood has an alkaline pH (range of 7.35 to 7.45) and a temperature slightly higher than core body temperature (38°C vs. 37°C, or 100.4°F vs. 98.6°F). Clinicians use the terms **hypovolemic** (hī-pō-vō-LĒ-mik), **normovolemic** (nor-mō-vō-LĒ-mik), and **hypervolemic** (hī-per-vō-LĒ-mik) to refer to low, normal, or excessive blood volumes, respectively. Low or high blood volumes are potentially dangerous—for example, a hypervolemic condition can place a severe stress on the heart (e.g., hypertension or "high blood pressure"), which must push the extra fluid around the circulatory system.

Plasma [FIGURE 20.1 AND TABLE 20.2]

Plasma contributes approximately 55% of the volume of whole blood, and water accounts for 92% of the plasma volume. These are average values, and the actual concentrations vary depending on the region of the cardiovascular system or area of the body sampled and the ongoing activity within that particular region. Information concerning the composition of plasma is summarized in Figure 20.1• and Table 20.2.

DIFFERENCES BETWEEN PLASMA AND INTERSTITIAL FLUID

In many respects, plasma resembles interstitial fluid. For example, the ion concentrations in plasma are similar to those of interstitial fluid but very different from the ion concentrations in cytoplasm. The principal differences between plasma and interstitial fluid involve the concentrations of dissolved gases and proteins.

1. *Concentrations of dissolved oxygen and carbon dioxide*: The dissolved oxygen concentration in plasma is higher than that of interstitial fluid. As a result, oxygen diffuses out of the bloodstream and into peripheral tissues. The carbon dioxide concentration in interstitial fluid is much higher than that of plasma, so carbon dioxide diffuses out of the tissues and into the bloodstream.
2. *Concentration of dissolved proteins*: The plasma contains significant quantities of dissolved proteins, whereas interstitial fluid does not. The large size and globular shapes of most plasma proteins prevent them from crossing capillary walls, and they remain trapped within the cardiovascular system.

TABLE 20.2 COMPOSITION OF WHOLE BLOOD

Component	Significance
PLASMA	
Water	Dissolves and transports organic and inorganic molecules, distributes blood cells, and transfers heat
Electrolytes	Normal extracellular fluid ion composition essential for vital cellular activities
Nutrients	Used for energy production, growth, and maintenance of cells
Organic wastes	Carried to sites of breakdown or excretion
Proteins	
Albumins	Major contributor to osmotic concentration of plasma; transport some lipids
Globulins	Transport ions, hormones, lipids
Fibrinogen	Essential component of clotting system; can be converted to insoluble fibrin
FORMED ELEMENTS	
Red blood cells	Transport gases (oxygen and carbon dioxide)
White blood cells	Defend body against pathogens; remove toxins, wastes, and damaged cells
Platelets	Participate in clotting response

THE PLASMA PROTEINS [FIGURE 20.1]

Plasma proteins account for approximately 7% of the plasma composition (Figure 20.1●). One hundred milliliters of human plasma normally contains 6–7.8 g of soluble proteins. There are three major classes of plasma proteins: *albumins* (al-BŪ-minz), *globulins* (GLOB-ū-linz), and *fibrinogen* (fī-BRIN-ō-jen).

1. **Albumins** constitute roughly 60% of the plasma proteins. As the most abundant proteins, they are major contributors to the osmotic pressure of the plasma. They are also important in the transport of fatty acids, steroid hormones, and other substances. Albumins are the smallest of the major plasma proteins.
2. **Globulins** account for approximately 35% of the plasma protein population. Globulins include both *immunoglobulins* and *transport globulins*. **Immunoglobulins** (i-mū-nō-GLOB-ū-linz), also called **antibodies**, attack foreign proteins and pathogens. **Transport globulins** bind small ions, hormones, or compounds that are either insoluble or that might be filtered out of the blood at the kidneys.
3. **Fibrinogen**, which accounts for about 4% of the plasma proteins, participates in the clotting reaction. Under certain conditions fibrinogen molecules interact, forming large, insoluble strands of *fibrin* (FĪ-brin). These fibers provide the basic framework for a blood clot. If steps are not taken to prevent clotting, the conversion of fibrinogen to fibrin will occur in a sample of plasma. This conversion removes the clotting proteins, leaving a fluid known as **serum**. Fibrinogen is the largest of the plasma proteins.

Both albumins and globulins can attach to lipids, such as triglycerides, fatty acids, or cholesterol, that are not water-soluble. These protein-lipid combinations, called **lipoproteins** (lī-pō-PRŌ-tēnz), readily dissolve in plasma, and this is how insoluble lipids are delivered to peripheral tissues.

The liver synthesizes and releases more than 90% of the plasma proteins. Because the liver is the primary source of plasma proteins, liver disorders can alter the composition and functional properties of the blood. For example, some forms of liver disease can lead to uncontrolled bleeding, caused by inadequate synthesis of fibrinogen and other plasma proteins involved in the clotting response.

CLINICAL BRIEF

PLASMA EXPANDERS

Plasma expanders are solutions that can be used to increase blood volume temporarily, over a period of hours, while preparing for a transfusion of whole blood. Plasma expanders contain large carbohydrate molecules, rather than dissolved proteins, to maintain proper osmolarity. Although these carbohydrates are not metabolized, they are gradually removed from circulation by phagocytes, and the blood volume steadily declines. Plasma expanders are easily stored, and their sterile preparation ensures that there are no problems with viral or bacterial contamination. Although they provide a temporary solution to hypovolemia (low blood volume), plasma expanders fail to increase the amount of oxygen delivered to peripheral tissues.

✓ CONCEPT CHECK

- How would a slow blood flow affect the stability of your body's temperature?
- If a person is diagnosed as being hypovolemic, how would blood pressure be affected?
- Why does whole blood have such a high viscosity?

Formed Elements [TABLE 20.3]

The major cellular components of blood are *red blood cells* and *white blood cells*. There are two major classes of white blood cells: *granular* (with granules) and *agranular* (without granules). In addition, blood contains noncellular formed elements called *platelets*, that function in the clotting response. Table 20.3 summarizes information concerning the formed elements of the blood.

■ Red Blood Cells (RBCs) [FIGURE 20.1]

Red blood cells (RBCs), or **erythrocytes** (e-RITH-rō-sīts; *erythros*, red) account for slightly less than half of the total blood volume (Figure 20.1●, p. 533). The **hematocrit** (hē-MA-tō-krit) value indicates the percentage of whole blood contributed by formed elements. The normal hematocrit in adult men averages 46 (range: 40–54); the average for adult women is 42 (range: 37–47). Because whole blood contains roughly 1000 red blood cells for each white blood cell, the hematocrit closely approximates the volume of erythrocytes. As a result, hematocrit values are often reported as the **volume of packed red cells (VPRC)**, or, simply, the **packed cell volume (PCV)**.

The number of erythrocytes in the blood of a normal individual staggers the imagination. One microliter (μl), or cubic millimeter (mm^3), of

TABLE 20.3

A REVIEW OF THE FORMED ELEMENTS OF THE BLOOD

Formed Elements	Abundance (per μl)[a]	Characteristics	Functions	Remarks
RED BLOOD CELLS	5.2 million (range: 4.4–6.0 million)	Biconcave disc without a nucleus, mitochondria, or ribosomes; red color due to presence of hemoglobin molecules	Transport oxygen from lungs to tissues, and carbon dioxide from tissues to lungs	120-day life expectancy; amino acids and iron recycled; produced in bone marrow
WHITE BLOOD CELLS	7000 (range: 6000–9000)			
Granulocytes				
Neutrophils	4150 (range: 1800–7300) differential count: 57%	Round cell; nucleus resembles a series of beads; cytoplasm contains large, pale inclusions	Phagocytic; engulf pathogens or debris in tissues	Survive minutes to days, depending on activity; produced in bone marrow
Eosinophils	165 (range: 0–700) differential count: 2.4%	Round cell; nucleus usually in two lobes; cytoplasm contains large granules that stain bright orange-red with acid dyes	Attack anything that is labeled with antibodies; important in fighting parasitic infections; suppress inflammation	Produced in bone marrow
Basophils	44 (range: 0–150) differential count: 0.6%	Round cell; nucleus usually cannot be seen, because of dense, purple-blue granules in cytoplasm	Enter damaged tissues and release histamine and other chemicals	Assist mast cells of tissues in producing inflammation; produced in bone marrow
Agranulocytes				
Monocytes	456 (range: 200–950) differential count: 6.5%	Very large, kidney bean-shaped nucleus; abundant pale cytoplasm	Enter tissues to become free macrophages; engulf pathogens or debris	Primarily produced in bone marrow
Lymphocytes	2185 (range: 1500–4000) differential count: 31%	Slightly larger than RBC; round nucleus; very little cytoplasm	Cells of lymphatic system, providing defense against specific pathogens or toxins	T cells attack directly; B cells form plasma cells that produce antibodies; produced in bone marrow and lymphoid tissues
PLATELETS	350,000 (range: 150,000–500,000)	Cytoplasmic fragments; contain enzymes and proenzymes; no nucleus	Hemostasis: clump together and stick to vessel wall (platelet phase); activate intrinsic pathway of coagulation phase	Produced by megakaryocytes in bone marrow

[a] Values reported are averages. Differential count: percentage of circulating white blood cells.

whole blood from a man contains roughly 5.4 million erythrocytes; a microliter of blood from a woman contains about 4.8 million erythrocytes. There are approximately 260 million red blood cells in a single drop of whole blood, and 25 trillion (2.5×10^{13}) RBCs in the blood of an average adult.

STRUCTURE OF RBCS [FIGURE 20.2]

Erythrocytes transport both oxygen and carbon dioxide within the bloodstream. They are among the most specialized cells of the body, and their anatomical specializations are apparent when RBCs are compared with "typical" body cells. p. 28 Figure 20.2a,c● indicates the significant differences detected using light and electron microscopy. Each red blood cell is a biconcave disc that has a thin central region and a thick outer margin (Figure 20.2d●). The diameter of a typical erythrocyte measured in a standard blood smear is 7.7 μm; it has a maximum thickness of about 2.85 μm, although the center narrows to about 0.8 μm.

This unusual shape, which provides strength and flexibility, also gives each RBC a disproportionately large surface area for a cell its size. The large surface area permits rapid diffusion between the RBC cytoplasm and surrounding plasma. As blood circulates from the capillaries of the lungs to the capillaries in peripheral tissues and then back to the lungs, respiratory

CLINICAL BRIEF

ANEMIA AND POLYCYTHEMIA

Anemia (a-NĒ-mē-a) exists when the oxygen-carrying capacity of the blood is reduced, diminishing the delivery of oxygen to peripheral tissues. Such a reduction causes a variety of symptoms, including premature muscle fatigue, weakness, lethargy, and a general lack of energy. Anemia may exist because the hematocrit is abnormally low or because the amount of hemoglobin in the RBCs is reduced. Standard laboratory tests can be used to differentiate between the various forms of anemia on the basis of the number, size, shape, and hemoglobin content of red blood cells. For a more detailed discussion of blood tests, see the Clinical Issues appendix. *Blood Tests and RBCs p. 797*

An elevated hematocrit with a normal blood volume constitutes **polycythemia** (po-lē-sī-THĒ-mē-ah). There are several different types of polycythemia. **Erythrocytosis** (e-rith-rō-sī-TŌ-sis), a polycythemia affecting only red blood cells, will be considered later in the chapter. **Polycythemia vera** ("true polycythemia") results from an increase in the numbers of all types of blood cells. Many if not all of the blood cells develop from one abnormal hematopoietic stem cell. The hematocrit may reach 80–90, at which point the tissues become oxygen-starved because red blood cells are blocking the smaller vessels. This condition seldom strikes young people; most cases involve persons aged 60–80. There are several treatment options, but none cures the condition. The cause of polycythemia vera is unknown, although there is some evidence that the condition is linked to radiation exposure.

gases are absorbed and released by RBCs. The total surface area of the red blood cells in the blood of a typical adult is roughly 3800 m^2, 2000 times the total surface area of the body.

The flattened shape also enables them to form stacks, like dinner plates. These stacks, called **rouleaux** (roo-LŌ; "little rolls," singular, *rouleau*), form and dissociate repeatedly without affecting the cells involved. One rouleau can pass along a blood vessel (Figure 20.2b●) little larger than the diameter of a single erythrocyte, whereas individual cells would bump the walls, band together, and form log jams that could block the vessel. Finally, the slender profile of an erythrocyte gives the cell considerable flexibility; erythrocytes can bend and flex with apparent ease, and by changing their shape, individual red blood cells can squeeze through small diameter, distorted, or compressed capillaries.

RBC LIFE SPAN AND CIRCULATION

During their differentiation and maturation, red blood cells lose most of their organelles, retaining only an extensive cytoskeleton. As a result, circulating RBCs lack mitochondria, endoplasmic reticulum, ribosomes, and nuclei. (The process of RBC formation will be described in a later section.) Without mitochondria, these cells can obtain energy only through anaerobic metabolism, and they rely on glucose obtained from the surrounding

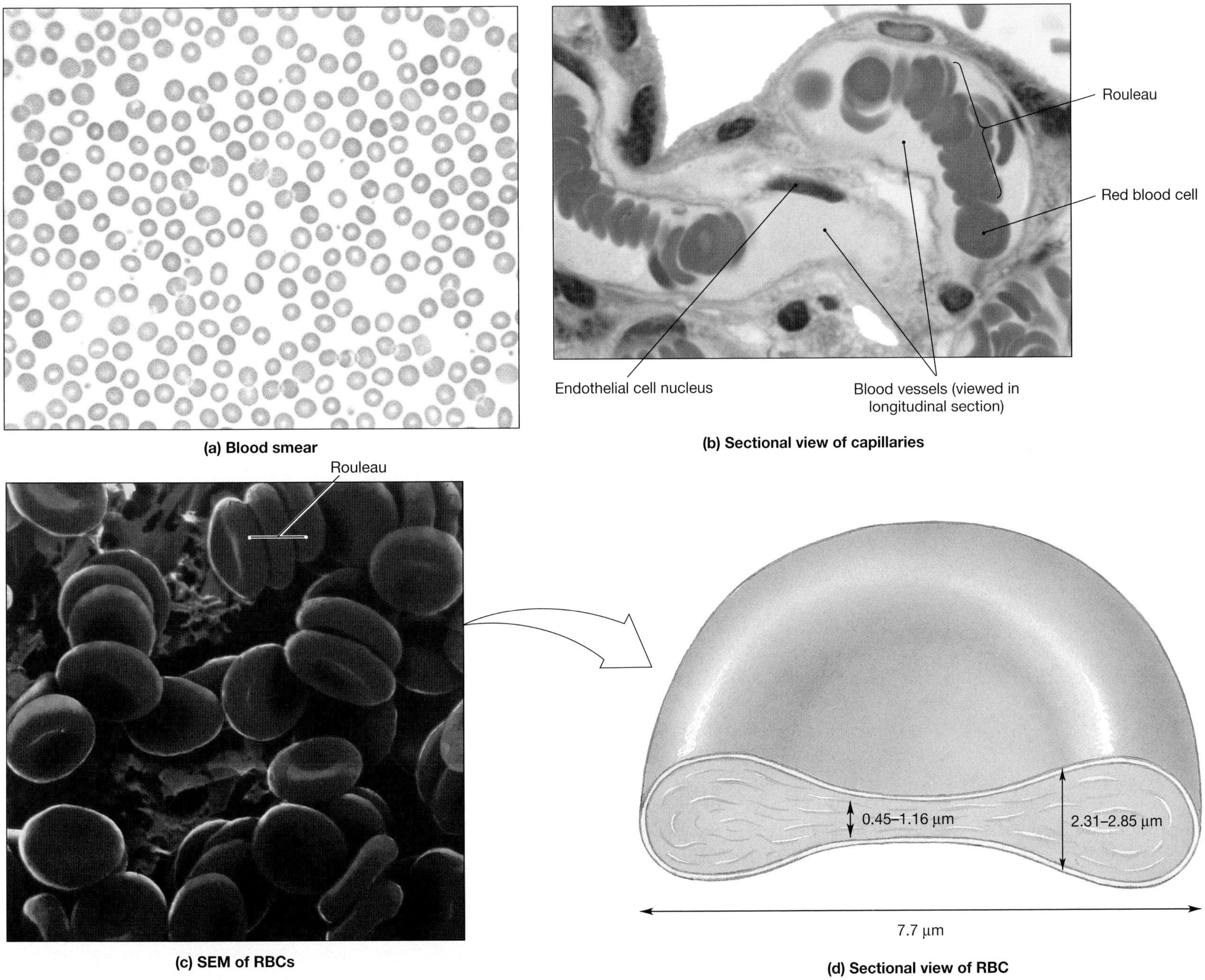

FIGURE 20.2 **Anatomy of Red Blood Cells**

(a) When viewed in a standard blood smear, red blood cells appear as two-dimensional objects because they are flattened against the surface of the slide. (LM × 477) **(b)** When traveling through relatively narrow capillaries, erythrocytes may stack like dinner plates, forming rouleaux. (LM × 1430) **(c)** A scanning electron micrograph of red blood cells reveals their three-dimensional structure quite clearly.(SEM × 1838) **(d)** A sectional view of a red blood cell.

plasma. This mechanism ensures that absorbed oxygen will be carried to peripheral tissues, not "stolen" by mitochondria in the RBCs. The lack of a nucleus or ribosomes means that protein synthesis cannot occur; thus, an RBC cannot replace damaged enzymes or structural proteins.

This is a serious problem because in its lifetime, an erythrocyte is exposed to severe stresses. A single circuit of the circulatory system usually takes less than 30 seconds. In that time it stacks in rouleaux, contorts and squeezes through capillaries, and then joins its comrades in a headlong rush back to the heart for another round. With all this wear and tear and no repair mechanisms, a typical red blood cell has a relatively short life span of about 120 days. After traveling about 700 miles in 120 days, either the cell membrane ruptures or the aged cell is detected and destroyed by phagocytic cells. About 1% of the circulating erythrocytes are replaced each day, and in the process approximately 3 million new erythrocytes enter the circulation *each second*!

RBCs AND HEMOGLOBIN [FIGURE 20.3]

A developing erythrocyte loses all intracellular components not directly associated with its primary functions: oxygen and carbon dioxide transport. A mature red blood cell consists of a cell membrane surrounding cytoplasm containing water (66%) and proteins (about 33%). **Hemoglobin** (HĒ-mō-glō-bin) (**Hb**) molecules account for over 95% of the erythrocyte's proteins. Hemoglobin is responsible for the cell's ability to transport oxygen and carbon dioxide. Hemoglobin is a red pigment; its presence gives blood its characteristic red color. Oxygenated hemoglobin has a bright red color, whereas deoxygenated hemoglobin has a deep red color. This accounts for the color difference between arterial (oxygen-rich) and venous (oxygen-poor) blood.

Four globular protein subunits combine to form a single molecule of hemoglobin. Each subunit contains a single molecule of **heme** (Figure 20.3●). Each heme unit holds an iron ion in such a way that it can interact with an oxygen molecule. The iron-oxygen interaction is very weak, and the two separate easily without damage to either the hemoglobin or the oxygen molecule. There are approximately 280 million molecules of hemoglobin in each red blood cell, and because a hemoglobin molecule contains four heme units, each erythrocyte potentially can carry more than a billion molecules of oxygen. Hemoglobin also carries about 23% of the carbon dioxide transported in the blood. Carbon dioxide binds to amino acids of the globin subunits rather than competing with oxygen for binding with iron. The binding of carbon dioxide to a globin subunit is just as reversible as the binding of oxygen to heme.

As red blood cells circulate through capillaries in the lungs, oxygen enters and carbon dioxide leaves the plasma by diffusion. As plasma oxygen levels climb, oxygen diffuses into RBCs and binds to hemoglobin; as plasma carbon dioxide levels fall, hemoglobin releases CO_2 that diffuses into the plasma. In other words, the red blood cells absorb oxygen and release carbon dioxide. In the peripheral tissues, the conditions reverse because active cells are consuming oxygen and producing carbon dioxide. As blood flows through these tissues, oxygen diffuses out of the plasma and carbon dioxide diffuses in. The RBCs then release oxygen and absorb CO_2. † *Sickle Cell Anemia p. 799*

BLOOD TYPES [FIGURE 20.4 AND TABLE 20.4]

An individual's **blood type** is determined by the presence or absence of specific components in erythrocyte cell membranes. A typical red blood cell membrane contains a number of **surface antigens**, or *agglutinogens* (a-gloo-TIN-ō-jenz), exposed to the plasma. These surface antigens are glycoproteins or glycolipids whose characteristics are genetically determined. At least 50 different kinds of surface antigens have been localized on the surfaces of RBCs. Three of particular importance have been designated surface antigens **A**, **B**, and **D** (**Rh**).

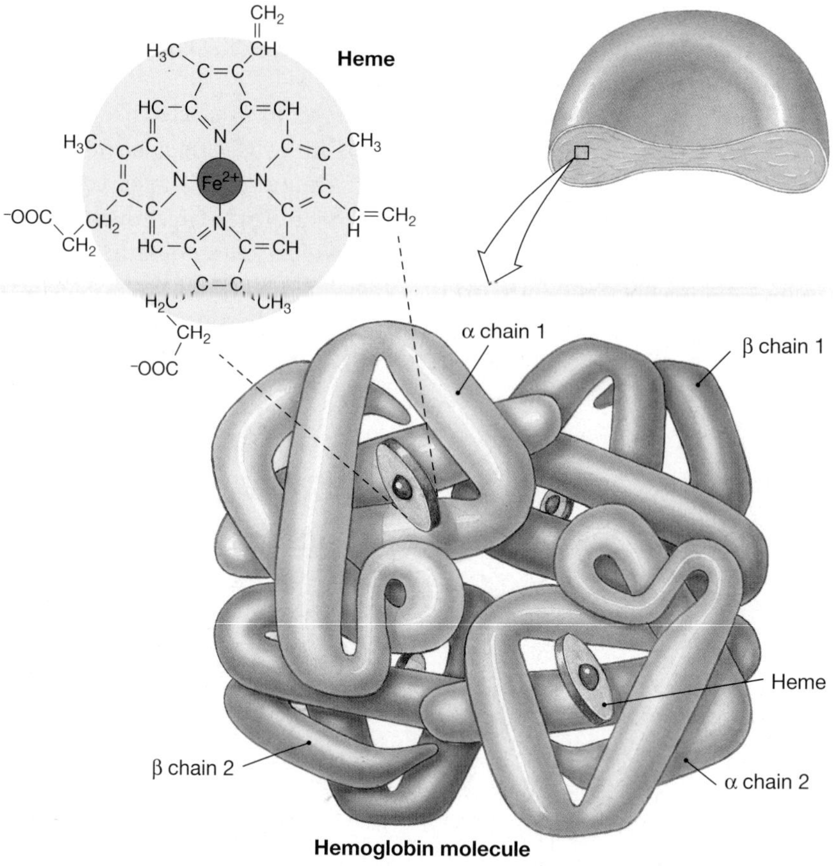

FIGURE 20.3 **THE STRUCTURE OF HEMOGLOBIN**

Hemoglobin consists of four globular protein subunits. Each subunit contains a single molecule of heme, a porphyrin ring surrounding a single ion of iron. It is the iron ion that reversibly binds to an oxygen molecule.

CLINICAL BRIEF

ERYTHROCYTOSIS

In **erythrocytosis** the blood contains abnormally large numbers of red blood cells. Erythrocytosis usually results from the massive release of erythropoietin by tissues deprived of oxygen. People moving to high altitudes usually experience erythrocytosis, because the air contains less oxygen than it does at sea level. The increased number of red blood cells compensates for the fact that individually each RBC is carrying less oxygen than it would at sea level. Mountaineers and those living at altitudes of 10,000–12,000 feet may have hematocrits as high as 65.

Individuals whose hearts or lungs are functioning inadequately may also develop erythrocytosis. For example, this condition is often seen in heart failure and emphysema, two conditions discussed in later chapters. Whether the blood fails to circulate efficiently or the lungs do not deliver enough oxygen to the blood, peripheral tissues remain oxygen-poor despite the rising hematocrit. Having a higher concentration of red blood cells increases the oxygen-carrying capacity of the blood, but it also makes the blood thicker and harder to push around the circulatory system. This change increases the workload on the heart, making a bad situation even worse.

The red blood cells of each person have a characteristic combination of surface antigens (Figure 20.4●). For example, **Type A** blood has antigen A, **Type B** has antigen B, **Type AB** has both, and **Type O** has neither. The average values for the U.S. population are: Type O 46%, Type A 40%, Type B 10%, and Type AB 4%. These values may differ among different racial and ethnic groups (Table 20.4).

The presence of the Rh antigen, sometimes called the *Rh factor,* is indicated by the terms **Rh-positive** (present) or **Rh-negative** (absent). In recording the complete blood type, the term *Rh* is usually omitted, and the data are reported as O-negative, A-positive, and so forth.

Antibodies and Cross-Reactions [Figure 20.4] You probably know that your blood type must be checked before you give or receive blood. The surface antigens on your own red blood cells are ignored by your immune system. (This ability to recognize your own body cells will be examined in Chapter 23.) However, your plasma contains antibodies (immunoglobulins), that will attack "foreign" surface antigens. These antibodies are known as *agglutinins* (a-GLOO-ti-ninz). The blood of a Type A, Type B, or Type O individual always contains antibodies that will attack foreign surface antigens. For example, if you have Type A blood, your plasma contains circulating anti-B antibodies that will attack Type B erythrocytes (Figure 20.4●). If you are Type B, your plasma contains anti-A antibodies. In a Type O individual the red blood cells lack surface antigens A and B, and the plasma contains both anti-A and anti-B antibodies. At the other extreme, the plasma of a Type AB individual contains neither anti-A nor anti-B antibodies.

Even if a Type A person has never been exposed to Type B blood, the individual will still have anti-B antibodies in the plasma. In contrast, the plasma of an Rh-negative individual does not always contain anti-Rh antibodies. These antibodies are present only if the individual has been **sensitized** by previous exposure to Rh-positive erythrocytes. Such exposure may occur accidentally, during a transfusion, but it also may accompany a seemingly normal pregnancy involving an Rh-negative mother and an Rh-positive fetus. *Hemolytic Disease of the Newborn p. 799*

When an antibody meets its specific surface antigen, a **cross-reaction** occurs. Initially the red blood cells clump together, a process called **agglutination** (a-gloo-ti-NĀ-shun), and they may also **hemolyze**, or rupture. Clumps and fragments of red blood cells under attack form drifting masses that can plug small vessels in the kidneys, lungs, heart, or brain, damaging or destroying the tissues deprived of circulation. Such reactions can be avoided by ensuring that the blood types of the donor and the recipient are **compatible**. In practice, this procedure involves choosing a donor whose blood cells will not undergo cross-reaction with the plasma of the recipient.

CLINICAL BRIEF

Elevating Hemoglobin Levels

Blood for transfusion is a limited resource, and transfusions are not without risk. In an attempt to develop a lower-risk alternative, genetic engineering procedures are now being used to synthesize one of the subunits of normal human hemoglobin, which can be introduced into the circulation to increase oxygen transport and total blood volume. An alternative strategy involves removing hemoglobin molecules from red cells and attaching them to inert carrier molecules that will prevent their loss at the kidneys. Still a third approach draws on natural but nonhuman sources. The FDA recently approved testing of a blood substitute called *Hemopure*™ that contains purified hemoglobin from cattle. Because the hemoglobin is removed from erythrocytes and infused without plasma, cross-reactions are not expected to occur.

TABLE 20.4 **Differences in Blood Group Distribution**

	Percentage with Each Blood Type				
POPULATION	**O**	**A**	**B**	**AB**	**Rh⁺**
U.S. (average)	46	40	10	4	85
Caucasian	45	40	11	4	85
African-American	49	27	20	4	95
Chinese	42	27	25	6	100
Japanese	31	39	21	10	100
Korean	32	28	30	10	100
Filipino	44	22	29	6	100
Hawaiian	46	46	5	3	100
Native North American	79	16	4	< 1	100
Native South American	100	0	0	0	100
Australian Aborigines	44	56	0	0	100

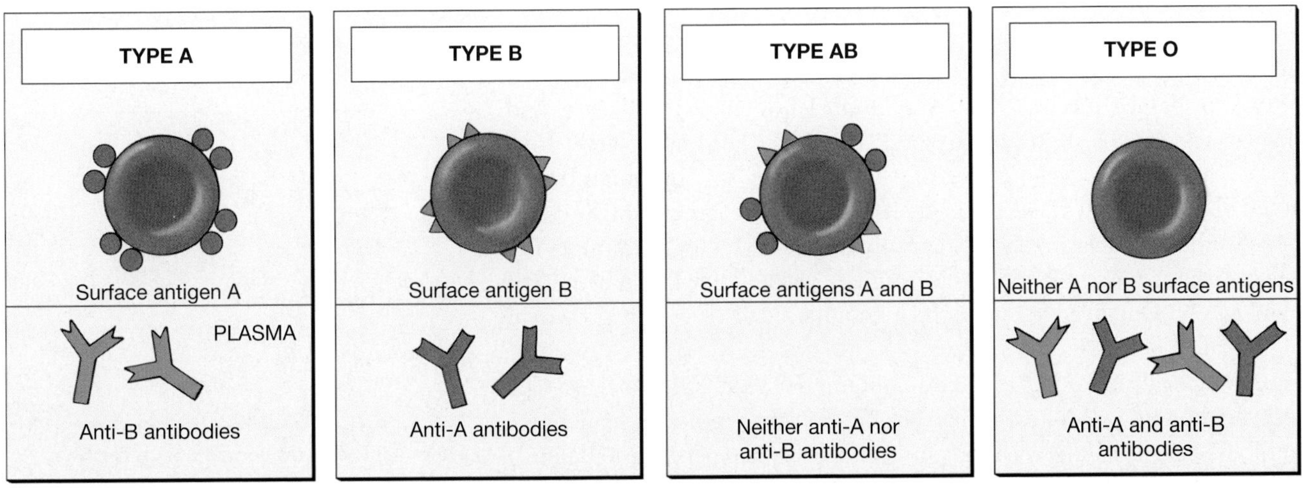

FIGURE 20.4 **Blood Typing**

The blood type depends on the presence of agglutinogens on RBC surfaces. The plasma contains agglutinins, antibodies that will react with foreign agglutinogens. The relative frequencies of each blood type in the U.S. population are indicated in *Table 20.4.*

CLINICAL BRIEF

Testing for Compatibility

Testing for compatibility normally involves two steps: a determination of blood type and a *cross-match test*. At least 50 surface antigens have been identified on red cell surfaces, but the standard test for blood type categorizes a blood sample on the basis of the three most likely to produce dangerous cross-reactions. The test involves taking drops of blood and mixing them with solutions containing anti-A, anti-B, and anti-Rh (anti-D) antibodies. Any cross-reactions are then recorded. For example, if the red blood cells clump together when exposed to anti-A and anti-B, the individual has Type AB blood. If no reactions occur, the person must be Type O. The presence or absence of the Rh surface antigen is also noted, and the individual is classified as Rh-positive or Rh-negative on that basis. Type O-positive is the most common blood type. The red blood cells of these individuals do not have surface antigens A and B, but they do have surface antigen D.

Standard blood typing can be completed in a matter of minutes, and Type O-negative blood can be safely administered to anyone of any blood type in an emergency. However, with at least 48 other possible surface antigens on the cell surface, cross-reactions can occur, even to Type O blood. Whenever time and facilities permit, further testing is performed to ensure complete compatibility.

Cross-match testing involves exposing the donor's red blood cells to a sample of the recipients plasma under controlled conditions. This procedure reveals the presence of significant cross-reactions involving other surface antigens and antibodies.

✓ CONCEPT CHECK

- If the hematocrit value of a woman is 42, what is the percentage of red blood cells present in her blood?
- How does the shape of red blood cells aid in blood flow and the diffusion of oxygen?
- Red blood cells have neither nuclei nor ribosomes. What effect does this have on their life span?
- A person with Type AB blood can receive blood from any other blood types. Why?

■ White Blood Cells (WBCs) [FIGURE 20.5 AND TABLE 20.3]

White blood cells (WBCs), or **leukocytes** (LOO-kō-sīts; *leukos*, white), are scattered throughout peripheral tissues. Circulating leukocytes represent only a small fraction of their total population; most of the white blood cells in the body are found in peripheral tissues. White blood cells help defend the body against invasion by pathogens and remove toxins, wastes, and abnormal or damaged cells. WBCs contain nuclei of characteristic sizes and shapes (Figure 20.5●). All white blood cells are as large as or larger than RBCs. There are two major classes of white blood cells: (1) **granular leukocytes**, or **granulocytes** (GRAN-ū-lō-sīts), which have large granular inclusions in their cytoplasm, and (2) **agranular leukocytes**, or **agranulocytes**, which do not possess cytoplasmic granules visible with the light microscope. Representative granular and agranular leukocytes are shown in Figure 20.5●.

A typical microliter of blood contains 6000–9000 leukocytes. The term **leukopenia** (loo-kō-PĒ-nē-a; *penia*, poverty) indicates inadequate numbers of white blood cells; a count of below 2500 per μl usually indicates a serious disorder. **Leukocytosis** (loo-kō-sī-TŌ-sis) refers to excessive numbers of white blood cells; a count of over 30,000 per μl usually indicates a serious disorder. A stained blood smear provides a **differential count** of the white blood cell population. The values obtained indicate the number of each type of cell encountered in a sample of 100 white blood cells. The normal range for each cell type is indicated in Table 20.3. The endings *-penia* and *-osis* can also be used to indicate low or high numbers, respectively, of specific types of white blood cells. For example, *lymphopenia* means too few lymphocytes, and *lymphocytosis* means an unusually high number.

Leukocytes have a very short life span, typically only a few days. In instances of injury or invasion of an area by a foreign organism, a leukocyte can migrate across the endothelial lining of a capillary by squeezing between adjacent endothelial cells. This process is known as **diapedesis**. The bloodstream provides rapid transportation for WBCs to these injured sites, where they are attracted to the chemical signs of inflammation or infection in the adjacent interstitial fluids. This attraction to specific chemical stimuli, called **chemotaxis**, draws them to invading pathogens, damaged tissues, and white blood cells already in the damaged tissues.

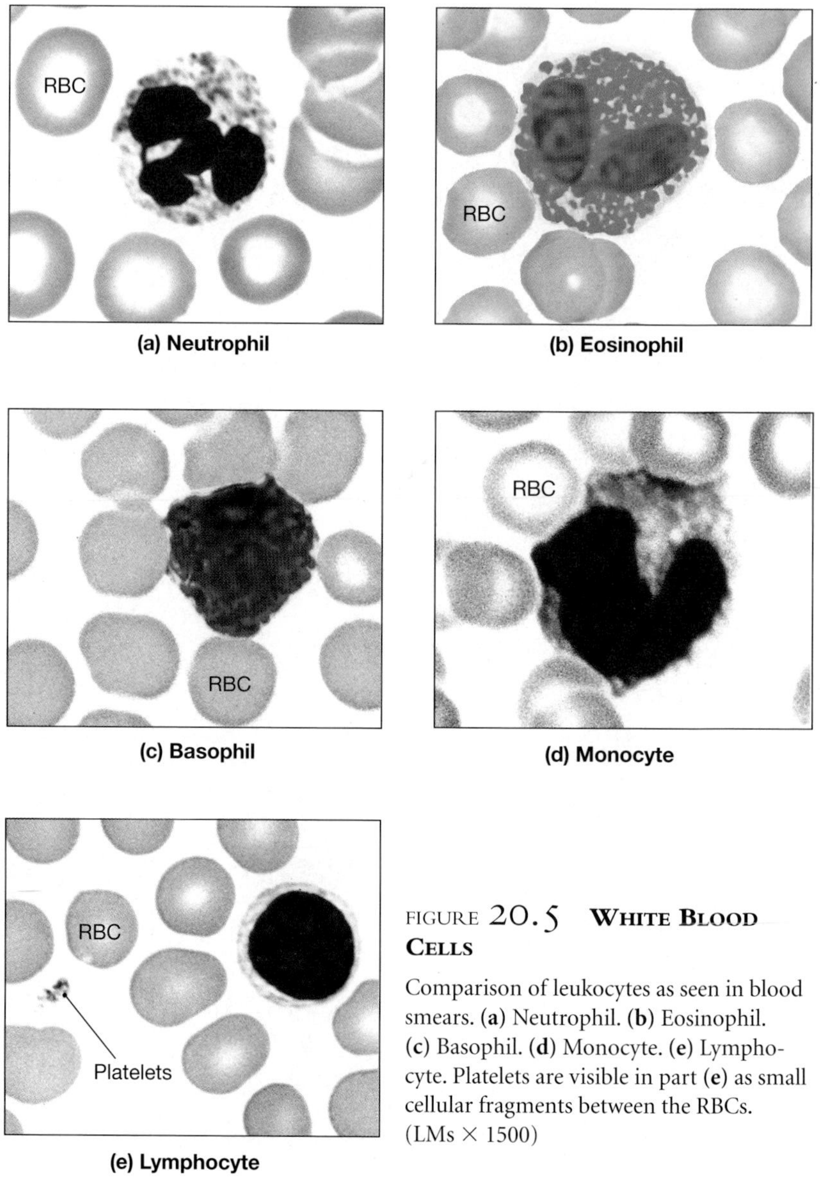

FIGURE 20.5 **White Blood Cells**

Comparison of leukocytes as seen in blood smears. **(a)** Neutrophil. **(b)** Eosinophil. **(c)** Basophil. **(d)** Monocyte. **(e)** Lymphocyte. Platelets are visible in part **(e)** as small cellular fragments between the RBCs. (LMs × 1500)

GRANULAR LEUKOCYTES

Granular leukocytes are subdivided on the basis of their staining characteristics into *neutrophils*, *eosinophils*, and *basophils*. Neutrophils and eosinophils are important phagocytic cells that participate in the immune response; they are sometimes called *microphages*, so that they will not be confused with the monocytes of the blood or the fixed and free macrophages found in peripheral tissues.

Neutrophils [FIGURE 20.5a] Approximately 70% of the circulating white blood cells are **neutrophils** (NOO-trō-fils). They are called neutrophils because their cytoplasm is packed with pale, neutral-staining granules containing lysosomal enzymes and bactericidal (bacteria-killing) compounds. Each mature neutrophil (Figure 20.5a●) has a diameter of 12–15 μm, nearly twice that of a red blood cell. A neutrophil has a very dense, contorted nucleus that may be condensed into a series of lobes like beads on a string. This attribute has given these cells another name, **polymorphonuclear leukocytes** (pol-ē-mor-fō-NŪ-klē-ar; *poly*, many + *morphe*, form), or **PMNs**.

Neutrophils are highly mobile and are usually the first of the WBCs to arrive at an injury site. They are very active phagocytes, specializing in attacking and digesting bacteria. Neutrophils usually have a short life span, surviving for about 12 hours. After actively engulfing debris or pathogens, a neutrophil dies, but its breakdown releases some chemicals that attract other neutrophils to the site, and others that have a broad antibiotic activity against the pathogens.

Eosinophils [FIGURE 20.5b] **Eosinophils** (ē-ō-SIN-ō-fils), also called **acidophils** (a-SID-ō-fils), are so named because their granules stain with *eosin*, an acidic red dye. Eosinophils are similar in size to neutrophils and represent 2–4% of the circulating WBCs. These cells have both deep red granules and a bilobed (two-lobed) nucleus, making an eosinophil relatively easy to identify (Figure 20.5b●). Eosinophils are phagocytic cells attracted to foreign compounds that have reacted with circulating antibodies. Eosinophil numbers increase dramatically during an allergic reaction or a parasitic infection. Eosinophils also are attracted to injury sites, where they release enzymes that reduce the degree of inflammation and control its spread to adjacent tissues.

Basophils [FIGURE 20.5c] **Basophils** (BĀ-sō-fils) are so named because they have numerous granules that stain with basic dyes. These inclusions stain a deep purple or blue with the stains used in a standard blood smear (Figure 20.5c●). Basophils are relatively rare, accounting for less than 1% of the leukocyte population. They migrate to sites of injury and cross the capillary endothelium to accumulate within the damaged tissues, where they discharge their granules into the interstitial fluids. The granules contain histamine; its release exaggerates the inflammation response at the injury site by increasing capillary permeability. Basophils also release chemicals that stimulate mast cells and attract basophils and other white blood cells to the area.[1]

AGRANULAR LEUKOCYTES

Circulating blood contains two types of agranular leukocytes: *monocytes* and *lymphocytes*. They differ both structurally and functionally.

Monocytes [FIGURE 20.5d] **Monocytes** (MON-ō-sīts) are the largest WBC, at 16–20 μm in diameter, 2–3 times the diameter of a typical RBC. These cells account for 2–8% of the WBC population. They are normally almost spherical, and, when flattened in a blood smear, they appear even larger; they are relatively easy to identify by their size and shape of their nucleus. Each cell has a large oval or kidney bean–shaped nucleus (Figure 20.5d●). Monocytes circulate for just a few days before entering peripheral tissues. Outside the bloodstream, monocytes are called *free macrophages*, to distinguish them from the immobile *fixed macrophages* found in many connective tissues. p. 64 Free macrophages are highly mobile, phagocytic cells. They usually arrive at the injury site shortly after the first neutrophils. While phagocytizing, free and fixed macrophages release chemicals that attract and stimulate other monocytes and other phagocytic cells. Active macrophages also secrete substances that lure fibroblasts into the region. The fibroblasts begin producing a dense network of collagen fibers around the site. This *scar tissue* may eventually wall off the injured area. Monocytes are one component of the **monocyte-macrophage system** that includes related cell types, such as fixed macrophages and more specialized cells such as the microglia of the CNS, the Langerhans cells of the skin, and phagocytic cells in the liver, spleen, and lymph nodes. This system will be discussed further in Chapter 23.

Lymphocytes [FIGURE 20.5e] Typical **lymphocytes** (LIM-fō-sīts) have very little cytoplasm, just a thin halo around a relatively large, round, purple-staining nucleus (Figure 20.5e●). Lymphocytes are usually slightly larger than RBCs and account for 20–30% of the WBC population. Blood lymphocytes represent a minute segment of the entire lymphocyte population, for lymphocytes are the primary cells of the **lymphatic system**, a network of special vessels and organs distinct from, but connected to, those of the cardiovascular system.

Lymphocytes are responsible for *specific immunity*: the ability of the body to mount a counterattack against invading pathogens or foreign proteins *on an individual basis*. Lymphocytes respond to such threats in three ways. One group of lymphocytes, called **T cells**, enter peripheral tissues and attack foreign cells directly. Another group of lymphocytes, the **B cells**, differentiate into plasma cells that secrete antibodies that attack foreign cells or proteins in distant portions of the body. T cells and B cells cannot be distinguished with the light microscope. **NK cells**, a third group, sometimes known as *large granular lymphocytes*, are responsible for *immune surveillance*, the destruction of abnormal tissue cells. These cells are important in preventing cancer. (The lymphatic system and immunity are discussed in Chapter 23.) *The Leukemias p. 799*

Platelets [FIGURES 20.5e/20.6/20.7]

Platelets are flattened, membrane-enclosed packets, which are round when viewed from above (Figure 20.5e●) and spindle-shaped in section. Platelets were once thought to be cells that had lost their nuclei, because similar functions in vertebrates other than mammals are performed by small nucleated blood cells. Histologists called all of these cells **thrombocytes** (THROM-bō-sīts; *thrombos*, clot). The term is still in use, although in mammals the term *platelet* is more suitable because these are membrane-enclosed enzyme packets, not individual cells.

Normal red bone marrow contains a number of very unusual cells, called **megakaryocytes** (meg-a-KAR-ē-ō-sīts; *mega-*, big + *karyon*, nucleus

[1] Histamine and other chemicals are also found in the granules of mast cells, connective tissue cells introduced in Chapter 3, p. 000 and the mast cells in damaged connective tissues also release their granules. However, the two cells are separate and distinct.

+ *cyte*, cell). As the name suggests, these are enormous cells (up to 160 μm in diameter) with large nuclei (Figure 20.6●). The dense nucleus may be lobed or ring-shaped, and the surrounding cytoplasm contains Golgi apparatus, ribosomes, and mitochondria in abundance. The cell membrane communicates with an extensive membrane network that radiates throughout the peripheral cytoplasm.

During their development and growth, megakaryocytes manufacture structural proteins, enzymes, and membranes. They then begin shedding cytoplasm in small membrane-enclosed packets. These packets are the platelets that enter the circulation. A mature megakaryocyte gradually loses all of its cytoplasm, producing around 4000 platelets before the nucleus is engulfed by phagocytes and broken down for recycling.

Platelets are continually replaced, and an individual platelet circulates for 10–12 days before being removed by phagocytes. A microliter of circulating blood contains an average of 350,000 platelets. An abnormally low platelet count (80,000 per μl, or less) is known as **thrombocytopenia** (throm-bō-sī-tō-PĒ-nē-a) and indicates excessive platelet destruction or inadequate platelet production. Symptoms include: bleeding along the digestive tract, bleeding within the skin, and occasional bleeding inside the CNS. Platelet counts in **thrombocytosis** (throm-bō-sī-TŌ-sis) may exceed 1,000,000 per μl, which usually results from accelerated platelet formation in response to infection, inflammation, or cancer.

Platelets are one participant in a vascular *clotting system* that also includes plasma proteins and the cells and tissues of the circulatory network. The process of **hemostasis** (*haima*, blood + *stasis*, halt) prevents the loss of blood through the walls of damaged vessels. In doing so, it both restricts blood loss and establishes a framework for tissue repairs. A portion of a blood clot is shown in Figure 20.7●.

Hemostasis involves a complex chain of events, and a disorder that affects any one step can disrupt the entire process. In addition, there are general requirements—for example, a deficiency of calcium ions or vitamin K will interfere with virtually all aspects of hemostasis.

The functions of platelets include:

1. *Transport of chemicals important to the clotting process.* By releasing enzymes and other factors at the appropriate times, platelets help initiate and control the clotting process.
2. *Formation of a temporary patch in the walls of damaged blood vessels.* Platelets clump together at an injury site, forming a *platelet plug* that can slow the rate of blood loss while clotting occurs.
3. *Active contraction after clot formation has occurred.* Platelets contain filaments of actin and myosin that can interact to produce contractions that shorten them. After a blood clot has formed, the contractions of platelets reduces the size of the clot and pulls together the cut edges of the vessel wall. *Problems with the Clotting Process p. 800*

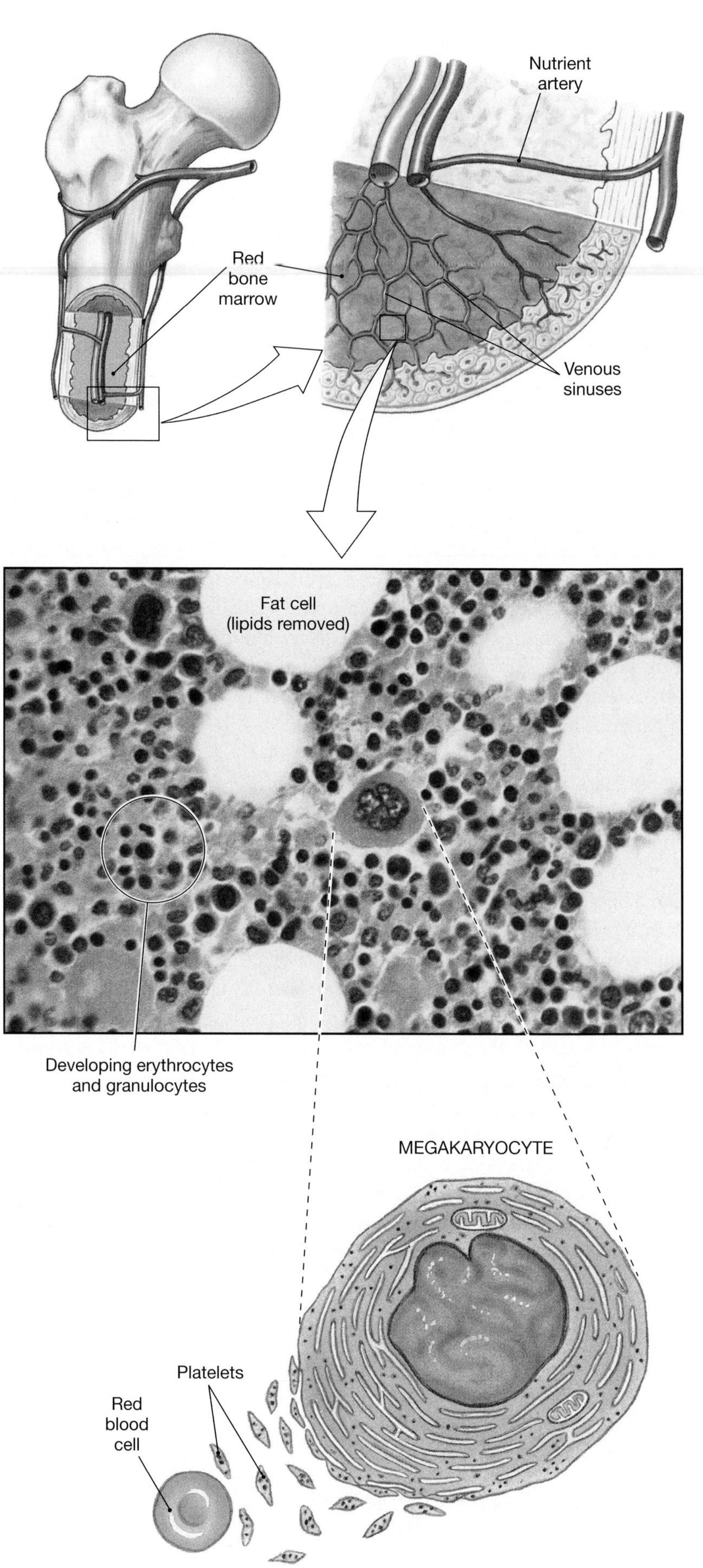

FIGURE 20.6 **MEGAKARYOCYTES AND PLATELET FORMATION**

Megakaryocytes stand out in bone marrow sections because of their enormous size and the unusual shape of their nuclei. These cells are continually shedding chunks of cytoplasm that enter the circulation as platelets. (LM × 673)

HEMOPHILIA

CLINICAL BRIEF

Hemophilia (hē-mō-FĒL-ē-ah) is one of many inherited disorders characterized by inadequate production of clotting factors. The incidence of this condition in the general population is about 1 in 10,000, with males accounting for 80–90% of those affected. In hemophilia A, production of a single clotting factor is reduced; the severity of the condition depends on the degree of reduction. In severe cases extensive bleeding accompanies the slightest mechanical stresses, and hemorrhages occur spontaneously at joints and around muscles.

Transfusions of clotting factors can often reduce or control the symptoms of hemophilia, but plasma samples from many individuals must be pooled (combined) to obtain adequate amounts of clotting factors. Pooling is very expensive and increases the risk of infection with blood-borne infections such as hepatitis or AIDS. Gene-splicing techniques have been used to manufacture the clotting factor most often involved (factor VIII). Although supplies are now limited, this procedure should eventually provide a safer method of treatment.

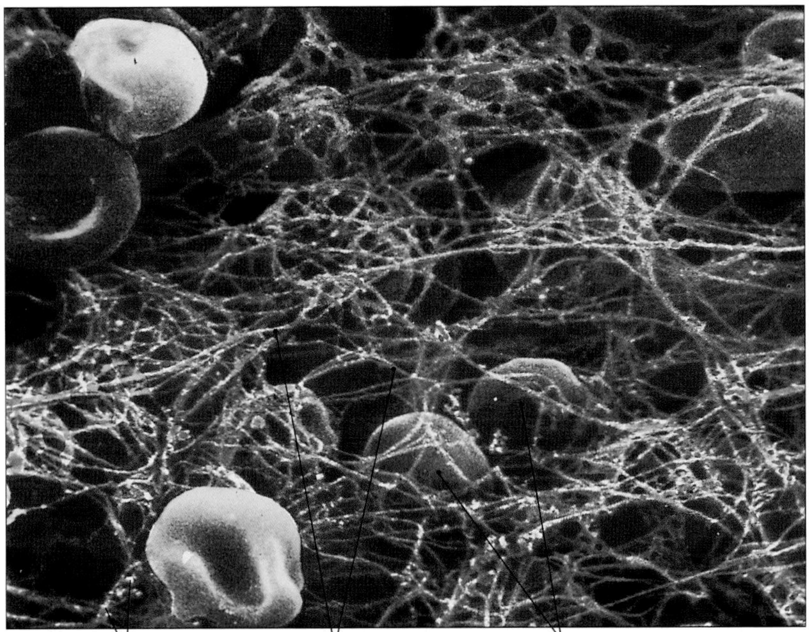

FIGURE 20.7 **STRUCTURE OF A BLOOD CLOT**

A colorized scanning electron micrograph showing the network of fibers that forms the framework of the clot. Red blood cells trapped in the clot add to its mass and give it a red color. (SEM × 4625)

✓ CONCEPT CHECK

- What type of white blood cells would you expect to find in the greatest number in an infected cut?
- What is thrombocytosis, and when does it occur?
- If you have an allergic reaction, which white blood cell type would increase dramatically?
- What is the function of the granules in basophils?

Hemopoiesis [FIGURE 20.8]

The process of blood cell formation is called **hemopoiesis** (hēm-ō-poi-Ē-sis). Blood cells appear in the circulation during the third week of embryonic development. These cells divide repeatedly, increasing their numbers. As other organ systems appear, some of the embryonic blood cells move out of the circulation and into the liver, spleen, thymus, and bone marrow (Figure 20.8●). These embryonic cells differentiate into stem cells that produce blood cells by their divisions. As the skeleton enlarges, the bone marrow becomes increasingly important; in the adult it is the primary site of blood cell formation.

Stem cells, called **hemocytoblasts**, produce all blood cells, but the process occurs in a series of separate steps. Hemocytoblast divisions produce at least four different types of stem cells with relatively restricted functions. Three of these stem cell lines are responsible for the production of red blood cells and megakaryocytes; these are the **myeloid stem cells** (Figure 20.8●). The daughter cells produced by the divisions of myeloid stem cells are called **progenitor cells**, and they have more restricted fates. For example, one type of progenitor cell produces daughter cells that mature into red blood cells; another gives rise to granulocytes or monocytes. Figure 20.8● includes important details concerning the formation of the various cellular components of the blood.

SYNTHETIC BLOOD

CLINICAL BRIEF

Despite improvements in transfusion technology, shortages of blood and anxieties over safety persist. In addition, some people may be unable or unwilling to accept transfusions for medical or religious reasons. Thus there has been widespread interest in a number of recent attempts to develop synthetic blood components.

Whole blood substitutes are highly experimental solutions still undergoing clinical evaluation. In addition to the osmotic agents found in plasma expanders, these solutions contain small clusters of synthetic molecules built of carbon and fluorine atoms. The mixtures, known as **perfluorochemical (PFC) emulsions**, can carry roughly 70% of the oxygen of whole blood. Animals have been kept alive after an exchange transfusion that completely replaced their circulating blood with a PFC emulsion.

PFC solutions have the same advantages of other plasma expanders, plus the added benefits of transporting oxygen. Because there are no RBCs involved, the PFC emulsions can carry oxygen to regions whose capillaries have been partially blocked by fatty deposits or blood clots. Unfortunately, PFCs do not absorb oxygen as effectively as normal blood. To ensure that they deliver adequate oxygen to peripheral tissues, the individual must breathe air rich in oxygen, usually through an oxygen mask. In addition, phagocytes appear to engulf the PFC clusters. These problems have limited the use of PFC emulsions on humans. However, one PFC solution, *Fluosol*, has been used to enhance oxygen delivery to cardiac muscle during heart surgery.

Another approach involves the manufacture of miniature erythrocytes by enclosing small bundles of hemoglobin in a lipid membrane. These **neohematocytes** (nē-ō-hēm-AT-ō-sīts) are spherical, with a diameter of under 1 mm, and they can easily pass through narrow or partially blocked vessels. The major problem with this technique is that phagocytes treat neohematocytes like fragments of normal erythrocytes, so they remain in circulation for only about 5 hours.

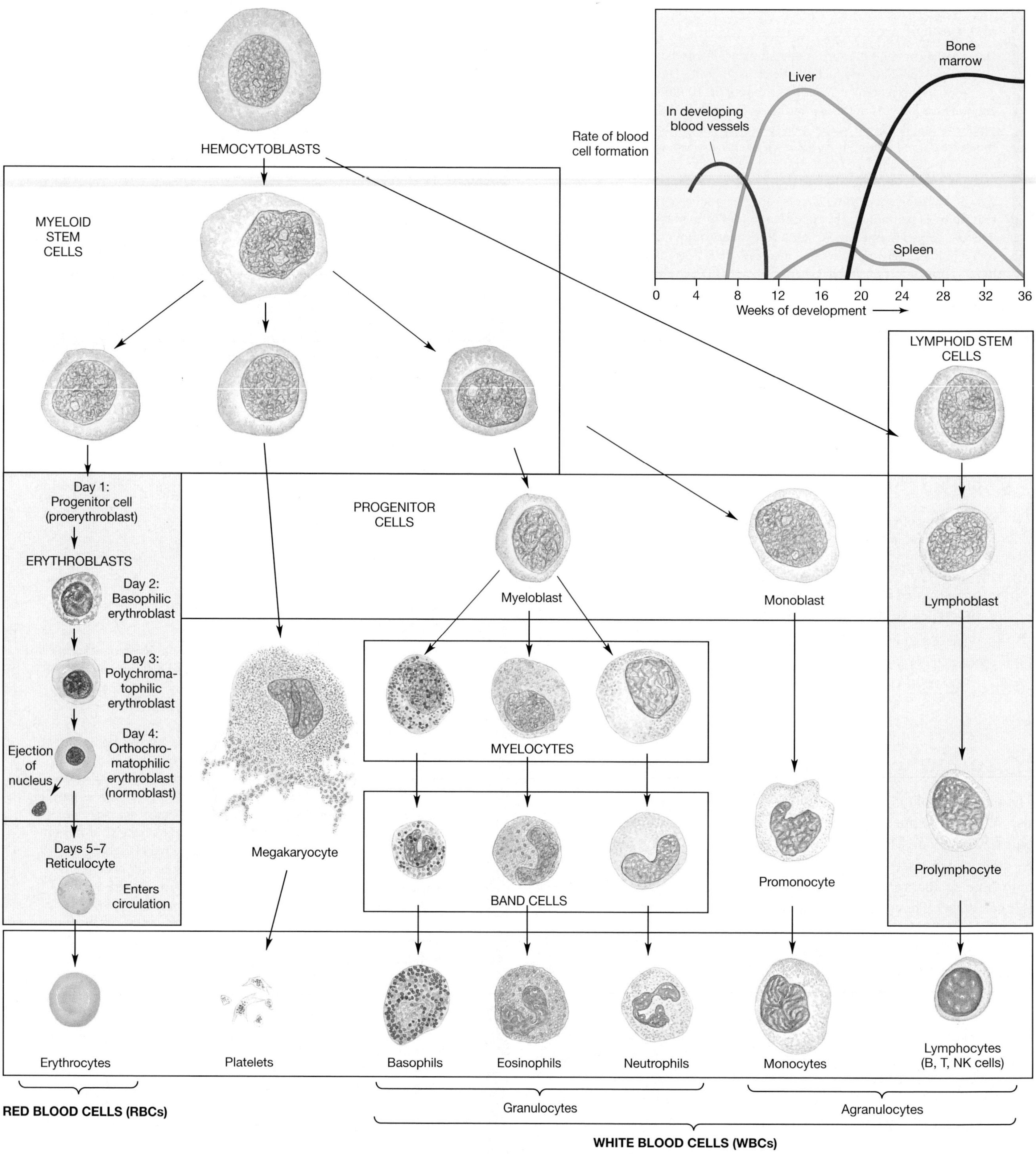

FIGURE 20.8 **The Origins and Differentiation of Formed Elements**

Hemocytoblast stem cells give rise to both myeloid and lymphoid stem cells. Myeloid stem cells produce progenitor cells, which divide to produce the various classes of blood cells. The graph indicates the primary locations of blood cell formation during embryonic and fetal development.

Erythropoiesis [FIGURE 20.8]

Erythropoiesis (e-rith-rō-poy-Ē-sis) refers specifically to the formation of erythrocytes (Figure 20.8●). The red bone marrow, or **myeloid tissue** (MĪ-e-loyd; *myelos*, marrow), is the primary site of blood cell formation in the adult. Blood cells are produced in areas of **red marrow** (Figure 20.6●). p. 114 Red marrow is found in portions of the vertebrae, sternum, ribs, skull, scapulae, pelvis, and proximal limb bones. Under extreme conditions the fatty **yellow marrow** found in other bones can be converted to red marrow. For example, this conversion may occur after a severe and sustained blood loss, thereby increasing the rate of red blood cell formation. For erythropoiesis to proceed normally, the myeloid tissues must receive adequate supplies of amino acids, iron, and **vitamin B_{12}**, a vitamin obtained from dairy products and meat.

Erythropoiesis is regulated by the hormone **erythropoiesis-stimulating hormone**, or **erythropoietin (EPO)**, introduced in Chapter 19. p. 519 Erythropoietin is produced and secreted under hypoxic (low-oxygen) conditions, primarily in the kidneys. Erythropoietin has two major effects:

- It stimulates increased rates of cell division in erythroblasts and in the stem cells that produce erythroblasts; and
- It speeds up the maturation of RBCs, primarily by accelerating the rate of hemoglobin synthesis. Under maximum EPO stimulation, the bone marrow can increase the rate of red blood cell formation tenfold, to around 30 million per second.

CLINICAL BRIEF

Blood Doping

The practice of **blood doping** in various forms has become widespread among competitive athletes involved with endurance sports such as cycling. One procedure entails removing whole blood from the athlete in the weeks before an event. The packed red cells are separated from the plasma and stored. By the time of the race, the competitor's bone marrow will have replaced the lost blood. Immediately before the event, the packed red cells are reinfused, increasing the hematocrit. The objective is to elevate the oxygen-carrying capacity of the blood, and so increase endurance. The consequence is that the athlete's heart is placed under a tremendous strain. The long-term effects are unknown, but the practice carries a significant risk of stroke or heart attack; it has recently been banned in amateur sports. Training at high altitudes is a safer and currently acceptable alternative.

Because it is now being synthesized using recombinant DNA techniques, EPO can be obtained by individuals attempting to circumvent these rules. Although use of this drug will be difficult for officials to detect, individuals using it are taking the same risks as those who practice blood doping. Over the last few years, the deaths of 18 European cyclists and the disqualification of several Olympic competitors during the 2002 Winter Olympics apparently resulted from blood doping, EPO abuse, or both.

STAGES IN RBC MATURATION [FIGURE 20.8]

A maturing red blood cell passes through a series of developmental stages. **Hematologists** (hē-ma-TOL-ō-jists), specialists in blood formation and function, have given specific names to key stages (Figure 20.8●). **Erythroblasts** are very immature red blood cells that are actively synthesizing hemoglobin. **Reticulocytes** (re-TIK-ū-lō-sīts) represent the last step in the maturation process. After shedding their nuclei, these cells will develop the appearance of mature erythrocytes and enter the circulation. These immature red blood cells normally account for about 0.8% of the RBC population.

Leukopoiesis [FIGURE 20.8]

Stem cells responsible for the production of white blood cells (**leukopoiesis**) originate in the bone marrow (Figure 20.8●). Granulocytes complete their development in myeloid tissue; monocytes begin their differentiation in the bone marrow, enter the circulation, and complete their development when they become free macrophages in peripheral tissues. Stem cells responsible for the production of lymphocytes, a process called **lymphopoiesis**, also originate in the bone marrow, but many of them subsequently migrate to the thymus. The bone marrow and thymus are called *primary lymphoid organs* because the divisions of undifferentiated stem cells at these sites produce daughter cells destined to become specialized lymphocytes. Immature B cells and NK cells are produced in the bone marrow, and immature T cells are produced in the thymus. These cells may subsequently migrate to *secondary lymphoid organs*, such as the spleen, tonsils, or lymph nodes. Although they retain the ability to divide, their divisions always produce cells of the same type—a dividing T cell produces daughter T cells and not NK or B cells. We will consider formation of lymphocytes in more detail in Chapter 23.

Factors that regulate lymphocyte maturation are as yet incompletely understood; however, prior to maturity, hormones of the thymus gland promote the differentiation and maintenance of T-cell populations. Several hormones, collectively called **colony-stimulating factors (CSFs)**, are involved in the regulation of other white blood cell populations. Commercially available CSFs are now used to stimulate the production of WBCs in individuals undergoing cancer chemotherapy.

✓ CONCEPT CHECK

- What are the two main effects of erythropoietin?
- What is the function of hemocytoblasts?
- The ejection of the nucleus marks which stage of red blood cell maturation?
- What cell sheds cytoplasmic packets to produce platelets?

RELATED CLINICAL TERMS

anemia (a-NE-me-ah): A condition in which the oxygen-carrying capacity of the blood is reduced, because of low hematocrit or low blood hemoglobin concentrations. *p. 536*

aplastic anemia: Anemia caused by failure of the bone marrow, leading to a low hematocrit and a low reticulocyte count. *Blood Tests and RBCs p. 797*

bone marrow transplant: A transfusion of bone marrow cells, including stem cells, that can be used to repopulate the bone marrow after radiation exposure, chemotherapy, or aplastic anemia. *The Leukemias p. 799*

embolism: A condition in which a drifting blood clot becomes stuck in a blood vessel, blocking circulation to the area downstream. *Problems with the Clotting Process p. 800*

hemolytic disease of the newborn (HDN): An anemia in the newborn usually caused by an incompatibility between the maternal (Rh^-) and fetal (Rh^+) blood types. *Hemolytic Disease of the Newborn p. 799*

hemophilia: One of many inherited disorders characterized by inadequate production of clotting factors. *p. 543*

hemorrhagic anemia: Anemia caused by severe bleeding, typified by a low hematocrit and a low hemoglobin but normal RBCs. *Blood Tests and RBCs p. 797*

normochromic: The condition in which red blood cells contain normal amounts of hemoglobin. *Blood Tests and RBCs p. 797*

normocytic: A term referring to red blood cells of normal size. *Blood Tests and RBCs p. 797*

normovolemic (nor-mo-vo-LĒ-mik): The condition of having normal blood volume. *p. 534*

packed red blood cells (PRBCs): Red blood cells from which most of the plasma has been removed. *p. 534*

plaque: An abnormal area within a blood vessel where large quantities of lipids accumulate. *Problems with the Clotting Process p. 800*

polycythemia (po-lē-sī-THĒ-mē-ah): A blood condition showing an elevated hematocrit with a normal blood volume. *p. 536*

thrombus: A blood clot. *Problems with the Clotting Process p. 800*

transfusion: A procedure in which blood components are given to someone whose blood volume has been reduced or whose blood is deficient in some components. *p. 534*

Additional Clinical Terms Discussed in Appendix I (pp. 797–800)

acute leukemia; aplastic anemia; autologous marrow transplant; chronic leukemia; coumadin; embolus; erythroblast; erythroblastosis fetalis; hemoglobin (Hb) concentration; hemorrhagic anemia; heparin; heterologous marrow transplant; hyperchromic; hypochromic; iron deficiency anemia; leukemia; lymphoid; macrocytic; malaria; mean corpuscular hemoglobin concentration (MCHC); mean corpuscular volume (MCV); microcytic; myeloid; pernicious anemia; RBC count; reticulocyte count; *RhoGam*™; sickle cell anemia; sickling trait; streptokinase; tissue plasminogen activator (t-PA); urokinase

STUDY OUTLINE & CHAPTER REVIEW

Introduction 533

1. The cardiovascular system provides a mechanism for the rapid transport of nutrients, waste products, and cells within the body.

Functions of the Blood 533

1. **Blood** is a specialized connective tissue. Its functions include: (1) transporting dissolved gases; (2) transporting and distributing nutrients; (3) transporting metabolic wastes; (4) transporting and delivering enzymes and hormones; (5) stabilizing the pH and electrolyte composition of interstitial fluids; (6) restricting fluid losses through damaged vessels or injuries via the **clotting reaction**; (7) defending the body against toxins and pathogens; and (8) stabilizing body temperature by absorbing and redistributing heat. *(see Table 20.1)*

Composition of the Blood 533

1. Blood consists of two components: **plasma**, the liquid matrix of blood, and **formed elements**, which include: **red blood cells (RBCs)**, **white blood cells (WBCs)**, and **platelets**. The plasma and **formed elements** constitute **whole blood**, which can be **fractionated** for analytical or clinical purposes. *(see Figure 20.1 and Table 20.2)*
2. There are 4–6 liters of whole blood in an average adult. The terms hypovolemic, normovolemic, and hypervolemic refer to low, normal, or excessive blood volume, respectively.

Plasma 534

3. Plasma accounts for about 55% of the volume of blood; roughly 92% of plasma is water. *(see Figure 20.1 and Table 20.2)*
4. Plasma differs from interstitial fluid because it has a higher dissolved oxygen concentration and large numbers of dissolved proteins. There are three classes of plasma proteins: *albumins*, *globulins*, and *fibrinogen*. *(see Table 20.2)*
5. **Albumins** constitute about 60% of plasma proteins. **Globulins** constitute roughly 35% of plasma proteins: they include **immunoglobulins** (**antibodies**), which attack foreign proteins and pathogens, and **transport globulins**, which bind ions, hormones, and other compounds. **Fibrinogen** molecules function in the clotting reaction by interacting to form *fibrin*; removing fibrinogen from plasma leaves a fluid called **serum**. *(see Table 20.2)* When albumins or globulins become attached to lipids, they form lipoproteins, which are carried in the circulatory system until the lipids are delivered to the tissues.

Formed Elements 535

Red Blood Cells (RBCs) 535

1. Red blood cells (RBCs), or **erythrocytes**, account for slightly less than half the blood volume. The **hematocrit** value indicates the percentage of whole blood occupied by cellular elements. Since blood contains about 1000 RBCs for each WBC, this value closely approximates the volume of RBCs. *(see Figures 20.2 to 20.4)*
2. RBCs transport oxygen and carbon dioxide within the bloodstream. They are highly specialized cells with large surface-to-volume ratios. Each RBC is a biconcave disc. This shape gives them a large surface area, allowing for rapid diffusion of gases and the ability to form stacks (called **rouleaux**), that can pass easily through small vessels.
3. Because RBCs lack mitochondria, ribosomes, and nuclei, they are unable to perform normal maintenance operations, so they usually degenerate after about 120 days in the circulation. Damaged or dead RBCs are recycled by phagocytes. *(see Figure 20.3 and Table 20.3)*
4. Molecules of **hemoglobin (Hb)** account for over 95% of the RBCs' proteins. Hemoglobin gives RBCs the ability to transport oxygen and carbon dioxide. Hemoglobin is a globular protein formed from four subunits. Each subunit contains a single molecule of **heme**, which holds an iron ion that can reversibly bind an oxygen molecule At the lungs, carbon dioxide diffuses out of the **blood** and oxygen diffuses into the blood. In the peripheral tissues, the opposite occurs: Oxygen diffuses out of the blood and carbon dioxide diffuses into the blood.
5. One's **blood type** is determined by the presence or absence of specific **surface antigens** *(agglutinogens)*, in the RBC cell membranes: **A**, **B**, and **D (Rh)**. Type A

blood has surface antigen A, Type B blood has surface antigen B, Type AB has both, and Type O has neither. Rh-positive blood has the Rh surface antigen, and Rh-negative does not. **Antibodies** specific to these surface antigens are called *agglutinins*. Antibodies within a person's plasma will react with RBCs bearing foreign surface antigens, causing a cross-reaction. *(see Figure 20.4 and Table 20.4)*

White Blood Cells (WBCs) 540

6. White blood cells (WBCs), or **leukocytes**, defend the body against pathogens and remove toxins, wastes, and abnormal or damaged cells *(see Figure 20.5)*. The two classes of WBCs are granular leukocytes (granulocytes) and agranular leukocytes (agranulocytes).
7. A stained blood smear provides a differential count of the white blood cell population. The word endings *-penia* and *-osis* are used to indicate low or high numbers, respectively, of specific types of white blood cells.
8. Leukocytes show **chemotaxis** (the attraction to specific chemicals) and **diapedesis** (the ability to move through vessel walls).
9. **Granular leukocytes (granulocytes)** are subdivided into **neutrophils**, **eosinophils (acidophils)**, and **basophils**. Fifty to 70% of circulating WBCs are neutrophils, which are highly mobile phagocytes. The much less common eosinophils are phagocytic cells, which are attracted to foreign compounds that have reacted with circulating antibodies. The relatively rare basophils migrate to damaged tissues and release histamines, aiding the inflammation response. *(see Figure 20.5 and Table 20.3)*
10. **Agranular leukocytes** are subdivided into **monocytes** and **lymphocytes**. Monocytes migrating into peripheral tissues become free macrophages, which are highly mobile, phagocytic cells. Lymphocytes, the primary cells of the **lymphatic system**, include **T cells** (which enter peripheral tissues and attack foreign cells directly), **B cells** (which produce antibodies), and **NK cells** (which destroy abnormal tissue cells). *(see Figure 20.5 and Table 20.3)*

Platelets 541

11. Platelets are sometimes called **thrombocytes**; they are not cells but are membrane-enclosed packets of cytoplasm.
12. **Megakaryocytes** are enormous cells in the bone marrow that release packets of cytoplasm (platelets) into the circulating blood. The functions of platelets include: (1) transporting chemicals important to the clotting process; (2) forming a temporary patch in the walls of damaged blood vessels; and (3) causing contraction after a clot has formed in order to reduce the size of the break in the vessel wall. *(see Figures 20.6/20.7 and Table 20.3)*

Hemopoiesis 543

1. **Hemopoiesis** is the process of blood cell formation. Circulating **stem cells** called **hemocytoblasts** divide to form all of the blood cells. *(see Figure 20.8)*

Erythropoiesis 545

2. **Erythropoiesis**, the formation of erythrocytes, occurs mainly within the **myeloid tissue** (bone marrow) in adults. RBC formation increases under **erythropoiesis-stimulating hormone (erythropoietin, EPO)** stimulation. Stages in RBC development include **erythroblasts** and **reticulocytes**. *(see Figure 20.8)*

Leukopoiesis 545

3. **Leukopoiesis**, the formation of white blood cells, occurs in bone marrow. Granulocytes and monocytes are produced by stem cells in the bone marrow. Stem cells responsible for **lymphopoiesis** (production of lymphocytes) also originate in the bone marrow, but many migrate to peripheral lymphoid tissues *(see Figure 20.8)*.
4. The bone marrow and the thymus are called *primary lymphoid organs. Secondary lymphoid organs*, such as the spleen, tonsils, or lymph nodes, contain white blood cells that divide to produce cells of the same type.
5. Factors that regulate lymphocyte maturation are not completely understood. Several **colony-stimulating factors (CSFs)** are involved in regulating other WBC populations and coordinating RBC and WBC production.

LEVEL 1 REVIEWING FACTS AND TERMS

Match each numbered item with the most closely related lettered item. Use letters for answers in the spaces provided.

Column A	Column B
_____ 1. basophils	a. stem cell source of all blood cells
_____ 2. embolism	b. low blood volume
_____ 3. lymphocyte	c. granular white blood cell
_____ 4. leukopoiesis	d. clotting protein
_____ 5. hemostasis	e. agranular white blood cell
_____ 6. erythroblasts	f. process of white blood cell production
_____ 7. fibrinogen	g. process of preventing blood loss
_____ 8. immunoglobulins	h. clot in a blood vessel
_____ 9. hemocytoblasts	i. antibodies
_____10. hypovolemic	j. immature red blood cells

11. Functions of the blood include
    (a) transport of nutrients and waste
    (b) regulation of pH and electrolyte concentrations
    (c) restricting fluid loss
    (d) all of the above
12. The formed elements of the blood include
    (a) red blood cells (b) defense proteins
    (c) clotting proteins (d) plasma proteins
13. The most abundant proteins in blood are
    (a) globulins
    (b) albumins
    (c) fibrinogens
    (d) lipoproteins
14. Plasma proteins that are important in body defense are the
    (a) albumins (b) fibrinogens
    (c) immunoglobulins (d) metalloproteins
15. Stem cells responsible for the production of white blood cells originate in the
    (a) liver (b) thymus
    (c) spleen (d) bone marrow
16. Each of the following statements concerning red blood cells (RBCs) is true except
    (a) RBCs are biconcave discs (b) RBCs lack mitochondria
    (c) RBCs have a large nucleus (d) RBCs can form stacks called rouleaux
17. The primary function of hemoglobin is to
    (a) store iron (b) transport glucose
    (c) give RBCs their color (d) carry oxygen to peripheral tissues

18. A person with Type A blood has
    (a) A surface antigens on their red blood cells
    (b) B surface antigens in their plasma
    (c) anti-A antibodies in their plasma
    (d) anti-O antibodies in their plasma
19. The white blood cells that increase in number during an allergic reaction or in response to parasitic infections are the
    (a) neutrophils (b) eosinophils
    (c) basophils (d) monocytes
20. Platelets are
    (a) large cells that lack a nucleus
    (b) small cells that lack a nucleus
    (c) fragments of cells
    (d) small cells with a many-shaped nucleus

LEVEL 2 REVIEWING CONCEPTS

1. A reduction in the number of oxygen-binding sites on hemoglobin would result in which of the following?
    (a) anemia (b) polycythemia
    (c) hypoxia (d) anoxia
2. A difference between the A, B, O blood types, and the Rh factor is that
    (a) Rh surface antigens are not found on the surface of RBCs
    (b) Rh surface antigens do not produce a cross-reaction
    (c) individuals who are Rh^- do not carry antibodies to Rh surface antigens unless they have been previously sensitized
    (d) Rh^+ antibodies are present on all RBCs
3. A sample of tissue from an injury shows a large number of basophils. This would indicate that the tissue was
    (a) normal (b) inflamed
    (c) being rejected (d) infected by multicellular parasites
4. Iron deficiency would result in which of the following?
    (a) decreased leukocyte count
    (b) decreased monocyte count
    (c) anemia
    (d) polycythemia
5. What is the volume of packed red cells, and why is it sometimes called "packed cell volume"?
6. What is the function of the clotting reaction?
7. What is the fate of megakaryocytes?
8. Give some examples of secondary lymphoid organs.
9. What is the function of the plasma proteins called lipoproteins?
10. Can a person with Type O blood receive Type AB blood? Why or why not?

LEVEL 3 CRITICAL THINKING AND CLINICAL APPLICATIONS

1. Children who suffer from protein starvation have bloated bellies because of ascites (an accumulation of fluid in the tissue spaces of the abdomen). How would a shortage of protein in the diet cause this condition?
2. Mononucleosis is a disease that can cause an enlarged spleen because of increased numbers of phagocytic and other cells. Common symptoms include pale complexion, a tired feeling, and a lack of energy sometimes to the point of not being able to get out of bed. What might cause these symptoms?
3. Almost half of our vitamin K is synthesized by bacteria that inhabit our large intestine. Based on this information, why would taking a broad-spectrum antibiotic (antibiotics are chemical agents that selectively kill pathogenic microorganisms) produce frequent nosebleeds?

✓ ANSWERS TO CONCEPT CHECK QUESTIONS

p. 535 **1.** Blood carries heat away from areas that are warm and distributes it either to the skin when the body is too warm, or to vital organs when the body is cold; marked slowing of flow would disrupt the body's ability to cool or warm itself properly. **2.** A hypovolemic individual has a low blood volume and would therefore have abnormally low blood pressure. **3.** Whole blood contains significant numbers of formed elements, such as RBCs, WBCs, and platelets. These components of the blood make it thicker and more resistant to flow.
p. 540 **1.** The hematocrit value closely approximates the percentage of red blood cells; thus, red blood cells account for 42% of her blood volume. **2.** Red blood cells have the ability to stack and create a rouleau, which can pass through a tiny blood vessel more easily than could many separate red blood cells. In addition, red blood cells are flexible, which allows them to squeeze through small capillaries. **3.** Because RBCs do not have a nucleus or ribosomes, protein synthesis for repair and replacement cannot take place. As a result they have a shorter life span that other cells; RBCs survive for only around 120 days in the circulation. **4.** Type AB blood has A and B surface antigens, so it has neither anti-A nor anti-B antibodies because the body's immune system ignores its own antigens.
p. 543 **1.** We would expect to find a large number of neutrophils in an infected cut. **2.** Thrombocytosis refers to unusually high numbers of platelets; this probably occurs in response to infection, inflammation, or cancer. **3.** Eosinophils. **4.** The granules contain histamine; its release exaggerates the inflammation response at the injury site.
p. 545 **1.** Erythropoietin increases the rate of both erythroblast cell division and stem cell division, and it speeds up the maturation of RBCs. **2.** Hemocytoblasts are stem cells, which produce all formed elements. **3.** The ejection of the nucleus marks the final stage of red blood cell maturation, at which point it is called a reticulocyte. **4.** Megakaryocytes, which are derived from hemocytoblasts, produce platelets.

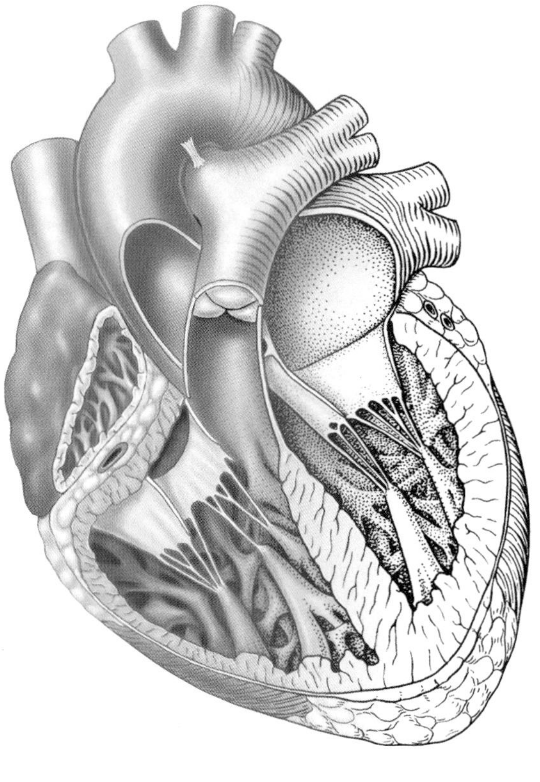

21

THE CARDIOVASCULAR SYSTEM

The Heart

Every living cell relies on the surrounding interstitial fluid as a source of oxygen and nutrients and as a place for the disposal of wastes. Levels of gases, nutrients, and waste products in the interstitial fluid are kept stable through continuous exchange between the interstitial fluid and the circulating blood. The blood must stay in motion to maintain homeostasis. If blood stops flowing through a tissue, its oxygen and nutrient supplies are exhausted quickly, its capacity to absorb wastes is soon reached, and neither hormones nor white blood cells can get to their intended targets. Thus, all of the functions of the cardiovascular system ultimately depend on the heart, because it is the heart that keeps blood moving. This muscular organ beats approximately 100,000 times each day, propelling blood through the blood vessels. Each year the heart pumps over 1.5 million gallons of blood, enough to fill 200 train tank cars.

For a practical demonstration of the heart's pumping abilities, turn on the faucet in the kitchen and open it all the way. To deliver an amount of water equal to the volume of blood pumped by the heart in an average lifetime (65 years), that faucet would have to be left on for at least 45 years. Equally remarkable, the volume of blood pumped by the heart can vary widely, between 5 and 30 liters per minute. The performance of the heart is closely monitored and finely regulated by the nervous system to ensure that gas, nutrient, and waste levels in the peripheral tissues remain within normal limits, whether one is sleeping peacefully, reading a book, or involved in a vigorous racquetball game.

We begin this chapter by examining the structural features that enable the heart to perform so reliably, even in the face of widely varying physical demands. We will then consider the mechanisms that regulate cardiac activity to meet the body's ever changing needs.

An Overview of the Cardiovascular System [FIGURE 21.1]

Despite its impressive workload, the heart is a small organ; your heart is roughly the size of your clenched fist. The heart's four muscular chambers, right and left **atria** (A-tre-a; singular, *atrium*; "chamber") and right and left **ventricles** (VEN-tri-k'ls; "little belly"), work together to pump blood through a network of blood vessels between the heart and the peripheral tissues. The network can be subdivided into two circuits: the **pulmonary circuit**, which carries carbon dioxide–rich blood from the heart to the gas-exchange surfaces of the lungs and returns oxygen-rich blood to the heart; and the **systemic circuit**, which transports oxygen-rich blood from the heart to the rest of the body's cells, returning carbon dioxide–rich blood back to the heart. The right atrium receives blood from the systemic circuit, and the right ventricle discharges blood into the pulmonary circuit. The left atrium collects blood from the pulmonary circuit, and the left ventricle ejects blood into the systemic circuit. When the heart beats, the atria contract first, followed by the ventricles. The two ventricles contract at the same time and eject equal volumes of blood into the pulmonary and systemic circuits.

Each circuit begins and ends at the heart. **Arteries** transport blood away from the heart; **veins** return blood to the heart (Figure 21.1•). Blood travels through these circuits in sequence. For example, blood returning to the heart in the systemic veins must complete the pulmonary circuit before reentering the systemic arteries. **Capillaries** are small thin-walled vessels that interconnect the smallest arteries and veins. Capillaries are called **exchange vessels** because their thin walls permit exchange of nutrients, dissolved gases, and waste products between the blood and surrounding tissues.

The Pericardium [FIGURE 21.2]

The heart is located near the anterior chest wall, directly posterior to the sternum in the **pericardial** (per-i-KAR-dē-al) **cavity** (Figure 21.2a•), a portion of the ventral body cavity. The pericardial cavity is situated between the pleural cavities, in the mediastinum, which also contains the thymus, esophagus, and trachea. p. 19 The position of the heart relative to other structures in the mediastinum is shown in Figure 21.2d•.

The **pericardium** is the serous membrane lining the pericardial cavity. To visualize the relationship between the heart and the pericardial cavity, imagine pushing your fist toward the center of a large balloon (Figure 21.2b•). The wall of the balloon represents the pericardium, and your fist is the heart. The wall of the balloon corresponds to the pericardium. The pericardium is divided into the **visceral pericardium** (the part of the balloon in contact with your fist) and the **parietal pericardium** (the rest of the

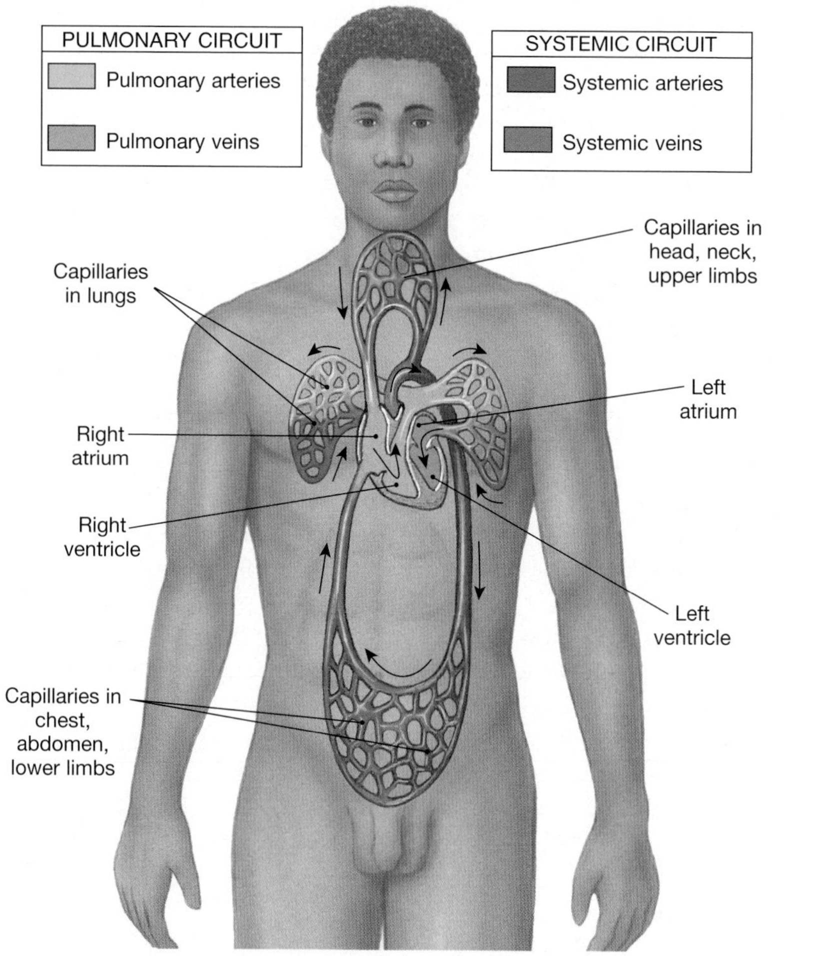

FIGURE 21.1 **A Generalized View of the Pulmonary and Systemic Circuits**

Blood flows through separate pulmonary and systemic circuits, driven by the pumping of the heart. Each circuit begins and ends at the heart and contains arteries, capillaries, and veins. Arrows indicate the direction of blood flow within each circuit.

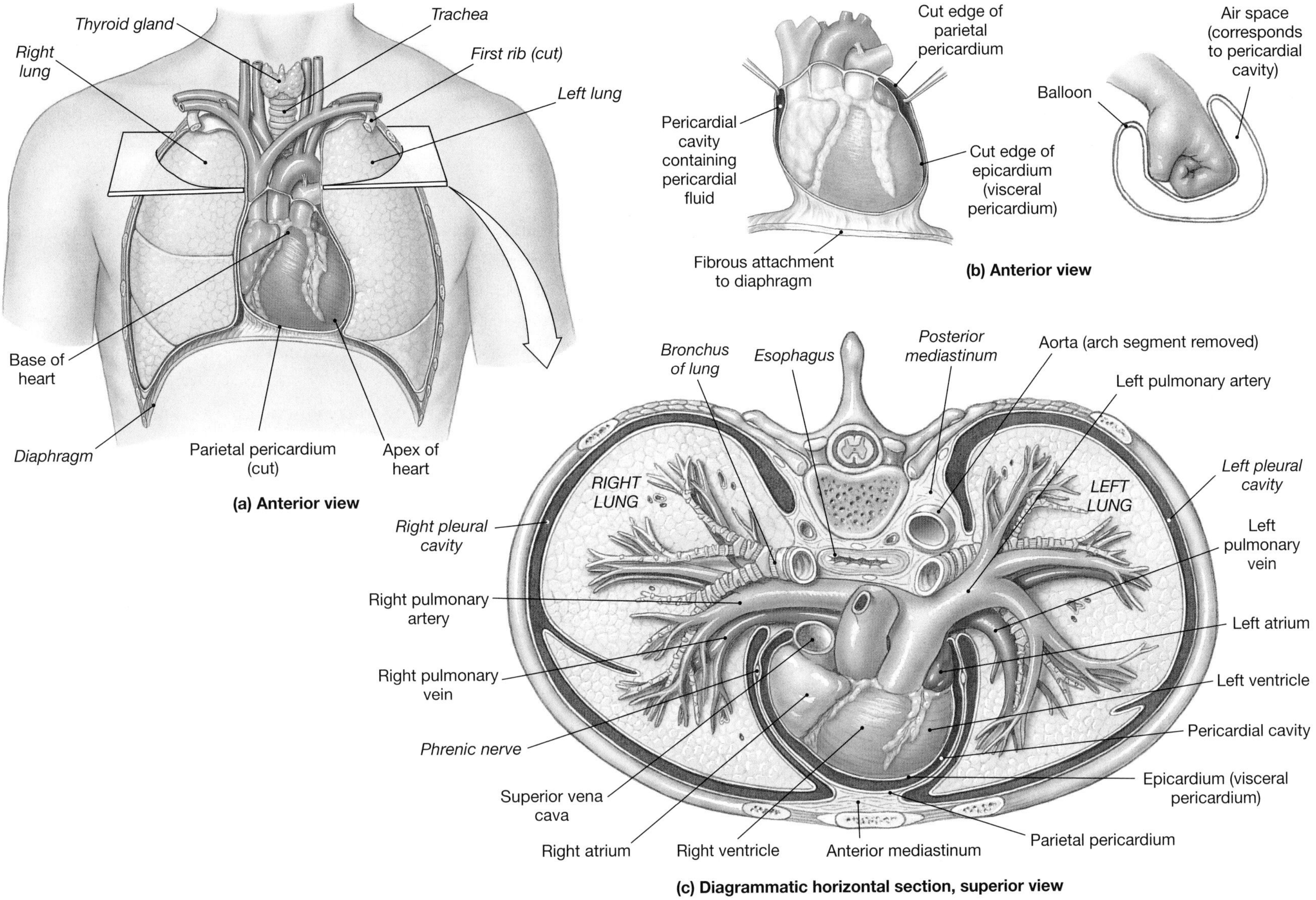

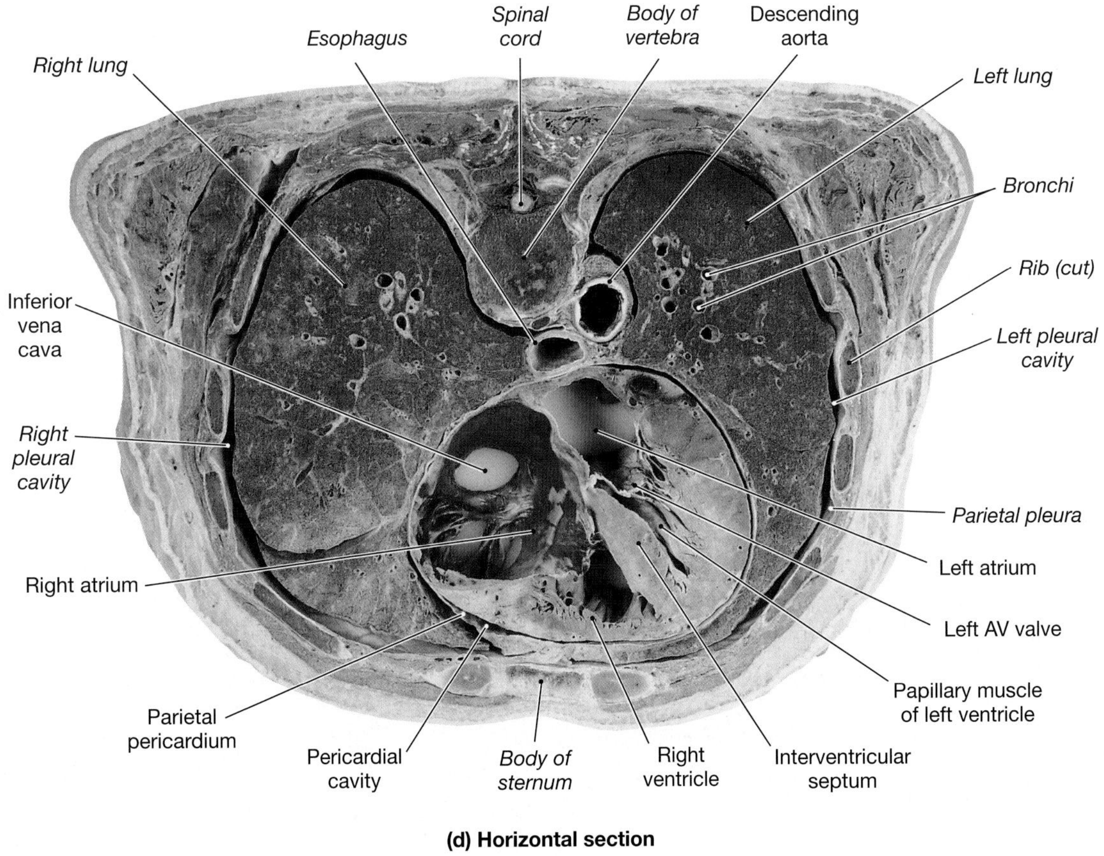

FIGURE 21.2 **LOCATION OF THE HEART IN THE THORACIC CAVITY**

The heart is situated within the anterior portion of the mediastinum, immediately posterior to the sternum. (**a**) Anterior view of the open chest cavity, showing the position of the heart and major vessels relative to the lungs. The sectional plane indicates the orientation of (c). (**b**) Relationships between the heart and the pericardial cavity. The pericardial cavity surrounds the heart like the balloon surrounds the fist (right). (**c**) Diagrammatic view showing the position of the heart and the location of other organs within the mediastinum. In this sectional view, the heart is shown intact so you can see the orientation of the major vessels. (**d**) A superior view of a horizontal section through the trunk at the level of vertebra T_8.

balloon). Your wrist, where the balloon folds back upon itself, corresponds to the *base* of the heart (so named because it is where the heart is attached to the major vessels and bound to the mediastinum).

The loose connective tissue of the visceral pericardium, or *epicardium*, is bound to the cardiac muscle tissue of the heart. The serous membrane of the parietal pericardium is reinforced by an outer layer of dense, irregular connective tissue containing abundant collagen fibers. This reinforcing layer is known as the **fibrous pericardium**. Together, the parietal pericardium and the fibrous pericardium form the tough **pericardial sac**. At the base of the heart, the collagen fibers of the fibrous pericardium stabilize the positions of the pericardium, heart, and associated vessels in the mediastinum. The slender gap between the opposing parietal and visceral surfaces is the pericardial cavity. This cavity normally contains 10–20 ml of **pericardial fluid** secreted by the pericardial membranes. Pericardial fluid acts as a lubricant, reducing friction between the opposing surfaces. The moist pericardial lining prevents friction as the heart beats, and the collagen fibers binding the base of the heart to the mediastinum limit movement of the major vessels during a contraction.

Structure of the Heart Wall [FIGURE 21.3]

A section through the wall of the heart (Figure 21.3a,b●) reveals three distinct layers: (1) an outer *epicardium* (visceral pericardium), (2) a middle *myocardium*, and (3) an inner *endocardium*.

1. The **epicardium** is the visceral pericardium, and it forms the external surface of the heart. This serous membrane is superficial to the myocardium.
2. The **myocardium** consists of multiple, interlocking layers of cardiac muscle tissue, with associated connective tissues, blood vessels, and nerves. The relatively thin atrial myocardium contains layers that form figure eights, as they pass from atrium to atrium. The ventricular myocardium is much thicker, and the muscle orientation changes from layer to layer. Superficial ventricular muscles wrap around both ventricles; deeper muscle layers spiral around and between the ventricles from the attached *base* toward the free tip, or *apex*, of the heart (Figure 21.3a-c●).
3. The inner surfaces of the heart, including the valves, are covered by a simple squamous epithelium, known as the **endocardium** (en-dō-KAR-dē-um; *endo-*, inside). The endocardium is continuous with the endothelium of the attached blood vessels.

Cardiac Muscle Tissue [FIGURE 21.3b–e]

The unusual histological characteristics of cardiac muscle tissue give the myocardium its unique functional properties. Cardiac muscle tissue was introduced in Chapter 3, and its properties were briefly compared with those of other muscle types. p. 77 Cardiac muscle cells, or *cardiocytes*, are relatively small, averaging 10–20 μm in diameter and 50–100 μm in length. A typical cardiocyte has a single, centrally placed nucleus (Figure 21.3b–d●).

Although they are much smaller than skeletal muscle fibers, cardiac muscle cells resemble skeletal muscle fibers in that each cardiac muscle cell contains organized myofibrils, and the alignment of their sarcomeres produces striations. However, cardiac muscle cells differ from skeletal muscle fibers in several important respects:

1. Cardiac muscle cells are almost totally dependent on aerobic respiration to obtain the energy needed to continue contracting. The sarcoplasm of a cardiac muscle cell thus contains hundreds of mitochondria and abundant reserves of myoglobin (to store oxygen). Energy reserves are maintained in the form of glycogen and lipid inclusions.
2. The relatively short T-tubules of cardiac muscle cells do not form triads with the sarcoplasmic reticulum.
3. The circulatory supply of cardiac muscle tissue is more extensive even than that of red skeletal muscle tissue. p. 256
4. Cardiac muscle cells contract without instructions from the nervous system; their contractions will be discussed later in this chapter.
5. Cardiac muscle cells are interconnected by specialized cell junctions called *intercalated discs* (Figure 21.3c–e●).

THE INTERCALATED DISCS [FIGURE 21.3d,e]

Cardiac muscle cells are connected to neighboring cells at specialized cell junctions known as **intercalated** (in-TER-ka-lā-ted) **discs**. Intercalated discs are unique to cardiac muscle tissue. The jagged appearance is due to the extensive interlocking of opposing sarcolemmal membranes (Figure 21.3d,e●). At an intercalated disc:

1. The cell membranes of two cardiac muscle cells are bound together by desmosomes. p. 44 This locks the cells together and helps to maintain the three-dimensional structure of the tissue.
2. Myofibrils in these muscle cells anchor firmly to the membrane at the intercalated disc. The intercalated disc thus ties together the myofibrils of adjacent cells. As a result, the two muscle cells "pull together" with maximum efficiency.
3. Cardiac muscle cells are also connected by gap junctions. pp. 43, 77 Ions and small molecules can move between cells at gap junctions, thereby creating a direct electrical connection between the two muscle cells. As a result, the stimulus for contraction—an action potential—can move from one cardiac muscle cell to another as if the membranes were continuous.

Because cardiac muscle cells are mechanically, chemically, and electrically connected to one another, cardiac muscle tissue functions like a single, enormous muscle cell. The contraction of any one cell will trigger the contraction of several others, and the contraction will spread throughout the myocardium. For this reason, cardiac muscle has been called a *functional syncytium* (sin-SIT-ē-um; "fused mass of cells").

The Fibrous Skeleton [FIGURE 21.3b/21.7b]

The connective tissues of the heart include large numbers of collagen and elastic fibers. Each cardiac muscle cell is wrapped in a strong but elastic sheath, and adjacent cells are tied together by fibrous cross-links, or "struts." In turn, each muscle layer has a fibrous wrapping, and fibrous sheets separate the superficial and deep muscle layers. These connective tissue layers are continuous with dense bands of fibroelastic tissue that encircle (1) the bases of the pulmonary trunk and aorta and (2) the valves of the heart. This extensive connective tissue network is called the **fibrous skeleton** of the heart (Figures 21.3b and 21.7b●).

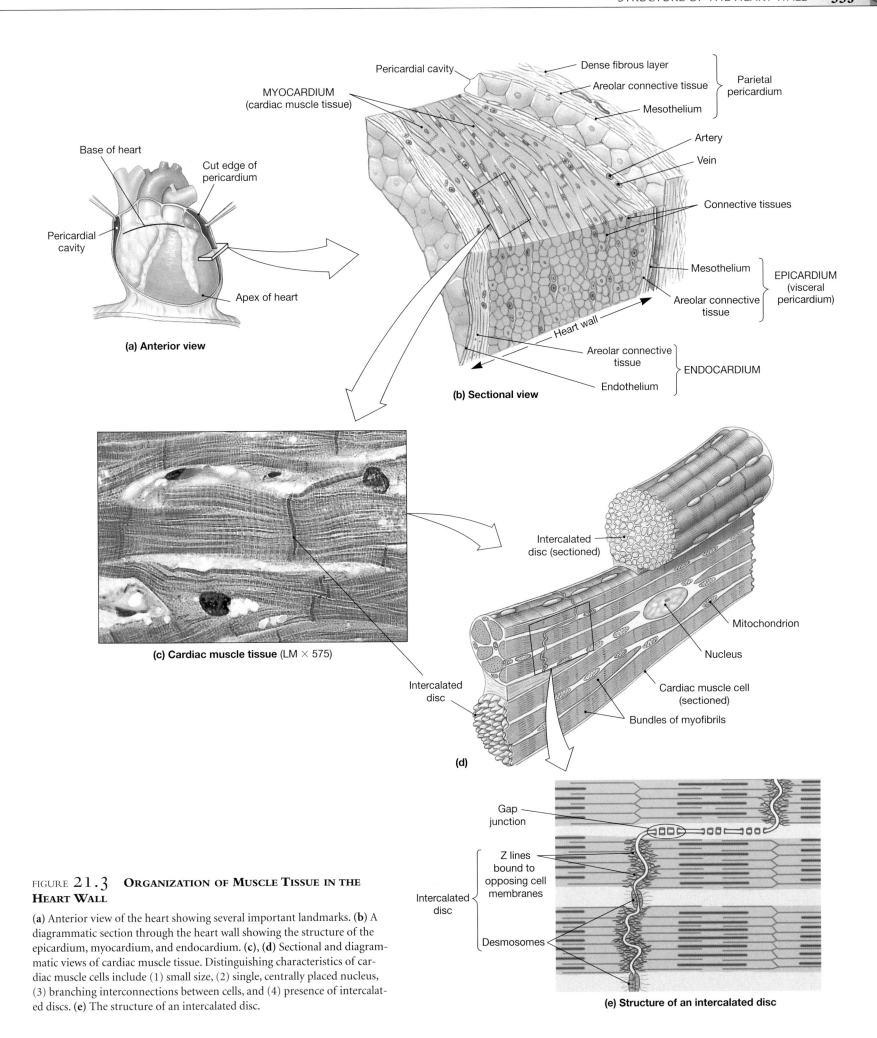

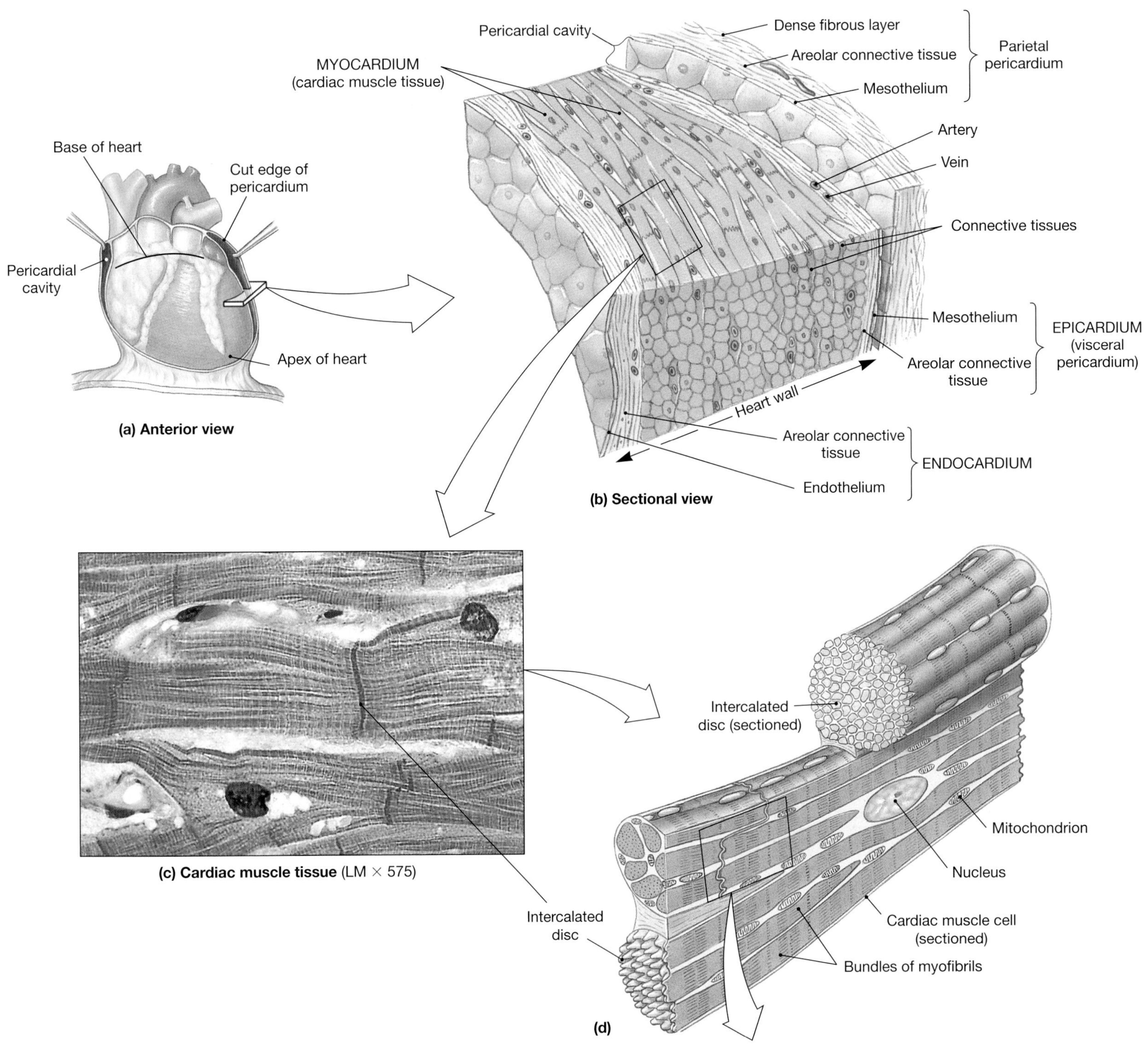

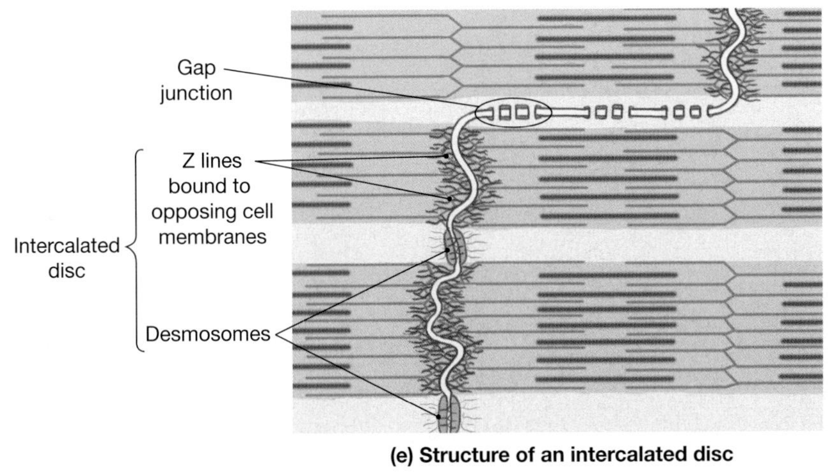

FIGURE 21.3 **ORGANIZATION OF MUSCLE TISSUE IN THE HEART WALL**

(**a**) Anterior view of the heart showing several important landmarks. (**b**) A diagrammatic section through the heart wall showing the structure of the epicardium, myocardium, and endocardium. (**c**), (**d**) Sectional and diagrammatic views of cardiac muscle tissue. Distinguishing characteristics of cardiac muscle cells include (1) small size, (2) single, centrally placed nucleus, (3) branching interconnections between cells, and (4) presence of intercalated discs. (**e**) The structure of an intercalated disc.

The fibrous skeleton has the following functions:

1. Stabilizing the positions of the muscle cells and valves in the heart.
2. Providing physical support for the cardiac muscle cells and for the blood vessels and nerves in the myocardium.
3. Distributing the forces of contraction.
4. Reinforcing the valves and helping to prevent over-expansion of the heart.
5. Providing elasticity that helps return the heart to its original shape after each contraction.
6. Physically isolating the atrial muscle cells from the ventricular muscle cells; as you will see in a later section, this isolation is vital for the coordination of cardiac contractions.

✓ CONCEPT CHECK

- How could you distinguish a sample of cardiac muscle tissue from a sample of skeletal muscle tissue?
- What is the pericardial cavity?
- How are cardiac muscle cells connected to their neighbors?
- Why is a cardiac muscle called a functional syncytium?

Orientation and Superficial Anatomy of the Heart [FIGURES 21.2/21.4/21.5]

Although advertisements and cartoons often show the heart at the center of the chest, a midsagittal section would not cut the heart in half. This is because the heart (1) lies slightly to the left of the midline, (2) sits at an angle to the longitudinal axis of the body, and (3) is rotated toward the left side.

1. *The heart lies slightly to the left of the midline.* The heart is located within the mediastinum, between the two lungs. Because the heart lies slightly to the left of the midline, the notch within the medial surface of the left lung is considerably deeper than the corresponding notch in the medial surface of the right lung. The **base** is the broad superior portion of the heart, where the heart is attached to the major arteries and veins of the systemic and pulmonary circuits. The base of the heart includes both the origins of the major vessels and the superior surfaces of the two atria. In terms of our balloon analogy, the base corresponds to the wrist (Figure 21.2b●). The base sits posterior to the sternum at the level of the third costal cartilage, centered about 1.2 cm (0.5 in.) to the left side (Figure 21.4●). The **apex** (Ā-peks) is the inferior, rounded tip of the heart, which points laterally at an oblique angle. A typical adult heart measures approximately 12.5 cm (5 in.) from the attached base to the apex. The apex reaches the fifth intercostal space approximately 7.5 cm (3 in.) to the left of the midline.
2. *The heart sits at an oblique angle to the longitudinal axis of the body.* The base forms the **superior border** of the heart. The **right border** of the heart is formed by the right atrium; the **left border** is formed by the left ventricle and a small portion of the left atrium. The left border extends to the apex, where it meets the **inferior border**. The inferior border is formed mainly by the inferior wall of the right ventricle.
3. *The heart is rotated slightly toward the left.* As a result of this rotation, the anterior surface, or **sternocostal** (ster-nō-KOS-tal) **surface**, consists primarily of the right atrium and right ventricle (Figure 21.5a●). The posterior and inferior wall of the left ventricle forms much of the sloping posterior surface, or **diaphragmatic surface**, that extends between the base and the apex of the heart (Figure 21.5b●).

The atria and the ventricles have very different functions—the atria receive venous blood that must continue on to the ventricles, whereas the ventricles must propel blood around the systemic and pulmonary circuits. These functional differences are of course linked to external and internal structural differences. Examine Figure 21.5●, which details the superficial anatomy of the heart, and note the distinguishing characteristics of the atria and ventricles. The atria have relatively thin muscular walls, and, as a

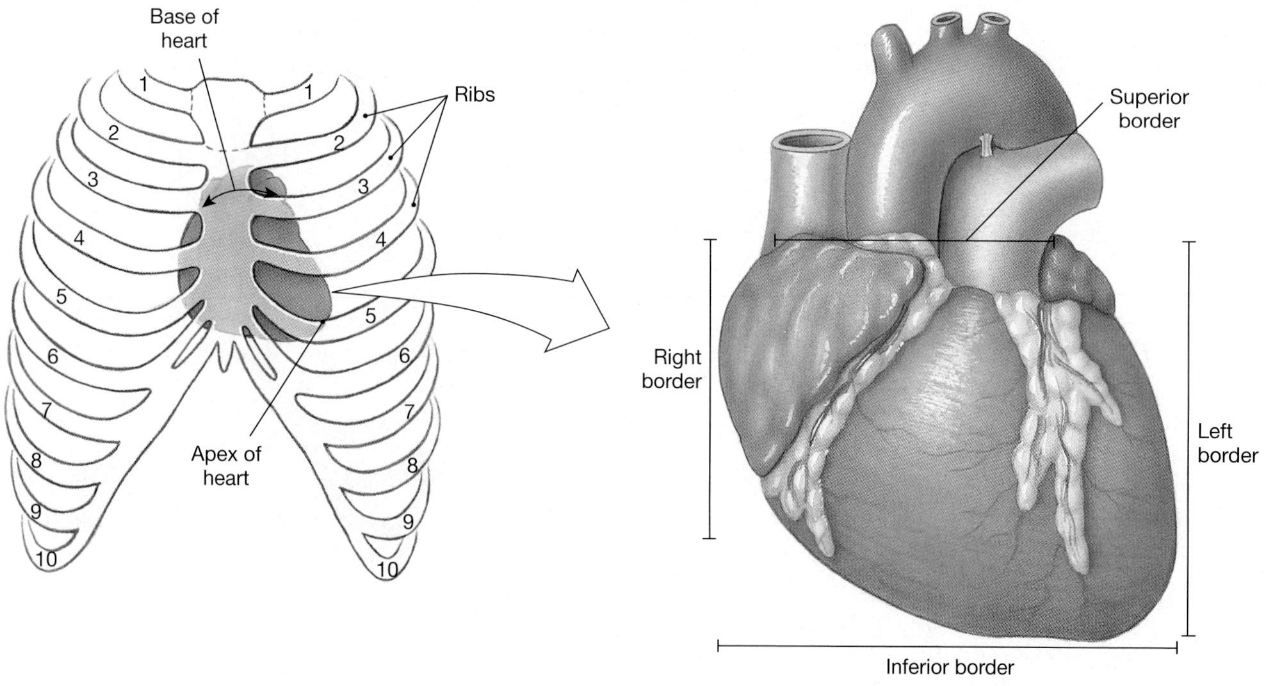

FIGURE 21.4 **POSITION AND ORIENTATION OF THE HEART**

The location of the heart within the thoracic cavity and the borders of the heart.

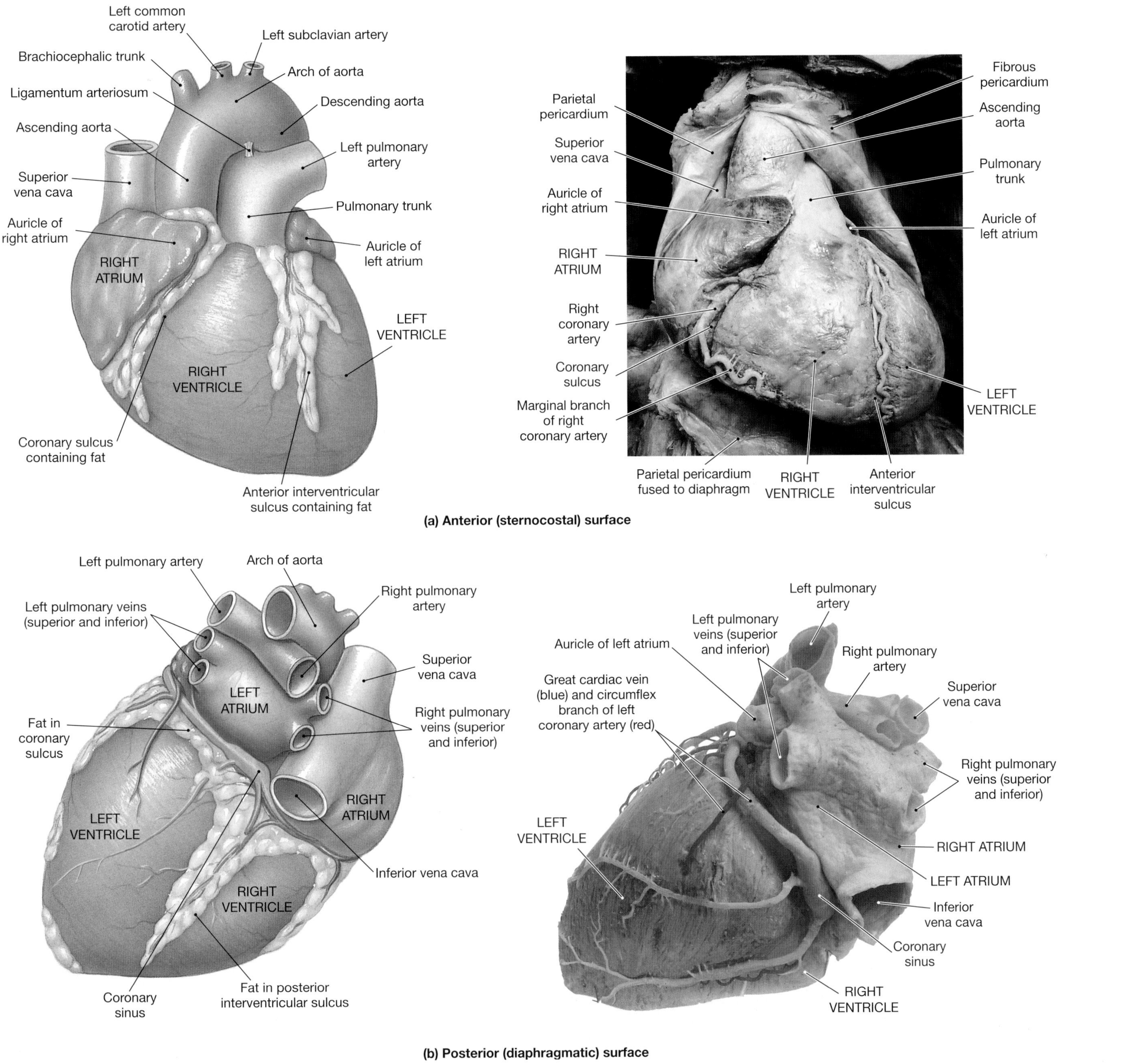

FIGURE 21.5 **Superficial Anatomy of the Heart**

(**a**) Anterior view of the heart and great vessels. In the photo, the pericardial sac has been cut and reflected to expose the heart and great vessels. (**b**) Posterior view of the heart and great vessels.

result, they are highly distensible. When not filled with blood, the outer portion of each atrium deflates and becomes a rather lumpy and wrinkled flap. This expandable extension of an atrium is called an **auricle** (AW-ri-kel; *auris,* ear) because it reminded early anatomists of the external ear. The auricle is also known as an *atrial appendage.* A deep groove, the **coronary sulcus**, marks the border between the atria and the ventricles. The boundary line between the left and right ventricles is marked by shallower depressions on the anterior surface (the **anterior interventricular sulcus**) and the posterior surface (the **posterior interventricular sulcus**). The connective tissue of the epicardium at the coronary and interventricular sulci usually contains substantial amounts of fat that must be removed to expose the underlying grooves. These sulci also contain the arteries and veins that supply blood to the cardiac muscle of the heart.

Internal Anatomy and Organization of the Heart [FIGURE 21.6]

Figure 21.6● details the internal anatomy and functional organization of the atria and ventricles. The atria are separated by the *interatrial septum* (*septum,* a wall), and the *interventricular septum* divides the ventricles (Figure 21.6a,c●). Each atrium communicates with the ventricle of the same side. **Valves** are folds of endocardium that extend into the openings between the atria and ventricles. These valves open and close to prevent backflow, thereby maintaining a one-way flow of blood from the atria into the ventricles. (Valve structure and function will be described under a separate heading.)

An atrium functions to collect blood returning to the heart and deliver it to the attached ventricle. The functional demands placed on the right and left atria are very similar, and the two chambers look almost identical. The demands placed on the right and left ventricles are very different, and there are significant structural differences between the two.

■ The Right Atrium [FIGURES 21.5/21.6a,c]

The right atrium receives oxygen-poor venous blood from the systemic circuit through the **superior vena cava** (VĒ-na CĀ-va) and the **inferior vena cava** (Figures 21.5 and 21.6a,c●). The superior vena cava, which opens into the posterior, superior portion of the right atrium, delivers venous blood from the head, neck, upper limbs, and chest. The inferior vena cava, which opens into the posterior and inferior portion of the right atrium, delivers venous blood from the rest of the trunk, the viscera, and the lower limbs. The veins of the heart itself, called *coronary veins,* collect blood from the heart wall and deliver it to the *coronary sinus* (Figure 21.5b●). This collecting vessel opens into the right atrium, slightly inferior to the opening of the inferior vena cava. (The coronary blood vessels will be described under a separate heading.)

Prominent muscular ridges, the **pectinate muscles** (*pectin,* comb), or *musculi pectinati,* extend along the inner surface of the right auricle and across the adjacent anterior atrial wall. The **interatrial septum** separates the right and left atria. From the fifth week of embryonic development until birth, there is an oval opening, the **foramen ovale**, in this septum. (*See Embryology Summaries* on p. 568 and pp. 604–605.) The foramen ovale permits blood flow directly from the right atrium to the left atrium while the lungs are developing and nonfunctional. At birth the lungs begin functioning and the foramen ovale closes; after 48 hours it is permanently sealed. A small depression, the **fossa ovalis**, persists at this site in the adult heart. Occasionally the foramen ovale does not close, and it remains *patent* (open). As a result, blood recirculates into the pulmonary circuit, reducing the efficiency of systemic circulation and elevating blood pressure in the pulmonary vessels. This can lead to cardiac enlargement, fluid build-up in the lungs, and eventual heart failure.

■ The Right Ventricle [FIGURES 21.5/21.6]

Oxygen-poor venous blood travels from the right atrium into the right ventricle through a broad opening bounded by three fibrous flaps. These flaps, or **cusps**, form the **right atrioventricular (AV) valve**, called the **tricuspid valve** (trī-KUS-pid; *tri,* three) (Figure 21.6●). The free edges of the cusps are attached to bundles of collagen fibers, the chordae tendineae (KOR-dē TEN-di-nē-ē; "tendinous cords"). These bundles arise from the **papillary** (PAP-i-ler-ē) **muscles**, cone-shaped muscular projections of the inner ventricular surface. The chordae tendineae limit the movement of the cusps and prevent backflow of blood from the right ventricle into the right atrium; the mechanism will be detailed in a later section.

The internal surface of the ventricle contains a series of irregular muscular folds, the **trabeculae carneae** (tra-BEK-ū-lē CAR-nē-ē; *carneus,* fleshy). The **moderator band** is a band of ventricular muscle that extends from the **interventricular septum**, a thick, muscular partition that separates the two ventricles, to the anterior wall of the right ventricle and the bases of the papillary muscles.

The superior end of the right ventricle tapers to a smooth-walled, cone-shaped pouch, the **conus arteriosus**, which ends at the **pulmonary semilunar valve** *(pulmonary valve).* This valve consists of three thick semilunar (half moon–shaped) cusps. As blood is ejected from the right ventricle, it passes through this valve to enter the **pulmonary trunk**, the start of the pulmonary circuit. The arrangement of cusps in this valve prevents the backflow of blood into the right ventricle, when that chamber relaxes. From the pulmonary trunk, blood flows into both the **left** and **right pulmonary arteries** (Figures 21.5 and 21.6●). These vessels branch repeatedly within the lungs before supplying the pulmonary capillaries where gas exchange occurs.

■ The Left Atrium [FIGURES 21.5/21.6a]

From the pulmonary capillaries, the blood, now oxygen-rich, flows into small veins that ultimately unite to form four pulmonary veins, two from each lung. These **left** and **right pulmonary veins** empty into the posterior portion of the left atrium (Figures 21.5 and 21.6a●). The left atrium lacks pectinate muscles, but it has an auricle. Blood flowing from the left atrium into the left ventricle passes through the **left atrioventricular (AV) valve**, or **mitral** (also termed the *bicuspid*) (bī-KUS-pid) **valve**. As the name bicuspid implies, this valve contains a pair of cusps (*bi-,* two) rather than a trio (*tri-,* three). Clinicians, however, often prefer the term *mitral* (MĪ-tral; *mitre,* a bishop's hat) *valve.* The left atrioventricular valve permits the flow of oxygen-rich blood from the left atrium into the left ventricle, but prevents bloodflow in the reverse direction.

FIGURE 21.6 **SECTIONAL ANATOMY OF THE HEART**

(**a**) A diagrammatic frontal section through the relaxed heart, showing major landmarks and the path of blood flow through the atria and ventricles (arrows). (**b**) Photograph of papillary muscles and chordae tendineae supporting the right AV valve. The picture was taken inside the right ventricle, looking toward a light shining from the right atrium. (**c**) Anterior view of a frontally sectioned heart, showing internal features and valves. The cardiac arteries and veins have been injected with latex; the arteries are red, the veins blue. (**d**) horizontal section through the heart at the level of vertebra T_8.

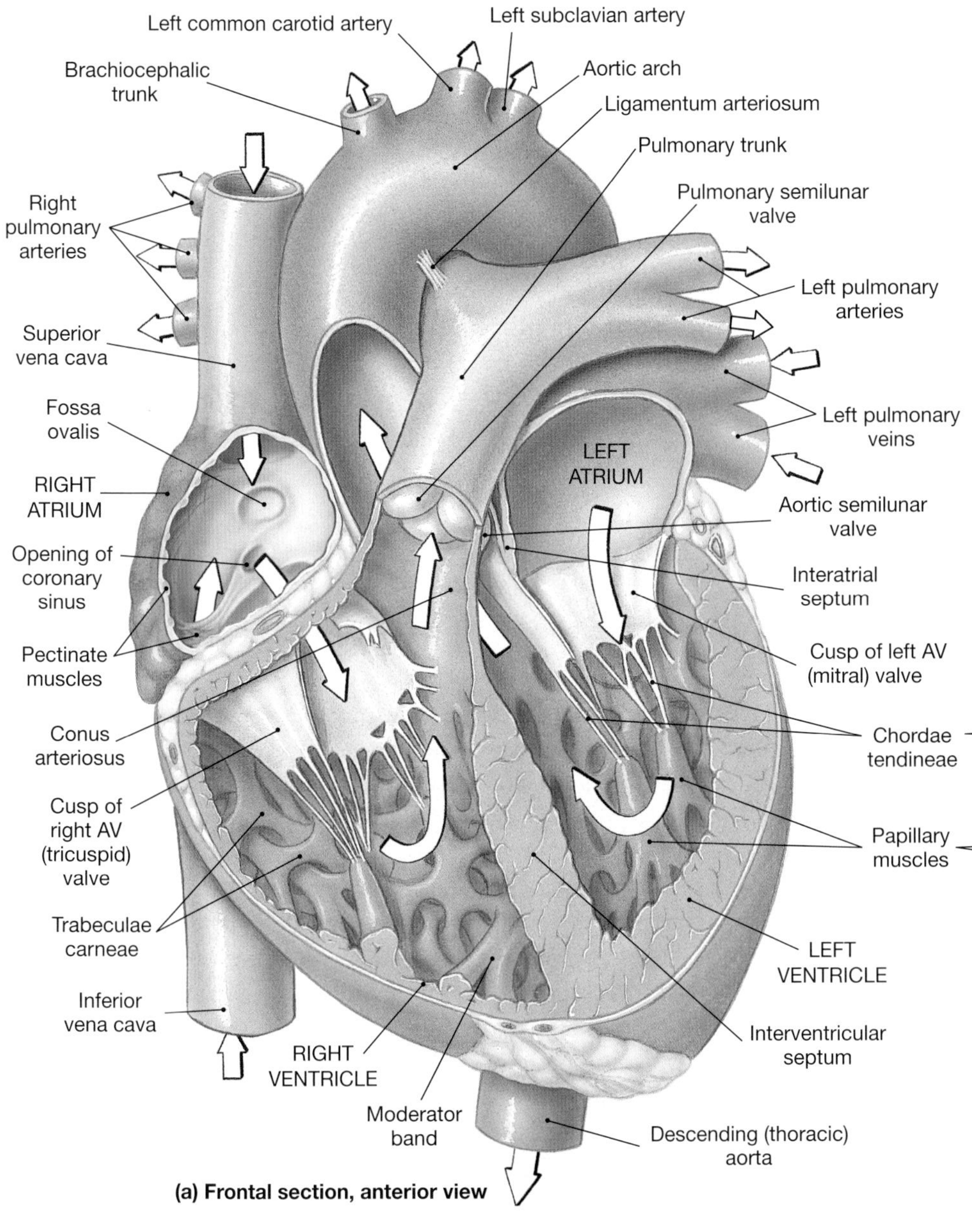

(a) Frontal section, anterior view

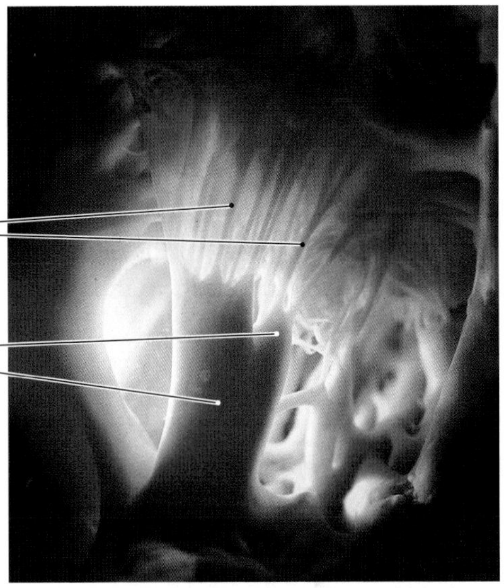

(b) Interior view, right ventricle

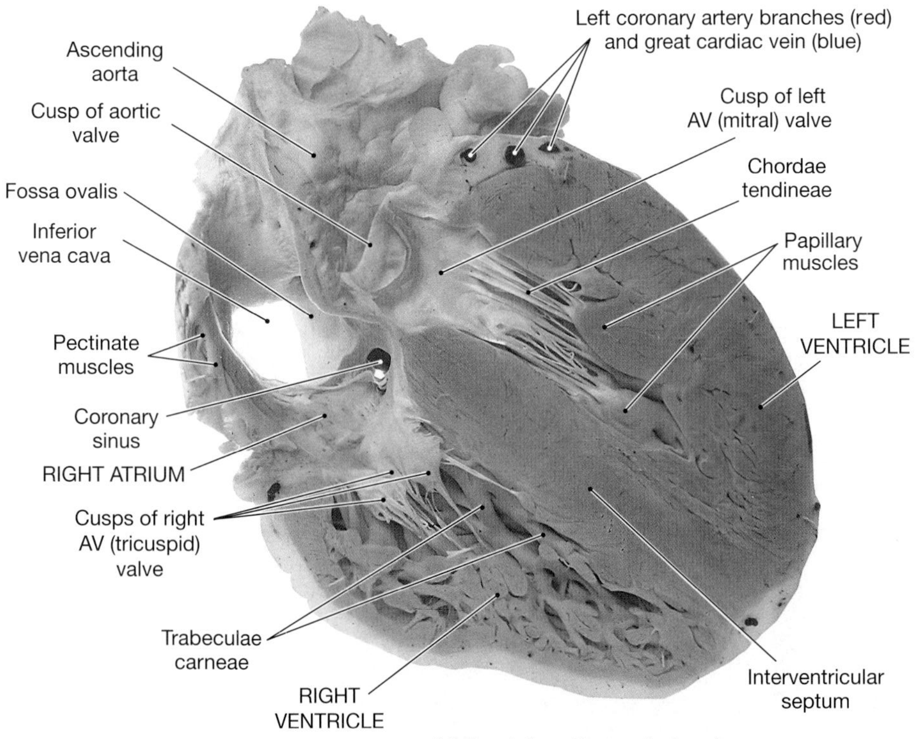

(c) Frontal section, anterior view

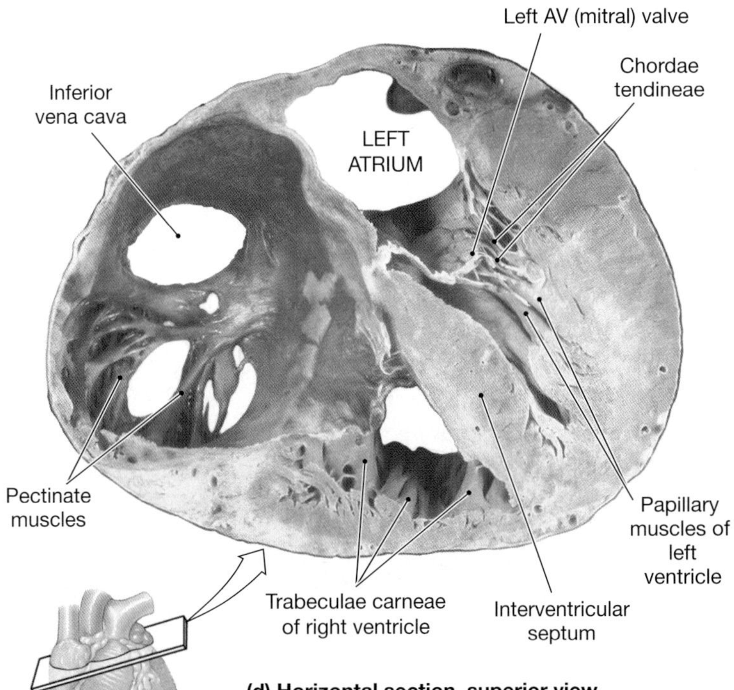

(d) Horizontal section, superior view

The Left Ventricle [FIGURES 21.5/21.6a,c,d]

The left ventricle has the thickest wall of any heart chamber. The extra-thick myocardium enables it to develop enough pressure to force blood around the entire systemic circuit; by comparison the right ventricle, which has a relatively thin wall, must push blood to the lungs and then back to the heart, a distance of only about 30 cm (1 ft). The internal organization of the left ventricle resembles that of the right ventricle (Figure 21.6a,c,d●). However, the trabeculae carneae are more prominent than they are in the right ventricle, and since the mitral valve has two cusps, there are two large papillary muscles rather than three.

Blood leaves the left ventricle by passing through the **aortic semilunar valve** *(aortic valve)* into the **ascending aorta.** The arrangement of cusps in the aortic semilunar valve is the same as in the pulmonary semilunar valve. Saclike dilations of the base of the ascending aorta occur adjacent to each cusp. These sacs, called **aortic sinuses,** prevent the individual cusps from sticking to the wall of the aorta when the valve opens. The *right* and *left coronary arteries,* which deliver blood to the myocardium, originate at the aortic sinus. The aortic semilunar valve prevents the backflow of blood into the left ventricle once it has been pumped out of the heart and into the systemic circuit. From the ascending aorta, blood flows on through the **aortic arch** and into the **descending aorta** (Figures 21.5 and 21.6●). The pulmonary trunk is attached to the aortic arch by the *ligamentum arteriosum,* a fibrous band that is the remnant of an important fetal blood vessel. Circulatory changes that occur at birth are described in Chapter 22.

Structural Differences between the Left and Right Ventricles [FIGURE 21.6a,c,d]

Anatomical differences between left and right ventricles are best seen in three-dimensional or sectional views (Figure 21.6a,c,d●). The lungs partially enclose the pericardial cavity, and the base of the heart lies between the left and right lungs. As a result, the pulmonary arteries and veins are relatively short and wide, and the right ventricle normally does not need to push very hard to propel blood through the pulmonary circuit. The wall of the right ventricle is relatively thin, and in sectional view it resembles a pouch attached to the massive wall of the left ventricle. When the right ventricle contracts, it moves toward the massive wall of the left ventricle. This compresses the blood within the right ventricle, and the rising pressure forces the blood through the pulmonary semilunar valve and into the pulmonary trunk. This mechanism moves blood very efficiently at relatively low pressures, which are all that one needs to move blood around the pulmonary circuit. Higher pressures would actually be dangerous, because the pulmonary capillaries are very delicate. Pressures as high as those found in systemic capillaries would both damage the pulmonary vessels and force fluid into the alveoli of the lungs.

A comparable pumping arrangement would not be suitable for the left ventricle, because six to seven times as much force must be exerted to propel blood through the systemic circuit. The left ventricle, which has an extremely thick muscular wall, is round in cross section. When the left ventricle contracts, two things happen: The distance between the base and apex decreases, and the diameter of the ventricular chamber decreases. If you imagine the effects of simultaneously squeezing and rolling up the end of a toothpaste tube, you will get the idea. The forces generated are quite powerful, more than enough to force open the aortic semilunar valve and eject blood into the ascending aorta. As the powerful left ventricle contracts, it also bulges into the right ventricular cavity. This intrusion improves the efficiency of the right ventricle's efforts. Individuals whose right ventricular musculature has been severely damaged may continue to survive because of the extra push provided by the contraction of the left ventricle. *The Cardiomyopathies p. 800*

✓ CONCEPT CHECK

- What is the name of the groove separating the atria from the ventricles?
- What are some external characteristics that distinguish the atria from the ventricles?

The Structure and Function of Heart Valves [FIGURES 21.7/21.8]

Details of the structure and function of the heart valves are shown in Figure 21.7a,b●. The chordae tendineae and papillary muscles play an important role in the normal function of the AV valves. During the period of ventricular relaxation (*ventricular diastole*) the ventricles are filling with blood, the papillary muscles are relaxed, and the open AV valve offers no resistance to the flow of blood from atrium to ventricle. Over this period the semilunar valves are closed; the semilunar valves do not need chordae tendineae because the relative positions of the cusps are stable and the three symmetrical cusps support one another like the legs of a tripod.

When the period of ventricular contraction (*ventricular systole*) begins, blood moving back toward the atria swings the cusps of the AV valves together. Tension in the papillary muscles and chordae tendineae keeps the cusps from swinging farther and opening into the atria. Thus, the chordae tendineae and papillary muscles are essential to prevent the backflow, or *regurgitation*, of blood into the atria each time the ventricles contract.

Serious valvular abnormalities can interfere with cardiac function; the timing and intensity of the related heart sounds can provide useful diagnostic information. Physicians use an instrument called a **stethoscope** (STETH-ō-scōp) to listen to normal and abnormal heart sounds. Valve sounds may be muffled as they pass through the pericardium, surrounding

CLINICAL BRIEF

Mitral Valve Prolapse

Minor abnormalities in valve shape are relatively common. For example, an estimated 10% of normal individuals age 14–30 have some degree of **mitral valve prolapse (MVP)**. In this condition the mitral valve cusps do not close properly. The problem may involve abnormally long (or short) chordae tendineae or malfunctioning papillary muscles. Because the valve does not work perfectly, some regurgitation may occur during left ventricular systole. The surges, swirls, and eddies that occur during regurgitation create a rushing, gurgling sound known as a **heart murmur**. Most of these individuals are completely asymptomatic, and they live normal, healthy lives unaware of any circulatory malfunction. However, regurgitation may increase the risk of valve infection after dental (or some medical) procedures.

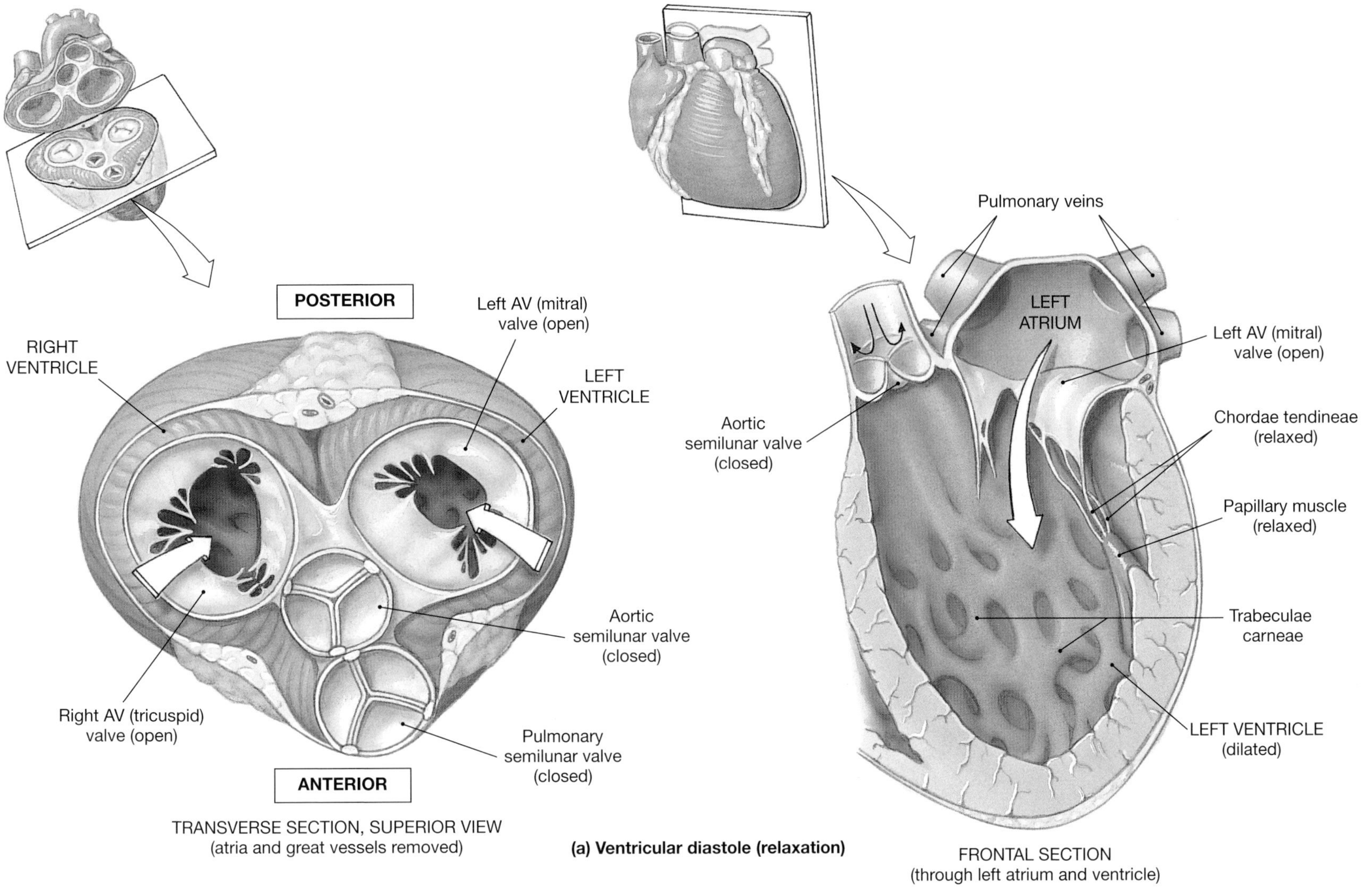

(a) Ventricular diastole (relaxation)

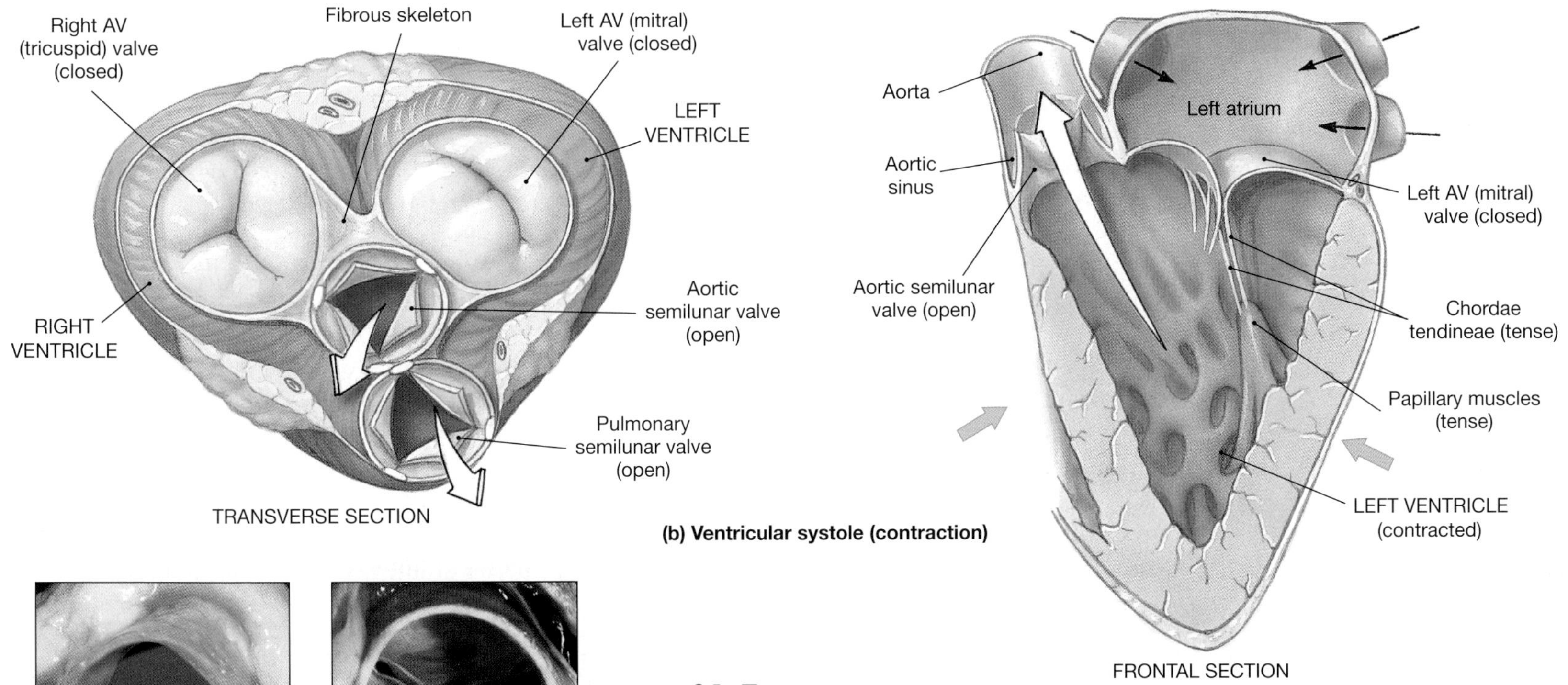

(b) Ventricular systole (contraction)

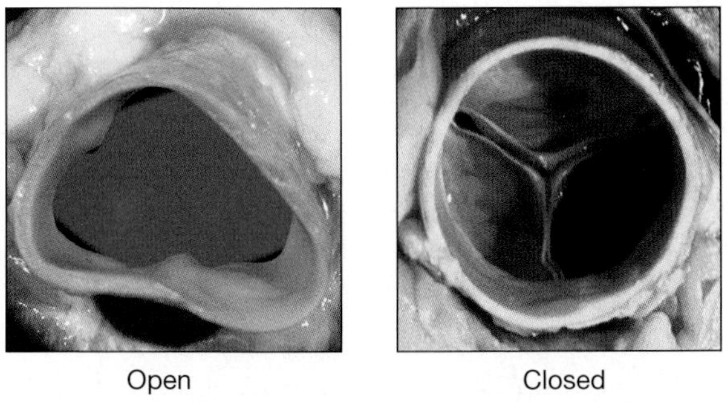

(c) Semilunar valve function

FIGURE 21.7 **VALVES OF THE HEART**

White arrows indicate blood flow into or out of a ventricle; black arrows, blood flow into an atrium; and green arrows, ventricular contraction. (**a**) When the ventricles are relaxed, the AV valves are open and the semilunar valves are closed. The chordae tendineae are loose, and the papillary muscles are relaxed. (**b**) When the ventricles are contracting, the AV valves are closed and the semilunar valves are open. In the frontal section notice the attachment of the left AV valve to the chordae tendineae and papillary muscles. (**c**) The aortic semilunar valve in the open (left) and closed (right) positions. The individual cusps brace one another in the closed position.

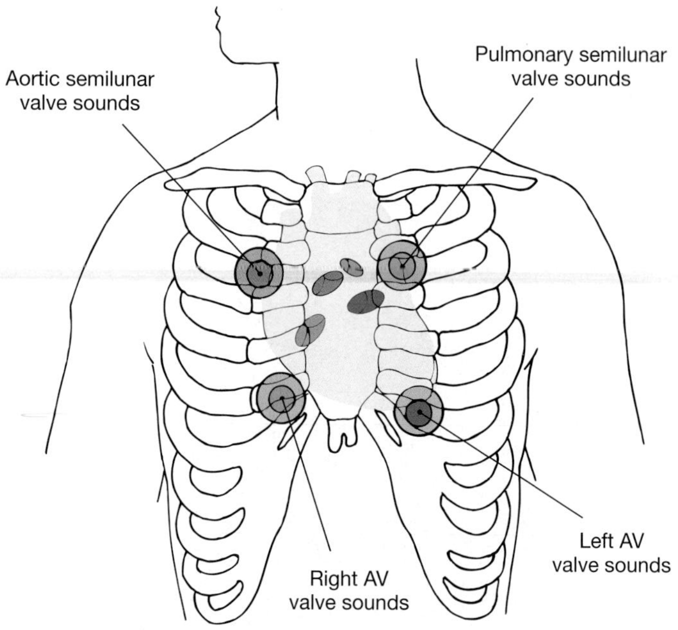

FIGURE 21.8 **HEART VALVES AND HEART SOUNDS**

The proper placement of a stethoscope varies depending on which heart sounds and valves are of interest.

tissues, and the chest wall. As a result, the stethoscope placement does not always correspond to the position of the valve under review (Figure 21.8●).

HEART SOUNDS [FIGURE 21.8]

Figure 21.8● shows the placement of a stethoscope to listen to the sounds generated by the action of individual valves, the familiar "lubb-dupp" **heart sounds** that accompany each heartbeat. This technique, called *auscultation*, is a simple and effective method of cardiac diagnosis.

When you listen to your own heart, you usually hear the first and second heart sounds. These sounds accompany the action of the heart valves. The first heart sound, the "lubb," lasts a little longer than the second. It is produced as the AV valves close at the start of ventricular systole. The second heart sound, "dupp," occurs at the beginning of ventricular diastole, when the semilunar valves close.

Third and fourth heart sounds may be audible as well, but even with a stethoscope they are usually very faint and are seldom detectable in healthy adults. These sounds are not related to valve action; they are associated with (1) blood flowing into the ventricles and (2) atrial contraction. More sophisticated procedures can be used to obtain a detailed assessment of cardiac function. One of the most important techniques involves the analysis of the electrical signals generated by the heart; this technique, the *electrocardiogram*, will be described in a later section. † *Valvular Heart Disease p. 800, Infection and Inflammation of the Heart p. 800, Rheumatic Heart Disease and Valvular Stenosis p. 801*

■ Coronary Blood Vessels [FIGURE 21.9]

The heart works continuously, and cardiac muscle cells require reliable supplies of oxygen and nutrients. The **coronary circulation** supplies blood to the muscle tissue of the heart. During maximum exertion, the oxygen demand rises considerably, and the blood flow to the heart may increase to nine times above resting levels.

The coronary circulation (Figure 21.9●) includes an extensive network of coronary blood vessels. The left and right **coronary arteries** originate at the base of the ascending aorta, within the aortic sinus, as the first branches of this vessel. Blood pressure here is the highest found anywhere in the systemic circuit, and this pressure ensures a continuous flow of blood to meet the demands of active cardiac muscle tissue.

THE RIGHT CORONARY ARTERY [FIGURE 21.9]

The **right coronary artery** follows the coronary sulcus around the heart, and supplies blood to (1) the right atrium, (2) portions of both ventricles, and (3) portions of the conducting system of the heart. The major branches are shown in (Figure 21.9●).

1. *Atrial branches*: As it curves across the anterior surface of the heart, the right coronary artery gives rise to **atrial arteries** that supply the myocardium of the right atrium.
2. *Ventricular branches*: Near the right border of the heart, the right coronary artery usually gives rise to one or more **marginal arteries** that extend across the ventricular surface. It then continues across the posterior surface of the heart, supplying the **posterior interventricular artery**, which runs toward the apex within the posterior interventricular sulcus. This branch supplies blood to the interventricular septum and adjacent portions of the ventricles.
3. *Branches to the conducting system*: A small branch near the base of the right coronary artery penetrates the atrial wall to reach the *sinoatrial (SA) node*, also known as the *cardiac pacemaker*. A small artery to the *atrioventricular (AV) node*, another part of the conducting system of the heart, originates from the right coronary artery near the posterior interventricular branch. These nodes and their role in the regulation of the heartbeat will be the topic of a later section.

THE LEFT CORONARY ARTERY [FIGURE 21.9]

The **left coronary artery** supplies blood to the left ventricle, left atrium, and the interventricular septum. As it reaches the anterior surface of the heart, it gives rise to a *circumflex branch* and an *anterior interventricular branch* (Figure 21.9●). The **circumflex artery** curves to the left within the coronary sulcus, eventually reaching the posterior surface of the heart, where it meets and fuses with small branches of the right coronary artery. The much larger **anterior interventricular artery**, or *left anterior descending artery*, runs along the anterior surface within the anterior interventricular sulcus. This artery supplies the anterior ventricular myocardium and the interventricular septum. Small branches from the anterior interventricular branch of the left coronary artery are continuous with those of the posterior interventricular branch of the right coronary artery. Such interconnections between arteries are called **anastomoses** (a-nas-tō-MŌ-ses; *anastomosis*, outlet). Because the arteries are interconnected in this way, the blood supply to the ventricular muscle remains relatively constant, regardless of pressure fluctuations within the left and right coronary arteries.

THE CARDIAC VEINS [FIGURE 21.9a,b]

The **great cardiac vein** and **middle cardiac vein** collect blood from smaller veins draining the myocardial capillaries; they deliver this venous blood to the **coronary sinus**, a large thin-walled vein that lies in the posterior

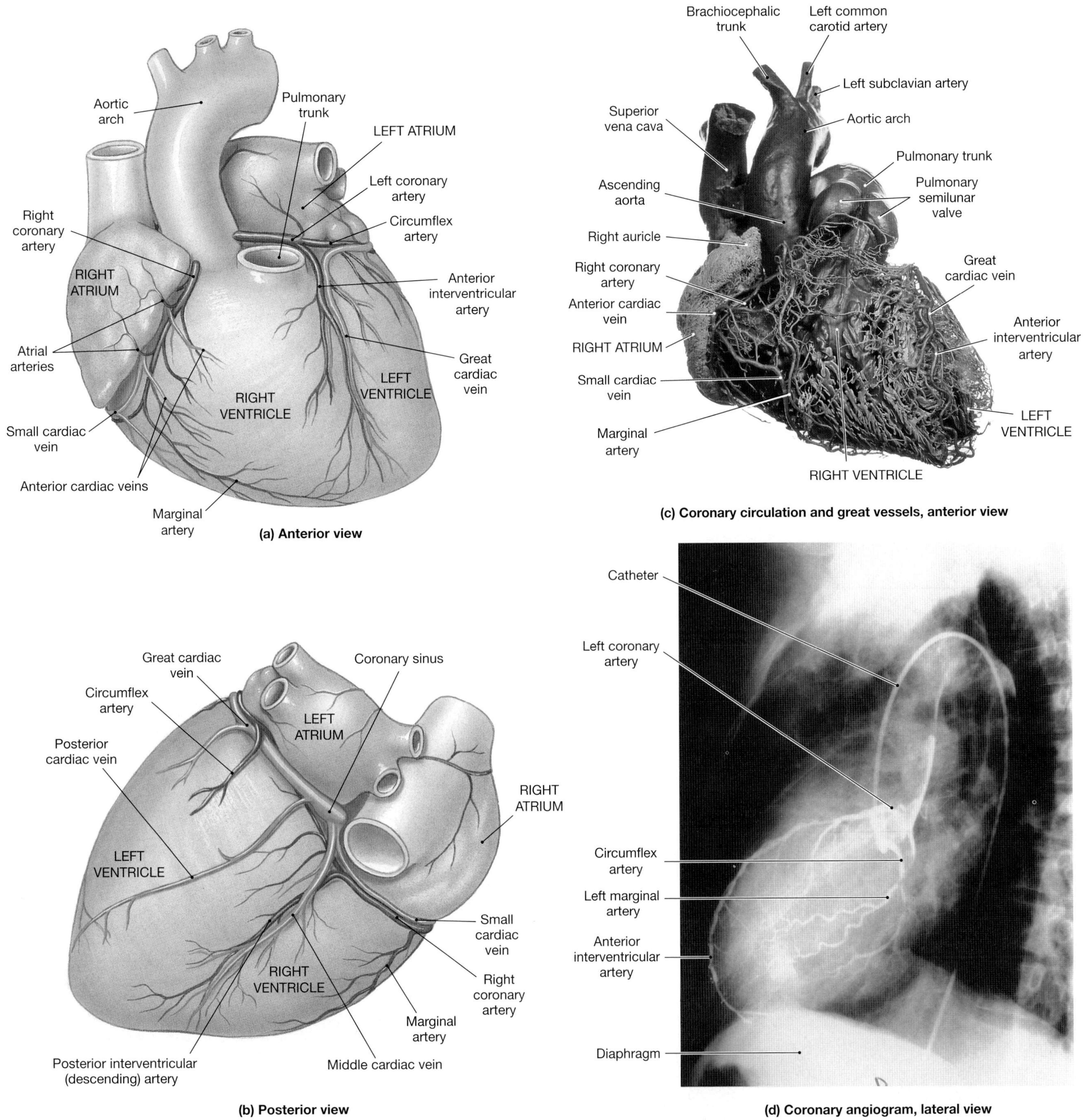

(a) Anterior view

(c) Coronary circulation and great vessels, anterior view

(b) Posterior view

(d) Coronary angiogram, lateral view

FIGURE 21.9 **Coronary Circulation**

(**a**) Coronary vessels supplying the anterior surface of the heart. (**b**) Coronary vessels supplying the posterior surface of the heart. (**c**) A cast of the coronary vessels, showing the complexity and extent of the coronary circulation. Coronary vessels are also seen in *Figure 21.5*. (**d**) Coronary angiogram showing the coronary arteries, left lateral projection.

CLINICAL DISCUSSION

Coronary Artery Disease

The term **coronary artery disease (CAD)** refers to degenerative changes in the coronary circulation. Cardiac muscle fibers need a constant supply of oxygen and nutrients, and any reduction in coronary circulation produces a corresponding reduction in cardiac performance. Such reduced circulatory supply, known as **coronary ischemia** (is-KĒ-mē-a), usually results from partial or complete blockage of the coronary arteries. The usual cause is the formation of a fatty deposit, or *plaque*, in the wall of a coronary vessel. The plaque, or an associated thrombus, then narrows the passageway and reduces or stops blood flow. Spasms in the smooth muscles of the vessel wall can further decrease blood flow or even stop it altogether. A variety of imaging procedures can be used to visualize coronary circulation, including **coronary angiography** (Figure 21.9d●) and DSA (Digital Subtraction Angiography) scans (Figure 21.10a,b●).

One of the first symptoms of CAD is often **angina pectoris** (an-JĪ-na PEK-tor-is; *angina*, pain spasm + *pectoris*, of the chest). In the most common form of angina, temporary insufficiency of oxygen delivery and ischemia develop when the workload of the heart increases. Although the individual may feel comfortable at rest, any unusual exertion or emotional stress can produce a sensation of pressure, chest constriction, and pain that may radiate from the sternal area to the arms, back, and neck.

Angina can often be controlled by a combination of drug treatment and changes in lifestyle. Lifestyle changes to combat angina include (1) limiting activities known to trigger angina attacks, such as strenuous exercise, and avoiding stressful situations while doing moderate, regular exercise within tolerated limits; (2) stopping smoking; and (3) modifying the diet to lower fat consumption. Medications useful for controlling angina include drugs that block sympathetic stimulation (*propranolol* or *metoprolol*); vasodilators, such as *nitroglycerin* (nī-trō-GLIS-er-in) or *atrial natriuretic peptide*; or drugs that block calcium movement into the cardiac muscle fibers (*calcium channel blockers*). Drugs that lower cholesterol and lipid levels in the blood may prevent plaque growth or even cause plaque regression.

Angina can also be treated surgically. A single, soft plaque may be removed with the aid of a slender, elongated **catheter** (KATH-e-ter). The catheter, a small-diameter tube, is inserted into a large artery (generally the femoral) and guided to the plaque in a coronary artery. A variety of different surgical tools can be slid into the catheter, and the plaque can then be removed with laser beams or chewed to pieces by a miniature version of the Rotor-Rooter™ machine. Debris created during plaque destruction is sucked up by the catheter, preventing blockage of smaller vessels.

In **balloon angioplasty** (AN-jē-ō-plas-tē; *angeion*, vessel), the catheter tip contains an inflatable balloon. Once in position, the balloon is inflated, compressing the plaque against the vessel walls (Figure 21.10c●). This procedure works best on small (under 10 mm), soft plaques. Because *restenosis*, or repeated narrowing, may develop, metal *stents*, or sleeves, can often be put into the artery to hold it open.

Coronary bypass surgery involves taking a small section from either a small artery (often the *internal thoracic artery*) or a peripheral vein, such as a branch of the femoral vein, and using it to create a detour around the obstructed portion of a coronary artery. As many as four coronary arteries can be rerouted this way during a single operation. The procedures are named according to the number of vessels repaired, so one speaks of single, double, triple, or quadruple coronary bypass operations. Current recommendations are that coronary bypass surgery should be reserved for cases of severe angina that do not respond to other treatment.

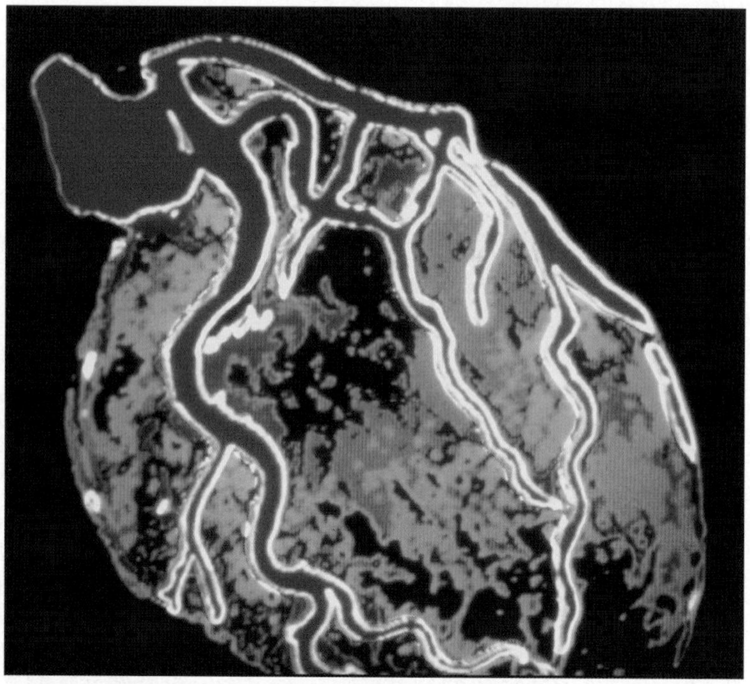

(a) Normal circulation

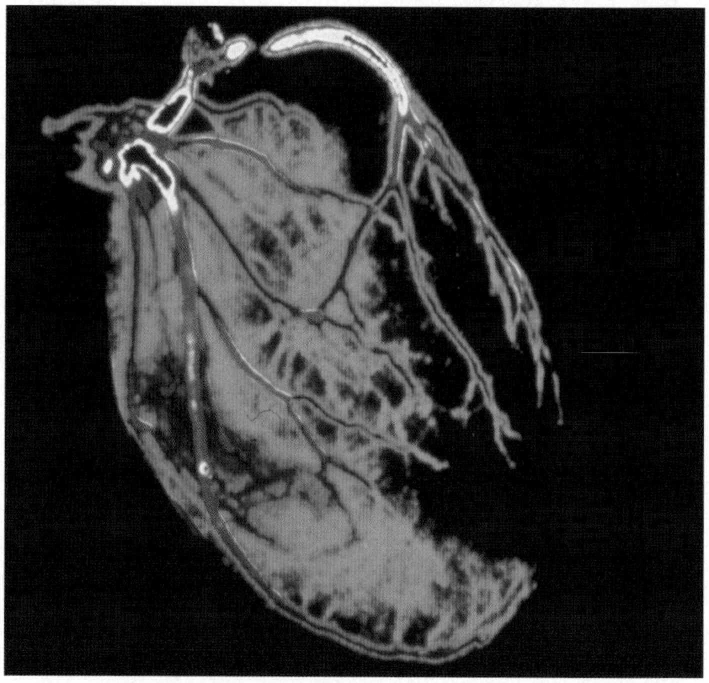

(b) Restricted circulation

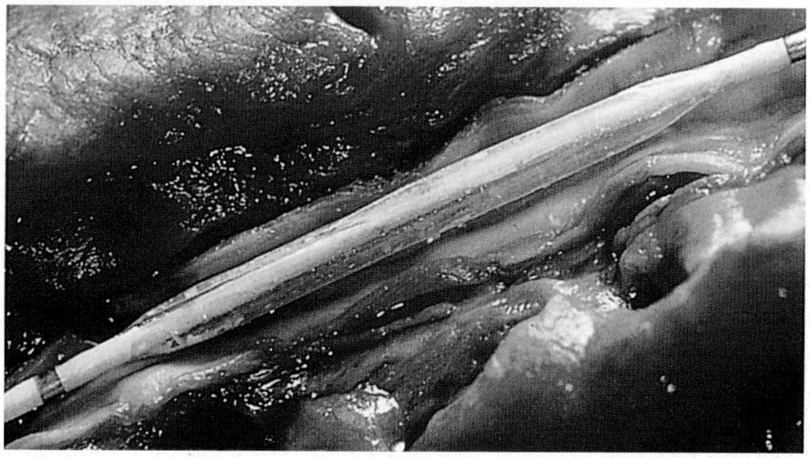

(c)

FIGURE 21.10 **Coronary Circulation and Clinical Testing**

(a) A color-enhanced DSA image of a healthy heart. The ventricular walls have an extensive circulatory supply. (The atria are not shown.) **(b)** A color-enhanced DSA image of a damaged heart. Most of the ventricular myocardium is deprived of circulation. **(c)** Balloon angioplasty can sometimes be used to remove a circulatory blockage. The catheter is guided through the coronary arteries to the site of blockage and inflated to press the soft plaque against the vessel wall.

portion of the coronary sulcus (Figure 21.9a,b●). As noted earlier in the chapter (p. 556), the coronary sinus communicates with the right atrium inferior to the opening of the inferior vena cava.

Other cardiac veins that empty into the great cardiac vein or the coronary sinus include (1) the **posterior cardiac vein**, draining the area served by the circumflex artery; (2) the **middle cardiac vein**, draining the area supplied by the posterior interventricular artery; and (3) the **small cardiac vein**, which receives blood from the posterior surfaces of the right atrium and ventricle. The **anterior cardiac veins**, which drain the anterior surface of the right ventricle, empty directly into the right atrium.

✓ CONCEPT CHECK

- What would happen if there were no valves between the atria and ventricles?
- What three major veins open into the right atrium?
- Trace the path of blood from the left ventricle to the respiratory surfaces of the lungs.
- What prevents the AV valves from opening back into the atria?
- What causes the familiar "lubb-dupp" heart sounds heard with a stethoscope? Please be specific.

The Cardiac Cycle [FIGURE 21.11]

The period between the start of one heartbeat and the beginning of the next is a single cardiac cycle. The cardiac cycle therefore includes alternate periods of contraction and relaxation. For any one chamber in the heart, the cardiac cycle can be divided into two phases. During contraction, or **systole** (SIS-tō-lē), a chamber ejects blood either into another heart chamber or into an arterial trunk. Systole is followed by the second phase, one of relaxation, or **diastole** (dī-AS-tō-lē). During diastole a chamber fills with blood and prepares for the start of the next cardiac cycle. The events of the cardiac cycle are summarized in Figure 21.11●.

■ The Coordination of Cardiac Contractions [FIGURE 21.12]

The function of any pump is to develop pressure and move a particular volume of fluid in a specific direction at an acceptable speed. The heart works in cycles of contraction and relaxation, and the pressure within each chamber alternately rises and falls. The AV and semilunar valves help to ensure a one-way flow of blood despite these pressure oscillations. Blood will flow out of an atrium only as long as the AV valve is open and atrial pressure exceeds ventricular pressure. Similarly, blood will flow from a ventricle into an arterial trunk only as long as the semilunar valve is open and ventricular pressure exceeds the arterial pressure. The proper functioning of the heart

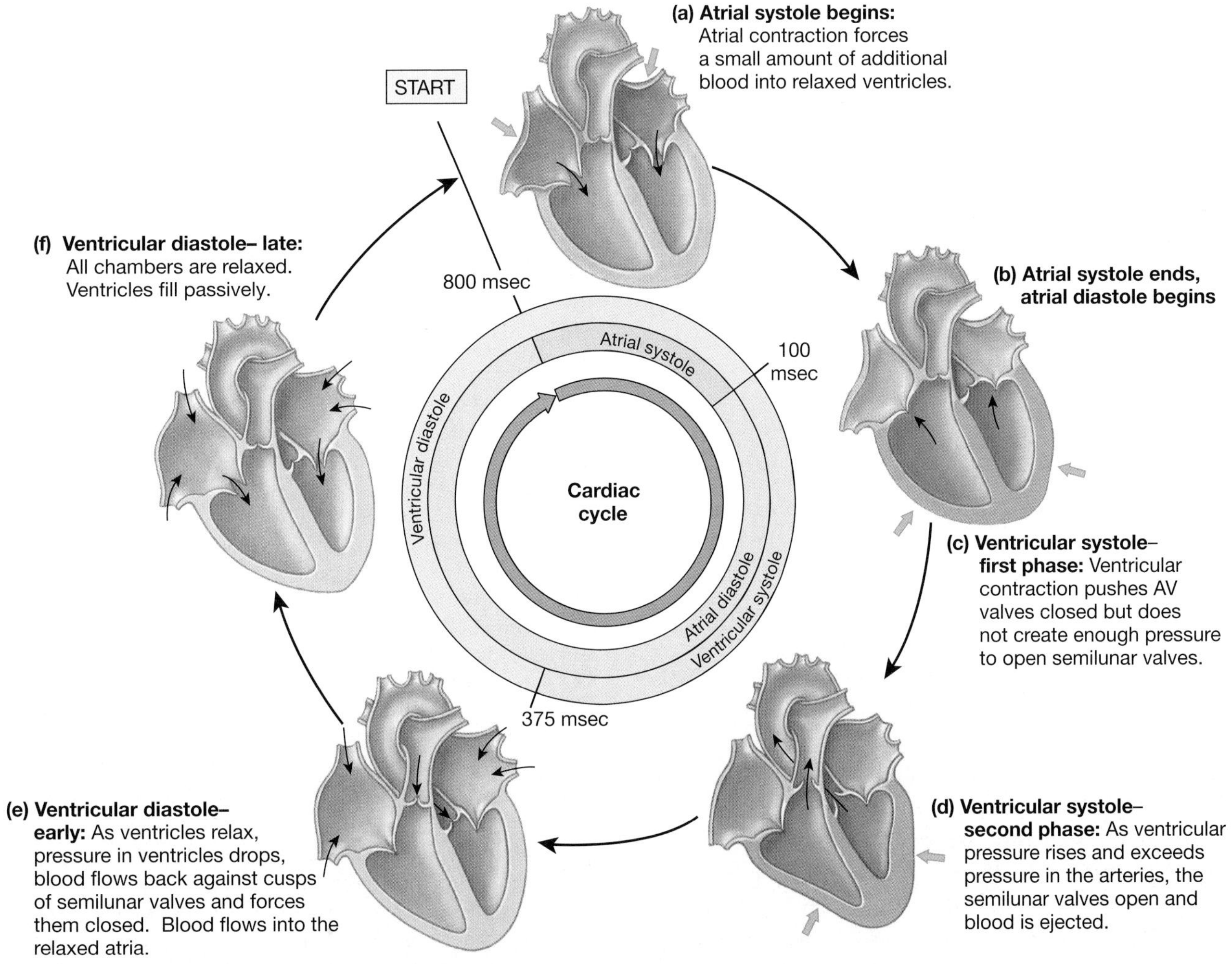

FIGURE 21.11 **The Cardiac Cycle**

Black arrows indicate movement of blood or valves; green arrows indicate myocardial contraction.

thus depends on proper timing of atrial and ventricular contractions. The elaborate pacemaking and conduction systems normally provide the required timing.

Unlike skeletal muscle, cardiac muscle tissue contracts on its own, in the absence of neural or hormonal stimulation. This inherent ability to generate and conduct impulses is called *automaticity*, or *autorhythmicity*. (Automaticity is also characteristic of some types of smooth muscle tissue discussed in Chapter 25). Neural or hormonal stimuli can alter the basic rhythm of contraction, but even a heart removed for a heart transplant will continue to beat unless steps are taken to prevent it.

Each contraction follows a precise sequence: The atria contract first and then the ventricles. If the contractions follow another sequence, the normal pattern of blood flow is disturbed. For example, if the atria and ventricles contract at the same time, the closing of the AV valves prevents blood flow between the atria and ventricles. Cardiac contractions are coordinated by specialized *conducting cells*, cardiac muscle cells that are incapable of undergoing powerful contractions. There are two distinct populations of these cells. **Nodal cells** are responsible for establishing the rate of cardiac contraction, and **conducting fibers** distribute the contractile stimulus to the general myocardium (Figure 21.12●).

The Sinoatrial and Atrioventricular Nodes [FIGURE 21.12]

Nodal cells are unusual because their cell membranes spontaneously depolarize to threshold. p. 346 Nodal cells are electrically coupled to one another, to conducting fibers, and to normal cardiac muscle cells. As a result, when an action potential appears in a nodal cell, it sweeps through the conducting system, reaching all of the cardiac muscle tissue and causing a contraction. In this way, nodal cells determine the heart rate.

Not all nodal cells depolarize at the same rate, and the normal rate of contraction is established by the nodal cells that reach threshold first; the impulse they produce will bring all other nodal cells to threshold. These rapidly depolarizing cells are called **pacemaker cells**. They are found in the **sinoatrial** (sī-nō-Ā-trē-al) **node (SA node)**, or **cardiac pacemaker**. The SA node is embedded in the posterior wall of the right atrium, near the entrance of the superior vena cava (Figure 21.12●). Isolated pacemaker cells depolarize rapidly and spontaneously, generating 80–100 action potentials per minute.

Each time the SA node generates an impulse, it produces a heartbeat, so theoretically the resting heart rate would be 80–100 beats per minute (bpm). However, any factor that changes either the resting potential or the rate of spontaneous depolarization at the SA node will alter the heart rate. For example, nodal cell activity is affected by the activity of the autonomic nervous system. When acetylcholine (ACh) is released by parasympathetic neurons, it slows the rate of spontaneous depolarization and lowers the heart rate. In contrast, when norepinephrine (NE) is released by sympathetic neurons, the rate of depolarization increases, and the heart rate accelerates. Under normal resting conditions, parasympathetic activity reduces the heart rate from the inherent nodal rate of 80–100 impulses per minute to a more leisurely 70–80 beats per minute.

A number of clinical problems are the result of abnormal pacemaker function. **Bradycardia** (brā-dē-KAR-dē-a; *bradys*, slow) is the term used to indicate a heart rate that is slower than normal, whereas a faster-than-normal heart rate is termed **tachycardia** (tak-ē-KAR-dē-a; *tachys*, swift). Both terms are relative, and in clinical practice the definition varies depending on the normal resting heart rate and conditioning of the individual.

THE CONDUCTING SYSTEM OF THE HEART [FIGURE 21.12]

The cells of the SA node are electrically connected to those of the larger **atrioventricular** (ā-trē-ō-ven-TRIK-ū-lar) **node (AV node)** through conducting fibers in the atrial walls (Figure 21.12●). As the signal for contraction passes from the SA node to the AV node via the *internodal pathways*, the conducting fibers also pass the contractile stimulus to cardiac muscle cells of both atria. The action potential then spreads across the atrial surfaces through cell-to-cell contact. The stimulus affects only the atria, because the fibrous skeleton electrically isolates the atrial myocardium from the ventricular myocardium.

The AV node sits within the floor of the right atrium near the opening of the coronary sinus. Due to differences in the shape of the nodal cells, the impulse slows as it passes through the AV node. From there, the impulse travels to the **AV bundle**, also known as the *bundle of His* (HISS). This rather massive bundle of conducting fibers travels

CLINICAL BRIEF

Myocardial Infarctions

In a **myocardial** (mī-ō-KAR-dē-al) **infarction (MI)**, or *heart attack*, the coronary circulation becomes blocked and the cardiac muscle cells die from lack of oxygen. The affected tissue then degenerates, creating a nonfunctional area known as an *infarct*. Heart attacks most often result from severe coronary artery disease. The consequences depend on the site and nature of the circulatory blockage. If it occurs near the base of one of the coronary arteries, the damage will be widespread and the heart will probably stop beating. If the blockage involves one of the smaller arterial branches, the individual may survive the immediate crisis, but there are many potential complications, all unpleasant. As scar tissue forms in the damaged area, the heartbeat may become irregular and less effective as a pump, and other vessels can become constricted, creating additional cardiovascular problems such as angina.

Myocardial infarctions are most often associated with fixed blockages, such as those seen in CAD. When the crisis develops because of *thrombus* (stationary clot) formation at an area of plaque, the condition is called **coronary thrombosis**. A vessel already narrowed by plaque formation may also become blocked by a sudden spasm in the smooth muscles of the vascular wall. The individual then experiences intense pain, similar to that of an angina attack but persisting even at rest.

Roughly 25% of MI patients die before obtaining medical assistance, and 65% of MI deaths among those under age 50 occur within an hour after the initial infarct. The goals of treatment are to limit the size of the infarct and prevent additional complications by preventing irregular contractions, improving circulation with vasodilators, providing additional oxygen, reducing the cardiac workload, and, if possible, eliminating the cause of the circulatory blockage. Anticoagulants may help prevent the formation of additional thrombi; chewing an aspirin early in the course of an MI is helpful, and clot-dissolving enzymes may reduce the extent of the damage if they are administered within 6 hours after the MI has occurred.

There are roughly 1.3 million MIs in the United States each year, and half of the victims die within a year of the incident. A number of factors have been identified that increase the risk of a heart attack. They include smoking, high blood pressure, high blood cholesterol levels, diabetes, and obesity. There are also hereditary factors that may predispose an individual to coronary artery disease. The presence of two risk factors more than doubles the risk, so eliminating as many risk factors as possible will improve one's chances of preventing or surviving a heart attack. For example, changes in eating habits to limit dietary cholesterol, exercise to lower weight, and seeking treatment for high blood pressure are relatively easy steps in the right direction; and the benefits are considerable.

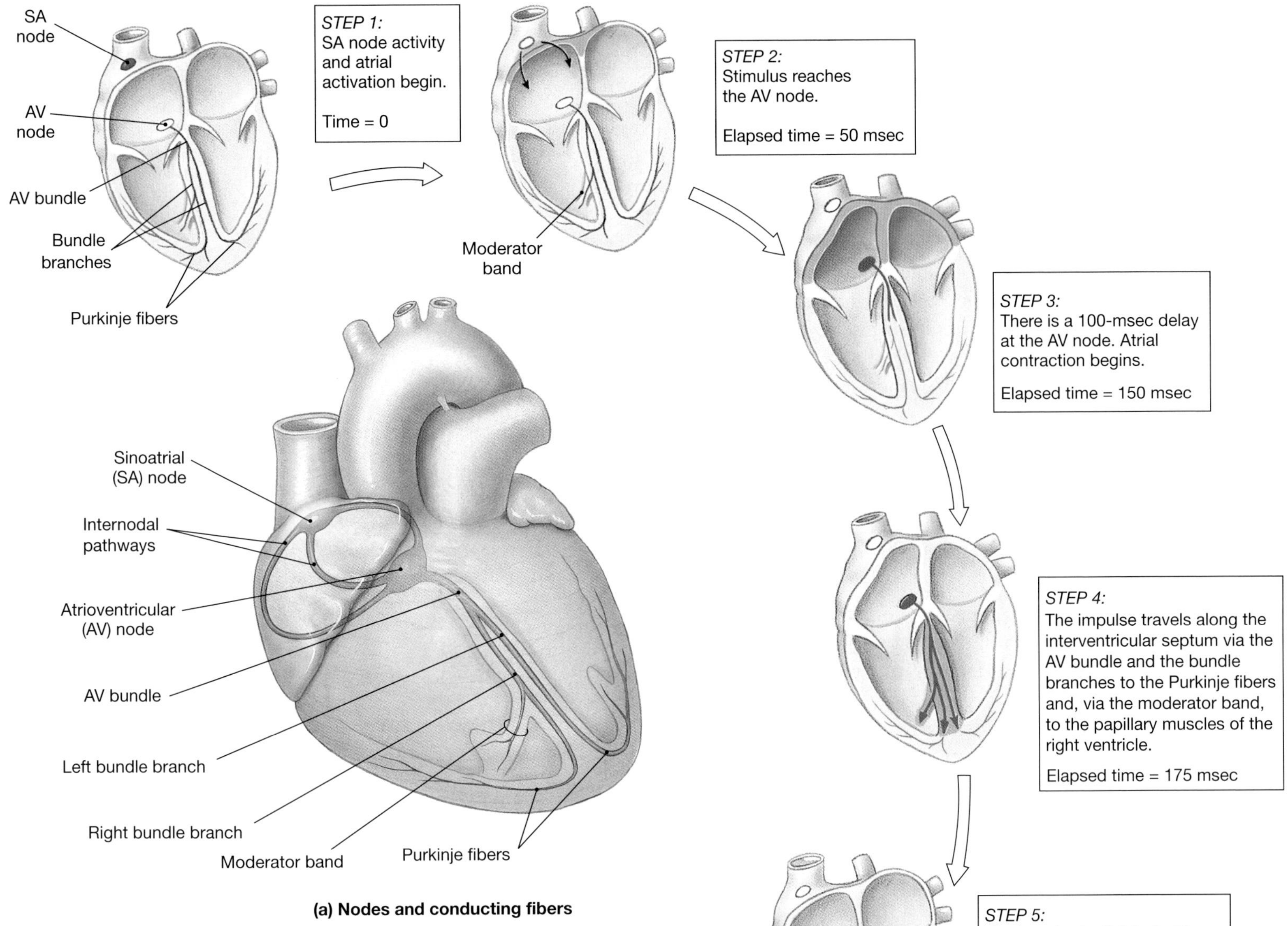

SA node activity and atrial activation begin.

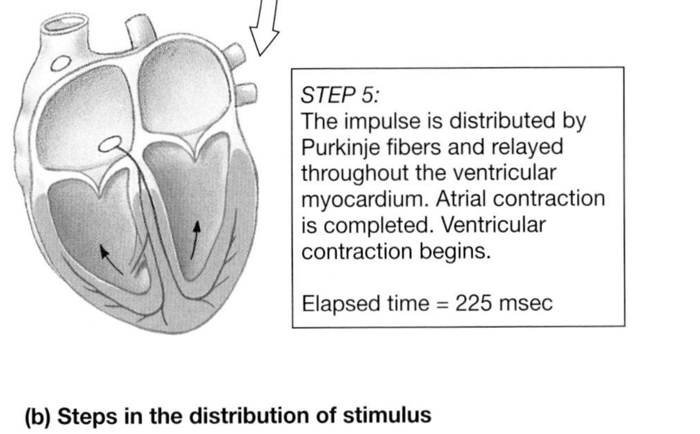

STEP 5:
The impulse is distributed by Purkinje fibers and relayed throughout the ventricular myocardium. Atrial contraction is completed. Ventricular contraction begins.
Elapsed time = 225 msec
(b) Steps in the distribution of stimulus

FIGURE 21.12 **The Conducting System of the Heart**

(**a**) The stimulus for contraction is generated by pacemaker cells at the SA node. From there, impulses follow three different paths through the atrial walls to reach the AV node. After a brief delay, the impulses are conducted to the bundle of His (AV bundle), and then on to the bundle branches, the Purkinje cells, and the ventricular myocardial cells. (**b**) The movement of the contractile stimulus through the heart is shown in *Steps 1 to 5*.

along the interventricular septum a short distance before dividing into a **right bundle branch** and a **left bundle branch** that extend toward the apex and then radiate across the inner surfaces of both ventricles. At this point, **Purkinje** (pur-KIN-jē) **cells** (*Purkinje fibers*) convey the impulses very rapidly to the contractile cells of the ventricular myocardium. The conducting fibers of the moderator band relay the stimulus to the papillary muscles, which tense the chordae tendineae before the ventricles contract.

The stimulus for a contraction is generated at the SA node, and the anatomical relationships among the contracting cells, the nodal cells, and the conducting fibers distribute the impulse so that (1) the atria contract together, before the ventricles, and (2) the ventricles contract together in a wave that begins at the apex and spreads toward the base. When the ventricles contract in this way, blood is pushed toward the base of the heart and out into the aortic and pulmonary trunks. *Problems with Pacemaker Function p. 801*

✓ CONCEPT CHECK

- If the cells of the SA node were not functioning, what effect would this have on heart rate?
- If norepinephrine is released at the heart, what is the effect on heart rate?
- How do nodal cells coordinate cardiac muscle contractions?

■ The Electrocardiogram (ECG) [FIGURES 21.12/ 21.13]

The electrical events associated with the depolarization and repolarization of the heart are powerful enough to be detected by electrodes placed on the body surface. A recording of these electrical activities constitutes an **electrocardiogram** (ē-lek-trō-KAR-dē-ō-gram), also called an **ECG** or **EKG**. During each cardiac cycle, a wave of depolarization radiates through the atria, reaches the AV node, travels down the interventricular septum to the apex, turns, and spreads through the ventricular myocardium toward the base (Figure 21.12●). This electrical activity can be monitored from the body surface. By comparing the information obtained from electrodes placed at different locations, one can monitor the performance of specific nodal, conducting, and contractile components. For example, when a portion of the heart has been damaged, as after a MI, these cardiac muscle cells can no longer conduct action potentials, so an ECG will reveal an abnormal pattern of electrical conduction.

The appearance of the ECG tracing varies, depending on the placement of the monitoring electrodes, or *leads*. Figure 21.13● shows the important features of a representative electrocardiogram. The **P wave** accompanies the depolarization of the atria. The **QRS complex** appears as the ventricles depolarize. This electrical signal is relatively strong because the mass of the ventricular muscle is much larger than that of the atria. The smaller **T wave** indicates ventricular repolarization. You do not see a deflection corresponding to atrial repolarization because it occurs while the ventricles are depolarizing, and the electrical events are masked by the QRS complex.

Analyzing an ECG usually focuses on the size and duration of the P wave, the QRS complex, and the ST segment (between the QRS and T waves). For example, a smaller-than-normal electrical signal may mean that the mass of the heart muscle has decreased; in contrast, excessively large signals may mean that the heart muscle has become enlarged. Changes in the size and shape of the T wave may indicate a condition that slows ventricular repolarization.

Despite the variety of sophisticated equipment available to assess or visualize cardiac function, in the vast majority of cases, the electrocardiogram provides the first and most important diagnostic information. ECG analysis is especially useful in detecting and diagnosing **cardiac arrhythmias** (a-RITH-mē-az), abnormal patterns of cardiac activity. Clinical problems appear when the arrhythmias reduce the pumping efficiency of the heart. Serious arrhythmias may indicate damage to the myocardial musculature, injuries to the pacemakers or conduction pathways, or other factors.

■ Autonomic Control of Heart Rate [FIGURE 21.14]

The basic heart rate is established by the pacemaker cells of the SA node, but this intrinsic rate can be modified by the autonomic nervous system (ANS). The sympathetic and parasympathetic divisions of the ANS provide innervation to the heart through the cardiac plexus (anatomical details were presented in Chapter 17). ∞ p. 459 Both ANS divisions innervate the SA and AV nodes as well as the atrial and ventricular cardiac muscle cells and smooth muscle in the walls of the cardiac blood vessels (Figure 21.14●).

The effects of NE and ACh on nodal tissues were detailed earlier in this chapter, and may be summarized as:

- NE release produces an increase in both heart rate and force of contractions through the stimulation of beta receptors on nodal cells and contractile cells;

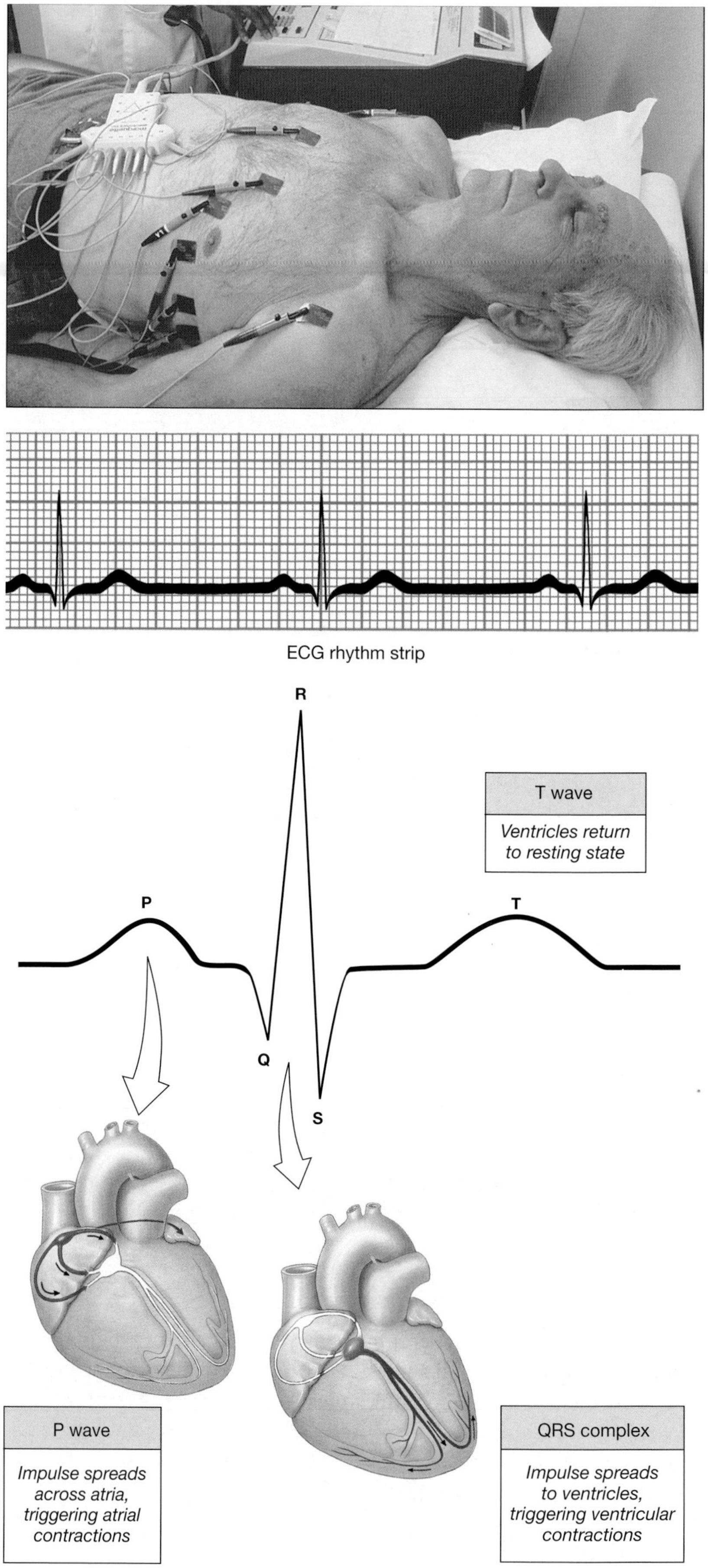

FIGURE 21.13 **AN ELECTROCARDIOGRAM**

An ECG printout is a strip of graph paper containing a record of the electrical events monitored by electrodes attached to the body surface. The photo shows electrode locations for a standard ECG. The enlarged section indicates the major components of the ECG.

- ACh release produces a decrease in both heart rate and force of contractions through the stimulation of muscarinic receptors of nodal cells and contractile cells.

The cardiac centers of the medulla oblongata contain the autonomic centers for cardiac control. Stimulation of the **cardioacceleratory center** activates the necessary sympathetic neurons; the nearby **cardioinhibitory center** governs the activities of the parasympathetic neurons. The cardiac centers receive inputs from higher centers, especially from the parasympathetic and sympathetic headquarters in the hypothalamus. (These centers were described in Chapter 15.) p. 400

Information concerning the status of the cardiovascular system arrives at the cardiac centers from visceral sensory fibers that monitor baroreceptors sensitive to blood pressure and chemoreceptors sensitive to dissolved gas concentrations. These receptors are innervated by the glossopharyngeal (N IX) and vagus (N X) nerves. In response to this information, the cardiac centers adjust the cardiac performance to maintain adequate circulation to vital organs, such as the brain. These centers respond very quickly to changes in blood pressure and to the amount of dissolved oxygen and carbon dioxide in arterial blood. For example, a drop in blood pressure or an increase in carbon dioxide concentration usually indicates that the heart must work harder to meet the demands of peripheral tissues. The cardiac centers then respond by increasing the heart rate and force of contraction by activating the sympathetic nervous system. *Monitoring the Living Heart p. 801*

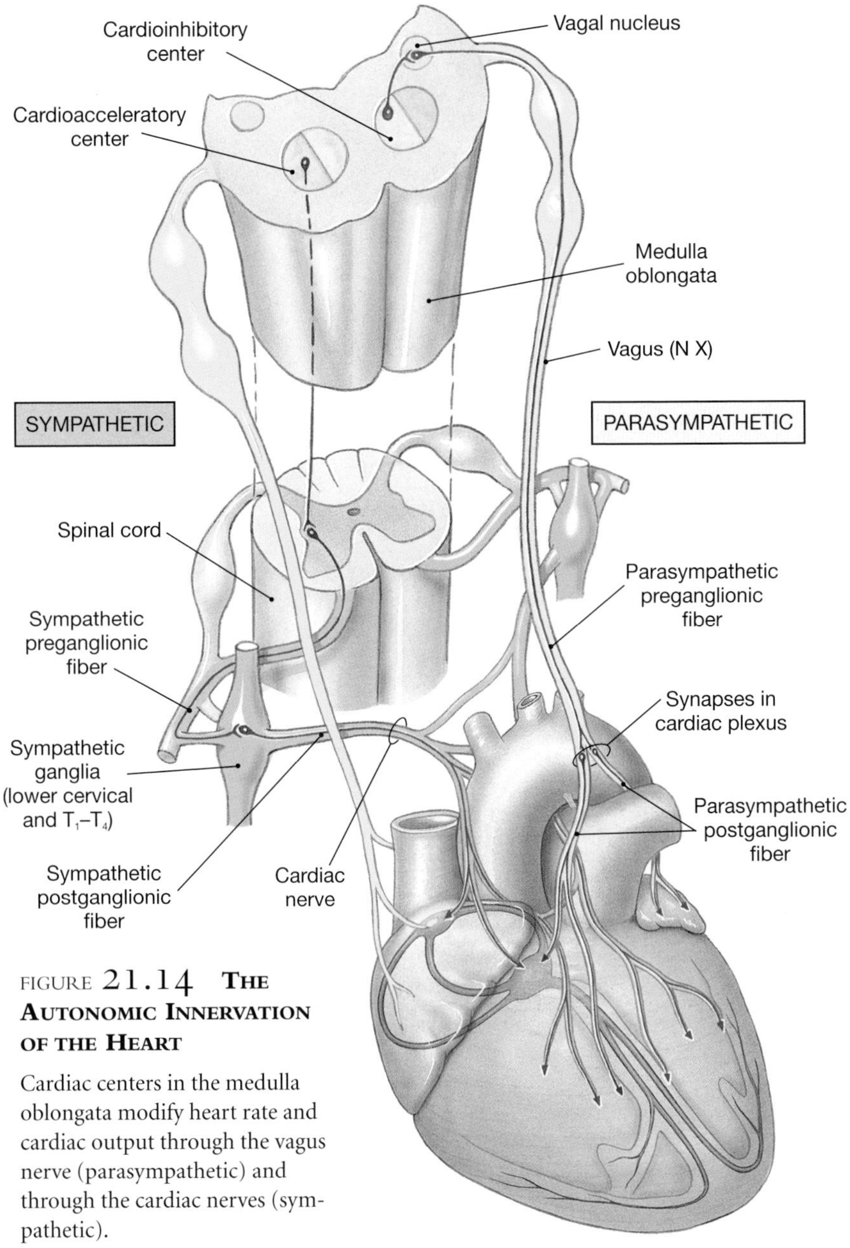

FIGURE 21.14 **The Autonomic Innervation of the Heart**

Cardiac centers in the medulla oblongata modify heart rate and cardiac output through the vagus nerve (parasympathetic) and through the cardiac nerves (sympathetic).

✓ CONCEPT CHECK

- If the pressure in the ventricle remained equal to that in the atrium, what would happen to the blood flow?
- What might be The cause of an abnormal P wave on an ECG?

RELATED CLINICAL TERMS

angina pectoris (an-JĪ-na PEK-tor-is): A condition in which exertion or stress can produce severe chest pain, resulting from temporary circulatory insufficiency and ischemia when the heart's workload increases. *p. 562*

bradycardia (brā-dē-KAR-dē-a): A heart rate that is slower than normal. *p. 564*

cardiac arrhythmias (a-RITH-mē-az): Abnormal patterns of cardiac contraction. *p. 566*

cardiac tamponade: A condition resulting from pericardial irritation and inflammation, in which fluid collects in the pericardial sac and restricts cardiac output. *Infection and Inflammation of the Heart p. 800*

cardiomyopathies (kar-dē-ō-mī-OP-a-thēz): A group of diseases characterized by the progressive, irreversible degeneration of the myocardium. *The Cardiomyopathies p. 800*

carditis (kar-DĪ-tis): A general term indicating inflammation of the heart. *Infection and Inflammation of the Heart p. 800*

coronary artery disease (CAD): Degenerative changes in the coronary circulation. *p. 562*

coronary thrombosis: A blockage due to the formation of a clot (thrombus) at a plaque in a coronary artery. *p. 564*

heart failure: A condition in which the heart weakens and peripheral tissues suffer from oxygen and nutrient deprivation. *Rheumatic Heart Disease and Valvular Stenosis p. 801*

heart murmur: A rushing, gurgling sound caused by blood regurgitation back through faulty heart valves. *p. 558*

mitral valve prolapse: A condition in which the mitral valve cusps do not close properly because of abnormally long (or short) chordae tendineae or malfunctioning papillary muscles. *p. 558*

myocardial (mī-ō-KAR-dē-al) **infarction (MI):** A condition in which the coronary circulation becomes blocked and the cardiac muscle cells die from oxygen starvation; also called a heart attack. *p. 564*

rheumatic heart disease (RHD): A disorder in which the heart valves become thickened and stiffen into a partially closed position, affecting the efficiency of the pumping action of the heart. *Rheumatic Heart Disease and Valvular Stenosis p. 801*

tachycardia (tak-e-KAR-dē-a): A heart rate that is faster than normal. *p. 564*

valvular stenosis (ste-NŌ-sis): A condition in which the opening between the heart valves is narrower than normal. *Valvular Heart Disease p. 800*

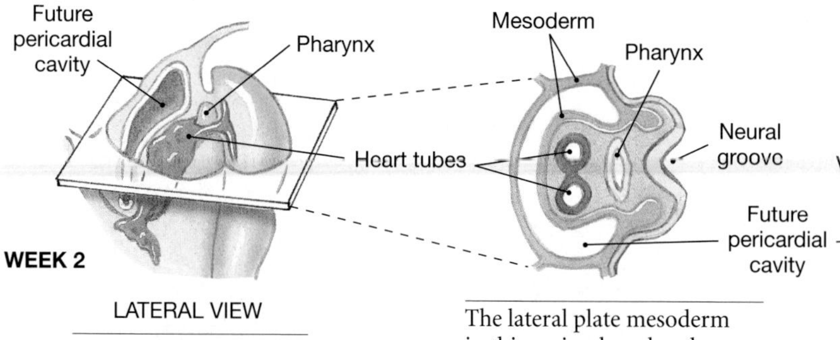

During the second developmental week, the heart consists of a pair of thin-walled, muscular tubes beneath the floor of the pharynx.

The lateral plate mesoderm in this region has already split into parietal and visceral layers, creating a space that will eventually form the pericardial cavity.

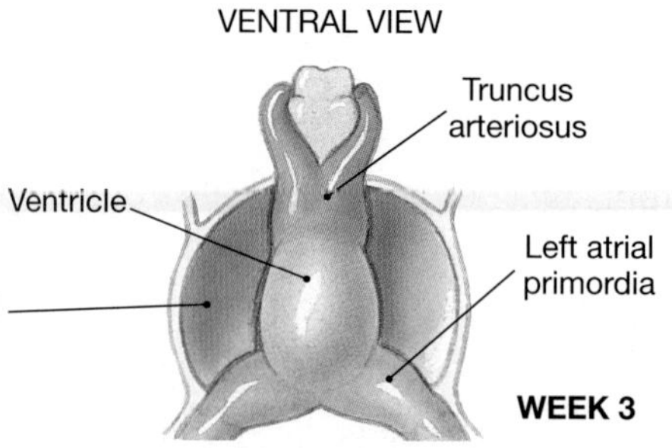

By the third week, the heart is pumping, and circulating blood. The cardiac tubes have fused, producing a heart with a single central chamber. Two large veins bring blood to the heart, and a single large artery, **the truncus arteriosus**, carries blood to the general circulation.

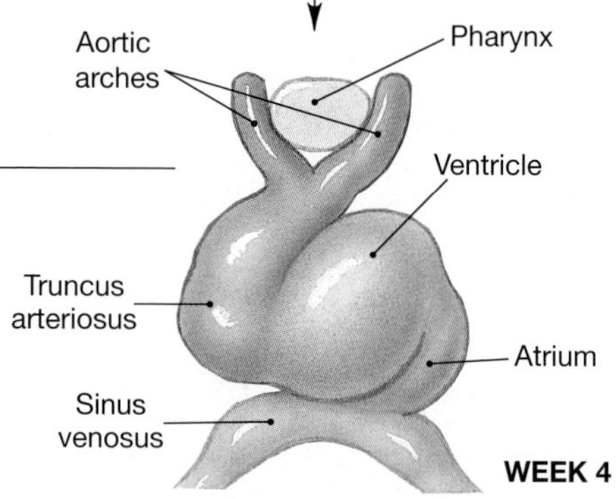

The heart elongates as the embryo grows larger. It curves back upon itself, forming an S-curve that gradually becomes more pronounced. The atrial and ventricular regions already differ in thickness.

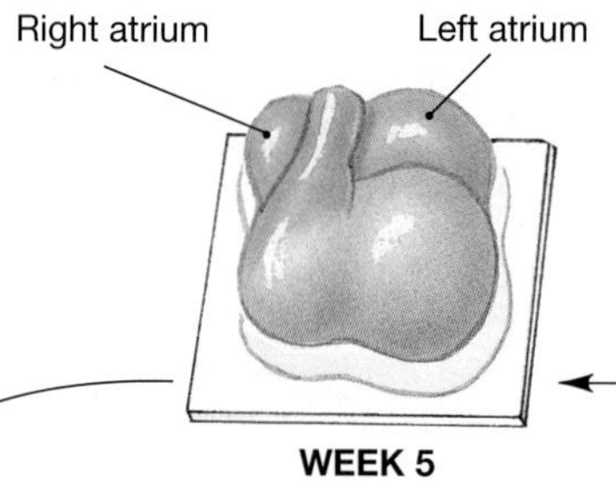

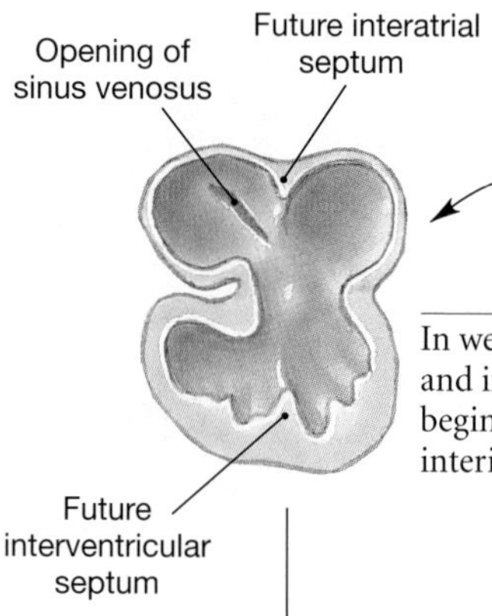

In week 5, the interatrial and interventricular septa begin to subdivide the interior of the heart.

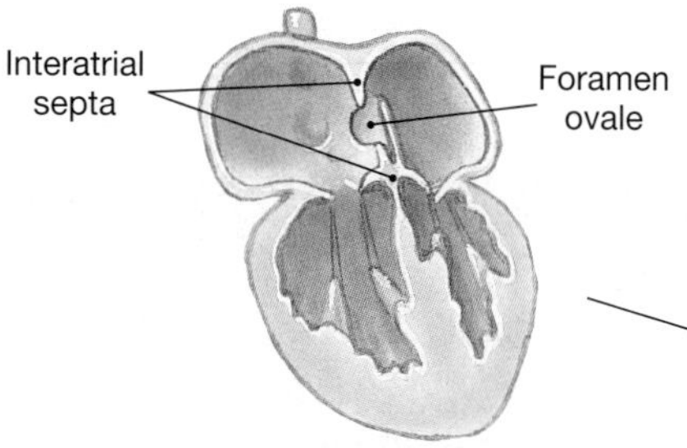

Two interatrial septa develop, one overlapping the other. A gap between the two, called the **foramen ovale**, permits blood flow from the right atrium to the left atrium. Back flow from left to right is prevented by a flap that acts as a one-way valve. Until birth, this atrial short circuit diverts blood from the pulmonary circuit.

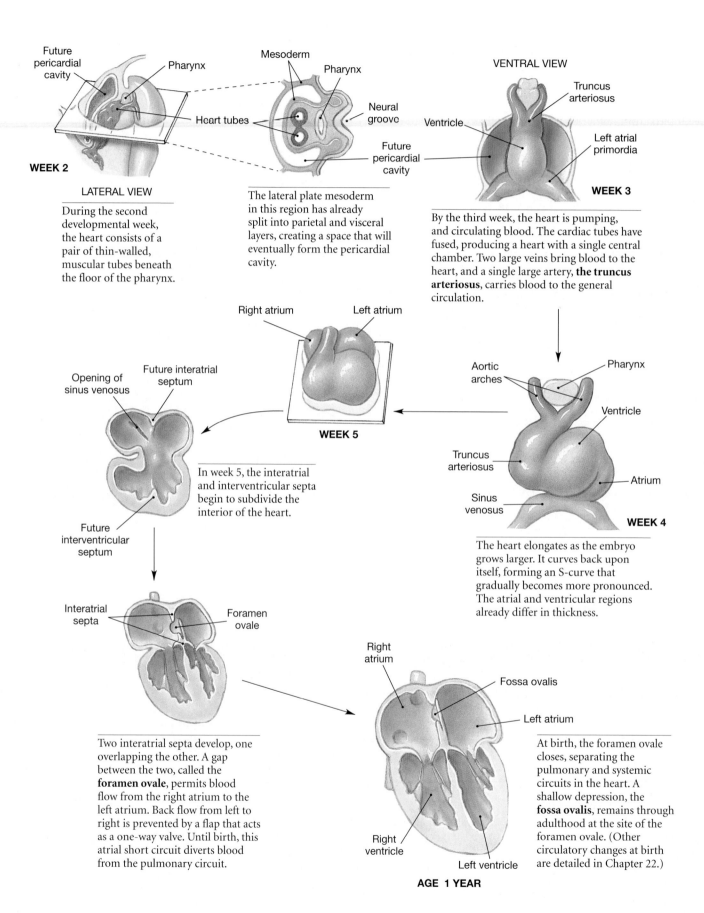

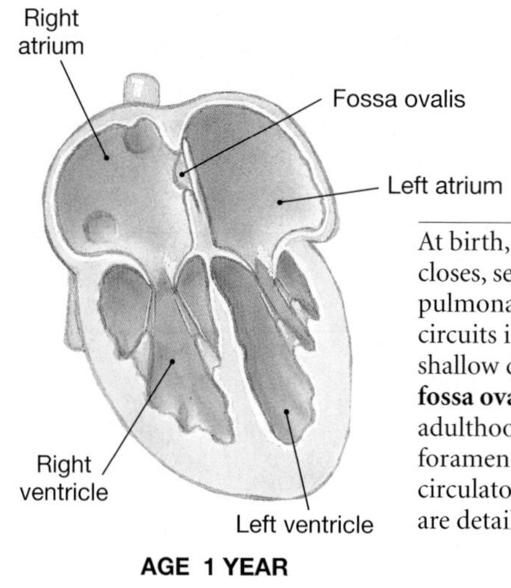

At birth, the foramen ovale closes, separating the pulmonary and systemic circuits in the heart. A shallow depression, the **fossa ovalis**, remains through adulthood at the site of the foramen ovale. (Other circulatory changes at birth are detailed in Chapter 22.)

Additional Clinical Terms Discussed in Appendix I (p. 800-801)
aortic stenosis; echocardiography; endocarditis; heart disease; heart transplants; mitral stenosis; myocarditis; pacemakers; pericarditis; valvular stenosis

STUDY OUTLINE & CHAPTER REVIEW

Introduction 550

1. All of the tissues and fluids in the body rely on the cardiovascular system to maintain homeostasis. The proper functioning of the cardiovascular system depends on the activity of the **heart**, which can vary its pumping capacity depending on the needs of the peripheral tissues.

An Overview of the Cardiovascular System 550

1. The circulatory system can be subdivided into two closed circuits that occur in series. Each circuit functions individually in series, while the two circuits together function in parallel. The **pulmonary circuit** carries oxygen-poor blood from the heart to the lungs and back, and the **systemic circuit** transports oxygen-rich blood from the heart to the rest of the body and back. **Arteries** carry blood away from the heart; **veins** return blood to the heart. **Capillaries** are tiny vessels between the smallest arteries and veins. *(see Figure 21.1)*
2. The heart contains four chambers, the **right atrium** and **ventricle**, and the **left atrium** and **ventricle**. Atria collect blood returning to the heart, and the ventricles discharge blood into vessels to leave the heart.

The Pericardium 550

1. The heart is surrounded by the **pericardial cavity**, which is lined by the **pericardium** and contains a small amount of lubricating fluid, called the **pericardial fluid**. The **visceral pericardium (epicardium)** covers the heart's outer surface, and the **parietal pericardium** lines the inner surface of the **pericardial sac** that surrounds the heart. The heart lies in the anterior portion of the **mediastinum.** *(see Figure 21.2)*

Structure of the Heart Wall 552

1. The heart wall contains three layers: the **epicardium** (the visceral pericardium), the **myocardium** (muscular wall of the heart), and the **endocardium** (the epithelium covering the inner surfaces of the heart). *(see Figure 21.3)*

Cardiac Muscle Tissue 552

2. The bulk of the heart consists of the muscular **myocardium**. Cardiac muscle cells (**cardiocytes**), which are smaller than skeletal muscle cells, are almost totally dependent on aerobic respiration. *(see Figure 21.3)*
3. Cardiocytes are interconnected by **intercalated discs**, which both convey the force of contraction from cell to cell and conduct action potentials. Intercalated discs join cardiac muscle cells through desmosomes, myofibrils, and gap junctions. Because cardiac muscle cells are connected in this way, they function like a single, enormous cell. *(see Figure 21.3d,e)*

The Fibrous Skeleton 552

4. The internal connective tissue of the heart is called the **fibrous skeleton**. *(see Figures 21.3b/21.7)*
5. The fibrous skeleton of the heart functions to stabilize the heart's contractile cells and valves; support the muscle cells, blood vessels, and nerves; distribute the forces of contraction; add strength and elasticity; and physically isolate the atria from the ventricles.

Orientation and Superficial Anatomy of the Heart 554

1. The **auricle** (atrial appendage) is an expandable extension of the atrium. The **coronary sulcus** is the deep groove between the atria and the ventricles. Other shallower depressions include the **anterior interventricular sulcus** and the **posterior interventricular sulcus.**
2. The great veins are connected to the superior end of the heart at the **base**. The inferior, pointed tip of the heart is the **apex**. *(see Figures 21.2a/21.4)*
3. The heart sits at an angle to the longitudinal axis of the body and presents the following **borders**: **superior**, **inferior**, **left**, and **right**. *(see Figure 21.4)*
4. The heart has the following surfaces: the **sternocostal surface** is formed by the anterior surfaces of the right atrium and ventricle; the **diaphragmatic surface** is formed primarily by the posterior, inferior wall of the left ventricle. *(see Figure 21.5)*

Internal Anatomy and Organization of the Heart 556

1. The atria are separated by the **interatrial septum**, and the ventricles are divided by the **interventricular septum.** The openings between the atria and ventricles contain folds of connective tissue covered by endocardium; these *valves* maintain a one-way flow of blood. *(see Figure 21.6)*

The Right Atrium 556

2. The right atrium receives blood from the systemic circuit through two great veins, the **superior vena cava** and **inferior vena cava**. The atrial walls contain prominent muscular ridges, the **pectinate muscles**. The **coronary veins** return blood to the *coronary sinus*. During embryonic development an opening, called the **foramen ovale**, penetrates the interatrial septum. This opening closes after birth, leaving a depression, termed the **fossa ovalis**. *(see Figure 21.6)*

The Right Ventricle 556

3. Blood flows from the right atrium into the right ventricle through the **right atrioventricular (AV) (tricuspid) valve**. This opening is bounded by three **cusps** of fibrous tissue braced by the tendinous **chordae tendineae** that are connected to **papillary muscles.** *(see Figure 21.6a,c)*
4. Blood leaving the right ventricle enters the **pulmonary trunk** after passing through the **pulmonary semilunar valve**. The pulmonary trunk divides to form the **left** and **right pulmonary arteries.** *(see Figure 21.6)*

The Left Atrium 556

5. The left atrium receives oxygenated blood from the **left** and **right pulmonary veins**; it has thicker walls than that of the right atrium. *(see Figure 21.6a,c)*
6. Blood leaving the left atrium flows into the left ventricle through the **left atrioventricular (AV) (mitral) valve**, or *bicuspid valve.*

The Left Ventricle 558

7. The left ventricle is the largest and thickest of the four chambers because it must pump blood to the entire body. Blood leaving the left ventricle passes through the **aortic semilunar valve** and into the systemic circuit via the **ascending aorta**. Blood passes from the ascending aorta through the **aortic arch** and into the **descending aorta**. *(see Figure 21.6a,c)*

Structural Differences between the Left and Right Ventricles 558

8. The right ventricle has thin walls and develops low pressure when pumping into the pulmonary circuit to and from the adjacent lungs. Functionally, low pressure is necessary because the pulmonary capillaries at the gas-exchange surfaces of the lungs are very delicate. The left ventricle has a thick wall because it pumps blood throughout the systemic circuit. Anatomical differences between the left and right ventricles are shown in *Figure 21.6.*

The Structure and Function of Heart Valves 558

9. Valves normally permit blood flow in only one direction, preventing the **regurgitation** (backflow) of blood. The valves of the heart are detailed in *Figure 21.7.*
10. The closure of valves and rushing of blood through the heart cause characteristic **heart sounds** that can be heard during auscultation. *(see Figure 21.8)*

Coronary Blood Vessels 560

11. The **coronary circulation** supplies blood to the muscles of the heart, to meet the high oxygen and nutrient demands of cardiac muscle cells. The **coronary arteries** originate at the base of the ascending aorta, and each gives rise to two branches. The right coronary artery gives rise to both a **right marginal artery** and a **posterior interventricular artery**. The **left coronary artery** gives rise to both a **circumflex artery** and a **left marginal artery**. Interconnections between arteries called **anastomoses** ensure a constant blood supply.
12. The **great** and **middle cardiac veins** carry blood from the coronary capillaries to the **coronary sinus**. *(see Figure 21.9)*
13. Other cardiac veins that empty into the great cardiac vein or the coronary sinus are the: **posterior cardiac vein**, draining the areas served by the circumflex

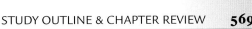

artery; **middle cardiac vein**, draining the areas supplied by the posterior interventricular artery; **small cardiac vein**, draining blood from the posterior surfaces of the right atrium and ventricle.

14. The **anterior cardiac veins** drain the anterior surface of the right ventricle and empty directly into the right atrium.

The Cardiac Cycle 563

1. The **cardiac cycle** consists of periods of **atrial** and **ventricular systole** (contraction) and atrial and ventricular **diastole** (relaxation/filling). *(see Figure 21.11)*

The Coordination of Cardiac Contractions 563

2. Cardiac muscle tissue contracts on its own, without neural or hormonal stimulation. This is called *automaticity* or *autorhythmicity*.
3. **Nodal cells** establish the rate of cardiac contraction, and **conducting fibers** distribute the contractile stimulus to the general myocardium. *(see Figure 21.12)*

The Sinoatrial and Atrioventricular Nodes 564

4. Nodal cells depolarize spontaneously and determine the heart rate.
5. **Pacemaker cells** found in the **sinoatrial (SA) node (cardiac pacemaker)** normally establish the rate of contraction. *(see Figure 21.12)*
6. From the SA node, the stimulus travels over the **internodal pathways** to the **atrioventricular (AV) node**, then to the **AV bundle**, which divides into a **right** and **left bundle branch**. From here **Purkinje cells** convey the impulses to the ventricular myocardium. *(see Figure 21.12)*

The Electrocardiogram (ECG) 566

7. A recording of electrical activities in the heart is an **electrocardiogram (ECG or EKG)**. Important landmarks of an ECG include the **P wave** (atrial depolarization), **QRS complex** (ventricular depolarization), and **T wave** (ventricular repolarization). ECG analysis can detect cardiac arrhythmias, which are abnormal patterns of cardiac activity. *(see Figure 21.13)*

Autonomic Control of Heart Rate 566

8. The basic heart rate is established by the pacemaker cells, but it can be modified by the ANS. Norepinephrine produces an increase in heart rate and force of contraction, while acetylcholine produces a decrease in heart rate and contraction.
9. The **cardioacceleratory center** in the medulla oblongata activates sympathetic neurons; the **cardioinhibitory center** governs the activities of the parasympathetic neurons. The cardiac centers receive inputs from higher centers and from receptors monitoring blood pressure and the concentrations of dissolved gases in the blood. *(see Figure 21.14)*

LEVEL 1 REVIEWING FACTS AND TERMS

Match each numbered item with the most closely related lettered item. Use letters for answers in the spaces provided.

Column A	Column B
_____ 1. cardiocytes	a. vein to the left atrium
_____ 2. bradycardia	b. covers the outer surface of heart
_____ 3. diastole	c. supplies blood to heart muscles
_____ 4. coronary circulation	d. lines inner surface of pericardial sac
_____ 5. visceral pericardium	e. slow heart rate
_____ 6. systole	f. cardiac muscle cells
_____ 7. myocardium	g. muscular wall of the heart
_____ 8. right pulmonary vein	h. relaxation phase of the cardiac cycle
_____ 9. superior vena cava	i. vein to the right atrium
_____10. parietal pericardium	j. contraction phase of the cardiac cycle

11. The heart lies in the
    (a) pleural cavity (b) peritoneal cavity
    (c) abdominopelvic cavity (d) pericardial cavity
12. The atrioventricular valve that is located on the side of the heart that receives blood from the superior vena cava is the
    (a) mitral valve (b) bicuspid valve
    (c) tricuspid valve (d) aortic semilunar valve
13. The functions of the pericardium include
    (a) returning blood to the atria
    (b) pumping blood into circulation
    (c) anchoring the heart to surrounding structures
    (d) preventing expansion of the heart
14. The heart wall is composed of ___ layers of tissue.
    (a) 2 (b) 3
    (c) 4 (d) 5
15. The heart is innervated by
    (a) parasympathetic nerves
    (b) sympathetic nerves
    (c) both sympathetic and parasympathetic nerves
    (d) splanchnic nerves
16. The pacemaker cells of the heart are located in the
    (a) SA node
    (b) wall of the left ventricle
    (c) Purkinje fibers
    (d) both the left and right ventricles
17. Analysis of an electrocardiogram can reveal
    (a) the sequence of valves opening and closing
    (b) the volume of blood passing through each chamber
    (c) the condition of the conducting system
    (d) the damage to specific valves
18. The first heart sound is heard when the
    (a) AV valves open
    (b) AV valves close
    (c) semilunar valves open
    (d) atria contract
19. The fibrous skeleton of the heart functions to
    (a) maintain the normal shape of the heart
    (b) strengthen and help overexpansion of the heart
    (c) help distribute the forces of cardiac contraction
    (d) all of the above
20. The mitral or bicuspid valve is located
    (a) in the opening of the aorta
    (b) between the left atrium and left ventricle
    (c) between the right atrium and right ventricle
    (d) in the opening of the pulmonary trunk

LEVEL 2 REVIEWING CONCEPTS

1. When a clot forms in a coronary vessel and obstructs blood flow to the muscle, the condition is referred to as a(n)
   (a) coronary thrombosis (b) pulmonary embolism
   (c) angina pectoris (d) myocardial infarction
2. If the papillary muscles fail to contract,
   (a) blood will not enter the atria
   (b) the ventricles will not pump blood
   (c) the AV valves will not close properly
   (d) the semilunar valves will not open
3. During ventricular systole, the
   (a) atria are contracting
   (b) ventricles are relaxed
   (c) AV valves are closed
   (d) pressure in the ventricles declines
4. How is cardiac muscle similar to skeletal muscle?
5. Why do semilunar valves lack muscular braces like those found in AV valves?
6. Define and describe the endocardium.
7. What is the function of the pericardial fluid?
8. Which chamber of the heart is the largest and has the thickest walls? Why are its walls so thick?
9. Why are nodal cells unique? What is their function?
10. What is the effect of NE release on cardiac function?

LEVEL 3 CRITICAL THINKING AND CLINICAL APPLICATIONS

1. Harvey has a heart murmur in his left ventricle that produces a loud "gurgling" sound at the beginning of systole. What do you suspect to be the cause of this sound?
2. Lee is brought to the emergency room of a hospital suffering from a cardiac arrhythmia. In the emergency room he begins to exhibit tachycardia and as a result loses consciousness. His wife asks you why he lost consciousness. What would you tell her?
3. If the cardiac centers detect an abundance of oxygen in the blood, what chemical is likely to be released?
4. If a newborn baby becomes cyanotic, what is this condition, and why might this occur?

✓ ANSWERS TO CONCEPT CHECK QUESTIONS

p. 554 **1.** The most obvious characteristic that differentiates cardiac muscle tissue from skeletal muscle tissue is that cardiac muscle cells are small with a centrally placed nucleus. Additionally, they are mechanically, chemically, and electrically connected to neighboring cells at intercalated discs. **2.** The pericardial cavity is a small space between the visceral surface of the heart and the parietal surface of the pericardial sac. It contains pericardial fluid, which acts as a lubricant, reducing friction between opposing surfaces. **3.** Cardiac muscle cells are connected to their neighbors at specialized junctional sites termed intercalated discs. At an intercalated disc, the membranes are bound together by desmosomes, myofibrils of the cells are anchored to the membrane, and gap junctions connect the cells. **4.** A syncitium is a single giant cell with multiple nuclei; it forms through the fusion of many small individual cells. Cardiac muscle tissue functions much like a single muscle cell. The contraction of one cardiac muscle cell triggers the contractions of several others, and so on, spreading throughout the myocardium.
p. 558 **1.** The groove between the atria and ventricles is the coronary sulcus. **2.** The atria have relatively thin, highly distensible muscular walls, and when not distended, the outer portion looks like a deflated, wrinkled flap (the auricle).
p. 563 **1.** The valves prevent backflow from the ventricles into the atria when the ventricles contract, so without them, the blood would rush back into the atria. **2.** The superior vena cava, the inferior vena cava, and the coronary sinus. **3.** Blood from the left ventricle passes through the aortic semilular valve, travels through the systemic circuit, and enters the right atrium. From there it passes through the tricuspid valve and enters the right ventricle. Blood leaving the right ventricle passes through the pulmonary semilunar valve and enters the pulmonary trunk. Then, it flows into the pulmonary arteries before arriving at the capillaries in the respiratory surfaces of the lungs. **4.** As the ventricles begin to contract, they force the AV valves to close, which in turn pull on the chordae tendineae. The chordae tendineae pull on the papillary muscles. The papillary muscles respond by contracting, opposing the force that is pushing the valves toward the atria. **5.** The "lubb" is caused by the closing of the AV valves. The "dupp" is caused by the closing of the semilunar valves.
p. 565 **1.** If these cells were not functioning, the heart would still continue to beat, but at a slower rate, following the pace set by the AV node. **2.** The heart rate is increased in response to norepinephrine. **3.** Because the cell membranes of nodal cells depolarize spontaneously and nodal cells are electrically connected to one another, to conducting fibers, and to cardiac muscle cells, contractions are able to sweep through the conducting system and trigger the contractions of cardiac muscle cells. The timing of contraction is controlled by the rate of propagation and the distribution of the contractile stimulus to the cardiac muscle tissue in the atria versus the ventricles.
p. 567 **1.** The flow of blood from one chamber to the next occurs only if the pressure is higher in the first chamber than in the second, so equal pressures would mean that the blood would not flow from the atrium to the ventricle. **2.** The P wave accompanies atrial depolarization, so an abnormality suggests that the atria may be damaged.

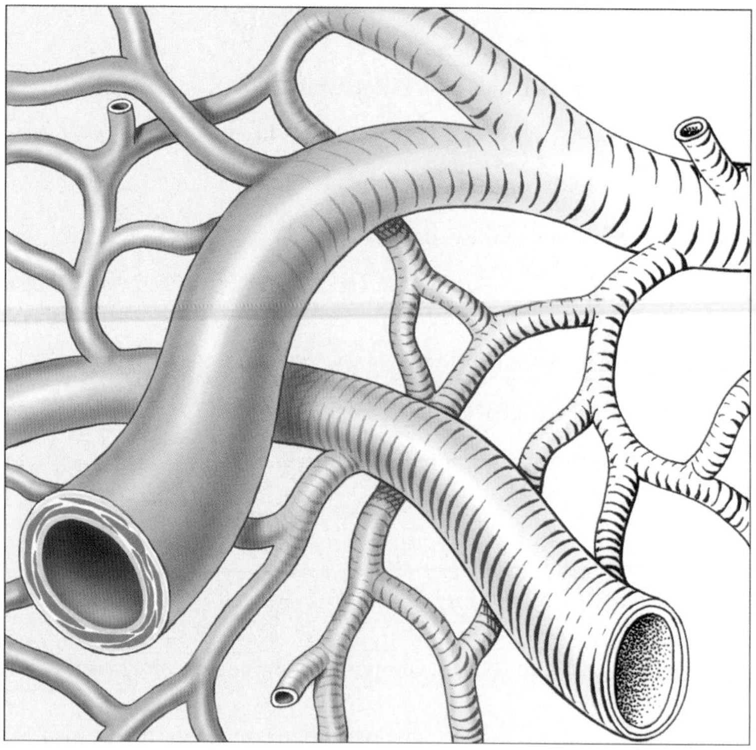

22

THE CARDIOVASCULAR SYSTEM

Vessels and Circulation

The cardiovascular system is a closed system that circulates blood throughout the body. There are two groups of blood vessels: one supplies the lungs (the *pulmonary circuit*), and the other supplies the rest of the body (the *systemic circuit*). Blood is pumped from the heart into both the pulmonary and systemic (aortic) trunks simultaneously. The relatively small pulmonary circuit begins at the pulmonary semilunar valve and ends at the entrance to the left atrium. Pulmonary arteries that branch from the pulmonary trunk carry blood to the lungs for gas exchange. The systemic circuit begins at the aortic semilunar valve and ends at the entrance to the right atrium. Systemic arteries branch from the aorta and distribute blood to all other organs for nutrient, gas, and waste exchange. The pulmonary trunk and the aorta each have a lumenal diameter of around 2.5 cm (1 in.). These vessels branch to form numerous smaller vessels that supply individual regions and organs.

After entering the organs, further branching occurs, creating several hundred million tiny arteries that provide blood to more than 10 billion capillaries barely the diameter of a single red blood cell. These capillaries form extensive branching networks; estimates of the combined length of all of the capillaries in the body (placed end to end) range from 5000 to 25,000 miles. In other words, the capillaries in your body could at least cross the continental United States and perhaps circle the globe. All chemical and gaseous exchange between the blood and interstitial fluid takes place across capillary walls. Tissue cells rely on capillary diffusion to obtain nutrients and oxygen and to remove metabolic waste products. Blood leaving capillary networks enters a network of small veins that gradually merge to form larger vessels that ultimately supply either one of the pulmonary veins (pulmonary circuit) or the inferior or superior vena cava (systemic circuit).

Our discussion early in this chapter will focus attention on the histological and anatomical organization of arteries, capillaries, and veins. We will then proceed to identify the major blood vessels and circulatory routes of the cardiovascular system.

Histological Organization of Blood Vessels [FIGURE 22.1]

The walls of arteries and veins contain three distinct layers: (1) an inner *tunica intima* or *tunica interna* (in-TER-na), (2) a middle *tunica media*; and (3) an outer *tunica externa*, or *tunica adventitia*. View Figure 22.1● as we examine the histological structure of arteries and veins.

- The **tunica intima**, or *tunica interna*, is the innermost layer of a blood vessel. This layer includes the endothelial lining of the vessel and an underlying layer of connective tissue containing variable amounts of elastic fibers. In arteries the outer margin of the tunica intima contains a thick layer of elastic fibers called the **internal elastic membrane**. In the largest arteries, the connective tissue is more extensive, and the tunica intima is thicker than in smaller arteries.
- The **tunica media** is the middle layer, and it contains concentric sheets of smooth muscle tissue in a framework of loose connective tissue. The smooth muscle fibers of the tunica media encircle the lumen of the blood vessel. When stimulated by sympathetic activation, these smooth muscles constrict and reduce the diameter of the blood vessel, a process called **vasoconstriction**. Relaxation of the smooth muscles increases the diameter of the lumen, a process called **vasodilation** (vaz-ō-dī-LĀ-shun). These smooth muscles may contract or relax in response to local stimuli or under control of the sympathetic division of the ANS. Any resulting change in vessel diameter affects both blood pressure and blood flow through this tissue. Collagen fibers bind the tunica media to both the tunica intima and tunica externa. Arteries have a thin band of elastic fibers, the **external elastic membrane**, located between the tunica media and tunica externa.

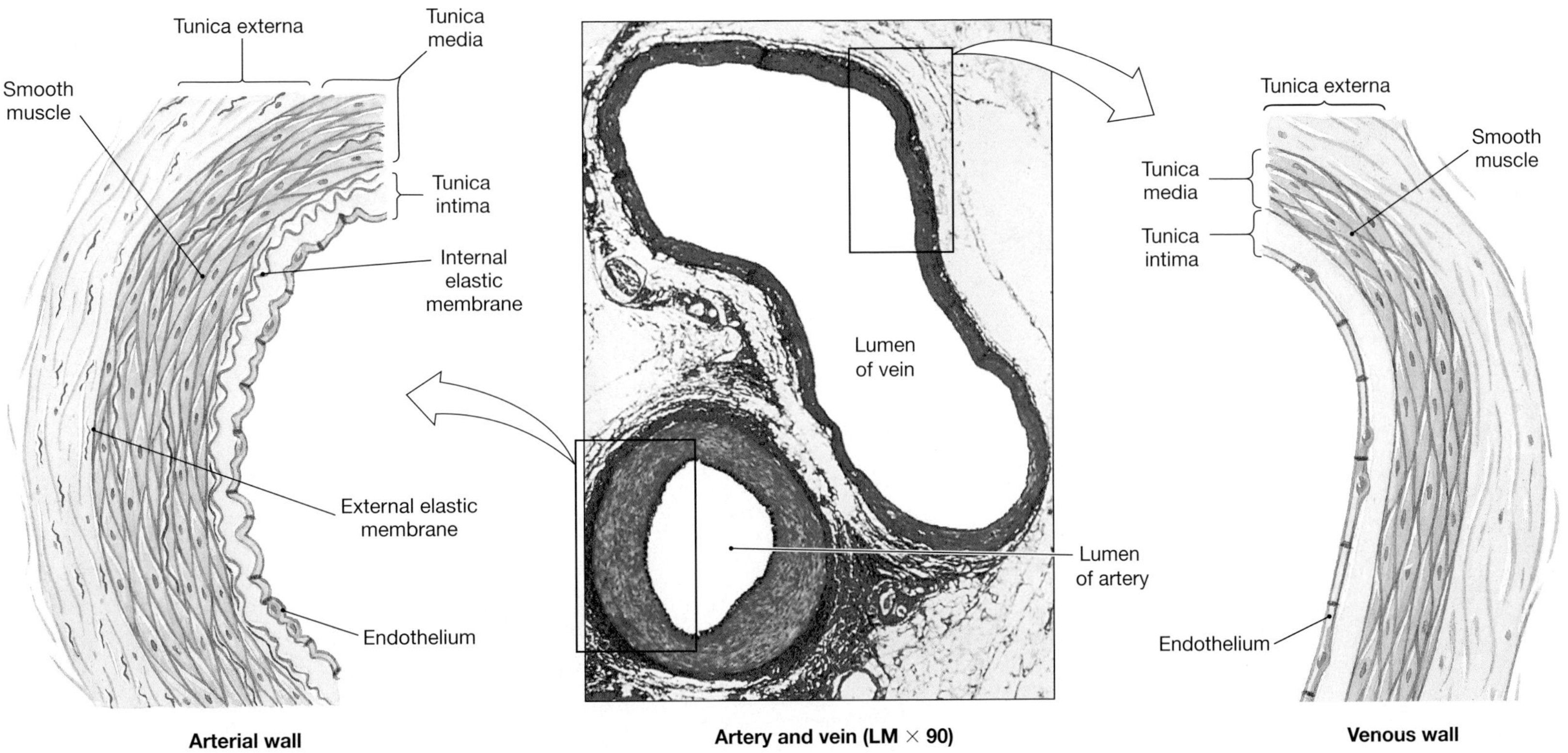

FIGURE 22.1 **A Comparison of Typical Arteries and Veins**
Light micrograph of an artery and vein.

- The outer **tunica externa** (eks-TER-na), or *tunica adventitia* (ad-ven-TISH-ē-a), forms a connective tissue sheath around the vessel. This layer is very thick, composed chiefly of collagen fibers, with scattered bands of elastic fibers. The fibers of the tunica externa typically blend into those of adjacent tissues, stabilizing and anchoring the blood vessel. In veins this layer is usually thicker than the tunica media.

Their layered walls give arteries and veins considerable strength. The combination of muscular and elastic components permits controlled alterations in diameter as blood pressure or blood volume changes. However, the vessel walls are too thick to allow diffusion between the bloodstream and surrounding tissues, or even between the blood and the tissues of the vessel itself. Instead, the walls of large vessels contain small arteries and veins that supply the smooth muscle fibers and fibroblasts of the tunica media and tunica externa. These blood vessels are called the **vasa vasorum** ("vessels of vessels"). *Aneurysms p. 801*

Distinguishing Arteries from Veins [FIGURE 22.1]

Arteries supplying and veins draining the same region typically lie side by side within a narrow band of connective tissue (Figure 22.1●). Arteries and veins may be distinguished in histological sections by the following characteristics:

1. In general, when comparing two adjacent vessels the walls of arteries are thicker than those of veins. The tunica media of an artery contains more smooth muscle and elastic fibers than does that of a vein. These contractile and elastic components resist the pressure generated by the heart as it forces blood into the circuit.
2. When not opposed by blood pressure, arterial walls contract. Thus, when seen on dissection or in sectional view (Figure 22.1●), arteries appear smaller than the corresponding veins. Because the walls of arteries are relatively thick and strong, they retain their circular shape

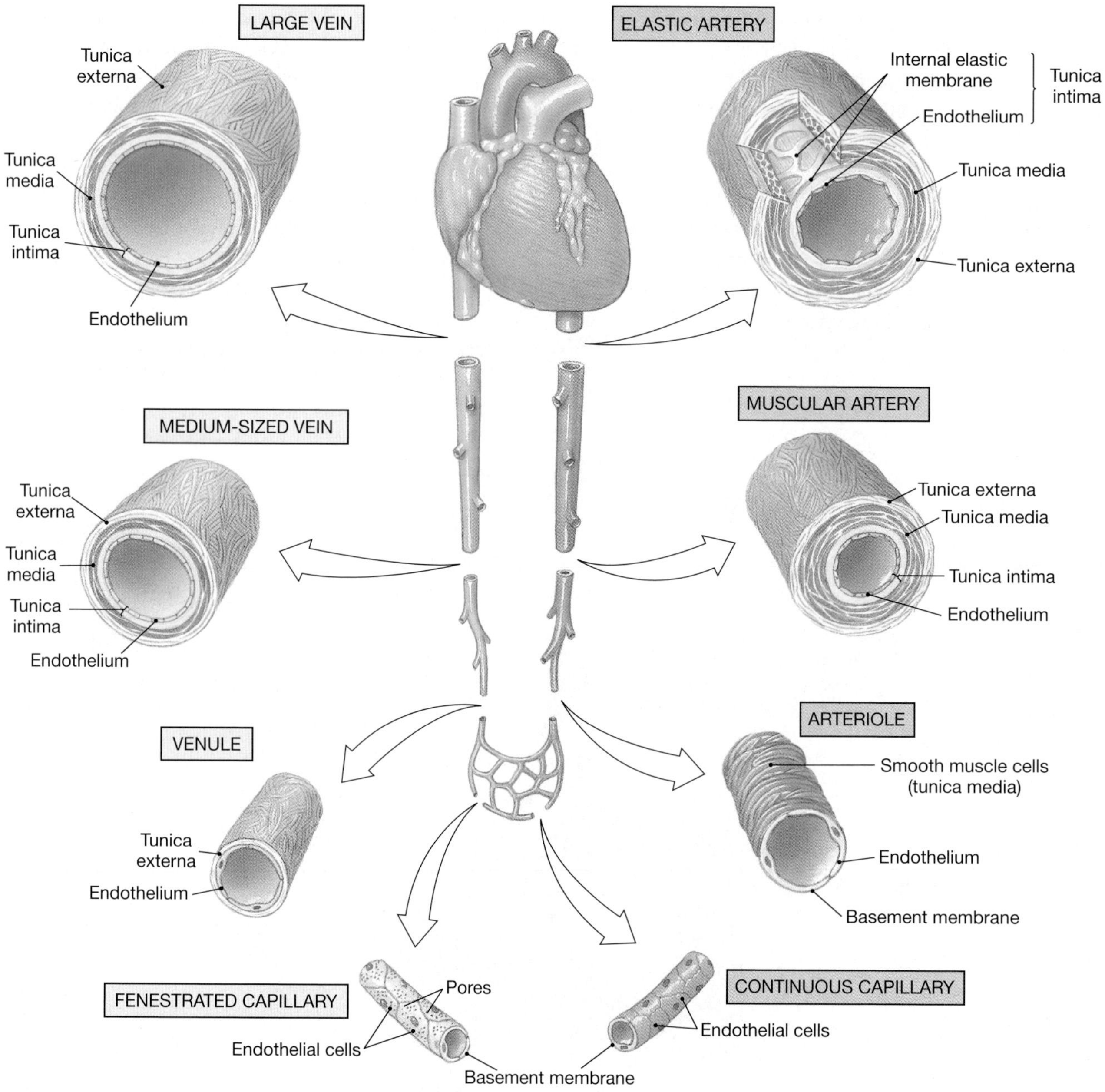

FIGURE 22.2 **Histological Structure of Blood Vessels**

in section. Cut veins tend to collapse, and in section they often look flattened or grossly distorted.

3. The endothelial lining of an artery cannot contract, so when an artery constricts, the endothelium is thrown into folds that give arterial sections a pleated appearance. The lining of a vein lacks these folds.

Arteries [FIGURE 22.2]

In traveling from the heart to peripheral capillaries, blood passes through a series of arteries of ever smaller diameter: *elastic arteries*, *muscular arteries*, and *arterioles* (Figure 22.2●).

ELASTIC ARTERIES [FIGURE 22.2]

Elastic arteries, or *conducting arteries*, are large vessels with diameters of up to 2.5 cm (1 in.). They transport large volumes of blood away from the heart. The pulmonary and aortic trunks and their major branches (the *pulmonary*, *common carotid*, *subclavian*, and *common iliac arteries*) are examples of elastic arteries. The walls of elastic arteries are not very thick relative to the vessel diameter, but they are extremely resilient. The tunica media of these vessels contains a high density of elastic fibers and relatively few smooth muscle fibers (Figure 22.2●). As a result, elastic arteries are able to tolerate the pressure changes that occur during the cardiac cycle. During ventricular systole, pressures rise rapidly and the elastic arteries are stretched, whereas during ventricular diastole, blood pressure within the arterial system falls, and the elastic fibers recoil to their original dimensions. Their stretching cushions the sudden rise in pressure during ventricular systole, and their recoiling both slows the decline in pressure during ventricular diastole and forces blood onward toward the capillaries.

MUSCULAR ARTERIES [FIGURES 22.1/22.2]

Muscular arteries, or *distribution arteries* (also known as *medium-sized arteries*), transport blood to the body's skeletal muscle and internal organs. A typical muscular artery has a diameter of approximately 0.4 cm (0.15 in.). Muscular arteries have a thicker tunica media with a greater percentage of smooth muscle fibers than one finds in elastic arteries (Figures 22.1 and 22.2●). The *external carotid artery* of the neck, the *brachial arteries* of the arms, the *femoral arteries* of the thighs, and the *mesenteric arteries* of the abdomen are examples of muscular arteries. The sympathetic division of the ANS can control the diameter of each of these arteries. By constricting *(vasoconstriction)* or relaxing *(vasodilation)* the smooth muscle in the tunica media, the ANS can regulate the blood flow to each organ independently.

ARTERIOLES [FIGURE 22.2]

Arterioles (ar-TĒ-rē-ōlz) are considerably smaller than muscular arteries. Arterioles have an average diameter of about 30 μm. They have a poorly defined tunica externa, and their tunica media consists of scattered smooth muscle fibers that do not form a complete layer (Figure 22.2●). The smaller muscular arteries and arterioles change their diameter in response to local conditions or to sympathetic or endocrine stimulation. For example, arterioles in most tissues vasoconstrict under sympathetic stimulation. p. 454 Arterioles control the blood flow between arteries and capillaries.

Elastic and muscular arteries are seamlessly interconnected, and vessel characteristics change gradually as the vessels get farther away from the heart. For example, the largest of the muscular arteries contain a considerable amount of elastic tissue, while the smallest resemble heavily muscled arterioles.

Capillaries [FIGURES 22.2/22.3]

Capillaries are the smallest and most delicate blood vessels (Figure 22.2●). They are important functionally because they are the only blood vessels whose walls permit exchange between the blood and the surrounding interstitial fluids. Because the walls are relatively thin, the diffusion distances are small, and exchange can occur quickly. In addition, blood flows slowly through capillaries, allowing sufficient time for diffusion or active transport of materials across the capillary walls. Some substances cross the capillary walls by diffusing across the endothelial cell lining; other substances pass through gaps between adjacent endothelial cells. The fine structure of each capillary determines its ability to regulate the exchange of substances between the blood and the interstitial fluid.

A typical capillary consists of an endothelial tube enclosed within a delicate basement membrane. The average internal diameter of a capillary is a mere 8 μm, very close to that of a single red blood cell.

Continuous capillaries are found in most regions of the body. In these capillaries, the endothelium is a complete lining, and the endothelial cells are connected by tight junctions and desmosomes (Figure 22.3a,b●).

Fenestrated capillaries (FEN-es-trā-ted; *fenestra*, window) are capillaries that contain "windows," or pores in their walls, due to an incomplete or perforated endothelial lining (Figure 22.3c,d●). The walls of a fenestrated capillary have a "Swiss cheese" appearance, and the pores allow molecules as large as peptides and small proteins to pass into or out of the circulation. This type of capillary permits very rapid exchange of fluids and solutes. Examples of fenestrated capillaries noted in earlier chapters include the choroid plexus of the brain, and the capillaries in a variety of endocrine organs, including the hypothalamus, the pituitary, the pineal, the adrenals, and the thyroid gland. pp. 389, 503 Fenestrated capillaries are also found at filtration sites in the kidneys.

Sinusoids (SĪ-nus-oydz) resemble fenestrated capillaries except they have larger pores and a thinner basement membrane. (In some organs, such as the liver, there is no basement membrane.) Sinusoids are flattened and irregular and follow the internal contours of complex organs. They permit an extensive exchange of fluids and large solutes, including suspended proteins, between blood and interstitial fluid. Blood moves through sinusoids relatively slowly, maximizing the time available for absorption and secretion across the sinusoidal walls. Sinusoids are found in the liver, bone marrow, and adrenal glands.

There are four basic mechanisms responsible for the exchange of materials across the walls of capillaries and sinusoids:

1. Diffusion across the capillary endothelial cells (lipid-soluble materials, gases, and water by osmosis).
2. Diffusion through gaps between adjacent endothelial cells (water and small solutes; larger solutes in the case of sinusoids).
3. Diffusion through the pores in fenestrated capillaries (water and solutes).
4. Vesicular transport by endothelial cells (endocytosis at lumenal side, exocytosis at basal side), water, and specific bound and unbound solutes. pp. 32–33, 40

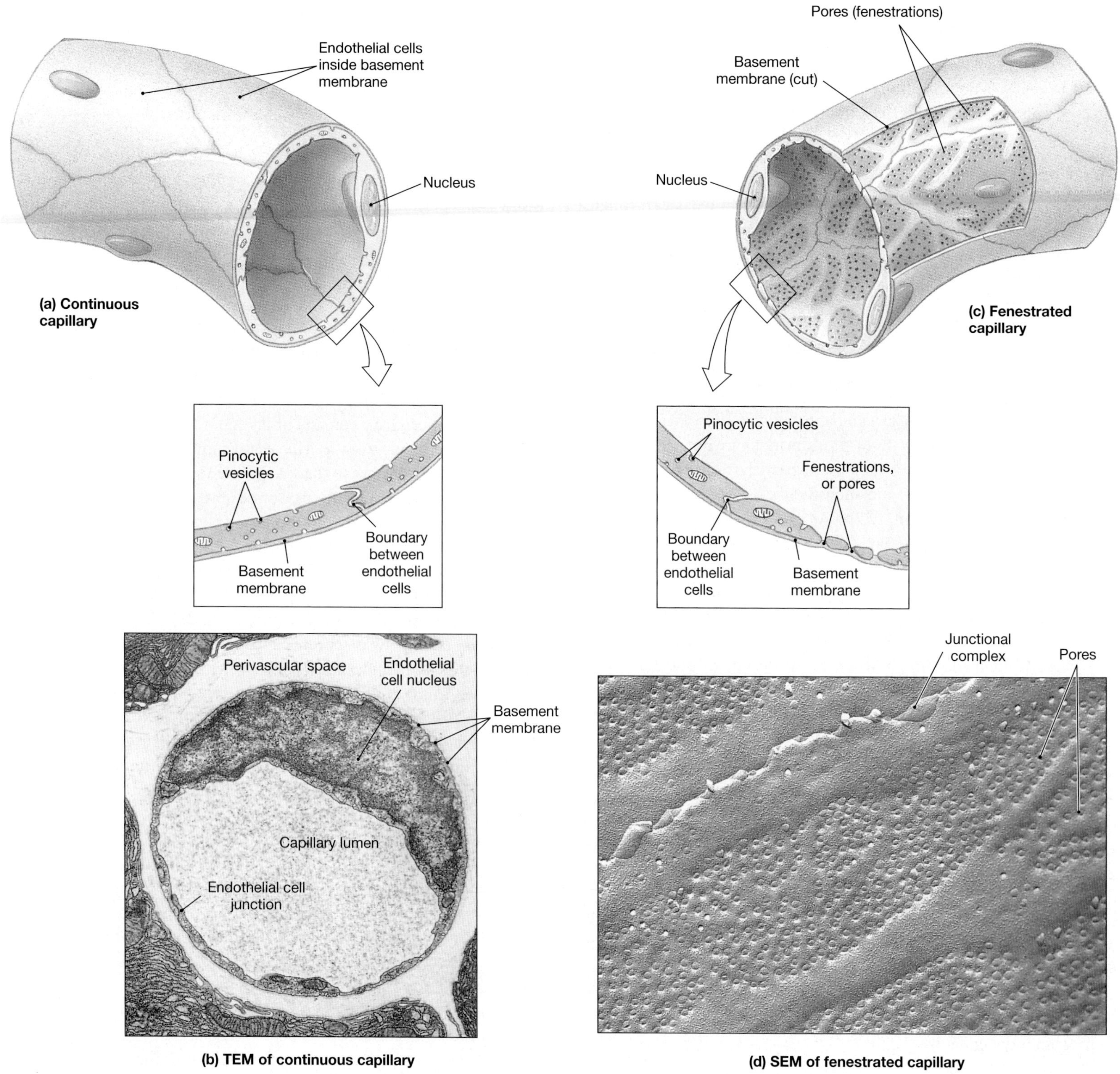

FIGURE 22.3 **STRUCTURE OF CAPILLARIES**

(a,b) Continuous capillaries. The diagrammatic view of a continuous capillary shows the structure of its wall. The TEM shows a cross section through a continuous capillary. A single endothelial cell forms a complete lining around this portion of the capillary. **(c,d)** Fenestrated capillaries. The diagrammatic view of a fenestrated capillary details the structure of the wall. The SEM shows the wall of a fenestrated capillary. The pores are gaps in the endothelial lining that permit the passage of large volumes of fluid and solutes. (SEM × 12,425)

CAPILLARY BEDS [FIGURE 22.4]

Capillaries do not function as individual units. Each capillary is a part of an interconnected network called a **capillary bed**, or **capillary plexus** (Figure 22.4●). A single arteriole usually gives rise to dozens of capillaries that empty into several venules. The entrance to each capillary is guarded by a band of smooth muscle, termed a **precapillary sphincter**. Contraction of the smooth muscle fibers constricts and narrows the diameter of the capillary entrance, thereby reducing or stopping the flow of blood. Relaxation of the sphincter dilates the opening, allowing blood to enter the capillary at a faster rate. Precapillary sphincters open when carbon dioxide levels rise; such a rise indicates that the tissue needs oxygen and nutrients. The sphincters close when carbon dioxide levels decline, or under sympathetic stimulation.

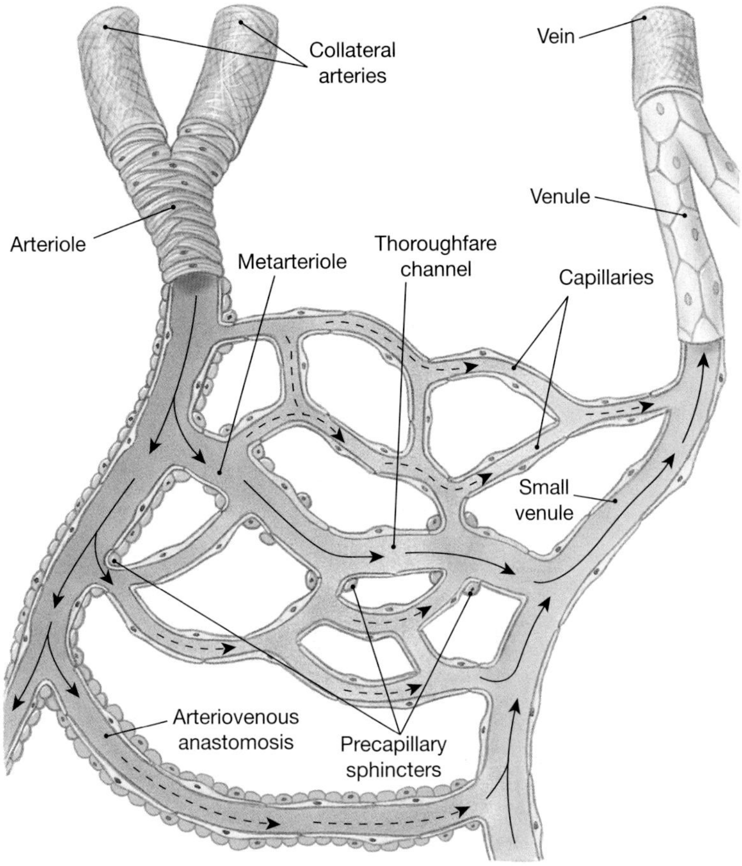

(a) Capillary bed

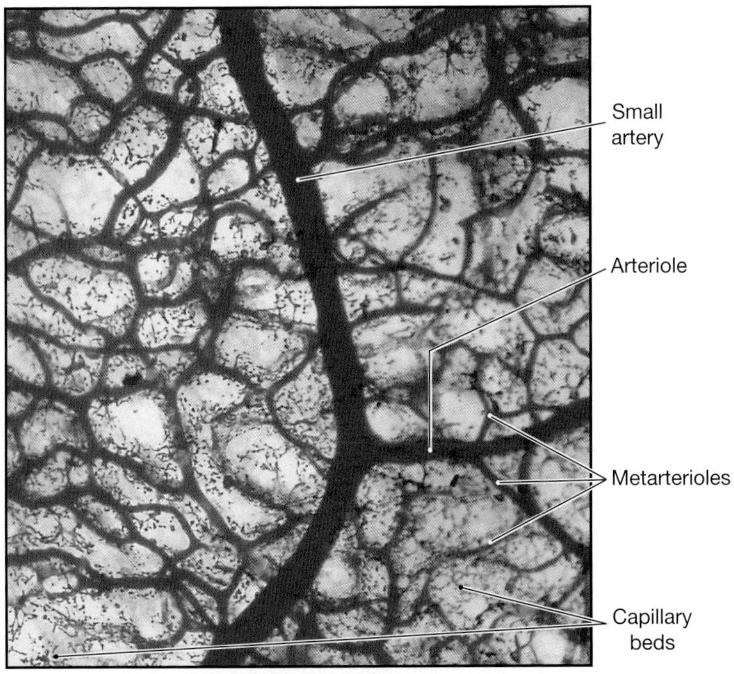

(b) Micrograph of capillary bed

FIGURE 22.4 **ORGANIZATION OF A CAPILLARY BED**

(**a**) Basic organization of a typical capillary bed. The pattern of blood flow changes continually in response to regional alterations in tissue oxygen demand. (**b**) Capillary bed as seen in living tissue.

A capillary bed contains several relatively direct connections between arterioles and venules. The arteriolar segment of such a passageway contains smooth muscle cells capable of altering its diameter, and this region is often called a **metarteriole** (met-ar-TĒ-rē-ōl) (Figure 22.4a●). It is somewhat intermediate in structure between arterioles and capillaries. The rest of the passageway resembles a typical capillary, and it is called a **thoroughfare channel**.

Blood usually flows from the arterioles to the venules at a constant rate, but the blood flow within a single capillary can be quite variable. Each precapillary sphincter goes through cycles of alternately contracting and relaxing, perhaps a dozen times each minute. As a result, the blood flow within any one capillary occurs in a series of pulses rather than as a steady and constant stream. The net effect is that blood may reach the venules by one route now and by quite a different route later. This process, which is controlled at the tissue level, is called capillary **autoregulation**.

There are also mechanisms to modify the circulatory supply to the entire capillary complex. The capillary networks within an area are often supplied by more than one artery. The arteries, called **collaterals**, enter the region and fuse together rather than ending in a series of arterioles. The interconnection is an **arterial anastomosis**. Arterial anastomoses are common in the brain and heart. Such an arrangement guarantees a reliable blood supply to the tissues, for if one arterial supply should become blocked, the other will supply blood to the capillary bed. **Arteriovenous** (ar-tē-rē-ō-VĒ-nus) **anastomoses** are direct connections between arterioles and venules (Figure 22.4a●). Anastomoses are abundant in visceral organs and joints, especially where changes in body position could hinder blood flow through one vessel or another. Smooth muscles in the walls of these vessels can contract or relax to regulate the amount of blood reaching the capillary bed. For example, when the arteriovenous anastomoses are dilated, blood will bypass the capillary bed and flow directly into the venous circulation. Collateral blood flow is regulated primarily by sympathetic innervation, under the control of cardiovascular centers in the medulla oblongata. p. 408

Veins [FIGURES 22.1/22.2]

Veins collect blood from all tissues and organs and return it to the heart. Following the pattern of blood flow, veins are discussed from smallest to largest (venule to medium-sized vein to large vein), whereas arteries were discussed from largest to smallest (elastic artery to muscular artery to arteriole). The walls of veins are thinner and less elastic than those of corresponding arteries because the blood pressure in veins is lower than that in arteries. Veins are classified on the basis of their size, and, in general, veins are larger in diameter than their corresponding arteries. Review Figures 22.1, p. 573, and 22.2●, p. 574, to compare typical arteries and veins.

VENULES

Venules, the smallest veins, collect blood from capillaries. They vary widely in size and character, but an average venule has a lumenal diameter of roughly 20 μm. The smallest venules resemble expanded capillaries, and venules smaller than 50 μm in total diameter lack a tunica media altogether.

MEDIUM-SIZED VEINS

Medium-sized veins range from 2 to 9 mm in internal diameter and correspond in general size to medium-sized arteries. In these veins, the tunica media is thin, and it contains relatively few smooth muscle fibers. The thickest layer of a medium-sized vein is the tunica externa, which contains longitudinal bundles of elastic and collagen fibers.

CLINICAL DISCUSSION

ARTERIOSCLEROSIS

Arteriosclerosis (ar-tē-rē-ō-skle-RŌ-sis) is a thickening and toughening of arterial walls. Complications related to arteriosclerosis account for roughly one-half of all deaths in the United States. There are many different forms of arteriosclerosis; one example is coronary artery disease (CAD), which was described in Chapter 21. p. 562

Arteriosclerosis takes two major forms, focal calcification and atherosclerosis:

1. **Focal calcification** is the gradual degeneration of smooth muscle in the tunica media and the subsequent deposition of calcium salts. This process typically involves arteries of the limbs and genital organs. Some focal calcification occurs as part of the aging process, and it may develop in association with atherosclerosis. Rapid and severe calcification may occur as a complication of diabetes mellitus, an endocrine disorder.

2. **Atherosclerosis** (ath-er-ō-skle-RŌ-sis) is associated with damage to the endothelial lining and the formation of lipid deposits in the tunica media. This is the most common form of arteriosclerosis.

Many factors may be involved in the development of atherosclerosis. One major factor is lipid levels in the blood. Atherosclerosis tends to develop in persons whose blood contains elevated levels of plasma lipids, specifically cholesterol. Circulating cholesterol is transported to peripheral tissues in *lipoproteins,* protein–lipid complexes.

When cholesterol-rich lipoproteins remain in circulation for an extended period, circulating monocytes begin removing them from the bloodstream. Eventually, the monocytes become filled with lipid droplets. Now called *foam cells,* they attach themselves to the endothelial walls of blood vessels, where they release growth factors. These cytokines stimulate the divisions of smooth muscle cells near the tunica interna, thickening the vessel wall.

Other monocytes then invade the area, migrating between the endothelial cells. As these changes occur, the monocytes, smooth muscle cells, and endothelial cells begin phagocytizing lipids as well. The result is a **plaque**, a fatty mass of tissue that projects into the lumen of the vessel. At this point, the plaque has a relatively simple structure, and evidence suggests that the process can be reversed if appropriate dietary adjustments are made.

If the conditions persist, the endothelial cells become swollen with lipids, and gaps appear in the endothelial lining. Platelets now begin sticking to the exposed collagen fibers. The combination of platelet adhesion and aggregation leads to the formation of a localized blood clot, which will further restrict blood flow through the artery. The structure of the plaque is now relatively complex. Plaque growth can be halted, but the structural changes are generally permanent.

Typical plaques are shown in Figure 22-5●. Elderly individuals, especially elderly men, are most likely to develop atherosclerotic plaques. Evidence suggests that estrogens may slow plaque formation; this may account for the lower incidence of CAD, myocardial infarctions (MIs), and strokes in women. After menopause, when estrogen production declines, the risk of CAD, MIs, and strokes in women increases markedly.

In addition to advanced age and male gender, other important risk factors include high blood cholesterol levels, high blood pressure, and cigarette smoking. Roughly 20 percent of middle-aged men have all three of these risk factors; these individuals are four times more likely to experience an MI or a cardiac arrest than are other men in their age group. Although fewer women develop atherosclerotic plaques, elderly women smokers with high blood cholesterol and high blood pressure are at much greater risk than are other women. Factors that can promote the development of atherosclerosis in both men and women include diabetes mellitus, obesity, and stress. Evidence also indicates that at least some forms of atherosclerosis may be linked to chronic infection with *Chlamydia pneumoniae,* a bacterium responsible for several types of respiratory infections, including some forms of pneumonia.

Potential treatments for atherosclerotic plaques include *catheterization* and *balloon angioplasty* and *stents*. p. 562 and Scan 10d in the Companion Atlas However, the best approach is to try to avoid atherosclerosis by eliminating or reducing risk factors. Suggestions include: (1) reducing the amount of dietary cholesterol and saturated fats by restricting consumption of fatty meats (such as beef, lamb, and pork), egg yolks, and cream, (2) giving up smoking, (3) checking your blood pressure and taking steps to lower it if necessary, (4) having your blood cholesterol levels monitored regularly and treated if necessary, (5) controlling your weight, and (6) exercising regularly.

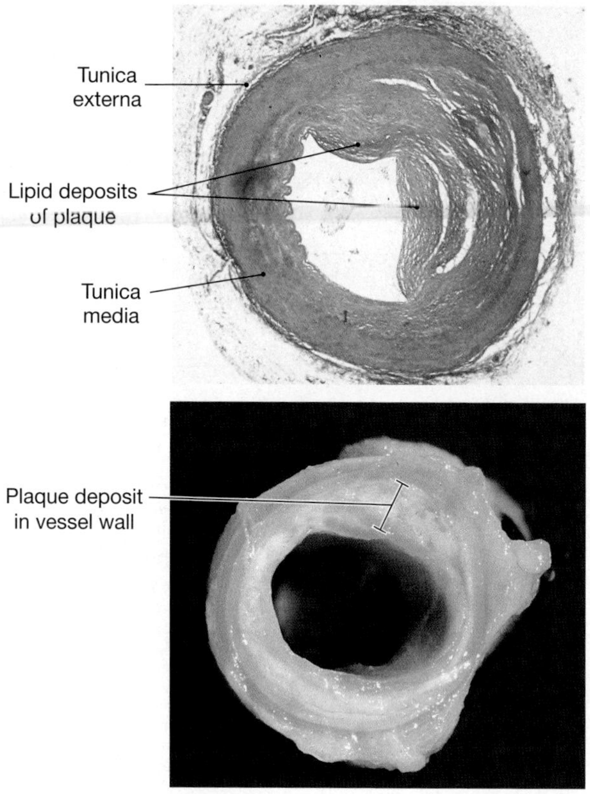

FIGURE 22.5 **A PLAQUE BLOCKING A PERIPHERAL ARTERY**

The fatty mass extends into the vessel lumen and restricts blood flow. (LM × 28)

LARGE VEINS

Large veins include the *great veins,* the *superior* and *inferior venae cavae,* and their tributaries within the abdominopelvic and thoracic cavities. In veins, all of the tunica layers are thickest in the large veins. The slender tunica media is surrounded by a thick tunica externa composed of a mixture of elastic and collagenous fibers.

VENOUS VALVES [FIGURE 22.6]

The blood pressure in venules and medium-sized veins is too low to oppose the force of gravity. In the limbs, veins of this size contain one-way **valves** that are formed from infoldings of the tunica intima (Figure 22.6●). These valves act like the valves in the heart, preventing the backflow of blood. As long as the valves function normally, any movement that distorts or com-

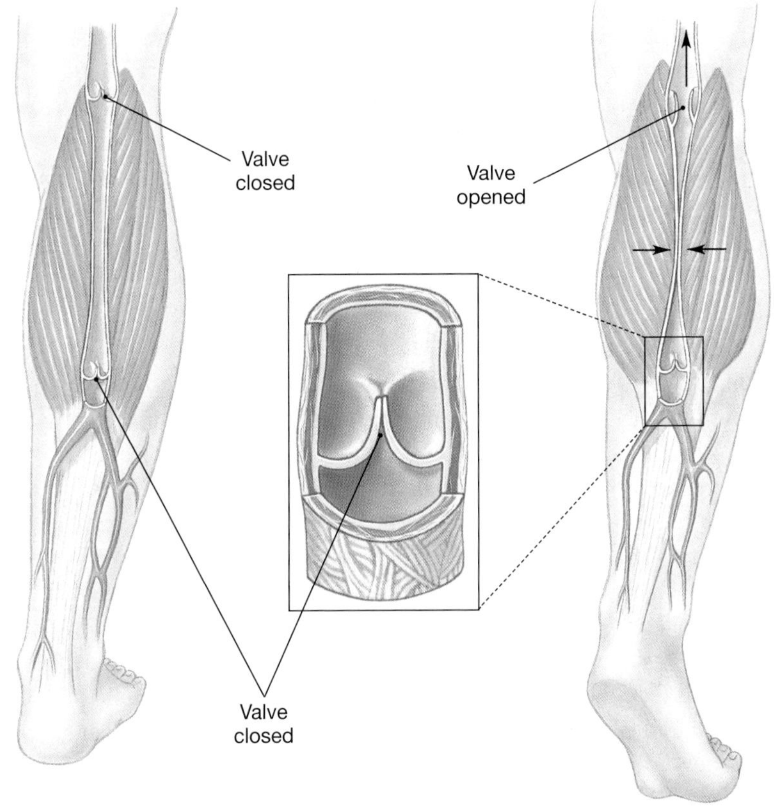

FIGURE 22.6 **Function of Valves in the Venous System**

Valves in the walls of medium-sized veins prevent the backflow of blood. Venous compression caused by the contraction of adjacent skeletal muscles creates pressure (shown by arrows) which assists in maintaining venous blood flow. Changes in body position and the thoracoabdominal pump may provide additional assistance.

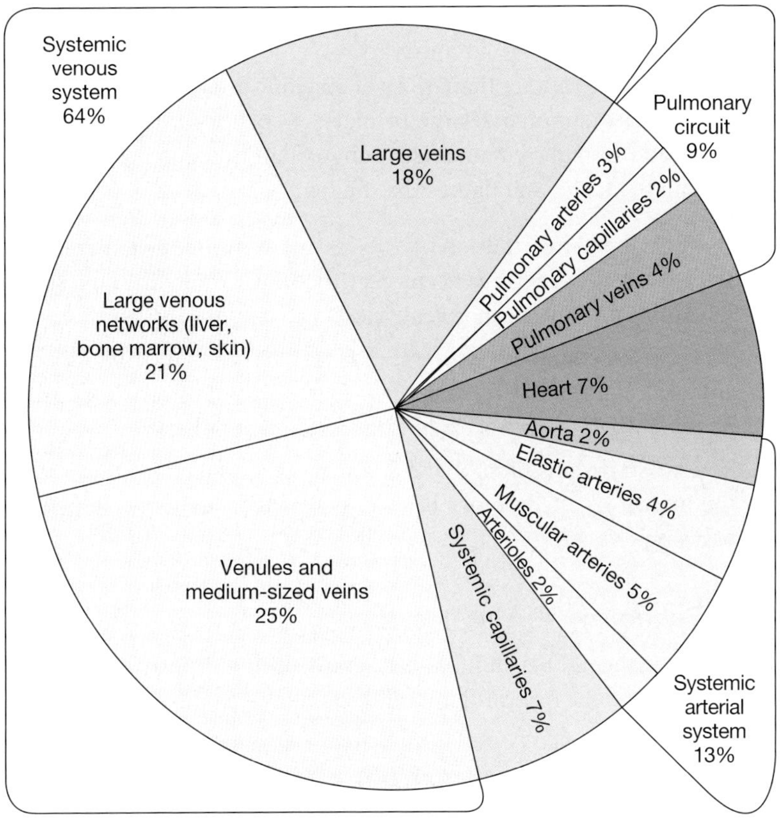

FIGURE 22.7 **The Distribution of Blood in the Cardiovascular System**

presses a vein will push blood toward the heart. For example, when you are standing, blood returning from the foot must overcome the pull of gravity to ascend to the heart. Valves compartmentalize the blood within the veins, thereby dividing the weight of the blood between the compartments. Any movement in the surrounding skeletal muscles squeezes the blood toward the heart. This mechanism is called a *skeletal muscle pump*. Large veins such as the vena cava do not have valves, but changes of pressure in the thoracic cavity assist in moving blood toward the heart. This mechanism, called the *thoracoabdominal pump*, will be considered further in Chapter 24. † *Problems with Venous Valve Function p. 801*

The Distribution of Blood [FIGURE 22.7]

The total blood volume is unevenly distributed among arteries, veins, and capillaries (Figure 22.7●). The heart, arteries, and capillaries normally contain 30–35% of the blood volume (roughly 1.5 l of whole blood), and the venous system contains the rest (65–70% or around 3.5 l).

Because their walls are thinner and contain a lower proportion of smooth muscle, veins are much more distensible than arteries. For a given rise in pressure, a typical vein will stretch about eight times as much as a corresponding artery. If the blood volume rises or falls, the elastic walls stretch or recoil, changing the volume of blood in the venous system.

If serious hemorrhaging occurs, the vasomotor center of the medulla oblongata stimulates sympathetic nerves innervating smooth muscle cells in the walls of medium-sized veins. When the smooth muscles in the walls contract, this **venoconstriction** (vē-nō-kon-STRIK-shun) reduces the volume of the venous system. In addition, blood enters the general circulation from venous networks in the liver, bone marrow, and skin. Reducing the amount of blood in the venous system can maintain the volume within the arterial system at near-normal levels despite a significant blood loss. The venous system therefore acts as a **blood reservoir**, with the liver acting as the primary reservoir; the change in volume constitutes the **venous reserve**. The venous reserve normally amounts to just over 1 liter, 21% of the total blood volume.

✓ CONCEPT CHECK

- Examination of a section of tissue shows several small, thin-walled vessels with very little smooth muscle tissue in the tunica media. What type of vessels are these?
- Why are valves found in veins but not in arteries?
- The femoral artery is an example of which type of artery?
- Does gas exchange occur between the blood and surrounding tissues in arterioles?

Blood Vessel Distribution [FIGURE 22.8]

The blood vessels of the body can be divided into the pulmonary circuit and the systemic circuit. The pulmonary circuit is composed of arteries and veins that transport blood between the heart and the lungs, a relatively short distance. The arteries and veins of the systemic circuit transport oxygenated blood between the heart and all other tissues, a round-trip that involves much longer distances. There are some functional and structural differences between the vessels in these circuits. For example, blood pressure within the pulmonary circuit is relatively low, and the walls of pulmonary arteries are thinner than those of systemic arteries.

Figure 22.8• summarizes the primary circulatory routes within the pulmonary and systemic circuits. Three important functional patterns will emerge from the tables and figures that follow:

1. The peripheral distribution of arteries and veins on the left and right sides is usually identical except near the heart, where the largest vessels connect to the atria or ventricles.
2. A single vessel may have several different names as it crosses specific anatomical boundaries, making accurate anatomical descriptions possible when the vessel extends far into the periphery.
3. Arteries and veins often make anastomotic connections that reduce the impact of a temporary or even permanent occlusion (blockage) of a single vessel.

The Pulmonary Circuit [FIGURE 22.9]

Blood entering the right atrium has just returned from capillary beds in peripheral tissues and the myocardium. While blood was in those capillary beds, oxygen was released and carbon dioxide absorbed. After traveling through the right atrium and right ventricle, blood enters the pulmonary trunk, the start of the pulmonary circuit. This circuit, which at any given moment contains about 9% of the total blood volume, begins at the pulmonary semilunar valve and ends at the entrance to the left atrium. In the pulmonary circuit (Figure 22.9a•), oxygen will be replenished, carbon dioxide excreted, and the oxygenated blood returned to the heart for distribution to the body within the systemic circuit. Compared with the systemic circuit, the pulmonary circuit is relatively short; the base of the pulmonary trunk and lungs are only about 15 cm (6 in.) apart.

The arteries of the pulmonary circuit differ from those of the systemic circuit in that they carry deoxygenated blood. (For this reason, color-coded diagrams usually show the pulmonary arteries in blue, the same color as systemic veins.) As the pulmonary trunk curves over the superior border of the heart, it gives rise to the **left** and **right pulmonary arteries**. These large arteries enter the lungs before branching repeatedly, giving rise to smaller and smaller arteries. The smallest branches, the *pulmonary arterioles*, provide blood to capillary networks that surround small air pockets, or **alveoli** (al-VĒ-ō-lī; *alveolus*, sac) in the lungs. The walls of alveoli are thin enough for gas exchange to occur between the capillary blood and inspired air. (Alveolar structure is described in Chapter 24.) As it leaves the alveolar capillaries, oxygenated blood enters venules that in turn merge to form larger vessels carrying blood toward the **pulmonary veins**. These four veins, two from each lung, empty into the left atrium, completing the pulmonary circuit (Figure 22.9a•). Figure 22.9b• is a coronary angiogram showing the vessels of the pulmonary circuit and their relationship to the heart and lungs in a living subject.

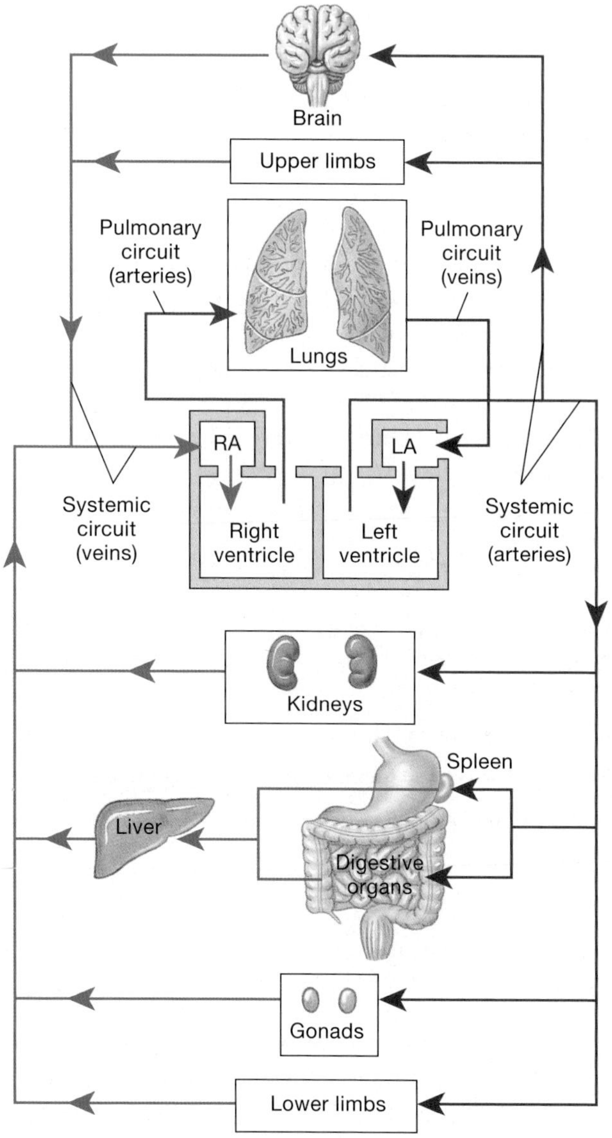

FIGURE 22.8
AN OVERVIEW OF THE GENERAL PATTERN OF CIRCULATION

The Systemic Circuit

The systemic circuit begins at the aortic semilunar valve and ends at the entrance to the right atrium. It supplies the capillary beds in all parts of the body not supplied by the pulmonary circuit and, at any given moment, contains about 84% of the total blood volume.

SYSTEMIC ARTERIES [FIGURES 22.10 TO 22.20]

Figure 22.10• is an overview of the arterial system. This figure indicates the relative locations of major systemic arteries. The detailed distribution of these vessels and their branches will be found in Figures 22.11 to 22.20•. Because we have separate figures for arteries versus veins, we have not included the terms "artery" or "vein" in the labels. By convention, several large arteries are called *trunks*; examples include the *pulmonary trunk*, *brachiocephalic trunk*, *thyrocervical trunk*, and *celiac trunk*. To avoid confusion, these names will be shown in their entirety. Because the descriptions that follow focus attention on major branches found on both sides of the body, for clarity the terms "right" or "left" will not be used in the accompanying figures unless both vessels are labeled.

THE ASCENDING AORTA [FIGURES 21.6a/22.9] The **ascending aorta** begins at the aortic semilunar valve of the left ventricle (Figures 21.6a, p. 557, and 22.9•). The left and right coronary arteries originate at the base of the ascending aorta, just superior to the aortic semilunar valve. The distribution of coronary vessels was described in Chapter 21 and illustrated in Figure 21.9. p. 561

THE AORTIC ARCH [FIGURES 22.10 TO 22.13] The **aortic arch** curves like a cane handle across the superior surface of the heart, connecting the ascending aorta with the *descending aorta*. Three elastic arteries originate along the aortic arch (Figures 22.10, 22.11, and 22.12•). These arteries,

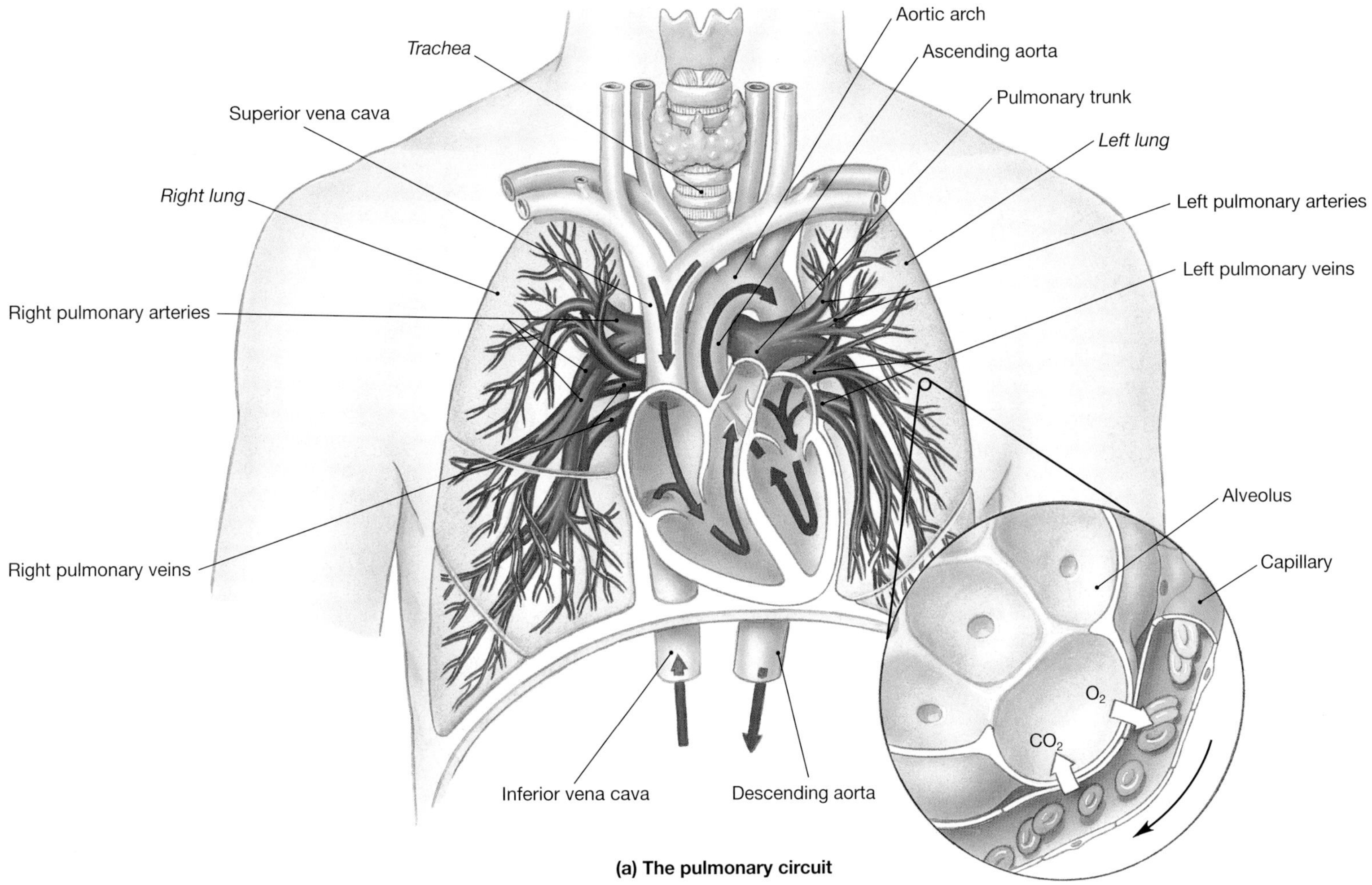

(a) The pulmonary circuit

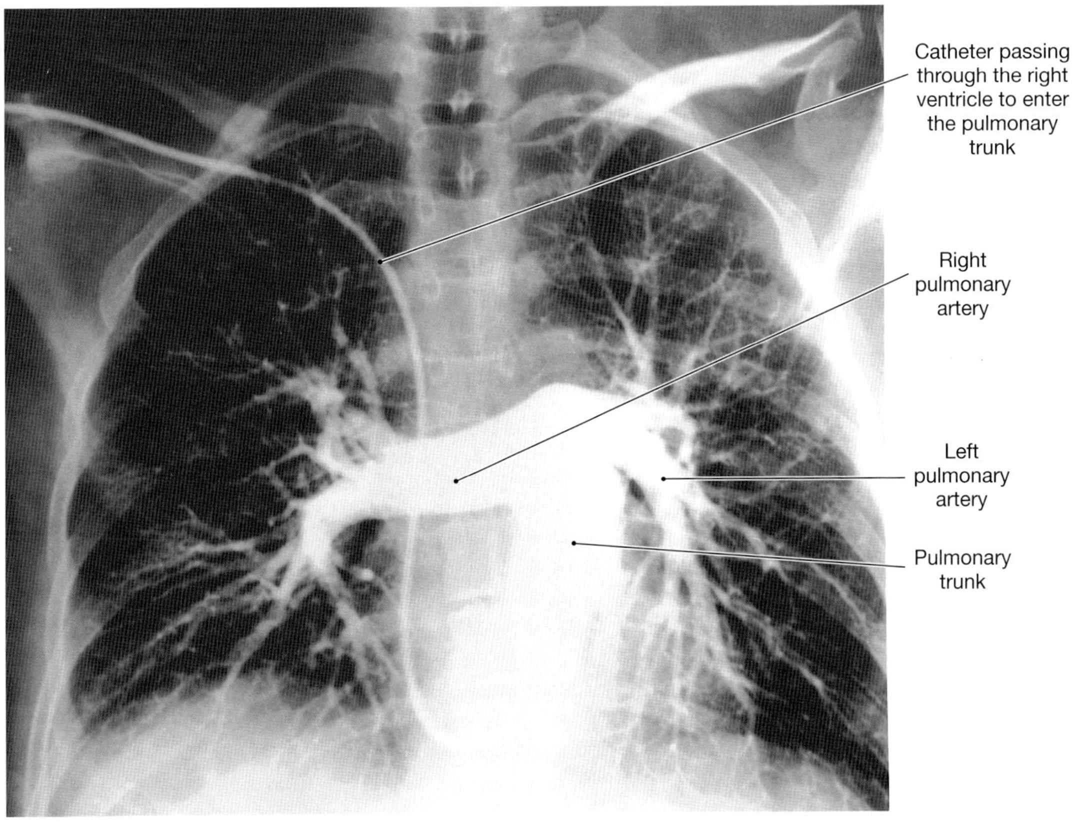

(b) Angiogram

FIGURE 22.9 **The Pulmonary Circuit**

(**a**) Anatomy of the pulmonary circuit. Blue arrows indicate the flow of deoxygenated blood; red arrows the flow of oxygenated blood. The breakout shows the alveoli of the lung and the routes of gas diffusion into the bloodstream across the walls of the alveolar capillaries. (**b**) Coronary angiogram, showing heart, vessels of the pulmonary circuit, diaphragm, and lungs.

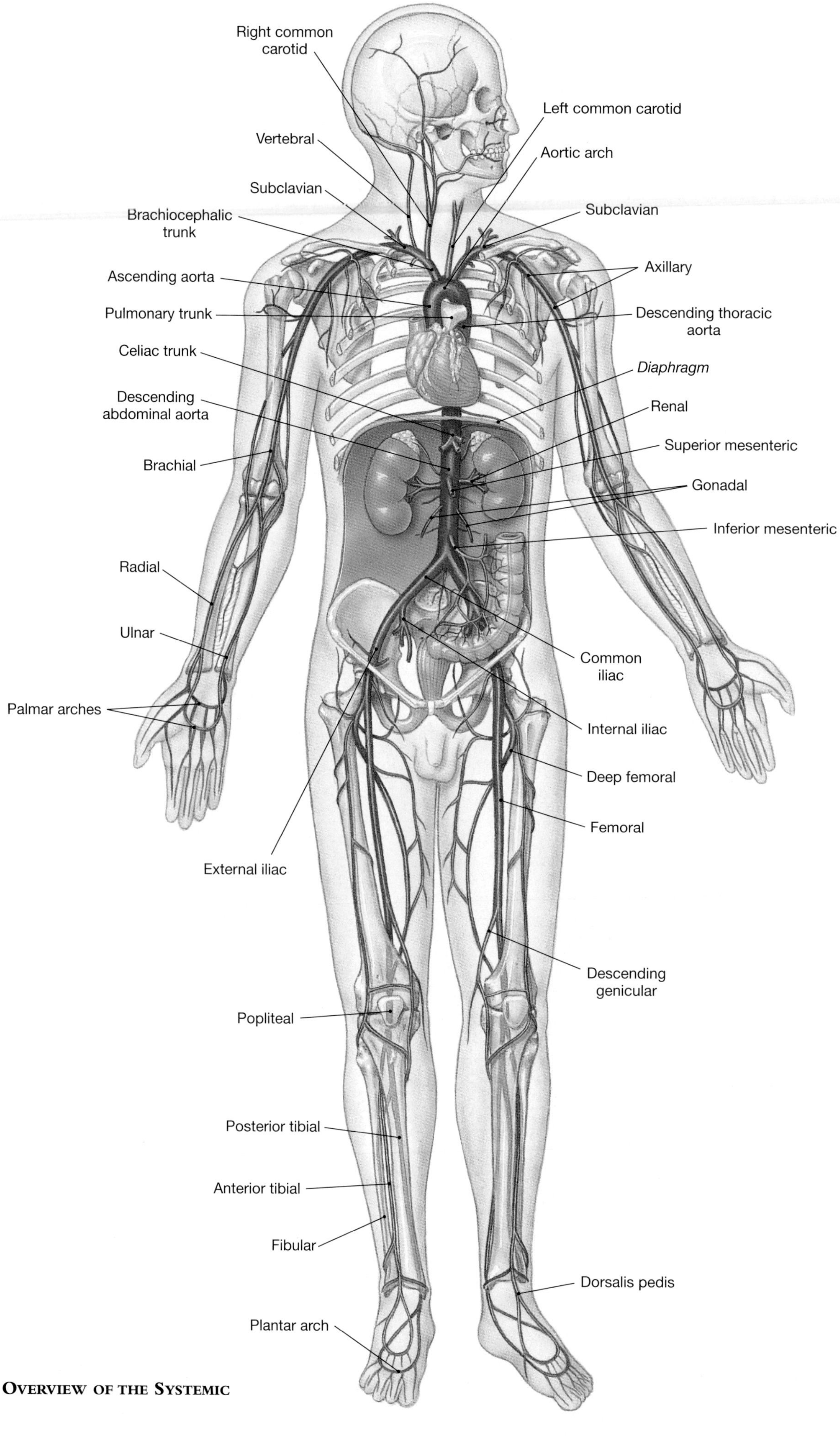

FIGURE 22.10 **An Overview of the Systemic Arterial System**

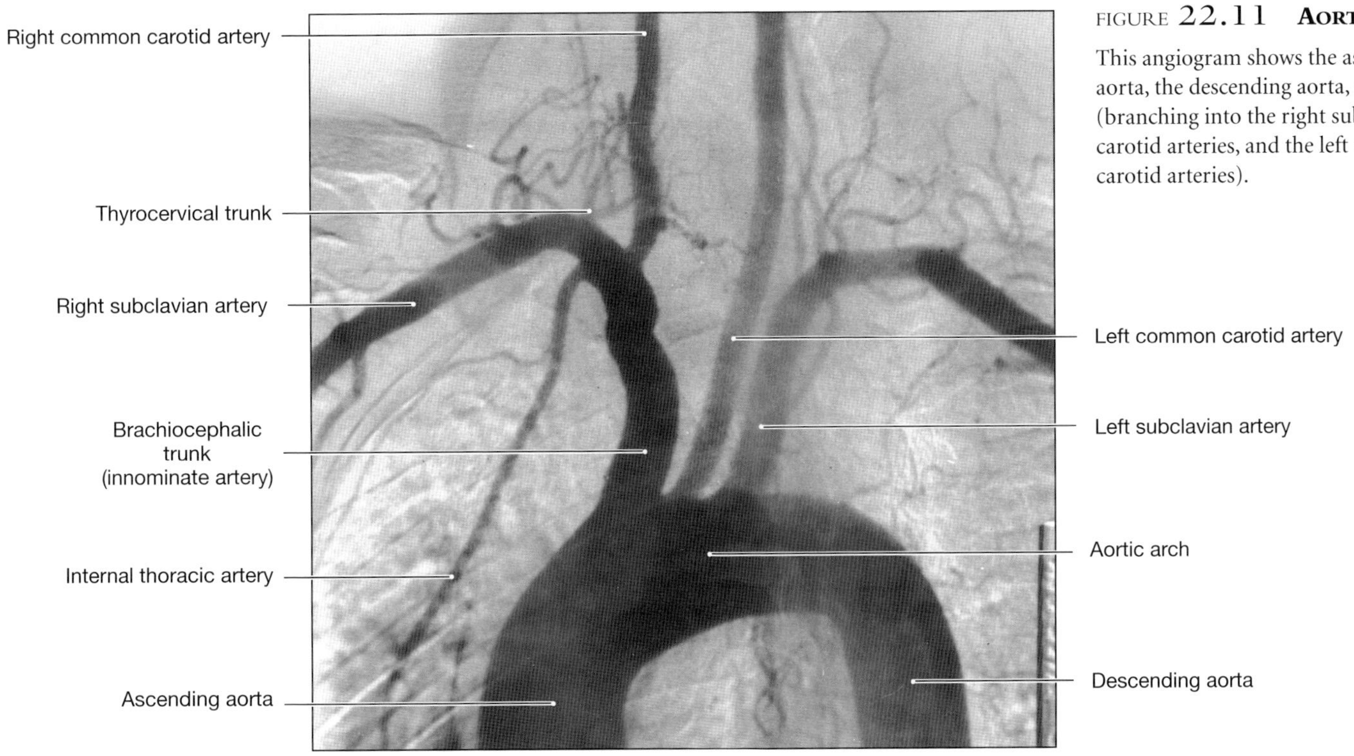

FIGURE 22.11 **Aortic Angiogram**

This angiogram shows the ascending aorta, the arch of the aorta, the descending aorta, and the brachiocephalic trunk (branching into the right subclavian and right common carotid arteries, and the left subclavian and left common carotid arteries).

the **brachiocephalic** (brā-kē-ō-se-FAL-ik) **trunk**, the **left common carotid**, and the **left subclavian** (sub-CLĀ-vē-an), deliver blood to the head, neck, shoulders, and upper limbs. The brachiocephalic trunk, also called the *innominate artery* (i-NOM-i-nāt; "unnamed"), ascends for a short distance before branching to form the **right subclavian artery** and the **right common carotid artery**. There is only one brachiocephalic trunk, and the left common carotid and left subclavian arteries arise separately from the aortic arch. However, in terms of their peripheral distribution, the vessels on the left side are mirror images of those on the right side. Figures 22.12 and 22.13● illustrate the major branches of these arteries.

The Subclavian Arteries [FIGURES 22.10 TO 22.12] The subclavian arteries supply blood to the upper limbs, chest wall, shoulders, back, brain, and spinal cord (Figures 22,10 and 22.12●). Three major branches arise before a subclavian artery leaves the thoracic cavity: (1) the **thyrocervical trunk**, which provides blood to muscles and other tissues of the neck, shoulder, and upper back; (2) an **internal thoracic artery**, supplying the pericardium and anterior wall of the chest; and (3) a **vertebral artery**, which provides blood to the brain and spinal cord.

After leaving the thoracic cavity and passing over the outer border of the first rib, the subclavian artery becomes the **axillary artery**, which supplies blood to the muscles of the pectoral region and axilla. The axillary artery crosses the axilla to enter the arm, where it gives rise to the *humeral circumflex arteries,* which supply structures near the head of the humerus. Distally the axillary artery becomes the **brachial artery**, which supplies blood to the upper limb.

The brachial artery first gives rise to the **deep brachial artery**, which supplies deep structures along the posterior surface of the arm. It then supplies blood to *ulnar collateral arteries,* which, together with the *ulnar recurrent arteries,* supply the area around the elbow. At the cubital fossa, the brachial artery divides into the **radial artery**, which follows the radius, and the **ulnar artery**, which follows the ulna to the wrist. These arteries supply blood to the forearm. At the wrist, these arteries anastomose to form a **superficial palmar arch** and a **deep palmar arch** that respectively supply blood to the palm and to the **digital arteries** of the thumb and fingers.

The Carotid Arteries and the Blood Supply to the Brain [FIGURES 22.13 TO 22.15] The common carotid arteries ascend deep in the tissues of the neck. A common carotid artery can usually be located by pressing gently along either side of the windpipe (trachea) until a strong pulse is felt. Each common carotid artery divides at the level of the larynx into an **external carotid artery** and an **internal carotid artery**. The **carotid sinus**, located at the base of the internal carotid artery, may extend along a portion of the common carotid artery (Figure 22.13●). The carotid sinus contains baroreceptors and chemoreceptors involved in cardiovascular regulation. ∞ p. 472 The external carotid arteries supply blood to the structures of the neck, pharynx, esophagus, larynx, lower jaw, and face. The internal carotid arteries enter the skull through the *carotid canals* of the temporal bones, delivering blood to the brain. ∞ p. 139 Each internal carotid artery ascends to the level of the optic nerves, where it divides into three branches: (1) an **ophthalmic artery**, which supplies the eyes; (2) an **anterior cerebral artery**, which supplies the frontal and parietal lobes of the brain; and (3) a **middle cerebral artery**, which supplies the midbrain and lateral surfaces of the cerebral hemispheres (Figures 22.13 and 22.15●, p. 588).

The brain is extremely sensitive to changes in its circulatory supply. An interruption of circulation for several seconds will produce unconsciousness, and after 4 minutes there may be some permanent neural damage. Such circulatory crises are rare, because blood reaches the brain through the vertebral arteries as well as the internal carotid arteries. The vertebral arteries arise from the subclavian arteries and ascend within the *transverse foramina* of the cervical vertebrae. ∞ pp. 166–168 The vertebral arteries enter the cranium at the foramen magnum, where they fuse along the ventral surface of the medulla oblongata to form the **basilar artery**. The basilar artery continues on the ventral surface of the brain along the pons, branching many times before dividing into the **posterior cerebral arteries**. The **posterior communicating arteries** branch off the posterior cerebral arteries (Figures 22.13a and 22.15a,b●).

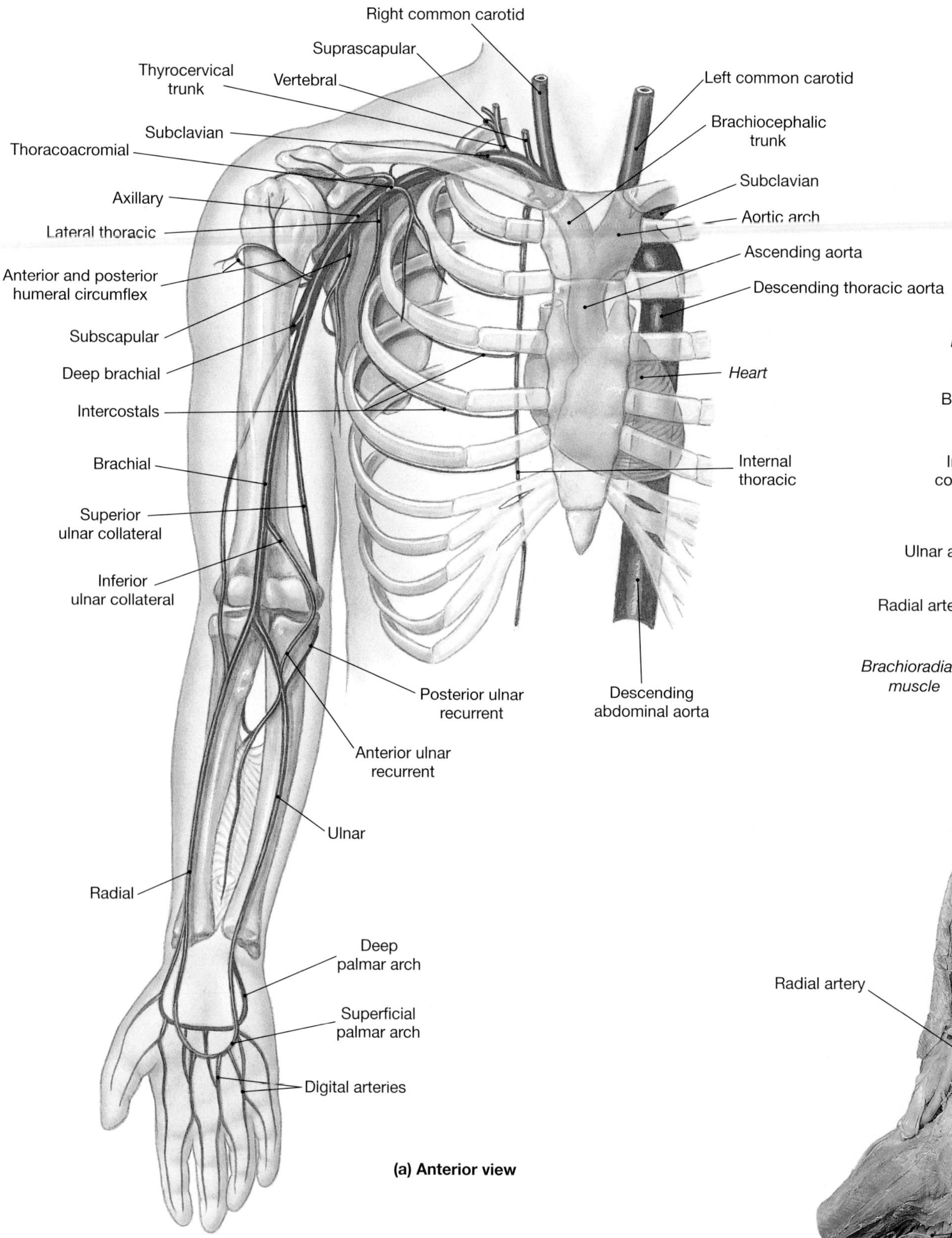

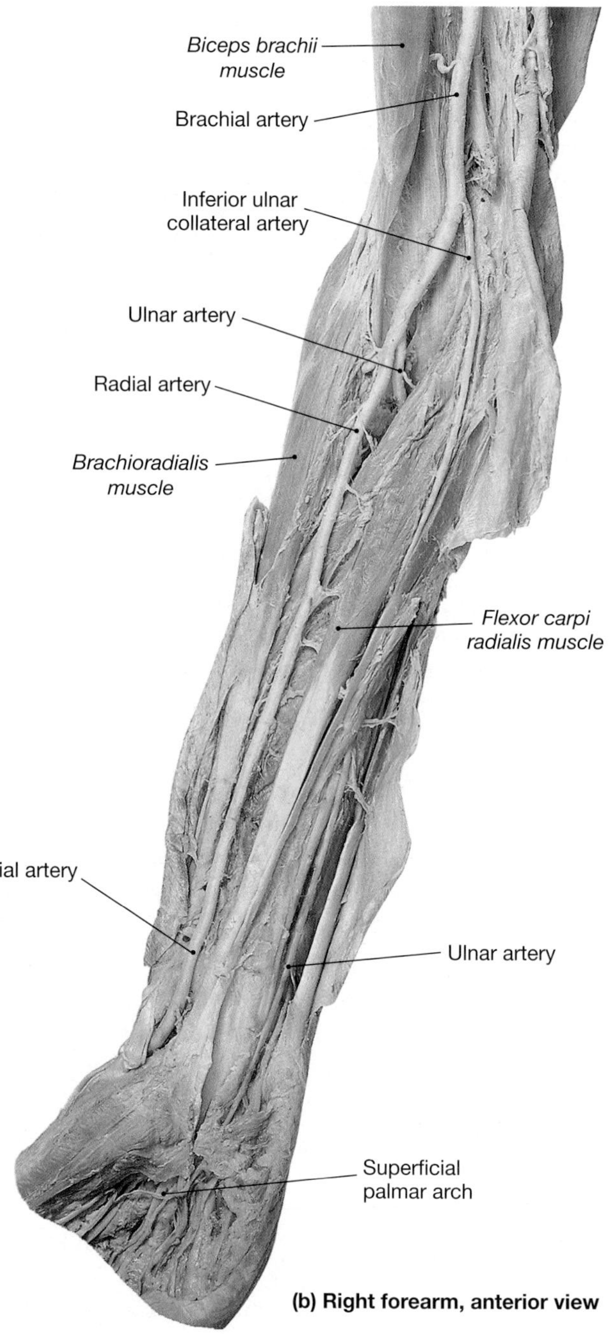

FIGURE 22.12 **Arteries of the Chest and Upper Limb**

(**a**) Arteries originating along the aortic arch shown branching into the chest and right upper limb. (**b**) Anterior view of the right forearm, dissected to show the main arteries. (**c**) Anterior view of the right axillary region dissected to show blood vessels and nerves in this region. (**d**) A flow chart showing the arterial distribution from the aortic arch. White arrows show major pathways of blood flow; black arrows show distribution to secondary or terminal pathways. *See MRI Scan 10c in the Companion Atlas.*

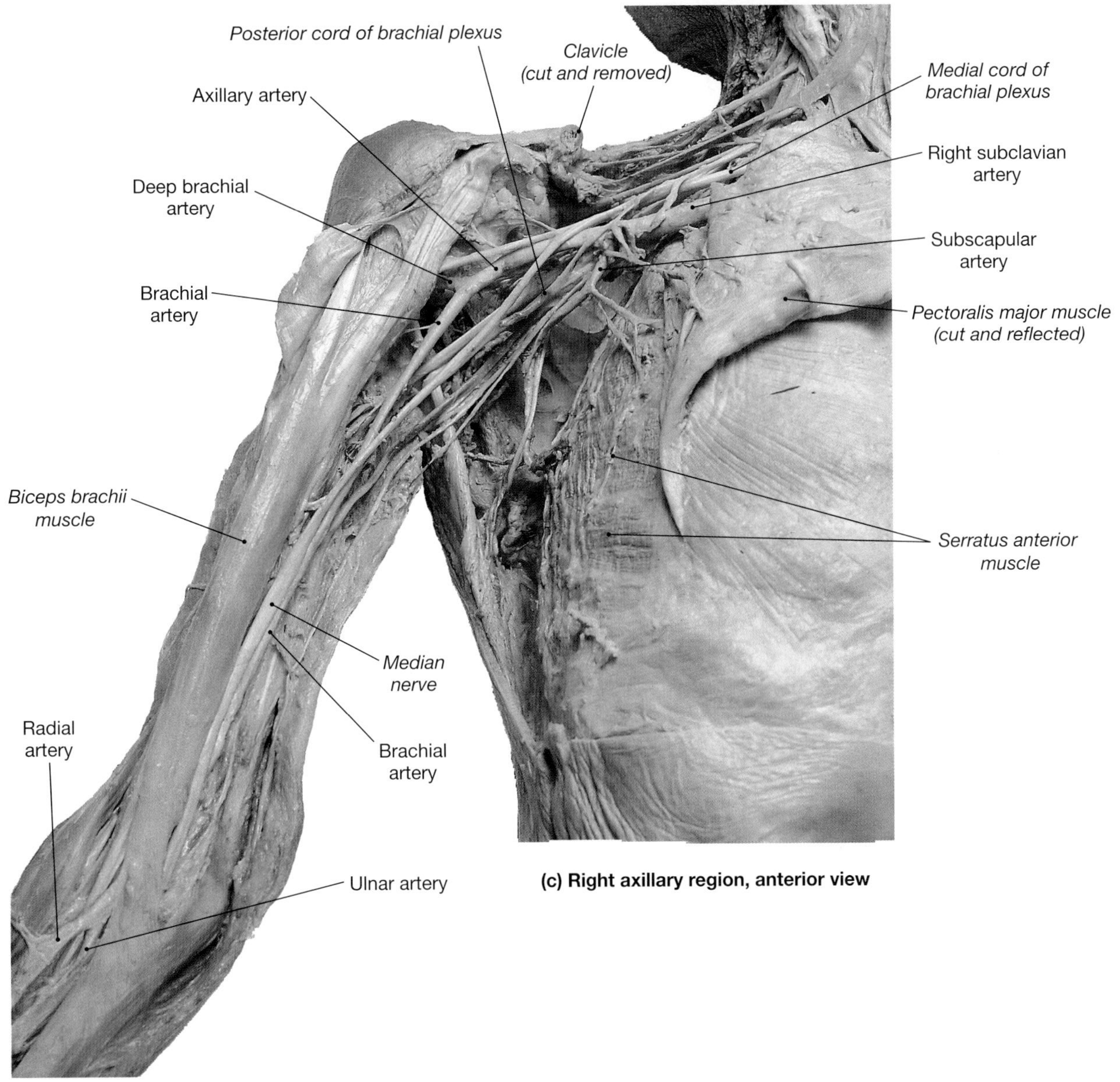

Posterior cord of brachial plexus
Clavicle (cut and removed)
Axillary artery
Medial cord of brachial plexus
Right subclavian artery
Deep brachial artery
Subscapular artery
Brachial artery
Pectoralis major muscle (cut and reflected)
Biceps brachii muscle
Serratus anterior muscle
Median nerve
Radial artery
Brachial artery
Ulnar artery

(c) Right axillary region, anterior view

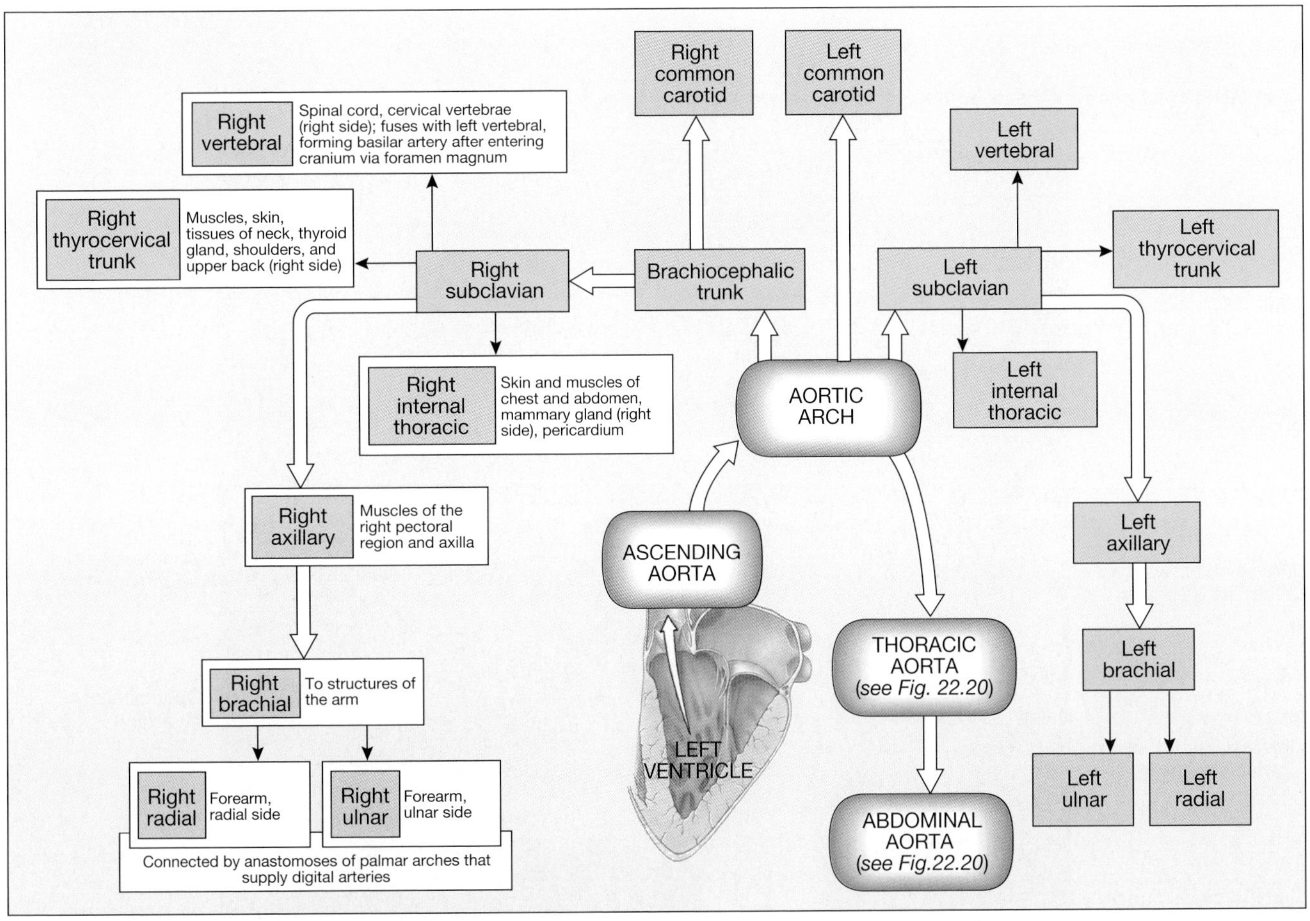

Right common carotid
Left common carotid
Right vertebral
Spinal cord, cervical vertebrae (right side); fuses with left vertebral, forming basilar artery after entering cranium via foramen magnum
Left vertebral
Right thyrocervical trunk
Muscles, skin, tissues of neck, thyroid gland, shoulders, and upper back (right side)
Right subclavian
Brachiocephalic trunk
Left subclavian
Left thyrocervical trunk
Right internal thoracic
Skin and muscles of chest and abdomen, mammary gland (right side), pericardium
AORTIC ARCH
Left internal thoracic
Right axillary
Muscles of the right pectoral region and axilla
ASCENDING AORTA
Left axillary
THORACIC AORTA (see Fig. 22.20)
Left brachial
Right brachial
To structures of the arm
LEFT VENTRICLE
Right radial
Forearm, radial side
Right ulnar
Forearm, ulnar side
Left ulnar
Left radial
Connected by anastomoses of palmar arches that supply digital arteries
ABDOMINAL AORTA (see Fig.22.20)

(d)

FIGURE 22.12 *(continued)*

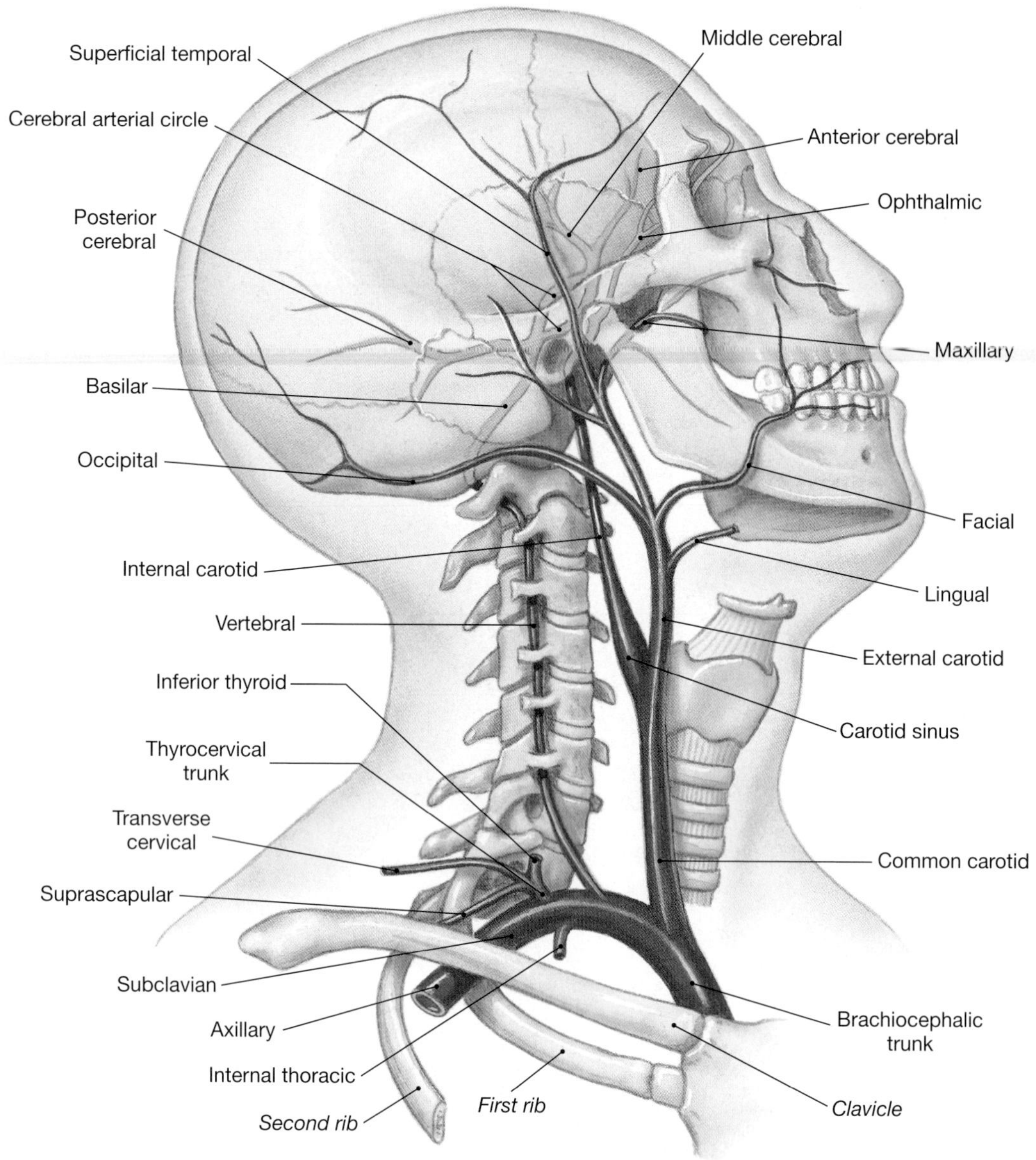

(a) Arteries of neck and head, an oblique lateral view from the right side

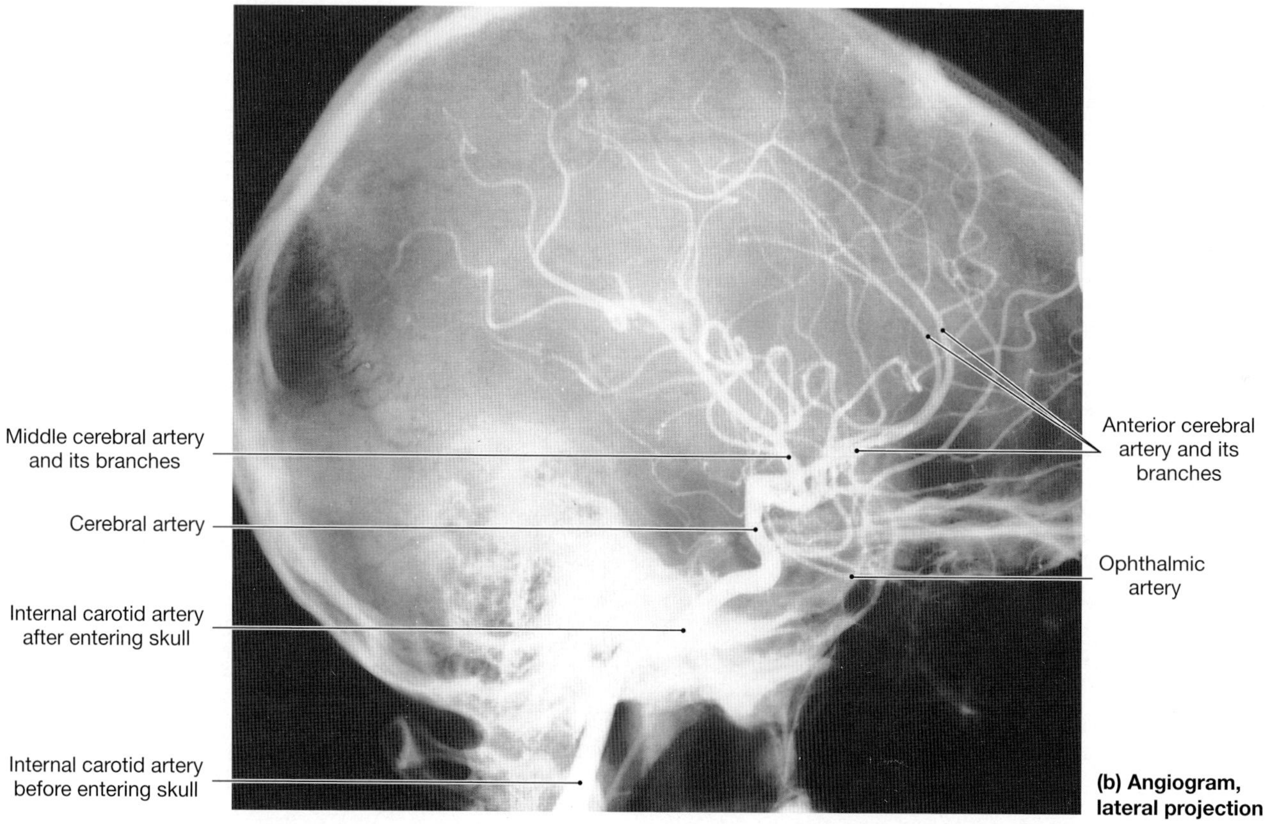

(b) Angiogram, lateral projection

FIGURE 22.13 **Arteries of the Neck and Head**

(**a**) General circulation pattern of arteries supplying the neck and superficial structures of the head; this is an oblique lateral view from the right side. (**b**) Angiogram showing the internal carotid arteries and its branches within the cranium.

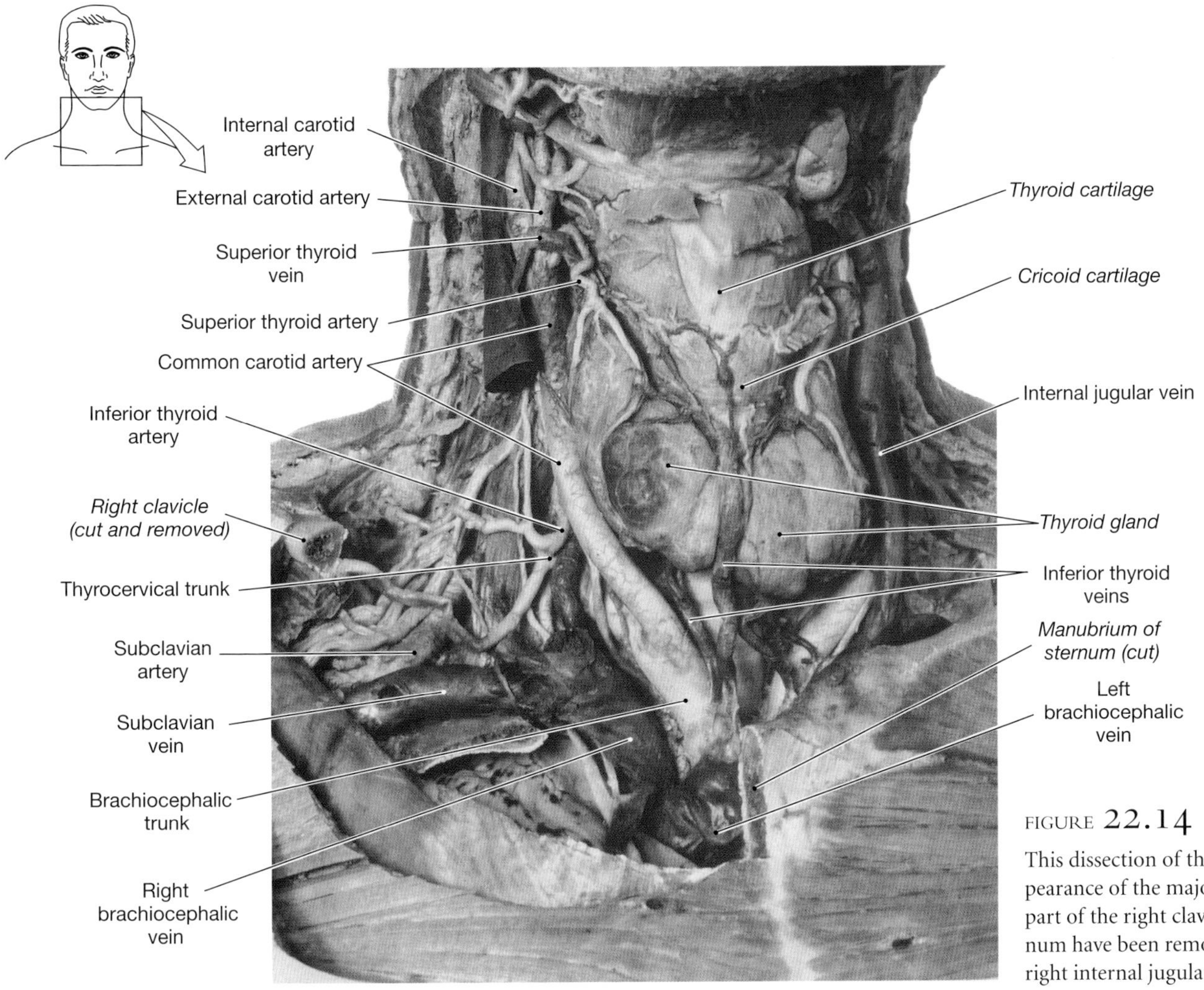

FIGURE 22.14 **MAJOR ARTERIES OF THE NECK**

This dissection of the anterior neck shows the position and appearance of the major arteries in this region. In this dissection, part of the right clavicle, first rib, and manubrium of the sternum have been removed, along with the inferior portion of the right internal jugular vein.

The internal carotid arteries normally supply the arteries of the anterior half of the cerebrum, and the rest of the brain receives blood from the vertebral arteries. But this circulatory pattern can easily change, because the internal carotid arteries and the basilar artery are interconnected in a ring-shaped anastomosis, the **cerebral arterial circle**, or *circle of Willis*, that encircles the infundibulum of the pituitary gland (Figure 22.15a,b●). With this arrangement, the brain can receive blood from either or both arterial sources, the carotids or the vertebrals, and the chances for a serious interruption of circulation are reduced.

THE DESCENDING AORTA [FIGURES 22.10/22.16/22.20] The **descending aorta** is continuous with the aortic arch. The diaphragm divides the descending aorta into a superior **thoracic aorta** and an inferior **abdominal aorta** (Figure 22.10●, p. 582). A summary of the distribution of blood from the descending aorta is presented in Figure 22.20●, p. 594. The tributaries of the thoracic aorta are shown in Figure 22.16●.

The Thoracic Aorta [FIGURE 22.16] The thoracic aorta begins at the level of vertebra T_5 and penetrates the diaphragm at the level of vertebra T_{12} (Figure 22.16●). The thoracic aorta travels within the mediastinum, on the dorsal thoracic wall, slightly to the left of the vertebral column. It supplies blood to branches servicing the viscera of the thorax, the muscles of the chest and the diaphragm, and the thoracic portion of the spinal cord. The branches of the thoracic aorta are grouped anatomically as either *visceral branches* or *parietal branches*. Visceral branches supply the organs of the chest: the **bronchial arteries** supply the conducting passageways of the lungs, **pericardial arteries** supply the pericardium, **mediastinal arteries** supply general mediastinal structures, and **esophageal arteries** supply the esophagus. The parietal branches supply the chest wall: the **intercostal arteries** supply the chest muscles and the vertebral column area, and the **superior phrenic** (FREN-ik) **arteries** deliver blood to the superior surface of the muscular diaphragm that separates the thoracic and abdominopelvic cavities.

The Abdominal Aorta [FIGURES 22.16/22.17] The abdominal aorta begins immediately inferior to the diaphragm (Figures 22.16 and 22.17●, pp. 589–592). The abdominal aorta descends slightly to the left of the vertebral column, but posterior to the peritoneal cavity; it is often surrounded by a cushion of adipose tissue. At the level of vertebra L_4, the abdominal aorta splits into the *right* and *left common iliac arteries* that supply deep pelvic structures and the lower limbs. The region where the aorta splits is called the **terminal segment of the aorta**.

The abdominal aorta delivers blood to all of the abdominopelvic organs and structures. The major branches to visceral organs are unpaired, and they arise on the anterior surface of the abdominal aorta and extend into the mesenteries to reach the visceral organs. Branches to the body wall, the kidneys, and other structures outside of the abdominopelvic cavity are paired, and they originate along the lateral surfaces of the abdominal aorta. Figure 22.16● shows the major arteries of the trunk, with the thoracic and abdominal organs removed.

FIGURE 22.15 **THE ARTERIAL SUPPLY TO THE BRAIN**

(**a**) An inferior view of the brain, showing the distribution of arteries. *See Figure 22.22b* for a comparable view of the veins on the inferior surface of the brain. (**b**) The arteries on the inferior surface of the brain; the vessels have been injected with red latex, making them easier to see. (**c**) A lateral view of the arteries supplying the brain. This is a corrosion cast: the vessels have been injected with latex, and then the brain tissue was dissolved and removed in an acid bath.

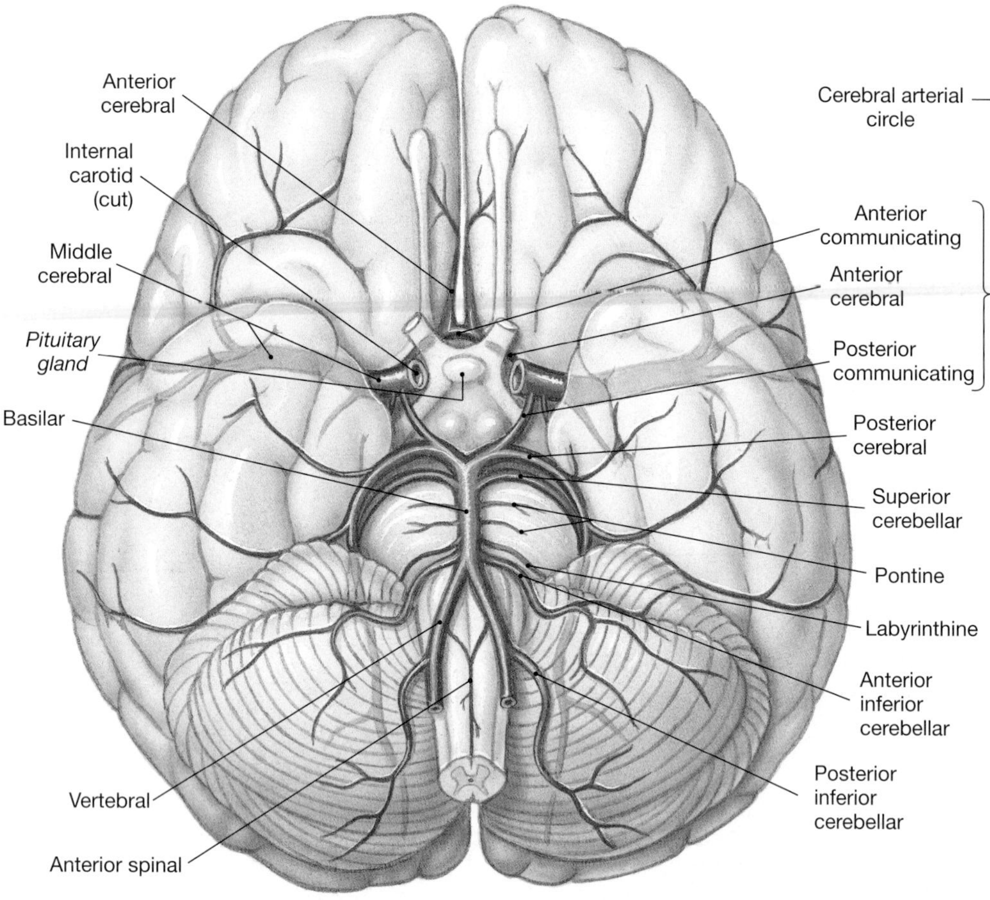

(a) Arteries of the brain, inferior view

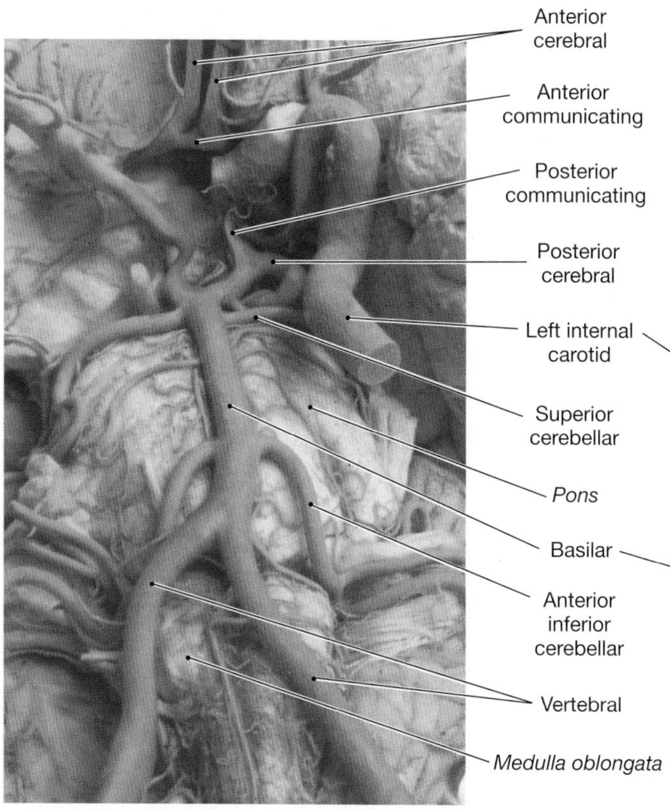

(b) Arteries injected to show cerebral arterial circle

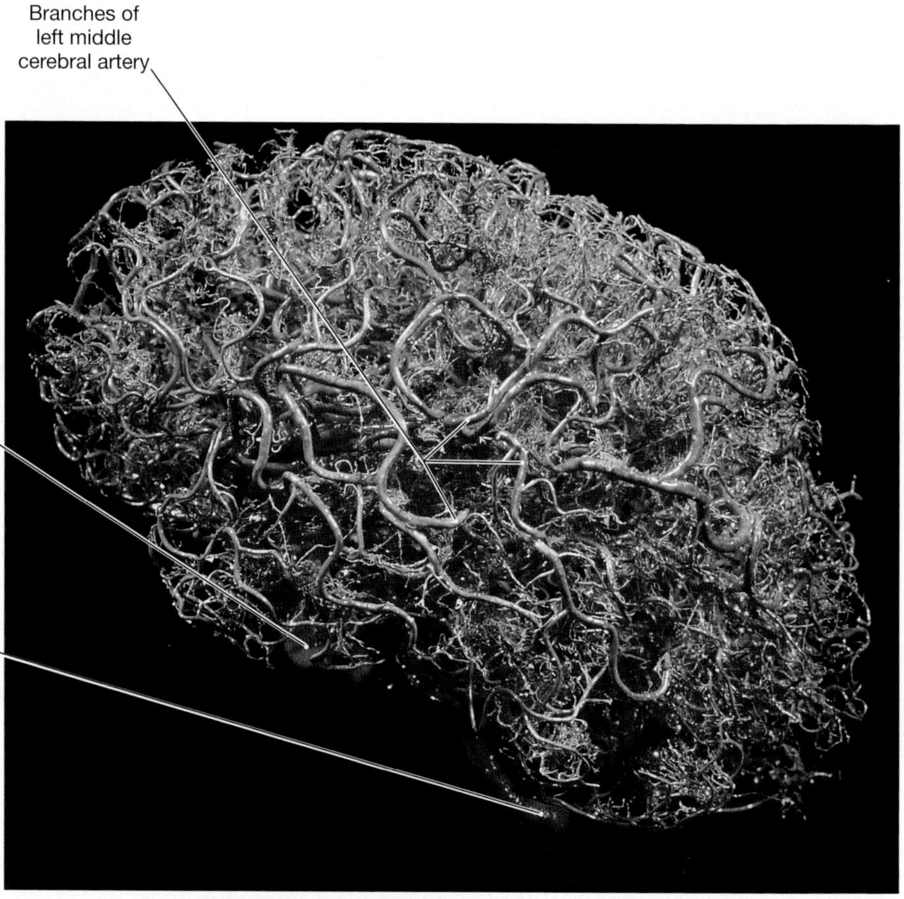

(c) Corrosion cast of cerebral arteries, left cerebral hemisphere

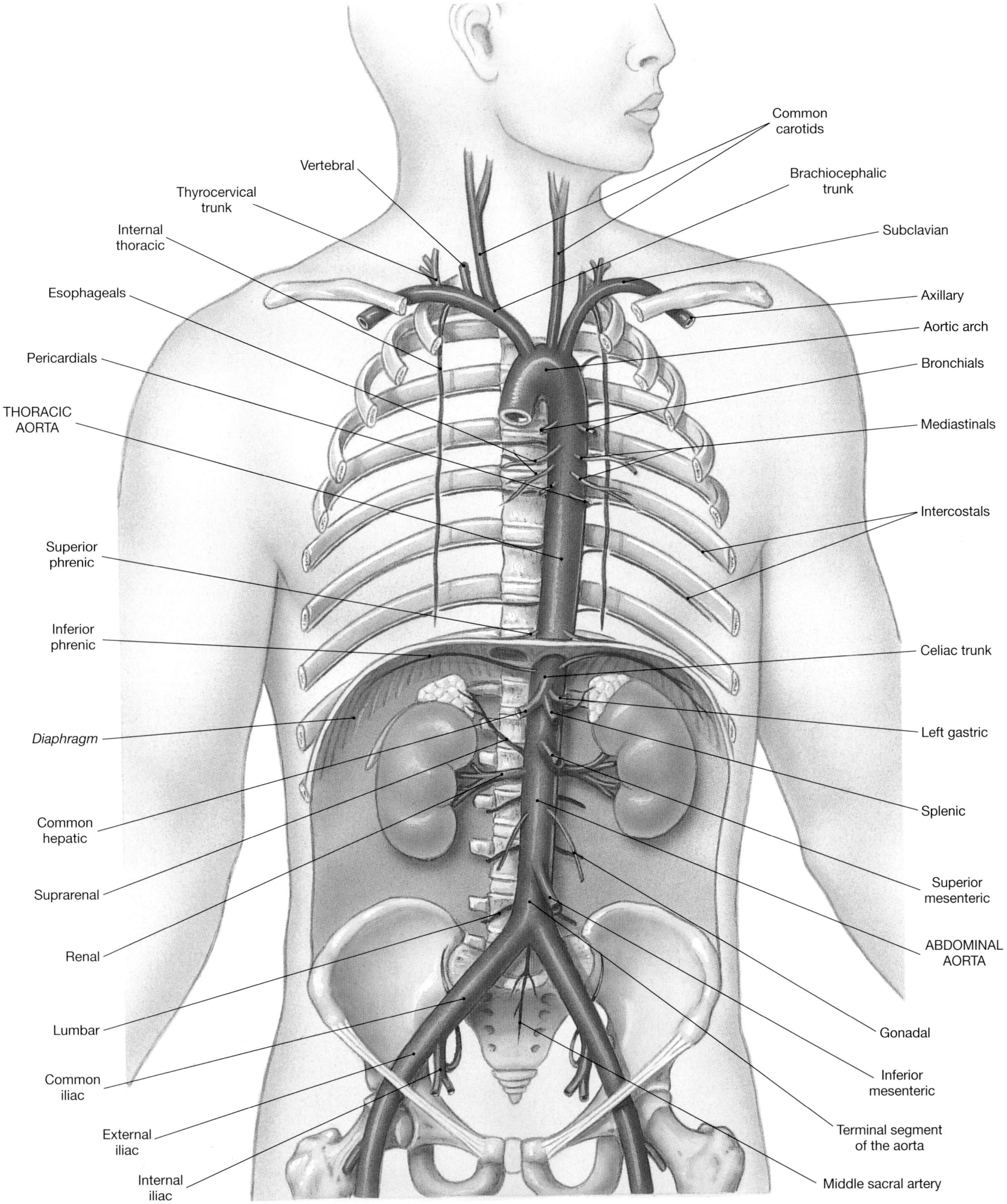

FIGURE 22.16 **MAJOR ARTERIES OF THE TRUNK.** *See MRI Scan 10d in the Companion Atlas.*

There are three unpaired arteries: (1) the *celiac trunk*, (2) the *superior mesenteric artery*, and (3) the *inferior mesenteric artery* (Figures 22.16 and 22.17●).

1. The **celiac** (SĒ-lē-ak) **trunk** delivers blood to the liver, stomach, esophagus, gallbladder, duodenum, pancreas, and spleen. The celiac artery divides into three branches:
   - the **left gastric artery**, which supplies the stomach and the inferior portion of the esophagus;
   - the **splenic artery**, which supplies the spleen and arteries to the stomach (*left gastroepiploic artery*) and pancreas (*pancreatic arteries*);
   - the **common hepatic artery**, which supplies arteries to the liver (*hepatic artery proper*), stomach (*right gastric artery*), gallbladder (*cystic artery*), and duodenal area (*gastroduodenal*, *right gastroepiploic*, and *superior pancreaticoduodenal arteries*).
2. The **superior mesenteric** (mez-en-TER-ik) **artery** arises about 2.5 cm inferior to the celiac trunk to supply arteries to the pancreas and duodenum (*inferior pancreaticoduodenal artery*), small intestine (*intestinal arteries*), and most of the large intestine (*right colic*, *middle colic*, and *ileocolic arteries*).
3. The **inferior mesenteric artery** arises about 5 cm superior to the terminal segment of the aorta and delivers blood to the terminal portions of the colon (*left colic* and *sigmoid arteries*) and the rectum (*rectal arteries*).

There are five paired arteries: (1) the *inferior phrenics*, (2) the *suprarenals*, (3) the *renals*, (4) the *lumbars*, and (5) the *gonadals*.

1. The **inferior phrenic arteries** supply the inferior surface of the diaphragm and the inferior portion of the esophagus.
2. The **suprarenal arteries** originate on either side of the aorta near the base of the superior mesenteric artery. Each suprarenal artery supplies an adrenal gland, which caps the superior part of a kidney.
3. The short (about 7.5 cm) **renal arteries** arise along the posterolateral surface of the abdominal aorta, about 2.5 cm inferior to the superior mesenteric artery, and travel posterior to the peritoneal lining to reach the adrenal glands and kidneys. We shall discuss branches of the renal arteries in Chapter 26.
4. **Gonadal** (gō-NAD-al) **arteries** originate between the superior and inferior mesenteric arteries. In males, they are called *testicular arteries* and are long, thin arteries that supply blood to the testes and scrotum. In females, they are termed *ovarian arteries* and supply blood to the ovaries, uterine tubes, and uterus. The distribution of gonadal vessels (both arteries and veins) differs in males and females; the differences will be described in Chapter 27.
5. Small **lumbar arteries** arise on the posterior surface of the aorta and supply the vertebrae, spinal cord, and the abdominal wall.

ARTERIES OF THE PELVIS AND LOWER LIMBS [FIGURES 22.16/22.18/22.19] Near the level of vertebra L_4, the terminal segment of the abdominal aorta divides to form a pair of muscular arteries, the **right** and **left common iliac** (IL-ē-ak) **arteries** and the small **middle sacral artery**. These arteries carry blood to the pelvis and lower limbs (Figures 22.16, 22.18, and 22.19●). As these arteries travel along the inner surface of the ilium, they descend posterior to the cecum and sigmoid colon, and, at the level of the lumbosacral joint, each common iliac divides to form an **internal iliac artery** and an **external iliac artery**. The internal iliac arteries enter the pelvic cavity to supply the urinary bladder, the internal and external walls of the pelvis, the

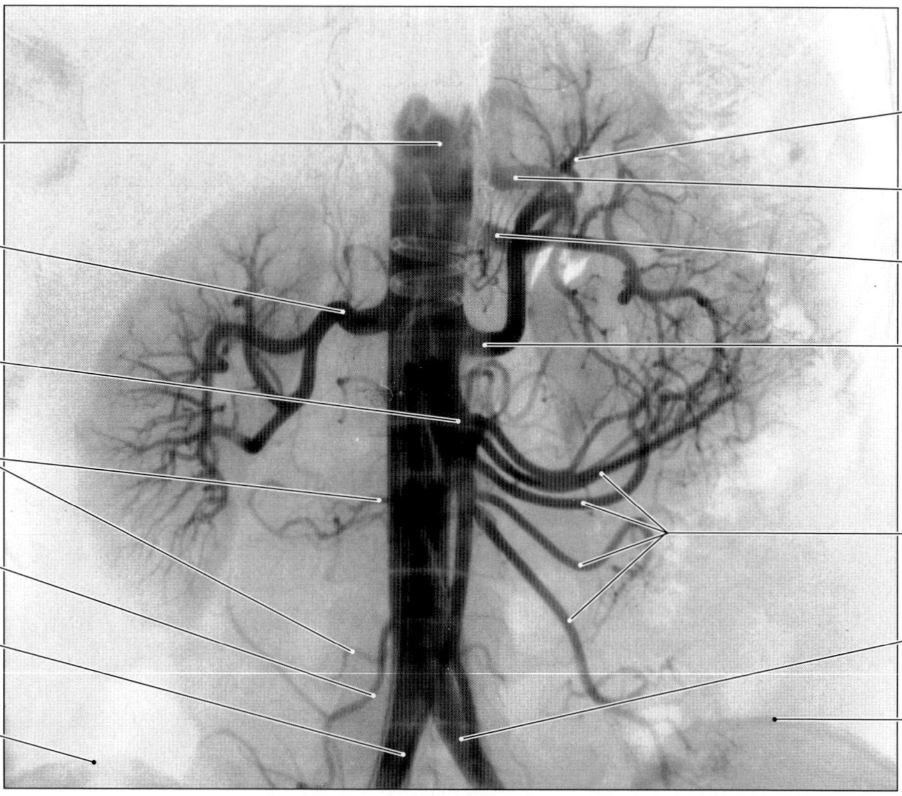

FIGURE 22.17 **ARTERIES OF THE ABDOMEN**

(**a**) Angiogram of the abdominal aorta, showing arteries to the viscera and to the kidneys (anterior/posterior projection). (**b**) Major arteries supplying the abdominal viscera. *See Scan 10d in the Companion Atlas.*

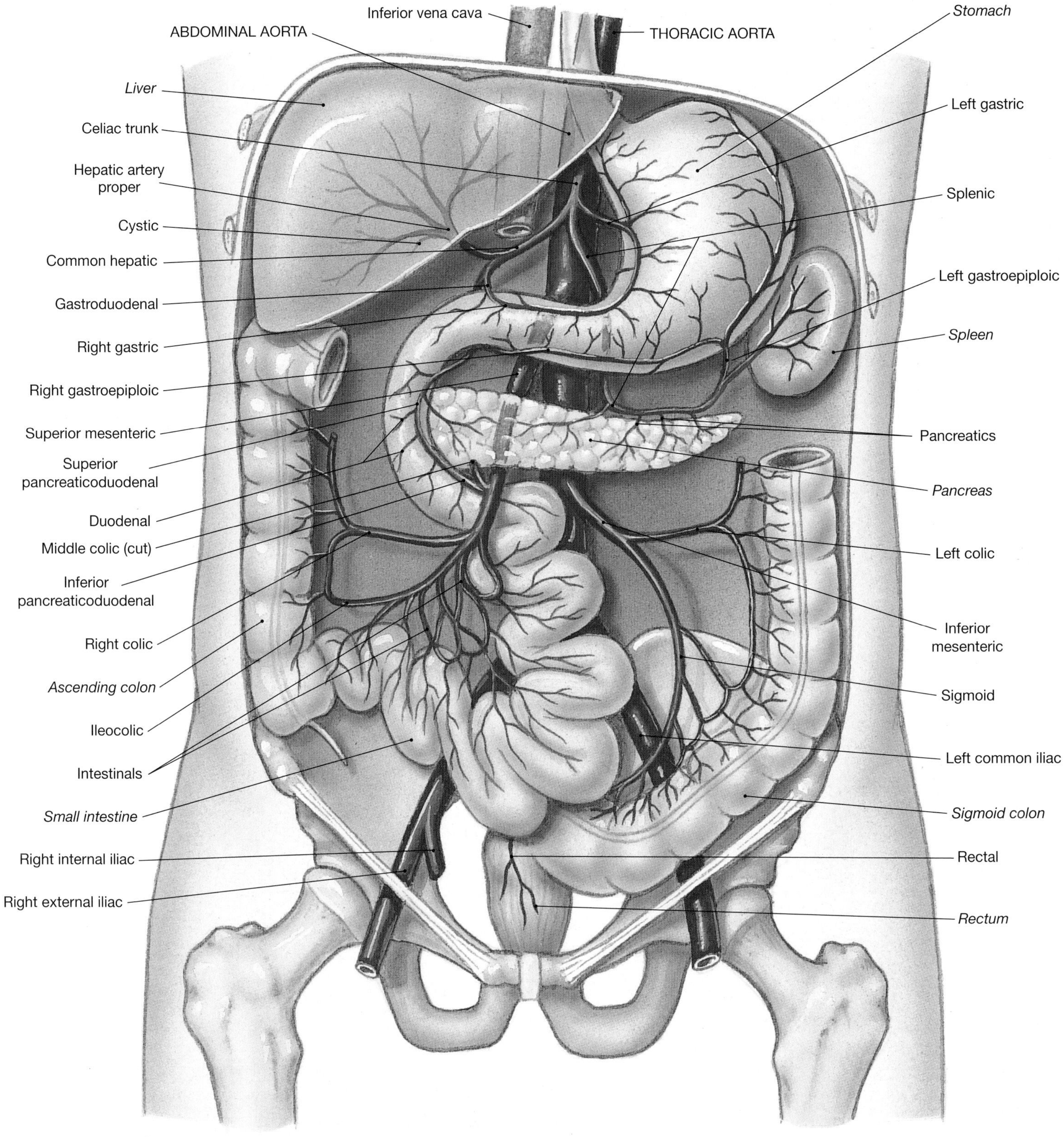

FIGURE 22.17 *(continued)* **(b) Arteries supplying abdominal organs, anterior view**

external genitalia, and the medial side of the thigh. The major tributaries of the internal iliac artery are the *superior gluteal, internal pudendal, obturator,* and *lateral sacral arteries.* In females, these vessels also supply the uterus and vagina. The external iliac arteries supply blood to the lower limbs, and they are much larger in diameter than the internal iliac arteries.

Arteries of the Thigh and Leg [FIGURES 22.18/22.19] The external iliac artery crosses the surface of the iliopsoas muscle and penetrates the abdominal wall midway between the anterior superior iliac spine and the symphysis pubis. It emerges on the anteromedial surface of the thigh as the **femoral artery** and roughly 5 cm distal to its emergence, the **deep femoral artery**

branches off its lateral surface (Figure 22.18●). The deep femoral artery, which gives rise to the *medial* and *lateral circumflex arteries*, supplies blood to the ventral and lateral regions of the skin and deep muscles of the thigh.

The femoral artery continues inferiorly and posterior to the femur. As it reaches the popliteal fossa, it gives off a branch, the **descending genicular artery**, which supplies the medial aspect of the knee. The femoral artery continues and as it passes through the adductor magnus muscle it becomes the **popliteal** (pop-LIT-ē-al) **artery** (Figure 22.19●). The popliteal

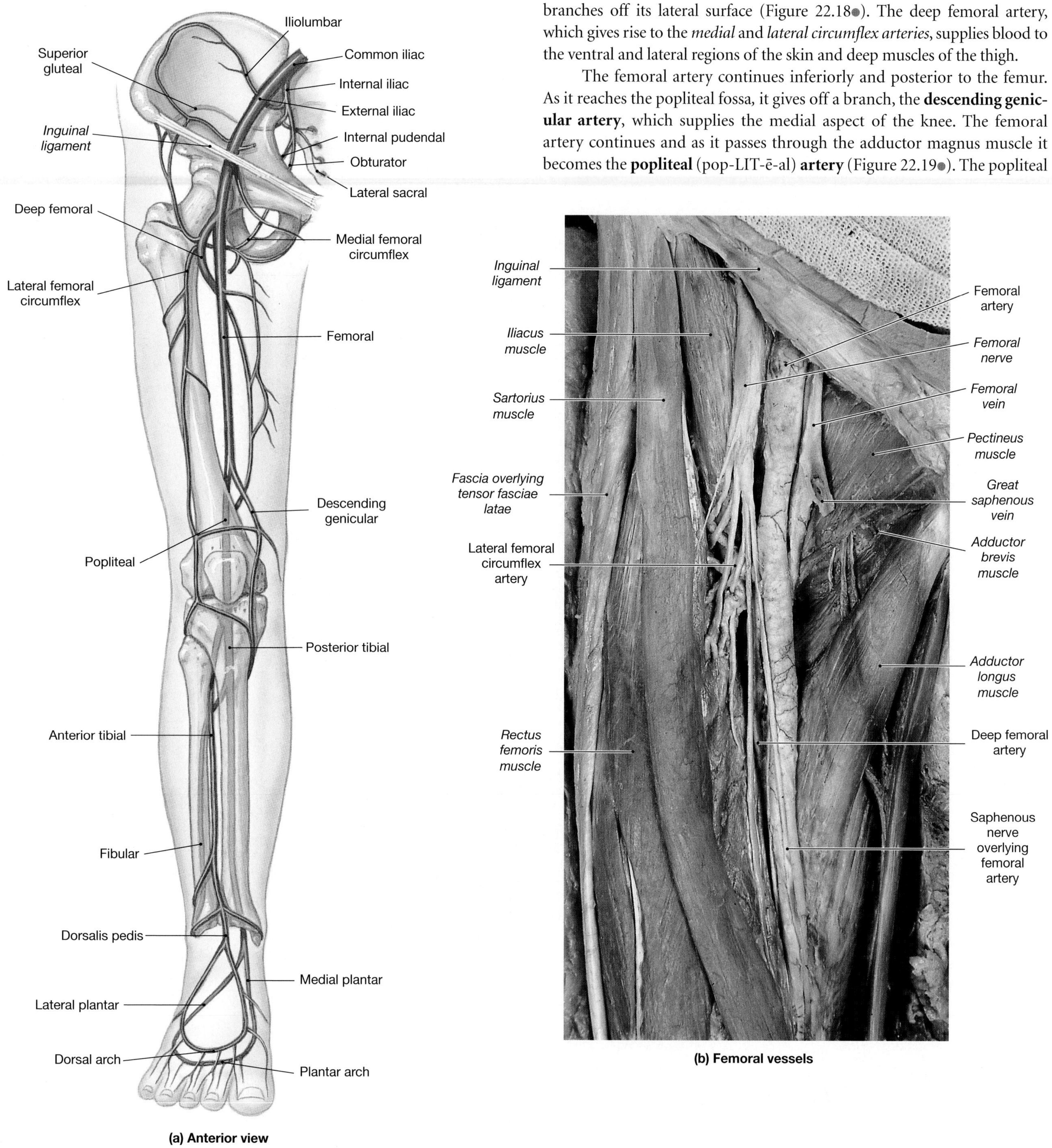

FIGURE 22.18 **Major Arteries of the Lower Limb, Part I**

(**a**) Anterior view of the arteries supplying the right lower limb. (**b**) Major arteries of the right thigh. *See MRI Scans 5, 6b, and 7 in the Companion Atlas.*

artery crosses the popliteal fossa before branching to form the **posterior tibial artery** and the **anterior tibial artery**. The posterior tibial artery gives rise to the **fibular artery**, or *peroneal artery*, and continues inferiorly along the posterior surface of the tibia. The anterior tibial artery passes between the tibia and fibula, emerging on the anterior surface of the tibia. As it descends toward the foot, the anterior tibial artery provides blood to the skin and muscles of the anterior portion of the leg.

Arteries of the Foot [FIGURES 22.18/22.20] When the anterior tibial artery reaches the ankle, it becomes the **dorsalis pedis artery**. The dorsalis pedis branches repeatedly, supplying the ankle and dorsal portion of the foot (Figure 22.18●).

As it reaches the ankle, the posterior tibial artery divides to form the **medial** and **lateral plantar arteries**, which supply blood to the plantar surface of the foot. The medial and lateral plantar arteries are connected to the dorsalis pedis artery by a pair of anastomoses. This connection links the **dorsal arch** (*arcuate arch*) to the **plantar arch**. Small arteries branching off these arches supply the distal portions of the foot and the toes.

Before proceeding, review Figure 22.20●, which summarizes the distribution of blood from the thoracic, abdominal, and terminal portions of the aorta.

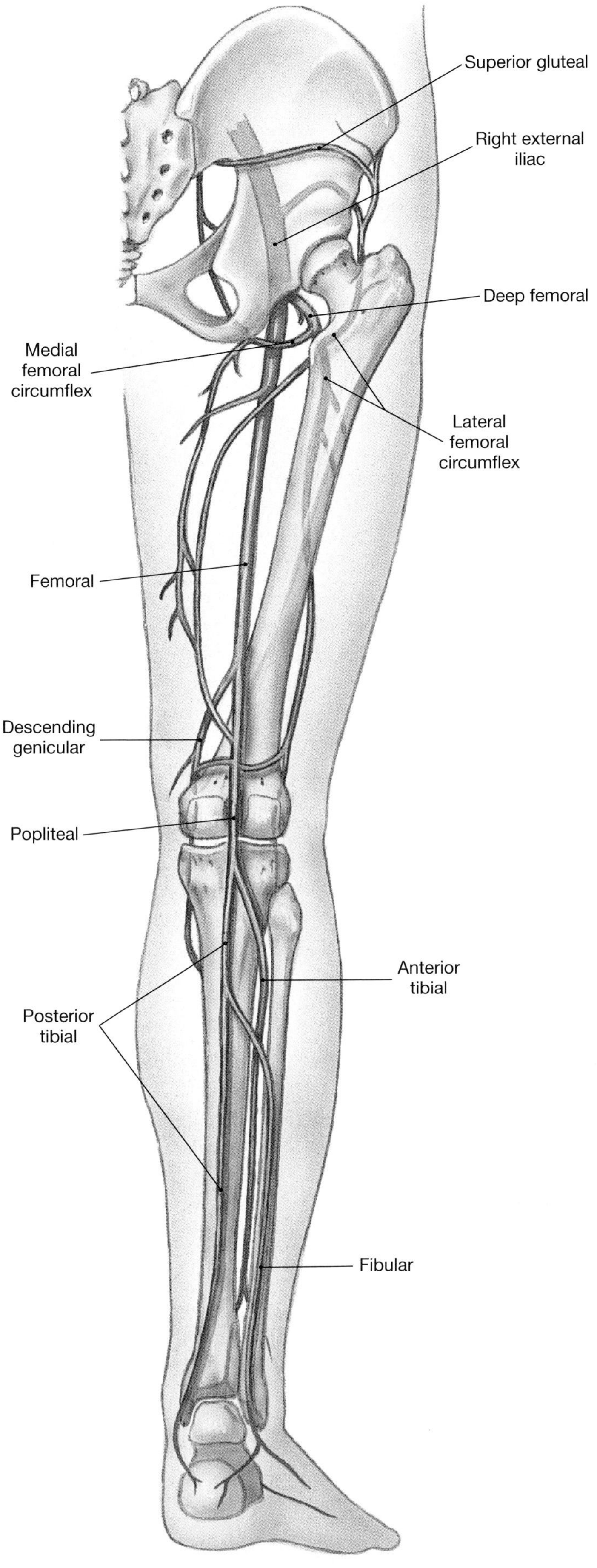

(a) Posterior view

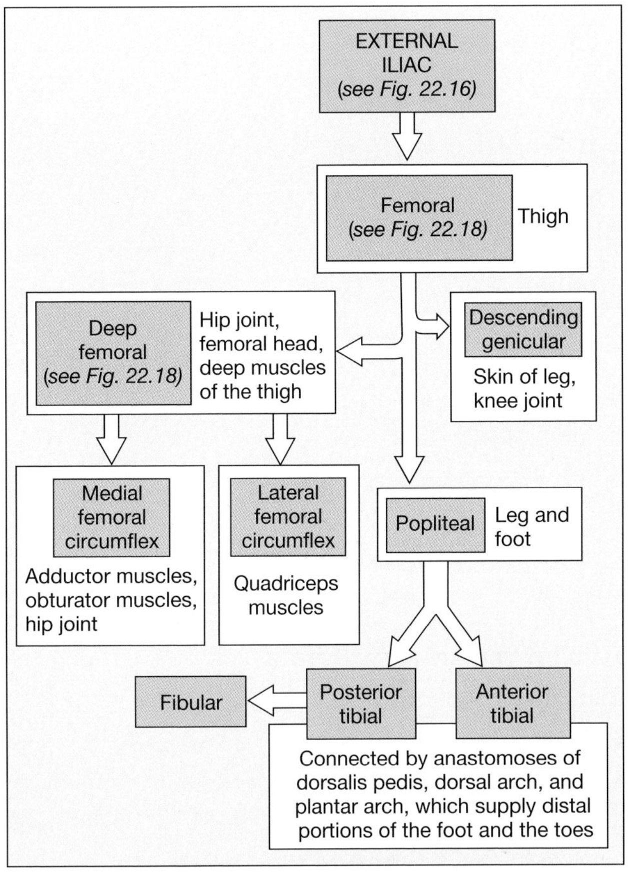

(b) Major arteries of the lower limb

FIGURE 22.19 **Major Arteries of the Lower Limb, Part II**

(**a**) Posterior view of the arteries supplying the right lower limb. (**b**) A summary of the major arteries of the lower limb. *See MRI Scans 5, 6b, and 7 in the Companion Atlas.*

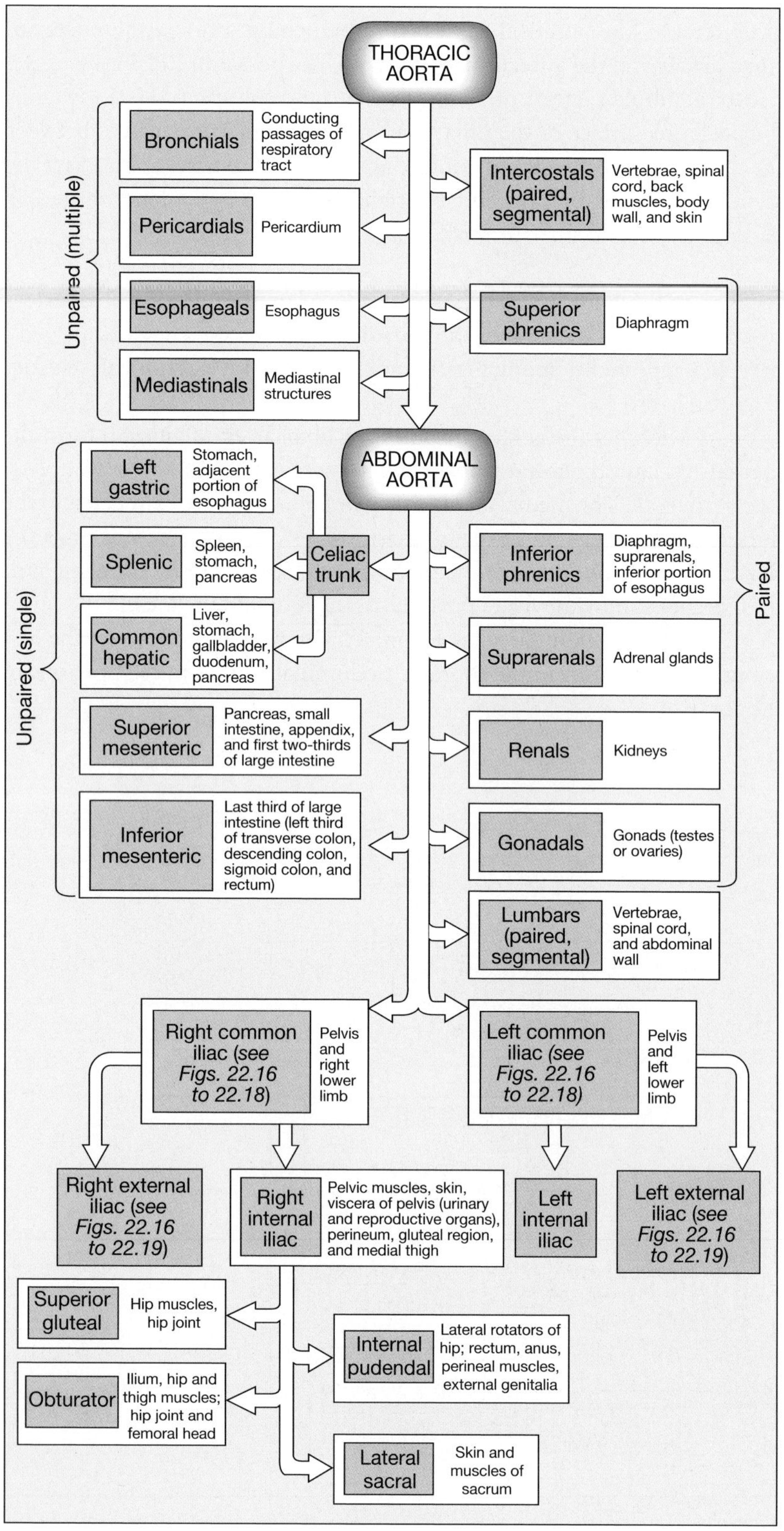

FIGURE 22.20 **A Summary of the Arterial System**
The distribution of blood from the aorta.

✓ **CONCEPT CHECK**

- What regions of the body receive their blood from the carotid arteries?
- Which artery is found at the biceps region of the right arm?
- What artery does the external iliac become after leaving the abdominal cavity?
- Would damage to the internal carotid arteries always result in brain damage? Why or why not?

SYSTEMIC VEINS [FIGURES 22.10/22.21]

Veins collect blood from the body's tissues and organs in an elaborate venous network that drains into the right atrium of the heart through the *superior* and *inferior venae cavae* (Figure 22.21●). A comparison between Figure 22.21 and Figure 22.10● reveals that arteries and veins typically run side by side, and in many cases they have comparable names. For example, the axillary arteries run alongside the axillary veins. In addition, arteries and veins often travel in the company of peripheral nerves that have the same names and innervate the same structures. *p. 367*

One significant difference between the arterial and venous systems concerns the distribution of major veins in the neck and limbs. Arteries in these areas are not found at the body surface; instead, they are deep, protected by bones and surrounding soft tissues. In contrast, the neck and limbs usually have two sets of peripheral veins, one superficial and the other deep. The superficial veins are so close to the surface that they can be seen quite easily. Because they are so close to the surface, they are easy targets for obtaining blood samples, and most blood tests are performed on venous blood collected from the superficial veins of the upper limb (usually the antecubital surface).

This dual venous drainage plays an important role in the control of body temperature. When body temperature becomes abnormally low, the arterial blood supply to the skin is reduced, and the superficial veins are bypassed. Blood enters the limbs, then returns to the trunk in the deep veins. When overheating occurs, the blood supply to the skin increases, and the superficial veins dilate. This mechanism is one reason why superficial veins in the arms and legs become prominent during periods of heavy exercise or when sitting in a sauna, hot tub, or steam bath.

The branching pattern of peripheral veins is much more variable than that of arteries. Arterial pathways are usually direct, because developing arteries grow toward active tissues. By the time blood reaches the venous system, pressures are low, and routing variations make little functional difference. The discussion that follows is based on the most common arrangement of veins.

The Superior Vena Cava [FIGURES 22.21/22.24a] All of the systemic veins (except the cardiac veins, which drain into the coronary sinus) drain into either the *superior vena cava* or the *inferior vena cava*. The **superior vena cava (SVC)** receives blood from the tissues and organs of the head, neck, chest, shoulders, and upper limbs (Figures 22.21 and 22.24a●, p. 598).

Venous Return from the Cranium [FIGURE 22.22] Numerous *superficial cerebral veins* and *internal cerebral veins* drain the cerebral hemispheres. The **superficial cerebral veins** empty into a network of dural sinuses, including the *superior* and *inferior sagittal sinuses*, the *petrosal sinuses*, the *occipital sinus*, the *left* and *right transverse sinuses*, and the *straight sinus* (Figure 22.22●). The largest sinus, the **superior sagittal sinus**, is in the falx cerebri. *p. 387* The majority of the **internal cerebral veins** collect inside the brain to form the **great cerebral vein**, which collects blood from the interior of the cerebral hemispheres and the choroid plexus and delivers it to the **straight sinus**. Other cerebral veins drain into the **cavernous sinus** in company with numerous small veins from the orbit. Blood from the cavernous sinus reaches the internal jugular vein through the petrosal sinuses.

The venous sinuses converge within the dura mater in the region of the lambdoid suture. The left and right transverse sinuses converge at the base of the petrous portion of the temporal bone, forming the **sigmoid sinus**, which penetrates the jugular foramen and leaves the skull as the **internal jugular vein**. The internal jugular vein descends parallel to the common carotid artery in the neck.

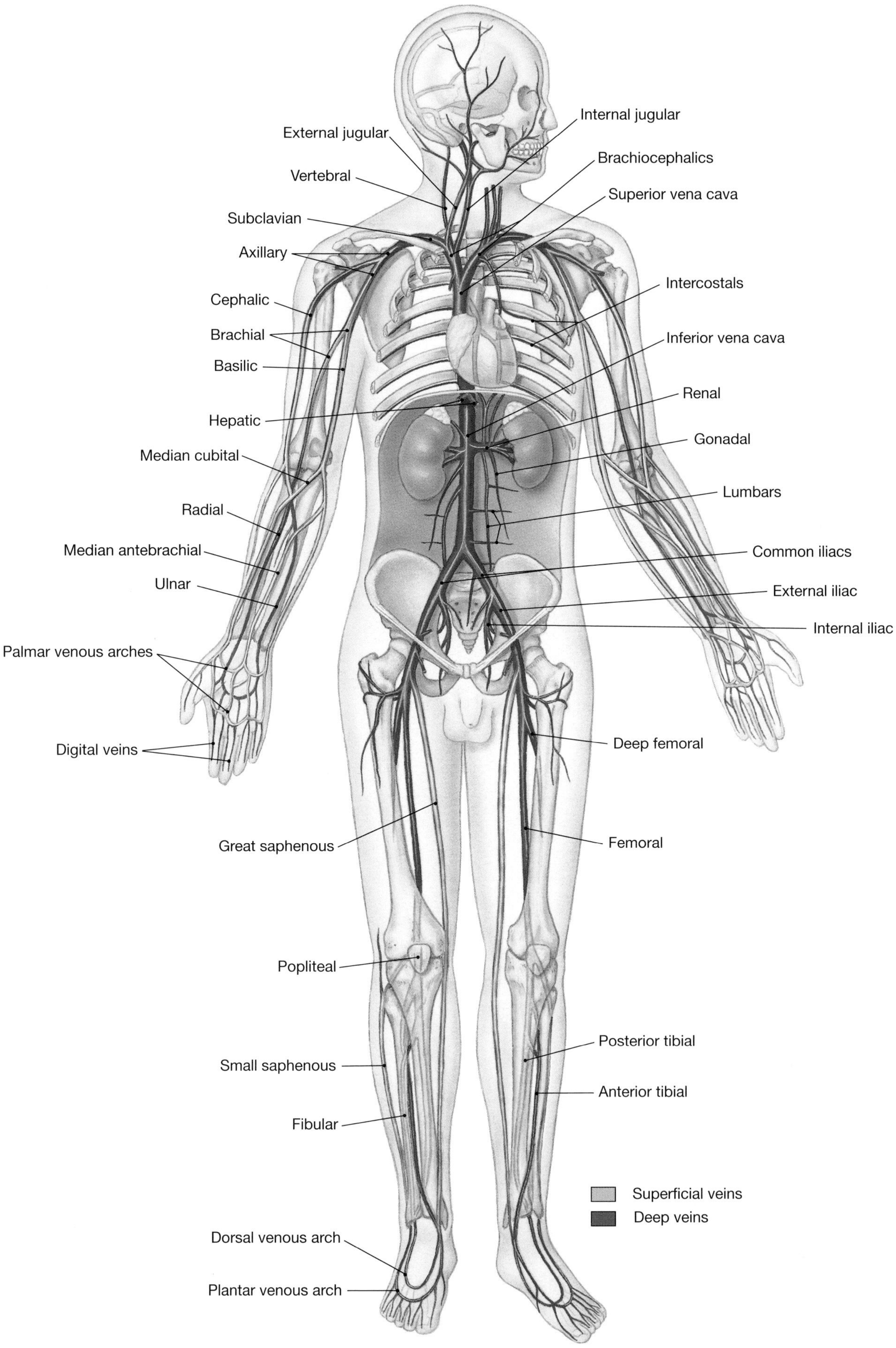

External jugular
Vertebral
Subclavian
Axillary
Cephalic
Brachial
Basilic
Hepatic
Median cubital
Radial
Median antebrachial
Ulnar
Palmar venous arches
Digital veins
Great saphenous
Popliteal
Small saphenous
Fibular
Dorsal venous arch
Plantar venous arch
Internal jugular
Brachiocephalics
Superior vena cava
Intercostals
Inferior vena cava
Renal
Gonadal
Lumbars
Common iliacs
External iliac
Internal iliac
Deep femoral
Femoral
Posterior tibial
Anterior tibial
Superficial veins
Deep veins

FIGURE 22.21 **AN OVERVIEW OF THE SYSTEMIC VENOUS SYSTEM**

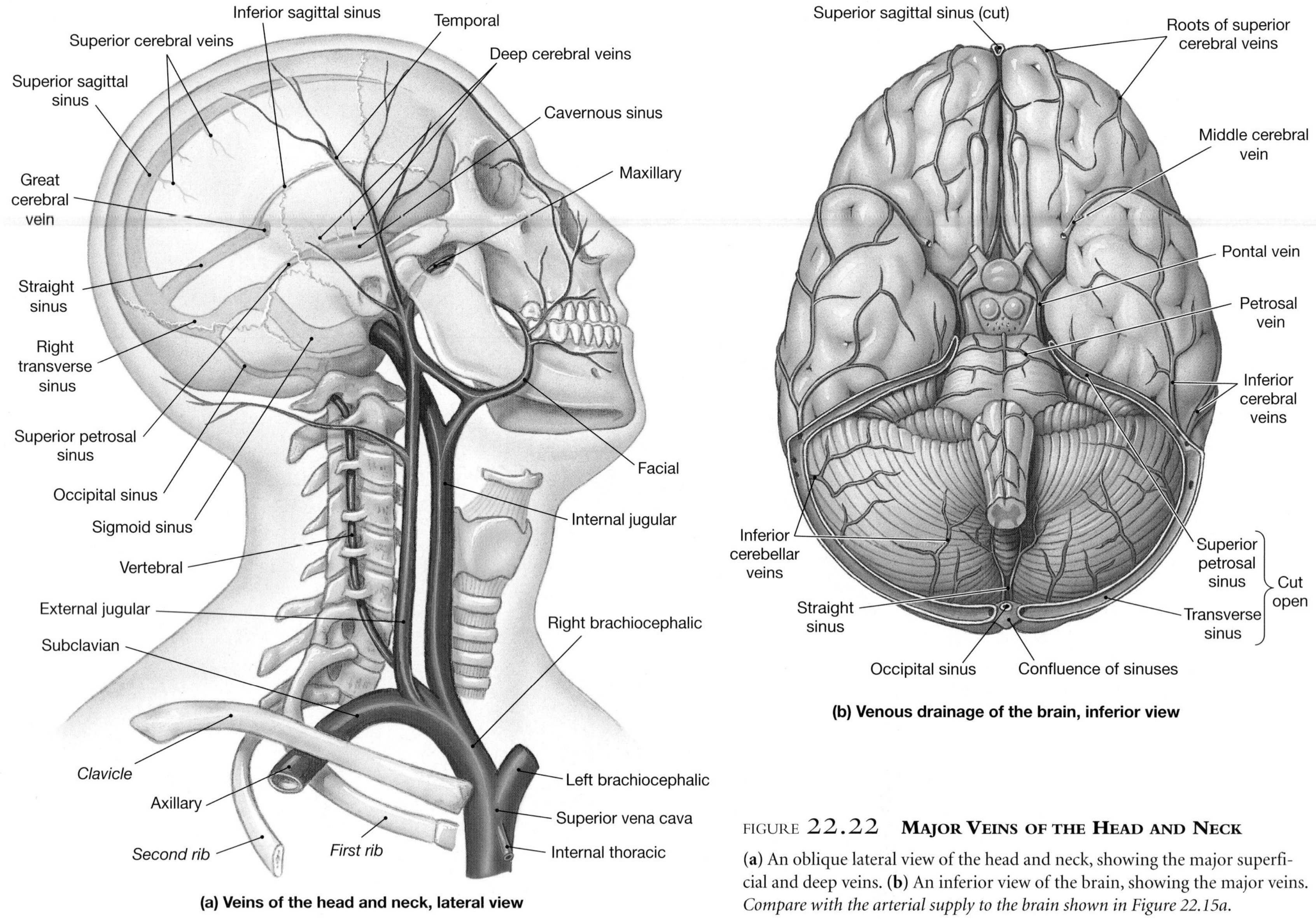

FIGURE 22.22 **MAJOR VEINS OF THE HEAD AND NECK**

(**a**) An oblique lateral view of the head and neck, showing the major superficial and deep veins. (**b**) An inferior view of the brain, showing the major veins. *Compare with the arterial supply to the brain shown in Figure 22.15a.*

Vertebral veins drain the cervical spinal cord and the posterior surface of the skull. These vessels descend within the transverse foramina of the cervical vertebrae, in company with the vertebral arteries. The vertebral veins empty into the *brachiocephalic veins* of the chest.

Superficial Veins of the Head and Neck [FIGURE 22.22] Superficial veins of the head converge to form the **temporal**, **facial**, and **maxillary veins** (Figure 22.22●). The temporal and maxillary veins drain into the **external jugular vein**. The facial vein drains into the internal jugular vein; a broad anastomosis between the external and internal jugular veins at the angle of the mandible provides dual venous drainage of the face, scalp, and cranium. The external jugular vein descends superficial to the sternocleidomastoid muscle. Posterior to the clavicle, the external jugular empties into the *subclavian vein*. In healthy individuals, the external jugular vein is easily palpable, and a *jugular venous pulse* (*JVP*) can sometimes be seen at the base of the neck.

Venous Return from the Upper Limb [FIGURE 22.23] The **digital veins** empty into the **superficial** and **deep palmar veins** of the hand, which interconnect to form the **palmar venous arches** (Figure 22.23●). The superficial arch empties into the **cephalic vein**, which ascends along the radial side of the forearm, the **median antebrachial vein**, and the **basilic vein**, which ascends on the ulnar side. Anterior to the elbow is the superficial **median cubital vein**, which interconnects the cephalic and basilic veins. Venous blood samples are typically collected from the median cubital vein.

From the elbow, the basilic vein passes superiorly along the medial surface of the biceps brachii muscle. As it approaches the axilla, the basilic vein joins the brachial vein to form the **axillary vein** (Figure 22.23●).

The deep palmar veins drain into the **radial vein** and the **ulnar vein**. After crossing the elbow, these veins fuse to form the **brachial vein**. The brachial vein lies parallel to the brachial artery. As the brachial vein continues toward the trunk, it receives blood from the basilic vein before entering the axilla as the axillary vein.

The Formation of the Superior Vena Cava [FIGURES 22.23/22.24] The cephalic vein joins the axillary vein on the outer surface of the first rib, forming the **subclavian vein**, which continues onto the chest. The subclavian vein passes over the superior surface of the first rib, along the clavicle, and into the thoracic cavity. The subclavian then merges with the external and internal jugular veins of that side. This creates the **brachiocephalic vein**, or *innominate vein* (Figure 22.23●). The brachiocephalic vein receives blood from the **vertebral**

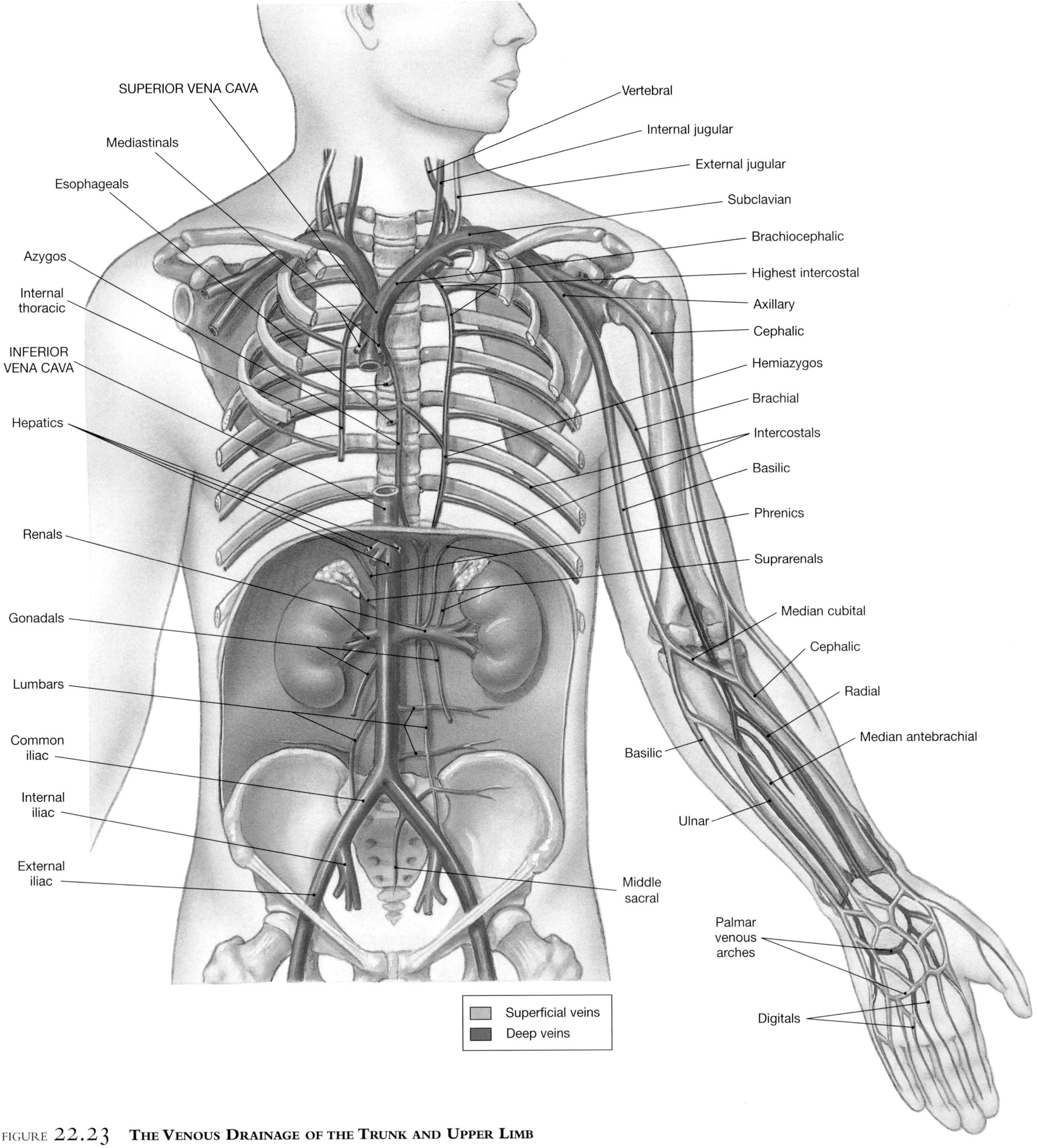

FIGURE 22.23 **The Venous Drainage of the Trunk and Upper Limb**

FIGURE 22.24 **A Summary Flow Chart of the Venous System**

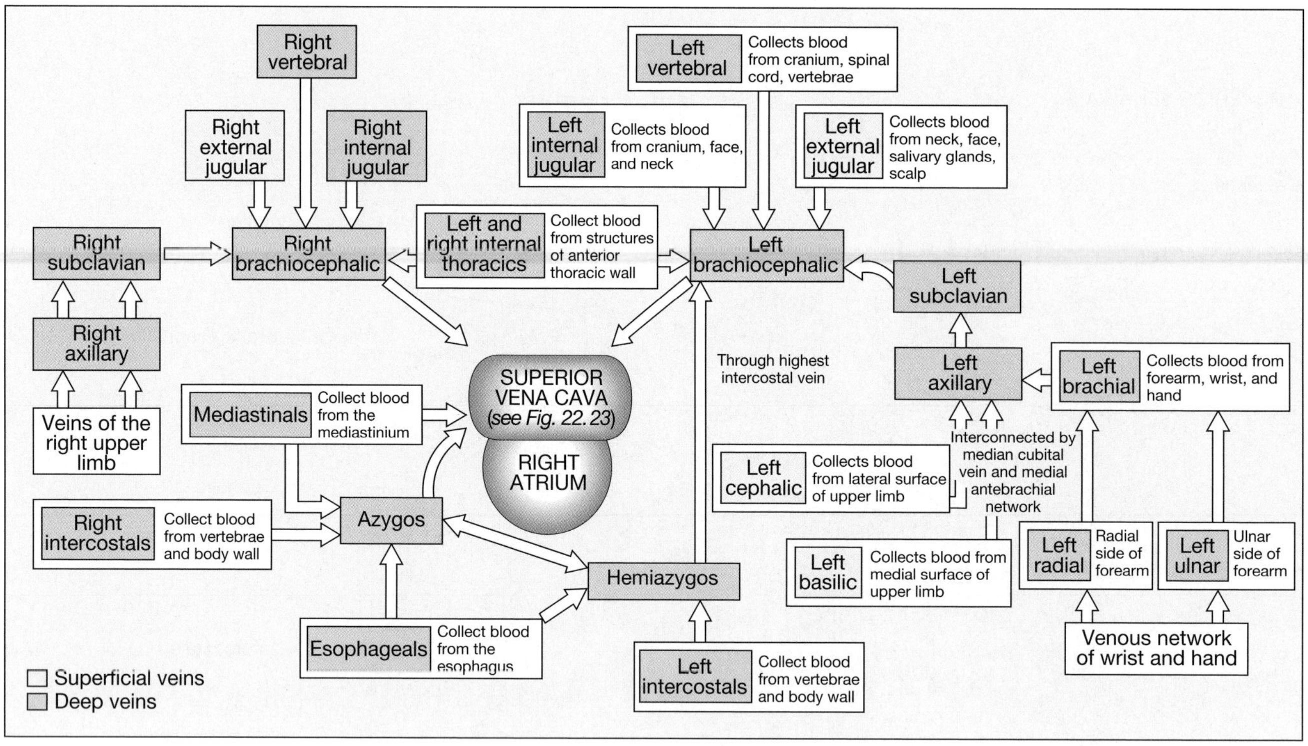

(**a**) Tributaries of the superior vena cava.

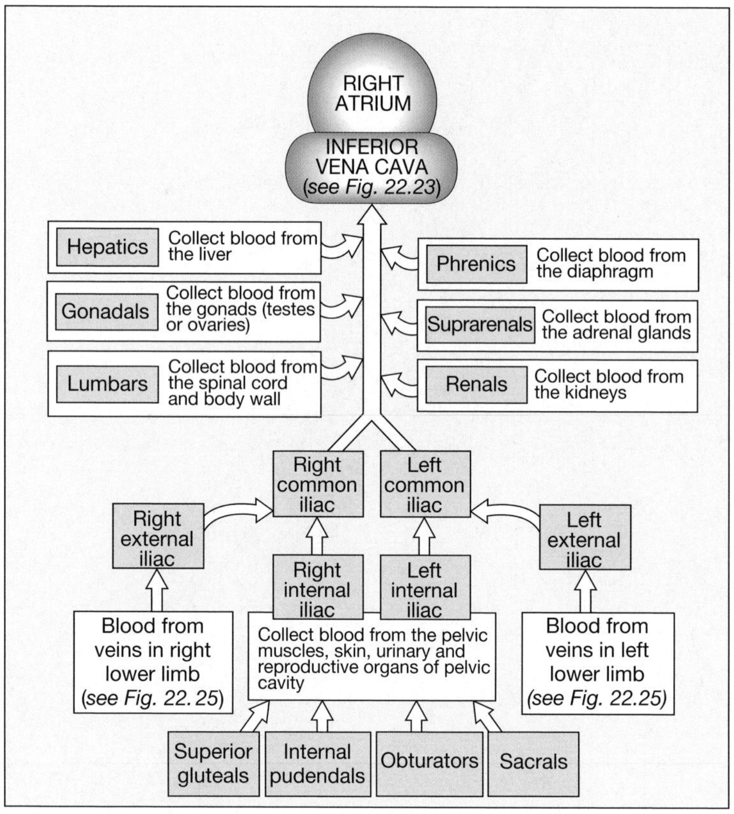

(**b**) Tributaries of the inferior vena cava.

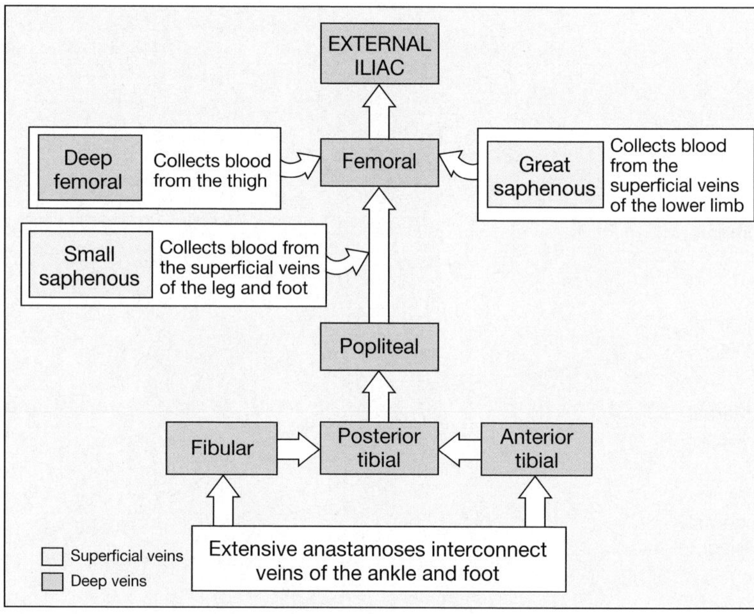

(**c**) Summary of the veins of the lower limb.

vein, which drains the posterior portion of the skull and the spinal cord. At the level of the first and second ribs, the left and right brachiocephalic veins merge to form the superior vena cava (SVC). Close to the point of fusion, the **internal thoracic vein** empties into the left brachiocephalic vein.

The **azygos** (AZ-i-gos) **vein** is the major tributary of the superior vena cava. This vessel ascends from the lumbar region over the right side of the vertebral column to invade the thoracic cavity through the diaphragm. The azygos joins the superior vena cava at the level of vertebra T_2. The azygos receives blood from the smaller **hemiazygos vein**. The hemiazygos vein may also drain into the **highest intercostal vein**, a tributary of the left brachiocephalic vein. The azygos and hemiazygos veins are the chief collecting vessels of the thorax. They receive blood from: (1) numerous **intercostal veins**, which receive blood from the chest muscles; (2) **esophageal veins**, which drain blood from the esophagus; and (3) smaller veins draining other mediastinal structures.

Figure 22.24a● summarizes the venous tributaries of the superior vena cava.

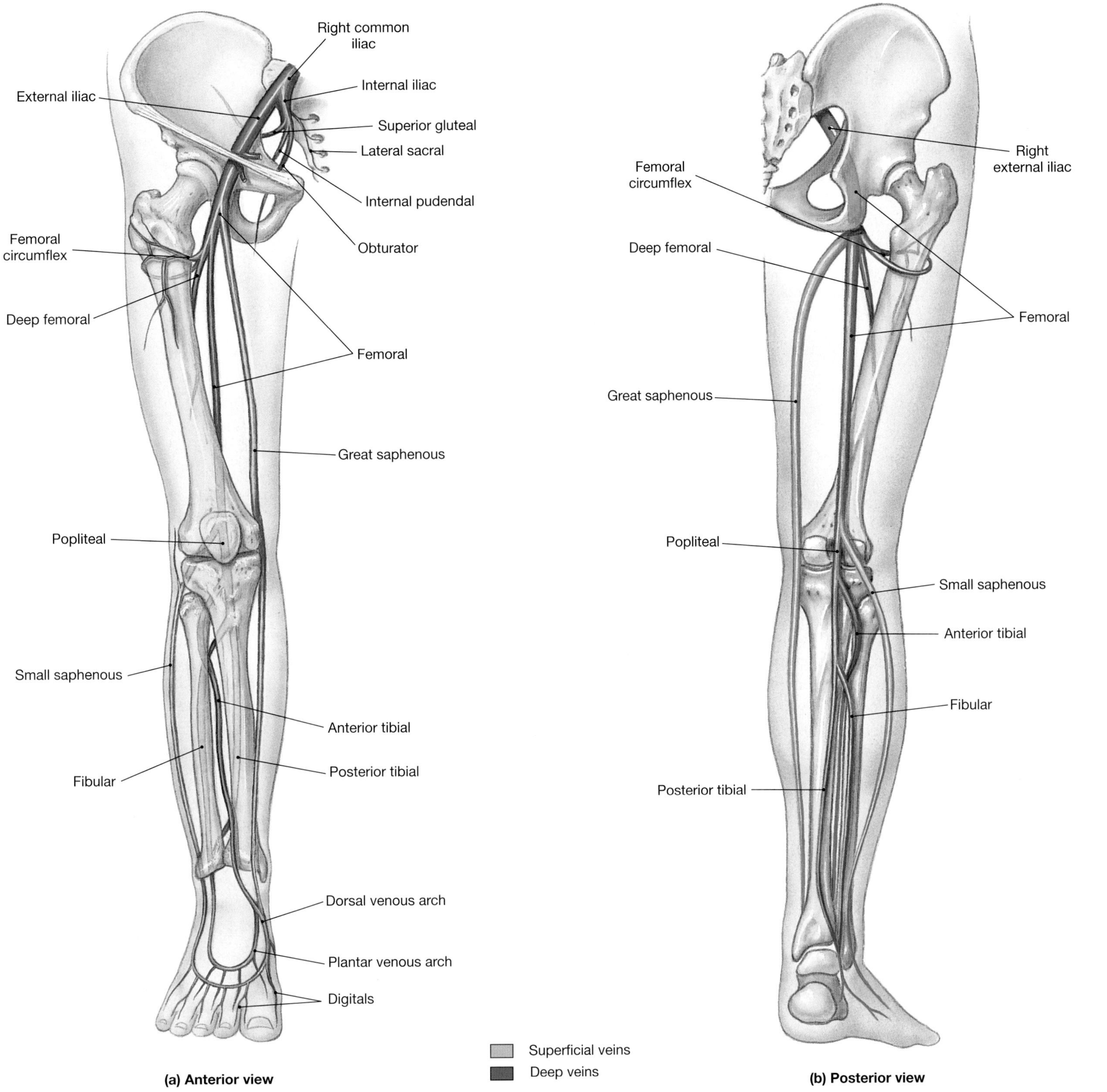

FIGURE 22.25 **The Venous Drainage of the Lower Limb**

(**a**) Anterior view, showing the veins of the right lower limb. (**b**) Posterior view, showing the veins of the right lower limb. *See MRI Scans 5, 6, and 7 in the Companion Atlas.*

The Inferior Vena Cava The **inferior vena cava (IVC)** collects most of the venous blood from organs inferior to the diaphragm (a small amount reaches the superior vena cava via the azygos and hemiazygos veins). Figure 22.24b,c● summarizes the venous tributaries of the inferior vena cava.

Veins Draining the Lower Limb [FIGURES 22.24C/22.25] Blood leaving capillaries in the sole of each foot collects into a network of **plantar veins**. The plantar network provides blood to the deep veins of the leg: the **anterior tibial vein**, the **posterior tibial vein**, and the **fibular vein,** or *peroneal vein* (Figures 22.24c and 22.25a●). The **dorsal venous arch**

collects blood from capillaries on the superior surface of the foot and the **digital veins** of the foot. There are extensive interconnections between the plantar arch and the dorsal arch, and the path of blood flow can easily shift from superficial to deep veins.

The dorsal venous arch is drained by two superficial veins, the **great saphenous vein** (sa-FĒ-nus; *saphenes*, prominent) and the **small saphenous vein**. The great saphenous vein is used in coronary bypass operations to replace blocked coronary vessels. It is the longest vein in the body; it ascends along the medial aspect of the leg and thigh, draining into the *femoral vein* near the hip joint. The small saphenous vein arises from the dorsal venous arch and ascends along the posterior and lateral aspect of the calf. It then enters the popliteal fossa, where it meets the **popliteal vein** formed by the union of the tibial and fibular veins. The popliteal vein may be easily palpated in the popliteal fossa adjacent to the adductor magnus muscle (Figure 22.25b,c●). Once it reaches the femur, the popliteal vein becomes the **femoral vein**, which ascends along the thigh next to the femoral artery. Immediately before penetrating the abdominal wall, the femoral vein receives blood from (1) the great saphenous vein; (2) the **deep femoral vein**, which collects blood from deeper structures of the thigh, and (3) the **femoral circumflex vein**, which drains the area around the neck and head of the femur. The large vein that results penetrates the body wall and emerges in the pelvic cavity as the **external iliac vein**.

Veins Draining the Pelvis [FIGURE 22.23] The external iliac veins receive blood from the lower limbs, pelvis, and lower abdomen. As each external iliac vein travels across the inner surface of the ilium, it merges with the **internal iliac vein**, which drains the pelvic organs on that side. The internal iliac veins are formed by the fusion of the superior gluteal, internal pudendal, obturator, and lateral sacral veins. The union of the external and internal iliac veins forms the **common iliac vein**. The left and right common iliacs ascend at an oblique angle. Anterior to vertebra L_5, they unite to form the inferior vena cava (IVC) (Figure 22.23●).

Veins Draining the Abdomen [FIGURES 22.23/22.24] The inferior vena cava ascends posterior to the peritoneal cavity, parallel to the aorta. Blood from the inferior vena cava flows into the right atrium, where it mixes with venous blood from the superior vena cava. This blood then enters the right ventricle and is pumped into the pulmonary circulation for oxygenation at the lungs. The abdominal portion of the inferior vena cava collects blood from six major veins (Figures 22.23 and 22.24●).

1. **Lumbar veins** drain the lumbar portion of the abdomen. Superior branches of these veins are connected to the azygos vein (right side) and hemiazygos vein (left side), which empty into the superior vena cava.
2. **Gonadal** *(ovarian* or *testicular)* **veins** drain the ovaries or testes. The right gonadal vein empties into the inferior vena cava; the left gonadal usually drains into the left renal vein.
3. **Hepatic veins** leave the liver and empty into the inferior vena cava at the level of vertebra T_{10}.
4. **Renal veins** collect blood from the kidneys. These are the largest tributaries of the inferior vena cava.
5. **Suprarenal veins** drain the adrenal glands. Usually only the right suprarenal vein drains into the inferior vena cava, and the left drains into the left renal vein.
6. **Phrenic veins** drain the diaphragm. Only the right phrenic vein drains into the inferior vena cava; the left drains into the left renal vein.

THE HEPATIC PORTAL SYSTEM [FIGURE 22.26] The liver is the only digestive organ drained by the inferior vena cava. Instead of traveling directly to the inferior vena cava, blood leaving the capillary beds supplied by the celiac, superior mesenteric, and inferior mesenteric arteries flows into the veins of the **hepatic portal system**. You may recall from Chapter 19 that a blood vessel connecting two capillary beds is called a *portal vessel*, and the network is a *portal system*. p. 512 In the hepatic portal system, venous blood which absorbs nutrients from the small intestine, parts of the large intestine, stomach, and pancreas, flows directly to the liver for processing and storage.

Blood flowing in the hepatic portal system is therefore quite different from that in other systemic veins. For example, levels of blood glucose and amino acids in the hepatic portal vein often exceed those found anywhere else in the circulatory system. Additionally, at the same time that this oxygen-poor and nutrient-rich blood arrives at the liver from the digestive organs, the liver also receives oxygen-rich and nutrient-poor blood from the systemic circuit through the hepatic artery proper. This means that the liver receives mixed blood with respect to nutrients and oxygen.

The liver regulates the concentrations of nutrients, such as glucose or amino acids, in the circulating blood. During digestion, the stomach and intestines absorb high concentrations of nutrients, along with various wastes and even toxins. The hepatic portal system delivers these compounds directly to the liver for storage, metabolic conversion, or excretion by liver cells. After passing through the liver sinusoids, blood collects into the hepatic veins, which empty into the inferior vena cava (Figure 22.26●). Because blood goes to the liver first, the composition of the blood in the general systemic circuit remains relatively stable, regardless of the digestive activities under way.

The hepatic portal system begins in the capillaries of the digestive organs and ends as the hepatic portal vein discharges blood into sinusoids in the liver. The tributaries of the hepatic portal vein include:

- The **inferior mesenteric vein** collects blood from capillaries along the inferior portion of the large intestine. Its tributaries include the **left colic vein** and the **superior rectal veins**, which drain the descending colon, sigmoid colon, and rectum.
- The **splenic vein** is formed by the union of the inferior mesenteric vein and veins from the spleen, the lateral border of the stomach (*left gastroepiploic vein*), and the pancreas (*pancreatic veins*).
- The **superior mesenteric vein** collects blood from veins draining the stomach (*right gastroepiploic vein*), the small intestine (*intestinal* and *pancreaticoduodenal veins*), and two-thirds of the large intestine (*ileocolic, right colic,* and *middle colic veins*).

The hepatic portal vein forms through the fusion of the superior mesenteric and splenic veins. Of the two, the superior mesenteric normally contributes the greater volume of blood and most of the nutrients. As it proceeds toward the liver, the hepatic portal receives blood from the **gastric veins**, which drain the medial border of the stomach and from the **cystic veins** from the gallbladder.

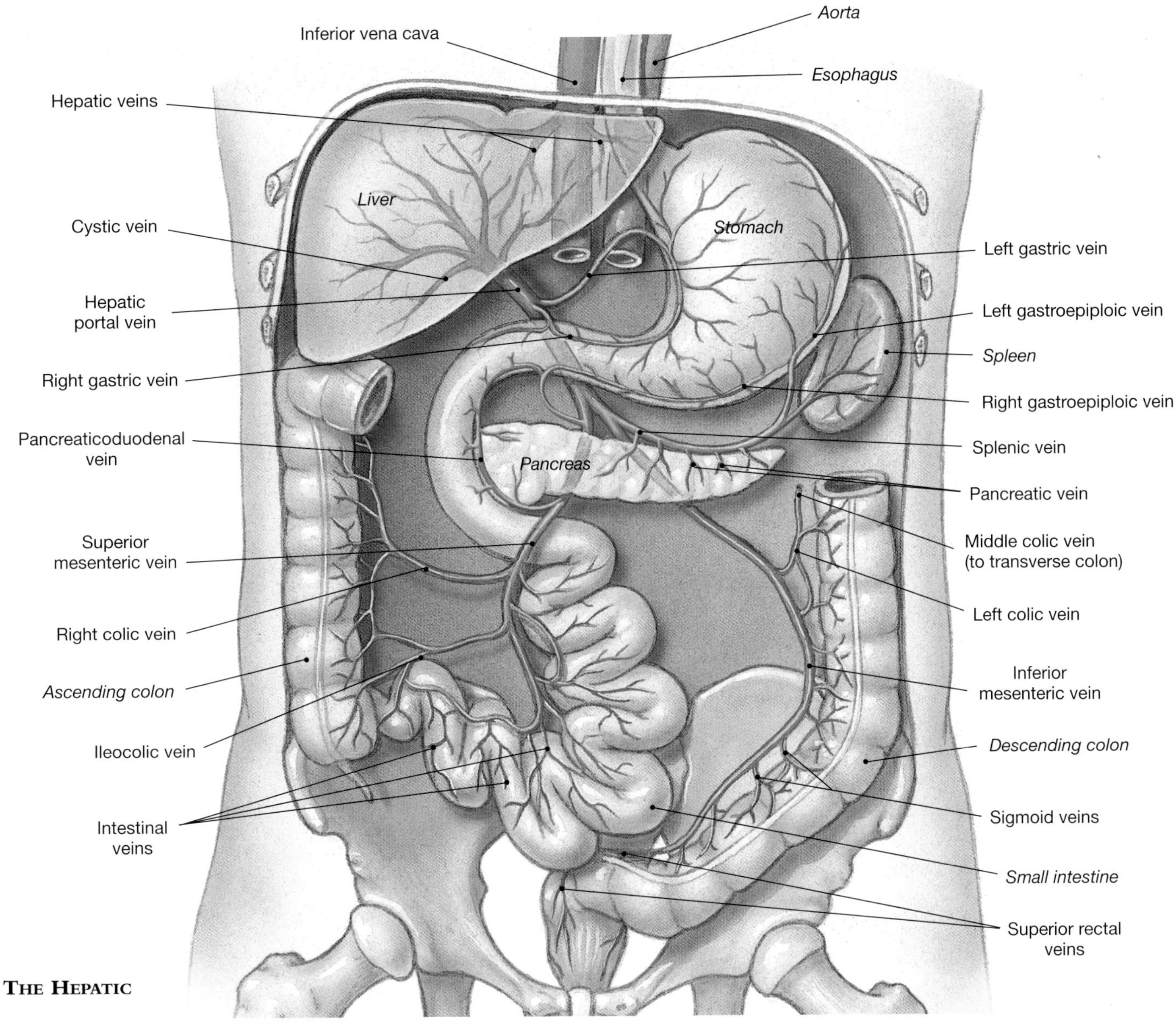

FIGURE 22.26 **The Hepatic Portal System**

✓ CONCEPT CHECK

- Diane is in an automobile accident and ruptures her celiac artery. What organs would be affected most directly by this injury?
- It is 110°F outside, and you are very hot. What changes have occurred in your veins and why?
- Which major vein receives the blood from the head, neck, chest, shoulders, and upper limbs?
- Why does blood leaving the intestines first go to the liver?

Cardiovascular Changes at Birth

[FIGURE 22.27]

There are significant differences between the fetal and adult cardiovascular systems that reflect differing sources of respiratory and nutritional support. The embryonic lungs are collapsed and nonfunctional, and the digestive tract has nothing to digest. All of the fetal nutritional and respiratory needs are provided by diffusion across the *placenta*, a complex organ that regulates exchange between the fetal and maternal bloodstreams. (We will discuss the structure of the placenta in Chapter 28.) Two **umbilical arteries** leave the internal iliac arteries of the fetus, enter the umbilical cord, and deliver blood to the placenta. Blood returns to the fetus from the placenta in the single **umbilical vein**, bringing oxygen and nutrients to the developing fetus. The umbilical vein drains into the **ductus venosus**, which is connected to an intricate network of veins within the developing liver. The ductus venosus collects the blood from the veins of the liver and from the umbilical vein and delivers it to the inferior vena cava (Figure 22.27a,c●). When the placental connection is broken at birth, blood flow ceases along the umbilical vessels, and they soon degenerate.

One of the most interesting aspects of circulatory development reflects the differences between the life of an embryo or fetus and that of a newborn infant. Throughout embryonic and fetal life, the lungs are collapsed; yet after delivery, the newborn infant must be able to extract oxygen from inspired air rather than across the placenta.

Although the interatrial and interventricular septa develop early in fetal life, the interatrial partition remains functionally incomplete up to the time of birth. The interatrial opening, or **foramen ovale**, is associated with an elongated flap that acts as a valve. Blood can flow freely from the right atrium to the left atrium, but any backflow closes the valve and isolates the two chambers. Thus, blood can enter the heart at the right atrium and bypass the pulmonary circuit altogether. A second short circuit exists between the pulmonary and aortic trunks. This connection, the **ductus arteriosus**, consists of a short, muscular vessel.

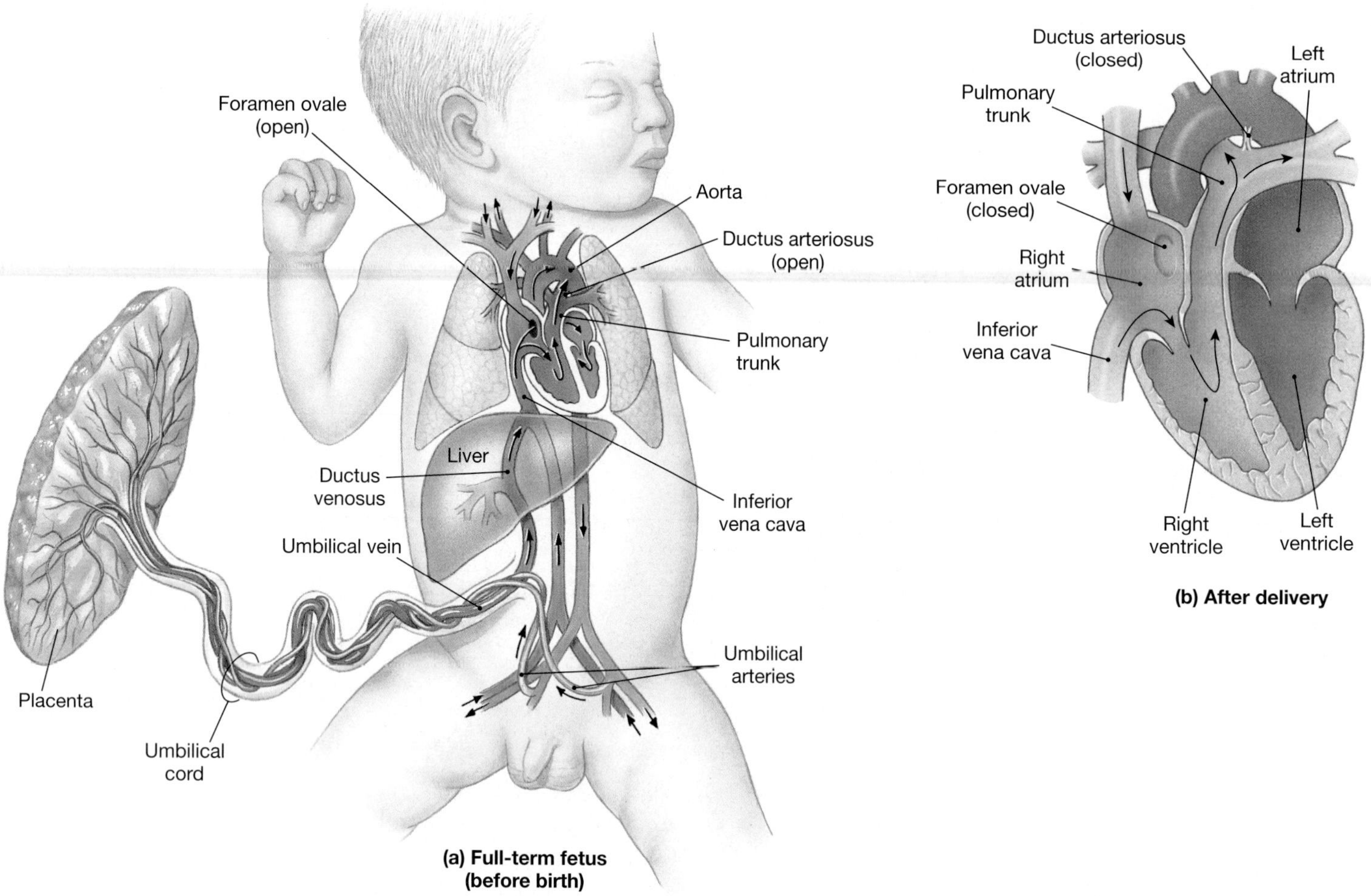

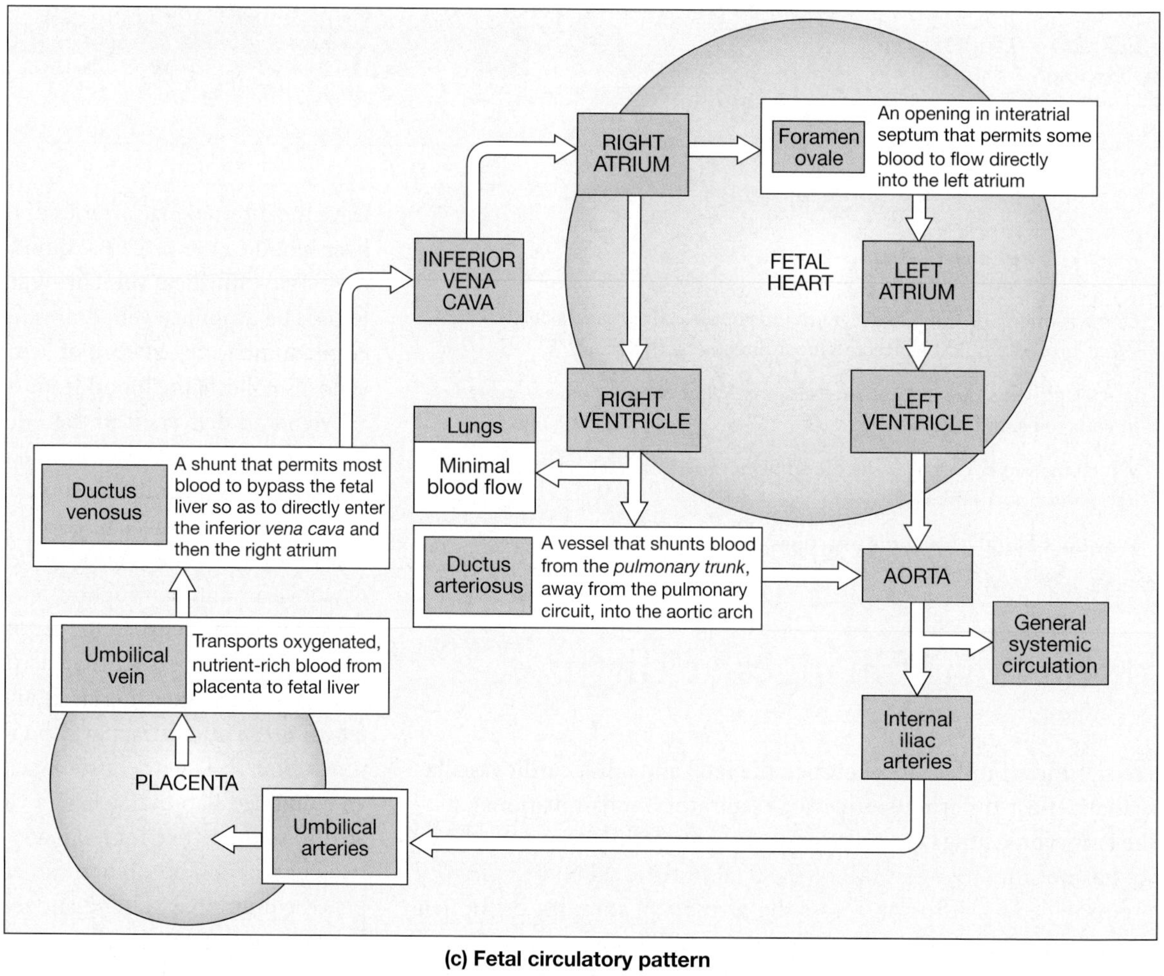

FIGURE 22.27 **CHANGES IN FETAL CIRCULATION AT BIRTH**

(**a**) Circulation pathways in a full-term fetus. (**b**) Blood flow through the heart of the newborn. (**c**) Flow chart for circulatory patterns in the fetus and newborn infant.

With the lungs collapsed, the capillaries are compressed and little blood flows through the lungs. During diastole, blood enters the right atrium and flows into the right ventricle, but some also passes into the left atrium via the foramen ovale. About 25% of the blood arriving at the right atrium bypasses the pulmonary circuit in this way. In addition, over 90% of the blood leaving the right ventricle passes through the ductus arteriosus and enters the systemic circuit, rather than continuing to the lungs.

At birth, dramatic changes occur. When the infant takes its first breath, the lungs expand, and so do the pulmonary vessels. The smooth muscles in the ductus arteriosus contract, isolating the pulmonary and aortic trunks, and blood begins flowing through the pulmonary circuit. As pressures rise in the left atrium, the valvular flap closes the foramen ovale and completes the cardiovascular remodeling. These alterations are diagrammed and summarized in Figure 22.27a,b●. In the adult, the interatrial septum bears a shallow depression, the **fossa ovalis**, that marks the site of the original foramen ovale. The remnants of the ductus arteriosus persist as a fibrous cord, the **ligamentum arteriosum**. ∞ *p. 558*

If the proper circulatory changes do not occur at birth or shortly thereafter, problems will eventually develop. The severity of the problem varies depending on which connection remains open and the size of the opening. Treatment may involve surgical closure of the foramen ovale or the ductus arteriosus or both. Other forms of congenital heart defects result from abnormal cardiac development or inappropriate connections between the heart and major arteries and veins. (For further discussion of cardiovascular changes during development, see the *Embryology Summary* on pp. 604–605.)

Exercise and Cardiovascular Disease

Regular exercise has several beneficial effects. Even a moderate exercise routine (power walking or bicycling, for example) can lower total blood cholesterol levels. High cholesterol is one of the major risk factors for atherosclerosis, leading to cardiovascular disease and strokes. In addition, a healthy lifestyle with regular exercise, a balanced diet, weight control, and no smoking reduces stress, lowers blood pressure, slows plaque formation, and reduces the chances of plaque formation.

Large-scale statistical studies indicate that regular moderate exercise may cut the incidence of heart attacks almost in half. Exercise is also beneficial in accelerating recovery after a heart attack. Regular light to moderate exercise, coupled with a low-fat diet and low-stress lifestyle, not only reduces symptoms, such as angina, but also improves both mood and the overall quality of life. However, exercise does not remove an underlying medical problem.

There is no evidence that intense athletic training lowers the incidence of cardiovascular disease. On the contrary, the strains placed on the cardiovascular system during intense athletic training can be severe. Healthy individuals can develop acute disorders, such as kidney failure, after extreme exercise.

Aging and the Cardiovascular System

The capabilities of the cardiovascular system gradually decline with age. The major changes are listed and summarized below, in the same sequence as the cardiovascular chapters: blood, heart, and vessels.

1. *Age-related changes in the blood* may include: (1) decreased hematocrit; (2) constriction or blockage of peripheral veins by a **thrombus** (stationary blood clot); the thrombus can become detached, pass through the heart, and become wedged in a small artery, most often in the lungs, causing a **pulmonary embolism**; and (3) pooling of blood in the veins of the legs because valves are not working effectively.
2. *Age-related changes in the heart* include: (1) a reduction in the maximum cardiac output, (2) changes in the activities of the nodal and conducting fibers, (3) a reduction in the elasticity of the fibrous skeleton, (4) a progressive atherosclerosis that can restrict coronary circulation, and (5) replacement of damaged cardiac muscle fibers by scar tissue.
3. *Age-related changes in blood vessels* are often related to arteriosclerosis and include: (1) inelastic walls of arteries become less tolerant of sudden increases in pressure, which may lead to an **aneurysm** (AN-ū-rizm), causing a stroke, infarct, or massive blood loss, depending on the vessel involved; (2) calcium salts can deposit on weakened vascular walls, increasing the risk of a stroke or infarct; and (3) thrombi can form at atherosclerotic plaques.

✓ CONCEPT CHECK

- What major changes occur in the heart and major vessels of a newborn at birth?
- What causes varicose veins?
- Why is a reduction in elasticity of the arteries with age dangerous?

RELATED CLINICAL TERMS

aneurysm (AN-Ū-rizm): A bulge in the weakened wall of a blood vessel, usually an artery. *Aneurysms p. 801*

arteriosclerosis (ar-tē-rē-ō-skle-RŌ-sis): A thickening and toughening of arterial walls. *p. 578*

atherosclerosis (ath-er-ō-skle-RŌ-sis): A type of arteriosclerosis characterized by changes in the endothelial lining and the formation of plaques. *p. 578*

hemorrhoids (HEM-ō-roidz): Varicose veins in the walls of the rectum and/or anus, often associated with pregnancy or frequent straining to force bowel movements. *Problems with Venous Valve Function p. 801*

pulmonary embolism: Circulatory blockage caused by the trapping of a freed thrombus in a pulmonary artery. *p. 603*

thrombus: A stationary blood clot within a blood vessel. *p. 603*

varicose (VAR-i-kōs) **veins:** Sagging, swollen veins distorted by gravity and the failure of the venous valves. *Problems with Venous Valve Function p. 801*

Additional Clinical Term Discussed in Appendix I (p. 789)
Marfan's syndrome

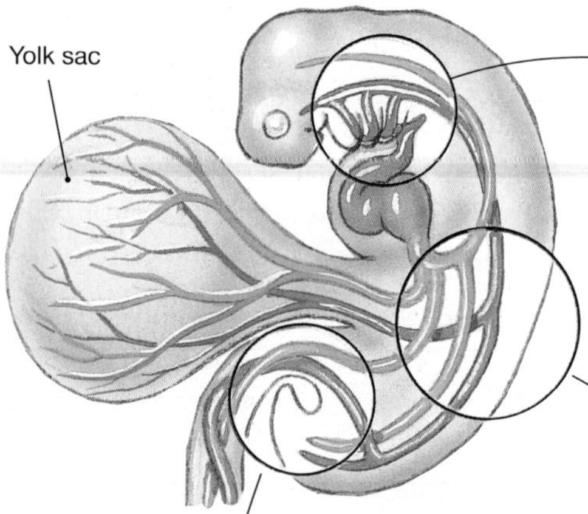

We will follow the development of three major vessel complexes: the aortic arch, the venae cavae, and the hepatic portal and umbilical systems. (Arteries are shown in red and veins in blue regardless of the oxygenation of the blood they carry.)

THE AORTIC ARCHES

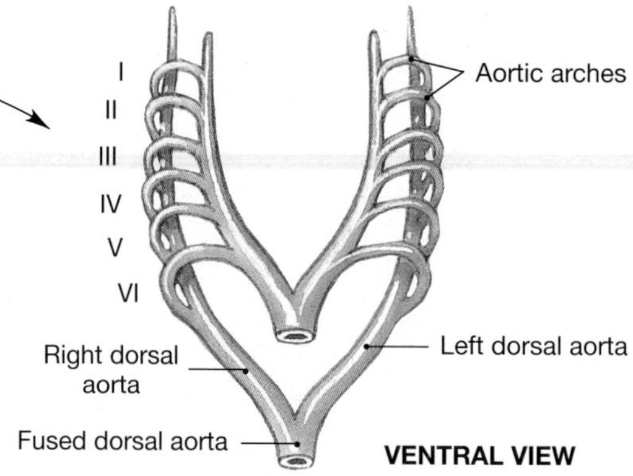

VENTRAL VIEW

An **aortic arch** carries arterial blood through each of the *pharyngeal arches* (p. 000). In the dorsal pharyngeal wall, these vessels fuse to create the **dorsal aorta**, which distributes blood throughout the body. The arches are usually numbered from I to VI, corresponding to the pharyngeal arches.

THE VENAE CAVAE

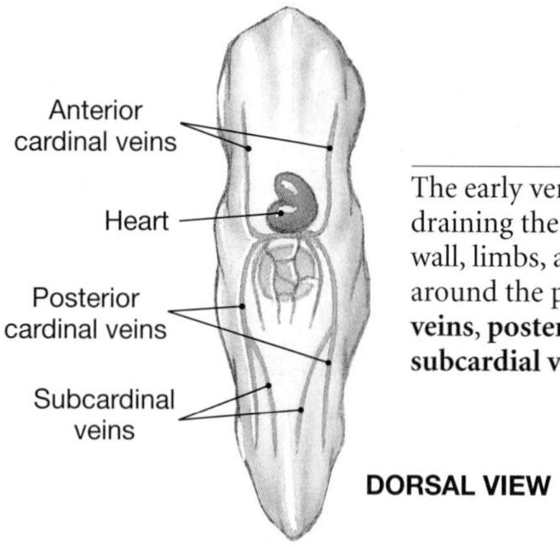

DORSAL VIEW

The early venous circulation draining the tissues of the body wall, limbs, and head centers around the paired **anterior cardinal veins**, **posterior cardinal veins**, and **subcardinal veins**.

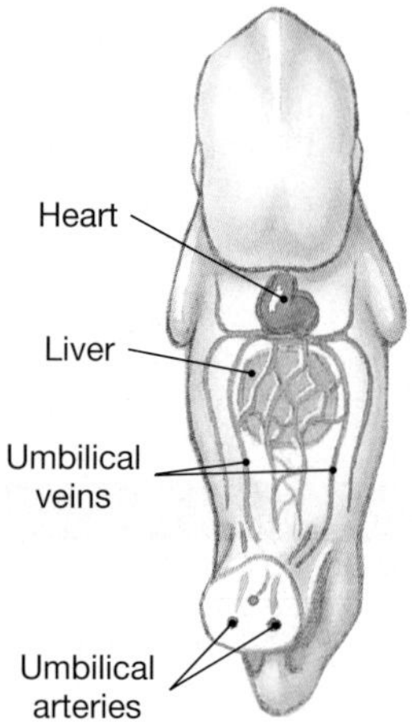

4 WEEKS

Paired **umbilical arteries** deliver blood to the placenta. At 4 weeks, paired **umbilical veins** return blood to capillary networks in the liver. Veins running along the length of the digestive tract have extensive interconnections.

THE HEPATIC PORTAL AND UMBILICAL VESSELS

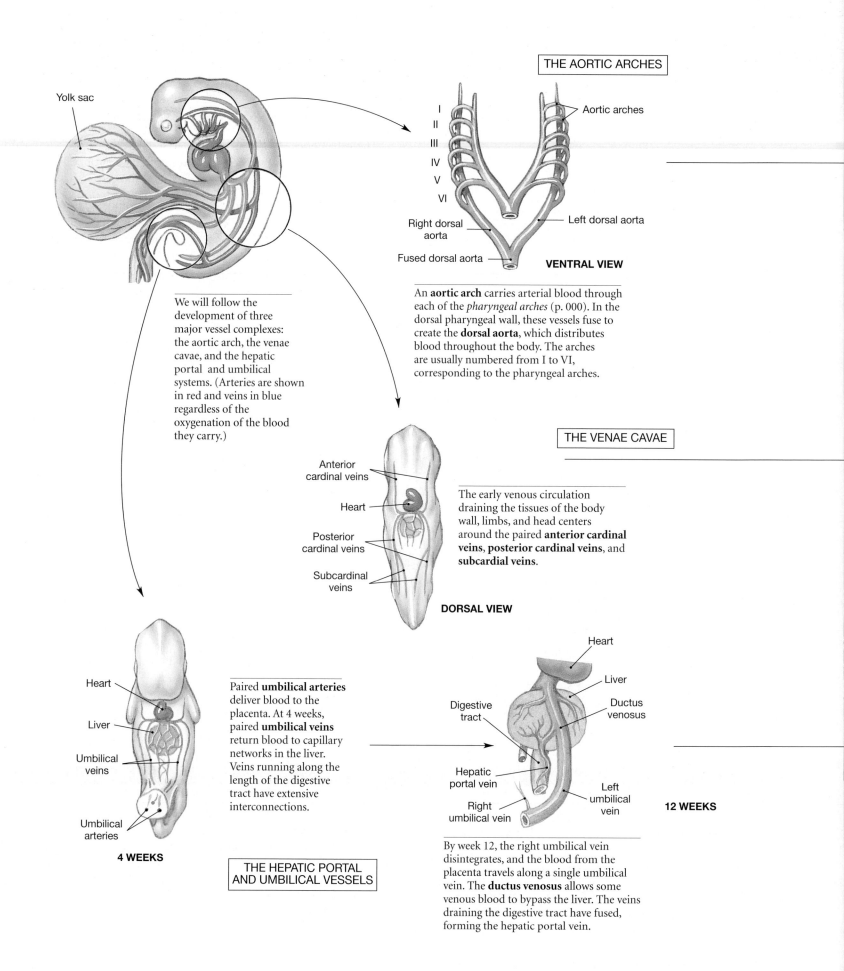

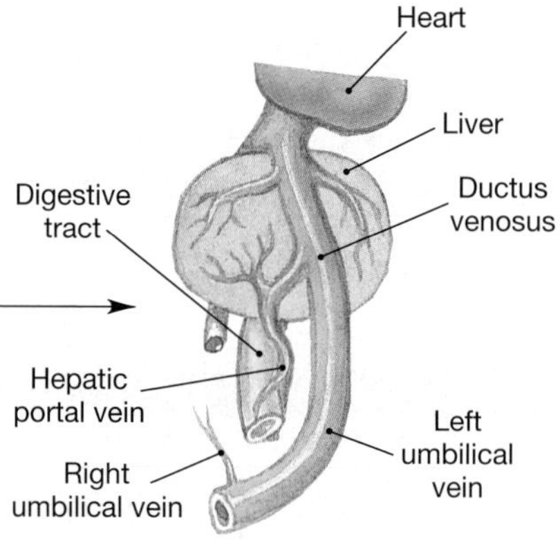

12 WEEKS

By week 12, the right umbilical vein disintegrates, and the blood from the placenta travels along a single umbilical vein. The **ductus venosus** allows some venous blood to bypass the liver. The veins draining the digestive tract have fused, forming the hepatic portal vein.

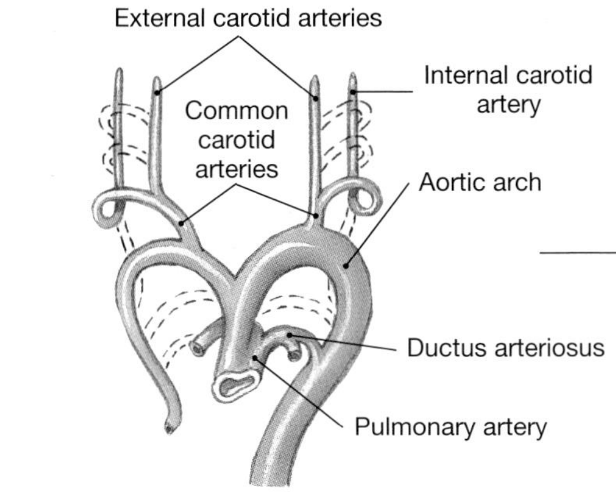

As development proceeds, some of these arches disintegrate. The **ductus arteriosus** provides an external short-circuit between the pulmonary and systemic circuits. Most of the blood entering the right atrium bypasses the lungs, passing instead through the ductus arteriosus or the **foramen ovale** in the heart.

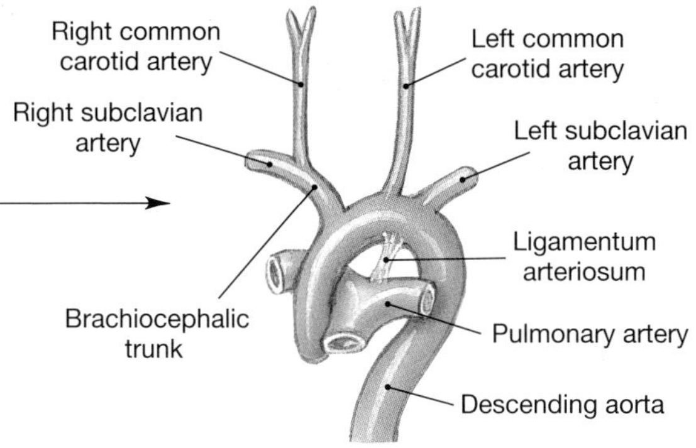

The left half of arch IV ultimately becomes the aortic arch, which carries blood away from the left ventricle.

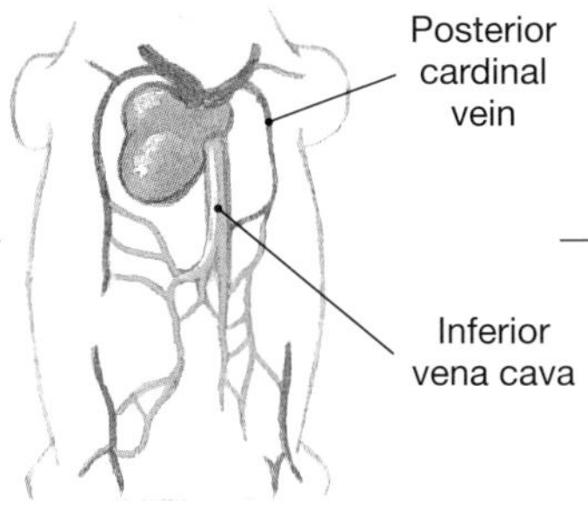

Interconnections form among these veins, and a combination of fusion and disintegration produces more-direct, larger-diameter connections to the right atrium.

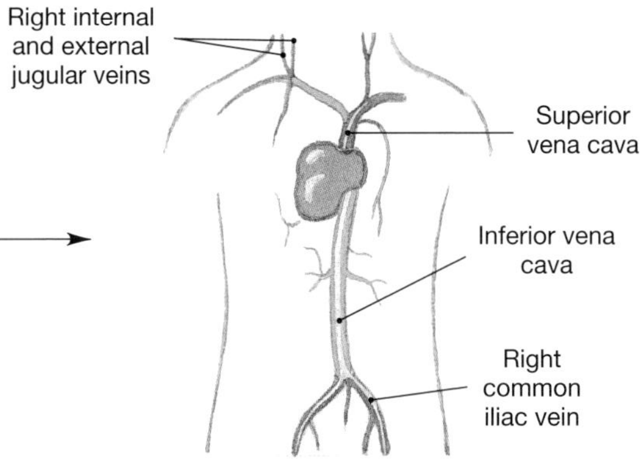

This process continues, ultimately producing the superior and inferior venae cavae.

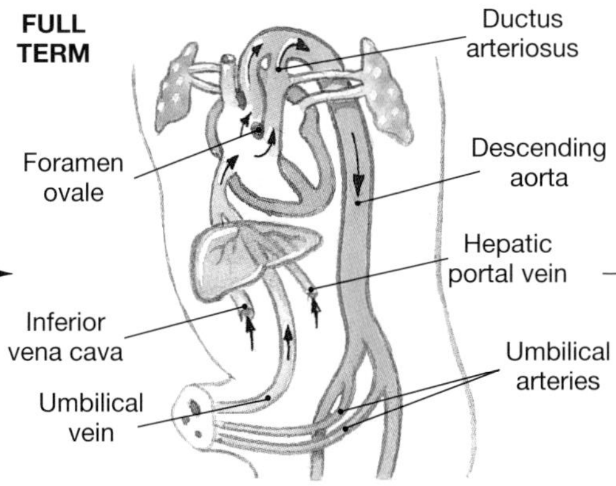

Shortly before birth, blood returning from the placenta travels through the liver in the ductus venosus to reach the inferior vena cava. Much of the blood delivered by the venae cavae bypasses the lungs by traveling through the foramen ovale and the ductus arteriosus.

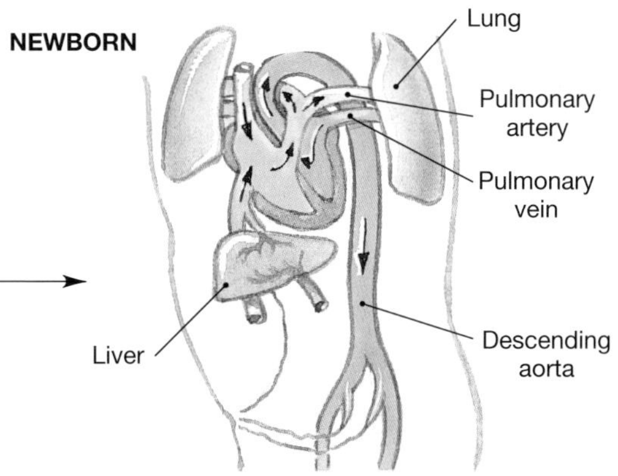

At birth, pressures drop in the pleural cavities as the chest expands and the infant takes its first breath. The pulmonary vessels dilate, and blood flow to the lungs increases. Pressure falls in the right atrium, and the higher left atrial pressures close the valve that guards the foramen ovale. Smooth muscles contract the ductus arteriosus, which ultimately converts to the **ligamentum arteriosum**, a fibrous strand.

CLINICAL DISCUSSION

Congenital Cardiovascular Problems

Congenital cardiovascular problems serious enough to represent a threat to homeostasis are relatively rare. They usually reflect abnormal formation of the heart or problems with the interconnections between the heart and the great vessels. Several examples of congenital cardiovascular defects are illustrated in Figure 22.28●. Most of these conditions can be surgically corrected, although multiple surgeries may be required and life expectancy may be shortened in more severe defects.

The incomplete closure of the foramen ovale or ductus arteriosus (Figure 22.28a●) results in the recirculation of blood into the pulmonary circuit. Pressures rise in the pulmonary circuit, leading to pulmonary edema and cardiac enlargement that can ultimately lead to heart failure.

Ventricular septal defects (Figure 22.28b●) are the most common congenital heart problems, affecting 0.12% of newborn infants. The opening between the left and right ventricles has the reverse effect of a connection between the atria: When it beats, the more powerful left ventricle ejects blood into the right ventricle and pulmonary circuit. Pulmonary hypertension, pulmonary edema, and cardiac enlargement may result.

The *tetralogy of Fallot* (fa-LŌ) (Figure 22.28c●) is a complex group of heart and circulatory defects that affect 0.10% of newborn infants. In this condition (1) the pulmonary trunk is abnormally narrow, (2) the interventricular septum is incomplete, (3) the aorta originates where the interventricular septum normally ends, and (4) the right ventricle is enlarged. Because normal blood oxygenation does not occur, the circulating blood has a deep red color. The skin then develops the blue tones typical of cyanosis, a condition noted in Chapter 4, and the infant is known as a "blue baby."

In the *transposition of great vessels* (Figure 22.28d●) the aorta is connected to the right ventricle and the pulmonary artery is connected to the left ventricle. This malformation affects 0.05% of newborn infants.

In an *atrioventricular septal defect* (Figure 22.28e●) the atria and ventricles are incompletely separated. The results are quite variable, depending on the extent of the defect and the effects on the atrioventricular valves. This type of defect most often affects infants with *Down syndrome*, a disorder caused by the presence of an extra copy of chromosome 21.

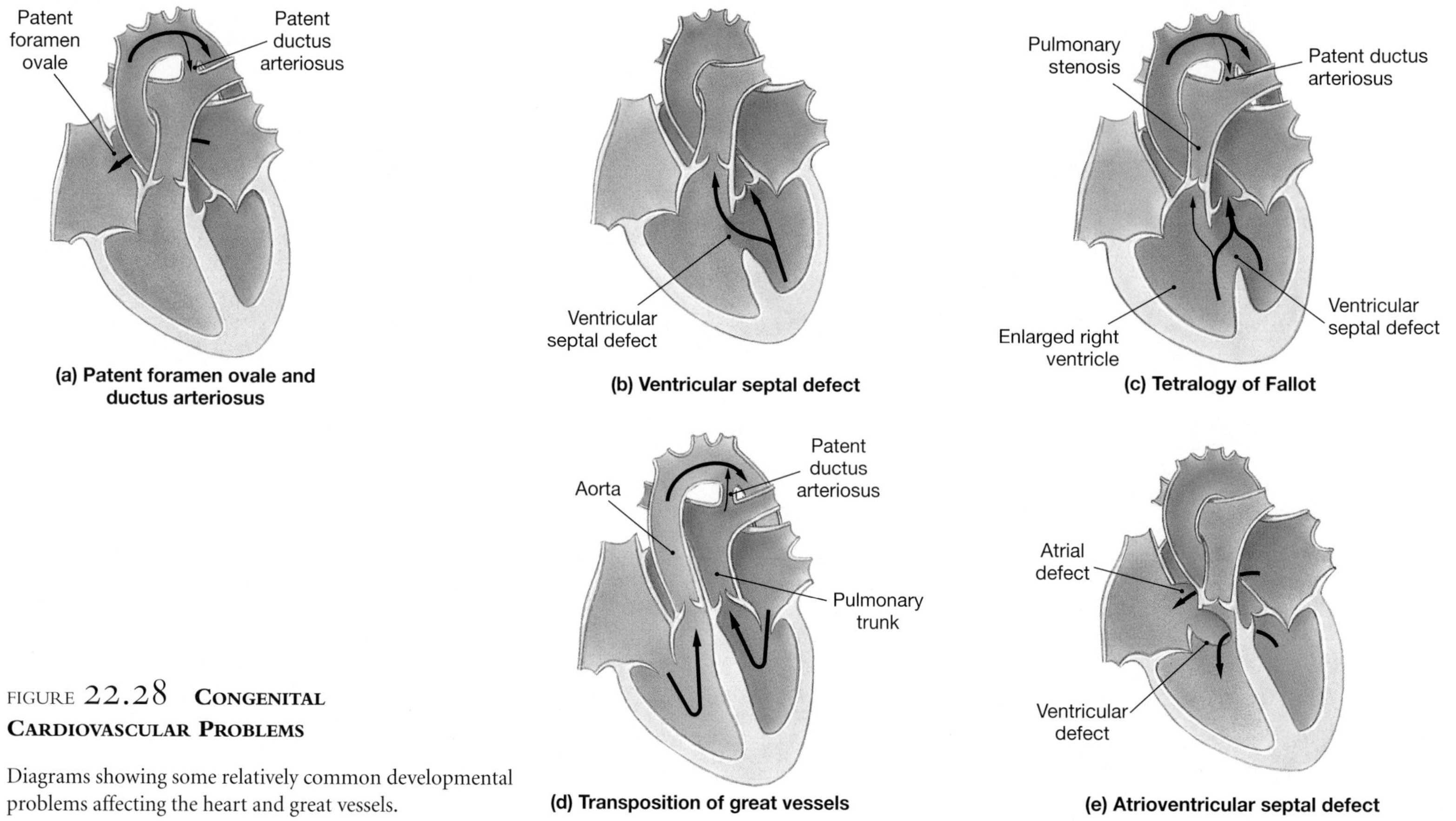

FIGURE 22.28 **Congenital Cardiovascular Problems**

Diagrams showing some relatively common developmental problems affecting the heart and great vessels.

STUDY OUTLINE & CHAPTER REVIEW

Introduction 573

1. The cardiovascular system is a closed system with two circulatory patterns: a *pulmonary circuit* and a *systemic circuit.*
2. Blood flows through a network of arteries, veins, and capillaries. All chemical and gaseous exchange between the blood and interstitial fluid takes place across capillary walls.

Histological Organization of Blood Vessels 573

1. The walls of arteries and veins contain three layers: the **tunica intima** (the innermost layer), **tunica media** (the middle layer), and **tunica externa** (the connective tissue sheath around the vessel). *(see Figure 22.1)*

Distinguishing Arteries from Veins 574

2. In general, the walls of arteries are thicker than those of veins. The endothelial lining of an artery cannot contract, so it is thrown into folds. Arteries constrict when blood pressure does not distend them; veins constrict very little. *(see Figure 22.1)*

Arteries 575

3. The arterial system includes the large **elastic arteries**, **muscular** or **medium-sized arteries**, and smaller **arterioles.** Elastic arteries, or *conducting arteries*, transport large volumes of blood away from the heart. They are able to stretch and recoil with pressure changes. Muscular arteries *(distribution arteries)* distribute blood to skeletal muscles and organs. Arterioles can change their diameter **(vasoconstriction or vasodilation)** in response to conditions in the body. As we proceed toward the capillaries, the number of vessels increase, but the diameters of the individual vessels decrease and the walls become thinner. *(see Figures 22.1 and 22.2)*

Capillaries 575

4. Capillaries are the smallest blood vessels and the only blood vessels whose walls permit exchange between blood and interstitial fluid. Capillaries may be **continuous** (the endothelium is a complete lining) or **fenestrated** (the endothelium contains "windows"). **Sinusoids** are specialized fenestrated capillaries found in selected tissues (e.g., liver) that allow very slow blood flow. *(see Figure 22.3)*
5. Capillaries form interconnected networks called **capillary beds (capillary plexuses)**. A **precapillary sphincter** (a band of smooth muscle) adjusts the blood flow into each capillary. **Central**, or **preferred**, **channels** provide the means of arteriole-venule communication. A **metarteriole** is the arteriolar segment of the channel. The entire capillary plexus may be bypassed by blood flow through **arteriovenous anastomoses** or via central channels within the capillary plexus. *(see Figure 22.4)*

Veins 577

6. **Veins** collect deoxygenated blood from the tissues and organs and return it to the heart. **Venules** collect blood from the capillaries and merge into **medium-sized veins** and then **large veins**. The arterial system is a high-pressure system; blood pressure in veins is much lower. **Valves** in veins prevent the backflow of blood. *(see Figures 22.1/22.2/22.6)*

The Distribution of Blood 579

7. While the heart, arteries, and capillaries usually contain about 30–35% of the blood volume, most of the blood volume is in the venous system (65–70%). Peripheral **venoconstriction** helps maintain adequate blood volume in the arterial system after a hemorrhage. The **venous reserve,** which is the extra blood in the venous system that can be distributed within the arterial system, normally accounts for up to 20% of the total blood volume. *(see Figure 22.7)*

Blood Vessel Distribution 579

1. The blood vessels of the body can be divided into those of the **pulmonary circuit** (between the heart and lungs) and the **systemic circuit** (from the heart to all organs and tissues). *(see Figure 22.8)*

The Pulmonary Circuit 580

2. The arteries of the pulmonary circuit carry deoxygenated blood. The pulmonary circuit includes the **pulmonary trunk**, the **left** and **right pulmonary arteries**, and the **pulmonary veins**, which empty into the left atrium. *(see Figure 22.9)*

The Systemic Circuit 580

3. The **ascending aorta** gives rise to the coronary circulation. The **aortic arch** continues as the **descending aorta**. Three large arteries arise from the aortic arch to collectively supply the head, neck, shoulder, and upper limbs: **brachiocephalic trunk, left common carotid**, and **left subclavian** arteries. The detailed distribution of these blood vessels and their branches will be found in *Figures 22.10 to 22.20.*
4. The brachiocephalic trunk gives rise to the **right subclavian** and the **right common carotid arteries**. These arteries supply the right upper limb and portions of the right shoulder, neck, and head. *(see Figures 22.12/22.13/22.14)*
5. Each subclavian artery exits the thoracic cavity to become the **axillary artery**, which enters the arm to become the **brachial artery**. The brachial arteries and their branches supply blood to the upper limbs. *(see Figure 22.12)*
6. Each common carotid artery divides into an **external carotid artery** and **internal carotid artery**. The external carotids and their branches supply blood to structures in the neck and face. The internal carotids and their branches enter the skull to supply blood to the brain and eyes. The brain also receives blood from the **vertebral arteries**. The vertebral arteries and the internal carotids form the **cerebral arterial circle**, (*or circle of Willis*) which ensures the blood supply to the brain. *(see Figures 22.13 to 22.15)*
7. The descending aorta superior to the diaphragm is termed the **thoracic aorta** and inferior to it the **abdominal aorta**. The thoracic aorta and its branches supply blood to the thorax and thoracic viscera. The abdominal aorta and its branches supply blood to the abdominal wall, abdominal viscera, pelvic structures, and lower limbs. The three unpaired arteries are the **celiac trunk**, the **superior mesenteric artery**, and the **inferior mesenteric artery**. The celiac trunk divides into the **left gastric artery**, the **common hepatic artery**, and the **splenic artery**. Paired arteries include the **suprarenal arteries**, the **renal arteries**, the **lumbar arteries**, and the **gonadal arteries.** The detailed distribution of these blood vessels and their branches will be found in *Figures 22.10 to 22.18/22.20.*
8. Arteries in the neck and limbs are deep beneath the skin; in contrast, there are usually two sets of peripheral veins—one superficial and one deep. This dual venous drainage is important for controlling body temperature *(see Figures 22.10/22.21)*. Arteries of the pelvis and lower limbs include the **right** and **left common iliac** arteries, which branch to form **external** and **internal iliac** arteries. The **femoral** and **deep femoral** arteries supply the lower limb *(see Figure 22.18)*. The arteries of the foot can be seen in *Figures 22.18 and 22.20.*
9. The **superior vena cava (SVC)** receives blood from the head, neck, chest, shoulders, and upper limbs. The detailed distribution of these collecting vessels and their branches may be seen in *Figures 22.21 to 22.24*. The **inferior vena cava (IVC)** collects most of the venous blood from organs below the diaphragm. The detailed distribution of these collecting vessels and their branches may be seen in *Figures 22.23 to 22.26.*
10. Any blood vessel connecting two capillary beds is called a *portal vessel*, and the network of blood vessels comprises a *portal system.*
11. Blood leaving the capillaries supplied by the celiac, superior, and inferior mesenteric arteries flows into the **hepatic portal system**. Blood in the hepatic portal system is unique compared to that of the other systemic veins, because portal blood contains high concentrations of nutrients. These substances are collected from the digestive organs through the vessels of the portal system and are transported directly to the liver for processing. *(see Figure 22.26b)*
12. The vessels that form the hepatic portal system are shown in *Figure 22.26.*

Cardiovascular Changes at Birth 601

1. During fetal development, the **umbilical arteries** carry blood to the placenta. It returns via the **umbilical vein** and enters a network of vascular sinuses in the liver. The **ductus venosus** collects this blood and returns it to the inferior vena cava *(see Figure 22.27 and Embryology Summary).*
2. At this time, the interatrial septum is incomplete and the **foramen ovale** allows the passage of blood from the right atrium to the left atrium. The **ductus arteriosus** also permits the flow of blood between the pulmonary trunk and the aortic arch. At birth or shortly thereafter, as the pulmonary circuit becomes functional, these connections normally close, forming a depression called the **fossa ovalis,** where the foramen ovale was, and the **ligamentum arteriosum**, where the ductus arteriosus used to be. *(see Figure 22.27 and Embryology Summary)*

Exercise and Cardiovascular Disease 603

1. Regular moderate exercise can cut the risk of heart attack almost in half and lower total blood cholesterol levels (one of the major risk factors for atherosclerosis, leading to cardiovascular disease and strokes). In addition, regular exercise reduces stress, lowers blood pressure, and slows plaque formation.

Aging and the Cardiovascular System 603

1. Age-related changes in the blood can include: (1) decreased hematocrit, (2) constriction or blockage of peripheral veins by a **thrombus** (stationary blood clot), and (3) pooling of blood in the veins of the lower legs because valves are not working effectively.

2. Age-related changes in the heart include: (1) a reduction in the maximum cardiac output, (2) changes in the activities of the nodal and conducting fibers, (3) a reduction in the elasticity of the fibrous skeleton, (4) a progressive atherosclerosis that can restrict coronary circulation, and (5) replacement of damaged cardiac muscle fibers by scar tissue.

3. Age-related changes in blood vessels are often related to arteriosclerosis and include: (1) inelastic walls of arteries being less tolerant of sudden increases in pressure, which may lead to an **aneurysm**, (2) calcium salts, which can deposit on weakened vascular walls, increasing the risk of a stroke or infarct, and (3) thrombi that form at atherosclerotic plaques.

LEVEL 1 REVIEWING FACTS AND TERMS

Match each numbered item with the most closely related lettered item. Use letters for answers in the spaces provided.

Column A	Column B
____ 1. elastic arteries	a. delivers blood to the head
____ 2. thrombus	b. distribution arteries
____ 3. collaterals	c. conducting arteries
____ 4. renal veins	d. stationary blood clot
____ 5. iliac arteries	e. network of capillaries
____ 6. alveoli	f. arteries that supply a capillary network
____ 7. carotid arteries	g. supply blood to the upper limbs
____ 8. subclavian arteries	h. small air sacs
____ 9. capillary plexus	i. supply blood to the lower limbs
____ 10. muscular arteries	j. collect blood from the kidneys

11. Compared to arteries, veins
    (a) are more elastic
    (b) have thinner walls
    (c) have more smooth muscle in their tunica media
    (d) have a pleated endothelium

12. Capillaries that have a complete lining are called
    (a) continuous capillaries (b) fenestrated capillaries
    (c) sinusoidal capillaries (d) sinusoids

13. The only blood vessels whose walls permit exchange between the blood and the surrounding interstitial fluids are the
    (a) arteries (b) arterioles
    (c) veins (d) capillaries

14. Blood flow through the capillaries is regulated by the
    (a) arterial anastomosis (b) central channel
    (c) vasa vasorum (d) precapillary sphincter

15. Blood from the brain returns to the heart by way of the
    (a) vertebral vein (b) internal jugular vein
    (c) external jugular vein (d) azygos vein

16. Branches off the aortic arch include
    (a) the left subclavian artery (b) the right subclavian artery
    (c) the right axillary artery (d) the right common carotid artery

17. During increased exercise
    (a) stroke volume decreases
    (b) cardiac output decreases
    (c) venous return increases
    (d) vasoconstriction occurs at the active skeletal muscles

18. In the leg, the femoral artery becomes the
    (a) popliteal artery (b) deep femoral artery
    (c) tibial artery (d) iliac artery

19. The fusion of the brachiocephalic veins forms the
    (a) azygos vein (b) superior vena cava
    (c) inferior vena cava (d) subclavian vein

20. Elderly individuals usually have
    (a) elevated hematocrits (b) stiff inelastic arteries
    (c) increased venous return (d) decreased blood pressure

LEVEL 2 REVIEWING CONCEPTS

1. A major difference between the arterial and venous systems is that
    (a) arteries are usually more superficial than veins
    (b) in the limbs there is dual venous drainage
    (c) veins are usually less branched than the arteries
    (d) veins exhibit a more orderly pattern of branching in the limbs

2. You would expect to find fenestrated capillaries in
    (a) the pancreas (b) skeletal muscles
    (c) cardiac muscle (d) the spleen

3. Under normal conditions, is the blood volume of the venous system greater than, equal to, or less than that of the arterial system? Why?

4. What are some examples of elastic arteries?

5. Where are sinusoids found?

6. What are arteriovenous anastomoses?

7. What are the functions of venous valves in the limbs?

8. What three elastic arteries originate along the aortic arch?

9. From which regions of the body does the superior vena cava receive blood?

10. What is the function of the foramen ovale in a fetus?

LEVEL 3 CRITICAL THINKING AND CLINICAL APPLICATIONS

1. Paul falls on an axe and cuts several major blood vessels in his leg. What would you expect to observe as a result of this accident?
2. John loves to soak in hot tubs and whirlpools. One day he decides to raise the temperature in his hot tub as high as it will go. After a few minutes in the water, he feels faint, passes out, and nearly drowns. Luckily he is saved by a bystander. Explain what happened.
3. Millie's grandfather suffers from congestive heart failure. When she visits him, she notices that his ankles and feet appear swollen. She asks you why this occurs. What would you tell her?

✓ ANSWERS TO CONCEPT CHECK QUESTIONS

p. 579 **1.** The blood vessels are veins. Arteries and arterioles have a relatively large amount of smooth muscle tissue in a thick, well-developed tunica media. **2.** Blood pressure in the arterial system pushes blood into the capillaries. Blood pressure on the venous side is very low, and other forces help keep the blood moving. Valves in the walls of venules and medium-sized veins permit blood flow in only one direction, toward the heart, preventing the backflow of blood toward the capillaries. **3.** The femoral artery is a muscular artery. **4.** No gas exchange occurs in arterioles.

p. 594 **1.** The carotid arteries supply blood to structures of the head and the neck, including the brain. **2.** The right brachial artery is the artery at the biceps region. **3.** The external iliac artery gives rise to the femoral artery in the thigh. **4.** Damage to the internal carotid arteries would not always result in brain damage, because the vertebral arteries also supply blood to the brain.

p. 601 **1.** Organs served by the celiac artery include the stomach, spleen, liver, and pancreas. **2.** The superficial veins are dilated to promote heat loss through the skin. **3.** The superior vena cava receives blood from the head, neck, chest, shoulders, and upper limbs. **4.** Blood from the intestines goes to the liver first. This blood contains high amounts of glucose, amino acids, and other nutrients and toxins absorbed from the digestive tract. These are processed by the liver before the blood goes to the general systemic circuit in order to keep the composition of the blood in the body relatively stable.

p. 603 **1.** The major changes in the heart and the major vessels that occur at birth are: (1) the pulmonary vessels expand, (2) the ductus arteriosus contracts, forcing blood to flow through the pulmonary circuit, and (3) the valvular flap closes the foramen ovale. **2.** The blood pools in the veins of the legs because the valves are not working effectively. **3.** The arteries must be able to expand with a sudden increase in pressure. A reduced ability to do so could result in a bulge or tear in the wall of the artery.

REVIEW IT

Challenge 1

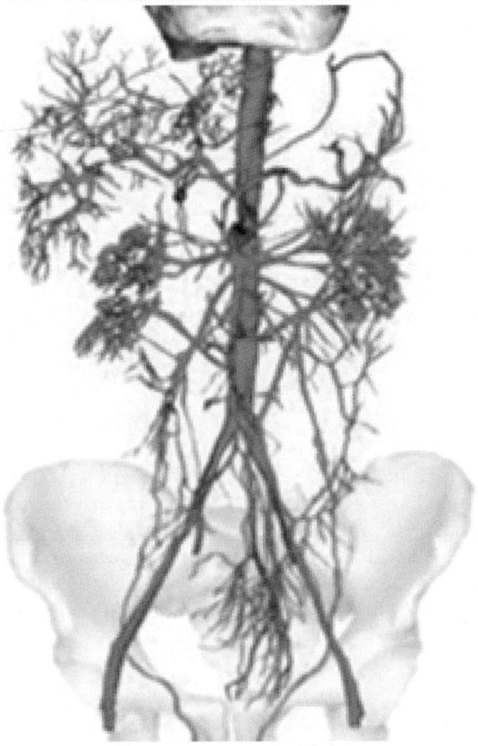

A study of the blood vessels usually begins with a discussion of the arteries, due to the relatively consistent placement of arteries as opposed to veins. Here you see an image taken from the Interactive CD/Cardiovascular system/Abdomen and pelvis. The organs have been stripped away, along with the more superficial veins. What remains are the major arteries, as they were found in the individual used to prepare the Visible Human data set. Compare what you see here with Figure 22.17●, pp. 590–591. List the vessels you see. Are they in the same location on the individual as in the figure? Would you expect the same level of consistency when comparing the venous structures?

Go to the Interactive CD, Animations/Cardiovascular system/Abdomen and pelvis, and compare the CD image to that in the text.

Challenge 2

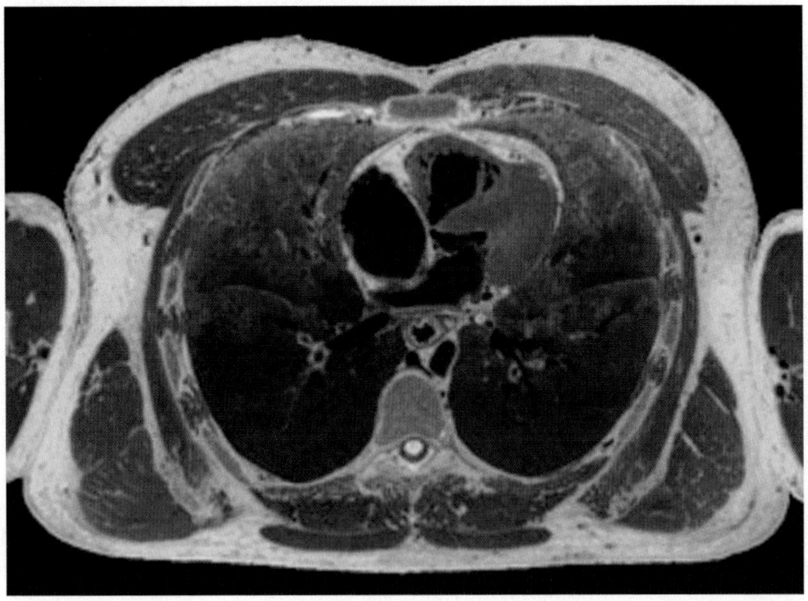

As you know, the center of the cardiovascular system is the heart. Here you see a transverse section through the thoracic cavity at the level of the heart. Visible are structures of the cavity and mediastinum as well as internal anatomy of the heart. Can you label the following in this view? Spinal cord, rib, intercostals, left lung, pulmonary arteries, esophagus, pleural cavity, pericardial cavity, ventricle, atrium, pericardial fat. Approximately how far from the top of the heart is this section? What clues did you use to identify the heart structures through which the section passes? Compare this image and your answers with Figure 21.2d●, p. 551.

Images provided by the Digital Cadaver™ project, courtesy of Visible Productions, Inc.

APPLY IT

Below are two separate exercises that provide more information on a topic presented in the preceding chapters. Each one is designed to take approximately 10 minutes, and will help you gain a better appreciation for the material presented in Chapters 20–22.

Critical Linking 1

Heart disease is listed as the number one killer of American men and women in the United States. Recently coronary health came to national attention as the current United States Vice President Cheney was treated for coronary artery disease. He required both angioplasty and the surgical implantation of an automatic defibrillator. To understand the effectiveness of these procedures requires a firm base in cardiovascular anatomy. Review the figures in your text that illustrate cardiac blood supply and the conduction system of the heart. Using Figure 21.9●, p.561, as a guide, indicate which you feel is most important to healthy blood supply to the heart. Move to Figure 21.12●, p. 565, which depicts the electrical system of the heart, and indicate the most appropriate site for implantation of a device designed to monitor and control heart rate through electrical stimulation.

To check your choices, go to the Companion Website, Critical Linking portion and choose the key word "Angioplasty." Read the articles on Vice President Cheney from CNN November 22, 2000, and June 30, 2001. Compare these descriptions with your original thoughts. Did your choices match the clinical operation? Would your choices serve just as well, if they were not those used in VP Cheney's surgeries?

Critical Linking 2

A thorough understanding of the cardiovascular system includes being able to identify the arteries. Look back at Figures 22.10, 22.12, 22.13, 22.15, 22.16, 22.18, and 22.19●. Can you name these arteries?

Test yourself by visiting the Companion Website and clicking on the keyword "arteries." This will take you to the Lumen "learn-em" site for arteries. Here you will find illustrations similar to those listed above with no labels. Moving your cursor over the circled numbers on these illustrations will provide the names. Try to match each figure listed above with the appropriate illustration on the Website. Name the arteries depicted on the Website, then check your answers by moving your cursor.

FURTHER STUDIES

"Blood is thicker than water." This saying has seemingly been a part of American culture forever. Just how long we have understood the properties of blood is an interesting study in medical history. You may be surprised at how recently we have discovered some of these facts! The circulation of blood through the body was not described accurately until 1628 when it was reported by English physician William Harvey. Shortly after this, transfusions were attempted. The first transfusions were attempted in animals other than humans, with some success in dogs as early as 1665. For a brief period of time, transfusions were carried out between humans and animals, but (not surprisingly!) this practice was banned in 1667 due to violent and often fatal cross-reactions. The first recorded human blood transfusion is credited to James Blundell in 1818. He was an obstetrician, and transfused a patient to treat postpartum hemorrhage. The donor blood was taken from the husband's arm and placed directly into her vein with no thought of contamination, cross reaction, or compatibility. Dr. Blundell performed a total of 10 transfusions in the years 1825 to 1830, which he carefully recorded and published to further this branch of medicine. The ABO blood groups were first described by Karl Landsteiner in 1900, with type AB described shortly thereafter by his colleagues.

Landsteiner received the Nobel Prize for Medicine for this discovery in 1930. The first blood bank in the United States was established at Cook County Hospital in Chicago in 1935. Five years later, Karl Landsteiner, Alex Weiner, Philip Levine, and R. E. Stetson described the Rh factor, paving the way for modern understanding of blood transfusion and rejection. It wasn't until 1943 that blood-born diseases were uncovered, with the description of transfusion-transmitted hepatitis by P. Beeson. In 1985 the first blood screening test to detect HIV was licensed and implemented in blood banks, followed closely by tests for hepatitis C.

More information on the history of our understanding of blood can be obtained through visiting *http://www.aabb.org/All_About_Blood/FAQs/aabb_faqs.htm#highlights* or *http://smfbc.org/about/blood_banking.htm*.

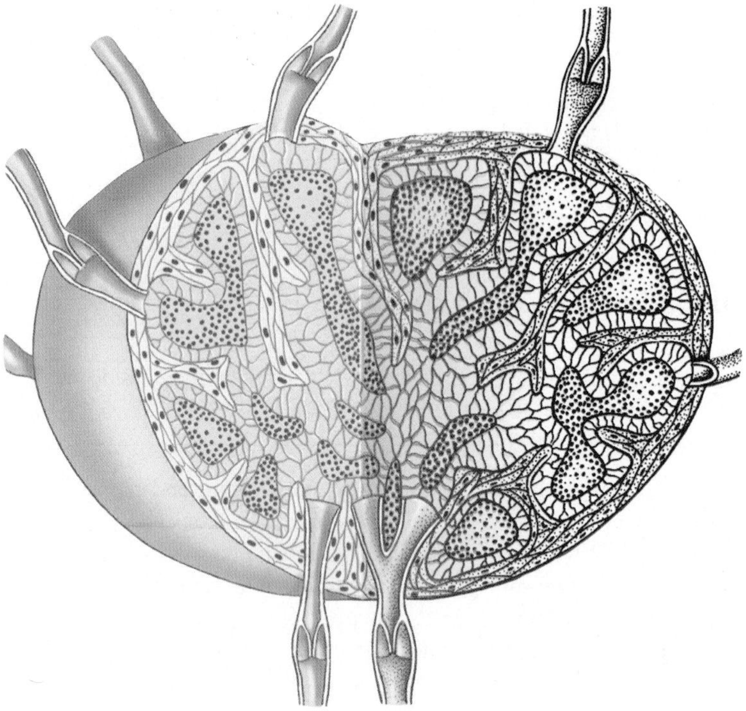

23

THE LYMPHATIC SYSTEM

The world is not always kind to the human body. Accidental collisions and interactions with objects in our environment produce bumps, cuts, and burns. The effects of an injury may be compounded by assorted viruses, bacteria, and other microorganisms that thrive in our environment. Some of these microorganisms normally live on the surface of and inside our bodies, but all have the potential to cause us great harm. Remaining alive and healthy involves a massive, combined effort involving many different organs and systems. In this ongoing struggle, the lymphatic system plays the primary role.

In this chapter we will describe the anatomical organization of the lymphatic system and consider how this system interacts with other systems and tissues to defend the body against infection and disease.

An Overview of the Lymphatic System

[FIGURE 23.1]

The **lymphatic system** (Figure 23.1●) consists of a network of *lymphatic vessels, lymph,* the fluid connective tissue transported in these vessels, and *lymphoid tissues* and *lymphoid organs,* which monitor and alter composition of the lymph. Lymphatic vessels originate in peripheral tissues and deliver lymph to the venous system. Lymph consists of (1) interstitial fluid, which resembles blood plasma, but with a lower concentration of proteins, (2) lymphocytes, cells responsible for the immune response, and (3) macrophages of various types. pp. 64–65 Lymphatic vessels often begin within or pass through lymphoid tissues and lymphoid organs, structures that contain large numbers of lymphocytes, macrophages, and (in many cases) lymphoid stem cells.

Functions of the Lymphatic System [FIGURE 23.2]

The primary function of the lymphatic system are:

1. *The production, maintenance, and distribution of lymphocytes.*

   Lymphocytes, which are essential to the normal defense mechanisms of the body, are produced and stored within lymphoid organs, such as the spleen, thymus, and bone marrow. Lymphoid tissues and organs are classified either as *primary* or *secondary. Primary lymphoid structures* contain stem cells that divide to produce daughter cells that differentiate into B, T, or NK cells. p. 541 The bone marrow and thymus of the adult are primary lymphoid structures. However, most immune responses begin in *secondary lymphoid structures,* where immature or activated lymphocytes divide to produce additional lymphocytes of the same type. For example, the divisions of activated B cells can produce the additional B cells needed to fight off an infection. Secondary lymphoid structures are located "at the front lines," where invading bacteria are first encountered. Examples include lymph nodes and tonsils.

2. *Maintain normal blood volume* and *eliminate local variations in the chemical composition of the interstitial fluid.*

   The blood pressure at the beginning of a systemic capillary is approximately 35 mm Hg. The blood pressure tends to force water and solutes out of the plasma and into the interstitial fluid (Figure 23.2b●). This movement is opposed by the *colloid osmotic pressure* of the blood, the osmotic pressure resulting from the presence of suspended plasma proteins. The colloid osmotic pressure at the start of a systemic capillary is usually about 25 mm Hg. Consequently, there is a net outward pressure of roughly 10 mm Hg that forces water and solutes out of the plasma and into the interstitial fluid. Blood pressure declines along the capillary due to a combination of friction and fluid

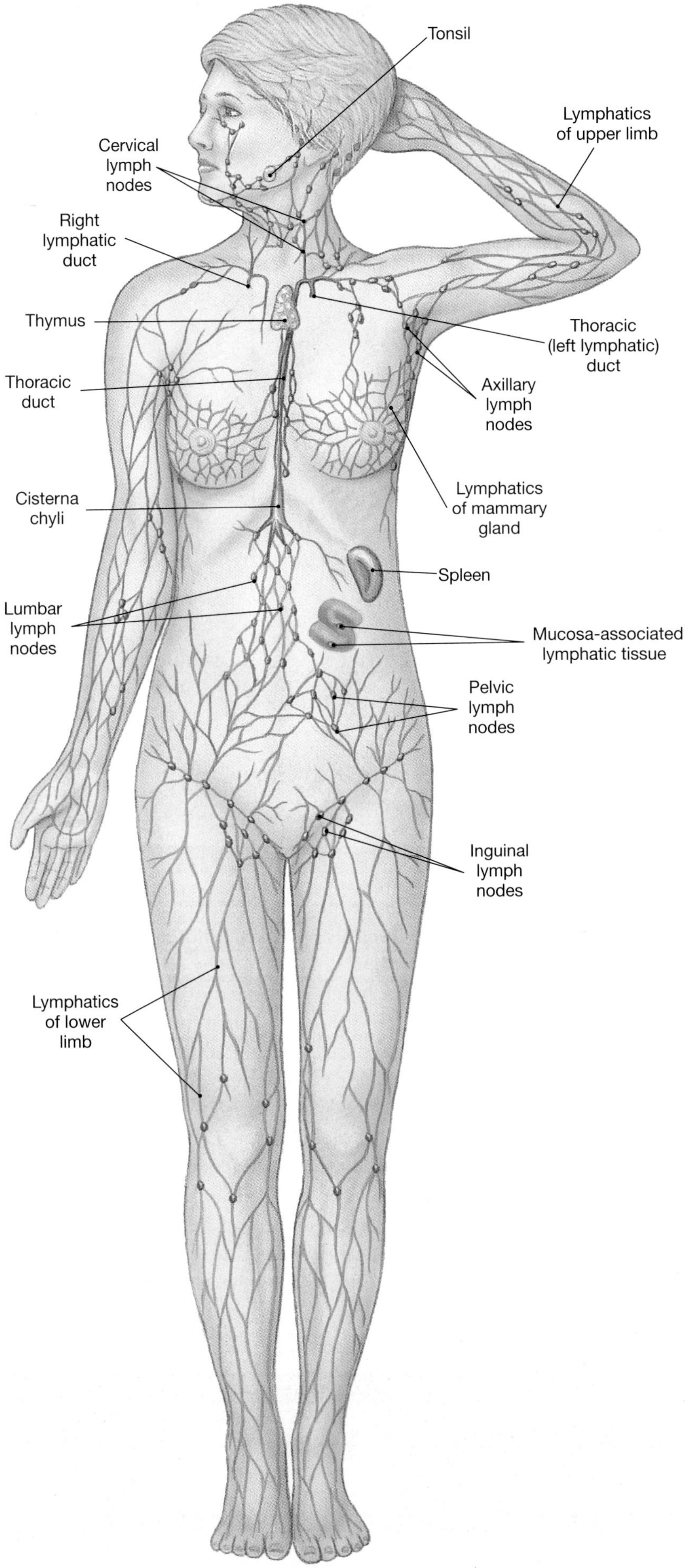

FIGURE 23.1 **LYMPHATIC SYSTEM**
An overview of the arrangement of lymphatic vessels, lymph nodes, and lymphoid organs.

FIGURE 23.2 **Lymphatic Capillaries**

Lymphatic capillaries are blind vessels that begin in areas of loose connective tissue. (**a**) A three-dimensional view of the association of blood capillaries, tissue, interstitial fluid, and lymphatic capillaries. Interstitial fluid enters these capillaries where it now is called lymph, by passing between adjacent endothelial cells or through pores, when the endothelial cells are fenestrated. Arrows show the direction of blood, interstitial fluid, and lymph movement. (**b**) Sectional view through a cluster of lymphatic capillaries.

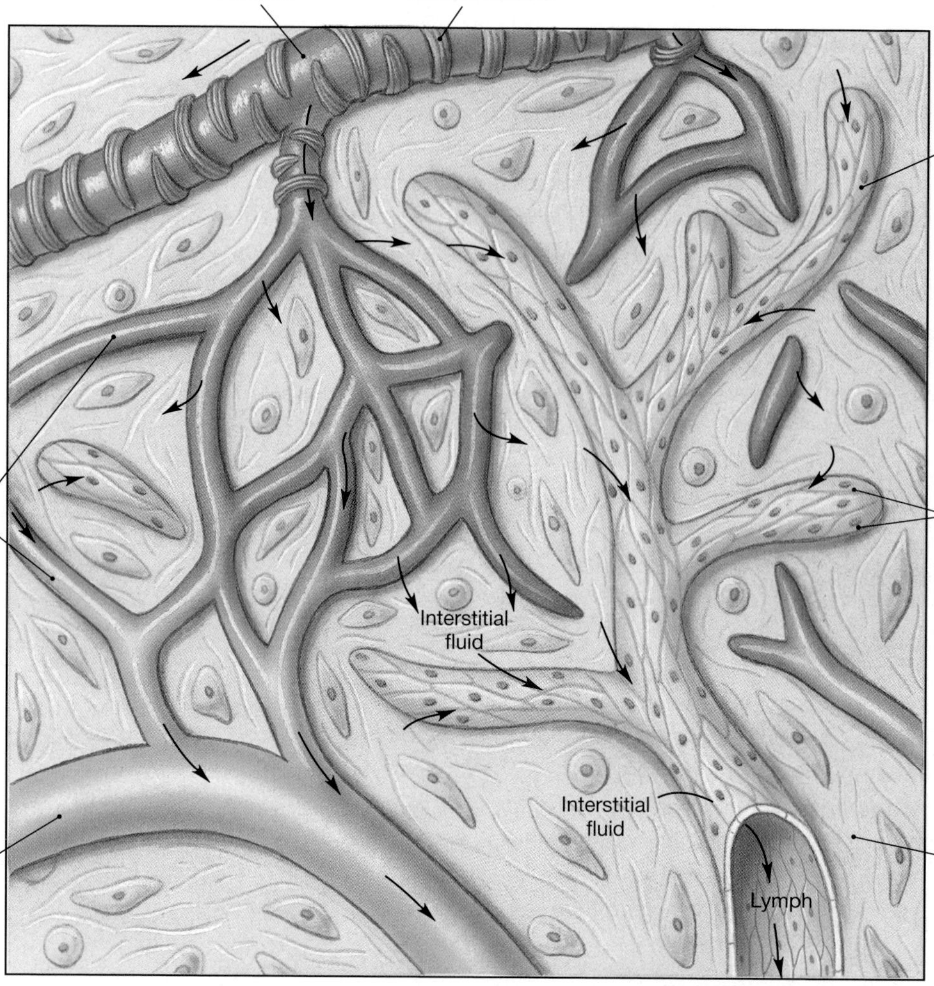

(a) Association of blood capillaries, tissue, and lymphatic capillaries

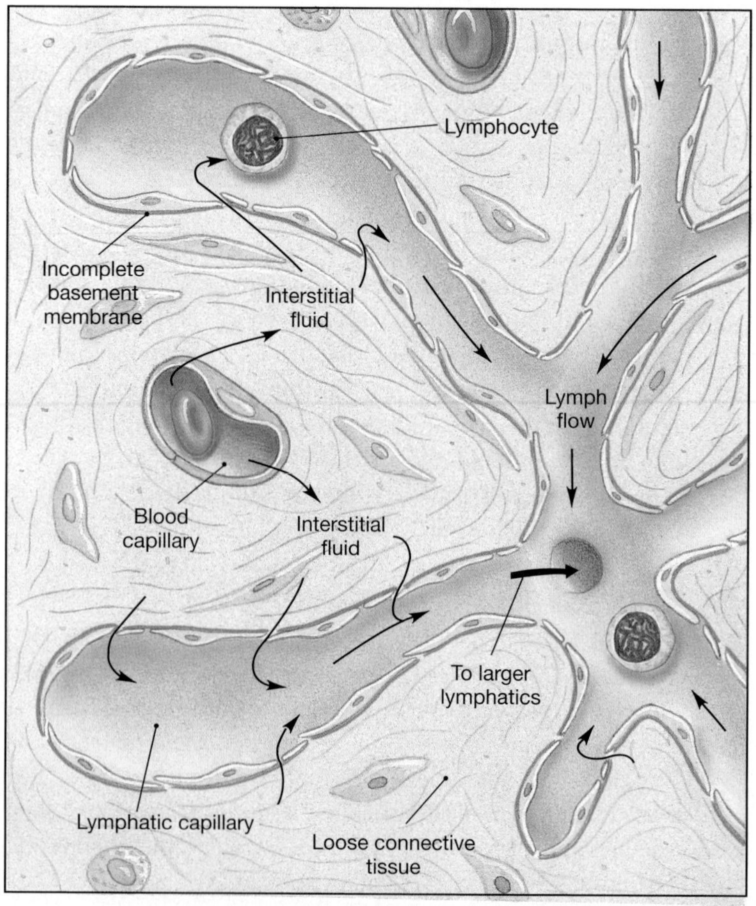

(b) Sectional view

loss. As blood pressure falls below 25 mm Hg, water and solutes begin to be reabsorbed. However, there is a small net movement of fluid from the plasma into the interstitial fluid along every systemic capillary. The total volume is substantial—approximately 3.6 l, or 72% of the total blood volume, enters the interstitial fluid each day. Under normal circumstances this movement goes unnoticed, because each day the vessels of the lymphatic system return an equal volume of interstitial fluid to the bloodstream. Consequently, there is a continual movement of fluid from the bloodstream into the tissues and then back to the bloodstream through lymphatic vessels. This circulation of fluid helps to eliminate regional differences in the composition of interstitial fluid. Because so much fluid moves through the lymphatic system each day, a break in a major lymphatic vessel can cause a rapid and potentially fatal decline in blood volume.

3. *Provide an alternative route for the transport of hormones, nutrients, and waste products.*

   For example, certain lipids absorbed by the digestive tract are carried to the bloodstream by lymphatic vessels rather than by absorption across capillary walls.

Structure of Lymphatic Vessels

Lymphatic vessels, often called *lymphatics*, carry lymph from peripheral tissues to the venous system. As with blood vessels, the lymphatic vessels range in size from small-diameter *lymphatic capillaries* to large-diameter collecting vessels, called *lymphatic ducts*.

Lymphatic Capillaries [FIGURE 23.2]

The lymphatic network begins with the **lymphatic capillaries**, or *terminal lymphatics*, which form a complex network within peripheral tissues. Lymphatic capillaries differ from vascular capillaries in several ways: (1) they are larger both in diameter and in sectional view; (2) they have thinner walls because their endothelial cells lack a continuous basement membrane; (3) they often have a flat or irregular outline, in part because the walls are too thin to hold their shape when the already low lymph pressure disappears entirely; and (4) their endothelial cells overlap instead of being tightly bound to one another (Figure 23.2●). The region of endothelial cell overlap acts as a one-way valve, permitting passage of interstitial fluid into the lymphatic capillary but preventing its escape (Figure 23.2b●). The endothelial cells are often fenestrated, and with pores in the cells and gaps between cells there is little in the interstitial fluid that cannot find its way into a lymphatic capillary. The gaps between endothelial cells are large enough that the lymphatic capillaries absorb not only interstitial fluid and dissolved solutes but any viruses or other abnormal items, such as cell debris or bacteria, that are present in damaged or infected tissues. As a result, a lymphatic capillary contains chemical and physical evidence about the health of the surrounding tissues.

Lymphatic capillaries are present in almost every tissue and organ in the body. Prominent lymphatic capillaries in the small intestine, called *lacteals*, transport lipids absorbed by the digestive tract. Lymphatic capillaries are absent in areas that lack a blood supply, such as the matrix of cartilage and the cornea of the eye, and there are no lymphatics in either the bone marrow or the central nervous system.

Valves of Lymphatic Vessels [FIGURE 23.3]

From the lymphatic capillaries, lymph flows into larger lymphatic vessels that lead toward the lymphatic trunks in the abdominopelvic and thoracic cavities. The larger lymphatics are comparable to veins both in the layers in their walls and in the presence of internal valves. The valves are quite close together, and at each valve the lymphatic vessel bulges noticeably. This configuration gives large lymphatics a beaded appearance (Figure 23.3●). Pressures within the lymphatic system are minimal; in fact, interstitial fluid pressure is lower than that of the venous system. The valves prevent the backflow of lymph within lymph vessels, especially those of the limbs. The larger lymphatic vessels have layers of smooth muscle in their walls. Rhythmic contractions of these vessels propel lymph toward the lymphatic ducts. Skeletal muscle contraction and respiratory movements work together to help move lymph through the lymphatic vessels. Contractions of skeletal muscles in the limbs compress the lymphatics and squeeze lymph toward the trunk; a comparable mechanism assists venous return. With each inhalation, pressure decreases inside the thoracic cavity, and lymph is pulled from smaller lymphatic vessels into the lymphatic ducts.

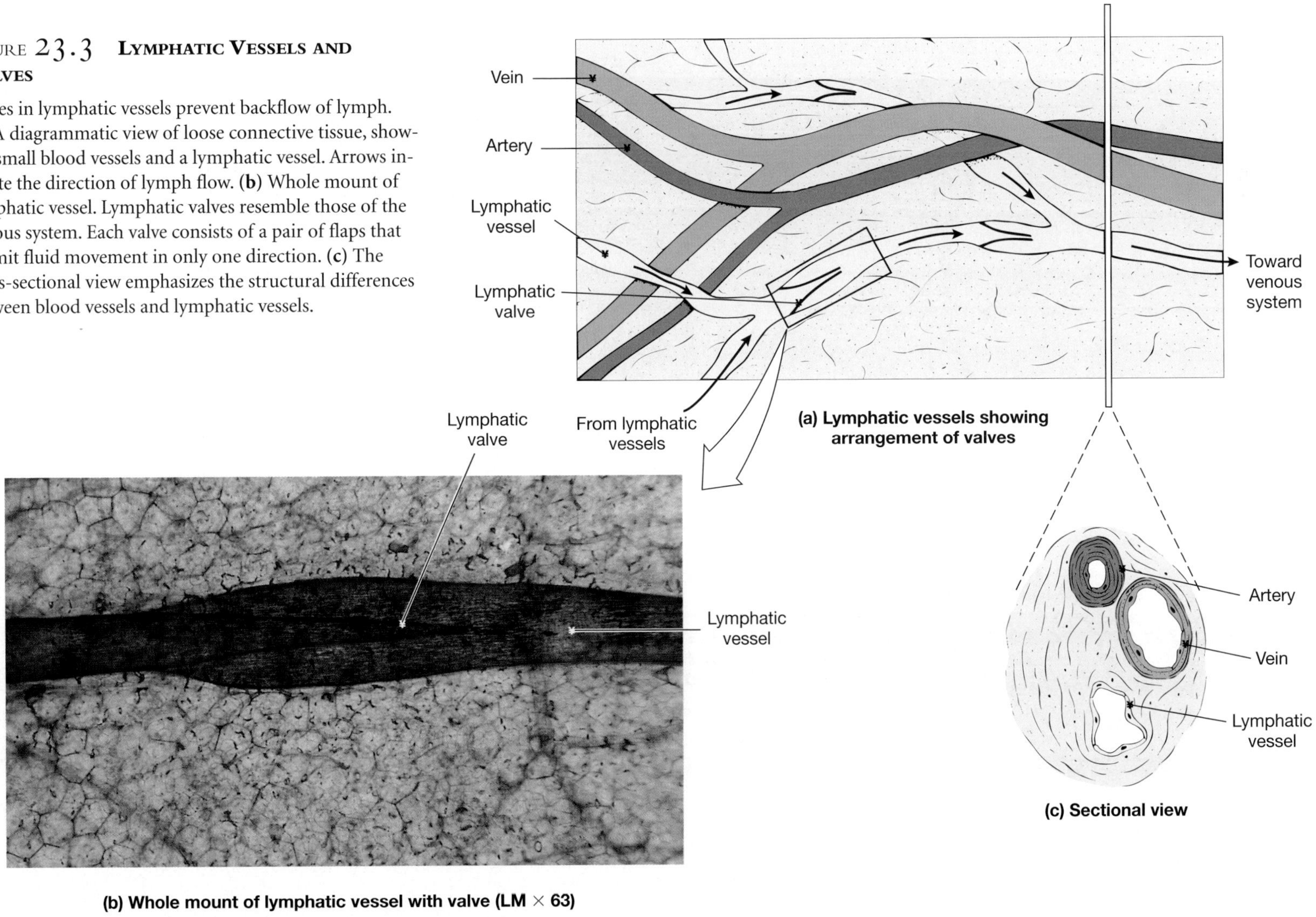

FIGURE 23.3 **LYMPHATIC VESSELS AND VALVES**

Valves in lymphatic vessels prevent backflow of lymph. (**a**) A diagrammatic view of loose connective tissue, showing small blood vessels and a lymphatic vessel. Arrows indicate the direction of lymph flow. (**b**) Whole mount of lymphatic vessel. Lymphatic valves resemble those of the venous system. Each valve consists of a pair of flaps that permit fluid movement in only one direction. (**c**) The cross-sectional view emphasizes the structural differences between blood vessels and lymphatic vessels.

If a lymphatic vessel is compressed or blocked or its valves are damaged, lymph drainage slows or ceases in the affected area. When fluid continues to leave the vascular capillaries in that region but the lymphatic system is no longer able to remove it, the interstitial fluid volume and pressure gradually increase. The affected tissues become distended and swollen, and this condition is called *lymphedema*.

Lymphatic vessels are typically found in association with blood vessels. Note the differences in relative size, general appearance, and branching pattern that distinguish lymphatic vessels from arteries and veins (Figure 23.3a●). There are also characteristic color differences that are apparent on examining living tissues. Arteries are usually a bright red, veins a dark red, and lymphatics a pale golden color.

■ Major Lymph-Collecting Vessels [FIGURE 23.4]

Two sets of lymphatic vessels, the superficial lymphatics and the deep lymphatics, collect lymph from the lymphatic capillaries. **Superficial lymphatics** travel with superficial veins and are found in:

- The subcutaneous layer next to the skin.

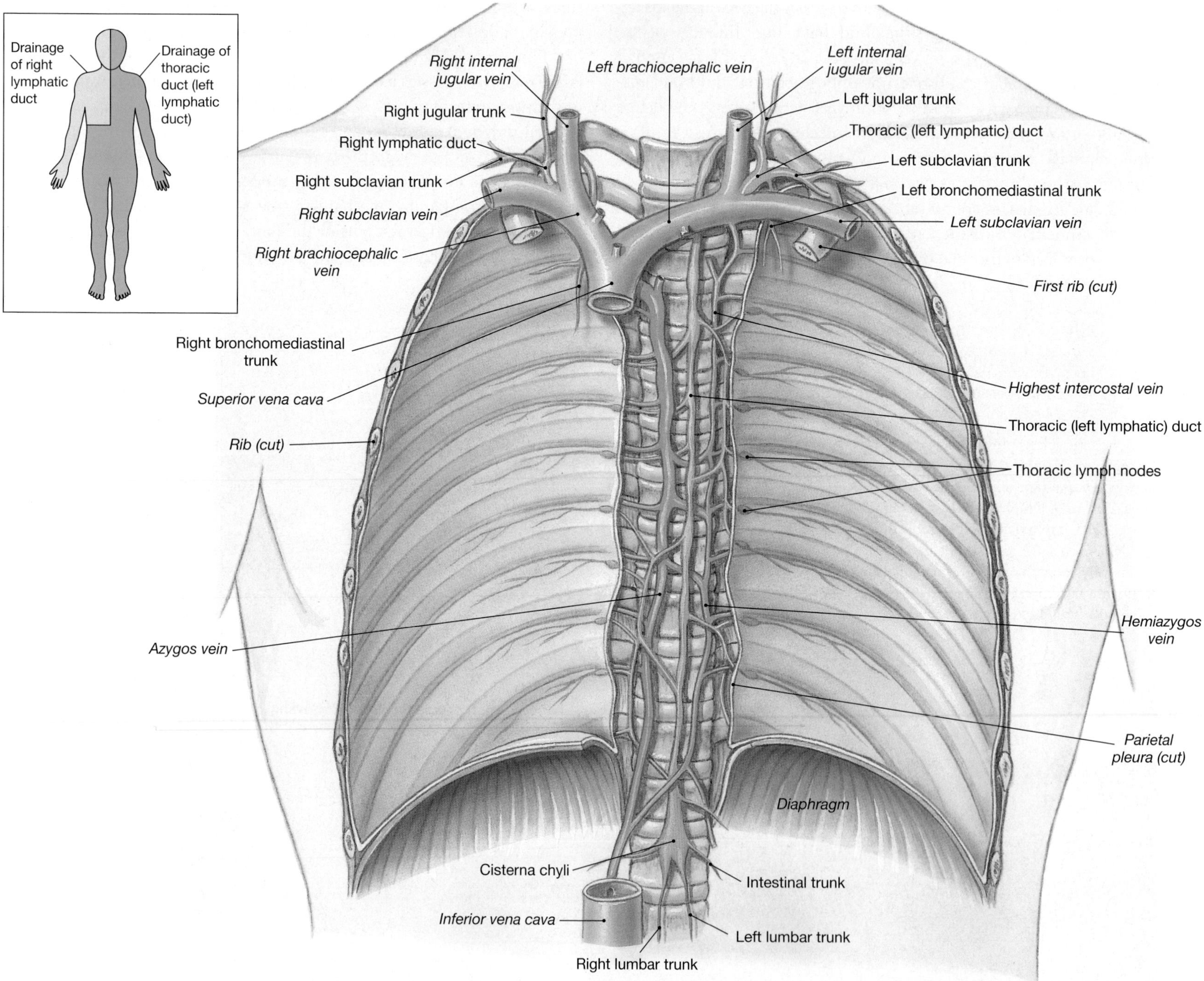

FIGURE 23.4 **LYMPHATIC DUCTS AND LYMPHATIC DRAINAGE**

The collecting system of lymph vessels, nodes, major lymphatic collecting ducts, and their relationship to the brachiocephalic veins. The thoracic duct collects lymph from tissues inferior to the diaphragm and from the left side of the upper body. The right lymphatic duct drains the right half of the body superior to the diaphragm.

- The loose connective tissues of the mucous membranes lining the digestive, respiratory, urinary, and reproductive tracts.
- The loose connective tissues of the serous membranes lining the pleural, pericardial, and peritoneal cavities.

Deep lymphatics are large lymph vessels that accompany the deep arteries and veins. These lymphatic vessels collect lymph from skeletal muscles and other organs of the neck, limbs, and trunk, as well as visceral organs in the thoracic and abdominopelvic cavities.

Within the trunk, superficial and deep lymphatics converge to form larger vessels called **lymphatic trunks**. The lymphatic trunks include the (1) *lumbar trunks*, (2) *intestinal trunks*, (3) *bronchomediastinal trunks*, (4) *subclavian trunks*, and (5) *jugular trunks* (Figure 23.4●). The lymphatic trunks in turn empty into two large collecting vessels, the **lymphatic ducts**, that deliver lymph to the venous circulation.

THE THORACIC DUCT [FIGURES 23.4/23.5]

The **thoracic duct** collects lymph from both sides of the body inferior to the diaphragm and from the left side of the body superior to the diaphragm. The thoracic duct begins inferior to the diaphragm at the level of vertebra L_2. The base of the thoracic duct is an expanded, saclike chamber, the **cisterna chyli** (KĪ-lē) (Figures 23.4 and 23.5●). The cisterna chyli receives lymph from the inferior region of the abdomen, pelvis, and lower limbs through the *right* and *left lumbar trunks* and the *intestinal trunk*.

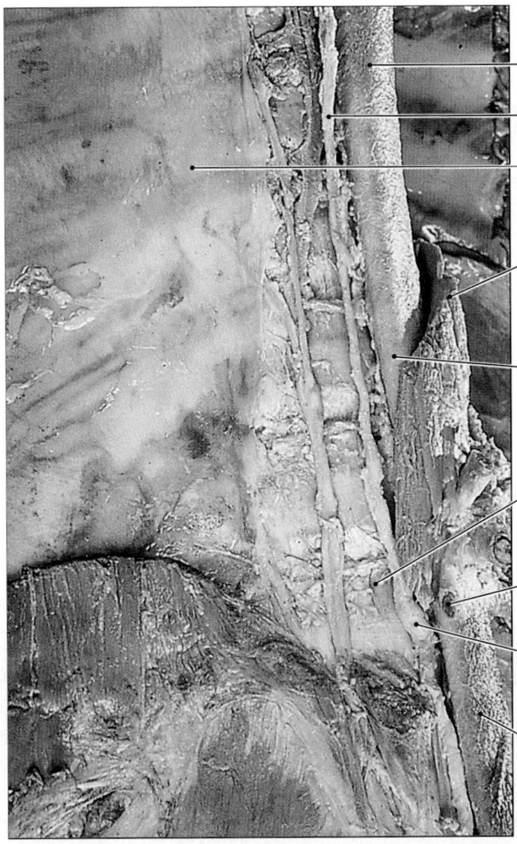

FIGURE 23.5 **MAJOR LYMPHATIC VESSELS OF THE TRUNK**

Anterior view of a dissection of the thoracic duct and adjacent blood vessels. The thoracic and abdominopelvic organs have been removed.

The inferior segment of the thoracic duct lies anterior to the vertebral column. From its origin anterior to the second lumbar vertebra, it penetrates the diaphragm with the aorta, at an opening known as the *aortic hiatus*, and ascends along the left side of the vertebral column to the level of the left clavicle. After collecting lymph from the *left bronchomediastinal trunk*, the *left subclavian trunk*, and the *left jugular trunk*, it empties into the left subclavian vein near the base of the left internal jugular vein (Figure 23.4●). Lymph collected from the left side of the head, neck, and thorax as well as lymph from the entire body inferior to the diaphragm reenters the venous system in this way.

THE RIGHT LYMPHATIC DUCT [FIGURE 23.4]

The relatively small **right lymphatic duct** collects lymph from the right side of the body superior to the diaphragm. The right lymphatic duct receives lymph from smaller lymphatic vessels that converge in the region of the right clavicle. This duct empties into the venous system at or near the junction of the right internal jugular vein and right subclavian vein (Figure 23.4●). *Lymphedema p. 801*

✓ CONCEPT CHECK

- What is the main function of the lymphatic system?
- You are looking through a microscope at a cross section of two capillaries. Capillary number one is larger than number two and has a thinner wall. Capillary number two has a round shape, whereas capillary one has an irregular outline. Which capillary is probably a lymphatic capillary?
- Would the rupture of a major lymphatic vessel be fatal? Why or why not?
- Which lymphoid structures contain stem cells?

Lymphocytes

Lymphocytes are the primary cells of the lymphatic system, and they are responsible for specific immunity. pp. 703, 541 They respond to the presence of: (1) invading organisms, such as bacteria or viruses; (2) abnormal body cells, such as virus-infected cells or cancer cells; and (3) foreign proteins, such as the toxins released by some bacteria. Lymphocytes attempt to eliminate these threats or render them harmless by a combination of physical and chemical attack. They travel throughout the body, circulating in the bloodstream, then moving through peripheral tissues and eventually returning to the bloodstream through the lymphatic system. The time spent within the lymphatic system varies; a lymphocyte may remain within a lymph node or other lymphatic organ for hours, days, or even years. When in peripheral tissues, lymphocytes may encounter invading pathogens or foreign proteins; while in the lymphatic system, they may be exposed to pathogens or proteins carried by lymph. Regardless of the source, lymphocytes respond by initiating an immune response.

Types of Lymphocytes

There are three different classes of lymphocytes in the blood: **T cells** (thymus-dependent), **B cells** (**b**one marrow–derived), and **NK cells** (**n**atural **k**iller). Each type has distinctive biochemical and functional characteristics.

T CELLS

Approximately 80% of circulating lymphocytes are classified as T cells. There are several types of T cells. **Cytotoxic T cells** attack foreign cells or body cells infected by viruses. Their attack often involves direct contact. These lymphocytes are responsible for providing **cell-mediated immunity. Helper T cells** and **suppressor T cells** assist in the regulation and coordination of the immune response; for this reason they are also called *regulatory T cells*. Regulatory T cells control both the activation and the activity of B cells. **Memory T cells** are produced by the division of activated T cells following exposure to a particular antigen. They are called memory cells because they remain "on reserve," becoming activated only if the same antigen appears in the body at a later date. This is not a complete list, as there are several other types of specialized T cells in the body.

B CELLS

B cells account for 10–15% of circulating lymphocytes. When stimulated by exposure to an antigen, a B cell can differentiate into **plasma cells**. Plasma cells are responsible for the production and secretion of **antibodies.** p. 539 These soluble proteins react with specific chemical targets, called **antigens**. Antigens are usually associated with pathogens, parts or products of pathogens, or other foreign compounds. Most antigens are short peptide chains or short amino-acid sequences along a complex protein, but some lipids, polysaccharides, and nucleic acids can also stimulate antibody production. When an antigen-antibody complex forms, it starts a chain of events leading to the destruction, neutralization, or elimination of the antigen. Antibodies are also known as **immunoglobulins** (i-mū-nō-GLOB-ū-lins). Because the blood is the primary distribution route for immunoglobulins, B cells are said to be responsible for **antibody-mediated immunity**, or *humoral* ("liquid") *immunity*. **Memory B cells** are produced by the division of activated B cells; activation occurs following exposure to a particular antigen. Memory B cells become activated only if the antigen appears in the body at a later date.

Helper T cells promote the differentiation of plasma cells and accelerate the production of antibodies. Suppressor T cells inhibit the formation of plasma cells and reduce the production of antibodies by existing plasma cells.

NK CELLS

The remaining 5–10% of circulating lymphocytes are NK cells, also known as *large granular lymphocytes*. These lymphocytes attack foreign cells, normal cells infected with viruses, and cancer cells that appear in normal tissues. The continual policing of peripheral tissues by NK cells and activated macrophages has been called **immunological surveillance**.

Lymphocytes and the Immune Response [FIGURE 23.6]

The goal of the **immune response** is the destruction or inactivation of pathogens, abnormal cells, and foreign molecules such as toxins. The body has two different ways to do this:

1. Direct attack by activated T cells *(cell-mediated immunity)*;
2. Attack by circulating antibodies released by the *plasma cells* derived from activated B cells *(antibody-mediated immunity)*.

Figure 23.6● provides an overview of the immune response to bacterial infection and viral infection. After the appearance of an antigen, the first step in the immune response is often the phagocytosis of the antigen

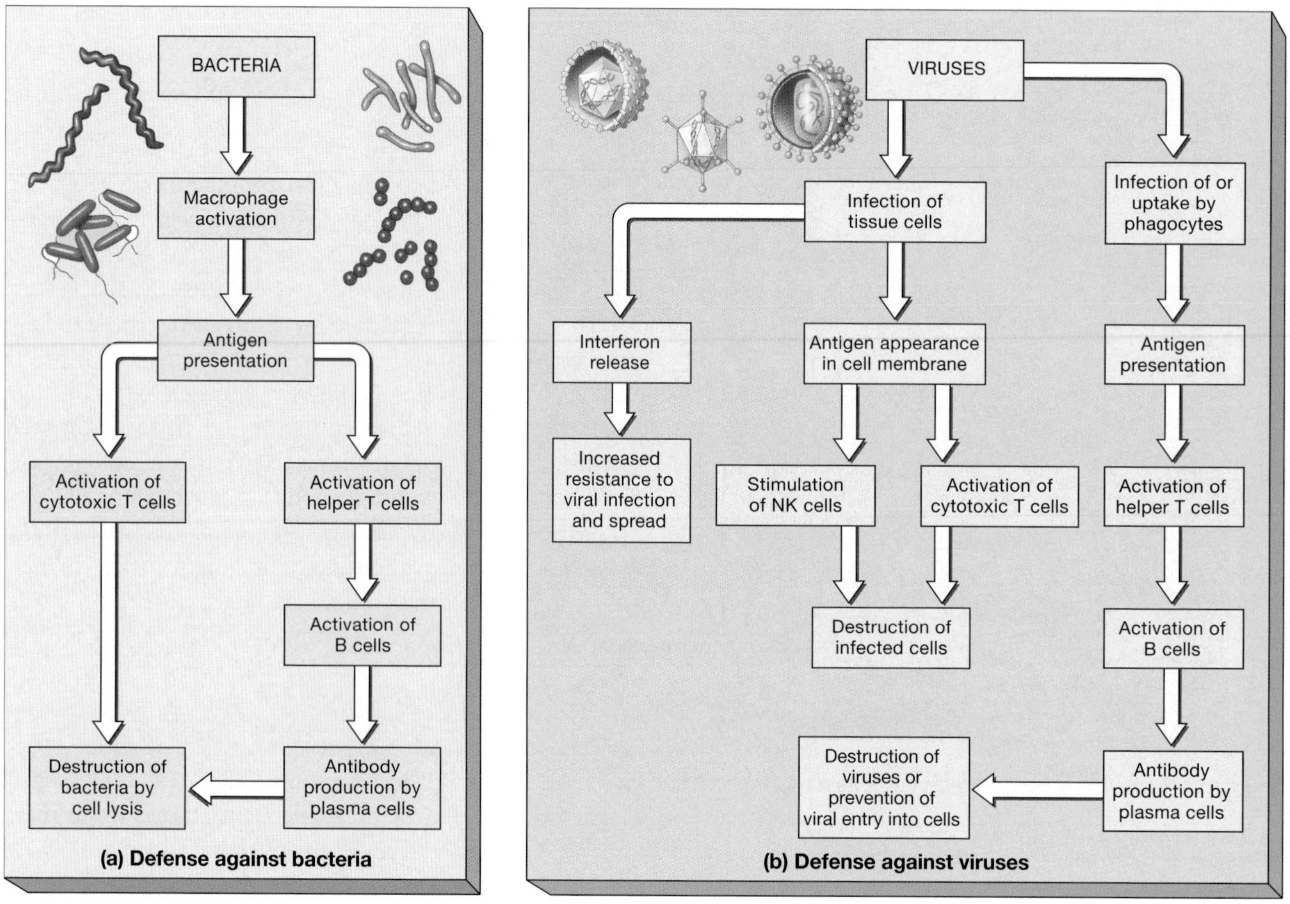

FIGURE 23.6
LYMPHOCYTES AND THE IMMUNE RESPONSE

(**a**) Defenses against bacterial pathogens are usually initiated by active macrophages. (**b**) Defenses against viruses are usually activated after the infection of normal cells. In each instance, B cells and T cells cooperate to produce a coordinated chemical and physical attack.

by a macrophage. The macrophage then displays pieces of the antigen in its cell membrane. In this way the macrophage "presents" the antigen to T cells, and the process is called *antigen presentation*. The T cells that respond are sensitive to that particular antigen and none other. These lymphocytes respond because their cell membranes contain receptors capable of binding that specific antigen. When binding occurs, the T cells undergo activation, and begin to divide. Some daughter cells differentiate into cytotoxic T cells, others into helper T cells that will in turn activate B cells, and some become **memory T cells** that will differentiate further only if it encounters this antigen at a later date.

Your immune system has no way of anticipating which antigens it will actually encounter. Its protective strategy is to prepare for *any* antigen that might appear. During development, differentiation of cells in the lymphatic system produces an enormous number of lymphocytes with varied antigen sensitivities. The ability of a lymphocyte to recognize a specific antigen is called *immunocompetence*. Among the trillion or so lymphocytes in the human body are millions of different lymphocyte populations. Each population consists of several thousand cells that are prepared to recognize a specific antigen. When one of these lymphocytes binds an antigen, it becomes activated, dividing to produce more lymphocytes sensitive to that particular antigen. Some of the lymphocytes function immediately to eliminate the antigen, and others (the memory cells) will be ready if the antigen reappears at a later date. This mechanism provides an immediate defense and ensures that there will be an even more massive and rapid response if the antigen appears in the body at some later date.

Distribution and Life Span of Lymphocytes

The ratio of B cells to T cells varies depending on the tissue or organ considered. For example, B cells are seldom found in the thymus, and in the blood, T cells outnumber B cells by a ratio of 8:1. This ratio changes to 1:1 in the spleen and 1:3 in the bone marrow.

The lymphocytes within these organs are visitors, not residents. Lymphocytes continually move throughout the body; they wander through a tissue and then enter a blood vessel or lymphatic vessel for transport to another site. T cells move relatively quickly. For example, a wandering T cell may spend about 30 minutes in the blood and 15–20 hours in a lymph node. B cells move more slowly; a typical B cell spends around 30 hours in a lymph node before moving to another location.

In general, lymphocytes have relatively long life spans, far longer than other formed elements in the blood. Roughly 80% survive for four years, and some last 20 years or more. Throughout life, normal lymphocyte populations are maintained by the process of *lymphopoiesis*.

Lymphopoiesis: Lymphocyte Production [FIGURE 23.7]

Lymphopoiesis occurs in the bone marrow and thymus. Figure 23.7● shows the relationships among bone marrow, thymus, and peripheral tissues with respect to lymphocyte production, maturation, and distribution.

Hemocytoblasts in the bone marrow produce lymphocytic stem cells with two distinct fates. One group remains in the bone marrow. These stem cells divide to produce NK cells and B cells, which gain immunocompetence and migrate into peripheral tissues. The NK cells continuously circulate through peripheral tissues, whereas the B cells take up residence in lymph nodes, the spleen, and lymphatic tissues. The second group of stem cells migrates to the thymus. Under the influence of the thymic hormones (thymosin-1, thymopentin, thymulin, and others) these stem cells divide repeatedly, producing daughter cells that undergo functional maturation into T cells. These T cells subsequently migrate to the spleen, other lymphoid organs, and the bone marrow.

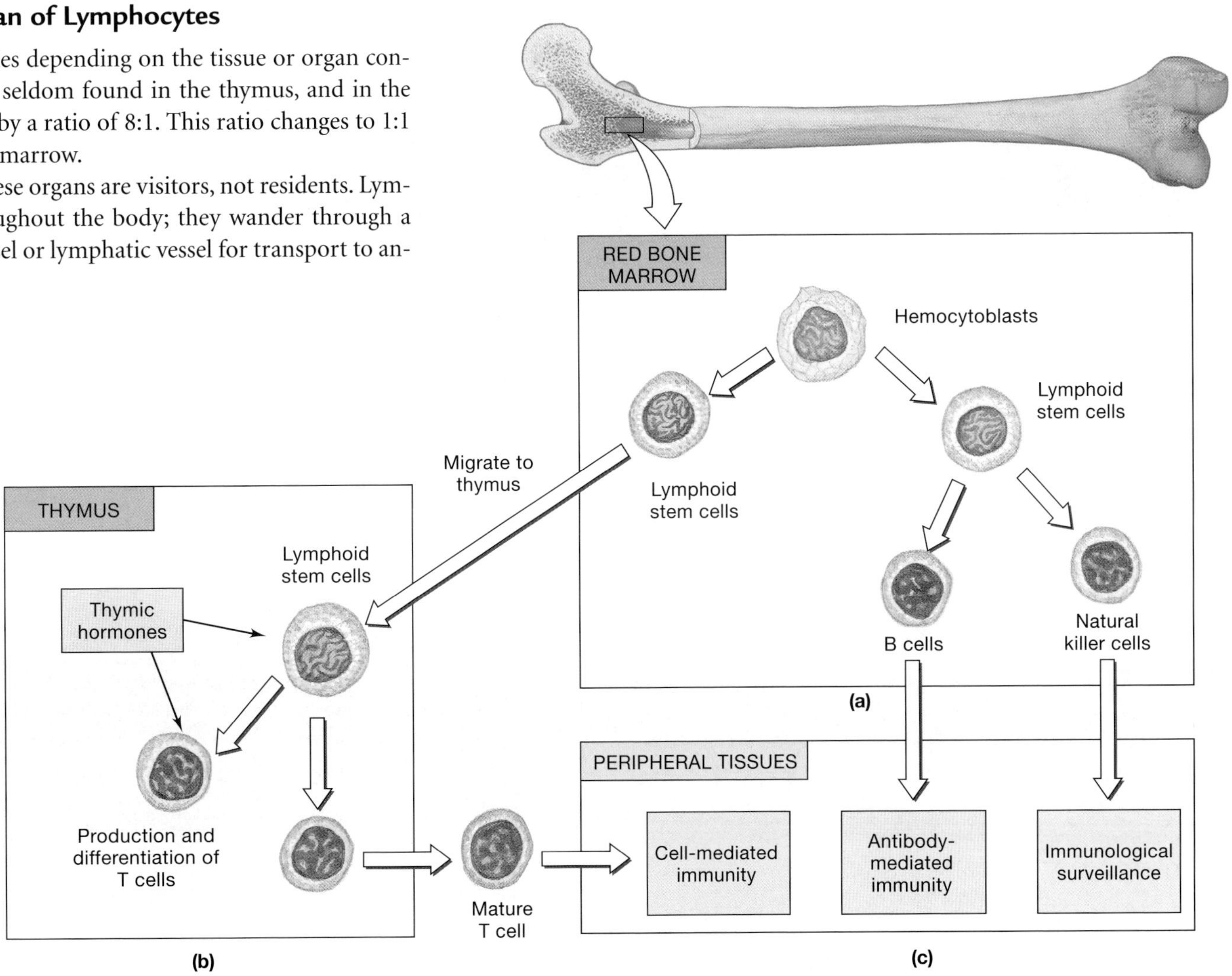

FIGURE 23.7
DERIVATION AND DISTRIBUTION OF LYMPHOCYTES

Hemocytoblast divisions produce lymphoid stem cells with two different fates. **(a)** One group remains in the bone marrow, producing daughter cells that mature into B cells and NK cells that enter peripheral tissues. **(b)** The second group of stem cells migrates to the thymus, where subsequent divisions produce daughter cells that mature into T cells. **(c)** Mature T cells leave the circulation to take temporary residence in peripheral tissues. All three types of lymphocyte circulate throughout the body in the bloodstream.

As a lymphocyte migrates through peripheral tissues, it retains the ability to divide. Its divisions produce daughter cells of the same type and with sensitivity to the same specific antigen. For example, a dividing B cell produces other B cells, not T cells or NK cells. The ability to increase the number of lymphocytes of a specific type is important to the success of the immune response. If that ability is compromised, the individual will be unable to mount an effective defense against infection and disease. For example, the disease *AIDS* (*acquired immune deficiency syndrome*) results from infection with a virus that selectively destroys T cells. Individuals with AIDS are likely to be killed by bacterial or viral infections that would be overcome easily by a normal immune system. For a detailed discussion of AIDS, see the Clinical Issues appendix. *SCID* and *AIDS p. 802*

Lymphoid Tissue [FIGURE 23.8]

Lymphoid tissues are connective tissues dominated by lymphocytes. In a **lymphoid nodule**, the lymphocytes are densely packed within the loose connective tissue of the mucous membranes lining the respiratory, digestive, urinary, and reproductive tracts (Figure 23.8●). Typical lymphoid nodules average around a millimeter in diameter, but the boundaries are indistinct because no fibrous capsule surrounds them. They often have a pale, central zone, called a **germinal center**, which contains activated, dividing lymphocytes (Figure 23.8a●).

The digestive tract has an extensive array of lymphoid nodules collectively known as the **mucosa-associated lymphatic tissue (MALT)**. Large nodules in the wall of the pharynx are called **tonsils** (Figure 23.8b●). The lymphocytes aggregated in tonsils gather and remove pathogens that enter the pharynx in either inspired air or food. There are usually five tonsils:

- a single **pharyngeal tonsil**, often called the *adenoids*, located in the posterior superior wall of the nasopharynx;
- a pair of **palatine tonsils**, located at the posterior margin of the oral cavity along the boundary of the pharynx to the soft palate; and
- a pair of **lingual tonsils**, which are not visible, because they are located at the base of the tongue.

Clusters of lymphoid nodules in the mucosal lining of the small intestine are know as **aggregated lymphoid nodules**, or *Peyer's patches*. In addition, the walls of the *vermiform appendix*, a blind pouch that originates near the junction between the small and large intestines, contain a mass of fused lymphoid nodules.

The lymphocytes within these lymphoid nodules are not always able to destroy bacterial or viral invaders that have crossed the epithelium of the digestive tract. An infection may then develop; familiar examples include *tonsillitis* and *appendicitis*. *Infected Lymphoid Nodules p. 802*

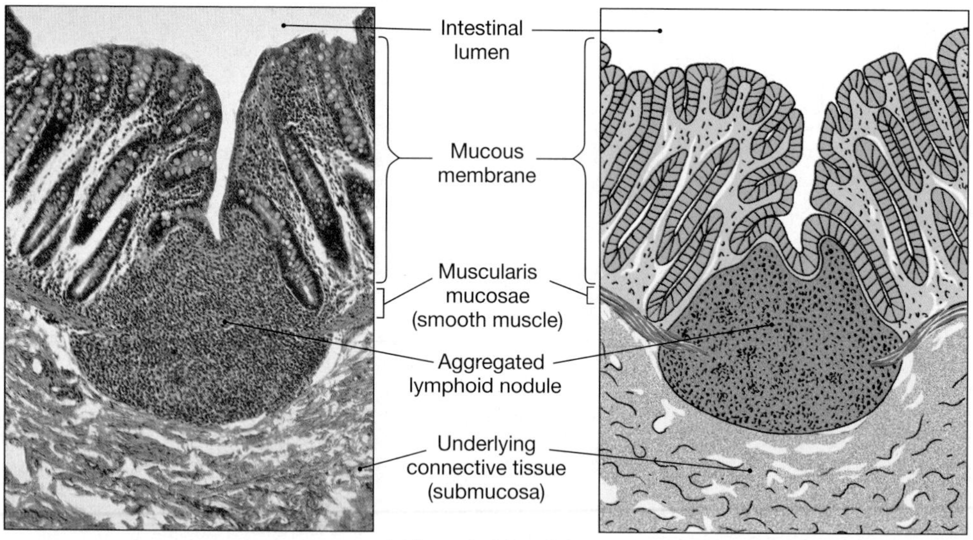

(a) Lymphoid nodule

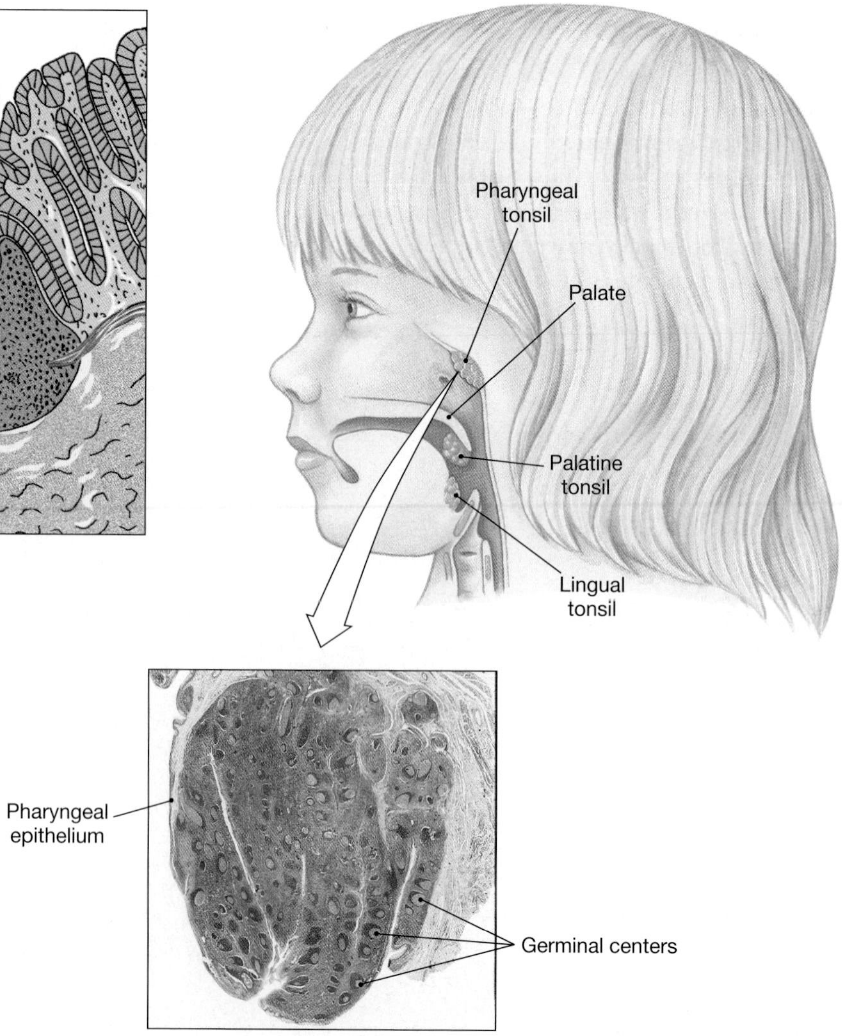

(b) Pharyngeal tonsil

FIGURE 23.8 **LYMPHOID TISSUES**

(a) Appearance of an isolated lymphoid nodule in the large intestine. Note the relatively pale pink germinal center, where lymphocyte cell divisions occur. (LM × 51).
(b) The location of the tonsils and a histological organization of a single tonsil.

✓ CONCEPT CHECK

- Which type of lymphocyte is the most common?
- John becomes infected with a pathogen. A few months later, while relatively healthy, he is exposed to the same pathogen. Will he definitely get sick again? Why or why not?
- Circulating lymphocytes retain the ability to divide. Why is this important?
- What is the name of the lymphoid nodules found internal to the epithelium of the intestine?

Lymphoid Organs

Lymphoid organs are separated from surrounding tissues by a fibrous connective tissue capsule. These organs include the *lymph nodes*, the *thymus*, and the *spleen*.

Lymph Nodes [FIGURES 23.1/23.4/23.9 TO 23.14]

Lymph nodes are small, oval lymphoid organs ranging in diameter from 1 to 25 mm (up to around 1 in.). The general pattern of lymph node distribution in the body can be seen in Figure 23.1●, p. 613. Each lymph node is covered by a dense fibrous connective tissue capsule. Fibrous extensions from the capsule extend partway into the interior of the node (Figure 23.9●). These fibrous partitions are called *trabeculae*.

The shape of a typical lymph node resembles that of a lima bean. Blood vessels and nerves attach to the lymph node at the indentation, or **hilus** (Figure 23.9●). Two sets of lymphatic vessels are connected to each lymph node: *afferent lymphatics* and *efferent lymphatics*. The afferent lymphatic vessels, which bring lymph to the node from peripheral tissues, penetrate the capsule on the side opposite the hilus. Lymph then flows slowly through the lymph node within a network of sinuses, which are open passageways with incomplete walls. Upon arriving at the node, lymph first enters the *subcapsular sinus* that contains a meshwork of branching reticular fibers, macrophages, and *dendritic cells*. **Dendritic cells** collect antigens from the lymph and present them in their cell membranes. T cells encountering these bound antigens become activated, thus initiating an immune response. After passing through the subcapsular sinus, lymph flows through the **outer cortex** of the node. The outer cortex contains aggregated B cells with germinal centers similar to those of lymphoid nodules.

Lymph flow continues through lymph sinuses in the **deep cortex** *(paracortical area)*. Here, circulating lymphocytes leave the bloodstream and enter the lymph node by crossing the walls of blood vessels within the deep cortex. The deep cortical area is dominated by T cells.

After flowing through the sinuses of the deep cortex, lymph continues into the core, or **medulla**, of the lymph node. The medulla contains B cells and plasma cells organized into elongate masses known as **medullary cords**. Lymph enters the efferent lymphatics at the hilus after passing through a network of sinuses in the medulla.

The lymph nodes function like a kitchen water filter: They filter and purify lymph before it reaches the venous system. As lymph flows through a

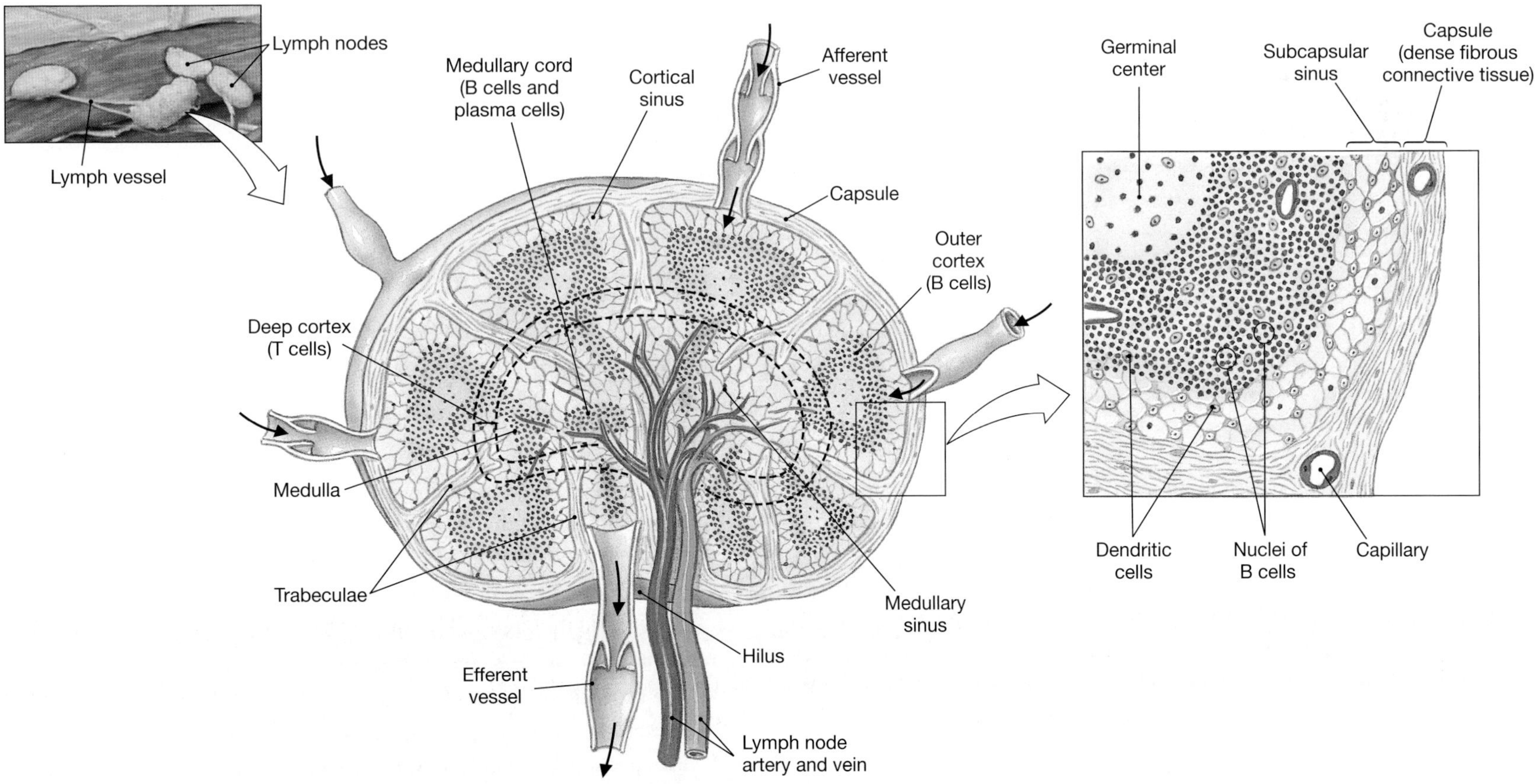

FIGURE 23.9 **STRUCTURE OF A LYMPH NODE**

Lymph nodes are covered by a dense, fibrous, connective tissue capsule. Lymphatic vessels and blood vessels penetrate the capsule to reach the lymphoid tissue within. Note that there are several afferent lymphatic vessels and only one efferent vessel.

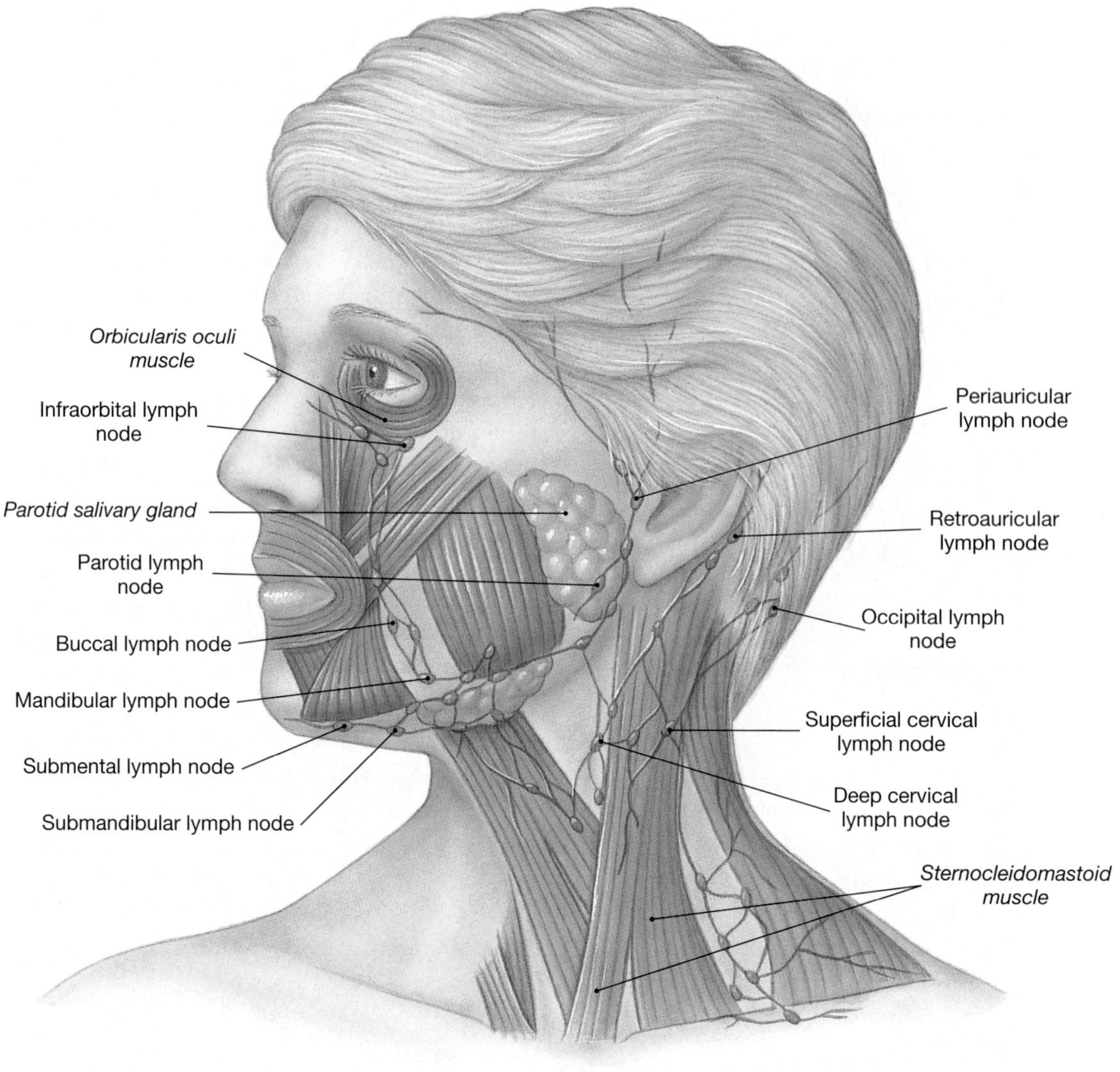

FIGURE 23.10 **Lymphatic Drainage of the Head and Neck**
Position of the lymphatic vessels and nodes that drain the head and neck regions.

lymph node, at least 99% of the antigens present in the arriving lymph are removed. Fixed macrophages in the walls of the lymphatic sinuses engulf debris or pathogens in the lymph as it flows past. Antigens removed in this way are then processed by the macrophages and "presented" to nearby T cells. Other antigens stick to the surfaces of dendritic cells, where they can stimulate T cell activity.

The largest lymph nodes are found where peripheral lymphatics connect with the trunk (Figure 23.4●, p. 616), in regions such as the base of the neck (Figure 23.10●), the axillae (Figure 23.11●), and the groin (Figures 23.12 to 23.14●). These nodes are often called *lymph glands.* "Swollen glands" usually indicate inflammation or infection of peripheral structures. Dense collections of lymph nodes also exist within the mesenteries of the gut, near the trachea and passageways leading to the lungs, and in association with the thoracic duct.

DISTRIBUTION OF LYMPHOID TISSUES AND LYMPH NODES [FIGURES 23.4/23.10 TO 23.15]

Lymphoid tissues and lymph nodes are distributed in areas particularly susceptible to injury or invasion. If we wanted to protect a house against intrusion, we might guard all doors and windows and perhaps keep a pitbull indoors. The distribution of lymphoid tissues and lymph nodes is based on a similar strategy.

1. The **cervical lymph nodes** monitor lymph originating in the head and neck (Figure 23.10●).
2. The **axillary lymph nodes** filter lymph arriving at the trunk from the upper limbs (Figure 23.11a●). In women, the axillary nodes also drain lymph from the mammary glands (Figure 23.11b●).
3. The **popliteal lymph nodes** filter lymph arriving at the thigh from the leg, and the **inguinal lymph nodes** monitor lymph arriving at the trunk from the lower limbs (Figures 23.12 to 23.14●).
4. The **thoracic lymph nodes** receive lymph from the lungs, respiratory passageways, and mediastinal structures (Figure 23.4●, p. 616).
5. The **abdominal lymph nodes** filter lymph arriving from the urinary and reproductive systems.
6. The lymphoid tissue of Peyer's patches, the **intestinal lymph nodes**, and the **mesenterial lymph nodes** receive lymph originating from the digestive tract (Figure 23.15●, p. 625).

FIGURE 23.11 **Lymphatic Drainage of the Upper Limb**

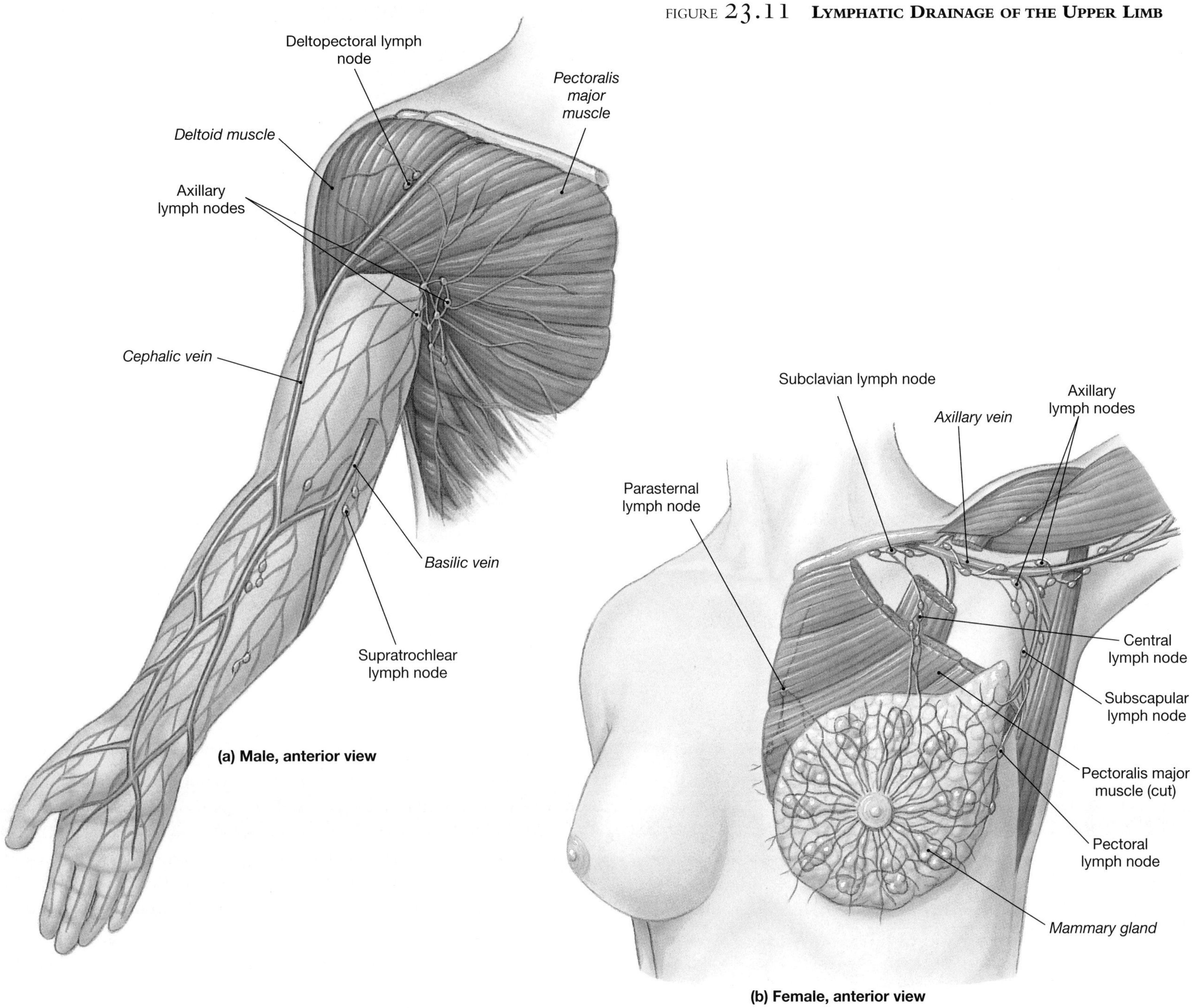

CLINICAL BRIEF

Lymphadenopathy and Metastatic Cancer

A minor injury normally produces a slight enlargement of the nodes along the lymphatic draining the region. The enlargement usually results from an increase in the number of lymphocytes and phagocytes in the node, in response to a minor, localized infection. Chronic or excessive enlargement of lymph nodes constitutes **lymphadenopathy** (lim-fad-e-NOP-a-thē). This condition may occur in response to scarring of damaged lymphatic ducts, bacterial, viral or parasitic infections, or cancer.

Lymphatics are found in most regions of the body, and lymphatic capillaries offer little resistance to the passage of cancer cells. As a result, metastasizing cancer cells often spread along the lymphatics. Under these circumstances the lymph nodes serve as way stations for migrating cancer cells. Thus, an analysis of lymph nodes can provide information on the spread of the cancer cells, and such information has a direct influence on the selection of appropriate therapies. One example is the classification of breast cancer or lymphomas by the degree of lymph node involvement.
Lymphomas p. 802 and *Breast Cancer p. 803*.

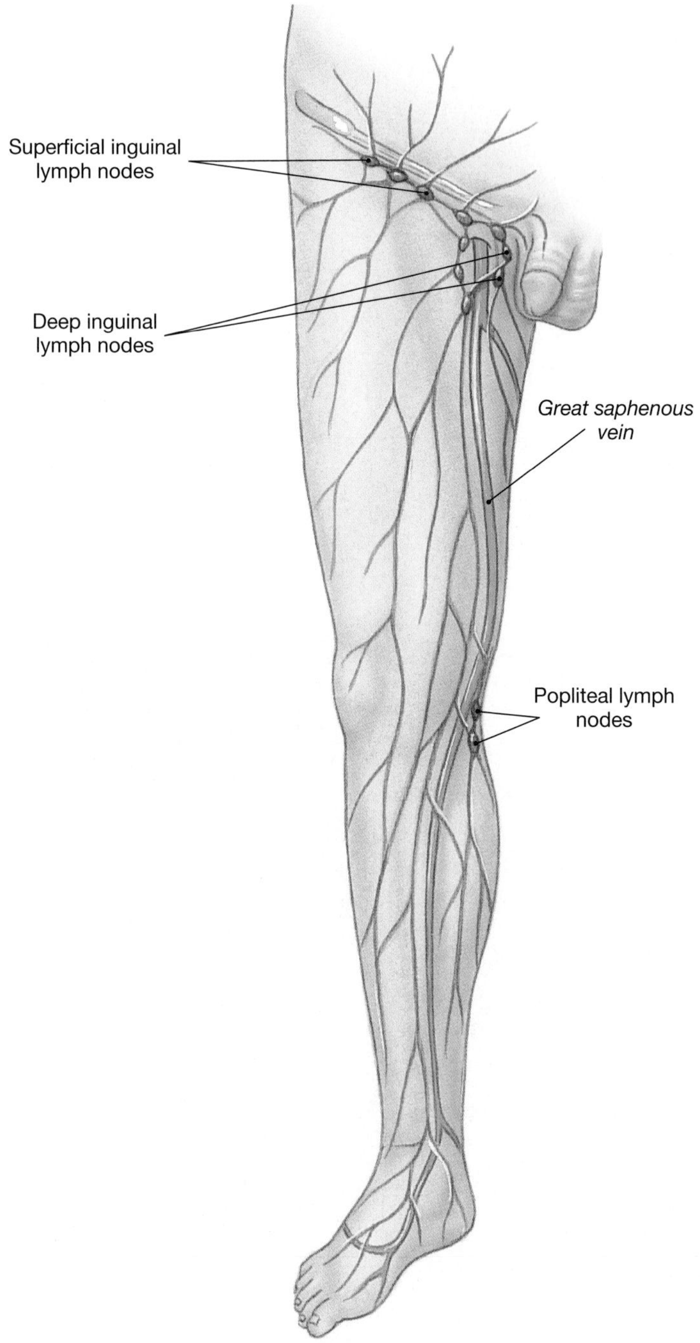

FIGURE 23.12 **Lymphatic Drainage of the Lower Limb**

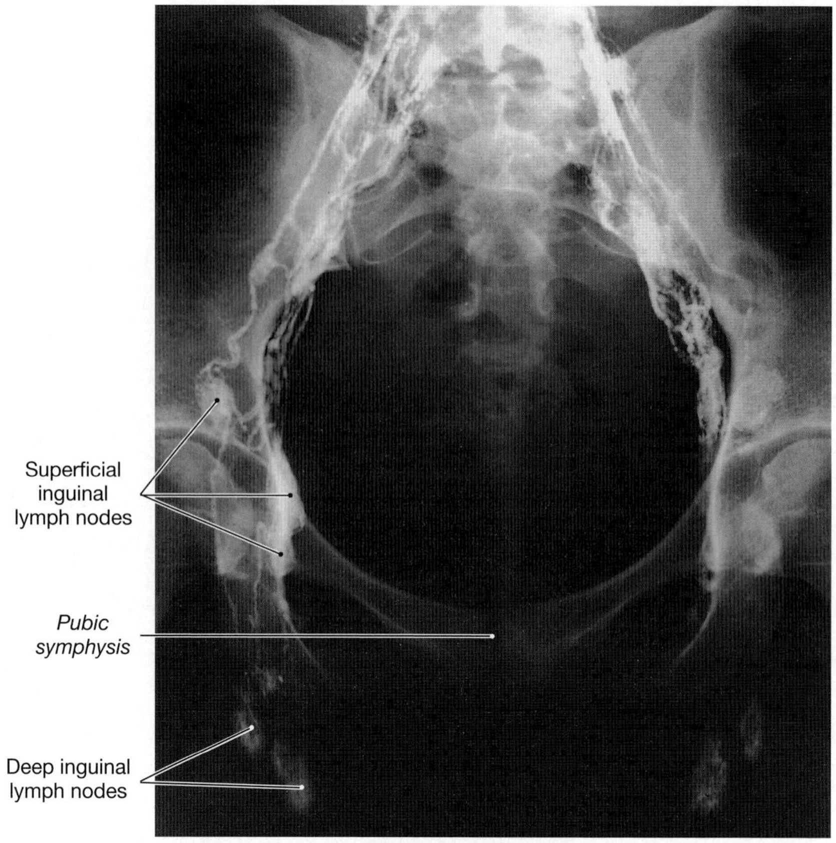

FIGURE 23.13 **A Pelvic Lymphangiogram**

The lymph vessels and lymph nodes can be visualized in a *lymphangiogram*, an X-ray taken after the introduction of a radiopaque dye into the lymphatic system.

The Thymus [FIGURE 23.16]

The **thymus** lies posterior to the sternum, in the anterior portion of the mediastinum. It has a nodular consistency and a pinkish coloration. The thymus reaches its greatest size (relative to body size) in the first year or two after birth, and its maximum absolute size during puberty, when it weighs between 30 and 40 g. Thereafter, the thymus gradually decreases in size and the functional cells are replaced by connective tissue fibers. This degenerative process is called *involution*.

The capsule that covers the thymus divides it into two **thymic lobes** (Figure 23.16a,b●). Fibrous partitions, or **septae**, extend from the capsule to divide the lobes into **lobules** averaging 2 mm in width (Figure 23.16b,c●, p. 626). Each lobule consists of a dense outer **cortex** and a somewhat diffuse, paler central **medulla**. The cortex contains lymphoid stem cells that divide rapidly, producing daughter cells that mature into T cells and migrate into the medulla. During the maturation process, any T cells that are sensitive to normal tissue antigens are destroyed. The surviving T cells eventually enter one of the specialized blood vessels in that region. While they are within the thymus, T cells do not participate in the immune response; they remain inactive until they enter the general circulation. The capillaries of the thymus resemble those of the CNS in that they do not permit free exchange between the interstitial fluid and the circulation. This **blood-thymus barrier** prevents premature stimulation of the developing T cells by circulating antigens.

Epithelial cells are scattered among the lymphocytes of the thymus. These cells are responsible for the production of thymic hormones that promote the differentiation of functional T cells. In the medulla, these cells cluster together in concentric layers, forming distinctive structures known as **Hassall's corpuscles** (Figure 23.16d●), whose function remains unknown.

The Spleen [FIGURE 23.17a]

The **spleen** is the largest lymphoid organ in the body. It is around 12 cm (5 in.) long and weighs up to 160 g (5.6 oz). The spleen lies along the curving lateral border of the stomach, extending between the ninth and eleventh ribs on the left side. It is attached to the lateral border of the stomach by a broad mesenteric band, the **gastrosplenic ligament** (Figure 23.17a●).

On gross dissection the spleen has a deep red color because of the blood it contains. The spleen performs functions for the blood comparable to those performed by the lymph nodes for lymph. The functions of the spleen include (1) the removal of abnormal blood cells and other blood components through phagocytosis, (2) the storage of iron recycled from broken down red blood cells, and (3) the initiation of immune responses by B cells and T cells in response to antigens in the circulating blood.

FIGURE 23.14 **Lymphatic Drainage of the Inguinal Region**

(**a**) A superficial and deeper view of the inguinal region of a male, showing the distribution of superficial lymph nodes and lymphatics. (**b**) An anterior view of a dissection of the inguinal lymph nodes and vessels.

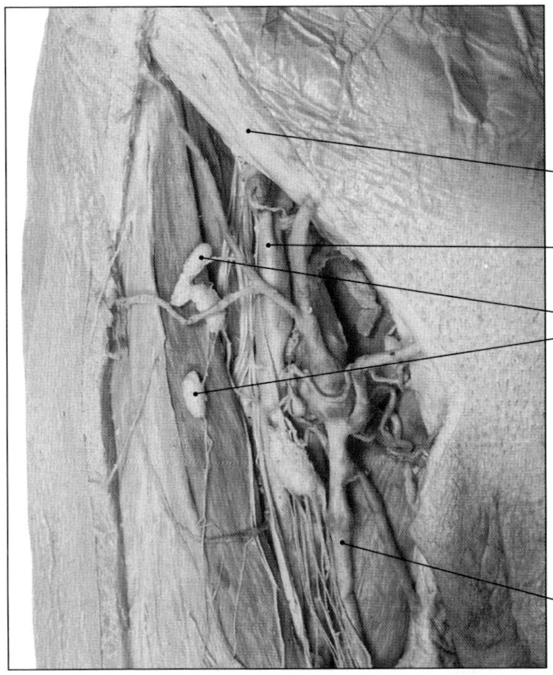

(a) Dissection of the upper right inguinal region in a male

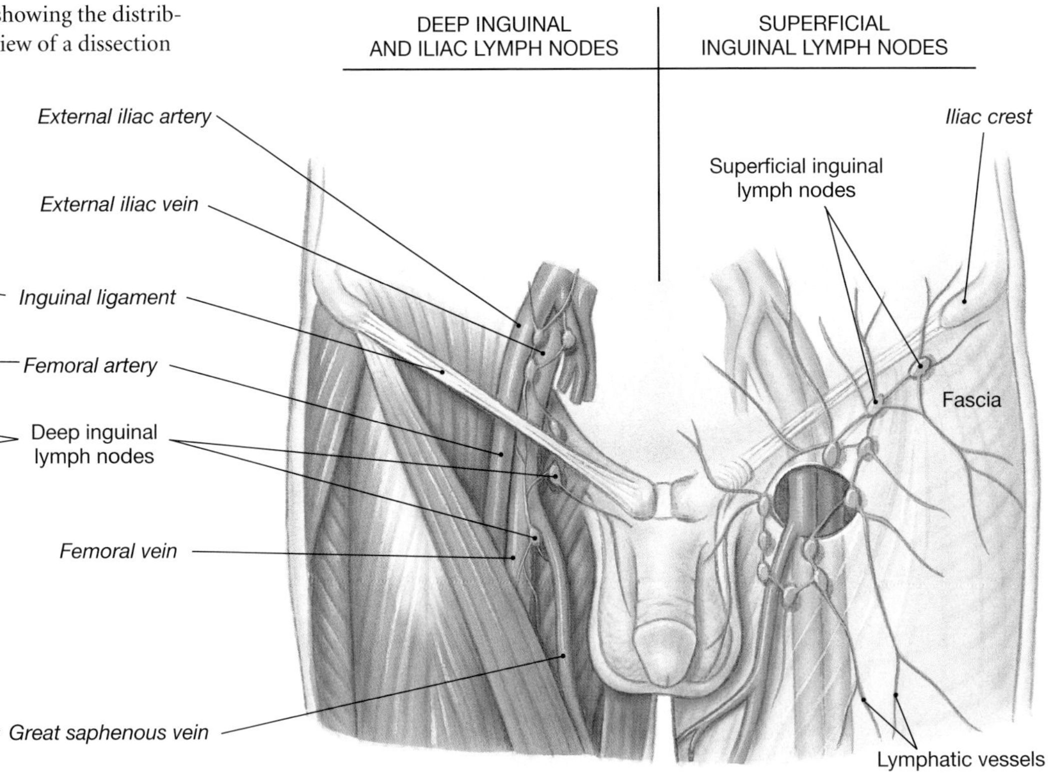

(b) Inguinal lymph nodes and vessels

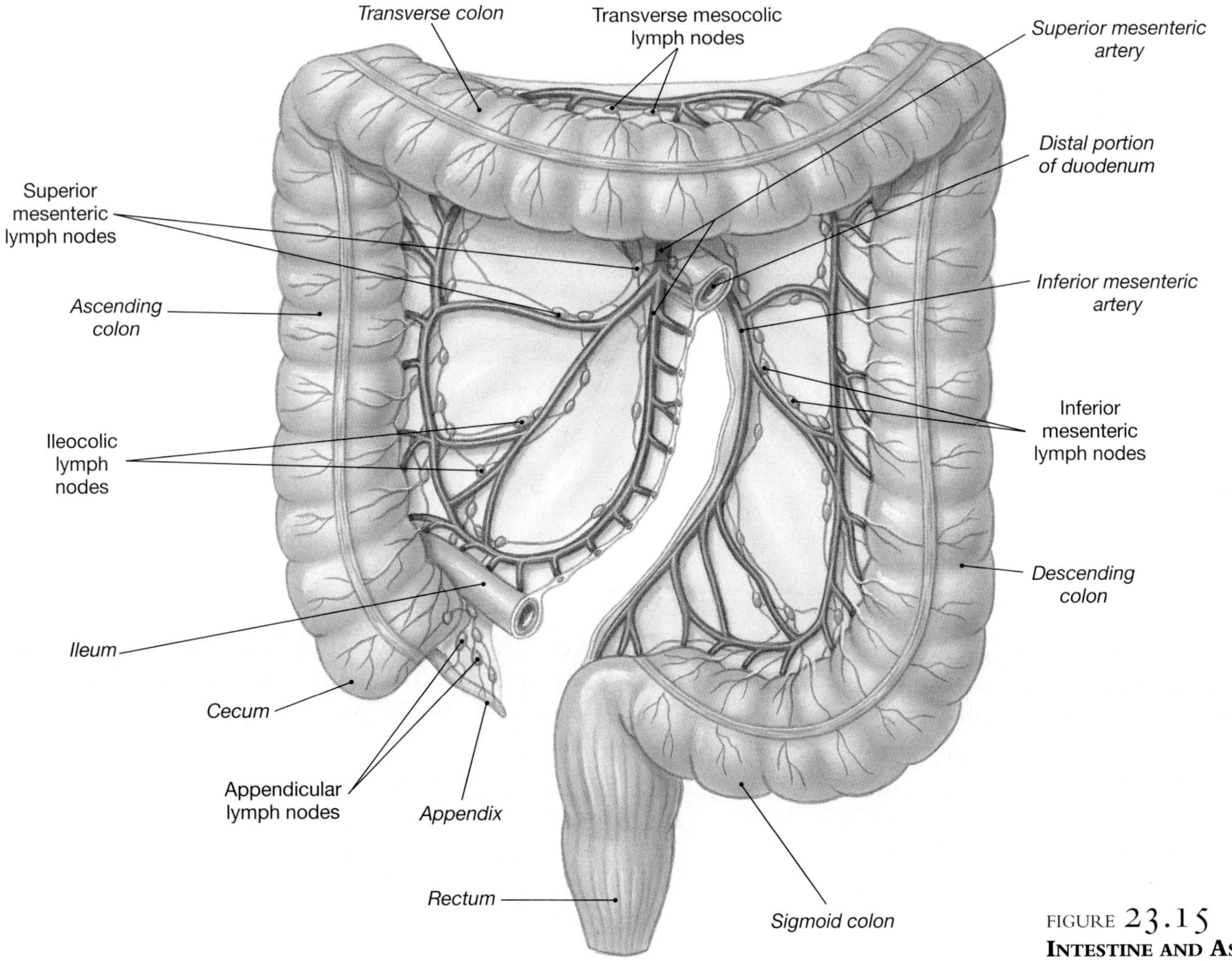

FIGURE 23.15 **Lymph Nodes in the Large Intestine and Associated Mesenteries**

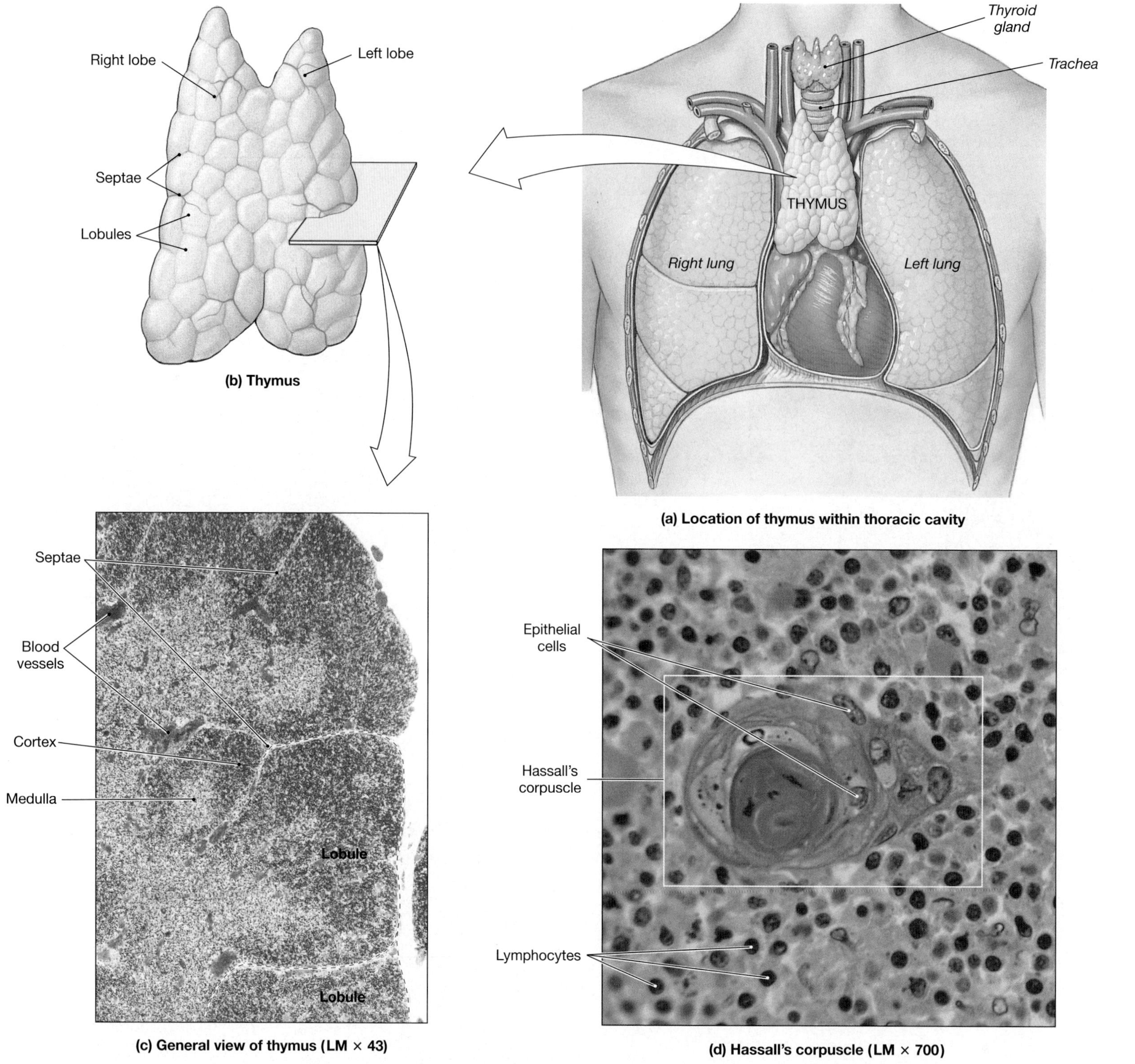

FIGURE 23.16 **THE THYMUS**

(**a**) The location of the thymus on gross dissection; note the relationship to other organs in the chest. (**b**) Anatomical landmarks on the thymus. (**c**) A low-power light micrograph of the thymus. Note the fibrous septae that divide the thymic tissue into lobules resembling interconnected lymphoid nodules. The dashed line follows the edges of a single lobule. (**d**) At higher magnification the unusual structure of Hassall's corpuscles can be examined. The small cells in view are lymphocytes in various stages of development.

SURFACES OF THE SPLEEN [FIGURE 23.17]

The spleen has a soft consistency, and its shape primarily reflects its association with the structures around it. It lies wedged between the stomach, the left kidney, and the muscular diaphragm. The **diaphragmatic surface** (Figure 23.17a●) is smooth and convex, conforming to the shape of the diaphragm and body wall. The **visceral surface** (Figure 23.17b●) contains indentations that follow the shapes of the stomach (the **gastric area**) and kidney (the **renal area**). Splenic blood vessels and lymphatics communicate with the spleen on the visceral surface at the **hilus**, a groove marking the border between the gastric and renal areas. The **splenic artery**, **splenic vein**, and the lymphatics draining the spleen are attached at the hilus.

HISTOLOGY OF THE SPLEEN [FIGURE 23.17C]

The spleen is surrounded by a capsule containing collagen and elastic fibers. The cellular components within constitute the **pulp** of the spleen (Figure 23.17c●). Areas of **red pulp** contain large quantities of red blood cells, whereas areas of **white pulp** resemble lymphatic nodules. The splenic artery enters at the hilus and branches to produce a number of arteries that radi-

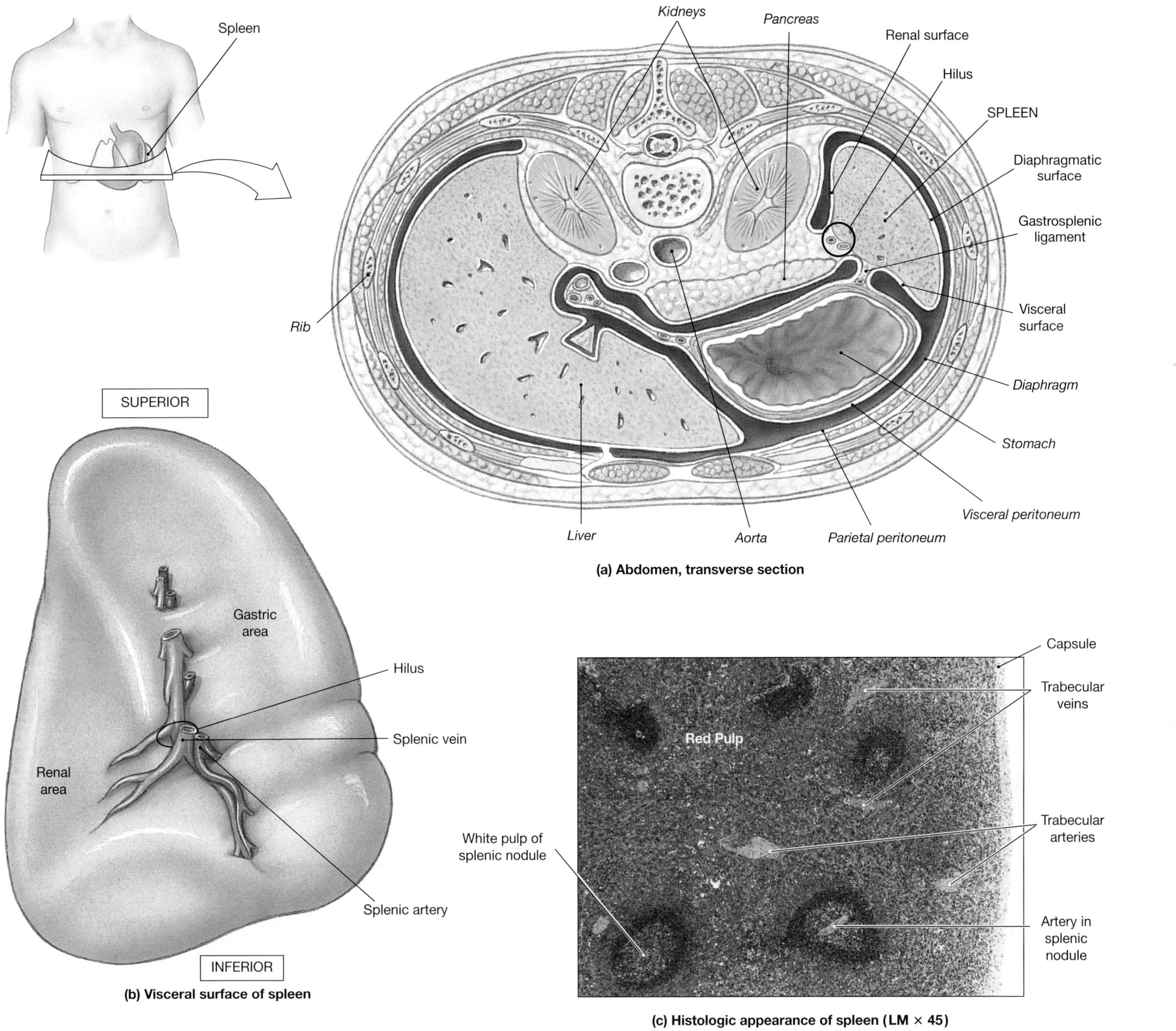

FIGURE 23.17 **THE SPLEEN**

(**a**) The shape of the spleen roughly conforms to the shapes of adjacent organs. This transverse section through the trunk shows the typical position of the spleen within the abdominopelvic cavity (superior view). (**b**) External appearance of the visceral surface of the intact spleen, showing major anatomical landmarks. This view should be compared with that of (a). (**c**) Histological appearance of the spleen. Areas of white pulp are dominated by lymphocytes; they appear blue or purple because the nuclei of the lymphocytes stain very darkly. Areas of red pulp contain a preponderance of red blood cells.

ate outward toward the capsule. These **trabecular arteries** branch extensively, and their finer branches are surrounded by areas of white pulp. Capillaries then discharge the blood into the venous sinuses of the red pulp.

The cell population of the red pulp includes all of the normal components of the circulating blood, plus fixed and free macrophages. The structural framework of the red pulp consists of a network of reticular fibers. The blood passes through this meshwork and enters large sinusoids, also lined by fixed macrophages. The sinusoids empty into small veins, and these ultimately merge to form **trabecular veins**, which continue toward the hilus.

This circulatory arrangement gives the phagocytes of the spleen an opportunity to identify and engulf any damaged or infected cells in the circulating blood. Lymphocytes are scattered throughout the red pulp, and the region surrounding each area of white pulp has a high concentration of macrophages. Thus, any microorganisms or abnormal plasma components will quickly come to the attention of the splenic lymphocytes. *Disorders of the Spleen, p. 803, Immune Disorders, p. 804,* and *Systemic Lupus Erythematosus p. 804*

✓ CONCEPT CHECK

- Why is it important that lymph encounters T cells before B cells?
- What is important about the placement of lymph nodes?
- Why is there a blood-thymus barrier in the capillaries of the thymus?
- Why do lymph nodes often enlarge during an infection?

Aging and the Lymphatic System

With advancing age, the lymphatic system becomes less effective at combating disease. T cells become less responsive to antigens; as a result, fewer cytotoxic T cells respond to an infection. Because the number of helper T cells is also reduced, B cells are less responsive, and antibody levels do not rise as quickly after antigen exposure. The net result is an increased susceptibility to viral and bacterial infection. For this reason, vaccinations for acute viral diseases, such as the flu (influenza), are strongly recommended for elderly individuals. The increased incidence of cancer in the elderly reflects the fact that surveillance by the lymphatic system declines, and tumor cells are not eliminated as effectively.

RELATED CLINICAL TERMS

acquired immune deficiency syndrome (AIDS): A disorder that develops after HIV infection, characterized by reduced T cell populations and depressed cellular immunity. *AIDS p. 802*

allergy: An inappropriate or excessive immune response to antigens. *Immune Disorders p. 804*

appendicitis: Inflammation of the appendix, often requiring treatment in the form of an **appendectomy**. *Infected Lymphoid Nodules p. 802*

autoimmune disorder: A disorder that develops when the immune response mistakenly targets normal body cells and tissues. *Immune Disorders p. 804*

immunodeficiency disease: A disease in which either the immune system fails to develop normally or the immune response is somehow blocked. *Immune Disorders p. 804*

lymphadenopathy (lim-fad-e-NOP-a-thē): Chronic or excessive enlargement of lymph nodes. *p. 623*

lymphomas: Malignant cancers consisting of abnormal lymphocytes or lymphocytic stem cells; includes *Hodgkin's disease* and *non-Hodgkin's lymphoma*. *Lymphomas p. 802*

severe combined immunodeficiency disease (SCID): A congenital condition in which an individual fails to develop either cellular or humoral immunity because of a lack of normal B and T cells. *SCID p. 802*

splenectomy (splē-NEK-to-mē): Surgical removal of the spleen, usually after splenic rupture. *Disorders of the Spleen p. 803*

splenomegaly (splē-nō-MEG-a-lē): Enlargement of the spleen; often caused by infection, inflammation, or cancer. *Disorders of the Spleen p. 803*

systemic lupus erythematosus (LOO-pus e-rith-ē-ma-TŌ-sis) **(SLE):** A condition resulting from a generalized breakdown in the antigen-recognition mechanism. *Systemic Lupus Erythematosus p. 804*

tonsillectomy: The removal of an infected tonsil to remove symptoms of **tonsillitis**. *Infected Lymphoid Nodules p. 802*

Additional Clinical Terms Discussed in Appendix I (p. 801-804)

acquired immune deficiency syndrome (AIDS); allergens; allergies; autoantibodies; autoimmune disorders; breast cancer; Burkitt's lymphoma; elephantiasis; Epstein-Barr virus (EBV); filariasis; graft versus host (GVH); disease; Hodgkin's disease (HD); non-Hodgkin's lymphoma (NHL); human immunodeficiency virus (HIV); hypersplenism; hyposplenism; Kaposi's sarcoma; lymphedema; mammography; modified radical mastectomy; mononucleosis; opportunistic infections; radical mastectomy; segmental mastectomy; severe combined immunodeficiency disease (SCID); systemic lupus erythematosus (SLE); thermography; total mastectomy

STUDY OUTLINE & CHAPTER REVIEW

Introduction 613

1. The cells, tissues, and organs of the *lymphatic system* play a central role in the body's defenses against viruses, bacteria, and other microorganisms.

An Overview of the Lymphatic System 613

1. The **lymphatic system** includes a network of lymphatic vessels that carry **lymph** (a fluid similar to plasma but with a lower concentration of proteins). A series of lymphoid organs and lymphoid tissues are interconnected by the lymphatic vessels. *(see Figures 23.1/23.2)*

Functions of the Lymphatic System 613

2. The lymphatic system produces, maintains, and distributes lymphocytes (cells that attack invading organisms, abnormal cells, and foreign proteins). The system also helps maintain blood volume and eliminate local variations in the composition of the interstitial fluid. Lymphoid structures can be classified as *primary* (containing stem cells) or *secondary* (containing immature or activated lymphocytes).

Structure of Lymphatic Vessels 614

Lymphatic Capillaries 615

1. **Lymphatic vessels**, or *lymphatics*, carry lymph from peripheral tissues to the venous system. Lymph flows along a network of lymphatics that originate in the **lymphatic capillaries** (*terminal lymphatics*). The endothelial cells of a lymphatic capillary overlap to act as a one-way valve, preventing fluid from returning to the intercellular spaces. *(see Figure 23.2b)*

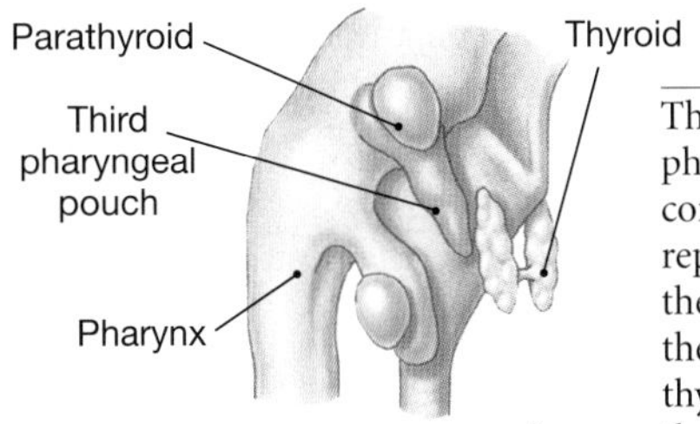

6 WEEKS

The thymus forms from cells of the third pharyngeal pouch. These cells lose their connection with the epithelium and divide repeatedly. As the embryo changes shape, the thymic lobes are brought together near the midline of the chest. At birth, the thymus is relatively large, filling much of the anterior mediastinum.

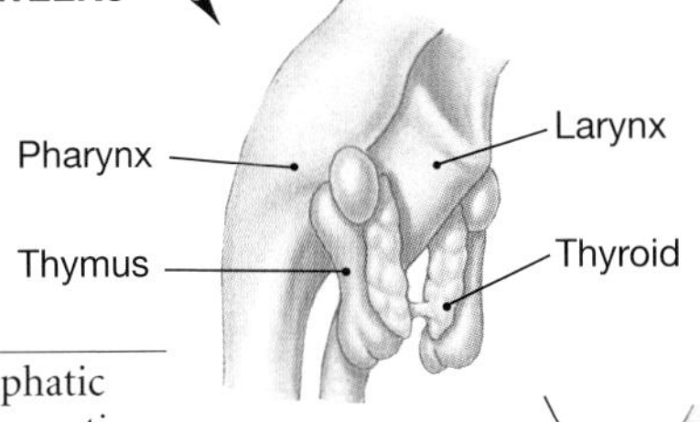

7 WEEKS

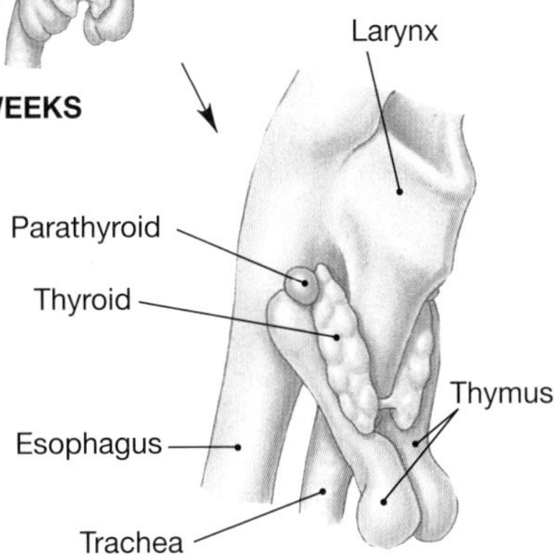

8 WEEKS

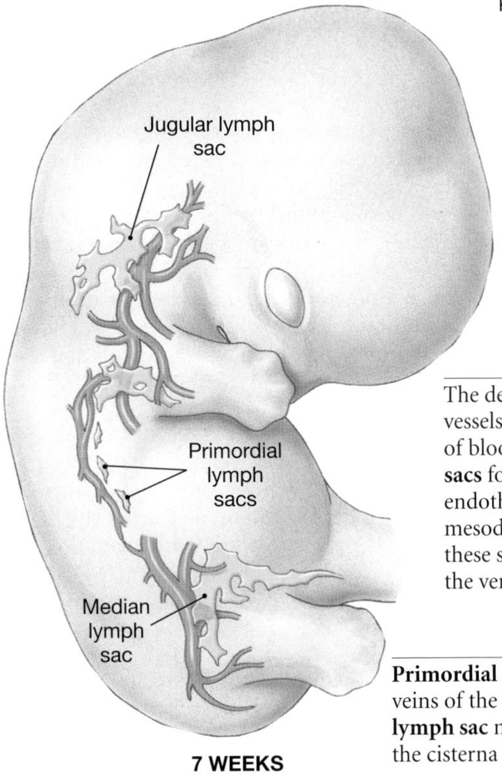

7 WEEKS

The development of the lymphatic vessels is closely tied to the formation of blood vessels. Paired **jugular lymph sacs** form from the fusion of small, endothelium-lined pockets in the mesoderm of the neck. By week 7, these sacs become connected to the venous system.

Primordial lymph sacs form parallel with veins of the trunk, and a large **median lymph sac** marks the future location of the cisterna chyli.

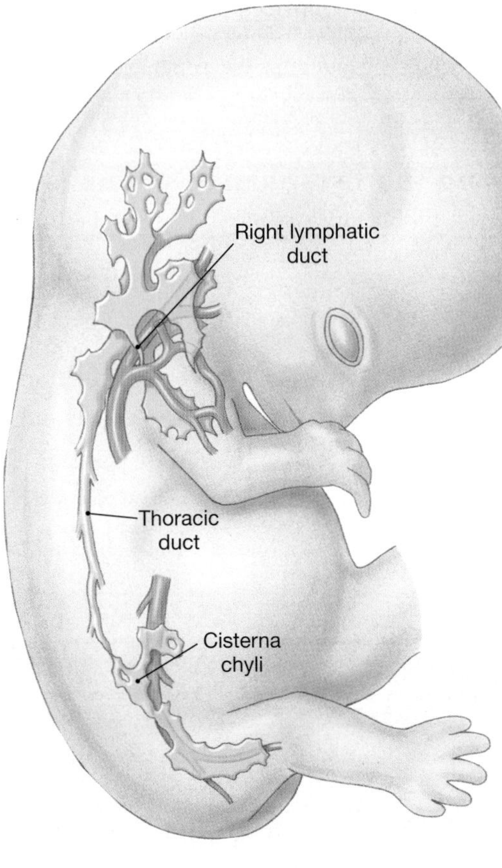

8 WEEKS

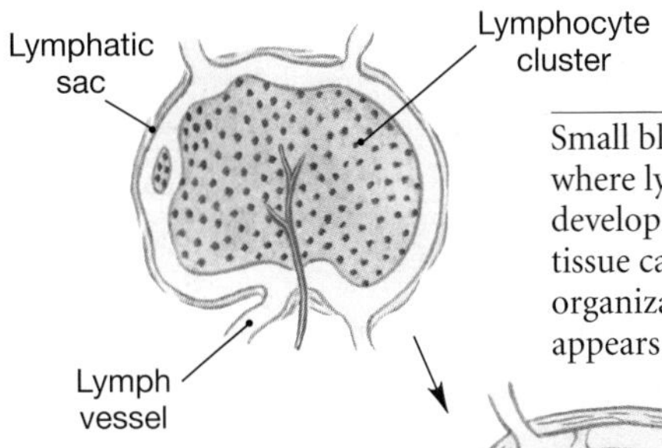

Small blood vessels grow into areas where lymphocytes cluster within developing lymphatic sacs. Connective tissue capsules form, and the internal organization of a lymph node gradually appears.

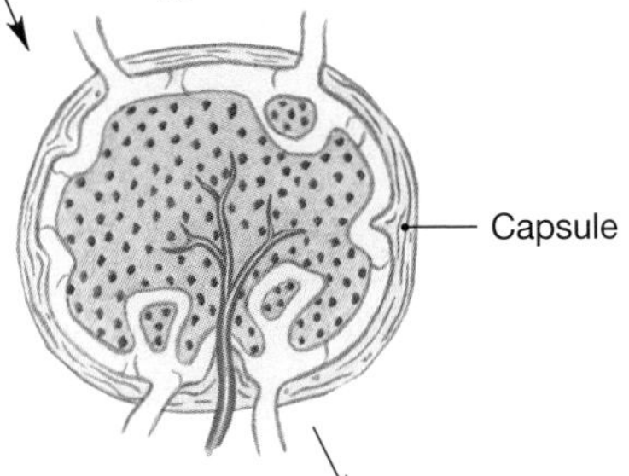

As growth continues, the isolated lymphatic sacs fuse, forming the thoracic duct and right lymphatic duct. As the limb buds enlarge, lymphatic vessels grow into the area along with developing arteries and veins.

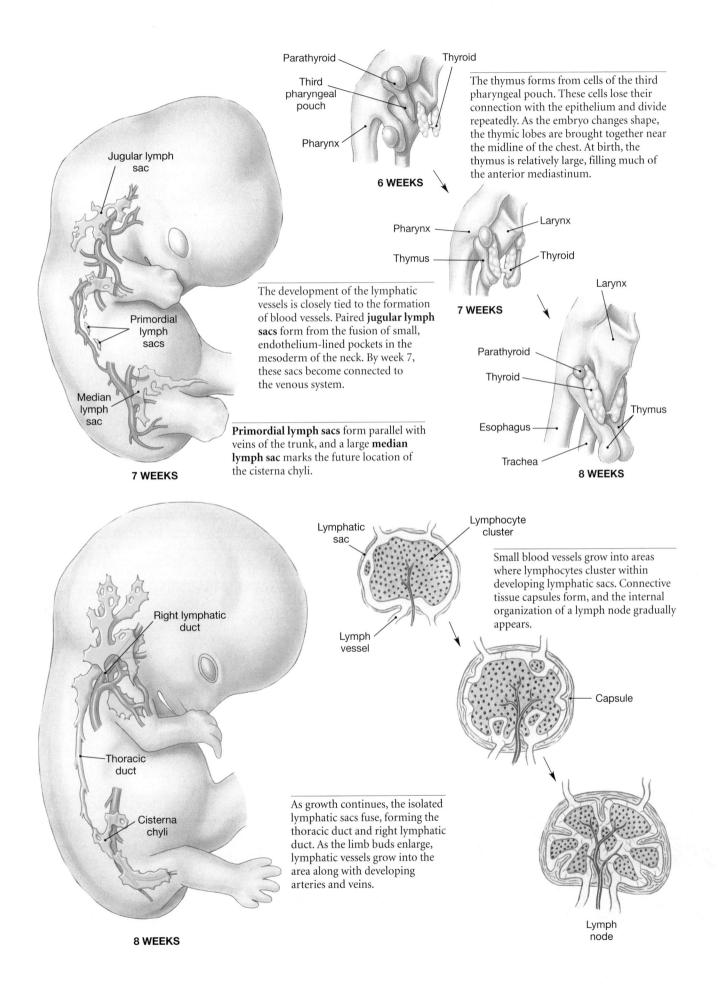

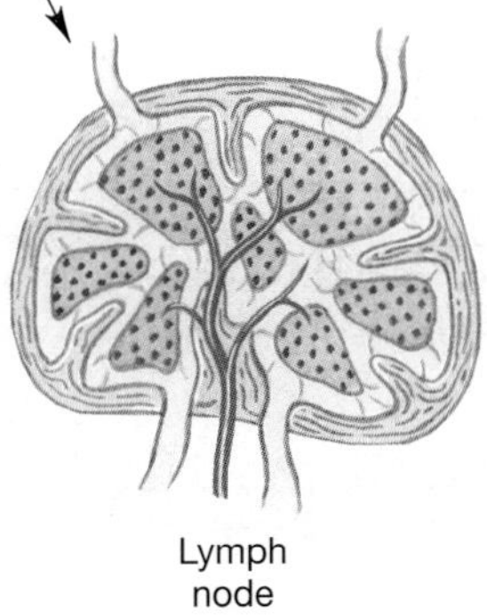

Valves of Lymphatic Vessels 615

2. Lymphatics contain numerous internal valves to prevent backflow of lymph.

Major Lymph-Collecting Vessels 616

3. Two sets of lymphatic vessels collect blood from the lymphatic capillaries: the **superficial lymphatics** and the **deep lymphatics**. The lymphatic vessels empty into the **thoracic duct** and the **right lymphatic duct**. *(see Figures 23.2 to 23.5)*

Lymphocytes 617

Types of Lymphocytes 617

1. There are three different classes of lymphocytes: **T cells** (thymus-dependent), **B cells** (bone marrow–derived), and **NK cells** (natural killer). *(see Figures 23.6/23.7)*

Lymphocytes and the Immune Response 618

2. **Cytotoxic T cells** attack foreign cells or body cells infected by viruses; they provide **cell-mediated immunity**. *Regulatory T cells* (**helper** and **suppressor**) regulate and coordinate the immune response, while **memory T cells** remain "on reserve." *(see Figure 23.6)*
3. B cells can differentiate into **plasma cells**, which produce and secrete *antibodies* that react with specific chemical targets, or **antigens**. Antibodies in body fluids are called **immunoglobulins**. B cells are responsible for **antibody-mediated immunity**. **Memory B cells** will become activated if the antigen appears again at a later date. *(see Figure 23.6)*
4. NK cells (also called *large granular lymphocytes*) attack foreign cells, normal cells infected with viruses, and cancer cells. They provide **immunological surveillance**.
5. The goal of the **immune response** is the destruction or inactivation of pathogens, abnormal cells, and foreign molecules such as toxins. Antigens are engulfed by macrophages, which then present the antigen to T cells so they can begin differentiating. The millions of different lymphocytes, which retain the ability to divide, allow the body to be prepared for any antigen. The process of acquiring the ability to recognize antigens is called "gaining immunocompetence." *(see Figure 23.6)*

Distribution and Life Span of Lymphocytes 619

Lymphopoiesis: Lymphocyte Production 619

6. Lymphocytes continually migrate in and out of the blood through the lymphoid tissues and organs and, in general, have relatively long life spans. **Lymphopoiesis** (lymphocyte production) involves the bone marrow, thymus, and peripheral lymphatic tissues. *(see Figure 23.7)*

Lymphoid Tissues 620

1. **Lymphoid tissues** are connective tissues dominated by lymphocytes. In a **lymphoid nodule**, the lymphocytes are densely packed in an area of loose connective tissue. Important lymphoid nodules are **aggregated lymphoid nodules** beneath the lining of the intestine, the *vermiform appendix*, and the **tonsils** in the walls of the pharynx. *(see Figure 23.8)*

Lymphoid Organs 621

1. Important **lymphoid organs** include the *lymph nodes*, the *thymus*, and the *spleen*. Lymphoid tissues and lymph nodes are distributed in areas especially vulnerable to injury or invasion. *(see Figures 23.1/23.9 to 23.15)*

Lymph Nodes 621

2. **Lymph nodes** are encapsulated masses of lymphoid tissue. The **deep cortex** is dominated by T cells; the **outer cortex** and **medulla** contain B cells arranged into **medullary cords. Lymph glands** are the largest lymph nodes, found where peripheral lymphatics connect with the trunk. *(see Figures 23.1/23.9 to 23.15)*
3. Lymphoid tissues and nodes are located in areas particularly susceptible to injury or invasion by microorganisms.
4. The **cervical lymph nodes, axillary lymph nodes, inguinal lymph nodes, popliteal lymph nodes, thoracic lymph nodes, abdominal lymph nodes, intestinal lymph nodes**, and **mesenterial lymph nodes** serve to protect the vulnerable areas of the body. *(see Figures 23.4/23.5/23.10 to 23.15)*

The Thymus 624

5. The **thymus** lies behind the sternum, in the anterior mediastinum. **Epithelial cells** scattered among the lymphocytes produce thymic hormones. These hormones promote the differentiation of T cells. The **blood-thymus barrier** does not allow free exchange between the interstitial fluid and the circulation, protecting the T cells from being prematurely activated. After puberty the thymus gradually decreases in size, a process called *involution*. *(see Figure 23.16)*

The Spleen 624

6. The adult **spleen** contains the largest mass of lymphoid tissue in the body. The spleen performs the same functions for the blood that lymph nodes perform for the lymph. The **diaphragmatic surface** of the spleen lies against the diaphragm; the **visceral surface** is against the stomach and kidney and contains a groove called the **hilus**. The cellular components form the **pulp** of the spleen. **Red pulp** contains large numbers of red blood cells, and areas of **white pulp** resemble lymphatic nodules. Lymphocytes are scattered throughout the red pulp, and the region surrounding the white pulp has a high concentration of macrophages. *(see Figure 23.17)*

Aging and the Lymphatic System 628

1. With aging, the immune system becomes less effective at combating disease.

LEVEL 1 REVIEWING FACTS AND TERMS

Match each numbered item with the most closely related lettered item. Use letters for answers in the spaces provided.

Column A	Column B
_____ 1. plasma cells	a. terminal lymphatics
_____ 2. spleen	b. responsible for cell-mediated immunity
_____ 3. thymus	c. produce antibodies
_____ 4. cytotoxic T cells	d. aid in immunological surveillance
_____ 5. antibodies	e. contains developing T cells
_____ 6. NK cells	f. immunoglobulins
_____ 7. lymphatic capillaries	g. responsible for antibody-mediated immunity
_____ 8. cisterna chyli	h. production of lymphocytes
_____ 9. lymphopoiesis	i. saclike chamber of the thoracic duct
_____10. B cells	j. largest lymphoid organ in the body

11. The lymphatic system is composed of
    (a) lymphatic vessels
    (b) the spleen
    (c) lymph nodes
    (d) all of the above

12. Compared to blood capillaries, lymph capillaries
    (a) have a basement membrane
    (b) are smaller in diameter
    (c) have walls of a smooth endothelial lining
    (d) are frequently irregular in shape

13. Most of the lymph returns to the venous circulation by way of the
(a) right lymphatic duct (b) cisterna chyli
(c) hepatic portal vein (d) thoracic duct
14. Some cells known as lymphocytes
(a) are actively phagocytic
(b) destroy red blood cells
(c) produce proteins called antibodies
(d) are primarily found in red bone marrow
15. ___ are large lymphatic nodules that are located in the walls of the pharynx.
(a) Tonsils (b) Lymph nodes
(c) Thymus gland (d) Hassall's corpuscles
16. Areas of the spleen that contain large numbers of lymphocytes are known as
(a) white pulp (b) red pulp
(c) adenoids (d) lymph nodes
17. The red pulp of the spleen contains large numbers of
(a) macrophages (b) antibodies
(c) neutrophils (d) lymphocytes
18. The cells responsible for the production of circulating antibodies are
(a) NK cells (b) plasma cells
(c) helper T cells (d) cytotoxic T cells
19. The medullary cords of a lymph node contain
(a) cytotoxic T cells (b) suppressor T cells
(c) NK cells (d) B cells
20. Lymphocytes that attack foreign cells or body cells infected with viruses are
(a) B cells (b) helper T cells
(c) cytotoxic T cells (d) suppressor T cells

LEVEL 2 REVIEWING CONCEPTS

1. If the thymus failed to produce the hormone thymosin, we would expect to see a decrease in the number of
(a) B lymphocytes (b) NK cells
(c) cytotoxic T cells (d) neutrophils
2. The human immunodeficiency virus (HIV) that causes the disease known as AIDS selectively infects
(a) helper T cells (b) plasma cells
(c) cytotoxic T cells (d) suppressor T cells
3. Blocking the antigen receptors on the surface of lymphocytes would interfere with
(a) phagocytosis of the antigen
(b) that lymphocyte's ability to produce antibodies
(c) antigen recognition
(d) the ability of the lymphocyte to present antigen
4. What is the function of the blood-thymus barrier?
5. What major artery and vein pass through the hilus of the spleen?
6. From what areas of the body does the thoracic duct collect lymph?
7. Which type of lymphocyte is most common?
8. What is lymphedema?
9. What occurs in secondary lymphoid structures?
10. Where are Peyer's patches found?

LEVEL 3 CRITICAL THINKING AND CLINICAL APPLICATIONS

1. Tom has just been exposed to the measles virus, and since he can't remember if he has had measles before, he wonders if he is going to come down with the disease or not. He asks you if there is any way he can tell if he has been previously exposed or if he is going to get sick before it actually happens. What would you tell him?
2. Willy is allergic to ragweed pollen and tells you that he read about a medication that can help his condition by suppressing his immune response. Do you think that this treatment could help Willy? Explain!
3. Paula's grandfather is diagnosed as having lung cancer. His physician orders biopsies of several lymph nodes from neighboring regions of the body, and Paula wonders why, since the cancer is in his lungs. What would you tell her?

✓ ANSWERS TO CONCEPT CHECK QUESTIONS

p. 617 **1.** The lymphatic system produces, maintains, and distributes lymphocytes, which are essential to the defense of the body. **2.** Capillary number one. **3.** Because lymph delivers so much of the body's fluid back to the bloodstream, a break in a major lymphatic vessel could mean a potentially fatal decline in blood volume. **4.** Primary lymphoid structures carry stem cells.
p. 621 **1.** T cells account for about 80% of the body's lymphocytes. **2.** Some B cells differentiate into memory B cells, which will become activated when the antigen appears again. Therefore, John's body will be able to mount an immune response faster and more effectively, possibly warding off any symptoms of illness. **3.** The body must be able to mount a rapid and powerful response to antigens, requiring the production of certain types of lymphocytes at certain times. Without this ability, the body's immune response would be slower and weaker and could fall victim to infection or disease. **4.** Peyer's patches.
p. 628 **1.** T cells regulate the activation or suppression of B cells and therefore manage the extent of the immune response. **2.** Lymph nodes are strategically placed throughout the body in areas susceptible to injury or invasion. **3.** The capillaries in the thymus do not allow free exchange between the interstitial fluid and the circulation. If they did, circulating antigens would prematurely stimulate the developing T cells. **4.** Lymph nodes enlarge during an infection due to increased numbers of lymphocytes and phagocytes within the active nodes.

REVIEW IT

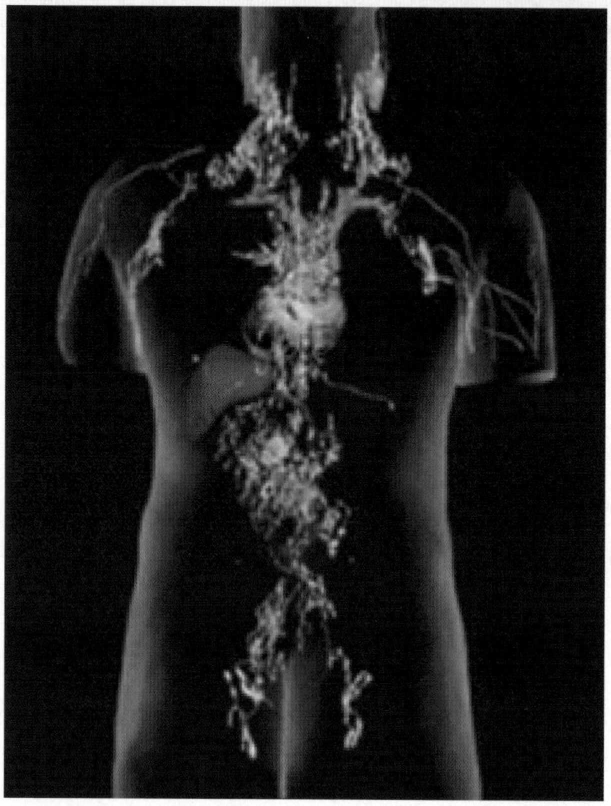

Challenge 1

As you have learned, the lymphatic system is composed of lymphatic vessels, lymphoid tissues and lymphoid organs. In illustrations, these structures appear relatively robust. However, in reality the lymphatic vessels are quite thin and delicate with many flimsy valves. This difference can be seen when comparing the image you see here with Figure 23.1●, p. 613.

This image is taken from the Interactive CD/Lymphatic System/Whole body view. Note that this image is an anterior view. Compare this image directly to Figure 23.1● locating the same structures. Are all the structures labeled on Figure 23.1● identifiable here? If not, which are missing? Why do you suppose structures shown in illustrations are not visible in the reconstructed image? Check your answers by going to the Interactive CD/Animations/Lymphatic system and studying the labeled structures.

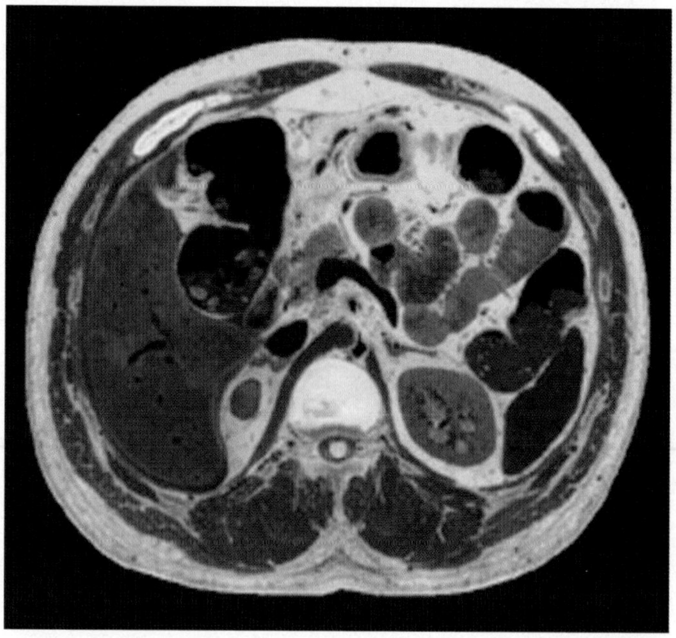

Challenge 2

This section is taken from the Visible Human Data Set. It represents an area of the abdominal cavity at the level of the spleen. Label the surfaces of the spleen and identify the adjacent organs, bones, muscles and major blood vessels. Use Figure 23.17●, p. 627 as a guide to recognizing this section. What conclusions can you draw about the location and texture of the spleen? How do you explain the fact that traumatic injuries to the spleen are both relatively common and very dangerous?

Images provided by the Digital Cadaver™ project, courtesy of Visible Productions, Inc.

APPLY IT

Below are two separate exercises that provide more information on a topic presented in the preceding chapters. Each one is designed to take approximately 10 minutes, and will help you gain a better appreciation for the material presented in this chapter.

Critical Linking 1

The lymphatic system is responsible for maintaining the health of the body. You have learned the various parts to the system, and been given an overview of its functions. It is often helpful to supplement your understanding of complicated processes by reviewing another author's interpretation of the same process. Using the text as a guide, prepare a study outline of the lymphatic system with the following headings: Thoracic duct, Lymph nodes, Thymus, and Spleen. Try to look at these structures from a regional viewpoint, linking the structures of the head and neck, the upper limbs, the lower limbs and the abdomen and pelvis. Include in this outline the location of each organ, its general function and how it relates to the flow of lymph through the body.

Expand on your outline by visiting the companion website and clicking on the key word "lymphatic system". Here you will find a description of the lymphatic system as presented in 1918 by Henry Gray. You may recognize this as an electronic version of the classic anatomical text "Gray's Anatomy". It is a good idea to bookmark this site for future reference.

Critical Linking 2

The thymus plays an important role in establishing immune competence and providing functional T cells. It reaches its maximum relative size early in life, and involutes significantly after maturity. The effect of involution on immune function is uncertain. Several companies are selling extracts from crushed cow thymuses as expensive dietary supplements, claiming that they improve immune function. Another company sells thymic extracts in shampoos and skin creams. Do these products make sense to you?

Attempts have been made to improve immune function through the transplantation of thymic tissues. Do you think this approach has greater potential for success? Write a summary of your opinions and discuss them with a study partner. You can read about this project by visiting the Companion Website's Critical Linking section, and choosing the keyword "thymus." Use the information on this site to expand your summary.

FURTHER STUDIES

The lymphatic system is designed to protect the body from foreign pathogens. Occasionally, this system can turn against the body, targeting normal tissues and cells. While the exact mechanisms behind this are not completely understood, it is thought that bacteria, viruses, toxins, and some drugs may play a role in triggering an autoimmune process in someone who already has a genetic predisposition to develop such a disorder. The inflammation initiated by these agents somehow provokes in the body a "sensitization" (autoimmune reaction) in the involved tissues. The resulting signs and symptoms are referred to as autoimmune diseases. Autoimmune diseases affect nearly 50 million Americans, and a full 70% of these are female. While a genetic link has been established for predisposition to autoimmune disease, it remains unclear why females are more susceptible. Symptoms of autoimmune disease include inflammation and damage to body tissues. Autoimmunity can lead to destruction of one type of tissue, excessive growth of an organ, or interference with organ function. These diseases can be either organ-specific, or systemic.

A full listing of autoimmune diseases along with specific patient information can be obtained by visiting *http://www.aarda.org/indexf.html* on the Internet. Reviewing these diseases in some detail will help to reinforce the concepts of homeostasis presented through out this text, as well as the principles of immunity discussed in the preceding chapter.

Two of the most common systemic autoimmune diseases are rheumatoid arthritis and Systemic Lupus Erythematosus (SLE). Rheumatoid arthritis is an autoimmune disease most prevalent in older people, first occurring between ages 20 and 50. This disease affects the joints, most usually the small joints of the hands. It can lead to destruction of bone in severe cases. Type *http://www.duq.edu/PT/RA/RA.html* into your browser to read a thorough discussion of this disease prepared by the Rangos School of Health Sciences at Duquesne University.

A less well known autoimmune disease, Systemic Lupus Erythematosus, has recently begun to appear in the news. This disease primarily affects women, resulting in a series of characteristic symptoms including a characteristic reddened "butterfly" rash on the face. SLE, or Lupus, can affect many parts of the body, including the joints, skin, kidneys, heart, lungs, blood vessels, and brain. Although people with the disease may have many different symptoms, some of the most common ones include extreme fatigue, painful or swollen joints (arthritis), unexplained fever, skin rashes, and kidney problems. A good discussion of this disorder can be found at: *http://www.niams.nih.gov/hi/topics/lupus/slehandout/index.htm*.

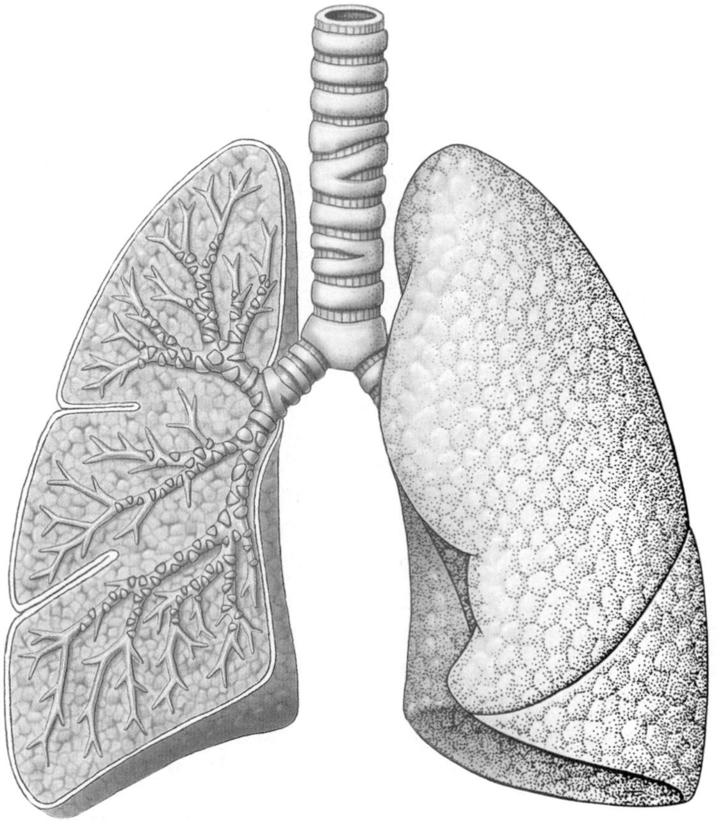

24

THE RESPIRATORY SYSTEM

Cells obtain energy primarily through aerobic metabolism, a process that requires oxygen and produces carbon dioxide. For cells to survive, they must have a way to obtain that oxygen and eliminate the carbon dioxide. The cardiovascular system provides the link between the interstitial fluids around peripheral cells and the gas exchange surfaces of the lungs. The respiratory system facilitates the exchange of gases between the air and the blood. As it circulates, blood carries oxygen from the lungs to peripheral tissues; it also accepts the carbon dioxide produced by these tissues and transports it to the lungs for elimination.

We begin our discussion of the respiratory system by describing the anatomical structures that conduct air from the external environment to the gas-exchange surfaces in the lungs. We will then discuss the mechanics of breathing and the neural control of respiration.

An Overview of the Respiratory System [FIGURE 24.1]

The **respiratory system** includes the nose, nasal cavity and sinuses, the pharynx, the larynx (voice box), the trachea (windpipe), and smaller conducting passageways leading to the gas-exchange surfaces of the lungs. These structures are illustrated in Figure 24.1●. The **respiratory tract** consists of the airways that carry air to and from these surfaces. The respiratory tract can be divided into a *conducting portion* and a *respiratory portion*. The conducting portion extends from the entrance to the nasal cavity to the smallest *bronchioles* of the lungs. The respiratory portion of the tract includes the delicate *respiratory bronchioles* and the delicate air sacs, or **alveoli** (al-VĒ-ō-lī), where gas exchange occurs.

The respiratory system includes the respiratory tract and the associated tissues, organs, and supporting structures. The **upper respiratory system** consists of the nose, nasal cavity, paranasal sinuses, and pharynx. These passageways filter, warm, and humidify the air, protecting the more delicate conduction and exchange surfaces of the **lower respiratory system** from debris, pathogens, and environmental extremes. The lower respiratory system includes the larynx, trachea, bronchi, and lungs.

Filtering, warming, and humidification of the inhaled air begin at the entrance to the upper respiratory system and continue throughout the rest of the conducting system. By the time the air reaches lung alveoli, most foreign particles and pathogens have been removed, and the humidity and temperature are within acceptable limits. The success of this "conditioning process" is due primarily to the properties of the *respiratory epithelium*, discussed in a later section.

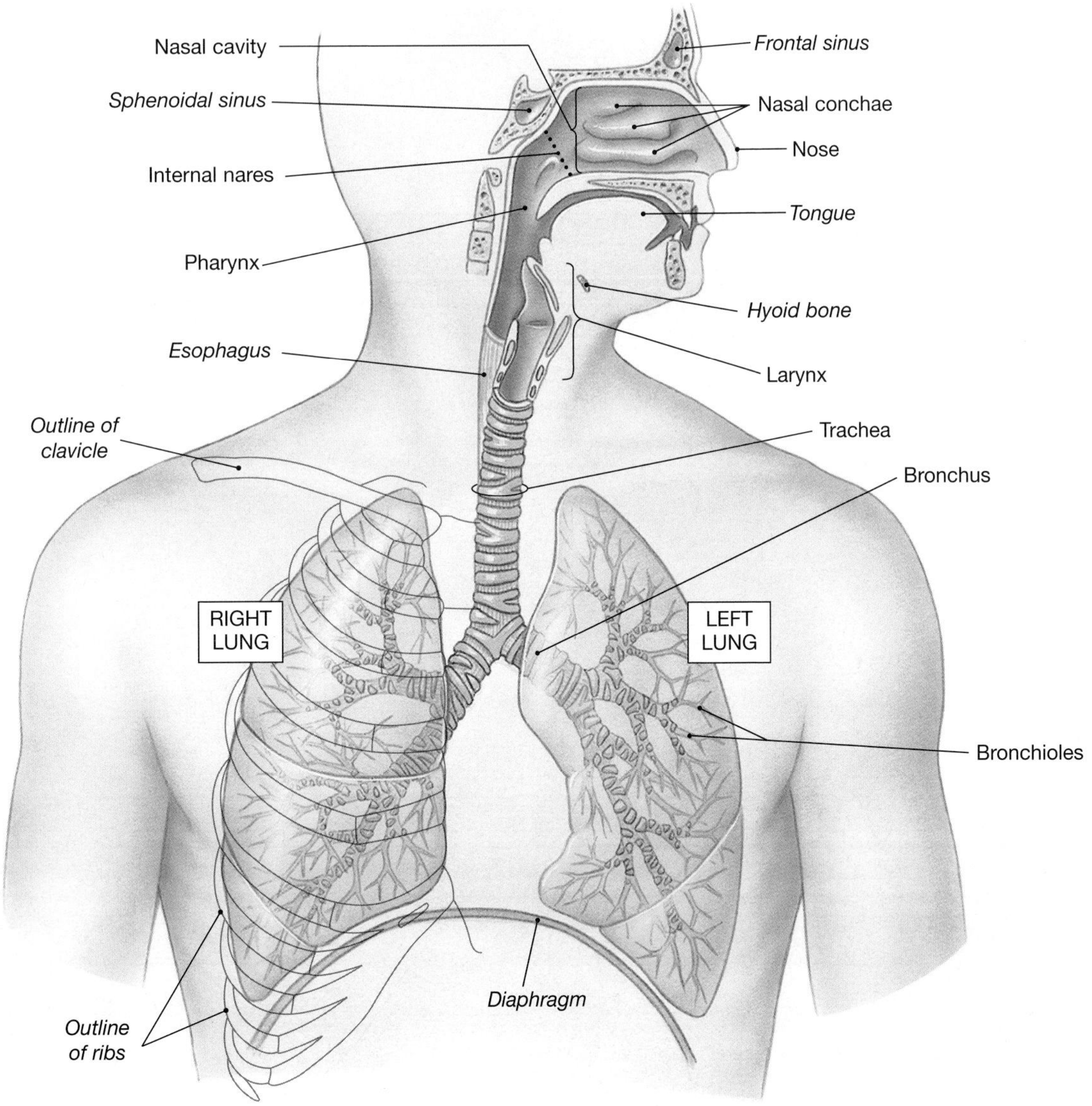

FIGURE 24.1 **STRUCTURES OF THE RESPIRATORY SYSTEM**

Functions of the Respiratory System

The functions of the respiratory system are:

1. Providing an extensive area for gas exchange between air and circulating blood;
2. Moving air to and from the exchange surfaces of the lungs;
3. Protecting respiratory surfaces from dehydration, temperature changes, and other environmental variations;
4. Defending the respiratory system and other tissues from invasion by pathogenic microorganisms;
5. Producing sounds involved in speaking, singing, or nonverbal communication;
6. Assisting in the regulation of blood volume, blood pressure, and the control of body fluid pH.

The respiratory system performs these functions with the cooperation of the circulatory system, selected skeletal muscles, and the nervous system.

The Respiratory Epithelium [FIGURE 24.2]

The **respiratory epithelium** consists of a pseudostratified, ciliated, columnar epithelium with numerous goblet cells (Figure 24.2●). The respiratory epithelium lines the entire respiratory tract except for the inferior portions of the pharynx and the finest conducting portions and alveoli. Goblet cells in the epithelium and mucous glands deep to the epithelium produce a sticky mucus that bathes the exposed surfaces. In the nasal cavity, cilia sweep any debris trapped in mucus or microorganisms toward the pharynx, where it will be swallowed and exposed to the acids and enzymes of the stomach. In lower portions of the respiratory tract, the cilia also beat

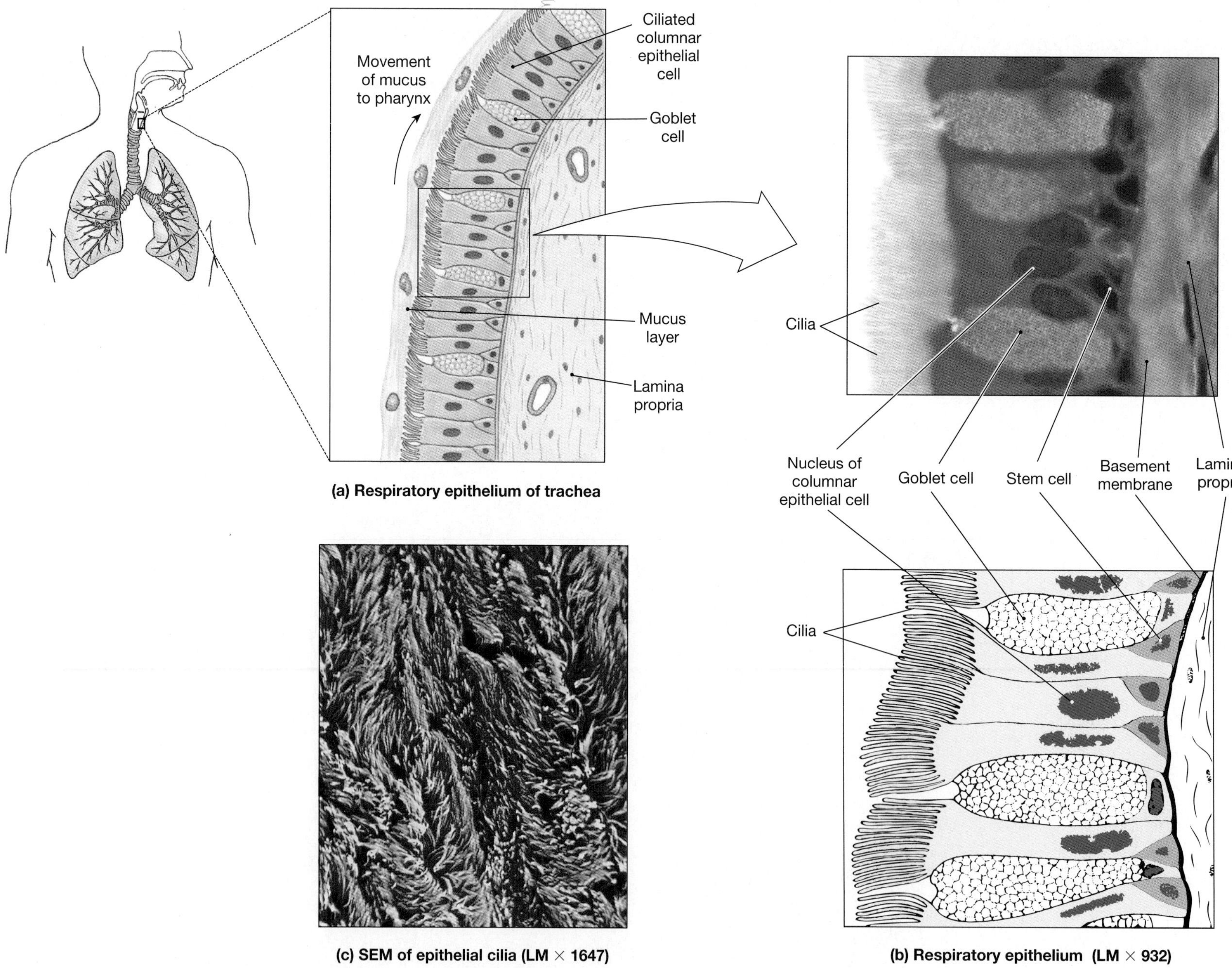

FIGURE 24.2 **The Respiratory Epithelium**

(**a**) Diagrammatic view of the respiratory epithelium. (**b**) Sketch and light micrograph showing the sectional appearance of the respiratory epithelium. (**c**) A surface view of the epithelium, as seen with the scanning electron microscope. In this colorized image, the cilia of the epithelial cells form a dense layer that resembles a shag carpet. The movement of these cilia propels mucus across the epithelial surface.

toward the pharynx, creating a *mucus escalator* that cleans the respiratory passageways. The delicate surfaces of the respiratory system can be severely damaged if the inspired air becomes contaminated with debris or pathogens. The respiratory mechanisms for filtration are responsible for the **respiratory defense system.** *Overloading the Respiratory Defenses p. 804*

In the nasal cavity, this filtering mechanism removes virtually all particles larger than around 10 μm from the inspired air. Smaller particles may be trapped by the mucus of the nasopharynx or secretions of the pharynx before proceeding farther along the conducting system. Exposure to unpleasant stimuli, such as noxious vapors, large quantities of dust and debris, allergens, or pathogens, usually causes a rapid increase in the rate of mucus production. (The familiar symptoms of the "common cold" result from the invasion of the respiratory epithelium by one of over 200 viruses.)

Filtration, warming, and humidification of inhaled air occurs throughout the conducting portion of the respiratory system, but the greatest changes occur within the nasal cavity. Breathing through the mouth eliminates much of the preliminary filtration, heating, and humidifying of the inspired air. Patients breathing on a respirator, which utilizes a tube to provide air directly into the trachea, must receive air that has been externally filtered and humidified, or risk alveolar damage.

CLINICAL BRIEF

Cystic Fibrosis

Cystic fibrosis (CF) is the most common lethal inherited disease in the Caucasian population, occurring at a frequency of 1 birth in 3000. Each year approximately 2000 babies are born with this condition in the United States alone. Even with improved therapies only 34% reach adulthood and less than 10% survive past age 30. Death is usually the result of a massive bacterial infection of the lungs and associated lung and heart failure.

The underlying problem involves an abnormality in a membrane transport protein responsible for the active transport of chloride ions. This membrane protein is abundant in exocrine cells that produce watery secretions. In persons with CF, these cells cannot transport salts and water effectively, and the secretions produced are thick and gooey. Mucous glands of the respiratory tract and secretory cells of the pancreas, salivary glands, digestive tract, and reproductive tract are affected.

The most serious symptoms appear because the respiratory defense system cannot transport such dense mucus. The mucus escalator stops working, and mucus plugs block the smaller respiratory passageways. This blockage reduces the diameter of the airways, and the inactivation of the normal respiratory defenses leads to frequent bacterial infections.

The gene responsible for CF has been identified, and the structure of the membrane protein determined. Now that the structure of the gene is understood, research continues with the goal of correcting the defect by the insertion of normal genes.

The Upper Respiratory System

The Nose and Nasal Cavity [FIGURE 24.3]

The nose is the primary passageway for air entering the respiratory system. The bones, cartilages, and sinuses associated with the nose were introduced in Chapter 6. pp. 138, 153 Air normally enters the respiratory system through the paired **external nares** (NA-rēs) that open into the **nasal cavity**. The **vestibule** (VES-ti-būl) is the portion of the nasal cavity enclosed by the flexible tissues of the nose (Figure 24.3a, d●). The vestibule is supported by thin, paired *lateral cartilages* and two pairs of *alar cartilages*. The epithelium of the vestibule contains coarse hairs that extend across the external nares. Large airborne particles such as sand, sawdust, or even insects are trapped in these hairs and are prevented from entering the nasal cavity.

The *nasal septum* separates right and left portions of the nasal cavity. The bony portion of the nasal septum is formed by the fusion of the perpendicular plate of the ethmoid and the plate of the vomer. The anterior portion of the nasal septum is formed of hyaline cartilage. This cartilaginous plate supports the bridge, or **dorsum nasi** (DOR-sum NĀ-zī), and **apex** (tip) of the nose.

The maxillae, nasal and frontal bones, ethmoid, and sphenoid form the lateral and superior walls of the nasal cavity. The mucous secretions produced in the associated *paranasal sinuses* (p. 153), aided by the tears draining through the nasolacrimal ducts, help keep the surface of the nasal cavity moist and clean. The superior portion, or *olfactory region*, of the nasal cavity includes the area lined by olfactory epithelium: the inferior surface of the cribriform plate and superior nasal conchae of the ethmoid and the superior portion of the nasal septum. p. 148

The superior, middle, and inferior nasal conchae, or *turbinate bones*, project toward the nasal septum from the lateral walls of the nasal cavity. To pass from the vestibule to the internal nares, or *choanae* (kō-Ā-nē), air tends to flow between adjacent conchae, through the **superior, middle**, or **inferior meatuses** (mē-Ā-tus-es; *meatus*, passage) (Figure 24.3b,d●). These are narrow grooves rather than open passageways, and the incoming air bounces off the conchal surfaces and churns around like water flowing over rapids. This turbulence serves a purpose: As the air eddies and swirls, small airborne particles are likely to come in contact with the mucus that coats the lining of the nasal cavity. In addition to promoting filtration, the turbulence allows extra time for warming and humidifying the incoming air. *Nosebleeds p. 804*

A bony **hard palate**, formed by the maxillary and palatine bones, forms the floor of the nasal cavity and separates the oral and nasal cavities. p. 150 A fleshy **soft palate** extends posterior to the hard palate, marking the boundary line between the superior *nasopharynx* and the rest of the pharynx (Figure 24.3c,d●). The nasal cavity opens into the nasopharynx at the **internal nares.**

✓ CONCEPT CHECK

- If it is very cold outside, why is it hard on the lower respiratory system if you breathe only through your mouth?
- What is the mucus escalator?
- Why does the nasal cavity have nasal conchae instead of having smooth, straight walls? How does this affect the lower respiratory system?
- With cystic fibrosis, the mucus produced is much thicker than normal. How would this affect the defense of the respiratory system?

The Pharynx [FIGURE 24.3c,d]

The nose, mouth, and throat connect to each other by a common passageway or chamber called the **pharynx** (FAR-inks). The pharynx is shared by the digestive and respiratory systems. It extends between the internal nares and the entrances to the larynx and esophagus. The curving superior and posterior walls are closely bound to the axial skeleton, but the lateral walls are quite

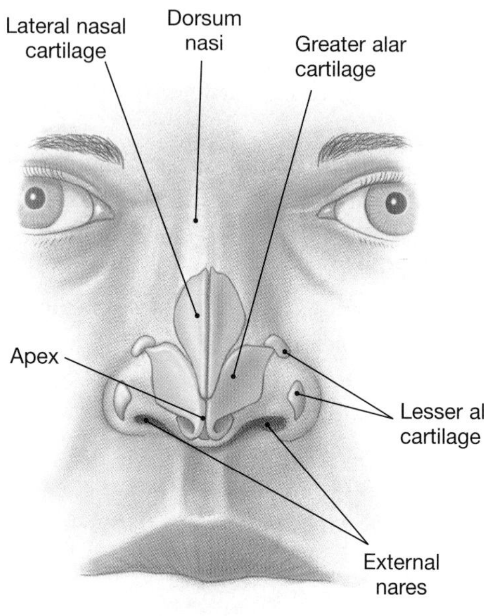

(a) Anterior view

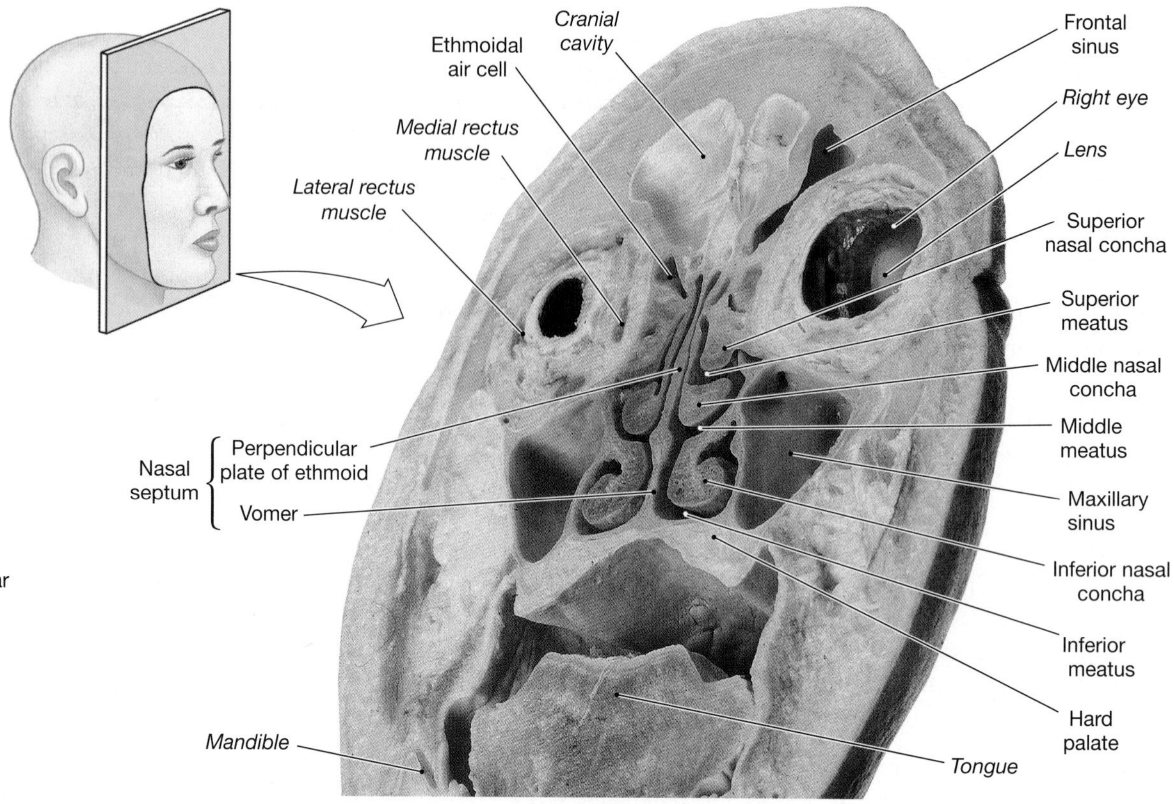

(b) Head, coronal section

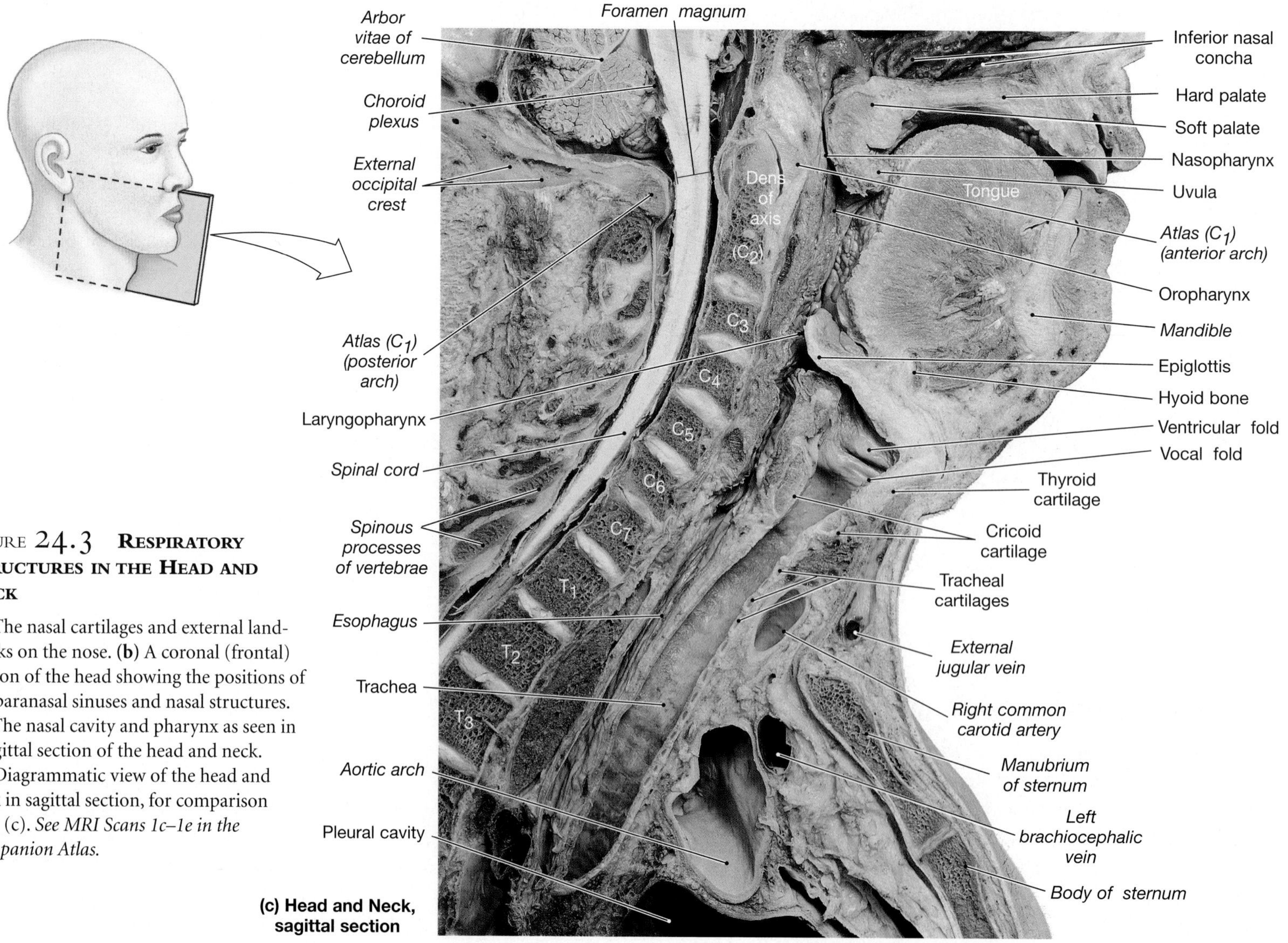

(c) Head and Neck, sagittal section

FIGURE 24.3 **RESPIRATORY STRUCTURES IN THE HEAD AND NECK**

(**a**) The nasal cartilages and external landmarks on the nose. (**b**) A coronal (frontal) section of the head showing the positions of the paranasal sinuses and nasal structures. (**c**) The nasal cavity and pharynx as seen in a sagittal section of the head and neck. (**d**) Diagrammatic view of the head and neck in sagittal section, for comparison with (c). *See MRI Scans 1c–1e in the Companion Atlas.*

flexible and muscular. The pharynx is divided into three regions (Figure 24.3c,d●): the *nasopharynx*, the *oropharynx*, and the *laryngopharynx*.

THE NASOPHARYNX [FIGURE 24.3c,d]

The **nasopharynx** (nā-zō-FAR-inks) is the superior portion of the pharynx. It is connected to the posterior portion of the nasal cavity via the internal nares and is separated from the oral cavity by the soft palate (Figure 24.3c,d●).

The nasopharynx, like the nasal cavities, is lined by typical respiratory epithelium. The *pharyngeal tonsil* is located on the posterior wall of the nasopharynx; the lateral walls contain the openings of the *auditory tubes* (see Figure 25.5a●). p. 669

THE OROPHARYNX [FIGURE 24.3c,d]

The **oropharynx** (ōr-ō-FAR-inks, *oris*, mouth) extends between the soft palate and the base of the tongue at the level of the hyoid bone. The posterior portion of the oral cavity communicates directly with the oropharynx, as do the posterior and inferior portions of the nasopharynx (Figure 24.3c,d●). At the boundary between the nasopharynx and oropharynx, the epithelium changes from a typical respiratory epithelium to a stratified squamous epithelium similar to that of the oral cavity.

The posterior margin of the soft palate supports the dangling **uvula** (Ū-vū-la) and two pairs of muscular **pharyngeal arches**. On either side a palatine tonsil lies between an anterior **palatoglossal** (pal-a-tō-GLOS-al) **arch** and a posterior **palatopharyngeal** (pal-a-tō-fa-RIN-jē-al) **arch** (see Figure 25.5a●). A curving line that connects the palatoglossal arches and uvula forms the boundaries of the **fauces** (FAW-sēz), the passageway between the oral cavity and the oropharynx.

THE LARYNGOPHARYNX [FIGURE 24.3c,d]

The narrow **laryngopharynx** (la-RING-gō-far-inks) includes that region of the pharynx lying between the hyoid bone and the entrance to the esophagus (Figure 24.3c,d●). The laryngopharynx is the most inferior part of the pharynx, and like the oropharynx it is lined by a stratified squamous epithelium that can resist mechanical abrasion, chemical attack, and pathogenic invasion.

The Lower Respiratory System

The Larynx [FIGURES 24.3d/24.4]

Inspired (inhaled) air leaves the pharynx by passing through a narrow opening, the **glottis** (GLOT-is) (Figure 24.3d●). The **larynx** (LAR-inks) surrounds and protects the glottis. The larynx begins at the level of vertebra C_4 or C_5 and ends at the level of vertebra C_7. The larynx is essentially a cylinder whose cartilaginous walls are stabilized by ligaments or skeletal muscles or both.

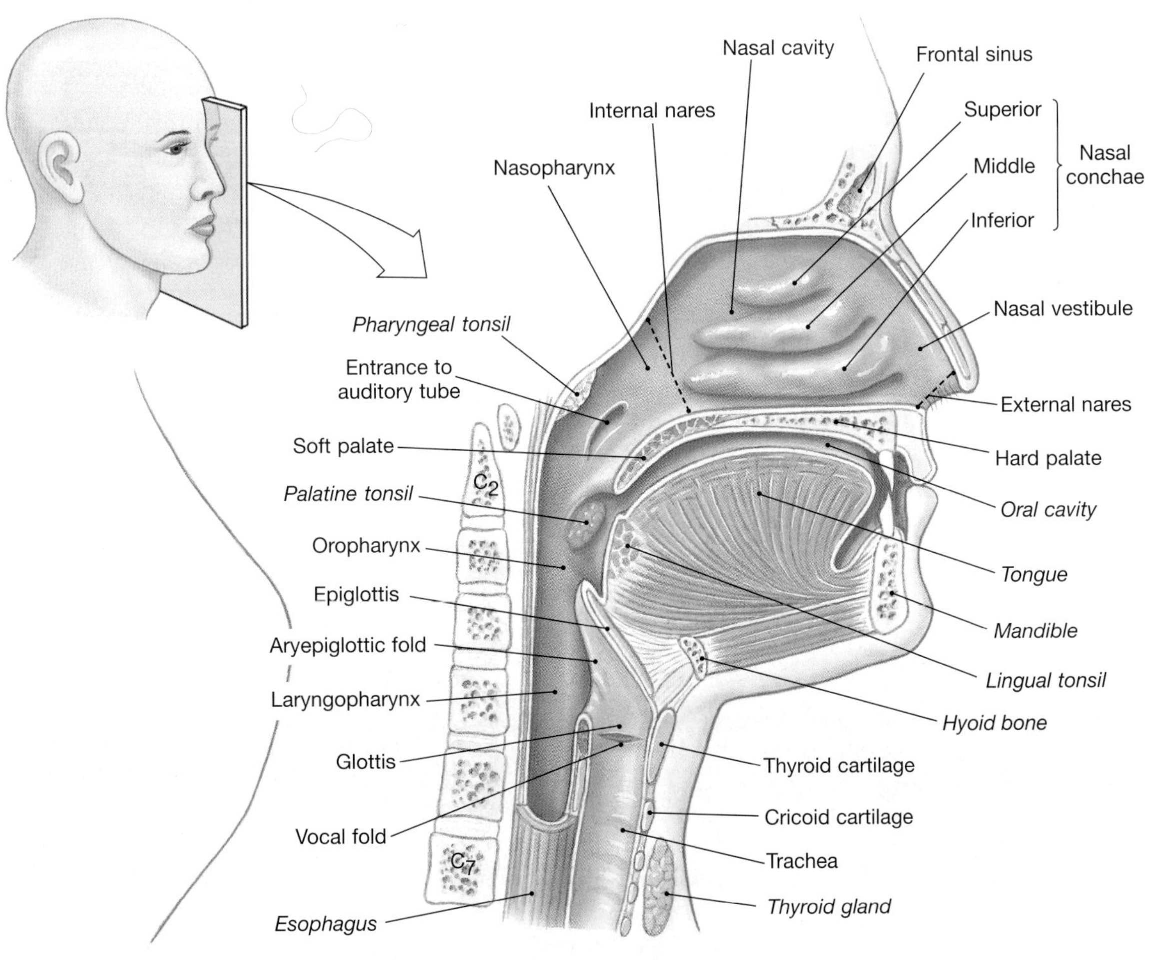

(d) Sagittal section

FIGURE 24.3 *(continued)*

CARTILAGES OF THE LARYNX [FIGURE 24.4]

Three large unpaired cartilages form the body of the larynx: the *thyroid cartilage*, the *cricoid cartilage*, and the *epiglottis*. View Figure 24.4● as we describe the cartilages of the larynx.

THE THYROID CARTILAGE [FIGURE 24.4a,b] The **thyroid** ("shield-shaped") **cartilage** is the largest laryngeal cartilage, and it forms most of the anterior and lateral walls of the larynx (Figure 24.4a,b●). The thyroid cartilage, when viewed in sagittal section, is incomplete posteriorly. The anterior surface of this cartilage bears a thick ridge, the **laryngeal prominence**. This ridge is easily seen and felt, and the thyroid cartilage is commonly called the *Adam's apple*. During embryological development, the thyroid cartilage is formed by two pieces of cartilage that meet in the anterior midline to form the laryngeal prominence.

The inferior surface of the thyroid cartilage articulates with the cricoid cartilage; the superior surface has ligamentous attachments to the epiglottis and smaller laryngeal cartilages.

THE CRICOID CARTILAGE [FIGURE 24.4a,c] The thyroid cartilage sits superior to the **cricoid cartilage** (KRĪ-koyd; "ring-shaped"). It is a complete ring, whose posterior portion is greatly expanded, providing support in the absence of the thyroid cartilage. The cricoid and thyroid cartilages protect the glottis and

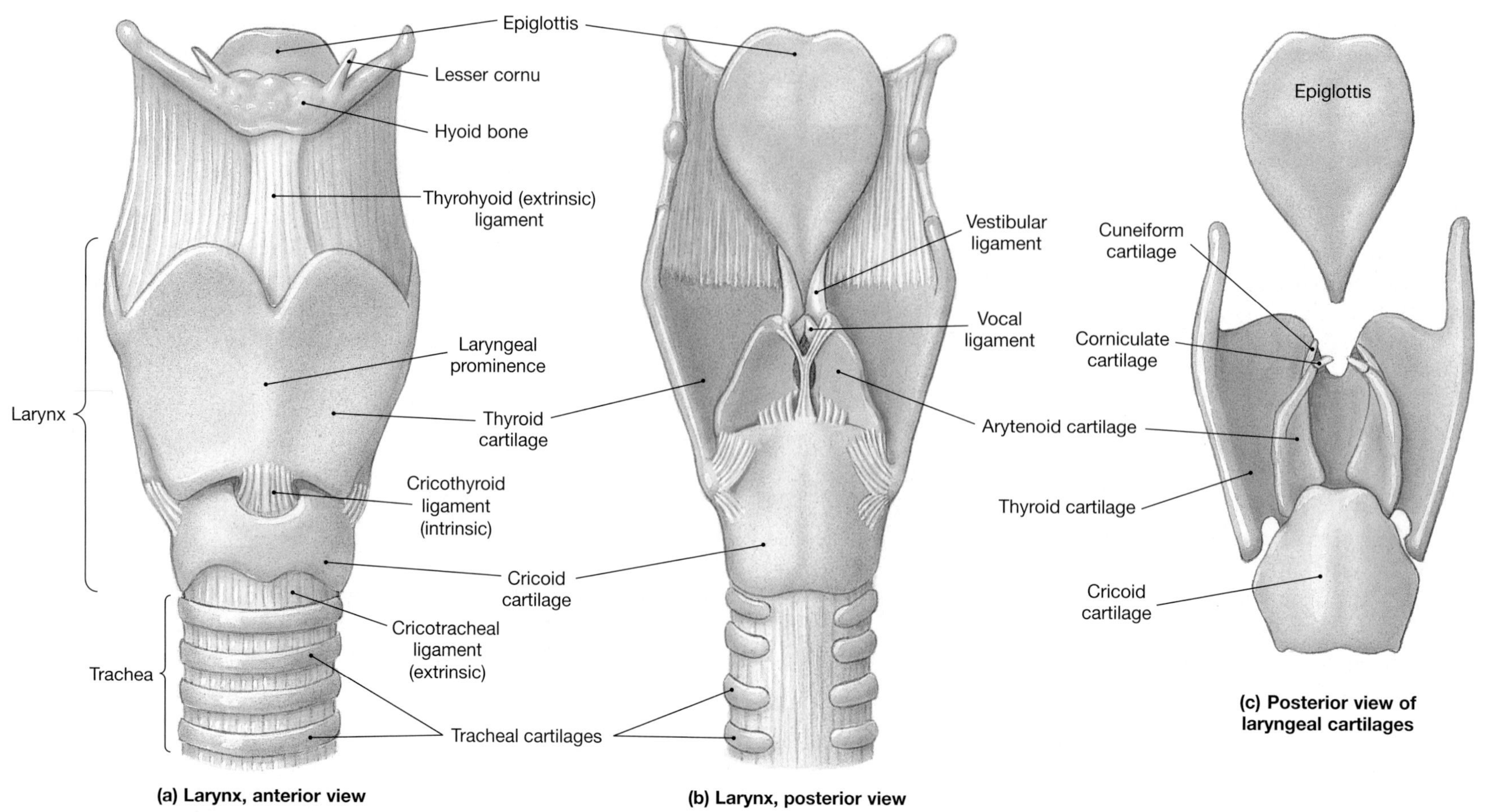

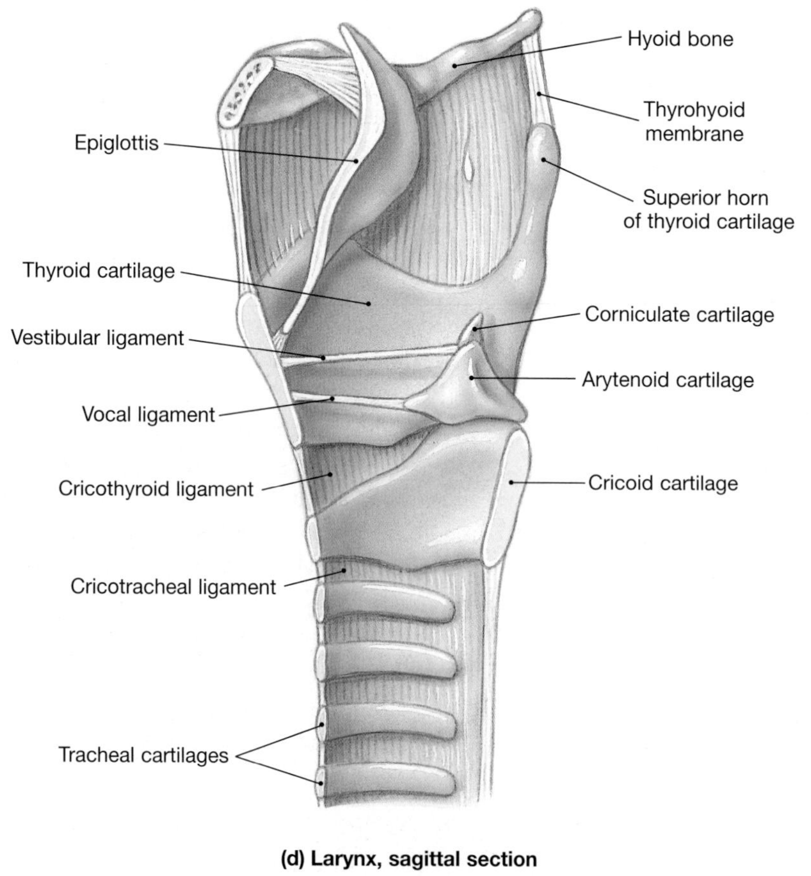

FIGURE 24.4 **ANATOMY OF THE LARYNX**

(**a**) Anterior view of the intact larynx. (**b**) Posterior view of the intact larynx. (**c**) Posterior view, showing the relationships among the individual laryngeal cartilages. (**d**) Sagittal section of the intact larynx.

the entrance to the trachea, and their broad surfaces provide sites for the attachment of important laryngeal muscles and ligaments. Ligaments attach the inferior surface of the cricoid cartilage to the first cartilage of the trachea (Figure 24.4a,c●). The superior surface of the cricoid cartilage articulates with the small paired *arytenoid cartilages.*

THE EPIGLOTTIS [FIGURES 24.3c,d/24.4b,c] The shoehorn-shaped **epiglottis** (ep-i-GLOT-is) projects superior to the glottis (Figures 24.3c,d and 24.4b,c●). The *epiglottal cartilage* that supports it has ligamentous attachments to the anterior and superior borders of the thyroid cartilage and the hyoid bone. During swallowing, the larynx is elevated, and the epiglottis folds back over the glottis, preventing the entry of liquids or solid food into the respiratory passageways.

PAIRED LARYNGEAL CARTILAGES [FIGURES 24.4c,d/24.5] The larynx also contains three pairs of smaller cartilages: the *arytenoid, corniculate,* and *cuneiform cartilages.* The anytenoids and corniculates are hyaline cartilages; the cuneiforms are elastic cartilages.

- The paired **arytenoid cartilages** (ar-i-TĒ-noyd; "ladle-shaped") articulate with the superior border of the enlarged portion of the cricoid cartilage (Figure 24.4c●).
- The **corniculate cartilages** (kor-NIK-ū-lāt; "horn-shaped") articulate with the arytenoid cartilages (Figure 24.4c●). The corniculate and arytenoid cartilages are involved with the opening and closing of the glottis and the production of sound.
- Elongate, curving **cuneiform cartilages** (kū-NĒ-i-form; "wedge-shaped") lie within the *aryepiglottic fold* that extends between the

lateral aspect of each arytenoid cartilage and the epiglottis (Figures 24.4d and 24.5●).

LARYNGEAL LIGAMENTS [FIGURES 24.4a,b/24.5]

A series of **intrinsic ligaments** binds all nine cartilages together to form the larynx (Figure 24.4a,b●). **Extrinsic ligaments** attach the thyroid cartilage to the hyoid bone and the cricoid cartilage to the trachea. The **vestibular ligaments** and the **vocal ligaments** extend between the thyroid cartilage and the arytenoids.

The vestibular and vocal ligaments are covered by folds of laryngeal epithelium that project into the glottis. The vestibular ligaments lie within the superior pair of folds, known as the **vestibular folds** (Figures 24.4b and 24.5●). The vestibular folds, which are relatively inelastic, help to prevent foreign objects from entering the glottis and provide protection for the more delicate **vocal folds**.

The vocal folds are highly elastic, because the vocal ligament is a band of elastic tissue. The vocal folds are involved with the production of sounds, and for this reason they are known as the **true vocal cords**. Because the vestibular folds play no part in sound production, they are often called the **false vocal cords**.

Sound Production Air passing through the glottis vibrates the vocal folds and produces sound waves. The pitch of the sound produced depends on the diameter, length, and tension in the vocal folds. The diameter and length are directly related to the size of the larynx. The tension is controlled by the contraction of voluntary muscles that change the relative positions of the thyroid and arytenoid cartilages. When the distance increases, the vocal folds tense and the pitch rises; when the distance decreases, the vocal folds relax and the pitch falls.

Children have slender, short vocal folds, and their voices tend to be high-pitched. At puberty the larynx of a male enlarges considerably more than that of a female. The true vocal cords of an adult male are thicker and longer, and they produce lower tones than those of an adult female.

The entire larynx is involved in sound production, because its walls vibrate, creating a composite sound. Amplification and echoing of the sound occur within the pharynx, the oral cavity, the nasal cavity, and the paranasal sinuses. The final production of distinct sounds depends on voluntary movements of the tongue, lips, and cheeks.

THE LARYNGEAL MUSCULATURE [FIGURE 24.6]

The larynx is associated with two different groups of muscles, the *intrinsic laryngeal muscles* and the *extrinsic laryngeal muscles*. The **intrinsic laryngeal muscles** have two major functions. One group regulates tension in the vocal folds, while a second set opens and closes the glottis. Those involved with the vocal folds insert upon the thyroid, arytenoid, and corniculate cartilages. Opening or closing the glottis involves rotational movements of the arytenoids that move the vocal folds apart or together.

The **extrinsic laryngeal musculature** positions and stabilizes the larynx. These muscles were considered in Chapter 10. p. 276

During swallowing, both extrinsic and intrinsic muscles cooperate to prevent food or drink from entering the glottis. Before you swallow, the material is crushed and chewed into a pasty mass known as a *bolus*. Extrinsic muscles then elevate the larynx, bending the epiglottis over the entrance to the glottis, so that the bolus can glide across the epiglottis, rather than falling into the larynx (Figure 24.6●). While this movement is under way, intrinsic muscles close the glottis. Should any food particles or liquids manage to touch the surfaces of the vestibular or vocal folds, the coughing reflex will be triggered. Coughing usually prevents the material from entering the glottis. *Disorders of the Larynx p. 804*

✓ CONCEPT CHECK

- What are the functions of the thyroid cartilage?
- What are the functions of the larynx?
- Laurel uses voluntary muscle contraction to shorten the distance between her thyroid and arytenoid cartilages. What is happening to the pitch of her voice?
- How would the absence of intrinsic laryngeal muscles affect swallowing?

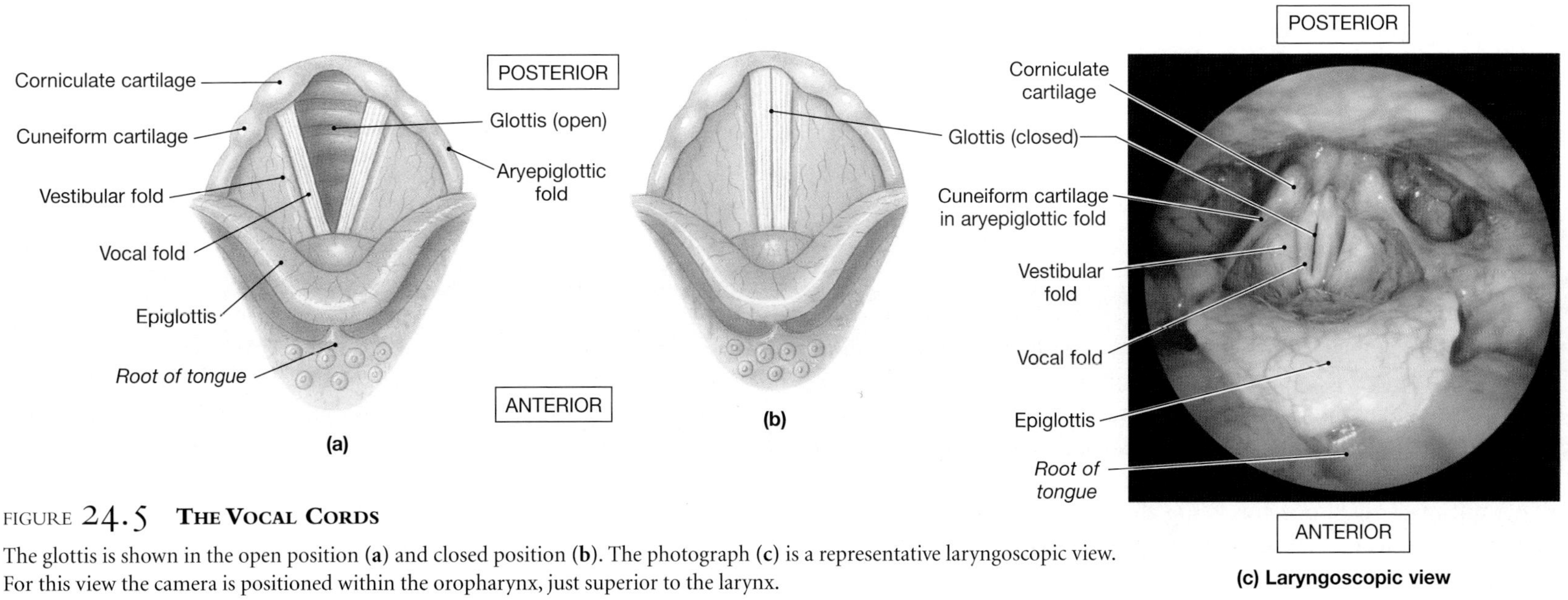

FIGURE 24.5 **The Vocal Cords**

The glottis is shown in the open position (**a**) and closed position (**b**). The photograph (**c**) is a representative laryngoscopic view. For this view the camera is positioned within the oropharynx, just superior to the larynx.

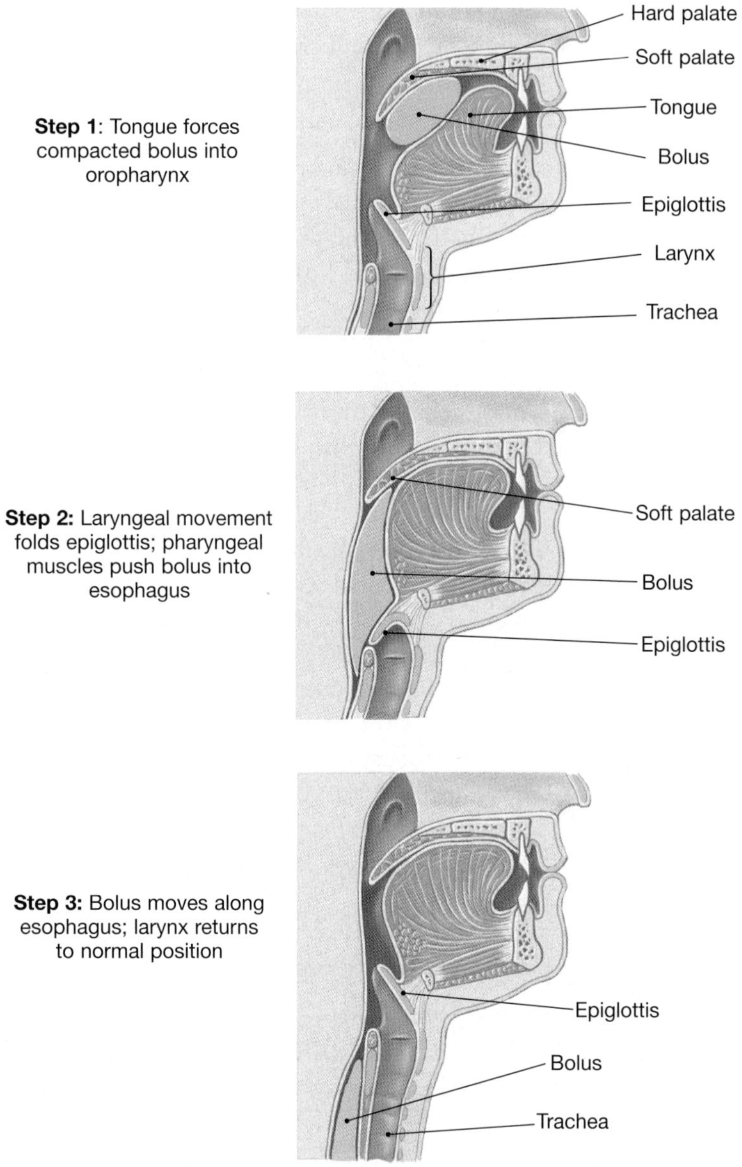
Step 1: Tongue forces compacted bolus into oropharynx
Hard palate
Soft palate
Tongue
Bolus
Epiglottis
Larynx
Trachea
Step 2: Laryngeal movement folds epiglottis; pharyngeal muscles push bolus into esophagus
Soft palate
Bolus
Epiglottis
Step 3: Bolus moves along esophagus; larynx returns to normal position
Epiglottis
Bolus
Trachea

FIGURE 24.6 **MOVEMENTS OF THE LARYNX DURING SWALLOWING**

During swallowing the elevation of the larynx folds the epiglottis over the glottis, steering materials into the esophagus.

The Trachea [FIGURES 24.2a/24.3c/24.7]

The epithelium of the larynx is continuous with that of the **trachea** (TRĀ-kē-a), or "windpipe." The trachea is a tough, flexible tube with a diameter of around 2.5 cm (1 in.) and a length of approximately 11 cm (4.25 in.) (Figures 24.3c, p. 638 and 24.7●). The trachea begins anterior to vertebra C_6 in a ligamentous attachment to the cricoid cartilage; it ends in the mediastinum, at the level of vertebra T_5, where it branches to form the *right* and *left primary bronchi.*

The lining of the trachea consists of respiratory epithelium overlying a layer of loose connective tissue called the **lamina propria** (LA-mi-na PRŌ-prē-a) (Figure 24.2a●, p. 636). The lamina propria separates the respiratory epithelium from underlying cartilages. The epithelium and lamina propria are interdependent, and the combination is an example of a *mucous membrane*, or **mucosa** (mū-KŌ-sa). ∞ p. 75

A thick layer of connective tissue, the **submucosa** (sub-mū-KŌ-sa), surrounds the mucosa. The submucosa contains mucous glands that communicate with the epithelial surface through a number of secretory ducts.

Superficial to the submucosa, the trachea contains 15–20 **tracheal cartilages** (Figure 24.7●). Each tracheal cartilage is bound to neighboring cartilages by elastic **anular ligaments**. The tracheal cartilages stiffen the tracheal walls and protect the airway. They also prevent its collapse or overexpansion as pressures change in the respiratory system.

Each tracheal cartilage is C-shaped. The closed portion of the C protects the anterior and lateral surfaces of the trachea. The open portions of the tracheal cartilages face posteriorly, toward the esophagus (Figure 24.7b●). Because the cartilages do not continue around the trachea, the posterior tracheal wall can easily distort during swallowing, permitting the passage of large masses of food along the esophagus.

An elastic ligament and a band of smooth muscle, the **trachealis**, connect the ends of each tracheal cartilage (Figure 24.7b●). Contraction of the trachealis muscle alters the diameter of the tracheal lumen, changing the resistance to air-flow. The normal diameter of the trachea changes from moment to moment, primarily under the control of the sympathetic division of the autonomic nervous system. Sympathetic stimulation increases the diameter of the trachea and makes it easier to move large volumes of air along the respiratory passageways.

CLINICAL BRIEF

TRACHEAL BLOCKAGE

Foreign objects, usually food, that become lodged in the larynx or trachea are usually expelled by coughing. If the individual can speak or make a sound, the airway is still open, and no emergency measures should be taken. If the victim can neither breathe nor speak, an immediate threat to life exists. Unfortunately, many victims become acutely embarrassed by this situation, fail to realize the danger, and rather than seek assistance, they run to the nearest restroom and quietly die there.

In the **Heimlich** (HĪM-lik) **maneuver**, or *abdominal thrust*, a rescuer applies compression to the abdomen just inferior to the diaphragm. Compression elevates the diaphragm forcefully and may generate enough pressure to remove the blockage. This maneuver must be performed properly to avoid damage to internal organs. Organizations such as the Red Cross, the local fire department, or other charitable groups usually hold brief training sessions throughout the year.

If the blockage remains, professionally qualified rescuers may perform a **tracheostomy** (trā-kē-OS-tō-mē). In this procedure an incision is made through the anterior tracheal wall, and a tube is inserted. The tube bypasses the larynx and permits air-flow directly into the trachea. A tracheostomy may also be required when the larynx becomes blocked by a foreign object, inflammation, or sustained laryngeal spasms when a portion of the trachea has been crushed, or when a portion of the trachea has been removed during treatment of laryngeal cancer.

The Primary Bronchi [FIGURE 24.7a]

The trachea branches within the mediastinum, giving rise to the **right** and **left primary bronchi** (BRONG-kī). The left and right primary bronchi are outside the lungs and are called the **extrapulmonary bronchi**. An internal ridge, the **carina** (ka-RĪ-na), lies between the entrances to the two primary bronchi (Figure 24.7a●). The histological organization of the primary bronchi is the same as the trachea, with cartilaginous C-shaped supporting rings. The right primary bronchus supplies the right lung, and the left supplies the left lung. The right primary bronchus has a larger diameter than

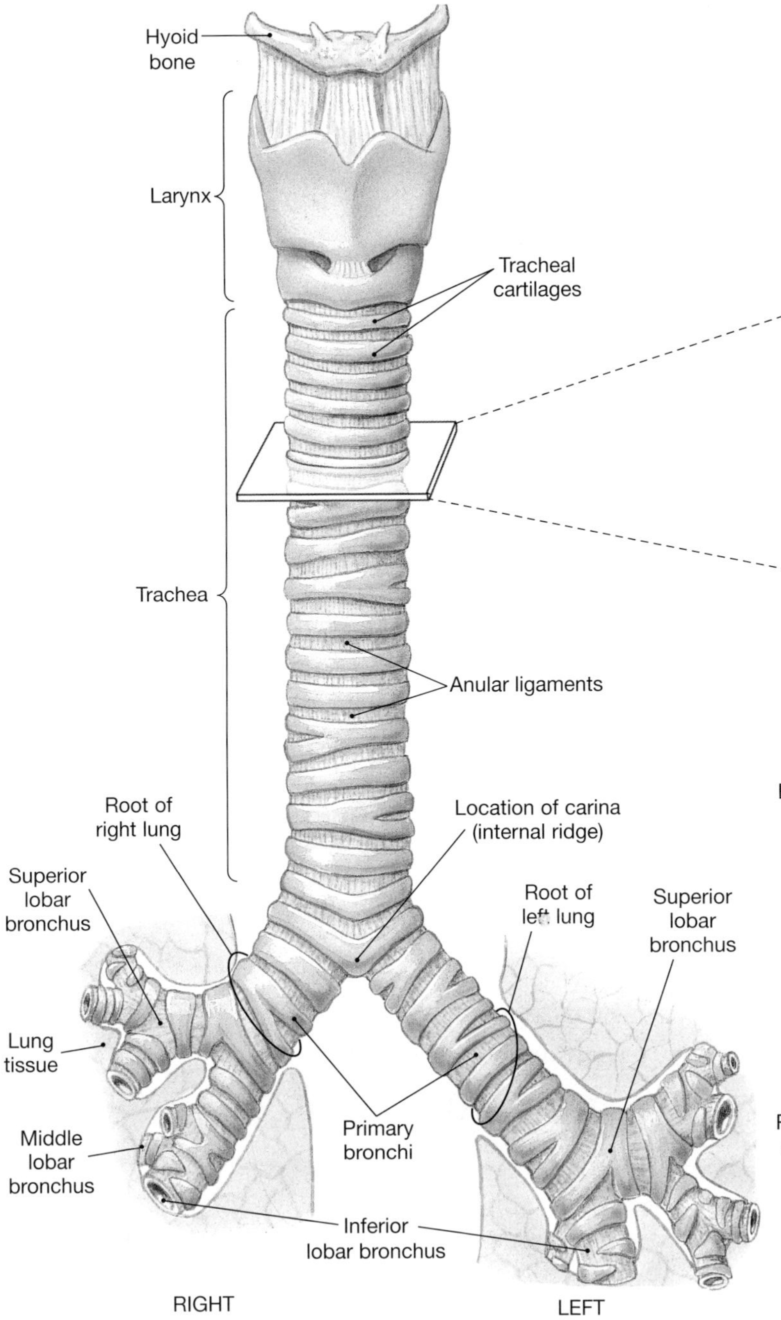

(a) Trachea and bronchi, anterior view

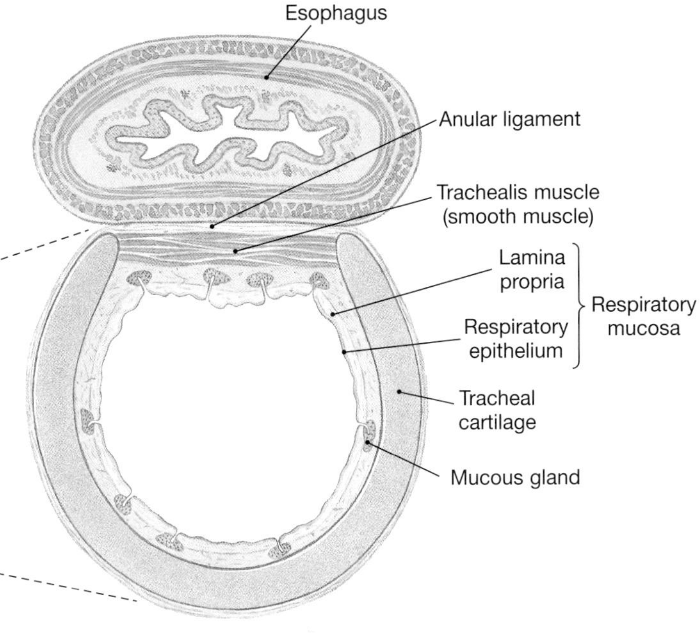

(b) Trachea and esophagus, transverse section

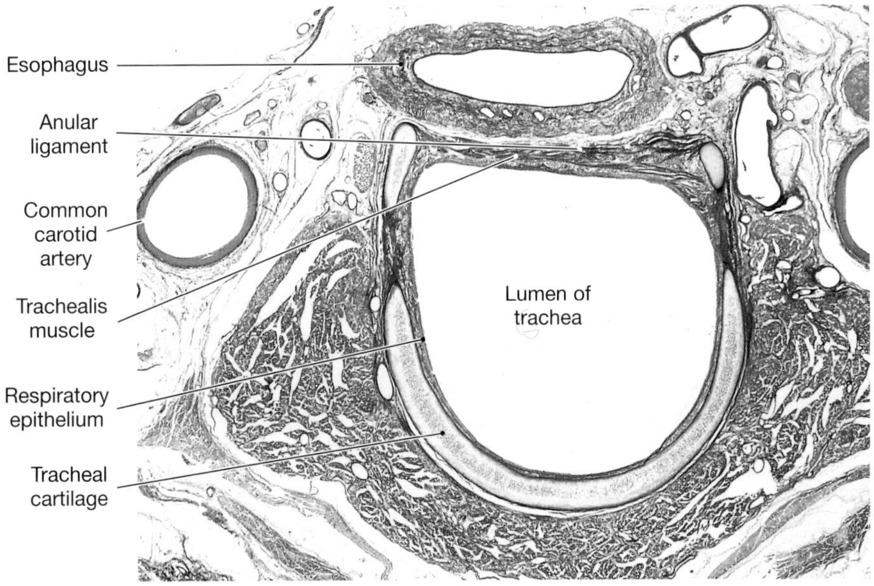

(c) Trachea, transverse-sectional view (LM × 60)

FIGURE 24.7 **ANATOMY OF THE TRACHEA AND PRIMARY BRONCHI**

(**a**) Anterior view on dissection, showing the plane of section for (b). (**b,c**) Cross-sectional views of the trachea, showing its relationship to surrounding structures.

the left, and it descends toward the lung at a steeper angle. For these reasons foreign objects that enter the trachea usually become lodged in the right bronchus rather than the left.

Each primary bronchus travels to a groove along the medial surface of its lung before branching further. This groove, the **hilus** (HĪ-lus), also provides access for entry to pulmonary vessels and nerves. The entire array is firmly anchored in a meshwork of dense connective tissue. This complex, known as the **root** of the lung, attaches it to the mediastinum and fixes the positions of the major nerves, vessels, and lymphatics. The roots of the lungs are located anterior to vertebrae T_5 (right) and T_6 (left).

✓ CONCEPT CHECK

- Why are the cartilages that reinforce the trachea C-shaped instead of complete rings?
- What type of epithelium can be observed in the trachea?
- How are tracheal cartilages involved in respiration?
- How can you identify the right primary bronchus from the left primary bronchus?

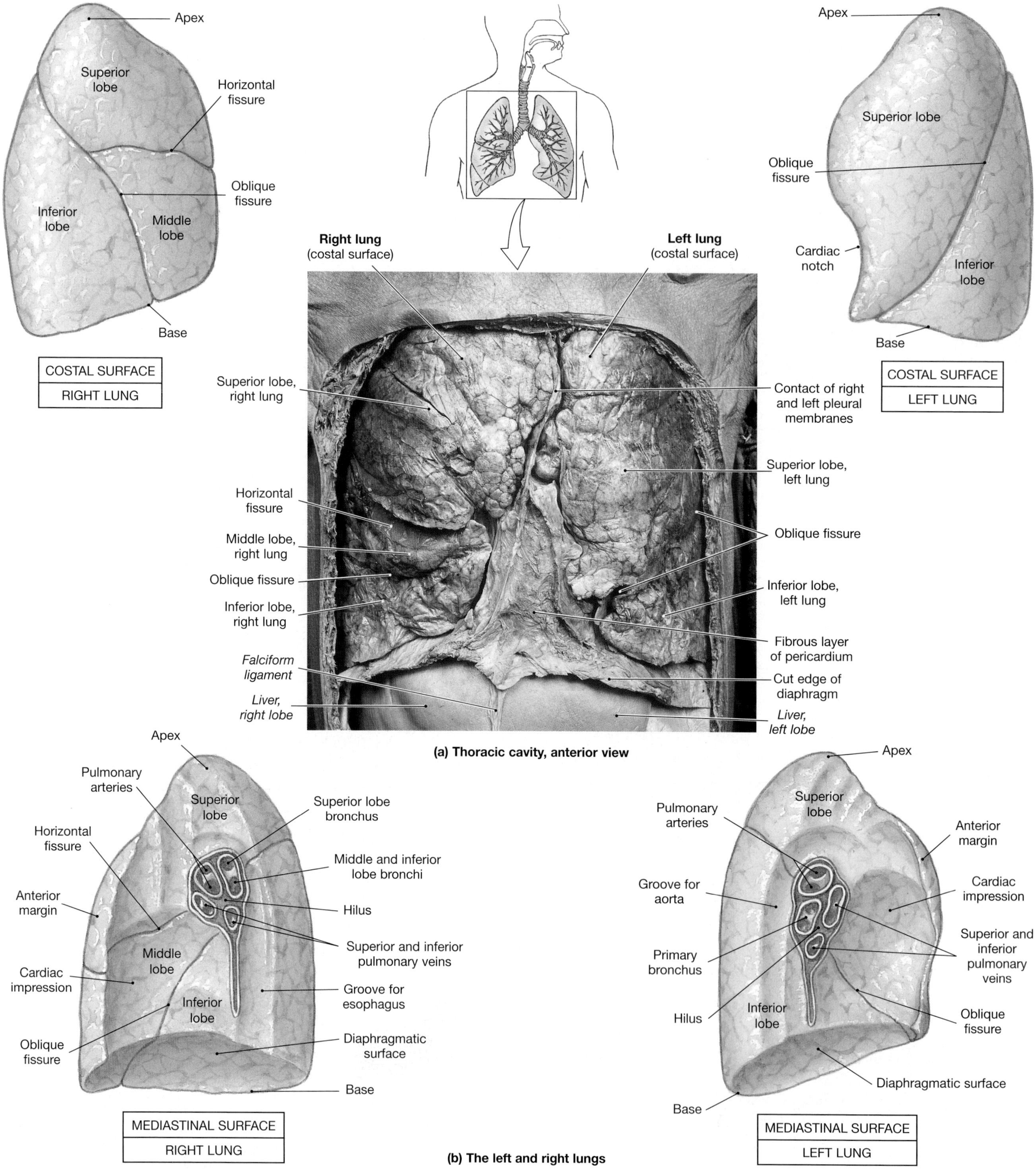

FIGURE 24.8 **SUPERFICIAL ANATOMY OF THE LUNGS**

(**a**) Anterior view of the opened chest, showing relative positions of lungs and heart. (**b**) Diagrammatic views of the costal (lateral) and mediastinal (medial) surfaces of the isolated right and left lungs.

The Lungs [FIGURE 24.8]

The left and right lungs (Figure 24.8●) are situated in the left and right pleural cavities. Each lung is a blunt cone with the tip, or **apex**, pointing superiorly. The apex on each side extends into the base of the neck above the first rib. The broad concave inferior portion, or **base**, of each lung rests on the superior surface of the diaphragm.

Lobes of the Lungs [FIGURE 24.8]

The lungs have distinct **lobes** separated by deep fissures. The **right lung** has three lobes: **superior, middle**, and **inferior**. The *horizontal fissure* separates the superior and middle lobes. The *oblique fissure* separates the superior and inferior lobes. The **left lung** has only two lobes, **superior** and **inferior**, separated by the *oblique fissure* (Figure 24.8●). The right lung is broader than the left because most of the heart and great vessels project into the left pleural cavity. However, the left lung is longer than the right lung, because the diaphragm rises on the right side to accommodate the mass of the liver.

Lung Surfaces [FIGURE 24.8]

The curving anterior portion of the lung that follows the inner contours of the rib cage is the **costal surface** or *anterior surface* (Figure 24.8a●). The **mediastinal surface**, or *medial surface*, contains the hilus and has a more irregular shape (Figure 24.8b●). The mediastinal surfaces of both lungs bear grooves that mark the positions of the great vessels and the heart. The heart is located to the left of the midline, and the left lung has a large *cardiac impression*. In anterior view, the medial margin of the right lung forms a vertical line, whereas the medial margin of the left lung bears a concavity, the **cardiac notch**.

The connective tissues of the root of each lung extend into its substance, or **parenchyma** (par-ENG-ki-ma). These fibrous partitions, or **trabeculae**, contain elastic fibers, smooth muscles, and lymphatics. They branch repeatedly, dividing the lobes into smaller and smaller compartments. The branches of the conducting passageways, pulmonary vessels, and nerves of the lungs follow these trabeculae to reach their peripheral destinations. The terminal partitions, or **septa**, divide the lung into **lobules** (LOB-ūlz), each supplied by tributaries of the pulmonary arteries, pulmonary veins, and respiratory passageways. The connective tissues of the septa are in turn continuous with those of the visceral pleura. We will now follow the branching pattern of the bronchi from the hilus to the alveoli of each lung.

The Pulmonary Bronchi [FIGURES 24.7/24.9/24.10]

The primary bronchi and their branches form the *bronchial tree*. Because the left and right bronchi are outside the lungs, they are called *extrapulmonary bronchi*. As the primary bronchi enter the lungs, they divide to form smaller passageways (Figures 24.7, 24.9, and 24.10●). Those branches are collectively called the **intrapulmonary bronchi**.

Each primary bronchus divides to form **secondary bronchi**, also known as **lobar bronchi**. Secondary bronchi in turn branch to form **tertiary bronchi**, or **segmental bronchi**. The branching pattern differs depending on the lung considered; details are provided below. Each tertiary bronchus supplies air to a single *bronchopulmonary segment*, a specific region of one lung (Figure 24.10a,b●). There are 10 tertiary bronchi (and 10 bronchopulmonary segments) in the right lung. The left lung also has 10 segments during development, but subsequent fusion usually reduces that number to eight or nine. The walls of the primary, secondary, and tertiary bronchi contain progressively lesser amounts of cartilage. The walls of secondary and tertiary bronchi contain cartilage plates arranged around the lumen. These cartilages serve the same purpose as the rings of cartilage in the trachea and primary bronchi. *Bronchitis p. 804*

BRANCHES OF THE RIGHT PRIMARY BRONCHUS [FIGURES 24.7/24.10]

The right lung has three lobes, and the right primary bronchus divides into three secondary bronchi: a **superior lobar bronchus**, a **middle lobar bronchus**, and an **inferior lobar bronchus**. The middle and inferior lobar bronchi branch from the right primary bronchus almost as soon as it enters the lung at the hilus (Figure 24.7●). Each lobar branch delivers air to one of the lobes of the right lung (Figure 24.10●).

BRANCHES OF THE LEFT PRIMARY BRONCHUS [FIGURES 24.7/24.9/24.10]

The left lung has two lobes, and the left primary bronchus divides into two secondary bronchi: a **superior lobar bronchus** and an **inferior lobar bronchus** (Figures 24.7, 24.9, and 24.10●).

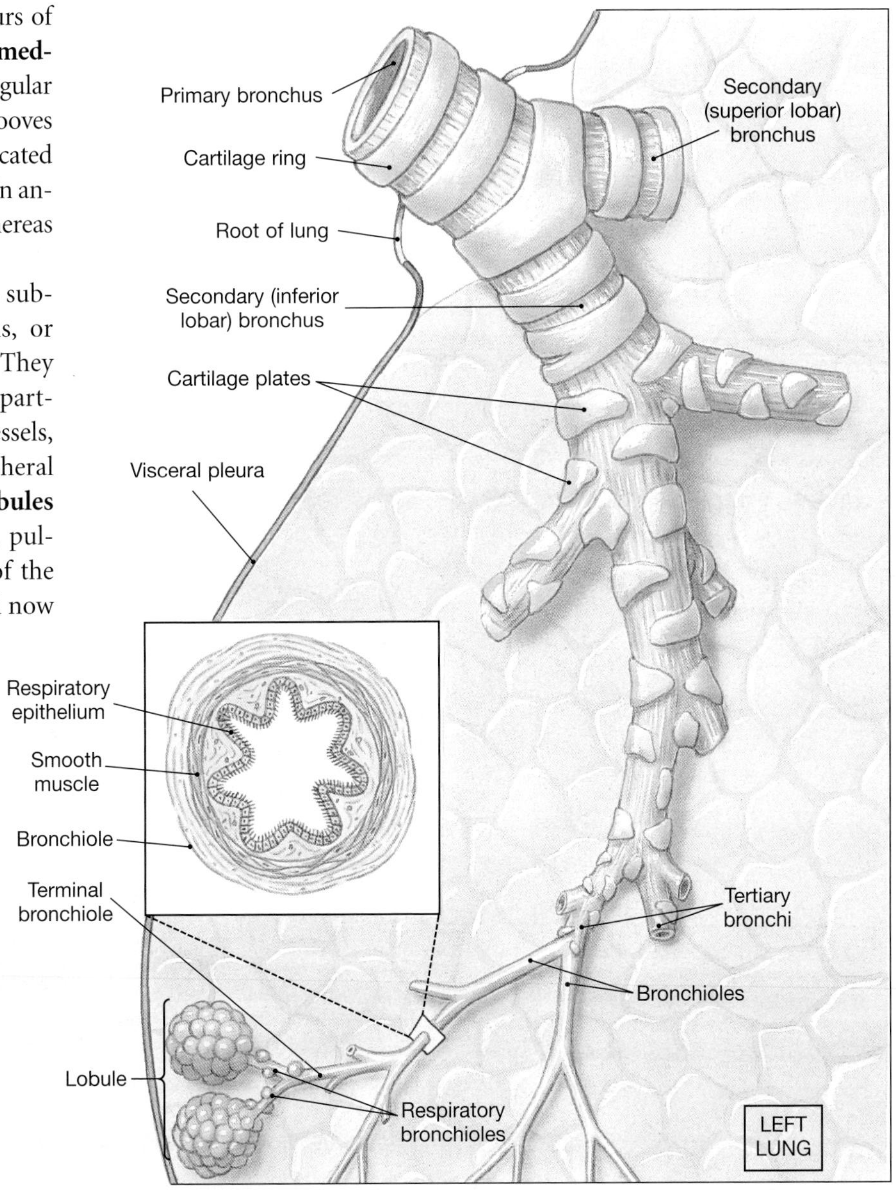

FIGURE 24.9 **BRONCHI AND BRONCHIOLES**

For clarity, the degree of branching has been reduced; an airway branches approximately 23 times before reaching the level of a lobule.

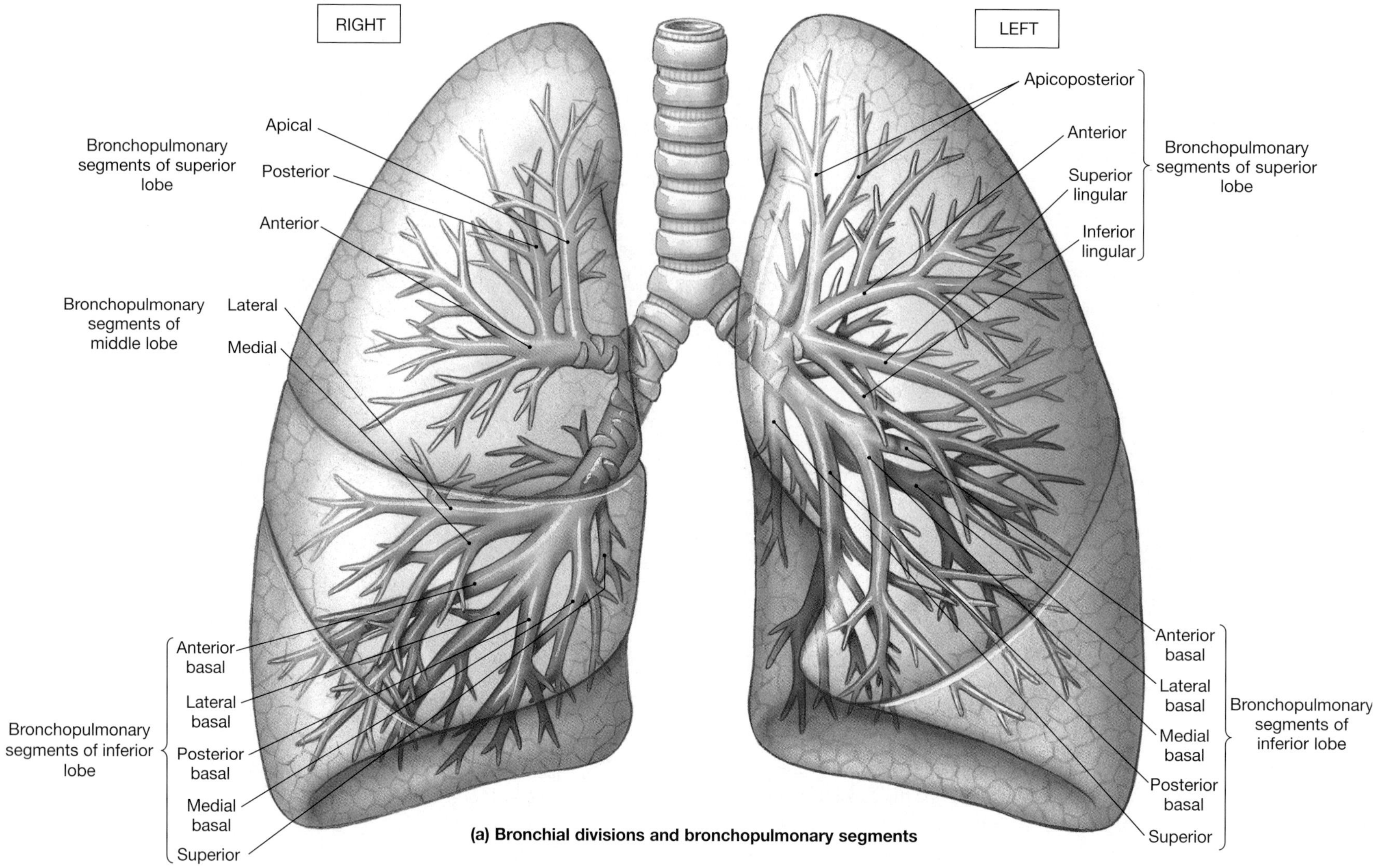

(a) Bronchial divisions and bronchopulmonary segments

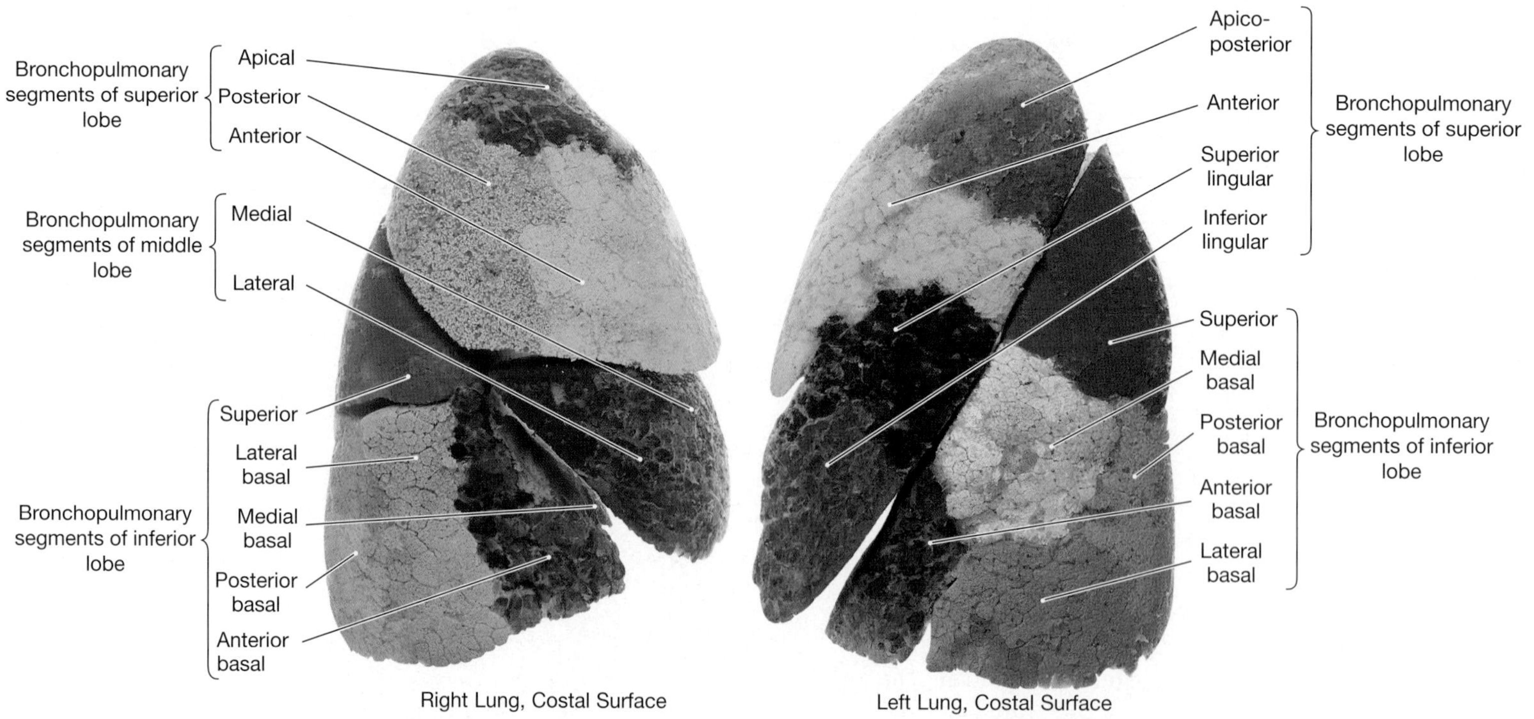

(b) Bronchopulmonary segments of left and right lungs, anterior view

FIGURE 24.10 **The Bronchial Tree and Divisions of the Lungs, Anterior View**

(**a**) Gross anatomy of the lungs, showing the bronchial tree and its divisions. (**b**) Isolated left and right lungs have been colored to show the distribution of the bronchopulmonary segments. (**c**) Bronchogram of the right and left bronchial tree, slightly oblique, posteroanterior view. (**d**) Plastic cast of the adult bronchial tree. All of the branches in a given bronchiopulmonary segment have been painted the same color.

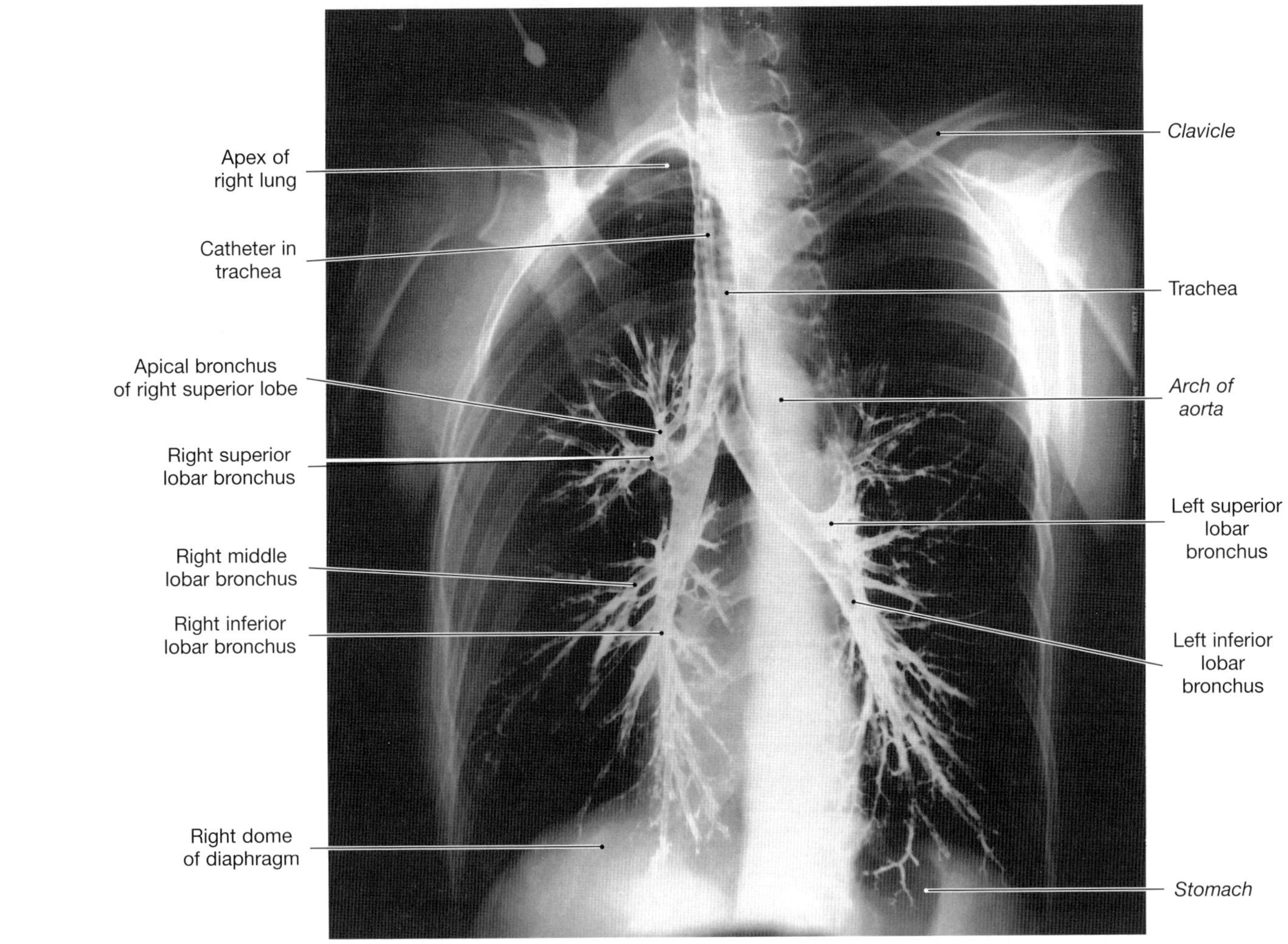
Apex of right lung
Catheter in trachea
Apical bronchus of right superior lobe
Right superior lobar bronchus
Right middle lobar bronchus
Right inferior lobar bronchus
Right dome of diaphragm
Clavicle
Trachea
Arch of aorta
Left superior lobar bronchus
Left inferior lobar bronchus
Stomach

(c) Bronchogram

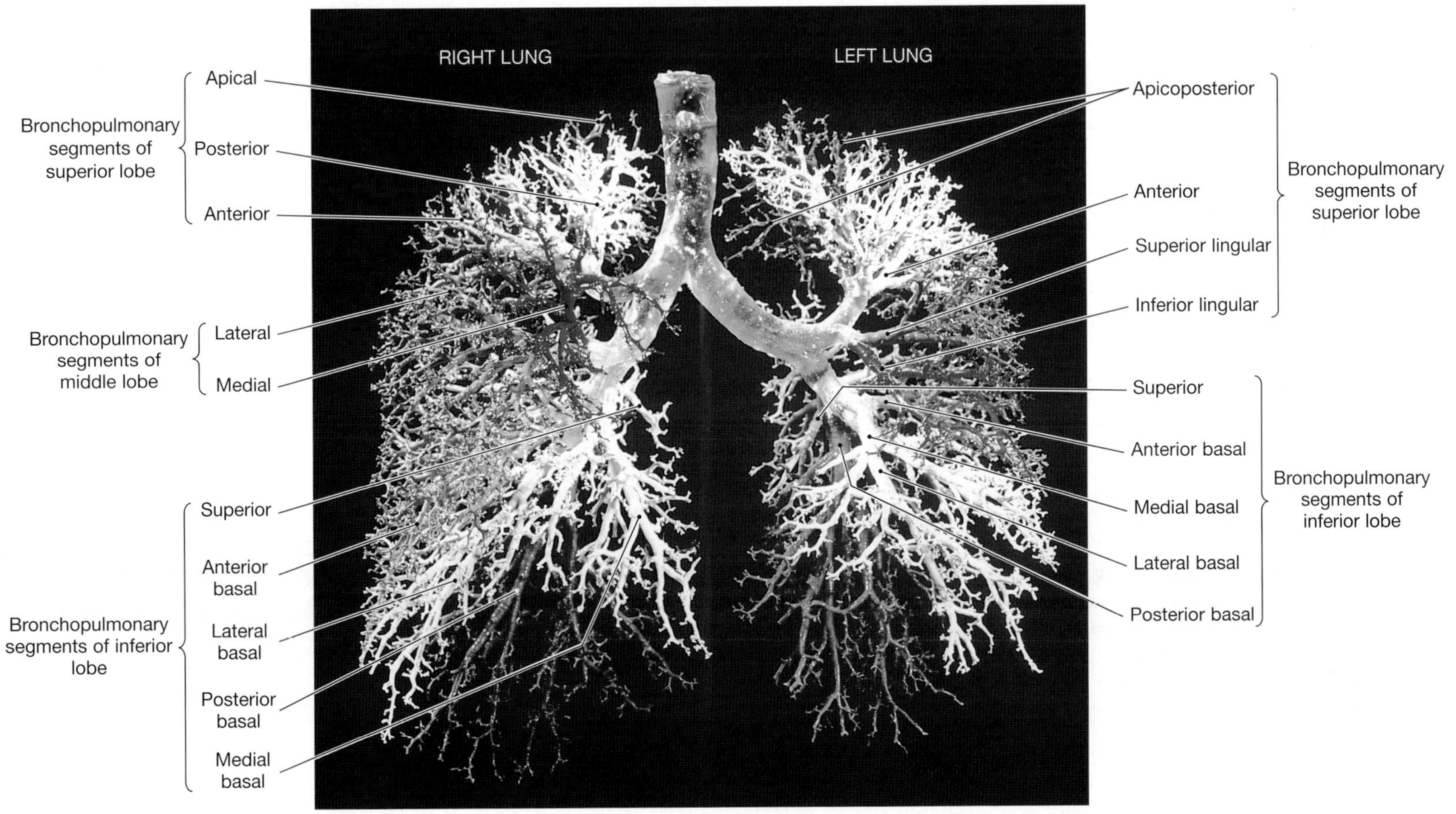
RIGHT LUNG
LEFT LUNG
Bronchopulmonary segments of superior lobe
Apical
Posterior
Anterior
Bronchopulmonary segments of middle lobe
Lateral
Medial
Bronchopulmonary segments of inferior lobe
Superior
Anterior basal
Lateral basal
Posterior basal
Medial basal
Apicoposterior
Anterior
Superior lingular
Inferior lingular
Bronchopulmonary segments of superior lobe
Superior
Anterior basal
Medial basal
Lateral basal
Posterior basal
Bronchopulmonary segments of inferior lobe

(d) The bronchial tree

FIGURE 24.10 *(continued)*

BRANCHES OF THE SECONDARY BRONCHI [FIGURE 24.10a,d]

The secondary bronchi in each lung divide to form tertiary bronchi. In the right lung three tertiary bronchi supply the superior lobe, two enter the middle lobe, and five supply the inferior lobe. The superior lobe of the left lung usually contains four tertiary bronchi, whereas the inferior lobe has five (Figure 24.10a,d●). The tertiary bronchi deliver air to the bronchopulmonary segments of the lungs.

THE BRONCHOPULMONARY SEGMENTS [FIGURE 24.10a,b,d]

Each lobe of the lung can be divided into smaller units called **bronchopulmonary segments.** Each segment consists of the lung tissue associated with a single tertiary bronchus. The bronchopulmonary segments have names that correspond to those of the associated tertiary bronchi (Figure 24.10a,b,d●). *Examining the Living Lung p. 804*

The Bronchioles [FIGURES 24.9/24.11b]

Each tertiary bronchus branches several times within the bronchopulmonary segment, ultimately giving rise to 6500 smaller **terminal bronchioles.** Terminal bronchioles have a lumenal diameter of 0.3–0.5 mm. The walls of terminal bronchioles, which lack cartilaginous supports, are dominated by smooth muscle tissue (Figures 24.9 and 24.11b●). The autonomic nervous system regulates the activity in the smooth muscle layer of the terminal bronchioles and thereby controls the diameter. Sympathetic activation leads to enlargement of the airway, or **bronchodilation.** Parasympathetic stimulation leads to **bronchoconstriction.** These changes alter the resistance to air-flow toward or away from the respiratory exchange surfaces. Tension in the smooth muscles often throws the bronchiolar mucosa into a series of folds, and excessive stimulation, as in *asthma*, can almost completely prevent airflow along the terminal bronchioles.

CLINICAL BRIEF

Asthma

Asthma (AZ-ma) affects an estimated 4–6% of the U.S. population. For unknown reasons, possibly environmental, more children have been getting asthma over the last 20 years. There are several different forms of asthma, but all are characterized by unusually sensitive and irritable conducting passageways. In many cases the trigger appears to be an immediate hypersensitivity reaction to an allergen in the inspired air. Drug reactions, air pollution, chronic respiratory infections, exercise, or emotional stress can also cause an asthmatic attack in sensitive individuals.

The most obvious and potentially dangerous symptoms include (1) the constriction of smooth muscles all along the bronchial tree, (2) edema and inflammation of the walls of the respiratory passageways, and (3) accelerated production of mucus. The combination makes breathing very difficult. Exhalation is affected more than inhalation; the narrowed passageways often collapse before exhalation is completed. Although mucus production increases, mucus transport slows, and fluids accumulate along the passageways. Coughing and wheezing then develop.

Severe asthmatic attacks reduce the functional capabilities of the respiratory system, and the peripheral tissues become oxygen-starved. This condition can prove fatal, and asthma fatalities have been increasing in recent years. The annual death rate from asthma in the United States is approximately four deaths per million population (for ages 5–34). Mortality among asthmatic blacks is higher than that of whites.

Each terminal bronchiole delivers air to a single pulmonary lobule. Within the lobule, the terminal bronchiole branches to form several **respiratory bronchioles.** These are the thinnest and most delicate branches of the bronchial tree, and they deliver air to the exchange surfaces of the lungs. The preliminary filtration and humidification of the incoming air are completed before air leaves the terminal bronchioles. The epithelial cells of the respiratory bronchioles and the smaller terminal bronchioles are cuboidal. Cilia are rare, and there are no goblet cells or underlying mucous glands.

CLINICAL BRIEF

Emphysema

Emphysema (em-fi-ZĒ-ma) is a chronic, progressive condition characterized by shortness of breath and an inability to tolerate physical exertion. The underlying problem is destruction of respiratory-exchange surfaces in the lungs. In essence, respiratory bronchioles and alveoli are functionally eliminated. The alveoli gradually expand, capillaries deteriorate, and gas exchange in the region comes to a halt.

Emphysema has been linked to the inhalation of air containing fine particulate matter or toxic vapors, such as those found in cigarette smoke. As the condition progresses, the reduction in exchange surface limits the ability to provide adequate oxygen. The condition is widespread. Some degree of emphysema is a normal consequence of aging, and an estimated 66% of adult males and 25% of adult females have detectable areas of emphysema in their lungs.

Clinical symptoms typically fail to appear unless the damage is extensive. Most people requiring treatment for emphysema are adult smokers. Unfortunately, the loss of alveoli and bronchioles in emphysema is permanent and irreversible. Further progression can be limited by cessation of smoking; a partially effective treatment for severe cases is administration of oxygen. Asthma medications and prevention of respiratory infections (flu and pneumonia vaccinations) may help as well.

Alveolar Ducts and Alveoli [FIGURES 24.11/24.12]

Respiratory bronchioles are connected to individual alveoli and to multiple alveoli along regions called **alveolar ducts.** These passageways end at **alveolar sacs,** common chambers connected to several individual alveoli (Figures 24.11 and 24.12a-c●). Each lung contains approximately 150 million alveoli, and their abundance gives the lung an open, spongy appearance. An extensive network of capillaries is associated with each alveolus (Figure 24.12b●); the capillaries are surrounded by a network of elastic fibers. This elastic tissue helps maintain the relative positions of the alveoli and respiratory bronchioles. Recoil of these fibers during expiration reduces the size of the alveoli and assists in the process of expiration.

THE ALVEOLUS AND THE RESPIRATORY MEMBRANE [FIGURE 24.12c,d]

The alveolar epithelium consists primarily of simple squamous epithelium (Figure 24.12c●). The squamous epithelial cells, called *Type I cells,* or *respiratory epitheliocytes,* are unusually thin and delicate. **Septal cells,** also called **surfactant** (sur-FAK-tant) **cells** or *Type II cells,* are scattered among the squamous cells. These large cells produce an oily secretion containing a mixture of phospholipids. This secretion, termed **surfactant,** coats the inner surface of each alveolus. Surfactant reduces surface tension in the fluid coating the alveolar surface; without surfactant, the alveoli would collapse. Roaming **alveolar macrophages** (*dust cells*) patrol the epithelium,

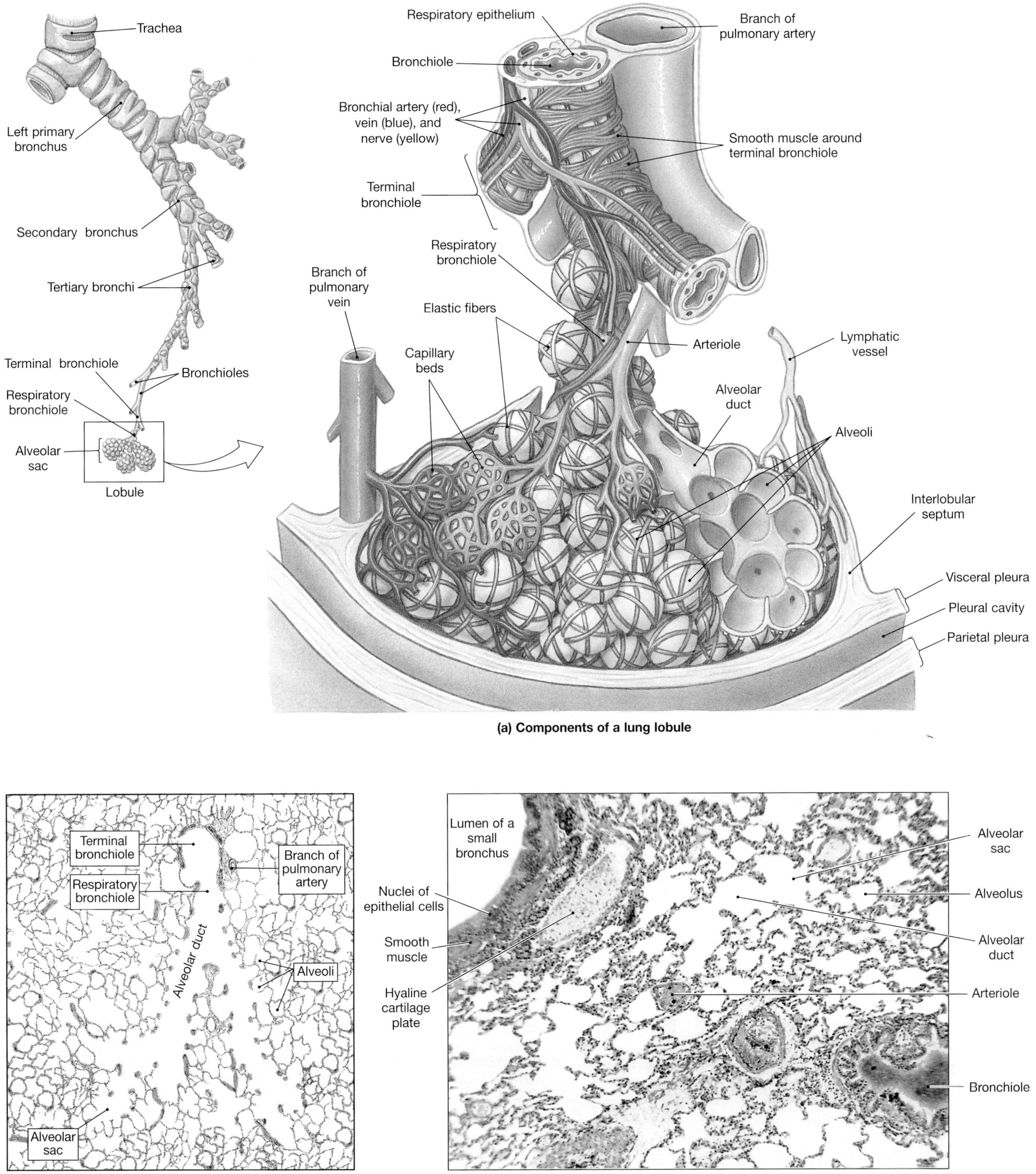

(a) Components of a lung lobule

(b) Lung tissue

(c) Lung (LM x 62)

FIGURE 24.11 **BRONCHI AND BRONCHIOLES**

(**a**) The structure of one portion of a single lobule. (**b**) Diagrammatic view of lung tissue. (**c**) Light micrograph of a section of the lung.

FIGURE 24.12 **Alveolar Organization**

(a) Basic structure of a lobule, cut to reveal the arrangement between the alveolar ducts and alveoli. A network of capillaries surrounds each alveolus. These capillaries are surrounded by elastic fibers. **(b)** SEM of lung tissue, showing the appearance and organization of the alveoli. **(c)** Diagrammatic sectional view of alveolar structure and the respiratory membrane. **(d)** The respiratory membrane.

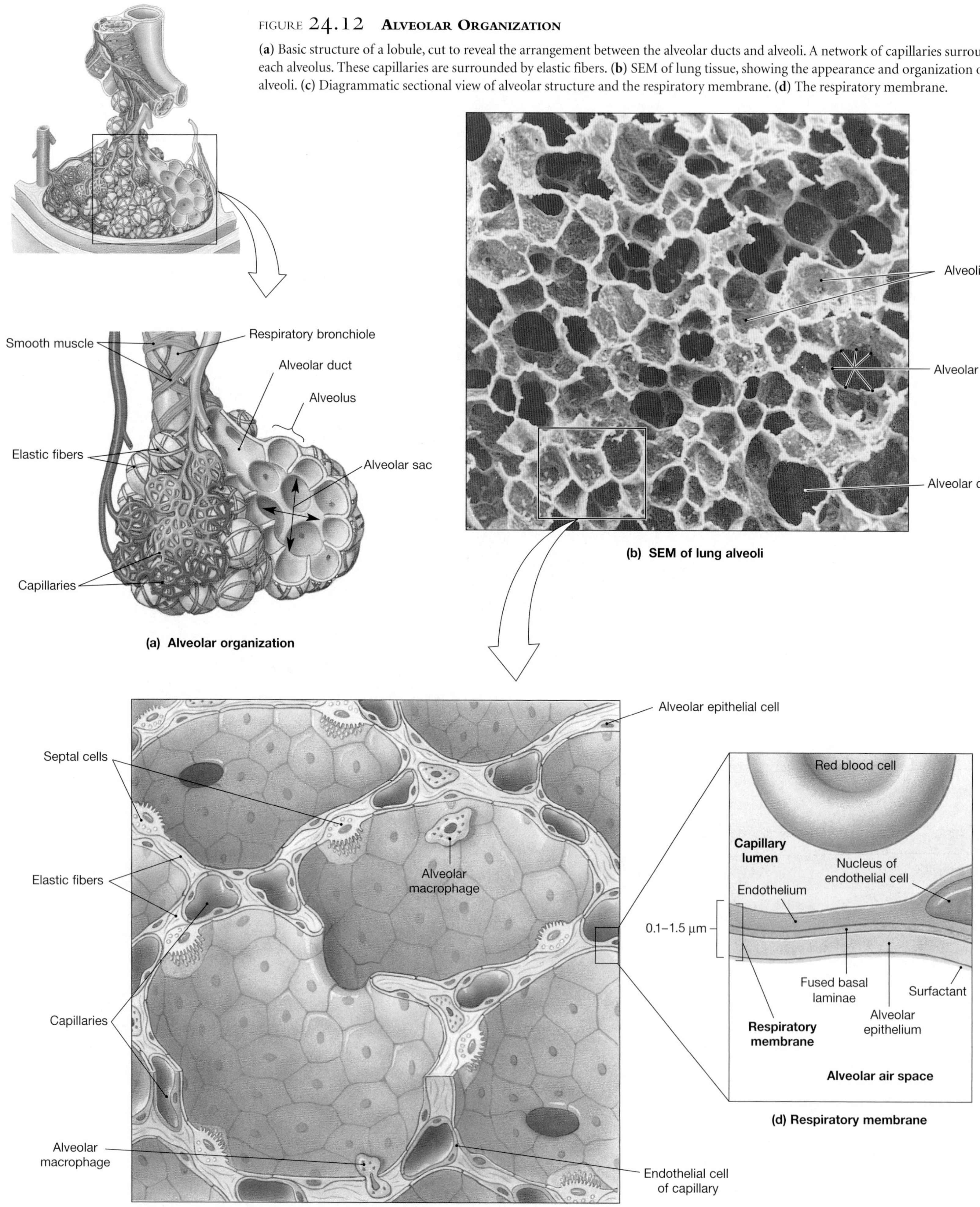

(a) Alveolar organization

(b) SEM of lung alveoli

(c) Alveoli, sectional view

(d) Respiratory membrane

phagocytizing any particulate matter that has eluded the respiratory defenses and reached the alveolar surfaces. *Respiratory Distress Syndrome p. 804*

Gas exchange occurs in areas where the *basal laminae*, or basement membranes, of the alveolar epithelium and adjacent capillaries have fused (Figure 24.12d●). In these areas the total distance separating the respiratory and circulatory systems can be as little as 0.1 μm. Diffusion across this **respiratory membrane** proceeds very rapidly, because (1) the distance is small and (2) the gases are lipid-soluble. The membranes of the epithelial and endothelial cells thus do not pose a barrier to the movement of oxygen and carbon dioxide between the blood and alveolar airspaces. *Tuberculosis p. 805* and *Pneumonia p. 805*

✓ CONCEPT CHECK

- Why do chronic smokers develop a hacking "smoker's" cough?
- Chronic bronchitis involves the overproduction of mucus. Over time, how could this affect respiration?
- Why are there almost no cilia and no goblet cells or mucous glands in the respiratory bronchiole?
- Why do the cells of the alveolus produce surfactant?

The Blood Supply to the Lungs

The respiratory-exchange surfaces receive blood from arteries of the pulmonary circuit. p. 580 The pulmonary arteries enter the lungs at the hilus and branch with the bronchi as they approach the lobules. Each lobule receives an arteriole and a venule, and a network of capillaries surrounds each alveolus directly beneath the respiratory membrane. In addition to providing a mechanism for gas exchange, the alveolar capillaries are the primary source of *angiotensin converting enzyme*, which converts circulating angiotensin I to angiotensin II, a hormone involved with the regulation of blood volume and blood pressure. p. 519

Blood from the alveolar capillaries passes through the pulmonary venules, and then enters the pulmonary veins that deliver it to the left atrium. The conducting portions of the respiratory tract receive blood from the *external carotid arteries* (nasal passages and larynx), the *thyrocervical trunk* (branches of the subclavian arteries that supply the inferior larynx and trachea), and the *bronchial arteries*. p. 583 The capillaries supplied by the bronchial arteries provide oxygen and nutrients to the conducting passageways of the lungs. The venous blood flows into the pulmonary veins, bypassing the rest of the systemic circuit and diluting the oxygenated blood leaving the alveoli. *Pulmonary Embolism p. 805*

The Pleural Cavities And Pleural Membranes [FIGURES 24.8a/24.13]

The thoracic cavity has the shape of a broad cone. Its walls are the rib cage, and the muscular diaphragm forms the floor. The two *pleural cavities* are separated by the mediastinum (Figure 24.13●). Each lung occupies a single pleural cavity, lined by a serous membrane, or **pleura** (PLOO-ra; pleural *pleurae*). The **parietal pleura** covers the inner surface of the thoracic wall and extends over the diaphragm and mediastinum. The **visceral pleura** covers the outer surfaces of the lungs, extending into the fissures between the lobes. Each **pleural cavity** actually represents a potential space rather than an open chamber, for the parietal and visceral layers are usually in close contact. A small amount of **pleural fluid** is secreted by both pleural membranes. Pleural fluid gives a moist, slippery coating that provides lubrication, thereby reducing friction between the parietal and visceral surfaces during breathing. *Pneumothorax p. 805*

Inflammation of the pleurae, a condition called **pleurisy**, may cause the membranes to produce and secrete excess amounts of pleural fluid, or the inflamed pleura may adhere to one another, limiting relative movement. In either form of this disorder, breathing becomes difficult, and prompt medical attention is required. *Thoracentesis p. 805*

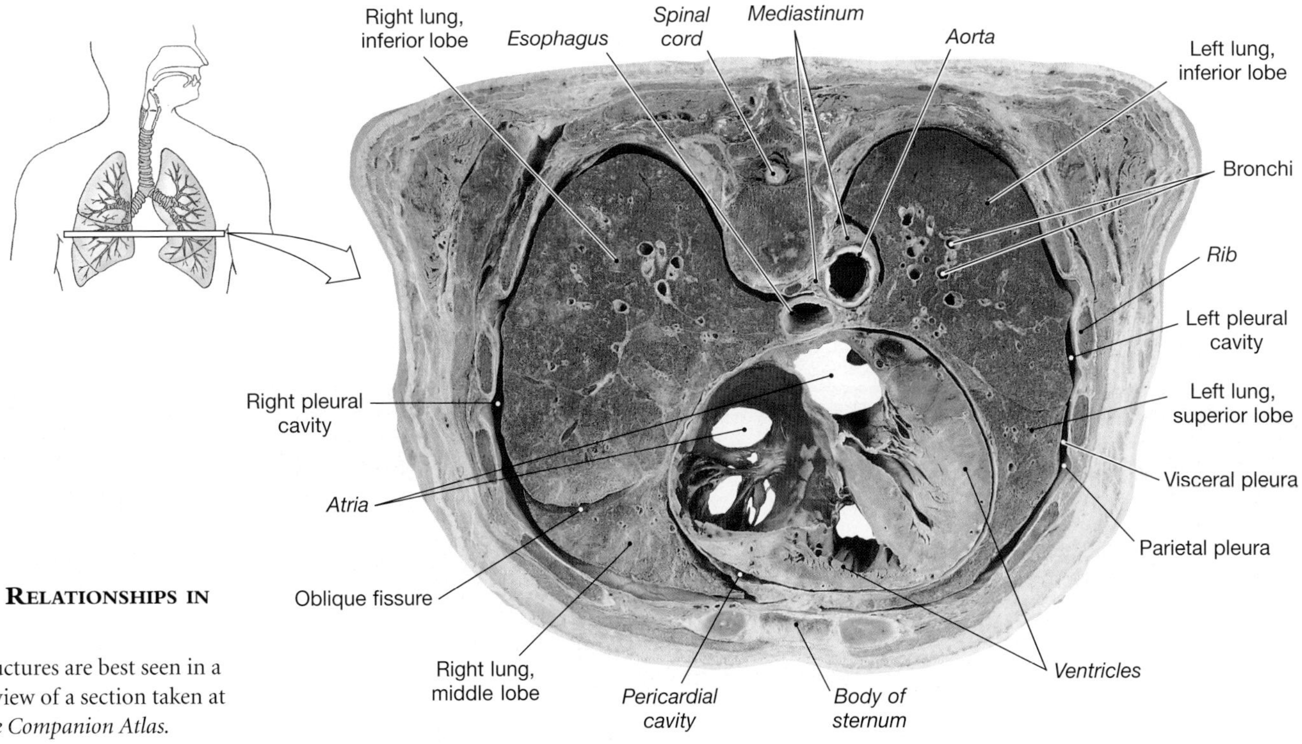

FIGURE 24.13 **ANATOMICAL RELATIONSHIPS IN THE THORACIC CAVITY**

The relationships among thoracic structures are best seen in a horizontal section. This is a superior view of a section taken at the level of T_8. *See MRI Scan 9b in the Companion Atlas.*

THE LUNGS

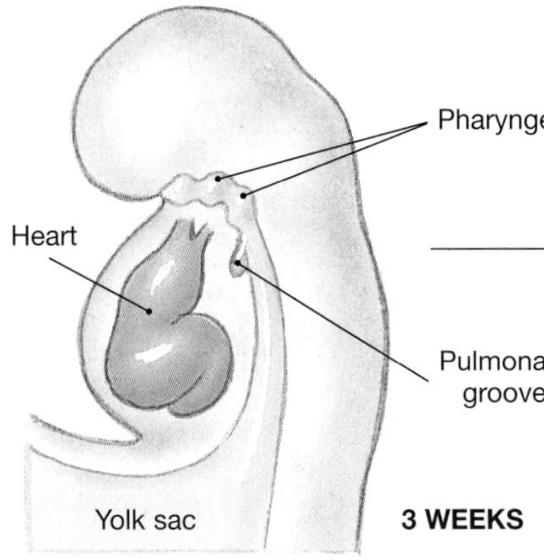

A shallow **pulmonary groove** appears in the midventral floor of the pharynx after roughly $3^{1}/_{2}$ weeks of development. This groove, which lies near the level of the last pharyngeal arch, gradually deepens.

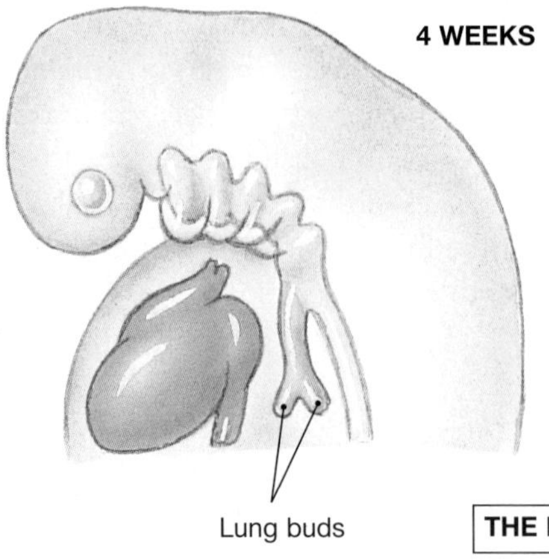

By week 4, the groove has become a blind pocket that extends caudally, anterior to the esophagus. This tube will become the trachea. At its tip, the tube branches, forming a pair of **lung buds**.

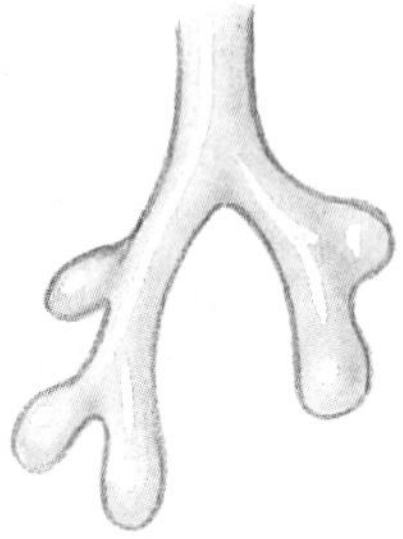

The lung buds continue to elongate and branch repeatedly.

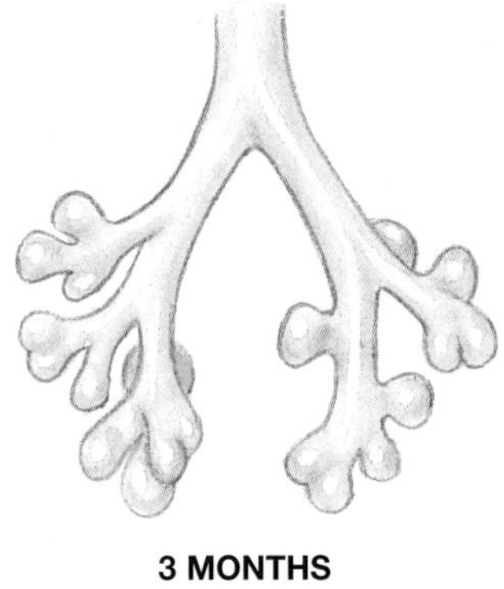

By the end of the sixth fetal month, there are around a million terminal branches, and the conducting passageways are complete to the level of the bronchioles.

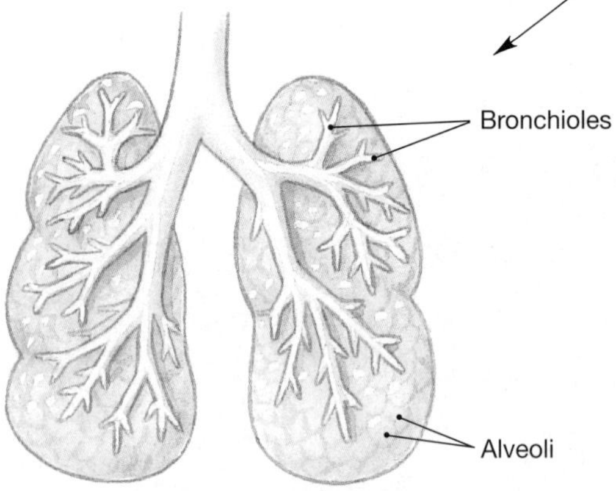

Over the next three months, each of the bronchioles gives rise to several hundred alveoli. This process continues for a variable period after birth.

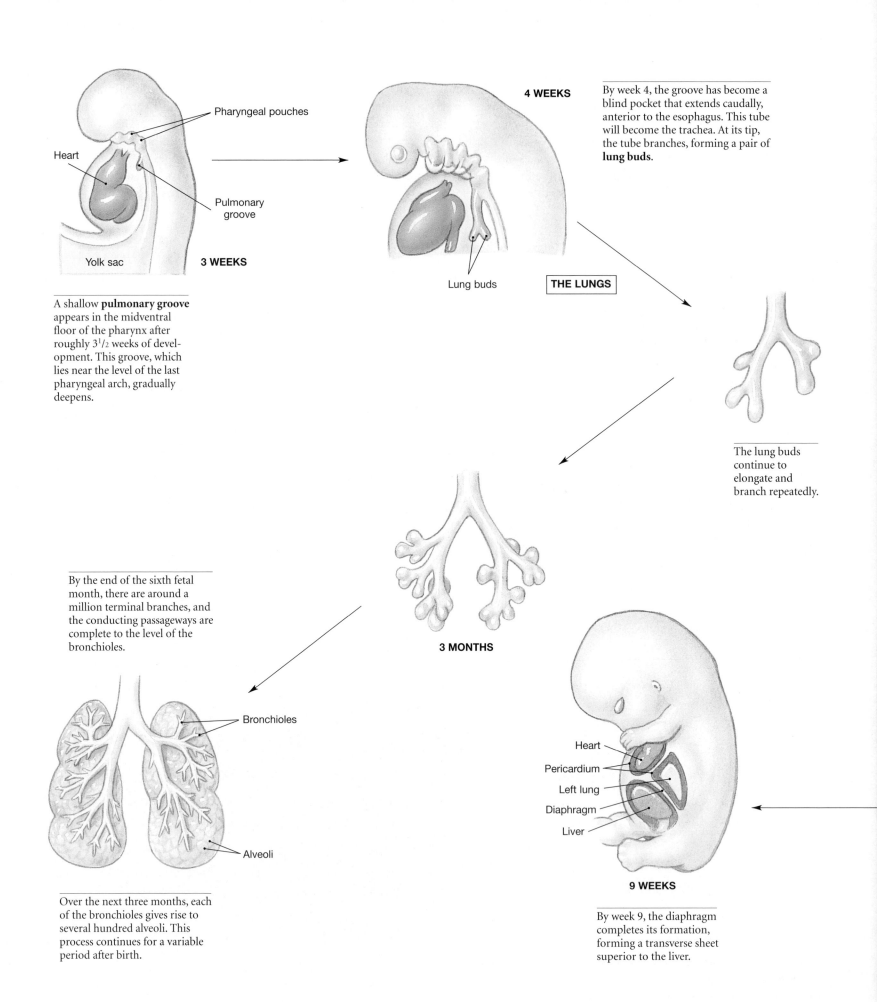

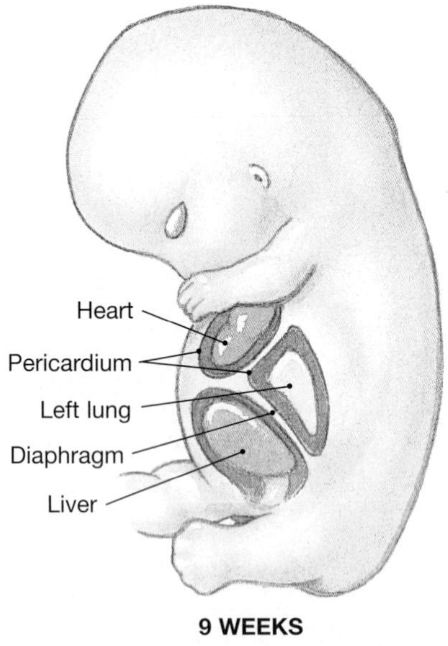

By week 9, the diaphragm completes its formation, forming a transverse sheet superior to the liver.

THE PLEURAL CAVITIES

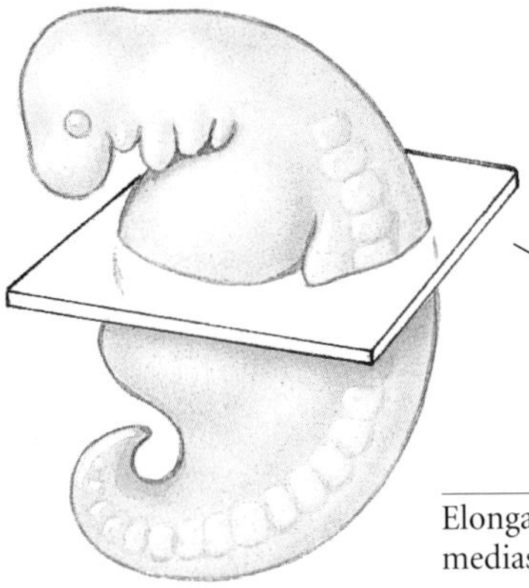

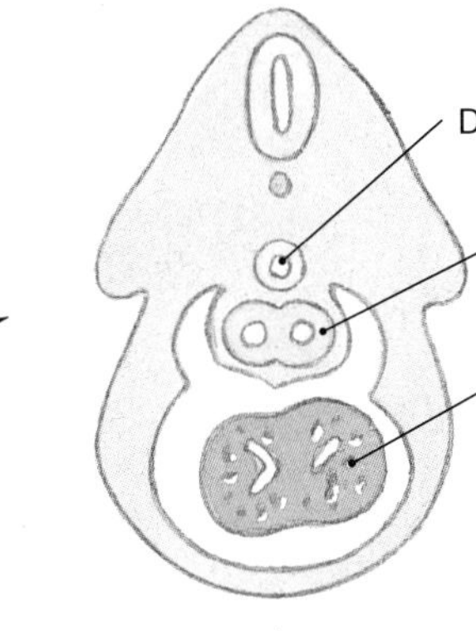

Elongation of the tube carries it into the mediastinum, and as the branching proceeds, the lung buds project into the ventral cavity dorsal to the developing heart.

4 WEEKS

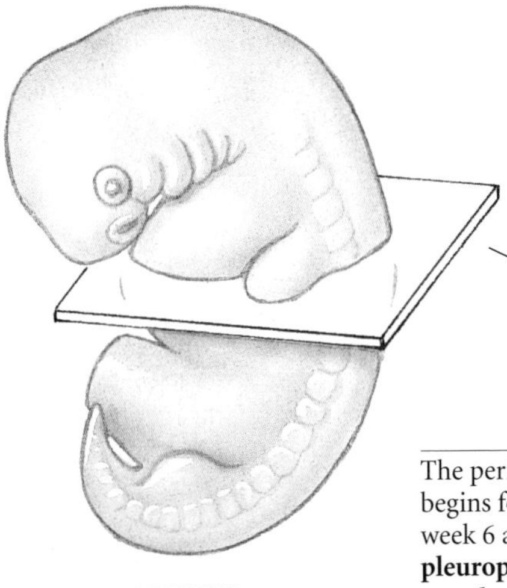

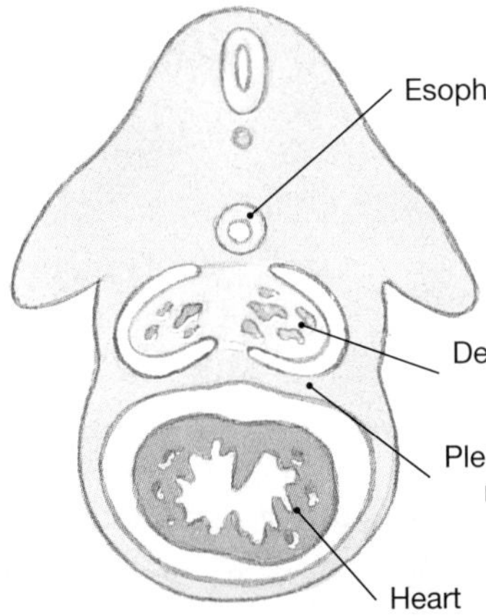

The pericardial sac begins forming in week 6 as a thin **pleuropericardial membrane** forms between the heart and the developing lungs.

6 WEEKS

By week 8, the pericardial sac is complete and the pericardial cavity is isolated from the rest of the ventral body cavity. The diaphragm then completes its formation, attaching to the pericardial sac and tissues of the mediastinum. This attachment separates the abdominopelvic cavity from the pleural cavities.

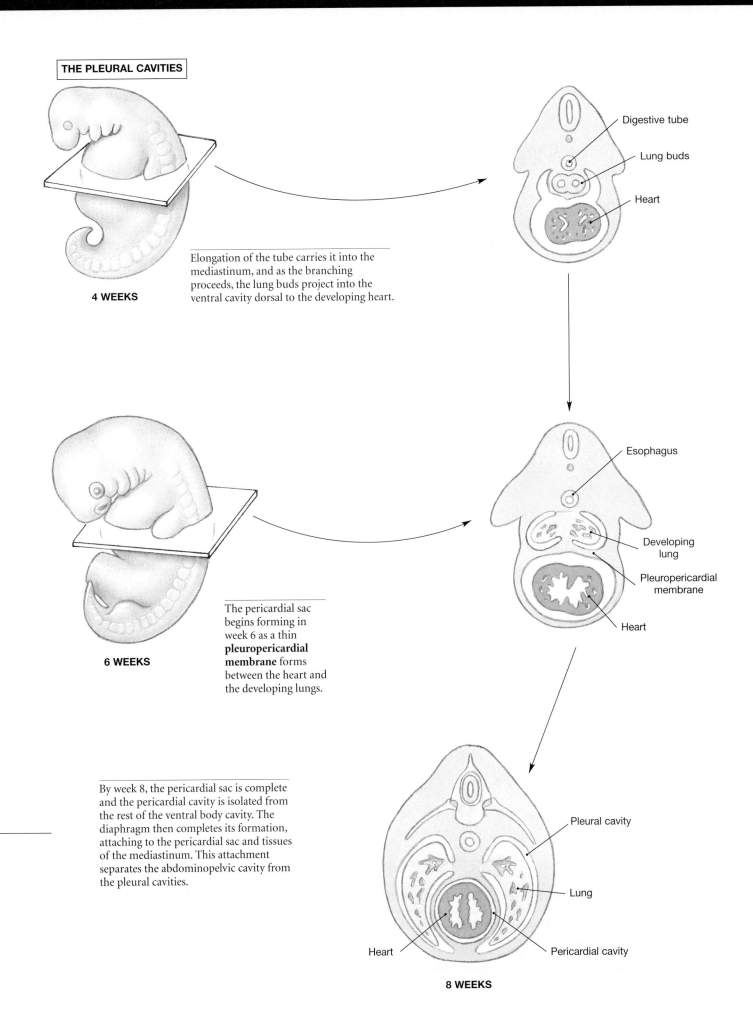

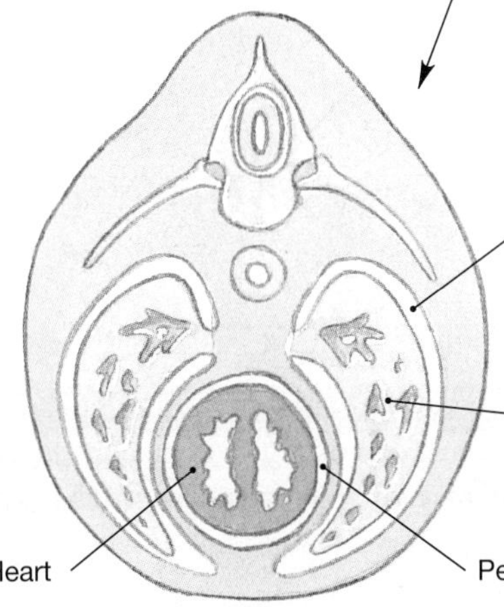

8 WEEKS

✓ CONCEPT CHECK

- If a pulmonary embolism involves a major pulmonary vessel, why can this cause heart failure?
- What is the function of the pleural fluid?
- What vessels supply the respiratory tract?

Respiratory Muscles And Pulmonary Ventilation

Pulmonary ventilation, or breathing, refers to the physical movement of air into and out of the bronchial tree. The function of pulmonary ventilation is to maintain adequate *alveolar ventilation*, the movement of air into and out of the alveoli. Alveolar ventilation prevents the buildup of carbon dioxide in the alveoli and ensures a continual supply of oxygen that keeps pace with absorption by the bloodstream.

Respiratory Muscles [FIGURE 24.14]

The skeletal muscles involved in respiratory movements were introduced in Chapters 10 and 11. p. 278 Of these, the most important are the **diaphragm** and the **external** and **internal intercostal muscles**. Contraction of the diaphragm increases the volume of the thoracic cavity by tensing and flattening its floor, and this increase draws air into the lungs. The external intercostal muscles may assist in inspiration by elevating the ribs. When the ribs are elevated they swing anteriorly, and the width of the thoracic cage increases along its anterior-posterior axis. The internal intercostal muscles depress the ribs and reduce the width of the thoracic cavity, thereby contributing to expiration. These muscles and their actions are diagrammed in Figure 24.14●.

The **accessory respiratory muscles** become active when the depth and frequency of respiration must be increased markedly. The sternocleidomastoid, serratus anterior, pectoralis minor, and scalene muscles assist the external intercostal muscles in elevating the ribs and performing inspiration. The transversus thoracis, oblique, and rectus abdominis muscles assist the internal intercostal muscles in expiration by compressing the abdominal contents, forcing the diaphragm upward, and further reducing the volume of the thoracic cavity.

RESPIRATORY MOVEMENTS

The respiratory muscles may be used in various combinations, depending on the volume of air that must be moved in or out of the lungs. Respiratory movements may be classified as *eupnea* or *hyperpnea* on the basis of whether expiration is passive or active.

EUPNEA In **eupnea** (ūp-NĒ-a), or **quiet breathing**, inspiration involves muscular contractions, but expiration is a passive process. During quiet breathing, expansion of the lungs stretches their elastic fibers. In addition, elevation of the rib cage stretches opposing skeletal muscles and elastic fibers in the connective tissues of the body wall. When the inspiratory muscles relax, these elastic structures contract, returning the diaphragm or rib cage or both to their original positions.

Eupnea may involve *diaphragmatic breathing* or *costal breathing*.

- During **diaphragmatic breathing**, or **deep breathing**, contraction of the diaphragm provides the necessary change in thoracic volume. Air is drawn into the lungs as the diaphragm contracts, and exhalation occurs when the diaphragm relaxes.
- In **costal breathing**, or **shallow breathing**, the thoracic volume changes because the rib cage changes shape. Inhalation occurs when contraction of the external intercostal muscles elevates the ribs and enlarges the thoracic cavity. Exhalation occurs when these muscles relax. During pregnancy women increasingly rely on costal breathing as the uterus enlarges and pushes the abdominal viscera against the diaphragm.

HYPERPNEA **Hyperpnea** (hī-perp-NĒ-a), or **forced breathing**, involves active inspiratory and expiratory movements. Forced breathing calls upon the accessory muscles to assist with inspiration, and expiration involves contraction of the transversus thoracis and internal intercostal muscles. When a person is breathing at absolute maximum levels, such as during vigorous exercise, the abdominal muscles are used in exhalation. Their contraction compresses the abdominal contents, pushing them up against the diaphragm and further reducing the volume of the thoracic cavity.

Respiratory Changes at Birth

There are several important differences between the respiratory system of a fetus and that of a newborn infant. Before delivery, pulmonary arterial resistance is high, because the pulmonary vessels are collapsed. The rib cage is compressed, and the lungs and conducting passageways contain only small amounts of fluid and no air. At birth the newborn infant takes a truly heroic first breath through powerful contractions of the diaphragmatic and external intercostal muscles. The inspired air enters the passageways with enough force to push the contained fluids out of the way and inflate the entire bronchial tree and most of the alveoli. The same drop in pressure that pulls air into the lungs pulls blood into the pulmonary circulation; the changes in blood flow that occur lead to the closure of the *foramen ovale*, the embryonic interatrial connection, and the *ductus arteriosus*, the fetal connection between the pulmonary trunk and the aorta. p. 605

The exhalation that follows fails to completely empty the lungs, for the rib cage does not return to its former, fully compressed state. Cartilages and connective tissues keep the conducting passageways open, and the surfactant covering the alveolar surfaces prevents their collapse. Subsequent breaths complete the inflation of the alveoli.

✓ CONCEPT CHECK

- John breaks a rib that punctures the thoracic cavity on his left side. Which structures are potentially damaged, and what do you predict will happen to the lung as a result?
- In emphysema, alveoli are replaced by large air spaces and elastic fibrous connective tissue. How do these changes affect the lungs?
- Elise just had a baby, and he takes his very first breath. Summarize the changes that occur in his body due to this first breath.

FIGURE 24.14 **RESPIRATORY MUSCLES**

(a) As the ribs are elevated or the diaphragm is depressed, the volume of the thoracic cavity increases and air moves into the lungs. The outward movement of the ribs as they are elevated resembles the outward swing of a raised bucket handle. **(b)** Anterior view at rest, with no air movement. **(c) Inhalation**, showing the primary and accessory respiratory muscles that elevate the ribs and flatten the diaphragm. **(d) Exhalation**, showing the primary and accessory respiratory muscles that depress the ribs and elevate the diaphragm.

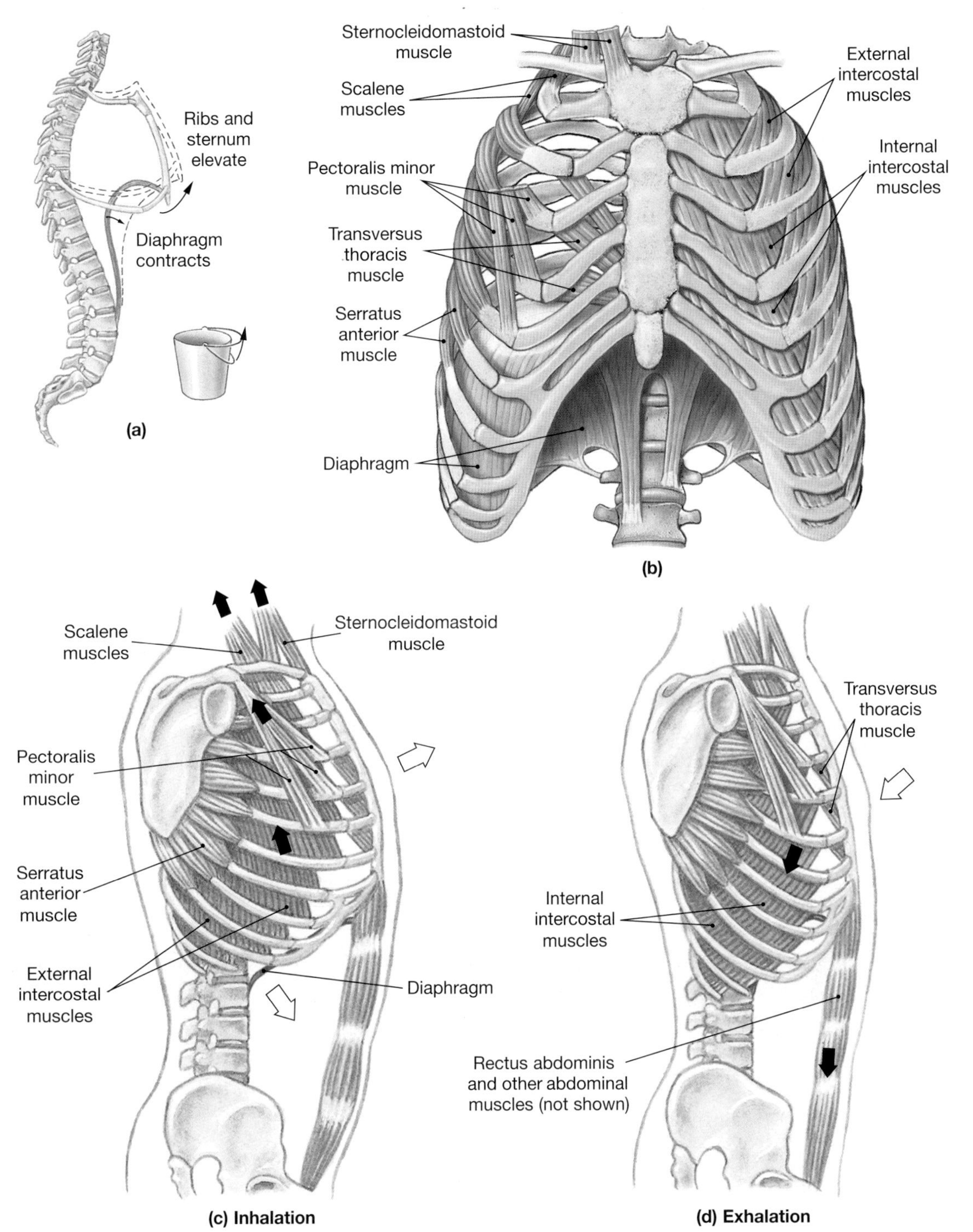

Respiratory Centers of the Brain

[FIGURE 24.15]

Under normal conditions cellular rates of oxygen absorption and carbon dioxide generation are matched by the capillary rates of delivery and removal. When adjustments by the cardiovascular and respiratory systems are needed to meet the body's ever changing demands for oxygen, these systems must be coordinated. The regulatory centers that integrate the responses by these systems are located in the pons and medulla oblongata.

The **respiratory centers** are three pairs of loosely organized nuclei in the reticular formation of the pons and medulla oblongata. p. 408 These nuclei regulate the activities of the respiratory muscles by adjusting the frequency and depth of pulmonary ventilation.

The **respiratory rhythmicity center** sets the basic pace and depth of respiration. It can be subdivided into a **dorsal respiratory group (DRG)** and a **ventral respiratory group (VRG)**. The dorsal respiratory group, or *inspiratory center*, controls motor neurons innervating the external intercostal muscles and the diaphragm. This group functions in every respiratory cycle, whether quiet or forced. The ventral respiratory group functions only during forced respiration. It innervates motor neurons controlling accessory muscles involved in active exhalation and maximal inhalation. The neurons involved with active exhalation are sometimes said to form an *expiratory center.*

The **apneustic** (ap-NŪ-stik) and **pneumotaxic** (nū-mō-TAKS-ik) **centers** of the pons are paired nuclei that adjust the output of the rhythmicity center, thereby modifying the pace of respiration. Their activities adjust the respiratory rate and the depth of respiration in response to sensory stimuli or instructions from higher centers. The locations of the apneustic, pneumotaxic, and respiratory rhythmicity centers are reviewed in Figure 24.15●.

Normal breathing occurs automatically, without conscious control. Three different reflexes are involved in the regulation of respiration: (1) *mechanoreceptor reflexes* that respond to changes in the volume of the lungs or to changes in arterial blood pressure, (2) *chemoreceptor reflexes* that respond to changes in the P_{CO_2}, pH, and P_{O_2} of the blood and cerebrospinal fluid, and (3) *protective reflexes* that respond to physical injury or irritation of the respiratory tract. p. 472

Higher centers influence respiration via inputs to the pneumotaxic center and by their direct influence on respiratory muscles. The higher centers involved are found in the cerebrum, especially the cerebral cortex, and in the hypothalamus. Although pyramidal output provides conscious control over the respiratory muscles, these muscles most often receive instructions via extrapyramidal pathways. In addition, the respiratory centers are embedded in the reticular formation, and almost every sensory and motor nucleus has some connection with this complex. As a result, emotional and autonomic activities often affect the pace and depth of respiration.

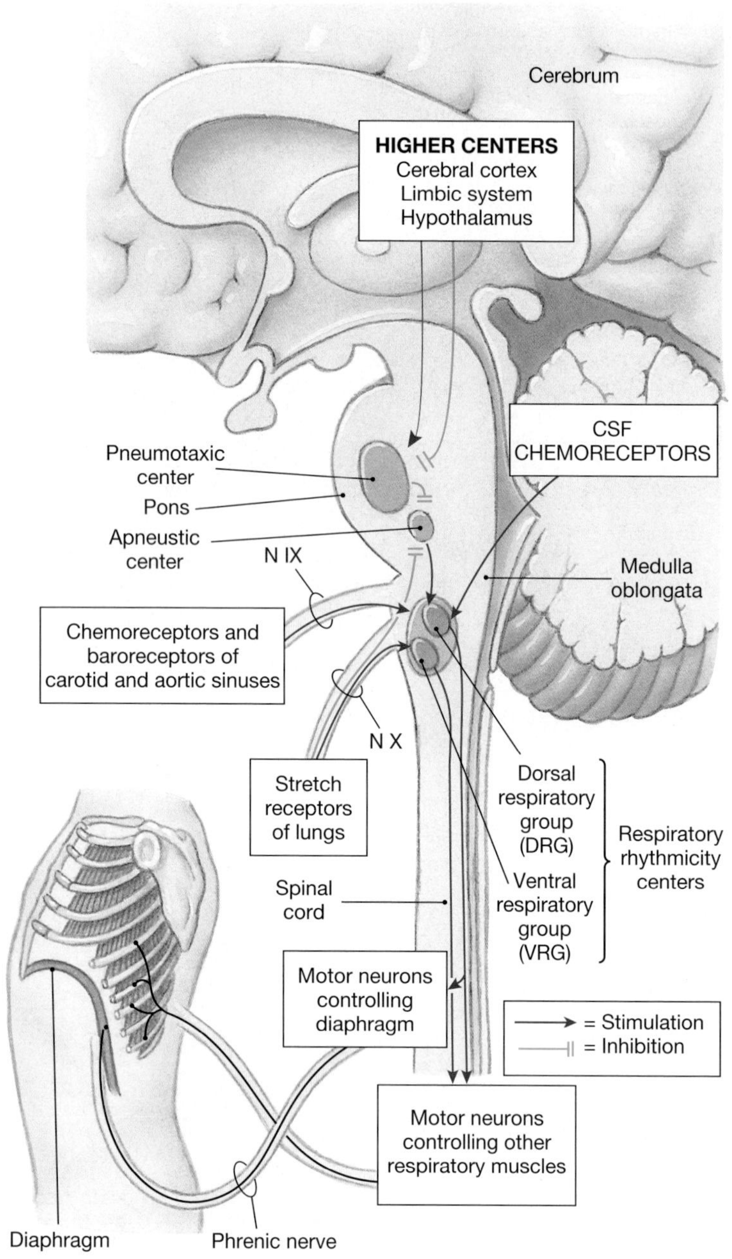

FIGURE 24.15 **Respiratory Centers and Reflex Controls**
The positions and relationships between the major respiratory centers and other factors important to respiratory control. Pathways for conscious control over respirating muscles are not shown.

CLINICAL BRIEF

Artificial Respiration and CPR

Artificial respiration provides air to an individual whose cardiovascular system continues to function normally but whose respiratory muscles are no longer operating. In **mouth-to-mouth resuscitation**, a rescuer provides ventilation by exhaling into the mouth or mouth and nose of the victim. After each breath, contact is broken to permit passive exhalation by the victim. Air provided in this way contains adequate oxygen to meet the needs of the victim. *Mechanical ventilation*, using machines, can support life for years following neurologic or muscular respiratory failure.

Cardiopulmonary resuscitation, or **CPR**, restores some circulation and ventilation to an individual whose heart has stopped beating. Compression applied to the rib cage over the sternum reduces the volume of the thoracic cavity, squeezing the heart and propelling blood into the aorta and pulmonary trunk. When the pressure is removed, the thorax expands, and blood moves into the great veins. Cycles of compression are interspersed with cycles of mouth-to-mouth breathing that maintain pulmonary ventilation. Basic CPR techniques can be mastered in about eight hours of intensive training, using special equipment. Training is available at minimal cost through charitable organizations such as the Red Cross and the American Heart Association.

Aging and the Respiratory System

Many factors interact to reduce the efficiency of the respiratory system in elderly individuals. Three examples are particularly noteworthy:

1. With increasing age, elastic tissue deteriorates throughout the body. This deterioration reduces the lungs' ability to inflate and deflate.
2. Movements of the rib cage are restricted by arthritic changes in the rib articulations and by decreased flexibility at the costal cartilages. In combination with the changes noted in (1), the stiffening and reduction in chest movement effectively limit the respiratory volume. This restriction contributes to the reduction in exercise performance and capabilities with increasing age.
3. Some degree of emphysema is normally found in individuals aged 50–70. On average, roughly 1 square foot of respiratory membrane is lost each year after age 30. However, the extent varies widely depending on the lifetime exposure to cigarette smoke and other respiratory irritants. *Lung Cancer, Smoking, and Diet p. 805*

RELATED CLINICAL TERMS

asthma (AZ-ma): A condition characterized by unusually sensitive, irritable, inflamed conducting passageways. *p. 648*

atelectasis (at-e-LEK-ta-sis): A partially or completely collapsed lung. *Pneumothorax p. 805*

bronchitis (brong-KĪ-tis): An inflammation of the bronchial lining. *Bronchitis p. 804*

cardiopulmonary resuscitation (CPR): Applying cycles of compression to the rib cage and mouth-to-mouth breathing to maintain circulatory and respiratory function. *p. 656*

cystic fibrosis (CF): A relatively common, lethal inherited disease in which mucous secretions in the lungs become too thick to be transported easily. *p. 637*

emphysema (em-fi-ZĒ-ma): A chronic, progressive condition characterized by shortness of breath and resulting from the destruction of respiratory exchange surfaces. *p. 648*

epistaxis (ep-i-STAK-sis): A nosebleed caused by trauma, infection, allergies, hypertension, or other factors. *Nosebleeds p. 804*

Heimlich (HĪM-lik) **maneuver:** A method of applying abdominal pressure to force the expulsion of foreign objects lodged in the trachea or larynx. *p. 642*

laryngitis (lar-in-JĪ-tis): Infection or inflammation of the larynx. *Disorders of the Larynx p. 804*

lung cancer (pleuropulmonary neoplasm): A class of aggressive malignancies originating in the bronchial passageways or alveoli. *Lung Cancer, Smoking, and Diet p. 805*

pleural effusion: An abnormal accumulation of fluid within the pleural cavities. *Thoracentesis p. 805*

pneumonia (nū-MŌ-nē-a): A condition caused by an infection of the lobules of the lung and characterized by a decline in respiratory function due to fluid leakage into the alveoli and/or swelling and constriction of the respiratory bronchioles. *Pneumonia p. 805*

pneumothorax (nū-mō-THŌ-raks): The entry of air into the pleural cavity. *Pneumothorax p. 805*

pulmonary embolism: Blockage of a pulmonary artery by a blood clot, fat mass, or air bubble. *Pulmonary Embolism p. 805*

respiratory distress syndrome (RDS): A condition resulting from inadequate surfactant production; characterized by collapse of the alveoli and an inability to maintain adequate levels of gas exchange at the lungs. *Respiratory Distress Syndrome (RDS) p. 804*

silicosis (sil-i-KŌ-sis), **asbestosis** (as-bes-TŌ-sis), and **anthracosis** (an-thra-KŌ-sis): Serious clinical conditions caused by the inhalation of dust or other particulate matter in sufficient quantities to overload the respiratory defense system, resulting in lung scarring and a reduction in respiratory function. *Overloading the Respiratory Defenses p. 804*

thoracentesis: Removal of a sample of pleural fluid for diagnostic evaluation. *Thoracentesis p. 805*

tracheostomy (trā-kē-OS-tō-mē): Insertion of a tube through an incision in the anterior tracheal wall to bypass a foreign body or damaged larynx. *p. 642*

tuberculosis (tū-ber-kū-LŌ-sis) **(TB):** Infection of the lungs by the bacteria *Mycobacterium tuberculosis*. Symptoms are variable but usually include coughing and chest pain, with fever, night sweats, fatigue, and weight loss. *Tuberculosis p. 805*

Additional Clinical Terms Discussed in Appendix I (pp. 804–806)

acute epiglottitis; adult respiratory distress syndrome (ARDS); bronchography; bronchoscope; bronchoscopy; chronic airways obstruction or chronic obstructive pulmonary disease (COPD); laryngotracheobronchitis or croup; nebulization; neonatal respiratory distress syndrome (NRDS) or hyaline membrane disease (HMD); positive end-expiratory pressure (PEEP); spontaneous pneumothorax

STUDY OUTLINE & CHAPTER REVIEW

Introduction 635

An Overview of the Respiratory System 635

1. The **respiratory system** includes the nose, nasal cavity and sinuses, pharynx, larynx, trachea, and conducting passageways leading to the exchange surfaces of the lungs. *(see Figure 24.1)*
2. The **upper respiratory system** includes the nose, nasal cavity, and pharynx. These structures begin the process of filtration and humidification of the incoming air. The **lower respiratory system** includes the larynx, trachea, bronchi, bronchioles, and **alveoli**.
3. The **respiratory tract** consists of a *conducting portion*, which extends from the entrance to the nasal cavity to the bronchioles; and a *respiratory portion*, which includes the *respiratory bronchioles* and the **alveoli**. *(see Figure 24.1)*

Functions of the Respiratory System 636

4. The respiratory system: (1) provides an area for gas exchange between air and circulating blood; (2) moves air to and from exchange surfaces; (3) protects respiratory surfaces, (4) defends the respiratory system and other tissues from pathogens, (5) permits vocal communication, and (6) helps regulate blood volume and pressure and body fluid pH.

The Respiratory Epithelium 636

5. The **respiratory epithelium** lines the conducting portions of the respiratory system down to the level of the smallest bronchioles.
6. The respiratory epithelium consists of a pseudostratified, ciliated, columnar epithelium with goblet cells. *(see Figure 24.2)*
7. The respiratory epithelium produces mucus that traps incoming particles. Underneath is the **lamina propria** (a layer of connective tissue); the combined respiratory epithelium and lamina propria form a **mucosa** (mucous membrane). *(see Figure 24.2)*
8. The **respiratory defense system** includes the *mucus escalator* (which washes particles toward the stomach), alveolar macrophages, hairs, and cilia.

The Upper Respiratory System 637

The Nose and Nasal Cavity 637

1. Air normally enters the respiratory system via the **external nares**, which open into the **nasal cavity**. The **vestibule** (entryway) of the nose is guarded by hairs that screen out large particles. *(see Figure 24.3)*
2. Incoming air flows through the **superior**, **middle**, or **inferior meatuses** (narrow grooves) and bounces off the conchal surfaces. *(see Figure 24.3b,d)*
3. The **hard palate** separates the oral and nasal cavities. The **soft palate** separates the superior *nasopharynx* from the oral cavity. The connections between the nasal cavity and nasopharynx represent the **internal nares**. *(see Figure 24.3)*

The Pharynx 637

4. The **pharynx** is a chamber shared by the digestive and respiratory systems. The **nasopharynx** is the superior part of the pharynx. The **oropharynx** is continuous with the oral cavity; the **laryngopharynx** includes the narrow zone between the hyoid and the entrance to the esophagus. *(see Figure 24.3c,d)*

The Lower Respiratory System 639

The Larynx 639

1. Inhaled air passes through the **glottis** en route to the lungs; the **larynx** surrounds and protects the glottis. The **epiglottis** projects into the pharynx. *(see Figures 24.3c,d/24.4/24.5)*
2. Two pairs of folds span the glottal opening; the relatively inelastic **vestibular folds** and the more delicate **vocal folds.** Air passing through the glottis vibrates the vocal folds and produces sound. *(see Figure 24.5)*
3. The **intrinsic laryngeal muscles** regulate tension in the vocal folds and open and close the glottis. The **extrinsic laryngeal musculature** positions and stabilizes the larynx. During swallowing, both sets of muscles help to prevent particles from entering the glottis. *(see Figure 24.6)*

The Trachea 642

1. The **trachea** ("windpipe") extends from the sixth cervical vertebra to the fifth thoracic vertebra. The **submucosa** contains C-shaped **tracheal cartilages** that stiffen the tracheal walls and protect the airway. The posterior tracheal wall can distort to permit large masses of food to move along the esophagus. *(see Figures 24.2a/24.3c/24.9)*

The Primary Bronchi 642

1. The trachea branches within the mediastinum to form the **right** and **left primary bronchi**. The primary bronchi and their branches form the *bronchial tree*. Each bronchus enters a lung at the **hilus**. The **root** of the lung is a connective tissue mass including the bronchus, pulmonary vessels, and nerves. *(see Figures 24.7/24.8)*

The Lungs 645

Lobes of the Lungs 645

1. The **lobes** of the lungs are separated by fissures; the **right lung** has three lobes and three secondary bronchi: **superior lobar**, **middle lobar**, and **inferior lobar bronchi**. The **left lung** has two lobes and two secondary bronchi: **superior lobar** and **inferior lobar bronchi.** *(see Figure 24.8)*

Lung Surfaces 645

2. The anterior portion of the lung, termed the **costal surface**, follows the inner contours of the rib cage. The **mediastinal surface** contains a hilus, and the left lung bears the **cardiac notch** *(see Figure 24.8)*

3. The connective tissues of the root extend into the **parenchyma** of the lung as a series of **trabeculae** (partitions). These branches form **septa** that divide the lung into **lobules**. *(see Figure 24.9)*

The Pulmonary Bronchi 645

4. *Extrapulmonary bronchi* (**left** and **right primary bronchi**) are outside of the lung tissue. *Intrapulmonary bronchi* (branches within the lung) are surrounded by bands of smooth muscle. *(see Figures 24.7/24.9/24.10/24.11)*

The Bronchioles 648

5. Each lung is further divided into smaller units called **bronchopulmonary segments**. These segments are named according to the associated tertiary bronchi. The right lung contains 10 and the left lung usually contains 8–9 bronchopulmonary segments. *(see Figures 24.7/24.9/24.10)*
6. Within the bronchopulmonary segments each tertiary bronchus ultimately gives rise to 50–80 **terminal bronchioles** that supply individual lobules. *(see Figures 24.9/24.11)*

Alveolar Ducts and Alveoli 648

7. The **respiratory bronchioles** open into **alveolar ducts**; many alveoli are interconnected at each duct. *(see Figures 24.11/24.12)*
8. The **respiratory membrane** (alveolar lining) consists of a simple squamous epithelium of *Type I cells*; **septal cells** (*Type II cells*) scattered in it produce an oily secretion (**surfactant**) that keeps the alveoli from collapsing. **Alveolar macrophages** (*dust cells*) patrol the epithelium and engulf foreign particles. *(see Figure 24.12)*

The Blood Supply to the Lungs 651

9. The respiratory exchange surfaces are extensively connected to the circulatory system via the vessels of the pulmonary circuit.

The Pleural Cavities And Pleural Membranes 651

1. Each lung occupies a single **pleural cavity** lined by a **pleura** (serous membrane). The two types of pleurae are the **parietal pleura**, covering the inner surface of the thoracic wall, and the **visceral pleura**, covering the lungs. **Pleural fluid** is secreted by both pleural membranes. *(see Figure 24.13)*

Respiratory Muscles And Pulmonary Ventilation 654

Respiratory Muscles 654

1. **Pulmonary ventilation** is the movement of air into and out of the lungs. The most important respiratory muscles are the **diaphragm** and the **external** and **internal intercostal muscles**. Contraction of the diaphragm increases the volume of the thoracic cavity; the external intercostals may assist in inspiration by elevating the ribs; the internal intercostals depress the ribs and reduce the width of the thoracic cavity, thereby contributing to expiration. The **accessory respiratory muscles** become active when the depth and frequency of respiration must be increased markedly. Accessory muscles include the sternocleidomastoid, serratus anterior, transversus thoracis, scalene, pectoralis minor, oblique, and rectus abdominis muscles. *(see Figure 24.14)*

Respiratory Changes at Birth 654

2. Before delivery the fetal lungs are fluid-filled and collapsed. At the first breath, the lungs inflate and never collapse completely thereafter.

Respiratory Centers of the Brain 655

3. The **respiratory centers** include three pairs of nuclei in the reticular formation of the pons and medulla. The **respiratory rhythmicity center** sets the pace for respiration. The **apneustic center** causes strong, sustained inspiratory movements, and the **pneumotaxic center** inhibits the apneustic center and the inspiratory center in the medulla oblongata. *(see Figure 24.15)*
4. Three different reflexes are involved in the regulation of respiration: (1) *Mechanoreceptor reflexes* respond to changes in the volume of the lungs or to changes in arterial blood pressure; (2) *chemoreceptor reflexes* respond to changes in the Pco_2 pH, and Po_2 of the blood and cerebrospinal fluid; and (3) *protective reflexes* respond to physical injury or irritation of the respiratory tract. *(see Figure 24.15)*
5. Conscious and unconscious thought processes can also control respiratory activity by affecting the respiratory centers or controlling the respiratory muscles.

Aging And The Respiratory System 656

1. The respiratory system is generally less efficient in the elderly because: (1) elastic tissue deteriorates, lowering the vital capacity of the lungs; (2) movements of the chest cage are restricted by arthritic changes and decreased flexibility of costal cartilages; and (3) some degree of emphysema is normal in the elderly.

LEVEL 1 REVIEWING FACTS AND TERMS

Match each numbered item with the most closely related lettered item. Use letters for answers in the spaces provided.

Column A	Column B
____ 1. extrinsic laryngeal musculature	a. superior portion of the pharynx
____ 2. tertiary bronchi	b. bridge of the nose
____ 3. nasopharynx	c. nose, nasal cavity, paranasal sinuses
____ 4. vestibular folds	d. windpipe
____ 5. lower respiratory system	e. lobar bronchi
____ 6. trachea	f. surrounds and protects the glottis
____ 7. dorsum nasi	g. larynx, trachea, bronchi, lungs
____ 8. secondary bronchi	h. provide protection for the vocal folds
____ 9. upper respiratory tract	i. positions and stabilizes the larynx
____10. larynx	j. segmental bronchi

11. Air entering the body is filtered, warmed, and humidified by the
(a) upper respiratory tract (b) lower respiratory tract
(c) lungs (d) alveoli

12. Surfactant
(a) protects the surfaces of the lungs
(b) phagocytizes small particulates
(c) replaces mucus in the alveoli
(d) helps prevent the alveoli from collapsing

13. The portion of the pharynx that receives both air and food is the
(a) nasopharynx (b) oropharynx
(c) vestibule (d) laryngopharynx

14. The portion of the nasal cavity contained within the flexible tissues of the external nose is the
(a) nasopharynx (b) vestibule
(c) internal chamber (d) glottis

15. The ______ is lined by squamous epithelium.
(a) nasopharynx (b) oropharynx
(c) laryngopharynx (d) larynx

16. The larynx is composed of ______ cartilages.
(a) 2 (b) 3
(c) 6 (d) 9

17. The cartilage that serves as a base for the larynx is the
    (a) thyroid cartilage
    (b) cuneiform cartilage
    (c) corniculate cartilage
    (d) cricoid cartilage
18. The vocal folds are located in the
    (a) nasopharynx (b) oropharynx
    (c) larynx (d) trachea
19. The trachea
    (a) is lined by simple squamous epithelium
    (b) is reinforced with C-shaped cartilages
    (c) contains no mucous glands
    (d) does not alter its diameter
20. The septa divide the lungs into
    (a) lobes (b) lobules
    (c) alveoli (d) bronchi

LEVEL 2 REVIEWING CONCEPTS

1. Which of the following would have a higher-pitched voice?
   (a) person with thick vocal folds (b) person with thin vocal folds
2. At birth, an infant takes its first breaths, and
   (a) the pressure in the lungs decreases
   (b) the resistance in the pulmonary arteries increases
   (c) changes in blood flow cause the foramen ovale to open
   (d) very little air enters the alveoli
3. Damage to the septal cells of the lungs would result in
   (a) respiratory membrane thickening (b) increased gas exchange
   (c) alveolar rupture (d) alveolar collapse
4. How many lobes does the right lung have?
5. What is bronchodilation?
6. What is the function of the alveolar macrophages?
7. What is the function of the septa?
8. What do the intrinsic laryngeal muscles do?
9. Which paired laryngeal cartilages are involved with the opening and closing of the glottis?
10. What portion of the pharynx does the laryngopharynx include?

LEVEL 3 CRITICAL THINKING AND CLINICAL APPLICATIONS

1. During winter, Mitch sleeps in a dorm room that lacks any humidifier for the heated air. In the mornings he notices that his nose is "stuffy" similar to when he has a cold. After showering and drinking some water, the stuffiness disappears until the next morning. What might be the cause of Mitch's nasal condition?
2. A newborn infant was found dead, abandoned by the road. Among the many questions that the police would like to have answered is whether the infant was born dead or alive. After an autopsy the medical examiner tells them that the infant was dead at birth. How could the medical examiner determine this?

✓ ANSWERS TO CONCEPT CHECK QUESTIONS

p. 637 **1.** The upper respiratory system warms and humidifies the air, which protects the lower respiratory system from extreme temperatures and dry conditions. Breathing through the mouth eliminates much of the filtration, humidification, and warming that must occur before the air reaches the sensitive tissues of the lungs. **2.** The mucus escalator is present in the lower regions of the respiratory tract, where the cilia beat toward the pharynx, cleaning the respiratory passageways. **3.** The conchae cause turbulence in the inspired air. This slows air movement and brings the air into contact with the moist, warm walls of the nasal cavity. Turbulent air-flow is essential to the filtration, humidification, and warming of air; these processes protect more delicate regions of the lower respiratory system. If the nasal cavity were a tubular passageway with straight walls, turbulence would be minimal. **4.** This would affect the mucus escalator, which cannot transport such dense mucus. Respiratory passageways would become clogged and reduce the diameter of the airways, leading to infection.
p. 641 **1.** The thyroid cartilage protects the glottis and the opening to the trachea. **2.** During swallowing, the epiglottis folds over the glottis, preventing food or liquids from entering the respiratory passageways. **3.** The pitch of her voice is getting lower. **4.** The glottis could not close without the intrinsic laryngeal muscles, so food could enter the respiratory passageways.
p. 643 **1.** The tracheal cartilages are C-shaped to allow room for esophageal expansion when large portions of food or liquid are swallowed. **2.** The trachea has a typical respiratory epithelium, which is pseudostratified, ciliated, columnar epithelial cells. **3.** Tracheal cartilages prevent the overexpansion or collapse of the airways during respiration, thereby keeping the airway open and functional. **4.** The right primary bronchus has a larger diameter, and it extends toward the lung at a steeper angle.
p. 651 **1.** Chronic smoking damages the lining of the air passageways. Cilia are seared off the surface of the cells by the heat, and the large number of particles that escape filtering are trapped in the excess mucus that is secreted to protect the irritated lining. This combination of circumstances creates a situation in which there is a large amount of thick mucus that is difficult to clear from the passages. The cough reflex is an attempt to remove this material from the airways. **2.** The overproduction of mucus can lead to the obstruction of smaller airways, causing a decrease in respiratory efficiency. **3.** Filtration and humidification are complete by the time air reaches this point, so the need for those structures is eliminated. **4.** The surfactant coats the inner surface of each alveolus and helps to reduce surface tension and avoid the collapse of the alveoli.
p. 654 **1.** This would place a great strain on the right ventricle, as it continues to try to force blood through the blocked vessel. Over time, the strain can cause heart failure. **2.** Pleural fluid provides lubrication between the parietal and visceral surfaces during breathing. **3.** The respiratory tract receives blood from the external carotid arteries, the thyrocervical trunk, and the bronchial arteries.
p. 654 **1.** Since the rib penetrates the chest wall, the thoracic cavity will be damaged, as well as the inner pleura. Atmospheric air will then enter the pleural cavity. This space is normally at a lower pressure than the outside air, so when the air enters, the natural elasticity of the lungs will not be compensated and the lung will collapse. This condition is called a pneumothorax. **2.** As a result of emphysema, the larger air spaces and lack of elasticity will reduce the efficiency of capillary exchange and pulmonary ventilation. **3.** During a baby's first breath, air is forced into the lungs due to the change in pressure. Fluids are pushed out of the way of the conducting passageways, and the alveoli immediately inflate with air. Pulmonary circulation becomes activated, and this closes the foramen ovale and the ductus arteriosus.

REVIEW IT

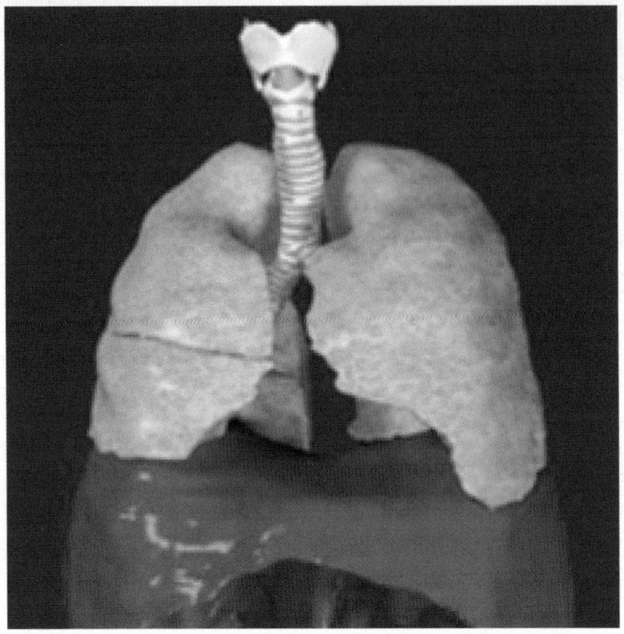

Challenge 1

Test your understanding of the respiratory tract and associated structures by labeling every respiratory structure you see on this illustration. This was taken from the *Interactive CD/Animations/Respiratory system*. It is tilted slightly to left of center. Compare this illustration to the figures found in your text. Which one most closely matches this view? Are all the structures present in the text figure visible on the CD reconstruction?

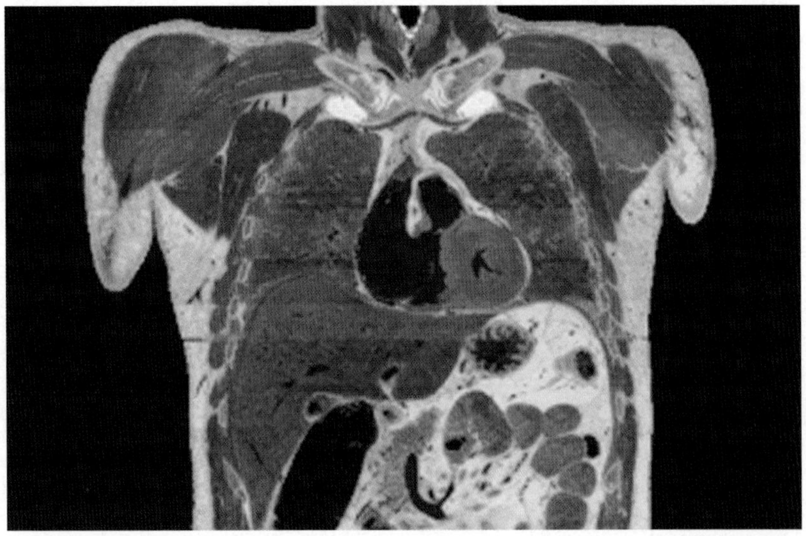

Challenge 2

In the previous media spread, you were challenged to identify structures from the thoracic cavity based on a transverse section from the Visible Human data set. Here you see a reconstruction of the same region of the body in coronal section. What structures are visible to you here? Compare this to Figure 24.1●, p. 635 for orientation purposes. Identify the structures of the respiratory system including trachea, larynx, left and right lungs, (superior and inferior lobes), bronchioles, pulmonary arteries and veins, pleural cavity, and diaphragm. What other structures are visible to you in this view? You should be able to identify every structure you see here!

Images provided by the Digital Cadaver™ project, courtesy of Visible Productions, Inc.

APPLY IT

Below are two separate exercises that provide more information on the preceding topics. These exercises differ from the usual critical linking exercises found in the media spreads in that these are designed to help you understand the interactions between systems, rather than hone your knowledge of one particular system. As before, each activity is designed to take approximately 10 minutes, and will help you gain a better appreciation for the integration of the material presented in this chapter.

Critical Linking 1

"Air goes in and out, blood goes 'round and 'round." As you know, the cardiovascular system and the respiratory system are slightly more complicated than that, however the description does indicate the simplistic view of their activities. Perhaps not quite so obvious is the interplay between these two. As heart rate increases, respirations increase to keep pace. This means that as you increase your level of exercise, you must increase both the rate of blood flow to the muscles and the rate of oxygen intake to meet the demands of active muscle. Is there a limit to this dual increase? Yes, that limit is referred to as the VO_2 max, and is used to calculate aerobic efficiency and fitness. This is the maximum amount of oxygen you can take in and use measured in ml per kilogram per minute.

To understand this interplay, and to get a rough idea of your personal VO_2 max, go to the Companion Website and click on the keyword "VO_2max". Read the information presented and then click on the link to the endurance test page under "measuring VO_2 max." Perform the Rockport test and obtain your score. Compare your score to others in your class, making predictions first as to who will have the highest VO_2 max, and who will score in the average range. Include in your predictions your reasons for your predictions.

Critical Linking 2

Chronic Obstructive Pulmonary Disease (COPD) is the medical term used to describe a series of conditions characterized by abnormalities in the lungs that make exhalation difficult. The two diseases most often associated with COPD are emphysema and chronic bronchitis. Both cause excessive inflammation of the respiratory structures, leading to problems with air-flow. The cardiovascular system is integrally tied to the respiratory system, so that diseases such as this affect both.

Think about this interaction in light of the effects of COPD. Using what you know of these two systems, prepare a summary of possible side effects from these two diseases on the cardiovascular system. To check your answers, and perhaps uncover new relationships between the two, visit the Companion Website and click the key word "COPD." Upgrade your summary sheet to include any new findings.

FURTHER STUDIES

Poetry it is said, is the language of the heart. There are innumerable references to the heart in this branch of literary work. Why is that? Perhaps the relationship between heart activity and muscular or emotional activity spurred this link. We are all aware that our heart beats faster when we are excited, nervous or active. This beating can be interpreted as the heart's being the center of our emotional life, even though anatomists will argue the limbic system of the brain is ultimately responsible for emotional thought. Anatomical fact aside, poets have written "from the heart" for ages.

Tour the poetry of the previous centuries by visiting *http://poetry.about.com/mbody.htm* and choosing an era. You will find references to the heart in all the categories listed here. Contemporary poets write of the heart just as often as their ancient predecessors. The latest collection of contemporary poetry can be found at either *http://poetry.about.com/cs/contemporarypoets/index.htm* or *http://www.cprw.com/*. Occasionally anatomy has inspired poetry! The Franklin Institute in Philadelphia has a large scientifically correct heart display that allows participants to walk through taking the same path as blood cells. This has inspired a collection of poems entitled "The Poetry of the Heart" published on line at *http://sln.fi.edu/biosci/history/database.poems.html*. A short perusal of poetry may just provide the brief intellectual diversion every anatomist should take in order to fully appreciate their field of study!

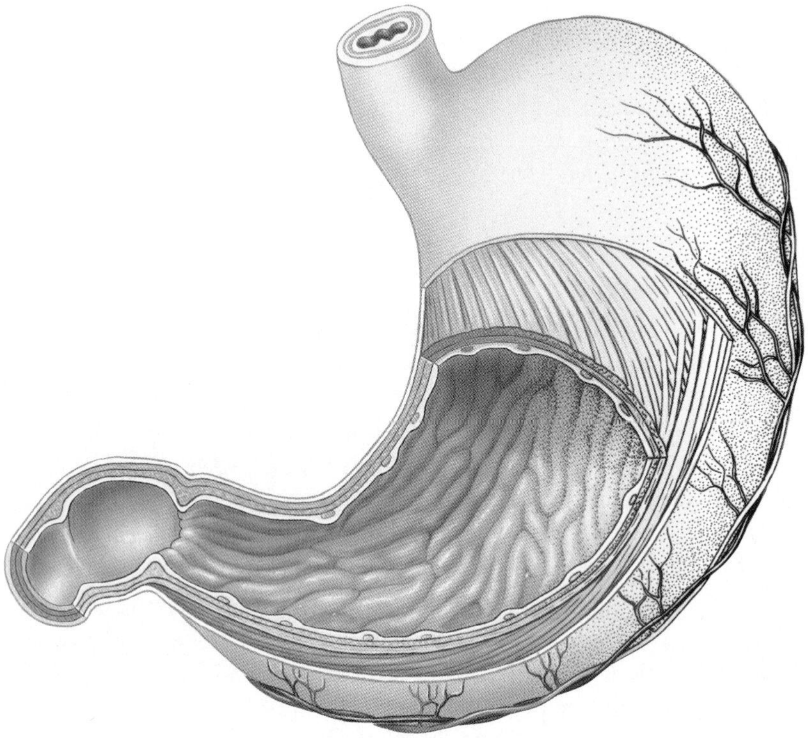

25

THE DIGESTIVE SYSTEM

Few of us give any serious thought to the digestive system unless it malfunctions. Yet each day we spend hours of conscious effort filling and emptying it. References to this system are part of our everyday language. Think of how often we use expressions relating to the digestive system: We may "have a gut feeling," "want to chew on" something, or find someone's opinions "hard to swallow." When something does go wrong with the digestive system, even something minor, most of us seek treatment immediately. For this reason, television advertisements promote toothpaste and mouthwash, diet supplements, antacids, and laxatives.

The digestive system consists of a muscular tube, called the **digestive tract**, and various **accessory organs**. The *oral cavity* (mouth), *pharynx, esophagus, stomach, small intestine,* and *large intestine* make up the digestive tract. Accessory digestive organs include the teeth, tongue, and various *glandular organs,* such as the salivary glands, liver, and pancreas, which secrete into ducts emptying into the digestive tract. Food enters the digestive tract and passes along its length. On the way, the secretions of the glandular organs, which contain water, enzymes, buffers, and other components, assist in preparing organic and inorganic nutrients for absorption across the epithelium of the digestive tract. The digestive tract and accessory organs work together to perform the following functions:

1. *Ingestion*: Ingestion occurs when foods and liquids enter the digestive tract via the mouth.
2. *Mechanical processing*: Most ingested solids must undergo mechanical processing before they are swallowed. Squashing with the tongue and tearing and crushing with the teeth are examples of mechanical processing that occur before ingestion. Swirling, mixing, churning, and propulsive motions of the digestive tract continue to provide mechanical processing after swallowing.
3. *Digestion*: Digestion is the chemical and enzymatic breakdown of complex sugars, lipids, and proteins into small organic molecules that can be absorbed by the digestive epithelium.
4. *Secretion*: Digestion usually involves the action of acids, enzymes, and buffers produced by active secretion. Some of these secretions are produced by the lining of the digestive tract, but most are provided by the accessory organs, such as the pancreas.
5. *Absorption*: Absorption is the movement of organic molecules, electrolytes, vitamins, and water across the digestive epithelium and into the interstitial fluid of the digestive tract.
6. *Compaction*: Compaction is the progressive dehydration of indigestible materials and organic wastes prior to elimination from the body.
7. *Excretion*: Waste products are secreted into the digestive tract, primarily by the accessory glands (especially the liver). **Defecation** (def-e-KĀ-shun) is the elimination of fecal material from the body.

The lining of the digestive tract also plays a defensive role by protecting surrounding tissues against (1) the corrosive effects of digestive acids and enzymes, (2) mechanical stresses, such as abrasion, and (3) pathogens that are either swallowed with food or residing within the digestive tract.

In summary, the organs of the digestive system mechanically and chemically process food that is introduced into the mouth and passed along the digestive tract. The purpose of these activities is to reduce the solid, complex chemical structures of food into small molecules that can be absorbed by the epithelium lining the digestive tract, for transfer to the circulating blood.

An Overview of the Digestive System

[FIGURE 25.1]

The major components of the digestive system are shown in Figure 25.1●. Although several of these structures have overlapping functions, each has certain areas of specialization and shows distinctive histological characteristics. Before discussing these specializations and distinctions, we will consider the basic structural characteristics common to all portions of the digestive tract.

Histological Organization of the Digestive Tract

[FIGURE 25.2]

The major layers of the digestive tract include: (1) the *mucosa*, (2) the *submucosa*, (3) the *muscularis externa*, and (4) the *serosa*. Variations in the structure of these four layers occur along the digestive tract. These variations are related to the specific functions of each organ and region. Sectional and diagrammatic views of a representative region of the digestive tract are presented in Figure 25.2●.

THE MUCOSA [FIGURE 25.2]

The inner lining, or **mucosa**, of the digestive tract is an example of a **mucous membrane**. Mucous membranes, introduced in Chapter 3, consist of a layer of loose connective tissue covered by an epithelium moistened by glandular secretions. p. 75 The **mucosal epithelium** may be simple or stratified, depending on the location and the stresses involved. For example, the oral cavity and esophagus are lined by a stratified squamous epithelium that can resist stress and abrasion, whereas the stomach, small intestine, and almost the entire large intestine have a simple columnar epithelium specialized for secretion and absorption. The mucosa of the digestive tract is often organized in transverse or longitudinal folds (Figure 25.2●). The folds, called *plicae* (PLĪ-sē; singular, *plica* PLĪ-ka), resemble pleats. Plicae dramatically increase the surface area available for absorption. In some regions of the digestive tract, plicae are permanent features that involve both the mucosa and submucosa. In other regions, plicae are temporary features that disappear as the lumen fills. They thus permit expansion of the lumen after a large meal. Ducts opening onto the epithelial surfaces carry the secretions of gland cells located in either the mucosa and submucosa or within accessory organs.

The underlying layer of areolar tissue, called the **lamina propria**, contains blood vessels, sensory nerve endings, lymphatic vessels, smooth muscle fibers, and scattered areas of lymphoid tissue. The latter is part of the network of mucosa-associated lymphoid tissue, or MALT, introduced in Chapter 23. p. 620 In most regions of the digestive tract the outer portion of the mucosa, external to the lamina propria, is a narrow band of smooth muscle and elastic fibers. This band is called the **muscularis** (mus-kū-LAR-is) **mucosae**. The smooth muscle fibers in the muscularis mucosae are arranged in two thin concentric layers (Figure 25.2a●). The inner layer encircles the lumen (the *circular muscle*), and the outer layer contains muscle fibers oriented parallel to the long axis of the tract (the *longitudinal layer*). Contraction of these layers alters the shape of the lumen and moves the epithelial pleats and folds.

THE SUBMUCOSA [FIGURE 25.2a]

The **submucosa** (sub-myū-KŌ-sa) is a layer of areolar tissue that surrounds the muscularis mucosae. Large blood vessels and lymphatics are found in this layer, and in some regions of the digestive tract the submucosa also contains exocrine glands that secrete buffers and enzymes into the lumen of the digestive tract. Along its outer margin, the submucosa contains a network of nerve fibers and scattered neuron cell bodies. This **submucosal plexus** (*plexus of Meissner*) innervates the mucosa; it contains sensory neurons, parasympathetic ganglia, and sympathetic postganglionic fibers (Figure 25.2a●).

THE MUSCULARIS EXTERNA [FIGURE 25.2]

The **muscularis externa,** which surrounds the submucosa, is a region dominated by smooth muscle fibers. These fibers are arranged in both circular (inner) and longitudinal (outer) layers (Figure 25.2●). These layers of smooth muscle play an essential role in mechanical processing and in the propulsion of materials along the digestive tract. These movements are coordinated primarily by neurons of the **myenteric plexus** (mī-en-TER-ik; *mys*, muscle + *enteron*, intestine), or *plexus of Auerbach*. This network of parasympathetic ganglia and sympathetic postganglionic fibers lies sandwiched between the circular and longitudinal muscle layers. Parasympathetic stimulation increases muscular tone and stimulates contractions, and sympathetic stimulation promotes inhibition of muscular activity and relaxation.

Additionally, at specific locations along the digestive tract, this layer forms sphincters, or *valves*, that help to prevent materials from moving along the tract at an inappropriate time or in the wrong direction. Valves or sphincters have thickened areas of the circular smooth muscle layer that constricts the lumen to restrict the flow of digestive material through the lumen.

THE SEROSA [FIGURE 25.2]

Along most regions of the digestive tract within the peritoneal cavity, the muscularis externa is covered by a *serous membrane* known as the **serosa** (Figure 25.2●). There is no serosa, however, surrounding the muscularis externa of the oral cavity, pharynx, esophagus, and rectum. Instead, the muscularis externa is wrapped by a dense network of collagen fibers that firmly attaches the digestive tract to adjacent structures. This fibrous sheath is called the **adventitia** (ad-ven-TISH-a).

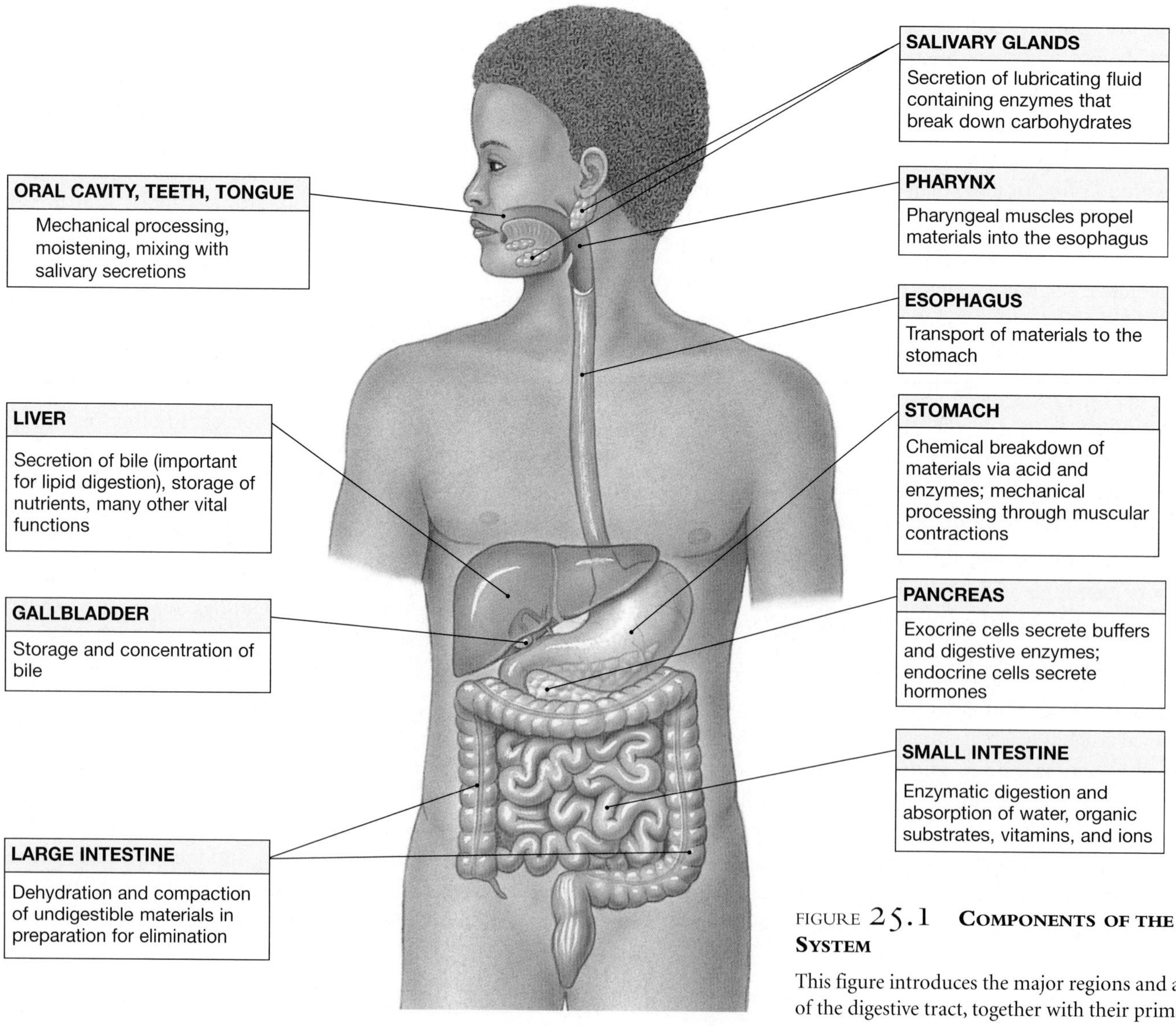

SALIVARY GLANDS
Secretion of lubricating fluid containing enzymes that break down carbohydrates
ORAL CAVITY, TEETH, TONGUE
Mechanical processing, moistening, mixing with salivary secretions
PHARYNX
Pharyngeal muscles propel materials into the esophagus
ESOPHAGUS
Transport of materials to the stomach
LIVER
Secretion of bile (important for lipid digestion), storage of nutrients, many other vital functions
STOMACH
Chemical breakdown of materials via acid and enzymes; mechanical processing through muscular contractions
GALLBLADDER
Storage and concentration of bile
PANCREAS
Exocrine cells secrete buffers and digestive enzymes; endocrine cells secrete hormones
SMALL INTESTINE
Enzymatic digestion and absorption of water, organic substrates, vitamins, and ions
LARGE INTESTINE
Dehydration and compaction of undigestible materials in preparation for elimination

FIGURE 25.1 **Components of the Digestive System**

This figure introduces the major regions and accessory organs of the digestive tract, together with their primary functions.

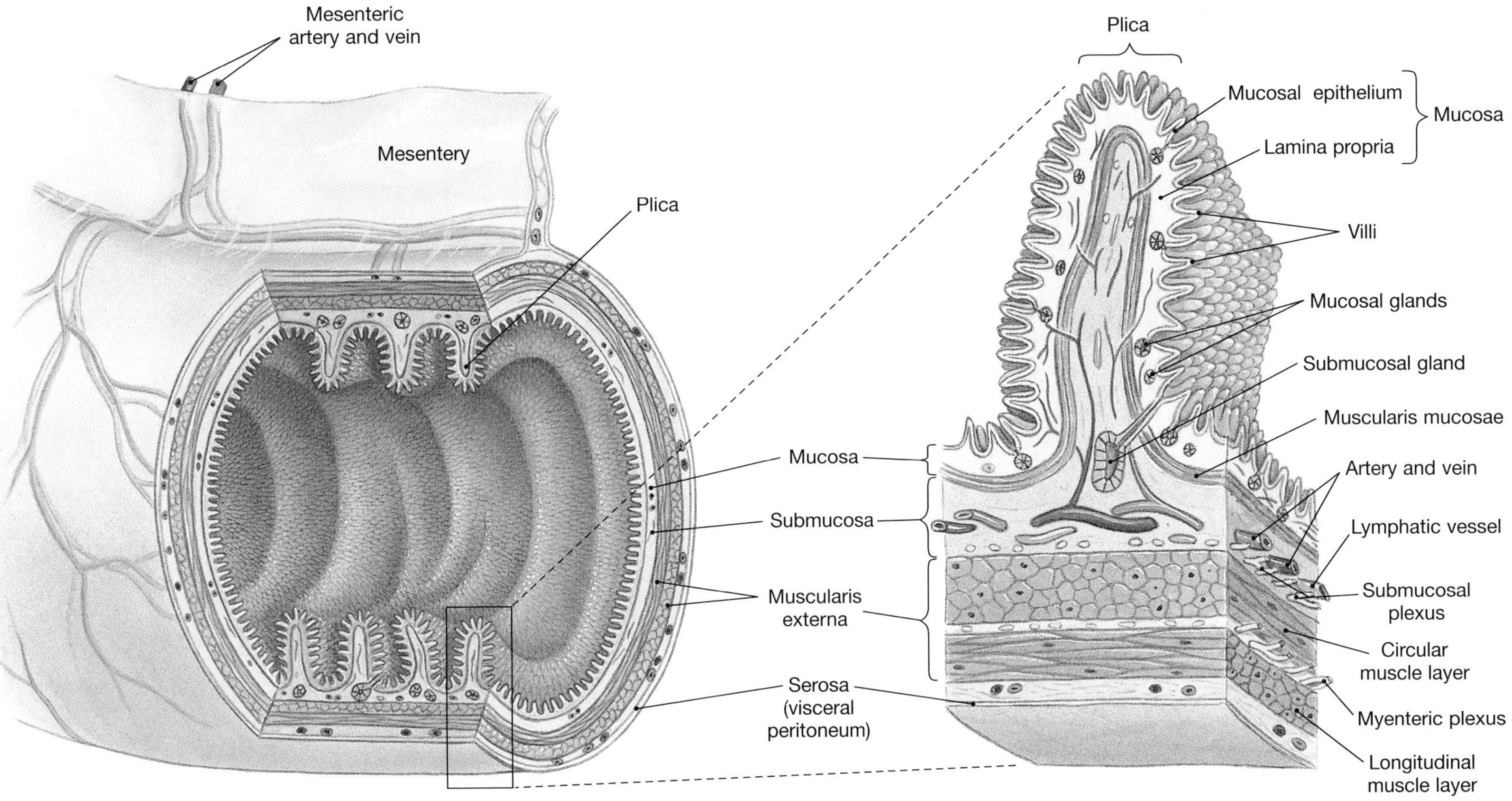

(a) Histological organization of the digestive tract

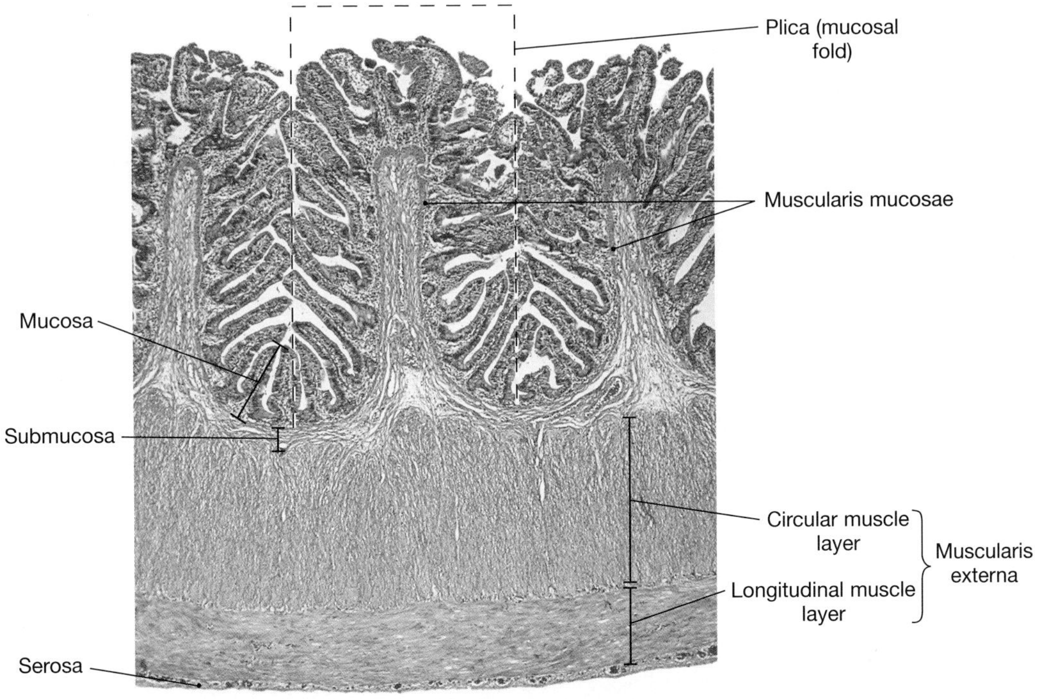

(b) Photomicrograph of ileum (LM X 160)

FIGURE 25.2 **HISTOLOGICAL STRUCTURE OF THE DIGESTIVE TRACT**

(**a**) Three-dimensional view of the histological organization of the general digestive tube. (**b**) Photomicrograph of the ileum, showing aspects of the histological organization of the small intestine.

Muscularis Layers and the Movement of Digestive Materials

The digestive tract contains **visceral smooth muscle tissue**. A single smooth muscle cell ranges from 5 to 10 μm in diameter and from 30 to 200 μm in length. They are surrounded by connective tissue, but the collagen fibers do not form tendons or aponeuroses. The contractile proteins of these smooth muscle cells are not organized into sarcomeres, so the muscle cells are nonstriated, involuntary muscle. p. 77 Although they are nonstriated, their contractions are as strong as those of skeletal or cardiac muscle cells. In visceral smooth muscle tissue, many of the muscle cells have no motor innervation. The muscle cells are arranged in sheets or layers, and adjacent muscle cells are electrically connected by gap junctions. When one visceral smooth muscle cell contracts, the contraction spreads in a wave that travels throughout the tissue. The initial stimulus may be the activation of a motor neuron that contacts one of the smooth muscle cells in the region. It may also be a local response to chemicals, hormones, the concentrations of oxygen or carbon dioxide, or physical factors such as extreme stretching or irritation.

Because the contractile filaments of smooth muscle cells are not rigidly organized, a stretched smooth muscle cell soon adapts to its new length and

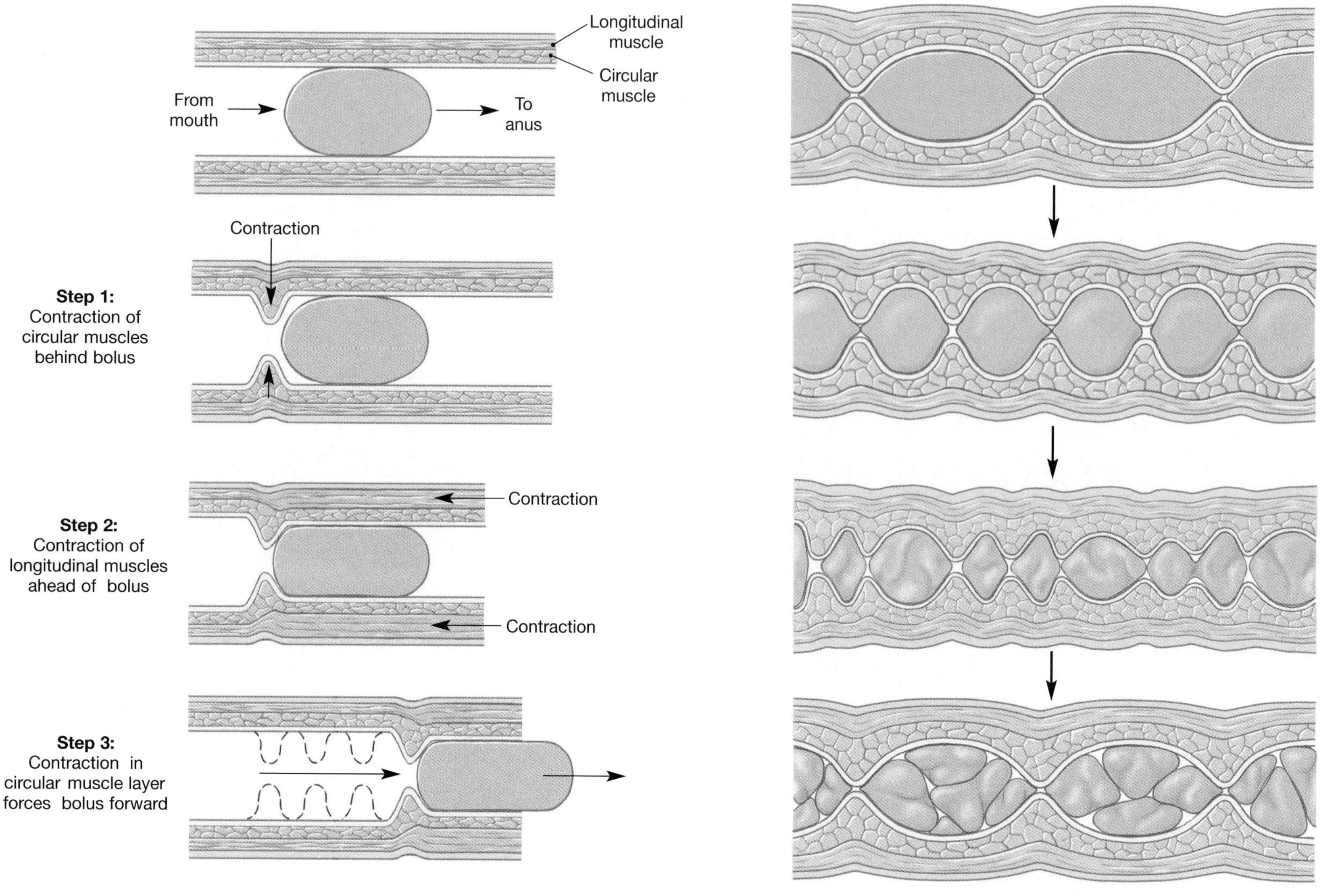
Longitudinal muscle
Circular muscle
From mouth
To anus
Contraction
Step 1: Contraction of circular muscles behind bolus
Contraction
Contraction
Step 2: Contraction of longitudinal muscles ahead of bolus
Step 3: Contraction in circular muscle layer forces bolus forward

(a) Peristalsis

(b) Segmentation

FIGURE 25.3 **Peristalsis and Segmentation**

(**a**) Peristalsis propels materials along the length of the digestive tract by coordinated contractions of the circular and longitudinal layers. (**b**) Segmentation movements involve primarily the circular muscle layers. These activities churn and mix the contents of the digestive tract, but do not produce net movement in a particular direction.

retains the ability to contract on demand. This ability to tolerate extreme stretching is called *plasticity*. Plasticity is especially important for digestive organs that undergo great changes in volume, such as the stomach.

Smooth muscle in the digestive tract shows rhythmic cycles of activity because of the presence of **pacesetter cells**. Pacesetter cells are found in both the muscularis mucosae and muscularis externa. These smooth muscle cells undergo spontaneous depolarization, which triggers contractions leading to two types of movement events, *peristalsis* and *segmentation*. These wavelike contractions spread through the entire muscular sheet and facilitate the propulsion and mixing of contents along the digestive tract.

PERISTALSIS [FIGURE 25.3a]

The muscularis externa propels materials from one region of the digestive tract to another through the contractions of **peristalsis** (per-i-STAL-sis). Peristalsis consists of waves of muscular contractions that move a **bolus** (a small oval mass of food) along the length of the digestive tract. During a **peristaltic wave**, the circular muscles contract behind the digestive contents. Longitudinal muscles contract next, shortening adjacent segments. A wave of contraction in the circular muscles then forces the materials in the desired direction (Figure 25.3a●).

SEGMENTATION [FIGURE 25.3b]

Most areas of the small intestine and some regions of the large intestine undergo contractions that produce **segmentation** (Figure 25.3b●). These movements churn and fragment the digestive materials, mixing the contents with intestinal secretions. They do not produce net movement in any particular direction.

Segmentation and peristalsis may be triggered by pacesetter cells, hormones, chemicals, and physical stimulation. Peristaltic waves also can be initiated by afferent and efferent fibers within the glossopharyngeal, vagus, or pelvic nerves. Local peristaltic movements limited to a few centimeters of the digestive tract are triggered by sensory receptors in the walls of the digestive tract. These afferent fibers synapse within the myenteric plexus to produce localized **myenteric reflexes**. These are short reflexes that do not involve the CNS. p. 462 The term *enteric nervous system* is often used to refer to the neural network that coordinates these reflexes.

In general, short reflexes control activities in one region of the digestive tract. This control may involve coordinating local peristalsis and triggering the secretion of digestive glands. Many neurons are involved—the enteric nervous system has roughly as many neurons and neurotransmitters as the spinal cord has.

Sensory information from receptors in the digestive tract is also distributed to the CNS, where it can trigger *long reflexes.* p. 462 Long reflexes, which involve interneurons and motor neurons in the CNS, provide a higher level of control over digestive and glandular activities. These reflexes generally control large-scale peristaltic waves that move materials from one region of the digestive tract to another. Long reflexes may involve motor fibers in the glossopharyngeal, vagus, or pelvic nerves that synapse in the myenteric plexus.

The Peritoneum

The serosa, or **visceral peritoneum**, is continuous with the **parietal peritoneum** that lines the inner surfaces of the body wall. p. 20 The peritoneal lining continually produces a watery peritoneal fluid that lubricates the peritoneal surfaces. About 7 liters of fluid are secreted and reabsorbed each day, although the volume within the peritoneal cavity at any one time is very small. Under unusual conditions, such as liver disease, heart failure, or disturbances in electrolyte balance, the volume of peritoneal fluid can increase markedly. This results in a dangerous reduction in blood volume and distortion of visceral organs.

CLINICAL BRIEF

PERITONITIS

Inflammation of the peritoneal membrane produces symptoms of **peritonitis** (per-i-tō-NĪ-tis), a painful condition that interferes with the normal functioning of the affected organs. Physical damage, chemical irritation, or bacterial invasion of the peritoneum can lead to severe and even fatal cases of peritonitis. Peritonitis due to bacterial infection is a potential complication of any surgery that involves opening the peritoneal cavity or any disease that perforates the walls of the stomach or intestines. Liver disease, kidney disease, or heart failure can cause an increase in the rate of fluid movement through the peritoneal lining. The accumulation of fluid, called **ascites** (a-SĪ-tēz), creates a characteristic abdominal swelling. Distortion of internal organs by the contained fluid can result in a variety of symptoms; heartburn, indigestion, and low back pain are common complaints.

MESENTERIES [FIGURES 25.2a/25.4/25.10a/25.11b]

Within the peritoneal cavity, most regions of the digestive tract are suspended by sheets of serous membrane that connect the parietal peritoneum with the visceral peritoneum. These **mesenteries** (MEZ-en-ter-ēz) are fused, double sheets of peritoneal membrane (Figure 25.2a●). The areolar connective tissue between the mesothelial surfaces provides an access route for the passage of the blood vessels, nerves, and lymphatics to and from the digestive tract. Mesenteries also stabilize the relative positions of the attached organs and prevent the entanglement of intestines during digestive movements or sudden changes in body position.

During development, the digestive tract and accessory organs are suspended within the peritoneal cavity by dorsal and ventral mesenteries (Figure 25.4a,b●). The ventral mesentery later disappears along most of the digestive tract, persisting only on the ventral surface of the stomach, between the stomach and liver (the **lesser omentum**), and between the liver and the anterior abdominal wall and diaphragm (the *falciform ligament*). (Although this peritoneal sheet is called a "ligament," it is not comparable to the ligaments that interconnect bones.) For additional information concerning the development of the digestive tract, accessory organs, and associated mesenteries, see the Embryology Summary on pp. 670–671.

As the digestive tract elongates, it twists and turns within the relatively crowded peritoneal cavity. The dorsal mesentery of the stomach becomes greatly enlarged, and it forms a pouch that extends inferiorly, between the body wall and the anterior surface of the small intestine. This pouch is the **greater omentum** (ō-MEN-tum; *omentum*, fat skin) (see Figure 25.4b● and look ahead to Figures 25.10a, and 25.11b●). The loose connective tissue within the mesentery of the greater omentum usually contains a thick layer of adipose tissue. It also contains numerous lymph nodes. All but the first 25 cm of the small intestine is suspended by a thick *mesenterial sheet*, termed the **mesentery proper**, that provides stability but permits a degree of independent movement. The mesentery associated with the initial portion of the small intestine (the *duodenum*) and the pancreas fuses with the posterior abdominal wall, locking these structures in position. Their anterior surfaces are covered by peritoneum, but the rest of the organs lie outside of the peritoneal cavity. This position is called **retroperitoneal** (re-trō-per-i-tō-NĒ-al; *retro*, behind) because part or all of these organs lie posterior to the peritoneum.

The **mesocolon** is a mesentery attached to the large intestine. The middle portion of the large intestine (the *transverse colon*) is suspended by the **transverse mesocolon**. The *sigmoid colon*, which leads to the rectum and anus, is suspended by the **sigmoid mesocolon**. The mesocolon of the *ascending colon*, the *descending colon*, and the *rectum* of the large intestine usually fuses to the body wall, fixing them in position. These organs are now retroperitoneal, and visceral peritoneum covers only their anterior surfaces and portions of their lateral surfaces (Figure 25.4b–d●).

✓ CONCEPT CHECK

- What are the components and functions of the mucosa of the digestive tract?
- What are the functions of mesenteries?
- What is the functional difference between peristalsis and segmentation?
- What is important about the lack of organization in the contractile filaments of smooth muscle cells?

The Oral Cavity

Our exploration of the digestive tract will follow the path of food from the mouth to the anus. The mouth opens into the *oral cavity*. The functions of the oral cavity include: (1) *analysis* of material before swallowing; (2) *mechanical processing* through the actions of the teeth, tongue, and palatal surfaces; (3) *lubrication* by mixing with mucous and salivary secretions; and (4) limited *digestion* of carbohydrates by a salivary enzyme.

Anatomy of the Oral Cavity [FIGURE 25.5]

The **oral cavity**, or *buccal* (BUK-al) *cavity* (Figure 25.5●), is lined by the **oral mucosa**, which has a stratified squamous epithelium that protects the

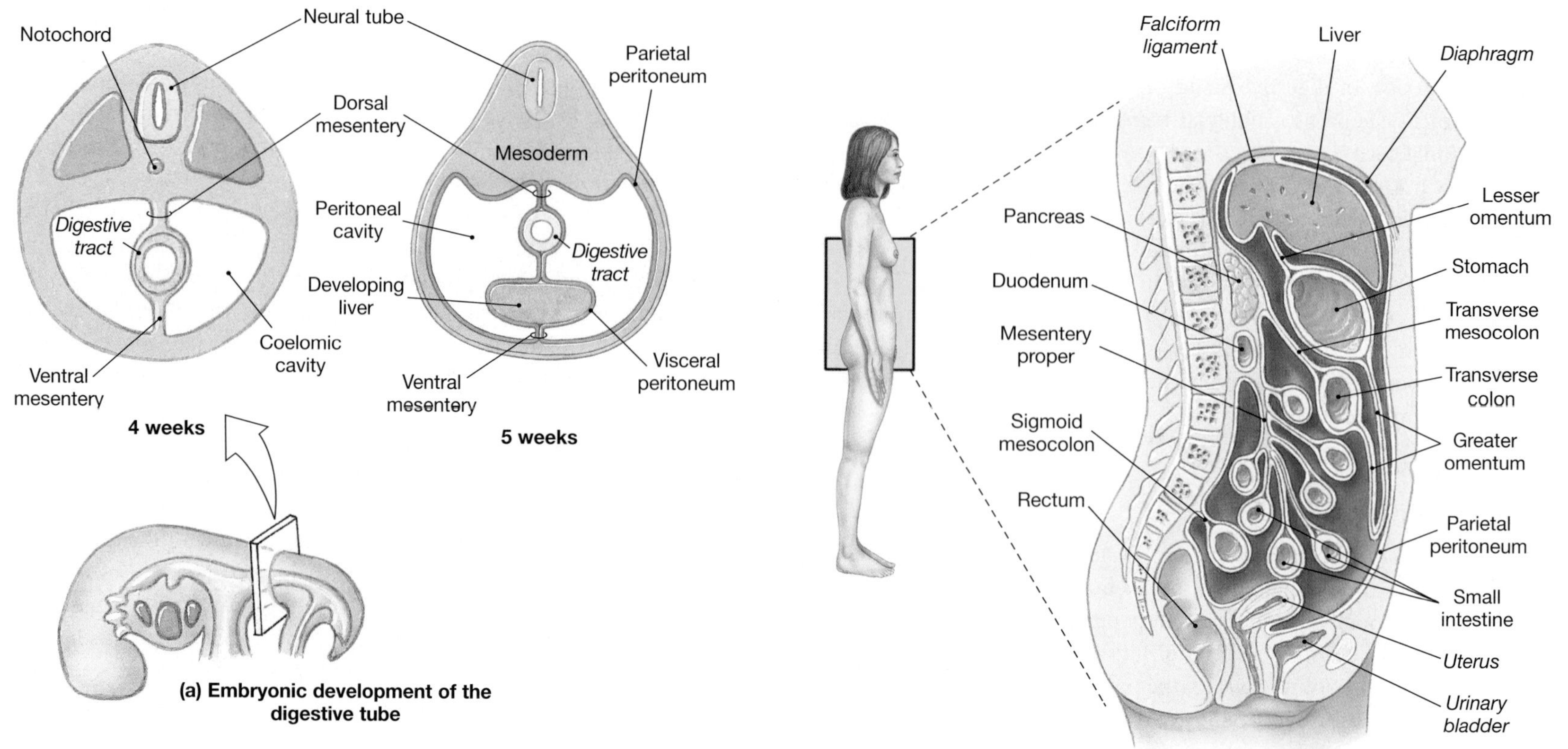

(a) Embryonic development of the digestive tube

(b) Sagittal section, showing adult mesenteries

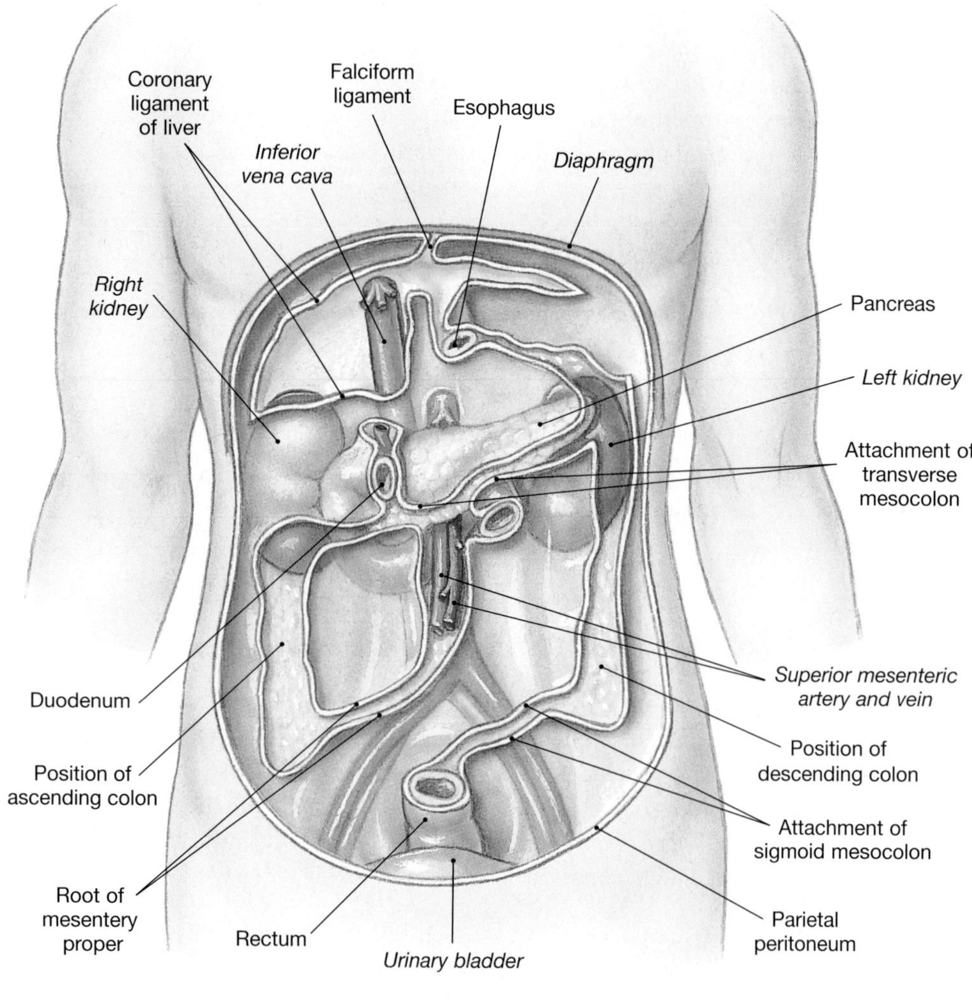

(c) Mesentery attachments, anterior view

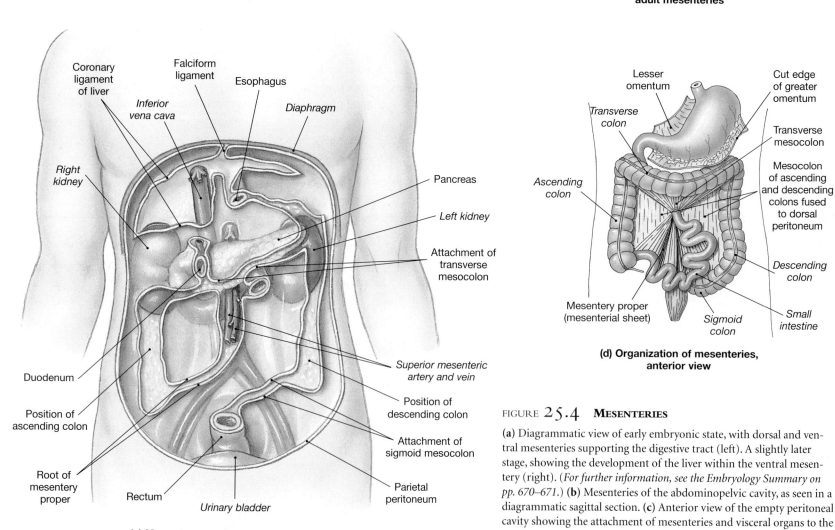

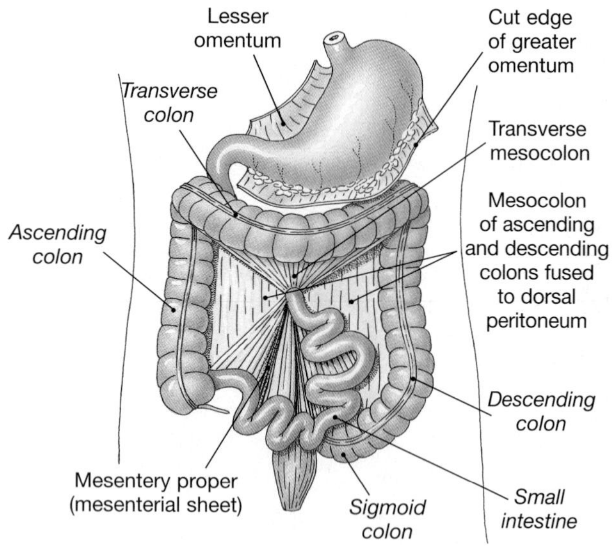

(d) Organization of mesenteries, anterior view

FIGURE 25.4 **MESENTERIES**

(**a**) Diagrammatic view of early embryonic state, with dorsal and ventral mesenteries supporting the digestive tract (left). A slightly later stage, showing the development of the liver within the ventral mesentery (right). (*For further information, see the Embryology Summary on pp. 670–671.*) (**b**) Mesenteries of the abdominopelvic cavity, as seen in a diagrammatic sagittal section. (**c**) Anterior view of the empty peritoneal cavity showing the attachment of mesenteries and visceral organs to the posterior wall of the abdominal cavity. (**d**) The organization of mesenteries in the adult. This is a simplified view; the length of the small intestine has been greatly reduced.

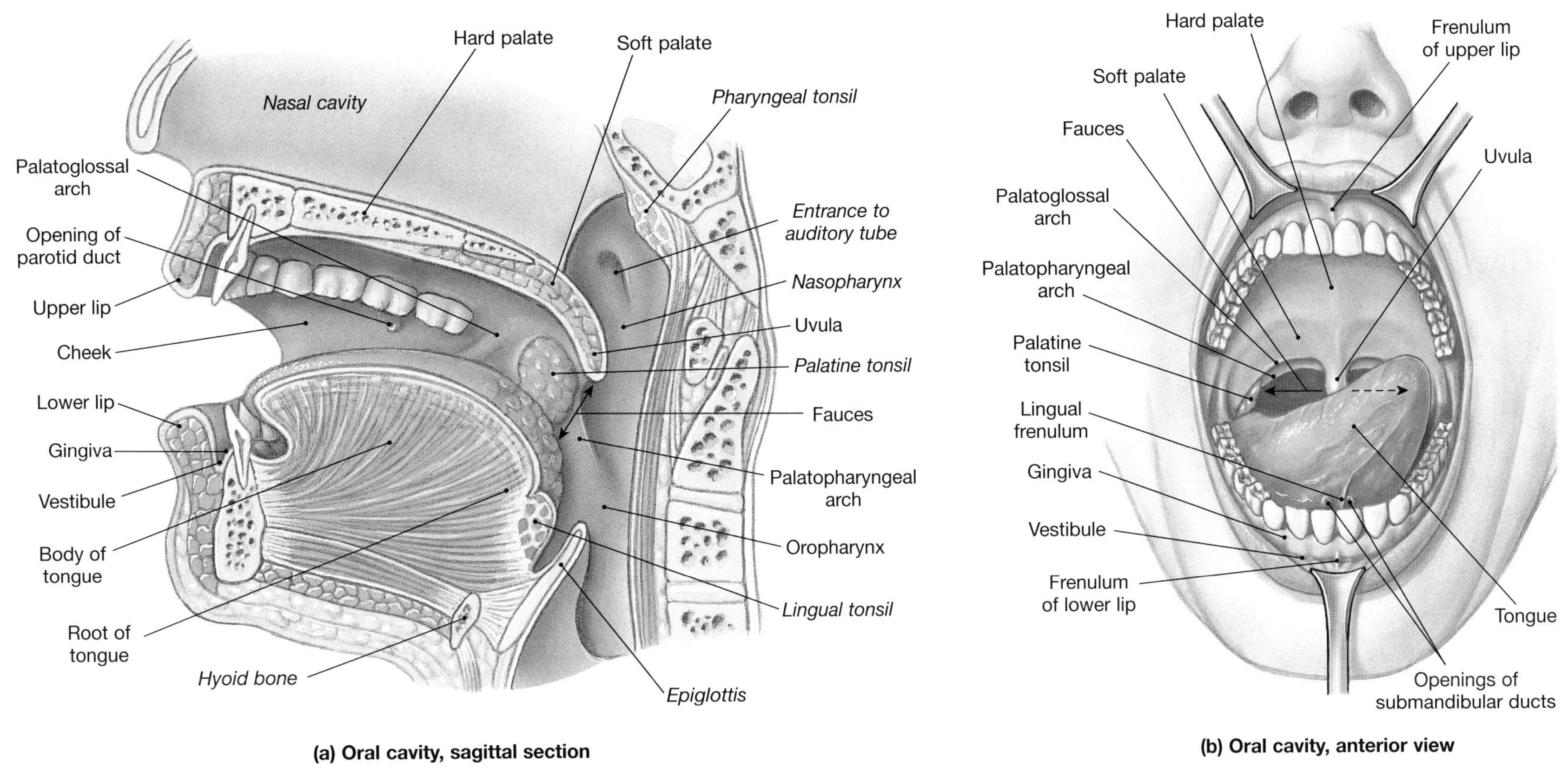

FIGURE 25.5 **The Oral Cavity**

(a) The oral cavity as seen in sagittal section. **(b)** An anterior view of the oral cavity, as seen through the open mouth.

mouth from abrasion during the ingestion of food. Unlike the stratified squamous epithelium of the skin, the oral epithelium does not undergo keratinization. The mucosa of the **cheeks**, or lateral walls of the oral cavity, are supported and formed by *buccal fat pads* and the *buccinator muscles.* p. 269 Anteriorly, the mucosa of the cheeks are continuous with the lips, or **labia** (LĀ-bē-a). The **vestibule** is the space between the cheeks, lips, and the teeth. A ridge of oral mucosa, the gums, or **gingivae** (JIN-ji-vē), surrounds the base of each tooth on the alveolar surfaces of the maxillary bone and mandible. pp. 151–152

The roof of the oral cavity is formed by the **hard** and **soft palates,** while the tongue dominates its floor. The hard palate separates the oral cavity from the nasal cavity. The soft palate separates the oral cavity from the nasopharynx and closes off the nasopharynx during swallowing. The finger-shaped **uvula** (ū-vū-la), which dangles from the center of the posterior margin of the soft palate, helps to prevent food from entering the pharynx prematurely. Inferior to the tongue, the floor receives additional support from the *mylohyoid muscle.* p.276 The hard palate is formed by the palatine process of the maxilla and the palatine bone. The soft palate lies posterior to the hard palate. The posterior margin of the soft palate supports the dangling uvula and two pairs of muscular *pharyngeal arches.*

1. The *palatoglossal arches* extend between the soft palate and the base of the tongue. Each arch consists of a mucous membrane and the underlying *palatoglossus muscle* and associated tissues. p. 274
2. The *palatopharyngeal arches* extend from the soft palate to the side of the pharynx. Each arch consists of a mucous membrane and the underlying *palatopharyngeus muscle* and associated tissues. p. 275

The palatine tonsils lie between the palatoglossal and palatopharyngeal arches. The posterior margin of the soft palate, including the uvula, the palatopharyngeal arches, and the base of the tongue, frame the **fauces** (FAW-sēz), the entrance to the oropharynx.

THE TONGUE [FIGURES 18.7/25.5/25.6]

The **tongue** (Figure 25.5●) manipulates materials inside the mouth and may occasionally be used to bring foods (such as ice cream or pudding) into the oral cavity. The primary functions of the tongue are: (1) mechanical processing by compression, abrasion, and distortion; (2) manipulation to assist in chewing and to prepare the material for swallowing; (3) sensory analysis by touch, temperature, and taste receptors; and (4) secretion of mucins and an enzyme that aids in fat digestion.

The tongue can be divided into an anterior **body**, or *oral portion*, and a posterior **root**, or *pharyngeal portion*. The superior surface, or **dorsum**, of the body contains numerous fine projections called *papillae*. The thickened epithelium covering each papilla produces additional friction that assists in the movement of materials by the tongue. Additionally, taste buds are found along the edges of many papillae. Surface features and histological details of the tongue can be found in Figure 18.7●. p. 475 A V-shaped line of circumvallate papillae roughly indicates the boundary between the body and the root of the tongue, which is situated in the pharynx.

The tongues epithelium is flushed by the secretions of small glands that extend into the lamina propria of the tongue. These secretions contain water mucins, and the enzyme *lingual lipase.* Lingual lipase begins the enzymatic breakdown of lipids, specifically triglycerides. The epithelium covering the inferior surface of the tongue is thinner and more delicate than

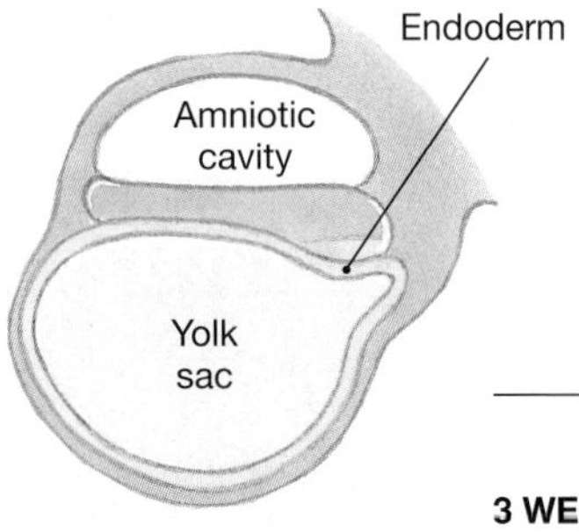

By week 3, endodermal cells have migrated around the inside of the *blastocyst* (p. 00), completing a pouch known as the **yolk sac**.

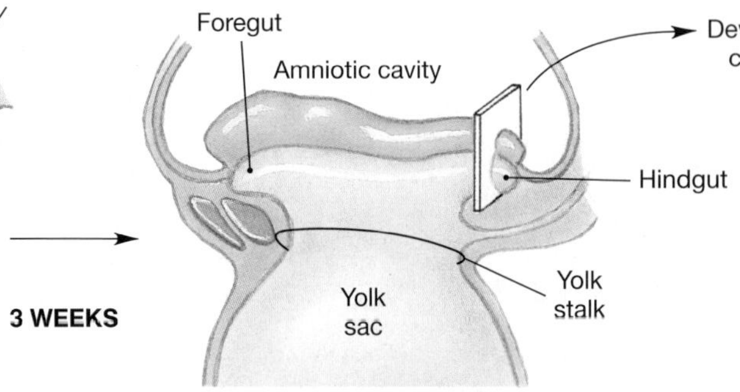

3 WEEKS

As the embryo forms on the *embryonic shield*, two pockets of endoderm are created: the **foregut** and **hindgut**. A broad connection between these pockets and the yolk sac remains within the **yolk stalk**.

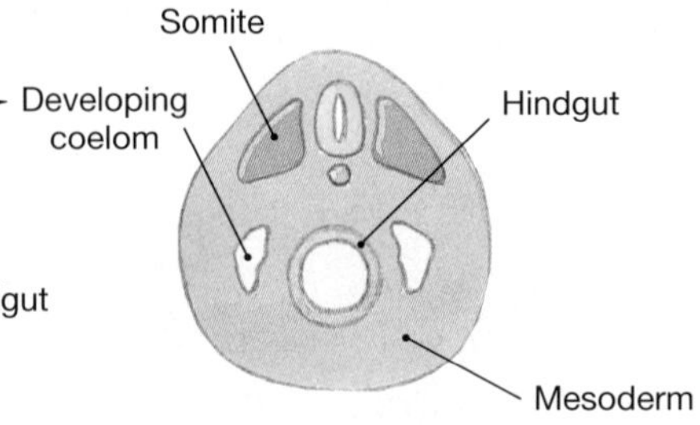

In sectional view, the embryonic gut is a simple endodermal tube surrounded by mesoderm. Cavities appearing within the mesoderm create the *coelom* (ventral body cavity).

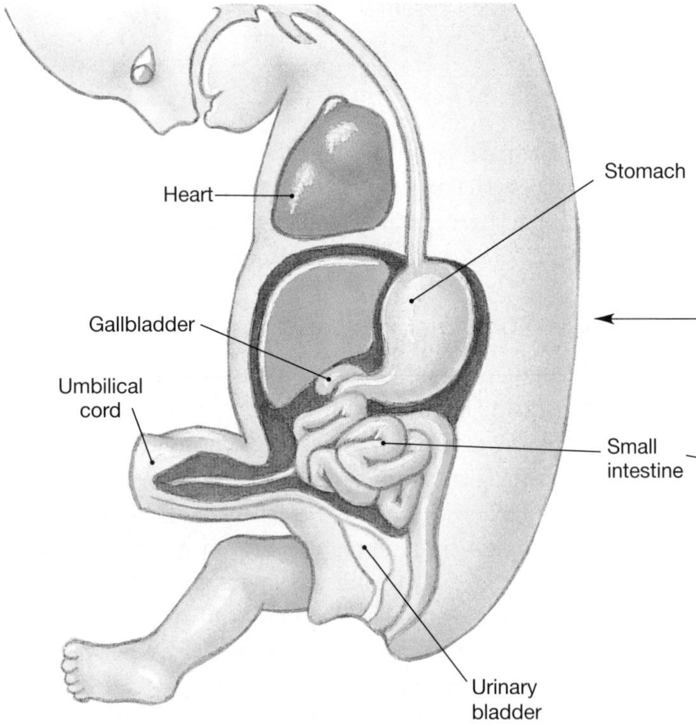

10 WEEKS

By week 10, the intestines have begun moving back into the coelomic cavity, although they continue to grow longer.

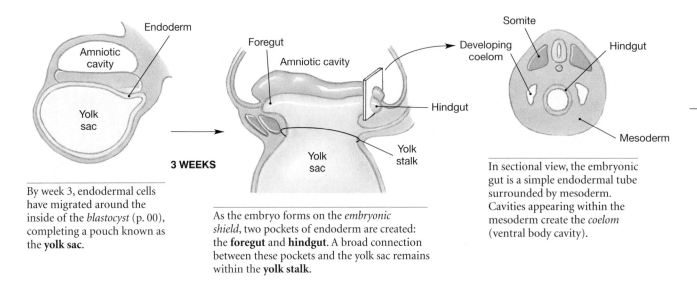

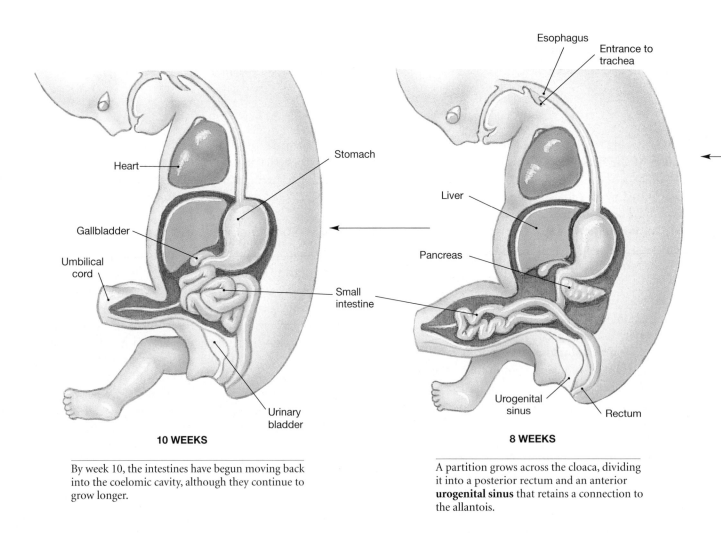

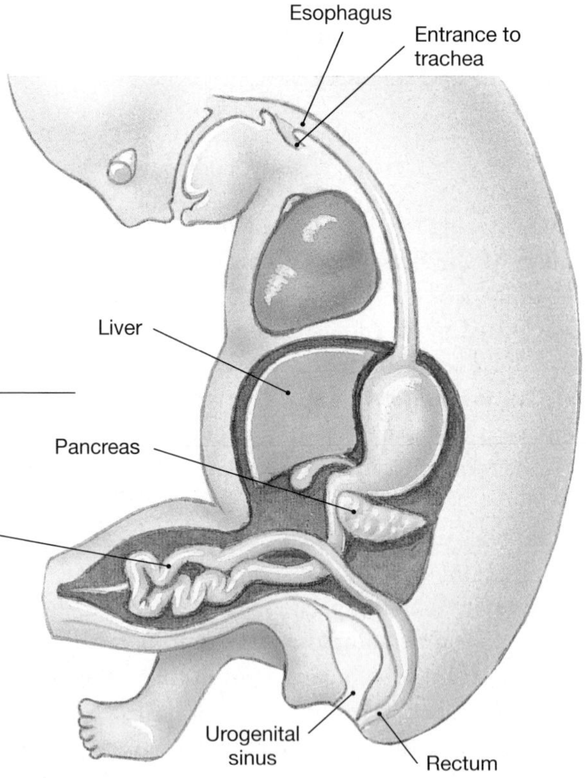

8 WEEKS

A partition grows across the cloaca, dividing it into a posterior rectum and an anterior **urogenital sinus** that retains a connection to the allantois.

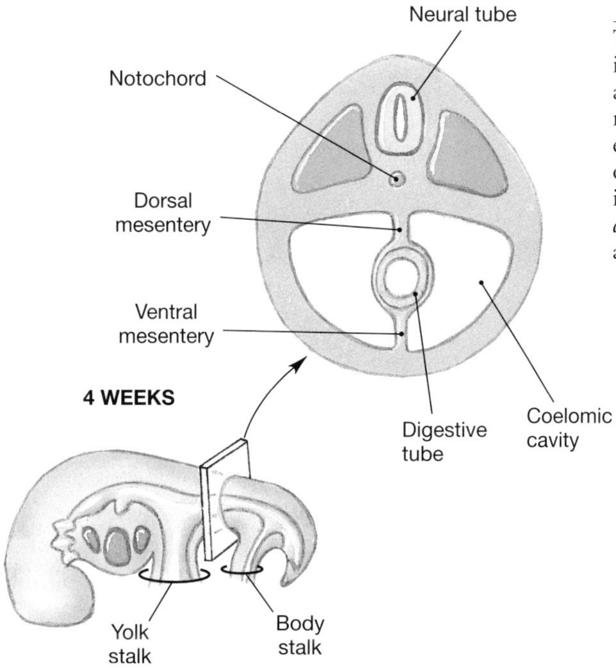

The digestive tube remains suspended in the coelom by a **dorsal mesentery** and a **ventral mesentery**. The ventral mesentery disintegrates everywhere except where major vessels or visceral organs have grown into it. It remains intact along the path of the *umbilical arteries* and where the *umbilical vein* and liver develop.

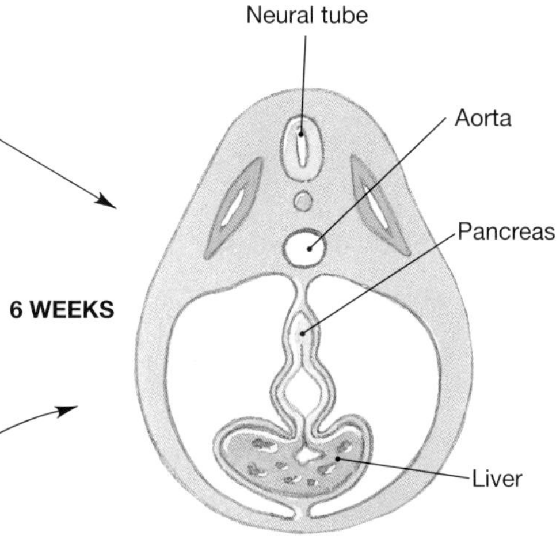

The pancreas and liver begin as epithelial pockets that grow away from the digestive tract and into the dorsal and ventral mesenteries, respectively.

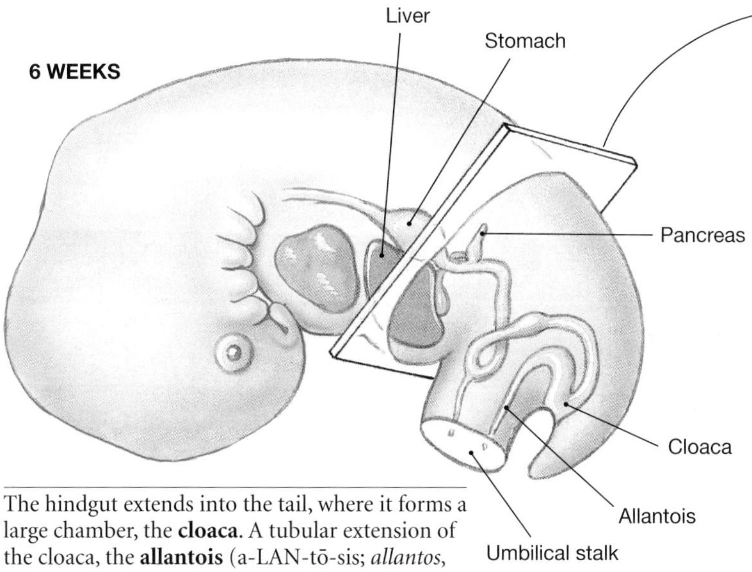

The intestines begin to elongate, and with the breakdown of the ventral mesentery, they push outward into the umbilical stalk. Further elongation and coiling occur ouside the body of the embryo.

The hindgut extends into the tail, where it forms a large chamber, the **cloaca**. A tubular extension of the cloaca, the **allantois** (a-LAN-tō-sis; *allantos*, sausage) projects away from the body and into the **body stalk**. Fusion of the yolk stalk and body stalk will create the **umbilical stalk**, also known as the *umbilical cord*.

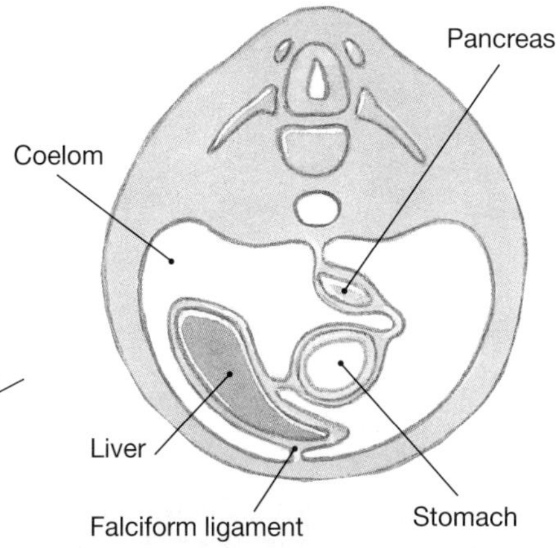

As the embryo enlarges, the stomach and liver rotate toward the right, creating two pockets: The mesenteries that form these pockets are the greater omentum and the lesser omentum.

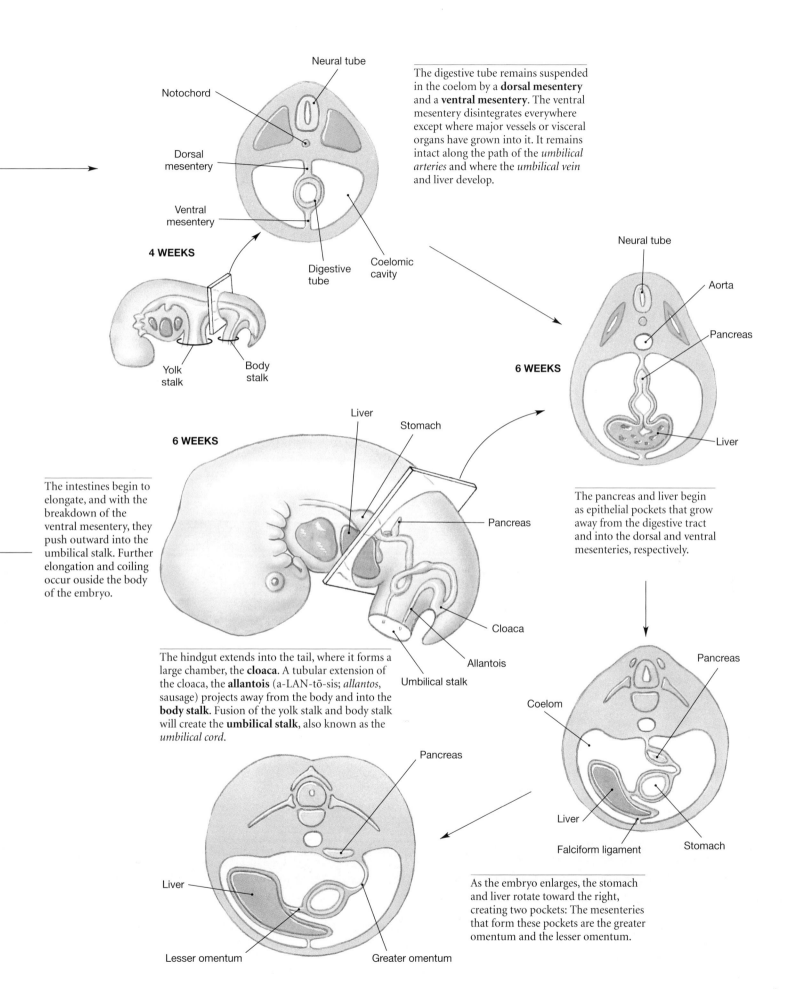

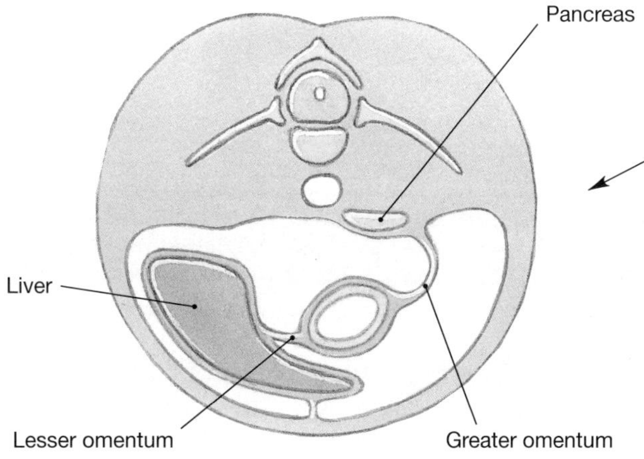

that of the dorsum. Along the inferior midline is a thin fold of mucous membrane, the **lingual frenulum** (FREN-ū-lum; *frenulum*, small bridle), which connects the body of the tongue to the mucosa of the oral floor. Ducts from one of the salivary glands are visible as they open on either side of the lingual frenulum (Figure 25.5b●).

The lingual frenulum prevents extreme movements of the tongue. However, if the lingual frenulum is *too* restrictive, the individual cannot eat or speak normally. When properly diagnosed, this condition, called **ankyloglossia** (ang-ki-lō-GLOS-ē-a), can be corrected surgically.

The tongue contains two different groups of muscles, **intrinsic tongue muscles** and **extrinsic tongue muscles**. Both intrinsic and extrinsic tongue muscles are under the control of the hypoglossal nerve (N XII). The extrinsic muscles, discussed in Chapter 10, include the *hyoglossus*, *styloglossus*, *genioglossus*, and *palatoglossus muscles*. p. 274 All gross movements of the tongue are performed by the extrinsic muscles. The less massive intrinsic muscles alter the shape of the tongue and assist the extrinsic muscles during precise movements, as in speech.

SALIVARY GLANDS [FIGURES 25.5/25.6]

Three pairs of salivary glands secrete into the oral cavity (Figure 25.6●). Each salivary gland is covered by a fibrous capsule. The saliva produced by the secretory cells of the gland is transported through a network of fine ducts to a single large drainage duct. This main duct penetrates the capsule and opens onto the surface of the oral mucosa. These salivary glands are:

1. The **parotid** (pa-ROT-id) **salivary glands** are the largest salivary glands, with an average weight of roughly 20 g. They are irregularly shaped, extending between the inferior surface of the zygomatic arch and the anterior margin of the sternocleidomastoid muscle, and from the mastoid process of the temporal bone anteriorly across the superficial surface of the masseter muscle. The secretions of each gland are drained by a **parotid duct**, or *Stensen's duct*, which empties into the vestibule at the level of the second upper molar (Figure 25.5a●).
2. The **sublingual** (sub-LING-gwal) **salivary glands** are covered by the mucous membrane of the floor of the mouth. Numerous **sublingual ducts**, or *ducts of Rivinus*, open along either side of the lingual frenulum (Figure 25.6a●).
3. The **submandibular salivary glands** are found in the floor of the mouth along the medial surfaces of the mandible inferior of the *mylohyoid line*. p. 152 The **submandibular ducts**, or *Wharton's ducts*, open into the mouth on either side of the lingual frenulum immediately posterior to the teeth (Figure 25.5b●). The histological appearance of the submandibular gland can be seen in Figure 25.6b●.

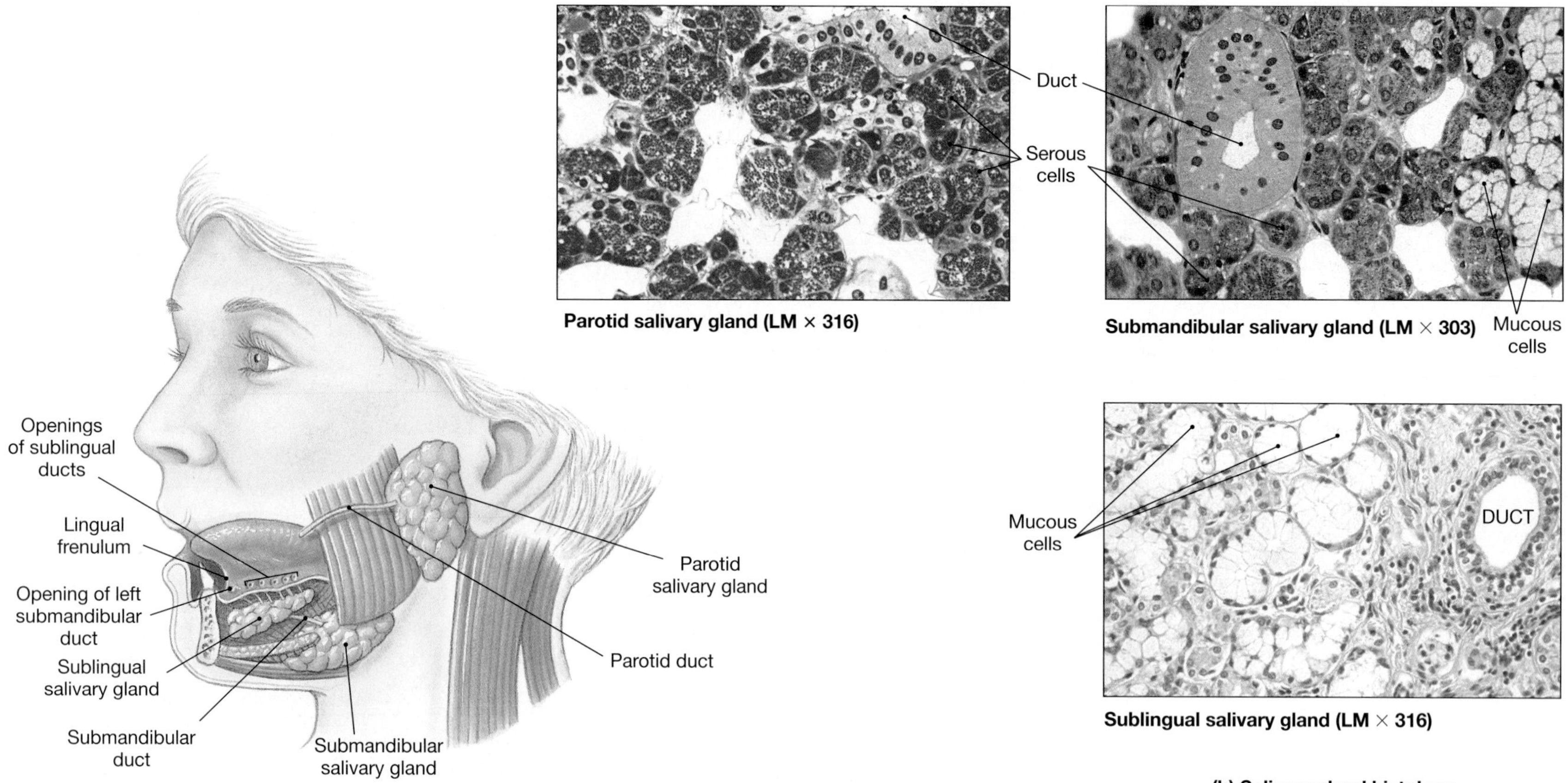

(b) Salivary gland histology

(a) Lateral view with left mandibular body and ramus removed

FIGURE 25.6 **THE SALIVARY GLANDS.**

(a) Lateral view, showing the relative positions of the salivary glands and ducts on the left side of the head. Much of the left half of the body and the left ramus of the mandible have been removed. For the positions of the ducts inside the oral cavity, *see Figure 25.5*.
(b) Photomicrographs showing histological detail of the parotid, submandibular, and sublingual salivary glands. The parotid salivary gland produces saliva rich in enzymes. The gland is dominated by serous secretory cells. The submandibular salivary gland produces saliva containing enzymes and mucins, and it contains both serous and mucous secretory cells. The sublingual salivary gland produces saliva rich in mucins. This gland is dominated by mucous secretory cells.

Each of the salivary glands has a distinctive cellular organization and produces saliva with slightly different properties. For example, the parotid salivary glands produce a thick, serous secretion containing the digestive enzyme **salivary amylase**, which begins the chemical breakdown of complex carbohydrates. The saliva in the mouth is a mixture of glandular secretions; about 70% of the saliva originates in the submandibular salivary glands, 25% from the parotid salivary glands, and 5% from the sublingual salivary glands. Collectively the salivary glands produce 1.0–1.5 l of saliva each day, with a composition of 99.4% water, plus an assortment of ions, buffers, metabolites, and enzymes. Glycoproteins called *mucins* are primarily responsible for the lubricating effects of saliva. p. 30

At mealtimes the production of large quantities of saliva lubricates the mouth, moistens the food, and dissolves chemicals that stimulate the taste buds. A continual background level of secretion flushes the oral surfaces and helps to control populations of oral bacteria. A reduction or elimination of salivary secretions triggers a bacterial population explosion in the oral cavity. This proliferation rapidly leads to recurring infections and the progressive erosion of the teeth and gums.

Salivary secretions are usually controlled by the autonomic nervous system. Each salivary gland receives parasympathetic and sympathetic innervation. Any object placed within the mouth can trigger a salivary reflex by stimulating receptors monitored by the trigeminal nerve or by stimulating taste buds innervated by N VII, N IX, or N X. pp. 413, 475 Parasympathetic stimulation accelerates secretion by all of the salivary glands, resulting in the production of large amounts of watery saliva. In contrast, sympathetic activation results in the secretion of a small volume of viscous saliva containing high enzyme concentrations. The reduced volume produces the sensation of a dry mouth.

CLINICAL BRIEF

Mumps

The **mumps** virus preferentially targets the salivary glands, most often the parotids, although other organs may also become infected. Infection most often occurs at 5–9 years of age. The first exposure stimulates antibody production and usually confers permanent immunity; active immunity can be conferred by immunization. In postadolescent males, the mumps virus may also infect the testes and cause sterility. Infection of the pancreas by the mumps virus may produce temporary or permanent diabetes; other organ systems, including the CNS, may be affected in severe cases. An effective mumps vaccine is available; widespread distribution has almost eliminated this disease in the United States.

THE TEETH [FIGURE 25.7a]

The movements of the tongue are important in passing food across the surfaces of the teeth. Teeth perform chewing, or **mastication** (mas-ti-KĀ-shun), of food. Mastication breaks down tough connective tissues and plant fibers and helps saturate the materials with salivary secretions and enzymes.

Figure 25.7a● is a sectional view through an adult tooth. The bulk of each tooth consists of a mineralized matrix similar to bone. This material, called **dentin** (DEN-tin), differs from bone because it does not contain living cells. Instead, cytoplasmic processes extend into the dentin from cells in the central **pulp cavity**. The pulp cavity is spongy and highly vascular. It receives blood vessels and nerves via a narrow tunnel, the **root canal**, which is located at the base, or **root**, of the tooth. The **dental artery**, **dental vein**, and **dental nerve** enter the root canal through the **apical foramen** to service the pulp cavity.

The root of the tooth is anchored into a bony socket, or *alveolus*. Collagen fibers of the **periodontal** (per-ē-ō-DON-tal) **ligament** extend from the dentin of the root to the alveolar bone, creating a strong articulation known as a *gomphosis*. p. 214 A layer of **cementum** (se-MEN-tum) covers the dentin of the root, providing protection and firmly anchoring the periodontal ligament. Cementum is very similar in histological structure to bone, and less resistant to erosion than dentin.

The **neck** of the tooth marks the boundary between the root and the **crown**. The crown is the visible portion of the tooth that projects above the soft tissue of the gingiva. Epithelial cells of the **gingival** (JIN-ji-val) **sulcus** form tight attachments to the tooth above the neck, preventing bacterial access to the lamina propria of the gingiva or the relatively soft cementum of the root. If this attachment breaks down, bacterial infection of the gingiva may occur. This condition is called *gingivitis*.

The dentin of the crown is covered by a layer of **enamel**. Enamel contains densely packed crystals of calcium phosphate and is the hardest biologically manufactured substance. Adequate amounts of calcium, phosphates, and vitamin D during childhood are essential if the enamel coating is to be complete and resistant to decay.

Types of Teeth [FIGURE 25.7b,c] There are four types of teeth, each with specific functions (Figure 25.7b,c●):

1. **Incisors** (in-SĪ-zerz) are blade-shaped teeth found at the front of the mouth. Incisors are useful for clipping or cutting, as when nipping off the tip of a carrot stick.
2. The **cuspids** (KUS-pidz), or *canines*, are conical with a sharp ridgeline and a pointed tip. They are used for tearing or slashing. A tough piece of celery might be weakened by the clipping action of the incisors, but then moved to one side to take advantage of the shearing action provided by the cuspids. Incisors and cuspids each have a single root.
3. **Bicuspids** (bī-KUS-pidz), or *premolars*, have one or two roots. Premolars have flattened crowns with prominent ridges. They are used for crushing, mashing, and grinding.
4. **Molars** have very large flattened crowns with prominent ridges and typically have three or more roots. Molars also have flattened crowns for crushing and grinding.

Dental Succession [FIGURE 25.7d,e] During development, two sets of teeth begin to form. The first to appear are the **deciduous teeth** (dē-SID-ū-us; *deciduus*, falling off), also called the *primary teeth*, *milk teeth*, or *baby teeth* (Figure 25.7d,e●). There are usually 20 deciduous teeth, five on each side of the upper and lower jaws. These teeth will be replaced by the adult **secondary dentition**, or *permanent dentition*. The larger adult jaws can accommodate more than 20 permanent teeth, and three additional molars appear on each side of the upper and lower jaws as the individual ages. These teeth extend the length of the tooth rows posteriorly and bring the permanent tooth count to 32.

On each side of the upper or lower jaw, the primary dentition consists of two incisors, one cuspid, and a pair of deciduous molars. These are gradually replaced by the permanent dentition.

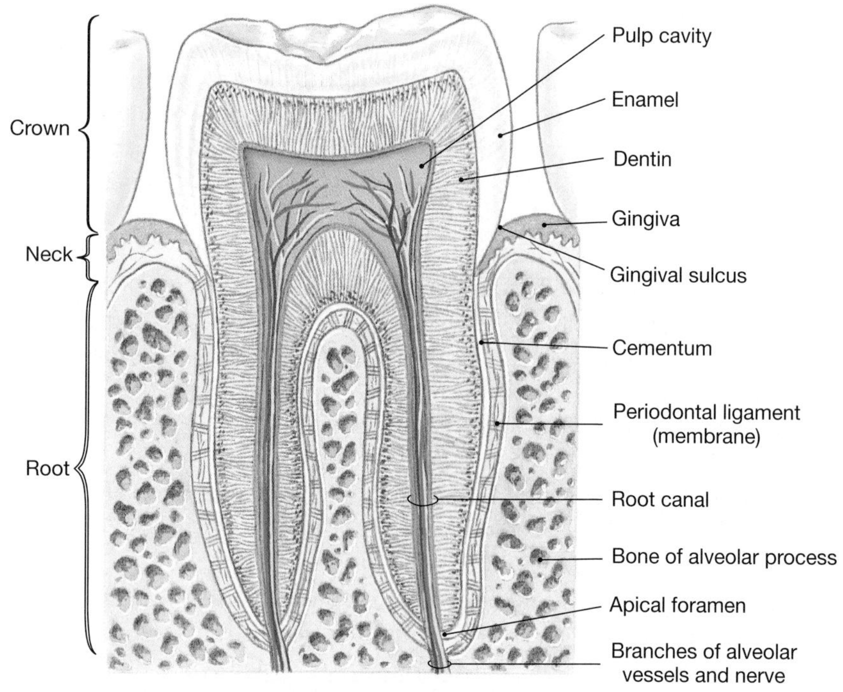

(a) Tooth, sectional view

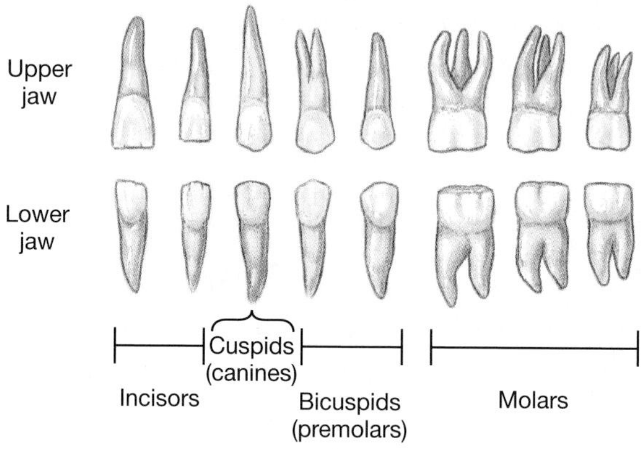

(b) Adult upper and lower teeth

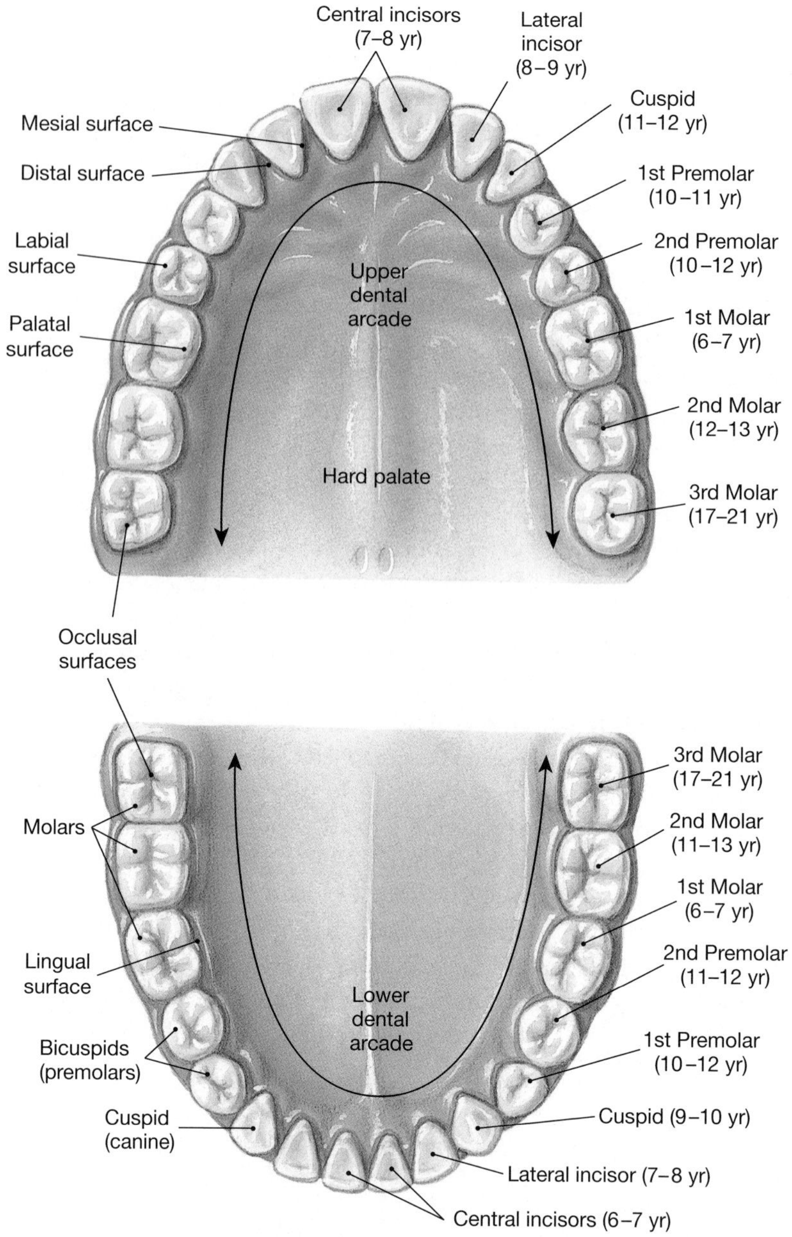

(c) Adult teeth, upper and lower jaws

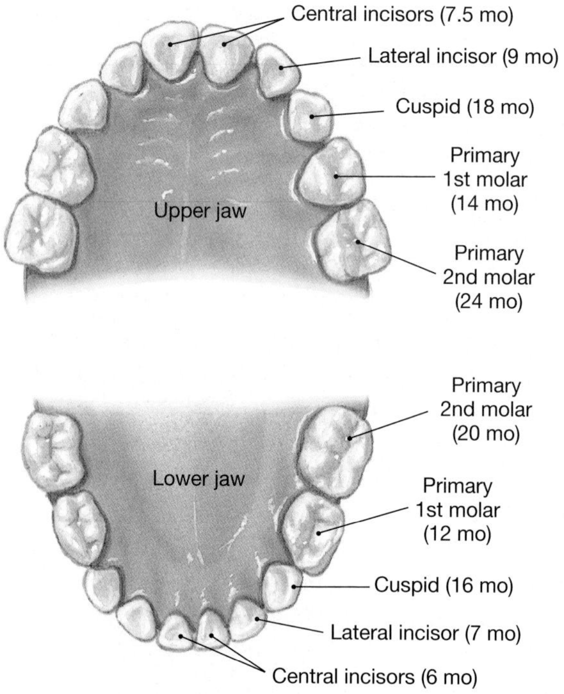

(d) The primary teeth

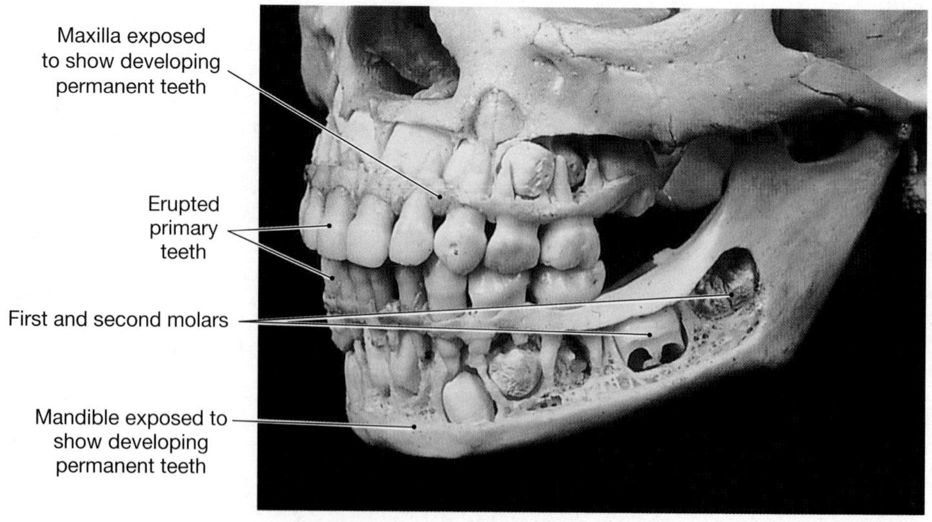

(e) Mandible and maxillary bones with unerupted teeth exposed

FIGURE 25.7 **TEETH**

Teeth perform chewing or mastication of food. **(a)** Diagrammatic section through a typical adult tooth. **(b)** The adult teeth. **(c)** Dental reference terms and the normal orientation of adult teeth. The normal range of ages at eruption for each tooth is shown in parentheses. **(d)** The primary teeth with the age at eruption given in months. **(e)** The skull of a 4-year-old child, with the maxillae and mandible cut away to expose the unerupted permanent teeth.

Figure 25.7d● indicates the sequence of eruption of the primary dentition and the approximate ages at their replacement. In this process the periodontal ligaments and roots of the primary teeth are eroded away, until they fall out or are pushed aside by the emergence, or **eruption**, of the secondary teeth. The adult premolars take the place of the deciduous molars, and the definitive adult molars extend the tooth row as the jaw enlarges. The third molars, or *wisdom teeth*, may not erupt before age 21, if they appear at all. Wisdom teeth often develop in inappropriate positions, and they may be unable to erupt properly.

A Dental Frame Of Reference [Figure 25.7b,c] The upper and lower rows of teeth form a curving **dental arcade**. Relative positions along the arch are indicated by the use of special terms (Figure 25.7b,c●). The terms **labial** or **buccal** refer to the outer surface of the dental arch, adjacent to the lips or cheeks. **Palatal** (upper) or **lingual** (lower) refers to the inner surface of the dental arch. **Mesial** (MĒ-zē-al) or **distal** refers to the opposing surfaces between the teeth in a single dental arch. The mesial surface is the side of a tooth that faces away from the last molar; the distal surface faces away from the first incisor. For example, the mesial surface of each canine faces the distal surface of the second incisor. The **occlusal surfaces** (o-KLOO-zal; *occlusio*, closed) of the teeth face their counterparts on the opposing dental arch. The occlusal surfaces perform the actual clipping, tearing, crushing, and grinding actions of the teeth.

Mastication The muscles of mastication close the jaws and slide or rock the lower jaw from side to side. ∞ p. 273 During mastication, food is forced back and forth between the vestibule and the rest of the oral cavity, crossing and recrossing the occlusal surfaces. This movement results in part from the action of the masticatory muscles, but control would be impossible without the aid of the buccal, labial, and lingual muscles. Once the material has been shredded or torn to a satisfactory consistency and moistened with salivary secretions, the tongue begins compacting the debris into a small oval mass, or **bolus**, that can be swallowed relatively easily.

CLINICAL BRIEF

Dental Problems

Most common dental problems result from the action of oral bacteria. Bacteria adhering to the surfaces of the teeth produce a sticky matrix that traps food particles and creates deposits of **plaque**. The mass of the plaque deposit protects the bacteria from salivary secretions, and as they digest nutrients, the bacteria generate acids that erode the structure of the tooth. The results are *dental caries*, otherwise known as cavities. Brushing the exposed surfaces of the teeth after meals helps to prevent the settling of bacteria and the entrapment of food particles, but bacteria between the teeth and within the gingival sulcus may elude the brush. Dentists therefore recommend the daily use of dental floss to clean these spaces.

If the bacteria remain within the gingival sulcus, the acids generated begin eroding the connections between the neck of the tooth and the gingiva. The gums appear to recede from the teeth, and **periodontal disease** develops. As it progresses, the bacteria attack the cementum, progressively destroying the periodontal ligament and eroding the alveolar bone. This deterioration loosens the tooth, and periodontal disease is the most common cause for the loss of teeth.

If teeth are broken or they must be removed because of disease, the usual treatment involves replacing them with "false teeth" attached to a plate or frame inserted into the mouth. Over the past 10 years, an alternative has been developed, using dental implants. A ridged titanium cylinder is inserted into the alveolus, and osteoblasts lock the ridges into the surrounding bone. After 4–6 months, an artificial tooth is screwed into the cylinder. Roughly 42% of individuals over age 65 have lost all of their teeth; the rest have lost an average of 10 teeth.

✓ CONCEPT CHECK

- What type of epithelium lines the oral cavity?
- What are the functions of saliva?
- What nutrient begins its chemical breakdown in the mouth?
- Pretend you are eating an apple. Summarize the actions of the teeth involved.

The Pharynx

The pharynx serves as a common passageway for food, liquids, and air. The epithelial lining and divisions of the pharynx, the nasopharynx, the oropharynx, and the laryngopharynx were described and illustrated in Chapter 24. ∞ pp. 636, 639 Deep to the lamina propria of the mucosa lies a dense layer of elastic fibers, bound to the underlying skeletal muscles. The specific pharyngeal muscles involved in swallowing, summarized below, were detailed in Chapter 10. ∞ p. 275

- The *pharyngeal constrictors* (superior, middle, and inferior) push the bolus toward the esophagus.
- The *palatopharyngeus* and *stylopharyngeus muscles* elevate the larynx.
- The *palatal muscles* raise the soft palate and adjacent portions of the pharyngeal wall.

The pharyngeal muscles cooperate with muscles of the oral cavity and esophagus to initiate the swallowing process, or **deglutition** (dē-gloo-TISH-un).

The Swallowing Process [Figure 25.8]

Swallowing is a complex process whose initiation is voluntarily controlled, but it proceeds involuntarily once initiated. It can be divided into *buccal*, *pharyngeal*, and *esophageal phases*. Key aspects of each phase are illustrated in Figure 25.8●.

1. The **buccal phase** begins with the compression of the bolus against the hard palate. Subsequent retraction of the tongue then forces the bolus into the pharynx and assists in the elevation of the soft palate by the palatal muscles, thereby isolating the nasopharynx (Figure 25.8a,b●). The buccal phase is strictly voluntary; once the bolus enters the oropharynx, involuntary reflexes are initiated, and the bolus is moved toward the stomach.
2. The **pharyngeal phase** begins as the bolus comes in contact with the palatal arches or the posterior pharyngeal wall or both (Figure 25.8c,d●). Elevation of the larynx (by the palatopharyngeus and stylopharyngeus muscles) and folding of the epiglottis direct the bolus past the closed glottis. In less than a second, the pharyngeal constrictor muscles have propelled the bolus into the esophagus. During the

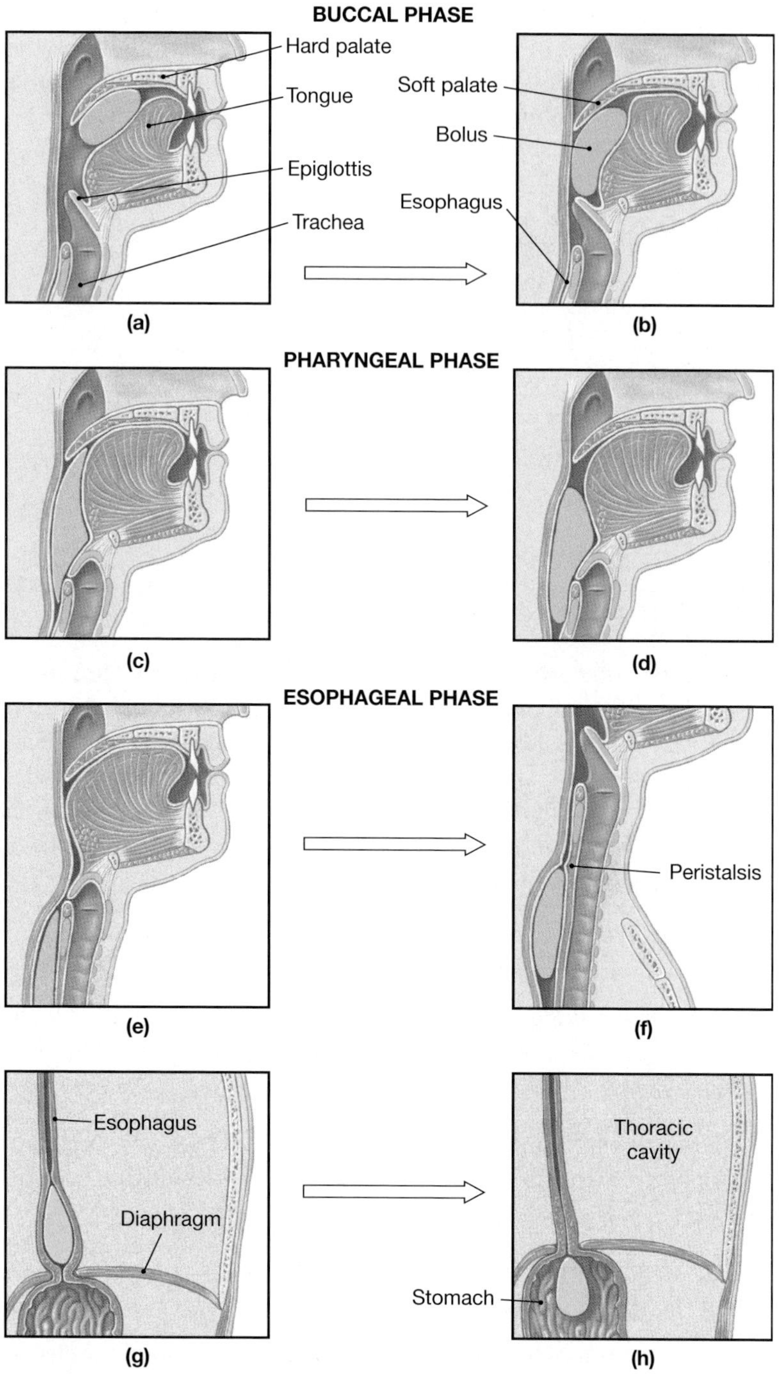

FIGURE 25.8 **The Swallowing Process**

This sequence, based on a series of X-rays, shows the stages of swallowing and the movement of materials from the mouth to the stomach. **(a,b)** buccal phase, **(c,d)** pharyngeal phase, and **(e–h)** esophageal phase.

time it takes for the bolus to travel through the pharynx and into the esophagus, the respiratory centers are inhibited and breathing ceases. p. 656

3. The **esophageal phase** of swallowing (Figure 25.8e–g●) starts with the opening of the **upper esophageal sphincter**. After passing through the open sphincter, the bolus is pushed along the length of the esophagus by peristaltic waves. The approach of the bolus triggers the opening of the weak **lower esophageal sphincter**, and the bolus then continues into the stomach (Figure 25.8g,h●). *Achalasia and Esophagitis p. 806*

The Esophagus [FIGURES 25.1/25.8]

The **esophagus** (Figure 25.1●, p. 664) is a hollow muscular tube that transports foods and liquids to the stomach. It is located posterior to the trachea (Figure 25.8●) and passes along the dorsal wall of the mediastinum in the thoracic cavity and enters the peritoneal cavity through an opening in the diaphragm, the **esophageal hiatus** (hī-Ā-tus), before emptying into the stomach. p. 282 The esophagus is approximately 25 cm (1 ft) long and about 2 cm (0.75 in.) in diameter. It begins at the level of the cricoid cartilage anterior to vertebra C_6 and ends anterior to vertebra T_7.

The esophagus receives blood from the esophageal arteries and branches of (1) the *thyrocervical trunk* and *external carotid arteries* of the neck, (2) the *bronchial arteries* and *esophageal arteries* of the mediastinum, and (3) the *inferior phrenic artery* and *left gastric artery* of the abdomen. Venous blood from the esophageal capillaries collects into the *esophageal, inferior thyroid, azygos,* and *gastric veins*. The esophagus is innervated by the vagus and sympathetic trunks via the *esophageal plexus*. pp. 458, 589, 591 Neither the upper nor the lower portion of the esophagus has a well-defined sphincter muscle comparable to those located elsewhere along the digestive tract. Nevertheless, the terms *upper esophageal sphincter* and *lower esophageal sphincter (cardiac sphincter)* are used to describe these regions because they are similar in function to other sphincters.

Histology of the Esophageal Wall [FIGURES 25.2/25.9]

The wall of the esophagus has mucosal, submucosal, and muscularis layers comparable to those described above (Figure 25.2●, p. 665). There are several distinctive features of the esophageal wall, shown in Figure 25.9●:

- The mucosa of the esophagus contains an abrasion-resistant stratified squamous epithelium.
- The mucosa and submucosa are thrown into large folds that run the length of the esophagus. These folds permit expansion during the passage of a large bolus; except during swallowing, muscle tone in the walls keeps the lumen closed.
- The smooth muscle layer of the muscularis mucosae may be very thin or absent near the pharynx, but it gradually thickens to 200–400μm as it approaches the stomach.
- The submucosa contains scattered *esophageal glands*. These simple, branched, tubular glands produce a mucous secretion that lubricates the bolus and protects the epithelial surface.
- The muscularis externa has inner circular and outer longitudinal muscle layers. In the superior third of the esophagus both layers contain skeletal muscle fibers; in the middle third there is a mixture of skeletal and smooth muscle tissue; along the inferior third only smooth muscles are found. The skeletal muscle and smooth muscle in the esophagus are controlled by visceral reflexes, and you do not have voluntary control over these contractions.
- There is no serosa. A layer of connective tissue outside the muscularis externa anchors the esophagus in position against the dorsal body wall. This outer fibrous layer is called the *adventitia*. Over the 1-2 cm between the diaphragm and the stomach the esophagus is retroperitoneal, with peritoneum covering the anterior and left lateral surfaces.

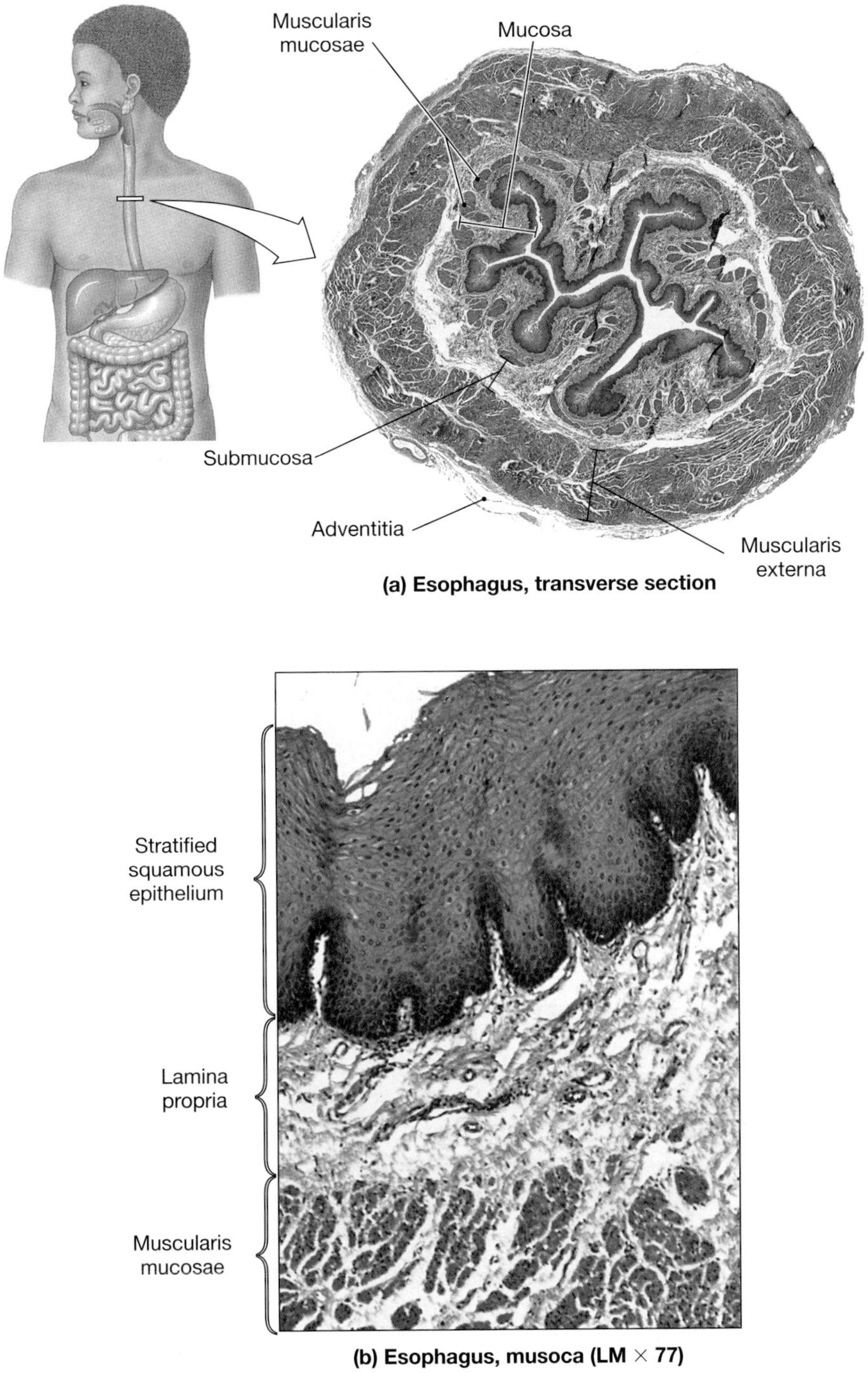

FIGURE 25.9 **THE ESOPHAGUS**

(**a**) Low-power view of a section through the esophagus. (**b**) The esophageal mucosa.

✓ CONCEPT CHECK

- Where is the fauces?
- What is occurring when the soft palate and larynx are elevated and the glottis closes?
- Which stage of swallowing is voluntary?
- Emily is experiencing heartburn. What is happening in her esophagus?

The Stomach

The stomach performs three major functions: (1) the bulk storage of ingested food, (2) the mechanical breakdown of ingested food, and (3) chemical digestion of ingested food through the disruption of chemical bonds by acids and enzymes. The mixing of ingested substances with the gastric juices secreted by the glands of the stomach produces a viscous, soupy mixture called **chyme** (KĪM).

Anatomy of the Stomach [FIGURES 25.10/25.11/25.12/25.13a]

The stomach has the shape of an expanded J (Figures 25.10 and 25.11●). It occupies the *left hypochondriac, epigastric*, and portions of the *umbilical* and *left lumbar regions* (Figure 25.12●, p. 680). The shape and size of the stomach are extremely variable from individual to individual and from one meal to the next.

The J-shaped stomach has a short **lesser curvature**, forming the **medial surface** of the organ, and a long **greater curvature**, which forms the **lateral surface**. In an average stomach the lesser curvature has a length of approximately 10 cm (4 in.) and the greater curvature measures about 40 cm (16 in.). The **anterior** and **posterior surfaces** are smoothly rounded. The stomach typically extends between the levels of vertebrae T_7 and L_3.

The stomach is divided into four regions (Figures 25.10 and 25.11●):

1. The esophagus contacts the medial surface of the stomach at the **cardia** (KAR-dē-a), so named because of its proximity to the heart. It consists of the superior, medial portion of the stomach within 3 cm (1.2 in.) of the junction between the stomach and the esophagus. The esophageal lumen opens into the cardia at the **cardiac orifice**.
2. The region of the stomach superior to the gastroesophageal junction is the **fundus** (FUN-dus). The fundus contacts the inferior and posterior surface of the diaphragm.
3. The area between the fundus and the curve of the J is the **body** of the stomach. The body is the largest region of the stomach; it functions as a mixing tank for ingested food and gastric secretions.
4. The **pylorus** (pī-LŌR-us) is the curve of the J. The pyloris is divided into the *pyloric antrum*, which is connected to the body of the stomach, and the *pyloric canal*, which is connected to the *duodenum*, the proximal segment of the small intestine. As mixing movements occur during digestion, the pylorus frequently changes shape. A muscular **pyloric sphincter** regulates the release of chyme from the **pyloric outlet** into the duodenum.

The volume of the stomach increases at mealtimes and decreases as chyme leaves the stomach and enters the small intestine. In the relaxed (empty) stomach, the mucosa is thrown into a number of prominent longitudinal folds, called **rugae** (ROO-gē; "wrinkles") (Figure 25.11a●). Rugae permit expansion of the gastric lumen. As expansion occurs, the epithelial lining, which cannot stretch, flattens out, and the rugae become less prominent. In a full stomach, the rugae almost disappear.

MESENTERIES OF THE STOMACH [FIGURES 25.4/25.10a]

The visceral peritoneum covering the outer surface of the stomach is continuous with a pair of prominent mesenteries. The **greater omentum** forms an enormous pouch that hangs like an apron from the greater curvature of the stomach. The greater omentum lies posterior to the abdominal

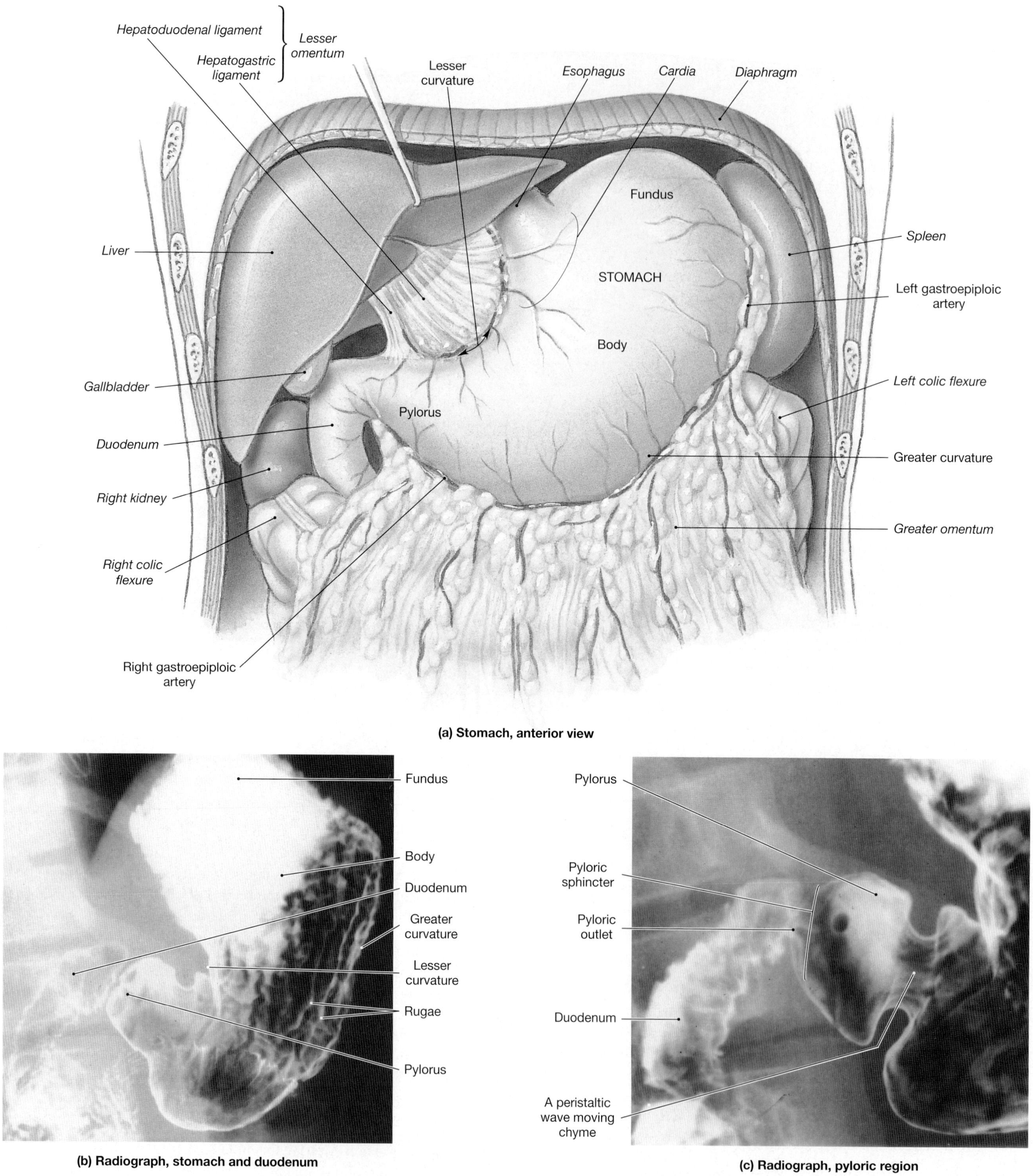

FIGURE 25.10 **The Stomach and Omenta**

(**a**) Surface anatomy of the stomach, showing blood vessels and relation to liver and intestines. (**b**) Radiograph of the stomach and duodenum, after swallowing a barium solution to increase contrast. (**c**) Radiograph of the pyloric region, pyloric valve, and duodenum.

FIGURE 25.11 **Gross Anatomy of the Stomach**

(**a**) External and internal anatomy of the stomach. (**b**) Anterior view of the superior portion of abdominal cavity after removal of the left lobe of the liver and the lesser omentum. Note the position and orientation of the stomach.

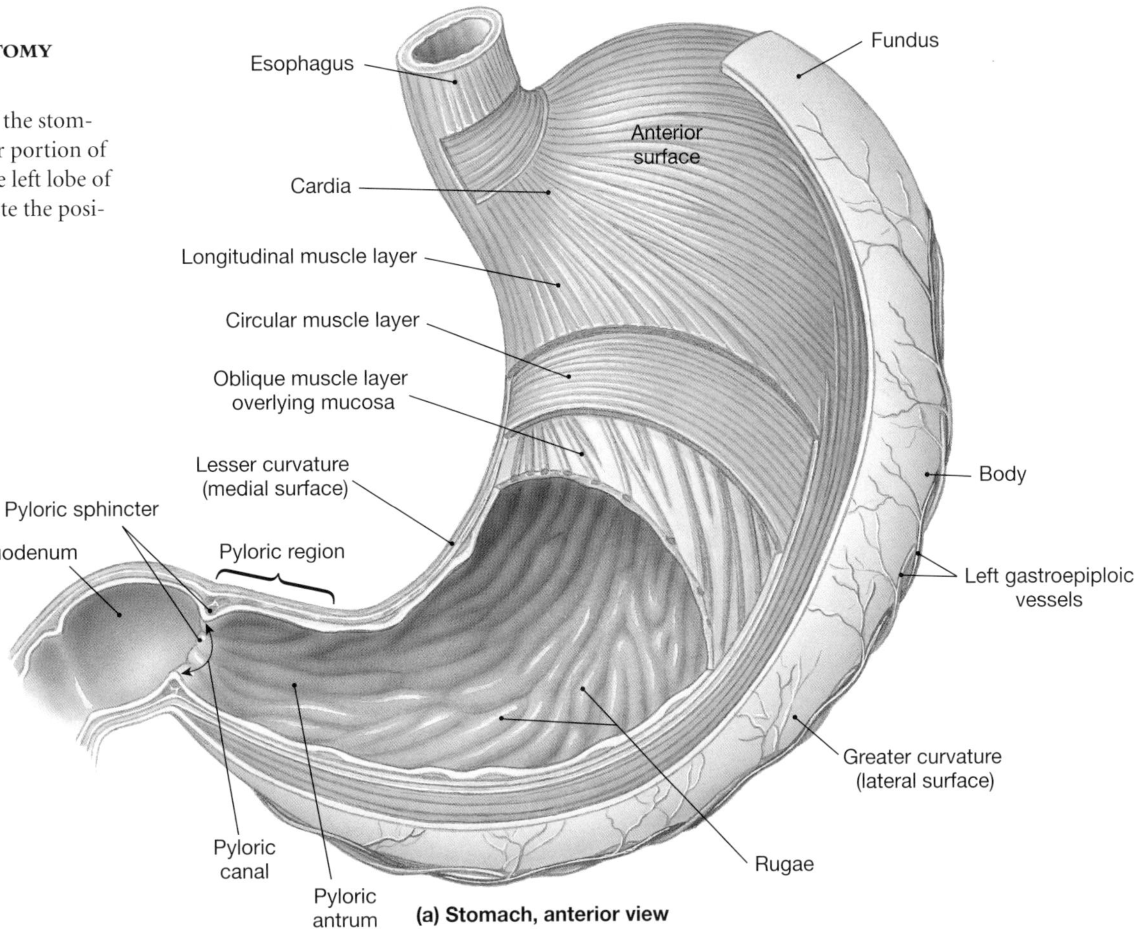

(a) Stomach, anterior view

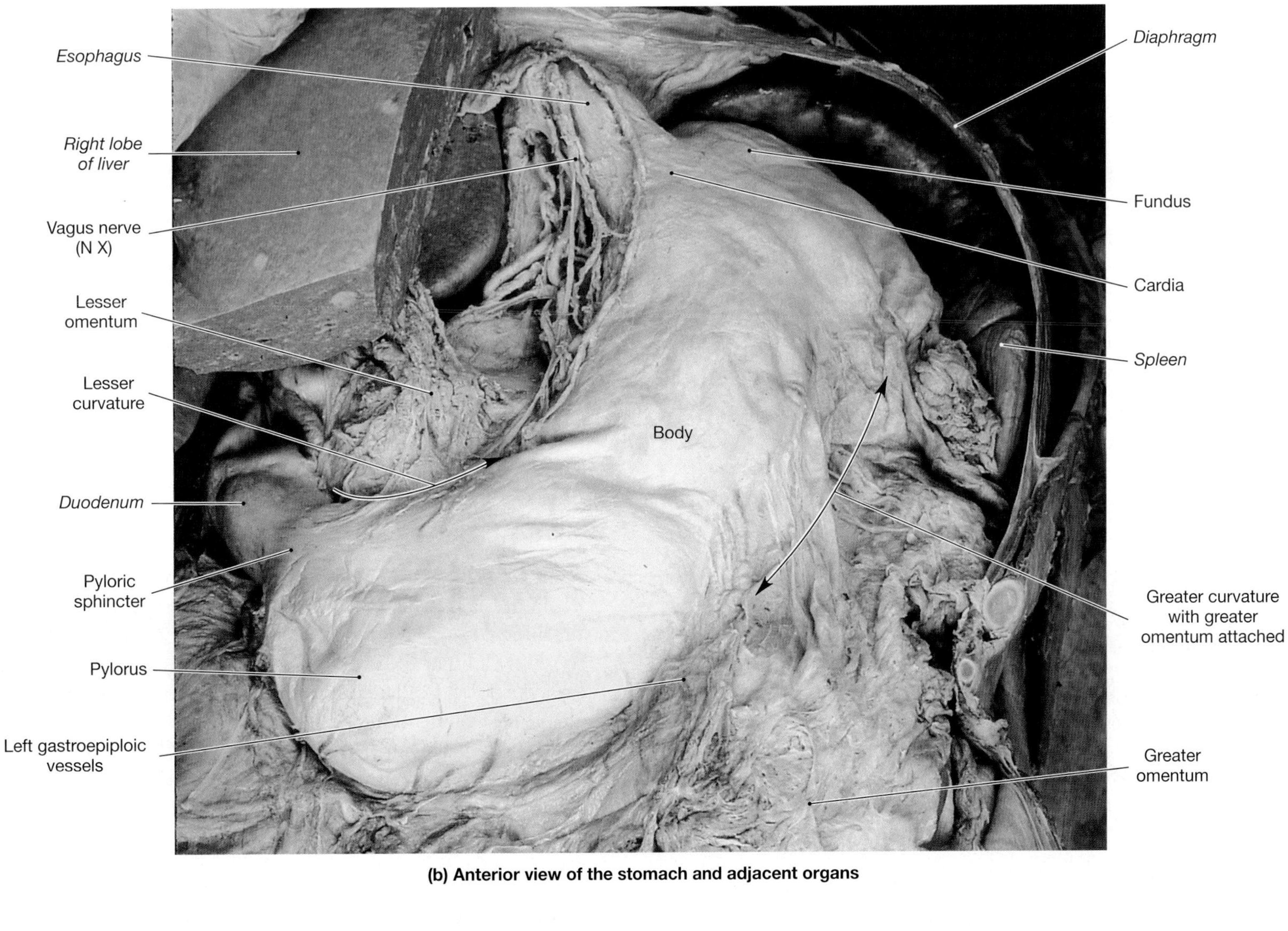

(b) Anterior view of the stomach and adjacent organs

FIGURE 25.12
ABDOMINAL REGIONS AND PLANES

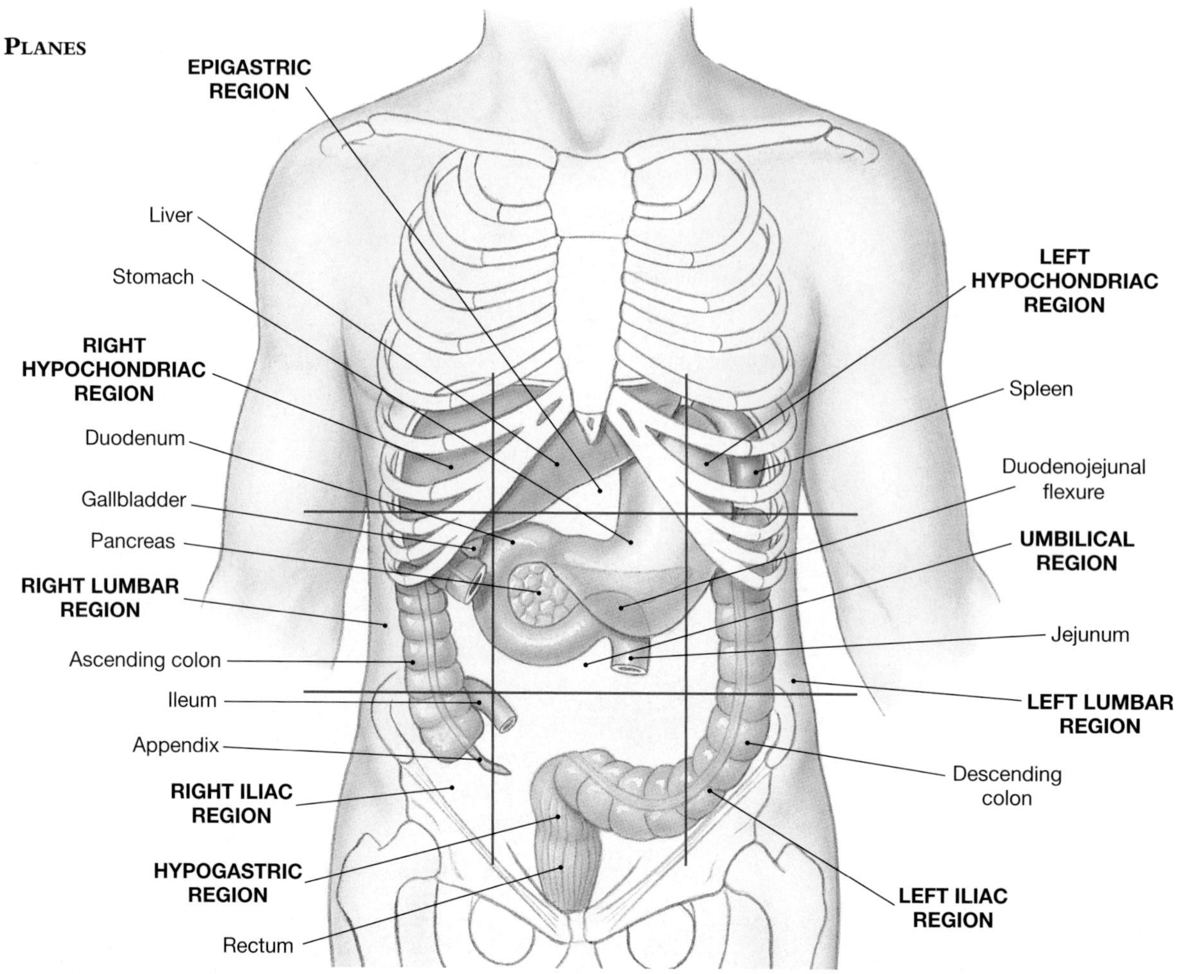

wall and anterior to the abdominal viscera (Figures 25.4, p. 668, and 25.10a●). Adipose tissue in the greater omentum conforms to the shapes of the surrounding organs; it provides padding and protects the anterior and lateral surfaces of the abdomen. The lipids in the adipose tissue represent an important energy reserve, and the greater omentum provides insulation that reduces heat loss across the anterior abdominal wall. The *lesser omentum* is a much smaller pocket in the ventral mesentery between the lesser curvature of the stomach and the liver. The lesser omentum stabilizes the position of the stomach and provides an access route for blood vessels and other structures entering or leaving the liver.

BLOOD SUPPLY TO THE STOMACH [FIGURES 22.17/22.26/25.10a]

The three branches of the celiac artery supply blood to the stomach (Figures 22.17, p. 591 and 25.10a●):

- The *left gastric artery* supplies blood to the lesser curvature and cardia.
- The *splenic artery* supplies the fundus and greater curvature through the *left gastroepiploic artery*.
- The *common hepatic artery* supplies blood to the lesser and greater curvatures of the pylorus through the *right gastric, right gastroepiploic artery* and the *gastroduodenal artery*. *Gastric* and *gastroepiploic veins* drain blood from the stomach into the hepatic portal vein (Figure 22.26●, p. 601).

MUSCULATURE OF THE STOMACH [FIGURE 25.11a]

The muscularis mucosae and muscularis externa of the stomach contain extra layers of smooth muscle in addition to the usual *circular* and *longitudinal layers*. The muscularis mucosae usually contains an additional outer, circular layer of muscle fibers. The muscularis externa has an extra inner, *oblique layer* of smooth muscle (Figure 25.11a●). The extra layers of smooth muscle strengthen the stomach wall, and they perform the mixing and churning activities essential to the formation of chyme.

Histology of the Stomach [FIGURE 25.13]

A simple columnar epithelium lines all regions of the stomach. The epithelium is a *secretory sheet* that produces a carpet of mucous that covers the lumenal surfaces of the stomach. p. 75 The mucous layer protects the epithelium against the acids and enzymes in the gastric lumen. Shallow depressions, called **gastric pits**, open onto the gastric surface (Figure 25.13●). The mucous cells at the base, or *neck*, of each gastric pit actively divide to replace superficial cells that are shed continuously into the chyme. The continual replacement of epithelial cells provides an additional defense against the gastric contents. If stomach acid and digestive enzymes penetrate the mucous layers, any damaged epithelial cells are quickly replaced.

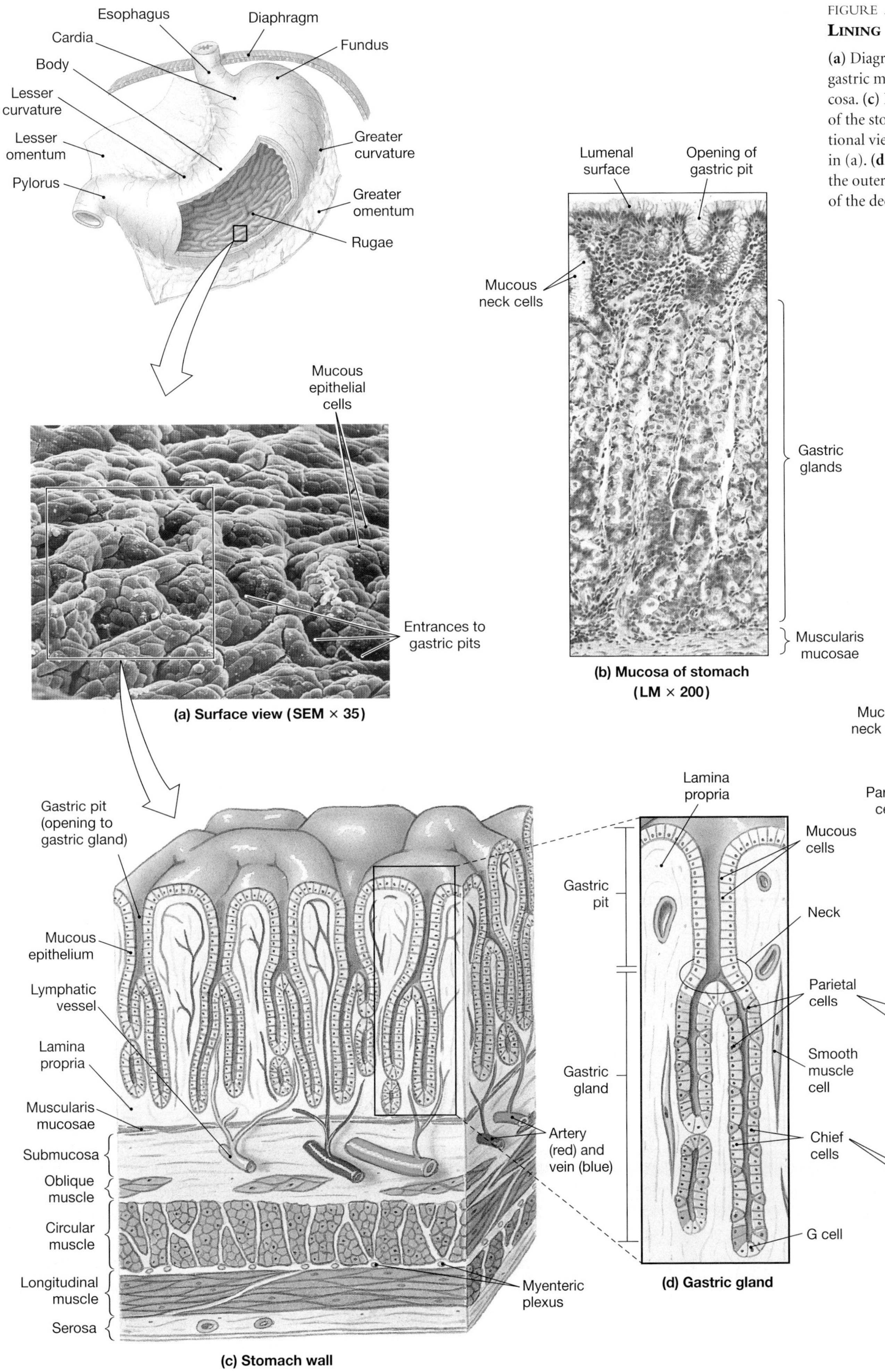

(a) Surface view (SEM × 35)

(b) Mucosa of stomach (LM × 200)

(c) Stomach wall

(d) Gastric gland

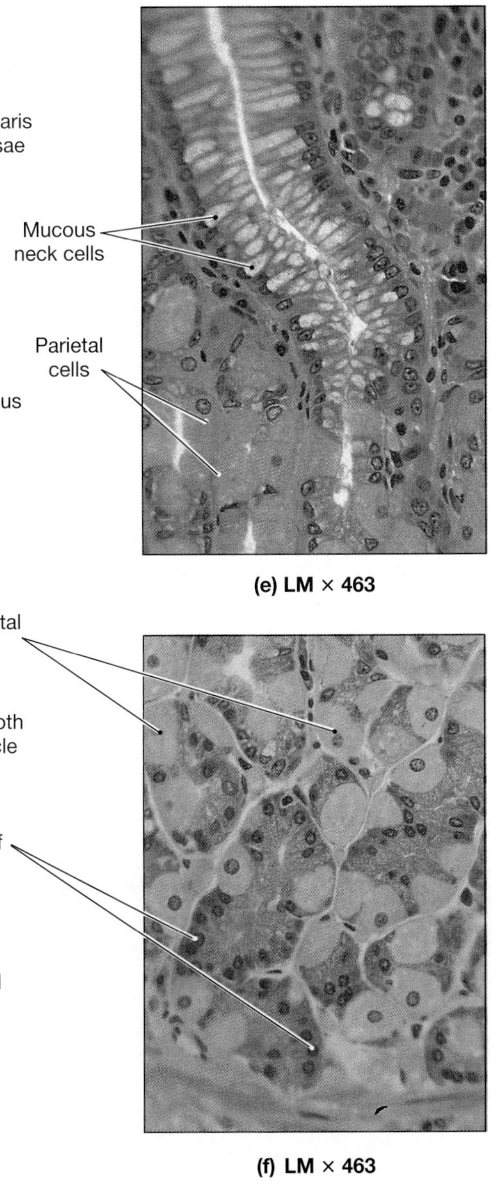

(e) LM × 463

(f) LM × 463

FIGURE 25.13 **THE STOMACH LINING**

(**a**) Diagrammatic view and colorized SEM of the gastric mucosa. (**b**) Micrograph of the gastric mucosa. (**c**) Diagrammatic view of the organization of the stomach wall. This corresponds to a sectional view through the area indicated by the box in (a). (**d**) A gastric gland. (**e**) The parietal cells of the outer portion of a gastric gland. (**f**) Chief cells of the deepest portions of a gastric gland.

GASTRIC SECRETORY CELLS

In the fundus and body of the stomach, each gastric pit communicates with several **gastric glands** that extend deep into the underlying lamina propria. Gastric glands (Figure 25.13b●) are simple branched tubular glands dominated by three types of secretory cells: *parietal cells*, *chief cells*, and *enteroendocrine cells* scattered between the other two cell types (Figure 25.12c–f●). The parietal and chief cells work together to secrete about 1500 ml of **gastric juice** each day.

PARIETAL CELLS **Parietal cells**, or *oxyntic cells*, are especially common along the proximal portions of each gastric gland. These cells secrete *intrinsic factor* and *hydrochloric acid* (HCl). **Intrinsic factor** facilitates the absorption of vitamin B_{12} across the intestinal lining. Vitamin B_{12} is necessary for normal erythropoiesis. p. 545 Hydrochloric acid lowers the pH of the gastric juice, kills microorganisms, breaks down cell walls and connective tissues in food, and activates the secretions of the chief cells.

CHIEF CELLS **Chief cells**, or *zymogen cells*, are most abundant near the base of a gastric gland. These cells secrete **pepsinogen** (pep-SIN-ō-jen), which is converted by the acids in the gastric lumen to an active proteolytic enzyme, **pepsin**. The stomachs of newborn infants (but not adults) also produce **rennin** and **gastric lipase**, enzymes important for the digestion of milk. Rennin coagulates milk proteins, and gastric lipase initiates the digestion of milk fats.

ENTEROENDOCRINE CELLS **Enteroendocrine** (en-ter-ō-EN-dō-krin) **cells** are scattered among the parietal and chief cells. These cells produce at least seven different secretions. **G cells** are enteroendocrine cells that are most abundant in the gastric pits of the pyloric region. They secrete the hormone **gastrin** (GAS-trin). Gastrin, which is released when food enters the stomach, stimulates the secretory activity of both parietal and chief cells. It also promotes smooth muscle activity in the stomach wall. This enhances mixing and churning activity.

CLINICAL BRIEF

GASTRITIS AND PEPTIC ULCERS

Inflammation of the gastric mucosa causes **gastritis** (gas-TRĪ-tis), a superficial inflammation of the stomach lining. This condition may develop after swallowing drugs, including alcohol and aspirin. Gastritis may also appear after severe emotional or physical stress, bacterial infection of the gastric wall, or the ingestion of strong acid or alkaline chemicals. Infections involving the bacterium *Helicobacter pylori* are an important factor in ulcer formation; an *ulcer* is a distinct, localized erosion of the stomach lining. These bacteria are able to survive long enough to penetrate the mucus that coats the epithelium. Once within the protective layer of mucus, they are safe from the action of gastric acids and enzymes. Over time, the infection damages the epithelial lining, with two major results: (1) erosion of the lamina propria by gastric juices and (2) entry and spread of the bacteria through the gastric wall and into the bloodstream.

A **peptic ulcer** develops when the digestive acids and enzymes manage to erode their way through the defenses of the stomach lining or proximal portions of the small intestine. The locations may be indicated by using the terms **gastric ulcer** (stomach) or **duodenal ulcer** (duodenum). Peptic ulcers result from the excessive production of acid or the inadequate production of the alkaline mucus that poses an epithelial defense. An estimated 50–80% of gastric ulcers and 95% of duodenal ulcers involve people infected with *H. pylori*.

Once gastric juices have destroyed the epithelial layers, the virtually defenseless lamina propria will be exposed to digestive attack. Sharp abdominal pain results, and bleeding can develop. The administration of antacids can often control peptic ulcers by neutralizing the acids and allowing time for the mucosa to regenerate. Several drugs inhibit acid production and secretion. Treatment with an intense 2-week course of multiple antibiotics can eliminate *H. pylori* and cure most ulcers, although reinfection may occur. Dietary restrictions to limit the intake of acidic beverages and eliminate foods that promote acid production (caffeine) or damage unprotected mucosal cells (alcohol) also help. In severe cases the damage may provoke significant bleeding, and the acids may even erode their way through the wall of the digestive tract and enter the peritoneal cavity. This condition, called a **perforated ulcer**, requires immediate surgical correction.

Regulation of the Stomach

The production of acid and enzymes by the gastric mucosa can be directly controlled by the central nervous system and indirectly regulated by local hormones. CNS regulation involves both the vagus nerve (parasympathetic innervation) and branches of the celiac plexus (sympathetic innervation). The sight or thought of food triggers motor output in the vagus nerve. Postganglionic parasympathetic fibers innervate parietal cells, chief cells, and mucous cells of the stomach. Stimulation causes an increase in the production of acids, enzymes, and mucus. The arrival of food in the stomach stimulates stretch receptors in the stomach wall and chemoreceptors in the mucosa. Reflexive contractions occur in the muscularis layers of the stomach wall, and gastrin is released by enteroendocrine cells. Both parietal and chief cells respond to the presence of gastrin by accelerating their secretory activities: parietal cells are especially sensitive to gastrin, so the rate of acid production increases more dramatically than the rate of enzyme secretion.

Sympathetic activation leads to the inhibition of gastric activity. In addition, two hormones released by the small intestine inhibit gastric secretion. The release of these hormones, **secretin** (se-KRĒ-tin) and **cholecystokinin** (kō-lē-sis-tō-KĪ-nin), stimulates secretion by both the pancreas and liver; the depression of gastric activity is a secondary but complementary effect. *Stomach Cancer p. 806*

✓ CONCEPT CHECK

- What are the functions of the greater omentum?
- How do the cells that line the stomach keep from becoming damaged in such an acidic environment?
- What do chief cells secrete?
- What hormone stimulates the secretion of parietal and chief cells?

The Small Intestine

[FIGURES 25.2a,b/25.10/25.12/25.14 TO 25.16]

The **small intestine** plays the primary role in the digestion and absorption of nutrients. The small intestine averages 6 m (20 ft) in length (range, 15–25 ft) and has a diameter ranging from 4 cm (1.6 in.) at the stomach to about 2.5 cm (1 in.) at the junction with the large intestine. It occupies all abdominal regions except the left hypochondriac and epigastric regions (Figures 25.12, p. 680, and 25.14●). Ninety percent of nutrient absorption occurs in the small intestine, and most of the rest occurs in the large intestine.

The small intestine fills much of the peritoneal cavity: its position is stabilized by mesenteries attached to the dorsal body wall. Movement of

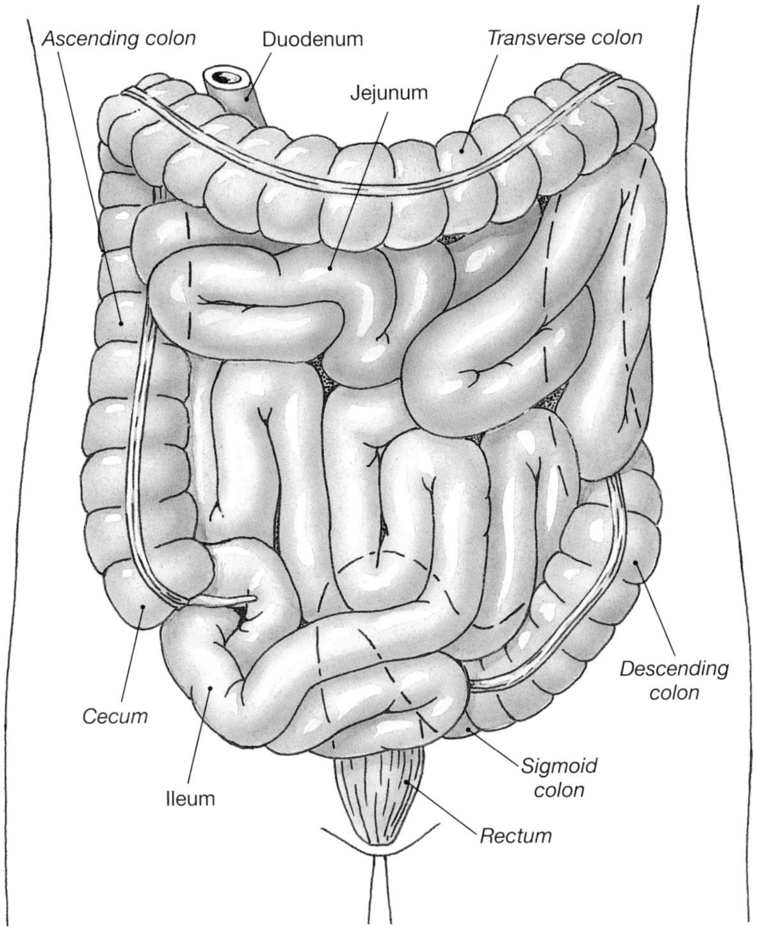

FIGURE 25.14 **REGIONS OF THE SMALL INTESTINE**

The color coding indicates the relative sizes and positions of the duodenum, jejunum, and ileum.

> CLINICAL BRIEF
>
> **DRASTIC WEIGHT LOSS TECHNIQUES**
>
> At any given moment, an estimated 30% of the U.S. population is overweight and most are dieting to promote weight loss. "Fat farms" and exercise clubs have sprung up across the country, and there has been an increase in the use of surgery to promote weight loss. Many of these techniques involve surgically remodeling the gastrointestinal tract. **Gastric stapling** attempts to correct an overeating problem by reducing the size of the stomach. A large portion of the gastric lumen is stapled closed, leaving only a small pouch in communication with the esophagus and duodenum. After this surgery the individual will be able to eat only a small amount before the stretch receptors in the gastric wall become stimulated and the individual feels "full." Gastric stapling is a major surgical procedure, and in addition, the smooth muscle in the wall of the functional portion of the stomach gradually becomes increasingly tolerant of distension, and the operation may have to be repeated.
>
> In *gastric bypass surgery,* much of the stomach is bypassed, with only a small remnant pouch connected to the jejunum. Gastric stapling and gastric bypass surgery both produce weight loss. One contributing factor is that the anatomic changes force modification in the amounts eaten and frequency of meals. However, vitamin deficiencies as well as iron and other mineral deficiencies may occur, and nutritional supplements are recommended. Surgical and metabolic complications for these procedures are less than those of *intestinal bypass surgery.* This surgery practice, now avoided, involved the removal of large portions of the jejunum.

the small intestine during digestion is restricted by the stomach, the large intestine, the abdominal wall, and the pelvic girdle. Figures 25.10, p. 678, 25.12, and 25.14● show the position of the small intestine relative to the other segments of the digestive tract.

The intestinal lining bears a series of transverse folds called **plicae circulares** (sir-kū-LAR-ēs) (Figures 25.2a,b, p. 665, 25.15a, and 25.16b●). Unlike the rugae in the stomach, each plica is a permanent feature of the intestinal lining; plicae do not disappear as the small intestine fills. Roughly 800 plicae are found along the length of the small intestine, and their presence greatly increases the surface area available for absorption.

Regions of the Small Intestine [FIGURE 25.14]

The small intestine has three anatomical subdivisions: the *duodenum*, the *jejunum*, and the *ileum* (Figure 25.14●).

THE DUODENUM [FIGURES 25.4b/25.10b,c/25.12/25.14]

The **duodenum** (dū-ō-DĒ-num) (Figures 25.10b,c, p. 678, 25.12, and 25.14●) is the shortest and widest segment of the small intestine; it is approximately 25 cm (10 in.) long. The duodenum is connected to the pylorus of the stomach, and the interconnection is guarded by the pyloric sphincter. From its start at the pyloric sphincter, the duodenum curves in a C that encloses the pancreas. Except for the proximal 2.5 cm (1 in.), the duodenum is in a retroperitoneal position between vertebrae L_1 and L_4 (Figure 25.4b●, p. 668).

The duodenum is a "mixing bowl" that receives chyme from the stomach and digestive secretions from the pancreas and liver. Almost all essential digestive enzymes enter the small intestine from the pancreas.

THE JEJUNUM

A rather abrupt bend, the *duodenojejunal flexure,* marks the boundary between the duodenum and the **jejunum** (je-JOO-num) (Figure 25.12●). At this junction the small intestine reenters the peritoneal cavity, supported by a sheet of mesentery. The jejunum is about 2.5 m (8 ft) long. The bulk of chemical digestion and nutrient absorption occurs in the jejunum.

THE ILEUM

The **ileum** (IL-ē-um) is the third and last segment of the small intestine. It is also the longest, averaging 3.5 m (12 ft) in length. The ileum ends at a sphincter, the *ileocecal valve*, which controls the flow of materials from the ileum into the *cecum* of the large intestine. The ileocecal valve protrudes into the cecum.

Support of the Small Intestine [FIGURES 22.17/22.26/25.4]

The duodenum has no supporting mesentery. The proximal 2.5 cm is movable, but the rest is retroperitoneal and fixed in position. The jejunum and ileum are supported by an extensive, fan-shaped mesentery known as the **mesentery proper** (Figure 25.4●, p. 668). Blood vessels, lymphatics, and nerves reach these segments of the small intestine by passing through the connective tissue of the mesentery. The blood vessels involved are *intestinal arteries,* branches of the *superior mesenteric artery* and *superior mesentric*

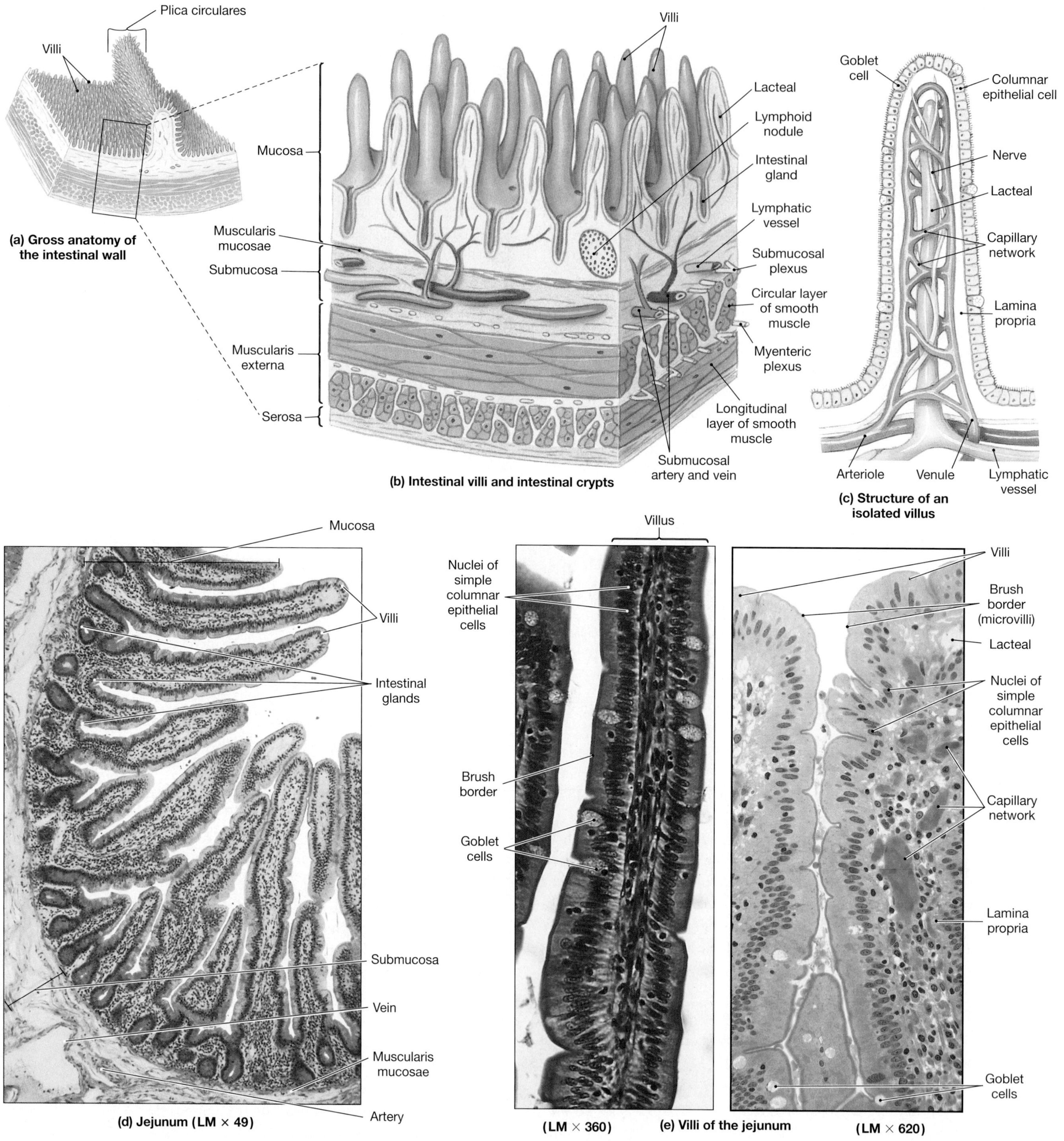

FIGURE 25.15 **The Intestinal Wall**

(**a**) Characteristic features of the intestinal lining. (**b**) The organization of villi and the intestinal crypts. (**c**) Diagrammatic view of a single villus, showing the capillary and lymphatic supply. (**d**) Panoramic view of the wall of the small intestine showing mucosa with characteristic villi, plica, submucosa, and muscularis layers. (**e**) Photomicrographs of villi from the jejunum.

vein (Figures 22.17 and 22.26●). pp. 591, 601 Parasympathetic innervation is provided by the vagus nerve; sympathetic innervation involves postganglionic fibers from the superior mesenteric ganglion. p. 452

Histology of the Small Intestine

THE INTESTINAL EPITHELIUM [FIGURES 25.15/25.16]

The mucosa of the small intestine forms a series of fingerlike projections, the **intestinal villi** (Figures 25.15 and 25.16●), that project into the lumen. Each villus is covered by a simple columnar epithelium. The apical surfaces of the epithelial cells are carpeted with microvilli, and the cells are often said to have a "brush border." If the small intestine were a simple tube with smooth walls, it would have a total absorptive area of roughly 0.33 m^2. Instead, the epithelium contains plicae circulares, each plica supports a forest of villi, and each villus is covered by epithelial cells whose exposed surfaces contain microvilli. This arrangement increases the total area for absorption to over 200 m^2.

INTESTINAL CRYPTS [FIGURE 25.15b,d]

Between the columnar epithelial cells, goblet cells eject mucins onto the intestinal surfaces. At the bases of the villi are found the entrances to the **intestinal glands**, or *crypts* (KRIPTS) *of Lieberkühn*. These pockets extend deep into the underlying lamina propria (Figure 25.15b,d●). Near the base of each gland, stem cell divisions continually produce new generations of epithelial cells. These new cells are continually displaced toward the intestinal surface, and within a few days they will have reached the tip of a villus, where they are shed into the intestinal lumen. This ongoing process renews the epithelial surface and adds intracellular enzymes to the chyme.

Intestinal glands also contain enteroendocrine cells responsible for the production of several intestinal hormones, including cholecystokinin and secretin, and enzymes with antibacterial activity.

THE LAMINA PROPRIA [FIGURE 25.15b,c,e]

The lamina propria of each villus contains an extensive network of capillaries that absorbs and carries nutrients to the hepatic portal circulation. In addition to capillaries and nerve endings, each villus contains a terminal lymphatic called a **lacteal** (LAK-tē-al; *lacteus*, milky) (Figure 25.15b,c,e●). Lacteals transport materials that cannot enter local capillaries. These materials, such as large lipid-protein complexes, ultimately reach the venous circulation by way of the thoracic duct. The name *lacteal* refers to the pale, cloudy, milky appearance of lymph containing large quantities of lipids.

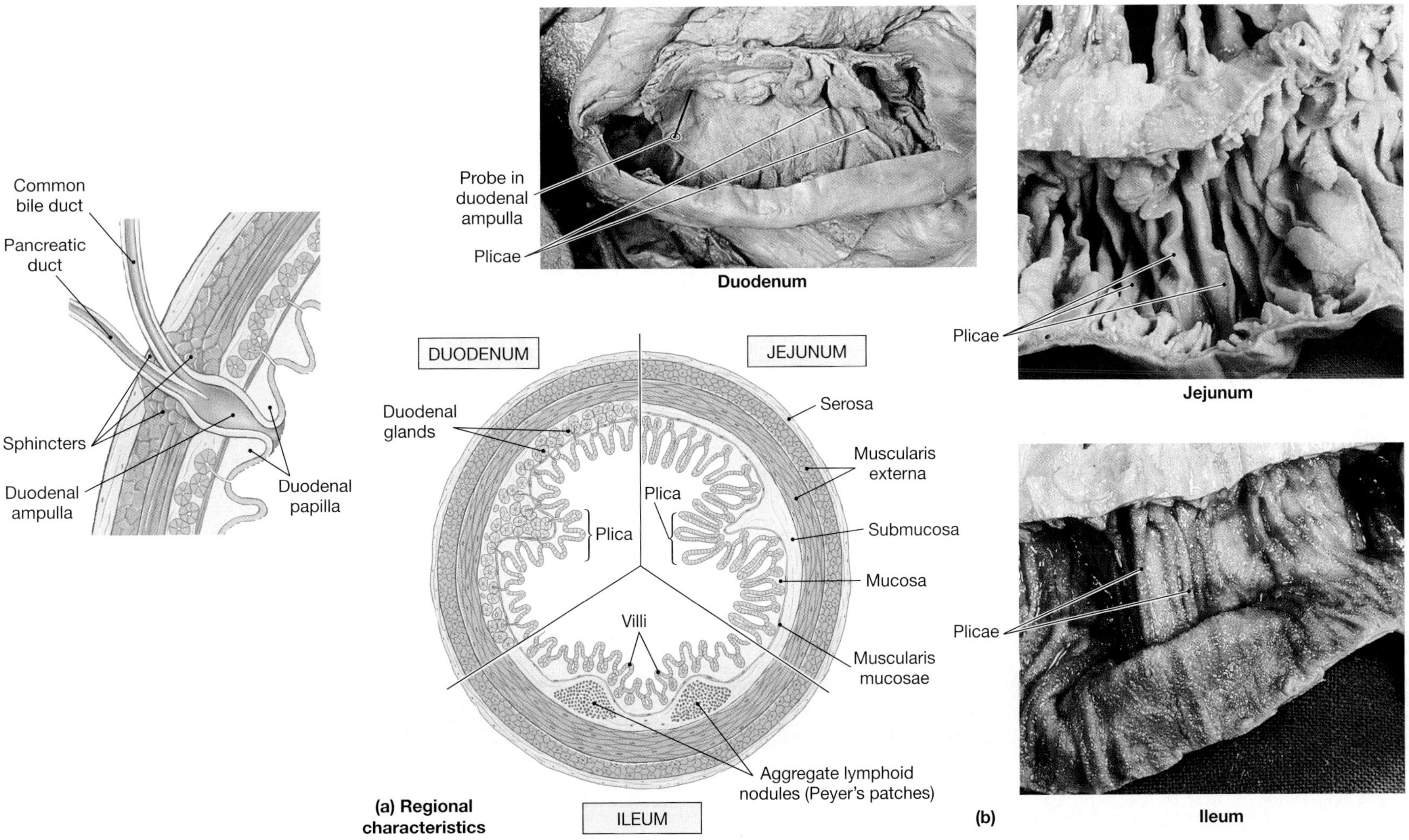

FIGURE 25.16 **Regions of the Small Intestine**

(a) Diagrammatic view highlighting the distinguishing features of each region of the small intestine. The detail shows a cross section through the ampulla of the duodenum. **(b)** The photographs show the gross anatomy of the intestinal lining in each region of the small intestine.

REGIONAL SPECIALIZATIONS [FIGURE 25.16]

The regions of the small intestine have histological specializations related to their primary functions. Representative sections from each region of the small intestine are presented in Figure 25.16●.

The Duodenum [FIGURES 25.16/25.22b] The duodenum contains numerous mucous glands. In addition to the intestinal glands, the submucosa contains **duodenal glands**, also known as *Brunner's glands*, that produce copious quantities of mucus (Figure 25.16a●). The mucus produced by the intestinal and submucosal glands protects the epithelium from the acid chyme arriving from the stomach. The mucus also contains buffers that help elevate the pH of the chyme. Submucosal glands are most abundant in the proximal portion of the duodenum, and their numbers decrease approaching the jejunum. Over this distance the pH of the intestinal contents changes from 1–2 to 7–8; by the beginning of the jejunum, the extra mucous production is no longer needed.

Buffers and enzymes from the pancreas and bile from the liver enter the duodenum roughly halfway along its length. Within the duodenal wall, the *common bile duct* from the liver and *pancreatic duct* from the pancreas come together at a muscular chamber called the **duodenal ampulla** (am-PYŪL-la), or **hepatopancreatic ampulla**. This chamber opens into the duodenal lumen at a small mound known as the **duodenal papilla**, or, when an accessory pancreatic duct is present, the *greater duodenal papilla*. (Figures 25.16 and 25.22b●, p. 693).

The Jejunum and Ileum [FIGURES 25.15d,e/25.16] Plicae and villi remain prominent over the proximal half of the jejunum (Figures 25.15d,e and 25.16●). Most nutrient absorption occurs here. As one approaches the ileum, the plicae and villi become smaller and continue to diminish in size to the end of the ileum. This reduction parallels the reduction in absorptive activity; most nutrient absorption has occurred before materials reach the terminal portion of the ileum. The region adjacent to the large intestine lacks plicae altogether, and the scattered villi are stumpy and conical.

Bacteria, especially *E. coli*, are normal inhabitants within the lumen of the large intestine. These bacteria are nourished by the surrounding mucosa. The epithelial barriers (cells, mucus, and digestive juices) and underlying cells of the immune system protect the small intestine from bacteria migrating from the large intestine. Small, isolated, individual lymphoid nodules are present in the lamina propria of the jejunum. In the ileum the lymphoid nodules become more numerous, and they fuse together to form large masses of lymphoid tissue (Figure 25.16a●). These lymphoid centers, called **aggregated lymphoid nodules**, or *Peyer's patches*, may reach the size of cherries. ∞ p. 620 They are most abundant in the terminal portion of the ileum, near the entrance to the large intestine, which normally contains large numbers of potentially harmful bacteria.

Regulation of the Small Intestine

As absorption occurs, weak peristaltic contractions slowly move materials along the length of the small intestine. Movements of the small intestine are controlled primarily by neural reflexes involving the submucosal and myenteric plexuses. Stimulation of the parasympathetic system increases the sensitivity of these reflexes and accelerates peristaltic contractions and segmentation movements. These contractions and movements, which promote mixing of the intestinal contents, are usually limited to within a few centimeters of the original stimulus. Coordinated intestinal movements occur when food enters the stomach; these movements tend to move materials away from the duodenum and toward the large intestine. During these periods the ileocecal valve permits the passage of material into the large intestine. † *Gastroenteritis p. 806*

Hormonal and CNS controls regulate the secretory output of the small intestine and accessory glands. The secretions of the small intestine are collectively called **intestinal juice**. Secretory activities are triggered by local reflexes and parasympathetic (vagal) stimulation. Sympathetic stimulation inhibits secretion. Duodenal enteroendocrine cells produce secretin and cholecystokinin, hormones that coordinate the secretory activities of the stomach, duodenum, liver, and pancreas.

✓ CONCEPT CHECK

- Which histological features of the small intestine facilitate the digestion and absorption of nutrients?
- What is the function of plicae?
- What are the functions of intestinal glands?
- Which section of the small intestine serves as a "mixing bowl"?

The Large Intestine [FIGURES 25.2/25.12/25.14/25.17]

The horseshoe-shaped **large intestine** begins at the end of the ileum and ends at the anus. The large intestine lies inferior to the stomach and liver and almost completely frames the small intestine (Figures 25.2, p. 665, 25.12, p. 680, and 25.14●).

The large intestine, often called the **large bowel**, has an average length of about 1.5 m (5 ft) and a width of 7.5 cm (3 in.). It can be divided into three parts: (1) the *cecum*, the first portion of the large intestine, which appears as a pouch; (2) the *colon*, the largest portion of the large intestine; and (3) the *rectum*, the last 15 cm (6 in.) of the large intestine and the end of the digestive tract (Figure 25.17●).

The major functions of the large intestine are: (1) the *reabsorption of water and electrolytes, and compaction* of the intestinal contents into feces, (2) the *absorption of important vitamins* produced by bacterial action, and (3) the *storing of fecal material* before defecation.

The large intestine receives blood from tributaries of the *superior mesenteric* and *inferior mesenteric arteries*. Venous blood is collected from the large intestine by the *superior mesenteric* and *inferior mesenteric veins*. ∞ pp. 591, 601

The Cecum [FIGURE 25.17]

Materials arriving from the ileum first enter an expanded pouch called the **cecum** (SĒ-kum). The ileum attaches to the medial surface of the cecum and opens into the cecum at the **ileal papilla**. Muscles encircling the opening form the **ileocecal** (il-ē-ō-SĒ-kal) **valve** (Figure 25.17●), which regulates the passage of materials into the large intestine. The cecum collects and stores the arriving materials and begins the process of compaction. The slender, hollow **vermiform appendix**, or *appendix*, is attached to the posteromedial surface of the cecum. The appendix usually is approximately 9 cm (3.5 in.) long, but its size and shape are quite variable. A band of mesentery, the **mesoappendix**, connects the appendix to the ileum and cecum. The mucosa and submucosa of the appendix are dominated by lymphoid nodules, and its primary function is as an organ of the lymphatic system

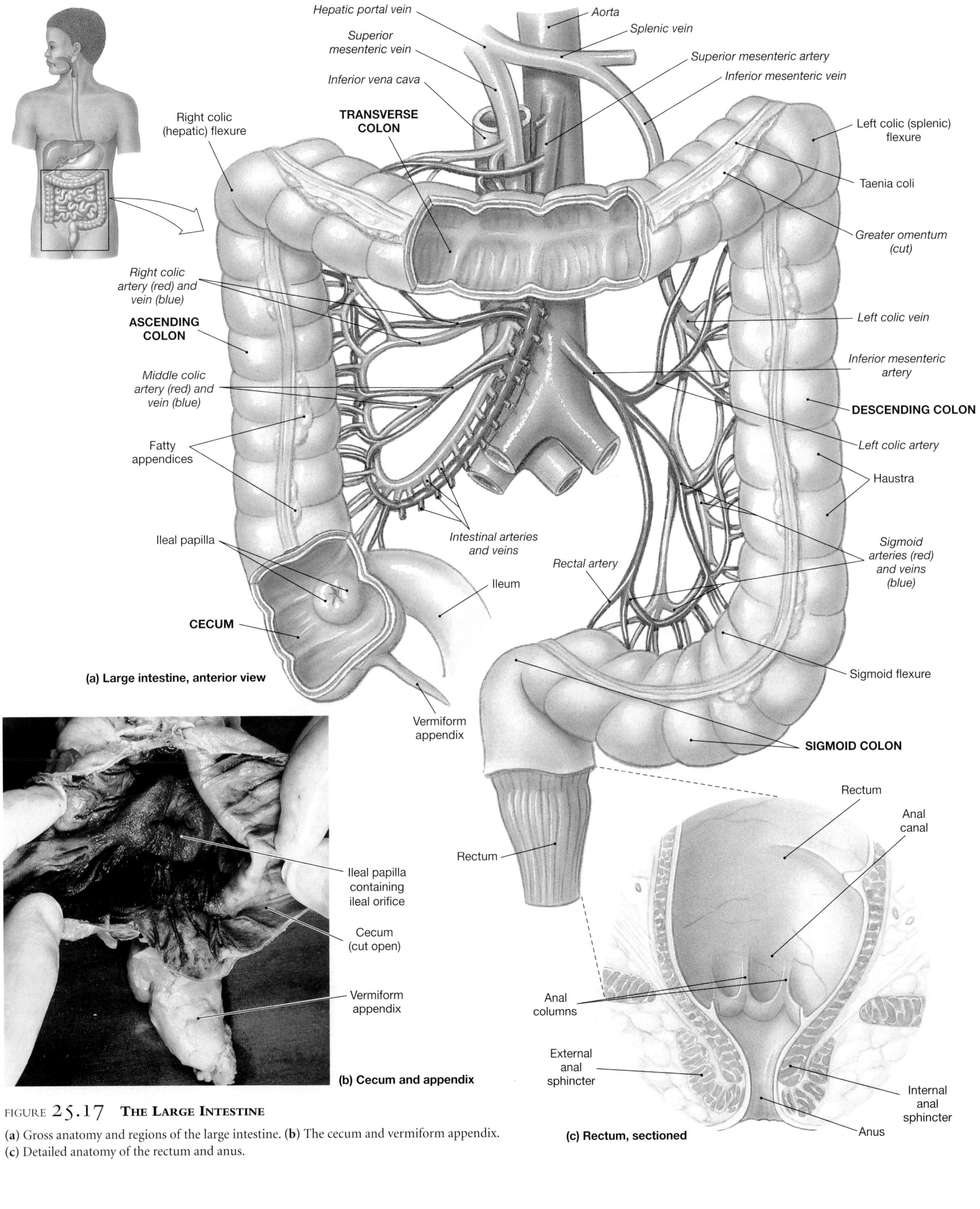

FIGURE 25.17 **The Large Intestine**

(**a**) Gross anatomy and regions of the large intestine. (**b**) The cecum and vermiform appendix. (**c**) Detailed anatomy of the rectum and anus.

comparable to one of the tonsils. Inflammation of the appendix produces the symptoms of *appendicitis.* *Infected Lymphoid Nodules p. 802*

The Colon [FIGURE 25.17]

The **colon** has a larger diameter and a thinner wall than the small intestine. Refer to Figure 25.17• as we describe the colon. There are several distinctive features of the colon:

1. The wall of the colon forms a series of pouches, or **haustra** (HAWS-tra; singular *haustrum*), that permit considerable distension and elongation. Cutting into the intestinal lumen reveals that the creases between the haustra extend into the mucosal lining, producing a series of internal folds.
2. Three separate longitudinal ribbons of smooth muscle, the **taeniae coli** (TĒ-nē-ā KŌ-lī), are visible on the outer surfaces of the colon just beneath the serosa. Muscle tone within these bands creates the haustra.
3. The serosa of the colon contains numerous tear drop–shaped sacs of fat, called the **fatty appendices of the colon**, or *epiploic appendages* (Figure 25.17a•).

REGIONS OF THE COLON [FIGURES 25.17a/25.18]

The colon is subdivided into four regions: the *ascending colon*, the *transverse colon*, the *descending colon*, and the *sigmoid colon* (Figure 25.17a•). The regions of the colon may be seen clearly in the radiograph in Figure 25.18•.

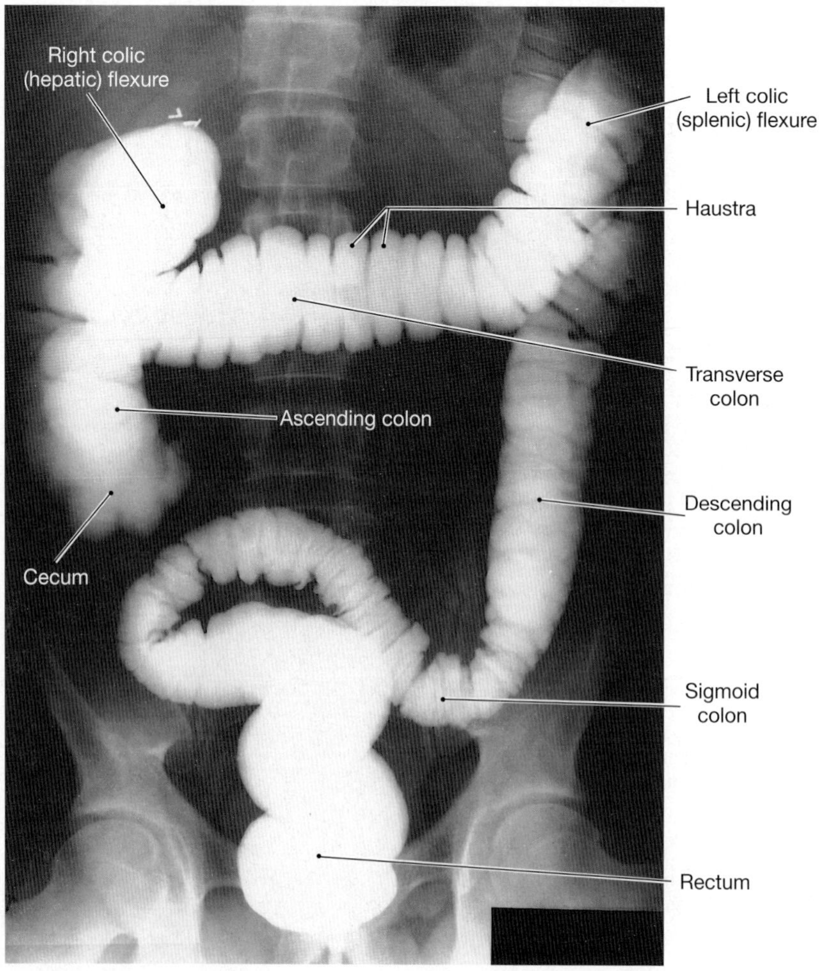

FIGURE 25.18 **ANTERIOR/POSTERIOR RADIOGRAPH OF THE COLON**

THE ASCENDING COLON [FIGURE 25.4] The **ascending colon** begins at the superior border of the cecum and ascends along the right lateral and posterior wall of the peritoneal cavity to the inferior surface of the liver. At this point the colon turns to the left at the **right colic flexure**, or *hepatic flexure*. The right colic flexure marks the end of the ascending colon and the beginning of the *transverse colon*. The lateral and anterior surfaces of the ascending colon are covered by visceral peritoneum. There is normally no mesentery, and the ascending colon is retroperitoneal (Figure 25.4c,d, p. 668•).

THE TRANSVERSE COLON The **transverse colon** begins at the right colic flexure. It curves anteriorly and crosses the abdomen from right to left. It is supported by the *transverse mesocolon* and is separated from the anterior abdominal wall by the layers of the greater omentum. As the transverse colon reaches the left side, it passes inferior to the greater curvature of the stomach. The **gastrocolic ligament** attaches the transverse colon to the greater curvature of the stomach. Near the spleen, the colon makes a right-angle bend, termed the **left colic flexure**, or *splenic flexure*, and then proceeds caudally.

THE DESCENDING COLON The **descending colon** proceeds inferiorly along the left side until reaching the iliac fossa. The descending colon, which is retroperitoneal, is firmly attached to the abdominal wall. At the iliac fossa, the descending colon enters an S-shaped segment, the *sigmoid colon*, at the **sigmoid flexure** (SIG-moid; *sigmoides*, the Greek letter S).

THE SIGMOID COLON [FIGURE 25.4] The sigmoid flexure begins the **sigmoid colon**, an S-shaped segment of the large intestine that is only about 15 cm (6 in.) long. It lies posterior to the urinary bladder, suspended from the *sigmoid mesocolon* (Figure 25.4•, p. 668). The sigmoid colon empties into the *rectum*.

The Rectum [FIGURES 25.12/25.17/25.18]

The **rectum** (REK-tum) forms the last 15 cm (6 in.) of the digestive tract (Figures 25.12, p. 680, 25.17, and 25.18•). The rectum is an expandable organ for the temporary storage of fecal material. Movement of fecal materials into the rectum triggers the urge to defecate.

The last portion of the rectum, the **anal canal**, contains small longitudinal folds, the **anal columns**. The distal margins of the anal columns are joined by transverse folds that mark the boundary between the columnar epithelium of the proximal rectum and a stratified squamous epithelium similar to that found in the oral cavity. The anal canal ends at the **anus**, or *anal orifice*. Very close to the anus, the epidermis becomes keratinized and identical to the surface of the skin.

Veins in the lamina propria and submucosa of the anal canal occasionally become distended, producing *hemorrhoids*. The circular muscle layer of the muscularis externa in this region forms the **internal anal sphincter**. The smooth muscle fibers of the internal anal sphincter are not under voluntary control. The **external anal sphincter** encircles the distal portion of the anal canal. This sphincter, which consists of a ring of skeletal muscle fibers, is under voluntary control.

Histology of the Large Intestine [FIGURE 25.19]

The histological characteristics that distinguish the large intestine from the small intestine are as follows:

- The wall of the large intestine is relatively thin. Although the diameter of the colon is roughly three times that of the small intestine, the wall is much thinner.

- The large intestine lacks villi, which are characteristic of the small intestine.
- Goblet cells are much more abundant than they are in the small intestine.
- The large intestine has distinctive intestinal glands (Figure 25.19●). The glands of the large intestine are deeper than those of the small intestine, and they are dominated by goblet cells. Secretion occurs as local stimuli trigger reflexes involving the local nerve plexuses, resulting in the production of copious amounts of mucus to promote lubrication as undigested waste is compacted.
- Large lymphoid nodules are scattered throughout the lamina propria and extend into the submucosa (see Figure 23.8a●).
- The muscularis externa differs from that of other intestinal regions because the longitudinal layer has been reduced to the muscular bands of the taeniae coli. However, the mixing and propulsive contractions of the colon resemble those of the small intestine. *Diverticulitis and Colitis p. 806*

Regulation of the Large Intestine

Movement of ingested materials from the cecum to the transverse colon occurs very slowly. It involves both peristaltic activity and *haustral churning*, the segmentation movements of the large intestine. The slow passage of materials along the large intestine allows time for the fecal material to be converted into a sludgy paste. Movement from the transverse colon through the rest of the large intestine results from powerful peristaltic contractions, called **mass movements**, that occur a few times each day. The stimulus is distension of the stomach and duodenum, and the commands are relayed over the intestinal nerve plexuses. The contractions force fecal materials into the rectum and produce the conscious urge to defecate.

The rectal chamber is usually empty except when one of those powerful mass movements forces fecal materials out of the sigmoid colon into the rectum. Distension of the rectal wall then stimulates the conscious urge to defecate. It also leads to the relaxation of the internal sphincter through the *defecation reflex*, and fecal material moves into the anal canal. When the external anal sphincter is voluntarily relaxed, defecation can occur.

Accessory Digestive Organs

The major accessory organs of the digestive tract are the *salivary glands*, the *liver*, the *gallbladder*, and the *pancreas*. The salivary glands, liver, and pancreas have exocrine functions. The accessory digestive organs produce and store enzymes and buffers that are essential to normal digestive function. The liver and pancreas have other vital functions in addition to their roles in digestion.

The Liver [FIGURE 25.12 AND TABLE 25.1]

The **liver** is the largest visceral organ, and it is one of the most versatile organs in the body. Most of its mass lies within the right hypochondriac and epigastric regions (Figure 25.12●, p. 680). The liver weighs about 1.5 kg (3.3 lb). This large, firm, reddish-brown organ provides essential metabolic and synthetic services that fall into three basic categories: *metabolic regulation*, *hematological regulation*, and *bile production*.

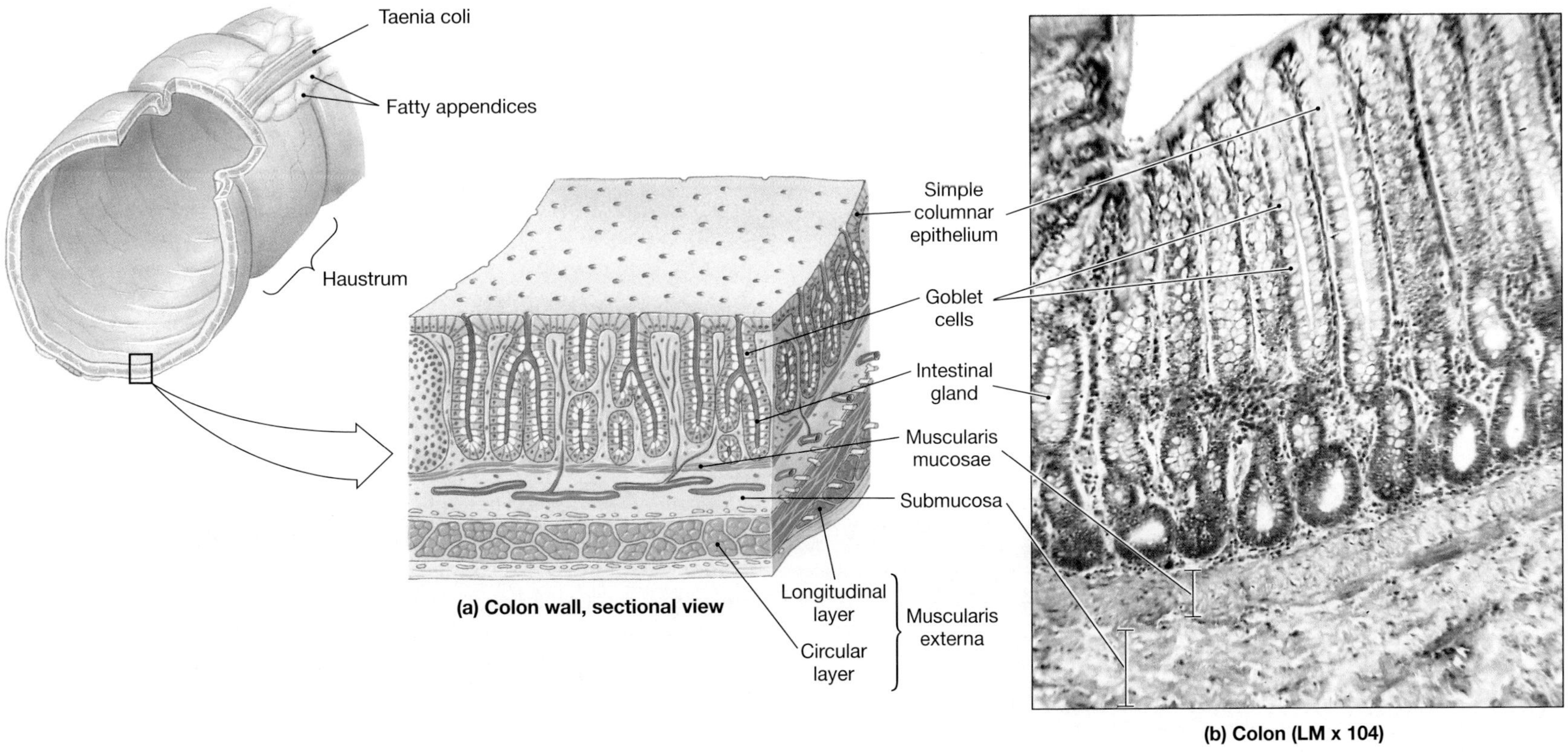

FIGURE 25.19 **THE WALL OF THE LARGE INTESTINE**

(**a**) Three-dimensional view of the colon wall. (**b**) Photomicrograph of colon, showing detail of mucosal and submucosa layers.

1. *Metabolic Regulation:* The liver represents the central clearinghouse for metabolic regulation in the body. Circulating levels of carbohydrates, lipids, and amino acids are regulated by the liver. All blood leaving the absorptive surfaces of the digestive tract enters the hepatic portal system and flows into the liver. This arrangement gives liver cells the opportunity to extract absorbed nutrients or toxins from the blood before it reaches the systemic circulation through the hepatic veins. Liver cells, or *hepatocytes* (he-PAT-ō-sīts), monitor the circulating levels of metabolites and adjust them as necessary. Excess nutrients are removed and stored, and deficiencies are corrected by mobilizing stored reserves or performing appropriate synthetic activities. Circulating toxins and metabolic waste products are also removed, for subsequent inactivation, storage, or excretion. Finally, fat-soluble vitamins (A, D, K, and E) are absorbed and stored in the liver.
2. *Hematological Regulation:* The liver is the largest blood reservoir in the body, and it receives about 25% of the cardiac output. As blood passes through the liver sinusoids, (1) phagocytic cells in the liver remove old or damaged RBCs, cellular debris, and pathogens from the circulation; and (2) liver cells synthesize plasma proteins that contribute to the osmotic concentration of the blood, transport nutrients, and establish the clotting and complement systems.
3. *Synthesis and Secretion of Bile:* **Bile** is synthesized by liver cells, stored in the gallbladder, and excreted into the lumen of the duodenum. Bile consists mostly of water, with minor amounts of ions, *bilirubin* (a pigment derived from hemoglobin), and an assortment of lipids collectively known as the *bile salts.* The water and ions assist in the dilution and buffering of acids in chyme as it enters the small intestine. Bile salts associate with lipids in the chyme and make it possible for enzymes to break down those lipids into fatty acids suitable for absorption.

To date, over 200 different functions have been assigned to the liver. A partial listing of these functions is presented in Table 25.1. Any condition that severely damages the liver represents a serious threat to life. The liver has a limited ability to regenerate after injury, but liver function will not fully recover unless the normal vascular pattern returns.

TABLE 25.1 MAJOR FUNCTIONS OF THE LIVER

DIGESTIVE AND METABOLIC FUNCTIONS
Synthesis and secretion of bile
Storage of glycogen and lipid reserves
Maintenance of normal blood glucose, amino acid, and fatty acid concentrations
Synthesis and interconversion of nutrient types (e.g., transamination of amino acids, or conversion of carbohydrates to lipids)
Synthesis and release of cholesterol bound to transport proteins
Inactivation of toxins
Storage of iron reserves
Storage of fat-soluble vitamins
OTHER MAJOR FUNCTIONS
Synthesis of plasma proteins
Synthesis of clotting factors
Synthesis of the inactive hormone, angiotensinogen
Phagocytosis of damaged red blood cells (by Kupffer cells)
Blood storage (major contributor to venous reserve)
Absorption and breakdown of circulating hormones (including insulin and epinephrine) and immunoglobulins
Absorption and inactivation of lipid-soluble drugs

ANATOMY OF THE LIVER [FIGURE 25.20]

The liver is wrapped in a tough fibrous capsule and covered by a layer of visceral peritoneum. On the anterior surface, a ventral mesentery, the **falciform** (FAL-si-form) **ligament**, marks the division between the **left lobe** and **right lobe** of the liver (Figure 25.20a–c●). A thickening in the inferior margin of the falciform ligament is the **round ligament**, or *ligamentum teres*, a fibrous band that marks the path of the degenerated fetal umbilical vein. The liver is suspended from the inferior surface of the diaphragm by the **coronary ligament**.

The shape of the liver conforms to its surroundings. The **anterior surface**, or *parietal surface*, follows the smooth curve of the body wall (Figure 25.20c●). The **posterior surface**, or *visceral surface*, bears the impressions of the stomach, small intestine, right kidney, and large intestine (Figure 25.20d●). The impression left by the inferior vena cava marks the division between the right lobe and the small **caudate** (KAW-dāt) **lobe**. Inferior to the caudate lies the **quadrate lobe** sandwiched between the left lobe and the gallbladder. Afferent blood vessels and other structures reach the liver by traveling within the connective tissue of the lesser omentum. They converge at the **hilus** of the liver, a region known as the *porta hepatis* ("doorway to the liver").

The Blood Supply to the Liver [FIGURES 22.26/25.17a/25.20d] The circulation to the liver was detailed in Chapter 22 and summarized in Figures 22.17 and 22.26●. pp. 591, 601 Two blood vessels deliver blood to the liver, the **hepatic artery proper** and the **hepatic portal vein** (Figures 25.17a and 25.20d●). Roughly one-third of the normal hepatic blood flow arrives via the hepatic artery, and the rest is provided by the hepatic portal vein. Blood returns to the systemic circuit through the **hepatic veins** that open into the inferior vena cava. The arterial supply provides oxygenated blood to the liver and the hepatic portal vein supplies nutrients and other chemicals absorbed from the intestine.

HISTOLOGICAL ORGANIZATION OF THE LIVER [FIGURE 25.21]

Each lobe of the liver is divided by connective tissue into approximately 100,000 **liver lobules**, the basic functional units of the liver. The histological organization and structure of a typical liver lobule are shown in Figure 25.21●.

The Liver Lobule [FIGURE 25.21] The liver cells, or **hepatocytes** (he-PAT-ō-sīts), in a liver lobule form a series of irregular plates arranged like the spokes of a wheel (Figure 25.21a,c●). The plates are only one cell thick, and exposed hepatocyte surfaces are covered with short microvilli. Sinusoids between adjacent plates empty into the **central vein** (Figure 25.21b●). The fenestrated walls of the sinusoids contain large openings that allow substances to pass out of the circulation and into the spaces surrounding the hepatocytes. In addition to typical endothelial cells, the sinusoidal lining includes a large number of **Kupffer** (KOOP-fer) **cells**, also known as *stellate reticuloendothelial cells.* These phagocytic cells are part of the monocyte-macrophage system, and they engulf pathogens, cell debris, and damaged blood cells. Kupffer cells also engulf and retain any heavy metals, such as tin or mercury, that are absorbed by the digestive tract.

Blood enters the liver sinusoids from small branches of the portal vein and hepatic artery. A typical lobule is hexagonal in cross section (Figure 25.21a,b●). There are six **portal areas**, or *hepatic triads*, one at each of the six corners of the lobule. A portal area (Figure 25.21c●) contains three structures: (1) a branch of the hepatic portal vein, (2) a branch of the hepatic artery proper, and (3) a small branch of the bile duct.

FIGURE 25.20 **ANATOMY OF THE LIVER**

(**a**) Horizontal section through the upper abdomen, showing the position of the liver relative to other visceral organs. (**b**) Horizontal section through the upper abdomen, showing structures illustrated in (a). (**c**) Anatomical landmarks on the anterior surface of the liver. (**d**) The posterior surface of the liver.

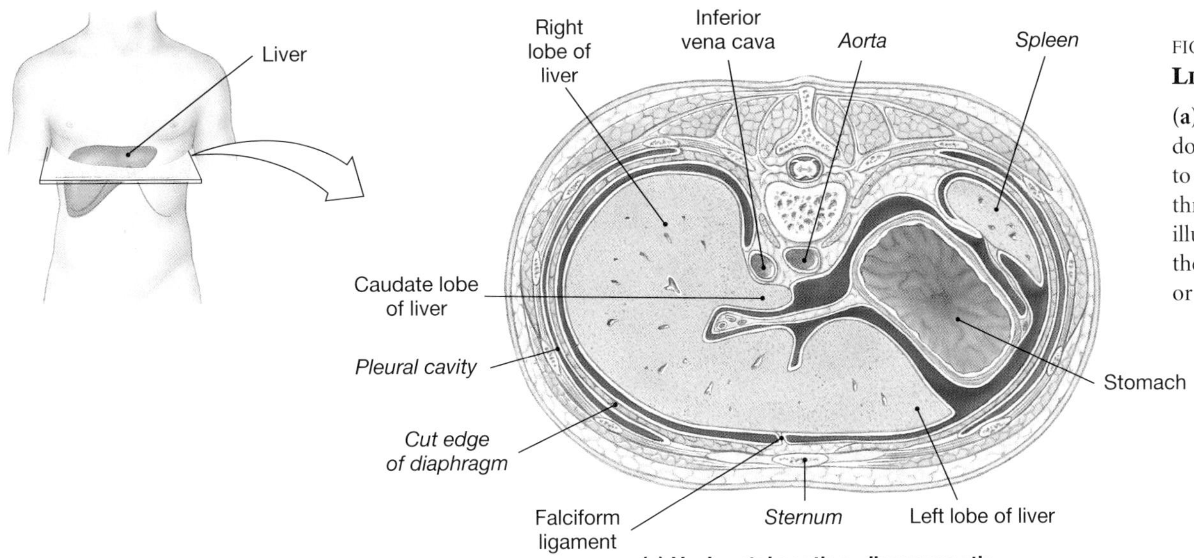

(a) Horizontal section, diagrammatic

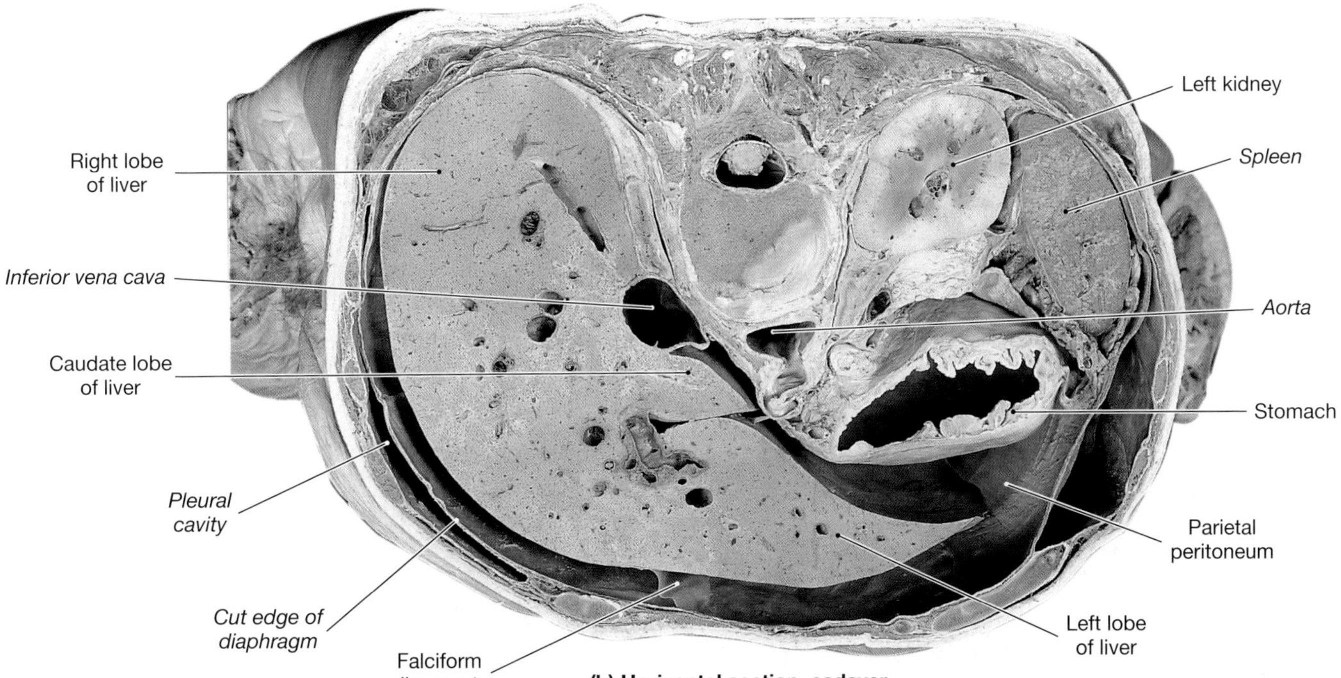

(b) Horizontal section, cadaver

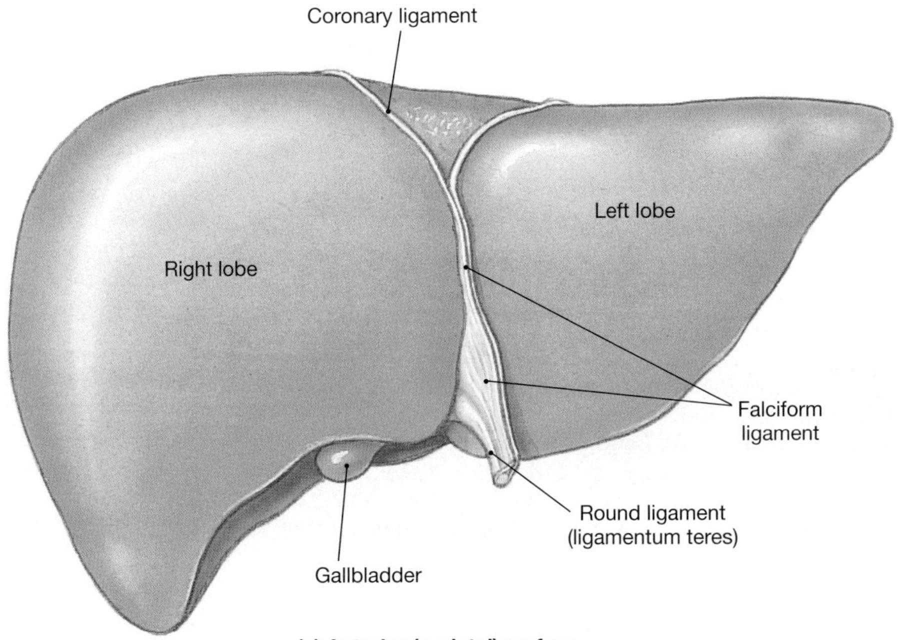

(c) Anterior (parietal) surface

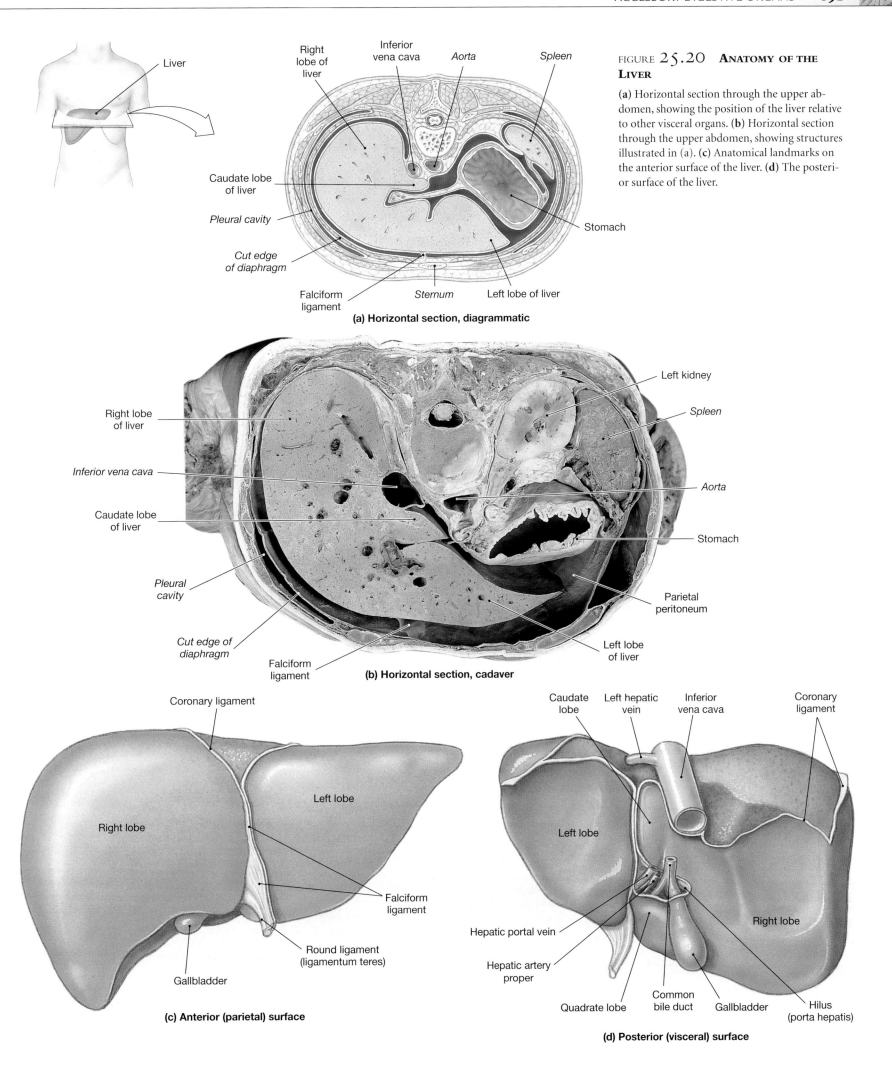

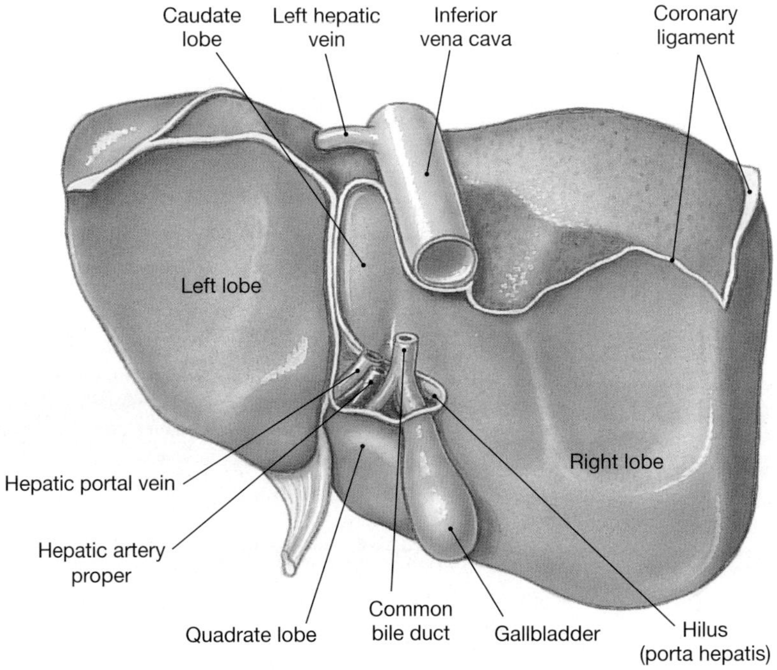

(d) Posterior (visceral) surface

Branches from the arteries and veins deliver blood to the sinusoids of adjacent lobules (Figure 25.21a●). As blood flows through the sinusoids, hepatocytes absorb and secrete materials into the bloodstream across their exposed surfaces. Blood then leaves the sinusoids and enters the central vein of the lobule. The central veins ultimately merge to form the hepatic veins that empty into the inferior vena cava. † *Cirrhosis p. 806*

Bile Secretion And Transport [Figures 25.21a and 25.22] Bile is secreted into a network of narrow channels between the opposing membranes of adjacent liver cells. These small passageways are called **bile canaliculi**. The bile canaliculi extend outward through the liver lobule, away from the central vein. These eventually connect with fine **bile ductules** (DUK-tūlz) that carry bile to a bile duct in the nearest portal area (Figure 25.21a●). The **right** and **left hepatic**

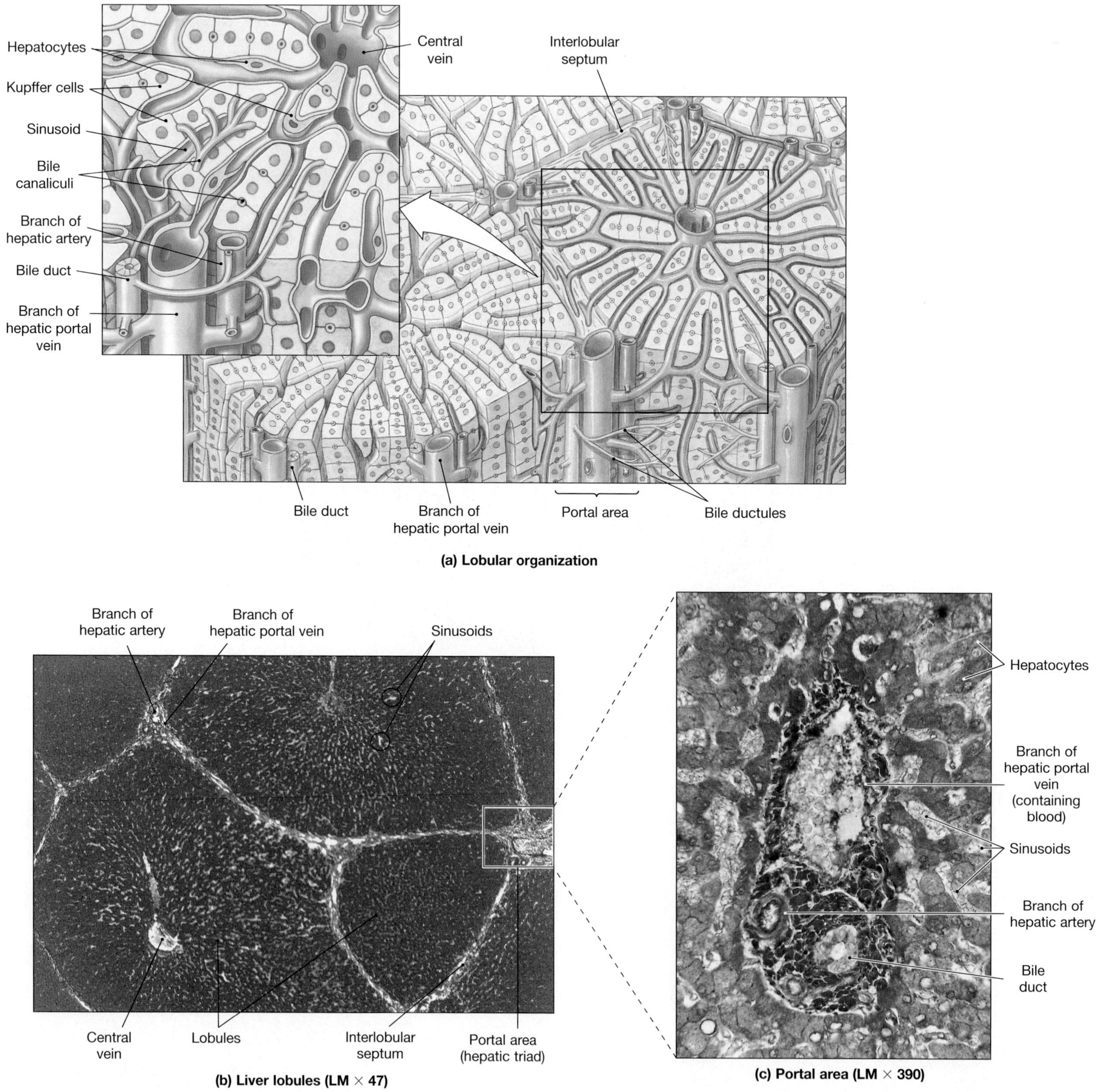

FIGURE 25.21 **Liver Histology**

(**a**) Diagrammatic view of lobular organization. (**b**) Light micrograph showing representative mammalian liver lobules. Human liver lobules lack a distinct connective tissue boundary, making them difficult to distinguish in histological section. (**c**) Light micrograph, showing detail of portal area (hepatic triad).

ducts collect bile from all of the bile ducts of the liver lobes. These ducts unite to form the **common hepatic duct** that leaves the liver. The bile within the common hepatic duct may either (1) flow into the *common bile duct* that empties into the duodenum or (2) enter the *cystic duct* that leads to the gallbladder. These structures are illustrated and shown in a radiograph in Figure 25.22●.

The Gallbladder [FIGURES 25.16a/25.22]

The **gallbladder** is a hollow, pear-shaped, muscular organ. The gallbladder is a muscular sac that stores and concentrates bile before its excretion into the small intestine. The gallbladder is located in a recess, or fossa, in the visceral surface of the right lobe. The gallbladder is divided into three regions: the **fundus**, the **body**, and the **neck** (Figure 25.22a,c●). The **cystic duct** leads from the gallbladder toward the hilus of the liver. At the porta hepatis, the common hepatic duct and the cystic duct unite to create the **common bile duct** (Figure 25.22a●). At the duodenum, a muscular **hepatopancreatic sphincter**, or *sphincter of Oddi*, surrounds the lumen of the common bile duct and the duodenal ampulla (Figures 25.16, p. 685 and 25.22b●). Contraction of this sphincter seals off the passageway and prevents bile from entering the small intestine.

The gallbladder has two major functions, *bile storage* and *bile modification*. When the hepatopancreatic sphincter is closed, bile enters the cystic duct. Over the interim, when bile cannot flow along the common bile duct, it enters the cystic duct for storage within the expandable gallbladder. When filled to capacity, the gallbladder contains 40–70 ml of bile. As bile remains in the gallbladder, its composition gradually changes. Water is absorbed from the bile, and the bile salts and other components of bile become increasingly concentrated.

Bile ejection occurs under stimulation of the hormone *cholecystokinin*, or *CCK*. Cholecystokinin is released into the bloodstream at the duodenum, when chyme arrives containing large amounts of lipids and partially digested proteins. CCK causes relaxation of the hepatopancreatic sphincter and contraction of the gallbladder.

The Pancreas [FIGURES 25.22a/25.23]

The **pancreas** lies posterior to the stomach, extending laterally from the duodenum toward the spleen (Figures 25.22a and 25.23●). The pancreas is an elongate, pinkish-gray organ, approximately 15 cm (6 in.) long and around 80 g (3 oz). The broad **head** of the pancreas lies within the loop formed by the duodenum as it leaves the pylorus. The slender **body** extends transversely toward the spleen, and the **tail** is short and bluntly rounded. The pancreas is retroperitoneal, and it is firmly bound to the posterior wall of the abdominal cavity.

The surface of the pancreas has a lumpy, nodular texture. A thin, transparent connective tissue capsule wraps the pancreas. The pancreatic lobules, associated blood vessels,

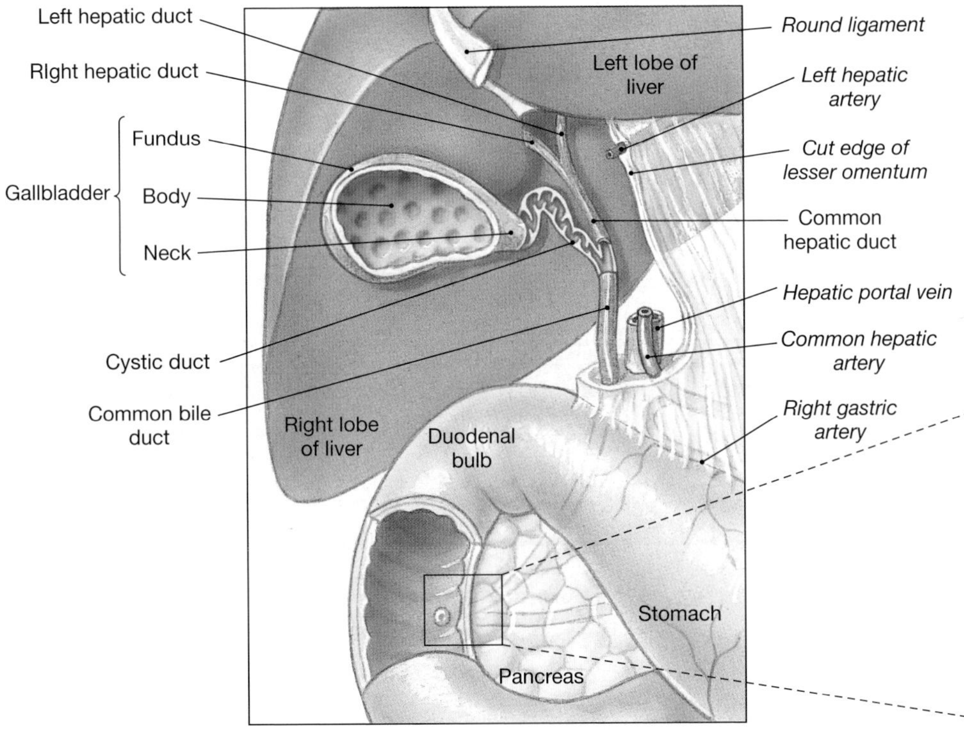

(a) Gallbladder and associated ducts

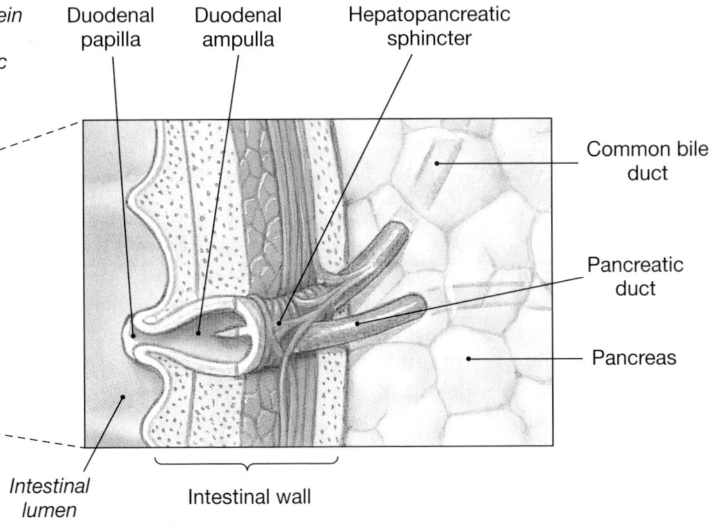

(b) Hepatopancreatic sphincter

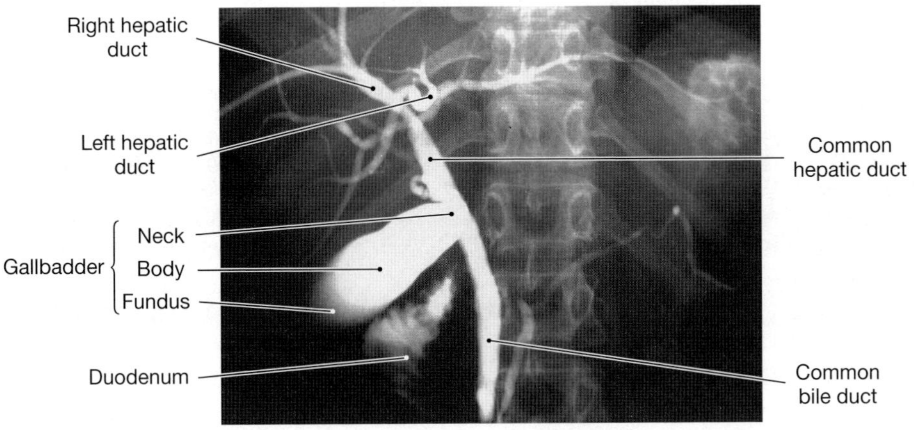

(c) Cholangiopancreatogram

FIGURE 25.22 **THE GALLBLADDER AND ASSOCIATED BILE DUCTS**

(**a**) A view of the inferior surface of the liver, showing the position of the gallbladder and ducts that transport bile from the liver to the gallbladder and duodenum. (**b**) A portion of the lesser omentum has been cut away to make it easier to see the relationships among the common bile duct, the hepatic duct, and the cystic duct. (**c**) Radiograph (cholangiopancreatogram, anterior-posterior view) of the gallbladder, biliary ducts, and pancreatic ducts.

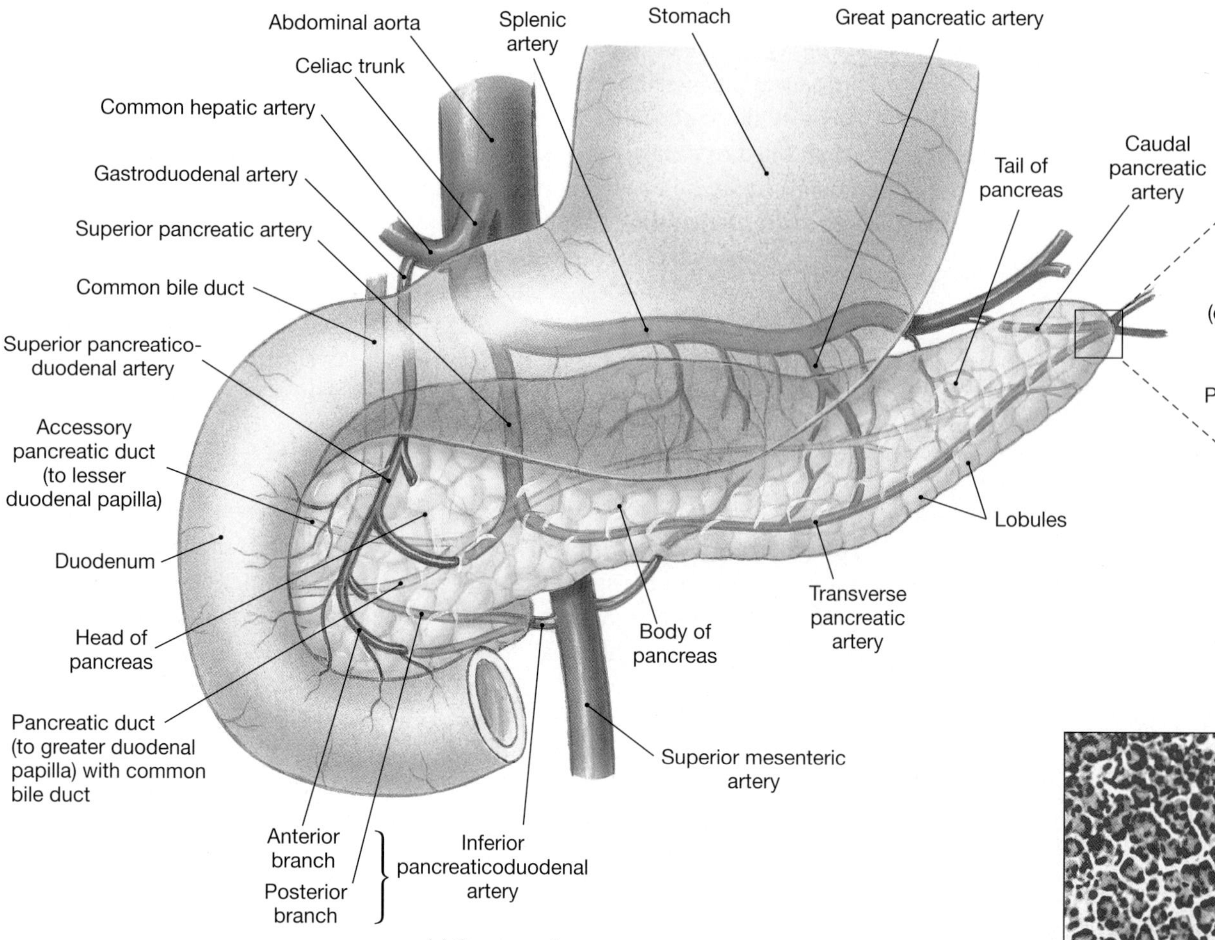

(a) Gross anatomy

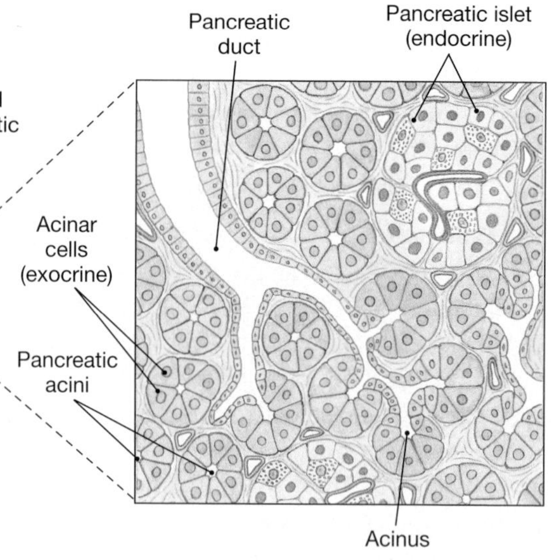

(b) Pancreatic histology

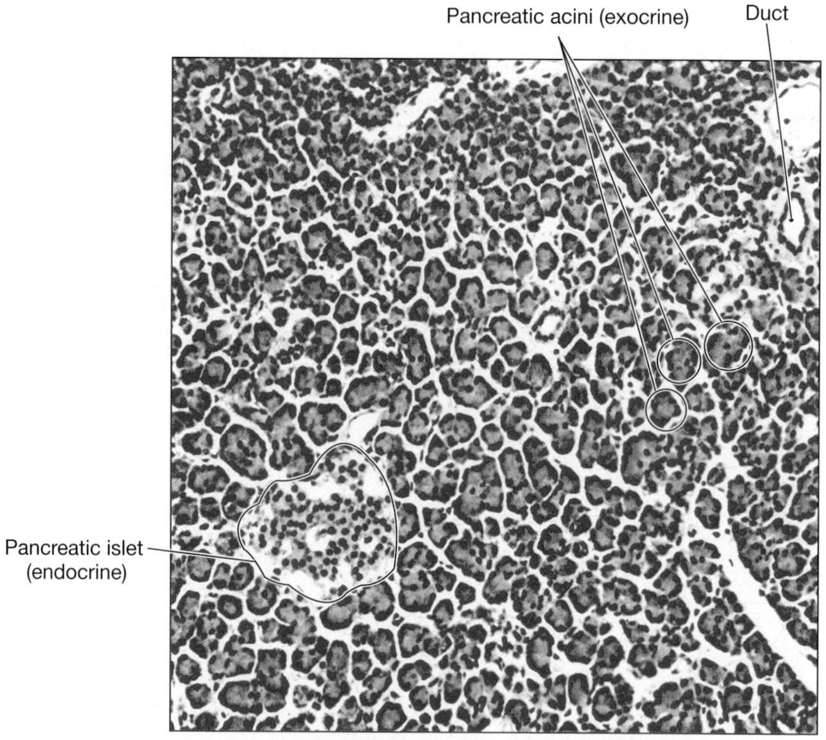

(c) Exocrine and endocrine cells (LM x 120)

FIGURE 25.23 **THE PANCREAS**

(a) Gross anatomy of the pancreas. The head of the pancreas is tucked into a curve of the duodenum that begins at the pylorus of the stomach. **(b)** Diagrammatic view of the cellular organization of the pancreas, showing exocrine and endocrine regions. **(c)** Photomicrograph of the pancreas, showing exocrine and endocrine cells.

CLINICAL BRIEF

PROBLEMS WITH BILE STORAGE AND SECRETION

If bile becomes too concentrated, crystals of insoluble minerals and salts begin to appear. These deposits are called **gallstones**. Merely having them, a condition termed **cholelithiasis** (kō-lē-li-THĪ-a-sis; *chole*, bile), does not represent a problem as long as the stones remain very small. Small stones are normally flushed down the bile duct and excreted.

If gallstones enter and jam in the cystic or bile duct, the painful symptoms of **cholecystitis** (kō-lē-sis-TĪ-tis) appear. The gallbladder becomes swollen and inflamed, infections may develop, and if the blockage does not work its way down the common bile duct to the duodenum, it must be removed or destroyed. Small gallstones can be chemically dissolved. One chemical now under clinical review is *methyl tert-butyl ether* (MTBE). When introduced into the gallbladder, it dissolves gallstones in a matter of hours. Unfortunately, stones frequently recur. One's health and digestion seem totally unaffected by the removal of the gallbladder, so even asymptomatic individuals with gallstones are often advised to have their gallbladders removed. Surgery is usually required to remove large gallstones, and the gallbladder is removed at the same time, to prevent recurrence. The procedure can be performed with a laparoscope inserted through several small incisions.

Another therapy for cholecystitis involves placing the individual in a water bath on a soft cushion and shattering the stones with focused sound waves. The apparatus used is called a *lithotripter*. The particles produced are then small enough to pass through the bile duct without difficulty.

and excretory ducts can be seen through the anterior capsule and the overlying layer of peritoneum.

The pancreas is primarily an exocrine organ producing digestive enzymes and buffers, although it does serve an endocrine function as discussed in Chapter 19. p. 519 The large **pancreatic duct** (*duct of Wirsung*) delivers these secretions to the duodenal ampulla. A small **accessory pancreatic duct**, or *duct of Santorini*, may branch from the pancreatic duct before it leaves the pancreas (Figure 25.23a●). When present, the accessory pancreatic duct typically empties into the duodenum at a separate papilla, the *lesser duodenal papilla*. This papilla lies a few centimeters proximal to the greater duodenal papilla.

Arterial blood reaches the pancreas through branches of the *splenic, superior mesenteric,* and *common hepatic arteries* (Figures 22.17 and 25.23a●). p. 694 The **pancreatic arteries** and **pancreaticoduodenal arteries** (**superior** and **inferior**) are the major branches from these vessels. The *splenic vein* and its branches drain the pancreas (Figure 22.26●). p. 601

HISTOLOGICAL ORGANIZATION OF THE PANCREAS [FIGURE 25.23b,c]

Partitions of connective tissue divide the pancreatic tissue into distinct lobules (Figure 25.23b,c●). The blood vessels and tributaries of the pancreatic ducts are found within these connective tissue septa. The pancreas is an example of a compound tubuloacinar gland. Within each lobule, the ducts branch repeatedly before ending in blind pockets, the **pancreatic acini** (AS-i-nī). Each pancreatic acinus is lined by a simple cuboidal epithelium. *Pancreatic islets* are scattered between the acini, but they account for only around 1% of the cellular population of the pancreas.

The pancreatic acini secrete a mixture of water, ions, and pancreatic digestive enzymes into the duodenum. This secretion is called **pancreatic juice.** Pancreatic enzymes do most of the digestive work in the small intestine, breaking down ingested materials into small molecules suitable for absorption. The pancreatic ducts secrete buffers (primarily sodium bicarbonate) in a watery solution. These secretions are important in neutralizing the acid in chyme and stabilizing the pH of the intestinal contents.

PANCREATIC ENZYMES

Pancreatic enzymes are classified according to their intended targets. **Lipases** (LĪ-pā-zez) digest lipids, **carbohydrases** (kar-bō-HĪ-drā-zez) digest sugars and starches, **nucleases** attack nucleic acids, and **proteolytic** (prō-tē-ō-LIT-ik) **enzymes** break proteins apart. Proteolytic enzymes include **proteinases** and **peptidases**; proteinases break apart large protein complexes, whereas peptidases break small peptide chains into individual amino acids.

THE REGULATION OF PANCREATIC SECRETION

Secretion of pancreatic juice occurs primarily in response to hormonal instructions from the duodenum. When acid chyme arrives in the small intestine, secretin is released. This hormone triggers the production of watery pancreatic juice containing buffers, especially sodium bicarbonate. Another duodenal hormone, cholecystokinin, stimulates the production and secretion of pancreatic enzymes.

✓ CONCEPT CHECK

- In cystic fibrosis, the pancreatic duct can become blocked with thickened secretions. This condition would interfere with the proper digestion of which group(s) of nutrients?
- What are the three major functions of the liver?
- What does contraction of the hepatopancreatic sphincter accomplish?
- What are the functions of the pancreas?

CLINICAL BRIEF

PANCREATITIS

Pancreatitis (pan-krē-a-TĪ-tis) is an inflammation of the pancreas. Blockage of the excretory ducts (such as by gallstones), bacterial or viral infections, and drug reactions, especially those involving alcohol, are among the factors that may produce this condition. These stimuli provoke a crisis by injuring exocrine cells in at least a portion of the organ. Lysosomes within the damaged cells then activate the proenzymes, and autodigestion begins. The proteolytic enzymes digest the surrounding, undamaged cells, activating their enzymes and starting a chain reaction.

In most cases only a portion of the pancreas will be affected, and the condition subsides in a few days. In 10–15% of pancreatitis cases, the process does not subside, and the enzymes may ultimately destroy the pancreas. They are also likely to enter the systemic circulation, via the pancreatic blood vessels, and digest their way through the pancreatic capsule and enter the abdominopelvic cavity. Roughly two-thirds of these patients survive, but they often suffer recurrent painful episodes of pancreatitis. If the islet cells are damaged, diabetes mellitus may occur. *Diabetes mellitus, p. 797*

Aging And The Digestive System

Essentially, normal digestion and absorption occur in elderly individuals. However, there are many changes in the digestive system that parallel age-related changes already described for other systems.

1. *The rate of epithelial stem cell division declines.* The digestive epithelium becomes more susceptible to damage by abrasion, acids, or enzymes. Peptic ulcers therefore become more likely. In the mouth, esophagus, and anus the stratified epithelium becomes thinner and more fragile.
2. *Smooth muscle tone decreases.* General motility decreases, and peristaltic contractions are weaker. This change slows the rate of chyme movement and promotes constipation. Sagging of the walls of haustra in the colon can produce symptoms of *diverticulitis.* Straining to eliminate compacted fecal materials can stress the less resilient walls of blood vessels, producing *hemorrhoids.* Weakening of the cardiac sphincter can lead to *esophageal reflux* and frequent bouts of "heartburn."
3. *The effects of cumulative damage become apparent.* A familiar example would be the gradual loss of teeth because of tooth decay or gingivitis. Cumulative damage can involve internal organs as well. Toxins such as alcohol, heavy metals, and other injurious chemicals that are absorbed by the digestive tract are transported to the liver for processing or storage. The liver cells are not immune to these compounds, and chronic exposure can lead to *cirrhosis* or other types of liver disease.
4. *Cancer rates increase.* Cancers are most common in organs where stem cells divide to maintain epithelial cell populations. p. 82 Rates of colon cancer and stomach cancer rise in the elderly; oral and pharyngeal cancers are particularly common in elderly smokers.
5. *Changes in other systems have direct or indirect effects on the digestive system.* For example, the reduction in bone mass and calcium content in the skeleton is associated with erosion of the tooth sockets and eventual tooth loss. The decline in olfactory and gustatory sensitivity with age can lead to dietary changes that affect the entire body.

RELATED CLINICAL TERMS

achalasia (ak-a-LĀ-zē-a): Blockage of the lower part of the esophagus due to weak peristalsis and malfunction of the lower esophageal sphincter. *Achalasia and Esophagitis p. 806*

cholecystitis (kō-lē-sis-TĪ-tis): A painful condition caused by blockage of the cystic or common bile duct by gallstones. *p. 694*

cholelithiasis (kō-lē-li-THĪ-a-sis): Presence of gallstones in the gallbladder. *p. 694*

cirrhosis: A condition caused by scarring of the liver following destruction of hepatocytes by drug exposure, viral infection, ischemia, or other factors. *Cirrhosis p. 806*

colitis: Irritation of the colon, leading to abnormal bowel function. *Diverticulitis and Colitis p. 806*

colostomy: Attachment and opening of the colon to the abdominal wall, bypassing the distal portion of the large intestine. *Diverticulitis and Colitis p. 806*

diverticulosis (dī-ver-tik-ū-LŌ-sis): A condition in which pockets (*diverticula*) form in the mucosa of the colon, usually in the sigmoid colon. *Diverticulitis and Colitis p. 806*

enteritis (en-ter-Ī-tis): Irritation of the small intestine by toxins or other irritants; causes diarrhea due to frequent peristalsis along the small intestine. *Gastroenteritis p. 806*

esophagitis (ē-sof-a-JĪ-tis): Inflammation of the esophagus due to erosion by gastric juices. *Achalasia and Esophagitis p. 806*

gastrectomy: Surgical removal of the stomach; a potential treatment for stomach cancer. *Stomach Cancer p. 806*

gastric stapling and gastric bypass surgery: Surgical procedures to promote weight loss by blocking off a large portion of the gastric lumen. *p. 683*

gastritis (gas-TRĪ-tis): Inflammation of the gastric mucosa. *p. 682*

gastroenteritis: Vomiting and diarrhea caused by an extremely powerful irritating stimulus. *Gastroenteritis p. 806*

gastroscope: A fiberoptic instrument used to visualize the interior of the stomach. *Stomach Cancer p. 806*

mumps: A viral infection that most often affects the parotid salivary glands between ages 5–9. *p. 673*

pancreatitis (pan-krē-a-TĪ-tis): An inflammation of the pancreas due to blockage of the pancreatic ducts, bacterial or viral infections, or drug reactions. *p. 695*

peptic ulcer: A localized erosion of the gastric or duodenal mucosa by acids and enzymes in chyme. *p. 682*

periodontal disease: A progressive condition resulting from erosion of the connections between the necks of teeth and the gingiva. *p. 675*

peritonitis (per-i-tō-NĪ-tis): A painful condition resulting from inflammation of the peritoneal membrane. *p. 667*

plaque: A dense deposit of food particles and bacterial secretions on the surfaces of teeth. *p. 675*

Additional Clinical Terms Discussed in Appendix I (p. 806)
colectomy; diverticulitis; ileostomy; irritable bowel syndrome; spastic colon (or spastic colitis); total gastrectomy

STUDY OUTLINE & CHAPTER REVIEW

Introduction 663

1. The **digestive system** consists of the muscular **digestive tract** and various **accessory organs.** *(see Figure 25.1)*
2. Digestive functions include: ingestion, mechanical processing, digestion, secretion, absorption, compaction, and excretion.

An Overview of the Digestive System 663

Histological Organization of the Digestive Tract 663

1. The **lamina propria** and epithelium form the **mucosa** (a **mucous membrane**) of the digestive tract. The major layers of the digestive tract are the **submucosa** (areolar tissue), the **muscularis externa** (a region of smooth muscle fibers), and (in the peritoneal cavity) a serous membrane called the **serosa.** *(see Figure 25.2)*
2. The digestive tract is lined by a **mucosal epithelium** moistened by glandular secretions of the epithelial and accessory organs. The mucosal epithelium may be simple or stratified. The lining of the digestive tract contains pleats which allow for expansion. *(see Figure 25.2)*

Muscularis Layers and the Movement of Digestive Materials 665

3. The smooth muscle cells of the digestive tract are capable of plasticity, which is the ability to tolerate extreme stretching. The digestive system contains visceral smooth muscle tissue, in which the muscle cells are arranged in sheets and contain no motor innervation. The presence of **pacesetter cells** allows for rhythmic waves of contraction that spread through the entire muscular sheet.
4. The muscularis externa propels materials through the digestive tract through the contractions of **peristalsis. Segmentation** movements in areas of the small intestine churn digestive materials. *(see Figure 25.3)*

The Peritoneum 667

5. The serosa, also known as the **visceral peritoneum,** is continuous with the **parietal peritoneum** that lines the inner surfaces of the body wall.
6. Fused double sheets of peritoneal membrane called **mesenteries** suspend portions of the digestive tract. *(see Figure 25.2a)*
7. Important **mesenteries** include the **greater omentum, lesser omentum, mesentery proper, transverse mesentery,** and **sigmoid mesocolon.** The ascending colon, descending colon, duodenum, and pancreas are attached to the posterior wall of the abdominopelvic cavity; they are **retroperitoneal.** *(see Figures 25.4/25.10/25.11)*

The Oral Cavity 667

1. The functions of the oral cavity include: (1) **analysis** of potential foods; (2) **mechanical processing** using the teeth, tongue, and palatal surfaces; (3) **lubrication** by mixing with mucus and salivary secretions; and (4) **digestion** by salivary enzymes. Structures of the oral cavity include the tongue, salivary glands, and teeth. *(see Figure 25.5)*

Anatomy of the Oral Cavity 667

2. The **buccal cavity** (oral cavity) is lined by a stratified squamous epithelium. The **hard** and **soft palates** form the roof of the oral cavity. Other important features can be seen in *Figure 25.5.*
3. The **tongue** aids in mechanical processing and manipulation of food as well as sensory analysis. The superior surface of the **body** of the tongue is covered with *papillae.* The inferior portion of the tongue contains a thin fold of mucous membrane called the **lingual frenulum. Intrinsic** and **extrinsic tongue muscles** are controlled by the hypoglossal nerve. *(see Figure 25.5)*
4. The **parotid, sublingual,** and **submandibular salivary glands** discharge their secretions into the oral cavity. The parotid salivary glands release **salivary amylase,** which begins the breakdown of carbohydrates. *(see Figures 25.5/25.6)*
5. **Saliva** lubricates the mouth, solubilizes food, dissolves chemicals, flushes the oral surfaces, and helps control bacteria. Salivation is usually controlled by the autonomic nervous system.
6. **Dentin** forms the basic structure of a tooth. The **crown** is coated with **enamel,** and the **root** with **cementum.** The **neck** marks the boundary between the root and the crown. The **periodontal ligament** anchors the tooth in an alveolar socket. **Mastication** (chewing) occurs through the contact of the opposing **occlusal surfaces** of the teeth. *(see Figure 25.7)*

7. There are four types of teeth, each with specific functions: **incisors**, for cutting; **cuspids** (canines) for tearing; **bicuspids** (premolars), for crushing; and **molars**, for grinding. *(see Figure 25.7b,c)*
8. The first set of teeth to appear are called **deciduous teeth** (*primary, milk*, or *baby teeth*), followed by the adult **secondary dentition** (*permanent dentition*). The sequence of tooth eruption is presented in *Figure 25.7d.*
9. The upper rows of teeth form a **dental arcade** with **labial**, **palatal** (upper teeth) or **lingual** (lower teeth), **mesial**, and **distal surfaces**. *(Figure 25.7c)*
10. **Mastication** forces the food across the surfaces of the teeth until it forms a **bolus** that can be swallowed easily.

The Pharynx 675

1. Skeletal muscles involved with swallowing include the *pharyngeal constrictor muscles* and the *palatopharyngeus, stylopharyngeus*, and *palatal muscles.*

The Swallowing Process 675

2. **Deglutition** (swallowing) has three phases. The **buccal phase** begins with the compaction of a **bolus** and its movement into the pharynx. The **pharyngeal phase** involves the elevation of the larynx, reflection of the epiglottis, and closure of the glottis. Finally, the **esophageal phase** involves the opening of the **upper esophageal sphincter** and peristalsis moving the bolus down the esophagus to the **lower esophageal sphincter**. *(see Figure 25.8)*

The Esophagus 676

1. The **esophagus** is a hollow muscular tube that transports food and liquid to the stomach, through the **esophageal hiatus**, an opening in the diaphragm.

Histology of the Esophageal Wall 676

2. The wall of the esophagus is formed by *mucosa, submucosa, muscularis*, and *adventitia* layers. *(see Figures 25.1/25.2/25.9)*

The Stomach 677

1. The **stomach** has three major functions: (1) *bulk storage of ingested matter*, (2) *mechanical breakdown of resistant materials*, and (3) *chemical digestion* through the disruption of chemical bonds using acids and enzymes.

Anatomy of the Stomach 677

2. The stomach is divided into four regions: the **cardia**, the **fundus**, the **body**, and the **pylorus**. The **pyloric sphincter** guards the exit from the stomach. The mucosa and submucosa are thrown into longitudinal folds, called **rugae**. The muscularis layer is formed of three bands of smooth muscle: a *longitudinal layer*, a *circular layer*, and an inner *oblique layer. (see Figures 25.4/25.10 to 25.13)*
3. The mesenteries of the stomach are the **greater omentum**, which hangs from the greater curvature, and the **lesser omentum**, which is attached to the lesser curvature.
4. Three branches of the **celiac trunk** supply blood to the stomach: the *left gastric artery, splenic artery*, and the *common hepatic artery. (see Figures 22.17/22.26/25.10a)*

Histology of the Stomach 680

5. Simple columnar epithelium line all portions of the stomach. Shallow depressions, called **gastric pits**, contain the **gastric glands** of the fundus and body. **Parietal cells** secrete **intrinsic factor** and hydrochloric acid. **Chief cells** secrete **pepsinogen**, which acids in the gastric lumen convert to the enzyme **pepsin**. **G cells** of the stomach secrete the hormone **gastrin**. *(see Figure 25.13)*

Regulation of the Stomach 682

6. The production and secretion of **gastric juices** are directly controlled by the CNS (the vagus nerve, parasympathetic innervation, and the celiac plexus, sympathetic innervation). The release of the local hormones **secretin** and **cholecystokinin** inhibits gastric secretion but stimulates secretion by the pancreas and liver.

The Small Intestine 682

Regions of the Small Intestine 683

1. The small intestine includes the **duodenum**, the **jejunum**, and the **ileum**. The intestinal mucosa bears transverse folds, called **plicae circulares**, and small projections, called **intestinal villi**, that increase the surface area for absorption. Each villus contains a terminal lymphatic called a **lacteal**. Pockets called **intestinal glands** house enteroendocrine, goblet, and stem cells. *(see Figures 25.4/25.10/25.12/25.14 to 25.16)*

Support of the Small Intestine 683

2. The *superior mesenteric artery* and *superior mesenteric vein* supply numerous branches to the segments of the small intestine. *(see Figures 22.17/22.26)*
3. The **mesentery proper** supports the branches of the superior mesenteric artery and vein, lymphatics, and nerves that supply the jejunum and ileum. *(see Figure 25.4)*

Histology of the Small Intestine 685

4. The regions of the small intestine have histological specializations that determine their primary functions. The duodenum (1) contains **duodenal** *(Brunner's)* **glands** that aid the crypts in producing mucus and (2) receives the secretions of the common bile duct and pancreatic duct. The jejunum and ileum contain large groups of **aggregated lymphoid nodules** (*Peyer's patches*) within the lamina propria. *(see Figures 25.15/ 25.16)*

Regulation of the Small Intestine 686

5. **Intestinal juice** moistens the chyme, helps to buffer acids, and dissolves digestive enzymes and the products of digestion.
6. *Secretin* and *cholecystokinin* (*CCK*) are two hormones important in the coordination of digestive activities. Parasympathetic (vagal) innervation stimulates digestive function; sympathetic stimulation inhibits activity along the digestive tract.

The Large Intestine 686

1. The **large intestine (large bowel),** begins as a pouch inferior to the terminal portion of the ileum and ends at the anus. The main functions of the large intestine are to: (1) *reabsorb water and compact feces*, (2) *absorb vitamins by bacteria*, and (3) *store fecal material* prior to defecation. *(see Figures 25.4/25.12/25.14/25.17 to 25.19)*
2. The large intestine is divided into three parts: the **cecum**, the **colon**, and the **rectum**.

The Cecum 686

3. The **cecum** collects and stores materials arriving from the ileum. The **ileocecal valve** opens into the cecum. The **vermiform appendix** is attached to the cecum, and it functions as part of the lymphatic system. *(see Figure 25.17)*

The Colon 688

4. The **colon** has a larger diameter and a thinner wall than the small intestine. It bears **haustra** (pouches), the **taeniae coli** (longitudinal bands of muscle), and **fatty appendices** (fat aggregations within the serosa). *(see Figures 25.17/25.18)*
5. The colon is subdivided into four regions: **ascending**, **transverse**, **descending**, and **sigmoid**. *(see Figures 25.4/25.17/25.18)*

The Rectum 688

6. The rectum terminates in the **anal canal** leading to the **anus**. **Internal** and **external anal sphincters** control the passage of fecal material to the anus. Distension of the rectal wall triggers the **defecation reflex**. *(see Figures 25.12/25.17/25.18)*

Histology of the Large Intestine 688

7. The major histological features of the colon are: lack of villi, abundance of goblet cells, and distinctive mucous-secreting intestinal glands. *(see Figure 25.19)*

Regulation of the Large Intestine 689

8. Movement from the cecum to the transverse colon occurs slowly via peristalsis and *haustral churning*. Movement from the transverse to the sigmoid colon occurs several times each day via **mass movements**.
9. Distention of the rectal wall from a mass movement may stimulate the conscious desire to relax internal and external anal sphincters to defecate.

Accessory Digestive Organs 689

The Liver 689

1. The **liver** performs metabolic and hematological regulation and produces **bile.** Its metabolic role is to regulate the concentrations of wastes and nutrients in the blood, and its hematological role is as a blood reservoir. *(see Figures 25.12/25.20/25.21 and Table 25.1)*
2. The liver is divided into four lobes: **left**, **right**, **quadrate**, and **caudate**. The gallbladder is located in a fossa within the **posterior surface** of the right lobe. *(see Figure 25.20)*
3. The **hepatic artery proper** and **hepatic portal vein** supply blood to the liver. **Hepatic veins** drain blood from the liver and return it to the systemic circuit via the inferior vena cava. *(see Figures 22.17/22.26/25.20d)*
4. Liver cells are specialized epithelial cells, termed **hepatocytes. Kupffer cells**, or *stellate reticuloendothelial cells*, are phagocytic cells that reside in the sinusoidal lining. The **liver lobule** is the basic functional unit of the liver. Each lobule is hexagonal in cross section and contains six **portal areas**, or *hepatic triads*. A portal area consists of a branch of the hepatic portal vein, a branch of the hepatic artery proper, and a branch of the hepatic (bile) duct. **Bile canaliculi** carry bile to **bile ductules** that lead to portal areas. The bile ducts from each lobule unite to form the **left** and **right hepatic ducts**, which merge to form the **common hepatic duct**. *(see Figures 25.20 to 25.22)*

The Gallbladder 693

5. The **gallbladder** is a hollow muscular organ that stores and concentrates bile before excretion in the small intestine. **Bile salts** break apart large drops of lipids and make them accessible to digestive enzymes. Bile ejection occurs under stimulation of cholecytstokinin (CCK).
6. The gallbladder is divided into: **fundus**, **body**, and **neck** regions. The **cystic duct** leads from the gallbladder to merge with the common hepatic duct to form the **common bile duct**. *(see Figures 25.16a/25.20d/25.22)*

The Pancreas 693

7. The **pancreas** is divided into **head**, **body**, and **tail** regions. The **pancreatic duct** penetrates the wall of the duodenum. Within each lobule, ducts branch repeatedly before ending in the **pancreatic acini** (blind pockets). The **accessory pancreatic duct** (if present) and pancreatic duct perforate the wall of the duodenum to discharge **pancreatic juice** at the *lesser duodenal papilla* and *greater duodenal papilla*, respectively. *(see Figures 25.22/25.23)*
8. Pancreatic tissue consists of exocrine and endocrine portions. The bulk of the organ is exocrine in function, as the pancreatic acini secrete water, ions, and digestive enzymes into the small intestine. Pancreatic enzymes include **lipases**, **carbohydrases**, **nucleases**, and **proteolytic enzymes**. The major hormones produced by the endocrine portion are insulin and glucagon.
9. Regulation of the production of pancreatic juice is via the hormones cholecystokinin and secretin.

Aging and the Digestive System 695

1. Normal digestion and absorption occur in elderly individuals; however, changes in the digestive system reflect age-related changes in other body systems. These include a slowed rate of epithelial stem cell division, a decrease in smooth muscle tone, the appearance of cumulative damage, an increase in cancer rates, as well as numerous changes in other systems.

LEVEL 1 REVIEWING FACTS AND TERMS

Match each numbered item with the most closely related lettered item. Use letters for answers in the spaces provided.

Column A	Column B
_____ 1. segmentation	a. mesentery sheet suspending small intestine
_____ 2. mesentery proper	b. propels materials through the digestive tract
_____ 3. cuspids	c. churn and fragment digestive materials
_____ 4. serosa	d. canines
_____ 5. buccal fat pads	e. mechanical/chemical digestion of food
_____ 6. mastication	f. serous membrane covering muscularis externa
_____ 7. bicuspids	g. chewing
_____ 8. digestive tract	h. premolars
_____ 9. mesocolon	i. mesentery associated with the large intestine
_____ 10. peristalsis	j. form the cheeks

11. Which of the following is not part of the digestive tract?
(a) stomach
(b) pharynx
(c) esophagus
(d) spleen

12. Digestion refers to
(a) the progressive dehydration of indigestible residue
(b) the input of food into the digestive tract
(c) the chemical breakdown of food
(d) the absorption of nutrients into the gut

13. Most of the digestive tract is lined by
(a) pseudostratified ciliated columnar epithelium
(b) cuboidal epithelium
(c) stratified squamous epithelium
(d) simple columnar epithelium

14. The ___ are double sheets of peritoneal membrane that hold some of the visceral organs in their proper position.
(a) serosa (b) adventitia
(c) mesenteries (d) fibrosa

15. The activities of the digestive system are regulated by
(a) hormones
(b) parasympathetic neurons
(c) sympathetic neurons
(d) all of the above

16. Sandwiched between the layer of circular and longitudinal muscle in the muscularis externa is the
(a) mucosa
(b) submucosa
(c) muscularis mucosa
(d) myenteric plexus

17. The bulk of each tooth has a mineralized matrix similar to that of bone called
(a) enamel (b) cementum
(c) dentin (d) pulp

18. The passageway between the oral cavity and the pharynx is the
(a) uvula
(b) fauces
(c) palatoglossal arch
(d) palatopharyngeal arch

19. Plicae and intestinal villi
    (a) increase the surface area of the mucosa of the small intestine
    (b) carry digestion products that do not enter blood capillaries
    (c) produce new cells for the mucosa of the small intestine
    (d) secrete digestive enzymes

20. Chief cells secrete
    (a) pepsinogen
    (b) gastrin
    (c) mucus
    (d) hydrochloric acid

LEVEL 2 REVIEWING CONCEPTS

1. The lining of the stomach
   (a) is composed of simple columnar epithelium
   (b) is covered by a thick, viscous mucus
   (c) is constantly being replaced
   (d) all of the above
2. A blockage of the ducts from the parotid salivary glands would
   (a) result in the production of more viscous saliva
   (b) impair the lubricating properties of saliva
   (c) interfere with carbohydrate digestion in the mouth
   (d) eliminate the sense of taste
3. In response to the hormone cholecystokinin, the pancreas secretes
   (a) a fluid rich in enzymes
   (b) a fluid rich in bicarbonate
   (c) a fluid rich in bile
   (d) a fluid that contains only proteinases
4. What is the function of the lipase from the pancreas?
5. What is the function of the hepatopancreatic sphincter?
6. What does the gallbladder do with bile?
7. What is the function of Kupffer cells?
8. What is the last region of the colon before it reaches the rectum?
9. What is the function of the lacteals in the small intestine?
10. What triggers the release of gastrin?

LEVEL 3 CRITICAL THINKING AND CLINICAL APPLICATIONS

1. Jack has just had some dental work done, at which time the dentist used a topical anesthetic to deaden sensation in the oral cavity. Some of the anesthetic has numbed the pharynx, and the dentist cautions Jack about eating anything until the numbness wears off. Jack is hungry and ignores the warning and eats a hamburger for lunch. As he tries to swallow the bolus, he begins to gag. Can you explain why this occurs?
2. Leon suffers from gallstones. His doctor puts him on a diet low in fat. Why?
3. A condition known as lactose intolerance is characterized by painful abdominal cramping, gas, and diarrhea. The cause of the problem is an inability to digest the milk sugar, lactose. How would this cause the observed symptoms?

✓ ANSWERS TO CONCEPT CHECK QUESTIONS

p. 667 **1.** The components of the mucosa are: (1) mucosal epithelium (depending upon location, may be simple columnar or stratified squamous); (2) lamina propria, areolar tissue underlying the epithelium; and (3) muscularis mucosae, bands of smooth muscle fibers arranged in concentric layers. The mucosa of the digestive tract is an example of a mucous membrane, serving both absorptive and secretory functions. **2.** Mesenteries provide an access route for the passage of blood vessels, nerves, and lymphatics to and from the digestive tract. They also stabilize the relative positions of the attached organs. **3.** Peristalsis is waves of muscular contractions that move substances the length of the digestive tube. Segmentation activities churn and mix the contents of the small and large intestines but do not produce net movement in a particular direction. **4.** This allows the stretched smooth muscle to adapt to its new shape and still have the ability to contract when necessary.
p. 675 **1.** The oral cavity is lined by the oral mucosa, which is composed of nonkeratinized stratified squamous epithelium. **2.** At mealtime, saliva lubricates the mouth and dissolves chemicals that stimulate the taste buds. Saliva contains the digestive enzyme salivary amylase, which begins the chemical breakdown of carbohydrates. **3.** Carbohydrates are broken down by salivary amylase in the mouth. **4.** The incisors cut away a section of the apple, which then enters the mouth. The cuspids tear at the rough skin and pulp of the apple. The apple then moves to the bicuspids and molars for thorough mashing and grinding before finally being swallowed.
p. 677 **1.** The fauces is the opening between the oral cavity and the pharynx. **2.** The process that is being described is deglutition, or swallowing. **3.** The buccal phase is the only voluntary phase of swallowing. **4.** Her lower esophageal sphincter may not have closed completely, and powerful stomach acids are entering her lower esophagus, causing uncomfortable acid reflux.
p. 682 **1.** The greater omentum provides support to the surrounding organs, pads the organs from the surfaces of the abdomen, provides an important energy reserve, and provides insulation. **2.** The epithelium produces a carpet of mucus that covers the interior surfaces of the stomach, providing protection against the powerful acids and enzymes. Any cells that do become damaged are quickly replaced. **3.** Chief cells secrete pepsinogen. In infants, they also secrete rennin and gastric lipase. **4.** Gastrin, produced by enteroendocrine cells, stimulates the secretion of parietal and chief cells.
p. 686 **1.** The characteristic lining of the small intestine contains plicae circulares, which support intestinal villi. The villi are covered by a simple columnar epithelium whose apical surface is covered by microvilli. This arrangement increases the total area for digestion and absorption to over 200 m^2. **2.** Plicae are folds in the lining of the intestine, and they greatly increase the surface area available for absorption. **3.** Intestinal glands house the stem cells that produce new epithelial cells, which renew the epithelial surface and add intracellular enzymes to the chyme. In addition, intestinal glands contain cells that produce several intestinal hormones. **4.** The duodenum acts as a mixing bowl for the chyme entering from the stomach.
p. 695 **1.** The condition of cystic fibrosis interferes with the digestion of sugars, starches, lipids, nucleic acids, and proteins. **2.** The liver acts as a metabolic regulator by extracting nutrients and toxins from the blood before it enters the bloodstream. It also regulates the blood, serving as a blood reservoir, phagocytizing old or damaged RBCs, and synthesizing plasma proteins. Finally, the liver synthesizes and secretes bile. **3.** Contraction of the hepatopancreatic sphincter seals off the passageway between the gallbladder and the small intestine and keeps bile from entering the small intestine. **4.** The pancreas produces digestive enzymes and buffers (exocrine functions) and hormones (endocrine functions).

REVIEW IT

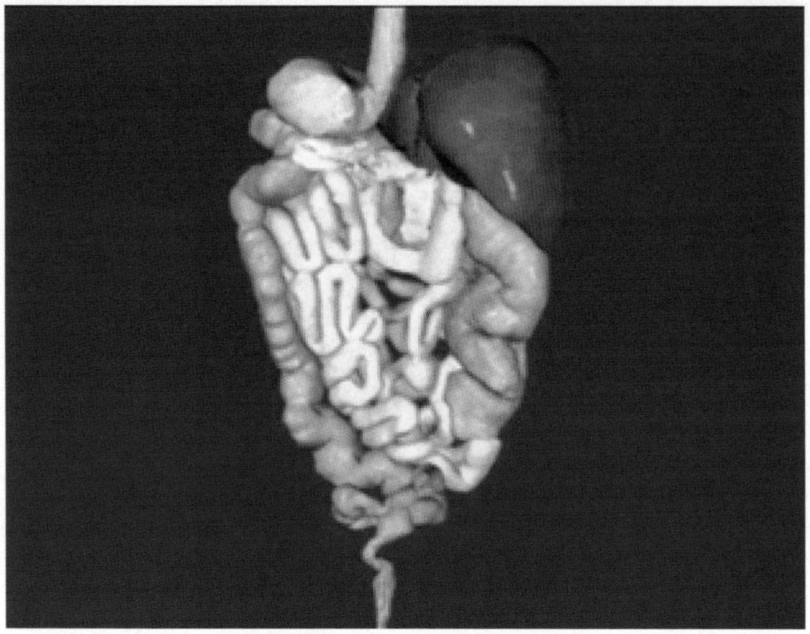

Challenge 1

Although the relationship of form to function is seen throughout this text, it is most apparent in the digestive system. The basic structure of the digestive tract is a tube, with accessory organs periodically adding contents to that tube. Each accessory organ of the system develops in association with a specific part of the digestive tract.. The image presented is taken from the *Interactive CD/Animations/Digestive system*. What structures can you identify in this view? You should be able to list at least 5 major organs, and 5 or 6 landmarks or regions of those organs. Is it obvious from this view that these structures are modifications of one long convoluted tube? How many accessory organs can you identify in this view? Using the figures in the text as a guide, determine whether this is an anterior, posterior or lateral view. Label the left and right sides of the body.

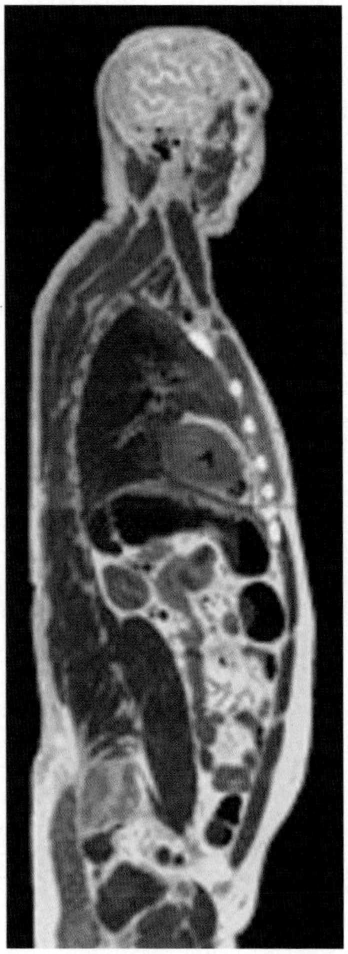

Challenge 2

Understanding the spatial relationship between structures of the digestive system and the abdominal cavity is sometimes difficult. The organs twist and move with respect to one anther as food is processed. How does this occur without displacing organs or causing potentially fatal twists and closures of the lumen? The accompanying image is a sagittal section reconstructed from cross-sectional data provided by the Visible Human project. Study this image carefully. What structures of the digestive system are visible? What tissue type(s) is (are) present between these organs? How much freedom of movement does this allow the organs of the digestive system? If necessary, review Chapter 3 to help identify these tissues.

Images provided by the Digital Cadaver™ project, courtesy of Visible Productions, Inc.

APPLY IT

Below are two separate exercises that provide more information on a topic presented in the preceding chapters. Each one is designed to take approximately 10 minutes, and will help you gain a better appreciation for the material presented in this chapter.

Critical Linking 1

Normally when the digestive system is mentioned, food processing and nutrient absorption are discussed. Thinking back to the first chapter of this text will remind you that all 11 body systems work in concert to maintain homeostasis. This means what is detrimental to one system will directly or indirectly affect other systems as well. We are all aware of the effects of smoking on the respiratory system. Recent studies have indicated a similar relationship between smoking and the digestive system. Using your knowledge of the digestive system, homeostasis and the respiratory system, predict what these effects might be. Prepare a short outline of your predictions, listing what organs are affected and what the effects might be.

The government has prepared a short essay "Smoking and your digestive system" designed for the lay person. Go to the Companion Website, Critical Linking portion and choose the key word "Smoking." Use the article presented to modify your outline. Go beyond what is presented in the article to explain how these effects might be related to smoking.

Critical Linking 2

Identifying the organs of the GI tract is an important skill, however many practicing physicians do not see the digestive organs in the views presented in either the text figures or the Interactive CD/Animations. Endoscopy is a technique that provides a noninvasive means of investigating the GI tract without the health risks of exploratory surgery. However, identifying structures of the tract from the lumen requires a new set of skills. Cues must be taken from the structure of the walls and the overall shape of the organ.

To practice this skill yourself, visit the Companion Website, Critical Linking portion. Click on the key word "endoscopy," and choose "normal findings in upper endoscopy," "stomach," "duodenum," "colon," "sigmoid colon," and "rectum." In each case you will be presented with labeled endoscopic views of the organ chosen. Click through a few to discover identifying characteristics of each organ. Prepare a list of these characteristics, relating them to the information presented in the text. Go to the *Interactive CD/Animations/Regional view/Panoramic/Digestive System.* Practice your endoscopic identification skills on these views.

FURTHER STUDIES

A consideration of the digestive system in a social context leads one to reflect on the historical significance of famine. What happens both sociologically and biologically when nutrients become scarce? The Irish potato famine provides an excellent look at both the biology of an event such as this and the social implications. The "Great Famine" lasted from 1845 to 1848, and is attributed to a fungal blight which caused potato plants to rot to the ground. The island of Ireland was at that time dependant on the potato crop for the survival of its 8 million plus inhabitants. In 1845 half the crop was lost to the blight, and by 1846 the potato crop was a complete failure. Governmental assistance came too late for most people, and starvation was rampant. The winter of 1846 brought excessive rain and cold, causing outbreaks of typhus and recurring fever, commonly referred to as "famine fever". Scurvy and dysentery added to the health troubles of the Irish, and an outbreak of cholera in the major cities in 1849 decimated the population. There are newspaper articles that chronicle the cause of death related to starvation. These range from poisoning due to eating seaweeds to death by overtaxing the digestive system with a large meal after weeks of starvation. In many cases the suffering individuals merely fell over. Post-mortem investigation usually revealed some small scrap of food in the stomach such as turnip peels or wheat paste.

This harsh lesson on single crop dependence and health hazards associated with starvation has been preserved in many articles, biographies and works of fiction. Take a minute to explore the information presented on this topic via the Internet by visiting: *http://www.irelandseye.com/articles/history/events/dates/famine.shtm* first for a general description of the famine. Newspaper reports can be obtained by visiting: *http://www.people.virginia.edu/~eas5e/Irish/Wexford.html*.

To gain a personal perspective, look at Gerald Keegan's diary online: *http://www.people.virginia.edu/~eas5e/Irish/Keegan.html*.

Drawing depicting this event can be obtained through *http://vassun.vassar.edu/~sttaylor/FAMINE/PT/PT.html*.

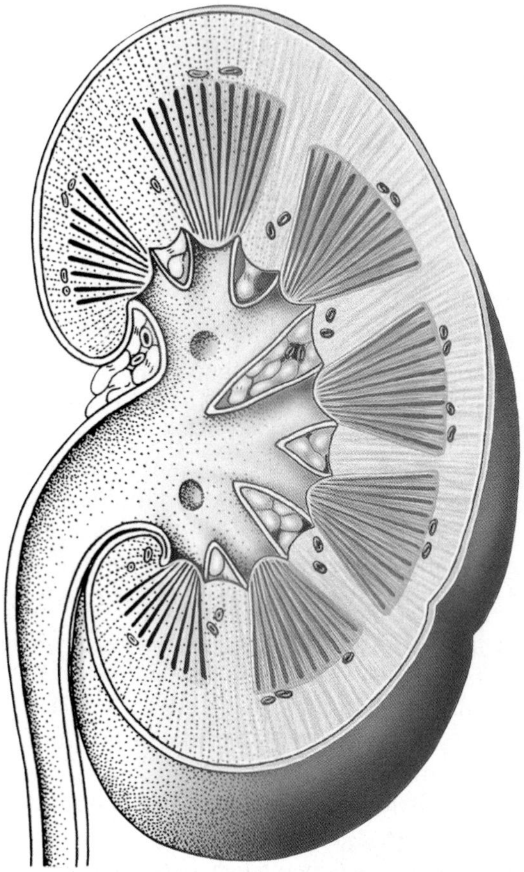

26 THE URINARY SYSTEM

The coordinated activities of the digestive, cardiovascular, respiratory, and urinary systems prevent the development of "pollution" problems inside the body. The digestive tract absorbs nutrients from food, and the liver adjusts the nutrient concentration of the circulating blood. The cardiovascular system delivers these nutrients, along with oxygen from the respiratory system, to peripheral tissues. As blood leaves these tissues it carries the carbon dioxide and other waste products generated by active cells to sites of excretion. The carbon dioxide is eliminated at the lungs. Most of the organic waste products, along with excess water and electrolytes, are removed and excreted by the urinary system, the focus of this chapter. Other systems assist the kidneys in the excretion of water and solutes; for example, the sweat produced by sweat glands contains water and solutes, and various digestive organs secrete waste product into the lumen of the digestive tract. However, the contribution of these organs is minor compared to that of the kidneys

The urinary system performs vital excretory functions and eliminates the organic waste products generated by cells throughout the body. It also has a number of other essential functions that are often overlooked. A more complete list of urinary system functions includes:

1. *regulating plasma concentrations of sodium, potassium, chloride, calcium, and other ions* by controlling the quantities lost in the urine;
2. *regulating blood volume and blood pressure* by: (a) adjusting the volume of water lost in the urine, (b) releasing erythropoietin, and (c) releasing renin; p. 519
3. *contributing to the stabilization of blood pH;*
4. *conserving valuable nutrients* by preventing their excretion in the urine;
5. *eliminating organic waste products*, especially nitrogenous wastes such as *urea* and *uric acid*, toxic substances, and drugs;
6. *synthesizing calcitriol*, a hormone derivative of vitamin D_3 that stimulates calcium ion absorption by the intestinal epithelium; and
7. *assisting the liver* in detoxifying poisons and, during starvation, deaminating amino acids so that other tissues can break them down.

All urinary system activities are carefully regulated to maintain the solute composition and concentration of the blood within acceptable limits. A disruption of any one of these functions will have immediate and potentially fatal consequences.

This chapter considers the functional organization of the urinary system and describes the major regulatory mechanisms that control urine production and concentration. The urinary system (Figure 26.1a●) includes: the *kidneys*, *ureters*, *urinary bladder*, and *urethra*. The excretory functions of this system are performed by a pair of **kidneys**. These organs produce **urine**, a fluid waste product containing water, ions, and small soluble compounds. Urine leaving the kidneys travels along the **urinary tract**, which consists of the paired **ureters** (ū-RĒ-terz) and the **urinary bladder** where urine is temporarily stored. When **urination**, or **micturition** (mik-tū-RISH-un) occurs, contraction of the muscular urinary bladder forces urine through the **urethra** (ū-RĒTH-ra) and out of the body.

The Kidneys [FIGURES 26.1/26.2]

The kidneys are located lateral to the vertebral column between the last thoracic and third lumbar vertebrae on each side (Figure 26.2a●). The superior surface of the right kidney is often situated inferior to the superior surface of the left kidney (Figures 26.1 and 26.2a●).

On gross dissection, the anterior surface of the right kidney is covered by the liver, the hepatic flexure of the colon, and the duodenum. The anterior surface of the left kidney is covered by the spleen, stomach, pancreas, jejunum, and splenic flexure of the colon. The superior surface of each kidney is capped by an adrenal gland (Figures 26.1a and 26.2●). The kidneys, adrenal glands, and ureters lie between the muscles of the dorsal body wall and the parietal peritoneum in a retroperitoneal position (Figures 26.1c and 26.2●).

The position of the kidneys in the abdominal cavity is maintained by: (1) the overlying peritoneum, (2) contact with adjacent visceral organs, and (3) supporting connective tissues. Each kidney is protected and stabilized by three concentric layers of connective tissue (Figure 26.1c●):

1. A layer of collagen fibers covers the outer surface of the entire organ. This layer, the **renal capsule**, is also known as the *fibrous tunic* of the kidney. It maintains the shape of the kidney and provides mechanical protection.
2. A layer of adipose tissue, the **adipose capsule**, or *perirenal fat* (*peri-*, around + *renes*, kidneys), surrounds the renal capsule. This layer can be quite thick, and on dissection the adipose capsule usually obscures the outline of the kidney.
3. Collagen fibers extend outward from the inner renal capsule through the adipose capsule to a dense outer layer known as the **renal fascia.** The renal fascia anchors the kidney to surrounding structures. Posteriorly the renal fascia is bound to the deep fascia surrounding the muscles of the body wall. A layer of *pararenal* (*para*, near) *fat* separates the posterior and lateral portions of the renal fascia from the body wall. Anteriorly the renal fascia is attached to the peritoneum and to the anterior renal fascia of the opposite side.

In effect, the kidney is suspended by collagen fibers from the renal fascia and is packed in a soft cushion of adipose tissue. This arrangement prevents the jolts and shocks of day-to-day existence from disturbing normal kidney function. If the suspensory fibers stretch, or body fat content declines, reducing the amount of adipose tissue padding, the kidneys may become more vulnerable to traumatic injury.

Superficial Anatomy of the Kidney [FIGURES 26.2/26.3]

Each brownish-red kidney has the shape of a kidney bean. A typical adult kidney (Figures 26.2 and 26.3●) measures approximately 10 cm (4 in.) in length, 5.5 cm (2.2 in.) in width, and 3 cm (1.2 in.) in thickness. A single kidney averages around 150 g (5.25 oz). A prominent medial indentation, the **hilus**, is the entry point for the *renal artery* and the exit for the *renal vein* and *ureter*.

The fibrous renal capsule has inner and outer layers. In sectional view (Figure 26.3a●), the inner layer folds inward at the hilus and lines the **renal sinus**, an internal cavity within the kidney. Renal blood vessels, lymphatic vessels, renal nerves, and the ureter that drains the kidney pass through the hilus and branch within the renal sinus. The thick, outer layer of the capsule extends across the hilus and stabilizes the position of these structures.

Sectional Anatomy of the Kidney [FIGURE 26.3]

The interior of the each kidney contains a renal cortex, renal medulla, and renal sinus. The renal **cortex** is the outer layer of the kidney in contact with the capsule (Figure 26.3a●). The cortex is reddish-brown and granular. The

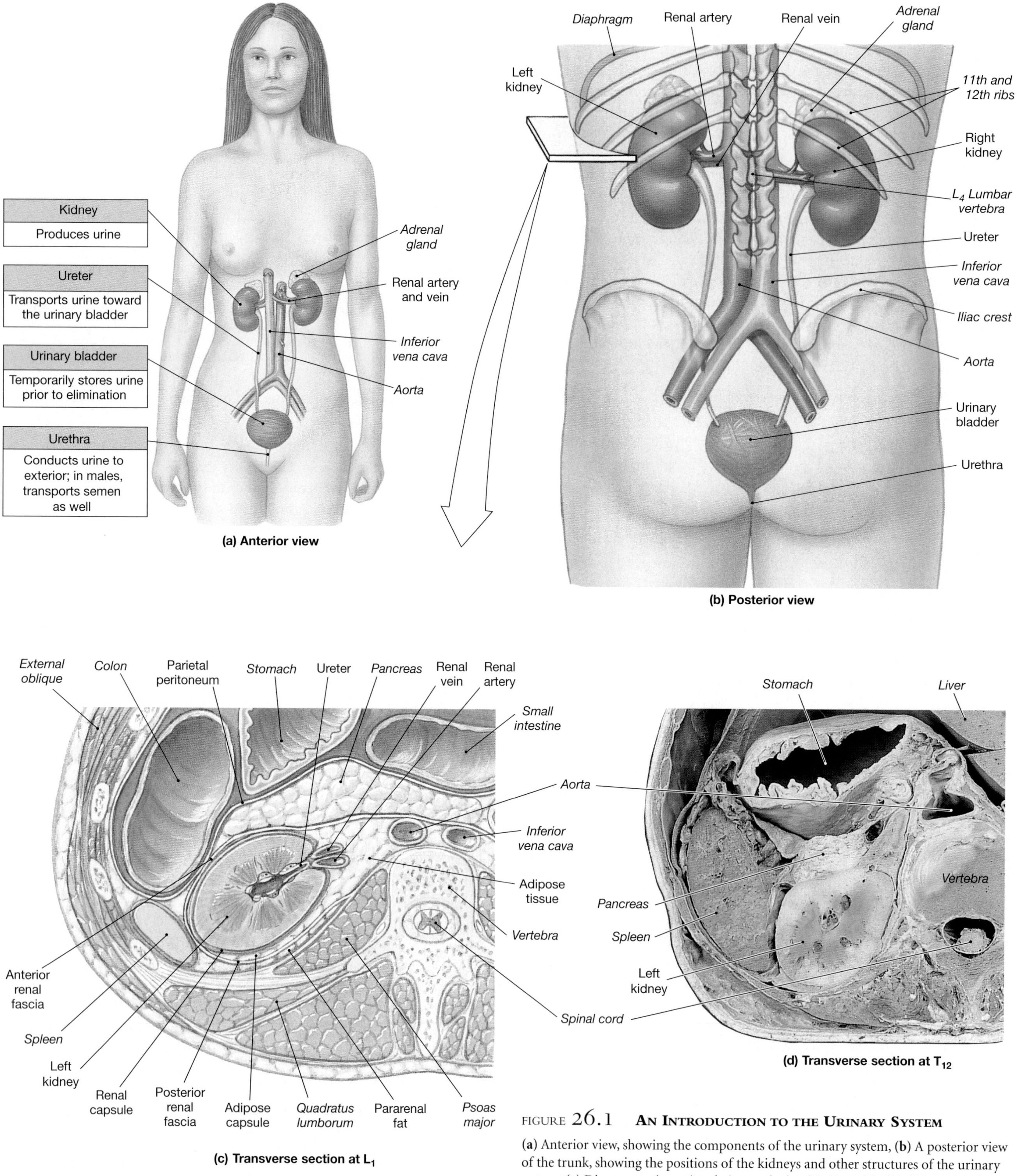

FIGURE 26.1 **AN INTRODUCTION TO THE URINARY SYSTEM**

(a) Anterior view, showing the components of the urinary system, **(b)** A posterior view of the trunk, showing the positions of the kidneys and other structures of the urinary system. **(c)** Diagrammatic sectional view at the level indicated in (b). **(d)** Sectional view slightly superior to the plane of (c).

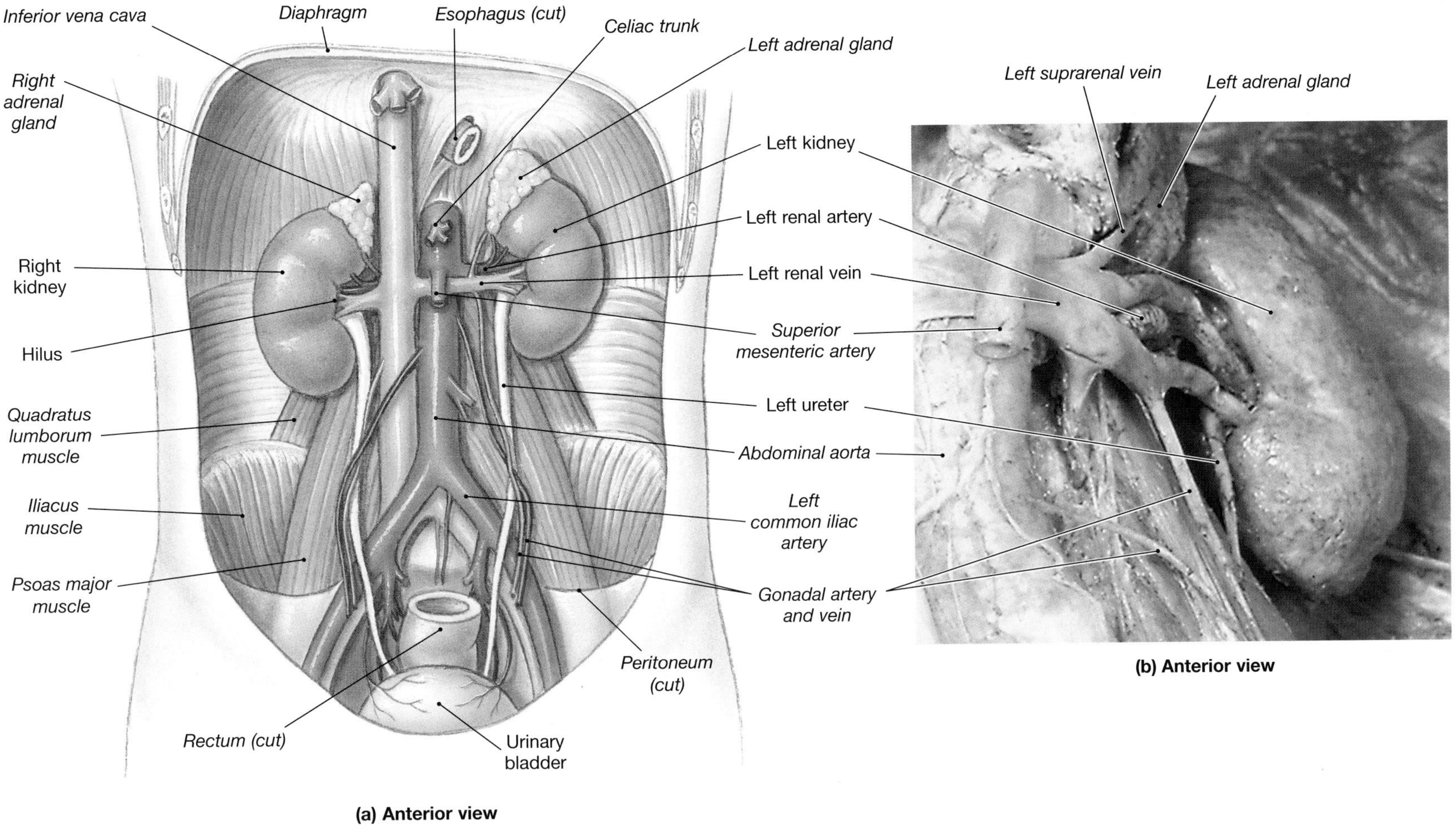

FIGURE 26.2 **The Urinary System in Gross Dissection**

(**a**) Diagrammatic anterior view of the abdominopelvic cavity showing the kidneys, adrenal glands, ureters, urinary bladder, and the blood supply to the kidneys. (**b**) Anterior view of the left kidney and associated structures.

medulla is internal to the cortex and consists of 6 to 18 distinct conical or triangular structures, called **renal pyramids**, or *medullary pyramids.* The base of each pyramid faces the cortex, and the tip, or **renal papilla**, projects into the renal sinus. Each pyramid has a series of fine grooves that converge at the papilla. Adjacent renal pyramids are separated by bands of cortical tissue, called **renal columns**. The columns have a distinctly granular texture, similar to that of the cortex. A **renal lobe** contains a renal pyramid, the overlying area of renal cortex, and adjacent tissues of the renal columns.

Urine production occurs in the renal lobes, and ducts within each renal papilla discharge urine into a cup-shaped drain, called a **minor calyx** (KĀ-liks). Four or five minor calyces (KĀL-i-sēz) merge to form a **major calyx**, and the major calyces combine to form a large, funnel-shaped chamber, the **renal pelvis**. The renal pelvis, which fills most of the renal sinus, is connected to the ureter at the hilus of the kidney.

Urine production begins in microscopic, tubular structures called **nephrons** (NEF-ronz) in the cortex of each renal lobe. Each kidney has roughly 1.25 million nephrons, with a combined length of about 145 km (85 miles).

The Blood Supply to the Kidneys [FIGURES 22.16/26.4/26.5/26.9b]

The kidneys receive 20–25% of the total cardiac output. In normal individuals, about 1200 ml of blood flows through the kidneys each minute. Each kidney receives blood from a **renal artery** that originates along the lateral surface of the abdominal aorta near the level of the superior mesenteric artery. p. 589 As the renal artery enters the renal sinus, it branches into the **segmental arteries** (Figure 26.4●). Segmental arteries further divide into a series of **interlobar arteries** that radiate outward, penetrating the renal capsule and extending through the renal columns between the renal pyramids. The interlobar arteries supply blood to the **arcuate** (AR-kū-āt) **arteries** that parallel the boundary between the cortex and medulla of the kidney. Each arcuate artery gives rise to a number of **interlobular arteries** supplying portions of the adjacent renal lobe. Numerous **afferent arterioles** branch from each interlobular artery to supply individual nephrons.

From the nephrons, blood enters a network of venules and small veins that converge on the **interlobular veins**. In a mirror image of the arterial distribution, the interlobular veins deliver blood to **arcuate veins** that empty into **interlobar veins**. The interlobar veins merge to form the **renal vein**. Many of the blood vessels just described are visible in the corrosion cast of the kidneys and in the renal angiogram shown in Figures 26.5b and 26.9b●, respectively.

Innervation of the Kidneys

Urine production in the kidneys is regulated in part through *autoregulation*, which involves reflexive changes in the diameters of the arterioles supplying the nephrons, thereby altering blood flow and filtration rates. Both hormonal and neural mechanisms can supplement or adjust the local responses. The kidneys and ureters are innervated by **renal nerves**.

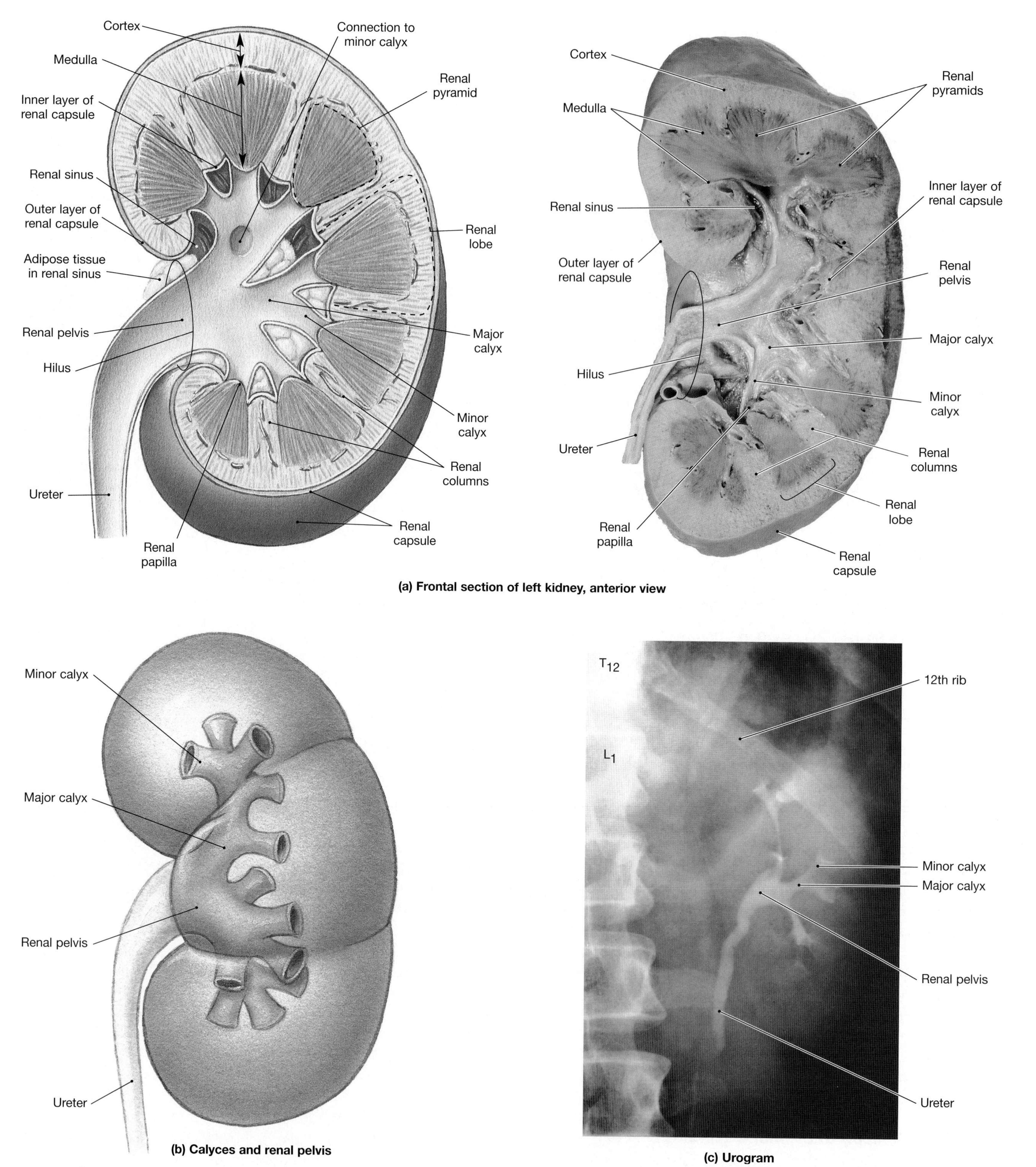

(a) Frontal section of left kidney, anterior view

(b) Calyces and renal pelvis

(c) Urogram

FIGURE 26.3 **STRUCTURE OF THE KIDNEY**

(**a**) Frontal section through the left kidney, showing major structures. The outlines of a renal lobe and a renal pyramid are indicated by dotted lines. (**b**) Shadow drawing to show the arrangement of the calyces and renal pelvis within the kidney. (**c**) Urogram of left kidney showing calyces, renal pelvis, and ureter.

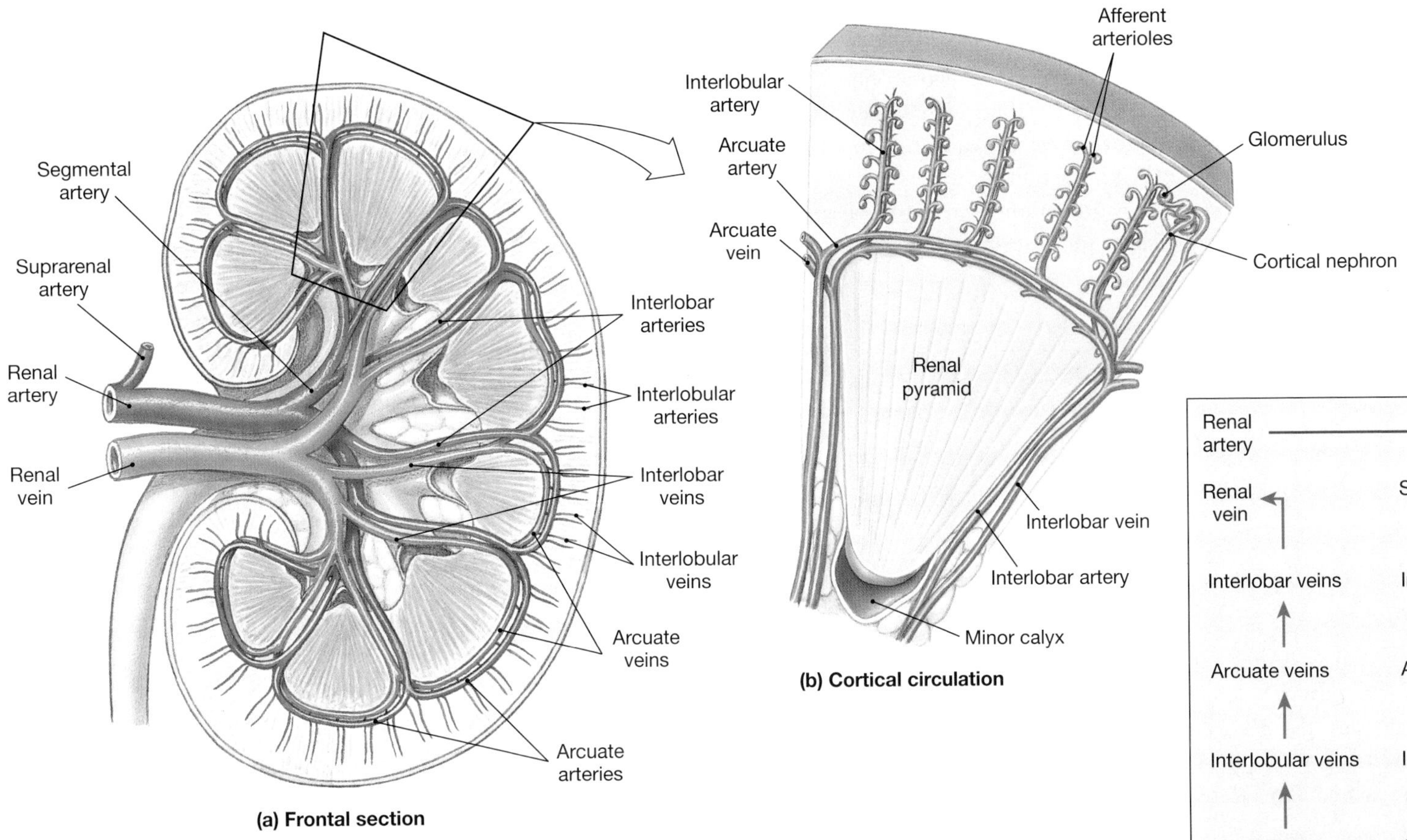

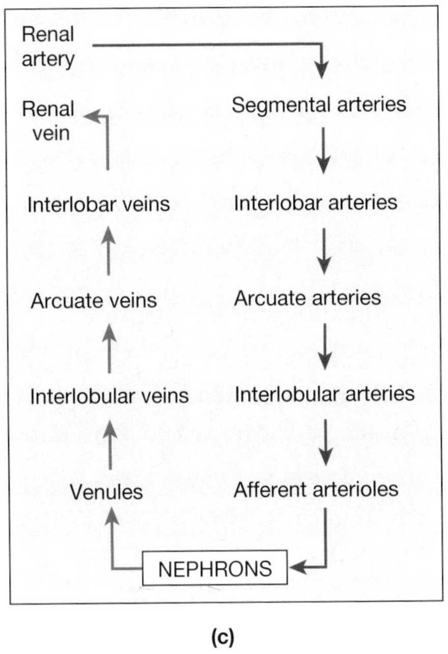

FIGURE 26.4 **Blood Supply to the Kidneys**

The flow chart summarizes the pattern of renal circulation. (**a**) Sectional view, showing major arteries and veins; compare with *Figures 26.3* and *26.8*. (**b**) Circulation in the cortex. (**c**) Flow chart of renal circulation.

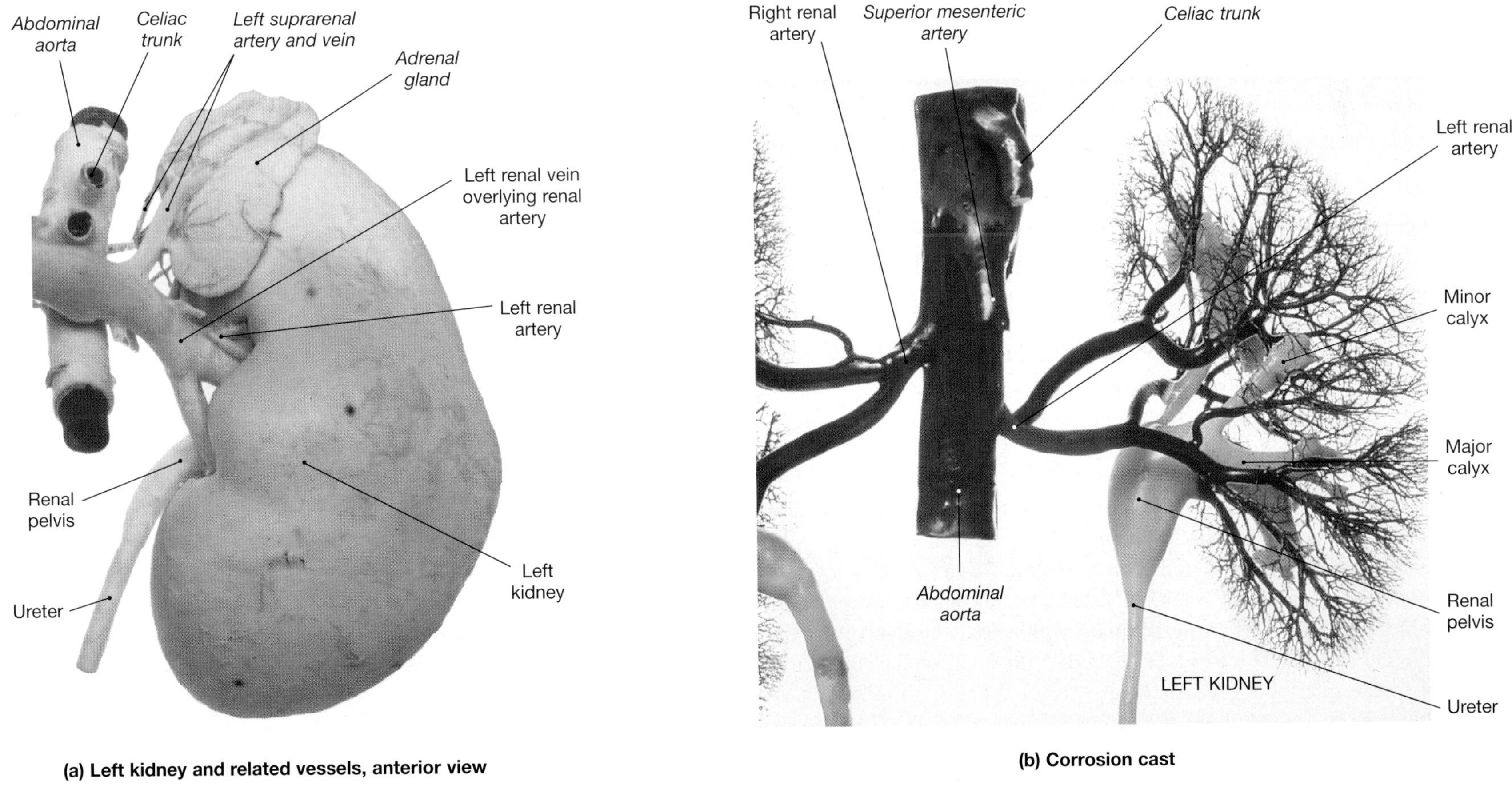

FIGURE 26.5 **Renal Vessels and Blood Flow**

(**a**) The left kidney, ureter, and associated vessels. The vessels have been injected with latex to make them easier to see. (**b**) Corrosion cast of the circulation and conducting passageways of the kidneys.

Most of the nerve fibers involved are sympathetic postganglionic fibers from the superior mesenteric ganglion. A renal nerve enters each kidney at the hilus and follows the branches of the renal artery to reach individual nephrons. Known functions of sympathetic innervation include: (1) regulation of renal blood flow and pressure; (2) stimulation of renin release; and (3) direct stimulation of water and sodium ion reabsorption.

✓ CONCEPT CHECK

- If your blood pressure is low, what changes will you see in the kidneys?
- After leaving the kidneys, where does the urine go?
- What is the function of calcitriol?

Histology of the Kidney [FIGURE 26.6]

The **nephron** (NEF-ron), the basic structural and functional unit of the kidney, can be viewed only with a microscope. For clarity, the nephron diagrammed in Figure 26.6● has been shortened and straightened out.

AN INTRODUCTION TO THE STRUCTURE AND FUNCTION OF THE NEPHRON [FIGURES 26.4c/26.6/26.7]

The **renal tubule**, a long tubular passageway, begins at the **renal corpuscle** (KOR-pus-ul), a cup-shaped chamber. The renal corpuscle is approximately 200 μm (0.2 mm) in diameter. It contains a capillary network called the **glomerulus** (glō-MER-ū-lus; pleural, *glomeruli*), which consists of about 50 intertwining capillaries. Blood arrives at the glomerulus by way of an **afferent arteriole** and departs in an **efferent arteriole**. These structures of the nephon are shown in Figure 26.6●.

Filtration across the walls of the glomerulus produces a protein-free solution known as **glomerular filtrate**, or simply *filtrate*. From the renal corpuscle, the filtrate enters a long tubular passageway that is subdivided into regions with varied structural and functional characteristics. Major subdivisions include the **proximal convoluted tubule (PCT)**, the **loop of Henle** (HEN-lē), or *nephron loop*, and the **distal convoluted tubule (DCT)**.

Each nephron empties into the **collecting system**. A **connecting tubule** connected to the distal convoluted tubule carries the filtrate toward a nearby **collecting duct**. The collecting duct leaves the cortex and descends into the medulla, carrying fluid toward a **papillary duct** that drains into the renal pelvis.

Nephrons from different locations differ slightly in structure. Roughly 85% of all nephrons are **cortical nephrons**, located almost entirely within the superficial cortex of the kidney (Figure 26.7a,e●). In a cortical nephron, the loop of Henle is relatively short, and the efferent arteriole delivers blood to a network of **peritubular capillaries**, which surround the entire renal tubules. These capillaries drain into small venules that carry blood to the interlobular veins (Figure 26.4c●). The remaining 15% of nephrons, termed **juxtamedullary nephrons** (juks-ta-MED-ū-lar-ē; *juxta*, near) are located closer to the medulla, and they have long loops of Henle that extend deep into the renal pyramids (Figure 26.7a,f●). Because they are more numerous than juxtamedullary nephrons, cortical nephrons perform most of the reabsorptive and secretory functions of the kidneys. However, it is the juxtaglomerular nephrons that create the conditions necessary for the production of a concentrated urine.

The urine arriving at the renal pelvis is very different from the filtrate produced at the renal corpuscle. Filtration is a passive process that promotes movement across a barrier based solely on solute size. A filter with pores large enough to permit the passage of organic waste products is unable to prevent the passage of water, ions, and other organic molecules, such as glucose, fatty acids, or amino acids. The other distal segments of the nephron are responsible for:

- reabsorbing all of the useful organic substrates from the filtrate,
- reabsorbing over 80% of the water in the filtrate, and
- secreting into the filtrate waste products that were missed by the filtration process.

We will now examine each of the segments of a juxtamedullary nephron in greater detail.

THE RENAL CORPUSCLE [FIGURE 26.6/26.8c–e]

The renal corpuscle has a diameter averaging 150–250 μm. It includes: (1) the capillary knot of the glomerulus and (2) the expanded initial segment of the renal tubule, a region known as the **Bowman's capsule**. The glomerulus projects into the Bowman's capsule like the heart projects into the pericardial cavity (Figure 26.8c●). The outer wall of the capsule is lined by a simple squamous **parietal *(capsular)* epithelium**, and this layer is continuous with the **visceral *(glomerular)* epithelium** that covers the glomerular capillaries. The visceral epithelium consists of large cells with complex processes, or "feet," that wrap around the glomerular capillaries. These specialized cells are called **podocytes** (PŌD-ō-sīts; *podos*, foot + *cyte*, cell) as seen in Figure 26.8c–e●. The **capsular space** separates the parietal and visceral epithelia. The connection between the parietal and visceral epithelia lies at the **vascular pole** of the renal corpuscle. At the vascular pole, the glomerular capillaries are connected to the bloodstream. Blood arrives at these capillaries through the afferent arteriole and departs in the relatively smaller diameter efferent arteriole (Figure 26.8c●). (This unusual circulatory arrangement will be discussed further in a later section.)

Filtration occurs as blood pressure forces fluid and dissolved solutes out of the glomerulus and into the capsular space. The filtrate produced in this way is very similar to plasma with the blood proteins removed. The filtration process involves passage across three physical barriers (Figure 26.8d●):

1. *The Capillary Endothelium*: The glomerular capillaries are *fenestrated capillaries* with pores ranging from 60 to 100 nm (0.06–0.1 μm) in diameter. p. 575 These openings are small enough to prevent the passage of blood cells, but they are too large to restrict the diffusion of solutes, even those the size of plasma proteins.
2. *The Basement Membrane*: The basement membrane that surrounds the capillary endothelium has several times the density and thickness of a typical basement membrane. This layer, called the **lamina densa**, restricts the passage of the larger plasma proteins but permits the movement of smaller plasma proteins, nutrients, and ions. Unlike basement membranes elsewhere in the body, the lamina densa may encircle two or more capillaries. When it does, **mesangial cells** are sit-

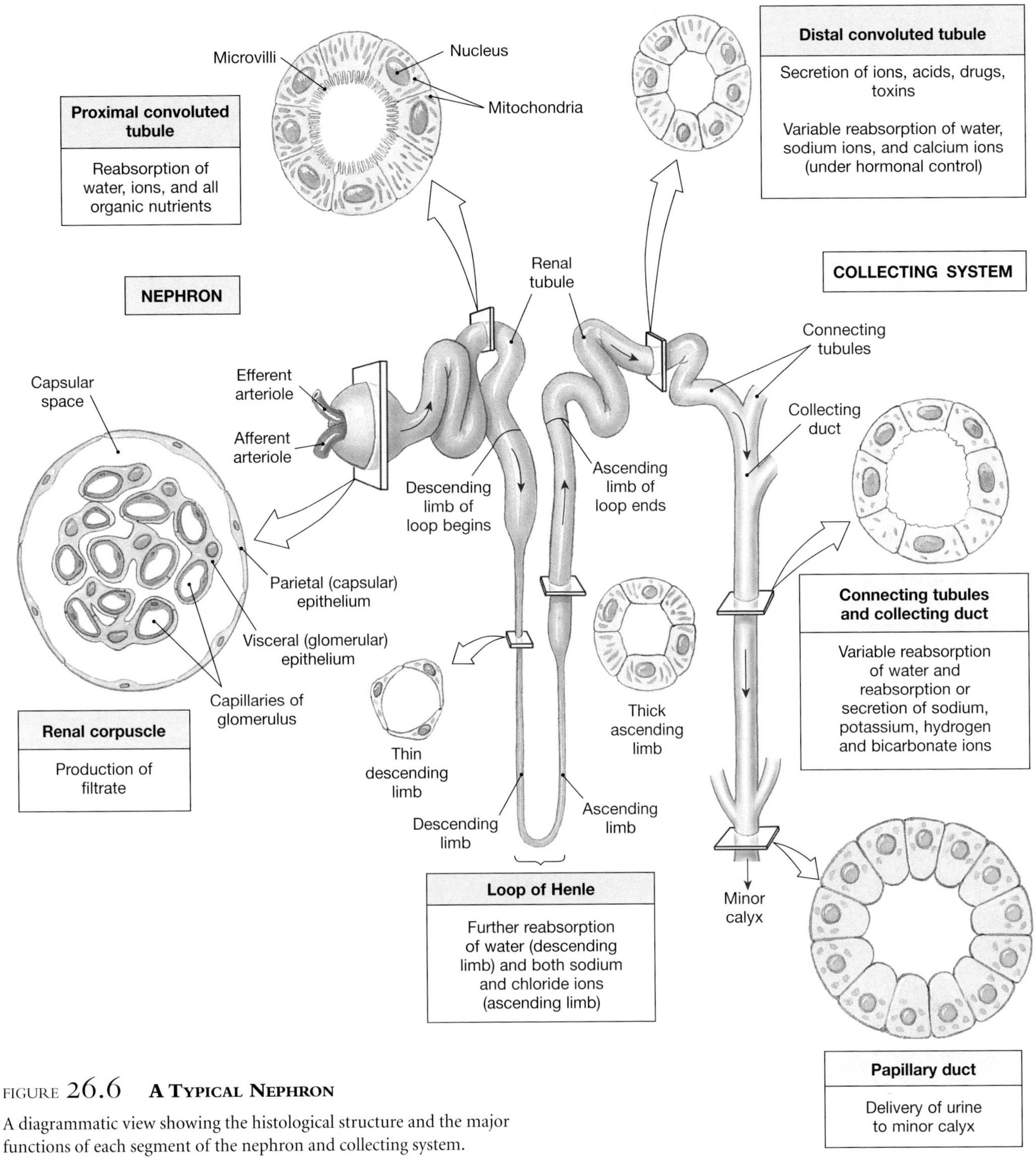

FIGURE 26.6 **A Typical Nephron**

A diagrammatic view showing the histological structure and the major functions of each segment of the nephron and collecting system.

uated between the endothelial cells of adjacent capillaries. Mesangial cells (1) provide physical support for the capillaries, (2) engulf organic materials that might otherwise clog the lamina densa, and (3) regulate the diameters of the glomerular capillaries, and thus play a role in the regulation of glomerular blood flow and filtration.

3. *The Glomerular Epithelium*: The podocytes have long cellular processes that wrap around the outer surfaces of the basement membrane. These delicate "feet," or **pedicels** (PED-i-selz) (Figure 26.8d,e●), are separated by narrow gaps called **filtration slits**. Because the filtration slits are very narrow, the filtrate entering the capsular space consists of water with dissolved ions, small organic molecules, and few if any plasma proteins.

In addition to metabolic wastes, the filtrate contains other organic compounds such as glucose, free fatty acids, amino acids, and vitamins. These potentially useful materials are reabsorbed by the proximal convoluted tubule.

THE PROXIMAL CONVOLUTED TUBULE [FIGURES 26.6/26.7a,d/26.8c]

The proximal convoluted tubule (PCT) is the first part of the renal tubule. Entry into the PCT lies almost directly opposite the vascular pole of the renal corpuscle, at the **tubular pole** of the renal corpuscle (Figure 26.8c●). The lining of the PCT consists of a simple cuboidal epithelium whose apical surfaces

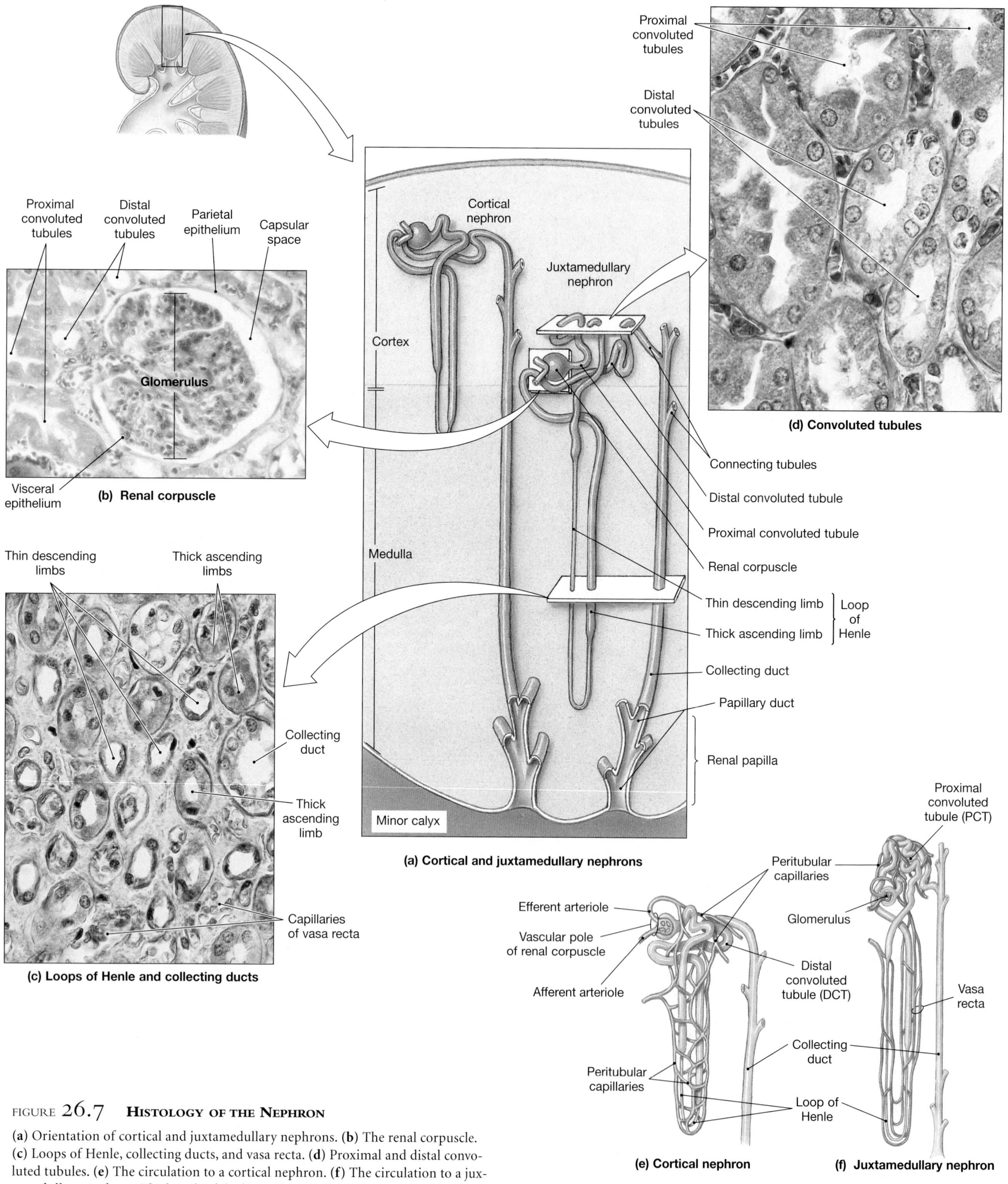

FIGURE 26.7 **Histology of the Nephron**

(**a**) Orientation of cortical and juxtamedullary nephrons. (**b**) The renal corpuscle. (**c**) Loops of Henle, collecting ducts, and vasa recta. (**d**) Proximal and distal convoluted tubules. (**e**) The circulation to a cortical nephron. (**f**) The circulation to a juxtamedullary nephron. The length of the loop of Henle is not drawn to scale.

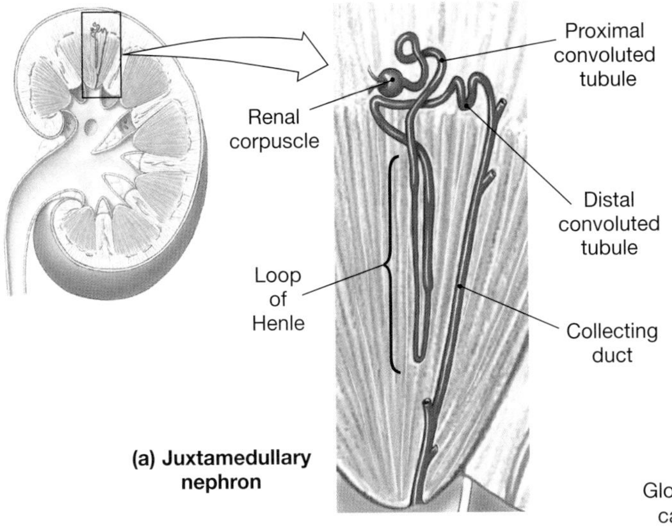

(a) Juxtamedullary nephron

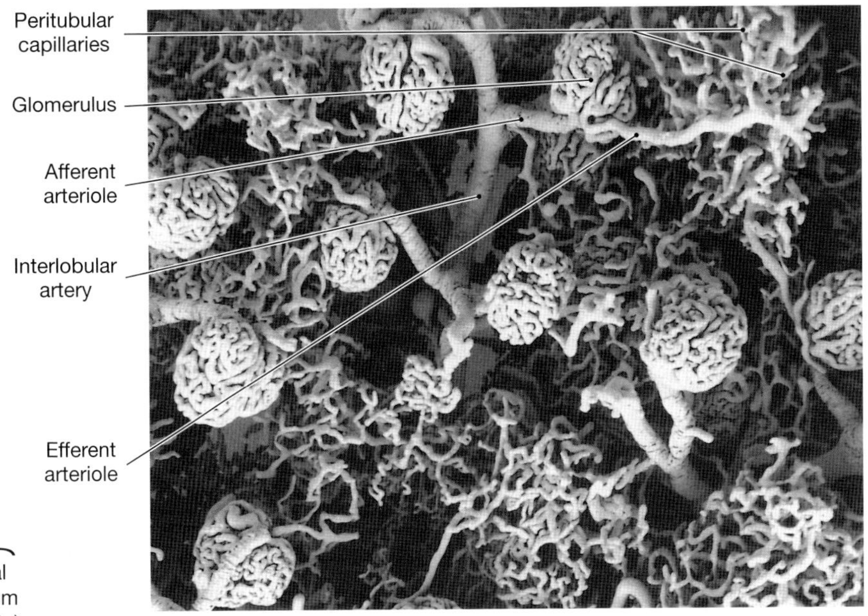

(b) Glomeruli and associated blood vessels (SEM x 94)

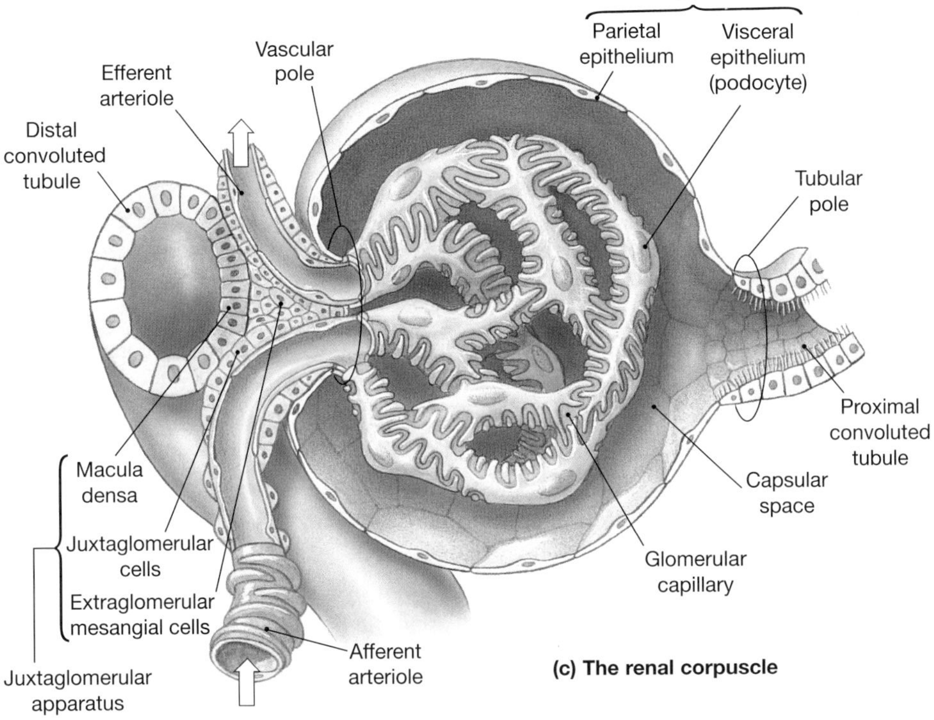

(c) The renal corpuscle

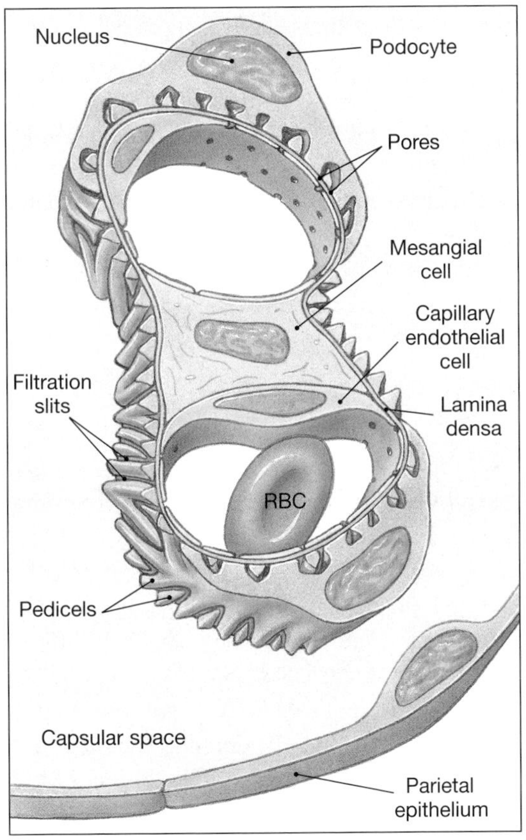

(d) The filtration apparatus

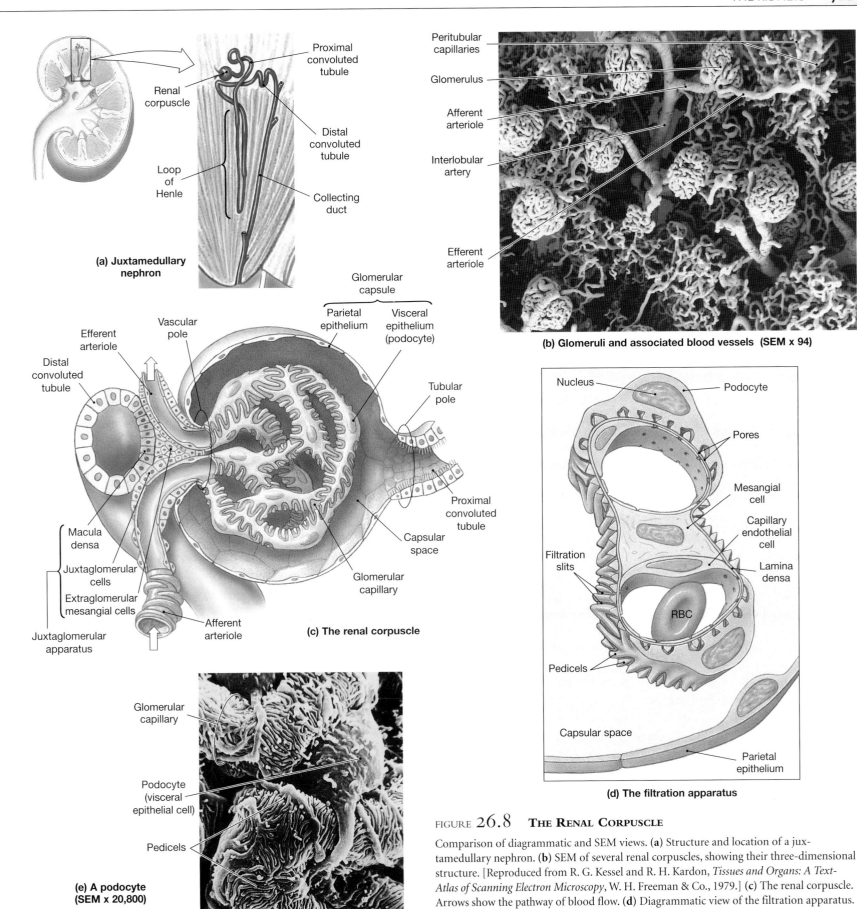

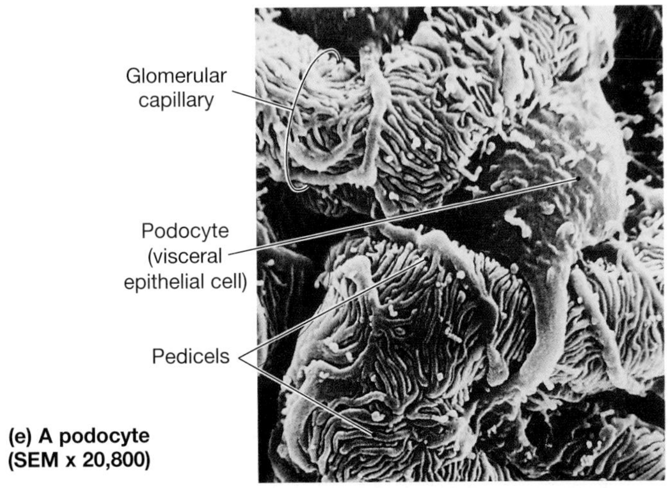

(e) A podocyte (SEM x 20,800)

FIGURE 26.8 **The Renal Corpuscle**

Comparison of diagrammatic and SEM views. (**a**) Structure and location of a juxtamedullary nephron. (**b**) SEM of several renal corpuscles, showing their three-dimensional structure. [Reproduced from R. G. Kessel and R. H. Kardon, *Tissues and Organs: A Text-Atlas of Scanning Electron Microscopy*, W. H. Freeman & Co., 1979.] (**c**) The renal corpuscle. Arrows show the pathway of blood flow. (**d**) Diagrammatic view of the filtration apparatus. (**e**) Colorized SEM of the electron micrograph of the glomerular surface, showing individual podocytes and their foot processes and pedicels.

are blanketed with microvilli, increasing the surface area for reabsorption (Figures 26.6 and 26.7d●). These cells actively absorb organic nutrients, ions, and plasma proteins (if any) from the filtrate as it flows along the PCT. As these solutes are absorbed, osmotic forces pull water across the wall of the PCT and into the surrounding interstitial fluid, or *peritubular fluid*. Absorption represents the primary function of the PCT; as tubular fluid passes along the PCT, the epithelial cells reabsorb virtually all of the organic nutrients and plasma proteins and 60% of the sodium ions, chloride ions, and water. The PCT also actively reabsorbs potassium, calcium, magnesium, bicarbonate, phosphate, and sulfate ions.

THE LOOP OF HENLE [FIGURES 26.6/26.7a,c]

The proximal convoluted tubule ends at an acute bend that turns the renal tubule toward the medulla. This bend marks the start of the loop of Henle (Figure 26.7a,c●). The loop of Henle can be divided into a **descending limb** and an **ascending limb**. The descending limb travels in the medulla toward the renal pelvis, and the ascending limb returns toward the cortex. Each limb contains a **thick segment** and a **thin segment** (Figures 26.6 and 26.7a,c●). (The terms *thick* and *thin* refer to the thickness of the surrounding epithelium, not to the diameter of the lumen.)

Thick segments are found closest to the cortex, whereas a thin squamous epithelium lines the deeper medullary portions. The thick ascending limb, which begins deep in the medulla, contains active transport mechanisms that pump sodium and chloride ions out of the tubular fluid. As a result of these transport activities, the medullary interstitial fluid contains an unusually high concentration of solutes. Solute concentration is usually expressed in terms of milliosmoles (mOsml). Near the base of the loop, in the deepest part of the medulla, the solute concentration of the interstitial fluid is roughly four times that of plasma (1200 mOsml vs. 300 mOsml). The thin descending and ascending limbs are freely permeable to water, but relatively impermeable to ions and other solutes. The high osmotic concentration of their surroundings results in an osmotic flow of water out of the nephron.

The net effect is that the loop of Henle reabsorbs an additional 25% of the water from the tubular fluid and an even higher percentage of the sodium and chloride ions. Reabsorption in the PCT and loop of Henle normally reclaims all of the organic nutrients, 85% of the water, and over 90% of the sodium and chloride ions. The remaining water, ions, and all of the organic wastes filtered at the glomerulus remain in the loop of Henle and enter the distal convoluted tubule.

THE DISTAL CONVOLUTED TUBULE [FIGURES 26.7a,b,d/26.8c]

The ascending limb of the loop of Henle ends where it forms a sharp angle that places the tubular wall in close contact with the glomerulus and its accompanying vessels. The distal convoluted tubule (DCT) begins at that bend. The initial portion of the DCT crosses the vascular pole of the renal corpuscle, passing between the afferent and efferent arterioles (Figure 26.8a,c,d●).

In sectional view (Figure 26.7b,d●), the DCT differs from the PCT in that (1) the DCT has a smaller diameter, (2) the epithelial cells of the DCT lack microvilli, and (3) the boundaries between the epithelial cells in the DCT are distinct. These characteristics reflect the major differences in function between the two regions of convoluted tubules: the PCT is primarily involved in reabsorption, whereas the DCT is primarily involved in secretion.

The distal convoluted tubule is an important site for (1) the active secretion of ions, acids, and other materials, (2) the selective reabsorption of sodium and calcium ions from the tubular fluid, and (3) the selective reabsorption of water, which helps concentrate the tubular fluid. The sodium transport activities of the DCT are controlled by circulating levels of *aldosterone* secreted by the adrenal cortex. p. 518

The Juxtaglomerular Apparatus [FIGURE 26.8c] The epithelial cells in the distal convoluted tubule immediately adjacent to the afferent arteriole at the vascular pole of the glomerulus are taller than those seen elsewhere along the DCT. This region of the DCT, detailed in Figure 26.8c●, is called the **macula densa** (MAK-ū-la DEN-sa). These cells monitor electrolyte concentration (specifically sodium and chloride ions) in the tubular fluid. The cells of the macula densa are closely associated with unusual smooth muscle fibers in the wall of the afferent arteriole. These muscle fibers are known as **juxtaglomerular cells** (*juxta*, near). **Extraglomerular mesangial cells** occupy the space between the glomerulus, the afferent and efferent arterioles, and the DCT. Together the macula densa, juxtaglomerular cells, and extraglomerular mesangial cells form the **juxtaglomerular apparatus**, an endocrine structure that secretes two hormones, *renin* and *erythropoietin*, that were described in Chapter 19. p. 519 These hormones, released when renal blood pressure, blood flow, or local oxygen levels decrease, elevate blood volume, hemoglobin levels, and blood pressure, and restore normal rates of filtrate production.

THE COLLECTING SYSTEM [FIGURES 26.6/26.7]

The distal convoluted tubule, the last segment of the nephron, opens into the collecting system which consists of *connecting tubules*, *collecting ducts*, and *papillary ducts* (Figure 26.7a,c●). Individual connecting tubules connect each nephron to a nearby collecting duct (Figure 26.7a,e,f●). Each collecting duct receives fluid from many connecting tubules, draining both cortical and juxtamedullary nephrons. Several collecting ducts converge to empty into the larger papillary duct that empties into a minor calyx in the renal pelvis. The epithelium lining the collecting system begins as simple cuboidal cells in the connecting tubules and changes to a columnar epithelium in the collecting and papillary ducts (Figure 26.6●).

In addition to transporting tubular fluid from the nephron to the renal pelvis, the collecting system makes final adjustments to its osmotic concentration and volume. The regulatory mechanism involves changing the permeability of the collecting ducts to water. This change is significant because the collecting ducts pass through the medulla, where the loop of Henle has established very high solute concentrations in the interstitial fluids. If collecting duct permeability is low, most of the tubular fluid reaching the collecting duct will flow into the renal pelvis and the urine will be dilute. In contrast, if collecting duct permeability is high, the osmotic flow of water out of the duct into the medulla will be promoted. This results in a small amount of highly concentrated urine. ADH (antidiuretic hormone) is the hormone responsible for controlling the permeability of the collecting system. p. 511 The higher the levels of circulating ADH, the greater the amount of water reabsorbed, and the more concentrated the urine. *Urine Composition* and *Hemodialysis p. 806*

✓ CONCEPT CHECK

- Trace the path of a drop of blood from the renal artery to a glomerulus and back to a renal vein.
- Trace the path taken by filtrate in traveling from a glomerulus to a minor calyx.
- Explain why filtration alone is not sufficient for urine production.
- What is the function of the loop of Henle?

Structures for Urine Transport, Storage, and Elimination [FIGURE 26.9c]

Filtrate modification and urine production end when the fluid enters the minor calyx. The remaining parts of the urinary system (the *ureters*, *urinary bladder*, and *urethra*) are responsible for the transport, storage, and elimination of the urine. Figure 26.9c●, a pyelogram of the urinary system, provides an orientation to the relative sizes and positions of these organs.

The minor and major calyces, the renal pelvis, the ureters, the urinary bladder, and the proximal portion of the urethra are lined by a *transitional epithelium* that can tolerate cycles of distension and contraction without damage. p. 57

CLINICAL DISCUSSION

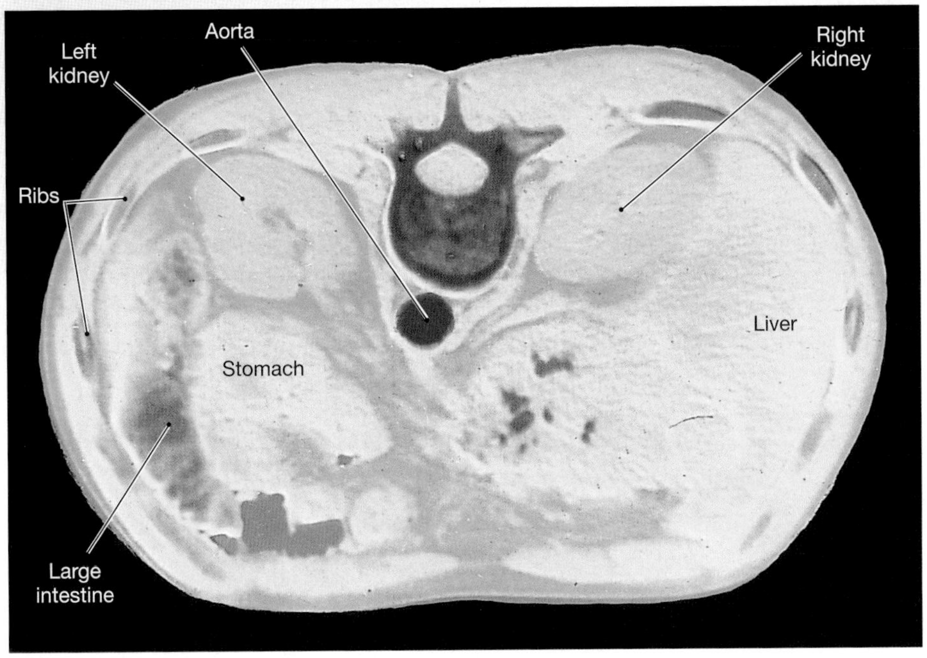

(a) Color-enhanced CT scan

Examination of the Kidneys and Urinary System

The anatomy of the urinary system can be examined using a variety of sophisticated procedures. **Computerized tomography (CT)** scans can provide useful information concerning localized abnormalities (Figure 26.9a●). The blood flow through the kidney can be checked by **arteriography** (Figure 26.9b●). Administering a radiopaque compound intravenously that will enter the urine and taking X-rays produces a **pyelogram** (PĪ-e-lō-gram; *pyelos*, pelvis), or IVP, or of the entire urinary system (Figure 26.9c●). Pyelography can detect anatomical abnormalities of the kidney, ureter, or urinary bladder.

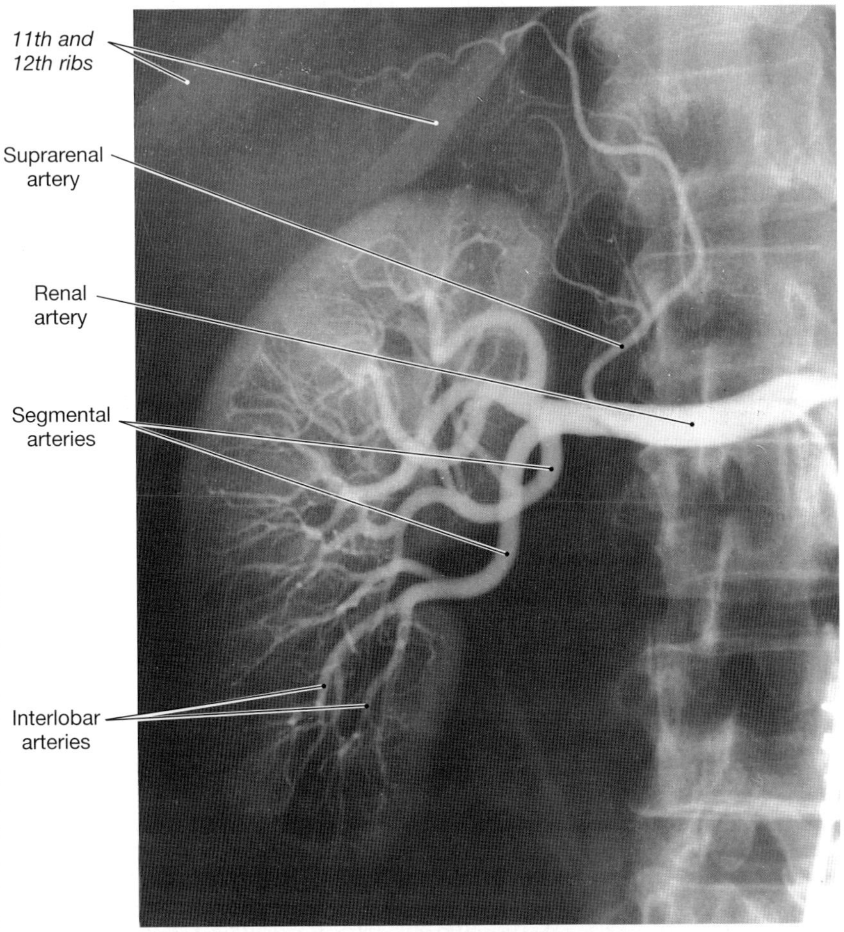

(b) Renal angiogram

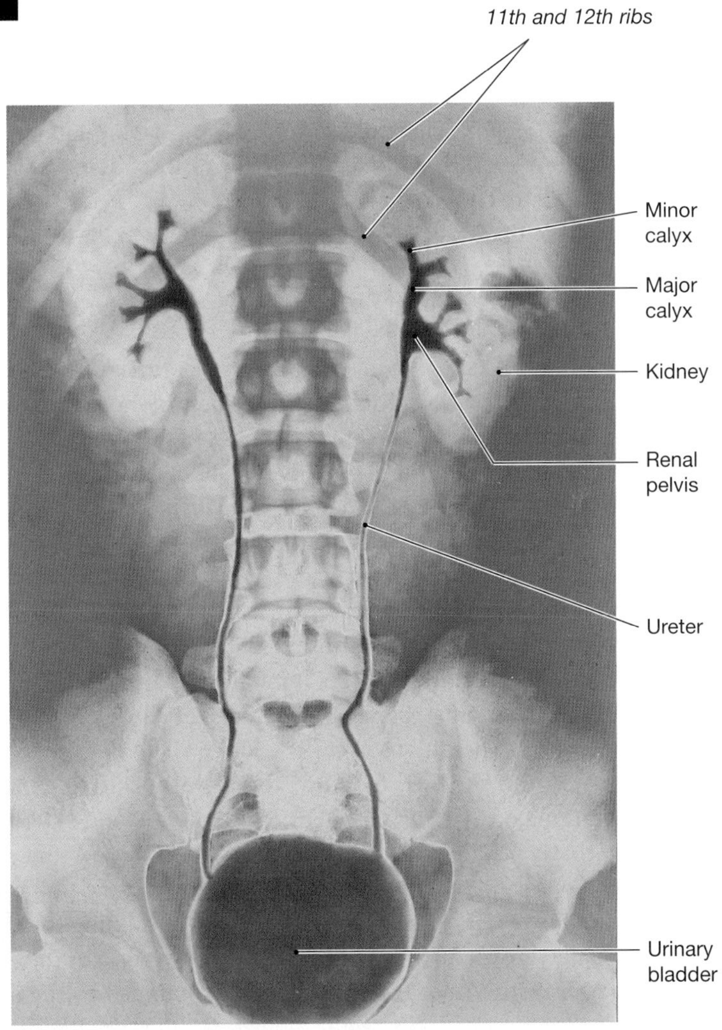

(c) Normal pyelogram

FIGURE 26.9 **Images of the Urinary System**

(**a**) A CT scan showing the position of the kidneys in a transverse section through the trunk as viewed from below. (**b**) An arteriogram of the right kidney. (**c**) This colorized X-ray was taken after the injection of a radiopaque dye that was filtered into the urine.

The Ureters [FIGURES 26.1/26.2/26.3/26.9c/26.10]

The ureters are a pair of muscular tubes that extend inferiorly from the kidneys for about 30 cm (12 in.) before reaching the urinary bladder (Figure 26.1a,b●). Each ureter begins as a continuation of the funnel-shaped renal pelvis through the hilus (Figures 26.3 and 26.9c●). As the ureters extend to the urinary bladder, they pass inferiorly and medially over the psoas major muscles. The ureters are retroperitoneal, and they are firmly attached to the posterior abdominal wall (Figure 26.2●). The paths taken by the ureters as they approach the wall of the urinary bladder differ in men and women because of variations in the nature, size, and position of the reproductive organs (Figure 26.10a,b●).

The ureters penetrate the posterior wall of the urinary bladder without entering the peritoneal cavity. They pass through the bladder wall at an oblique angle, and the **ureteral opening** is slitlike rather than rounded (Figure 26.10c●). This shape helps to prevent backflow of urine toward the ureter and kidneys when the urinary bladder contracts.

HISTOLOGY OF THE URETERS [FIGURE 26.11a]

The wall of each ureter consists of three layers: (1) an inner mucosa lined by a transitional epithelium, (2) a middle muscular layer made up of longitudinal (inner) and circular (outer) bands of smooth muscle, and (3) an outer connective tissue layer (adventitia) that is continuous with the fibrous renal capsule and peritoneum (Figure 26.11a●). Starting at the kidney, about every half-minute, peristaltic contractions of the muscular wall are triggered by stimulation of stretch receptors in the ureteral wall. These contractions "milk" urine out of the renal pelvis and along the ureter to the bladder.

The Urinary Bladder [FIGURE 26.10c,d]

The urinary bladder is a hollow muscular organ that functions as a temporary storage reservoir for urine. In the male, the base of the urinary bladder lies between the rectum and the symphysis pubis; in the female, the base of the urinary bladder sits inferior to the uterus and anterior to the vagina. The dimensions of the urinary bladder vary, depending on the state of distension, but the full urinary bladder can contain about a liter of urine.

The superior surfaces of the urinary bladder are covered by a layer of peritoneum; several peritoneal folds assist in stabilizing its position. The **median umbilical ligament**, or **urachus** (ū-ra-kus), extends from the anterior and superior border toward the umbilicus (Figure 26.10c●). The **lateral umbilical ligaments** pass along the sides of the bladder and also reach the umbilicus. These fibrous cords contain the vestiges of the two *umbilical arteries* that supplied blood to the placenta during embryonic and fetal development. p. 601 The posterior, inferior, and anterior surfaces of the urinary bladder lie outside the peritoneal cavity. In these areas, tough ligamentous bands anchor the bladder to the pelvic and pubic bones.

In sectional view (Figure 26.10c,d●), the mucosa lining the urinary bladder is usually thrown into folds, or **rugae**, that disappear as the bladder stretches and fills with urine. The triangular area bounded by the ureteral openings and the entrance to the urethra constitutes the **trigone** (TRĪ-gōn) of the urinary bladder. The mucosa here lacks rugae; it is smooth and very thick, and the trigone acts as a funnel that channels urine into the urethra when the urinary bladder contracts.

The urethral entrance lies at the apex of the trigone, at the most inferior point in the bladder. The region surrounding the urethral opening, known as the **neck** of the urinary bladder, contains a muscular **internal urethral sphincter**, or *sphincter vesicae*. The smooth muscle of the internal urethral sphincter provides involuntary control over the discharge of urine from the bladder. The urinary bladder is innervated by postganglionic fibers from ganglia in the *hypogastric plexus* and by parasympathetic fibers from intramural ganglia that are controlled by branches of the pelvic nerves. p. 459

HISTOLOGY OF THE URINARY BLADDER [FIGURE 26.11b]

The wall of the bladder contains mucosa, submucosa, and muscularis layers (Figure 26.11b●). The muscularis layer consists of inner and outer longitudinal smooth muscle layers, with a layer of circular muscle sandwiched between. Collectively, these layers form the powerful **detrusor** (dē-TROO-sor) muscle of the urinary bladder. Contraction of this muscle compresses the urinary bladder and expels its contents into the urethra. A layer of serosa covers the superior surface of the urinary bladder.

CLINICAL BRIEF

PROBLEMS WITH THE CONDUCTING SYSTEM

Local blockages of the collecting tubules, collecting ducts, or ureters may result from the formation of small blood clots, epithelial cells, lipids, or other materials, collectively called **casts**. Casts are often excreted in the urine and visible in microscopic analysis of urine samples. **Calculi** (KAL-kū-lī), or "kidney stones," form from calcium deposits, magnesium salts, or crystals of uric acid. This condition is called **nephrolithiasis** (nef-rō-li-THĪ-a-sis). The blockage of the urinary passage by a stone or other factors, such as external compression, results in **urinary obstruction**. Urinary obstruction is a painful and serious problem because it will reduce or eliminate filtration in the affected kidney.

Kidney stones are usually visible on an X-ray, and if peristalsis and fluid pressures are insufficient to dislodge them, they must be surgically removed or destroyed. One interesting nonsurgical procedure involves breaking kidney stones apart with a *lithotripter*, similar in principle to the one used to destroy gallstones. p. 694

The Urethra [FIGURE 26.10]

The urethra extends from the neck of the urinary bladder (Figure 26.10c●) to the exterior. The female and male urethra differ in length and in function. In the female, the urethra is very short, extending 3–5 cm (1–1.5 in.) from the bladder to the vestibule (Figure 26.10b●). The external urethral opening, or **external urethral meatus**, is situated near the anterior wall of the vagina.

In the male, the urethra extends from the neck of the urinary bladder to the tip of the penis, a distance that may be 18–20 cm (7–8 in.). The male urethra can be subdivided into three portions (Figure 26.10a,c,d●): (1) the *prostatic urethra*, (2) the *membranous urethra*, and (3) the *penile urethra*.

The **prostatic urethra** passes through the center of the prostate gland (Figure 26.10c●). The **membranous urethra** includes the short segment that penetrates the urogenital diaphragm, the muscular floor of the pelvic cavity. The **penile** (PĒ-nīl) **urethra** extends from the distal border of the urogenital diaphragm to the external urethral meatus at the tip of the penis (Figure 26.10a●). The functional differences between these regions will be considered in Chapter 27.

In both sexes, as the urethra passes through the urogenital diaphragm, a circular band of skeletal muscle forms the **external urethral sphincter,** or *sphincter urethrae*. p. 284–285 The contractions of both the external and internal urethral sphincters are controlled by branches of the hypogastric plexus. Only the external urethral sphincter is under voluntary control, through the perineal branch of the pudendal nerve. p. 370 The sphincter has a resting muscle tone and usually must be voluntarily relaxed to permit urination.The autonomic innervation of the external sphincter

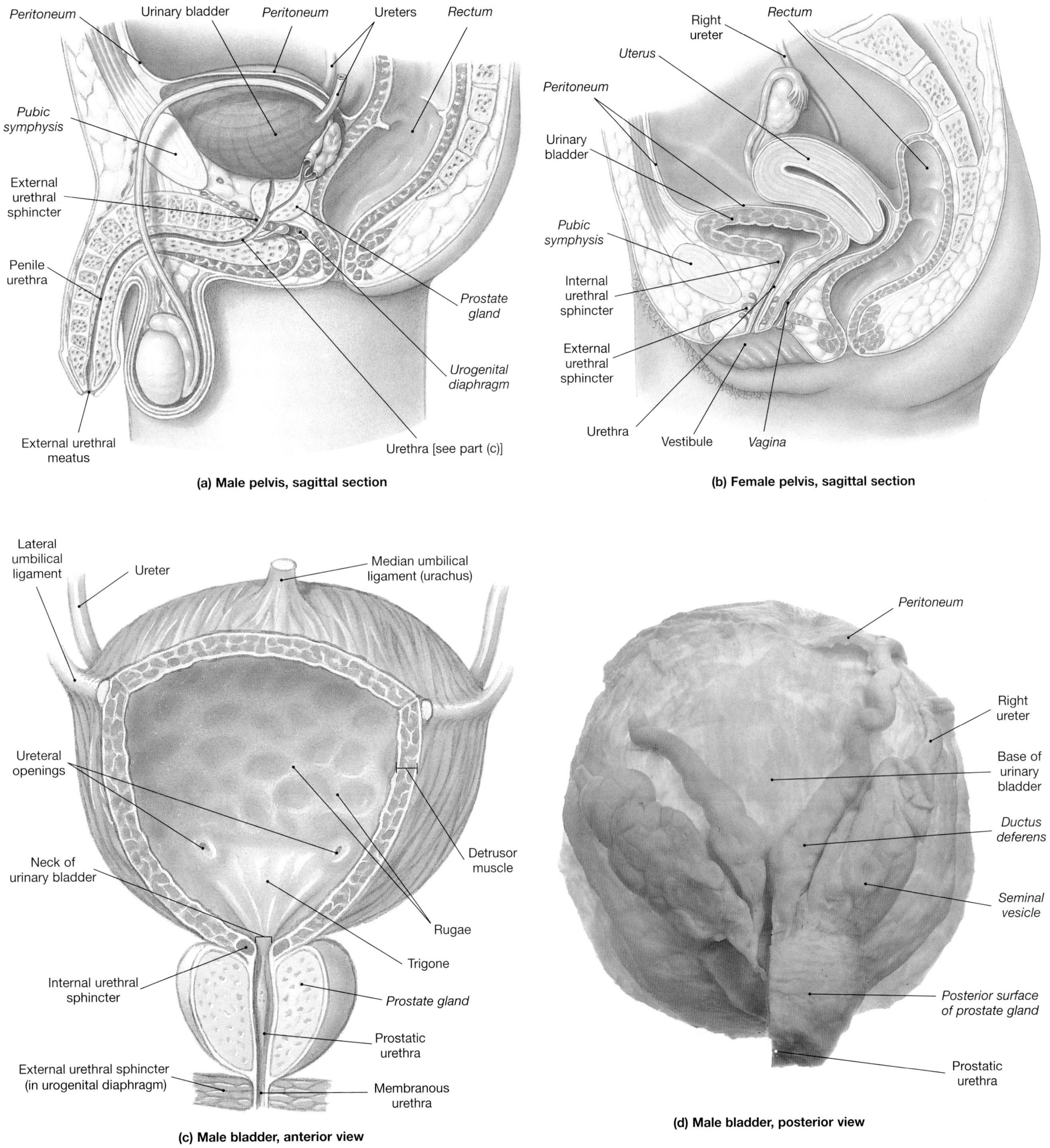

(a) Male pelvis, sagittal section

(b) Female pelvis, sagittal section

(c) Male bladder, anterior view

(d) Male bladder, posterior view

FIGURE 26.10 **Organs Responsible for the Conduction and Storage of Urine**

(**a**) Position of the ureter, urinary bladder, and urethra in the male. (**b**) Position of these organs in the female. (**c**) Anatomy of the urinary bladder in a male. (**d**) The male urinary bladder and accessory reproductive structures as seen in posterior view.

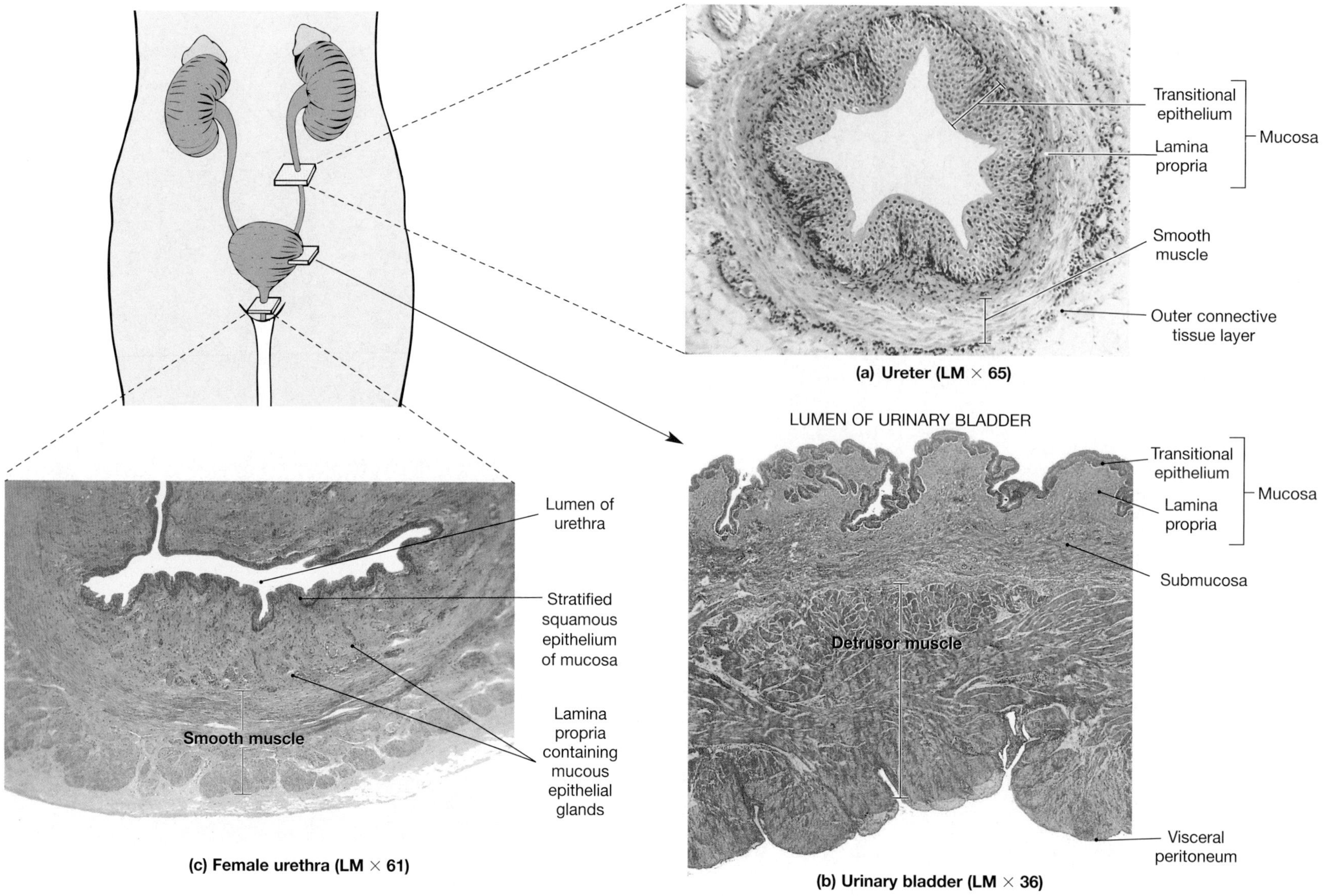

FIGURE 26.11 **HISTOLOGY OF THE COLLECTING AND TRANSPORT ORGANS**

(**a**) A ureter seen in transverse section. Note the thick layer of smooth muscle surrounding the lumen. (*See also Figure 3.5c.*) (**b**) The wall of the urinary bladder. (**c**) A transverse section through the female urethra.

becomes important only if voluntary control is lacking, as in infants or adults after spinal cord injuries. (See the section on the micturition reflex.)

HISTOLOGY OF THE URETHRA [FIGURE 26.11c]

In females, the urethral lining is usually a transitional epithelium near the urinary bladder. The rest of the urethra is typically lined by a stratified squamous epithelium (Figure 26.11c●). The lamina propria contains an extensive network of veins, and the entire complex is surrounded by concentric layers of smooth muscle.

In males, the histological organization of the urethra varies along its length. As one proceeds from the neck of the urinary bladder to the external urethral meatus the epithelium changes from transitional to pseudostratified columnar or stratified columnar, and then to stratified squamous. The lamina propria is thick and elastic; the mucous membrane is thrown into longitudinal folds. Mucous-secreting cells are found in the epithelial pockets, and in the male the epithelial mucous glands may form tubules that extend into the lamina propria. Connective tissues of the lamina propria anchor the urethra to surrounding structures.

CLINICAL BRIEF

URINARY TRACT INFECTIONS

Urinary tract infections, or **UTIs**, result from the colonization of the urinary tract by bacterial or fungal invaders. The intestinal bacterium *Escherichia coli* is most often involved, and women are particularly susceptible to urinary tract infections because of the close proximity of the external urethral orifice to the anus. Sexual intercourse may also push bacteria into the urethra and, since the female urethra is relatively short, toward the urinary bladder.

The condition may be asymptomatic (without symptoms), but it can be detected by the presence of bacteria and blood cells in the urine. If inflammation of the urethral wall occurs, the condition may be termed **urethritis**, while inflammation of the lining of the urinary bladder represents **cystitis**. Many infections affect both regions to some degree. Urination becomes painful, a symptom known as **dysuria** (dis-ū-rē-a), and the bladder becomes tender and sensitive to pressure. Despite the discomfort produced, the individual feels the urge to urinate frequently. Urinary tract infections usually respond to antibiotic therapies, although subsequent reinfections may occur.

In untreated cases, the bacteria may proceed along the ureters to the renal pelvis. The resulting inflammation of the walls of the renal pelvis produces *pyelitis* (pī-e-LĪ-tis). If the bacteria invade the renal cortex and medulla as well, **pyelonephritis** (pī-e-lō-nef-RĪ-tis) results. Signs and symptoms of pyelonephritis include a high fever, intense pain on the affected side, vomiting, diarrhea, and the presence of blood cells and pus in the urine.

The Micturition Reflex and Urination

Urine reaches the urinary bladder by the peristaltic contractions of the ureters. The process of urination, which empties the urinary bladder, is coordinated by the **micturition reflex**. Stretch receptors in the wall of the urinary bladder are stimulated as it fills with urine. Afferent fibers in the pelvic nerves carry the impulses generated to the sacral spinal cord. Their increased level of activity: (1) facilitates parasympathetic motor neurons in the sacral spinal cord, (2) stimulates contraction of the urinary bladder, and (3) stimulates interneurons that relay sensations to the cerebral cortex. As a result, we become consciously aware of the fluid pressure in the urinary bladder. The urge to urinate first develops when the urinary bladder contains approximately 200 ml of urine. Voluntary urination involves conscious relaxation of the external sphincter and subconscious facilitation of the micturition reflex. When the external urethral sphincter relaxes, feedback through the autonomic nervous system relaxes the internal urethral sphincter. Tensing of the abdominal and expiratory muscles increases abdominal pressures and assists in compressing the urinary bladder. In the absence of voluntary relaxation of the external urethral sphincter, reflexive relaxation of both sphincters will eventually occur as the urinary bladder nears capacity. *Problems with the Micturition Reflex p. 807*

✓ CONCEPT CHECK

- An obstruction of a ureter by a kidney stone would interfere with the flow of urine between what two points?
- Explain how the lining of the urinary bladder allows the bladder to become distended.
- Why are females more susceptible to urinary tract infections than males?
- How is the urinary bladder held in place?

Aging and the Urinary System

In general, aging is associated with an increased incidence of kidney problems. Age-related changes in the urinary system include

1. *A decline in the number of functional nephrons*: The total number of kidney nephrons drops by 30–40% between ages 25 and 85.
2. *A reduction in glomerular filtration*: This reduction results from decreased numbers of glomeruli, cumulative damage to the filtration apparatus in the remaining glomeruli, and reductions in renal blood flow.
3. *Reduced sensitivity to ADH*: With age the distal portions of the nephron and the entire collecting system become less responsive to ADH. Less reabsorption of water and sodium ions occurs as a result; urination becomes more frequent, and daily fluid requirements increase.
4. *Problems with the micturition reflex*: Several factors are involved in such problems:
   a. The sphincter muscles lose muscle tone and become less effective at voluntarily retaining urine. This loss of tone leads to *incontinence*, a slow leakage of urine.
   b. The ability to control micturition is often lost after a stroke, Alzheimer's disease, or other CNS problems affecting the cerebral cortex or hypothalamus.
   c. In males, **urinary retention** may develop secondary to chronic inflammation of the prostate gland. In this condition swelling and distortion of prostatic tissues compress the prostatic urethra, restricting or preventing the flow of urine.

RELATED CLINICAL TERMS

automatic bladder: A condition in which the micturition reflex remains intact but the individual cannot prevent the reflexive emptying of the urinary bladder. *Problems with the Micturition Reflex p. 807*

cystitis: Inflammation of the urinary bladder lining, usually from infection. *p. 716*

dysuria (dis-ū-rē-a): Painful urination. *p. 716*

hemodialysis (hē-mō-dī-AL-i-sis): A technique in which an artificial membrane is used to regulate the composition of the blood and remove waste products. *Hemodialysis p. 806*

incontinence (in-KON-ti-nens): An inability to voluntarily control urination. *Problems with the Micturition Reflex p. 807*

nephrolithiasis (nef-rō-li-THĪ-a-sis): The presence of kidney stones. *p. 714*

pyelogram (PĪ-e-lō-gram): An image obtained by taking an X-ray of the kidneys after a radiopaque compound has been administered. *p. 713*

renal calculi (KAL-kū-lī): "Kidney stones" formed from calcium deposits, magnesium salts, or crystals of uric acid. *p. 714*

urethritis: Inflammation of the urethral wall. *p. 716*

urinary obstruction: Blockage of the conducting system by a calculus or other factors. *p. 714*

urinary tract infection (UTI): Urinary tract inflammation caused by bacteria or fungal infection. *p. 716*

Additional Clinical Terms Discussed in Appendix I(pp. 806–807)
dialysis; shunts; stress incontinence

STUDY OUTLINE & CHAPTER REVIEW

Introduction 703

1. The functions of the urinary system include: (1) regulating plasma concentrations of ions; (2) regulating blood volume and pressure by adjusting the volume of water lost and releasing erythropoietin and renin; (3) helping to stabilize blood pH; (4) conserving nutrients; (5) eliminating organic wastes; and (6) synthesizing **calcitriol**.
2. The **urinary system** includes the **kidneys**, the **ureters**, the **urinary bladder**, and the **urethra**. The kidneys produce **urine** (a fluid containing water, ions, and soluble compounds); during **urination (micturition)** urine is forced out of the body. *(see Figure 26.1)*

The Kidneys 703

Superficial Anatomy of the Kidney 703

1. The kidneys are located on either side of the vertebral column between the last thoracic and third lumbar vertebrae. *(see Figures 26.1 and 26.2)*

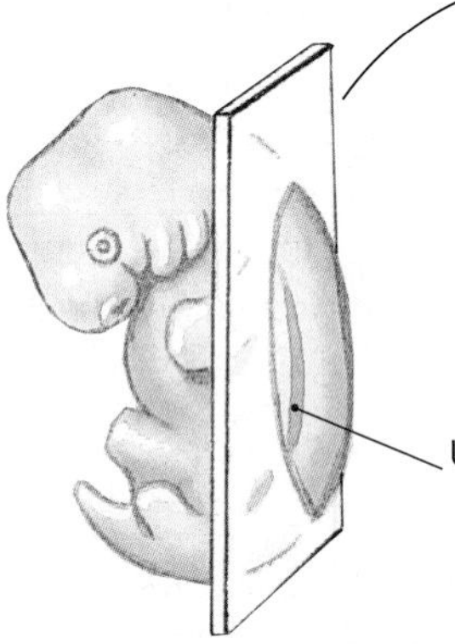

The kidneys develop in stages along the axis of the **urogenital ridge**, a thickened area beneath the dorsolateral wall of the coelomic cavity.

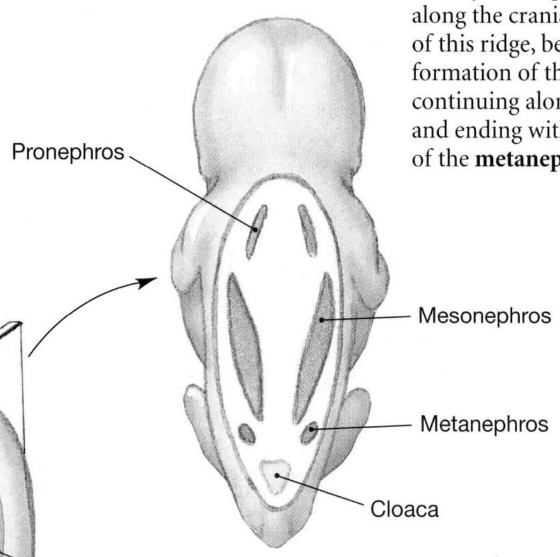

Kidney development proceeds along the cranial/caudal axis of this ridge, beginning with the formation of the **pronephros**, continuing along the **mesonephros**, and ending with the development of the **metanephros**.

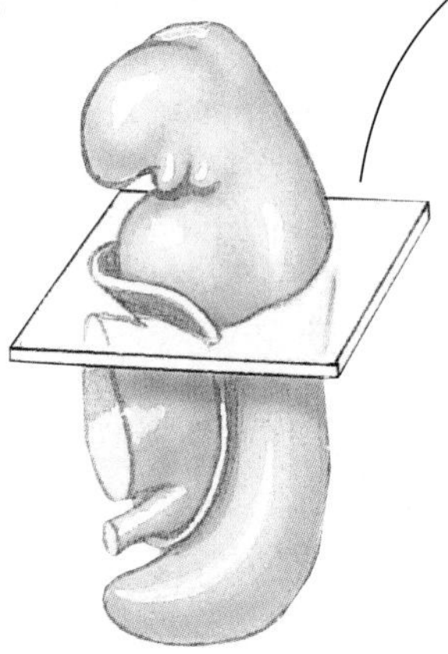

3½ WEEKS

The pronephros consists of a series of tubules (generally 7 pairs) that appears within the **nepthrotome**, the narrow band of mesoderm between the somites and the lateral plate.

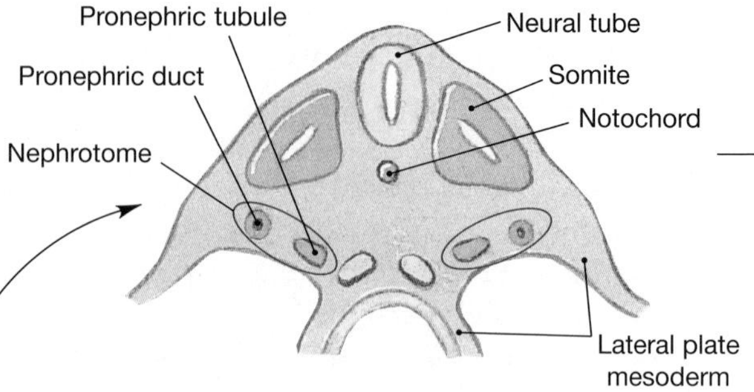

The pronephric tubules are very small and nonfunctional, and they disintegrate almost at once. The only significant contribution of the pronephros is the formation of a pair of **pronephric ducts** that grow caudally until they connect to the *cloaca* (p. 671).

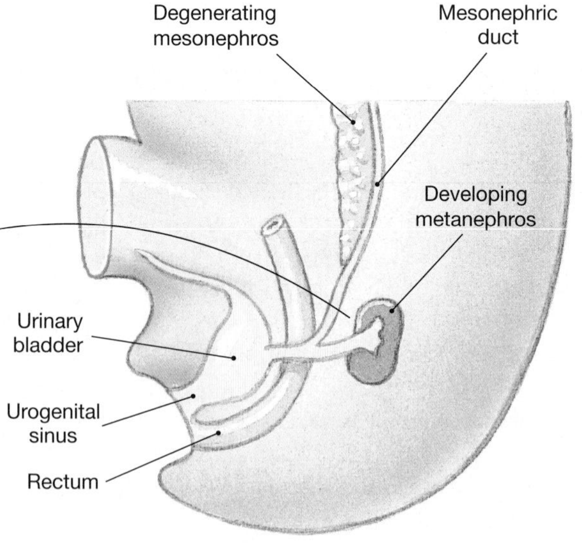

Near the end of the second developmental month, the cloaca is subdivided into a dorsal rectum and a ventral **urogenital sinus**. The proximal portions of the allantois persists as the **urinary bladder**, and the connection between the bladder and an opening on the body surface will form the **urethra**.

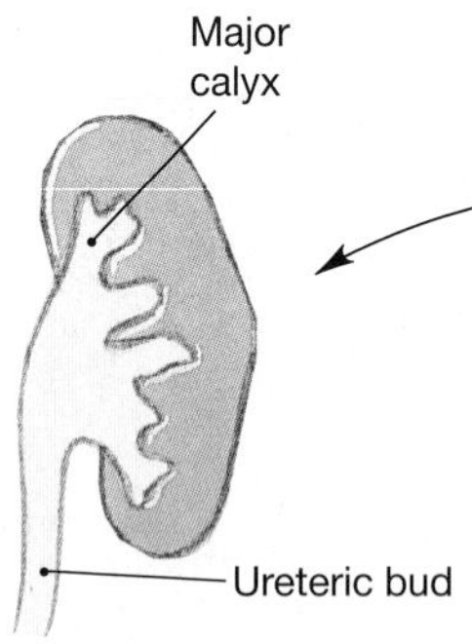

The ureteric bud branches within the metanephros, creating the calyces and the collecting system. The nephrons, which form within the mesoderm of the metanephros, tap into the collecting tubules.

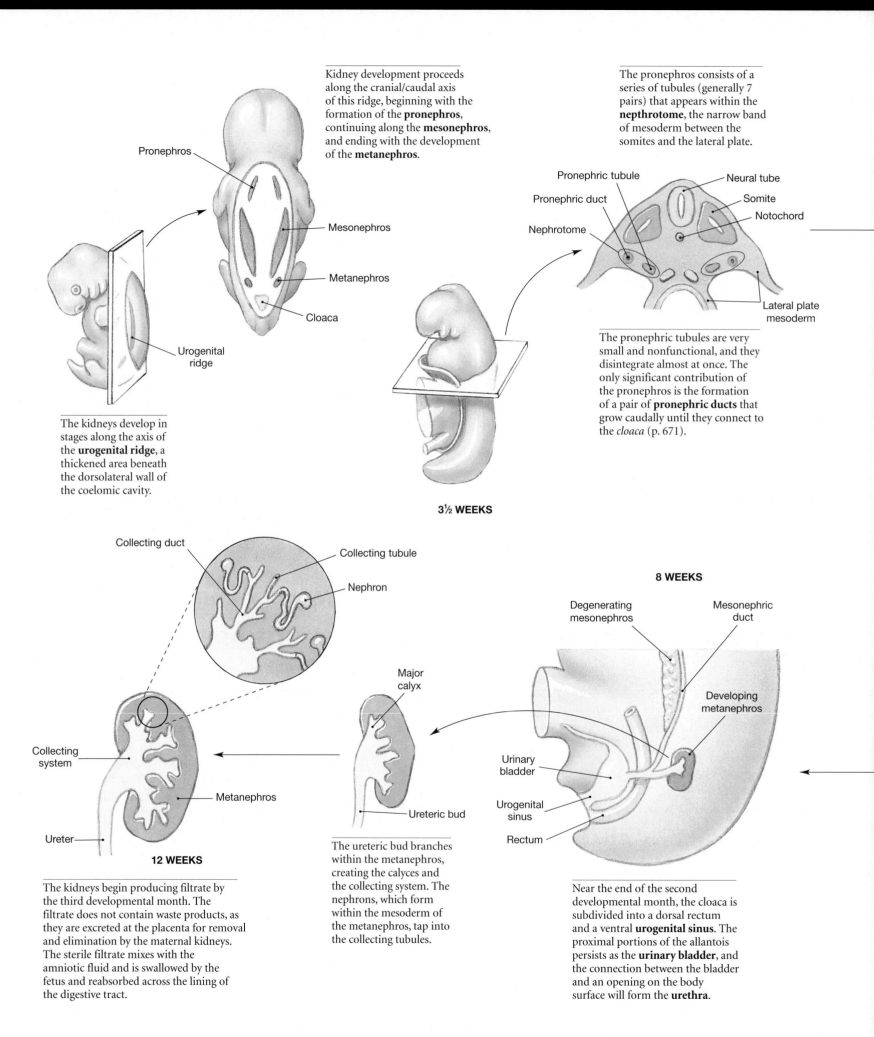

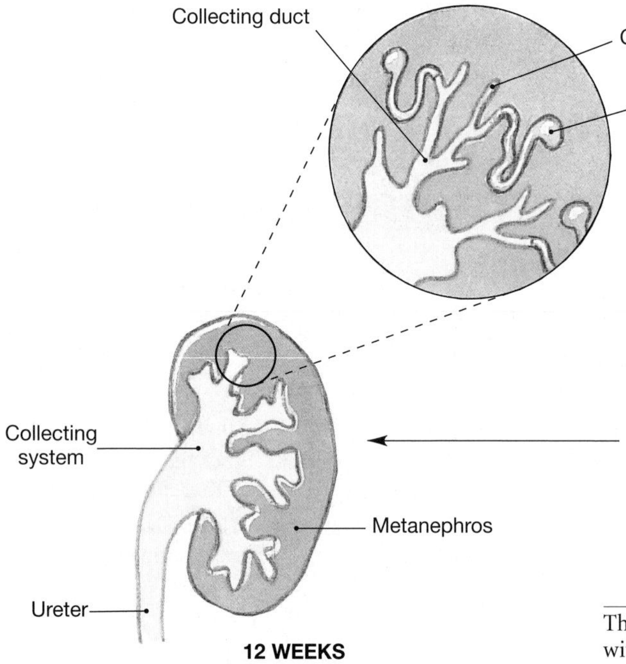

12 WEEKS

The kidneys begin producing filtrate by the third developmental month. The filtrate does not contain waste products, as they are excreted at the placenta for removal and elimination by the maternal kidneys. The sterile filtrate mixes with the amniotic fluid and is swallowed by the fetus and reabsorbed across the lining of the digestive tract.

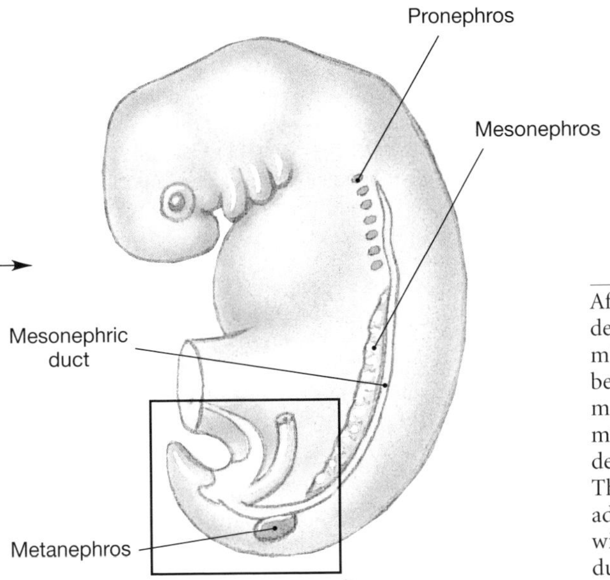

4 WEEKS

After approximately 4 weeks of development, the mesoderm midway along the urogenital ridge begins organizing into the mesonephros. On either side of the midline, approximately 70 tubules develop within these segments. These tubules grow toward the adjacent pronephric duct and fuse with it. From this moment on, the duct is called the **mesonephric duct**.

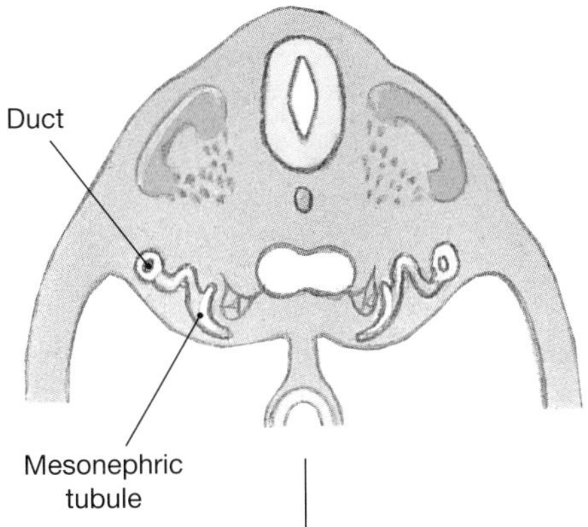

4 WEEKS

Nephrotomal mesoderm of the metanephros forms a dense mass without a trace of segmental organization. This will become the functional adult kidney.

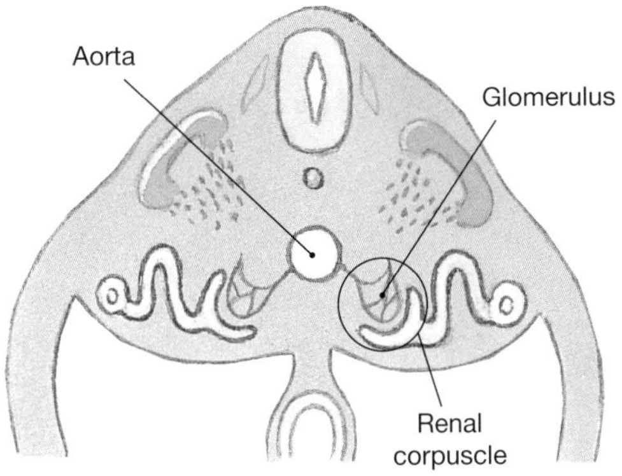

In each segment, a branch of the aorta grows toward the nephrotome, and the tubules form large nephrons with enormous glomeruli. Like the pronephros, the mesonephros does not persist, and when the last segments of the mesonephros are forming, the first are already beginning to degenerate.

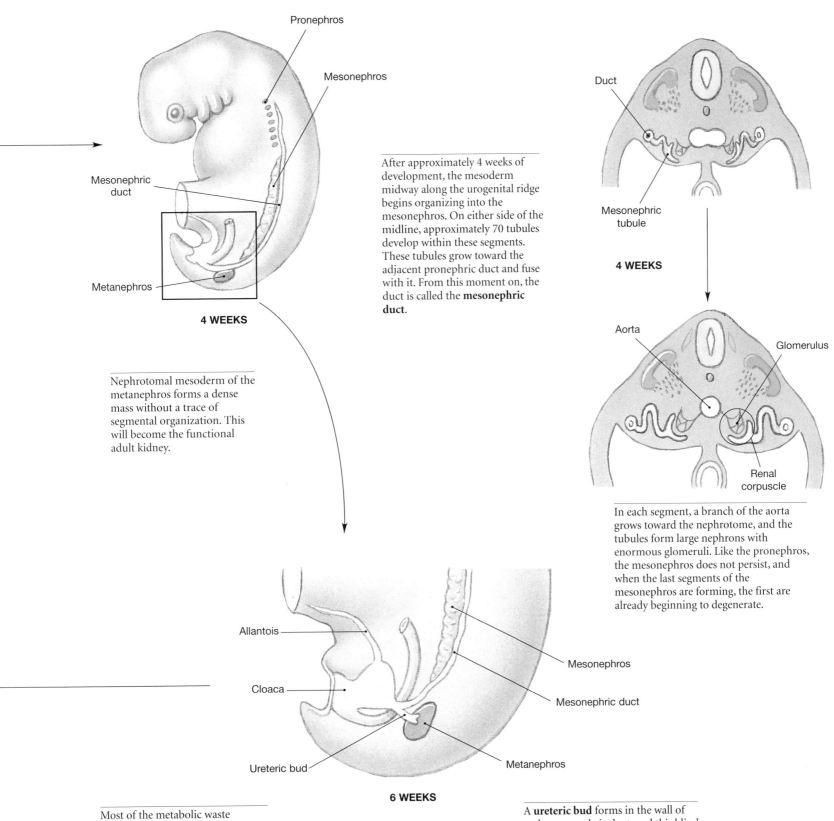

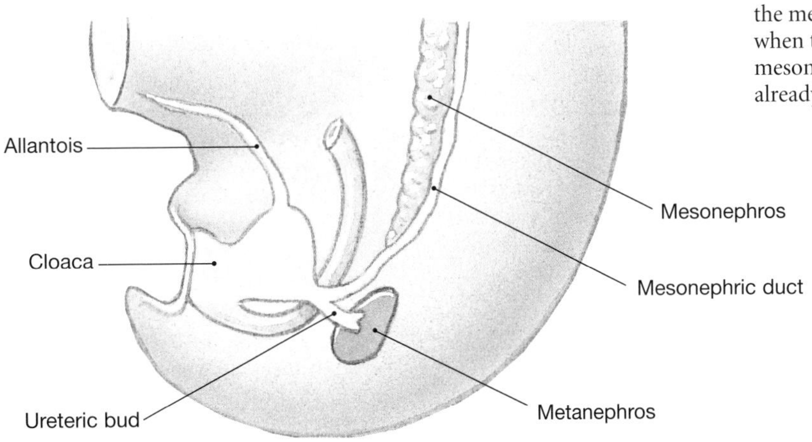

6 WEEKS

Most of the metabolic waste produced by the developing embryo are passed across the placenta to enter the maternal circulation. The small amount of urine produced by the kidneys accumulates within the cloaca and the *allantois* (p. 671), an endoderm-lined sac that extends into the umbilical stalk.

A **ureteric bud** forms in the wall of each mesonephric duct, and this blind tube elongates and branches within the adjacent metanephros. Tubules developing within the metanephros then connect to the terminal branches.

2. The position of the kidneys in the abdominal cavity is maintained by: (1) the overlying peritoneum, (2) contact with adjacent visceral organs, and (3) supporting connective tissues. The three concentric layers of connective tissue are the **renal capsule**, which covers the outer surface of the organ; the **adipose capsule**, which surrounds the renal capsule; and the **renal fascia**, which anchors the kidney to surrounding structures. The inner layer of the renal capsule lines the **renal sinus**. *(see Figure 26.3)*
3. The ureter and renal blood vessels are attached to the **hilus** of the kidney. *(see Figure 26.3)*

Sectional Anatomy of the Kidney 703

4. The kidney is divided into an outer **renal cortex**, a central **renal medulla**, and an inner **renal sinus**. The medulla contains 6–18 **renal pyramids**, whose tips, or **renal papillae**, project into the **renal sinus**. **Renal columns** composed of cortex separate adjacent pyramids. A **renal lobe** contains a renal pyramid, the overlying area of renal cortex, and adjacent tissues of the renal columns. *(see Figure 26.3a)*
5. The **minor calyces** are continuous with the **major calyces**. These spaces lead into the **renal sinus**, which is continuous with the **ureter**. *(see Figure 26.3a–c)*

The Blood Supply to the Kidneys 705

6. The vasculature of the kidneys includes the **renal**, **segmental**, **interlobar**, **arcuate**, and **interlobular arteries** to the **afferent arterioles** that supply the nephrons. From the nephrons, blood flows into the **interlobular**, **arcuate**, **interlobar**, and **renal veins**. *(see Figures 26.4 and 26.5)*

Innervation of the Kidneys 705

7. The kidneys and ureters are innervated by **renal nerves**. Sympathetic activation regulates glomerular blood flow and pressure, stimulates renin release, and accelerates sodium ion and water reabsorption.

Histology of the Kidney 707

8. The **nephron** (the basic functional unit in the kidney) consists of a **renal tubule** that empties into the **collecting system**. From the **renal corpuscle**, the **tubular fluid** travels through the **proximal convoluted tubule (PCT)**, the **loop of Henle (nephron loop)**, and the **distal convoluted tubule (DCT)**. It then flows through the **connecting tubule**, **collecting duct**, and **papillary duct** to reach the renal minor calyx. *(see Figure 26.6)*
9. Roughly 85% of the nephrons are **cortical nephrons** found within the cortex. The loops of Henle are short and the **efferent arteriole** provides blood to the **peritubular capillaries** that surround the renal tubules. The **juxtamedullary nephrons** are closer to the medulla, with their loops of Henle extending deep into the renal pyramids. *(see Figures 26.4c,d/26.7a)*
10. Nephrons are responsible for: (1) production of filtrate, (2) reabsorption of organic nutrients, and (3) reabsorption of water and ions. The **parietal epithelium** lines the outer wall of the renal corpuscle. Blood arrives via the relatively large **afferent arteriole** and departs in the relatively small **efferent arteriole**. *(see Figure 26.8c)*
11. The renal corpuscle contains the capillary knot of the **glomerulus** and **Bowman's capsule** *(glomerular capsule)*. At the glomerulus, **podocytes** wrap their "feet" around the capillaries. The **pedicels** of the **podocytes** are separated by narrow **filtration slits**. The **capsular space** separates the **parietal** and **visceral epithelia**. The glomerular capillaries are *fenestrated capillaries.* The basement membrane is unusually thick and is called the **lamina densa**. Blood arrives at the **vascular pole** of the renal corpuscle via the afferent arteriole and departs in the **efferent arteriole**. From the efferent arteriole blood enters the peritubular capillaries and the **vasa recta** that follow the loops of Henle in the medulla. *(see Figure 26.8)*
12. The proximal convoluted tubule (PCT) actively reabsorbs nutrients, ions, plasma proteins, and electrolytes from the tubular fluid. The loop of Henle includes a **descending limb** and an **ascending limb**; each limb contains a **thick segment** and a **thin segment**. The ascending limb delivers fluid to the distal convoluted tubule (DCT), which actively secretes ions and reabsorbs sodium ions from the urine. Reabsorption in the PCT and the loop of Henle reclaims all of the organic nutrients, 85% of the water, and over 90% of the Na and Cl ions. *(see Figures 26.6 to 26.8)*
13. The **distal convoluted tubule** is an important site for the secretion of ions and other materials, and the reabsorption of sodium ions.
14. The **juxtaglomerular apparatus** is composed of the **macula densa, juxtaglomerular cells**, and the **extraglomerular mesangial cells.** The juxtaglomerular apparatus secretes the hormones *renin* and *erythropoietin.*
15. The DCT opens into the **collecting system**. The collecting system consists of *connecting tubules, collecting ducts,* and *papillary ducts.* In addition to transporting fluid from the nephron to the renal pelvis, the collecting system makes adjustments to the osmotic concentrations and volume.

Structures for Urine Transport, Storage, and Elimination 712

1. Tubular fluid modification and urine production ends when the fluid enters the minor calyx in the renal sinus. The rest of the urinary system (the *ureters, urinary bladder,* and *urethra)* is responsible for transporting, storing, and eliminating the urine. *(see Figure 26.10)*

The Ureters 714

2. The ureters extend from the renal pelvis to the urinary bladder and are responsible for transporting urine to the bladder. The wall of each ureter consists of an inner mucosa layer, a middle muscular layer, and an outer connective tissue layer (adventitia). *(see Figures 26.2/26.3/26.10/26.11)*

The Urinary Bladder 714

3. The **urinary bladder** is a hollow muscular organ that serves as a storage reservoir for urine. The bladder is stabilized by the **urachus** or **middle umbilical ligament** and the **lateral umbilical ligaments**. Internal features include the **trigone**, the **neck**, and the **internal urethral sphincter**. The mucosal lining contains prominent **rugae** (folds). Contraction of the **detrusor** muscle compresses the urinary bladder and expels the urine into the urethra. *(see Figures 26.10/26.11)*

The Urethra 714

4. The urethra extends from the neck of the urinary bladder to the exterior. In the female, the urethra is short and ends in the **external urethral meatus** (external urethral opening), and in the male, the urethra has **prostatic**, **membranous**, and **penile** sections; the penile urethra ends at the external urethral meatus. In both sexes, as the urethra passes through the urogenital diaphragm, a circular band of skeletal muscles forms the **external urethral sphincter**, which is under voluntary control.
5. The female urethral lining is usually a transitional epithelium near the urinary bladder; the rest is usually a stratified squamous epithelium. The urethral lining of males varies from a transitional epithelium at the urinary bladder, to a stratified columnar or a pseudostratified epithelium, and then to stratified squamous epithelium near the external urethral meatus. *(see Figures 26.10/26.11)*

The Micturition Reflex and Urination 717

6. The process of urination is coordinated by the **micturition reflex**, which is initiated by stretch receptors in the bladder wall. Voluntary urination involves coupling this reflex with the voluntary relaxation of the external urethral sphincter.

Aging and the Urinary System 717

1. Aging is usually associated with increased kidney problems. Age-related changes in the urinary system include: (1) a declining number of functional nephrons, (2) reduced glomerular filtration, (3) reduced sensitivity to ADH, and (4) problems with the micturition reflex (**urinary retention** may develop in men whose prostate glands are enlarged).

LEVEL 1 REVIEWING FACTS AND TERMS

Match each numbered item with the most closely related lettered item. Use letters for answers in the spaces provided.

Column A	Column B
____ 1. urination	a. muscle of the urinary bladder
____ 2. detrusor muscle	b. basic functional unit of a kidney
____ 3. macula densa	c. micturition
____ 4. medulla	d. fibrous tunic of the kidney
____ 5. hilus	e. kidney stones
____ 6. nephron	f. consists of 16 to 18 renal pyramids
____ 7. calculi	g. dense outer layer of the kidney

_____ 8. vasa recta
_____ 9. renal capsule
_____ 10. renal fascia

h. series of capillaries
i. region of the juxtaglomerular apparatus
j. site of the entry/exit for the renal artery and vein

11. Urine is carried to the urinary bladder by
(a) blood vessels (b) lymphatics
(c) the ureters (d) the urethra

12. The urinary system does all of the following except
(a) secretes excess glucose molecules
(b) regulates blood volume
(c) contributes to stabilizing blood pH
(d) eliminates organic waste products

13. The renal sinus is
(a) the innermost layer of kidney tissue
(b) a conical-shaped structure that is located in the renal medulla
(c) an internal cavity lined by the fibrous capsule: located inside hilus
(d) a large branch of the renal pelvis

14. The glomerular capsule and the glomerulus make up the
(a) renal pyramid (b) loop of Henle
(c) renal corpuscle (d) renal papilla

15. The process of filtration occurs at the
(a) proximal convoluted tubule (b) renal corpuscle
(c) collecting duct (d) loop of Henle

16. The ability to form a concentrated urine depends on the functions of the
(a) proximal convoluted tubule (b) distal convoluted tubule
(c) collecting duct (d) loop of Henle

17. The ureters and urinary bladder are lined by ______ epithelium.
(a) stratified squamous (b) pseudostratified columnar
(c) simple cuboidal (d) transitional

18. Each of the following is a normal component of urine except
(a) hydrogen ions (b) urea
(c) large proteins (d) salts

19. The expanded end of a nephron is the
(a) glomerulus (b) renal corpuscle
(c) proximal convoluted tubule (d) distal convoluted tubule

20. The portion of the nephron that attaches to the collecting duct is the
(a) loop of Henle
(b) proximal convoluted tubule
(c) distal convoluted tubule
(d) collecting duct

LEVEL 2 REVIEWING CONCEPTS

1. An obstruction in the glomerulus would affect the flow of blood into the
(a) renal artery (b) efferent arteriole
(c) afferent arteriole (d) intralobular artery

2. Damage to the renal medulla would interfere first with the functioning of the
(a) glomerular capsule (b) distal convoluted tubule
(c) proximal convoluted tubule (d) collecting ducts

3. In the loop of Henle
(a) water performs osmosis into the descending limb
(b) sodium ions are passively transported out of the ascending limb
(c) the ascending limb is very permeable to water
(d) the filtrate in the descending limb becomes more and more hypotonic

4. Where is the glomerulus located in a nephron?
5. What is unique about the glomerular epithelium?
6. The juxtaglomerular apparatus secretes what product?
7. What is the trigone of the urinary bladder?
8. Which urethral sphincter is under voluntary control?
9. What is the primary function of the proximal convoluted tubule?
10. Identify the purpose of rugae in the urinary bladder.

LEVEL 3 CRITICAL THINKING AND CLINICAL APPLICATIONS

1. Sylvia is suffering from severe edema in both her hands and ankles. Her physician prescribes a diuretic (a substance that will increase the volume of urine produced). Why might this help to alleviate Sylvia's problem?
2. Frank has been fasting for several days and begins to notice a strong smell of ammonia in his urine. He begins to worry and borrows some pH paper from the chemistry lab and finds that his urine is very acidic. What could account for these observations?
3. David's grandfather suffers from hypertension. His doctor tells him that part of the problem stems from renal arteriosclerosis. Why would this cause hypertension?

✓ ANSWERS TO CONCEPT CHECK QUESTIONS

p. 708 **1.** The kidneys will absorb more water, release erythropoietin, and release renin. **2.** The urine goes through the ureters to the urinary bladder, where it is stored until urination occurs. From there, it travels through the urethra before exiting the body. **3.** Calcitriol, released by the kidneys in response to low calcium levels in the blood, causes increased calcium absorption by the intestinal epithelium.
p. 712 **1.** The route blood must take from the renal artery to the glomerulus is: Renal artery → Segmental arteries → Interlobar arteries → Arcuate arteries → Interlobular arteries → Afferent arterioles → Glomerulus. The route blood must take from the glomerulus to the renal vein is: Glomerulus → Efferent arterioles → Peritubular capillaries/vasa recta → Venules → Interlobular veins → Arcuate veins → Interlobar veins → Renal vein. **2.** Filtrate flows from glomerulus to a minor calyx via the following route: Glomerulus → Capsular space → PCT → Descending limb of loop → Loop of Henle → Ascending limb of loop → DCT → Connecting tubule → Collecting duct → Papillary duct → Minor calyx. **3.** Filtration, which is a passive process, allows ions and molecules to pass based on their size. This means that if the pores are big enough to allow the passage of organic wastes, they are also big enough to allow the passage of water, ions, and other important organic molecules. This is why some elements of the filtrate must be actively reabsorbed before the production of urine is complete. **4.** The loop absorbs additional water from the tubular fluid and an even higher percentage of the sodium and chloride ions. This results from the high osmotic concentration of its surroundings in the medulla.
p. 717 **1.** An obstruction of the ureters would interfere with the passage of urine from the renal pelvis to the urinary bladder. **2.** The mucosa lining of the urinary bladder is thrown into folds, called rugae, that allow the bladder to stretch when it is full. **3.** Females experience urinary tract infections, which are caused by bacteria, more often than do males because of the proximity of the urethra to the anus. In addition, the urethra is shorter in females, so the bacteria can easily migrate to the urinary bladder. **4.** The urinary bladder is held in place by the median umbilical ligament extending from the anterior and superior border, and the lateral umbilical ligaments passing along the sides of the bladder.

REVIEW IT

Challenge 1

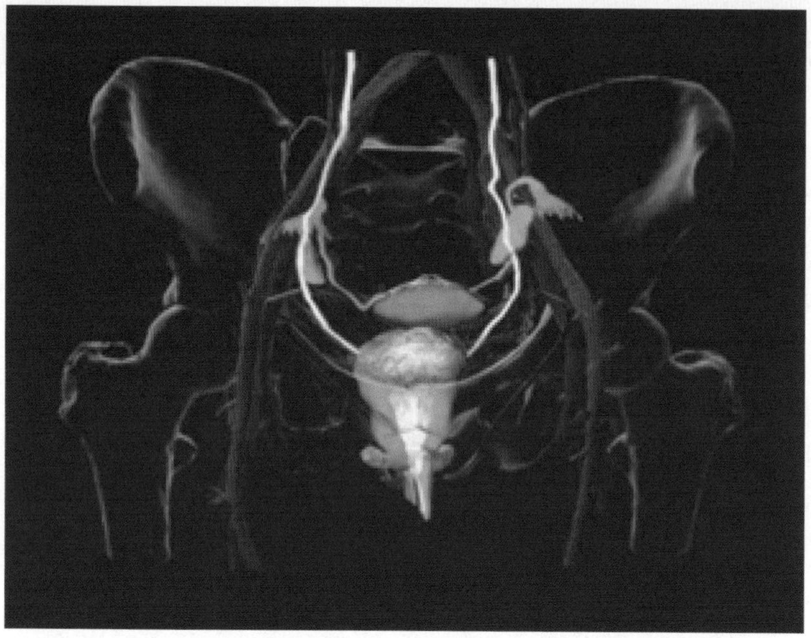

The urinary system is unique in many ways. One such way is that the anatomy of this system is straight forward and relatively simple. There are after all only four organs to identify. The placement of these organs is quite interesting, however. The Interactive CD has two different views of the urinary system—one using the male Visible Human data set found under Animations/Systemic/Urinary system, and the other using the female data set. The female images on the Interactive CD demonstrate the placement of the organs of the urinary system very well, with the exception of the kidney, which lie superior to the pelvis, and are not included in the pelvic image. The image presented here is taken from the regional view of the pelvis on your CD. List *every* structure you see, noting the relationship of each of these organs to the urinary system. At this point, all but the structures of the reproductive system should be familiar to you. What figure from the text is most helpful to you in this task?

Challenge 2

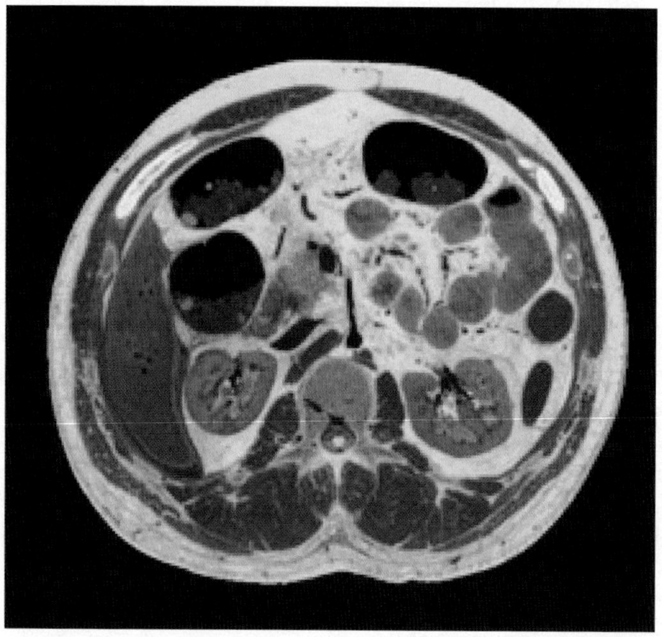

You have learned that the kidneys are retroperitoneal, embedded in adipose tissue. Identify the left and right kidneys on this transverse section from the Visible Human data set. You should be able to identify the spinal cord, vertebral body, liver, mesenteries, intercostals muscles, rib, small intestine, colon, vena cava, spleen, psoas major, quadratus lumorum, external oblique, and latissimus dorsi muscles as well as the ureters and the internal structures of the kidney. What figures in the text are most helpful to you as you identify EVERYTHING in this section? Is there anything in this section that does not easily match a figure from the text? If so, why do you suppose that to be the case?

Images provided by the Digital Cadaver™ project, courtesy of Visible Productions, Inc.

APPLY IT

Below are two separate exercises that provide more information on a topic presented in the preceding chapters. Each one is designed to take approximately 10 minutes, and will help you gain a better appreciation for the material presented in this chapter.

Critical Linking 1

Arguably, the really interesting part of the urinary system is not the gross anatomy but the functional histology. An appreciation for the complex and intricate monitoring of the internal environment performed by this system requires an understanding of the microscopic anatomy involved. Your text provides a few photomicrographs to introduce you to this topic. Review these images, then prepare two line drawings, one of a nephron and the other of the entire system. These do not have to be recognizable to anyone other than you. Next indicate on your drawings where you expect there to be histological differences related to structure. Sketch the histology of the areas you indicated which are provided in the text.

Move to the Companion Website, Critical Linking and click the keyword "urinary histology." There you will find a series of 32 additional histological photographs. Add to your sketches using this database. When completed, you will have a wonderful reference relating the structures, histology and functions of the urinary system.

Critical Linking 2

Practice, practice, practice! The best way to reinforce knowledge is to continually challenge your understanding. It is helpful to consider many different views of the same structures to gain a three dimensional understanding of anatomical relationships.

One of the best anatomy quizzing sites on the Internet can be found by visiting the Companion Website and clicking the key word "urinary quizzes." This site was created by Murray Jensen, University of Minnesota, and has won many awards for excellence in teaching online. Test your knowledge of the urinary system by attempting the quizzes found there. If you are unsuccessful, review the specific topic you missed in the text, and try the quizzes again. If you can sweep through the entire set, you will be well prepared for your exams!

FURTHER STUDIES

Although the urinary system is not usually in the news, we do in fact hear quite a bit about it indirectly. There is an entire sports drink market built around the science of fluid balance and hydration. As you are aware, the underlying system in these campaigns is the urinary system. It is worth taking a look at these marketing schemes forearmed with knowledge of the urinary system anatomy and physiology. Amazingly, a quick search of the Internet reveals hundreds of sites designed to sell fluids purporting restoration of fluid balance.

Many of these are scientifically well-respected companies. Powerade, for example, is owned by The Coca-Cola Company. This company is known to conduct nutrition research on all its products. The Gatorade Sports Science Institute, located in Barrington Illinois (*http://www.gssiweb.com/*) carries out scientific experiments to determine what compounds are lost from the body during exercise, and in what concentrations. Topics currently being researched by these companies include: (1) assessing the physiological, metabolic, and hormonal response to fluid and nutrient intake before, during, and after exercise; (2) researching the psychosensory responses to food and fluid intake in the exercise setting in an effort to understand how taste characteristics influence voluntary intake and physiological response; and (3) evaluating emerging science in sports nutrition as it relates to physiological and performance responses. To test your personal fluid balance and loss due to exercise, you can visit the Internet at: *http://www.rice.edu/~jenky/sports/fluid.balance.html*. Here you will find a short fluid balance test that will help you determine how much exercise affects your fluid volumes. Take some time from your studies to get some exercise, and to further your appreciation for the urinary system!

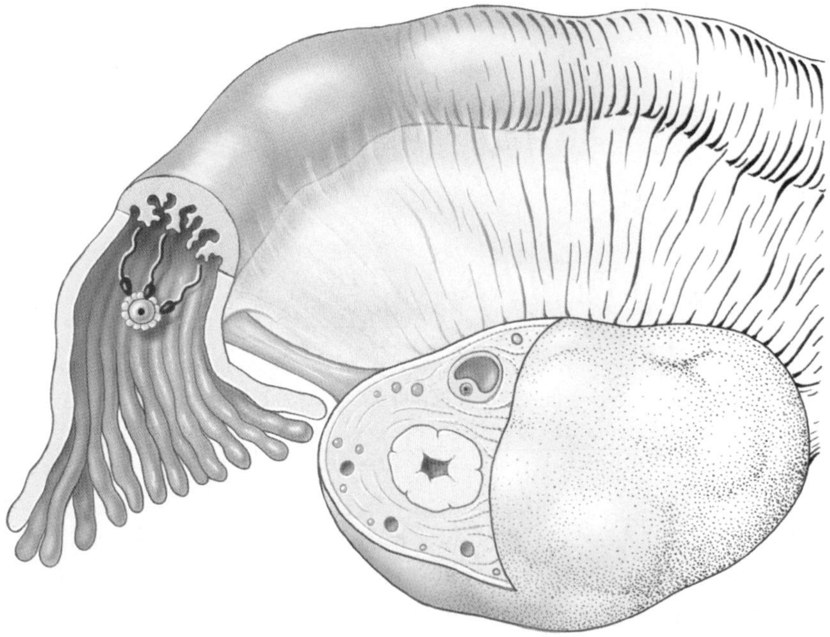

27

THE REPRODUCTIVE SYSTEM

An individual's life span can be measured in decades, but the human species has survived for hundreds of thousands of years through the activities of the reproductive system. The human reproductive system produces, stores, nourishes, and transports functional male and female reproductive cells, or **gametes** (GAM-ēts). The combination of the genetic material provided by a **sperm** from the father and an immature **ovum** (Ō-vum) from the mother occurs shortly after **fertilization**, or *conception*. Fertilization produces a **zygote** (ZĪ-gōt), a single cell whose growth, development, and repeated divisions will, in approximately 9 months, produce an infant who will grow and mature as part of the next generation. The reproductive system also produces sex hormones that affect the structure and function of all other systems.

This chapter will describe the structures and mechanisms involved in the production and maintenance of gametes and, in the female, the development and support of the developing embryo and fetus. Chapter 28, the last chapter in the text, will begin at fertilization and consider the process of development.

Organization of the Reproductive System

The reproductive system includes:

- Reproductive organs, or **gonads** (GŌ-nadz), which produce gametes and hormones,
- A reproductive tract, consisting of ducts that receive, store, and transport the gametes,
- Accessory glands and organs that secrete fluids into the ducts of the reproductive system or into other excretory ducts, and
- Perineal structures associated with the reproductive system. These perineal structures are collectively known as the **external genitalia** (jen-i-TĀ-lē-a).

The male and female reproductive systems are functionally quite different. In the adult male, the gonads, called **testes** (TES-tēz; singular *testis*), secrete sex hormones called *androgens*, principally *testosterone*, and produce one-half billion sperm per day. p. 521 After storage, mature sperm travel along a lengthy duct system where they are mixed with the secretions of accessory glands, forming the mixture known as **semen** (SĒ-men). During **ejaculation** (e-jak-ū-LĀ-shun), the semen is expelled from the body.

In adult females the gonads, or **ovaries**, typically produce only one immature gamete, or *oocyte*, per month. The oocyte travels along short **uterine tubes** (*oviducts*) that open into the muscular **uterus** (ū-ter-us). A short passageway, the **vagina** (va-JĪ-na), connects the uterus with the exterior. During intercourse, male ejaculation introduces semen into the vagina, and the sperm cells may ascend the female reproductive tract, where they may encounter an oocyte and begin the process of fertilization.

Anatomy of the Male Reproductive System [FIGURE 27.1]

The principal structures of the male reproductive system are shown in Figure 27.1●. Sperm cells, or **spermatozoa** (sper-ma-tō-ZŌ-a), leave the testes and travel along a duct system that includes: the **epididymis** (ep-i-DID-i-mus), the **ductus deferens** (DUK-tus DEF-e-renz), or *vas deferens*, the **ejaculatory** (ē-JAK-ū-la-tō-rē) **duct**, and the **urethra** before leaving the body. Accessory organs, notably the **seminal** (SEM-i-nal) **vesicles**, the **prostate** (PROS-tāt) **gland**, and the **bulbourethral** (bul-bō-ū-RĒ-thral) **glands** secrete into the ejaculatory ducts and urethra. The external genitalia include the **scrotum** (SKRŌ-tum), which encloses the testes, and the **penis** (PĒ-nis), an erectile organ through which the distal portion of the urethra passes.

The Testes [FIGURES 27.1/27.3]

Each testis has the shape of a flattened egg roughly 5 cm (2 in.) long, 3 cm (1.2 in.) wide, and 2.5 cm (1 in.) thick. Each has a weight of 10–15 g (0.35–0.53 oz). The testes hang within the **scrotum**, a pouch of skin suspended inferior to the perineum and anterior to the anus. Note the location and relation of the testes in sagittal section (Figure 27.1●) and in frontal section (Figure 27.3●).

DESCENT OF THE TESTES [FIGURE 27.2]

During development, the testes form inside the body cavity adjacent to the kidneys. The relative positions of these organs change as the fetus enlarges, resulting in a gradual movement of the testes inferiorly and anteriorly toward the anterior abdominal wall. The **gubernaculum testis** is a cord of connective tissue and muscle fibers that extends from the inferior part of each testis to the posterior wall of a small, inferior pocket of the peritoneum. As growth proceeds, the gubernacula do not elongate, and the testes are held in position (Figure 27.2●). During the seventh developmental month (1) growth continues at a rapid pace, and (2) circulating hormones stimulate contraction of the gubernaculum testis. Over this period the relative position of the testes changes further, and they move through the abdominal musculature, accompanied by small pockets of the peritoneal cavity. This process is known as the **descent of the testes**.

As it moves through the body wall, each testis is accompanied by the ductus deferens, the testicular blood vessels, nerves, and lymphatics. Once the testes have descended, the ducts, vessels, nerves, and lymphatics remain bundled together within the *spermatic cords*.

THE SPERMATIC CORD [FIGURE 27.3]

The **spermatic cords** consist of layers of fascia, tough connective tissue, and muscle enclosing the blood vessels, nerves, and lymphatics supplying the testes. Each spermatic cord begins at the deep inguinal ring, extends through the inguinal canal, exits at the superficial inguinal ring, and descends to the testes within the scrotum (Figure 27.3●). The spermatic cords form during the descent of the testes. *Testicular Torsion p. 807*

Each spermatic cord contains the ductus deferens, the **testicular artery**, the **pampiniform plexus** (pam-PIN-i-form; *pampinus*, tendril + *forma*, form) of the **testicular vein**, and the **ilioinguinal** and **genitofemoral nerves** from the lumbar plexus. p. 371, 594

The narrow canals linking the scrotal chambers with the peritoneal cavity are called the *inguinal canals*. These passageways usually close, but the persistence of the spermatic cords creates weak points in the abdominal wall that remain throughout life. As a result, *inguinal hernias*, discussed in Chapter 10, are relatively common in males. p. 278 The inguinal canals

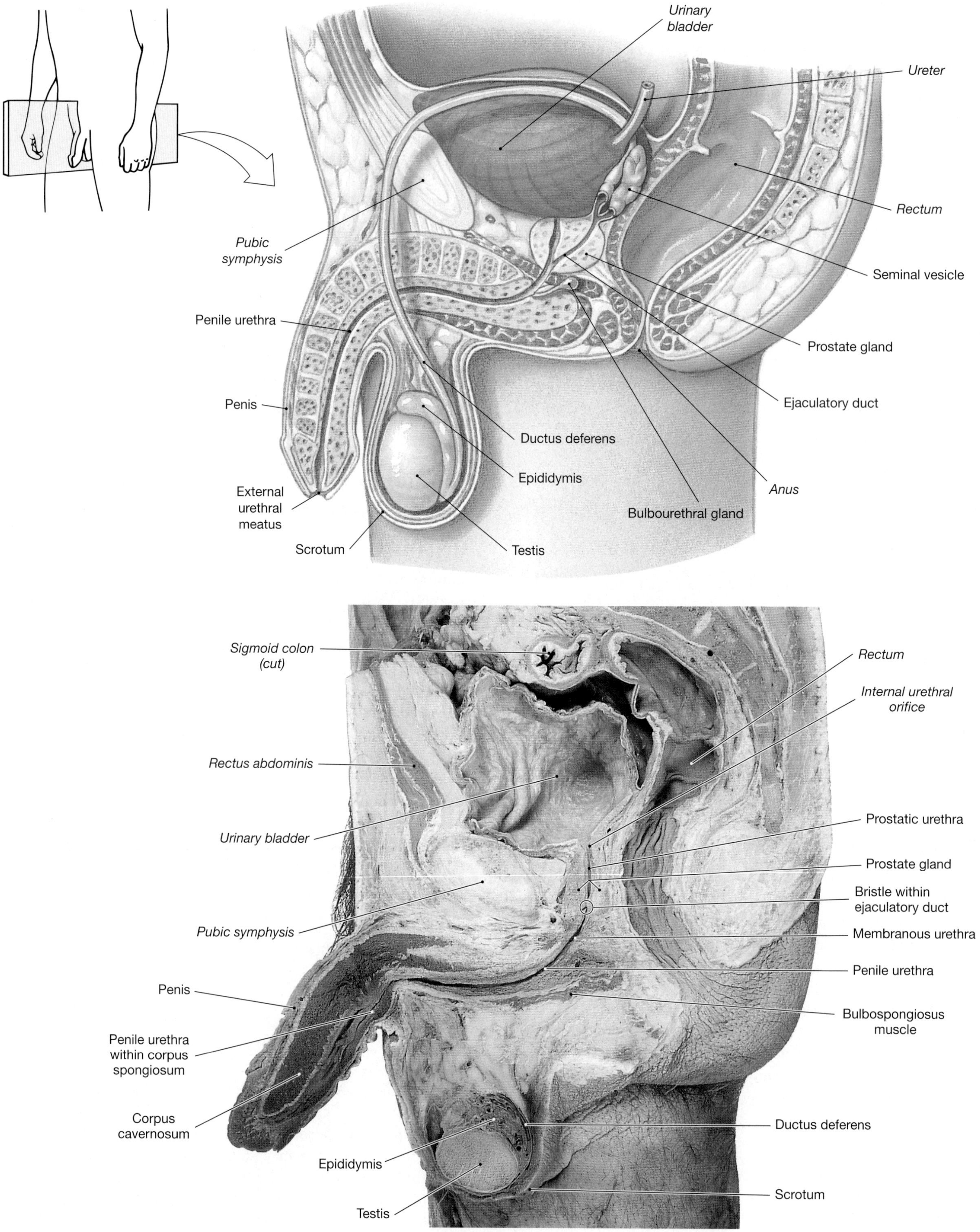

FIGURE 27.1 **The Male Reproductive System, Part I**

The male reproductive system as seen in sagittal section. The diagrammatic view shows several intact organs on the left side; this should help you to interpret the cadaver section.

FIGURE 27.2 **The Descent of the Testes**

(**a**) Diagrammatic sectional view at representative stages in the descent of the testes. As the fetus enlarges, the gubernaculum testis does not. Note the scale bar to the left of each image. (**b**) Anterior view of the opened abdomen at representative stages, showing the descent of the testes and origin of the spermatic cord.

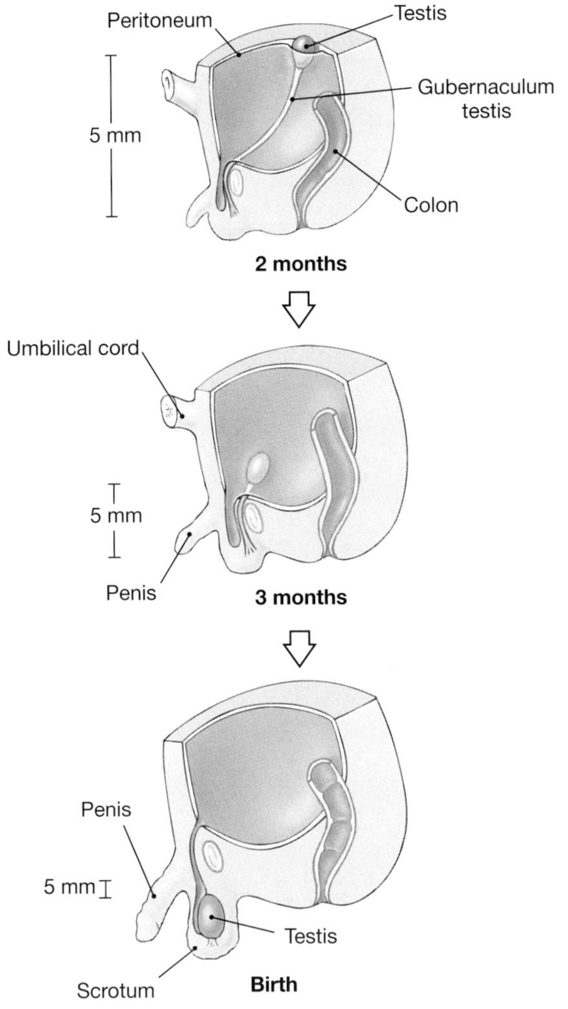

(a) Descent of testis, lateral view

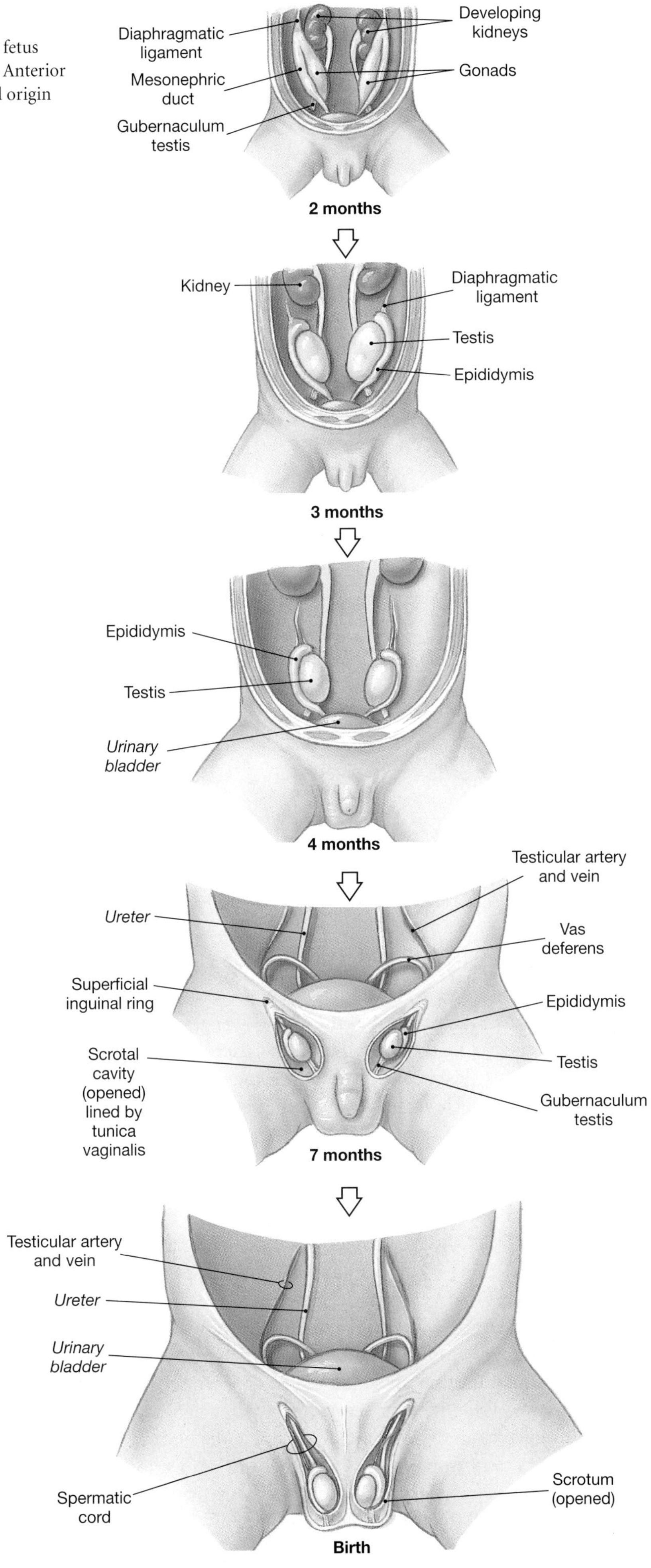

(b) Descent of testis, anterior view

in females are very small, containing only the ilioinguinal nerves and the round ligaments of the uterus. The abdominal wall is nearly intact, and inguinal hernias in women are rare.

CLINICAL BRIEF

Cryptorchidism

In **cryptorchidism** (kript-OR-ki-dizm; *crypto*, hidden), one or both of the testes have not descended into the scrotum by the time of birth. Typically the testes are lodged in the abdominal cavity or within the inguinal canal. This condition occurs in about 3% of full-term deliveries and in roughly 30% of premature births. In most instances normal descent occurs a few weeks later, but the condition can be surgically corrected if it persists. Corrective measures are usually taken before puberty because cryptorchid (abdominal) testes will not produce sperm, as the temperature inside the peritoneal cavity is too high for sperm production (1–2°C higher than the temperature within the scrotal cavity). The individual will therefore be sterile (infertile). If the testes cannot be moved into the scrotum, they will usually be removed, because about 10% of those with uncorrected cryptorchid testes eventually develop testicular cancer.

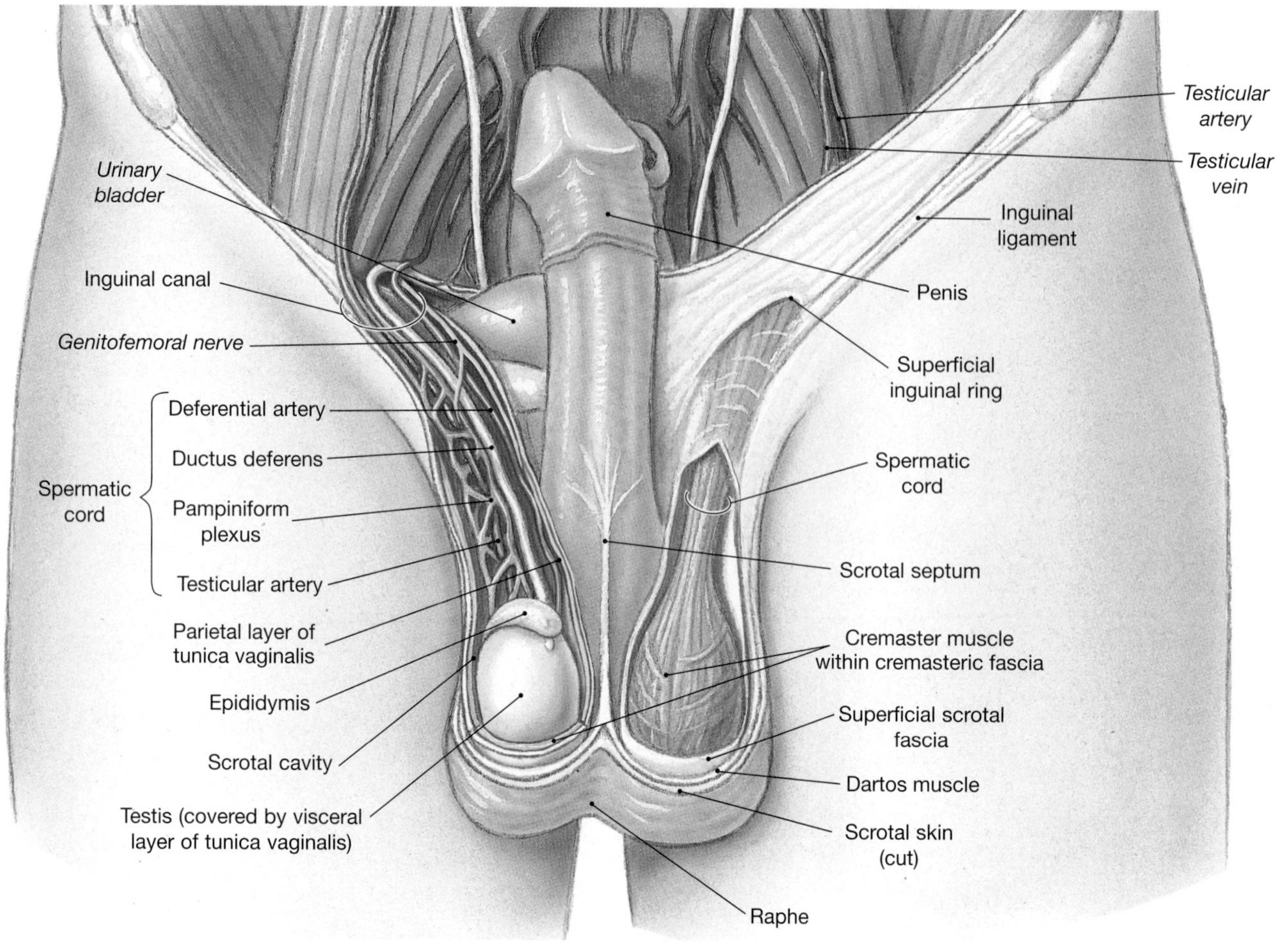

FIGURE 27.3 **The Male Reproductive System, Part II**
Diagrammatic anterior view of the gonads, external genitalia, and associated structures in a male.

THE SCROTUM AND THE POSITION OF THE TESTES

[FIGURES 27.3/27.4a]

The scrotum is divided internally into two separate chambers, and the partition between the two is marked externally by a raised thickening in the scrotal surface known as the **perineal raphe** (RĀ-fē) (Figures 27.3 and 27.4a●). Each testis occupies a separate compartment, or **scrotal cavity**, with a narrow space separating the inner surface of the scrotum from the outer surface of the testis. The **tunica vaginalis** (TOO-nĭk-a vaj-i-NAL-is) is a serous membrane which covers the outside of each testis and lines the scrotal cavity. This membrane reduces friction between the opposing surfaces.

The scrotum consists of a thin layer of skin and the underlying superficial fascia. The dermis of the scrotum contains a layer of smooth muscle, the **dartos** (DAR-tōs) **muscle**. Its tonic contraction causes the characteristic wrinkling of the scrotal surface. A layer of skeletal muscle, the **cremaster** (kre-MAS-ter) **muscle**, lies deep to the dermis. Contraction of the cremaster tenses the scrotum and pulls the testes closer to the body. These contractions are controlled by the *cremasteric reflex*. Contraction occurs during sexual arousal and in response to changes in temperature. Normal sperm development in the testes requires temperatures around 1.1°C (2°F) lower than those found elsewhere in the body. The cremaster moves the testes away from or toward the body as needed to maintain acceptable testicular temperatures. When air or body temperatures rise, the cremaster relaxes and the testes then move away from the body. Cooling the scrotum, as when entering a cold swimming pool, results in cremasteric contractions that pull the testes closer to the body and keep testicular temperatures from falling.

The scrotum is richly supplied with sensory and motor nerves from the *hypogastric plexus* and branches of the *ilioinguinal nerves*, the *genitofemoral nerves*, and the *pudendal nerves*. p. 370 The vascular supply to the scrotum includes the **internal pudendal arteries** (supplied by the internal iliac arteries), the **external pudendal arteries** (supplied by the femoral arteries), and the *cremasteric branch* of the **inferior epigastric arteries** (supplied by the external iliac arteries). p. 592 The names and distributions of the veins follow those of the arteries.

STRUCTURE OF THE TESTES [FIGURE 27.4]

The **tunica albuginea** (TŪ-ni-ka al-bū-JIN-ē-a) is a dense fibrous layer that surrounds the testis and is covered by the tunica vaginalis. This fibrous capsule is rich in collagen fibers that are continuous with those surrounding the adjacent epididymis. The collagen fibers of the tunica albuginea also extend into the interior of the testis, forming fibrous partitions, or *septa* (Figure 27.4●). These septa converge toward the **mediastinum** of the testis. The mediastinum supports the blood vessels and lymphatics supplying the testis and the ducts that collect sperm and transport them into the epididymis.

HISTOLOGY OF THE TESTES [FIGURES 27.4/27.5c,d/27.7a]

The septa partition the testis into compartments called **lobules**. Roughly 800 slender, tightly coiled **seminiferous** (sem-in-IF-er-us) **tubules** are distributed among the lobules (Figure 27.4●). Each tubule averages around 80 cm (31 in.) in length, and a typical testis contains nearly one-half mile of seminiferous tubules. Sperm production occurs within these tubules.

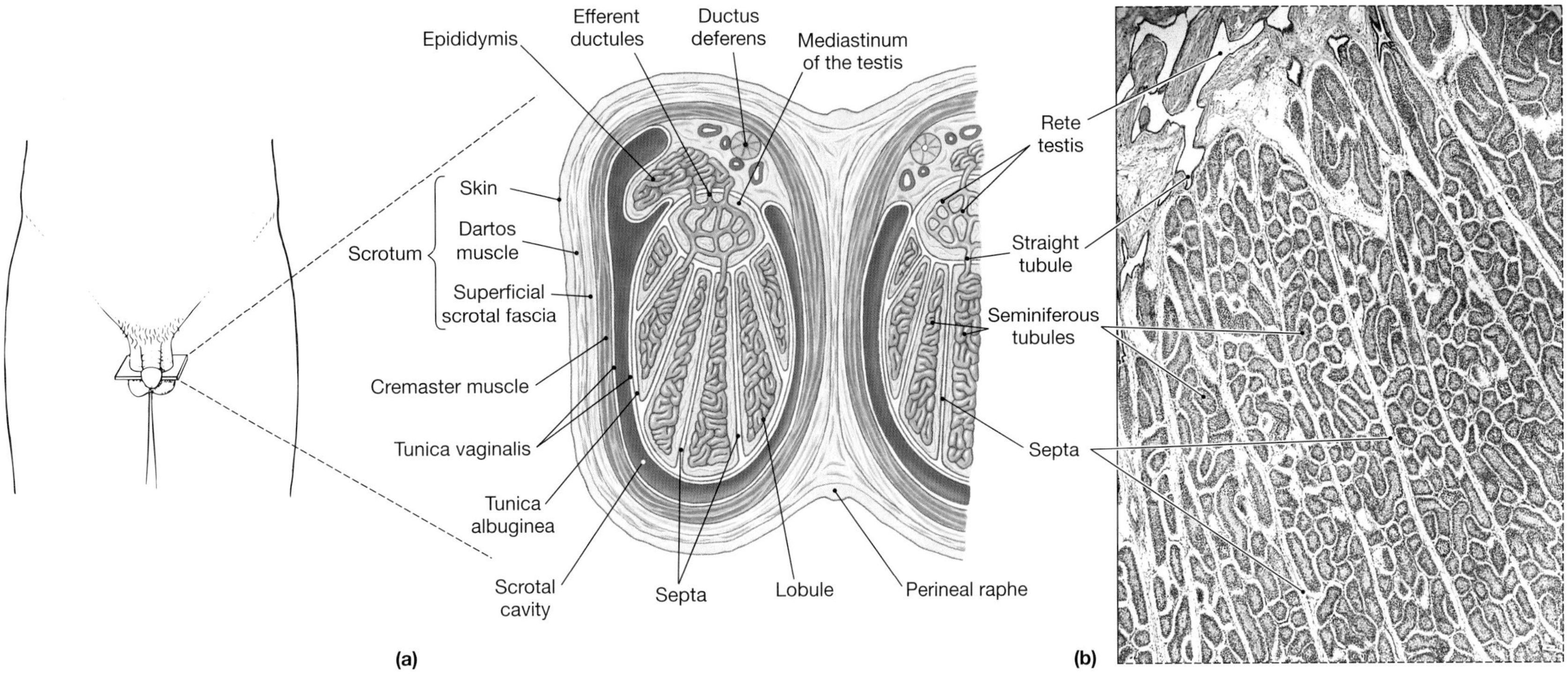

FIGURE 27.4 **STRUCTURE OF THE TESTES**

(a) Diagrammatic horizontal section showing the anatomical relationships of the testes within the scrotal cavities. The connective tissues surrounding the seminiferous tubules and the rete testis are not shown. **(b)** Photomicrograph showing general view of septa separating seminiferous tubules. (LM × 25)

Each seminiferous tubule is U-shaped and connected to a single **straight tubule** that enters the mediastinum of the testis (Figures 27.4a and 27.7a●). Within the mediastinum, these tubules are extensively interconnected, forming a maze of passageways known as the **rete testis** (RĒ-tē; *rete,* net). Fifteen to 20 large **efferent ductules** connect the rete testis to the epididymis.

Because the seminiferous tubules are tightly coiled, histological preparations most often show them in transverse section. A delicate capsule surrounds each tubule, and loose connective tissue fills the external spaces between the tubules. Within those spaces are found numerous blood vessels and large **interstitial cells** (*interstitial endocrinocytes,* or *cells of Leydig*) (Figure 27.5c,d●). Interstitial cells produce male sex hormones, called *androgens. Testosterone* is the most important androgen. Testosterone functions to: (1) stimulate spermatogenesis; (2) promote the physical and functional maturation of spermatozoa; (3) maintain the accessory organs of the male reproductive tract; (4) cause development of secondary sexual characteristics by influencing the development and maturation of nonreproductive structures such as the distribution of facial hair and adipose tissue, muscle mass, and total body size; (5) stimulate growth and metabolism throughout the body; and (6) influence brain development by stimulating sexual behaviors and sexual drive.

Testosterone production accelerates markedly at puberty, initiating sexual maturation and the appearance of secondary sexual characteristics.

SPERMATOGENESIS AND MEIOSIS [FIGURE 27.5]

Sperm cells, or spermatozoa, are produced through the process of **spermatogenesis** (sper-ma-tō-JEN-e-sis). Spermatogenesis begins at the outermost layer of cells in the seminiferous tubules. Stem cells called **spermatogonia** (sper-mat-ō-GŌ-nē-a) form during embryonic development but remain dormant until puberty. Beginning at sexual maturation, spermatogonia divide throughout the individual's reproductive years. As each division occurs, one of the daughter cells remains in the outer layer of the seminiferous tubule, as an undifferentiated stem cell, while the other is pushed toward the lumen. The latter cell differentiates into a **primary spermatocyte** (sper-MAT-ō-sīt) that prepares to begin *meiosis,* a form of cell division that produces gametes containing one-half the normal number of chromosomes. Because they contain only one member of each pair of chromosomes, gametes are called **haploid** (HAP-loid; *haplo,* single).

Meiosis (mī-Ō-sis) is a special form of cell division that produces gametes. Mitosis and meiosis differ significantly in terms of the nuclear events. In mitosis there is a single division that produces two identical daughter cells, each containing 23 pairs of chromosomes. In meiosis there is a pair of divisions that produces four different, haploid gametes, each containing 23 individual chromosomes.

In the testes, the first step in meiosis is the division of a primary spermatocyte to produce a pair of **secondary spermatocytes**. Each secondary spermatocyte divides, to produce a pair of **spermatids** (SPER-ma-tidz). As a result, four spermatids are produced for every primary spermatocyte that enters meiosis (Figure 27.5b●). Spermatogonia, spermatocytes undergoing meiosis, and spermatids are depicted in Figure 27.5c,d●.

Spermatogenesis is directly stimulated by testosterone and indirectly stimulated by FSH (follicle-stimulating hormone), as detailed below. Testosterone is produced by interstitial cells of the testis in response to LH (luteinizing hormone). p. 513

SPERMIOGENESIS [FIGURE 27.5b–d]

Each spermatid matures into a single **spermatozoon** (sper-ma-tō-ZŌ-on), or sperm cell, by a maturation process called **spermiogenesis** (Figure 27.5b–d●). During spermiogenesis, spermatids are embedded within the cytoplasm of large **sustentacular** (sus-ten-TAK-ū-lar) **cells** (*Sertoli cells*). Sustentacular cells are attached to the basement membrane at the tubular capsule and extend toward the lumen between the spermatocytes undergoing meoisis. As spermiogenesis proceeds, the spermatids gradually develop the appearance of mature spermatozoa. At *spermiation,* a spermatozoon loses its attachment to the sustentacular cell and enters the lumen of the

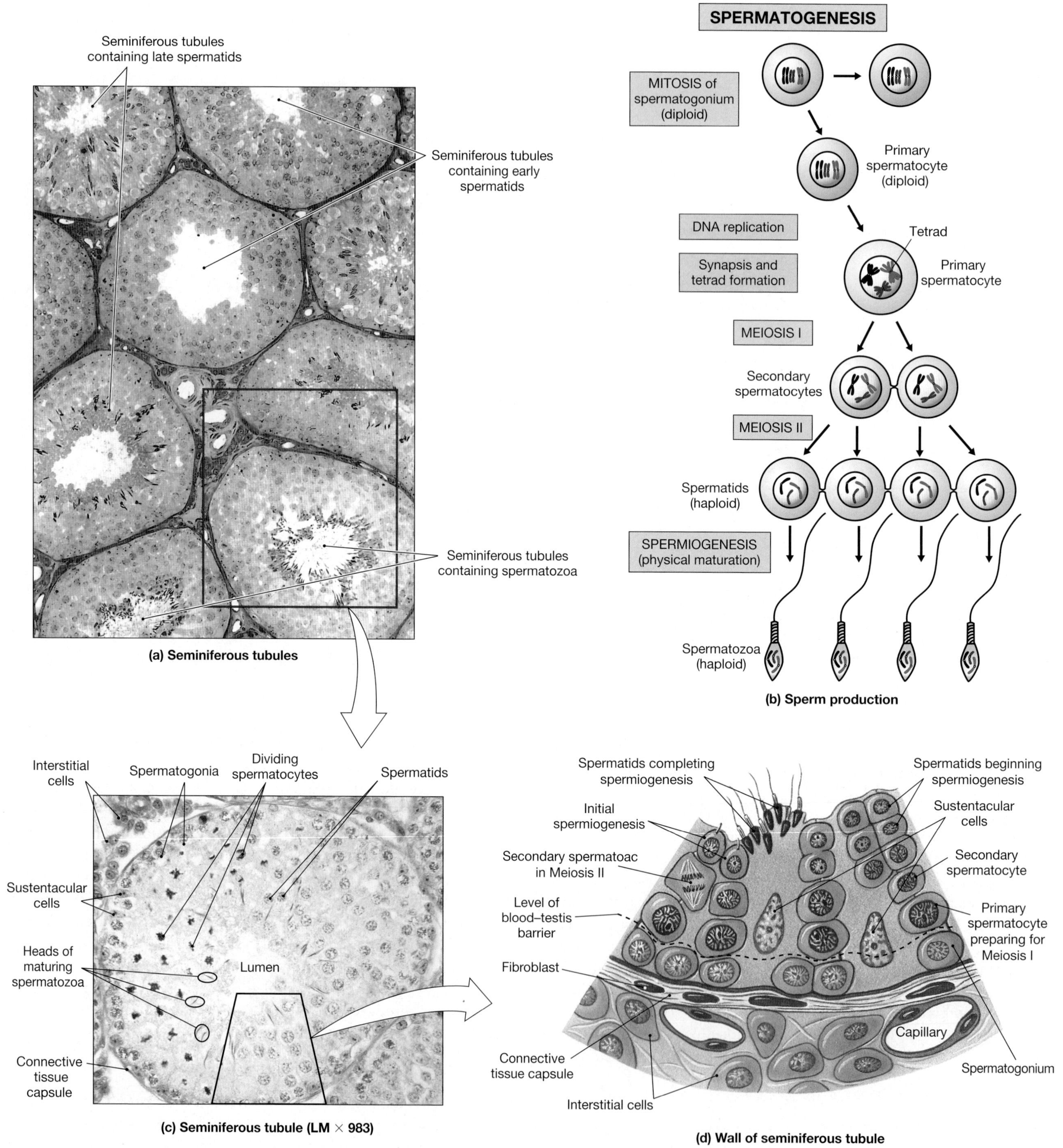

FIGURE 27.5 **The Seminiferous Tubules**

(**a**) Seminiferous tubules in sectional view. (**b**) Meiosis in the testes. (**c**) Spermatogenesis within one segment of a seminiferous tubule. (**d**) The blood-testis barrier and the structure of the wall of a seminiferous tubule.

seminiferous tubule. The entire process, from spermatogonial division to spermiation, takes approximately 9 weeks.

SUSTENTACULAR CELLS [FIGURE 27.5c,d]

Sustentacular cells have five important functions:

1. *Maintenance of the blood-testis barrier*: The seminiferous tubules are isolated from the general circulation by a **blood-testis barrier** (Figure 27.5c,d●) comparable to the blood-brain barrier. p. 388 Tight junctions between extensions of sustentacular cells isolate the lumenal portion of the seminiferous tubule from the surrounding interstitial fluid. Transport of materials across the sustentacular cells is tightly regulated so that the environment surrounding the spermatocytes and spermatids remains very stable. The lumen of a seminiferous tubule contains a fluid very different from interstitial fluid; for example, tubular fluid is high in androgens, estrogens, potassium, and amino acids. The blood-testis barrier is essential to preserving these differences. In addition, developing spermatozoa contain sperm-specific antigens in their cell membranes. These antigens, not found in somatic cell membranes, would be attacked by the immune system if the blood-testis barrier did not prevent their detection.
2. *Support of spermatogenesis*: Spermatogenesis depends on the stimulation of sustentacular cells by circulating FSH and testosterone. Stimulated sustentacular cells then support the division of spermatogonia and the meiotic divisions of spermatocytes.
3. *Support of spermiogenesis*: Spermiogenesis requires the presence of sustentacular cells. These cells surround and enfold the spermatids, providing nutrients and chemical stimuli that promote their development.
4. *Secretion of inhibin*: Sustentacular cells secrete a hormone called **inhibin** (in-HIB-in). Inhibin depresses the pituitary production of follicle-stimulating hormone (FSH) and gonadotropin-releasing hormone (GnRH). p. 513 The faster the rate of sperm production, the greater the amount of inhibin secreted.
5. *Secretion of androgen-binding protein*: **Androgen-binding protein (ABP)** binds androgens (primarily testosterone) in the fluid contents of the seminiferous tubules. This protein is thought to be important in elevating the concentration of androgens within the tubules and stimulating spermiogenesis.

Anatomy of a Spermatozoon [FIGURE 27.6]

There are three distinct regions to each spermatozoon: the *head*, the *middle piece*, and the *tail* (Figure 27.6●).

- The **head** is a flattened oval containing densely packed chromosomes. The tip contains the **acrosomal** (ak-rō-SŌ-mal) **cap**, a membrane-bound, vesicular compartment containing enzymes involved in the preliminary steps of fertilization.
- A short **neck** attaches the head to the **middle piece**. The neck contains both centrioles of the original spermatid. The microtubules of the distal centriole are continuous with those of the middle piece and

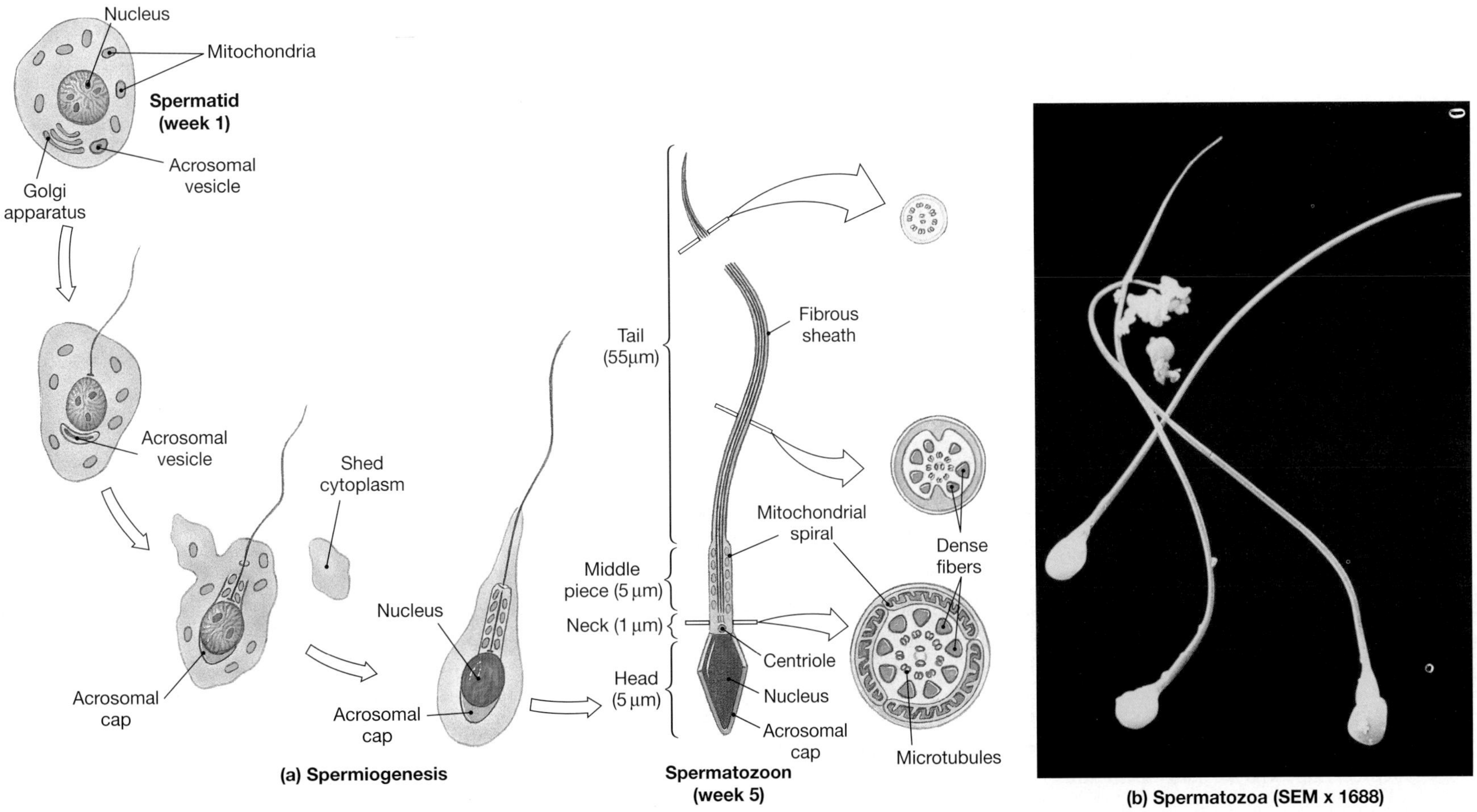

FIGURE 27.6 **SPERMIOGENESIS AND SPERMATOZOON STRUCTURE**

(a) Differentiation of a spermatid into a spermatozoon. **(b)** Human spermatozoa.

tail. Mitochondria arranged in a spiral around the microtubules provide the energy needed to move the tail.

- The **tail** is the only example of a flagellum in the human body. p. 37 A **flagellum** moves a cell from one place to another. In contrast to cilia, which beat in a predictable, waving fashion, the flagellum of a spermatozoon has a complex, corkscrew motion. The microtubules of the flagellum are surrounded by a dense, fibrous sheath.

Unlike other, less specialized cells, a mature spermatozoon lacks an endoplasmic reticulum, Golgi apparatus, lysosomes, peroxisomes, inclusions, and many other intracellular structures. Because the cell does not contain glycogen or other energy reserves, it must absorb nutrients (primarily fructose) from the surrounding fluid.

✓ CONCEPT CHECK

- What structures make up the body of the spermatic cord?
- Why are inguinal hernias relatively common in males?
- How is the location of the testes (outside the peritoneal cavity) important to the production of viable sperm?
- What is the function of the blood-testis barrier?

The Male Reproductive Tract

The testes produce physically mature spermatozoa that are not yet capable of successful fertilization because they are not yet motile. The other regions of the male reproductive system are concerned with the functional maturation, nourishment, storage, and transport of spermatozoa.

THE EPIDIDYMIS [FIGURES 27.1/27.3/27.4/27.7]

Late in their development, the spermatozoa become detached from the sustentacular cells and are free within the lumen of the seminiferous tubule. Although they have most of the physical characteristics of mature sperm cells, they are still functionally immature and incapable of coordinated locomotion or fertilization. Fluid currents then transport the spermatozoa along the straight tubule, through the rete testis (Figure 27.7a●), and into the epididymis. The lumen of the epididymis is lined by a distinctive simple columnar epithelium with long stereocilia (Figure 27.7b,c●).

The epididymis lies along the posterior border of the testis (Figures 27.1, 27.3, 27.4, and 27.7a●). It has a firm texture and can be felt through the skin of the scrotum. The epididymis consists of a tubule almost 7 m (23 ft) long, coiled and twisted so as to take up very little space. The epididymis has a *head*, a *body*, and a *tail*.

- The superior **head** receives spermatozoa via the efferent ducts of the mediastinum of the testis.
- The **body** begins distal to the last efferent duct and extends inferiorly along the posterior margin of the testis.
- Near the inferior border of the testis, the number of convolutions decreases, marking the start of the **tail**. The tail reverses direction, and as it ascends, the tubular epithelium changes. The stereocilia disappear and the epithelium becomes indistinguishable from that of the attached ductus deferens. The tail of the epididymis is the principal region involved with sperm storage.

The epididymis has three major functions:

1. It *monitors and adjusts the composition of the fluid produced by the seminiferous tubules*. The columnar epithelial lining of the epididymis bears distinctive *stereocilia* (Figure 27.7b,c●) that increase the surface area available for absorption and secretion into the tubular fluid. p. 57
2. It *acts as a recycling center for damaged spermatozoa*. Cellular debris and damaged spermatozoa are absorbed, and the products of enzymatic breakdown are released into the surrounding interstitial fluids for pickup by the epididymal circulation.
3. It *stores spermatozoa and facilitates their functional maturation*. It takes about 2 weeks for a spermatozoon to pass through the epididymis, and during this period it completes its functional maturation in a protected environment. Although spermatozoa leaving the epididymis are mature, they remain immobile. To become active, motile, and fully functional, the spermatozoa must undergo **capacitation**; a secretory product of the epididymis prevents premature capacitation. Capacitation normally occurs in two steps: Spermatozoa become motile when mixed with secretions of the seminal vesicles, and they become capable of successful fertilization when, upon exposure to conditions inside the female reproductive tract, the permeability of the sperm cell membrane changes.

Transport along the epididymis involves some combination of fluid movement and peristaltic contractions of smooth muscle. Spermatozoa eventually leave the tail of the epididymis and enter the ductus deferens.

THE DUCTUS DEFERENS [FIGURES 27.1/27.3/27.7a/27.8a,b]

The ductus deferens, also called the *deferential duct* or *vas deferens*, is 40–45 cm (16–18 in.) long. It begins at the end of the tail of the epididymis (Figure 27.7a●) and ascends into the abdominopelvic cavity through the inguinal canal as part of the spermatic cord (Figure 27.3●). Inside the abdominal cavity, the ductus deferens passes posteriorly, curving inferiorly along the lateral surface of the urinary bladder toward the superior and posterior margin of the prostate gland (Figure 27.1●). Just before it reaches the prostate, the ductus deferens becomes enlarged, and the expanded portion is known as the **ampulla** (am-PŪL-la) (Figure 27.8a●).

The wall of the ductus deferens contains a thick layer of smooth muscle (Figure 27.8b●). Peristaltic contractions in this layer propel spermatozoa and fluid along the duct, which is lined by a columnar epithelium with stereocilia. In addition to transporting sperm, the ductus deferens can store spermatozoa for several months. During this time the spermatozoa remain in a state of suspended animation, with low metabolic rates.

The junction of each ampulla with the base of a seminal vesicle marks the start of an **ejaculatory duct**. This relatively short passageway (2 cm, or less than 1 in.) penetrates the muscular wall of the prostate gland and empties into the urethra (Figure 27.1●) near the ejaculatory duct from the other side.

THE URETHRA [FIGURES 27.1/27.9]

The urethra of the male extends from the urinary bladder to the tip of the penis, a distance of 15–20 cm (6–8 in.). It is divided into *prostatic*,

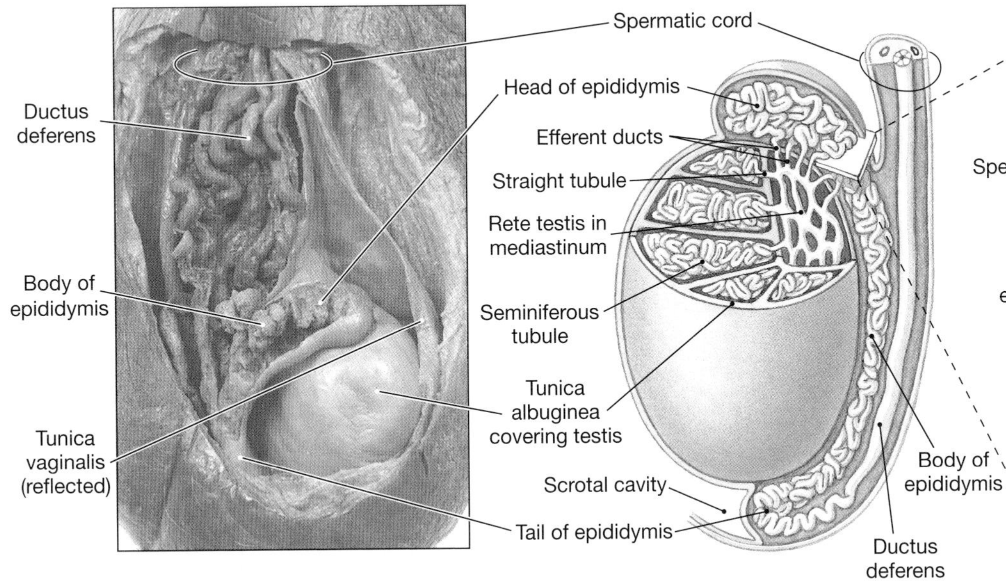

(a) Testis and epididymis

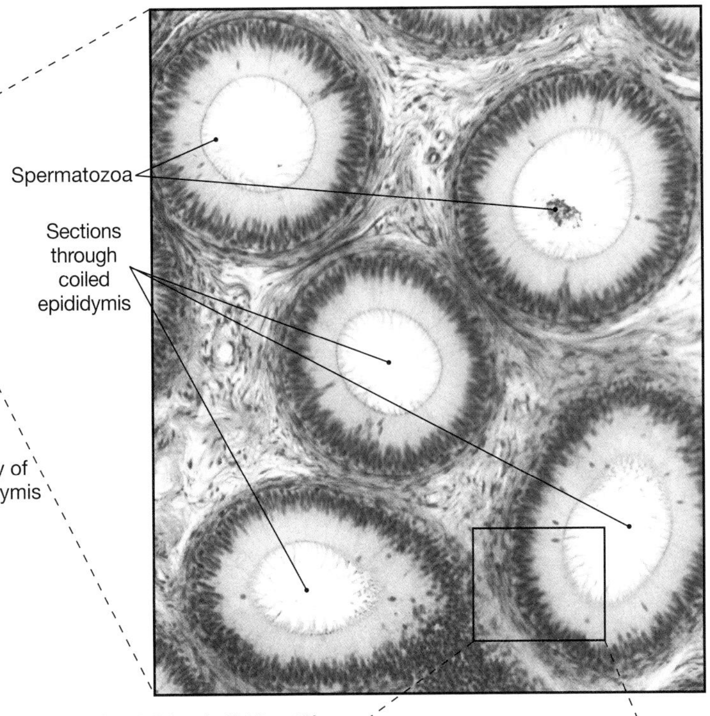

(b) Epididymis (LM × 49)

FIGURE 27.7 **The Epididymis**

(**a**) (Left) Appearance of the testis and epididymis on gross dissection. (Right) Diagrammatic view of the testis and epididymis, showing the sectional plane of (b). (**b**) A light micrograph showing the organization of tubules and the surrounding connective tissues. (**c**) A micrograph showing epithelial characteristics, especially the elongate stereocilia.

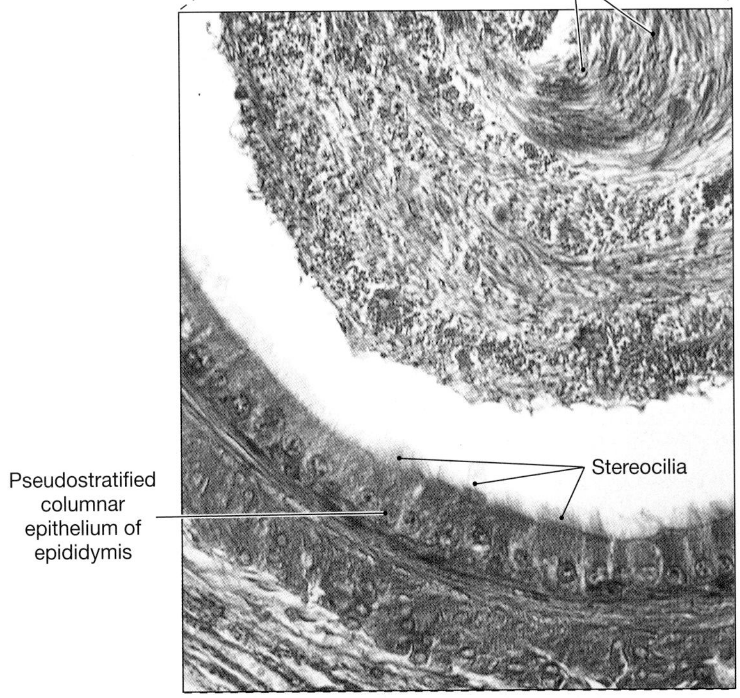

(c) Epididymis (LM × 1304)

membranous, and *penile regions* (Figures 27.1 and 27.9●). These subdivisions were considered in Chapter 26. p. 714 The urethra in the male is a passageway used by both the urinary and reproductive systems.

The Accessory Glands [FIGURES 27.1/27.8a]

The fluids contributed by the seminiferous tubules and the epididymis account for only about 5% of the final volume of semen. The seminal fluid is a mixture of the secretions from many different glands, each with distinctive biochemical characteristics. These glands include: the *seminal vesicles*, the *prostate gland*, and the *bulbourethral glands* (Figures 27.1, p. 726, and 27.8a●). Major functions of these glands include: (1) activating the spermatozoa, (2) providing the nutrients spermatozoa need for motility, and (3) producing buffers that counteract the acidity of the urethral and vaginal contents.

THE SEMINAL VESICLES [FIGURES 27.1a/27.8a,c/27.9a]

The ductus deferens on each side ends at the junction between the ampulla and the duct draining the seminal vesicle. The **seminal vesicles** are embedded in connective tissue on either side of the midline, sandwiched between the posterior wall of the urinary bladder and the anterior wall of the rectum. Each seminal vesicle is a tubular gland, with a total length of around 15 cm (6 in.). Each tubule has many short side branches, and the entire assemblage is coiled and folded into a compact, tapered mass roughly 5 cm by 2.5 cm (2 in. × 1 in.). The location of the seminal vesicles can be seen in Figures 27.1a, 27.8a, and 27.9a●.

The seminal vesicles are extremely active secretory glands, with a pseudostratified, columnar or cuboidal epithelial lining (Figure 27.8c●). These glands contribute about 60% of the volume of semen, and although the vesicular fluid usually has the same osmotic concentration as blood plasma, the composition is quite different. In particular, the secretion of the seminal vesicles contains prostaglandins, clotting proteins, and relatively high concentrations of fructose, which is easily metabolized by spermatozoa to produce ATP. Seminal fluid is discharged into the ductus deferens at *emission*, when peristaltic contractions are under way in the ductus deferens, seminal vesicles, and prostate gland. These contractions are under the

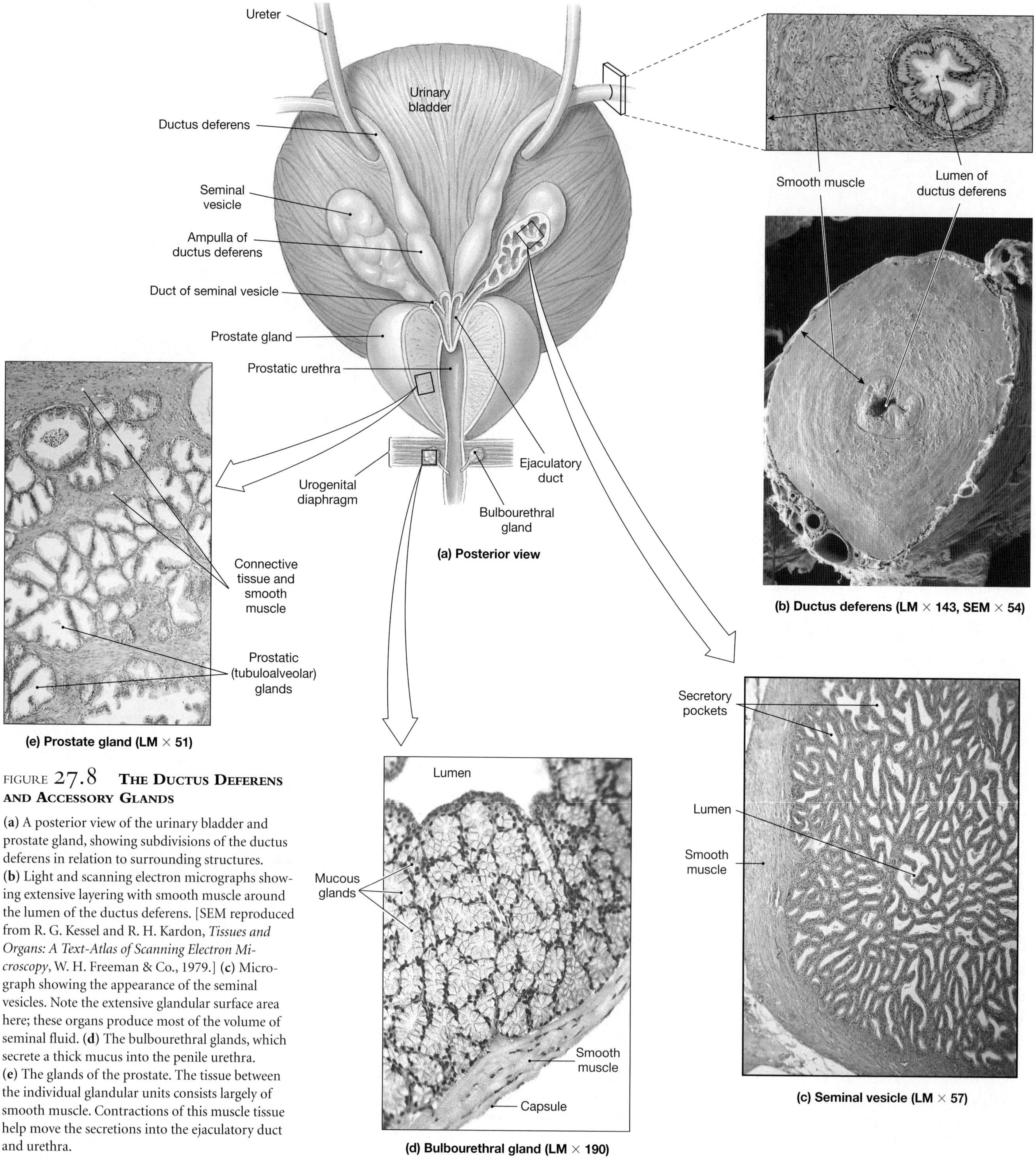

FIGURE 27.8 **THE DUCTUS DEFERENS AND ACCESSORY GLANDS**

(a) A posterior view of the urinary bladder and prostate gland, showing subdivisions of the ductus deferens in relation to surrounding structures. **(b)** Light and scanning electron micrographs showing extensive layering with smooth muscle around the lumen of the ductus deferens. [SEM reproduced from R. G. Kessel and R. H. Kardon, *Tissues and Organs: A Text-Atlas of Scanning Electron Microscopy*, W. H. Freeman & Co., 1979.] **(c)** Micrograph showing the appearance of the seminal vesicles. Note the extensive glandular surface area here; these organs produce most of the volume of seminal fluid. **(d)** The bulbourethral glands, which secrete a thick mucus into the penile urethra. **(e)** The glands of the prostate. The tissue between the individual glandular units consists largely of smooth muscle. Contractions of this muscle tissue help move the secretions into the ejaculatory duct and urethra.

control of the sympathetic nervous system. p. 453 When mixed with the secretions of the seminal vesicles, previously inactive but functional spermatozoa begin beating their flagella, becoming highly mobile.

THE PROSTATE GLAND [FIGURES 27.1/27.8a,e/27.9a]

The **prostate gland** is a small, muscular, rounded organ with a diameter of about 4 cm (1.6 in.). The prostate gland encircles the prostatic urethra as it leaves the urinary bladder (Figures 27.1, 27.8a, and 27.9a●). The glandular tissue of the prostate consists of a cluster of 30–50 *compound tubuloalveolar glands* (Figure 27.8e●). p. 60 These glands are surrounded and wrapped in a thick blanket of smooth muscle fibers. The epithelial lining typically varies from a simple to a pseudostratified columnar epithelium.

The prostatic glands produce weakly acidic secretion, **prostatic fluid**, that contributes 20–30% of the volume of semen. In addition to several other compounds of uncertain significance, prostatic secretions contain **seminalplasmin** (sem-i-nal-PLAZ-min), an antibiotic that may help prevent urinary tract infections in males. These secretions are ejected into the prostatic urethra by peristaltic contractions of the muscular wall. *Prostate Cancer p. 807*

CLINICAL BRIEF

PROSTATITIS

Prostatic inflammation, or **prostatitis** (pros-ta-TĪ-tis), can occur at any age, but it most often afflicts older men. Prostatitis usually results from bacterial infections, but the condition may also develop in the apparent absence of known pathogens. Individuals with prostatitis complain of pain in the lower back, perineum, or rectum, sometimes accompanied by painful urination and the discharge of mucous secretions containing numerous white blood cells (leukocytes) from the external urethral meatus. Antibiotic therapy is usually effective in treating cases resulting from bacterial infection, but in other cases, antibiotics may not provide relief.

Prostatic enlargement, or **benign prostatic hypertrophy**, usually occurs spontaneously in men over age 50. The increase in size occurs at the same time that hormonal changes are under way within the testes. Androgen production by the interstitial cells decreases over this period, and at the same time, these endocrine cells begin releasing small quantities of estrogens into the circulation. The combination of lower testosterone levels and the presence of estrogen probably stimulates prostatic growth. Blood pressure drugs that relax smooth muscle can relieve symptoms while vasoconstrictors, such as cold decongestants, can increase obstruction. Drugs and herbs that inhibit formation of testosterone are also effective in reducing symptoms. In severe cases, prostatic swelling can constrict and block the urethra and even the rectum. The urinary obstruction can cause permanent kidney damage if not corrected. Partial surgical removal is the most effective treatment. In the procedure known as a **TURP** (transurethral prostatectomy), an instrument pushed along the urethra restores normal function by cutting away the swollen prostatic tissue. Most of the prostate remains in place, and there are no external scars.

THE BULBOURETHRAL GLANDS [FIGURES 27.1/27.8a,d/27.9a]

The paired **bulbourethral glands**, or *Cowper's glands*, are located at the base of the penis, covered by the fascia of the uro-genital diaphragm (Figures 27.1, 27.8a, and 27.9a●). The bulbourethral glands are round, with diameters approaching 10 mm (less than 0.5 in.). The duct of each gland parallels the penile urethra for 3–4 cm (1.2–1.6 in.) before emptying into the urethral lumen. The glands and ducts are lined by a simple columnar epithelium. These are compound, tubuloalveolar mucous glands (Figure 27.8d●) that secrete a thick, sticky, alkaline mucus. This secretion helps to neutralize any urinary acids that may remain in the urethra and provides lubrication for the tip of the penis.

Semen

A typical ejaculation releases 2–5 ml of semen. This volume of fluid, called an **ejaculate**, contains:

1. *Spermatozoa*: A normal **sperm count** ranges from 20 to 100 million spermatozoa per cubic milliliter of semen.
2. *Seminal fluid*: **Seminal fluid**, the fluid component of semen, is a mixture of glandular secretions with a distinctive ionic and nutrient composition. In terms of total volume, seminal fluid contains the combined secretions of the seminal vesicles (60%), the prostate (30%), the sustentacular cells and epididymis (5%), and the bulbourethral glands (5%).
3. *Enzymes*: Several important enzymes are present in the seminal fluid, including: (1) a protease that may help to dissolve mucous secretions in the vagina, and (2) seminalplasmin, an antibiotic enzyme that kills a variety of bacteria including *Escherichia coli*.

The Penis [FIGURES 27.1/27.9]

The **penis** is a tubular organ that contains the distal portion of the urethra (Figure 27.1●). It conducts urine to the exterior and introduces semen into the female vagina during sexual intercourse. The penis (Figure 27.9a,c ●) is divided into three regions:

- The **root** of the penis is the fixed portion that attaches the penis to the rami of the ischium. This connection occurs within the urogenital triangle immediately inferior to the pubic symphysis.
- The **body (shaft)** of the penis is the tubular, movable portion. Masses of *erectile tissue* are found within the body.
- The **glans** of the penis is the expanded distal end that surrounds the *external urethral meatus*.

The skin overlying the penis is generally hairless and is generally more darkly pigmented than skin elsewhere on the body. The dermis contains a layer of smooth muscle, and the underlying loose connective tissue allows the thin skin to move without distorting underlying structures. The subcutaneous layer also contains superficial arteries, veins, and lymphatics but relatively few fat cells.

A fold of skin, the **prepuce** (PRĒ-pūs), or *foreskin*, surrounds the tip of the penis. The prepuce attaches to the relatively narrow **neck** of the penis and continues over the glans that surrounds the **external urethral meatus**. There are no hair follicles on the opposing surfaces, but **preputial** (prē-PŪ-shal) **glands** in the skin of the neck and the inner surface of the prepuce secrete a waxy material known as **smegma** (SMEG-ma). Unfortunately, smegma can be an excellent nutrient source for bacteria. Mild inflammation and infections in this region are common, especially if the area is not washed thoroughly and frequently. One way of avoiding trouble is to perform a **circumcision** (ser-kum-SIZH-un) and surgically remove the prepuce. In Western societies (especially in the United States),

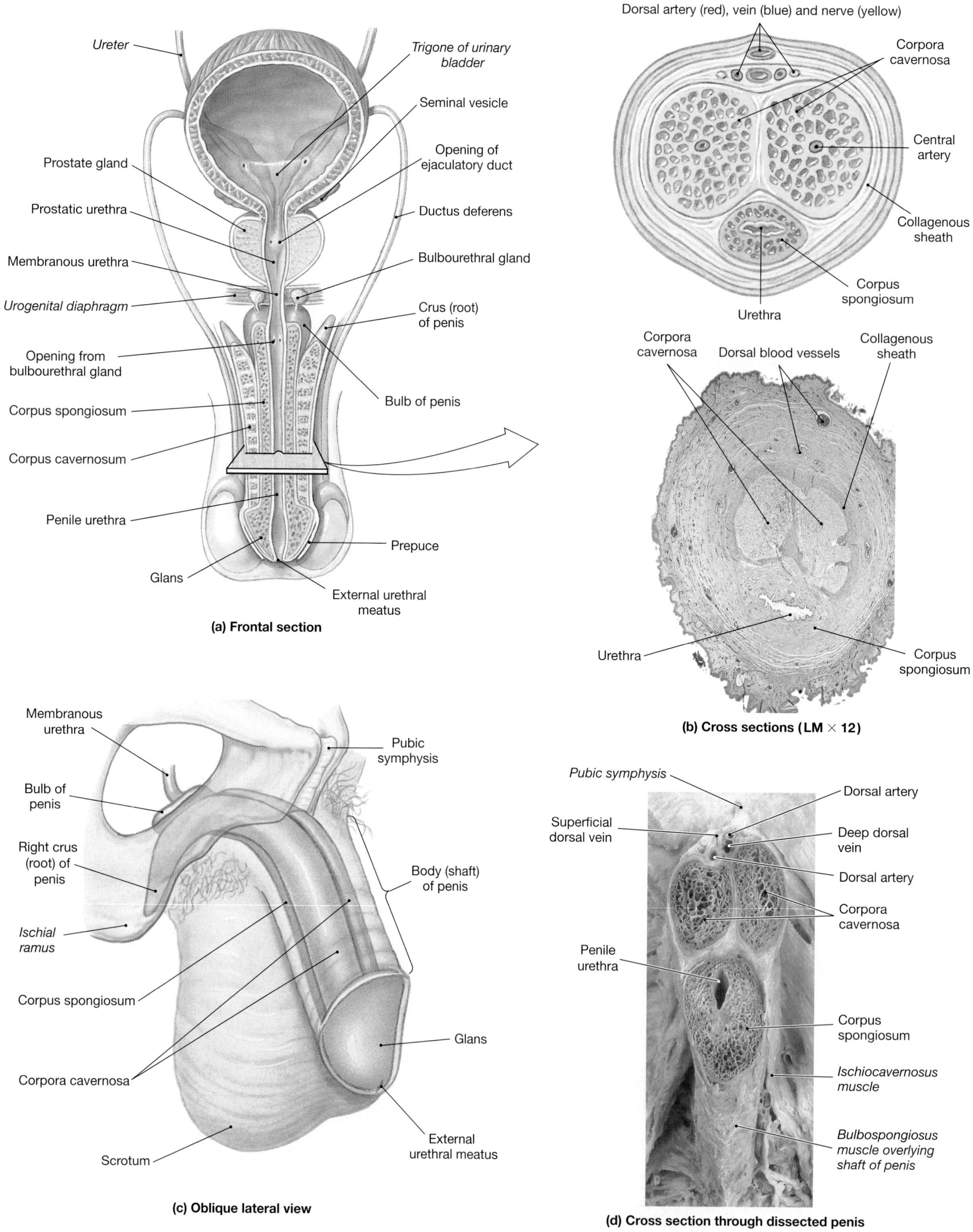

FIGURE 27.9 **The Penis**

(**a**) Frontal section showing the structures of the penis. (**b**) Cross section of the penis, showing the relation of the urethra and three masses of erectile tissue. (**c**) Lateral and oblique view of the penis, showing the orientation of the erectile tissues. (**d**) A cross section through the shaft of the penis.

this procedure is usually performed shortly after birth. Although controversial, continuation of this practice is supported by strong cultural biases and epidemiological evidence. Uncircumcised males have a higher incidence of urinary tract infections and are at greater risk for penile cancer than circumcised males.

Deep to the loose connective tissue that underlies the dermis, a dense network of elastic fibers encircles the internal structures of the penis. Most of the body of the penis consists of three parallel cylindrical columns of **erectile tissue** (Figure 27.9a–c ●). Erectile tissue consists of a three-dimensional maze of vascular channels incompletely separated by partitions of elastic connective tissue and smooth muscle fibers. In the resting state, the arterial branches are constricted and the muscular partitions are tense. This combination reduces blood flow into the erectile tissue. The smooth muscles in the arterial walls relax under parasympathetic stimulation. When the muscles relax: (1) the vessels dilate, (2) blood flow increases, (3) the vascular channels become engorged with blood, and (4) **erection** of the penis occurs. The flaccid (non-erect) penis hangs inferior to the pelvic symphysis anterior to the scrotum, but during erection the penis stiffens and assumes a more upright position.

On the anterior surface of the flaccid penis, the two cylindrical **corpora cavernosa** (KOR-pō-ra ka-ver-NŌ-sa) are separated by a thin septum and encircled by a dense collagenous sheath. At their bases the corpora cavernosa diverge, forming the **crura** (*crura*, legs; singular *crus*) of the penis. Each crus is bound to the ramus of the ischium by tough connective tissue ligaments. The corpora cavernosa extend distally along the length of the penis as far as the glans, and each corpus cavernosum contains a central artery (Figure 27.9a–c●).

The **corpus spongiosum** (spon-jē-Ō-sum) surrounds the penile urethra. This erectile body extends from the superficial fascia of the urogenital diaphragm to the tip of the penis, where it expands to form the glans. The sheath surrounding the corpus spongiosum contains more elastic fibers than does that of the corpora cavernosa, and the erectile tissue contains a pair of arteries. The **bulb** of the penis is the thickened, proximal end of the corpus spongiosum.

After erection has occurred, semen release involves a two-step process. During **emission** the sympathetic nervous system coordinates peristaltic contractions that in sequence sweep along the ductus deferens, the seminal vesicles, the prostate gland, and the bulbourethral glands. These contractions mix the fluid components of the semen within the male reproductive tract. **Ejaculation** then occurs, as powerful, rhythmic contractions begin in the *ischiocavernosus* and *bulbospongiosus muscles* of the pelvic floor. p. 283, 284 The ischiocavernosus muscles insert along the sides of the penis, and their contractions primarily serve to stiffen the organ. The bulbospongiosus wraps around the base of the penis, and its contraction pushes semen toward the external urethral orifice. These contractions are controlled by reflexes involving the inferior lumbar and superior sacral segments of the spinal cord. p. 462

✓ CONCEPT CHECK

- What are the two steps of capacitation and where do they occur?
- Trace the path of a developing sperm from the time it becomes detached from the sustentacular cells until it is released from the body.

Anatomy of the Female Reproductive System [FIGURES 27.10/27.11/27.15]

A woman's reproductive system must produce functional gametes, protect and support a developing embryo, and nourish the newborn infant. The principal structures of the female reproductive system are shown in Figure 27.10●. Female gametes leave the **ovaries**, travel along the **uterine tubes** (*Fallopian tubes* or *oviducts*), where fertilization may occur, and eventually reach the **uterus**. The uterus opens into the **vagina**; the external opening of the vagina is surrounded by the female external genitalia. As in the male, a variety of accessory glands secrete into the reproductive tract.

The ovaries, uterine tubes, and uterus are enclosed within an extensive mesentery known as the **broad ligament** (Figures 27.11 and 27.15●). The uterine tubes extend along the superior border of the broad ligament and open into the pelvic cavity lateral to the ovaries. The free edge of the broad ligament that attaches to each uterine tube is known as the **mesosalpinx** (mez-ō-SAL-pinks). A thickened fold of the broad ligament, the **mesovarium** (mez-ō-VAR-ē-um), supports and stabilizes the position of each ovary.

The broad ligament attaches to the sides and floor of the pelvic cavity, where it becomes continuous with the parietal peritoneum. The broad ligament thus subdivides the pelvic cavity. The pocket formed between the posterior wall of the uterus and the anterior surface of the colon is the **rectouterine** (rek-tō-Ū-te-rin) **pouch**, while the pocket between the anterior wall of the uterus and the posterior wall of the urinary bladder is the **vesicouterine** (ves-i-kō-Ū-ter-in) **pouch**. These subdivisions are most apparent in sagittal section (Figures 27.10 and 27.11c●).

Several other ligaments assist the broad ligament in supporting and stabilizing the position of the uterus and associated reproductive organs. These ligaments travel within the mesentery sheet of the broad ligament on the way to the ovaries or uterus. The broad ligament limits side-to-side movement and rotation, and the other ligaments (detailed in the next section) prevent superior-inferior movement.

■ The Ovaries [FIGURES 27.10/27.11/27.15]

The **ovaries** are small, paired organs located near the lateral walls of the pelvic cavity (Figures 27.10, 27.11, and 27.15●). A typical ovary is a flattened oval that measures approximately 5 cm in length, by 2.5 cm in width, and 8 mm in thickness (2 in $\times$ 1 in $\times$ 0.33 in) Each ovary weighs 6–8 g (roughly 0.25 oz). These organs are responsible for the production of ova and the secretion of hormones. The position of each ovary is stabilized by the mesovarium and by a pair of supporting ligaments: the *ovarian ligament* and the *suspensory ligament*. The **ovarian ligament** extends from the lateral wall of the uterus, near the attachment of the uterine tube, to the medial surface of the ovary. The **suspensory ligament** extends from the lateral surface of the ovary past the open end of the uterine tube to the pelvic wall. The major blood vessels, the **ovarian artery** and **ovarian vein**, travel to and from the ovary within the suspensory ligament. They extend through the mesovarium, along with nerves and lymphatic vessels, and are connected to the ovary at the **ovarian hilum** (Figure 27.11●).

Each ovary has a pink or yellowish coloration and a nodular consistency that resembles cottage cheese or lumpy oatmeal. The visceral peritoneum covering the surface of each ovary is a single layer of cuboidal

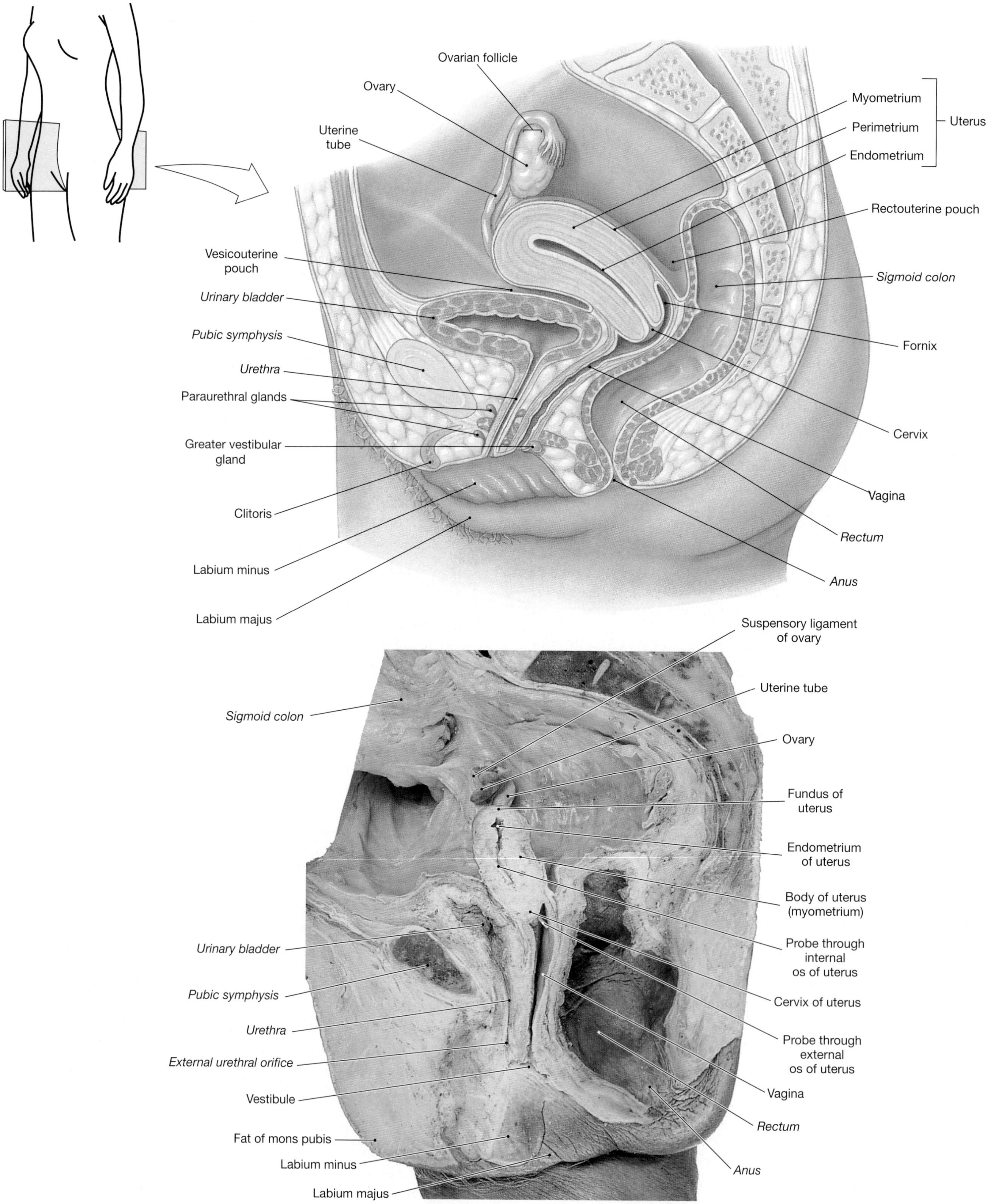

FIGURE 27.10 **The Female Reproductive System**

The female pelvis and perineum in sagittal section.

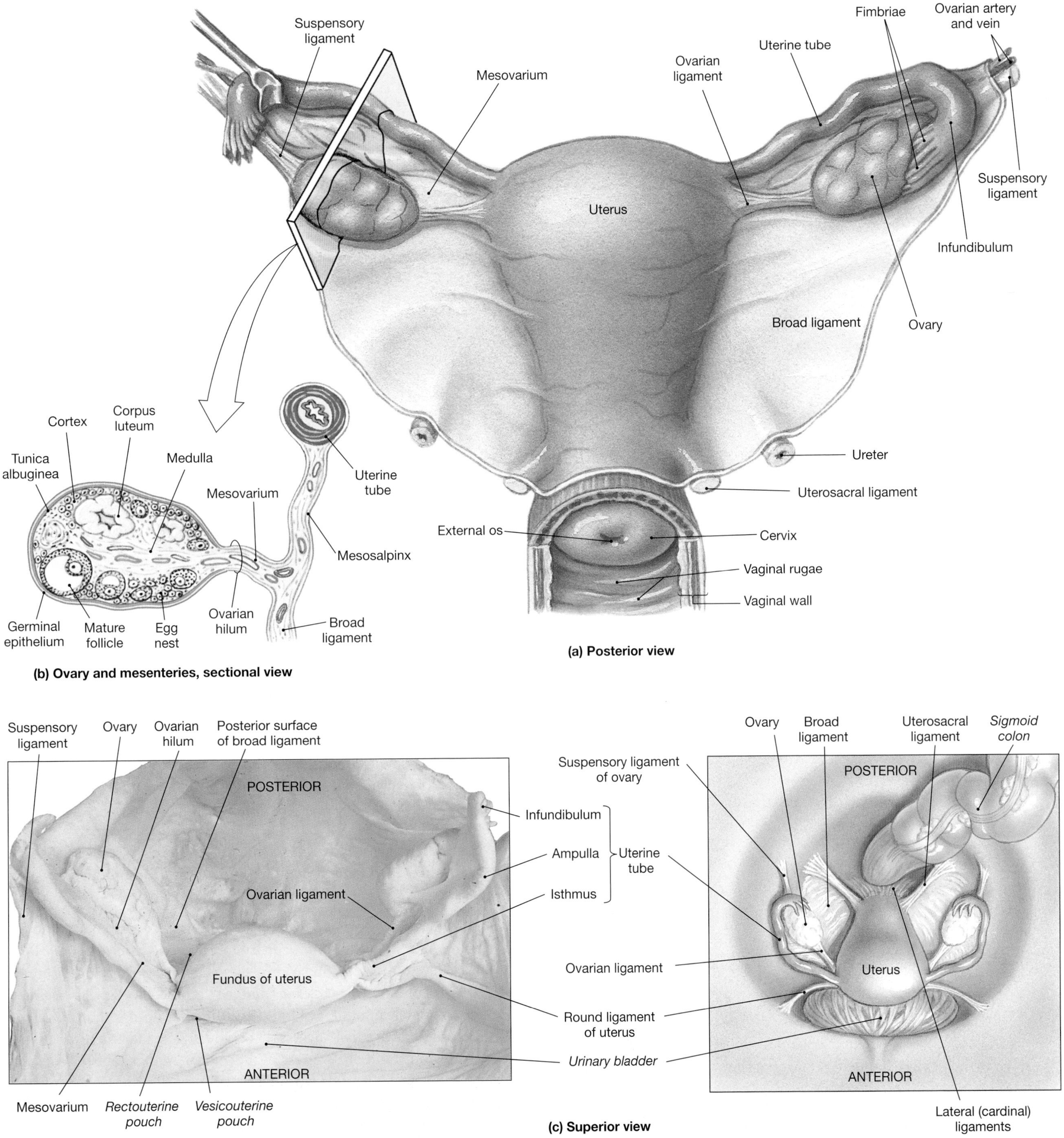

FIGURE 27.11 **THE OVARIES, UTERINE TUBES, AND UTERUS**

(**a**) Posterior view of the ovaries, uterine tubes, and uterus along with their supporting ligaments. (**b**) The ovary and associated mesenteries in sectional view. (**c**) Superior view of the pelvic cavity in a female.

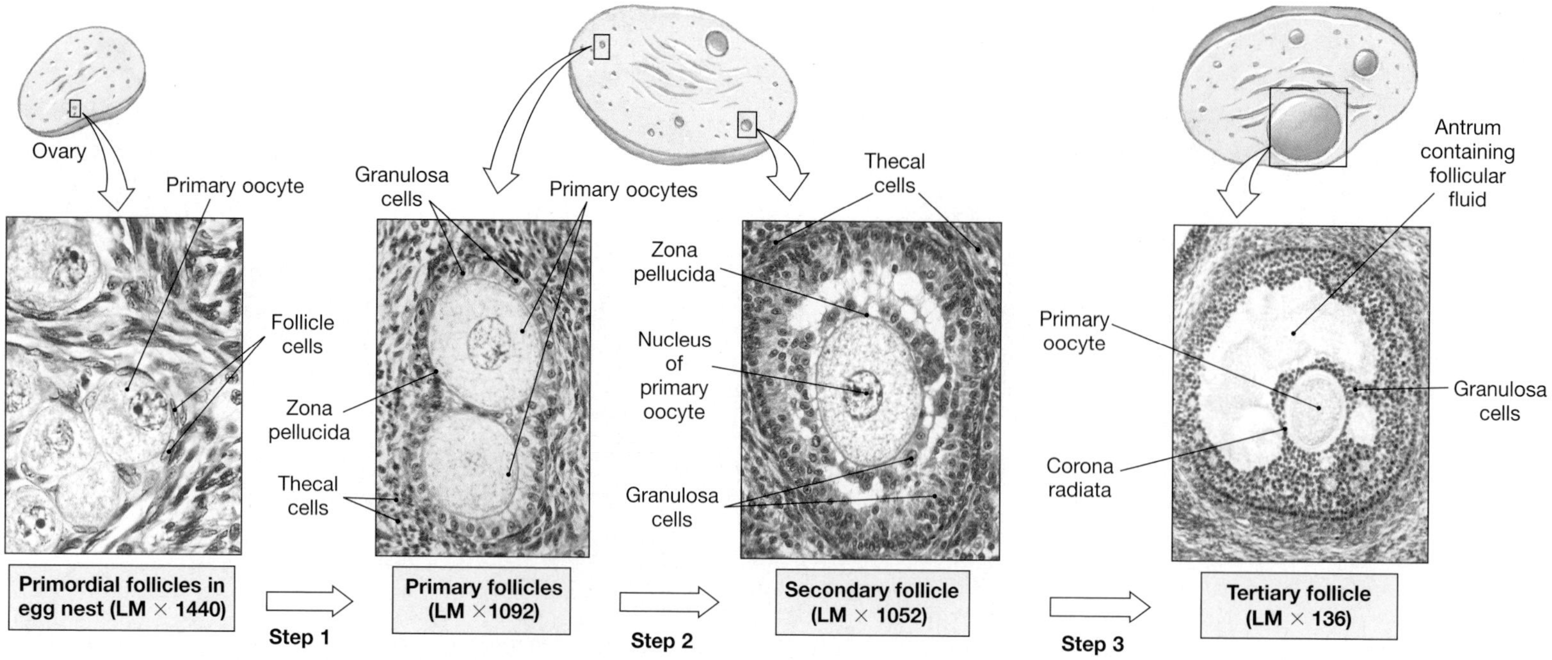

FIGURE 27.12 **THE OVARIAN CYCLE**

Follicular development during the ovarian cycle. The nuclear events that occur in ovum production (oogenesis) are illustrated in *Figure 27.13*.

epithelium called the **germinal epithelium**. It overlies a layer of dense connective tissue called the **tunica albuginea**. The interior of the ovary can be divided into a superficial *cortex* and a deep *medulla* (Figure 27.11b●). The production of gametes occurs in the ovarian cortex.

THE OVARIAN CYCLE AND OOGENESIS [FIGURES 27.12/27.13]

The production of female gametes, a process called **oogenesis** (ō-ō-JEN-e-sis), usually occurs on a monthly basis, as part of the **ovarian cycle**. Gamete development occurs in specialized structures called **ovarian follicles** (ō-VAR-ē-an FOL-i-klz). Unlike the situation in the male gonads, the stem cells, or **oogonia** (Ō-ō-GŌ-nē-a), of the female complete their mitotic divisions before birth. There are roughly 2 million **primary oocytes** (Ō-ō-sīts) in the ovaries at birth; by the time of puberty, that number has declined to around 400,000 due to degeneration. This degenerative process is called **atresia** (a-TRĒ-zē-a), and forms *atretic follicles*. The remaining primary oocytes are located in the outer portion of the ovarian cortex near the tunica albuginea, in clusters called **egg nests**. Each primary oocyte within an egg nest is surrounded by a layer of simple squamous follicular cells. Together, the primary oocyte and follicle cells constitute a **primordial** (prī-MOR-dē-al) **follicle**.

At puberty, rising levels of FSH trigger the start of the ovarian cycle, and each month thereafter some of the primordial follicles are stimulated to undergo further development. Important steps of this cycle are shown in Figure 27.12● and summarized below.

STEP 1. ***Formation of Primary Follicles.*** The ovarian cycle begins as activated primordial follicles develop into **primary follicles**. In a primary follicle, the follicular cells enlarge and undergo repeated cell divisions. As a result of these divisions, several layers of follicular cells develop around the oocyte. As the wall of the follicle thickens further, a space opens up between the developing oocyte and the innermost follicular cells. This region, which contains interdigitating microvilli from the follicle cells and the oocyte, is called the **zona pellucida** (ZŌ-na pel-OO-si-da). Follicular cells continually provide the developing oocyte with nutrients. These follicle cells are called **granulosa cells**.

The development from primordial to primary follicles and subsequent follicular maturation occurs under FSH stimulation. As the follicular cells enlarge and multiply, adjacent cells in the ovarian stroma form a layer of **thecal cells** around the follicle. Thecal cells and granulosa cells work together to release steroid hormones called *estrogens*. The hormone **estradiol** (es-tra-DĪ-ol) is the most important estrogen, and it is the dominant hormone prior to ovulation. Estrogens have several important functions, including: (1) stimulating bone and muscle growth, (2) maintaining female secondary sex characteristics, (3) affecting CNS activity, including sex-related behaviors and drives, (4) maintaining the function of the reproductive glands and organs, and (5) initiating repair and growth of the uterine lining.

STEP 2. ***Formation of Secondary Follicles.*** Although many primordial follicles develop into primary follicles, only a few primary follicles develop into secondary follicles. The transformation begins as the wall of the follicle thickens and the deeper follicular cells begin secreting small amounts of fluid. This **follicular fluid**, or *liquor folliculi*, accumulates in small pockets that gradually expand and separate cells of the inner and outer layers of the follicle. At this stage the complex is known as a **secondary follicle**. Although the oocyte continues to grow at a slow pace, the follicle as a whole now enlarges rapidly because of this accumulation of fluid.

STEP 3. ***Formation of a Tertiary Follicle.*** Eight to ten days after the start of the ovarian cycle, the ovaries usually contain only a single secondary follicle destined for further development. By the tenth to the fourteenth day of the cycle, it has developed into a **tertiary follicle**, or **mature Graafian** (GRAF-ē-an) **follicle**, roughly 15 mm in diameter.

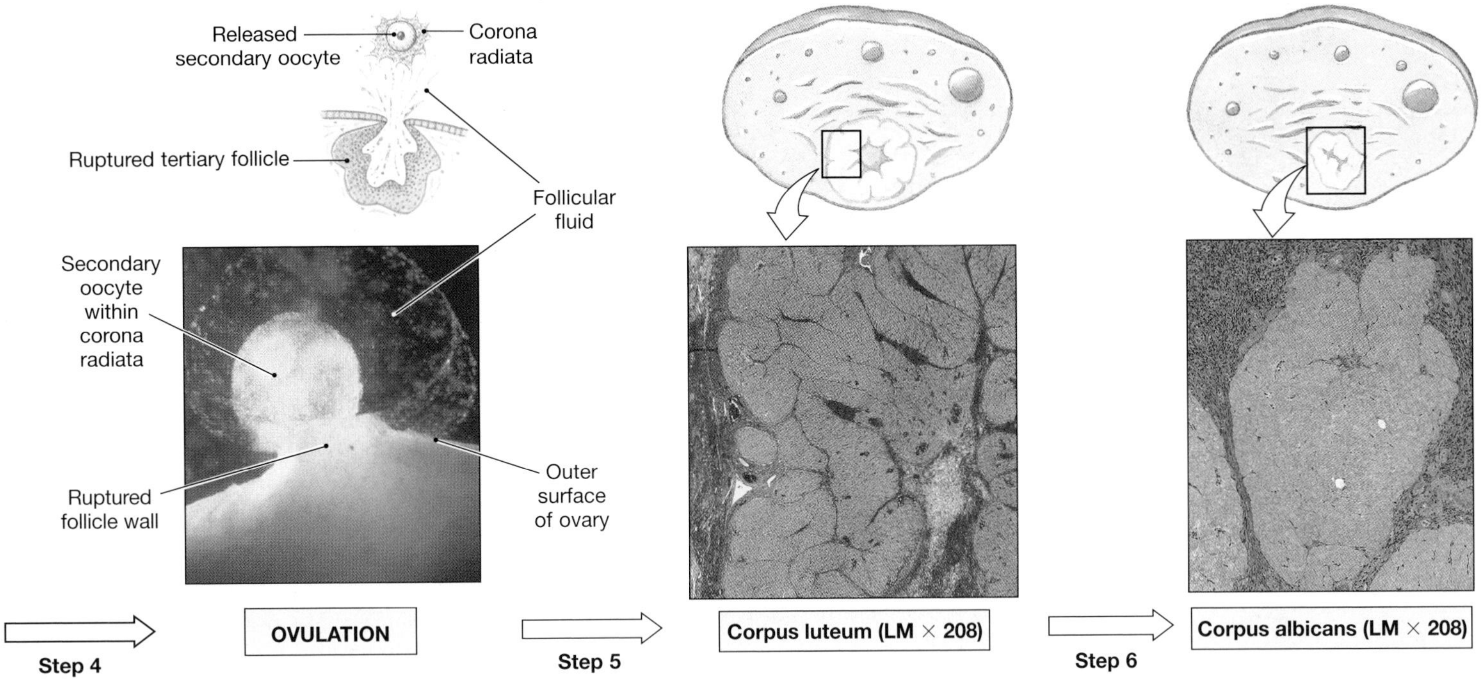

This complex spans the entire width of the cortex and stretches the ovarian wall, creating a prominent bulge in the surface of the ovary. The oocyte projects into the expanded central chamber, or **antrum** (AN-trum), surrounded by a mass of follicular cells.

Until this time, the primary oocyte has been suspended in prophase of the first meiotic division. That division is now completed. Although the nuclear events during oogenesis are identical to those in spermatogenesis, the cytoplasm of the primary oocyte is not evenly distributed (Figure 27.13●). Instead of producing two secondary oocytes, the first meiotic division yields a **secondary oocyte** and a small, nonfunctional **polar body**. The secondary oocyte then proceeds to the metaphase stage of the second meiotic division; that division will not be completed unless fertilization occurs. If fertilization does occur, the second meiotic division will be completed, producing an ovum and a nonfunctional polar body. Thus instead of producing four gametes, each of approximately equal size, oogenesis produces a single ovum containing most of the cytoplasm of the primary oocyte and polar bodies that are simply containers for the "extra" chromosomes.

STEP 4. ***Ovulation.*** As the time of gamete release, or **ovulation** (ov-ū-LĀ-shun), approaches, the secondary oocyte and the surrounding follicular cells lose their connections with the follicular wall and drift free within the antrum. This event ideally occurs on day 14 of a 28-day cycle. The follicular cells surrounding the oocyte are now known as the **corona radiata** (ko-RŌ-na rā-dē-A-ta). The distended follicular wall then ruptures, releasing the follicular contents, including the oocyte, into the peritoneal cavity. The sticky follicular fluid usually keeps the corona radiata attached to the surface of the ovary, where either direct contact with the entrance to the uterine tube or fluid currents move the oocyte into the uterine tube.

The stimulus for ovulation is a sudden rise in LH levels that weakens the follicular wall. The surge in LH release coincides with, and is triggered by, a peak in estrogen levels as the tertiary follicle matures. The **follicular phase** of the ovarian cycle is the period between the start of the cycle and the completion of ovulation. The duration of this phase commonly varies from 7–28 days.

STEP 5. ***Formation of the Corpus Luteum.*** The empty follicle initially collapses, and ruptured vessels bleed into its lumen. The remaining follicular cells then invade the area, and proliferate to form a relatively short-lived endocrine structure known as the **corpus luteum** (LŪ-tē-um), named for its yellow color (*lutea*, yellow). The formation of the corpus luteum occurs under LH stimulation.

The lipids contained in the corpus luteum are used to synthesize steroid hormones known as **progestins** (prō-JES-tinz); the principal progestin is **progesterone** (prō-JES-ter-ōn). Although moderate amounts of estrogens are also secreted by the corpus luteum, progesterone is the principal hormone of the postovulatory period. Its primary function is to continue the preparation of the uterus for pregnancy.

STEP 6. ***Formation of the Corpus Albicans.*** Unless pregnancy occurs, the corpus luteum begins to degenerate roughly 12 days after ovulation. Progesterone and estrogen levels then fall markedly. Fibroblasts invade the nonfunctional corpus luteum, producing a knot of pale scar tissue called a **corpus albicans** (AL-bi-kanz). The disintegration, or *involution*, of the corpus luteum marks the end of the ovarian cycle. The **luteal phase** of the ovarian cycle begins at ovulation and ends with the involution of the corpus luteum. The duration of this phase is usually 14 days. Because of the wide variation in the length of the follicular phase, the entire ovarian cycle may range from 21–35 days.

Another ovarian cycle begins immediately, because the decline in progesterone and estrogen levels that occur as one cycle ends stimulates **gonadotropin-releasing hormone** (**GnRH**) production at the hypothalamus. This hormone triggers a rise in FSH and LH production by the anterior lobe of the pituitary gland, and this rise stimulates another period of follicle development.

The hormonal changes involved with the ovarian cycle in turn affect the activities of other reproductive tissues and organs. In the uterus, the hormonal changes are responsible for the maintenance of the *uterine cycle*, discussed in a later section.

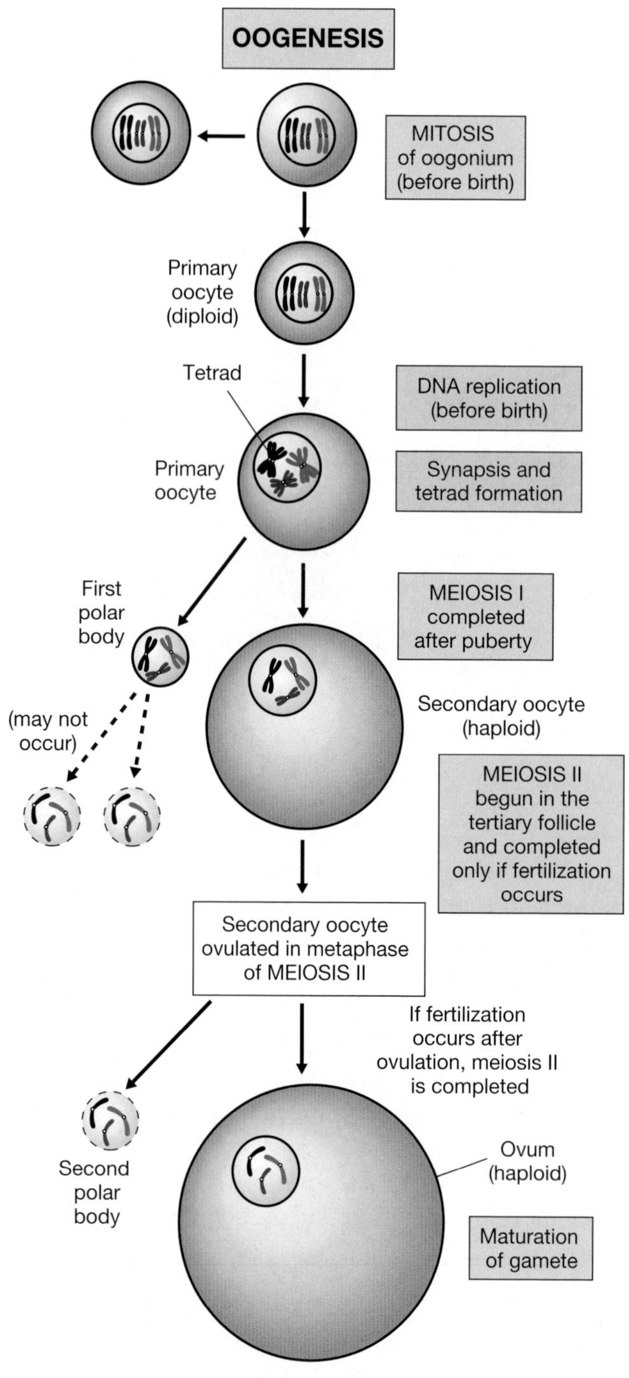

FIGURE 27.13 **MEIOSIS AND OVUM PRODUCTION**

AGE AND OOGENESIS

Although many primordial follicles may have developed into primary follicles, and several primary follicles matured into secondary follicles, usually only a single secondary oocyte is released into the pelvic cavity at ovulation. The rest undergo atresia. At puberty there are approximately 200,000 primordial follicles in each ovary. Forty years later, few if any follicles remain, although only around 500 will have been ovulated over the interim.

The Uterine Tubes [FIGURES 27.10/27.11/27.14/27.15a]

Each **uterine tube** is a hollow, muscular tube measuring roughly 13 cm (5 in.) in length (Figures 27.10, 27.11, 27.14, and 27.15a●). Each uterine tube is divided into three regions:

1. *The Infundibulum*: The end closest to the ovary forms an expanded funnel, or **infundibulum** (in-fun-DIB-ū-lum), with numerous fingerlike projections that extend into the pelvic cavity. The projections are called **fimbriae** (FIM-brē-ē). The cells lining the inner surface of the infundibulum have cilia that beat toward the middle segment of the uterine tube, the *ampulla*.
2. *The Ampulla*: The **ampulla** is the intermediate portion of the uterine tube. The thickness of the smooth muscle layers in the wall of the ampulla greatly increases as it approaches the uterus.
3. *The Isthmus*: The ampulla leads to the **isthmus** (IS-mus), a short segment adjacent to the uterine wall. The isthmus is continuous with the *intramural* (in-tra-MŪ-ral) *portion*, or *uterine part*, of the uterine tube that opens into the uterine cavity.

HISTOLOGY OF THE UTERINE TUBE [FIGURE 27.14]

The epithelium lining the uterine tube has both ciliated and nonciliated simple columnar cells (Figure 27.14c●). The mucosa is surrounded by concentric layers of smooth muscle (Figure 27.14b●). Transport of the materials along the uterine tube involves a combination of ciliary movement and peristaltic contractions in the walls of the uterine tube. A few hours before ovulation, sympathetic and parasympathetic nerves from the hypogastric plexus "turn on" this beating pattern. The uterine tubes transport a secondary oocyte for final maturation and fertilization. It normally takes 3–4 days for an oocyte to travel from the infundibulum to the uterine chamber. *If fertilization is to occur, the secondary oocyte must encounter spermatozoa during the first 12–24 hours of its passage.* Fertilization typically occurs in the ampulla of the uterine tube.

Along with its transport function, the uterine tube also provides a rich, nutritive environment containing lipids and glycogen. This mixture provides nutrients to both spermatozoa and a developing pre-embryo. Unfertilized oocytes degenerate in the terminal portions of the uterine tubes or within the uterus.

CLINICAL BRIEF

Pelvic Inflammatory Disease (PID)

Pelvic inflammatory disease (PID) is a major cause of sterility in women. This condition, an infection of the uterine tubes, affects an estimated 850,000 women each year in the United States alone. Sexually transmitted pathogens are often involved, and as many as 50–80% of all first cases may be due to infection by *Neisseria gonorrhoeae*, the organism responsible for symptoms of **gonorrhea** (gon-ō-RĒ-a), a sexually transmitted disease discussed in the Clinical Issues appendix. PID may also result from invasion of the region by bacteria normally found within the vagina. Symptoms of pelvic inflammatory disease include fever, lower abdominal pain, and elevated white blood cell counts. In severe cases the infection may spread to other visceral organs or produce a generalized peritonitis.

Sexually active women in the 15–24 age group have the highest incidence of PID. Although oral contraceptive use decreases the risk of infection, the presence of an intrauterine device (IUD) may increase the risk by 1.4–7.3 times. Treatment with antibiotics may control the condition, but chronic abdominal pain may persist. In addition, damage and scarring of the uterine tubes may cause infertility by preventing the passage of a zygote to the uterus. Another sexually transmitted bacterium, *Chlamydia*, has been identified as the probable cause of up to 50% of all cases of PID. Despite the fact that women with this infection may develop few if any symptoms, scarring of the uterine tubes may still produce infertility.

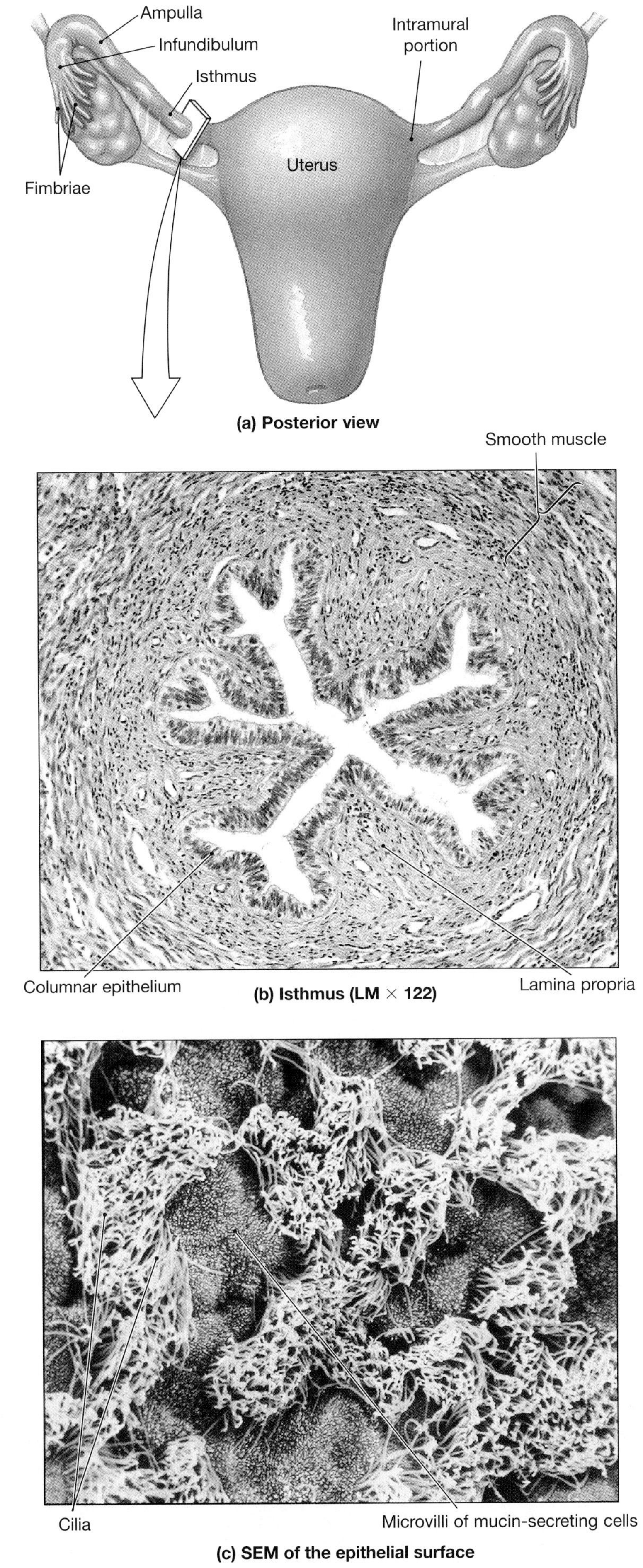

FIGURE 27.14 **The Uterine Tubes**

(**a**) Regions of the uterine tubes. (**b**) A sectional view of the isthmus. (**c**) A colorized SEM of the ciliated lining of the uterine tube. *Professor P. Motta, Dept. of Anatomy, University "La Sapienza," Rome/Science Photo Library/Photo Researchers, Inc.*

✓ CONCEPT CHECK

- What are the functions of the follicular cells?
- How could scarring of the uterine tubes cause infertility?

The Uterus [FIGURES 27.10/27.11/27.15]

The **uterus** (Ū-ter-us) provides mechanical protection, nutritional support, and waste removal for the developing embryo (weeks 1–8) and fetus (from week 9 to delivery). In addition, contractions in the muscular wall of the uterus are important in ejecting the fetus at the time of birth. The position of the uterus within the pelvic cavity and its relation to other pelvic organs can be seen in different views in Figures 27.10, 27.11, and 27.15●.

The uterus is a small, pear-shaped organ about 7.5 cm (3 in.) long with a maximum diameter of 5 cm (2 in.). It weighs 30–40 g (1–1.4 oz). In its normal position, the uterus bends anteriorly near its base, a condition known as **anteflexion** (an-tē-FLEK-shun). In this position the body of the uterus lies across the superior and posterior surfaces of the urinary bladder (Figure 27.10●). If instead the uterus bends backward toward the sacrum, the condition is termed **retroflexion** (re-trō-FLEK-shun). Retroflexion, which occurs in about 20% of adult women, has no clinical significance.

SUSPENSORY LIGAMENTS OF THE UTERUS [FIGURE 27.15a]

In addition to the mesenteric sheet of the broad ligament, three pairs of suspensory ligaments stabilize the position of the uterus and limit its range of movement (Figure 27.15a●). The **uterosacral** (ū-te-rō-SĀ-kral) **ligaments** extend from the lateral surfaces of the uterus to the anterior face of the sacrum, keeping the *body* of the uterus from moving inferiorly and anteriorly. The **round ligaments** arise on the lateral margins of the uterus just inferior to the bases of the uterine tubes. They extend anteriorly, passing through the inguinal canal before ending in the connective tissues of the external genitalia. These ligaments primarily restrict posterior movement of the uterus. The **lateral** (*cardinal*) **ligaments** extend from the base of the uterus and vagina to the lateral walls of the pelvis. These ligaments also tend to prevent the inferior movement of the uterus. Additional mechanical support is provided by the skeletal muscles and fascia of the pelvic floor. p. 283, 284

INTERNAL ANATOMY OF THE UTERUS [FIGURE 27.15]

The uterine **body**, or *corpus*, is the largest region of the uterus (Figure 27.15a●). The **fundus** is the rounded portion of the body superior to the attachment of the uterine tubes. The body ends at a constriction known as the **isthmus** of the uterus. The **cervix** (SER-viks) of the uterus is the inferior portion that extends from the isthmus to the vagina.

The tubular cervix projects about 1.25 cm (0.5 in.) into the vagina. Within the vagina, the distal end of the cervix forms a curving surface surrounding the **external os** (*os*, opening or mouth), or *external orifice* of the uterus. The external os leads into the **cervical canal**, a constricted passageway that opens into the **uterine cavity** of the body at the **internal os**, or *internal orifice* (Figure 27.15a,b●). The mucus that fills the cervical canal and covers the external orifice helps prevent the passage of bacteria from

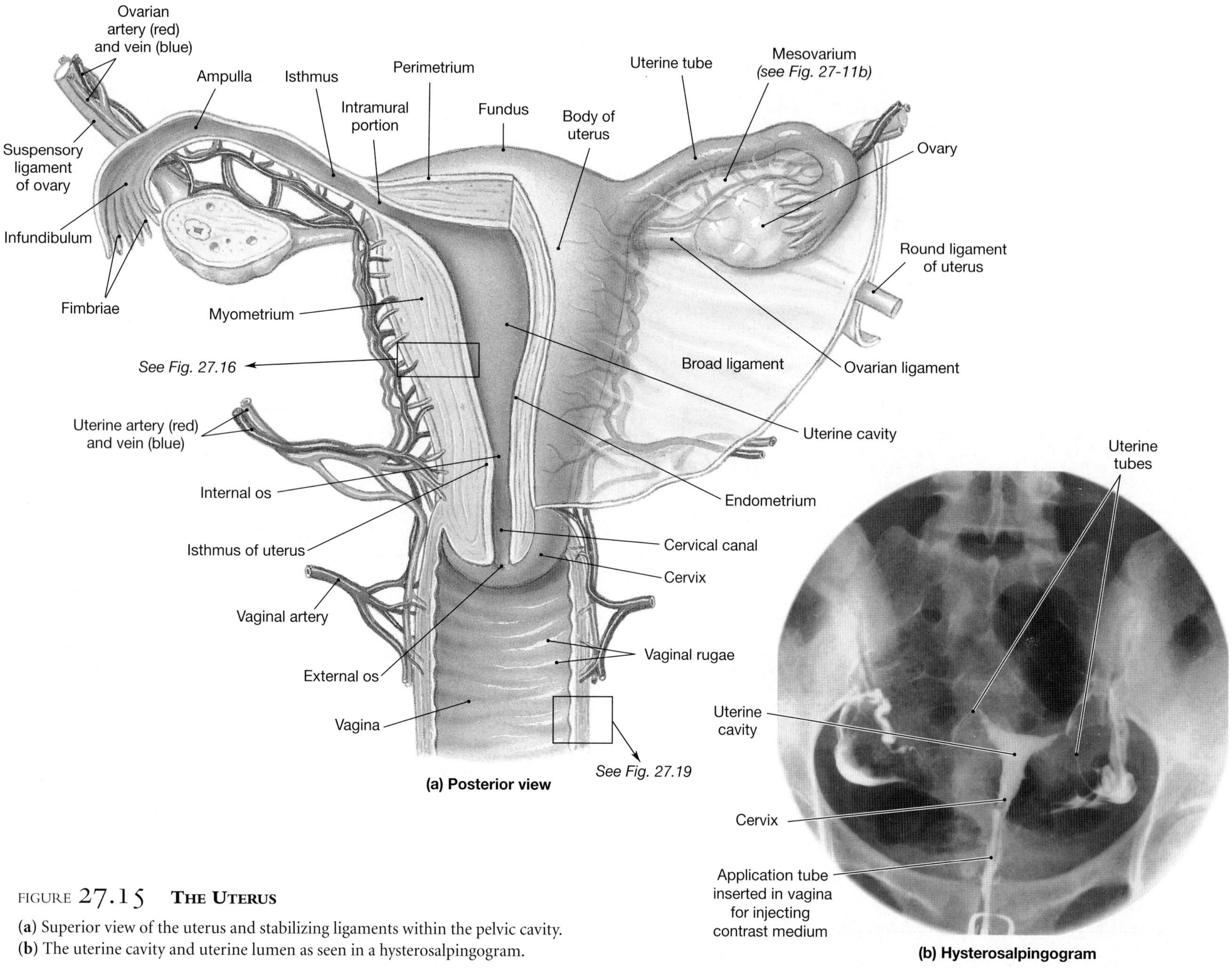

FIGURE 27.15 **THE UTERUS**

(**a**) Superior view of the uterus and stabilizing ligaments within the pelvic cavity.
(**b**) The uterine cavity and uterine lumen as seen in a hysterosalpingogram.

the vagina into the cervical canal. As ovulation approaches, the mucus changes its consistency, becoming more watery. If the mucus remains viscous, sperm have a difficult time entering the uterus, and they may be unable to ascend the uterine tubes. This reduces the odds for successful fertilization. For this reason, treatment of female infertility may include drugs that reduce the viscosity of the cervical mucus.

The uterus is supplied with branches of the uterine and ovarian arteries and veins. Numerous lymphatic vessels also supply each portion of the uterus. The uterus is innervated by autonomic fibers from the hypogastric plexus (sympathetic nervous system) and sacral segments S_3 and S_4 (parasympathetic nervous system). Sensory afferents from the uterus enter the spinal cord in the dorsal roots of spinal nerves T_{11} and T_{12}.

THE UTERINE WALL [FIGURES 27.15a, 27.16]

The dimensions of the uterus are highly variable. In adult women of reproductive age who have not borne children, the uterine wall is about 1.5 cm (0.5 in.) thick. The uterine wall has an outer, muscular **myometrium** (mī-ō-MĒ-trē-um; *myo-*, muscle + *metra* , uterus) and an inner glandular **endometrium** (en-dō-MĒ-trē-um), or *mucosa.* The fundus and the anterior and posterior surfaces of the uterine body are covered by a serous membrane continuous with the peritoneal lining. This incomplete serosal layer is called the **perimetrium** (Figures 27.15a and 27.16●).

The endometrium contributes about 10% to the mass of the uterus. The glandular and vascular tissues of the endometrium support the physiological demands of the growing fetus. Vast numbers of uterine glands open onto the endometrial surface. These glands extend deep into the lamina propria almost all the way to the myometrium. Under the influence of estrogen, the uterine glands, blood vessels, and endothelium change with the various phases of the monthly *uterine cycle.*

The myometrium is the thickest portion of the uterine wall, and it forms almost 90% of the mass of the uterus. Smooth muscle in the myometrium is arranged into longitudinal, circular, and oblique layers. The smooth muscle tissue of the myometrium provides much of the force needed to push a large fetus out of the uterus and into the vagina.

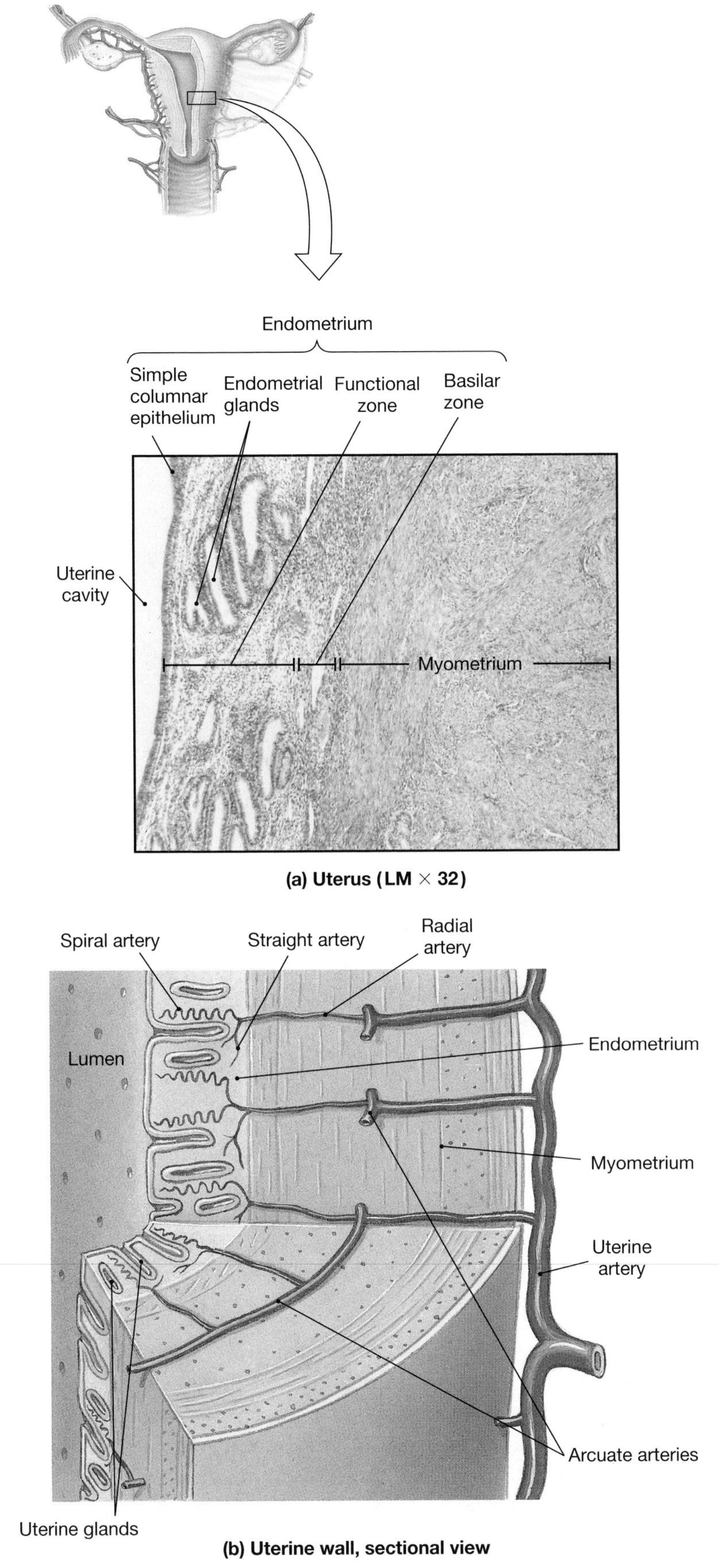

(a) Uterus (LM × 32)

(b) Uterine wall, sectional view

FIGURE 27.16 **THE UTERINE WALL**

(**a**) Basic histology of the uterine wall. (**b**) A diagrammatic sectional view of the uterine wall, showing the endometrial regions and the circulatory supply to the endometrium.

CLINICAL BRIEF

CERVICAL CANCER

Cervical cancer is the most common reproductive system cancer in women age 15–34. Roughly 13,000 new cases of invasive cervical cancer are diagnosed each year in the United States, and approximately 40% of them will eventually die of this condition. Another 50,000 patients are diagnosed with a less aggressive form of cervical cancer. (Other uterine tumors and cancers are discussed in the Clinical Issues appendix.) *Uterine Tumors and Cancers p. 807*

Most women with cervical cancer fail to develop symptoms until late in the disease. At that stage, vaginal bleeding, especially after intercourse, pelvic pain, and vaginal discharge may appear. Early detection is the key to reducing the mortality rate for cervical cancer. The standard screening test is the *Pap smear*, named for Dr. George Papanicolaou, an anatomist and cytologist. The cervical epithelium normally sheds its superficial cells, and a sample of cells scraped or brushed from the epithelial surface can be examined for abnormal or cancerous cells. The American Cancer Society recommends yearly Pap tests at ages 20 and 21, followed by smears at 1- to 3-year intervals until age 65.

The primary risk factor for this cancer is a history of multiple sexual partners. It appears likely that these cancers develop after viral infection by one of several different human papillomaviruses (HPV) that can be transmitted through sexual contact. *Sexually Transmitted Diseases p. 807*

Treatment of localized, noninvasive cervical cancer involves the removal of the affected portion of the uterus. Treatment of more advanced cancers typically involves a combination of radiation therapy, **hysterectomy** (complete or partial removal of the uterus), lymph node removal, and chemotherapy.

BLOOD SUPPLY TO THE UTERUS [FIGURE 27.15a]

The uterus receives blood from branches of the **uterine arteries** (Figure 27.15a●), which arise from branches of the internal iliac arteries, and the ovarian arteries, which arise from the abdominal aorta inferior to the renal arteries. p. 589 There are extensive interconnections among the arteries to the uterus. This arrangement helps ensure a reliable flow of blood to the organ, despite changes in position and the changes in uterine shape that accompany pregnancy.

HISTOLOGY OF THE UTERUS [FIGURE 27.16]

The endometrium can be subdivided into an inner **functional zone**, the layer closest to the uterine cavity, and an outer **basilar zone** adjacent to the myometrium. The functional zone contains most of the uterine glands and contributes most of the endometrial thickness. The basilar zone attaches the endometrium to the myometrium and contains the terminal branches of the tubular glands (Figure 27.16a●).

Within the myometrium, branches of the uterine arteries form **arcuate arteries** that encircle the endometrium. **Radial arteries** branch from the arcuate arteries and supply both **straight arteries** that deliver blood to the basilar zone of the endometrium and **spiral arteries** that supply the functional zone (Figure 27.16b●).

The structure of the basilar zone remains relatively constant over time, but that of the functional zone undergoes cyclical changes in response to sexual hormone levels. These alterations produce the characteristic histological features of the *uterine cycle.* *Endometriosis p. 808*

THE UTERINE CYCLE [FIGURES 27.17/27.18]

The **uterine cycle**, or *menstrual* (MEN-stroo-al) *cycle,* averages 28 days in length, but it can range from 21 to 35 days in normal individuals. The three phases of the uterine cycle are: (1) *menses*, (2) the *proliferative phase*, and

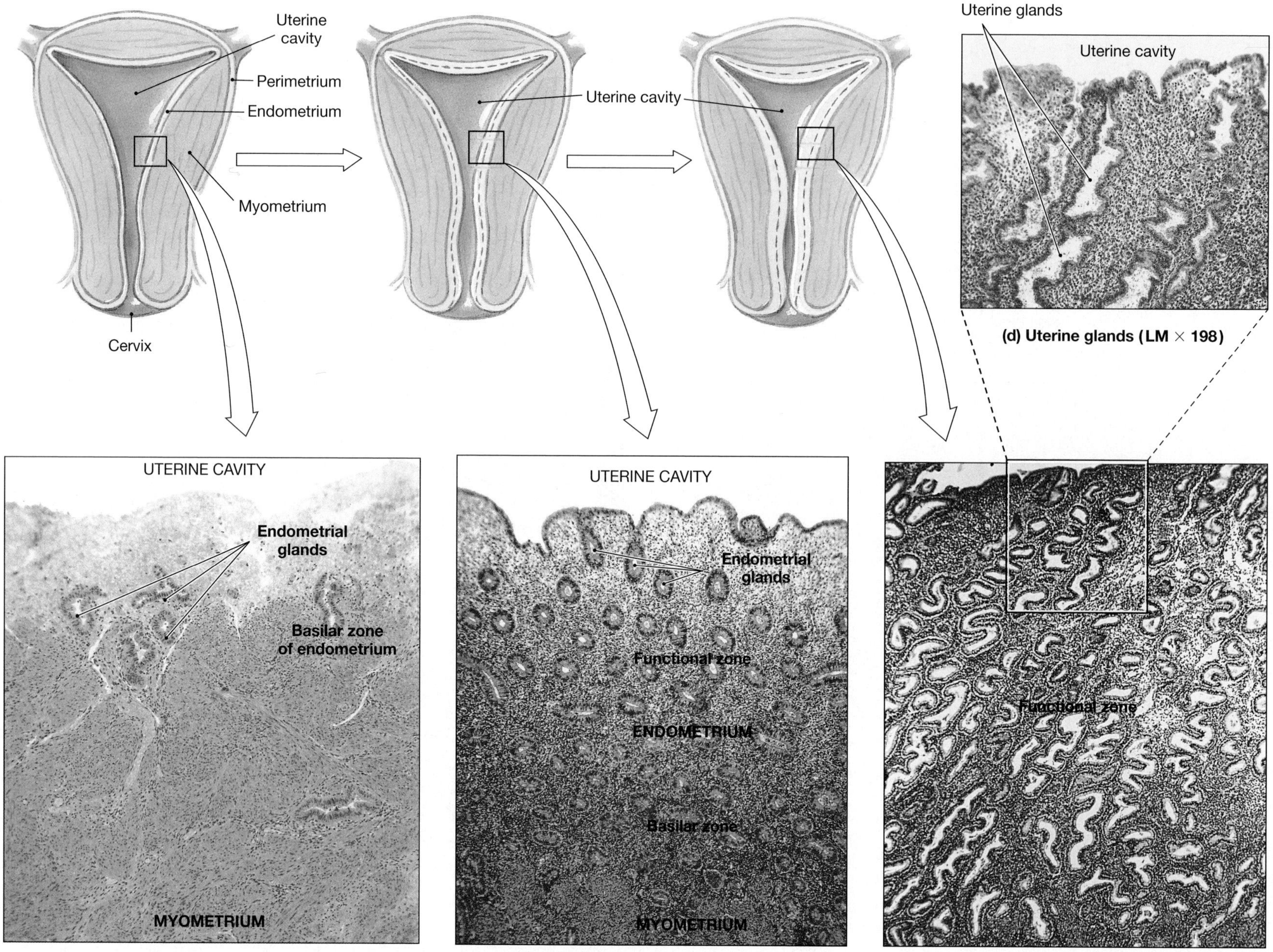

(a) Menses (LM × 83)

(b) Proliferative phase (LM × 87)

(c) Secretory phase (LM × 69)

FIGURE 27.17 **HISTOLOGICAL CHANGES IN THE UTERINE CYCLE**

(**a**) Menses. (**b**) Proliferative phase. (**c**) Secretory phase. The functional zone is now so thick that at a magnification comparable to that of (a) or (b) you cannot capture the entire width of the endometrium in one image. (**d**) Detail of uterine glands.

(3) the *secretory phase*. The histological appearance of the endometrium during each phase is shown in Figure 27.17●. The phases occur in response to hormones associated with the regulation of the ovarian cycle (Figure 27.18●).

MENSES [FIGURES 27.17a/27.18] The uterine cycle begins with the onset of **menses** (MEN-sēz), a period marked by the wholesale destruction of the functional zone of the endometrium. The arteries begin constricting, reducing blood flow to this region, and the secretory glands and tissues of the functional zone begin to die. Eventually the weakened arterial walls rupture, and blood pours into the connective tissues of the functional zone. Blood cells and degenerating tissues break away from the endometrium and enter the uterine lumen, to be lost by passage through the external os and into the vagina. This sloughing of tissue, which continues until the entire functional zone has been lost (Figure 27.17a●) is called **menstruation** (men-stroo-Ā-shun). Menstruation usually lasts from 1 to 7 days, and over this period roughly 35–50 ml, of blood are lost. Painful menstruation, or **dysmenorrhea**, may result either from uterine inflammation and contraction or from conditions involving adjacent pelvic structures.

Menses occurs when progestin and estrogen concentrations decrease at the end of the ovarian cycle. It continues until the next group of follicles has developed to the point where estrogen levels rise once again (Figure 27.18●).

THE PROLIFERATIVE PHASE [FIGURES 27.17b/27.18] The basilar zone, including the basal portions of the uterine glands, survives menses because its circulatory supply remains constant. In the days following the completion of menses, and under the influence of circulating estrogens, the epithelial cells of the glands multiply and spread across the endometrial surface, restoring the integrity of the uterine epithelium (Figure 27.17b●). Further growth and vascularization results in the complete restoration of the functional zone. As this reorganiza-

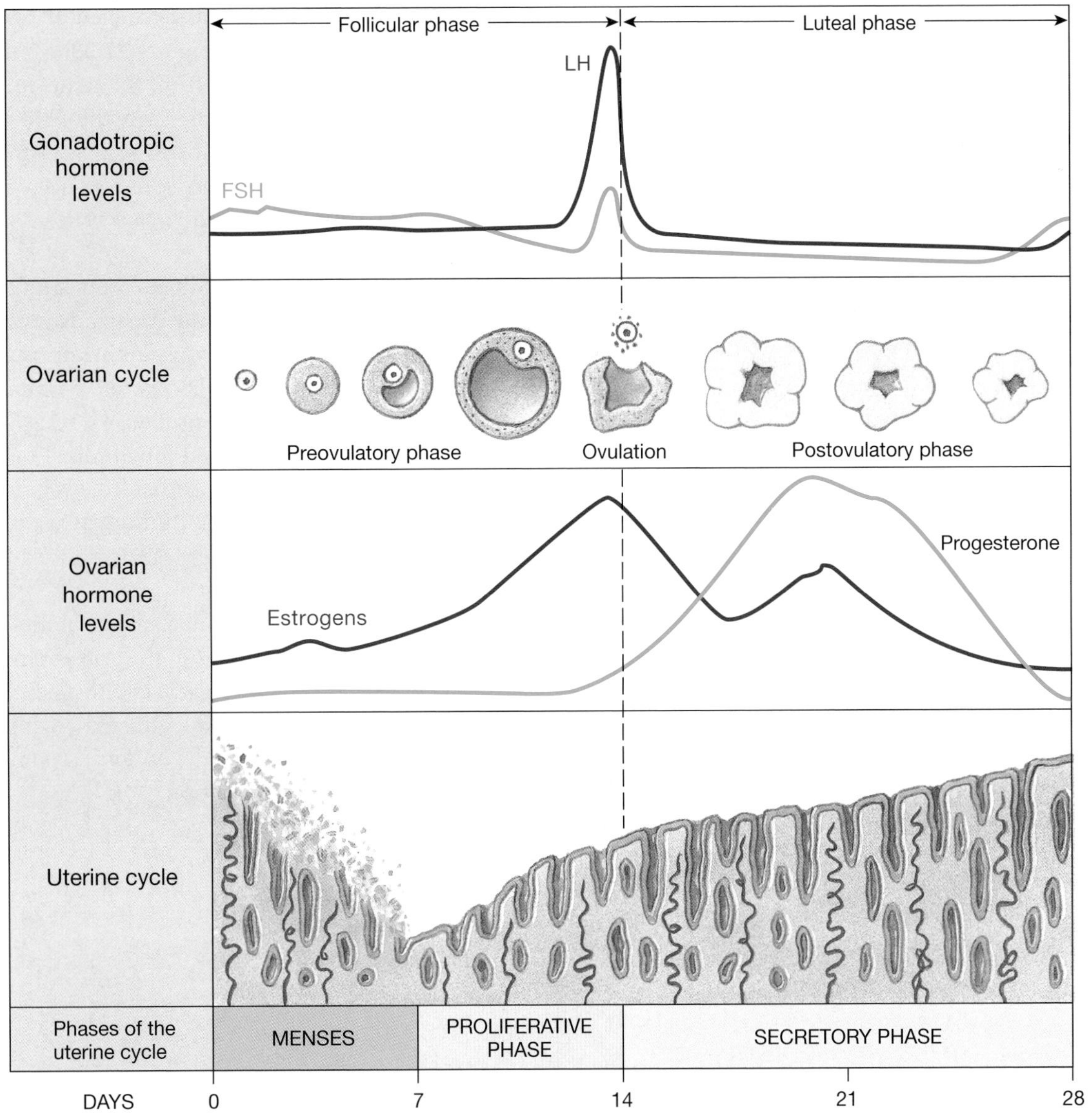

FIGURE 27.18 **The Hormonal Regulation of Female Reproductive Function**

tion proceeds, the endometrium is said to be in the **proliferative phase**. This restoration occurs at the same time as the enlargement of primary and secondary follicles in the ovary. The proliferative phase is stimulated and sustained by estrogen secreted by the developing follicles (Figure 27.18●).

By the time ovulation occurs, the functional zone is several millimeters thick, and prominent mucous glands extend to the border with the basilar zone. At this time the endometrial glands are manufacturing a mucus rich in glycogen. The entire functional zone is highly vascularized, with small arteries spiraling toward the inner surface from larger trunks in the myometrium.

The Secretory Phase [FIGURES 27.17c/27.18] During the secretory phase of the uterine cycle, the endometrial glands enlarge, accelerating their rates of secretion, and the arteries elongate and spiral through the tissues of the functional zone (Figure 27.17c●). This activity occurs under the combined stimulatory effects of progestins and estrogens from the corpus luteum (Figure 27.18●). This phase begins at the time of ovulation and it persists as long as the corpus luteum remains intact.

Secretory activities peak about 12 days after ovulation. Over the next day or two, the glandular activity declines, and the uterine cycle ends as the corpus luteum stops producing stimulatory hormones. A new cycle then begins with the onset of menses and the disintegration of the functional zone. The secretory phase usually lasts 14 days. As a result, the date of ovulation can be determined after the fact by counting backward 14 days from the first day of menses.

Menarche and Menopause The uterine cycle of events begins with the **menarche** (me-NAR-kē), or first uterine cycle at puberty, typically at age 11–12. The cycles continue until age 45–55, when **menopause** (MEN-ō-paws), the last uterine cycle, occurs. Over the intervening decades, the regular appearance of uterine cycles is interrupted only by unusual circumstances such as illness, stress, starvation, or pregnancy. Typically, irregular cycles are characteristic of the first two years after menarche and the last 2 years before menopause.

■ The Vagina [FIGURES 27.10/27.11a]

The **vagina** is an elastic, muscular tube extending from the cervix of the uterus to the *vestibule*, a space bounded by the external genitalia (Figures 27.10, p. 738, and 27.11a●, p. 739). The vagina has an average length of 7.5–9 cm (3–3.5 in.), but because the vagina is highly distensible, its length and width are quite variable.

At the proximal end of the vagina, the cervix projects into the **vaginal canal**. The shallow recess surrounding the cervical protrusion is known as the **fornix** (FOR-niks). The vagina lies parallel to the rectum, and the two are in close contact posteriorly. Anteriorly, the urethra travels along the superior wall of the vagina as it travels from the urinary bladder to its opening into the vestibule. The primary blood supply of the vagina is via the **vaginal branches** of the internal iliac (or uterine) arteries and veins. Innervation is from the hypogastric plexus, sacral nerves S_2–S_4, and branches of the pudendal nerve.

The vagina has three major functions:

1. It serves as a passageway for the elimination of menstrual fluids;
2. It receives the penis during sexual intercourse and holds spermatozoa before they pass into the uterus;
3. In childbirth, it forms the lower portion of the birth canal through which the fetus passes during delivery.

CLINICAL BRIEF

VAGINITIS

An infection of the vaginal canal, known as **vaginitis** (va-jin-ī-tis), may be caused by fungal, bacterial, or parasitic organisms. In addition to any discomfort that may result, the condition may affect the survival of sperm and thereby reduce fertility. There are several different forms of vaginitis, and minor cases are relatively common. **Candidiasis** (kan-di-DĪ-a-sis) results from a fungal (yeast) infection. The organism responsible appears to be a normal component of the vaginal microbial population in 30–80% of normal women. Symptoms include itching and burning sensations, and a lumpy white discharge may also be produced. Antifungal medications are used to treat this condition.

Bacterial (nonspecific) vaginitis results from the combined action of several bacteria in extremely high numbers. The bacteria involved are normally present in small numbers in about 30% of adult women. In this form of vaginitis, the vaginal discharge contains epithelial cells and large numbers of bacteria. Antibiotics are often effective in controlling this condition.

Trichomoniasis (trik-ō-mō-NĪ-a-sis) involves infection by a parasite, *Trichomonas vaginalis*, introduced by sexual contact with a carrier. Because it is a sexually transmitted disease, both partners must be treated to prevent reinfection.

A more serious vaginal infection by *Staphylococcus* bacteria is responsible for symptoms of **toxic shock syndrome (TSS)**. Symptoms include high fever, sore throat, vomiting and diarrhea, and a generalized rash. As the condition progresses, shock, respiratory distress, and kidney or liver failure may develop, and 10–15% of all cases prove fatal. Women have developed this condition while using tampons or vaginal sponges, and people of either gender may develop TSS after abrasion or burn injuries that promote bacterial infection.

HISTOLOGY OF THE VAGINA [FIGURES 27.15a/27.19/27.20b]

In sectional view, the lumen of the vagina appears constricted, roughly forming the shape of an H. The vaginal walls contain a network of blood vessels and layers of smooth muscle genitalia (Figures 27.19●), and the lining is moistened by the secretions of the cervical glands and by the movement of water across the permeable epithelium. The vagina and vestibule are separated by an elastic epithelial fold, the **hymen** (HĪ-men), which may partially or completely block the entrance to the vagina. The two bulbospongiosus muscles pass on either side of the vaginal orifice, and their contractions constrict the entrance. p. 283, 284 These muscles cover the *vestibular bulbs*, masses of erectile tissue that pass on either side of the vaginal entrance (Figure 27.20b●). During development, the vestibular bulbs are derived from the same embryonic tissues as the corpus spongiosum of the male (see the Embryology Summary on pp. 751–753). The vestibular bulbs and corpus spongiosum are called *homologous* (ho-MOL-ō-gus; *homo*, same + *logos*, relation) because they are similar in structure and origin; however, homologous structures can have very different functions.

The vaginal lumen is lined by a stratified squamous epithelium (Figure 27.19●) that in the relaxed state is thrown into folds, called *rugae* (Figure 27.15a●). The underlying lamina propria is thick and elastic, and it contains small blood vessels, nerves, and lymph nodes. The vaginal mucosa is surrounded by an elastic **muscularis layer**, with layers of smooth muscle fibers arranged in circular and longitudinal bundles continuous with the uterine myometrium. The portion of the vagina adjacent to the uterus has a serosal covering continuous with the pelvic peritoneum; over the rest of the vagina the muscularis layer is surrounded by an *adventitia* of fibrous connective tissue.

The vagina contains a normal population of resident bacteria, supported by the nutrients found in the cervical mucus. The metabolic activity of these bacteria creates an acid environment, which restricts the growth of many pathogenic organisms. An acid environment also inhibits sperm motility, and for this reason the buffers contained in seminal fluid are important to successful fertilization.

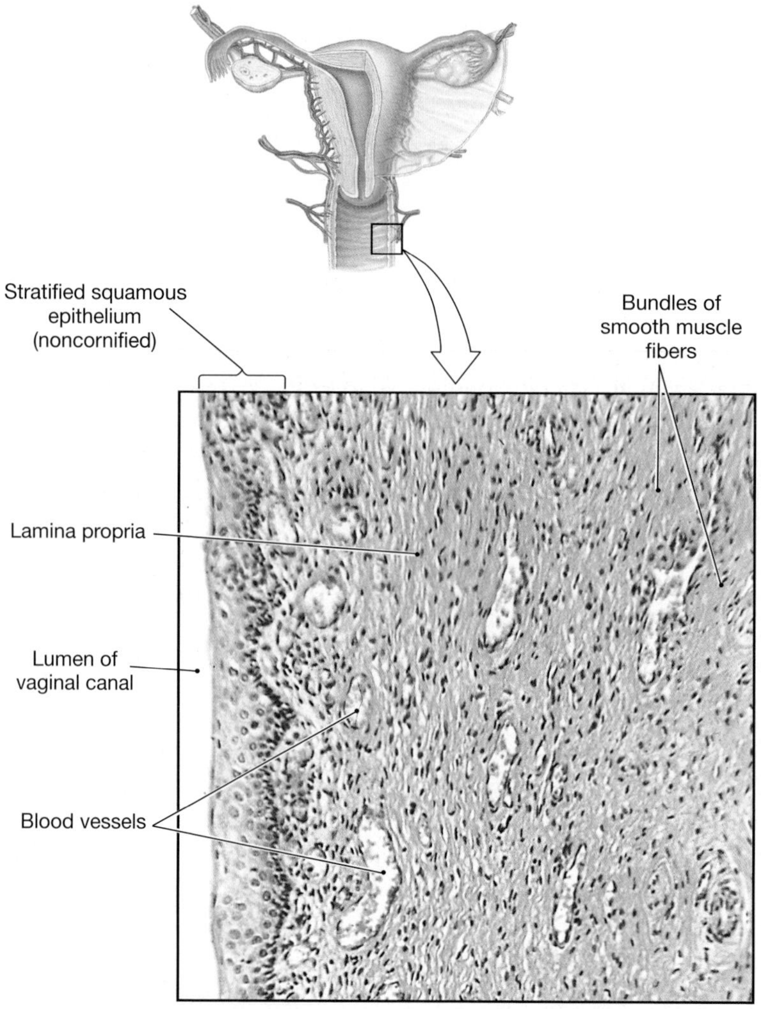

FIGURE 27.19 **HISTOLOGY OF THE VAGINAL WALL** (LM × 36)

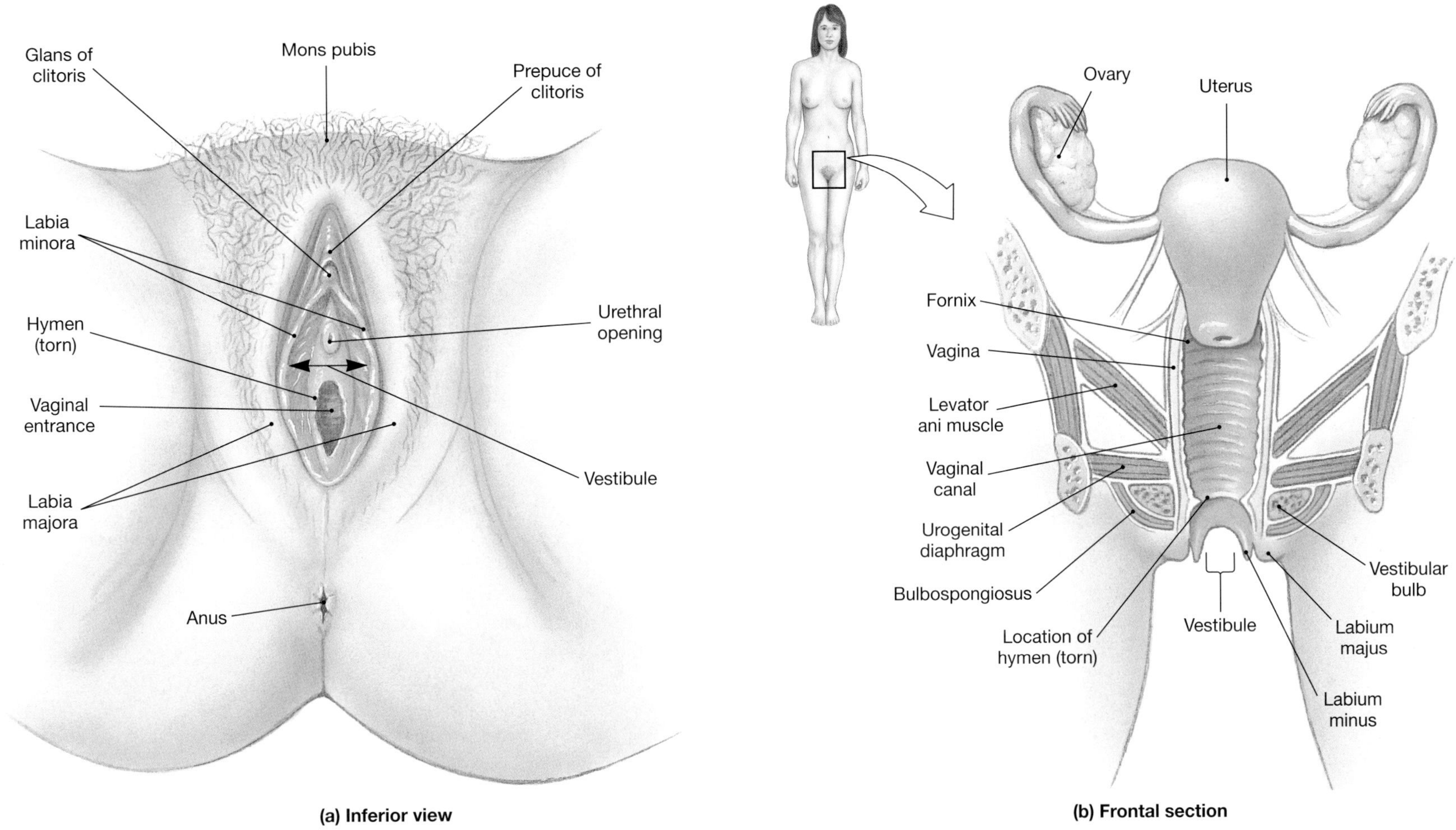

FIGURE 27.20 **THE FEMALE EXTERNAL GENITALIA**

(a) An inferior view of the female perineum. **(b)** A diagrammatic frontal section, showing the relative positions of the internal and external reproductive structures.

The External Genitalia [FIGURES 27.10a/27.20]

The region enclosing the female external genitalia is called the **vulva** (VUL-va), or *pudendum* (pū-DEN-dum) (Figure 27.20a●). The vagina opens into the **vestibule**, a central space bounded by the **labia minora** (LĀ-bē-a mi-NOR-a; singular *labium minus*). The labia minora are covered with a smooth, hairless skin. The urethra opens into the vestibule just anterior to the vaginal entrance. The **paraurethral glands**, or *Skene's glands*, discharge into the urethra near the external urethral orifice. Anterior to the urethral opening, the **clitoris** (KLI-tō-ris) projects into the vestibule. Internally the clitoris contains erectile tissue homologous with the corpora cavernosa in males (see the Embryological Summary on pp. 752–753). The clitoris becomes engorged with blood during sexual arousal. A small erectile *glans* sits atop the organ, and extensions of the labia minora encircle the body of the clitoris, forming the **prepuce**, or *hood*, of the clitoris.

A variable number of small **lesser vestibular glands** discharge their secretions onto the exposed surface of the vestibule, keeping it moistened. During arousal, a pair of ducts discharges the secretions of the **greater vestibular glands** (*Bartholin's glands*) into the vestibule near the posterolateral margins of the vaginal entrance (Figure 27.10a●). These mucous glands resemble the bulbourethral glands of the male.

The outer limits of the vulva are established by the *mons pubis* and the *labia majora*. The prominent bulge of the **mons pubis** is formed by adipose tissue beneath the skin anterior to the pubic symphysis. Adipose tissue also accumulates within the fleshy **labia majora** (singular, *labium majus*), which are homologous to the scrotum of males. The labia majora encircle and partially conceal the labia minora and vestibular structures. The outer margins of the labia majora are covered with the same coarse hair that covers the mons pubis, but the inner surfaces are relatively hairless. Sebaceous glands and scattered apocrine sweat glands secrete onto the inner surface of the labia majora, moistening them and providing lubrication.

The Mammary Glands [FIGURE 27.21]

At birth, the newborn infant cannot fend for itself, and several key systems have yet to complete their development. Over the initial period of adjustment to an independent existence, the infant gains nourishment from the milk secreted by the maternal **mammary glands**. Milk production, or **lactation** (lak-TĀ-shun), occurs in the mammary glands of the breasts, specialized accessory organs of the female reproductive system.

The mammary glands lie on either side of the chest in the subcutaneous tissue of the **pectoral fat pad** deep to the skin (Figure 27.21a,b●). Each breast bears a small conical projection, the **nipple**, where the ducts of underlying mammary glands open onto the body surface. The region of reddish-brown skin surrounding each nipple is known as the **areola** (a-RĒ-ō-la). The presence of large sebaceous glands in the underlyng dermis gives the areolar surface a granular texture.

The glandular tissue of the breast consists of a number of separate lobes, each containing several secretory lobules (Figure 27.21a●). Ducts

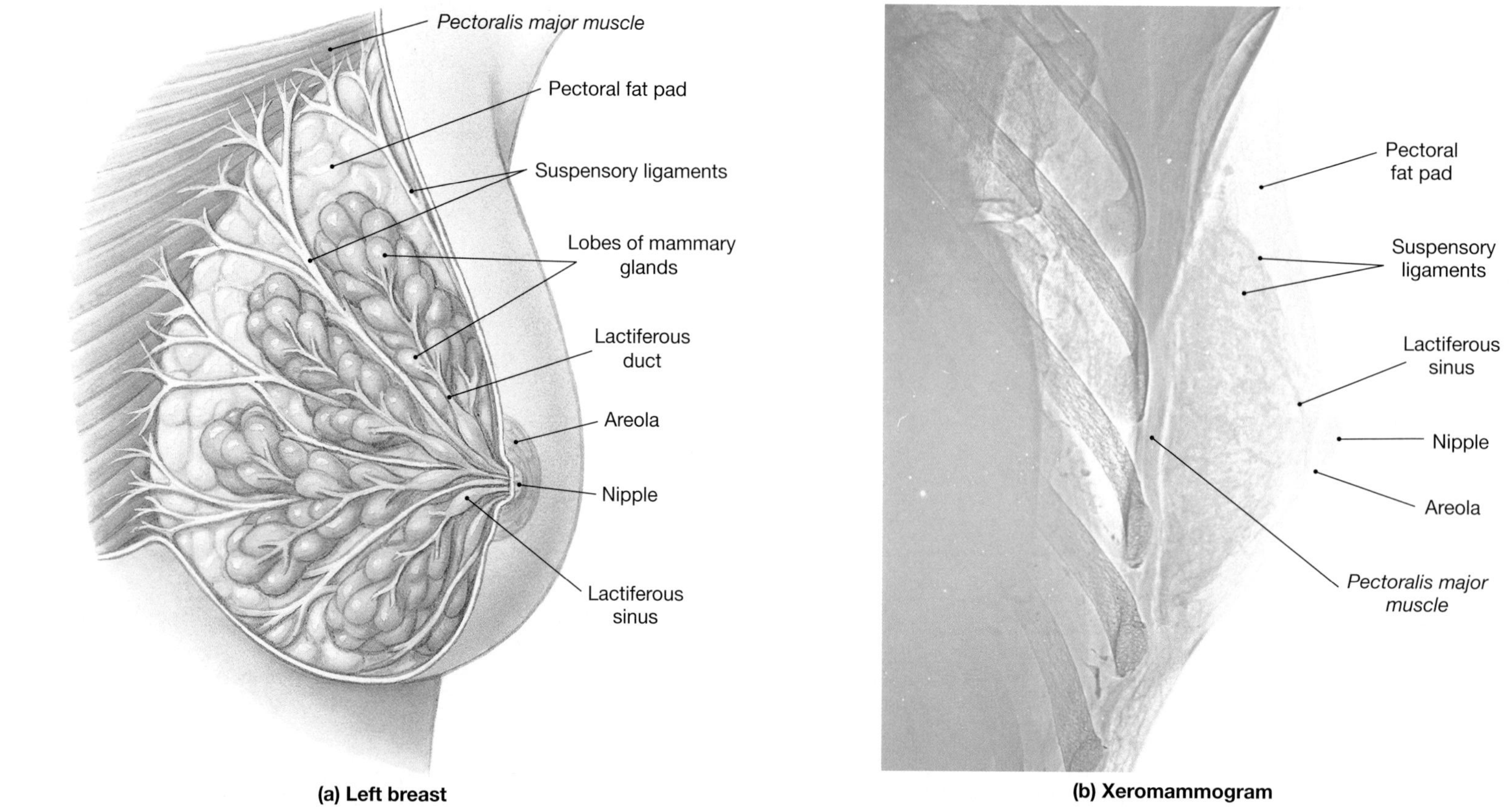

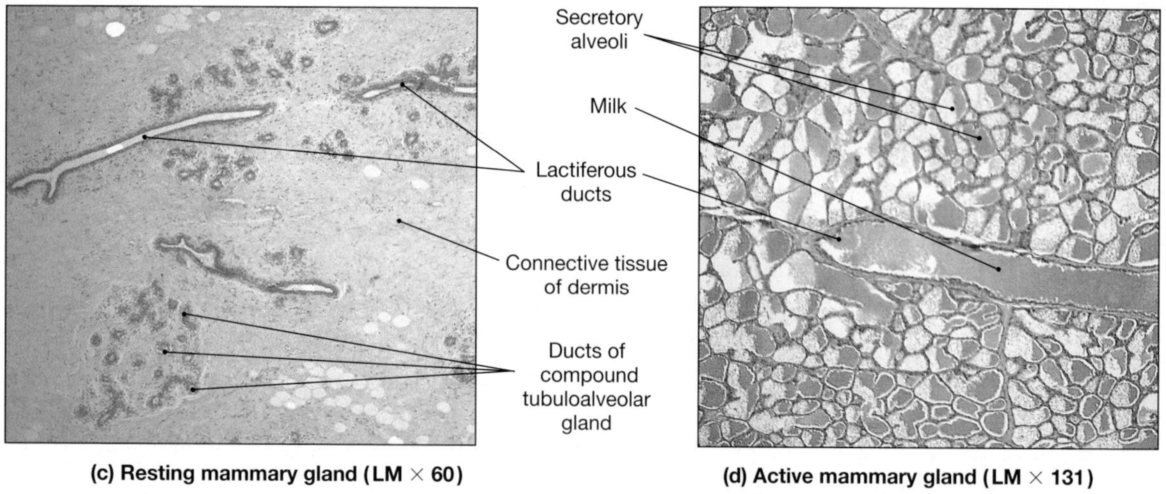

FIGURE 27.21 **THE MAMMARY GLANDS**

(a) Gross anatomy of the breast. **(b)** Xeromammogram, a radiographic technique designed to show the tissue detail of the breast, mediolateral projection. **(c,d)** Micrographs comparing the histological organization of the resting and active mammary glands.

leaving the lobules converge, giving rise to a single **lactiferous** (lak-TIF-e-rus) **duct** in each lobe (Figure 27.21a,c,d●). Near the nipple, that lactiferous duct expands, forming an expanded chamber called a **lactiferous sinus**. Usually 15–20 lactiferous sinuses open onto the surface of each nipple. Dense connective tissue surrounds the duct system and forms partitions that extend between the lobes and lobules. These bands of connective tissue, known as the **suspensory ligaments of the breast**, originate in the dermis of the overlying skin. A layer of loose connective tissue separates the mammary complex from the underlying pectoralis muscles. Branches of the *internal thoracic artery* supply blood to each mammary gland. p. 583 The lymphatic drainage of the mammary gland was detailed in Chapter 23. p. 622

DEVELOPMENT OF THE MAMMARY GLANDS DURING PREGNANCY [FIGURE 27.21c,d]

Figure 27.21c,d● compares the histological organization of the inactive and active mammary glands. The resting mammary gland is dominated by a duct system, rather than by active glandular cells. The size of the breast in a nonpregnant woman reflects primarily the amount of adipose tissue in the breast, rather than the amount of glandular tissue. The secretory apparatus does not develop until pregnancy occurs.

Further mammary gland development requires a combination of hormones, including *prolactin (PRL)* and *growth hormone (GH)* from the anterior pituitary gland. p. 513 Under stimulation by these hormones, aided by **human placental lactogen** (LAK-tō-jen) (**HPL**) from the placenta, the mammary gland ducts become mitotically active, and gland cells begin to appear. By the end of the sixth month of pregnancy, the mammary glands are fully developed, and the gland begins producing secretions that are stored in the duct system. Milk is released when the infant begins to suck on the nipple. This stimulation causes the release of oxytocin by the posterior lobe of the pituitary gland, and oxytocin triggers the contraction of smooth muscles in the walls of the lactiferous ducts and sinuses, ejecting milk.

GENDER-INDIFFERENT STAGES (WEEKS 3–6)

DEVELOPMENT OF THE GONADS

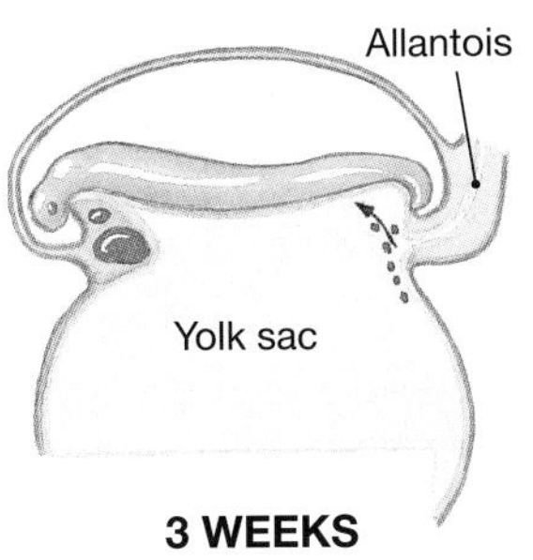

During the third week, endodermal cells migrate from the wall of the yolk sac near the allantois to the dorsal wall of the abdominal cavity. These primordial germ cells enter the **genital ridges** that parallel the mesonephros.

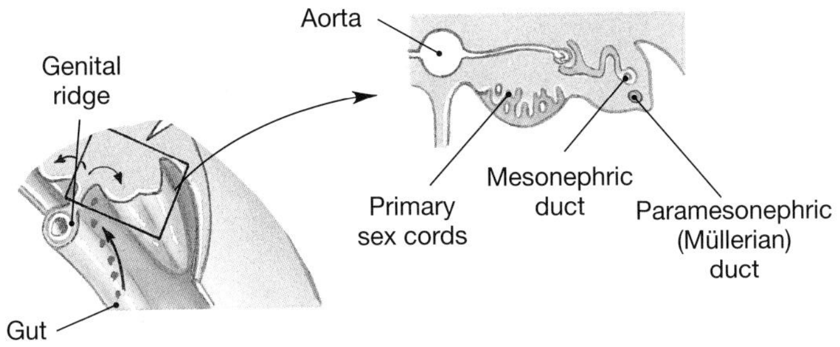

Each ridge has a thick epithelium continuous with columns of cells, the **primary sex cords**, that extend into the center (medulla) of the ridge. Anterior to each mesonephric duct, a duct forms that has no connection to the kidneys. This is the **paramesonephric** (Müllerian) **duct**; it extends along the genital ridge and continues toward the cloaca. At this gender-indifferent stage, male embryos cannot be distinguished from female embryos.

DEVELOPMENT OF DUCTS AND ACCESSORY ORGANS

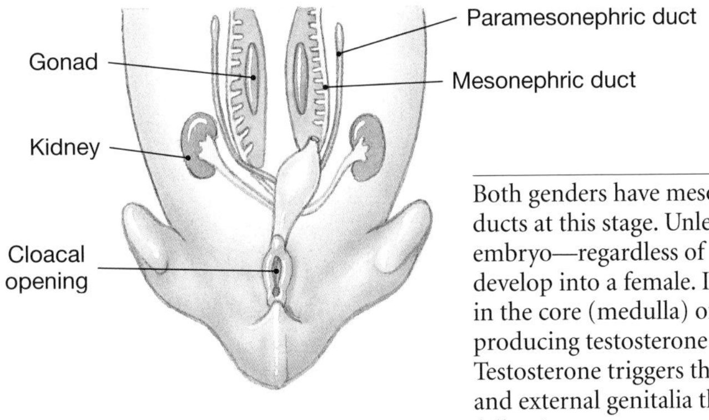

Both genders have mesonephric and paramesonephric ducts at this stage. Unless exposed to androgens, the embryo—regardless of its genetic gender—will develop into a female. In a normal male embryo, cells in the core (medulla) of the genital ridge begin producing testosterone sometime after week 6. Testosterone triggers the changes in the duct system and external genitalia that are detailed on the following page.

DEVELOPMENT OF EXTERNAL GENITALIA

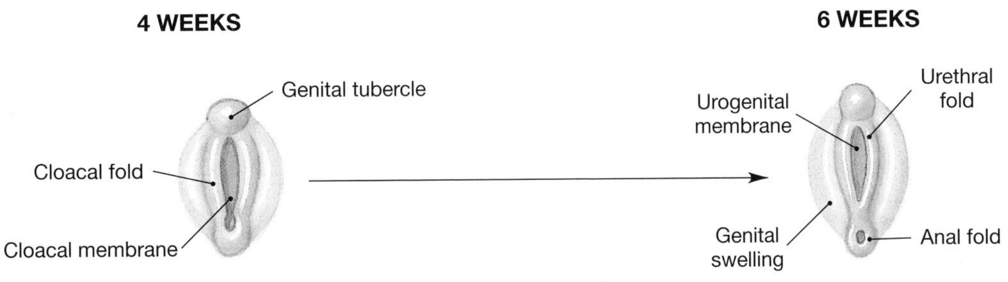

After 4 weeks of development, there are mesenchymal swellings called **cloacal folds** around the **cloacal membrane** (the cloaca does not open to the exterior). The **genital tubercle** forms the glans of the penis in males and the clitoris in females.

Two weeks later, the cloaca has been subdivided, separating the cloacal membrane into a posterior *anal membrane*, bounded by the *anal folds*, and an anterior **urogenital membrane**, bounded by the **urethral folds**. A prominent **genital swelling** forms lateral to each urethral fold.

DEVELOPMENT OF THE MALE REPRODUCTIVE SYSTEM

DEVELOPMENT OF THE TESTES

7 WEEKS | **12 WEEKS**

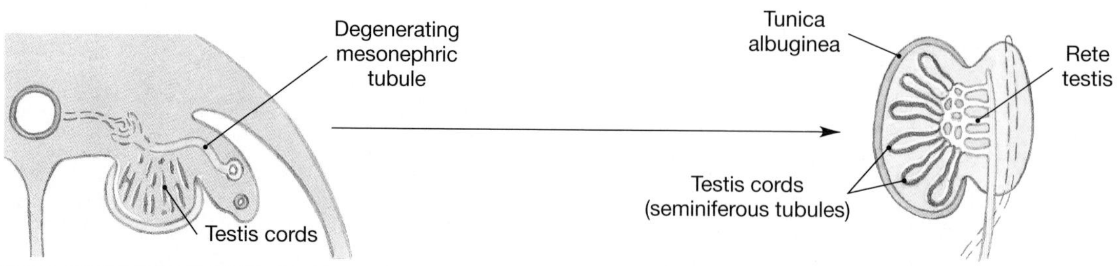

In the male, the primary sex cords proliferate and the germ cells migrate into the sex cords. The resulting **testis cords** will form the seminiferous tubules.

Connections form between the arching testis cords and the adjacent mesonephric nephrons. Although these nephrons later degenerate, the seminiferous tubules remain connected to the mesonephric duct.

DEVELOPMENT OF MALE DUCTS AND ACCESSORY ORGANS

4 MONTHS | **7 MONTHS**

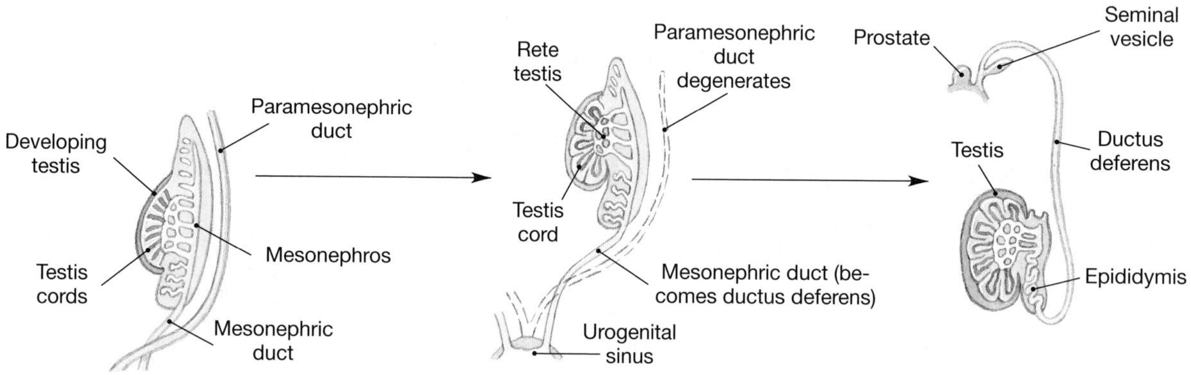

A view of the testis and ducts of the right side as seen in frontal section. Note the location and orientation of the mesonephros relative to the developing testis.

After 4 months of development, the testis cords are connected to the remnants of the mesonephric tubules by the rete testis. The paramesonephric (Müllerian) duct has degenerated.

Definitive organization after the testis has descended into the scrotum (*see Figure 27.2*). Note the relationships between the definitive sex organs and the embryonic structures.

DEVELOPMENT OF MALE EXTERNAL GENITALIA

10 WEEKS | **BIRTH**

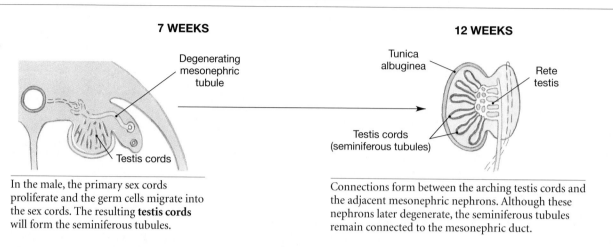

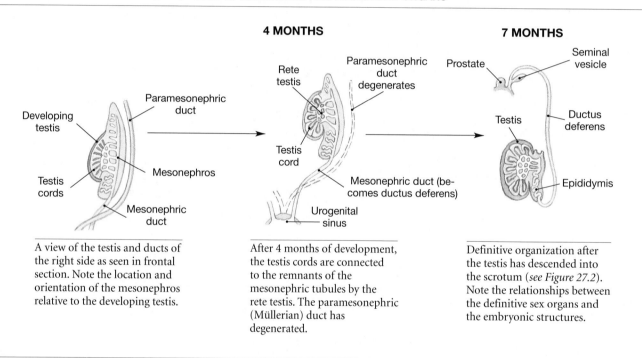

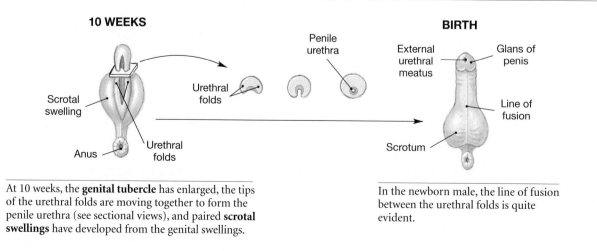

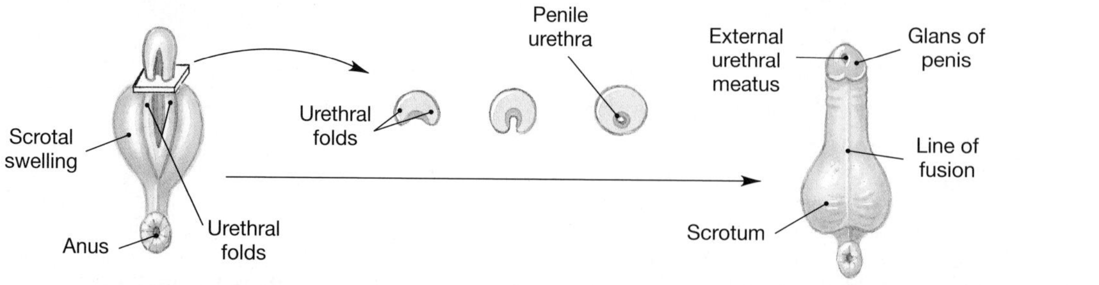

At 10 weeks, the **genital tubercle** has enlarged, the tips of the urethral folds are moving together to form the penile urethra (see sectional views), and paired **scrotal swellings** have developed from the genital swellings.

In the newborn male, the line of fusion between the urethral folds is quite evident.

DEVELOPMENT OF THE FEMALE REPRODUCTIVE SYSTEM

DEVELOPMENT OF THE OVARIES

7 WEEKS

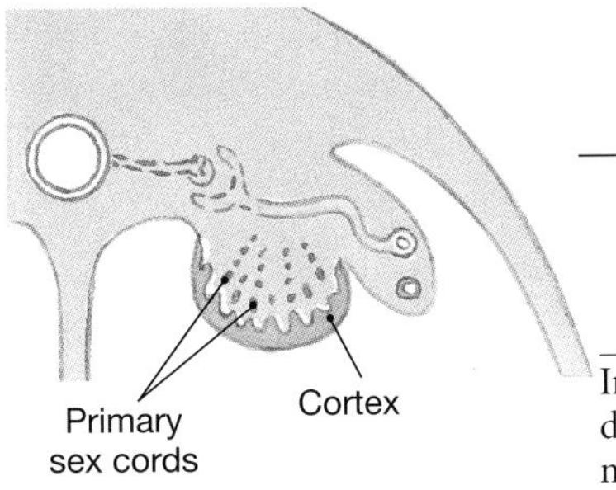

12 WEEKS

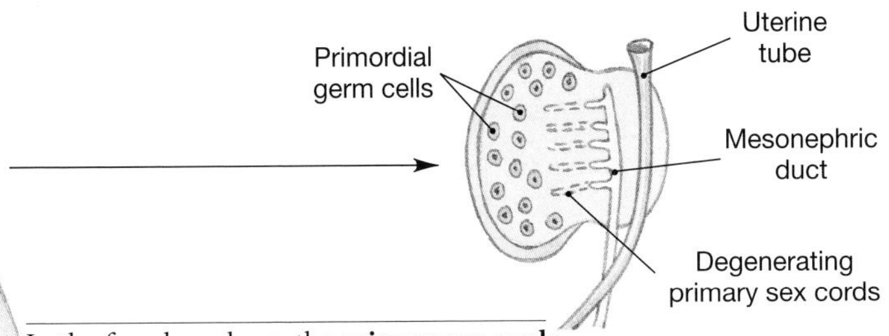

In the female embryo, the **primary sex cords** degenerate and the **primordial germ cells** migrate into the outer region (cortex) of the genital ridge.

DEVELOPMENT OF FEMALE DUCTS AND ACCESSORY ORGANS

7 WEEKS

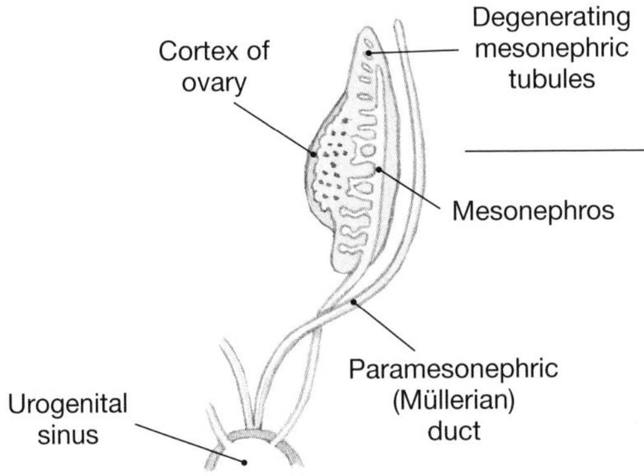

10 WEEKS / BIRTH

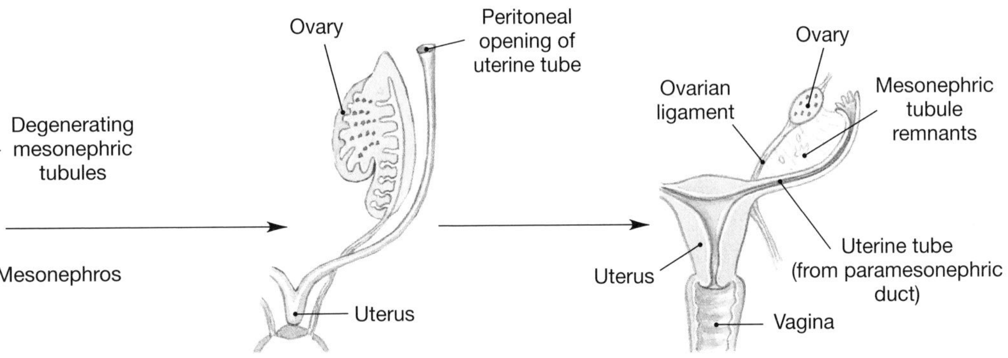

The mesonephric tubules and duct degenerate; the paramesonephric (Müllerian) duct develops a broad opening into the peritoneal cavity. Note the fusion of the ducts and the separation of the common chamber, which will form the uterus, from the urogenital sinus.

DEVELOPMENT OF FEMALE EXTERNAL GENITALIA

Comparison of Male and Female External Genitalia

Males	Females
Penis	Clitoris
Corpora cavernosa	Erectile tissue
Corpus spongiosum	Vestibular bulbs
Proximal shaft of penis	Labia minora
Penile urethra	Vestibule
Bulbourethral glands	Greater vestibular glands
Scrotum	Labia majora

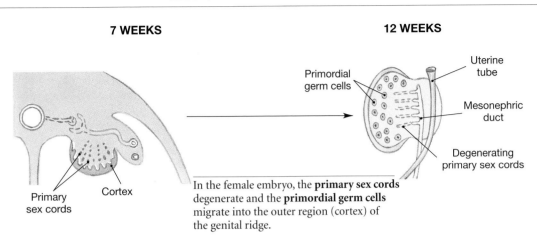

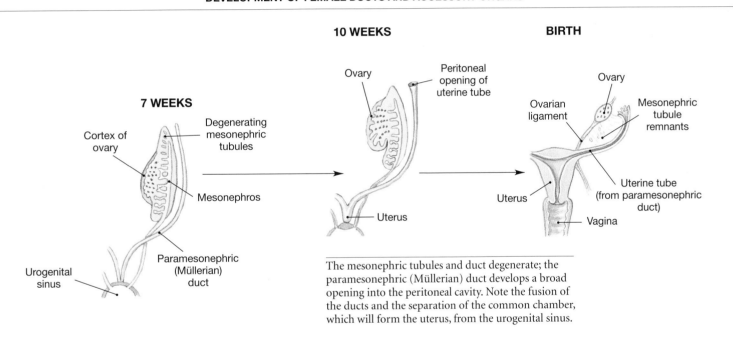

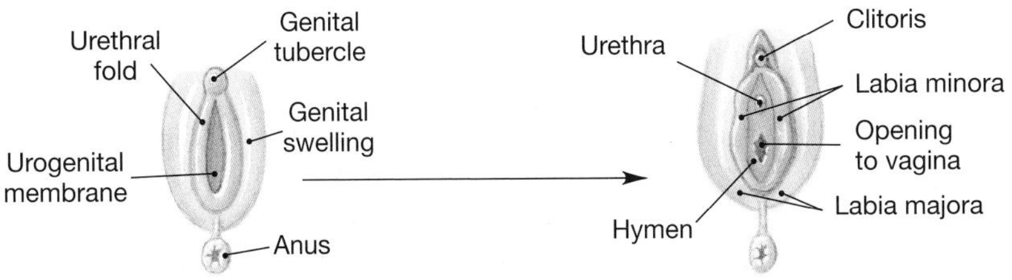

In the female, the urethral folds do not fuse; they develop into the labia minora. The genital swellings will form the labia majora. The genital tubercle develops into the clitoris. The urethra opens to the exterior immediately posterior to the clitoris. The hymen remains as an elaboration of the urogenital membrane.

✓ CONCEPT CHECK

- During the proliferative phase of the uterine cycle, what activities are occurring in the ovarian cycle?
- What marks the beginning of the secretory phase?
- Would blockage of a single lactiferous sinus interfere with delivery of milk to the nipple? Explain.
- What hormones stimulate lactation?

Pregnancy and the Female Reproductive System

If fertilization occurs, the zygote (fertilized egg) undergoes a series of cell divisions, forming a hollow ball of cells known as a **blastocyst** (BLAS-tō-sist). Upon arrival in the uterine cavity, the blastocyst initially obtains nutrients by absorbing the secretions of the uterine glands. Within a few days, it contacts the endometrial wall, erodes the epithelium, and buries itself in the endometrium. This process, known as **implantation** (im-plan-TĀ-shun), initiates the chain of events leading to the formation of a special organ, the **placenta** (pla-SEN-ta), that will support embryonic and fetal development over the next 9 months.

The placenta provides a medium for the transfer of dissolved gases, nutrients, and waste products between the fetal and maternal bloodstreams. It also acts as an endocrine organ, producing hormones. The hormone **human chorionic** (kō-rē-ON-ik) **gonadotropin (HCG)** appears in the maternal bloodstream soon after implantation has occurred. The presence of HCG in blood or urine samples provides a reliable indication of pregnancy. In function, HCG resembles LH, for in the presence of HCG, the corpus luteum does not degenerate. If it did, the pregnancy would end, for the functional zone of the endometrium would disintegrate.

In the presence of HCG, the corpus luteum persists for about 3 months before degenerating. Its departure does not trigger the return of menstrual periods, because by then the placenta is actively secreting both estrogen and progesterone. Over the following months, the placenta also synthesizes two additional hormones: **relaxin**, which increases the flexibility of the pelvis during delivery and causes dilation of the cervix during birth, and human placental lactogen, which helps prepare the mammary glands for milk production. *Conception Control p. 808*

Aging and the Reproductive System

The aging process affects the reproductive systems of men and women. The most striking age-related changes in the female reproductive system occur at menopause, while changes in the male reproductive system occur more gradually and over a longer period of time.

Menopause

Menopause is usually defined as the time that ovulation and menstruation cease. Menopause typically occurs at age 45–55, but in the years preceding it, the regularity of the ovarian and menstrual cycles gradually fades. **Premature menopause** occurs before age 40. A shortage of primordial follicles is the underlying cause of these developments. Menopause is accompanied by a sharp and sustained rise in the production of GnRH, FSH, and LH, while circulating concentrations of estrogen and progesterone decline. The decline in estrogen levels leads to reductions in the size of the uterus and breasts, accompanied by a thinning of the urethral and vaginal walls. The reduced estrogen concentrations have also been linked to the development of osteoporosis and a variety of cardiovascular and neural effects, including "hot flashes," anxiety, and depression.

The Male Climacteric

Changes in the male reproductive system occur more gradually, over a period known as the **male climacteric**. Circulating testosterone levels begin to decline between ages 50 and 60, coupled with increases in circulating levels of FSH and LH. Although sperm production continues (men can father children well into their eighties), there is a gradual reduction in sexual activity in older men, which may be linked to declining testosterone levels. Some clinicians are now tentatively suggesting the use of testosterone replacement therapy to enhance libido (sexual drive) in elderly men.

RELATED CLINICAL TERMS

cryptorchidism (kript-OR-ki-dizm): Failure of one or both testes to descend into the scrotum by the time of birth. *p. 727*

endometrial polyps: Benign epithelial tumors of the uterine lining. *Uterine Tumors and Cancers p. 807*

endometriosis (en-dō-mē-trē-Ō-sis): Growth of endometrial tissue outside of the uterus. *Endometriosis p. 808*

gonorrhea (gon-ō-RĒ-a): A sexually transmitted disease of the reproductive tract. *p. 742*

leiomyomas (lī-ō-mī-Ō-maz), or **fibroids:** Benign myometrial tumors that are the most common reproductive system tumors in women. *Uterine Tumors and Cancers p. 807*

orchiectomy (or-kē-EK-tō-mē): Surgical removal of a testis. *Testicular Torsion p. 807*

pelvic inflammatory disease (PID): An infection of the uterine tubes. *p. 742*

prostate cancer: A malignant, metastasizing cancer that is the second most common cause of cancer deaths in males. *Prostate Cancer p. 807*

prostatectomy (pros-ta-TEK-tō-mē): Surgical removal of the prostate gland. *Prostate Cancer p. 807*

prostate-specific antigen (PSA): An antigen whose concentration in the blood increases in individuals with prostate cancer or other prostate diseases. *Prostate Cancer p. 807*

sexually transmitted diseases (STDs): Diseases transferred from one individual to another primarily or exclusively through sexual contact. Examples include gonorrhea, syphilis, herpes, and AIDS. *Sexually Transmitted Diseases p. 807*

testicular torsion: Twisting of the spermatic cord and reduction in testicular blood flow resulting from rotation of the testis inside the scrotal cavity. *Testicular Torsion p. 807*

vaginitis (va-jin-ī-tis): Infection of the vaginal canal by fungal or bacterial pathogens. *p. 748*

vasectomy (vaz-EK-to-mē): Surgical removal of a segment of the ductus deferens, making it impossible for spermatozoa to reach the distal portions of the reproductive tract. *Conception Control p. 808*

Additional Clinical Terms Discussed in Appendix I (pp. 807-808)

castration; cervical cancer; cervical cap; chancre; condom (prophylactic); congenital syphilis; diaphragm; dysuria; endometrial cancer; Herpes Virus genital herpes; hysterectomy; intrauterine device (IUD); Norplant system; oral contraceptives; primary syphilis; secondary syphilis; spermicide; sterilization; syphilis; tabes dorsalis; tertiary syphilis; tubal ligation; vaginal rings; vaginal sponge; vasectomy

STUDY OUTLINE & CHAPTER REVIEW

Introduction 725

1. The human reproductive system produces, stores, nourishes, and transports functional **gametes** (reproductive cells). **Fertilization** is the fusion of a **sperm** from the male and an immature **ovum** from the female to form a **zygote** (fertilized egg).

Organization of the Reproductive System 725

1. The reproductive system includes **gonads**, ducts, accessory glands and organs, and the **external genitalia**.
2. In the male, the **testes** produce sperm, which are expelled from the body in **semen** during **ejaculation**. The **ovaries** (gonads) of a sexually mature female produce an egg that travels along **uterine tubes** to reach the **uterus**. The **vagina** connects the uterus with the exterior.

Anatomy of the Male Reproductive System 725

1. The **spermatozoa** travel along the **epididymis**, the **ductus deferens**, the **ejaculatory duct**, and the **urethra** before leaving the body. Accessory organs (notably the **seminal vesicles**, **prostate gland**, and **bulbourethral glands**) secrete into the ejaculatory ducts and urethra. The **scrotum** encloses the testes, and the **penis** is an erectile organ. *(see Figures 27.1 to 27.9)*

The Testes 725

2. The testes hang within the scrotum, and each measures about 2 in. long and 1 in. in diameter.
3. The **descent of the testes** through the inguinal canals occurs during development. Before this time, the testes are held in place by the **gubernaculum testis**. During the seventh developmental month, differential growth and contraction of the gubernaculum testis causes the testes to descend. *(see Figure 27.2)*
4. Layers of fascia, connective tissue, and muscle collectively form a sheath, the spermatic cord, which encloses the ductus deferens, testicular artery and vein, **pampiniform plexus**, and the **ilioinguinal** and **genitofemoral nerves**. *(see Figure 27.3)*
5. The testes remain connected to the abdominal cavity through the **spermatic cords**. The **perineal raphe** marks the boundary between the two chambers in the scrotum. Each testis lies in its own **scrotal cavity**. *(see Figures 27.1/27.2/27.4)*
6. Contraction of the **dartos muscle** gives the scrotum a wrinkled appearance; the **cremaster muscle** pulls the testes closer to the body. The **tunica vaginalis** is a serous membrane that covers the **tunica albuginea**, the fibrous capsule that surrounds each testis. Septa extend from the tunica albuginea to the **mediastinum**, creating a series of **lobules**. **Seminiferous tubules** within each lobule are the sites of sperm production. From there, sperm pass through a **straight tubule** to the **rete testis**. **Efferent ducts** connect the rete testis to the epididymis. Between the seminiferous tubules, there are **interstitial cells** that secrete male sex hormones, *androgens*. Seminiferous tubules contain **spermatogonia**, stem cells involved in **spermatogenesis** (the production of sperm). *(see Figures 27.1 to 27.5)*
7. The spermatogonia produce **primary spermatocytes**, **diploid** cells ready to undergo **meiosis**. Four spermatids are produced for every primary spermatocyte. The spermatids remain embedded within **sustentacular cells** while they mature into a **spermatozoon**, a process called **spermiogenesis**. The sustentacular cells function to maintain the **blood-testis barrier**, support spermatogenesis, support spermiogenesis, secrete **inhibin**, and secrete **androgen-binding protein**.

Anatomy of a Spermatozoon 731

8. Each spermatozoon has a **head**, **neck**, **middle piece**, and **tail**. The tip of the head contains the **acrosomal cap**. The tail consists of a single **flagellum**. Lacking most intracellular structures, the spermatozoon must absorb nutrients from the environment. *(see Figure 27.6)*

The Male Reproductive Tract 732

9. After detaching from the sustentacular cells, the spermatozoa are carried along fluid currents into the epididymis, an elongate tubule with **head**, **body**, and **tail** regions. The epididymis monitors and adjusts the composition of the tubular fluid, serves as a recycling center for damaged spermatozoa, stores spermatozoa, and facilitates their functional maturation (a process called **capacitation**). *(see Figures 27.1/27.4/27.7)*
10. The **ductus deferens** (*vas deferens*) begins at the epididymis and passes through the inguinal canal as one component of the spermatic cord. Near the prostate, it enlarges to form the **ampulla**. The junction of the base of the seminal vesicle and the ampulla creates the **ejaculatory duct**, which empties into the urethra. The ductus deferens functions to transport and store spermatozoa. *(see Figures 27.1/27.7/27.8)*
11. The urethra extends from the urinary bladder to the tip of the penis. It can be divided into three regions: the *prostatic urethra, membranous urethra*, and *penile urethra. (see Figures 27.1/27.9)*

The Accessory Glands 733

12. The accessory glands function to activate and provide nutrients to the spermatozoa and to produce buffers to neutralize the acidity of the urethra and vagina.
13. Each **seminal vesicle** is an active secretory gland that contributes about 60% of the volume of semen; its secretions are high in fructose, which is easily used to produce ATP by spermatozoa. The spermatozoa become highly active after mixing with the secretions of the seminal vesicles. *(see Figures 27.1/27.8/27.9)*
14. The **prostate gland** secretes a weakly acidic fluid (**prostatic fluid**) that accounts for 20–30% of the volume of semen. These secretions contain an antibiotic, **seminalplasmin**, which may help prevent urinary tract infections in men. *(see Figures 27.1/27.8/27.9)*
15. Alkaline mucus secreted by the **bulbourethral glands** (*Cowper's glands*) has lubricating properties. *(see Figures 27.1/27.8/27.9)*

Semen 735

16. A typical ejaculation releases 2–5 ml, of semen (an **ejaculate**), which contains a **sperm count** of 20 to 100 million sperm per milliliter. The **seminal fluid** is a specific mixture of secretions of the accessory glands and contains important enzymes.

The Penis 735

17. The **penis** can be divided into a **root**, **body (shaft)**, and **glans**. The skin overlying the penis resembles that of the scrotum. The **prepuce (foreskin)** surrounds the tip of the penis. **Preputial glands** on the inner surface of the prepuce secrete **smegma**. Most of the body of the penis consists of three masses of **erectile tissue**. Beneath the superficial fascia, there are two **corpora cavernosa** and a single **corpus spongiosum** that surrounds the urethra. When the smooth muscles in the arterial walls relax, the erectile tissue becomes engorged with blood, producing an **erection.** *(see Figures 27.1/27.2/27.9)*

Anatomy of the Female Reproductive System 737

1. Principal structures of the female reproductive system include the **ovaries**, **uterine tubes**, **uterus**, **vagina**, and external genitalia. *(see Figures 27.10 to 27.21)*
2. The ovaries, uterine tubes, and uterus are enclosed within the **broad ligament** (an extensive mesentery). The **mesovarium** supports and stabilizes each ovary. *(see Figures 27.10/27.11/27.15)*

The Ovaries 737

3. The ovaries are held in position by the **ovarian ligament** and the **suspensory ligament**. The **ovarian artery** and **vein** enter the ovary at the **ovarian hilum**. Each ovary is covered by a **tunica albuginea**. *(see Figures 27.10/27.11/27.15)*

4. **Oogenesis** (gamete production) occurs monthly in **ovarian follicles** as part of the **ovarian cycle**. As **primordial follicles** develop into **primary follicles**, **thecal granulosa cells** surrounding the oocyte release estrogens, most importantly **estradiol**. **Follicular fluid** encourages rapid growth during the formation of only a few **secondary follicles**. Finally, one **tertiary (mature Graffian) follicle** develops. The primary oocyte undergoes meiotic division, producing a secondary oocyte. At **ovulation**, a **secondary oocyte** surrounded by follicular cells (the **corona radiata**) is released through the ruptured ovarian wall. *(see Figure 27.12)*
5. The follicular cells remaining within the ovary form the **corpus luteum**, which produces **progestins**, mainly **progesterone**. If pregnancy does not occur, it degenerates into a **corpus albicans** of scar tissue. *(see Figure 27.12)*
6. The decline in progesterone and estrogen triggers the secretion of **GnRH**, which in turn triggers a rise in FSH and LH production, and the entire cycle begins again. *(see Figure 27.18)*

The Uterine Tubes 742

7. Each **uterine tube** has an expanded funnel, the **infundibulum**, with **fimbriae** (projections), an **ampulla**, an **isthmus**, and an **intramural portion** that opens into the **uterine cavity**. *(see Figures 27.10/27.11/27.14/27.15)*
8. The uterine tube is lined with ciliated and nonciliated simple columnar epithelial cells, which aid in the transport of materials. For fertilization to occur, the ovum must encounter spermatozoa during the first 12–24 hours of its passage from the infundibulum to the uterus.

The Uterus 743

9. The **uterus** provides mechanical protection and nutritional support to the developing embryo. Normally the uterus bends anteriorly near its base *(anteflexion)*. It is stabilized by the broad ligament, **uterosacral ligaments**, **round ligaments**, and the **lateral ligaments**. *(see Figures 27.10/27.11/27.15)*
10. The gross divisions of the uterus include the **body** (the largest portion), **fundus**, **isthmus**, **cervix**, **external os**, **uterine cavity**, **cervical canal**, and **internal os**. The uterine wall can be divided into an inner **endometrium**, a muscular **myometrium**, and a superficial **perimetrium**. *(see Figures 27.15/27.16)*
11. The uterus receives blood from the **uterine arteries**, which then branch and form extensive interconnections.
12. A typical 28-day **uterine cycle** *(menstrual cycle)* begins with the onset of **menses** and the destruction of the functional zone of the endometrium. This process of **menstruation** continues from 1 to 7 days. *(see Figures 27.17/27.18)*
13. After menses, the **proliferative phase** begins and the functional zone undergoes repair and thickens. Menstrual activity begins at **menarche** (first uterine cycle) and continues until **menopause**. *(see Figures 27.17/27.18)*

The Vagina 747

14. The **vagina** is an elastic, muscular tube extending between the uterus and external genitalia. The vagina serves as a passageway for menstrual fluids, receives the penis during sexual intercourse, and forms the lower portion of the birth canal. A thin epithelial fold, the **hymen**, partially blocks the entrance to the vagina. The vagina is lined by a stratified squamous epithelium which, when relaxed, forms *rugae* (folds). *(see Figures 27.10/27.11/27.19)*

The External Genitalia 749

15. The structures of the **vulva** *(pudendum)* include the **vestibule**, **labia minora**, **clitoris**, **prepuce** (hood), and **labia majora**. The **lesser** and **greater vestibular glands** keep the area moistened in and around the vestibule. The fatty **mons pubis** creates the outer limit of the vulva. *(see Figures 27.10/27.20)*

The Mammary Glands 749

16. The **mammary glands** lie in the subcutaneous layer beneath the skin of the chest and are the site of milk production, or **lactation**. The glandular tissue of the breast consists of secretory lobules. Ducts leaving the lobules converge into a single **lactiferous duct** and expand near the nipple, forming a **lactiferous sinus**. The ducts of underlying mammary glands open onto the body surface at the **nipple**. *(see Figure 27.21)*
17. Branches of the *internal thoracic artery* supply blood to each breast.
18. Mammary glands develop during pregnancy under the influence of *prolactin (PRL)* and *growth hormone (GH)* from the anterior pituitary, as well as **human placental lactogen (HPL)** from the placenta.

Pregnancy and the Female Reproductive System 754

1. If fertilization occurs, **implantation** of the **blastocyst** occurs in the endometrial wall. The **placenta** that develops functions as a temporary endocrine organ, producing several important hormones. **Human chorionic gonadotropin (HCG)** maintains the corpus luteum for several months. By the time the corpus luteum degenerates, the placenta is actively secreting both estrogen and progesterone. The placenta also produces **relaxin**, which is important for delivery, and human placental lactogen (HPL).

Aging and the Reproductive System 754

Menopause 754

1. **Menopause** (the time that ovulation and menstruation cease) typically occurs around age 50. **Premature menopause** occurs before age 40. Production of GnRH, FSH, and LH rises, while circulating concentrations of estrogen and progestins decline.

The Male Climacteric 754

2. The **male climacteric**, which occurs between ages 50 and 60, involves a decline in circulating testosterone levels and a rise in FSH and LH levels.

LEVEL 1 REVIEWING FACTS AND TERMS

Match each numbered item with the most closely related lettered item. Use letters for answers in the spaces provided.

Column A	Column B
____ 1. interstitial cells	a. link from scrotal chamber to peritoneal cavity
____ 2. inguinal canals	b. stem cells that produce spermatozoa
____ 3. mammary glands	c. compartment of the scrotum containing a testis
____ 4. endometrium	d. site of lactation
____ 5. dartos	e. flagellum
____ 6. scrotal cavity	f. inner layer of the uterine wall
____ 7. spermatogonia	g. enlarged portion of the uterine tube
____ 8. estradiol	h. important estrogen hormone
____ 9. ampulla	i. responsible for the production of androgens
____10. spermatozoon tail	j. layer of smooth muscle in dermis of the scrotum

11. The reproductive system includes
(a) gonads and external genitalia
(b) ducts that receive and transport the gametes
(c) accessory glands and organs that secrete fluids
(d) all of the above

12. Straight tubules originate at the seminiferous tubules and form a maze of passageways called the
(a) epididymis (b) ductus deferens
(c) rete testis (d) efferent ducts

13. Sperm production occurs in the
    (a) ductus deferens (b) seminiferous tubules
    (c) epididymis (d) seminal vesicles
14. The structure that carries sperm from the epididymis to the urethra is the
    (a) ductus deferens (b) epididymis
    (c) seminal vesicle (d) ejaculatory duct
15. Semen contains all of the following except
    (a) spermatozoons (b) seminal fluid
    (c) prostaglandins (d) spermatocytes
16. The structure that transports the ovum to the uterus is the
    (a) uterosacral ligament (b) vagina
    (c) uterine tube (d) infundibulum
17. The muscular layer of the uterus is the
    (a) endometrium (b) perimetrium
    (c) myometrium (d) eterometrium
18. During the proliferative phase of the menstrual cycle
    (a) ovulation occurs
    (b) a new functional layer is formed in the uterus
    (c) secretory glands and blood vessels develop in the endometrium
    (d) the old functional layer is sloughed off
19. The vagina is
    (a) a central space surrounded by the labia minora
    (b) the inner lining of the uterus
    (c) the inferior portion of the uterus
    (d) a muscular tube extending between the uterus and the external genitalia
20. The generally dark, pigmented skin that surrounds the nipple is called the
    (a) clitoris
    (b) fornix
    (c) areola
    (d) hymen

LEVEL 2 REVIEWING CONCEPTS

1. Contraction of the cremaster muscle
   (a) moves the testis closer to the body cavity
   (b) produces an erection
   (c) propels sperm through the urethra
   (d) moves sperm through the ductus deferens
2. In the follicular phase of the ovarian cycle, the ovary is
   (a) undergoing atresia (b) forming a corpus luteum
   (c) developing a mature follicle (d) secreting progesterone
3. Major functions of the accessory glands of the male reproductive system include all of the following except
   (a) production of spermatozoa
   (b) providing nutrients that spermatozoa need for motility
   (c) propelling spermatozoa and fluids along the reproductive tract
   (d) activating the spermatozoa
4. Which region of the sperm cell contains chromosomes?
5. What is the function of the acrosomal cap?
6. Where is seminalplasmin produced, and what is its function?
7. How does follicular fluid benefit the development of a follicle?
8. Menstruation results in the loss of which layer of the endometrium?
9. What is the hormone that can be detected soon after implantation is complete?
10. When does descent of the testes normally begin?

LEVEL 3 CRITICAL THINKING AND CLINICAL APPLICATIONS

1. Jerry is in an automobile accident that severs his spinal cord at the L_3 level. After his recovery, he wonders if he will still be able to have an erection. What would you tell him?
2. In a condition known as endometriosis, endometrial cells proliferate within a uterine tube or within the peritoneal cavity. A major symptom of endometriosis is periodic pain. Why do you think this occurs?
3. Contraceptive pills contain estradiol alone or estradiol and progesterone that are given at programmed doses during the ovarian cycle to prevent follicle maturation and ovulation. How would this happen?

✓ ANSWERS TO CONCEPT CHECK QUESTIONS

p. 732 **1.** The ductus deferens, testicular blood vessels, nerves, and lymphatics make up the body of the spermatic cord. **2.** The inguinal canals, which are narrow canals linking the scrotal chambers with the peritoneal cavity, usually close, but the presence of the spermatic cords leaves weak points in the abdominal wall. **3.** The temperature inside the peritoneal cavity is too high for the production of sperm cells, so the testes are located within the scrotal cavity, outside the peritoneal cavity, where temperatures are cooler. **4.** The blood-testis barrier isolates the inner portions of the seminiferous tubule from the surrounding interstitial fluid. Transport across the sustentacular cells is tightly regulated to maintain a very stable environment inside the tubule.
p. 737 **1.** Capacitation involves sperm becoming active, motile, and fully functional. It occurs when sperm are mixed with secretions of the seminal vesicles, and upon exposure to conditions within the female reproductive tract. **2.** After becoming detached from the sustentacular cells, the sperm lies within the lumen of the seminiferous tubule. It will then travel in fluid currents along the straight tubule, through the rete testis, and into the epididymis, where it will remain for about 2 weeks, completing its functional maturation. Upon leaving the epididymis, the sperm will enter the ductus deferens, where it can be stored for several months. The sperm then enters the ejaculatory duct, which enters the urethra for passage out of the body.
p. 743 **1.** The follicular cells provide the developing oocyte with nutrients and release estrogens. They stimulate the growth of the follicle by secreting follicular fluid. Once ovulation occurs, the remaining follicular cells in the empty follicle create the corpus luteum. **2.** Scarring in the uterine tubes can cause infertility by preventing the passage of a zygote to the uterus. This can be caused by pelvic inflammatory disease, which is an infection of the uterine tubes.
p. 754 **1.** Enlargement of the primary and secondary follicles in the ovary happens at the same time as the proliferative phase of the uterine cycle. **2.** Ovulation occurs at the beginning of the secretory phase. **3.** Blockage of a single lactiferous sinus would not interfere with milk moving to the nipple because each breast usually has between 15 and 20 lactiferous sinuses. **4.** Prolactin, growth hormone, and human placental lactogen stimulate the mammary glands to become mitotically active, producing the gland cells, which are necessary for lactation.

REVIEW IT

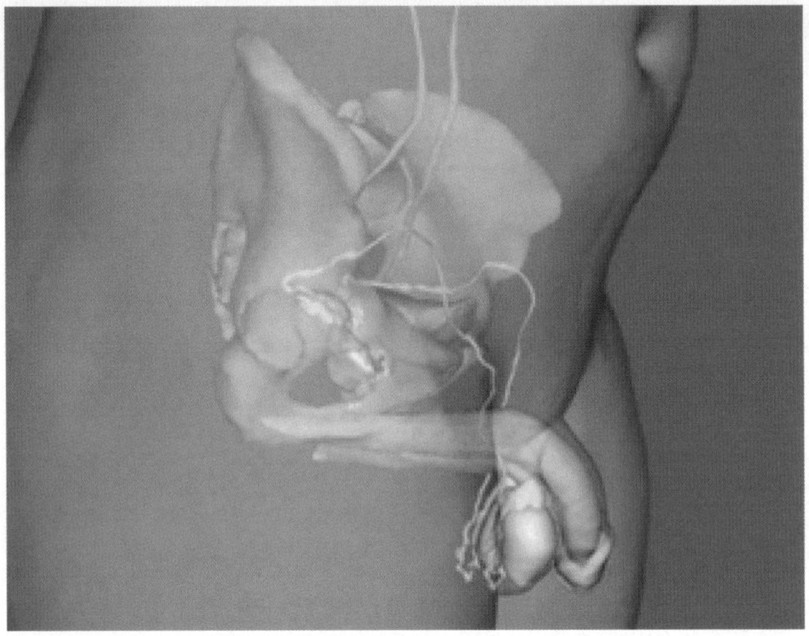

Challenge 1

The reproductive system is challenging because males and females have different structures. While the female reproductive system is more complex in terms of hormonal control, the male reproductive system is more complex anatomically. Review the structures of the male reproductive system by labeling all structures visible on this image. Compare the images in the text to this view to gain an appreciation of the spatial relationships between these structures.

If you have access to a computer, use the Interactive CD to help with this task. The image you see is taken from the Interactive *CD/Reproductive system/Male*. Why does the ductus deferens appear convoluted in this image, but appears straight in the figures in the text? Are there other anomalies such as this that you can identify? Do they have the same origin?

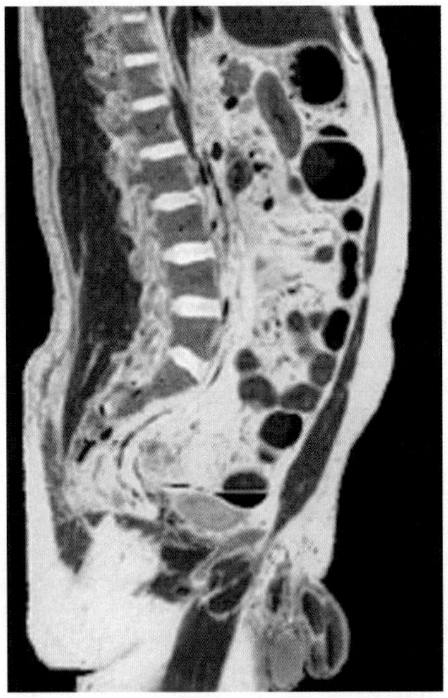

Challenge 2

Looking at reconstructed data from the Visible Human data set presents quite different views of the reproductive system than the figures presented in the text. Part of this is due to the method of preparation of the sections in the Visible Human. These are not created to demonstrate organ continuity, but rather to represent the structures visible in one 1-mm-thick cross-sections of the body. Three-dimensional composites such as the one in Challenge 1 are rendered by computer software. This image is a computer-generated sagittal section based on the original cross-sectional data. Label as many structures as you can identify, including muscles, bones and visceral organs, paying particular attention to the reproductive structures. Where are the remaining reproductive structures in relation to this section? Which of the figures in the text most closely illustrates this view of the male? Are you able to learn more about the physical relationships of the reproductive system from the illustration or this sagittal section?

Images provided by the Digital Cadaver™ project, courtesy of Visible Productions, Inc.

APPLY IT

Below are two separate exercises that provide more information on a topic presented in the preceding chapters. Each one is designed to take approximately 10 minutes, and will help you gain a better appreciation for the material presented in this chapter.

Critical Linking 1

To further enhance your understanding of the reproductive system, it is helpful to read many different presentations of the material. Two such presentations, complete with quizzes and illustrations can be found by moving to the Critical Linking section of the Companion Website and choose the keyword "Reproductive Quizzes 1" and "Reproductive Quizzes 2." The first will take you to a presentation created by D. Jay Cervino in 1997 for "Maternal and Child Health." This has some interesting information, without excessive detail. The second is designed for a higher level student and deals with the histology of the reproductive system. Associated with this site, when you click on "home," is a review section allowing you to review what you missed on line.

Critical Linking 2

Most medical students will eventually perform a cadaver dissection. To preview this experience or to review your current lab dissections visit the Companion Website and click on the keyword "cadaver." Here you will find a site prepared by Donal Stranahan, Ph.D. of the LTSN Center for Bioscience. This site walks you through a cadaver dissection. Choose the dissection of the female pelvis to review the organs of the female reproductive system. The reproductive organs are discussed on the right, and demonstrated via dissection photos on the left. Clicking on highlighted terms will provide a photograph of the cadaver with the chosen structure pointed out for you. Test your knowledge by first trying to identify the structures yourself, before clicking highlighted terms. This site is graphic and not for the faint of heart, but well worth the time should you be planning on a medical career.

FURTHER STUDIES

One of the most interesting aspects of the reproductive system is that society's treat males and females differently. There are obvious biological differences, with which you are now quite familiar. Less obvious is the impact these differences have had on society and the history of mankind. Simply because women bear children, their role in society differs from males. In most cases, but notably not all, this has largely prevented women from holding positions of power in government and society. An interesting historical perspective on this can be gained by studying Ancient Egypt as it moved from a Matriarchal to a Patriarchal society.

Enter *http://www.geocities.com/jywanza1/Matriarcahldecline.html* into your web browser to read a brief history of this switch. Many Asian societies still have patriarchal regimes.

The article found at *http://www.dartmouth.edu/~ames21/classwork/yim-3-women.html* is an excellent review of the problems facing Korean women today. In the United States, women's rights groups have fought for decades to improve the status of women in government and society. See the Website *http://equalrightsamendment.org/* to read about the current status of this movement.

This battle has even crept into our daily conversations through such popular books as "Men are from Mars; Women are from Venus" by John Gray, Ph. D. (*http://www.goodbiz.com/johngray/menhome.html*). Notably, as societies move into the information age, gender is becoming less important than knowledge and the ability to gather relevant information.

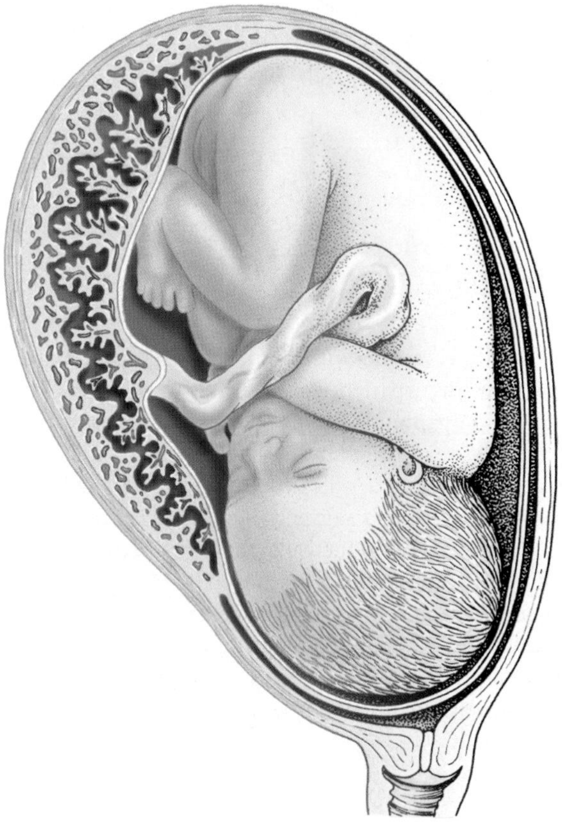

28 HUMAN DEVELOPMENT

The process of **development** is the gradual modification of anatomical structures during the period from conception to maturity. The changes are truly remarkable—what begins as a single cell slightly larger than the period at the end of this sentence becomes a human body containing trillions of cells organized into tissues, organs, and organ systems. The formation of specialized cell types during development is called **differentiation**. Differentiation occurs through selective changes in genetic activity. A basic appreciation of human development provides a framework for enhancing the understanding of anatomical structures. This discussion will focus on highlights of the developmental process; the Embryology Summaries in earlier chapters described the development of specific systems.

An Overview Of Development

Development involves (1) the division and differentiation of cells, resulting in the formation of diverse cell types, and (2) reorganization of those cell types to produce or modify anatomical structures. Development produces a mature individual capable of reproduction. The process is a continuum that begins at fertilization, or **conception**, and can be separated into periods characterized by specific anatomical changes. **Prenatal development** occurs in the period from conception to delivery; this period will be the primary focus of this chapter. The term **embryology** (em-brē-OL-ō-jē) refers to the study of the developmental events that occur during prenatal development. **Postnatal development** commences at birth and continues to maturity. We will briefly consider the *neonatal period* that immediately follows delivery, but other aspects of childhood and adolescent development were considered in earlier chapters dealing with specific systems.

The period of prenatal development can be further subdivided. **Pre-embryonic development** begins at fertilization and continues through *cleavage* (an initial series of cell divisions) and *implantation* (the movement of the pre-embryo into the uterine lining).

Pre-embryonic development is followed by **embryonic development**, which extends from implantation, typically occurring on the ninth or tenth day after fertilization, and continuing to the end of the eighth developmental week. **Fetal development** begins at the start of the ninth developmental week and continues up to the time of birth. We will now look at each of these processes in greater detail.

Fertilization

Fertilization involves the fusion of two haploid gametes, producing a diploid zygote containing the normal somatic number of chromosomes (46). pp. 729, 740 The functional roles and contributions of the spermatozoon and the ovum are very different. The spermatozoon simply delivers the paternal chromosomes to the site of fertilization, but the ovum must provide all of the nourishment and genetic programming to support the embryonic development for nearly a week after conception. The volume of the ovum is therefore much greater than that of the spermatozoon.

Normal fertilization occurs in the ampulla of the uterine tube, usually within a day of ovulation. Over this period of time, the secondary oocyte has traveled a few centimeters, but the spermatozoa must cover the distance between the vagina and the ampulla. The sperm arriving in the vagina are already motile, but they cannot fertilize an oocyte until they have undergone capacitation within the female reproductive tract.

Contractions of the uterine musculature and ciliary currents in the uterine tubes have been proposed as likely mechanisms for accelerating the movement of spermatozoa from the vagina to the fertilization site. The passage time may be from 30 minutes to 2 hours. Even with transport assistance and available nutrients, this is not an easy passage. Of the 200 million spermatozoa introduced into the vagina from a typical ejaculate, only around 10,000 enter the uterine tube, and fewer than 100 actually reach the ampulla. A male with a sperm count below 20 million per milliliter is functionally **sterile** because too few spermatozoa will survive to reach the secondary oocyte. One or two spermatozoa cannot accomplish fertilization, because of the condition of the secondary oocyte at ovulation.

The Oocyte at Ovulation [FIGURE 28.1]

Ovulation occurs before the completion of oocyte maturation, and the secondary oocyte leaving the follicle is in metaphase of the second meiotic division (meiosis II). Metabolic operations have also been discontinued, and the secondary oocyte drifts in a sort of suspended animation, awaiting the stimulus for further development. If fertilization does not occur, it will disintegrate without completing meiosis.

Fertilization is complicated by the fact that when the secondary oocyte is ejected from the ovary, it is surrounded by a layer of follicle cells, the *corona radiata*. p. 741 The events that follow are diagrammed in Figure 28.1●. The corona radiata protects the secondary oocyte as it passes through the ruptured follicular wall and into the infundibulum of the uterine tube. Although the physical process of fertilization requires only a single sperm in contact with the oocyte membrane, that spermatozoon must first penetrate the corona radiata. The acrosomal cap of the sperm contains **hyaluronidase** (hī-a-lūr-ON-a-dāz), an enzyme that breaks down the intercellular cement between adjacent follicle cells in the corona radiata. Dozens of spermatozoa must release hyaluronidase before the connections between the follicular cells break down enough to permit fertilization. No matter how many spermatozoa slip through the gap, only a single spermatozoon will accomplish fertilization and activate the oocyte. When that spermatozoon passes through the zona pellucida and contacts the secondary oocyte, their plasma membranes fuse, and the sperm enters the **ooplasm**, or cytoplasm of the oocyte. The process of membrane fusion is the trigger for **oocyte activation**. Oocyte activation involves a series of changes in the metabolic activity of the secondary oocyte. The metabolic rate of the oocyte rises suddenly, and immediate changes in the cell membrane prevent fertilization by additional sperm. (If more than one sperm does penetrate the oocyte membrane, an event called *polyspermy*, normal development cannot occur.) Perhaps the most dramatic change in the oocyte is the completion of meiosis.

Pronucleus Formation and Amphimixis [FIGURE 28.1]

After oocyte activation and the completion of meiosis, the nuclear material remaining within the ovum reorganizes as the **female pronucleus** (Figure 28.1●). While these changes are under way, the nucleus of the spermatozoon swells, becoming the **male pronucleus**. The male pronucleus then migrates toward the center of the cell, and the two pronuclei fuse in a process called **amphimixis** (am-fi-MIK-sis). Fertilization is now complete, with the formation of a **zygote** containing the normal complement of 46 chromosomes. The zygote now prepares to begin dividing; these mitotic divisions will ultimately produce billions of specialized cells. *Chromosomal Analysis p. 809*

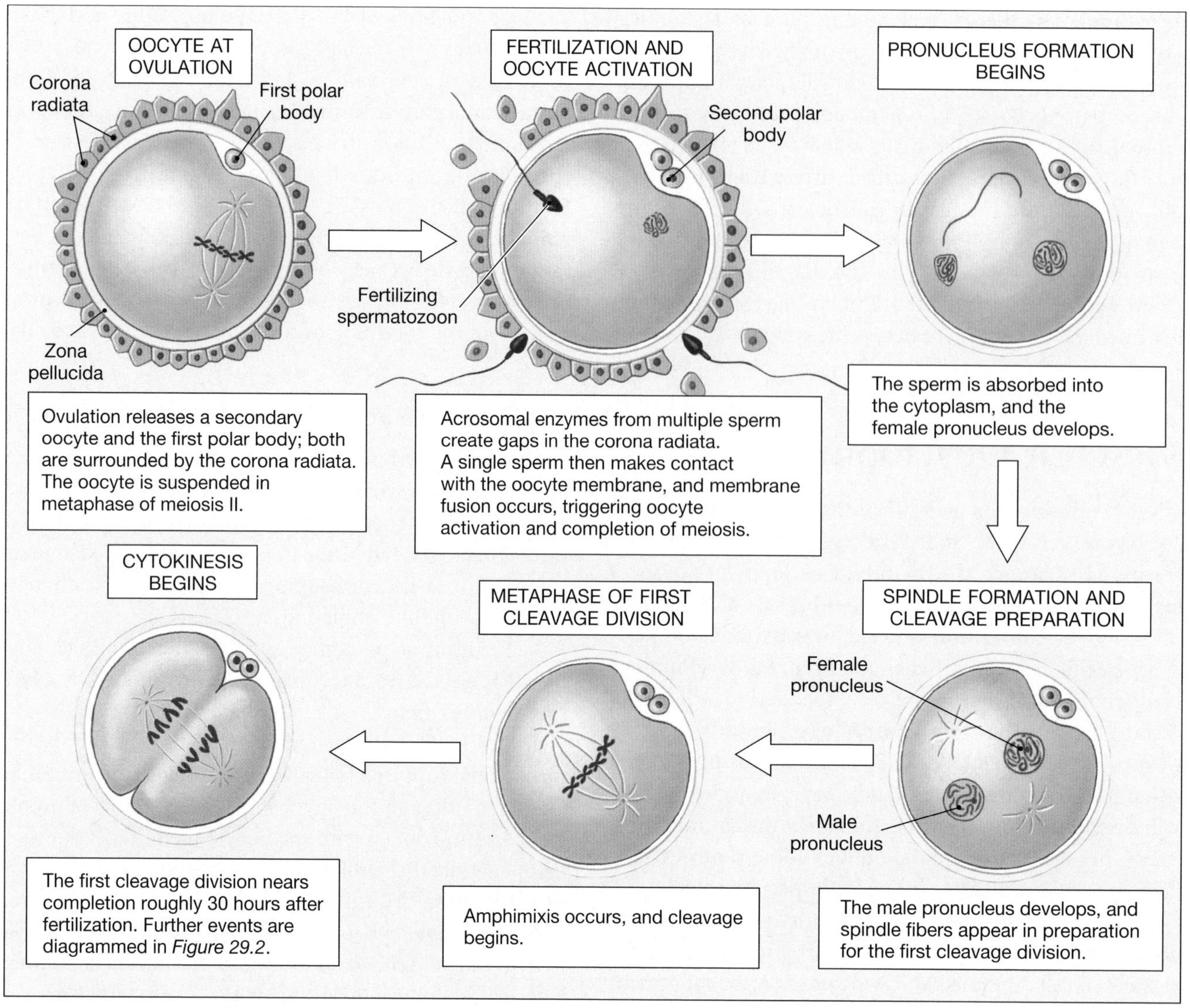

FIGURE 28.1 **FERTILIZATION AND PREPARATION FOR CLEAVAGE**

Events at fertilization and immediately thereafter. During human fertilization the entire sperm enters the cytoplasm. In many other mammals and invertebrates, only the head of the sperm enters the ooplasm.

Prenatal Development

The time spent in prenatal development is known as the period of **gestation** (jes-TĀ-shun). For convenience, the gestation period is usually considered as three integrated **trimesters**, each 3 months in duration:

- The **first trimester** is the period of embryonic and early fetal development. During this period the rudiments of all the major organ systems appear.
- In the **second trimester** the organs and organ systems complete most of their development. The body proportions change, and by the end of the second trimester the fetus looks distinctively human.
- The **third trimester** is characterized by rapid fetal growth. Early in the third trimester most of the major organ systems become fully functional, and an infant born 1 month or even 2 months prematurely has a reasonable chance of survival.

✓ **CONCEPT CHECK**

- Why must large numbers of sperm reach the secondary oocyte to accomplish fertilization?
- Can sperm arriving in the vagina perform fertilization immediately? Explain.
- As soon as the sperm enters the ooplasm, what happens to the secondary oocyte?
- What developments characterize the first trimester?

The First Trimester

By the end of the first trimester (twelfth developmental week), the fetus is almost 75 mm (3 in.) long and weighs perhaps 14 g (0.5 oz). The events that occur in the first trimester are complex, and the first trimester is the most dangerous period in prenatal life. Only about 40% of conceptions produce embryos that survive the first trimester. For this reason pregnant women are usually warned to take great care to avoid drugs or other disruptive stresses during the first trimester, in the hopes of preventing an error in the delicate processes under way.

Many important and complex developmental events occur during the first trimester. We will focus attention on four general processes: *cleavage*, *implantation*, *placentation*, and *embryogenesis*.

1. **Cleavage** (KLĒ-vej) is a sequence of cell divisions that begins immediately after fertilization and ends at the first contact with the uterine wall. Over this period the zygote becomes a **pre-embryo** that develops into a multicellular complex known as a **blastocyst**. (Cleavage and blastocyst formation were introduced in the Embryology Summary in Chapter 3.) p. 53
2. **Implantation** begins with the attachment of the blastocyst to the endometrium and continues as the blastocyst invades the uterine wall. As implantation proceeds, a number of other important events take place that set the stage for the formation of vital embryonic structures.
3. **Placentation** (pla-sen-TĀ-shun) begins as blood vessels form around the edges of the blastocyst. This is the first step in the formation of the **placenta**. The placenta provides the link between maternal and embryonic systems; it provides respiratory and nutritional support essential for further prenatal development.
4. **Embryogenesis** (em-brē-ō-JEN-e-sis) is the formation of a viable embryo. This process forms the body of the embryo and its internal organs.

CLEAVAGE AND BLASTOCYST FORMATION [FIGURE 28.2]

Cleavage (Figure 28.2●) is a series of cell divisions that subdivides the cytoplasm of the zygote into smaller cells called **blastomeres** (BLAS-tō-mērz). The first cleavage division produces a pre-embryo consisting of two identical blastomeres. The first division is completed roughly 30 hours after fertilization, and subsequent cleavage divisions occur at intervals of 10–12 hours. During the initial cleavage divisions, all of the blastomeres undergo mitosis simultaneously, but as the number of blastomeres increases, the timing becomes less predictable.

At this point in time, the pre-embryo is a solid ball of cells, resembling a mulberry. This stage is called the **morula** (MOR-ū-la; "mulberry"). After 5 days of cleavage, the blastomeres form a hollow ball, the **blastocyst**, with an inner cavity known as the **blastocoele** (BLAS-tō-sēl). At this stage you can begin to see differences between the cells of the blastocyst. The outer layer of cells, separating the external environment from the blastocoele, is called the **trophoblast** (TRŌ-fō-blast). The function is implied by

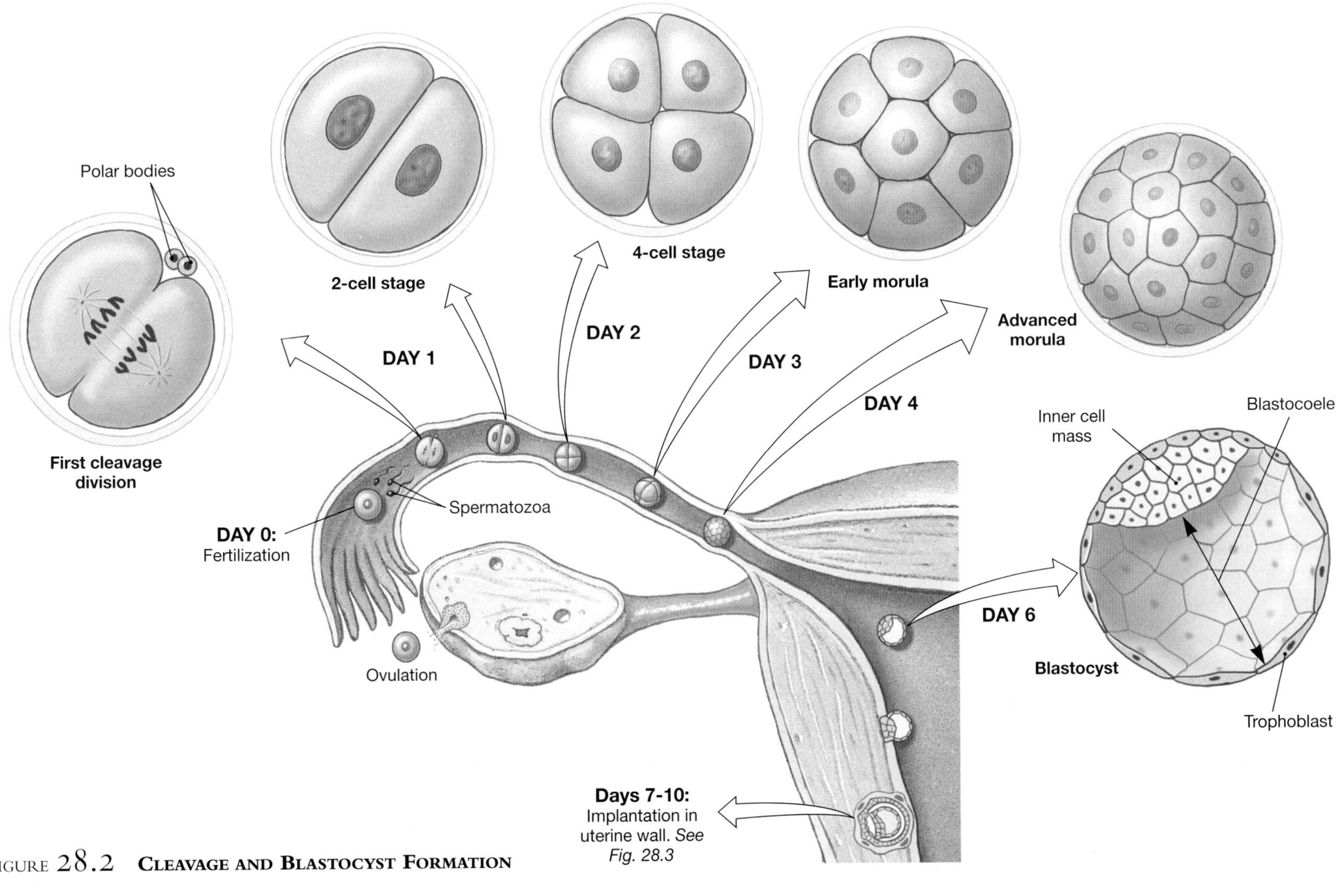

FIGURE 28.2 **Cleavage and Blastocyst Formation**

the name: *trophos,* food + *blast*, precursor. These cells will be responsible for providing food to the developing embryo. Trophoblast cells are the only cells of the pre-embryo that contact the uterine wall. A second group of cells, the **inner cell mass**, lies clustered at one end of the blastocyst. These cells are exposed to the blastocoele but are insulated from contact with the outside environment by the trophoblast. The cells of the inner cell mass are the stem cells responsible for producing all of the cells and cell types of the body. *Technology and the Treatment of Infertility p. 809*

IMPLANTATION [FIGURE 28.3]

At fertilization the zygote is still 4 days away from the uterus. It arrives in the uterine cavity as a morula, and over the next 2–3 days, blastocyst formation occurs. During this period the cells are absorbing nutrients from the fluid within the uterine cavity. This fluid, rich in glycogen, is secreted by the endometrial glands. When fully formed, the blastocyst contacts the endometrium, usually in the fundus or body of the uterus, and implantation occurs. Stages in the implantation process are illustrated in Figure 28.3●.

Implantation begins as the surface of the blastocyst closest to the inner cell mass touches and adheres to the uterine lining (see *Day 7*, Figure 28.3●). At the point of contact, the trophoblast cells divide rapidly, making the trophoblast several layers thick. Near the endometrial wall, the cell membranes separating the trophoblast cells disappear, forming a layer of cytoplasm containing multiple nuclei (*Day 8*). This outer layer, called the **syncytial** (sin-SISH-al) **trophoblast**, begins to erode a path through the uterine epithelium, by secreting the enzyme *hyaluronidase.* This enzyme breaks down the intercellular cement between adjacent epithelial cells, just as the hyaluronidase released by spermatozoa dissolved the connections between cells of the corona radiata. At first this erosion creates a gap in the uterine lining, but the migration and divisions of adjacent epithelial cells soon repair the surface. When the repairs are completed, the blastocyst loses contact with the uterine cavity and is completely embedded within the endometrium. Development hereafter occurs entirely within the functional zone of the endometrium.

As implantation proceeds, the syncytial trophoblast continues to enlarge and spread into the surrounding endometrium (*Day 9*). This results in the disruption and enzymatic digestion of uterine glands. The nutrients released are absorbed by the syncytial trophoblast and distributed by diffusion across the underlying **cellular trophoblast** to the inner cell mass. These nutrients provide the energy needed to support the early stages of embryo formation. Trophoblastic extensions grow around endometrial capillaries, and as the capillary walls are destroyed, maternal blood begins to percolate through trophoblastic channels, called **lacunae**. Fingerlike **primary villi** extend away from the trophoblast into the surrounding endometrium; each primary villus consists of an extension of syncytial trophoblast with a core of cellular trophoblast. Over the next few days, the trophoblast begins breaking down larger endometrial veins and arteries, and blood flow through the lacunae increases. *Problems with the Implantation Process p. 809*

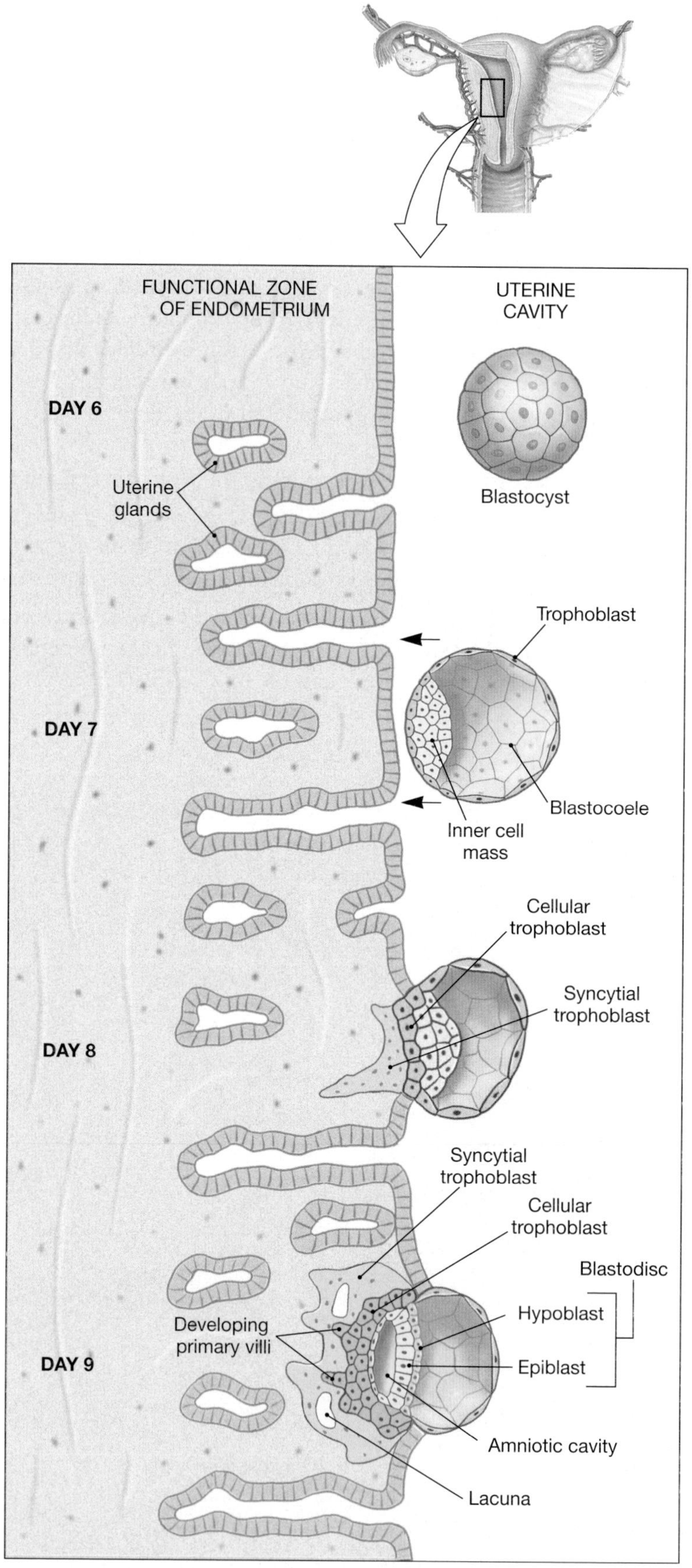

FIGURE 28.3 **STAGES IN THE IMPLANTATION PROCESS**

FORMATION OF THE BLASTODISC [FIGURES 28.3/28.4] In the early blastocyst stage, the inner cell mass has little visible organization. But, by the time of implantation, the inner cell mass has already started separating from the trophoblast. The separation gradually increases, creating a fluid-filled chamber called the **amniotic** (am-nē-OT-ik) **cavity**. The amniotic cavity can be seen in *Day 9* of Figure 28.3●; additional details from *Days 10–12* are shown in Figure 28.4●. At this stage the cells of the inner cell mass are organized into an oval sheet that is two cell layers thick. This oval, called a **blastodisc** (BLAS-tō-disk), initially consists of two epithelial layers: the **epiblast** (EP-i-blast), which faces the amniotic cavity, and the **hypoblast** (HĪ-pō-blast), which is exposed to the fluid contents of the blastocoele.

Gastrulation And Germ Layer Formation [FIGURE 28.4 AND TABLE 28.1] A few days later, a third layer begins forming through the process of **gastrulation** (gas-troo-LĀ-shun) (*Day 12*, Figure 28.4●). During gastrulation, cells in specific areas of the epiblast move toward the center of the blastodisc, toward a line known as the **primitive streak.** Once they arrive at the primitive streak, the migrating cells leave the surface and move between the epiblast and hypoblast. This movement creates three distinct embryonic layers with markedly different fates. Once gastrulation begins, the layer remaining in contact with the amniotic cavity is called the **ectoderm**, the hypoblast is known as the **endoderm**, and the new intervening layer is the **mesoderm.** The formation of mesoderm and the developmental fates of these **germ layers** were introduced in an Embryology Summary in Chapter 3. p. 53 Table 28.1 contains a more comprehensive listing of the contributions each germ layer makes to the body systems described in earlier chapters.

The Formation Of Extraembryonic Membranes [FIGURE 28.5] In addition to forming body structures and organs, germ layers also form four structures that extend outside of the embryonic body. These structures, known as **extraembryonic membranes**, are: (1) the *yolk sac* (endoderm and mesoderm), (2) the *amnion* (ectoderm and mesoderm), (3) the *allantois* (endoderm and mesoderm), and (4) the *chorion* (mesoderm and trophoblast). These membranes support embryonic and fetal development by maintaining a consistent, stable environment and by providing access to the oxygen and nutrients carried by the maternal bloodstream. Despite their importance during prenatal development, they leave few traces of their existence in adult systems. Figure 28.5● shows representative stages in the development of the extraembryonic membranes.

The Yolk Sac [FIGURES 28.4/28.5a–c] The first of the extraembryonic membranes to appear is the **yolk sac** (Figures 28.4 and 28.5●). The yolk sac begins as migrating hypoblast cells spread out around the outer edges of the blastocoele to form a complete pouch suspended below the blastodisc. This pouch is already visible 10 days after fertilization (Figure 28.4●). As

TABLE 28.1 **The Fates of the Primary Germ Layers**

Ectodermal Contributions
Integumentary system: epidermis, hair follicles and hairs, nails, and glands communicating with the skin (apocrine and merocrine sweat glands, mammary glands, and sebaceous glands)
Skeletal system: pharyngeal cartilages and their derivatives in the adult (portion of sphenoid, the auditory ossicles, the styloid processes of the temporal bones, the cornu and superior rim of the hyoid bone)*
Nervous system: all neural tissue, including brain and spinal cord
Endocrine system: pituitary gland and adrenal medullae
Respiratory system: mucous epithelium of nasal passageways
Digestive system: mucous epithelium of mouth and anus, salivary glands
Mesodermal Contributions
Skeletal system: all components except some pharyngeal derivatives
Muscular system: all components
Endocrine system: adrenal cortex and endocrine tissues of heart, kidneys, and gonads
Cardiovascular system: all components, including bone marrow
Lymphatic system: all components
Urinary system: the kidneys, including the nephrons and the initial portions of the collecting system
Reproductive system: the gonads and the adjacent portions of the duct systems
Miscellaneous: the lining of the body cavities (thoracic, pericardial, peritoneal) and the connective tissues supporting all organ systems
Endodermal Contributions
Endocrine system: thymus, thyroid, and pancreas
Respiratory system: respiratory epithelium (except nasal passageways) and associated mucous glands
Digestive system: mucous epithelium (except mouth and anus); exocrine glands (except salivary glands); liver, and pancreas
Urinary system: urinary bladder and distal portions of the duct system
Reproductive system: distal portions of the duct system; stem cells that produce gametes

*The neural crest is derived from ectoderm and contributes to the formation of the skull and the skeletal derivatives of the embryonic pharyngeal arches.

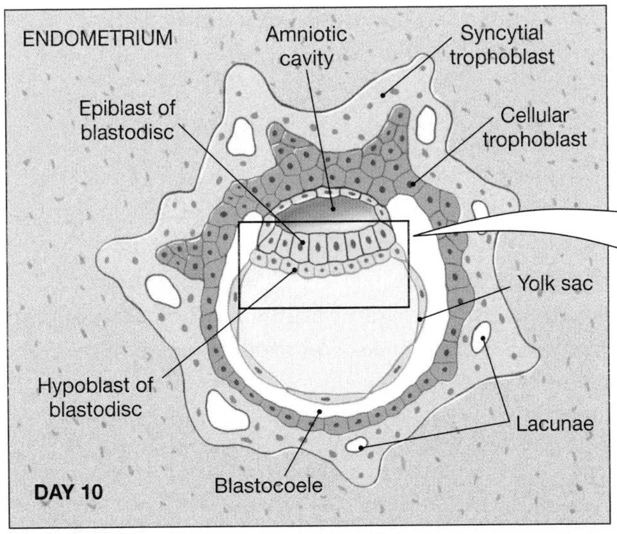

The blastodisc begins as two layers: the *epiblast*, facing the amniotic cavity, and the *hypoblast*, exposed to the blastocoele. Migration of epiblast cells around the amniotic cavity is the first step in the formation of the amnion. Migration of hypoblast cells creates a sac that hangs below the blastodisc. This is the first step in yolk sac formation.

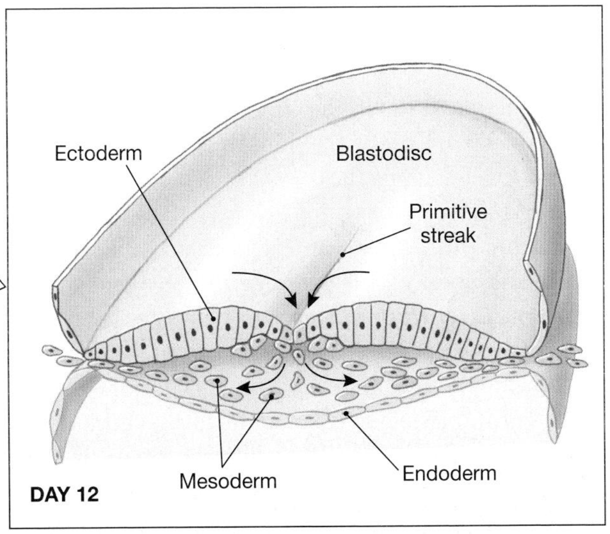

Migration of epiblast cells into the region between epiblast and hypoblast gives the blastodisc a third layer. From the time this process (gastrulation) begins, the epiblast is called *ectoderm*, the hypoblast *endoderm*, and the migrating cells *mesoderm*.

FIGURE 28.4 **Blastodisc Organization and Gastrulation**

gastrulation proceeds, mesodermal cells migrate around this pouch and complete the formation of the yolk sac. Blood vessels soon appear within the mesoderm, and the yolk sac becomes an important site of early blood cell formation.

The Amnion [FIGURE 28.5a,b,e] The ectodermal layer also undergoes an expansion, and ectodermal cells spread over the inner surface of the amniotic cavity. Mesodermal cells soon follow, creating a second, outer layer. This combination of ectoderm and mesoderm is the **amnion** (AM-nē-on) (Figure 28.5a,b●). As the embryo and later the fetus enlarges, this membrane continues to expand, increasing the size of the amniotic cavity. The amniontic cavity contains **amniotic fluid** which surrounds and cushions the developing embryo or fetus (Figure 28.5b,e●).

The Allantois [FIGURE 28.5b] The third extraembryonic membrane begins as an outpocketing of the endoderm near the base of the yolk sac (Figure 28.5b●). The free endodermal tip then grows toward the wall of the blastocyst, surrounded by a mass of mesodermal cells. This sac of endoderm and mesoderm is the **allantois** (a-LAN-tō-is); its base later gives rise to the urinary bladder. (The formation of the allantois and its relationship to the urinary bladder was detailed in the Embryology Summary in Chapter 26.) *p. 718–719*

The Chorion [FIGURE 28.5a,b] The mesoderm associated with the allantois spreads until it extends completely around the inside of the trophoblast, forming a mesodermal layer underneath the trophoblast. This combination of mesoderm and trophoblast is the **chorion** (KOR-ē-on) (Figure 28.5a,b●).

When implantation first occurs, the nutrients absorbed by the trophoblast can easily reach the blastodisc by simple diffusion. But as the embryo and the trophoblastic complex enlarge, the distance between the two increases and diffusion alone can no longer keep pace with the demands of the developing embryo. The chorion solves this problem, for blood vessels developing within the mesoderm provide a rapid transit system linking the embryo with the trophoblast. Circulation through those chorionic vessels begins early in the third week of development, when the heart starts beating.

PLACENTATION [FIGURE 28.5]

The appearance of blood vessels in the chorion is the first step in the formation of a functional placenta. By the third week of development (Figure 28.5b●), mesoderm extends along the core of each of the trophoblastic villi, forming **chorionic villi** in contact with maternal tissues. These villi continue to enlarge and branch, forming an intricate network within the endometrium. Maternal blood vessels continue to be eroded, and maternal blood slowly percolates through lacunae lined by syncytial trophoblast. Diffusion occurs between the maternal blood flowing through the lacunae and fetal blood flowing through vessels within the chorionic villi.

At first the entire blastocyst is surrounded by chorionic villi. The chorion continues to enlarge, expanding like a balloon within the endometrium, and by the fourth week, the embryo, amnion, and yolk sac are suspended within an expansive, fluid-filled chamber (Figure 28.5c●). The connection between the embryo and chorion, known as the **body stalk**, contains the distal portions of the allantois and blood vessels carrying blood to and from the placenta. The narrow connection between the endoderm of the embryo and the yolk sac is called the **yolk stalk**. (The formation of the yolk stalk and body stalk was detailed in the Embryology Summary in Chapter 25.) *pp. 670–671*

The placenta does not continue to enlarge indefinitely. Regional differences in placental organization begin to develop as placental expansion forms a prominent bulge in the endometrial surface. The relatively thin portion of the endometrium that covers the embryo and separates it from the uterine cavity is called the **decidua capsularis** (dē-SID-ū-a kap-sū-LA-ris). This layer no longer participates in nutrient exchange, and the chorionic villi disappear in this region (Figure 28.5d●). Placental functions are now concentrated in a disc-shaped area situated in the deepest portion of the endometrium, a region called the **decidua basalis** (ba-SA-lis). The rest of the endometrium, which has no contact with the chorion, is called the **decidua parietalis** (pa-rī-e-TAL-is). As the end of the first trimester approaches, the fetus moves farther away from the placenta (Figure 28.5d,e●). It remains connected by the **umbilical cord**, or *umbilical stalk,* which contains the allantois, placental blood vessels, and the yolk stalk.

The developing fetus is totally dependent on maternal organ systems for nourishment, respiration, and waste removal. These functions must be performed by maternal systems in addition to their normal operations. For example, the mother must absorb enough oxygen, nutrients, and vitamins for herself and her fetus, and she must eliminate all of the generated wastes. Although this is not a burden over the initial weeks of gestation, the demands placed upon the mother become significant in subsequent trimesters, as the fetus grows larger. In practical terms the mother must breathe, eat, and excrete for two.

PLACENTAL CIRCULATION [FIGURE 28.6] Figure 28.6a● diagrams circulation at the placenta near the end of the first trimester. Blood flows from the fetus to the placenta through the paired **umbilical arteries** and returns in a single **umbilical vein**. The chorionic villi (Figure 28.6b●) provide the surface area for active and passive exchange between the fetal and maternal bloodstreams. As noted in Chapter 27, the placenta also synthesizes important hormones that affect maternal as well as embryonic tissues. *p. 754* Human chorionic gonadotropin (HCG) production begins within a few days of implantation; it stimulates the corpus luteum so that it continues to produce progesterone throughout the early stages of the pregnancy. During the second and third trimesters the placenta also releases progesterone, estrogens, human placental lactogen (HPL), and relaxin. These hormones are synthesized and released into the maternal circulation by the trophoblast. *Problems with Placentation p. 809*

EMBRYOGENESIS [FIGURES 28.5/28.7 AND TABLE 28.2]

Shortly after gastrulation begins, folding and differential growth of the embryonic disc produce a bulge that projects into the amniotic cavity (Figure 28.5b●). This projection is known as the **head fold**; similar movements lead to the formation of a **tail fold** (Figure 28.5c●). The **embryo** is now physically as well as developmentally separated from the rest of the blastodisc and the extraembryonic membranes. The definitive orientation of the embryo can now be seen, complete with dorsal and ventral surfaces and left and right sides.

Many of these changes in proportions and appearance that occur between the fourth developmental week and the end of the first trimester are seen in Figure 28.7●.

The first trimester is a critical period for development because events in the first 12 weeks establish the basis for organ formation, a process called **organogenesis**. Embryology Summaries in earlier chapters described major features of organogenesis in each organ system. Important developmental milestones are indicated in Table 28.2; those interested in additional details should refer to the Embryology Summaries cross-referenced in the table. *Teratogens and Abnormal Development p. 809*

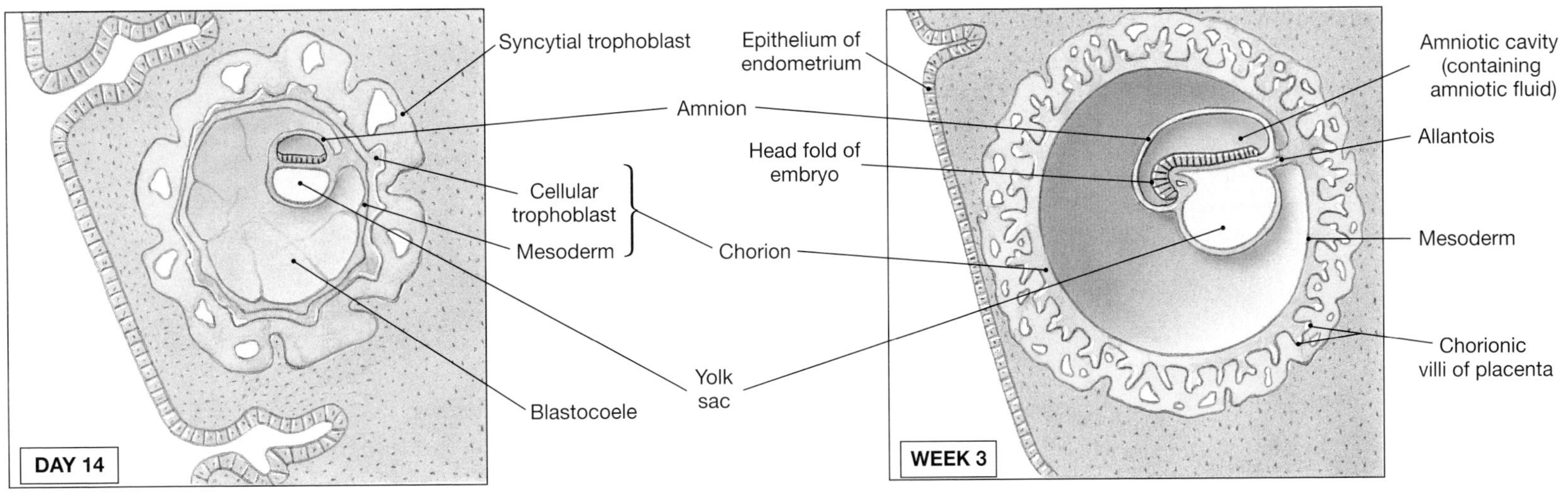

(a) Migration of mesoderm around the inner surface of the trophoblast forms the chorion. Mesodermal migration around the outside of the amniotic cavity, between the ectodermal cells and the trophoblast, forms the amnion. Mesodermal migration around the endodermal pouch below the blastodisc forms the definitive yolk sac.

(b) The embryonic disc bulges into the amniotic cavity at the head fold. The allantois, an endodermal extension surrounded by mesoderm, extends toward the trophoblast.

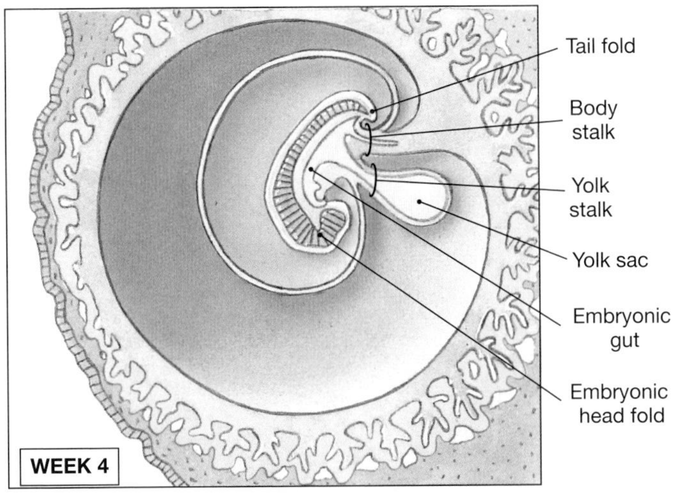

(c) The embryo now has a head fold and a tail fold. Constriction of the connection between the embryo and the surrounding trophoblast constricts the yolk stalk and body stalk.

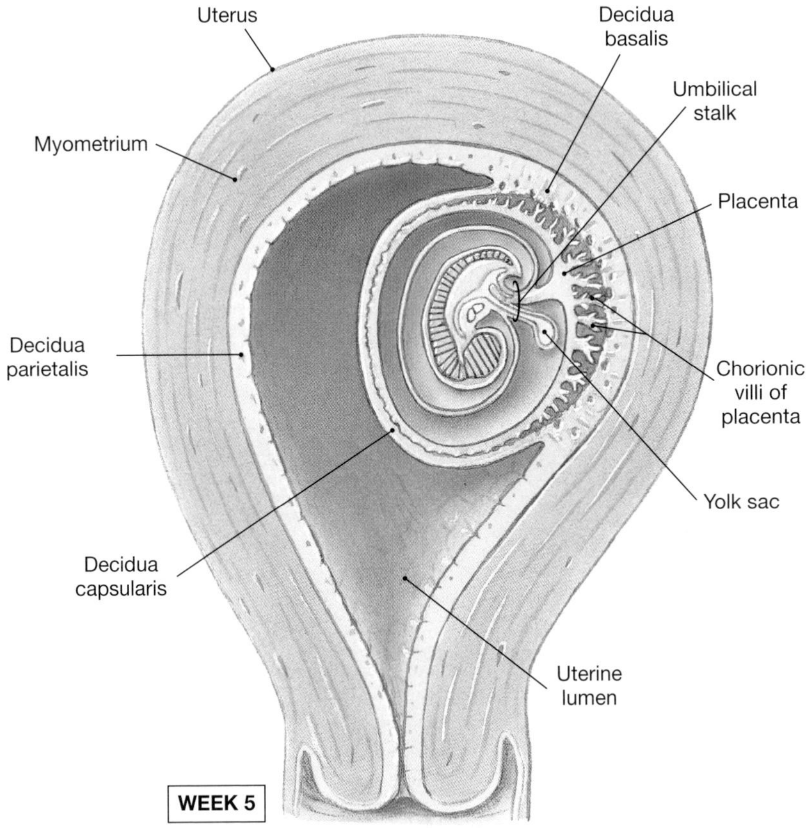

(d) The developing embryo and extraembryonic membranes bulge into the uterine cavity. The trophoblast pushing out into the uterine lumen remains covered by endometrium, but no longer participates in nutrient absorption and embryo support. The embryo moves away from the placenta, and the body stalk and yolk stalk fuse to form an umbilical stalk.

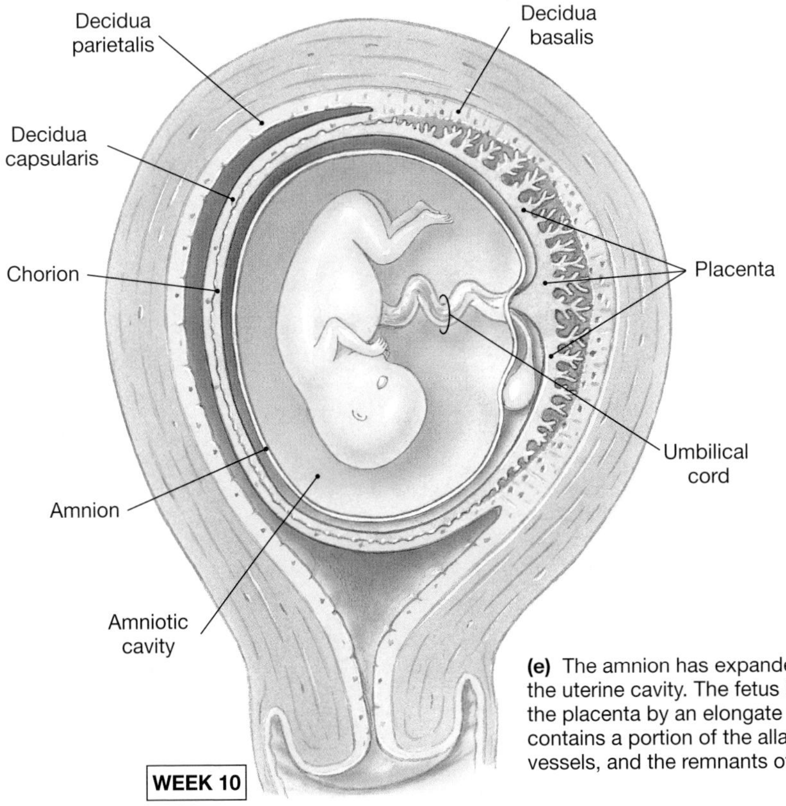

(e) The amnion has expanded greatly, filling the uterine cavity. The fetus is connected to the placenta by an elongate umbilical cord that contains a portion of the allantois, blood vessels, and the remnants of the yolk stalk.

FIGURE 28.5 **The Embryonic Membranes and Placenta Formation**

FIGURE 28.6 **A Three-Dimensional View of Placental Structure**

(**a**) For clarity, the uterus is shown after the embryo has been removed and the umbilical cord cut. Blood flows into the placenta through ruptured maternal blood arteries. It then flows around chorionic villi, which contain fetal blood vessels. Fetal blood arrives over paired umbilical arteries and leaves over a single umbilical vein. Maternal blood reenters the venous system of the mother through the broken walls of small uterine veins. Maternal blood flow is shown by arrows; note that no actual mixing of maternal and fetal blood occurs. (**b**) A cross section through a chorionic villus, showing the syncytial trophoblast exposed to the maternal blood space.

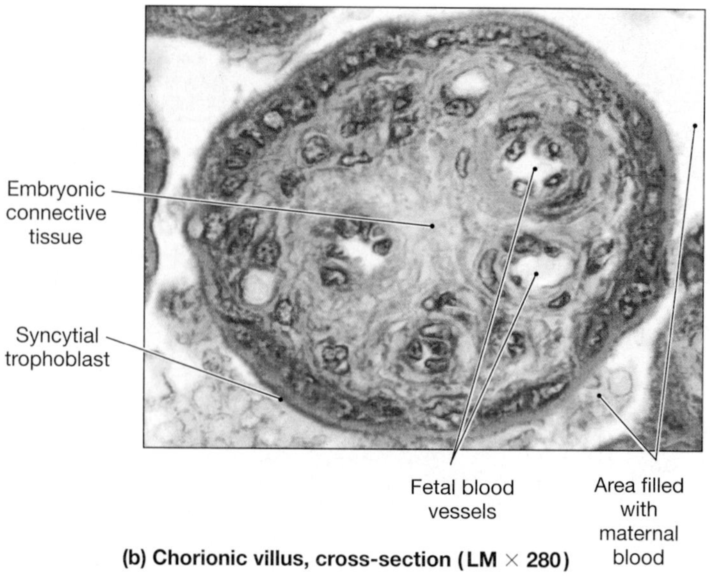

(b) Chorionic villus, cross-section (LM × 280)

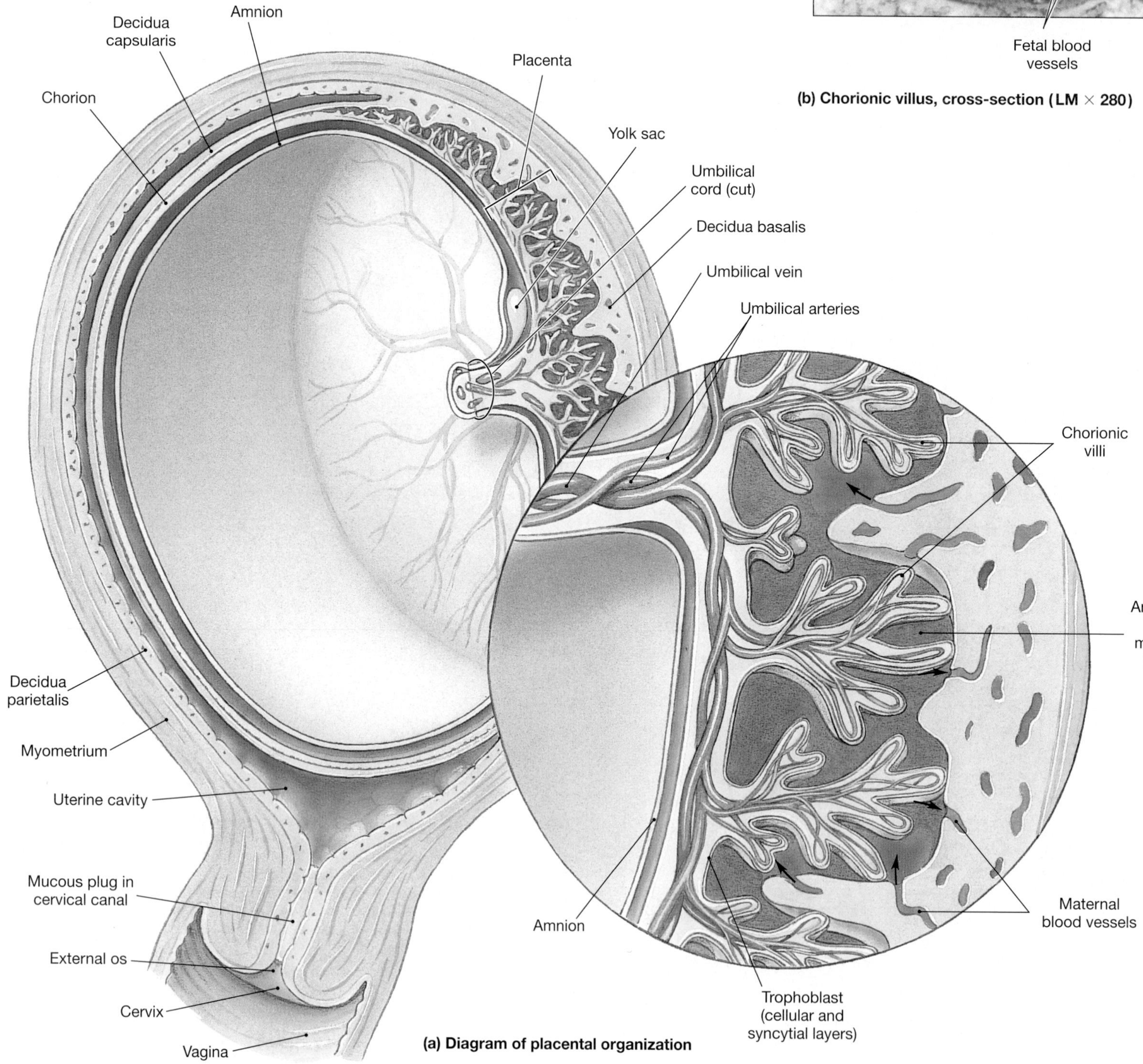

(a) Diagram of placental organization

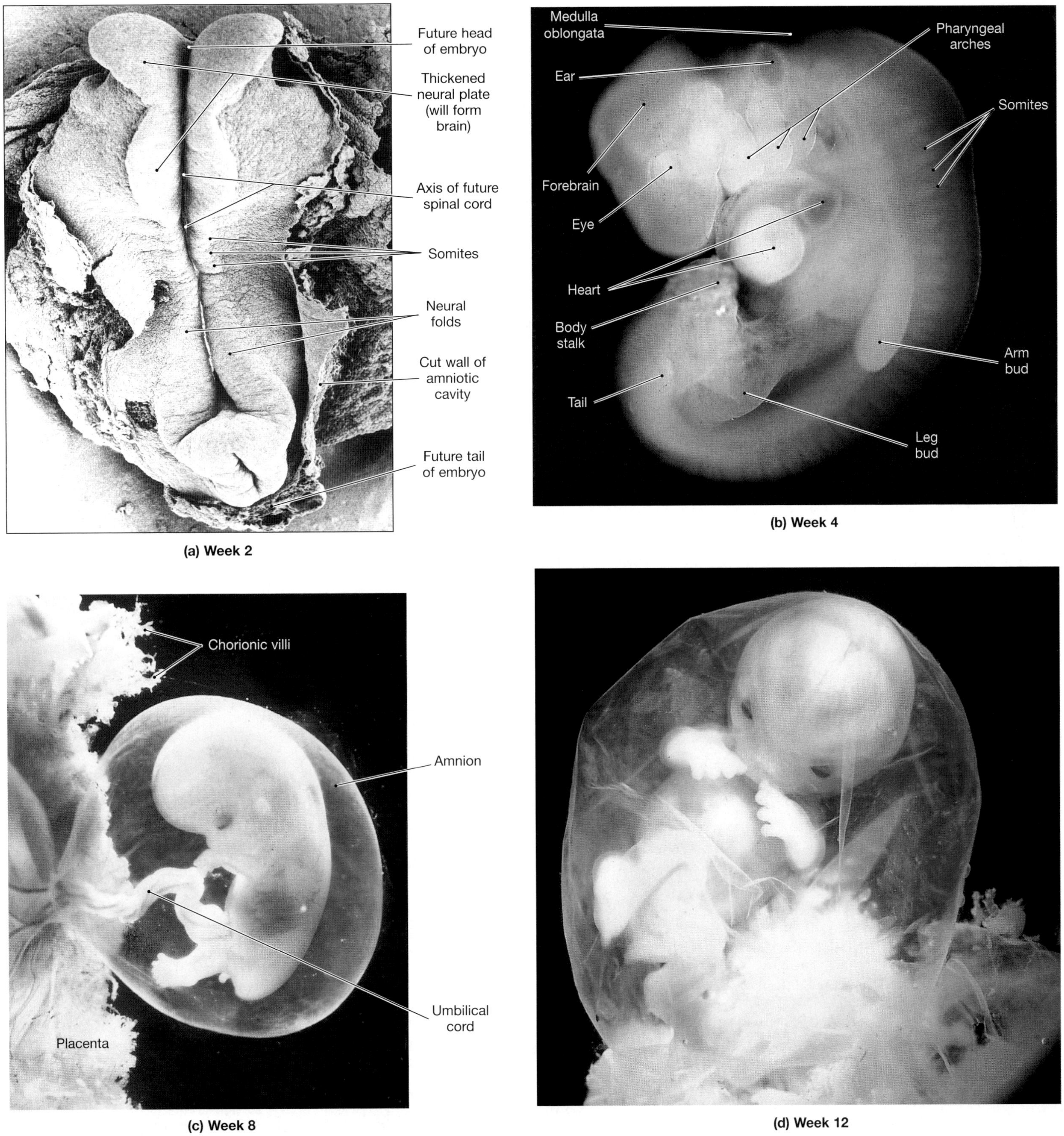

FIGURE 28.7 **The First Trimester**

(a) An SEM of the superior surface of a 2-week embryo; neurulation (neural tube formation) is under way. **(b–d)** fiber-optic views of human development in the first trimester.

TABLE 28.2

An Overview of Prenatal Development

Background Material:
Ch. 3: Formation of Tissues (p. 53)
Development of Epithelia (p. 63)
Origins of Connective Tissues (p. 74)
Development of Organ Systems (pp. 80–81)

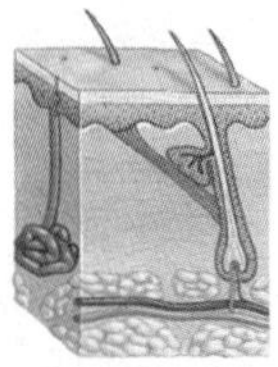

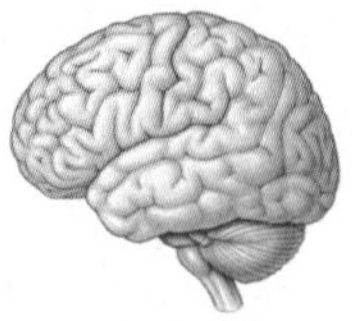
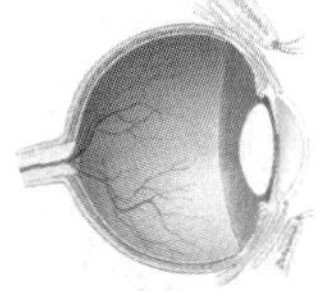

Gestational Age (Months)	Size and Weight	Integumentary System	Skeletal System	Muscular System	Nervous System	Special Sense Organs
1	5 mm 0.02 g		(b) Somite formation	(b) Somite formation	(b) Neural tube formation	(b) Eye and ear formation
2	28 mm 2.7 g	(b) Formation of nail beds, hair follicles, sweat glands	(b) Formation of axial and appendicular cartilages	(c) Rudiments of axial musculature	(b) CNS, PNS organization, growth of cerebrum	(b) Formation of taste buds, olfactory epithelium
3	78 mm 26 g	(b) Epidermal layers appear	(b) Ossification centers spreading	(c) Rudiments of appendicular musculature	(c) Basic spinal cord and brain structure	
4	133 mm 150 g	(b) Formation of hair, sebaceous glands (c) Sweat glands	(b) Articulations (c) Facial and palatal organization	Fetus starts moving	(b) Rapid expansion of cerebrum	(c) Basic eye and ear structure (b) Peripheral receptor formation
5	185 mm 460 g	(b) Keratin production, nail production			(b) Myelination of spinal cord	
6	230 mm 823 g			(c) Perineal muscles	(b) CNS tract formation (c) Layering of cortex	
7	270 mm 1492 g	(b) Keratinization, nail formation, hair formation				(c) Eyelids open, retina sensitive to light
8	310 mm 2274 g		(b) Epiphyseal cartilage formation			Taste receptors functional
9	346 mm 2912 g					
Postnatal development		Hair changes in consistency and distribution	Formation and growth of epiphyseal cartilages continue	Muscle mass and control increase	Myelination, layering, CNS tract formation continue	
Chapter containing relevant Embryology Summary		4: Development of the Integumentary System (pp. 102–103)	6: Development of the Skull (pp. 156–157), Development of the Vertebral Column (pp. 172–173) 7: Development of the Appendicular Skeleton (pp. 192–193)	10: Development of the Muscular System (pp. 286–287)	13: Introduction to the Development of the Nervous System (p. 349) 14: Development of the Spinal Cord and Spinal Nerves (pp. 376–377) 15: Development of the Brain and Cranial Nerves (pp. 420–421)	18: Development of Special Sense Organs (pp. 498–499)

Note: (b) = begin formation; (c) = complete formation.

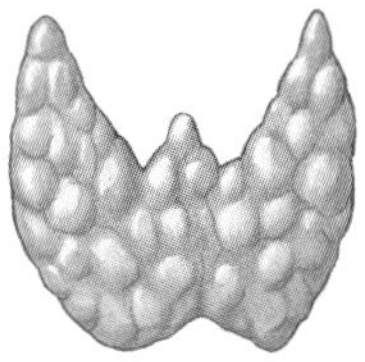 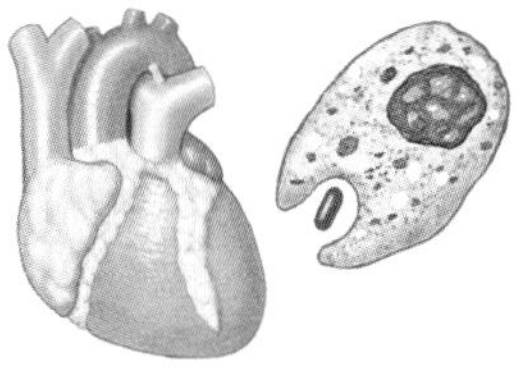 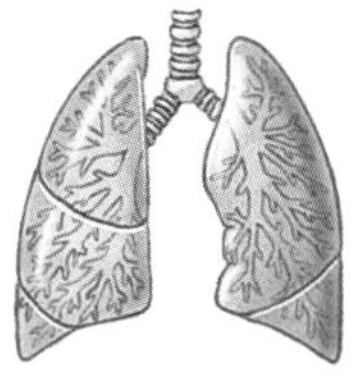 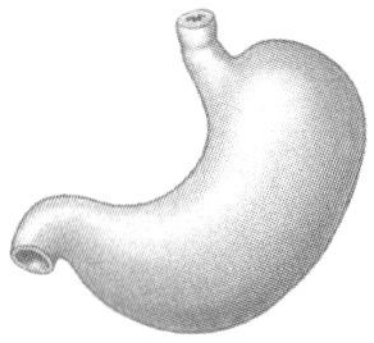 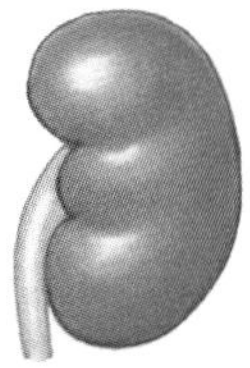 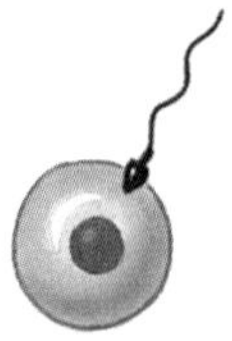

Endocrine System	Cardiovascular and Lymphatic Systems	Respiratory System	Digestive System	Urinary System	Reproductive System
	(b) Heartbeat	(b) Trachea and lung formation	(b) Formation of intestinal tract, liver, pancreas (c) Yolk sac	(c) Allantois	
(b) Formation of thymus, thyroid, pituitary, adrenal glands	(c) Basic heart structure, major blood vessels, lymph nodes and ducts (b) Blood formation in liver	(b) Extensive bronchial branching into mediastinum (c) Diaphragm	(b) Formation of intestinal subdivisions, villi, salivary glands	(b) Kidney formation (adult form)	(b) Formation of mammary glands
(c) Thymus, thyroid gland	(b) Tonsils, blood formation in bone marrow		(c) Gallbladder, pancreas		(b) Formation of definitive gonads, ducts, genitalia
	(b) Migration of lymphocytes to lymphoid organs, blood formation in spleen (c) Tonsils	(c) Nostrils open	(c) Intestinal subdivisions	(b) Degeneration of embryonic kidneys	
(c) Adrenal glands	(c) Spleen, liver, bone marrow	(b) Alveolar formation	(c) Epithelial organization, glands		
(c) Pituitary gland			(c) Intestinal plicae		(b) Descent of testes
		Complete pulmonary branching and alveolar formation		Complete nephron formation at birth	Descent of testes complete at or near time of birth
	Cardiovascular changes at birth; immune system gradually becomes fully operational				
19: Development of the Endocrine System (pp. 522–523)	21: Development of the Heart (p. 568) 22: Development of the Circulatory System (pp. 604–605) 23: Development of Lymphatic System (p. 629)	24: Development of the Respiratory System (pp. 652–653)	25: Development of the Digestive System (pp. 670–671)	26: Development of the Urinary System (pp. 718–719)	27: Development of the Reproductive System (pp. 751–753)

✓ **Concept Check**

- What is the fate of the inner cell mass of the blastocyst?
- What is the function of the syncytial trophoblast?
- What systems does the mesodermal layer give rise to?
- What are the functions of the placenta?

■ The Second and Third Trimesters [FIGURES 28.7d/28.8/28.9 AND TABLE 28.2]

By the end of the first trimester (Figure 28.7d●), the rudiments of all of the major organ systems have formed. Over the next 3 months, these systems will complete their functional development, and by the end of the second trimester, the fetus weighs around 0.64 kg (1.4 lb). During the second trimester the fetus, encircled by the amnion, grows faster than the surrounding placenta. Soon the mesodermal outer covering of the amnion fuses with the inner lining of the chorion. Figure 28.8● shows a 4-month fetus as viewed with a fiber-optic endoscope and a 6-month fetus as seen in ultrasound.

During the third trimester, all of the organ systems become functional. The rate of growth begins to decrease, but in absolute terms this trimester sees the largest weight gain. In 3 months the fetus puts on around 2.6 kg (5.7 lb), reaching a full-term weight of somewhere near 3.2 kg (7 lb). Important events in organ system development in the second and third trimesters are detailed in the Embryology Summaries in earlier chapters, and highlights are noted in Table 28.2.

At the end of gestation, a typical uterus will have undergone a tremendous increase in size. It will grow from 7.5 cm (3 in.) to 30 cm (12 in.) long and will contain almost 5 l of fluid. The uterus and its contents weigh roughly 10 kg (22 lb). This remarkable expansion occurs through the enlargement and elongation of existing smooth muscle fibers. Figure 28.9● shows the position of the uterus, fetus, and placenta from 16 weeks to *full term* (9 months). When the pregnancy is at term, the uterus and fetus push many of the abdominal organs out of their normal positions (Figure 28.9c●).

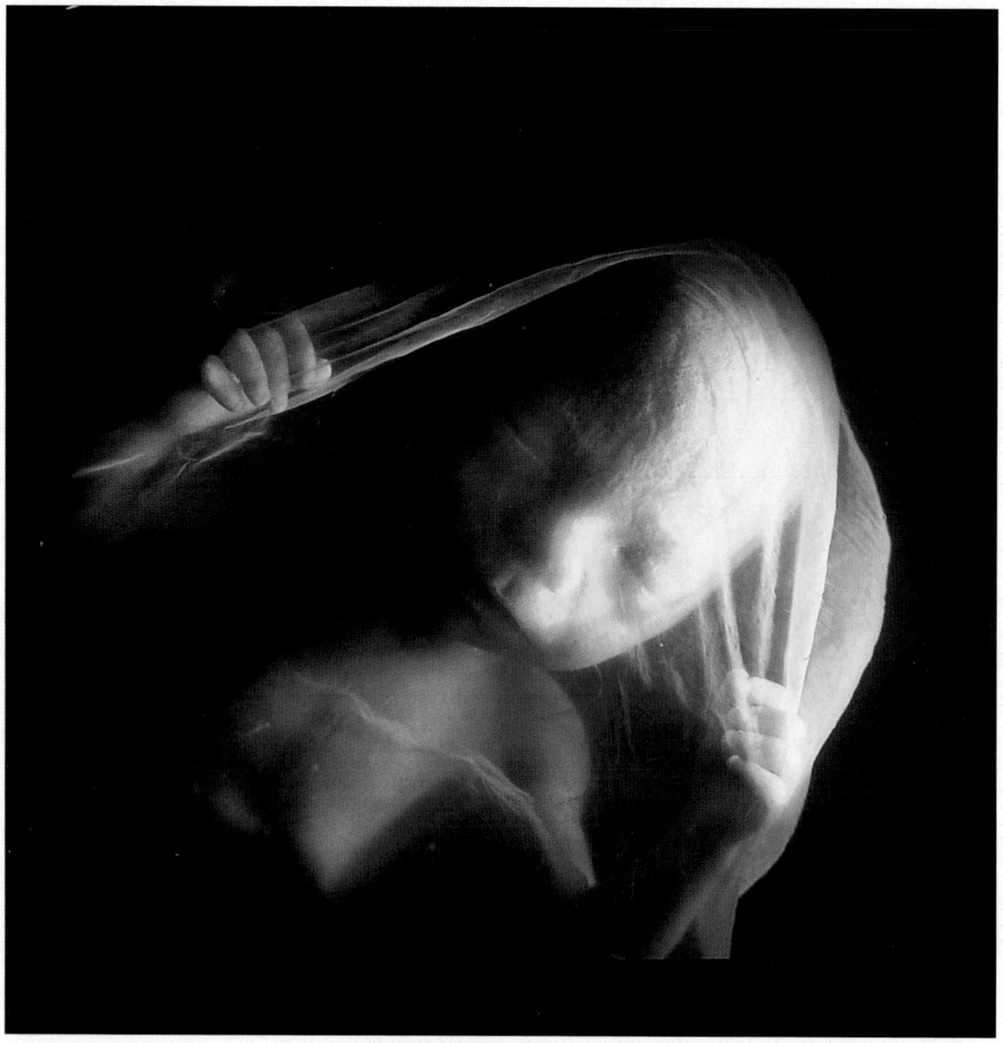

(a) 4-month fetus (endoscopic view)

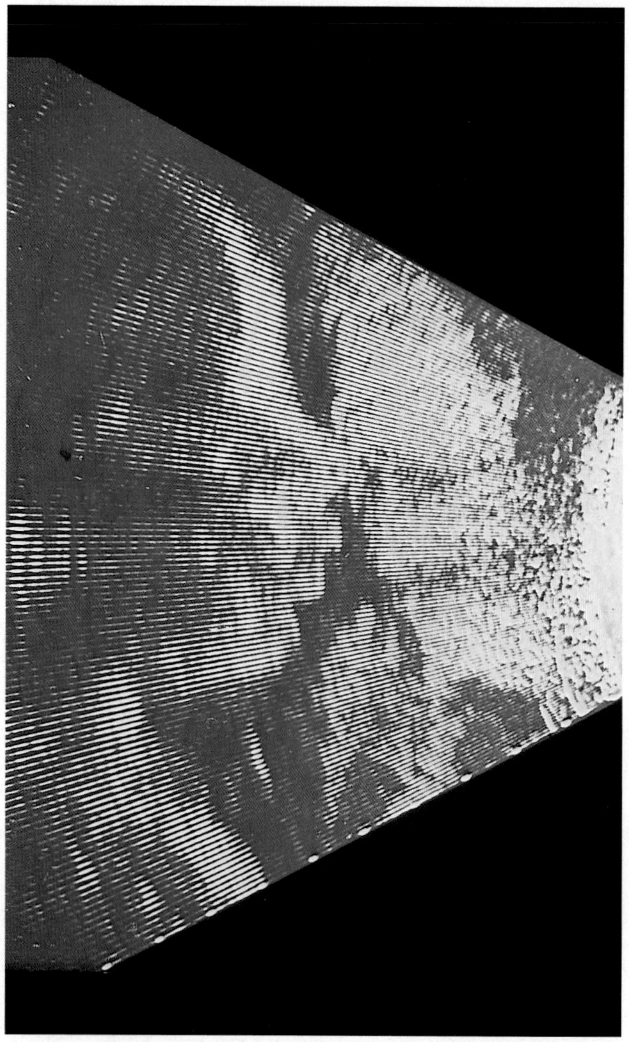

(b) 6-month fetus (ultrasound image)

FIGURE 28.8 **THE SECOND AND THIRD TRIMESTERS**
(a) A 4-month fetus seen through a fiber-optic endoscope. **(b)** Head of a 6-month fetus as seen in an ultrasound scan.

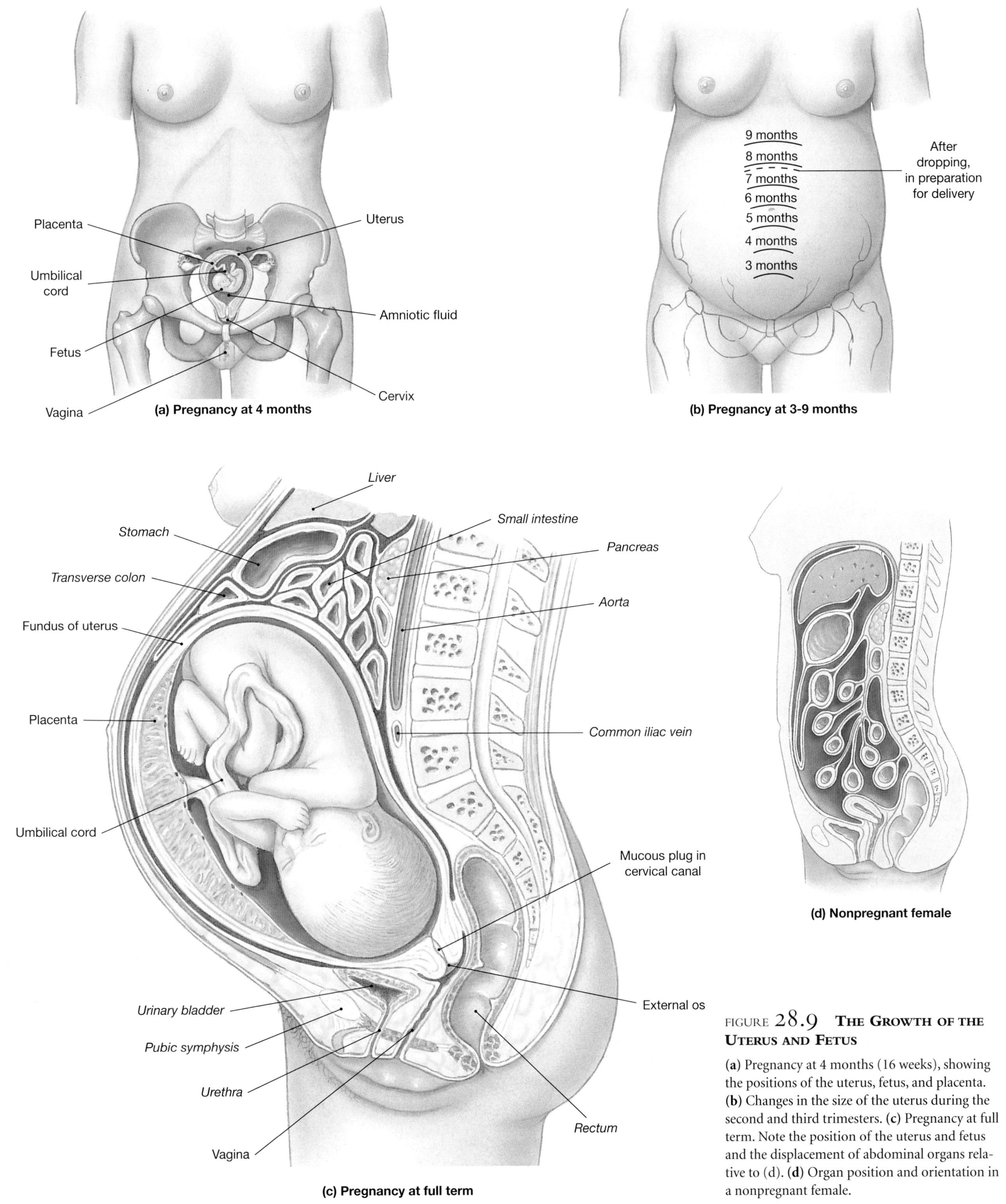

FIGURE 28.9 **THE GROWTH OF THE UTERUS AND FETUS**

(**a**) Pregnancy at 4 months (16 weeks), showing the positions of the uterus, fetus, and placenta. (**b**) Changes in the size of the uterus during the second and third trimesters. (**c**) Pregnancy at full term. Note the position of the uterus and fetus and the displacement of abdominal organs relative to (d). (**d**) Organ position and orientation in a nonpregnant female.

Labor And Delivery

The goal of labor is the expulsion of the fetus, a process known as **parturition** (par-tū-RISH-un), or birth. Labor is triggered by the combination of increased oxytocin levels and increased uterine sensitivity to oxytocin. This combination stimulates *labor contractions* in the myometrium. During *true labor* as opposed to the occasional uterine spasms of *false labor*, each labor contraction begins near the fundus of the uterus and sweeps in a wave toward the cervix. These contractions are strong and occur at regular intervals. As parturition approaches, the contractions increase in force and frequency, changing the position of the fetus and moving it toward the cervical canal.

Stages of Labor [FIGURE 28.10]

Labor is usually divided into three stages (Figure 28.10●): the dilation stage, the expulsion stage, and the placental stage.

THE DILATION STAGE [FIGURE 28.10a]

The **dilation stage** (Figure 28.10a●) begins with the onset of true labor, as the cervix dilates and the fetus begins to move down the cervical canal. This stage typically lasts 8 or more hours, but during this period the labor contractions occur at intervals of once every 10–30 minutes. Late in the process, the amnion usually ruptures, an event sometimes referred to as "having the water break."

THE EXPULSION STAGE [FIGURE 28.10b]

The **expulsion stage** (Figure 28.10b●) begins as the cervix dilates completely, pushed open by the approaching fetus. Expulsion continues until the fetus has completed its emergence from the vagina, a period usually lasting less than 2 hours. The arrival of the newborn infant into the outside world represents birth, or **delivery**.

If the vaginal canal is too small to permit the passage of the fetus, and there is acute danger of perineal tearing, the passageway may be temporarily enlarged by making an incision through the perineal musculature. After delivery this **episiotomy** (e-pē-zē-OT-ō-mē) can be repaired with sutures, a much simpler procedure than dealing with the bleeding and tissue damage associated with a potentially extensive perineal tear.

The relative sizes of the fetal skull and the maternal pelvic outlet affect the ease and success of delivery. If progress is slow or complications arise during the dilation or expulsion stages, the infant may be removed by **cesarean section**, or *"c-section."* In such cases an incision is made through the abdominal wall, and the uterus is opened just enough to allow passage of the infant's head. This procedure is performed during 15–25% of the deliveries in the United States. Efforts are now being made to reduce the frequency of both episiotomies and cesarean sections.

THE PLACENTAL STAGE [FIGURE 28.10c]

During the **placental stage** of labor (Figure 28.10c●), the muscle tension builds in the walls of the partially empty uterus, and the organ gradually decreases in size. This uterine contraction tears the connections between the endometrium and the placenta. Usually within an hour after delivery, the placental stage ends with the ejection of the placenta, or *afterbirth*. The disruption of the placenta is accompanied by a loss of blood (as much as 500–600 ml), but because the maternal blood volume has increased during pregnancy, the loss can be tolerated.

CLINICAL BRIEF

FORCEPS DELIVERIES AND BREECH BIRTHS

In most pregnancies, by the end of gestation, the fetus has rotated within the uterus to transit the birth canal head first, facing the mother's sacrum. In about 6% of deliveries, the fetus faces the mother's pubis instead. These infants can be delivered normally, given enough time, but risks to infant and mother are reduced by a *forceps delivery*. The forceps resemble large, curved salad tongs that can be separated for insertion into the vaginal canal one side at a time. Once in place, they are reunited and used to grasp the head of the fetus. An intermittent pull is applied, so that the forces on the head resemble those of normal delivery.

In 3–4% of deliveries, the legs or buttocks of the fetus enter the vaginal canal first. Such deliveries are known as **breech births**. Risks to the infant are relatively higher in breech births because the umbilical cord may become constricted, and placental circulation cut off. Because the head is normally the widest part of the fetus, the cervix may dilate enough to pass the legs and body but not the head. Entrapment of the fetal head compresses the umbilical cord, prolongs delivery, and subjects the fetus to severe distress and potential damage. If the fetus cannot be repositioned manually, a cesarean section is usually performed.

Premature Labor

Premature labor occurs when true labor begins before the fetus has completed normal development. The chances of newborn survival are directly related to body weight at delivery. Even with massive supportive efforts, infants born weighing less than 400 g (14 oz) will not survive, primarily because the respiratory, cardiovascular, and urinary systems are unable to support life without the aid of maternal systems. As a result, the dividing line between *spontaneous abortion* and **immature delivery** is usually set at 500 g (17.6 oz), the normal weight near the end of the second trimester.

Infants delivered before completing 7 months of gestation (weight under 1 kg) have less than a 50:50 chance of survival, and many survivors suffer from severe developmental abnormalities. A **premature delivery** produces a newborn weighing over 1 kg (35.2 oz), and its chances of survival range from fair to excellent, depending on the individual circumstances. *Complexity and Perfection: An Improbable Dream p. 810*

The Neonatal Period

Developmental processes do not cease at delivery, for the newborn infant has few of the anatomical, functional, or physiological characteristics of the mature adult. The **neonatal period** extends from the moment of birth to 1 month thereafter. A variety of physiological and anatomical alterations occurs as the fetus completes the transition to the status of a newborn infant, or **neonate**. Before delivery, transfer of dissolved gases, nutrients, waste products, hormones, and immunoglobulins occurred across the placental interface. At birth, the newborn infant must become relatively self-sufficient, with the processes of respiration, digestion, and excretion performed by its own specialized organs and organ systems. The transition from fetus to neonate may be summarized as:

1. The lungs at birth are collapsed and filled with fluid, and filling them with air involves a massive and powerful inhalation.

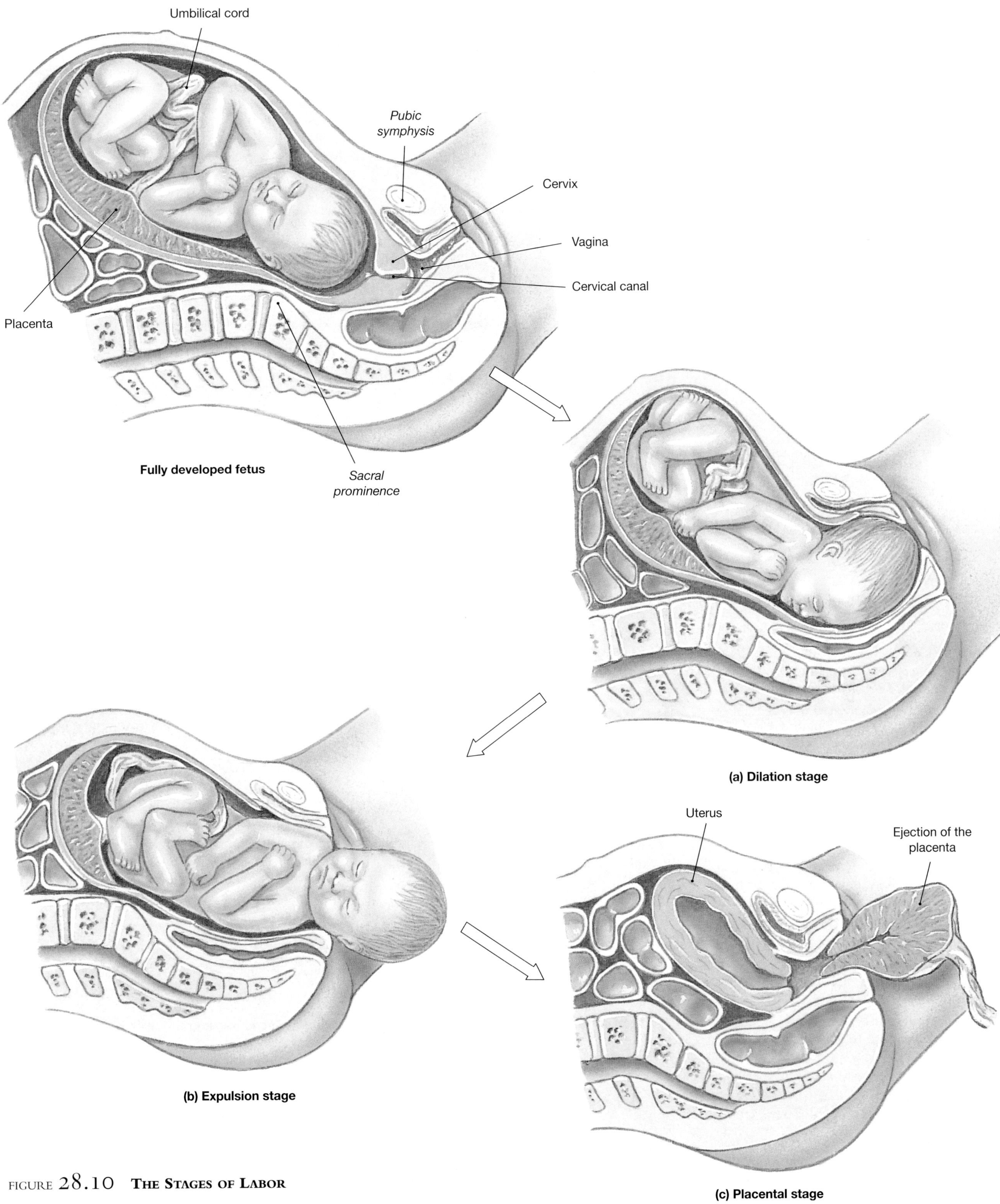

FIGURE 28.10 **THE STAGES OF LABOR**

2. When the lungs expand, the pattern of cardiovascular circulation changes because of alterations in blood pressure and flow rates. The ductus arteriosus closes, isolating the pulmonary and systemic trunks, and the closure of the foramen ovale separates the atria of the heart, completing separation of the pulmonary and systemic circuits. These cardiovascular changes were discussed in Chapters 21 and 22.
3. Typical heart rates of 120–140 beats per minute and respiratory rates of 30 breaths per minute in neonates are normal, and considerably higher than those of adults.
4. Prior to birth, the digestive system remains relatively inactive, although it does accumulate a mixture of bile secretions, mucus, and epithelial cells. This collection of debris is excreted in the first few days of life. Over that period the newborn infant begins to nurse.
5. As waste products build up in the arterial blood, they are filtered into the urine at the kidneys. Glomerular filtration is normal, but the urine cannot be concentrated to any significant degree. As a result, urinary water losses are high, and neonatal fluid requirements are much greater than those of adults.
6. The neonate has little ability to control body temperature, particularly in the first few days after delivery. As the infant grows larger and increases the thickness of its insulating subcutaneous adipose "blanket," its metabolic rate also rises. Daily and even hourly alterations in body temperature continue throughout childhood.

CLINICAL BRIEF

Evaluating the Newborn Infant

Each newborn infant gets closely scrutinized after delivery. This procedure checks for the presence of anatomical and physiological abnormalities. It also provides baseline information useful in assessing postnatal development. In addition to general appearance, the pulse, respiratory rate, weight, length, and other physical dimensions are noted. Newborn infants are also screened for genetic and metabolic disorders, such as *phenylketonuria (PKU)* or congenital *hypothyroidism*.

The **Apgar rating** considers heart rate, respiratory rate, muscle tone, response to stimulation, and color at 1 and 5 minutes after birth. In each category the infant receives a score ranging from 0 (poor) to 2 (excellent), and the scores are then totaled. An infant's Apgar rating (0–10) has been shown to be an accurate predictor of newborn survival and the presence of neurological damage. For example, newborn infants with *cerebral palsy* usually have a low Apgar rating.

✓ CONCEPT CHECK

- In general, what embryonic changes are seen in the second trimester?
- Which stage of labor usually takes the longest?
- Why does a newborn infant have a much higher relative fluid intake than do adults?
- What triggers the expulsion of the placenta?

RELATED CLINICAL TERMS

abruptio placentae (a-BRUP-shē-ō pla-SEN-tē): Tearing of the placenta sometime after the fifth gestational month. *Problems with Placentation p. 809*

amniocentesis: An analysis of fetal cells taken from a sample of amniotic fluid. *Chromosomal Analysis p. 809*

Apgar rating: A method of evaluating newborn infants; a test for developmental problems and neurological damage. *p. 776*

breech birth: A delivery wherein the legs or buttocks of the fetus enter the vaginal canal first. *p. 774*

chorionic villi sampling: An analysis of cells collected from the chorionic villi during the first trimester. *Chromosomal Analysis p. 809*

congenital malformation: A severe structural abnormality, present at birth, that affects major systems. *Complexity and Perfection: An Improbable Dream p. 810*

ectopic pregnancy: A pregnancy in which implantation occurs somewhere other than the uterus. *Problems with the Implantation Process p. 809*

fetal alcohol syndrome (FAS): A neonatal condition resulting from maternal alcohol consumption; characterized by developmental defects often involving the skeletal, nervous, and/or cardiovascular systems. *Teratogens and Abnormal Development p. 809*

gestational neoplasm: A tumor formed by undifferentiated, rapid growth of the syncytial trophoblast; if untreated, the neoplasm may become malignant. *Problems with the Implantation Process p. 809*

in vitro fertilization: Fertilization outside the body, usually in a test tube or petri dish. *Technology and the Treatment of Infertility p. 809*

infertility: The inability to have children. *Technology and the Treatment of Infertility p. 809*

placenta previa: A condition resulting from implantation in or near the cervix. *Problems with Placentation p. 809*

teratogens (ter-AT-ō-jens): Stimuli that disrupt normal development by damaging cells, altering chromosome structure, or altering the chemical environment of the embryo. *Teratogens and Abnormal Development p. 809*

Additional Clinical Terms Discussed in Appendix I (pp. 809–810)

abdominal pregnancies; artificial insemination; GIFT (gamete intrafallopian tube transfer); in vitro fertilization (IVF); spontaneous mutations; teratology; ZIFT (zygote intrafallopian tube transfer)

STUDY OUTLINE & CHAPTER REVIEW

Introduction 761

1. **Development** is the gradual modification of physical and physiological characteristics from conception to maturity. The creation of different cell types during development is called **differentiation**.

An Overview Of Development 761

1. Development begins at **conception** (fertilization) and can be divided into prenatal (before birth) and postnatal (birth to maturity) development. **Prenatal development** occurs before birth; **postnatal development** begins at birth and continues to maturity.

Fertilization 761

1. Fertilization normally occurs in the ampulla of the uterine tube within a day after ovulation. Sperm cannot fertilize an egg until they have undergone *capacitation*. Around 200 million sperm are ejaculated into the vagina, and only around 100 will reach the ampulla. A man with fewer than 20 million sperm per milliliter is functionally **sterile**.

The Oocyte at Ovulation 761

2. The acrosomal caps of the spermatozoa release **hyaluronidase**, an enzyme that separates cells of the corona radiata and exposes the oocyte membrane. When a single spermatozoon contacts that membrane, fertilization occurs and **oocyte activation** follows. *(see Figure 28.1)*

Pronucleus Formation and Amphimixis 761

3. During activation the secondary oocyte completes meiosis. The **female pronucleus** then fuses with the **male pronucleus**, a process called **amphimixis**. *(see Figure 28.1)*

Prenatal Development 762

1. The 9-month **gestation** period can be divided into three **trimesters**, each 3 months in duration.

The First Trimester 762

2. The **first trimester** is the most critical period in prenatal life. **Cleavage** subdivides the cytoplasm of the zygote in a series of mitotic divisions; the zygote becomes a **blastocyst**. During **implantation** the blastocyst burrows into the uterine endometrium. **Placentation** occurs as blood vessels form around the blastocyst and the **placenta** appears. **Embryogenesis** is the formation of a viable embryo.
3. The blastocyst consists of an outer **trophoblast** and an **inner cell mass**, which is clustered at one end next to a hollow cavity (the **blastocoele**). *(see Figure 28.2 and Table 28.1)*
4. When the blastocyst adheres to the uterine lining, the trophoblast next to the endometrium undergoes changes and becomes a **syncytial trophoblast**, which then erodes a path through the uterine epithelium. As the trophoblast enlarges and spreads, maternal blood flows through trophoblastic channels, or **lacunae**. The blastocyst organizes into layers, and becomes a **blastodisc**. After **gastrulation** the blastodisc contains an embryo composed of **endoderm**, **ectoderm**, and an intervening **mesoderm**. These **germ layers** help form four **extraembryonic membranes**: the yolk sac, amnion, allantois, and chorion. *(see Figures 28.3 to 28.5 and Table 28.1)*
5. The **yolk sac** is an important site of blood cell formation. The **amnion** encloses fluid that surrounds and cushions the developing embryo. The base of the **allantois** later gives rise to the urinary bladder. Circulation within the vessels of the **chorion** provides a rapid transit system linking the embryo with the trophoblast. *(see Figures 28.4/28.5)*
6. **Chorionic villi** extend outward into the maternal tissues, forming an intricate, branching network through which maternal blood flows. The **body stalk** and the **yolk stalk** connect the embryo with the chorion and the yolk sac, respectively. As development proceeds, the **umbilical cord** connects the fetus to the placenta. Blood flow occurs through the umbilical arteries and the umbilical vein, and exchange occurs at the chorionic villi. The placenta synthesizes HCG, estrogens, progestins, HPL, and relaxin. *(see Figures 28.5/28.6)*
7. The **embryo**, which has a **head fold** and a **tail fold**, will undergo critical changes in the first trimester. Events in the first 12 weeks establish the basis for **organogenesis** (organ formation). *(see Figures 28.5b,c/28.7 and Table 28.2)*

The Second and Third Trimesters 772

8. In the **second trimester**, the organ systems near functional completion and the fetus grows rapidly. During the **third trimester**, the organ systems become functional and the fetus undergoes its largest weight gain. *(see Figures 28.7 to 28.9 and Table 28.2)*
9. By the end of gestation, the uterus will be about 12 inches long and contain almost 5 liters of fluid. *(see Figure 28.9)*

Labor And Delivery 774

1. The goal of *true labor* is **parturition**, the forcible expulsion of the fetus.

Stages of Labor 774

2. Labor can be divided into three stages: **dilation stage**, **expulsion stage**, and **placental stage**. The dilation stage, in which the cervix dilates, usually lasts 8 or more hours. The expulsion stage involves the birth (**delivery**) of the fetus. The placenta is ejected during the placental stage. *(see Figure 28.10)*

Premature Labor 774

3. **Premature labor** occurs before the fetus has completely developed. A **premature delivery** produces a newborn weighing over 1 kg, which may or may not survive.

The Neonatal Period 774

1. The **neonatal period** extends from birth to 1 month of age.
2. In the transition from fetus to **neonate**, the respiratory, circulatory, digestive, and urinary systems undergo tremendous changes as they begin functioning independently. The newborn must also begin thermoregulating.

LEVEL 1 REVIEWING FACTS AND TERMS

Match each numbered item with the most closely related lettered item. Use letters for answers in the spaces provided.

Column A	Column B
_____ 1. corona radiata	a. begins with the onset of true labor
_____ 2. hyaluronidase	b. begins as the cervix dilates completely
_____ 3. second trimester	c. birth
_____ 4. expulsion stage	d. characterized by rapid fetal growth
_____ 5. ooplasm	e. incision through the perineal musculature
_____ 6. third trimester	f. layer of follicular cells around the oocyte
_____ 7. first trimester	g. enzyme in the acrosomal cap
_____ 8. parturition	h. cytoplasm of the oocyte
_____ 9. dilation stage	i. time of embryonic/early fetal development
_____10. episiotomy	j. characterized by the continued development of organ systems

11. Sperm cannot perform fertilization until they
    (a) undergo capacitation (b) undergo activation
    (c) lose their acrosome (d) are in the vagina for 3 days
12. During amphimixis
    (a) sperm become capacitated
    (b) the ovum finishes meiosis II
    (c) the male and female pronuclei fuse
    (d) meiosis occurs
13. Which trimester is the most dangerous in prenatal life?
    (a) first (b) second
    (c) third
14. The process of cell division that occurs after fertilization is called
    (a) cleavage (b) implantation
    (c) placentation (d) embryogenesis
15. A blastocyst is a(n)
    (a) extraembryonic membrane that forms blood vessels
    (b) solid ball of cells
    (c) hollow ball of cells
    (d) portion of the placenta
16. The extraembryonic membrane that forms a fluid-filled sac is the
    (a) yolk sac (b) amnion
    (c) allantois (d) chorion
17. The hormone that is the basis for pregnancy tests is
    (a) LH (b) progesterone
    (c) human chorionic gonadotropin (HCG) (d) human placental lactogen (HPL)
18. The first stage of labor is the ___ stage.
    (a) dilation (b) expulsion
    (c) placental (d) decidual
19. The extraembryonic membrane that forms the urinary bladder is the
    (a) yolk sac (b) allantois
    (c) amnion (d) chorion
20. During implantation, the
    (a) syncytial trophoblast erodes a path through the endometrium
    (b) inner cell mass begins to form the placenta
    (c) maternal blood vessels in the endometrium are walled off from the blastocyst
    (d) entire trophoblast becomes syncytial

LEVEL 2 REVIEWING CONCEPTS

1. If a sperm lacked sufficient quantities of hyaluronidase, it would not be able to
   (a) move its flagellum (b) penetrate the corona radiata
   (c) become capacitated (d) survive the environment
2. Problems with the formation of the chorion would affect
   (a) the embryo's ability to produce blood cells
   (b) formation of the limbs
   (c) the embryo's ability to derive nutrition from the mother
   (d) lung formation
3. During implantation, why doesn't the hyaluronidase erode the trophoblast as well as the endometrium?
   (a) the trophoblast is coated with anti-hyaluronidase
   (b) the outer layer of the trophoblast is syncytial, with no cell junctions for the enzyme to attack
   (c) the trophoblast secretes an enzyme that breaks down the hyaluronidase
   (d) the endometrium inactivates the hyaluronidase before it can act on the trophoblast
4. What is cleavage and in which trimester does it occur?
5. What is the primitive streak?
6. Describe the effects of human chorionic gonadotropin. Where and when is it produced?
7. What occurs during amphimixis?
8. Which extraembryonic membrane is important for blood cell formation?
9. Describe the process of organogenesis?
10. Why is the first trimester considered a critical period?

LEVEL 3 CRITICAL THINKING AND CLINICAL APPLICATIONS

1. Joe and Jane desperately want to have children, and although they have tried for two years, they have not been successful. Finally, each of them consults a physician, and it turns out that Joe suffers from oligospermia (low sperm count). He confides to you that he doesn't understand why this would interfere with his ability to have children since he remembers from biology class that it only takes one sperm to fertilize an egg. What would you tell him?
2. Cathy has just given birth to a little girl. When the nurses take the infant back to the nursery and try to feed her, she becomes cyanotic (turns blue). The episode passes, but when the infant is bathed, she becomes cyanotic again. Blood gas levels are taken and they show the arterial blood is only 60% saturated. Physical examinations show that there is nothing structurally wrong with the lungs. What might be causing the problem?

✓ ANSWERS TO CONCEPT CHECK QUESTIONS

p. 762 **1.** The oocyte is surrounded by a layer of follicular cells. The acrosomal cap of the spermatozoa contains hyaluronidase to break down the connections between adjacent follicular cells. One or two spermatozoa is not enough, however, because at least a hundred must release hyaluronidase in order for the connections to break down enough to allow fertilization. **2.** The sperm is not ready for fertilization until it undergoes capacitation, which occurs in the female reproductive tract. **3.** The oocyte undergoes oocyte activation, which includes a sudden rise in metabolism, a change in the cell membrane to prevent fertilization by other sperm, and the completion of meiosis. **4.** The first trimester involves the development of all the major organ systems.

p. 772 **1.** The inner cell mass of the blastocyst eventually develops into the embryo. **2.** The syncytial trophoblast erodes a path through the uterine epithelium. There is some digestion of uterine glands, which release nutrients that are absorbed by the syncytial trophoblast and distributed by diffusion to the inner cell mass. **3.** The mesodermal layer gives rise to the skeletal, muscular, endocrine, cardiovascular, lymphatic, urinary, reproductive, lining of body cavities, and forms connective tissues. **4.** The placenta contains the maternal arteries and single umbilical vein, permits nutrient exchange at the chorionic villi, and it synthesizes hormones, which are important to the mother and to the embryo.

p. 776 **1.** The organ systems continue development, and the embryo grows rapidly, though more notably in height than in weight. **2.** The dilation stage, in which the cervix dilates and the fetus slides down the cervical canal, usually takes 8 or more hours. **3.** The kidneys, while able to filter wastes, cannot yet concentrate the urine. This means that the newborn loses a large amount of water through urination and must take in a large amount of fluid to make up for the loss. **4.** During the placental stage, the empty uterus contracts and decreases in size, which breaks the connection between the endometrium and the placenta, causing the expulsion of the placenta.

REVIEW IT

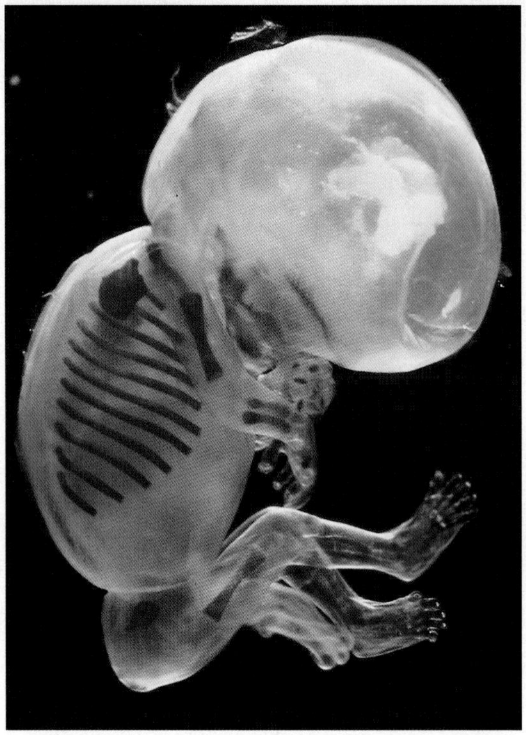

Challenge 1

The figure seen here is a photograph of a 10-week old human fetus. The fetus has been stained with alizarin red, indicating developing skeletal elements. The rest of the specimen was then cleared to allow maximum observation of the developing skeletal system. What portions of the adult skeleton can you recognize in this fetus? Make a list indicating the names of the adult bones you see.

To check your answers, go to the Interactive CD/Animations/Skeletal System/Whole Skeleton. Compare the red elements seen here with the entire adult skeleton. Correct your list if need be. Why do you suppose these bones develop first? Can you identify areas that begin as separate ossification centers in the fetus, but become one skeletal element in the adult? Why do you suppose these areas exist?

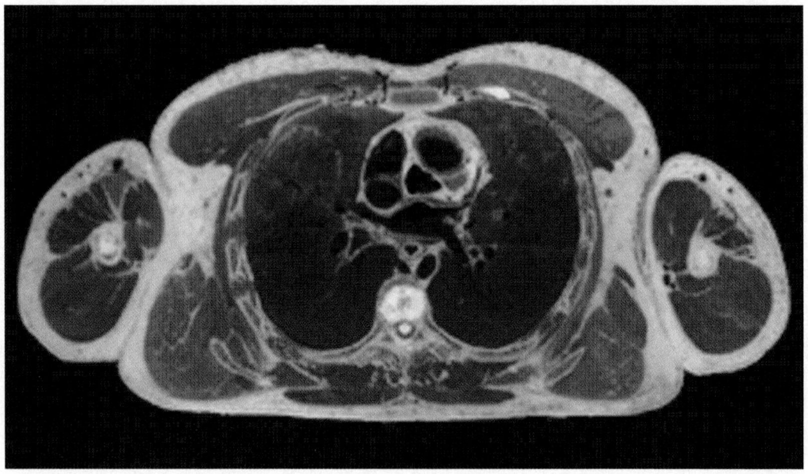

Challenge 2

This image comes from the male Visible Human data set. Identify the approximate location of this section. Now consider this section from a developmental standpoint. What structures developed first in the embryo? Can you list the visible structures in order of development? Title that list "Embryologic Development". Prepare a second list titled "Post-natal Development." This list will identify the changes in these organs during adult life; enlargement of the organs is one such change. Can you come up with others?

Images provided by the Digital Cadaver™ project, courtesy of Visible Productions, Inc.

APPLY IT

Below are two separate exercises that provide more information on the preceding topics. These exercises differ from the usual critical linking exercises found in the media spreads in that these are designed to help you understand the interactions among systems, rather than hone your knowledge of one particular system. As before, each activity is designed to take approximately 10 minutes, and will help you gain a better appreciation for the material presented in the text.

Critical Linking 1

Pregnancy is a time of wonder for many couples. As you have learned, much is occurring that will shape the health of the future infant. How much of that development is genetically controlled, and how much is affected by the environment of the developing fetus?

The US Food and Drug Administration closely monitors the nutritional health of pregnant women through research into the various claims both for and against food products and supplements during pregnancy. We are usually clear on what is detrimental to the health of the developing fetus, but less certain of what is beneficial. List at least 10 *food* items (not pharmaceuticals of any sort) that you think are helpful and 10 that you think are harmful to embryological development. Do you know why you considered these to be helpful or harmful? What specific developmental process is facilitated or hindered? To verify your conclusions and expand your list, visit the Companion Website, Critical Linking and choose the key word "USFDA."

Critical Linking 2

The definition of congenital defect is one that is present at birth. These are usually due either to teratogens introduced during development, or to a genetic anomaly present at conception. Congenital heart defects are among the most well studied, affecting nearly 1% of the births in the United States. These defects commonly include such troubles as: abnormalities impeding the flow of blood through the vessels, malformed or missing valves, structural problems that allow blood to flow from one side of the heart to the other or problems with the connection of the heart to the major vessels. Reflecting on what you have learned of the heart and its role in homeostasis, predict the symptoms of two of these defects: patent (open) ductus arteriosus and transposition of the great vessels. How are these defects detected? What can be done to correct these defects? Prepare a short essay based on your understanding of the cardiovascular system and development.

Refine your thoughts by visiting the Companion Website, Critical Linking section and choosing the key word "congenital defects." Read the general description of cardiac congenital defects, then click on the terms "Patent ductus arteriosus" and "Transposition of the great arteries." These will lead you to short descriptions of each defect. How well did your predictions match the information presented? Refine your answers if needed.

FURTHER STUDIES

You have just completed a thorough study of the anatomy of the human body. Although the majority of the information presented in this text was scientific in nature, each of the 10 preceding Connections has included one section such as this that dealt with information outside the usual boundaries of anatomy. In this final media spread, you are presented with a topic that every scientist should consider, specifically how will you put your scientific knowledge to use? There is growing concern among the general population that the goals of science and scientific knowledge are moving beyond societal control. Do we know too much? Are we pushing the boundaries of morality?

As a case in point, cloning through the manipulation of pluripotent cells has been discussed in scientific circles since Hans Spemann proposed taking an adult nucleus and implanting it in an enucleated pluripotent stem cell in 1938. *(http://www.genesage.com/professionals/geneletter/archives/historyofcloning.html)* Since that time, scientists have been working toward an understanding of differentiation and cloning. By 1983, sheep and mice were successfully cloned. Although there was some question as to the validity of the mouse experiments, by 1986 a University of Wisconsin team (Neal First, Randall Prather, Willard Eysestone) irrefutably cloned a cow from an embryonic cell. Stem cell research was a natural offshoot of these experiments. If this could be done for humans, might there be tremendous medical gains? The push to discover ways in which to do this has led to controversy as well as experimental success. Frozen human embryos, destined to be implanted in females unable to conceive naturally, have come under scrutiny. The research community would like to use these embryos as a source of human stem cells for further study. However, in the year 2001 the United States passed laws restricting the use of stem cells in research in the United States. An exodus of leading geneticists and embryological scientists ensued. Are we using our knowledge wisely? To read a primer on the stem cell debate prepared by the NIH, type *http://www.nih.gov/news/stemcell/primer.htm* into your Internet browser. For a review of current issues in science, ethics, and religion, visit the American Association for the Advancement of Science, Dialog on Science, Ethics, and Religion Webpage *(http://www.aaas.org/spp/dser/)*. As you continue in your scientific education, please keep these issues in mind. Our future requires thoughtful consideration of science, ethics and religion.

A

APPENDIXES

CLINICAL ISSUES

1 AN INTRODUCTION TO ANATOMY

Anatomy and Observation PAGE 14

Technical innovation has affected modern medicine in many ways, but some of the most important diagnostic procedures have not changed significantly in a thousand years. The next time you visit a physician, pay attention to the way information is obtained. First, your chart is reviewed, giving the doctor information about past or preexisting problems. Then you are asked to describe the reasons for your visit.

As you talk, the doctor not only listens carefully, but also watches closely. This is really the first stage in physical assessment, a time for general questions such as, "Is the patient moving, speaking, and thinking normally?" The answers will later be integrated with the results of more precise observations.

A physical examination typically follows. There are four basic components to a physical examination:

1. **Inspection:** Inspection is careful observation. A general inspection involves examining body proportions, posture, and patterns of movements. Local inspection is the examination of sites or regions of suspected injury. Of the four components of the physical exam, inspection is often the most important because it provides the largest amount of useful information. Many diagnostic conclusions can be made on the basis of inspection alone; most skin conditions, for example, are identified in this way. A number of endocrine problems and inherited metabolic disorders can produce subtle changes in body proportions that are easily overlooked by the untrained eye.
2. **Palpation:** In palpation the physician uses hands and fingers to feel the body. This procedure provides information on skin texture and temperature, the presence and nature of abnormal masses, the pattern of the pulse, and the location of tender spots. Once again, the procedure relies on an understanding of normal anatomy. A small, soft, lumpy mass in one spot is a salivary gland; in another location it could be a tumor. The importance of a tender spot can be recognized in diagnosis only if the observer knows what organs lie beneath it.
3. **Percussion:** Percussion is tapping with the fingers or hand to obtain information about the densities of underlying tissues. For example, the chest normally produces a hollow sound, because the lungs are filled with air. That sound changes in pneumonia, when the lungs contain large amounts of fluid. Of course, to get the clearest chest percussions, the fingers must be placed on the right spots.
4. **Auscultation:** Auscultation (aws-kul-TĀ-shun; *auscultare*, to listen) is listening to body sounds, usually using a stethoscope. This technique is particularly useful for checking the condition of the lungs during breathing. The wheezing sound heard in asthma is caused by constriction of the airways, and pneumonia produces a gurgling sound, indicating that fluid has accumulated in the lungs. Auscultation is also important in diagnosing heart conditions. Many cardiac problems affect the sound of the heartbeat or produce abnormal swirling sounds (heart murmurs) during blood flow.

The entire process of physical examination relies on one fact: The doctor already knows the superficial and deep anatomy of the human body. To detect the abnormal, you must first understand the normal.

2 THE CELL

Causes of Cancer PAGE 45

A relatively small number of cancers are actually inherited; 18 different types have been identified to date, including two forms of leukemia. Most cancers develop through the interaction of genetic and environmental factors. They will be discussed individually, although in practice it is difficult to separate the two completely.

GENETIC FACTORS Two related genetic factors are involved in the development of cancer: *hereditary predisposition* and *oncogene activation*. An individual born with genes that increase the likelihood of cancer has a **hereditary predisposition** for the disease. Under these conditions a cancer is not guaranteed, but it is highly likely. The inherited genes usually affect tissue abilities to metabolize toxins, control mitosis and growth, perform repairs after injury, or identify and destroy abnormal tissue cells. As a result, body cells become more sensitive to local or environmental factors that would have little effect on normal tissues.

Cancers may also result from idiopathic (of no known cause) somatic mutations that modify genes involved with cell growth, differentiation, or mitosis. As a result, an ordinary cell converts into a cancer cell. The modified genes are called **oncogenes** (ONG-kō-jēnz); the normal genes are called **proto-oncogenes.** Oncogene activation occurs through the alteration of normal somatic genes. Because these mutations do not affect reproductive cells, the cancers caused by active oncogenes are not inherited.

A proto-oncogene, like other genes, has a regulatory component (gene ON/OFF) and a structural component that contains the triplets that determine protein structure. Mutations in either portion of the gene may convert it to an active oncogene. A small mutation can accomplish this conversion; changing one nucleotide out of a chain of 5000 can convert a normal proto-oncogene to an active oncogene. In some cases, a viral infection can trigger activation of an oncogene. For example, one of the papilloma viruses appears to be responsible for many cancers of the cervix.

More than 50 oncogenes have now been identified. In addition, a group of anticancer genes has now been identified. These genes, called **tumor-suppressing genes** (TSG), or **anti-oncogenes,** suppress mitosis and growth in normal cells. Mutations that alter these genes make oncogene activation more likely. TSG mutation has been suggested as important in promoting several cancers, including several blood cell cancers, breast cancer, and ovarian cancer.

ENVIRONMENTAL FACTORS Many cancers can be directly or indirectly attributed to environmental factors called **carcinogens** (kar-SIN-ō-jenz). Carcinogens are substances that stimulate the conversion of a normal cell to a cancer cell. Some carcinogens are **mutagens,** compounds that damage DNA strands and sometimes cause chromosomal breakage. Radiation is an example of a mutagen that has carcinogenic effects.

There are many different carcinogens in the environment. Plants manufacture poisons that protect them from insects and other predators, and although their carcinogenic activities are relatively weak, many common spices, vegetables, and beverages contain compounds that can be carcinogenic if consumed in large quantities. Animal tissues may also store or concentrate toxins, and these may be swallowed in contaminated meals. Cosmic radiation, X-rays, and other radiation sources can also cause cancer.

Specific carcinogens will affect only those cells capable of responding to that particular physical or chemical stimulus. The responses vary because differentiation produces cell types with specific sensitivities. For example, benzene can produce a cancer of the blood, cigarette smoke a lung cancer, and vinyl chloride a liver cancer. Very few stimuli can produce cancers throughout the body; radiation exposure is a notable exception. In general, cells undergoing mitosis are more likely to respond to chemical or radiational carcinogens than are cells in interphase. As a result, cancer rates are highest in epithelial tissues, where frequent stem cell divisions keep pace with the replacement of cells damaged or lost at the surface, and relatively low in neural and muscle tissues, where stem cell divisions do not normally occur.

3 THE TISSUE LEVEL OF ORGANIZATION

Infants and Brown Fat PAGE 68

Infants' body temperatures are more unstable than those of adults. They cannot shiver, and they lose heat more rapidly because of their relatively large surface-to-volume ratios. The adipose tissue between the shoulder blades, around the neck, and possibly elsewhere in the upper body is different from most of the adipose tissue in the adult. It is highly vascularized, and the individual adipocytes contain numerous mitochondria. Together these characteristics give the tissue a deep, rich color responsible for the name **brown fat.** When stimulated by the nervous system, lipid breakdown accelerates; the cells do not capture the energy released, and it radiates into the surrounding tissues as heat. The circulating blood is warmed by this heat and also distributes it throughout the body. In this way an infant can accelerate metabolic heat generation by 100% very quickly. There is little if any brown fat in the adult; with larger body size, skeletal muscle mass, and insulation, shivering is significantly more effective in elevating body temperature.

Liposuction PAGE 68

Liposuction is a surgical procedure for the removal of unwanted adipose tissue. In liposuction, a small incision is made through the skin, and a tube is inserted into the underlying adipose tissue. Suction is then applied, and chunks of tissue containing adipocytes, other cells, fibers, blood, lymph, and ground substance are removed.

This practice has received a lot of attention in the news, and many advertisements praise the technique as easy, safe, and effective. In fact, it is (1) not always easy, (2) sometimes dangerous, and (3) of limited effectiveness. The density of adipose tissue varies according to body location and from individual to individual, and adipose tissue is not always easy to suck through a tube. An anesthetic must be used to control pain, and this poses risks. Blood vessels are stretched and torn, and extensive bleeding and fluid loss can occur. One serious complication is the potential for infection after the liposuction treatment.

Finally, adipose tissue can repair itself, and adipocyte populations recover over time. The only way to ensure that fat lost through liposuction will not return is to adopt a lifestyle that includes a proper diet and adequate exercise. In fact, such a lifestyle could produce the same weight loss *without* liposuction, eliminating the surgical expense and risk.

Cartilages and Knee Injuries PAGE 71

The knee is an extremely complex joint that contains both hyaline cartilage and fibrocartilage. The hyaline cartilage caps bony surfaces, while pads of fibrocartilage within the joint prevent bone-to-bone contact when movements are under way. Many sports injuries involve tearing of the fibrocartilage pads or supporting ligaments. The loss of support and cushioning places more strain on the hyaline cartilages within joints and leads to further joint damage. Articular cartilages are not only avascular, but they also lack a perichondrium. As a result, they heal even more slowly than other cartilages. Surgery usually produces only a temporary or incomplete repair. For this reason, most competitive sports have rules designed to reduce the number of knee injuries. For example, in football "clipping" is outlawed because it produces stresses that can tear the fibrocartilages and the supporting ligaments at the knee.

Recent advances in tissue culture have enabled researchers to grow fibrocartilage in the laboratory. Chondrocytes removed from the knees of injured dogs are cultured in an artificial framework of collagen fibers. They eventually produce masses of fibrocartilage that can be inserted into the damaged joints. Over time the pads change shape and grow, restoring normal joint function. In the future this technique may be used to treat severe knee injuries in humans.

Problems with Serous Membranes PAGE 76

If serous membranes are damaged, the production of transudate fluid may cease, and the parietal and visceral mesothelia will rub against one another, producing an abrasion. This mesothelial damage attracts fibroblasts, which migrate into the damaged area and bind the opposing membranes together with a network of collagen fibers. Although this binding reduces friction by limiting movement, the collagen fibers may compress blood vessels, nerves, or other vital structures in the region. These restrictive fibrous connections are called **adhesions**. Adhesions may occur as a result of surgery, infection, or other injuries that damage serous membranes.

Several clinical conditions, including infection and chronic irritation, can cause the abnormal buildup of fluid within one of the ventral body cavities. **Pleurisy**, or **pleuritis**, is an inflammation of the pleural cavities. At first the membranes become inflamed and rough, and the opposing membranes may scratch against one another. This scratching produces a characteristic sound known as a **pleural rub**. Adhesions occasionally form between the serous membranes of the pleural cavities. More frequently, continued inflammation and rubbing lead to a gradual increase in the production of fluid to levels well above normal. Fluid then accumulates in the pleural cavities, producing a condition known as **pleural effusion**. Pleural effusion can also be caused by heart conditions that elevate the pressure in blood vessels of the lungs. As fluids build up in the pleural cavities, the lungs are compressed, making it difficult to breathe. The combination of large pleural effusions and heart disease can be lethal.

Pericarditis is an inflammation of the pericardium. This condition often leads to a **pericardial effusion**: an abnormal accumulation of fluid in the pericardial cavity. When sudden or severe, the fluid buildup can seriously reduce the efficiency of the heart and reduce blood flow through major vessels. Pericarditis can be infectious or a complication of a heart attack or heart surgery.

Peritonitis, an inflammation of the peritoneum, can occur after an infection of or injury to the peritoneal lining. Peritonitis is a potential complication of any surgical procedure that requires opening the peritoneal cavity. Liver disease, kidney disease, or heart failure can cause an increase in the rate of fluid movement through the peritoneal lining. The accumulation of fluid, called **ascites** (a-SĪ-tēz), creates a characteristic abdominal swelling. Distortion of internal organs by the contained fluid can result in a variety of symptoms; heartburn, indigestion, and low back pain are common complaints. The condition of heartburn will be discussed in greater detail in the digestive system.

Tissue Structure and Disease PAGE 79

Pathologists (pa-THOL-ō-jists) are physicians who specialize in the anatomical diagnosis of disease processes. In their analyses they integrate anatomical and histological observations to determine the nature and extent of the disease.

Disease processes affect the histological organization of tissues and organs. As an example, consider the histological changes induced in the respiratory epithelium by one relatively common irritating stimulus, cigarette smoke. The first abnormality to be observed is **dysplasia** (dis-PLĀ-zē-a), a change in the normal shape, size, and organization of tissue cells. It is usually a response to chronic irritation or inflammation, and the changes are reversible if the individual stops smoking.

Epithelia and connective tissues may undergo more radical changes in structure, caused by the division and differentiation of stem cells. **Metaplasia** (me-ta-PLĀ-zē-a) is a structural change that dramatically alters the character of the tissue. Metaplasia is also a reversible condition. During **anaplasia** (a-na-PLĀ-zē-a) tissue cells change size and shape, often becoming unusually large or abnormally small. In anaplasia tissue organization breaks down, and a tumor forms. Unlike dysplasia and metaplasia, anaplasia is irreversible.

Cancer Treatment and Statistics PAGE 82

There is no single, universally effective cure or preventive measure for cancer; there are too many separate causes, possible mechanisms, and individual differences. The goal of cancer treatment is to achieve **remission**. A tumor in remission either ceases to grow or decreases in size. Basically, the treatment of malignant tumors must accomplish one of the following to produce remission.

1. **Surgical removal or destruction of individual tumors:** Tumors that contain malignant cells can be surgically removed or destroyed by radiation, heat, or freezing. These techniques are very effective if the treatment is undertaken before metastasis has occurred. For this reason early detection is important in improving survival rates for all forms of cancer.
2. **Killing metastatic cells throughout the body:** This treatment is much more difficult and potentially dangerous, because healthy tissues are likely to be damaged at the same time. At present the most widely approved treatments are chemotherapy and radiation.

Chemotherapy involves the administration of drugs that will either kill the cancerous tissues or prevent mitotic divisions. These drugs often affect stem cells in normal tissues, and the side effects are often unpleasant. For example, because chemotherapy slows the regeneration and maintenance of epithelia of the skin and digestive tract, patients lose their hair and experience nausea and vomiting. Several drugs are often administered simultaneously or in sequence, because cancer cells can develop a resistance to a single drug. This strategy may also reduce the severity of any side effects. Chemotherapy is often used in the treatment of many kinds of metastatic cancer.

Hormone manipulation may offer a less toxic form of chemotherapy. The compounds used may suppress tumor growth or hormones that stimulate tumor growth. Examples include *Tamoxifen*, used to treat breast cancer, and *antigonadotropins* or *estrogens* used to treat prostate cancer.

Localized radiation is often used to destroy cancers confined to a small region of the body. Massive doses of radiation are necessary if the cancer has metastasized widely. For example, in advanced cases of lymphoma, a cancer of the immune system, enough radiation is administered to kill all of the blood-forming cells in the body. After treatment, new blood cells must be provided by a *bone marrow transplant*.

An understanding of molecular mechanisms and cell biology is leading to new approaches that may revolutionize the treatment of cancer. One approach focuses on the fact that cancer cells are usually ignored by the immune system. In **immunotherapy**, substances are administered that help the immune system recognize and attack the cancer cells. More elaborate experimental procedures involve the creation of customized antibodies using gene-splicing techniques. The resulting antibodies are specifically designed to attack the tumor cells in one particular patient. Although this technique shows promise, it remains difficult, costly, and very labor-intensive.

Another promising approach is based on the identification of specific oncogenes and their corresponding proteins. One cancer, chronic myeloid leukemia, is caused, in its initial chronic state, by genetic changes that lead to the production of an abnormal

TABLE A.1

CANCER INCIDENCE AND SURVIVAL RATES IN THE UNITED STATES

Site	Estimated New Cases (2002)	Estimated Deaths (2002)	Five-Year Survival Rates Diagnosis Date 1960–1963	Diagnosis Date 1992–1997
DIGESTIVE TRACT				
Esophagus	13,100	12,600	2.5%	14%
Stomach	21,600	12,400	9.5%	21%
Colon and rectum	148,300	56,600	36.0%	61%
RESPIRATORY TRACT				
Lung and bronchi	169,400	154,900	6.5%	15%
URINARY TRACT				
Kidney and other urinary structures	34,200	12,300	37.5%	62%
Urinary bladder	54,306	12,600	38.5%	81%
REPRODUCTIVE SYSTEM				
Breast	203,500	39,600	54.0%	86%
Ovary	23,300	13,900	32.0%	52%
Testis	7,500	400	63.0%	95%
Prostate gland	189,100	30,200	42.5%	96%
NERVOUS SYSTEM	17,000	13,100	18.5%	>26%
CARDIOVASCULAR SYSTEM[B]	114,600	62,200	14.0%[a]	>46%[a]
SKIN (MELANOMA ONLY)	53,600	7,400	60.0%	89%

Data courtesy of the American Cancer Society.
[a]Acute lymphocytic leukemia only.
[b]Includes cancers of blood forming tissues.

enzyme, *tyrosine kinase*. A chemical has been found that inhibits this abnormal enzyme. In preliminary trials, this chemical, brand name *Gleevec*, produced complete remission in some patients and generally reversed the progression of the disease.

INCIDENCE AND SURVIVAL RATES A statistical profile of cancer incidence and survival rates in the United States is presented in Table A-1. Interestingly, the profile would be different for other countries, reflecting different environmental conditions, hereditary factors, exposure to infectious diseases, and cultural dietary preferences. Bladder cancer is common in Egypt, stomach cancer in Japan, and liver cancer in Africa.

Advances in chemotherapy, radiation procedures, and molecular biology have produced significant improvements in the survival rates for several types of cancer. The improved survival rates indicated in Table A-1, however, reflect advances not only in therapy, but also in early detection. Much of the credit goes to increased public awareness and concern about cancer. In general, the odds of survival increase markedly if the cancer is detected early, especially before it undergoes metastasis. Despite the variety of possible cancers, the American Cancer Society has identified seven "warning signs" that mean it's time to consult a physician. These warning signs are presented in Table A-2. Screening people for pre-cancerous lesions and early cancer reduces the death rate from cervical, breast and colon cancer.

TABLE A.2

SEVEN WARNING SIGNS OF CANCER

- Change in bowel or bladder habits
- A sore that does not heal
- Unusual bleeding or discharge
- Thickening or lump in breast or elsewhere
- Indigestion or difficulty in swallowing
- Obvious change in wart or mole
- Nagging cough or hoarseness

4 THE INTEGUMENTARY SYSTEM

Psoriasis and Xerosis PAGE 91

Excessive production of keratin is called **hyperkeratosis** (hī-per-ker-a-TŌ-sis). The effects are easily observed as calluses and corns. Calluses appear on thick-skinned areas exposed to mechanical stress, such as the palms of the hands or the heels. Corns are more localized and form on or between the toes.

In **psoriasis** (sō-RĪ-a-sis) the stratum germinativum becomes unusually active in specific areas, including the scalp, elbows, palms, soles, groin, and nails. Normally an individual stem cell divides once every 20 days, but in psoriasis it may divide every day-and-a-half. Keratinization is abnormal and often incomplete by the time the outer layers are shed. The affected areas are red, and covered with small white scales that continually flake away. Most cases are painless and controllable.

Psoriasis appears in 20–30% of the individuals with an inherited tendency for the condition. Roughly 5% of the general U.S. population has psoriasis to some degree, often triggered by stress and anxiety.

Xerosis (zē-RŌ-sis), or "dry skin," is a common complaint of older persons and almost anyone living in an arid climate. Under these conditions the cell membranes deteriorate, and the stratum corneum becomes more a collection of scales than a unified covering. This condition can increase the rate of water loss by insensible perspiration by 75 times.

Skin Cancers PAGE 93

Almost everyone has several benign lesions of the skin; freckles and moles are examples. Skin cancers are the most common form of cancer, and the most common skin cancers are caused by prolonged exposure to sunlight.

A **basal cell carcinoma** is a malignant cancer that originates in the germinativum (basal) layer. This is the most common skin cancer, and roughly two-thirds of these cancers appear in areas subjected to chronic UV exposure. These carcinomas have recently been linked to an inherited gene.

Squamous cell carcinomas are less common, but almost totally restricted to areas of sun-exposed skin. Metastasis seldom occurs in squamous cell carcinomas and almost never in basal cell carcinomas, and most people survive these cancers. The usual treatment involves surgical removal of the tumor, and at least 95% of patients survive 5 years or more after treatment. (This statistic, the 5-year survival rate, is a common method of reporting long-term prognoses.)

Compared with these common and seldom life-threatening cancers, **malignant melanomas** (mel-a-NŌ-maz) are extremely dangerous. In this condition, cancerous melanocytes grow rapidly and metastasize throughout the lymphatic system. The outlook for long-term survival is dramatically different, depending on when the condition is diagnosed. If the condition is localized, the 5-year survival rate is 90%; if widespread, the survival rate drops to 14%.

Fair-skinned individuals who live in the tropics are most susceptible to all forms of skin cancer, because their melanocytes are unable to shield them from the ultraviolet radiation. Sun damage can be prevented by avoiding exposure to the sun during the middle hours of the day and by using clothing, a hat, and a sunblock (not a tanning oil or sunscreen)—a practice that also delays the cosmetic problems of sagging and wrinkling. Everyone who expects to be out in the sun for any length of time should choose a broad-spectrum sunblock with a sun protection factor (SPF) of at least 15; blonds, redheads, and people with very fair skin are better off with a sun protection factor of 20 to 30. (One should also remember the risks before spending time in a tanning salon or tanning bed.)

The use of sunscreens has now become even more important as the ozone gas in the upper atmosphere is destroyed by our industrial emissions. Ozone absorbs UV before it reaches the earth's surface, and in doing so, it assists the melanocytes in preventing skin cancer. Australia, which is most affected by the depletion of the ozone layer near the South Pole (the "ozone hole"), is already reporting an increased incidence of skin cancers.

Dermatitis PAGE 94

Dermatitis is an inflammation of the skin that involves primarily the papillary region of the dermis. The inflammation usually begins in a portion of the skin exposed to infection or irritated by chemicals, radiation, or mechanical stimuli. Dermatitis may cause no physical discomfort, or it may produce an annoying itch, as in poison ivy. Other forms of this condition can be quite painful, and the inflammation may spread rapidly across the entire integument. Some common forms of dermatitis are:

1. **Contact dermatitis** usually occurs in response to strong chemical irritants or in an allergic response to a foreign compound. It produces an itchy rash that may spread to other areas as scratching distributes the chemical agent; poison ivy is an example.
2. **Eczema** (EK-se-ma), or *ectopic dermatitis,* is a dermatitis that can be triggered by temperature changes, fungus, chemical irritants, greases, detergents, or stress. Hereditary or environmental factors or both can encourage the development of eczema.
3. **Diaper rash** is a localized dermatitis caused by a combination of moisture, irritating chemicals from fecal or urinary wastes, and flourishing microorganisms.
4. **Urticaria** (ur-ti-KAR-ē-a), also known as **hives**, is an extensive allergic response to food, drugs, an insect bite, infection, stress, or other stimuli.

Tumors in the Dermis PAGE 95

Tumors seldom develop in the dermis, and those that do appear are usually benign. Two forms of benign tumors called **hemangiomas** may appear among dermal blood vessels during embryonic development. Viewed from the surface, these form prominent *birthmarks*. A **capillary hemangioma** involves capillaries of the papillary layer. It usually enlarges after birth, but subsequently fades and disappears. **Cavernous hemangiomas**, or "port-wine stains," affect larger vessels in the dermis. Such birthmarks usually last a lifetime unless treated.

Baldness and Hirsutism PAGE 99

Two factors interact to determine baldness. A bald individual has a genetic susceptibility, but this tendency must be triggered by large quantities of male sex hormones. Testosterone is converted to a related compound within hair follicles. In genetically predisposed men, this compound causes conversion of terminal hairs to vellus hairs. This results in baldness. A medicine called *Propecia* (finasteride) blocks the conversion of testosterone and in one study reduced hair loss by up to 75%. Many women carry the genetic background for baldness, but unless major hormonal abnormalities develop, as in certain endocrine tumors, nothing happens to their hair.

Male pattern baldness affects the top of the head first, only later reducing the hair density along the sides. Thus hair follicles can be removed from the sides and implanted on the top or front of the head, temporarily delaying a receding hairline. This procedure is rather expensive (thousands of dollars), and not every transplant is successful.

Alopecia areata (al-ō-PĒ-shē-ah ar-ē-A-ta) is a localized hair loss that can affect either sex. The cause is not known, and the severity of hair loss varies from case to case. This condition is associated with several disorders of the immune system; it has also been suggested that periods of stress may promote alopecia areata in individuals already genetically prone to baldness.

Skin conditions that affect follicles can contribute to hair loss. Baldness can also result from exposure to radiation or to many of the toxic (poisonous) drugs used in cancer therapy. Both radiation and anticancer drugs have their greatest effects on rapidly dividing cells, and thus tend to damage matrix cells.

Hairs are dead, keratinized structures, and no amount of oiling, shampooing, or dousing with kelp extracts, vitamins, or nutrients will influence either the exposed hair or the follicle buried in the dermis. Untested treatments for baldness were banned by the Food and Drug Administration in 1984. Minoxidil, a drug originally marketed for the control of blood pressure, appears to stimulate hair follicles when rubbed onto the scalp. It is now available on a nonprescription basis. Treatment involves applying a 2–4% solution to the scalp twice daily; after 4 months, over one-third of patients reported satisfactory results.

Hirsutism (HER-sūt-izm; *hirsutus*, bristly) refers to the growth of hair on women in patterns usually characteristic of men. Because considerable overlap exists between the two sexes in terms of normal hair distribution, and there are significant racial and genetic differences, the precise definition is more often a matter of personal taste than objective analysis. Age and sex hormones may play a role, for hairiness increases late in pregnancy, and menopause produces a change in body hair patterns. Severe hirsutism may be associated with abnormal androgen (male sex hormone) production, either in the ovaries or adrenal glands. A cream (*Vaniqa*) which reduces cell division and differentiation has been shown to reduce facial hair growth in some women.

Acne PAGE 100

Individuals with a genetic tendency toward acne have larger than average sebaceous glands, and when the ducts become blocked the secretions accumulate. Inflammation develops, and bacterial infection may occur. The condition usually surfaces at puberty, as production of sex hormones accelerates. The secretory output of the sebaceous glands may be further encouraged by anxiety, stress, physical exertion, certain foods, and drugs.

The visible signs of acne are called **comedos** (ko-MĒ-dŪz). "Whiteheads" contain accumulated, stagnant secretions. "Blackheads" contain more solid material that has been invaded by bacteria. Although neither condition indicates the presence of dirt in the pores, washing may help to reduce superficial oiliness.

Acne usually fades after sex hormone concentrations stabilize. Peeling agents may help reduce inflammation and minimize scarring. In cases of severe acne, the most effective treatment usually involves the discouragement of bacteria by the administration of topical or systemic antibiotic drugs. Topical or systemic vitamin A derivatives, such as tretinoin (*Retin-A*), increase the rate of epithelial all turnover and reduce sebaceous gland activity. These medications can be very effective.

Inflammation Of The Skin PAGE 104

The epidermis provides significant protection from mechanical and chemical hazards because of its thick cell layers and keratin covering. Although the surface of the skin normally contains a variety of microorganisms, most of them are harmless as long as they remain outside the stratum corneum. Penetration of the superficial layers may produce familiar conditions such as "athlete's foot" (a fungal infection), and warts and cold sores (both viral infections). To reach the underlying connective tissues, a bacterium must survive the bacteriocidal ingredients of sebum, avoid being flushed from the surface by the sweat gland secretions, penetrate the stratum corneum, squeeze between the junctional complexes of deeper layers, escape the Langerhans cells, and cross the basement membrane. Unfortunately, once there, the papillary layer of the dermis provides all the elements for the growth of microorganisms: warmth, darkness, moisture, and nutrients.

If the protective barriers are crossed, or if an injury breaks through the epidermis, foreign materials or pathogens reach the underlying tissues. Mast cells within the dermis then respond by triggering a powerful inflammatory response. Inflammation begins immediately after an injury and produces swelling, redness, heat, and pain. Inflammation in the skin is important in defending the body against injury and disease. Inflammation and regeneration are controlled at the tissue level. The two phases overlap; inflammation establishes a framework that guides the cells responsible for reconstruction, and repairs are under way well before cleanup operations have ended.

INFLAMMATION Many stimuli can produce inflammation, including impact, abrasion, distortion, chemical irritation, infection by pathogenic organisms (such as bacteria or viruses), or extreme temperatures (hot or cold). The common factor is that each of these stimuli either kills cells, damages fibers, or injures the tissue in some other way. These changes alter the chemical composition of the interstitial fluid: Damaged cells release prostaglandins, proteins, and potassium ions, and the injury itself may have introduced foreign proteins or pathogens.

Immediately after the injury, tissue conditions become even more abnormal. **Necrosis** (ne-KRŌ-sis) is the tissue degeneration that occurs after cells have been injured or destroyed. This process begins several hours after the initial event, and the damage is caused by lysosomal enzymes. Lysosomes break down through autolysis, releasing digestive enzymes that first destroy the injured cells and then attack surrounding tissues. The accumulation of debris, fluid, dead and dying cells, and other necrotic tissue components that may result is known as **pus**. Pus often forms at an infection site in the dermis. An accumulation of pus in an enclosed tissue space is called an **abscess**.

These tissue changes trigger the inflammatory response by releasing chemicals that stimulate mast cells. When an injury occurs that damages fibers and cells, the stimulated mast cells release chemicals (**histamine** and **heparin**). Activation of mast cells results in three major changes:

1. Histamine relaxes the smooth muscle tissue in the vessel walls, and the vessels enlarge, or **dilate**. This dilation increases the blood flow through the tissue, giving the region a reddish color and making it warm to the touch. The increased blood flow accelerates the delivery of nutrients and oxygen and the removal of waste products and toxic chemicals. It also brings white blood cells to the region. These cells migrate into the injury site and assist in the defense and cleanup operations.
2. Histamine makes the endothelial cells of local capillaries more permeable. Plasma, including blood proteins, diffuses into the injured tissue, and the area becomes swollen. Once in the tissue, some of the blood proteins combine to form large, insoluble fibers, known as **fibrin**. The heparin released by mast cells prevents the formation of fibrin at the injury site, but a dense fibrous meshwork, called a **clot**, surrounds the damaged area. The clot walls off the inflamed region, isolating it and slowing the spread of cellular debris or bacteria into adjacent tissues.
3. Stimulated macrophages and newly arrived white blood cells defend the tissue and perform cleanup operations. Fixed macrophages, free macrophages, and microphages engulf debris and bacteria. Lymphocytes convert to plasma cells, producing antibodies that attack pathogens and foreign proteins. Usually the combination of physical attack, through phagocytosis, and chemical attack, through antibodies, succeeds in cleaning up the region and eliminating the inflammatory stimulus.

The combination of abnormal tissue conditions and chemicals released by mast cells also stimulate neurons which generate pain sensations. These sensations trigger immediate reflexive action (such as pulling the affected part of the body away from the source of the injury) that reduces the chances for further damage.

REGENERATION When damage extends through the epidermis and into the dermis, bleeding usually occurs. The fibrin clot, or **scab**, that forms at the surface temporarily restores the integrity of the epidermis and restricts the entry of additional microorganisms and further blood loss. Cells of the stratum germinativum undergo rapid divisions and begin to migrate along the sides of the wound in an attempt to replace the missing epidermal cells.

If the wound covers an extensive area or involves a region covered by thin skin, dermal repairs must be under way before epithelial cells can cover the surface. Fibroblast and mesenchymal cell divisions produce mobile cells that invade the deeper area of injury, migrating along fibrin strands. Endothelial cells of damaged blood vessels also begin to divide, and capillaries follow the fibroblasts, eventually uniting and providing a circulatory supply. The combination of fibrin clot, fibroblasts, and an extensive capillary network is called **granulation tissue**. Over time, the fibrin dissolves, and the number of capillaries declines. Fibroblast activity leads to the appearance of collagen fibers and typical ground substance.

While dermal repairs are in progress, **contraction** pulls the edges of the wound closer together. The mechanism of contraction is uncertain, but it is an essential part of the healing process when damage has been extensive. For example, after amputation of a finger or even the distal portion of a limb, over 90% of the exposed surface is covered by contraction of the wound edges, rather than by the divisions and migration of epithelial cells. Contraction distorts the adjacent surface, as if the skin were being stretched to cover the injury site. If contraction and epithelial cell migration cannot cover the wound, **skin grafts** may be required.

Complications Of Inflammation PAGE 104

When pus accumulates in an enclosed tissue space, the result is an abscess. In the skin, an abscess can form as pus builds up inside the fibrin clot that surrounds the injury site. If the cellular defenses succeed in destroying the invaders, the pus will be either absorbed or surrounded by a fibrous capsule, creating a cyst.

Erysipelas (er-i-SIP-e-las; *erythros*, red + *pella*, skin) is a widespread inflammation of the dermis caused by bacterial infection. If the inflammation spreads into the subcutaneous layer and deeper tissues, the condition is called **cellulitis** (sel-ū-LĪ-tis). Erysipelas and cellulitis develop when bacterial invaders break through the fibrin wall. The bacteria involved produce large quantities of hyaluronidase, an enzyme that liquefies the ground substance, and fibrinolysin, an enzyme that breaks down fibrin and prevents clot and abscess formation. These are serious conditions that require prompt antibiotic therapy.

An **ulcer** is a localized shedding of an epithelium. **Decubitus ulcers**, also known as "bedsores," may afflict bedridden or mobile patients with circulatory restrictions, especially when splints, casts, or bedding continually press against superficial blood vessels. Such sores most often affect the skin near joints or projecting bones, where the vessels are constricted by pressure against hard underlying structures. The chronic lack of circulation kills epidermal cells, removing a barrier to bacterial infection, and eventually the dermal tissues deteriorate as well. (A comparable necrosis will occur in any tissues deprived of adequate circulation.) Bedsores can be prevented or treated by frequent changes in body position that vary the pressures applied to vulnerable skin sites.

Burns and Grafts PAGE 104

Burns result from exposure of the skin to heat, radiation, electrical shock, or strong chemical agents. The severity of the burn reflects the depth of penetration and the total area affected. Table A-3 summarizes a classification of burns based on the depth of involvement.

First- and second-degree burns are also called **partial-thickness** burns because damage is restricted to the superficial layers of the skin. Accessory structures such as hair follicles and glands are usually unaffected. **Full-thickness burns**, or **third-degree burns**, destroy the epidermis and dermis, extending into subcutaneous tissues. These burns are actually less painful than second-degree burns, because sensory nerves are destroyed along with accessory structures, blood vessels, and other dermal components. Extensive third-degree burns cannot repair themselves, and the site remains exposed to potential infection.

Roughly 10,000 people die from burns each year in the United States. The larger the area burned, the more significant the effects on integumentary function. Burns that cover more than 20% of the skin surface represent serious threats to life because they affect the following functions:

FLUID AND ELECTROLYTE BALANCE Even areas with partial-thickness burns lose their effectiveness as barriers to fluid and electrolyte losses. In full-thickness burns, the rate of fluid loss through the skin may reach five times the normal level.

THERMOREGULATION Increased fluid loss means increased evaporative cooling. More energy must be expended to keep body temperature within acceptable limits.

PROTECTION FROM ATTACK The burned epidermal surface, oozing tissue fluid, encourages bacterial growth. If the skin is broken at a blister or the site of a third-degree burn, infection is likely. Widespread bacterial infection, or **sepsis** (*septikos*, rotting), is the leading cause of death in burn victims.

Because full-thickness burns cannot heal without aid, surgical procedures are necessary to encourage healing. In a **skin graft**, areas of intact skin are transplanted to cover the burn site. A **split-thickness graft** takes a shaving of the epidermis and superficial portions of the dermis. A **full-thickness graft** involves the epidermis and both layers of the dermis.

With the development of fluid replacement therapies, infection control methods, and grafting techniques, the recovery rate for severe burns has improved dramatically. At present, young patients with burns over 80% of the body have an approximately 50% chance of recovery.

TABLE A.3 **A CLASSIFICATION OF BURNS**

Classification	Damage Report	Appearance and Sensation
First-degree burn	*Killed*: superficial cells of epidermis. *Injured*: deeper layers of epidermis, papillary dermis.	Inflamed, tender. (example: mild sunburn)
Second-degree burn	*Killed*: superficial and deeper cells of epidermis; dermis may be affected. *Injured*: damage may extend into reticular layer of the dermis, but many accessory structures unaffected.	Blisters, pain. (example: touching a hot iron)
Third-degree burn	*Killed*: all epidermal and dermal cells. *Injured*: hypodermal and deeper tissues and organs.	Charred, no sensation at all.

Recent advances in cell culture techniques may improve survival rates further. It is now possible to remove a small section of skin and grow it under controlled laboratory conditions. Over time, the germinative cell divisions produce large sheets of epidermal cells that can then be used to cover the burn area. From initial samples the size of postage stamps, square yards of epidermis have been grown and transplanted onto body surfaces. Although questions remain concerning the strength and flexibility of the repairs, skin cultivation represents a substantial advance in the treatment of extensive burns.

Scar Tissue Formation PAGE 104

What regulates the extent of scar tissue formation is not known. In some individuals, most often dark-skinned people, scar tissue formation may continue beyond the requirements of tissue repair. The result is a flattened mass of scar tissue that begins at the injury site and grows into the surrounding dermis. This thickened area of scar tissue, called a **keloid** (KĒ-loyd), is covered by a shiny, smooth epidermal surface. Keloids most often develop on the upper back, shoulders, anterior chest, and earlobes.

Synthetic Skin PAGE 104

Epidermal culturing can produce a new epithelial layer to cover the burn site. A second new procedure provides a model for dermal repairs that takes the place of normal granulation tissue. A special synthetic skin is used. The imitation has a plastic (silastic) epidermis and a dermis composed of collagen fibers and ground cartilage. The collagen fibers are taken from cow skin, and the cartilage from sharks. Over time, fibroblasts migrate among the collagen fibers and gradually replace the model framework with their own. The silastic epidermis is intended only as a temporary cover that will be replaced by either skin grafts or a cultured epidermal layer.

5 THE SKELETAL SYSTEM Osseous Tissue and Skeletal Structure

Inherited Abnormalities In Skeletal Development PAGE 121

There are several inherited conditions that result in abnormal bone formation. Three examples are *osteogenesis imperfecta, Marfan's syndrome*, and *achondroplasia*.

Osteogenesis imperfecta (im-per-FEK-ta) is an inherited condition appearing in 1 individual in about 20,000, that affects the organization of collagen fibers. Osteoblast function is impaired, growth is abnormal, and the bones are very fragile, leading to progressive skeletal deformation and repeated fractures. The ligaments and tendons are very "loose," thereby permitting excessive movement at the joints.

Marfan's syndrome is also linked to defective collagen fiber production. This condition, which affects approximately 1 individual in 10,000, is due to a genetic defect that causes the production of an abnormal form of fibrillin, a protein important to normal connective tissue strength and elasticity. Because connective tissues are found in most organs, the effects of this defect are widespread. The most visible sign of Marfan's syndrome involves the skeleton; individuals with Marfan's syndrome are usually tall, with abnormally long limbs and fingers. But the most serious consequences involve the cardiovascular system. Roughly 90% of the people with Marfan's syndrome have abnormal cardiovascular systems. Inside the heart, connective tissue supports the valves that ensure a one-way flow of blood. In Marfan's syndrome, these valves are often defective, and the heart works at reduced efficiency. Outside the heart, connective tissue that reinforces the walls of the aorta, a large blood vessel leaving the heart, may be too weak to resist the blood pressure. As a result, a bubble (aneurysm) forms in the vessel wall. If the bubble breaks, there is a sudden, fatal loss of blood.

Achondroplasia (a-kon-drō-PLĀ-zhē-a) is a condition resulting from abnormal epiphyseal activity. In this case the epiphyseal plates grow unusually slowly, and the individual develops short, stocky limbs. Although there are other skeletal abnormalities, the trunk is normal in size, and other areas of growth and development remain unaffected. The adult will be an *achondroplastic dwarf*.

Hyperostosis PAGE 121

The excessive formation of bone is termed **hyperostosis** (hī-per-os-TŌ-sis). In **osteopetrosis** (os-tē-ō-pe-TRŌ-sis; *petros*, stone) the total mass of the skeleton gradually increases because of a decrease in osteoclast activity. Remodeling stops, and the shapes of the bones gradually change. Osteopetrosis in children produces a variety of skeletal deformities. The primary cause for this relatively rare condition is unknown.

In **acromegaly** (*akron*, extremity + *megale* great) an excessive amount of growth hormone is released after puberty, when most of the epiphyseal plates have already closed. Cartilages and small bones respond to the hormone, however, resulting in abnormal growth at the hands, feet, lower jaw, skull, and clavicles.

Osteomalacia PAGE 123

In the condition of **osteomalacia** (os-tē-ō-ma-LĀ-shē-ah; *malakia*, softness) bone size remains the same but the mineral content decreases, thereby softening the bones. In this condition the osteoblasts are working hard, but the matrix is not accumulating enough calcium salts. This condition can occur in adults or children whose diet contains inadequate levels of calcium or vitamin D.

Osteoporosis and Other Skeletal Abnormalities Associated with Aging PAGE 125

Osteoporosis (os-tē-ō-pōr-Ō-sis) exists when a reduction in bone mass has proceeded to the point that normal function is compromised. In typical osteoporosis, trabecular bone and the endosteal surfaces are eroded, so bone sizes and shapes remain normal but bone mass and strength decrease. Current statistics suggest that 29% of women between the ages of 45 and 79 can be considered osteoporotic. The increase in incidence after menopause has been linked to decreases in the production of the female sex hormones (estrogens). The incidence of osteoporosis in men of the same age is estimated at 18%.

Osteoporotic bones become very fragile. This frequently leads to fractures, and subsequent healing occurs very slowly. Vertebrae may collapse (*compression fractures*), distorting the normal alignment of vertebrae and putting pressure on spinal nerves. The hunched-over posture (*kyphosis*) that develops in older women with severe osteoporosis has been called a "widow's hump." Therapies that restore estrogen levels, increased dietary calcium intake, and exercise that stresses bones and stimulates osteoblast activity appear to slow but not completely prevent the development of osteoporosis. Treatment with calcitonin (by injection or nasal spray) has been also shown to slow the development of osteoporosis. Drugs that inhibit osteoclasts (biphosphonates such as *fosama*) have been shown to reduce fracture rates and increase bone mass. Some combination of these therapies may reduce disability in the elderly.

Osteoporosis can also develop as a secondary effect of many cancers. Cancers of the bone marrow, breast, or other tissues release a chemical known as **osteoclast-activating factor**. This compound increases both the number and activity of osteoclasts and produces a severe osteoporosis.

Infectious diseases that affect the skeletal system also become more common in older individuals.

- **Osteomyelitis** (os-tē-ō-mī-e-LĪ-tis; *myelos*, marrow) is a painful infection of a bone most often caused by bacteria. This condition, most common in people over 50 years of age, can lead to dangerous systemic infections.
- **Paget's disease**, also known as **osteitis deformans** (os-tē-Ī-tis de-FŌR-manz), appears to be the result of a viral infection. This condition affects roughly 10% of the population over 70. Osteoclast activity accelerates, producing areas of acute osteoporosis, and osteoblasts produce abnormal matrix proteins. The result is a gradual deformation of the entire skeleton that is particularly noticeable in the skull. Calcitonin and/or biphosphonate therapy reduces bone pain, normalizes calcium levels, and may improve the x-ray appearance of the bones.

6 THE SKELETAL SYSTEM Axial Division

The Thoracic Cage and Surgical Procedures PAGE 174

Surgery on the heart, lungs, or other organs in the thorax often involves entering the thoracic cavity. The mobility of the ribs and the cartilaginous connections with the sternum allow the ribs to be temporarily moved out of the way. Special rib-spreaders are used, which push them apart in much the same way that a jack lifts a car off the ground for a tire change. If more extensive access is required, the sternal cartilages can be cut and the entire sternum can be folded out of the way. Once replaced, the cartilages are reunited by scar tissue, and the ribs heal fairly rapidly.

After thoracic surgery, chest tubes may penetrate the thoracic wall to permit drainage of fluids. To install a **chest tube** or obtain a sample of pleural fluid, the wall of the thorax must be penetrated. This process, called **thoracentesis** (thō-ra-sen-TĒ-sis), involves the penetration of the thoracic wall along the superior border of one of the ribs. Penetration at this location avoids damaging vessels and nerves within the costal groove.

8 THE SKELETAL SYSTEM: ARTICULATIONS

Rheumatism, Arthritis, and Synovial Function PAGE 215

Proper synovial function depends on healthy articular cartilages. When an articular cartilage has been damaged, the matrix begins to break down, and the exposed cartilage changes from a slick, smooth gliding surface to a rough feltwork of bristly collagen fibers. This feltwork drastically increases friction, damaging the cartilage further. Eventually the central area of the articular cartilage may completely disappear, exposing the underlying bone. Fibroblasts are attracted to areas of friction, and they begin tying the opposing bones together with a network of collagen fibers. This network may later be converted to bone, locking the articulating elements into position. Such a bony fusion, called **ankylosis** (an-ke-LŌ-sis), eliminates the friction by making movement impossible. Degenerative changes may be caused by simply immobilizing a joint. When motion ceases, so does circulation of synovial fluid, and the cartilages begin to suffer. **Continuous passive motion (CPM)** of any injured joint appears to encourage the repair process by improving the circulation of synovial fluid. The movement is often performed by a physical therapist or machine during the recovery period following joint surgery.

Rheumatism (ROO-ma-tizm) is a general term that indicates pain and stiffness affecting the skeletal system, the muscular system, or both. There are several major forms of rheumatism. **Arthritis** (ar-THRĪ-tis) includes all of the rheumatic diseases that affect synovial joints. Arthritis always involves damage to the articular cartilages, but the specific cause may vary. For example, arthritis can result from bacterial or viral infection, mechanical injury to the joint, metabolic problems, or immune disorders.

The diseases of arthritis are usually considered as either **degenerative** or **inflammatory** in nature. Degenerative diseases begin at the articular cartilages, and modification of the underlying bone and inflammation of the joint occur secondarily. Inflammatory diseases start with the inflammation of synovial tissues, and damage later spreads to the articular surfaces. We will consider a single example of each type.

Osteoarthritis (os-tē-ō-ar-THRĪ-tis), also known as **degenerative arthritis**, or **degenerative joint disease (DJD)**, usually affects older individuals. In the U.S. population 25% of women and 15% of men over 60 years of age show signs of this disease. The condition seems to result from cumulative wear and tear on the joint surfaces. Some individuals, however, may have a genetic predisposition to develop osteoarthritis, for researchers have recently isolated a gene linked to the disease. This gene codes for an abnormal form of collagen that differs from the normal protein in only 1 of its 1000 amino acids.

Rheumatoid arthritis is an inflammatory condition that affects roughly 2.5% of the adult population. The cause is uncertain, although misdirected immune responses, bacteria, viruses, and genetic factors have all been proposed. The synovial membrane becomes swollen and inflamed, a condition known as **synovitis** (sī-nō-VĪ-tis). The cartilaginous matrix begins to break down, and the process accelerates as dying cartilage cells release lysosomal enzymes.

In their later stages, inflammatory and degenerative forms of arthritis produce an inflammation that spreads into the surrounding area. Ankylosis (joint fusion), common in the past when complete rest was routinely prescribed for arthritis patients, is rarely seen today. Regular exercise, physical therapy, drugs that reduce inflammation and pain, such as aspirin, and low doses of the immunosuppressant *methotrexate*, may slow the progress of the disease. Surgical procedures can realign or redesign the affected joint, and in extreme cases involving the hip, knee, elbow, or shoulder, the defective joint can be replaced by an artificial one. Joint replacement has the advantage of eliminating the pain and restoring full range of motion. Prosthetic (artificial) joints are weaker than natural ones, but elderly people seldom stress them to their limits.

Bursitis PAGE 227

When bursae become inflamed, causing pain in the affected area whenever the tendon or ligament moves, such a condition is termed **bursitis**. Inflammation can result from the friction associated with repetitive motion, pressure over the joint, irritation by chemical stimuli, infection, or trauma. Bursitis associated with repetitive motion often occurs at the shoulder; for example, golfers, pitchers, ceiling painters, and tennis players may develop bursitis, usually at the subscapular bursa. The most common pressure-related bursitis is a **bunion**. Bunions form over the base of the big toe as a result of the friction and distortion of the joint, frequently caused by tight shoes, frequently with pointed toes. There is chronic inflammation of the region, and as the wall of the bursa thickens, fluid builds up in the surrounding tissues. The result is a firm, tender nodule. There are special names for bursitis at other locations; the names of these conditions indicate the occupations most often associated with them. In "housemaid's knee," which accompanies prolonged kneeling, the affected bursa lies between the patella (*kneecap*) and the skin. The condition of "student's elbow" is a form of bursitis that can result from propping your head above a desk while reading your anatomy textbook.

Most of the symptoms of bursitis subside if the stimulus is removed, and a variety of anti-inflammatory drugs may help. In extreme cases the affected bursae can be surgically removed in a **bursectomy**.

Hip Fractures and the Aging Process PAGE 231

Hip fractures are most often suffered by individuals over 60 years of age, when osteoporosis has weakened the femora. These injuries may be accompanied by dislocation of the hip or pelvic fractures. For individuals with osteoporosis, healing proceeds very slowly. In addition, spasms of the powerful muscles that surround the joint can easily prevent proper alignment of the bone fragments. Trochanteric fractures can heal well, if the joint can be stabilized; steel frames, pins, screws, or some combination of those devices may be needed to preserve alignment and permit healing to proceed normally.

Severe fractures of the femoral neck have the highest complication rate of any fracture because the blood supply to the region is relatively delicate. The surgical procedures used to treat trochanteric fractures are often unsuccessful in stabilizing femoral neck fractures. If more complex pinning operations fail, the entire joint can be replaced. In this "total hip" procedure, the damaged portion of the femur is removed. An artificial portion of the femur, including the head and neck, is attached by a spike that extends into the marrow cavity of the shaft. Special cement may be used to anchor it in place and to attach a new articular surface to the acetabulum.

A Case Study: Avascular Necrosis PAGE 231

Everyone agreed that Bo Jackson was an athletic phenomenon. At age 28 he was playing professional football with the L.A. Raiders and professional baseball for the Kansas City Royals—and starring in both sports. But on January 13, 1991, things changed dramatically when Bo was tackled near the sidelines in an NFL playoff game. The combination of pressure and twisting applied by the tackler and the tremendous power of Bo's thigh muscles produced an unusual injury, a fracture-dislocation of the hip. Although he experienced severe pain, the immediate damage to the femur and hip was limited; however, the complications of the injury were more damaging than the injury itself. The dislocation tore blood vessels in the capsule and along the femoral neck, where the capsular fibers attach. Two problems gradually developed as a result:

1. The mineral deposits in the bone of the pelvis and femur are turned over very rapidly, and osteocytes have high energy demands. A reduction in blood flow first injures and then kills them. When bone maintenance stops in the affected region, the matrix begins to break down. This process, called *avascular necrosis*, weakened and eroded Bo's femoral head.
2. The initial impact and damage to the articular cartilages was followed by joint immobility, and the articular chondrocytes became nutrient-starved. The combination resulted in the gradual loss of the articular cartilages of the femur and acetabulum.

After more than a year of rehabilitation, Bo Jackson had surgery to replace the damaged joint with an artificial hip. Joint replacement eliminates the pain and restores full range of motion. However, prosthetic (artificial) joints are weaker than natural ones, and his professional sports career ended soon after.

Knee Injuries PAGE 232

Athletes place tremendous stresses on their knees. Ordinarily, the medial and lateral menisci move as the femoral position changes. Placing a lot of weight on the knee while it is partially flexed can trap a meniscus between the tibia and femur, resulting in a break or tear in the cartilage. In the most common injury, the lateral surface of the lower leg is driven medially, tearing the medial meniscus. In addition to being quite painful, the torn cartilage may restrict movement at the joint. It can also lead to chronic problems and the development of a "trick knee," a knee that feels rather unstable. Sometimes the meniscus can be heard popping in and out of position when the knee is extended.

An **arthroscope** uses fiber optics to permit exploration of a joint without major surgery. Optical fibers are thin threads of glass or plastic that conduct light. The fibers can be bent around corners, so they can be introduced into a knee or other joint and moved around, enabling the physician to see what is going on inside the joint. If necessary, the apparatus can be modified to perform surgical modification of the joint at the same time. This procedure, called **arthroscopic surgery**, has greatly simplified the treatment of knee and other joint injuries. Ideally, small torn pieces of cartilage can be removed and the meniscus surgically trimmed. Large tears may require **meniscectomy**, the removal of the affected cartilage. New tissue-culturing techniques may someday permit the replacement of the meniscus.

An arthroscope cannot show the physician soft tissue details outside the joint cavity. Magnetic resonance imaging is a cost-effective and noninvasive method of viewing and examining, without injury, soft tissues around the joint prior to surgery.

Other knee injuries involve tearing one or more stabilizing ligaments or damaging the patella. *Torn ligaments* are often difficult to correct surgically, and healing is slow. Impacts to the anterior surface of the knee may shatter the patella. Treatment involves surgical removal of fragments and repair of the tendons and ligaments. Total knee replacements are rarely performed on young people, but they are becoming increasingly common among elderly patients with severe arthritis.

Problems with the Ankle and Foot PAGE 236

The ankle and foot are subjected to a variety of stresses during normal daily activities. In a **sprain**, a ligament is stretched to the point where some of the collagen fibers are torn. The ligament remains functional, and the structure and stability of the joint is not affected. The most common cause of a sprained ankle is a forceful inversion of the foot that stretches the lateral ligament. An ice pack is usually required to reduce swelling, and with rest and support the ankle should heal in about 3 weeks. The mnemonic PRICE applies to the first aid involved: Protect, Rest, Ice, Compression, and Elevation.

In more serious incidents, the entire ligament may be torn apart, simply termed a *torn ligament*, or the connection between the ligament and the malleolus may be so strong that the bone breaks before the ligament. In general, a broken bone heals more quickly and effectively than does a torn ligament. A dislocation often accompanies such injuries.

In a **dancer's fracture** the proximal portion of the fifth metatarsal is broken. This fracture usually occurs while the body weight is being supported by the longitudinal arch. A sudden shift in weight from the medial portion of the arch to the lateral, less elastic border breaks the fifth metatarsal close to its articulation.

Individuals with abnormal arch develoment are more likely to suffer metatarsal injuries than individuals with normal arch development. In the condition of **flat feet**, one loses or never develops the longitudinal arch. *Fallen arches* may develop as tendons and ligaments stretch and become less elastic. Obese individuals or those who must constantly stand or walk on the job are likely candidates. Children have very mobile articulations and elastic ligaments, and they often have *flexible flat feet*. Their feet look flat only while they are standing, and the arch reappears when they stand on their toes or sit down. This condition usually disappears as growth continues.

Congenital talipes equinovarus (**clubfoot**), results from an inherited developmental abnormality that affects 2 in 1000 births. Boys are affected roughly twice as often as girls. One or both feet may be involved, and the condition may be mild, moderate, or severe. The underlying problem is abnormal muscle development (often related to abnormal innervation) that distorts growing bones and joints. Usually the tibia, ankle, and foot are affected, and the feet are turned medially and inverted. The longitudinal arch is exaggerated, and if both feet are involved, the soles face one another. Prompt treatment with casts or other supports in infancy helps to alleviate the problem, and fewer than half of the cases require surgery. Kristi Yamaguchi, Olympic Gold Medalist in figure skating, was born with this condition.

9 THE MUSCULAR SYSTEM Skeletal Muscle Tissue

Duchenne's Muscular Dystrophy PAGE 254

The **muscular dystrophies** (DIS-trō-fēz) are congenital diseases that produce progressive muscle weakness and deterioration. One of the most common and best understood conditions is **Duchenne's muscular dystrophy** (DMD).

This form of muscular dystrophy appears in childhood, often between the ages of 3 and 7. The condition generally affects only males. A progressive muscular weakness develops, and the individual usually dies before age 20 because of respiratory paralysis. Skeletal muscles are primarily affected, initially in the limbs. In later stages of the disease the facial muscles and cardiac muscle tissue may also become involved.

The skeletal muscle fibers in a DMD patient are structurally different from those of normal individuals. DMD sufferers also lack a protein, called *dystrophin*, found in normal muscle fibers. Although the functions of this protein remain uncertain, researchers have recently identified and cloned the gene for dystrophin. Rats with DMD have been cured by insertion of this gene into their muscle fibers—a technique that has been tried on human patients. It appears, based on studies on rats, that the loss of membrane proteins normally associated with dystrophin may be the initial cause of the muscle degeneration.

The inheritance of DMD is sex-linked: Women carrying the defective genes are unaffected, but each of their male children will have a 50% chance of developing DMD. Now that the specific location of the gene has been identified, it is possible to determine whether a woman is carrying the defective gene or not.

Myasthenia Gravis PAGE 254

Myasthenia gravis (mī-as-THĒ-nē-a GRA-vis) is characterized by muscular weakness that is often most pronounced in muscles innervated by the cranial nerves. The first symptom is usually a weakness of the eye muscles and drooping eyelids. Facial muscles are often weak as well, and the individual develops a peculiar smile known as the "myasthenic snarl." As the disease progresses, pharyngeal weakness leads to problems with chewing and swallowing, and it becomes difficult to hold the head upright.

The muscles of the upper chest and arms are next to be affected. All of the voluntary muscles of the body may ultimately be involved. Severe myasthenia gravis produces respiratory paralysis, with a mortality rate of 5–10%. However, the disease does not always progress to such a life-threatening stage. For example, roughly 20% of patients experience eye problems but no other symptoms.

The condition results from a decrease in the number of ACh receptors on the motor end plate. Before the remaining receptors can be stimulated enough to trigger a strong contraction, the ACh molecules are destroyed by cholinesterase. As a result, muscular weakness develops.

The primary cause of myasthenia gravis appears to be a malfunction in the immune system. The body attacks the ACh receptors of the motor end plate as if they were foreign proteins. For unknown reasons, women are affected twice as often as men. Estimates of the incidence of this disease in the United States range from 2 to 10 cases per 100,000 population.

One approach to therapy involves the administration of drugs, such as *neostigmine*, that are termed *cholinesterase inhibitors*. These compounds are enzyme inhibitors that prevent the enzymatic breakdown of ACh. With cholinesterase activity reduced, the concentration of ACh at the synapse can rise enough to stimulate the surviving receptors and produce muscle contraction.

Polio PAGE 255

Because skeletal muscles depend on their motor neurons for stimulation, disorders that affect the nervous system have an indirect effect on the muscular system. The poliovirus attacks and kills motor neurons in the spinal cord and brain. As the neurons die, the dependent motor units first become paralyzed and then undergo atrophy. The resulting condition is called **polio**. In severe cases, paralysis of the respiratory muscles makes breathing impossible without the assistance of an "iron lung" or respirator that provides mechanical ventilation.

Polio has been almost completely eliminated from the U.S. population through a successful immunization program. In 1954 there were 18,000 new cases; there were 8 in 1976. Unfortunately, many parents refuse to immunize their children against the poliovirus on the assumption that the disease has been "conquered." This assumption is a mistake because (1) there is no cure for polio, (2) the virus remains in the environment, and (3) approximately 38% of children ages 1–4 have not been immunized. Although recent efforts by the World Health Organization for universal childhood immunization have eliminated polio from the Western Hemisphere, it remains a threat in the rest of the world. With the rate of international travel increasing, a major epidemic could still occur.

10 THE MUSCULAR SYSTEM Muscle Organization and the Axial Musculature

Hernias PAGE 278

When the abdominal muscles contract forcefully, pressure in the abdominopelvic cavity can rise dramatically, and those pressures are applied to internal organs. If the individual exhales at the same time, the pressure is relieved, because the diaphragm can move upward as the lungs collapse. But during vigorous periods of isometric exercises or when lifting a weight while holding one's breath, pressure in the abdominopelvic cavity can rise to 1500 pounds per square inch, roughly 100 times normal pressures. Pressures this high can cause a variety of problems, among them the development of a *hernia*. A **hernia** develops when an organ protrudes through an abnormal opening. There are many locations where hernias may develop; at this time we will consider only *inguinal* (groin) *hernias* and *diaphragmatic hernias*.

During the sixth month of embryonic development, the male testes descend into the scrotum by passing through the abdominal wall at the **inguinal** (ING-gwuh-null) **canals**. Abdominal pressure actually assists in the descent process. The spermatic ducts and associated blood vessels penetrate the abdominal musculature at the inguinal canals on their way to the abdominal reproductive organs. In an **inguinal**

hernia, the inguinal canal enlarges and abdominal contents such as a portion of the intestine (or more rarely the urinary bladder) are forced into the inguinal canal. If the herniated structures become trapped or twisted within the inguinal sac, surgery may be required to prevent serious complications. Inguinal hernias are not always caused by unusually high abdominal pressures. Injuries to the abdomen or inherited weakness or distensibility of the canal may have the same effect. Although a weak spot does exist at the site of the inguinal canal, females are much less likely to develop inguinal hernias.

The esophagus and major blood vessels pass through an opening in the muscular diaphragm. In a **diaphragmatic hernia** or *hiatal hernia* (hī-Ā-tal; *hiatus*, a gap or opening), abdominal organs slide into the thoracic cavity, most often through the **esophageal hiatus**, the opening used by the esophagus. The severity of the condition will depend on the size of the opening and when it occurs. Small hiatal hernias are actually very common in adults and most go unnoticed. Radiologists see them in about 30% of patients examined with barium contrast techniques. When clinical complications develop, they usually occur because abdominal organs that have pushed into the thoracic cavity are exerting pressure on structures or organs there. As is the case with inguinal hernias, a diaphragmatic hernia may result from congenital factors or from an injury that weakens or tears the diaphragmatic muscle. A large diaphragmatic hernia sometimes forms during embryological development. In this condition the abdominal organs can move into the thoracic cavity, compressing and restricting the developing lungs. After birth the infant may die from lung insufficiency. If dragnosed before birth, prenatal surgery can be performed to improve the chances for survival.

11 THE MUSCULAR SYSTEM The Appendicular Musculature

Sports Injuries PAGE 295

Many Americans participate in exercise programs and sports on a regular basis; more than 30 million Americans go jogging, and millions more participate in various amateur and professional sports. As a result, sports injuries are very common, and sports medicine has become an active area of professional and academic research interest.

Sports injuries affect amateurs and professionals alike. For example, in 1986 the approximately 1500 players in the National Football League reported 1500 injuries. At the amateur level, a 5-year study of college football players indicated that 73.5% experienced mild injuries, 21.5% moderate injuries, and 11.6% severe injuries in the course of their playing careers. Contact sports are not the only activities that show a significant injury rate; a study of 1650 joggers running at least 27 miles per week reported 1819 injuries in a single year.

Muscles and bones respond to increased use by enlarging and strengthening. Poorly conditioned individuals are therefore more likely to subject their bones and muscles to intolerable stresses. Training is also important in minimizing overuse injuries and keeping joint movements within the intended ranges of motion. Planned warm-up exercises before athletic events stimulate circulation, improve muscular performance and control, and help prevent injuries to muscles, joints, and ligaments. Stretching exercises stimulate muscle circulation and help keep ligaments and joint capsules supple. Such conditioning extends the range of motion and may prevent sprains and strains when sudden loads are applied.

Dietary planning can also be important in preventing injuries to muscles during endurance events, such as marathon running. Emphasis has often been placed on the importance of carbohydrates, leading to the practice of "carbohydrate loading" before a marathon. But muscles also utilize amino acids extensively while operating within aerobic limits, and an adequate diet must include both carbohydrates and proteins.

Improved playing conditions, equipment, and regulations also play a role in reducing the incidence of sports injuries. Jogging shoes, ankle or knee braces, helmets, and body padding are examples of equipment that can be effective. The substantial penalties now earned for "personal fouls" in contact sports have reduced the numbers of neck and knee injuries.

Several injuries common to those engaged in active sports may also affect nonathletes, although the primary causes may differ considerably. A partial listing of activity-related conditions would include the following:

Bone bruise: bleeding within the periosteum of a bone

Bursitis: inflammation of the bursae around one or more joints

Stress fractures: cracks or breaks in bones subjected to repeated stresses or trauma

Muscle cramps: prolonged, involuntary, and painful muscular contractions

Sprains: tears or breaks in ligaments or tendons

Strains: tears or breaks in muscles

Tendinitis: inflammation of the connective tissue surrounding a tendon

13 THE NERVOUS SYSTEM Neural Tissue

Axoplasmic Transport and Disease PAGE 343

Each synaptic knob contains rough endoplasmic reticulums, ribosomes, and mitochondria needed to synthesize the neurotransmitters released at the synapse. Much synthetic activity also occurs at the cell body: Additional neurotransmitter, enzymes, mitochondria, and lysosomes all form within the perikaryon. Many of these products are transported to the synaptic knobs along the length of the axon. This **axoplasmic transport** occurs along neurotubules, and the process requires adenosine triphosphate (ATP).

While some materials are traveling from the cell body to the periphery of the neuron, other substances are being transported to the cell body. This reverse transport process is called *retrograde flow*. Retrograde flow may deliver chemicals or debris from damage to the axon or synaptic knobs. This flow will trigger an immediate change in the cell's functions.

Rabies is a viral disease of the central nervous system (CNS). The rabies virus can infect any mammal, wild or domestic, and with few exceptions the result is death within 3 weeks. Rabies is usually transmitted to people through the bite of a rabid animal. For unknown reasons, bats can survive rabies infection for prolonged periods, and may act as a reservoir for the disease in the wild. There are an estimated 15,000 cases of rabies each year worldwide, the majority of them the result of dog bites. Only about five of those cases, however, are diagnosed in the United States, and since dogs and cats in the United States are vaccinated against rabies, cases here are most commonly caused by the bites of raccoons, foxes, skunks, and bats.

Although these bites usually involve peripheral sites, such as the hand or foot, the symptoms are caused by CNS damage. The virus present at the injury site is absorbed by synaptic knobs in the region. It then gets a free ride to the CNS, courtesy of retrograde flow. Over the first few days following exposure, the individual experiences headache, fever, muscle pain, nausea, and vomiting. The victim then enters a phase marked by extreme excitability, with hallucinations, muscle spasms, and disorientation. There is difficulty in swallowing, and the accumulation of saliva makes the individual appear to be foaming at the mouth. Coma and death soon follow.

Preventive treatment after exposure to the rabies virus consists of separate injections of a vaccine to promote antibody formation and manufactured antibodies against the rabies virus. This postexposure treatment may not be sufficient following massive infection, which can lead to death in as little as 4 days. Individuals such as veterinarians or field biologists who are at high risk for exposure often take a preexposure series of immunizations. These injections bolster the immune defenses and improve the effectiveness of the postexposure treatment. Without treatment, human rabies infection is always fatal.

Rabies is perhaps the most dramatic example of a clinical condition directly related to axoplasmic flow. Many toxins, however, including heavy metals, some pathogenic bacteria, and other viruses, use this mechanism to enter the CNS.

Demyelination Disorders PAGE 346

Demyelination is the progressive destruction of myelin sheaths in the CNS and peripheral nervous system (PNS). The result is a gradual loss of sensation and motor control that leaves affected regions numb and paralyzed. Many unrelated conditions can cause demyelination; four important examples of demyelinating disorders are heavy metal poisoning, diphtheria, multiple sclerosis, and Guillain-Barré syndrome. Multiple sclerosis is discussed in the appendix for Chapter 14; other demyelination disorders are considered here.

HEAVY METAL POISONING Chronic exposure to heavy metal ions, such as arsenic, lead, or mercury, can lead to glial cell damage and demyelination. As demyelination occurs, the affected axons deteriorate, and the condition becomes irreversible. Historians note several interesting examples of heavy metal poisoning with widespread impact. For example, lead contamination of drinking water has been cited as one factor in the decline of the Roman Empire. In the seventeenth century, the great physicist Sir Isaac Newton is thought to have suffered several episodes of physical illness and mental instability brought on by his use of mercury in chemical experiments. Well into the nineteenth century, mercury used in the preparation of felt presented a serious occupational hazard for those employed in the manufacture of stylish hats. Over time, mercury absorbed through the skin and across the lungs accumulated in the CNS, producing neurological damage that affected both physical and mental function. (This effect is the source of the expression "mad as a hatter.") More recently, Japanese fishermen working in Minamata Bay, Japan, collected and consumed seafood contaminated with mercury discharged from a nearby chemical plant. Levels of mercury in their systems gradually rose to the point that clinical symptoms appeared in hundreds of people. Making matters worse, babies born to pregnant women exposed to mercury had severe, crippling birth defects.

DIPHTHERIA Diphtheria (dif-THĒ-rē-a; *diphthera, membrane* + *-ia*, disease) is a disease that results from a bacterial infection of the respiratory tract. In addition to restricting airflow and sometimes damaging the respiratory surfaces, the bacteria produce a powerful toxin that injures the kidneys and adrenal glands, among other tissues. In the nervous system, diphtheria toxin damages Schwann cells and destroys myelin sheaths in the PNS. This demyelination leads to sensory and motor problems that may ultimately produce a fatal paralysis. The toxin also affects cardiac muscle cells, and heart enlargement and failure may occur. The fatality rate for untreated cases ranges from 35 to 90%, depending on the site of infection and the subspecies of bacterium. Because an effective vaccine exists, cases are relatively rare in countries with adequate health care. With political change and poverty causing a drop in vaccine availability and immunization rates, Russia is experiencing a diphtheria epidemic that began early in 1993.

GUILLAIN-BARRÉ SYNDROME Guillain-Barré syndrome is characterized by a progressive demyelination. Symptoms initially involve pain and weakness of the legs, which spreads rapidly to muscles of the trunk and arms. These symptoms usually increase in intensity for 1–2 weeks before subsiding. The mortality rate is low (under 5%), but there may be some permanent loss of motor function. The cause is unknown, but because roughly two-thirds of Guillain-Barré patients develop symptoms within 2 months after a viral infection, it is suspected that the condition may result from a malfunction of the immune system. In severe cases, *plasmapheresis* has proven helpful. In this procedure the patient's blood is filtered, and the plasma and antibodies are discarded and replaced by donor plasma. High intravenous doses of gamma globulin have also helped.

Growth and Myelination of the Nervous System PAGE 350

The development of the nervous system continues for many years after birth. Nerve cells increase in number for the first year after delivery, and most of the important interconnections between neurons occur after birth, rather than before. Growth of the brain is completed by age 4, but the neurons have yet to be interconnected extensively. The myelination of axons may not be completed until early adolescence. The level of nervous system development limits mental and physical performance. For example, the degree of interconnection between neurons in the CNS affects intellectual abilities, and the myelination of axons improves coordination and control by decreasing the time between reception of a sensation and completion of a response. Because the early years are crucial for the development of normal interconnections, malnutrition in children under 5 years of age can lead to mental retardation and disability.

14 THE NERVOUS SYSTEM The Spinal Cord and Spinal Nerves

Spinal Meningitis PAGE 359

Meningitis is the inflammation of the meningeal membranes. Meningitis often disrupts the normal circulatory and cerebrospinal fluid supplies, damaging or killing neurons and glial cells in the affected areas. Although the initial diagnosis may specify the meninges of the spinal cord (spinal meningitis) or brain (cerebral meningitis), in later stages the entire meningeal system is usually affected.

The warm, dark, nutrient-rich environment of the meninges provides ideal conditions for a variety of bacteria and viruses. Microorganisms that cause meningitis include those associated with middle ear infections, pneumonococcal, streptococcal ("strep"), staphylococcal ("staph"), or meningococcal infections, and tuberculosis. The malaria parasite can also cause a severe form of meningitis. These pathogens may gain access to the meninges by traveling within blood vessels or by entering at sites of vertebral or cranial injury. Headache, chills, high fever, disorientation, and rapid heart and respiratory rates appear as higher centers are affected. Without treatment, delirium, coma, convulsions, and death may follow within hours.

The mortality rate for viral meningitis ranges from 1 to 50% or higher, depending on the type of virus, the age and health of the patient, and other factors. There is no effective treatment for most forms of viral meningitis. Bacterial meningitis can be combated with antibiotics and the maintenance of proper fluid and electrolyte balance, given early diagnosis. The incidence of one previously common form of meningitis caused by the bacteria *Hemophilus influenza* has been much reduced by the vaccination of infants.

The most common clinical assessment involves checking for a "stiff neck" by asking the patient to touch chin to chest. Meningitis affecting the cervical portion of the spinal cord results in an increase in the muscle tone of the extensor muscles of the neck. So many motor units become activated that voluntary or involuntary flexion of the neck becomes painfully difficult if not impossible.

Spinal Anesthesia PAGE 359

Local anesthetics introduced into the subarachnoid space of the spinal cord produce a temporary sensory and motor paralysis. The effects spread as the anesthetic diffuses along the cord, and precise control of the regional effects can be difficult to achieve. Problems with overdosing are seldom serious, because the diaphragmatic breathing muscles are controlled by upper cervical spinal nerves and most spinal anesthesia involves the lumbar region. Thus respiration continues even when the thoracic and abdominal segments have been paralyzed. Sometimes more precise control can be obtained by injecting the anesthetic into the epidural space, producing an **epidural block**. This technique has the advantage of affecting only the spinal nerves in the immediate area of the injection. Epidural anesthesia may be difficult to achieve, however, in the upper cervical, midthoracic, and lumbar regions, where the epidural space is extremely narrow. **Caudal anesthesia** involves the introduction of anesthetics into the epidural space of the sacrum. Injection at this site paralyzes lower abdominal and perineal structures. Epidural blocks in the lower lumbar or sacral regions or caudal anesthesia are typically used to control pain during childbirth.

Technology and Motor Paralysis PAGE 362

If a peripheral axon is damaged but not displaced, normal function may eventually return as the cut axon stump grows across the injury site away from the soma and along its former path. The mechanics of this process were detailed at the close of Chapter 13. p. 345

In order for normal function to be restored, several things must happen. The severed ends must be relatively close together (1–2 mm), they must remain in proper alignment, and there must be no physical obstacles between them, such as the dense fibers of scar tissue. These conditions can be created in the laboratory, using experimental animals and individual axons or small axonal bundles. But in accidental injuries to peripheral nerves the edges are likely to be jagged, intervening segments may be lost entirely, and elastic contraction in the surrounding connective tissues may pull the cut ends apart and misalign them.

Until recently the surgical response would involve trimming the injured nerve ends, neatly sewing them together, and hoping for the best. This procedure was often unsuccessful, in part because scalpels do not produce a smoothly cut surface and the thousands of broken axons would never be perfectly aligned. Moreover, nerve axons are not highly elastic, so if a large segment of the nerve was removed, crushed, or otherwise destroyed there would be no way to bring the intact ends close enough to permit any regeneration. In such instances a **nerve graft** could be inserted, using a section from some other, less important peripheral nerve. The functional results were even less likely to be wholly satisfactory, for the growing axonal tips had to find their way across not one but two gaps, and the chances for successful alignment were proportionately smaller. Nevertheless, any return of function was certainly better than none at all!

Two very different research strategies are now being pursued. One focuses on the physical and biochemical control of nerve regeneration, while the other looks for a more "high-tech" solution. An example of a biological approach that focuses on physical factors is the use of a synthetic sleeve to guide nerve growth. The sleeve is a tube with an outer layer of silicone around an inner layer of cowhide collagen bound to the proteoglycans from shark cartilage. Using this sleeve as a guide, axons can grow across gaps as large as 20 mm (0.75 in.). The procedure has yet to be tried on humans, and functional restoration is incomplete because proper alignment does not always occur. Experiments on other mammals have successfully reunited the cut sections of large peripheral nerves.

A second biological line of investigation involves the biochemical control of nerve growth and regeneration. Neurons are influenced by a combination of growth promoters and growth inhibitors. Damaged myelin sheaths apparently release an inhibitory factor that slows the repair process. Researchers have made a monoclonal antibody, *IN-1*, that will inactivate the inhibitory factor released in the damaged spinal cords of rats. They are also experimenting with the application of *nerve growth factor*, a protein in the CNS that is required for normal growth and development of many neural components. These treatments stimulate repairs, even in severed spinal cords. Prior to this research, regeneration of nerves in the CNS was thought to be impossible or highly unlikely. The recent discovery of stem cells in the ependymal layer of the brain has opened new treatment possibilities, and the regulation of neural stem cell activity is now an active line of research.

In the meantime, other research teams are experimenting with the use of computers to stimulate specific muscles and muscle groups electrically. The technique is called **functional electrical stimulation**, or **FES**. This approach often involves implanting a network of wires beneath the skin with their tips in skeletal muscle tissue. The wires are connected to a computer that may be small enough to be worn at the waist. The wires deliver minute electrical stimuli to the muscles, depolarizing their membranes and causing contractions. With this equipment and lightweight braces, quadriplegics have walked several hundred yards and paraplegics several thousand.

Even more impressive results have been obtained using a network of wires woven into the fabric of close-fitting garments. This network provides the necessary stimulation without the complications and maintenance problems that accompany implanted wires. A paraplegic woman in a set of electronic "hot pants" completed several miles of the 1985 Honolulu Marathon, and more recently a paraplegic woman walked down the aisle at her wedding. A commercial version of the system may be on the market in the near future.

Such technological solutions can provide only a degree of motor control without accompanying sensation. Everyone would prefer a biological procedure that would restore the functional integrity of the nervous system. For now, however, computer-assisted programs such as FES can improve the quality of life for thousands of individuals. The 1995 spinal cord injury of actor Christopher Reeves has brought publicity and additional funds to this area of research.

Multiple Sclerosis PAGE 363

Multiple sclerosis (skler-Ō-sis; *sklerosis*, hardness), or **MS**, is a disease characterized by recurrent incidents of demyelination affecting axons in the optic nerve, and/or multiple sites in the brain or spinal cord. Common symptoms include partial loss of vision and problems with speech, balance, and general motor coordination. The initial symptoms appear as the result of myelin degeneration within the white matter of the lateral and posterior columns of the spinal cord or along tracts within the brain. For example, spinal cord involvement may produce weakness, tingling sensations, and a loss of "position sense" for the arms or legs. During subsequent attacks the effects become more widespread, and the cumulative sensory and motor losses may eventually lead to a generalized muscular paralysis. MRI imaging shows plaques in the affected areas of the CNS.

The time between incidents and the degree of recovery vary from case to case. In about one-third of all cases the disorder is progressive, and each incident leaves a greater degree of functional impairment. The average age at the first attack is 30–40; the incidence in women is 1.5 times that among men.

Recent evidence suggests that this condition may be linked to a defect in the immune system that causes it to attack myelin sheaths. MS patients have lymphocytes that do not respond normally to foreign proteins, and because several viral proteins have amino acid sequences similar to those of normal myelin, it has been proposed that MS results from a case of mistaken identity. For unknown reasons MS appears to be associated with cold and temperate climates. It has been suggested that individuals developing MS may have an inherited susceptibility to the virus that is exaggerated by environmental conditions. The yearly incidence in the United States averages around 50 cases for every 100,000 in the population. Treatment with two types of interferon have slowed the progress of the disease in some patients.

Reflexes and Diagnostic Testing PAGE 374

Many reflexes can be assessed through careful observation and the use of simple tools. The procedures are easy to perform, and the results can provide valuable information about location and extent of damage to the spinal cord or spinal nerves. By testing a series of spinal and cranial reflexes, a physician can assess the function of sensory pathways and motor centers throughout the spinal cord and brain.

Neurologists test many different reflexes; only a few are so generally useful that physicians make them part of a standard physical examination. The *knee jerk* (patellar), *ankle jerk*, and *biceps* and *triceps reflexes* are stretch reflexes controlled by specific segments of the spinal cord. Testing these reflexes provides information about the corresponding spinal segments.

Stroking an infant's foot on the side of the sole produces a fanning of the toes known as the **Babinski sign**, or *positive Babinski reflex*. This response disappears as descending inhibitory synapses develop. In the adult, the same stimulus produces a curling of the toes, called a **plantar reflex** or *negative Babinski reflex*, after about a 1-second delay. If either the higher centers or the descending tracts are damaged, the Babinski sign will reappear.

The **abdominal reflex**, present in the normal adult, results from descending spinal facilitation. In this reflex, a light stroking of the skin produces a reflexive twitch in the abdominal muscles that moves the navel toward the stimulus. This reflex disappears following damage to descending tracts.

In **hyporeflexia** normal reflexes are present but weak. In **areflexia** (ā-re-FLEK-sē-a; *a*-, without) normal reflexes cannot be detected. Hyporeflexia or areflexia may indicate temporary or permanent damage to skeletal muscles, dorsal or ventral nerve roots, spinal nerves, the spinal cord, or the brain.

Hyperreflexia occurs when higher centers maintain a high degree of stimulation along the spinal cord. Under these conditions reflexes are easily triggered, and the responses may be grossly exaggerated. This effect can also result from spinal cord compression or diseases that target higher centers or descending tracts. One potential result of hyperreflexia is the appearance of alternating contractions in opposing muscles. When one muscle contracts, it stimulates the stretch receptors in the other. The stretch reflex then triggers a contraction in that muscle, and this contraction stretches receptors in the original muscle. This self-perpetuating sequence can be repeated indefinitely; it is called **clonus** (KLŌ-nus).

A more extreme hyperreflexia develops if the motor neurons of the spinal cord lose contact with higher centers. Often, following a severe spinal injury, the individual first experiences a temporary period of areflexia known as *spinal shock*. p. 362 When the reflexes return, they respond in an exaggerated fashion, even to mild stimuli. The reflex contractions may occur in a series of intense muscle spasms potentially strong enough to break bones. In the **mass reflex** the entire spinal cord becomes hyperactive for several minutes, issuing exaggerated skeletal muscle and visceral motor commands.

15 THE NERVOUS SYSTEM The Brain and Cranial Nerves

Cranial Trauma PAGE 386

Cranial trauma is a head injury resulting from violent contact with another object. Head injuries account for over half of the deaths attributed to trauma each year. There are roughly 8 million cases of cranial trauma annually, and over a million of these are major incidents involving intracranial hemorrhaging, **concussion**, contusion, or laceration of the brain. The characteristics of spinal concussion, contusion, and laceration are introduced in the Clinical Brief on *Epidural and Subdural Hemorrhages* on p. 386 of the text, and the same descriptions can be applied to injuries of the brain.

Concussions usually accompany even minor head injuries. A concussion involves an immediate and temporary loss of consciousness and some degree of amnesia or change in other mental functions. Physicians examine any concussed individual quite closely and may request X-ray or CT scans of the head to check for skull fractures or cranial bleeding. Mild concussions produce a brief interruption of consciousness and little memory loss. Severe concussions produce extended periods of unconsciousness and abnormal neurological functions. Severe concussions are often associated with contusions (bruises) or lacerations (tears); the possibilities for recovery vary depending on the areas affected. Extensive damage to the reticular formation may produce a permanent state of unconsciousness, while damage to the lower brain stem will usually prove fatal. Repeated concussions can permanently affect the brain, and football quarterbacks, and boxers have retired early to reduce this risk.

Cerebellar Dysfunction PAGE 406

Cerebellar function may be permanently altered by trauma or a stroke, or temporarily by drugs such as alcohol. Such alterations can produce severe disturbances in motor control. **Ataxia** (*ataxia*, a lack of order) refers to the disturbance of balance that in severe cases leaves the individual unable to stand without assistance. Less severe conditions cause an obvious unsteadiness and irregular patterns of movement. The individual often watches his or her feet to see where they are going and controls ongoing movements by intense concentration and voluntary effort. Reaching for something becomes a major exertion, for the only information available must be gathered by sight or touch while the movement is taking place. Without the cerebellar ability to adjust movements while they are occurring, the individual becomes unable to anticipate the time course of a movement. Most often, a reaching movement ends with the hand overshooting the target. This inability to anticipate and stop a movement precisely is called **dysmetria** (dis-MET-rē-a; *dys*-, bad + *metron*, measure). When the person attempts to correct the situation, the hand usually overshoots again in the opposite direction, and then again. The hand continues to oscillate back and forth until either the object can be grasped or the attempt is abandoned. This oscillatory movement is known as an **intention tremor**.

Clinicians check for ataxia by watching an individual walk in a straight line; a test for dysmetria involves touching the tip of the index finger to the tip of the nose. Because many drugs impair cerebellar performance, the same tests are used by police officers to check drivers suspected of alcohol or other drug abuse.

16 THE NERVOUS SYSTEM Pathways and Higher-Order Functions

Tay-Sachs Disease PAGE 434

Tay-Sachs disease is a genetic abnormality involving the metabolism of *gangliosides*, important components of nerve cell membranes. Victims lack an enzyme needed to break down one particular ganglioside, which accumulates within the lysosomes of CNS neurons and causes them to deteriorate. Affected infants seem normal at birth, but within 6 months neurological problems begin to appear. The progress of symptoms typically includes muscular weakness, blindness, seizures, and death, usually before age 4. No effective treatment exists, but prospective parents can be tested to determine whether they are carrying the gene responsible for this condition. The disorder is most prevalent in one ethnic group, the Ashkenazi Jews of Eastern Europe.

Huntington's Disease PAGE 438

Huntington's disease (*Huntington's chorea*) is an inherited disease marked by a progressive deterioration of mental abilities and pronounced movement disorders. There are racial differences in the incidence of this condition; Caucasians have by far the highest incidence, 3–4 cases per million population.

In Huntington's disease the cerebral nuclei show degenerative changes, as do the frontal lobes of the cerebral cortex. The basic problem is the destruction of neurons secreting ACh and GABA (an inhibitory neurotransmitter) in the cerebral nuclei. The first signs of the disease usually appear in early adulthood. As you would expect in view of the areas affected, the symptoms involve difficulties in performing voluntary and involuntary patterns of movement and a gradual decline in intellectual abilities leading eventually to dementia. Screening tests can now detect the presence of the gene for Huntington's disease, which is located on chromosome 4. However, no effective treatment is available.

18 THE NERVOUS SYSTEM General and Special Senses

The Control of Pain PAGE 469

Pain management poses a number of problems for clinicians. Painful sensations can result from tissue damage or sensory nerve irritation; it may originate where it is perceived, be referred from another location, or represent a "false" signal generated along the sensory pathway. The treatment differs in each case, and an accurate diagnosis is an essential first step.

When pain results from tissue damage, the most effective solution is to stop the damage and end the stimulation. This solution is not always possible. Alternatively, the painful sensations can be suppressed at the injury site. Topical or locally injected anesthetics inactivate nociceptors in the immediate area. Aspirin and related analgesics reduce inflammation and suppress the release of irritating chemicals, such as enzymes or prostaglandins, in damaged tissues.

Pain can also be suppressed by inhibition of the pain pathway. Analgesics related to morphine reduce pain by mimicking the action of *endorphins*, pain-relieving chemicals released inside the CNS. Surgical steps can be taken to control severe pain, including: (1) the sensory innervation of an area can be destroyed by an electric current; (2) the dorsal roots carrying the painful sensations can be cut (a **rhizotomy**); (3) the ascending tracts can be severed (a **tractotomy**); or (4) thalamic or limbic centers can be stimulated or destroyed. These options, listed in order of increasing degree of effect, surgical complexity, and associated risk, are used only when other methods of pain control have failed to provide relief.

Many aspects of pain generation and control remain a mystery. Some patients experience a significant reduction in pain after receiving a nonfunctional medication. It has been suggested that this "placebo effect" results from endorphin release triggered by the expectation of pain relief. The Chinese technique of acupuncture to control pain has recently received considerable attention. Fine needles are inserted at specific locations and are either heated or twirled by the therapist. Several theories have been proposed to account for the positive effects, but none is widely accepted. The acupuncture points do not correspond to the distribution of any of the major peripheral nerves, and pain relief may come from the central release of endorphins. Developing a second source of pain or sensation may reduce the perception of pre-existing pain elsewhere in the body. Perhaps acupuncture and old-fashioned "mustard plasters" work in this way.

Assessment of Tactile Sensitivities PAGE 472

Because tactile sensitivities can be affected by various pathological conditions or damage to neural pathways, mapping tactile responses can sometimes aid clinical assessment. Sensory losses with clear regional boundaries indicate trauma to spinal nerves. For example, sensory loss along a dermatomal boundary (see Figure 14–8●, p. 364) can permit a reasonably precise determination of the affected spinal nerve(s). More widespread sensory loss may indicate damage to ascending tracts in the spinal cord. **Anesthesia** implies a total loss of sensation; **hypesthesia**, a reduction in sensitivity; and **paresthesia**, abnormal sensations, such as the pins-and-needles sensation when an arm or leg "falls asleep" because of pressure on a peripheral nerve.

Regional sensitivity to light touch can be checked by gentle contact with a fingertip or a slender wisp of cotton. The **two-point discrimination test** provides a more detailed sensory map for tactile receptors. Two fine points of a drawing compass, bent paper clip, or other object are applied to the skin surface simultaneously. The subject then describes the contact. When the points fall within a single receptive field, the individual will report only one point of contact. A normal individual loses two-point discrimination at 1 mm (0.04 in.) on the surface of the tongue, at 2–3 mm (0.08–0.12 in.) on the lips, at 3–5 mm (0.12–0.20 in.) on the backs of the hands and feet, and at 4–7 cm (1.6–2.75 in.) over the general body surface.

Vibration receptors are tested by applying the base of a tuning fork to the skin overlying bony prominences. Damage to an individual spinal nerve produces insensitivity to vibration along the paths of the related sensory nerves. If the sensory loss results from spinal cord damage, the injury site can often be located by walking the tuning fork down the spinal column, and resting its base on the vertebral spines.

Vertigo, Dizziness, and Motion Sickness PAGE 483

The term **vertigo** describes an inappropriate sense of motion. This description distinguishes it from "**dizziness**," a sensation of light-headedness and disorientation that often precedes a fainting spell. Vertigo can result from disturbances in central processing or abnormal conditions at the peripheral receptors.

Any event that sets endolymph into motion can stimulate the equilibrium receptors. Placing an ice pack in contact with the temporal bone or flushing the external auditory canal with cold water may chill the endolymph in the outermost portions of the labyrinth and establish a temperature-related circulation of fluid. A mild and temporary vertigo is the result. Consumption of excessive quantities of alcohol and exposure to certain drugs can also produce vertigo by changing the composition of the endolymph or disturbing the hair cells.

In **Ménière's disease**, distortion of the membranous labyrinth by high fluid pressures may rupture the membranous wall and mix endolymph and perilymph together. The receptors in the vestibule and semicircular canals then become highly stimulated, and the individual may be unable to start a voluntary movement because of the intense spinning or rolling sensations experienced. In addition to the vertigo, the victim may "hear" unusual sounds as the cochlear receptors are activated. Other causes of vertigo include viral infection of the vestibular nerve and damage to the vestibular nucleus or its tracts.

The exceedingly unpleasant symptoms of **motion sickness** include headache, sweating, flushing of the face, nausea, vomiting, and various changes in mental perspective. (Sufferers may go from a state of giddy excitement to almost suicidal despair in a matter of moments.) It has been suggested that the condition results when central processing stations, such as the mesencephalic tectum, receive conflicting sensory information. Sitting belowdecks on a moving boat or reading a magazine in a car or airplane often provides the necessary conditions. Your eyes report that your position in space is not changing, but your labyrinthine receptors detect every bump and roll. As a result, "seasick" sailors watch the horizon rather than their immediate surroundings, so that the eyes will provide visual confirmation of the movements detected by the inner ear. It is not known why some individuals are almost immune to motion sickness, whereas others find travel by boat or plane almost impossible. Even space shuttle astronauts are not immune. Roughly half of them develop acute space-sickness that can limit their activities for 1–2 days. NASA research continues and may soon help sailors as well as astronauts.

Testing and Treating Hearing Deficits PAGE 486

In the most common hearing test, a subject listens to sounds of varying frequency and intensity generated at irregular intervals. A record is kept of the responses, and the graphed record, or **audiogram**, is compared with results from a subject with "normal" hearing.

Bone conduction tests are used to discriminate between conductive and nerve deafness. If you put your fingers in your ears and talk quietly, you can still hear yourself because the bones of the skull conduct the sound waves to the cochlea, bypassing the middle ear. In a bone conduction test the physician places a vibrating tuning fork against the skull. If the subject hears the sound of the tuning fork in contact with the skull, but not as well when held next to the external auditory canal, the problem must lie within the external or middle ear. If the subject remains unresponsive to either stimulus, the problem must be at the receptors or along the auditory pathway.

Several effective treatments exist for conductive deafness. A hearing aid overcomes the loss in sensitivity by simply increasing the intensity of stimulation. Surgery may repair the tympanic membrane or free damaged or immobilized ossicles. Artificial ossicles may also be implanted if the originals are damaged beyond repair.

There are few possible treatments for nerve deafness. Mild conditions may be overcome by the use of a hearing aid if some functional hair cells remain. In a **cochlear implant** a small battery-powered device is inserted beneath the skin behind the mastoid process. Small wires run through the round window to reach the cochlear nerve, and when the implant "hears" a sound it stimulates the nerve directly. Increasing the number of wires and varying their implantation sites make it possible to create a number of different frequency sensations. Those sensations do not approximate normal hearing because there is as yet no way to target the specific afferent fibers responsible for the perception of a particular sound. Instead, a random assortment of afferent fibers are stimulated, and the individual must learn to recognize the meaning and probable origin of the perceived sound. Congenitally deaf children have developed normal speech with multi-channel implants, and adults with acquired nerve deafness have recovered functional hearing using these devices. Recently, scientists have been able to stimulate regrowth of damaged hair cells in chickens and hamsters.

Scotomas and Floaters PAGE 493

Abnormal blind spots, or **scotomas**, that are fixed in position may result from compression of the optic nerve, damage to retinal photoreceptors, or central damage along the visual pathway. Scotomas are abnormal, permanent features of the visual field. Most readers will probably be more familiar with floaters, small spots that drift across the field of vision. Floaters are common, temporary phenomena that result from blood cells or cellular debris within the vitreous body. They can be detected by staring at a blank wall or a white sheet of paper.

19 THE ENDOCRINE SYSTEM

Diabetes Insipidus PAGE 511

There are several different forms of **diabetes**, all characterized by excessive urine production (**polyuria**). Although diabetes can be caused by physical damage to the kidneys, most forms are the result of endocrine abnormalities. The two most important forms are *diabetes insipidus*, considered here, and *diabetes mellitus*, considered later.

Diabetes insipidus develops when the posterior pituitary no longer releases adequate amounts of antidiuretic hormone (ADH). Water conservation at the kidneys is impaired, and excessive amounts of water are lost in the urine. As a result, an individual with diabetes insipidus is constantly thirsty, but the fluids consumed are not retained by the body. Mild cases may not require treatment, as long as fluid and electrolyte intake keep pace with urinary losses. In severe diabetes insipidus the fluid losses can reach 10 liters, per day, and a fatal dehydration will occur unless treatment is provided. Administering a synthetic form of ADH, *desmopressin acetate* (DDAVP), in a nasal spray concentrates the urine and reduces urine volume. The drug enters the bloodstream after diffusing through the nasal epithelium. It is an effective treatment for bed-wetting if used at bed-time.

Thyroid Gland Disorders PAGE 515

Normal production of thyroid hormones establishes the background rates of cellular metabolism. These hormones exert their primary effects on large, metabolically active tissues and organs, including skeletal muscles, the liver, and the kidneys. Inadequate production of thyroid hormones, or **hypothyroidism**, in an infant is marked by inadequate skeletal and nervous development and a metabolic rate as much as 40% below normal levels. This condition affects approximately 1 birth out of every 5000. Hypothyroidism beginning later in childhood will retard growth and delay puberty. Adults with this condition are lethargic and unable to adjust to cold temperatures. The symptoms, collectively known as **myxedema** (miks-e-DĒ-ma), include subcutaneous swelling, dry skin, hair loss, low body temperature, muscular weakness, and slowed reflexes.

Hypothyroidism usually results from some problem involving the thyroid gland rather than with pituitary production of thyroid-stimulating hormone (TSH). One useful testing procedure involves monitoring the rate of iodine uptake by the thyroid gland, which provides an indication of its functional capabilities. Thyroid hormone and TSH levels may also be analyzed. Treatment involves the administration of synthetic thyroid hormones to maintain normal blood concentrations.

A **goiter** is an enlargement of the thyroid gland. That enlargement may not necessarily indicate increased production of thyroid hormones, but merely an increase in follicular size. One form of goiter occurs if the thyroid fails to obtain enough iodine to meet its synthetic requirements. Under TSH stimulation, follicle cells produce large quantities of thyroglobulin, but they are unable to provide the iodine needed to create functional thyroid hormones. As a result, TSH levels rise further, and the thyroid gland begins to enlarge. Administering iodine may not solve the problem, for the sudden availability of iodine may produce symptoms of *hyperthyroidism* (see below) as the stored thyroglobulin becomes activated. The usual therapy involves the ingestion of thyroxine, which feeds back on the hypothalamus and pituitary to inhibit the production of TSH. Over time the resting thyroid may return to its normal size and functional capabilities.

Thyrotoxicosis, or **hyperthyroidism**, occurs when thyroid hormones are produced in excessive quantities. The metabolic rate climbs, and the skin becomes flushed and moist with perspiration. Blood pressure and heart rate increase, and the heartbeat may become irregular as circulatory demands escalate. The effects on the CNS make the individual restless, excitable, and subject to shifts in mood and emotional states. Despite the drive for increased activity, the subject has limited energy reserves and fatigues easily.

In **Graves' disease** excessive thyroid activity leads to goiter and the symptoms of hyperthyroidism. Protrusion of the eyes, or **exophthalmos** (eks-ahf-THAL-mos) may also appear, for unknown reasons. Graves' disease has a genetic basis and seems to be an autoimmune disorder. It affects women much more often than men. Treatment may involve the use of antithyroid drugs, surgical removal of portions of the glandular mass, or destruction of part of the gland by exposure to radioactive iodine. President and Mrs. George H. Bush both had Graves' disease while in the White House. Mr. Bush developed heart arrhythmias, and Mrs. Bush suffered exophthalmos. No environmental cause was found despite an intensive investigation.

Hyperthyroidism may also result from thyroid tumors, inflammation, or immune system disorders. In extreme cases the individual's metabolic processes accelerate out of control. During a **thyrotoxic crisis** the subject experiences an extremely high fever, rapid heart rate, and the malfunctioning of a variety of physiological systems. *p. 525*

Disorders of Parathyroid Function PAGE 515

When the parathyroid gland secretes inadequate or excessive amounts of parathyroid thormone (PTH), calcium concentrations move outside of normal homeostatic limits. Inadequate PTH production, a condition called **hypoparathyroidism**, leads to low calcium concentrations in body fluids. The most obvious symptoms involve neural and muscle tissues, where calcium ions have important functions. The nervous system becomes more excitable, and the affected individual may experience hypocalcemic tetany, characterized by spasms in the muscles of the arms, hands, and face. Hypoparathyroidism may develop after neck surgery, especially for thyroid diseases if the blood supply to these glands is restricted. In many other cases the primary cause of the condition is uncertain. Treatment with PTH is not practical, due to problems with cost and availability. As an alternative, a dietary combination of supplemental vitamin D and calcium can be used to elevate body fluid calcium concentrations. (Vitamin D stimulates the absorption of calcium ions across the lining of the digestive tract.)

Disorders of the Adrenal Cortex PAGE 518

Clinical problems related to the adrenal gland vary depending on which of the adrenal zones becomes involved. The conditions may result from changes in the functional capabilities of the adrenal cells (primary conditions) or disorders affecting the regulatory mechanisms (secondary conditions). In **hypoaldosteronism** the zona glomerulosa fails to produce enough aldosterone, usually because the kidneys are not releasing adequate amounts of renin. Affected individuals lose excessive amounts of water and sodium ions at the kidneys, and the water loss leads to low blood volume and pressure. Changes in electrolyte concentrations affect transmembrane potentials, eventually causing dysfunctions in neural and muscular tissues.

Hypersecretion of aldosterone results in the condition of **aldosteronism**. Under continued aldosterone stimulation, the kidneys retain sodium ions very effectively, but potassium ions are lost in large quantities. In response, potassium ions move out of the cells and into the interstitial fluids, only to be lost in turn. A crisis eventually develops when low extracellular potassium concentrations disrupt normal cardiac, neural, and kidney cell functions.

An equally great variety of pathogens can infect the pericardium, producing a condition called **pericarditis**. The inflamed pericardial layers rub against one another as the heart beats, producing a characteristic scratching sound. In addition, pericardial irritation and inflammation often results in an increased production of pericardial fluid. Fluid then collects in the pericardial sac, potentially restricting the movement of the heart. This condition is called **cardiac tamponade** (tam-po-NĀD). Treatment involves draining excess pericardial fluid and cutting a window in the restricting pericardium.

■ Rheumatic Heart Disease and Valvular Stenosis PAGE 560

Mitral stenosis and *aortic stenosis* are the most common forms of valvular heart disease. In any form of valvular stenosis, the valves have a narrower than normal opening, and they don't close completely. They thus restrict blood flow when the associated chamber contracts, and permit backflow when it relaxes. About 40% of patients with **rheumatic heart disease** (**RHD**) develop mitral stenosis. Roughly two-thirds of them are women, although the reason for the correlation between female gender and mitral stenosis is unknown. In **mitral stenosis** blood enters the left ventricle at a slower than normal rate, and when the ventricle contracts, blood flows back into the left atrium as well as into the aortic trunk. As a result, the left ventricle has to work much harder to maintain adequate systemic circulation. In severe cases of mitral stenosis the ventricular musculature is not up to the task. The heart weakens, and peripheral tissues begin to suffer from oxygen and nutrient deprivation. This weakened condition of the heart leads to **heart failure**.

Symptoms of **aortic stenosis** develop in roughly 25% of patients with RHD; 80% of these individuals are males. Symptoms of aortic stenosis are initially less severe than those of mitral stenosis. In this condition the left ventricle works harder to maintain normal circulatory function, because of the narrow valvular opening. Clinical problems develop when the opening narrows to the point that adequate blood flow cannot occur. Symptoms then resemble those of mitral stenosis.

■ Problems with Pacemaker Function PAGE 565

Symptoms of severe bradycardia (below 50 beats per minute) include weakness, fatigue, confusion, and loss of consciousness. Drug therapies are seldom helpful, but artificial pacemakers can be used with considerable success. Wires run to the atria, the ventricles, or both, depending on the nature of the problem, and the unit delivers small electrical pulses to stimulate the myocardium. Internal pacemakers are surgically implanted, batteries and all. These units last 7–8 years or more before another operation is required to change the battery. External pacemakers are used for temporary emergencies, such as immediately after cardiac surgery. Only the wires are implanted, and an external control box is worn on the belt.

There are over 50,000 artificial pacemakers in use at present. The simplest provide constant stimulation to the ventricles at rates of 70–80 per minute. More sophisticated pacemakers vary their rates to adjust to changing circulatory demands, as during exercise. Others are able to monitor cardiac activity and respond whenever the heart begins to function abnormally.

Tachycardia, usually defined as a heart rate of over 100 beats per minute, increases the workload on the heart. At very high heart rates cardiac performance suffers because the ventricles do not have enough time to refill with blood before the next contraction occurs. Chronic or acute incidents of tachycardia may be controlled by drugs that affect the permeability of pacemaker membranes or block the effects of sympathetic stimulation.

■ Monitoring the Living Heart PAGE 567

Many techniques can be used to examine the structure and performance of the living heart. No single diagnostic procedure can provide the complete picture, so the tests used will vary depending on the suspected nature of the problem. A standard chest X-ray will show the basic size, shape, and orientation of the heart. Additional details require more specialized procedures to enhance the clarity of the images.

Because the heart is constantly moving, ordinary computerized tomography (CT) and magnetic resonance imaging (MRI) scans create blurred images. Special instruments and computers that generate images at high speed can be used to develop three-dimensional still or moving pictures of the heart as it beats. These procedures produce dramatic images, but the cost and complexity of the equipment have so far limited their use to major research institutions. Although PET scans can be used to diagnose disorders of coronary circulation (see Figure 21–9a,b●, p. 561), the cost limits the clinical use of this technology too.

Ultrasound analysis, called **echocardiography** (ek-ō-kar-dē-OG-ra-fē), provides images that lack the clarity of CT or MRI scans, but the equipment is relatively inexpensive and portable. Recent advances in data processing have made the images suitable for following details of cardiac contractions, and echocardiography is now an important diagnostic tool. Another relatively widespread approach involves a **coronary angiogram** (see Figure 21–9c●, p. 561). In this procedure, radiopaque dyes are injected into the coronary circulation, permitting X-ray analysis of coronary blood flow.

22 THE CARDIOVASCULAR SYSTEM: VESSELS AND CIRCULATION

■ Aneurysms PAGE 574

An **aneurysm** (AN-yū-rizm) is a bulge in the weakened wall of a blood vessel, usually an artery. This bulge resembles a bubble in the wall of a tire, and like a bad tire, the affected artery may suffer a catastrophic blowout. The most dangerous aneurysms are those involving arteries of the brain, where they cause strokes, and of the aorta, where a blowout will cause fatal bleeding in a matter of minutes.

Aortic aneurysms are most often caused by chronic high blood pressure. With age, the vessel walls become less elastic, and when a weak point develops the high arterial pressures distort the wall, creating an aneurysm. Unfortunately, because they are often painless, they are likely to go undetected. When aneurysms are found by ultrasound or other scanning procedures, the risk of rupture can sometimes be estimated on the basis of their size. For example, an aortic aneurysm larger than 6 cm has a 50:50 chance of rupturing in the next 10 years. Treatment often begins with the reduction of blood pressure by means of vasodilators or beta-blockers. An aneurysm in an accessible area, such as the abdomen, may be surgically removed and the vessel repaired.

Although high blood pressure is most often responsible for aneurysm formation, any trauma or infection that weakens vessel walls can lead to an aneurysm. In addition, at least some aortic aneurysms have been linked to inherited disorders, such as Marfan's syndrome (p. 789), that weaken connective tissues. Affected individuals have weak arterial walls, making them more likely to develop aortic aneurysms. It is not known whether genetic factors are involved in the development of other aneurysms.

■ Problems with Venous Valve Function PAGE 579

One of the consequences of aging is a loss of elasticity and resilience in connective tissues throughout the body. Blood vessels are no exception, and with age the walls of veins begin to sag. This change usually affects the superficial veins of the arms and legs first, because at these locations gravity opposes blood flow. The situation is aggravated by a lack of exercise or an occupation requiring long hours standing or sitting. Because there is no muscular activity to help keep the blood moving, venous blood pools on the proximal (heart) side of each valve. As the venous walls are distorted, the valves become leaky, and gravity can then pull blood back toward the capillaries. Normal blood flow is thereby further impeded, and the veins become grossly distended. These sagging, swollen vessels are called **varicose** (VĀR-i-kōs) **veins**. Varicose veins of the leg are relatively harmless but unsightly; surgical procedures are sometimes used to remove or constrict the offending vessels.

Varicose veins are not limited to the extremities, and another common site involves a network of veins in the walls of the anus. Pressures within the abdominopelvic cavity rise dramatically when the abdominal muscles are tensed. Straining to force defecation can force blood into these veins, and repeated incidents leave them permanently distended. These distended veins, known as **hemorrhoids** (HEM-ō-roydz), can bleed and in severe cases be extremely painful.

Hemorrhoids are often associated with pregnancy, because of changes in circulation and abdominal pressures. Minor cases can be treated by the topical application of drugs that promote contraction of smooth muscles within the venous walls. More severe cases may require the surgical removal or destruction of the distended veins.

23 THE LYMPHATIC SYSTEM

■ Lymphedema PAGE 617

Blockage of the lymphatic drainage from a limb produces **lymphedema** (lim-fe-DĒ-ma). In this condition, interstitial fluids accumulate, and the limb gradually becomes swollen and grossly distended. If the condition persists the connective tissues lose their elasticity, and the swelling becomes permanent. Lymphedema is painless, and the condition by itself does not pose a major threat to life. The danger comes from the continual risk of an uncontrolled infection developing in the affected area. Because the interstitial fluids are essentially stagnant, toxins and pathogens can accumulate and overwhelm the local defenses without fully activating the immune system in the rest of the body.

Temporary lymphedema of the feet and ankles may be caused by tight clothing constricting the lymphatic circulation or by prolonged standing or sitting. Elevating the feet and loosening clothing may eliminate the problem. Chronic lymphedema usually results as scar tissue forms in an area of injury. Trauma, infections, and surgical procedures are often implicated. In **filariasis** (fil-a-RĪ-a-sis), a parasitic nematode (roundworm) carried by mosquitoes forms massive colonies within lymphatic channels. Repeated scarring of the passageways eventually blocks lymphatic drainage and produces the extreme lymphedema with permanent distension of tissues known as **elephantiasis** (el-e-fan-TĪ-a-sis).

Therapy for chronic lymphedema consists of treating infections by the administration of antibiotics, and (when possible) reducing the swelling. One possible treatment involves the application of elastic wrappings that squeeze the tissue. This external compression elevates the hydrostatic pressure of the interstitial fluids and opposes the entry of additional fluid from the capillaries.

Scid Page 620

In **severe combined immunodeficiency disease (SCID)**, the individual fails to develop either cellular or humoral immunity. Lymphocyte populations are reduced, and normal B and T cells are not present. Patients with SCID are unable to provide an immune defense, and even a mild infection can prove fatal.

Total isolation offers protection at great cost and with severe restrictions on the individual's lifestyle. Bone marrow transplants from compatible donors, usually a close relative, have been used to colonize lymphatic tissues with functional lymphocytes.

The most famous SCID patient was the "bubble boy," David. David was kept in physical isolation until age 12, when he received a bone marrow transplant. Before the donor marrow cells established a functional immune system, David died of cancer. Although technology had protected him from external pathogens, without immune surveillance he had no defense against threats from within.

For these patients, whose SCID reflects the lack of a specific enzyme (adenosine deaminase), treatment with the enzyme itself has helped. In 1990, genetic engineering techniques were used to insert normal genes for this enzyme into the lymphocytes of two children suffering from this form of SCID. The experiment was successful, and the children have regained and (thus far) retained normal immune function with periodic infusions of the gene.

AIDS Page 620

Acquired immune deficiency syndrome, or **AIDS**, is caused by a virus known as **human immunodeficiency virus** (**HIV**), which selectively infects helper T cells. Infection of T cells by itself impairs the immune response, because these cells play a central role in coordinating cellular and humoral responses to antigens. To make matters worse, suppressor T cells are relatively unaffected by the virus, and over time the excess of suppressing factors "turns off" the normal immune response. Circulating antibody levels decline, cellular immunity is reduced, and the body is left without defenses against a wide variety of microbial invaders.

Because the immune function is so reduced, ordinarily harmless pathogens can initiate lethal infections, known as *opportunistic infections*. AIDS patients are especially prone to lung infections and pneumonia, often caused by *Pneumocystis carinii* or other fungi, and to a wide variety of bacterial, viral, and protozoan diseases. Because immune surveillance is also depressed, the risk of cancer increases. One of the most common cancers seen in AIDS patients, though very rare in normal individuals, is **Kaposi's sarcoma**, characterized by rapid cell division in endothelial cells of cutaneous blood vessels. This cancer has been linked to repeated infection with a herpes virus, *Herpes type 8*.

Infection with HIV occurs through intimate contact with the body fluids of infected individuals. Although all body fluids carry the virus, the major routes of transmission involve contact with blood, semen, or vaginal secretions. Most AIDS patients become infected through sexual contact with an HIV-infected person (who may *not necessarily* be suffering from the clinical symptoms of AIDS). The next largest group of patients consists of intravenous drug users who share blood-contaminated needles. A relatively small number of individuals have become infected with the virus after receiving a transfusion of contaminated blood or blood products. Screening of donated blood has now reduced the risk of HIV infection from donated blood to perhaps 1 in 40,000 transfusions. Finally, infants are born with the disease, having acquired it from infected mothers prior to or at delivery. Treatment of pregnant HIV-infected women with antiviral drugs can reduce the risk of maternal-fetal transmission of the virus.

The best defense against AIDS consists of avoiding sexual contact with infected individuals. All forms of sexual intercourse carry the potential risk of viral transmission. The use of synthetic condoms greatly reduces the chance of infection (although it does not completely eliminate it). Condoms that are not made of synthetic materials are effective in preventing pregnancy but do not block the passage of viruses.

Clinical symptoms of AIDS may not appear for 5–10 years or more after infection, and when they do appear they are often mild, consisting of *lymphadenopathy* and chronic but nonfatal infections. After a variable period of time, full-blown AIDS develops. AIDS is almost invariably fatal, but the recent use of multiple antiviral drugs in combination has dramatically improved survival in up to 50 percent of patients. Drug resistance is a concern, but hope exists that HIV may be treatable as a chronic infection, if not cured outright. Deaths in the United States have already climbed above 450,000, and estimates of the number of infected individuals range up to 1 million. The numbers worldwide are even more frightening: the World Health Organization (2000) estimates that over 36 million people may be infected, with 3 million deaths each year.

Despite intensive efforts, a vaccine has yet to be developed that will provide immunity from HIV infection. However, the survival rate for AIDS patients has been steadily increasing because new drugs are available that slow the progress of the disease, and improved antibiotic therapies help combat secondary infections. This combination is extending the life span of patients while the search for more-effective treatment continues.

Infected Lymphoid Nodules Page 620

The lymphocytes in a lymphoid nodule are not always able to destroy bacterial or viral invaders that have crossed the adjacent epithelium. When such pathogens become established in a lymphoid nodule, an infection develops. **Tonsillitis** is an infection of one of the tonsils, most often the pharyngeal tonsil. An individual with bacterial tonsillitis develops a high fever and leukocytosis (an abnormally high white blood cell count). The affected tonsil becomes swollen and inflamed, sometimes enlarging enough to partially block the entrance to the trachea. Breathing then becomes difficult, and in severe cases impossible. Historians believe that George Washington died this way. If the infection proceeds, abscesses develop within the tonsillar tissues, and the bacteria may enter the bloodstream by passing through the lymphatic capillaries and vessels to the venous system.

In the early stages, antibiotics may control the infection, but once abscesses have formed, the best treatment involves surgical drainage of the abscesses. **Tonsillectomy**, the removal of the tonsil, was once highly recommended and frequently performed to prevent recurring tonsillar infections. The procedure does reduce the incidence and severity of subsequent infections, but questions have been raised concerning the overall cost to the individual. The tonsils represent a first line of defense against bacterial invasion of the pharyngeal walls. If they are removed, bacteria may not be detected until a truly severe infection is well under way.

Appendicitis usually follows an erosion of the epithelial lining of the appendix. Several factors may be responsible for the initial ulceration, notably bacterial or viral pathogens. Bacteria that normally inhabit the lumen of the large intestine then cross the epithelium and enter the underlying tissues. Inflammation occurs, and the opening between the appendix and the rest of the intestinal tract may become constricted. Mucus secretion accelerates, and the organ becomes increasingly distended. Eventually the swollen and inflamed appendix may rupture or perforate. If it does, the bacteria will be released into the warm, dark, moist confines of the abdominopelvic cavity, where they can cause a life-threatening infection (peritonitis). The most effective treatment for appendicitis is the surgical removal of the organ, a procedure known as an **appendectomy**.

Lymphomas Page 623

Lymphomas are malignant cancers consisting of abnormal lymphocytes or lymphocytic stem cells. Over 60,000 cases of lymphoma are diagnosed in the United States each year, and that number has been steadily increasing. There are many different types of lymphoma. One form, called **Hodgkin's disease (HD)**, accounts for roughly 12% of all lymphoma cases. Hodgkin's disease most often strikes individuals at ages 15–35 or those over age 50. The reason for this pattern of incidence is unknown; although the cause of the disease is uncertain, an infectious agent (probably a virus) is suspected. Other types are usually lumped together under the heading of **non-Hodgkin's lymphoma (NHL)**. They are extremely diverse, and in most cases the primary cause remains a mystery. At least some forms reflect a combination of inherited and environmental factors. For example, one form, called *Burkitt's lymphoma*, most often affects male children in Africa and New Guinea. The affected children have been infected with the *Epstein-Barr virus* (EBV).[1] The EBV infects B cells, but under normal circumstances the infected cells are destroyed by the immune system. EBV is widespread in the environment, and childhood exposure usually produces lasting immunity. Children developing Burkitt's lymphoma may have a genetic susceptibility to EBV infection; in addition, presence of another illness, such as malaria, may weaken their immune

[1]This highly variable virus is also responsible for *infectious mononucleosis* (discussed in *Disorders of the Spleen, p. 803*)

systems to the point that a lymphoma can develop. As improved antibiotic treatment prolongs AIDS patients' lives, more of them are developing NHL.

The first symptom associated with any lymphoma is usually a painless enlargement of lymph nodes. The involved nodes have a firm, rubbery texture. Because the nodes are painless, the condition is often overlooked until it has progressed to the point that secondary symptoms appear. For example, patients seeking help for recurrent fever, night sweats, gastrointestinal or respiratory problems, or weight loss may be unaware of any underlying lymph node changes. In the late stages of the disease, symptoms can include liver or spleen enlargement, CNS dysfunction, pneumonia, a variety of skin conditions, and anemia.

In planning treatment, clinicians consider the histological structure of the nodes and the stage of the disease. In examination of a biopsy, the structure of the node is described as nodular or diffuse. A nodular node retains a semblance of normal structure, with follicles and germinal centers. In a diffuse node the interior of the node has changed, and follicular structure has broken down. In general, the nodular lymphomas progress more slowly than the diffuse forms, which tend to be more aggressive. On the other hand, the nodular lymphomas are more difficult to treat and are more likely to recur even after remission has been achieved.

The most important factor influencing treatment selection is the stage of the disease. Table A-6 includes a simplified staging classification for lymphomas. When a lymphoma is diagnosed early (stage I or II), localized therapies may be effective. For example, the cancerous node(s) may be surgically removed and the region(s) irradiated to kill residual cancer cells. Success rates are very high when a lymphoma is detected in these early stages. For Hodgkin's disease, localized radiation can produce remission lasting 10 years or more in over 90% of patients. Treatment of localized NHL is somewhat less effective. The 5-year remission rates average 60–80% for all types; success rates are higher in nodular forms than for diffuse forms.

Although these are encouraging results, it should be noted that few lymphoma patients are diagnosed while in the early stages of the disease. For example, only 10–15% of NHL patients are diagnosed at stages I or II. For lymphomas at stages III and IV, treatment most often involves chemotherapy. Combination chemotherapy, in which two or more drugs are administered simultaneously, is the most effective treatment. For Hodgkin's disease, a four-drug combination with the acronym MOPP (nitrogen mustard, Oncovin [vincristine], prednisone, and procarbazine) produces lasting remission in 80% of patients.

Bone marrow transplantation is a treatment option for acute, late-stage lymphoma. When suitable donor marrow is available, the patient receives whole-body irradiation, chemotherapy, or some combination of the two sufficient to kill tumor cells throughout the body. This treatment also destroys normal bone marrow cells. Donor bone marrow is then infused, and over the next 2 weeks the donor cells colonize the bone marrow and begin producing red blood cells, granulocytes, monocytes, and lymphocytes.

Potential complications of this treatment include the risk of infection and bleeding while the donor marrow is becoming established. The immune cells of the donor marrow may also attack the tissues of the recipient, a response called **graft versus host**, or **GVH, disease**. For a patient with stage I or II lymphomas, without bone marrow involvement, bone marrow or stem cells can be removed and stored (frozen) for over 10 years. If other treatment options fail, or the patient comes out of remission at a later date, an **autologous marrow transplant** can be performed. This option eliminates the need for donor typing and the risk of GVH disease.

TABLE A.6 **CANCER STAGING IN LYMPHOMA**

Stage I: Involvement of a single node or region (or a single extranodal site)

Typical treatment: surgical removal and/or localized irradiation; in slowly progressing forms of NHL, treatment may be postponed indefinitely.

Stage II: Involvement of nodes in two or more regions (or an extranodal site and nodes in one or more regions) on the same side of the diaphragm

Typical treatment: surgical removal and localized irradiation that includes an extended area around the cancer site (the extended field).

Stage III: Involvement of lymph node regions on both sides of the diaphragm. This is a large category that is subdivided on the basis of the organs or regions involved. For example, in stage III the spleen contains cancer cells.

Typical treatment: combination chemotherapy, with or without radiation; radiation treatment may involve irradiating all of the thoracic and abdominal nodes plus the spleen (total axial nodal irradiation, or TANI).

Stage IV: Widespread involvement of extranodal tissues above and below the diaphragm

Treatment is highly variable, depending on the circumstances. Combination chemotherapy is always used; it may be combined with whole-body irradiation. The "last resort" treatment involves massive chemotherapy followed by a bone marrow transplant.

Breast Cancer PAGE 623

Breast cancer is the primary cause of death for women between the ages of 35 and 45, but the disease actually becomes more common after age 50. There will be an estimated 39,600 deaths in the United States from breast cancer in 2002, and roughly 203,500 new cases reported. An estimated 12% of women in the United States will develop breast cancer at some point in their lifetimes. The incidence is highest among Caucasians, somewhat lower in African Americans, and the lowest in Asians and Native Americans. Notable risk factors include (1) a family history of breast cancer, (2) a pregnancy after age 30, and (3) early menarche (first menstrual period) or late menopause (last menstrual period). Despite repeated studies (and rumors), there are no proven links between oral contraceptive use, estrogen therapy, breast feeding, fat consumption, or alcohol use and breast cancer. It appears likely that multiple factors are involved; most women never develop breast cancer, even women in families with a history of this disease.

Early detection of breast cancer is the key to reducing the death rate. *Most breast cancers are found through self-examination*, but the use of clinical screening techniques has increased in recent years. **Mammography** involves the use of X-rays to examine breast tissues; the radiation dosage can be restricted because only soft tissues must be penetrated. This procedure gives the clearest picture of conditions within the breast tissues. Ultrasound can provide some information, but the images lack the detail of standard mammograms.

For treatment to be successful the cancer must be identified while it is still relatively small and localized. Once it has grown larger than 2 cm (0.78 in.) the chances for long-term survival worsen. A poor prognosis also follows if the cancer cells have spread through the lymphatic system to the axillary lymph nodes. If the nodes are not yet involved, the chances of 5-year survival are about 82%, but if four or more nodes are involved the survival rate drops to 21%.

Treatment of breast cancer begins with the removal of the tumors and sampling of the axillary lymph nodes. Because the cancer cells usually begin spreading before the condition is diagnosed, surgical treatment involves the removal of part or all of the affected breast.

- In a **segmental mastectomy**, or "lumpectomy," only a portion of the breast is removed. This, combined with irradiation, is the preferred treatment for small, localized tumors.
- In a **total mastectomy**, the entire breast is removed, but other tissues are left intact.
- In a **modified radical mastectomy**, the most common operation, the breast and nodes are removed but the muscular tissue remains intact.

A combination of chemotherapy, radiation treatments, and hormone treatments may be used to supplement the surgical procedures.

Disorders of the Spleen PAGE 628

An impact to the left side of the abdomen can distort or damage the spleen. Such injuries are known risks of automobile accidents, contact sports, such as football or hockey, and more solitary athletic activities, such as skiing or sledding. The spleen tears so easily, however, that a seemingly minor blow to the side may rupture the capsule. The result is serious internal bleeding and eventual circulatory shock.

Because the spleen is relatively fragile, it is very difficult to repair surgically. (Sutures usually tear out before they have been tensed enough to stop the bleeding.) Treatment for a severely ruptured spleen involves its complete removal, a process called a **splenectomy** (splē-NEK-tō-mē).

The spleen responds like a lymph node to infection, inflammation, or invasion by cancer cells. The enlargement that follows is called **splenomegaly** (splē-nō-MEG-a-lē; *megas*, large), and splenic rupture may also occur under these conditions. One relatively common condition causing splenomegaly is **mononucleosis.** This condition, also known as the "kissing disease" because it can be spread by kissing, results from infection by the Epstein-Barr virus (EBV). In addition to splenic enlargement, symptoms of acute mononucleosis include fever, sore throat, widespread swelling of lymph nodes, increased numbers of lymphocytes in the blood, and the presence of circulating antibodies to the virus. The condition most often affects young adults (age 15–25) in the spring or fall. Only the symptoms are treated, because no drugs are effective against this virus. The most dangerous aspect of the disease is the risk of rupturing the enlarged spleen, which becomes fragile. Patients are therefore cautioned against heavy exercise or other activities that increase abdominal pressures. If the spleen does rupture, severe hemorrhaging may occur; death will follow unless transfusion and an immediate splenectomy can be performed.

An individual whose spleen is missing (following splenectomy) or nonfunctional has **hyposplenism** (hī-pō-SPLĒN-ism). Hyposplenism usually does not pose a serious problem, but such individuals are prone to pneumococcal bacterial infections, and pneumococcal vaccination is recommended. In **hypersplenism** the spleen becomes overactive, and the increased phagocytic activities lead to anemia (low numbers of RBCs), leukopenia (low numbers of WBCs), and thrombocytopenia (low numbers of platelets). Splenectomy may be required for hypersplenism.

Immune Disorders PAGE 628

Because the immune response is so complex, there are many opportunities for things to go wrong. A great variety of clinical conditions may result from disorders of the immune functions associated with the lymphatic system. In an **immunodeficiency disease** either the immune system fails to develop normally or the immune response is blocked in some way. Examples of immunodeficiency diseases include **severe combined immunodeficiency disease (SCID)** and *acquired immune deficiency syndrome* (AIDS). SCID and AIDS are discussed in this appendix on page 802.

Autoimmune disorders develop when the lymphatic system mistakenly targets normal body cells and tissues. Lymphocytes usually recognize and ignore the antigens normally found in the body. The recognition system can malfunction, however, and when it does the activated B cells begin to manufacture antibodies against other cells and tissues. The trigger may be a reduction in suppressor T cell activity, excessive stimulation of helper T cells, tissue damage that releases large quantities of antigens, viral or bacterial toxins, or some combination of factors.

The symptoms produced depend on the identity of the antigen attacked by these misguided antibodies, called **autoantibodies**. Several conditions described in earlier chapters are autoimmune disorders. For example, *rheumatoid arthritis* occurs when autoantibodies form immune complexes within connective tissues, especially around the joints. p. 790 Many other autoimmune disorders appear to be cases of mistaken identity. For example, proteins associated with the measles, Epstein-Barr, influenza, and other viruses contain amino acid sequences that are similar to those of myelin proteins. As a result, antibodies that target these viruses may also attack myelin sheaths. This mechanism accounts for the neurologic complications that sometimes follow a vaccination or viral infection.

Allergies are inappropriate or excessive immune responses to antigens. The sudden increase in cellular activity or antibody titers can have a number of unpleasant side effects. For example, neutrophils or cytotoxic T cells may destroy normal cells while attacking the antigen, or the antibody-antigen complex may trigger a massive inflammatory response. Antigens that trigger allergic reactions are often called **allergens**.

Systemic Lupus Erythematosus PAGE 628

Systemic lupus erythematosus (LOO-pus e-rith-ē-ma-TŌ-sis), or **SLE**, appears to result from a generalized breakdown in the antigen recognition mechanism. An individual with SLE manufactures autoantibodies against nucleic acids, ribosomes, clotting factors, blood cells, platelets, and lymphocytes. The immune complexes form deposits in tissues, producing anemia, kidney damage, arthritis, and vascular inflammation. CNS function deteriorates with CNS vasculitis.

The most obvious sign of this condition is the presence of a butterfly-shaped rash of the face, centered over the bridge of the nose. SLE affects women nine times as often as men, and the U.S. incidence averages 2–3 cases per 100,000 population. There is no known cure, but almost 80% of SLE patients survive 10 years and up to 75% survive 20 years after diagnosis. Treatment consists of controlling the symptoms and depressing the immune response through administration of specialized drugs or corticosteroids.

24 THE RESPIRATORY SYSTEM

Overloading the Respiratory Defenses PAGE 637

Large quantities of airborne particles may overload the respiratory defenses and produce a variety of different illnesses. The presence of irritants in the lining of the conducting passageways may provoke the formation of abscesses, and damage to the epithelium in the affected areas may allow the irritants to enter the surrounding tissues of the lung. The scar tissue that forms reduces the elasticity of the lung and may restrict airflow along the passageways. Irritants or foreign particles may also enter the lymphatics of the lung, producing inflammation of the regional lymph nodes. Chronic irritation and stimulation of the epithelium and its defenses cause changes in the epithelium that increase the likelihood of lung cancer. Severe symptoms of such disorders develop slowly and may take 20 years or more to appear. **Silicosis** (sil-i-KŌ-sis), produced by the inhalation of silica dust, **asbestosis** (as-bes-TŌ-sis), from the inhalation of asbestos fibers, and **anthracosis** (an-thra-KŌ-sis), the "black lung disease" of coal miners, are examples of conditions caused by overloading the respiratory defenses. Tobacco smoke accelerates the course of these diseases while it causes disease on its own (p. 805).

Nosebleeds PAGE 637

The extensive vascularization of the nasal cavity and the relatively vulnerable position of the nose make a nosebleed, or **epistaxis** (ep-i-STAK-sis), relatively common. Bleeding usually involves vessels of the mucosa covering the anterior cartilaginous portion of the septum. Packing the vestibule with gauze or pinching the external nares together to squeeze the vessels against the septum will often control the bleeding until clotting occurs. More severe bleeding originating elsewhere in the nasal cavity may require packing the posterior portion of the cavity via the internal nares.

Epistaxis can result from any factor affecting the integrity of the epithelium or the underlying vessels. Examples would include trauma, such as a punch in the nose, drying, infections, allergies, or clotting disorders. Hypertension may also provoke a nosebleed by rupturing small vessels of the lamina propria.

Disorders of the Larynx PAGE 641

Infection or inflammation of the larynx is known as **laryngitis** (lar-in-JĪ-tis). This condition often affects the vibrational qualities of the vocal cords; hoarseness is the most familiar symptom. Mild cases are temporary and seldom serious, but bacterial or viral infection of the epiglottis or upper trachea in children can be very dangerous because the swelling may close the glottis or trachea and cause suffocation. **Acute epiglottitis** (ep-i-glot-TĪ-tis) can develop relatively rapidly following a bacterial infection of the throat. Although most common in children, it does occur in adults, especially those with *Hodgkin's disease* or *leukemia*, two cancers described earlier. Serious inflammation and edema of the trachea can also occur in small children following infection with one of the parainfluenza viruses. The condition, **laryngotracheobronchitis**, is more commonly called **croup** (kroop).

Bronchitis PAGE 645

Bronchitis (brong-KĪ-tis) is an inflammation of the bronchial lining. The most characteristic symptom is the overproduction of mucus, which leads to frequent coughing. An estimated 20% of adult males have chronic bronchitis. This condition is most often related to cigarette smoking, but it can also result from other environmental irritants, such as chemical vapors. Over time the increased mucus production can block smaller airways and reduce respiratory efficiency. This condition is called **chronic airways obstruction** or *chronic obstructive pulmonary disease (COPD)*.

Examining the Living Lung PAGE 648

A chest X-ray remains the standard diagnostic screening test for chest conditions. This procedure can detect abnormalities in lung structure including scar tissue formation, fluid accumulation, or distortion of the conducting passageways. CT scans show much greater definition of internal structures and can clarify the nature of abnormalities initially spotted on a chest X-ray. CT scans are particularly helpful in diagnosing and tracking the progression of lung cancers. Lung scans are made when radioactive tracers are injected or radioactive gases are inhaled. These procedures can detect abnormalities in airflow or pulmonary blood circulation.

Bronchoscopy (brong-KOS-kō-pē) involves the insertion of a fiberoptic bundle, or *bronchoscope*, a few millimeters in diameter into the trachea. The bundle, once inserted, can be steered along the conducting passageways to the level of the smaller bronchi. In addition to permitting direct visualization of bronchial structures, the bronchoscope can collect tissue or mucus samples from the respiratory tract. In **bronchography** (brong-KOG-ra-fē) a bronchoscope or catheter introduces a radiopaque material into the bronchi. This technique can permit detailed analysis of bronchial masses, such as tumors, or other obstructions along the bronchial tree (see Figure 24-10b●, p. 646).

Respiratory Distress Syndrome (RDS) PAGE 651

Surfactant cells begin producing surfactants at the end of the sixth fetal month. By the eighth month surfactant production has risen to the level required for normal respiratory function. **Neonatal respiratory distress syndrome (NRDS)**, also known as *hyaline membrane disease (HMD)*, develops when surfactant amounts are inadequate. Although there are inherited forms of NRDS, the condition most often accompanies premature delivery.

In the absence of surfactants, the alveoli tend to collapse during exhalation, and although the conducting passageways remain open, the newborn infant must then inhale with extra force to reopen the alveoli on the next breath. In effect, every breath must approach the power of the first, and the infant rapidly becomes exhausted. Respiratory movements become progressively weaker, eventually the alveoli fail to expand, and gas exchange ceases.

One method of treatment involves assisting the infant by administering air under pressure, so that the alveoli are held open. This procedure, known as **positive end-expiratory pressure (PEEP)**, can keep the newborn alive until surfactant production increases to normal levels. Surfactant from other sources can also be provided; suitable surfactants can be extracted from cow lungs (*Survanta*), obtained from the fluids that surround full-term infants, or synthesized using gene-splicing techniques (*Exosurf*). These preparations are usually administered in the form of a fine mist of surfactant droplets. This method of delivery is called **nebulization**.

Surfactant abnormalities may also develop in adults, as the result of severe respiratory or systemic infections or other sources of pulmonary injury. Alveolar collapse follows, producing a condition known as **acute respiratory distress syndrome (ARDS)**. PEEP is often used in an attempt to maintain life until the underlying problem can be corrected, but at least 50–60% of ARDS cases result in fatalities.

Tuberculosis PAGE 651

Tuberculosis (tū-ber-kū-LŌ-sis), or **TB**, results from a bacterial infection of the lungs, although other organs may be invaded as well. The bacteria, *Mycobacterium tuberculosis*, may colonize the respiratory passageways, the interstitial spaces, and/or the alveoli. Symptoms are variable, but usually include coughing and chest pain, with fever, night sweats, fatigue, and weight loss.

At the site of infection, macrophages and fibroblasts proceed to wall off the area, forming an abscess. If the scar tissue barricade fails, the bacteria move into the surrounding tissues and the process repeats itself. The resulting masses of fibrous tissue distort the conducting passageways, increasing resistance and decreasing airflow. In the alveoli, the attacked surfaces are destroyed. The combination reduces the vital capacity and the area available for gas exchange.

Treatment for TB is complex, because (1) the bacteria can spread to many different tissues, (2) they can develop a resistance to standard antibiotics relatively quickly, and (3) to be effective, treatment is prolonged. As a result, several drugs are used in combination over a period of 6–9 months. The most effective drugs now available include *isoniazid*, which interferes with bacterial replication, and *rifampin*, which blocks bacterial protein synthesis.

Tuberculosis is a major health problem throughout the world, but especially in underdeveloped countries. An estimated 2 billion people are infected at this time. There are 7–9 million cases diagnosed each year and 3 million deaths annually due to tuberculosis infection. The problem is much less severe in developed nations, such as the United States. However, tuberculosis was extremely common in the United States earlier in this century. An estimated 80% of the people born around the turn of the century became infected with tuberculosis during their lives. Although many were able to meet the bacterial challenge, it was still the number one cause of death in 1906. These statistics have been drastically altered with the arrival of antibiotics and techniques for early detection of infection. Between 1906 and 1984 the death rate fell from 2 deaths per 1000 population to 1.5 deaths per 100,000 population.

Tuberculosis today is unevenly distributed through the U.S. population, and is resurging as a public health problem. Several groups are at relatively high risk for infection. For example, Hispanics, African Americans, prison inmates, new immigrants, hospital employees, and individuals with immune disorders, such as AIDS patients, are more likely to be infected than other members of the population. Although at present only 2–5% of young American adults have been infected, the incidence and death rate have increased since 1984. Recently, strains resistant to many antibiotics have appeared, further complicating efforts to control the disease. An infectious disease that is airborne (spread through the air), potentially fatal, and threatens to become resistant to all current treatments has led to calls for supervised treatment of infected persons up to and including institutionalization, quarantine, and other forms of confinement until the patient is noninfectious. Interest in developing a more effective vaccine than the BCG vaccine currently used in Europe is increasing.

Pneumonia PAGE 651

Pneumonia (nū-MŌ-nē-a) is an infection of the lobules of the lung. As inflammation occurs within the lobules, respiratory function deteriorates as a result of fluid leakage into the alveoli and/or swelling and constriction of the respiratory bronchioles. When bacteria are involved, they are frequently species normally found in the mouth and pharynx that have somehow managed to evade the respiratory defenses. As a result, pneumonia becomes more likely when the respiratory defenses have been compromised by other factors, such as epithelial damage from smoking or the breakdown of the immune system in AIDS. The most common pneumonia that develops in AIDS patients results from infection by the fungus *Pneumocystis carinii*. These organisms are normally found in the alveoli, but in healthy individuals the respiratory defenses are able to prevent infection and tissue damage.

Pulmonary Embolism PAGE 651

The lungs are the only organs that receive the entire cardiac output. Blood pressure in the pulmonary circuit is usually relatively low. Pulmonary vessels can easily become blocked by blood clots, fat masses, or air bubbles in the pulmonary arteries. Blockage of a vessel stops blood flow to all of the alveoli serviced by the obstructed vessel. This condition is called a **pulmonary embolism**. A condition described earlier, *venous thrombosis*, can promote development of a pulmonary embolism. If the blockage remains in place for several hours, the alveoli will permanently collapse. If the blockage occurs in a major pulmonary vessel, rather than a minor tributary, pulmonary resistance increases. This resistance places extra strain on the right ventricle, which may be unable to maintain cardiac output. Congestive heart failure may be the result. Thrombolytic therapy with heparin or other drugs can be life saving.

Pneumothorax PAGE 651

Any injury to the thorax that penetrates the parietal pleura or damages the alveoli and the visceral pleura can allow air into the pleural cavity. This condition, called **pneumothorax** (nū-mō-THŌR-aks), breaks the fluid bond between the pleurae and allows the elastic fibers to contract. The result is a partially collapsed lung, a condition termed **atelectasis** (at-e-LEK-ta-sis; *ateles*, imperfect + *ektasis* expansion).

A pneumothorax may develop because of imperfections and air leakage in the walls of superficial alveoli. These problems often have a congenital basis. A sudden lung collapse, called a **spontaneous pneumothorax**, may be triggered by heavy exercise, coughing, or other activities that increase the pressures inside the alveoli of the lungs.

Thoracentesis PAGE 651

The delicate pleural membranes can become inflamed because of chronic irritation or infection. Inflammation produces symptoms of pleuritis, or pleurisy. Acute and chronic inflammation often causes a change in the permeability of the pleura, leading to a **pleural effusion** (an abnormal accumulation of fluid within the pleural cavities). When a pleural effusion is detected on an X-ray, samples of pleural fluid may be obtained, using a long needle inserted between the ribs. This sampling procedure is called **thoracentesis** (or *thoracocentesis*). The fluid collected is usually checked for the presence of blood, white blood cells, bacteria, proteins, and glucose.

Lung Cancer, Smoking, and Diet PAGE 656

Lung cancer, or *pleuropulmonary neoplasm*, is an aggressive class of malignancies originating in the bronchial passageways or alveoli. These cancers affect the epithelial cells lining conducting passageways, mucous glands, or alveoli. Symptoms usually do not appear until the condition has progressed to the point that the tumor masses are restricting airflow or compressing adjacent mediastinal structures. Chest pain, shortness of breath, a cough or wheeze, and weight loss are common symptoms. Treatment programs vary depending on the cellular organization of the tumor and whether metastasis (cancer cell migration) has occurred, but surgery, radiation, or chemotherapy may be involved.

Deaths from lung cancer in the U. S. were rare in 1900, but there were 29,000 in 1956, 105,000 in 1978, and an estimated 154,900 in 2002. Each year 22% of new cancers detected are lung cancers, and in 2002, roughly 89,000 men and 65,000 women will be diagnosed with this condition. Lung cancers now account for 35% of all cancer deaths, making this condition the primary cause of cancer death in the U.S. population. Despite advances in the treatment of other forms of cancer, the survival statistics for lung cancer have not changed significantly. Even with early detection the 5-year survival rates are only 30% (men) to 50% (women), and most lung cancer patients die within a year of diagnosis.

Detailed statistical and experimental evidence has shown that *85–90%* of all lung cancers are the direct result of cigarette smoking. Claims to the contrary are simply unjustified and insupportable. The data are far too extensive to detail here, but the incidence of lung cancer for nonsmokers is 3.4 per 100,000 population, while the incidence for smokers ranges from 59.3 per 100,000 for those burning between a half-pack and a pack per day, to 217.3 per 100,000 for those smoking one to two packs per day. Before 1970, this disease affected primarily middle-aged men, but as the number of women smokers has increased so has the number of women dying from lung cancer.

Smoking changes the quality of the inspired air, making it drier and contaminated with several carcinogenic compounds and particulate matter. The combination overloads the respiratory defenses and damages the epithelial cells throughout the respiratory system. The histological changes that follow were described in Chapter 3. p. 82. Whether lung cancer develops appears to be related to the total cumulative exposure to the carcinogenic stimuli. The more cigarettes smoked, the greater the risk, whether those cigarettes are smoked over a period of weeks or years. The histological changes induced by smoking are reversible, and a normal epithelium will return if the stimulus is removed. At the same time the statistical risks decline to significantly lower levels. Ten years after quitting, a former smoker stands only a 10% greater chance of developing lung cancer than a nonsmoker.

The fact that cigarette smoking often causes cancer is not surprising in view of the toxic chemicals contained in the smoke. What is surprising is that more smokers do not develop lung cancer. There is evidence that some smokers have a genetic predisposition for developing one form of lung cancer. Dietary factors may also play a role in preventing lung cancer, although the details are controversial. In terms of their influence on the risk of lung cancer, there is general agreement that (1) vitamins C and A have no effect, (2) foods containing beta carotene, an orange pigment, and other vegetable components reduce the risk, and (3) a high-cholesterol, high-fat diet increases the risk.

25 THE DIGESTIVE SYSTEM

Achalasia and Esophagitis PAGE 676

In the condition known as **achalasia** (ak-a-LĀ-zē-a), the smooth muscle of the esophagus and lower esophageal sphincter has increased muscle tone. This reduces or eliminates the lumen. A bolus of food from the mouth descends the esophagus relatively slowly and its arrival does not cause the opening of the lower esophageal sphincter. Materials then accumulate at the base of the esophagus like cars at a stop light. Secondary peristaltic waves may occur repeatedly, adding to the individual's discomfort. The most successful treatment involves cutting the circular muscle layer at the base of the esophagus or expanding a balloon in the lower esophagus until the muscle layer tears.

A weakened or permanently relaxed sphincter can cause inflammation of the esophagus, or **esophagitis** (e-sof-a-JĪ-tis), as powerful gastric acids enter the lower esophagus. The esophageal epithelium has few defenses from acid and enzyme attack, and inflammation, epithelial erosion, and intense discomfort are the result. Occasional incidents of reflux, or backflow, from the stomach are responsible for the symptoms of "heartburn." This relatively common problem supports a multimillion dollar industry devoted to producing and promoting antacids and acid inhibitors.

Stomach Cancer PAGE 681

Stomach cancer is one of the most common lethal cancers, responsible for roughly 21,000 deaths in the United States each year. However, the incidence has declined over the last 100 years. (Dietary changes–fewer preservatives and less smoking and salting of food, for example) have been linked to the decline.) Because the symptoms may resemble those of gastric ulcers, the condition may not be discovered in its early stages. Diagnosis usually involves X-rays of the stomach at various degrees of distension. The mucosa can also be visually inspected using a flexible instrument called a **gastroscope**. Attachments permit the collection of tissue samples for histological analysis.

Treatment of gastric cancer involves the surgical removal of part or all of the stomach. Even a **total gastrectomy** (gas-TREK-tō-mē) can be tolerated, because the stomach provides no essential digestive services other than the secretion of intrinsic factor for vitamin B_{12} absorption.

Gastroenteritis PAGE 686

An irritation of the small intestine may lead to a series of powerful peristaltic contractions that eject the contents of the small intestine into the large intestine. An extremely powerful irritating stimulus will produce a "clean sweep" of the absorptive areas of the digestive tract. Vomiting clears the stomach, duodenum, and proximal jejunum, and peristaltic contractions evacuate the distal jejunum and ileum. Bacterial toxins, viral infections, and various poisons will sometimes produce these extensive gastrointestinal responses. Conditions affecting primarily the small intestine are usually referred to as **enteritis** (en-ter-Ī-tis) of one kind or another. If both vomiting and diarrhea are present, the term **gastroenteritis** (gas-trō-en-ter-Ī-tis) may be used instead.

Diverticulitis and Colitis PAGE 689

In **diverticulosis** (dī-ver-tik-ū-LŌ-sis) pockets (*diverticula*) form in the mucosa, usually in the sigmoid colon. These get forced outward, probably by the pressures generated during defecation. If the pockets push through weak points in the muscularis externa, they form semi-isolated chambers that are subject to recurrent infection and inflammation. The infections cause pain and the occasional bleeding characteristic of **diverticulitis** (dī-ver-tik-ū-LĪ-tis). In severe cases the diverticula may perforate, setting the bacteria loose in the peritoneal cavity.

The general term **colitis** (kō-LĪ-tis) may be used to indicate a condition characterized by inflammation of the colon. **Irritable bowel disease** involves abdominal pain and variations in bowel activity without detectable anatomical abnormalities. This condition is accompanied by ANS dysfunction, with motor and sensory components. Individuals with irritable bowel disease have alternating constipation and diarrhea, (one may predominate), with symptoms worsened by psychological stress. **Inflammatory bowel disease** is characterized by chronic overactivity of the MALT system. Mucosal ulcerations, with diarrhea, pain, and bleeding indicate a condition know as *ulcerative colitis*. In *Crohn's disease* the inflammation extends through the entire intestinal wall, all the way to the serosa. Treatment of severe inflammatory bowel disease may involve a **colectomy** (kō-LEK-tō-mē), the removal of all or a portion of the colon. If a large part or even all of the colon must be removed, normal connection with the anus cannot be maintained. Instead, the end of the intact digestive tube is sutured to the abdominal wall, and wastes then accumulate in a plastic pouch or sac attached to the opening. If the attachment involves the colon, the procedure is a **colostomy** (kō-LOS-tō-mē); if the ileum is involved it is an **ileostomy** (il-ē-OS-tō-mē).

Cirrhosis PAGE 692

The underlying problem in **cirrhosis** (sir-Ō-sis) appears to be the widespread destruction of hepatocytes by exposure to drugs (especially alcohol), viral infection, ischemia, or blockage of the hepatic ducts. Two processes are involved in producing the symptoms. Initially the damage to hepatocytes leads to the formation of extensive areas of scar tissue that branch throughout the liver. The surviving hepatocytes then undergo repeated cell divisions, but the fibrous tissue prevents the new hepatocytes from achieving a normal lobular arrangement. So the liver gradually converts from an organized assemblage of lobules to a fibrous aggregation of poorly functioning cell clusters.

Cirrhotic blockage of blood flow along the hepatic portal vein eventually leads to *portal hypertension*. Clinical problems include *esophageal varices* (distended varicose veins), and ascites. Damage to liver cells leads to a variety of metabolic problems and clotting system abnormalities. Symptomatic treatment can slow disease progression. Liver transplant surgery is an option, with a 60% 5-year survival for recipients.

26 THE URINARY SYSTEM

Urine Composition PAGE 712

Table A-7 presents typical values for the most important components of normal urine.

Hemodialysis PAGE 717

In kidney failure, if drugs, infusions, and dietary controls cannot stabilize the composition of the blood, more drastic measures are taken. In one technique, called **hemodialysis** (hē-mō-dī-AL-i-sis), an artificial membrane is used to regulate the composition of the blood. The basic principle involved in this process, called **dialysis**, involves passive diffusion across a semipermeable membrane. The patient's blood flows across an artificial *dialysis membrane* that contains pores large enough to permit the diffusion of small ions, but small enough to prevent the loss of plasma proteins. On the other side of the membrane flows a special dialysis fluid.

TABLE A.7 GENERAL CHARACTERISTICS OF NORMAL URINE

pH:	6.0 (RANGE: 4.5–8)
Specific gravity:	1.003–1.030
Osmolarity:	855–1335 mOsm
Water content:	93–97%
Volume:	500–1500 ml/day
Color:	clear yellow
Odor:	varies depending on composition
Bacterial content:	sterile

In effect, diffusion across the dialysis membrane takes the place of normal glomerular filtration, and the characteristics of the dialysis fluid ensure that important metabolites remain in the circulation, rather than diffusing across the membrane.

In practice, silastic tubes, called *shunts*, are inserted into a medium-sized artery and vein. (The usual location is in the forearm, although the lower leg is sometimes used.) When connected to the dialysis machine, the individual sits quietly while blood circulates from the arterial shunt, through the machine, and back via the venous shunt. Inside the machine, the blood flows across a dialyzing membrane, where diffusion occurs. For chronic dialysis, an artery is surgically connected to a vein in the forearm creating an arterio-venous anastomosis that serves as a biological shunt. During dialysis, two needles are inserted into the shunt and connected to the machine.

Problems With the Micturition Reflex PAGE 717

Incontinence (in-KON-ti-nens) refers to an inability to voluntarily control urination. This condition may reflect damage to the CNS, the spinal cord, or the nerve supply to the bladder or external sphincter. Incontinence often accompanies the dementia of Alzheimer's disease, and it may also result from a stroke or spinal cord injury. In many cases the individual develops an **automatic bladder**. The micturition reflex remains intact, but voluntary control of the external sphincter is lost and the individual cannot prevent the reflexive emptying of the bladder.

Infants lack voluntary control over urination because the necessary corticospinal connections have yet to be established. Toilet training before age 2 usually involves training the parent to anticipate the timing of the reflex, rather than training the child to exert conscious control.

Childbirth can stretch and damage the sphincter muscles, and some women then develop **stress incontinence**. In this condition elevated intraabdominal pressures, such as during a cough or sneeze, can overwhelm the sphincter muscles, causing urine to leak out.

27 THE REPRODUCTIVE SYSTEM

Testicular Torsion PAGE 725

Because the testes are only loosely attached to the scrotal walls, they may become twisted within the scrotal cavity. This condition, called **testicular torsion**, usually occurs in children or adolescents. Symptoms include pain in the groin and inguinal region, local inflammation, and swelling of the scrotum on the affected side. Treatment involves prompt external or surgical manipulation of the testis to relieve the twisting and securing the testes to the scrotal wall. When unilateral torsion is found, both testes are at risk and both treated. Because any kinks in the spermatic cord severely restrict the arterial supply to the testis, corrective measures must be taken within 4–6 hours, before the testicular tissues become permanently damaged. If the tissues are deprived of circulation for longer periods, the damage will be irreversible, and the affected testis may have to be removed. This surgical procedure is called an **orchidectomy** (ōr-ki-DEK-tō-mē); *orchis*, testis); one testis produces sufficient testosterone and sperm for normal male reproductive function. The common term *castration* indicates that both testes have been removed.

Prostate Cancer PAGE 735

Prostate cancer is the second most common cancer in men, and it is the second most common cause of cancer deaths in males. Each year almost 200,000 cases are diagnosed in the United States, and there are approximately 30,000 deaths. Most patients are over age 65. There are racial differences in susceptibility that are poorly understood; the incidence is relatively high among black Americans and low among Asians.

Prostate cancer usually originates in one of the secretory glands, and as it progresses it produces a nodular lump or swelling on the prostatic surface. Palpation of the prostate gland through the rectal wall is the easiest diagnostic screening procedure, but transrectal prostatic ultrasound provides more detailed information.

If the condition is detected before the cancer cells have spread to other organs, the usual treatment is either localized radiation or the surgical removal of the prostate gland. This operation, called a **prostatectomy** (pros-ta-TEK-tō-mē), is often effective in controlling the condition, but undesirable side effects from either radiation or surgery may include a loss of erectile function and urinary incontinence. Modified surgical procedures can reduce these risks and maintain normal sexual function in roughly three out of four patients. The medication *viagra* improves blood flow into the corpora cavernosa, improving and prolonging erection in patients with ercetile dysfunction.

One recent screening method involves a blood test for **prostate-specific antigen (PSA)**. Elevated levels of this antigen, normally present in low concentrations, may indicate the presence of prostate cancer. This test is more sensitive than the serum enzyme assay previously used for screening purposes. The enzyme test for prostatic acid phosphatase detects prostate cancer in comparatively late stages of development.

Once metastasis is under way, and the lymphatic system, lungs, bone marrow, liver, or adrenal glands are involved, the survival rates are significantly lower. Potential treatments for metastatic prostate cancer include more intensive radiation dosage, hormonal manipulation, lymph node removal, and aggressive chemotherapy. Because the cancer cells are stimulated by testosterone, treatment may involve castration or hormones that depress gonadotropin-releasing hormone (GnRH) or luteinizing hormone (LH) production. Until recently the usual hormone selected was diethylstilbestrol (DES), an estrogen. There are two other drug options: (1) Drugs that mimic GnRH are given in high doses, producing a surge in LH production followed by a sharp decline to very low levels, presumably as the endocrine cells adapt to the excessive stimulation. (2) Drugs that block the action of androgens can be given. Several new drugs, including Flutamide, block the cytoplasmic receptors for testosterone and prevent stimulation of the cancer cells. Despite these theoretically interesting advances in treatment, however, the average survival time for patients diagnosed with advanced prostatic cancer is only 2.5 years.

Uterine Tumors and Cancers PAGE 745

Uterine tumors are the most common tumors in women. It has been estimated that 40% of women over age 50 have benign uterine tumors involving smooth muscle and connective tissue cells. These **leiomyomas** (lē-ō-mĪ-Ō-maz), or *fibroids*, are stimulated by estrogens, and they can grow quite large, reaching weights as great as 13.6 kg (30 lb) if left untreated. Occlusion of the uterine tubes, pressure on adjacent organs, especially the urinary bladder, and compression of blood vessels may lead to a variety of complications, including heavy, prolonged, and painful menstruation. In young women, observation or conservative treatment with drugs or restricted surgery to preserve fertility may be utilized. In older women, a decision may be made to remove the entire uterus, a procedure termed a *hysterectomy*.

Benign epithelial tumors in the uterus are called **endometrial polyps**. Roughly 10% of women probably have polyps, but because of their small size and lack of symptoms the condition passes unnoticed. If bleeding occurs or if the polyps become excessively enlarged, they can be surgically removed.

Uterine cancers are less common, affecting approximately 11.9 per 100,000 women. In 2002 there will be roughly 52,000 new cases reported and 10,700 deaths in the United States. There are two types of uterine cancers, *endometrial* and *cervical*.

Endometrial cancer is an invasive cancer of the endometrial lining. Roughly 39,000 cases are reported each year in the United States, with approximately 6600 deaths. The condition most often affects women age 50–70. Estrogen therapy, used to treat osteoporosis in postmenopausal women (see Osteoporosis and Other Skeletal Abnormalities Associated with Aging on Page 779) increases the risk of endometrial cancer by 2–10 times. Adding cyclic progesterone therapy to the estrogen therapy seems to reduce or eliminate this risk.

There is no satisfactory screening test for endometrial cancer. The most common symptom is irregular bleeding, and diagnosis typically involves examination of a biopsy of the endometrial lining by suction or scraping. Treatment of early-stage endometrial cancer involves a hysterectomy followed by localized radiation therapy. In advanced stages, more aggressive radiation treatment is recommended. Chemotherapy has not proven to be very successful in treating endometrial cancers, and only 30–40% of patients with advanced endometrial cancer benefit from this approach.

Cervical cancer can be detected early by the Pap smear, and depending on how advanced it is, can be treated by local surgery, hysterectomy, and/or irradiation. In 2002 there were estimates of 13,000 new cases and 4,100 deaths in the United States. Many cases are associated with infection by *human papilloma virus*, the cause of genital warts.

Sexually Transmitted Diseases PAGE 745

Sexually transmitted diseases, or **STDs**, are transferred from individual to individual, usually or exclusively by sexual intercourse. A variety of bacterial, viral, and fungal infections are included in this category. At least two dozen different STDs are currently recognized. All are unpleasant. *Chlamydia* can cause pelvic inflammatory disease (PID) and infertility. p. 754 Other types of STDs are quite dangerous, and a few, including AIDS, are deadly. Here we will discuss three of the most common sexually transmitted diseases: *gonorrhea*, *syphilis*, and *herpes*. AIDS was discussed earlier in Chapter 23 on Page 802.

GONORRHEA The bacterium *Neisseria gonorrhoeae* is responsible for gonorrhea, one of the most common sexually transmitted diseases in the United States. Nearly 360,000 cases are reported each year. These bacteria usually invade epithelial cells lining the male or female reproductive tract. In relatively rare cases they will also colonize the pharyngeal or rectal epithelium.

The symptoms of genital infection vary, depending on the sex of the individual concerned. It has been estimated that up to 80% of women infected with gonorrhea experience no symptoms, or symptoms so minor that medical treatment is thought to be unnecessary. As a result these individuals act as carriers, spreading the infection through their sexual contacts. An estimated 10–15% of women infected with gonorrhea experience more acute symptoms, because the bacteria invade the epithelia of the uterine tubes. This type of infection probably accounts for many of the cases of PID in the U.S. population; tens of thousands of women may become infertile each year as the result of scar tissue formation along the uterine tubes after gonorrheal infections.

Diagnosis in males seldom poses as great a problem, for all but 20–30% of infected males develop painful symptoms requiring immediate medical attention. The urethral invasion is accompanied by pain on urination (*dysuria*) and often a viscous urethral discharge. A sample of the discharge can be cultured to permit positive identification of the organism involved. Antibiotic treatment cures gonorrhea infections.

SYPHILIS Syphilis (SIF-i-lis) results from infection by the bacterium *Treponema pallidum*. Untreated syphilis can cause serious cardiovascular and neurological illness years after infection, or it can be spread to the fetus during pregnancy to produce congenital malformations. The annual reported incidence of new infections in 2000 was 2.2 cases per 100,000 population.

Primary syphilis begins as the bacteria cross the mucous epithelium and enter the lymphatics and bloodstream. At the invasion site the bacteria multiply, and after an incubation period ranging from 1.5 to 6 weeks their activities produce a raised lesion, or **chancre** (SHANG-ker). This lesion remains for several weeks before fading away, even without treatment. In heterosexual men the chancre usually appears on the penis; in women it may develop on the labia, vagina, or cervix. Lymph nodes in the region usually enlarge and remain swollen even after the chancre has disappeared.

Symptoms of **secondary syphilis** appear roughly 6 weeks later. Secondary syphilis usually involves a diffuse, reddish skin rash. Like the chancre, the rash fades over a period of 2–6 weeks. These symptoms may be accompanied by fever, headaches, and uneasiness. The combination is so vague that the disease may easily be overlooked or diagnosed as something else entirely. In a few instances more serious complications such as meningitis, hepatitis, or arthritis may develop.

The individual then enters the **latent phase**. The duration of the latent phase varies widely. Fifty to seventy percent of untreated individuals with latent syphilis fail to develop the symptoms of **tertiary syphilis**, or late syphilis, although the bacterial pathogens remain within their tissues. Those destined to develop tertiary syphilis may do so 10 or more years after infection.

The most severe symptoms of tertiary syphilis involve the CNS and the cardiovascular system. Neurosyphilis may result from bacterial infection of the meninges or the tissues of the brain or spinal cord or both. **Tabes dorsalis** (TĀ-bēz dor-SAL-is) results from the invasion and demyelination of the posterior columns of the spinal cord and the sensory ganglia and nerves. In the cardiovascular system the disease affects the major vessels, leading to aortic stenosis, aneurysms, or calcification.

Equally disturbing are the effects of transmission from mother to fetus across the placenta. These cases of congenital syphilis are marked by infections of the developing bones and cartilages of the skeleton and progressive damage to the spleen, liver, bone marrow, and kidneys. The risk of transmission may be as high as 80–95%, so maternal blood testing is recommended early in pregnancy.

Treatment of syphilis involves the administration of penicillin or other antibiotics.

HERPES VIRUS **Genital herpes** results from infection by herpesviruses. Two different viruses are involved. Eighty to ninety percent of genital herpes cases are caused by a specific virus known as *HV-2* (herpes simplex virus Type 2), a virus usually associated with the genitalia. The remaining cases are caused by *HV-1*, the same virus responsible for cold sores on the mouth. Typically within a week of the initial infection the individual develops a number of painful, ulcerated lesions on the external genitalia. In women ulcers may also appear on the cervix. In adults with normal immune systems, these gradually heal over the next 2–3 weeks. Recurring infections are common, although subsequent incidents are less severe. However, infection of the newborn infant during delivery with herpesviruses present in the vagina can lead to serious illness because the infant has few immunological defenses. Recent development of the antiviral agent *acyclovir* has helped treatment of initial infections and chronic treatment reduces the frequency of recurrences.

■ Endometriosis PAGE 745

In **endometriosis** (en-dō-mē-trē-Ō-sis) an area of endometrial tissue reattaches elsewhere in the body and begins to grow outside the uterine body. The cause is unknown, though retrograde menstruation through the uterine tubes is suspected. The severity of the condition depends on the size of the abnormal mass and its location. Abdominal pain, bleeding, pressure on adjacent structures, and infertility are common symptoms. If the island of endometrial tissue enlarges, the symptoms become more severe.

Diagnosis can usually be made by using a laparoscope inserted through a small opening in the abdominal wall. Using this device a physician can see the outer surfaces of the uterus and uterine tubes, the ovaries, and the lining of the pelvic cavity. Treatment may involve surgical removal of the endometrial masses or hormonal therapies that suppress menstruation.

■ Conception Control PAGE 754

For physiological, logistical, financial, or emotional reasons most adults practice some form of conception control during their reproductive years. When the simplest and most obvious method, sexual abstinence, is unsatisfactory for some reason, another method of contraception must be used to avoid unwanted pregnancies.

Well over 50% of U.S. women age 15–44 are practicing some method of contraception; there are many different methods of contraception; only a few will be considered here.

Sterilization makes one unable to provide functional gametes for fertilization. Either sexual partner may be sterilized with the same net result. In a **vasectomy** (vaz-EK-tō-mē) a segment of the ductus deferens is removed, making it impossible for spermatozoa to pass from the epididymis to the distal portions of the reproductive tract. The surgery can be performed in a physician's office in a matter of minutes. After a 1-cm section is removed, the cut ends are usually separated and tied off, with scar tissue forming a permanent seal. Alternatively, the cut ends of the ductus deferens can be blocked with silicone plugs that can later be removed to restore fertility. After a vasectomy the man experiences normal sexual function, for the epididymal and testicular secretions normally account for only around 5% of the volume of the semen. Spermatozoa continue to develop, but they remain within the epididymis until they degenerate.

In the female the uterine tubes can be blocked through a surgical procedure known as a **tubal ligation**. Because the surgery involves general anesthesia and entering the abdominopelvic cavity (usually with a laparascope), complications are more likely than with the vasectomy. As in a vasectomy, attempts may be made to restore fertility after a tubal ligation.

Oral contraceptives manipulate the female hormonal cycle so that ovulation does not occur. Most contraceptive pills use a combination of progestins and estrogens that suppresses production of FSH and LH and the resulting stimulation of ovarian follicles. There are now at least 20 brands of combination oral contraceptives available, and over 200 million women are using them worldwide. In the United States, 25% of women under age 45 use the combination pill to prevent conception.

The **condom**, also called a *prophylactic*, or "rubber," covers the body of the penis during intercourse and keeps spermatozoa from reaching the female reproductive tract. **Vaginal barriers** such as the *diaphragm* and *cervical cap* rely on similar principles. A diaphragm, the most popular form of vaginal barrier in use at the moment, consists of a dome of latex rubber with a small metal hoop supporting the rim. Because vaginas vary in size, women choosing this method must be individually fitted. Before intercourse the diaphragm is inserted so that it covers the cervical os, and it is usually coated with a small amount of spermicidal jelly or cream, for more effective contraception. The cervical cap is smaller and lacks the metal rim. It, too, must be fitted carefully, but unlike the diaphragm it may be left in place for several days.

An **intrauterine device (IUD)** consists of a small plastic and copper T that is inserted into the uterine cavity. The mechanism of action remains uncertain, but it is known that IUDs stimulate prostaglandin production in the uterus. The net result is an alteration in the chemical composition of uterine secretions, and the changes in the intrauterine environment lower the chances for fertilization and subsequent implantation. IUDs are in limited use today in the United States, but they remain popular in many other countries.

Vaginal rings and the **Norplant™ system** involve silicone rubber structures impregnated with progesterone hormones that diffuse at a constant rate. The vaginal rings are placed in the vagina, while the Norplant system involves the insertion of small rods beneath the skin. Both methods have undergone clinical trials, and the Norplant system has shown long-term effectiveness comparable to that of combined oral contraceptives, at lower progesterone levels. Although the Norplant system is now available by prescription, a relatively high cost has to date limited the use of this contraceptive method. An injectable progesterone that provides contraception for three months, and a progesterone-only pill that is taken daily are also available.

28 HUMAN DEVELOPMENT

Chromosomal Analysis PAGE 761

In **amniocentesis** a sample of amniotic fluid is removed and the fetal cells that it contains are analyzed. This procedure permits the identification of at least 20 different congenital conditions. The needle inserted to obtain a fluid sample is guided into position using ultrasound. Unfortunately, amniocentesis has two major drawbacks:

1. Because the sampling procedure represents a potential threat to the health of the fetus and mother, amniocentesis is performed only when there are known risk factors present. Examples of risk factors would include a family history for specific conditions, or in the case of Down syndrome, a maternal age over 35.
2. Sampling cannot safely be performed until the volume of amniotic fluid is large enough that the fetus will not be injured during the sampling process. The usual time for amniocentesis is at a gestational age of 14–15 weeks so that if abnormalities are present the option of a therapeutic abortion remains available.

A procedure called *chorionic villi sampling* analyzes cells collected from the villi during the first trimester. This procedure has a higher risk of miscarriage and misdiagnosis.

Technology and the Treatment of Infertility PAGE 764

Infertility, or the inability to have children, has been the focus of media attention for the past 10 years. An estimated 10–15% of U.S. marriages are infertile, and another 10% are unable to have as many children as desired. An infertile, or sterile, woman is unable to produce functional eggs and/or support a developing embryo. An infertile man is incapable of providing a sufficient number of motile spermatozoa for successful fertilization. Because sterility of either sexual partner will have the same result, diagnosis and treatment of infertility must involve evaluation of both sexual partners. Approximately 60% of infertility cases can be attributed to problems with the female reproductive system.

The term *Assisted Reproductive Technologies*, or *ART*, encompasses the ever-increasing types of treatment than currently help roughly 30% of infertile couples "take home a baby." In cases of male infertility due to low sperm counts, semen from several ejaculates can be pooled, concentrated, and introduced into the female reproductive tract. This technique, known as *artificial insemination*, may lead to normal fertilization and pregnancy. If the husband cannot produce functional spermatozoa, viable spermatozoa can be obtained from a "sperm bank" that stores donor sperm cells. Insertion of a single sperm into a harvested oocyte has produced normal babies.

When there are problems with the transport of the egg from the ovary to the uterine tube, due to scarring of the fimbriae or other problems, a procedure called **GIFT** (*gamete intrafallopian tube transfer*) can be used. In this process, mature oocytes are removed from the ovaries, placed in the uterine tubes, and exposed to high concentrations of spermatozoa from the husband or donor.

In the GIFT procedure, fertilization occurs in its normal location, within the uterine tube. This site is not essential, and fertilization can also take place in a test tube or Petri dish. This process is called **in vitro fertilization** (*vitro*, glass), or **IVF**. If a carefully controlled fluid environment is provided, early development will proceed normally. One variation on the GIFT procedure, called **ZIFT** (*zygote intrafallopian tube transfer*), exposes selected oocytes to spermatozoa outside the body and inserts zygotes or early pre-embryos, rather than mature oocytes, into the uterine tubes. Alternatively, the zygote can be maintained in an artificial environment through the first 2–3 days of development. This procedure is often selected if the uterine tubes are damaged or blocked. The pre-embryo is then placed directly into the uterus, rather than into one of the uterine tubes.

Problems with the Implantation Process PAGE 764

The trophoblast undergoes repeated nuclear divisions, shows extensive and rapid growth, has a very high demand for energy, invades and spreads through adjacent tissues, and fails to activate the maternal immune system. In short, the trophoblast has many of the characteristics of cancer cells. In an estimated 0.1% of pregnancies, a definitive placenta does not develop, and instead the syncytial trophoblast runs wild, forming a **gestational neoplasm** or **hydatidiform** (hī-da-TID-i-fōrm) **mole**. Prompt surgical removal of the mass is essential, sometimes followed by chemotherapy, for about 20% of hydatidiform moles will metastasize, invading other tissues with potentially fatal results.

Implantation usually occurs at the endometrial surface lining the uterine cavity. The precise location within the uterus varies, although most often implantation occurs in the body of the uterus. This is not an ironclad rule, and in an **ectopic pregnancy** implantation occurs somewhere other than within the uterus.

The incidence of ectopic pregnancies is approximately 0.6%. Women who douche regularly have a **4.4** times higher risk of experiencing an ectopic pregnancy than those who don't, presumably because the flushing action pushes the zygote away from the uterus. If the uterine tube has been scarred by a previous episode of pelvic inflammatory disease, there is also an increased risk of an ectopic pregnancy. Although implantation may occur within the peritoneal cavity, in the ovarian wall, or in the cervix, 95% of ectopic pregnancies involve implantation within a uterine tube. The tube cannot expand enough to accommodate the developing embryo, and it usually ruptures during the first trimester. At this time the hemorrhaging that occurs in the peritoneal cavity may be severe enough to pose a threat to the woman's life.

In a few instances the ruptured uterine tube releases the embryo with an intact umbilical cord, and further development can occur. About 5% of these **abdominal pregnancies** actually complete full-term development; normal birth cannot occur, but the infant can be surgically removed from the abdominopelvic cavity. Because abdominal pregnancies are possible, it has been suggested that men as well as women could act as surrogate mothers if a zygote were surgically implanted in the peritoneal wall. It is not clear how the endocrine, cardiovascular, nervous, and other systems of a man would respond to the stresses of pregnancy. The procedure has been tried successfully in mice, however, and experiments continue.

Problems with Placentation PAGE 766

In a **placenta previa** (PRĒ-vē-a; "in the way") implantation occurs in or near the internal cervical os. This condition causes problems as the growing placenta approaches the internal cervical os. In a total placenta previa the placenta actually extends across the internal os, while a partial placenta previa only partially blocks the os. The placenta is characterized by a rich fetal blood supply, with erosion of maternal blood vessels within the endometrium. In placenta previa, the placenta passes across the internal os, so the delicate complex hangs like an unsupported water balloon. As the pregnancy advances, even minor mechanical stresses can be enough to tear the placental tissues, leading to massive fetal and maternal hemorrhaging. The only treatment is immediate emergency C-section and delivery of the baby and placenta.

Prior to ultrasound scanning, most cases were not diagnosed until the seventh month of pregnancy, when the placenta reaches its full size. At this time the dilation of the cervical canal and the weight of the uterine contents are pushing against the placenta where it bridges the internal os. Minor, painless hemorrhaging usually appears as the first sign of the condition. The diagnosis can usually be detected early by ultrasound scanning. Treatment in cases of total placenta previa usually involves bed rest for the mother until the fetus reaches a size at which cesarean delivery can be performed with a reasonable chance of neonatal (newborn) survival.

In an **abruptio placentae** (ab-RUP-shē-ō pla-SEN-tē) part or all of the placenta tears away from the uterine wall sometime after the fifth month of gestation. The bleeding into the uterine cavity and the pain that follows usually will be noted and reported, although in some cases the shifting placenta may block the passage of blood through the cervical canal. In severe cases the hemorrhaging leads to maternal anemia, shock, and kidney failure. Although maternal mortality is low, the fetal mortality rate from this condition ranges from 30 to 100%, depending on the severity of the hemorrhaging.

Teratogens and Abnormal Development PAGE 766

Teratogens (ter-AT-ō-jens) are stimuli that disrupt normal development by damaging cells, altering chromosome structure, or acting as abnormal inducers. **Teratology** (ter-a-TOL-ō-jē) is literally the "study of monsters," and it considers extensive departures from the pathways of normal development. Teratogens that affect the embryo in the first trimester will potentially disrupt cleavage, gastrulation, or neurulation. The embryonic survival rate will be low, and the survivors will usually have severe anatomical and physiological defects affecting all of the major organ systems. Errors introduced into the developmental process during the second and third trimesters will be more likely to affect specific organs or organ systems, for the major organizational patterns are already established. Nevertheless, the alterations reduce the chances for long-term survival.

Many powerful teratogens are encountered in everyday life. The location and severity of the resulting defects vary depending on the nature of the stimulus and the time of exposure. Radiation is a powerful teratogen that can affect all living cells. Even the X-rays used in diagnostic procedures can break chromosomes and produce developmental errors; thus nonionizing procedures such as ultrasound are used to track embryonic and fetal development. Fetal exposure to the microorganisms responsible for syphilis or rubella ("German measles") can also produce serious developmental abnormalities, including congenital heart defects, mental retardation, and deafness.

Even practices usually considered socially acceptable for adults may not be acceptable to the fetus. Two examples are consumption of alcohol and smoking. **Fetal alcohol syndrome (FAS)** occurs when maternal alcohol consumption produces developmental defects such as skeletal deformation, cardiovascular defects, and neurological disorders. Mortality rates can be as high as 17%, and the survivors are plagued by problems in later development. The most severe cases involve mothers who consume the alcohol content of at least 7 ounces of hard liquor, 10 beers, or several bottles of wine each day. But because the effects produced are directly related to the degree of exposure, there is probably no level of alcohol consumption that can be considered completely safe. Fetal alcohol syndrome is the number one cause of mental retardation in the United States today, affecting roughly 7500 infants each year.

Smoking presents another major risk to the developing fetus. In addition to introducing potentially harmful chemicals, such as nicotine, smoking lowers the oxygen content of maternal blood and reduces the amount of oxygen arriving at the placenta. The fetus carried by a smoking mother will not grow as rapidly as one carried by a nonsmoker, and smoking increases the risks of spontaneous abortion, prematurity, and fetal death. There is also a higher rate of infant mortality after delivery, and postnatal development can be adversely affected.

Complexity and Perfection: An Improbable Dream PAGE 774

The expectation of prospective parents that every pregnancy will be idyllic and every baby a perfect specimen reflects deep-seated misconceptions about the nature of the developmental process. These misconceptions lead many to believe that when serious developmental errors occur someone or something is at fault, and blame must be assigned to maternal habits, such as smoking, alcohol consumption, improper diet, maternal exposure to toxins or prescription drugs, or the presence of other disruptive stimuli in the environment. The prosecution of women giving birth to severely impaired infants for "fetal abuse" (exposing a fetus to known or suspected risk factors) represents an extreme example of this philosophy.

Although environmental stimuli may indeed lead to developmental problems, such factors are only one component of a complex system normally subject to considerable variation. Even if every pregnant woman were packed in cotton fluff and locked in her room from conception to delivery, developmental accidents and errors would continue to appear with statistical regularity.

Spontaneous mutations are the result of random errors in the replication process, and such incidents are relatively common. At least 10% of fertilizations produce zygotes with abnormal chromosomes, and because spontaneous mutations usually fail to produce visible defects, the actual number of mutations must be far larger. Most of the severely affected zygotes die before completing development, and only about 0.5% of newborn infants show chromosomal abnormalities resulting from spontaneous mutations.

Because of the nature of the regulatory mechanisms, prenatal development does not follow precise, predetermined pathways. For example, considerable variation exists in the pathways of blood vessels and nerves, because it doesn't matter how the blood or neural impulses get to their destination, as long as they do get there. If the variations fall outside acceptable limits, however, the embryo or fetus fails to complete development. Very small changes in heart structure may result in the death of a fetus, while large variations in venous distribution are extremely common and relatively harmless. Virtually everyone can be considered abnormal to some degree, because no one has characteristics that are statistically "average" in every respect. An estimated 20% of your genes differ from those found in the majority of the population, and minor recognized defects such as extra nipples or birthmarks are quite common.

Current evidence suggests that as many as half of all conceptions produce zygotes that do not survive the cleavage stage. These disintegrate within the uterine tubes or uterine cavity, and because implantation never occurs there are no obvious signs of pregnancy. These instances of preimplantation mortality are often associated with chromosomal abnormalities. Of those embryos that implant, roughly 20% fail to complete 5 months of development, with an average survival time of 8 weeks. Severe problems affecting early embryogenesis or placenta formation are usually responsible.

Prenatal mortality tends to eliminate the most severely affected fetuses. Those with less extensive defects may survive, completing full-term gestation or arriving via premature delivery. **Congenital malformations** are structural abnormalities, present at birth, that affect major systems. Spina bifida, hydrocephaly, and Down syndrome are among the most common congenital malformations; these conditions were described in earlier chapters of the text and this appendix. The incidence of congenital malformations at birth averages around 6%, but only one-third of them are categorized as severe. Only 10% of these congenital problems can be attributed to environmental factors. The rest probably reflect chromosomal abnormalities or genetic factors, including a family history of similar or related defects.

Medical technology continues to improve our abilities to understand and manipulate physiological processes. Genetic analysis of potential parents may now provide estimates concerning the likelihood of specific problems, although the problems themselves remain outside our control. But even with a better understanding of the genetic mechanisms involved, it will probably never be possible to control every aspect of development and thereby prevent spontaneous abortions and congenital malformations. There are simply too many complex, interdependent steps involved in prenatal development, and malfunctions of some kind are statistically inevitable.

WEIGHTS AND MEASURES

Accurate descriptions of physical objects would be impossible without a precise method of reporting the pertinent data. Dimensions such as length and width are reported in standardized units of measurement, such as centimeters or inches. These values can be used to calculate the volume of an object, a measurement of the amount of space it fills. **Mass** is another important physical property. The mass of an object is determined by the amount of matter it contains; on earth the mass of an object determines its weight.

Most U.S. readers describe length and width in terms of inches, feet, or yards; volumes in pints, quarts, or gallons; and weights in ounces, pounds, or tons. These are units of the **U.S. system** of measurement. Table A-8 summarizes the familiar and unfamiliar terms used in the U.S. system. For reference purposes, this table also includes a definition of the "household units," popular in recipes and cookbooks. The U.S. system can be very difficult to work with, because there is no logical relationship between the various units. For example, there are 12 inches in a foot, 3 feet in a yard, and 1760 yards in a mile. Without a clear pattern of organization, converting feet to inches or miles to feet can be confusing and time-consuming. The relationships between ounces, pints, quarts, and gallons, or ounces, pounds, and tons are no more logical.

TABLE A-8 THE U.S. SYSTEM OF MEASUREMENT

Physical Property	Unit	Relationship to Other U.S. Units	Relationship to Household Units
Length	inch (in.)	1 in. = 0.083 ft	
	foot (ft)	1 ft = 12 in.	
		= 0.33 yd	
	yard (yd)	1 yd = 36 in.	
		= 3 ft	
	mile (mi)	1 mi = 5,280 ft	
		= 1,760 yd	
Volume	fluidram (fl dr)	1 fl dr = 0.125 fl oz	
	fluid ounce (fl oz)	1 fl oz = 8 fl dr	= 6 teaspoons (tsp)
		= 0.0625 pt	= 2 tablespoons (tbsp)
	pint (pt)	1 pt = 128 fl dr	= 32 tbsp
		= 16 fl oz	= 2 cups (c)
		= 0.5 qt	
	quart (qt)	1 qt = 256 fl dr	= 4 c
		= 32 fl oz	
		= 2 pt	
		= 0.25 gal	
	gallon (gal)	1 gal = 128 fl oz	
		= 8 pt	
		= 4 qt	
Mass	grain (gr)	1 gr = 0.002 oz	
	dram (dr)	1 dr = 27.3 gr	
		= 0.063 oz	
	ounce (oz)	1 oz = 437.5 gr	
		= 16 dr	
	pound (lb)	1 lb = 7000 gr	
		= 256 dr	
		= 16 oz	
	ton (t)	1 t = 2000 lb	

In contrast, the **metric system** has a logical organization based on powers of 10, as indicated in Table A-9. For example, **a meter** (m) represents the basic unit for the measurement of size. When measuring larger objects, data can be reported in terms of **dekameters** (*deka*, ten), **hectometers** (*hekaton*, hundred), or **kilometers** (**km**; *chilioi*, thousand); for smaller objects, data can be reported in **decimeters** (**dm**; 0.1 m; *decem*, ten), **centimeters** (**cm** = 0.01 m *centum*, hundred), **millimeters** (**mm** = 0.001 m *mille*, thousand), and so forth. Notice that the same prefixes are used to report weights, based on the **gram** (g), and volumes, based on the **liter** (l). This text reports data in metric units, usually with U.S. equivalents. You should use this opportunity to become familiar with the metric system, because most technical sources report data only in metric units, and most of the rest of the world uses the metric system exclusively. Conversion factors are included in Table A-9.

Pharmacies at one time used the **apothecary system**, a relatively specialized system of measurement borrowed from England when America was still a colony. This system has been largely replaced by the metric system, and we will ignore apothecary units in this text. The apothecary system deals only with volumes and weights, and the volumetric units are comparable to those of the U.S. system. The two systems differ, however, in terms of the definitions of mass units.

The U.S. and metric systems also differ in their methods of reporting temperatures; in the United States, temperatures are usually reported in degrees Fahrenheit (°F) whereas scientific literature and individuals in most other countries report temperatures in degrees Centigrade or Celsius (°C) The relationship between temperatures in degrees Fahrenheit and those in degrees Centigrade has been indicated at the bottom of Table A-9.

TABLE A-9

THE METRIC SYSTEM OF MEASUREMENT

Physical Property	Unit	Relationship to Standard Metric Units	Conversion to U.S. Units	
Length	nanometer (nm)	1 nm = 0.000000001 m (10^{-9})	= 4 × 10^{-8} in.	25,000,000 nm = 1 in.
	micrometer (µm)	1 µm = 0.000001 m (10^{-6})	= 4 × 10^{-5} in.	25,000 µm = 1 in.
	millimeter (mm)	1 mm = 0.001 m (10^{-3})	= 0.0394 in.	25.4 mm = 1 in.
	centimeter (cm)	1 cm = 0.01 m (10^{-2})	= 0.394 in.	2.54 cm = 1 in.
	decimeter (dm)	1 dm = 0.1 m (10^{-1})	= 3.94 in.	0.25 dm = 1 in.
	meter (m)	standard unit of length	= 39.4 in.	0.0254 m = 1 in.
			= 3.28 ft	0.3048 m = 1 ft
			= 1.09 yd	0.914 m = 1 yd
	dekameter (dam)	1 dam = 10 m		
	hectometer (hm)	1 hm = 100 m		
	kilometer (km)	1 km = 1000 m	= 3280 ft	
			= 1093 yd	
			= 0.62 mi	1.609 km = 1 mi
Volume	microliter (µl)	1 ml = 0.000001 l (10^{-6})		
		= 1 cubic millimeter (mm^3)		
	milliliter (ml)	1 ml = 0.001 l (10^{-3})	= 0.03 fl oz	5 ml = 1 tsp
		= 1 cubic centimeter (cm^3 or cc)		15 ml = 1 tbsp
				30 ml = 1 fl oz
	centiliter (cl)	1 cl = 0.01 l (10^{-2})	= 0.34 fl oz	3 cl = 1 fl oz
	deciliter (dl)	1 dl = 0.1 l (10^{-1})	= 3.38 fl oz	0.29 dl = 1 fl oz
	liter (l)	standard unit of volume	= 33.8 fl oz	0.0295 l = 1 fl oz
			= 2.11 pt	0.473 l = 1 pt
			= 1.06 qt	0.946 l = 1 qt
Mass	picogram (pg)	1 pg = 0.000000000001 g (10^{-12})		
	nanogram (ng)	1 ng = 0.000000001 g (10^{-9})		
	microgram (µg)	1 µg = 0.000001 g (10^{-6})	= 0.000015 gr	66,666 µg = 1 gr
	milligram (mg)	1 mg = 0.001 g (10^{-3})	= 0.015 gr	66.7 mg = 1 gr
	centigram (cg)	1 cg = 0.01 g (10^{-2})	= 0.15 gr	6.7 cg = 1 gr
	decigram (dg)	1 dg = 0.1 g (10^{-1})	= 1.5 gr	0.67 dg = 1 gr
	gram (g)	standard unit of mass	= 0.035 oz	28.35 g = 1 oz
			= 0.0022 lb	453.6 g = 1 lb
	dekagram (dag)	1 dag = 10 g		
	hectogram (hg)	1 hg = 100 g		
	kilogram (kg)	1 kg = 1000 g	= 2.2 lb	0.453 kg = 1 lb
	metric ton (kt)	1 kt = 1000 kg	= 1.1 t	
			= 2205 lb	0.907 kt = 1 t

Temperature	Centigrade		Fahrenheit
Freezing point of pure water	0°		32°
Normal body temperature	36.8°		98.6°
Boiling point of pure water	100°		212°
Conversion °C → °F:	°F = (1.8 × °C) + 32	°F → °C:	°C = (°F − 32) × 0.56

ANSWERS TO REVIEW QUESTIONS

CHAPTER 1

Level 1

1. F	2. E	3. G	4. H
5. B	6. D	7. C	8. A
9. C	10. C	11. D	12. B
13. D	14. A	15. B	

Level 2

1. A
2. All living organisms have the same basic functions: responsiveness, growth and differentiation, reproduction, movement, metabolism and excretion.
3. In the anatomical position, a person stands with the legs together and the feet flat on the floor. The hands are at the sides, and the palms face forward.
4. B 5. B
6. Serous membranes provide a slippery cover for the inside of the ventral body cavities and the outside of most organs located within these cavities. This slippery lining prevents friction between moving organs and the body wall.

Level 3

1. A chemical imbalance in a heart muscle cell would disrupt tissue function, and this could affect the ability of the heart to pump blood. This compromises the cardiovascular system, and other tissues and organs will soon stop functioning, with potentially fatal results. This is a good example of how all levels of organization within an organism are interdependent. A disruption at the cellular level affects the tissue, the organ that contains the tissue, the body system that contains the organ, and finally the organism as a whole.
2. The body systems affected would include: digestive, respiratory, and skeletal systems. The anatomical specialties involved include surface anatomy, regional anatomy studies, systemic anatomy, comparative anatomy, and developmental anatomy.

CHAPTER 2

Level 1

1. E	2. G	3. H	4. B
5. D	6. I	7. C	8. F
9. A	10. B	11. A	12. B
13. B	14. C	15. D	16. D
17. A	18. C	19. A	20. C

Level 2

1. C
2. The three basic concepts of the cell theory are:
   (a) Cells are the structural "building blocks" of all plants and animals.
   (b) Cells are produced by the division of preexisting cells.
   (c) Cells are the smallest structural units that perform all vital functions.
3. The four passive processes by which substances get into and out of cells are: diffusion, osmosis, filtration, and facilitated diffusion.
4. Three important differences between cytosol and extracellular fluid are:
   (a) The cytosol is high in potassium ions; the extracellular fluid is high in sodium ions.
   (b) The cytosol has a relatively high concentration of dissolved and suspended proteins.
   (c) The cytosol contains relatively small quantities of carbohydrates and large reserves of amino acids and lipids.
5. Three major factors that determine whether a substance can diffuse across a cell membrane are: size of the molecule, the concentration gradient across the membrane, and the solubility of the molecule.
6. Organelles are structures that perform specific functions within the cell. Organelles can be divided into two broad categories: (1) nonmembranous organelles, which are always in contact with the cytoplasm, and (2) membranous organelles, which are surrounded by membranes that isolate their contents from the cytosol.
7. Cytokinesis is the process by which daughter cells complete their physical separation at the end of mitosis. The completion of cytokinesis marks the end of cell division.
8. The stages of mitosis are:
   (a) Prophase—chromosomes (called chromatids) become visible: paired centrioles move apart; spindle fibers extend between centriole pairs; nuclear membrane disappears.
   (b) Metaphase—chromatids move to metaphase plate.
   (c) Anaphase—chromatid pairs separate; daughter chromosomes move toward opposite ends of the cell.
   (d) Telophase—'reverse of prophase': nuclear membranes form; nuclei enlarge; chromosomes gradually uncoil.
9. The general functions of the cell membrane include:
   (a. physical isolation
   (b) regulation of exchange with the environment
   (c) sensitivity
   (d) structural support
10. Microfilaments have two major functions:
    (a) microfilaments anchor the cytoskeleton to integral proteins of the cell membrane.
    (b) microfilaments can interact with other microfilaments or thick filaments to produce active movement of a portion of a cell.

Level 3

1. Your body tissues have a higher solute concentration than fresh water, so over time osmosis draws water into your skin, producing swelling and distortion.
2. Side A had the higher solute concentration, as osmosis is drawing water into this solution from Solution B.
3. The isolation of the internal contents of membranous organelles allows them to manufacture or store secretions, enzymes, or toxins that could adversely affect the cytoplasm in general. Another benefit is the increased efficiency of having specialized enzyme systems concentrated in one place. For example, mitochondria contain the concentration of enzymes necessary for energy production in the cell.
4. Since the transport of the molecule is against the concentration gradient (that is, from lower to higher concentration), and it requires energy to move across the membrane, this type of transport is called "active" transport.

CHAPTER 3

Level 1

1. J	2. H	3. A	4. C
5. F	6. G	7. I	8. B
9. E	10. D	11. B	12. C
13. C	14. D	15. D	16. A
17. D	18. B	19. A	20. A

Level 2

1. A single cell is the unit of structure and function, whereas groups of cells work together as units called tissues (collections of special cells and cell products that perform limited functions).
2. B 3. D
4. Cartilage has a matrix that is equivalent to a firm gel; in contrast, the matrix of bone is hardened and impermeable.
5. If the epithelium is one cell layer thick (simple), it provides no mechanical protection, and is found only in protected areas; this epithelium is found where absorption and secretion occur. If the epithelium is multilayered (stratified), it is found where there is mechanical and/or chemical stress. It provides protection.
6. Exocrine secretions occur in ducts that lead onto surfaces; endocrine secretions occur in the interstitial fluid, and then enter into blood vessels.
7. The presence of the cilia at the surface of the epithelium means that the mucous layer, which rests on the epithelium surface, will be kept moving and the respiratory surface will be kept clean (or clear).
8. Dense irregular connective tissue is found in the dermis where it gives the skin strength and allows it to resist stresses from many directions.
9. To ensure that absorption and secretion occur through the lining epithelium and not between the cells.
10. Germinative cells are stem cells usually found in the deepest layer of the epithelium. They divide to replace cells lost or destroyed at the epithelial surface.

Level 3

1. The presence of DNA, RNA, and membrane components suggest that the cells were destroyed during the process of secretion. This is consistent with a holocrine type of secretion.
2. Skeletal muscle tissue would be made up of densely packed fibers running in the same direction, but since these fibers are composed of cells, they would have many nuclei and mitochondria. Skeletal muscle also has an obvious banding pattern or striations due to the arrangement of the actin and myosin filaments within the cell. The student is probably looking at a slide of tendon (dense regular connective tissue).
3. Since air must diffuse from the alveoli into the bloodstream, you would expect to find very thin cells, or squamous epithelium. Thicker types of epithelial cells would slow the process of gas diffusion to and from the blood.
4. The first step would be to determine if the muscle was striated. Two categories would be found: striated would contain skeletal and cardiac muscle, and nonstriated would contain smooth muscle. The second step would be to determine voluntary versus involuntary. The striated category from step one has skeletal muscle, which is voluntary, and cardiac muscle, which is involuntary. The nonstriated category from step one has smooth muscle, which is involuntary. Thus, the three muscle types are: (1) skeletal: striated/voluntary, (2) cardiac: striated/involuntary, and (3) smooth: nonstriated/involuntary.

CHAPTER 4

Level 1

1. E	2. F	3. D	4. G
5. C	6. H	7. B	8. I
9. A	10. B	11. A	12. C
13. C	14. B	15. B	16. B
17. D	18. C	19. A	20. B

Level 2

1. B
2. Sebum is secreted by sebaceous glands. It provides lubrication on the surface of the skin and inhibits the growth of bacteria.
3. Sensible perspiration is a clear secretion produced by eccrine glands, termed sweat. Insensible perspiration occurs when water from interstitial fluids slowly penetrates through the stratum corneum to be evaporated into the surrounding air.
4. If the skin stretches and then does not contract to its original size, then it wrinkles and creases, creating a network of stretch marks.
5. Keratin is produced in large amounts in the stratum granulosum. Fibers of keratin interlock in the cells in this layer during the process of keratinization. The cells become thinner and flatter; subsequently, they dehydrate. The keratin is important in maintaining the structure of the outer part of the epidermis and participating in its water-resistance. Additionally, keratin forms the basic structural component of hair and nails.
6. It has no vital organs.
7. These activities either reduce the amount of potentially odorous secretion on the surface of the skin or inhibit the activity of bacteria on these secretions.
8. An individual who is cyanotic has a sustained reduction in circulatory supply to the skin. As a result, the skin takes on a bluish coloration, called cyanosis.
9. D 10. D

Level 3

1. When the body temperature increases, more blood flow is directed to the vessels of the skin. The red pigment in the blood gives the skin a redder than usual color and accounts for the victim's flushed appearance. The skin is dry because the sweat glands are not producing sweat (avoids further dehydration). Without evaporation cooling, not enough heat is dissipated from the skin, the skin is warm, and the body temperature rises.
2. Under the action of ultraviolet irradiation from sunlight, Vitamin D_3 (cholecalciferol) is formed from a steroid precursor. This vitamin is converted in the liver into an intermediate used by the kidneys to synthesize calcitriol, which is essential for normal calcium absorption by the small intestine. An inadequate supply of calcitriol leads to impaired bone maintenance and growth.
3. The palms of the hands and the soles of the feet have a thicker epidermis and an extra layer in the epidermis, the stratum lucidum. This thicker layer slows down the rate of diffusion of the medication and significantly decreases its effectiveness.

CHAPTER 5

Level 1

1. A	2. B	3. C	4. B
5. B	6. A	7. A	8. C
9. A	10. B		

Level 2

1. C 2. A
3. In intramembranous ossification, bone develops from mesenchyme or fibrous connective tissue. In endochondral ossification, bone replaces an existing cartilage model.
4. Compact bone.
5. Spongy bone is found where bones receive stresses from many directions. In the expanded ends of the long bones, the epiphyses, the trabeculae in the spongy bone are extensively cross-braced to withstand stress applied from different directions.
6. The periosteum (1) isolates and protects the bone from surrounding tissues, (2) stabilizes the positions of blood vessels and nerves that supply the bone, and (3) actively participates in bone growth and repair.
7. The final repair is slighty thicker and stronger than the original bone.
8. Epiphyseal cartilage is required for continued growth in length of a developing bone. Its presence indicates that growth in length still occurs in a bone.
9. A sesamoid bone is usually small, round, and flat. It develops inside tendons and is most often encountered near joints at the knee, the hands, and the feet. Wormian (sutural) bones are small, flat, oddly shaped bones found between the flat bones of the skull in the suture line.
10. Ossification is the process of replacing other tissues with bone. Calcification refers to the deposition of calcium salts within a tissue.

Level 3

1. At the fracture point, bleeding into the area creates a fracture hematoma. An internal callus forms as a network of spongy bone unites the inner surfaces, and an external callus of cartilage and bone stabilizes the outer edges. Together, these are seen as the enlargement. A swelling initially marks the location of the fracture. Over time this region will be remodeled, and little evidence of the fracture will remain.
2. Assuming that there is no other disease process involved, John's problem is probably related to his poor diet. Good nutrition is important for promoting the healing process. Insufficient quantities of protein, vitamin C, vitamin A, and vitamin D would cause the normal healing process to occur at a much slower rate. John's inactivity may also contribute to his slower rate of healing.
3. The patient most likely suffered the fracture due to osteoporosis, which has an increased incidence in post-menopausal women. Osteoporosis is a reduction in bone mass that compromises normal bone function. Increased activity of osteoclasts is responsible for the reduction in bone mass. Therapy may include estrogen replacement treatment, dietary changes to elevate calcium levels in the blood, and exercise that stresses bones and stimulates osteoblast activity.

CHAPTER 6

Level 1

1. B	2. E	3. H	4. J
5. I	6. A	7. D	8. G
9. F	10. C	11. D	12. C
13. A	14. A	15. C	16. A
17. A	18. C	19. A	20. B

Level 2

1. D 2. B
3. A prominent central depression between the wings of the sphenoid bone cradles the pituitary gland just inferior to the brain. This depression is called the hypophyseal fossa, and the bony enclosure is called the sella turcica.
4. Sutures are immovable joints that are boundaries between skull bones. At a suture, the bones are tied firmly together with dense fibrous connective tissue.
5. The ligamentum nuchae is a large elastic ligament that begins at the vertebra prominens and extends cranially to an insertion along the external occipital crest. Along the way, it attaches to the spinous processes of the other cervical vertebrae. When the head is held upright, this ligament is like the string on a bow, maintaining the cervical curvature without muscular effort.
6. The mucous membrane of the paranasal sinuses responds to environmental stress by accelerating the production of mucus. The mucus flushes irritants off the walls of

the nasal cavities. A variety of stimuli produce this result, including sudden changes in temperature or humidity, irritating vapors, and bacterial or viral infections.
7. The first cervical vertebra is the atlas. Its articulation between the occipital condyles permits nodding (indicating "yes"), but prevents twisting. The second cervical vertebra is the axis. A transverse ligament binds the odontoid process of the axis to the inner surface of the atlas, forming a pivot for rotation of the atlas and skull.
8. The thick petrous portion of the temporal bone houses the inner ear structures that provide information about hearing and balance.
9. The foramina in the cribriform plate permit passage of the olfactory nerves, providing the sense of smell.
10. The body or centrum of the lumbar vertebrae is largest because these vertebrae bear the most weight.

Level 3

1. Women in later stages of pregnancy develop lower back pain because of changes in the lumbar curvature of the spine. The increased mass of the pregnant uterus shifts the center of gravity, and to compensate for this the lumbar curvature is exaggerated and more of the body weight supported by the lumbar region than normal. This results in sore muscles and lower back pain.
2. Jeff probably has a deviated septum as result of his broken nose. In a deviated septum, the cartilaginous portion of the septum is bent where it joins the bone. This condition often blocks the drainage of one or more sinuses, with resulting sinus headaches, infections, and sinusitis.
3. As a result of a cold or the flu, the teeth in the maxillae ache and the front of the head feels heavy. An inflammation of the paranasal sinuses mucous membranes increases the production of mucus. The maxillary sinuses are often involved, because gravity does little to assist mucus drainage from these sinuses. Congestion increases and the patient experiences headaches, a feeling of pressure in facial bones, and an ache in the teeth rooted in the maxillae.
4. Prior to birth, the brain enlarges rapidly. Although the bones of the skull are also growing, they fail to keep pace with the growing brain. At birth, the cranial bones are connected by areas of fibrous connective tissue called fontanels. These connections are quite flexible, and the skull can be distorted without damage. Such distortion normally occurs during delivery and eases the passage of the infant through the birth canal. The distortion is only temporary.

CHAPTER 7

Level 1

1. B	2. E	3. G	4. J
5. C	6. I	7. F	8. A
9. H	10. D	11. B	12. B
13. C	14. D	15. D	16. B
17. B	18. C	19. C	20. C

Level 2

1. B
2. D
3. C
4. To determine the age of a skeleton, one would consider some or all of the following: the fusion of the epiphyseal plates, the amount of mineral content, the size of roughness of bone markings, teeth, bone mass of the mandible, and intervertebral disc size.
5. Weight transfer occurs along the longitudinal arch of the foot. Ligaments and tendons maintain this arch by tying the calcaneus to the distal portions of the metatarsal bones. The lateral, calcaneal side of the foot carries most of the weight of the body while standing normally. This portion of the arch has less curvature than the medial, talar portion.
6. Fractures of the medial portion of the clavicle are common because a fall on the palm of the hand of an outstretched arm produces compressive forces that are conducted to the clavicle and its articulation with the manubrium.
7. The tibia is part of the knee joint and is involved in the transfer of weight to the ankle and foot. The fibula is excluded from the knee joint and does not transfer weight to the ankle and foot.
8. The olecranon process of the ulna is the point of the elbow. During extreme extension, this process swings into the olecranon fossa on the posterior surface of the humerus to prevent overextension of the forearm relative to the arm.
9. Body weight is passed to the metatarsal bones through the cuboid bone and the cuneiform bones.

Level 3

1. In osteoporosis, a decrease in the calcium content of the body leads to bones that are weak and brittle. Since the hip joint and leg bones must support the weight of the body, any weakening of these bones may result in insufficient strength to support the body mass, and as a result the bone will break under the great weight. The shoulder joint is not a load-bearing joint and is not subject to the same great stresses or strong muscle contractions as the hip joint. As a result, breaks in the bones of this joint occur less frequently.
2. The general appearance of the pelvis, the shape of the pelvic inlet, the depth of the iliac fossa, the characteristics of the ilium, the angle inferior to the pubic symphysis, the position of the acetabulum, the shape of the obturator foramen, and the characteristics of the ischium are all important in determining an individual's sex from a skeleton. Age can be determined by the size, degree of mineralization, and various markings on the bone. The individual's general appearance can be reconstructed by looking at the markings where muscles attach to the bones. This can indicate the size and shape of the muscles and thus the individual.
3. As a result of the lack of clavicles, we might expect that Joe would have an increased range of motion at the shoulder joint.
4. The condition known as "flat feet" is due to a lower than normal longitudinal arch in the foot. A weakness in the ligaments and tendons that attach the calcaneus to the distal ends of the metatarsals would most likely contribute to this condition.

CHAPTER 8

Level 1

1. I	2. D	3. H	4. C
5. G	6. B	7. F	8. A
9. E	10. B	11. A	12. B
13. D	14. D	15. B	16. A
17. C	18. D	19. A	20. A

Level 2

1. C
2. A joint cannot be both highly mobile and very strong. The greater the range of motion at a joint, the weaker it becomes, and vice versa. For example, a synarthrosis, which is the strongest type of joint, does not permit any movement.
3. The large biceps brachii muscle covers the anterior surface of the arm. Its tendon is attached to the radius at the radial tuberosity. Contraction of this muscle produces supination of the forearm and flexion of the elbow.
4. The tibiotalar joint, or ankle joint, involves the distal articular surface of the tibia, including the medial malleolus, the lateral malleolus of the fibula, and the trochlea and lateral articular facets of the talus. The malleoli, supported by ligaments of the ankle joint (medial deltoid ligament and the three lateral ligaments) and associated fat pads, prevent the ankle bones from sliding from side to side.
5. Articular cartilages cover articulating surfaces of bones. They resemble hyaline cartilages elsewhere in the body, but they have no perichondrium, and the matrix contains more water than other cartilages have.
6. Locking the knee allows you to stand for prolonged periods without using (and tiring) the extensor muscles of the leg. The knee joint can "lock" in the extended position by a slight external rotation of the tibia, which tightens the anterior cruciate ligament and jams the meniscus between the tibia and femur.
7. The joint capsule that surrounds the entire synovial joint is continuous with the periostea of the articulating bones. Accessory ligaments are localized thickenings of the capsule. Extracapsular ligaments are on the outside of the capsule; intracapsular ligaments are found inside the capsule. In the humeroulnar joint, the capsule is reinforced by strong ligaments. The radial collateral ligament stabilizes the lateral surface of the joint. The annular ligament binds the proximal radial head to the ulna. The medial surface of the joint is stabilized by the ulnar collateral ligament.
8. The edges of the bones are interlocked and bound together at the suture by dense connective tissue. A different type of synarthrosis binds each tooth to the surrounding bony socket. This fibrous connection is the periodontal ligament.
9. The movement of the wrist and hand from palm-facing-front to palm-facing-back is called pronation. Circumduction is a special type of angular motion that encompasses all types of angular motion: flexion, extension, adduction, and abduction.
10. As one ages, the water content of the nucleus pulposus within each disc decreases. Loss of water by the discs causes shortening of the vertebral column.

LEVEL 3

1. The term whiplash is used to describe an injury wherein the body suddenly changes position, as in a fall or during rapid acceleration or deceleration. The balancing muscles are not strong enough to stabilize the head. A dangerous partial or complete dislocation of the cervical vertebrae can result, with injury to muscles and ligaments and potential injury to the spinal cord. It is called whiplash because the movement of the head resembles the cracking of a whip.
2. In a sprain, a ligament is stretched to the point where some of the collagen fibers are torn. The ligament remains functional, and the structure and stability of the joint are not affected. In a more serious incident, the entire ligament may be torn apart, simply termed a torn ligament, or the connection between the ligament and the malleolus may be so strong that the bone breaks before the ligament. In general, a broken bone heals more quickly and effectively than a torn ligament does. A dislocation often accompanies such injuries.
3. Cartilage does not have any blood vessels, so the chondrocytes rely on diffusion to gain nutrients and eliminate wastes. The synovial fluid supplies nutrients to the articular cartilages that it bathes, removing wastes. If the circulation of the synovial fluid is impaired or stopped, the cells will not get enough nutrients or be able to get rid of their waste products. This combination of factors can lead to the death of the chondrocytes and the breakdown of the cartilage.

CHAPTER 9

LEVEL 1

1. C	2. A	3. B	4. D
5. C	6. A	7. C	8. B
9. B	10. C		

LEVEL 2

1. D
2. C
3. C
4. B
5. Rectus means "straight": These muscles are parallel muscles whose fibers generally run along the long axis of the body. Muscles that are visible at the body surface are often called externus. A muscle whose name includes flexor indicates that flexion is a primary function of the muscle. The name trapezius indicates the shape of the muscle.
6. A nerve impulse arrives at the synaptic knob of the neuromuscular junction and causes acetylcholine to be released into the synaptic cleft. The acetylcholine released then binds to receptors on the sarcolemma surface, initiating a change in the local transmembrane potential. This change results in the generation of electrical signals that sweep over the surface of the sarcolemma.
7. Connective tissues bind and attach skeletal muscles to other structures. There are three concentric layers of connective tissue. The outer layer, or epimysium, surrounds the entire muscle. The middle layer, or perimysium, divides the muscle into a series of internal compartments, each containing a bundle of muscle fibers. Each compartment is called a fascicle. The inner layer, or endomysium, surrounds each skeletal muscle fiber and binds each fiber to its neighbors.
8. A motor unit with 1500 fibers would be involved in powerful, gross movements. The greater the number of fibers in a motor unit, the more powerful the contraction, and the less fine control exhibited by the motor unit.
9. In the zone of overlap, the thin filaments pass between the thick filaments. It is in this region that interaction between thick and thin filaments occurs to form cross-bridges so that contraction may occur and tension generated.

LEVEL 3

1. If a muscle is not stimulated by a motor neuron on a regular basis, the muscle will lose tone and mass and become weak (atrophy). During the time that his leg was immobilized, it did not receive sufficient stimulation to maintain proper tone. It will take a while for the muscle to build back up to support Tom's weight.
2. Weight lifting requires anaerobic endurance. The students would want to develop fast fibers for short-term maximum strength. This could be achieved by engaging in activities that involve frequent, brief but intensive workouts, such as with progressive-resistance machines. Repeated exhaustive stimulation will help the fast fibers develop more mitochondria and a higher concentration of glycolytic enzymes as well as increase the size and strength of the muscle (hypertrophy).
3. The drug would cause flaccid paralysis. In this case, the muscles are relaxed and unable to contract.

CHAPTER 10

LEVEL 1

1. H	2. D	3. A	4. I
5. F	6. B	7. J	8. E
9. C	10. G	11. C	12. B
13. A	14. B	15. C	16. A
17. A	18. C	19. A	20. D

LEVEL 2

1. C
2. A
3. C
4. C
5. In addition to using the flexor muscles, the trunk can be assisted to move forward by gravity. The bulk of the mass of the body is anterior to the vertebral column, making the movement toward the anterior easier to make.
6. The urogenital triangle is the anterior half of the perineum. The superficial muscles in this region are those of the external genitalia, which overlie the deeper muscles that strengthen the pelvic floor and encircle the urethra. These muscles form sphincters around the urinary openings in both males and females, and the vaginal opening in females.
7. The contraction of the internal oblique effects the following activities: compresses the abdomen; depresses the ribs; flexes, bends to the side, or rotates the spine.
8. The anterior muscles of the neck control the position of the larynx, depress the mandible, tense the floor of the mouth and provide a stable foundation for muscles of the tongue and pharynx.
9. The diaphragm is a major muscle of respiration. It is included in the axial musculature because it is developmentally linked to the other muscles of the chest wall.
10. The muscles of mastication move the lower jaw during chewing and grinding of the food. The muscles of the tongue manipulate food within the mouth in preparation for swallowing. The pharyngeal muscles are important in the initiation of the swallowing of the food.

LEVEL 3

1. The muscles of the anal triangle form the posterior aspect of the perineum, a structure that has boundaries established by the inferior margins of the pelvis. A muscular sheet, the pelvic diaphragm, forms the foundation of the anal triangle and extends anteriorly superior to the urogenital diaphragm to attach anteriorly to the posterior aspect of the pubic symphysis. A sphincter muscle in this region surrounds the opening to the anus.
2. The contraction of the frontalis and procerus muscles cause Mary to raise her eyebrows and wrinkle her forehead (frontalis), and to move her nose and change the position and shape of the nostrils. The raising of the eyebrows and flaring of the nostrils show that she has some concern with this meeting.
3. The muscles involved in controlling the position of the head on the vertebral column involve the muscles of the erector spinae, sacrospinalis and multifidus groups, as well as the sternocleidomastoid. These muscles control the position of the head, both maintaining posture involuntarily as well as moving the head voluntarily.

CHAPTER 11

LEVEL 1

1. C	2. K	3. I	4. G
5. J	6. D	7. F	8. B
9. E	10. H	11. A	12. D
13. B	14. B	15. A	16. D

LEVEL 2

1. C
2. C
3. D
4. B
5. The appendicular muscle that makes it possible for a person to do a push-up is the serratus anterior.
6. The muscle that becomes greatly enlarged in ballet dancers because of the need to flex and abduct the hip and which supports the knee laterally is the tensor fascia lata.
7. The most important muscle involved in sitting cross-legged is the sartorius muscle.
8. The intrinsic muscles of the hand are involved in fine control of hand and finger movement.

9. Both the tensor fasciae latae and the gluteus maximus pull on the iliotibial tract, a band of collagen fibers that provides a lateral brace for the knee. It is especially important when a person balances on one foot.
10. These sheets of connective tissue that encircle the wrist and ankle joint act as a bracelet and anklet and allow the tendons on the long hand and foot extensors and flexors to pass internal to them. These structures hold the tendons close to the surface of the limb and effectively change the position from which the tendons act to change the position of the hand and the foot.

Level 3

1. As with writing, the intrinsic muscles of the hand and fingers permit the small, precise movements needed to hit specific keys of the typewriter or computer keyboard. However, the large muscles of the forearm hold the hand in the proper position and provide extension and flexion of the fingers.
2. Jerry probably injured either the semimembranosus and/or semitendinosus, since both of these muscles are primarily involved in the action with which he is having difficulty.
3. Although the pectoralis muscle is located across the chest, it inserts on the greater tubercle joint of the humerus, the large bone of the arm. When the muscle contracts, it contributes to flexion, adduction, and medial rotation of the humerus at the shoulder joint. All of these arm movements would be in part impaired if the muscle were damaged.

CHAPTER 13

Level 1

1. C	2. J	3. H	4. E
5. B	6. I	7. F	8. D
9. A	10. G	11. B	12. D
13. B	14. A	15. C	
16. B	17. D	18. B	
19. C	20. B		

Level 2

1. C   2. B   3. D
4. Collaterals enable a single neuron to communicate with several other cells at the same time.
5. Exteroceptors provide information about the external environment in the form of touch, temperature, pressure, sight, smell, hearing, and taste. Interoceptors monitor the digestive, respiratory, cardiovascular, urinary, and reproductive systems and provide sensations of taste, deep pressure, and pain.
6. The blood-brain barrier is needed to isolate neural tissue from the general circulation because hormones or other chemicals normally present in the blood could have disruptive effects on neuron function.
7. The CNS is responsible for integrating, processing, and coordinating sensory data and motor commands. It is also the seat of higher functions, such as intelligence, memory, learning, and emotion. The PNS provides sensory information to the CNS and carries motor commands to peripheral tissues and systems.
8. The somatic nervous system controls skeletal muscle contractions, which may be voluntary or involuntary. The autonomic nervous system regulates smooth muscle, cardiac muscle, and glandular activity, usually outside of our conscious awareness or control.
9. An electrical synapse is a more efficient carrier of impulses than is a chemical synapse because the two cells are linked by gap junctions, and they function as if they shared a common cell membrane. However, chemical synapses are more versatile because the neuron membrane can be influenced by excitatory and inhibitory stimuli simultaneously. The cell membrane at an electrical synapse simply passes the signal from one cell to another, whereas a chemical synapse integrates information arriving across multiple synapses.
10. In serial processing, information may be relayed in a stepwise sequence, from one neuron to another or from one neuronal pool to the next. Parallel processing occurs when several neurons or neuronal pools are processing the same information at one time.

Level 3

1. Action potentials travel faster along fibers that are myelinated than fibers that are unmyelinated. Destruction of the myelin sheath slows the time it takes for motor neurons to communicate with their effector muscles. This delay in response results in varying degrees of uncoordinated muscle activity. The situation is very similar to a newborn, where the infant cannot control its arms and legs very well because the myelin sheaths are still being laid down for the first year. Since not all motor neurons to the same muscle may be demyelinated to the same degree, there would be some fibers that are slow to respond while others are responding normally, producing contractions that are erratic and poorly controlled.
2. In the process known as Wallerian degeneration, the axons distal to the injury site deteriorate, and macrophages migrate in to phagocytize the debris. The Schwann cells in the area divide and form a solid cellular cord that follows the path of the original axon. As the neuron recovers, its axon grows into the injury site, and the Schwann cells wrap around it. If the axon continues to grow into the periphery alongside the appropriate cord of Schwann cells, it may eventually reestablish normal synaptic contacts. If it stops growing or wanders off in some new direction, normal function will not return.
3. The cell type that would be found in increased numbers in the area of the brain that is affected by the stroke would be microglial cells.

CHAPTER 14

Level 1

1. D	2. I	3. A	4. G
5. B	6. C	7. J	8. F
9. H	10. E	11. B	12. C
13. A	14. B	15. D	16. B
17. B	18. D	19. A	20. B

Level 2

1. B   2. D
3. 1 walls of vertebral canal; 4 subdural space; 7 pia mater; 6 subarachnoid space; 3 dura mater; 2 epidural space; 5 arachnoid membrane; 8 spinal cord
4. The meninges provide tough protective covering, longitudinal physical stability, and a space for shock-absorbing fluid.
5. A reflex is an immediate involuntary response, whereas voluntary motor movement is under conscious control and is voluntary.
6. Incoming sensory information would be disrupted.
7. Transmission of information between neurons at synapses takes a finite amount of time. Thus, the more synapses in a reflex, the longer the delay between the stimulus and responses. In a monosynaptic reflex, there is only one synapse. It has the most rapid response time. In a polysynaptic reflex, there could be many synapses, each contributing to the overall delay.
8. In the cervical region, the first pair of spinal nerves, C_1 exits between the skull and the first cervical vertebra. Thereafter, a numbered cervical spinal nerve exits after each cervical vertebra. For instance, nerve C_2 exits after vertebra C_1 nerve C_3 exits after vertebra C_2 and so on until nerve C_8 exits after vertebra C_7.
9. The denticulate ligaments prevent side-to-side movements of the spinal cord.
10. The adult spinal cord extends only as far as vertebra L_1 or L_2 Inferior to this point in the vertebral foramen, the meningeal layers enclose the relatively sturdy components of the cauda equina and a significant quantity of CSF.

Level 3

1. The person would still exhibit a defecation (bowel) and micturition (bladder) reflex because the spinal reflex is processed at the level of the spinal cord. Efferent impulses from the organs would stimulate specific interneurons in the sacral region that would synapse with the motor neurons controlling the sphincters, thus bringing about emptying when organs began to fill. This is the same situation that exists in a newborn infant who has not yet fully developed the descending tracts necessary for conscious control. The individual with the spinal cord transection would lose voluntary control of the bowel and bladder because these functions rely on impulses carried by motor neurons in the brain that must travel down the cord and synapse with the interneurons and motor neurons that are involved in the reflex.
2. The part of the cord that is most likely compressed is an ascending tract.
3. The anterior horn cells of the spinal cord are somatic motor neurons that direct the activity of skeletal muscles. The lumbar region of the spinal cord controls the skeletal muscles that are involved with the control of the muscles of the hip, leg, and foot. As a result of the injury, Karen would have poor control of most leg muscles, a problem with walking if she could walk at all, and if she could stand, problems maintaining balance.

CHAPTER 15

Level 1

1. C	2. H	3. J	4. F
5. E	6. A	7. D	8. B
9. G	10. I	11. C	12. A
13. C	14. D	15. A	16. B
17. D	18. C	19. D	20. C

Level 2

1. B 2. C 3. D
4. Damage would have occurred in the premotor cortex of the frontal lobe.
5. Impulses from proprioceptors must pass through the olivary nuclei on their way to the cerebellum.
6. The nuclei involved in the coordinate movement of the head in the direction of a loud noise are the inferior colliculi.
7. The cranial nerves that collectively participate in eye function are: II, III, IV, V, and VI.
8. The person might have a lesion in the limbic system.
9. The pons provides links between the cerebellar hemispheres and the mesencephalon, diencephalon, cerebrum, and spinal cord.
10. A less intact blood-brain barrier suggests that the endothelium is extremely permeable. This permeability exposes hypothalamic nuclei to circulating hormones and permits the diffusion of hypothalamic hormones into the circulation.

Level 3

1. The cerebrospinal fluid is accumulating inside the skull of the child, causing the bones of the skull, which have not yet fused, to expand somewhat at the sutures. Because the arachnoid villi, which usually serve to drain the CSF, do not develop until children reach three years of age, there must be a drain installed to reduce the amount of fluid retained within the skull. This drain is called a shunt and most commonly empties into the jugular veins.
2. The condition is Bell's palsy, and it is caused by inflammation of the facial nerve. The problem is easily distinguished from tic doloureux, as it is usually painless and disappears on its own within a few days to weeks.
3. Since the right arm is impaired as a result of the stroke, the damaged area would likely be in the left frontal lobe.

CHAPTER 16

Level 1

1. H	2. E	3. A	4. K
5. B	6. F	7. C	8. I
9. J	10. D	11. G	12. A
13. B	14. D	15. B	16. B
17. D	18. C	19. C	20. B

Level 2

1. C 2. A 3. A 4. C
5. The first-order neuron is the sensory neuron that delivers the sensations to the CNS.
6. The sensory homunculus is distorted because the area of sensory cortex devoted to a particular region is proportional not to its absolute size but rather to the number of sensory receptors the region contains.
7. The cerebral nuclei are processing centers that provide background patterns of movement involved in the performance of voluntary motor activities.
8. The pyramidal and extrapyramidal systems function effectively when their outputs are continually readjusted as movement occurs. The complex processing and integration of neural information from peripheral structures, visual information from the eyes, and equilibrium-related sensations from the inner ear is performed by the cerebellum.
9. The speech center, or Broca's area, lies along the edge of the premotor cortex in the same hemisphere (left) as the general interpretive area.
10. The vestibulospinal tracts direct the involuntary regulation of balance in response to sensations from the inner ear. The reticulospinal tracts direct the involuntary regulation of reflex activity and autonomic functions.

Level 3

1. The problem with Cindy's spinal cord is probably located in the lateral spinothalamic tract on the right side, somewhere around the level of spinal segment L_2 To figure this out, you would need to determine (1) what tract carries sensory information from the lower limb, (2) where it decussates, and (3) what spinal segments innervate the hip and lower limb, as detailed in the dermatome illustration in Chapter 14 (Figure 14-8).
2. Injuries to the motor cortex eliminate the ability to produce fine control of motor units. However, as long as the cerebral nuclei are functional, gross movements would still be possible. Evan should still be able to walk, maintain his balance, and perform voluntary and involuntary movements primarily by using the reticulospinal tracts and rubrospinal tracts rather than the corticospinal tracts. Although these movements may be awkward or difficult, they will still be able to take place.

CHAPTER 17

Level 1

1. C	2. G	3. H	4. D
5. I	6. A	7. F	8. J
9. B	10. E	11. B	12. B
13. C	14. D	15. D	16. C
17. B	18. C	19. B	20. C

Level 2

1. A 2. A
3. B 4. D
5. There are no enzymes to break down epinephrine and norepinephrine in the blood and very little in peripheral tissues.
6. The parasympathetic division innervates only visceral structures served by some cranial nerves or lying within the thoracic and/or abdominopelvic cavities. The sympathetic division has widespread impact due to extensive collateral branching of preganglionic fibers, which reach visceral organs and tissues throughout the body.
7. The distribution of sympathetic ganglionic cholinergic fibers provides a method for regulating sweat gland secretion in the skin.
8. Sympathetic chain ganglia are innervated by preganglionic fibers from the thoracolumbar regions of the spinal cord, and interconnected by preganglionic fibers and axons from each ganglion in the chain innervating a particular body segment. The collateral ganglia are part of the abdominal autonomic plexuses anterior to the vertebral column. Preganglionic sympathetic fibers innervate the collateral ganglia as splanchnic nerves. Intramural ganglia (also termed terminal ganglia) are part of the parasympathetic division. They are located near or within the tissues of the visceral organs.
9. The sympathetic division of the ANS stimulates metabolism, increases alertness, and prepares for emergency in "fight or flight." The parasympathetic division promotes relaxation, nutrient uptake, energy storage, and "rest and repose."
10. Visceral motor neurons, called preganglionic neurons, send their axons, called preganglionic fibers, from the CNS to synapse on ganglionic neurons, whose cell bodies are located in ganglia outside the CNS.

Level 3

1. Cutting off autonomic nervous system stimulation to the stomach through the vagus nerve decreases stimulation of digestive glands, thus reducing their secretion. This may diminish ulcers in the wall of the stomach.
2. The described events in the patient indicate a "fight or flight" response and suggest that sympathetic activation has occurred.
3. Kassie should be treated with epinephrine. This would mimic sympathetic activation, which dilates air passageways in the lungs. The constriction of her respiratory passages would be alleviated.

CHAPTER 18

Level 1

1. B	2. D	3. E	4. H
5. J	6. C	7. I	8. F
9. G	10. A	11. A	12. B
13. D	14. A	15. C	16. B
17. A	18. B	19. C	20. C

Level 2

1. C 2. B 3. C
4. Each receptor has a characteristic sensitivity. For example, a touch receptor is very sensitive to pressure but relatively insensitive to chemical stimuli.

5. There is an increased quantity of neurotransmitter released when the stereocilia of the hair cell are displaced toward the kinocilium.
6. The hair cells in the inner ear act as sensory receptors.
7. Sensory adaptation is a reduction in sensitivity in the presence of a constant stimulus. The receptor responds strongly at first, but thereafter the activity along the afferent fiber gradually declines, in part because of synaptic fatigue.
8. Sensory coding provides information about the strength, duration, variation, and movement of the stimulus.
9. An individual suffering from damage to the Pacinian corpuscles would have trouble feeling direct pressure, like a pinch.
10. The bony labyrinth is a shell of dense bone. It surrounds and protects fluid-filled tubes and chambers known as the membranous labyrinth.

Level 3

1. In removing the polyps, some of the olfactory epithelium was probably damaged or destroyed. This would decrease the surface area available for the detection of odor molecules and thus the intensity of the stimulus. As a result, it would take a larger stimulus to provide the same level of smell after the surgery than before the surgery.
2. Jared has an infection of the middle ear, most often of bacterial origin, most commonly found in children and infants. Pathogens usually gain access to the middle ear cavity through the pharyngotympanic tube, which is shorter and more horizontally oriented in infants and children than in adults, usually during an upper respiratory infection. As the infection progresses, the middle ear cavity can fill with pus. The increase in pressure in the middle ear cavity can eventually rupture the tympanum. This condition can be treated with antibiotics.
3. Ruth has a condition that results from irritation or damage to the conjunctiva of the eye, most commonly from the dilation of blood vessels beneath the conjunctival surface. Bacteria, fungi, and viruses can cause this problem, or it can be caused by chemical or physical irritation.

CHAPTER 19

Level 1

1. C	2. I	3. F	4. B
5. H	6. J	7. A	8. G
9. E	10. D	11. D	12. C
13. C	14. D	15. D	16. C
17. B	18. A	19. C	20. B

Level 2

1. C
2. C
3. A
4. The nervous system has localized, immediate, short-term effects on neurons, gland cells, muscle cells, and fat cells. The endocrine system has widespread, gradual, long-term effects on all tissues.
5. The three types of chemical structure of hormones are amino acid derivatives, peptide hormones, and steroid hormones.
6. The primary targets are most cells in the body. The effects include supporting functional maturation of sperm, protein synthesis in skeletal muscles, male secondary sex characteristics, and associated behaviors.
7. Thyroid hormones increase energy utilization, oxygen consumption, growth and development of cells.
8. Parathyroid glands produce parathyroid hormone (PTH) in response to low calcium concentrations. PTH increases calcium ion concentrations in body fluids by stimulating osteoclasts, inhibiting osteoblasts, reducing urinary excretion of calcium ions, and promoting intestinal absorption of calcium (through stimulation of calcitriol production by the kidneys) until blood calcium ion concentrations return to normal.
9. Melatonin slows the maturation of sperm, eggs, and reproductive organs by inhibiting the production of a hypothalamic-releasing hormone that stimulates FSH and LH secretion.
10. The capillary network is part of the hypophyseal portal system. Capillaries in the hypothalamus absorb regulatory secretions from hypothalamic nuclei, then unite as portal vessels, which proceed to the anterior pituitary gland. Here, the portal vessels branch into a second capillary network where regulatory factors leave the vessels and stimulate endocrine cells in the anterior pituitary gland.

Level 3

1. One would expect to observe acromegaly from the overproduction of growth hormone after the epiphyseal plates have fused. Additionally, it might be expected that hyperglycemia would be apparent in the patient.
2. The two disorders are: (1) diabetes insipidus, when the posterior pituitary no longer produces ADH. Consequently, dehydration occurs and increased urination is an outcome: (2) diabetes mellitus, when there is an inadequate production of insulin with resulting elevation of blood glucose levels and an increase in urine production. TEST–look for the presence of glucose in the urine.
3. The physician could order blood tests to determine the hormone levels involved with secretory control. Either (1) the hypothalamus isn't secreting enough releasing hormone to stimulate adequate production of TSH by the anterior pituitary gland; (2) the anterior pituitary gland cannot produce normal levels of TSH under normal stimulation; or (3) the thyroid gland is unable to respond normally to TSH stimulation.
4. One benefit of a portal system is that it ensures that the controlling hormones will be delivered directly to the target cells. In addition, since the hormones go directly to their target cells without first passing through the general circulation, they are not diluted. The hypothalamus can control the cells of the pituitary with much smaller amounts of releasing and inhibiting hormones than would be necessary if the hormones had to first go through the circulatory pathway before reaching the anterior pituitary gland.

CHAPTER 20

Level 1

1. C	2. H	3. E	4. F
5. G	6. J	7. D	8. I
9. A	10. B	11. D	12. A
13. B	14. C	15. D	16. C
17. D	18. A	19. B	20. C

Level 2

1. C
2. C
3. B
4. C
5. The volume of packed cells is a hematocrit. It is expressed as a percentage and closely approximates the volume of erythrocytes in the blood sample. As a result, the hematocrit value is often called the volume of packed red cells or simply the packed cell volume.
6. The clotting reaction seals the breaks in blood vessel walls, preventing changes in blood volume that could seriously affect blood pressure and cardiovascular function.
7. A mature megakaryocyte begins to shed its cytoplasm in small membrane-enclosed packets called platelets.
8. Secondary lymphoid organs include the spleen, tonsils, and lymph nodes.
9. Lipoproteins are protein-lipid molecules that readily dissolve in plasma. Some lipoproteins function to transport insoluble lipids to peripheral tissues.
10. People with type O blood have anti-A and anti-B antibodies in their plasma. Thus, they could not receive blood from an AB donor, because the RBCs in this blood type contains surface antigens A and B on their surface. A cross-reaction would occur.

Level 3

1. Protein is needed by the body for proper growth, maintenance, and reproduction. If there is a shortage of protein in the diet, the body must turn to itself for a source of amino acids. One of the first proteins to be dismantled for amino acids is plasma albumin. As the level of albumin in the plasma decreases, so does the osmotic pressure. Fluid leaks to the interstitial spaces because of the decreased osmotic pressure, and because of the large vascular network in the abdominal viscera, the majority pools in this region.
2. A major function of the spleen is to destroy old, defective, and worn-out red blood cells. As the spleen increases in size, so does its capacity to eliminate red blood cells, and this produces anemia. The decreased number of red blood cells decreases the blood's ability to deliver oxygen to the tissues and thus the metabolism is slowed down. This would account for the tired feeling and lack of energy. Because there are fewer red blood cells than normal, the blood circulating through the skin is not as red and so the person has a pale or white skin coloration.
3. Broad-spectrum antibiotics act to kill a wide range of bacteria, both pathogenic and nonpathogenic. When an individual takes such an antibiotic, it kills a large number of bacteria normally found in the intestine. Because these bacteria produce vitamin K, their elimination substantially decreases the amount of vitamin K that is available to the liver for the production of prothrombin, a procoagulant

that is vital to clotting reactions. With decreased amounts of prothrombin in the blood, normal daily injuries such as breaks in the vessels in the nasal passageways, which are normally sealed off quickly by coagulation, do not seal off as quickly, producing the effect of nosebleeds.

CHAPTER 21

Level 1

1. F	2. E	3. H	4. C
5. B	6. J	7. G	8. A
9. I	10. D	11. D	12. C
13. C	14. B	15. C	16. A
17. C	18. B	19. D	20. B

Level 2

1. A 2. C 3. C
4. Cardiac muscle cells are like skeletal muscle fibers in that each cardiac muscle cell contains organized myofibrils, and the alignment of their sarcomeres gives the cardiocyte a striated appearance.
5. The semilunar valves do not require muscular braces because the arterial walls do not contract, and the relative position of the cusps are stable.
6. The endocardium is the simple squamous epithelium that covers the inner surfaces of the heart, including the valves. It is continuous with the endothelium of the attached blood vessels.
7. Pericardial fluid is secreted by pericardial membranes, and it acts as a lubricant, reducing friction between the opposing visceral and parietal pericardium surfaces.
8. The left ventricle is the largest heart chamber, and it has the thickest walls. The walls of the left ventricle are the thickest because it needs to exert so much force to push blood around the systemic circuit.
9. Nodal cells are unusual because their cell membranes depolarize spontaneously. Nodal cells are responsible for establishing the rate of cardiac contraction.
10. Norepinephrine release produces an increase in both heart rate and force of contractions through the stimulation of beta receptors on nodal cells and contractile cells.

Level 3

1. It would appear that Harvey has a regurgitating mitral valve. When an AV valve fails to close properly, the blood flowing back into the atrium produces the abnormal heart sound or murmur. If the sound is heard at the beginning of the systole, this would indicate the AV valve because this is the period when the valve is just closed and the blood in the ventricle is under increasing pressure; thus, the likelihood of backflow is the greatest. If the sound were heard at the end of systole or the beginning of diastole, it would indicate a regurgitating semilunar valve—in this case, the aortic semilunar valve.
2. During tachycardia, the heart beats at an abnormally fast rate. The faster the heart beats, the less time there is in between contractions for it to fill with blood again. As a result, over a period of time, the heart fills with less and less blood and thus pumps less blood out. The stroke volume decreases, as does the cardiac output. When the cardiac output decreases to the point where not enough blood reaches the central nervous system, loss of consciousness occurs.
3. Acetylcholine would be released, resulting in a decreased heart rate.
4. Cyanosis is a bluish coloration of the skin due to the presence of deoxygenated blood in vessels near the body surface. It develops in infants in which the foramen ovale remains patent (open).

CHAPTER 22

Level 1

1. C	2. D	3. F	4. J
5. I	6. H	7. A	8. G
9. E	10. B	11. B	12. A
13. D	14. D	15. B	16. A
17. C	18. A	19. B	20. B

Level 2

1. B 2. A
3. The blood volume is greater in the venous system because it acts as a reservoir for blood.
4. The pulmonary, common carotid, subclavian, and common iliac arteries are examples of elastic arteries.
5. Sinusoids are found in liver, bone marrow, and adrenal glands.
6. They are a direct connection between arterioles and venules.
7. They prevent backflow and aid in the flow of blood back to the heart by compartmentalizing it.
8. The brachiocephalic, left common carotid, and left subclavian are elastic arteries that originate on the aortic arch.
9. The superior vena cava receives blood from the head, neck, shoulders, and upper extremities.
10. Blood can flow directly from the right atrium to the left atrium, bypassing the pulmonary circuit.

Level 3

1. This injury would cause an increased heart rate, an increased secretion of renin by the kidneys, a decreased secretion of atrial natriuretic hormone, and an increased total peripheral resistance.
2. In response to the high temperature of the water, John's body shunted more blood to the superficial veins to decrease body temperature. The dilation of the superficial veins caused a shift in blood to the arms and legs and resulted in a decreased venous return. Because of the decreased venous return, the cardiac output decreased and less blood with oxygen was delivered to the brain. This caused John to feel light-headed and faint and nearly caused his demise.
3. In heart failure, the heart is not able to produce enough force to circulate the blood properly. The blood tends to pool in the extremities and as more and more fluid accumulates in the capillaries, the blood hydrostatic pressure increases and the blood osmotic pressure decreases. The fluid accumulation exceeds the ability of the lymphatics to drain it, and, as a result, edema occurs and produces obvious swelling.

CHAPTER 23

Level 1

1. C	2. J	3. E	4. B
5. F	6. D	7. A	8. I
9. H	10. G	11. D	12. D
13. D	14. C	15. A	16. A
17. A	18. B	19. D	20. C

Level 2

1. C
2. A
3. C
4. The blood-thymus barrier prevents premature stimulation of developing T cells by circulating antigens.
5. The splenic artery and the splenic vein.
6. The thoracic duct collects lymph from areas of the body inferior to the diaphragm and from the left side of the body superior to the diaphragm.
7. T cells
8. Lymphedema is the swelling in the tissues as a result of damaged valves in lymphatic vessels or blocked lymphatic vessels.
9. Immature or activated lymphocytes divide to produce additional lymphocytes of the same type.
10. In the lamina propria, which lies internal to the epithelium that lines part of the small intestine.

Level 3

1. If Tom has previously had the measles, there should be a significant amount of IgG antibody in his blood shortly after the exposure, the result of a secondary humoral immune response. If he has not previously had the disease and is in the early stages of a primary response, his blood might show an elevated level of antibodies.
2. Allergies occur when antigens called allergens bind to specific types of antibodies that are bound to the surface of mast cells and basophils. A person becomes allergic when they develop antibodies for a specific allergen. Theoretically, at least, a molecule that would bind to the specific antibodies for ragweed and prevent the allergen from binding should help to relieve the allergy.
3. A key characteristic of cancer cells is their ability to break free from a tumor and migrate to other tissues of the body forming new tumors. This process is called metastasis. The primary route for the spread of cancer cells is the lymphatic system, and cancer cells may remain in a lymph node for a period of time before moving on to other tissues. Examination of regional lymph nodes for the presence of cancer cells can help the physician determine if the cancer was caught in an early stage or whether it has started to spread to other tissues. It can also give the physician an idea of what other tissues may be affected by the cancer, which would help to decide upon the proper treatment for the patient.

CHAPTER 24

Level 1

1. I	2. J	3. A	4. H
5. G	6. D	7. B	8. E
9. C	10. F	11. A	12. D
13. B	14. B	15. C	16. D
17. D	18. C	19. B	20. B

Level 2

1. B 2. A 3. D
4. The right lung has three lobes.
5. Bronchodilation is the enlargement of the airway.
6. They phagocytize particulate matter.
7. The septa divides the lung into lobules.
8. One group regulates tension in the vocal folds and the second group opens and closes the glottis.
9. The paired laryngeal cartilages involved with the opening and closing of the glottis are the corniculate and arytenoid cartilages
10. It includes the portion lying between the hyoid bone and the entrance to the esophagus.

Level 3

1. Since the air that Mitch is breathing is not humidified (thus dry) large amounts of moisture are leaving the mucus to humidify the air that is being respired. This makes the mucus tacky and difficult for the cilia to move. As more mucus is produced, it builds up forming the nasal congestion in the morning. As Mitch showers and drinks fluid, the moisture is replaced and the mucus loosens up and is moved along the proper route as usual. The reason this happens mostly at night is because Mitch is probably not getting up frequently to drink water to replace what is being lost to humidify the air.
2. Unless the infant was suffocated immediately when it was born, the first breath that it took would start to inflate the lungs and some of the air would be trapped in the lungs. By placing the lungs in water to see if they would float or not, the medical examiner can determine whether or not there is any air in the lungs. Other measurements and tests could also be used to determine if the infant had breathed at all (air in the lungs) or was dead at birth (lungs collapsed and a small amount of flui(d).

CHAPTER 25

Level 1

1. C	2. A	3. D	4. F
5. J	6. G	7. H	8. E
9. I	10. B	11. D	12. C
13. D	14. C	15. D	16. D
17. C	18. B	19. A	20. A

Level 2

1. D 2. C 3. A
4. Lipase attacks lipids.
5. The hepatopancreatic sphincter seals off the passageway between the gallbladder and the small intestine and prevents bile from entering the small intestine.
6. The gallbladder stores bile and concentrates it.
7. Kupffer cells engulf pathogens, cell debris, and damaged blood cells in the liver.
8. The last region of the colon before the rectum is the sigmoid colon.
9. Lacteals transport materials that could not enter local capillaries. These materials eventually reach the circulation via the thoracic duct.
10. Gastrin release is triggered by food entering the stomach.

Level 3

1. The deglutition or swallowing reflex is initiated when a bolus of food is positioned in such a way as to stimulate tactile receptors on the palatal arches and uvula. If these areas are still unresponsive as a result of the anesthetic, the bolus can be pushed back by the tongue, but no muscle contractions will occur to move the food into the pharynx or elevate the larynx. Some of the saliva or parts of the bolus are likely to fall into the laryngopharynx, stimulating a gag reflex (assuming that these receptors are not affected by the anesthetic). If this is the case, it is unlikely that the pharynx will be able to move the bolus back to the oral cavity, and Jack will temporarily be in a very uncomfortable position.
2. The gallbladder functions to concentrate and store bile produced by the liver. In this capacity, the gallbladder reabsorbs water from the bile. Since bile salts are produced from cholesterol, these salts will be precipitated as cholesterol-like gallstones if too much water is reaborbed. Whenever bile is released by the gallbladder in response to the presence of fats in the duodenum, the smooth muscle in the wall of the gallbladder must contract. This generates pain and a burning sensation. To minimize or prevent this pain, Leon's doctor prescribes a diet low or absent in fat: no fat in the diet, no contraction of the gallbladder, and no pain.
3. If an individual cannot digest lactose, then the sugar will pass through to the large intestine in an undigested form. The presence of the extra sugar in the chyme increases the solute concentration of the chyme, resulting in less water being reabsorbed by the intestinal mucosa. The bacteria that inhabit the large intestine can metabolize the lactose, and in the process they produce large amounts of carbon dioxide. The gas overstretches the intestine, which stimulates local reflexes that increase peristalsis. The combination of more fluid content and increased peristalsis produces the symptoms of diarrhea. The overexpansion of the intestine by gas causes the severe pain and abdominal cramping, and, of course, the increase in intestinal gas is directly related to increased gas production by the bacteria.

CHAPTER 26

Level 1

1. C	2. A	3. I	4. F
5. J	6. B	7. E	8. H
9. D	10. G	11. C	12. A
13. C	14. C	15. B	16. D
17. D	18. C	19. B	20. C

Level 2

1. B 2. D 3. B
4. The glomerulus is contained within the expanded chamber of the nephron (Bowman's capsule).
5. It consists of large cells (podocytes) with "feet" that wrap around the glomerular capillaries.
6. The juxtaglomerular apparatus secretes two hormones—renin and erythropoietin.
7. The trigone is the triangular area of the urinary bladder bounded by the ureteral openings and the entrance to the urethra.
8. Voluntary control is under the control of the external urethral sphincter.
9. The primary function of the proximal convoluted tubule is absorption.
10. The rugae in the urinary bladder allow it to expand as it fills with urine.

Level 3

1. Increasing the volume of urine produced would decrease the total blood volume of the body. This would lead to a decreased blood hydrostatic pressure. Edema is frequently the result of hydrostatic pressure of the blood exceeding the opposing forces at the capillaries in the affected area. Depending on the actual cause of the edema, decreasing the blood hydrostatic pressure would decrease edema formation and possibly cause some of the fluid to move from the interstitial spaces back to the blood.
2. Frank's fasting has resulted in his body's utilizing proteins and fats as a source of energy since carbohydrate sources of energy are not readily available. Metabolism of these nutrients produces acidic products such as lactic acid and ketone bodies. The amount of these substances, as well as mobilized fatty acids, exceeds their tubular maximum, and the excess is secreted from the urine. The presence of these substances in the plasma also shifts the pH to be lower, prompting the kidneys to secrete more hydrogen ions. Since hydrogen ion secretion is dependant on the pH of the filtrate (among other things), the filtrate must be buffered to accept the extra ions. The body conserves base in the form of bicarbonate, so it produces ammonia, which can accept hydrogen ions in the filtrate to form ammonia and thus buffer the filtrate while conserving bicarbonate. This accounts for the heavy ammonia smell of Frank's urine.
3. Renal hypertension restricts blood flow to the kidneys and produces renal ischemia. Decreased blood flow and ischemia triggers the juxtaglomerular apparatus to produce more renin, which leads to elevated levels of angiotensin II and then aldosterone. Angiotensin II causes vasoconstriction, increased peripheral resistance, and thus increased blood pressure. The aldosterone promotes sodium retention. This leads to more water retained by the body and in increase in blood volume. This too contributes to a higher blood pressure. Another factor to consider is the release of more erythropoietin in response to tissue hypoxia. The erythropoietin stimulates the formation of red blood cells, which leads to increased blood viscosity and again contributes to the hypertension.

CHAPTER 27

Level 1

1. I	2. A	3. D	4. F
5. J	6. C	7. B	8. H
9. G	10. E	11. D	12. C
13. B	14. A	15. D	16. C
17. C	18. B	19. D	20. C

Level 2

1. A 2. C 3. A
4. The head of the sperm cell contains a nucleus with the chromosomes.
5. The acrosomal cap contains enzymes involved in the primary steps of fertilization.
6. Seminalplasmin is produced by the prostate gland, it is an antibiotic that may help prevent UTIs in males.
7. Follicular fluid causes the follicle to enlarge rapidly.
8. During menstruation, the functional zone of the endometrium is lost.
9. Soon after implantation, human chorionic gonadotropin (hCG) can be detected.
10. The testes normally begin to descend at the seventh developmental month.

Level 3

1. Yes, he would still be able to have an erection. Forming an erection is a parasympathetic reflex that is controlled by the sacral region of the spinal cord (inferior to the injury). Tactile stimulation of the penis would initiate the parasympathetic reflex that controls erection. He would also be able to experience an erection by sympathetic route, since this would be controlled in the T_{12} to L_2 area of the cord (superior to the injury). Stimulation by higher centers could produce a decreased sympathetic tone in the vessels to the penis, resulting in an erection.
2. The endometrial cells have receptors for the hormone estrogen and progesterone and respond to these hormones the same as they would if they were in the body of the uterus. Under the influence of estrogen, they proliferate at the beginning of the menstrual cycle and begin to develop glands and blood vessels, which then further develop under the control of progesterone. This dramatic change in tissue size and characteristics interferes with neighboring tissues by pressing on them or interrupting their functions in other ways. It is this interference that causes the periodic painful sensations.
3. Slightly elevated levels of estradiol or estradiol and progesterone inhibits both GnRH at the hypothalamus and the release of FSH and LH from the pituitary. Without FSH, primordial follicles do not initiate development, and the endogenous levels of estrogen remain low. An LH surge is necessary for ovulation to occur, and the LH surge is triggered by the peaking of estradiol. If the level of estradiol is not allowed to rise above the critical level, the LH surge and ovulation will not occur, even if a follicle managed to develop to a stage that it could ovulate. Any mature follicles would ultimately degenerate, and no new follicles would mature to take their place. Although the ovarian cycle is interrupted, the level of hormones is still adequate to regulate a normal menstrual cycle.

CHAPTER 28

Level 1

1. F	2. G	3. J	4. B
5. H	6. D	7. I	8. C
9. A	10. E	11. A	12. C
13. B	14. A	15. C	16. B
17. C	18. B	19. B	20. A

Level 2

1. B 2. B 3. B
4. Cleavage is a sequence of cell divisions immediately after fertilization in the first trimester.
5. The primitive streak is the center line of the blastodisc where cells migrate and begin separating into germ layers.
6. Human chorionic gonadotropin is produced in the trophoblast cells shortly after implantation. It signals the corpus luteum to produce more progesterone.
7. Amphimixis is the fusion of the male and female pronuclei.
8. The yolk sac functions in the production of blood cells.
9. Organogenesis is the process of organ formation.
10. Because events in the first 12 weeks establish the basis for organ formation.

Level 3

1. Although technically what Joe says is true, it only takes one sperm to fertilize an egg, the probability of this occurring if not enough sperm are deposited is very slim. Of the millions of sperm that enter the female reproductive tract, most are killed or disabled before they reach the uterus. The acid environment, temperature, and presence of immunoglobulins in the vaginal secretions are just a few of the factors responsible for the demise of so many sperm. Many sperm are not capable of making the complete trip. Once at the secondary oocyte, the sperm must penetrate the corona radiata, and this requires the combined enzyme contributions of one hundred or more sperm. If the ejaculate contains few sperm, it is likely that none will reach the oocyte, and fertilization will be impossible.
2. The most obvious possibility is that there is a problem with the cardiovascular supply to the lungs. A good guess would be a patent ductus arteriosus (the ductus arteriosus has failed to completely close off). The baby seems normal at rest. But there is not enough oxygen in the bloodstream to meet the demands when the baby becomes more active, as during the excitement of a bath or a meal, so the infant then becomes cyanotic.

FOREIGN WORD ROOTS, PREFIXES, SUFFIXES, AND COMBINING FORMS

Many of the words we use in everyday English have their roots in other languages, particularly Greek and Latin. This is especially true for anatomical terms, many of which were introduced into the anatomical literature by Greek and Roman anatomists. This list includes some of the foreign word roots, prefixes, suffixes, and combining forms that are part of many of the biological and anatomical terms you will see in this text.

Each entry starts with the commonly encountered form or forms of the prefix, suffix, or combining form followed by the word root (shown in italics) with its English translation. One example is given to illustrate the use of the prefix, suffix, or combining form, but there are many others, and you will see them as you progress through the text.

a-, *a-*, without: avascular
ab-, ab, from: abduct
-ac, *-akos,* pertaining to: cardiac
ad-, *ad,* to, toward: adduct
aden-, adeno-, *adenos,* gland: adenoid
af-, *ad,* toward: afferent
-al, *-alis,* pertaining to: brachial
-algia, *algos,* pain: neuralgia
ana-, *ana,* up, back: anaphase
andro-, *andros,* male: androgen
angio-, *angeion,* vessel: angiogram
anti-, ant-, *anti,* against: antibiotic
apo-, apo, from: apocrine
arachn-, *arachne,* spider: arachnoid
arthro-, *arthros,* joint: arthroscopy
-asis, -asia, state, condition: homeostasis
astro-, *aster,* star: astrocyte
atel-, *ateles,* imperfect: atelectasis
baro-, *baros,* pressure: baroreceptor
bi-, bi-, two: bifurcate
blast-, -blast, *blastos,* precursor: blastocyst
brachi-, *brachium,* arm: brachiocephalic
brady-, *bradys,* slow: bradycardia
bronch-, *bronchus,* windpipe, airway: bronchial
cardi-, cardio-, -cardia, *kardia,* heart: cardiac
-centesis, *kentesis,* puncture: thoracocentesis
cerebro-, *cerebrum,* brain: cerebrospinal
chole-, *chole,* bile: cholecystitis
chondro-, *chondros,* cartilage: chondrocyte
chrom-, chromo-, *chroma,* color: chromatin
circum-, *circum,* around: circumduction
-clast, *klastos,* broken: osteoclast
coel-, -coel, *koila,* cavity: coelom
contra-, *contra,* against: contralateral
cranio-, *cranium,* skull: craniosacral
cribr-, *cribrum,* sieve: cribriform
-crine, *krinein,* to separate: endocrine
cyst-, -cyst, *kystis,* sac: blastocyst
desmo-, *desmos,* band: desmosome
di-, *dis,* twice: disaccharide
dia-, *dia,* through: diameter
diure-, *diourein,* to urinate: diuresis
dys-, dys-, painful: dysmenorrhea
-ectasis, *ektasis,* expansion: atelectasis
ecto-, *ektos,* outside: ectoderm
ef-, *ex,* away from: efferent
emmetro-, *emmetros,* in proper measure: emmetropia
encephalo-, *enkephalos,* brain: encephalitis
end-, endo-, *endos,* inside: endometrium
entero-, *enteron,* intestine: enteric
epi-, *epi,* on: epimysium
erythema-, *erythema,* flushed (skin): erythematosis
erythro-, *erythros,* red: erythrocyte
ex-, *ex,* out, away from: exocytosis
ferr-, *ferrum,* iron: transferrin
-gen, -genic, *gennan,* to produce: mutagen
genicula-, *geniculum,* knee-like structure: geniculate
genio-, *geneion,* chin: geniohyoid
glosso-, -glossus, *glossus,* tongue: hypoglossal
glyco-, *glykys,* sugar: glycogen
-gram, *gramma,* record: myogram
-graph, -graphia, *graphein,* to write, record: electroencephalograph
gyne-, gyno-, *gynaikos,* woman: gynecologist
hem-, hemato-, *haima,* blood: hemopoiesis
hemi-, hemi-, half: hemisphere
hepato-, *hepaticus,* liver: hepatocyte
hetero-, *heteros,* other: heterosexual
histo-, *histos,* tissue: histology
holo-, *holos,* entire: holocrine
homeo-, homo-, *homos,* same: homeostasis
hyal-, hyalo-, *hyalos,* glass: hyaline
hydro-, *hydros,* water: hydrolysis
hyo-, *hyoeides,* U-shaped: hyoid
hyper-, *hyper,* above: hyperpolarization
ili-, ilio-, *ilium:* iliac
infra-, *infra,* beneath: infraorbital
inter-, *inter,* between: interventricular
intra-, *intra,* within: intracapsular
ipsi-, *ipse,* itself: ipsilateral
iso-, *isos,* equal: isotonic
-itis, -itis, inflammation: dermatitis
karyo-, *karyon,* body: megakaryocyte
kerato-, *keros,* horn: keratin
kino-, -kinin, *kinein,* to move: bradykinin
lact-, lacto-, -lactin, *lac,* milk: prolactin
-lemma, *lemma,* husk: plasmalemma
leuko-, *leukos,* white: leukocyte
liga-, *ligare,* to bind together: ligase
lip-, lipo-, *lipos,* fat: lipoid
lyso-, -lysis, -lyze, *lysis,* dissolution: hydrolysis
mal-, *mal,* abnormal: malabsorption
mamilla-, *mamilla,* little breast: mamillary
mast-, masto-, *mastos,* breast: mastoid
mega-, *megas,* big: megakaryocyte
mero-, *meros,* part: merocrine
meso-, *mesos,* middle: mesoderm
meta-, *meta,* after, beyond: metaphase
mono-, *monos,* single: monocyte
morpho-, *morphe,* form: morphology
-mural, *murus,* wall: intramural
myelo-, *myelos,* marrow: myeloblast
myo-, *mys,* muscle: myofilament
natri-, *natrium,* sodium: natriuretic
neur-, neuro-, *neuron,* nerve: neuromuscular
oculo-, *oculus,* eye: oculomotor
oligo-, *oligos,* little, few: oligopeptide
-ology, *logos,* the study of: physiology
-oma, -oma, swelling: carcinoma
onco-, *onkos,* mass, tumor: oncology
-opia, *ops,* eye: optic
-osis, -osis, state, condition: neurosis
osteon, osteo-, *os,* bone: osteocyte
oto-, *otikos,* ear: otoconia
para-, *para,* beyond: paraplegia
patho-, -path, -pathy, *pathos,* disease: pathology
pedia-, *paidos,* child: pediatrician
peri-, *peri,* around: perineurium
-phasia, *phasis,* speech: aphasia
-phil, -philia, *philus,* love: hydrophilic
-phobe, -phobia, *phobos,* fear: hydrophobic
-phylaxis, *phylax,* a guard: prophylaxis
physio-, *physis,* nature: physiology
-plasia, *plasis,* formation: dysplasia
platy-, *platys,* flat: platysma
-plegia, *plege,* a blow, paralysis: paraplegia
-plexy, *plessein,* to strike: apoplexy
podo-, *podon,* foot: podocyte
-poiesis, *poiesis,* making: hemopoiesis
poly-, *polys,* many: polysaccharide
presby-, *presbys,* old: presbyopia
pro-, *pro,* before: prophase
pterygo-, *pteryx,* wing: pterygoid
pulp-, *pulpa,* flesh: pulpitis
retro-, *retro,* backward: retroperitoneal
-rrhea, *rhein,* flow, discharge: amenorrhea
sarco-, *sarkos,* flesh: sarcomere
scler-, sclero-, *skleros,* hard: sclera
semi-, *semis,* half: semitendinosus
-septic, *septikos,* putrid: antiseptic
-sis, state or condition: metastasis
som-, -some, *soma,* body: somatic
spino-, *spina,* spine, vertebral column: spinodeltoid
-stomy, *stoma,* mouth, opening: colostomy
stylo-, *stylus,* stake, pole: styloid
sub-, *sub,* below; subcutaneous
syn-, *syn,* together: synthesis
tachy-, *tachys,* swift: tachycardia
telo-, *telos,* end: telophase
therm-, thermo-, *therme,* heat: thermoregulation
-tomy, *temnein,* to cut: appendectomy
trans-, *trans,* through: transudate
-trophic, -trophin, -trophy, *trophikos,* nourishing: adrenocorticotropic
tropho-, *trophe,* nutrition: trophoblast
tropo-, *tropikos,* turning: troponin
uro-, -uria, *ouron,* urine: glycosuria

EPONYMS IN COMMON USE

Eponym	Equivalent Terms	Individual Referenced
The Cellular Level Of Organization (Chapter 2)		
Golgi apparatus		Camillo Golgi (1844–1926), Italian histologist; shared Nobel Prize in 1906
Krebs cycle	Tricarboxylic or citric acid cycle	Hans Adolph Krebs (1900–1981), British biochemist; shared Nobel Prize in 1953
The Skeletal System (Chapters 5–8)		
Colles' fracture		Abraham Colles (1773–1843), Irish surgeon
Haversian canals	Central canals	Clopton Havers (1650–1702), English anatomist and microscopist
Haversian systems		Osteons
Pott's fracture		Percivall Pott (1714–1788), English surgeon
Volkmann's canals	Perforating canals	Alfred Wilhelm Volkmann (1800–1877), German surgeon
Wormian bones	Sutural bones	Olas Worm (1588–1654), Danish anatomist
The Muscular System (Chapters 9–11)		
Achilles tendon	Calcaneal tendon	Achilles, hero of Greek mythology
Cori cycle		Carl Ferdinand Cori (1896–1984) and Gerty Theresa Cori (1896–1957), American biochemists; shared Nobel Prize in 1947
The Nervous System (Chapters 13–17)		
Broca's center	Speech center	Pierre Paul Broca (1824–1880), French surgeon
Foramina of Luschka	Lateral foramina	Hubert von Luschka (1820–1875), German anatomist
Foramen of Magendie	Median foramen	François Magendie (1783–1855), French physiologist
Foramen of Munro	Interventricular foramen	John Cummings Munro (1858–1910), American surgeon
Nissl bodies		Franz Nissl (1860–1919), German neurologist
Purkinje cells		Johannes E. Purkinje (1787–1869), Czechoslovakian physiologist
Nodes of Ranvier		Louis Antoine Ranvier (1835–1922), French physiologist
Island of Reil	Insula	Johann Christian Reil (1759–1813), German anatomist
Fissure of Rolando	Central sulcus	Luigi Rolando (1773–1831), Italian anatomist
Schwann cells		Theodor Schwann (1810–1882), German anatomist
Aqueduct of Sylvius	Mesencephalic aqueduct	Jacobus Sylvius (Jacques Dubois, 1478–1555), French anatomist
Sylvian fissure	Lateral sulcus	Franciscus Sylvius (Franz de le Boë, 1614–1672), Dutch anatomist
Pons varolii	Pons	Costanzo Varolio (1543–1575), Italian anatomist
Sensory Function (Chapter 18)		
Organ of Corti		Alfonso Corti (1822–1888), Italian anatomist
Eustachian tube	Auditory (pharyngotympanic) tube	Bartolomeo Eustachio (1520–1574), Italian anatomist
Golgi tendon organs	Tendon organs	*See* Golgi apparatus *under* The Cellular Level (Chapter 2)
Hertz (Hz)		Heinrich Hertz (1857–1894), German physicist
Meibomian glands		Heinrich Meibom (1638–1700), German anatomist
Corpuscles of Meissner	Tarsal glands	Georg Meissner (1829–1905), German physiologist
Merkel's discs		Friedrich Siegismund Merkel (1845–1919), German anatomist
Pacinian corpuscles	Tactile discs	Filippo Pacini (1812–1883), Italian anatomist
Ruffini's corpuscles		Angelo Ruffini (1864–1929), Italian anatomist
Canal of Schlemm	Scleral venous sinus	Friedrich S. Schlemm (1795–1858), German anatomist
Glands of Zeis		Edward Zeis (1807–1868), German ophthalmologist
The Endocrine System (Chapter 19)		
Islets of Langerhans	Pancreatic islets	Paul Langerhans (1847–1888), German pathologist
Interstitial cells of Leydig	Interstitial cells	Franz von Leydig (1821–1908), German anatomist

Eponym	Equivalent Terms	Individual Referenced
The Cardiovascular System (Chapters 20–22)		
Bundle of His		Wilhelm His (1863–1934), German physician
Purkinje cells		*See under* The Nervous System (Chapters 13–17)
Starling's law		Ernest Henry Starling (1866–1927), English physiologist
Circle of Willis	Cerebral arterial circle	Thomas Willis (1621–1675), English physician
The Lymphatic System (Chapter 23)		
Hassall's corpuscles		Arthur Hill Hassall (1817–1894), English physician
Kupffer cells	Stellate reticuloendothelial cells	Karl Wilhelm Kupffer (1829–1902), German anatomist
Langerhans cells		*See Islets* of Langerhans under The Endocrine System (Chapter 19)
Peyer's patches	Aggregated lymphoid nodules	Johann Conrad Peyer (1653–1712), Swiss anatomist
The Respiratory System (Chapter 24)		
Adam's apple	Laryngeal prominence of thyroid cartilage	Biblical reference
Bohr effect		Niels Bohr (1885–1962), Danish physicist; won 1922 Nobel Prize
Boyle's law		Robert Boyle (1621–1691), English physicist
Charles' law		Jacques Alexandre César Charles (1746–1823), French physicist
Dalton's law		John Dalton (1766–1844), English physicist
Henry's law		William Henry (1775–1837), English chemist
The Digestive System (Chapter 25)		
Plexus of Auerbach	Myenteric plexus	Leopold Auerbach (1827–1897), German anatomist
Brunner's glands	Duodenal glands	Johann Conrad Brunner (1653–1727), Swiss anatomist
Kupffer cells	Stellate reticuloendothelial cells	*See under* The Lymphatic System (Chapter 23)
Crypts of Lieberkuhn	Intestinal glands	Johann Nathaniel Lieberkuhn (1711–1756), German anatomist
Plexus of Meissner	Submucosal plexus	See Corpuscles of Meissner under Sensory Function (Chapter 18)
Sphincter of Oddi	Hepatopancreatic sphincter	Ruggero Oddi (1864–1913), Italian physician
Peyer's patches	Aggregated lymphoid nodules	*See under* The Lymphatic System (Chapter 23)
Duct of Santorini	Accessory pancreatic duct	Giovanni Domenico Santorini (1681–1737), Italian anatomist
Stensen's duct	Parotid duct	Niels Stensen (1638–1686), Danish physician/priest
Ampulla of Vater	Duodenal ampulla	Abraham Vater (1684–1751), German anatomist
Wharton's duct	Submandibular duct	Thomas Wharton (1614–1673), English physician
Foramen of Winslow	Epiploic foramen	Jacob Benignus Winslow (1669–1760), French anatomist
Duct of Wirsung	Pancreatic duct	Johann Georg Wirsung (1600–1643), German physician
The Urinary System (Chapter 26)		
Bowman's capsule	Glomerular capsule	Sir William Bowman (1816–1892), English physician
Loop of Henle	Nephron loop	Friedrich Gustav Jakob Henle (1809–1885), German histologist
The Reproductive System (Chapters 27–28)		
Bartholin's glands	Greater vestibular glands	Casper Bartholin, Jr. (1655–1738), Danish anatomist
Cowper's glands	Bulbourethral glands	William Cowper (1666–1709), English surgeon
Fallopian tube	Uterine tube/oviduct	Gabriele Falloppio (1523–1562), Italian anatomist
Graafian follicle	Tertiary follicle	Reijnier de Graaf (1641–1673), Dutch physician
Interstitial cells of Leydig	Interstitial cells	*See under* The Endocrine System (Chapter 19)
Glands of Littre	Lesser vestibular glands	Alexis Littre (1658–1726), French surgeon
Sertoli cells	Sustentacular cells	Enrico Sertoli (1842–1910), Italian histologist

GLOSSARY OF KEY TERMS

abdomen: Region of trunk bounded by the diaphragm and pelvis.
abdominopelvic cavity: Portion of the ventral body cavity that contains abdominal and pelvic subdivisions.
abducens (ab-DŪ-senz): Cranial nerve VI; innervates the lateral rectus muscle of the eye.
abduction: Movement away from the midline.
abortion: Premature loss or expulsion of an embryo or fetus.
abscess: A localized collection of pus within a damaged tissue.
absorption: The active or passive uptake of gases, fluids, or solutes.
accommodation: Alteration in the curvature of the lens to focus an image on the retina; decrease in receptor sensitivity or perception following chronic stimulation.
acetabulum (a-se-TAB-ū-lum): Fossa on lateral aspect of pelvis that accommodates the head of the femur.
acetylcholine (ACh) (as-e-til-KŌ-lēn): Chemical neurotransmitter in the brain and PNS; dominant neurotransmitter in the PNS, released at neuromuscular junctions and synapses of the parasympathetic division.
acetylcholinesterase (AChE): Enzyme found in the synaptic cleft, bound to the postsynaptic membrane, and in tissue fluids; breaks down and inactivates ACh molecules.
achalasia (ak-a-LĀ-zē-a): Condition that develops when the lower esophageal sphincter fails to dilate, and ingested materials cannot enter the stomach.
Achilles tendon: Calcaneal tendon.
acid: A compound whose dissociation in solution releases a hydrogen ion and an anion; an acid solution has a pH below 7.0 and contains an excess of hydrogen ions.
acinus/acini (AS-i-nī): Histological term referring to a blind pocket, pouch, or sac.
acne: Condition characterized by inflammation of sebaceous glands and follicles; commonly affects adolescents and most often involves the face.
acoustic: Pertaining to sound or the sense of hearing.
acquired immune deficiency syndrome (AIDS): A disease caused by the **human immunodeficiency virus (HIV)**, characterized by destruction of helper T cells and a resulting severe impairment of the immune response.
acromegaly: Condition caused by overproduction of growth hormone in the adult, characterized by thickening of bones and enlargement of cartilages and other soft tissues.
acromion (a-KRŌ-mē-on): Continuation of the scapular spine that projects superior to the capsule of the shoulder joint.
acrosomal cap (ak-rō-SŌ-mal): Membranous sac at the tip of a sperm cell that contains hyaluronic acid.
actin: Protein component of microfilaments; forms thin filaments in skeletal muscles and produces contractions of all muscles through interaction with thick (myosin) filaments; *see* **sliding filament theory**.
action potential: A conducted change in the transmembrane potential of excitable cells, initiated by a change in the membrane permeability to sodium ions: *see also* **nerve impulse**.
active transport: The ATP-dependent absorption or excretion of solutes across a cell membrane.
acute: Sudden in onset, severe in intensity, and brief in duration.
adaptation: Alteration of pupillary size in response to changes in light intensity; in CNS often used as a synonym for accommodation: physiological responses that produce acclimatization.
Addison's disease: Condition resulting from hyposecretion of glucocorticoids, characterized by lethargy, weakness, hypotension, and increased skin pigmentation.
adduction: Movement toward the axis or midline of the body as viewed in the anatomical position.
adenine: A purine, one of the nitrogen bases in the nucleic acids RNA and DNA.
adenohypophysis (ad-e-nō-hī-POF-i-sis): The anterior lobe of the pituitary gland, also called the *anterior pituitary* or the **pars distalis**.
adenoid: The pharyngeal tonsil.
adenosine triphosphate (ATP): A high-energy compound consisting of adenosine with three phosphate groups attached; the third is attached by a high-energy bond.
adhesion: Fusion of two mesenteric layers following damage or irritation of their opposing surfaces.
adipocyte (AD-i-pō-sīt): A fat cell.
adipose tissue: Loose connective tissue dominated by adipocytes.
adrenal cortex: Superficial portion of adrenal gland that produces steroid hormones.
adrenal gland: Small endocrine gland secreting steroids and catecholamines, located superior to each kidney.
adrenal medulla: Core of the adrenal gland; a modified sympathetic ganglion that secretes catecholamines into the blood following sympathetic activation.
adrenocortical hormone: Any of the steroids produced by the adrenal cortex.
adrenocorticotropic hormone (ACTH): Hormone that stimulates the production and secretion of glucocorticoids by the zona fasciculata of the adrenal cortex; released by the anterior pituitary in response to CRF.
adventitia (ad-ven-TISH-a): Superficial layer of connective tissue surrounding an internal organ; fibers are continuous with those of surrounding tissues, providing support and stabilization.
afferent: Toward.
afferent arteriole: An arteriole bringing blood to the glomerulus of the kidney.
afferent fiber: Axons carrying sensory information to the CNS.
agglutination (a-gloo-ti-NĀ-shun): Aggregation of red blood cells due to interactions between surface agglutinogens and plasma agglutinins.
agglutinins (a-GLOO-ti-ninz): Immunoglobulins in plasma that react with antigens on the surfaces of foreign red blood cells when donor and recipient differ in blood type.
agglutinogens (a-GLOO-tin-ō-jenz): Antigens on the surfaces of red blood cells whose presence and structure are genetically determined.
agonist: A muscle responsible for a specific movement.
agranular: Without granules; **agranular leukocytes** are monocytes and lymphocytes; the **agranular reticulum** is an intracellular organelle that synthesizes and stores carbohydrates and lipids.
AIDS: *See* **acquired immune deficiency syndrome**.
AIDS-related complex (ARC): Early symptoms of HIV infection, consisting chiefly of lymphadenopathy, fevers, and chronic nonfatal infections.
alba, albicans, albuginea (AL-bi-kanz) (al-bū-JIN-ē-a): White.
albinism: Absence of pigment in hair and skin caused by inability of body to produce melanin.
albumins (al-BŪ-mins): The smallest of the plasma proteins; function as transport proteins and important in contributing to plasma oncotic pressure.
aldosterone: A mineralocorticoid (steroid) produced by the zona glomerulosa of the adrenal cortex that stimulates sodium and water conservation at the kidneys; secreted in response to the presence of angiotensin II.
aldosteronism: Condition caused by the oversecretion of aldosterone, characterized by fluid retention, edema, and hypertension.
allantois (a-LAN-tō-is): One of the extraembryonic membranes; it provides vascularity to the chorion and is therefore essential to placenta formation; the proximal portion becomes the urinary bladder.
alpha cells: Cells in the pancreatic islets that secrete glucagon.
alpha receptors: Membrane receptors sensitive to norepinephrine or epinephrine; stimulation usually results in excitation of the target cell.
alveolar sac: An air-filled chamber that supplies air to several alveoli.
alveolus/alveoli (al-VĒ-ō-lī): Blind pockets at the end of the respiratory tree, lined by a simple squamous epithelium and surrounded by a capillary network; gas exchange with the blood occurs here.
Alzheimer's disease: Disorder resulting from degenerative changes in populations of neurons in the cerebrum, causing dementia characterized by problems with attention, short-term memory, and emotions.
amacrine cells (AM-a-krīn): Modified neurons in the retina that facilitate or inhibit communication between bipolar and ganglion cells.
amino acids: Organic compounds whose chemical structure can be summarized as $R—CHNH_2COOH$
amnesia: Temporary or permanent memory loss.
amniocentesis: Sampling of amniotic fluid for analytical purposes; used to detect certain forms of genetic abnormalities.
amnion (AM-nē-on): One of the extraembryonic membranes; surrounds the developing embryo/fetus.
amniotic fluid (am-nē-OT-ik): Fluid that fills the amniotic cavity; provides cushioning and support for the embryo/fetus.
amphiarthrosis (am-fē-ar-THRŌ-sis): An articulation that permits a small degree of independent movement.
amphicytes (AM-fi-sīts): Supporting cells that surround neurons in the PNS; also called satellite cells.
amphimixis (am-fi-MIK-sis): The fusion of male and female pronuclei following fertilization.
ampulla/ampullae (am-PYŪL-la): A localized dilation in the lumen of a canal or passageway.
amygdala/amygdaloid nucleus (ah-MIG-da-loid): A basal nucleus that is a component of the limbic system and acts as an interface between that system, the cerebrum, and sensory systems.
amylase: An enzyme that breaks down polysaccharides, produced by the salivary glands and pancreas.

anabolism (a-NAB-ō-lizm): The synthesis of complex organic compounds from simpler precursors.
analgesia: Relief from pain.
anal canal: The distal portion of the rectum that contains the anal columns and ends at the anus.
anal triangle: The posterior subdivision of the perineum.
anaphase (AN-a-fāz): Mitotic stage in which the paired chromatids separate and move toward opposite ends of the spindle apparatus.
anastomosis (a-nas-tō-MŌ-sis): The joining of two tubes, usually referring to a connection between two peripheral vessels without an intervening capillary bed.
anatomical position: An anatomical reference position, the body viewed from the anterior surface with the palms facing forward; supine.
anatomy (a-NAT-ō-mē): The study of the structure of the body.
anaxonic neuron (an-ak-SON-ik): A CNS neuron that has many processes but no apparent axon.
androgen (AN-drō-jen): A steroid sex hormone produced primarily by the interstitial cells of the testis, and manufactured in small quantities by the adrenal cortex in either sex.
anemia (a-NĒ-mē-ah): Condition marked by a reduction in the hematocrit and/or hemoglobin content of the blood.
anencephaly (an-en-SEF-a-lē): Development defect characterized by incomplete development of cerebral hemispheres and cranium.
anesthesia: Total or partial loss of sensation from a region of the body.
aneurysm (AN-yū-rizm): A weakening and localized dilation in the wall of a blood vessel.
angiogram (AN-jē-ō-gram): An X-ray image of circulatory pathways.
angiography: X-ray examination of vessel distribution following the introduction of radiopaque substances into the bloodstream.
angiotensin I, II: Angiotensin II is a hormone that causes an elevation in systemic blood pressure, stimulates secretion of aldosterone, promotes thirst, and causes the release of ADH; a converting enzyme in the pulmonary capillaries, converts angiotensin I to angiotensin II.
angiotensinogen: Blood protein produced by the liver that is converted to angiotensin I by the enzyme renin.
ankyloglossia (ang-ki-lō-GLOS-ē-a): Condition characterized by an overly robust and restrictive lingual frenulum.
annulus (AN-ū-lus): A cartilage or bone shaped like a ring.
anorexia nervosa: An eating disorder marked by a loss of appetite and pronounced weight loss.
anoxia (an-OK-sē-a): Tissue oxygen deprivation.
antagonist: A muscle that opposes the movement of an agonist.
antebrachium: The forearm.
anteflexion (an-te-FLEK-shun): Normal position of the uterus, with the superior surface bent forward.
anterior: On or near the front or ventral surface of the body.
antibiotic: Chemical agent that selectively kills pathogenic microorganisms.
antibody (AN-ti-bod-ē): A globular protein produced by plasma cells that will bind to specific antigens and promote their destruction or removal from the body.
anticoagulant: Compound that slows or prevents clot formation by interfering with the clotting system.
antidiuretic hormone (ADH) (an-tī-dī-ū-RET-ik): Hormone synthesized in the hypothalamus and secreted at the posterior pituitary; causes water retention at the kidneys, and an elevation of blood pressure.
antigen: A substance capable of inducing the production of antibodies.
antrum (AN-trum): A chamber or pocket.
anuria (a-NŪ-rē-a): Cessation of urine production.
anus: External opening of the anal canal.
aorta: Large, elastic artery that carries blood away from the left ventricle, and into the systemic circuit.
aortic reflex: Baroreceptor reflex triggered by increased aortic pressures; leads to a reduction in cardiac output and a fall in systemic pressure.
apex (Ā-peks): A pointed tip, usually referring to a triangular object and positioned opposite a broad base.
Apgar: A test used to assess the neurological status of a newborn infant.
aphasia: Inability to speak.
apnea (AP-nē-a): Cessation of breathing.
apneustic center (ap-NŪ-stik): Respiratory center whose chronic activation would lead to apnea at full inspiration.
apocrine secretion: Mode of secretion where the glandular cell sheds portions of its cytoplasm.
aponeurosis/aponeuroses (ap-ō-nū-RŌ-sēz): A broad tendinous sheet that may serve as the origin or insertion of a skeletal muscle.
appendicitis: Inflammation of the appendix.
appendicular: Pertaining to the upper or lower limbs.
appendix: A blind tube connected to the cecum of the large intestine.
appositional growth: Enlargement by the addition of cartilage or bony matrix to the outer surface.
aqueous humor: Fluid similar to perilymph or CSF that fills the anterior chamber of the eye.
arachnoid (a-RAK-noyd): The middle meninges that encloses CSF and protects the central nervous system.
arachnoid granulations: Processes of the arachnoid that project into the superior sagittal sinus; sites where CSF enters the venous circulation.
arbor vitae: Central, branching mass of white matter inside the cerebellum.
arcuate (AR-kū-āt): Curving.
areflexia (ā-re-FLEK-sē-a): Absence of normal reflex responses to stimulation.
areola (a-RĒ-ō-la): Pigmented area that surrounds the nipple of a breast.
areolar: Containing minute spaces, as in areolar connective tissue.
arrector pili (ar-REK-tor PĪ-lī): Smooth muscles whose contractions cause piloerection.
arrhythmias (a-RITH-mē-az): Abnormal patterns of cardiac contractions.
arteriole (ar-TĒ-rē-ōl): A small arterial branch that delivers blood to a capillary network.
artery: A blood vessel that carries blood away from the heart and toward a peripheral capillary.
arthritis (ar-THRĪ-tis): Inflammation of a joint.
arthroscope: Fiber-optic device intended for visualizing the interior of joints; may also be used for certain forms of joint surgery.
articular: Pertaining to a joint.
articular capsule: Dense collagen fiber sleeve that surrounds a joint and provides protection and stabilization.
articular cartilage: Cartilage pad that covers the surface of a bone inside a joint cavity.
articulation (ar-tik-ū-LĀ-shun): A joint; formation of words.
arytenoid cartilages (ar-i-TĒ-noid): A pair of small cartilages in the larynx.
ascending tract: A tract carrying information from the spinal cord to the brain.
ascites (a-SĪ-tēz): Overproduction and accumulation of peritoneal fluid.
aseptic: Free from pathogenic contamination.
asphyxia: Unconsciousness due to oxygen deprivation at the CNS.
aspirate: To remove or obtain by suction; to inhale.
association areas: Cortical areas of the cerebrum responsible for integration of sensory inputs and/or motor commands.
association neuron: *See* **interneuron**.
asthma (AZ-ma): Reversible constriction of smooth muscles around respiratory passageways frequently caused by an allergic response.
astigmatism: Visual disturbance due to an irregularity in the shape of the cornea.
astrocyte (AS-trō-sīt): One of the glial cells in the CNS.
atelectasis (at-e-LEK-ta-sis): Collapse of a lung or a portion of a lung.
atherosclerosis (ath-er-ō-skle-RŌ-sis): Formation of fatty plaques in the walls of arteries, leading to circulatory impairment.
atresia (a-TRĒ-zē-a): Closing of a cavity, or its incomplete development; used in the reproductive system to refer to the degeneration of developing ovarian follicles.
atria: Thin-walled chambers of the heart that receive venous blood from the pulmonary or systemic circuits.
atrial natriuretic peptide (nā-tre-ū-RET-ik): Hormone released by specialized atrial cardiocytes when they are stretched by an abnormally large venous return; promotes fluid loss and reductions in blood pressure and venous return.
atrioventricular (AV) node (ā-trē-ō-ven-TRIK-ū-lar): Specialized cardiocytes that relay the contractile stimulus to the bundle of His, the bundle branches, the Purkinje fibers, and the ventricular myocardium; located at the boundary between the atria and ventricles.
atrioventricular valve: One of the valves that prevent backflow into the atria during ventricular systole.
atrophy (AT-rō-fē): Wasting away of tissues from lack of use or nutritional abnormalities.
auditory ossicles: The bones of the middle ear: malleus, incus, and stapes.
auditory tube: A passageway that connects the nasopharynx with the middle ear cavity; *also called the Eustachian or pharyngotympanic tube.*
autoimmunity: Immune system sensitivity to normal cells and tissues, resulting in the production of autoantibodies.
autolysis: Destruction of a cell due to the rupture of lysosomal membranes in its cytoplasm.
automatic bladder: Reflex micturition following stimulation of stretch receptors in the bladder wall; seen in patients who have lost motor control of the lower body.
automaticity: Spontaneous depolarization to threshold, a characteristic of cardiac pacemaker cells.
autonomic ganglion: A collection of visceral motor neurons outside the CNS.
autonomic nerve: A peripheral nerve consisting of preganglionic or postganglionic autonomic fibers.
autonomic nervous system (ANS): Centers, nuclei, tracts, ganglia, and nerves involved in the unconscious regulation of visceral functions; includes components of the CNS and PNS.

autopsy: Detailed examination of a body after death, usually performed by a pathologist.
autoregulation: Alterations in activity that maintain homeostasis in direct response to changes in the local environment; does not require neural or endocrine control.
autosomal (aw-tō-SŌ-mal): Chromosomes other than the X or Y chromosomes that determine the genetic sex of an individual.
avascular (ā-VAS-kū-lar): Without blood vessels.
avulsion: An injury involving the violent tearing away of body tissues.
axilla: The armpit.
axolemma: The cell membrane of an axon, continuous with the cell membrane of the soma and dendrites and distinct from any glial cell coverings.
axon: Elongate extension of a neuron that conducts an action potential away from the soma and toward the synaptic terminals.
axon hillock: Portion of the neural cell body adjacent to the initial segment.
axoplasm (AK-sō-plazm): Cytoplasm within an axon.
Babinski sign: Reflexive dorsiflexion of the toes following stroking of the plantar surface of the foot; positive reflex (Babinski sign) is normal up to age 1.5 years; thereafter a positive reflex indicates damage to descending tracts.
bacteria: Single-celled microorganisms, some pathogenic, that are common in the environment.
baroreception: Ability to detect changes in pressure.
baroreceptor reflex: A reflexive change in cardiac activity in response to changes in blood pressure.
baroreceptors (bar-ō-rē-SEP-torz): Receptors responsible for baroreception.
basal nuclei: Nuclei of the cerebrum that are involved in the regulation of somatic motor activity at the subconcious level.
base: A compound whose dissociation releases a hydroxide ion (OH^-) or removes a hydrogen ion from the solution.
basement membrane: A layer of filaments and fibers that attach an epithe lium to the underlying connective tissue.
basilar membrane: Membrane that supports the organ of Corti and separates the cochlear duct from the scala tympani in the inner ear.
basophils (BĀ-sō-filz): Circulating granulocytes (WBCs) similar in size and function to tissue mast cells.
B cells: Lymphocytes responsible for the production of antibodies, following their conversion to plasma cells.
benign: Not malignant.
beta cells: Cells of the pancreatic islets that secrete insulin in response to elevated blood sugar concentrations.
beta receptors: Membrane receptors sensitive to epinephrine; stimulation may result in excitation or inhibition of the target cell.
bicarbonate ions: HCO_3^-; anion components of the bicarbonate buffer system.
bicuspid (bī-KUS-pid): A sharp, conical tooth, also called a canine tooth.
bicuspid valve: The left AV valve, also known as the **mitral valve**.
bifurcate: To branch into two parts.
bile: Exocrine secretion of the liver that is stored in the gallbladder and ejected into the duodenum.
bile salts: Steroid derivatives in the bile, responsible for the emulsification of ingested lipids.
bilirubin (bil-ē-ROO-bin): A reddish pigment, a product of hemoglobin catabolism.
biopsy: The removal of a small sample of tissue for pathological analysis.
bipennate muscle: A muscle whose fibers are arranged on either side of a common tendon.
bladder: A muscular sac that distends as fluid is stored and whose contraction ejects the fluid at an appropriate time; used alone, the term usually refers to the urinary bladder.
blastocoele (BLAS-tō-sēl): Fluid-filled cavity within a blastocyst.
blastocyst (BLAS-tō-sist): Early stage in the developing embryo, consisting of an outer trophoblast and an inner cell mass.
blastodisc (BLAS-tō-disk): Later stage in the development of the inner cell mass; it includes the cells that will form the embryo.
blastomere (BLAS-tō-mēr): One of the cells in the morula, a collection of cells produced by the division of the zygote.
blood-brain barrier: Isolation of the CNS from the general circulation; primarily the result of astrocyte regulation of capillary permeabilities.
blood clot: A network of fibrin fibers and trapped blood cells.
blood pressure: A force exerted against the vascular walls by the blood, as the result of the push exerted by cardiac contraction and the elasticity of the vessel walls. It is usually measured along one of the muscular arteries, with systolic pressure measured during ventricular systole, and diastolic pressure during ventricular diastole.
blood-testis barrier: Isolation of the seminiferous tubules from the general circulation, due to the activities of the sustentacular (Sertoli) cells.
boil: An abscess of the skin, usually involving a sebaceous gland.
bolus: A compact mass; usually refers to compacted ingested material on its way to the stomach.
bone: *See* **osseous tissue**.
bowel: The intestinal tract.
brachial: Pertaining to the arm.
brachial plexus: Network formed by branches of spinal nerves $C_5 - T_1$ en route to innervate the upper limb.
brachium: The arm.
bradycardia (brā-dē-KAR-dē-a): An abnormally slow heart rate.
brain stem: The brain minus the cerebrum and cerebellum.
brevis: Short.
Broca's center: The speech center of the brain, usually found on the neural cortex of the left cerebral hemisphere.
bronchial tree: The trachea, bronchi, and bronchioles.
bronchitis (brong-KĪ-tis): Inflammation of the bronchial passageways.
bronchoscope: A fiber-optic instrument used to examine the bronchial passageways.
bronchus/bronchi: One of the branches of the bronchial tree between the trachea and bronchioles.
buccal (BUK-al): Pertaining to the cheeks.
buffer: A compound that stabilizes the pH of a solution by removing or releasing hydrogen ions.
bulbar: Pertaining to the brain stem.
bulbourethral glands (bul-bō-ū-RĒ-thral): Mucous glands at the base of the penis that secrete into the penile urethra; also called Cowper's glands.
bundle branches: Specialized conducting cells in the ventricles that carry the contractile stimulus from the bundle of His to the Purkinje fibers.
bundle of His (hiss): Specialized conducting cells in the interventricular septum that carry the contracting stimulus from the AV node to the bundle branches and thence to the Purkinje fibers.
bursa: A small sac filled with synovial fluid that cushions adjacent structures and reduces friction.
bursectomy: The surgical removal of an inflamed bursa.
bursitis: Painful inflammation of one or more bursae.
cesarean section: Surgical delivery of an infant via an incision through the lower abdominal wall and uterus.
calcaneal tendon: Large tendon that inserts on the calcaneus; tension on this tendon produces plantar flexion of the foot; also called the *Achilles tendon.*
calcaneus (kal-KĀ-nē-us): The heelbone, the largest of the tarsal bones.
calcification: The deposition of calcium salts within a tissue.
calcitonin (kal-si-TŌ-nin): Hormone secreted by C cells of the thyroid when calcium ion concentrations are abnormally high; restores homeo-stasis by increasing the rate of bone deposition and the renal rate of calcium loss, and inhibiting calcium uptake at the digestive tract.
calculus/calculi (KAL-kū-lī): Concretions of insoluble materials that form within body fluids, especially the gallbladder, kidneys, or urinary bladder.
callus: A localized thickening of the epidermis due to chronic mechanical stresses; a thickened area that forms at the site of a bone break as part of the repair process.
calvaria (kal-VAR-ē-a): The skullcap, formed of the frontal, parietal, and occipital bones.
calyx/calyces (KĀL-i-sēz): A cup-shaped division of the renal pelvis.
canaliculi (kan-a-LIK-ū-lī): Microscopic passageways between cells; bile canaliculi carry bile to bile ducts in the liver; in bone, canaliculi permit the diffusion of nutrients and wastes to and from osteocytes.
cancellous bone (KAN-sel-us): Spongy bone, composed of a network of bony struts.
cancer: A malignant tumor that tends to undergo metastasis.
cannula: A tube that can be inserted into the body; often placed in blood vessels prior to transfusion or dialysis.
canthus, medial and lateral (KAN-thus): The angles formed at either corner of the eye between the upper and lower eyelids.
capacitation (ka-pas-i-TĀ-shun): Activation process that must occur before a spermatozoon can successfully fertilize an egg; occurs in the vagina following ejaculation.
capillaries: Small blood vessels, interposed between arterioles and venules, whose thin walls permit the diffusion of gases, nutrients, and wastes between the plasma and interstitial fluids.
capitulum (ka-PIT-ū-lum): General term for a small, elevated articular process; used to refer to the rounded distal surface of the humerus that articulates with the radial head.
caput: The head.
carbohydrase (kar-bō-HĪ-drāz): An enzyme that breaks down carbohydrate molecules.
carbohydrate (kar-bō-HĪ-drāt): Organic compound containing carbon, hydrogen, and oxygen in a ratio that approximates 1:2:1.
carbon dioxide: CO_2, a compound produced by the decarboxylation reactions of aerobic glycolysis.
carbonic anhydrase: An enzyme that catalyzes the reaction $H_2O + CO_2 \rightarrow H_2CO_3$; important in carbon dioxide transport, gastric acid secretion, and renal pH regulation.

carboxypeptidase (kar-bok-sē-PEP-ti-dāz): A protease that breaks down proteins and releases amino acids.
carcinogenic (kar-sin-ō-JEN-ik): Stimulating cancer formation in affected tissues.
cardia (KAR-dē-a): The area of the stomach surrounding its connection with the esophagus.
cardiac: Pertaining to the heart.
cardiac cycle: One complete heartbeat, including atrial and ventricular systole and diastole.
cardiac glands: Mucous glands characteristic of the cardia of the stomach.
cardiac output: The amount of blood ejected by the left ventricle each minute; normally about 5 liters.
cardiac tamponade: Compression of the heart due to fluid accumulation in the pericardial cavity.
cardiocyte (KAR-dē-ō-sīt): A cardiac muscle cell.
cardiomyopathy (kar-dē-ō-mī-OP-a-thē): A progressive disease characterized by damage to the cardiac muscle tissue.
cardiopulmonary resuscitation: Method of artificially maintaining respiratory and circulatory function.
cardiovascular: Pertaining to the heart, blood, and blood vessels.
cardiovascular centers: Poorly localized centers in the reticular formation of the medulla of the brain; includes cardioacceleratory, cardioinhibitory, and vasomotor centers.
cardium: The heart.
carina (ka-RĪ-na): A ridge on the inner surface of the base of the trachea that runs anteroposteriorly, between the two primary bronchi.
carotene (KAR-ō-tēn): A yellow-orange pigment found in carrots and in green and orange leafy vegetables; a compound that the body can convert to vitamin A.
carotid artery: The principal artery of the neck, servicing cervical and cranial structures; one branch, the internal carotid, represents a major blood supply for the brain.
carotid body: A group of receptors adjacent to the carotid sinus that are sensitive to changes in the carbon dioxide levels, pH, and oxygen concentrations of the arterial blood.
carotid sinus: A dilated segment of the internal carotid artery whose walls contain baroreceptors sensitive to changes in blood pressure.
carotid sinus reflex: Reflexive changes in blood pressure that maintain homeostatic pressures at the carotid sinus, stabilizing blood flow to the brain.
carpus/carpal: The wrist.
cartilage: A connective tissue with a gelatinous matrix and an abundance of fibers.
castration: Removal of the testes, also called bilateral orchiectomy.
catabolism (ka-TAB-ō-lizm): The breakdown of complex organic molecules into simpler components, accompanied by the release of energy.
cataract: A reduction in lens transparency that causes visual impairment.
catecholamines (kat-e-KŌL-am-inz): Epinephrine, norepinephrine, and related compounds.
catheter (KATH-e-ter): Surgical instrument, a tube inserted into a body cavity or along a blood vessel or excretory passageway for the collection of body fluids, blood pressure monitoring, or the introduction of medications or radiographic dyes.
cauda equina (KAW-da ek-WĪ-na): Spinal nerve roots distal to the tip of the adult spinal cord; they extend caudally inside the vertebral canal en route to lumbar and sacral segments.
caudal/caudally: Closest to or toward the tail (coccyx).
caudate nucleus (KAW-dāt): One of the basal nuclei, involved with the subconscious control of muscular activity.
cavernous tissue: Erectile tissue that can be engorged with blood; found in the penis and clitoris.
cecum (SĒ-kum): An expanded pouch at the start of the large intestine.
cell: The smallest living unit in the human body.
cell-mediated immunity: Resistance to disease through the activities of sensitized T cells that destroy antigen-bearing cells by direct contact or through the release of lymphotoxins; also called cellular immunity.
cellulitis (sel-ū-LĪ-tis): Diffuse inflammation, usually involving areas of loose connective tissue, such as the subcutaneous layer.
cementum (se-MEN-tum): Bony material covering the root of a tooth, not shielded by a layer of enamel.
center of ossification: Site in a connective tissue where bone formation begins.
central canal: Longitudinal canal in the center of an osteon that contains blood vessels and nerves, also called the Haversian canal; a passageway along the longitudinal axis of the spinal cord that contains cerebrospinal fluid.
central nervous system (CNS): The brain and spinal cord.
central sulcus: Groove in the surface of a cerebral hemisphere, between the primary sensory and primary motor areas of the cortex.
centriole: A cylindrical intracellular organelle composed of nine groups of microtubules, three in each group; functions in mitosis or meiosis by forming the basis of the spindle apparatus.
centromere (SEN-trō-mēr): Localized region where two chromatids remain connected following chromosome replication; site of spindle fiber attachment.
centrosome: Region of cytoplasm containing a pair of centrioles oriented at right angles to one another.
centrum: The vertebral body.
cephalic: Pertaining to the head.
cerebellum (ser-e-BEL-um): Posterior portion of the metencephalon, containing the cerebellar hemispheres; includes the arbor vitae, cerebellar nuclei, and cerebellar cortex.
cerebral cortex: An extensive area of neural cortex covering the surfaces of the cerebral hemispheres.
cerebral hemispheres: Expanded portions of the cerebrum covered in neu ral cortex.
cerebral palsy: Chronic condition resulting from damage to motor areas of the brain during development or at delivery.
cerebral peduncle: Mass of nerve fibers on the ventrolateral surface of the mesencephalon; contains ascending tracts that terminate in the thalamus and descending tracts that originate in the cerebral hemispheres.
cerebrospinal fluid: Fluid bathing the internal and external surfaces of the CNS; secreted by the choroid plexus.
cerebrovascular accident (CVA): A stroke; occlusion of a blood vessel supplying a portion of the brain, resulting in damage to the dependent neurons.
cerebrum (SER-e-brum or se-RĒ-brum): The largest portion of the brain, composed of the cerebral hemispheres; includes the cerebral cortex, the basal nuclei, and the internal capsule.
cerumen: Waxy secretion of integumentary glands along the external acoustic meatus.
ceruminous glands (se-RŪ-mi-nus): Integumentary glands that secrete cerumen.
cervical enlargement: Relative enlargement of the cervical portion of the spinal cord due to the abundance of CNS neurons involved with motor control of the arms.
cervix: The inferior part of the uterus.
chalazion (kah-LĀ-zē-on): An inflammation and distension of a Meibomian gland on the eyelid; also called a sty.
chancre (SHANG-ker): A skin lesion that develops at the primary site of a syphilis infection.
charley horse: Soreness and stiffness in a strained muscle, usually involving the quadriceps group.
chemoreception: Detection of alterations in the concentrations of dissolved compounds or gases.
chemotaxis (kē-mō-TAK-sis): The attraction of phagocytic cells to the source of abnormal chemicals in tissue fluids.
chloride shift: Movement of plasma chloride ions into RBCs in exchange for bicarbonate ions generated by the intracellular dissociation of carbonic acid.
cholecystitis (kō-lē-sis-TĪ-tis): Inflammation of the gallbladder.
cholecystokinin (CCK) (kō-lē-sis-tō-KĪ-nin): Duodenal hormone that stimulates the contraction of the gallbladder and the secretion of enzymes by the exocrine pancreas; also called pancreozymin.
cholelithiasis (kō-lē-li-THĪ-a-sis): The formation or presence of gallstones.
cholesterol: A steroid component of cell membranes and a substrate for the synthesis of steroid hormones and bile salts.
choline: Chemical compound; a breakdown product or precursor of acetylcholine.
cholinergic synapse (kō-lin-ER-jik): Synapse where the presynaptic membrane releases ACh on stimulation.
cholinesterase (kō-li-NES-te-rās): Enzyme that breaks down and inactivates ACh.
chondrocyte (KON-drō-sīt): Cartilage cell.
chondroitin sulfate (kon-DROY-tin): The predominant proteoglycan in cartilage, responsible for the gelatinous consistency of the matrix.
chordae tendineae (KOR-dē TEN-di-nē-ē): Fibrous cords that brace the AV valves in the heart, stabilizing their position and preventing backflow during ventricular systole.
chorion/chorionic (KOR-ē-on) (kō-rē-ON-ik): An extraembryonic membrane, consisting of the trophoblast and underlying mesoderm, that forms the placenta.
choroid: Middle, vascular layer in the wall of the eye.
choroid plexus: The vascular complex in the roof of the third and fourth ventricles of the brain, responsible for CSF production.
chromatid (KRŌ-ma-tid): One complete copy of a DNA strand.
chromatin (KRŌ-ma-tin): Histological term referring to the grainy material visible in cell nuclei during interphase; the appearance of the DNA content of the nucleus when the chromosomes are uncoiled.
chromosomes: Dense structures, composed of tightly coiled DNA strands and associated histones, that become visible in the nucleus when a cell prepares to undergo mitosis or meiosis; normal human somatic cells contain 46 chromosomes apiece.
chronic: Habitual or long-term.
chylomicrons (kī-lō-MĪ-kronz): Relatively large droplets that may contain triglycerides, phospholipids, and cholesterol in association with proteins; synthesized and released by intestinal cells and transported to the venous blood via the lymphatic system.
chyme (kīm): A semifluid mixture of ingested food and digestive secretions that is found in the stomach as digestion proceeds.
chymotrypsin (kī-mō-TRIP-sin): A protease found in the small intestine.

chymotrypsinogen: Inactive proenzyme secreted by the pancreas that is subsequently converted to chymotrypsin.
ciliary body: A thickened region of the choroid that encircles the lens of the eye; it includes the ciliary muscle and the ciliary processes that support the suspensory ligaments of the lens.
cilium/cilia: A slender organelle that extends above the free surface of an epithelial cell, and usually undergoes cycles of movement; composed of a basal body and microtubules in a 9×2 array.
circulatory system: The network of lymphatic and blood vessels involved in the circulation and recirculation of extracellular fluid.
circumduction (sir-kum-DUK-shun): A movement at a synovial joint where the distal end of the bone describes a circle, but the shaft does not rotate.
circumvallate papilla (sir-kum-VAL-āt pa-PIL-la): One of the large, dome-shaped papillae on the dorsum of the tongue that form the V that separates the body of the tongue from the root.
cirrhosis (sir-RŌ-sis): A liver disorder characterized by the degeneration of hepatocytes and their replacement by connective tissue.
cisterna (sis-TUR-na): An expanded chamber.
clavicle (KLAV-i-kul): The collarbone.
cleavage (KLĒV-ij): Mitotic divisions that follow fertilization of the ovum and lead to the formation of a blastocyst.
cleavage lines: Stress lines in the skin that follow the orientation of major bundles of collagen fibers in the dermis.
climacteric: Age-related cessation of gametogenesis in the male or female due to reduced sex hormone production.
clitoris (KLI-to-ris): A small erectile organ of the female that is the developmental equivalent of the male penis.
clone: The production of genetically identical cells.
clonus (KLŌ-nus): Rapid cycles of muscular contraction and relaxation.
clot: A network of fibrin fibers and trapped blood cells; also called a thrombus.
clotting factors: Plasma proteins synthesized by the liver that are essential to the clotting response.
clotting response: Series of events that result in the formation of a clot.
coccygeal ligament: Fibrous extension of the dura mater and filum terminale; provides longitudinal stabilization to the spinal cord.
coccyx (KOK-siks): Terminal portion of the vertebral column, consisting of relatively tiny, fused vertebrae.
cochlea (KOK-lē-a): Spiral portion of the bony labyrinth of the inner ear that surrounds the organ of hearing.
cochlear duct (KOK-lē-ar): Membranous tube within the cochlea that is filled with endolymph and contains the organ of Corti; also called the **scala media**.
codon (KŌ-don): A sequence of three nitrogen bases along an mRNA strand that will specify the location of a single amino acid in a peptide chain.
coelom (SĒ-lom): The ventral body cavity, lined by a serous membrane and subdivided during development into the pleural, pericardial, and abdominopelvic (peritoneal) cavities.
coenzymes (kō-EN-zīmz): Complex organic cofactors, usually structurally related to vitamins.
cofactors: Ions or molecules that must be attached to the active site before an enzyme can function; examples include mineral ions and several vitamins.
colectomy (kō-LEK-tō-mē): Surgical removal of part or all of the colon.
colitis: Inflammation of the colon.
collagen: Strong, insoluble protein fiber common in connective tissues.
collagenous tissues: (ko-LA-jin-us): Dense connective tissues in which collagen fibers are the dominant fiber type; includes dense regular and dense irregular connective tissues.
collateral ganglion (kō-LAT-er-al): A sympathetic ganglion situated in front of the spinal column and separate from the sympathetic chain.
Colles' fracture (KOL-lēz): Fracture of the distal end of the radius and possibly the ulna, with posterior and dorsal displacement of the distal bone fragments.
colliculus/colliculi (ko-LIK-ū-lus): A little mound; in the brain, used to refer to one of the cortical thickenings in the roof of the mesencephalon; the superior colliculus is associated with the visual system, and the inferior colliculi with the auditory system.
colon: The large intestine.
colonoscope (kō-LON-ō-skōp): A fiber-optic device for examining the interior of the colon.
colostomy (kō-LOS-tō-mē): The surgical connection of a portion of the colon to the body wall, sometimes performed after a colectomy to permit the discharge of fecal materials.
colostrum (kō-LOS-trum): Secretion of the mammary glands at the time of childbirth and for a few days thereafter; contains more protein and less fat than the milk secreted later.
coma (kō-ma): An unconscious state from which the individual cannot be aroused, even by strong stimuli.
comedo (kō-MĒ-dō): An inflamed sebaceous gland.
comminuted: Broken or crushed into small pieces.
commissure: A crossing over from one side to another.
common bile duct: Duct formed by the union of the cystic duct from the gallbladder and the bile ducts from the liver; terminates at the duodenal ampulla, where it meets the pancreatic duct.
common pathway: In the clotting response, the events that begin with the appearance of thromboplastin and end with the formation of a clot.
compact bone: Dense bone containing parallel osteons.
compensation curves: The cervical and lumbar curves that develop to center the body weight over the legs.
complement: 11 plasma proteins that interact in a chain reaction following exposure to activated antibodies on the surfaces of certain pathogens, and which promote cell lysis, phagocytosis, and other defense mechanisms.
compliance: The ability of certain organs to tolerate changes in volume; a property that reflects the presence of elastic fibers and smooth muscles.
compound: A molecule containing two or more elements in combination.
concentration: Amount (in grams) or number of atoms, ions, or molecules (in moles) per unit volume.
conception: Fertilization.
concha/conchae (KONG-kē): Three pairs of thin, scroll-like bones that project into the nasal cavities; the superior and medial conchae are part of the ethmoid, and the inferior are separate bones.
concussion: A violent blow or shock; loss of consciousness due to a violent blow to the head.
condyle: A rounded articular projection on the surface of a bone.
cone: Retinal photoreceptor responsible for color vision.
congenital (kon-JEN-i-tal): Already present at the birth of an individual.
congestive heart failure (CHF): Failure to maintain adequate cardiac output due to circulatory problems or myocardial damage.
conjunctiva (kon-junk-TĪ-va): A layer of stratified squamous epithelium that covers the inner surfaces of the lids and the anterior surface of the eye to the edges of the cornea.
conjunctivitis: Inflammation of the conjunctiva.
connective tissue: One of the four primary tissue types; provides a struc tural framework for the body that stabilizes the relative positions of the other tissue types; includes connective tissue proper, cartilage, bone, and blood; always has cell products, cells, and ground substance.
contractility: The ability to contract, possessed by skeletal, smooth, and cardiac muscle cells.
contracture: A permanent contraction of an entire muscle following the atrophy of individual muscle cells.
contralateral reflex: A reflex that affects the side of the body opposite to the stimulus.
conus medullaris: Conical tip of the spinal cord that gives rise to the filum terminale.
convergence: In the nervous system, the innervation of a single neuron by the axons from several neurons; this is most common along motor pathways.
coracoid process (KOR-a-koid): A hook-shaped process of the scapula that projects above the anterior surface of the capsule of the shoulder joint.
Cori cycle: Metabolic exchange of lactic acid from skeletal muscle for glucose from the liver; performed during the recovery period following muscular exertion.
cornea (KOR-nē-a): Transparent portion of the fibrous tunic of the anterior surface of the eye.
corniculate cartilages (kor-NIK-ū-lāt): A pair of small laryngeal cartilages.
cornification: The production of keratin by a stratified squamous epithelium; also called **keratinization**.
cornu: A horn.
corona radiata (ko-RŌ-na rā-dē-A-ta): A layer of follicle cells surrounding an oocyte at ovulation.
coronoid (KOR-ō-noid): Hooked or curved.
corpora quadrigemina (KOR-pō-ra quad-ri-JEM-i-na): The superior and inferior colliculi of the mesencephalic tectum (roof) in the brain.
corpus/corpora: Body.
corpus albicans: The scar tissue that remains after degeneration of the corpus luteum at the end of a uterine cycle.
corpus callosum: Bundle of axons linking centers in the left and right cerebral hemispheres.
corpus cavernosum (KOR-pus ka-ver-NŌ-sum): Masses of erectile tissue within the body of the penis (male) or clitoris (female).
corpus luteum (LOO-tē-um): Progestin-secreting mass of follicle cells that develops in the ovary after ovulation.
corpus spongiosum (spon-jē-Ō-sum): Mass of erectile tissue that surrounds the urethra in the penis, and expands distally to form the glans.
cortex: Outer layer or portion of an organ.
Corti, organ of: The spiral organ, a receptor complex in the scala media of the cochlea that includes the inner and outer hair cells, supporting cells and structures, and the tectorial membrane; provides the sensation of hearing.

corticobulbar tracts (kor-ti-kō-BUL-bar): Descending tracts that carry information/commands from the cerebral cortex to nuclei and centers in the brain stem.
corticospinal tracts: Descending tracts that carry motor commands from the cerebral cortex to the anterior gray horns of the spinal cord.
corticosteroid: A steroid hormone produced by the adrenal cortex.
corticosterone (kor-ti-KOS-te-rōn): One of the corticosteroids secreted by the zona fasciculata of the adrenal cortex; a glucocorticoid.
corticotropin: *See* **adrenocorticotropic hormone (ACTH)**.
corticotropin-releasing hormone (CRH): Releasing hormone secreted by the hypothalamus that stimulates secretion of ACTH by the anterior pituitary.
cortisol (KOR-ti-sol): One of the corticosteroids secreted by the zona fasciculata of the adrenal cortex; a glucocorticoid.
costa/costae: A rib.
cotransport: Membrane transport of a nutrient, such as glucose, in company with the movement of an ion, usually sodium; transport requires a carrier protein but does not involve direct ATP expenditure and can occur regard less of the concentration gradient for the nutrient.
countercurrent multiplication: Active transport between two limbs of a loop that contains a fluid moving in one direction; responsible for the concentration of the urine in the kidney tubules.
cranial: Pertaining to the head.
cranial nerves: Peripheral nerves originating at the brain.
craniosacral division (krā-nē-ō-SAK-ral): *See* **parasympathetic division**.
craniostenosis (krā-nē-ō-sten-Ō-sis): Skull deformity caused by premature closure of the cranial sutures.
cranium: The braincase; the skull bones that surround the brain.
creatine: A nitrogenous compound synthesized in the body that can bind a high-energy phosphate and serve as an energy reserve.
creatine phosphate: A high-energy compound present in muscle cells; during muscular activity the phosphate group is donated to ADP, regenerating ATP.
creatinine: A breakdown product of creatine metabolism.
crenation: Cellular shrinkage due to an osmotic movement of water out of the cytoplasm.
cribriform plate: Portion of the ethmoid of the skull that contains the foramina used by the axons of olfactory receptors en route to the olfactory bulbs of the cerebrum.
cricoid cartilage (KRĪ-koid): Ring-shaped cartilage forming the inferior margin of the larynx.
crista/cristae: A ridge-shaped collection of hair cells in the ampulla of a semicircular canal; the crista and cupula form a receptor complex sensitive to movement along the plane of the canal.
cross-bridge: Myosin head that projects from the surface of a thick filament, and that can bind to an active site of a thin filament in the presence of calcium ions.
cruciate ligaments: A pair of intracapsular ligaments (anterior and poste rior) in the knee.
cryptorchidism (kript-OR-ki-dizm): The failure of the testes to descend into the inguinal canal during late fetal development.
cryptorchid testis: An undescended testis that is in the abdominopelvic cavity rather than the scrotum.
cuneiform cartilages (kū-NĒ-i-form): A pair of small cartilages in the larynx.
cupula (KŪ-pū-la): A gelatinous mass that sits in the ampulla of a semicircular canal in the inner ear, and whose movement stimulates the hair cells of the crista.
Cushing's disease: Condition caused by oversecretion of adrenal steroids.
cuspids (KUS-pids): Conical upper teeth with sharp ridges, located in the upper jaw on either side posterior to the second incisor.
cutaneous membrane: The epidermis and papillary layer of the dermis.
cuticle: Layer of dead, cornified cells surrounding the shaft of a hair; for nails, *see* **eponychium**.
cutis: The skin.
cyanosis: Bluish coloration of the skin due to the presence of deoxygenated blood in *vessels* near the body surface.
cyst: A fibrous capsule containing fluid or other material.
cystic duct: A duct that carries bile between the gallbladder and the common bile duct.
cystitis: Inflammation of the urinary bladder.
cytokinesis (sī-tō-ki-NĒ-sis): The cytoplasmic movement that separates two daughter cells at the completion of mitosis.
cytology (sī-TOL-ō-jē): The study of cells.
cytoplasm: The material between the cell membrane and the nuclear membrane.
cytoskeleton: A network of microtubules and microfilaments in the cytoplasm.
cytosol: The fluid portion of the cytoplasm.
cytotoxic: Poisonous to living cells.
cytotoxic T cells: Lymphocytes of the cellular immune response that kill target cells by direct contact or through the secretion of lymphotoxins; also called killer T cells.
daughter cells: Genetically identical cells produced by mitosis.
decerebrate: Lacking a cerebrum.
decomposition reaction: A chemical reaction that breaks a molecule into smaller fragments.
decubitis ulcers: Ulcers that form where chronic pressure interrupts circulation to a portion of the skin.
decussate: To cross over to the opposite side, usually referring to the crossover of the pyramidal tracts on the ventral surface of the medulla oblongata.
defecation (def-e-KĀ-shun): The elimination of fecal wastes.
deglutition (deg-loo-TISH-un): Swallowing.
delta cell: A pancreatic islet cell that secretes somatostatin.
dementia: Loss of mental abilities.
demyelination: The loss of the myelin sheath of an axon, usually due to chemical or physical damage to Schwann cells or oligodendrocytes.
dendrite (DEN-drīt): A sensory process of a neuron.
denticulate ligaments: Supporting fibers that extend laterally from the surface of the spinal cord, tying the pia mater to the dura mater and providing lateral support for the spinal cord.
dentin (DEN-tin): Bonelike material that forms the body of a tooth; it differs from bone in lacking osteocytes and osteons.
deoxyribonucleic acid (DNA) (dē-ok-sē-rī-bō-nū-KLĀ-ik): DNA strand: a nucleic acid consisting of a chain of nucleotides containing the sugar deoxyribose and the nitrogen bases adenine, guanine, cytosine, and thymine. DNA molecule: two DNA strands wound in a double helix and held together by weak bonds between complementary nitrogen base pairs.
depolarization: A change in the transmembrane potential that moves it from a negative value toward 0 mV.
depression: Inferior (downward) movement of a body part.
dermatitis: Inflammation of the skin.
dermatome: A sensory region monitored by the dorsal rami of a single spinal segment.
dermis: The connective tissue layer beneath the epidermis of the skin.
desmosomes (DEZ-mō-sōmz): A cell junction consisting of a thin proteoglycan layer reinforced by a network of intermediate filaments that lock the two cells together.
detrusor muscle (dē-TROO-sor): Smooth muscle in the wall of the urinary bladder.
detumescence (dē-tū-MES-ens): Loss of a penile erection in the male.
development: Growth and the acquisition of increasing structural and functional complexity; includes the period from conception to maturity.
diabetes insipidus: Polyuria due to inadequate production of ADH.
diabetes mellitus (mel-LĪ-tus): Polyuria and glycosuria, most often due to inadequate production of insulin with resulting elevation of blood glucose levels.
dialysis: Diffusion between two solutions of differing solute concentrations across a semipermeable membrane containing pores that permit the passage of some solutes and not others.
diapedesis (dī-a-pe-DĒ-sis): Movement of white blood cells through the walls of blood vessels by migration between adjacent endothelial cells.
diaphragm (DĪ-a-fram): Any muscular partition; often used to refer to the respiratory muscle that separates the thoracic from the abdominopelvic cavities.
diaphysis (dī-AF-i-sis): The shaft of a long bone.
diarrhea (dī-a-RĒ-a): Abnormally frequent defecation, associated with the production of unusually fluid feces.
diarthrosis (dī-ar-THRŌ-sis): A synovial joint.
diastole (dī-AS-tō-lē): A period of relaxation; may refer to either the atria or the ventricles.
diastolic pressure: Pressure measured in the walls of a muscular artery when the left ventricle is in diastole.
diencephalon (dī-en-SEF-a-lon): A division of the brain that includes the epithalamus, thalamus, and hypothalamus.
differential count: The determination of the relative abundance of each type of white blood cell, based on a random sampling of 100 WBCs.
differentiation: The gradual appearance of characteristic cellular specializations during development, as the result of gene activation or repression.
diffusion: Passive molecular movement from an area of relatively high concentration to an area of relatively low concentration.
digestion: The chemical breakdown of ingested materials into simple molecules that can be absorbed by the cells of the digestive tract.
digestive system: The digestive tract and associated glands.
digestive tract: An internal passageway that begins at the mouth and ends at the anus.
dilate: To increase in diameter; to enlarge or expand.
diploē (DIP-lō-ē): A layer of spongy bone between the internal and external tables of a flat bone.
dislocation: Forceful displacement of an articulating bone to an abnormal position, usually accompanied by damage to tendons, ligaments, the articular capsule, or other structures.
distal: Movement away from the point of attachment or origin; for a limb, away from its attachment to the trunk.
distal convoluted tubule: Portion of the nephron closest to the collecting tubule and duct; an important site of active secretion.

diuresis: Fluid loss at the kidneys; the production of urine.
divergence: In neural tissue, the spread of excitation from one neuron to many neurons; an organizational pattern common along sensory pathways of the CNS.
diverticulitis (dī-ver-tik-ū-LĪ-tis): Inflammation of a diverticulum.
diverticulosis (dī-ver-tik-ū-LŌ-sis): The formation of diverticula.
diverticulum: A sac or pouch in the wall of the colon or other organ.
dizygotic twins (dī-zī-GOT-ik): Twins that result from the fertilization of two different ova.
dopamine (DŌ-pah-mēn): An important neurotransmitter in the CNS.
dorsal: Toward the back, posterior.
dorsal root ganglion: PNS ganglion containing the cell bodies of sensory neurons.
dorsiflexion: Elevation of the superior surface of the foot.
Down syndrome: A genetic abnormality resulting from the presence of three copies of chromosome 21; individuals with this condition have characteristic physical and intellectual deficits.
duct: A passageway that delivers exocrine secretions to an epithelial surface.
ductus arteriosus (ar-tē-rē-Ō-sus): Vascular connection between the pulmonary trunk and the aorta that functions throughout fetal life; normally closes at birth or shortly thereafter, and persists as the ligamentum arteriosum.
ductus deferens (DUK-tus DEF-e-renz): A passageway that carries sperm from the epididymis to the ejaculatory duct.
duodenal ampulla: Chamber that receives bile from the common bile duct and pancreatic secretions from the pancreatic duct.
duodenal glands: *See* **submucosal glands.**
duodenal papilla: Conical projection from the inner surface of the duodenum that contains the opening of the duodenal ampulla.
duodenum (dū-OD-e-num): The proximal 1 ft of the small intestine that contains short villi and submucosal glands.
dura mater (DŪ-ra MĀ-ter): Outermost component of the meninges that surround the brain and spinal cord.
dynamic equilibrium: Maintenance of normal body orientation as sudden changes in position (rotation, acceleration, etc.) occur.
dyslexia: Impaired ability to comprehend written words.
dysmenorrhea: Painful menstruation.
dysuria (dis-Ū-rē-a): Painful urination.
eccrine glands (EK-rin): Sweat glands of the skin that produce a watery secretion.
echocardiography (ek-ō-kar-dē-OG-ra-fē): Examination of the heart using modified ultrasound techniques.
ectoderm: One of the three primary germ layers; covers the surface of the embryo and gives rise to the nervous system, the epidermis and associated glands, and a variety of other structures.
ectopic (ek-TOP-ik): Outside of its normal location.
effector: A peripheral gland or muscle cell innervated by a motor neuron.
efferent: Away from.
efferent arteriole: An arteriole carrying blood away from the glomerulus of the kidney.
efferent fiber: An axon that carries impulses away from the CNS.
ejaculation (ē-jak-ū-LĀ-shun): The ejection of semen from the penis as the result of muscular contractions of the bulbocavernosus and ischiocavernosus muscles.
ejaculatory duct (ē-JAK-ū-la-tō-rē): Short ducts that pass within the walls of the prostate and connect the ductus deferens with the prostatic urethra.
elastase (ē-LAS-tāz): A pancreatic enzyme that breaks down elastin fibers.
elastin: Connective tissue fibers that stretch and rebound, providing elasticity to connective tissues.
electrocardiogram (ECG, EKG) (ē-lek-trō-KAR-dē-ō-gram): Graphic record of the electrical activities of the heart, as monitored at specific locations on the body surface.
electroencephalogram (EEG): Graphic record of the electrical activities of the brain.
electrolytes (ē-LEK-trō-līts): Soluble inorganic compounds whose ions will conduct an electric current in solution.
electron: One of the three fundamental particles; a subatomic particle that bears a negative charge and normally orbits around the protons of the nucleus.
elephantiasis (el-e-fan-TĪ-a-sis): A lymphedema caused by infection and blockage of lymphatics by mosquito-borne parasites.
elevation: Movement in a superior, or upward, direction.
embolism (EM-bō-lizm): Obstruction or closure of a vessel by an embolus.
embolus (EM-bō-lus): An air bubble, fat globule, or blood clot drifting in the circulation.
embryo (EM-brē-ō): Developmental stage beginning at fertilization and ending at the start of the third developmental month.
embryogenesis (em-brē-ō-JEN-e-sis): The process of embryo formation.
embryology (em-brē-OL-ō-jē): The study of embryonic development, focusing on the first 2 months after fertilization.
emesis (EM-e-sis): Vomiting.
emmetropia: Normal vision.
emulsification (ē-mul-si-fi-KĀ-shun): The physical breakup of fats in the digestive tract, forming smaller droplets accessible to digestive enzymes; normally the result of mixing with bile salts.
enamel: Crystalline material similar in mineral composition to bone, but harder and without osteocytes, that covers the exposed surfaces of the teeth.
encephalitis: Inflammation of the brain.
endocarditis: Inflammation of the endocardium of the heart.
endocardium (en-dō-KAR-dē-um): The simple squamous epithelium that lines the heart and is continuous with the endothelium of the great vessels.
endochondral ossification (en-dō-KON-dral): The conversion of a cartilaginous model to bone, the characteristic mode of formation for skeletal elements other than the bones of the cranium, the clavicles, and sesamoid bones.
endocrine gland: A gland that secretes hormones into the blood.
endocrine system: The endocrine glands of the body.
endocytosis (en-dō-sī-TŌ-sis): The movement of relatively large volumes of extracellular material into the cytoplasm via the formation of a membranous vesicle at the cell surface; includes pinocytosis and phagocytosis.
endoderm: One of the three primary germ layers; the layer on the undersurface of the embryonic disc, that gives rise to the epithelia and glands of the digestive system, the respiratory system, and portions of the urinary system.
endogenous: Produced within the body.
endolymph (EN-dō-limf): Fluid contents of the membranous labyrinth (the saccule, utricle, semicircular ducts and cochlear duct) of the inner ear.
endometrial glands: Secretory glands of the endometrium.
endometrium (en-dō-MĒ-trē-um): The mucous membrane lining the uterus.
endomysium (en-dō-MĪS-ē-um): A delicate network of connective tissue fibers that surrounds individual muscle cells.
endoneurium: A delicate network of connective tissue fibers that surrounds individual nerve fibers.
endoplasmic reticulum (en-dō-PLAZ-mik re-TIK-ū-lum): A network of membranous channels in the cytoplasm of a cell that function in intracellular transport, synthesis, storage, packaging, and secretion.
endorphins (en-DOR-finz): Neuromodulators produced in the CNS that inhibit activity along pain pathways.
endosteum: An incomplete cellular lining found on the inner (medullary) surfaces of bones.
endothelium (en-dō-THĒ-lē-um): The simple squamous epithelium that lines blood and lymphatic vessels.
enkephalins (en-KEF-a-linz): Neuromodulators produced in the CNS that inhibit activity along pain pathways.
enteritis (en-ter-Ī-tis): Inflammation of the intestinal tract.
enterocrinin: A hormone secreted by the duodenal lining when exposed to acid chyme; stimulates the secretion of the duodenal glands.
enteroendocrine cells (en-ter-ō-EN-dō-krin): Endocrine cells scattered among the epithelial cells lining the digestive tract.
enterogastric reflex: Reflexive inhibition of gastric secretion initiated by the arrival of acid chyme in the small intestine.
enterohepatic circulation: Excretion of bile salts by the liver, followed by absorption of bile salts by intestinal cells for return to the liver via the hepatic portal vein.
enterokinase: An enzyme in the lumen of the small intestine that activates the proenzymes secreted by the pancreas.
enzyme: A protein that catalyzes a specific biochemical reaction.
eosinophils (ē-ō-SIN-ō-filz): A granulocyte (WBC) with a lobed nucleus and red-staining granules; participates in the immune response, and is especially important during allergic reactions.
ependyma (ep-EN-di-mah): Layer of cells lining the ventricles and central canal of the CNS.
epiblast (EP-i-blast): The layer of the inner cell mass facing the amniotic cavity prior to gastrulation.
epicardium: Serous membrane covering the outer surface of the heart; also called the visceral pericardium.
epidermis: The epithelium covering the surface of the skin.
epididymis (ep-i-DID-i-mus): Coiled duct that connects the rete testis to the ductus deferens; site of functional maturation of spermatozoa.
epidural block: Anesthesia caused by the elimination of sensory inputs from dorsal nerve roots following the introduction of drugs into appropriate regions of the epidural space.
epidural space: Space between the spinal dura mater and the walls of the vertebral foramen; contains blood vessels and adipose tissue; a frequent site of injection for regional anesthesia.
epiglottis (ep-i-GLOT-is): Blade-shaped flap of tissue, reinforced by cartilage, that is attached to the dorsal and superior surface of the thyroid cartilage; it folds over the entrance to the larynx during swallowing.
epimysium (ep-i-MĪS-ē-um): A dense investment of collagen fibers that surrounds a skeletal muscle, and is continuous with the tendons/aponeuroses of the muscle, and with the perimysium.

epineurium: A dense investment of collagen fibers that surrounds a peripheral nerve.
epiphyseal cartilage (e-pi-FI-sē-al): Cartilaginous region between the epiphysis and diaphysis of a growing bone.
epiphysis (e-PIF-i-sis): The end of a long bone.
epistaxis (ep-i-STAK-sis): Nosebleed.
epithelium (e-pi-THĒ-lē-um): One of the four primary tissue types; a layer of cells that forms a superficial covering or an internal lining of a body cavity or vessel.
eponychium (ep-ō-NIK-ē-um): A narrow zone of stratum corneum that extends across the surface of a nail at its exposed base; also called the **cuticle**.
equational division: The second meiotic division.
equilibrium (ē-kwi-LIB-rē-um): A dynamic state, where two opposing forces or processes are in balance.
erection: Stiffening of the penis prior to copulation due to the engorgement of the erectile tissues of the corpora cavernosa and the corpus spongiosum.
erythema (er-i-THĒ-ma): Redness and inflammation at the surface of the skin.
erythrocyte (e-RITH-rō-sīt): A red blood cell; an anucleate blood cell containing large quantities of hemoglobin.
erythrocytosis (e-rith-rō-sī-TŌ-sis): An abnormally large number of erythrocytes in the circulating blood.
erythropoiesis (e-rith-rō-poy-Ē-sis): Red blood cell formation.
erythropoietin (e-rith-rō-POI-e-tin): Hormone released by tissues, especially the kidneys, exposed to low oxygen concentrations; stimulates hematopoiesis in bone marrow.
Escherichia coli: Normal bacterial resident of the large intestine.
esophagus: A muscular tube that connects the pharynx to the stomach.
estradiol (es-tra-DĪ-ol): The primary estrogen secreted by ovarian follicles.
estrogens (ES-trō-jenz): The dominant sex hormones in females; notably estradiol.
eupnea (ŪP-nē-a): Normal quiet breathing.
eversion (ē-VER-shun): A turning outward.
excitable membranes: Membranes that conduct action potentials, a characteristic of muscle and nerve cells.
excretion: Elimination from the body.
exocrine gland: A gland that secretes onto the body surface or into a passageway connected to the exterior.
exocytosis (ek-sō-sī-TŌ-sis): The ejection of cytoplasmic materials by fusion of a membranous vesicle with the cell membrane.
expiration: Exhalation; breathing out.
expiratory reserve: The amount of additional air that can be voluntarily moved out of the respiratory tract after a normal tidal expiration.
extension: An increase in the angle between two articulating bones; the opposite of flexion.
extensor retinaculum (ret-i-NAK-ū-lum): A thickening of the fascia of the forearm at the wrist or the leg at the ankle, forming a band of dense connective tissue that holds extensor muscle tendons in place.
external acoustic meatus: Passageway in the temporal bone that leads to the tympanic membrane.
external ear: The auricle, external acoustic meatus, and tympanic membrane.
external nares: The nostrils; the external openings into the nasal cavity.
exteroceptors: Sensory receptors in the skin, mucous membranes, and special sense organs that provide information about the external environment and our position within it.
extracellular fluid: All body fluid other than that contained within cells; includes plasma and interstitial fluid.
extraembryonic membranes: The yolk sac, amnion, chorion, and allantois.
extrafusal fibers: Contractile muscle fibers, as opposed to the sensory intrafusal fibers (muscle spindles).
extrinsic pathway: Clotting pathway that begins with damage to blood vessels or surrounding tissues and ends with the formation of tissue thromboplastin.
fabella: A sesamoid bone often found in the tendon of the lateral head of the gastrocnemius muscle.
facilitated diffusion: Passive movement of a substance across a cell membrane via a protein carrier.
facilitation: Depolarization of a neuron cell membrane toward threshold, or making the cell more sensitive to depolarizing stimuli.
falciform ligament (FAL-si-form): A sheet of mesentery that contains the ligamentum teres, the fibrous remains of the umbilical vein of the fetus.
falx (FALKS): Sickle-shaped.
falx cerebri (FALKS ser-Ē-brē): Curving sheet of dura mater that extends between the two cerebral hemispheres; encloses the superior sagittal sinus.
fascia (FASH-a): Connective tissue fibers, primarily collagenous, that form sheets or bands beneath the skin to attach, stabilize, enclose, and separate muscles and other internal organs.
fasciculus (fa-SIK-ū-lus): A small bundle, usually referring to a collection of nerve axons or muscle fibers.
fatty acids: Hydrocarbon chains ending in a carboxyl group.
fauces (FAW-sēz): The passage from the mouth to the pharynx, bounded by the palatal arches, the soft palate, and the uvula.
feces: Waste products eliminated by the digestive tract at the anus; contain indigestible residue, bacteria, mucus, and epithelial cells.
fenestra: An opening.
fenestrated (FEN-es-trāt-ed): Having multiple openings; used when referring to very permeable capillaries whose endothelial cells are penetrated by pores of varying sizes.
fertilization: Fusion of egg and sperm to form a zygote.
fetus: Developmental stage lasting from the start of the third developmental month to delivery.
fibrillation (fi-bril-Ā-shun): Uncoordinated contractions of individual muscle cells that impair or prevent normal function.
fibrin (FĪ-brin): Insoluble protein fibers that form the basic framework of a blood clot.
fibrinogen (fī-BRIN-ō-jen): Plasma proteins that can be converted, by the action of enzymes, into insoluble strands of fibrin that form the basis for a blood clot.
fibrinolysis (fī-brin-OL-i-sis): The breakdown of the fibrin strands of a blood clot by a proteolytic enzyme.
fibroblasts (FĪ-brō-blasts): Cells of connective tissue proper that are responsible for the production of extracellular fibers and the secretion of the organic compounds of the extracellular matrix.
fibrocartilage: Cartilage containing an abundance of collagen fibers; found around the edges of joints, in the intervertebral discs, the menisci of the knee, etc.
fibrous tunic: The outermost layer of the eye, composed of the sclera and cornea.
fibula (FIB-ū-la): The lateral, relatively small bone of the lower leg.
filariasis (fil-a-RĪ-a-sis): Condition resulting from infection by mosquitoborne parasites; may cause elephantiasis.
filiform papillae: Slender conical projections from the dorsal surface of the anterior two-thirds of the tongue.
filtrate: Fluid produced by filtration at a glomerulus in the kidney.
filtration: Movement of a fluid across a membrane whose pores restrict the passage of solutes on the basis of size.
filum terminale: A fibrous extension of the spinal cord that extends from the conus medullaris to the coccygeal ligament.
fimbriae (FIM-brē-ē): A fringe; used to describe the fingerlike processes that surround the entrance to the uterine tube.
fissure: An elongated groove or opening.
fistula: An abnormal passageway between two organs or from an internal organ or space to the body surface.
flaccid: Limp, soft, flabby; a muscle without muscle tone.
flagellum/flagella (fla-JEL-ah): An organelle structurally similar to a cilium, but used to propel a cell through a fluid.
flatus: Intestinal gas.
flexion (FLEK-shun): A movement at a joint that reduces the angle between two articulating bones; the opposite of extension.
flexor: A muscle that produces flexion.
flexor reflex: A reflex contraction of the flexor muscles of a limb in response to an unpleasant stimulus.
flexure: A bending.
fluoroscope: An instrument that permits the examination of the body with X-rays in real time, rather than via fixed images on photographic plates.
folia (FŌ-lē-ah): Leaflike folds; used in reference to the slender folds in the surface of the cerebellar cortex.
follicle (FOL-i-kl): A small secretory sac or gland.
follicle-stimulating hormone (FSH): A hormone secreted by the anterior pituitary; stimulates oogenesis (female) and spermatogenesis (male).
folliculitis (fō-lik-ū-LĪ-tis): Inflammation of a follicle, such as a hair follicle of the skin.
fontanel (fon-tah-NEL): A relatively soft, flexible, fibrous region between two flat bones in the developing skull.
foramen: An opening or passage through a bone.
forearm: Distal portion of the upper limb between the elbow and wrist.
forebrain: The cerebrum.
fornix (FOR-niks): An arch, or the space bounded by an arch; in the brain, an arching tract that connects the hippocampus with the mamillary bodies; in the eye, a slender pocket found where the epithelium of the ocular conjunctiva folds back upon itself as the palpebral conjunctiva.
fossa: A shallow depression or furrow in the surface of a bone.
fourth ventricle: An elongate ventricle of the metencephalon (pons and cerebellum) and the myelencephalon (medulla) of the brain; the roof contains a region of choroid plexus.
fovea (FŌ-vē-a): Portion of the retina providing the sharpest vision, with the highest concentration of cones; also called the **macula lutea**.
fracture: A break or crack in a bone.
frenulum (FREN-ū-lum): A bridle; *see* **lingual frenulum**.

frontal plane: A sectional plane that divides the body into anterior and posterior portions.
fructose: A hexose (simple sugar containing six carbons) found in foods and in semen.
fundus (FUN-dus): The base of an organ.
fungiform papillae: Mushroom-shaped papillae on the dorsal and dorsolateral surfaces of the tongue.
furuncle (FŪ-rung-kl): A boil, resulting from the invasion and inflammation of a hair follicle or sebaceous gland.
gallbladder: Pear-shaped reservoir for the bile secreted by the liver.
gametes (GAM-ēts): Reproductive cells (sperm or eggs) that contain one-half of the normal chromosome complement.
gametogenesis (ga-mē-tō-JEN-e-sis): The formation of gametes.
gamma aminobutyric acid (GABA) (GAM-ma a-MĒ-nō-bū-TIR-ik): A neurotransmitter of the CNS whose effects are usually inhibitory.
gamma motor neurons: Motor neurons that adjust the sensitivities of muscle spindles (intrafusal fibers).
ganglion/ganglia: A collection of nerve cell bodies outside of the CNS.
gap junctions: Connections between cells that permit electrical coupling.
gaster (GAS-ter): The stomach; the body or belly of a skeletal muscle.
gastrectomy (gas-TREK-tō-mē): Partial or total surgical removal of the stomach.
gastric: Pertaining to the stomach.
gastric glands: Tubular glands of the stomach whose cells produce acid, enzymes, intrinsic factor, and hormones.
gastrin (GAS-trin): Hormone produced by enteroendocrine cells of the stomach, when exposed to mechanical stimuli or vagal stimulation, and the duodenum, when exposed to chyme containing undigested proteins.
gastritis (gas-TRĪ-tis): Inflammation of the stomach.
gastroenteric reflex (gas-trō-en-TER-ik): An increase in peristalsis along the small intestine triggered by the arrival of food in the stomach.
gastroileal reflex (gas-trō-IL-ē-al): Peristaltic movements that shift materials from the ileum to the colon, triggered by the arrival of food in the stomach.
gastrointestinal (GI) tract: An internal passageway that begins at the mouth, ends at the anus, and is lined by a mucous membrane; also known as the digestive tract.
gastroscope: A fiber-optic instrument that permits visual inspection of the stomach lining.
gastrulation (gas-troo-LĀ-shun): The movement of cells of the inner cell mass that creates the three primary germ layers of the embryo.
gene: A portion of a DNA strand that functions as a hereditary unit and is found at a particular locus on a specific chromosome.
genetic engineering: Research and experiments involving the manipulation of the genetic makeup of an organism.
genetics: The study of mechanisms of heredity.
geniculate (je-NIK-ū-lāt): Like a little knee; the medial geniculate nuclei and the lateral geniculate nuclei are thalamic nuclei in the walls of the thalamus of the brain.
genitalia (jen-i-TĀ-lē-a): Reproductive organs.
genotype (JĒN-ō-tīp): The genetic complement of a particular individual.
germinal centers: Pale regions in the interior of lymphatic tissues or nodules, where mitoses are under way.
gestation (jes-TĀ-shun): The period of intrauterine development.
gingivae (JIN-ji-vē): The gums.
gingivitis: Inflammation of the gums.
gland: Cells that produce exocrine or endocrine secretions, derived from epithelia.
glans penis: Expanded tip of the penis that surrounds the urethra meatus; continuous with the corpus spongiosum.
glaucoma: Eye disorder characterized by rising intraocular pressures due to inadequate drainage of aqueous humor at the canal of Schlemm.
glenoid cavity: A rounded depression that forms the articular surface of the scapula at the shoulder joint.
glial cells (GLĒ-al): Supporting cells in the neural tissue of the CNS and PNS.
globular proteins: Proteins whose tertiary structure makes them rounded and compact.
globulins (GLOB-ū-lins): Globular plasma proteins with a variety of important functions.
glomerular capsule: Expanded initial portion of the nephron that surrounds the glomerulus.
glomerulonephritis (glō-mer-ū-lō-nef-RĪ-tis): Inflammation of the glomeruli of the kidneys.
glomerulus (glō-MER-ū-lus): A ball or knot; in the kidneys, a knot of capillaries that projects into the enlarged, proximal end of a nephron; the site where filtration occurs, the first step in the production of urine.
glossopharyngeal nerve (glos-ō-fa-RIN-jē-al): Cranial nerve IX.
glottis (GLOT-is): The passage from the pharynx to the larynx.
glucagon (GLOO-ka-gon): Hormone secreted by the alpha cells of the pancreatic islets; elevates blood glucose concentrations.
glucocorticoids (gloo-kō-KŌR-ti-koyds): Hormones secreted by the zona fasciculata of the adrenal cortex to modify glucose metabolism; cortisol and corticosterone are important examples.
glucose (GLOO-kōs): A 6-carbon sugar, $C_6H_{12}O_6$, the preferred energy source for most cells and the only energy source for neurons under normal conditions.
glucose-dependent insulinotropic hormone (GIP): A duodenal hormone released when the arriving chyme contains large quantities of carbohydrates; triggers the secretion of insulin and a slowdown in gastric activity.
glycerides: Lipids composed of glycerol bound to 1–3 fatty acids.
glycogen (GLI-kō-jen): A polysaccharide that represents an important energy reserve; a polymer consisting of a long chain of glucose molecules.
glycolipids (glī-cō-LIP-idz): Compounds created by the combination of carbohydrate and lipid components.
glycoprotein (glī-kō-PRŌ-tēn): A compound containing a relatively small carbohydrate group attached to a large protein.
glycosuria (glī-cō-SŪ-rē-a): The presence of glucose in the urine.
goblet cell: A goblet-shaped, mucus-producing, unicellular gland found in certain epithelia of the digestive and respiratory tracts.
goiter: Enlargement of the thyroid gland.
Golgi apparatus (GOL-jē): Cellular organelle consisting of a series of membranous plates that give rise to lysosomes and secretory vesicles.
Golgi tendon organ (GOL-jē): *See* **tendon organ**.
gomphosis (gom-FŌ-sis): A fibrous synarthrosis that binds a tooth to the bone of the jaw; *see* **periodontal ligament**.
gonadotropic hormones: FSH and LH, hormones that stimulate gamete development and sex hormone secretion.
gonadotropin-releasing hormone (GnRH) (gō-nad-ō-TRŌ-pin): Hypothalamic releasing hormone that causes the secretion of FSH and LH by the anterior pituitary gland.
gonadotropins (gō-nad-ō-TRŌ-pins): Hormones that stimulate the gonads (testes or ovaries).
gonads (GŌ-nadz): Organs that produce gametes and hormones.
gout: Clinical condition resulting from elevated uric acid concentrations in the blood and peripheral tissues.
granulocytes (GRAN-ū-lō-sīts): White blood cells containing granules visible with the light microscope; includes eosinophils, basophils, and neutrophils; also called granular leukocytes.
gray matter: Areas in the central nervous system dominated by nerve cell bodies, glial cells, and unmyelinated axons.
gray ramus: A bundle of postganglionic sympathetic nerve fibers that go to a spinal nerve for distribution to effectors in the body wall, skin, and extremities.
greater omentum: A large fold of the dorsal mesentery of the stomach that hangs in front of the intestines.
greater vestibular glands: Mucous glands in the vaginal walls that secrete into the vestibule; the equivalent of the bulbourethral glands of the male.
greenstick fracture: A fracture most often affecting the long bones of young children.
groin: The inguinal region.
gross anatomy: The study of the structural features of the human body without the aid of a microscope.
growth hormone (GH): Anterior pituitary hormone that stimulates tissue growth and anabolism when nutrients are abundant, and restricts tissue glucose dependence when nutrients are in short supply.
gustation (gus-TĀ-shun): Taste.
gynecologists (gī-ne-KOL-ō-jists): Physicians specializing in the pathology of the female reproductive system.
gyrus (JĪ-rus): A prominent fold or ridge of neural cortex on the surfaces of the cerebral hemispheres.
hair: A keratinous strand produced by epithelial cells of the hair follicle.
hair cells: Sensory cells of the inner ear.
hair follicle: An accessory structure of the integument; a tube lined by a stratified squamous epithelium that begins at the surface of the skin and ends at the hair papilla.
hair root: A thickened, conical structure consisting of a connective tissue papilla and the overlying matrix, a layer of epithelial cells that produces the hair shaft.
hallux: The great toe.
haploid (HAP-loid): Possessing one-half of the normal number of chromosomes; a characteristic of gametes.
hard palate: The bony roof of the oral cavity, formed by the maxillary and palatine bones.
Hassall's corpuscles: Aggregations of epithelial cells in the thymus whose functions are unknown.
haustra (HAWS-tra): Saclike pouches along the length of the large intestine that result from tension in the taenia coli.
Haversian system: *See* **osteon**.
heart block: A cardiac arrhythmia due to conduction delays that affect communication between the atria and ventricles.

Heimlich maneuver (HĪM-lik): A technique for removing an airway blockage by external compression of the abdomen and forceful elevation of the diaphragm.
helper T cells: Lymphocytes (T cells) whose secretions and other activities coordinate the cellular and humoral immune responses.
hematocrit (hē-MAT-ō-krit): Percentage of the volume of whole blood contributed by cells; also called the packed cell volume (PCV) or the volume of packed red cells (VPRC).
hematologists (hē-ma-TOL-ō-jists): Specialists in disorders of the blood and blood-forming tissues.
hematoma: A tumor or swelling filled with blood.
hematuria (hē-ma-TŪ-rē-a): The presence of abnormal numbers of red blood cells in the urine.
heme (HĒM): A porphyrin ring containing a central iron atom that can reversibly bind oxygen molecules; a component of the hemoglobin molecule.
hemiplegia: Paralysis affecting one side of the body (arm, trunk, and leg).
hemocytoblasts: Stem cells whose divisions produce all of the various populations of blood cells.
hemodialysis (hē-mō-dī-AL-i-sis): Dialysis of the blood.
hemoglobin (hē-mō-GLŌ-bin): Protein composed of four globular subunits, each bound to a single molecule of heme; the protein found in red blood cells that gives them the ability to transport oxygen in the blood.
hemolysis: Breakdown (lysis) of red blood cells.
hemophilia (hē-mō-FIL-ē-a): A congenital condition resulting from the inadequate synthesis of one of the clotting factors.
hemopoiesis (hē-mō-poi-Ē-sis): Blood cell formation and differentiation.
hemorrhage: Blood loss.
hemorrhoids (HEM-ō-roidz): Swollen, varicose veins that protrude from the walls of the rectum and/or anal canal.
hemostasis: The cessation of bleeding.
hemothorax: The entry of blood into one of the pleural cavities.
heparin (HEP-a-rin): An anticoagulant released by activated basophils and mast cells.
hepatic duct: Duct carrying bile away from the liver lobes and toward the union with the cystic duct.
hepatic portal vein: Vessel that carries blood between the intestinal capillaries and the sinusoids of the liver.
hepatitis (hep-a-TĪ-tis): Inflammation of the liver, resulting from exposure to toxic chemicals, drugs, or viruses.
hepatocyte (he-PAT-ō-sīt): A liver cell.
hernia: The protrusion of a loop or portion of a visceral organ through the abdominopelvic wall or into the thoracic cavity.
herniated disc: Rupture of the connective tissue sheath of the nucleus pulposus of an intervertebral disc.
heterotopic: Ectopic; outside of its normal location.
heterozygous (het-er-ō-ZĪ-gus): Possessing two different alleles at corresponding loci on a chromosome pair; the individual's phenotype may be determined by one or both of the alleles.
hexose: A 6-carbon simple sugar.
hiatus (hī-Ā-tus): A gap, cleft, or opening.
hilus (HĪ-lus) **or hilum** (HĪ-lum): A localized region where blood vessels, lymphatics, nerves, and/or other anatomical structures are attached to an organ.
hippocampus: A portion of the limbic system that is concerned with the organization and storage of memories.
hirsutism (HER-sut-izm): Excessive hair growth in women that follows the distribution pattern typical of adult males; sometimes caused by the overproduction of androgens.
histamine (HIS-ta-mēn): Chemical released by stimulated mast cells or basophils to initiate or enhance an inflammatory response.
histology (his-TOL-ō-jē): The study of tissues.
histones: Proteins associated with the DNA of the nucleus, and around which the DNA strands are wound.
holocrine secretion (HŌL-ō-krin): Form of exocrine secretion where the secretory cell becomes swollen with vesicles and then ruptures.
homeostasis (hō-mē-ō-STĀ-sis): The maintenance of a relatively constant internal environment.
homologous chromosomes (hō-MOL-ō-gus): The members of a chromosome pair, each containing the same gene loci.
homozygous (hō-mō-ZĪ-gus): Having the same gene for a particular character on two homologous chromosomes.
hormone: A compound secreted by one cell that travels through the circulatory system to affect the activities of cells in another portion of the body.
human chorionic gonadotropin (HCG): Placental hormone that maintains the corpus luteum for the first 3 months of pregnancy.
human immunodeficiency virus (HIV): The infectious agent that causes **acquired immune deficiency syndrome** (**AIDS**).
human leukocyte antigen (HLA): Antigens on cell surfaces important to foreign antigen recognition and that play a role in the coordination and activation of the immune response.
human placental lactogen (HPL): Placental hormone that stimulates the functional development of the mammary glands.
humoral immunity: Immunity resulting from the presence of circulating antibodies produced by plasma cells.
hyaline cartilage (HĪ-a-lin): The most common type of cartilage; the matrix contains collagen fibers. Examples include the connections between the ribs and sternum, the tracheal and bronchial cartilages, and synovial cartilages.
hyaluronic acid or hyaluronan: A proteoglycan in the matrix of many connective tissues that gives the matrix a viscous consistency; also functions as intercellular cement.
hyaluronidase (hī-a-lūr-ON-a-dāz): An enzyme that breaks down hyaluronic acid; produced by some bacteria and found in the acrosomal cap of a sperm cell.
hydrocephalus: Condition resulting from excessive production or inadequate drainage of cerebrospinal fluid.
hydrostatic pressure: Fluid pressure.
hymen: A membrane that forms during development, covering the entrance to the vagina.
hypercapnia (hī-per-KAP-nē-a): High plasma carbon dioxide concentrations, often the result of hypoventilation or inadequate tissue perfusion.
hyperglycemia: Elevated plasma glucose concentrations.
hyperopia: The farsighted condition, characterized by an inability to focus on objects close by.
hyperplasia: Abnormal enlargement of an organ due to an increase in the number of cells.
hyperpnea (hī-perp-NĒ-a): Abnormal increases in the rate and depth of respiration.
hyperreflexia: Abnormally exaggerated reflex responses to stimulation.
hypersecretion: Overactivity of glands that produce exocrine or endocrine secretions.
hypertension: Abnormally high blood pressure.
hyperthermia: Excessively high body temperature.
hyperthyroidism: Excessive production of thyroid hormones.
hypertonic: When comparing two solutions, used to refer to the solution with the higher osmolarity.
hypertrophy (HĪ-per-trō-fē): Increase in the size of tissue without cell division.
hyperventilation (hī-per-ven-ti-LĀ-shun): A rate of respiration sufficient to reduce the plasma P_{CO_2} to levels below normal.
hypoblast (HĪ-pō-blast): The undersurface of the inner cell mass that faces the blastocoele of the early embryo.
hypocapnia: Abnormally low plasma P_{CO_2}, usually the result of hyperventilation.
hypodermic needle: A needle inserted through the skin to introduce drugs into the subcutaneous layer.
hypodermis: The subcutaneous layer, a region of loose connective tissue also called the **superficial fascia**.
hypoesthesia: Abnormally decreased sensitivity to stimuli.
hypoglossal nerve: N XII, the cranial nerve responsible for the control of the muscles that move the tongue.
hyponychium (hī-pō-NIK-ē-um): A thickening in the epidermis beneath the free edge of a nail.
hypophyseal portal system (hī-pō-FIS-ē-al): Network of vessels that carry blood from capillaries in the hypothalamus to capillaries in the anterior lobe of the pituitary gland (hypophysis).
hypophysis (hī-POF-i-sis): The anterior lobe of the pituitary gland, which can be further subdivided into the pars distalis and the pars intermedia.
hyporeflexia: Abnormally depressed reflex responses to stimuli.
hyposecretion: Abnormally low rates of exocrine or endocrine secretion.
hypothalamus: The floor of the diencephalon; region of the brain containing centers involved with the unconscious regulation of visceral functions, emotions, drives, and the coordination of neural and endocrine functions.
hypotonic: When comparing two solutions, used to refer to the one with the lower osmolarity.
hypoventilation: A respiratory rate insufficient to keep plasma P_{CO_2} within normal levels.
hypovolemic (hī-pō-vō-LĒ-mik): An abnormally low blood volume.
hypoxia (hī-POKS-ē-a): Low tissue oxygen concentrations.
ileocecal valve (il-ē-ō-SĒ-kal): A fold of mucous membrane that guards the connection between the ileum and the cecum.
ileostomy (il-ē-OS-tō-mē): Surgical creation of an opening into the ileum; the opening created when the ileum is surgically attached to the abdominal wall.
ileum (IL-ē-um): The last 8 ft of the small intestine.
ilium (IL-ē-um): The largest of the three bones whose fusion creates a coxa.
immunity: Resistance to injuries and diseases caused by foreign compounds, toxins, and pathogens.

immunization: Developing immunity by the deliberate exposure to antigens under conditions that prevent the development of illness but stimulate the production of memory B cells.
immunoglobulin (i-mū-nō-GLOB-ū-lin): A circulating antibody.
implantation (im-plan-TĀ-shun): The erosion of a blastocyst into the uterine wall.
impotence: Inability to obtain or maintain an erection in the male.
incisors (in-SĪ-zerz): Two pairs of flattened, bladelike teeth located at the front of the dental arches in both the upper and lower jaws.
inclusions: Aggregations of insoluble pigments, nutrients, or other materials in the cytoplasm.
incontinence (in-KON-ti-nens): Inability to voluntarily control micturition (or defecation).
incus (IN-kus): The central auditory ossicle, situated between the malleus and the stapes in the middle ear cavity.
infarct: An area of dead cells resulting from an interruption of circulation.
infection: Invasion and colonization of body tissues by pathogenic organisms.
inferior: A directional reference meaning below.
inferior vena cava: The vein that carries blood from the parts of the body below the heart to the right auricle.
infertility: Inability to conceive.
inflammation: A nonspecific defense mechanism that operates at the tissue level, characterized by swelling, redness, warmth, pain, and some loss of function.
inflation reflex: A reflex mediated by the vagus nerve that prevents overexpansion of the lungs.
infundibulum (in-fun-DIB-ū-lum): A tapering, funnel-shaped structure; in the nervous system, refers to the connection between the pituitary gland and the hypothalamus; the infundibulum of the uterine tube is the entrance bounded by fimbriae that receives the ova at ovulation.
ingestion: The introduction of materials into the digestive tract via the mouth.
inguinal canal: A passage through the abdominal wall that marks the path of testicular descent, and that contains the testicular arteries, veins, and ductus deferens.
inguinal region: The area near the junction of the trunk and the thighs that contains the external genitalia.
inhibin (in-HIB-in): A hormone produced by the sustentacular cells that inhibits the pituitary secretion of FSH.
initial segment: The proximal portion of the axon, adjacent to the axon hillock, where an action potential first appears.
injection: Forcing of fluid into a body part or organ.
inner cell mass: Cells of the blastocyst that will form the body of the embryo.
inner ear: *See* **internal ear.**
innervation: The distribution of sensory and motor nerves to a specific region or organ.
insertion: Point of attachment of a muscle that is more movable.
inspiration: Inhalation; the movement of air into the respiratory system.
inspiratory reserve: The maximum amount of air that can be drawn into the lungs over and above the normal tidal volume.
insoluble: Incapable of dissolving in solution.
insomnia: Sleep disorder characterized by an inability to fall asleep.
insula (IN-sū-la): A region of the temporal lobe that is visible only after opening the lateral sulcus.
insulin (IN-sū-lin): Hormone secreted by the beta cells of the pancreatic islets; causes a reduction in plasma glucose concentrations.
integument (in-TEG-ū-ment): The skin and associated organs (hair, glands, receptors, etc.).
intercalated discs (in-TER-ka-lā-ted): Regions where adjacent cardiocytes interlock and where gap junctions permit electrical coupling between the cells.
intercellular cement: Proteoglycans, containing the polysaccharide hyaluronic acid, found between adjacent epithelial cells.
intercellular fluid: *See* **interstitial fluid.**
interdigitate: To interlock.
interferons (in-ter-FĒR-ons): Peptides released by virally infected cells, especially lymphocytes, that make other cells more resistant to viral infection and slow viral replication.
interleukins (in-ter-LOO-kins): Peptides released by activated monocytes and lymphocytes that assist in the coordination of the cellular and humoral immune responses.
internal capsule: Term given to the appearance of the white matter of the cerebral hemispheres on gross dissection of the brain.
internal ear: The membranous labyrinth that contains the organs of hearing and equilibrium.
internal nares: The entrance to the nasopharynx from the nasal cavity.
internal respiration: Diffusion of gases between the blood and interstitial fluid.
interneuron: An association neuron; neurons inside the CNS that are interposed between sensory and motor neurons.
interoceptors: Sensory receptors monitoring the functions and status of internal organs and systems.
interosseous membrane: Fibrous connective tissue membrane between the shafts of the tibia and fibula or the radius and ulna; an example of a fibrous amphiarthrosis.
interphase: Stage in the life of a cell during which the chromosomes are uncoiled and all normal cellular functions except mitosis are under way.
intersegmental reflex: A reflex that involves several segments of the spinal cord.
interstitial cell-stimulating hormone: An alternative name for LH in the male; stimulates androgen production by the interstitial cells of the testes.
interstitial fluid (in-ter-STISH-al): Fluid in the tissues that fills the spaces between cells.
interstitial growth: Form of cartilage growth through the growth, mitosis, and secretion of chondrocytes inside the matrix.
interventricular foramen: The opening that permits fluid movement between the lateral and third ventricles.
intervertebral disc: Fibrocartilage pad between the centra of successive vertebrae that acts as a shock absorber.
intestinal crypt: A tubular epithelial pocket lined by secretory cells and opening into the lumen of the digestive tract; also called an intestinal gland.
intestine: Tubular organ of the digestive tract.
intracellular fluid: The cytosol.
intrafusal fibers: Muscle spindle fibers.
intramembranous ossification (in-tra-MEM-bra-nus): The formation of bone within a connective tissue without the prior development of a cartilaginous model.
intramuscular injection: Injection of medication into the bulk of a skeletal muscle.
intraocular pressure: The hydrostatic pressure exerted by the aqueous humor of the eye.
intrapleural pressure: The pressure measured in a pleural cavity; also called the intrathoracic pressure.
intrapulmonary pressure (in-tra-PUL-mō-ner-ē): The pressure measured in an alveolus; also called the intra-alveolar pressure.
intrauterine: Within the uterus; used to refer to the period of prenatal development.
intrinsic factor: Glycoprotein secreted by the parietal cells of the stomach that facilitates the intestinal absorption of vitamin B_{12}.
inversion: A turning inward.
in vitro: Outside of the body, in an artificial environment.
in vivo: In the living body.
involuntary: Not under conscious control.
ion: An atom or molecule bearing a positive or negative charge due to the acceptance or donation of an electron.
ipsilateral: A reflex response affecting the same side as the stimulus.
iris: A contractile structure made up of smooth muscle that forms the colored portion of the eye.
ischemia (is-KĒ-mē-a): Inadequate blood supply to a region of the body.
ischium (IS-kē-um): One of the three bones whose fusion creates the coxa.
islets of Langerhans: *See* **pancreatic islets.**
isometric contraction: A muscular contraction characterized by rising tension production but no change in length.
isotonic: A solution having an osmolarity that does not result in water movement across cell membranes; of the same contractive strength.
isotonic contraction: A muscular contraction during which tension climbs and then remains stable as the muscle shortens.
isthmus (IS-mus): A narrow band of tissue connecting two larger masses.
jaundice (JAWN-dis): Condition characterized by yellowing of connective tissues due to elevated tissue bilirubin levels; usually associated with damage to the liver or biliary system.
jejunum (je-JŪ-num): The middle portion of the small intestine.
joint: An area where adjacent bones interact; an articulation.
juxtaglomerular apparatus: The macula densa and the juxtaglomerular cells; a complex responsible for the release of renin and erythropoietin.
juxtaglomerular cells: Modified smooth muscle cells in the walls of the afferent and efferent arterioles adjacent to the glomerulus and the macula densa.
juxtamedullary nephrons: The 15% of nephrons whose loops of Henle extend into the medulla; these nephrons are responsible for creating the osmotic gradient within the medulla.
karyotyping (KAR-ē-ō-tī-ping): The determination of the chromosomal characteristics of an individual or cell.
keratin (KER-a-tin): Tough, fibrous protein component of nails, hair, calluses, and the general integumentary surface.
keratinization (KER-a-tin-ī-zā-shun): The production of keratin by epithelial cells.
keratinized (KER-a-tin-īzd): Containing large quantities of keratin.
keratohyalin (ker-a-tō-HĪ-a-lin): A protein within maturing keratinocytes.
ketone bodies: Keto acids produced during the catabolism of lipids and ketogenic amino acids; specifically acetone, acetoacetate, and beta-hydroxybutyrate.

kidney: A component of the urinary system; an organ functioning in the regulation of plasma composition, including the excretion of wastes and the maintenance of normal fluid and electrolyte balance.
killer T cells: *See* **cytotoxic T cells**.
Kupffer cells (KOOP-fer): Stellate cells of the liver; phagocytic cells of the liver sinusoids.
kyphosis (kī-FŌ-sis): Exaggerated thoracic curvature.
labia (LĀ-bē-a): Lips; labia majora and minora are components of the female external genitalia.
labrum: A lip or rim.
labyrinth: A maze of passageways; usually refers to the structures of the inner ear.
lacrimal gland (LAK-ri-mal): Tear gland on the dorsolateral surface of the eye.
lactase: An enzyme that breaks down a disaccharide (lactose) in milk.
lactation (lak-TĀ-shun): The production of milk by the mammary glands.
lacteal (LAK-tē-al): A terminal lymphatic within an intestinal villus.
lactic acid: Compound produced from pyruvic acid during glycolysis.
lactiferous duct (lak-TIF-e-rus): Duct draining one lobe of the mammary gland.
lactiferous sinus: An expanded portion of a lactiferous duct adjacent to the nipple of a breast.
lacuna (la-KŪ-nē): A small pit or cavity.
lambdoid suture (lam-DOID-al): Synarthrotic articulation between the parietal and occipital bones of the cranium.
lamellae (la-MEL-lē): Concentric layers of bone within an osteon.
lamellated corpuscle: Receptor sensitive to vibration; also called a *Pacinian corpuscle*.
lamina (LA-min-a): A thin sheet or layer.
lamina propria (PRŌ-prē-a): A layer of loose connective tissue situated immediately beneath the epithelium of a mucous membrane.
laminectomy: Removal of the spinous processes of a vertebra to gain access and treat a herniated disc.
Langerhans cells (LAN-ger-hanz): Cells in the epithelium of the skin and digestive tract that participate in the immune response by presenting antigens to T cells.
laparoscope (LAP-a-rō-skōp): Fiber-optic instrument used to visualize the contents of the abdominopelvic cavity.
large intestine: The terminal portions of the intestinal tract, consisting of the colon, the rectum, and the anal canal.
laryngopharynx (la-ring-gō-FAR-inks): Division of the pharynx inferior to the epiglottis and superior to the esophagus.
larynx (LAR-inks): A complex cartilaginous structure that surrounds and protects the glottis and vocal cords; the superior margin is bound to the hyoid bone and the inferior margin is bound to the trachea.
lateral: Pertaining to the side.
lateral apertures: Openings in the roof of the fourth ventricle that permit the circulation of CSF into the subarachnoid space.
lateral ventricle: Fluid-filled chamber within one of the cerebral hemispheres.
laxatives: Compounds that promote defecation via increased peristalsis or an increase in the water content and volume of the feces.
lens: The transparent body lying behind the iris and pupil and in front of the vitreous humor.
lesion: A localized abnormality in tissue organization.
lesser omentum: A small pocket in the mesentery that connects the lesser curvature of the stomach to the liver.
leukemia (loo-KĒ-mē-ah): A malignant disease of the blood-forming tissues.
leukocyte (LOO-kō-sīt): A white blood cell.
leukocytosis (loo-kō-sī-TŌ-sis): Abnormally high numbers of circulating white blood cells.
leukopenia (loo-kō-PĒ-nē-ah): Abnormally low numbers of circulating white blood cells.
leukopoiesis (loo-kō-poy-Ē-sis): White blood cell formation.
ligament (LIG-a-ment): Dense band of connective tissue fibers that attach one bone to another.
ligamentum arteriosum: The fibrous strand found in the adult that represents the remains of the ductus arteriosus of the fetus.
ligamentum nuchae (NOO-kē): An elastic ligament that extends between the vertebra prominens and the external occipital crest.
ligamentum teres: The fibrous strand in the falciform ligament that represents the remains of the umbilical vein of the fetus.
ligate: To tie off.
limbic system (LIM-bik): Group of nuclei and centers in the cerebrum and diencephalon that are involved with emotional states, memories, and behavioral drives.
limbus (LIM-bus): The edge of the cornea, marked by the transition from the corneal epithelium to the ocular conjunctiva.
liminal stimulus: A stimulus sufficient to depolarize the transmembrane potential of an excitable membrane to threshold, and produce an action potential.
linea alba: Tendinous band that runs along the midline of the rectus abdominis.
lingual: Pertaining to the tongue.
lingual frenulum: An epithelial fold that attaches the inferior surface of the tongue to the floor of the mouth.
lipase (LI-pāz): A pancreatic enzyme that breaks down triglycerides.
lipemia (lip-Ē-mē-a): Elevated concentration of lipids in the circulation.
lipid: An organic compound containing carbons, hydrogens, and oxygens in a ratio that does not approximate 1:2:1; includes fats, oils, and waxes.
lipogenesis (lī-pō-JEN-e-sis): Synthesis of lipids from nonlipid precursors.
lipolysis: The catabolism of lipids as a source of energy.
lipoprotein (lī-pō-PRŌ-tēn): A compound containing a relatively small lipid bound to a protein.
liver: An organ of the digestive system with varied and vital functions that include the production of plasma proteins, the excretion of bile, the storage of energy reserves, the detoxification of poisons, and the interconversion of nutrients.
lobule (LOB-ūl): The basic organizational unit of the liver at the histological level.
loose connective tissue: A loosely organized, easily distorted connective tissue containing several different fiber types, a varied population of cells, and a viscous ground substance.
lordosis (lor-DŌ-sis): An exaggeration of the lumbar curvature.
lumbar: Pertaining to the lower back.
lumen: The central space within a duct or other internal passageway.
lungs: Paired organs of respiration, situated in the left and right pleural cavities.
luteinizing hormone (LH) (LOO-tē-in-ī-zing): Anterior pituitary hormone that in the female assists FSH in follicle stimulation, triggers ovulation, and promotes the maintenance and secretion of the endometrial glands; in the male, stimulates spermatogenesis; also known as **interstitial cell-stimulating hormone**.
luxation (luks-Ā-shun): Dislocation of a joint.
lymph: Fluid contents of lymphatic vessels, similar in composition to interstitial fluid.
lymphadenopathy (lim-fad-e-NOP-a-thē): Pathological enlargement of the lymph nodes.
lymphatics: Vessels of the lymphatic system.
lymphedema (lim-fe-DĒ-ma): Swelling of peripheral tissues due to excessive lymph production or inadequate drainage.
lymph nodes: Lymphoid organs that monitor the composition of lymph.
lymphocyte (LIM-fō-sīt): A cell of the lymphatic system that participates in the immune response.
lymphokines: Chemicals secreted by activated lymphocytes.
lymphopoiesis: The production of lymphocytes.
lymphotoxin (lim-fō-TOK-sin): A secretion of lymphocytes that kills the target cells.
lysis (LĪ-sis): The destruction of a cell through the rupture of its cell membrane.
lysosome (LĪ-sō-sōm): Intracellular vesicle containing digestive enzymes.
lysozyme: An enzyme present in some exocrine secretions that has antibiotic properties.
macrophage: A phagocytic cell of the monocyte-macrophage system.
macula (MAK-ū-la): A receptor complex in the saccule or utricle that responds to linear acceleration or gravity.
macula densa (MAK-ū-la DEN-sa): A group of specialized secretory cells in a portion of the distal convoluted tubule adjacent to the glomerulus and the juxtaglomerular cells; a component of the juxtaglomerular apparatus.
macula lutea (LOO-tē-a): The fovea.
malignant cancer: A form of cancer characterized by rapid cellular growth and the spread of cancer cells throughout the body.
malleus (MAL-ē-us): The first auditory ossicle, bound to the tympanic membrane and the incus.
mamillary bodies (MAM-i-lar-ē): Nuclei in the hypothalamus concerned with feeding reflexes and behaviors; a component of the limbic system.
mammary glands: Milk-producing glands of the female breast.
manubrium: The broad, roughly triangular, superior element of the sternum.
manus: The hand.
marrow: A tissue that fills the internal cavities in a bone; may be dominated by hemopoietic cells (red marrow) or adipose tissue (yellow marrow).
mass peristalsis: Powerful peristaltic contraction that moves fecal materials along the colon and into the rectum.
mass reflex: Hyperreflexia in an area innervated by spinal cord segments distal to an area of injury.
mast cell: A connective tissue cell that when stimulated releases histamine, serotonin, and heparin, initiating the inflammatory response.
mastectomy: Surgical removal of part or all of a mammary gland.
mastication (mas-ti-KĀ-shun): Chewing.
mastoid sinus: Air-filled spaces in the mastoid process of the temporal bone.
matrix: The ground substance of a connective tissue.
maxillary sinus (MAK-si-ler-ē): One of the paranasal sinuses; an air-filled chamber lined by a respiratory epithelium that is located in a maxilla and opens into the nasal cavity.

meatus (mē-Ā-tus): An opening or entrance into a passageway.
mechanoreception: Detection of mechanical stimuli, such as touch, pressure, or vibration.
medial: Toward the midline of the body.
mediastinum (mē-dē-as-TĪ-num): Central tissue mass that divides the thoracic cavity into two pleural cavities; includes the aorta and other great vessels, the esophagus, trachea, thymus, the pericardial cavity and heart, and a host of nerves, small vessels, and lymphatics.
medulla: Inner layer or core of an organ.
medulla oblongata: The most caudal of the five brain regions, also known as the **myelencephalon.**
medullary cavity: The space within a bone that contains the marrow.
medullary rhythmicity center: Center in the medulla oblongata that sets the background pace of respiration; includes inspiratory and expiratory centers.
megakaryocytes (meg-a-KAR-ē-ō-sīts): Bone marrow cells responsible for the formation of platelets.
meiosis (mī-Ō-sis): Cell division that produces gametes with half of the normal somatic chromosome complement.
melanin (MEL-a-nin): Yellow-brown pigment produced by the melanocytes of the skin.
melanocyte (me-LAN-ō-sīt): Specialized cell found in the deeper layers of the stratified squamous epithelium of the skin, responsible for the production of melanin.
melanocyte-stimulating hormone (MSH): Hormone of the pars intermedia of the anterior pituitary that stimulates melanin production.
melanomas (mel-a-NŌ-maz): Dangerous malignant skin cancers that involve melanocytes.
melatonin (mel-a-TŌ-nin): Hormone secreted by the pineal gland; inhibits secretion of MSH and gonadotropins.
membrane: Any sheet or partition; a layer consisting of an epithelium and the underlying connective tissue.
membrane flow: The movement of sections of membrane surface to and from the cell surface and components of the endoplasmic reticulum, the Golgi apparatus, and vesicles.
membranous labyrinth: Endolymph-filled tubes of the inner ear that enclose the receptors of the inner ear.
menarche (me-NAR-kē): The beginning of menstrual function.
meninges (men-IN-jēz): Three membranes that surround the surfaces of the CNS; the dura mater, the pia mater, and the arachnoid.
meningitis: Inflammation of the spinal or cranial meninges.
meniscectomy: Removal of a meniscus.
meniscus (men-IS-kus): A fibrocartilage pad between opposing surfaces in a joint.
menopause (MEN-ō-paws): The cessation of uterine cycles as a consequence of the aging process and exhaustion of viable follicles.
menses (MEN-sēz): The first menstrual period that normally occurs at puberty.
menstrual (MEN-stroo-al) **cycle:** *See* **uterine cycle.**
menstruation (men-stroo-Ā-shun): The sloughing of blood and endometrial tissue at menses.
merocrine (MER-ō-krin): A method of secretion where the cell ejects materials through exocytosis.
mesencephalic aqueduct: Passageway that connects the third ventricle (diencephalon) with the fourth ventricle (metencephalon).
mesencephalon (mez-en-SEF-a-lon): The midbrain.
mesenchyme (MEZ-en-kīm): Embryonic/fetal connective tissue.
mesentery (MEZ-en-ter-ē): A double layer of serous membrane that supports and stabilizes the position of an organ in the abdominopelvic cavity and provides a route for the associated blood vessels, nerves, and lymphatics.
mesoderm: The middle germ layer that lies between the ectoderm and endoderm of the embryo.
mesothelium (mez-ō-THĒ-lē-um): A simple squamous epithelium that lines one of the divisions of the ventral body cavity.
messenger RNA (mRNA): RNA formed at transcription to direct protein synthesis in the cytoplasm.
metabolism (me-TAB-ō-lizm): The sum of all of the biochemical processes under way within the human body at a given moment; includes anabolism and catabolism.
metabolites (me-TAB-ō-līts): Compounds produced in the body as the result of metabolic reactions.
metacarpals (met-a-KAR-pals): The five bones of the palm of the hand.
metalloproteins (me-tal-ō-PRŌ-tēnz): Plasma proteins that transport metal ions.
metaphase (MET-a-fāz): A stage of mitosis wherein the chromosomes line up along the equatorial plane of the cell.
metaphysis (me-TAF-i-sis): The region of a long bone between the epiphysis and diaphysis, corresponding to the location of the epiphyseal cartilage of the developing bone.
metarteriole (met-ar-TĒ-rē-ōl): A vessel that connects an arteriole to a venule and that provides blood to a capillary plexus.
metastasis (me-TAS-ta-sis): The spread of a disease from one organ to another.
metatarsal: One of the five bones of the foot that articulate with the tarsals (proximally) and the phalanges (distally).
metencephalon (met-en-SEF-a-lon): The pons and cerebellum of the brain.
micelle (mī-SEL): A spherical aggregation of bile salts, monoglycerides, and fatty acids in the lumen of the intestinal tract.
microcephaly (mī-krō-SEF-a-lē): An abnormally small cranium, due to premature closure of one or more fontanels.
microfilaments: Fine protein filaments visible with the electron microscope; components of the cytoskeleton.
microglia (mī-KROG-lē-a): Phagocytic glial cells in the CNS, derived from the monocytes of the blood.
microphages: Neutrophils and eosinophils.
microtubules: Microscopic tubules that are part of the cytoskeleton, and are found in cilia, flagella, the centrioles, and spindle fibers.
microvilli: Small, fingerlike extensions of the exposed cell membrane of an epithelial cell.
micturition (mik-tū-RI-shun): Urination.
midbrain: The mesencephalon.
middle ear: Space between the external and internal ear that contains auditory ossicles.
midsagittal plane: A plane passing through the midline of the body that divides it into left and right halves.
mineralocorticoids: Corticosteroids produced by the zona glomerulosa of the adrenal cortex; steroids such as aldosterone, that affect mineral metabolism.
miscarriage: Spontaneous abortion.
mitochondrion (mī-tō-KON-drē-on): An intracellular organelle responsible for generating most of the ATP required for cellular operations.
mitosis (mī-TŌ-sis): The division of a single cell that produces two identical daughter cells; the primary mechanism of tissue growth.
mitral valve (MĪ-tral): The left AV, or bicuspid, valve of the heart.
mixed gland: A gland that contains exocrine and endocrine cells, or an exocrine gland that produces serous and mucous secretions.
mixed nerve: A peripheral nerve that contains sensory and motor fibers.
modiolus (mō-DĪ-ō-lus): The bony central hub of the cochlea.
mole: A quantity of an element or compound having a mass in grams equal to its atomic or molecular weight.
molecular weight: The sum of the atomic weights of the atoms in a molecule.
molecule: A compound containing two or more atoms that are held together by chemical bonds.
monocytes (MON-ō-sīts): Phagocytic agranulocytes (white blood cells) in the circulating blood.
monoglyceride (mon-ō-GLI-se-rīd): A lipid consisting of a single fatty acid bound to a molecule of glycerol.
monokines: Secretions released by activated cells of the monocyte-macrophage system to coordinate various aspects of the immune response.
monosaccharide (mon-ō-SAK-ah-rīd): A simple sugar, such as glucose or ribose.
monosynaptic reflex: A reflex where the sensory afferent synapses directly on the motor efferent.
monozygotic twins: Twins produced through the splitting of a single fertilized egg (zygote).
morula (MOR-ū-la): A mulberry-shaped collection of cells produced through the mitotic divisions of a zygote.
motor unit: All of the muscle cells controlled by a single motor neuron.
mucins (MŪ-sins): Proteoglycans responsible for the lubricating properties of mucus.
mucosa (mū-KŌ-sa): A mucous membrane; the epithelium plus the lamina propria.
mucous: An adjective referring to the presence or production of mucus.
mucous membrane: *See* **mucosa.**
mucus: Lubricating secretion produced by unicellular and multicellular glands along the digestive, respiratory, urinary, and reproductive tracts.
multipennate muscle: A muscle whose internal fibers are organized around several different tendons.
multipolar neuron: A neuron with many dendrites and a single axon, the typical form of a motor neuron.
multiunit smooth muscle: Smooth muscle tissue whose muscle cells are innervated in motor units.
muriatic acid: Hydrochloric acid (HCl).
muscarinic receptors (mus-kar-IN-ik): Membrane receptors sensitive to acetylcholine (ACh) and to muscarine, a toxin produced by certain mushrooms; found at all parasympathetic neuroeffector junctions and at a few sympathetic neuroeffector junctions.
muscle: A contractile organ composed of muscle tissue, blood vessels, nerves, connective tissues, and lymphatics.
muscle tissue: A tissue characterized by the presence of cells capable of contraction; includes skeletal, cardiac, and smooth muscle tissue.

muscularis externa (mus-kū-LAR-is): Concentric layers of smooth muscle responsible for peristalsis.
muscularis mucosae: Layer of smooth muscle beneath the lamina propria responsible for moving the mucosal surface.
mutagens (MŪ-ta-jenz): Chemical agents that induce mutations and may be carcinogenic.
myalgia (mī-AL-jē-a): Muscle pain.
myasthenia gravis (mī-as-THĒ-nē-a GRA-vis): Muscular weakness due to a reduction in the number of ACh receptor sites on the sarcolemmal surface; suspected to be an autoimmune disorder.
myelencephalon (mī-el-en-SEF-a-lon): The medulla oblongata.
myelin (MĪ-e-lin): Insulating sheath around an axon consisting of multiple layers of glial cell membrane; significantly increases conduction rate along the axon.
myelination: The formation of myelin.
myeloid tissue: Tissue responsible for the production of red blood cells, granulocytes, monocytes, and platelets.
myenteric plexus (mī-en-TER-ik): Parasympathetic motor neurons and sympathetic postganglionic fibers located between the circular and longitudinal layers of the muscularis externa.
myocardial infarction (mī-ō-KAR-dē-al): Heart attack; damage to the heart muscle due to an interruption of regional coronary circulation.
myocarditis: Inflammation of the myocardium.
myocardium: The cardiac muscle tissue of the heart.
myofibrils: Organized collections of myofilaments in skeletal and cardiac muscle cells.
myofilaments: Fine protein filaments, composed of the proteins actin (thin filaments) and myosin (thick filaments).
myoglobin (MĪ-ō-GLŌ-bin): An oxygen-binding pigment especially common in slow skeletal and cardiac muscle fibers.
myogram: A recording of the tension produced by muscle fibers on stimulation.
myometrium (mī-ō-MĒ-trē-um): The thick layer of smooth muscle in the wall of the uterus.
myopia: Nearsightedness, an inability to accommodate for distant vision.
myosepta: Connective tissue partitions that separate adjacent skeletal muscles.
myosin: Protein component of the thick myofilaments.
myositis (mī-ō-SĪ-tis): Inflammation of muscle tissue.
nail: Keratinous structure produced by epithelial cells of the nail root.
narcolepsy: A sleep disorder characterized by falling asleep at inappropriate moments.
nares, external (NA-rēz): The entrance from the exterior to the nasal cavity.
nares, internal: The entrance from the nasal cavity to the nasopharynx.
nasal cavity: A chamber in the skull bounded by the internal and external nares.
nasolacrimal duct: Passageway that transports tears from the nasolacrimal sac to the nasal cavity.
nasolacrimal sac: Chamber that receives tears from the nasolacrimal ducts.
nasopharynx (nā-zō-FAR-inks): Region posterior to the internal nares, superior to the soft palate, and ending at the oropharynx.
necrosis (NEK-rō-sis): Death of cells or tissues from disease or injury.
negative feedback: Corrective mechanism that opposes or negates a variation from normal limits.
neonate: A newborn infant, or baby.
neoplasm: A tumor, or mass of abnormal tissue.
nephritis (nef-RĪ-tis): Inflammation of the kidney.
nephrolithiasis (nef-rō-li-THĪ-a-sis): Condition resulting from the formation of kidney stones.
nephron (NEF-ron): Basic functional unit of the kidney.
nerve impulse: An action potential in a nerve cell membrane.
neural cortex: An area where gray matter is found at the surface of the CNS.
neurilemma (nū-ri-LEM-ma): The outer surface of a glial cell that encircles an axon.
neuroeffector junction: A synapse between a motor neuron and a peripheral effector, such as a muscle cell or gland cell.
neurofibrils: Microfibrils in the cytoplasm of a neuron.
neurofilaments: Microfilaments in the cytoplasm of a neuron.
neuroglandular junction: A specific type of neuroeffector junction.
neuroglia (nū-ROG-lē-a): Nonneural cells of the CNS and PNS that support and protect the neurons.
neurohypophysis (nū-rō-hī-POF-i-sis): The posterior lobe of the pituitary gland or *pars nervosa*.
neuromuscular junction: A specific type of neuroeffector junction.
neuron (NŪ-ron): A nerve cell.
neurotransmitter: Chemical compound released by one neuron to affect the transmembrane potential of another.
neurotubules: Microtubules in the cytoplasm of a neuron.
neurulation: The embryological process responsible for the formation of the CNS.
neutropenia: An abnormally low number of neutrophils in the circulating blood.
neutrophil (NŪ-trō-fil): A phagocytic microphage (granulocyte, WBC) that is very numerous and usually the first of the mobile phagocytic cells to arrive at an area of injury or infection.
nicotinic receptors (nik-ō-TIN-ik): ACh receptors found on the surfaces of sympathetic and parasympathetic ganglion cells, that will also respond to the compound nicotine.
nipple: An elevated epithelial projection on the surface of the breast, containing the openings of the lactiferous sinuses.
Nissl bodies: The ribosomes, Golgi, RER, and mitochondria of the perikar yon of a typical nerve cell.
nitrogenous wastes: Organic waste products of metabolism that contain nitrogen, such as urea, uric acid, and creatinine.
NK cells (natural killer cells): Lymphocytes responsible for immune surveillance, the detection and destruction of cancer cells.
nociception (nō-sē-SEP-shun): Pain perception.
node of Ranvier: Area between adjacent glial cells where the myelin covering of an axon is incomplete.
nodose ganglion (NŌ-dōs): A sensory ganglion of cranial nerve X.
noradrenaline: Catecholamine secreted by the adrenal medulla, released at most sympathetic neuroeffector junctions, and at certain synapses inside the CNS; also called **norepinephrine**.
norepinephrine (nor-ep-i-NEF-rin): A catecholamine neurotransmitter in the PNS and CNS, and a hormone secreted by the adrenal medulla; also called **noradrenaline**.
normovolemic (nor-mō-vō-LĒ-mik): Having a normal blood volume.
nucleic acid (nū-KLĒ-ik): A polymer of nucleotides containing a pentose sugar, a phosphate group, and one of four nitrogenous bases that regulate the synthesis of proteins and make up the genetic material in cells.
nucleolus (nū-KLĒ-ō-lus): Dense region in the nucleus that represents the site of RNA synthesis.
nucleoplasm: Fluid content of the nucleus.
nucleoproteins: Proteins of the nucleus that are generally associated with the DNA.
nucleus: Cellular organelle that contains DNA, RNA, and proteins; a mass of gray matter in the CNS.
nucleus pulposus (pul-PŌ-sus): The gelatinous core of an intervertebral disc.
nutrient: An organic compound that can be broken down in the body to produce energy.
nystagmus: Involuntary, continual movement of the eyes as if to adjust to constant motion.
obesity: Body weight 10–20 percent above standard values, as the result of body fat accumulation.
occlusal surface (o-KLŪ-sal): The opposing surfaces of the teeth that come into contact when processing food.
ocular: Pertaining to the eye.
oculomotor nerve (ok-ū-lō-MŌ-ter): Cranial nerve III, that controls the extra-ocular muscles other than the superior oblique and the lateral rectus.
olecranon: The proximal end of the ulna that forms the prominent point of the elbow.
olfaction: The sense of smell.
olfactory bulb (ol-FAK-tor-ē): Two olfactory nerves that lie beneath the frontal lobe of the cerebrum.
olfactory tract: Tract over which nerve impulses from the retina are transmitted between the optic chiasma and the thalamus.
oligodendrocytes (ol-i-gō-DEN-drō-sīts): CNS glial cells responsible for maintaining cellular organization in the gray matter and providing a myelin sheath in areas of white matter.
oncogene (ON-kō-jēn): A gene that can turn a normal cell into a cancer cell.
oncologists (on-KOL-ō-jists): Physicians specializing in the study and treatment of tumors.
oocyte (Ō-ō-sīt): A cell whose meiotic divisions will produce a single ovum and three polar bodies.
oogenesis (ō-ō-JEN-e-sis): Ovum production.
oogonia (ō-ō-GŌ-nē-a): Stem cells in the ovaries whose divisions give rise to oocytes.
oophorectomy (ō-of-ō-REK-tō-mē): Surgical removal of the ovaries.
oophoritis (ō-of-ō-RĪ-tis): Inflammation of the ovaries.
ooplasm: The cytoplasm of the ovum.
opsin: A protein, one structural component of the visual pigment rhodopsin.
opsonization: An effect of coating an object with antibodies; the attraction and enhancement of phagocytosis.
optic chiasma (OP-tik kī-AZ-ma): Crossing point of the optic nerves.
optic nerve: Nerve that carries signals from the eye to the optic chiasma.
ora serrata (Ō-ra ser-RA-ta): The anterior edge of the neural retina.
orbit: Bony cavity of the skull that contains the eyeball.
orchiectomy (or-kē-EK-tō-mē): Surgical removal of one or both testes.
orchitis: Inflammation of the testes.

organelle (or-gan-EL): An intracellular structure that performs a specific function or group of functions.
organic compound: A compound containing carbon, hydrogen, and usually oxygen.
organogenesis: The formation of organs during embryological and fetal development.
organs: Combinations of tissues that perform complex functions.
origin: Point of attachment of a muscle that is less movable.
oropharynx (or-ō-FAR-inks): The middle portion of the pharynx, bounded superiorly by the nasopharynx, anteriorly by the oral cavity, and inferiorly by the laryngopharynx.
os coxa/ossa coxae: The bones of the hip.
osmolarity (oz-mō-LAR-i-tē): The total concentration of dissolved materials in a solution, regardless of their specific identities, expressed in terms of moles.
osmoreceptor: A receptor sensitive to changes in the osmolarity of the plasma.
osmosis (oz-MŌ-sis): The movement of water across a semipermeable membrane toward a solution containing a relatively high solute concentration.
osmotic pressure: The force of osmotic water movement; the pressure that must be applied to prevent osmotic movement across a membrane.
osseous tissue: A strong connective tissue containing specialized cells and a mineralized matrix of crystalline calcium phosphate and calcium carbonate.
ossicles: Small bones.
ossification: The formation of bone.
osteoblasts (OS-tē-ō-blasts): Cells that produce bone within connective tissue (intramembranous ossification) or cartilage (endochondral ossification); may differentiate into osteocytes.
osteoclast (OS-tē-ō-klast): A cell that dissolves the fibers and matrix of bone.
osteocyte (OS-tē-ō-sīt): A bone cell responsible for the maintenance and turnover of the mineral content of the surrounding bone.
osteogenesis (os-tē-ō-JEN-e-sis): Bone production.
osteogenic layer (os-tē-ō-JEN-ik): The inner, cellular layer of the periosteum that participates in bone growth and repair.
osteoid (OS-tē-oyd): The organic components of the bone matrix, produced by osteoblasts and osteocytes.
osteolysis (os-tē-OL-i-sis): The breakdown of the mineral matrix of bone.
osteon (OS-tē-on): The basic histological unit of compact bone, consisting of osteocytes organized around a central canal and separated by concentric lamellae.
osteopenia (os-tē-ō-PĒ-nē-a): The condition of inadequate bone production in the adult, leading to a loss in bone mass and strength.
osteoporosis (os-tē-ō-pōr-Ō-sis): A reduction in bone mass and strength sufficient to compromise normal bone function.
osteoprogenitor cells: Stem cells that give rise to osteoblasts.
otic: Pertaining to the ear.
otitis media: Inflammation of the middle ear cavity.
oval window: Opening in the bony labyrinth where the stapes attaches to the membranous wall of the scala vestibuli.
ovarian cycle (ō-VAR-ē-an): Monthly cycle of gamete development in the ovaries, associated with cyclical changes in the production of sex hormones (estrogens and progestins).
ovary: Female reproductive gland.
ovulation (ōv-ū-LĀ-shun): The release of a secondary oocyte, surrounded by cells of the corona radiata, following the rupture of the wall of a tertiary follicle.
ovum/ova (Ō-vum): A gamete produced by the reproductive system of a female; an egg.
oxytocin (oks-i-TŌ-sin): Hormone produced by hypothalamic cells and secreted into capillaries at the posterior pituitary; stimulates smooth muscle contractions of the uterus or mammary glands in the female, but has no known function in males.
pacemaker cells: Cells of the SA node that set the pace of cardiac contraction.
palate: Horizontal partition separating the oral cavity from the nasal cavity and nasopharynx; can be divided into an anterior bony (hard) palate and a posterior fleshy (soft) palate.
palatine: Pertaining to the palate.
palpate: To examine by touch.
palpebrae (pal-PĒ-brē): Eyelids.
pancreas: Digestive organ containing exocrine and endocrine tissues; exocrine portion secretes pancreatic juice; endocrine portion secretes hormones, including insulin and glucagon.
pancreatic duct: A tubular duct that carries pancreatic juice from the pancreas to the duodenum.
pancreatic islets: Aggregations of endocrine cells in the pancreas.
pancreatic juice: A mixture of buffers and digestive enzymes that is discharged into the duodenum under the stimulation of the enzymes secretin and cholecystokinin.
pancreatitis (pan-krē-a-TĪ-tis): Inflammation of the pancreas.
Papanicolaou (Pap) test: Test for the detection of malignancies of the female reproductive tract, especially the cervix and uterus.
papilla (pa-PIL-la): A small, conical projection.
paralysis: Loss of voluntary motor control over a portion of the body.
paranasal sinuses: Bony chambers lined by respiratory epithelium that open into the nasal cavity; includes the frontal, ethmoidal, sphenoid, and maxillary sinuses.
parasagittal: A section or plane that parallels the midsagittal plane but that does not pass along the midline.
parasympathetic division: One of the two divisions of the autonomic nervous system; also known as the craniosacral division; generally responsible for activities that conserve energy and lower the metabolic rate.
parathyroid glands: Four small glands embedded in the posterior surface of the thyroid; responsible for parathyroid hormone secretion.
parathyroid hormone: Hormone secreted by the parathyroid gland when plasma calcium levels fall below the normal range; causes increased osteoclast activity, increased intestinal calcium uptake, and decreased calcium ion loss at the kidneys.
parenchyma (par-ENG-ki-ma): The cells of a tissue or organ that are responsible for fulfilling its functional role.
paresthesia: Sensory abnormality that produces a tingling sensation.
parietal: Referring to the body wall or outer layer.
parietal cell: Cells of the gastric glands that secrete HCl and intrinsic factor.
Parkinson's disease: Progressive motor disorder due to degeneration of the cerebral nuclei.
parotid glands (pa-ROT-id): Large salivary glands that secrete a saliva containing high concentrations of salivary (alpha) amylase.
pars distalis (dis-TAL-is): The large, anterior portion of the anterior pituitary gland.
pars intermedia (in-ter-MĒ-dē-a): The portion of the anterior lobe of the pituitary gland immediately adjacent to the posterior pituitary and the infundibulum.
pars nervosa: The posterior pituitary gland.
parturition (par-tū-RISH-un): Childbirth, delivery.
patella (pa-TEL-la): The sesamoid bone of the kneecap.
pathogenic: Disease-causing.
pathologist (pa-THOL-ō-jist): An M.D. specializing in the identification of diseases based on characteristic structural and functional changes in tissues and organs.
pedicel (PED-i-sel): A slender process of a podocyte that forms part of the filtration apparatus of the kidney glomerulus.
pedicles (PE-di-kls): Thick bony struts that connect the vertebral body with the articular and spinous processes.
pelvic cavity: Inferior subdivision of the abdominopelvic (peritoneal) cavity; encloses the urinary bladder, the sigmoid colon and rectum, and male or female reproductive organs.
pelvis: A bony complex created by the articulations between the coxae, the sacrum, and the coccyx.
penis (PĒ-nis): Component of the male external genitalia; a copulatory organ that surrounds the urethra, and that serves to introduce semen into the female vagina.
pepsin: Proteolytic enzyme secreted by the chief cells of the gastric glands in the stomach.
pepsinogen (pep-SIN-ō-jen): The inactive proenzyme that is secreted by chief cells of the gastric pits; after secretion it is converted to the proteolytic enzyme pepsin.
peptidases: Enzymes that split peptide bonds and release amino acids.
perforating canal: A passageway in compact bone that runs at right angles to the axes of the osteons, between the periosteum and endosteum.
perfusion: The blood flow through a tissue.
pericardial cavity (per-i-KAR-dē-al): The space between the parietal pericardium and the epicardium (visceral pericardium) that covers the outer surface of the heart.
pericarditis: Inflammation of the pericardium.
pericardium (per-i-KAR-dē-um): The fibrous sac that surrounds the heart, and whose inner, serous lining is continuous with the epicardium.
perichondrium (per-i-KON-drē-um): Layer that surrounds a cartilage, consisting of an outer fibrous and an inner cellular region.
perikaryon (per-i-KAR-ē-on): The cytoplasm that surrounds the nucleus in the soma of a nerve cell.
perilymph (PER-ē-limf): A fluid similar in composition to cerebrospinal fluid; found in the spaces between the bony labyrinth and the membranous labyrinth of the inner ear.
perimysium (per-i-MĪS-ē-um): Connective tissue partition that separates adjacent fasciculi in a skeletal muscle.
perineum (per-i-NĒ-um): The pelvic floor and associated structures.
perineurium: Connective tissue partition that separates adjacent bundles of nerve fibers in a peripheral nerve.
periodontal ligament (per-ē-ō-DON-tal): Collagen fibers that bind the cementum of a tooth to the periosteum of the surrounding alveolus.
periosteum (per-ē-OS-tē-um): Layer that surrounds a bone, consisting of an outer fibrous and inner cellular region.
peripheral nervous system (PNS): All neural tissue outside of the CNS.
peristalsis (per-i-STAL-sis): A wave of smooth muscle contractions that propels materials along the axis of a tube such as the digestive tract, the ureters, or the ductus deferens.

peritoneal cavity: *See* **abdominopelvic cavity.**
peritoneum (per-i-tō-NĒ-um): The serous membrane that lines the peritoneal (abdominopelvic) cavity.
peritonitis (per-i-tō-NĪ-tis): Inflammation of the peritoneum.
peritubular capillaries: A network of capillaries that surrounds the proximal and distal convoluted tubules of the kidneys.
permeability: Ease with which dissolved materials can cross a membrane; if freely permeable, any molecule can cross the membrane; if impermeable, nothing can cross; most biological membranes are selectively permeable.
peroxisome: A membranous vesicle containing enzymes that break down hydrogen peroxide (H_2O_2).
pes: The foot.
petrosal ganglion: Sensory ganglion of the glossopharyngeal nerve (N IX).
petrous: Stony, usually used to refer to the thickened portion of the temporal bone that encloses the inner ear.
Peyer's patches (PĪ-erz): Lymphatic nodules beneath the epithelium of the small intestine.
pH: The negative exponent of the hydrogen ion concentration, in moles per liter.
phagocyte: A cell that performs phagocytosis.
phagocytosis (fa-gō-sī-TŌ-sis): The engulfing of extracellular materials or pathogens; movement of extracellular materials into the cytoplasm by enclosure in a membranous vesicle.
phalanx/phalanges (fa-LAN-jēz): Digits; the bones of the fingers and toes.
pharmacology: The study of drugs, their physiological effects, and their clinical uses.
pharynx (FAR-inks): The throat; a muscular passageway shared by the digestive and respiratory tracts.
phasic response: A pattern of response to stimulation by sensory neurons that are normally inactive; stimulation causes a burst of neural activity that ends when the stimulus either stops or stops changing in intensity.
phenotype (FĒN-ō-tīp): Physical characteristics that are genetically determined.
phonation (fō-NĀ-shun): Sound production at the larynx.
phosphate group: PO_4^{3-}.
phospholipid (fos-fō-LIP-id): An important membrane lipid whose structure includes hydrophilic and hydrophobic regions.
phosphorylation (fos-for-i-LĀ-shun): The addition of a phosphate group to a molecule.
photoreception: Sensitivity to light.
physiology (fiz-ē-OL-ō-jē): Literally the study of function; considers the ways living organisms perform vital activities.
pia mater: The delicate inner meningeal layer that is in direct contact with the neural tissue of the CNS.
pigment: A compound with a characteristic color.
piloerection: "Goosebumps" effect produced by the contraction of the arrector pili muscles of the skin.
pineal gland: Neural tissue in the posterior portion of the roof of the diencephalon, responsible for the secretion of melatonin.
pinealocytes (PĪN-ē-a-lō-sīts): Secretory cells of the pineal gland.
pinna: The expanded, projecting portion of the external ear that surrounds the external auditory meatus.
pinocytosis (pin-ō-sī-TŌ-sis): The introduction of fluids into the cytoplasm by enclosing them in membranous vesicles at the cell surface.
pituitary gland: The "master gland," situated in the sella turcica of the sphenoid bone and connected to the hypothalamus by the infundibulum; includes the posterior lobe (pars nervosa) and the anterior lobe (pars intermedia and pars distalis).
placenta (pla-SENT-a): A complex structure in the uterine wall that permits diffusion between the fetal and maternal circulatory systems; also called the afterbirth.
placentation (pla-sen-TĀ-shun): Formation of a functional placenta following implantation of a blastocyst in the endometrium.
plantar: Referring to the sole of the foot.
plasma (PLAZ-mah): The fluid ground substance of whole blood; what remains after the cells have been removed from a sample of whole blood.
plasma cell: Activated B cells that secrete antibodies.
plasmalemma (plaz-ma-LEM-a): Cell membrane.
platelets (PLĀT-lets): Small packets of cytoplasm that contain enzymes important in the clotting response; manufactured in the bone marrow by cells called **megakaryocytes.**
pleura (PLOO-ra): The serous membrane lining the pleural cavities.
pleural cavities: Subdivisions of the thoracic cavity that contain the lungs.
pleuritis (ploor-Ī-tis): Inflammation of the pleura.
plexus (PLEK-sus): A complex interwoven network of peripheral nerves or blood vessels.
plica (PLĪ-ka): A permanent transverse fold in the wall of the small intestine.
pneumotaxic center (nū-mō-TAKS-ik): A center in the reticular formation of the pons that regulates the activities of the apneustic and respiratory rhythmicity centers to adjust the pace of respiration.
pneumothorax (nū-mō-THŌ-raks): The introduction of air into the pleural cavity.
podocyte (POD-ō-sīt): A cell whose processes surround the glomerular capillaries and assist in the filtration process.
polar body: A nonfunctional packet of cytoplasm containing chromosomes eliminated from an oocyte during meiosis.
pollex (POL-eks): The thumb.
polycythemia (po-lē-sī-THĒ-mē-a): An unusually high hematocrit due to the presence of excess numbers of formed elements, especially RBCs.
polymorph: Polymorphonuclear leukocyte; a neutrophil.
polypeptide: A chain of amino acids strung together by peptide bonds; those containing over 100 peptides are called proteins.
polysaccharide (pol-ē-SAK-ah-rīd): A complex sugar, such as glycogen or a starch.
polysynaptic reflex: A reflex with interneurons interposed between the sensory fiber and the motor neuron(s).
polyuria (pol-ē-Ū-rē-a): Excessive urine production.
pons: The portion of the metencephalon anterior to the cerebellum.
popliteal (pop-lit-Ē-al): Pertaining to the back of the knee.
porphyrins (POR-fi-rinz): Ring-shaped molecules that form the basis for important respiratory and metabolic pigments, including heme and the cytochromes.
porta hepatis: A region of mesentery between the duodenum and liver that contains the hepatic artery, the hepatic portal vein, and the common bile duct.
positive feedback: Mechanism that increases a deviation from normal limits following an initial stimulus.
postcentral gyrus: The primary sensory cortex, where touch, vibration, pain, temperature, and taste sensations arrive and are consciously perceived.
posterior: Toward the back; dorsal.
postganglionic neuron: An autonomic neuron in a peripheral ganglion, whose activities control peripheral effectors.
postovulatory phase: The secretory phase of the menstrual cycle.
precentral gyrus: The primary motor cortex on a cerebral hemisphere, located rostral to the central sulcus.
prefrontal cortex: Rostral portion of each cerebral hemisphere thought to be involved with higher intellectual functions, predictions, calculations, and so forth.
preganglionic neuron: Visceral motor neuron inside the CNS whose output controls one or more ganglionic motor neurons in the PNS.
premolars: Bicuspids; teeth with flattened occlusal surfaces located anterior to the molar teeth.
premotor cortex: Motor association area between the precentral gyrus and the prefrontal area.
preoptic nucleus: Hypothalamic nucleus that coordinates thermoregulatory activities.
preovulatory phase: A portion of the menstrual cycle; period of estrogen-induced repair of the functional zone of the endometrium through the growth and proliferation of epithelial cells in the glands not lost during menses.
prepuce (PRĒ-pūs): Loose fold of skin that surrounds the glans penis (males) or the clitoris (females).
preputial glands (prē-PŪ-shal): Glands on the inner surface of the prepuce that produce a viscous, odorous secretion, called **smegma.**
presbyopia: Farsightedness; an inability to accommodate for near vision.
prevertebral ganglion: *See* **collateral ganglion.**
prime mover: A muscle that performs a specific action.
proenzyme: An inactive enzyme secreted by an epithelial cell.
progesterone (prō-JES-ter-ōn): The most important progestin secreted by the corpus luteum following ovulation.
progestins (prō-JES-tinz): Steroid hormones structurally related to cholesterol.
prognosis: A prediction concerning the possibility or time course of recovery from a specific disease.
projection fibers: Axons carrying information from the thalamus to the cerebral cortex.
prolactin (prō-LAK-tin): Hormone that stimulates functional development of the mammary gland in females; secreted by the anterior lobe of the pituitary gland.
prolapse: The abnormal descent or protrusion of a portion of an organ, such as the vagina or anorectal canal.
proliferative phase: *See* **preovulatory phase.**
pronation (prō-NĀ-shun): Rotation of the forearm that makes the palm face posteriorly.
pronucleus: Enlarged egg or sperm nucleus that forms after fertilization but before amphimixis.
properdin: Complement factor that prolongs and enhances non-antibody- dependent complement binding to bacterial cell walls.
prophase (PRŌ-fāz): The initial phase of mitosis, characterized by the appearance of chromosomes, breakdown of the nuclear membrane, and formation of the spindle apparatus.
proprioception (prō-prē-ō-SEP-shun): Awareness of the positions of bones, joints, and muscles.
prostaglandin (pros-tah-GLAN-din): Lipoid secreted by one cell that alters the metabolic activities or sensitivities of adjacent cells; sometimes called "local hormones."

prostate gland (PROS-tāt): Accessory gland of the male reproductive tract, contributing roughly one-third of the volume of semen.
prostatectomy (pros-ta-TEK-tō-mē): Surgical removal of the prostate.
prostatitis (pros-ta-TĪ-tis): Inflammation of the prostate.
prosthesis: An artificial substitute for a body part.
protease: *See* **proteinase**.
protein: A large polypeptide with a complex structure.
proteinase: An enzyme that breaks down proteins into peptides and amino acids.
proteinuria (pro-ten-Ū R-ē-a): Abnormal amounts of protein in the urine.
proteoglycan (prō-tē-ō-GLĪ-kan): Compound containing a large polysaccharide complex attached to a relatively small protein; examples include hyaluronic acid and chondroitin sulfate.
prothrombin: Circulating proenzyme of the common pathway of the clotting system; converted to thrombin by the enzyme thromboplastin.
proton: A fundamental particle bearing a positive charge.
protraction: Movement anteriorly in the horizontal plane.
proximal: Toward the attached base of an organ or structure.
proximal convoluted tubule: The portion of the nephron between Bowman's capsule and the loop of Henle; the major site of active reabsorption from the filtrate.
pruritis (prū-RĪ-tus): Itching.
pseudopodia (sū-dō-PŌ-dē-a): Temporary cytoplasmic extensions typical of mobile or phagocytic cells.
pseudostratified epithelium: An epithelium containing several layers of nuclei, but whose cells are all in contact with the underlying basement membrane.
psoriasis (sō-RĪ-a-sis): Skin condition characterized by excessive keratin production and the formation of dry, scaly patches on the body surface.
psychosomatic condition: An abnormal physiological state with a psychological origin.
puberty: Period of rapid growth, sexual maturation, and the appearance of secondary sexual characteristics; usually occurs at ages 10–15.
pubic symphysis: Fibrocartilaginous amphiarthrosis between the pubic bones of the coxae.
pubis (PŪ-bis): The anterior, inferior component of the coxa.
pudendum (pū-DEN-dum): The external genitalia.
pulmonary circuit: Blood vessels between the pulmonary semilunar valve of the right ventricle and the entrance to the left atrium; the blood circulation through the lungs.
pulmonary ventilation: Movement of air in and out of the lungs.
pulp cavity: Internal chamber in a tooth, containing blood vessels, lymphatics, nerves, and the cells that maintain the dentin.
pulpitis (pul-PĪ-tis): Inflammation of the tissues of the pulp cavity.
pupil: The opening in the center of the iris through which light enters the eye.
purine: An N compound with a ring-shaped structure; examples include adenine and guanine, two nitrogen bases common in nucleic acids.
Purkinje cell (pur-KIN-jē): Large, branching neuron of the cerebellar cortex.
Purkinje fibers: Specialized conducting cardiocytes in the ventricles.
pus: An accumulation of debris, fluid, dead and dying cells, and necrotic tissue.
putamen (pū-TĀ-men): Thalamic nucleus involved in the integration of sensory information prior to projection to the cerebral hemispheres.
P wave: Deflection of the ECG corresponding to atrial depolarization.
pyelogram (PĪ-el-ō-gram): A radiographic image of the kidneys and ureters.
pyelonephritis (pī-e-lō-nef-RĪ-tis): Inflammation of the kidneys.
pyloric sphincter (pī-LOR-ic): Sphincter of smooth muscle that regulates the passage of chyme from the stomach to the duodenum.
pylorus (pī-LOR-us): Gastric region between the body of the stomach and the duodenum; includes the pyloric sphincter.
pyrexia (pī-REK-se-a): A fever.
pyrimidine: An N compound with a ring-shaped structure; examples include cytosine, thymine, and uracil, nitrogen bases common in nucleic acids.
pyruvic acid (pī-RŪ-vik): 3-carbon compound produced by glycolysis.
quadriplegia: Paralysis of the upper and lower limbs.
radiodensity: Relative resistance to the passage of X-rays.
radiographic techniques: Methods of visualizing internal structures using various forms of radiational energy.
radiopaque: Having a relatively high radiodensity.
rami communicantes: Axon bundles that link the spinal nerves with the ganglia of the sympathetic chain.
ramus: A branch.
raphe (RĀ-fē): A seam.
receptor field: The area monitored by a single sensory receptor.
recessive gene: An allele that will affect the phenotype only when the individual is homozygous for that trait.
rectal columns: Longitudinal folds in the walls of the anorectal canal.
rectouterine pouch (rek-tō-Ū-te-rin): Peritoneal pocket between the anterior surface of the rectum and the posterior surface of the uterus.
rectum (REK-tum): The last 15 cm (6 in.) of the digestive tract.
rectus: Straight.
red blood cell: *See* **erythrocyte**.
reduction: The gain of hydrogen atoms or electrons, or the loss of an oxygen molecule.
reductional division: The first meiotic division, which reduces the chromosome number from 46 to 23.
reflex: A rapid, automatic response to a stimulus.
reflex arc: The receptor, sensory neuron, motor neuron, and effector involved in a particular reflex; interneurons may or may not be present, depending on the reflex considered.
refraction: The bending of light rays as they pass from one medium to another.
refractory period: Period between the initiation of an action potential and the restoration of the normal resting potential; over this period the membrane will not respond normally to stimulation.
relaxation phase: The period following a contraction when the tension in the muscle fiber returns to resting levels.
relaxin: Hormone that loosens the pubic symphysis; a hormone secreted by the placenta.
renal: Pertaining to the kidneys.
renal corpuscle: The initial portion of the nephron, consisting of an expanded chamber that encloses the glomerulus.
renin: Enzyme released by the juxtaglomerular cells when renal blood pressure or pO_2 declines; converts angiotensinogen to angiotensin I.
rennin: Gastric enzyme that breaks down milk proteins.
replication: Duplication.
repolarization: Movement of the transmembrane potential away from $+$ mV values and toward the resting potential.
residual volume: Amount of air remaining in the lungs after maximum forced expiration.
respiration: Exchange of gases between living cells and the environment; includes pulmonary ventilation, external respiration, internal respiration, and cellular respiration.
respiratory minute volume: The amount of air moved in and out of the respiratory system each minute.
resting potential: The transmembrane potential of a normal cell under homeostatic conditions.
rete (RĒ-tē): An interwoven network of blood vessels or passageways.
reticular activating center: Mesencephalic portion of the reticular formation responsible for arousal and the maintenance of consciousness.
reticular formation: Diffuse network of gray matter that extends the entire length of the brain stem.
reticulocytes (re-TIK-ū-lō-sīts): The last stage in the maturation of red blood cells; normally the youngest red blood cells present in the blood.
reticulospinal tracts: Descending tracts that carry involuntary motor commands issued by neurons of the reticular formation.
retina: The innermost layer of the eye, lining the vitreous chamber; also known as the neural tunic.
retinene (RET-i-nēn): Visual pigment derived from vitamin A.
retraction: Movement posteriorly in the horizontal plane.
retroflexion (ret-rō-FLEK-shun): A posterior tilting of the uterus that has no clinical significance.
retrograde flow (RET-rō-grād): Transport of materials from the telodendria to the soma of a neuron.
retroperitoneal (re-trō-per-i-tō-NĒ-al): Situated behind or outside of the peritoneal cavity.
reverberation: Positive feedback along a chain of neurons, so that they remain active once stimulated.
rheumatism (ROO-ma-tizm): A condition characterized by pain in muscles, tendons, bones, or joints.
Rh factor: Agglutinogen that may be present (Rh-positive) or absent (Rh-negative) from the surfaces of red blood cells.
rhizotomy: Surgical transection of a dorsal root, usually performed to relieve pain.
rhodopsin (rō-DOP-sin): The visual pigment found in the membrane discs of the distal segments of rods.
rhythmicity center: Medullary center responsible for the basic pace of respiration; includes inspiratory and expiratory centers.
ribonucleic acid (RNA) (rī-bō-nū-KLĀ-ik): A nucleic acid consisting of a chain of nucleotides that contain the sugar ribose and the nitrogen bases adenine, guanine, cytosine, and uracil.
ribosome: An organelle containing rRNA and proteins, that is essential to mRNA translation and protein synthesis.
right lymphatic duct: Lymphatic vessel delivering lymph from the right side of the head, neck, and chest to the venous system via the right subclavian vein.
rigor mortis: Extended muscular contraction and rigidity that occurs after death, as the result of calcium ion release from the SR and the exhaustion of cytoplasmic ATP reserves.

rod: Photoreceptor responsible for vision under dimly lit conditions.
rostral: Toward the nose; used when referring to relative position inside the skull.
rough endoplasmic reticulum (RER): A membranous organelle that is a site of protein synthesis and storage.
rouleau/rouleaux (roo-LŌ): A stack of red blood cells.
round window: An opening in the bony labyrinth of the inner ear that exposes the membranous wall of the scala tympani to the air of the middle ear cavity.
rubrospinal tracts: Descending tracts that carry involuntary motor commands issued by the red nucleus of the mesencephalon.
Ruffini corpuscles (ru-FĒ-nē): Receptors sensitive to tension and stretch in the dermis of the skin.
rugae (ROO-gē): Mucosal folds in the lining of the empty stomach that disappear as gastric distension occurs.
saccule (SAK-ūl): A portion of the vestibular apparatus of the inner ear, responsible for static equilibrium.
sagittal plane: Sectional plane that divides the body into left and right portions.
salivatory nucleus (SAL-i-va-tōr-ē): Medullary nucleus that controls the secretory activities of the salivary glands.
saltatory conduction: Relatively rapid conduction of a nerve impulse between successive nodes of a myelinated axon.
sarcolemma: The cell membrane of a muscle cell.
sarcoma (sar-KŌ-ma): A tumor of connective tissues.
sarcomere: The smallest contractile unit of a striated muscle cell.
sarcoplasm: The cytoplasm of a muscle cell.
scala media: The central, endolymph-filled chamber of the inner ear; *see* cochlear duct.
scala tympani: The perilymph-filled chamber of the inner ear below the basilar membrane; pressure changes here distort the round window.
scala vestibuli: The perilymph-filled chamber of the inner ear above the vestibular membrane; pressure changes here result from distortions of the oval window.
scapula (SKAP-ū-la): The shoulder blade.
scar tissue: Thick, collagenous tissue that forms at an injury site.
Schlemm, canal of: Passageway that delivers aqueous humor from the anterior chamber of the eye to the venous circulation.
Schwann cells: Glial cells responsible for the neurilemma that surrounds axons in the PNS.
sciatica (sī-AT-i-ka): Pain resulting from compression of the roots of the sciatic nerve.
sciatic nerve (sī-AT-ik): Nerve innervating the posteromedial portions of the thigh and lower leg.
sclera (SKLER-a): The fibrous, outer layer of the eye forming the white area of the anterior surface; a portion of the fibrous tunic of the eye.
sclerosis: A hardening and thickening that often occurs secondary to tissue inflammation.
scoliosis (skō-lē-Ō-sis): An abnormal, exaggerated lateral curvature of the spine.
scrotum (SKRŌ-tum): Loose-fitting, fleshy pouch that encloses the testes of the male.
sebaceous glands (sē-BĀ-shus): Glands that secrete sebum, usually associated with hair follicles.
sebum (SĒ-bum): A waxy secretion that coats the surfaces of hairs.
secondary sex characteristics: Physical characteristics that appear at puberty in response to sex hormones, but that are not involved in the production of gametes.
secretin (sē-KRĒ-tin): Duodenal hormone that stimulates pancreatic buffer secretion and inhibits gastric activity.
semen (SĒ-men): Fluid ejaculate containing spermatozoa and the secretions of accessory glands of the male reproductive tract.
semicircular ducts: Tubular components of the vestibular apparatus responsible for dynamic equilibrium.
semilunar valve: A three-cusped valve guarding the exit from one of the cardiac ventricles; includes the pulmonary and aortic valves.
seminal vesicles (SEM-i-nal): Glands of the male reproductive tract that produce roughly 60 percent of the volume of semen.
seminiferous tubules (se-mi-NIF-e-rus): Coiled tubules where sperm production occurs in the testis.
senescence: Aging.
septae (SEP-tē): Partitions that subdivide an organ.
serosa: *See* **serous membrane**.
serotonin (ser-ō-TŌ-nin): A neurotransmitter in the CNS; a compound that enhances inflammation, released by activated mast cells and basophils.
serous cell: A cell that produces a watery secretion containing high concentrations of enzymes.
serous membrane: A squamous epithelium and the underlying loose connective tissue; the lining of the pericardial, pleural, and peritoneal cavities.
serum: Blood plasma from which clotting agents have been removed.
sesamoid bone: A bone that forms in a tendon.
sigmoid colon (SIG-moid): The S-shaped 8-inch portion of the colon between the descending colon and the rectum.
sign: A clinical term for objective evidence of the presence of a disease.
simple epithelium: An epithelium containing a single layer of cells above the basement membrane.
sinus: A chamber or hollow in a tissue; a large, dilated vein.
sinusitis: Inflammation of a nasal sinus.
sinusoid (SĪ-nus-oid): An extensive network of vessels found in the liver, adrenal cortex, spleen, and pancreas; similar in histological structure to capillaries.
skeletal muscle: A contractile organ of the muscular system.
skeletal muscle tissue: Contractile tissue dominated by skeletal muscle fibers; characterized as striated, voluntary muscle.
sliding filament theory: The concept that a sarcomere shortens as the thick and thin filaments slide past one another.
small intestine: The duodenum, jejunum, and ileum; the digestive tract between the stomach and large intestine.
smegma (SMEG-ma): Secretion of the preputial glands of the penis or clitoris.
smooth endoplasmic reticulum: Membranous organelle where lipid and carbohydrate synthesis and storage occur.
smooth muscle tissue: Muscle tissue found in the walls of many visceral organs; characterized as nonstriated, involuntary muscle.
soft palate: Fleshy posterior extension of the hard palate, separating the nasopharynx from the oral cavity.
sole: The inferior surface of the foot.
solute: Material dissolved in a solution.
solution: A fluid containing dissolved materials.
solvent: The fluid component of a solution.
soma (SŌ-ma): Body.
somatic (sō-MAT-ik): Pertaining to the body.
somatic nervous system: System of nerve fibers that run from the central nervous system to the muscles of the skeleton.
somatomedins: Compounds stimulating tissue growth, released by the liver following GH secretion.
somatostatin: GH-IH, a hypothalamic regulatory hormone that inhibits GH secretion by the anterior pituitary.
somatotropin: Growth hormone, produced by the anterior pituitary in response to GH-RH.
sperm: *See* **spermatozoa**.
spermatic cord: Spermatic vessels, nerves, lymphatics, and the ductus deferens, extending between the testes and the proximal end of the inguinal canal.
spermatids (SPER-ma-tidz): The product of meiosis in the male, cells that differentiate into spermatozoa.
spermatocyte (sper-MAT-ō-sīt): Cells of the seminiferous tubules that are engaged in meiosis.
spermatogenesis (sper-ma-tō-JEN-e-sis): Sperm production.
spermatogonia (sper-ma-tō-GŌ-nē-a): Stem cells whose mitotic divisions give rise to other stem cells and spermatocytes.
spermatozoon/spermatozoa (sper-ma-tō-ZŌ-on): A sperm cell, the male gamete.
spermicide: Compound toxic to sperm cells, sometimes used as a contraceptive method.
spermiogenesis: The process of spermatid differentiation that leads to the formation of physically mature spermatozoa.
sphincter (SFINK-ter): Muscular ring that contracts to close the entrance or exit of an internal passageway.
spina bifida (SPĪ-na BĪ-fi-da): A developmental abnormality in which the vertebral laminae fail to unite at the midline; the entire vertebral column and skull may be affected in severe cases.
spinal meninges (men-IN-jēz): Specialized membranes that line the vertebral canal and provide protection, stabilization, nutrition, and shock absorption to the spinal cord.
spinal nerve: One of 31 pairs of nerves that originate on the spinal cord from anterior and posterior roots.
spindle apparatus: A muscle spindle (intrafusal fiber) and its sensory and motor innervation.
spinocerebellar tracts: Ascending tracts carrying sensory information to the cerebellum.
spinothalamic tracts: Ascending tracts carrying poorly localized touch, pressure, pain, vibration, and temperature sensations to the thalamus.
spinous process: Prominent posterior projection of a vertebra, formed by the fusion of two laminae.
splanchnic nerves: Preganglionic (myelinated) sympathetic nerves that end in one of the collateral ganglia.
spleen: Lymphatic organ important for red blood cell phagocytosis, immune response, and lymphocyte production.
splenectomy (splē-NEK-tō-mē): Surgical removal of the spleen.
sprain: Forceful distortion of an articulation that produces damage to the capsule, ligaments, or tendons but not dislocation.

sputum (SPŪ-tum): Viscous mucus ejected from the mouth after transport to the pharynx by the mucus escalator of the respiratory tract.
squama: A broad, flat surface.
squamous (SKWĀ-mus): Flattened.
squamous epithelium: An epithelium whose superficial cells are flattened and platelike.
stapedius (stā-PĒ-dē-us): A muscle of the middle ear whose contraction tenses the auditory ossicles and reduces the forces transmitted to the oval window.
stapes (STĀ-pēz): The auditory ossicle attached to the tympanic membrane.
statoconia (otoliths) (statō-KŌ-nē-a): Aggregations of calcium carbonate crystals in a gelatinous membrane that sits above one of the maculae of the vestibular apparatus.
stenosis (ste-NŌ-sis): A constriction or narrowing of a passageway.
stereocilia: Elongate microvilli characteristic of the epithelium of the epididymis and portions of the ductus deferens.
steroid: A ring-shaped lipid structurally related to cholesterol.
stimulus: An environmental alteration that produces a change in cellular activities; often used to refer to events that alter the transmembrane potentials of excitable cells.
stratified: Containing several layers.
stratum (STRĀ-tum): Layer.
stratum corneum (KŌR-nē-um): Layers of flattened, dead, keratinized cells covering the epidermis of the skin.
stretch receptors: Sensory receptors that respond to stretching of the surrounding tissues.
stroma: The connective tissue framework of an organ, as distinguished from the functional cells (parenchyma) of that organ.
subarachnoid space: Meningeal space containing CSF; the area between the arachnoid membrane and the pia mater.
subclavian (sub-CLĀ-vē-an): Pertaining to the region under the clavicle.
subcutaneous layer: The layer of loose connective tissue below the dermis; also called the **hypodermis** or **superficial fascia**.
sublingual salivary glands (sub-LING-gwal): Mucus-secreting salivary glands situated under the tongue.
subluxation (sub-luks-Ā-shun): A partial dislocation of a joint.
submandibular salivary glands: Salivary glands nestled in depressions on the medial surfaces of the mandible; salivary glands that produce a mixture of mucins and enzymes (salivary amylase).
submucosa (sub-mū-KŌ-sa): Region between the muscularis mucosae and the muscularis externa.
submucosal glands: Mucous glands in the submucosa of the duodenum.
subserous fascia: Loose connective tissue layer beneath the serous membrane lining the ventral body cavity.
substantia nigra: A nucleus in the midbrain that is responsible for negative feedback control of the basal nuclei.
substrate: A participant (product or reactant) in an enzyme-catalyzed reaction.
sulcus (SUL-kus): A groove or furrow.
summation: Temporal or spatial addition of stimuli.
superficial fascia: *See* **subcutaneous layer**.
superior: Directional reference meaning above.
superior vena cava: The vein that carries blood from the parts of the body above the heart to the right atrium.
supination (su-pin-Ā-shun): Rotation of the forearm so that the palm faces anteriorly.
supine (sū-PĪN): Lying face up, with palms facing anteriorly.
suppressor T cells: Lymphocytes that inhibit B cell activation and plasma cell secretion of antibodies.
suprarenal gland (sū-pra-RĒ-nal): *See* **adrenal gland**.
surfactant (sur-FAK-tant): Lipid secretion that coats alveolar surfaces and prevents their collapse.
sustentacular cells (sus-ten-TAK-ū-lar): Supporting cells of the seminiferous tubules of the testis, responsible for the differentiation of spermatids, the maintenance of the blood-testis barrier, and the secretion of inhibin.
sutural bones: Irregular bones that form in fibrous tissue between the flat bones of the developing cranium; also called **Wormian bones**.
suture: Fibrous joint between flat bones of the skull.
sympathectomy (sim-path-EK-tō-mē): Transection of the sympathetic innervation to a region.
sympathetic division: Division of the autonomic nervous system responsible for "fight or flight" reactions; concerned primarily with the elevation of metabolic rate and increased alertness.
symphysis: A fibrous amphiarthrosis, such as those between adjacent vertebrae or between the pubic bones of the coxae.
symptom: Clinical term for an abnormality of function due to the presence of disease.
synapse (SIN-aps): Site of communication between a nerve cell and some other cell; if the other cell is not a neuron, the term neuroeffector junction is often used.
synarthrosis (sin-ar-THRŌ-sis): A joint that does not permit relative movement between the articulating elements.
synchondrosis (sin-kon-DRŌ-sis): A cartilaginous synarthrosis, such as the articulation between the epiphysis and diaphysis of a growing bone.
syncope: A sudden, transient loss of consciousness; a faint.
syncytial trophoblast (sin-SISH-al): Multinucleate cytoplasmic layer that covers the blastocyst; the layer responsible for uterine erosion and implantation.
syncytium: A multinucleate mass of cytoplasm, produced by the fusion of cells or repeated mitoses without cytokinesis.
syndesmosis (sin-dez-MŌ-sis): A fibrous amphiarthrosis.
syndrome: A discrete set of symptoms that occur together.
syneresis (si-NER-ē-sis): Clot retraction.
synergist (SIN-er-jist): A muscle that assists a prime mover in performing its primary action.
synostosis (sin-os-TŌ-sis): A synarthrosis formed through the fusion of the articulating elements.
synovial cavity (si-NŌ-vē-ul): Fluid-filled chamber in a diarthrodial joint.
synovial fluid: Substance secreted by synovial membranes that lubricates joints.
synovial membrane: An incomplete layer of fibroblasts confronting the synovial cavity, plus the underlying loose connective tissue.
synthesis (SIN-the-sis): Manufacture; anabolism.
system: An interacting group of organs that performs one or more specific functions.
systemic circuit: Vessels between the aortic semilunar valve and the entrance to the right atrium; the circulatory system other than vessels of the pulmonary circuit.
systole (SIS-tō-lē): The period of cardiac contraction.
systolic pressure: Peak arterial pressure measured during ventricular systole.
tachycardia (tak-ē-KAR-dē-a): An abnormally rapid heart rate.
tactile: Pertaining to the sense of touch.
tactile corpuscles: Touch receptors located within dermal papillae adjacent to the basement membrane of the epidermis; also called *Meisner's corpuscles*.
tactile discs: Sensory nerve endings that contact special receptors called Merkel cells, located within the deeper layers of the epidermis; also called *Merkel's discs*.
taenia coli (TĒ-nē-a KŌ-lī): Three longitudinal bands of smooth muscle in the muscularis externa of the colon.
tarsus: The ankle.
T cells: Lymphocytes responsible for cellular immunity, and for the coordination and regulation of the immune response; includes regulatory T cells (helpers and suppressors) and cytotoxic (killer) T cells.
tears: Fluid secretions of the lacrimal glands that bathe the anterior surfaces of the eyes.
tectorial membrane (tek-TŌR-ē-al): Gelatinous membrane suspended over the hair cells of the organ of Corti.
tectospinal tracts: Descending extrapyramidal tracts carrying involuntary motor commands issued by the colliculi.
tectum: The roof of the mesencephalon of the brain.
telencephalon (tel-en-SEF-a-lon): The forebrain or cerebrum, including the cerebral hemispheres, the internal capsule, and the cerebral nuclei.
telodendria (te-lō-DEN-drē-a): Terminal axonal branches that end in synaptic knobs.
telophase (TEL-ō-fāz): The final stage of mitosis, characterized by the disappearance of the spindle apparatus, the reappearance of the nuclear membrane and the disappearance of the chromosomes, and the completion of cytokinesis.
temporal: Pertaining to time (temporal summation) or pertaining to the temples (temporal bone).
tendinitis: Painful inflammation of a tendon.
tendon: A collagenous band that connects a skeletal muscle to an element of the skeleton.
tendon organ: Receptor sensitive to tension in a tendon.
tentorium cerebelli (ten-TŌR-ē-um ser-e-BEL-ē): Dural partition that separates the cerebral hemispheres from the cerebellum.
teratogen (ter-AT-ō-jen): Stimulus that causes developmental defects.
teres: Long and round.
terminal: Toward the end.
tertiary follicle: A mature ovarian follicle, containing a large, fluid-filled chamber.
testes (TES-tēz): The male gonads, sites of gamete production and hormone secretion.
testosterone (tes-TOS-te-rōn): The principal androgen produced by the interstitial cells of the testes.
tetanic contraction: Sustained skeletal muscle contraction due to repeated stimulation at a frequency that prevents muscle relaxation.
tetanus: A tetanic contraction; also used to refer to a disease state resulting from the stimulation of muscle cells by bacterial toxins.
tetrad (TET-rad): Paired, duplicated chromosomes visible at the start of meiosis I.
tetraiodothyronine (tet-ra-ī-ō-dō-THĪ-rō-nēn): T_4, or thyroxine, a thyroid hormone.
thalamus: The walls of the diencephalon.
thalassemia (thal-ah-SĒ-mē-ah): A hereditary disorder affecting hemoglobin synthesis and producing anemia.

theory: A hypothesis that makes valid predictions, as demonstrated by evidence that is testable, unbiased, and repeatable.
therapy: Treatment of disease.
thermogenesis (ther-mō-JEN-e-sis): Heat production.
thermography: Diagnostic procedure involving the production of an infrared image.
thermoreception: Sensitivity to temperature changes.
thermoregulation: Homeostatic maintenance of body temperature.
thick filament: A myosin filament in a skeletal or cardiac muscle cell.
thin filament: An actin filament in a skeletal or cardiac muscle cell.
thoracoabdominal pump (thō-ra-kō-ab-DOM-i-nal): Changes in the intrapleural pressures during the respiratory cycle that assist the venous return to the heart.
thoracolumbar division (thō-ra-kō-LUM-bar): The sympathetic division of the ANS.
thorax: The chest.
threshold: The transmembrane potential at which an action potential begins.
thrombin (THROM-bin): Enzyme that converts fibrinogen to fibrin.
thrombocytes (THROM-bō-sīts): *See* **platelets.**
thrombocytopenia (throm-bō-sī-tō-PĒ-nē-ah): Abnormally low platelet count in the circulating blood.
thromboembolism (throm-bō-EM-bō-lizm): Occlusion of a blood vessel by a drifting blood clot.
thromboplastin: Enzyme that converts prothrombin to thrombin; enzyme formed by the intrinsic or extrinsic clotting pathways.
thrombus: A blood clot.
thymine: A pyrimidine found in DNA.
thymosin (THĪ-mō-sin): Thymic hormone essential to the development and differentiation of T cells.
thymus: Lymphatic organ, site of T cell formation.
thyroglobulin (thī-rō-GLOB-ū-lin): Circulating transport globulin that binds thyroid hormones.
thyroid gland: Endocrine gland whose lobes sit lateral to the thyroid cartilage of the larynx.
thyroid hormones: Thyroxine (T_4) and triiodothyronine (T_3), hormones of the thyroid gland; hormones that stimulate tissue metabolism, energy utilization, and growth.
thyroid-stimulating hormone (TSH): Anterior pituitary hormone that triggers the secretion of thyroid hormones by the thyroid gland.
thyroxine (TX) (thī-ROKS-in): A thyroid hormone (T_4).
tibia (TIB-ē-a): The large, medial bone of the leg.
tidal volume: The volume of air moved in and out of the lungs during a normal quiet respiratory cycle.
tissue: A collection of specialized cells and cell products that perform a specific function.
tonsil: A lymphatic nodule beneath the epithelium of the pharynx; includes the palatine, pharyngeal, and lingual tonsils.
topical: Applied to the body surface.
trabecula (tra-BEK-ū-la): A connective tissue partition that subdivides an organ.
trabeculae carneae (tra-BEK-ū-lē CAR-nē-ē): Muscular ridges projecting from the walls of the ventricles of the heart.
trachea (TRĀ-kē-a): The windpipe, an airway extending from the larynx to the primary bronchi.
tracheal ring: C-shaped supporting cartilage of the trachea.
tracheostomy (trā-kē-OS-tō-mē): Surgical opening of the anterior tracheal wall to permit airflow.
trachoma: An infectious disease of the conjunctiva and cornea.
tract: A bundle of axons inside the CNS.
tractotomy: The surgical transection of a tract, sometimes used to relieve pain.
transcription: The encoding of genetic instructions on a strand of mRNA.
transdermal medication: Administration of medication by absorption through the skin.
transection: To sever or cut in the transverse plane.
transfusion: Transfer of blood from a donor directly into the bloodstream of another person.
transient ischemic attack: A temporary loss of consciousness due to the occlusion of a small blood vessel in the brain.
translation: The process of peptide formation using the instructions carried by an mRNA strand.
transmembrane potential: The potential difference, in millivolts, measured across the cell membrane; a potential difference that results from the uneven distribution of positive and negative ions across a cell membrane.
transudate (TRANS-ū-dāt): Fluid that diffuses across a serous membrane and lubricates opposing surfaces.
treppe (TREP-ē): "Staircase" increase in tension production following repeated stimulation of a muscle, even though the muscle is allowed to complete each relaxation phase.
triad (liver): The combination of branches of the hepatic duct, the hepatic portal vein, and the hepatic artery, found at each corner of a liver lobule.
triad (muscle cell): The combination of a T tubule and two cisternae of the sarcoplasmic reticulum.
tricuspid valve (trī-KUS-pid): The right atrioventricular valve that prevents backflow of blood into the right atrium during ventricular systole.
trigeminal nerve (trī-JEM-i-nal): Cranial nerve V, responsible for providing sensory information from the lower portions of the face, including the upper and lower jaws, and delivering motor commands to the muscles of mastication.
triglyceride (trī-GLIS-e-rīd): A lipid composed of a molecule of glycerol attached to three fatty acids.
trigone (TRĪ-gōn): Triangular region of the bladder bounded by the exits of the ureters and the entrance to the urethra.
triiodothyronine: T_3, one of the thyroid hormones.
trisomy: The abnormal possession of three copies of a chromosome; trisomy 21 is responsible for the Down syndrome.
trochanters (trō-KAN-terz): Large processes near the head of the femur.
trochlea (TROK-lē-a): A pulley.
trochlear nerve (TROK-lē-ar): Cranial nerve IV, controlling the superior oblique muscle of the eye.
trophoblast (TRŌ-fō-blast): Superficial layer of the blastocyst that will be involved with implantation, hormone production, and placenta formation.
troponin/tropomyosin (trō-PŌ-nin) (trō-pō-MI-ō-sin): Proteins on the thin filaments that mask the active sites in the absence of free calcium ions.
trunk: The thoracic and abdominopelvic regions.
trypsin (TRIP-sin): One of the pancreatic proteases.
trypsinogen: The inactive proenzyme secreted by the pancreas and converted to trypsin in the duodenum.
T tubules: Transverse, tubular extensions of the sarcolemma that extend deep into the sarcoplasm to contact cisternae of the sarcoplasmic reticulum.
tuberculum (tū-BER-kū-lum): A small, localized elevation on a bony surface.
tuberosity: A large, roughened elevation on a bony surface.
tubulin: Protein subunit of microtubules.
tumor: A tissue mass formed by the abnormal growth and replication of cells.
tunica (TŪ-ni-ka): A layer or covering; in blood vessels: t. externa, the outermost layer of connective tissue fibers that stabilizes the position of the vessel; t. intima, the innermost layer, consisting of the endothelium plus an underlying elastic membrane; t. media, a middle layer containing collagen, elastin, and smooth muscle fibers in varying proportions.
turbinates: *See* **conchae.**
T wave: Deflection of the ECG corresponding to ventricular repolarization.
twitch: A single contraction/relaxation cycle in a skeletal muscle.
tympanic membrane (tim-PAN-ik): Membrane that separates the external acoustic meatus from the middle ear; membrane whose vibrations are transferred to the auditory ossicles and ultimately to the oval window; the "eardrum."
ulcer: An area of epithelial sloughing associated with damage to the underlying connective tissues and vasculature.
ultrasound: Diagnostic visualization procedure that uses high-frequency sound waves.
umbilical cord (um-BIL-i-kal): Connecting stalk between the fetus and the placenta; contains the allantois, the umbilical arteries, and the umbilical vein.
umbilicus: The navel.
unicellular gland: Goblet cell.
unipennate muscle: A muscle whose fibers are all arranged on one side of the tendon.
unipolar neuron: A sensory neuron whose soma lies in a dorsal root ganglion or a sensory ganglion of a cranial nerve.
unmyelinated axon: Axon whose neurilemma does not contain myelin, and where continuous conduction occurs.
urachus (Ū-ra-kus): The middle umbilical ligament.
uracil: One of the pyrimidines characteristic of RNA.
uremia (ū-RĒ-mē-a): Abnormal condition caused by impaired kidney function, characterized by the retention of wastes and the disruption of many other organ systems.
ureters (ū-RĒ-terz): Muscular tubes, lined by transitional epithelium, that carry urine from the renal pelvis to the urinary bladder.
urethra (ū-RĒ-thra): A muscular tube that carries urine from the urinary bladder to the exterior.
urethritis: Inflammation of the urethra.
urinalysis: Analysis of the physical and chemical characteristics of the urine.
urinary bladder: Muscular, distensible sac that stores urine prior to micturition.
urination: The voiding of urine; micturition.
uterine (menstrual) cycle: Cyclical changes in the uterine lining that occur in reproductive-age women. Each uterine cycle, which occurs in response to circulating hormones (*see* **ovarian cycle**), lasts 21–35 days.
uterus (Ū-ter-us): Muscular organ of the female reproductive tract where implantation, placenta formation, and fetal development occur.
utricle (Ū-tri-kl): The largest chamber of the vestibular apparatus; contains a macula important for static equilibrium.

uvea: The vascular tunic of the eye.
uvula (Ū-vū-la): A dangling, fleshy extension of the soft palate.
vagina (va-JI-na): A muscular tube extending between the uterus and the vestibule.
vagus nerve: N X, the cranial nerve responsible for most (75%) of the parasympathetic preganglionic output from the CNS.
varicose veins (VAR-i-kōs): Distended superficial veins.
vasa vasorum: Blood vessels that supply the walls of large arteries and veins.
vascular: Pertaining to blood vessels.
vascularity: The blood vessels in a tissue.
vascular spasm: Contraction of the wall of a blood vessel at an injury site, a process that may slow the rate of blood loss.
vasoconstriction: A reduction in the diameter of arterioles due to contraction of smooth muscles in the tunica media; an event that elevates peripheral resistance, and that may occur in response to local factors, through the action of hormones, or from stimulation of the vasomotor center.
vasodilation (vaz-ō-dī-LĀ-shun): An increase in the diameter of arterioles due to the relaxation of smooth muscles in the tunica media; an event that reduces peripheral resistance, and that may occur in response to local factors, through the action of hormones, or following decreased stimulation of the vasomotor center.
vasomotion: Alterations in the pattern of blood flow through a capillary bed in response to changes in the local environment.
vasomotor center: Medullary center whose stimulation produces vasoconstriction and an elevation in peripheral resistance.
vein: Blood vessel carrying blood from a capillary bed toward the heart.
venae cavae (VĒ-nē CĀ-vē): The major veins delivering systemic blood to the right atrium.
ventilation: Air movement in and out of the lungs.
ventilatory rate: The respiratory rate.
ventral: Pertaining to the anterior surface.
ventricle (VEN-tri-kl): One of the large, muscular pumping chambers of the heart that discharges blood into the pulmonary or systemic circuits.
venules (VEN-ūlz): Thin-walled veins that receive blood from capillaries.
vermis (VER-mis): Midsagittal band of neural cortex on the surface of the cerebellum.
vertebral canal: Passageway that encloses the spinal cord, a tunnel bounded by the neural arches of adjacent vertebrae.
vertebral column: The cervical, thoracic, and lumbar vertebrae, the sacrum, and the coccyx.
vertebrochondral ribs: Ribs 8–10, false ribs connected to the sternum by shared cartilaginous bars.
vertebrosternal ribs: Ribs 1–7, true ribs connected to the sternum by individual cartilaginous bars.
vertigo: Dizziness.
vesicle: A membranous sac in the cytoplasm of a cell.
vestibular folds: Mucosal folds in the laryngeal walls that do not play a role in sound production; the false vocal cords.
vestibular membrane: The membrane that separates the scala media from the scala vestibuli of the inner ear.
vestibular nucleus: Processing center for sensations arriving from the vestibular apparatus; located near the border between the pons and medulla oblongata.
vestibule (VES-ti-būl): A chamber; in the inner ear, the term refers to the utricle, saccule, and semicircular ducts; also refers to (1) a region of the female external genitalia, (2) the space within the fleshy portion of the nose between the nostrils and the external nares, and (3) the space between the ventricular folds and the vocal folds of the larynx.
vestibulospinal tracts: Descending tracts of the extrapyramidal system, carrying involuntary motor commands issued by the vestibular nucleus to stabilize the position of the head.
villus: A slender projection of the mucous membrane of the small intestine.
virus: A pathogenic microorganism.
viscera: Organs in the ventral body cavity.
visceral: Pertaining to viscera or their outer coverings.
visceral smooth muscle tissue: Smooth muscle tissue forming sheets or layers in the walls of visceral organs; the cells may not be innervated, and the layers often show automaticity (rhythmic contractions).
viscosity: The resistance to flow exhibited by a fluid, due to molecular interactions within the fluid.
viscous: Thick, syrupy.
vital capacity: The maximum amount of air that can be moved in or out of the respiratory system; the sum of the inspiratory reserve, the expiratory reserve, and the tidal volume.
vitamin: An essential organic nutrient that functions as a coenzyme in vital enzymatic reactions.
vitreous humor: Gelatinous mass in the vitreous chamber of the eye.
vocal folds: Folds in the laryngeal wall containing elastic ligaments whose tension can be voluntarily adjusted; the true vocal cords, responsible for phonation.
voluntary: Controlled by conscious thought processes.
vulva (VUL-va): The female pudendum (external genitalia).
Wallerian degeneration: Disintegration of an axon and its myelin sheath distal to an injury site.
white blood cells: Leukocytes; the granulocytes and agranulocytes of the blood.
white matter: Regions inside the CNS that are dominated by myelinated axons.
white ramus: A nerve bundle containing the myelinated preganglionic axons of sympathetic motor neurons en route to the sympathetic chain or a collateral ganglion.
Wormian bones: *See* **sutural bones**.
xiphoid process (ZĪ-foid): Slender, inferior extension of the sternum.
Y chromosome: The sex chromosome whose presence indicates that the individual is a genetic male.
yolk sac: One of the three extraembryonic membranes, composed of an inner layer of endoderm and an outer layer of mesoderm.
Zeis, glands of (ZĪS): Enlarged sebaceous glands on the free edges of the eyelids.
zona fasciculata (ZŌ-na fa-sik-ū-LA-ta): Region of the adrenal cortex responsible for glucocorticoid secretion.
zona glomerulosa (glō-mer-ū-LŌ-sa): Region of the adrenal cortex responsible for mineralocorticoid secretion.
zona pellucida (pel-LŪ-si-da): Region between a developing oocyte and the surrounding follicular cells of the ovary.
zona reticularis (re-tik-ū-LAR-is): Region of the adrenal cortex responsible for androgen secretion.
zygote (ZĪ-gōt): The fertilized ovum prior to the start of cleavage.

INDEX

PHOTO CREDITS

Chapter 1
1.7 ©The New Yorker Collection 1990 Ed Fisher from cartoonbank.com. All Rights Reserved.
1.9a,1.9b, 1.9c Custom Medical Stock Photo, Inc.
1.14d Ralph T. Hutchings
1.15aL, 1.15aR, 1.15bL, Science Source/Photo Researchers, Inc.
1.15bR Custom Medical Stock Photo, Inc.
1.16b,c,d Dr. Kathleen Welch
1.16b CNRI/Photo Researchers, Inc.
1.16d Photo Researchers, Inc.
1.17a Picker International
1.17b Alexander Tsiaras/Science Source/Photo Researchers, Inc.

Chapter 2
2.1a Todd Derksen
2.1b David M. Phillips/Visuals Unlimited
2.1c Todd Derksen
2.7a M. Sahliwa/Visuals Unlimited
2.9b Fawcett/Hirokawa/Heuser/Science Source/Photo Researchers, Inc.
2.9c M. Schliwa/Visuals Unlimited
2.10a Fawcett/de Harven/Kalnins/Photo Researchers, Inc.
2.12 CNRI/Science Source/Photo Researchers, Inc.
2.13a Don W. Fawcett, M.D., Harvard Medical School
2.13b, 2.16a Biophoto Associates/Photo Researchers, Inc.
2.17b Dr. Birgit H. Satir
2.22a, b Ed Reschke/Peter Arnold, Inc.
2.22c James Solliday/Biological Photo Service
2.22d,e,f Ed Reschke/Peter Arnold, Inc.

Chapter 3
3.2b Custom Medical Stock Photo, Inc.
3.3c Dr. C.P. Leblond and A. Rambourg, McGill University
3.4a Ward's Natural Science Establishment, Inc.
3.4b Frederic H. Martini
3.5a Pearson Education/PH College
3.5b Courtesy of Gregory N. Fuller, M.D., Ph.D., Chief, Section of Neuropathology, M.D. Anderson Cancer Center,Houston, Texas
3.5c,d; 3.06a,b Frederic H. Martini
3.6c Courtesy of Gregory N. Fuller, M.D., Ph.D., Chief, Section of Neuropathology, M.D. Anderson Cancer Center,Houston, Texas
3.7a S. Elem/Visuals Unlimited
3.7b Frederic H. Martini
3.9a Phototake/Carolina Biological Supply Company
3.11b Ward's Natural Science Establishment, Inc.
3.12a,b Project Masters, Inc./The Bergman Collection
3.13a Science Source/Photo Researchers, Inc.
3.13b Frederic H. Martini
3.13c Ward's Natural Science Establishment, Inc.
3.14a John D. Cunningham/Visuals Unlimited
3.14b Bruce Iverson/Visuals Unlimited
3.14c Frederic H. Martini
3.17a Robert Brons/Biological Photo Service
3.17b Science Source/Photo Researchers, Inc.
3.17c Ed Reschke/Peter Arnold, Inc.
3.18 Frederic H. Martini
3.21a G. W. Willis, MD/Biological Photo Service
3.21b Phototake NYC
3.21c,3.22b Pearson Education/PH College

Chapter 4
4.3 John D. Cunningham/Visuals Unlimited
4.4b,c Frederic H. Martini
4.5 © R. G. Kessel and R. H. Kardon, *"Tissues and Organs: A Text-Atlas of Scanning Electroni Microscopy,"* W.H. Freeman & Co., 1979. All Rights Reserved.
4.6 Pearson Education/PH College
4.7a © R. G. Kessel and R. H. Kardon, *"Tissues and Organs: A Text-Atlas of Scanning Electroni Microscopy,"* W.H. Freeman & Co., 1979. All Rights Reserved.
4.7b David Scharf Photography
4.7c Prof. P. Motta, Dept. of Anatomy, University *"La Sapienza,"* Rome/Science Photo Library/Photo Researchers, Inc.
4.9b Manfred Kage/Peter Arnold, Inc.
4.10b John D. Cunningham/Visuals Unlimited
4.13, 4.14a,b Frederic H. Martini
4.16 D.Falconer/Getty Images, Inc/PhotoDisc, Inc.
CD 4.1 The New England Journal of Medicine

Chapter 5
5.1b,c,d © R. G. Kessel and R. H. Kardon, *"Tissues and Organs: A Text-Atlas of Scanning Electroni Microscopy,"* W.H. Freeman & Co., 1979. All Rights Reserved.
5.2d, 5.3a,b,c, Ralph T. Hutchings
5.4c, 5.5L,R Frederic H. Martini
5.6a,b Ralph T. Hutchings
5.7b Pearson Education/PH College
5.8a,b Project Masters, Inc./The Bergman Collection
5.12a,b Prof. P. Motta, Dept. of Anatomy, University *"La Sapienza,"* Rome/Science Photo Library/Photo Researchers, Inc.
F05.1,F05.2 Southern Illinois University/Visuals Unlimited
F05.3 Grace Moore/Medichrome/The Stock Shop, Inc.
F05.4 Southern Illinois University/Peter Arnold, Inc.
F05.5 Custom Medical Stock Photo, Inc.
F05.6 Scott Camazine/Photo Researchers, Inc.
F05.7 Patricia Barber, RBP/Custom Medical Stock Photo, Inc.
F05.8 Southern Illinois University/Visuals Unlimited
F05.9 Project Masters, Inc./The Bergman Collection

Chapter 6
6.1bB,b 6.3a,b,c,d,e; 6.4; 6.5 6.6a,b,c,; 6.7a,b,c; 6.8a Ralph T. Hutchings
6.8b Michael J. Timmons
6.8c; 6.9a,b; 6.10a,c; 6.11b; 6.12a,b; 6.13b,c; 6.14a,b; 6.15; 6.16a; 6.17b Ralph T. Hutchings
6.18c Michael J. Timmons
6.18d; 6.19b Ralph T. Hutchings
6.19c Siemens Medical Systems, Inc.
6.20a Science Photo Library/Custom Medical Stock Photo, Inc.
6.20b National Medical Slide/Custom Medical Stock Photo, Inc.
6.20c Princess Margaret Rose Orthopaedic Hospital/ Science Photo Library/Photo Researchers, Inc.
6.22a,b,c,d; 6.23a,b,c,d,e; 6.24a,b,c,d,; 6.25a,b; 6.26a,b,c; 6.27a,b,c,d; Ralph T. Hutchings

Chapter 7
7.1a,b; 7.2a Ralph T. Hutchings
7.2b Bates/Custom Medical Stock Photo, Inc.
7.3a,b; 7.4a; 7.5d,e,f; 7.6a,b,c,d; 7.7a,b,c,d,e; 7.8a,b,c; 7.9a Ralph T. Hutchings
7.9b Bates/Custom Medical Stock Photo, Inc.
7.10a,b; 7.11a,b; 7.14a,b,c,d,e,f; 7.15a,b; 7.16a,b,c,d; 7.17a,b; 7.18a,b Ralph T. Hutchings

Chapter 8
8.3a,b,c,d; 8.4L,R; 8.5a,b,c,d,e,f; 8.9; 8.10, 8.11d; 8.12b Ralph T. Hutchings
8.12d Patrick M. Timmons/ Michael J. Timmons
8.12e,f Ralph T. Hutchings
8.13d Patrick M. Timmons/Michael J. Timmons
8.17b,c; 8.18a Ralph T. Hutchings
8.18b Courtesy of Dr. Eugene C. Wasson, III and staff of Maui Radiology Consultants, Maui Memorial Hospital
8.19b,e Ralph T. Hutchings

Chapter 9
9.2a Fred Hossler/Visuals Unlimited
9.2b Don W. Fawcett/Science Source/Photo Researchers, Inc.
9.3b Ward's Natural Science Establishment, Inc.
9.4b Don W. Fawcett/Photo Researchers, Inc.
9.7B,T J.J. Head/Carolina Biological Supply Company/Phototake NYC
9.13a Lippincott Williams & Wilkins
9.13b Frederic H. Martini

Chapter 10
10.04b; 10.11a Ralph T. Hutchings
10.11b,c Mentor Networks Inc.
10.13d; 10.14c Ralph T. Hutchings

Chapter 11
11.6a Custom Medical Stock Photo, Inc.
11.6d Ralph T. Hutchings
11.7a Mentor Networks Inc.
11.7d; 11.9f; 11.1b,c; 11.12b; 11.13b; 11.14c Ralph T. Hutchings
11.14d Mentor Networks Inc.
11.15b; 11.16c; 11.17c; 11.18a Ralph T. Hutchings

Chapter 12
12.1a,b,c; 12.2a,b Mentor Networks Inc.
12.3a Custom Medical Stock Photo, Inc.
12.3b; 12.4a,b; 12.5a,b Mentor Networks Inc.
12.6a Custom Medical Stock Photo, Inc.
12.6b,c; 12.7a,b Mentor Networks Inc.
12.7c Custom Medical Stock Photo, Inc.
12.7d Mentor Networks Inc.

Chapter 13
13.6a Frederic H. Martini
13.6b Dr. Richard Kessel and Dr. Randy H. Kardon
13.7 John D. Cunningham/Visuals Unlimited
13.8a Biophoto Associates/Photo Researchers, Inc.
13.8b Photo Researchers, Inc.
13.9a Pearson Education/PH College
13.13a David Scott/Phototake NYC

Chapter 14
14.1b,c; 14.2a Ralph T. Hutchings/Ralph T. Hutchings
14.3 Patrick M. Timmons/Michael J. Timmons
14.4a Ralph T. Hutchings
14.4b Hinerfeld/Custom Medical Stock Photo, Inc.
14.5a Michael J. Timmons
14.6a /Dr. Richard Kessel and Dr. Randy H. Kardon
14.12; 14.14a,b; 15.2b; 15.3c; 15.4a Ralph T. Hutchings

Chapter 15
15.7 Visuals Unlimited
15.8a,b,c; 15.9a Ralph T. Hutchings
15.11c Michael J. Timmons
15.11d Pat Lynch/Photo Researchers, Inc.
15.13a,b; 15.15a; 15.16b,d Ralph T. Hutchings
15.17a /Daniel P. Perl, M.D.
15.17b; 15.19a,b Ralph T. Hutchings
15.19b Ward's Natural Science Establishment, Inc.
15.21a,c Ralph T. Hutchings

Chapter 17
17.5b Ward's Natural Science Establishment, Inc.

Chapter 18
18.3d,f; 18.5 Frederic H. Martini
18.7b Pearson Education/PH College
18.7c G. W. Willis/Terraphotographics/Biological Photo Service
18.10c Ralph T. Hutchings/Ralph T. Hutchings
18.10d; 18.14b Lennart Nilsson/Albert Bonniers Forlag AB
18.16c Michael J. Timmons
18.16e Ward's Natural Science Establishment, Inc.
18.16f P. Motta/Science Photo Library/Photo Researchers, Inc.
18.18a; 18.19 Ralph T. Hutchings
18.20d Michael J. Timmons
18.20f Ralph T. Hutchings
18.22a Ed Reschke/Peter Arnold, Inc.
18.22c Custom Medical Stock Photo, Inc.
18.24; 18.25a Ralph T. Hutchings

Chapter 19
19.4b Manfred Kage/Peter Arnold, Inc.
19.7b,c; 19.9b,c Frederic H. Martini
19.10c; 19.11b Ward's Natural Science Establishment, Inc.
19.11c,d Michael S. Ballo, M.D., Duke University Medical Center
19.12a,b Project Masters, Inc./The Bergman Collection
19.12c John Paul Kay/Peter Arnold, Inc.
19.12d Custom Medical Stock Photo, Inc.
19.12e Biophoto Associates/Science Source/Photo Researchers, Inc.

Chapter 20
20.2a David Scharf/Peter Arnold, Inc.
20.2b Ed Reschke/Peter Arnold, Inc.
20.2c Dennis Kunkel/CNRI/Phototake NYC
20.5a,b,c,d,e Ed Reschke/Peter Arnold, Inc.
20.6 Frederic H. Martini
20.7 Custom Medical Stock Photo, Inc.

Chapter 21
21.2d Ralph T. Hutchings
21.3c Peter Arnold, Inc.
21.5a,b Ralph T. Hutchings
21.6b Lennart Nilsson/Albert Bonniers Forlag AB
21.6c Ralph T. Hutchings
21.6cL Science Photo Library/Photo Researchers, Inc.
21.6cR Biophoto Associates/Science Source/Photo Researchers, Inc.
21.6d; 21.9c Ralph T. Hutchings/Ralph T. Hutchings
21.10a,b Howard Sochurek/Medichrome/The Stock Shop, Inc.
21.10c Peter Arnold, Inc.
21.13 Larry Mulvehill/Photo Researchers, Inc.

Chapter 22
22.1 Biophoto Associates/Photo Researchers, Inc.
22.3b,d *Bailey's Textbook of Microscopic Anatomy by Kelly, Wood, & Enders. Copyright 1984. Williams & Wilkins.*
22.4b Biophoto Associates/Photo Researchers, Inc.
22.5B B & B Photos/Custom Medical Stock Photo, Inc.
22.5T William Ober/Visuals Unlimited
22.12b,c; 22.14; 22.15b,c; 22.18b Ralph T. Hutchings

Chapter 23
23.3b Frederic H. Martini
23.5 Ralph T. Hutchings
23.8a David M. Phillips/Visuals Unlimited
23.8b Biophoto Associates/Photo Researchers, Inc.
23.9U Ralph T. Hutchings
23.13 Frederic H. Martini
23.14a Ralph T. Hutchings
23.16c,d; 23.17c Frederic H. Martini

Chapter 24
24.2a Frederic H. Martini
24.2c Photo Researchers, Inc.
24.3b,c Ralph T. Hutchings
24.5c Phototake NYC
24.7c John D. Cunningham/Visuals Unlimited
24.8a; 24.10b,d Ralph T. Hutchings
24.11c Ward's Natural Science Establishment, Inc.
24.12b Micrograph by P. Gehr, from Bloom & Fawcett, *"Textboook of Histology,"* W.B. Saunders Co.
24.13 Ralph T. Hutchings

Chapter 25
25.2b G. W. Willis, MD/Biological Photo Service
25.6b Frederic H. Martini
25.7e Ralph T. Hutchings
25.9a Alfred Pasieka/Peter Arnold, Inc.
25.9b Astrid and Hanns.Frieder Michler/Science Photo Library/Photo Researchers, Inc.
25.11b Ralph T. Hutchings
25.13a Prof. P. Motta, Dept. of Anatomy, University *"La Sapienza,"* Rome/Science Photo Library/Photo Researchers, Inc.
25.13b John D. Cunningham/Visuals Unlimited
25.13e,f Frederic H. Martini
25.15d John D. Cunningham/Visuals Unlimited
25.15e Michael J. Timmons
25.15e G. W. Willis, MD/Biological Photo Service
25.16,b; 25.17b Ralph T. Hutchings
25.19b Ward's Natural Science Establishment, Inc.
25.20b Ralph T. Hutchings
25.21b Ward's Natural Science Establishment, Inc.
25.21c Michael J. Timmons
25.23c Frederic H. Martini

Chapter 26
26.2b; 26.3a Ralph T. Hutchings
26.3c Mentor Networks Inc.
26.5a,b Ralph T. Hutchings
26.6c,d Pearson Education/PH College
26.7b © R. G. Kessel and R. H. Kardon, *"Tissues and Organs: A Text-Atlas of Scanning Electroni Microscopy,"* W.H. Freeman & Co., 1979. All Rights Reserved.
26.7e David M. Phillips/Visuals Unlimited
26.8b Pearson Education/PH College
26.9a,c Photo Researchers, Inc.
26.10d Ralph T. Hutchings
26.11a Ward's Natural Science Establishment, Inc.
26.11b,c Frederic H. Martini

Chapter 27
27.1 Ralph T. Hutchings
27.4b Frederic H. Martini
27.5a Don W. Fawcett ,M.D., Harvard Medical School
27.5c Ward's Natural Science Establishment, Inc.
27.6b David M. Phillips/Visuals Unlimited
27.7a Ralph T. Hutchings
27.7b,c Frederic H. Martini
27.8a Ward's Natural Science Establishment, Inc.
27.8b © R. G. Kessel and R. H. Kardon, *"Tissues and Organs: A Text-Atlas of Scanning Electroni Microscopy,"* W.H. Freeman & Co., 1979. All Rights Reserved.
27.8c,d,e Frederic H. Martini
27.9b Ward's Natural Science Establishment, Inc.
27.9d; 27.10; 27.11c Ralph T. Hutchings
27.12a,b,c,d Frederic H. Martini
27.12e C. Edelmann/La Villete/Photo Researchers, Inc.
27.12f G.W. Willis, MD/Biological Photo Service
27.12g G. W. Willis/Terraphotographics/Biological Photo Service
27.14b Frederic H. Martini
27.14c Custom Medical Stock Photo, Inc.
27.16a Ward's Natural Science Establishment, Inc.
27.17a,b Frederic H. Martini
27.17c Michael J. Timmons
27.17d Frederic H. Martini
27.19 Michael J. Timmons
27.21b Ralph T. Hutchings
27.21c Fred E. Hossler/Visuals Unlimited
27.21d Frederic H. Martini

Chapter 28
28.6b Frederic H. Martini
28.7a Dr. Arnold Tamarin
28.7b,c,d; 28.8a Lennart Nilsson/Albert Bonniers Forlag AB
28.8b Photo Researchers, Inc.